Ranking the '80s

Ranking the '80s

**A complete catalog
of all the acts and records
from pop's video decade**

Bill Carroll

Published by Carroll Applied Science, LLC
Dallas, TX

Cover Art by Lighthouse 24

Ranking the '80s
A complete catalog of all the acts and records from pop's video decade

Copyright © 2021, William F. Carroll, Jr.

Published by Carroll Applied Science, LLC
Dallas, TX

All rights reserved. No part of this book may be reproduced (except for inclusion in reviews) disseminated or utilized in any form or by any means, electronic or mechanical, including photocopying, recording or in any information storage or retrieval system or the Internet/World Wide Web without written permission from the author or publisher.

Please visit the website at **http://ranking.rocks** for more information or to contact the author.

Ranking the '80s
Bill Carroll

1. Title 2. Author 3. Popular Music/History

Library of Congress Control Number: 2021905801

ISBN-13: 978-0-578-88109-6

Praise for Ranking the '80s:

"Video might have killed the radio star…but it gave birth to one of the most vibrant decades in pop history. This is the definitive overview of pop's most visual decade…the stars, songs and albums. If only the book came with a pair of parachute pants….."

Lou Simon, VP/Music Programming, host of The Diner, Sirius XM Radio

"From a-ha to ZZ Top…from 'Down Under' to 'Straight Up'…it's the ultimate 'Celebration' of the music of the '80s. If you find yourself digging out your Walkman and Rubik's Cube, don't say I didn't warn you!"

Rich Appel, Host of the worldwide-syndicated radio show *That Thing with Rich Appel*

"Bill Carroll has done it again."

"After successfully ranking The Songwriters, The Albums and (with more than a little help from Dann Isbell) The '70's, this time around he's Ranking The '80's, in a brand-new, meticulously researched book that ranks every charted single of the 1980's based on a scientifically developed formula to determine the comparative values of each title charted. (And he's got the charts and graphs to prove it!)"

"In addition, this book has some very interesting and intriguing new perspectives to share ... Additional tidbits such as comparing a given artist's Singles Chart Performance against their Album Track Record. (By the 1980's, the preferred choice in purchasing music was the LP format rather than the pop singles of the past ... and this book shows you, side by side, how an artist performed in each medium. You may be surprised by some of these results!)"

The new book also incorporates Bill's Songwriters research AND throws in the additional bonus of Producer information for these chart hits. It's about as "All-In-One" as any chartaholic could ever dream of.
Here's hoping he'll eventually find the time to add similar updates to his 1970's volume and partner with Isbell again to update their outstanding Ranking The '60's volume ... the book that launched this series. (Could a Ranking the '50's book ALSO be in our future???)

"A more complete analysis simply isn't available ... and you just can't argue with the research (no matter what your own 20/20 hindsight may THINK was the case.) These books present the facts AS THEY ACTUALLY HAPPENED ... regardless of the "pick-and-choose" airplay we hear today."

"Highly recommended as a necessary addition to any Music Chart Reference Library."

Kent Kotal--forgottenhits60s.blogspot.com, http://www.forgottenhits.com

To Dann Isbell, Lou Simon and
"The Diner" crowd
Mentors, friends and fellow chart-hounds.

"Video killed the radio star...
We can't rewind, we've gone too far"
--Geoff Downes, Trevor Horn, Bruce Woolley

CONTENTS

Foreword: Dann Isbell xiii
How to Use This Book/FAQ xv
Twenty (Trivia) Questions (and a couple of other challenges) xvii

The Singles
Ranking The Singles 3
   Entry, Peak, and Exit Date; Peak Rank, Weeks at Peak, Total Weeks; Album Source and Producer
 Singles are listed in order of chart strength

Singles, Alphabetical Order with Rank 93

Singles Special Lists 119
  Singles: 25 Weeks or More on Chart 121
  Singles: 9 Weeks or More in Top 10 122
  Singles: 16 Weeks or More in Top 40 123
  Top 50 Singles Each Year 124
  Number 1 Singles by Weeks 129
    Chart Entries and Number 1s by year 130
    Graph of Average Weeks at No. 1, Number 1s Per Year and Number 1 Scores by Peak Date 131
  Singles: Weakest Follow Ups to Number 1s 132
  Singles: Consecutive Number 1s 133
  50 Highest Charting Singles to Miss the Weekly Top 5 134
  50 Lowest Scoring Number 1 Singles 134
  Singles: No Proximate Album 135
  80's Covers 136

The Albums
Ranking The Albums by Album Charts 143
   Entry, Peak, and Exit Date; Peak Rank, Weeks at Peak, Total Weeks
 Albums are listed in order of chart strength

Albums, Alphabetical Index with Rank 197

Albums Special Lists 227
  75 Weeks or More on Chart 229
  13 Weeks or More in Top 10 230
  31 Weeks or More in Top 40 231
  Top 50 By Year of Peak 232
    Number 1 Albums by Weeks 237
    Graph of Average Weeks at No. 1 By Year 237
  Albums: Highest Scoring Missing the Weekly Top 5 238
  Albums: Lowest Scoring Number 1s 238
  Highest Scoring Albums with No Hot 100 Singles 238
  Evaluating Albums Graphically Two Ways: Album Chart History vs. Score of Derived Singles 239
  Albums: Ranked By Derived Singles 242
  Albums with Three or More Hot 100 Singles 243

Cont'd

The Acts

The Acts with Their Singles 257
All 1,399 Acts: Act Rank; Highest Peak; Top 10, Top 40 and Total Chart Entries; Act Score
All 4,168 Singles: Year of Entry; Peak, Wks at Peak; Wks at Top 10, Top 40 & Total
Acts are listed in Alphabetical Order; Singles for each Act are listed in order of chart strength

The Singles Acts, Ranked 291
Highest Peak; Top 10s, Top 10 Wks, Top 40s, Top 40 Wks, Total Entries and Total Chart Weeks
Singles are listed in order of chart strength

The Singles Acts Special Lists 305
50 Highest Charting Acts Within Each Year 307
Yearly Top 50 Acts Through the Decade 309
Acts: 100 or More Total Chart Weeks 315
Acts: 6 or More Chart Appearances 316
Acts: 3 or More Top 10 Entries 317
Acts: 4 or More Top 40 Entries 318
Acts: More Than 40 Consecutive Chart Weeks 319
Acts: 3 Singles on Chart Simultaneously 320
Acts: 2 Singles in Top 10 Simultaneously 320
Acts: Highest Average Record Score 321
Acts: Number 1 Records 322

The Acts with Their Albums 323
All 1,927 Acts: Act Rank; Highest Peak; Top 10, Top 40 and Total Chart Entries; Act Score
All 5,155 Albums: Year of Entry; Peak, Wks at Peak; Wks at Top 10, Top 40 & Total
Acts are listed in Alphabetical Order; Singles for each Act are listed in order of chart strength

The Albums Acts, Ranked
Highest Peak; Top 10s, Top 10 Wks., Top 40s, Top 40 Wks., Total Entries and Total Chart Wks. ... 365
Albums are listed in order of chart strength

The Album Acts Special Lists 383
50 Highest Charting Acts Within Each Year 385
Yearly Top 50s Through the Decade 387
Acts: 150 or More Total Chart Weeks 392
Acts: 6 or More Chart Appearances 393
Acts: All Top 10 Entries 394
Acts: 3 or More Top 40 Entries 395
Acts: More Than 70 Consecutive Weeks on the Chart 396
Acts: Simultaneously-Most Entries, 2 Albums in Top 10, 3 Albums in Top 40 397
Acts: Number 1 Albums 398

The Acts Alphabetical Index, Singles and Albums, with Ranks 399

Evaluating Acts Two Ways: Singles-Centric vs. Album-Centric Acts 411

cont'd

Chronology Graphs of 20 Top Singles Acts and 20 Top Album Acts 417

Michael Jackson	420	Diana Ross	450
Madonna	422	Prince	452
Hall & Oates	424	Elton John	454
Lionel Richie	426	Cyndi Lauper	456
Phil Collins	428	Billy Ocean	458
Billy Joel	430	Bruce Springsteen	462
Duran Duran	432	Journey	464
Whitney Houston	434	Def Leppard	466
Kenny Rogers	436	U2	468
John Mellencamp	438	The Police	470
Air Supply	440	Van Halen	472
Huey Lewis & The News	442	Bon Jovi	474
Kool & The Gang	444	Pat Benatar	476
Olivia Newton-John	446	Pink Floyd	478
Chicago	448	Rolling Stones	480

The Writers

Ranking the Writers and Writer Teams 485
    Top 200 Writers 487
    Top 150 Writer Teams 488
    Top 50 Writer Teams with Their Acts and Singles 489
    Top 50 Singles Acts and Their Writers 493
    Writers: 50 or More Consecutive Weeks in the Hot 100 496
    Writers: Four or More Songs on the Hot 100 Simultaneously 497

The Producers

Ranking the Producers 499
    Top 150 Producer Teams 501
    Top 50 Producer Teams, Their Acts and Singles 502

The Appendix

Methodology for *Ranking the '80s* 509
    Singles with Credit Allocated to the Underlying Acts 511
Answers to Ranking the '80s Twenty (Trivia) Questions
    (and a couple of other challenges) 513
Acknowledgements 515
About the Author 517

FOREWORD

By Dann Isbell
Author of *Ranking the '60s* and *Ranking the '70s*

This is the third volume in the Ranking the Decades series. I began ranking the popular songs for a decade more than ten years ago with the sixties compilation, *Ranking the '60s: A Comprehensive Listing of the Top Songs and Acts from Pop's Golden Decade*. What originally started out as a personal project to assign more meaning to the songs I loved expanded into a more ambitious endeavor to meet research requirements set by my university. The academic version gave birth to a commercial version in 2013. Known popularly as *Ranking the '60s*, the title has since been whittled down further in informal discourse to *RT60s*.

The second volume, published two years later in 2015, *Ranking the '70s: A Complete Compilation of the Chart Songs and Acts from Pop's Eclectic Decade*, was a collaborative effort with Bill Carroll. I extended an invitation to Bill as co-author for several reasons: First and foremost, his love of chronicling pop music matched my own. That was critically important because our shared intense commitment was the means by which we were able to overcome the challenge of working together while separated by the Pacific Ocean, he in Texas, myself in Taiwan. Second, Bill's application of scientific data complemented my aesthetic approach carried over from *RT60s*. Third, with Bill on board, I concluded that with his industriousness we could deliver *RT70s* in rapid time.

In looking over Bill's books, *Ranking the Rock Writers,* published in 2018 (886 pages!) and *Ranking the Albums,* released in 2020, I must have felt much as Tommy Lasorda did at Mike Piazza's baseball Hall of Fame induction. And so with full confidence I untethered Bill to assume sole authorship of *Ranking the '80s*.

For my long-standing interest in the ranking of pop music, the '80s decade opened with a bang. On New Year's Eve and on into New Year's Day, I found myself listening to and recording reel-to-reel the Top 1000 of the '70s over a radio station serving the Coachella Valley in Southern California, long before the locale became famous for its annual twin music festivals, the Coachella and the Stagecoach. What made this especially satisfying was the fact that the station had agreed to play my set of 45s while adhering to a countdown I provided based on the national charts.

Over the course of the new decade, this wonderfully cathartic moment set in motion an irresistible urge to record and preserve radio countdowns, first while living in the Southern California area from 1980-84, then in Taiwan, 1984-89. Shows kept to this day include all American Top 40 year-end countdowns with Casey Kasem, a KRLA oldies top 300, and a pair of KHTZ broadcasts: the top 97 for the year 1982 and a listener's choice top 970 from the summer of 1984.

Detailing why I would claim a fascination over countdowns and rankings over a twenty-five year stretch is beyond the limitations of this note. Let's just say that I can agree with many of you who tune in to a countdown show that there is this primal "need to know" feature to our brains—we are naturally inclined to ordering and list-making because it reflects a drive to make sense of one part of our world.

Offbeat as this may sound, as you leaf through the pages of *Ranking the '80s*, I would hope that you will feel moments of empowerment as you make sense of how the songs of the '80s stacked up relative to each other.

March, 2021

How To Use This Book/FAQs

The information used to compile this book consists of Title, Act and Chart Position week by week for all those records peaking on the Billboard Hot 100 Singles or Top 200 albums between January 1, 1980 and December 31, 1989. All the other quantities are derived from these data. Descriptive information about albums, writers, producers and acts is found from various sources, including act websites, Wikipedia, discogs.com and 45cat.com. Much of this information was assembled for Ranking the Rock Writers and Ranking the Albums.

There are six major sections of this book and they interoperate: Acts With Their Singles, Acts With Their Albums, then sections for Acts, Singles, Albums, and Writers and Producers. There are complete alphabetical indexes for singles and albums and a concordance list between singles acts and album acts. Additionally, there are specialized lists. The Chronologies section contains a two-page graphical snapshot of the decade for the twenty most important singles acts and album acts. More detail can be found in the Table of Contents.

Q: What was the rank of *Bette Davis Eyes* by Kim Carnes?
A: Section is listed at the top outside page border. Go to the Singles, Alphabetical Order list, p 96. It's number 4.

Q: Who wrote it?
A: That, as well as the album it came from is found in the Ranking the Singles list, p. 5. Jackie DeShannon and Donna Weiss wrote it; it came from *Mistaken Identity,* produced by Val Garay. There are other statistics there.

Q: What was the number 10 single, then?
A: Ranking the Singles list again, p. 5. *Waiting For A Girl Like You* by Foreigner.

Q: I hated that record. Why did you rank it so high?
A: Remember, this is not about what is good or bad, it's about what charted strongly.

Q: OK. How many singles did The Police chart?
A: Go to the Acts With Their Singles Section, p. 280 and look up The Police (under P—the words a, an and the are not used for alphabetical order). There you will find The POLICE along with their Act ranking, (31) their highest peak with any single, (1) and the number of their singles that made the Top 10, Top 40, and the chart itself [6|8|10].

Q. What was their biggest single?
A. First record on the list under The POLICE is *Every Breath You Take*, and you will find its year of entry, (83), Peak and Weeks at Peak, 1(8), and weeks in the Top 10, Top 40 and in Total: [13|20|22]. Finally, there is the score.

Q: Let's try one more. Where did Night Ranger rank as a group?
A: Do you mean for singles or albums?

Q: Well, both actually
A: Start with the Acts, Alphabetical Index p. 406 because the list has act rank for both singles and albums.

Q: Right. I see they're number 108 for albums and 144 for singles. Who was 145 for singles?
A: Go To the Singles Acts, Ranked, p. 294. Peter Gabriel.

Q: There's so much stuff here. Are there any mistakes?
A: Probably. Typos are most likely in manual transcription as in the album sources and producer lists. I hope you look hard enough to find them and let me know where they are. Contact me through http://ranking.rocks.

Ranking the '80s: Twenty (Trivia) Questions

1. Which four albums that peaked in the '80s also charted seven singles on the Hot 100 *in the '80s*?

2. Which single spent the most weeks at number 2 but never made number 1?
   a. Open Arms-Journey   b. Electric Avenue-Eddy Grant   c. Waiting For A Girl Like You-Foreigner
   d. Rosanna-Toto   e. Looking For A New Love-Jody Watley

3. Which two acts charted three of their singles in the top 40 simultaneously?

   a. Diana Ross   b. Prince   c. New Kids On The Block   d. Cyndi Lauper   e. John Lennon

4. This 1983 hit was the last country record to top the Hot 100 for nearly 17 years. Name it.

5. Name the three James Bond themes that charted in the '80s. Which was the first to go to number 1?

6. Which act had seven consecutive released singles go to number 1?

7. Which act scored its eighth consecutive number 1 album of new material in the '80s?

8. Name the only '80s charted single and the only charting singles act beginning with the letter X.

9. Name the highest-ranking male and female solo singles acts having only one name.

10. What was the highest-ranking singles duo of the '80s? What was the highest-ranking singles group?

11. Which act had seven albums simultaneously on the Top 200 chart during the '80s?

    a. Led Zeppelin   b. U2   c. Beatles   d. Julio Iglesias   e. Bob Seger

12. What was the highest charting album with no Hot 100 singles?

13. What was the highest ranked '80s cover of a previous Hot 100 single?

14. Name the three covers of previous number 1 records that also went to number 1 in the '80s.

15. Name the three writers to have songs simultaneously occupy number 1 and 2 on the Hot 100.

16. Who was the highest charting producer of the '80s, measured by singles chart scores?

17. Which act charted the most albums (alone or in collaboration) that peaked in the '80s?

18. Which act charted the most albums (alone or in collaboration) including those holdovers or recurrent from previous decades?

    a. Rolling Stones   b. Elton John   c. Pink Floyd   d. Kenny Rogers   e. Elvis Presley

19. Which writer, alone or as part of a team, charted the most songs on the Hot 100 in the '80s? Which writer team charted the most?

20. What are the top three soundtracks of the '80s?

Bonus Questions? OK. Here You Go.

Bonus 1: What was the highest charting writing team, who were not primarily performers?

Bonus 2: The titles of The Police's two biggest hits both started with "Every." Name them.

Bonus 3: One act ranked number 1 in three consecutive years: 1985, 1986, 1987. Name that act.

Bonus 4: Three solo male acts, three solo female acts, and three duos of the '80s each having one name, charted number 1 singles in the '80s. Name them.

Bonus 5: These three '80s acts, which were not one-off star collaborations (i.e. USA For Africa)—had one single ever in the Hot 100 and it peaked at number 1. Two were instrumental and one was a cappella. Name them.

Bonus Bonus: One other non-US-based '80s act peaked at number 2 and never made the Hot 100 again. Who was it?

Other Challenges: Who was bigger?

Based on their '80s singles, pick which of the pair ranked higher.

Go-Gos or Belinda Carlisle

Genesis or Phil Collins

George Michael or Wham!

Stevie Nicks or Fleetwood Mac

Peter Cetera or Chicago

Bobby Brown or New Edition

Don Henley or Glenn Frey

Neil Diamond or Barbra Streisand

Rick Astley or Bananarama

The Fixx or A-Ha

The Pretenders or The Motels

Pat Benatar or Debbie Gibson

Billy Joel or Elton John

Answers in the Appendix

Ranking the '80s:

The Singles

The Singles

Ranking the Singles

Singles 1-2000

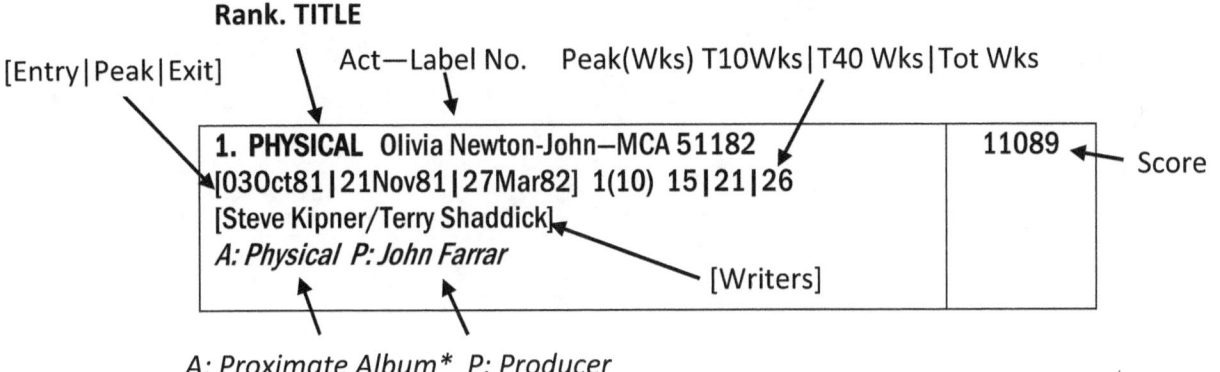

Singles 2001-4168

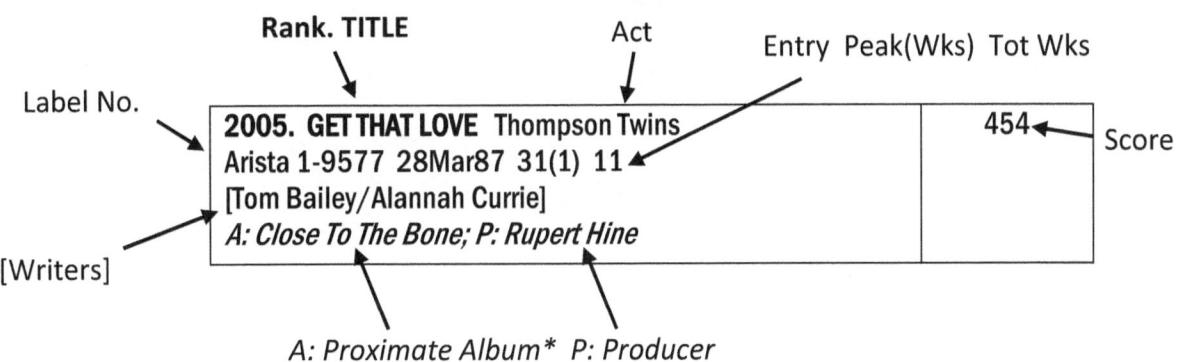

*The "Proximate Album" is the album from which the single is most likely taken. In many cases that could be a Soundtrack and that Soundtrack is cited even if the song later appears on an album by the act; in some duet cases the song appeared on albums for both acts. See Appendix.

Ranking the Singles

Singles 1-2000

Rank. TITLE Artist Label Number
[Entry Date | Peak Date | Exit Date] Peak(Peak Wks) Top10 | Top40 | Tot Wks
[Writers] *Album*; Producers — Score

1. PHYSICAL Olivia Newton-John MCA 51182 — 11089
[03Oct81 | 21Nov81 | 27Mar82] 1(10) 15 | 21 | 26
[Steve Kipner/Terry Shaddick] *A: Physical*; P: John Farrar

2. ENDLESS LOVE Diana Ross & Lionel Richie Motown 1519 — 10386
[11Jul81 | 15Aug81 | 09Jan82] 1(9) 13 | 19 | 27
[Lionel Richie] *A: Endless Love Soundtrack*; P: Lionel Richie

3. SAY SAY SAY Paul McCartney and Michael Jackson Columbia 04168 — 10336
[15Oct83 | 10Dec83 | 10Mar84] 1(6) 13 | 18 | 22
[Michael Jackson/Paul McCartney] *A: Pipes Of Peace*; P: George Martin

4. BETTE DAVIS EYES Kim Carnes EMI America 8077 — 10255
[28Mar81 | 16May81 | 19Sep81] 1(9) 14 | 20 | 26
[Jackie DeShannon/Donna Weiss] *A: Mistaken Identity*; P: Val Garay

5. EVERY BREATH YOU TAKE The Police A&M 2542 — 10182
[04Jun83 | 09Jul83 | 29Oct83] 1(8) 13 | 20 | 22
[Sting] *A: Synchronicity*; P: Hugh Padgham/Police

6. CALL ME (Theme From...American Gigolo) Blondie Chrysalis 2414 — 9931
[16Feb80 | 19Apr80 | 02Aug80] 1(6) 12 | 19 | 25
[Clement A. Bozewski/Deborah Harry/Giorgio Moroder]
A: American Gigolo Soundtrack; P: Giorgio Moroder

7. FLASHDANCE...WHAT A FEELING Irene Cara Casablanca 811440 — 9800
[02Apr83 | 28May83 | 17Sep83] 1(6) 14 | 20 | 25
[Irene Cara/Keith Forsey/Giorgio Moroder]
A: Flashdance Soundtrack; P: Giorgio Moroder

8. EYE OF THE TIGER Survivor Scotti Brothers 02912 — 9773
[05Jun82 | 24Jul82 | 20Nov82] 1(6) 15 | 18 | 25
[Jim Peterik/Frankie Sullivan] *A: Eye Of The Tiger*; P: Jim Peterik/Frankie Sullivan

9. ANOTHER ONE BITES THE DUST Queen Elektra 47031 — 9572
[16Aug80 | 04Oct80 | 14Mar81] 1(3) 15 | 21 | 31
[John Deacon] *A: The Game*; P: Queen

10. WAITING FOR A GIRL LIKE YOU Foreigner Atlantic 3868 — 9492
[10Oct81 | 28Nov81 | 13Mar82] 2(10) 15 | 19 | 23
[Lou Gramm/Michael Leslie Jones] *A: 4*; P: Mick Jones/Mutt Lange

11. DO THAT TO ME ONE MORE TIME The Captain & Tennille Casablanca 2215 — 8662
[20Oct79 | 16Feb80 | 19Apr80] 1(1) 14 | 22 | 27
[Toni Tennille] *A: Make Your Move*; P: Daryl Dragon

12. EBONY AND IVORY Paul McCartney Columbia 02860 — 8625
[10Apr82 | 15May82 | 14Aug82] 1(7) 12 | 15 | 19
[Paul McCartney] *A: Tug Of War*; P: George Martin

13. ABRACADABRA Steve Miller Band Capitol 5126 — 8547
[29May82 | 04Sep82 | 13Nov82] 1(2) 14 | 19 | 25
[Steve Miller] *A: Abracadabra*; P: Gary Mallaber/Steve Miller

14. UPSIDE DOWN Diana Ross Motown 1494 — 8529
[12Jul80 | 06Sep80 | 24Jan81] 1(4) 14 | 17 | 29
[Bernard Edwards/Nile Rodgers] *A: Diana*; P: Bernard Edwards/Nile Rodgers

15. ALL NIGHT LONG (All Night) Lionel Richie Motown 1698 — 8437
[17Sep83 | 12Nov83 | 25Feb84] 1(4) 13 | 17 | 24
[Lionel Richie] *A: Can't Slow Down*; P: James Anthony Carmichael/Lionel Richie

16. LADY Kenny Rogers Liberty 1380 — 8423
[04Oct80 | 15Nov80 | 21Mar81] 1(6) 13 | 19 | 25
[Lionel Richie] *A: Kenny Rogers' Greatest Hits*; P: Lionel Richie

17. BILLIE JEAN Michael Jackson Epic 03509 — 8360
[22Jan83 | 05Mar83 | 02Jul83] 1(7) 11 | 17 | 24
[Michael Jackson] *A: Thriller*; P: Quincy Jones

18. CRAZY LITTLE THING CALLED LOVE Queen Elektra 46579 — 8343
[22Dec79 | 23Feb80 | 17May80] 1(4) 12 | 17 | 22
[Freddie Mercury] *A: The Game*; P: Queen

19. I LOVE ROCK 'N ROLL Joan Jett & The Blackhearts Boardwalk 7-11-135 — 8293
[06Feb82 | 20Mar82 | 19Jun82] 1(7) 12 | 16 | 20
[Jake Hooker/Alan Merrill]
A: I Love Rock 'N Roll; P: Ritchie Cordell/Kenny Laguna

20. CENTERFOLD The J. Geils Band EMI America 8102 — 8247
[07Nov81 | 06Feb82 | 24Apr82] 1(6) 12 | 20 | 25
[Seth Justman] *A: Freeze-Frame*; P: Seth Justman

21. ANOTHER BRICK IN THE WALL (Part II) Pink Floyd Columbia 11187 — 8171
[19Jan80 | 22Mar80 | 05Jul80] 1(4) 12 | 19 | 25
[Roger Waters] *A: The Wall*; P: Bob Ezrin/David Gilmour/Roger Waters

22. COMING UP (Live At Glasgow) Paul McCartney & Wings Columbia 11263 — 8092
[26Apr80 | 28Jun80 | 13Sep80] 1(3) 11 | 16 | 21
[Paul McCartney] *A: McCartney II**; P: Paul McCartney

23. TOTAL ECLIPSE OF THE HEART Bonnie Tyler Columbia 03906 — 7997
[16Jul83 | 01Oct83 | 28Jan84] 1(4) 11 | 18 | 29
[Jim Steinman] *A: Faster Than The Speed Of Night*; P: Jim Steinman

24. (Just Like) STARTING OVER John Lennon Geffen 49604 — 7930
[01Nov80 | 27Dec80 | 28Mar81] 1(5) 14 | 19 | 22
[John Lennon] *A: Double Fantasy*; P: Jack Douglas/John Lennon/Yoko Ono

25. WOMAN IN LOVE Barbra Streisand Columbia 11364 — 7930
[06Sep80 | 25Oct80 | 14Feb81] 1(3) 11 | 19 | 24
[Barry Gibb/Robin Gibb]
A: Guilty; P: Albhy Galuten/Barry Gibb/Karl Richardson

26. HURTS SO GOOD John Cougar Riva 209 — 7863
[24Apr82 | 07Aug82 | 30Oct82] 2(4) 16 | 22 | 28
[George Green/John Mellencamp]
A: American Fool; P: Don Gehman/John Mellencamp

27. WHEN DOVES CRY Prince Warner Bros. 29286 — 7679
[02Jun84 | 07Jul84 | 20Oct84] 1(5) 11 | 16 | 21
[Prince] *A: Purple Rain (Soundtrack)*; P: Prince

28. JUMP Van Halen Warner Bros. 29384 — 7537
[14Jan84 | 25Feb84 | 02Jun84] 1(5) 10 | 15 | 21
[Michael Anthony/David Lee Roth/Alex Van Halen/Eddie Van Halen]
A: 1984 (MCMLXXXIV); P: Ted Templeman

29. FUNKYTOWN Lipps, Inc. Casablanca 2233 — 7303
[29Mar80 | 31May80 | 30Aug80] 1(4) 9 | 15 | 23
[Steve Greenberg] *A: Mouth To Mouth*; P: Steve Greenberg

30. ROCK WITH YOU Michael Jackson Epic 50797 — 7256
[03Nov79 | 19Jan80 | 12Apr80] 1(4) 9 | 19 | 24
[Rod Temperton] *A: Off The Wall*; P: Quincy Jones

31. JACK & DIANE John Cougar Riva 210 — 7190
[24Jul82 | 02Oct82 | 18Dec82] 1(4) 10 | 17 | 22
[John Mellencamp] *A: American Fool*; P: Don Gehman/John Mellencamp

32. DOWN UNDER Men At Work Columbia 03303 — 7153
[06Nov82 | 15Jan83 | 23Apr83] 1(4) 10 | 19 | 25
[Colin James Hay/Ron Strykert] *A: Business As Usual*; P: Peter Mclan

33. AGAINST ALL ODDS (Take A Look At Me Now) Phil Collins Atlantic 89700 A — 7041
[25Feb84 | 21Apr84 | 04Aug84] 1(3) 10 | 16 | 24
[Phil Collins] *A: Against All Odds Soundtrack*; P: Arif Mardin

34. IT'S STILL ROCK AND ROLL TO ME Billy Joel Columbia 11276 — 6941
[24May80 | 19Jul80 | 11Oct80] 1(2) 11 | 19 | 21
[Billy Joel] *A: Glass Houses*; P: Phil Ramone

35. BEAT IT Michael Jackson Epic 03759 — 6892
[26Feb83 | 30Apr83 | 13Aug83] 1(3) 10 | 18 | 25
[Michael Jackson] *A: Thriller*; P: Quincy Jones

36. I CAN'T GO FOR THAT (No Can Do) Daryl Hall & John Oates RCA 12357 — 6855
[14Nov81 | 30Jan82 | 03Apr82] 1(1) 12 | 18 | 24
[Sara Allen/Daryl Hall/John Oates] *A: Private Eyes*; P: Daryl Hall/John Oates

37. HARD TO SAY I'M SORRY Chicago Full Moon 29979 — 6847
[05Jun82 | 11Sep82 | 13Nov82] 1(2) 12 | 18 | 24
[Peter Cetera/David Foster] *A: Chicago 16*; P: David Foster

38. ISLANDS IN THE STREAM Kenny Rogers Duet with Dolly Parton RCA 13615 — 6841
[27Aug83 | 29Oct83 | 11Feb84] 1(2) 12 | 18 | 25
[Barry Gibb/Maurice Gibb/Robin Gibb]
A: Eyes That See In The Dark; P: Albhy Galuten/Barry Gibb/Karl Richardson

39. HELLO Lionel Richie Motown 1722 — 6840
[25Feb84 | 12May84 | 04Aug84] 1(2) 10 | 17 | 24
[Lionel Richie] *A: Can't Slow Down*; P: James Anthony Carmichael/Lionel Richie

40. WHAT'S LOVE GOT TO DO WITH IT Tina Turner Capitol 5354 — 6823
[19May84 | 01Sep84 | 24Nov84] 1(3) 10 | 18 | 28
[Terry Britten/Graham Lyle] *A: Private Dancer*; P: Terry Britten

41. DON'T YOU WANT ME The Human League A&M/Virgin 2397 — 6816
[06May82 | 03Jul82 | 11Sep82] 1(3) 12 | 21 | 28
[John William Callis/Philip Oakey/Philip Wright]
A: Dare; P: Human League/Martin Rushent

Ranking the Singles

Rank. TITLE Artist Label Number
[Entry Date | Peak Date | Exit Date] Peak(Peak Wks) Top10 | Top40 | Tot Wks
[Writers] *Album;* Producers — Score

42. **MANEATER** Daryl Hall & John Oates RCA Victor 13354 — 6774
[16Oct82 | 18Dec82 | 19Mar83] 1(4) 13 | 17 | 23
[Sara Allen/Daryl Hall/John Oates] A: *H2O*; P: Daryl Hall/John Oates

43. **FOOTLOOSE** Kenny Loggins Columbia 04310 — 6733
[28Jan84 | 31Mar84 | 30Jun84] 1(3) 11 | 16 | 23
[Kenny Loggins/Dean Pitchford]
A: *Footloose Soundtrack*; P: Lee DeCarlo/Kenny Loggins

44. **SWEET DREAMS (Are Made Of This)** Eurythmics RCA 13533 — 6665
[14May83 | 03Sep83 | 05Nov83] 1(1) 9 | 17 | 26
[Annie Lennox/David A. Stewart]
A: *Sweet Dreams (Are Made Of This)*; P: David A. Stewart

45. **MAGIC** Olivia Newton-John MCA 41247 — 6621
[24May80 | 02Aug80 | 25Oct80] 1(4) 9 | 16 | 23
[John Farrar] A: *Xanadu (Soundtrack)*; P: John Farrar

46. **OWNER OF A LONELY HEART** Yes Atco 99817 — 6601
[05Nov83 | 21Jan84 | 07Apr84] 1(2) 10 | 17 | 23
[Jon Anderson/Trevor Horn/Trevor Rabin/Chris Squire]
A: *90125*; P: Trevor Horn

47. **LIKE A VIRGIN** Madonna Sire 29210 — 6531
[17Nov84 | 22Dec84 | 23Mar85] 1(6) 9 | 14 | 19
[Thomas F. Kelly/Billy Steinberg] A: *Like A Virgin*; P: Nile Rodgers

48. **JESSIE'S GIRL** Rick Springfield RCA 12201 — 6460
[28Mar81 | 01Aug81 | 31Oct81] 1(2) 12 | 22 | 32
[Rick Springfield] A: *Working Class Dog*; P: Keith Olsen

49. **GHOSTBUSTERS** Ray Parker Jr. Arista 1-9212 — 6445
[16Jun84 | 11Aug84 | 03Nov84] 1(3) 10 | 14 | 21
[Ray Parker Jr.] A: *Ghostbusters Soundtrack*; P: Ray Parker Jr.

50. **ARTHUR'S THEME (Best That You Can Do)** Christopher Cross Warner Bros. 49787 — 6437
[15Aug81 | 17Oct81 | 23Jan82] 1(3) 12 | 17 | 24
[Peter Allen/Burt Bacharach/Christopher Cross/Carole Bayer Sager]
A: *Arthur (The Album)*; P: Michael Omartian

51. **ROSANNA** Toto Columbia 02811 — 6431
[17Apr82 | 03Jul82 | 18Sep82] 2(5) 11 | 18 | 23
[David Paich] A: *Toto IV*; P: Toto

52. **KARMA CHAMELEON** Culture Club Virgin/Epic 04221 — 6404
[03Dec83 | 04Feb84 | 28Apr84] 1(3) 9 | 16 | 22
[Michael Craig/Roy Hay/Jon Moss/George O'Dowd/Phil Pickett]
A: *Colour By Numbers*; P: Steve Levine

53. **LET'S DANCE** David Bowie EMI America 8158 — 6341
[26Mar83 | 21May83 | 06Aug83] 1(1) 10 | 14 | 20
[David Bowie] A: *Let's Dance*; P: David Bowie/Nile Rodgers

54. **I JUST CALLED TO SAY I LOVE YOU** Stevie Wonder Motown 1745 — 6339
[18Aug84 | 13Oct84 | 09Feb85] 1(3) 10 | 15 | 26
[Stevie Wonder] A: *The Woman In Red (Soundtrack)*; P: Stevie Wonder

55. **PLEASE DON'T GO** K.C. And The Sunshine Band TK 1035 — 6303
[25Aug79 | 05Jan80 | 16Feb80] 1(1) 11 | 18 | 26
[Harry Wayne Casey/Rick Finch]
A: *Do You Wanna Go Party*; P: Harry Wayne Casey/Richard Finch

56. **SAY YOU, SAY ME** Lionel Richie Motown 1819 — 6296
[09Nov85 | 21Dec85 | 22Mar86] 1(4) 9 | 16 | 20
[Lionel Richie]
A: *Dancing On The Ceiling*; P: James Anthony Carmichael/Lionel Richie

57. **GLORIA** Laura Branigan Atlantic 4048 — 6294
[10Jul82 | 27Nov82 | 12Mar83] 2(3) 10 | 22 | 36
[Giancarlo Bigazzi/Umberto Tozzi/Trevor Veitch]
A: *Branigan*; P: Jack White

58. **WOMAN** John Lennon Geffen 49644 — 6131
[17Jan81 | 21Mar81 | 30May81] 2(3) 12 | 17 | 20
[John Lennon] A: *Double Fantasy*; P: Jack Douglas/John Lennon/Yoko Ono

59. **9 TO 5** Dolly Parton RCA 12133 — 6099
[29Nov80 | 21Feb81 | 23May81] 1(2) 9 | 18 | 26
[Dolly Parton] A: *9 To 5 And Odd Jobs*; P: Gregg Perry

60. **ALL OUT OF LOVE** Air Supply Arista 0520 — 6080
[14Jun80 | 13Sep80 | 13Dec80] 2(4) 10 | 17 | 27
[Clive Davis/Graham Russell] A: *Lost In Love*; P: Robie Porter

61. **MANIAC** Michael Sembello Casablanca 812516 — 6058
[04Jun83 | 10Sep83 | 29Oct83] 1(2) 9 | 16 | 22
[Dennis Matkosky/Michael Sembello]
A: *Flashdance Soundtrack*; P: Phil Ramone/Michael Sembello

62. **BEING WITH YOU** Smokey Robinson Tamla 54321 — 5997
[14Feb81 | 23May81 | 01Aug81] 2(3) 10 | 16 | 25
[Smokey Robinson] A: *Being With You*; P: George Tobin

63. **I LOVE A RAINY NIGHT** Eddie Rabbitt Elektra 47066 — 5924
[08Nov80 | 28Feb81 | 16May81] 1(2) 9 | 18 | 28
[David Malloy/Eddie Rabbitt/Even Stevens] A: *Horizon*; P: David Malloy

64. **DON'T TALK TO STRANGERS** Rick Springfield RCA 13070 — 5887
[06Mar82 | 22May82 | 24Jul82] 2(4) 11 | 16 | 21
[Rick Springfield] A: *Success Hasn't Spoiled Me Yet*; P: Keith Olsen

65. **THAT'S WHAT FRIENDS ARE FOR** Dionne & Friends Arista 1-9422 — 5879
[09Nov85 | 18Jan86 | 12Apr86] 1(4) 10 | 17 | 23
[Burt Bacharach/Carole Bayer Sager]
A: *Friends*; P: Burt Bacharach/Carole Bayer Sager

66. **CARELESS WHISPER** Wham! Featuring George Michael Columbia 04691 — 5877
[22Dec84 | 16Feb85 | 11May85] 1(3) 9 | 17 | 21
[George Michael/Andrew Ridgeley] A: *Make It Big*; P: George Michael

67. **ELECTRIC AVENUE** Eddy Grant Portrait 37-03793 — 5865
[16Apr83 | 02Jul83 | 10Sep83] 2(5) 8 | 15 | 22
[Eddy Grant] A: *Killer On The Rampage*; P: Eddy Grant

68. **ANOTHER DAY IN PARADISE** Phil Collins Atlantic 88774 — 5854
[04Nov89 | 23Dec89 | 03Mar90] 1(4) 10 | 14 | 18
[Phil Collins] A: *...But Seriously*; P: Phil Collins/Hugh Padgham

69. **MICKEY** Toni Basil Chrysalis 2638 — 5851
[04Sep82 | 11Dec82 | 05Mar83] 1(1) 10 | 18 | 27
[Mike Chapman/Nicky Chinn]
A: *Word Of Mouth*; P: Greg Mathieson/Trevor Veitch

70. **RIDE LIKE THE WIND** Christopher Cross Warner Bros. 49184 — 5822
[16Feb80 | 26Apr80 | 05Jul80] 2(4) 9 | 17 | 21
[Christopher Cross] A: *Christopher Cross*; P: Michael Omartian

71. **LET'S HEAR IT FOR THE BOY** Deniece Williams Columbia 04417 — 5783
[07Apr84 | 26May84 | 11Aug84] 1(2) 9 | 14 | 19
[Dean Pitchford/Tom Snow] A: *Let's Hear It For The Boy*; P: George Duke

72. **MISSING YOU** John Waite EMI America 8212 — 5752
[23Jun84 | 22Sep84 | 01Dec84] 1(1) 9 | 16 | 22
[Mark Leonard/Chas Sandford/John Waite]
A: *No Brakes*; P: Gary Gersh/David Thoener/John Waite

73. **OPEN ARMS** Journey Columbia 02687 — 5744
[16Jan82 | 27Feb82 | 15May82] 2(6) 10 | 14 | 18
[Jonathan Cain/Steve Perry] A: *Escape*; P: Kevin Elson/Mike Stone

74. **SAY IT ISN'T SO** Daryl Hall & John Oates RCA Victor 13654 — 5735
[29Oct83 | 17Dec83 | 25Feb84] 2(4) 10 | 15 | 18
[Daryl Hall] A: *Rock 'N Soul Part 1*; P: Daryl Hall/John Oates

75. **CRAZY FOR YOU** Madonna Geffen 29051 — 5654
[02Mar85 | 11May85 | 20Jul85] 1(1) 9 | 14 | 19
[John Bettis/Jon Lind] A: *Vision Quest Soundtrack*; P: Jellybean Benitez

76. **SHAME ON THE MOON** Bob Seger & The Silver Bullet Band Capitol 5187 — 5590
[18Dec82 | 26Feb83 | 07May83] 2(4) 8 | 19 | 21
[Rodney Crowell] A: *The Distance*; P: Jimmy Iovine

77. **WALK LIKE AN EGYPTIAN** Bangles Columbia 06257 — 5581
[27Sep86 | 20Dec86 | 28Feb87] 1(4) 8 | 15 | 23
[Liam Sternberg] A: *Different Light*; P: David Kahne

78. **DANCING IN THE DARK** Bruce Springsteen Columbia 04463 — 5565
[26May84 | 30Jun84 | 13Oct84] 2(4) 9 | 15 | 21
[Bruce Springsteen] A: *Born In The U.S.A.*;
P: Jon Landau/Chuck Plotkin/Bruce Springsteen/Steven Van Zandt

79. **WE ARE THE WORLD** USA for Africa Columbia 7-04839 — 5554
[23Mar85 | 13Apr85 | 20Jul85] 1(4) 8 | 12 | 13
[Michael Jackson/Lionel Richie] A: *We Are The World*; P: Quincy Jones

80. **CHARIOTS OF FIRE - TITLES** Vangelis Polydor 2189 — 5496
[12Dec81 | 08May82 | 19Jun82] 1(1) 9 | 15 | 28
[Vangelis] A: *Chariots Of Fire (Soundtrack)*; P: Vangelis

81. **START ME UP** The Rolling Stones Rolling Stones 21003 — 5491
[22Aug81 | 31Oct81 | 30Jan82] 2(3) 11 | 19 | 24
[Mick Jagger/Keith Richards] A: *Tattoo You*; P: Glimmer Twins

82. **WHO CAN IT BE NOW?** Men At Work Columbia 02888 — 5487
[10Jul82 | 30Oct82 | 08Jan83] 1(1) 9 | 17 | 27
[Colin James Hay] A: *Business As Usual*; P: Peter Mclan

83. **WAKE ME UP BEFORE YOU GO-GO** Wham! Columbia 04552 — 5473
[08Sep84 | 17Nov84 | 16Feb85] 1(3) 8 | 14 | 24
[George Michael] A: *Make It Big*; P: George Michael

Ranking the Singles

Rank. TITLE Artist Label Number [Entry Date | Peak Date | Exit Date] Peak(Peak Wks) Top10 | Top40 | Tot Wks [Writers] *Album*; Producers	Score
84. CELEBRATION Kool & The Gang De-Lite 807 [25Oct80 | 07Feb81 | 16May81] 1(2) 7 | 21 | 30 [Robert Bell/Ronald Bell/George M. Brown/Eumir Deodato/Robert Mickens/Claydes Smith/J.T. Taylor/Dennis Thomas/Earl Toon] *A: Celebrate!*; P: Eumir Deodato	5462
85. LOVE ON THE ROCKS Neil Diamond Capitol 4939 [01Nov80 | 10Jan81 | 14Mar81] 2(3) 10 | 17 | 20 [Gilbert Bécaud/Neil Diamond] *A: The Jazz Singer (Soundtrack)*; P: Bob Gaudio	5450
86. PRIVATE EYES Daryl Hall & John Oates RCA 12296 [29Aug81 | 07Nov81 | 30Jan82] 1(2) 9 | 17 | 25 [Janna Allen/Sara Allen/Daryl Hall/Warren Pash] *A: Private Eyes*; P: Daryl Hall/John Oates	5434
87. BABY, COME TO ME Patti Austin (A duet with James Ingram) Qwest 50036 [24Apr82 | 19Feb83 | 23Apr83] 1(2) 9 | 18 | 32 [Rod Temperton] *A: Every Home Should Have One*; P: Quincy Jones	5417
88. JUST THE TWO OF US Grover Washington, Jr. Elektra 47103 [14Feb81 | 02May81 | 25Jul81] 2(3) 11 | 16 | 24 [Ralph MacDonald/William Salter/Bill Withers] *A: Winelight*; P: Ralph MacDonald/Grover Washington Jr.	5413
89. TIME AFTER TIME Cyndi Lauper Portrait 37-04432 [14Apr84 | 09Jun84 | 25Aug84] 1(2) 9 | 14 | 20 [Rob Hyman/Cyndi Lauper] *A: She's So Unusual*; P: Rick Chertoff	5392
90. UPTOWN GIRL Billy Joel Columbia 04149 [24Sep83 | 12Nov83 | 18Feb84] 3(5) 10 | 16 | 22 [Billy Joel] *A: An Innocent Man*; P: Phil Ramone	5362
91. THE GIRL IS MINE Michael Jackson/Paul McCartney Epic 03288 [06Nov82 | 08Jan83 | 05Mar83] 2(3) 10 | 14 | 18 [Michael Jackson] *A: Thriller*; P: Quincy Jones	5346
92. OUT OF TOUCH Daryl Hall & John Oates RCA Victor 13916 [29Sep84 | 08Dec84 | 02Mar85] 1(2) 9 | 16 | 23 [Daryl Hall/John Oates] *A: Big Bam Boom*; P: Bob Clearmountain/Daryl Hall/John Oates	5335
93. DO YOU REALLY WANT TO HURT ME Culture Club Virgin/Epic 03368 [04Dec82 | 26Mar83 | 21May83] 2(3) 9 | 18 | 25 [Michael Craig/Roy Hay/Jon Moss/George O'Dowd] *A: Kissing To Be Clever*; P: Steve Levine	5296
94. ON MY OWN Patti LaBelle And Michael McDonald MCA 52770 [22Mar86 | 14Jun86 | 23Aug86] 1(3) 7 | 15 | 23 [Burt Bacharach/Carole Bayer Sager] *A: Winner In You*; P: Burt Bacharach/Carole Bayer Sager	5294
95. THE TIDE IS HIGH Blondie Chrysalis 2465 [15Nov80 | 31Jan81 | 09May81] 1(1) 10 | 17 | 26 [John Holt/Duke Reid] *A: Autoamerican*; P: Mike Chapman	5289
96. I WANT TO KNOW WHAT LOVE IS Foreigner Atlantic 89596 [08Dec84 | 02Feb85 | 27Apr85] 1(2) 8 | 16 | 21 [Michael Leslie Jones] *A: Agent Provocateur*; P: Mick Jones/Alex Sadkin	5285
97. THEME FROM "THE GREATEST AMERICAN HERO" (Believe It Or Not) Joey Scarbury Elektra 47147 [09May81 | 15Aug81 | 31Oct81] 2(2) 10 | 18 | 26 [Stephen Geyer/Mike Post] *A: America's Greatest Hero*; P: Mike Post	5261
98. LITTLE JEANNIE Elton John MCA 41236 [03May80 | 19Jul80 | 20Sep80] 3(4) 11 | 17 | 21 [Elton John/Gary Osborne] *A: 21 At 33*; P: Clive Franks/Elton John	5217
99. CARIBBEAN QUEEN (No More Love On The Run) Billy Ocean Jive 9199 [11Aug84 | 03Nov84 | 02Feb85] 1(2) 7 | 15 | 26 [Keith Diamond/Billy Ocean] *A: Suddenly*; P: Keith Diamond	5211
100. LET'S GO CRAZY Prince And The Revolution Warner Bros. 29216 [04Aug84 | 29Sep84 | 08Dec84] 1(2) 9 | 14 | 19 [Prince] *A: Purple Rain (Soundtrack)*; P: Prince And The Revolution	5200
101. MORE THAN I CAN SAY Leo Sayer Warner Bros. 49565 [27Sep80 | 06Dec80 | 28Feb81] 2(5) 9 | 15 | 23 [Jerry Allison/Sonny Curtis] *A: Living In A Fantasy*; P: Alan Tarney	5195
102. FAITH George Michael Columbia 07623 [24Oct87 | 12Dec87 | 05Mar88] 1(4) 9 | 15 | 20 [George Michael] *A: Faith*; P: George Michael	5187
103. CAN'T FIGHT THIS FEELING REO Speedwagon Epic 04713 [19Jan85 | 09Mar85 | 18May85] 1(3) 8 | 14 | 18 [Kevin Cronin] *A: Wheels Are Turnin'*; P: Kevin Cronin/Gary Richrath	5170
104. SOMEBODY'S WATCHING ME Rockwell Motown 1702 [28Jan84 | 24Mar84 | 02Jun84] 2(3) 8 | 14 | 19 [Kennedy Gordy] *A: Somebody's Watching Me*; P: Kennedy Gordy/Curtis Anthony Nolen	5142
105. ALONE Heart Capitol 44002 [16May87 | 11Jul87 | 03Oct87] 1(3) 8 | 15 | 21 [Thomas F. Kelly/Billy Steinberg] *A: Bad Animals*; P: Ron Nevison	5121
106. THE REFLEX Duran Duran Capitol 5345 [21Apr84 | 23Jun84 | 08Sep84] 1(2) 8 | 15 | 21 [Simon Le Bon/Nick Rhodes/Andy Taylor/Nigel John Taylor/Roger Taylor] *A: Seven And The Ragged Tiger*; P: Duran Duran/Ian Little/Alex Sadkin	5085
107. MAKING LOVE OUT OF NOTHING AT ALL Air Supply Arista 1-9056 [30Jul83 | 08Oct83 | 14Jan84] 2(3) 9 | 17 | 25 [Jim Steinman] *A: Greatest Hits*; P: Jim Steinman	5057
108. UP WHERE WE BELONG Joe Cocker and Jennifer Warnes Island 99996 [21Aug82 | 06Nov82 | 22Jan83] 1(3) 7 | 15 | 23 [Will Jennings/Jack Nitzsche/Buffy Sainte-Marie] *A: An Officer And A Gentleman Soundtrack*; P: Stewart Levine	5023
109. MISS YOU MUCH Janet Jackson A&M 1445 [02Sep89 | 07Oct89 | 13Jan90] 1(4) 8 | 13 | 20 [James Samuel Harris/Terry Lewis] *A: Janet Jackson's Rhythm Nation 1814*; P: Jimmy Jam Harris/Terry Lewis	5007
110. HARDEN MY HEART Quarterflash Geffen 49824 [17Oct81 | 13Feb82 | 27Mar82] 3(2) 12 | 19 | 24 [Marv Ross] *A: Quarterflash*; P: John Boylan	4972
111. LIVIN' ON A PRAYER Bon Jovi Mercury 888184 [13Dec86 | 14Feb87 | 02May87] 1(4) 7 | 13 | 21 [Jon Bon Jovi/Desmond Child/Richie Sambora] *A: Slippery When Wet*; P: Bruce Fairbairn	4968
112. STOP DRAGGIN' MY HEART AROUND Stevie Nicks (with Tom Petty And The Heartbreakers) Modern 7336 [25Jul81 | 05Sep81 | 12Dec81] 3(6) 10 | 15 | 21 [Mike Campbell/Tom Petty] *A: Bella Donna*; P: Jimmy Iovine/Tom Petty	4960
113. TRULY Lionel Richie Motown 1644 [09Oct82 | 27Nov82 | 05Feb83] 1(2) 8 | 13 | 18 [Lionel Richie] *A: Lionel Richie*; P: James Anthony Carmichael/Lionel Richie	4947
114. THE WILD BOYS Duran Duran Capitol 5417 [03Nov84 | 15Dec84 | 02Mar85] 2(4) 8 | 14 | 18 [Simon Le Bon/Nick Rhodes/Andy Taylor/Nigel John Taylor/Roger Taylor] *A: Arena*; P: Duran Duran/Nile Rodgers	4921
115. I WANNA DANCE WITH SOMEBODY (Who Loves Me) Whitney Houston Arista 1-9598 [16May87 | 27Jun87 | 12Sep87] 1(2) 9 | 14 | 18 [George Merrill/Shannon Rubicam] *A: Whitney*; P: Narada Michael Walden	4899
116. MONEY FOR NOTHING Dire Straits Warner Bros. 28950 [13Jul85 | 21Sep85 | 07Dec85] 1(3) 8 | 13 | 22 [Mark Knopfler/Sting] *A: Brothers In Arms*; P: Neil Dorfsman/Mark Knopfler	4894
117. GIRLS JUST WANT TO HAVE FUN Cyndi Lauper Portrait 37-04120 [17Dec83 | 10Mar84 | 02Jun84] 2(2) 8 | 14 | 25 [Robert Hazard] *A: She's So Unusual*; P: Rick Chertoff	4890
118. BROKEN WINGS Mr. Mister RCA 14136 [21Sep85 | 07Dec85 | 15Feb86] 1(2) 9 | 15 | 22 [Steve George/John Lang/Richard Page] *A: Welcome To The Real World*; P: Paul DeVilliers/Mr. Mister	4890
119. WE GOT THE BEAT Go-Go's I.R.S. 9903 [30Jan82 | 10Apr82 | 05Jun82] 2(3) 9 | 15 | 19 [Charlotte Caffey] *A: Beauty And The Beat*; P: Rob Freeman/Richard Gottehrer	4878
120. KISS ON MY LIST Daryl Hall & John Oates RCA 12142 [24Jan81 | 11Apr81 | 27Jun81] 1(3) 8 | 17 | 23 [Janna Allen/Daryl Hall] *A: Voices*; P: Daryl Hall/John Oates	4872
121. KEEP ON LOVING YOU REO Speedwagon Epic 50953 [29Nov80 | 21Mar81 | 06Jun81] 1(1) 9 | 20 | 28 [Kevin Cronin] *A: Hi Infidelity*; P: Kevin Beamish/Kevin Cronin/Gary Richrath	4866
122. SHOUT Tears For Fears Mercury 880294 [15Jun85 | 03Aug85 | 19Oct85] 1(3) 7 | 13 | 19 [Roland Orzabal/Ian Stanley] *A: Songs From The Big Chair*; P: Chris Hughes	4854
123. MEDLEY Stars On 45 Radio 3810 [11Apr81 | 20Jun81 | 29Aug81] 1(1) 8 | 14 | 21 *A: Stars On Long Play*; P: Jaap Eggermont	4837
124. EVERY ROSE HAS ITS THORN Poison Enigma 44203 [29Oct88 | 24Dec88 | 18Mar89] 1(3) 8 | 14 | 21 [Bobby Dall/C.C. DeVille/Bret Michaels/Rikki Rockett] *A: Open Up And Say...Ahh!*; P: Tom Werman	4836
125. LET'S GROOVE Earth, Wind & Fire ARC Columbia 18-02536 [03Oct81 | 19Dec81 | 13Mar82] 3(5) 9 | 16 | 24 [Wayne Vaughn/Maurice White] *A: Raise!*; P: Maurice White	4813

Ranking the Singles

Rank. TITLE Artist Label Number
[Entry Date | Peak Date | Exit Date] Peak(Peak Wks) Top10 | Top40 | Tot Wks
[Writers] Album; Producers — Score

126. COME ON EILEEN Dexys Midnight Runners Mercury 76189 — 4804
[22Jan83 | 23Apr83 | 25Jun83] 1(1) 6 | 14 | 23
[Kevin Adams/Jimmy Paterson/Kevin Rowland]
A: Too-Rye-Ay; P: Clive Langer/Alan Winstanley

127. RAPTURE Blondie Chrysalis 2485 — 4803
[31Jan81 | 28Mar81 | 13Jun81] 1(2) 8 | 14 | 20
[Clement A. Bozewski/Deborah Harry/Christopher Stein]
A: Autoamerican; P: Mike Chapman

128. SLOW HAND Pointer Sisters Planet 47929 — 4799
[30May81 | 29Aug81 | 07Nov81] 2(3) 11 | 16 | 24
[John Bettis/Michael Clark] A: Black & White; P: Richard Perry

129. SAILING Christopher Cross Warner Bros. 49507 — 4791
[14Jun80 | 30Aug80 | 01Nov80] 1(1) 7 | 13 | 21
[Christopher Cross] A: Christopher Cross; P: Michael Omartian

130. EVERYBODY WANTS TO RULE THE WORLD Tears For Fears Mercury 880659 — 4785
[16Mar85 | 08Jun85 | 24Aug85] 1(2) 8 | 14 | 24
[Chris Hughes/Roland Orzabal/Ian Stanley]
A: Songs From The Big Chair; P: Chris Hughes

131. STRAIGHT UP Paula Abdul Virgin 99256 — 4776
[03Dec88 | 11Feb89 | 20May89] 1(3) 7 | 16 | 25
[Elliot Wolff] A: Forever Your Girl; P: Elliot Wolff

132. SEPARATE LIVES (Love Theme From White Nights) Phil Collins and Marilyn Martin Atlantic 89498 — 4751
[05Oct85 | 30Nov85 | 22Feb86] 1(1) 9 | 16 | 21
[Stephen Bishop]
A: White Nights Soundtrack; P: Phil Collins/Arif Mardin/Hugh Padgham

133. LA BAMBA Los Lobos Slash 28336 — 4741
[27Jun87 | 29Aug87 | 14Nov87] 1(3) 7 | 14 | 21
[Traditional] A: La Bamba (Soundtrack); P: Mitchell Froom

134. NOTHING'S GONNA STOP US NOW Starship Grunt 5109-7 — 4720
[31Jan87 | 04Apr87 | 27Jun87] 1(2) 8 | 15 | 22
[Albert Hammond/Diane Warren]
A: No Protection; P: Narada Michael Walden

135. WORKING MY WAY BACK TO YOU/FORGIVE ME GIRL (Medley) Spinners Atlantic 3637 — 4689
[15Dec79 | 29Mar80 | 31May80] 2(2) 8 | 16 | 25
[Sandy Linzer/Denny Randell//Michael Zager]
A: Dancin' And Lovin'; P: Michael Zager

136. ONE MORE TRY George Michael Columbia 07773 — 4687
[16Apr88 | 28May88 | 13Aug88] 1(3) 7 | 14 | 18
[George Michael] A: Faith; P: George Michael

137. MY PREROGATIVE Bobby Brown MCA 53383 — 4660
[22Oct88 | 14Jan89 | 01Apr89] 1(1) 7 | 15 | 24
[Bobby Brown/Gene Griffin] A: Don't Be Cruel; P: Gene Griffin

138. THE ROSE Bette Midler Atlantic 3656 — 4647
[22Mar80 | 28Jun80 | 06Sep80] 3(3) 8 | 16 | 25
[Amanda McBroom] A: The Rose (Soundtrack); P: Paul A. Rothchild

139. PARTY ALL THE TIME Eddie Murphy Columbia 05609 — 4645
[05Oct85 | 28Dec85 | 01Mar86] 2(3) 9 | 14 | 22
[Rick James] A: How Could It Be; P: Rick James

140. THE POWER OF LOVE Huey Lewis And The News Chrysalis 42876 — 4636
[29Jun85 | 24Aug85 | 02Nov85] 1(2) 8 | 15 | 19
[Johnny Colla/Chris Hayes/Huey Lewis]
A: Back To The Future Soundtrack; P: Huey Lewis And The News

141. LOOK AWAY Chicago Reprise/Full Moon 27766 — 4616
[24Sep88 | 10Dec88 | 04Mar89] 1(2) 8 | 16 | 24
[Diane Warren] A: Chicago 19; P: Ron Nevison

142. RIGHT HERE WAITING Richard Marx EMI 50219 — 4616
[08Jul89 | 12Aug89 | 25Nov89] 1(3) 7 | 13 | 21
[Richard Marx] A: Repeat Offender; P: David Cole/Richard Marx

143. SHAKEDOWN Bob Seger MCA 53094 — 4610
[23May87 | 01Aug87 | 19Sep87] 1(1) 7 | 14 | 18
[Harold Faltermeyer/Keith Forsey/Bob Seger]
A: Beverly Hills Cop II Soundtrack; P: Harold Faltermeyer/Keith Forsey

144. DIRTY LAUNDRY Don Henley Asylum 69894 — 4595
[30Oct82 | 08Jan83 | 05Mar83] 3(3) 10 | 14 | 19
[Don Henley/Danny Kortchmar]
A: I Can't Stand Still; P: Don Henley/Danny Kortchmar/Greg Ladanyi

145. WE DIDN'T START THE FIRE Billy Joel Columbia 73021 — 4586
[14Oct89 | 09Dec89 | 17Feb90] 1(2) 8 | 15 | 19
[Billy Joel] A: Storm Front; P: Billy Joel/Mick Jones

146. GREATEST LOVE OF ALL Whitney Houston Arista 1-9466 — 4572
[29Mar86 | 17May86 | 26Jul86] 1(3) 7 | 14 | 18
[Linda Creed/Michael Masser] A: Whitney Houston; P: Michael Masser

147. SEXUAL HEALING Marvin Gaye Columbia 03302 — 4570
[30Oct82 | 29Jan83 | 19Mar83] 3(3) 10 | 15 | 21
[Odell Brown/Marvin Gaye/David Ritz] A: Midnight Love; P: Marvin Gaye

148. MORNING TRAIN (Nine To Five) Sheena Easton EMI America 8071 — 4553
[14Feb81 | 02May81 | 04Jul81] 1(2) 6 | 15 | 21
[Florrie Palmer] A: Sheena Easton; P: Christopher Neil

149. HUNGRY LIKE THE WOLF Duran Duran Harvest 5195 — 4539
[25Dec82 | 26Mar83 | 28May83] 3(3) 9 | 16 | 23
[Simon Le Bon/Nick Rhodes/Andy Taylor/Nigel John Taylor/Roger Taylor]
A: Rio; P: Colin Thurston

150. ROLL WITH IT Steve Winwood Virgin 99326 — 4536
[11Jun88 | 30Jul88 | 08Oct88] 1(4) 7 | 14 | 18
[Lamont Dozier/Brian Holland/Eddie Holland/Will Jennings/Steve Winwood]
A: Roll With It; P: Tom Lord-Alge/Steve Winwood

151. EASY LOVER (Duet With Phil Collins) Philip Bailey Columbia 04679 — 4519
[24Nov84 | 02Feb85 | 27Apr85] 2(2) 7 | 16 | 23
[Philip Bailey/Phil Collins/Nathan East] A: Chinese Wall; P: Phil Collins

152. LOST IN LOVE Air Supply Arista 0479 — 4516
[09Feb80 | 03May80 | 12Jul80] 3(4) 6 | 17 | 23
[Graham Russell] A: Lost In Love; P: Rick Chertoff/Robie Porter

153. TALKING IN YOUR SLEEP The Romantics Nemperor 04135 — 4501
[08Oct83 | 28Jan84 | 31Mar84] 3(3) 7 | 15 | 26
[Coz Canler/Jimmy Marinos/Wally Palmar/Mike Skill/Pete Solley]
A: In Heat; P: Peter Solley

154. KISS Prince And The Revolution Paisley Park 7-28751 — 4499
[22Feb86 | 19Apr86 | 21Jun86] 1(2) 7 | 13 | 18
[Prince] A: Parade: Music From The Motion Picture Under The Cherry Moon (Soundtrack); P: Prince And The Revolution

155. LIKE A PRAYER Madonna Sire 27539 — 4497
[18Mar89 | 22Apr89 | 01Jul89] 1(3) 7 | 12 | 16
[Madonna L. Ciccone/Patrick Raymond Leonard]
A: Like A Prayer; P: Patrick Leonard/Madonna

156. I FEEL FOR YOU Chaka Khan Warner Bros. 29195 — 4486
[08Sep84 | 24Nov84 | 02Mar85] 3(3) 9 | 17 | 26
[Prince] A: I Feel For You; P: Arif Mardin

157. WE BUILT THIS CITY Starship Grunt 14170 — 4486
[07Sep85 | 16Nov85 | 15Feb86] 1(2) 7 | 15 | 24
[Dennis Lambert/Martin Page/Bernie Taupin/Peter F. Wolf]
A: Knee Deep In The Hoopla; P: Jeremy Smith/Peter Wolf

158. WITH OR WITHOUT YOU U2 Island 99469 — 4480
[21Mar87 | 16May87 | 18Jul87] 1(3) 8 | 13 | 18
[Adam Clayton/Dave Evans/Paul Hewson/Larry Mullen]
A: The Joshua Tree; P: Brian Eno/Daniel Lanois

159. DON'T YOU (Forget About Me) Simple Minds A&M 2703 — 4475
[23Feb85 | 18May85 | 20Jul85] 1(1) 8 | 14 | 22
[Keith Forsey/Steve Schiff] A: The Breakfast Club Soundtrack; P: Keith Forsey

160. HOW WILL I KNOW Whitney Houston Arista 1-9434 — 4469
[07Dec85 | 15Feb86 | 10May86] 1(2) 6 | 16 | 23
[George Merrill/Shannon Rubicam/Narada Michael Walden]
A: Whitney Houston; P: Narada Michael Walden

161. KYRIE Mr. Mister RCA 14258 — 4458
[21Dec85 | 01Mar86 | 03May86] 1(2) 7 | 13 | 20
[Steve George/John Lang/Richard Page]
A: Welcome To The Real World; P: Paul DeVilliers/Mr. Mister

162. DON'T KNOW MUCH Linda Ronstadt Featuring Aaron Neville Elektra 69261 — 4448
[30Sep89 | 23Dec89 | 24Mar90] 2(2) 8 | 16 | 26
[Barry Mann/Tom Snow/Cynthia Weil]
A: Cry Like A Rainstorm, Howl Like The Wind; P: Peter Asher

163. ST. ELMO'S FIRE (Man In Motion) John Parr Atlantic 89541 — 4434
[22Jun85 | 07Sep85 | 16Nov85] 1(2) 7 | 14 | 22
[David Foster/John Parr] A: St. Elmo's Fire Soundtrack; P: David Foster

164. THE ONE THAT YOU LOVE Air Supply Arista 0604 — 4433
[16May81 | 25Jul81 | 19Sep81] 1(1) 8 | 14 | 19
[Graham Russell] A: The One That You Love; P: Harry Maslin

165. ONE MORE NIGHT Phil Collins Atlantic 89588 — 4427
[09Feb85 | 30Mar85 | 08Jun85] 1(2) 6 | 12 | 18
[Phil Collins]
A: No Jacket Required; P: Phil Collins/Hugh Padgham

Ranking the Singles

Rank. TITLE Artist Label Number
[Entry Date | Peak Date | Exit Date] Peak(Peak Wks) Top10 | Top40 | Tot Wks
[Writers] *Album*; Producers   Score

166. SWEET CHILD O' MINE Guns N' Roses Geffen 27963   4427
[25Jun88 | 10Sep88 | 03Dec88] 1(2) 7 | 14 | 24
[Steven Adler/Saul Hudson/Duff McKagan/Axl Rose/Izzy Stradlin]
A: Appetite For Destruction; P: Mike Clink

167. COWARD OF THE COUNTY Kenny Rogers United Artists 1327 4425
[17Nov79 | 26Jan80 | 22Mar80] 3(4) 8 | 15 | 19
[Roger Bowling/Billy Edd Wheeler] *A: Kenny*; P: Larry Butler

168. BLAME IT ON THE RAIN Milli Vanilli Arista 1-9904   4409
[07Oct89 | 25Nov89 | 10Mar90] 1(2) 6 | 14 | 23
[Diane Warren] *A: Girl You Know It's True*; P: Frank Farian

169. SHE WORKS HARD FOR THE MONEY Donna Summer Mercury 812370 4407
[28May83 | 06Aug83 | 15Oct83] 3(3) 8 | 17 | 21
[Michael Omartian/Donna Summer]
A: She Works Hard For The Money; P: Michael Omartian

170. HOLD ME Fleetwood Mac Warner Bros. 29966   4397
[19Jun82 | 24Jul82 | 09Oct82] 4(7) 10 | 15 | 17
[Christine McVie/Robbie Patton]
A: Mirage; P: Lindsey Buckingham/Ken Caillat/Richard Dashut/Fleetwood Mac

171. COULD'VE BEEN Tiffany MCA 53231   4391
[28Nov87 | 06Feb88 | 09Apr88] 1(2) 6 | 14 | 20
[Lois Ann Blaisch] *A: Tiffany*; P: George Tobin

172. TIME (Clock Of The Heart) Culture Club Virgin 03796   4390
[16Apr83 | 18Jun83 | 13Aug83] 2(2) 9 | 13 | 18
[Michael Craig/Roy Hay/Jon Moss/George O'Dowd]
A: Kissing To Be Clever; P: Steve Levine

173. LOST IN YOUR EYES Debbie Gibson Atlantic 88970   4389
[21Jan89 | 04Mar89 | 27May89] 1(3) 7 | 12 | 19
[Debbie Gibson] *A: Electric Youth*; P: Debbie Gibson

174. THE BEST OF TIMES Styx A&M 2300   4383
[24Jan81 | 21Mar81 | 30May81] 3(4) 10 | 15 | 19
[Dennis DeYoung] *A: Paradise Theater*; P: Styx

175. TELL HER ABOUT IT Billy Joel Columbia 04012   4357
[30Jul83 | 24Sep83 | 26Nov83] 1(1) 7 | 15 | 18
[Billy Joel] *A: An Innocent Man*; P: Phil Ramone

176. YES, I'M READY Teri DeSario Casablanca 2227   4354
[17Nov79 | 01Mar80 | 19Apr80] 2(2) 7 | 16 | 23
[Barbara Mason] *A: Moonlight Madness*; P: Harry Wayne Casey

177. PART-TIME LOVER Stevie Wonder Tamla 1808   4348
[07Sep85 | 02Nov85 | 25Jan86] 1(1) 8 | 14 | 21
[Stevie Wonder] *A: In Square Circle*; P: Stevie Wonder

178. PAPA DON'T PREACH Madonna Sire 28660   4347
[28Jun86 | 16Aug86 | 25Oct86] 1(2) 7 | 13 | 18
[Madonna L. Ciccone/Brian Elliot] *A: True Blue*; P: Stephen Bray/Madonna

179. SO EMOTIONAL Whitney Houston Arista 1-9642   4343
[31Oct87 | 09Jan88 | 05Mar88] 1(1) 8 | 14 | 19
[Thomas F. Kelly/Billy Steinberg] *A: Whitney*; P: Narada Michael Walden

180. HERE I GO AGAIN Whitesnake Geffen 28339   4336
[04Jul87 | 10Oct87 | 09Jan88] 1(1) 7 | 14 | 28
[David Coverdale/Bernie Marsden] *A: Whitesnake*; P: Keith Olsen

181. WEST END GIRLS Pet Shop Boys EMI America 8307   4327
[01Mar86 | 10May86 | 12Jul86] 1(2) 7 | 14 | 20
[Chris Lowe/Neil Tennant] *A: Please*; P: Stephen Hague

182. THE SAFETY DANCE Men Without Hats Backstreet 52232   4322
[25Jun83 | 10Sep83 | 03Dec83] 3(4) 7 | 16 | 24
[Ivan Doroschuk] *A: Rhythm Of Youth*; P: Marc Durand

183. EYE IN THE SKY Alan Parsons Project Arista 0696   4321
[03Jul82 | 16Oct82 | 18Dec82] 3(3) 8 | 17 | 25
[Alan Parsons/Eric Woolfson] *A: Eye In The Sky*; P: Alan Parsons

184. NEVER GONNA GIVE YOU UP Rick Astley RCA 5347   4313
[19Dec87 | 12Mar88 | 28May88] 1(2) 7 | 14 | 24
[Matt Aitken/Mike Stock/Pete Waterman]
A: Whenever You Need Somebody; P: Matt Aitken/Mike Stock/Pete Waterman

185. QUEEN OF HEARTS Juice Newton Capitol 4997   4309
[30May81 | 19Sep81 | 28Nov81] 2(2) 10 | 19 | 27
[Hank DeVito] *A: Juice*; P: Richard Landis

186. STUCK WITH YOU Huey Lewis And The News Chrysalis 43019 4301
[02Aug86 | 20Sep86 | 06Dec86] 1(3) 7 | 13 | 19
[Chris Hayes/Huey Lewis] *A: Fore!*; P: Huey Lewis And The News

187. A VIEW TO A KILL Duran Duran Capitol 5475   4298
[18May85 | 13Jul85 | 07Sep85] 1(2) 6 | 13 | 17 [John Barry/Simon Le Bon/
Nick Rhodes/Andy Taylor/Nigel John Taylor/Roger Taylor]
A: A View To A Kill Soundtrack; P: Jason Corsaro/Duran Duran/Bernard Edwards

188. JEOPARDY Greg Kihn Band Beserkley 69847   4294
[29Jan83 | 07May83 | 25Jun83] 2(1) 7 | 14 | 22
[Greg Kihn/Steve Wright] *A: Kihnspiracy*; P: Matthew King Kaufman

189. SHAKE YOU DOWN Gregory Abbott Columbia 06191   4286
[18Oct86 | 17Jan87 | 14Mar87] 1(1) 8 | 16 | 22
[Gregory Abbott] *A: Shake You Down*; P: Gregory Abbott

190. ROCK ME AMADEUS Falco A&M 2821   4285
[08Feb86 | 29Mar86 | 31May86] 1(3) 7 | 13 | 17
[Ferdi Bolland/Rob Bolland/Falco]
A: Falco 3; P: Ferdi Bolland/Rob Bolland

191. EVERYTIME YOU GO AWAY Paul Young Columbia 04867   4283
[11May85 | 27Jul85 | 12Oct85] 1(1) 8 | 15 | 23
[Daryl Hall] *A: The Secret Of Association*;
P: Laurie Latham

192. 99 LUFTBALLONS Nena Epic 04108   4276
[10Dec83 | 03Mar84 | 12May84] 2(1) 6 | 13 | 23
[Jörn Fahrenkrog-Petersen/Carlo Karges]
A: 99 Luftballons; P: Reinhold Heil/Manne Praeker

193. LIVE TO TELL Madonna Sire 28717   4244
[12Apr86 | 07Jun86 | 09Aug86] 1(1) 6 | 13 | 18
[Madonna L. Ciccone/Patrick Raymond Leonard]
A: True Blue; P: Patrick Leonard/Madonna

194. LONGER Dan Fogelberg Full Moon/Epic 50824   4243
[15Dec79 | 15Mar80 | 10May80] 2(2) 7 | 13 | 22
[Dan Fogelberg] *A: Phoenix*; P: Dan Fogelberg/Marty Lewis/Norbert Putnam

195. CRUISIN' Smokey Robinson Tamla 54306   4240
[06Oct79 | 02Feb80 | 22Mar80] 4(4) 7 | 17 | 25
[Smokey Robinson/Marvin Tarplin]
A: Where There's Smoke; P: Smokey Robinson

196. GIRL I'M GONNA MISS YOU Milli Vanilli Arista 1-9870   4238
[05Aug89 | 23Sep89 | 30Dec89] 1(2) 6 | 14 | 22
[Peter Bischof-Fallenstein/Frank Farian/Dietmar Kawohl]
A: Girl You Know It's True; P: Frank Farian

197. WILD, WILD WEST The Escape Club Atlantic 89048   4208
[20Aug88 | 12Nov88 | 18Feb89] 1(1) 7 | 16 | 27
[Johnnie Christo/John Holliday/Amos Pizzey/Trevor Steel/Milan Zekavica]
A: Wild, Wild West; P: Chris Kimsey

198. COLD HEARTED Paula Abdul Virgin 99196   4158
[24Jun89 | 02Sep89 | 11Nov89] 1(1) 8 | 15 | 21
[Elliot Wolff] *A: Forever Your Girl*; P: Elliot Wolff

199. SARA Starship Grunt 14253   4156
[28Dec85 | 15Mar86 | 10May86] 1(1) 7 | 13 | 20
[Ina Wolf/Peter F. Wolf] *A: Knee Deep In The Hoopla*;
P: Jeremy Smith/Peter Wolf

200. HEAD TO TOE Lisa Lisa And Cult Jam Columbia 07008   4139
[11Apr87 | 20Jun87 | 22Aug87] 1(1) 7 | 17 | 20
[Curtis Bedeau/Gerard Charles/Hugh Clarke/Brian George/
Lucien George/Paul George] *A: Spanish Fly*; P: Full Force

201. HEAVEN Bryan Adams A&M 2729   4136
[20Apr85 | 22Jun85 | 24Aug85] 1(2) 6 | 14 | 19
[Bryan Adams/Jim Vallance] *A: Reckless*; P: Bryan Adams/Bob Clearmountain

202. LOOKING FOR A NEW LOVE Jody Watley MCA 52956   4128
[07Mar87 | 02May87 | 11Jul87] 2(4) 6 | 14 | 19
[André Cymone/Jody Watley] *A: Jody Watley*; P: André Cymone/David Rivkin

203. THERE'LL BE SAD SONGS (To Make You Cry) Billy Ocean Jive 9465 4121
[19Apr86 | 05Jul86 | 06Sep86] 1(1) 7 | 14 | 21
[Wayne Brathwaite/Barry Eastmond/Billy Ocean]
A: Love Zone; P: Wayne Brathwaite/Barry Eastmond

204. AFRICA Toto Columbia 03335   4117
[30Oct82 | 05Feb83 | 19Mar83] 1(1) 6 | 16 | 21
[David Paich/Jeff Porcaro] *A: Toto IV*; P: Toto

205. (I've Had) THE TIME OF MY LIFE Bill Medley And Jennifer Warnes 4114
RCA 5224 [28Sep87 | 28Nov87 | 13Feb88] 1(1) 6 | 15 | 21
[John DeNicola/Donald Markowitz/Franke Previte]
A: Dirty Dancing Soundtrack; P: Michael Lloyd

206. HEAVEN IS A PLACE ON EARTH Belinda Carlisle MCA 53181 4109
[26Sep87 | 05Dec87 | 13Feb88] 1(1) 6 | 15 | 21
[Rick Nowels/Ellen Shipley] *A: Heaven On Earth*; P: Rick Nowels

207. LEAN ON ME Club Nouveau King Jay/Warner 28430   4095
[14Feb87 | 21Mar87 | 06Jun87] 1(2) 6 | 12 | 17
[Bill Withers] *A: Life, Love & Pain*;
P: Denzil Foster/Jay King/Thomas McElroy

Ranking the Singles

Rank. TITLE Artist Label Number [Entry Date | Peak Date | Exit Date] Peak(Peak Wks) Top10 | Top40 | Tot Wks [Writers] Album; Producers	Score
208. ALWAYS Atlantic Starr Warner Bros. 28455 [28Mar87 | 13Jun87 | 22Aug87] 1(1) 7 | 14 | 22 [David E. Lewis/Jonathan Lewis/Wayne I. Lewis] A: All In The Name Of Love; P: David Lewis/Wayne Lewis	4081
209. ADDICTED TO LOVE Robert Palmer Island 99570 [08Feb86 | 03May86 | 05Jul86] 1(1) 7 | 14 | 22 [Robert Palmer] A: Riptide; P: Bernard Edwards	4061
210. GLORY OF LOVE (Theme From The Karate Kid Part II) Peter Cetera Full Moon 28662 [07Jun86 | 02Aug86 | 25Oct86] 1(2) 6 | 14 | 21 [Peter Cetera/David Foster] A: The Karate Kid Part II Soundtrack; P: Michael Omartian	4056
211. EVERYTHING SHE WANTS (Remix) Wham! Columbia 04840 [23Mar85 | 25May85 | 03Aug85] 1(2) 6 | 14 | 20 [George Michael] A: Make It Big; P: George Michael	4050
212. HUMAN The Human League A&M/Virgin 2861 [13Sep86 | 22Nov86 | 24Jan87] 1(1) 7 | 15 | 20 [James Samuel Harris/Terry Lewis] A: Crash; P: Jimmy Jam Harris/Terry Lewis	4045
213. SUSSUDIO Phil Collins Atlantic 89560 [11May85 | 06Jul85 | 31Aug85] 1(1) 6 | 14 | 17 [Phil Collins] A: No Jacket Required; P: Phil Collins/Hugh Padgham	4040
214. HOLD ME NOW Thompson Twins Arista 1-9164 [11Feb84 | 05May84 | 30Jun84] 3(2) 7 | 15 | 21 [Tom Bailey/Alannah Currie/Joe Leeway] A: Into The Gap; P: Tom Bailey/Alex Sadkin	4030
215. HEART ATTACK Olivia Newton-John MCA 52100 [04Sep82 | 06Nov82 | 22Jan83] 3(4) 7 | 13 | 21 [Paul Steven Bliss/Steve Kipner] A: Olivia's Greatest Hits, Vol. 2; P: John Farrar	4023
216. THE WAY IT IS Bruce Hornsby And The Range RCA Victor 5023 [20Sep86 | 13Dec86 | 14Feb87] 1(1) 8 | 15 | 22 [Bruce Hornsby] A: The Way It Is; P: Bruce Hornsby/Elliot Scheiner	4013
217. EMOTIONAL RESCUE The Rolling Stones Rolling Stones 20001 [05Jul80 | 06Sep80 | 08Nov80] 3(2) 7 | 14 | 19 [Mick Jagger/Keith Richards] A: Emotional Rescue; P: Glimmer Twins	4012
218. DIDN'T WE ALMOST HAVE IT ALL Whitney Houston Arista 1-9616 [01Aug87 | 26Sep87 | 21Nov87] 1(2) 7 | 13 | 17 [Will Jennings/Michael Masser] A: Whitney; P: Michael Masser	4007
219. I STILL HAVEN'T FOUND WHAT I'M LOOKING FOR U2 Island 99430 [13Jun87 | 08Aug87 | 03Oct87] 1(2) 7 | 13 | 17 [Adam Clayton/Dave Evans/Paul Hewson/Larry Mullen] A: The Joshua Tree; P: Brian Eno/Daniel Lanois	4006
220. GOT MY MIND SET ON YOU George Harrison Dark Horse 28178 [24Oct87 | 16Jan88 | 19Mar88] 1(1) 8 | 15 | 22 [Rudy Clark] A: Cloud Nine; P: George Harrison/Jeff Lynne	4004
221. BIGGEST PART OF ME Ambrosia Warner Bros. 49225 [05Apr80 | 07Jun80 | 09Aug80] 3(3) 8 | 14 | 19 [David Pack] A: One Eighty; P: Ambrosia/Freddie Piro	4001
222. GET OUTTA MY DREAMS, GET INTO MY CAR Billy Ocean Jive 9678 [13Feb88 | 09Apr88 | 25Jun88] 1(2) 7 | 14 | 20 [Mutt Lange/Billy Ocean] A: Tear Down These Walls; P: Mutt Lange	3998
223. NEED YOU TONIGHT INXS Atlantic 89188 [24Oct87 | 30Jan88 | 09Apr88] 1(1) 8 | 17 | 25 [Andy Farriss/Michael Hutchence] A: Kick; P: Chris Thomas	3984
224. FRIENDS AND LOVERS Gloria Loring & Carl Anderson USA Carrere 4-06122 [05Jul86 | 27Sep86 | 22Nov86] 2(2) 7 | 14 | 21 [Paul Gordon/Jay Gruska] A: Gloria Loring; P: Yves Dessca	3983
225. THE LOOK Roxette EMI 50190 [11Feb89 | 08Apr89 | 17Jun89] 1(1) 7 | 13 | 19 [Per Gessle] A: Look Sharp!; P: Clarence Ofwerman	3981
226. ANYTHING FOR YOU Gloria Estefan and Miami Sound Machine Epic 07759 [12Mar88 | 14May88 | 13Aug88] 1(2) 7 | 14 | 23 [Gloria Estefan] A: Let It Loose; P: Emilio Estefan	3981
227. MAN IN THE MIRROR Michael Jackson Epic 07668 [06Feb88 | 26Mar88 | 28May88] 1(2) 7 | 13 | 17 [Glen Ballard/Siedah Garrett] A: Bad; P: Quincy Jones	3974
228. PURPLE RAIN Prince And The Revolution Warner Bros. 29174 [06Oct84 | 17Nov84 | 19Jan85] 2(2) 7 | 11 | 16 [Prince] A: Purple Rain (Soundtrack); P: Prince	3959

Rank. TITLE Artist Label Number [Entry Date | Peak Date | Exit Date] Peak(Peak Wks) Top10 | Top40 | Tot Wks [Writers] Album; Producers	Score
229. CHERISH Kool & The Gang De-Lite 880869 [06Jul85 | 21Sep85 | 21Dec85] 2(3) 7 | 15 | 25 [Robert Bell/Ronald Bell/James Bonneford/George M. Brown/ Claydes Smith/J.T. Taylor/Curtis Fitzgerald Williams] A: Emergency; P: Ronald Bell/Jim Bonneford/Kool & The Gang	3953
230. TAKE ON ME A-Ha Warner Bros. 29011 [13Jul85 | 19Oct85 | 11Jan86] 1(1) 7 | 15 | 27 [Magne Furuholmen/Morten Harket/Pal Waaktaar] A: Hunting High And Low; P: Alan Tarney	3952
231. SUKIYAKI A Taste Of Honey Capitol 4953 [07Mar81 | 13Jun81 | 15Aug81] 3(3) 8 | 16 | 24 [Janice-Marie Johnson/Hachidai Nakamura] A: Twice As Sweet; P: George Duke	3949
232. AT THIS MOMENT (Live) ('86) Billy Vera & The Beaters Rhino 74403 [08Nov86 | 24Jan87 | 28Mar87] 1(2) 6 | 15 | 21 [Billy Vera] A: By Request (The Best of Billy Vera & The Beaters); P: Jeff Baxter	3948
233. SAVING ALL MY LOVE FOR YOU Whitney Houston Arista 1-9381 [17Aug85 | 26Oct85 | 11Jan86] 1(1) 7 | 15 | 22 [Gerry Goffin/Michael Masser] A: Whitney Houston; P: Michael Masser	3944
234. NEVER GONNA LET YOU GO Sergio Mendes A&M 2540 [16Apr83 | 09Jul83 | 17Sep83] 4(4) 8 | 16 | 23 [Barry Mann/Cynthia Weil] A: Sergio Mendes(2); P: Sergio Mendes	3938
235. TWO HEARTS Phil Collins Atlantic 88980 [19Nov88 | 21Jan89 | 18Mar89] 1(2) 7 | 13 | 18 [Phil Collins/Lamont Dozier] A: Buster Soundtrack; P: Phil Collins/Lamont Dozier	3934
236. OVERKILL Men At Work Columbia 03795 [09Apr83 | 04Jun83 | 23Jul83] 3(1) 8 | 13 | 16 [Colin James Hay] A: Cargo; P: Peter Mclan	3933
237. GROOVY KIND OF LOVE Phil Collins Atlantic 89017 [03Sep88 | 22Oct88 | 18Feb89] 1(2) 6 | 13 | 25 [Carole Bayer Sager/Toni Wine] A: Buster Soundtrack; P: Phil Collins/Anne Dudley	3922
238. MIAMI VICE THEME Jan Hammer MCA 52666 [07Sep85 | 09Nov85 | 01Feb86] 1(1) 7 | 13 | 22 [Jan Hammer] A: Miami Vice Soundtrack; P: Jan Hammer	3913
239. UNION OF THE SNAKE Duran Duran Capitol 5290 [05Nov83 | 24Dec83 | 25Feb84] 3(3) 6 | 12 | 17 [Simon Le Bon/Nick Rhodes/Andy Taylor/Nigel John Taylor/Roger Taylor] A: Seven And The Ragged Tiger; P: Alex Sadkin	3910
240. WHEN I THINK OF YOU Janet Jackson A&M 2855 [09Aug86 | 11Oct86 | 13Dec86] 1(2) 6 | 13 | 19 [James Samuel Harris/Terry Lewis] A: Control; P: Jimmy Jam Harris/Terry Lewis	3899
241. THESE DREAMS Heart Capitol 5541 [18Jan86 | 22Mar86 | 31May86] 1(1) 6 | 13 | 20 [Martin Page/Bernie Taupin] A: Heart; P: Ron Nevison	3896
242. I THINK WE'RE ALONE NOW Tiffany MCA 53167 [29Aug87 | 07Nov87 | 06Feb88] 1(2) 6 | 13 | 24 [Ritchie Cordell] A: Tiffany; P: George Tobin	3894
243. I'LL BE THERE FOR YOU Bon Jovi Mercury 872564 [04Mar89 | 13May89 | 29Jul89] 1(1) 6 | 13 | 22 [Jon Bon Jovi/Richie Sambora] A: New Jersey; P: Bruce Fairbairn	3890
244. SLEDGEHAMMER Peter Gabriel Geffen 28718 [10May86 | 26Jul86 | 27Sep86] 1(1) 7 | 14 | 21 [Peter Gabriel] A: So; P: Peter Gabriel/Daniel Lanois	3883
245. C'EST LA VIE Robbie Nevil EMI 50047 [11Oct86 | 17Jan87 | 14Mar87] 2(2) 7 | 16 | 23 [Mark Holding/Robbie Nevil/Duncan Pain] A: Robbie Nevil; P: Alex Sadkin	3882
246. MONY MONY "LIVE" Billy Idol Chrysalis 43161 [05Sep87 | 21Nov87 | 30Jan88] 1(1) 6 | 12 | 22 [Bobby Bloom/Ritchie Cordell/Bo Gentry/Tommy James] A: Vital Idol; P: Keith Forsey	3879
247. JOANNA Kool & The Gang De-Lite 829 [05Nov83 | 11Feb84 | 14Apr84] 2(1) 6 | 16 | 24 [Clifford Adams/Robert Bell/Ronald Bell/James Bonneford/George M. Brown/Claydes Smith/J.T. Taylor/Curtis Fitzgerald Williams] A: In The Heart; P: Ronald Bell/Jim Bonneford/Kool & The Gang	3878

Rank. TITLE Artist Label Number [Entry Date | Peak Date | Exit Date] Peak(Peak Wks) Top10 | Top40 | Tot Wks [Writers] *Album; Producers* Score	Rank. TITLE Artist Label Number [Entry Date | Peak Date | Exit Date] Peak(Peak Wks) Top10 | Top40 | Tot Wks [Writers] *Album; Producers* Score
248. DANCING ON THE CEILING Lionel Richie Motown 1843 3868 [19Jul86 | 13Sep86 | 08Nov86] 2(2) 8 | 14 | 17 [Michael Frenchik/Lionel Richie/Carlos Rios] A: *Dancing On The Ceiling*; P: *James Anthony Carmichael/Lionel Richie*	**268. WITH YOU I'M BORN AGAIN** Billy Preston & Syreeta Motown 1477 3728 [08Dec79 | 08Mar80 | 21Jun80] 4(4) 6 | 15 | 29 [Carol Connors/David Shire] A: *Fast Break Soundtrack*; P: *James Di Pasquale/David Shire*
249. EVERYBODY HAVE FUN TONIGHT Wang Chung Geffen 28562 3858 [04Oct86 | 27Dec86 | 21Feb87] 2(2) 7 | 15 | 21 [Nick Feldman/Jack Hues/Peter F. Wolf] A: *Mosaic*; P: *Peter Wolf*	**269. KOKOMO** The Beach Boys Elektra 69385 3725 [03Sep88 | 05Nov88 | 11Mar89] 1(1) 5 | 15 | 28 [Mike Love/Scott McKenzie/Terry Melcher/John Phillips] A: *Cocktail Soundtrack*; P: *Terry Melcher*
250. FATHER FIGURE George Michael Columbia 07682 3850 [16Jan88 | 27Feb88 | 07May88] 1(2) 6 | 13 | 17 [George Michael] A: *Faith*; P: *George Michael*	**270. DON'T DREAM IT'S OVER** Crowded House Capitol 5614 3721 [17Jan87 | 25Apr87 | 27Jun87] 2(1) 7 | 15 | 24 [Neil Finn] A: *Crowded House*; P: *Mitchell Froom*
251. HE'S SO SHY Pointer Sisters Planet 47916 3840 [26Jul80 | 25Oct80 | 17Jan81] 3(3) 5 | 17 | 26 [Tom Snow/Cynthia Weil] A: *Special Things*; P: *Richard Perry*	**271. MR. ROBOTO** Styx A&M 2525 3714 [12Feb83 | 16Apr83 | 11Jun83] 3(2) 8 | 16 | 18 [Dennis DeYoung] A: *Kilroy Was Here*; P: *Styx*
252. WIND BENEATH MY WINGS Bette Midler Atlantic 88972 3826 [04Mar89 | 10Jun89 | 16Sep89] 1(1) 7 | 15 | 29 [Larry Henley/Jeff Silbar] A: *Beaches (Soundtrack)*; P: *Arif Mardin*	**272. HARD HABIT TO BREAK** Chicago Warner/Full Moon 29214 3712 [04Aug84 | 20Oct84 | 19Jan85] 3(2) 6 | 15 | 25 [Steve Kipner/John Lewis Parker] A: *Chicago 17*; P: *David Foster*
253. ON OUR OWN Bobby Brown MCA 53662 3811 [10Jun89 | 05Aug89 | 21Oct89] 2(3) 6 | 13 | 20 [Kenneth Edmonds/Antonio Reid/Daryl Simmons] A: *Dance!...Ya Know It!*; P: *Kenneth Edmonds/L.A. Reid*	**273. THE LIVING YEARS** Mike + The Mechanics Atlantic 88964 3709 [07Jan89 | 25Mar89 | 20May89] 1(1) 6 | 14 | 20 [Brian A. Robertson/Mike Rutherford] A: *Living Years*; P: *Christopher Neil/Mike Rutherford*
254. BURNING HEART Survivor Scotti Brothers 05663 3808 [02Nov85 | 01Feb86 | 29Mar86] 2(2) 6 | 16 | 22 [Jim Peterik/Frankie Sullivan] A: *Rocky IV Soundtrack*; P: *Jim Peterik/Frankie Sullivan*	**274. GIRL YOU KNOW IT'S TRUE** Milli Vanilli Arista 1-9781 3708 [07Jan89 | 01Apr89 | 01Jul89] 2(1) 7 | 15 | 26 [Kayode Adeyemo/Rodney Hollaman/Kevin Lyles/William Pettaway/ Sean Spencer] A: *Girl You Know It's True*; P: *Frank Farian*
255. AMANDA Boston MCA 52756 3807 [27Sep86 | 08Nov86 | 24Jan87] 1(2) 6 | 12 | 18 [Tom Scholz] A: *Third Stage*; P: *Tom Scholz*	**275. YOU GIVE LOVE A BAD NAME** Bon Jovi Mercury 884953 3706 [06Sep86 | 29Nov86 | 14Feb87] 1(1) 6 | 14 | 24 [Jon Bon Jovi/Desmond Child/Richie Sambora] A: *Slippery When Wet*; P: *Bruce Fairbairn*
256. THE FLAME Cheap Trick Epic 07745 3801 [09Apr88 | 09Jul88 | 08Oct88] 1(2) 6 | 14 | 27 [Alan Nicholas Graham/Bobby Mitchell] A: *Lap Of Luxury*; P: *Richie Zito*	**276. VENUS** Bananarama London 886056 3701 [28Jun86 | 06Sep86 | 01Nov86] 1(1) 7 | 12 | 19 [Robbie Van Leeuwen] A: *True Confessions*; P: *Matt Aitken/Mike Stock/Pete Waterman*
257. WHEN I SEE YOU SMILE Bad English Epic 69082 3797 [16Sep89 | 11Nov89 | 10Feb90] 1(2) 6 | 15 | 22 [Diane Warren] A: *Bad English*; P: *Richie Zito*	**277. DON'T WANNA LOSE YOU** Gloria Estefan Epic 68959 3701 [08Jul89 | 16Sep89 | 04Nov89] 1(1) 8 | 13 | 18 [Gloria Estefan] A: *Cuts Both Ways*; P: *Jorge Casas/Emilio Estefan/Clay Ostwald*
258. RED RED WINE ('88) UB40 A&M 1244 3793 [13Aug88 | 15Oct88 | 28Jan89] 1(1) 6 | 12 | 25 [Neil Diamond] A: *Labour Of Love*; P: *Ray Falconer/UB40*	**278. JUMP (For My Love)** Pointer Sisters Planet 13780 3698 [28Apr84 | 07Jul84 | 06Oct84] 3(2) 8 | 15 | 24 [Stephen Mitchell/Marti Sharron/Gary Skardina] A: *Break Out*; P: *Richard Perry*
259. (A) BABY, I LOVE YOUR WAY/(B) FREEBIRD MEDLEY (FREE BABY) 3786 Will To Power Epic 08034 [10Sep88 | 03Dec88 | 18Feb89] 1(1) 6 | 15 | 24 [Peter Frampton//Allen Collins/Ronnie Van Zant] A: *Will To Power*; P: *Bob Rosenberg*	**279. I'LL BE LOVING YOU (Forever)** New Kids On The Block Columbia 68671 3690 [01Apr89 | 17Jun89 | 19Aug89] 1(1) 6 | 14 | 21 [Maurice Starr] A: *Hangin' Tough*; P: *Michael Jonzun/Maurice Starr*
260. MATERIAL GIRL Madonna Sire 29083 3774 [09Feb85 | 23Mar85 | 01Jun85] 2(2) 6 | 12 | 17 [Peter H. Brown/Robert Rans] A: *Like A Virgin*; P: *Nile Rodgers*	**280. YOU KEEP ME HANGIN' ON** Kim Wilde MCA 53024 3690 [28Mar87 | 06Jun87 | 15Aug87] 1(1) 6 | 13 | 21 [Lamont Dozier/Brian Holland/Eddie Holland] A: *Another Step*; P: *Ricky Wilde*
261. SHE BOP Cyndi Lauper Portrait 37-04516 3772 [21Jul84 | 08Sep84 | 17Nov84] 3(3) 8 | 14 | 18 [Rick Chertoff/Gary Corbett/Cyndi Lauper/Stephen Lunt] A: *She's So Unusual*; P: *Rick Chertoff*	**281. (I Just) DIED IN YOUR ARMS** Cutting Crew Virgin 99481 3669 [07Mar87 | 02May87 | 11Jul87] 1(2) 6 | 15 | 25 [Nick Van Eede] A: *Broadcast*; P: *Terry Brown/Cutting Crew/John Jansen*
262. WISHING WELL Terence Trent D'Arby Columbia 07675 3764 [16Jan88 | 07May88 | 02Jul88] 1(1) 6 | 15 | 25 [Terence Trent D'Arby/Sean Oliver] A: *Introducing The Hardline According To Terence Trent D'Arby*; P: *Terence Trent D'Arby/Martyn Ware*	**282. FOREVER YOUR GIRL** Paula Abdul Virgin 99230 3654 [11Mar89 | 20May89 | 05Aug89] 1(2) 6 | 14 | 22 [Oliver Leiber] A: *Forever Your Girl*; P: *Oliver Leiber*
263. HEAVEN Warrant Columbia 68985 3755 [22Jul89 | 23Sep89 | 25Nov89] 2(2) 7 | 14 | 19 [Joseph Cagle/Steven Chamberlin/Jerry Dixon/John Oswald/Eric Turner] A: *Dirty Rotten Filthy Stinking Rich*; P: *Beau Hill*	**283. TRUE COLORS** Cyndi Lauper Portrait 37-06247 3650 [30Aug86 | 25Oct86 | 10Jan87] 1(2) 6 | 12 | 20 [Thomas F. Kelly/Billy Steinberg] A: *True Colors*; P: *Cyndi Lauper/Lennie Petze*
264. HIGHER LOVE Steve Winwood Island 28710 3755 [14Jun86 | 30Aug86 | 08Nov86] 1(1) 6 | 14 | 22 [Will Jennings/Steve Winwood] A: *Back In The High Life*; P: *Russ Titelman/Steve Winwood*	**284. NOTORIOUS** Duran Duran Capitol 5648 3641 [01Nov86 | 10Jan87 | 21Feb87] 2(1) 6 | 13 | 17 [Simon Le Bon/Nick Rhodes/Nigel John Taylor] A: *Notorious*; P: *Duran Duran/Nile Rodgers*
265. DON'T WORRY BE HAPPY (Edit) Bobby McFerrin EMI-Manhattan 50146 3746 [30Jul88 | 24Sep88 | 21Jan89] 1(2) 6 | 13 | 26 [Bobby McFerrin] A: *Simple Pleasures*; P: *Linda Goldstein*	**285. SHE DRIVES ME CRAZY** Fine Young Cannibals I.R.S./MCA 53483 3641 [28Jan89 | 15Apr89 | 01Jul89] 1(1) 7 | 14 | 23 [Roland Gift/David Steele] A: *The Raw & The Cooked*; P: *Fine Young Cannibals/David Rivkin*
266. THE HEAT IS ON Glenn Frey MCA 52512 3741 [08Dec84 | 16Mar85 | 18May85] 2(1) 6 | 13 | 24 [Harold Faltermeyer/Keith Forsey] A: *Beverly Hills Cop Soundtrack*; P: *Harold Faltermeyer/Keith Forsey*	**286. DESIRE** Andy Gibb RSO 1019 3634 [26Jan80 | 08Mar80 | 03May80] 4(4) 8 | 12 | 15 [Barry Gibb/Maurice Gibb/Robin Gibb] A: *After Dark*; P: *Albhy Galuten/Barry Gibb/Karl Richardson*
267. LOVE BITES Def Leppard Mercury 870402 3728 [13Aug88 | 08Oct88 | 14Jan89] 1(1) 6 | 13 | 23 [Rick Allen/Steve Clark/Phil Collen/Joe Elliott/Mutt Lange/Rick Savage] A: *Hysteria*; P: *Mutt Lange*	**287. SHATTERED DREAMS** Johnny Hates Jazz Virgin 99383 3615 [19Mar88 | 14May88 | 23Jul88] 2(3) 6 | 13 | 19 [Clark Datchler] A: *Turn Back The Clock*; P: *Calvin Hayes/Mike Nocito*

Ranking the Singles

Rank. TITLE Artist Label Number
[Entry Date | Peak Date | Exit Date] Peak(Peak Wks) Top10 | Top40 | Tot Wks
[Writers] *Album; Producers*   Score

288. ALL THOSE YEARS AGO George Harrison Dark Horse 49725   3615
[23May81 | 04Jul81 | 05Sep81] 2(3) 6 | 11 | 16
[George Harrison] *A: Somewhere In England; P: Ray Cooper/George Harrison*

289. LISTEN TO YOUR HEART Roxette EMI 50223   3612
[26Aug89 | 04Nov89 | 20Jan90] 1(1) 6 | 14 | 22
[Per Gessle/Mats Persson] *A: Look Sharp!; P: Clarence Ofwerman*

290. YOU BELONG TO THE CITY Glenn Frey MCA 52651   3607
[14Sep85 | 16Nov85 | 01Feb86] 2(2) 7 | 13 | 21
[Glenn Frey/Jack Tempchin] *A: Miami Vice Soundtrack; P: Glenn Frey*

291. ETERNAL FLAME Bangles Columbia 68533   3603
[04Feb89 | 01Apr89 | 10Jun89] 1(1) 6 | 14 | 19
[Susanna Hoffs/Thomas F. Kelly/Billy Steinberg]
A: Everything; P: Davitt Sigerson

292. CUPID/I'VE LOVED YOU FOR A LONG TIME (Medley) Spinners   3595
Atlantic 3664
[17May80 | 19Jul80 | 20Sep80] 4(3) 7 | 14 | 19
[Sam Cooke//Michael Zager] *A: Love Trippin'; P: Michael Zager*

293. ALIVE & KICKING Simple Minds A&M 2783   3576
[19Oct85 | 28Dec85 | 01Mar86] 3(2) 6 | 16 | 20
[Charles Burchill/Jim Kerr/Michael MacNeil]
A: Once Upon A Time; P: Bob Clearmountain/Jimmy Iovine

294. OPEN YOUR HEART Madonna Sire 28508   3575
[06Dec86 | 07Feb87 | 04Apr87] 1(1) 6 | 14 | 18
[Madonna L. Ciccone/Gardner Cole/Peter Rafelson]
A: True Blue; P: Patrick Leonard/Madonna

295. TOY SOLDIERS Martika Columbia 68747   3573
[20May89 | 22Jul89 | 30Sep89] 1(2) 6 | 13 | 20
[Michael Jay/Martika] *A: Martika; P: Michael Jay*

296. WHERE DO BROKEN HEARTS GO Whitney Houston Arista 1-9674   3563
[27Feb88 | 23Apr88 | 25Jun88] 1(2) 6 | 13 | 18
[Charles Henry Jackson Jr./Frank Wildhorn]
A: Whitney; P: Narada Michael Walden

297. TAKE MY BREATH AWAY (Love Theme From Top Gun) Berlin   3558
Columbia 05903
[21Jun86 | 13Sep86 | 08Nov86] 1(1) 7 | 13 | 21
[Giorgio Moroder/Tom Whitlock] *A: Top Gun Soundtrack; P: Giorgio Moroder*

298. SEASONS CHANGE Exposé Arista 1-9640   3558
[28Nov87 | 20Feb88 | 09Apr88] 1(1) 6 | 16 | 20
[Lewis Martineé] *A: Exposure; P: Lewis A. Martineé*

299. I DON'T NEED YOU Kenny Rogers Liberty 1415   3554
[13Jun81 | 15Aug81 | 10Oct81] 3(2) 8 | 14 | 18
[Ricky Lynn Christian] *A: Share Your Love; P: Lionel Richie*

300. STUCK ON YOU Lionel Richie Motown 1746   3553
[23Jun84 | 25Aug84 | 27Oct84] 3(2) 7 | 14 | 19
[Lionel Richie]
A: Can't Slow Down; P: James Anthony Carmichael/Lionel Richie

301. SHAKE IT UP The Cars Elektra 47250   3549
[21Nov81 | 27Feb82 | 17Apr82] 4(3) 7 | 17 | 22
[Ric Ocasek] *A: Shake It Up; P: Roy Thomas Baker*

302. TAKE YOUR TIME (Do It Right) Part 1 The S.O.S. Band Tabu 5522   3545
[31May80 | 16Aug80 | 18Oct80] 3(2) 7 | 14 | 21
[Sigidi Abdullah/Harold Clayton] *A: S.O.S.; P: Sigidi Abdulla*

303. GUILTY Barbra Streisand & Barry Gibb Columbia 11390   3545
[01Nov80 | 10Jan81 | 28Mar81] 3(2) 7 | 15 | 22
[Barry Gibb/Maurice Gibb/Robin Gibb]
A: Guilty; P: Albhy Galuten/Barry Gibb/Karl Richardson

304. DRIVE The Cars Elektra 69706   3543
[04Aug84 | 29Sep84 | 08Dec84] 3(3) 7 | 14 | 19
[Ric Ocasek] *A: Heartbeat City; P: Cars/Mutt Lange*

305. WHO'S THAT GIRL Madonna Sire 28341   3537
[11Jul87 | 22Aug87 | 24Oct87] 1(1) 6 | 11 | 16
[Madonna L. Ciccone/Patrick Raymond Leonard]
A: Who's That Girl (Soundtrack); P: Patrick Leonard/Madonna

306. JACOB'S LADDER Huey Lewis And The News Chrysalis 43097   3536
[17Jan87 | 14Mar87 | 25Apr87] 1(1) 6 | 12 | 15
[Bruce Hornsby/John Hornsby] *A: Fore!; P: Huey Lewis And The News*

307. RASPBERRY BERET Prince & The Revolution Paisley Park 7-28972   3528
[18May85 | 20Jul85 | 07Sep85] 2(1) 6 | 14 | 17
[Prince] *A: Around The World In A Day; P: Prince And The Revolution*

308. THE WANDERER Donna Summer Geffen 49563   3526
[20Sep80 | 15Nov80 | 31Jan81] 3(3) 6 | 13 | 20
[Giorgio Moroder/Donna Summer]
A: The Wanderer; P: Pete Bellotte/Giorgio Moroder

309. HANDS TO HEAVEN Breathe A&M 2991   3503
[16Apr88 | 06Aug88 | 29Oct88] 2(2) 6 | 16 | 29
[Michael Delahunty/David Glasper/Marcus Lillington/Ian Spice]
A: All That Jazz; P: Bob Sargeant

310. IS THIS LOVE Whitesnake Geffen 28233   3497
[24Oct87 | 19Dec87 | 27Feb88] 2(1) 7 | 13 | 19
[David Coverdale/John Sykes] *A: Whitesnake; P: Keith Olsen/Mike Stone*

311. SIMPLY IRRESISTIBLE Robert Palmer EMI-Manhattan 50133   3496
[02Jul88 | 10Sep88 | 12Nov88] 2(2) 6 | 14 | 20
[Robert Palmer] *A: Heavy Nova; P: Robert Palmer*

312. LOST IN EMOTION Lisa Lisa And Cult Jam Columbia 07267   3495
[01Aug87 | 17Oct87 | 12Dec87] 1(1) 6 | 13 | 20
[Curtis Bedeau/Gerard Charles/Hugh Clarke/Brian George/
Lucien George/Paul George] *A: Spanish Fly; P: Full Force*

313. HOLD ON TO THE NIGHTS Richard Marx EMI-Manhattan 50106   3490
[21May88 | 23Jul88 | 08Oct88] 1(1) 6 | 14 | 21
[Richard Marx] *A: Richard Marx; P: David Cole/Richard Marx*

314. MONKEY George Michael Columbia 07941   3473
[09Jul88 | 27Aug88 | 22Oct88] 1(2) 6 | 12 | 16
[George Michael] *A: Faith;*
P: Jimmy Jam Harris/Terry Lewis/George Michael

315. I KNEW YOU WERE WAITING (For Me) Aretha Franklin & George Michael   3467
Arista 1-9559
[21Feb87 | 18Apr87 | 13Jun87] 1(2) 7 | 12 | 17
[Simon Climie/Dennis W. Morgan] *A: Aretha (II); P: Narada Michael Walden*

316. 867-5309/JENNY Tommy Tutone Columbia 02646   3466
[23Jan82 | 22May82 | 24Jul82] 4(3) 8 | 16 | 27
[Alex Call/Jim Keller]
A: Tommy Tutone-2; P: Jim Keller/Chuck Plotkin/Tommy Tutone

317. WHEN I'M WITH YOU ('88) Sheriff Capitol 44302   3465
[26Nov88 | 04Feb89 | 15Apr89] 1(1) 5 | 13 | 21
[Arnold Lanni] *A: Sheriff; P: Stacy Heydon*

318. BATDANCE Prince Warner Bros. 22924   3450
[17Jun89 | 05Aug89 | 14Oct89] 1(1) 6 | 11 | 18
[Prince] *A: Batman (Soundtrack); P: Prince*

319. SECRET LOVERS Atlantic Starr A&M 2788   3447
[28Dec85 | 22Mar86 | 31May86] 3(2) 6 | 14 | 23
[David E. Lewis/Wayne I. Lewis] *A: As The Band Turns; P: David Lewis/Wayne Lewis*

320. ANGEL OF THE MORNING Juice Newton Capitol 4976   3447
[21Feb81 | 02May81 | 18Jul81] 4(4) 6 | 16 | 22
[Chip Taylor] *A: Juice; P: Richard Landis*

321. SEA OF LOVE The Honeydrippers Es Paranza 7-99701   3446
[13Oct84 | 05Jan85 | 23Feb85] 3(1) 6 | 14 | 20
[Phil Baptiste/George Khoury]
A: Volume One; P: Phil Carson/Ahmet Ertegun/Robert Plant

322. LOVERBOY Billy Ocean Jive 9284   3446
[01Dec84 | 23Feb85 | 20Apr85] 2(1) 6 | 15 | 21
[Keith Diamond/Mutt Lange/Billy Ocean] *A: Suddenly; P: Keith Diamond*

323. POUR SOME SUGAR ON ME Def Leppard Mercury 870298   3419
[23Apr88 | 23Jul88 | 01Oct88] 2(1) 7 | 15 | 24
[Rick Allen/Steve Clark/Phil Collen/Joe Elliott/Mutt Lange/Rick Savage]
A: Hysteria; P: Mutt Lange

324. HOLDING BACK THE YEARS Simply Red Elektra 69564   3419
[05Apr86 | 12Jul86 | 06Sep86] 1(1) 6 | 14 | 23
[Mick Hucknall/Neil Moss] *A: Picture Book; P: Stewart Levine*

325. URGENT Foreigner Atlantic 3831   3417
[04Jul81 | 05Sep81 | 05Dec81] 4(4) 7 | 17 | 23
[Michael Leslie Jones] *A: 4; P: Mick Jones/Mutt Lange*

326. MASTER BLASTER (Jammin') Stevie Wonder Tamla 54317   3415
[20Sep80 | 06Dec80 | 21Feb81] 5(3) 8 | 16 | 23
[Stevie Wonder] *A: Hotter Than July; P: Stevie Wonder*

327. BABY DON'T FORGET MY NUMBER Milli Vanilli Arista 1-9832   3415
[29Apr89 | 01Jul89 | 16Sep89] 1(1) 6 | 14 | 21
[Bernd Berwanger/Roger Dalton/Frank Farian/Brad Howell]
A: Girl You Know It's True; P: Frank Farian

Rank. TITLE Artist Label Number [Entry Date | Peak Date | Exit Date] Peak(Peak Wks) Top10 | Top40 | Tot Wks [Writers] *Album; Producers*	Score
328. INVISIBLE TOUCH Genesis Atlantic 89407 [31May86 | 19Jul86 | 20Sep86] 1(1) 6 | 12 | 17 [Tony Banks/Phil Collins/Mike Rutherford] *A: Invisible Touch; P: Genesis/Hugh Padgham*	3412
329. IF YOU DON'T KNOW ME BY NOW Simply Red Elektra 69297 [06May89 | 15Jul89 | 30Sep89] 1(1) 6 | 15 | 22 [Kenneth Gamble/Leon Huff] *A: A New Flame; P: Stewart Levine*	3412
330. STRAY CAT STRUT Stray Cats EMI America 8122 [25Dec82 | 26Feb83 | 30Apr83] 3(3) 5 | 14 | 19 [Brian Setzer] *A: Built For Speed; P: Dave Edmunds*	3396
331. I JUST CAN'T STOP LOVING YOU Michael Jackson Epic 07253 [08Aug87 | 19Sep87 | 07Nov87] 1(1) 6 | 11 | 14 [Michael Jackson] *A: Bad; P: Quincy Jones*	3395
332. FOOLISH BEAT Debbie Gibson Atlantic 89109 [23Apr88 | 25Jun88 | 03Sep88] 1(1) 6 | 14 | 20 [Debbie Gibson] *A: Out Of The Blue; P: Debbie Gibson*	3393
333. THAT GIRL Stevie Wonder Tamla 1602 [16Jan82 | 20Mar82 | 15May82] 4(3) 9 | 13 | 18 [Stevie Wonder] *A: Stevie Wonder's Original Musiquarium I; P: Stevie Wonder*	3383
334. REAL LOVE Jody Watley MCA 53484 [18Mar89 | 20May89 | 15Jul89] 2(2) 6 | 12 | 18 [André Cymone/Jody Watley] *A: Larger Than Life; P: André Cymone*	3382
335. OH SHERRIE Steve Perry Columbia 04391 [07Apr84 | 09Jun84 | 18Aug84] 3(3) 7 | 13 | 20 [Bill Cuomo/Randy Goodrum/Craig Krampf/Steve Perry] *A: Street Talk; P: Steve Perry*	3379
336. SHE BLINDED ME WITH SCIENCE Thomas Dolby Capitol 5204 [19Feb83 | 14May83 | 16Jul83] 5(4) 8 | 15 | 22 [Thomas Dolby/Jo Kerr] *A: The Golden Age Of Wireless; P: Thomas Dolby/Tim Friese-Greene*	3379
337. OH SHEILA Ready For The World MCA 52636 [03Aug85 | 12Oct85 | 21Dec85] 1(1) 6 | 13 | 21 [Melvin Riley/Gordon Strozier/Gerald Valentine] *A: Ready For The World; P: Ready For The World*	3376
338. GOOD THING Fine Young Cannibals I.R.S./MCA 53639 [06May89 | 08Jul89 | 26Aug89] 1(1) 6 | 13 | 17 [Roland Gift/David Steele] *A: The Raw & The Cooked; P: Andy Cox/Roland Gift/David Steele*	3365
339. TRUE Spandau Ballet Chrysalis 42720 [06Aug83 | 08Oct83 | 03Dec83] 4(4) 6 | 13 | 18 [Gary Kemp] *A: True; P: Steve Jolley/Spandau Ballet/Tony Swain*	3350
340. BAD MEDICINE Bon Jovi Mercury 870657 [24Sep88 | 19Nov88 | 04Feb89] 1(2) 6 | 12 | 20 [Jon Bon Jovi/Desmond Child/Richie Sambora] *A: New Jersey; P: Bruce Fairbairn*	3342
341. TYPICAL MALE Tina Turner Capitol 5615 [30Aug86 | 18Oct86 | 13Dec86] 2(1) 6 | 12 | 16 [Terry Britten/Graham Lyle] *A: Break Every Rule; P: Terry Britten*	3330
342. RHYTHM OF THE NIGHT DeBarge Gordy 1770 [16Feb85 | 27Apr85 | 13Jul85] 3(2) 7 | 14 | 22 [Diane Warren] *A: Rhythm Of The Night; P: Richard Perry*	3325
343. EVERY WOMAN IN THE WORLD Air Supply Arista 0564 [25Oct80 | 31Jan81 | 21Mar81] 5(1) 8 | 17 | 22 [Dominic Bugatti/Frank Musker] *A: Lost In Love; P: Harry Maslin/Robie Porter*	3319
344. GIVING YOU THE BEST THAT I GOT Anita Baker Elektra 69371 [24Sep88 | 17Dec88 | 18Feb89] 3(3) 7 | 15 | 22 [Anita Baker/Randy Holland/Skip Scarborough] *A: Giving You The Best That I Got; P: Michael J. Powell*	3304
345. HUNGRY HEART Bruce Springsteen Columbia 11391 [08Nov80 | 27Dec80 | 07Mar81] 5(5) 8 | 14 | 18 [Bruce Springsteen] *A: The River; P: Jon Landau/Bruce Springsteen/Steven Van Zandt*	3294
346. I WANT YOUR SEX George Michael Columbia 07164 [06Jun87 | 08Aug87 | 17Oct87] 2(1) 6 | 14 | 20 [George Michael] *A: Faith; P: George Michael*	3293
347. HANGIN' TOUGH New Kids On The Block Columbia 68960 [15Jul89 | 09Sep89 | 04Nov89] 1(1) 6 | 12 | 17 [Maurice Starr] *A: Hangin' Tough; P: Maurice Starr*	3272
348. SHINING STAR The Manhattans Columbia 11222 [26Apr80 | 19Jul80 | 11Oct80] 5(3) 5 | 14 | 25 [Leo Graham/Paul Richmond] *A: After Midnight; P: Leo Graham*	3269
349. I MISS YOU Klymaxx Constellation 52606 [14Sep85 | 28Dec85 | 29Mar86] 5(4) 7 | 17 | 29 [Lynn Malsby] *A: Meeting In The Ladies Room; P: Klymaxx*	3259
350. WILD THING Tone Loc Delicious Vinyl 102 [03Dec88 | 18Feb89 | 20May89] 2(1) 6 | 14 | 25 [Matt Dike/Michael Ross/Anthony Terrell Smith/Marvin Young] *A: Loc-ed After Dark; P: Matt Dike/Michael Ross*	3259
351. BREAK MY STRIDE Matthew Wilder Private I 4-04113 [17Sep83 | 21Jan84 | 31Mar84] 5(3) 7 | 14 | 29 [Greg Prestopino/Matthew Wilder] *A: I Don't Speak The Language; P: Peter Bunetta/Rick Chudacoff/Bill Elliott*	3257
352. GIVE ME THE NIGHT George Benson Warner Bros. 49505 [05Jul80 | 27Sep80 | 06Dec80] 4(2) 7 | 14 | 23 [Rod Temperton] *A: Give Me The Night; P: Quincy Jones*	3251
353. SELF CONTROL Laura Branigan Atlantic 89676 [14Apr84 | 30Jun84 | 29Sep84] 4(2) 6 | 15 | 25 [Giancarlo Bigazzi/Steve Piccolo/Raffaele Riefoli] *A: Self Control; P: Robbie Buchanan/Jack White*	3250
354. THE LADY IN RED Chris DeBurgh A&M 2848 [14Feb87 | 23May87 | 08Aug87] 3(2) 6 | 14 | 26 [Chris de Burgh] *A: Into The Light; P: Paul Hardiman*	3242
355. YOU'RE THE INSPIRATION Chicago Warner/Full Moon 29126 [17Nov84 | 19Jan85 | 13Apr85] 3(2) 6 | 14 | 22 [Peter Cetera/David Foster] *A: Chicago 17; P: David Foster*	3237
356. FREEZE-FRAME The J. Geils Band EMI America 8108 [20Feb82 | 10Apr82 | 05Jun82] 4(4) 8 | 12 | 16 [Seth Justman/Peter Wolf] *A: Freeze-Frame; P: Seth Justman*	3237
357. FAME Irene Cara RSO 1034 [14Jun80 | 13Sep80 | 06Dec80] 4(2) 6 | 12 | 26 [Michael Gore/Dean Pitchford] *A: Fame Soundtrack; P: Michael Gore*	3221
358. ALL I NEED Jack Wagner Qwest 29238 [20Oct84 | 12Jan85 | 16Mar85] 2(2) 6 | 14 | 22 [Glen Ballard/Clif Magness/David Pack] *A: All I Need; P: Glen Ballard/Clif Magness*	3215
359. A WOMAN NEEDS LOVE (Just Like You Do) Ray Parker Jr. & Raydio Arista 0592 [07Mar81 | 20Jun81 | 05Sep81] 4(2) 6 | 15 | 27 [Ray Parker Jr.] *A: A Woman Needs Love; P: Ray Parker Jr.*	3214
360. LOVE IS A BATTLEFIELD Pat Benatar Chrysalis 42732 [24Sep83 | 10Dec83 | 18Feb84] 5(1) 7 | 14 | 22 [Mike Chapman/Holly Knight] *A: Live From Earth; P: Peter Coleman/Neil Geraldo*	3213
361. (There's) NO GETTIN' OVER ME Ronnie Milsap RCA 12264 [27Jun81 | 05Sep81 | 07Nov81] 5(5) 8 | 15 | 20 [Walt Aldridge/Tom Brasfield] *A: There's No Gettin' Over Me; P: Tom Collins/Ronnie Milsap*	3210
362. FOR YOUR EYES ONLY Sheena Easton Liberty 1418 [25Jul81 | 17Oct81 | 09Jan82] 4(4) 5 | 14 | 25 [Bill Conti/Mike Leeson] *A: For Your Eyes Only Soundtrack; P: Christopher Neil*	3209
363. HERE COMES THE RAIN AGAIN Eurythmics RCA 13725 [28Jan84 | 31Mar84 | 09Jun84] 4(2) 7 | 14 | 20 [Annie Lennox/David A. Stewart] *A: Touch; P: David A. Stewart*	3206
364. THE WAY YOU MAKE ME FEEL Michael Jackson Epic 07645 [21Nov87 | 23Jan88 | 19Mar88] 1(1) 6 | 13 | 18 [Michael Jackson] *A: Bad; P: Quincy Jones*	3201
365. THE OTHER WOMAN Ray Parker Jr. Arista 0669 [20Mar82 | 12Jun82 | 07Aug82] 4(2) 7 | 14 | 21 [Ray Parker Jr.] *A: The Other Woman; P: Ray Parker Jr.*	3200
366. TOGETHER FOREVER Rick Astley RCA 8319 [16Apr88 | 18Jun88 | 13Aug88] 1(1) 6 | 12 | 18 [Matt Aitken/Mike Stock/Pete Waterman] *A: Whenever You Need Somebody; P: Matt Aitken/Mike Stock/Pete Waterman*	3198
367. DON'T FALL IN LOVE WITH A DREAMER Kenny Rogers with Kim Carnes United Artists 1345 [29Mar80 | 24May80 | 02Aug80] 4(3) 5 | 14 | 19 [Kim Carnes/Dave Ellingson] *A: Gideon; P: Larry Butler/Kenny Rogers*	3194
368. STATE OF SHOCK Jacksons Epic 04503 [30Jun84 | 04Aug84 | 06Oct84] 3(3) 6 | 11 | 15 [Randy Hansen/Michael Jackson] *A: Victory; P: Michael Jackson*	3191
369. WHO'S CRYING NOW Journey Columbia 02241 [18Jul81 | 03Oct81 | 05Dec81] 4(4) 8 | 14 | 21 [Jonathan Cain/Steve Perry] *A: Escape; P: Kevin Elson/Mike Stone*	3180

Ranking the Singles

Rank. TITLE Artist Label Number
[Entry Date | Peak Date | Exit Date] Peak(Peak Wks) | Top10 | Top40 | Tot Wks
[Writers] *Album; Producers* Score

370. THE NEXT TIME I FALL Peter Cetera w/Amy Grant Full Moon 28597 3174
[20Sep86 | 06Dec86 | 07Feb87] 1(1) | 6 | 15 | 21
[Bobby Caldwell/Paul Gordon]
A: *Solitude/Solitaire*; P: Michael Omartian

371. EXPRESS YOURSELF Madonna Sire 22948 3158
[03Jun89 | 15Jul89 | 16Sep89] 2(2) | 5 | 11 | 16
[Stephen Bray/Madonna L. Ciccone]
A: *Like A Prayer*; P: Stephen Bray/Madonna

372. WE DON'T NEED ANOTHER HERO (Thunderdome) Tina Turner Capitol 5491 3151
[06Jul85 | 14Sep85 | 02Nov85] 2(1) | 6 | 12 | 18
[Terry Britten/Graham Lyle]
A: *Mad Max Beyond Thunderdome Soundtrack*; P: Terry Britten

373. HEAT OF THE MOMENT Asia Geffen 50040 3149
[17Apr82 | 26Jun82 | 14Aug82] 4(3) | 6 | 12 | 18
[Geoffrey Downes/John Wetton] A: *Asia*; P: Mike Stone

374. YOU SHOULD HEAR HOW SHE TALKS ABOUT YOU Melissa Manchester Arista 0676 3133
[22May82 | 18Sep82 | 06Nov82] 5(3) | 5 | 15 | 25
[Dean Pitchford/Tom Snow] A: *Hey Ricky*; P: Arif Mardin

375. ROCK ON Michael Damian Cypress 1420 3129
[18Mar89 | 03Jun89 | 05Aug89] 1(1) | 5 | 13 | 21
[David Essex]
A: *Where Do We Go From Here*; P: Michael Damian/Larry Weir/Tom Weir

376. ELVIRA The Oak Ridge Boys MCA 51084 3123
[16May81 | 25Jul81 | 10Oct81] 5(4) | 8 | 14 | 22
[Dallas Frazier] A: *Fancy Free*; P: Ron Chancey

377. I'VE NEVER BEEN TO ME Charlene Motown 1611 3113
[06Mar82 | 22May82 | 17Jul82] 3(3) | 6 | 14 | 20
[Ken Hirsch/Ronald Norman Miller]
A: *I've Never Been To Me*; P: Don Costa/Berry Gordy/Ronald Norman Miller

378. ENDLESS SUMMER NIGHTS Richard Marx EMI 50113 3112
[23Jan88 | 26Mar88 | 11Jun88] 2(2) | 6 | 15 | 21
[Richard Marx] A: *Richard Marx*; P: Humberto Gatica

379. PUTTIN' ON THE RITZ Taco RCA 13574 3108
[25Jun83 | 03Sep83 | 12Nov83] 4(2) | 6 | 14 | 21
[Irving Berlin] A: *After Eight*; P: David Parker

380. (It's Just) THE WAY THAT YOU LOVE ME ('89) Paula Abdul Virgin 99282 3106
[23Sep89 | 02Dec89 | 03Feb90] 3(1) | 6 | 14 | 22
[Oliver Leiber] A: *Forever Your Girl*; P: Oliver Leiber

381. CAUSING A COMMOTION Madonna Sire 28224 3098
[12Sep87 | 24Oct87 | 09Jan88] 2(3) | 5 | 11 | 18
[Stephen Bray/Madonna L. Ciccone]
A: *Who's That Girl (Soundtrack)*; P: Stephen Bray/Madonna

382. IF YOU LOVE SOMEBODY SET THEM FREE Sting A&M 2738 3098
[08Jun85 | 03Aug85 | 05Oct85] 3(2) | 6 | 14 | 18
[Sting] A: *The Dream Of The Blue Turtles*; P: Pete Smith/Sting

383. WHEN THE GOING GETS TOUGH, THE TOUGH GET GOING Billy Ocean Jive 9432 3075
[30Nov85 | 15Feb86 | 03May86] 2(1) | 5 | 14 | 23
[Wayne Brathwaite/Barry Eastmond/Mutt Lange/Billy Ocean]
A: *Jewel Of The Nile Soundtrack*; P: Wayne Brathwaite/Barry Eastmond

384. EYES WITHOUT A FACE Billy Idol Chrysalis 42786 3061
[05May84 | 14Jul84 | 29Sep84] 4(2) | 6 | 14 | 22
[Billy Idol/Steve Stevens] A: *Rebel Yell*; P: Missing

385. DEVIL INSIDE INXS Atlantic 89144 3056
[13Feb88 | 16Apr88 | 04Jun88] 2(2) | 5 | 12 | 17
[Andy Farriss/Michael Hutchence] A: *Kick*; P: Chris Thomas

386. SEXY EYES Dr. Hook Capitol 4831 3047
[16Feb80 | 24May80 | 05Jul80] 5(2) | 6 | 15 | 21
[Robert Mather/Keith Stegall/Chris Waters]
A: *Sometimes You Win*; P: Ron Haffkine

387. LET'S WAIT AWHILE Janet Jackson A&M 2906 3037
[17Jan87 | 21Mar87 | 23May87] 2(1) | 5 | 11 | 19
[Melanie Andrews/James Samuel Harris/Janet Jackson/Terry Lewis]
A: *Control*; P: Jimmy Jam Harris/Janet Jackson/Terry Lewis

388. I'LL ALWAYS LOVE YOU Taylor Dayne Arista 1-9700 3037
[11Jun88 | 24Sep88 | 31Dec88] 3(2) | 6 | 16 | 30
[Jimmy George] A: *Tell It To My Heart*; P: Ric Wake

389. BACK TO LIFE Soul II Soul Virgin 99171 3034
[23Sep89 | 16Dec89 | 31Mar90] 4(1) | 7 | 18 | 28
[Nellee Hooper/Simon Law/Beresford Romeo/Caron Wheeler]
A: *Keep On Movin'*; P: Nellee Hooper/Beresford Romeo

390. EVEN THE NIGHTS ARE BETTER Air Supply Arista 0692 3028
[12Jun82 | 04Sep82 | 09Oct82] 5(2) | 8 | 13 | 18
[Kenneth Edward Bell/Terry Skinner/J.L. Wallace]
A: *Now And Forever*; P: Harry Maslin

391. ALWAYS ON MY MIND Willie Nelson Columbia 02741 3027
[06Mar82 | 12May82 | 07Aug82] 5(3) | 6 | 15 | 23
[Johnny Christopher/Mark James/Wayne Carson Thompson]
A: *Always On My Mind*; P: Chips Moman

392. AXEL F Harold Faltermeyer MCA 52536 3024
[30Mar85 | 01Jun85 | 03Aug85] 3(3) | 5 | 12 | 19
[Harold Faltermeyer] A: *Beverly Hills Cop Soundtrack*; P: Harold Faltermeyer

393. LOVE SHACK The B-52's Reprise 22817 3009
[02Sep89 | 18Nov89 | 03Mar90] 3(2) | 6 | 17 | 27
[Kate Pierson/Fred Schneider/Keith Strickland/Cindy Wilson]
A: *Cosmic Thing*; P: Don Was

394. NEVER SURRENDER Corey Hart EMI America 8268 3008
[08Jun85 | 17Aug85 | 19Oct85] 3(2) | 6 | 14 | 20
[Corey Hart] A: *Boy In The Box*; P: Jon Astley/Phil Chapman/Corey Hart

395. STEPPIN' OUT Joe Jackson A&M 2428 3003
[21Aug82 | 11Dec82 | 19Feb83] 6(2) | 5 | 17 | 27
[Joe Jackson] A: *Night And Day*; P: Joe Jackson/David Kershenbaum

396. PASSION Rod Stewart Warner Bros. 49617 3000
[22Nov80 | 07Feb81 | 04Apr81] 5(2) | 6 | 17 | 20
[Phil Chen/Jim Cregan/Gary Grainger/Kevin Savigar/Rod Stewart]
A: *Foolish Behaviour*; P: Rod Stewart

397. I KEEP FORGETTIN' Michael McDonald Warner Bros. 29933 2999
[07Aug82 | 23Oct82 | 11Dec82] 4(3) | 7 | 13 | 19
[Jerry Leiber/Michael McDonald/Ed Sanford/Mike Stoller]
A: *If That's What It Takes*; P: Ted Templeman/Lenny Waronker

398. BAD Michael Jackson Epic 07418 2998
[19Sep87 | 24Oct87 | 19Dec87] 1(2) | 5 | 11 | 14
[Michael Jackson] A: *Bad*; P: Quincy Jones

399. ONE THING LEADS TO ANOTHER The Fixx MCA 52264 2996
[27Aug83 | 05Nov83 | 31Dec83] 4(1) | 7 | 13 | 19
[Alfie Agius/Cy Curnin/Peter John Greenall/Jamie West-Oram/Adam Woods]
A: *Reach The Beach*; P: Rupert Hine

400. EVERY LITTLE THING SHE DOES IS MAGIC The Police A&M 2371 2994
[26Sep81 | 05Dec81 | 30Jan82] 3(2) | 4 | 15 | 19
[Sting] A: *Ghost In The Machine*; P: Hugh Padgham/Police

401. YOU AND I Eddie Rabbitt with Crystal Gayle Elektra 69936 2991
[09Oct82 | 12Feb83 | 23Apr83] 7(4) | 6 | 21 | 29
[Frank Myers] A: *Radio Romance*; P: David Malloy

402. STEAL AWAY Robbie Dupree Elektra 46621 2990
[12Apr80 | 12Jul80 | 13Sep80] 6(2) | 7 | 15 | 23
[Rick Chudacoff/Robbie Dupree]
A: *Robbie Dupree*; P: Peter Bunetta/Rick Chudacoff

403. DANGER ZONE Kenny Loggins Columbia 05893 2987
[10May86 | 26Jul86 | 27Sep86] 2(1) | 6 | 13 | 21
[Giorgio Moroder/Tom Whitlock] A: *Top Gun Soundtrack*; P: Giorgio Moroder

404. LET IT WHIP Dazz Band Motown 1609 2985
[24Aug82 | 17Jul82 | 25Sep82] 5(2) | 6 | 14 | 23
[Reggie Andrews/Leon Chancler] A: *Keep It Live*; P: Reggie Andrews

405. HAZY SHADE OF WINTER Bangles Def Jam 07630 2983
[14Nov87 | 06Feb88 | 02Apr88] 2(1) | 5 | 14 | 21
[Paul Simon] A: *Less Than Zero Soundtrack*; P: Bangles/Bill Drescher/David White

406. DON'T FORGET ME (When I'm Gone) Glass Tiger Manhattan 50037 2981
[12Jul86 | 11Oct86 | 20Dec86] 2(1) | 5 | 14 | 24
[Alan Frew/Sam Reid/Jim Vallance] A: *The Thin Red Line*; P: Jim Vallance

407. I GUESS THAT'S WHY THEY CALL IT THE BLUES Elton John Geffen 29460 2975
[29Oct83 | 28Jan84 | 31Mar84] 4(1) | 5 | 15 | 23
[Elton John/Davey Johnstone/Bernie Taupin]
A: *Too Low For Zero*; P: Chris Thomas

408. SWEET DREAMS Air Supply Arista 0655 2974
[12Dec81 | 20Mar82 | 24Apr82] 5(2) | 7 | 15 | 20
[Graham Russell] A: *The One That You Love*; P: Harry Maslin

Rank. TITLE Artist Label Number
[Entry Date | Peak Date | Exit Date] Peak(Peak Wks) | Top10 | Top40 | Tot Wks
[Writers] Album; Producers — Score

409. YOUNG TURKS Rod Stewart Warner Bros. 49843 — 2970
[17Oct81 | 19Dec81 | 19Feb82] 5(4) 6 | 15 | 19
[Carmine Appice/Duane Hitchings/Kevin Savigar/Rod Stewart]
A: Tonight I'm Yours; P: Rod Stewart

410. KING OF PAIN The Police A&M 2569 — 2967
[27Aug83 | 08Oct83 | 10Dec83] 3(2) 5 | 13 | 16
[Sting] A: Synchronicity; P: Hugh Padgham/Police

411. DIRTY DIANA Michael Jackson Epic 07739 — 2958
[07May88 | 02Jul88 | 06Aug88] 1(1) 5 | 11 | 14
[Michael Jackson] A: Bad; P: Quincy Jones

412. HOW CAN I FALL? Breathe A&M 1224 — 2953
[10Sep88 | 03Dec88 | 04Feb89] 3(2) 5 | 16 | 22
[Michael Delahunty/David Glasper/Marcus Lillington/Ian Spice]
A: All That Jazz; P: Bob Sargeant

413. MANIC MONDAY Bangles Columbia 05757 — 2950
[25Jan86 | 19Apr86 | 07Jun86] 2(1) 5 | 14 | 20
[Prince] A: Different Light; P: David Kahne

414. MAD ABOUT YOU Belinda Carlisle I.R.S. 52815 — 2942
[17May86 | 09Aug86 | 04Oct86] 3(2) 6 | 14 | 21
[Paula Brown/Mitchel Evans/James Whelan] A: Belinda; P: Michael Lloyd

415. I DIDN'T MEAN TO TURN YOU ON Robert Palmer Island 99537 — 2941
[16Aug86 | 08Nov86 | 10Jan87] 2(1) 5 | 13 | 22
[James Samuel Harris/Terry Lewis] A: Riptide; P: Bernard Edwards

416. SHE'S LIKE THE WIND Patrick Swayze (featuring Wendy Fraser) — 2939
RCA 5363
[19Dec87 | 27Feb88 | 07May88] 3(3) 6 | 13 | 21
[Patrick Swayze/Stacey Widelitz]
A: Dirty Dancing Soundtrack; P: Michael Lloyd

417. KEEP YOUR HANDS TO YOURSELF Georgia Satellites Elektra 69502 — 2929
[22Nov86 | 21Feb87 | 04Apr87] 2(1) 5 | 14 | 20
[Dan Baird] A: Georgia Satellites; P: Jeff Glixman

418. U GOT THE LOOK Prince Paisley Park 7-28289 — 2927
[01Aug87 | 17Oct87 | 16Jan88] 2(1) 6 | 13 | 25
[Prince] A: Sign 'O' The Times; P: Prince

419. THE LOCO-MOTION Kylie Minogue Geffen 27752 — 2927
[27Aug88 | 12Nov88 | 25Feb89] 3(2) 6 | 13 | 27
[Gerry Goffin/Carole King] A: Kylie; P: Matt Aitken/Mike Stock/Pete Waterman

420. COOL IT NOW New Edition MCA 52455 — 2926
[22Sep84 | 05Jan85 | 09Mar85] 4(1) 6 | 14 | 25
[Vincent Brantley/Ricky Timas]
A: New Edition; P: Vincent Brantley/Rick Timas

421. I CAN'T WAIT Nu Shooz Atlantic 89446 — 2921
[08Mar86 | 14Jun86 | 09Aug86] 3(1) 5 | 15 | 23
[John Robert Smith] A: Poolside; P: John Smith/Rick Waritz

422. DRIVIN' MY LIFE AWAY Eddie Rabbitt Elektra 46656 — 2919
[21Jun80 | 04Oct80 | 06Dec80] 5(2) 6 | 15 | 25
[David Malloy/Eddie Rabbitt/Even Stevens] A: Horizon; P: David Malloy

423. WHAT HAVE I DONE TO DESERVE THIS? Pet Shop Boys And Dusty — 2918
Springfield EMI-Manhattan 50107
[12Dec87 | 20Feb88 | 09Apr88] 2(2) 5 | 13 | 18
[Chris Lowe/Neil Tennant/Allee Willis] A: Actually; P: Stephen Hague

424. YOU ARE Lionel Richie Motown 1657 — 2911
[15Jan83 | 26Mar83 | 14May83] 4(2) 4 | 16 | 18
[Brenda Richie/Lionel Richie] A: Lionel Richie; P: James Anthony
Carmichael/Lionel Richie

425. OH NO Commodores Motown 1527 — 2910
[26Sep81 | 05Dec81 | 06Feb82] 4(3) 4 | 15 | 20
[Lionel Richie] A: In The Pocket; P: James Anthony Carmichael/Commodores

426. DON'T RUSH ME Taylor Dayne Arista 1-9722 — 2909
[05Nov88 | 21Jan89 | 18Mar89] 2(1) 7 | 13 | 20
[Alexandra Forbes/Jeffrey Franzel] A: Tell It To My Heart; P: Ric Wake

427. FREEWAY OF LOVE Aretha Franklin Arista 1-9354 — 2899
[22Jun85 | 31Aug85 | 26Oct85] 3(1) 6 | 13 | 19
[Jeffrey E. Cohen/Narada Michael Walden]
A: Who's Zoomin' Who?; P: Narada Michael Walden

428. BUFFALO STANCE Neneh Cherry Virgin 99231 — 2898
[01Apr89 | 24Jun89 | 09Sep89] 3(1) 6 | 14 | 24
[Neneh Cherry/Cameron McVey/Jamie Morgan/Phil Ramacon]
A: Raw Like Sushi; P: Bomb The Bass/Mark Saunders/Tim Simenon

429. R.O.C.K. IN THE U.S.A. (A Salute To 60's Rock) John Cougar Mellencamp — 2895
Riva 884 455
[01Feb86 | 05Apr86 | 24May86] 2(1) 5 | 11 | 17
[John Mellencamp] A: Scarecrow; P: Don Gehman/John Mellencamp

430. SOMEWHERE OUT THERE Linda Ronstadt And James Ingram — 2890
MCA 52973
[20Dec86 | 14Mar87 | 16May87] 2(1) 5 | 12 | 22
[James Horner/Barry Mann/Cynthia Weil]
A: An American Tail Soundtrack; P: Peter Asher

431. TRUE BLUE Madonna Sire 28591 — 2882
[04Oct86 | 15Nov86 | 17Jan87] 3(3) 5 | 12 | 16
[Stephen Bray/Madonna L. Ciccone] A: True Blue; P: Stephen Bray/Madonna

432. SHAKE YOUR LOVE Debbie Gibson Atlantic 89187 — 2868
[03Oct87 | 19Dec87 | 27Feb88] 4(1) 7 | 15 | 22
[Debbie Gibson] A: Out Of The Blue; P: Fred Zarr

433. CRYING Don McLean Millennium 11799 — 2868
[24Jan81 | 21Mar81 | 23May81] 5(3) 6 | 15 | 18
[Joe Melson/Roy Orbison] A: Chain Lightning; P: Larry Butler

434. HEAD OVER HEELS Tears For Fears Mercury 880899 — 2862
[14Sep85 | 09Nov85 | 25Jan86] 3(1) 6 | 12 | 20
[Roland Orzabal/Curt Smith] A: Songs From The Big Chair; P: Chris Hughes

435. SOLDIER OF LOVE Donny Osmond Capitol 44369 — 2861
[25Mar89 | 03Jun89 | 22Jul89] 2(1) 6 | 11 | 18
[Evan Rogers/Carl Sturken] A: Donny Osmond; P: Evan Rogers/Carl Sturken

436. SATISFIED Richard Marx EMI 50189 — 2851
[06May89 | 24Jun89 | 12Aug89] 1(1) 5 | 13 | 15
[Richard Marx] A: Repeat Offender; P: David Cole/Richard Marx

437. HERE I AM (Just When I Thought I Was Over You) Air Supply Arista 0626 — 2848
[26Sep81 | 21Nov81 | 06Feb82] 5(3) 6 | 15 | 20
[Norman Sallitt] A: The One That You Love; P: Harry Maslin

438. STEP BY STEP Eddie Rabbitt Elektra 47174 — 2841
[25Jul81 | 17Oct81 | 19Dec81] 5(2) 8 | 15 | 22
[David Malloy/Eddie Rabbitt/Even Stevens] A: Step By Step; P: David Malloy

439. NAUGHTY GIRLS (Need Love Too) Samantha Fox Jive 1089 — 2837
[27Feb88 | 04Jun88 | 27Aug88] 3(1) 6 | 14 | 27
[Curtis Bedeau/Gerard Charles/Hugh Clarke/Brian George/
Lucien George/Paul George] A: Samantha Fox; P: Full Force

440. WAITING FOR A STAR TO FALL Boy Meets Girl RCA 8691 — 2834
[10Sep88 | 17Dec88 | 25Feb89] 5(3) 6 | 16 | 25
[George Merrill/Shannon Rubicam] A: Reel Life; P: Arif Mardin

441. TWIST OF FATE Olivia Newton-John MCA 52284 — 2833
[05Nov83 | 07Jan84 | 03Mar84] 5(2) 7 | 14 | 18
[Peter Beckett/Steve Kipner] A: Two Of A Kind (Soundtrack); P: David Foster

442. NIGHTSHIFT Commodores Motown 1773 — 2832
[26Jan85 | 20Apr85 | 22Jun85] 3(1) 6 | 13 | 22
[Franne Golde/Dennis Lambert/Walter Orange] A: Nightshift; P: Dennis Lambert

443. I'M COMING OUT Diana Ross Motown 1491 — 2830
[06Sep80 | 15Nov80 | 07Feb81] 5(3) 6 | 14 | 23
[Bernard Edwards/Nile Rodgers] A: Diana; P: Bernard Edwards/Nile Rodgers

444. THE SWEETEST THING (I've Ever Known) Juice Newton Capitol 5046 — 2829
[17Oct81 | 13Feb82 | 27Mar82] 7(2) 7 | 18 | 24
[Otha Young] A: Juice; P: Richard Landis

445. ONE NIGHT IN BANGKOK Murray Head RCA 13988 — 2828
[23Feb85 | 18May85 | 06Jul85] 3(1) 6 | 13 | 20
[Benny Andersson/Tim Rice/Björn Ulvaeus]
A: Chess; P: Benny Andersson/Tim Rice/Björn Ulvaeus

446. LEATHER AND LACE Stevie Nicks (with Don Henley) Modern 7341 — 2825
[24Oct81 | 23Jan82 | 27Feb82] 6(3) 8 | 15 | 19
[Stevie Nicks] A: Bella Donna; P: Jimmy Iovine

447. CUM ON FEEL THE NOIZE Quiet Riot Pasha 04005 — 2822
[17Sep83 | 19Nov83 | 04Feb84] 5(2) 5 | 14 | 21
[Noddy Holder/Jim Lea] A: Metal Health; P: Spencer Proffer

448. IF I COULD TURN BACK TIME Cher Geffen 22886 — 2821
[08Jul89 | 23Sep89 | 09Dec89] 3(2) 5 | 14 | 23
[Diane Warren] A: Heart Of Stone; P: Guy Roche/Diane Warren

449. TURN YOUR LOVE AROUND George Benson Warner Bros. 49846 — 2821
[24Oct81 | 06Feb82 | 20Mar82] 5(2) 6 | 16 | 22
[Bill Champlin/Jay Graydon/Steve Lukather]
A: Give Me The Night; P: Jay Graydon

Ranking the Singles

Rank. **TITLE** Artist Label Number
[Entry Date | Peak Date | Exit Date] Peak(Peak Wks) Top10 | Top40 | Tot Wks
[Writers] *Album; Producers* — Score

450. I DON'T WANNA GO ON WITH YOU LIKE THAT Elton John MCA 53345 — 2818
[18Jun88 | 27Aug88 | 15Oct88] 2(1) 6 | 13 | 18
[Elton John/Bernie Taupin]
A: Reg Strikes Back; P: Chris Thomas

451. IS THERE SOMETHING I SHOULD KNOW Duran Duran Capitol 5233 — 2818
[04Jun83 | 06Aug83 | 24Sep83] 4(1) 6 | 12 | 17
[Simon Le Bon/Nick Rhodes/Andy Taylor/Nigel John Taylor/Roger Taylor]
A: Duran Duran; P: Duran Duran/Ian Little

452. AGAINST THE WIND Bob Seger Capitol 4863 — 2801
[03May80 | 14Jun80 | 23Aug80] 5(3) 6 | 11 | 17
[Bob Seger] *A: Against The Wind; P: Bill Szymczyk*

453. CHERISH Madonna Sire 22883 — 2798
[19Aug89 | 07Oct89 | 25Nov89] 2(2) 5 | 12 | 15
[Madonna L. Ciccone/Patrick Raymond Leonard]
A: Like A Prayer; P: Patrick Leonard/Madonna

454. BORN TO BE MY BABY Bon Jovi Mercury 872156 — 2796
[26Nov88 | 18Feb89 | 08Apr89] 3(1) 5 | 13 | 20
[Jon Bon Jovi/Desmond Child/Richie Sambora]
A: New Jersey; P: Bruce Fairbairn

455. ONLY IN MY DREAMS Debbie Gibson Atlantic 89322 — 2792
[09May87 | 05Sep87 | 14Nov87] 4(1) 5 | 16 | 28
[Debbie Gibson] *A: Out Of The Blue; P: Fred Zarr*

456. ANGEL Aerosmith Geffen 28249 — 2787
[30Jan88 | 30Apr88 | 16Jul88] 3(2) 5 | 15 | 25
[Desmond Child/Steven Tyler] *A: Permanent Vacation; P: Bruce Fairbairn*

457. YOU GIVE GOOD LOVE Whitney Houston Arista 1-9274 — 2786
[11May85 | 27Jul85 | 28Sep85] 3(1) 6 | 13 | 21
[LaForrest Cope] *A: Whitney Houston; P: Kashif*

458. MAKE ME LOSE CONTROL Eric Carmen Arista 1-9686 — 2785
[21May88 | 13Aug88 | 01Oct88] 3(1) 6 | 13 | 20
[Eric Carmen/Dean Pitchford] *A: No Album; P: Jimmy Ienner*

459. BACK ON THE CHAIN GANG The Pretenders Sire 29840 — 2775
[11Dec82 | 19Mar83 | 21May83] 5(3) 5 | 14 | 24
[Chrissie Hynde] *A: No Album; P: Chris Thomas*

460. THE LOVER IN ME Sheena Easton MCA 53416 — 2769
[05Nov88 | 04Mar89 | 22Apr89] 2(1) 5 | 14 | 25
[Kenneth Edmonds/Antonio Reid/Daryl Simmons]
A: The Lover In Me; P: Kenneth Edmonds/L.A. Reid

461. TOO HOT Kool & The Gang De-Lite 802 — 2769
[19Jan80 | 05Apr80 | 17May80] 5(2) 5 | 13 | 18
[George M. Brown] *A: Ladies' Night; P: Eumir Deodato*

462. NEVER Heart Capitol 5512 — 2769
[14Sep85 | 07Dec85 | 22Feb86] 4(1) 5 | 14 | 24
[Gene Black/Holly Knight/Ann Wilson/Nancy Wilson] *A: Heart; P: Ron Nevison*

463. NASTY Janet Jackson A&M 2830 — 2753
[17May86 | 19Jul86 | 20Sep86] 3(1) 6 | 11 | 19
[James Samuel Harris/Terry Lewis] *A: Control; P: Jimmy Jam Harris/Terry Lewis*

464. ON THE RADIO Donna Summer Casablanca 2236 — 2750
[12Jan80 | 08Mar80 | 03May80] 5(2) 6 | 12 | 17
[Giorgio Moroder/Donna Summer]
A: On The Radio: Greatest Hits: Volumes I & II; P: Giorgio Moroder

465. BUST A MOVE Young M.C. Delicious Vinyl 105 — 2743
[29Jul89 | 14Oct89 | 21Apr90] 7(1) 4 | 20 | 39
[Matt Dike/Luther Rabb/Michael Ross/Marvin Young]
A: Stone Cold Rhymin'; P: Matt Dike/Michael Ross

466. EVERY LITTLE STEP Bobby Brown MCA 53618 — 2738
[25Mar89 | 10Jun89 | 12Aug89] 3(2) 6 | 13 | 21
[Kenneth Edmonds/Antonio Reid]
A: Don't Be Cruel; P: Kenneth Edmonds/L.A. Reid

467. TAKE IT ON THE RUN REO Speedwagon Epic 01054 — 2737
[21Mar81 | 30May81 | 01Aug81] 5(2) 6 | 15 | 20
[Gary Richrath] *A: Hi Infidelity; P: Kevin Beamish/Kevin Cronin/Gary Richrath*

468. LOVERGIRL Teena Marie Epic 04619 — 2721
[15Dec84 | 30Mar85 | 25May85] 4(1) 5 | 13 | 24
[Teena Marie] *A: Starchild; P: Teena Marie*

469. WHY CAN'T THIS BE LOVE Van Halen Warner Bros. 28740 — 2706
[15Mar86 | 17May86 | 28Jun86] 3(1) 5 | 11 | 16
[Michael Anthony/Sammy Hagar/Alex Van Halen/Eddie Van Halen]
A: 5150; P: Mick Jones/Donn Landee/Van Halen

470. SPECIAL LADY Ray, Goodman & Brown Polydor 2033 — 2690
[26Jan80 | 19Apr80 | 24May80] 5(2) 4 | 14 | 18
[Al Goodman/Walter Lee Morris/Harry Ray]
A: Ray, Goodman & Brown; P: Vincent Castellano

471. SOWING THE SEEDS OF LOVE Tears For Fears Fontana 874 710 — 2677
[02Sep89 | 28Oct89 | 09Dec89] 2(1) 5 | 12 | 15
[Roland Orzabal/Curt Smith]
A: The Seeds Of Love; P: David Bascombe/Tears For Fears

472. AUTOMATIC Pointer Sisters Planet 13730 — 2676
[28Jan84 | 14Apr84 | 09Jun84] 5(1) 6 | 14 | 20
[Mark Goldenberg/Brock Walsh] *A: Break Out; P: Richard Perry*

473. YOU MAKE MY DREAMS Daryl Hall & John Oates RCA Victor 12217 — 2669
[02May81 | 04Jul81 | 19Sep81] 5(3) 6 | 14 | 21
[Sara Allen/Daryl Hall/John Oates] *A: Voices; P: Daryl Hall/John Oates*

474. DER KOMMISSAR After The Fire Epic 03559 — 2668
[12Feb83 | 30Apr83 | 02Jul83] 5(2) 5 | 14 | 21
[Falco/Andrew Piercy/Robert Ponger] *A: ATF; P: After The Fire/John Eden*

475. SO ALIVE Love and Rockets RCA 8956 — 2663
[20May89 | 05Aug89 | 30Sep89] 3(1) 6 | 12 | 20
[Daniel Ash/David Haskins/Kevin Haskins]
A: Love And Rockets; P: John Fryer/Love And Rockets

476. WE BELONG Pat Benatar Chrysalis 42826 — 2660
[27Oct84 | 05Jan85 | 09Mar85] 5(2) 7 | 14 | 20
[Eric Lowen/Dan Navarro] *A: Tropico; P: Peter Coleman/Neil Geraldo*

477. CRUSH ON YOU The Jets MCA 52774 — 2659
[12Apr86 | 21Jun86 | 23Aug86] 3(2) 5 | 13 | 20
[Jerry Knight/Aaron Zigman]
A: The Jets; P: Jerry Knight/Don Powell/David Rivkin/Aaron Zigman

478. HUNGRY EYES Eric Carmen RCA 5315 — 2648
[07Nov87 | 13Feb88 | 23Apr88] 4(1) 5 | 16 | 25
[John DeNicola/Franke Previte] *A: Dirty Dancing Soundtrack; P: Eric Carmen*

479. HIP TO BE SQUARE Huey Lewis And The News Chrysalis 43065 — 2647
[18Oct86 | 06Dec86 | 31Jan87] 3(2) 5 | 12 | 16
[Bill Gibson/Sean Hopper/Huey Lewis] *A: Fore!; P: Huey Lewis And The News*

480. LIVING INSIDE MYSELF Gino Vannelli Arista 0588 — 2640
[21Mar81 | 30May81 | 01Aug81] 6(3) 7 | 14 | 20
[Gino Vannelli] *A: Nightwalker; P: Gino Vannelli/Joe Vannelli/Ross Vannelli*

481. STAND BACK Stevie Nicks Modern 99863 — 2639
[04Jun83 | 20Aug83 | 08Oct83] 5(1) 6 | 13 | 19
[Stevie Nicks/Prince] *A: The Wild Heart; P: Jimmy Iovine*

482. HEARTLIGHT Neil Diamond Columbia 03219 — 2634
[11Sep82 | 13Nov82 | 15Jan83] 5(4) 6 | 11 | 19
[Burt Bacharach/Neil Diamond/Carole Bayer Sager]
A: Heartlight; P: Burt Bacharach/Neil Diamond/Carole Bayer Sager

483. IN TOO DEEP Genesis Atlantic 89316 — 2632
[25Apr87 | 27Jun87 | 15Aug87] 3(1) 5 | 12 | 17
[Tony Banks/Phil Collins/Mike Rutherford]
A: Invisible Touch; P: Genesis/Hugh Padgham

484. WE'VE GOT TONIGHT Kenny Rogers And Sheena Easton Liberty 1492 — 2632
[29Jan83 | 26Mar83 | 28May83] 6(3) 7 | 15 | 18
[Bob Seger] *A: We've Got Tonight; P: David Foster/Kenny Rogers*

485. SHOULD'VE KNOWN BETTER Richard Marx Manhattan 50083 — 2631
[26Sep87 | 12Dec87 | 13Feb88] 3(1) 5 | 13 | 21
[Richard Marx] *A: Richard Marx; P: Humberto Gatica*

486. MERCEDES BOY Pebbles MCA 53279 — 2616
[07May88 | 09Jul88 | 03Sep88] 2(2) 5 | 11 | 18
[Perri Arlette Reid] *A: Pebbles; P: Charlie Wilson*

487. LUKA Suzanne Vega A&M 2937 — 2614
[06Jun87 | 22Aug87 | 10Oct87] 3(1) 5 | 12 | 19
[Suzanne Vega] *A: Solitude Standing; P: Steve Addabbo/Lenny Kaye*

488. THE HEART OF ROCK & ROLL Huey Lewis And The News Chrysalis 42782 — 2607
[21Apr84 | 09Jun84 | 01Sep84] 6(4) 7 | 14 | 20
[Johnny Colla/Huey Lewis] *A: Sports; P: Huey Lewis And The News*

489. I'M YOUR MAN Wham! Columbia 05721 — 2606
[30Nov85 | 01Feb86 | 29Mar86] 3(2) 4 | 12 | 18
[George Michael] *A: Music From The Edge Of Heaven; P: George Michael*

490. NEVER KNEW LOVE LIKE THIS BEFORE Stephanie Mills — 2602
20th Century 2460 [09Aug80 | 15Nov80 | 24Jan81] 6(2) 5 | 16 | 25
[Reggie Lucas/James Mtume] *A: Sweet Sensation; P: Reggie Lucas/James Mtume*

Ranking the Singles

Rank. TITLE Artist Label Number	
[Entry Date | Peak Date | Exit Date) Peak(Peak Wks) Top10 | Top40 | Tot Wks	
[Writers] Album; Producers	Score

491. I GET WEAK Belinda Carlisle MCA 53242   2582
[16Jan88 | 19Mar88 | 30Apr88] 2(1) 5 | 13 | 16
[Diane Warren]
A: Heaven On Earth; P: Rick Nowels

492. CARS Gary Numan Atco 7211   2581
[16Feb80 | 07Jun80 | 02Aug80] 9(3) 5 | 17 | 25
[Gary Numan]   A: The Pleasure Principle; P: Gary Numan

493. LATE IN THE EVENING Paul Simon Warner Bros. 49511   2579
[09Aug80 | 27Sep80 | 22Nov80] 6(3) 7 | 12 | 16
[Paul Simon]   A: One-Trick Pony; P: Phil Ramone/Paul Simon

494. LITTLE RED CORVETTE Prince Warner Bros. 29746   2573
[26Feb83 | 21May83 | 23Jul83] 6(2) 6 | 15 | 22
[Prince]   A: 1999; P: Prince

495. ARMAGEDDON IT Def Leppard Mercury 870692   2569
[19Nov88 | 21Jan89 | 18Mar89] 3(2) 5 | 12 | 18
[Rick Allen/Steve Clark/Phil Collen/Joe Elliott/Mutt Lange/Rick Savage]
A: Hysteria; P: Mutt Lange

496. HEART AND SOUL T'Pau Virgin 99466   2549
[02May87 | 08Aug87 | 31Oct87] 4(4) 5 | 16 | 27
[Carol Decker/Ron Rogers]   A: T'Pau; P: Roy Thomas Baker

497. OUT OF THE BLUE Debbie Gibson Atlantic 89129   2548
[30Jan88 | 09Apr88 | 21May88] 3(1) 5 | 13 | 17
[Debbie Gibson]   A: Out Of The Blue; P: Debbie Gibson/Fred Zarr

498. NEUTRON DANCE Pointer Sisters Planet 13951   2532
[24Nov84 | 16Feb85 | 27Apr85] 6(3) 6 | 14 | 23
[Danny Sembello/Allee Willis]   A: Break Out; P: Richard Perry

499. WHAT HAVE YOU DONE FOR ME LATELY Janet Jackson A&M 2812   2532
[22Feb86 | 17May86 | 12Jul86] 4(1) 6 | 13 | 21
[James Samuel Harris/Terry Lewis]
A: Control; P: Jimmy Jam Harris/Terry Lewis

500. THE NIGHT OWLS Little River Band Capitol 5033   2525
[22Aug81 | 07Nov81 | 09Jan82] 6(2) 6 | 14 | 21
[Graham Goble]   A: Time Exposure; P: George Martin

501. WANNA BE STARTIN' SOMETHIN' Michael Jackson Epic 03914   2524
[28May83 | 16Jul83 | 03Sep83] 5(2) 6 | 11 | 15
[Michael Jackson]   A: Thriller; P: Quincy Jones

502. YOU GOT IT ALL The Jets MCA 52968   2523
[15Nov86 | 07Mar87 | 09May87] 3(1) 4 | 12 | 26
[Rupert Holmes]   A: The Jets; P: Don Powell/David Rivkin

503. SUDDENLY Billy Ocean Jive 9323   2519
[23May85 | 08Jun85 | 17Aug85] 4(1) 4 | 13 | 22
[Keith Diamond/Billy Ocean]   A: Suddenly; P: Keith Diamond

504. COVER GIRL New Kids On The Block Columbia 69088   2518
[16Sep89 | 04Nov89 | 13Jan90] 2(1) 4 | 10 | 18
[Maurice Starr]   A: Hangin' Tough; P: Maurice Starr

505. WITH EVERY BEAT OF MY HEART Taylor Dayne Arista 1-9895   2496
[21Oct89 | 16Dec89 | 17Feb90] 5(4) 6 | 13 | 18
[Arthur Baker/Tommy Faragher/Lotti Golden]   A: Can't Fight Fate; P: Ric Wake

506. THRILLER Michael Jackson Epic 04363   2493
[11Feb84 | 03Mar84 | 12May84] 4(2) 5 | 9 | 14
[Rod Temperton]   A: Thriller; P: Quincy Jones

507. KEY LARGO Bertie Higgins Kat Family 02524   2483
[14Nov81 | 17Apr82 | 29May82] 8(2) 4 | 17 | 29
[Bertie Higgins/Sonny Limbo]
A: Just Another Day In Paradise; P: Sonny Limbo/Scott MacLellan

508. TO ALL THE GIRLS I'VE LOVED BEFORE Julio Iglesias & Willie Nelson   2482
Columbia 04217
[03Mar84 | 19May84 | 21Jul84] 5(1) 6 | 12 | 21
[Hal David/Albert Hammond]   A: Swept Away; P: Richard Perry

509. WHAT'S ON YOUR MIND (Pure Energy) Information Society   2481
Tommy Boy 27826
[16Jul88 | 22Oct88 | 31Dec88] 3(1) 6 | 14 | 25
[Paul Robb/Kurt Valaquen]   A: Information Society; P: Fred Maher

510. 65 LOVE AFFAIR Paul Davis Arista 0661   2478
[27Feb82 | 22May82 | 10Jul82] 6(2) 7 | 13 | 20
[Paul Davis]   A: Cool Night; P: Paul Davis/Ed Seay

511. CARRIE Europe Epic 07282   2472
[01Aug87 | 10Oct87 | 05Dec87] 3(2) 5 | 12 | 19
[Mic Michaeli/Joey Tempest]   A: The Final Countdown; P: Kevin Elson

512. WILL YOU STILL LOVE ME? Chicago Warner/Full Moon 28512   2470
[15Nov86 | 21Feb87 | 18Apr87] 3(1) 4 | 13 | 23
[Richard Baskin/David Foster/Tom Keane]   A: Chicago 18; P: David Foster

513. SEPARATE WAYS (Worlds Apart) Journey Columbia 03513   2468
[05Feb83 | 19Mar83 | 28May83] 8(6) 7 | 16 | 17
[Jonathan Cain/Steve Perry]   A: Frontiers; P: Kevin Elson/Mike Stone

514. LOOKIN' FOR LOVE Johnny Lee Full Moon/Asylum 47004   2465
[12Jul80 | 20Sep80 | 29Nov80] 5(2) 5 | 13 | 21
[Wanda Mallette/Bob Morrison/Patti Ryan]
A: Urban Cowboy Soundtrack; P: John Boylan

515. YOU CAN DO MAGIC America Capitol 5412   2458
[31Jul82 | 16Oct82 | 11Dec82] 8(5) 6 | 15 | 20
[Russ Ballard]   A: View From The Ground; P: Russ Ballard

516. THE SEARCH IS OVER Survivor Scotti Brothers 04871   2456
[20Apr85 | 13Jul85 | 07Sep85] 4(1) 5 | 14 | 21
[Jim Peterik/Frankie Sullivan]   A: Vital Signs; P: Ron Nevison

517. SIGN O' THE TIMES Prince Paisley Park 7-28399   2446
[07Mar87 | 25Apr87 | 06Jun87] 3(1) 5 | 11 | 14
[Prince]   A: Sign 'O' The Times; P: Prince

518. CONTROL Janet Jackson A&M 2877   2445
[01Nov86 | 24Jan87 | 28Feb87] 5(1) 6 | 13 | 18
[James Samuel Harris/Terry Lewis]   A: Control; P: Jimmy Jam Harris/Terry Lewis

519. LA ISLA BONITA Madonna Sire 28425   2445
[21Mar87 | 02May87 | 11Jul87] 4(3) 5 | 12 | 17
[Madonna L. Ciccone/Bruce Gaitsch/Patrick Raymond Leonard]
A: True Blue; P: Patrick Leonard/Madonna

520. TALK TO ME Stevie Nicks Modern 99582   2443
[16Nov85 | 25Jan86 | 15Mar86] 4(2) 5 | 13 | 18
[Chas Sandford]   A: Rock A Little; P: Jimmy Iovine/Chas Sandford

521. LADIES NIGHT Kool & The Gang De-Lite 801   2426
[06Oct79 | 12Jan80 | 15Mar80] 8(2) 5 | 14 | 24
[Robert Bell/Ronald Bell/George M. Brown/Meekaaeel Muhammed/
Claydes Smith/J.T. Taylor/Dennis Thomas/Earl Toon]
A: Ladies' Night; P: Eumir Deodato

522. LUCKY STAR Madonna Sire 29177   2425
[25Aug84 | 20Oct84 | 08Dec84] 4(1) 5 | 12 | 16
[Madonna L. Ciccone]   A: Madonna; P: Reggie Lucas

523. DESIRE U2 Island 99250   2423
[01Oct88 | 26Nov88 | 21Jan89] 3(1) 5 | 13 | 17
[Adam Clayton/Dave Evans/Paul Hewson/Larry Mullen]
A: Rattle And Hum (Soundtrack); P: Jimmy Iovine

524. I DON'T WANNA LIVE WITHOUT YOUR LOVE Chicago   2418
Reprise/Full Moon 27855
[04Jun88 | 27Aug88 | 22Oct88] 3(1) 5 | 13 | 21
[Albert Hammond/Diane Warren]   A: Chicago 19; P: Ron Nevison

525. CALIFORNIA GIRLS David Lee Roth Warner Bros. 29102   2418
[19Jan85 | 02Mar85 | 04May85] 3(1) 5 | 11 | 16
[Mike Love/Brian Wilson]   A: Crazy From The Heat; P: Ted Templeman

526. YOU GOT IT (The Right Stuff) New Kids On The Block Columbia 08092   2405
[19Nov88 | 11Mar89 | 13May89] 3(1) 5 | 13 | 21
[Maurice Starr]   A: Hangin' Tough; P: Michael Jonzun/Maurice Starr

527. ONCE BITTEN TWICE SHY Great White Capitol 44366   2400
[13May89 | 12Aug89 | 04Nov89] 5(2) 5 | 14 | 26
[Ian Hunter]   A: ...Twice Shy; P: Michael Lardie/Alan Niven

528. MY HEART CAN'T TELL YOU NO Rod Stewart Warner Bros. 27729   2400
[10Dec88 | 01Apr89 | 27May89] 4(1) 5 | 13 | 25
[Simon Climie/Dennis W. Morgan]
A: Out Of Order; P: Bernard Evans/Rod Stewart/Andy Taylor

529. I HEARD A RUMOUR Bananarama London 886165   2398
[18Jul87 | 26Sep87 | 21Nov87] 4(3) 6 | 12 | 19
[Matt Aitken/Sarah Dallin/Siobhan Fahey/Mike Stock/Pete Waterman/
Keren Woodward]   A: Wow!; P: Matt Aitken/Mike Stock/Pete Waterman

530. DON'T MEAN NOTHING Richard Marx Manhattan 50079   2396
[13Jun87 | 29Aug87 | 31Oct87] 3(1) 4 | 12 | 21
[Bruce Gaitsch/Richard Marx]   A: Richard Marx; P: David Cole/Richard Marx

531. NEW SENSATION INXS Atlantic 89080   2395
[14May88 | 23Jul88 | 03Sep88] 3(1) 5 | 12 | 17
[Andy Farriss/Michael Hutchence]   A: Kick; P: Chris Thomas

532. I CAN DREAM ABOUT YOU Dan Hartman MCA 52378   2393
[05May84 | 18Aug84 | 20Oct84] 6(2) 4 | 16 | 25
[Dan Hartman]   A: I Can Dream About You; P: Dan Hartman/Jimmy Iovine

533. FUNKY COLD MEDINA Tone Loc Delicious Vinyl 104   2392
[04Mar89 | 29Apr89 | 01Jul89] 3(1) 5 | 11 | 18
[Matt Dike/Michael Ross/Marvin Young]
A: Loc-ed After Dark; P: Matt Dike/Michael Ross

Ranking the Singles

Rank. TITLE Artist Label Number
[Entry Date | Peak Date | Exit Date] Peak(Peak Wks) Top10 | Top40 | Tot Wks
[Writers] *Album; Producers* Score

534. PERFECT WORLD Huey Lewis And The News Chrysalis 43265 2388
[16Jul88 | 10Sep88 | 22Oct88] 3(2) 5 | 12 | 15
[Alex Call] *A: Small World; P: Huey Lewis And The News*

535. MAKE IT REAL The Jets MCA 53311 2388
[16Apr88 | 25Jun88 | 27Aug88] 4(2) 5 | 13 | 20
[Rick Kelly/Linda Mallah/Don Powell]
A: Magic; P: Rick Kelly/Don Powell/Michael Verdick

536. SISTER CHRISTIAN Night Ranger Camel/MCA 52350 2385
[10Mar84 | 09Jun84 | 18Aug84] 5(2) 4 | 12 | 24
[Kelly Keagy] *A: Midnight Madness; P: Pat Glasser*

537. THE WINNER TAKES IT ALL ABBA Atlantic 3776 2383
[22Nov80 | 14Mar81 | 16May81] 8(2) 4 | 16 | 26
[Benny Andersson/Björn Ulvaeus]
A: Super Trouper; P: Benny Andersson/Björn Ulvaeus

538. I WANT A NEW DRUG Huey Lewis And The News Chrysalis 42766 2382
[14Jan84 | 24Mar84 | 19May84] 6(2) 5 | 13 | 19
[Chris Hayes/Huey Lewis] *A: Sports; P: Huey Lewis And The News*

539. ONE ON ONE Daryl Hall & John Oates RCA Victor 13421 2378
[29Jan83 | 09Apr83 | 28May83] 7(3) 6 | 15 | 18
[Daryl Hall] *A: H2O; P: Daryl Hall/John Oates*

540. LAND OF CONFUSION Genesis Atlantic 89336 2370
[01Nov86 | 31Jan87 | 21Mar87] 4(1) 4 | 15 | 21
[Tony Banks/Phil Collins/Mike Rutherford]
A: Invisible Touch; P: Genesis/Hugh Padgham

541. TOUCH ME (I Want Your Body) Samantha Fox Jive 1006 2369
[01Nov86 | 14Feb87 | 04Apr87] 4(1) 4 | 13 | 23
[Jon Astrop/Peter Brian Harris/Mark Shreeve]
A: Touch Me; P: Jon Astrop/Pete Q. Harris

542. THE GLAMOROUS LIFE Sheila E. Warner Bros. 29285 2352
[16Jun84 | 06Oct84 | 08Dec84] 7(1) 5 | 16 | 26
[Prince] *A: The Glamorous Life; P: Prince/Sheila E.*

543. TRYIN' TO LIVE MY LIFE WITHOUT YOU (Live) Bob Seger Capitol 5042 2351
[12Sep81 | 07Nov81 | 16Jan82] 5(2) 4 | 12 | 19
[Eugene Williams] *A: Nine Tonight; P: Ed Andrews/Bob Seger*

544. BETTER BE GOOD TO ME Tina Turner Capitol 5387 2351
[15Sep84 | 24Nov84 | 02Feb85] 5(2) 4 | 13 | 21
[Mike Chapman/Nicky Chinn/Holly Knight] *A: Private Dancer; P: Rupert Hine*

545. WHAT YOU NEED INXS Atlantic 89460 2349
[18Jan86 | 12Apr86 | 31May86] 5(1) 4 | 14 | 22
[Andy Farriss/Michael Hutchence] *A: Listen Like Thieves; P: Chris Thomas*

546. SMOOTH OPERATOR Sade Portrait 37-04807 2347
[02Mar85 | 18May85 | 13Jul85] 5(2) 5 | 13 | 20
[Helen Adu/Ray St. John] *A: Diamond Life; P: Robin Millar*

547. TAINTED LOVE Soft Cell Sire 49855 2345
[16Jan82 | 17Jul82 | 06Nov82] 8(2) 3 | 15 | 43
[Ed Cobb] *A: Non-Stop Erotic Cabaret; P: Mike Thorne*

548. TWO OF HEARTS Stacey Q Atlantic 89381 2345
[12Jul86 | 11Oct86 | 06Dec86] 3(1) 4 | 13 | 22
[Sue Gatlin/Tim Greene/John Mitchell]
A: Better Than Heaven; P: Jon St. James

549. WE DON'T TALK ANYMORE Cliff Richard EMI America 8025 2342
[20Oct79 | 19Jan80 | 01Mar80] 7(2) 4 | 14 | 20
[Alan Tarney] *A: We Don't Talk Anymore; P: Bruce Welch*

550. WHO'S JOHNNY (Short Circuit Theme) El DeBarge Gordy 1842 2341
[26Apr86 | 05Jul86 | 30Aug86] 3(1) 5 | 13 | 19
[Ina Wolf/Peter F. Wolf] *A: El DeBarge; P: Peter Wolf*

551. MISS ME BLIND Culture Club Virgin/Epic 04388 2336
[03Mar84 | 21Apr84 | 16Jun84] 5(2) 6 | 12 | 16
[Michael Craig/Roy Hay/Jon Moss/George O'Dowd]
A: Colour By Numbers; P: Steve Levine

552. WHY DO FOOLS FALL IN LOVE Diana Ross RCA 12349 2330
[17Oct81 | 19Dec81 | 27Feb82] 7(3) 6 | 14 | 20
[George Goldner/Morris Levy/Frankie Lymon/Jimmy Merchant/
Herman Santiago] *A: Why Do Fools Fall In Love; P: Diana Ross*

553. THINGS CAN ONLY GET BETTER Howard Jones Elektra 69651 2329
[23Mar85 | 15Jun85 | 24Aug85] 5(1) 6 | 14 | 23
[Howard Jones] *A: Dream Into Action; P: Rupert Hine*

554. HIT ME WITH YOUR BEST SHOT Pat Benatar Chrysalis 2464 2328
[04Oct80 | 20Dec80 | 14Mar81] 9(3) 4 | 15 | 24
[Eddie Schwartz] *A: Crimes Of Passion; P: Keith Olsen*

555. CHANGE OF HEART Cyndi Lauper Portrait 37-06431 2326
[29Nov86 | 14Feb87 | 21Mar87] 3(2) 5 | 13 | 17
[Cyndi Lauper/Essra Mohawk]
A: True Colors; P: Cyndi Lauper/Lennie Petze

556. IF YOU LEAVE Orchestral Manoeuvres In The Dark A&M 2811 2321
[08Mar86 | 31May86 | 19Jul86] 4(1) 5 | 13 | 20
[Martin H. Cooper/Paul Humphreys/Andrew McCluskey]
A: Pretty In Pink Soundtrack; P: Tom Lord-Alge/O.M.D.

557. TONIGHT, TONIGHT, TONIGHT Genesis Atlantic 89290 2320
[14Feb87 | 04Apr87 | 23May87] 3(1) 4 | 10 | 15
[Tony Banks/Phil Collins/Mike Rutherford]
A: Invisible Touch; P: Genesis/Hugh Padgham

558. NO ONE IS TO BLAME Howard Jones Elektra 69549 2312
[12Apr86 | 05Jul86 | 13Sep86] 4(1) 5 | 14 | 23
[Howard Jones] *A: Dream Into Action; P: Phil Collins/Hugh Padgham*

559. LOVE SOMEBODY Rick Springfield RCA 13738 2297
[10Mar84 | 05May84 | 23Jun84] 5(2) 5 | 12 | 16
[Rick Springfield]
A: Hard To Hold (Soundtrack); P: Bill Drescher/Rick Springfield

560. BOY FROM NEW YORK CITY The Manhattan Transfer Atlantic 3816 2291
[23May81 | 08Aug81 | 10Oct81] 7(3) 6 | 13 | 21
[George Davis/John Issac Taylor] *A: Mecca For Moderns; P: Jay Graydon*

561. SONGBIRD Kenny G Arista 1-9588 2290
[04Apr87 | 11Jul87 | 29Aug87] 4(1) 4 | 12 | 22
[Kenny G] *A: Duotones; P: Preston Glass*

562. ROCK THE CASBAH The Clash Epic 03245 2285
[02Oct82 | 22Jan83 | 12Mar83] 8(4) 5 | 15 | 24
[Nicky Headon/Michael Geoffrey Jones/Paul Simonon/Joe Strummer]
A: Combat Rock; P: Clash

563. THE WARRIOR Scandal Featuring Patty Smyth Columbia 04424 2284
[30Jun84 | 22Sep84 | 17Nov84] 7(2) 5 | 15 | 21
[Nick Gilder/Holly Knight] *A: Warrior; P: Mike Chapman*

564. LITTLE LIES Fleetwood Mac Warner Bros. 28291 2284
[29Aug87 | 07Nov87 | 16Jan88] 4(1) 5 | 13 | 21
[Christine McVie/Eddy Quintela]
A: Tango In The Night; P: Lindsey Buckingham/Richard Dashut

565. WHEN THE CHILDREN CRY White Lion Atlantic 89015 2283
[05Nov88 | 04Feb89 | 08Apr89] 3(1) 4 | 12 | 23
[Vito Bratta/Mike Tramp] *A: Pride; P: Michael Wagener*

566. HIM Rupert Holmes MCA 41173 2280
[19Jan80 | 29Mar80 | 10May80] 6(2) 5 | 12 | 17
[Rupert Holmes] *A: Partners In Crime; P: Jim Boyer/Rupert Holmes*

567. ONLY THE LONELY The Motels Capitol 5114 2280
[24Apr82 | 17Jul82 | 25Sep82] 9(4) 4 | 15 | 23
[Martha Davis] *A: All Four One; P: Val Garay*

568. FIRE LAKE Bob Seger Capitol 4836 2278
[23Feb80 | 03May80 | 07Jun80] 6(2) 4 | 12 | 16
[Bob Seger] *A: Against The Wind; P: Muscle Shoals Rhythm Section/Bob Seger*

569. I'M ALRIGHT Kenny Loggins Columbia 11317 2264
[12Jul80 | 11Oct80 | 06Dec80] 7(2) 5 | 12 | 22
[Kenny Loggins] *A: Caddyshack Soundtrack; P: Bruce Botnick/Kenny Loggins*

570. LADY (You Bring Me Up) Commodores Motown 1514 2260
[20Jun81 | 05Sep81 | 14Nov81] 8(3) 6 | 15 | 22
[Harold Hudson/Shirley King/William King]
A: In The Pocket; P: James Anthony Carmichael/Commodores

571. DON'T LOSE MY NUMBER Phil Collins Atlantic 89536 2260
[20Jul85 | 28Sep85 | 16Nov85] 4(1) 4 | 13 | 18
[Phil Collins] *A: No Jacket Required; P: Phil Collins/Hugh Padgham*

572. SIGN YOUR NAME Terence Trent D'Arby Columbia 07911 2254
[28May88 | 13Aug88 | 15Oct88] 4(1) 5 | 13 | 21
[Terence Trent D'Arby]
A: Introducing The Hardline According To Terence Trent D'Arby;
P: Terence Trent D'Arby/Martyn Ware

573. THAT'S ALL Genesis Atlantic 89724 2253
[26Nov83 | 11Feb84 | 07Apr84] 6(2) 4 | 14 | 20
[Tony Banks/Phil Collins/Mike Rutherford]
A: Genesis; P: Genesis/Hugh Padgham

574. SAD SONGS (Say So Much) Elton John Geffen 29292 2244
[09Jun84 | 11Aug84 | 13Oct84] 5(1) 5 | 13 | 19
[Elton John/Bernie Taupin] *A: Breaking Hearts; P: Chris Thomas*

Rank. TITLE Artist Label Number	
[Entry Date | Peak Date | Exit Date] Peak(Peak Wks) | Top10 | Top40 | Tot Wks	
[Writers] *Album*; Producers	Score
575. I WANT TO BE YOUR MAN Roger Reprise 28229	2242
[14Nov87 | 13Feb88 | 02Apr88] 3(1) 5 | 13 | 21	
[Larry Troutman/Roger Troutman] *A: Unlimited*; P: Roger Troutman	
576. RUNNING WITH THE NIGHT Lionel Richie Motown 1710	2232
[26Nov83 | 04Feb84 | 31Mar84] 7(2) 5 | 14 | 19	
[Lionel Richie/Cynthia Weil]	
A: Can't Slow Down; P: James Anthony Carmichael/Lionel Richie	
577. SOMEBODY'S BABY Jackson Browne Asylum 69982	2217
[31Jul82 | 16Oct82 | 04Dec82] 7(3) 6 | 12 | 19	
[Jackson Browne/Danny Kortchmar]	
A: Fast Times At Ridgemont High Soundtrack; P: Jackson Browne	
578. LOVE SONG The Cure Elektra 69280	2215
[12Aug89 | 21Oct89 | 02Dec89] 2(1) 4 | 12 | 17	
[Simon Gallup/Roger O'Donnell/Robert James Smith/	
Porl Thompson/Laurence Tolhurst/Boris Williams]	
A: Disintegration; P: Dave Allen/Robert Smith	
579. 1-2-3 Gloria Estefan and Miami Sound Machine Epic 07921	2215
[04Jun88 | 20Aug88 | 08Oct88] 3(1) 5 | 13 | 19	
[Gloria Estefan/Enrique Garcia]	
A: Let It Loose; P: Lawrence Dermer/Emilio Estefan/Joe Galdo/Rafael Vigil	
580. TAKE ME HOME TONIGHT Eddie Money Columbia 06231	2214
[16Aug86 | 15Nov86 | 17Jan87] 4(1) 4 | 12 | 23	
[Jeff Barry/Ellie Greenwich/Mike Leeson/Phil Spector/Peter Vale]	
A: Can't Hold Back; P: Eddie Money/Richie Zito	
581. I DON'T WANT YOUR LOVE Duran Duran Capitol 44237	2212
[15Oct88 | 03Dec88 | 28Jan89] 4(2) 4 | 13 | 16	
[Simon Le Bon/Nick Rhodes/Nigel John Taylor]	
A: Big Thing; P: Daniel Abraham/Duran Duran/Jonathan Elias	
582. WORD UP Cameo Atlanta Artists 884933	2212
[13Sep86 | 22Nov86 | 31Jan87] 6(3) 5 | 14 | 21	
[Larry Blackmon/Tomi Jenkins] *A: Word Up!*; P: Larry Blackmon	
583. SUNGLASSES AT NIGHT Corey Hart EMI America 8203	2211
[26May84 | 01Sep84 | 27Oct84] 7(1) 5 | 15 | 23	
[Corey Hart] *A: First Offense*; P: Jon Astley/Phil Chapman	
584. DON'T LET IT END Styx A&M 2543	2206
[30Apr83 | 02Jul83 | 13Aug83] 6(1) 5 | 13 | 16	
[Dennis DeYoung] *A: Kilroy Was Here*; P: Styx	
585. THE BOYS OF SUMMER Don Henley Geffen 29141	2201
[10Nov84 | 09Feb85 | 06Apr85] 5(1) 4 | 14 | 22	
[Mike Campbell/Don Henley]	
A: Building The Perfect Beast;	
P: Mike Campbell/Don Henley/Danny Kortchmar/Greg Ladanyi	
586. DON'T DISTURB THIS GROOVE The System Atlantic 89320	2199
[11Apr87 | 18Jul87 | 29Aug87] 4(1) 4 | 13 | 21	
[David Frank/Mic Murphy] *A: Don't Disturb This Groove*; P: System	
587. MAKE A MOVE ON ME Olivia Newton-John MCA 52000	2194
[13Feb82 | 03Apr82 | 15May82] 5(3) 5 | 10 | 14	
[John Farrar/Tom Snow] *A: Physical*; P: John Farrar	
588. (She's) SEXY + 17 Stray Cats EMI America 8168	2192
[06Aug83 | 01Oct83 | 12Nov83] 5(2) 5 | 14 | 15	
[Brian Setzer]	
A: Rant 'N' Rave With The Stray Cats; P: Dave Edmunds	
589. ALL THROUGH THE NIGHT Cyndi Lauper Portrait 37-04639	2186
[06Oct84 | 08Dec84 | 09Feb85] 5(1) 5 | 14 | 19	
[Jules Shear] *A: She's So Unusual*; P: Rick Chertoff	
590. DON'T YOU WANT ME Jody Watley MCA 53162	2173
[03Oct87 | 19Dec87 | 05Mar88] 6(3) 6 | 14 | 23	
[David Paul Bryant/Franne Golde/Jody Watley]	
A: Jody Watley; P: Bernard Edwards	
591. REAL LOVE The Doobie Brothers Warner Bros. 49503	2173
[06Sep80 | 25Oct80 | 20Dec80] 5(2) 5 | 11 | 16	
[Patrick Henderson/Michael McDonald]	
A: One Step Closer; P: Ted Templeman	
592. TOO SHY Kajagoogoo EMI America 8161	2160
[23Apr83 | 09Jul83 | 27Aug83] 5(1) 4 | 12 | 19	
[Steven Askew/Nick Beggs/Christopher Hamill/Stuart Neale/Jeremy Strode]	
A: White Feathers; P: Nick Rhodes/Colin Thurston	
593. KEEP THE FIRE BURNIN' REO Speedwagon Epic 02967	2157
[12Jun82 | 14Aug82 | 25Sep82] 7(3) 6 | 13 | 16	
[Kevin Cronin]	
A: Good Trouble; P: Kevin Beamish/Kevin Cronin/Alan Gratzer/Gary Richrath	

Rank. TITLE Artist Label Number	
[Entry Date | Peak Date | Exit Date] Peak(Peak Wks) | Top10 | Top40 | Tot Wks	
[Writers] *Album*; Producers	Score
594. MANDOLIN RAIN Bruce Hornsby And The Range RCA 5087	2155
[17Jan87 | 21Mar87 | 16May87] 4(1) 5 | 12 | 18	
[Bruce Hornsby/John Hornsby]	
A: The Way It Is; P: Bruce Hornsby/Elliot Scheiner	
595. FREEDOM Wham! Columbia 05409	2153
[27Jul85 | 28Sep85 | 23Nov85] 3(1) 4 | 12 | 18	
[George Michael] *A: Make It Big*; P: George Michael	
596. MY LOVE Lionel Richie Motown 1677	2148
[09Apr83 | 11Jun83 | 23Jul83] 5(1) 5 | 12 | 16	
[Lionel Richie] *A: Lionel Richie*; P: James Anthony Carmichael/Lionel Richie	
597. LIVING IN AMERICA James Brown Scotti Brothers 05682	2143
[07Dec85 | 01Mar86 | 12Apr86] 4(1) 5 | 11 | 19	
[Dan Hartman/Charles E. Kaufman] *A: Rocky IV Soundtrack*; P: Dan Hartman	
598. NO MORE LONELY NIGHTS Paul McCartney Columbia 04581	2137
[13Oct84 | 08Dec84 | 09Feb85] 6(2) 5 | 14 | 18	
[Paul McCartney] *A: Give My Regards To Broad Street*; P: George Martin	
599. OBSESSION Animotion Mercury 880266	2137
[26Jan85 | 04May85 | 06Jul85] 6(1) 5 | 14 | 24	
[Michael Des Barres/Holly Knight] *A: Animotion*; P: John Ryan	
600. EVERYTHING YOUR HEART DESIRES Daryl Hall & John Oates	2134
Arista 1-9684 [16Apr88 | 11Jun88 | 30Jul88] 3(1) 5 | 11 | 16	
[Daryl Hall] *A: Ooh Yeah!*; P: Daryl Hall/John Oates/Tom Wolk	
601. SMALL TOWN John Cougar Mellencamp Riva 884 202	2127
[02Nov85 | 28Dec85 | 01Mar86] 6(4) 5 | 13 | 18	
[John Mellencamp] *A: Scarecrow*; P: Don Gehman/John Mellencamp	
602. GLORY DAYS Bruce Springsteen Columbia 04924	2124
[01Jun85 | 03Aug85 | 28Sep85] 5(1) 5 | 13 | 18	
[Bruce Springsteen] *A: Born In The U.S.A.*;	
P: Jon Landau/Chuck Plotkin/Bruce Springsteen/Steven Van Zandt	
603. DON'T BE CRUEL Cheap Trick Epic 07965	2120
[30Jul88 | 08Oct88 | 19Nov88] 4(2) 5 | 12 | 17	
[Otis Blackwell/Elvis Presley] *A: Lap Of Luxury*; P: Richie Zito	
604. RUN TO YOU Bryan Adams A&M 2686	2116
[03Nov84 | 19Jan85 | 09Mar85] 6(1) 4 | 12 | 19	
[Bryan Adams/Jim Vallance] *A: Reckless*; P: Bryan Adams/Bob Clearmountain	
605. THROWING IT ALL AWAY Genesis Atlantic 89372	2113
[16Aug86 | 11Oct86 | 29Nov86] 4(2) 4 | 12 | 16	
[Tony Banks/Phil Collins/Mike Rutherford]	
A: Invisible Touch; P: Genesis/Hugh Padgham	
606. HEARTS Marty Balin EMI America 8084	2113
[23May81 | 08Aug81 | 10Oct81] 8(2) 4 | 13 | 21	
[Jesse Barish] *A: Balin*; P: John Hug	
607. STRUT Sheena Easton EMI America 8227	2111
[25Aug84 | 24Nov84 | 09Feb85] 7(1) 3 | 15 | 25	
[Charlie Dore/Julian Littman] *A: A Private Heaven*; P: Greg Mathieson	
608. IF THIS IS IT Huey Lewis And The News Chrysalis 42803	2110
[21Jul84 | 15Sep84 | 10Nov84] 6(2) 4 | 13 | 17	
[Johnny Colla/Huey Lewis] *A: Sports*; P: Huey Lewis And The News	
609. I CAN'T TELL YOU WHY Eagles Asylum 46608	2108
[23Feb80 | 19Apr80 | 07Jun80] 8(3) 4 | 12 | 16	
[Glenn Frey/Don Henley/Timothy B. Schmit] *A: The Long Run*; P: Bill Szymczyk	
610. (Keep Feeling) FASCINATION The Human League A&M/Virgin 2547	2106
[28May83 | 20Aug83 | 08Oct83] 8(2) 3 | 13 | 20	
[John William Callis/Philip Oakey]	
A: Fascination!; P: Human League/Martin Rushent	
611. ROCK THIS TOWN Stray Cats EMI America 8132	2104
[18Sep82 | 11Dec82 | 05Feb83] 9(5) 5 | 13 | 21	
[Brian Setzer] *A: Built For Speed*; P: Dave Edmunds	
612. THE SECOND TIME AROUND Shalamar Solar 11709	2101
[08Dec79 | 22Mar80 | 10May80] 8(2) 4 | 13 | 23	
[William Shelby/Leon Sylvers] *A: Big Fun*; P: Leon Sylvers	
613. PATIENCE Guns N' Roses Geffen 22996	2099
[08Apr89 | 03Jun89 | 05Aug89] 4(1) 5 | 12 | 18	
[Steven Adler/Saul Hudson/Duff McKagan/Axl Rose/Izzy Stradlin]	
A: G N' R Lies; P: Mike Clink	
614. TELEFONE (Long Distance Love Affair) Sheena Easton EMI America 8172	2099
[20Aug83 | 29Oct83 | 14Jan84] 9(2) 3 | 14 | 22	
[Greg Mathieson/Trevor Veitch] *A: Best Kept Secret*; P: Greg Mathieson	
615. 18 AND LIFE Skid Row Atlantic 88883	2097
[08Jul89 | 23Sep89 | 18Nov89] 4(1) 5 | 13 | 20	
[Rachel Bolan/Dave Sabo] *A: Skid Row*; P: Michael Wagener	

Ranking the Singles

Rank. TITLE Artist Label Number
[Entry Date	Peak Date	Exit Date] Peak(Peak Wks) Top10	Top40	Tot Wks
[Writers] *Album; Producers* — Score

616. ALMOST PARADISE Mike Reno And Ann Wilson Columbia 04418 — 2095
[12May84 | 14Jul84 | 22Sep84] 7(2) 4 | 13 | 20
[Eric Carmen/Dean Pitchford] *A: Footloose Soundtrack; P: Keith Olsen*

617. IT'S MY TURN Diana Ross Motown 1496 — 2085
[25Oct80 | 24Jan81 | 14Mar81] 9(3) 3 | 15 | 21
[Michael Masser/Carole Bayer Sager]
A: It's My Turn Soundtrack; P: Michael Masser

618. TELL IT TO MY HEART Taylor Dayne Arista 1-9612 — 2084
[10Oct87 | 23Jan88 | 26Mar88] 7(1) 4 | 15 | 25
[Ernie Gold/Seth Swirsky] *A: Tell It To My Heart; P: Ric Wake*

619. INFATUATION Rod Stewart Warner Bros. 29256 — 2076
[26May84 | 28Jul84 | 22Sep84] 6(2) 4 | 13 | 18
[Duane Hitchings/Roland Robinson/Rod Stewart]
A: Camouflage; P: Michael Omartian

620. THE SWEETEST TABOO Sade Portrait 37-05713 — 2075
[23Nov85 | 01Mar86 | 19Apr86] 5(1) 4 | 13 | 22
[Helen Adu/Martin Ditcham] *A: Promise; P: Robin Millar*

621. IT'S A MISTAKE Men At Work Columbia 03959 — 2066
[02Jul83 | 20Aug83 | 08Oct83] 6(2) 4 | 12 | 15
[Colin James Hay] *A: Cargo; P: Peter Mclan*

622. TOO LATE FOR GOODBYES Julian Lennon Atlantic 89589 — 2063
[26Jan85 | 23Mar85 | 18May85] 5(1) 5 | 12 | 17
[Julian Lennon] *A: Valotte; P: Phil Ramone*

623. LET THE MUSIC PLAY Shannon Mirage/Emergency 99810 — 2058
[12Nov83 | 25Feb84 | 21Apr84] 8(1) 3 | 12 | 24
[Chris Barbosa/Ed Chisolm]
A: Let The Music Play; P: Chris Barbosa/Rod Hui/Mark Liggett

624. WE DON'T HAVE TO TAKE OUR CLOTHES OFF Jermaine Stewart — 2052
Arista 1-9424
[17May86 | 09Aug86 | 11Oct86] 5(2) 4 | 13 | 22
[Preston Glass/Narada Michael Walden]
A: Frantic Romantic; P: Narada Michael Walden

625. IN MY HOUSE Mary Jane Girls Gordy 1741 — 2049
[09Mar85 | 08Jun85 | 03Aug85] 7(3) 5 | 12 | 22
[Rick James] *A: Only Four You; P: Rick James*

626. COME GO WITH ME Exposé Arista 1-9555 — 2047
[24Jan87 | 04Apr87 | 30May87] 5(1) 5 | 12 | 19
[Lewis Martineé] *A: Exposure; P: Lewis A. Martineé*

627. LEADER OF THE BAND Dan Fogelberg Full Moon 02647 — 2045
[28Nov81 | 06Mar82 | 10Apr82] 9(2) 4 | 16 | 20
[Dan Fogelberg] *A: The Innocent Age; P: Dan Fogelberg/Marty Lewis*

628. HELLO AGAIN Neil Diamond Capitol 4960 — 2040
[31Jan81 | 28Mar81 | 16May81] 6(2) 5 | 12 | 16
[Neil Diamond/Alan Lindgren] *A: The Jazz Singer (Soundtrack); P: Bob Gaudio*

629. LET'S GET SERIOUS Jermaine Jackson Motown 1469 — 2039
[29Mar80 | 12Jul80 | 30Aug80] 9(2) 3 | 14 | 23
[Lee Garrett/Stevie Wonder] *A: Let's Get Serious; P: Stevie Wonder*

630. GIRLFRIEND Pebbles MCA 53185 — 2038
[30Jan88 | 23Apr88 | 11Jun88] 5(1) 5 | 12 | 20
[Kenneth Edmonds/Antonio Reid] *A: Pebbles; P: Kenneth Edmonds/L.A. Reid*

631. LOVE'S BEEN A LITTLE BIT HARD ON ME Juice Newton Capitol 5120 — 2037
[08May82 | 10Jul82 | 28Aug82] 7(2) 4 | 13 | 17
[Gary Burr] *A: Quiet Lies; P: Richard Landis*

632. SOMETHING ABOUT YOU Level 42 Polydor 883362 — 2029
[15Feb86 | 31May86 | 16Aug86] 7(2) 3 | 14 | 27
[Wally Badarou/Boon Gould/Phil Gould/Mark King/Mike Lindup]
A: World Machine; P: Wally Badarou/Level 42

633. GIVING IT UP FOR YOUR LOVE Delbert McClinton Capitol/MSS 4948 — 2028
[06Dec80 | 21Feb81 | 11Apr81] 8(3) 5 | 14 | 19
[Jerry Lynn Williams]
A: The Jealous Kind; P: Barry Beckett/Muscle Shoals Rhythm Section

634. I LIKE IT Dino 4th & Broadway 7483 — 2013
[13May89 | 12Aug89 | 28Oct89] 7(2) 4 | 14 | 25
[Dino Esposito] *A: 24/7; P: Dino Esposito*

635. TROUBLE Lindsey Buckingham Asylum 47223 — 2011
[24Oct81 | 16Jan82 | 27Feb82] 9(2) 5 | 14 | 19
[Lindsey Buckingham]
A: Law And Order; P: Lindsey Buckingham/Richard Dashut

636. WHILE YOU SEE A CHANCE Steve Winwood Island 49656 — 2011
[07Feb81 | 18Apr81 | 06Jun81] 7(2) 5 | 12 | 18
[Will Jennings/Steve Winwood] *A: Arc Of A Diver; P: Steve Winwood*

637. THIS IS IT Kenny Loggins Columbia 11109 — 2011
[20Oct79 | 09Feb80 | 22Mar80] 11(2) 0 | 16 | 23
[Kenny Loggins/Michael McDonald] *A: Keep The Fire; P: Tom Dowd*

638. BORDERLINE Madonna Sire 29354 — 2009
[10Mar84 | 16Jun84 | 29Sep84] 10(1) 1 | 15 | 30
[Reggie Lucas] *A: Madonna; P: Jellybean Benitez/Reggie Lucas*

639. SUMMER OF '69 Bryan Adams A&M 2739 — 2007
[29Jun85 | 31Aug85 | 19Oct85] 5(2) 4 | 12 | 17
[Bryan Adams/Jim Vallance] *A: Reckless; P: Bryan Adams/Bob Clearmountain*

640. LET'S GO ALL THE WAY Sly Fox Capitol 5552 — 2000
[28Dec85 | 12Apr86 | 14Jun86] 7(2) 4 | 14 | 25
[Gary Lee Cooper] *A: Let's Go All The Way; P: Ted Currier*

641. TWILIGHT ZONE Golden Earring 21 Records 103 — 2000
[27Nov82 | 26Mar83 | 28May83] 10(2) 2 | 15 | 27
[George Kooymans] *A: Cut; P: Shell Schellekens*

642. ELECTION DAY Arcadia Capitol 5501 — 1991
[26Oct85 | 14Dec85 | 08Feb86] 6(2) 6 | 12 | 16
[Simon Le Bon/Nick Rhodes/Roger Taylor]
A: So Red The Rose; P: Arcadia/Alex Sadkin

643. ANGELIA Richard Marx EMI 50218 — 1991
[07Oct89 | 02Dec89 | 27Jan90] 4(1) 4 | 11 | 17
[Richard Marx] *A: Repeat Offender; P: David Cole/Richard Marx*

644. COME DANCING The Kinks Arista 1054 — 1990
[07May83 | 16Jul83 | 27Aug83] 6(2) 3 | 12 | 17
[Ray Davies] *A: State Of Confusion; P: Ray Davies*

645. RONI Bobby Brown MCA 53463 — 1982
[07Jan89 | 18Mar89 | 29Apr89] 3(1) 4 | 11 | 17
[Darnell Bristol/Kenneth Edmonds]
A: Don't Be Cruel; P: Kenneth Edmonds/L.A. Reid

646. SOME LIKE IT HOT The Power Station Capitol 5444 — 1980
[16Mar85 | 11May85 | 13Jul85] 6(2) 5 | 12 | 18
[Robert Palmer/Andy Taylor/Nigel John Taylor]
A: The Power Station; P: Bernard Edwards

647. ONE GOOD WOMAN Peter Cetera Full Moon 27824 — 1979
[23Jul88 | 01Oct88 | 19Nov88] 4(1) 3 | 13 | 18
[Peter Cetera/Patrick Raymond Leonard]
A: One More Story; P: Peter Cetera/Patrick Leonard

648. YOUR LOVE The Outfield Columbia 05796 — 1977
[15Feb86 | 10May86 | 12Jul86] 6(2) 4 | 12 | 22
[John Spinks] *A: Play Deep; P: William Wittman*

649. YOU MIGHT THINK The Cars Elektra 69744 — 1977
[10Mar84 | 28Apr84 | 30Jun84] 7(3) 4 | 11 | 17
[Ric Ocasek] *A: Heartbeat City; P: Cars/Mutt Lange*

650. THE LONG RUN Eagles Asylum 46569 — 1975
[08Dec79 | 02Feb80 | 15Mar80] 8(2) 3 | 12 | 15
[Glenn Frey/Don Henley]
A: The Long Run; P: Bill Szymczyk

651. YOU CAN'T HURRY LOVE Phil Collins Atlantic 89933 — 1974
[06Nov82 | 05Feb83 | 26Mar83] 10(3) 3 | 16 | 21
[Lamont Dozier/Brian Holland/Eddie Holland]
A: Hello, I Must Be Going!; P: Phil Collins

652. YOU MAY BE RIGHT Billy Joel Columbia 11231 — 1973
[15Mar80 | 03May80 | 21Jun80] 7(3) 4 | 11 | 15
[Billy Joel] *A: Glass Houses; P: Phil Ramone*

653. WORDS GET IN THE WAY Miami Sound Machine Epic 06120 — 1973
[14Jun86 | 20Sep86 | 22Nov86] 5(1) 3 | 13 | 24
[Gloria Estefan] *A: Primitive Love; P: Emilio Estefan*

654. WALK THIS WAY Run-D.M.C. Profile 5112 — 1958
[26Jul86 | 27Sep86 | 08Nov86] 4(1) 5 | 10 | 16
[Joe Perry/Steven Tyler] *A: Raising Hell; P: Rick Rubin/Russell Simmons*

655. SOLITAIRE Laura Branigan Atlantic 89868 — 1953
[19Mar83 | 21May83 | 09Jul83] 7(2) 4 | 13 | 17
[Martine Clémenceau/Diane Warren] *A: Branigan 2; P: Jack White*

656. LONELY OL' NIGHT John Cougar Mellencamp Riva 880 984 — 1951
[24Aug85 | 12Oct85 | 04Jan86] 6(2) 5 | 13 | 20
[John Mellencamp] *A: Scarecrow; P: Don Gehman/John Mellencamp*

657. OUR HOUSE Madness Geffen 29668 — 1950
[07May83 | 23Jul83 | 10Sep83] 7(1) 4 | 13 | 19
[Michael Barson/Mark Bedford/Christopher Foreman/
Graham McPherson/Charles Smyth/Lee Jay Thompson/Daniel Woodgate]
A: Madness; P: Clive Langer/Alan Winstanley

Ranking the Singles

Rank. TITLE Artist Label Number	
[Entry Date	Peak Date	Exit Date] Peak(Peak Wks) Top10	Top40	Tot Wks	
[Writers] *Album; Producers*	Score

658. MORE LOVE Kim Carnes EMI America 8045 — 1950
[31May80 | 16Aug80 | 04Oct80] 10(3) 3 | 15 | 19
[Smokey Robinson] *A: Romance Dance; P: George Tobin*

659. WALK OF LIFE Dire Straits Warner Bros. 28878 — 1949
[02Nov85 | 25Jan86 | 22Mar86] 7(1) 4 | 15 | 21
[Mark Knopfler] *A: Brothers In Arms; P: Neil Dorfsman/Mark Knopfler*

660. SECOND CHANCE Thirty Eight Special A&M 1273 — 1948
[11Feb89 | 06May89 | 01Jul89] 6(1) 4 | 14 | 21
[Max Carl/Jeff Carlisi/Cal Curtis] *A: Rock & Roll Strategy; P: Rodney Mills*

661. IN YOUR ROOM Bangles Columbia 08090 — 1943
[15Oct88 | 07Jan89 | 25Feb89] 5(1) 4 | 12 | 20
[Susanna Hoffs/Thomas F. Kelly/Billy Steinberg]
A: Everything; P: Davitt Sigerson

662. CAN'T WE TRY Dan Hill (Duet with Vonda Shepard) Columbia 07050 — 1939
[06Jun87 | 12Sep87 | 14Nov87] 6(1) 4 | 13 | 24
[Beverly Hill/Dan Hill] *A: Dan Hill(2); P: John Capek/Hank Medress*

663. I'VE DONE EVERYTHING FOR YOU Rick Springfield RCA 12166 — 1929
[22Aug81 | 07Nov81 | 16Jan82] 8(2) 4 | 12 | 22
[Sammy Hagar] *A: Working Class Dog; P: Keith Olsen*

664. WOULD I LIE TO YOU? Eurythmics RCA 14078 — 1926
[27Apr85 | 13Jul85 | 31Aug85] 5(1) 4 | 13 | 19
[Annie Lennox/David A. Stewart]
A: Be Yourself Tonight; P: David A. Stewart

665. IF EVER YOU'RE IN MY ARMS AGAIN Peabo Bryson Elektra 69728 — 1922
[12May84 | 18Aug84 | 27Oct84] 10(3) 3 | 13 | 25
[Michael Masser/Tom Snow/Cynthia Weil]
A: Straight From The Heart; P: Michael Masser

666. SARA Fleetwood Mac Warner Bros. 49150 — 1920
[15Dec79 | 02Feb80 | 15Mar80] 7(3) 5 | 11 | 14
[Stevie Nicks] *A: Tusk; P: Fleetwood Mac*

667. NOBODY TOLD ME John Lennon Polydor 817254 — 1920
[21Jan84 | 03Mar84 | 21Apr84] 5(1) 4 | 11 | 14
[John Lennon] *A: Milk And Honey; P: MISSING*

668. METHOD OF MODERN LOVE Daryl Hall & John Oates RCA 13970 — 1918
[15Dec84 | 16Feb85 | 20Apr85] 5(1) 4 | 11 | 19
[Janna Allen/Daryl Hall]
A: Big Bam Boom; P: Bob Clearmountain/Daryl Hall/John Oates

669. LEGS ZZ Top Warner Bros. 29272 — 1913
[19May84 | 21Jul84 | 22Sep84] 8(2) 4 | 12 | 19
[Frank Beard/Billy Gibbons/Dusty Hill] *A: Eliminator; P: Bill Ham*

670. DELIRIOUS Prince Warner Bros. 29503 — 1912
[03Sep83 | 22Oct83 | 31Dec83] 8(4) 4 | 11 | 18
[Prince] *A: 1999; P: Prince*

671. BREAKOUT Swing Out Sister Mercury 888016 — 1911
[15Aug87 | 14Nov87 | 16Jan88] 6(2) 4 | 11 | 23
[Andy Connell/Corinne Drewery/Martin Jackson]
A: It's Better To Travel; P: Paul Staveley O'Duffy

672. LAY YOUR HANDS ON ME Thompson Twins Arista 1-9396 — 1908
[21Sep85 | 23Nov85 | 01Feb86] 6(2) 4 | 14 | 20
[Tom Bailey/Alannah Currie/Joe Leeway]
A: Here's To Future Days; P: Tom Bailey/Nile Rodgers/Alex Sadkin

673. RHYTHM IS GONNA GET YOU Gloria Estefan and Miami Sound Machine — 1904
Epic 07059 [30May87 | 01Aug87 | 19Sep87] 5(1) 5 | 12 | 17
[Gloria Estefan/Enrique Garcia]
A: Let It Loose; P: Lawrence Dermer/Emilio Estefan/Joe Galdo/Rafael Vigil

674. WHEN SMOKEY SINGS ABC Mercury 888604 — 1899
[04Jul87 | 19Sep87 | 07Nov87] 5(1) 4 | 12 | 19
[Martin Fry/Mark White]
A: Alphabet City; P: Bernard Edwards/Martin Fry/Mark White

675. I'M ON FIRE Bruce Springsteen Columbia 04772 — 1895
[16Feb85 | 13Apr85 | 29Jun85] 6(2) 4 | 12 | 20
[Bruce Springsteen] *A: Born In The U.S.A.;*
P: Jon Landau/Chuck Plotkin/Bruce Springsteen/Steven Van Zandt

676. HURT SO BAD Linda Ronstadt Asylum 46624 — 1882
[12Apr80 | 24May80 | 12Jul80] 8(3) 4 | 11 | 14
[Bobby Hart/Teddy Randazzo/Bobby Weinstein] *A: Mad Love; P: Peter Asher*

677. COOL CHANGE Little River Band Capitol 4789 — 1881
[20Oct79 | 19Jan80 | 16Feb80] 10(1) 1 | 13 | 18
[Glenn Shorrock] *A: First Under The Wire; P: John Boylan/Little River Band*

678. ANGEL Madonna Sire 29008 — 1880
[27Apr85 | 29Jun85 | 17Aug85] 5(1) 4 | 12 | 17
[Stephen Bray/Madonna L. Ciccone] *A: Like A Virgin; P: Nile Rodgers*

679. TO BE A LOVER Billy Idol Chrysalis 43024 — 1879
[04Oct86 | 20Dec86 | 31Jan87] 6(1) 3 | 13 | 18
[William Bell/Booker T. Jones Jr.] *A: Whiplash Smile; P: Keith Forsey*

680. TONIGHT, I CELEBRATE MY LOVE Peabo Bryson/Roberta Flack — 1875
Capitol 5242 [09Jul83 | 05Nov83 | 21Jan84] 16(2) 0 | 15 | 29
[Gerry Goffin/Michael Masser] *A: Born To Love; P: Michael Masser*

681. THE WAY YOU LOVE ME Karyn White Warner Bros. 27773 — 1874
[15Oct88 | 04Feb89 | 01Apr89] 7(1) 4 | 14 | 25
[Kenneth Edmonds/Antonio Reid/Daryl Simmons]
A: Karyn White; P: Kenneth Edmonds/L.A. Reid

682. AMERICA Neil Diamond Capitol 4994 — 1871
[25Apr81 | 13Jun81 | 15Aug81] 8(3) 4 | 13 | 17
[Neil Diamond] *A: The Jazz Singer (Soundtrack); P: Bob Gaudio*

683. I KNOW THERE'S SOMETHING GOING ON Frida Atlantic 89984 — 1869
[06Nov82 | 26Mar83 | 21May83] 13(3) 0 | 12 | 29
[Russ Ballard] *A: Something's Going On; P: Phil Collins*

684. STOMP! The Brothers Johnson A&M 2216 — 1868
[15Mar80 | 24May80 | 19Jul80] 7(2) 2 | 13 | 19
[George Johnson/Louis E. Johnson/Valerie Johnson/Rod Temperton]
A: Light Up The Night; P: Quincy Jones

685. ANGEL EYES The Jeff Healey Band Arista 1-9808 — 1866
[17Jun89 | 02Sep89 | 11Nov89] 5(1) 4 | 13 | 22
[John Hiatt/Fred Koller] *A: See The Light; P: Greg Ladanyi*

686. SWEET FREEDOM Michael McDonald MCA 52857 — 1864
[14Jun86 | 30Aug86 | 25Oct86] 7(3) 3 | 13 | 20
[Rod Temperton] *A: Running Scared Soundtrack;*
P: Richard Rudolph/Bruce Swedien/Rod Temperton

687. ALL I NEED IS A MIRACLE Mike + The Mechanics Atlantic 89450 — 1864
[22Mar86 | 07Jun86 | 26Jul86] 5(1) 4 | 12 | 19
[Christopher Neil/Mike Rutherford]
A: Mike + The Mechanics; P: Christopher Neil

688. ROCKET 2 U The Jets MCA 53254 — 1861
[23Jan88 | 02Apr88 | 18Jul88] 6(2) 4 | 13 | 22
[Bobby Nunn] *A: Magic; P: Bobby Nunn*

689. WANTED DEAD OR ALIVE Bon Jovi Mercury 888467 — 1860
[11Apr87 | 06Jun87 | 01Aug87] 7(3) 5 | 12 | 17
[Jon Bon Jovi/Richie Sambora] *A: Slippery When Wet; P: Bruce Fairbairn*

690. PINK CADILLAC Natalie Cole EMI-Manhattan 50117 — 1860
[05Mar88 | 07May88 | 25Jun88] 5(2) 4 | 12 | 17
[Bruce Springsteen] *A: Everlasting; P: Dennis Lambert*

691. CATCH ME (I'm Falling) Pretty Poison Virgin 99416 — 1859
[26Sep87 | 19Dec87 | 27Feb88] 8(3) 3 | 14 | 23
[Vincent Corea/Jeanne Roberts]
A: Catch Me, I'm Falling; P: Kurt Shore/Kae Williams Jr.

692. WASTED ON THE WAY Crosby, Stills & Nash Atlantic 4058 — 1857
[26Jun82 | 21Aug82 | 02Oct82] 9(4) 5 | 12 | 15
[Graham Nash]
A: Daylight Again;
P: Crosby, Stills & Nash/Steve Gursky/Stanley Johnston

693. COVER ME Bruce Springsteen Columbia 04561 — 1853
[11Aug84 | 20Oct84 | 08Dec84] 7(1) 4 | 13 | 18
[Bruce Springsteen] *A: Born In The U.S.A.; P: Jon Landau/Chuck Plotkin/Bruce*
Springsteen/Steven Van Zandt

694. ALWAYS SOMETHING THERE TO REMIND ME Naked Eyes — 1852
EMI-America 8155 [12Mar83 | 11Jun83 | 06Aug83] 8(2) 2 | 13 | 22
[Burt Bacharach/Hal David] *A: Naked Eyes; P: Tony Mansfield*

695. SHOWER ME WITH YOUR LOVE Surface Columbia 68746 — 1852
[01Jul89 | 16Sep89 | 04Nov89] 5(1) 4 | 11 | 19
[Bernard Leon Jackson]
A: 2nd Wave; P: David Conley/Bernard Jackson/David Townsend

696. FAMILY MAN Daryl Hall & John Oates RCA Victor 13507 — 1851
[30Apr83 | 25Jun83 | 13Aug83] 6(1) 3 | 12 | 16
[Tim Cross/Richard Fenn/Mike Frye/Mike Oldfield/Morris Pert/Maggie Reilly]
A: H2O; P: Daryl Hall/John Oates

697. BIG LOVE Fleetwood Mac Warner Bros. 28398 — 1847
[28Mar87 | 30May87 | 11Jul87] 5(1) 5 | 11 | 16
[Lindsey Buckingham]
A: Tango In The Night; P: Lindsey Buckingham/Richard Dashut

698. SUDDENLY LAST SUMMER The Motels Capitol 5271 — 1846
[03Sep83 | 19Nov83 | 14Jan84] 9(1) 2 | 13 | 20
[Martha Davis] *A: Little Robbers; P: Val Garay*

Ranking the Singles

Rank. TITLE Artist Label Number
[Entry Date | Peak Date | Exit Date] Peak(Peak Wks) Top10 | Top40 | Tot Wks
[Writers] *Album;* Producers — Score

699. HUMAN NATURE Michael Jackson Epic 04026 — 1844
[23Jul83 | 17Sep83 | 22Oct83] 7(2) 4 | 11 | 14
[John Bettis/Steve Porcaro] A: *Thriller;* P: Quincy Jones

700. I WANT HER Keith Sweat Vintertainment 7-69431 — 1842
[16Jan88 | 02Apr88 | 28May88] 5(1) 5 | 13 | 20
[Teddy Riley/Keith Sweat] A: *Make It Last Forever;* P: Keith Sweat

701. TIRED OF TOEIN' THE LINE Rocky Burnette EMI America 8043 — 1840
[10May80 | 26Jul80 | 13Sep80] 8(2) 2 | 12 | 19
[Rocky Burnette/Ron Coleman] A: *The Son Of Rock And Roll;* P: Bill House/Jim Seiter

702. DIAMONDS Herb Alpert A&M 2929 — 1839
[11Apr87 | 20Jun87 | 15Aug87] 5(1) 4 | 12 | 19
[James Samuel Harris/Terry Lewis] A: *Keep Your Eye On Me;* P: Jimmy Jam Harris/Terry Lewis

703. STAND BY ME Ben E. King Atlantic 89361 — 1837
[04Oct86 | 20Dec86 | 21Feb87] 9(3) 4 | 13 | 21
[Ben E. King/Jerry Leiber/Mike Stoller] A: *Stand By Me Soundtrack;* P: Jerry Leiber/Mike Stoller

704. WHIP IT Devo Warner Bros. 49550 — 1837
[30Aug80 | 15Nov80 | 14Feb81] 14(3) 0 | 15 | 25
[Jerry Casale/Mark Mothersbaugh] A: *Freedom Of Choice;* P: Devo/Robert Margouleff

705. SILENT RUNNING (ON DANGEROUS GROUND) Mike + The Mechanics Atlantic 89488 — 1835
[23Nov85 | 08Mar86 | 03May86] 6(1) 4 | 11 | 24
[Brian A. Robertson/Mike Rutherford] A: *Mike + The Mechanics;* P: Christopher Neil

706. SAY YOU WILL Foreigner Atlantic 89169 — 1833
[05Dec87 | 20Feb88 | 09Apr88] 6(1) 4 | 12 | 19
[Lou Gramm/Michael Leslie Jones] A: *Inside Information;* P: Frank Filipetti/Mick Jones

707. CONGA Miami Sound Machine Epic 05457 — 1829
[19Oct85 | 08Feb86 | 19Apr86] 10(2) 2 | 16 | 27
[Enrique Garcia] A: *Primitive Love;* P: Emilio Estefan

708. AFFAIR OF THE HEART Rick Springfield RCA 13497 — 1827
[16Apr83 | 18Jun83 | 13Aug83] 9(2) 3 | 13 | 18
[Rick Springfield/Danny Tate/Blaise Tosti] A: *Living In Oz;* P: Bill Drescher/Rick Springfield

709. HEART AND SOUL Huey Lewis And The News Chrysalis 42726 — 1827
[10Sep83 | 26Nov83 | 04Feb84] 8(1) 4 | 13 | 21
[Mike Chapman/Nicky Chinn] A: *Sports;* P: Huey Lewis And The News

710. ADULT EDUCATION Daryl Hall & John Oates RCA 13714 — 1822
[18Feb84 | 07Apr84 | 09Jun84] 8(1) 4 | 11 | 17
[Sara Allen/Daryl Hall/John Oates] A: *Rock 'N Soul Part 1;* P: Bob Clearmountain/Daryl Hall/John Oates

711. HEARTBREAKER Dionne Warwick Arista 1015 — 1820
[09Oct82 | 15Jan83 | 05Mar83] 10(2) 2 | 13 | 22
[Barry Gibb/Maurice Gibb/Robin Gibb] A: *Heartbreaker;* P: Albhy Galuten/Barry Gibb/Karl Richardson

712. PENNY LOVER Lionel Richie Motown 1762 — 1819
[06Oct84 | 01Dec84 | 02Feb85] 8(2) 4 | 13 | 18
[Brenda Richie/Lionel Richie] A: *Can't Slow Down;* P: James Anthony Carmichael/Lionel Richie

713. CASANOVA Levert Atlantic 89217 — 1814
[15Aug87 | 31Oct87 | 12Dec87] 5(1) 4 | 12 | 18
[Reggie Calloway] A: *The Big Throwdown;* P: Reggie Calloway

714. DREAMING Cliff Richard EMI America 8057 — 1812
[13Sep80 | 22Nov80 | 07Feb81] 10(3) 3 | 13 | 22
[Leo Sayer/Alan Tarney] A: *I'm No Hero;* P: Alan Tarney

715. THE VALLEY ROAD Bruce Hornsby & The Range RCA 7645 — 1810
[30Apr88 | 02Jul88 | 13Aug88] 5(1) 5 | 11 | 16
[Bruce Hornsby/John Hornsby] A: *Scenes From The Southside;* P: Neil Dorfsman/Bruce Hornsby

716. IF IT ISN'T LOVE New Edition MCA 53264 — 1807
[02Jul88 | 17Sep88 | 19Nov88] 7(1) 4 | 13 | 21
[James Samuel Harris/Terry Lewis] A: *Heart Break;* P: Jimmy Jam Harris/Terry Lewis

717. SOMEDAY Glass Tiger Manhattan 50048 — 1807
[01Nov86 | 24Jan87 | 21Mar87] 7(2) 3 | 13 | 21
[Al Connelly/Alan Frew/Jim Vallance] A: *The Thin Red Line;* P: Jim Vallance

718. SECRET RENDEZVOUS Karyn White Warner Bros. 27863 — 1805
[27May89 | 26Aug89 | 14Oct89] 6(1) 3 | 13 | 21
[Kenneth Edmonds/Antonio Reid/Daryl Simmons] A: *Karyn White;* P: Kenneth Edmonds/L.A. Reid

719. ALWAYS ON MY MIND Pet Shop Boys EMI-Manhattan 50123 — 1802
[26Mar88 | 21May88 | 02Jul88] 4(1) 4 | 10 | 15
[Johnny Christopher/Mark James/Wayne Carson Thompson] A: *Introspective;* P: Julian Mendelsohn/Pet Shop Boys

720. ALL CRIED OUT Lisa Lisa And Cult Jam With Full Force Featuring Paul Anthony And Bow Legged Lou Columbia 05844 — 1799
[26Jul86 | 25Oct86 | 17Jan87] 8(1) 3 | 13 | 26
[Curtis Bedeau/Gerard Charles/Hugh Clarke/Brian George/Lucien George/Paul George]
A: *Lisa Lisa & Cult Jam With Full Force;* P: Full Force

721. TAKE ME HOME Phil Collins Atlantic 89472 — 1797
[15Mar86 | 10May86 | 28Jun86] 7(3) 5 | 11 | 16
[Phil Collins] A: *No Jacket Required;* P: Phil Collins/Hugh Padgham

722. BREAKDANCE Irene Cara Geffen 29328 — 1793
[24Mar84 | 09Jun84 | 28Jul84] 8(1) 3 | 11 | 19
[Irene Cara/Bunny Hull/Giorgio Moroder] A: *What A Feelin';* P: Giorgio Moroder

723. POINT OF NO RETURN Exposé Arista 1-9579 — 1791
[09May87 | 18Jul87 | 29Aug87] 5(1) 4 | 11 | 17
[Lewis Martineé] A: *Exposure;* P: Lewis A. Martineé

724. DE DO DO DO, DE DA DA DA The Police A&M 2275 — 1789
[25Oct80 | 17Jan81 | 14Mar81] 10(2) 2 | 13 | 21
[Sting] A: *Zenyatta Mondatta;* P: Nigel Gray/Police

725. FAST CAR Tracy Chapman Elektra 69412 — 1788
[04Jun88 | 27Aug88 | 22Oct88] 6(2) 4 | 12 | 21
[Tracy Chapman] A: *Tracy Chapman;* P: David Kershenbaum

726. HOW DO I MAKE YOU Linda Ronstadt Asylum 46602 — 1783
[02Feb80 | 22Mar80 | 17May80] 10(3) 3 | 12 | 16
[Billy Steinberg] A: *Mad Love;* P: Peter Asher

727. ROCK STEADY The Whispers Solar 70006 — 1781
[06Jun87 | 29Aug87 | 07Nov87] 7(1) 3 | 12 | 23
[Kenneth Edmonds/Antonio Reid/Bo Watson] A: *Just Gets Better With Time;* P: Kenneth Edmonds/L.A. Reid

728. THE FINER THINGS Steve Winwood Island 28498 — 1776
[07Feb87 | 18Apr87 | 11Jul87] 8(3) 4 | 12 | 23
[Will Jennings/Steve Winwood] A: *Back In The High Life;* P: Russ Titelman/Steve Winwood

729. WHO'S ZOOMIN' WHO Aretha Franklin Arista 1-9410 — 1775
[28Sep85 | 30Nov85 | 01Feb86] 7(1) 3 | 13 | 19
[Aretha Franklin/Preston Glass/Narada Michael Walden] A: *Who's Zoomin' Who?;* P: Narada Michael Walden

730. BIG TIME Peter Gabriel Geffen 28503 — 1775
[29Nov86 | 07Mar87 | 02May87] 8(2) 4 | 11 | 23
[Peter Gabriel] A: *So;* P: Peter Gabriel/Daniel Lanois

731. DO YOU BELIEVE IN LOVE Huey Lewis And The News Chrysalis 2589 — 1774
[06Feb82 | 17Apr82 | 29May82] 7(3) 4 | 13 | 17
[Mutt Lange] A: *Picture This;* P: Huey Lewis And The News

732. HARLEM SHUFFLE The Rolling Stones Rolling Stones 38-05802 — 1773
[15Mar86 | 03May86 | 07Jun86] 5(1) 5 | 10 | 13
[Earl Nelson/Bob Relf] A: *Dirty Work;* P: Glimmer Twins/Steve Lillywhite

733. MISSING YOU Diana Ross RCA 13966 — 1770
[01Dec84 | 13Apr85 | 01Jun85] 10(2) 2 | 9 | 27
[Lionel Richie] A: *Swept Away;* P: James Anthony Carmichael/Lionel Richie

734. AFTER ALL (Love Theme From Chances Are) Cher and Peter Cetera Geffen 27529 — 1766
[11Mar89 | 13May89 | 22Jul89] 6(1) 4 | 11 | 20
[Dean Pitchford/Tom Snow] A: *Chances Are Soundtrack;* P: Peter Asher

735. LOVE TOUCH Rod Stewart Warner Bros. 28668 — 1765
[31May86 | 09Aug86 | 27Sep86] 6(1) 4 | 12 | 18
[Gene Black/Mike Chapman/Holly Knight] A: *Rod Stewart;* P: Mike Chapman

736. CANDLE IN THE WIND (Live) Elton John MCA 53196 — 1764
[07Nov87 | 23Jan88 | 26Mar88] 6(1) 3 | 12 | 21
[Elton John/Bernie Taupin]
A: *Live In Australia (With The Melbourne Symphony Orchestra);* P: Gus Dudgeon

Rank. TITLE Artist Label Number	
[Entry Date | Peak Date | Exit Date] Peak(Peak Wks) | Top10 | Top40 | Tot Wks	
[Writers] *Album; Producers*	Score

737. CHURCH OF THE POISON MIND Culture Club Virgin/Epic 04144 — 1763
[22Oct83 | 03Dec83 | 11Feb84] 10(3) 3 | 12 | 17
[Michael Craig/Roy Hay/Jon Moss/George O'Dowd]
A: *Colour By Numbers*; P: Steve Levine

738. MISS YOU LIKE CRAZY Natalie Cole EMI 50185 — 1760
[15Apr89 | 08Jul89 | 19Aug89] 7(2) 4 | 13 | 19
[Preston Glass/Gerry Goffin/Michael Masser]
A: *Good To Be Back*; P: Michael Masser

739. TAKE IT EASY ON ME Little River Band Capitol 5057 — 1760
[05Dec81 | 06Mar82 | 10Apr82] 10(2) 2 | 15 | 19
[Graham Goble] A: *Time Exposure*; P: George Martin

740. HEAT OF THE NIGHT Bryan Adams A&M 2921 — 1759
[28Mar87 | 16May87 | 11Jul87] 6(2) 4 | 10 | 16
[Bryan Adams/Jim Vallance]
A: *Into The Fire*; P: Bryan Adams/Bob Clearmountain

741. HARD TO SAY Dan Fogelberg Full Moon 02488 — 1758
[29Aug81 | 31Oct81 | 02Jan82] 7(2) 4 | 10 | 19
[Dan Fogelberg] A: *The Innocent Age*; P: Dan Fogelberg/Marty Lewis

742. STAND R.E.M. Warner Bros. 27688 — 1753
[21Jan89 | 08Apr89 | 27May89] 6(3) 4 | 11 | 19
[Bill Berry/Peter Buck/Mike Mills/Michael Stipe]
A: *Green*; P: Scott Litt/R.E.M.

743. MIDNIGHT BLUE Lou Gramm Atlantic 89304 — 1752
[31Jan87 | 18Apr87 | 13Jun87] 5(1) 4 | 11 | 19
[Lou Gramm/Bruce Turgon] A: *Ready Or Not*; P: Lou Gramm/Pat Moran

744. WELCOME TO THE JUNGLE Guns N' Roses Geffen 27759 — 1749
[22Oct88 | 24Dec88 | 11Feb89] 7(2) 4 | 12 | 17
[Steven Adler/Saul Hudson/Duff McKagan/Axl Rose/Izzy Stradlin]
A: *Appetite For Destruction*; P: Mike Clink

745. HEARTBEAT Don Johnson Epic 06285 — 1742
[23Aug86 | 18Oct86 | 29Nov86] 5(2) 3 | 10 | 15
[Eric Kaz/Wendy Waldman] A: *Heartbeat*; P: Chas Sandford

746. BRILLIANT DISGUISE Bruce Springsteen Columbia 07595 — 1737
[03Oct87 | 21Nov87 | 16Jan88] 5(1) 5 | 11 | 16
[Bruce Springsteen]
A: *Tunnel Of Love*; P: Jon Landau/Chuck Plotkin/Bruce Springsteen

747. YOU'VE LOST THAT LOVIN' FEELING Daryl Hall & John Oates RCA 12103 — 1734
[27Sep80 | 29Nov80 | 07Feb81] 12(3) 0 | 14 | 20
[Barry Mann/Phil Spector/Cynthia Weil] A: *Voices*; P: Daryl Hall/John Oates

748. DREAMTIME Daryl Hall RCA 14387 — 1729
[02Aug86 | 04Oct86 | 08Nov86] 5(1) 4 | 11 | 15
[John Beeby/Daryl Hall] A: *Three Hearts In The Happy Ending Machine*;
P: Daryl Hall/David A. Stewart/Tom Wolk

749. I LOVE YOU Climax Blues Band Warner Bros. 49669 — 1722
[21Feb81 | 20Jun81 | 22Aug81] 12(1) 0 | 17 | 27
[Derek Holt] A: *Flying The Flag*; P: John Ryan

750. THEY DON'T KNOW Tracey Ullman MCA/Stiff 52347 — 1719
[25Feb84 | 28Apr84 | 16Jun84] 8(2) 4 | 11 | 20
[Kirsty MacColl] A: *You Broke My Heart In 17 Places*; P: Peter Collins

751. TELL IT LIKE IT IS Heart Epic 50950 — 1716
[22Nov80 | 10Jan81 | 07Mar81] 8(2) 4 | 11 | 16
[George Richard Davis Jr./Lee Diamond] A: *Greatest Hits/Live*; P: Heart

752. CLOSE MY EYES FOREVER (remix) Lita Ford (duet with Ozzy Osbourne) — 1716
RCA 8899 [04Mar89 | 17Jun89 | 19Aug89] 8(1) 4 | 12 | 25
[Lita Ford/Ozzy Osbourne] A: *Lita*; P: Mike Chapman

753. I'LL TUMBLE 4 YA Culture Club Virgin/Epic 03912 — 1713
[02Jul83 | 27Aug83 | 15Oct83] 9(2) 4 | 12 | 16
[Michael Craig/Roy Hay/Jon Moss/George O'Dowd]
A: *Kissing To Be Clever*; P: Steve Levine

754. DRESS YOU UP Madonna Sire 28919 — 1711
[17Aug85 | 05Oct85 | 30Nov85] 5(1) 4 | 11 | 16
[Andrea LaRusso/Peggy Stanziale] A: *Like A Virgin*; P: Nile Rodgers

755. TAKE IT AWAY Paul McCartney Columbia 03018 — 1711
[10Jul82 | 21Aug82 | 23Oct82] 10(5) 5 | 11 | 16
[Paul McCartney] A: *Tug Of War*; P: George Martin

756. WHEN IT'S LOVE Van Halen Warner Bros. 27827 — 1708
[02Jul88 | 10Sep88 | 05Nov88] 5(1) 3 | 12 | 19
[Michael Anthony/Sammy Hagar/Alex Van Halen/Eddie Van Halen]
A: *OU812*; P: Donn Landee

757. MUSCLES Diana Ross RCA 13348 — 1707
[02Oct82 | 13Nov82 | 22Jan83] 10(6) 6 | 10 | 17
[Michael Jackson] A: *Silk Electric*; P: Michael Jackson

758. RUMORS Timex Social Club Jay 7001 — 1707
[14Jun86 | 16Aug86 | 18Oct86] 8(2) 4 | 12 | 19
[Alex Hill/Michael Marshall/Marcus Thompson]
A: *Rumors*; P: Denzil Foster/Jay King

759. LET ME BE THE ONE Exposé Arista 1-9617 — 1707
[15Aug87 | 31Oct87 | 09Jan88] 7(1) 3 | 13 | 22
[Lewis Martineé] A: *Exposure*; P: Lewis A. Martineé

760. BETTER LOVE NEXT TIME Dr. Hook Capitol 4785 — 1704
[13Oct79 | 19Jan80 | 16Feb80] 12(2) 0 | 14 | 19
[Larry Keith/Steve Pippin/Johnny Slate]
A: *Sometimes You Win*; P: Ron Haffkine

761. SHE'S OUT OF MY LIFE Michael Jackson Epic 50871 — 1704
[19Apr80 | 21Jun80 | 02Aug80] 10(2) 2 | 11 | 16
[Tom Bahler] A: *Off The Wall*; P: Quincy Jones

762. IS THIS LOVE Survivor Scotti Brothers 06381 — 1703
[25Oct86 | 17Jan87 | 28Feb87] 9(2) 3 | 13 | 19
[Jim Peterik/Frankie Sullivan]
A: *When Seconds Count*; P: Ron Nevison/Frankie Sullivan

763. THE OTHER GUY Little River Band Capitol 5185 — 1703
[20Nov82 | 05Feb83 | 19Mar83] 11(3) 0 | 13 | 18
[Graham Goble] A: *Greatest Hits*; P: Little River Band/Ernie Rose

764. VOICES CARRY 'til tuesday Epic 04795 — 1701
[13Apr85 | 13Jul85 | 31Aug85] 8(1) 2 | 13 | 21
[Michael Hausman/Robert Holmes/Aimee Mann/Joey Pesce]
A: *Voices Carry*; P: Mike Thorne

765. HEY NINETEEN Steely Dan MCA 51036 — 1701
[29Nov80 | 14Feb81 | 04Apr81] 10(2) 2 | 13 | 19
[Walter Becker/Donald Fagen] A: *Gaucho*; P: Gary Katz

766. SOMETHING SO STRONG Crowded House Capitol 5695 — 1699
[02May87 | 25Jul87 | 19Sep87] 7(1) 3 | 11 | 21
[Neil Finn/Mitchell Froom] A: *Crowded House*; P: Mitchell Froom

767. LOVE IN AN ELEVATOR Aerosmith Geffen 22845 — 1697
[02Sep89 | 28Oct89 | 16Dec89] 5(1) 3 | 11 | 16
[Joe Perry/Steven Tyler] A: *Pump*; P: Bruce Fairbairn

768. BAD BOY Miami Sound Machine Epic 05805 — 1695
[08Mar86 | 10May86 | 12Jul86] 8(3) 3 | 12 | 19
[Larry Dermer/Joe Galdo/Rafael Vigil] A: *Primitive Love*; P: Emilio Estefan

769. TOO MUCH TIME ON MY HANDS Styx A&M 2323 — 1694
[21Mar81 | 23May81 | 25Jul81] 9(2) 3 | 13 | 19
[Tommy Shaw] A: *Paradise Theater*; P: Styx

770. DON'T BE CRUEL Bobby Brown MCA 53327 — 1692
[23Jul88 | 15Oct88 | 14Jan89] 8(2) 2 | 14 | 26
[Kenneth Edmonds/Antonio Reid/Daryl Simmons]
A: *Don't Be Cruel*; P: Kenneth Edmonds/L.A. Reid

771. ELECTRIC BLUE Icehouse Chrysalis 43201 — 1689
[13Feb88 | 21May88 | 02Jul88] 7(1) 3 | 13 | 21
[Iva Davies/John Oates] A: *Man Of Colours*; P: David Lord

772. RESPECT YOURSELF Bruce Willis Motown 1876 — 1683
[17Jan87 | 07Mar87 | 18Apr87] 5(1) 4 | 10 | 14
[Luther Ingram/Bonny Rice] A: *The Return Of Bruno*; P: Robert Kraft

773. JUST LIKE JESSE JAMES Cher Geffen 22844 — 1680
[21Oct89 | 23Dec89 | 17Feb90] 8(4) 5 | 11 | 18
[Desmond Child/Diane Warren] A: *Heart Of Stone*; P: Desmond Child

774. SLEEPING BAG ZZ Top Warner Bros. 28884 — 1676
[19Oct85 | 14Dec85 | 08Feb86] 8(1) 4 | 13 | 17
[Frank Beard/Billy Gibbons/Dusty Hill] A: *Afterburner*; P: Bill Ham

775. ON THE DARK SIDE John Cafferty And The Beaver Brown Band Scotti — 1673
Brothers 04594
[18Aug84 | 27Oct84 | 15Dec84] 7(2) 3 | 11 | 18
[John Cafferty] A: *Eddie & The Cruisers (Soundtrack)*; P: Kenny Vance

776. NIKITA Elton John Geffen 28800 — 1673
[18Jan86 | 22Mar86 | 17May86] 7(2) 4 | 11 | 18
[Elton John/Bernie Taupin] A: *Ice On Fire*; P: Gus Dudgeon

777. DON'T DO ME LIKE THAT Tom Petty And The Heartbreakers — 1671
Backstreet 41138 [17Nov79 | 02Feb80 | 15Mar80] 10(2) 2 | 13 | 18
[Tom Petty] A: *Damn The Torpedoes*; P: Jimmy Iovine/Tom Petty

778. XANADU Olivia Newton-John/Electric Light Orchestra MCA 41285 — 1669
[09Aug80 | 11Oct80 | 29Nov80] 8(2) 3 | 10 | 17
[Jeff Lynne] A: *Xanadu (Soundtrack)*; P: Jeff Lynne

779. VALOTTE Julian Lennon Atlantic 89609 — 1668
[20Oct84 | 12Jan85 | 23Feb85] 9(1) 4 | 12 | 19
[Justin Clayton/Julian Lennon/Carlton Morales] A: *Valotte*; P: Phil Ramone

Ranking the Singles

Rank. TITLE Artist Label Number
[Entry Date | Peak Date | Exit Date] Peak(Peak Wks) | Top10 | Top40 | Tot Wks
[Writers] *Album; Producers* Score

780. CRUMBLIN' DOWN John Cougar Mellencamp Riva 214 1668
[15Oct83 | 26Nov83 | 28Jan84] 9(3) 3 | 11 | 16
[George Green/John Mellencamp]
A: Uh-Huh; P: Don Gehman/John Mellencamp

781. TONIGHT SHE COMES The Cars Elektra 69589 1667
[02Nov85 | 11Jan86 | 22Feb86] 7(1) 4 | 12 | 17
[Ric Ocasek] *A: Greatest Hits; P: Cars/Mike Shipley*

782. GOODY TWO SHOES Adam Ant Epic 03367 1667
[13Nov82 | 12Feb83 | 02Apr83] 12(3) 0 | 14 | 21
[Adam Ant/Marco Pirroni] *A: Friend Or Foe; P: Adam Ant/Marco Pirroni*

783. ONE MOMENT IN TIME Whitney Houston Arista 1-9743 1661
[10Sep88 | 12Nov88 | 31Dec88] 5(1) 4 | 11 | 17
[John Bettis/Albert Hammond]
A: 1988 Summer Olympics-One Moment In Time; P: Narada Michael Walden

784. BRASS IN POCKET (I'm Special) The Pretenders Sire 49181 1661
[16Feb80 | 31May80 | 12Jul80] 14(2) 0 | 12 | 22
[James Honeyman-Scott/Chrissie Hynde] *A: Pretenders; P: Chris Thomas*

785. LAY YOUR HANDS ON ME Bon Jovi Mercury 874452 1661
[03Jun89 | 29Jul89 | 16Sep89] 7(2) 4 | 11 | 16
[Jon Bon Jovi/Richie Sambora] *A: New Jersey; P: Bruce Fairbairn*

786. PAC-MAN FEVER Buckner & Garcia Columbia 02673 1661
[09Jan82 | 27Mar82 | 15May82] 9(2) 3 | 14 | 19
[Jerry Buckner/Gary Garcia] *A: Pac-Man Fever; P: Jerry Buckner/Gary Garcia*

787. I'M SO EXCITED ('84) Pointer Sisters Planet 13857 1658
[04Aug84 | 27Oct84 | 12Jan85] 9(1) 2 | 12 | 24
[Trevor Lawrence/Anita Pointer/June Pointer/Ruth Pointer]
A: Break Out; P: Richard Perry

788. UNDERCOVER OF THE NIGHT The Rolling Stones Rolling Stones 99813 1657
[12Nov83 | 24Dec83 | 11Feb84] 9(3) 3 | 10 | 14
[Mick Jagger/Keith Richards] *A: Undercover; P: Glimmer Twins/Chris Kimsey*

789. PARADISE CITY Guns N' Roses Geffen 27570 1657
[14Jan89 | 11Mar89 | 06May89] 5(2) 3 | 11 | 16
[Steven Adler/Saul Hudson/Duff McKagan/Axl Rose/Izzy Stradlin]
A: Appetite For Destruction; P: Mike Clink

790. WALK ON WATER Eddie Money Columbia 08060 1653
[01Oct88 | 24Dec88 | 18Feb89] 9(2) 4 | 13 | 21
[Jesse Harms] *A: Nothing To Lose; P: Eddie Money/Richie Zito*

791. I DON'T WANT TO LIVE WITHOUT YOU Foreigner Atlantic 89101 1652
[19Mar88 | 28May88 | 09Jul88] 5(1) 4 | 11 | 17
[Michael Leslie Jones] *A: Inside Information; P: Mick Jones*

792. DAYDREAM BELIEVER Anne Murray Capitol 4813 1648
[22Dec79 | 23Feb80 | 12Apr80] 12(3) 0 | 11 | 17
[John Stewart] *A: I'll Always Love You; P: Jim Ed Norman*

793. DON'T YOU KNOW WHAT THE NIGHT CAN DO? Steve Winwood
Virgin 99290 1648
[20Aug88 | 29Oct88 | 10Dec88] 6(1) 3 | 11 | 17
[Will Jennings/Steve Winwood]
A: Roll With It; P: Tom Lord-Alge/Steve Winwood

794. I RAN (SO FAR AWAY) A Flock Of Seagulls Jive 102 1647
[10Jul82 | 23Oct82 | 04Dec82] 9(2) 4 | 10 | 22
[Frank Maudsley/Paul Reynolds/Ali Score/Mike Score]
A: A Flock Of Seagulls; P: Mike Howlett

795. BE NEAR ME ABC Mercury 880626 1646
[24Aug85 | 09Nov85 | 18Jan86] 9(2) 3 | 11 | 22
[Martin Fry/Mark White]
A: How To Be A...Zillionaire!; P: Martin Fry/Mark White

796. LIFE IN A NORTHERN TOWN The Dream Academy Warner Bros. 28841 1645
[30Nov85 | 22Feb86 | 19Apr86] 7(3) 3 | 11 | 21
[Gilbert Gabriel/Nick Laird-Clowes]
A: The Dream Academy; P: David Gilmour/Nick Laird-Clowes/George Nicholson

797. WE'LL BE TOGETHER Sting A&M 2983 1644
[10Oct87 | 05Dec87 | 06Feb88] 7(1) 4 | 12 | 18
[Sting] *A: ...Nothing Like The Sun; P: Bryan Loren/Sting*

798. LET MY LOVE OPEN THE DOOR Pete Townshend Atco 7217 1643
[14Jun80 | 16Aug80 | 18Oct80] 9(3) 3 | 12 | 19
[Pete Townshend] *A: Empty Glass; P: Chris Thomas*

799. VACATION Go-Go's I.R.S. 9907 1638
[03Jul82 | 21Aug82 | 02Oct82] 8(3) 4 | 9 | 14
[Charlotte Caffey/Kathy Valentine/Jane Wiedlin]
A: Vacation; P: Richard Gottehrer

800. SWEETHEART Franke And The Knockouts Millennium 11801 1638
[07Mar81 | 06Jun81 | 11Jul81] 10(2) 2 | 14 | 19
[Billy Elworthy/Franke Previte] *A: Franke & The Knockouts; P: Steve Verroca*

801. DID IT IN A MINUTE Daryl Hall & John Oates RCA 13065 1635
[20Mar82 | 22May82 | 03Jul82] 9(2) 4 | 11 | 16
[Janna Allen/Sara Allen/Daryl Hall] *A: Private Eyes; P: Daryl Hall/John Oates*

802. DON'T SHED A TEAR Paul Carrack Chrysalis 43164 1635
[14Nov87 | 13Feb88 | 23Apr88] 9(3) 3 | 13 | 24
[Rob Friedman/Eddie Schwartz] *A: One Good Reason; P: Christopher Neil*

803. I HATE MYSELF FOR LOVING YOU Joan Jett And The Blackhearts 1633
Blackheart 4-07919
[25Jun88 | 01Oct88 | 17Dec88] 8(2) 3 | 12 | 26
[Desmond Child/Joan Jett]
A: Up Your Alley; P: Desmond Child/Kenny Laguna

804. 1999 Prince Warner Bros. 29896 1633
[30Oct82 | 23Jul83 | 10Sep83] 12(2) 0 | 10 | 27
[Prince] *A: 1999; P: Prince*

805. WHO'S HOLDING DONNA NOW DeBarge Gordy 1793 1631
[01Jun85 | 10Aug85 | 05Oct85] 6(1) 3 | 12 | 19
[David Foster/Randy Goodrum/Jay Graydon]
A: Rhythm Of The Night; P: Jay Graydon

806. JUST TO SEE HER Smokey Robinson Motown 1877 1629
[28Mar87 | 04Jul87 | 15Aug87] 8(1) 4 | 12 | 21
[Jimmy George/Louis Pardini]
A: One Heartbeat; P: Peter Bunetta/Rick Chudacoff

807. HOOKED ON CLASSICS The Royal Philharmonic Orchestra RCA 12304 1627
[31Oct81 | 30Jan82 | 13Mar82] 10(2) 2 | 12 | 20
A: Hooked On Classics; P: Jeff Jarrett/Don Reedman

808. COMIN' IN AND OUT OF YOUR LIFE Barbra Streisand Columbia 02621 1626
[14Nov81 | 30Jan82 | 27Feb82] 11(1) 0 | 11 | 16
[Bobby Whiteside/Richard Lee Wold] *A: Memories; P: Andrew Lloyd Webber*

809. LET ME LOVE YOU TONIGHT Pure Prairie League Casablanca 2266 1625
[10May80 | 12Jul80 | 30Aug80] 10(1) 4 | 11 | 17
[George Daniel Greer/Jeff Wilson/Steve Woodard]
A: Firin' Up; P: John Ryan

810. PRIVATE DANCER Tina Turner Capitol 5433 1623
[19Jan85 | 23Mar85 | 18May85] 7(2) 3 | 12 | 18
[Mark Knopfler] *A: Private Dancer; P: John S. Carter*

811. ALL THIS TIME Tiffany MCA 53371 1617
[05Nov88 | 11Feb89 | 25Mar89] 6(1) 3 | 14 | 21
[Tim James/Steve McClintock] *A: Hold An Old Friend's Hand; P: George Tobin*

812. FORTRESS AROUND YOUR HEART Sting A&M 2767 1614
[24Aug85 | 26Oct85 | 04Jan86] 8(2) 3 | 11 | 20
[Sting] *A: The Dream Of The Blue Turtles; P: Pete Smith/Sting*

813. WALKING ON SUNSHINE Katrina And The Waves Capitol 5466 1612
[23Mar85 | 22Jun85 | 10Aug85] 9(1) 3 | 13 | 21
[Kimberly Rew] *A: Katrina And The Waves; P: Pat Collier/Katrina And The Waves*

814. PIANO IN THE DARK Brenda Russell Feat. Joe Esposito A&M 3003 1611
[20Feb88 | 04Jun88 | 06Aug88] 6(1) 2 | 13 | 25
[Scott Cutler/Jeff Hull/Brenda Russell]
A: Get Here; P: Andre Fischer/Jeff Hull/Brenda Russell

815. HOLD ON TIGHT ELO Jet 02408 1611
[25Jul81 | 03Oct81 | 28Nov81] 10(2) 2 | 13 | 19
[Jeff Lynne] *A: Time; P: Jeff Lynne*

816. CRIMSON AND CLOVER Joan Jett & The Blackhearts 1605
Boardwalk 7-11-144
[01May82 | 19Jun82 | 07Aug82] 7(2) 3 | 10 | 15
[Tommy James/Peter Lucia]
A: I Love Rock 'N Roll; P: Ritchie Cordell/Kenny Laguna

817. MISLED Kool & The Gang De-Lite 880431 1600
[24Nov84 | 09Mar85 | 04May85] 10(1) 1 | 13 | 24
[Robert Bell/Ronald Bell/James Bonneford/George M. Brown/
Claydes Smith/J.T. Taylor/Curtis Fitzgerald Williams]
A: Emergency; P: Ronald Bell/Jim Bonneford/Kool & The Gang

818. I CAN'T STAND IT Eric Clapton And His Band RSO 1060 1600
[28Feb81 | 02May81 | 20Jun81] 10(2) 2 | 12 | 17
[Eric Clapton] *A: Another Ticket; P: Tom Dowd*

819. COOL NIGHT Paul Davis Arista 0645 1596
[07Nov81 | 06Feb82 | 13Mar82] 11(2) 0 | 13 | 19
[Paul Davis] *A: Cool Night; P: Paul Davis/Ed Seay*

Ranking the Singles

Rank. TITLE Artist Label Number
[Entry Date | Peak Date | Exit Date] Peak(Peak Wks) Top10 | Top40 | Tot Wks
[Writers] *Album; Producers*   Score

820. PROMISES, PROMISES Naked Eyes EMI-America 8170   1595
[16Jul83 | 08Oct83 | 26Nov83] 11(1) 0 | 12 | 20
[Pete Byrne/Rob Fisher] *A: Naked Eyes; P: Tony Mansfield*

821. JESSE Carly Simon Warner Bros. 49518   1594
[02Aug80 | 01Nov80 | 03Jan81] 11(2) 0 | 13 | 23
[Mike Mainieri/Carly Simon] *A: Come Upstairs; P: Mike Mainieri*

822. ALLENTOWN Billy Joel Columbia 03413   1591
[27Nov82 | 05Feb83 | 23Apr83] 17(6) 0 | 16 | 22
[Billy Joel] *A: The Nylon Curtain; P: Phil Ramone*

823. BORN IN THE U.S.A. Bruce Springsteen Columbia 04680   1590
[10Nov84 | 19Jan85 | 02Mar85] 9(1) 2 | 11 | 17
[Bruce Springsteen] *A: Born In The U.S.A.; P: Jon Landau/Bruce Springsteen*

824. DON'T STOP BELIEVIN' Journey Columbia 02567   1589
[31Oct81 | 19Dec81 | 13Feb82] 9(3) 4 | 13 | 16
[Jonathan Cain/Steve Perry/Neal Schon]
A: Escape; P: Kevin Elson/Mike Stone

825. WRAPPED AROUND YOUR FINGER The Police A&M 2614   1588
[07Jan84 | 03Mar84 | 21Apr84] 8(1) 2 | 10 | 16
[Sting] *A: Synchronicity; P: Hugh Padgham/Police*

826. NEVER TEAR US APART INXS Atlantic 89038   1587
[13Aug88 | 05Nov88 | 14Jan89] 7(3) 3 | 11 | 23
[Andy Farriss/Michael Hutchence] *A: Kick; P: Chris Thomas*

827. PERFECT WAY Scritti Politti Warner Bros. 28949   1583
[07Sep85 | 21Dec85 | 22Feb86] 11(1) 0 | 13 | 25
[David Gamson/Green Gartside]
A: Cupid And Psyche 85; P: David Gamson/Green Gartside/Fred Maher

828. THE THEME FROM HILL STREET BLUES Mike Post Featuring Larry Carlton   1579
Elektra 47186
[22Aug81 | 14Nov81 | 16Jan82] 10(2) 2 | 10 | 22
[Mike Post] *A: Television Theme Songs; P: Mike Post*

829. DON'T STAND SO CLOSE TO ME ('81) The Police A&M 2301   1578
[07Feb81 | 11Apr81 | 06Jun81] 10(3) 3 | 13 | 18
[Sting] *A: Zenyatta Mondatta; P: Nigel Gray/Police*

830. WHO WILL YOU RUN TO Heart Capitol 44040   1573
[15Aug87 | 03Oct87 | 09Jan88] 7(3) 3 | 11 | 22
[Diane Warren] *A: Bad Animals; P: Ron Nevison*

831. CAUGHT UP IN YOU 38 Special A&M 2412   1571
[01May82 | 03Jul82 | 21Aug82] 10(3) 3 | 12 | 17
[Don Barnes/Jeff Carlisi/Jim Peterik/Frankie Sullivan]
A: Special Forces; P: Rodney Mills

832. LOVE IS IN CONTROL (Finger On The Trigger) Donna Summer   1571
Geffen 29982 [26Jun82 | 25Sep82 | 23Oct82] 10(1) 1 | 11 | 18
[Quincy Jones/Merria Ross/Rod Temperton]
A: Donna Summer; P: Quincy Jones

833. CAN'T STAY AWAY FROM YOU Gloria Estefan and Miami Sound Machine   1570
Epic 07641 [21Nov87 | 05Mar88 | 23Apr88] 6(1) 3 | 11 | 23
[Gloria Estefan]
A: Let It Loose; P: Lawrence Dermer/Emilio Estefan/Joe Galdo/Rafael Vigil

834. THIS LITTLE GIRL Gary U.S. Bonds EMI America 8079   1565
[25Apr81 | 20Jun81 | 22Aug81] 11(3) 0 | 13 | 18
[Bruce Springsteen] *A: Dedication; P: Bruce Springsteen/Steven Van Zandt*

835. I CAN'T HOLD BACK Survivor Scotti Brothers 04603   1564
[15Sep84 | 08Dec84 | 16Feb85] 13(2) 0 | 13 | 23
[Jim Peterik/Frankie Sullivan] *A: Vital Signs; P: Ron Nevison*

836. ALL RIGHT Christopher Cross Warner Bros. 29843   1564
[22Jan83 | 05Mar83 | 07May83] 12(3) 0 | 13 | 16
[Christopher Cross] *A: Another Page; P: Michael Omartian*

837. SAME OLD LANG SYNE Dan Fogelberg Full Moon/Epic 50961   1561
[13Dec80 | 21Feb81 | 11Apr81] 9(2) 2 | 13 | 18
[Dan Fogelberg] *A: The Innocent Age; P: Dan Fogelberg/Marty Lewis*

838. WHAT I AM Edie Brickell & The New Bohemians Geffen 27696   1561
[26Nov88 | 04Mar89 | 01Apr89] 7(1) 3 | 10 | 19
[Alan Aly/Edie Brickell/John Bush/John Houser/Ken Withrow]
A: Shooting Rubberbands At The Stars; P: Pat Moran

839. MAJOR TOM (Coming Home) Peter Schilling Elektra 69811   1551
[24Sep83 | 24Dec83 | 18Feb84] 14(3) 0 | 10 | 22
[David Lodge/Peter Schilling]
A: Error In The System; P: Armin Sabol/Peter Schilling

840. THINK OF LAURA Christopher Cross Warner Bros. 29658   1548
[10Dec83 | 04Feb84 | 31Mar84] 9(2) 2 | 11 | 17
[Christopher Cross] *A: Another Page; P: Michael Omartian*

841. NOTHING'S GONNA CHANGE MY LOVE FOR YOU Glenn Medeiros   1547
Amherst 311
[14Feb87 | 06Jun87 | 01Aug87] 12(1) 0 | 13 | 25
[Gerry Goffin/Michael Masser] *A: Glenn Medeiros; P: Jay Stone*

842. A DIFFERENT CORNER George Michael Columbia 05888   1547
[26Apr86 | 14Jun86 | 09Aug86] 7(3) 4 | 10 | 16
[George Michael] *A: Music From The Edge Of Heaven; P: George Michael*

843. SHE WANTS TO DANCE WITH ME Rick Astley RCA 8838   1547
[17Dec88 | 25Feb89 | 15Apr89] 6(1) 3 | 10 | 18
[Rick Astley] *A: Hold Me In Your Arms; P: Rick Astley/Ian Curnow/Phil Harding*

844. I WON'T HOLD YOU BACK Toto Columbia 03597   1544
[12Mar83 | 07May83 | 02Jul83] 10(1) 1 | 12 | 17
[Steve Lukather] *A: Toto IV; P: Toto*

845. THE LOOK OF LOVE (Part One) ABC Mercury 76168   1544
[11Sep82 | 08Jan83 | 26Feb83] 18(3) 0 | 13 | 25
[Martin Fry/David Gerald Palmer/Stephen Singleton/Mark White]
A: The Lexicon Of Love; P: Trevor Horn

846. WHEN SHE WAS MY GIRL Four Tops Casablanca 2338   1544
[15Aug81 | 07Nov81 | 09Jan82] 11(2) 0 | 11 | 22
[Marc Blatte/Larry Gottlieb] *A: Tonight!; P: David Wolfert*

847. AN INNOCENT MAN Billy Joel Columbia 04259   1543
[17Dec83 | 25Feb84 | 14Apr84] 10(1) 1 | 11 | 18
[Billy Joel] *A: An Innocent Man; P: Phil Ramone*

848. HEART TO HEART Kenny Loggins Columbia 03377   1542
[27Nov82 | 29Jan83 | 19Mar83] 15(5) 0 | 13 | 17
[David Foster/Kenny Loggins/Michael McDonald]
A: High Adventure; P: Bruce Botnick/Kenny Loggins

849. HEAVEN HELP ME Deon Estus Mika 871 538-7   1539
[25Feb89 | 29Apr89 | 10Jun89] 5(1) 2 | 11 | 16
[Deon Estus/George Michael] *A: Spell; P: George Michael*

850. LOVE WILL CONQUER ALL Lionel Richie Motown 1866   1534
[04Oct86 | 29Nov86 | 31Jan87] 9(2) 3 | 10 | 18
[Greg Phillinganes/Lionel Richie/Cynthia Weil]
A: Dancing On The Ceiling; P: James Anthony Carmichael/Lionel Richie

851. ALL SHE WANTS TO DO IS DANCE Don Henley Geffen 29065   1534
[23Feb85 | 04May85 | 29Jun85] 9(1) 2 | 11 | 19
[Danny Kortchmar]
A: Building The Perfect Beast; P: Don Henley/Danny Kortchmar/Greg Ladanyi

852. CHINA GIRL David Bowie EMI America 8165   1531
[04Jun83 | 27Aug83 | 01Oct83] 10(1) 1 | 11 | 18
[David Bowie/Iggy Pop] *A: Let's Dance; P: David Bowie/Nile Rodgers*

853. KISSING A FOOL George Michael Columbia 08050   1530
[08Oct88 | 26Nov88 | 14Jan89] 5(1) 4 | 10 | 15
[George Michael] *A: Faith; P: George Michael*

854. SPIES LIKE US Paul McCartney Capitol 5537   1530
[23Nov85 | 08Feb86 | 15Mar86] 7(1) 3 | 11 | 17
[Paul McCartney]
A: Spies Like Us Soundtrack; P: Paul McCartney/Hugh Padgham/Phil Ramone

855. THE PROMISE When In Rome Virgin 99323   1528
[03Sep88 | 10Dec88 | 18Feb89] 11(2) 0 | 13 | 25
[Clive Farrington/Mike Floreale/Andrew Mann]
A: When In Rome; P: Ben Rogan

856. I AIN'T GONNA STAND FOR IT Stevie Wonder Tamla 54320   1526
[13Dec80 | 07Mar81 | 18Apr81] 11(3) 0 | 11 | 19
[Stevie Wonder] *A: Hotter Than July; P: Stevie Wonder*

857. I FOUND SOMEONE Cher Geffen 28191   1525
[21Nov87 | 05Mar88 | 14May88] 10(1) 1 | 12 | 26
[Michael Bolton/Mark Mangold] *A: Cher(2); P: Michael Bolton*

858. CROSS MY BROKEN HEART The Jets MCA 53123   1523
[06Jun87 | 01Aug87 | 19Sep87] 7(1) 3 | 11 | 16
[Stephen Bray/Tony Pierce] *A: Magic; P: Stephen Bray/Michael Verdick*

859. BURNING DOWN THE HOUSE Talking Heads Sire 29565   1522
[30Jul83 | 22Oct83 | 10Dec83] 9(1) 1 | 11 | 20
[David Byrne/Chris Frantz/Jerry Harrison/Tina Weymouth]
A: Speaking In Tongues; P: Talking Heads

860. NITE AND DAY Al B. Sure! Warner Bros. 28192   1515
[09Apr88 | 16Jul88 | 27Aug88] 7(1) 2 | 13 | 21
[Albert Joseph Brown/Kyle West] *A: In Effect Mode; P: Kyle West*

861. MIXED EMOTIONS The Rolling Stones Columbia 69008   1515
[02Sep89 | 14Oct89 | 18Nov89] 5(2) 3 | 9 | 12
[Mick Jagger/Keith Richards] *A: Steel Wheels; P: Glimmer Twins/Chris Kimsey*

Ranking the Singles

Rank. TITLE Artist Label Number [Entry Date	Peak Date	Exit Date] Peak(Peak Wks) Top10	Top40	Tot Wks [Writers] Album; Producers	Score
862. I REMEMBER HOLDING YOU Boys Club MCA 53430 [22Oct88	14Jan89	11Mar89] 8(1) 1	12	21 [Joe Pasquale] A: Boys Club; P: David Cole/Joe Pasquale	1515
863. CHERRY BOMB John Cougar Mellencamp Mercury 888 934 [24Oct87	09Jan88	12Mar88] 8(1) 3	12	21 [John Mellencamp] A: The Lonesome Jubilee; P: Don Gehman/John Mellencamp	1512
864. JUST LIKE PARADISE David Lee Roth Warner Bros. 28119 [16Jan88	12Mar88	30Apr88] 6(1) 3	11	16 [David Lee Roth/Brett Tuggle] A: Skyscraper; P: David Lee Roth	1511
865. MY HOMETOWN Bruce Springsteen Columbia 05728 [07Dec85	25Jan86	15Mar86] 6(2) 2	9	15 [Bruce Springsteen] A: Born In The U.S.A.; P: Jon Landau/Chuck Plotkin/Bruce Springsteen/Steven Van Zandt	1511
866. SUGAR WALLS Sheena Easton EMI America 8253 [22Dec84	02Mar85	13Apr85] 9(2) 3	9	17 [Prince] A: A Private Heaven; P: Greg Mathieson/Prince	1509
867. KING FOR A DAY Thompson Twins Arista 1-9450 [18Jan86	22Mar86	03May86] 8(1) 3	11	16 [Tom Bailey/Alannah Currie/Joe Leeway] A: Here's To Future Days; P: Tom Bailey/Nile Rodgers	1507
868. NOTHIN' BUT A GOOD TIME Poison Enigma 44145 [23Apr88	09Jul88	27Aug88] 6(1) 3	11	19 [Bobby Dall/C.C. DeVille/Bret Michaels/Rikki Rockett] A: Open Up And Say...Ahh!; P: Tom Werman	1507
869. SHE'S A BEAUTY The Tubes Capitol 5217 [09Apr83	02Jul83	20Aug83] 10(1) 1	12	20 [David Foster/Steve Lukather/Fee Waybill] A: Outside Inside; P: David Foster	1506
870. BABY LOVE Regina Atlantic 89417 [21Jun86	13Sep86	01Nov86] 10(1) 2	12	20 [Stephen Bray/Mary E. Kessler/Regina Richards] A: Curiosity; P: Stephen Bray	1504
871. DESERT MOON Dennis DeYoung A&M 2666 [08Sep84	10Nov84	02Feb85] 10(1) 1	12	22 [Dennis DeYoung] A: Desert Moon; P: Dennis DeYoung	1501
872. RIGHT ON TRACK Breakfast Club MCA 52954 [14Mar87	30May87	18Jul87] 7(1) 1	11	19 [Stephen Bray/Dan Gilroy] A: The Breakfast Club; P: Jimmy Iovine	1500
873. DON'T CLOSE YOUR EYES Kix Atlantic 88902 [09Sep89	16Dec89	10Feb90] 11(1) 0	13	23 [Bob Halligan/John Palumbo/Donnie Purnell] A: Blow My Fuse; P: Tom Werman	1498
874. OFF THE WALL Michael Jackson Epic 50838 [16Feb80	12Apr80	07Jun80] 10(2) 2	11	17 [Rod Temperton] A: Off The Wall; P: Quincy Jones	1498
875. ROMEO'S TUNE Steve Forbert Nemperor 7525 [01Dec79	23Feb80	05Apr80] 11(2) 0	12	19 [Steve Forbert] A: Jackrabbit Slim; P: John Simon	1497
876. THIS TIME I KNOW IT'S FOR REAL Donna Summer Atlantic 88899 [22Apr89	24Jun89	12Aug89] 7(2) 3	10	17 [Matt Aitken/Mike Stock/Donna Summer/Pete Waterman] A: Another Place And Time; P: Matt Aitken/Mike Stock/Pete Waterman	1494
877. YESTERDAY'S SONGS Neil Diamond Columbia 02604 [07Nov81	09Jan82	13Feb82] 11(2) 0	12	15 [Neil Diamond] A: On The Way To The Sky; P: Neil Diamond	1494
878. POISON Alice Cooper Epic 68958 [23Sep89	25Nov89	27Jan90] 7(1) 3	10	19 [Desmond Child/Alice Cooper/John McCurry] A: Trash; P: Desmond Child	1494
879. HOW AM I SUPPOSED TO LIVE WITHOUT YOU Laura Branigan Atlantic 89805 [02Jul83	08Oct83	12Nov83] 12(1) 0	12	20 [Michael Bolton/Doug James] A: Branigan 2; P: Jack White	1493
880. IT'S NO CRIME Babyface Solar 68966 [12Aug89	28Oct89	09Dec89] 7(1) 4	10	18 [Kenneth Edmonds/Antonio Reid/Daryl Simmons] A: Tender Lover; P: Kenneth Edmonds/L.A. Reid	1490
881. BLUE JEAN David Bowie EMI America 8231 [15Sep84	03Nov84	12Jan85] 8(2) 2	10	19 [David Bowie] A: Tonight; P: David Bowie/Derek Bramble/Hugh Padgham	1487
882. DON'T CRY Asia Geffen 29571 [30Jul83	17Sep83	22Oct83] 10(2) 2	11	13 [Geoffrey Downes/John Wetton] A: Alpha; P: Mike Stone	1486

Rank. TITLE Artist Label Number [Entry Date	Peak Date	Exit Date] Peak(Peak Wks) Top10	Top40	Tot Wks [Writers] Album; Producers	Score
883. VALERIE ('87) Steve Winwood Island 28231 [10Oct87	19Dec87	20Feb88] 9(1) 1	12	20 [Will Jennings/Steve Winwood] A: Chronicles; P: Steve Winwood	1485
884. FRESH Kool & The Gang De-Lite 880623 [23Mar85	08Jun85	27Jul85] 9(1) 2	11	19 [Robert Bell/Ronald Bell/James Bonneford/George M. Brown/ Sandy Linzer/Claydes Smith/J.T. Taylor/Curtis Fitzgerald Williams] A: Emergency; P: Ronald Bell/Jim Bonneford/Kool & The Gang	1484
885. PINK HOUSES John Cougar Mellencamp Riva 215 [10Dec83	11Feb84	24Mar84] 8(1) 2	11	16 [John Mellencamp] A: Uh-Huh; P: Don Gehman/John Mellencamp	1484
886. INTO THE NIGHT ('80) Benny Mardones Polydor 2091 [14Jun80	06Sep80	25Oct80] 11(2) 0	12	20 [Benny Mardones/Robert Tepper] A: Never Run Never Hide; P: Barry Mraz	1483
887. HOW 'BOUT US Champaign Columbia 11433 [14Feb81	06Jun81	18Jul81] 12(1) 0	13	23 [Dana Walden] A: How 'Bout Us; P: Leo Graham	1482
888. LIVING IN SIN Bon Jovi Mercury 876070 [07Oct89	16Dec89	10Feb90] 9(1) 3	12	19 [Jon Bon Jovi] A: New Jersey; P: Bruce Fairbairn	1480
889. DOING IT ALL FOR MY BABY Huey Lewis And The News Chrysalis 43143 [18Jul87	19Sep87	31Oct87] 6(1) 3	11	16 [Phil Cody/Mike Duke] A: Fore!; P: Huey Lewis And The News	1480
890. SENTIMENTAL STREET Night Ranger Camel/MCA 52591 [25May85	27Jul85	14Sep85] 8(2) 2	11	17 [Jack Blades/Alan Fitzgerald] A: 7 Wishes; P: Pat Glasser	1480
891. ONE FINE DAY Carole King Capitol 4864 [17May80	26Jul80	06Sep80] 12(2) 0	10	17 [Gerry Goffin/Carole King] A: Pearls-Songs Of Goffin And King; P: Mark Hallman/Carole King	1479
892. BALLERINA GIRL Lionel Richie Motown 1873 [06Dec86	21Feb87	04Apr87] 7(1) 3	10	18 [Lionel Richie] A: Dancing On The Ceiling; P: James Anthony Carmichael/Lionel Richie	1473
893. SMOOTH CRIMINAL Michael Jackson Epic 08044 [12Nov88	14Jan89	18Feb89] 7(2) 3	11	15 [Michael Jackson] A: Bad; P: Quincy Jones	1470
894. FAR FROM OVER Frank Stallone RSO 815023 [30Jul83	01Oct83	12Nov83] 10(2) 2	11	16 [Vince DiCola/Frank Stallone] A: Frank Stallone; P: Johnny Mandel	1468
895. DREAMIN' Vanessa Williams Wing 871 078 [07Jan89	08Apr89	20May89] 8(1) 2	11	20 [Michael Forte/Lisa Montgomery/Geneva Paschal] A: The Right Stuff; P: Donald R. Robinson	1466
896. CRUEL SUMMER Bananarama London 810127 [21Jul84	29Sep84	17Nov84] 9(1) 2	11	18 [Sarah Dallin/Siobhan Fahey/Steve Jolley/Tony Swain/Keren Woodward] A: Bananarama; P: Steve Jolley/Tony Swain	1465
897. ROCK WIT'CHA Bobby Brown MCA 53652 [26Aug89	04Nov89	13Jan90] 7(1) 3	11	21 [Kenneth Edmonds/Daryl Simmons] A: Don't Be Cruel; P: Kenneth Edmonds/L.A. Reid	1465
898. STRAIGHT FROM THE HEART Bryan Adams A&M 2536 [12Mar83	28May83	16Jul83] 10(2) 2	11	19 [Bryan Adams/Eric Kagna] A: Cuts Like A Knife; P: Bryan Adams/Bob Clearmountain	1464
899. I WANNA HAVE SOME FUN Samantha Fox Jive 1154 [12Nov88	11Feb89	15Apr89] 8(1) 1	12	23 [Curtis Bedeau/Gerard Charles/Hugh Clarke/Brian George/ Lucien George/Paul George] A: I Wanna Have Some Fun; P: Full Force	1459
900. HEAD OVER HEELS Go-Go's I.R.S. 9926 [17Mar84	19May84	30Jun84] 11(2) 0	10	16 [Charlotte Caffey/Kathy Valentine] A: Talk Show; P: Martin Rushent	1455
901. I STILL BELIEVE Brenda K. Starr MCA 53288 [02Apr88	02Jul88	24Sep88] 13(1) 0	12	26 [Antonina Armato/Giuseppe Cantarelli] A: Brenda K. Starr; P: Eumir Deodato	1454
902. FAITHFULLY Journey Columbia 03840 [16Apr83	11Jun83	30Jul83] 12(3) 0	11	16 [Jonathan Cain] A: Frontiers; P: Kevin Elson/Mike Stone	1452
903. SURRENDER TO ME Ann Wilson and Robin Zander Capitol 44288 [24Dec88	11Mar89	29Apr89] 6(1) 2	10	19 [Richard Marx/Ross Vannelli] A: Tequila Sunrise Soundtrack; P: Richie Zito	1452

Rank. TITLE Artist Label Number	
[Entry Date | Peak Date | Exit Date] Peak(Peak Wks) | Top10 | Top40 | Tot Wks	
[Writers] *Album*; Producers	Score
904. DON'T GET ME WRONG The Pretenders Sire 28630	1450
[11Oct86 | 27Dec86 | 07Feb87] 10(2) 2 | 12 | 18	
[Chrissie Hynde] A: *Get Close*; P: Bob Clearmountain/Jimmy Iovine	
905. AN AMERICAN DREAM The Dirt Band United Artists 1330	1448
[08Dec79 | 01Mar80 | 12Apr80] 13(2) 0 | 11 | 19	
[Rodney Crowell] A: *An American Dream*; P: Bob Edwards/Jeff Hanna	
906. MEET ME HALF WAY Kenny Loggins Columbia 06690	1448
[07Mar87 | 13Jun87 | 22Aug87] 11(1) 0 | 12 | 25	
[Giorgio Moroder/Tom Whitlock]	
A: *Over The Top Soundtrack*; P: Giorgio Moroder	
907. SOLID Ashford & Simpson Capitol 5397	1448
[10Nov84 | 09Feb85 | 20Apr85] 12(2) 0 | 11 | 24	
[Nick Ashford/Valerie Simpson] A: *Solid*; P: Nick Ashford/Valerie Simpson	
908. PROVE YOUR LOVE Taylor Dayne Arista 1-9676	1445
[20Feb88 | 07May88 | 18Jun88] 7(1) 3 | 11 | 18	
[Arnie Roman/Seth Swirsky] A: *Tell It To My Heart*; P: Ric Wake	
909. VICTORY Kool & The Gang Mercury 888074	1442
[01Nov86 | 24Jan87 | 28Feb87] 10(1) 1 | 12 | 18	
[Robert Bell/Ronald Bell/George M. Brown/Claydes Smith/J.T. Taylor/	
Curtis Fitzgerald Williams] A: *Forever*; P: Kool & The Gang	
910. SWEET LOVE Anita Baker Elektra 69557	1441
[16Aug86 | 01Nov86 | 10Jan87] 8(1) 2 | 11 | 22	
[Anita Baker/Gary Bias/Louis A. Johnson] A: *Rapture*; P: Michael J. Powell	
911. OUR LIPS ARE SEALED Go-Go's I.R.S. 9901	1438
[29Aug81 | 12Dec81 | 20Mar82] 20(2) 0 | 13 | 30	
[Terry Hall/Jane Wiedlin]	
A: *Beauty And The Beat*; P: Rob Freeman/Richard Gottehrer	
912. MIRROR, MIRROR Diana Ross RCA 13021	1437
[09Jan82 | 06Mar82 | 10Apr82] 8(3) 3 | 10 | 14	
[Dennis Matkosky/Michael Sembello]	
A: *Why Do Fools Fall In Love*; P: Diana Ross	
913. GOT A HOLD ON ME Christine McVie Warner Bros. 29372	1435
[28Jan84 | 24Mar84 | 12May84] 10(1) 1 | 11 | 16	
[Christine McVie/Todd Sharp] A: *Christine McVie*; P: Russ Titelman	
914. WE'RE IN THIS LOVE TOGETHER Al Jarreau Warner Bros. 49746	1433
[01Aug81 | 07Nov81 | 09Jan82] 15(2) 0 | 11 | 24	
[Roger Murrah/Keith Stegall] A: *Breakin' Away*; P: Jay Graydon	
915. WHAT ARE WE DOIN' IN LOVE Dottie West Liberty 1404	1433
[28Mar81 | 27Jun81 | 08Aug81] 14(1) 0 | 12 | 20	
[Randy Goodrum] A: *Wild West*; P: Randy Goodrum/Brent Maher	
916. FUNKYTOWN Pseudo Echo RCA 5217	1431
[16May87 | 18Jul87 | 22Aug87] 6(1) 3 | 10 | 15	
[Steve Greenberg] A: *Love An Adventure*; P: Brian Canham	
917. WAIT White Lion Atlantic 89126	1431
[27Feb88 | 21May88 | 16Jul88] 8(1) 2 | 11 | 21	
[Vito Bratta/Mike Tramp] A: *Pride*; P: Michael Wagener	
918. I'M STILL STANDING Elton John Geffen 29639	1429
[07May83 | 09Jul83 | 20Aug83] 12(1) 0 | 12 | 19	
[Elton John/Bernie Taupin] A: *Too Low For Zero*; P: Chris Thomas	
919. (You Gotta) FIGHT FOR YOUR RIGHT (To Party!) Beastie Boys	1427
Def Jam 06595	
[20Dec86 | 07Mar87 | 18Apr87] 7(1) 2 | 10 | 18	
[Adam Horovitz/Rick Rubin/Adam Yauch]	
A: *Licensed To Ill*; P: Beastie Boys/Rick Rubin	
920. HOT ROD HEARTS Robbie Dupree Elektra 47005	1426
[19Jul80 | 11Oct80 | 15Nov80] 15(2) 0 | 12 | 18	
[Stephen Geyer/Bill LaBounty]	
A: *Robbie Dupree*; P: Peter Bunetta/Rick Chudacoff	
921. ONE IN A MILLION YOU Larry Graham Warner Bros. 49221	1425
[28Jun80 | 20Sep80 | 08Nov80] 9(2) 2 | 9 | 20	
[Sam Dees] A: *One In A Million You*; P: Larry Graham	
922. LOVIN' EVERY MINUTE OF IT Loverboy Columbia 05569	1425
[24Aug85 | 02Nov85 | 11Jan86] 9(1) 2 | 9 | 21	
[Mutt Lange] A: *Lovin' Every Minute Of It*; P: Tom Allom/Paul Dean	
923. BREAKIN'.. THERE'S NO STOPPING US	1422
Ollie And Jerry Polydor 821708	
[02Jun84 | 04Aug84 | 29Sep84] 9(1) 1 | 11 | 18	
[Ollie Brown/Jerry Knight] A: *No Album*; P: Ollie Brown	
924. HOT GIRLS IN LOVE Loverboy Columbia 03941	1420
[11Jun83 | 20Aug83 | 24Sep83] 11(1) 0 | 11 | 16	
[Paul Dean/Bruce Fairbairn] A: *Keep It Up*; P: Paul Dean/Bruce Fairbairn	

Rank. TITLE Artist Label Number	
[Entry Date | Peak Date | Exit Date] Peak(Peak Wks) | Top10 | Top40 | Tot Wks	
[Writers] *Album*; Producers	Score
925. DO WHAT YOU DO Jermaine Jackson Arista 1-9279	1417
[27Oct84 | 05Jan85 | 09Mar85] 13(1) 0 | 12 | 20	
[Ralph Dino/Larry Di Tomaso] A: *Jermaine Jackson*; P: Jermaine Jackson	
926. YOUR WILDEST DREAMS The Moody Blues Polydor 883906	1412
[19Apr86 | 12Jul86 | 06Sep86] 9(2) 2 | 12 | 21	
[Justin Hayward] A: *The Other Side Of Life*; P: Tony Visconti	
927. PILOT OF THE AIRWAVES Charlie Dore Island 49166	1411
[23Feb80 | 03May80 | 14Jun80] 13(2) 0 | 10 | 17	
[Charlie Dore] A: *Where To Now*; P: Alan Tarney/Bruce Welch	
928. THINKING OF YOU Sa-Fire Cutting 872 502-7	1406
[04Feb89 | 06May89 | 15Jul89] 12(2) 0 | 12 | 24	
[Wilma Cosmé/Russ DeSalvo/Robert Steele]	
A: *Sa-Fire*; P: Aldo Marin/Carlos Rodgers	
929. GET DOWN ON IT Kool & The Gang De-Lite 818	1405
[27Feb82 | 22May82 | 19Jun82] 10(2) 2 | 9 | 17	
[Robert Bell/Ronald Bell/George M. Brown/Eumir Deodato/	
Robert Mickens/Claydes Smith/J.T. Taylor]	
A: *Something Special*; P: Eumir Deodato/Kool & The Gang	
930. WHAT ABOUT LOVE? Heart Capitol 5481	1405
[01Jun85 | 24Aug85 | 19Oct85] 10(1) 1 | 12 | 21	
[Brian Allen/Sheron Alton/Jim Vallance] A: *Heart*; P: Ron Nevison	
931. PASS THE DUTCHIE Musical Youth MCA 52149	1404
[11Dec82 | 26Feb83 | 09Apr83] 10(2) 2 | 10 | 19	
[Headley Bennett/Huford Benjamin Brown/Lloyd Ferguson/Robert Lyn/Jackie	
Mittoo/Leroy Sibblis/Fitzroy Simpson] A: *The Youth Of Today*; P: Peter Collins	
932. PLEASE DON'T GO GIRL New Kids On The Block Columbia 07700	1400
[25Jun88 | 08Oct88 | 31Dec88] 10(1) 1 | 12 | 28	
[Maurice Starr] A: *Hangin' Tough*; P: Maurice Starr	
933. IT'S RAINING AGAIN Supertramp A&M 2502	1397
[30Oct82 | 11Dec82 | 22Jan83] 11(4) 0 | 11 | 13	
[Rick Davies/Roger Hodgson]	
A: *"...Famous Last Words..."*; P: Peter Henderson/Supertramp	
934. THE ONE I LOVE R.E.M. I.R.S. 53171	1394
[19Sep87 | 05Dec87 | 30Jan88] 9(1) 1 | 10 | 20	
[Bill Berry/Peter Buck/Mike Mills/Michael Stipe]	
A: *Document*; P: Scott Litt/R.E.M.	
935. I WANNA BE YOUR LOVER Prince Warner Bros. 49050	1389
[24Nov79 | 26Jan80 | 08Mar80] 11(2) 0 | 12 | 16	
[Prince] A: *Prince*; P: Prince	
936. THE FINAL COUNTDOWN Europe Epic 06416	1387
[24Jan87 | 28Mar87 | 23May87] 8(2) 2 | 9 | 18	
[Joey Tempest] A: *The Final Countdown*; P: Kevin Elson	
937. NEW MOON ON MONDAY Duran Duran Capitol 5309	1387
[14Jan84 | 17Mar84 | 28Apr84] 10(1) 1 | 10 | 16	
[Simon Le Bon/Nick Rhodes/Andy Taylor/Nigel John Taylor/Roger Taylor]	
A: *Seven And The Ragged Tiger*; P: Duran Duran/Ian Little/Alex Sadkin	
938. THE END OF THE INNOCENCE Don Henley Geffen 22925	1386
[24Jun89 | 26Aug89 | 21Oct89] 8(1) 2 | 12 | 18	
[Don Henley/Bruce Hornsby]	
A: *The End Of The Innocence*; P: Don Henley/Bruce Hornsby	
939. LOVE WILL TURN YOU AROUND Kenny Rogers Liberty 1471	1386
[03Jul82 | 28Aug82 | 23Oct82] 13(5) 0 | 10 | 17	
[David Malloy/Kenny Rogers/Thom Schuyler/Even Stevens]	
A: *Love Will Turn You Around*; P: David Malloy/Kenny Rogers	
940. WHAT YOU DON'T KNOW Exposé Arista 1-9836	1384
[20May89 | 15Jul89 | 26Aug89] 8(3) 3 | 11 | 17	
[Lewis Martineé] A: *What You Don't Know*; P: Lewis A. Martineé	
941. IT'S GONNA TAKE A MIRACLE Deniece Williams ARC 02812	1384
[03Apr82 | 12Jun82 | 24Jul82] 10(2) 2 | 9 | 17	
[Teddy Randazzo/Lou Stallman/Bobby Weinstein]	
A: *Niecy*; P: Thom Bell/Deniece Williams	
942. HOLD ON TO MY LOVE Jimmy Ruffin RSO 1021	1383
[01Mar80 | 03May80 | 31May80] 10(2) 2 | 9 | 14	
[Robin Gibb/Blue Weaver] A: *Sunrise*; P: Robin Gibb/Blue Weaver	
943. I'VE BEEN IN LOVE BEFORE Cutting Crew Virgin 99425	1383
[05Sep87 | 21Nov87 | 23Jan88] 9(2) 2 | 11 | 21	
[Nick Van Eede] A: *Broadcast*; P: Michael Barbiero/Steve Thompson	
944. DANCING IN THE STREET Mick Jagger & David Bowie EMI America 8288	1381
[31Aug85 | 12Oct85 | 30Nov85] 7(1) 3 | 9 | 14	
[Marvin Gaye/Ivy Jo Hunter/William Stevenson]	
A: *No Album*; P: Clive Langer/Alan Winstanley	

Ranking the Singles

Rank. TITLE Artist Label Number [Entry Date | Peak Date | Exit Date] Peak(Peak Wks) Top10 | Top40 | Tot Wks [Writers] Album; Producers	Score
945. I DROVE ALL NIGHT Cyndi Lauper Epic 68759 [06May89 | 08Jul89 | 12Aug89] 6(1) 2 | 10 | 15 [Thomas F. Kelly/Billy Steinberg] A: A Night To Remember; P: Cyndi Lauper/Lennie Petze	1380
946. NOBODY'S FOOL Kenny Loggins Columbia 07971 [09Jul88 | 17Sep88 | 05Nov88] 8(1) 3 | 11 | 18 [Kenny Loggins/Mike Towers] A: Back To Avalon; P: Dennis Lambert	1379
947. POP LIFE Prince & The Revolution Paisley Park 7-28998 [27Jul85 | 21Sep85 | 26Oct85] 7(1) 3 | 10 | 14 [Prince] A: Around The World In A Day; P: Prince And The Revolution	1376
948. LAWYERS IN LOVE Jackson Browne Asylum 69826 [09Jul83 | 10Sep83 | 15Oct83] 13(2) 0 | 12 | 15 [Jackson Browne] A: Lawyers In Love; P: Jackson Browne/Greg Ladanyi	1375
949. I MADE IT THROUGH THE RAIN Barry Manilow Arista 0566 [22Nov80 | 31Jan81 | 07Mar81] 1(1) 1 | 11 | 16 [Jack Feldman/Gerard Kenny/Barry Manilow/Drey Shepperd/Bruce Sussman] A: Barry; P: Ron Dante/Barry Manilow	1373
950. I STILL CAN'T GET OVER LOVING YOU Ray Parker Jr. Arista 1-9116 [12Nov83 | 04Feb84 | 17Mar84] 12(1) 0 | 11 | 19 [Ray Parker Jr.] A: Woman Out Of Control; P: Ray Parker Jr.	1373
951. I'LL BE OVER YOU Toto Columbia 06280 [30Aug86 | 22Nov86 | 31Jan87] 11(1) 0 | 12 | 23 [Randy Goodrum/Steve Lukather] A: Fahrenheit; P: Toto	1372
952. TOGETHER Tierra Boardwalk 5702 [08Nov80 | 21Feb81 | 28Mar81] 18(2) 0 | 15 | 21 [Kenneth Gamble/Leon Huff] A: City Nights; P: Rudy Salas	1369
953. SOMEBODY'S KNOCKIN' Terri Gibbs MCA 41309 [17Jan81 | 25Apr81 | 13Jun81] 13(2) 0 | 12 | 22 [Jerry Gillespie/Ed Penney] A: Somebody's Knockin'; P: Ed Penney	1367
954. WAR (Live) Bruce Springsteen & The E Street Band Columbia 06432 [22Nov86 | 27Dec86 | 07Feb87] 8(3) 3 | 9 | 12 [Barrett Strong/Norman Whitfield] A: Bruce Springsteen & The E Street Band Live 1975-1985; P: Jon Landau/Chuck Plotkin/Bruce Springsteen	1362
955. CIRCLE IN THE SAND Belinda Carlisle MCA 53308 [16Apr88 | 18Jun88 | 06Aug88] 7(1) 2 | 10 | 17 [Rick Nowels/Ellen Shipley] A: Heaven On Earth; P: Rick Nowels	1361
956. HIGH ON YOU Survivor Scotti Brothers 04685 [26Jan85 | 23Mar85 | 18May85] 8(2) 2 | 11 | 17 [Jim Peterik/Frankie Sullivan] A: Vital Signs; P: Ron Nevison	1361
957. MAKING LOVE Roberta Flack Atlantic 4005 [06Mar82 | 12Jun82 | 24Jul82] 13(3) 0 | 11 | 21 [Burt Bacharach/Bruce Roberts/Carole Bayer Sager] A: I'm The One; P: Burt Bacharach/Carole Bayer Sager	1361
958. WHAT KIND OF FOOL Barbra Streisand & Barry Gibb Columbia 11430 [31Jan81 | 21Mar81 | 16May81] 10(3) 3 | 10 | 16 [Albhy Galuten/Barry Gibb] A: Guilty; P: Albhy Galuten/Barry Gibb/Karl Richardson	1360
959. TIME Alan Parsons Project Arista 0598 [18Apr81 | 01Aug81 | 19Sep81] 15(3) 0 | 12 | 23 [Alan Parsons/Eric Woolfson] A: The Turn Of A Friendly Card; P: Alan Parsons	1359
960. MAGIC The Cars Elektra 69724 [19May84 | 07Jul84 | 08Sep84] 12(3) 0 | 11 | 17 [Ric Ocasek] A: Heartbeat City; P: Mutt Lange	1357
961. YOUR LOVE IS DRIVING ME CRAZY Sammy Hagar Geffen 29816 [11Dec82 | 26Feb83 | 16Apr83] 13(2) 0 | 13 | 19 [Sammy Hagar] A: Three Lock Box; P: Keith Olsen	1353
962. TARZAN BOY Baltimora Manhattan 50018 [19Oct85 | 01Mar86 | 12Apr86] 13(1) 0 | 10 | 26 [Maurizio Bassi/Naimy Hackett] A: Living In The Background; P: Maurizio Bassi	1350
963. I WOULD DIE 4 U Prince And The Revolution Warner Bros. 29121 [15Dec84 | 02Feb85 | 23Mar85] 8(1) 3 | 10 | 15 [Prince] A: Purple Rain (Soundtrack); P: Prince	1349
964. DR. FEELGOOD Mötley Crüe Elektra 69271 [02Sep89 | 28Oct89 | 16Dec89] 6(1) 2 | 9 | 16 [Mick Mars/Nikki Sixx] A: Dr. Feelgood; P: Bob Rock	1349
965. KEEP ON MOVIN' Soul II Soul Virgin 99205 [24Jun89 | 09Sep89 | 04Nov89] 11(1) 0 | 10 | 20 [Beresford Romeo] A: Keep On Movin'; P: Nellee Hooper/Beresford Romeo	1348
966. THE BREAKUP SONG (They Don't Write 'Em) Greg Kihn Band Beserkley 47149 [23May81 | 05Sep81 | 24Oct81] 15(2) 0 | 13 | 23 [Greg Kihn/Gary Philippet/Steve Wright] A: Rockihnroll; P: Matthew King Kaufman	1346
967. GET IT ON The Power Station Capitol 5479 [08Jun85 | 03Aug85 | 14Sep85] 9(2) 3 | 10 | 15 [Marc Bolan] A: The Power Station; P: Bernard Edwards	1345
968. JUST ONCE Quincy Jones Featuring James Ingram A&M 2357 [15Aug81 | 14Nov81 | 16Jan82] 17(2) 0 | 13 | 23 [Barry Mann/Cynthia Weil] A: The Dude; P: Quincy Jones	1343
969. YOU ARE MY LADY Freddie Jackson Capitol 5495 [07Sep85 | 16Nov85 | 18Jan86] 12(2) 0 | 11 | 20 [Barry Eastmond] A: Rock Me Tonight; P: Barry Eastmond	1340
970. BLUE EYES Elton John Geffen 29954 [10Jul82 | 02Oct82 | 06Nov82] 12(3) 0 | 10 | 18 [Elton John/Gary Osborne] A: Jump Up!; P: Chris Thomas	1339
971. LOVE YOU DOWN Ready For The World MCA 52947 [29Nov86 | 21Feb87 | 04Apr87] 9(1) 1 | 12 | 19 [Melvin Riley] A: Long Time Coming; P: Ready For The World/Gary Spaniola	1339
972. WALK THE DINOSAUR Was (Not Was) Chrysalis 43331 [28Jan89 | 01Apr89 | 13May89] 7(1) 3 | 9 | 16 [Randy Jacobs/David Was/Don Was] A: What Up, Dog?; P: David Was/Don Was	1333
973. WHY ME? Irene Cara Geffen 29464 [22Oct83 | 03Dec83 | 28Jan84] 13(3) 0 | 10 | 15 [Irene Cara/Keith Forsey/Giorgio Moroder] A: What A Feelin'; P: Giorgio Moroder	1330
974. WHERE ARE YOU NOW? ('89) Jimmy Harnen w/Synch WTG 31-68625 [25Feb89 | 10Jun89 | 05Aug89] 10(1) 1 | 11 | 24 [Richard Congdon/Jimmy Harnen] A: No Album; P: Jerry G. Hludzik/Bill Kelly	1326
975. THE LONGEST TIME Billy Joel Columbia 04400 [24Mar84 | 12May84 | 21Jul84] 14(3) 1 | 11 | 18 [Billy Joel] A: An Innocent Man; P: Phil Ramone	1325
976. FRIENDS Jody Watley with Eric B. & Rakim MCA 53660 [17Jun89 | 26Aug89 | 14Oct89] 9(1) 2 | 11 | 18 [Eric Barrier/André Cymone/William Griffin/Jody Watley] A: Larger Than Life; P: André Cymone	1323
977. P.Y.T. (Pretty Young Thing) Michael Jackson Epic 04165 [08Oct83 | 26Nov83 | 21Jan84] 10(1) 1 | 9 | 16 [James Ingram/Quincy Jones] A: Thriller; P: Quincy Jones	1323
978. A LITTLE IN LOVE Cliff Richard EMI America 8068 [13Dec80 | 14Mar81 | 09May81] 17(2) 0 | 11 | 22 [Alan Tarney] A: I'm No Hero; P: Alan Tarney	1321
979. BREAK IT TO ME GENTLY Juice Newton Capitol 5148 [21Aug82 | 23Oct82 | 11Dec82] 11(3) 0 | 10 | 17 [Diane Lampert/Joe Seneca] A: Quiet Lies; P: Richard Landis	1321
980. YOU'RE ONLY HUMAN (Second Wind) Billy Joel Columbia 05417 [13Jul85 | 31Aug85 | 26Oct85] 9(2) 2 | 11 | 16 [Billy Joel] A: Greatest Hits Vol. I & II; P: Phil Ramone	1320
981. A LOVE BIZARRE Sheila E. Paisley Park 7-28890 [16Nov85 | 01Mar86 | 19Apr86] 11(1) 0 | 12 | 23 [Prince/Sheila E.] A: Romance 1600; P: Prince/Sheila E.	1320
982. ROUND AND ROUND Ratt Atlantic 89693 [16Jun84 | 25Aug84 | 13Oct84] 12(2) 0 | 10 | 18 [Robbin Crosby/Warren DeMartini/Stephen Pearcy] A: Out Of The Cellar; P: Beau Hill	1317
983. THE RAIN Oran "Juice" Jones Def Jam 06209 [13Sep86 | 15Nov86 | 17Jan87] 9(1) 2 | 9 | 19 [Vincent Bell] A: Juice; P: Vincent Bell/Russell Simmons	1316
984. WATCHING THE WHEELS John Lennon Geffen 49695 [28Mar81 | 23May81 | 18Jul81] 10(2) 2 | 10 | 17 [John Lennon] A: Double Fantasy; P: Jack Douglas/John Lennon/Yoko Ono	1316
985. PAPER IN FIRE John Cougar Mellencamp Mercury 888763 [15Aug87 | 03Oct87 | 28Nov87] 9(3) 3 | 10 | 16 [John Mellencamp] A: The Lonesome Jubilee; P: Don Gehman/John Mellencamp	1315
986. GO HOME Stevie Wonder Tamla 1817 [23Nov85 | 01Feb86 | 15Mar86] 10(1) 1 | 11 | 17 [Stevie Wonder] A: In Square Circle; P: Stevie Wonder	1315
987. YOU'RE THE ONLY WOMAN (You And I) Ambrosia Warner Bros. 49508 [12Jul80 | 20Sep80 | 08Nov80] 13(2) 0 | 10 | 18 [David Pack] A: One Eighty; P: Ambrosia/Freddie Piro	1314

Rank. TITLE Artist Label Number	
[Entry Date | Peak Date | Exit Date] Peak(Peak Wks) Top10 | Top40 | Tot Wks	
[Writers] Album; Producers	Score
988. DON'T LET GO Isaac Hayes Polydor 2011	1314
[27Oct79 | 02Feb80 | 15Mar80] 18(2) 0 | 12 | 21	
[Jesse Stone] A: Don't Let Go; P: Isaac Hayes	
989. JUST GOT PAID Johnny Kemp Columbia 07744	1314
[14May88 | 13Aug88 | 01Oct88] 10(1) 1 | 11 | 21	
[Gene Griffin/Johnny Kemp] A: Secrets Of Flying; P: Johnny Kemp/Teddy Riley	
990. THE OLD MAN DOWN THE ROAD John Fogerty Warner Bros. 29100	1313
[22Dec84 | 02Mar85 | 20Apr85] 10(1) 1 | 9 | 18	
[John Fogerty] A: Centerfield; P: John Fogerty	
991. DIDN'T I (Blow Your Mind) New Kids On The Block Columbia 68960	1312
[16Sep89 | 18Nov89 | 20Jan90] 8(1) 2 | 10 | 19	
[Thom Bell/William Hart] A: New Kids On The Block; P: Maurice Starr	
992. TENDER LOVE Force M.D.'s Warner Bros. 28818	1312
[01Feb86 | 12Apr86 | 07Jun86] 10(2) 2 | 11 | 19	
[James Samuel Harris/Terry Lewis] A: Chillin'; P: Jimmy Jam Harris/Terry Lewis/Cheryl Lynn	
993. DOCTOR! DOCTOR! Thompson Twins Arista 1-9209	1310
[26May84 | 21Jul84 | 08Sep84] 11(1) 0 | 9 | 16	
[Tom Bailey/Alannah Currie/Joe Leeway] A: Into The Gap; P: Tom Bailey/Alex Sadkin	
994. INVINCIBLE (Theme From The Legend Of Billie Jean) Pat Benatar Chrysalis 41507	1310
[06Jul85 | 14Sep85 | 26Oct85] 10(1) 1 | 11 | 17	
[Simon Climie/Holly Knight] A: Seven The Hard Way; P: Mike Chapman	
995. DARE ME Pointer Sisters Planet 14126	1309
[13Jul85 | 21Sep85 | 09Nov85] 11(1) 0 | 13 | 18	
[David Innis/Sam Lorber] A: Contact; P: Richard Perry	
996. PHOTOGRAPH Def Leppard Mercury 811215	1309
[12Mar83 | 21May83 | 02Jul83] 12(1) 0 | 9 | 17	
[Rick Allen/Steve Clark/Joe Elliott/Mutt Lange/Rick Savage/Pete Willis] A: Pyromania; P: Mutt Lange	
997. THINK I'M IN LOVE Eddie Money Columbia 02964	1308
[03Jul82 | 18Sep82 | 23Oct82] 16(3) 0 | 12 | 17	
[Eddie Money/Randy Oda] A: No Control; P: Tom Dowd	
998. TONIGHT Kool & The Gang De-Lite 830	1305
[25Feb84 | 05May84 | 23Jun84] 13(2) 0 | 10 | 18	
[Clifford Adams/Robert Bell/Ronald Bell/James Bonneford/George M. Brown/Michael Ray/Claydes Smith/J.T. Taylor/Curtis Fitzgerald Williams] A: In The Heart; P: Ronald Bell/Jim Bonneford/Kool & The Gang	
999. LET'S GO Wang Chung Geffen 28531	1304
[24Jan87 | 11Apr87 | 23May87] 9(1) 2 | 11 | 18	
[Nick Feldman/Jack Hues] A: Mosaic; P: Peter Wolf	
1000. PUMP UP THE VOLUME M/A/R/R/S 4th & Broadway 7452	1300
[28Nov87 | 20Feb88 | 30Apr88] 13(3) 0 | 11 | 23	
[Andrew Martyn Biggs/Steven Anthony Biggs] A: No Album; P: Martyn Young	
1001. IS IT LOVE Mr. Mister RCA 14313	1297
[29Mar86 | 31May86 | 19Jul86] 8(1) 1 | 11 | 17	
[Steve George/John Lang/Pat Mastelotto/Richard Page] A: Welcome To The Real World; P: Paul DeVilliers/Mr. Mister	
1002. DON'T KNOW WHAT YOU GOT (Till It's Gone) Cinderella Mercury 870644	1292
[03Sep88 | 19Nov88 | 28Jan89] 12(1) 0 | 11 | 22	
[Tom Keifer] A: Long Cold Winter; P: Eric Brittingham/Andy Johns/Tom Keifer	
1003. LOVE ZONE Billy Ocean Jive 9510	1291
[26Jul86 | 27Sep86 | 08Nov86] 10(1) 0 | 11 | 16	
[Wayne Brathwaite/Barry Eastmond/Billy Ocean] A: Love Zone; P: Wayne Brathwaite/Barry Eastmond	
1004. MISUNDERSTANDING Genesis Atlantic 3662	1289
[24May80 | 16Aug80 | 20Sep80] 14(2) 0 | 11 | 18	
[Phil Collins] A: Duke; P: Genesis/David Hentschel	
1005. WAITING ON A FRIEND The Rolling Stones Rolling Stones 21004	1289
[05Dec81 | 06Feb82 | 13Mar82] 13(3) 0 | 12 | 15	
[Mick Jagger/Keith Richards] A: Tattoo You; P: Glimmer Twins	
1006. RUSH HOUR Jane Wiedlin EMI 50118	1286
[07May88 | 30Jul88 | 10Sep88] 9(1) 2 | 10 | 19	
[Peter Rafelson/Jane Wiedlin] A: Fur; P: Stephen Hague	
1007. TWO OCCASIONS The Deele Solar 70015	1285
[27Feb88 | 21May88 | 16Jul88] 10(1) 1 | 12 | 21	
[Darnell Bristol/Kenneth Edmonds/Sid Johnson] A: Eyes Of A Stranger; P: Kenneth Edmonds/L.A. Reid	
1008. IT'S A SIN Pet Shop Boys EMI America 43027	1284
[05Sep87 | 14Nov87 | 09Jan88] 9(1) 2 | 10 | 19	
[Chris Lowe/Neil Tennant] A: Actually; P: Julian Mendelsohn	
1009. WALKING AWAY Information Society Tommy Boy 27736	1284
[26Nov88 | 18Feb89 | 01Apr89] 9(1) 3 | 10 | 19	
[Paul Robb] A: Information Society; P: Fred Maher	
1010. I SAW HIM STANDING THERE Tiffany MCA 53285	1282
[27Feb88 | 23Apr88 | 28May88] 7(1) 3 | 9 | 14	
[John Lennon/Paul McCartney] A: Tiffany; P: George Tobin	
1011. FOREVER YOUNG Rod Stewart Warner Bros. 27796	1281
[06Aug88 | 15Oct88 | 14Jan89] 12(1) 0 | 10 | 24	
[Jim Cregan/Bob Dylan/Kevin Savigar/Rod Stewart] A: Out Of Order; P: Bernard Edwards/Rod Stewart/Andy Taylor	
1012. SHADOWS OF THE NIGHT Pat Benatar Chrysalis 2647	1280
[16Oct82 | 11Dec82 | 29Jan83] 13(4) 0 | 10 | 16	
[David Leigh Byron] A: Get Nervous; P: Peter Coleman/Neil Geraldo	
1013. YOU COULD HAVE BEEN WITH ME Sheena Easton EMI America 8101	1280
[28Nov81 | 20Feb82 | 27Mar82] 15(2) 0 | 12 | 18	
[Lea Maalfrid] A: You Could Have Been With Me; P: Christopher Neil	
1014. THE BEACH BOYS MEDLEY The Beach Boys Capitol 5030	1280
[25Jul81 | 03Oct81 | 21Nov81] 12(2) 0 | 11 | 18	
A: No Album; P: Missing	
1015. THIS COULD BE THE NIGHT Loverboy Columbia 05765	1278
[18Jan86 | 29Mar86 | 17May86] 10(1) 1 | 10 | 18	
[Jonathan Cain/Paul Dean/Mike Reno/Bill Wray] A: Lovin' Every Minute Of It; P: Tom Allom/Paul Dean	
1016. SMUGGLER'S BLUES Glenn Frey MCA 52546	1274
[06Apr85 | 22Jun85 | 10Aug85] 12(1) 0 | 11 | 19	
[Glenn Frey/Jack Tempchin] A: The Allnighter; P: Allan Blazek/Glenn Frey	
1017. LOVE IN THE FIRST DEGREE Alabama RCA 12288	1264
[14Nov81 | 06Mar82 | 03Apr82] 15(2) 0 | 10 | 21	
[Tim DuBois/Jim Hurt] A: Feels So Right; P: Alabama/Larry McBride/Harold Shedd	
1018. SOME GUYS HAVE ALL THE LUCK Rod Stewart Warner Bros. 29215	1263
[25Aug84 | 27Oct84 | 15Dec84] 10(1) 1 | 10 | 17	
[Jeff Fortgang] A: Camouflage; P: Michael Omartian	
1019. WIPEOUT Fat Boys and The Beach Boys Tin Pan Apple 885 960	1261
[11Jul87 | 19Sep87 | 14Nov87] 12(2) 0 | 11 | 19	
[Bob Berryhill/Pat Connolly/Jim Fuller/Ron Wilson] A: Crushin'; P: Albert Cabrera/Tony Moran	
1020. DUDE (Looks Like A Lady) Aerosmith Geffen 28240	1261
[03Oct87 | 12Dec87 | 13Feb88] 14(4) 0 | 10 | 20	
[Desmond Child/Joe Perry/Steven Tyler] A: Permanent Vacation; P: Bruce Fairbairn	
1021. JANE Jefferson Starship Grunt 11750	1259
[03Nov79 | 19Jan80 | 09Feb80] 14(1) 0 | 10 | 15	
[Craig Chaquico/David Freiberg/Paul Kantner/Jim McPherson] A: Freedom At Point Zero; P: Ron Nevison	
1022. IN AMERICA Charlie Daniels Band Epic 50888	1256
[31May80 | 02Aug80 | 06Sep80] 11(2) 0 | 8 | 15	
[Tom Crain/Charlie Daniels/Joel DiGregorio/Fred Edwards/Charles Hayward/James W. Marshall] A: Full Moon; P: John Boylan	
1023. WHEN I LOOKED AT HIM Exposé Arista 1-9868	1254
[19Aug89 | 21Oct89 | 30Dec89] 10(1) 1 | 12 | 20	
[Lewis Martineé] A: What You Don't Know; P: Lewis A. Martineé	
1024. SUPER FREAK (Part I) Rick James Gordy 7205	1254
[08Aug81 | 24Oct81 | 16Jan82] 16(2) 0 | 10 | 24	
[Rick James/Alonzo H. Miller] A: Street Songs; P: Rick James	
1025. LIGHTS OUT Peter Wolf EMI America 8208	1249
[14Jul84 | 08Sep84 | 13Oct84] 12(2) 0 | 10 | 14	
[Don Covay/Peter Wolf] A: Lights Out; P: Michael Jonzun/Peter Wolf	
1026. CRY Waterfront Polydor 871110	1249
[08Apr89 | 17Jun89 | 29Jul89] 10(2) 2 | 10 | 17	
[Phil Cilia/Chris Duffy] A: Waterfront; P: Glenn Skinner	
1027. NEVER BE THE SAME Christopher Cross Warner Bros. 49580	1246
[11Oct80 | 29Nov80 | 14Feb81] 15(3) 0 | 12 | 19	
[Christopher Cross] A: Christopher Cross; P: Michael Omartian	
1028. YOU GOT IT Roy Orbison Virgin 99245	1245
[21Jan89 | 15Apr89 | 20May89] 9(1) 1 | 11 | 18	
[Jeff Lynne/Roy Orbison/Tom Petty] A: Mystery Girl; P: Jeff Lynne	

Ranking the Singles

Rank. TITLE Artist Label Number
[Entry Date | Peak Date | Exit Date] Peak(Peak Wks) Top10 | Top40 | Tot Wks
[Writers] *A: Album; P: Producers* Score

1029. LOOK WHAT YOU'VE DONE TO ME Boz Scaggs Columbia 11349   1237
[23Aug80 | 25Oct80 | 13Dec80] 14(2) 0 | 10 | 17
[David Foster/Boz Scaggs]
A: Urban Cowboy Soundtrack; P: David Foster/Bill Schnee

1030. COOL LOVE Pablo Cruise A&M 2349   1232
[04Jul81 | 05Sep81 | 24Oct81] 13(2) 0 | 11 | 17
[David Jenkins/Cory Lerios/John Pierce] *A: Reflector; P: Tom Dowd*

1031. TAKE MY HEART (You Can Have It If You Want It) Kool & The Gang De-Lite 815   1228
[17Oct81 | 26Dec81 | 06Feb82] 17(3) 0 | 12 | 17
[Robert Bell/Ronald Bell/George M. Brown/Eumir Deodato/Claydes Smith/J.T. Taylor] *A: Something Special; P: Eumir Deodato*

1032. I COULD NEVER TAKE THE PLACE OF YOUR MAN Prince Paisley Park 7-28288   1224
[14Nov87 | 06Feb88 | 05Mar88] 10(1) 1 | 12 | 17
[Prince] *A: Sign 'O' The Times; P: Prince*

1033. IT WOULD TAKE A STRONG STRONG MAN Rick Astley RCA 8663   1222
[16Jul88 | 17Sep88 | 29Oct88] 10(1) 1 | 10 | 16
[Matt Aitken/Mike Stock/Pete Waterman]
A: Whenever You Need Somebody; P: Matt Aitken/Mike Stock/Pete Waterman

1034. WHAT ABOUT ME ('82) Moving Pictures Network 69952   1221
[18Sep82 | 12Feb83 | 12Mar83] 29(2) 0 | 13 | 26
[Frances Frost/Garry Frost] *A: Days Of Innocence; P: Charles Fisher*

1035. HOLIDAY Madonna London-Sire 7-29478   1219
[29Oct83 | 28Jan84 | 17Mar84] 16(2) 0 | 11 | 21
[Curtis Hudson/Lisa Stevens] *A: Madonna; P: Jellybean Benitez*

1036. TIME WILL REVEAL DeBarge Gordy 1705   1215
[15Oct83 | 21Jan84 | 03Mar84] 18(1) 0 | 11 | 21
[Bunny DeBarge/El DeBarge] *A: In A Special Way; P: El DeBarge*

1037. SOMEBODY Bryan Adams A&M 2701   1214
[02Feb85 | 06Apr85 | 25May85] 11(1) 0 | 10 | 17
[Bryan Adams/Jim Vallance] *A: Reckless; P: Bryan Adams/Bob Clearmountain*

1038. KISS ME DEADLY Lita Ford RCA 6866   1212
[02Apr88 | 18Jun88 | 03Sep88] 12(2) 0 | 10 | 23
[Mick Smiley] *A: Lita; P: Mike Chapman*

1039. A MATTER OF TRUST Billy Joel Columbia 06108   1210
[09Aug86 | 18Oct86 | 06Dec86] 10(1) 1 | 10 | 18
[Billy Joel] *A: The Bridge; P: Phil Ramone*

1040. TAKE ME TO HEART Quarterflash Geffen 29603   1206
[18Jun83 | 20Aug83 | 01Oct83] 14(1) 0 | 11 | 17
[Marv Ross] *A: Take Another Picture; P: John Boylan*

1041. ALL OVER THE WORLD Electric Light Orchestra MCA 41289   1205
[02Aug80 | 04Oct80 | 15Nov80] 13(2) 0 | 9 | 16
[Jeff Lynne] *A: Xanadu (Soundtrack); P: Jeff Lynne*

1042. PUT A LITTLE LOVE IN YOUR HEART Annie Lennox & Al Green A&M 1255   1205
[05Nov88 | 14Jan89 | 25Feb89] 9(1) 2 | 10 | 17
[Jackie DeShannon/Jimmy Holiday/Randy Myers]
A: Scrooged Soundtrack; P: David A. Stewart

1043. AIN'T EVEN DONE WITH THE NIGHT John Cougar Riva 207   1205
[31Jan81 | 09May81 | 20Jun81] 17(2) 0 | 12 | 21
[John Mellencamp] *A: Nothin' Matters And What If It Did; P: Steve Cropper*

1044. ROOM TO MOVE Animotion Polydor 871418   1205
[18Feb89 | 06May89 | 17Jun89] 9(1) 1 | 11 | 18
[Simon Climie/Rob Fisher/Dennis W. Morgan]
A: Animotion(2); P: Steve Barri/Tony Peluso

1045. SOME KIND OF LOVER Jody Watley MCA 53235   1205
[30Jan88 | 16Apr88 | 21May88] 10(1) 1 | 10 | 17
[André Cymone/Jody Watley] *A: Jody Watley; P: André Cymone/David Rivkin*

1046. TUFF ENUFF The Fabulous Thunderbirds CBS Associated 05838   1204
[19Apr86 | 12Jul86 | 23Aug86] 10(1) 1 | 10 | 19
[Kim Wilson] *A: Tuff Enuff; P: Dave Edmunds*

1047. GAMES PEOPLE PLAY Alan Parsons Project Arista 0573   1204
[06Dec80 | 14Mar81 | 09May81] 16(2) 0 | 10 | 23
[Alan Parsons/Eric Woolfson] *A: The Turn Of A Friendly Card; P: Alan Parsons*

1048. ANOTHER LOVER Giant Steps A&M 1226   1204
[20Aug88 | 12Nov88 | 14Jan89] 13(1) 0 | 10 | 22
[Colin Campsie/Gardner Cole/George McFarlane]
A: The Book Of Pride; P: Gardner Cole

1049. EMPTY GARDEN (Hey Hey Johnny) Elton John Geffen 50049   1202
[20Mar82 | 29May82 | 10Jul82] 13(2) 0 | 10 | 17
[Elton John/Bernie Taupin] *A: Jump Up!; P: Chris Thomas*

1050. DANCE HALL DAYS Wang Chung Geffen 29310   1199
[21Apr84 | 07Jul84 | 15Sep84] 16(1) 0 | 10 | 22
[Darren Costin/Nick Feldman/Jack Hues]
A: Points On The Curve; P: Ross Cullum/Chris Hughes

1051. YOU'LL ACCOMP'NY ME Bob Seger Capitol 4904   1197
[26Jul80 | 20Sep80 | 08Nov80] 14(3) 0 | 9 | 16
[Bob Seger] *A: Against The Wind; P: Ed Andrews/Bob Seger*

1052. ALPHABET ST. Prince Paisley Park 7-27900   1192
[30Apr88 | 25Jun88 | 23Jul88] 8(1) 2 | 9 | 13
[Prince] *A: Lovesexy; P: Prince*

1053. DEJA VU Dionne Warwick Arista 0459   1191
[10Nov79 | 02Feb80 | 15Mar80] 15(2) 0 | 11 | 19
[Adrienne Anderson/Isaac Hayes] *A: Dionne(2); P: Barry Manilow*

1054. I CAN'T HELP IT Andy Gibb & Olivia Newton-John RSO 1026   1190
[29Mar80 | 24May80 | 21Jun80] 12(2) 0 | 8 | 13
[Barry Gibb] *A: After Dark; P: Albhy Galuten/Barry Gibb/Karl Richardson*

1055. PARENTS JUST DON'T UNDERSTAND D.J. Jazzy Jeff & The Fresh Prince Jive 1099   1188
[21May88 | 23Jul88 | 24Sep88] 12(2) 0 | 10 | 19
[Peter Brian Harris/Will Smith/Jeff Townes] *A: He's The D.J., I'm The Rapper; P: Pete Q. Harris/Bryan New/Will Smith/Jeff Townes*

1056. ALL THIS LOVE DeBarge Gordy 1660   1188
[23Apr83 | 09Jul83 | 27Aug83] 17(3) 0 | 10 | 19
[El DeBarge] *A: All This Love; P: El DeBarge/Iris Gordy*

1057. LESSONS IN LOVE Level 42 Polydor 883956   1188
[04Apr87 | 27Jun87 | 01Aug87] 12(1) 0 | 10 | 18
[Wally Badarou/Boon Gould/Mark King]
A: Running In The Family; P: Wally Badarou/Level 42

1058. BREAKDOWN DEAD AHEAD Boz Scaggs Columbia 11241   1188
[29Mar80 | 17May80 | 28Jun80] 15(3) 0 | 9 | 14
[David Foster/Boz Scaggs] *A: Middle Man; P: Bill Schnee*

1059. I LIVE FOR YOUR LOVE Natalie Cole Manhattan 50094   1187
[07Nov87 | 06Feb88 | 02Apr88] 13(1) 0 | 11 | 22
[Pam Reswick/Allan Rich/Steve Werfel] *A: Everlasting; P: Dennis Lambert*

1060. DO I DO Stevie Wonder Tamla 1612   1185
[29May82 | 10Jul82 | 28Aug82] 13(3) 0 | 9 | 14
[Stevie Wonder] *A: Stevie Wonder's Original Musiquarium I; P: Stevie Wonder*

1061. SUPERWOMAN Karyn White Warner Bros. 27783   1183
[28Jan89 | 15Apr89 | 27May89] 8(1) 1 | 10 | 18
[Kenneth Edmonds/Antonio Reid/Daryl Simmons]
A: Karyn White; P: Kenneth Edmonds/L.A. Reid

1062. LOVE WILL SAVE THE DAY Whitney Houston Arista 1-9720   1182
[02Jul88 | 27Aug88 | 15Oct88] 9(1) 1 | 11 | 16
[Toni Colandreo] *A: Whitney; P: Jellybean Benitez*

1063. LOVELY ONE The Jacksons Epic 50938   1177
[27Sep80 | 15Nov80 | 24Jan81] 12(2) 0 | 9 | 18
[Michael Jackson/Steven Randall Jackson] *A: Triumph; P: Jacksons*

1064. REFUGEE Tom Petty And The Heartbreakers Backstreet 41169   1176
[26Jan80 | 15Mar80 | 26Apr80] 15(2) 0 | 10 | 14
[Mike Campbell/Tom Petty]
A: Damn The Torpedoes; P: Jimmy Iovine/Tom Petty

1065. MAN ON YOUR MIND Little River Band Capitol 5061   1174
[03Apr82 | 29May82 | 17Jul82] 14(3) 0 | 8 | 16
[Glenn Shorrock/Kerryn Tolhurst] *A: Time Exposure; P: George Martin*

1066. EDGE OF SEVENTEEN (Just Like The White Winged Dove) Stevie Nicks Modern 7401   1172
[20Feb82 | 17Apr82 | 22May82] 11(2) 0 | 10 | 14
[Stevie Nicks] *A: Bella Donna; P: Jimmy Iovine*

1067. TOUCH OF GREY Grateful Dead Arista 1-9606   1171
[25Jul87 | 26Sep87 | 31Oct87] 9(1) 2 | 9 | 15
[Jerry Garcia/Robert Hunter] *A: In The Dark; P: John Cutler/Jerry Garcia*

1068. EVERLASTING LOVE Howard Jones Elektra 69308   1171
[18May89 | 03Jun89 | 22Jul89] 12(1) 0 | 11 | 19
[Howard Jones] *A: Cross That Line; P: Ross Cullum/Chris Hughes/Ian Stanley*

1069. YOU SHOULD BE MINE (The Woo Woo Song) Jeffrey Osborne A&M 2814   1168
[24May86 | 23Aug86 | 27Sep86] 13(1) 0 | 11 | 19
[Andy Goldmark/Bruce Roberts] *A: Emotional; P: Richard Perry*

1070. GIVE IT UP KC Meca 1001   1166
[24Dec83 | 17Mar84 | 12May84] 18(1) 0 | 10 | 21
[Deborah Carter/Harry Wayne Casey]
A: KC Ten; P: Harry Wayne Casey/Richard Finch

Rank. TITLE Artist Label Number	
[Entry Date | Peak Date | Exit Date] Peak(Peak Wks) Top10 | Top40 | Tot Wks	
[Writers] *Album; Producers*	Score
1071. **RELAX (Mix) ('85)** Frankie Goes To Hollywood ZTT/Island 7-99805	1165
[19Jan85 | 16Mar85 | 04May85] 10(2) 2 | 10 | 16	
[Peter Gill/Holly Johnson/Mark O'Toole]	
A: *Welcome To The Pleasure Dome*; P: *Trevor Horn*	
1072. **TALK DIRTY TO ME** Poison Enigma 5686	1164
[14Mar87 | 16May87 | 27Jun87] 9(1) 2 | 9 | 16	
[Bobby Dall/C.C. DeVille/Bret Michaels/Rikki Rockett]	
A: *Look What The Cat Dragged In*; P: *Ric Browde*	
1073. **ONLY THE YOUNG** Journey Geffen 29090	1163
[26Jan85 | 23Mar85 | 11May85] 9(1) 1 | 10 | 16	
[Jonathan Cain/Steve Perry/Neal Schon]	
A: *Vision Quest Soundtrack*; P: *Kevin Elson/Mike Stone*	
1074. **PRETTY WOMAN** Van Halen Warner Bros. 50003	1163
[06Feb82 | 17Apr82 | 22May82] 12(2) 0 | 9 | 16	
[William Dees/Roy Orbison] A: *Diver Down*; P: *Ted Templeman*	
1075. **PEOPLE ARE PEOPLE** Depeche Mode Sire 29221	1163
[25May85 | 03Aug85 | 21Sep85] 13(2) 0 | 10 | 18	
[Martin L. Gore] A: *Some Great Reward*; P: *Depeche Mode/Daniel Miller*	
1076. **I KNOW WHAT I LIKE** Huey Lewis And The News Chrysalis 43108	1162
[04Apr87 | 30May87 | 04Jul87] 9(1) 1 | 10 | 14	
[Chris Hayes/Huey Lewis] A: *Fore!*; P: *Huey Lewis And The News*	
1077. **DANCING IN THE SHEETS** Shalamar Columbia 04372	1162
[17Mar84 | 26May84 | 14Jul84] 17(2) 0 | 10 | 18	
[Dean Pitchford/Bill Wolfer] A: *Heart Break*; P: *Bill Wolfer*	
1078. **HER TOWN TOO** James Taylor And J.D. Souther Columbia 60514	1160
[14Mar81 | 02May81 | 13Jun81] 11(2) 0 | 10 | 14	
[J.D. Souther/James Taylor/Waddy Wachtel]	
A: *Dad Loves His Work*; P: *Peter Asher*	
1079. **SEPTEMBER MORN'** Neil Diamond Columbia 11175	1159
[22Dec79 | 01Mar80 | 05Apr80] 17(1) 0 | 10 | 16	
[Gilbert Bécaud/Neil Diamond] A: *September Morn*; P: *Bob Gaudio*	
1080. **SHARE YOUR LOVE WITH ME** Kenny Rogers Liberty 1430	1159
[05Sep81 | 24Oct81 | 12Dec81] 14(2) 0 | 10 | 15	
[Al Braggs/Deadric Malone] A: *Share Your Love*; P: *Lionel Richie*	
1081. **I WANNA BE A COWBOY** Boys Don't Cry Profile 5084	1158
[05Apr86 | 21Jun86 | 09Aug86] 12(1) 0 | 9 | 19	
[Brian Chatton/Nico Ramsden/Nick Richards/Jeff Seopardie]	
A: *Boys Don't Cry*; P: *Boys Don't Cry*	
1082. **WHAT'S FOREVER FOR** Michael Murphey Liberty 1466	1157
[24Jul82 | 09Oct82 | 04Dec82] 19(5) 0 | 11 | 20	
[Rafe Van Hoy] A: *Michael Martin Murphey*; P: *Jim Ed Norman*	
1083. **STONE LOVE** Kool & The Gang Mercury 888292	1157
[07Feb87 | 02May87 | 06Jun87] 10(1) 1 | 10 | 18	
[Robert Bell/Ronald Bell/George M. Brown/Claydes Smith/J.T. Taylor/	
Dennis Thomas/Curtis Fitzgerald Williams]	
A: *Forever*; P: *Ronald Bell/I.B.M.C./Kool & The Gang*	
1084. **NOTHIN' AT ALL (Remix)** Heart Capitol 5572	1156
[19Apr86 | 21Jun86 | 02Aug86] 10(1) 1 | 10 | 16	
[Mark Mueller] A: *Heart*; P: *Ron Nevison*	
1085. **THERE'S THE GIRL** Heart Capitol 44089	1156
[07Nov87 | 23Jan88 | 12Mar88] 12(2) 0 | 11 | 19	
[Holly Knight/Nancy Wilson] A: *Bad Animals*; P: *Ron Nevison*	
1086. **PUSH IT** Salt-N-Pepa Next Plateau 315	1154
[21Nov87 | 20Feb88 | 07May88] 19(1) 0 | 13 | 25	
[Hurby Azor/Ray Davies] A: *Hot, Cool And Vicious*; P: *Hurby Azor*	
1087. **OPPORTUNITIES (Let's Make Lots Of Money)** Pet Shop Boys EMI America 8330	1154
[31May86 | 02Aug86 | 13Sep86] 10(1) 0 | 9 | 16	
[Chris Lowe/Neil Tennant] A: *Please*; P: *Nicholas Froome/J.J. Jeczalik*	
1088. **ANY DAY NOW** Ronnie Milsap RCA 13216	1152
[01May82 | 10Jul82 | 14Aug82] 14(2) 0 | 9 | 16	
[Burt Bacharach/Bob Hilliard]	
A: *Inside Ronnie Milsap*; P: *Tom Collins/Ronnie Milsap*	
1089. **NOBODY'S FOOL** Cinderella Mercury 884851	1151
[08Nov86 | 14Feb87 | 28Mar87] 13(1) 0 | 8 | 21	
[Tom Keifer] A: *Night Songs*; P: *Andy Johns*	
1090. **CRAZY ABOUT HER** Rod Stewart Warner Bros. 27657	1151
[13May89 | 29Jul89 | 02Sep89] 11(1) 0 | 10 | 17	
[Jim Cregan/Duane Hitchings/Rod Stewart]	
A: *Out Of Order*; P: *Bernard Edwards/Rod Stewart/Andy Taylor*	

Rank. TITLE Artist Label Number	
[Entry Date | Peak Date | Exit Date] Peak(Peak Wks) Top10 | Top40 | Tot Wks	
[Writers] *Album; Producers*	Score
1091. **MISS SUN** Boz Scaggs Columbia 11406	1150
[29Nov80 | 07Feb81 | 21Mar81] 14(2) 0 | 9 | 17	
[David Paich] A: *Hits!*; P: *Bill Schnee*	
1092. **THE LANGUAGE OF LOVE** Dan Fogelberg Full Moon 04314	1149
[04Feb84 | 24Mar84 | 05May84] 13(1) 0 | 10 | 14	
[Dan Fogelberg] A: *Windows And Walls*; P: *Dan Fogelberg/Marty Lewis*	
1093. **IT'S NOW OR NEVER** John Schneider Scotti Brothers 02105	1149
[30May81 | 15Aug81 | 03Oct81] 14(2) 0 | 11 | 19	
[Eduardo di Capua/Wally Gold/Aaron Schroeder]	
A: *Now Or Never*; P: *John D'Andrea/Tony Scotti*	
1094. **IT'S NOT OVER ('TIL IT'S OVER)** Starship RCA 5225	1147
[27Jun87 | 29Aug87 | 10Oct87] 9(1) 1 | 10 | 16	
[Phil Galdston/Robbie Nevil/John Van Tongeren]	
A: *No Protection*; P: *Keith Olsen*	
1095. **TUNNEL OF LOVE** Bruce Springsteen Columbia 07663	1146
[05Dec87 | 06Feb88 | 19Mar88] 9(1) 1 | 11 | 16	
[Bruce Springsteen]	
A: *Tunnel Of Love*; P: *Jon Landau/Chuck Plotkin/Bruce Springsteen*	
1096. **(Sittin' On) THE DOCK OF THE BAY** Michael Bolton Columbia 07680	1144
[23Jan88 | 26Mar88 | 14May88] 11(2) 0 | 10 | 17	
[Steve Cropper/Otis Redding] A: *The Hunger*; P: *Jonathan Cain*	
1097. **THE OLD SONGS** Barry Manilow Arista 0633	1143
[10Oct81 | 28Nov81 | 23Jan82] 15(2) 0 | 10 | 16	
[Buddy Kaye/David Pomeranz] A: *If I Should Love Again*; P: *Barry Manilow*	
1098. **WE'RE READY** Boston MCA 52985	1142
[06Dec86 | 14Feb87 | 14Mar87] 9(1) 2 | 10 | 15	
[Tom Scholz] A: *Third Stage*; P: *Tom Scholz*	
1099. **ROCK ME TONITE** Billy Squier Capitol 5370	1141
[07Jul84 | 08Sep84 | 20Oct84] 15(1) 0 | 12 | 16	
[Billy Squier] A: *Signs Of Life*; P: *Billy Squier/Jim Steinman*	
1100. **EVEN NOW** Bob Seger & The Silver Bullet Band Capitol 5213	1141
[12Mar83 | 07May83 | 28May83] 12(2) 0 | 9 | 12	
[Bob Seger] A: *The Distance*; P: *Jimmy Iovine*	
1101. **ONE HEARTBEAT** Smokey Robinson Motown 1897	1141
[18Jul87 | 03Oct87 | 21Nov87] 10(1) 1 | 11 | 19	
[Steve LeGassick/Brian Ray]	
A: *One Heartbeat*; P: *Peter Bunetta/Rick Chudacoff*	
1102. **GIRLS, GIRLS, GIRLS** Mötley Crüe Elektra 69465	1139
[30May87 | 25Jul87 | 05Sep87] 12(2) 0 | 9 | 15	
[Tommy Lee/Mick Mars/Nikki Sixx] A: *Girls, Girls, Girls*; P: *Tom Werman*	
1103. **THE PLEASURE PRINCIPLE** Janet Jackson A&M 2927	1139
[23May87 | 01Aug87 | 19Sep87] 14(1) 0 | 10 | 18	
[Monte Moir] A: *Control*; P: *Monte Moir*	
1104. **EVERYBODY'S GOT TO LEARN SOMETIME** The Korgis Asylum 47055	1136
[11Oct80 | 27Dec80 | 14Feb81] 18(2) 0 | 11 | 19	
[James Edward Warren] A: *Dumb Waiters*; P: *Korgis/David Lord*	
1105. **ONE** Bee Gees Warner Bros. 22899	1136
[29Jul89 | 30Sep89 | 28Oct89] 7(1) 1 | 10 | 14	
[Barry Gibb/Maurice Gibb/Robin Gibb] A: *One*; P: *Bee Gees/Brian Tench*	
1106. **NOBODY** Sylvia (2) RCA 13223	1134
[28Aug82 | 20Nov82 | 08Jan83] 15(3) 0 | 9 | 20	
[Kye Fleming/Dennis W. Morgan] A: *Just Sylvia*; P: *Tom Collins*	
1107. **LOVE THEME FROM ST. ELMO'S FIRE** David Foster Atlantic 89528	1134
[24Aug85 | 16Nov85 | 18Jan86] 15(2) 0 | 10 | 22	
[David Foster] A: *St. Elmo's Fire Soundtrack*; P: *David Foster/Humberto Gatica*	
1108. **YOU SPIN ME ROUND (Like A Record)** Dead Or Alive Epic 04894	1133
[01Jun85 | 17Aug85 | 28Sep85] 11(1) 0 | 11 | 18	
[Pete Burns/Steve Coy/Wayne Hussey/Tim Lever/Mike Percy]	
A: *Youthquake*; P: *Pete Waterman*	
1109. **ONE HUNDRED WAYS** Quincy Jones Featuring James Ingram A&M 2387	1132
[19Dec81 | 17Apr82 | 08May82] 14(1) 0 | 11 | 21	
[Tony Coleman/Kathy Wakefield/Ben Wright] A: *The Dude*; P: *Quincy Jones*	
1110. **HEAVEN IN YOUR EYES** Loverboy Columbia 06178	1132
[02Aug86 | 11Oct86 | 22Nov86] 12(2) 0 | 11 | 17	
[Paul Dean/John Dexter/Debra Mae Moore/Mike Reno]	
A: *Top Gun Soundtrack*; P: *Paul Dean/John Dexter*	
1111. **WALKING DOWN YOUR STREET** Bangles Columbia 06674	1131
[14Feb87 | 18Apr87 | 30May87] 11(3) 0 | 9 | 16	
[Louis Gutierrez/Susanna Hoffs/David Kahne]	
A: *Different Light*; P: *David Kahne*	

Ranking the Singles

Rank. TITLE Artist Label Number [Entry Date | Peak Date | Exit Date] Peak(Peak Wks) Top10 | Top40 | Tot Wks [Writers] Album; Producers	Score
1112. I'LL BE ALRIGHT WITHOUT YOU Journey Columbia 06301 [06Dec86 | 28Feb87 | 25Apr87] 14(2) 0 | 9 | 21 [Jonathan Cain/Steve Perry/Neal Schon] A: Raised On Radio; P: Steve Perry	1131
1113. I'LL WAIT Van Halen Warner Bros. 29307 [14Apr84 | 02Jun84 | 14Jul84] 13(2) 0 | 10 | 14 [Michael Anthony/David Lee Roth/Alex Van Halen/Eddie Van Halen] A: 1984 (MCMLXXXIV); P: Ted Templeman	1128
1114. IT'S A MIRACLE Culture Club Virgin/Epic 04457 [12May84 | 16Jun84 | 04Aug84] 13(2) 0 | 8 | 13 [Michael Craig/Roy Hay/Jon Moss/George O'Dowd/Phil Pickett] A: Colour By Numbers; P: Steve Levine	1127
1115. LEAVE A LIGHT ON Belinda Carlisle MCA 53706 [30Sep89 | 09Dec89 | 27Jan90] 11(1) 0 | 10 | 18 [Rick Nowels/Ellen Shipley] A: Runaway Horses; P: Rick Nowels	1127
1116. BABY JANE Rod Stewart Warner Bros. 29608 [28May83 | 30Jul83 | 27Aug83] 14(1) 0 | 9 | 14 [Jay Davis/Rod Stewart] A: Body Wishes; P: Tom Dowd/Rod Stewart	1124
1117. JUNGLE LOVE The Time Warner Bros. 29181 [27Oct84 | 23Feb85 | 13Apr85] 20(1) 0 | 10 | 25 [Morris Day/Jesse Woods Johnson/Prince] A: Ice Cream Castle; P: Morris Day/Prince	1122
1118. JAMIE Ray Parker Jr. Arista 1-9293 [17Nov84 | 26Jan85 | 09Mar85] 14(1) 0 | 11 | 17 [Ray Parker Jr.] A: Chartbusters; P: Ray Parker Jr.	1122
1119. PERSONALLY Karla Bonoff Columbia 02805 [01May82 | 07Aug82 | 28Aug82] 19(2) 0 | 12 | 18 [Paul Kelly] A: Wild Heart Of The Young; P: Kenny Edwards	1122
1120. HYSTERIA Def Leppard Mercury 870004 [23Jan88 | 26Mar88 | 07May88] 10(1) 1 | 10 | 16 [Rick Allen/Steve Clark/Phil Collen/Joe Elliott/Mutt Lange/Rick Savage] A: Hysteria; P: Mutt Lange	1121
1121. GEMINI DREAM The Moody Blues Threshold 601 [06Jun81 | 01Aug81 | 12Sep81] 12(1) 0 | 9 | 15 [Justin Hayward/John Lodge] A: Long Distance Voyager; P: Pip Williams	1121
1122. THE STROKE Billy Squier Capitol 5005 [16May81 | 15Aug81 | 26Sep81] 17(2) 0 | 11 | 20 [Billy Squier] A: Don't Say No; P: Reinhold Mack/Billy Squier	1121
1123. BE GOOD TO YOURSELF Journey Columbia 05869 [12Apr86 | 31May86 | 19Jul86] 9(1) 1 | 10 | 16 [Jonathan Cain/Steve Perry/Neal Schon] A: Raised On Radio; P: Steve Perry	1119
1124. SYNCHRONICITY II The Police A&M 2571 [05Nov83 | 10Dec83 | 04Feb84] 16(4) 0 | 9 | 14 [Sting] A: Synchronicity; P: Hugh Padgham/Police	1118
1125. CHAINS OF LOVE Erasure Sire 27844 [30Jul88 | 29Oct88 | 10Dec88] 12(1) 0 | 11 | 20 [Andy Bell/Vince Clarke] A: The Innocents; P: Stephen Hague	1118
1126. WHEN YOU CLOSE YOUR EYES Night Ranger Camel/MCA 52420 [14Jul84 | 29Sep84 | 03Nov84] 14(1) 0 | 11 | 17 [Jack Blades/Alan Fitzgerald/Brad Gillis] A: Midnight Madness; P: Pat Glasser	1118
1127. ME SO HORNY 2 Live Crew Skywalker 130 [09Sep89 | 18Nov89 | 31Mar90] 26(1) 0 | 9 | 30 [Luther Campbell/David Hobbs/Mark Ross/Ricardo Williams/Chris Wong Won] A: As Nasty As They Wanna Be; P: Luther Campbell/2 Live Crew	1116
1128. KISS HIM GOODBYE The Nylons Open Air 0022 [16May87 | 01Aug87 | 05Sep87] 12(1) 0 | 10 | 17 [Gary DeCarlo/Dale Frashuer/Paul Leka] A: Happy Together; P: Bill Henderson	1116
1129. YAH MO B THERE James Ingram (With Michael McDonald) Qwest 29394 [10Dec83 | 03Mar84 | 07Apr84] 19(1) 0 | 10 | 18 [James Ingram/Quincy Jones/Michael McDonald/Rod Temperton] A: It's Your Night; P: Quincy Jones	1115
1130. FINISH WHAT YA STARTED Van Halen Warner Bros. 27746 [01Oct88 | 10Dec88 | 11Feb89] 13(1) 0 | 10 | 20 [Michael Anthony/Sammy Hagar/Alex Van Halen/Eddie Van Halen] A: OU812; P: Donn Landee	1114
1131. I'M GOIN' DOWN Bruce Springsteen Columbia 05603 [07Sep85 | 26Oct85 | 30Nov85] 9(1) 1 | 9 | 13 [Bruce Springsteen] A: Born In The U.S.A.; P: Jon Landau/Chuck Plotkin/Bruce Springsteen/Steven Van Zandt	1114
1132. DYNAMITE Jermaine Jackson Arista 1-9190 [21Jul84 | 15Sep84 | 10Nov84] 15(2) 0 | 10 | 16 [Andy Goldmark/Bruce Roberts] A: Jermaine Jackson; P: Jermaine Jackson	1109
1133. SHOULD I DO IT The Pointer Sisters Planet 47960 [23Jan82 | 03Apr82 | 08May82] 13(2) 0 | 10 | 16 [Layng Martine Jr.] A: Black & White; P: Richard Perry	1109
1134. STOP TO LOVE Luther Vandross Epic 06523 [15Nov86 | 14Feb87 | 21Mar87] 15(1) 0 | 11 | 19 [Nat Adderley Jr./Luther Vandross] A: Give Me The Reason; P: Marcus Miller/Luther Vandross	1104
1135. THROUGH THE YEARS Kenny Rogers Liberty 1444 [26Dec81 | 06Mar82 | 03Apr82] 13(2) 0 | 11 | 15 [Steve Dorff/Marty Panzer] A: Share Your Love; P: Lionel Richie	1103
1136. IN A BIG COUNTRY Big Country Mercury 814467 [22Oct83 | 03Dec83 | 28Jan84] 17(3) 0 | 9 | 15 [William Stuart Adamson/Mark Brzezicki/Anthony Earl Butler/ Bruce William Watson] A: The Crossing; P: Steve Lillywhite	1101
1137. JUST ANOTHER NIGHT Mick Jagger Columbia 04743 [09Feb85 | 30Mar85 | 11May85] 12(2) 0 | 10 | 14 [Mick Jagger] A: She's The Boss; P: Mick Jagger/Bill Laswell/Material	1101
1138. HOLD ON Santana Columbia 03160 [14Aug82 | 23Oct82 | 13Nov82] 15(2) 0 | 10 | 14 [Ian Thomas] A: Shango; P: John Ryan	1100
1139. TIME IS TIME Andy Gibb RSO 1059 [22Nov80 | 24Jan81 | 14Mar81] 15(2) 0 | 11 | 17 [Andy Gibb/Barry Gibb] A: Andy Gibb's Greatest Hits; P: Albhy Galuten/Barry Gibb/Karl Richardson	1100
1140. DON'T TELL ME LIES Breathe A&M 1267 [14Jan89 | 18Mar89 | 29Apr89] 10(1) 0 | 10 | 16 [Michael Delahunty/David Glasper/Marcus Lillington/Ian Spice] A: All That Jazz; P: Bob Sargeant	1099
1141. YOU'RE NOT ALONE Chicago Reprise/Full Moon 27757 [21Jan89 | 25Mar89 | 13May89] 10(1) 1 | 10 | 17 [James Scott] A: Chicago 19; P: Ron Nevison	1099
1142. BACK IN THE HIGH LIFE AGAIN Steve Winwood Island 28472 [30May87 | 15Aug87 | 17Oct87] 13(1) 0 | 10 | 21 [Will Jennings/Steve Winwood] A: Back In The High Life; P: Russ Titelman/Steve Winwood	1098
1143. DIAL MY HEART The Boys Motown 53301 [10Dec88 | 25Feb89 | 15Apr89] 13(1) 0 | 9 | 19 [Kenneth Edmonds/Antonio Reid/Daryl Simmons] A: Messages From The Boys; P: Kenneth Edmonds/L.A. Reid	1097
1144. I'VE GOT A ROCK N' ROLL HEART Eric Clapton Duck/Warner 29780 [29Jan83 | 26Mar83 | 14May83] 18(3) 0 | 10 | 16 [Steve Diamond/Troy Seals/Eddie Setser] A: Money And Cigarettes; P: Tom Dowd	1097
1145. FALL IN LOVE WITH ME Earth, Wind & Fire Columbia 03375 [22Jan83 | 19Mar83 | 07May83] 17(3) 0 | 10 | 16 [Wanda Hutchinson/Wayne Vaughn/Maurice White] A: Powerlight; P: Maurice White	1096
1146. WAIT FOR ME Daryl Hall & John Oates RCA 11747 [27Oct79 | 26Jan80 | 01Mar80] 18(1) 0 | 10 | 19 [Daryl Hall] A: X-Static; P: David Foster	1094
1147. THE VOICE The Moody Blues Threshold 602 [08Aug81 | 03Oct81 | 28Nov81] 15(2) 0 | 11 | 17 [Justin Hayward] A: Long Distance Voyager; P: Pip Williams	1093
1148. CRAZY Icehouse Chrysalis 43156 [17Oct87 | 23Jan88 | 05Mar88] 14(1) 0 | 11 | 21 [Iva Davies/Robert Kretschmer/Andy Qunta] A: Man Of Colours; P: David Lord	1091
1149. DON'T COME AROUND HERE NO MORE Tom Petty and The Heartbreakers MCA 52496 [16Mar85 | 18May85 | 15Jun85] 13(1) 0 | 9 | 14 [Tom Petty/David A. Stewart] A: Southern Accents; P: Jimmy Iovine/Tom Petty/David A. Stewart	1091
1150. SHOULD'VE NEVER LET YOU GO Neil Sedaka & Dara Sedaka Elektra 46615 [29Mar80 | 28Jun80 | 02Aug80] 19(2) 0 | 10 | 19 [Phil Cody/Neil Sedaka] A: In The Pocket; P: Robert Appere/Neil Sedaka	1089
1151. SILHOUETTE Kenny G Arista 1-9751 [29Oct88 | 07Jan89 | 18Feb89] 13(1) 0 | 10 | 17 [Kenny G] A: Silhouette; P: Kenny G	1088
1152. THE GOONIES 'R' GOOD ENOUGH Cyndi Lauper Portrait 34-04918 [18May85 | 13Jul85 | 24Aug85] 10(1) 1 | 9 | 15 [Cyndi Lauper/Stephen Lunt/Arthur Stead] A: The Goonies (Soundtrack); P: Cyndi Lauper/Lennie Petze	1086

Ranking the Singles

Rank. TITLE Artist Label Number
[Entry Date | Peak Date | Exit Date] Peak(Peak Wks) Top10 | Top40 | Tot Wks
[Writers] *Album; Producers* Score

1153. I WON'T FORGET YOU Poison Enigma 44038 1085
[05Sep87 | 21Nov87 | 23Jan88] 13(1) 0 | 9 | 21
[Bobby Dall/C.C. DeVille/Bret Michaels/Rikki Rockett]
A: Look What The Cat Dragged In; P: Ric Browde

1154. THE DOCTOR The Doobie Brothers Capitol 44376 1079
[20May89 | 15Jul89 | 19Aug89] 9(1) 1 | 9 | 14
[Tom Johnston/Charles E. Kaufman/Eddie Schwartz]
A: Cycles; P: Charlie Midnight/Eddie Schwartz

1155. I WANNA GO BACK Eddie Money Columbia 06569 1077
[20Dec86 | 14Mar87 | 09May87] 14(1) 0 | 10 | 21
[Monty Byrom/Danny Chauncey/Ira Walker]
A: Can't Hold Back; P: Eddie Money/Richie Zito

1156. JUST A GIGOLO/I AIN'T GOT NOBODY David Lee Roth 1077
Warner Bros. 29040
[23Mar85 | 01Jun85 | 13Jul85] 12(1) 0 | 10 | 17
[Irving Caesar/ Leonello Casucci/ Roger Graham/Spencer Williams]
A: Crazy From The Heat; P: Ted Templeman

1157. DON'T SHUT ME OUT Kevin Paige Chrysalis 23389 1076
[19Aug89 | 25Nov89 | 27Jan90] 18(2) 0 | 10 | 24
[Kevin Paige] *A: Kevin Paige; P: Kevin Paige*

1158. THE EDGE OF HEAVEN Wham! Columbia 06182 1071
[05Jul86 | 16Aug86 | 27Sep86] 10(2) 2 | 8 | 13
[George Michael] *A: Music From The Edge Of Heaven; P: George Michael*

1159. SPIRITS IN THE MATERIAL WORLD The Police A&M 2390 1069
[16Jan82 | 13Mar82 | 10Apr82] 11(2) 0 | 10 | 13
[Sting] *A: Ghost In The Machine; P: Hugh Padgham/Police*

1160. YOU GOT LUCKY Tom Petty and The Heartbreakers Backstreet 52144 1068
[13Nov82 | 29Jan83 | 12Mar83] 20(2) 0 | 11 | 18
[Mike Campbell/Tom Petty] *A: Long After Dark; P: Jimmy Iovine*

1161. HAND TO HOLD ON TO John Cougar Riva 211 1066
[06Nov82 | 22Jan83 | 05Mar83] 19(2) 0 | 11 | 18
[John Mellencamp] *A: American Fool; P: Don Gehman/John Mellencamp*

1162. IF ANYONE FALLS Stevie Nicks Modern 99832 1066
[10Sep83 | 05Nov83 | 10Dec83] 14(1) 0 | 9 | 14
[Stevie Nicks/Sandy Stewart] *A: The Wild Heart; P: Jimmy Iovine*

1163. GYPSY Fleetwood Mac Warner Bros. 29918 1065
[04Sep82 | 23Oct82 | 04Dec82] 12(3) 0 | 8 | 14
[Stevie Nicks]
A: Mirage; P: Lindsey Buckingham/Ken Caillat/Richard Dashut/Fleetwood Mac

1164. AUTHORITY SONG John Cougar Mellencamp Riva 216 1065
[17Mar84 | 19May84 | 23Jun84] 15(1) 0 | 9 | 15
[John Mellencamp] *A: Uh-Huh; P: Don Gehman/John Mellencamp*

1165. IN MY DREAMS REO Speedwagon Epic 07255 1064
[18Jul87 | 24Oct87 | 06Feb88] 19(1) 0 | 8 | 30
[Kevin Cronin/Thomas F. Kelly] *A: Life As We Know It;*
P: Kevin Cronin/David DeVore/Alan Gratzer/Gary Richrath

1166. WOT'S IT TO YA Robbie Nevil EMI 50075 1064
[30May87 | 01Aug87 | 12Sep87] 10(2) 2 | 9 | 16
[Robbie Nevil/Brock Walsh] *A: Robbie Nevil; P: Alex Sadkin/Philip Thornalley*

1167. FOOLISH HEART Steve Perry Columbia 04693 1063
[24Nov84 | 16Feb85 | 30Mar85] 18(1) 0 | 11 | 19
[Randy Goodrum/Steve Perry] *A: Street Talk; P: Steve Perry*

1168. BRAND NEW LOVER Dead Or Alive Epic 06374 1062
[29Nov86 | 14Mar87 | 25Apr87] 15(1) 0 | 9 | 22
[Pete Burns/Steve Coy/Wayne Hussey/Tim Lever/Mike Percy]
A: Mad, Bad And Dangerous To Know; P: Matt Aitken/Mike Stock/Pete Waterman

1169. I PLEDGE MY LOVE Peaches & Herb Polydor/MVP 2053 1062
[19Jan80 | 26Apr80 | 24May80] 19(1) 0 | 8 | 19
[Dino Fekaris/Freddie Perren] *A: Twice The Fire; P: Freddie Perren*

1170. KISSES ON THE WIND Neneh Cherry Virgin 99183 1059
[22Jul89 | 30Sep89 | 21Oct89] 8(1) 1 | 8 | 14
[Neneh Cherry/Cameron McVey]
A: Raw Like Sushi; P: Dynamik Duo/Nick Plytas

1171. LOVE POWER Dionne Warwick & Jeffrey Osborne Arista 1-9567 1059
[11Jul87 | 29Aug87 | 10Oct87] 12(2) 0 | 9 | 14
[Burt Bacharach/Carole Bayer Sager]
A: Reservations For Two; P: Burt Bacharach/Carole Bayer Sager

1172. TAKE A LITTLE RHYTHM Ali Thomson A&M 2243 1058
[14Jun80 | 23Aug80 | 04Oct80] 15(2) 0 | 9 | 17
[Ali Thomson] *A: Take A Little Rhythm; P: Jon Kelly/Ali Thomson*

1173. IKO IKO The Belle Stars Capitol 44343 1058
[04Mar89 | 13May89 | 01Jul89] 14(2) 0 | 10 | 18
[James Crawford/Barbara Anne Hawkins/Rosa Lee Hawkins/
Joan Marie Johnson/Traditional] *A: Rain Man Soundtrack; P: Brian Tench*

1174. THAT'S WHAT LOVE IS ALL ABOUT Michael Bolton Columbia 7322 1057
[05Sep87 | 12Dec87 | 20Feb88] 19(2) 0 | 10 | 25
[Michael Bolton/Eric Kaz] *A: The Hunger; P: Keith Diamond*

1175. SOMEONE COULD LOSE A HEART TONIGHT Eddie Rabbitt 1056
Elektra 47235
[14Nov81 | 23Jan82 | 20Feb82] 15(2) 0 | 10 | 15
[David Malloy/Eddie Rabbitt/Even Stevens] *A: Step By Step; P: David Malloy*

1176. YOU DON'T WANT ME ANYMORE Steel Breeze RCA 13283 1056
[28Aug82 | 13Nov82 | 08Jan83] 16(2) 0 | 11 | 20
[Ken Goorabian] *A: Steel Breeze; P: Kim Fowley*

1177. YOUR MAMA DON'T DANCE Poison Enigma 44293 1054
[18Feb89 | 15Apr89 | 20May89] 10(1) 1 | 10 | 14
[Kenny Loggins/Jim Messina] *A: Open Up And Say...Ahh!; P: Tom Werman*

1178. STAY THE NIGHT Chicago Warner/Full Moon 29306 1052
[05May84 | 23Jun84 | 25Aug84] 16(2) 0 | 10 | 17
[Peter Cetera/David Foster] *A: Chicago 17; P: David Foster*

1179. MR. TELEPHONE MAN New Edition MCA 52484 1051
[22Dec84 | 23Feb85 | 06Apr85] 12(2) 0 | 8 | 16
[Ray Parker Jr.] *A: New Edition; P: Ray Parker Jr.*

1180. LOVE IS FOREVER Billy Ocean Jive 9540 1049
[25Oct86 | 27Dec86 | 07Feb87] 16(3) 0 | 11 | 16
[Wayne Brathwaite/Barry Eastmond/Billy Ocean]
A: Love Zone; P: Wayne Brathwaite/Barry Eastmond

1181. LET ME BE YOUR ANGEL Stacy Lattisaw Cotillion 46001 1047
[09Aug80 | 22Nov80 | 17Jan81] 21(2) 0 | 10 | 24
[Bunny Hull/Narada Michael Walden]
A: Let Me Be Your Angel; P: Narada Michael Walden

1182. PANAMA Van Halen Warner Bros. 29250 1047
[23Jun84 | 18Aug84 | 29Sep84] 13(1) 0 | 10 | 15
[Michael Anthony/David Lee Roth/Alex Van Halen/Eddie Van Halen]
A: 1984 (MCMLXXXIV); P: Ted Templeman

1183. ONE NIGHT LOVE AFFAIR Bryan Adams A&M 2770 1047
[14Sep85 | 09Nov85 | 21Dec85] 13(1) 0 | 9 | 15
[Bryan Adams/Jim Vallance] *A: Reckless; P: Bryan Adams/Bob Clearmountain*

1184. TREAT ME RIGHT Pat Benatar Chrysalis 2487 1046
[17Jan81 | 14Mar81 | 16May81] 18(2) 0 | 10 | 18
[Pat Benatar/Doug Lubahn] *A: Crimes Of Passion; P: Keith Olsen*

1185. EVERYWHERE Fleetwood Mac Warner Bros. 28143 1045
[28Nov87 | 06Feb88 | 26Mar88] 14(2) 0 | 10 | 18
[Christine McVie] *A: Tango In The Night; P: Lindsey Buckingham/Richard Dashut*

1186. YOU CAN'T GET WHAT YOU WANT (Till You Know What You Want) 1045
Joe Jackson A&M 2628 [21Apr84 | 23Jun84 | 04Aug84] 15(1) 0 | 9 | 16
[Joe Jackson] *A: Body And Soul; P: Joe Jackson/David Kershenbaum*

1187. HOLDING ON Steve Winwood Virgin 99261 1045
[26Nov88 | 28Jan89 | 18Mar89] 11(1) 0 | 11 | 17
[Will Jennings/Steve Winwood]
A: Roll With It; P: Tom Lord-Alge/Steve Winwood

1188. I'D STILL SAY YES Klymaxx Constellation 53028 1042
[02May87 | 25Jul87 | 12Sep87] 18(1) 0 | 9 | 20
[Kenneth Edmonds/Joyce Irby/Greg Scelsa] *A: Klymaxx; P: Joyce Irby*

1189. CALL TO THE HEART Giuffria MCA 52497 1039
[10Nov84 | 02Feb85 | 16Mar85] 15(2) 0 | 7 | 19
[David Glen Eisley/Gregg Giuffria] *A: Giuffria; P: Gregg Giuffria*

1190. THEME FROM "THE DUKES OF HAZZARD" (Good Ol' Boys) Waylon 1039
RCA 12067 [13Sep80 | 13Dec80 | 14Feb81] 21(2) 0 | 10 | 23
[Waylon Jennings] *A: Music Man; P: Richie Albright*

1191. OUT HERE ON MY OWN Irene Cara RSO 1048 1039
[16Aug80 | 15Nov80 | 17Jan81] 19(2) 0 | 9 | 23
[Lesley Gore/Michael Gore] *A: Fame Soundtrack; P: Michael Gore*

1192. THE ONE YOU LOVE Glenn Frey Asylum 69974 1039
[21Aug82 | 06Nov82 | 11Dec82] 15(2) 0 | 11 | 17
[Glenn Frey/Jack Tempchin]
A: No Fun Aloud; P: Allan Blazek/Glenn Frey/Jim Ed Norman

1193. YOU CAN CALL ME AL Paul Simon Warner Bros. 28667 1039
[09Aug86 | 23May87 | 04Jul87] 23(2) 0 | 7 | 29
[Paul Simon] *A: Graceland; P: Paul Simon*

Ranking the Singles

Rank. TITLE Artist Label Number
[Entry Date | Peak Date | Exit Date] Peak(Peak Wks) | Top10 | Top40 | Tot Wks
[Writers] *Album; Producers* Score

1194. SOUTHERN CROSS Crosby, Stills & Nash Atlantic 89969 1037
[18Sep82 | 20Nov82 | 08Jan83] 18(3) 0 | 9 | 17
[Michael Curtis/Richard Curtis/Stephen Stills]
A: Daylight Again; P: Crosby, Stills & Nash

1195. IT'S NOT ENOUGH Starship RCA 9032 1036
[05Aug89 | 07Oct89 | 18Nov89] 12(1) 0 | 9 | 16
[Tommy Funderburk/Martin Page]
A: Love Among The Cannibals; P: Larry Klein/Mike Shipley

1196. I WON'T BACK DOWN Tom Petty MCA 53369 1035
[29Apr89 | 01Jul89 | 05Aug89] 12(1) 0 | 9 | 15 [Jeff Lynne/Tom Petty]
A: Full Moon Fever; P: Mike Campbell/Jeff Lynne/Tom Petty

1197. MODERN WOMAN (FROM "RUTHLESS PEOPLE") Billy Joel Epic 06118 1035
[07Jun86 | 26Jul86 | 13Sep86] 10(3) 1 | 9 | 15
[Billy Joel] *A: The Bridge; P: Phil Ramone*

1198. WHEN THE HEART RULES THE MIND GTR Arista 1-9470 1035
[10May86 | 12Jul86 | 23Aug86] 14(2) 0 | 10 | 16
[Steve Hackett/Steve Howe] *A: GTR; P: Geoffrey Downes*

1199. GIVE IT ALL YOU GOT Chuck Mangione A&M 2211 1035
[19Jan80 | 22Mar80 | 03May80] 18(2) 0 | 9 | 16
[Chuck Mangione] *A: Fun And Games; P: Chuck Mangione*

1200. NEW ATTITUDE Patti LaBelle MCA 52517 1035
[16Feb85 | 11May85 | 06Jul85] 17(1) 0 | 9 | 21
[Jonathan Gilutin/Bunny Hull/Sharon Robinson]
A: Beverly Hills Cop Soundtrack; P: Peter Bunetta/Rick Chudacoff/Howie Rice

1201. YOU KNOW I LOVE YOU...DON'T YOU? Howard Jones Elektra 69512 1034
[18Oct86 | 20Dec86 | 31Jan87] 17(3) 0 | 10 | 16
[Howard Jones] *A: One To One; P: Arif Mardin*

1202. BREAKING US IN TWO Joe Jackson A&M 2510 1034
[15Jan83 | 19Mar83 | 30Apr83] 18(1) 0 | 10 | 16
[Joe Jackson] *A: Night And Day; P: Joe Jackson/David Kershenbaum*

1203. I COULD NEVER MISS YOU (More Than I Do) Lulu Alfa 7006 1032
[01Aug81 | 10Oct81 | 28Nov81] 18(2) 0 | 10 | 18
[Neil Harrison] *A: Lulu; P: Mark London*

1204. MISSIONARY MAN Eurythmics RCA 14414 1031
[26Jul86 | 11Oct86 | 08Nov86] 14(1) 0 | 9 | 16
[Annie Lennox/David A. Stewart] *A: Revenge; P: David A. Stewart*

1205. MODERN LOVE David Bowie EMI America 8177 1029
[17Sep83 | 12Nov83 | 10Dec83] 14(1) 0 | 9 | 13
[David Bowie] *A: Let's Dance; P: David Bowie/Nile Rodgers*

1206. THE BEATLES' MOVIE MEDLEY The Beatles Capitol 5107 1029
[27Mar82 | 08May82 | 19Jun82] 12(3) 0 | 8 | 11
A: No Album; P: George Martin

1207. JUMP START Natalie Cole Manhattan 50073 1027
[25Jul87 | 03Oct87 | 21Nov87] 13(2) 0 | 10 | 18
[Reggie Calloway/Vincent Calloway]
A: Everlasting; P: Reggie Calloway/Vincent Calloway

1208. GET ON YOUR FEET Gloria Estefan Epic 69064 1027
[30Sep89 | 25Nov89 | 20Jan90] 11(1) 0 | 8 | 17
[Jorge Casas/John DeFaria/Clay Ostwald]
A: Cuts Both Ways; P: Jorge Casas/Emilio Estefan/Clay Ostwald

1209. SUDDENLY Olivia Newton-John & Cliff Richard MCA 51007 1025
[25Oct80 | 17Jan81 | 28Feb81] 20(1) 0 | 11 | 19
[John Farrar] *A: Xanadu (Soundtrack); P: John Farrar*

1210. FEELS SO RIGHT Alabama RCA 12236 1024
[06Jun81 | 05Sep81 | 31Oct81] 20(2) 0 | 8 | 22
[Randy Owen]
A: Feels So Right; P: Alabama/Larry McBride/Harold Shedd

1211. JOJO Boz Scaggs Columbia 11281 1024
[14Jun80 | 23Aug80 | 04Oct80] 17(2) 0 | 9 | 16
[David Foster/David Lasley/Boz Scaggs] *A: Middle Man; P: Bill Schnee*

1212. THAT WAS YESTERDAY Foreigner Atlantic 89571 1021
[16Mar85 | 04May85 | 22Jun85] 12(1) 0 | 10 | 15
[Lou Gramm/Michael Leslie Jones]
A: Agent Provocateur; P: Mick Jones/Alex Sadkin

1213. ALONG COMES A WOMAN Chicago Warner/Full Moon 29082 1018
[23Feb85 | 20Apr85 | 08Jun85] 14(2) 0 | 10 | 16
[Peter Cetera/Mark Goldenberg] *A: Chicago 17; P: David Foster*

1214. GIRLS Dwight Twilley EMI America 8196 1018
[18Feb84 | 21Apr84 | 02Jun84] 16(1) 0 | 10 | 16
[Dwight Twilley] *A: Jungle; P: Noah Shark/Mark Smith*

1215. DEAD GIVEAWAY Shalamar Solar 69819 1018
[25Jun83 | 24Sep83 | 05Nov83] 22(2) 0 | 10 | 20
[Marcus Dare/Joey Gallo/Leon Sylvers] *A: The Look; P: Leon Sylvers*

1216. BOBBIE SUE Oak Ridge Boys MCA 52006 1017
[16Jan82 | 20Mar82 | 17Apr82] 12(2) 0 | 9 | 14
[Jerry Leiber/Wood Newton/Mike Stoller/Adele Tyler/Dan Tyler]
A: Bobbie Sue; P: Ron Chancey

1217. BODY LANGUAGE Queen Elektra 47452 1016
[01May82 | 19Jun82 | 31Jul82] 11(2) 0 | 8 | 14
[Freddie Mercury] *A: Hot Space; P: Reinhold Mack/Queen*

1218. SOMEONE THAT I USED TO LOVE Natalie Cole Capitol 4869 1016
[21Jun80 | 20Sep80 | 08Nov80] 21(2) 0 | 9 | 21
[Gerry Goffin/Michael Masser] *A: Don't Look Back; P: Michael Masser*

1219. ROCK OF AGES Def Leppard Mercury 812604 1015
[11Jun83 | 13Aug83 | 17Sep83] 16(1) 0 | 9 | 15
[Rick Allen/Steve Clark/Joe Elliott/Mutt Lange/Rick Savage/Pete Willis]
A: Pyromania; P: Mutt Lange

1220. LOST IN YOU Rod Stewart Warner Bros. 27927 1015
[07May88 | 16Jul88 | 03Sep88] 12(1) 0 | 9 | 18
[Rod Stewart/Andy Taylor]
A: Out Of Order; P: Bernard Edwards/Rod Stewart/Andy Taylor

1221. BEAT'S SO LONELY Charlie Sexton MCA 52715 1010
[14Dec85 | 22Mar86 | 26Apr86] 17(3) 0 | 10 | 20
[Keith Forsey/Charlie Sexton] *A: Pictures For Pleasure; P: Keith Forsey*

1222. JUST BECAUSE Anita Baker Elektra 69327 1009
[21Jan89 | 01Apr89 | 06May89] 14(1) 0 | 11 | 16
[Alex Brown/Sami McKinney/Michael O'Hara]
A: Giving You The Best That I Got; P: Michael J. Powell

1223. GOIN' DOWN Greg Guidry Columbia 02691 1008
[13Feb82 | 01May82 | 29May82] 17(3) 0 | 10 | 16
[Greg Guidry/David C. Martin] *A: Over The Line; P: Greg Guidry/John Ryan*

1224. HEARTBREAKER Pat Benatar Chrysalis 2395 1008
[22Dec79 | 15Mar80 | 19Apr80] 23(1) 0 | 9 | 17
[Geoff Gill/Cliff Wade] *A: In The Heat Of The Night; P: Peter Coleman*

1225. THE CAPTAIN OF HER HEART Double A&M 2838 1006
[28Jun86 | 13Sep86 | 25Oct86] 16(2) 0 | 9 | 18
[Felix Haug/Kurt Maloo] *A: Blue; P: Double*

1226. MOVE AWAY Culture Club Virgin/Epic 05847 1004
[05Apr86 | 31May86 | 05Jul86] 12(2) 0 | 10 | 14
[Michael Craig/Roy Hay/Jon Moss/George O'Dowd/Phil Pickett]
A: From Luxury To Heartache; P: Lew Hahn/Arif Mardin

1227. SWEET BABY Stanley Clarke/George Duke Epic 01052 1003
[02May81 | 01Aug81 | 12Sep81] 19(2) 0 | 9 | 20
[George Duke] *A: The Clarke/Duke Project; P: Stanley Clarke/George Duke*

1228. DIGGING YOUR SCENE The Blow Monkeys RCA 14325 1002
[03May86 | 02Aug86 | 06Sep86] 14(1) 0 | 10 | 19
[Bruce Robert Howard] *A: Animal Magic; P: Peter Wilson*

1229. IF SHE WOULD HAVE BEEN FAITHFUL... Chicago 1001
Warner/Full Moon 28424
[21Mar87 | 30May87 | 25Jul87] 17(1) 0 | 8 | 19
[Randy Goodrum/Steve Kipner] *A: Chicago 18; P: David Foster*

1230. YOU'RE A FRIEND OF MINE Clarence Clemons and Jackson Browne 1001
Columbia 05660 [26Oct85 | 18Jan86 | 01Mar86] 18(1) 0 | 12 | 19
[Jeffrey E. Cohen/Narada Michael Walden] *A: Hero; P: Narada Michael Walden*

1231. A LITTLE RESPECT Erasure Sire 27738 1000
[10Dec88 | 04Mar89 | 01Apr89] 14(1) 0 | 9 | 17
[Andy Bell/Vince Clarke] *A: The Innocents; P: Stephen Hague*

1232. ELECTRIC YOUTH Debbie Gibson Atlantic 88919 1000
[01Apr89 | 13May89 | 24Jun89] 11(3) 0 | 8 | 13
[Debbie Gibson] *A: Electric Youth; P: Fred Zarr*

1233. TRUE LOVE Glenn Frey MCA 53363 1000
[20Aug88 | 15Oct88 | 26Nov88] 13(3) 0 | 9 | 15
[Glenn Frey/Jack Tempchin] *A: Soul Searching; P: Glenn Frey/Elliot Scheiner*

1234. SACRED EMOTION Donny Osmond Capitol 44379 1000
[17Jun89 | 26Aug89 | 30Sep89] 13(1) 0 | 9 | 16
[Evan Rogers/Carl Sturken] *A: Donny Osmond; P: Evan Rogers/Carl Sturken*

1235. OBJECT OF MY DESIRE Starpoint Elektra 69621 1000
[28Sep85 | 14Dec85 | 08Mar86] 25(1) 0 | 9 | 24
[Kayode Adeyemo/Keith Diamond/Ernesto Phillips]
A: Restless; P: Keith Diamond

Rank. TITLE Artist Label Number [Entry Date | Peak Date | Exit Date] Peak(Peak Wks) Top10 | Top40 | Tot Wks [Writers] *Album; Producers*	Score
1236. **AMERICAN STORM** Bob Seger & The Silver Bullet Band Capitol 5532 [15Mar86 | 03May86 | 14Jun86] 13(1) 0 | 9 | 14 [Bob Seger] *A: Like A Rock; P: Ed Andrews/Bob Seger*	999
1237. **I'M GONNA TEAR YOUR PLAYHOUSE DOWN** Paul Young Columbia 05577 [07Sep85 | 02Nov85 | 07Dec85] 13(1) 0 | 9 | 14 [Earl Randle] *A: The Secret Of Association; P: Laurie Latham*	999
1238. **LOVE THE WORLD AWAY** Kenny Rogers United Artists 1359 [21Jun80 | 02Aug80 | 06Sep80] 14(2) 0 | 8 | 12 [Bob Morrison/Johnny Wilson] *A: Urban Cowboy Soundtrack; P: Larry Butler*	998
1239. **HUMAN TOUCH** Rick Springfield RCA 13576 [09Jul83 | 10Sep83 | 15Oct83] 18(1) 0 | 11 | 15 [Rick Springfield] *A: Living In Oz; P: Bill Drescher/Rick Springfield*	997
1240. **THREE TIMES IN LOVE** Tommy James Millennium 11785 [26Jan80 | 29Mar80 | 10May80] 19(2) 0 | 9 | 16 [Tommy James/Ronald Serota] *A: Three Times In Love; P: Tommy James*	997
1241. **SINCERELY YOURS** Sweet Sensation Atco 99246 [04Feb89 | 29Apr89 | 03Jun89] 14(1) 0 | 8 | 19 [Joseph E. Malloy/Ricardo Pagan] *A: Take It While It's Hot; P: Steve Peck*	997
1242. **I BEG YOUR PARDON** Kon Kan Atlantic 88969 [24Dec88 | 11Mar89 | 22Apr89] 15(2) 0 | 9 | 18 [Barry Harris/Joe South] *A: Move To Move; P: Barry Harris*	995
1243. **DON'T ANSWER ME** Alan Parsons Project Arista 1-9160 [03Mar84 | 05May84 | 09Jun84] 15(1) 0 | 8 | 15 [Alan Parsons/Eric Woolfson] *A: Ammonia Avenue; P: Alan Parsons*	995
1244. **STEAL THE NIGHT** Stevie Woods Cotillion 46016 [12Sep81 | 26Dec81 | 30Jan82] 25(3) 0 | 10 | 21 [Bill Bowersock/Trevor Veitch/Matt Vernon] *A: Take Me To Your Heaven; P: Jack White*	994
1245. **WINNING** Santana Columbia 01050 [11Apr81 | 18Jul81 | 08Aug81] 17(1) 0 | 11 | 18 [Russ Ballard] *A: Zebop!; P: Keith Olsen/Carlos Santana*	994
1246. **CULT OF PERSONALITY** Living Colour Epic 68611 [11Mar89 | 06May89 | 17Jun89] 13(2) 0 | 9 | 15 [William Calhoun/Corey Glover/Vernon Reid/Muzz Skillings] *A: Vivid; P: Ed Stasium*	989
1247. **MY GIRL (Gone, Gone, Gone)** Chilliwack Millennium 11813 [26Sep81 | 12Dec81 | 30Jan82] 22(2) 0 | 11 | 19 [William A. Henderson/Brian MacLeod] *A: Wanna Be A Star; P: Bill Henderson/Brian MacLeod*	989
1248. **SANCTIFY YOURSELF** Simple Minds A&M 2810 [25Jan86 | 15Mar86 | 26Apr86] 14(3) 0 | 9 | 14 [Charles Burchill/Jim Kerr/Michael MacNeil] *A: Once Upon A Time; P: Bob Clearmountain/Jimmy Iovine*	986
1249. **BOULEVARD** Jackson Browne Asylum 47003 [05Jul80 | 06Sep80 | 18Oct80] 19(2) 0 | 10 | 16 [Jackson Browne] *A: Hold Out; P: Jackson Browne*	984
1250. **CUTS LIKE A KNIFE** Bryan Adams A&M 2553 [11Jun83 | 06Aug83 | 10Sep83] 15(1) 0 | 8 | 18 [Bryan Adams/Jim Vallance] *A: Cuts Like A Knife; P: Bryan Adams/Bob Clearmountain*	982
1251. **CRY** Godley & Creme Polydor 881786 [20Jul85 | 05Oct85 | 09Nov85] 16(1) 0 | 10 | 17 [Lol Crème/Kevin Godley] *A: The History Mix Vol. I; P: Godley & Crème/Trevor Horn*	981
1252. **READ 'EM AND WEEP** Barry Manilow Arista 1-9101 [19Nov83 | 07Jan84 | 18Feb84] 18(2) 0 | 10 | 14 [Jim Steinman] *A: Greatest Hits-Vol. II; P: Jim Steinman*	981
1253. **DREAMER** Supertramp A&M 2269 [20Sep80 | 08Nov80 | 20Dec80] 15(2) 0 | 8 | 14 [Rick Davies/Roger Hodgson] *A: Paris; P: Peter Henderson/Russel Pope*	981
1254. **IT AIN'T ENOUGH** Corey Hart EMI America 8236 [29Sep84 | 01Dec84 | 19Jan85] 17(1) 0 | 9 | 16 [Corey Hart] *A: First Offense; P: Jon Astley/Phil Chapman*	980
1255. **TRY AGAIN** Champaign Columbia 03563 [02Apr83 | 11Jun83 | 13Aug83] 23(3) 0 | 9 | 20 [Michael Day/Rocky Maffit/Dana Walden] *A: Modern Heart; P: Champaign/George Massenburg*	979
1256. **OLD-FASHION LOVE** Commodores Motown 1489 [21Jun80 | 30Aug80 | 04Oct80] 20(2) 0 | 11 | 16 [Milan Williams] *A: Heroes; P: James Anthony Carmichael/Commodores*	973
1257. **ON THE ROAD AGAIN** Willie Nelson Columbia 11351 [06Sep80 | 08Nov80 | 17Jan81] 20(2) 0 | 10 | 20 [Willie Nelson] *A: Honeysuckle Rose (Soundtrack); P: Willie Nelson*	973
1258. **TALK TO ME** Chico DeBarge Motown 1858 [08Nov86 | 21Feb87 | 21Mar87] 21(1) 0 | 11 | 20 [Paul Fox/Franne Golde/Nick Mundy] *A: Chico DeBarge; P: Skip Drinkwater*	972
1259. **MODERN GIRL** Sheena Easton EMI America 8080 [09May81 | 18Jul81 | 05Sep81] 18(2) 0 | 9 | 18 [Dominic Bugatti/Frank Musker] *A: Sheena Easton; P: Christopher Neil*	971
1260. **MIDDLE OF THE ROAD** The Pretenders Sire 29444 [17Dec83 | 11Feb84 | 17Mar84] 19(2) 0 | 9 | 14 [Chrissie Hynde] *A: Learning To Crawl; P: Chris Thomas*	968
1261. **IF I'D BEEN THE ONE** 38 Special A&M 2584 [12Nov83 | 21Jan84 | 25Feb84] 19(1) 0 | 9 | 16 [Don Barnes/Jeff Carlisi/Larry Steele/Donnie Van Zant] *A: Tour De Force; P: Rodney Mills*	966
1262. **WHO DO YOU GIVE YOUR LOVE TO?** Michael Morales Wing 887 743 [29Apr89 | 29Jul89 | 02Sep89] 15(1) 0 | 10 | 19 [Michael Morales] *A: Michael Morales; P: Michael Morales*	965
1263. **REALLY WANNA KNOW YOU** Gary Wright Warner Bros. 49769 [04Jul81 | 05Sep81 | 24Oct81] 16(2) 0 | 10 | 17 [Ali Thomson/Gary Wright] *A: The Right Place; P: Dean Parks/Gary Wright*	964
1264. **RIO** Duran Duran Capitol 5215 [02Apr83 | 14May83 | 25Jun83] 14(2) 0 | 9 | 13 [Simon Le Bon/Nick Rhodes/Andy Taylor/Nigel John Taylor/Roger Taylor] *A: Rio; P: Colin Thurston*	963
1265. **EAT IT** "Weird Al" Yankovic Rock 'n' Roll 04374 [10Mar84 | 14Apr84 | 26May84] 12(1) 0 | 7 | 12 [Michael Jackson/Al Yankovic] *A: "Weird Al" Yankovic In 3-D; P: Rick Derringer*	963
1266. **EVERY LITTLE KISS ('87)** Bruce Hornsby And The Range RCA 5165 [16May87 | 11Jul87 | 22Aug87] 14(1) 0 | 9 | 15 [Bruce Hornsby] *A: The Way It Is; P: Bruce Hornsby/Elliot Scheiner*	961
1267. **LITTLE LIAR** Joan Jett And The Blackhearts Blackheart 4-08095 [29Oct88 | 21Jan89 | 11Mar89] 19(1) 0 | 10 | 20 [Desmond Child/Joan Jett] *A: Up Your Alley; P: Desmond Child/Kenny Laguna*	961
1268. **WHAT'S GOING ON** Cyndi Lauper Portrait 37-06970 [14Mar87 | 09May87 | 06Jun87] 12(1) 0 | 10 | 13 [Renaldo Benson/Al Cleveland/Marvin Gaye] *A: True Colors; P: Cyndi Lauper/Lennie Petze*	961
1269. **STAND BY ME** Mickey Gilley Full Moon/Asylum 46640 [17May80 | 02Aug80 | 13Sep80] 22(2) 0 | 9 | 18 [Ben E. King/Jerry Leiber/Mike Stoller] *A: Urban Cowboy Soundtrack; P: Jim Ed Norman*	960
1270. **WHAT ABOUT ME?** Kenny Rogers RCA 13899 [15Sep84 | 17Nov84 | 19Jan85] 15(1) 0 | 9 | 19 [David Foster/Richard Marx/Kenny Rogers] *A: What About Me?; P: David Foster/Kenny Rogers*	959
1271. **ALL NIGHT LONG** Joe Walsh Full Moon/Asylum 46639 [17May80 | 26Jul80 | 30Aug80] 19(2) 0 | 8 | 16 [Joe Walsh] *A: Urban Cowboy Soundtrack; P: Joe Walsh*	959
1272. **DO YOU LOVE ME** The Contours Motown Yesteryear 448 [04Jun88 | 06Aug88 | 17Sep88] 11(1) 0 | 8 | 16 [Berry Gordy] *A: More Dirty Dancing; P: Berry Gordy*	958
1273. **LIKE A ROCK** Bob Seger & The Silver Bullet Band Capitol 5592 [24May86 | 12Jul86 | 16Aug86] 12(1) 0 | 9 | 13 [Bob Seger] *A: Like A Rock; P: Ed Andrews/Bob Seger*	958
1274. **HOURGLASS** Squeeze A&M 2967 [19Sep87 | 05Dec87 | 23Jan88] 15(1) 0 | 9 | 19 [Chris Difford/Glenn Tilbrook] *A: Babylon And On; P: Eric Thorngren/Glenn Tilbrook*	958
1275. **GOODBYE** Night Ranger Camel/MCA 52729 [09Nov85 | 01Feb86 | 08Mar86] 17(1) 0 | 10 | 18 [Jack Blades/Jeff Watson] *A: 7 Wishes; P: Pat Glasser*	956
1276. **UNDERSTANDING** Bob Seger & The Silver Bullet Band Capitol 5413 [10Nov84 | 05Jan85 | 16Feb85] 17(2) 0 | 8 | 15 [Bob Seger] *A: Teachers Soundtrack; P: Ed Andrews/Bob Seger*	952
1277. **AMERICAN MUSIC** Pointer Sisters Planet 13254 [26Jun82 | 28Aug82 | 25Sep82] 16(3) 0 | 8 | 14 [Parker McGee] *A: So Excited!; P: Richard Perry*	952
1278. **DON'T MAKE ME OVER** Sybil Next Plateau 325 [23Sep89 | 02Dec89 | 24Feb90] 20(1) 0 | 9 | 23 [Burt Bacharach/Hal David] *A: Sybil; P: James Bratten/Dolores Drewry*	950

Ranking the Singles

Rank. TITLE Artist Label Number
[Entry Date | Peak Date | Exit Date] Peak(Peak Wks) Top10 | Top40 | Tot Wks
[Writers] Album; Producers   Score

1279. BABY I LIED Deborah Allen RCA 13600   950
[15Oct83 | 21Jan84 | 03Mar84] 26(3) 0 | 7 | 21
[Deborah Allen/Rory Bourke/Rafe Van Hoy]
A: Cheat The Night; P: Charles Callelo

1280. AIN'T NOBODY Rufus And Chaka Khan Warner Bros. 29555   947
[01Oct83 | 03Dec83 | 04Feb84] 22(3) 0 | 8 | 19
[David Wolinski] A: Live-Stompin' At The Savoy; P: Russ Titelman

1281. IT'S ONLY LOVE Bryan Adams/Tina Turner A&M 2791   947
[23Nov85 | 18Jan86 | 22Feb86] 15(1) 0 | 9 | 14
[Bryan Adams/Jim Vallance] A: Reckless; P: Bryan Adams/Bob Clearmountain

1282. SMOKY MOUNTAIN RAIN Ronnie Milsap RCA 12084   945
[29Nov80 | 28Feb81 | 18Apr81] 24(2) 0 | 9 | 21
[Kye Fleming/Dennis W. Morgan]
A: Greatest Hits; P: Tom Collins/Ronnie Milsap

1283. ANOTHER PART OF ME Michael Jackson Epic 07962   945
[23Jul88 | 10Sep88 | 15Oct88] 11(1) 0 | 8 | 13
[Michael Jackson] A: Bad; P: Quincy Jones

1284. SAVED BY ZERO The Fixx MCA 52213   941
[28May83 | 13Aug83 | 10Sep83] 20(1) 0 | 8 | 16
[Alfie Agius/Cy Curnin/Peter John Greenall/Jamie West-Oram/Adam Woods]
A: Reach The Beach; P: Rupert Hine

1285. I WOULDN'T HAVE MISSED IT FOR THE WORLD Ronnie Milsap   941
RCA 12342
[24Oct81 | 16Jan82 | 13Feb82] 20(2) 0 | 11 | 17
[Kye Fleming/Dennis W. Morgan/Charles Quillen]
A: There's No Gettin' Over Me; P: Tom Collins/Ronnie Milsap

1286. CHECK IT OUT John Cougar Mellencamp Mercury 870 126   941
[06Feb88 | 16Apr88 | 14May88] 14(1) 0 | 8 | 15
[John Mellencamp]
A: The Lonesome Jubilee; P: Don Gehman/John Mellencamp

1287. IS IT YOU Lee Ritenour Elektra 47124   941
[25Apr81 | 27Jun81 | 08Aug81] 15(2) 0 | 9 | 16
[Bill Champlin/Lee Ritenour/Eric Tagg]
A: Rit; P: David Foster/Harvey Mason/Lee Ritenour

1288. ONE OF THE LIVING Tina Turner Capitol 5518   940
[05Oct85 | 23Nov85 | 01Feb86] 15(2) 0 | 10 | 18
[Holly Knight]
A: Mad Max Beyond Thunderdome Soundtrack; P: Mike Chapman

1289. OPERATOR Midnight Star Solar 69684   939
[01Dec84 | 02Feb85 | 23Mar85] 18(2) 0 | 8 | 17
[Reggie Calloway/Vincent Calloway/Belinda Lipscomb/Bo Watson]
A: Planetary Invasion; P: Reggie Calloway

1290. THIRD TIME LUCKY (First Time I Was A Fool) Foghat Bearsville 49125 938
[17Nov79 | 26Jan80 | 23Feb80] 23(2) 0 | 10 | 15
[Dave Peverett] A: Boogie Motel; P: Foghat/Tony Outeda

1291. EMOTION IN MOTION Ric Ocasek Geffen 28617   937
[06Sep86 | 15Nov86 | 10Jan87] 15(1) 0 | 8 | 19
[Ric Ocasek]
A: This Side Of Paradise; P: Ross Cullum/Chris Hughes/Ric Ocasek

1292. STAY WITH ME TONIGHT Jeffrey Osborne A&M 2591   936
[15Oct83 | 14Jan84 | 03Mar84] 30(2) 0 | 8 | 21
[Raymond C. Jones] A: Stay With Me Tonight; P: George Duke

1293. SIDEWALK TALK Jellybean EMI America 8297   935
[16Nov85 | 01Feb86 | 15Mar86] 18(1) 0 | 9 | 18
[Madonna L. Ciccone] A: Wotupski?; P: Jellybean Benitez

1294. TAKE ME DOWN Alabama RCA 13210   934
[22May82 | 03Jul82 | 14Aug82] 18(4) 0 | 8 | 13
[Mark Eugene Gray/James P. Pennington]
A: Mountain Music; P: Alabama/Harold Shedd

1295. WITHOUT YOUR LOVE Roger Daltrey Polydor 2121   933
[13Sep80 | 29Nov80 | 17Jan81] 20(2) 0 | 8 | 19
[William Morris Nicholls] A: McVicar (Soundtrack); P: Jeff Wayne

1296. WALKING ON A THIN LINE Huey Lewis And The News Chrysalis 42825 933
[20Oct84 | 08Dec84 | 26Jan85] 18(2) 0 | 10 | 15
[André Pessis/Kevin Wells] A: Sports; P: Huey Lewis And The News

1297. EMPIRE STRIKES BACK (Medley) Meco RSO 1038   933
[14Jun80 | 09Aug80 | 13Sep80] 18(2) 0 | 8 | 14
[John Williams] A: Meco Plays Music From The Empire Strikes Back;
P: Tony Bongiovi/Meco Monardo/Lance Quinn

1298. FALLEN ANGEL Poison Enigma 44191   930
[30Jul88 | 08Oct88 | 12Nov88] 12(1) 0 | 9 | 16
[Bobby Dall/C.C. DeVille/Bret Michaels/Rikki Rockett]
A: Open Up And Say...Ahh!; P: Tom Werman

1299. A PENNY FOR YOUR THOUGHTS Tavares RCA 13292   930
[18Sep82 | 25Dec82 | 05Feb83] 33(4) 0 | 9 | 21
[Kenny Nolan] A: New Directions; P: Kenny Nolan/Jay Senter

1300. TOUCH ME WHEN WE'RE DANCING Carpenters A&M 2344   929
[20Jun81 | 01Aug81 | 19Sep81] 16(4) 0 | 8 | 14
[Kenneth Edward Bell/Terry Skinner/J.L. Wallace]
A: Made In America; P: Richard Carpenter

1301. WHERE THE STREETS HAVE NO NAME U2 Island 99408   927
[12Sep87 | 07Nov87 | 12Dec87] 13(2) 0 | 9 | 14
[Adam Clayton/Dave Evans/Paul Hewson/Larry Mullen]
A: The Joshua Tree; P: Brian Eno/Daniel Lanois

1302. I'M ALIVE Electric Light Orchestra MCA 41246   926
[24May80 | 12Jul80 | 30Aug80] 16(2) 0 | 8 | 15
[Jeff Lynne] A: Xanadu (Soundtrack); P: Jeff Lynne

1303. NEVER ENDING STORY Limahl EMI America 8230   925
[23Mar85 | 15Jun85 | 27Jul85] 17(1) 0 | 8 | 19
[Keith Forsey/Giorgio Moroder] A: Don't Suppose; P: Giorgio Moroder

1304. 19 Paul Hardcastle Chrysalis 42860   925
[01Jun85 | 20Jul85 | 31Aug85] 15(2) 0 | 8 | 14
[William Couturie/Paul Hardcastle/Jonas McCord/Mike Oldfield]
A: Paul Hardcastle; P: Paul Hardcastle

1305. ONE STEP UP Bruce Springsteen Columbia 07726   924
[27Feb88 | 23Apr88 | 04Jun88] 13(1) 0 | 8 | 15
[Bruce Springsteen]
A: Tunnel Of Love; P: Jon Landau/Chuck Plotkin/Bruce Springsteen

1306. BIG LOG Robert Plant Atlantic 99844   921
[06Aug83 | 15Oct83 | 19Nov83] 20(2) 0 | 9 | 16
[Robbie Blunt/Robert Plant/Jezz Woodroffe]
A: The Principle Of Moments; P: Benji Lefevre/Robert Plant

1307. BEDS ARE BURNING Midnight Oil Columbia 07433   921
[02Apr88 | 02Jul88 | 27Aug88] 17(1) 0 | 9 | 22
[Peter Garrett/Peter Gifford/Rob Hirst/James Moginie/Martin Rotsey]
A: Diesel And Dust; P: Warne Livesey/Midnight Oil

1308. DOMINOES Robbie Nevil EMI 50053   919
[14Feb87 | 25Apr87 | 30May87] 14(1) 0 | 9 | 16
[Richard Eastman/Bobby Hart/Robbie Nevil]
A: Robbie Nevil; P: Alex Sadkin/Philip Thornalley

1309. ROCKET Def Leppard Mercury 872614   918
[04Mar89 | 29Apr89 | 27May89] 12(1) 0 | 9 | 13
[Rick Allen/Steve Clark/Phil Collen/Joe Elliott/Mutt Lange/Rick Savage]
A: Hysteria; P: Mutt Lange

1310. AND THE BEAT GOES ON The Whispers Solar 11894   917
[09Feb80 | 12Apr80 | 17May80] 19(2) 0 | 8 | 15
[William Shelby/Stephen Shockley/Leon Sylvers]
A: The Whispers; P: Dick Griffey/Whispers

1311. AND WE DANCED Hooters Columbia 05568   914
[10Aug85 | 26Oct85 | 21Dec85] 21(1) 0 | 8 | 20
[Eric Bazilian/Rob Hyman] A: Nervous Night; P: Rick Chertoff

1312. GIMME SOME LOVIN' Blues Brothers Atlantic 3666   911
[31May80 | 26Jul80 | 30Aug80] 18(2) 0 | 8 | 14
[Spencer Davis/Muff Winwood/Steve Winwood]
A: The Blues Brothers (Soundtrack); P: Bob Tischler

1313. MODERN DAY DELILAH Van Stephenson MCA 52376   910
[21Apr84 | 30Jun84 | 11Aug84] 22(2) 0 | 10 | 17
[Jan Buckingham/Van Stephenson] A: Righteous Anger; P: Richard Landis

1314. PLAY THE GAME TONIGHT Kansas Kirshner 2903   910
[08May82 | 03Jul82 | 14Aug82] 17(3) 0 | 9 | 15
[Phil Ehart/Daniel Fleishour/Robert Frazier/Kerry Livgren/
Richard John Williams] A: Vinyl Confessions; P: Kansas/Ken Scott

1315. SHE'S A BAD MAMA JAMA (She's Built, She's Stacked) Carl Carlton 909
20th Century 2488 [22Aug81 | 24Oct81 | 09Jan82] 22(2) 0 | 7 | 21
[Leon Haywood] A: Carl Carlton; P: Leon Haywood

1316. STILL THEY RIDE Journey Columbia 02883   907
[22May82 | 17Jul82 | 21Aug82] 19(3) 0 | 9 | 14
[Jonathan Cain/Steve Perry/Neal Schon] A: Escape; P: Kevin Elson/Mike Stone

Rank. TITLE Artist Label Number	
[Entry Date | Peak Date | Exit Date] Peak(Peak Wks) Top10 | Top40 | Tot Wks	
[Writers] *Album;* Producers	Score
1317. SMOKIN' IN THE BOYS ROOM Mötley Crüe Elektra 69625 [13Jul85 | 07Sep85 | 19Oct85] 16(2) 0 | 9 | 15 [Cub Koda/Michael Lutz] A: *Theatre Of Pain;* P: Tom Werman	907
1318. THE CURLY SHUFFLE Jump 'N The Saddle Atlantic 89718 [03Dec83 | 21Jan84 | 03Mar84] 15(1) 0 | 7 | 14 [Peter Quinn] A: *Jump 'N The Saddle Band;* P: T.C. Furlong/Mike Rasfield/Barney Schwartz	905
1319. ARE WE OURSELVES? The Fixx MCA 52444 [18Aug84 | 20Oct84 | 24Nov84] 15(1) 0 | 8 | 15 [Alfie Agius/Dan Brown/Cy Curnin/Peter John Greenall/Jamie West-Oram/Adam Woods] A: *Phantoms;* P: Rupert Hine	905
1320. EMERGENCY Kool & The Gang De-Lite 884199 [26Oct85 | 28Dec85 | 08Feb86] 18(2) 0 | 8 | 16 [Clifford Adams/Robert Bell/Ronald Bell/James Bonneford/George M. Brown/Robert Mickens/Michael Ray/Claydes Smith/J.T. Taylor/Dennis Thomas/Curtis Fitzgerald Williams] A: *Emergency;* P: Ronald Bell/Jim Bonneford/Kool & The Gang	904
1321. SAY GOODBYE TO HOLLYWOOD (Live) Billy Joel Columbia 02518 [12Sep81 | 07Nov81 | 19Dec81] 17(1) 0 | 8 | 15 [Billy Joel] A: *Songs In The Attic;* P: Phil Ramone/Brian Ruggles	904
1322. DON'T ASK ME WHY Billy Joel Columbia 11331 [02Aug80 | 20Sep80 | 08Nov80] 19(3) 0 | 9 | 15 [Billy Joel] A: *Glass Houses;* P: Phil Ramone	903
1323. LOVE LIGHT IN FLIGHT Stevie Wonder Tamla 1769 [01Dec84 | 02Feb85 | 16Mar85] 17(1) 0 | 10 | 16 [Stevie Wonder] A: *The Woman In Red (Soundtrack);* P: Stevie Wonder	903
1324. WONDERING WHERE THE LIONS ARE Bruce Cockburn Millennium 11786 [22Mar80 | 07Jun80 | 12Jul80] 21(2) 0 | 9 | 17 [Bruce Cockburn] A: *Dancing In The Dragon's Jaws;* P: Gene Martynec	901
1325. I BELIEVE IN YOU Don Williams MCA 41304 [27Sep80 | 20Dec80 | 07Feb81] 24(3) 0 | 9 | 20 [Roger Cook/Sam Hogin] A: *I Believe In You;* P: Garth Fundis/Don Williams	901
1326. IT MIGHT BE YOU (Theme From Tootsie) Stephen Bishop Warner Bros. 29791 [29Jan83 | 07May83 | 11Jun83] 25(1) 0 | 8 | 20 [Alan Bergman/Marilyn Bergman/Dave Grusin] A: *Tootsie Soundtrack;* P: Dave Grusin	901
1327. FIRE AND ICE Pat Benatar Chrysalis 2529 [18Jul81 | 05Sep81 | 24Oct81] 17(1) 0 | 9 | 15 [Pat Benatar/Thomas F. Kelly/Scott Sheets] A: *Precious Time;* P: Neil Geraldo/Keith Olsen	901
1328. ROCK 'N' ROLL IS KING ELO Jet 03964 [25Jun83 | 20Aug83 | 17Sep83] 19(2) 0 | 9 | 13 [Jeff Lynne] A: *Secret Messages;* P: Jeff Lynne	901
1329. HEALING HANDS Elton John MCA 53692 [26Aug89 | 28Oct89 | 02Dec89] 13(1) 0 | 9 | 15 [Elton John/Bernie Taupin] A: *Sleeping With The Past;* P: Chris Thomas	900
1330. POP GOES THE WORLD Men Without Hats Mercury 888859 [31Oct87 | 13Feb88 | 19Mar88] 20(0) 0 | 10 | 21 [Ivan Doroschuk] A: *Pop Goes The World;* P: Bernd Held/Men Without Hats	898
1331. ONLY TIME WILL TELL Asia Geffen 29970 [24Jul82 | 18Sep82 | 23Oct82] 17(3) 0 | 8 | 14 [Geoffrey Downes/John Wetton] A: *Asia;* P: Mike Stone	897
1332. HEARTS ON FIRE Randy Meisner Epic 50964 [24Jan81 | 14Mar81 | 02May81] 19(3) 0 | 9 | 15 [Eric Kaz/Randy Meisner] A: *One More Song;* P: Val Garay	895
1333. THE POLITICS OF DANCING Re-Flex Capitol 5301 [26Nov83 | 17Mar84 | 14Apr84] 24(1) 0 | 5 | 21 [Paul Fishman] A: *The Politics Of Dancing;* P: John Punter	893
1334. DRESSED FOR SUCCESS Roxette EMI 50204 [27May89 | 29Jul89 | 23Sep89] 14(1) 0 | 9 | 18 [Per Gessle] A: *Look Sharp!;* P: Clarence Ofwerman	891
1335. YOU BETTER YOU BET The Who Warner Bros. 49698 [21Mar81 | 09May81 | 27Jun81] 18(3) 0 | 10 | 15 [Pete Townshend] A: *Face Dances;* P: Bill Szymczyk	891
1336. I'M THAT TYPE OF GUY LL Cool J Def Jam 68902 [10Jun89 | 05Aug89 | 23Sep89] 15(2) 0 | 8 | 16 [Steve Ettinger/Dwayne Simon/James Todd Smith] A: *Walking With A Panther;* P: Dwayne Simon/James Todd Smith	890

Rank. TITLE Artist Label Number	
[Entry Date | Peak Date | Exit Date] Peak(Peak Wks) Top10 | Top40 | Tot Wks	
[Writers] *Album;* Producers	Score
1337. ANIMAL Def Leppard Mercury 888832 [10Oct87 | 26Dec87 | 13Feb88] 19(2) 0 | 9 | 19 [Rick Allen/Steve Clark/Phil Collen/Joe Elliott/Mutt Lange/Rick Savage] A: *Hysteria;* P: Mutt Lange	890
1338. WE ALL SLEEP ALONE Cher Geffen 27986 [09Apr88 | 11Jun88 | 16Jul88] 14(1) 0 | 9 | 15 [Jon Bon Jovi/Desmond Child/Richie Sambora] A: *Cher(2);* P: Jon Bon Jovi/Desmond Child/Richie Sambora	889
1339. ALMOST OVER YOU Sheena Easton EMI America 8186 [10Dec83 | 10Mar84 | 21Apr84] 25(1) 0 | 6 | 20 [Jennifer Kimball/Cindy Richardson] A: *Best Kept Secret;* P: Greg Mathieson	889
1340. WHAT YOU GET IS WHAT YOU SEE Tina Turner Capitol 5668 [07Feb87 | 04Apr87 | 09May87] 13(1) 0 | 7 | 14 [Terry Britten/Graham Lyle] A: *Break Every Rule;* P: Terry Britten	887
1341. WHO WEARS THESE SHOES? Elton John Geffen 29189 [08Sep84 | 03Nov84 | 08Dec84] 16(1) 0 | 10 | 14 [Elton John/Bernie Taupin] A: *Breaking Hearts;* P: Chris Thomas	887
1342. MAN SIZE LOVE Klymaxx MCA 52841 [05Jul86 | 13Sep86 | 11Oct86] 15(1) 0 | 8 | 15 [Rod Temperton] A: *Klymaxx;* P: Richard Rudolph/Bruce Swedien/Rod Temperton	887
1343. ALL I WANTED Kansas MCA 52958 [01Nov86 | 17Jan87 | 28Feb87] 19(1) 0 | 10 | 18 [Steve Morse/Steve Walsh] A: *Power;* P: Andrew Powell	885
1344. DON'T MAKE ME WAIT FOR LOVE Kenny G Vocal by Lenny Williams Arista 1-9625 [29Aug87 | 07Nov87 | 02Jan88] 15(1) 0 | 9 | 19 [Walter Afanasieff/Preston Glass/Narada Michael Walden] A: *Duotones;* P: Preston Glass	884
1345. DON'T LOOK BACK Fine Young Cannibals MCA 53695 [12Aug89 | 07Oct89 | 28Oct89] 11(1) 0 | 8 | 12 [Roland Gift/David Steele] A: *The Raw & The Cooked;* P: Andy Cox/Roland Gift/David Steele	884
1346. HELLO AGAIN The Cars Elektra 69681 [27Oct84 | 22Dec84 | 02Feb85] 20(2) 0 | 10 | 15 [Ric Ocasek] A: *Heartbeat City;* P: Cars/Mutt Lange	882
1347. COME AS YOU ARE Peter Wolf EMI America 8350 [28Feb87 | 25Apr87 | 06Jun87] 15(2) 0 | 9 | 15 [Tim Mayer/Peter Wolf] A: *Come As You Are;* P: Eric Thorngren/Peter Wolf	881
1348. ANGEL OF HARLEM U2 Island 99254 [17Dec88 | 11Feb89 | 25Mar89] 14(2) 0 | 8 | 15 [Adam Clayton/Dave Evans/Paul Hewson/Larry Mullen] A: *Rattle And Hum (Soundtrack);* P: Jimmy Iovine	878
1349. A NIGHTMARE ON MY STREET DJ Jazzy Jeff & The Fresh Prince Jive 1124 [30Jul88 | 24Sep88 | 12Nov88] 15(1) 0 | 9 | 16 [Peter Brian Harris/Will Smith/Jeff Townes] A: *He's The D.J., I'm The Rapper;* P: Pete Q. Harris/Bryan New/Will Smith/Jeff Townes	878
1350. KEEPING THE FAITH Billy Joel Columbia 04681 [26Jan85 | 23Mar85 | 11May85] 18(2) 0 | 10 | 16 [Billy Joel] A: *An Innocent Man;* P: Phil Ramone	876
1351. LIKE NO OTHER NIGHT 38 Special A&M 2831 [03May86 | 05Jul86 | 16Aug86] 14(1) 0 | 9 | 16 [Don Barnes/John Bettis/Jeff Carlisi/Jim Vallance] A: *Strength In Numbers;* P: Keith Olsen	874
1352. HEADED FOR A HEARTBREAK Winger Atlantic 88922 [03Jun89 | 19Aug89 | 30Sep89] 19(1) 0 | 9 | 18 [Kip Winger] A: *Winger;* P: Beau Hill	872
1353. SPY IN THE HOUSE OF LOVE Was (Not Was) Chrysalis 43266 [01Oct88 | 17Dec88 | 21Jan89] 16(1) 0 | 10 | 17 [David Was/Don Was] A: *What Up, Dog?;* P: Paul Staveley O'Duffy	869
1354. HEY BABY Henry Lee Summer CBS Associated 68891 [20May89 | 05Aug89 | 16Sep89] 18(8) 0 | 8 | 18 [Henry Lee Summer] A: *I've Got Everything;* P: Henry Lee Summer	867
1355. DREAMING Orchestral Manoeuvres In The Dark A&M 3002 [12Mar88 | 21May88 | 02Jul88] 16(2) 0 | 9 | 17 [Paul Humphreys/Andrew McCluskey] A: *The Best Of OMD;* P: O.M.D.	865
1356. HOT IN THE CITY ('82) Billy Idol Chrysalis 2605 [03Jul82 | 11Sep82 | 23Oct82] 23(4) 0 | 9 | 17 [Billy Idol] A: *Billy Idol;* P: Keith Forsey	864

Ranking the Singles

Rank. TITLE Artist Label Number
[Entry Date | Peak Date | Exit Date] Peak(Peak Wks) Top10 | Top40 | Tot Wks
[Writers] *Album*; Producers — Score

1357. I THINK IT'S LOVE Jermaine Jackson Arista 1-9444 861
[22Feb86 | 26Apr86 | 31May86] 16(1) 0 | 9 | 15
[Jermaine Jackson/Michael Omartian/Stevie Wonder]
A: *Precious Moments*; P: Michael Omartian

1358. MISSING YOU Dan Fogelberg Full Moon 03289 860
[09Oct82 | 04Dec82 | 22Jan83] 23(5) 0 | 9 | 16
[Dan Fogelberg]
A: *Dan Fogelberg/Greatest Hits*; P: Dan Fogelberg/Marty Lewis

1359. DAY BY DAY Hooters Columbia 05730 860
[14Dec85 | 22Feb86 | 12Apr86] 18(3) 0 | 7 | 18
[Eric Bazilian/Rick Chertoff/Rob Hyman] A: *Nervous Night*; P: Rick Chertoff

1360. 99 Toto Columbia 11173 859
[22Dec79 | 15Mar80 | 12Apr80] 26(1) 0 | 8 | 17
[David Paich] A: *Hydra*; P: Tom Knox/Toto

1361. THE FUTURE'S SO BRIGHT, I GOTTA WEAR SHADES Timbuk 3 859
I.R.S. 52940
[25Oct86 | 27Dec86 | 07Feb87] 19(2) 0 | 9 | 16
[Pat MacDonald] A: *Greetings From Timbuk 3*; P: Dennis Herring

1362. WHY NOT ME Fred Knoblock Scotti Brothers 600 859
[28Jun80 | 23Aug80 | 27Sep80] 18(2) 0 | 7 | 14
[Fred Knoblock/Carson Whitsett] A: *Why Not Me*; P: James Stroud

1363. PRESSURE Billy Joel Columbia 03244 859
[25Sep82 | 20Nov82 | 15Jan83] 20(3) 0 | 8 | 17
[Billy Joel] A: *The Nylon Curtain*; P: Phil Ramone

1364. KIDS IN AMERICA Kim Wilde EMI America 8110 858
[22May82 | 14Aug82 | 18Sep82] 25(4) 0 | 8 | 18
[Marty Wilde/Ricky Wilde] A: *Kim Wilde*; P: Ricky Wilde

1365. EVERYBODY DANCE Ta Mara & The Seen A&M 2768 856
[12Oct85 | 18Jan86 | 01Mar86] 24(1) 0 | 10 | 21
[Jesse Woods Johnson] A: *Ta Mara & The Seen*; P: Jesse Johnson

1366. WHEN I WANTED YOU Barry Manilow Arista 0481 855
[15Dec79 | 01Mar80 | 29Mar80] 20(2) 0 | 7 | 14
[Gino Cunico] A: *One Voice*; P: Ron Dante/Barry Manilow

1367. CRAZY IN THE NIGHT (Barking At Airplanes) Kim Carnes 855
EMI America 8267
[11May85 | 13Jul85 | 24Aug85] 15(1) 0 | 9 | 16
[Kim Carnes] A: *Barking At Airplanes*; P: Kim Carnes/Bill Cuomo

1368. ROCK ME TONIGHT (For Old Times Sake) Freddie Jackson Capitol 5459 854
[25May85 | 10Aug85 | 28Sep85] 18(1) 0 | 8 | 19
[Paul Laurence] A: *Rock Me Tonight*; P: Paul Laurence

1369. TAKE OFF Bob & Doug McKenzie Mercury 76134 854
[30Jan82 | 27Mar82 | 01May82] 16(2) 0 | 9 | 14
[Kerry Crawford/Marc Giacomelli/Jonathan Goldsmith/Rick Moranis/
Dave Thomas] A: *Great White North*; P: Marc Giacomelli

1370. NO WAY OUT Jefferson Starship Grunt 13811 851
[12May84 | 21Jul84 | 25Aug84] 23(1) 0 | 8 | 16
[Ina Wolf/Peter F. Wolf] A: *Nuclear Furniture*; P: Ron Nevison

1371. THIS IS THE TIME Billy Joel Columbia 06526 850
[15Nov86 | 31Jan87 | 07Mar87] 18(1) 0 | 9 | 17
[Billy Joel] A: *The Bridge*; P: Phil Ramone

1372. WHO FOUND WHO Jellybean (Elisa Fiorillo Vocals) Chrysalis 43120 849
[11Jul87 | 19Sep87 | 17Oct87] 16(1) 0 | 8 | 15
[Paul Gurvitz] A: *Just Visiting This Planet*; P: Jellybean Benitez

1373. STAY THE NIGHT Benjamin Orr Elektra 69506 849
[08Nov86 | 14Feb87 | 21May87] 24(1) 0 | 6 | 19
[Benjamin Orr/Diane Page] A: *The Lace*; P: Larry Klein/Benjamin Orr/Mike Shipley

1374. LOVE IS ALRIGHT TONITE Rick Springfield RCA 13008 848
[05Dec81 | 13Feb82 | 20Mar82] 20(2) 0 | 10 | 16
[Rick Springfield] A: *Working Class Dog*; P: Bill Drescher/Rick Springfield

1375. I MISSED AGAIN Phil Collins Atlantic 3790 847
[21Mar81 | 23May81 | 04Jul81] 19(2) 0 | 9 | 16
[Phil Collins] A: *Face Value*; P: Phil Collins

1376. SEXY GIRL Glenn Frey MCA 52413 845
[30Jun84 | 18Aug84 | 06Oct84] 20(2) 0 | 9 | 15
[Glenn Frey/Jack Tempchin]
A: *The Allnighter*; P: Barry Beckett/Allan Blazek/Glenn Frey

1377. LOVE IS THE SEVENTH WAVE Sting A&M 2787 844
[09Nov85 | 28Dec85 | 01Feb86] 17(3) 0 | 9 | 13
[Sting] A: *The Dream Of The Blue Turtles*; P: Pete Smith/Sting

1378. ALL OF YOU Julio Iglesias & Diana Ross Columbia 04507 844
[07Jul84 | 01Sep84 | 20Oct84] 19(2) 0 | 8 | 16
[Julio Iglesias/Tony Renis/Cynthia Weil]
A: *Swept Away*; P: Ramon Acusa/Richard Perry

1379. COME GO WITH ME The Beach Boys Caribou 02633 843
[21Nov81 | 30Jan82 | 27Feb82] 18(2) 0 | 8 | 15
[Clarence Quick] A: *Ten Years Of Harmony (1970-1980)*; P: Alan Jardine

1380. IN THE AIR TONIGHT Phil Collins Atlantic 3824 842
[30May81 | 15Aug81 | 19Sep81] 19(2) 0 | 8 | 17
[Phil Collins] A: *Face Value*; P: Phil Collins

1381. SHOCK THE MONKEY Peter Gabriel Geffen 29883 841
[23Oct82 | 29Jan83 | 19Feb83] 29(2) 0 | 10 | 18
[Peter Gabriel] A: *Peter Gabriel (Security)*; P: Peter Gabriel/David Lord

1382. RUN RUNAWAY Slade CBS Associated 04398 840
[07Apr84 | 16Jun84 | 28Jul84] 20(1) 0 | 8 | 17 [Noddy Holder/Jim Lea]
A: *Keep Your Hands Off My Power Supply*; P: John Punter

1383. HERE WITH ME REO Speedwagon Epic 07901 840
[25Jun88 | 10Sep88 | 29Oct88] 20(1) 0 | 9 | 19
[Rick Braun/Kevin Cronin] A: *The Hits*; P: Keith Olsen

1384. YANKEE ROSE David Lee Roth Warner Bros. 28656 840
[05Jul86 | 30Aug86 | 11Oct86] 16(1) 0 | 8 | 15
[David Lee Roth/Steve Vai] A: *Eat 'Em And Smile*; P: Ted Templeman

1385. RUN FOR THE ROSES Dan Fogelberg Full Moon 02821 840
[03Apr82 | 29May82 | 03Jul82] 18(1) 0 | 8 | 14
[Dan Fogelberg]
A: *The Innocent Age*; P: Dan Fogelberg/Marty Lewis

1386. RAG DOLL Aerosmith Geffen 27915 840
[04Jun88 | 20Aug88 | 24Sep88] 17(1) 0 | 8 | 17
[Holly Knight/Joe Perry/Steven Tyler/Jim Vallance]
A: *Permanent Vacation*; P: Bruce Fairbairn

1387. AMERICAN HEARTBEAT Survivor Scotti Brothers 03213 839
[25Sep82 | 20Nov82 | 08Jan83] 17(2) 0 | 7 | 16
[Jim Peterik/Frankie Sullivan]
A: *Eye Of The Tiger*; P: Jim Peterik/Frankie Sullivan

1388. I NEED LOVE LL Cool J Def Jam 07350 838
[01Aug87 | 12Sep87 | 24Oct87] 14(2) 0 | 8 | 13
[Bobby Ervin/Steve Ettinger/Darryl Pierce/Dwayne Simon/James Todd Smith]
A: *Bigger And Deffer*;
P: Bobby Ervin/Darryl Pierce/Dwayne Simon/James Todd Smith

1389. BE STILL MY BEATING HEART Sting A&M 2992 838
[16Jan88 | 12Mar88 | 16Apr88] 15(2) 0 | 8 | 14
[Sting] A: *...Nothing Like The Sun*; P: Neil Dorfsman/Sting

1390. WORKING FOR THE WEEKEND Loverboy Columbia 02589 837
[14Nov81 | 13Feb82 | 27Mar82] 29(2) 0 | 8 | 20
[Paul Dean/Matt Frenette/Mike Reno] A: *Get Lucky*;
P: Paul Dean/Bruce Fairbairn

1391. SWEPT AWAY Diana Ross RCA 13864 837
[01Sep84 | 27Oct84 | 01Dec84] 19(2) 0 | 8 | 14
[Sara Allen/Daryl Hall] A: *Swept Away*; P: Arthur Baker/Daryl Hall

1392. ONE LONELY NIGHT REO Speedwagon Epic 04848 834
[30Mar85 | 01Jun85 | 13Jul85] 19(2) 0 | 9 | 16
[Neal Doughty]
A: *Wheels Are Turnin'*; P: Kevin Crouch/Alan Gratzer/Gary Richrath

1393. THE COLOUR OF LOVE Billy Ocean Jive 9707 833
[28May88 | 30Jul88 | 10Sep88] 19(2) 0 | 9 | 16
[Wayne Brathwaite/Barry Eastmond/Billy Ocean/Jolyon Skinner]
A: *Tear Down These Walls*; P: Wayne Brathwaite/Barry Eastmond

1394. OUT OF WORK Gary U.S. Bonds EMI America 8117 833
[12Jun82 | 21Aug82 | 25Sep82] 21(2) 0 | 9 | 16
[Bruce Springsteen] A: *On The Line*; P: Bruce Springsteen/Steven Van Zandt

1395. MORE THAN YOU KNOW Martika Columbia 08103 832
[24Dec88 | 01Apr89 | 06May89] 18(1) 0 | 8 | 19
[Michael Jay/Martika/Marvin Morrow] A: *Martika*; P: Michael Jay

1396. NAUGHTY NAUGHTY John Parr Atlantic 89612 824
[15Dec84 | 09Mar85 | 27Apr85] 23(1) 0 | 8 | 20
[John Parr] A: *John Parr*; P: Peter Solley

1397. LOVE COME DOWN Evelyn King RCA 13273 822
[28Aug82 | 06Nov82 | 11Dec82] 17(2) 0 | 8 | 16
[Michael Jones] A: *Get Loose*; P: Morrie Brown

Ranking the Singles

Rank. TITLE Artist Label Number
[Entry Date | Peak Date | Exit Date] Peak(Peak Wks) Top10 | Top40 | Tot Wks
[Writers] *Album; Producers* — Score

1398. COMING AROUND AGAIN Carly Simon Arista 1-9525 — 821
[01Nov86 | 24Jan87 | 21Feb87] 18(1) 0 | 9 | 17
[Carly Simon] *A: Coming Around Again;*
P: Russ Kunkel/George Massenburg/Bill Payne/Paul Samwell-Smith

1399. SAVE A PRAYER Duran Duran Capitol 5438 — 820
[02Feb85 | 16Mar85 | 04May85] 16(2) 0 | 8 | 14
[Simon Le Bon/Nick Rhodes/Andy Taylor/Nigel John Taylor/Roger Taylor]
A: Arena; P: Colin Thurston

1400. POINT OF NO RETURN Nu Shooz Atlantic 89392 — 820
[05Jul86 | 11Oct86 | 29Nov86] 28(1) 0 | 8 | 22
[Valerie Day/John Robert Smith] *A: Poolside; P: John Smith/Rick Waritz*

1401. LOVE OVERBOARD Gladys Knight And The Pips MCA 53210 — 818
[09Jan88 | 12Mar88 | 09Apr88] 13(1) 0 | 9 | 14
[Reggie Calloway] *A: All Our Love; P: Reggie Calloway/Vincent Calloway*

1402. YOU ARE THE GIRL The Cars Elektra 69446 — 818
[29Aug87 | 24Oct87 | 28Nov87] 17(1) 0 | 9 | 14
[Ric Ocasek] *A: Door To Door; P: Ric Ocasek*

1403. JUST AS I AM Air Supply Arista 1-9353 — 817
[25May85 | 13Jul85 | 31Aug85] 19(2) 0 | 10 | 15
[Rob Hegel/Dick Wagner] *A: Air Supply; P: Bob Ezrin*

1404. LET ME TICKLE YOUR FANCY Jermaine Jackson Motown 1628 — 813
[24Jul82 | 25Sep82 | 30Oct82] 18(2) 0 | 7 | 15
[Jermaine Jackson/Paul M. Jackson Jr./Marilyn McLeod/Pam Sawyer]
A: Let Me Tickle Your Fancy; P: Berry Gordy/Jermaine Jackson

1405. ALIBIS Sergio Mendes A&M 2639 — 812
[26May84 | 11Aug84 | 29Sep84] 29(2) 0 | 7 | 19
[Tony Macaulay/Tom Snow] *A: Confetti; P: Robbie Buchanan/Sergio Mendes*

1406. BECAUSE OF YOU The Cover Girls Fever 1914 — 811
[28Nov87 | 27Feb88 | 09Apr88] 27(2) 0 | 8 | 20
[David Cole] *A: Show Me; P: Robert Clivillés/Louie Vega*

1407. HEART OF THE NIGHT Juice Newton Capitol 5192 — 809
[27Nov82 | 29Jan83 | 12Mar83] 25(3) 0 | 10 | 16
[John Bettis/Michael Clark] *A: Quiet Lies; P: Richard Landis*

1408. FOOL IN THE RAIN Led Zeppelin Swan Song 71003 — 807
[22Dec79 | 16Feb80 | 15Mar80] 21(2) 0 | 8 | 13
[John Paul Jones/Jimmy Page/Robert Plant]
A: In Through The Out Door; P: Jimmy Page

1409. SOUL PROVIDER Michael Bolton Columbia 68909 — 807
[01Jul89 | 16Sep89 | 21Oct89] 17(1) 0 | 7 | 17
[Michael Bolton/Andy Goldmark]
A: Soul Provider; P: Peter Bunetta/Rick Chudacoff

1410. DON'T STOP THE MUSIC Yarbrough & Peoples Mercury 76085 — 806
[07Feb81 | 11Apr81 | 23May81] 19(2) 0 | 7 | 16
[Jonah Ellis/Alisa Peoples/Lonnie Simmons]
A: The Two Of Us; P: Jonah Ellis/Lonnie Simmons

1411. COME BACK AND STAY Paul Young Columbia 04313 — 806
[04Feb84 | 07Apr84 | 12May84] 22(1) 0 | 8 | 15
[Jack Lee] *A: No Parlez; P: Laurie Latham*

1412. RADIO GA-GA Queen Capitol 5317 — 805
[18Feb84 | 07Apr84 | 12May84] 16(1) 0 | 8 | 13
[Roger Taylor] *A: The Works; P: Reinhold Mack/Queen*

1413. ALL FIRED UP Pat Benatar Chrysalis 43268 — 805
[02Jul88 | 27Aug88 | 22Oct88] 19(2) 0 | 8 | 17
[Pat Benatar/Myron Grombacher/Kerryn Tolhurst]
A: Wide Awake In Dreamland; P: Keith Forsey/Neil Geraldo

1414. MAKE A LITTLE MAGIC The Dirt Band United Artists 1356 — 804
[21Jun80 | 16Aug80 | 04Oct80] 25(3) 0 | 9 | 16
[Robert Carpenter/Jeff Hanna/Richard Hathaway]
A: Make A Little Magic; P: Bob Edwards/Jeff Hanna

1415. YOU'RE MY ONE AND ONLY (True Love) Seduction Vendetta 1433 — 804
[08Jul89 | 07Oct89 | 25Nov89] 23(1) 0 | 8 | 21
[Robert Clivillés/David Cole/Frederick B. Williams]
A: Nothing Matters Without Love; P: Robert Clivillés/David Cole

1416. SEVEN YEAR ACHE Rosanne Cash Columbia 11426 — 802
[25Apr81 | 18Jul81 | 05Sep81] 22(2) 0 | 7 | 20
[Rosanne Cash] *A: Seven Year Ache; P: Rodney Crowell*

1417. THE BEST Tina Turner Capitol 44442 — 800
[02Sep89 | 04Nov89 | 02Dec89] 15(1) 0 | 8 | 14
[Mike Chapman/Holly Knight] *A: Foreign Affair; P: Dan Hartman/Tina Turner*

1418. STRANGER IN MY HOUSE Ronnie Milsap RCA 13470 — 799
[26Mar83 | 28May83 | 09Jul83] 23(1) 0 | 8 | 16
[Mike Reid] *A: Keyed Up; P: Tom Collins/Ronnie Milsap*

1419. ON THE LOOSE Saga Portrait 37-03359 — 798
[04Dec82 | 26Feb83 | 02Apr83] 26(3) 0 | 8 | 18
[Ian Stevenson Crichton/Jim Crichton/Jim Gilmour/Steve Negus/
Michael Sadler] *A: Worlds Apart; P: Rupert Hine*

1420. SERIOUS Donna Allen 21 Records 99497 — 797
[14Feb87 | 09May87 | 13Jun87] 21(1) 0 | 9 | 18
[Donna Allen/Emridge Jones/Gary King/Harry J. King/Louis Pace/
Reggie White] *A: Perfect Timing; P: Lou Pace*

1421. I CAN'T WAIT Stevie Nicks Modern 99565 — 796
[22Feb86 | 12Apr86 | 17May86] 16(2) 0 | 8 | 13
[Stevie Nicks/Rick Nowels/Eric Pressly]
A: Rock A Little; P: Jimmy Iovine/Rick Nowels

1422. EARLY IN THE MORNING Robert Palmer EMI-Manhattan 50157 — 795
[22Oct88 | 17Dec88 | 28Jan89] 19(1) 0 | 9 | 15
[Lonnie Simmons/Rudy Taylor/Charles K. Wilson]
A: Heavy Nova; P: Robert Palmer

1423. THE LUCKY ONE Laura Branigan Atlantic 89636 — 795
[04Aug84 | 29Sep84 | 10Nov84] 20(2) 0 | 8 | 15
[Bruce Roberts]
A: Self Control; P: Robbie Buchanan/Jack White

1424. DO YOU WANNA TOUCH ME (Oh Yeah) Joan Jett & The Blackhearts Boardwalk 11-150-7 — 794
[31Jul82 | 25Sep82 | 30Oct82] 20(1) 0 | 7 | 14
[Gary Glitter/Mike Leander] *A: Bad Reputation; P: Ritchie Cordell/Kenny Laguna*

1425. BREAKING AWAY Balance Portrait 24-02177 — 793
[11Jul81 | 26Sep81 | 31Oct81] 22(2) 0 | 9 | 17
[Peppy Castro] *A: Balance; P: Balance*

1426. LITTLE TOO LATE Pat Benatar Chrysalis 03536 — 792
[05Feb83 | 26Mar83 | 07May83] 20(3) 0 | 7 | 14
[Alex Call] *A: Get Nervous; P: Peter Coleman/Neil Geraldo*

1427. SOULS Rick Springfield RCA 13650 — 792
[15Oct83 | 03Dec83 | 21Jan84] 23(3) 0 | 6 | 15
[Rick Springfield] *A: Living In Oz; P: Bill Drescher/Rick Springfield*

1428. ON THE WINGS OF LOVE Jeffrey Osborne A&M 2434 — 790
[25Sep82 | 18Dec82 | 22Jan83] 29(3) 0 | 7 | 18
[Jeffrey Osborne/Peter Schless] *A: Jeffrey Osborne; P: George Duke*

1429. ENDLESS NIGHTS Eddie Money Columbia 07035 — 790
[11Apr87 | 27Jun87 | 15Aug87] 21(1) 0 | 7 | 19
[John Cesario/Michele Collyer/Steve Mullen]
A: Can't Hold Back; P: Eddie Money/Richie Zito

1430. THE SUN ALWAYS SHINES ON T.V. A-Ha Warner Bros. 28846 — 789
[30Nov85 | 22Feb86 | 22Mar86] 20(1) 0 | 8 | 17
[Pal Waaktaar] *A: Hunting High And Low; P: Alan Tarney*

1431. SEND ME AN ANGEL ('83) Real Life Curb/MCA 52287 — 789
[12Nov83 | 11Feb84 | 17Mar84] 29(2) 0 | 6 | 19
[David Sterry/Richard Zatorski] *A: Heart Land; P: Ross Cockle/Glenn Wheatley*

1432. WRAP HER UP Elton John Geffen 28873 — 786
[26Oct85 | 07Dec85 | 25Jan86] 20(2) 0 | 10 | 14
[Elton John/Davey Johnstone/Fred Mandel/Charlie Morgan/
Bernie Taupin/Paul Westwood] *A: Ice On Fire; P: Gus Dudgeon*

1433. GO INSANE Lindsey Buckingham Elektra 69714 — 783
[28Jul84 | 13Oct84 | 10Nov84] 23(1) 0 | 9 | 16
[Lindsey Buckingham] *A: Go Insane; P: Lindsey Buckingham/Gordon Fordyce*

1434. HEARTBREAK HOTEL The Jacksons Epic 50959 — 782
[06Dec80 | 14Feb81 | 21Mar81] 22(2) 0 | 8 | 16
[Michael Jackson] *A: Triumph; P: Jacksons*

1435. JUST BETWEEN YOU AND ME April Wine Capitol 4975 — 782
[07Feb81 | 25Apr81 | 23May81] 21(1) 0 | 7 | 16
[Myles Goodwyn]
A: The Nature Of The Beast; P: Myles Goodwyn/Mike Stone

1436. WHAT KIND OF FOOL AM I Rick Springfield RCA 13245 — 782
[05Jun82 | 03Jul82 | 21Aug82] 21(6) 0 | 9 | 12
[Rick Springfield] *A: Success Hasn't Spoiled Me Yet; P: Keith Olsen*

1437. MARY'S PRAYER Danny Wilson Virgin 99465 — 782
[06Jun87 | 05Sep87 | 17Oct87] 23(1) 0 | 8 | 20
[Gary Clark] *A: Meet Danny Wilson; P: David Bascombe*

1438. CENTIPEDE Rebbie Jackson Columbia 04547 — 781
[06Oct84 | 22Dec84 | 09Feb85] 24(2) 0 | 8 | 19
[Michael Jackson] *A: Centipede; P: Michael Jackson*

1439. FIND ANOTHER FOOL Quarterflash Geffen 50006 — 781
[13Feb82 | 17Apr82 | 08May82] 16(2) 0 | 7 | 13
[Marv Ross] *A: Quarterflash; P: John Boylan*

Ranking the Singles

Rank. TITLE Artist Label Number	
[Entry Date	Peak Date	Exit Date] Peak(Peak Wks) Top10	Top40	Tot Wks	
[Writers] Album; Producers	Score

1440. THE TWIST (YO, TWIST!) Fat Boys Tin Pan Apple 887 571 — 781
[18Jun88 | 06Aug88 | 24Sep88] 16(2) 0 | 8 | 15
[Hank Ballard] A: Coming Back Hard Again; P: Albert Cabrera/Tony Moran

1441. NO MORE WORDS Berlin Geffen 29360 — 778
[10Mar84 | 12May84 | 30Jun84] 23(2) 0 | 8 | 17
[John Crawford] A: Love Life; P: Giorgio Moroder/Richie Zito

1442. GIRL CAN'T HELP IT Journey Columbia 06302 — 777
[30Aug86 | 01Nov86 | 29Nov86] 17(1) 0 | 8 | 14
[Jonathan Cain/Steve Perry/Neal Schon] A: Raised On Radio; P: Steve Perry

1443. WILD WILD LIFE Talking Heads Sire 28629 — 777
[06Sep86 | 06Dec86 | 24Jan87] 25(1) 0 | 7 | 21
[David Byrne/Chris Frantz/Jerry Harrison/Tina Weymouth] A: True Stories: A Film By David Byrne, The Complete Soundtrack; P: Talking Heads

1444. THE LAST WORTHLESS EVENING Don Henley Geffen 22771 — 775
[07Oct89 | 09Dec89 | 03Feb90] 21(1) 0 | 8 | 18
[John Corey/Don Henley/Stan Lynch]
A: The End Of The Innocence; P: John Corey/Don Henley/Stan Lynch

1445. BOP 'TIL YOU DROP Rick Springfield RCA 13861 — 775
[18Aug84 | 20Oct84 | 24Nov84] 20(1) 0 | 9 | 15
[Rick Springfield] A: Hard To Hold (Soundtrack); P: Bill Drescher/Rick Springfield

1446. RUSSIANS Sting A&M 2799 — 774
[18Jan86 | 01Mar86 | 12Apr86] 16(2) 0 | 8 | 13
[Sergei Prokofiev/Sting]
A: The Dream Of The Blue Turtles; P: Pete Smith/Sting

1447. POISON ARROW ABC Mercury 810340 — 772
[29Jan83 | 26Mar83 | 07May83] 25(3) 0 | 8 | 15
[Martin Fry/Mark Lickley/Stephen Singleton/Mark White]
A: The Lexicon Of Love; P: Trevor Horn

1448. INTO THE NIGHT ('89) Benny Mardones Polydor 889368 — 771
[06May89 | 01Jul89 | 26Aug89] 20(1) 0 | 7 | 17
[Benny Mardones/Robert Tepper] A: Benny Mardones; P: Barry Mraz

1449. AS WE LAY Shirley Murdock Elektra 69518 — 769
[17Jan87 | 28Mar87 | 16May87] 23(2) 0 | 7 | 18
[William Beck/Larry Troutman] A: Shirley Murdock!; P: Roger Troutman

1450. CALL IT LOVE Poco RCA 9038 — 769
[26Aug89 | 04Nov89 | 09Dec89] 18(1) 0 | 8 | 16
[Billy Crain/Ronnie Guilbeau/Richard Lonow] A: Legacy; P: David Cole

1451. LOVE ME TOMORROW Chicago Full Moon 29911 — 766
[25Sep82 | 04Dec82 | 01Jan83] 22(2) 0 | 8 | 15
[Peter Cetera/David Foster] A: Chicago 16; P: David Foster

1452. TORTURE Jacksons Epic 04575 — 765
[18Aug84 | 29Sep84 | 03Nov84] 17(2) 0 | 8 | 12
[Jackie Jackson/Kathy Wakefield] A: Victory; P: Jackie Jackson

1453. TALK IT OVER Grayson Hugh RCA 8802 — 765
[24Jun89 | 09Sep89 | 21Oct89] 19(1) 0 | 8 | 18
[Irwin Levine/Sandy Linzer]
A: Blind To Reason; P: Michael Baker/Axel Kroell

1454. LIFE IN ONE DAY Howard Jones Elektra 69631 — 765
[06Jul85 | 07Sep85 | 19Oct85] 19(1) 0 | 8 | 16
[Howard Jones] A: Dream Into Action; P: Rupert Hine

1455. RUNNER Manfred Mann's Earth Band Arista 1-9143 — 764
[21Jan84 | 24Mar84 | 28Apr84] 22(1) 0 | 8 | 15
[Ian Thomas] A: No Album; P: Manfred Mann

1456. FASCINATED Company B Atlantic 89294 — 763
[21Mar87 | 13Jun87 | 18Jul87] 21(1) 0 | 8 | 18
[Ish Ledesma] A: Company B; P: Ish Ledesma

1457. NO REPLY AT ALL Genesis Atlantic 3858 — 763
[26Sep81 | 28Nov81 | 23Jan82] 29(2) 0 | 6 | 18
[Tony Banks/Phil Collins/Mike Rutherford] A: Abacab; P: Genesis

1458. I LIKE IT DeBarge Gordy 1645 — 761
[05Feb83 | 09Apr83 | 28May83] 31(2) 0 | 6 | 17
[Bunny DeBarge/El DeBarge/Randy DeBarge]
A: All This Love; P: El DeBarge/Iris Gordy

1459. THIS TIME John Cougar Riva 205 — 760
[27Sep80 | 06Dec80 | 17Jan81] 27(2) 0 | 7 | 17
[John Mellencamp] A: Nothin' Matters And What If It Did; P: Steve Cropper

1460. THEME FROM MAGNUM P.I. Mike Post Elektra 47400 — 760
[06Feb82 | 08May82 | 29May82] 25(2) 0 | 7 | 17
[Pete Carpenter/Mike Post]
A: Television Theme Songs; P: Mike Post

1461. SISTERS ARE DOIN' IT FOR THEMSELVES — 759
Eurythmics And Aretha Franklin RCA 14214
[19Oct85 | 07Dec85 | 25Jan86] 18(1) 0 | 8 | 15
[Annie Lennox/David A. Stewart] A: Who's Zoomin' Who/Aretha Franklin; Be Yourself Tonight/Eurythmics; P: David A. Stewart

1462. THINK ABOUT ME Fleetwood Mac Warner Bros. 49196 — 759
[08Mar80 | 26Apr80 | 31May80] 20(2) 0 | 7 | 13
[Christine McVie] A: Tusk; P: Fleetwood Mac

1463. SO BAD Paul McCartney Columbia 04296 — 759
[24Dec83 | 11Feb84 | 24Mar84] 23(2) 0 | 8 | 14
[Paul McCartney] A: Pipes Of Peace; P: George Martin

1464. NO NIGHT SO LONG Dionne Warwick Arista 0527 — 758
[26Jul80 | 04Oct80 | 08Nov80] 23(2) 0 | 6 | 16
[Will Jennings/Richard Kerr] A: No Night So Long; P: Steve Buckingham

1465. COMING HOME Cinderella Mercury 872982 — 758
[08Apr89 | 24Jun89 | 29Jul89] 20(1) 0 | 7 | 17
[Tom Keifer] A: Long Cold Winter; P: Eric Brittingham/Andy Johns/Tom Keifer

1466. PARADISE Sade Epic 07904 — 758
[14May88 | 23Jul88 | 20Aug88] 16(1) 0 | 8 | 15
[Helen Adu/Paul Denman/Andrew Hale/Stuart Matthewman]
A: Stronger Than Pride; P: Helen Adu

1467. A FINE FINE DAY Tony Carey MCA 52343 — 755
[03Mar84 | 05May84 | 09Jun84] 22(1) 0 | 8 | 15
[Tony Carey] A: Some Tough City; P: Peter Hauke

1468. YOU DON'T KNOW Scarlett & Black Virgin 99405 — 755
[30Jan88 | 16Apr88 | 28May88] 20(1) 0 | 8 | 18
[Robin Hild] A: Scarlett & Black;
P: Skip Drinkwater/Paul Fox/Phil Harding/Daize Washbourn

1469. LIVING IN A BOX Living In A Box Chrysalis 43104 — 754
[20Jun87 | 22Aug87 | 26Sep87] 17(1) 0 | 8 | 15
[Steve Piggott/Marcus Vere] A: Living In A Box; P: Richard James Burgess

1470. LOVE WALKS IN Van Halen Warner Bros. 28626 — 753
[09Aug86 | 04Oct86 | 15Nov86] 22(2) 0 | 9 | 15
[Michael Anthony/Sammy Hagar/Alex Van Halen/Eddie Van Halen]
A: 5150; P: Mick Jones/Donn Landee/Van Halen

1471. (Forever) LIVE AND DIE Orchestral Manoeuvres In The Dark A&M 2872 — 751
[27Sep86 | 06Dec86 | 17Jan87] 19(1) 0 | 7 | 17
[Paul Humphreys/Graham Weir/Neil Weir]
A: The Pacific Age; P: Stephen Hague

1472. ALL THE LOVE IN THE WORLD The Outfield Columbia 05894 — 749
[07Jun86 | 16Aug86 | 20Sep86] 19(1) 0 | 7 | 16
[John Spinks] A: Play Deep; P: William Wittman

1473. WE ARE THE YOUNG Dan Hartman MCA 52471 — 749
[06Oct84 | 15Dec84 | 26Jan85] 25(1) 0 | 9 | 17
[Dan Hartman/Charles E. Kaufman]
A: I Can Dream About You; P: Dan Hartman/Jimmy Iovine

1474. SKELETONS Stevie Wonder Motown 1907 — 748
[17Oct87 | 05Dec87 | 30Jan88] 19(1) 0 | 7 | 16
[Stevie Wonder] A: Characters; P: Stevie Wonder

1475. SOME THINGS ARE BETTER LEFT UNSAID Daryl Hall & John Oates — 748
RCA 14035
[16Mar85 | 04May85 | 08Jun85] 18(1) 0 | 8 | 13
[Daryl Hall] A: Big Bam Boom; P: Bob Clearmountain/Daryl Hall/John Oates

1476. I WISH I HAD A GIRL Henry Lee Summer CBS Associated 07720 — 747
[13Feb88 | 30Apr88 | 11Jun88] 20(1) 0 | 7 | 18
[Henry Lee Summer] A: Henry Lee Summer; P: Michael Frondelli

1477. TOUGH ALL OVER John Cafferty And The Beaver Brown Band — 747
Scotti Brothers 04891
[11May85 | 06Jul85 | 17Aug85] 22(1) 0 | 8 | 15
[John Cafferty] A: Tough All Over; P: Kenny Vance

1478. BACK WHERE YOU BELONG 38 Special A&M 2615 — 746
[04Feb84 | 24Mar84 | 28Apr84] 20(1) 0 | 8 | 13
[Gary O'Connor] A: Tour De Force; P: Rodney Mills

1479. C-I-T-Y John Cafferty And The Beaver Brown Band — 745
Scotti Brothers 05452
[10Aug85 | 05Oct85 | 16Nov85] 18(1) 0 | 8 | 15
[John Cafferty] A: Tough All Over; P: Kenny Vance

1480. WHO'S THAT GIRL? Eurythmics RCA 13800 — 745
[05May84 | 23Jun84 | 28Jul84] 21(2) 0 | 7 | 13
[Annie Lennox/David A. Stewart] A: Touch; P: David A. Stewart

Rank. TITLE Artist Label Number	
[Entry Date	Peak Date	Exit Date] Peak(Peak Wks) Top10	Top40	Tot Wks	
[Writers] Album; Producers	Score

1481. WE'RE NOT GONNA TAKE IT Twisted Sister Atlantic 89641 — 745
[28Jul84 | 22Sep84 | 03Nov84] 21(2) 0 | 7 | 15
[Dee Snider] A: Stay Hungry; P: Tom Werman

1482. MORNIN' Jarreau Warner Bros. 29720 — 743
[19Mar83 | 14May83 | 25Jun83] 21(2) 0 | 6 | 15
[David Foster/Jay Graydon/Al Jarreau] A: Jarreau; P: Jay Graydon

1483. FORGET ME NOTS Patrice Rushen Elektra 47427 — 743
[01May82 | 03Jul82 | 14Aug82] 23(3) 0 | 7 | 16
[Theresa McFaddin/Patrice Rushen/Freddie Washington]
A: Straight From The Heart; P: Charles Mims Jr./Patrice Rushen

1484. VIENNA CALLING (The New '86 Edit) Falco A&M 2832 — 742
[26Apr86 | 21Jun86 | 26Jul86] 18(1) 0 | 8 | 14
[Ferdi Bolland/Rob Bolland/Falco] A: Falco 3; P: Ferdi Bolland/Rob Bolland

1485. PAMELA Toto Columbia 07715 — 740
[20Feb88 | 07May88 | 25Jun88] 22(2) 0 | 8 | 19
[David Paich/Joseph Stanley Williams]
A: The Seventh One; P: George Massenburg/Bill Payne/Toto

1486. TENDER IS THE NIGHT Jackson Browne Asylum 69791 — 740
[24Sep83 | 19Nov83 | 14Jan84] 25(1) 0 | 7 | 17
[Jackson Browne/Danny Kortchmar/Russ Kunkel]
A: Lawyers In Love; P: Jackson Browne/Greg Ladanyi

1487. SPACE AGE LOVE SONG A Flock Of Seagulls Jive 2003 — 740
[13Nov82 | 12Feb83 | 12Mar83] 30(2) 0 | 7 | 18
[Frank Maudsley/Paul Reynolds/Ali Score/Mike Score]
A: A Flock Of Seagulls; P: Mike Howlett

1488. KILLIN' TIME Fred Knoblock & Susan Anton Scotti Brothers 609 — 739
[22Nov80 | 07Feb81 | 21Mar81] 28(1) 0 | 9 | 18
[Jeff Harrington/Jeff Pennig] A: Killin' Time; P: James Stroud

1489. LOVE ON A TWO WAY STREET Stacy Lattisaw Cotillion 46015 — 739
[20Jun81 | 05Sep81 | 10Oct81] 26(2) 0 | 7 | 17
[Bert Keyes/Sylvia Robinson] A: With You; P: Narada Michael Walden

1490. DOUBLE DUTCH BUS Frankie Smith WMOT 5356 — 738
[16May81 | 15Aug81 | 19Sep81] 30(1) 0 | 7 | 19
[William A. Bloom/Frankie Smith]
A: Children Of Tomorrow; P: Bill Bloom/Frankie Smith

1491. HAPPY Surface Columbia 06611 — 738
[23May87 | 18Jul87 | 22Aug87] 20(2) 0 | 8 | 14
[Dave Conley/Bernard Leon Jackson/David Townsend]
A: Surface; P: David Conley/Bernard Jackson/David Townsend

1492. TONIGHT I'M YOURS (Don't Hurt Me) Rod Stewart Warner Bros. 49886 — 735
[23Jan82 | 20Mar82 | 24Apr82] 20(2) 0 | 8 | 14
[Jim Cregan/Kevin Savigar/Rod Stewart] A: Tonight I'm Yours; P: Rod Stewart

1493. NO MORE RHYME Debbie Gibson Atlantic 88885 — 734
[17Jun89 | 12Aug89 | 16Sep89] 17(1) 0 | 7 | 14
[Debbie Gibson] A: Electric Youth; P: Fred Zarr

1494. HOW DO I SURVIVE Amy Holland Capitol 4884 — 733
[09Aug80 | 11Oct80 | 22Nov80] 22(2) 0 | 6 | 16
[Paul Steven Bliss] A: Amy Holland; P: Patrick Henderson/Michael McDonald

1495. PRECIOUS TO ME Phil Seymour Boardwalk 8-5703 — 731
[24Jan81 | 28Mar81 | 09May81] 22(2) 0 | 7 | 16
[Phil Seymour] A: Phil Seymour; P: Richard Podolor

1496. LET'S STAY TOGETHER Tina Turner Capitol 5322 — 730
[21Jan84 | 24Mar84 | 28Apr84] 26(1) 0 | 7 | 15
[Al Green/Al Jackson/Willie Mitchell]
A: Private Dancer; P: Greg Walsh/Martyn Ware

1497. ROOMS ON FIRE Stevie Nicks Modern 99216 — 730
[06May89 | 01Jul89 | 05Aug89] 16(1) 0 | 7 | 14
[Stevie Nicks/Rick Nowels] A: The Other Side Of The Mirror; P: Rupert Hine

1498. LEAVE IT Yes Atco 99787 — 729
[03Mar84 | 21Apr84 | 09Jun84] 24(2) 0 | 7 | 15
[Trevor Horn/Trevor Rabin/Chris Squire] A: 90125; P: Trevor Horn

1499. THERE GOES MY BABY Donna Summer Geffen 29291 — 729
[11Aug84 | 13Oct84 | 10Nov84] 21(1) 0 | 8 | 14
[Ben E. King/Jerry Leiber/Lover Patterson/Mike Stoller/George Treadwell]
A: Cats Without Claws; P: Michael Omartian

1500. POWER OF LOVE Laura Branigan Atlantic 89191 — 727
[24Oct87 | 09Jan88 | 20Feb88] 26(2) 0 | 9 | 18
[Mary Applegate/Candy DeRouge/Gunther Mende/Jennifer Rush]
A: Touch; P: David Kershenbaum

1501. TRAIN IN VAIN (Stand By Me) The Clash Epic 50851 — 723
[22Mar80 | 24May80 | 21Jun80] 23(2) 0 | 7 | 14
[Nicky Headon/Michael Geoffrey Jones/Paul Simonon/Joe Strummer]
A: London Calling; P: Guy Stevens

1502. SAY YOU'LL BE MINE Christopher Cross Warner Bros. 49705 — 722
[28May81 | 23May81 | 27Jun81] 20(2) 0 | 7 | 14
[Christopher Cross] A: Christopher Cross; P: Michael Omartian

1503. ROMANCING THE STONE Eddy Grant Portrait 37-04433 — 722
[19May84 | 21Jul84 | 08Sep84] 26(2) 0 | 6 | 17
[Eddy Grant] A: Going For Broke; P: Eddy Grant

1504. SOMEWHERE DOWN THE ROAD Barry Manilow Arista 0658 — 720
[19Dec81 | 20Feb82 | 27Mar82] 21(2) 0 | 7 | 15
[Tom Snow/Cynthia Weil] A: If I Should Love Again; P: Barry Manilow

1505. SAUSALITO SUMMERNIGHT Diesel Regency 7339 — 719
[12Sep81 | 21Nov81 | 09Jan82] 25(1) 0 | 6 | 18
[Mark Boon/Rob Vunderink] A: Watts In A Tank; P: Pim Koopman

1506. THAT AIN'T LOVE REO Speedwagon Epic 06656 — 719
[31Jan87 | 04Apr87 | 02May87] 16(1) 0 | 7 | 14
[Kevin Cronin] A: Life As We Know It;
P: Kevin Cronin/David DeVore/Alan Gratzer/Gary Richrath

1507. TENDERNESS General Public I.R.S. 9934 — 718
[17Nov84 | 16Feb85 | 16Mar85] 27(1) 0 | 5 | 18
[Micky Billingham/Roger Charlery/Dave Wakeling]
A: ...All The Rage; P: Colin Fairley/General Public/Gavin McKillop

1508. SEND HER MY LOVE Journey Columbia 04151 — 718
[24Sep83 | 19Nov83 | 31Dec83] 23(2) 0 | 7 | 15
[Jonathan Cain/Steve Perry] A: Frontiers; P: Kevin Elson/Mike Stone

1509. HONESTLY Stryper Enigma 75009 — 718
[07Nov87 | 30Jan88 | 12Mar88] 23(1) 0 | 8 | 19
[Michael Sweet] A: To Hell With The Devil;
P: Oz Fox/Stephan Galfas/Michael Sweet/Robert Sweet

1510. (How To Be A) MILLIONAIRE ABC Mercury 884382 — 717
[18Jan86 | 22Mar86 | 19Apr86] 20(1) 0 | 7 | 14
[Martin Fry/Mark White] A: How To Be A...Zillionaire!; P: Martin Fry/Mark White

1511. SOME KIND OF FRIEND Barry Manilow Arista 1046 — 717
[26Feb83 | 30Apr83 | 11Jun83] 26(2) 0 | 7 | 16
[Adrienne Anderson/Barry Manilow]
A: Here Comes The Night; P: Barry Manilow

1512. DIGITAL DISPLAY Ready For The World MCA 52734 — 717
[07Dec85 | 22Feb86 | 05Apr86] 21(1) 0 | 6 | 18
[Greg Potts] A: Ready For the World; P: Ready For The World

1513. SHE'S GOT A WAY Billy Joel Columbia 02628 — 715
[21Nov81 | 23Jan82 | 20Feb82] 23(2) 0 | 9 | 14
[Billy Joel] A: Songs In The Attic; P: Phil Ramone

1514. THE WAR SONG Culture Club Virgin/Epic 04638 — 715
[06Oct84 | 17Nov84 | 29Dec84] 17(1) 0 | 7 | 13
[Michael Craig/Roy Hay/Jon Moss/George O'Dowd]
A: Waking Up With The House On Fire; P: Steve Levine

1515. ANY WAY YOU WANT IT Journey Columbia 11213 — 715
[01Mar80 | 26Apr80 | 07Jun80] 23(2) 0 | 6 | 15
[Steve Perry/Neal Schon] A: Departure; P: Kevin Elson/Geoffrey Workman

1516. NOBODY SAID IT WAS EASY (Lookin' For The Lights) Le Roux RCA 13059 — 714
[13Feb82 | 17Apr82 | 08May82] 18(2) 0 | 6 | 13
[Tony Haselden] A: Last Safe Place; P: Leon Medica

1517. SEVEN BRIDGES ROAD Eagles Asylum 47100 — 714
[20Dec80 | 07Feb81 | 21Mar81] 21(1) 0 | 7 | 14
[Steve Young] A: Eagles Live; P: Bill Szymczyk

1518. I'M FREE (Heaven Helps The Man) Kenny Loggins Columbia 04452 — 714
[16Jun84 | 28Jul84 | 15Sep84] 22(2) 0 | 8 | 14
[Kenny Loggins/Dean Pitchford]
A: Footloose Soundtrack; P: David Foster/Kenny Loggins

1519. YOU BE ILLIN' Run-D.M.C. Profile 5119 — 713
[25Oct86 | 20Dec86 | 21Feb87] 29(3) 0 | 7 | 18
[Jason Mizell/Joseph Ward Simmons/Ray White]
A: Raising Hell; P: Rick Rubin/Russell Simmons

1520. WONDERLAND Commodores Motown 1479 — 713
[15Dec79 | 23Feb80 | 22Mar80] 25(2) 0 | 6 | 15
[Milan Williams] A: Midnight Magic; P: James Anthony Carmichael/Commodores

Ranking the Singles

Rank. TITLE Artist Label Number
[Entry Date | Peak Date | Exit Date] Peak(Peak Wks) Top10 | Top40 | Tot Wks
[Writers] *Album; Producers*   Score

1521. SUPERSONIC J.J. Fad Ruthless 7-99328   712
[23Apr88 | 18Jun88 | 17Sep88] 30(1) 0 | 4 | 22
[Dania Birks/Juana Burns/Juanita Lee/Kim Nazel/Fatimah Shaheed]
A: Supersonic--The Album; P: Antoine Carraby/Dr. Dre/Mik Lezan

1522. ROCK YOU LIKE A HURRICANE Scorpions Mercury 818440   712
[24Mar84 | 26May84 | 07Jul84] 25(1) 0 | 7 | 16
[Klaus Meine/Herman Rarebell/Rudolf Schenker]
A: Love At First Sting; P: Dieter Dierks

1523. POP SINGER John Cougar Mellencamp Mercury 874 012   712
[29Apr89 | 17Jun89 | 15Jul89] 15(1) 0 | 7 | 12
[John Mellencamp] *A: Big Daddy; P: John Mellencamp*

1524. LOVE IN STORE Fleetwood Mac Warner Bros. 29848   711
[27Nov82 | 15Jan83 | 26Feb83] 22(3) 0 | 8 | 14
[Christine McVie/Jim Recor]
A: Mirage; P: Lindsey Buckingham/Ken Caillat/Richard Dashut/Fleetwood Mac

1525. FOR TONIGHT Nancy Martinez Atlantic 89371   711
[04Oct86 | 27Dec86 | 21Feb87] 32(3) 0 | 7 | 21
[Phil George/Donna Pacifici]
A: Not Just The Girl Next Door; P: Teneen Ali/Sergio Munzibai

1526. GIVE TO LIVE Sammy Hagar Geffen 28314   710
[20Jun87 | 29Aug87 | 10Oct87] 23(1) 0 | 7 | 17
[Sammy Hagar] *A: I Never Said Goodbye; P: Sammy Hagar/Edward Van Halen*

1527. WISHING (If I Had A Photograph Of You) A Flock Of Seagulls Jive 2006   709
[14May83 | 09Jul83 | 13Aug83] 26(2) 0 | 7 | 15
[Frank Maudsley/Paul Reynolds/Ali Score/Mike Score]
A: Listen; P: Mike Howlett

1528. WILD WORLD Maxi Priest Virgin 99269   708
[29Oct88 | 07Jan89 | 25Feb89] 25(1) 0 | 7 | 18
[Cat Stevens] *A: Maxi Priest; P: Sly Dunbar/Willie Lindo/Robbie Shakespeare*

1529. THE HONEYTHIEF Hipsway Columbia 06579   708
[24Jan87 | 04Apr87 | 16May87] 19(1) 0 | 6 | 15
[Johnny McElhone/Ali McLeod/Grahame Skinner/Harry Travers]
A: Hipsway; P: Gary Langan

1530. I WONDER IF I TAKE YOU HOME Lisa Lisa and Cult Jam with Full Force Columbia 04886   707
[08Jun85 | 17Aug85 | 26Oct85] 34(1) 0 | 6 | 21
[Curtis Bedeau/Gerard Charles/Hugh Clarke/Brian George/Lucien George/Paul George] *A: Lisa Lisa & Cult Jam With Full Force; P: Full Force*

1531. SEVEN WONDERS Fleetwood Mac Warner Bros. 28317   704
[20Jun87 | 15Aug87 | 12Sep87] 19(1) 0 | 8 | 13
[Stevie Nicks/Sandy Stewart]
A: Tango In The Night; P: Lindsey Buckingham/Richard Dashut

1532. DO RIGHT Paul Davis Bang 4808   704
[08Mar80 | 17May80 | 07Jun80] 23(1) 0 | 6 | 14
[Paul Davis] *A: Paul Davis; P: Paul Davis/Ed Seay*

1533. ONE STEP CLOSER The Doobie Brothers Warner Bros. 49622   704
[22Nov80 | 10Jan81 | 21Feb81] 24(2) 0 | 7 | 14
[Carlene Carter/Keith Knudsen/John McFee]
A: One Step Closer; P: Ted Templeman

1534. THE WAITING Tom Petty And The Heartbreakers Backstreet 51100   703
[02May81 | 13Jun81 | 25Jul81] 19(2) 0 | 7 | 13
[Tom Petty] *A: Hard Promises; P: Jimmy Iovine/Tom Petty*

1535. SO FAR AWAY Dire Straits Warner Bros. 28789   703
[01Mar85 | 26Apr86 | 31May86] 19(1) 0 | 7 | 14
[Mark Knopfler] *A: Brothers In Arms; P: Neil Dorfsman/Mark Knopfler*

1536. I AM BY YOUR SIDE Corey Hart EMI America 8348   703
[20Sep86 | 15Nov86 | 13Dec86] 18(1) 0 | 7 | 13
[Corey Hart] *A: Fields Of Fire; P: Phil Chapman/Corey Hart*

1537. CANDY Cameo Atlanta Artists 888193   702
[27Dec86 | 21Mar87 | 18Apr87] 21(1) 0 | 7 | 17
[Larry Blackmon/Tomi Jenkins] *A: Word Up!; P: Larry Blackmon*

1538. STRANGE BUT TRUE Times Two Reprise 27998   702
[12Mar88 | 21May88 | 02Jul88] 21(2) 0 | 8 | 17
[Gardner Cole/Shanti Jones]
A: X2; P: Steve Barri/Gardner Cole/Shanti Jones/Tony Peluso

1539. DON'T FIGHT IT Kenny Loggins with Steve Perry Columbia 03192   700
[28Aug82 | 23Oct82 | 13Nov82] 17(2) 0 | 6 | 12
[Kenny Loggins/Steve Perry/Dean Pitchford]
A: High Adventure; P: Bruce Botnick/Kenny Loggins

1540. DEEP INSIDE MY HEART Randy Meisner Epic 50939   699
[18Oct80 | 06Dec80 | 31Jan81] 22(1) 0 | 7 | 16
[Eric Kaz/Randy Meisner] *A: One More Song; P: Val Garay*

1541. FALLING IN LOVE (Uh-Oh) Miami Sound Machine Epic 06352   697
[01Nov86 | 17Jan87 | 14Feb87] 25(1) 0 | 8 | 16
[Larry Dermer/Joe Galdo/Rafael Vigil] *A: Primitive Love; P: Emilio Estefan*

1542. I DO (LIVE VERSION) The J. Geils Band EMI America 8148   696
[20Nov82 | 08Jan83 | 19Feb83] 24(3) 0 | 7 | 14
[Melvin Mason/Frank Paden/Johnny Paden/Jesse Smith/Willie Stephenson]
A: Showtime; P: Seth Justman

1543. SUZANNE Journey Columbia 06134   696
[21Jun86 | 16Aug86 | 13Sep86] 17(1) 0 | 7 | 13
[Jonathan Cain/Steve Perry] *A: Raised On Radio; P: Steve Perry*

1544. SE LA Lionel Richie Motown 1883   696
[28Mar87 | 16May87 | 20Jun87] 20(2) 0 | 7 | 13
[Greg Phillinganes/Lionel Richie]
A: Dancing On The Ceiling; P: James Anthony Carmichael/Lionel Richie

1545. SUNSET GRILL Don Henley Geffen 28906   696
[31Aug85 | 19Oct85 | 30Nov85] 22(2) 0 | 8 | 14
[Don Henley/Danny Kortchmar/Benmont Tench]
A: Building The Perfect Beast; P: Don Henley/Danny Kortchmar/Greg Ladanyi

1546. TIME AND TIDE Basia Epic 07730   693
[23Jul88 | 29Oct88 | 03Dec88] 26(1) 0 | 7 | 20
[Basia Trzetrzelewska/Danny White]
A: Time And Tide; P: Basia Trzetrzelewska/Danny White

1547. I WANNA BE THE ONE Stevie B LMR 74003   693
[18Feb89 | 15Apr89 | 01Jul89] 32(1) 0 | 5 | 20
[Dadgel Atabay/Stevie B] *A: In My Eyes; P: Stevie B*

1548. HOOKED ON YOU ('89) Sweet Sensation Atco 99210   692
[03Jun89 | 12Aug89 | 16Sep89] 23(1) 0 | 7 | 16
[Joseph E. Malloy/David Sanchez]
A: Take It While It's Hot; P: Ted Currier/David Sanchez

1549. FOOL IN LOVE WITH YOU Jim Photoglo 20th Century 2487   692
[18Apr81 | 04Jul81 | 01Aug81] 25(2) 0 | 7 | 16
[Brian Neary/Jim Photoglo] *A: Fool In Love With You; P: Brian Neary*

1550. AFTER THE FALL Journey Columbia 04004   692
[09Jul83 | 27Aug83 | 24Sep83] 23(1) 0 | 8 | 12
[Jonathan Cain/Steve Perry] *A: Frontiers; P: Kevin Elson/Mike Stone*

1551. TAKE IT EASY Andy Taylor Atlantic 89414   690
[31May86 | 02Aug86 | 20Sep86] 24(1) 0 | 7 | 17
[Stephen Jones/Andy Taylor]
A: American Anthem Soundtrack; P: Roy Thomas Baker

1552. STATE OF THE HEART Rick Springfield RCA 14120   689
[08Jun85 | 17Aug85 | 14Sep85] 22(1) 0 | 7 | 15
[Eric McCusker/Tim Pierce/Rick Springfield]
A: Tao; P: Bill Drescher/Rick Springfield

1553. IN YOUR LETTER REO Speedwagon Epic 02457   689
[08Aug81 | 26Sep81 | 31Oct81] 20(2) 0 | 7 | 13
[Gary Richrath] *A: Hi Infidelity; P: Kevin Beamish/Kevin Cronin/Gary Richrath*

1554. LEAVE A TENDER MOMENT ALONE Billy Joel Columbia 04514   688
[07Jul84 | 25Aug84 | 13Oct84] 27(2) 0 | 7 | 15
[Billy Joel] *A: An Innocent Man; P: Phil Ramone*

1555. THIS WOMAN Kenny Rogers RCA 13710   688
[14Jan84 | 03Mar84 | 07Apr84] 23(2) 0 | 6 | 13
[Albhy Galuten/Barry Gibb]
A: Eyes That See In The Dark; P: Albhy Galuten/Barry Gibb/Karl Richardson

1556. CALLING AMERICA Electric Light Orchestra CBS Associated 05766   687
[01Feb86 | 05Apr86 | 10May86] 18(1) 0 | 7 | 15
[Jeff Lynne] *A: Balance Of Power; P: Jeff Lynne*

1557. NO EASY WAY OUT Robert Tepper Scotti Brothers 05750 687
[25Jan86 | 29Mar86 | 10May86] 22(1) 0 | 7 | 16
[Robert Tepper] *A: No Easy Way Out; P: Joe Chiccarelli*

1558. SHE'S MINE Steve Perry Columbia 04496   686
[30Jun84 | 11Aug84 | 22Sep84] 21(2) 0 | 8 | 13
[Randy Goodrum/Steve Perry] *A: Street Talk; P: Steve Perry*

1559. ORINOCO FLOW (Sail Away) Enya Geffen 27633   683
[21Jan89 | 15Apr89 | 13May89] 24(1) 0 | 8 | 17
[Enya/Roma Ryan] *A: Watermark; P: Nicky Ryan*

1560. WHITE HORSE Laid Back Sire 29346   683
[25Feb84 | 12May84 | 23Jun84] 26(2) 0 | 4 | 18
[John Guldberg/Tim Stahl] *A: ...Keep Smiling; P: Laid Back/Seven Dwarfs*

1561. WITHOUT YOU (Not Another Lonely Night) Franke And The Knockouts Millennium 13105   681
[03Apr82 | 12Jun82 | 10Jul82] 24(2) 0 | 7 | 15
[Billy Elworthy/Blake Levinsohn/Franke Previte]
A: Below The Belt; P: Knockouts/Franke Previte/Peter Solley

Rank. TITLE Artist Label Number	
[Entry Date | Peak Date | Exit Date] Peak(Peak Wks) Top10 | Top40 | Tot Wks	
[Writers] *Album*; Producers	Score

1562. ALL OUR TOMORROWS Eddie Schwartz Atco 7342 — 680
[12Dec81 | 20Feb82 | 20Mar82] 28(2) 0 | 7 | 15
[Eddie Schwartz/David Tyson] *A: No Refuge; P: Eddie Schwartz/David Tyson*

1563. LIES Thompson Twins Arista 1024 — 680
[22Jan83 | 26Mar83 | 07May83] 30(3) 0 | 5 | 16
[Tom Bailey/Alannah Currie/Joe Leeway] *A: Side Kicks; P: Alex Sadkin*

1564. SEQUEL Harry Chapin Boardwalk 5700 — 679
[01Nov80 | 13Dec80 | 31Jan81] 23(2) 0 | 7 | 14
[Harry Chapin] *A: Sequel; P: Howard Albert/Ron Albert*

1565. I DO YOU The Jets MCA 53193 — 679
[17Oct87 | 12Dec87 | 23Jan88] 20(1) 0 | 6 | 15
[Rick Kelly/Linda Mallah]
A: Magic; P: Rick Kelly/Michael Verdick

1566. ONE STEP CLOSER TO YOU Gavin Christopher Manhattan 50028 — 679
[24May86 | 16Aug86 | 13Sep86] 22(1) 0 | 7 | 17
[David Grant/Jeff Pescetto/Evan Rogers/Carl Sturken]
A: One Step Closer; P: Evan Rogers/Carl Sturken

1567. THE LOVE IN YOUR EYES Eddie Money Columbia 68532 — 679
[14Jan89 | 18Mar89 | 13May89] 24(1) 0 | 7 | 18
[David Paul Bryant/Steven Dubin/Adrian Gurvitz] *A*
: Nothing To Lose; P: Eddie Money/Richie Zito

1568. DON'T YOU GET SO MAD Jeffrey Osborne A&M 2561 — 678
[16Jul83 | 17Sep83 | 15Oct83] 25(1) 0 | 6 | 14
[Don Freeman/Jeffrey Osborne/Michael Sembello]
A: Stay With Me Tonight; P: George Duke

1569. UNDER PRESSURE Queen & David Bowie Elektra 47235 — 678
[07Nov81 | 09Jan82 | 13Feb82] 29(2) 0 | 8 | 15
[David Bowie/John Deacon/Brian May/Freddie Mercury/Roger Taylor]
A: Hot Space; P: David Bowie/Queen

1570. FOUR IN THE MORNING (I Can't Take Any More) Night Ranger — 677
Camel/MCA 52661
[24Aug85 | 12Oct85 | 16Nov85] 19(1) 0 | 6 | 13
[Jack Blades] *A: 7 Wishes; P: Pat Glasser*

1571. HEART HOTELS Dan Fogelberg Full Moon/Epic 50862 — 677
[22Mar80 | 24May80 | 14Jun80] 21(1) 0 | 6 | 13
[Dan Fogelberg] *A: Phoenix; P: Dan Fogelberg/Marty Lewis/Norbert Putnam*

1572. MYSTERY LADY Billy Ocean Jive 9374 — 677
[06Jul85 | 31Aug85 | 12Oct85] 24(1) 0 | 8 | 15
[Keith Diamond/Billy Ocean/James Woodley] *A: Suddenly; P: Keith Diamond*

1573. WELCOME TO HEARTLIGHT Kenny Loggins Columbia 03555 — 677
[12Mar83 | 30Apr83 | 11Jun83] 24(2) 0 | 7 | 14
[Kenny Loggins] *A: High Adventure; P: Bruce Botnick/Kenny Loggins*

1574. CHANGE OF HEART Tom Petty and The Heartbreakers — 676
Backstreet 52181
[26Feb83 | 02Apr83 | 07May83] 21(3) 0 | 7 | 11
[Tom Petty] *A: Long After Dark; P: Jimmy Iovine/Tom Petty*

1575. SEND ME AN ANGEL ('89) Real Life Curb 10531 — 675
[13May89 | 22Jul89 | 26Aug89] 26(1) 0 | 8 | 16
[David Sterry/Richard Zatorski]
A: Send Me An Angel '89; P: Ross Cockle/Glenn Wheatley

1576. HOLD ON LOOSELY .38 Special A&M 2316 — 675
[28Feb81 | 16May81 | 20Jun81] 27(2) 0 | 6 | 17
[Don Barnes/Jeff Carlisi/Jim Peterik]
A: Wild-Eyed Southern Boys; P: Rodney Mills

1577. FANTASY Aldo Nova Portrait 24-02799 — 672
[27Mar82 | 29May82 | 10Jul82] 23(2) 0 | 7 | 16
[Aldo Nova] *A: Aldo Nova; P: Aldo Nova*

1578. FACE THE FACE Pete Townshend Atco 99590 — 672
[09Nov85 | 18Jan86 | 22Feb86] 26(1) 0 | 7 | 16
[Pete Townshend] *A: White City - A Novel; P: Chris Thomas*

1579. SOUL KISS Olivia Newton-John MCA 52686 — 671
[05Oct85 | 23Nov85 | 11Jan86] 20(1) 0 | 7 | 15
[Mark Goldenberg] *A: Soul Kiss; P: John Farrar*

1580. IF THE LOVE FITS WEAR IT Leslie Pearl RCA 13235 — 671
[22May82 | 14Aug82 | 04Sep82] 28(2) 0 | 7 | 16
[Leslie Pearl/Phil Redrow] *A: Words And Music; P: Leslie Pearl*

1581. SO IN LOVE Orchestral Manoeuvres In The Dark A&M 2746 — 671
[31Aug85 | 09Nov85 | 21Dec85] 26(2) 0 | 7 | 17
[Stephen Hague/Paul Humphreys/Andrew McCluskey]
A: Crush; P: Stephen Hague

1582. A SHOULDER TO CRY ON Tommy Page Sire 27645 — 670
[11Feb89 | 06May89 | 24Jun89] 29(1) 0 | 6 | 20
[Tommy Page] *A: Tommy Page; P: Arif Mardin/Joe Mardin*

1583. KISS THE BRIDE Elton John Geffen 29568 — 670
[06Aug83 | 01Oct83 | 22Oct83] 25(1) 0 | 8 | 12
[Elton John/Bernie Taupin] *A: Too Low For Zero; P: Chris Thomas*

1584. TURN UP THE RADIO Autograph RCA 13953 — 670
[22Dec84 | 16Mar85 | 27Apr85] 29(1) 0 | 5 | 19
[Steve Isham/Steve Lynch/Steve Plunkett/Randy Rand/Keni Richards]
A: Sign In Please; P: Neil Kernon

1585. STILL IN SAIGON Charlie Daniels Band Epic 02828 — 668
[27Mar82 | 29May82 | 12Jun82] 22(2) 0 | 8 | 12
[Dan Daley] *A: Windows; P: John Boylan*

1586. THAT WAS THEN, THIS IS NOW — 667
Micky Dolenz And Peter Tork (Of The Monkees) Arista 1-9505
[05Jul86 | 30Aug86 | 04Oct86] 20(1) 0 | 7 | 14
[Vance Brescia] *A: Then & Now...The Best Of The Monkees; P: Michael Lloyd*

1587. ANOTHER NIGHT Aretha Franklin Arista 1-9453 — 667
[18Jan86 | 22Mar86 | 19Apr86] 22(1) 0 | 7 | 14
[Giuseppe Cantarelli/Roy Freeland]
A: Who's Zoomin' Who?; P: Narada Michael Walden

1588. FREEDOM OVERSPILL Steve Winwood Island 28595 — 666
[27Sep86 | 22Nov86 | 03Jan87] 20(2) 0 | 7 | 15
[George Fleming/James Hooker/Will Jennings/Steve Winwood]
A: Back In The High Life; P: Russ Titelman/Steve Winwood

1589. THAT OLD SONG Ray Parker Jr. & Raydio Arista 0616 — 665
[11Jul81 | 05Sep81 | 17Oct81] 21(2) 0 | 6 | 15
[Ray Parker Jr.] *A: A Woman Needs Love; P: Ray Parker Jr.*

1590. TRUE FAITH New Order Qwest 28271 — 665
[24Oct87 | 26Dec87 | 20Feb88] 32(2) 0 | 8 | 18
[Gillian Gilbert/Stephen Hague/Peter Hook/Stephen Morris/Bernard Sumner]
A: Substance; P: Stephen Hague/New Order

1591. NEW SONG Howard Jones Elektra 69766 — 661
[21Jan84 | 31Mar84 | 28Apr84] 27(1) 0 | 6 | 15
[Howard Jones] *A: Human's Lib; P: Colin Thurston*

1592. HEART LIKE A WHEEL Steve Miller Band Capitol 5068 — 661
[31Oct81 | 12Dec81 | 30Jan82] 24(1) 0 | 9 | 14
[Steve Miller] *A: Circle Of Love; P: Steve Miller*

1593. SECRET SEPARATION The Fixx MCA 52832 — 661
[24May86 | 26Jul86 | 23Aug86] 19(1) 0 | 6 | 14
[Dan Brown/Cy Curnin/Peter John Greenall/Jeannette Obstoj/
Jamie West-Oram/Adam Woods] *A: Walkabout; P: Rupert Hine*

1594. AH! LEAH! Donnie Iris MCA/Carousel 51025 — 660
[13Dec80 | 28Feb81 | 11Apr81] 29(1) 0 | 6 | 18
[Mark Avsec/Donnie Iris] *A: Back On The Streets; P: Mark Avsec*

1595. DO THEY KNOW IT'S CHRISTMAS? Band Aid Columbia 04749 — 660
[22Dec84 | 19Jan85 | 16Feb85] 13(1) 0 | 4 | 9
[Bob Geldof/Midge Ure] *A: No Album; P: Trevor Horn/Midge Ure*

1596. BE MY LADY Jefferson Starship Grunt 13350 — 659
[09Oct82 | 11Dec82 | 22Jan83] 28(2) 0 | 6 | 16
[Jeannette Sears/Pete Sears] *A: Winds Of Change; P: Kevin Beamish*

1597. WE TWO Little River Band Capitol 5231 — 659
[07May83 | 18Jun83 | 23Jul83] 22(2) 0 | 6 | 12
[Graham Goble] *A: The Net; P: Little River Band/Ernie Rose*

1598. TALL COOL ONE Robert Plant Es Paranza 7-99348 — 658
[16Apr88 | 02Jul88 | 13Aug88] 25(1) 0 | 7 | 18
[Phil Johnstone/Robert Plant]
A: Now And Zen; P: Phil Johnstone/Tim Palmer/Robert Plant

1599. MIDNIGHT ROCKS Al Stewart Arista 0552 — 658
[30Aug80 | 25Oct80 | 22Nov80] 24(1) 0 | 6 | 13
[Al Stewart/Peter White] *A: 24 Carrots; P: Chris Desmond/Al Stewart*

1600. THE WOMAN IN ME Donna Summer Geffen 29805 — 657
[18Dec82 | 26Feb83 | 02Apr83] 33(2) 0 | 6 | 16
[John Bettis/Michael Clark] *A: Donna Summer; P: Quincy Jones*

1601. LOVE CHANGES (Everything) Climie Fisher Capitol 44137 — 655
[14May88 | 30Jul88 | 10Sep88] 23(1) 0 | 6 | 18
[Simon Climie/Rob Fisher/Dennis W. Morgan] *A: Everything; P: Stephen Hague*

1602. MY GIRL Suavé Capitol 44124 — 655
[12Mar88 | 14May88 | 18Jun88] 20(2) 0 | 7 | 15
[Smokey Robinson/Ronnie White] *A: I'm Your Playmate; P: Suave*

Ranking the Singles

Rank. TITLE Artist Label Number
[Entry Date | Peak Date | Exit Date] Peak(Peak Wks) Top10 | Top40 | Tot Wks
[Writers] *Album; Producers*   Score

1603. EVERYBODY WANTS YOU Billy Squier Capitol 5163   654
[02Oct82 | 18Dec82 | 22Jan83] 32(3) 0 | 6 | 17
[Billy Squier] *A: Emotions In Motion; P: Reinhold Mack/Billy Squier*

1604. WHIRLY GIRL Oxo Geffen 29765   654
[19Feb83 | 23Apr83 | 21May83] 28(1) 0 | 6 | 14
[Ish Ledesma] *A: Oxo; P: Ish Ledesma/Ken Mansfield*

1605. AND I AM TELLING YOU I'M NOT GOING Jennifer Holliday   654
Geffen 29983
[03Jul82 | 28Aug82 | 02Oct82] 22(3) 0 | 7 | 14
[Tom Eyen/Henry Krieger] *A: Dreamgirls Original Cast; P: David Foster*

1606. DOMINO DANCING Pet Shop Boys EMI-Manhattan 50161   653
[08Oct88 | 03Dec88 | 07Jan89] 18(1) 0 | 6 | 14
[Chris Lowe/Neil Tennant] *A: Introspective; P: Lewis A. Martineé*

1607. EDGE OF A BROKEN HEART Vixen EMI 50141   651
[17Sep88 | 19Nov88 | 04Feb89] 26(1) 0 | 5 | 21
[Richard Marx/Fee Waybill] *A: Vixen; P: Richard Marx*

1608. THIS TIME Bryan Adams A&M 2574   649
[03Sep83 | 29Oct83 | 19Nov83] 24(1) 0 | 6 | 12
[Bryan Adams/Jim Vallance]
A: Cuts Like A Knife; P: Bryan Adams/Bob Clearmountain

1609. THE ANGEL SONG Great White Capitol 44449   648
[30Sep89 | 02Dec89 | 17Feb90] 30(1) 0 | 4 | 17
[Mark Kendall/Alan Niven] *A: ...Twice Shy; P: Michael Lardie/Alan Niven*

1610. NIGHT MOVES Marilyn Martin Atlantic 89465   647
[18Jan86 | 22Mar86 | 17May86] 28(1) 0 | 6 | 18
[Marilyn Martin/John Parr] *A: Marilyn Martin; P: Jon Astley/Phil Chapman*

1611. NOT JUST ANOTHER GIRL Ivan Neville Polydor 887814   647
[08Oct88 | 10Dec88 | 11Feb89] 26(1) 0 | 6 | 19
[Ivan Neville] *A: If My Ancestors Could See Me Now; P: Danny Kortchmar*

1612. EARLY IN THE MORNING The Gap Band Total Experience 8201   647
[22May82 | 24Jul82 | 21Aug82] 24(2) 0 | 6 | 14
[Lonnie Simmons/Rudy Taylor/Charles K. Wilson]
A: Gap Band IV; P: Lonnie Simmons

1613. LOVE IS A STRANGER Eurythmics RCA 13618   646
[17Sep83 | 12Nov83 | 10Dec83] 23(1) 0 | 6 | 13
[Annie Lennox/David A. Stewart]
A: Sweet Dreams (Are Made Of This); P: David A. Stewart/Adam Williams

1614. TWO PLACES AT THE SAME TIME Ray Parker Jr. & Raydio Arista 0494   645
[19Apr80 | 28Jun80 | 19Jul80] 30(2) 0 | 6 | 14
[Ray Parker Jr.] *A: Two Places At The Same Time; P: Ray Parker Jr.*

1615. WE LIVE FOR LOVE Pat Benatar Chrysalis 2419   643
[05Apr80 | 14Jun80 | 05Jul80] 27(2) 0 | 6 | 14
[Neil Geraldo] *A: In The Heat Of The Night; P: Peter Coleman*

1616. DADDY'S HOME Cliff Richard EMI America 8103   642
[16Jan82 | 20Mar82 | 10Apr82] 23(1) 0 | 8 | 13
[Charles Baskerville/Clarence Bassett/William Henry Miller/James Sheppard]
A: Wired For Sound; P: Cliff Richard

1617. ROCK AND A HARD PLACE The Rolling Stones Columbia 73057   641
[04Nov89 | 23Dec89 | 03Feb90] 23(3) 0 | 8 | 14
[Mick Jagger/Keith Richards]
A: Steel Wheels; P: Glimmer Twins/Chris Kimsey

1618. NOBODY WINS Elton John Geffen 49722   641
[09May81 | 20Jun81 | 01Aug81] 21(2) 0 | 6 | 13
[Jean-Paul Dreau/Gary Osborne] *A: The Fox; P: Clive Franks/Elton John*

1619. FADE AWAY Bruce Springsteen Columbia 11431   640
[07Feb81 | 14Mar81 | 25Apr81] 20(2) 0 | 6 | 12
[Bruce Springsteen]
A: The River; P: Jon Landau/Bruce Springsteen/Steven Van Zandt

1620. SEVENTEEN Winger Atlantic 88958   640
[25Feb89 | 06May89 | 10Jun89] 26(1) 0 | 6 | 16
[Reb Beach/Beau Hill/Kip Winger] *A: Winger; P: Beau Hill*

1621. TURN ME LOOSE Loverboy Columbia 11421   639
[31Jan81 | 25Apr81 | 23May81] 35(1) 0 | 6 | 17
[Paul Dean/Mike Reno] *A: Loverboy; P: Bruce Fairbairn*

1622. NEW WORLD MAN Rush Mercury 76179   636
[18Sep82 | 30Oct82 | 04Dec82] 21(3) 0 | 6 | 12
[Geddy Lee/Alex Lifeson/Neil Peart] *A: Signals; P: Terry Brown/Rush*

1623. THROUGH THE STORM Aretha Franklin And Elton John Arista 1-9809   636
[15Apr89 | 27May89 | 24Jun89] 16(1) 0 | 7 | 11
[Albert Hammond/Diane Warren]
A: Through The Storm; P: Narada Michael Walden

1624. INVISIBLE Alison Moyet Columbia 04781   635
[09Mar85 | 01Jun85 | 29Jun85] 31(1) 0 | 6 | 17
[Lamont Dozier] *A: Alf; P: Steve Jolley/Tony Swain*

1625. STAGES ZZ Top Warner Bros. 28810   634
[18Jan86 | 08Mar86 | 05Apr86] 21(1) 0 | 7 | 12
[Frank Beard/Billy Gibbons/Dusty Hill] *A: Afterburner; P: Bill Ham*

1626. SAY YOU'RE WRONG Julian Lennon Atlantic 89567   634
[20Apr85 | 01Jun85 | 06Jul85] 21(2) 0 | 8 | 12
[Julian Lennon] *A: Valotte; P: Phil Ramone*

1627. MY GUY Sister Sledge Cotillion 47000   634
[30Jan82 | 03Apr82 | 08May82] 23(2) 0 | 6 | 15
[Smokey Robinson] *A: The Sisters; P: Sister Sledge*

1628. MAKE LOVE STAY Dan Fogelberg Full Moon 03525   633
[05Feb83 | 19Mar83 | 21May83] 29(3) 0 | 6 | 16
[Dan Fogelberg] *A: Dan Fogelberg/Greatest Hits; P: Dan Fogelberg/Marty Lewis*

1629. NEVER TOO MUCH Luther Vandross Epic 02409   633
[10Oct81 | 28Nov81 | 16Jan82] 33(2) 0 | 4 | 15
[Luther Vandross] *A: Never Too Much; P: Luther Vandross*

1630. ROCK AND ROLL GIRLS John Fogerty Warner Bros. 29053   632
[16Mar85 | 27Apr85 | 01Jun85] 20(1) 0 | 6 | 12
[John Fogerty] *A: Centerfield; P: John Fogerty*

1631. LET'S GO DANCIN' (Ooh La, La, La) Kool & The Gang De-Lite 824   632
[30Oct82 | 08Jan83 | 05Feb83] 30(2) 0 | 7 | 15
[Clifford Adams/Amir Bayyan/Robert Bell/Ronald Bell/George M. Brown/
Eumir Deodato/Robert Mickens/Michael Ray/Claydes Smith/
J.T. Taylor/Curtis Fitzgerald Williams] *A: As One; P: Eumir Deodato*

1632. JUST LIKE HEAVEN The Cure Elektra 69443   632
[10Oct87 | 09Jan88 | 13Feb88] 40(1) 0 | 1 | 19
[Simon Gallup/Robert James Smith/Porl Thompson/Laurence Tolhurst/
Boris Williams]
A: Kiss Me, Kiss Me, Kiss Me; P: Dave Allen/Robert Smith

1633. SWEET SIXTEEN Billy Idol Chrysalis 43114   632
[25Apr87 | 27Jun87 | 25Jul87] 20(1) 0 | 7 | 14
[Billy Idol] *A: Whiplash Smile; P: Keith Forsey*

1634. FOREVER MINE The O'Jays Philadelphia Int'l 3727   632
[24Nov79 | 02Feb80 | 16Feb80] 28(1) 0 | 5 | 13
[Kenneth Gamble/Leon Huff] *A: Identify Yourself; P: Kenny Gamble/Leon Huff*

1635. SHAME The Motels Capitol 5497   632
[20Jul85 | 14Sep85 | 12Oct85] 21(1) 0 | 7 | 13
[Martha Davis] *A: Shock; P: Richie Zito*

1636. DREAMS Van Halen Warner Bros. 28702   631
[24May86 | 19Jul86 | 23Aug86] 22(1) 0 | 7 | 14
[Michael Anthony/Sammy Hagar/Alex Van Halen/Eddie Van Halen] *A: 5150;
P: Mick Jones/Donn Landee/Van Halen*

1637. SINCE I DON'T HAVE YOU Don McLean Millennium 11804   631
[11Apr81 | 30May81 | 11Jul81] 23(2) 0 | 6 | 14
[James Beaumont/Walter Lester/Lennie Martin/Joseph Rock/John L.
Taylor/Joseph Verscharen/Janet Vogel] *A: Chain Lightning; P: Larry Butler*

1638. VERONICA Elvis Costello Warner Bros. 22981   630
[22Apr89 | 24Jun89 | 22Jul89] 19(1) 0 | 6 | 14
[Elvis Costello/Paul McCartney]
A: Spike; P: T-Bone Burnett/Elvis Costello/Kevin Killen

1639. EARTH ANGEL New Edition MCA 52905   629
[23Aug86 | 11Oct86 | 22Nov86] 21(1) 0 | 6 | 14
[Jesse Belvin/Gaynel Hodge/Curtis Edward Williams]
A: Under The Blue Moon; P: Freddie Perren

1640. TWIST AND SHOUT The Beatles Capitol 5624   628
[09Aug86 | 27Sep86 | 15Nov86] 23(1) 0 | 7 | 15
[Bert Berns/Phil Medley] *A: No Album; P: George Martin*

1641. I.G.Y. (What A Beautiful World) Donald Fagen Warner Bros. 29900   626
[09Oct82 | 27Nov82 | 08Jan83] 26(3) 0 | 7 | 14
[Donald Fagen] *A: The Nightfly; P: Gary Katz*

1642. CAN'T SHAKE LOOSE Agnetha Faltskog Polydor 815230   626
[27Aug83 | 05Nov83 | 03Dec83] 29(2) 0 | 5 | 15
[Russ Ballard] *A: Wrap Your Arms Around Me; P: Mike Chapman*

1643. SUNSHINE Dino 4th & Broadway 7489   626
[09Sep89 | 11Nov89 | 16Dec89] 23(1) 0 | 6 | 15
[Dino Esposito] *A: 24/7; P: Dino Esposito*

1644. ROCK OF LIFE Rick Springfield RCA 6853   624
[06Feb88 | 02Apr88 | 14May88] 22(1) 0 | 6 | 15
[Rick Springfield] *A: Rock Of Life; P: Keith Olsen/Rick Springfield*

Ranking the Singles

Rank. TITLE Artist Label Number
[Entry Date | Peak Date | Exit Date] Peak(Peak Wks) Top10 | Top40 | Tot Wks
[Writers] *Album; Producers* Score

1645. SATURDAY LOVE Cherrelle with Alexander O'Neal Tabu 4-05767   623
[15Feb86 | 19Apr86 | 07Jun86] 26(2) 0 | 6 | 17
[James Samuel Harris/Terry Lewis]
A: High Priority; P: Jimmy Jam Harris/Terry Lewis

1646. DAY-IN DAY-OUT David Bowie EMI America 8380   622
[04Apr87 | 23May87 | 20Jun87] 21(2) 0 | 7 | 12
[David Bowie] *A: Never Let Me Down; P: David Bowie/David Richards*

1647. ROUGH BOY ZZ Top Warner Bros. 28733   622
[29Mar86 | 17May86 | 21Jun86] 22(1) 0 | 7 | 13
[Frank Beard/Billy Gibbons/Dusty Hill] *A: Afterburner; P: Bill Ham*

1648. LOVE YOU LIKE I NEVER LOVED BEFORE John O'Banion Elektra 47125   622
[28Mar81 | 16May81 | 20Jun81] 24(2) 0 | 7 | 13
[Joey Carbone/Kathi Pinto/Richie Zito]
A: John O'Banion; P: Joey Carbone/Richie Zito

1649. BREAK IT UP Foreigner Atlantic 4044   621
[15May82 | 26Jun82 | 07Aug82] 26(2) 0 | 6 | 13
[Michael Leslie Jones] *A: 4; P: Mick Jones/Mutt Lange*

1650. DREAMIN' IS EASY Steel Breeze RCA 13427   621
[15Jan83 | 05Mar83 | 09Apr83] 30(3) 0 | 6 | 13
[Ken Goorabian] *A: Steel Breeze; P: Kim Fowley*

1651. SWEET TIME REO Speedwagon Epic 03264   620
[28Aug82 | 23Oct82 | 27Nov82] 26(2) 0 | 6 | 14
[Kevin Cronin]
A: Good Trouble; P: Kevin Beamish/Kevin Cronin/Alan Gratzer/Gary Richrath

1652. I'M SO EXCITED ('82) Pointer Sisters Planet 13327   620
[18Sep82 | 27Nov82 | 01Jan83] 30(2) 0 | 6 | 16
[Trevor Lawrence/Anita Pointer/June Pointer/Ruth Pointer]
A: So Excited!; P: Richard Perry

1653. A WORD IN SPANISH Elton John MCA 53408   618
[17Sep88 | 12Nov88 | 10Dec88] 19(1) 0 | 6 | 13
[Elton John/Bernie Taupin] *A: Reg Strikes Back; P: Chris Thomas*

1654. JAMMIN' ME Tom Petty and The Heartbreakers MCA 53065   617
[25Apr87 | 20Jun87 | 11Jul87] 18(1) 0 | 6 | 12
[Mike Campbell/Bob Dylan/Tom Petty]
A: Let Me Up (I've Had Enough); P: Mike Campbell/Tom Petty

1655. WHEN ALL IS SAID AND DONE ABBA Atlantic 3889   615
[09Jan82 | 13Mar82 | 10Apr82] 27(2) 0 | 8 | 14
[Benny Andersson/Björn Ulvaeus]
A: The Visitors; P: Benny Andersson/Björn Ulvaeus

1656. BIG FUN Kool & The Gang De-Lite 822   615
[28Aug82 | 16Oct82 | 06Nov82] 21(2) 0 | 7 | 11
[Clifford Adams/Robert Bell/Ronald Bell/George M. Brown/Eumir Deodato/
Robert Mickens/Michael Ray/Claydes Smith/J.T. Taylor/
Curtis Fitzgerald Williams] *A: As One; P: Eumir Deodato*

1657. GENERAL HOSPI-TALE The Afternoon Delights MCA 51148   615
[25Jul81 | 03Oct81 | 07Nov81] 33(2) 0 | 5 | 16
[Harry Francis King/Lisa Tedesco] *A: General Hospi-tale; P: Harry King*

1658. SOME DAYS ARE DIAMONDS (Some Days Are Stone) John Denver   614
RCA 12246
[13Jun81 | 19Sep81 | 24Oct81] 36(1) 0 | 4 | 20
[Dick Feller] *A: Some Days Are Diamonds; P: Larry Butler*

1659. DON'T LET HIM GO REO Speedwagon Epic 02127   614
[13Jun81 | 01Aug81 | 12Sep81] 24(2) 0 | 6 | 14
[Kevin Cronin] *A: Hi Infidelity; P: Kevin Beamish/Kevin Cronin/Gary Richrath*

1660. THE RIGHT THING Simply Red Elektra 69487   614
[28Feb87 | 16May87 | 06Jun87] 27(2) 0 | 6 | 15
[Mick Hucknall] *A: Men And Women; P: Alex Sadkin*

1661. LOVE PLUS ONE Haircut One Hundred Arista 0672   614
[15May82 | 07Aug82 | 04Sep82] 37(1) 0 | 4 | 17
[Nicholas Heyward] *A: Pelican West; P: Bob Sargeant*

1662. GENIUS OF LOVE Tom Tom Club Sire 49882   614
[23Jan82 | 24Apr82 | 15May82] 31(2) 0 | 6 | 17
[Adrian Belew/Christopher Frantz/Steven J.C. Stanley/Tina Weymouth]
A: Tom Tom Club; P: Tom Tom Club

1663. THAT'S THE WAY Katrina And The Waves SBK 07303   613
[22Jul89 | 16Sep89 | 07Oct89] 16(1) 0 | 6 | 12
[Katrina Leskanich/Kimberly Rew] *A: Break Of Hearts;
P: Alex Cooper/Vince De La Cruz/Katrina Leskanich/Kimberley Rew*

1664. LADY LOVE ME (One More Time) George Benson Warner Bros. 29563   612
[23Jul83 | 10Sep83 | 15Oct83] 30(1) 0 | 6 | 13
[James Newton Howard/David Paich] *A: In Your Eyes; P: Arif Mardin*

1665. WHY ME Styx A&M 2206   612
[15Dec79 | 09Feb80 | 08Mar80] 26(1) 0 | 5 | 13
[Dennis DeYoung] *A: Cornerstone; P: Styx*

1666. STRANGER IN TOWN Toto Columbia 04672   611
[27Oct84 | 22Dec84 | 02Feb85] 30(2) 0 | 6 | 15
[David Paich/Jeff Porcaro] *A: Isolation; P: Toto*

1667. ATHENA The Who Warner Bros. 29905   611
[04Sep82 | 30Oct82 | 04Dec82] 28(3) 0 | 6 | 14
[Pete Townshend] *A: It's Hard; P: Glyn Johns*

1668. WHEN IT'S OVER Loverboy Columbia 02814   611
[10Apr82 | 12Jun82 | 17Jul82] 26(2) 0 | 6 | 15
[Paul Dean/Mike Reno] *A: Get Lucky; P: Paul Dean/Bruce Fairbairn*

1669. COME ON, LET'S GO Los Lobos Slash 28186   610
[12Sep87 | 07Nov87 | 12Dec87] 21(2) 0 | 7 | 14
[Ritchie Valens] *A: La Bamba (Soundtrack); P: Steve Berlin*

1670. IT'S A LOVE THING The Whispers Solar 12154   610
[14Feb81 | 18Apr81 | 23May81] 28(2) 0 | 5 | 15
[Dana Meyers/William Shelby] *A: Imagination; P: Dick Griffey/Whispers*

1671. HANG FIRE The Rolling Stones Rolling Stones 21300   608
[20Mar82 | 08May82 | 29May82] 20(2) 0 | 6 | 11
[Mick Jagger/Keith Richards] *A: Tattoo You; P: Glimmer Twins*

1672. ROCK AND ROLL DREAMS COME THROUGH Jim Steinman   608
Cleveland International 19-02111
[30May81 | 15Aug81 | 12Sep81] 32(1) 0 | 6 | 16
[Jim Steinman] *A: Bad For Good; P: Jimmy Iovine/John Jansen/Jim Steinman*

1673. TURNING JAPANESE The Vapors United Artists 1364   608
[27Sep80 | 29Nov80 | 17Jan81] 36(1) 0 | 3 | 17
[David Fenton] *A: New Clear Days; P: Vic Coppersmith-Heaven*

1674. HE CAN'T LOVE YOU Michael Stanley Band EMI America 8063   607
[22Nov80 | 31Jan81 | 07Mar81] 33(2) 0 | 5 | 16
[Kevin Raleigh] *A: Heartland; P: Michael Stanley Band*

1675. MONEY$ TOO TIGHT (To Mention) Simply Red Elektra 69528   607
[19Jul86 | 04Oct86 | 25Oct86] 28(1) 0 | 6 | 15
[John Valentine/William Valentine/Caroline Wiggins]
A: Picture Book; P: Stewart Levine

1676. THE WOMAN IN YOU Bee Gees RSO 813173   605
[21May83 | 18Jun83 | 30Jul83] 24(3) 0 | 6 | 11
[Barry Gibb/Maurice Gibb/Robin Gibb] *A: Staying Alive (Soundtrack);
P: Albhy Galuten/Barry Gibb/Maurice Gibb/Robin Gibb/Karl Richardson*

1677. LOVE WILL FIND A WAY Yes Atco 99449   605
[03Oct87 | 28Nov87 | 06Feb88] 30(1) 0 | 6 | 19 [Trevor Rabin]
A: Big Generator; P: Paul DeVilliers/Trevor Horn/Trevor Rabin/Yes

1678. THAT GIRL COULD SING Jackson Browne Asylum 47036   604
[20Sep80 | 08Nov80 | 13Dec80] 22(2) 0 | 5 | 13
[Jackson Browne] *A: Hold Out; P: Jackson Browne/Greg Ladanyi*

1679. ABACAB Genesis Atlantic 3891   603
[26Dec81 | 20Feb82 | 27Mar82] 26(2) 0 | 6 | 14
[Tony Banks/Phil Collins/Mike Rutherford] *A: Abacab; P: Genesis*

1680. RED RED WINE ('84) UB40 A&M 2600   603
[28Jan84 | 31Mar84 | 05May84] 34(2) 0 | 4 | 15
[Neil Diamond] *A: Labour Of Love; P: Ray Falconer/UB40*

1681. YOU SHOOK ME ALL NIGHT LONG AC/DC Atlantic 3761   603
[06Sep80 | 08Nov80 | 27Dec80] 35(1) 0 | 3 | 16
[Brian Johnson/Angus Young/Malcolm Young] *A: Back In Black; P: Mutt Lange*

1682. FOOLIN' Def Leppard Mercury 814178   602
[03Sep83 | 05Nov83 | 03Dec83] 28(1) 0 | 5 | 14
[Rick Allen/Steve Clark/Joe Elliott/Mutt Lange/Rick Savage/Pete Willis]
A: Pyromania; P: Mutt Lange

1683. RUNNING UP THAT HILL Kate Bush EMI America 8285   602
[07Sep85 | 30Nov85 | 18Jan86] 30(1) 0 | 4 | 20
[Kate Bush] *A: Hounds Of Love; P: Kate Bush*

1684. ROTATION Herb Alpert A&M 2202   602
[17Nov79 | 19Jan80 | 09Feb80] 30(2) 0 | 6 | 13
[Andy Armer/Randy Badazz] *A: Rise; P: Herb Alpert/Randy Badazz*

1685. GOLD Spandau Ballet Chrysalis 42743   601
[19Nov83 | 14Jan84 | 04Feb84] 29(2) 0 | 6 | 12
[Gary Kemp] *A: True; P: Steve Jolley/Spandau Ballet/Tony Swain*

1686. CAN'TCHA SAY (You Believe In Me)/STILL IN LOVE Boston MCA 53029   598
[07Mar87 | 18Apr87 | 30May87] 20(2) 0 | 5 | 13
[Brad Delp/Gerald Green/Tom Scholz// Brad Delp/Tom Scholz]
A: Third Stage; P: Tom Scholz

Ranking the Singles

Rank. TITLE Artist Label Number [Entry Date	Peak Date	Exit Date] Peak(Peak Wks) Top10	Top40	Tot Wks [Writers] *Album; Producers*	Score
1687. **WHAT YOU SEE IS WHAT YOU GET** Brenda K. Starr MCA 53367 [06Aug88	01Oct88	29Oct88] 24(2) 0	7	13 [Stephen Lunt/Arthur Stead] *A: Brenda K. Starr; P: Stephen Broughton Lunt*	597
1688. **NEVER AS GOOD AS THE FIRST TIME** Sade Portrait 37-05761 [29Mar86	17May86	14Jun86] 20(1) 0	7	12 [Helen Adu/Stuart Matthewman] *A: Promise; P: Mike Pela/Ben Rogan*	596
1689. **THE SEDUCTION (LOVE THEME)** The James Last Band Polydor 2071 [29Mar80	24May80	21Jun80] 28(2) 0	6	13 [Giorgio Moroder] *A: American Gigolo Sountrack; P: James Last/Ron Last*	595
1690. **LIVE IS LIFE** Opus Polydor 883730 [25Jan86	29Mar86	10May86] 32(2) 0	5	16 [Guenter Grasmuck/Niki Gruber/Ewald Pfleger/Kurt Plisnier/Herwig Rudisser] *A: Up And Down; P: Peter J. Muller*	594
1691. **SILENT MORNING** Noel 4th & Broadway 7439 [22Aug87	14Nov87	16Jan88] 47(1) 0	0	22 [Noel Pagan] *A: Noel; P: Roman Ricardo/Paul Robb*	594
1692. **WE CONNECT** Stacey Q Atlantic 89331 [13Dec86	28Feb87	18Apr87] 35(2) 0	4	19 [John Griffitt Wilcox] *A: Better Than Heaven; P: Jon St. James*	594
1693. **IT'S MY LIFE** Talk Talk EMI America 8195 [24Mar84	19May84	23Jun84] 31(2) 0	6	14 [Tim Friese-Greene/Mark Hollis] *A: It's My Life; P: Tim Friese-Greene*	593
1694. **BABY MAKES HER BLUE JEANS TALK** Dr. Hook Casablanca 2347 [27Feb82	17Apr82	15May82] 25(3) 0	6	17 [Ron Haffkine/Dennis Locorriere/Sam Weedman] *A: Players In The Dark; P: Ron Haffkine*	592
1695. **SMOKING GUN** The Robert Cray Band Mercury 888343 [07Feb87	18Apr87	09May87] 22(1) 0	6	14 [Bruce Bromberg/Richard Cousins/Robert Cray] *A: Strong Persuader; P: Bruce Bromberg/Dennis Walker*	592
1696. **EVERYTHING IN MY HEART** Corey Hart EMI America 8300 [30Nov85	01Feb86	08Mar86] 30(1) 0	7	15 [Corey Hart] *A: Boy In The Box; P: Jon Astley/Phil Chapman/Corey Hart*	591
1697. **DOWN BOYS** Warrant Columbia 68606 [29Apr89	08Jul89	12Aug89] 27(1) 0	6	16 [Joseph Cagle/Steven Chamberlin/Jerry Dixon/John Oswald/Eric Turner] *A: Dirty Rotten Filthy Stinking Rich; P: Beau Hill*	591
1698. **A NITE AT THE APOLLO LIVE: THE WAY YOU DO THE THINGS YOU DO/ MY GIRL** Daryl Hall & John Oates Featuring David Ruffin & Eddie Kendricks RCA 14178 [31Aug85	12Oct85	09Nov85] 20(1) 0	7	11 [Smokey Robinson/Robert Rogers//Smokey Robinson/Ronald White] *A: Hall & Oates Live At The Apollo With David Ruffin & Eddie Kendrick;* *P: Bob Clearmountain/Daryl Hall/John Oates*	591
1699. **RUNNIN' DOWN A DREAM** Tom Petty MCA 53682 [29Jul89	23Sep89	28Oct89] 23(1) 0	7	14 [Mike Campbell/Jeff Lynne/Tom Petty] *A: Full Moon Fever; P: Mike Campbell/Jeff Lynne/Tom Petty*	590
1700. **RADIOACTIVE** The Firm Atlantic 89586 [09Feb85	13Apr85	18May85] 28(2) 0	6	15 [Paul Rodgers] *A: The Firm; P: Jimmy Page/Paul Rodgers*	590
1701. **TULSA TIME//COCAINE [TSW]** Eric Clapton And His Band RSO 1039 [05Jul80	16Aug80	20Sep80] 30(2) 0	5	12 [Danny Flowers// J. J. Cale] *A: Just One Night; P: Jon Astley*	589
1702. **SHE'S SO COLD** The Rolling Stones Rolling Stones 21001 [27Sep80	08Nov80	20Dec80] 26(2) 0	5	13 [Mick Jagger/Keith Richards] *A: Emotional Rescue; P: Glimmer Twins*	588
1703. **CHIQUITITA** ABBA Atlantic 3629 [10Nov79	12Jan80	26Jan80] 29(1) 0	6	12 [Benny Andersson/Björn Ulvaeus] *A: Voulez-Vous; P: Benny Andersson/Björn Ulvaeus*	587
1704. **DON'T WALK AWAY** Rick Springfield RCA 13813 [26May84	07Jul84	11Aug84] 26(2) 0	6	12 [Rick Springfield] *A: Hard To Hold (Soundtrack); P: Bill Drescher/Rick Springfield*	587
1705. **TO LIVE AND DIE IN L.A.** Wang Chung Geffen 28891 [12Oct85	14Dec85	08Feb86] 41(2) 0	0	18 [Nick Feldman/Jack Hues] *A: To Live And Die In L.A. (Soundtrack); P: Steve Jolley/Tony Swain*	587

Rank. TITLE Artist Label Number [Entry Date	Peak Date	Exit Date] Peak(Peak Wks) Top10	Top40	Tot Wks [Writers] *Album; Producers*	Score
1706. **THE SIGN OF FIRE** The Fixx MCA 52316 [26Nov83	14Jan84	18Feb84] 32(2) 0	7	13 [Alfie Agius/Cy Curnin/Peter John Greenall/Jamie West-Oram/Adam Woods] *A: Reach The Beach; P: Rupert Hine*	586
1707. **STILL RIGHT HERE IN MY HEART** Pure Prairie League Casablanca 2332 [18Apr81	20Jun81	18Jul81] 28(2) 0	7	14 [George Daniel Greer/Jeff Wilson] *A: Something In The Night; P: Rob Fraboni*	586
1708. **TAKE ME WITH U** Prince And The Revolution Warner Bros. 29079 [09Feb85	23Mar85	27Apr85] 25(2) 0	6	12 [Prince] *A: Purple Rain (Soundtrack); P: Prince And The Revolution*	585
1709. **CAN'T HELP FALLING IN LOVE** Corey Hart EMI America 8368 [13Dec86	21Feb87	14Mar87] 24(1) 0	5	14 [Luigi Creatore/Hugo Peretti/George David Weiss] *A: Fields Of Fire; P: Phil Chapman/Corey Hart*	585
1710. **ATLANTA LADY (Something About Your Love)** Marty Balin EMI America 8093 [19Sep81	24Oct81	12Dec81] 27(3) 0	5	13 [Jesse Barish] *A: Balin; P: John Hug*	584
1711. **HOW DO YOU KEEP THE MUSIC PLAYING (Theme From "Best Friends")** James Ingram And Patti Austin Qwest 29618 [14May83	09Jul83	03Sep83] 45(1) 0	0	17 [Alan Bergman/Marilyn Bergman/Michel Legrand] *A: It's Your Night; P: Quincy Jones/Johnny Mandel*	584
1712. **I'LL STILL BE LOVING YOU** Restless Heart RCA 5065 [11Apr87	13Jun87	08Aug87] 33(2) 0	5	18 [Pat Bunch/Todd Cerney/Mary Ann Kennedy/Pam Rose] *A: Wheels; P: Tim DuBois/Scott Hendricks/Restless Heart*	583
1713. **PARTYMAN** Prince Warner Bros. 22814 [26Aug89	07Oct89	28Oct89] 18(1) 0	7	10 [Prince] *A: Batman (Soundtrack); P: Prince*	582
1714. **TIL MY BABY COMES HOME** Luther Vandross Epic 04760 [16Mar85	25May85	29Jun85] 29(1) 0	6	16 [Marcus Miller/Luther Vandross] *A: The Night I Fell In Love; P: Marcus Miller/Luther Vandross*	582
1715. **FIND A WAY** Amy Grant A&M 2734 [18May85	27Jul85	31Aug85] 29(1) 0	6	16 [Amy Grant/Michael W. Smith] *A: Unguarded; P: Brown Bannister*	581
1716. **WASN'T THAT A PARTY** The Rovers Epic 51007 [21Feb81	02May81	13Jun81] 37(2) 0	4	17 [Tom Paxton] *A: Wasn't That A Party; P: Jack Richardson*	579
1717. **LIES** Jonathan Butler Jive 1038 [27Jun87	29Aug87	26Sep87] 27(2) 0	5	14 [Jonathan Butler/Jolyon Skinner] *A: Jonathan Butler; P: Barry Eastmond*	579
1718. **SHOW ME** The Pretenders Sire 29317 [17Mar84	05May84	09Jun84] 28(1) 0	6	13 [Chrissie Hynde] *A: Learning To Crawl; P: Chris Thomas*	579
1719. **GIRLS CAN GET IT** Dr. Hook Casablanca 2314 [01Nov80	13Dec80	31Jan81] 34(2) 0	6	14 [Leslie Pearl] *A: Rising; P: Ron Haffkine*	579
1720. **I CAN'T DRIVE 55** Sammy Hagar Geffen 29173 [29Sep84	24Nov84	12Jan85] 26(1) 0	6	16 [Sammy Hagar] *A: VOA; P: Ted Templeman*	578
1721. **DO YOU LOVE WHAT YOU FEEL** Rufus and Chaka MCA 41131 [24Nov79	02Feb80	01Mar80] 30(2) 0	4	15 [David Wolinski] *A: Masterjam; P: Quincy Jones*	578
1722. **HEARTBREAK BEAT** Psychedelic Furs Columbia 06420 [14Mar87	23May87	13Jun87] 26(1) 0	5	14 [John Ashton/Richard Butler/Tim Butler] *A: Midnight To Midnight; P: Chris Kimsey*	577
1723. **THE SMILE HAS LEFT YOUR EYES** Asia Geffen 29475 [15Oct83	26Nov83	07Jan84] 34(2) 0	5	13 [John Wetton] *A: Alpha; P: Mike Stone*	577
1724. **THE BELLE OF ST. MARK** Sheila E. Warner Bros. 29180 [27Oct84	15Dec84	02Feb85] 34(4) 0	5	15 [Prince] *A: The Glamorous Life; P: Prince/Sheila E.*	576
1725. **SHINE ON** L.T.D. A&M 2283 [15Nov80	31Jan81	28Feb81] 40(1) 0	1	16 [Richard Kerr/Billy Osborne/Jeffrey Osborne] *A: Shine On; P: Bobby Martin*	576

Rank. TITLE Artist Label Number [Entry Date | Peak Date | Exit Date] Peak(Peak Wks) Top10|Top40|Tot Wks [Writers] *Album*; Producers	Score
1726. HOLIDAY The Other Ones Virgin 99428 [01Aug87|17Oct87|21Nov87] 29(1) 0|4|17 [Stephan Gottwald/Uwe Hoffmann/Alf Klimek/Jayney Klimek/Johnny Klimek/Andreas Schwarz-Ruszczynski] *A: The Other Ones*; P: Christopher Neil	575
1727. SET ME FREE Utopia Bearsville 49180 [23Feb80|19Apr80|10May80] 27(1) 0|5|12 [Roger Powell/Todd Rundgren/Kasim Sulton/John Griffitt Wilcox] *A: Adventures In Utopia*; P: Todd Rundgren/Utopia	574
1728. FIRE IN THE MORNING Melissa Manchester Arista 0485 [23Feb80|26Apr80|17May80] 32(2) 0|5|13 [Steve Dorff/Gary Harju/Larry Herbstritt] *A: Melissa Manchester*; P: Steve Buckingham	573
1729. WHO'LL BE THE FOOL TONIGHT Larsen-Feiten Band Warner Bros. 49282 [16Aug80|11Oct80|15Nov80] 29(2) 0|6|14 [Buzz Feiten] *A: Larsen-Feiten Band*; P: Tommy LiPuma	573
1730. WHEN HE SHINES Sheena Easton EMI America 8113 [03Apr82|05Jun82|10Jul82] 30(2) 0|6|15 [Dominic Bugatti/Florrie Palmer] *A: You Could Have Been With Me*; P: Christopher Neil	572
1731. WE WERE MEANT TO BE LOVERS Photoglo 20th Century 2446 [29Mar80|14Jun80|02Aug80] 31(1) 0|6|15 [Brian Neary/Jim Photoglo] *A: Photoglo*; P: Brian Neary	572
1732. SEX AS A WEAPON Pat Benatar Chrysalis 42927 [23Nov85|18Jan86|15Feb86] 28(1) 0|7|13 [Thomas F. Kelly/Billy Steinberg] *A: Seven The Hard Way*; P: Neil Geraldo	571
1733. WHY CAN'T I HAVE YOU The Cars Elektra 69657 [26Jan85|30Mar85|18May85] 33(1) 0|5|17 [Ric Ocasek] *A: Heartbeat City*; P: Cars/Mutt Lange	571
1734. IN YOUR EYES ('86) Peter Gabriel Geffen 28622 [30Aug86|25Oct86|29Nov86] 26(2) 0|7|14 [Peter Gabriel] *A: So*; P: Peter Gabriel/Daniel Lanois/Bill Laswell	570
1735. DON'T PAY THE FERRYMAN Chris DeBurgh A&M 2511 [30Apr83|02Jul83|30Jul83] 34(1) 0|4|14 [Chris de Burgh] *A: The Getaway*; P: Rupert Hine	569
1736. TIME OUT OF MIND Steely Dan MCA 51082 [14Mar81|25Apr81|23May81] 22(2) 0|7|11 [Walter Becker/Donald Fagen] *A: Gaucho*; P: Gary Katz	568
1737. SPRING LOVE (Come Back To Me) Stevie B LMR 74002 [09Jul88|13Aug88|05Nov88] 43(1) 0|0|18 [Stevie B] *A: Party Your Body*; P: Tolga Katas/Stevie B	566
1738. SINCE YOU'VE BEEN GONE The Outfield Columbia 07170 [13Jun87|15Aug87|19Sep87] 31(1) 0|5|15 [John Spinks] *A: Bangin'*; P: William Wittman	565
1739. STOP IN THE NAME OF LOVE The Hollies Atlantic 89819 [04Jun83|30Jul83|20Aug83] 29(1) 0|6|12 [Lamont Dozier/Brian Holland/Eddie Holland] *A: What Goes Around*; P: Paul Bliss/Hollies/Stanley Johnston/Graham Nash	565
1740. ALIEN Atlanta Rhythm Section Columbia 02471 [29Aug81|24Oct81|05Dec81] 29(2) 0|4|15 [Buddy Buie/Randy Lewis/Steve McRay] *A: Quinella*; P: Buddy Buie	564
1741. MORE THAN JUST THE TWO OF US Sneaker Handshake 02557 [31Oct81|23Jan82|06Feb82] 34(1) 0|6|15 [Mitch Crane/Michael Schneider] *A: Sneaker*; P: Jeff Baxter	564
1742. NEVER THOUGHT (That I Could Love) Dan Hill Columbia 07618 [12Dec87|20Feb88|23Apr88] 43(1) 0|0|20 [Dan Hill] *A: Dan Hill(2)*; P: John Capek/Hank Medress	563
1743. NOT ENOUGH LOVE IN THE WORLD Don Henley Geffen 29012 [25May85|27Jul85|14Sep85] 34(1) 0|5|17 [Don Henley/Danny Kortchmar/Benmont Tench] *A: Building The Perfect Beast*; P: Don Henley/Danny Kortchmar/Greg Ladanyi	563
1744. PRESS Paul McCartney Capitol 5597 [02Aug86|13Sep86|11Oct86] 21(2) 0|6|11 [Paul McCartney] *A: Press To Play*; P: Paul McCartney/Hugh Padgham	563
1745. SAYIN' SORRY (Don't Make It Right) Denise Lopez Vendetta 7200 [11Jun88|20Aug88|01Oct88] 31(1) 0|5|17 [David Bowler/Eric Li] *A: Truth In Disguise*; P: John Morales/Sergio Munzibai	563
1746. IN MY EYES Stevie B LMR 74004 [27May89|29Jul89|16Sep89] 37(1) 0|2|17 [Stevie B] *A: In My Eyes*; P: Stevie B	562
1747. THE DREAM (Hold On To Your Dream) Irene Cara Geffen/Network 29396 [10Dec83|11Feb84|10Mar84] 37(1) 0|3|14 [Pete Bellotte/Irene Cara/Giorgio Moroder] *A: What A Feelin'*; P: Giorgio Moroder	562
1748. THEME FROM NEW YORK, NEW YORK Frank Sinatra Reprise 49233 [03May80|14Jun80|19Jul80] 32(2) 0|6|12 [Fred Ebb/John Kander] *A: Trilogy: Past, Present And Future*; P: Sonny Burke	561
1749. POOR MAN'S SON Survivor Scotti Brothers 02560 [17Oct81|12Dec81|16Jan82] 33(1) 0|4|14 [Jim Peterik/Frankie Sullivan] *A: Premonition*; P: Jim Peterik/Frankie Sullivan	561
1750. GETCHA BACK The Beach Boys Caribou 04913 [25May85|29Jun85|10Aug85] 26(2) 0|7|12 [Mike Love/Terry Melcher] *A: The Beach Boys*; P: Steve Levine	560
1751. VOICES OF BABYLON The Outfield Columbia 68601 [25Mar89|27May89|24Jun89] 25(1) 0|6|14 [John Spinks] *A: Voices Of Babylon*; P: David Kahne/David Leonard/John Spinks	560
1752. YOU DROPPED A BOMB ON ME The Gap Band Total Experience 8203 [14Aug82|25Sep82|06Nov82] 31(5) 0|7|13 [Lonnie Simmons/Rudy Taylor/Charles K. Wilson] *A: Gap Band IV*; P: Lonnie Simmons	559
1753. FOREVER Kenny Loggins Columbia 04931 [25May85|20Jul85|19Oct85] 40(1) 0|1|22 [David Foster/E. Ein Loggins/Kenny Loggins] *A: Vox Humana*; P: David Foster/Kenny Loggins	559
1754. THAT'S LOVE Jim Capaldi Atlantic 89849 [30Apr83|18Jun83|23Jul83] 28(2) 0|5|13 [Jim Capaldi] *A: Fierce Heart*; P: Jim Capaldi/Steve Winwood	559
1755. EVERY STEP OF THE WAY John Waite EMI America 8282 [10Aug85|28Sep85|26Oct85] 25(2) 0|6|12 [Ivan Kral/Mark Sidgwick/John Waite] *A: Mask Of Smiles*; P: Stephan Galfas/John Waite	559
1756. PRIDE (In The Name Of Love) U2 Island 99704 [27Oct84|15Dec84|02Feb85] 33(1) 0|5|15 [Adam Clayton/Dave Evans/Paul Hewson/Larry Mullen] *A: The Unforgettable Fire*; P: Brian Eno/Daniel Lanois	559
1757. PROMISE ME The Cover Girls Fever 1917 [26Mar88|21May88|30Jul88] 40(1) 0|1|19 [Albert Cabrera/Tony Moran/Andy Tripoli] *A: Show Me*; P: Albert Cabrera/Tony Moran/Andy Tripoli	559
1758. FAKE Alexander O'Neal Tabu 4-07100 [25Jul87|26Sep87|31Oct87] 25(1) 0|6|15 [James Samuel Harris/Terry Lewis] *A: Hearsay*; P: Jimmy Jam Harris/Terry Lewis	559
1759. TENDER YEARS ('84a) John Cafferty And The Beaver Brown Band Scotti Brothers 04682 [17Nov84|12Jan85|16Feb85] 31(1) 0|7|14 [John Cafferty] *A: Eddie & The Cruisers (Soundtrack)*; P: Kenny Vance	559a
1760. DRAW OF THE CARDS Kim Carnes EMI America 8087 [08Aug81|19Sep81|24Oct81] 28(2) 0|6|12 [Kim Carnes/Bill Cuomo/Dave Ellingson/Val Garay] *A: Mistaken Identity*; P: Val Garay	558
1761. JUST GOT LUCKY JoBoxers RCA 13601 [10Sep83|19Nov83|17Dec83] 36(2) 0|4|15 [Chris Bostock/Dig Wayne] *A: Like Gangbusters*; P: Alan Shacklock	558
1762. HE'LL NEVER LOVE YOU (Like I Do) Freddie Jackson Capitol 5535 [14Dec85|22Feb86|22Mar86] 25(1) 0|6|15 [Keith Diamond/Barry Eastmond] *A: Rock Me Tonight*; P: Barry Eastmond	557
1763. MOONLIGHTING (Theme) Al Jarreau MCA 53124 [30May87|18Jul87|22Aug87] 23(1) 0|5|13 [Lee Holdridge/Al Jarreau] *A: Moonlighting Soundtrack*; P: Nile Rodgers	557
1764. TAKEN IN Mike + The Mechanics Atlantic 89404 [28Jun86|09Aug86|04Oct86] 32(4) 0|5|15 [Christopher Neil/Mike Rutherford] *A: Mike + The Mechanics*; P: Christopher Neil	557
1765. MY EVER CHANGING MOODS The Style Council Geffen 29359 [07Apr84|09Jun84|07Jul84] 29(1) 0|6|14 [Paul Weller] *A: My Ever Changing Moods*; P: Paul Weller/Peter Wilson	557
1766. THE ONE THING INXS Atco 99905 [26Mar83|28May83|25Jun83] 30(2) 0|5|14 [Andy Farriss/Michael Hutchence] *A: Shabooh Shoobah*; P: Mark Opitz	557

Ranking the Singles

Rank. TITLE Artist Label Number	
[Entry Date | Peak Date | Exit Date] Peak(Peak Wks) Top10 | Top40 | Tot Wks	
[Writers] *Album; Producers*	Score
1767. TWO LESS LONELY PEOPLE IN THE WORLD Air Supply Arista 1004	557
[13Nov82 | 08Jan83 | 12Feb83] 38(3) 0 | 5 | 14	
[Howard Greenfield/Ken Hirsch] *A: Now And Forever; P: Harry Maslin*	
1768. OVERJOYED Stevie Wonder Tamla 1832	556
[22Feb86 | 12Apr86 | 17May86] 24(2) 0 | 6 | 13	
[Stevie Wonder] *A: In Square Circle; P: Stevie Wonder*	
1769. COME BACK The J. Geils Band EMI America 8032	555
[02Feb80 | 22Mar80 | 19Apr80] 32(2) 0 | 5 | 12	
[Seth Justman/Peter Wolf] *A: Love Stinks; P: Seth Justman*	
1770. THERE MUST BE AN ANGEL (Playing With My Heart) Eurythmics RCA 14160	554
[03Aug85 | 21Sep85 | 12Oct85] 22(1) 0 | 7 | 11	
[Annie Lennox/David A. Stewart] *A: Be Yourself Tonight; P: David A. Stewart*	
1771. EVERYDAY I WRITE THE BOOK Elvis Costello And The Attractions Columbia 04045	554
[20Aug83 | 22Oct83 | 19Nov83] 36(1) 0 | 2 | 14	
[Elvis Costello] *A: Punch The Clock; P: Clive Langer/Alan Winstanley*	
1772. RAIN ON THE SCARECROW John Cougar Mellencamp Riva 884 635	553
[26Apr86 | 14Jun86 | 12Jul86] 21(1) 0 | 6 | 12	
[George Green/John Mellencamp] *A: Scarecrow; P: Don Gehman/John Mellencamp*	
1773. FOREVER MAN Eric Clapton Duck/Warner 29081	553
[09Mar85 | 27Apr85 | 25May85] 26(1) 0 | 6 | 12	
[Eric Clapton/Jerry Lynn Williams] *A: Behind The Sun; P: Ted Templeman/Lenny Waronker*	
1774. LET ME BE THE CLOCK Smokey Robinson Tamla 54311	553
[15Mar80 | 17May80 | 14Jun80] 31(2) 0 | 4 | 14	
[Smokey Robinson] *A: Warm Thoughts; P: Smokey Robinson*	
1775. UNDER THE MILKY WAY The Church Arista 1-9673	552
[09Apr88 | 18Jun88 | 16Jul88] 24(1) 0 | 5 | 15	
[Karin Jansson/Steve Kilbey] *A: Starfish; P: Church/Greg Ladanyi/Waddy Wachtel*	
1776. THE BORDER America Capitol 5236	552
[18Jun83 | 06Aug83 | 03Sep83] 33(3) 0 | 6 | 12	
[Russ Ballard/Dewey Bunnell] *A: Your Move; P: Russ Ballard*	
1777. GEE WHIZ Bernadette Peters MCA 41210	552
[29Mar80 | 31May80 | 21Jun80] 31(2) 0 | 5 | 13	
[Carla Thomas] *A: Bernadette Peters; P: Brooks Arthur*	
1778. THE WAY HE MAKES ME FEEL Barbra Streisand Columbia 04177	551
[22Oct83 | 10Dec83 | 28Jan84] 40(2) 0 | 2 | 15	
[Alan Bergman/Marilyn Bergman/Michel Legrand] *A: Yentl (Soundtrack); P: Dave Grusin/Phil Ramone*	
1779. SHOW ME The Cover Girls Fever 1911	551
[28Feb87 | 09May87 | 27Jun87] 44(1) 0 | 0 | 18	
[Albert Cabrera/Bob Khozouri/Tony Moran/Andy Tripoli] *A: Show Me; P: Albert Cabrera/Tony Moran/Andy Tripoli*	
1780. WHITE WEDDING Billy Idol Chrysalis 2648	551
[21May83 | 02Jul83 | 13Aug83] 36(2) 0 | 3 | 13	
[Billy Idol] *A: Billy Idol; P: Keith Forsey*	
1781. STAYING TOGETHER Debbie Gibson Atlantic 89034	551
[06Aug88 | 24Sep88 | 22Oct88] 22(1) 0 | 6 | 12	
[Debbie Gibson] *A: Out Of The Blue; P: Debbie Gibson/Fred Zarr*	
1782. REMEMBER THE NIGHTS The Motels Capitol 5246	551
[03Dec83 | 28Jan84 | 18Feb84] 36(1) 0 | 3 | 12	
[Martha Davis/Scott Thurston] *A: Little Robbers; P: Val Garay*	
1783. FISHNET Morris Day Warner Bros. 28201	550
[20Feb88 | 23Apr88 | 14May88] 23(1) 0 | 6 | 13	
[Morris Day/James Samuel Harris/Terry Lewis] *A: Daydreaming; P: Jimmy Jam Harris/Terry Lewis*	
1784. DO IT FOR LOVE Sheena Easton EMI America 8295	550
[26Oct85 | 14Dec85 | 25Jan86] 29(1) 0 | 4 | 14	
[Adele Bertei/Mary Alice Kessler] *A: Do You; P: Nile Rodgers*	
1785. ALL TIME HIGH Rita Coolidge A&M 2551	549
[02Jul83 | 27Aug83 | 24Sep83] 36(1) 0 | 4 | 13	
[John Barry/Tim Rice] *A: Octopussy Soundtrack; P: John Barry*	
1786. WHEN THE LIGHTS GO OUT Naked Eyes EMI-America 8183	549
[22Oct83 | 24Dec83 | 21Jan84] 37(2) 0 | 3 | 14	
[Pete Byrne/Rob Fisher] *A: Naked Eyes; P: Tony Mansfield*	
1787. BRUCE Rick Springfield Mercury 880405	549
[17Nov84 | 12Jan85 | 09Feb85] 27(1) 0 | 6 | 13	
[Rick Springfield] *A: Beautiful Feelings; P: Tom Perry/Joey D. Vieira*	
1788. MONEY CHANGES EVERYTHING Cyndi Lauper Portrait 37-04737	549
[22Dec84 | 09Feb85 | 16Mar85] 27(1) 0 | 6 | 13	
[Thomas Gray] *A: She's So Unusual; P: Cyndi Lauper/Lennie Petze*	
1789. FACTS OF LOVE Jeff Lorber Featuring Karyn White Warner Bros. 28588	548
[06Dec86 | 21Feb87 | 21Mar87] 27(1) 0 | 5 | 16	
[Evan Rogers/Carl Sturken] *A: Private Passion; P: Jeff Lorber/Evan Rogers/Carl Sturken*	
1790. IF SHE KNEW WHAT SHE WANTS Bangles Columbia 05886	548
[10May86 | 12Jul86 | 09Aug86] 29(1) 0 | 5 | 14	
[Jules Shear] *A: Different Light; P: David Kahne*	
1791. NIGHTBIRD Stevie Nicks (with Sandy Stewart) Modern 99799	548
[17Dec83 | 28Jan84 | 03Mar84] 33(1) 0 | 4 | 12	
[Stevie Nicks/Sandy Stewart] *A: The Wild Heart; P: Jimmy Iovine*	
1792. HOLD ME 'TIL THE MORNIN' COMES Paul Anka Columbia 03897	548
[18Jun83 | 03Sep83 | 01Oct83] 40(2) 0 | 2 | 16	
[Paul Anka/David Foster] *A: Walk A Fine Line; P: Denny Diante*	
1793. GUITAR MAN (Remix) Elvis Presley RCA 12158	547
[24Jan81 | 21Mar81 | 25Apr81] 28(2) 0 | 5 | 14	
[Jerry Reed] *A: Guitar Man; P: Felton Jarvis*	
1794. BACK IN BLACK AC/DC Atlantic 3787	545
[20Dec80 | 21Feb81 | 28Mar81] 37(2) 0 | 5 | 15	
[Brian Johnson/Angus Young/Malcolm Young] *A: Back In Black; P: Mutt Lange*	
1795. CRYIN' Vixen EMI 50167	544
[28Jan89 | 25Mar89 | 22Apr89] 22(1) 0 | 6 | 13	
[Jeff Paris/Gregg Tripp] *A: Vixen; P: David Cole/Rick Neigher*	
1796. THE LEGEND OF WOOLEY SWAMP Charlie Daniels Band Epic 50921	544
[16Aug80 | 18Oct80 | 15Nov80] 31(1) 0 | 4 | 14	
[Tom Crain/Charlie Daniels/Joel DiGregorio/Fred Edwards/Charles Hayward/James W. Marshall] *A: Full Moon; P: John Boylan*	
1797. HEARTS ON FIRE Bryan Adams A&M 2948	544
[13Jun87 | 08Aug87 | 05Sep87] 26(1) 0 | 6 | 13	
[Bryan Adams/Jim Vallance] *A: Into The Fire; P: Bryan Adams/Bob Clearmountain*	
1798. COULD I HAVE THIS DANCE Anne Murray Capitol 4920	544
[06Sep80 | 01Nov80 | 06Dec80] 33(2) 0 | 4 | 14	
[Wayland Holyfield/Bob House] *A: Urban Cowboy Soundtrack; P: Jim Ed Norman*	
1799. BOY IN THE BOX Corey Hart EMI America 8287	544
[14Sep85 | 02Nov85 | 30Nov85] 26(1) 0 | 6 | 12	
[Corey Hart] *A: Boy In The Box; P: Jon Astley/Phil Chapman/Corey Hart*	
1800. THE SALT IN MY TEARS Martin Briley Mercury 812165	543
[21May83 | 30Jul83 | 27Aug83] 36(1) 0 | 3 | 15	
[Martin Briley] *A: One Night With A Stranger; P: Peter Coleman*	
1801. WHAT ABOUT LOVE 'Til Tuesday Epic 06289	543
[20Sep86 | 22Nov86 | 20Dec86] 26(1) 0 | 5 | 14	
[Aimee Mann] *A: Welcome Home; P: Rhett Davies*	
1802. TWILIGHT ZONE/TWILIGHT TONE The Manhattan Transfer Atlantic 3649	543
[19Apr80 | 14Jun80 | 05Jul80] 30(2) 0 | 4 | 12	
[Bernard Hermann//Jay Graydon,/Alan Paul] *A: Extensions; P: Jay Graydon*	
1803. MOUNTAINS Prince And The Revolution Paisley Park 7-28711	543
[24May86 | 05Jul86 | 02Aug86] 23(1) 0 | 6 | 11	
[Lisa Coleman/Wendy Melvoin/Prince] *A: Parade: Music From The Motion Picture Under The Cherry Moon (Soundtrack); P: Prince And The Revolution*	
1804. SHANGHAI BREEZES John Denver RCA 13071	541
[06Mar82 | 22May82 | 05Jun82] 31(1) 0 | 5 | 14	
[John Denver] *A: Seasons Of The Heart; P: John Denver/Barney Wyckoff*	
1805. I DON'T MIND AT ALL Bourgeois Tagg Island 99409	541
[10Oct87 | 05Dec87 | 30Jan88] 38(1) 0 | 2 | 17	
[Brent Bourgeois/Lyle Workman] *A: YoYo; P: Todd Rundgren*	
1806. MY KIND OF LADY Supertramp A&M 2517	541
[29Jan83 | 19Mar83 | 16Apr83] 31(1) 0 | 5 | 12	
[Rick Davies/Roger Hodgson] *A: "...Famous Last Words..."; P: Peter Henderson/Supertramp*	
1807. GOING TO A GO-GO The Rolling Stones Rolling Stones 21301	541
[12Jun82 | 17Jul82 | 21Aug82] 25(3) 0 | 5 | 11	
[Warren Moore/Smokey Robinson/Robert Rogers/Marvin Tarplin] *A: Still Life (American Concert 1981); P: Glimmer Twins*	
1808. I DO WHAT I DO (Theme For 9 1/2 Weeks) John Taylor Capitol 5551	540
[08Mar87 | 26Apr87 | 23May87] 23(1) 0 | 6 | 12	
[Michael Des Barres/Jonathan Elias/Nigel John Taylor] *A: 9 1/2 Weeks Soundtrack; P: Jason Corsaro/Jonathan Elias/John Taylor*	

Rank. TITLE Artist Label Number	
[Entry Date	Peak Date	Exit Date] Peak(Peak Wks) Top10	Top40	Tot Wks	
[Writers] A: Album; P: Producers	Score
1809. WALKS LIKE A LADY Journey Columbia 11275	540
[24May80	19Jul80	16Aug80] 32(2) 0	4	13	
[Steve Perry] A: Departure; P: Kevin Elson/Geoffrey Workman	
1810. EVEN IT UP Heart Epic 50847	540
[09Feb80	29Mar80	26Apr80] 33(2) 0	4	12	
[Sue Ennis/Ann Wilson/Nancy Wilson]	
A: Bebe Le Strange; P: Mike Flicker/Howard Leese/Ann Wilson/Nancy Wilson	
1811. LOST HER IN THE SUN John Stewart RSO 1016	539
[08Dec79	09Feb80	01Mar80] 34(1) 0	4	13	
[John Stewart] A: Bombs Away Dream Babies; P: John Stewart	
1812. TOMORROW DOESN'T MATTER TONIGHT Starship Grunt 14332	539
[05Apr86	24May86	28Jun86] 26(1) 0	6	13	
[Steven Cristol/Robin Randall]	
A: Knee Deep In The Hoopla; P: Jeremy Smith/Peter Wolf	
1813. HEAVY METAL (Takin' A Ride) Don Felder Full Moon 47175	539
[25Jul81	17Oct81	14Nov81] 43(1) 0	0	17	
[Don Felder] A: Heavy Metal Soundtrack; P: Don Felder	
1814. MIRACLES Stacy Lattisaw Cotillion 7-99855	538
[13Aug83	22Oct83	26Nov83] 40(1) 0	1	16	
[Gary Benson/Frank Wildhorn] A: Sixteen; P: Narada Michael Walden	
1815. MISTAKE NO. 3 Culture Club Virgin/Epic 04727	537
[15Dec84	02Feb85	09Mar85] 33(2) 0	5	13	
[Michael Craig/Roy Hay/Jon Moss/George O'Dowd]	
A: Waking Up With The House On Fire; P: Steve Levine	
1816. IT MUST BE LOVE Madness Geffen 29562	537
[20Aug83	08Oct83	05Nov83] 33(2) 0	5	12	
[Labi Siffre] A: Madness; P: Clive Langer/Alan Winstanley	
1817. SLIPPING AWAY Dave Edmunds Columbia 03877	536
[14May83	30Jul83	20Aug83] 39(1) 0	1	15	
[Jeff Lynne] A: Information; P: Jeff Lynne	
1818. (Ghost) RIDERS IN THE SKY The Outlaws Arista 0582	536
[27Dec80	07Mar81	04Apr81] 31(1) 0	4	15	
[Stan Jones] A: Ghost Riders; P: Gary Lyons	
1819. CAUGHT UP IN THE RAPTURE Anita Baker Elektra 69511	536
[29Nov86	14Feb87	28Mar87] 37(1) 0	2	18	
[Garry Glenn/Diane Quander] A: Rapture; P: Michael J. Powell	
1820. MY GIRL Donnie Iris MCA/Carousel 52031	536
[27Mar82	29May82	26Jun82] 25(2) 0	6	14	
[Mark Avsec/Donnie Iris] A: King Cool; P: Mark Avsec	
1821. DON'T GIVE IT UP Robbie Patton Liberty 1420	536
[11Jul81	22Aug81	03Oct81] 26(2) 0	6	13	
[David Adelstein/Buzz Cason/Mac Gayden/Robbie Patton]	
A: Distant Shores; P: Ken Caillat/Christine McVie/Robbie Patton	
1822. TELL HER NO Juice Newton Capitol 5265	534
[13Aug83	01Oct83	22Oct83] 27(1) 0	5	11	
[Rod Argent] A: Dirty Looks; P: Richard Landis	
1823. CASTLES IN THE AIR Don McLean Millennium 11819	534
[31Oct81	26Dec81	30Jan82] 36(3) 0	5	14	
[Don McLean] A: Believers; P: Larry Butler	
1824. JUMPIN' JACK FLASH Aretha Franklin Arista 1-9528	534
[27Sep86	08Nov86	06Dec86] 21(1) 0	6	11	
[Mick Jagger/Keith Richards] A: Aretha (II); P: Keith Richards	
1825. KING OF THE HILL Rick Pinette And Oak Mercury 76049	534
[10May80	19Jul80	09Aug80] 36(2) 0	3	14	
[Rick Pinette] A: Oak; P: Holden Raphael	
1826. JUMP TO IT Aretha Franklin Arista 0699	533
[21Aug82	09Oct82	06Nov82] 24(2) 0	6	12	
[Marcus Miller/Luther Vandross] A: Jump To It; P: Luther Vandross	
1827. OBSCENE PHONE CALLER Rockwell Motown 1731	533
[05May84	30Jun84	04Aug84] 35(1) 0	2	14	
[Kennedy Gordy]	
A: Somebody's Watching Me; P: Kennedy Gordy/Curtis Anthony Nolen	
1828. YEARS Wayne Newton Aries II 108	533
[02Feb80	05Apr80	26Apr80] 35(1) 0	3	13	
[Kye Fleming/Dennis W. Morgan] A: No Album; P: Tom Collins	
1829. SOMEBODY'S OUT THERE Triumph MCA 52898	532
[30Aug86	08Nov86	06Dec86] 27(1) 0	5	15	
[Rik Emmett/Mike Levine/Gil Moore] A: The Sport Of Kings; P: Mike Clink	
1830. TURN TO YOU Go-Go's I.R.S. 9928	532
[16Jun84	04Aug84	15Sep84] 32(2) 0	5	14	
[Charlotte Caffey/Jane Wiedlin] A: Talk Show; P: Martin Rushent	

Rank. TITLE Artist Label Number	
[Entry Date	Peak Date	Exit Date] Peak(Peak Wks) Top10	Top40	Tot Wks	
[Writers] A: Album; P: Producers	Score
1831. ALL SHE WANTS IS Duran Duran Capitol 44287	532
[24Dec88	18Feb89	18Mar89] 22(1) 0	6	13	
[Simon Le Bon/Nick Rhodes/Nigel John Taylor]	
A: Big Thing; P: Daniel Abraham/Duran Duran/Jonathan Elias	
1832. WAKE UP LITTLE SUSIE Simon and Garfunkel Warner Bros. 50053	532
[03Apr82	29May82	12Jun82] 27(2) 0	6	11	
[Boudleaux Bryant/Felice Bryant]	
A: The Concert In Central Park; P: Art Garfunkel/Paul Simon	
1833. PAPERLATE Genesis Atlantic 4053	530
[05Jun82	07Aug82	04Sep82] 32(3) 0	5	14	
[Tony Banks/Phil Collins/Mike Rutherford] A: Three Sides Live; P: Genesis	
1834. STRONGER THAN BEFORE Carole Bayer Sager Boardwalk WS8 02054	530
[16May81	11Jul81	08Aug81] 30(2) 0	7	13	
[Burt Bacharach/Bruce Roberts/Carole Bayer Sager]	
A: Sometimes Late At Night; P: Brooks Arthur/Burt Bacharach	
1835. DOES IT MAKE YOU REMEMBER Kim Carnes EMI America 8147	530
[06Nov82	08Jan83	29Jan83] 36(2) 0	4	13	
[Kim Carnes/Dave Ellingson] A: Voyeur; P: Val Garay	
1836. LADY The Whispers Solar 11928	530
[19Apr80	07Jun80	28Jun80] 28(2) 0	4	11	
[Nicholas Caldwell] A: The Whispers; P: Dick Griffey/Whispers	
1837. SO WRONG Patrick Simmons Elektra 69839	529
[19Mar83	07May83	11Jun83] 30(1) 0	5	13	
[Patrick Simmons/Chris Thompson] A: Arcade; P: John Ryan	
1838. NEVER KNEW LOVE LIKE THIS Alexander O'Neal featuring Cherrelle	529
Tabu 4-07646	
[23Jan88	02Apr88	23Apr88] 28(1) 0	6	14	
[James Samuel Harris/Terry Lewis]	
A: Hearsay; P: Jimmy Jam Harris/Terry Lewis	
1839. FLESH FOR FANTASY Billy Idol Chrysalis 42809	528
[25Aug84	06Oct84	10Nov84] 29(2) 0	6	12	
[Billy Idol/Steve Stevens] A: Rebel Yell; P: Keith Forsey	
1840. ME MYSELF AND I De La Soul Tommy Boy 7926	528
[03Jun89	22Jul89	23Sep89] 34(1) 0	3	17	
[Edwin Birdsong/George Clinton/Paul Huston/David Jolicoeur/	
Vincent Mason/Kelvin Mercer/Abrim Tilmon/Philippe Wynn]	
A: 3 Feet High And Rising; P: Paul Huston	
1841. MAKE BELIEVE Toto Columbia 03143	528
[07Aug82	25Sep82	30Oct82] 30(2) 0	5	13	
[David Paich] A: Toto IV; P: Toto	
1842. I'M ALMOST READY Pure Prairie League Casablanca 2294	527
[23Aug80	18Oct80	15Nov80] 34(2) 0	4	13	
[Vince Gill] A: Firin' Up; P: John Ryan	
1843. THIS MAN IS MINE Heart Epic 02925	527
[15May82	03Jul82	07Aug82] 33(2) 0	4	13	
[Sue Ennis/Ann Wilson/Nancy Wilson]	
A: Private Audition; P: Howard Leese/Ann Wilson/Nancy Wilson	
1844. ANYONE CAN SEE Irene Cara Network 47950	527
[28Nov81	20Mar82	03Apr82] 42(2) 0	0	18	
[Irene Cara/Bruce Roberts] A: Anyone Can See; P: Ron Dante	
1845. I'M HAPPY THAT LOVE HAS FOUND YOU Jimmy Hall Epic 50931	526
[27Sep80	22Nov80	17Jan81] 27(1) 0	4	17	
[Ellison Chase/William Haberman/Arthur Jacobson]	
A: Touch You; P: Norbert Putnam	
1846. YOU KNOW THAT I LOVE YOU Santana Columbia 11144	525
[24Nov79	02Feb80	16Feb80] 35(1) 0	3	13	
[Alex Ligertwood/Alan Pasqua/Carlos Santana/Chris Solberg]	
A: Marathon; P: Keith Olsen	
1847. FIND YOUR WAY BACK Jefferson Starship Grunt 12211	523
[04Apr81	23May81	27Jun81] 29(2) 0	6	13	
[Thomas Borsdorf/Craig Chaquico] A: Modern Times; P: Ron Nevison	
1848. I WON'T STAND IN YOUR WAY Stray Cats EMI America 8185	522
[29Oct83	17Dec83	21Jan84] 35(1) 0	3	13	
[Brian Setzer] A: Rant 'N' Rave With The Stray Cats; P: Dave Edmunds	
1849. I DO'WANNA KNOW REO Speedwagon Epic 04659	521
[27Oct84	08Dec84	19Jan85] 29(1) 0	5	13	
[Kevin Cronin]	
A: Wheels Are Turnin'; P: Kevin Cronin/Alan Gratzer/Gary Richrath	
1850. 10-9-8 Face To Face Epic 04430	521
[02Jun84	21Jul84	08Sep84] 38(3) 0	3	15	
[Angelo Petraglia] A: Face To Face; P: Arthur Baker	

Ranking the Singles

Rank. TITLE Artist Label Number
[Entry Date | Peak Date | Exit Date] Peak(Peak Wks) Top10 | Top40 | Tot Wks
[Writers] *Album; Producers* Score

1851. SHAKE FOR THE SHEIK The Escape Club Atlantic 88983 520
[10Dec88 | 28Jan89 | 11Mar89] 28(3) 0 | 5 | 14
[Johnnie Christo/John Holliday/Trevor Steel/Milan Zekavica]
A: Wild, Wild West; P: Chris Kimsey

1852. TEACHER, TEACHER 38 Special Capitol 5405 520
[29Sep84 | 24Nov84 | 15Dec84] 25(1) 0 | 5 | 12
[Bryan Adams/Jim Vallance]
A: Teachers Soundtrack; P: Rodney Mills/38 Special

1853. SOUL CITY Partland Brothers Manhattan 50065 519
[02May87 | 27Jun87 | 25Jul87] 27(1) 0 | 5 | 13
[Chris Partland/George Partland] *A: Electric Honey; P: Vini Poncia*

1854. SHIP OF FOOLS (Save Me From Tomorrow) World Party Chrysalis 43052 519
[14Feb87 | 25Apr87 | 23May87] 27(1) 0 | 5 | 15
[Karl Wallinger] *A: Private Revolution; P: Karl Wallinger*

1855. MIRROR MAN The Human League A&M 2587 518
[01Oct83 | 19Nov83 | 17Dec83] 30(1) 0 | 5 | 12
[Ian Burden/John William Callis/Philip Oakey]
A: Fascination!; P: Human League/Martin Rushent

1856. RIGHT BY YOUR SIDE Eurythmics RCA 13695 518
[21Jul84 | 08Sep84 | 06Oct84] 29(1) 0 | 5 | 12
[Annie Lennox/David A. Stewart] *A: Touch; P: David A. Stewart*

1857. TRAGEDY John Hunter Private I 4-04643 518
[08Dec84 | 16Feb85 | 23Mar85] 39(2) 0 | 2 | 16
[John Martin Hunter Jr.]
A: Famous At Night; P: Phil Bonanno/John Hunter

1858. YOU'RE SUPPOSED TO KEEP YOUR LOVE FOR ME Jermaine Jackson Motown 1490 518
[12Jul80 | 20Sep80 | 04Oct80] 34(1) 0 | 4 | 13
[Stevie Wonder] *A: Let's Get Serious; P: Stevie Wonder*

1859. SITTING AT THE WHEEL The Moody Blues Threshold 604 518
[03Sep83 | 15Oct83 | 05Nov83] 27(1) 0 | 6 | 10
[John Lodge] *A: The Present; P: Pip Williams*

1860. ROLL ME AWAY Bob Seger And The Silver Bullet Band Capitol 5235 516
[28May83 | 02Jul83 | 30Jul83] 27(2) 0 | 6 | 10
[Bob Seger] *A: The Distance; P: Jimmy Iovine*

1861. THE KID'S AMERICAN Matthew Wilder Private I 4-04363 516
[18Feb84 | 07Apr84 | 12May84] 33(2) 0 | 4 | 13
[Matthew Wilder]
A: I Don't Speak The Language; P: Peter Bunetta/Rick Chudacoff/Bill Elliott

1862. MOTORTOWN The Kane Gang Capitol 44062 515
[17Oct87 | 19Dec87 | 30Jan88] 36(1) 0 | 3 | 16
[Martin Brammer/David Brewis/Paul Woods]
A: Miracle; P: Kane Gang/Pete Wingfield

1863. THANKS FOR MY CHILD Cheryl "Pepsii" Riley Columbia 07996 514
[29Oct88 | 24Dec88 | 21Jan89] 32(2) 0 | 5 | 13
[Curtis Bedeau/Gerard Charles/Hugh Clarke/Brian George/
Lucien George/Paul George] *A: Me, Myself And I; P: Full Force*

1864. WHAT IS LOVE? Howard Jones Elektra 514
[21Apr84 | 09Jun84 | 14Jul84] 33(3) 0 | 4 | 13
[William Bryant/Howard Jones] *A: Human's Lib; P: Rupert Hine*

1865. HOLDING OUT FOR A HERO Bonnie Tyler Columbia 04370 512
[25Feb84 | 14Apr84 | 19May84] 34(3) 0 | 4 | 13
[Dean Pitchford/Jim Steinman] *A: Footloose Soundtrack; P: Jim Steinman*

1866. WHEN YOUR HEART IS WEAK Cock Robin Columbia 04875 511
[15Jun85 | 31Aug85 | 28Sep85] 35(1) 0 | 3 | 16
[Peter Kingsbery] *A: Cock Robin; P: Steve Hillage*

1867. ROCKIN' AT MIDNIGHT The Honeydrippers Es Paranza 7-99686 511
[05Jan85 | 23Feb85 | 16Mar85] 25(1) 0 | 6 | 11
[Roy Brown] *A: Volume One; P: Phil Carson/Ahmet Ertegun/Robert Plant*

1868. I FOUND SOMEBODY Glenn Frey Asylum 47466 510
[05Jun82 | 07Aug82 | 28Aug82] 31(2) 0 | 5 | 13
[Glenn Frey/Jack Tempchin]
A: No Fun Aloud; P: Allan Blazek/Glenn Frey/Jim Ed Norman

1869. CUDDLY TOY (Feel For Me) Roachford Epic 68549 510
[15Apr89 | 10Jun89 | 15Jul89] 25(1) 0 | 5 | 14
[Andrew Roachford] *A: Roachford; P: Michael H. Brauer*

1870. NEVER BEEN IN LOVE Randy Meisner Epic 03032 509
[31Jul82 | 18Sep82 | 09Oct82] 28(3) 0 | 6 | 11
[Craig Bickhardt] *A: Randy Meisner; P: Mike Flicker*

1871. IT TAKES TWO Rob Base & D.J. E-Z Rock Profile 5186 509
[20Aug88 | 29Oct88 | 03Dec88] 36(1) 0 | 5 | 16
[James Brown/Robert Ginyard] *A: It Takes Two; P: Rob Base/William Hamilton*

1872. RUMBLESEAT John Cougar Mellencamp Riva 884 856 508
[28Jun86 | 16Aug86 | 20Sep86] 28(1) 0 | 4 | 13
[John Mellencamp] *A: Scarecrow; P: Don Gehman/John Mellencamp*

1873. YEAH, YEAH, YEAH Judson Spence Atlantic 88999 507
[15Oct88 | 10Dec88 | 14Jan89] 32(1) 0 | 4 | 14
[Monroe Jones/Judson Spence]
A: Judson Spence; P: Monroe Jones/Judson Spence/David Tickle

1874. 17 Rick James Gordy 1730 507
[14Jul84 | 18Aug84 | 13Oct84] 36(3) 0 | 3 | 14
[Rick James] *A: Reflections; P: Rick James*

1875. ALL THE THINGS SHE SAID Simple Minds A&M 2828 506
[05Apr86 | 31May86 | 28Jun86] 28(1) 0 | 6 | 13
[Charles Burchill/Jim Kerr/Michael MacNeil]
A: Once Upon A Time; P: Bob Clearmountain/Jimmy Iovine

1876. A LOVER'S HOLIDAY Change Warner Bros./RFC 49208 506
[17May80 | 19Jul80 | 09Aug80] 40(1) 0 | 1 | 13
[Davide Romani/Tanyayette Willoughby]
A: The Glow Of Love; P: Jacques Fred Petrus

1877. IN THE MOOD Robert Plant Atlantic 99820 506
[19Nov83 | 14Jan84 | 04Feb84] 39(2) 0 | 5 | 12
[Robbie Blunt/Paul Martinez/Robert Plant]
A: The Principle Of Moments; P: Benji Lefevre/Robert Plant

1878. JUKE BOX HERO Foreigner Atlantic 4017 505
[13Feb82 | 03Apr82 | 08May82] 26(2) 0 | 6 | 13
[Lou Gramm/Michael Leslie Jones] *A: 4; P: Mick Jones/Mutt Lange*

1879. WHAT I LIKE ABOUT YOU Michael Morales Wing 889 678 505
[12Aug89 | 14Oct89 | 04Nov89] 28(1) 0 | 6 | 13
[Jimmy Marinos/Wally Palmar/Mike Skill]
A: Michael Morales; P: Roy Thomas Baker

1880. HOLD ME Teddy Pendergrass Asylum 69720 505
[09Jun84 | 28Jul84 | 06Oct84] 46(2) 0 | 0 | 18
[Linda Creed/Michael Masser] *A: Love Language; P: Michael Masser*

1881. HOW DOES IT FEEL TO BE BACK Daryl Hall & John Oates RCA 12048 504
[19Jul80 | 13Sep80 | 11Oct80] 30(1) 0 | 4 | 13
[John Oates] *A: Voices; P: Daryl Hall/John Oates*

1882. ONLY WHEN YOU LEAVE Spandau Ballet Chrysalis 42792 503
[28Jul84 | 15Sep84 | 13Oct84] 34(2) 0 | 4 | 12
[Gary Kemp] *A: Parade; P: Steve Jolley/Spandau Ballet/Tony Swain*

1883. I SHOULD BE SO LUCKY Kylie Minogue Geffen 27922 503
[07May88 | 16Jul88 | 06Aug88] 28(1) 0 | 4 | 14
[Matt Aitken/Mike Stock/Pete Waterman]
A: Kylie; P: Matt Aitken/Mike Stock/Pete Waterman

1884. BACK ON MY FEET AGAIN The Babys Chrysalis 2398 502
[19Jan80 | 15Mar80 | 05Apr80] 33(1) 0 | 3 | 12
[Dominic Bugatti/Frank Musker/John Waite] *A: Union Jacks; P: Keith Olsen*

1885. HANGING ON A HEART ATTACK Device Chrysalis 42996 502
[14Jun86 | 09Aug86 | 13Sep86] 35(3) 0 | 4 | 13
[Mike Chapman/Holly Knight] *A: 22B3; P: Mike Chapman*

1886. EVERLASTING LOVE Rex Smith/Rachel Sweet Columbia 02169 501
[27Jun81 | 22Aug81 | 19Sep81] 32(2) 0 | 4 | 13
[Buzz Cason/Mac Gayden]
*A: Everlasting Love/Rex Smith; And Then He Kissed Me/Rachel Sweet;
P: Rick Chertoff*

1887. BOYS DO FALL IN LOVE Robin Gibb Mirage 99743 500
[02Jun84 | 21Jul84 | 18Aug84] 37(1) 0 | 4 | 12
[Maurice Gibb/Robin Gibb] *A: Secret Agent; P: Maurice Gibb/Robin Gibb*

1888. I GOT YOU BABE UB40 With Chrissie Hynde A&M 2758 500
[27Jul85 | 21Sep85 | 26Oct85] 28(1) 0 | 4 | 14
[Sonny Bono] *A: Little Baggariddim; P: Ray Falconer/UB40*

1889. YOU'RE MY GIRL Franke And The Knockouts Millennium 11808 500
[04Jul81 | 22Aug81 | 26Sep81] 27(2) 0 | 5 | 13
[Billy Elworthy/Benny Harrison/Franke Previte]
A: Franke & The Knockouts; P: Steve Verroca

1890. GHOST TOWN Cheap Trick Epic 08089 499
[05Nov88 | 24Dec88 | 04Feb89] 33(3) 0 | 5 | 14
[Rick Nielsen/Diane Warren] *A: Lap Of Luxury; P: Richie Zito*

1891. WHEN WE WAS FAB George Harrison Dark Horse 28131 499
[06Feb88 | 26Mar88 | 16Apr88] 23(1) 0 | 6 | 11
[George Harrison/Jeff Lynne]
A: Cloud Nine; P: George Harrison/Jeff Lynne

1892. MOTHERS TALK Tears For Fears Mercury 884638 499
[12Apr86 | 24May86 | 28Jun86] 27(2) 0 | 4 | 13
[Roland Orzabal/Ian Stanley] *A: Songs From The Big Chair; P: Chris Hughes*

Ranking the Singles

Rank. TITLE Artist Label Number [Entry Date | Peak Date | Exit Date] Peak(Peak Wks) Top10|Top40|Tot Wks [Writers] *Album;* Producers	Score
1893. **VOYEUR** Kim Carnes EMI America 8127 [21Aug82|25Sep82|06Nov82] 29(4) 0|6|12 [Kim Carnes/Dave Ellingson/Duane Hitchings] *A: Voyeur;* P: Val Garay	498
1894. **IN NEON** Elton John Geffen 29111 [01Dec84|12Jan85|23Feb85] 38(3) 0|3|13 [Elton John/Bernie Taupin] *A: Breaking Hearts;* P: Chris Thomas	498
1895. **HOW MANY TIMES CAN WE SAY GOODBYE** Dionne Warwick And Luther Vandross Arista 1-9073 [08Oct83|12Nov83|31Dec83] 27(2) 0|5|13 [Steve Goldman] *A: How Many Times Can We Say Goodbye;* P: Luther Vandross	498
1896. **SYMPTOMS OF TRUE LOVE** Tracie Spencer Capitol 44198 [01Oct88|03Dec88|14Jan89] 38(1) 0|3|16 [Irmgard Klarmann/Felix Weber] *A: Tracie Spencer;* P: Ron Kersey	497
1897. **NICE GIRLS** Eye To Eye Warner Bros. 50050 [22May82|24Jul82|14Aug82] 37(2) 0|3|13 [Deborah Berg/Julian Marshall] *A: Eye To Eye;* P: Gary Katz	497
1898. **CANNONBALL** Supertramp A&M 2731 [25May85|06Jul85|10Aug85] 28(2) 0|7|12 [Rick Davies] *A: Brother Where You Bound;* P: David Kershenbaum/Supertramp	497
1899. **THIS IS NOT AMERICA** David Bowie/Pat Metheny Group EMI America 8251 [02Feb85|23Mar85|20Apr85] 32(2) 0|4|12 [David Bowie/Lyle Mays/Pat Metheny] *A: The Falcon & The Snowman (Soundtrack);* P: David Bowie/Pat Metheny	497
1900. **VALLEY GIRL** Frank Zappa Barking Pumpkin 9-02972 [17Jul82|11Sep82|02Oct82] 32(2) 0|3|12 [Frank Zappa/Moon Unit Zappa] *A: A Ship Arriving Too Late To Save A Drowning Witch;* P: Frank Zappa	496
1901. **I LIVE BY THE GROOVE** Paul Carrack Chrysalis 23427 [28Oct89|16Dec89|20Jan90] 31(1) 0|4|13 [Paul Carrack/Eddie Schwartz] *A: Groove Approved;* P: Paul Carrack/Eddie Schwartz/Tom Wolk	494
1902. **GOING BACK TO CALI** LL Cool J Def Jam 07679 [20Feb88|02Apr88|21May88] 31(2) 0|5|14 [Rick Rubin/James Todd Smith] *A: Less Than Zero Soundtrack;* P: Rick Rubin	494
1903. **WE CLOSE OUR EYES** Go West Chrysalis 42850 [23Feb85|27Apr85|01Jun85] 41(1) 0|0|15 [Peter Cox/Richard Drummie] *A: Go West;* P: Gary Stevenson	493
1904. **LOST IN LOVE** New Edition MCA 52553 [30Mar85|11May85|29Jun85] 35(2) 0|4|14 [Russ Kramer] *A: New Edition;* P: Richard Rudolph/Michael Sembello	493
1905. **YOUR IMAGINATION** Daryl Hall & John Oates RCA 13252 [19Jun82|14Aug82|28Aug82] 33(2) 0|5|11 [Daryl Hall] *A: Private Eyes;* P: Daryl Hall/John Oates	492
1906. **WHAT AM I GONNA DO (I'm So In Love With You)** Rod Stewart Warner Bros. 29564 [27Aug83|08Oct83|12Nov83] 35(1) 0|3|12 [Tony Brock/Jay Davis/Rod Stewart] *A: Body Wishes;* P: Tom Dowd/Jimmy Iovine/Rod Stewart	491
1907. **IF YOU SHOULD SAIL** Nielsen/Pearson Capitol 4910 [13Sep80|22Nov80|13Dec80] 38(1) 0|2|14 [Reed Nielsen/Mark Pearson] *A: Nielson Pearson;* P: Richard Landis	491
1908. **THE CLAPPING SONG** Pia Zadora Elektra/Curb 69889 [11Dec82|19Feb83|19Mar83] 36(2) 0|3|15 [Lincoln Chase/Kay Werner/Sue Werner] *A: Rock It Out;* P: Charles Callelo	491
1909. **BOYS NIGHT OUT** Timothy B. Schmit MCA 53137 [19Sep87|07Nov87|12Dec87] 25(1) 0|5|13 [Bruce Gaitsch/Will Jennings/Timothy B. Schmit] *A: Timothy B.;* P: Richard Rudolph	491
1910. **WHATCHA GONNA DO** Chilliwack Millennium 13110 [23Oct82|04Dec82|15Jan83] 41(3) 0|0|13 [William A. Henderson/Brian MacLeod] *A: Opus X;* P: Bill Henderson/Brian MacLeod	491
1911. **DR. HECKYLL & MR. JIVE** Men At Work Columbia 04111 [17Sep83|22Oct83|26Nov83] 28(2) 0|5|11 [Colin James Hay] *A: Cargo;* P: Peter Mclan	490
1912. **CELEBRATE YOUTH** Rick Springfield RCA 14047 [06Apr85|18May85|15Jun85] 26(1) 0|6|11 [Rick Springfield] *A: Tao;* P: Bill Drescher/Rick Springfield	490
1913. **COLD LOVE** Donna Summer Geffen 49634 [29Nov80|17Jan81|14Feb81] 33(2) 0|1|13 [Pete Bellotte/Harold Faltermeyer/Keith Forsey] *A: The Wanderer;* P: Pete Bellotte/Giorgio Moroder	489
1914. **I NEED YOUR LOVIN'** Teena Marie Gordy 7189 [22Nov80|31Jan81|21Feb81] 37(1) 0|3|14 [Teena Marie] *A: Irons In The Fire;* P: Teena Marie	489
1915. **TWO PEOPLE** Tina Turner Capitol 5644 [22Nov86|10Jan87|07Feb87] 30(2) 0|5|12 [Terry Britten/Graham Lyle] *A: Break Every Rule;* P: Terry Britten	489
1916. **BOP** Dan Seals EMI America 8289 [25Jan86|15Mar86|03May86] 42(1) 0|0|15 [Paul Davis/Jennifer Kimball] *A: Won't Be Blue Anymore;* P: Kyle Lehning	488
1917. **SUN CITY** Artists United Against Apartheid Manhattan 50017 [02Nov85|14Dec85|25Jan86] 38(1) 0|3|13 [Steven Van Zandt] *A: Sun City;* P: Arthur Baker/Steven Van Zandt	488
1918. **LIVING IN A FANTASY** Leo Sayer Warner Bros. 49657 [24Jan81|07Mar81|11Apr81] 23(2) 0|6|12 [Leo Sayer/Alan Tarney] *A: Living In A Fantasy;* P: Alan Tarney	488
1919. **I KNOW YOU'RE OUT THERE SOMEWHERE** The Moody Blues Polydor 887600 [04Jun88|06Aug88|17Sep88] 30(1) 0|4|16 [Justin Hayward] *A: Sur La Mer;* P: Tony Visconti	487
1920. **VOICES** Cheap Trick Epic 50814 [08Dec79|02Feb80|16Feb80] 32(1) 0|3|11 [Rick Nielsen] *A: Dream Police;* P: Tom Werman	487
1921. **SUGAR DON'T BITE** Sam Harris Motown 1743 [15Sep84|10Nov84|15Dec84] 36(1) 0|3|14 [Bruce Roberts/Donna Weiss] *A: Sam Harris;* P: Steve Barri/Tony Belluso	487
1922. **BAD BOY** Ray Parker Jr. Arista 1030 [04Dec82|29Jan83|19Feb83] 35(1) 0|4|12 [Ray Parker Jr.] *A: Greatest Hits;* P: Ray Parker Jr.	487
1923. **TWILIGHT WORLD** Swing Out Sister Mercury 888484 [19Dec87|27Feb88|26Mar88] 31(1) 0|3|15 [Andy Connell/Corinne Drewery/Martin Jackson] *A: It's Better To Travel;* P: Paul Staveley O'Duffy	486
1924. **ONE GOOD REASON** Paul Carrack Chrysalis 43204 [19Mar88|14May88|11Jun88] 28(2) 0|5|13 [Paul Carrack/Chris Difford] *A: One Good Reason;* P: Christopher Neil	486
1925. **BACK ON HOLIDAY** Robbie Nevil EMI 50152 [12Nov88|21Jan89|11Feb89] 34(1) 0|3|14 [David Paul Bryant/Steven Dubin/Robbie Nevil] *A: A Place Like This;* P: Robbie Nevil	485
1926. **I COULDN'T SAY NO** Robert Ellis Orrall With Carlene Carter RCA 13431 [26Mar83|21May83|11Jun83] 32(1) 0|3|12 [Robert Ellis Orrall] *A: Special Pain;* P: Roger Bechirian	485
1927. **I CAN'T LET GO** Linda Ronstadt Asylum 46654 [28Jun80|09Aug80|13Sep80] 31(1) 0|4|12 [Al Gorgoni/Chip Taylor] *A: Mad Love;* P: Peter Asher	484
1928. **BIRTHDAY SUIT** Johnny Kemp Columbia 68569 [25Feb89|15Apr89|27May89] 36(2) 0|3|14 [Rhett Lawrence/Dean Pitchford] *A: Sing Soundtrack;* P: Rhett Lawrence	483
1929. **HEY THERE LONELY GIRL** Robert John EMI America 8049 [19Jul80|13Sep80|11Oct80] 31(1) 0|4|13 [Leon Carr/Earl Shuman] *A: Back On The Street;* P: Mike Piccirillo/George Tobin	483
1930. **BURNIN' FOR YOU** Blue Öyster Cult Columbia 02415 [15Aug81|03Oct81|14Nov81] 40(3) 0|3|14 [Rich Meltzer/Donald Roeser] *A: Fire Of Unknown Origin;* P: Martin Birch	483
1931. **SMALL WORLD** Huey Lewis And The News Chrysalis 43306 [08Oct88|26Nov88|17Dec88] 25(1) 0|6|11 [Chris Hayes/Huey Lewis] *A: Small World;* P: Huey Lewis And The News	482
1932. **WHAT ABOUT ME ('89)** Moving Pictures Geffen 22859 [12Aug89|30Sep89|02Dec89] 46(1) 0|0|17 [Frances Frost/Garry Frost] *A: The Last Picture Show;* P: Charles Fisher	482
1933. **SHOULD I SAY YES?** Nu Shooz Atlantic 89108 [16Apr88|18Jun88|30Jul88] 41(1) 0|0|16 [Valerie Day/John Robert Smith] *A: Told U So;* P: John Smith/Rick Waritz	482
1934. **TWO TRIBES** Frankie Goes To Hollywood ZTT/Island 7-99695 [20Oct84|15Dec84|26Jan85] 43(1) 0|0|15 [Peter Gill/Holly Johnson/Mark O'Toole] *A: Welcome To The Pleasure Dome;* P: Trevor Horn	482
1935. **THE BIRD** The Time Warner Bros. 29094 [23Feb85|13Apr85|18May85] 36(2) 0|2|13 [Morris Day/Prince] *A: Ice Cream Castle;* P: Morris Day/Prince	482

Ranking the Singles

Rank. TITLE Artist Label Number
[Entry Date | Peak Date | Exit Date] Peak(Peak Wks) Top10 | Top40 | Tot Wks
[Writers] *Album; Producers* Score

1936. ONE Metallica Elektra 69329 — 481
[18Feb89 | 08Apr89 | 27May89] 35(1) 0 | 4 | 15
[James Hetfield/Lars Ulrich]
A: *...And Justice For All*; P: Metallica/Flemming Rasmussen

1937. QUEEN OF THE BROKEN HEARTS Loverboy Columbia 04096 — 481
[17Sep83 | 12Nov83 | 03Dec83] 34(1) 0 | 3 | 12
[Paul Dean/Mike Reno] A: *Keep It Up*; P: Paul Dean/Bruce Fairbairn

1938. KEEP THE FIRE Kenny Loggins Columbia 11215 — 480
[23Feb80 | 12Apr80 | 17May80] 36(1) 0 | 2 | 13
[E. Ein Loggins/Kenny Loggins] A: *Keep The Fire*; P: Tom Dowd

1939. I'M ALIVE Neil Diamond Columbia 03503 — 479
[15Jan83 | 19Feb83 | 02Apr83] 35(4) 0 | 4 | 12
[Neil Diamond/David Foster]
A: *Heartlight*; P: Burt Bacharach/Neil Diamond/Carole Bayer Sager

1940. BE WITH YOU Bangles Columbia 68744 — 478
[06May89 | 24Jun89 | 22Jul89] 30(1) 0 | 5 | 12
[Walker Igleheart/Debbi Peterson] A: *Everything*; P: Davitt Sigerson

1941. THE WORD IS OUT Jermaine Stewart Arista 1-9256 — 478
[02Feb85 | 30Mar85 | 11May85] 41(1) 0 | 0 | 15
[Gregory Craig/Jermaine Stewart] A: *The Word Is Out*; P: Peter Collins

1942. ROCK THE NIGHT Europe Epic 07091 — 477
[02May87 | 20Jun87 | 01Aug87] 30(1) 0 | 4 | 13
[Joey Tempest] A: *The Final Countdown*; P: Kevin Elson

1943. PUT IT IN A MAGAZINE Sonny Charles Highrise 2001 — 476
[13Nov82 | 22Jan83 | 12Feb83] 40(2) 0 | 2 | 14
[Sonny Charles/Bobby Paris] A: *The Sun Still Shines*; P: Bobby Paris

1944. YOU COULD TAKE MY HEART AWAY Silver Condor Columbia 02268 — 476
[25Jul81 | 19Sep81 | 17Oct81] 32(1) 0 | 4 | 13
[John Corey] A: *Silver Condor*; P: Mike Flicker

1945. THE PARTY'S OVER (Hopelessly In Love) Journey Columbia 60505 — 475
[28Feb81 | 25Apr81 | 23May81] 34(1) 0 | 4 | 13
[Steve Perry] A: *Captured*; P: Kevin Elson

1946. LOVE WILL SHOW US HOW Christine McVie Warner Bros. 29313 — 474
[28Apr84 | 09Jun84 | 30Jun84] 30(1) 0 | 6 | 10
[Christine McVie/Todd Sharp] A: *Christine McVie*; P: Russ Titelman

1947. FEEL IT AGAIN Honeymoon Suite Warner Bros. 28779 — 472
[08Mar86 | 10May86 | 21Jun86] 34(1) 0 | 3 | 16
[Ray Coburn] A: *The Big Prize*; P: Bruce Fairbairn

1948. SAVANNAH NIGHTS Tom Johnston Warner Bros. 49096 — 472
[17Nov79 | 12Jan80 | 02Feb80] 34(2) 0 | 2 | 12
[Tom Johnston] A: *Everything You've Heard Is True*; P: Ted Templeman

1949. THE PRISONER Howard Jones Elektra 69288 — 472
[01Jul89 | 26Aug89 | 23Sep89] 30(1) 0 | 4 | 13
[Howard Jones] A: *Cross That Line*; P: Chris Hughes

1950. LUCKY Greg Kihn EMI America 8255 — 472
[16Feb85 | 06Apr85 | 04May85] 30(1) 0 | 4 | 12
[Greg Kihn/Steve Wright] A: *Citizen Kihn*; P: Matthew King Kaufman

1951. RUNAWAY Bon Jovi Mercury 818309 — 472
[25Feb84 | 21Apr84 | 19May84] 39(1) 0 | 1 | 13
[Jon Bon Jovi/George Karak] A: *Bon Jovi*; P: Tony Bongiovi/Lance Quinn

1952. WELCOME TO THE BOOMTOWN David & David A&M 2857 — 471
[04Oct86 | 29Nov86 | 17Jan87] 37(2) 0 | 3 | 16
[David Baerwald/David Ricketts] A: *Boomtown*; P: Davitt Sigerson

1953. STICK AROUND Julian Lennon Atlantic 89437 — 471
[22Mar86 | 10May86 | 14Jun86] 32(1) 0 | 4 | 13
[Julian Lennon] A: *The Secret Value Of DayDreaming*; P: Phil Ramone

1954. CALL ME Skyy Salsoul 2152 — 471
[16Jan82 | 06Mar82 | 27Mar82] 26(2) 0 | 4 | 11
[Randy Muller] A: *Skyy Line*; P: Randy Muller/Solomon Roberts Jr.

1955. BURNING HEART Vandenberg Atco 99947 — 471
[08Jan83 | 12Mar83 | 09Apr83] 39(2) 0 | 2 | 13
[Adrian Vandenberg] A: *Vandenberg*; P: Stuart Epps/Vandenburg

1956. THE ARMS OF ORION Prince With Sheena Easton Warner Bros. 22757 — 470
[21Oct89 | 16Dec89 | 20Jan90] 36(1) 0 | 3 | 14
[Sheena Easton/Prince] A: *Batman (Soundtrack)*; P: Prince

1957. A GIRL IN TROUBLE (Is A Temporary Thing) Romeo Void
Columbia 04534 — 470
[01Sep84 | 27Oct84 | 24Nov84] 35(1) 0 | 2 | 13
[Debora Iyall/David Kahne/Peter Woods/Frank Zincavage]
A: *Instincts*; P: David Kahne

1958. 853-5937 Squeeze A&M 2994 — 470
[19Dec87 | 13Feb88 | 05Mar88] 32(1) 0 | 5 | 12
[Chris Difford/Glenn Tilbrook] A: *Babylon And On*; P: Eric Thorngren/Glenn Tilbrook

1959. MY MOTHER'S EYES Bette Midler Atlantic 3771 — 470
[22Nov80 | 17Jan81 | 14Feb81] 39(1) 0 | 2 | 13
[Tom Jans] A: *Divine Madness (Soundtrack)*; P: Dennis Kirk

1960. JIMMY LEE Aretha Franklin Arista 1-9546 — 469
[06Dec86 | 07Feb87 | 28Feb87] 28(1) 0 | 4 | 13
[Jeffrey E. Cohen/Preston Glass/Lisa Walden/Narada Michael Walden]
A: *Aretha (II)*; P: Narada Michael Walden

1961. I DON'T WANT TO BE A HERO Johnny Hates Jazz Virgin 99304 — 469
[09Jul88 | 27Aug88 | 24Sep88] 31(1) 0 | 5 | 12
[Clark Datchler] A: *Turn Back The Clock*; P: Calvin Hayes/Mike Nocito

1962. MY HEART SKIPS A BEAT The Cover Girls Capitol 44436 — 469
[09Sep89 | 28Oct89 | 09Dec89] 38(1) 0 | 2 | 13
[David Cole] A: *We Can't Go Wrong*; P: Robert Clivillés/David Cole

1963. FOR AMERICA Jackson Browne Asylum 69566 — 467
[01Mar86 | 19Apr86 | 17May86] 30(1) 0 | 5 | 12
[Jackson Browne] A: *Lives In The Balance*; P: Jackson Browne

1964. CARRIE Cliff Richard EMI America 8035 — 467
[23Feb80 | 19Apr80 | 03May80] 34(1) 0 | 3 | 11
[Terry Britten/Brian A. Robertson]
A: *We Don't Talk Anymore*; P: Terry Britten/Cliff Richard

1965. I WANT YOU, I NEED YOU Chris Christian Boardwalk 7-11-126 — 467
[03Oct81 | 21Nov81 | 02Jan82] 37(1) 0 | 3 | 14
[Chris Christian/J.C. Crowley/Shanon Smith] A: *Chris Christian*; P: Bob Gaudio

1966. BLACK CARS Gino Vannelli HME 04889 — 467
[04May85 | 29Jun85 | 17Aug85] 42(1) 0 | 0 | 16
[Roy Freeland/Gino Vannelli]
A: *Black Cars*; P: Gino Vannelli/Joe Vannelli/Ross Vannelli

1967. AI NO CORRIDA (I-No-Ko-Ree-Da) Quincy Jones A&M 2309 — 466
[11Apr81 | 30May81 | 27Jun81] 28(2) 0 | 5 | 12
[Chaz Jankel/Kenny Young] A: *The Dude*; P: Quincy Jones

1968. FOOL FOR YOUR LOVING ('89) Whitesnake Geffen 22715 — 465
[04Nov89 | 23Dec89 | 03Feb90] 37(2) 0 | 3 | 14
[David Coverdale/Bernie Marsden/Micky Moody]
A: *Slip Of The Tongue*; P: Mike Clink/Keith Olsen

1969. TILL I LOVED YOU (The Love Theme From Goya) — 465
Barbra Streisand And Don Johnson Columbia 08062
[22Oct88 | 03Dec88 | 07Jan89] 25(1) 0 | 5 | 12
[Maury Yeston] A: *Till I Loved You*; P: Phil Ramone

1970. GOLDMINE Pointer Sisters RCA 5062 — 465
[01Nov86 | 20Dec86 | 24Jan87] 33(1) 0 | 3 | 13
[Andy Goldmark/Bruce Roberts] A: *Hot Together*; P: Richard Perry

1971. FLAMES OF PARADISE — 465
Jennifer Rush (Duet With Elton John) Epic 07119
[16May87 | 11Jul87 | 08Aug87] 36(1) 0 | 3 | 13
[Andy Goldmark/Bruce Roberts]
A: *Heart Over Mind*; P: Andy Goldmark/Bruce Roberts

1972. TAKE THE SHORT WAY HOME Dionne Warwick Arista 1040 — 464
[26Feb83 | 02Apr83 | 21May83] 41(1) 0 | 0 | 13
[Albhy Galuten/Barry Gibb]
A: *Heartbreaker*; P: Albhy Galuten/Barry Gibb/Karl Richardson

1973. SPICE OF LIFE The Manhattan Transfer Atlantic 89786 — 464
[10Sep83 | 05Nov83 | 03Dec83] 40(2) 0 | 2 | 13
[Derek Bramble/Rod Temperton] A: *Bodies And Souls*; P: Richard Rudolph

1974. SWINGIN' John Anderson Warner Bros. 29788 — 464
[05Mar83 | 16Apr83 | 28May83] 43(2) 0 | 0 | 13
[John David Anderson/Lionel Delmore]
A: *Wild & Blue*; P: John David Anderson/Frank Jones

1975. DON'T FORGET TO DANCE The Kinks Arista 1-9075 — 464
[20Aug83 | 08Oct83 | 22Oct83] 29(1) 0 | 4 | 10
[Ray Davies] A: *State Of Confusion*; P: Ray Davies

1976. I ONLY WANNA BE WITH YOU Samantha Fox Jive 1192 — 463
[18Mar89 | 20May89 | 10Jun89] 31(1) 0 | 4 | 13
[Mike Hawker/Ivor Raymonde]
A: *I Wanna Have Some Fun*; P: Matt Aitken/Mike Stock/Pete Waterman

1977. I GET EXCITED Rick Springfield RCA 13303 — 463
[11Sep82 | 30Oct82 | 27Nov82] 32(2) 0 | 5 | 13
[Rick Springfield] A: *Success Hasn't Spoiled Me Yet*; P: Keith Olsen

Rank. TITLE Artist Label Number	
[Entry Date | Peak Date | Exit Date] Peak(Peak Wks) Top10 | Top40 | Tot Wks	
[Writers] *Album; Producers*	Score
1978. POSSESSION OBSESSION Daryl Hall & John Oates RCA 14098	463
[01Jun85 | 06Jul85 | 17Aug85] 30(1) 0 | 6 | 12	
[Sara Allen/Daryl Hall/John Oates]	
A: Big Bam Boom; P: Arthur Baker/Bob Clearmountain/Daryl Hall/John Oates	
1979. MEMORY Barry Manilow Arista 1025	463
[20Nov82 | 15Jan83 | 19Feb83] 39(2) 0 | 2 | 14	
[T.S. Eliot/Trevor Nunn/Andrew Lloyd Webber]	
A: Here Comes The Night; P: Barry Manilow	
1980. TEARS John Waite EMI America 8238	462
[20Oct84 | 24Nov84 | 12Jan85] 37(1) 0 | 4 | 13	
[Vince Cusano/Adam Mitchell]	
A: No Brakes; P: Gary Gersh/David Thoener/John Waite	
1981. AFTER THE GLITTER FADES Stevie Nicks Modern 7405	462
[15May82 | 03Jul82 | 24Jul82] 32(1) 0 | 4 | 11	
[Stevie Nicks] *A: Bella Donna; P: Jimmy Iovine*	
1982. WHISPER TO A SCREAM (Birds Fly) Icicle Works Arista 1-9155	461
[21Apr84 | 09Jun84 | 07Jul84] 37(2) 0 | 4 | 12	
[Robert Ian McNabb] *A: Icicle Works; P: Hugh Jones*	
1983. SMALLTOWN BOY Bronski Beat MCA 52494	461
[22Dec84 | 09Mar85 | 06Apr85] 48(1) 0 | 0 | 16	
[Steve Bronski/Jimmy Somerville/Larry Steinbachek]	
A: The Age Of Consent; P: Mike Thorne	
1984. NEW ROMANCE (It's A Mystery) Spider Dreamland 100	461
[19Apr80 | 07Jun80 | 28Jun80] 39(2) 0 | 2 | 11	
[Anton Fig/Holly Knight] *A: Spider; P: Peter Coleman*	
1985. A LIFE OF ILLUSION Joe Walsh Asylum 47144	460
[23May81 | 11Jul81 | 08Aug81] 34(2) 0 | 4 | 12	
[Kenny Passarelli/Joe Walsh]	
A: There Goes The Neighborhood; P: Joe Walsh	
1986. LIVIN' IN DESPERATE TIMES Olivia Newton-John MCA 52341	460
[11Feb84 | 17Mar84 | 14Apr84] 31(2) 0 | 5 | 10	
[Barry Alfonso/Tom Snow] *A: Two Of A Kind (Soundtrack); P: David Foster*	
1987. ONE IN A MILLION The Romantics Nemperor 04373	460
[25Feb84 | 07Apr84 | 12May84] 37(2) 0 | 3 | 12	
[Coz Canler/Jimmy Marinos/Wally Palmar/Mike Skill]	
A: In Heat; P: Peter Solley	
1988. TROUBLE Nia Peeples Mercury 870154	460
[14May88 | 09Jul88 | 20Aug88] 35(2) 0 | 3 | 15	
[Stephen Laurence Harvey]	
A: Nothin' But Trouble; P: Steve (2) Harvey	
1989. NICOLE Point Blank MCA 51132	459
[27Jun81 | 05Sep81 | 26Sep81] 39(1) 0 | 2 | 14	
[Rusty Burns/Kim Davis/Peter Gruen/Mike Hamilton/Bubba Keith/ Bill Randolph] *A: American Excess; P: Bill Ham*	

Rank. TITLE Artist Label Number	
[Entry Date | Peak Date | Exit Date] Peak(Peak Wks) Top10 | Top40 | Tot Wks	
[Writers] *Album; Producers*	Score
1990. YOU'RE DRIVING ME OUT OF MY MIND Little River Band Capitol 5256	459
[23Jul83 | 10Sep83 | 01Oct83] 35(1) 0 | 3 | 11	
[Beeb Birtles/Graham Goble] *A: The Net; P: Little River Band/Ernie Rose*	
1991. NEVER LET ME DOWN David Bowie EMI America 43031	458
[08Aug87 | 26Sep87 | 17Oct87] 27(1) 0 | 5 | 11	
[Carlos Alomar/David Bowie]	
A: Never Let Me Down; P: David Bowie/David Richards	
1992. TONIGHT IT'S YOU Cheap Trick Epic 05431	457
[27Jul85 | 12Oct85 | 16Nov85] 44(1) 0 | 0 | 17	
[Jon Brant/Rick Nielsen/Mark Radice/Robin Zander]	
A: Standing On The Edge; P: Jack Douglas	
1993. BANG YOUR HEAD (Metal Health) Quiet Riot Pasha 04267	457
[07Jan84 | 11Feb84 | 24Mar84] 31(2) 0 | 4 | 12	
[Frankie Banali/Carlos Cavazo/Tony Cavazo/Kevin Dubrow]	
A: Metal Health; P: Spencer Proffer	
1994. REBEL YELL Billy Idol Chrysalis 42762	457
[28Jan84 | 24Mar84 | 28Apr84] 46(2) 0 | 0 | 14	
[Billy Idol/Steve Stevens] *A: Rebel Yell; P: Keith Forsey*	
1995. I THANK YOU ZZ Top Warner Bros. 49163	456
[19Jan80 | 15Mar80 | 29Mar80] 34(2) 0 | 3 | 11	
[Isaac Hayes/David Porter] *A: Deguello; P: Bill Ham*	
1996. HEART OF MINE Boz Scaggs Columbia 07780	456
[30Apr88 | 18Jun88 | 30Jul88] 35(2) 0 | 4 | 14	
[Bobby Caldwell/Dennis Matkosky/Jason Scheff]	
A: Other Roads; P: Stewart Levine	
1997. HOOKED ON SWING Larry Elgart And His Manhattan Swing Orchestra RCA 13219	456
[05Jun82 | 24Jul82 | 21Aug82] 31(2) 0 | 5 | 12	
A: Hooked On Swing; P: Larry Elgart	
1998. FOOLISH PRIDE Daryl Hall RCA 5038	456
[18Oct86 | 06Dec86 | 10Jan87] 33(1) 0 | 5 | 13	
[Daryl Hall]	
A: Three Hearts In The Happy Ending Machine; *P: Daryl Hall/David A. Stewart/Tom Wolk*	
1999. VOX HUMANA Kenny Loggins Columbia 04849	456
[23Mar85 | 04May85 | 25May85] 29(1) 0 | 4 | 10	
[E. Ein Loggins/Kenny Loggins] *A: Vox Humana; P: Kenny Loggins*	
2000. I LOVED 'EM EVERY ONE T.G. Sheppard Warner/Curb 49690	456
[14May81 | 16May81 | 13Jun81] 37(2) 0 | 2 | 14	
[Phil Sampson] *A: I Love 'Em All; P: Buddy Killen*	

Ranking the Singles

Singles 2001-4168

Rank. TITLE Act Label Number Entry Date Peak(Wks) Tot Wks	
[Writers] A: Proximate Album; P: Producer	Score

2001. TALK TO MYSELF Christopher Williams Geffen 22936 — 455
09Sep89 49(2) 18 [Timmy Gatling/Alton Stewart]
A: Adventures In Paradise; P: Timmy Gatling/Alton Stewart

2002. INVISIBLE HANDS Kim Carnes EMI America 8181 15Oct83 40(2) 13 — 455
[Brian Fairweather/Martin Page] A: Cafe Racers; P: Keith Olsen

2003. SUPERSTITIOUS Europe Epic 07979 13Aug88 31(1) 13 — 455
[Joey Tempest] A: Out Of This World; P: Ron Nevison

2004. FAREWELL MY SUMMER LOVE Michael Jackson Motown 1739 — 454
26May84 38(2) 12 [Keni St. Lewis]
A: Farewell My Summer Love; P: Fonce Mizell/Freddie Perren

2005. GET THAT LOVE Thompson Twins Arista 1-9577 28Mar87 31(1) 11 — 454
[Tom Bailey/Alannah Currie] A: Close To The Bone; P: Rupert Hine

2006. BLESSED ARE THE BELIEVERS Anne Murray Capitol 4987 28Mar81 — 454
34(2) 13 [Charlie Black/Rory Bourke/Sandy Pinkard]
A: Where Do You Go When You Dream; P: Jim Ed Norman

2007. IN YOUR EYES ('89) Peter Gabriel WTG 31-68936 20May89 41(1) 14 — 453
[Peter Gabriel] A: Say Anything; P: Peter Gabriel/Daniel Lanois

2008. JAM TONIGHT Freddie Jackson Capitol 44037 27Jun87 32(1) 13 — 453
[Freddie Jackson/Paul Laurence] A: Just Like The First Time; P: Paul Laurence

2009. TWO SIDES OF LOVE Sammy Hagar Geffen 29446 14Jul84 38(3) 12 — 453
[Sammy Hagar] A: VOA; P: Ted Templeman

2010. ONLY A LONELY HEART SEES Felix Cavaliere Epic 50829 — 452
01Mar80 36(1) 11 [Felix Cavaliere/Jay Tran]
A: Castles In The Air; P: Felix Cavaliere/Cengiz Yaltkaya

2011. MY BRAVE FACE Paul McCartney Capitol 44367 27May89 25(1) 10 — 452
[Elvis Costello/Paul McCartney]
A: Flowers In The Dirt; P: Neil Dorfsman/Mitchell Froom/Paul McCartney

2012. TELL ME TOMORROW - PART 1 Smokey Robinson Tamla 1601 16Jan82 — 451
33(1) 12 [Gary Goetzman/Mike Piccirillo] A: Yes It's You Lady; P: George Tobin

2013. I DON'T WANT TO WALK WITHOUT YOU Barry Manilow Arista 0501 — 451
12Apr80 36(3) 11 [Frank Loesser/Jule Styne]
A: One Voice; P: Ron Dante/Barry Manilow

2014. I CAN'T HELP MYSELF (Sugar Pie, Honey Bunch) Bonnie Pointer — 451
Motown 1478 22Dec79 40(2) 13
[Lamont Dozier/Brian Holland/Eddie Holland] A: Bonnie Pointer (II); P: Jeffrey Bowen

2015. NO LOOKIN' BACK Michael McDonald Warner Bros. 28960 — 451
27Jul85 34(4) 12 [Kenny Loggins/Michael McDonald/Ed Sanford]
A: No Lookin' Back; P: Michael McDonald/Ted Templeman

2016. GIRLS ARE MORE FUN Ray Parker Jr. Arista 1-9352 05Oct85 34(1) 15 — 451
[Ray Parker Jr.] A: Sex And The Single Man; P: Ray Parker Jr.

2017. I KNEW YOU WHEN Linda Ronstadt Asylum 69853 11Dec82 37(1) 12 — 451
[Joe South] A: Get Closer; P: Peter Asher

2018. BE MINE TONIGHT Neil Diamond Columbia 02928 22May82 35(2) 11 — 450
[Neil Diamond] A: On The Way To The Sky; P: Neil Diamond

2019. GET CLOSER Linda Ronstadt Asylum 69948 02Oct82 29(2) 12 — 450
[Jonathan Carroll] A: Get Closer; P: Peter Asher

2020. ONE HIT (To The Body) The Rolling Stones Rolling Stones 38-05906 — 450
17May86 28(1) 11 [Mick Jagger/Keith Richards/Ron Wood]
A: Dirty Work; P: Glimmer Twins/Steve Lillywhite

2021. PUT YOUR MOUTH ON ME Eddie Murphy Columbia 68897 — 450
29Jul89 27(1) 12 [Jeffrey E. Cohen/Eddie Murphy/Narada Michael Walden]
A: So Happy; P: Narada Michael Walden

2022. GIMME ALL YOUR LOVIN ZZ Top Warner Bros. 29693 02Apr83 — 449
37(1) 12 [Frank Beard/Billy Gibbons/Dusty Hill] A: Eliminator; P: Bill Ham

2023. VELCRO FLY ZZ Top Warner Bros. 28650 26Jul86 35(2) 12 — 449
[Frank Beard/Billy Gibbons/Dusty Hill] A: Afterburner; P: Bill Ham

2024. LONELY EYES Robert John EMI America 8030 08Dec79 41(1) 11 — 447
[Mike Piccirillo] A: Robert John; P: George Tobin

2025. SUGAR DADDY Thompson Twins Warner Bros. 22819 23Sep89 — 447
28(1) 12 [Tom Bailey/Alannah Currie] A: Big Trash; P: Tom Bailey/Alannah Currie

2026. LOVE STINKS The J. Geils Band EMI America 8039 12Apr80 38(2) 12 — 447
[Seth Justman/Peter Wolf] A: Love Stinks; P: Seth Justman

2027. SEA OF LOVE Del Shannon Network 47951 12Dec81 33(2) 12 — 447
[Phil Baptiste/George Khoury] A: Drop Down And Get Me; P: Tom Petty

2028. MAMA USED TO SAY Junior Mercury 76132 13Feb82 30(1) 13 — 446
[Robert Ashley Carter/Junior Giscombe] A: "Ji"; P: Bob Carter

2029. MISTER SANDMAN Emmylou Harris Warner Bros. 49684 — 446
28Feb81 37(1) 13 [Pat Ballard] A: Evangeline; P: Brian Ahern

2030. ANSWERING MACHINE Rupert Holmes MCA 41235 03May80 — 445
32(1) 11 [Rupert Holmes] A: Partners In Crime; P: Jim Boyer/Rupert Holmes

2031. I'M NOT THE ONE The Cars Elektra 69569 01Feb86 32(1) 11 — 445
[Ric Ocasek] A: Shake It Up; P: Roy Thomas Baker

2032. SOMETHING REAL (Inside Me/Inside You) Mr. Mister RCA 5273 — 445
22Aug87 29(1) 11 [Steve George/John Lang/Richard Page]
A: Go On...; P: Kevin Killen/Mr. Mister

2033. HIGH ON EMOTION Chris DeBurgh A&M 2643 30Jun84 44(2) 13 — 444
[Chris de Burgh] A: Man On The Line; P: Rupert Hine

2034. I'M IN LOVE Evelyn King RCA 12243 25Jul81 40(2) 14 — 444
[Michael Jones] A: I'm In Love; P: Morrie Brown

2035. RAPPER'S DELIGHT Sugarhill Gang Sugar Hill 542 10Nov79 36(1) 12 — 444
[Bernard Edwards/Curtis Fisher/Henry Jackson/Guy O'Brien/Sylvia Robinson/
Nile Rodgers/Michael Wright] A: The Sugar Hill Gang; P: Sylvia, Inc.

2036. FEELS SO GOOD Van Halen Warner Bros. 27565 28Jan89 35(1) 14 — 443
[Michael Anthony/Sammy Hagar/Alex Van Halen/Eddie Van Halen]
A: OU812; P: Donn Landee

2037. GIRLS WITH GUNS Tommy Shaw A&M 2676 29Sep84 33(1) 12 — 443
[Tommy Shaw] A: Girls With Guns; P: Mike Stone

2038. ON THE WAY TO THE SKY Neil Diamond Columbia 02712 — 443
13Feb82 27(2) 10 [Neil Diamond/Carole Bayer Sager]
A: On The Way To The Sky; P: Neil Diamond/Dennis St. John

2039. I REALLY DON'T NEED NO LIGHT Jeffrey Osborne A&M 2410 05Jun82 — 443
39(2) 15 [Jeffrey Osborne/David Wolinski] A: Jeffrey Osborne; P: George Duke

2040. MINIMUM LOVE Mac McAnally Geffen 29736 05Mar83 41(1) 12 — 442
[Mac McAnally/Jerry Wexler]
A: Nothing But The Truth; P: Clayton Ivey/Terry Woodford

2041. CRAZAY Jesse Johnson (Featuring Sly Stone) A&M 2878 — 442
25Oct86 53(3) 16 [Jesse Woods Johnson] A: Shockadelica; P: Jesse Johnson

2042. SHE WON'T TALK TO ME Luther Vandross Epic 08513 — 442
28Jan89 30(1) 12 [Hubert Eaves III/Luther Vandross]
A: Any Love; P: Marcus Miller/Luther Vandross

2043. SAY IT AGAIN Jermaine Stewart Arista 1-9636 19Mar88 27(1) 12 — 442
[Carol Davis/Bunny Sigler] A: Say It Again; P: Jerry Knight/Aaron Zigman

2044. PIECES OF ICE Diana Ross RCA 13549 25Jun83 31(1) 10 [John — 441
Capek/Marc Jordan] A: Ross (II); P: Gary Katz

2045. I DON'T KNOW WHERE TO START Eddie Rabbitt Elektra 47435 — 440
10Apr82 35(1) 13 [Thom Schuyler] A: Step By Step; P: David Malloy

2046. MAKING LOVE IN THE RAIN Herb Alpert A&M 2949 18Jul87 35(1) 14 — 439
[James Samuel Harris/Terry Lewis]
A: Keep Your Eye On Me; P: Jimmy Jam Harris/Terry Lewis

2047. OUT OF MIND OUT OF SIGHT Models Geffen 28762 26Apr86 37(1) 13 — 438
[James Freud] A: Out Of Mind Out Of Sight; P: Mark Opitz

2048. EBONY EYES Rick James featuring Smokey Robinson Gordy 1714 — 438
10Dec83 43(1) 11 [Rick James] A: Cold Blooded; P: Rick James

2049. CHLOE Elton John Geffen 49788 25Jul81 34(1) 13 [Elton John/Gary — 437
Osborne] A: The Fox; P: Clive Franks/Elton John

2050. (What) IN THE NAME OF LOVE Naked Eyes EMI-America 8219 11Aug84 — 437
39(2) 12 [Pete Byrne/Rob Fisher] A: Fuel For The Fire; P: Arthur Baker

2051. COLD BLOODED Rick James Gordy 1687 30Jul83 40(1) 12 — 437
[Rick James] A: Cold Blooded; P: Rick James

2052. DOWNTOWN One 2 Many A&M 1272 25Mar89 37(1) 13 — 437
[Don Black/Dag Kolsrud/Jan Ovland] A: Mirror; P: John Hudson/Dag Kolsrud

2053. WHEN A MAN LOVES A WOMAN Bette Midler Atlantic 3643 — 437
19Jan80 35(2) 10 [Calvin Lewis/Andrew Wright]
A: The Rose (Soundtrack); P: Paul A. Rothchild

2054. LOVE T.K.O. Teddy Pendergrass Philadelphia Int'l 3116 — 437
29Nov80 44(1) 13 [Gip Noble/Cecil Womack/Linda Womack]
A: TP; P: Cynthia Biggs/Dexter Wansel/Cecil Womack

2055. COVER OF LOVE Michael Damian Cypress 1430 17Jun89 31(1) 12 — 436
[Janine Best/Michael Damian/Troy Johnson/Larry Weir]
A: Where Do We Go From Here;
P: Michael Damian/Troy Johnson/Larry Weir/Tom Weir

2056. SAY WHAT Jesse Winchester Bearsville 49711 25Apr81 32(2) 12 — 435
[Jesse Winchester] A: Talk Memphis; P: Willie Mitchell

2057. FAKE FRIENDS Joan Jett & The Blackhearts Blackheart/MCA 52256 — 434
09Jul83 35(2) 10 [Joan Jett/Kenny Laguna]
A: Album; P: Ritchie Cordell/Joan Jett/Kenny Laguna

2058. WINDS OF CHANGE Jefferson Starship Grunt 13439 29Jan83 — 434
38(2) 11 [Jeannette Sears/Pete Sears] A: Winds Of Change; P: Kevin Beamish

2059. 24/7 Dino 4th & Broadway 7471 04Feb89 42(2) 14 — 434
[Dino Esposito] A: 24/7; P: Dino Esposito

Ranking the Singles

Rank. TITLE Act Label Number Entry Date Peak(Wks) Tot Wks [Writers] A: Proximate Album; P: Producer	Score
2060. VICTIM OF LOVE Bryan Adams A&M 2964 22Aug87 32(1) 12 [Bryan Adams/Jim Vallance] A: Into The Fire; P: Bryan Adams/Bob Clearmountain	434
2061. I GOTTA TRY Michael McDonald Warner Bros. 29862 13Nov82 44(4) 11 [Kenny Loggins/Michael McDonald] A: If That's What It Takes; P: Ted Templeman/Lenny Waronker	433
2062. DON'T STOP Jeffrey Osborne A&M 2687 13Oct84 44(1) 15 [David Batteau/Danny Sembello] A: Don't Stop; P: George Duke	433
2063. LET ME GO, LOVE Nicolette Larson Warner Bros. 49130 12Jan80 35(2) 11 [B.J. Cook Foster/Michael McDonald] A: In The Nick Of Time; P: Ted Templeman	432
2064. WHEN WE KISS Bardeux Enigma 75018 16Apr88 36(1) 13 [Stacy Smith/Jon St. James] A: Bold As Love; P: Jon St. James	431
2065. TOUCH AND GO The Cars Elektra 47039 06Sep80 37(2) 11 [Ric Ocasek] A: Panorama; P: Roy Thomas Baker	431
2066. GIVE ME TONIGHT Shannon Mirage 99775 31Mar84 46(2) 13 [Chris Barbosa/Ed Chisolm] A: Let The Music Play; P: Chris Barbosa/Mark Liggett	430
2067. CAN'T WAIT ANOTHER MINUTE Five Star RCA 14421 13Sep86 41(2) 14 [Paul Chiten/Sue Sheridan] A: Silk And Steel; P: Richard James Burgess	430
2068. DREAMIN' Will To Power Epic 07199 27Jun87 50(1) 16 [Bob Rosenberg] A: Will To Power; P: Bob Rosenberg	429
2069. LET HIM GO Animotion Mercury 880737 01Jun85 39(1) 13 [Bill Wadhams] A: Animotion; P: John Ryan	429
2070. (SARTORIAL ELOQUENCE) DON'T YA WANNA PLAY THIS GAME NO MORE Elton John MCA 41293 09Aug80 39(1) 12 [Elton John/Tom Robinson] A: 21 At 33; P: Clive Franks/Elton John	429
2071. HYPERACTIVE Robert Palmer Island 99545 07Jun86 33(1) 12 [Tony Haynes/Dennis Nelson/Robert Palmer] A: Riptide; P: Bernard Edwards	428
2072. LOOKS LIKE LOVE AGAIN Dann Rogers IA 500 15Dec79 41(2) 11 [Dana Merino] A: Hearts Under Fire; P: Ian Gardiner	428
2073. I NEED YOU Paul Carrack Epic 03146 04Sep82 37(2) 13 [Martin Belmont/Paul Carrack/Nick Lowe] A: Suburban Voodoo; P: Nick Lowe	428
2074. I'M STILL SEARCHING Glass Tiger EMI 50116 09Apr88 31(3) 11 [Alan Frew/Michael Hanson/Sam Reid] A: Diamond Sun; P: Jim Vallance	428
2075. KISS AND TELL Bryan Ferry Reprise 28117 27Feb88 31(1) 13 [Bryan Ferry] A: Bete Noire; P: Bryan Ferry/Chester Kamen/Patrick Leonard	427
2076. SHE'S IN LOVE WITH YOU Suzi Quatro RSO 1014 24Nov79 41(2) 11 [Mike Chapman/Nicky Chinn] A: Suzi...And Other Four Letter Words; P: Mike Chapman	427
2077. MIDNIGHT BLUE Louise Tucker Arista 1-9022 18Jun83 46(1) 13 [Charlie Skarbek/Tim Smit/Ludwig van Beethoven] A: Midnight Blue; P: Charlie Skarbek/Tim Smit	427
2078. SAVE THE LAST DANCE FOR ME Dolly Parton RCA 13703 10Dec83 45(1) 12 [Doc Pomus/Mort Shuman] A: The Great Pretender; P: Val Garay	427
2079. WHAT'S NEW Linda Ronstadt & The Nelson Riddle Orchestra Asylum 69780 29Oct83 53(3) 14 [Johnny Burke/Bob Haggart] A: What's New; P: Peter Asher	427
2080. AUTOMATIC MAN Michael Sembello Warner Bros. 29485 24Sep83 34(1) 10 [David Batteau/Danny Sembello/Michael Sembello] A: Bossa Nova Hotel; P: Phil Ramone	426
2081. SOMEWHERE (From "West Side Story") Barbra Streisand Columbia 05680 14Dec85 43(2) 14 [Leonard Bernstein/Stephen Sondheim] A: The Broadway Album; P: David Foster	425
2082. DANCING IN THE STREET Van Halen Warner Bros. 29986 22May82 38(2) 11 [Marvin Gaye/Ivy Jo Hunter/William Stevenson] A: Diver Down; P: Ted Templeman	425
2083. LET GO Sharon Bryant Wing 871 722 19Aug89 34(2) 13 [Darryl Duncan] A: Here I Am; P: Darryl Duncan	425
2084. SUGAR FREE Wa Wa Nee Epic 07283 12Sep87 35(1) 13 [Paul Gray] A: Wa Wa Nee; P: Paul Gray/Jim Taig	424
2085. TURNED AWAY Chuckii Booker Atlantic 88917 08Jul89 42(1) 14 [Chuckii Booker/Donnell Spencer] A: Chuckii; P: Chuckii Booker	424
2086. STARS ON 45 III (A Tribute To Stevie Wonder) (Medley) Stars On Radio 4019 27Mar82 28(2) 10 A: Stars On Long Play III; P: Jaap Eggermont	423
2087. MISSED OPPORTUNITY Daryl Hall & John Oates Arista 1-9727 09Jul88 29(1) 11 [Sara Allen/Daryl Hall/John Oates] A: Ooh Yeah!; P: Daryl Hall/John Oates/Tom Wolk	423
2088. KISS The Art Of Noise featuring Tom Jones China 871038 03Dec88 31(2) 11 [Prince] A: The Best Of The Art Of Noise; P: Art Of Noise	423
2089. A LITTLE BIT OF LOVE (Is All It Takes) New Edition MCA 52768 22Feb86 38(2) 15 [Richard Wyatt/Christine Yarian] A: All For Love; P: Richard Rudolph/Michael Sembello	423
2090. POP GOES THE MOVIES Part I Meco Arista 0660 13Feb82 35(2) 11 A: Pop Goes The Movies; P: Tony Bongiovi/Meco Monardo/Lance Quinn	422
2091. JUST CAN'T WIN 'EM ALL Stevie Woods Cotillion 46030 23Jan82 38(1) 12 [Bill Bowersock/Greg Mathieson/Trevor Veitch/Matt Vernon] A: Take Me To Your Heaven; P: Jack White	422
2092. LITTLE BY LITTLE Robert Plant Es Paranza 7-99644 18May85 36(1) 11 [Robert Plant/Jezz Woodroffee] A: Shaken 'N' Stirred; P: Benji Lefevre/Tim Palmer/Robert Plant	422
2093. NO ONE IN THE WORLD Anita Baker Elektra 69456 22Aug87 44(1) 17 [Ken Hirsch/Marti Sharron] A: Rapture; P: Marti Sharron/Gary Skardina	422
2094. LOVE IS LIKE A ROCK Donnie Iris MCA/Carousel 51223 19Dec81 37(1) 14 [Mark Avsec/Martin Lee Hoenes/Donnie Iris/ Albritton McClain/Kevin Valentine] A: King Cool; P: Mark Avsec	420
2095. MY OH MY Slade CBS Associated 04528 07Jul84 37(2) 11 [Noddy Holder/Jim Lea] A: Keep Your Hands Off My Power Supply; P: John Punter	420
2096. PRIME TIME Alan Parsons Project Arista 1-9208 19May84 34(2) 11 [Alan Parsons/Eric Woolfson] A: Ammonia Avenue; P: Alan Parsons	420
2097. GLAMOUR BOYS Living Colour Epic 68858 26Aug89 31(1) 13 [Vernon Reid] A: Vivid; P: Mick Jagger	420
2098. I FEEL THE EARTH MOVE Martika Columbia 68996 02Sep89 25(1) 11 [Carole King] A: Martika; P: Michael Jay	417
2099. THE FIRST DAY OF SUMMER Tony Carey MCA 52388 09Jun84 33(1) 11 [Tony Carey] A: Some Tough City; P: Peter Hauke	417
2100. IT'S NO SECRET Kylie Minogue Geffen 27651 10Dec88 37(1) 13 [Matt Aitken/Mike Stock/Pete Waterman] A: Kylie; P: Matt Aitken/Mike Stock/Pete Waterman	417
2101. IT'S INEVITABLE Charlie Mirage 99862 25Jun83 38(1) 11 [Terry Thomas] A: Charlie; P: Kevin Beamish/Terry Thomas	416
2102. WHY YOU TREAT ME SO BAD Club Nouveau King Jay/Warner 28360 30May87 39(1) 13 [Denzil Foster/Jay King/Thomas McElroy] A: Life, Love & Pain; P: Denzil Foster/Jay King/Thomas McElroy	416
2103. PROMISES IN THE DARK Pat Benatar Chrysalis 2555 03Oct81 38(2) 11 [Pat Benatar/Neil Geraldo] A: Precious Time; P: Neil Geraldo/Keith Olsen	416
2104. HAD A DREAM (Sleeping With The Enemy) Roger Hodgson A&M 2678 13Oct84 48(2) 15 [Roger Hodgson] A: In The Eye Of The Storm; P: Roger Hodgson	416
2105. NO TIME FOR TALK Christopher Cross Warner Bros. 29662 07May83 33(1) 10 [Christopher Cross] A: Another Page; P: Michael Omartian	416
2106. ROUTE 101 Herb Alpert A&M 2422 26Jun82 37(2) 10 [Juan Carlos Calderón] A: Fandango; P: Herb Alpert/Jose Quintana	415
2107. DA' BUTT E.U. EMI 50115 16Apr88 35(3) 12 [Marcus Miller/Mark Stevens] A: School Daze Soundtrack; P: Marcus Miller	415
2108. NOTORIOUS Loverboy Columbia 07324 22Aug87 38(2) 14 [Jon Bon Jovi/Todd Cerney/Paul Dean/Mike Reno/Richie Sambora] A: Wildside; P: Bruce Fairbairn	415
2109. SO CLOSE Diana Ross RCA 13424 05Feb83 40(2) 10 [Rob Mounsey/Diana Ross/Bill Wray] A: Silk Electric; P: Diana Ross	415
2110. DO ME BABY Meli'sa Morgan Capitol 5523 25Jan86 46(1) 14 [Prince] A: Do Me Baby; P: Paul Laurence	414
2111. STIR IT UP Patti LaBelle MCA 52610 15Jun85 41(1) 14 [Danny Sembello/Allee Willis] A: Beverly Hills Cop Soundtrack; P: Harold Faltermeyer/Keith Forsey	414
2112. I NEED YOU TONIGHT Peter Wolf EMI America 8241 13Oct84 36(1) 13 [Peter S. Bliss/Peter Wolf] A: Lights Out; P: Michael Jonzun/Peter Wolf	414
2113. HOPE YOU LOVE ME LIKE YOU SAY YOU DO Huey Lewis And The News Chrysalis 2604 15May82 36(2) 11 [Mike Duke] A: Picture This; P: Huey Lewis And The News	413
2114. KISS ME IN THE RAIN Barbra Streisand Columbia 11179 12Jan80 37(2) 11 [Sandy Farina/Lisa Ratner] A: Wet; P: Gary Klein	413
2115. THE LAST TIME I MADE LOVE Joyce Kennedy & Jeffrey Osborne A&M 2656 18Aug84 40(2) 12 [Jeff Barry/Barry Mann/Cynthia Weil] A: Lookin' For Trouble; P: Jeffrey Osborne	413
2116. OH, PEOPLE Patti LaBelle MCA 52877 19Jul86 29(1) 12 [Andy Goldmark/Bruce Roberts] A: Winner In You; P: Richard Perry	413
2117. TURN AND WALK AWAY The Babys Chrysalis 2467 15Nov80 42(2) 13 [Jonathan Cain/John Waite] A: On The Edge; P: Keith Olsen	413
2118. NIGHTIME Pretty Poison Virgin 99350 02Apr88 36(1) 12 [Vincent Corea/Jeanne Roberts] A: Catch Me, I'm Falling; P: Kae Williams Jr.	412
2119. I NEED YOU Pointer Sisters Planet 13639 08Oct83 48(2) 15 [John Alden Black/Richard Feldman/Nan O'Byrne] A: Break Out; P: Richard Perry	412
2120. LIVE EVERY MOMENT REO Speedwagon Epic 05412 13Jul85 34(1) 11 [Kevin Cronin] A: Wheels Are Turnin'; P: Kevin Cronin/Alan Gratzer/Gary Richrath	412

Ranking the Singles

Rank. TITLE Act Label Number Entry Date Peak(Wks) Tot Wks [Writers] A: Proximate Album; P: Producer	Score
2121. **WHAT DOES IT TAKE** Honeymoon Suite Warner Bros. 28670 12Jul86 52(1) 16 [Dermot Grehan] A: *The Big Prize*; P: Bruce Fairbairn	412
2122. **DANCE LITTLE SISTER (Part One)** Terence Trent D'Arby Columbia 08023 10Sep88 30(2) 11 [Terence Trent D'Arby] A: *Introducing The Hardline According To Terence Trent D'Arby*; P: Terence Trent D'Arby/Martyn Ware	411
2123. **MY TOWN** Michael Stanley Band EMI America 8178 01Oct83 39(1) 10 [Michael Stanley] A: *You Can't Fight Fashion*; P: Michael Stanley Band	410
2124. **TIED UP** Olivia Newton-John MCA 52155 15Jan83 38(3) 11 [John Farrar/Lee Ritenour] A: *Olivia's Greatest Hits, Vol. 2*; P: John Farrar	410
2125. **LITTLE JACKIE WANTS TO BE A STAR** Lisa Lisa And Cult Jam Columbia 68674 15Apr89 29(1) 11 [Curtis Bedeau/Gerard Charles/Hugh Clarke/Brian George/Lucien George/ Paul George] A: *Straight To The Sky*; P: Full Force	410
2126. **TREAT HER LIKE A LADY** The Temptations Gordy 1765 15Dec84 48(2) 11 [Otis Clayburne Williams/Ollie Woodson] A: *Truly For You*; P: Ralph Johnson/Al McKay	409
2127. **DIG THE GOLD** Joyce Cobb Cream 7939 24Nov79 42(1) 12 [Joyce Cobb/Lynn Lewis] A: *No Album*; P: Andy Black	409
2128. **TWO HEARTS** Stephanie Mills featuring Teddy Pendergrass 20th Century 2492 16May81 40(2) 13 [Tawatha Agee/Reggie Lucas/James Mtume] A: *Stephanie*; P: Reggie Lucas/James Mtume	409
2129. **I WANNA HEAR IT FROM YOUR LIPS** Eric Carmen Geffen 29118 19Jan85 35(1) 11 [Eric Carmen/Dean Pitchford] A: *Eric Carmen (II)*; P: Bob Gaudio	408
2130. **LAST TRAIN TO LONDON** Electric Light Orchestra Jet 5067 08Dec79 39(1) 11 [Jeff Lynne] A: *Discovery*; P: Jeff Lynne	408
2131. **STRUNG OUT** Steve Perry Columbia 04598 08Sep84 40(1) 13 [Craig Krampf/Steve Perry/Billy Steele] A: *Street Talk*; P: Steve Perry	408
2132. **TASTY LOVE** Freddie Jackson Capitol 5616 08Nov86 41(1) 12 [Freddie Jackson/Paul Laurence] A: *Just Like The First Time*; P: Paul Laurence	407
2133. **DESTINATION UNKNOWN** Missing Persons Capitol 5161 02Oct82 42(2) 14 [Dale Bozzio/Terry Bozzio/Warren Cuccurullo] A: *Missing Persons*; P: Ken Scott	406
2134. **PARANOIMIA** The Art Of Noise with Max Headroom China 43002 16Aug86 34(1) 12 [Anne Dudley/Jonathan Jeczalik/Gary Langan] A: *In Visible Silence*; P: Art Of Noise	406
2135. **IN THE DARK** Billy Squier Capitol 5040 12Sep81 35(2) 12 [Billy Squier] A: *Don't Say No*; P: Reinhold Mack/Billy Squier	405
2136. **CATCH ME I'M FALLING** Real Life Curb/MCA 52362 24Mar84 40(1) 11 [David Sterry/Richard Zatorski] A: *Heart Land*; P: Steve Hillage	405
2137. **BEAT OF A HEART** Scandal Featuring Patty Smyth Columbia 04750 26Jan85 41(2) 14 [Keith Mack/Zack Smith/Patty Smyth] A: *Warrior*; P: Mike Chapman	404
2138. **BOOM BOOM (Let's Go Back To My Room)** Paul Lekakis ZYX 1266 07Mar87 43(2) 13 [Riccardo Ballerini/Michele Chieregato/Thomas Hooker/ Stefano Montin/Roberto Turatti] A: *Tattoo It*; P: Michele Chieregato/Roberto Turatti	404
2139. **WALK AWAY** Donna Summer Casablanca 2300 13Sep80 36(2) 11 [Pete Bellotte/Harold Faltermeyer] A: *Walk Away - Collector's Edition (The Best Of 1977-1980)*; P: Pete Bellotte/Giorgio Moroder	404
2140. **HANDLE WITH CARE** Traveling Wilburys Wilbury 7-27732 29Oct88 45(1) 14 [Bob Dylan/George Harrison/Jeff Lynne/ Roy Orbison/Tom Petty] A: *Volume 1*; P: George Harrison/Jeff Lynne	404
2141. **THE LOVE PARADE** The Dream Academy Reprise 28750 19Apr86 36(2) 11 [Gilbert Gabriel/Nick Laird-Clowes] A: *The Dream Academy*; P: Alan Tarney	404
2142. **I BELIEVE** Chilliwack Millennium 13102 16Jan82 33(2) 11 [William A. Henderson] A: *Wanna Be Star*; P: Bill Henderson/Brian MacLeod	403
2143. **I'D RATHER LEAVE WHILE I'M IN LOVE** Rita Coolidge A&M 2199 17Nov79 38(1) 10 [Peter Allen/Carole Bayer Sager] A: *Satisfied*; P: David Anderle/Booker T. Jones	403
2144. **I DON'T CARE ANYMORE** Phil Collins Atlantic 89877 12Feb83 39(3) 11 [Phil Collins] A: *Hello, I Must Be Going!*; P: Phil Collins	403
2145. **SAME OLE LOVE (365 Days A Year)** Anita Baker Elektra 69484 21Mar87 44(2) 14 [Marilyn McLeod/Darryl Keith Roberts] A: *Rapture*; P: Michael J. Powell	403
2146. **WAKE UP (Next To You)** Graham Parker And The Shot Elektra 69654 04May85 39(1) 12 [Graham Parker] A: *Steady Nerves*; P: Graham Parker/William Wittman	402

Rank. TITLE Act Label Number Entry Date Peak(Wks) Tot Wks [Writers] A: Proximate Album; P: Producer	Score
2147. **SOMEDAY, SOMEWAY** Marshall Crenshaw Warner Bros. 29974 10Jul82 36(2) 11 [Marshall Crenshaw] A: *Marshall Crenshaw*; P: Marshall Crenshaw/Richard Gottehrer	401
2148. **FRIENDS IN LOVE** Dionne Warwick And Johnny Mathis Arista 0673 17Apr82 38(2) 13 [Bill Champlin/David Foster/Jay Graydon] A: *Friends In Love/Dionne Warwick; Friends In Love/Johnny Mathis*; P: Jay Graydon	400
2149. **WRACK MY BRAIN** Ringo Starr Boardwalk 7-11-130 07Nov81 38(1) 11 [George Harrison] A: *Stop And Smell The Roses*; P: George Harrison	400
2150. **ISLAND OF LOST SOULS** Blondie Chrysalis 2603 29May82 37(2) 10 [Clement A. Bozewski/Deborah Harry/Christopher Stein] A: *The Hunter*; P: Mike Chapman	399
2151. **STONE COLD** Rainbow Mercury 76146 24Apr82 40(1) 12 [Ritchie Blackmore/Roger Glover/Joe Lynn Turner] A: *Straight Between The Eyes*; P: Roger Glover	399
2152. **SHOULD I STAY OR SHOULD I GO ('82)** The Clash Epic 03006 17Jul82 45(2) 13 [Nicky Headon/Michael Geoffrey Jones/Paul Simonon/Joe Strummer] A: *Combat Rock*; P: Clash	399
2153. **ASHES BY NOW** Rodney Crowell Warner Bros. 49224 10May80 37(1) 11 [Rodney Crowell] A: *But What Will The Neighbors Think*; P: Rodney Crowell/Craig Leon	399
2154. **HYPNOTIZE ME** Wang Chung Geffen 28359 06Jun87 36(1) 12 [Nick Feldman/Jack Hues] A: *Mosaic*; P: Peter Wolf	399
2155. **DO YOU WANT CRYING** Katrina And The Waves Capitol 5450 27Jul85 37(2) 10 [Vince De La Cruz] A: *Katrina And The Waves*; P: Pat Collier/Katrina And The Waves	398
2156. **HIPPY HIPPY SHAKE** The Georgia Satellites Elektra 69366 22Oct88 45(2) 14 [Chan Romero] A: *Cocktail Soundtrack*; P: Georgia Satellites/Brenden O'Brien	398
2157. **CIRCLES** Atlantic Starr A&M 2392 27Mar82 38(2) 11 [David E. Lewis/Wayne I. Lewis] A: *Brilliance*; P: James Anthony Carmichael	398
2158. **SQUARE BIZ** Teena Marie Gordy 7202 25Jul81 50(1) 13 [Teena Marie/Allen McGrier] A: *It Must Be Magic*; P: Teena Marie	397
2159. **DON'T WANT TO WAIT ANYMORE** The Tubes Capitol 5007 20Jun81 35(2) 12 [Rick Anderson/Michael Cotten/David Foster/Charles Prince/ William Spooner/Roger Steen/Fee Waybill/Vince Welnick] A: *The Completion Backward Principle*; P: David Foster	397
2160. **ALL THOSE LIES** Glenn Frey Asylum 69857 11Dec82 41(3) 12 [Glenn Frey/Jack Tempchin] A: *No Fun Aloud*; P: Allan Blazek/Glenn Frey/Jim Ed Norman	397
2161. **MIDAS TOUCH** Midnight Star Solar 69525 20Sep86 42(1) 14 [Bo Watson/June Watson Williams] A: *Headlines*; P: Reggie Calloway	396
2162. **WHO'S MAKING LOVE** Blues Brothers Atlantic 3785 20Dec80 39(2) 11 [Homer Banks/Bettye Crutcher/Don Davis/ Raymond Earl Jackson] A: *Made In America*; P: Paul Shaffer/Bob Tischler	395
2163. **RIGHT BEFORE YOUR EYES** America Capitol 5177 27Nov82 45(2) 13 [Ian Thomas] A: *View From The Ground*; P: Bobby Colomby	395
2164. **SECOND NATURE** Dan Hartman MCA 52519 09Feb85 39(1) 12 [Dan Hartman/Charles E. Kaufman] A: *I Can Dream About You*; P: Dan Hartman/Jimmy Iovine	395
2165. **I WILL BE THERE** Glass Tiger Manhattan 50066 28Feb87 34(2) 11 [Al Connelly/Alan Frew/Michael Hanson/Wayne Parker/Sam Reid] A: *The Thin Red Line*; P: Jim Vallance	395
2166. **ALL MY LIFE** Kenny Rogers Liberty 1495 30Apr83 37(2) 11 [Charles David Robbins/Jeff Silbar/Van Stephenson] A: *We've Got Tonight*; P: David Foster/Kenny Rogers	395
2167. **ANGEL IN BLUE** The J. Geils Band EMI America 8100 22May82 40(1) 11 [Seth Justman] A: *Freeze-Frame*; P: Seth Justman	395
2168. **ONE FOR THE MOCKINGBIRD** Cutting Crew Virgin 99464 06Jun87 38(1) 11 [Nick Van Eede] A: *Broadcast*; P: Terry Brown/Cutting Crew	395
2169. **LOOK OUT ANY WINDOW** Bruce Hornsby & The Range RCA 8678 23Jul88 35(1) 12 [Bruce Hornsby/John Hornsby] A: *Scenes From The Southside*; P: Neil Dorfsman/Bruce Hornsby	395
2170. **OH GIRL** Boy Meets Girl A&M 2713 06Apr85 39(1) 11 [George Merrill/Shannon Rubicam] A: *Boy Meets Girl*; P: Tom Werman	393
2171. **DON'T LET GO** Wang Chung Geffen 29377 04Feb84 38(2) 11 [Nick Feldman/Jack Hues] A: *Points On The Curve*; P: Ross Cullum/Chris Hughes	393
2172. **LOOKING FOR A STRANGER** Pat Benatar Chrysalis 42688 23Apr83 39(2) 10 [Franne Golde/Peter Mclan] A: *Get Nervous*; P: Peter Coleman/Neil Geraldo	393
2173. **COUNT ME OUT** New Edition MCA 52703 02Nov85 51(1) 15 [Vincent Brantley/Ricky Timas] A: *All For Love*; P: Vincent Brantley/Rick Timas	392

Ranking the Singles

Rank. TITLE Act Label Number Entry Date Peak(Wks) Tot Wks [Writers] A: Proximate Album; P: Producer	Score
2174. WHAT A WONDERFUL WORLD Louis Armstrong A&M 3010 20Feb88 32(1) 11 [Bob Thiele/George David Weiss] A: Good Morning, Vietnam Soundtrack; P: David Anderle	392
2175. GIVE IT TO ME BABY Rick James Gordy 7179 30May81 40(2) 14 [Rick James] A: Street Songs; P: Rick James	390
2176. THROUGH THE FIRE Chaka Khan Warner Bros. 29025 27Apr85 60(1) 19 [David Foster/Tom Keane/Cynthia Weil] A: I Feel For You; P: David Foster/Humberto Gatica/Arif Mardin	390
2177. DON'T TELL ME YOU LOVE ME Night Ranger Boardwalk 15Jan83 40(3) 11 [Jack Blades] A: Dawn Patrol; P: Pat Glasser	390
2178. LIGHT OF DAY The Barbusters (Joan Jett & The Blackhearts) Blackheart 4-06692 21Feb87 33(1) 11 [Bruce Springsteen] A: Good Music; P: Jimmy Iovine	390
2179. COOL PLACES Sparks And Jane Wiedlin Atlantic 89866 16Apr83 49(2) 12 [Ron Mael/Russell Mael] A: In Outer Space; P: Ron Mael/Russell Mael	389
2180. WHERE DO THE CHILDREN GO Hooters Columbia 05854 05Apr86 38(1) 12 [Eric Bazilian/Rob Hyman] A: Nervous Night; P: Rick Chertoff	389
2181. AS LONG AS YOU FOLLOW Fleetwood Mac Warner Bros. 27644 03Dec88 43(1) 14 [Christine McVie/Eddy Quintela] A: Greatest Hits; P: Fleetwood Mac/Greg Ladanyi	389
2182. THE BIG MONEY Rush Mercury 884191 09Nov85 45(1) 14 [Geddy Lee/Alex Lifeson/Neil Peart] A: Power Windows; P: Peter Collins/Rush	388
2183. FLIRTIN' WITH DISASTER Molly Hatchet Epic 50822 05Jan80 42(1) 10 [Danny Joe Brown/Dave Hlubek/Banner Thomas] A: Flirtin' With Disaster; P: Tom Werman	388
2184. OH JULIE Barry Manilow Arista 0698 31Jul82 38(2) 11 [Shakin' Stevens] A: Oh, Julie!; P: Barry Manilow	388
2185. COMMUNICATION The Power Station Capitol 5511 07Sep85 34(2) 10 [Derek Bramble/Robert Palmer/Andy Taylor/Nigel John Taylor] A: The Power Station; P: Bernard Edwards	387
2186. NICE GIRLS Melissa Manchester Arista 1045 05Feb83 42(2) 11 [Jan Buckingham/Steve Buckingham/Mark Eugene Gray] A: Greatest Hits; P: Arif Mardin	387
2187. WOULDN'T IT BE GOOD Nik Kershaw MCA 52371 31Mar84 46(1) 13 [Nik Kershaw] A: Human Racing; P: Peter Collins	387
2188. TOO YOUNG Jack Wagner Qwest 28931 26Oct85 52(1) 14 [David Foster/Jay Graydon/Steve Kipner/Donny Osmond] A: Lighting Up The Night; P: Glen Ballard/Clif Magness	385
2189. COMING DOWN FROM LOVE Bobby Caldwell Clouds 21 19Apr80 42(1) 11 [Bobby Caldwell] A: Cat In The Hat; P: Bobby Caldwell/Steve Kimball	385
2190. LET IT BE ME Willie Nelson Columbia 03073 07Aug82 40(3) 12 [Gilbert Bécaud/Mann Curtis/Pierre Delanoë] A: Always On My Mind; P: Chips Moman	385
2191. RIGHT BETWEEN THE EYES Wax RCA 14306 15Mar86 43(2) 13 [Andrew Gold/Graham Gouldman] A: Magnetic Heaven; P: Philip Thornalley	384
2192. (You Can Still) ROCK IN AMERICA Night Ranger MCA 52305 03Dec83 51(1) 12 [Jack Blades/Brad Gillis] A: Midnight Madness; P: Pat Glasser	384
2193. DON'T LOOK DOWN - THE SEQUEL Go West Chrysalis 43141 01Aug87 39(1) 13 [Peter Cox/Richard Drummie] A: Dancing On The Couch-US Vers; P: Gary Stevenson	384
2194. LOVE MY WAY Psychedelic Furs Columbia 03340 05Mar83 44(2) 10 [John Ashton/Richard Butler/Tim Butler/Vince Ely] A: Forever Now; P: Todd Rundgren	383
2195. BOY, I'VE BEEN TOLD Sa-Fire Cutting 870 514-7 24Sep88 48(2) 16 [Mark Muniz] A: Sa-Fire; P: Carlos Rodgers/Peter Schwartz	382
2196. THE BORDERLINES Jeffrey Osborne A&M 2695 19Jan85 38(1) 11 [Raymond C. Jones] A: Don't Stop; P: George Duke	382
2197. USED TO BE Charlene & Stevie Wonder Motown 1650 30Oct82 46(3) 11 [Ken Hirsch/Ronald Norman Miller] A: Used To Be; P: Ronald Norman Miller	380
2198. TOM SAWYER Rush Mercury 76109 06Jun81 44(1) 13 [Pye Dubois/Geddy Lee/Alex Lifeson/Neil Peart] A: Moving Pictures; P: Terry Brown/Rush	380
2199. FIRST TIME LOVE Livingston Taylor Epic 50894 26Jul80 42(1) 10 [Patrick Alger/Peter Kaminsky] A: Man's Best Friend; P: Jeff Baxter/John Boylan	380
2200. ATOMIC Blondie Chrysalis 2410 17May80 39(1) 9 [Clement A. Bozewski/Jimmy Destri/Deborah Harry] A: Eat To The Beat; P: Mike Chapman	380
2201. GOODBYE IS FOREVER Arcadia Capitol 5542 01Feb86 33(1) 10 [Simon Le Bon/Nick Rhodes/Roger Taylor] A: So Red The Rose; P: Arcadia/Alex Sadkin	380
2202. YOU KEEP RUNNIN' AWAY 38 Special A&M 2431 21Aug82 38(2) 11 [Don Barnes/Jeff Carlisi/Jim Peterik] A: Special Forces; P: Rodney Mills	379
2203. BLAME IT ON LOVE Smokey Robinson & Barbara Mitchell Tamla 1684 02Jul83 48(1) 12 [Dave DeLuca/Ted Munda] A: Blame It On Love And All The Great Hits; P: George Tobin	379
2204. SHOW SOME RESPECT Tina Turner Capitol 5461 20Apr85 37(1) 10 [Terry Britten/Sue Shifrin] A: Private Dancer; P: Terry Britten	378
2205. LUCKY IN LOVE Mick Jagger Columbia 04893 27Apr85 38(2) 11 [Carlos Alomar/Mick Jagger] A: She's The Boss; P: Mick Jagger/Bill Laswell	378
2206. JUICY FRUIT Mtume Epic 03578 18Jun83 45(1) 12 [James Mtume] A: Juicy Fruit; P: James Mtume	377
2207. MURPHY'S LAW Cheri Venture 149 10Apr82 39(2) 12 [Geraldine Hunt/Daniel Joseph] A: Cheri; P: Geraldine Hunt/Freddie James	377
2208. I'M THE ONE Roberta Flack Atlantic 4068 24Jul82 42(1) 11 [William Eaton/Ralph MacDonald/William Salter] A: I'm The One; P: William Eaton/Roberta Flack/Ralph MacDonald/William Salter	377
2209. CENTERFIELD John Fogerty Warner Bros. 29053 25May85 44(2) 13 [John Fogerty] A: Centerfield; P: John Fogerty	377
2210. IF I WERE YOU Lulu Alfa 7011 21Nov81 44(1) 11 [Jerry Fuller/John Hobbs] A: Lulu; P: Mark London	376
2211. STARTING OVER AGAIN Dolly Parton RCA 11926 29Mar80 36(2) 10 [Bruce Sudano/Donna Summer] A: Dolly Dolly Dolly; P: Gary Klein	376
2212. THE FANATIC Felony Rock 'n' Roll 03497 12Feb83 42(2) 12 [Arthur Blea/Louis Ruiz/Dan Sands/Jeff Spry/Joe Spry] A: The Fanatic; P: Artie Kornfield/Don Rubin	375
2213. HANDS TIED Scandal Featuring Patty Smyth Columbia 04650 20Oct84 41(1) 13 [Mike Chapman/Holly Knight] A: Warrior; P: Mike Chapman	374
2214. TOMORROW PEOPLE Ziggy Marley And The Melody Makers Virgin 99347 14May88 39(1) 13 [Ziggy Marley] A: Conscious Party; P: Chris Frantz/Tina Weymouth	373
2215. STRAIGHT FROM THE HEART The Allman Brothers Band Arista 0618 08Aug81 39(2) 11 [Dickey Betts/Johnny Cobb] A: Brothers Of The Road; P: John Ryan	372
2216. WITHOUT YOUR LOVE Toto Columbia 06570 27Dec86 38(1) 11 [David Paich] A: Fahrenheit; P: Toto	372
2217. FOOL THAT I AM Rita Coolidge A&M 2281 13Dec80 46(2) 12 [Bruce Roberts/Carole Bayer Sager] A: Coast To Coast Soundtrack; P: David Anderle	372
2218. ALL NIGHT THING The Invisible Man's Band Mango 103 17May80 45(2) 10 [Clarence Burke Jr./Dean Gant/Alex Masucci] A: The Invisible Man's Band; P: Clarence Burke Jr./Alex Masucci	372
2219. GOOD TIMES INXS And Jimmy Barnes Atlantic 89237 20Jun87 47(1) 13 [Harry Vanda/George Young] A: Lost Boys Soundtrack; P: Mark Opitz	372
2220. SAY IT'S GONNA RAIN Will To Power Epic 07908 18Jun88 49(1) 14 [Bob Rosenberg] A: Will To Power; P: Bob Rosenberg	371
2221. STRIP Adam Ant Epic 04337 04Feb84 42(1) 13 [Adam Ant/Marco Pirroni] A: Strip; P: Phil Collins/Hugh Padgham	370
2222. MAN ON THE CORNER Genesis Atlantic 4025 20Mar82 40(2) 11 [Phil Collins] A: Abacab; P: Genesis	370
2223. I CAN TAKE CARE OF MYSELF Billy & The Beaters Alfa 7002 25Apr81 39(1) 11 [Billy Vera] A: Billy & The Beaters; P: Jeff Baxter	370
2224. THE CLOSER YOU GET Alabama RCA 13524 07May83 38(2) 11 [Mark Eugene Gray/James P. Pennington] A: The Closer You Get; P: Alabama/Harold Shedd	369
2225. ONLY YOU Commodores Motown 1694 17Sep83 54(2) 13 [Milan Williams] A: Commodores 13; P: Milan Williams	369
2226. SHINE SHINE Barry Gibb MCA 52443 01Sep84 37(2) 10 [George Bitzer/Barry Gibb/Maurice Gibb] A: Now Voyager; P: Barry Gibb/Karl Richardson	369
2227. I CAN'T HELP IT Bananarama London 886212 21Nov87 47(1) 13 [Matt Aitken/Sarah Dallin/Siobhan Fahey/Mike Stock/Pete Waterman/Keren Woodward] A: Wow!; P: Matt Aitken/Mike Stock/Pete Waterman	369
2228. WE'RE GOING ALL THE WAY Jeffrey Osborne A&M 2618 25Feb84 48(1) 12 [Barry Mann/Cynthia Weil] A: Stay With Me Tonight; P: George Duke	368
2229. IF I HAD MY WISH TONIGHT David Lasley EMI America 8111 13Mar82 36(2) 10 [Randy Goodrum/Dave Loggins] A: Missin' Twenty Grand; P: David Lasley	368
2230. SAY IT AGAIN Santana Columbia 04758 23Feb85 46(1) 11 [Val Garay/Steve Goldstein/Anthony La Peau] A: Beyond Appearances; P: Val Garay	367
2231. THE NIGHT IS STILL YOUNG Billy Joel Columbia 05657 05Oct85 34(2) 10 [Billy Joel] A: Greatest Hits Vol. I & II; P: Phil Ramone	367

Ranking the Singles

Rank. TITLE Act Label Number Entry Date Peak(Wks) Tot Wks [Writers] A: Proximate Album; P: Producer	Score
2232. MY HEROES HAVE ALWAYS BEEN COWBOYS Willie Nelson Columbia 11186 09Feb80 44(2) 10 [Sharon Vaughn] A: The Electric Horseman (Soundtrack); P: Willie Nelson/Sydney Pollack	366
2233. HEADED FOR A FALL Firefall Atlantic 3657 12Apr80 35(2) 9 [Rick Roberts] A: Undertow; P: Howard Albert/Ron Albert	366
2234. HOLD ON Kansas Kirshner 4291 20Sep80 40(1) 11 [Kerry Livgren] A: Audio-Visions; P: Kansas	366
2235. GOTTA HAVE MORE LOVE Climax Blues Band Warner Bros. 49605 15Nov80 47(2) 12 [Greg Guidry/Randy Guidry/Jeff Silbar] A: Flying The Flag; P: John Ryan	366
2236. YOU BETTER RUN Pat Benatar Chrysalis 2450 26Jul80 42(2) 11 [Eddie Brigati/Felix Cavaliere] A: Crimes Of Passion; P: Keith Olsen	366
2237. CANDY GIRL New Edition Streetwise 1108 07May83 46(1) 11 [Michael Jonzun/Maurice Starr] A: Candy Girl; P: Michael Jonzun/Maurice Starr	366
2238. BABY CAN I HOLD YOU Tracy Chapman Elektra 69356 05Nov88 48(3) 12 [Tracy Chapman] A: Tracy Chapman; P: David Kershenbaum	365
2239. LIVE EVERY MINUTE Ali Thomson A&M 2260 13Sep80 42(2) 11 [Ali Thomson] A: Take A Little Rhythm; P: Jon Kelly/Ali Thomson	365
2240. NEED YOUR LOVING TONIGHT Queen Elektra 47086 29Nov80 44(3) 11 [John Deacon] A: The Game; P: Queen	365
2241. DON'T SAY GOODNIGHT (It's Time For Love) (Parts 1 And 2) The Isley Brothers T-Neck 2290 19Apr80 39(2) 9 [Ernie Isley/Marvin Isley/O'Kelly Isley/Ronald Isley/Rudolph Isley/Chris Jasper] A: Go All The Way; P: Ernie Isley/Marvin Isley/O'Kelly Isley/Ronald Isley/Rudolph Isley/Chris Jasper	365
2242. THE FINEST The S.O.S. Band Tabu 4-05848 17May86 44(1) 13 [James Samuel Harris/Terry Lewis] A: Sands Of Time; P: Jimmy Jam Harris/Terry Lewis	364
2243. LOVE OR LET ME BE LONELY Paul Davis Arista 0697 17Jul82 40(2) 10 [Jerry Peters/Anita Poree/Skip Scarborough] A: Cool Night; P: Paul Davis/Ed Seay	364
2244. YOU Earth, Wind & Fire ARC Columbia 11-11407 22Nov80 48(3) 12 [David Foster/Brenda Russell/Maurice White] A: Faces; P: Maurice White	364
2245. SPANISH EDDIE Laura Branigan Atlantic 89531 27Jul85 40(2) 11 [Charles E. Cochran/David Palmer] A: Hold Me; P: Jack White	364
2246. RHYTHM OF LOVE Yes Atco 99419 19Dec87 40(1) 12 [Jon Anderson/Tony Kaye/Trevor Rabin/Chris Squire/Alan (2) White] A: Big Generator; P: Paul DeVilliers/Trevor Horn/Trevor Rabin/Yes	364
2247. DO YOU WANNA GET AWAY Shannon Mirage 99655 06Apr85 49(1) 15 [Chris Barbosa/Ann Godwin] A: Do You Wanna Get Away; P: Chris Barbosa/Mark Liggett	364
2248. LET THE FEELING FLOW Peabo Bryson Capitol 5065 09Jan82 42(1) 12 [Peabo Bryson] A: I Am Love; P: Peabo Bryson/Johnny Pate	362
2249. AFTER I CRY TONIGHT Lanier and Co. Larc 81010 04Dec82 48(2) 13 [Phillip Mitchell] A: Lanier & Co.; P: Gene Miller	361
2250. IN GOD'S COUNTRY U2 Island 99385 05Dec87 44(1) 12 [Adam Clayton/Dave Evans/Paul Hewson/Larry Mullen] A: The Joshua Tree; P: Brian Eno/Daniel Lanois	361
2251. IF ONLY YOU KNEW Patti LaBelle Philadelphia Int'l 4248 07Jan84 46(1) 13 [Cynthia Biggs/Kenneth Gamble/Dexter Wansel] A: I'm In Love Again; P: Kenny Gamble/Dexter Wansel	360
2252. MISFIT Curiosity Killed The Cat Mercury 888674 01Aug87 42(1) 13 [Tobias Andersen/Julian Brookhouse/Michael Drummond/Nick Thorp/Ben Volpeliere-Pierrot] A: Keep Your Distance; P: Stewart Levine	358
2253. AIN'T NOTHIN' GOIN' ON BUT THE RENT Gwen Guthrie Polydor 885106 02Aug86 42(1) 13 [Gwen Guthrie] A: Good To Go Lover; P: Gwen Guthrie	358
2254. SO FAR SO GOOD Sheena Easton EMI America 8332 02Aug86 43(1) 12 [Tom Snow/Cynthia Weil] A: About Last Night... Soundtrack; P: Narada Michael Walden	358
2255. CROSS MY HEART Eighth Wonder WTG 31-08036 19Nov88 56(2) 16 [Michael Jay] A: Fearless; P: Pete Hammond	358
2256. BAD BOY/HAVING A PARTY Luther Vandross Epic 03205 30Oct82 55(5) 12 [Marcus Miller/Luther Vandross//Sam Cooke] A: Forever, For Always, For Love; P: Luther Vandross	357
2257. LIVE MY LIFE Boy George Virgin 99390 26Dec87 40(1) 12 [Danny Sembello/Allee Willis] A: Hiding Out Soundtrack; P: John Robie	357
2258. KNOCKED OUT Paula Abdul Virgin 99329 18Jun88 41(1) 13 [Kenneth Edmonds/Antonio Reid/Daryl Simmons] A: Forever Your Girl; P: Kenneth Edmonds/L.A. Reid	357
2259. WILD HORSES Gino Vannelli CBS Associated 06699 18Apr87 55(2) 15 [Roy Freeland/Gino Vannelli] A: Big Dreamers Never Sleep; P: Gino Vannelli/Joe Vannelli/Ross Vannelli	357
2260. BACKFIRED Debbie Harry Chrysalis 2526 15Aug81 43(1) 10 [Bernard Edwards/Nile Rodgers] A: KooKoo; P: Bernard Edwards/Nile Rodgers	357
2261. BETCHA SAY THAT Gloria Estefan and Miami Sound Machine Epic 07371 05Sep87 36(2) 11 [Larry Dermer/Joe Galdo/Rafael Vigil] A: Let It Loose; P: Lawrence Dermer/Emilio Estefan/Joe Galdo/Rafael Vigil	356
2262. YOU SAVED MY SOUL Burton Cummings Alfa 7008 12Sep81 37(2) 11 [Burton Cummings] A: Sweet Sweet; P: Burton Cummings/Bruce Robb	356
2263. PLAY THE GAME Queen Elektra 46652 28Jun80 42(2) 9 [Freddie Mercury] A: The Game; P: Queen	355
2264. HOLDIN' ON Tané Cain RCA 13287 14Aug82 37(3) 11 [Pug Baker/Jonathan Cain] A: Tane Cain; P: Jonathan Cain/Keith Olsen	355
2265. HAVEN'T YOU HEARD Patrice Rushen Elektra 46551 26Jan80 42(2) 9 [Sheree Brown/Charles Mims/Patrice Rushen/Freddie Washington] A: Pizzazz; P: Reggie Andrews/Charles Mims Jr./Patrice Rushen	355
2266. CLONES (We're All) Alice Cooper Warner Bros. 49204 17May80 40(1) 9 [David Carron] A: Flush The Fashion; P: Roy Thomas Baker	355
2267. INDESTRUCTIBLE Four Tops Arista 1-9706 20Aug88 35(1) 11 [Harvey Price/Bobby Sandstrom] A: Indestructible; P: Bobby Sandstrom	354
2268. AND SHE WAS Talking Heads Sire 28917 07Sep85 54(1) 20 [David Byrne/Chris Frantz/Jerry Harrison/Tina Weymouth] A: Little Creatures; P: Talking Heads	353
2269. DON'T LEAVE ME THIS WAY Communards MCA/London 52928 27Dec86 40(1) 13 [Kenneth Gamble/Cary Gilbert/Leon Huff] A: The Communards; P: Mike Thorne	353
2270. DON'T NEED A GUN Billy Idol Chrysalis 43087 24Jan87 37(1) 9 [Billy Idol] A: Whiplash Smile; P: Keith Forsey	353
2271. SHELTER Lone Justice Geffen 28520 10Jan87 47(2) 12 [Maria McKee/Steven Van Zandt] A: Shelter; P: Jimmy Iovine/Lone Justice/Steven Van Zandt	353
2272. DON'T WASTE YOUR TIME Yarbrough & Peoples Total Experience 2400 14Apr84 48(1) 12 [Jonah Ellis] A: Be A Winner; P: Jonah Ellis	353
2273. WORKING IN THE COAL MINE Devo Full Moon 47204 05Sep81 43(1) 12 [Allen Toussaint] A: New Traditionalists; P: Devo	352
2274. EVERYDAY PEOPLE Joan Jett And The Blackhearts Blackheart 52272 10Sep83 37(1) 9 [Sylvester Stewart] A: Album; P: Ritchie Cordell/Joan Jett/Kenny Laguna	352
2275. LOVE'S GOT A LINE ON YOU Scandal Columbia 03615 02Apr83 59(2) 13 [Kathe Green/Zack Smith] A: Scandal; P: Vini Poncia	352
2276. CARAVAN OF LOVE Isley, Jasper, Isley CBS Associated 05611 07Dec85 51(2) 14 [Ernie Isley/Marvin Isley/Chris Jasper] A: Caravan Of Love; P: Ernie Isley/Marvin Isley/Chris Jasper	352
2277. THE NIGHT The Animals I.R.S. 9920 13Aug83 48(2) 10 [Eric Burdon/Don Evans/John Sterling] A: Ark; P: Eric Burdon/Chas Chandler/Stephen Lipson/Alan Price/John Steel/Hilton Valentine	351
2278. LET'S HANG ON Barry Manilow Arista 0675 20Mar82 32(1) 10 [Bob Crewe/Sandy Linzer/Denny Randell] A: If I Should Love Again; P: Barry Manilow	351
2279. PEEK-A-BOO Siouxsie & The Banshees Geffen 27760 15Oct88 53(2) 14 [Susan Ballion/Peter Clarke/Steven Severin] A: Peepshow; P: Mike Hedges/Siouxsie & The Banshees	351
2280. IT'S RAINING MEN The Weather Girls Columbia 03354 22Jan83 46(3) 11 [Paul Jabara/Paul Shaffer] A: Success; P: Bob Esty/Paul Jabara	350
2281. ALL OF ME FOR ALL OF YOU 9.9 RCA 14082 24Aug85 51(1) 13 [Richard "Dimples" Fields/Josh Sklair/Belinda Wilson] A: 9.9; P: Richard "Dimples" Fields	350
2282. LITTLE FIGHTER White Lion Atlantic 88874 24Jun89 52(1) 14 [Vito Bratta/Mike Tramp] A: Big Game; P: Michael Wagener	350
2283. TOO TIGHT Con Funk Shun Mercury 76089 24Jan81 40(1) 10 [Michael V. Cooper] A: Touch; P: Con Funk Shun	350
2284. LOVE ALWAYS El DeBarge Gordy 1857 09Aug86 43(1) 12 [Burt Bacharach/Bruce Roberts/Carole Bayer Sager] A: El DeBarge; P: Burt Bacharach/Carole Bayer Sager	349
2285. LET ME GO Ray Parker Jr. Arista 0695 17Jul82 38(2) 9 [Ray Parker Jr.] A: The Other Woman; P: Ray Parker Jr.	349
2286. HOW CAN I REFUSE Heart Epic 04047 13Aug83 44(2) 11 [Mark Andes/Denny Carmassi/Sue Ennis/Howard Leese/Ann Wilson/Nancy Wilson] A: Passionworks; P: Keith Olsen	348
2287. IF YOU WANT MY LOVE Cheap Trick Epic 02968 05Jun82 45(2) 11 [Rick Nielsen] A: One On One; P: Roy Thomas Baker	346
2288. FLY AWAY Blackfoot Atco 7331 20Jun81 42(1) 12 [Rick Medlocke/Jakson Spires] A: Marauder; P: Al Nalli/Henry Weck	345
2289. RESTLESS Starpoint Elektra 69561 22Mar86 46(2) 12 [Keith Diamond/Ernesto Phillips] A: Restless; P: Keith Diamond/Lionel Job	344
2290. FIRE WOMAN The Cult Sire 27543 27May89 46(2) 11 [Ian Astbury/Billy Duffy] A: Sonic Temple; P: Bob Rock	344

Rank. TITLE Act Label Number Entry Date Peak(Wks) Tot Wks [Writers] A: Proximate Album; P: Producer	Score
2291. TEACHER TEACHER Rockpile Columbia 11388 22Nov80 51(1) 12 [Eddie Phillips/Ken Pickett] A: Seconds Of Pleasure; P: Global Record Prodns	344
2292. BREAKIN' AWAY Al Jarreau Warner Bros. 49842 05Dec81 43(1) 10 [Tom Canning/Jay Graydon/Al Jarreau] A: Breakin' Away; P: Jay Graydon	343
2293. SHADDAP YOU FACE Joe Dolce MCA 51053 02May81 53(1) 14 [Joe Dolce] A: Shaddap You Face; P: Joe Dolce/Ian McKenzie	343
2294. 20/20 George Benson Warner Bros. 29120 15Dec84 48(2) 13 [Randy Goodrum/Steve Kipner] A: 20/20; P: Russ Titelman	343
2295. DON'T LET HIM KNOW Prism Capitol 5082 30Jan82 39(2) 10 [Bryan Adams/Jim Vallance] A: Small Change; P: Bob Carter	343
2296. MUSIC TIME Styx A&M 2625 05May84 40(2) 9 [Dennis DeYoung] A: Caught In The Act - Live; P: Styx	342
2297. STAND BY ME Maurice White Columbia 05571 31Aug85 50(2) 13 [Ben E. King/Jerry Leiber/Mike Stoller] A: Maurice White; P: Maurice White	342
2298. ONE-TRICK PONY Paul Simon Warner Bros. 49601 25Oct80 40(2) 11 [Paul Simon] A: One-Trick Pony; P: Phil Ramone/Paul Simon	342
2299. PRETTY IN PINK Psychedelic Furs A&M 2826 12Apr86 41(1) 11 [John Ashton/Richard Butler/Tim Butler/Vince Ely/Duncan Kilburn/Roger Morris] A: Pretty In Pink Soundtrack; P: Charles Harrowell/Psychedelic Furs	342
2300. WOMEN Foreigner Atlantic 3651 16Feb80 41(2) 9 [Michael Leslie Jones] A: Head Games; P: Roy Thomas Baker/Mick Jones/Ian McDonald	341
2301. WHO DO YOU THINK YOU'RE FOOLIN' Donna Summer Geffen 49664 21Feb81 40(2) 11 [Pete Bellotte/Sylvester Levay/Jerry Rix] A: The Wanderer; P: Pete Bellotte/Giorgio Moroder	341
2302. GIVE A LITTLE BIT MORE Cliff Richard EMI America 8076 25Apr81 41(2) 11 [Andy Hill/John Philip Hodge] A: I'm No Hero; P: Alan Tarney	340
2303. BABY COME AND GET IT Pointer Sisters Planet 14041 23Mar85 44(2) 11 [James Ingram/Barry Mann/Cynthia Weil] A: Break Out; P: Richard Perry	340
2304. SOMEBODY LIKE YOU 38 Special A&M 2854 19Jul86 48(1) 12 [Don Barnes/Jeff Carlisi/Larry Steele/Jim Vallance/Donnie Van Zant] A: Strength In Numbers; P: Keith Olsen	340
2305. JUST BE GOOD TO ME The S.O.S. Band Tabu 4-03955 27Aug83 55(1) 14 [James Samuel Harris/Terry Lewis] A: On The Rise; P: Jimmy Jam Harris/Terry Lewis	340
2306. (Baby Tell Me) CAN YOU DANCE Shanice Wilson A&M 2939 07Nov87 50(1) 13 [Bryan Loren] A: Discovery; P: Bryan Loren	339
2307. HOT HOT HOT (Radio Edit) Buster Poindexter And His Banshees Of Blue RCA 5357 12Dec87 45(1) 13 [Alphonsus Cassell] A: Buster Poindexter; P: Hank Medress	339
2308. JOHNNY CAN'T READ Don Henley Asylum 69971 21Aug82 42(2) 11 [Don Henley/Danny Kortchmar] A: I Can't Stand Still; P: Don Henley/Danny Kortchmar/Greg Ladanyi	339
2309. WORDS Missing Persons Capitol 5127 03Jul82 42(1) 11 [Terry Bozzio/Warren Cuccurullo] A: Missing Persons; P: Ken Scott	339
2310. RAIN FOREST Paul Hardcastle Profile 7059 12Jan85 57(1) 18 [Paul Hardcastle] A: Rain Forest; P: Paul Hardcastle	339
2311. YOU'RE MY LATEST, MY GREATEST INSPIRATION Teddy Pendergrass Philadelphia Int'l 2619 09Jan81 43(1) 11 [Kenneth Gamble/Leon Huff] A: It's Time For Love; P: Kenny Gamble/Leon Huff	338
2312. FOOL MOON FIRE Walter Egan Backstreet 52200 09Apr83 46(1) 10 [Walter Egan] A: Wild Exhibitions; P: Walter Egan/Duane Scott	338
2313. AUTOGRAPH John Denver RCA 11915 23Feb80 52(2) 10 [John Denver] A: Autograph; P: Milton Okun	338
2314. STAYING WITH IT Firefall Atlantic 3791 24Jan81 37(2) 9 [John Lewis Parker/Tom Snow] A: Clouds Across The Sun; P: Kyle Lehning	337
2315. GO Asia Geffen 07Dec85 46(2) 11 [Geoffrey Downes/John Wetton] A: Astra; P: Geoffrey Downes/Mike Stone	337
2316. THE HORIZONTAL BOP Bob Seger Capitol 4951 08Nov80 42(2) 12 [Bob Seger] A: Against The Wind; P: Ed Andrews/Bob Seger	336
2317. JONES VS. JONES Kool & The Gang De-Lite 813 16May81 39(2) 11 [Robert Bell/Ronald Bell/George M. Brown/Eumir Deodato/Robert Mickens/Claydes Smith/J.T. Taylor/Dennis Thomas/Earl Toon] A: Celebrate!; P: Eumir Deodato	336
2318. I CAN'T STAND STILL Don Henley Asylum 69931 15Jan83 48(1) 11 [Don Henley/Danny Kortchmar] A: I Can't Stand Still; P: Don Henley/Danny Kortchmar/Greg Ladanyi	336
2319. BODY ROCK Maria Vidal EMI America 8233 08Sep84 48(2) 12 [John Bettis/Sylvester Levay] A: No Album; P: Phil Galdston/Sylvester Levay	336
2320. LOVE ME IN A SPECIAL WAY DeBarge Gordy 1723 17Mar84 45(1) 11 [El DeBarge] A: In A Special Way; P: El DeBarge	336

Ranking the Singles

Rank. TITLE Act Label Number Entry Date Peak(Wks) Tot Wks [Writers] A: Proximate Album; P: Producer	Score
2321. LAY IT DOWN Ratt Atlantic 89546 06Jul85 40(1) 11 [Robbin Crosby/Juan Croucier/Warren DeMartini/Stephen Pearcy] A: Invasion Of Your Privacy; P: Beau Hill	335
2322. RITUAL Dan Reed Network Mercury 870183 12Mar88 38(1) 11 [Dan Edward Reed] A: Dan Reed Network; P: Bruce Fairbairn	335
2323. MY PRAYER Ray, Goodman & Brown Polydor 2116 23Aug80 47(2) 10 [Georges Boulanger/Jimmy Kennedy] A: Ray, Goodman & Brown II; P: Vincent Castellano	335
2324. OLD TIME ROCK & ROLL Bob Seger & The Silver Bullet Band Capitol 5276 17Sep83 48(1) 11 [George Henry Jackson/Thomas Earl Jones III] A: Risky Business Soundtrack; P: Muscle Shoals Rhythm Section/Bob Seger	334
2325. SOMETIMES A FANTASY Billy Joel Columbia 11379 11Oct80 36(2) 9 [Billy Joel] A: Glass Houses; P: Phil Ramone	334
2326. THE WAY TO YOUR HEART Soulsister EMI 50217 23Sep89 41(2) 10 [Jan Leyers/Paul Michiels] A: It Takes Two; P: Jan Leyers/Paul Michiels	333
2327. RED LIGHT Linda Clifford RSO 1041 09Aug80 41(1) 11 [Michael Gore/Dean Pitchford] A: I'm Yours; P: Gil Askey/Michael Gore	333
2328. A NIGHT TO REMEMBER Shalamar Solar 48005 10Apr82 44(2) 10 [Nidra Beard/Dana Meyers/Charmaine Sylvers] A: Friends; P: Leon Sylvers	333
2329. I NEED YOUR LOVING The Human League A&M 2893 06Dec86 44(1) 11 [Herman R. Davis/Dave Eiland/James Samuel Harris/Terry Lewis/Langston Richey/Danny Williams] A: Crash; P: Jimmy Jam Harris/Terry Lewis	333
2330. AFTER THE FIRE Roger Daltrey Atlantic 89491 14Sep85 48(1) 13 [Pete Townshend] A: Under A Raging Moon; P: Alan Shacklock	333
2331. TI AMO Laura Branigan Atlantic 89608 03Nov84 55(3) 12 [Giancarlo Bigazzi/Umberto Tozzi/Diane Warren] A: Self Control; P: Robbie Buchanan/Jack White	332
2332. WAR GAMES Crosby, Stills & Nash Atlantic 89912 25Jun83 45(1) 9 [Stephen Stills] A: Allies; P: Stanley Johnston/Graham Nash/Stephen Stills	331
2333. INSIDE LOVE (So Personal) George Benson Warner Bros. 29649 14May83 43(1) 10 [Michael Jones] A: In Your Eyes; P: Kashif/Arif Mardin	331
2334. WORKIN' FOR A LIVIN' Huey Lewis And The News Chrysalis 2630 14Aug82 41(2) 9 [Mario Cipollina/Johnny Colla/Bill Gibson/Chris Hayes/Sean Hopper/Huey Lewis] A: Picture This; P: Huey Lewis And The News	331
2335. ROCKIN' INTO THE NIGHT 38-Special A&M 2205 02Feb80 43(1) 9 [Jim Peterik/Gary Smith/Frankie Sullivan] A: Rockin' Into The Night; P: Rodney Mills	330
2336. MY KINDA LOVER Billy Squier Capitol 5037 28Nov81 45(1) 10 [Billy Squier] A: Don't Say No; P: Reinhold Mack/Billy Squier	330
2337. IF IT AIN'T ONE THING... IT'S ANOTHER Richard "Dimples" Fields Boardwalk 7-11-139 03Apr82 47(3) 10 [Richard "Dimples" Fields/Belinda Wilson] A: Mr. Look So Good!; P: Richard "Dimples" Fields/Belinda Wilson	330
2338. A KIND OF MAGIC Queen Capitol 5590 21Jun86 42(1) 11 [Roger Taylor] A: A Kind Of Magic; P: Queen/David Richards	330
2339. EVERYTHING WORKS IF YOU LET IT Cheap Trick Epic 50887 24May80 44(2) 10 [Rick Nielsen] A: Roadie Soundtrack; P: George Martin	330
2340. I'LL FALL IN LOVE AGAIN Sammy Hagar Geffen 49881 30Jan82 43(2) 10 [Sammy Hagar] A: Standing Hampton; P: Keith Olsen	329
2341. JAM ON IT Newcleus Sunnyview SUN 3010 02Jun84 56(2) 15 [Maurice Cenac] A: Jam On Revenge; P: Frank Fair/Joe Webb	329
2342. SHE'S STRANGE Cameo Atlanta Artists 818384 07Apr84 47(2) 11 [Larry Blackmon/Tomi Jenkins/Nathan Leftenant/Charles L. Singleton] A: She's Strange; P: Larry Blackmon	329
2343. SUMMERGIRLS Dino 4th & Broadway 7468 16Jul88 50(3) 12 [Dino Esposito] A: 24/7; P: Dino Esposito	329
2344. BACK TO PARADISE 38 Special A&M 2955 25Jul87 41(1) 11 [Bryan Adams/Pat Benatar/Jim Vallance] A: Flashback; P: Don Barnes/Jim Vallance	328
2345. FLASH'S THEME AKA FLASH Queen Elektra 47092 17Jan81 42(2) 10 [Brian May] A: Flash Gordon (Soundtrack); P: Reinhold Mack/Brian May	328
2346. YOUNG LOVE Air Supply Arista 1005 18Sep82 38(2) 9 [Graham Russell] A: Now And Forever; P: Harry Maslin	328
2347. HE'S A LIAR Bee Gees RSO 1066 26Sep81 30(1) 8 [Barry Gibb/Maurice Gibb/Robin Gibb] A: Living Eyes; P: Albhy Galuten/Barry Gibb/Maurice Gibb/Robin Gibb/Karl Richardson	328
2348. STATE OF INDEPENDENCE Donna Summer Geffen 29895 02Oct82 41(2) 10 [Jon Anderson/Vangelis] A: Donna Summer; P: Quincy Jones	328
2349. SHOOT FOR THE MOON Poco Atlantic 89919 18Dec82 50(1) 13 [Rusty Young] A: Ghost Town; P: John Mills/Poco	327
2350. MAMA WEER ALL CRAZEE NOW Quiet Riot Pasha 04505 07Jul84 51(2) 12 [Noddy Holder/Jim Lea] A: Condition Critical; P: Spencer Proffer	327

Ranking the Singles

Rank. TITLE Act Label Number Entry Date Peak(Wks) Tot Wks
[Writers] A: Proximate Album; P: Producer — Score

2351. I'LL TRY SOMETHING NEW A Taste Of Honey Capitol 5099
13Mar82 41(3) 10 [Smokey Robinson] A: Ladies Of The Eighties; P: Al McKay 327

2352. CRAZY (Keep On Falling) The John Hall Band EMI America 8096
26Dec81 42(1) 11 [Johanna Hall/John Hall/Bob Leinbach/Eric Parker]
A: All Of The Above; P: John Hall/Richard Sanford Orshoff 327

2353. TEMPTED Squeeze A&M 2345 01Aug81 49(2) 11
[Chris Difford/Glenn Tilbrook]
A: East Side Story; P: Roger Bechirian/Elvis Costello 327

2354. LOVE OF THE COMMON PEOPLE Paul Young Columbia 04453
19May84 45(1) 11 [John Hurley/Ronnie Wilkins]
A: No Parlez; P: Laurie Latham 326

2355. BLACK AND BLUE Van Halen Warner Bros. 27891 21May88 34(1) 10
[Michael Anthony/Sammy Hagar/Alex Van Halen/Eddie Van Halen]
A: OU812; P: Donn Landee 326

2356. FORGIVE ME FOR DREAMING Elisa Fiorillo Chrysalis 43237
07May88 49(1) 12 [Scott Cutler/Gerry Goffin] A: Elisa Fiorillo; P: Scott Cutler 326

2357. A LITTLE LOVE Juice Newton RCA 13823 02Jun84 44(2) 10
[Danny Douma/Richard Feldman/Todd Sharp]
A: Can't Wait All Night; P: Richard Landis 326

2358. ANY LOVE Luther Vandross Epic 08047 08Oct88 44(1) 13
[Marcus Miller/Luther Vandross] A: Any Love; P: Marcus Miller/Luther Vandross 326

2359. BABY TALKS DIRTY The Knack Capitol 4822 09Feb80 38(1) 8
[Berton Averre/Doug Fieger]
A: But The Little Girls Understand; P: Mike Chapman 326

2360. YOU TAKE ME UP Thompson Twins Arista 1-9244 25Aug84 44(1) 9
[Tom Bailey/Alannah Currie/Joe Leeway]
A: Into The Gap; P: Tom Bailey/Alex Sadkin 325

2361. NIGHTWALKER Gino Vannelli Arista 0613 04Jul81 41(2) 10
[Gino Vannelli] A: Nightwalker; P: Gino Vannelli/Joe Vannelli/Ross Vannelli 324

2362. TAKE ME BACK Bonnie Tyler Columbia 04246 03Dec83 46(1) 9
[Billy Cross] A: Faster Than The Speed Of Night; P: Jim Steinman 324

2363. THE LAST MILE Cinderella Mercury 872148 21Jan89 36(1) 10
[Tom Keifer] A: Long Cold Winter; P: Eric Brittingham/Andy Johns/Tom Keifer 324

2364. ONE TO ONE Carole King Atlantic 4026 27Mar82 45(2) 10
[Carole King/Cynthia Weil] A: One To One; P: Mark Hallman/Carole King 323

2365. LET ME LOVE YOU ONCE Greg Lake Chrysalis 2571 21Nov81 48(3) 10
[Steve Dorff/Molly Ann Leikin] A: Greg Lake; P: Greg Lake 323

2366. LICENCE TO CHILL Billy Ocean Jive 1283 07Oct89 32(1) 9
[Mutt Lange/Billy Ocean] A: Greatest Hits; P: Mutt Lange 323

2367. IT'S HARD TO BE HUMBLE Mac Davis Casablanca 2244
22Mar80 43(1) 12 [Mac Davis] A: It's Hard To Be Humble; P: Larry Butler 323

2368. TEARS ARE FALLING KISS Mercury 884141 19Oct85 51(1) 13
[Paul Stanley] A: Asylum; P: Gene Simmons/Paul Stanley 322

2369. LET ME TALK Earth, Wind & Fire ARC 11366 20Sep80 44(2) 9
[Philip Bailey/Lorenzo Dunn/Ralph Johnson/Albert Philip McKay/
Maurice White/Verdine White] A: Faces; P: Maurice White 322

2370. GIVING UP ON LOVE Rick Astley RCA 8872 15Apr89 38(1) 10
[Rick Astley] A: Hold Me In Your Arms; P: Rick Astley/Daize Washboum 321

2371. OOH OOH SONG Pat Benatar Chrysalis 42843 19Jan85 36(2) 9
[Pat Benatar/Neil Geraldo] A: Tropico; P: Peter Coleman/Neil Geraldo 321

2372. HEY LADIES Beastie Boys Capitol 44454 05Aug89 36(1) 10
[Barbarella Bishop/Mike Diamond/Matt Dike/Ronald Ford/Adam Horovitz/
John King/Garrett Clinton Shider/Garry Marshall Shider/
Mike Simpson/Larry Troutman/Roger Troutman/Adam Yauch]
A: Paul's Boutique; P: Beastie Boys/Dust Brothers 321

2373. CARS WITH THE BOOM L'Trimm Atlantic 89005 24Sep88 54(1) 15
[Elana Cager/Larry Davis/Rachel De Rougemont/Paul (2) Klein/
Joseph Louis Stone] A: Grab It!; P: Larry Davis/Paul Klein/Joe Stone 321

2374. SUPER TROUPER ABBA Atlantic 3806 04Apr81 45(1) 11
[Benny Andersson/Björn Ulvaeus]
A: Super Trouper; P: Benny Andersson/Björn Ulvaeus 320

2375. LIVING EYES Bee Gees RSO 1067 07Nov81 45(2) 10
[Barry Gibb/Maurice Gibb/Robin Gibb] A: Living Eyes;
P: Albhy Galuten/Barry Gibb/Maurice Gibb/Robin Gibb/Karl Richardson 320

2376. ALWAYS Firefall Atlantic 89916 15Jan83 59(5) 13
[Paul Crosta/John Sambataro] A: Break Of Dawn; P: Howard Albert/Ron Albert 320

2377. ARE YOU SURE So EMI 50109 20Feb88 41(1) 11
[Marcus Bell/Mark Long] A: Horseshoe In The Glove; P: Walter Turbitt 319

2378. BRING DOWN THE MOON Boy Meets Girl RCA 8807 28Jan89 49(1) 11
[George Merrill/Shannon Rubicam] A: Reel Life; P: Arif Mardin 319

2379. YOU WEAR IT WELL El DeBarge with DeBarge Gordy 1804 31Aug85
46(1) 10 [Chico DeBarge/El DeBarge] A: Rhythm Of The Night; P: El DeBarge 319

2380. GIRL I AM SEARCHING FOR YOU Stevie B LMR 74005
07Oct89 56(1) 15 [Stevie B] A: In My Eyes; P: Stevie B 318

2381. TROUBLE ME 10,000 Maniacs Elektra 69298 17Jun89 44(1) 12
[Dennis Drew/Natalie Merchant] A: Blind Man's Zoo; P: Peter Asher 317

2382. CAN YOU FEEL THE BEAT Lisa Lisa and Cult Jam with Full Force
Columbia 05669 16Nov85 69(1) 20
[Curtis Bedeau/Gerard Charles/Hugh Clarke/Brian George/Lucien George/
Paul George] A: Lisa Lisa & Cult Jam With Full Force; P: Full Force 317

2383. BORN IN EAST L.A. Cheech & Chong MCA 52655 21Sep85 48(2) 11
[Tommy Chong/Cheech Marin/Bruce Springsteen]
A: Get Out Of My Room; P: Jeff Eyrich 317

2384. TWILIGHT ELO Jet 02559 24Oct81 38(2) 11
[Jeff Lynne] A: Time; P: Jeff Lynne 316

2385. GIVE ME ALL YOUR LOVE Whitesnake Geffen 28103
13Feb88 48(1) 11 [David Coverdale/John Sykes]
A: Whitesnake; P: Keith Olsen/Mike Stone 316

2386. JUST ANOTHER DAY IN PARADISE Bertie Higgins Kat Family 02839
01May82 46(2) 10 [Bertie Higgins/Columbia Jones/Sonny Limbo]
A: Just Another Day In Paradise; P: Sonny Limbo 316

2387. DOWNTOWN LIFE Daryl Hall & John Oates Arista 1-9753
01Oct88 31(1) 9 [Sara Allen/Daryl Hall/Rick Iantosca/John Oates]
A: Ooh Yeah!; P: Daryl Hall/John Oates/Tom Wolk 316

2388. OVER AND OVER Pajama Party Atlantic 88799 11Nov89 59(1) 14
[Jim Klein/Peggy Sendars] A: Up All Night; P: Jim Klein 316

2389. APACHE Sugar Hill Gang Sugar Hill 774 13Feb82 53(2) 11
[Clifton Chase/Cheryl Cook/Jerry Lordan/Sylvia Robinson/Michael Wright]
A: 8th Wonder; P: Sylvia Robinson 315

2390. GREAT GOSH A'MIGHTY! Little Richard MCA 52780
08Mar86 42(2) 10 [Richard Penniman/John Schuller/Sylvia Smith]
A: Lifetime Friend; P: Dan Hartman/Billy Preston 315

2391. SWITCHIN' TO GLIDE The Kings Elektra 47006/47052 23Aug80
56(1) 13 [David Diamond/Aryan Zero] A: The Kings Are Here; P: Bob Ezrin 315

2392. SMILING ISLANDS Robbie Patton Atlantic 89955 12Mar83 52(1) 12
[David Adelstein/Robbie Patton]
A: Orders From Headquarters; P: Peter Coleman 315

2393. ILLEGAL ALIEN Genesis Atlantic 89698 10Mar84 44(1) 10
[Tony Banks/Phil Collins/Mike Rutherford]
A: Genesis; P: Genesis/Hugh Padgham 315

2394. SKIN TRADE Duran Duran Capitol 5670 31Jan87 39(2) 9
[Simon Le Bon/Nick Rhodes/Nigel John Taylor]
A: Notorious; P: Duran Duran/Nile Rodgers 315

2395. OFF ON YOUR OWN (Girl) Al B. Sure! Warner Bros. 27870 06Aug88
45(1) 12 [Albert Joseph Brown/Kyle West] A: In Effect Mode; P: Kyle West 315

2396. WHAT THE BIG GIRLS DO Van Stephenson MCA 52437
04Aug84 45(1) 9 [Jan Buckingham/Steve Buckingham/Van Stephenson]
A: Righteous Anger; P: Richard Landis 315

2397. BETCHA SHE DON'T LOVE YOU Evelyn King RCA 13380
15Jan83 49(3) 11 [Michael Jones] A: Get Loose; P: Morrie Brown 314

2398. NEEDLES AND PINS Tom Petty and The Heartbreakers with Stevie Nicks
MCA 52772 01Feb86 37(2) 9 [Sonny Bono/Jack Nitzsche]
A: Pack Up The Plantation - Live!; P: Mike Campbell/Tom Petty 314

2399. THERE'S NOTHING BETTER THAN LOVE Luther Vandross (Duet With
Gregory Hines) Epic 06978 14Mar87 50(1) 14
[John Vernon Anderson/Luther Vandross]
A: Give Me The Reason; P: Marcus Miller/Luther Vandross 313

2400. DON'T PUSH IT DON'T FORCE IT Leon Haywood 20th Century 2443
05Apr80 49(1) 11 [Leon Haywood] A: Naturally; P: Leon Haywood 313

2401. LOVE ALL THE HURT AWAY Aretha Franklin & George Benson
Arista 0624 29Aug81 46(2) 10 [Sam Dees]
A: Love All The Hurt Away; P: Arif Mardin 312

2402. PLANET ROCK Afrika Bambaataa And The Soul Sonic Force Tommy Boy
823 17Jul82 48(1) 11 [Arthur Baker/Afrika Bambaataa/John Robie]
A: Planet Rock-The Album; P: Arthur Baker 312

2403. IN YOUR SOUL Corey Hart EMI 50134 11Jun88 38(1) 10
[Corey Hart] A: Young Man Running; P: Corey Hart/Andy Richards 311

2404. COULD I BE DREAMING Pointer Sisters Planet 47920
08Nov80 52(2) 11 [Marlo Henderson/Trevor Lawrence/Anita Pointer]
A: Special Things; P: Richard Perry 311

2405. NAIL IT TO THE WALL Stacy Lattisaw Motown 1859 18Oct86 48(1) 13
[Stephen Lunt/Arnie Roman] A: Take Me All The Way; P: Jellybean Benitez 311

2406. I DON'T WANT TO TALK ABOUT IT Rod Stewart Warner Bros. 49138
22Dec79 46(2) 11 [Danny Whitten] A: Rod Stewart Greatest Hits; P: Tom Dowd 310

Rank. TITLE Act Label Number Entry Date Peak(Wks) Tot Wks [Writers] A: Proximate Album; P: Producer	Score
2407. THESE TIMES ARE HARD FOR LOVERS John Waite EMI America 43018 20Jun87 53(1) 16 [Desmond Child/John Waite] A: Rover's Return; P: Desmond Child/Frank Filipetti/John Waite	310
2408. LEFT IN THE DARK Barbra Streisand Columbia 04605 22Sep84 50(2) 12 [Jim Steinman] A: Emotion; P: Jim Steinman	310
2409. DON'T SAY YOU LOVE ME Billy Squier Capitol 44420 24Jun89 58(1) 15 [Billy Squier] A: Hear & Now; P: Jason Corsaro/Godfrey Diamond/Billy Squier	309
2410. BRASS MONKEY Beastie Boys Def Jam 07020 21Mar87 48(1) 10 [Mike Diamond/Adam Horovitz/Rick Rubin/Adam Yauch] A: Licensed To Ill; P: Beastie Boys/Rick Rubin	309
2411. LADY SOUL The Temptations Gordy 1856 11Oct86 47(2) 11 [Mark Holden] A: To Be Continued...; P: Peter Bunetta/Rick Chudacoff	309
2412. FASCINATION STREET The Cure Elektra 69300 13May89 46(2) 11 [Simon Gallup/Roger O'Donnell/Robert James Smith/Porl Thompson/Laurence Tolhurst/Boris Williams] A: Disintegration; P: Dave Allen/Robert Smith	309
2413. FULL OF FIRE Shalamar Solar 12152 20Dec80 55(1) 12 [Joey Gallo/Richard Randolph/Jody Watley] A: Three For Love; P: Leon Sylvers	309
2414. YOU DON'T BELIEVE Alan Parsons Project Arista 1-9108 19Nov83 54(3) 10 [Alan Parsons/Eric Woolfson] A: The Best Of The Alan Parsons Project; P: Alan Parsons	308
2415. SOMETHIN' 'BOUT YOU BABY I LIKE Glen Campbell and Rita Coolidge Capitol 4865 17May80 42(2) 10 [Richie Supa] A: Somethin' 'Bout You Baby I Like; P: Gary Klein	307
2416. FANTASY GIRL .38 Special A&M 2330 06Jun81 52(2) 10 [Jeff Carlisi/Jim Peterik] A: Wild-Eyed Southern Boys; P: Rodney Mills	307
2417. THE BIG CRASH Eddie Money Columbia 04199 19Nov83 54(1) 11 [Duane Hitchings/Eddie Money] A: Where's The Party?; P: Tom Dowd/Eddie Money	306
2418. FORGET ME NOT Bad English Epic 68946 22Jul89 45(1) 11 [Jonathan Cain/Mark Spiro/John Waite] A: Bad English; P: Richie Zito	305
2419. NEW YEAR'S DAY U2 Island 99915 02Apr83 53(2) 12 [Adam Clayton/Dave Evans/Paul Hewson/Larry Mullen] A: War; P: Steve Lillywhite	305
2420. I AM LOVE Jennifer Holliday Geffen 29522 22Oct83 49(2) 11 [David Foster/Maurice White/Allee Willis] A: Feel My Soul; P: Maurice White	305
2421. I WILL ALWAYS LOVE YOU Dolly Parton RCA 13260 31Jul82 53(1) 14 [Dolly Parton] A: Best Little Whorehouse In Texas Soundtrack; P: Dolly Parton/Gregg Perry	305
2422. WALK LIKE A MAN Mary Jane Girls Motown 1851 12Jul85 41(1) 10 [Bob Crewe/Bob Gaudio] A: A Fine Mess Soundtrack; P: Rick James	304
2423. SHANDI KISS Casablanca 2282 21Jun80 47(2) 10 [Vini Poncia/Paul Stanley] A: Kiss Unmasked; P: Vini Poncia	304
2424. COULD IT BE LOVE Jennifer Warnes Arista 0611 05Dec81 47(2) 10 [Randy Sharp] A: The Best Of Jennifer Warnes; P: Jim Ed Norman	304
2425. ANOTHER SLEEPLESS NIGHT Anne Murray Capitol 5083 30Jan82 44(1) 9 [Charlie Black/Rory Bourke] A: Where Do You Go When You Dream; P: Jim Ed Norman	303
2426. WE CAN LAST FOREVER Chicago Reprise/Full Moon 22985 13May89 55(2) 12 [John Dexter/Jason Scheff] A: Chicago 19; P: Ron Nevison	302
2427. WHERE DOES THE LOVIN' GO David Gates Elektra 46588 09Feb80 46(2) 8 [David Gates] A: Falling In Love Again; P: David Gates	302
2428. CAN'T WE TRY Teddy Pendergrass Philadelphia Int'l 3107 30Aug80 52(2) 12 [Ken Hirsch/Ronald Norman Miller] A: TP; P: John R. Faith/Teddy Pendergrass	301
2429. US AND LOVE (We Go Together) Kenny Nolan Casablanca 2234 02Feb80 44(1) 8 [Kenny Nolan] A: Night Miracles; P: Juergen Koppers/Kenny Nolan	301
2430. TAKING IT ALL TOO HARD Genesis Atlantic 89656 16Jun84 50(1) 12 [Tony Banks/Phil Collins/Mike Rutherford] A: Genesis; P: Genesis/Hugh Padgham	300
2431. SHE'S ON THE LEFT Jeffrey Osborne A&M 1227 20Aug88 48(3) 11 [Clinton Blanson/Robert Brookins/Tony Haynes/Jeffrey Osborne] A: One Love--One Dream; P: Robert Brookins/Jeffrey Osborne	300
2432. DO IT AGAIN The Kinks Arista 1-9309 22Dec84 41(1) 10 [Ray Davies] A: Word Of Mouth; P: Ray Davies	300
2433. CALL ME Go West Chrysalis 42865 01Jun85 54(1) 14 [Peter Cox/Richard Drummie] A: Go West; P: Gary Stevenson	300
2434. BABY COME BACK Billy Rankin A&M 2613 10Mar84 52(1) 11 [Billy Rankin] A: Growin' Up Too Fast; P: John Ryan	300
2435. LONELY TOGETHER Barry Manilow Arista 0596 14Mar81 45(2) 10 [Kenny Nolan] A: Barry; P: Barry Manilow	299
2436. SHE DON'T KNOW ME Bon Jovi Mercury 818958 26May84 48(1) 11 [Mark Avsec] A: Bon Jovi; P: Lance Quinn	298
2437. CAN YOU STAND THE RAIN New Edition MCA 53464 11Feb89 44(1) 13 [James Samuel Harris/Terry Lewis] A: Heart Break; P: Jimmy Jam Harris/Terry Lewis	298
2438. UNDER THE GUN Poco MCA 41269 19Jul80 48(1) 10 [Paul Cotton] A: Under The Gun; P: Mike Flicker	298
2439. I WISH I WAS EIGHTEEN AGAIN George Burns Mercury 57011 19Jan80 49(2) 10 [Sonny Throckmorton] A: I Wish I Was Eighteen Again; P: Jerry Kennedy	297
2440. WINNER TAKES IT ALL Sammy Hagar Columbia 06647 14Feb87 54(1) 14 [Giorgio Moroder/Tom Whitlock] A: Over The Top Soundtrack; P: Sammy Hagar/Giorgio Moroder/Edward Van Halen	297
2441. IT DIDN'T TAKE LONG Spider Dreamland 111 30May81 43(2) 10 [Holly Knight] A: Between The Lines; P: Peter Coleman	296
2442. JODY Jermaine Stewart Arista 1-9476 20Sep86 42(1) 9 [Jeffrey E. Cohen/Jermaine Stewart/Narada Michael Walden] A: Frantic Romantic; P: Narada Michael Walden	296
2443. YOU CAME Kim Wilde MCA 53370 17Sep88 41(1) 10 [Kim Wilde/Ricky Wilde] A: Close; P: Tony Swain/Ricky Wilde	295
2444. I.O.U. Lee Greenwood MCA 52199 28May83 53(2) 11 [Kerry Chater/Austin Roberts] A: Somebody's Gonna Love You; P: Jerry Crutchfield	295
2445. DINNER WITH GERSHWIN Donna Summer Geffen 28418 22Aug87 48(1) 11 [Brenda Russell] A: All Systems Go; P: Richard Perry	295
2446. LANDLORD Gladys Knight & The Pips Columbia 11239 14Jun80 46(1) 9 [Nick Ashford/Valerie Simpson] A: About Love; P: Nick Ashford/Valerie Simpson	295
2447. SINCE YOU'RE GONE The Cars Elektra 47433 27Mar82 41(1) 9 [Ric Ocasek] A: Shake It Up; P: Roy Thomas Baker	294
2448. STILL IN THE GAME Steve Winwood Island 29940 07Aug82 47(2) 10 [Will Jennings/Steve Winwood] A: Talking Back To The Night; P: Steve Winwood	294
2449. LOVE ON YOUR SIDE Thompson Twins Arista 1056 30Apr83 45(1) 9 [Tom Bailey/Alannah Currie/Joe Leeway] A: Side Kicks; P: Alex Sadkin	294
2450. WALKING ON THE CHINESE WALL Philip Bailey Columbia 04826 06Apr85 46(2) 12 [Bill Hughes/Marcy Levy/Roxanne Seeman] A: Chinese Wall; P: Phil Collins	294
2451. ME (Without You) Andy Gibb RSO 1056 14Mar81 40(1) 8 [Andy Gibb] A: Andy Gibb's Greatest Hits; P: Albhy Galuten/Barry Gibb/Karl Richardson	294
2452. HANGIN' ON A STRING (Contemplating) Loose Ends MCA 52570 20Jul85 43(1) 10 [Jane Eugene/Carl McIntosh/Steve Nichol] A: So Where Are You?; P: Nick Martinelli	294
2453. WEIRD SCIENCE Oingo Boingo MCA 52633 31Aug85 45(2) 12 [Danny Elfman] A: Dead Man's Party; P: Steve Bartek/Danny Elfman	294
2454. PEOPLE GET READY Jeff Beck And Rod Stewart Epic 05416 15Jun85 48(1) 11 [Curtis Mayfield] A: Flash; P: Jeff Beck	294
2455. WHY CAN'T I BE YOU? The Cure Elektra 69474 20Jun87 54(1) 12 [Simon Gallup/Robert James Smith/Porl Thompson/Laurence Tolhurst/Boris Williams] A: Kiss Me, Kiss Me, Kiss Me; P: Dave Allen/Robert Smith	294
2456. GAMES WITHOUT FRONTIERS Peter Gabriel Mercury 76063 16Aug80 48(2) 11 [Peter Gabriel] A: Peter Gabriel (III); P: Steve Lillywhite	293
2457. LET'S GET IT UP AC/DC Atlantic 3894 16Jan82 44(2) 9 [Brian Johnson/Angus Young/Malcolm Young] A: For Those About To Rock (We Salute You); P: Mutt Lange	293
2458. WET MY WHISTLE Midnight Star Solar 69790 26Nov83 61(2) 11 [Reggie Calloway] A: No Parking On The Dance Floor; P: Reggie Calloway	293
2459. KEEPING OUR LOVE ALIVE Henry Paul Band Atlantic 3883 12Dec81 50(2) 10 [Steve Grisham/Henry Paul/Jim Peterik] A: Anytime; P: Kevin Beamish	292
2460. YOU ARE MY HEAVEN Roberta Flack With Donny Hathaway Atlantic 3627 16Feb80 47(1) 11 [Eric Mercury/Stevie Wonder] A: Roberta Flack Featuring Donny Hathaway; P: Roberta Flack/Eric Mercury	292
2461. CIRCLE Edie Brickell & The New Bohemians Geffen 27580 08Apr89 48(2) 10 [Alan Aly/Edie Brickell/John Bush/John Houser/Ken Withrow] A: Shooting Rubberbands At The Stars; P: Pat Moran	292
2462. BUT YOU KNOW I LOVE YOU Dolly Parton RCA 12200 04Apr81 41(1) 10 [Mike Settle] A: 9 To 5 And Odd Jobs; P: Mike Post	292
2463. LOVE CRIES Stage Dolls Chrysalis 23366 29Jul89 46(1) 13 [Torstein Flakne/Bobby Icon] A: Stage Dolls; P: Bjøern Nessjø	291
2464. FRENCH KISS Lil' Louis Epic 73007 21Oct89 50(1) 13 [Marvin Louis Burns] A: From The Mind Of Lil Louis; P: Louis Burns	291

Ranking the Singles

Rank. TITLE Act Label Number Entry Date Peak(Wks) Tot Wks
[Writers] *A: Proximate Album; P: Producer* — Score

2465. ON THE WINGS OF A NIGHTINGALE The Everly Brothers Mercury 880213 01Sep84 50(1) 12 [Paul McCartney] *A: EB 84; P: Dave Edmunds* — 291

2466. POWER The Temptations Gordy 7183 10May80 43(1) 9 [Angelo Bond/Berry Gordy/Jean Mayer] *A: Power; P: Angelo Bond/Berry Gordy* — 290

2467. THE POWER OF LOVE Jennifer Rush Epic 05754 08Feb86 57(1) 13 [Mary Applegate/Candy DeRouge/Gunther Mende/Jennifer Rush] *A: Jennifer Rush; P: Candy DeRouge/Gunther Mende* — 290

2468. FOR A ROCKER Jackson Browne Asylum 69764 14Jan84 45(2) 9 [Jackson Browne] *A: Lawyers In Love; P: Jackson Browne/Greg Ladanyi* — 290

2469. DON QUICHOTTE Magazine 60 Baja 001 10May86 56(1) 11 [Jean-Luc Drion/Domenico Regiacorte] *A: Costa Del Sol; P: Jean-Luc Drion* — 290

2470. WILD AND CRAZY LOVE Mary Jane Girls Gordy 1798 20Jul85 42(1) 10 [Kenny Hawkins/Rick James] *A: Only Four You; P: Rick James* — 289

2471. LUCKY ME Anne Murray Capitol 4848 05Apr80 42(2) 8 [Charlie Black/Rory Bourke] *A: Somebody's Waiting; P: Jim Ed Norman* — 288

2472. I'LL BE GOOD René and Angela Mercury 884009 28Sep85 47(2) 10 [René Moore/Angela Winbush] *A: Street Called Desire; P: René and Angela/Bruce Swedien/Bobby Watson* — 288

2473. SOMETHING TO GRAB FOR Ric Ocasek Geffen 29784 12Feb83 47(2) 9 [Ric Ocasek] *A: Beatitude; P: Ric Ocasek* — 288

2474. SHOULD I STAY OR SHOULD I GO ('83) The Clash Epic 03547 19Feb83 50(2) 10 [Nicky Headon/Michael Geoffrey Jones/Paul Simonon/Joe Strummer] *A: Combat Rock; P: Clash* — 288

2475. CATCH MY FALL Billy Idol Chrysalis 42840 03Nov84 50(1) 11 [Billy Idol] *A: Rebel Yell; P: Keith Forsey* — 287

2476. PETER GUNN The Art Of Noise featuring Duane Eddy China 42986 17May86 50(1) 11 [Henry Mancini] *A: In Visible Silence; P: Art Of Noise* — 287

2477. LET ME BE Korona United Artists 1341 22Mar80 43(1) 8 [Bruce Blackman] *A: Let Me Be; P: Bruce Blackman/Mike Clark* — 287

2478. ANGEL SAY NO Tommy Tutone Columbia 11278 24May80 38(2) 8 [Tommy Heath/Jim Keller] *A: Tommy Tutone; P: Ed E. Thacker* — 286

2479. CRY LIKE A BABY Kim Carnes EMI America 8058 04Oct80 44(2) 8 [Spooner Oldham/Dan Penn] *A: Romance Dance; P: George Tobin* — 286

2480. BETTER BE HOME SOON Crowded House Capitol 44164 09Jul88 42(1) 11 [Neil Finn] *A: Temple Of Low Men; P: Mitchell Froom* — 286

2481. MY TOOT TOOT Jean Knight Mirage 99643 04May85 50(1) 15 [Sidney Simien] *A: My Toot Toot; P: Isaac Bolden* — 286

2482. EVERYTHING I NEED Men At Work Columbia 04929 25May85 47(1) 9 [Colin James Hay] *A: Two Hearts; P: Greg Ham/Colin Hay* — 286

2483. BEAT PATROL Starship RCA 5308 26Sep87 46(1) 10 [Johnny Warman] *A: No Protection; P: Peter Wolf* — 285

2484. SHINE ON George Duke Epic 02701 20Feb82 41(1) 9 [George Duke] *A: Dream On; P: George Duke* — 285

2485. YOUR LOVE IS KING Sade Portrait 37-05408 22Jun85 54(1) 11 [Helen Adu/Stuart Matthewman] *A: Diamond Life; P: Robin Millar* — 285

2486. EVERYTHING MUST CHANGE Paul Young Columbia 05712 23Nov85 56(2) 11 [Ian Kewley/Paul Young] *A: The Secret Of Association; P: Laurie Latham* — 285

2487. UNCONDITIONAL LOVE Donna Summer Mercury 814088 03Sep83 43(2) 8 [Michael Omartian/Donna Summer] *A: She Works Hard For The Money; P: Michael Omartian* — 284

2488. IF WE NEVER MEET AGAIN Tommy Conwell And The Young Rumblers Columbia 08505 17Dec88 48(1) 11 [Jules Shear] *A: Rumble; P: Rick Chertoff* — 284

2489. GIRL, DON'T LET IT GET YOU DOWN The O'Jays TSOP 4790 23Aug80 55(1) 11 [Kenneth Gamble/Leon Huff] *A: The Year 2000; P: Kenny Gamble/Leon Huff* — 284

2490. MONEY (That's What I Want) The Flying Lizards Virgin 67003 01Dec79 50(1) 10 [Janie Bradford/Berry Gordy] *A: The Flying Lizards; P: David Cunningham* — 284

2491. MAKE NO MISTAKE, HE'S MINE Barbra Streisand (Duet With Kim Carnes) Columbia 04695 15Dec84 51(1) 10 [Kim Carnes] *A: Emotion; P: Kim Carnes/Bill Cuomo* — 284

2492. SHE WAS HOT The Rolling Stones Rolling Stones 99788 04Feb84 44(2) 9 [Mick Jagger/Keith Richards] *A: Undercover; P: Glimmer Twins/Chris Kimsey* — 283

2493. I GOT THE FEELIN' (It's Over) Gregory Abbott Columbia 06632 21Feb87 56(2) 11 [Gregory Abbott] *A: Shake You Down; P: Gregory Abbott* — 283

2494. PRIVATE NUMBER The Jets MCA 52846 09Aug86 47(1) 11 [Jerry Knight/Aaron Zigman] *A: The Jets; P: Jerry Knight/Don Powell/David Rivkin/Aaron Zigman* — 283

2495. MAGIC POWER Triumph MCA 12298 03Oct81 51(1) 11 [Rik Emmett/Mike Levine/Gil Moore] *A: Allied Forces; P: Triumph* — 282

2496. RADIO ROMANCE Tiffany MCA 53623 25Feb89 35(1) 9 [John Duarte/Mark E. Paul] *A: Hold An Old Friend's Hand; P: George Tobin* — 282

2497. GLIDE Pleasure Fantasy 874 08Dec79 55(2) 10 [Nathaniel Phillips/Bruce A. Smith] *A: Future Now; P: Phil Kaffel/Marlon McClain/Pleasure* — 282

2498. DIRTY WATER The Inmates Polydor 2032 08Dec79 51(2) 10 [Ed Cobb] *A: First Offence; P: Vi Maile* — 282

2499. LET'S WORK Mick Jagger Columbia 07306 12Sep87 39(1) 9 [Mick Jagger/David A. Stewart] *A: Primitive Cool; P: Mick Jagger/David A. Stewart* — 281

2500. ANIMAL INSTINCT Commodores Motown 1788 25May85 43(1) 9 [Martin Page] *A: Nightshift; P: Dennis Lambert* — 281

2501. RIGHT NEXT TO ME Whistle Select 2005 03Jun89 60(1) 13 [Shiller Shaun Fequiere] *A: Transformation; P: Siller Shaun Fequiere/Howard Thompson* — 280

2502. A LOVE SONG Kenny Rogers Liberty 1485 16Oct82 47(3) 10 [Lee Greenwood] *A: Love Will Turn You Around; P: Kenny Rogers* — 280

2503. LOVE OF A LIFETIME Chaka Khan Warner Bros. 28671 12Jul86 53(1) 12 [David Gamson/Green Gartside] *A: Destiny; P: David Gamson/Green Gartside/Arif Mardin* — 280

2504. GAMES Phoebe Snow Mirage 3800 21Feb81 46(2) 10 [Andrea Farber/Vince Melamed] *A: Rock Away; P: Richie Cannata/Greg Ladanyi* — 280

2505. LITTLE DARLIN' Sheila Carrere 02564 05Dec81 49(1) 9 [Amanda Blue/Holly Knight] *A: Little Darlin'; P: Keith Olsen* — 279

2506. I WANT TO BREAK FREE Queen Capitol 5350 28Apr84 45(1) 8 [John Deacon] *A: The Works; P: Reinhold Mack/Queen* — 278

2507. BELIEVE IN ME Dan Fogelberg Full Moon 04447 28Apr84 48(1) 9 [Dan Fogelberg] *A: Windows And Walls; P: Dan Fogelberg/Marty Lewis* — 277

2508. IF LOOKS COULD KILL Player RCA 13006 23Jan82 48(2) 9 [Peter Beckett/Dennis Lambert] *A: Spies Of Life; P: Dennis Lambert* — 277

2509. YO' LITTLE BROTHER Nolan Thomas Mirage 99697 05Jan85 57(2) 13 [Ann Godwin/Curtis Josephs] *A: Yo' Little Brother; P: Chris Barbosa/Mark Liggett* — 277

2510. YES Merry Clayton RCA 6989 05Mar88 45(1) 11 [Neal Cavanaugh/Terry Fryer/Thomas Graf] *A: Dirty Dancing Soundtrack; P: Michael Lloyd* — 277

2511. STOP DOGGIN' ME AROUND Klique MCA 52250 08Oct83 50(1) 9 [Lena Agree] *A: Try It Out; P: Thomas McClary* — 277

2512. WITH YOU ALL THE WAY New Edition MCA 52829 14Jun86 51(1) 11 [Carl Wurtz] *A: All For Love; P: George Tobin* — 277

2513. WRAP IT UP The Fabulous Thunderbirds CBS Associated 06270 09Aug86 50(2) 10 [Isaac Hayes/David Porter] *A: Tuff Enuff; P: Dave Edmunds* — 277

2514. HEAVEN'S ON FIRE KISS Mercury 880205 13Oct84 49(1) 10 [Desmond Child/Paul Stanley] *A: Animalize; P: Paul Stanley* — 276

2515. HOT IN THE CITY ('87) Billy Idol Chrysalis 43203 12Dec87 48(1) 10 [Billy Idol] *A: Vital Idol; P: Keith Forsey* — 276

2516. I'LL FIND MY WAY HOME Jon & Vangelis Polydor 2205 22May82 51(2) 9 [Jon Anderson/Vangelis] *A: The Friends Of Mr. Cairo; P: Vangelis* — 275

2517. YOU AND ME TONIGHT Deja Virgin 99422 21Nov87 54(2) 12 [Eban Kelly/Jimi Randolph] *A: Serious; P: Eban Kelly/Jimi Randolph* — 275

2518. BABY TALK Alisha Vanguard 35262 21Dec85 68(3) 17 [Greg Brown/Logankoya] *A: Alisha; P: Mark S. Berry* — 275

2519. INSIDE OUTSIDE The Cover Girls Fever 1916 23Jul88 55(2) 13 [Albert Cabrera/Tony Moran/Andy Tripoli] *A: Show Me; P: Albert Cabrera/Tony Moran/Andy Tripoli* — 275

2520. IT ISN'T, IT WASN'T, IT AIN'T NEVER GONNA BE Aretha Franklin and Whitney Houston Arista 1-9850 01Jul89 41(1) 8 [Albert Hammond/Diane Warren] *A: Through The Storm; P: Narada Michael Walden* — 273

2521. WHEN THE FEELING COMES AROUND Jennifer Warnes Arista 0497 05Apr80 45(2) 8 [Rick Cunha] *A: Shot Through The Heart; P: Rob Fraboni* — 273

2522. CALL ME Dennis DeYoung A&M 2816 15Mar86 54(1) 11 [Dennis DeYoung] *A: Back To The World; P: Dennis DeYoung* — 273

2523. I GOT YOU Split Enz A&M 2252 23Aug80 53(1) 11 [Neil Finn] *A: True Colours; P: David Tickle* — 273

2524. (You're So Square) BABY, I DON'T CARE Joni Mitchell Geffen 29849 20Nov82 47(2) 9 [Jerry Leiber/Mike Stoller] *A: Wild Things Run Fast; P: Joni Mitchell* — 273

2525. HONEY, HONEY David Hudson Alston 3750 28Jun80 59(1) 11 [Earl Kenneth King Jr.] *A: To You Honey, Honey With Love; P: Willie Clarke* — 272

2526. FOREVER YOUNG ('88) Alphaville Atlantic 89013 29Oct88 65(2) 14 [Marian Gold/Bernhard Lloyd/Frank Mertens] *A: Forever Young; P: Wolfgang Loos/Colin Pearson* — 272

Rank. TITLE Act Label Number Entry Date Peak(Wks) Tot Wks [Writers] A: Proximate Album; P: Producer	Score
2527. GIRLS AIN'T NOTHING BUT TROUBLE D.J. Jazzy Jeff & The Fresh Prince Jive 1147 29Oct88 57(1) 12 [Buddy Kaye/Hugo Montenegro/Will Smith/Jeff Townes] A: Rock The House; P: Pete Q. Harris/Will Smith/Jeff Townes	272
2528. LOVER'S LANE Georgio Motown 1906 05Dec87 59(4) 12 [Georgio Allentini] A: Sexappeal; P: Georgio Allentini	271
2529. I'M IN LOVE AGAIN Pia Zadora Elektra/Curb 47428 27Mar82 45(1) 9 [Linda Laurie/Jacques Morali/Dan Schmidt] A: Pia; P: Jacques Morali	271
2530. THE RIGHT STUFF Vanessa Williams Wing 887 386 23Jul88 44(1) 10 [Kevin Jones/Rex Salas] A: The Right Stuff; P: Rex Salas	271
2531. YOU DON'T KNOW ME Mickey Gilley Epic 02172 11Jul81 55(1) 12 [Eddy Arnold/Cindy Walker] A: You Don't Know Me; P: Jim Ed Norman	269
2532. GET UP AND GO Go-Go's I.R.S. 9910 25Sep82 50(2) 9 [Charlotte Caffey/Jane Wiedlin] A: Vacation; P: Richard Gottehrer	269
2533. KEEP YOUR EYE ON ME Herb Alpert A&M 2915 28Feb87 46(1) 10 [James Samuel Harris/Terry Lewis] A: Keep Your Eye On Me; P: Jimmy Jam Harris/Terry Lewis/John McClain	269
2534. DANGEROUS Natalie Cole Modern 99648 04May85 57(2) 10 [Stephen Mitchell/Marti Sharron/Gary Skardina] A: Dangerous; P: Marti Sharron/Gary Skardina	269
2535. EASY FOR YOU TO SAY Linda Ronstadt Asylum 69838 23Apr83 54(1) 10 [Jimmy Webb] A: Get Closer; P: Peter Asher	269
2536. ALLERGIES Paul Simon Warner Bros. 29453 05Nov83 44(1) 10 [Paul Simon] A: Hearts And Bones; P: Roy Halee/Paul Simon/Russ Titelman	269
2537. ONE MORE TIME FOR LOVE Billy Preston & Syreeta Tamla 54312 14Jun80 52(2) 10 [Jerry Peters] A: Syreeta; P: Jerry Peters	268
2538. HEARTACHE AWAY Don Johnson Epic 06426 22Nov86 56(3) 11 [Steve Cochran] A: Heartbeat; P: Chas Sandford	268
2539. JUNGLE BOY John Eddie Columbia 05858 07Jun86 52(3) 10 [John Eddie] A: John Eddie; P: Bill Drescher	268
2540. NEVER GIVE UP Sammy Hagar Geffen 29718 26Mar83 46(1) 8 [Keith Olsen/Alan Pasqua] A: Three Lock Box; P: Keith Olsen	268
2541. SHATTERED GLASS Laura Branigan Atlantic 89245 04Jul87 48(1) 10 [Steve Coe/Bobby Mitchell] A: Touch; P: Matt Aitken/Mike Stock/Pete Waterman	268
2542. ONE LOVER AT A TIME Atlantic Starr Warner Bros. 28327 15Aug87 58(1) 13 [Richard Feldman/James Scott] A: All In The Name Of Love; P: David Lewis/Wayne Lewis	267
2543. PSYCHOBABBLE Alan Parsons Project Arista 1029 27Nov82 57(4) 10 [Alan Parsons/Eric Woolfson] A: Eye In The Sky; P: Alan Parsons	267
2544. THE BEST MAN IN THE WORLD Ann Wilson Capitol 5654 29Nov86 61(2) 12 [John Barry/Sue Ennis/Ann Wilson/Nancy Wilson] A: The Golden Child Soundtrack; P: Ron Nevison	267
2545. TRIBUTE (Right On) The Pasadenas Columbia 68575 25Feb89 52(1) 10 [Jon Banfield/Jeffrey Aaron Brown/David Milliner/Michael Milliner/Hammish Seelochan/Pete Wingfield] A: To Whom It May Concern; P: Pete Wingfield	266
2546. LOVE AND PRIDE King Epic 04917 20Jul85 55(1) 11 [Paul King/Mick Roberts] A: Steps In Time; P: Richard James Burgess	266
2547. STRANGER Jefferson Starship Grunt 12275 11Jul81 48(2) 11 [Jeannette Sears/Pete Sears] A: Modern Times; P: Ron Nevison	266
2548. LEADER OF THE PACK Twisted Sister Atlantic 89478 30Nov85 53(1) 10 [Jeff Barry/Ellie Greenwich/Shadow Morton] A: Come Out And Play; P: Dieter Dierks	266
2549. FREE ME Roger Daltrey Polydor 2105 05Jul80 53(1) 10 [Russ Ballard] A: McVicar (Soundtrack); P: Jeff Wayne	266
2550. CAN WE STILL BE FRIENDS Robert Palmer Island 49137 22Dec79 52(1) 9 [Todd Rundgren] A: Secrets; P: Robert Palmer	266
2551. GIVE ME THE REASON Luther Vandross Epic 06129 23Aug86 57(3) 11 [Nat Adderley Jr./Luther Vandross] A: Give Me The Reason; P: Luther Vandross	265
2552. FALLING IN LOVE Balance Portrait 24-02608 21Nov81 58(1) 11 [Peppy Castro] A: Balance; P: Balance/Tony Bongiovi	265
2553. READY OR NOT Lou Gramm Atlantic 89269 09May87 54(2) 12 [Lou Gramm/Bruce Turgon] A: Ready Or Not; P: Lou Gramm/Pat Moran	265
2554. SAVE YOUR LOVE Great White Capitol 44104 23Jan88 57(1) 12 [Jack Russell/Stephan Williams] A: Once Bitten; P: Mark Kendall/Michael Lardie/Alan Niven	265
2555. CHINA Red Rockers Columbia 03786 04Jun83 53(2) 10 [John Griffith/Darren Hill/David Kahne/James Singletary] A: Good As Gold; P: David Kahne	264
2556. THE BLUES Randy Newman and Paul Simon Warner Bros. 29803 22Jan83 51(3) 8 [Randy Newman] A: Trouble In Paradise; P: Russ Titelman/Lenny Waronker	264
2557. OH YEAH Yello Mercury 884930 08Aug87 51(1) 11 [Boris Blank] A: One Second; P: Yello	264
2558. IT'S FOR YOU Player Casablanca 2265 07Jun80 46(1) 8 [Peter Beckett] A: Room With A View; P: Peter Beckett/Tony Peluso	263
2559. LOVE THAT GOT AWAY Firefall Atlantic 3670 28Jun80 50(2) 9 [Rick Roberts] A: Undertow; P: Howard Albert/Ron Albert/Kyle Lehning	263
2560. SING ME AWAY Night Ranger Boardwalk 09Apr83 54(2) 9 [Jack Blades/Kelly Keagy] A: Dawn Patrol; P: Pat Glasser	262
2561. TAKE IT WHILE IT'S HOT Sweet Sensation Atco 99352 09Apr88 57(1) 13 [Joseph E. Malloy] A: Take It While It's Hot; P: Ted Currier	262
2562. LOVE IN THE FIRST DEGREE Bananarama London 886255 19Mar88 48(1) 10 [Matt Aitken/Sarah Dallin/Siobhan Fahey/Mike Stock/Pete Waterman/Keren Woodward] A: Wow!; P: Matt Aitken/Mike Stock/Pete Waterman	262
2563. CRY WOLF A-Ha Warner Bros. 28500 24Jan87 50(1) 10 [Magne Furuholmen/Pal Waaktaar] A: Scoundrel Days; P: Alan Tarney	262
2564. LIKE TO GET TO KNOW YOU WELL Howard Jones Elektra 69598 28Sep85 49(1) 9 [Howard Jones] A: The 12" Album; P: Rupert Hine	262
2565. OUTSTANDING The Gap Band Total Experience 8205 12Mar83 51(3) 8 [Ray Calhoun] A: Gap Band IV; P: Lonnie Simmons	262
2566. STOP THIS GAME Cheap Trick Epic 50942 08Nov80 48(2) 12 [Rick Nielsen/Robin Zander] A: All Shook Up; P: George Martin	261
2567. WHEN SHE DANCES Joey Scarbury Elektra 47201 10Oct81 49(1) 9 [Brian Blugerman] A: America's Greatest Hero; P: Mike Post	260
2568. JUST SO LONELY Get Wet Boardwalk 8-02018 25Apr81 39(1) 9 [Jose Augusto Esquibel] A: Get Wet; P: Phil Ramone	260
2569. TONIGHT David Bowie EMI America 8246 01Dec84 53(3) 9 [David Bowie/Iggy Pop] A: Tonight; P: David Bowie/Derek Bramble/Hugh Padgham	259
2570. SECRET Orchestral Manoeuvres In The Dark A&M 2794 14Dec85 63(2) 13 [Paul Humphreys] A: Crush; P: Stephen Hague	259
2571. SILLY Deniece Williams ARC 02406 15Aug81 53(2) 10 [Fritz Baskett/Clarence McDonald/Deniece Williams] A: My Melody; P: Thom Bell/Deniece Williams	259
2572. THE OAK TREE Morris Day Warner Bros. 28899 28Sep85 65(1) 12 [Morris Day] A: Color Of Success; P: Morris Day	259
2573. I NEED A MAN Eurythmics RCA 5361 19Dec87 46(1) 10 [Annie Lennox/David A. Stewart] A: Savage; P: David A. Stewart	258
2574. SOLD ME DOWN THE RIVER The Alarm I.R.S. 73002 07Oct89 50(1) 13 [Eddie MacDonald/Mike Peters] A: Change.; P: Tony Visconti	258
2575. THE METRO Berlin Geffen 29638 28May83 58(2) 10 [John Crawford] A: Pleasure Victim; P: Daniel R. Van Patten	258
2576. LET'S PRETEND WE'RE MARRIED//IRRESISTABLE BITCH [TSW] Prince Warner Bros. 29548 17Dec83 52(1) 10 [Prince//Prince] A: 1999; P: Prince	258
2577. DON'T ASK ME WHY Eurythmics Arista 1-9880 30Sep89 40(1) 9 [Annie Lennox/David A. Stewart] A: We Too Are One; P: Jimmy Iovine/David A. Stewart	257
2578. HURTS TO BE IN LOVE Gino Vannelli CBS Associated 05586 21Sep85 57(1) 12 [Gino Vannelli] A: Black Cars; P: Gino Vannelli/Joe Vannelli/Ross Vannelli	257
2579. WHAT I LIKE ABOUT YOU The Romantics Nemperor 7527 16Feb80 49(2) 8 [Jimmy Marinos/Wally Palmar/Mike Skill] A: The Romantics; P: Peter Solley	256
2580. LIVING ON VIDEO Trans-X Atco 99534 10May86 61(1) 12 [Pascal Languirand] A: Living On Video; P: Daniel Bernier	256
2581. I LIKE YOU (Special Mix) Phyllis Nelson Carrere 4-05719 08Feb86 61(2) 11 [Phyllis Nelson] A: I Like You; P: Yves Dessca	256
2582. THE SPIRIT OF RADIO Rush Mercury 76044 23Feb80 51(1) 8 [Geddy Lee/Alex Lifeson/Neil Peart] A: Permanent Waves; P: Terry Brown/Rush	256
2583. BAD TIMES Tavares Capitol 4811 05Jan80 47(2) 10 [Gerard McMahon] A: Supercharged; P: Bobby Colomby	256
2584. THE DEAD HEART Midnight Oil Columbia 07964 06Aug88 53(1) 10 [Peter Garrett/Rob Hirst/James Moginie] A: Diesel And Dust; P: Warne Livesey/Midnight Oil	255
2585. SWITCHIN' TO GLIDE//THIS BEAT GOES ON [TSW] The Kings Elektra 47006/47052 22Nov80 43(1) 10 [David Diamond/Aryan Zero// David Diamond/Aryan Zero] A: The Kings Are Here; P: Bob Ezrin	255
2586. SUPER BOWL SHUFFLE The Chicago Bears Shufflin' Crew Red Label 71012 11Jan86 41(2) 9 [Lloyd Barry/Bobby Daniels/Richard E. Meyer/Melvin Owens] A: No Album; P: Bobby Daniels/Rich Tufo	254

Ranking the Singles

Rank. TITLE Act Label Number Entry Date Peak(Wks) Tot Wks [Writers] A: Proximate Album; P: Producer	Score
2587. SHARP DRESSED MAN ZZ Top Warner Bros. 29576 23Jul83 56(1) 9 [Frank Beard/Billy Gibbons/Dusty Hill] A: Eliminator; P: Bill Ham	254
2588. DON'T STAND SO CLOSE TO ME '86 The Police A&M 2879 25Oct86 46(1) 9 [Sting] A: Every Breath You Take-The Singles; P: Laurie Latham/Police	254
2589. SYSTEM OF SURVIVAL Earth, Wind & Fire Columbia 07608 31Oct87 60(1) 13 [Skylark] A: Touch The World; P: Preston Glass/Maurice White	254
2590. BOOM! THERE SHE WAS Scritti Politti Featuring Roger Warner Bros. 27976 18Jun88 53(1) 11 [David Gamson/Green Gartside] A: Provision; P: David Gamson/Green Gartside	253
2591. EVERYDAY James Taylor Columbia 05681 09Nov85 61(1) 11 [Buddy Holly/Norman Petty] A: That's Why I'm Here; P: Peter Asher/Frank Filipetti/James Taylor	253
2592. IF YOUR HEART ISN'T IN IT Atlantic Starr A&M 2822 19Apr86 57(1) 12 [Hamish Stuart] A: As The Band Turns; P: David Lewis/Wayne Lewis	253
2593. LOVIN' THE NIGHT AWAY The Dillman Band RCA 12206 09May81 45(1) 9 [Patrick Frederick/Steve Seamans] A: Lovin' The Night Away; P: Rick Hall	253
2594. LET THE RIVER RUN Carly Simon Arista 1-9793 04Mar89 49(2) 10 [Carly Simon] A: Working Girl Soundtrack; P: Rob Mounsey/Carly Simon	252
2595. HEADED FOR THE FUTURE Neil Diamond Columbia 05889 24May86 53(2) 10 [Neil Diamond/Tom Hensley/Alan Lindgren] A: Headed For The Future; P: Neil Diamond/Tom Hensley/Alan Lindgren	251
2596. LOOKS THAT KILL Mötley Crüe Elektra 69756 04Feb84 54(2) 9 [Nikki Sixx] A: Shout At The Devil; P: Tom Werman	251
2597. HOOKED ON BIG BANDS Frank Barber Orchestra Victory 1001 08May82 61(3) 12 A: Hooked On Big Bands; P: Terry Brown	250
2598. WE'VE SAVED THE BEST FOR LAST Kenny G Vocal by Smokey Robinson Arista 1-9785 04Feb89 47(1) 9 [Paul Gordon/Dennis Matkosky/ Louis Pardini] A: Silhouette; P: Peter Bunetta/Rick Chudacoff	250
2599. FRENCH KISSIN Debbie Harry Geffen 28546 22Nov86 57(1) 11 [Chuck Lorre] A: Rockbird; P: Seth Justman	249
2600. DON'T MISUNDERSTAND ME Rossington Collins Band MCA 41284 26Jul80 55(2) 9 [Allen Collins/Barry Harwood/Dale Krantz] A: Anytime, Anyplace, Anywhere; P: Allen Collins/Barry Harwood/Gary Rossington	249
2601. COMPUTER GAME "THEME FROM THE CIRCUS" Yellow Magic Orchestra Horizon 127 02Feb80 60(2) 9 [Haruomi Hosono/Ryuichi Sakamoto/ Yukihiro Takahashi] A: Yellow Magic Orchestra; P: Haruomi Hosono	249
2602. YOU LOOK MARVELOUS Billy Crystal A&M 2764 27Jul85 58(1) 12 [Billy Crystal/Paul Shaffer] A: Mahvelous; P: Arthur Baker/Bob Tischler	249
2603. SECRET JOURNEY The Police A&M 2408 10Apr82 46(2) 8 [Sting] A: Ghost In The Machine; P: Hugh Padgham/Police	249
2604. ESCALATOR OF LIFE Robert Hazard RCA 13449 05Mar83 58(2) 9 [Robert Hazard] A: Robert Hazard; P: Robert Hazard	248
2605. JOY AND PAIN Rob Base & D.J. E-Z Rock Profile 5247 27May89 58(1) 13 [Robert Ginyard] A: It Takes Two; P: Rob Base/William Hamilton	248
2606. BET YOUR HEART ON ME Johnny Lee Full Moon/Asylum 10Oct81 54(1) 9 [Jimmy Ray McBride] A: Bet Your Heart On Me; P: Jim Ed Norman	248
2607. DRIVEN OUT The Fixx RCA 8837 25Feb89 55(1) 10 [Dan Brown/Cy Curnin/Peter John Greenall/Jamie West-Oram/Adam Woods] A: Calm Animals; P: William Wittman	248
2608. FEEL LIKE A NUMBER Bob Seger & The Silver Bullet Band Capitol 4581 19Dec81 48(2) 8 [Bob Seger] A: Nine Tonight; P: Ed Andrews/Bob Seger	248
2609. CALIFORNIA DREAMIN' The Beach Boys Capitol 5630 20Sep86 57(1) 10 [John Phillips/Michelle Phillips] A: Made In U.S.A.; P: Terry Melcher	248
2610. BABY STEP BACK Gordon Lightfoot Warner Bros. 50012 03Apr82 50(2) 8 [Gordon Lightfoot] A: Shadows; P: Ken Friesen/Gordon Lightfoot	248
2611. TALKING BACK TO THE NIGHT Steve Winwood Island 28122 13Feb88 57(1) 10 [Will Jennings/Steve Winwood] A: Chronicles; P: Tom Lord-Alge/Steve Winwood	247
2612. NEVER GIVE UP ON A GOOD THING George Benson Warner Bros. 50005 20Feb82 52(1) 9 [Michael Garvin/Tom Shapiro] A: No Album; P: Jay Graydon	247
2613. LITTLE WALTER Tony! Toni! Toné! Wing 887 385 28May88 47(1) 10 [Denzil Foster/Thomas McElroy/Clemon Timothy Riley/Raphael Saadiq/ Dwayne Wiggins] A: Who?; P: Denzil Foster/Thomas McElroy	247
2614. 25 OR 6 TO 4 Chicago Warner/Full Moon 27-28628 06Sep86 48(3) 8 [Robert Lamm] A: Chicago 18; P: David Foster	247
2615. NEW DAY FOR YOU Basia Epic 08112 10Dec88 53(1) 11 [Peter Ross/Basia Trzetrzelewska/Danny White] A: Time And Tide; P: Basia Trzetrzelewska/Danny White	247
2616. TELL ME White Lion Atlantic 89051 02Jul88 58(2) 11 [Vito Bratta/Mike Tramp] A: Pride; P: Michael Wagener	246
2617. TEXAS IN MY REAR VIEW MIRROR Mac Davis Casablanca 2305 18Oct80 51(2) 9 [Mac Davis] A: Texas In My Rear View Mirror; P: Rick Hall	246
2618. BOUNCIN' OFF THE WALLS Matthew Wilder Private I 4-04617 22Sep84 51(1) 9 [Matthew Wilder] A: Bouncin' Off The Walls; P: Peter Bunetta/Rick Chudacoff	246
2619. FIRE (Live) Bruce Springsteen & The E Street Band Columbia 06657 31Jan87 46(2) 8 [Bruce Springsteen] A: Bruce Springsteen & The E Street Band Live 1975-1985; P: Jon Landau/Chuck Plotkin/Bruce Springsteen	246
2620. I MUST BE DREAMING Giuffria MCA 52794 03May86 52(2) 10 [Willy DeVille] A: Silk + Steel; P: David Glen Eisley/Gregg Giuffria	245
2621. SOMEWHERE IN AMERICA Survivor Scotti Brothers 511 23Feb80 70(2) 12 [Jim Peterik] A: Survivor; P: MISSING	245
2622. PRETENDING Eric Clapton Duck/Reprise 22732 18Nov89 55(1) 11 [Jerry Lynn Williams] A: Journeyman; P: Russ Titelman	245
2623. PROMISES Barbra Streisand Columbia 02065 23May81 48(2) 9 [Barry Gibb/Robin Gibb] A: Guilty; P: Albhy Galuten/Barry Gibb/Karl Richardson	245
2624. AIN'T SO EASY David & David A&M 2905 24Jan87 51(1) 11 [David Baerwald/David Ricketts] A: Boomtown; P: Davitt Sigerson	245
2625. THEMES FROM E.T. (The Extra-Terrestrial) Walter Murphy MCA 52099 31Jul82 47(1) 9 [John Williams] A: Themes From E.T. (The Extra-Terrestrial); P: Walter Murphy	245
2626. WORK THAT BODY Diana Ross RCA 13201 10Apr82 44(2) 7 [Ray Chew/Paul Jabara/Diana Ross] A: Why Do Fools Fall In Love; P: Diana Ross	243
2627. WELCOME TO THE PLEASUREDOME (Trevor Horn Remix) Frankie Goes To Hollywood ZTT/Island 7-99653 06Apr85 48(2) 8 [Peter Gill/Holly Johnson/Brian Nash/Mark O'Toole] A: Welcome To The Pleasure Dome; P: Trevor Horn	242
2628. IT'S MONEY THAT MATTERS Randy Newman Reprise 27709 05Nov88 60(2) 12 [Randy Newman] A: Land Of Dreams; P: Mark Knopfler	242
2629. CHANGE John Waite Chrysalis 42606 02Mar85 54(1) 10 [Holly Knight] A: Vision Quest Soundtrack; P: Neil Geraldo	242
2630. NOTHING IN COMMON Thompson Twins Arista 1-9511 26Jul86 54(1) 10 [Tom Bailey/Alannah Currie] A: Nothing In Common Soundtrack; P: Tom Bailey/Geoffrey Downes	242
2631. ROCK ME Great White Capitol 44042 22Aug87 60(2) 14 [Mark Kendall/Michael Lardie/Alan Niven/Jack Russell] A: Once Bitten; P: Mark Kendall/Michael Lardie/Alan Niven	241
2632. BEYOND Herb Alpert A&M 2246 28Jun80 50(1) 8 [Richard Hewson] A: Beyond; P: Herb Alpert/Andy Armer/Randy Badazz	241
2633. THERE'S NO EASY WAY James Ingram Qwest 29318 07Apr84 58(1) 10 [Barry Mann] A: It's Your Night; P: Quincy Jones	241
2634. ARC OF A DIVER Steve Winwood Island 49726 09May81 48(2) 9 [Vivian Stanshall/Steve Winwood] A: Arc Of A Diver; P: Steve Winwood	240
2635. LICK IT UP KISS Mercury 814671 12Nov83 66(3) 11 [Vince Cusano/Paul Stanley] A: Lick It Up; P: Michael James Jackson/Gene Simmons/Paul Stanley	240
2636. GIRLFRIEND Bobby Brown MCA 52866 20Dec86 57(2) 9 [Kirk Crumpler/Lee Peters/Larry White] A: King Of Stage; P: Larry White	240
2637. JOYSTICK Dazz Band Motown 1701 11Feb84 61(1) 11 [Eric Fearman/Robert Lamont Harris] A: Joystick; P: Reggie Andrews	240
2638. WE ARE WHAT WE ARE The Other Ones Virgin 99473 18Apr87 53(1) 9 [Stephan Gottwald/Uwe Hoffmann/Alf Klimek/Jayney Klimek/ Johnny Klimek/Andreas Schwarz-Ruszczynski] A: The Other Ones; P: Christopher Neil	240
2639. I ONLY WANT TO BE WITH YOU Nicolette Larson Warner Bros. 29948 07Aug82 53(1) 9 [Mike Hawker/Ivor Raymonde] A: All Dressed Up & No Place To Go; P: Andrew Gold	239
2640. NEED A LITTLE TASTE OF LOVE The Doobie Brothers Capitol 44441 12Aug89 45(1) 9 [Ernie Isley/Marvin Isley/O'Kelly Isley/Ronald Isley/ Rudolph Isley/Chris Jasper] A: Cycles; P: Rodney Mills	239
2641. I'LL BE YOU The Replacements Sire 22992 08Apr89 51(1) 10 [Paul Westerberg] A: Don't Tell A Soul; P: Replacements/Matt Wallace	239
2642. I WANT ACTION Poison Enigma 44004 13Jun87 50(1) 10 [Bobby Dall/C.C. DeVille/Bret Michaels/Rikki Rockett] A: Look What The Cat Dragged In; P: Ric Browde	238

Ranking the Singles

Rank. TITLE Act Label Number Entry Date Peak(Wks) Tot Wks
[Writers] A: Proximate Album; P: Producer — Score

2643. LISTEN LIKE THIEVES INXS Atlantic 89429 10May86 54(2) 9
[Garry Beers/Andy Farriss/Michael Hutchence]
A: Listen Like Thieves; P: Chris Thomas — 238

2644. ANOTHER HEARTACHE Rod Stewart Warner Bros. 28631
30Aug86 52(1) 9 [Bryan Adams/Rod Stewart/Jim Vallance/Randy Wayne]
A: Rod Stewart; P: Bob Ezrin — 238

2645. ALL I HAVE TO DO IS DREAM Andy Gibb And Victoria Principal RSO 1065 15Aug81 51(2) 8 [Boudleaux Bryant]
A: No Album; P: Michael Barbiero/Andy Gibb — 238

2646. I'D DO IT ALL AGAIN Sam Harris Motown 1829 01Feb86 52(1) 9
[Sam Harris/Mary Unobsky] A: Sam-I-Am; P: Sam Harris/Lauren Wood — 237

2647. NEVER CAN SAY GOODBYE Communards MCA/London 53224
30Jan86 51(1) 9 [Clifton Davis] A: Red; P: Stephen Hague — 237

2648. HIGH TIME Styx A&M 2568 13Aug83 48(1) 7
[Dennis DeYoung] A: Kilroy Was Here; P: Styx — 237

2649. THEME FROM DYNASTY Bill Conti Arista 1021 06Nov82 52(2) 9
[Bill Conti] A: No Album; P: Bill Conti — 237

2650. DOCTORIN' THE TARDIS The Timelords TVT 4025 17Dec88 66(1) 13
[James Cauty/Mike Chapman/Nicky Chinn/William Ernest Drummond/Gary Glitter/Ron Granier/Mike Leander/Ian Richardson] A: No Album; P: Timelords — 237

2651. SOMEONE BELONGING TO SOMEONE Bee Gees RSO 815235
20Aug83 49(1) 6 [Barry Gibb/Maurice Gibb/Robin Gibb]
A: Staying Alive (Soundtrack); P: Albhy Galuten/Barry Gibb/Maurice Gibb/Robin Gibb/Karl Richardson — 236

2652. STREET OF DREAMS Rainbow Mercury 815660 05Nov83 60(2) 10
[Ritchie Blackmore/Joe Lynn Turner] A: Bent Out Of Shape; P: Roger Glover — 236

2653. BE YOUR MAN Jesse Johnson's Revue A&M 2702 16Mar85 61(1) 11
[Jesse Woods Johnson] A: Jesse Johnson's Revue; P: Jesse Johnson — 236

2654. HOW LONG Rod Stewart Warner Bros. 50051 24Apr82 49(2) 9
[Paul Carrack] A: Tonight I'm Yours; P: Jim Cregan/Rod Stewart — 236

2655. STRANGELOVE ('88) Depeche Mode Sire 27777 10Sep88 50(1) 9
[Martin L. Gore] A: Music For The Masses; P: David Bascombe/Depeche Mode — 236

2656. SAVIN' MYSELF Eria Fachin Critique 7-99356 27Feb88 50(1) 10
[David Lodge] A: My Name Is Eria Fachin; P: Vincent Degiorgio — 236

2657. BE THERE Pointer Sisters MCA 53120 08Aug87 42(1) 9
[Franne Golde/Allee Willis]
A: Beverly Hills Cop II Soundtrack; P: Narada Michael Walden — 236

2658. STAND BY Roman Holliday Jive 9036 18Jun83 54(1) 9
[Robert Lambert/Steve Lambert] A: Roman Holliday; P: Peter Collins — 235

2659. I'M NO ANGEL Gregg Allman Band Epic 06998 04Apr87 49(1) 10
[Tony Colton/Phil Palmer] A: I'm No Angel; P: Rodney Mills — 235

2660. GIVE ME THE KEYS (And I'll Drive You Crazy) Huey Lewis And The News Chrysalis 43335 21Jan89 47(1) 8
[Bill Gibson/Huey Lewis/Steve Lewis]
A: Small World; P: Huey Lewis And The News — 235

2661. I WANT YOU SO BAD Heart Capitol 44116 20Feb88 49(1) 9
[Thomas F. Kelly/Billy Steinberg] A: Bad Animals; P: Ron Nevison — 235

2662. MAKE IT LAST FOREVER Keith Sweat (Duet with Jacci McGhee)
Vintertainment 7-69386 23Jul88 59(2) 11
[Teddy Riley/Keith Sweat] A: Make It Last Forever; P: Keith Sweat — 235

2663. WHY CAN'T THIS NIGHT GO ON FOREVER Journey Columbia 07043 25Apr87 60(1) 12
[Jonathan Cain/Steve Perry] A: Raised On Radio; P: Steve Perry — 235

2664. STREET CORNER Ashford & Simpson Capitol 5109 05Jun82 56(1) 10
[Nick Ashford/Valerie Simpson]
A: Street Opera; P: Nick Ashford/Valerie Simpson — 234

2665. FOOL FOR YOUR LOVING ('80) Whitesnake Mirage 3672
02Aug80 53(1) 8 [David Coverdale/Bernie Marsden/Micky Moody]
A: Ready An' Willing; P: Martin Birch — 234

2666. WHERE DID YOUR HEART GO? Wham! Columbia 06294
11Oct86 50(2) 8 [David Was/Don Was]
A: Music From The Edge Of Heaven; P: George Michael — 234

2667. SPACE AGE WHIZ KIDS Joe Walsh Full Moon 29611 11Jun83 52(1) 8
[Joe Vitale/Joe Walsh] A: You Bought It-You Name It; P: Bill Szymczyk — 234

2668. RAINBOW'S END Sergio Mendes A&M 2563 13Aug83 52(2) 8
[David Batteau/Don Freeman] A: Sergio Mendes(2); P: Sergio Mendes — 234

2669. CAN'T STOP Rick James Gordy 1776 30Mar85 50(1) 8
[Rick James] A: Glow; P: Rick James — 233

2670. JOHNNY B Hooters Columbia 07241 18Jul87 61(1) 11
[Eric Bazilian/Rick Chertoff/Rob Hyman] A: One Way Home; P: Rick Chertoff — 233

2671. YEARS FROM NOW Dr. Hook Capitol 4885 05Jul80 51(2) 9
[Charles Lincoln Cochran/Roger Cook] A: Sometimes You Win; P: Ron Haffkine — 232

2672. LIPSTICK Suzi Quatro Dreamland 107 24Jan81 51(2) 9
[Mike Chapman/Nicky Chinn] A: Rock Hard; P: Mike Chapman — 232

2673. WATERFALL Wendy & Lisa Columbia 07243 19Sep87 56(1) 10
[Lisa Coleman/Wendy Melvoin/Robert Rivkin]
A: Wendy And Lisa; P: Lisa Coleman/Wendy Melvoin/Robert Rivkin — 232

2674. HOOKED ON YOU ('87) Sweet Sensation Next Plateau 308 17Jan87
64(1) 12 [Joseph E. Malloy/David Sanchez]
A: Take It While It's Hot; P: Ted Currier/David Sanchez — 232

2675. DREAMIN' John Schneider Scotti Brothers 02889 15May82 45(2) 8
[Barry De Vorzon/Ted Ellis] A: Quiet Man; P: John D'Andrea/Tony Scotti — 232

2676. CLOSER TO FINE Indigo Girls Epic 68912 22Jul89 52(1) 9
[Amy Ray/Emily Ann Saliers] A: Indigo Girls; P: Scott Litt — 232

2677. ALL YOU ZOMBIES Hooters Columbia 04854 18May85 58(1) 11
[Eric Bazilian/Rob Hyman] A: Nervous Night; P: Rick Chertoff — 232

2678. CLOSER THAN FRIENDS Surface Columbia 08537 15Apr89 57(1) 13
[Bernard Leon Jackson/David Townsend]
A: 2nd Wave; P: David Conley/Bernard Jackson/David Townsend — 232

2679. FIRST NIGHT Survivor Scotti Brothers 05579 17Aug85 53(1) 9
[Jim Peterik/Frankie Sullivan] A: Vital Signs; P: Ron Nevison — 231

2680. JUST THE WAY YOU LIKE IT The S.O.S. Band Tabu 4-04523
11Aug84 64(1) 10 [James Samuel Harris/Terry Lewis]
A: Just The Way You Like It; P: Jimmy Jam Harris/Terry Lewis — 230

2681. LET'S TALK ABOUT ME Alan Parsons Project Arista 1-9282 16Feb85
56(1) 10 [Alan Parsons/Eric Woolfson] A: Vulture Culture; P: Alan Parsons — 230

2682. IT'S TRICKY Run-D.M.C. Profile 5131 28Feb87 57(1) 10
[Darryl McDaniels/Jason Mizell/Rick Rubin/Joseph Ward Simmons]
A: Raising Hell; P: Rick Rubin/Russell Simmons — 230

2683. NICE 'N' SLOW Freddie Jackson Capitol 44171 23Jul88 61(1) 12
[Barry Eastmond/Jolyon Skinner] A: Don't Let Love Slip Away; P: Barry Eastmond — 229

2684. JUST TO SATISFY YOU Waylon & Willie RCA 13073 03Apr82 52(1) 9
[Don Bowman/Waylon Jennings] A: Black On Black; P: Chips Moman — 229

2685. SUMMERTIME GIRLS Y&T A&M 2748 13Jul85 55(1) 10
[Joey Alves/Leonard Haze/Philip Kennemore/Dave Meniketti]
A: Open Fire; P: Kevin Beamish — 229

2686. HEROES Commodores Motown 1495 20Sep80 54(2) 9
[Darrell Jones/Lionel Richie]
A: Heroes; P: James Anthony Carmichael/Commodores — 229

2687. OUTSIDE MY WINDOW Stevie Wonder Tamla 54308 01Mar80 52(2) 7
[Stevie Wonder] A: Journey Through The Secret Life Of Plants; P: Stevie Wonder — 228

2688. HOW MUCH LOVE Survivor Scotti Brothers 06705 21Feb87 51(1) 9
[Jim Peterik/Frankie Sullivan]
A: When Seconds Count; P: Ron Nevison/Frankie Sullivan — 227

2689. WHEN THE RAIN BEGINS TO FALL Jermaine Jackson And Pia Zadora Curb/MCA 52521 02Feb85 54(2) 11
[Michael Bradley/Peggy March/Steve Wittmack]
A: Voyage Of The Rock Aliens Soundtrack; P: Jack White — 227

2690. FIRST...BE A WOMAN Lenore O'Malley Polydor 2055 26Jul80 53(2) 8
[Michele Michaele/Lana Sebastian/Paul Sebastian]
A: First Be A Woman; P: Michaele/Lana Sebastian/Paul Sebastian — 227

2691. MEETING IN THE LADIES ROOM Klymaxx Constellation 52545
11May85 59(1) 11 [Reggie Calloway/Vincent Calloway/Bo Watson]
A: Meeting In The Ladies Room; P: Vincent Calloway/Bo Watson — 227

2692. WHAT DO ALL THE PEOPLE KNOW The Monroes Alfa 7119
29May82 59(1) 8 [Bob Monroe] A: The Monroes; P: Bruce Botnick — 227

2693. IT'S ONLY LOVE Simply Red Elektra 69317 25Feb89 57(1) 9
[Jimmie Cameron/Vella Cameron] A: A New Flame; P: Stewart Levine — 226

2694. LIKE A SURGEON "Weird Al" Yankovic Rock 'n' Roll 04937
22Jun85 47(1) 8 [Thomas F. Kelly/Billy Steinberg/Al Yankovic]
A: Dare To Be Stupid; P: Rick Derringer — 225

2695. LET'S GO 'ROUND AGAIN Average White Band Atlantic 3354
21Jun80 53(1) 8 [Alan Gorrie] A: Shine; P: David Foster — 225

2696. YOU CAN'T RUN FROM LOVE Eddie Rabbitt Warner Bros. 29712
23Apr83 55(1) 8
[David Malloy/Eddie Rabbitt/Even Stevens] A: Radio Romance; P: David Malloy — 225

2697. YOU'RE LOOKING LIKE LOVE TO ME Peabo Bryson/Roberta Flack
Capitol 5307 24Dec83 58(1) 11 [Jerry Corbetta/Bob Crewe/Bob Gaudio]
A: Born To Love; P: Bob Crewe/Bob Gaudio — 225

2698. HOLD ME Menudo RCA 14087 11May85 62(1) 11
[Howard Lee Rice/Allan Rich] A: Menudo; P: Howie Rice — 224

2699. ONE VISION Queen Capitol 5530 07Dec85 61(2) 10
[John Deacon/Brian May/Freddie Mercury/Roger Taylor]
A: Iron Eagle; P: Reinhold Mack/Queen — 224

Ranking the Singles

Rank. TITLE Act Label Number Entry Date Peak(Wks) Tot Wks
[Writers] A: *Proximate Album*; P: Producer — Score

2700. IT'S ALL I CAN DO Anne Murray Capitol 5023 26Sep81 53(2) 9
[Archie Jordan/Richard Leigh]
A: *Where Do You Go When You Dream*; P: Jim Ed Norman — 224

2701. BODY Jacksons Epic 04673 27Oct84 47(1) 7
[Marlon Jackson] A: *Victory*; P: Marlon Jackson — 222

2702. HEARTS ON FIRE Steve Winwood Virgin 99234 18Mar89 53(1) 9
[Jim Capaldi/Steve Winwood] A: *Roll With It*; P: Tom Lord-Alge/Steve Winwood — 222

2703. FOOL FOR A PRETTY FACE (Hurt By Love) Humble Pie Atco 7216 26Apr80 52(2) 7
[Steve Marriott/Jerry Shirley] A: *On To Victory*; P: Humble Pie/Johnny Wright — 222

2704. HEART DON'T LIE La Toya Jackson Private I 4-04439
05May84 56(1) 8 [Amir Bayyan/LaToya Jackson/Donna Johnson]
A: *Heart Don't Lie*; P: Amir Bayyan — 221

2705. BACK TOGETHER AGAIN Roberta Flack with Donny Hathaway Atlantic 3661 17May80 56(2) 8 [Reggie Lucas/James Mtume]
A: *Roberta Flack Featuring Donny Hathaway*; P: Roberta Flack/Eric Mercury — 221

2706. MOST OF ALL Jody Watley MCA 53258 30Apr88 60(1) 11
[Gardner Cole/Patrick Raymond Leonard] A: *Jody Watley*; P: Patrick Leonard — 221

2707. GIVE ME ALL NIGHT Carly Simon Arista 1-9587 23May87 61(1) 12
[Gerard McMahon/Carly Simon]
A: *Coming Around Again*; P: Paul Samwell-Smith — 221

2708. IT'S YOU Bob Seger & The Silver Bullet Band Capitol 5623
16Aug86 52(1) 9 [Bob Seger] A: *Like A Rock*; P: Ed Andrews/Bob Seger — 221

2709. THE GIGOLO O'Bryan Capitol 5067 27Mar82 57(2) 9
[O'Bryan Burnette/Don Cornelius] A: *Doin' Alright*; P: Don Cornelius — 221

2710. TAKE THE L. The Motels Capitol 5149 04Sep82 52(2) 9
[John S. Carter/Martha Davis/Marty Jourard] A: *All Four One*; P: Val Garay — 220

2711. ALL FALL DOWN Five Star RCA 14108 14Sep85 65(3) 11
[Barry Blue/Robin Smith] A: *Luxury Of Life*; P: Nick Martinelli — 220

2712. LANDSLIDE Olivia Newton-John MCA 52069 12Jun82 52(2) 8
[John Farrar] A: *Physical*; P: John Farrar — 220

2713. PLAYING WITH THE BOYS Kenny Loggins Columbia 05902
16Aug86 60(2) 12
[Kenny Loggins/Ina Wolf/Peter F. Wolf] A: *Top Gun Soundtrack*; P: Peter Wolf — 219

2714. I DON'T WANNA DANCE Eddy Grant Portrait 37-04039
13Aug83 53(1) 7 [Eddy Grant] A: *Killer On The Rampage*; P: Eddy Grant — 219

2715. BAD BOYS Wham! U.K. Columbia 03932 20Aug83 60(2) 9
[George Michael] A: *Fantastic*; P: Steve Brown — 219

2716. WHAT CHA' GONNA DO FOR ME Chaka Khan Warner Bros. 49692
16May81 53(1) 9 [Ned Doheny/Hamish Stuart]
A: *What Cha' Gonna Do For Me*; P: Arif Mardin/Jerry Wexler — 219

2717. MEMORIZE YOUR NUMBER Leif Garrett Scotti Brothers 510
15Dec79 60(1) 9 [Billy Kirkland] A: *Same Goes For You*; P: Michael Lloyd — 218

2718. COME GIVE YOUR LOVE TO ME Janet Jackson A&M 2552
05Feb83 58(3) 9 [Glen Barbee/Charmaine Sylvers]
A: *Janet Jackson*; P: Foster Sylvers/Jerry Weaver — 218

2719. THAT LOVIN' YOU FEELIN' AGAIN Roy Orbison & Emmylou Harris Warner Bros. 49262 28Jun80 55(2) 9
[Roy Orbison/Chris Price] A: *Roadie Soundtrack*; P: Brian Ahern — 217

2720. ONE MORE NIGHT Streek Columbia 02529 10Oct81 47(2) 7
[Bill DeMartines] A: *Streek*; P: Fred Ruppert/Bob Stringer — 217

2721. COMING UP CLOSE 'Til Tuesday Epic 06571 10Jan87 59(1) 10
[Aimee Mann] A: *Welcome Home*; P: Rhett Davies — 217

2722. THE HEAT OF HEAT Patti Austin Qwest 28788 03May86 55(1) 9
[James Samuel Harris/Terry Lewis] A: *Gettin' Away With Murder*;
P: Jimmy Jam Harris/Terry Lewis/Cheryl Lynn — 217

2723. PAPA WAS A ROLLIN' STONE Wolf Constellation 69849
11Dec82 55(2) 9 [Barrett Strong/Norman Whitfield] A: *Wolf*; P: Bill Wolfer — 217

2724. TIP OF MY TONGUE The Tubes Capitol 5258 23Jul83 52(1) 7
[Michael Snyder/William Spooner/Maurice White]
A: *Outside Inside*; P: David Foster — 217

2725. LE BEL AGE Pat Benatar Chrysalis 42968 15Feb86 54(2) 8
[Guy Marshall/Robert Tepper] A: *Seven The Hard Way*; P: Neil Geraldo — 217

2726. RIGHT KIND OF LOVE Quarterflash Geffen 29994 29May82 56(1) 8
[Marv Ross] A: *Quarterflash*; P: John Boylan — 216

2727. REACTION TO ACTION Foreigner Atlantic 89542 01Jun85 54(1) 8
[Lou Gramm/Michael Leslie Jones]
A: *Agent Provocateur*; P: Mick Jones/Alex Sadkin — 216

2728. LOVERS AFTER ALL Melissa Manchester And Peabo Bryson Arista 0587
28Feb81 54(1) 9
[Melissa Manchester/Leon Ware]
A: *For The Working Girl*; P: Steve Buckingham — 216

2729. ABSOLUTE BEGINNERS David Bowie EMI America 8308
29Mar86 53(1) 9 [David Bowie] A: *Absolute Beginners Soundtrack*;
P: David Bowie/Clive Langer/Alan Winstanley — 216

2730. WE'VE ONLY JUST BEGUN (The Romance Is Not Over) Glenn Jones
Jive 1049 31Oct87 66(1) 14
[Timmy Allen/Glenn Jones] A: *Glenn Jones*; P: Timmy Allen — 216

2731. MEXICAN RADIO Wall Of Voodoo I.R.S. 9912 19Mar83 58(1) 9
[Charles Gray/Marc Moreland/Joe Nanini/Stan Ridgway]
A: *Call Of The West*; P: Richard Mazda — 215

2732. THE GHOST IN YOU Psychedelic Furs Columbia 04416
12May84 59(1) 9 [Richard Butler/Tim Butler] A: *Mirror Moves*; P: Keith Forsey — 215

2733. ROUTE 66/BEHIND THE WHEEL Depeche Mode Sire 27991
07May88 61(2) 11 [Bobby Troup//Martin L. Gore]
A: *Music For The Masses*; P: David Bascombe/Depeche Mode — 215

2734. (Sittin' On) THE DOCK OF THE BAY The Reddings
Believe In a Dream 5-02836 12Jun82 55(1) 9
[Steve Cropper/Otis Redding] A: *Steamin' Hot*; P: Reddings/Russell Timmons — 215

2735. REASON TO LIVE KISS Mercury 870022 05Dec87 64(1) 12
[Desmond Child/Paul Stanley] A: *Crazy Nights*; P: Ron Nevison — 215

2736. YOU MAKE MY HEART BEAT FASTER (And That's All That Matters)
Kim Carnes EMI America 8191 21Jan84 54(1) 8
[Kim Carnes/Dave Ellingson/Brian Fairweather/Martin Page]
A: *Cafe Racers*; P: Keith Olsen — 214

2737. RUTHLESS PEOPLE Mick Jagger Epic 06211 02Aug86 51(2) 8
[Daryl Hall/Mick Jagger/David A. Stewart]
A: *Ruthless People Soundtrack*; P: Daryl Hall/Mick Jagger/David A. Stewart — 214

2738. CONGRATULATIONS Vesta A&M 1407 09Sep89 55(1) 8
[Tena Clark/Gary Prim/Vesta Williams] A: *Vesta 4 U*; P: Tena Clark — 214

2739. LIMELIGHT Rush Mercury 76095 14Mar81 55(2) 9
[Geddy Lee/Alex Lifeson/Neil Peart] A: *Moving Pictures*; P: Terry Brown/Rush — 214

2740. FREEDOM Pointer Sisters RCA 14224 02Nov85 59(2) 11
[David McHugh] A: *Contact*; P: Richard Perry — 214

2741. GOIN' TO THE BANK Commodores Polydor 885358
01Nov86 65(1) 12 [Franne Golde/Andy Goldmark/Dennis Lambert]
A: *United*; P: Dennis Lambert/Jeremy Smith — 213

2742. RADAR LOVE White Lion Atlantic 88836 23Sep89 59(1) 13
[Barry Hay/George Kooymans] A: *Big Game*; P: Michael Wagener — 213

2743. SAIL AWAY The Temptations Gordy 1720 07Apr84 54(1) 8
[Angelo Bond/Norman Whitfield] A: *Back To Basics*; P: Norman Whitfield — 213

2744. ALONE AGAIN Dokken Elektra 69650 04May85 64(2) 11
[Mick Brown/Don Dokken/George Lynch/Jeff Pilson]
A: *Tooth And Nail*; P: Tom Werman — 213

2745. ALL I WANT IS YOU Carly Simon Arista 1-9653 23Jan88 54(2) 9
[Jacob Brackman/Andy Goldmark/Carly Simon]
A: *Coming Around Again*; P: John Boylan — 212

2746. BACK AND FORTH Cameo Atlanta Artists 888385 25Apr87 50(1) 8
[Larry Blackmon/Tomi Jenkins/Kevin Kendrick/Nathan Leftenant]
A: *Word Up!*; P: Larry Blackmon — 212

2747. SUZI Randy Vanwarmer Bearsville 49752 20Jun81 55(1) 8
[Randy Vanwarmer] A: *Beat Of Love*; P: John Holbrook/Ian Kimmet — 212

2748. EWOK CELEBRATION Meco Arista 1-9045 02Jul83 60(1) 8
[Ben Burtt/John Williams/Joseph Stanley Williams]
A: *Ewok Celebration*; P: Tony Bongiovi/Meco Monardo/Lance Quinn — 212

2749. DOWN ON LOVE Foreigner Atlantic 89493 17Aug85 54(2) 8
[Lou Gramm/Michael Leslie Jones]
A: *Agent Provocateur*; P: Mick Jones/Alex Sadkin — 211

2750. I'VE BEEN WAITING FOR YOU ALL OF MY LIFE Paul Anka RCA 12225
18Apr81 48(1) 9 [Linda J. Kimball/Mark Sherrill]
A: *Both Sides Of Love*; P: Larry Butler — 211

2751. ENOUGH IS ENOUGH April Wine Capitol 5133 03Jul82 50(2) 8
[Myles Goodwyn] A: *Power Play*; P: Myles Goodwyn/Mike Stone — 211

2752. THE MORE YOU LIVE, THE MORE YOU LOVE A Flock Of Seagulls
Jive 9220 11Aug84 56(2) 9 [Frank Maudsley/Paul Reynolds/Ali Score/
Mike Score] A: *The Story Of A Young Heart*; P: Steve Lovell — 211

2753. SKATEAWAY Dire Straits Warner Bros. 49632 20Dec80 58(1) 10
[Mark Knopfler] A: *Making Movies*; P: Jimmy Iovine/Mark Knopfler — 211

2754. I'M A BELIEVER Giant A&M 1454 30Sep89 56(1) 10
[Dann Huff/David Lyndon Huff/Philip Naish/Alan Pasqua/Mark Spiro]
A: *Last Of The Runaways*; P: Terry Thomas — 211

2755. I WANT IT ALL Queen Capitol 44372 13May89 50(1) 10
[John Deacon/Brian May/Freddie Mercury/Roger Taylor]
A: *The Miracle*; P: Queen/David Richards — 210

Rank. TITLE Act Label Number Entry Date Peak(Wks) Tot Wks [Writers] A: Proximate Album; P: Producer	Score
2756. CUTIE PIE One Way MCA 52049 29May82 61(1) 10 [Theodore Dudley/Greg Green/Al Hudson/Glenda Hudson/Jon Meadows/Terry Morgan/Dave Roberson] A: Who's Foolin' Who; P: ADK/Irene Perkins	210
2757. HELP ME! Marcy Levy & Robin Gibb RSO 1047 08Nov80 50(2) 10 [Robin Gibb/Blue Weaver] A: Times Square Soundtrack; Robin Gibb/Blue Weaver	209
2758. IS THIS LOVE Pat Travers Band Polydor 2080 17May80 50(2) 7 [Bob Marley] A: Crash And Burn; P: Dennis Mackay/Pat Travers	209
2759. WHITE HOT Red Rider Capitol 4845 05Apr80 48(1) 7 [Tom Cochrane/Ken Greer] A: Don't Fight It; P: Michael James Jackson	209
2760. BREAKAWAY Big Pig A&M 3014 19Mar88 60(1) 10 [Mitch Bottler/Gary Zekley] A: Bonk; P: Nick Launay	208
2761. CALLING IT LOVE Animotion Polydor 889054 10Jun89 53(1) 9 [Desmond Child/Anton Fig] A: Animotion(2); P: Steve Barri/Tony Peluso	208
2762. THE DIFFERENT STORY (World Of Lust And Crime) Peter Schilling Elektra 69307 25Mar89 61(1) 10 [Hubert Kemmler/Susanne Müller-Pi/Peter Schilling] A: The Different Story (World Of Lust And Crime); P: Michael Cretu	208
2763. ON THE REBOUND Russ Ballard Epic 50883 14Jun80 58(1) 8 [Russ Ballard] A: Barnet Dogs; P: Russ Ballard/John Stanley	208
2764. THE GREAT COMMANDMENT Camouflage Atlantic 89031 24Dec88 59(1) 12 [Oliver Kreyssig/Heiko Maile/Marcus Meyn] A: Voices & Images; P: Camouflage	208
2765. HEART TURNS TO STONE Foreigner Atlantic 89046 16Jul88 56(2) 10 [Lou Gramm/Michael Leslie Jones] A: Inside Information; P: Mick Jones	208
2766. AMERICA Prince And The Revolution Paisley Park 7-28999 19Oct85 46(2) 7 [Mark Alton Brown/Lisa Coleman/Matt Fink/Wendy Melvoin/Prince/Robert Rivkin] A: Around The World In A Day; P: Prince And The Revolution	208
2767. LOVE IS CONTAGIOUS Taja Sevelle Reprise 28257 12Sep87 62(2) 10 [Taja Sevelle] A: Taja Sevelle; P: Chico Bennett	207
2768. WHY SHOULD I CRY? Nona Hendryx EMI America 8382 23May87 58(1) 9 [Nona Hendryx/Garry Johnson/Lisa Keith] A: Female Trouble; P: Spencer Bernard/Jellybean Johnson	207
2769. THE ONLY FLAME IN TOWN Elvis Costello & The Atractions Columbia 04502 28Jul84 56(1) 9 [Elvis Costello] A: Goodbye Cruel World; P: Clive Langer/Alan Winstanley	207
2770. WHO WERE YOU THINKIN' OF The Doolittle Band Columbia 11355 11Oct80 49(2) 7 [Paul Gauvin/James W. Glaser/Cathie Pelletier] A: No Album; P: Louis LoFredo	206
2771. WALKING IN MY SLEEP Roger Daltrey Atlantic 89704 18Feb84 62(1) 9 [Leslie Adey/Jack Green] A: Parting Should Be Painless; P: Mike Thorne	206
2772. HEARTACHE ALL OVER THE WORLD Elton John Geffen 28578 18Oct86 55(1) 8 [Elton John/Bernie Taupin] A: Leather Jackets; P: Gus Dudgeon	206
2773. WKRP IN CINCINNATI Steve Carlisle MCA 51205 21Nov81 65(4) 10 [Tom Wells/Hugh Wilson] A: Steve Carlisle Sings WKRP In Cincinnati; P: Jerry Buckner/Gary Garcia	206
2774. GOOD MORNING GIRL/STAY AWHILE Journey Columbia 11339 23Aug80 55(2) 8 [Steve Perry/Neal Schon//Steve Perry/Neal Schon] A: Departure; P: Kevin Elson/Geoffrey Workman	206
2775. BIG IN JAPAN Alphaville Atlantic 89665 24Nov84 66(2) 10 [Marian Gold/Bernhard Lloyd/Frank Mertens] A: Forever Young; P: Wolfgang Loos	205
2776. LOVE ME AGAIN The John Hall Band EMI America 8151 22Jan83 64(4) 10 [John Hall] A: Searchparty; P: John Hall	205
2777. WORDS F.R. David Carrere 101 23Jul83 62(1) 9 [Robert Fitoussi/Marty Sanders/Louis Yaguda] A: Words; P: Jean-Michel Gallois-Montbrun/Frederic Liebovitz	205
2778. MOVE YOUR BOOGIE BODY Bar-Kays Mercury 76015 08Dec79 57(2) 7 [James Edward Alexander/Charles Leonard Allen/Michael Beard/Mark Bynum/Larry Dodson/Sherman Guy/Harvey Henderson/Allen Alvoid Jones Jr./Lloyd Ed Smith/Winston Stewart/Frank Thompson] A: Injoy; P: Allen Jones	205
2779. ROCK LOBSTER The B-52's Warner Bros. 49173 19Apr80 56(1) 8 [Kate Pierson/Fred Schneider/Keith Strickland/Cindy Wilson/Ricky Wilson] A: The B-52's; P: Chris Blackwell	204
2780. THE ROYAL MILE (SWEET DARLIN') Gerry Rafferty United Artists 1366 19Jul80 54(2) 8 [Gerry Rafferty] A: Snakes And Ladders; P: Hugh Murphy/Gerry Rafferty	204
2781. UNITED TOGETHER Aretha Franklin Arista 0569 27Dec80 56(1) 8 [Charles Henry Jackson Jr./Phil Perry] A: Aretha; P: Chuck (2) Jackson	204
2782. SWEETHEART LIKE YOU Bob Dylan Columbia 04301 17Dec83 55(1) 9 [Bob Dylan] A: Infidels; P: Bob Dylan/Mark Knopfler	204
2783. FLY AWAY Peter Allen A&M 2288 10Jan81 55(1) 8 [Peter Allen/David Foster/Carole Bayer Sager] A: Bi-Costal; P: David Foster	203
2784. YOU MIGHT NEED SOMEBODY Turley Richards Atlantic 3645 26Jan80 54(2) 7 [Nan O'Byrne/Tom Snow] A: Therfu; P: Mick Fleetwood/Turley Richards	203
2785. WILD, WILD WEST Kool Moe Dee Jive 1086 30Apr88 62(2) 11 [Mohandas DeWese] A: How Ya Like Me Now; P: Mohandas DeWese/Pete Q. Harris/LaVaba Mallison/Bryan New/Teddy Riley	203
2786. THUNDER AND LIGHTNING Chicago Columbia 11345 23Aug80 56(2) 9 [Peter Cetera/Robert Lamm/Danny Seraphine] A: Chicago XIV; P: Tom Dowd	203
2787. HUNTERS OF THE NIGHT Mr. Mister RCA 13741 17Mar84 57(1) 8 [Steve George/George Ghiz/John Lang/Richard Page] A: I Wear The Face; P: Peter Mclan	203
2788. EASY LOVE Dionne Warwick Arista 0572 22Nov80 62(2) 10 [Randy Cate/Steve Dorff/Larry Herbstritt] A: No Night So Long; P: Steve Buckingham	202
2789. LONELY SCHOOL Tommy Shaw A&M 2696 15Dec84 60(2) 9 [Tommy Shaw] A: Girls With Guns; P: Mike Stone	202
2790. NEVER ENOUGH Patty Smyth Columbia 06643 28Feb87 61(1) 11 [Eric Bazilian/Rick Chertoff/Rob Hyman/David Kagan/Patty Smyth] A: Never Enough; P: Rick Chertoff/William Wittman	202
2791. GIVE IT UP Steve Miller Band Capitol 5194 11Dec82 60(2) 9 [Steve Miller] A: Abracadabra; P: Gary Mallaber/Steve Miller	202
2792. NEVER LET YOU GO Sweet Sensation Atco 99284 17Sep88 58(1) 10 [Joseph E. Malloy] A: Take It While It's Hot; P: Ted Currier	202
2793. OWN THE NIGHT Chaka Khan MCA 52730 21Dec85 57(2) 9 [Mary Dean/Franne Golde/Marti Sharron] A: Miami Vice Soundtrack; P: Arif Mardin/Joe Mardin	202
2794. MAKE IT BETTER (Forget About Me) Tom Petty and The Heartbreakers MCA 52605 08Jun85 54(1) 8 [Tom Petty/David A. Stewart] A: Southern Accents; P: Jimmy Iovine/Tom Petty/David A. Stewart	201
2795. SAY YOU REALLY WANT ME Kim Wilde MCA 53130 18Jul87 44(1) 8 [Richard Rudolph/Danny Sembello/Donnell Spencer] A: Another Step; P: Richard Rudolph/Bruce Swedien/Rod Temperton	201
2796. OLD FASHIONED LOVE Smokey Robinson Tamla 1615 17Apr82 60(1) 9 [Gary Goetzman/Mike Piccirillo] A: Yes It's You Lady; P: George Tobin	201
2797. ANOTHERLOVERHOLENYOHEAD Prince And The Revolution Paisley Park 7-28620 19Jul86 63(2) 10 [Prince] A: Parade: Music From The Motion Picture Under The Cherry Moon (Soundtrack); P: Prince And The Revolution	201
2798. STRAIGHT FROM THE HEART (Into Your Life) The Coyote Sisters Morocco 1742 28Jul84 66(2) 10 [Tony Berg/Leah Kunkel] A: The Coyote Sisters; P: David J. Holman/Roger Paglia	201
2799. KISS AND TELL Breakfast Club MCA 53128 11Jul87 48(1) 9 [Dan Gilroy] A: The Breakfast Club; P: Stephen Bray/Michael Verdick	200
2800. WALKING ON THIN ICE Yoko Ono Geffen 49683 07Mar81 58(1) 10 [Yoko Ono] A: No Album; P: Jack Douglas/John Lennon/Yoko Ono	200
2801. STRANGERS IN A STRANGE WORLD Jenny Burton & Patrick Jude Atlantic 89660 09Jun84 54(2) 7 [Jake Holmes] A: Beat Street Soundtrack; P: Jake Holmes	200
2802. SMALL TOWN GIRL John Cafferty And The Beaver Brown Band Scotti Brothers 05668 09Nov85 64(1) 10 [John Cafferty] A: Tough All Over; P: Kenny Vance	200
2803. AND THE CRADLE WILL ROCK... Van Halen Warner Bros. 49501 24May80 55(1) 7 [Michael Anthony/David Lee Roth/Alex Van Halen/Eddie Van Halen] A: Women And Children First; P: Ted Templeman	200
2804. RIGHT HERE AND NOW Bill Medley Planet 13317 02Oct82 58(3) 8 [Barry Mann/Cynthia Weil] A: Right Here And Now; P: Richard Perry	199
2805. PLAYING TO WIN LRB Capitol 5411 26Jan85 60(1) 8 [John Farnham/Graham Goble/David Hirschfelder/Stephen Housden/Wayne Nelson/Stephen Prestvich/Spencer Proffer] A: Playing To Win; P: Spencer Proffer	199
2806. ON THE DARK SIDE Eddie And The Cruisers Scotti Brothers 04107 08Oct83 64(1) 9 [John Cafferty] A: Eddie & The Cruisers (Soundtrack); P: Kenny Vance	197
2807. LOVELINE Dr. Hook Casablanca 2351 12Jun82 60(2) 10 [David Malloy/Eddie Rabbitt/Even Stevens] A: Players In The Dark; P: Ron Haffkine	197
2808. LOVE ON A SHOESTRING The Captain & Tennille Casablanca 2243 08Mar80 55(2) 7 [Kerry Chater/Douglas Foxworthy] A: Make Your Move; P: Daryl Dragon	196
2809. LIVING IN A DREAM Pseudo Echo RCA 5125 28Mar87 57(1) 9 [Brian Canham] A: Love An Adventure; P: Mark S. Berry	196

Ranking the Singles

Rank. TITLE Act Label Number Entry Date Peak(Wks) Tot Wks [Writers] A: Proximate Album; P: Producer	Score
2810. SPIRIT IN THE SKY Doctor And The Medics I.R.S. 52880 02Aug86 69(2) 11 [Norman Greenbaum] A: Laughing At The Pieces; P: Craig Leon	195
2811. TELL ME IF YOU STILL CARE The S.O.S. Band Tabu 4-04160 19Nov83 65(1) 11 [James Samuel Harris/Terry Lewis] A: On The Rise; P: Jimmy Jam Harris/Terry Lewis	195
2812. DIDN'T KNOW IT WAS LOVE Survivor Scotti Brothers 08067 08Oct88 61(1) 10 [Jim Peterik/Frankie Sullivan] A: Too Hot To Sleep; P: Frank Filipetti/Frankie Sullivan	195
2813. TURN IT ON AGAIN Genesis Atlantic 3751 06Sep80 58(1) 8 [Tony Banks/Phil Collins/Mike Rutherford] A: Duke; P: Genesis/David Hentschel	195
2814. FINALLY T.G. Sheppard Warner/Curb 50041 03Apr82 58(2) 8 [Gary Chapman] A: Finally!; P: Buddy Killen	195
2815. MORE STARS ON 45 (Medley) Stars On 45 Radio 3863 26Sep81 55(2) 7 A: Stars On Long Play II; P: Jaap Eggermont	195
2816. RUNNING SCARED The Fools EMI America 8072 07Mar81 50(2) 7 [Joe Melson/Roy Orbison] A: Heavy Mental; P: Vini Poncia	195
2817. STAY IN TIME Off Broadway usa Atlantic 3647 22Mar80 51(1) 7 [Cliff Johnson] A: On; P: Tom Werman	194
2818. GOODNIGHT SAIGON Billy Joel Columbia 03780 19Mar83 56(3) 7 [Billy Joel] A: The Nylon Curtain; P: Phil Ramone	194
2819. WHY ME? Planet P Geffen 29705 16Apr83 64(1) 9 [Tony Carey] A: Planet P; P: Peter Hauke	194
2820. FEELINGS OF FOREVER Tiffany MCA 53325 11Jun88 50(1) 9 [John Duarte/Mark E. Paul] A: Tiffany; P: George Tobin	194
2821. VARIETY TONIGHT REO Speedwagon Epic 07055 09May87 60(1) 9 [Neal Doughty] A: Life As We Know It; P: Kevin Crouch/David DeVore/Alan Gratzer/Gary Richrath	194
2822. SATELLITE Hooters Columbia 07607 03Oct87 61(1) 8 [Eric Bazilian/Rick Chertoff/Rob Hyman] A: One Way Home; P: Rick Chertoff	194
2823. FIELDS OF FIRE Big Country Mercury 811450 04Feb84 52(2) 6 [William Stuart Adamson/Mark Brzezicki/Anthony Earl Butler/Bruce William Watson] A: The Crossing; P: Steve Lillywhite	194
2824. JACKIE BROWN John Cougar Mellencamp Mercury 874 644 15Jul89 48(1) 8 [John Mellencamp] A: Big Daddy; P: John Mellencamp	193
2825. DIRTY WATER Rock And Hyde Capitol 5691 25Apr87 61(1) 10 [Paul Hyde/Bob Rock] A: Under The Volcano; P: Bruce Fairbairn/Paul Hyde/Bob Rock	193
2826. THE SUN AIN'T GONNA SHINE ANYMORE Nielsen/Pearson Capitol 5032 08Aug81 56(2) 8 [Bob Crewe/Bob Gaudio] A: Nielson Pearson; P: Richard Landis	193
2827. ALMOST SATURDAY NIGHT Dave Edmunds Swan Song 72000 09May81 54(2) 8 [John Fogerty] A: Twangin...; P: Dave Edmunds	193
2828. LET ME BE THE ONE Five Star RCA 14229 08Feb86 59(2) 9 [Ian Foster] A: Luxury Of Life; P: Nick Martinelli	193
2829. ALL TOUCH Rough Trade Boardwalk 167 18Dec82 58(2) 7 [Carole Pope/Kevan Staples] A: For Those Who Think Young; P: Gene Martynec/Kevan Staples	193
2830. IN THE NAME OF LOVE Ralph MacDonald (with Bill Withers) Polydor 881221 01Sep84 58(1) 10 [Ralph MacDonald/William Salter/Bill Withers] A: Universal Rhythm; P: William Eaton/Ralph MacDonald	192
2831. JACKIE Blue Zone U.K. Arista 1-9725 13Aug88 54(1) 9 [Thomas F. Kelly/Billy Steinberg] A: Big Thing; P: Paul Staveley O'Duffy	192
2832. BRINGIN' ON THE HEARTBREAK Def Leppard Mercury 818779 09Jun84 61(1) 8 [Rick Allen/Steve Clark/Joe Elliott/Rick Savage/Pete Willis] A: High 'N' Dry(2); P: Mutt Lange	192
2833. RIBBON IN THE SKY Stevie Wonder Tamla 1639 25Sep82 54(2) 7 [Stevie Wonder] A: Stevie Wonder's Original Musiquarium I; P: Stevie Wonder	192
2834. TUG OF WAR Paul McCartney Columbia 03235 02Oct82 53(2) 8 [Paul McCartney] A: Tug Of War; P: George Martin	191
2835. NIGHT SHIFT Quarterflash Warner Bros. 29932 14Aug82 60(3) 8 [Burt Bacharach/Marv Ross/Carole Bayer Sager] A: Night Shift Soundtrack; P: John Boylan	191
2836. BREAKFAST IN AMERICA Supertramp A&M 2292 13Dec80 62(3) 8 [Rick Davies/Roger Hodgson] A: Paris; P: Peter Henderson/Russel Pope	191
2837. IT'S MY JOB Jimmy Buffett MCA 51061 21Feb81 57(2) 8 [Mac McAnally] A: Coconut Telegraph; P: Norbert Putnam	190
2838. SO MUCH IN LOVE Timothy B. Schmit Full Moon 69939 02Oct82 59(2) 8 [William E. Jackson III/Roy Straigis/George R. Williams] A: Fast Times At Ridgemont High Soundtrack; P: Russ Titelman	190
2839. ONLY LONELY Bon Jovi Mercury 880736 20Apr85 54(2) 8 [Jon Bon Jovi/Dave Bryan] A: 7800 Degrees Fahrenheit; P: Lance Quinn	190
2840. GET IT RIGHT Aretha Franklin Arista 1-9034 30Jul83 61(1) 8 [Marcus Miller/Luther Vandross] A: Get It Right; P: Luther Vandross	190
2841. THE OTHER SIDE OF LIFE The Moody Blues Polydor 885201 16Aug86 58(1) 9 [Justin Hayward] A: The Other Side Of Life; P: Tony Visconti	189
2842. DARLIN' DANIELLE DON'T Henry Lee Summer CBS Associated 07909 28May88 57(1) 8 [Henry Lee Summer] A: Henry Lee Summer; P: Michael Frondelli	189
2843. LAY DOWN YOUR ARMS The Graces A&M 1440 12Aug89 56(1) 9 [Charlotte Caffey/Ralph Schuckett/Ellen Shipley] A: Perfect View; P: Ralph Schuckett/Ellen Shipley	189
2844. ON A CAROUSEL Glass Moon Radio 4022 13Mar82 50(1) 7 [Allan Clarke/Tony Hicks/Graham Nash] A: Growing In The Dark; P: John Pace/Raymond Silva	189
2845. STOP Sam Brown A&M 1234 15Apr89 65(1) 10 [Bruce Brody/Samantha Brown/Gregg Sutton] A: Stop!; P: Pete Brown/Sam Brown	189
2846. BEST OF TIMES Peter Cetera Full Moon 27712 29Oct88 59(1) 8 [Peter Cetera/Patrick Raymond Leonard] A: One More Story; P: Peter Cetera/Patrick Leonard	189
2847. SO YOU RAN Orion The Hunter Portrait 04483 02Jun84 58(2) 8 [Fran Cosmo/Barry Goudreau] A: Orion The Hunter; P: Barry Goudreau	188
2848. WHO'S BEHIND THE DOOR? Zebra Atlantic 89821 09Jul83 61(1) 8 [Randolph Lynch Jackson] A: Zebra; P: Jack Douglas	188
2849. CRITICIZE Alexander O'Neal Tabu 4-07600 14Nov87 70(1) 11 [Garry Johnson/Alexander O'Neal] A: Hearsay; P: Jellybean Johnson	188
2850. EVERYTIME YOU CRY The Outfield Columbia 06295 20Sep86 66(2) 10 [John Spinks] A: Play Deep; P: William Wittman	188
2851. SEXAPPEAL Georgio Motown 1882 28Feb87 58(1) 9 [Georgio Allentini] A: Sexappeal; P: Georgio Allentini	188
2852. GOODBYE TO YOU Scandal Columbia 03234 13Nov82 65(2) 11 [Zack Smith] A: Scandal; P: Vini Poncia	188
2853. SAY IT, SAY IT E.G. Daily A&M 2825 26Apr86 70(3) 10 [Stephen Bray/Toni Colandreo/E.G. Daily] A: Wild Child; P: Jellybean Benitez	188
2854. WEST COAST SUMMER NIGHTS Tony Carey Rocshire 95037 02Jul83 64(1) 9 [Tony Carey] A: Tony Carey [I Won't Be Home Tonight]; P: Peter Hauke	187
2855. IF LOOKS COULD KILL Heart Capitol 5605 19Jul86 54(1) 9 [Jack Conrad/Bob Garrett] A: Heart; P: Ron Nevison	187
2856. ONLY FOR LOVE Limahl EMI America 8277 20Jul85 51(2) 7 [Christopher Hamill] A: Don't Suppose; P: De Harris/Tim Palmer	187
2857. LONELY IN LOVE Giuffria MCA 52558 23Mar85 57(2) 8 [David Glen Eisley/Gregg Giuffria] A: Giuffria; P: Gregg Giuffria	187
2858. A WORLD WITHOUT HEROES KISS Casablanca 2343 12Dec81 56(2) 9 [Bob Ezrin/Lou Reed/Gene Simmons/Paul Stanley] A: Music From The Elder; P: Bob Ezrin	187
2859. YESTERDAY ONCE MORE/NOTHING REMAINS THE SAME (Medley) Spinners Atlantic 3798 14Feb81 52(2) 8 [John Bettis/Richard Carpenter//Michael Zager] A: Labor Of Love; P: Michael Zager	186
2860. SPECIAL WAY Kool & The Gang Mercury 888867 17Oct87 72(1) 14 [Ronald Bell/George M. Brown/Dwania Kyles/Kendal Stubbs/J.T. Taylor] A: Forever; P: Ronald Bell/I.B.M.C./Kool & The Gang	186
2861. ANGELINE The Allman Brothers Band Arista 0555 13Sep80 58(1) 8 [Dickey Betts/Johnny Cobb/Mike Lawler] A: Reach For The Sky; P: Allman Brothers Band/Johnny Cobb/Mike Lawler	184
2862. NOBODY'S PERFECT Mike + The Mechanics Atlantic 88990 05Nov88 63(1) 11 [Brian A. Robertson/Mike Rutherford] A: Living Years; P: Christopher Neil/Mike Rutherford	184
2863. YOU LIKE ME DON'T YOU Jermaine Jackson Motown 1503 18Apr81 50(1) 9 [Jermaine Jackson] A: Jermaine(2); P: Jermaine Jackson	184
2864. THIS MUST BE THE PLACE (NAÏVE MELODY) Talking Heads London-Sire 7-29451 26Nov83 62(1) 8 [David Byrne/Chris Frantz/Jerry Harrison/Tina Weymouth] A: Speaking In Tongues; P: Talking Heads	184
2865. THIS IS MY NIGHT Chaka Khan Warner Bros. 29097 19Jan85 60(2) 9 [David Frank/Mic Murphy] A: I Feel For You; P: Arif Mardin	184
2866. I COULD BE GOOD FOR YOU 707 Casablanca 2280 11Oct80 52(2) 9 [Jim McClarty/Duke McFadden] A: 707; P: Norman Ratner	184
2867. IT CAN HAPPEN Yes Atco 99745 23Jun84 51(1) 7 [Jon Anderson/Trevor Rabin/Chris Squire] A: 90125; P: Trevor Horn	183
2868. TONIGHT TONIGHT Bill Champlin Elektra 47240 26Dec81 55(2) 8 [Bill Champlin/David Foster/Ray Kennedy] A: Runaway; P: David Foster	183
2869. STAY AWAKE Ronnie Laws Liberty 1424 12Sep81 60(2) 9 [Ronnie Laws] A: Solid Ground; P: Ronnie Laws	183
2870. A CERTAIN GIRL Warren Zevon Asylum 46610 08Mar80 57(1) 7 [Allen Toussaint] A: Bad Luck Streak In Dancing School; P: Greg Ladanyi/Warren Zevon	182
2871. CONTROVERSY Prince Warner Bros. 49808 24Oct81 70(1) 11 [Prince] A: Controversy; P: Prince	182

Rank. TITLE Act Label Number Entry Date Peak(Wks) Tot Wks	
[Writers] A: Proximate Album; P: Producer	Score
2872. VOICE OF AMERICA'S SONS John Cafferty And The Beaver Brown Band Scotti Brothers 06048 14Jun86 62(2) 8 [John Cafferty] A: Tough All Over; P: Kenny Vance	182
2873. WANNA BE WITH YOU Earth, Wind & Fire ARC 02688 23Jan82 51(2) 7 [Wayne Vaughn/Maurice White] A: Raise!; P: Maurice White	182
2874. MAGNETIC Earth, Wind & Fire Columbia 04210 12Nov83 57(2) 9 [Martin Page] A: Electric Universe; P: Maurice White	182
2875. SOMEBODY SEND MY BABY HOME Lenny LeBlanc Capitol/MSS 4979 28Mar81 55(1) 7 [Ava Aldridge/Lenny LeBlanc] A: Breakthrough; P: Barry Beckett	182
2876. NOWHERE TO RUN Santana Columbia 03376 27Nov82 66(2) 8 [Russ Ballard] A: Shango; P: John Ryan	182
2877. DANCE WIT' ME - PART 1 Rick James Gordy 1619 29May82 64(2) 9 [Rick James] A: Throwin' Down; P: Rick James	181
2878. SOME PEOPLE Paul Young Columbia 06423 15Nov86 65(2) 10 [Ian Kewley/Paul Young] A: Between Two Fires; P: Ian Kewley/Hugh Padgham/Paul Young	181
2879. STRENGTH The Alarm I.R.S. 52736 28Dec85 61(2) 10 [Eddie MacDonald/Mike Peters/Dave Sharp/Nigel Twist] A: Strength; P: Mike Howlett	181
2880. I'M JUST TOO SHY Jermaine Jackson Motown 1525 31Oct81 60(1) 8 [Jermaine Jackson/Paul M. Jackson Jr.] A: I Like Your Style; P: Jermaine Jackson	181
2881. I LOVE TO BASS Bardeux Enigma 75047 16Sep89 68(2) 10 [Michael Eckart/Stacy Smith/Jon St. James] A: Shangri-La; P: Michael Eckart/Jon St. James	180
2882. SHE GOT THE GOLDMINE (I Got The Shaft) Jerry Reed RCA 13268 24Jul82 57(2) 9 [Tim DuBois] A: The Man With The Golden Thumb; P: Rick Hall	180
2883. IT'S NOT A WONDER (Live) Little River Band Capitol 4862 03May80 51(1) 6 [Graham Goble] A: Backstage Pass; P: Little River Band/Ernie Rose	180
2884. MACHINERY Sheena Easton EMI America 8131 04Sep82 57(2) 7 [Julia Downes] A: Madness, Money And Music; P: Christopher Neil	179
2885. TAXI DANCING Rick Springfield & Randy Crawford RCA 13861 17Nov84 59(1) 10 [Rick Springfield] A: Hard To Hold (Soundtrack); P: Bill Drescher/Rick Springfield	179
2886. BABY COME TO ME Regina Belle Columbia 68969 14Oct89 60(1) 9 [Jeffrey E. Cohen/Narada Michael Walden] A: Stay With Me; P: Narada Michael Walden	179
2887. MERCY, MERCY, MERCY Phoebe Snow Mirage 3818 02May81 52(2) 8 [Don Covay/Ronald Alonzo Miller] A: Rock Away; P: Richie Cannata/Greg Ladanyi	178
2888. A HEART IN NEW YORK Art Garfunkel Columbia 02307 08Aug81 66(1) 9 [Benny Gallagher/Graham Lyle] A: Scissors Cut; P: Art Garfunkel/Roy Halee	178
2889. RESTLESS HEART John Waite EMI America 8252 26Jan85 59(2) 8 [John Waite] A: No Brakes; P: Gary Gersh/David Thoener/John Waite	178
2890. SUMMER NIGHTS Survivor Scotti Brothers 02700 20Feb82 62(2) 8 [Jim Peterik/Frankie Sullivan] A: Premonition; P: Jim Peterik/Frankie Sullivan	178
2891. LET THE DAY BEGIN The Call MCA 53658 22Jul89 51(1) 9 [Michael Been] A: Let The Day Begin; P: Michael Been/Jim Goodwin	178
2892. HOLE IN MY HEART (All The Way To China) Cyndi Lauper Epic 07940 09Jul88 54(2) 8 [Richard Orange] A: Vibes Soundtrack; P: Cyndi Lauper/Lennie Petze	177
2893. ORIGINAL SIN INXS Atco 99766 28Apr84 58(1) 7 [Andy Farriss/Michael Hutchence] A: The Swing; P: Nile Rodgers	177
2894. SHAKIN' Eddie Money Columbia 03252 09Oct82 63(1) 9 [Ralph Carter/Eddie Money/Elizabeth Myers] A: No Control; P: Tom Dowd	177
2895. SOLITAIRE Peter Mclan ARC 11214 05Apr80 52(1) 7 [Anne Mclan/Peter Mclan] A: Playing Near The Edge; P: Peter Mclan	177
2896. COOL MAGIC Steve Miller Band Capitol 5162 16Oct82 57(2) 8 [Kenny Lee Lewis/Gary Mallaber] A: Abracadabra; P: Gary Mallaber/Steve Miller	177
2897. FOREVER MINE The Motels Capitol 5182 13Nov82 60(2) 8 [Martha Davis] A: All Four One; P: Val Garay	177
2898. FANTASTIC VOYAGE Lakeside Solar 12129 31Jan81 55(1) 8 [Fred Alexander/Norman Beavers/Marvin Craig/Bryan Dodds/Tiemeyer McCain/Thomas Shelby/Stephen Shockley/Otis Stokes/Mark Wood] A: Fantastic Voyage; P: Lakeside	176
2899. MUTUAL SURRENDER (What A Wonderful World) Bourgeois Tagg Island 99558 12Apr86 62(1) 10 [Larry Tagg] A: Bourgeois Tagg; P: Brent Bourgeois/David J. Holman/Larry Tagg	176
2900. SWEET SENSATION Stephanie Mills 20th Century 2449 14Jun80 52(2) 6 [Reggie Lucas/James Mtume] A: Sweet Sensation; P: Reggie Lucas/James Mtume	176

Rank. TITLE Act Label Number Entry Date Peak(Wks) Tot Wks	
[Writers] A: Proximate Album; P: Producer	Score
2901. WIND HIM UP Saga Portrait 37-03791 02Apr83 64(1) 8 [Ian Stevenson Crichton/Jim Crichton/Jim Gilmour/Steve Negus/ Michael Sadler] A: Worlds Apart; P: Rupert Hine	176
2902. MOONLIGHT ON WATER Kevin Raleigh Atlantic 88962 20May89 60(1) 9 [Andy Goldmark/Steve Kipner] A: Delusions Of Grandeur; P: Peter Coleman	175
2903. FADING AWAY Will To Power Epic 68543 04Feb89 65(1) 10 [Bob Rosenberg] A: Will To Power; P: Bob Rosenberg	175
2904. HERE COMES MY GIRL Tom Petty And The Heartbreakers Backstreet 41227 26Apr80 59(1) 7 [Mike Campbell/Tom Petty] A: Damn The Torpedoes; P: Jimmy Iovine/Tom Petty	175
2905. HOLD ME Sheila E. Paisley Park 7-28580 07Feb87 68(1) 10 [Connie Guzman/Eddie Mininfield/Sheila E.] A: Sheila E.; P: Sheila E.	175
2906. I WANT CANDY Bow Wow Wow RCA 13204 29May82 62(1) 7 [Bert Berns/Bob Feldman/Jerry Goldstein/Richard Gottehrer] A: I Want Candy; P: Kenny Laguna	175
2907. CH CH CHERIE Johnny Average Band Bearsville 49671 21Feb81 53(1) 7 [Johnny Average/Griffith McRee] A: Some People; P: Johnny Average/Griffith McRee	174
2908. FALLING IN LOVE AGAIN Michael Stanley Band EMI America 8090 08Aug81 64(2) 8 [Bob Pelander/Michael Stanley] A: North Coast; P: Eddie Kramer/Michael Stanley Band	174
2909. THE SENSITIVE KIND Santana Columbia 02178 01Aug81 56(1) 8 [J.J. Cale] A: Zebop!; P: Bill Graham/Carlos Santana	174
2910. HAPPY ENDING Joe Jackson A&M 2635 14Jul84 57(2) 8 [Joe Jackson] A: Body And Soul; P: Joe Jackson/David Kershenbaum	174
2911. HEART DON'T FAIL ME NOW Holly Knight Columbia 07932 03Sep88 59(1) 9 [Holly Knight] A: Holly Knight; P: Holly Knight/Chris Lord-Alge	173
2912. END OF THE LINE Traveling Wilburys Wilbury 7-27637 11Feb89 63(2) 9 [Bob Dylan/George Harrison/Jeff Lynne/Roy Orbison/Tom Petty] A: Volume 1; P: George Harrison/Jeff Lynne	173
2913. SARA Bill Champlin Elektra 47456 31Jul82 61(1) 8 [Bill Champlin/Alan Thicke] A: Runaway; P: David Foster	173
2914. DANCE Ratt Atlantic 89354 21Feb87 59(2) 9 [Robbin Crosby/Warren DeMartini/Beau Hill/Stephen Pearcy] A: Dancing Undercover; P: Beau Hill	173
2915. SOMEBODY LIKE YOU Robbie Nevil EMI 50176 18Mar89 63(1) 11 [Richard Feldman/Robbie Nevil/Jeff Pescetto] A: A Place Like This; P: Robbie Nevil/Chris Porter	173
2916. THE SECRET OF MY SUCCESS Night Ranger Camel/MCA 53013 28Mar87 64(3) 8 [Jack Blades/David Foster/Tom Keane/Mike Landau] A: Big Life; P: David Foster	173
2917. THE KID IS HOT TONITE Loverboy Columbia 02068 20Jun81 55(1) 7 [Bernard Aubin/Paul Dean] A: Loverboy; P: Bruce Fairbairn	173
2918. STARGAZER Peter Brown Drive 6281 15Dec79 59(1) 8 [Peter H. Brown] A: Stargazer; P: Peter Brown/Cory Wade	173
2919. SAVE THE OVERTIME (For Me) Gladys Knight & The Pips Columbia 03761 21May83 66(1) 10 [Sam Dees/Joey Gallo/Gladys Knight/Merald Knight/Ricky Darnell Smith] A: Visions; P: Edmund Sylvers/Leon Sylvers	173
2920. GOODNIGHT MY LOVE Mike Pinera Spector 00003 05Jan80 70(1) 8 [Mike Pinera] A: Forever, Mike Pinera; P: Mike Pinera	172
2921. RICKY "Weird Al" Yankovic Rock 'n' Roll 03849 30Apr83 63(2) 8 [Mike Chapman/Nicky Chinn/Eliot Daniel/Al Yankovic] A: "Weird Al" Yankovic; P: Rick Derringer	172
2922. DON'T YOU KNOW HOW MUCH I LOVE YOU Ronnie Milsap RCA 13564 13Aug83 58(1) 7 [Michael D. Stewart/Dan Edward Williams] A: Keyed Up; P: Tom Collins/Ronnie Milsap	172
2923. MY GUY/MY GIRL Amii Stewart & Johnny Bristol Handshake 5300 30Aug80 63(2) 8 [Smokey Robinson//Smokey Robinson/Ronnie White] A: I'm Gonna Get Your Love; P: Barry Long/Simon May	172
2924. GIRL WITH THE HUNGRY EYES Jefferson Starship Grunt 11921 23Feb80 55(1) 6 [Paul Kantner] A: Freedom At Point Zero; P: Ron Nevison	172
2925. MEMORY - THE THEME FROM ANDREW LLOYD WEBBER'S MUSICAL ''CATS'' Barbra Streisand Columbia 02717 20Feb82 52(1) 7 [T.S. Eliot/Trevor Nunn/Andrew Lloyd Webber] A: Memories; P: Andrew Lloyd Webber	172
2926. ROUND & ROUND New Order Qwest 27524 22Apr89 64(2) 9 [Gillian Gilbert/Peter Hook/Stephen Morris/Bernard Sumner] A: Technique; P: Stephen Hague/New Order	171
2927. SHOULD I SEE Frozen Ghost Atlantic 89279 11Apr87 69(3) 10 [Arnold Lanni] A: Frozen Ghost; P: Arnold Lanni	171

Ranking the Singles

Rank. TITLE Act Label Number Entry Date Peak(Wks) Tot Wks
[Writers] *A: Proximate Album; P: Producer*   Score

2928. LOVE COMES QUICKLY Pet Shop Boys EMI America 8338
30Aug86 62(1) 8 [Stephen Hague/Chris Lowe/Neil Tennant]
A: Please; P: Stephen Hague   170

2929. WISHING I WAS LUCKY Wet Wet Wet Uni 50000 28May88 58(2) 8
[Graeme Clark/Tommy Cunningham/Neil Mitchell/Marti Pellow]
A: Popped In Souled Out; P: Wet Wet Wet   170

2930. SHANGRI-LA Steve Miller Band Capitol 5407 06Oct84 57(2) 6
[Kenny Lee Lewis/Steve Miller]
A: Italian X Rays; P: Kenny Lee Lewis/Steve Miller   170

2931. HOT FUN IN THE SUMMERTIME Dayton Liberty 1468 24Jul82 58(2) 7
[Sylvester Stewart]   *A: Hot Fun; P: Rahni P. Harris Jr.*   170

2932. GYPSY ROAD Cinderella Mercury 874578 05Aug89 51(1) 7
[Tom Keifer]   *A: Long Cold Winter; P: Eric Brittingham/Andy Johns/Tom Keifer*   170

2933. HOT FOR TEACHER Van Halen Warner Bros. 29199 27Oct84 56(1) 7
[Michael Anthony/David Lee Roth/Alex Van Halen/Eddie Van Halen]
A: 1984 (MCMLXXXIV); P: Ted Templeman   170

2934. LOVE NEVER FAILS Greg Kihn Band Beserkley 69820 04Jun83 59(2) 6
[Greg Douglass/Greg Kihn/Larry Lynch/Gary Philippet/Steve Wright]
A: Kihnspiracy; P: Matthew King Kaufman   169

2935. DEDICATED TO THE ONE I LOVE Bernadette Peters MCA 51152
08Aug81 65(1) 8 [Ralph Bass/Lowman Pauling]
A: Now Playing; P: Brooks Arthur   169

2936. HAPPY TOGETHER (A Fantasy) The Captain & Tennille Casablanca 2264
10May80 53(1) 6 [Garry Bonner/Alan Gordon]
A: Make Your Move; P: Daryl Dragon   169

2937. STRANGER Stephen Stills Atlantic 89633 11Aug84 61(1) 8
[Christopher Stills/Stephen Stills]
A: Right By You; P: Howard Albert/Ron Albert/Stephen Stills   169

2938. WINDOWS Missing Persons Capitol 5200 15Jan83 63(2) 8
[Dale Bozzio/Terry Bozzio]   *A: Spring Session M; P: Ken Scott*   169

2939. ONE DAY IN YOUR LIFE Michael Jackson Motown 1512
18Apr81 55(1) 7 [Renée Armand/Samuel F. Brown III]
A: One Day In Your Life; P: Sam Brown III   168

2940. PROUD Joe Chemay Band Unicorn 95001 14Feb81 68(2) 8
[Bunny Hull/Andrew Woolfolk]
A: The Riper The Finer; P: Joe Chemay/John Guess   168

2941. CLOSE ENOUGH TO PERFECT Alabama RCA 13294 04Sep82 65(2) 12
[Carl Chambers]   *A: Mountain Music; P: Alabama/Larry McBride/Harold Shedd*   168

2942. LONELY WON'T LEAVE ME ALONE Glenn Medeiros Amherst 317
19Dec87 67(1) 11 [David Foster/Jermaine Jackson/Tom Keane/
Kathy Wakefield]   *A: Glenn Medeiros; P: Jay Stone*   168

2943. NEVER LET ME DOWN AGAIN Depeche Mode Sire 28189
26Dec87 63(1) 10 [Martin L. Gore]
A: Music For The Masses; P: David Bascombe/Depeche Mode   168

2944. FRONT PAGE STORY Neil Diamond Columbia 03801 23Apr83 65(1) 8
[Burt Bacharach/Neil Diamond/Carole Bayer Sager]
A: Heartlight; P: Burt Bacharach/Neil Diamond/Carole Bayer Sager   168

2945. STEADY Jules Shear EMI America 8259 06Apr85 57(1) 7
[Cyndi Lauper/Jules Shear]   *A: The Eternal Return; P: Bill Drescher/Jules Shear*   167

2946. WHEN I'M WITH YOU ('83) Sheriff Capitol 5199 14May83 61(1) 7
[Arnold Lanni]   *A: Sheriff; P: Stacy Heydon*   167

2947. HAVE YOU EVER LOVED SOMEBODY Freddie Jackson Capitol 5661
07Feb87 69(1) 9 [Barry Eastmond/Jolyon Skinner]
A: Just Like The First Time; P: Barry Eastmond   167

2948. WEAPONS OF LOVE The Truth I.R.S. 53084 09May87 65(1) 9
[Dennis Greaves/Mick Lister]   *A: Weapons Of Love; P: Dennis Herring*   167

2949. SOMEONE LIKE YOU Daryl Hall RCA 5105 24Jan87 57(1) 8
[Daryl Hall]   *A: Three Hearts In The Happy Ending Machine;
P: Daryl Hall/David A. Stewart/Tom Wolk*   166

2950. FREAK-A-ZOID Midnight Star Solar 69828 20Aug83 66(1) 8
[Reggie Calloway/Vincent Calloway/Bill Simmons]
A: No Parking On The Dance Floor; P: Reggie Calloway   166

2951. I'LL DRINK TO YOU Duke Jupiter Coast To Coast 02801
27Mar82 58(1) 7 [Marshall James Styler]   *A: Duke Jupiter 1; P: Glen Kolotkin*   166

2952. AND LOVE GOES ON Earth, Wind & Fire ARC 11434 07Feb81 59(2) 7
[Lorenzo Dunn/David Foster/Brenda Russell/Maurice White/Verdine White]
A: Faces; P: Maurice White   166

2953. VITAMIN L B. E. Taylor Group MCA 52311 28Jan84 66(1) 8
[Debbie Witkowski/Rick Witkowski]
A: Love Won The Fight; P: Joe Macre/Rick Witkowski   166

2954. IF I SAY YES Five Star RCA 5083 27Dec86 67(2) 11
[Michael Jay/Marvin Morrow]   *A: Silk And Steel; P: Michael Jay/Buster Pearson*   165

2955. YOU ARE IN MY SYSTEM The System Mirage 99937 05Mar83 64(2) 8
[David Frank/Mic Murphy]
A: Sweat; P: David Frank/Mic Murphy   165

2956. KNOCKING AT YOUR BACK DOOR Deep Purple Mercury 880477
05Jan85 61(2) 7 [Ritchie Blackmore/Ian Gillan/Roger Glover]
A: Perfect Strangers; P: Deep Purple/Roger Glover   164

2957. ALL THE KINGS HORSES The Firm Atlantic 89458 15Feb86 61(2) 8
[Paul Rodgers]   *A: Mean Business; P: Jimmy Page/Paul Rodgers*   164

2958. BABY DON'T GO Karla Bonoff Columbia 11206 01Mar80 69(1) 7
[Karla Bonoff/Kenny Edwards]   *A: Restless Nights; P: Kenny Edwards*   164

2959. FASHION David Bowie RCA 12134 06Dec80 70(1) 9
[David Bowie]   *A: Scary Monsters; P: David Bowie/Tony Visconti*   164

2960. OLYMPIA Sergio Mendes A&M 2623 07Apr84 58(1) 7
[Barry Mann/Cynthia Weil]   *A: Confetti; P: Sergio Mendes*   164

2961. HOT THING Prince Paisley Park 7-28288 06Feb88 63(2) 9
[Prince]   *A: Sign 'O' The Times; P: Prince*   163

2962. HOLD ON Badfinger Radio 3793 28Feb81 56(2) 8
[Tom Evans/Joseph Tansini]
A: Say No More; P: Jack Richardson/Steve Wittmack   163

2963. PUSS N BOOTS/THESE BOOTS (ARE MADE FOR WALKIN') Kon Kan
Atlantic 88828 02Sep89 58(1) 8
[Barry Harris/Jimmy Page/Robert Plant/Kevin Wynne//Lee Hazlewood]
A: Move To Move; P: Mark Goldenberg/Barry Harris   163

2964. FOREVER Little Steven And The Disciples Of Soul EMI America 8144
25Dec82 63(2) 9
[Steven Van Zandt]   *A: Men Without Women; P: Steven Van Zandt*   163

2965. DANGEROUS Loverboy Columbia 05711 16Nov85 65(1) 9
[Bryan Adams/Jim Vallance]
A: Lovin' Every Minute Of It; P: Tom Allom/Paul Dean   163

2966. KISS AND TELL Isley, Jasper, Isley CBS Associated 04741
16Feb85 63(2) 7 [Ernie Isley/Marvin Isley/Chris Jasper]
A: Broadway's Closer To Sunset Blvd.; P: Ernie Isley/Marvin Isley/Chris Jasper   163

2967. HOW CAN I FORGET YOU Elisa Fiorillo Chrysalis 43189
23Jan88 60(1) 8 [Gardner Cole]   *A: Elisa Fiorillo; P: Gardner Cole*   163

2968. STILL A THRILL Jody Watley MCA 53081 20Jun87 56(1) 7
[André Cymone/Jody Watley]   *A: Jody Watley; P: André Cymone/David Rivkin*   162

2969. WILLIE AND THE HAND JIVE George Thorogood & The Destroyers
EMI America 8270 15Jun85 63(2) 8
[Johnny Otis]   *A: Maverick; P: Delaware Destroyers/Terry Manning*   162

2970. LOVE X LOVE George Benson Warner Bros./Qwest 49570
18Oct80 61(3) 6 [Rod Temperton]   *A: Give Me The Night; P: Quincy Jones*   162

2971. SEX (I'm A...) Berlin Geffen 29747 05Mar83 62(2) 7
[John Crawford/David Samuel Diamond/Terri Nunn]
A: Pleasure Victim; P: Daniel R. Van Patten   162

2972. FUNNY HOW TIME SLIPS AWAY Spinners Atlantic 89922
11Dec82 67(2) 8 [Willie Nelson]   *A: Grand Slam; P: Freddie Perren*   161

2973. STANDING ON THE TOP PART 1 The Temptations Featuring Rick James
Gordy 1616 08May82 66(1) 8 [Rick James]   *A: Reunion; P: Rick James*   161

2974. RUN LIKE HELL Pink Floyd Columbia 11265 10May80 53(1) 6
[David Gilmour/Roger Waters]
A: The Wall; P: Bob Ezrin/David Gilmour/Roger Waters   160

2975. JESSE Julian Lennon Atlantic 89529 03Aug85 54(1) 6
[China Burton]   *A: Valotte; P: Phil Ramone*   159

2976. BON BON VIE (Gimme The Good Life) T.S. Monk Mirage 3790 21Feb81
63(1) 8 [L. Russell Brown/Sandy Linzer]   *A: House Of Music; P: Sandy Linzer*   159

2977. NOTHING EVER GOES AS PLANNED Styx A&M 2348 11Jul81 54(2) 8
[Dennis DeYoung]   *A: Paradise Theater; P: Styx*   159

2978. TURN ON YOUR RADAR Prism Capitol 5106 17Apr82 64(3) 7
[Morgan Walker]   *A: Small Change; P: John S. Carter*   159

2979. I DON'T NEED YOU Rupert Holmes MCA 51092 04Apr81 56(1) 7
[Rupert Holmes]   *A: Rupert Holmes; P: Rupert Holmes*   159

2980. THE ALLNIGHTER Glenn Frey MCA 52461 29Sep84 54(1) 6
[Glenn Frey/Jack Tempchin]
A: The Allnighter; P: Allan Blazek/Glenn Frey   159

2981. TOUCH AND GO Emerson, Lake & Powell Polydor 885101
21Jun86 60(1) 8 [Keith Emerson/Greg Lake]
A: Emerson, Lake, & Powell; P: Greg Lake/Tony Taverner   159

2982. WISE UP Amy Grant A&M 2762 17Aug85 66(2) 9
[Wayne Kirkpatrick/Billy Simon]
A: Unguarded; P: Brown Bannister   158

2983. BACK TO THE BULLET Saraya Polydor 889976 11Nov89 63(1) 9
[Sandy Linzer/Gregg Munier/Sandi Saraya]   *A: Saraya; P: Jeff Glixman*   158

Rank. TITLE Act Label Number Entry Date Peak(Wks) Tot Wks [Writers] A: Proximate Album; P: Producer	Score
2984. EVERY HOME SHOULD HAVE ONE ('81) Patti Austin Qwest 49854 12Dec81 62(2) 8 [Dominic Bugatti/Frank Musker] A: Every Home Should Have One; P: Quincy Jones	158
2985. MAKE THAT MOVE Shalamar Solar 12192 25Apr81 60(1) 8 [William Shelby/Ricky Darnell Smith/Kevin Spencer] A: Three For Love; P: Leon Sylvers	158
2986. TOUGH WORLD Donnie Iris MCA 52127 23Oct82 57(1) 6 [Mark Avsec/Martin Lee Hoenes/Donnie Iris] A: The High And The Mighty; P: Mark Avsec	158
2987. NEW GIRL NOW Honeymoon Suite Warner Bros. 29208 08Sep84 57(1) 7 [Dermot Grehan] A: Honeymoon Suite; P: Tom Treumuth	158
2988. I THINK I CAN BEAT MIKE TYSON D.J. Jazzy Jeff & The Fresh Prince Jive 1282 04Nov89 58(1) 9 [Peter Brian Harris/Will Smith/Jeff Townes] A: And In This Corner...; P: Nigel Green/Pete Q. Harris/Will Smith/Jeff Townes	158
2989. HE GOT YOU Ronnie Milsap RCA 13296 21Aug82 59(2) 7 [Ralph Murphy/Bobby Wood] A: Inside Ronnie Milsap; P: Tom Collins/Ronnie Milsap	158
2990. MEMORIES Tierra Boardwalk 70073 14Mar81 62(2) 8 [Rudy Salas] A: City Nights; P: Rudy Salas	158
2991. I JUST CAN'T WALK AWAY Four Tops Motown 1706 22Oct83 71(2) 9 [Lamont Dozier/Brian Holland/Eddie Holland] A: Back Where I Belong; P: Lamont Dozier/Brian Holland/Eddie Holland	157
2992. THEME FROM "DOCTOR DETROIT" Devo Backstreet 52215 21May83 59(1) 6 [Jerry Casale/Mark Mothersbaugh] A: "Doctor Detroit" Soundtrack; P: Devo	157
2993. WAY OUT J.J. Fad Ruthless 7-99285 17Sep88 61(2) 9 [Juana Burns] A: Supersonic--The Album; P: Antoine Carraby/Dr. Dre/Mik Lezan	157
2994. THE MOMENT OF TRUTH Survivor Casablanca 880053 16Jun84 63(2) 7 [Peter Beckett/Bill Conti/Dennis Lambert] A: The Karate Kid Soundtrack; P: Ron Nevison	157
2995. TAKE ANOTHER PICTURE Quarterflash Geffen 29523 01Oct83 58(1) 6 [Marv Ross] A: Take Another Picture; P: John Boylan	157
2996. INTO YOU Giant Steps A&M 1256 28Jan89 58(1) 8 [Colin Campsie/Gardner Cole/George McFarlane] A: The Book Of Pride; P: Gardner Cole	157
2997. LAND OF A THOUSAND DANCES The J. Geils Band EMI America 8156 26Feb83 60(2) 6 [Chris Kenner] A: Showtime; P: Seth Justman	157
2998. DON'T STOP ME BABY (I'm On Fire) The Boys Band Elektra 47406 06Mar82 61(1) 8 [Larry Keith/Steve Pippin/Austin Roberts/Johnny Slate] A: The Boys Band; P: Peter Granet	157
2999. THAT'S FREEDOM Tom Kimmel Mercury 888571 20Jun87 64(2) 8 [Jean Chapman/Tom Kimmel] A: 5 To 1; P: Bill Szymczyk	156
3000. SUBURBIA Pet Shop Boys EMI America 8355 06Dec86 70(1) 10 [Chris Lowe/Neil Tennant] A: Please; P: Julian Mendelsohn	156
3001. FIGHT FIRE WITH FIRE Kansas CBS Associated 04057 03Sep83 58(1) 7 [Dino Elefante/John Elefante] A: Drastic Measures; P: Kansas/Neil Kernon	156
3002. COMMUNICATION Spandau Ballet Chrysalis 42770 31Mar84 59(1) 7 [Gary Kemp] A: True; P: Steve Jolley/Spandau Ballet/Tony Swain	156
3003. CIRCLE OF LOVE Steve Miller Band Capitol 5086 23Jan82 55(2) 7 [Steve Miller] A: Circle Of Love; P: Steve Miller	155
3004. ONLY YOU Yaz Sire 29844 26Feb83 67(3) 8 [Vince Clarke] A: Upstairs At Eric's; P: Eric Radcliffe/Yaz	155
3005. HYPERACTIVE Thomas Dolby Capitol 5321 25Feb84 62(1) 7 [Thomas Dolby] A: The Flat Earth; P: Thomas Dolby	154
3006. TELL ME I'M NOT DREAMING Robert Palmer EMI 50206 01Jul89 60(1) 8 [Jay Gruska/Michael Omartian/Bruce Sudano] A: Heavy Nova; P: Robert Palmer	154
3007. I'M NOT PERFECT (But I'm Perfect For You) Grace Jones Manhattan 50052 29Nov86 69(1) 9 [Grace Jones/Bruce Woolley] A: Inside Story; P: Grace Jones/Nile Rodgers	154
3008. LATE AT NIGHT England Dan Seals Atlantic 3674 16Aug80 57(2) 6 [Dan Seals/Rafe Van Hoy] A: Stones; P: Kyle Lehning	154
3009. LONG AND LASTING LOVE (Once In A Lifetime) Glenn Medeiros Amherst 324 13Aug88 68(1) 10 [Gerry Goffin/Michael Masser] A: Not Me; P: Michael Masser	154
3010. LATELY Stevie Wonder Tamla 54323 11Apr81 64(2) 7 [Stevie Wonder] A: Hotter Than July; P: Stevie Wonder	154
3011. JUST ONE MORE TIME Headpins SGR 90001 24Dec83 70(1) 9 [Brian MacLeod/Darby Mills] A: Line Of Fire; P: Brian MacLeod	153
3012. BROOKLYN GIRLS Robbie Dupree Elektra 47145 23May81 54(2) 7 [Roy Freeland/Bill LaBounty] A: Street Corner Heroes; P: Peter Bunetta/Rick Chudacoff	153
3013. MORNING DESIRE Kenny Rogers RCA 14194 23Nov85 72(2) 9 [Dave Loggins] A: The Heart Of The Matter; P: George Martin	153
3014. RIGHT AWAY Hawks Columbia 60500 14Mar81 63(1) 7 [Dave Steen] A: Hawks; P: Tom Werman	153
3015. TAKE ME NOW David Gates Arista 0615 26Sep81 62(1) 7 [David Gates] A: Take Me Now; P: David Gates	153
3016. OH DADDY Adrian Belew Atlantic 88904 05Aug89 58(1) 8 [Adrian Belew] A: Mr. Music Head; P: Adrian Belew	152
3017. HIDE YOUR HEART KISS Mercury 876146 25Nov89 66(1) 10 [Desmond Child/Holly Knight/Paul Stanley] A: Hot In The Shade; P: Gene Simmons/Paul Stanley	152
3018. ONLY ONE YOU T.G. Sheppard Warner/Curb 49858 30Jan82 68(1) 8 [Michael Garvin/Bucky Jones] A: Finally!; P: Buddy Killen	152
3019. SHE'S TROUBLE Musical Youth MCA 52312 14Jan84 65(1) 7 [Terry Britten/Billy Livsey/Sue Shifrin] A: Different Style!; P: Peter Collins	151
3020. BLACK STATIONS/WHITE STATIONS M+M RCA 13802 30Jun84 63(1) 7 [Mark Gane/Martha Johnson] A: Mystery Walk; P: Daniel Lanois	151
3021. HEAT OF THE MOMENT After 7 Virgin 99204 21Oct89 74(1) 12 [Kenneth Edmonds/Antonio Reid] A: After 7; P: Kenneth Edmonds/L.A. Reid	151
3022. SIGN OF THE GYPSY QUEEN April Wine Capitol 5001 30May81 57(1) 8 [Lorence Hud] A: The Nature Of The Beast; P: Myles Goodwyn/Mike Stone	151
3023. UNDER THE INFLUENCE Vanity Motown 1833 19Apr86 56(1) 7 [Tommy Faragher/Tony Haynes/Robbie Nevil] A: Skin On Skin; P: Skip Drinkwater/Tommy Faragher	151
3024. I HEAR YOU NOW Jon & Vangelis Polydor 2098 16Aug80 58(1) 6 [Jon Anderson/Vangelis] A: Short Stories; P: Vangelis	151
3025. DESPERATE BUT NOT SERIOUS Adam Ant Epic 03688 12Mar83 66(3) 8 [Adam Ant/Marco Pirroni] A: Friend Or Foe; P: Adam Ant/Marco Pirroni	150
3026. LET ME IN Eddie Money Columbia 68739 22Apr89 60(1) 7 [Paul Gordon/Dennis Matkosky] A: Nothing To Lose; P: Eddie Money/Richie Zito	150
3027. HURRY UP AND WAIT The Isley Brothers T-Neck 02033 18Apr81 58(1) 7 [Ernie Isley/Marvin Isley/O'Kelly Isley/Ronald Isley/Rudolph Isley/Chris Jasper] A: Grand Slam; P: Ernie Isley/Marvin Isley/O'Kelly Isley/Ronald Isley/Rudolph Isley/Chris Jasper	150
3028. THROWAWAY Mick Jagger Columbia 07653 28Nov87 67(1) 9 [Mick Jagger] A: Primitive Cool; P: Mick Jagger/David A. Stewart	150
3029. YOU'RE SO EASY TO LOVE Tommy James Millennium 11802 09May81 58(2) 7 [Joey Greco/Tommy James] A: No Album; P: Tommy James	150
3030. I'LL BE AROUND What Is This MCA 52593 17Aug85 62(1) 6 [Thom Bell/Phil Hurtt] A: What Is This; P: Todd Rundgren	150
3031. PERHAPS LOVE Placido Domingo And John Denver Columbia 02679 16Jan82 59(1) 7 [John Denver] A: Perhaps Love; P: Milton Okun	150
3032. DON'T MAKE A FOOL OF YOURSELF Stacey Q Atlantic 89135 27Feb88 66(1) 8 [Skip Hahn/Jon St. James/Stacey Swain] A: Hard Machine; P: Jon St. James	150
3033. I WOULDN'T BEG FOR WATER Sheena Easton EMI America 8142 30Oct82 64(3) 7 [Mike Leeson/Peter Vale] A: Madness, Money And Music; P: Christopher Neil	149
3034. ROCKIT Herbie Hancock Columbia 04054 10Sep83 71(2) 9 [Michael Beinhorn/Herbie Hancock/Bill Laswell] A: Future Shock; P: Herbie Hancock/Material	149
3035. YOU ARE FOREVER Smokey Robinson Tamla 54327 20Jun81 59(1) 7 [Smokey Robinson] A: Being With You; P: George Tobin	149
3036. EVERY LITTLE KISS ('86) Bruce Hornsby And The Range RCA 5165 26Jul86 72(2) 9 [Bruce Hornsby] A: The Way It Is; P: Bruce Hornsby/Elliot Scheiner	149
3037. CHAIN REACTION (Special New Mix) Diana Ross RCA 14244 03May86 66(1) 8 [Barry Gibb/Maurice Gibb/Robin Gibb] A: No Album; P: Albhy Galuten/Barry Gibb/Karl Richardson	149
3038. HAS ANYONE EVER WRITTEN ANYTHING FOR YOU Stevie Nicks Modern 99532 17May86 60(2) 6 [Stevie Nicks/Keith Olsen] A: Rock A Little; P: Rick Nowels	149
3039. BLUE LIGHT David Gilmour Columbia 04378 07Apr84 62(1) 7 [David Gilmour] A: About Face; P: Bob Ezrin/David Gilmour	148
3040. CONCEALED WEAPONS The J. Geils Band EMI America 8242 03Nov84 63(1) 7 [Paul Justman/Seth Justman] A: You're Gettin' Even While I'm Gettin' Odd; P: Seth Justman	148
3041. THE ONLY WAY OUT Cliff Richard EMI America 8135 09Oct82 64(2) 7 [Ray Martinez] A: Now You See Me......Now You Don't; P: Craig Pruess/Cliff Richard	148

Ranking the Singles

Rank. TITLE Act Label Number Entry Date Peak(Wks) Tot Wks	
[Writers] A: Proximate Album; P: Producer	Score
3042. DEAR MR. JESUS PowerSource PowerVision 8603 19Dec87 61(1) 7 [Richard Klender] A: Shelter From The Storm; P: MISSING	148
3043. LOVE HAS TAKEN ITS TOLL Saraya Polydor 889292 08Jul89 64(1) 9 [Anthony Michael Bruno/Sandy Linzer/Sandi Saraya] A: Saraya; P: Jeff Glixman	147
3044. GOOD LIFE Inner City Virgin 99236 04Mar89 73(1) 11 [Paris Grey/Roy Holman/Kevin Saunderson] A: Big Fun; P: Kevin Saunderson	147
3045. BLAZE OF GLORY Kenny Rogers Liberty 1441 21Nov81 66(2) 9 [Larry Keith/Danny Morrison/Johnny Slate] A: Share Your Love; P: Lionel Richie	147
3046. TURN AROUND Neil Diamond Columbia 04541 18Aug84 62(1) 8 [Burt Bacharach/Neil Diamond/Carole Bayer Sager] A: Primitive; P: Denny Diante	146
3047. THE VISITORS ABBA Atlantic 4031 17Apr82 63(2) 8 [Benny Andersson/Björn Ulvaeus] A: The Visitors; P: Benny Andersson/Björn Ulvaeus	146
3048. HEAVEN (Must Be There) Eurogliders Columbia 64626 10Nov84 65(1) 6 [Bernie Lynch] A: This Island; P: Nigel Gray	146
3049. THINKING OF YOU Earth, Wind & Fire Columbia 07695 13Feb88 67(1) 8 [Wanda Hutchinson/Wayne Vaughn/Maurice White] A: Touch The World; P: Maurice White	146
3050. ARE YOU SERIOUS Tyrone Davis Highrise 2005 25Dec82 57(2) 6 [L.V. Johnson/Roger Miner] A: Tyrone Davis; P: Leo Graham	146
3051. RELAX ('84) Frankie Goes To Hollywood Island 99805 07Apr84 67(1) 7 [Peter Gill/Holly Johnson/Mark O'Toole] A: Welcome To The Pleasure Dome; P: Trevor Horn	146
3052. INSIDE A DREAM Jane Wiedlin EMI 50145 03Sep88 57(1) 7 [Gardner Cole/Jane Wiedlin] A: Fur; P: Stephen Hague	146
3053. FOLLOW YOU Glen Burtnick A&M 2968 26Sep87 65(1) 8 [Glen Burtnick/Jack Ponti] A: Heroes & Zeros; P: Glen Burtnick/David Prater	146
3054. DON'T LOOK ANY FURTHER The Kane Gang Capitol 44115 06Feb88 64(1) 8 [Franne Golde/Duane Hitchings/Dennis Lambert] A: Miracle; P: Kane Gang/Pete Wingfield	145
3055. ARE YOU GETTING ENOUGH HAPPINESS Hot Chocolate EMI America 8143 18Dec82 65(2) 7 [Errol Brown] A: Mystery; P: Mickie Most	145
3056. WAKE UP MY LOVE George Harrison Dark Horse 29864 20Nov82 53(2) 5 [George Harrison] A: Gone Troppo; P: Ray Cooper/George Harrison/Phil McDonald	145
3057. AMNESIA Shalamar Solar 69682 17Nov84 73(2) 9 [George Duke/Howard Hewett] A: Heart Break; P: George Duke	145
3058. CAN'T PUT A PRICE ON LOVE The Knack Capitol 4853 05Apr80 62(2) 6 [Berton Averre/Doug Fieger] A: But The Little Girls Understand; P: Mike Chapman	145
3059. DO YOU COMPUTE? Donnie Iris MCA 52230 02Jul83 64(1) 7 [Mark Avsec/Donnie Iris] A: Fortune 410; P: Mark Avsec	145
3060. I'M ALIVE Gamma Elektra 46555 05Jan80 60(1) 6 [Clint Ballard Jr.] A: Gamma 1; P: Ken Scott	145
3061. SHOW ME THE WAY Regina Belle Columbia 07080 18Jul87 68(1) 9 [Joey Gallo/Sue Pomerantz/Wardell Potts] A: All By Myself; P: Nick Martinelli	145
3062. DANCING UNDER A LATIN MOON Candi I.R.S. 53436 22Oct88 68(1) 7 [Michael Jay/Alan Roy Scott/Robbie Seidman] A: Candi; P: David Shaw	145
3063. THE MESSAGE Grandmaster Flash & The Furious Five Featuring: Melle Mel and Duke Bootee Sugar Hill 584 16Oct82 62(2) 7 [Clifton Chase/Edward Fletcher/Melvin Glover/Sylvia Robinson] A: The Message; P: Clifton Chase/Edward Fletcher/Sylvia Robinson	144
3064. THIS COULD BE THE RIGHT ONE April Wine Capitol 5319 11Feb84 58(1) 6 [Myles Goodwyn/David Henman/Jim Henman/Richie Henman] A: Animal Grace; P: Myles Goodwyn/Mike Stone	144
3065. STAND UP David Lee Roth Warner Bros. 28108 16Apr88 64(1) 8 [David Lee Roth/Brett Tuggle] A: Skyscraper; P: David Lee Roth	144
3066. WE CAN GET TOGETHER Icehouse Chrysalis 2530 01Aug81 62(2) 7 [Iva Davies] A: Icehouse; P: Cameron Allan/Iva Davies	144
3067. THE ELVIS MEDLEY Elvis Presley RCA 13351 27Nov82 71(4) 7 A: The Elvis Medley; P: David Briggs	143
3068. UNDER THE BOARDWALK Bruce Willis Motown 1896 13Jun87 59(1) 7 [Artie Resnick/Kenny Young] A: The Return Of Bruno; P: Robert Kraft	143
3069. HAPPY MAN Greg Kihn Band Beserkley 47463 22May82 62(1) 7 [Greg Kihn/Steve Wright] A: Kihntinued; P: Matthew King Kaufman	143
3070. CALLING ALL GIRLS Queen Elektra 69981 31Jul82 60(2) 6 [Roger Taylor] A: Hot Space; P: Reinhold Mack/Queen	143
3071. I PREDICT Sparks Atlantic 4030 15May82 60(1) 7 [Ron Mael/Russell Mael] A: Angst In My Pants; P: Reinhold Mack	143
3072. MY BABY The Pretenders Sire 28496 07Feb87 64(1) 7 [Chrissie Hynde] A: Get Close; P: Bob Clearmountain/Jimmy Iovine	143

Rank. TITLE Act Label Number Entry Date Peak(Wks) Tot Wks	
[Writers] A: Proximate Album; P: Producer	Score
3073. PLEASE BE THE ONE Karla Bonoff Columbia 03172 25Sep82 63(2) 7 [Karla Bonoff] A: Wild Heart Of The Young; P: Kenny Edwards	143
3074. MY FANTASY Teddy Riley featuring Guy Motown 1968 23Sep89 62(1) 6 [William Aquart/Gene Griffin/Teddy Riley] A: Do The Right Thing Soundtrack; P: Gene Griffin/Teddy Riley	142
3075. SHE'S TIGHT Cheap Trick Epic 03233 09Oct82 65(3) 7 [Rick Nielsen] A: One On One; P: Roy Thomas Baker	142
3076. ROOTY TOOT TOOT John Cougar Mellencamp Mercury 870 327 14May88 61(1) 8 [John Mellencamp] A: The Lonesome Jubilee; P: Don Gehman/John Mellencamp	142
3077. VOLCANO Jimmy Buffett MCA 41161 22Dec79 66(2) 7 [Jimmy Buffett/Harry Dailey/Keith Sykes] A: Volcano; P: Norbert Putnam	141
3078. SHOT IN THE DARK Ozzy Osbourne CBS Associated 05810 22Mar86 68(1) 9 [Ozzy Osbourne/Phil Soussan] A: The Ultimate Sin; P: Ron Nevison	141
3079. (Want You) BACK IN MY LIFE AGAIN Carpenters A&M 2370 12Sep81 72(2) 8 [Kerry Chater/Chris Christian] A: Made In America; P: Richard Carpenter	141
3080. IT'S LIKE WE NEVER SAID GOODBYE Crystal Gayle Columbia 11198 23Feb80 63(2) 6 [Roger Greenaway/Geoff Stephens] A: Miss The Mississippi; P: Allen Reynolds	140
3081. WHAT IF (I Said I Love You) Unipop Kat Family 03533 25Dec82 71(3) 8 [Sonny Limbo/Manny Loiacono/Phyllis Loiacono] A: Unilove; P: Sonny Limbo/Scott MacLellan	140
3082. HALF MOON SILVER Hotel MCA 41277 12Jul80 72(2) 7 [Lee Bargeron/Tom Calton/Marc Phillips] A: Half Moon Silver; P: Dain Eric/Hotel	140
3083. I HAVE THE SKILL Sherbs Atco 7325 07Mar81 61(1) 7 [Daryl Braithwaite/Tony Mitchell/Garth Porter] A: The Skill; P: Richard Lush/Sherbs	140
3084. AFTER YOU Dionne Warwick Arista 0498 29Mar80 65(2) 6 [Doug Frank/Doug James] A: Dionne(2); P: Barry Manilow	140
3085. DON'T MAKE ME OVER Jennifer Warnes Arista 0455 22Dec79 67(2) 7 [Burt Bacharach/Hal David] A: Shot Through The Heart; P: Rob Fraboni	139
3086. IT'S THE END OF THE WORLD AS WE KNOW IT (And I Feel Fine) R.E.M. I.R.S. 53220 30Jan88 69(1) 9 [Bill Berry/Peter Buck/Mike Mills/ Michael Stipe] A: Document; P: Scott Litt/R.E.M.	139
3087. I EAT CANNIBALS Total Coelo Chrysalis 42669 16Apr83 66(2) 6 [Barry Blue/Paul Greedus/Roy Nicolson] A: Man O' War; P: Barry Blue	138
3088. YEARNING FOR YOUR LOVE The Gap Band Mercury 76101 23May81 60(1) 7 [Oliver Scott/Ronnie Wilson] A: The Gap Band III; P: Lonnie Simmons	138
3089. EYE ON YOU Billy Squier Capitol 5416 08Dec84 71(1) 8 [Billy Squier] A: Signs Of Life; P: Billy Squier/Jim Steinman	138
3090. BLUE WORLD The Moody Blues Threshold 6054 12Nov83 62(2) 6 [Justin Hayward] A: The Present; P: Pip Williams	138
3091. STAR Earth, Wind & Fire ARC 11165 22Dec79 64(1) 6 [Eddie Del Barrio/Maurice White/Allee Willis] A: I Am; P: Maurice White	137
3092. THE COWBOY AND THE LADY John Denver RCA 12345 31Oct81 66(2) 7 [Bobby Goldsboro] A: Some Days Are Diamonds; P: Larry Butler	137
3093. A LUCKY GUY Rickie Lee Jones Warner Bros. 49816 03Oct81 64(2) 7 [Rickie Lee Jones] A: Pirates; P: Russ Titelman/Lenny Waronker	137
3094. ANCHORAGE Michelle Shocked Mercury 870611 10Dec88 66(1) 8 [Michelle Shocked] A: Short Sharp Shocked; P: Pete Anderson	137
3095. SMOOTH UP Bulletboys Warner Bros. 22876 22Jul89 71(1) 10 [Jimmy D'Anda/Mick Sweda/Marq Torien/Lonnie Vencent] A: Bulletboys; P: Ted Templeman	137
3096. GLORIA The Doors Elektra 69770 03Dec83 71(1) 7 [Van Morrison] A: Alive She Cried; P: Paul A. Rothchild	137
3097. YOUNG LOVE Janet Jackson A&M 2440 18Dec82 64(2) 6 [René Moore/Angela Winbush] A: Janet Jackson; P: René Moore/Bobby Watson/Angela Winbush	137
3098. POWERFUL STUFF The Fabulous Thunderbirds Elektra 69384 27Aug88 65(1) 7 [Robert Field/Michael James Henderson/Wally Wilson] A: Cocktail Soundtrack; P: Terry Manning	137
3099. YO NO SE' Pajama Party Atlantic 88984 20May89 75(1) 10 [Jim Klein/Peggy Sendars] A: Up All Night; P: Jim Klein	136
3100. I'M STEPPING OUT John Lennon Polydor 821107 31Mar84 55(1) 6 [John Lennon] A: Milk And Honey; P: MISSING	136
3101. TOO MUCH LOVE TO HIDE Crosby, Stills & Nash Atlantic 89888 29Jan83 69(3) 6 [Stephen Stills/Gerry Tolman] A: Daylight Again; P: Crosby, Stills & Nash/Steve Gursky/Stanley Johnston	136
3102. YOU HAVE PLACED A CHILL IN MY HEART Eurythmics RCA 8619 28May88 64(1) 7 [Annie Lennox/David A. Stewart] A: Savage; P: David A. Stewart	136
3103. GO FOR IT Kim Wilde MCA 52513 19Jan85 65(1) 7 [Marty Wilde/Ricky Wilde] A: Teases & Dares; P: Marty Wilde/Ricky Wilde	136

Ranking the Singles

Rank. TITLE Act Label Number Entry Date Peak(Wks) Tot Wks [Writers] A: Proximate Album; P: Producer	Score
3104. MISTAKEN IDENTITY Kim Carnes EMI America 8098 24Oct81 60(1) 6 [Kim Carnes] A: Mistaken Identity; P: Val Garay	136
3105. BLUE MONDAY 1988 New Order Qwest 27979 30Apr88 68(1) 10 [Gillian Gilbert/Peter Hook/Stephen Morris/Bernard Sumner] A: Substance; P: New Order	135
3106. LITTLE LADY Duke Jupiter Morocco 1736 12May84 68(2) 7 [Marshall James Styler] A: White Knuckle Ride; P: Glen Kolotkin	135
3107. I WANNA ROCK Twisted Sister Atlantic 89617 20Oct84 68(2) 7 [Dee Snider] A: Stay Hungry; P: Tom Werman	135
3108. HOT HOT HOT!!! The Cure Elektra 69424 05Mar88 65(1) 7 [Simon Gallup/Robert James Smith/Porl Thompson/Laurence Tolhurst/Boris Williams] A: Kiss Me, Kiss Me, Kiss Me; P: Dave Allen/Robert Smith	135
3109. WIND BENEATH MY WINGS Lou Rawls Epic 03758 26Mar83 65(2) 6 [Larry Henley/Jeff Silbar] A: When The Night Comes; P: Ron Haffkine	135
3110. BIG MISTAKE Peter Cetera Full Moon 28507 24Jan87 61(2) 6 [Peter Cetera/Amos Galpin] A: Solitude/Solitaire; P: Michael Omartian	134
3111. SITUATION Yazoo Sire 29953 18Sep82 73(2) 8 [Vince Clarke/Alison Moyet] A: Upstairs At Eric's; P: Vince Clarke/Daniel Miller/Eric Radcliffe	134
3112. HANDS ACROSS AMERICA Voices Of America EMI America 8319 12Apr86 65(1) 8 [Marc Blatte/John Carney/Larry Gottlieb] A: No Album; P: Humberto Gatica	134
3113. I CRY JUST A LITTLE BIT Shakin' Stevens Epic 04338 21Apr84 67(2) 6 [Bob Heatlie] A: The Bop Won't Stop; P: Christopher Neil	134
3114. TO DREAM THE DREAM Frankie Miller Capitol/MSS 5131 19Jun82 62(1) 6 [Francis John Miller] A: Standing On The Edge; P: Barry Beckett	134
3115. MIAMI Bob Seger & The Silver Bullet Band Capitol 5658 15Nov86 70(2) 9 [Bob Seger] A: Like A Rock; P: Ed Andrews/Bob Seger	133
3116. MORNING MAN Rupert Holmes MCA 51019 08Nov80 68(1) 7 [Rupert Holmes] A: Partners In Crime; P: Rupert Holmes	133
3117. WORLD WHERE YOU LIVE Crowded House Capitol 44033 08Aug87 65(1) 8 [Neil Finn] A: Crowded House; P: Mitchell Froom	133
3118. REMEMBER (Walking In The Sand) Aerosmith Columbia 11181 12Jan80 67(1) 6 [Shadow Morton] A: Night In The Ruts; P: Gary Lyons	133
3119. I CAN'T WAIT Deniece Williams Columbia 08014 24Sep88 66(1) 8 [Skylark] A: As Good As It Gets; P: George Duke	133
3120. REBELS ARE WE Chic Atlantic 3665 30Aug80 61(1) 6 [Bernard Edwards/Nile Rodgers] A: Real People; P: Bernard Edwards/Nile Rodgers	133
3121. LET'S BE LOVERS AGAIN Eddie Money with Valerie Carter Columbia 11377 18Oct80 65(2) 6 [Jimmy Lyon/Eddie Money] A: Playing For Keeps; P: Ron Nevison	132
3122. TALK TO ME Fiona Atlantic 89572 13Apr85 64(1) 7 [Beau Hill/Michael Pfeifer] A: Fiona; P: Peppi Marchello	132
3123. THE SCREAMS OF PASSION The Family Paisley Park 7-28953 28Sep85 63(1) 6 [Prince] A: The Family; P: Family/David Rivkin	132
3124. NO ONE LIKE YOU Scorpions Mercury 76153 19Jun82 65(2) 7 [Klaus Meine/Rudolf Schenker] A: Blackout; P: Dieter Dierks	132
3125. THOSE GOOD OLD DREAMS Carpenters A&M 2386 19Dec81 63(1) 6 [John Bettis/Richard Carpenter] A: Made In America; P: Richard Carpenter	132
3126. A LITTLE GOOD NEWS Anne Murray Capitol 5264 17Sep83 74(2) 9 [Charlie Black/Rory Bourke/Tommy Rocco] A: A Little Good News; P: Jim Ed Norman	132
3127. MAKE MY DAY T.G. Sheppard with Clint Eastwood Warner/Curb 29343A 18Feb84 62(1) 6 [Dewayne Blackwell] A: Slow Burn; P: Jim Ed Norman	132
3128. PAINTED PICTURE Commodores Motown 1651 04Dec82 70(1) 6 [Harold Hudson/Walter Orange] A: All The Great Hits; P: James Anthony Carmichael/Commodores	131
3129. WHAT SHE DOES TO ME (The Diana Song) The Producers Portrait 12-02092 13Jun81 61(1) 6 [Wayne Famous/Kyle Henderson/Bryan Holmes/Van Temple] A: The Producers; P: Tom Werman	131
3130. ALWAYS THERE FOR YOU Stryper Enigma 75019 23Jul88 71(1) 8 [Michael Sweet] A: In God We Trust; P: Michael Lloyd/Stryper	131
3131. KING OF SUEDE "Weird Al" Yankovic Rock 'n' Roll 04451 05May84 62(2) 6 [Sting/Al Yankovic] A: "Weird Al" Yankovic In 3-D; P: Rick Derringer	131
3132. PRIDE & PASSION John Cafferty And The Beaver Brown Band Scotti Brothers 68999 29Jul89 66(1) 7 [John Cafferty] A: Eddie & The Cruisers II (Soundtrack); P: John Cafferty/Kenny Vance	131
3133. WHEN LOVE COMES TO TOWN U2 With B. B. King Island 99225 01Apr89 68(1) 7 [Adam Clayton/Dave Evans/Paul Hewson/Larry Mullen] A: Rattle And Hum (Soundtrack); P: Jimmy Iovine	130
3134. KILLING ME SOFTLY Al B. Sure! Warner Bros. 27772 19Nov88 80(1) 11 [Charles Fox/Norman Gimbel] A: In Effect Mode; P: Kyle West	130
3135. ROCK & ROLL STRATEGY Thirty Eight Special A&M 1246 29Oct88 67(1) 8 [Max Carl/Donnie Van Zant] A: Rock & Roll Strategy; P: Rodney Mills	130
3136. I CALL YOUR NAME Switch Gordy 7175 03Nov79 83(2) 15 [Bobby DeBarge/Greg Williams] A: Switch II; P: Bobby DeBarge	130
3137. IS THAT IT? Katrina And The Waves Capitol 5566 05Apr86 70(1) 8 [Kimberly Rew] A: Waves; P: Pat Collier/Katrina And The Waves/Scott Litt	130
3138. TALKING OUT OF TURN The Moody Blues Threshold 603 07Nov81 65(2) 7 [John Lodge] A: Long Distance Voyager; P: Pip Williams	130
3139. CAT PEOPLE (Putting Out Fire) David Bowie Backstreet 52024 17Apr82 67(1) 10 [David Bowie/Giorgio Moroder] A: Cat People Soundtrack; P: Giorgio Moroder	129
3140. CROSS MY HEART Lee Ritenour Elektra 69892 04Dec82 69(3) 7 [Lee Ritenour/Eric Tagg] A: Rit/2; P: Harvey Mason/Lee Ritenour	129
3141. SET THE NIGHT ON FIRE Oak Mercury 76087 13Dec80 71(1) 6 [Jeff Silbar/Van Stephenson] A: Set The Night On Fire; P: Rick Hall	129
3142. STILL LOVING YOU Scorpions Mercury 880802 07Jul84 64(2) 6 [Klaus Meine/Rudolf Schenker] A: Love At First Sting; P: Dieter Dierks	128
3143. CLUB MICHELLE Eddie Money Columbia 04376 25Feb84 66(1) 7 [Ray Burton/Ralph Carter/Mitchell Froom/Eddie Money] A: Where's The Party?; P: Tom Dowd/Eddie Money	128
3144. NEW THING Enuff Z'Nuff Atco 99207 21Oct89 67(1) 7 [Ron Fajerstein/Donnie Vie/Chip Z'Nuff] A: Enuff Z'Nuff; P: Enuff Z'Nuff/Ron Fajerstein	128
3145. WHAT LOVE IS Marty Balin EMI America 8153 19Feb83 63(2) 6 [Greg Prestopino/Brock Walsh] A: Lucky; P: Val Garay	128
3146. BEFORE I GO Starship Grunt 14393 05Jul86 68(2) 7 [David Scott Roberts] A: Knee Deep In The Hoopla; P: Jeremy Smith/Peter Wolf	128
3147. BODY TALK The Deele Solar 69785 21Jan84 77(2) 8 [Stanley Burke/Melvin Gentry/Carl Greene/Antonio Reid/Bo Watson] A: Street Beat; P: Reggie Calloway	128
3148. WHISPER IN THE DARK Dionne Warwick Arista 1-9460 15Mar86 72(1) 9 [Edgar Bronfman/Bruce Roberts] A: Friends; P: Albhy Galuten	128
3149. LAYIN' IT ON THE LINE Jefferson Starship Grunt 13872 08Sep84 66(2) 6 [Craig Chaquico/Mickey Thomas] A: Nuclear Furniture; P: Ron Nevison	128
3150. I KNOW WHAT BOYS LIKE The Waitresses Polydor 2196 08May82 62(2) 6 [Chris Butler] A: Wasn't Tomorrow Wonderful?; P: Chris Butler/Kurt Munkacsi	128
3151. TOUCH ME TONIGHT Shooting Star Enigma 75054 04Nov89 67(1) 9 [Van McLain] A: Touch Me Tonight-The Best Of Shooting Star; P: Van McLain	127
3152. CASE OF YOU Frank Stallone Scotti Brothers 603 27Sep80 67(2) 6 [Joni Mitchell] A: Heart And Souls; P: Harry Nilsson	127
3153. COMIN' DOWN TONIGHT Thirty Eight Special A&M 1424 24Jun89 67(1) 7 [Max Carl/Jeff Carlisi/Robert White Johnson/Donnie Van Zant] A: Rock & Roll Strategy; P: Rodney Mills	127
3154. TEARS RUN RINGS Marc Almond Capitol 44240 21Jan89 67(1) 8 [Marc Almond] A: The Stars We Are; P: Marc Almond/Annie Hogan/Billy McGee/Pete Schwier	127
3155. KISS YOU (When It's Dangerous) Eight Seconds Polydor 885352 31Jan87 72(1) 8 [Marcello Cesare/Andres Del Castillo/Frank Lavigne/Scott Milks] A: Almacantar; P: Rupert Hine	127
3156. PERFECT COMBINATION Stacy Lattisaw & Johnny Gill Cotillion 10Mar84 75(1) 9 [Preston Glass/Narada Michael Walden] A: Perfect Combination; P: Narada Michael Walden	127
3157. SOLID ROCK Goanna Atco 99895 11Jun83 71(2) 7 [Shane Howard] A: Spirit Of Place; P: Trevor Lucas	127
3158. I SURRENDER Arlan Day Pasha 02480 17Oct81 71(2) 7 [Arlan Day] A: I Surrender; P: Larry Brown	126
3159. SOMEONE El DeBarge Gordy 1867 27Dec86 70(1) 9 [Jay Graydon/Mark Mueller/Robbie Nevil] A: El DeBarge; P: Jay Graydon	126
3160. MAMA Genesis Atlantic 89770 01Oct83 73(1) 9 [Tony Banks/Phil Collins/Mike Rutherford] A: Genesis; P: Genesis/Hugh Padgham	126
3161. I WANNA BE LOVED House Of Lords RCA 8805 07Jan89 58(1) 7 [Steve Johnstad/Michael Meyer] A: House Of Lords; P: Gregg Giuffria/Andy Johns	126
3162. RAIN IN THE SUMMERTIME The Alarm I.R.S. 53219 19Dec87 71(2) 8 [Eddie MacDonald/Mike Peters/Dave Sharp/Nigel Twist] A: Eye Of The Hurricane; P: Alarm/John Porter	126
3163. TODAY IS THE DAY Bar-Kays Mercury 76036 22Mar80 60(1) 5 [James Edward Alexander/Charles Leonard Allen/Michael Beard/Mark Bynum/Larry Dodson/Sherman Guy/Harvey Henderson/Allen Alvoid Jones Jr./Lloyd Ed Smith/Winston Stewart/Frank Thompson] A: Injoy; P: Allen Jones	125

Ranking the Singles

Rank. TITLE Act Label Number Entry Date Peak(Wks) Tot Wks [Writers] A: Proximate Album; P: Producer	Score
3164. I'M NOT THE MAN I USED TO BE Fine Young Cannibals MCA 53686 11Nov89 54(1) 6 [Roland Gift/David Steele] A: The Raw & The Cooked; P: Andy Cox/Roland Gift/David Steele	125
3165. COME HOME WITH ME BABY Dead Or Alive Epic 68885 01Jul89 69(1) 8 [Pete Burns/Steve Coy/Wayne Hussey/Tim Lever/Mike Percy] A: Nude; P: Pete Burns/Steve Coy	125
3166. THE WILD LIFE Bananarama London 882019 10Nov84 70(2) 8 [Sarah Dallin/Siobhan Fahey/Steve Jolley/Tony Swain/Keren Woodward] A: The Wild Life Soundtrack; P: Steve Jolley/Tony Swain	125
3167. DANCIN' IN THE STREETS Teri DeSario with K.C. Casablanca 2278 28Jun80 66(2) 6 [Marvin Gaye/Ivy Jo Hunter/William Stevenson] A: Moonlight Madness; P: Harry Wayne Casey	124
3168. SHAKE IT UP TONIGHT Cheryl Lynn Columbia 02102 08Aug81 70(1) 7 [Brenda Sutton/Mike Sutton] A: In The Night; P: Ray Parker Jr.	124
3169. STAND UP Underworld Sire 22852 26Aug89 67(1) 8 [Karl Hyde/Richard David Smith/Alfie Thomas] A: Change The Weather; P: Rick Smith	124
3170. OO-EE-DIDDLEY-BOP! Peter Wolf EMI America 8254 27Apr85 61(1) 5 [Michael Jonzun/Peter Wolf/Gordon Worthy] A: Lights Out; P: Michael Jonzun/Peter Wolf	124
3171. I DON'T THINK THAT MAN SHOULD SLEEP ALONE Ray Parker Jr. Geffen 28417 29Aug87 68(1) 7 [Ray Parker Jr.] A: After Dark; P: Ray Parker Jr.	124
3172. WHERE ARE YOU NOW? ('86) Synch Columbia 05788 01Mar86 77(1) 12 [Richard Congdon/Jimmy Harnen] A: No Album; P: Jerry G. Hludzik/Bill Kelly	124
3173. DEEP RIVER WOMAN Lionel Richie Motown 1873 TSW 17Jan87 71(1) 8 [Lionel Richie] A: Dancing On The Ceiling; P: James Anthony Carmichael/Lionel Richie	124
3174. GOIN' CRAZY! David Lee Roth Warner Bros. 28584 27Sep86 66(1) 7 [David Lee Roth/Steve Vai] A: Eat 'Em And Smile; P: Ted Templeman	124
3175. IF YOU LET ME STAY Terence Trent D'Arby Columbia 07398 24Oct87 68(1) 8 [Terence Trent D'Arby] A: Introducing The Hardline According To Terence Trent D'Arby; P: Howard Gray	124
3176. MEGA FORCE 707 Boardwalk 7-11-146 10Jul82 62(2) 6 [Jonathan Cain/Tod Howarth/Jim McClarty/Kevin Russell] A: Mega Force; P: Keith Olsen	124
3177. ENCORE Cheryl Lynn Columbia 04256 11Feb84 69(1) 8 [James Samuel Harris/Terry Lewis] A: Preppie; P: Jimmy Jam Harris/Terry Lewis/Cheryl Lynn	124
3178. AFTER ALL Al Jarreau Warner Bros. 29262 13Oct84 69(1) 9 [David Foster/Jay Graydon/Al Jarreau] A: High Crime; P: Jay Graydon	124
3179. YOU BETTER DANCE The Jets MCA 53673 29Jul89 59(1) 7 [Michael Jonzun] A: Believe; P: David Rivkin	124
3180. YOUR DADDY DON'T KNOW Toronto Network 69986 07Aug82 77(1) 8 [Geoff Iwamoto/Michael Roth] A: Get It On Credit; P: Steve Smith	123
3181. DON'T LET ME IN Sneaker Handshake 02714 27Feb82 63(2) 5 [Walter Becker/Donald Fagen] A: Sneaker; P: Jeff Baxter	123
3182. FIREFLIES Fleetwood Mac Warner Bros. 49660 07Feb81 60(1) 6 [Stevie Nicks] A: Fleetwood Mac Live; P: Ken Caillat/Richard Dashut/Fleetwood Mac	123
3183. TEACH ME TONIGHT Al Jarreau Warner Bros. 50032 03Apr82 70(2) 7 [Sammy Cahn/Gene DePaul] A: Breakin' Away; P: Jay Graydon	123
3184. YOU'VE GOT ANOTHER THING COMIN' Judas Priest Columbia 03168 06Nov82 67(2) 7 [Kenneth Downing/Rob Halford/Glenn Tipton] A: Screaming For Vengeance; P: Tom Allom	123
3185. SOMEBODY SAVE ME Cinderella Mercury 888483 25Apr87 66(1) 7 [Tom Keifer] A: Night Songs; P: Andy Johns	123
3186. EVER SINCE THE WORLD BEGAN Tommy Shaw Atlantic 89138 13Feb88 75(1) 9 [Jim Peterik/Frankie Sullivan] A: Ambition; P: Tommy Shaw/Terry Thomas	123
3187. BURNING DOWN ONE SIDE Robert Plant Swan Song 99979 11Sep82 64(1) 6 [Robbie Blunt/Robert Plant/Jezz Woodroffe] A: Pictures at Eleven; P: Robert Plant	122
3188. JANE'S GETTING SERIOUS Jon Astley Atlantic 89258 27Jun87 77(1) 10 [Jon Astley] A: Everybody Loves The Pilot (Except The Crew); P: Phil Chapman/Andy MacPherson	122
3189. TAKE YOU TONIGHT Ozark Mountain Daredevils Columbia 11247 24May80 67(2) 5 [Steve Cash/John Dillon/Larry Michael Lee] A: Ozark Mountain Daredevils; P: John Boylan	122
3190. THE MEDICINE SONG Stephanie Mills Casablanca 880180 13Oct84 65(2) 6 [David Wolinski] A: I've Got The Cure; P: David Wolinski	122
3191. TRUE LOVE WAYS Mickey Gilley Epic 50876 16Aug80 66(2) 7 [Buddy Holly/Norman Petty] A: That's All That Matters To Me; P: Jim Ed Norman	122
3192. ROCK IT Lipps, Inc. Casablanca 2281 02Aug80 64(1) 7 [Steve Greenberg] A: Mouth To Mouth; P: Steve Greenberg	121
3193. DON'T SAY NO TONIGHT Eugene Wilde Philly World 7-99608 07Dec85 76(1) 10 [Ronnie E. Broomfield/McKinley Horton] A: Serenade; P: Donald R. Robinson	121
3194. LEARNING TO FLY Pink Floyd Columbia 07363 10Oct87 70(1) 8 [Jon Carin/Bob Ezrin/David Gilmour/Anthony Moore] A: A Momentary Lapse Of Reason; P: Bob Ezrin/David Gilmour	121
3195. 96 TEARS Garland Jeffreys Epic 51008 14Mar81 66(2) 7 [Rudy Martinez] A: Rock & Roll Adult; P: Bob Clearmountain/Garland Jeffreys	121
3196. BORROWED TIME Styx A&M 2228 29Mar80 64(1) 6 [Dennis DeYoung/Tommy Shaw] A: Cornerstone; P: Styx	120
3197. FOOLIN' YOURSELF Aldo Nova Portrait 24-03001 17Jul82 65(2) 6 [Aldo Nova] A: Aldo Nova; P: Aldo Nova	120
3198. DANCIN' IN THE KEY OF LIFE Steve Arrington Atlantic 89535 17Aug85 68(2) 6 [India Arrington/Steve Arrington] A: Dancin' In The Key Of Life; P: Keg Johnson/Wilmer Raglin	120
3199. NOW OR NEVER Axe Atco 7408 24Jul82 64(2) 6 [Bobby Barth] A: Offering; P: Al Nalli	120
3200. WHEN WE MAKE LOVE Alabama RCA 13763 19May84 72(1) 10 [Troy Seals/Mentor Williams] A: Roll On; P: Alabama/Harold Shedd	120
3201. THORN IN MY SIDE Eurythmics RCA 5058 15Nov86 68(2) 9 [Annie Lennox/David A. Stewart] A: Revenge; P: David A. Stewart	120
3202. I CAN'T FACE THE FACT Gina Go-Go Capitol 44233 25Feb89 78(1) 11 [Desmond Foster/Gina Gomez/Nick Mundy] A: Sweet Surrender; P: Nick Mundy	120
3203. ALABAMA GETAWAY Grateful Dead Arista 0519 21Jun80 68(1) 6 [Jerry Garcia/Robert Hunter] A: Go To Heaven; P: Gary Lyons	120
3204. SOMEBODY ELSE'S GUY Jocelyn Brown Vinyl Dreams 71 16Jun84 75(2) 10 [Annette E. Brown/Jocelyn Brown] A: Somebody Else's Guy; P: Jocelyn Brown/Allen George/Fred McFarlane	120
3205. YOUR SMILE René and Angela Mercury 884271 22Mar86 62(2) 6 [René Moore/Angela Winbush] A: Street Called Desire; P: René and Angela/Bruce Swedien/Bobby Watson	120
3206. YOU'RE GONNA GET WHAT'S COMING Bonnie Raitt Warner Bros. 49116 08Dec79 73(1) 6 [Robert Palmer] A: The Glow; P: Peter Asher	120
3207. NATURAL LOVE Petula Clark Scotti Brothers 02676 06Feb82 66(1) 6 [Kim Espy/Phil Gernhard/Jeff Harrington/Jeff Pennig] A: No Album; P: Tony Scotti	119
3208. BRITE EYES Robbin Thompson Band Ovation 1157 18Oct80 66(1) 9 [Robbin Thompson] A: Two B's Please; P: Ken Brown	119
3209. PRAYING TO A NEW GOD Wang Chung Geffen 22969 27May89 63(1) 7 [David Chandler/Nick Feldman/Jack Hues] A: The Warmer Side Of Cool; P: Peter Wolf	119
3210. WHEREVER I LAY MY HAT (That's My Home) Paul Young Columbia 04071 01Oct83 70(1) 7 [Marvin Gaye/Barrett Strong/Norman Whitfield] A: No Parlez; P: Laurie Latham	119
3211. LYING Peter Frampton Atlantic 89463 01Feb86 74(2) 8 [Peter Frampton] A: Premonition; P: Peter Frampton/Peter Solley	119
3212. EMINENCE FRONT The Who Warner Bros. 29814 25Dec82 68(2) 6 [Pete Townshend] A: It's Hard; P: Glyn Johns	119
3213. EVERY HOME SHOULD HAVE ONE (Remix) ('83) Patti Austin Qwest 29727 19Mar83 69(1) 7 [Dominic Bugatti/Frank Musker] A: No Album; P: Quincy Jones	119
3214. KICK THE WALL Jimmy Davis & Junction QMI 53107 07Nov87 67(2) 6 [James O. Davis] A: Kick The Wall; P: Jack Holder/Don Smith	118
3215. SATISFY ME Billy Satellite Capitol 5326 18Aug84 64(1) 6 [Monty Byrom/Danny Chauncey/Ira Walker] A: Billy Satellite; P: Don Gehman	118
3216. GIVE Missing Persons Capitol 5326 17Mar84 67(1) 6 [Dale Bozzio/Terry Bozzio/Warren Cuccurullo/Patrick O'Hearn] A: Rhyme & Reason; P: Terry Bozzio/Missing Persons/Bruce Swedien	118
3217. SCHOOL'S OUT Krokus Arista 1-9468 07Jun86 67(1) 7 [Michael Bruce/Glen Buxton/Alice Cooper/Dennis Dunaway/Neal Smith] A: Change Of Address; P: Tom Werman	118
3218. THE RUMOUR Olivia Newton-John MCA 53294 20Aug88 62(1) 6 [Elton John/Bernie Taupin] A: The Rumour; P: James Newton Howard/Elton John	118
3219. RESERVATIONS FOR TWO Dionne & Kashif Arista 1-9638 31Oct87 62(1) 7 [Tena Clark/Nathan East/Gary Prim] A: Reservations For Two; P: Kashif	118
3220. CHEROKEE Europe Epic 07638 28Nov87 72(1) 10 [Joey Tempest] A: The Final Countdown; P: Kevin Elson	118
3221. IF YOU WANNA GET BACK YOUR LADY Pointer Sisters Planet 13430 26Mar83 67(1) 5 [John Lewis Parker/Brian Potter] A: So Excited!; P: Richard Perry	117

Ranking the Singles

Rank. TITLE Act Label Number Entry Date Peak(Wks) Tot Wks
[Writers] A: Proximate Album; P: Producer   Score

3222. CLOSER TO THE HEART Rush Mercury 76124 12Dec81 69(1) 7
[Geddy Lee/Alex Lifeson/Neil Peart/Pete Talbot]
A: Exit...Stage Left; P: Terry Brown/Rush   117

3223. KAYLEIGH Marillion Capitol 5493 05Oct85 74(1) 8
[Derek Dick/Mark Kelly/Ian Mosley/Steve Rothery/Pete Trewavas]
A: Misplaced Childhood; P: Chris Kimsey   117

3224. POSSE' ON BROADWAY Sir Mix-A-Lot Nastymix 75555
17Dec88 70(1) 9 [Anthony Ray]
A: Swass; P: Sir Mix-A-Lot   117

3225. WILD AGAIN Starship Elektra 69349 17Dec88 73(2) 8
[John Bettis/Michael Clark]   A: Cocktail Soundtrack; P: Phil Galdston/Starship   117

3226. DREAMIN' OF LOVE Stevie B LMR 74001 23Apr88 80(2) 10
[Stevie B]   A: Party Your Body; P: Stevie B   117

3227. CAN'T LET GO Stephen Stills Featuring Michael Finnigan
Atlantic 89611 06Oct84 67(2) 6
[Joe Esposito/Allee Willis]   A: Right By You; P: Steve Alaimo   116

3228. (Closest Thing To) PERFECT Jermaine Jackson Arista 1-9356
08Jun85 67(1) 7 [Jermaine Jackson/Michael Omartian/Bruce Sudano]
A: Perfect Soundtrack; P: Michael Omartian   116

3229. LIKE THE WEATHER 10,000 Maniacs Elektra 69418 07May88 68(2) 6
[Natalie Merchant]   A: In My Tribe; P: Peter Asher   116

3230. MAKE UP YOUR MIND Aurra Salsoul 7017 20Mar82 71(1) 7
[George Curtis Jones/Stephen C. Washington/Starleana Young]
A: A Little Love; P: Steve Washington   116

3231. QUE TE QUIERO Katrina And The Waves Capitol 5528
12Oct85 71(2) 6
[Kimberly Rew]   A: Katrina And The Waves; P: Pat Collier/Katrina And The Waves   116

3232. LOVING YOU WITH MY EYES Starland Vocal Band Windsong 11899
23Feb80 71(1) 6 [Taffy Danoff/Margot Kunkel]   A: 4 X 4; P: Barry Beckett   115

3233. THE GOOD LORD LOVES YOU Neil Diamond Columbia 11232
05Apr80 67(1) 6 [Richard Fagan]   A: September Morn; P: Bob Gaudio   115

3234. LOVE TRAIN Holly Johnson Uni 50023 17Jun89 65(1) 6
[Holly Johnson]   A: Blast; P: Steve Lovell/Andy Richards   115

3235. BURNING LIKE A FLAME Dokken Elektra 69435 19Dec87 72(1) 8
[Mick Brown/Don Dokken/George Lynch/Jeff Pilson]
A: Back For The Attack; P: Neil Kernon   114

3236. BURNING FLAME Vitamin Z Geffen 29039 15Jun85 73(1) 7
[Geoff Barradale/Philip Jesson/Nick Lockwood]
A: Rites Of Passage; P: Ross Cullum/Chris Hughes   114

3237. HARD TIMES FOR LOVERS Jennifer Holliday Geffen 28958
21Sep85 69(1) 7 [Lotti Golden/Richard C. Scher]
A: Say You Love Me; P: Arthur Baker/Lotti Golden/Richard Scher   114

3238. WHAT CAN YOU GET A WOOKIEE FOR CHRISTMAS (When He Already Owns A Comb?) Star Wars Intergalactic Droid Choir & Chorale RSO 1058
13Dec80 69(3) 6 [Maury Yeston]
A: Christmas In The Stars; P: Tony Bongiovi/Meco Monardo/Lance Quinn   114

3239. LEAD A DOUBLE LIFE Loverboy Columbia 05867 26Apr86 68(2) 7
[Paul Dean/Doug Johnson/Ted Johnson/Mike Reno/Davitt Sigerson/Bill Wray]
A: Lovin' Every Minute Of It; P: Tom Allom/Paul Dean   114

3240. RIGHT AWAY Kansas Kirshner 3084 21Aug82 73(3) 6
[Dino Elefante/John Elefante]   A: Vinyl Confessions; P: Kansas/Ken Scott   113

3241. JIMMY MACK Sheena Easton EMI America 8309 08Feb86 65(2) 6
[Lamont Dozier/Brian Holland/Eddie Holland]   A: Do You; P: Nile Rodgers   113

3242. TROUBLE IN PARADISE Jarreau Warner Bros. 29501 10Sep83 63(1) 7
[Jay Graydon/Greg Mathieson/Trevor Veitch]   A: Jarreau; P: Jay Graydon   113

3243. GOT TO BE THERE Chaka Khan Warner Bros. 29881 08Jan83 67(1) 7
[Elliot Willensky]   A: Chaka Khan; P: Arif Mardin   113

3244. LULLABY The Cure Elektra 69249 02Dec89 74(1) 8
[Simon Gallup/Roger O'Donnell/Robert James Smith/Porl Thompson/Laurence Tolhurst/Boris Williams]   A: Disintegration; P: Dave Allen/Robert Smith   113

3245. EDGE OF A DREAM Joe Cocker Capitol 5412 20Oct84 69(1) 7
[Bryan Adams/Jim Vallance]   A: Teachers Soundtrack; P: Keith Forsey   113

3246. PLAY THAT FUNKY MUSIC Roxanne Scotti Brothers 07724
12Mar88 63(1) 7 [Robert Parissi]   A: Roxanne; P: Geoffrey Workman   113

3247. WHO SHOT J.R.? Gary Burbank with Band McNally Ovation 1150
28Jun80 67(1) 5 [Gary Burbank/Ron Reed/Ed Vanover]   A: No Album; P: Ed Vanover   113

3248. I'M HAPPY JUST TO DANCE WITH YOU Anne Murray Capitol 4878
14Jun80 64(1) 6
[John Lennon/Paul McCartney]   A: Somebody's Waiting; P: Jim Ed Norman   113

3249. SOME CHANGES ARE FOR GOOD Dionne Warwick Arista 0602
20Jun81 65(2) 6 [Michael Masser/Carole Bayer Sager]
A: Hot! Live And Otherwise; P: Michael Masser   112

3250. ONE WAY LOVE TKA Tommy Boy 866 07Jun86 75(1) 9
[Jeff Mann/Marco Olivo]   A: Modern Girls Soundtrack; P: Jeff Mann/Marco Olivo   112

3251. SHE DID IT Michael Damian LEG 007 30May81 69(1) 6
[Eric Carmen]   A: No Album; P: Bruce Miller/Pete Moore   112

3252. GOT IT MADE Crosby, Stills, Nash & Young Atlantic 88966
04Feb89 69(1) 8 [Stephen Stills/Neil Young]
A: American Dream; P: Niko Bolas/Crosby, Stills, Nash & Young   112

3253. HOLIDAY Kool & The Gang Mercury 888712 27Jun87 66(1) 7
[Robert Bell/Ronald Bell/George M. Brown/Claydes Smith/J.T. Taylor/Curtis Fitzgerald Williams]   A: Forever; P: Ronald Bell/I.B.M.C./Kool & The Gang   112

3254. CATCHING THE SUN Spyro Gyra MCA 41180 19Apr80 68(1) 5
[Jay Beckenstein]   A: Catching The Sun; P: Jay Beckenstein/Richard Calandra   112

3255. I LIKE Guy MCA 53490 06May89 70(1) 7
[Timmy Gatling/Gene Griffin/Aaron Hall/Teddy Riley]
A: Guy; P: Gene Griffin/Teddy Riley   112

3256. HERE TO LOVE YOU The Doobie Brothers Warner Bros. 50001
06Feb82 65(2) 5 [Michael McDonald]
A: The Doobie Brothers Farewell Tour; P: Ted Templeman   112

3257. DESIRE Rockets RSO 1022 16Feb80 70(2) 6
[John Badanjek/Dennis Robbins]   A: No Ballads; P: Johnny Sandlin   112

3258. FEEL THE HEAT Jean Beauvoir Columbia 05904 14Jun86 73(1) 8
[Jean Beauvoir]   A: Drums Along The Mohawk; P: Jean Beauvoir   112

3259. WEATHERMAN SAYS Jack Wagner Qwest 28387 09May87 67(1) 6
[Nick Jameson/Kim O'Leary]
A: Don't Give Up Your Day Job; P: Steve Barri/Tony Peluso   112

3260. SHE LOOKS A LOT LIKE YOU Clocks Boulevard 03075
28Aug82 67(2) 5 [Steve Swaim]   A: Clocks; P: Mike Flicker   112

3261. WHAT'S TOO MUCH Smokey Robinson Motown 1911 14Nov87
79(1) 10 [Lonnie Kirtz/Smokey Robinson/Ivory Stone/Homer Talbert]
A: One Heartbeat; P: Peter Bunetta/Rick Chudacoff   111

3262. I SHOULDA LOVED YA Narada Michael Walden Atlantic 3631
09Feb80 66(1) 6 [Thomas M. Stevens/Narada Michael Walden/Allee Willis]
A: The Dance Of Life; P: Bob Clearmountain/Narada Michael Walden   111

3263. SEEING IS BELIEVING Mike + The Mechanics Atlantic 88921
22Apr89 62(1) 6 [Brian A. Robertson/Mike Rutherford]
A: Living Years; P: Christopher Neil/Mike Rutherford   111

3264. PLEASURE AND PAIN Divinyls Chrysalis 42916 25Jan86 76(1) 7
[Mike Chapman/Holly Knight]   A: What A Life!; P: Mike Chapman   111

3265. DON'T TRY TO STOP IT Roman Holliday Jive 9092 01Oct83 68(1) 6
[Brian Bonhomme]   A: Roman Holliday; P: Peter Collins   111

3266. SERIOUS KINDA GIRL Christopher Max EMI 50229 02Dec89 75(2) 8
[Christopher Max/Nile Rodgers]
A: More Than Physical; P: Ron Fair/Christopher Max/Nile Rodgers   111

3267. IN LOVE WITH LOVE Debbie Harry Geffen 28476 04Jul87 70(1) 7
[Deborah Harry/Christopher Stein]   A: Rockbird; P: Seth Justman   111

3268. EAGLES FLY Sammy Hagar Geffen 28185 24Oct87 82(1) 13
[Sammy Hagar]   A: I Never Said Goodbye; P: Sammy Hagar/Edward Van Halen   111

3269. FREAKSHOW ON THE DANCE FLOOR Bar-Kays Mercury 818631
26May84 73(1) 8 [James Edward Alexander/Michael Beard/Mark Bynum/Larry Dodson/Harvey Henderson/Allen Alvoid Jones Jr./Lloyd Ed Smith/Winston Stewart/Frank Thompson]   A: Dangerous; P: Allen Jones   110

3270. DON'T CRY FOR ME ARGENTINA Festival RSO 1020 08Mar80 72(1) 8
[Tim Rice/Andrew Lloyd Webber]   A: Evita; P: Boris Midney   110

3271. SHOTGUN RIDER Joe Sun Ovation 1141 31May80 71(1) 6
[Larry Henley/Jim Hurt/Johnny Slate]   A: Out Of Your Mind; P: Brien Fisher   110

3272. SHOOTING STAR Dollar Atco 7208/Carrere 7208 22Dec79 74(1) 6
[David Courtney]   A: Shooting Stars; P: Christopher Neil   110

3273. HOLDIN' ON FOR DEAR LOVE Lobo MCA/Curb 41152
22Dec79 75(2) 8 [Larry Henley/Steve Pippin/Johnny Slate]
A: Lobo; P: Bob Montgomery   110

3274. HOLD ON Donny Osmond Capitol 44423 30Sep89 73(1) 7
[Donny Osmond/Evan Rogers/Carl Sturken]
A: Donny Osmond; P: Evan Rogers/Carl Sturken   110

3275. (I Know) I'M LOSING YOU Uptown Oak Lawn 3810 27Dec86 80(2) 11
[Cornelius Grant/Eddie Holland/Norman Whitfield]
A: No Album; P: Jack Malken/Scott Yahney   110

3276. GOT TO LOVE SOMEBODY Sister Sledge Cotillion 45007
19Jan80 64(1) 5 [Bernard Edwards/Nile Rodgers]
A: Love Somebody Today; P: Bernard Edwards/Nile Rodgers   110

3277. IT'S MY PARTY Dave Stewart With Barbara Gaskin Platinum 4
19Dec81 72(2) 8 [Cy Crane/John Gluck/Wally Gold/Herb Weiner]
A: Up From The Dark; P: Dave Stewart   109

Ranking the Singles

Rank. TITLE Act Label Number Entry Date Peak(Wks) Tot Wks
[Writers] *A: Proximate Album; P: Producer* Score

3278. FEMALE INTUITION Mai Tai Critique 722 24May86 71(1) 7
[Jochem Fluitsma/Eric Van Tijn]
A: 1 Touch 2 Much; P: Jochem Fluitsma/Eric Van Tijn 109

3279. LOVE IS A HOUSE Force M.D.'s Tommy Boy 28300 29Aug87 78(1) 9
[Gina Foster/Geoff Gurd/Martin Lascelles]
A: Touch And Go; P: Geoff Gurd/Martin Lascelles 109

3280. LOVE KILLS Freddie Mercury Columbia 04606 29Sep84 69(1) 6
[Freddie Mercury/Giorgio Moroder]
A: Metropolis Soundtrack; P: Reinhold Mack/Freddie Mercury/Giorgio Moroder 109

3281. KEEP THIS TRAIN A-ROLLIN' The Doobie Brothers Warner Bros. 49670 14Feb81 62(1) 5
[Michael McDonald] *A: One Step Closer; P: Ted Templeman* 109

3282. CAN'T WAIT ALL NIGHT Juice Newton RCA 13863 11Aug84 66(1) 6
[Bryan Adams/Eric Kagna/Jim Vallance]
A: Can't Wait All Night; P: Richard Landis 109

3283. STILL John Schneider Scotti Brothers 02489 26Sep81 69(2) 5
[Lionel Richie] *A: Now Or Never; P: John D'Andrea/Tony Scotti* 109

3284. UNDERNEATH THE RADAR Underworld Sire 27968 16Apr88 74(1) 8
[Karl Hyde/Richard David Smith/Alfie Thomas]
A: Underneath The Radar; P: Rupert Hine 109

3285. 99 1/2 Carol Lynn Townes Polydor 881008 07Jul84 77(3) 9
[Maxi Anderson/John Footman] *A: Satisfaction Guaranteed; P: Rod Hui* 108

3286. I WANT TO BE YOUR PROPERTY Blue Mercedes MCA 53262 05Mar88 66(1) 6 [Duncan Millar/David Titlow]
A: Rich And Famous; P: Ian Curnow/Phil Harding 108

3287. BACK TO SCHOOL AGAIN Four Tops RSO 1069 15May82 71(1) 7
[Howard Greenfield/Louis St. Louis] *A: Grease 2 Soundtrack; P: Louis St. Louis* 108

3288. STOP YOUR SOBBING The Pretenders Sire 49506 21Jun80 65(2) 5
[Ray Davies] *A: Pretenders; P: Nick Lowe* 108

3289. MIDNITE MANIAC Krokus Arista 1-9248 15Sep84 71(1) 6
[Marc Storace/Fernando Von Arb] *A: The Blitz; P: Bruce Fairbairn* 108

3290. LONELY IS THE NIGHT Air Supply Arista 1-9521 09Aug86 76(1) 8
[Albert Hammond/Diane Warren] *A: Hearts In Motion; P: John Boylan* 108

3291. BANG THE DRUM ALL DAY Todd Rundgren Bearsville 29686 07May83 63(1) 5 [Todd Rundgren]
A: The Ever Popular Tortured Artist Effect; P: Todd Rundgren 108

3292. IN THE SHAPE OF A HEART Jackson Browne Asylum 69543 07Jun86 70(1) 7 [Jackson Browne] *A: Lives In The Balance; P: Jackson Browne* 108

3293. GUARANTEED FOR LIFE Millions Like Us Virgin 99412 14Nov87 69(2) 6 [Jonathan Hook/John O'Kane] *A: ...Millions Like Us; P: David Wolinski* 108

3294. MY FIRST NIGHT WITHOUT YOU Cyndi Lauper Epic 68945 05Aug89 62(1) 6 [Thomas F. Kelly/Cyndi Lauper/Billy Steinberg]
A: A Night To Remember; P: Cyndi Lauper/Lennie Petze 108

3295. THE BLUE SIDE Crystal Gayle Columbia 11270 07Jun80 81(2) 8
[David Lasley/Allee Willis] *A: Miss The Mississippi; P: Allen Reynolds* 108

3296. HE COULD BE THE ONE Josie Cotton Elektra 47481 21Aug82 74(3) 7
[Bobby Paine/Larson Paine]
A: Convertible Music; P: Bobby Paine/Larson Paine 108

3297. HOW CAN I LIVE WITHOUT HER Christopher Atkins Polydor 2210 07Aug82 71(2) 7 [Terry Britten/Sue Shifrin]
A: The Pirate Movie Soundtrack; P: Terry Britten 107

3298. VOICE ON THE RADIO Conductor Montage 1210 30Jan82 63(1) 5
[Franne Golde/Peter Mclan] *A: Conductor; P: Stuart Alan Love* 107

3299. CRAZY CRAZY NIGHTS KISS Mercury 888796 26Sep87 65(1) 7
[Adam Mitchell/Paul Stanley] *A: Crazy Nights; P: Ron Nevison* 107

3300. DON'T YOU KNOW WHAT LOVE IS Touch Atco 7311 31Jan81 69(1) 6
[Mark Mangold] *A: Touch; P: Tim Friese-Greene* 107

3301. WORKING ON IT Chris Rea Geffen 27535 25Mar89 73(2) 7
[Chris Rea] *A: New Light Through Old Windows; P: Jon Kelly/Chris Rea* 107

3302. WALKING IN L.A. Missing Persons Capitol 5212 12Mar83 70(2) 6
[Terry Bozzio] *A: Spring Session M; P: Ken Scott* 107

3303. YOU AIN'T SEEN NOTHING YET Figures On A Beach Sire 27628 08Apr89 67(1) 7 [Randy Bachman] *A: Figures On A Beach; P: Ivan Baker* 107

3304. TRUE TO YOU Ric Ocasek Geffen 28504 20Dec86 75(1) 8
[Ric Ocasek]
A: This Side Of Paradise; P: Ross Cullum/Chris Hughes/Ric Ocasek 107

3305. THE REAL THING The Brothers Johnson A&M 2343 27Jun81 67(1) 6
[George Johnson/Louis E. Johnson] *A: Winners; P: Brothers Johnson* 107

3306. THE LEBANON The Human League A&M/Virgin 2641 09Jun84 64(2) 5
[John William Callis/Philip Oakey]
A: Hysteria; P: Human League/Hugh Padgham/Chris Thomas 107

3307. WORKING CLASS MAN Jimmy Barnes Geffen 28749 22Mar86 74(1) 8
[Jonathan Cain] *A: Jimmy Barnes; P: Jonathan Cain* 106

3308. DON'T GIRLS GET LONELY Glenn Shorrock Capitol 5267 24Sep83 69(2) 6 [Stephen Allen Davis/Carson Whitsett]
A: Villain Of The Peace; P: John Boylan 106

3309. PRIMITIVE LOVE RITES Mondo Rock Columbia 06981 23May87 71(1) 6 [J.J. Hackett/Ross Wilson] *A: Boom Baby Boom; P: Bill Drescher* 106

3310. NEVER SAY DIE (Give A Little Bit More) Cliff Richard EMI America 8180 08Oct83 73(1) 7 [Terry Britten/Sue Shifrin]
A: Give A Little Bit More; P: Terry Britten 106

3311. PAY THE DEVIL (Ooo, Baby, Ooo) The Knack Capitol 5054 31Oct81 67(2) 5 [Berton Averre] *A: Round Trip; P: Jack Douglas* 106

3312. COLORS Ice-T Sire/Warner 27902 11Jun88 70(2) 7
[Charles Andre Glenn/Tracy Marrow]
A: Colors Soundtrack; P: Charles Glenn/Tracy Marrow 105

3313. LITTLE THING CALLED LOVE Neil Young Geffen 29887 29Jan83 71(2) 6 [Neil Young] *A: Trans; P: David Briggs/Tim Mulligan/Neil Young* 105

3314. ROCK 'N' ROLL TO THE RESCUE The Beach Boys Capitol 5595 28Jun86 68(1) 6
[Mike Love/Terry Melcher] *A: Made In U.S.A.; P: Terry Melcher* 105

3315. YOU GOT THE POWER War RCA 13061 03Apr82 66(2) 6
[Thomas Allen/Harold Brown/Jerry Goldstein/Lonnie Jordan/Lee Oskar/
Luther Rabb/Howard Scott] *A: Outlaw; P: Jerry Goldstein/Lonnie Jordan* 105

3316. YOUNG THING, WILD DREAMS (Rock Me) Red Rider Capitol 5335 16Jun84 71(2) 6 [Tom Cochrane]
A: Breaking Curfew; P: Tom Cochrane/Ken Greer 105

3317. TRICKLE TRICKLE The Manhattan Transfer Atlantic 3772 29Nov80 73(1) 8 [Clarence Bassett] *A: Extensions; P: Jay Graydon* 105

3318. DON'T WALK AWAY Toni Childs A&M 1237 13Aug88 72(1) 7
[Toni Childs/Phil Ramacon]
A: Union; P: Toni Childs/David Ricketts/David Tickle 105

3319. LIKE A CHILD Noel 4th & Broadway 7458 09Apr88 67(1) 8
[Mary E. Kessler/Noel Pagan/Roman Ricardo]
A: Noel; P: Roman Ricardo/Louie Vega 104

3320. PRETTY MESS Vanity Motown 1752 08Sep84 75(1) 7
[Denise Matthews/Bill Wolfer] *A: Wild Animal; P: Denise Matthews/Bill Wolfer* 104

3321. RUN TO ME Savoy Brown Town House 1055 10Oct81 68(2) 5
[Chris Norman/Pete Spencer] *A: Greatest Hits Live In Concert; P: Richie Wise* 104

3322. STAND OR FALL The Fixx MCA 52106 30Oct82 76(1) 8
[Charles Barrett/Cy Curnin/Peter John Greenall/John Leroy Kinyon/
Jamie West-Oram/Adam Woods] *A: Shuttered Room; P: Rupert Hine* 104

3323. NEW FRONTIER Donald Fagen Warner Bros. 29792 29Jan83 70(1) 6
[Donald Fagen] *A: The Nightfly; P: Gary Katz* 104

3324. REAP THE WILD WIND Ultravox Chrysalis 42682 09Apr83 71(1) 5
[Warren Cann/Chris Cross/Billy Currie/Midge Ure] *A: Quartet; P: George Martin* 104

3325. WE COULD BE TOGETHER Debbie Gibson Atlantic 88896 23Sep89 71(1) 6 [Debbie Gibson] *A: Electric Youth; P: Debbie Gibson/Fred Zarr* 104

3326. SAY HELLO TO RONNIE Janey Street Arista 1-9265 06Oct84 68(2) 5
[Dennis Pereca/Janey Street] *A: Heroes, Angels & Friends; P: Jimmy Ienner* 104

3327. MEDLEY: LOVE SONGS ARE BACK AGAIN Band Of Gold RCA 13866 13Oct84 64(1) 7 *A: The Band Of Gold Album; P: Paco Saval/Pete Wingfield* 103

3328. NO PROMISES Icehouse Chrysalis 42978 05Jul86 79(1) 9
[Iva Davies/Robert Kretschmer] *A: Measure For Measure; P: Rhett Davies* 103

3329. SLEEPWALK Larry Carlton Warner Bros. 50019 27Feb82 74(1) 8
[Ann Farina/Johnny Farina/Santo Farina] *A: Sleepwalk; P: Larry Carlton* 103

3330. MEMORIES OF DAYS GONE BY (Medley) Fred Parris & The Five Satins Elektra 47411 27Feb82 71(1) 5
A: Fred Parris And The Satins; P: Marty Markiewicz 103

3331. SOMEONE TO LOVE ME FOR ME Lisa Lisa & Cult Jam Featuring Full Force Columbia 07619 14Nov87 78(1) 10
[Curtis Bedeau/Gerard Charles/Hugh Clarke/Brian George/Lucien George/
Paul George/Lisa Velez] *A: Spanish Fly; P: Full Force* 103

3332. SAMANTHA (What You Gonna Do?) Cellarful Of Noise CBS Associated 07731 05Mar88 69(1) 7 [Mark Avsec]
A: Magnificent Obsession; P: Mark Avsec 103

3333. WHY YOU WANNA TRY ME Commodores Motown 1604 06Feb82 66(2) 5 [David Cochrane/Lionel Richie]
A: In The Pocket; P: James Anthony Carmichael/Commodores 103

3334. I WANT MY GIRL Jesse Johnson's Revue A&M 2749 20Jul85 76(1) 8
[Jesse Woods Johnson] *A: Jesse Johnson's Revue; P: Jesse Johnson* 103

3335. JUST BE MY LADY Larry Graham Warner Bros. 49744 05Sep81 67(1) 5 [Larry Graham] *A: Just Be My Lady; P: Larry Graham* 102

3336. THE GAP Thompson Twins Arista 1-9290 10Nov84 69(1) 6
[Tom Bailey/Alannah Currie/Joe Leeway]
A: Into The Gap; P: Tom Bailey/Alex Sadkin 102

Rank. TITLE Act Label Number Entry Date Peak(Wks) Tot Wks
[Writers] A: Proximate Album; P: Producer   Score

3337. CHANGE Tears For Fears Mercury 812677 06Aug83 73(1) 6
[Roland Orzabal] A: The Hurting; P: Ross Cullum/Chris Hughes   102

3338. MY MISTAKE The Kingbees RSO 1032 28Jun80 81(2) 8
[Jamie James] A: The Kingbees; P: Rich Fitzgerald/David J. Holman   102

3339. EUROPA AND THE PIRATE TWINS Thomas Dolby Capitol 5238
18Jun83 67(1) 5 [Thomas Dolby]
A: The Golden Age Of Wireless; P: Thomas Dolby   102

3340. BASKETBALL Kurtis Blow Polydor 881529 13Apr85 71(2) 6
[Curtis Bedeau/James Bralower/Gerard Charles/Hugh Clarke/
Robert Arthur Ford Jr./Brian George/Paul George/James B. Moore/
Kurt Walker/Shirley Walker/William Waring]
A: Ego Trip; P: Robert Ford/J.B. Moore   102

3341. EYE TO EYE Go West Chrysalis 42903 28Sep85 73(1) 7
[Peter Cox/Richard Drummie] A: Go West; P: Gary Stevenson   102

3342. CRAZY The Manhattans Columbia 44-03940 30Jul83 72(1) 6
[John Vernon Anderson/Steven Richard Williams]
A: Forever By Your Side; P: John Vernon Anderson/Steve Williams   102

3343. THE WOMAN IN ME Crystal Gayle Columbia 02523 05Dec81 76(2) 6
[Susan Thomas] A: Hollywood, Tennessee; P: Allen Reynolds   101

3344. LOOKING OVER MY SHOULDER 'til tuesday Epic 04936 24Aug85
61(1) 5 [Michael Hausman/Robert Holmes/Aimee Mann/Joey Pesce]
A: Voices Carry; P: Mike Thorne   101

3345. I SEND A MESSAGE INXS Atco 99731 21Jul84 77(2) 7
[Andy Farriss/Michael Hutchence] A: The Swing; P: Nick Launay   101

3346. ON THE LINE Tangier Atco 99208 12Aug89 67(1) 5
[Doug Gordon] A: Four Winds; P: Andy Johns   101

3347. ACTION Evelyn "Champagne" King RCA 13682 07Jan84 75(1) 7
[Dana Meyers/Leon Sylvers]
A: Face To Face; P: Joey Gallo/Foster Sylvers/Leon Sylvers   101

3348. TEST OF TIME The Romantics Nemperor 05587 31Aug85 71(1) 6
[Coz Canler/Wally Palmar/Mike Skill]
A: Rhythm Romance; P: Gordon Fordyce/Peter Solley   101

3349. VICTORY LINE Limited Warranty Atco 99541 28Jun86 79(2) 8
[Gerald Brunskill/Dale Goulett/Paul Hartwig/Jon Erik Newman/Greg Sotebeer]
A: Limited Warranty; P: Brian Tench   101

3350. GONNA MAKE IT Sa-Fire Cutting 874 278-7 24Jun89 71(1) 7
[David Harris] A: Sa-Fire; P: Angelo Cosme/David Harris   101

3351. FRANKIE Sister Sledge Atlantic 89547 15Jun85 75(1) 8
[Joy Denny] A: When The Boys Meet The Girls; P: Nile Rodgers   101

3352. DON'T SAY NO Billy Burnette Columbia 11380 01Nov80 68(2) 5
[Billy Burnette] A: Billy Burnette; P: Barry Seidel   100

3353. YOUNG BLOOD Bruce Willis Motown 1886 11Apr87 68(1) 5
[Jerry Leiber/Doc Pomus/Mike Stoller] A: The Return Of Bruno; P: Robert Kraft 100

3354. ALL RIGHT NOW Pepsi & Shirlie Polydor 887277 20Feb88 66(1) 6
[Andy Fraser/Paul Rodgers] A: All Right Now; P: Gary Langan   99.5

3355. BEAST OF BURDEN Bette Midler Atlantic 89712 11Feb84 71(1) 6
[Mick Jagger/Keith Richards] A: No Frills; P: Chuck Plotkin   99.2

3356. BREAD AND BUTTER Robert John Motown 1664 12Feb83 68(2) 4
[Larry Parks/Jay Turnbow] A: No Album; P: George Tobin   99.2

3357. SHIP TO SHORE Chris DeBurgh A&M 2565 20Aug83 71(1) 5
[Chris de Burgh] A: The Getaway; P: Rupert Hine   99.1

3358. ANYTHING CAN HAPPEN Was (Not Was) Chrysalis 43365
06May89 75(2) 6 [David Was/Don Was/Aaron Zigman]
A: What Up, Dog?; P: Paul Staveley O'Duffy   99.0

3359. DREAMER The Association Elektra 47094 31Jan81 66(2) 5
[Moon Martin] A: No Album; P: Bones Howe   99.0

3360. FIRE WITH FIRE Wild Blue Chrysalis 42985 17May86 71(1) 6
[Chas Sandford] A: No More Jinx; P: Chas Sandford   98.8

3361. PUT THIS LOVE TO THE TEST Jon Astley Atlantic 89027
15Oct88 74(1) 8 [Jon Astley/Richard King]
A: The Compleat Angler; P: Jon Astley/Andy MacPherson   98.8

3362. DON'T BREAK MY HEART Romeo's Daughter Jive 1140 15Oct88
73(1) 7 [Craig Joiner/Mutt Lange] A:Romeo's Daughter; P:Mutt Lange/John Parr 98.8

3363. DON'T LOOK ANY FURTHER Dennis Edwards (Featuring Siedah Garrett)
Gordy 1715 28Apr84 72(1) 6 [Franne Golde/Duane Hitchings/
Dennis Lambert] A: Don't Look Any Further; P: Dennis Lambert   98.6

3364. MEDLEY II (Medley) Stars On 45 Radio 3830 18Jul81 67(1) 6
A: Stars On Long Play; P: Jaap Eggermont   98.5

3365. AIMING AT YOUR HEART The Temptations Gordy 7208
19Sep81 67(2) 5 [Joseph Jefferson/Rich Roebuck/Charles Simmons]
A: The Temptations; P: Thom Bell   98.0

3366. LOVER Michael Stanley Band EMI America 8064 28Mar81 68(2) 6
[Michael Stanley] A: Heartland; P: Michael Stanley Band   97.8

3367. LADY OF MY HEART Jack Wagner Qwest 29085 25May85 76(1) 8
[Glen Ballard/David Foster/Jay Graydon]
A: All I Need; P: Glen Ballard/Clif Magness   97.8

3368. SHE'S FLY Tony Terry Epic 07417 28Nov87 80(1) 9
[George Dick/Gary Henry/David Sanchez/Tony Terry]
A: Forever Yours; P: Ted Currier   97.8

3369. WHAT YOU DO TO ME Carl Wilson Caribou 03590 14May83 72(1) 6
[Johanna Hall/John Hall] A: Youngblood; P: Jeff Baxter   97.8

3370. LEAVING L.A. Deliverance Columbia 11320 30Aug80 71(2) 5
[Ken Janz/Paul Janz] A: Tightrope; P: Deliverance/Peter Kirsten   97.7

3371. HEADLINES Midnight Star Solar 69547 14Jun86 69(2) 7
[Reggie Calloway/Vincent Calloway/Melvin Gentry/Belinda Lipscomb/
Bobby Lovelace/Bill Simmons] A: Headlines; P: Reggie Calloway/Midnight Star 97.7

3372. ALL NIGHT WITH ME Laura Branigan Atlantic 4023 20Mar82 69(1) 7
[Chris Montan] A: Branigan; P: Jack White   97.6

3373. SOME PEOPLE Belouis Some Capitol 5492 03Aug85 67(1) 6
[Belouis Some]
A: Some People; P: Michael Barbiero/Pete Schwier/Steve Thompson   97.5

3374. MY PARADISE The Outfield Columbia 68943 22Jul89 72(2) 6
[John Spinks]
A: Voices Of Babylon; P: David Kahne/David Leonard/John Spinks   97.4

3375. SHERRY Robert John EMI America 8061 25Oct80 70(1) 5
[Bob Gaudio] A: Back On The Street; P: George Tobin   97.0

3376. IN AND OUT OF LOVE Bon Jovi Mercury 880951 03Aug85 69(2) 6
[Jon Bon Jovi] A: 7800 Degrees Fahrenheit; P: Lance Quinn   96.8

3377. BABY GRAND Billy Joel featuring Ray Charles Columbia 06994
04Apr87 75(1) 7 [Billy Joel] A: The Bridge; P: Phil Ramone   96.5

3378. ELECTRIC KINGDOM Twilight 22 Vanguard 35241 17Dec83 79(1) 8
[Gordon Bahary/Errol Moore/Joe Saulter] A: Twilight 22; P: Gordon Bahary   96.5

3379. WAITING FOR YOUR LOVE Toto Columbia 03981 02Jul83 73(1) 6
[Bobby Kimball/David Paich] A: Toto IV; P: Toto   96.4

3380. IF I COULD GET YOU (Into My Life) Gene Cotton Knoll 5002
13Mar82 76(1) 8 [Gene Cotton] A: Eclipse Of The Blue Moon; P: Missing   96.0

3381. I DIDN'T MEAN TO TURN YOU ON Cherrelle Tabu 4-04406
07Jul84 79(1) 9 [James Samuel Harris/Terry Lewis]
A: Fragile; P: Jimmy Jam Harris/Terry Lewis   95.9

3382. INVITATION TO DANCE Kim Carnes EMI America 8250
19Jan85 68(1) 6 [Kim Carnes/Dave Ellingson/Brian Fairweather/Martin Page]
A: Barking At Airplanes; P: Nile Rodgers   95.9

3383. ALL OF MY LOVE Bobby Caldwell Polydor 2212 11Sep82 77(3) 6
[Bobby Caldwell] A: Carry On; P: Bobby Caldwell   95.9

3384. DANCING IN HEAVEN (Orbital Be-Bop) Q-Feel Jive 1220
17Jun89 75(1) 7 [Brian Fairweather/Martin Page]
A: Q-Feel; P: Brian Fairweather/Martin Page   95.2

3385. YOU WIN AGAIN Bee Gees Warner Bros. 28351 19Sep87 75(1) 6
[Barry Gibb/Maurice Gibb/Robin Gibb] A: E-S-P; P: Arif Mardin   95.0

3386. LA LA MEANS I LOVE YOU Tierra Boardwalk 129 24Oct81 73(1) 6
[Thom Bell/William Hart] A: Together Again; P: Rudy Salas   94.8

3387. EMOTIONS IN MOTION Billy Squier Capitol 5135 07Aug82 68(2) 6
[Billy Squier] A: Emotions In Motion; P: Reinhold Mack/Billy Squier   94.7

3388. I DON'T WANT TO KNOW YOUR NAME Glen Campbell Capitol 4959
24Jan81 65(1) 5 [Micheal Smotherman]
A: It's The World Gone Crazy; P: Gary Klein   94.7

3389. I KNEW THE BRIDE (When She Used To Rock 'N Roll)
Nick Lowe and His Cowboy Outfit Columbia 05570 30Nov85 77(1) 9
[Nick Lowe] A: The Rose Of England; P: Huey Lewis   94.5

3390. THE HEART IS NOT SO SMART El DeBarge with DeBarge Gordy 1822
07Dec85 75(1) 7 [Diane Warren] A: Rhythm Of The Night; P: Jay Graydon   94.5

3391. JOLE BLON Gary U.S. Bonds EMI America 8089 18Jul81 65(1) 6
[Moon Mullican] A: Dedication; P: Bruce Springsteen/Steven Van Zandt   94.5

3392. ALL RIGHT NOW Rod Stewart Warner Bros. 29122 15Dec84 72(1) 6
[Andy Fraser/Paul Rodgers] A: Camouflage; P: Michael Omartian   94.3

3393. DON'T GIVE UP Peter Gabriel/Kate Bush Geffen 28463
04Apr87 72(1) 6 [Peter Gabriel] A: So; P: Peter Gabriel/Daniel Lanois   94.1

3394. CRAZY WORLD Big Trouble Epic 07432 17Oct87 71(1) 7
[Giorgio Moroder/Tom Whitlock] A: Big Trouble; P: Giorgio Moroder   93.9

3395. SNAKE EYES Alan Parsons Project Arista 0635 17Oct81 67(1) 5
[Alan Parsons/Eric Woolfson] A: The Turn Of A Friendly Card; P: Alan Parsons 93.6

3396. PRIVATE IDAHO The B-52's Warner Bros. 49537 18Oct80 74(1) 5
[Kate Pierson/Fred Schneider/Keith Strickland/Cindy Wilson]
A: Wild Planet; P: B-52s/Rhett Davies   93.3

3397. AMERICAN MADE Oak Ridge Boys MCA 52179 19Mar83 72(2) 5
[Bob DiPiero/Patrick J. McManus] A: American Made; P: Ron Chancey   93.3

Ranking the Singles

Rank. TITLE Act Label Number Entry Date Peak(Wks) Tot Wks
[Writers] A: Proximate Album; P: Producer — Score

3398. WHY Carly Simon Mirage 4051 17Jul82 74(2) 6 [Bernard Edwards/Nile Rodgers] A: Soup For One Soundtrack; P: Bernard Edwards/Nile Rodgers — 92.9

3399. IT'S A NIGHT FOR BEAUTIFUL GIRLS The Fools EMI America 8036 19Apr80 67(1) 4 [Doug Forman/Mike Girard] A: Sold Out; P: Peter Solley — 92.9

3400. TIME FOR ME TO FLY REO Speedwagon Epic 50858 31May80 77(1) 6 [Kevin Cronin] A: You Can Tune A Piano But You Can't Tuna Fish; P: Kevin Cronin/Paul Grupp/Gary Richrath — 92.7

3401. NO PARKING (ON THE DANCE FLOOR) Midnight Star Solar 69753 03Mar84 81(1) 8 [Vincent Calloway/Bobby Lovelace/Bill Simmons] A: No Parking On The Dance Floor; P: Reggie Calloway — 92.5

3402. TIRED OF BEING BLONDE Carly Simon Epic 05419 29Jun85 70(1) 5 [Larry Raspberry] A: Spoiled Girl; P: Arthur Baker/Frank Filipetti/G.E. Smith/Tom Wolk — 92.5

3403. TELL HER Kenny Loggins Columbia 68531 04Feb89 76(1) 8 [Bert Berns] A: Back To Avalon; P: Richie Zito — 92.5

3404. LOVER COME BACK TO ME Dead Or Alive Epic 05607 21Sep85 75(2) 7 [Pete Burns/Steve Coy/Wayne Hussey/Tim Lever/Mike Percy] A: Youthquake; P: Matt Aitken/Mike Stock/Pete Waterman — 92.3

3405. STILL OF THE NIGHT Whitesnake Geffen 28331 13Jun87 79(1) 7 [David Coverdale/John Sykes] A: Whitesnake; P: Keith Olsen/Mike Stone — 91.9

3406. (Call Me) WHEN THE SPIRIT MOVES YOU Touch Atco 7222 26Jul80 65(1) 5 [Mark Mangold/Craig Sprovach] A: Touch; P: Tim Friese-Greene/Mark Mangold — 91.7

3407. LIVING ON THE EDGE Jim Capaldi Atlantic 89799 27Aug83 75(2) 5 [Jim Capaldi] A: Fierce Heart; P: Jim Capaldi/Steve Winwood — 91.6

3408. IF I WAS YOUR GIRLFRIEND Prince Paisley Park 7-28334 30May87 67(1) 6 [Prince] A: Sign 'O' The Times; P: Prince — 91.6

3409. EATEN ALIVE Diana Ross RCA 14181 21Sep85 77(1) 7 [Barry Gibb/Maurice Gibb/Michael Jackson] A: Eaten Alive; P: Albhy Galuten/Barry Gibb/Michael Jackson/Karl Richardson — 91.5

3410. A GOOD HEART Feargal Sharkey A&M 2804 15Mar86 74(2) 6 [Maria McKee] A: Feargal Sharkey; P: David A. Stewart — 91.3

3411. LOOK MY WAY The Vels Mercury 880547 23Feb85 72(1) 6 [Alice DeSoto/Charles Hanson/Chris Larkin] A: Velocity; P: Steven Stanley — 91.2

3412. LET'S GO UP Diana Ross RCA 13671 17Dec83 77(1) 6 [Franne Golde/Peter Ivers] A: Ross (II); P: Gary Katz — 91.1

3413. LOOK AT THAT CADILLAC Stray Cats EMI America 8194 28Jan84 68(1) 5 [Brian Setzer] A: Rant 'N' Rave With The Stray Cats; P: Dave Edmunds — 91.1

3414. THE ONE THAT REALLY MATTERS Survivor Scotti Brothers 03485 22Jan83 74(1) 6 [Jim Peterik] A: Eye Of The Tiger; P: Jim Peterik/Frankie Sullivan — 91.1

3415. FEELS SO REAL (Won't Let Go) Patrice Rushen Elektra 69742 30Jun84 78(2) 6 [Patrice Rushen/Freddie Washington] A: Now; P: Charles Mims Jr./Patrice Rushen — 91.0

3416. SIMPLE Johnny Mathis Columbia 04468 23Jun84 81(1) 8 [Marvin Morrow/Keith Stegall] A: A Special Part Of Me; P: Denny Diante — 91.0

3417. LADY DOWN ON LOVE Alabama RCA 13590 29Oct83 76(1) 6 [Randy Owen] A: The Closer You Get; P: Alabama/Harold Shedd — 91.0

3418. THE MAYOR OF SIMPLETON XTC Geffen 27552 29Apr89 72(1) 6 [Andy Partridge] A: Oranges And Lemons; P: Paul Fox — 90.9

3419. RAPPIN' RODNEY Rodney Dangerfield RCA 13656 03Dec83 83(1) 8 [Dennis Blair/Rodney Dangerfield/Robert Arthur Ford Jr./Scott Henry/Doug Hoyt/James B. Moore/Lawrence Smith] A: Rappin' Rodney; P: Robert Ford/J.B. Moore — 90.9

3420. BREAKING UP IS HARD ON YOU (A/K/A DON'T TAKE MA BELL AWAY FROM ME) The American Comedy Network Critique 704 04Feb84 70(1) 5 [Howard Greenfield/Neil Sedaka] A: No Album; P: Bob Rivers — 90.6

3421. WHAT'S THE MATTER HERE? 10,000 Maniacs Elektra 69388 03Sep88 80(1) 8 [Robert Buck/Natalie Merchant] A: In My Tribe; P: Peter Asher — 90.6

3422. THE MONKEY TIME The Tubes Capitol 5254 01Oct83 68(1) 4 [Curtis Mayfield] A: Outside Inside; P: David Foster — 90.1

3423. I DON'T LIKE MONDAYS The Boomtown Rats Columbia 11117 02Feb80 73(1) 6 [Bob Geldof] A: The Fine Art Of Surfacing; P: Phil Wainman — 89.7

3424. SHE'S A RUNNER Billy Squier Capitol 5202 05Feb83 75(2) 6 [Billy Squier] A: Emotions In Motion; P: Reinhold Mack/Billy Squier — 89.5

3425. ATTACK OF THE NAME GAME Stacy Lattisaw Cotillion 7-99968 16Oct82 70(2) 6 [Lincoln Chase/Jeffrey E. Cohen/Shirley Ellis/Narada Michael Walden] A: Sneakin' Out; P: Narada Michael Walden — 89.2

3426. BOOGIE DOWN Jarreau Warner Bros. 29624 18Jun83 77(1) 6 [Al Jarreau/Michael Omartian] A: Jarreau; P: Jay Graydon — 89.2

3427. DARLIN' Yipes!! Millennium 11791 02Aug80 68(2) 5 [Mike Love/Brian Wilson] A: A Bit Irrational; P: John Jansen — 89.0

3428. LUANNE Foreigner Atlantic 4072 31Jul82 75(2) 6 [Lou Gramm/Michael Leslie Jones] A: 4; P: Mick Jones/Mutt Lange — 88.8

3429. GO FOR SODA Kim Mitchell Bronze/Island 7-99652 18May85 86(4) 9 [Pye Dubois/Kim Mitchell] A: Akimbo Alogo; P: Nick Blagona/Kim Mitchell — 88.8

3430. HAPPY HOUR Deodato Warner Bros. 29984 19Jun82 70(2) 5 [Jerry Barnes/Katreese Barnes] A: Happy Hour; P: Eumir Deodato — 88.8

3431. GET IT ON Kingdom Come Polydor 887436 02Apr88 69(1) 6 [Lenny Wolf/Martin Wolff] A: Kingdom Come; P: Bob Rock/Lenny Wolf — 88.5

3432. LOLA (Live Version) The Kinks Arista 0541 30Aug80 81(2) 6 [Ray Davies] A: One For The Road; P: Ray Davies — 88.1

3433. SOUTHERN PACIFIC Neil Young & Crazy Horse Reprise 49870 26Dec81 70(1) 5 [Neil Young] A: Re-ac-tor; P: David Briggs/Tim Mulligan/Neil Young — 88.0

3434. DROP THE PILOT Joan Armatrading A&M 2538 28May83 78(1) 6 [Joan Armatrading] A: The Key; P: Val Garay — 87.7

3435. REMEMBER WHAT YOU LIKE Jenny Burton Atlantic 89748 28Jan84 81(2) 6 [John Robie] A: In Black And White; P: John Robie — 87.5

3436. YOU'RE MINE TONIGHT Pure Prairie League Casablanca 2337 25Jul81 68(2) 5 [Rafe Van Hoy] A: Something In The Night; P: Rob Fraboni — 87.4

3437. SUNSHINE IN THE SHADE The Fixx MCA 52498 17Nov84 69(1) 5 [Dan Brown/Cy Curnin/Peter John Greenall/Jamie West-Oram/Adam Woods] A: Phantoms; P: Rupert Hine — 86.9

3438. HIGH ON YOUR LOVE Debbie Jacobs MCA 41167 15Mar80 70(2) 4 [Paul Sabu] A: High On Your Love; P: Paul Sabu — 86.7

3439. WAYS TO BE WICKED Lone Justice Geffen 29023 11May85 71(1) 6 [Mike Campbell/Tom Petty] A: Lone Justice; P: Jimmy Iovine — 86.7

3440. TURN OFF THE LIGHTS The World Class Wreckin' Cru Kru-Cut 006 26Mar88 84(1) 11 [Alonzo Williams] A: Turn Off The Lights In The Fast Lane; P: Dr. Dre/Alonzo Williams — 86.5

3441. SOMEBODY SPECIAL Rod Stewart Warner Bros. 49686 21Mar81 71(2) 5 [Phil Chen/Jim Cregan/Gary Grainger/Steve Harley/Kevin Savigar/Rod Stewart] A: Foolish Behaviour; P: Rod Stewart — 86.5

3442. SHOTGUN RIDER Delbert McClinton Capitol/MSS 4984 28Mar81 70(2) 6 [Larry Henley/Jim Hurt/Johnny Slate] A: The Jealous Kind; P: Barry Beckett/Muscle Shoals Rhythm Section — 86.3

3443. A LESSON IN LEAVIN' Dottie West United Artists 1339 08Mar80 73(1) 5 [Randy Goodrum/Brent Maher] A: Special Delivery; P: Randy Goodrum/Brent Maher — 86.1

3444. SECRETS Mac Davis Casablanca 2336 11Jul81 76(2) 6 [Sam Lorber/Mike Noble/Jeff Silbar] A: Texas In My Rear View Mirror; P: Rick Hall — 86.0

3445. VOO DOO Rachel Sweet Columbia 03411 05Feb83 72(2) 5 [Marc Blatte/Larry Gottlieb/Rachel Sweet] A: Blame It On Love; P: Marc Blatte/Larry Gottlieb/Rachel Sweet — 86.0

3446. I DIDN'T MEAN TO STAY ALL NIGHT Starship RCA 9109 25Nov89 75(1) 8 [Mutt Lange] A: Love Among The Cannibals; P: Larry Klein/Mike Shipley — 85.8

3447. HALLELUIAH MAN Love and Money Mercury 870596 18Feb89 75(1) 7 [James (2) Grant] A: Strange Kind Of Love; P: Gary Katz — 85.7

3448. I HEARD IT THROUGH THE GRAPEVINE (Part 1) Roger Warner Bros. 49786 07Nov81 79(2) 7 [Barrett Strong/Norman Whitfield] A: The Many Facets Of Roger; P: Roger Troutman — 85.3

3449. SOUP FOR ONE Chic Mirage 4032 05Jun82 80(2) 6 [Bernard Edwards/Nile Rodgers] A: Soup For One Soundtrack; P: Bernard Edwards/Nile Rodgers — 85.0

3450. WATCHING OVER YOU Glenn Medeiros Amherst 314 01Aug87 80(1) 8 [Paul Gordon] A: Glenn Medeiros; P: Jay Stone — 85.0

3451. BEECHWOOD 4-5789 Carpenters A&M 2405 24Apr82 74(2) 4 [Marvin Gaye/George Gordy/William Stevenson] A: Made In America; P: Richard Carpenter — 84.7

3452. RED HOT Herb Alpert A&M 2593 10Dec83 77(2) 5 [Howard Massey] A: Blow Your Own Horn; P: Herb Alpert/Andy Armer/Randy Badazz — 84.7

3453. FEET DON'T FAIL ME NOW Utopia Network 69859 08Jan83 82(3) 6 [Doug Howard/Roger Powell/Todd Rundgren/John Griffitt Wilcox] A: Utopia; P: Todd Rundgren/Utopia — 84.5

3454. CHARM THE SNAKE Christopher Cross Warner Bros. 28864 26Oct85 68(1) 5 [Christopher Cross/Michael Omartian] A: Every Turn Of The World; P: Michael Omartian — 84.4

3455. NOT FADE AWAY Eric Hine Montage 1200 22Aug81 73(1) 5 [Buddy Holly/Norman Petty] A: No Album; P: Eric Hine — 84.3

3456. WATCHING YOU Slave Cotillion 46006 17Jan81 78(1) 6 [Mark Adams/Steve Arrington/Raye Turner/Stephen C. Washington/Danny Webster] A: Stone Jam; P: Jimmy Douglass/Steve Washington — 84.2

Ranking the Singles

Rank. TITLE Act Label Number Entry Date Peak(Wks) Tot Wks	
[Writers] A: Proximate Album; P: Producer	Score

3457. SUPERNATURAL LOVE Donna Summer Geffen 29142
10Nov84 75(2) 5 [Michael Omartian/Bruce Sudano/Donna Summer]
A: Cats Without Claws; P: Michael Omartian — 84.1

3458. IF I WERE YOU Toby Beau RCA 11964 05Jul80 70(2) 4
[Jerry Fuller/John Hobbs] A: If You Believe; P: Jerry Fuller — 84.0

3459. MEET EL PRESIDENTE Duran Duran Capitol 44001 02May87 70(1) 5
[Simon Le Bon/Nick Rhodes/Nigel John Taylor]
A: Notorious; P: Duran Duran/Nile Rodgers — 83.9

3460. YOU ARE IN MY SYSTEM Robert Palmer Island 99866
18Jun83 78(1) 6 [David Frank/Mic Murphy] A: Pride; P: Robert Palmer — 83.9

3461. THE POWER OF LOVE (You Are My Lady) Air Supply Arista 1-9391
10Aug85 68(1) 6 [Mary Applegate/Candy DeRouge/Gunther Mende/
Jennifer Rush] A: Air Supply; P: Peter Collins — 83.8

3462. I WON'T BE HOME TONIGHT Tony Carey Rocshire 95030
26Mar83 79(2) 7 [Tony Carey]
A: Tony Carey [I Won't Be Home Tonight]; P: Peter Hauke — 83.8

3463. SLIPSTREAM Allan Clarke Elektra/Curb 46617 24May80 70(2) 4
[Gary Benson/Allan Clarke] A: The Only One; P: Spencer Proffer — 83.7

3464. EENIE MEENIE Jeffrey Osborne A&M 2530 19Mar83 76(2) 5
[Raymond Pounds/Michael Sembello] A: Jeffrey Osborne; P: George Duke — 83.7

3465. SHINY SHINY Haysi Fantayzee RCA 13534 23Jul83 74(1) 5
[Paul Caplin/Kate Garner/Jeremy Healy]
A: Battle Hymns For Children Singing; P: Clive Langer/Alan Winstanley — 83.5

3466. NEVER HAD A LOT TO LOSE Cheap Trick Epic 68563 18Feb89 75(1) 6
[Thomas John Peterson/Robin Zander] A: Lap Of Luxury; P: Richie Zito — 83.4

3467. WAY COOL JR. Ratt Atlantic 88985 07Jan89 75(1) 7
[Warren DeMartini/Beau Hill/Stephen Pearcy]
A: Reach For The Sky; P: Beau Hill — 83.3

3468. THE SUN AND THE RAIN Madness Geffen 29350 03Mar84 72(1) 5
[Michael Barson] A: Madness; P: Clive Langer/Alan Winstanley — 83.0

3469. SOMEONE LIKE YOU Michael Stanley Band EMI America 8189
24Dec83 75(1) 5 [Kevin Raleigh]
A: You Can't Fight Fashion; P: Bob Clearmountain/Michael Stanley Band — 82.8

3470. WHEN THE RAIN COMES DOWN Andy Taylor MCA 52946
25Oct86 73(1) 6 [Stephen Jones/Andy Taylor]
A: Miami Vice II Soundtrack; P: Steve Jones/Andy Taylor — 82.8

3471. VALERIE ('82) Steve Winwood Island 29879 13Nov82 70(2) 4
[Will Jennings/Steve Winwood]
A: Talking Back To The Night; P: Tom Lord-Alge/Steve Winwood — 82.8

3472. PLAYING WITH LIGHTNING Shot In The Dark RSO 1061 04Apr81
71(1) 5 [Robin Lamble/Adam Yurman]
A: Shot In The Dark; P: Chris Desmond/Al Stewart — 82.7

3473. SEDUCED Leon Redbone Emerald City 7326 11Apr81 72(2) 6
[Gary Tigerman] A: From Branch To Branch; P: Beryl Handler — 82.5

3474. IF ANYBODY HAD A HEART John Waite EMI America 8315
28Jun86 76(1) 6 [Danny Kortchmar/J.D. Souther]
A: About Last Night...Soundtrack; P: Don Henley/Danny Kortchmar/J.D. Souther — 82.4

3475. IT HURTS TO BE IN LOVE Dan Hartman Blue Sky 02115
27Jun81 72(1) 5 [Howard Greenfield/Helen Miller]
A: It Hurts To Be In Love; P: Dan Hartman — 82.2

3476. PLEDGE PIN Robert Plant Swan Song 99952 13Nov82 74(2) 5
[Robbie Blunt/Robert Plant] A: Pictures At Eleven; P: Robert Plant — 81.9

3477. LET ME GO Heaven 17 Arista/Virgin 1050 12Mar83 74(2) 5
[Glenn Gregory/Ian Craig Marsh/Martyn Ware]
A: The Luxury Gap; P: British Electric Foundation — 81.9

3478. I'M NOT YOUR MAN Tommy Conwell And The Young Rumblers
Columbia 07980 24Sep88 74(1) 7
[Tommy Conwell/Robert Dewald/Marci Rauer] A: Rumble; P: Rick Chertoff — 81.9

3479. STILL TAKING CHANCES Michael Martin Murphey Liberty 1486
08Jan83 76(2) 7 [Michael Murphey]
A: Michael Martin Murphey; P: Jim Ed Norman — 81.8

3480. BLUE KISS Jane Wiedlin I.R.S. 52674 28Sep85 77(1) 9
[Randell Kirsch/Jane Wiedlin]
A: Jane Wiedlin; P: Russ Kunkel/George Massenburg/Bill Payne — 81.6

3481. I PRETEND Kim Carnes EMI America 8202 19May84 74(1) 5
[Brian Fairweather/Martin Page] A: Cafe Racers; P: Keith Olsen — 81.6

3482. STATE OF THE NATION Industry Capitol 5268 19Nov83 81(2) 8
[Jon Carin/Mercury Caronia] A: Industry; P: Rhett Davies — 81.3

3483. STRANGELOVE ('87) Depeche Mode Sire 28366 25Jul87 76(1) 6
[Martin L. Gore] A: Music For The Masses; P: David Bascombe/Depeche Mode — 81.2

3484. CRAZY Kenny Rogers RCA 13975 26Jan85 79(1) 8
[Richard Marx/Kenny Rogers] A: What About Me?; P: David Foster — 81.2

3485. TOO LATE Journey Columbia 11143 12Jan80 70(1) 4
[Steve Perry/Neal Schon] A: Evolution; P: Roy Thomas Baker — 80.9

3486. RIGHT THE FIRST TIME Gamma Elektra 47423 03Apr82 77(2) 5
[Mitchell Froom/Ronnie Montrose/Jerry Stahl]
A: Gamma 3; P: Ronnie Montrose — 80.8

3487. THE MEN ALL PAUSE Klymaxx Constellation 52486 15Feb86 80(1) 8
[Bernadette Cooper/Joyce Irby]
A: Meeting In The Ladies Room; P: Stephen Shockley — 80.6

3488. ACROSS THE MILES Survivor Scotti Brothers 68526 21Jan89 74(1) 6
[Jim Peterik/Frankie Sullivan]
A: Too Hot To Sleep; P: Frank Filipetti/Frankie Sullivan — 80.0

3489. REMOTE CONTROL The Reddings Believe In A Dream 5600
15Nov80 89(2) 13 [William Ray Beard/Chet Fortune/Nick Mann]
A: The Awakening; P: Nick Mann/Russell Timmons — 79.9

3490. REPETITION Information Society Tommy Boy 27659 01Apr89 76(1) 6
[Paul Robb] A: Information Society; P: Fred Maher — 79.6

3491. ABADABADANGO Kim Carnes EMI America 8281 03Aug85 67(1) 5
[Kim Carnes/Dave Ellingson/Duane Hitchings]
A: Barking At Airplanes; P: Kim Carnes/Duane Hitchings — 79.6

3492. IN THE DRIVER'S SEAT John Schneider Scotti Brothers 03062
14Aug82 72(1) 6
[Jeff Harrington/Jeff Pennig] A: Quiet Man; P: John D'Andrea/Tony Scotti — 79.5

3493. HAPPY TOGETHER The Nylons Open Air 0024 29Aug87 75(1) 7
[Garry Bonner/Alan Gordon] A: Happy Together; P: Val Garay — 79.3

3494. EAT MY SHORTS Rick Dees Atlantic 89601 15Dec84 75(2) 5
[Rick Dees/Mondo Fax]
A: Put It Where The Moon Don't Shine; P: Rick Dees/Augie Johnson — 79.3

3495. HEARTACHE Pepsi & Shirlie Polydor 885470 15Aug87 78(1) 8
[Wayne Brown/Iris Fernando/Tambi Fernando]
A: All Right Now; P: Phil Fearon/Tambi Fernando — 79.2

3496. BEG, BORROW OR STEAL Hughes/Thrall Boulevard 18Dec82 79(1) 5
[Glenn Hughes/Pat Thrall]
A: Hughes/Thrall; P: Rob Fraboni/Glenn Hughes/Andy Johns/Pat Thrall — 79.1

3497. RUNAWAY RITA Leif Garrett Scotti Brothers 02579 05Dec81 84(4) 6
[Jeff Harrington/Jeff Pennig/Shunichi Tokura]
A: My Movie Of You; P: John D'Andrea/Shunichi Tokura — 79.0

3498. YOU DON'T HAVE TO CRY René and Angela Mercury 884587
21Jun86 75(1) 7 [René Moore/Angela Winbush]
A: Street Called Desire; P: René and Angela/Bruce Swedien/Bobby Watson — 79.0

3499. WHEN THE RADIO IS ON Paul Shaffer Capitol 44413 12Aug89 81(1) 8
[Kevin Calhoun/Matt Noble]
A: Coast To Coast; P: Paul Shaffer/Russell Simmons/Larry Smith — 78.7

3500. HEART'S ON FIRE John Cafferty Scotti Brothers 05774
01Mar86 76(1) 6 [Vince DiCola/Joe Esposito/Ed Frugé]
A: Heart's On Fire; P: Vince DiCola/Ed Frugé — 78.4

3501. BREAK EVERY RULE Tina Turner Capitol 44003 09May87 74(2) 5
[Rupert Hine/Jeannette Obstoj] A: Break Every Rule; P: Rupert Hine — 78.4

3502. PAY YOU BACK WITH INTEREST Gary O' Capitol 5018 18Jul81 70(2) 5
[Allan Clarke/Tony Hicks/Graham Nash] A: Gary O'; P: Richard Landis — 78.3

3503. DANCIN' LIKE LOVERS Mary MacGregor RSO 1025 17May80 72(2) 4
[Larry Herbstritt/Doug Thiele] A: Mary MacGregor; P: David J. Holman — 78.3

3504. DO IT AGAIN/BILLIE JEAN (Medley) Club House Atlantic 89795
20Aug83 75(1) 5 [Walter Becker/Donald Fagen//Michael Jackson]
A: No Album; P: Michele Interlandi/Carmelo La Bionda/Stefano Scalera — 78.2

3505. MAGICAL John Parr Atlantic 89568 06Apr85 73(1) 5
[Marvin Lee Aday/John Parr] A: John Parr; P: Peter Solley — 78.1

3506. 8TH WONDER Sugarhill Gang Sugar Hill 553 07Feb81 82(1) 9
[Cheryl Cook/Ronald LaPread] A: 8th Wonder; P: Joey Robinson Jr. — 78.1

3507. I ENGINEER Animotion Casablanca 884433 08Mar86 76(1) 6
[Mike Chapman/Holly Knight/Bernie Taupin] A: Strange Behavior; P: Richie Zito — 78.1

3508. RAIN IN MAY Max Werner Radio 3821 16May81 74(2) 6
[Chris Meldon/Chris Pilgram] A: Seasons; P: Chris Pilgram — 78.0

3509. GRACELAND Paul Simon Warner Bros. 28522 06Dec86 81(1) 7
[Paul Simon] A: Graceland; P: Paul Simon — 77.6

3510. SLIP AWAY Pablo Cruise A&M 2373 17Oct81 75(1) 5
[David Jenkins/Chuck Lutz/John Pierce] A: Reflector; P: Tom Dowd — 77.6

3511. THE BEST OF ME David Foster And Olivia Newton-John Atlantic 89420
14Jun86 80(1) 8 [David Foster/Jeremy Lubbock/Richard Marx]
A: David Foster; P: David Foster/Humberto Gatica — 77.4

3512. IN MY DREAMS Dokken Elektra 69563 22Feb86 77(1) 7
[Mick Brown/Don Dokken/George Lynch/Jeff Pilson]
A: Under Lock And Key; P: Neil Kernon/Michael Wagener — 77.3

Ranking the Singles

Rank. TITLE Act Label Number Entry Date Peak(Wks) Tot Wks	
[Writers] A: Proximate Album; P: Producer	Score
3513. FOR THE LOVE OF MONEY Bulletboys Warner Bros. 27554 29Apr89 78(1) 6 [Kenneth Gamble/Leon Huff/Anthony Jackson] A: Bulletboys; P: Ted Templeman	77.2
3514. BURN RUBBER (Why You Wanna Hurt Me) The Gap Band Mercury 76091 28Feb81 84(1) 8 [Lonnie Simmons/Rudy Taylor/Charles K. Wilson] A: The Gap Band III; P: Lonnie Simmons	77.2
3515. WITHOUT YOU David Bowie EMI America 8190 10Mar84 73(1) 4 [David Bowie] A: Let's Dance; P: David Bowie/Nile Rodgers	77.1
3516. HOLLYWOOD Shooting Star Epic/Virgin 02755 27Mar82 70(1) 5 [Van McLain/Gary West] A: Hang On For Your Life; P: Dennis McKay	76.9
3517. HOLYANNA Toto Columbia 04752 09Feb85 71(1) 5 [David Paich/Jeff Porcaro] A: Isolation; P: Toto	76.8
3518. HARD TIMES James Taylor Columbia 02093 13Jun81 72(2) 5 [James Taylor] A: Dad Loves His Work; P: Peter Asher	76.7
3519. CAN'T GET STARTED Peter Wolf EMI 43012 23May87 75(2) 5 [Peter Wolf] A: Come As You Are; P: Eric Thorngren/Peter Wolf	76.1
3520. I MELT WITH YOU Modern English Sire 29775 02Apr83 78(1) 7 [Richard Ian Brown/Michael Conroy/Robbie Grey/Gary McDowell/ Stephen Walker] A: After The Snow; P: Hugh Jones	76.1
3521. DO YOU REMEMBER ME? Jermaine Jackson Arista 1-9502 05Jul86 71(2) 5 [Jermaine Jackson/Michael Omartian/Bruce Sudano] A: Precious Moments; P: Michael Omartian	76.0
3522. DON'T MAKE ME DO IT Patrick Simmons Elektra 69817 18Jun83 75(1) 5 [Mario Cipollina/Johnny Colla/Bill Gibson/Chris Hayes/ Sean Hopper/Huey Lewis] A: Arcade; P: John Ryan	75.8
3523. DAYS ARE NUMBERS (The Traveller) Alan Parsons Project Arista 1-9349 27Apr85 71(1) 5 [Alan Parsons/Eric Woolfson] A: Vulture Culture; P: Alan Parsons	75.7
3524. DON'T WANT NO-BODY J.D. Drews Unicorn 95000 27Dec80 79(1) 6 [Paul Delph/Daphna Edwards] A: J. D. Drews; P: Joe Chemay/Daphna Edwards	75.6
3525. LOVE THEME FROM SHOGUN (Mariko's Theme) Meco RSO 1052 11Oct80 70(2) 4 [Maurice Jarre] A: No Album; P: Tony Bongiovi/Meco Monardo/Lance Quinn	75.5
3526. PUT AWAY YOUR LOVE Alessi Qwest 50055 01May82 71(1) 4 [Billy Alessi/Bobby Alessi] A: Put Away Your Love; P: Christopher Cross/Michael Ostin	75.4
3527. SATISFACTION GUARANTEED The Firm Atlantic 89561 04May85 73(2) 5 [Jimmy Page/Paul Rodgers] A: The Firm; P: Jimmy Page/Paul Rodgers	75.4
3528. COMING UP YOU The Cars Elektra 69432 23Jan88 74(1) 5 [Ric Ocasek] A: Door To Door; P: Ric Ocasek	75.3
3529. SHOWING OUT (Get Fresh At The Weekend) Mel & Kim Atlantic 89329 21Feb87 78(1) 7 [Matt Aitken/Mike Stock/Pete Waterman] A: Showing Out; P: Matt Aitken/Mike Stock/Pete Waterman	75.3
3530. REBELS Tom Petty and The Heartbreakers MCA 52658 17Aug85 74(1) 5 [Tom Petty] A: Southern Accents; P: Mike Campbell/Jimmy Iovine/Tom Petty	75.2
3531. SWEAR Sheena Easton EMI America 8263 23Mar85 80(2) 6 [Tim Scott] A: A Private Heaven; P: Greg Mathieson	75.2
3532. SAVE ME Dave Mason Columbia 11289 12Jul80 71(1) 3 [Jim Krueger] A: Old Crest On A New Wave; P: Dave Mason/Joe Wissert	74.8
3533. MARY, MARY Run-D.M.C. Profile 5211 30Jul88 75(1) 6 [Michael Nesmith] A: Tougher Than Leather; P: Rick Rubin/Run-D.M.C.	74.6
3534. A TRICK OF THE NIGHT Bananarama London 886119 27Dec86 76(1) 7 [Steve Jolley/Tony Swain] A: True Confessions; P: Steve Jolley/Tony Swain	74.5
3535. EYES THAT SEE IN THE DARK Kenny Rogers RCA 13774 28Apr84 79(1) 5 [Barry Gibb/Maurice Gibb] A: Eyes That See In The Dark; P: Albhy Galuten/Barry Gibb/Karl Richardson	74.3
3536. SCIENTIFIC LOVE Midnight Star Solar 69659 02Mar85 80(1) 7 [Vincent Calloway/Kenneth Gant/Melvin Gentry/Belinda Lipscomb/Bo Watson] A: Planetary Invasion; P: Reggie Calloway	74.1
3537. CLEANIN' UP THE TOWN The Bus Boys Arista 1-9229 04Aug84 68(1) 5 [Brian O'Neal/Kevin O'Neal] A: Ghostbusters Soundtrack; P: John Hug/Brian O'Neal/Kevin O'Neal	74.0
3538. LOVE AND LONELINESS The Motors Virgin 67007 17May80 78(1) 5 [Nick Garvey/Gordon Hann] A: Tenement Steps; P: Jimmy Iovine/Motors	73.8
3539. ELECTRICLAND Bad Company Swan Song 99966 02Oct82 74(2) 4 [Paul Rodgers] A: Rough Diamonds; P: Bad Company	73.6
3540. THIS TIME INXS Atlantic 89497 16Nov85 81(1) 6 [Andy Farriss] A: Listen Like Thieves; P: Chris Thomas	73.6
3541. SOMETHING'S ON YOUR MIND "D" Train Prelude 8080 07Jan84 79(1) 6 [Hubert Eaves III/James N. Williams] A: Something's On Your Mind; P: Hubert Eaves III	73.4
3542. GOTTA GET YOU HOME TONIGHT Eugene Wilde Philly World 7-99710 12Jan85 83(1) 8 [Ronnie E. Broomfield/McKinley Horton] A: Eugene Wilde; P: Michael Forte/Donald R. Robinson	73.3
3543. IT'S A HARD LIFE Queen Capitol 5372 28Jul84 72(1) 4 [Freddie Mercury] A: The Works; P: Reinhold Mack/Queen	72.9
3544. THAT DIDN'T HURT TOO BAD Dr. Hook Casablanca 2325 11Apr81 69(1) 4 [Tom Brasfield/Robert Byrne] A: Rising; P: Ron Haffkine	72.9
3545. NATURE OF LOVE Waterfront Polydor 871414 29Jul89 70(1) 5 [Phil Cilia/Chris Duffy] A: Waterfront; P: Glenn Skinner	72.6
3546. LAY ALL YOUR LOVE ON ME Information Society Tommy Boy 27534 12Aug89 83(1) 8 [Benny Andersson/Björn Ulvaeus] A: Information Society; P: Fred Maher	72.6
3547. WHEN WE GET MARRIED Larry Graham Warner Bros. 49581 25Oct80 76(2) 4 [Donald Hogan] A: One In A Million You; P: Larry Graham	72.6
3548. MATHEMATICS Melissa Manchester MCA 52575 27Apr85 74(1) 5 [Melissa Manchester/Robbie Nevil/Brock Walsh] A: Mathematics; P: Brock Walsh	72.5
3549. SWEET RACHEL Beau Coup Amherst 318 31Oct87 80(1) 6 [Dennis Lewin] A: Born And Raised (On Rock & Roll); P: Duane Baron/Dennis Lewin/John Purdell	72.4
3550. OVER MY HEAD Toni Basil Chrysalis 42753 21Jan84 81(1) 6 [Franne Golde/Sue Shifrin] A: Toni Basil; P: Richie Zito	72.4
3551. SHY GIRL Stacey Q On The Spot 110 11Apr87 89(1) 11 [Skip Hahn/Jon St. James/Stacey Swain] A: Stacey Q (EP); P: Jon St. James	72.2
3552. CAUGHT IN THE GAME Survivor Scotti Brothers 04074 22Oct83 77(2) 5 [Jim Peterik/Frankie Sullivan] A: Caught In The Game; P: Frankie Sullivan	72.0
3553. YOU'RE THE BEST THING The Style Council Geffen 29248 14Jul84 76(2) 5 [Paul Weller] A: My Ever Changing Moods; P: Paul Weller/Peter Wilson	71.8
3554. IF YOU ASKED ME TO Patti LaBelle MCA 53358 07Oct89 79(2) 5 [Diane Warren] A: Be Yourself; P: Stewart Levine	71.8
3555. RUN TO PARADISE Choirboys WTG 31-68564 11Mar89 80(1) 7 [Brad Carr/Mark Gable] A: Big Bad Noise; P: Peter Blyton/Choirboys/Brian McGee	71.6
3556. THE WALLS CAME DOWN The Call Mercury 811487 07May83 74(1) 5 [Michael Been] A: Modern Romans; P: Michael Been/Call	71.4
3557. DON'T RUN MY LIFE Spys EMI America 8124 14Aug82 82(3) 5 [John Blanco/John DiGaudio/Ed Gagliardi/Al Greenwood/Billy Milne] A: S.P.Y.S.; P: Neil Kernon	71.2
3558. ANOTHER TICKET Eric Clapton And His Band RSO 1064 13Jun81 78(3) 5 [Eric Clapton] A: Another Ticket; P: Tom Dowd	71.0
3559. DON'T BE AFRAID OF THE DARK The Robert Cray Band Mercury 870596 24Sep88 74(1) [Dennis Walker] A: Don't Be Afraid Of The Dark; P: Bruce Bromberg/Dennis Walker	70.6
3560. A LITTLE IS ENOUGH Pete Townshend Atco 7312 11Oct80 72(2) 4 [Pete Townshend] A: Empty Glass; P: Chris Thomas	70.5
3561. IT'S ALRIGHT (Baby's Coming Back) Eurythmics RCA 14284 15Feb86 78(1) 6 [Annie Lennox/David A. Stewart] A: Be Yourself Tonight; P: David A. Stewart	70.4
3562. MAGIC CARPET RIDE Bardeux Enigma 75016 27Feb88 81(1) 7 [Stacy Smith/Jon St. James] A: Bold As Love; P: Jon St. James	70.4
3563. RHYTHM OF LOVE Scorpions Mercury 870323 04Jun88 75(1) 6 [Klaus Meine/Rudolf Schenker] A: Savage Amusement; P: Dieter Dierks	70.3
3564. LOVE STRUCK Jesse Johnson A&M 3020 23Apr88 78(1) 8 [Jesse Woods Johnson] A: Every Shade Of Love; P: Jesse Johnson	70.1
3565. TREASURE The Brothers Johnson A&M 2254 16Aug80 73(1) 4 [Rod Temperton] A: Light Up The Night; P: Quincy Jones	70.0
3566. A CHANCE FOR HEAVEN Christopher Cross Columbia 04492 16Jun84 76(1) 5 [Burt Bacharach/Christopher Cross/Carole Bayer Sager] A: The Official Music Of The XXIIIrd Olympiad-Los Angeles 1984; P: Michael Omartian	70.0
3567. WALK RIGHT NOW The Jacksons Epic 02132 27Jun81 73(2) 4 [Jackie Jackson/Michael Jackson/Steven Randall Jackson] A: Triumph; P: Jacksons	70.0
3568. I CAN'T STOP THE FEELIN' Pure Prairie League Casablanca 2319 06Dec80 77(1) 6 [Daniel Fleishour/James Sandefur] A: Firin' Up; P: John Ryan	70.0
3569. LET ME GO The Rings MCA 51069 07Mar81 75(2) 5 [Mike Baker] A: The Rings; P: Rings	69.9
3570. NOTHIN (That Compares 2 U) Jacksons Epic 68688 03Jun89 77(1) 7 [Kenneth Edmonds/Antonio Reid] A: 2300 Jackson St.; P: Kenneth Edmonds/L.A. Reid	69.9
3571. TIL YOU AND YOUR LOVER ARE LOVERS AGAIN Engelbert Humperdinck Epic 03817 16Jul83 77(1) 5 [Jan Buckingham/Mark Eugene Gray] A: You And Your Lover; P: Even Stevens	69.9

Rank. TITLE Act Label Number Entry Date Peak(Wks) Tot Wks	
[Writers] A: Proximate Album; P: Producer	Score
3572. DO YOU BELIEVE IN SHAME? Duran Duran Capitol 44337 18Mar89 72(1) 5 [Eleanor Broadwater/Robert Chaisson/Dale Hawkins/ Simon Le Bon/Stanley J. Lewis/Nick Rhodes/Nigel John Taylor] A: *Big Thing*; P: *Daniel Abraham/Duran Duran/Jonathan Elias*	69.6
3573. YOU'RE ALL I NEED Mötley Crüe Elektra 69429 28Nov87 83(1) 8 [Tommy Lee/Nikki Sixx] A: *Girls, Girls, Girls*; P: *Tom Werman*	69.5
3574. IN IT FOR LOVE England Dan & John Ford Coley Big Tree 17002 08Mar80 75(1) 4 [Greg Guidry/Denny Henson] A: *The Best Of England Dan & John Ford Coley*; P: *Kyle Lehning*	69.5
3575. DON'T TELL ME THE TIME Martha Davis Capitol 44057 21Nov87 80(1) 8 [Martha Davis] A: *Policy*; P: *Richie Zito*	69.3
3576. CAN YOU FEEL IT The Jacksons Epic 01032 02May81 77(2) 5 [Jackie Jackson/Michael Jackson] A: *Triumph*; P: *Jacksons*	69.1
3577. I WILL FOLLOW U2 Island 99789 14Jan84 81(2) 5 [Adam Clayton/Dave Evans/Paul Hewson/Larry Mullen] A: *Under A Blood Red Sky*; P: *Steve Lillywhite*	69.0
3578. HERE SHE COMES Bonnie Tyler Columbia 04548 11Aug84 76(1) 5 [Pete Bellotte/Giorgio Moroder] A: *Metropolis Soundtrack*; P: *Giorgio Moroder*	69.0
3579. TALK TALK Talk Talk EMI America 8136 16Oct82 75(1) 7 [Ed Hollis/Mark Hollis] A: *The Party's Over*; P: *Colin Thurston*	68.8
3580. SWEET, SWEET BABY (I'm Falling) Lone Justice Geffen 28965 27Jul85 73(1) 5 [Maria McKee/Benmont Tench/Steven Van Zandt] A: *Lone Justice*; P: *Jimmy Iovine*	68.7
3581. BOY BLUE Cyndi Lauper Portrait 37-07181 13Jun87 71(1) 4 [Jeff Bova/Cyndi Lauper/Stephen Lunt] A: *True Colors*; P: *Cyndi Lauper/Lennie Petze*	68.5
3582. WANT YOU FOR MY GIRLFRIEND 4 By Four Capitol 5690 13Jun87 79(1) 6 [Chris Dixon/Charles Henry Jackson Jr.] A: *4 By Four*; P: *Chris Dixon/Chuck (2) Jackson*	68.5
3583. TOUCH A FOUR LEAF CLOVER Atlantic Starr A&M 2580 10Dec83 87(3) 7 [David E. Lewis/Wayne I. Lewis] A: *Yours Forever*; P: *James Anthony Carmichael*	68.5
3584. WAIT ON LOVE Michael Bolton Columbia 07794 28May88 79(1) 6 [Michael Bolton/Jonathan Cain] A: *The Hunger*; P: *Jonathan Cain*	68.3
3585. PRESENCE OF LOVE The Alarm I.R.S. 53259 26Mar88 77(1) 6 [Eddie MacDonald/Mike Peters] A: *Eye Of The Hurricane*; P: *Alarm/John Porter*	68.2
3586. LOVE IS THE KEY Maze Featuring Frankie Beverly Capitol 5221 04Jun83 80(2) 5 [Frankie Beverly] A: *We Are One*; P: *Frankie Beverly*	68.1
3587. RADIO FREE EUROPE R.E.M. I.R.S. 9916 23Jul83 78(1) 5 [Bill Berry/Peter Buck/Mike Mills/Michael Stipe] A: *Murmur*; P: *Don Dixon/Mitch Easter*	68.0
3588. BLACK DOG Newcity Rockers Critique 728 18Apr87 80(1) 6 [John Paul Jones/Jimmy Page/Robert Plant] A: *Newcity Rockers*; P: *Cliff Goodwin/Bob Rivers*	68.0
3589. STACY Fortune MCA 52727 21Dec85 80(2) 6 [Roger Craig/Larry Alan Greene] A: *Fortune*; P: *Kevin Beamish*	67.7
3590. I GET OFF ON IT Tony Joe White Casablanca 2279 28Jun80 79(1) 5 [Leann White/Tony Joe White] A: *The Real Thang*; P: *Tony Joe White*	67.7
3591. (Between A) ROCK AND A HARD PLACE Cutting Crew Virgin 99215 13May89 77(1) 5 [Kevin MacMichael/Nick Van Eede] A: *The Scattering*; P: *Cutting Crew/Peter Vettesse*	67.5
3592. SAVE THE NIGHT FOR ME Maureen Steele Motown 1787 04May85 77(1) 5 [Harvey Price/Bobby Sandstrom] A: *Nature Of The Beast*; P: *Steve Barri/Bobby Sandstrom*	67.4
3593. UNDER THE COVERS Janis Ian Columbia 02176 11Jul81 71(2) 4 [Janis Ian] A: *Restless Eyes*; P: *Gary Klein*	67.4
3594. SWEET MERILEE Donnie Iris MCA/Carousel 51198 31Oct81 80(1) 6 [Mark Avsec/Donnie Iris] A: *King Cool*; P: *Mark Avsec*	67.1
3595. IF YOU WERE A WOMAN (And I Was A Man) Bonnie Tyler Columbia 05839 12Apr86 77(1) 6 [Desmond Child] A: *Secret Dreams & Forbidden Fire*; P: *Jim Steinman*	66.9
3596. A WOMAN IN LOVE (It's Not Me) Tom Petty And The Heartbreakers Backstreet 51136 01Aug81 79(1) 6 [Mike Campbell/Tom Petty] A: *Hard Promises*; P: *Jimmy Iovine/Tom Petty*	66.8
3597. WIRED FOR SOUND Cliff Richard EMI America 8095 10Oct81 71(1) 4 [Brian A. Robertson/Alan Tarney] A: *Wired For Sound*; P: *Alan Tarney*	66.7
3598. YOU CAN Madleen Kane Chalet 1225 06Feb82 77(1) 5 [Pete Bellotte/Giorgio Moroder] A: *Don't Wanna Lose You*; P: *Giorgio Moroder*	66.5
3599. WILL THE WOLF SURVIVE? Los Lobos Slash 29093 23Mar85 78(1) 5 [David Hidalgo/Louie Perez] A: *How Will The Wolf Survive*; P: *Steve Berlin/T-Bone Burnett*	66.4
3600. MAGIC MAN Herb Alpert A&M 2356 05Sep81 79(1) 5 [Herb Alpert/Melvin Ragin/Michael Lee Stokes] A: *Magic Man*; P: *Herb Alpert/Michael Stokes*	66.4
3601. WALKING THROUGH WALLS The Escape Club Atlantic 88951 25Mar89 81(1) 6 [Johnnie Christo/John Holliday/Trevor Steel/Milan Zekavica] A: *Wild, Wild West*; P: *Chris Kimsey*	66.1
3602. JOHNNY COME HOME Fine Young Cannibals I.R.S. 52760 05Apr86 76(1) 5 [Roland Gift/David Steele] A: *Fine Young Cannibals*; P: *Andy Cox/Roland Gift/David Steele*	66.1
3603. JOY Teddy Pendergrass Elektra 69401 18Jun88 77(2) 6 [Reggie Calloway/Vincent Calloway/Joel Davis] A: *Joy*; P: *Reggie Calloway/Vincent Calloway*	66.0
3604. I WANNA GO BACK Billy Satellite Capitol 5409 08Dec84 78(1) 5 [Monty Byrom/Danny Chauncey/Ira Walker] A: *Billy Satellite*; P: *Don Gehman*	65.9
3605. (Come On) SHOUT Alex Brown Mercury 880694 04May85 76(1) 6 [Marti Sharron/Gary Skardina] A: *Girls Just Want To Have Fun Soundtrack*; P: *Marti Sharron/Gary Skardina*	65.9
3606. YOU'VE GOT WHAT I NEED Shooting Star Virgin 67005 12Apr80 76(1) 4 [Van McLain/Gary West] A: *Shooting Star*; P: *Gus Dudgeon*	65.8
3607. FADE AWAY Loz Netto 21 Records 104 04Jun83 82(2) 6 [Loz Netto] A: *Loz Netto's Bzar*; P: *Colin Thurston*	65.4
3608. BIT BY BIT Stephanie Mills MCA 52617 06Jul85 78(1) 6 [Harold Faltermeyer/Franne Golde] A: *Fletch Soundtrack*; P: *Harold Faltermeyer*	65.3
3609. REAL LOVE The Cretones Planet 45911 03May80 79(1) 6 [Mark Goldenberg] A: *Thin Red Line*; P: *Peter Bernstein*	65.3
3610. I DID IT FOR LOVE Night Ranger Camel/MCA 53364 01Oct88 75(1) 5 [Russ Ballard] A: *Man In Motion*; P: *Brian Foraker*	65.2
3611. SIGN OF THE TIMES The Belle Stars Warner Bros. 29672 07May83 75(2) 4 [Stella Barker/Clare Hirst/Miranda Joyce/Theresa Matthias/Sarah Jane Owen/Judy Parsons/Lesley Shone] A: *The Belle Stars*; P: *Peter Collins*	65.1
3612. SEX SHOOTER Apollonia 6 Warner Bros. 29182 20Oct84 85(1) 6 [Prince] A: *Apollonia 6*; P: *Apollonia 6/Starr Company*	65.1
3613. BREAK-A-WAY Tracey Ullman MCA/Stiff 52385 16Jun84 70(1) 4 [Jackie DeShannon/Sharon Sheeley] A: *You Broke My Heart In 17 Places*; P: *Peter Collins*	65.0
3614. I WAS LOOKING FOR SOMEONE TO LOVE Leif Garrett Scotti Brothers 516 12Apr80 78(1) 5 [Howard Greenfield/Michael Lloyd] A: *Same Goes For You*; P: *Michael Lloyd*	64.9
3615. SHOULD I LOVE YOU Cee Farrow Rocshire 95032 24Sep83 82(1) 6 [Cee Farrow/Lothar Krell] A: *Red And Blue*; P: *Andy Lunn*	64.5
3616. WHO SAYS Device Chrysalis 43063 27Sep86 79(1) 6 [Mike Chapman/Holly Knight] A: *22B3*; P: *Mike Chapman*	64.4
3617. TAKE NO PRISONERS (In The Game Of Love) Peabo Bryson Elektra 69632 29Jun85 78(1) 6 [Billy Livsey/Sue Shifrin] A: *Take No Prisoners*; P: *Arif Mardin*	64.3
3618. WHEN I'M HOLDING YOU TIGHT Michael Stanley Band EMI America 8130 11Sep82 78(2) 4 [Kevin Raleigh] A: *MSB*; P: *Don Gehman/Michael Stanley Band*	64.2
3619. ALL I NEED TO KNOW Bette Midler Atlantic 89789 03Sep83 77(1) 4 [Barry Mann/Tom Snow/Cynthia Weil] A: *No Frills*; P: *Chuck Plotkin*	64.1
3620. NO ONE CAN LOVE YOU MORE THAN ME Melissa Manchester Arista 1-9087 29Oct83 78(1) 4 [Terry Britten/Billy Livsey] A: *Emergency*; P: *Arif Mardin*	64.0
3621. GOT TO ROCK ON Kansas Kirshner 4292 27Dec80 76(1) 5 [Steve Walsh] A: *Audio-Visions*; P: *Kansas*	64.0
3622. THE LAST SAFE PLACE ON EARTH Le Roux RCA 13224 29May82 77(1) 5 [Jeff Pollard] A: *Last Safe Place*; P: *Leon Medica*	63.9
3623. HE'S MY GIRL David Hallyday Scotti Brothers 07299 29Aug87 79(2) 6 [Cecelia Bullard/Kim Bullard/Richie Wise] A: *True Cool*; P: *Kim Bullard/Richie Wise*	63.6
3624. BABY, WHAT ABOUT YOU Crystal Gayle Warner Bros. 29582 10Sep83 83(1) 5 [Josh Leo/Wendy Waldman] A: *True Love*; P: *Jimmy Bowen*	63.6
3625. MIDNIGHT RENDEZVOUS The Babys Chrysalis 2425 03May80 72(1) 4 [Jonathan Cain/John Waite] A: *Union Jacks*; P: *Keith Olsen*	63.5
3626. THE GREATEST GIFT OF ALL Kenny Rogers & Dolly Parton RCA 13945 22Dec84 81(2) 4 [John Jarvis] A: *Once Upon A Christmas*; P: *David Foster/Kenny Rogers*	63.5
3627. THIS IS THE WORLD CALLING Bob Geldof Atlantic 89341 13Dec86 82(1) 6 [Bob Geldof] A: *Deep In The Heart Of Nowhere*; P: *Bob Geldof/David A. Stewart*	63.4
3628. NO TIME TO LOSE The Tarney/Spencer Band A&M 2366 19Sep81 74(1) 4 [Trevor Spencer/Alan Tarney] A: *Run For Your Life*; P: *David Kershenbaum*	63.4

Ranking the Singles

Rank. TITLE Act Label Number Entry Date Peak(Wks) Tot Wks
[Writers] A: Proximate Album; P: Producer   Score

3629. CAFE AMORE Spyro Gyra MCA 51035 07Feb81 77(1) 5
[Chet Catallo] A: Carnaval; P: Jay Beckenstein/Richard Calandra   63.3

3630. ALL I WANT Howard Jones Elektra 69494 24Jan87 76(1) 5
[Howard Jones] A: One To One; P: Arif Mardin   63.1

3631. SURVIVE Jimmy Buffett MCA 41199 08Mar80 77(1) 5
[Jimmy Buffett/Mike Utley] A: Volcano; P: Norbert Putnam   63.0

3632. CECILIA Times Two Reprise 27871 09Jul88 79(1) 6
[Paul Simon] A: X2; P: David Agent/Jay King/Benny Medina   62.8

3633. LOVE HAS A MIND OF ITS OWN Donna Summer with Matthew Ward
Mercury 814922 14Jan84 70(1) 4 [Michael Omartian/Donna Summer]
A: She Works Hard For The Money; P: Michael Omartian   62.7

3634. SHOPPIN' FROM A TO Z Toni Basil Chrysalis 03537 26Feb83 77(2) 4
[Toni Basil/Bruce Roberts/Allee Willis]
A: Word Of Mouth; P: Greg Mathieson/Trevor Veitch   62.3

3635. SEASONS OF THE HEART John Denver RCA 13270 31Jul82 78(1) 5
[John Denver] A: Seasons Of The Heart; P: John Denver/Barney Wyckoff   62.0

3636. I DON'T WANT A LOVER Texas Mercury 872350 09Sep89 77(1) 6
[Johnny McElhone/Sharleen Spiteri] A: Southside; P: Tim Palmer   61.9

3637. KEEP IT TIGHT Single Bullet Theory Nemperor 03300 05Mar83
78(2) 4 [Michael Garrett/Gary Holmes] A: Single Bullet Theory; P: Rob Freeman 61.8

3638. NOW IT'S MY TURN Berlin Geffen 29283 30Jun84 74(1) 4
[John Crawford] A: Love Life; P: Mike Howlett   61.8

3639. REMO'S THEME (What If) Tommy Shaw A&M 2773 05Oct85 81(2) 5
[Richie Cannata/Tommy Shaw] A: What If; P: Richie Cannata/Tommy Shaw   61.8

3640. INSIDE OF YOU Ray, Goodman & Brown Polydor 2077 10May80
76(1) 4 [Al Goodman/Henrietta Goodman/Walter Lee Morris/Harry Ray]
A: Ray, Goodman & Brown; P: Vincent Castellano   61.7

3641. VOICE OF FREEDOM Jim Kirk And The TM Singers Capitol 4834
16Feb80 71(1) 3 [Jim Kirk] A: No Album; P: TM Productions   61.5

3642. MAKE BELIEVE IT'S YOUR FIRST TIME Bobby Vinton Tapestry 002
05Jan80 78(2) 4 [Bob Morrison/Johnny Wilson] A: Encore; P: Jack Bielan   61.5

3643. NOTHING'S GONNA STOP ME NOW Samantha Fox Jive 1072
10Oct87 80(1) 5 [Matt Aitken/Mike Stock/Pete Waterman]
A: Samantha Fox; P: Matt Aitken/Mike Stock/Pete Waterman   61.5

3644. JUST CAN'T WAIT The J. Geils Band EMI America 8047
12Jul80 78(1) 5 [Seth Justman/Peter Wolf] A: Love Stinks; P: Seth Justman   61.5

3645. OWWWW! Chunky A MCA 53736 02Dec89 77(1) 6
[Zane Giles/Arsenio Hall] A: Large And In Charge; P: A.Z. Groove   61.4

3646. PIECE OF MY HEART Sammy Hagar Geffen 50059 15May82 73(1) 4
[Bert Berns/Jerry Ragovoy] A: Standing Hampton; P: Keith Olsen   61.4

3647. MORE THAN PHYSICAL Bananarama London 886080
18Oct86 73(1) 5 [Matt Aitken/Sarah Dallin/Siobhan Fahey/Mike Stock/
Pete Waterman/Keren Woodward]
A: True Confessions; P: Matt Aitken/Mike Stock/Pete Waterman   61.1

3648. LITTLE BIT OF LOVE Dwight Twilley EMI America 8206
19May84 77(1) 4 [Dwight Twilley] A: Jungle; P: Noah Shark/Mark Smith   61.0

3649. NEVER DIE YOUNG James Taylor Columbia 07616 02Apr88 80(1) 5
[James Taylor] A: Never Die Young; P: Don Grolnick   61.0

3650. SHAKE IT UP Bad Company Atlantic 88939 22Apr89 82(1) 8
[Brian Howe/Terry Thomas] A: Dangerous Age; P: Terry Thomas   60.9

3651. SHE DON'T LOOK BACK Dan Fogelberg Full Moon 07044
30May87 84(1) 6 [Dan Fogelberg] A: Exiles; P: Dan Fogelberg/Russ Kunkel   60.6

3652. NEXT TIME YOU'LL KNOW Sister Sledge Cotillion 46012
09May81 82(3) 5 [Narada Michael Walden/Allee Willis]
A: All American Girls; P: Narada Michael Walden   60.5

3653. INDUSTRIAL DISEASE Dire Straits Warner Bros. 29880
08Jan83 75(1) 4 [Mark Knopfler] A: Love Over Gold; P: Mark Knopfler   60.5

3654. THEME FROM "TERMS OF ENDEARMENT" Michael Gore Capitol 5334
14Apr84 84(1) 6 [Michael Gore]
A: Terms Of Endearment Soundtrack; P: Michael Gore   60.5

3655. WALKING INTO SUNSHINE Central Line Mercury 76126
14Nov81 84(1) 6 [Linton Beckles/Roy Carter/Lipson Francis]
A: Central Line; P: Roy Carter   60.4

3656. TONIGHT IS WHAT IT MEANS TO BE YOUNG Fire Inc. MCA 52377
02Jun84 80(2) 5
[Jim Steinman] A: Tonight Is What It Means To Be Young; P: Jim Steinman   60.2

3657. RIGHT NEXT DOOR (Because Of Me) The Robert Cray Band
Mercury 888327 16May87 80(1) 6
[Dennis Walker] A: Strong Persuader; P: Bruce Bromberg/Dennis Walker   59.9

3658. INNOCENT EYES Graham Nash Atlantic 89434 26Apr86 84(1) 7
[Paul Steven Bliss]
A: Innocent Eyes; P: Craig Doerge/Stanley Johnston/Graham Nash   59.9

3659. CRY JUST A LITTLE Paul Davis Bang 4811 19Jul80 78(1) 4
[Paul Davis] A: Paul Davis; P: Paul Davis/Ed Seay   59.9

3660. HE'S A PRETENDER High Inergy Gordy 1662 21May83 82(2) 5
[Gary Goetzman/Mike Piccirillo] A: Groove Patrol; P: George Tobin   59.9

3661. PERFECT Fairground Attraction RCA 8789 17Dec88 80(1) 6
[Mark Nevin]
A: The First Of A Million Kisses; P: Fairground Attraction/Kevin Maloney   59.7

3662. SUMMERTIME, SUMMERTIME Nocera Sleeping Bag 7LX-22
17Jan87 84(1) 7 [Floyd Fisher/Maria Nocera]
A: Over The Rainbow; P: Floyd Fisher   59.7

3663. DAYS GONE BY Poco Atlantic 89674 28Apr84 80(1) 5
[Paul Cotton] A: Inamorata; P: Paul Cotton/Rusty Young   59.6

3664. ROUTE 66 The Manhattan Transfer Atlantic 4034 29May82 78(1) 5
[Bobby Troup] A: Sharky's Machine Soundtrack; P: Snuff Garrett   59.6

3665. BLUES POWER Eric Clapton And His Band RSO 1051
08Nov80 76(1) 5 [Eric Clapton/Leon Russell] A: Just One Night; P: Jon Astley 59.6

3666. THE SOUND OF GOODBYE Crystal Gayle Warner Bros. 29452
10Dec83 84(2) 5 [Hugh Prestwood] A: Cage The Songbird; P: Jimmy Bowen   59.5

3667. LA-DI-DA Sad Café Swan Song 72002 15Aug81 78(1) 4
[John Stimpson/Paul Young] A: Sad Café; P: Eric Stewart   59.2

3668. FOREVER YOURS Tony Terry Epic 07900 25Jun88 80(1) 5
[Ted Currier/Gary Henry] A: Forever Yours; P: Ted Currier   59.1

3669. YOU WILL KNOW Stevie Wonder Motown 1919 06Feb88 77(1) 6
[Stevie Wonder] A: Characters; P: Stevie Wonder   59.0

3670. WHEN THE LADY SMILES Golden Earring 21 Records 112
24Mar84 76(1) 5 [Barry Hay/George Kooymans]
A: N.E.W.S.; P: Shell Schellekens   58.8

3671. STAND BACK The Fabulous Thunderbirds CBS Associated 07230
04Jul87 76(1) 5 [Kim Wilson] A: Hot Number; P: Dave Edmunds   58.8

3672. FIRE IN THE SKY The Dirt Band Liberty 1429 03Oct81 76(2) 4
[Robert Carpenter/Jeff Hanna]
A: Jealousy; P: Bob Edwards/Jeff Hanna   58.7

3673. LEILA ZZ Top Warner Bros. 49782 03Oct81 77(2) 4
[Frank Beard/Billy Gibbons/Dusty Hill] A: El Loco; P: Bill Ham   58.7

3674. SOMEDAY, SOMEWAY Robert Gordon RCA 12239 27Jun81 76(1) 4
[Marshall Crenshaw]
A: Are You Gonna Be The One; P: Robert Gordon/Scott Litt/Lance Quinn   58.6

3675. THIN LINE BETWEEN LOVE AND HATE The Pretenders Sire 29249
30Jun84 83(2) 5 [Jackie Members/Richard Poindexter/Robert Poindexter]
A: Learning To Crawl; P: Chris Thomas   58.5

3676. SOMEBODY SOMEWHERE Platinum Blonde Epic 05804 12Apr86
82(2) 5 [Mark Holmes] A: Platinum Blonde; P: Mark Holmes/Eddy Offord   58.5

3677. THE REAL THING Jellybean Featuring Stephen Dante Chrysalis 43167
07Nov87 82(2) 6 [Toni Colandreo/Arnie Roman]
A: Just Visiting This Planet; P: Jellybean Benitez   58.4

3678. YOU PUT THE BEAT IN MY HEART Eddie Rabbitt Warner Bros. 29512
10Sep83 81(1) 5 [Rick Giles/Don Pfrimmer]
A: Greatest Hits - Vol. II; P: David Malloy   58.3

3679. TAKE THE TIME Michael Stanley Band EMI America 8146 25Dec82
81(2) 5 [Michael Stanley] A: MSB; P: Don Gehman/Michael Stanley Band   58.0

3680. SO FINE Oak Ridge Boys MCA 52065 12Jun82 76(1) 4
[Johnny Otis] A: Bobbie Sue; P: Ron Chancey   57.9

3681. TENDER YEARS ('84) John Cafferty And The Beaver Brown Band
Scotti Brothers 04327 28Jan84 78(1) 5 [John Cafferty]
A: Eddie & The Cruisers (Soundtrack); P: Kenny Vance   57.8

3682. RUNNING BACK Eddie Money Columbia 11325 13Sep80 78(1) 4
[Radcliffe Bryan] A: Playing For Keeps; P: Ron Nevison   57.7

3683. AND THE NIGHT STOOD STILL Dion Arista 1-9797 05Aug89 75(1) 5
[Diane Warren] A: Yo Frankie; P: Dave Edmunds   57.6

3684. BABY I'M HOOKED (Right Into Your Love) Con Funk Shun
Mercury 814581 07Jan84 76(1) 5
[Cedric Martin/Van Ross Redding] A: Fever; P: Eumir Deodato   57.5

3685. GIMME YOUR GOOD LOVIN' Diving For Pearls Epic 69036 09Dec89
84(2) 6 [Dan Malone/Jack Moran] A: Diving For Pearls; P: David Prater   57.3

3686. TULSA TIME Eric Clapton And His Band RSO 1039 21Jun80 64(1) 2
[Danny Flowers] A: Just One Night; P: Jon Astley   57.2

3687. GET IT Stevie Wonder & Michael Jackson Motown 1930
07May88 80(1) 6 [Stevie Wonder] A: Characters; P: Stevie Wonder   57.1

3688. MIDNIGHT RAIN Poco MCA 41326 11Oct80 74(1) 4
[Paul Cotton] A: Under The Gun; P: Mike Flicker   57.1

3689. WHAT YOU'RE MISSING Chicago Full Moon 29798 29Jan83 81(2) 5
[Jay Gruska/Joseph Amos Williams] A: Chicago 16; P: David Foster   56.3

Ranking the Singles

Rank. TITLE Act Label Number Entry Date Peak(Wks) Tot Wks
[Writers] A: *Proximate Album*; P: Producer — Score

3690. DO YOU WANNA HOLD ME? Bow Wow Wow RCA 13467 23Apr83
77(1) 4 [Matthew Ashman/Dave Barbarossa/Leigh Gorman/Annabella Lwin/
Malcolm McLaren]
A: *When The Going Gets Tough, The Tough Get Going*; P: Mike Chapman — 56.2

3691. WHERE EVERYBODY KNOWS YOUR NAME (The Theme From "Cheers")
Gary Portnoy Applause 106 30Apr83 83(1) 4
[Judy Hart Angelo/Gary Portnoy] A: *No Album*; P: Judy Hart Angelo/Gary Portnoy — 56.2

3692. YOU'RE MY BLESSING Lou Rawls Philadelphia Int'l 3570
26Apr80 77(1) 3 [Kenneth Gamble/Leon Huff]
A: *Sit Down And Talk To Me*; P: Kenny Gamble/Leon Huff — 56.1

3693. SHIVER AND SHAKE The Silencers (2) Precision 9800 26Jul80 81(2) 5
[Cathi Capiola/Warren King]
A: *Rock N' Roll Enforcers*; P: Bob Clearmountain/Silencers — 56.1

3694. SIDE BY SIDE Earth, Wind & Fire Columbia 03814 14May83 76(1) 4
[Wanda Hutchinson/Wayne Vaughn/Maurice White]
A: *Powerlight*; P: Maurice White — 56.0

3695. SATISFIED MAN Molly Hatchet Epic 04648 20Oct84 81(1) 5
[Thomas DeLuca/Tom Jans] A: *The Deed Is Done*; P: Terry Manning — 56.0

3696. I FEEL THE MAGIC Belinda Carlisle I.R.S. 52889 20Sep86 82(1) 5
[Charlotte Caffey/Jonathan Segal] A: *Belinda*; P: Michael Lloyd — 56.0

3697. THE HOUSE OF THE RISING SUN Dolly Parton RCA 12282 1
9Sep81 77(2) 4 [Traditional] A: *9 To 5 And Odd Jobs*; P: Mike Post — 56.0

3698. ANYWHERE WITH YOU Rubber Rodeo Mercury 880175
25Aug84 86(2) 5 [Bob Holmes/Trish Milliken] A: *Scenic Views*; P: Hugh Jones — 55.9

3699. GOOD MUSIC Joan Jett And The Blackhearts Blackheart 4-06336
11Oct86 83(2) 6 [Joan Jett/Kenny Laguna]
A: *Good Music*; P: Kenny Laguna/Thom Panunzio — 55.7

3700. HANG ON NOW Kajagoogoo EMI America 8171 27Aug83 78(1) 4
[Steven Askew/Nick Beggs/Christopher Hamill/Stuart Neale/Jeremy Strode]
A: *White Feathers*; P: Nick Rhodes/Colin Thurston — 55.7

3701. HOLIDAY ROAD Lindsey Buckingham Warner Bros. 29570 06Aug83
82(1) 5 [Lindsey Buckingham] A: *National Lampoon's Vacation Soundtrack*;
P: Lindsey Buckingham/Richard Dashut — 55.6

3702. RUNNING BACK Urgent Manhattan 50005 10Aug85 79(1) 5
[Don Kehr/Michael Kehr/Steve Kehr]
A: *Cast The First Stone*; P: Ian Hunter/Mick Ronson — 55.6

3703. ONE FOOT BACK IN YOUR DOOR Roman Holliday Jive 9287
09Feb85 76(1) 5 [Mutt Lange] A: *Fire Me Up*; P: Nigel Green — 55.5

3704. ROXANNE, ROXANNE UTFO Select 1182 09Mar85 77(1) 5
[Curtis Bedeau/Jeff Campbell/Gerard Charles/Hugh Clarke/
Shiller Shaun Fequiere/Brian George/Lucien George/Paul George/
Frederick Douglass Reeves] A: *UTFO*; P: Full Force/Fred Munao — 55.5

3705. ALL AMERICAN GIRLS Sister Sledge Cotillion 46007
21Mar81 79(2) 5 [Joni Sledge/Lisa Walden/Narada Michael Walden/
Allee Willis] A: *All American Girls*; P: Narada Michael Walden — 55.5

3706. DIVIDED HEARTS Kim Carnes EMI America 8322 24May86 79(2) 5
[Kim Carnes/Collin Ellingson/Kathy Kurasch/Donna Weiss]
A: *Lighthouse*; P: Val Garay — 55.4

3707. GONE TOO FAR Eddie Rabbitt Elektra 46613 03May80 82(1) 4
[David Malloy/Eddie Rabbitt/Even Stevens] A: *Loveline*; P: David Malloy — 55.3

3708. FOOL FOR YOUR LOVE Jimmy Hall Epic 02857 01May82 77(2) 3
[Michael Omartian/Leo Sayer] A: *Cadillac Tracks*; P: Norbert Putnam — 55.3

3709. THE VERY LAST TIME Utopia Bearsville 49247 07Jun80 76(1) 3
[Roger Powell/Todd Rundgren/Kasim Sulton/John Griffitt Wilcox]
A: *Adventures In Utopia*; P: Todd Rundgren/Utopia — 55.1

3710. I WAS BORN TO LOVE YOU Freddie Mercury Columbia 04869
27Apr85 76(1) 4 [Freddie Mercury]
A: *Mr. Bad Guy*; P: Reinhold Mack/Freddie Mercury — 55.0

3711. LET IT ALL BLOW Dazz Band Motown 1760 01Dec84 84(1) 7
[Robert Lamont Harris/Keith Harrison]
A: *Jukebox*; P: Reggie Andrews/Bobby Harris — 54.7

3712. REALISTIC Shirley Lewis Vendetta 1448 28Oct89 84(2) 6
[Richard Darbyshire/Charles Mole] A: *Passion*; P: Shep Pettibone — 54.6

3713. DO YOU LOVE ME Andy Fraser Island 99784 03Mar84 75(1) 4
[Berry Gordy] A: *Fine Fine Line*; P: Andy Fraser — 54.5

3714. ONE MORE CHANCE Diana Ross Motown 1508 11Apr81 79(1) 5
[Gerry Goffin/Michael Masser] A: *To Love Again*; P: Michael Masser — 54.4

3715. TALK TO ME Quarterflash Geffen 28908 19Oct85 83(2) 6
[Marv Ross/Rindy Ross] A: *Back Into Blue*; P: Steve Levine — 54.2

3716. WYNKEN BLYNKEN AND NOD The Doobie Brothers Sesame Street
49642 24Jan81 76(1) 4 [Eugene Field/Lucy Simon]
A: *In Harmony - A Sesame Street Record*; P: David Levine/Lucy Simon — 54.2

3717. WHATEVER YOU DECIDE Randy Vanwarmer Bearsville 49258 02Aug80
77(2) 3 [Randy Vanwarmer] A: *Terraform*; P: John Holbrook/Ian Kimmet — 54.1

3718. FAVORITE WASTE OF TIME Bette Midler Atlantic 89761
22Oct83 78(1) 4 [Marshall Crenshaw] A: *No Frills*; P: Chuck Plotkin — 54.0

3719. LONG TIME LOVIN' YOU McGuffey Lane Atco 7319 17Jan81 85(1) 7
[John Schwab] A: *McGuffey Lane*; P: Gary Platt/John Schwab — 53.7

3720. I DON'T KNOW Michael Morales Wing 873 282 09Dec89 81(1) 6
[Michael Morales] A: *Michael Morales*; P: Michael Morales — 53.4

3721. I DON'T WANT TO BE LONELY Dana Valery Scotti Brothers 509
19Jan80 87(1) 5 [Mark Mueller] A: *No Album*; P: Kyle Lehning — 53.4

3722. SEASONS Charles Fox Handshake 5307 24Jan81 75(1) 4
[Charles Fox/Ed Newmark] A: *Seasons*; P: Charles Fox/Ed Newmark — 53.4

3723. YOU WERE MADE FOR ME Irene Cara Geffen 29257 28Jul84 78(1) 5
[Eddie Brown/Irene Cara] A: *What A Feelin'*; P: James Newton Howard — 53.4

3724. STRANGLEHOLD Paul McCartney Capitol 5636 15Nov86 81(2) 6
[Paul McCartney/Eric Stewart]
A: *Press To Play*; P: Paul McCartney/Hugh Padgham — 53.4

3725. DAYDREAM BELIEVER The Monkees Arista 1-9532 01Nov86 79(2) 4
[John Stewart] A: *Then & Now...The Best Of The Monkees*; P: Chip Douglas — 53.0

3726. YOU AND ME Rockie Robbins A&M 2231 19Jul80 80(2) 4
[James P. Pennington] A: *You And Me*; P: Bobby Martin — 52.8

3727. TOO GOOD TO TURN BACK NOW Rick Bowles Polydor 2209 03Jul82
77(2) 3 [Rick Bowles/Richard Putnam] A: *Free For The Evening*; P: Ted Daryll — 52.6

3728. THE HUNTER GTR Arista 1-9512 23Aug86 85(1) 6
[Geoffrey Downes] A: *GTR*; P: Geoffrey Downes — 52.6

3729. IT TAKES TIME The Marshall Tucker Band Warner Bros. 49215
26Apr80 79(2) 3 [Toy Caldwell] A: *Tenth*; P: Stewart Levine — 52.4

3730. SOMETHING JUST AIN'T RIGHT Keith Sweat Vintertainment 7-69411
14May88 79(1) 5 [Teddy Riley/Keith Sweat]
A: *Make It Last Forever*; P: Keith Sweat — 52.2

3731. JEALOUS GUY John Lennon and the Plastic Ono Band (With the Flux
Fiddlers) Capitol/EMI 44230 15Oct88 80(1) 4
[John Lennon] A: *Imagine: John Lennon*; P: John Lennon/Yoko Ono/Phil Spector — 52.1

3732. YOU, ME AND HE Mtume Epic 04504 15Sep84 83(1) 5
[James Mtume] A: *You, Me And He*; P: James Mtume — 52.0

3733. ALL NIGHT LONG Billy Squier Capitol 5422 27Oct84 75(1) 3
[Billy Squier] A: *Signs Of Life*; P: Billy Squier/Jim Steinman — 51.9

3734. WOMEN Def Leppard Mercury 888757 22Aug87 80(1) 5
[Steve Clark/Phil Collen/Joe Elliott/Mutt Lange/Rick Savage]
A: *Hysteria*; P: Mutt Lange — 51.9

3735. DISCIPLINE OF LOVE (Why Did You Do It) Robert Palmer Island 99597
16Nov85 82(2) 5
[David Batteau/Don Freeman] A: *Riptide*; P: Bernard Edwards — 51.7

3736. LET'S DANCE Chris Rea Motown 1900 29Aug87 81(2) 5
[Chris Rea] A: *Dancing With Strangers*; P: Stuart Eales/Chris Rea — 51.6

3737. I CANNOT BELIEVE IT'S TRUE Phil Collins Atlantic 89864
14May83 79(1) 4 [Phil Collins] A: *Hello, I Must Be Going!*; P: Phil Collins — 51.6

3738. SKIN DEEP Cher Geffen 27894 30Jul88 79(1) 4
[Mark Goldenberg/Jon Lind] A: *Cher(2)*; P: Jon Lind — 51.5

3739. LOVE IS THE HERO Billy Squier Capitol 5619 04Oct86 80(1) 5
[Billy Squier] A: *Enough Is Enough*; P: Peter Collins — 51.2

3740. THE VERY BEST IN YOU Change RFC/Atlantic 4027 29May82 84(3) 5
[Mauro Malavasi/Herb Smith]
A: *Sharing Your Love*; P: Mauro Malavasi/Jacques Fred Petrus — 51.1

3741. ONE NIGHT IN BANGKOK Robey Silver Blue 4-04774 02Mar85
77(1) 3 [Benny Andersson/Tim Rice/Björn Ulvaeus] A: *Robey*; P: Joel Diamond — 51.0

3742. DOWNTOWN Dolly Parton RCA 13756 14Apr84 80(1) 4
[Tony Hatch] A: *The Great Pretender*; P: Val Garay — 51.0

3743. GARDEN PARTY Herb Alpert A&M 2562 20Aug83 81(1) 4
[Eythor Gunnarsson] A: *Blow Your Own Horn*; P: Herb Alpert — 50.9

3744. EVERY BEAT OF MY HEART Rod Stewart Warner Bros. 28625
29Nov86 83(2) 6 [Kevin Savigar/Rod Stewart] A: *Rod Stewart*; P: Bob Ezrin — 50.8

3745. THIS TIME Kiara (Duet with Shanice Williams) Arista 1-9772
25Feb89 78(1) 5 [Charles L. Singleton]
A: *To Change And/Or Make A Difference*; P: Nick Martinelli — 50.6

3746. SQUARE ROOMS Al Corley Mercury 822241 11May85 80(1) 5
[Al Corley/Harold Faltermeyer/Peter John Wood]
A: *Square Rooms*; P: Harold Faltermeyer — 50.3

3747. BLAME IT ON THE RADIO John Parr Atlantic 89333 13Dec86 88(2) 6
[John Parr] A: *Running The Endless Mile*; P: John Parr — 50.3

3748. IT'S GONNA BE SPECIAL Patti Austin Qwest 29373 11Feb84 82(1) 4
[Glen Ballard/Clif Magness] A: *Patti Austin*; P: Quincy Jones — 50.1

Ranking the Singles

Rank. TITLE Act Label Number Entry Date Peak(Wks) Tot Wks [Writers] A: Proximate Album; P: Producer	Score
3749. THE DEVIL MADE ME DO IT Golden Earring 21 Records 108 23Apr83 79(1) 4 [Barry Hay/George Kooymans] A: Cut; P: Shell Schellekens	50.1
3750. NEXT LOVE Deniece Williams Columbia 04537 11Aug84 81(1) 4 [George Duke/Deniece Williams] A: Let's Hear It For The Boy; P: George Duke	49.9
3751. GYPSY SPIRIT Pendulum Venture 131 29Nov80 89(3) 7 [David Barrow/James Kenny/David Quintana] A: No Album; P: Pendulum	49.8
3752. TWISTIN' THE NIGHT AWAY Rod Stewart Geffen 28303 18Jul87 80(1) 4 [Sam Cooke] A: Innerspace Soundtrack; P: Rod Stewart	49.7
3753. GUNS FOR HIRE AC/DC Atlantic 89774 A 01Oct83 84(1) 5 [Brian Johnson/Angus Young/Malcolm Young] A: Flick Of The Switch; P: AC/DC	49.6
3754. LEGAL TENDER The B-52's Warner Bros. 29579 16Jul83 81(1) 4 [Kate Pierson/Fred Schneider/Keith Strickland/ Robert Waldrop/Cindy Wilson/Ricky Wilson] A: Whammy!; P: Steven Stanley	49.5
3755. I ONLY WANT TO BE WITH YOU The Tourists Epic 50850 17May80 83(1) 4 [Mike Hawker/Ivor Raymonde] A: Reality Effect; P: Tom Allom	49.5
3756. YOU KNOW WHAT TO DO Carly Simon Warner Bros. 29484 24Sep83 83(1) 4 [Jacob Brackman/Mike Mainieri/Carly Simon/Peter John Wood] A: Hello Big Man; P: Mike Mainieri	49.4
3757. SO. CENTRAL RAIN (I'M SORRY) R.E.M. I.R.S. 9927 23Jun84 85(1) 6 [Bill Berry/Peter Buck/Mike Mills/Michael Stipe] A: Reckoning; P: Don Dixon/Mitch Easter	49.3
3758. I'VE JUST BEGUN TO LOVE YOU Dynasty Solar 12021 13Sep80 87(1) 6 [William Shelby/Ricky Darnell Smith] A: Adventures In The Land Of Music; P: Leon Sylvers	49.1
3759. YOU'VE GOT A GOOD LOVE COMING Van Stephenson Handshake 02140 12Sep81 79(1) 4 [Danny Morrison/Jeff Silbar/ Van Stephenson] A: China Girl; P: Bob Montgomery/Jeff Silbar	49.1
3760. I DON'T NEED YOU ANYMORE Together? Featuring Jackie DeShannon RCA 11902 08Mar80 86(1) 5 [Paul Anka/Burt Bacharach] A: Together? Soundtrack; P: Paul Anka/Burt Bacharach	49.1
3761. MY LOVE Julio Iglesias Featuring Stevie Wonder Columbia 07781 21May88 80(1) 5 [Stevie Wonder] A: Non Stop; P: Humberto Gatica/Stevie Wonder	48.7
3762. PEANUT BUTTER Twennynine Featuring Lenny White Elektra 46552 02Feb80 83(1) 4 [Donald Blackman/Lenny White] A: Best Of Friends; P: Larry Dunn/Lenny White	48.7
3763. SHY BOY (Don't It Make You Feel Good) Bananarama London 810112 02Jul83 83(1) 4 [Steve Jolley/Tony Swain] A: Deep Sea Skiving; P: Steve Jolley/Tony Swain	48.7
3764. THIEF OF HEARTS Melissa Manchester Casablanca 880308 24Nov84 86(1) 6 [Keith Forsey/Melissa Manchester/Giorgio Moroder] A: Thief Of Hearts Soundtrack; P: Harold Faltermeyer/Giorgio Moroder	48.6
3765. SOMETHING IN MY HOUSE Dead Or Alive Epic 07022 25Apr87 85(2) 6 [Pete Burns/Steve Coy/Wayne Hussey/Tim Lever/Mike Percy] A: Mad, Bad And Dangerous To Know; P: Matt Aitken/Mike Stock/Pete Waterman	48.5
3766. IT'S GETTIN' LATE The Beach Boys Caribou 05433 03Aug85 82(1) 5 [Robert White Johnson/Myrna Smith-Schilling/Carl Wilson] A: The Beach Boys; P: Steve Levine	48.4
3767. COOL (Part 1) The Time Warner Bros. 49864 30Jan82 90(1) 7 [Dez Dickerson/Prince] A: The Time; P: Morris Day/Prince	48.4
3768. TALKIN' BOUT A REVOLUTION Tracy Chapman Elektra 39383 01Oct88 75(2) 4 [Tracy Chapman] A: Tracy Chapman; P: David Kershenbaum	48.2
3769. OPPOSITES DO ATTRACT All Sports Band Radio 3892 06Feb82 78(2) 3 [Cy Sulack/Michael Toste] A: All Sports Band; P: Joey Carbone/Richie Zito	47.8
3770. DON'T LOSE ANY SLEEP John Waite EMI America 43040 26Sep87 81(1) 4 [Diane Warren] A: Rover's Return; P: Frank Filipetti/Rick Nowels/John Waite	47.8
3771. PLEASE MR. POSTMAN Gentle Persuasion Capitol 5207 26Feb83 82(2) 4 [Robert Bateman/Georgia Dobbins/William Garrett/ Freddie Gorman/Brian Holland] A: No Album; P: Nate Chacker	47.6
3772. MERRY CHRISTMAS IN THE NFL Willis "The Guard" and Vigorish Handshake 5308 27Dec80 82(3) 3 [Jerry Buckner/Gary Garcia] A: No Album; P: Jerry Buckner/Gary Garcia	47.4
3773. PARADISE Change RFC/Atlantic 3809 06Jun81 80(2) 4 [Mauro Malavasi/Davide Romani/Tanyayette Willoughby] A: Miracles; P: Mauro Malavasi/Jacques Fred Petrus	47.1
3774. THE BREAKS (Part 1) Kurtis Blow Mercury 76075 06Sep80 87(1) 6 [Robert Arthur Ford Jr./James B. Moore/Russell Simmons/Lawrence Smith/ Kurt Walker] A: Kurtis Blow; P: Robert Ford/J.B. Moore	47.0
3775. BETTER THINGS The Kinks Arista 0649 28Nov81 92(2) 8 [Ray Davies] A: Give The People What They Want; P: Ray Davies	46.9
3776. FOOLS LIKE ME Lorenzo Lamas Scotti Brothers 04686 22Dec84 85(1) 5 [Phil Galdston/Andy Goldmark/Sylvester Levay] A: No Album; P: Gaylong J. Horton/Sylvester Levay/Phil Ramone	46.8
3777. HURT Re-Flex Capitol 5348 05May84 82(1) 4 [Paul Fishman] A: The Politics Of Dancing; P: John Punter	46.6
3778. DARLIN' Frank Stallone Polydor 821382 05May84 81(1) 4 [Vince DiCola/Mark Hudson/Frank Stallone] A: Frank Stallone; P: Vince DiCola/Frank Stallone	46.6
3779. RUNAWAY Luis Cardenas Allied Artists 72500 20Sep86 83(1) 5 [Max Crook/Del Shannon] A: Animal Instinct; P: Kim Richards	46.5
3780. I'M GONNA MISS YOU Kenny Loggins Columbia 08091 12Nov88 82(1) 5 [Jeff Pescetto/Pam Reswick/Steve Werfel] A: Back To Avalon; P: Peter Wolf	46.4
3781. LET'S PUT THE FUN BACK IN ROCK N ROLL Freddy Cannon & The Belmonts MiaSound 1002 26Sep81 81(2) 4 [Bob Feldman/Joseph Nicoletti] A: No Album; P: Bob Feldman	46.3
3782. CARRIE'S GONE Le Roux RCA 13456 19Mar83 81(2) 4 [Fergie Frederiksen/Jim Odom/Rod Roddy] A: So Fired Up; P: Leon Medica	46.2
3783. CENTURY'S END Donald Fagen Warner Bros. 27972 02Apr88 83(2) 5 [Donald Fagen/Tim Meher] A: Bright Lights, Big City Soundtrack; P: Donald Fagen/Gary Katz	46.1
3784. SEXCRIME (Nineteen Eighty-Four) Eurythmics RCA 13956 24Nov84 81(1) 4 [Annie Lennox/David A. Stewart] A: 1984 (For The Love Of Big Brother) (Soundtrack); P: David A. Stewart	46.1
3785. DON'T CHANGE INXS Atco 99874 16Jul83 80(1) 4 [Garry Beers/Andy Farriss/Jon Farriss/Tim Farriss/Michael Hutchence/Kirk Pengilly] A: Shabooh Shoobah; P: Mark Opitz	46.0
3786. HUNGRY Winger Atlantic 88859 23Sep89 85(1) 6 [Reb Beach/Kip Winger] A: Winger; P: Beau Hill	45.6
3787. HOLD ME Laura Branigan Atlantic 89496 19Oct85 82(2) 4 [Beth Andersen/Bill Bodine] A: Hold Me; P: Harold Faltermeyer/Jack White	45.6
3788. YOU BELONG TO ME The Doobie Brothers Warner Bros. 29552 30Jul83 79(1) 4 [Michael McDonald/Carly Simon] A: The Doobie Brothers Farewell Tour; P: Ted Templeman	45.5
3789. PAINTED MOON The Silencers RCA 5220 08Aug87 82(1) 6 [Charles Burns/James O'Neill] A: A Letter From St. Paul; P: David Bascombe	45.4
3790. JUST FOR THE MOMENT Ray Kennedy ARC 11242 03May80 82(2) 3 [Jack Conrad/Ray Kennedy] A: Ray Kennedy; P: David Foster	45.4
3791. A LITTLE BIT OF HEAVEN Natalie Cole Modern 99630 07Sep85 81(1) 6 [Richard Kerr/Graham Lyle] A: Dangerous; P: Marti Sharron/Gary Skardina	45.2
3792. REACH OUT Giorgio Moroder Columbia 04511 21Jul84 81(1) 4 [Paul Engemann/Giorgio Moroder/Richie Zito] A: The Official Music Of The XXIIIrd Olympiad-Los Angeles 1984; P: Giorgio Moroder	45.1
3793. TWIST MY ARM Pointer Sisters RCA 14197 01Mar86 83(2) 5 [Andy Goldmark/Bruce Roberts] A: Contact; P: Richard Perry	45.0
3794. THE HARDEST PART Blondie Chrysalis 2408 02Feb80 84(2) 3 [Clement A. Bozewski/Deborah Harry/Christopher Stein] A: Eat To The Beat; P: Mike Chapman	44.9
3795. YOU ARE MY EVERYTHING Surface Columbia 69016 04Nov89 84(1) 5 [Everett Collins/Dave Conley/Derrick Culler/David Townsend] A: 2nd Wave; P: David Conley/Bernard Jackson/David Townsend	44.8
3796. OPEN LETTER (To A Landlord) Living Colour Epic 68934 08Jul89 82(1) 5 [Tracie Morris/Vernon Reid] A: Vivid; P: Ed Stasium	44.8
3797. EYE OF THE ZOMBIE John Fogerty Warner Bros. 28657 06Sep86 81(1) 4 [John Fogerty] A: Eye Of The Zombie; P: John Fogerty	44.8
3798. DON'T LET GO THE COAT The Who Warner Bros. 49743 27Jun81 84(1) 5 [Pete Townshend] A: Face Dances; P: Bill Szymczyk	44.8
3799. MORE BOUNCE TO THE OUNCE PART 1 Zapp Warner Bros. 49534 04Oct80 86(1) 7 [Roger Troutman] A: Zapp; P: William Collins/Roger Troutman	44.7
3800. STEREOTOMY Alan Parsons Project Arista 1-9443 15Feb86 82(1) 4 [Alan Parsons/Eric Woolfson] A: Stereotomy; P: Alan Parsons	44.6
3801. LIKE FLAMES Berlin Geffen 28563 25Oct86 82(1) 5 [Rob Brill] A: Count Three And Pray; P: Bob Ezrin	44.6
3802. SHE LOVES MY CAR Ronnie Milsap RCA 13847 04Aug84 84(2) 4 [Roy Freeland/Bill LaBounty] A: One More Try For Love; P: Ron Galbraith/Ronnie Milsap	44.6

Ranking the Singles

Rank. TITLE Act Label Number Entry Date Peak(Wks) Tot Wks
[Writers] A: Proximate Album; P: Producer   Score

3803. IT HURTS TOO MUCH Eric Carmen Arista 0506 12Jul80 75(1) 2
[Eric Carmen] A: Tonight You're Mine; P: Harry Maslin   44.6

3804. BLACK KISSES (Never Make You Blue) Curtie And The Boombox
RCA 14103 27Jul85 81(1) 4 [Peter Koelewijn]
A: Curtie And The Boombox; P: Albert Boekholdt/Peter Koelewijn   44.5

3805. AT THIS MOMENT ('81) Billy & The Beaters Alfa 7005
19Sep81 79(1) 3 [Billy Vera] A: Billy & The Beaters; P: Jeff Baxter   44.5

3806. YOU SEND ME The Manhattans Columbia 04754 02Mar85 81(1) 5
[Sam Cooke] A: Too Hot To Stop It; P: Morrie Brown   44.4

3807. YOUNGER DAYS Joe Fagin Millennium 13107 31Jul82 80(2) 3
[Richie Supa] A: Why Don't We Spend The Night; P: David Mackay   44.4

3808. LOVE RESURRECTION Alison Moyet Columbia 05411 20Jul85 82(1) 4
[Steve Jolley/Alison Moyet/Tony Swain] A: Alf; P: Steve Jolley/Tony Swain 44.1

3809. WHAT'S YOUR HURRY DARLIN' Ironhorse Scotti Brothers 512
26Apr80 89(1) 6 [Randy Bachman/Carl Wilson]
A: Everything Is Grey; P: Randy Bachman/Dennis Mackay   44.0

3810. HE WANTS MY BODY Starpoint Elektra 69489 14Mar87 89(1) 7
[Preston Glass] A: Sensational; P: Preston Glass/Lionel Job   43.9

3811. ALL THE RIGHT MOVES Jennifer Warnes/Chris Thompson
Casablanca 814603 12Nov83 85(1) 4 [Barry Alfonso/Tom Snow]
A: All The Right Moves Soundtrack; P: Brooks Arthur/Tom Snow   43.9

3812. JUST ANOTHER DAY Oingo Boingo MCA 52726 25Jan86 85(1) 4
[Danny Elfman] A: Dead Man's Party; P: Steve Bartek/Danny Elfman   43.9

3813. BABY BABY Eighth Wonder WTG 31-68610 01Apr89 84(1) 5
[Sam Domingo/John Forte] A: Fearless; P: Pete Hammond   43.7

3814. LET ME SLEEP ALONE Cugini Scotti Brothers 503 22Dec79 88(1) 4
[Don Cugini/Andrew DiTaranto/Anthony Papa]
A: No Album; P: John D'Andrea/Andrew DiTaranto/Anthony Papa   43.6

3815. BOUNCE, ROCK, SKATE, ROLL PT. 1 Vaughan Mason And Crew
Brunswick 211 22Mar80 81(1) 3 [Jerome Bell/Gregory Bufford/
Vaughan Mason] A: Bounce, Rock, Skate, Roll; P: Vaughan Mason   43.6

3816. LOVE HAS FINALLY COME AT LAST Bobby Womack and Patti LaBelle
Beverly Glen 2012 24Mar84 88(2) 5 [Patrick L. Moten/Bobby Womack]
A: The Poet II; P: James Gadson/Andrew Loog Oldham/Bobby Womack   43.3

3817. NO BIG DEAL Love and Rockets Big Time 9045 23Sep89 82(1) 4
[Daniel Ash/David Haskins/Kevin Haskins]
A: Love And Rockets; P: John Fryer/Love And Rockets   43.2

3818. HEAVEN IN YOUR ARMS Dan Hartman Blue Sky 70053
11Apr81 86(2) 5 [Dan Hartman] A: It Hurts To Be In Love; P: Dan Hartman   43.2

3819. DEVIL IN A FAST CAR Sheena Easton EMI America 8201 14Apr84
79(1) 3 [Greg Mathieson/Trevor Veitch] A: Best Kept Secret; P: Greg Mathieson 43.2

3820. SHIP OF FOOLS Robert Plant Es Paranza 7-99333 27Aug88 84(1) 4
[Phil Johnstone/Robert Plant]
A: Now And Zen; P: Phil Johnstone/Tim Palmer/Robert Plant   43.2

3821. JOHNNY B. GOODE Peter Tosh EMI America 8159 09Jul83 84(1) 4
[Chuck Berry] A: Mama Africa; P: Chris Kimsey/Peter Tosh   43.2

3822. AMERICAN MEMORIES Shamus M'Cool Perspective 107
04Jul81 80(2) 3 [Shamus M'Cool] A: No Album; P: Perspective   43.1

3823. I WANT YOU Animotion Casablanca 884729 17May86 84(1) 4
[Rick Neigher/Bill Wadhams] A: Strange Behavior; P: Richie Zito   42.8

3824. TONIGHT The Whispers Solar 69842 30Apr83 84(1) 4
[Jerry Knight] A: Love For Love; P: Leon Sylvers   42.8

3825. WELCOME TO PARADISE John Waite EMI America 8278
19Oct85 85(1) 4 [John Waite] A: Mask Of Smiles; P: Stephan Galfas/John Waite 42.4

3826. OVER YOU Roxy Music Atco 7301 09Aug80 80(1) 4
[Bryan Ferry/Phil Manzanera] A: Flesh + Blood; P: Rhett Davies/Roxy Music 42.0

3827. MAN AGAINST THE WORLD Survivor Scotti Brothers 07070
09May87 86(1) 5 [Jimi Jamison/Jim Peterik/Frankie Sullivan]
A: When Seconds Count; P: Ron Nevison/Frankie Sullivan   42.0

3828. THE REAL END Rickie Lee Jones Warner Bros. 29191
29Sep84 83(2) 4 [Rickie Lee Jones]
A: The Magazine; P: James Newton Howard/Rickie Lee Jones   41.7

3829. SLOW DANCIN' Peabo Bryson Elektra 69699 29Sep84 82(2) 4
[Peabo Bryson] A: Straight From The Heart; P: Peabo Bryson   41.7

3830. SUMMER '81 MEDLEY The Cantina Band Millennium 11818
25Jul81 81(2) 3 A: No Album; P: Billy Civitella/Meco Monardo/Lance Quinn 41.7

3831. YOU'RE THE ONLY LOVE Paul Hyde And The Payolas A&M 2733
18May85 84(1) 4 [David Foster/Paul Hyde/Miriam Nelson/Bob Rock]
A: Here's The World For Ya; P: David Foster   41.6

3832. COME OUT FIGHTING Easterhouse Columbia 68552 01Apr89 82(1) 4
[Andy Perry] A: Waitng For The Redbird; P: Andy Perry   41.4

3833. HOW CAN YOU LOVE ME Ambrosia Warner Bros. 29996
05Jun82 86(2) 4 [David Pack/Joe Puerta] A: Road Island; P: James Guthrie 41.2

3834. THE FLYER Saga Portrait 37-04178 26Nov83 79(1) 3
[Jim Crichton/Michael Sadler] A: Heads Or Tales; P: Rupert Hine   41.2

3835. YOU MAKE ME WORK Cameo Atlanta Artists 870587
05Nov88 85(1) 5 [Larry Blackmon] A: Machismo; P: Larry Blackmon   41.1

3836. THIS LOVE Bad Company Atlantic 89355 18Oct86 85(2) 5
[Chris Fretwell/Brian Howe] A: Fame And Fortune; P: Keith Olsen   40.5

3837. I FOUND SOMEONE Laura Branigan Atlantic 89451 01Mar86 90(2) 6
[Michael Bolton/Mark Mangold] A: Hold Me; P: Harold Faltermeyer/Jack White 40.4

3838. OVER THE LINE Eddie Schwartz Atco 7402 20Mar82 91(4) 5
[Eddie Schwartz/David Tyson] A: No Refuge; P: Eddie Schwartz/David Tyson 40.2

3839. WHEN WILL I BE FAMOUS Bros Epic 07905 25Jun88 83(1) 5
[Nicky Graham/Tom Watkins] A: Push; P: Nicky Graham   40.2

3840. SO THE STORY GOES Living In A Box Chrysalis 43162 17Oct87 81(1) 4
[Richard Darbyshire/Marcus Vere] A: Living In A Box; P: Richard James Burgess 40.1

3841. HIGH ENERGY Evelyn Thomas TSR 106 29Sep84 85(1) 5
[Ian Levine/Fiachra Trench] A: High Energy; P: Ian Levine/Fiachra Trench 40.1

3842. OOO LA LA LA Teena Marie Epic 07708 12Mar88 85(1) 5
[Teena Marie/Allen McGrier] A: Naked To The World; P: Teena Marie   40.0

3843. I'M YOUR MAN Barry Manilow RCA 14397 12Jul86 86(2) 5
[Barry Manilow/Howard Lee Rice/Allan Rich]
A: Manilow; P: Barry Manilow/Howie Rice   40.0

3844. JIMMY LOVES MARYANN Josie Cotton Elektra 69748 07Apr84 82(1) 4
[Elliot Lurie] A: From The Hip; P: Bobby Paine/Larson Paine   39.9

3845. ROCK-A-LOTT Aretha Franklin Arista 1-9574 20Jun87 82(1) 4
[Preston Glass/Joe Johnson/Narada Michael Walden]
A: Aretha (II); P: Narada Michael Walden   39.8

3846. WALKIN' SHOES Tora Tora A&M 1425 12Aug89 86(1) 6
[Anthony Corder/Keith Douglas/Patrick Francis/John Patterson]
A: Surprise Attack; P: Paul Ebersold/Joe Hardy   39.7

3847. DON'T STOP TRYING Rodway Millennium 13111 11Dec82 83(1) 5
[Norman Dolph/Steve Rodway]
A: Horizontal Hold; P: Mark Liggett/Steve Rodway   39.7

3848. MAYBE THIS DAY Kissing The Pink Atlantic 89796 06Aug83 87(2) 5
[Peter Barnett/Stephen Cusack/Sylvia Griffin/Jon Kingsley Hall/Peter Stewart/
Josephine Wells/Nick Whitecross] A: Naked; P: Colin Thurston   39.5

3849. IS THIS THE END New Edition Streetwise 1111 15Oct83 85(1) 4
[Michael Jonzun/Maurice Starr]
A: Candy Girl; P: Arthur Baker/Michael Jonzun/Maurice Starr   39.4

3850. ALL I WANT IS YOU U2 Island 99199 01Jul89 83(1) 4
[Adam Clayton/Dave Evans/Paul Hewson/Larry Mullen]
A: Rattle And Hum (Soundtrack); P: Jimmy Iovine   39.2

3851. FOOL'S GAME Michael Bolton Columbia 03800 14May83 82(2) 3
[Michael Bolton/Mark Mangold/Craig Sprovach]
A: Michael Bolton; P: Gerry Block/Michael Bolton   39.1

3852. RICH MAN Terri Gibbs MCA 51119 20Jun81 89(2) 5
[Ed Mattson] A: Somebody's Knockin'; P: Ed Penney   39.1

3853. BABY COME BACK TO ME (The Morse Code Of Love)
The Manhattan Transfer Atlantic 89594 02Feb85 83(1) 3
[Nick Santamaria] A: Bop Doo-Wop; P: Tim Hauser   39.0

3854. GOT A NEW LOVE Good Question Paisley Park 7-27861
22Oct88 86(1) 5 [Marc Douglas/Sean Douglas/Rick Neigher]
A: Good Question; P: Rick Neigher   38.9

3855. LOVE AGAIN John Denver & Sylvie Vartan RCA 13931
10Nov84 85(1) 4 [John Denver] A: Sylvie Vartan; P: Milton Okun   38.8

3856. IF LOVE SHOULD GO Streets Atlantic 89760 03Dec83 87(1) 5
[Mike Slamer/Steve Walsh] A: 1st; P: Neil Kernon   38.7

3857. HOME SWEET HOME Mötley Crüe Elektra 69591 26Oct85 89(1) 6
[Tommy Lee/Nikki Sixx] A: Theatre Of Pain; P: Tom Werman   38.3

3858. GO DOWN EASY Dan Fogelberg Full Moon 04835 23Mar85 85(1) 4
[Jay Bolotin] A: High Country Snows; P: Dan Fogelberg/Marty Lewis   37.6

3859. THE CELTIC SOUL BROTHERS Dexys Midnight Runners Mercury 811142
28May83 86(1) 4 [Micky Billingham/Jimmy Paterson/Kevin Rowland]
A: Too-Rye-Ay; P: Clive Langer/Kevin Rowland/Alan Winstanley   37.6

3860. JAMMIN' Teena Marie Epic 04738 27Apr85 81(1) 3
[Teena Marie] A: Starchild; P: Teena Marie   37.4

3861. WHAT'S SHE GOT Liquid Gold Critique 701 03Sep83 86(1) 4
[George Stanley Alexander/Rob Davis]
A: No Album; P: Rob Davis/Tony Taverner   37.4

3862. (It's Just) THE WAY THAT YOU LOVE ME ('88) Paula Abdul Virgin 99282
12Nov88 88(1) 5 [Oliver Leiber] A: Forever Your Girl; P: Oliver Leiber   37.3

3863. SAD HEARTS Four Tops Casablanca 2353 28Aug82 84(2) 3
[Marc Blatte/Larry Gottlieb]
A: One More Mountain; P: David Wolfert   37.2

Ranking the Singles

Rank. TITLE Act Label Number Entry Date Peak(Wks) Tot Wks
[Writers] A: Proximate Album; P: Producer — Score

3864. DON'T KNOW MUCH Bill Medley Liberty 1402 28Mar81 88(1) 4
[Barry Mann/Tom Snow/Cynthia Weil] A: Sweet Thunder; P: Michael Lloyd — 37.0

3865. SOLSBURY HILL Peter Gabriel Geffen 29542 27Aug83 84(1) 3
[Peter Gabriel] A: Peter Gabriel/Plays Live; P: Peter Gabriel/Peter Walsh — 36.8

3866. BACK IN STRIDE Maze Featuring Frankie Beverly Capitol 5431
16Mar85 88(1) 6 [Frankie Beverly] A: Can't Stop The Love; P: Frankie Beverly — 36.8

3867. SEASONS OF GOLD (Four Seasons Medley) Gidea Park featuring
Adrian Baker Profile 5003 23Jan82 82(1) 3 A: Gidea Park; P: Adrian Baker — 36.7

3868. CAN'T HOLD BACK (Your Loving) Kano Mirage 3878 26Dec81 89(2) 5
[Barbara Addoms/Anna Cantù/Luciano Ninzatti/Stefano Pulga]
A: New York Cake; P: Matteo Bonsanto/Luciano Ninzatti/Stefano Pulga — 36.7

3869. LET'S DO SOMETHING CHEAP AND SUPERFICIAL Burt Reynolds
MCA 51004 18Oct80 88(1) 5
[Richard Levinson] A: Smokey And The Bandit 2 Soundtrack; P: Snuff Garrett — 36.7

3870. REAL PEOPLE//CHIP OFF THE OLD BLOCK [TSW] Chic Atlantic 3768
22Nov80 79(2) 2
[Bernard Edwards/Nile Rodgers //Bernard Edwards/Nile Rodgers]
A: Real People; P: Bernard Edwards/Nile Rodgers — 36.6

3871. ON AND ON AND ON ABBA Atlantic 3826 27Jun81 90(1) 6
[Benny Andersson/Björn Ulvaeus] A:Super Trouper; P:Benny Andersson/BjörnUlvaeus — 36.6

3872. GOIN' ON The Beach Boys Caribou 9032 12Apr80 83(1) 3
[Mike Love/Brian Wilson] A: Keepin' The Summer Alive; P: Bruce Johnston — 36.6

3873. I'M BAD LL Cool J Def Jam 07120 11Jul87 84(1) 4
[Bobby Ervin/Dwayne Simon/James Todd Smith]
A: Bigger And Deffer; P: Darryl Pierce/Dwayne Simon — 36.2

3874. MASQUERADE Berlin Geffen 29504 24Sep83 82(1) 3
[Chris Ruiz-Velasco] A: Pleasure Victim; P: Daniel R. Van Patten — 36.2

3875. ALLIES Heart Epic 04184 29Oct83 83(1) 4
[Jonathan Cain] A: Passionworks; P: Keith Olsen — 36.0

3876. SHOOTING SHARK Blue Öyster Cult Columbia 04298
11Feb84 83(1) 3 [Donald Roeser/Patti Smith]
A: The Revolution By Night; P: Bruce Fairbairn — 36.0

3877. SPEND THE NIGHT IN LOVE The Four Seasons Warner/Curb 49597
13Dec80 91(3) 5 [Bob Gaudio/Lenny Lee Goldsmith/Judy Parker]
A: Reunited Live; P: Charles Callelo/Bob Gaudio — 36.0

3878. SNAP SHOT Slave Cotillion 46022 24Oct81 91(2) 7
[Mark Adams/Steve Arrington/Charles Carter/Jimmy Douglass/Floyd Miller]
A: Show Time; P: Jimmy Douglass — 35.9

3879. THE SAME LOVE The Jets MCA 53734 21Oct89 87(1) 6
[Diane Warren] A: Believe; P: Don Powell/David Rivkin — 35.9

3880. I LIKE Men Without Hats MCA 52293 12Nov83 84(1) 3
[Ivan Doroschuk] A: Rhythm Of Youth; P: Marc Durand — 35.9

3881. ENGLISHMAN IN NEW YORK Sting A&M 1200 16Apr88 84(2) 4
[Sting] A: ...Nothing Like The Sun; P: Neil Dorfsman/Sting — 35.8

3882. THAT'S LIFE David Lee Roth Warner Bros. 28511 22Nov86 85(1) 4
[Kelly Gordon/Dean Kay] A: Eat 'Em And Smile; P: Ted Templeman — 35.7

3883. YOU CAN CALL ME BLUE Michael Johnson EMI America 8054
23Aug80 86(1) 3 [Lawrence Anthony Brown/David Morgan]
A: You Can Call Me Blue; P: Steve Gibson/Brent Maher — 35.7

3884. GOTTA GIVE A LITTLE LOVE (Ten Years After) Timmy Thomas Gold
Mountain 82004 02Jun84 80(1) 3
[Timmy Thomas] A: Gotta Give A Little Love (Ten Years After); P: Lou Pace — 35.6

3885. WANGO TANGO Ted Nugent Epic 50907 26Jul80 86(1) 4
[Ted Nugent] A: Scream Dream; P: Cliff Davies — 35.6

3886. RUNNING IN THE FAMILY Level 42 Polydor 885957 08Aug87 83(1) 4
[Wally Badarou/Phil Gould/Mark King]
A: Running In The Family; P: Wally Badarou/Level 42 — 35.6

3887. ROUGH BOYS Pete Townshend Atco 7318 15Nov80 89(2) 4
[Pete Townshend] A: Empty Glass; P: Chris Thomas — 35.4

3888. RIGHT BACK WHERE WE STARTED FROM Sinitta Atlantic 88807
23Sep89 84(2) 4 [J. Vincent Edwards/Pierre Tubbs]
A: Wicked; P: Pete Hammond — 35.4

3889. THEME FROM "RAGING BULL" (Cavalleria Rusticana) Joel Diamond
Motown 1504 21Feb81 82(1) 3 [Pietro Mascagni]
A: Raging Bull Soundtrack; P: Joel Diamond/Harold Wheeler — 35.4

3890. I LOST ON JEOPARDY "Weird Al" Yankovic Rock 'n' Roll 04469
30Jun84 81(1) 3 [Greg Kihn/Steve Wright/Al Yankovic]
A: "Weird Al" Yankovic In 3-D; P: Rick Derringer — 35.1

3891. BATTLESHIP CHAINS Georgia Satellites Elektra 69497
28Mar87 86(1) 5 [Terry Anderson] A: Georgia Satellites; P: Jeff Glixman — 35.1

3892. I HEARD IT THROUGH THE GRAPEVINE The California Raisins
Priority 9719 23Jan88 84(1) 4 [Barrett Strong/Norman Whitfield]
A: The California Raisins; P: Ross Vannelli — 35.0

3893. NOW AND FOREVER (You And Me) Anne Murray Capitol 5547
01Mar86 92(1) 6 [David Foster/Randy Goodrum/Jim Vallance]
A: Something To Talk About; P: David Foster — 34.9

3894. DANCING IN THE SHADOWS After The Fire Epic 03908
28May83 85(1) 3 [Peter Douglas Cameron Banks/Andrew Piercy]
A: ATF; P: John Eden/Andy Piercy — 34.9

3895. DON'T BE MY ENEMY Wang Chung Geffen 29193 22Sep84 86(2) 3
[Darren Costin/Nick Feldman/Jack Hues]
A: Points On The Curve; P: Ross Cullum/Chris Hughes — 34.9

3896. TAXI J. Blackfoot Sound Town 0004 03Mar84 90(1) 5
[Homer Banks/Chuck Brooks] A: City Slicker; P: Homer Banks/Chuck Brooks — 34.6

3897. COUNT YOUR BLESSINGS Ashford and Simpson Capitol 5598
27Sep86 84(1) 4 [Nick Ashford/Valerie Simpson]
A: Real Love; P: Nick Ashford/Valerie Simpson — 34.5

3898. THE JAM WAS MOVING Debbie Harry Chrysalis 2554 31Oct81 82(1) 3
[Bernard Edwards/Nile Rodgers] A: KooKoo; P: Bernard Edwards/Nile Rodgers — 34.4

3899. DO YA DO YA (Wanna Please Me) Samantha Fox Jive 1031
14Mar87 87(2) 5 [Michael Bissell/Graham Richardson]
A: Touch Me; P: Steve Lovell/Steve Power — 34.3

3900. EMOTION Barbra Streisand Columbia 04707 09Mar85 79(1) 2
[Peter S. Bliss] A: Emotion; P: Richard Perry — 34.3

3901. POWER Kansas MCA 53027 14Feb87 84(1) 4
[Randy Goodrum/Steve Morse/Steve Walsh] A: Power; P: Andrew Powell — 34.3

3902. MEMPHIS Joe Jackson A&M 2601 26Nov83 85(1) 4
[Joe Jackson] A: Mike's Murder (Soundtrack); P: Joe Jackson — 34.2

3903. DANCING IN MY SLEEP Secret Ties Night Wave 9201
13Dec86 91(2) 5 [Brian Soares] A: All Through The Night; P: Gerry Caples — 34.0

3904. SENDIN' ALL MY LOVE The Jets MCA 53380 13Aug88 88(1) 4
[Stephen Bray/Linda Mallah] A: Magic; P: Rick Kelly/Michael Verdick — 33.8

3905. SHE'S ONLY 20 Tami Show Chrysalis 43146 12Mar88 88(1) 4
[Tommy Gawenda/Cathy Massey/Claire Massey]
A: Tami Show; P: Mike Chapman — 33.5

3906. LET ME DOWN EASY Roger Daltrey Atlantic 89471 28Dec85 86(1) 4
[Bryan Adams/Jim Vallance] A: Under A Raging Moon; P: Alan Shacklock — 33.2

3907. GOOD FRIENDS Joni Mitchell Geffen 28840 28Dec85 85(1) 3
[Joni Mitchell]
A: Dog Eat Dog; P: Thomas Dolby/Larry Klein/Joni Mitchell/Mike Shipley — 33.2

3908. HEART AND SOUL The Monkees Rhino 74408 26Sep87 87(1) 4
[Simon Byrne/Andy Howell] A: Pool It!; P: Roger Bechirian — 33.2

3909. LOVE LIGHT Yutaka Alfa 7004 18Jul81 88(1) 3
[Marti McCall/Yutaka Yokokura] A: Love Light; P: Dave Grusin/Larry Rosen — 33.1

3910. VIDEO! Jeff Lynne Virgin/Epic 04570 18Aug84 85(1) 3
[Jeff Lynne] A: Electric Dreams Soundtrack; P: Jeff Lynne — 33.1

3911. RAIN Dragon Polydor 817292 18Aug84 88(2) 4
[Marc Hunter/Todd Hunter/Johanna Pigott]
A: Body And The Beat; P: Alan Mansfield — 33.1

3912. DESTROYER The Kinks Arista 0619 31Oct81 85(1) 4
[Ray Davies] A: Give The People What They Want; P: Ray Davies — 32.7

3913. SING A SIMPLE SONG West Street Mob Sugar Hill 576
17Apr82 89(1) 4 [Sylvester Stewart]
A: Break Dance-Electric Boogie; P: Cheryl Cook/Joey Robinson Jr. — 32.7

3914. IMAGINE Tracie Spencer Capitol 44268 11Feb89 85(1) 4
[John Lennon] A: Tracie Spencer; P: Ollie Brown — 32.7

3915. DON'T TALK Larry Lee Columbia 02740 26Jun82 81(1) 2
[Terry Britten/Sue Shifrin] A: Marooned; P: John Ryan — 32.6

3916. YES OR NO Go-Go's I.R.S. 9933 22Sep84 84(1) 3
[Ron Mael/Russell Mael/Jane Wiedlin] A: Talk Show; P: Martin Rushent — 32.4

3917. SISTERS OF THE MOON Fleetwood Mac Warner Bros. 49500 07Jun80
86(1) 3 [Stevie Nicks] A: Tusk; P: Ken Caillat/Richard Dashut/Fleetwood Mac — 32.4

3918. RUNNING Chubby Checker MCA 51233 20Feb82 91(3) 5
[Joe Russo] A: The Change Has Come; P: Evan Pace — 32.3

3919. STRANGER IN MY HOME TOWN Foghat Bearsville 49510
02Aug80 81(1) 3 [Dave Peverett]
A: Tight Shoes; P: Don Berman/Foghat/Tony Outeda — 32.3

3920. LOVE'S ONLY LOVE Engelbert Epic 50844 15Mar80 83(1) 2
[Paul Ryan] A: Love's Only Love; P: Joel Diamond — 32.3

3921. AIN'T NOTHING LIKE THE REAL THING/YOU'RE ALL I NEED TO GET BY
Chris Christian Boardwalk 7-11-149 28Aug82 88(3) 3
[Nick Ashford/Valerie Simpson// Nick Ashford/Valerie Simpson]
A: Chris Christian; P: Bob Gaudio — 32.2

3922. VERY SPECIAL Debra Laws Elektra 47142 15Aug81 90(1) 5
[William Jeffery/Lisa Peters]
A: Very Special; P: Hubert Laws/Ronnie Laws — 32.1

Rank. TITLE Act Label Number Entry Date Peak(Wks) Tot Wks [Writers] A: Proximate Album; P: Producer	Score
3923. COME TO ME Aretha Franklin Arista 0600 30May81 84(1) 3 [Willard Price] A: Aretha; P: Arif Mardin	32.0
3924. HOT WATER Level 42 Polydor 885155 26Jul86 87(2) 4 [Wally Badarou/Phil Gould/Mark King/Mike Lindup] A: A Physical Presence; P: Ken Scott	31.9
3925. IMAGINATION Belouis Some Capitol 5464 04May85 88(1) 5 [Belouis Some] A: Some People; P: Michael Barbiero/Steve Thompson	31.7
3926. EACH WORD'S A BEAT OF MY HEART Mink DeVille Atlantic 89750 11Feb84 89(1) 4 [Willy DeVille] A: Each Song Is A Beat OF My Heart; P: Howard Albert/Ron Albert	31.6
3927. FREAK-A-RISTIC Atlantic Starr A&M 2718 25May85 90(2) 6 [David E. Lewis] A: As The Band Turns; P: David Lewis/Wayne Lewis	31.5
3928. FOOLISH HEART Sharon Bryant Wing 889 878 16Dec89 90(1) 5 [Randy Goodrum/Steve Perry] A: Here I Am; P: Sharon Bryant/Rick Gallwey	31.4
3929. STRAP ME IN The Cars Elektra 69427 14Nov87 85(1) 4 [Ric Ocasek] A: Door To Door; P: Ric Ocasek	31.3
3930. JANET Commodores Motown 1802 21Sep85 87(1) 4 [Bobby Caldwell/Paul Fox/Franne Golde] A: Nightshift; P: Dennis Lambert	31.3
3931. UNCHAINED MELODY (Live) Heart Epic 51010 28Mar81 83(1) 3 [Alex North/Hy Zaret] A: Greatest Hits/Live; P: Heart	31.2
3932. A MILLION MILES AWAY The Plimsouls Geffen 29600 30Jul83 82(1) 3 [Joey Alkes/Peter Case/Chris Fradkin] A: Everywhere At Once; P: Jeff Eyrich	31.2
3933. WORLD SHUT YOUR MOUTH Julian Cope Island 99479 21Mar87 84(1) 4 [Julian Cope] A: St. Julian; P: Ed Stasium	31.2
3934. WHEN YOU WERE MINE Mitch Ryder Riva 213 16Jul83 87(1) 3 [Prince] A: Never Kick A Sleeping Dog; P: John Mellencamp	31.0
3935. DON'T WAIT FOR HEROES Dennis DeYoung A&M 2692 08Dec84 83(1) 4 [Dennis DeYoung] A: Desert Moon; P: Dennis DeYoung	30.9
3936. POP SONG 89 R.E.M. Warner Bros. 27640 10Jun89 86(1) 4 [Bill Berry/Peter Buck/Mike Mills/Michael Stipe] A: Green; P: Scott Litt/R.E.M.	30.8
3937. STAIRWAY TO HEAVEN Far Corporation Atco 99509 04Oct86 89(1) 4 [Jimmy Page/Robert Plant] A: Division One-The Album; P: Frank Farian	30.7
3938. THE BOY IN THE BUBBLE Paul Simon Warner Bros. 28460 07Mar87 86(1) 4 [Forere Motloheloa/Paul Simon] A: Graceland; P: Paul Simon	30.7
3939. YOU ARE THE ONE TKA Warner Bros. 22946 03Jun89 91(1) 7 [David Gaskins/Keath Lowry] A: Louder Than Love; P: Joey Gardner/Tony Moran	30.6
3940. WAITING GAME Swing Out Sister Fontana 874 190 03Jun89 86(1) 4 [Andy Connell/Corinne Drewery] A: Kaleidoscope World; P: Paul Staveley O'Duffy	30.6
3941. I BELIEVE IN YOU Stryper Enigma 75028 05Nov88 88(1) 5 [Michael Sweet] A: In God We Trust; P: Michael Lloyd/Stryper	30.4
3942. WHERE DID WE GO WRONG Frankie Valli Introducing Chris Forde MCA/Curb 41253 19Jul80 90(1) 4 [Richard Kerr/Marty Panzer] A: Heaven Above Me; P: Bob Gaudio	30.4
3943. NEVER TELL AN ANGEL (When Your Heart's On Fire) The Stompers Boardwalk 177 18Jun83 88(1) 4 [Sal Baglio] A: The Stompers; P: Ritchie Cordell/Glen Kolotkin	30.3
3944. SUPERSTAR/UNTIL YOU COME BACK TO ME (That's What I'm Gonna Do) Luther Vandross Epic 04969 28Apr84 87(1) 4 [Bonnie Bramlett/ Delaney Bramlett//Morris Broadnax/Clarence Paul/Stevie Wonder] A: Busy Body; P: Luther Vandross	30.3
3945. TOUCH THE FIRE Icehouse Chrysalis 23414 07Oct89 84(1) 4 [Iva Davies] A: Great Southern Land; P: David Lord	30.1
3946. SO MUCH FOR LOVE The Venetians Chrysalis 43056 28Feb87 88(1) 5 [Rik Swinn] A: Calling In The Lions; P: Mark Opitz	30.1
3947. WANTED MAN Ratt Atlantic 89618 06Oct84 91(1) 3 [Joe Cristofanilli/Robbin Crosby/Stephen Pearcy] A: Out Of The Cellar; P: Beau Hill	30.0
3948. HOLIDAY Nazareth A&M 2219 29Mar80 87(1) 3 [Pete Agnew/Jeff Baxter/Manny Charlton/Alistair Cleminson/ Dan McCafferty/Darrell Sweet] A: Malice In Wonderland; P: Jeff Baxter	29.9
3949. SMALL PARADISE John Cougar Riva 203 16Feb80 87(1) 3 [John Mellencamp] A: John Cougar; P: Howard Albert/Ron Albert	29.8
3950. I'M GONNA LOVE HER FOR BOTH OF US Meat Loaf Cleveland International/Epic 02490 19Sep81 84(1) 3 [Jim Steinman] A: Dead Ringer; P: Stephan Galfas/Meat Loaf	29.6
3951. BONGO BONGO Steve Miller Band Capitol 5442 16Feb85 84(1) 3 [Chris McCarty/Steve Miller] A: Italian X Rays; P: Byron Allred/ Kenny Lee Lewis/Gary Mallaber	29.6
3952. LITTLE SHEILA Slade CBS Associated 04865 04May85 86(1) 3 [Noddy Holder/Jim Lea] A: Rogues Gallery; P: John Punter	29.4
3953. MASTER AND SERVANT Depeche Mode Sire 28918 07Sep85 87(2) 3 [Martin L. Gore] A: Some Great Reward; P: Depeche Mode/Gareth Jones/Daniel Miller	29.3
3954. EVERY LOVE SONG Greg Kihn Band Beserkley 47441 17Jul82 82(1) 2 [Dave Carpender/Greg Kihn/Larry Lynch/Gary Philippet/Steve Wright] A: Kihntinued; P: Matthew King Kaufman	29.3
3955. SAY GOODBYE TO LITTLE JO Steve Forbert Nemperor 7529 12Apr80 85(1) 3 [Steve Forbert] A: Jackrabbit Slim; P: John Simon	29.2
3956. LIVING IN THE BACKGROUND Baltimora Manhattan 50029 12Apr86 87(1) 4 [Maurizio Bassi/Naimy Hackett] A: Living In The Background; P: Maurizio Bassi	29.2
3957. SHOCK The Motels Capitol 5529 26Oct85 84(1) 3 [Martha Davis/Scott Thurston] A: Shock; P: Richie Zito	28.9
3958. MAKE IT MEAN SOMETHING Rob Jungklas Manhattan 50054 07Feb87 86(1) 3 [Chad Cromwell/Jack Holder/Rob Jungklas] A: Closer To The Flame; P: William Wittman	28.9
3959. LOVE IN SIBERIA Laban Critique 725 08Nov86 88(2) 4 [Ivan Pedersen] A: Caught By Surprise; P: Cai Leitner	28.7
3960. LET'S DANCE (Make Your Body Move) West Street Mob Sugar Hill 559 12Sep81 88(1) 3 [Dan Brewster/Bruce Carter/Sherman Davis/Wayne Henderson/Donald Hepburn/Augie Johnson/Marlon McClain/Jerry Peters/Nathaniel Phillips/ Bruce A. Smith/Dennis Springer] A: West Street Mob; P: Joey Robinson Jr.	28.6
3961. I'M THROUGH WITH LOVE Eric Carmen Geffen 29032 20Apr85 87(2) 3 [Eric Carmen] A: Eric Carmen (II); P: Bob Gaudio	28.6
3962. I LIKE TO ROCK April Wine Capitol 4828 09Feb80 86(1) 3 [Myles Goodwyn] A: Harder...Faster; P: Nick Blagona/Myles Goodwyn	28.6
3963. WHEN THINGS GO WRONG Robin Lane & The Chartbusters Warner Bros. 49246 12Jul80 87(2) 3 [Joanne Cipolla/Robin Lane] A: Robin Lane & The Chartbusters; P: Joe Wissert	28.5
3964. I STILL WANT YOU The Del Fuegos Slash 28822 31May86 87(2) 4 [Tom Lloyd/Dan Zanes] A: Boston, Mass.; P: Mitchell Froom	28.2
3965. BACK OF MY HAND (I've Got Your Number) The Jags Island 49202 07Jun80 84(1) 2 [John Alder/Nicholas Watkinson] A: Evening Standards; P: Jon Astley/Phil Chapman	28.0
3966. IT'S JUST THE SUN Don McLean Millennium 11809 08Aug81 83(1) 2 [Don McLean] A: Chain Lightning; P: Larry Butler	27.8
3967. BRING IT ALL BACK Grayson Hugh RCA 9093 28Oct89 87(1) 4 [Grayson Hugh] A: Blind To Reason; P: Michael Baker/Axel Kroell	27.7
3968. MY OBSESSION Icehouse Chrysalis 43240 09Jul88 88(1) 4 [Iva Davies/Robert Kretschmer] A: Man Of Colours; P: David Lord	27.7
3969. SHELTER ME Joe Cocker Capitol 5557 08Mar86 91(2) 4 [Nicholas DiStefano] A: Cocker; P: Terry Manning	27.6
3970. TOCCATA Sky Arista 0568 10Jan81 83(1) 2 [Johann Sebastian Bach] A: Sky 2; P: Haydn Bendall/Tony Clarke/Sky	27.5
3971. THEME FROM S-EXPRESS S-Express Capitol 44181 18Jun88 91(1) 6 [Pascal Gabriel/Miles Gregory/Mark Moore/Matt Noble/Thomassina Smith] A: Original Soundtrack; P: Pascal Gabriel/Mark Moore	27.3
3972. COUNT ON ME Gerard McMahon Full Moon 29699 09Apr83 85(1) 3 [Gerard McMahon] A: No Looking Back; P: Gerard McMahon/Michael Ostin	27.3
3973. REV IT UP Newcity Rockers Critique 7-99437 12Sep87 86(1) 4 [Robert Ernlund/Anders Wikström/Gregg Winter] A: Newcity Rockers; P: Cliff Goodwin/Bob Rivers	27.2
3974. LOVE ON MY MIND TONIGHT The Temptations Gordy 1666 16Apr83 88(2) 3 [Peter Beckett/Dennis Lambert] A: Surface Thrills; P: Steve Barri/Dennis Lambert	27.0
3975. TRUST ME Cindy Bullens Casablanca 2217 12Jan80 90(1) 3 [Cindy Bullens] A: Steal The Night; P: Cindy Bullens/Mark Doyle	27.0
3976. CROSSROADS Tracy Chapman Elektra 69273 28Oct89 90(2) 4 [Tracy Chapman] A: Crossroads; P: David Kershenbaum	26.9
3977. I LOVE WOMEN Jim Hurt Scotti Brothers 605 11Oct80 90(2) 4 [Michael D. Stewart/Dan Edward Williams] A: No Album; P: Bob Montgomery/Johnny Slate	26.9
3978. YOUR PERSONAL TOUCH Evelyn "Champagne" King RCA 14201 11Jan86 86(1) 4 [Allen George/Fred McFarlane] A: A Long Time Coming; P: Allen George/Fred McFarlane	26.7
3979. ANGEL IN MY POCKET One To One Warner Bros. 28739 23Aug86 92(2) 4 [Leslie Howe/Louise Reny] A: Forward Your Emotions; P: Leslie Howe	26.7
3980. I'M FOR REAL Howard Hewett Elektra 69527 08Nov86 90(3) 3 [Stanley Clarke/Howard Hewett] A: I Commit To Love; P: Stanley Clarke/Howard Hewett	26.3

Ranking the Singles

Rank. TITLE Act Label Number Entry Date Peak(Wks) Tot Wks [Writers] A: Proximate Album; P: Producer	Score
3981. EVERYTHING IS ALRIGHT Spider Dreamland 103 02Aug80 86(1) 3 [Holly Knight] A: Spider; P: Peter Coleman	26.2
3982. LIFE'S WHAT YOU MAKE IT Talk Talk EMI America 8303 01Feb86 90(2) 4 [Tim Friese-Greene/Mark Hollis] A: The Colour Of Spring; P: Tim Friese-Greene	26.1
3983. BE MINE (Tonight) Grover Washington, Jr. Elektra 47246 06Feb82 92(2) 4 [William Eaton/Ralph MacDonald/William Salter] A: Come Morning; P: Grover Washington Jr.	25.9
3984. WINTER GAMES David Foster Atlantic 89140 27Feb88 85(1) 3 [David Foster] A: The Symphony Sessions; P: David Foster	25.9
3985. GO SEE THE DOCTOR Kool Moe Dee Jive/Rooftop 1041 18Apr87 89(1) 5 [Mohandas DeWese] A: Kool Moe Dee; P: Mohandas DeWese/LaVaba Mallison/Teddy Riley	25.9
3986. LAND OF LA LA Stevie Wonder Tamla 1846 14Jun86 86(1) 3 [Stevie Wonder] A: In Square Circle; P: Stevie Wonder	25.8
3987. SHARING THE LOVE Rufus With Chaka Khan MCA 51203 05Dec81 91(2) 5 [Kevin Murphy] A: Camouflage; P: Rufus	25.8
3988. DON'T WALK AWAY Robert Tepper Scotti Brothers 05879 10May86 85(1) 3 [Robert Tepper] A: No Easy Way Out; P: Joe Chiccarelli	25.8
3989. VANITY KILLS ABC Mercury 884714 17May86 91(2) 4 [Martin Fry/Mark White] A: How To Be A...Zillionaire!; P: Martin Fry/Mark White	25.7
3990. FLY AWAY Stevie Woods Cotillion 47006 15May82 84(1) 2 [Peter Allen/David Foster/Carole Bayer Sager] A: Take Me To Your Heaven; P: Jack White	25.6
3991. FLASHES Tiggi Clay Morocco 1716 25Feb84 86(1) 3 [Romeo McCall/Billy Peaches/Fizzy Qwick] A: Tiggi Clay; P: Tiggi Clay	25.2
3992. IT'S NOT YOU, IT'S NOT ME KBC Band Arista 1-9526 29Nov86 89(1) 4 [Phil Brown/Van Stephenson] A: KBC Band; P: John Boylan/Jim Gaines/KBC Band	25.0
3993. 777-9311 The Time Warner Bros. 29952 09Oct82 88(1) 3 [Prince] A: What Time Is It?; P: Morris Day/Prince	24.9
3994. LONELY NIGHTS Bryan Adams A&M 2359 13Mar82 84(1) 2 [Bryan Adams/Jim Vallance] A: You Want It, You Got It; P: Bryan Adams/Bob Clearmountain	24.8
3995. LOVING YOU Chris Rea Columbia 02727 10Apr82 88(2) 3 [Chris Rea] A: Chris Rea; P: Jon Kelly/Chris Rea	24.7
3996. COMING TO AMERICA (Part One) The System Atco 99320 09Jul88 91(1) 5 [Nancy Huang/Nile Rodgers] A: Coming To America Soundtrack; P: David Frank/Mic Murphy/Nile Rodgers	24.7
3997. AM I FORGIVEN Isle Of Man Pasha 05900 09Aug86 90(2) 4 [Ron Aniello/Robert Parlee] A: Isle Of Man; P: Isle Of Man/Spencer Proffer	24.6
3998. WONDERLAND Big Country Mercury 818834 02Jun84 86(1) 2 [William Stuart Adamson/Mark Brzezicki/Anthony Earl Butler/Bruce William Watson] A: Wonderland; P: Steve Lillywhite	24.6
3999. JUST LIKE PARADISE Larry John McNally ARC 02200 08Aug81 86(2) 2 [Larry John McNally] A: Larry John McNally; P: Jon Lind	24.6
4000. WHERE'S YOUR ANGEL? Lani Hall A&M 2305 21Mar81 88(1) 3 [Greg Phillinganes/Allee Willis] A: Blush; P: Greg Phillinganes/Allee Willis	24.5
4001. NO FRILLS LOVE Jennifer Holliday Geffen 28845 08Feb86 87(1) 3 [Arthur Baker/Tina Baker/Gary Henry] A: Say You Love Me; P: Arthur Baker	24.4
4002. FOUR LITTLE DIAMONDS ELO Jet 04130 01Oct83 86(1) 2 [Jeff Lynne] A: Secret Messages; P: Jeff Lynne	24.3
4003. LOVE'S ABOUT TO CHANGE MY HEART Donna Summer Atlantic 88840 16Sep89 85(1) 3 [Matt Aitken/Mike Stock/Pete Waterman] A: Another Place And Time; P: Matt Aitken/Mike Stock/Pete Waterman	24.1
4004. LOVE AND ROCK AND ROLL Greg Kihn EMI America 8306 29Mar86 92(1) 5 [Greg Kihn] A: Love And Rock And Roll; P: Matthew King Kaufman	24.1
4005. LOVE ON THE PHONE Suzanne Fellini Casablanca 2242 15Mar80 87(1) 2 [Johanna Deneroff/Suzanne Fellini/David Sonenberg/Jeff Waxman] A: Suzanne Fellini; P: Steve Burgh	24.0
4006. THAT WAS THEN BUT THIS IS NOW ABC Mercury 814631 04Feb84 89(1) 3 [Martin Fry/Stephen Singleton/Mark White] A: Beauty Stab; P: ABC/Gary Langan	24.0
4007. IS IT LOVE J.J. Fad Ruthless 7-99257 10Dec88 92(2) 5 [Juana Burns/Michelle Franklin] A: Supersonic--The Album; P: Antoine Carraby/Dr. Dre	24.0
4008. I FEEL FREE Belinda Carlisle MCA 53377 30Jul88 88(1) 4 [Peter Ronald Brown/Jack Bruce] A: Heaven On Earth; P: Rick Nowels	23.8
4009. PRETTY BOYS AND PRETTY GIRLS Book Of Love Sire 27858 03Sep88 90(2) 4 [Ted Ottaviano] A: Lullaby; P: Mark Ellis/Ted Ottaviano	23.8
4010. TELL THAT GIRL TO SHUT UP Transvision Vamp Uni 50001 01Oct88 87(1) 3 [Holly Beth Vincent] A: Pop Art; P: Bernd Held	23.7
4011. BEAT STREET BREAKDOWN-Part 1 Grandmaster Melle Mel & The Furious Five Atlantic 89659 04Aug84 86(1) 2 [Melvin Glover/Reggie Griffin] A: Beat Street Soundtrack; P: Melvin Glover/Sylvia Robinson	23.5
4012. I CAN'T SAY GOODBYE TO YOU Helen Reddy MCA 51106 23May81 88(1) 3 [Becky Hobbs] A: Play Me Out; P: Joel Diamond	23.3
4013. THE RAMBLER Molly Hatchet Epic 50965 07Mar81 91(2) 3 [Jimmy Farrar/Dave Hlubek] A: Beatin' The Odds; P: Tom Werman	23.3
4014. TAKIN' IT BACK Breathless EMI America 8020 12Jan80 92(2) 4 [Jonah Koslen] A: Breathless; P: Don Gehman	23.3
4015. TAKE AWAY Big Ric Scotti Brothers 04084 03Sep83 91(2) 3 [Kevin DiSimone/Bud Harner/John Pondel] A: Big Ric; P: John D'Andrea/Carmine Rubino	23.2
4016. UNFAITHFULLY YOURS (One Love) Stephen Bishop Warner Bros. 29345 31Mar84 87(1) 3 [Stephen Bishop] A: No Album; P: Greg Mathieson	23.0
4017. I'M NEVER GONNA SAY GOODBYE Billy Preston Motown 1625 25Sep82 88(1) 3 [Artie Butler/Molly Ann Leikin] A: Pressin' On; P: Artie Butler	22.9
4018. IF I HAD A ROCKET LAUNCHER Bruce Cockburn Gold Mountain 82013 09Feb85 88(1) 3 [Bruce Cockburn] A: Stealing Fire; P: Kerry Crawford/John Goldsmith	22.9
4019. THE MOTION OF LOVE Gene Loves Jezebel Geffen 28183 06Feb88 87(1) 3 [Jay Aston/Michael Aston/Peter Risingham/James Stevenson] A: The House Of Dolls; P: Jimmy Iovine	22.7
4020. FRIENDS//FIVE MINUTES OF FUNK [TSW] Whodini Jive 9276 05Jan85 87(1) 3 [Jalil Hutchins/Lawrence Smith//John Fletcher/Jalil Hutchins/Larry Smith] A: Escape; P: Larry Smith	22.6
4021. LOVE ON THE AIRWAVES Night Planet 47921 21Feb81 87(1) 3 [Chris Thompson/Robert Watson] A: Long Distance; P: Tim Friese-Greene	22.5
4022. LADY, LADY, LADY Joe "Bean" Esposito Casablanca 814430 22Oct83 86(1) 2 [Keith Forsey/Giorgio Moroder] A: Solitary Men; P: Giorgio Moroder	22.5
4023. LIVIN' RIGHT Glenn Frey MCA 53497 18Mar89 90(1) 4 [Glenn Frey/Jack Tempchin] A: Soul Searching; P: Glenn Frey/Elliot Scheiner	22.4
4024. ONCE IN A LIFETIME Talking Heads Sire 29163 19Apr86 91(1) 4 [David Byrne/Brian Eno/Chris Frantz/Jerry Harrison/Tina Weymouth] A: Stop Making Sense; P: Talking Heads	22.3
4025. LEFT TO MY OWN DEVICES Pet Shop Boys EMI-Manhattan 50171 28Jan89 84(1) 3 [Chris Lowe/Neil Tennant] A: Introspective; P: Trevor Horn/Stephen Lipson	22.3
4026. HANDS ON THE RADIO Henry Lee Summer CBS Associated 07986 17Sep88 85(1) 3 [Henry Lee Summer] A: Henry Lee Summer; P: Michael Frondelli	21.9
4027. ONE IN A MILLION Eddie And The Tide Atco 99617 21Sep85 85(1) 2 [Eddie Rice] A: Go Out And Get It; P: Bobby Corona	21.7
4028. LUCKY Eye To Eye Warner Bros. 29455 29Oct83 88(1) 2 [Deborah Berg/Julian Marshall] A: Shakespeare Stole My Baby; P: Gary Katz	21.6
4029. CHEAP SUNGLASSES ZZ Top Warner Bros. 49220 12Jul80 89(2) 2 [Frank Beard/Billy Gibbons/Dusty Hill] A: Deguello; P: Bill Ham	21.4
4030. BACK TO THE 60's (Medley) Tight Fit Arista 0638 10Oct81 89(2) 3 A: Back To The '60s; P: Ken Gold	21.4
4031. WOOD BEEZ (Pray Like Aretha Franklin) Scritti Politti Warner Bros. 28811 08Feb86 91(2) 4 [Green Gartside] A: Cupid And Psyche 85; P: Arif Mardin	21.2
4032. FALLING OUT OF LOVE Ivan Neville Polydor 871484 25Feb89 91(2) 3 [Ivan Neville] A: If My Ancestors Could See Me Now; P: Danny Kortchmar	21.0
4033. SUCH A SHAME Talk Talk EMI America 8215 30Jun84 89(1) 3 [Mark Hollis] A: It's My Life; P: Tim Friese-Greene	20.9
4034. THE LONGER YOU WAIT Gino Vannelli Arista 0664 06Mar82 89(1) 3 [Gino Vannelli] A: No Album; P: Gino Vannelli/Joe Vannelli/Ross Vannelli	20.8
4035. MONTEGO BAY Amazulu Mango 121 25Jul87 90(1) 4 [Jeff Barry/Bobby Bloom] A: Amazulu; P: Andy Hill	20.5
4036. LOVE, TRUTH AND HONESTY Bananarama London 886362 26Nov88 89(1) 3 [Matt Aitken/Sarah Dallin/Jacquie O'Sullivan/Mike Stock/Pete Waterman/Keren Woodward] A: Greatest Hits Collection; P: Matt Aitken/Mike Stock/Pete Waterman	20.4
4037. WITHOUT YOU (Love Theme From 'Leonard Part 6') Peabo Bryson & Regina Belle Elektra 69426 06Feb88 89(1) 3 [Lamont Dozier] A: Leonard Part 6 Soundtrack; P: Dean Gant/Michael J. Powell	20.4
4038. FOLLOW YOUR HEART Triumph MCA 52540 09Mar85 88(1) 2 [Rik Emmett/Mike Levine/Gil Moore] A: Thunder Seven; P: Eddie Kramer/Triumph	20.4

Ranking the Singles

Rank. TITLE Act Label Number Entry Date Peak(Wks) Tot Wks
[Writers] *A: Proximate Album; P: Producer* Score

4039. PIECE BY PIECE The Tubes Capitol 5443 09Mar85 87(1) 2
[Rick Anderson/Michael Cotten/Charles Prince/Todd Rundgren/
Tom Snow/William Spooner/Roger Steen/Fee Waybill/Vince Welnick]
A: Love Bomb; P: Todd Rundgren 20.4

4040. LOVERBOY Karen Kamon Columbia 04474 28Jul84 88(1) 2
[Billy Alessi/Bobby Alessi] *A: Heart Of You; P: Phil Ramone* 20.3

4041. SIMILAR FEATURES Melissa Etheridge Island 99251 08Apr89 94(3) 4
[Melissa Etheridge] *A: Melissa Etheridge;*
P: Niko Bolas/Melissa Etheridge/Craig Krampf/Kevin McCormick 20.1

4042. FAMILY MAN Fleetwood Mac Warner Bros. 28114 02Apr88 90(1) 4
[Lindsey Buckingham/Richard Dashut]
A: Tango In The Night; P: Lindsey Buckingham/Richard Dashut 20.1

4043. INFORMATION Eric Martin Capitol 5502 24Aug85 87(1) 2
[Tony Fanucchi/Randall Darnell Jackson/Michael Mani/Eric Martin]
A: Eric Martin; P: Danny Kortchmar/Greg Ladanyi 19.8

4044. WHEN YOU WALK IN THE ROOM Paul Carrack Chrysalis 43252
18Jun88 90(1) 3 [Jackie DeShannon] *A: One Good Reason; P: Christopher Neil* 19.7

4045. LET'S GO OUT TONIGHT Nile Rodgers Warner Bros. 29049
01Jun85 88(1) 3 [Nile Rodgers] *A: B-Movie Matinee;*
P: Tommy Jymi/Nile Rodgers 19.7

4046. STEPPIN' OUT Kool & The Gang De-Lite 816 13Feb82 89(2) 2
[Robert Bell/Ronald Bell/George M. Brown/Eumir Deodato/Amos Guider/
Robert Mickens/Claydes Smith/J.T. Taylor]
A: Something Special; P: Eumir Deodato/Kool & The Gang 19.7

4047. HOLD TIGHT Change RFC/Atlantic 3832 08Aug81 89(2) 2
[Mauro Malavasi/Davide Romani/Paul Slade]
A: Miracles; P: Mauro Malavasi/Jacques Fred Petrus 19.6

4048. (You're Puttin') A RUSH ON ME Stephanie Mills MCA 53151
10Oct87 85(1) 2 [Timmy Allen/Paul Laurence]
A: If I Were Your Woman; P: Paul Laurence 19.5

4049. REASON TO TRY Eric Carmen Arista 1-9746 08Oct88 87(1) 3
[Phil Galdston/Jon Lind]
A: 1988 Summer Olympics-One Moment In Time; P: Eric Carmen/Michael Lloyd 19.4

4050. BIG TALK Warrant Columbia 73035 04Nov89 93(2) 4
[Joseph Cagle/Steven Chamberlin/Jerry Dixon/John Oswald/Eric Turner]
A: Dirty Rotten Filthy Stinking Rich; P: Beau Hill 19.3

4051. LIVE IT UP Gardner Cole Warner Bros. 27793 15Oct88 91(2) 3
[Gardner Cole/Danny Sembello] *A: Triangles; P: Gardner Cole* 19.2

4052. LOUIE, LOUIE Fat Boys Tin Pan Apple 871 010 15Oct88 89(1) 3
[Richard Berry] *A: Coming Back Hard Again; P: Albert Cabrera/Tony Moran* 19.2

4053. DANCIN' WITH MY MIRROR Corey Hart EMI America 8385
28Mar87 88(1) 3 [Corey Hart] *A: Fields Of Fire; P: Phil Chapman/Corey Hart* 19.1

4054. BAD, BAD BILLY Snuff Warner/Curb 29615 20Aug83 88(1) 2
[Jim Bowling/Robbie House/Chuck Larson] *A: Night Fighter; P: Phil Gernhard* 19.1

4055. DEAR GOD Midge Ure Chrysalis 43319 11Mar89 95(2) 5
[Midge Ure] *A: Answers To Nothing; P: Midge Ure* 18.9

4056. DARLIN' I Vanessa Williams Wing 871 936 03Jun89 91(1) 4
[Kenny Harris/Rex Salas] *A: The Right Stuff; P: Rex Salas* 18.9

4057. EDIE (Ciao Baby) The Cult Sire 22873 30Sep89 93(2) 4
[Ian Astbury/Billy Duffy] *A: Sonic Temple; P: Bob Rock* 18.8

4058. TOO YOUNG TO FALL IN LOVE Mötley Crüe Elektra 69732
16Jun84 90(2) 2 [Nikki Sixx] *A: Shout At The Devil; P: Tom Werman* 18.6

4059. NIGHTRAIN Guns N' Roses Geffen 22869 29Jul89 93(1) 5
[Steven Adler/Saul Hudson/Duff McKagan/Axl Rose/Izzy Stradlin]
A: Appetite For Destruction; P: Mike Clink 18.5

4060. BULLISH Herb Alpert/Tijuana Brass A&M 2655 15Sep84 90(2) 2
[John Barnes/Jimmie Cameron] *A: Bullish; P: Herb Alpert/John Barnes* 18.2

4061. I'LL BE THERE Kenny Loggins Columbia 05625 12Oct85 88(1) 2
[David Foster/E. Ein Loggins/Kenny Loggins] *A: Vox Humana; P: Kenny Loggins* 18.1

4062. NIGHT PULSE Double Image CBS Associated 03832 02Jul83 92(1) 3
[Patrick Bolen/William Eugene Butler/Gabriel Katona]
A: No Album; P: Bob Gaudio 18.0

4063. SEE WHAT LOVE CAN DO Eric Clapton Duck/Warner 28986
29Jun85 89(1) 2 [Jerry Lynn Williams]
A: Behind The Sun; P: Ted Templeman/Lenny Waronker 18.0

4064. READY FOR LOVE Silverado Pavillion 02077 04Jul81 92(1) 3
[Buzz Goodwin/Carl Shillo] *A: Ready For Love; P: Don Oriolo* 17.9

4065. HEARTS AWAY Night Ranger Camel/MCA 53131 04Jul87 90(1) 3
[Jack Blades] *A: Big Life; P: Kevin Elson/Night Ranger* 17.8

4066. NIAGARA FALLS Chicago Warner/Full Moon 28283 11Jul87 90(1) 2
[Bobby Caldwell/Steve Kipner] *A: Chicago 18; P: David Foster* 17.7

4067. ONCE A NIGHT Jackie English Venture 135 20Dec80 94(1) 4
[Beverly Bremers/Jackie English] *A: No Album; P: Cecile Barker/Tony Camillo* 17.6

4068. I CAN SURVIVE Triumph RCA 11945 07Jun80 91(2) 2
[Rik Emmett/Mike Levine/Gil Moore] *A: Progressions Of Power; P: Triumph* 17.5

4069. AIN'T TOO PROUD TO BEG Rick Astley RCA 9030 19Aug89 89(1) 3
[Eddie Holland/Norman Whitfield]
A: Hold Me In Your Arms; P: Matt Aitken/Mike Stock/Pete Waterman 17.5

4070. HANDS ACROSS THE SEA Modern English Sire 29339
07Apr84 91(1) 3 [Richard Ian Brown/Michael Conroy/Robbie Grey/
Gary McDowell/Stephen Walker] *A: Riccochet Days; P: Hugh Jones* 17.0

4071. I WOULDN'T LIE Yarbrough & Peoples Total Experience 2437
28Jun86 93(1) 4 [George Adams/Jimmy Hamilton/Maurice Hayes/
Lonnie Simmons] *A: I Wouldn't Lie; P: Jimmy Hamilton/Lonnie Simmons* 16.8

4072. WE SHOULD BE SLEEPING Eddie Money Columbia 07359
19Sep87 90(1) 3 [Kevin Burns/Greg Lowry/Eddie Money/Glenn Thompson]
A: Can't Hold Back; P: Eddie Money/Richie Zito 16.7

4073. RONNIE'S RAPP Ron And The D.C. Crew Profile 5130
24Jan87 93(1) 4 [Inez J. Kitts/Paul (2) Klein/Mark Mosseley/
Joseph Louis Stone] *A: No Album; P: Mark Klein/Joe Stone* 16.6

4074. LITTLE SUZI Tesla Geffen 28353 16May87 91(1) 3
[James Diamond/Tony Hymas]
A: Mechanical Resonance; P: Michael Barbiero/Steve Thompson 16.6

4075. LOVE GRAMMAR John Parr Atlantic 89484 16Nov85 89(1) 2
[John Parr] *A: John Parr; P: John Parr* 16.4

4076. DIRTY LOOKS Juice Newton Capitol 5289 05Nov83 90(1) 3
[Charles David Robbins/Van Stephenson] *A: Dirty Looks; P: Richard Landis* 16.2

4077. MIRAGE Eric Troyer Chrysalis 2445 26Jul80 92(2) 2
[Eric Troyer] *A: No Album; P: Jack Douglas* 16.0

4078. ONLY THE LONELY (Have A Reason To Be Sad)
La Flavour Sweet City 7377 14Jun80 91(1) 2
[Mark Avsec] *A: Mandolay; P: Mark Avsec/Carl Maduri* 16.0

4079. ONLY A MEMORY The Smithereens Enigma 44150 21May88 92(1) 4
[Pat DiNizio] *A: Green Thoughts; P: Don Dixon* 16.0

4080. HIGH SCHOOL NIGHTS Dave Edmunds Columbia 04762
20Apr85 91(2) 2 [John David/Dave Edmunds/Steve Gould]
A: Porky's Revenge Soundtrack; P: Dave Edmunds 15.9

4081. STIMULATION Wa Wa Nee Epic 07671 30Jan88 86(1) 2
[Paul Gray] *A: Wa Wa Nee; P: Paul Gray/Jim Taig* 15.9

4082. DON'T DO ME Randy Bell Epic 04497 07Jul84 90(1) 3
[Randy Bell] *A: No Album; P: Richard Podolor* 15.9

4083. YOU'RE IN LOVE Ratt Atlantic 89502 12Oct85 89(1) 2
[Juan Croucier/Stephen Pearcy] *A: Invasion Of Your Privacy; P: Beau Hill* 15.7

4084. KEEP IT CONFIDENTIAL Nona Hendryx RCA 13438 04Jun83 91(1) 3
[Ellen Foley/Ellie Greenwich/Jeff Kent] *A: Nona; P: Nona Hendryx/Material* 15.2

4085. NOW YOU'RE IN HEAVEN Julian Lennon Atlantic 88925
13May89 93(1) 4 [Julian Lennon/John McCurry] *A: Mr. Jordan;P: Patrick Leonard* 15.0

4086. THIS ONE Paul McCartney Capitol 44438 09Sep89 94(1) 3
[Paul McCartney] *A: Flowers In The Dirt; P: Paul McCartney* 14.9

4087. INJURED IN THE GAME OF LOVE Donnie Iris HME 04734
23Mar85 91(1) 2 [Mark Avsec/Donnie Iris] *A: No Muss...No Fuss; P: Mark Avsec* 14.4

4088. FOREVER YOUNG ('85) Alphaville Atlantic 89578 23Mar85 93(1) 4
[Marian Gold/Bernhard Lloyd/Frank Mertens]
A: Forever Young; P: Wolfgang Loos/Colin Pearson 14.4

4089. ONE SIMPLE THING Stabilizers Columbia 06700 04Apr87 93(1) 3
[Dave Christenson/Rich Nevens] *A: Tyranny; P: Denny Diante* 14.4

4090. ONE LIFE TO LIVE Wayne Massey Polydor 2112 11Oct80 92(1) 2
[Ritchie Adams/Gloria Nissenson] *A: One Life To Live; P: Joel Diamond* 14.3

4091. FRIGHT NIGHT The J. Geils Band Private I 4-05462 10Aug85 91(1) 2
[Joe Lamont] *A: Fright Night Soundtrack; P: Seth Justman* 14.1

4092. INTO MY LOVE Greg Guidry Columbia 02984 17Jul82 92(1) 2
[Cindy Guidry/Greg Guidry] *A: Over The Line; P: John Ryan* 13.8

4093. LOVE CHANGES EVERYTHING Honeymoon Suite Warner Bros. 27935
30Apr88 91(2) 2 [Johnnie Dee/Dermot Grehan/Rob Preuss]
A: Racing After Midnight; P: Jeff Hendrickson/Ted Templeman 13.7

4094. HEARTLINE Robin George Bronze/Island 7-99658 13Apr85 92(1) 2
[Robin George] *A: Dangerous Music; P: John Ryan* 13.6

4095. THIS IS THE TIME Dennis DeYoung A&M 2839 28Jun86 93(1) 3
[Dennis DeYoung] *A: Back To The World; P: Dennis DeYoung* 13.6

4096. REAL LOVE Dolly Parton (Duet With Kenny Rogers) RCA 14058
08Jun85 91(1) 3 [Richard Brannan/David Malloy/Randy McCormick]
A: Real Love; P: David Malloy 13.5

4097. CANVAS OF LIFE Minor Detail Polydor 815 329 24Sep83 92(1) 2
[John Hughes/Willie Hughes] *A: Minor Detail; P: Billy Whelan* 13.2

4098. SAD GIRL GQ Arista 0659 13Mar82 93(2) 2
[Lloyd Smith/Jay Wiggins] *A: Face To Face; P: Jimmy Simpson* 13.2

Ranking the Singles

Rank. TITLE Act Label Number Entry Date Peak(Wks) Tot Wks
[Writers] A: Proximate Album; P: Producer   Score

4099. SOUNDS OF YOUR VOICE Jon Butcher Axis Capitol 5534
30Nov85 94(2) 3 [Jon Butcher/Thom Gimbel]
A: Along The Axis; P: Spencer Proffer   13.2

4100. 500 MILES Hooters Columbia 73013 09Dec89 97(3) 5
[Bobby Bare/Hedy West/Charlie Williams] A: Zig Zag; P: Rick Chertoff   13.0

4101. BUTTON OFF MY SHIRT Paul Carrack Chrysalis 43288 10Sep88
91(1) 3 [Billy Livsey/Graham Lyle] A: One Good Reason; P: Christopher Neil 13.0

4102. SOLITUDE STANDING Suzanne Vega A&M 2960 12Sep87 94(1) 3
[Stephen Ferrera/Anton Sanko/Marc Shulman/Suzanne Vega/
Michael Visceglia] A: Solitude Standing; P: Steve Addabbo/Lenny Kaye   12.9

4103. STILL CRUISIN' The Beach Boys Capitol 44445 26Aug89 93(1) 3
[Mike Love/Terry Melcher] A: Still Cruisin'; P: Terry Melcher   12.8

4104. I'M YOUR SUPERMAN All Sports Band Radio 3871 28Nov81 93(1) 2
[Michael Toste] A: All Sports Band; P: Joey Carbone/Richie Zito   12.8

4105. SCANDAL RCR Radio 711 05Apr80 94(2) 2
[Donna Rhodes/Perry Rhodes/Sandra Rhodes]
A: Scandal; P: Howard Albert/Ron Albert   12.7

4106. I LOVE MY TRUCK Glen Campbell Mirage 3845 22Aug81 94(1) 3
[Joe Rainey]
A: The Night The Lights Went Out In Georgia Soundtrack; P: Glen Campbell   12.6

4107. FALL ON ME R.E.M. I.R.S. 52883 04Oct86 94(1) 3
[Bill Berry/Peter Buck/Mike Mills/Michael Stipe]
A: Lifes Rich Pageant; P: Don Gehman   12.6

4108. HEY MAMBO Barry Manilow with Kid Creole And The Coconuts
Arista 1-9666 19Mar88 90(1) 2 [Jack Feldman/Thomas F. Kelly/
Barry Manilow/Bruce Sussman] A: Swing Street;
P: Eddie Arkin/Lawrence Dermer/Emilio Estefan/Joe Galdo/Barry Manilow   12.5

4109. SHY BOYS Ana Parc 07056 11Jul87 94(1) 3
[Sue Shifrin/Frank Wildhorn] A: Shy Boys; P: Karl Richardson/Frank Wildhorn 12.3

4110. I WASN'T THE ONE (Who Said Goodbye) Agnetha Faltskog and
Peter Cetera Atlantic 89145 23Apr88 93(1) 3
[Mark Mueller/Aaron Zigman] A: Agnetha Faltskog; P: Peter Cetera   12.2

4111. IF YOU FEEL IT Denise Lopez Vendetta 7213 12Nov88 94(1) 3
[Eric Li] A: Truth In Disguise; P: David Bowler/Howard Bowler/Eric Li   12.2

4112. (Believed You Were) LUCKY 'Til Tuesday Epic 08059 21Jan89 95(2) 3
[Aimee Mann/Jules Shear] A: Everything's Different Now; P: Rhett Davies   11.4

4113. TOO MUCH AIN'T ENOUGH LOVE Jimmy Barnes Geffen 27920
09Jul88 91(1) 2 [Jimmy Barnes/Tony Brock/Jonathan Cain/
Randall Darnell Jackson/Neal Schon]
A: Freight Train Heart; P: Jonathan Cain/Mike Stone   11.2

4114. PROVE ME WRONG David Pack Warner Bros. 28802 25Jan86 95(1) 3
[James Newton Howard/David Pack]
A: Anywhere You Go; P: James Newton Howard/David Pack   11.2

4115. WHITER SHADE OF PALE Hagar, Schon, Aaronson, Shrieve
Geffen 29280 19May84 94(1) 2 [Gary Brooker/Matthew Fisher/Keith Reid]
A: Through The Fire; P: Sammy Hagar/Neil Schon   11.2

4116. STAY TRUE Sly Fox Capitol 5581 07Jun86 94(2) 2
[Michael Camacho] A: Let's Go All The Way; P: Ted Currier   11.0

4117. YOU'RE NOT MY KIND OF GIRL New Edition MCA 53405
12Nov88 95(2) 4 [James Samuel Harris/Terry Lewis]
A: Heart Break; P: Jimmy Jam Harris/Terry Lewis   10.7

4118. LIFE GETS BETTER Graham Parker Arista 1-9065 24Sep83 94(1) 2
[Graham Parker] A: The Real Macaw; P: David Kershenbaum   10.6

4119. AMERICAN DREAM Simon F. Reprise 28237 24Oct87 91(1) 2
[Simon Fellowes] A: Never Never Land; P: Philip Thornalley   10.5

4120. OUTLAW War RCA 13238 10Jul82 94(1) 3
[Thomas Allen/Harold Brown/Jerry Goldstein/Ronnie Hammon/
Lonnie Jordan/Lee Oskar/Luther Rabb/Howard Scott]
A: Outlaw; P: Jerry Goldstein/Lonnie Jordan   10.1

4121. CHAIN REACTION Diana Ross RCA 14244 30Nov85 95(2) 3
[Barry Gibb/Maurice Gibb/Robin Gibb]
A: Eaten Alive; P: Albhy Galuten/Barry Gibb/Karl Richardson   10.1

4122. NEVER THOUGHT I'D FALL IN LOVE The Spinners Atlantic 4007
06Mar82 95(2) 2 [Dean Gant]
A: Can't Shake This Feelin'; P: Reggie Lucas/James Mtume   10.0

4123. BROKEN LAND The Adventures Elektra 69414 30Apr88 95(1) 3
[Pat Gribben] A: The Sea Of Love; P: Garry Bell   9.9

4124. REAL PEOPLE Chic Atlantic 3768 15Nov80 89(1) 1
[Bernard Edwards/Nile Rodgers] A: Real People; P: Bernard Edwards/Nile Rodgers 9.9

4125. WHATEVER HAPPENED TO OLD FASHIONED LOVE B.J. Thomas
Cleveland International 03492 21May83 93(1) 2
[Lewis J. Anderson] A: New Looks; P: Pete Drake   9.8

4126. GRAVITY James Brown Scotti Brothers 06275 18Oct86 93(1) 2
[Dan Hartman/Charles E. Kaufman] A: Gravity; P: Dan Hartman   9.5

4127. ROBERT De NIRO'S WAITING Bananarama London 820033
19May84 95(1) 2 [Sarah Dallin/Siobhan Fahey/Steve Jolley/
Tony Swain/Keren Woodward] A: Bananarama; P: Steve Jolley/Tony Swain   9.4

4128. I'M SO GLAD I'M STANDING HERE TODAY The Crusaders MCA 51177
26Sep81 97(2) 3 [Will Jennings/Joe Sample]
A: Standing Tall; P: Wilton Felder/Stix Hooper/Joe Sample   9.4

4129. WEATHERMAN Nick Jameson Motown 1853 23Aug86 95(2) 2
[Nick Jameson/Kim O'Leary] A: A Crowd Of One; P: Nick Jameson   9.4

4130. HEAVEN KNOWS When In Rome Virgin 99253 18Feb89 95(2) 2
[Clive Farrington/Mike Floreale/Andrew Mann]
A: When In Rome; P: Richard James Burgess   9.4

4131. MY ONE TEMPTATION Mica Paris Island 99252 24Jun89 97(2) 4
[Mike Leeson/Peter Vale/Miles Waters] A: So Good; P: Peter Vale/Miles Waters 9.2

4132. SOMEBODY'S GONNA LOVE YOU Lee Greenwood MCA 52257
01Oct83 96(2) 2 [Don Cook/Rafe Van Hoy]
A: Somebody's Gonna Love You; P: Jerry Crutchfield   9.0

4133. I THINK YOU'LL REMEMBER TONIGHT Axe Atco 99823
22Oct83 94(1) 2 [Bobby Barth/Michael Osborne] A: Nemesis; P: Al Nalli   9.0

4134. DANCING WITH THE MOUNTAINS John Denver RCA 12017
21Jun80 97(2) 3 [John Denver] A: Autograph; P: Milton Okun   8.9

4135. SWEET LIES Robert Palmer Island 99377 26Mar88 94(1) 2
[Frank Blair/Robert Palmer/Dony Wynn]
A: Sweet Lies Soundtrack; P: Robert Palmer   8.8

4136. ALL I KNOW IS THE WAY I FEEL Pointer Sisters RCA 5112 21Feb87
93(1) 2 [Estelle Levitt/Jerry Ragovoy] A: Hot Together; P: Richard Perry   8.6

4137. SEASONS Grace Slick RCA Victor 11939 19Apr80 95(1) 2
[Grace Slick] A: Dreams; P: Ron Frangipane   8.6

4138. ONE SUNNY DAY/DUELING BIKES FROM QUICKSILVER
Ray Parker Jr. And Helen Terry Atlantic 89456 15Feb86 96(2) 3
[Dean Pitchford/Bill Wolfer] A: Quicksilver Soundtrack; P: Ray Parker Jr.   8.6

4139. CERTAIN THINGS ARE LIKELY KTP Mercury 885727 13Jun87 97(1) 4
[Simon Aldridge/Stephen Cusack/Jon Kingsley Hall/Nick Whitecross]
A: Certain Things Are Likely; P: KTP/Peter Walsh   8.6

4140. PART OF ME THAT NEEDS YOU MOST Jay Black Midsong Int'l 72012
20Sep80 98(3) 4 [Mike Chapman/Nicky Chinn] A: No Album; P: Joel Diamond   8.6

4141. NO SOUVENIRS Melissa Etheridge Island 99176 04Nov89 95(1) 3
[Melissa Etheridge]
A: Brave And Crazy; P: Niko Bolas/Melissa Etheridge/Kevin McCormick   8.5

4142. NAME AND NUMBER Big Noise Atco 99168 04Nov89 97(2) 3
[Anthony Fennell/Paul Morais Johnson/Huw Lucas] A: Bang!; P: Elliot Wolff   8.5

4143. SCARLET FEVER Kenny Rogers Liberty 1503 20Aug83 94(1) 3
[Mike Dekle] A: We've Got Tonight; P: Kenny Rogers   8.2

4144. WALK AWAY RENEE Southside Johnny & The Jukes Atlantic 89394
16Aug86 98(2) 5 [Bob Calilli/Michael David Lookofsky/Tony Sansone]
A: At Least We Got Shoes; P: John Lyon/John Rollo   7.8

4145. START IT ALL OVER McGuffey Lane Atco 7345 06Feb82 97(1) 3
[Robert McNelley] A: Aqua Dream; P: Al Nalli/Henry Weck   7.3

4146. I WANT TO MAKE THE WORLD TURN AROUND Steve Miller Band
Capitol 5646 15Nov86 97(1) 3
[Steve Miller] A: Living In The 20th Century; P: Steve Miller   7.1

4147. THE ONLY WAY IS UP Yazz And The Plastic Population Elektra 69365
26Nov88 96(1) 4 [Johnny Henderson/George Henry Jackson]
A: No Album; P: Coldcut   6.8

4148. JACK THE LAD 3 Man Island Chrysalis 43231 02Apr88 94(1) 2
[Tim Cox/Nigel Swanston/Mike Whitford] A: No Album; P: Three Man Island   6.7

4149. POWER PLAY Molly Hatchet Epic 02680 06Feb82 96(1) 2
[Steve Holland] A: Take No Prisoners; P: Tom Werman   6.5

4150. BOY TOY Tia RCA 5107 07Mar87 97(2) 2
[Roy Bermingham/James M. Cohn/Charles Ibgui]
A: No Album; P: Roy Be/James Cohn/Charles Ibgui   6.3

4151. KAREN B.E. Taylor Group Epic 05851 31May86 94(1) 2
[Joe Macre/B.E. Taylor/Rick Witkowski] A: Karen; P: Rick Witkowski   6.3

4152. DOWNTOWN TRAIN Patty Smyth Columbia 07112 13Jun87 95(1) 2
[Tom Waits] A: Never Enough; P: Rick Chertoff/William Wittman   6.2

4153. LEAD ME ON Amy Grant A&M 1218 06Aug88 96(1) 2
[Amy Grant/Wayne Kirkpatrick/Michael W. Smith]
A: Lead Me On; P: Brown Bannister   5.4

4154. I NEED YOU Maurice White Columbia 05726 08Feb86 95(1) 1
[Priscilla Coolidge/William Daniel Smith/Mary Unobsky]
A: Maurice White; P: Robbie Buchanan/Maurice White   4.7

Rank. TITLE Act Label Number Entry Date Peak(Wks) Tot Wks [Writers] A: Proximate Album; P: Producer	Score
4155. WHEN I FALL IN LOVE Natalie Cole EMI-Manhattan 50138 27Aug88 95(1) 1 [Edward Heyman/Victor Young] A: *Everlasting*; P: *Marcus Miller*	4.7
4156. STOP! Erasure Sire 22879 22Jul89 97(1) 2 [Andy Bell/Vince Clarke] A: *Crackers International*; P: *Erasure*	4.6
4157. BLACK LEATHER Kings Of The Sun RCA 8646 30Jul88 98(2) 2 [Cliff Hoad/Jeff Hoad] A: *Kings Of The Sun*; P: *Eddie Kramer*	4.6
4158. LET'S PUT THE X IN SEX KISS Mercury 872246 14Jan89 97(1) 2 [Desmond Child/Paul Stanley] A: *Smashes, Thrashes & Hits*; P: *Paul Stanley*	4.6
4159. TINA CHERRY Georgio Motown 1892 29Aug87 96(1) 2 [Georgio Allentini] A: *Sexappeal*; P: *Georgio Allentini*	4.5
4160. YOUTH GONE WILD Skid Row Atlantic 88935 10Jun89 99(2) 2 [Rachel Bolan/Dave Sabo] A: *Skid Row*; P: *Michael Wagener*	3.2
4161. K.I.S.S.I.N.G. Siedah Garrett Qwest 27928 16Jul88 97(1) 1 [Guy Babylon/Dana Merino] A: *Kiss Of Life*; P: *Richard Rudolph/Rod Temperton*	3.1
4162. INTO MY SECRET Alisha RCA 5219 22Aug87 97(1) 1 [Mike Leeson/Peter Vale] A: *Nightwalkin'*; P: *Mark S. Berry*	3.1
4163. FAT "Weird Al" Yankovic Rock 'n' Roll 07769 21May88 99(1) 2 [Michael Jackson/Al Yankovic] A: *Even Worse*; P: *Rick Derringer*	2.3
4164. SPRING LOVE The Cover Girls Fever 1913 01Aug87 98(1) 1 [Rainy Davis/Pete Warner] A: *Show Me*; P: *Rainy Davis/Pete Warner*	2.3
4165. HOLD ME Colin James Hay Columbia 06580 07Mar87 99(1) 1 [Colin James Hay] A: *Looking For Jack*; P: *Robin Millar*	1.6
4166. IN BETWEEN DAYS (Without You) The Cure Elektra 69604 15Feb86 99(1) 1 [Robert James Smith] A: *The Head On The Door*; P: *Dave Allen/Robert Smith*	1.6
4167. LONG WAY TO LOVE Britny Fox Columbia 07926 01Oct88 100(2) 2 [Dean Davidson] A: *Britny Fox*; P: *John Jansen*	1.5
4168. THAT'S WHEN I THINK OF YOU 1927 Atlantic 88878 26Aug89 100(1) 1 [Garry Frost/Eric Weideman] A: *...Ish*; P: *Charles Fisher*	0.8

Ranking the Singles

Ranking the Singles Alphabetical Index

Rank **Title** *Act*

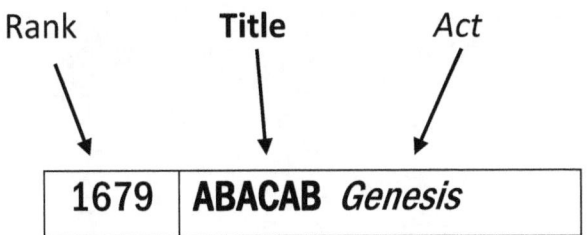

Articles A, An and The are omitted
[P] indicates Parody

Singles Alpha Index

A

1679 ABACAB *Genesis*
3491 ABADABADANGO *Kim Carnes*
  13 ABRACADABRA *Steve Miller Band*
2729 ABSOLUTE BEGINNERS *David Bowie*
3488 ACROSS THE MILES *Survivor*
3347 ACTION *Evelyn "Champagne" King*
 209 ADDICTED TO LOVE *Robert Palmer*
 710 ADULT EDUCATION *Daryl Hall & John Oates*
 708 AFFAIR OF THE HEART *Rick Springfield*
 204 AFRICA *Toto*
3178 AFTER ALL *Al Jarreau*
 734 AFTER ALL (Love Theme From Chances Are)
    Cher and Peter Cetera
2249 AFTER I CRY TONIGHT *Lanier and Co.*
1550 AFTER THE FALL *Journey*
2330 AFTER THE FIRE *Roger Daltrey*
1981 AFTER THE GLITTER FADES *Stevie Nicks*
3084 AFTER YOU *Dionne Warwick*
  33 AGAINST ALL ODDS (Take A Look At Me Now)
    Phil Collins
 452 AGAINST THE WIND *Bob Seger*
1594 AH! LEAH! *Donnie Iris*
1967 AI NO CORRIDA (I-No-Ko-Ree-Da) *Quincy Jones*
3365 AIMING AT YOUR HEART *The Temptations*
1043 AIN'T EVEN DONE WITH THE NIGHT *John Cougar*
1280 AIN'T NOBODY *Rufus And Chaka Khan*
2253 AIN'T NOTHIN' GOIN' ON BUT THE RENT
    Gwen Guthrie
3921 AIN'T NOTHING LIKE THE REAL THING/YOU'RE
    ALL I NEED TO GET BY *Chris Christian*
2624 AIN'T SO EASY *David & David*
4069 AIN'T TOO PROUD TO BEG *Rick Astley*
3203 ALABAMA GETAWAY *Grateful Dead*
1405 ALIBIS *Sergio Mendes*
1740 ALIEN *Atlanta Rhythm Section*
 293 ALIVE & KICKING *Simple Minds*
3705 ALL AMERICAN GIRLS *Sister Sledge*
 720 ALL CRIED OUT *Lisa Lisa And Cult Jam With
    Full Force Featuring Paul Anthony And
    Bow Legged Lou*
2711 ALL FALL DOWN *Five Star*
1413 ALL FIRED UP *Pat Benatar*
2645 ALL I HAVE TO DO IS DREAM
    Andy Gibb And Victoria Principal
4136 ALL I KNOW IS THE WAY I FEEL *Pointer Sisters*
 358 ALL I NEED *Jack Wagner*
    ALL I NEED see also YOU'RE ALL I NEED
 687 ALL I NEED IS A MIRACLE *Mike + The Mechanics*
3630 ALL I WANT *Howard Jones*
2745 ALL I WANT IS YOU *Carly Simon*
3850 ALL I WANT IS YOU *U2*
1343 ALL I WANTED *Kansas*
2166 ALL MY LIFE *Kenny Rogers*
1271 ALL NIGHT LONG *Joe Walsh*
3733 ALL NIGHT LONG *Billy Squier*
  15 ALL NIGHT LONG (All Night) *Lionel Richie*
2218 ALL NIGHT THING *The Invisible Man's Band*
3372 ALL NIGHT WITH ME *Laura Branigan*
2281 ALL OF ME FOR ALL OF YOU *9.9*
3383 ALL OF MY LOVE *Bobby Caldwell*
1378 ALL OF YOU *Julio Iglesias & Diana Ross*
1562 ALL OUR TOMORROWS *Eddie Schwartz*
  60 ALL OUT OF LOVE *Air Supply*

1041 ALL OVER THE WORLD *Electric Light Orchestra*
 836 ALL RIGHT *Christopher Cross*
    ALL RIGHT NOW
3354   *Pepsi & Shirlie*
3392   *Rod Stewart*
1831 ALL SHE WANTS IS *Duran Duran*
 851 ALL SHE WANTS TO DO IS DANCE *Don Henley*
2957 ALL THE KINGS HORSES *The Firm*
1472 ALL THE LOVE IN THE WORLD *The Outfield*
3811 ALL THE RIGHT MOVES
    Jennifer Warnes/Chris Thompson
1875 ALL THE THINGS SHE SAID *Simple Minds*
1056 ALL THIS LOVE *DeBarge*
 811 ALL THIS TIME *Tiffany*
2160 ALL THOSE LIES *Glenn Frey*
 288 ALL THOSE YEARS AGO *George Harrison*
 589 ALL THROUGH THE NIGHT *Cyndi Lauper*
1785 ALL TIME HIGH *Rita Coolidge*
2829 ALL TOUCH *Rough Trade*
2677 ALL YOU ZOMBIES *Hooters*
 822 ALLENTOWN *Billy Joel*
2536 ALLERGIES *Paul Simon*
3875 ALLIES *Heart*
2980 THE ALLNIGHTER *Glenn Frey*
1339 ALMOST OVER YOU *Sheena Easton*
 616 ALMOST PARADISE *Mike Reno And Ann Wilson*
2827 ALMOST SATURDAY NIGHT *Dave Edmunds*
 105 ALONE *Heart*
2744 ALONE AGAIN *Dokken*
1213 ALONG COMES A WOMAN *Chicago*
1052 ALPHABET ST. *Prince*
 208 ALWAYS *Atlantic Starr*
2376 ALWAYS *Firefall*
    ALWAYS ON MY MIND
 391   *Willie Nelson*
 719   *Pet Shop Boys*
 694 ALWAYS SOMETHING THERE TO REMIND ME
    Naked Eyes
3130 ALWAYS THERE FOR YOU *Stryper*
3997 AM I FORGIVEN *Isle Of Man*
 255 AMANDA *Boston*
 682 AMERICA *Neil Diamond*
2766 AMERICA *Prince And The Revolution*
 905 AN AMERICAN DREAM *The Dirt Band*
4119 AMERICAN DREAM *Simon F.*
1387 AMERICAN HEARTBEAT *Survivor*
3397 AMERICAN MADE *Oak Ridge Boys*
3822 AMERICAN MEMORIES *Shamus M'Cool*
1277 AMERICAN MUSIC *Pointer Sisters*
1236 AMERICAN STORM
    Bob Seger & The Silver Bullet Band
3057 AMNESIA *Shalamar*
3094 ANCHORAGE *Michelle Shocked*
1605 AND I AM TELLING YOU I'M NOT GOING
    Jennifer Holliday
2952 AND LOVE GOES ON *Earth, Wind & Fire*
2268 AND SHE WAS *Talking Heads*
1310 AND THE BEAT GOES ON *The Whispers*
2803 AND THE CRADLE WILL ROCK... *Van Halen*
3683 AND THE NIGHT STOOD STILL *Dion*
1311 AND WE DANCED *Hooters*
 456 ANGEL *Aerosmith*
 678 ANGEL *Madonna*
 685 ANGEL EYES *The Jeff Healey Band*
2167 ANGEL IN BLUE *The J. Geils Band*
3979 ANGEL IN MY POCKET *One To One*

1348 ANGEL OF HARLEM *U2*
 320 ANGEL OF THE MORNING *Juice Newton*
2478 ANGEL SAY NO *Tommy Tutone*
1609 THE ANGEL SONG *Great White*
 643 ANGELIA *Richard Marx*
2861 ANGELINE *The Allman Brothers Band*
1337 ANIMAL *Def Leppard*
2500 ANIMAL INSTINCT *Commodores*
  21 ANOTHER BRICK IN THE WALL (Part II) *Pink Floyd*
  68 ANOTHER DAY IN PARADISE *Phil Collins*
2644 ANOTHER HEARTACHE *Rod Stewart*
1048 ANOTHER LOVER *Giant Steps*
1587 ANOTHER NIGHT *Aretha Franklin*
   9 ANOTHER ONE BITES THE DUST *Queen*
1283 ANOTHER PART OF ME *Michael Jackson*
2425 ANOTHER SLEEPLESS NIGHT *Anne Murray*
3558 ANOTHER TICKET *Eric Clapton And His Band*
2797 ANOTHERLOVERHOLENYOHEAD
    Prince And The Revolution
2030 ANSWERING MACHINE *Rupert Holmes*
1088 ANY DAY NOW *Ronnie Milsap*
2358 ANY LOVE *Luther Vandross*
1515 ANY WAY YOU WANT IT *Journey*
1844 ANYONE CAN SEE *Irene Cara*
3358 ANYTHING CAN HAPPEN *Was (Not Was)*
 226 ANYTHING FOR YOU
    Gloria Estefan and Miami Sound Machine
3698 ANYWHERE WITH YOU *Rubber Rodeo*
2389 APACHE *Sugar Hill Gang*
2634 ARC OF A DIVER *Steve Winwood*
1319 ARE WE OURSELVES? *The Fixx*
3055 ARE YOU GETTING ENOUGH HAPPINESS
    Hot Chocolate
3050 ARE YOU SERIOUS *Tyrone Davis*
2377 ARE YOU SURE *So*
 495 ARMAGEDDON IT *Def Leppard*
1956 THE ARMS OF ORION
    Prince With Sheena Easton
  50 ARTHUR'S THEME (Best That You Can Do)
    Christopher Cross
2181 AS LONG AS YOU FOLLOW *Fleetwood Mac*
1449 AS WE LAY *Shirley Murdock*
2153 ASHES BY NOW *Rodney Crowell*
3805 AT THIS MOMENT ('81) *Billy & The Beaters*
 232 --AT THIS MOMENT (Live) ('86)
    Billy Vera & The Beaters
1667 ATHENA *The Who*
1710 ATLANTA LADY (Something About Your Love)
    Marty Balin
2200 ATOMIC *Blondie*
3425 ATTACK OF THE NAME GAME *Stacy Lattisaw*
1164 THE AUTHORITY SONG *John Cougar Mellencamp*
2313 AUTOGRAPH *John Denver*
 472 AUTOMATIC *Pointer Sisters*
2080 AUTOMATIC MAN *Michael Sembello*
 392 AXEL F *Harold Faltermeyer*

B

3813 BABY BABY *Eighth Wonder*
2238 BABY CAN I HOLD YOU *Tracy Chapman*
2303 BABY COME AND GET IT *Pointer Sisters*
2434 BABY COME BACK *Billy Rankin*
3853 BABY COME BACK TO ME
    (The Morse Code Of Love)
    The Manhattan Transfer
2886 BABY COME TO ME *Regina Belle*

Singles Alpha Index

#	Title	Artist
87	**BABY, COME TO ME**	*Patti Austin (A Duet With James Ingram)*
327	**BABY DON'T FORGET MY NUMBER**	*Milli Vanilli*
2958	**BABY DON'T GO**	*Karla Bonoff*
3377	**BABY GRAND**	*Billy Joel featuring Ray Charles*
1279	**BABY I LIED**	*Deborah Allen*
259	**(A) BABY, I LOVE YOUR WAY/(B) FREEBIRD MEDLEY (Free Baby)**	*Will To Power*
3684	**BABY I'M HOOKED (Right Into Your Love)**	*Con Funk Shun*
1116	**BABY JANE**	*Rod Stewart*
870	**BABY LOVE**	*Regina*
1694	**BABY MAKES HER BLUE JEANS TALK**	*Dr. Hook*
2610	**BABY STEP BACK**	*Gordon Lightfoot*
2518	**BABY TALK**	*Alisha*
2359	**BABY TALKS DIRTY**	*The Knack*
2306	**(Baby Tell Me) CAN YOU DANCE**	*Shanice Wilson*
3624	**BABY, WHAT ABOUT YOU**	*Crystal Gayle*
2746	**BACK AND FORTH**	*Cameo*
1794	**BACK IN BLACK**	*AC/DC*
3866	**BACK IN STRIDE**	*Maze Featuring Frankie Beverly*
1142	**BACK IN THE HIGH LIFE AGAIN**	*Steve Winwood*
3965	**BACK OF MY HAND (I've Got Your Number)**	*The Jags*
1925	**BACK ON HOLIDAY**	*Robbie Nevil*
1884	**BACK ON MY FEET AGAIN**	*The Babys*
459	**BACK ON THE CHAIN GANG**	*The Pretenders*
389	**BACK TO LIFE**	*Soul II Soul*
2344	**BACK TO PARADISE**	*38 Special*
3287	**BACK TO SCHOOL AGAIN**	*Four Tops*
2983	**BACK TO THE BULLET**	*Saraya*
4030	**BACK TO THE 60's (Medley)**	*Tight Fit*
2705	**BACK TOGETHER AGAIN**	*Roberta Flack with Donny Hathaway*
1478	**BACK WHERE YOU BELONG**	*38 Special*
2260	**BACKFIRED**	*Debbie Harry*
398	**BAD**	*Michael Jackson see also: FAT*
4054	**BAD, BAD BILLY**	*Snuff*
768	**BAD BOY**	*Miami Sound Machine*
1922	**BAD BOY**	*Ray Parker Jr.*
2256	**BAD BOY/HAVING A PARTY**	*Luther Vandross*
2715	**BAD BOYS**	*Wham! U.K.*
340	**BAD MEDICINE**	*Bon Jovi*
2583	**BAD TIMES**	*Tavares*
892	**BALLERINA GIRL**	*Lionel Richie*
3291	**BANG THE DRUM ALL DAY**	*Todd Rundgren*
1993	**BANG YOUR HEAD (Metal Health)**	*Quiet Riot*
3340	**BASKETBALL**	*Kurtis Blow*
318	**BATDANCE**	*Prince*
3891	**BATTLESHIP CHAINS**	*Georgia Satellites*
1123	**BE GOOD TO YOURSELF**	*Journey*
3983	**BE MINE (Tonight)**	*Grover Washington, Jr.*
2018	**BE MINE TONIGHT**	*Neil Diamond*
1596	**BE MY LADY**	*Jefferson Starship*
795	**BE NEAR ME**	*ABC*
1389	**BE STILL MY BEATING HEART**	*Sting*
2657	**BE THERE**	*Pointer Sisters*
1940	**BE WITH YOU**	*Bangles*
2653	**BE YOUR MAN**	*Jesse Johnson's Revue*
1014	**THE BEACH BOYS MEDLEY**	*The Beach Boys*
3355	**BEAST OF BURDEN**	*Bette Midler*
35	**BEAT IT**	*Michael Jackson see also: EAT IT*
2137	**BEAT OF A HEART**	*Scandal Featuring Patty Smyth*
2483	**BEAT PATROL**	*Starship*
4011	**BEAT STREET BREAKDOWN-Part 1**	*Grandmaster Melle Mel & The Furious Five*
1206	**THE BEATLES' MOVIE MEDLEY**	*The Beatles*
1221	**BEAT'S SO LONELY**	*Charlie Sexton*
1406	**BECAUSE OF YOU**	*The Cover Girls*
1307	**BEDS ARE BURNING**	*Midnight Oil*
3451	**BEECHWOOD 4-5789**	*Carpenters*
3146	**BEFORE I GO**	*Starship*
3496	**BEG, BORROW OR STEAL**	*Hughes/Thrall*
	Behind The Wheel...see: ROUTE 66	
62	**BEING WITH YOU**	*Smokey Robinson*
2507	**BELIEVE IN ME**	*Dan Fogelberg*
	Believe It Or Not...see: THEME FROM "THE GREATEST AMERICAN HERO"	
4112	**(Believed You Were) LUCKY**	*'Til Tuesday*
1724	**THE BELLE OF ST. MARK**	*Sheila E.*
1417	**THE BEST**	*Tina Turner*
2544	**THE BEST MAN IN THE WORLD**	*Ann Wilson*
3511	**THE BEST OF ME**	*David Foster And Olivia Newton-John*
2846	**BEST OF TIMES**	*Peter Cetera*
174	**THE BEST OF TIMES**	*Styx*
	Best That You Can Do...see: ARTHUR'S THEME	
2606	**BET YOUR HEART ON ME**	*Johnny Lee*
2261	**BETCHA SAY THAT**	*Gloria Estefan and Miami Sound Machine*
2397	**BETCHA SHE DON'T LOVE YOU**	*Evelyn King*
4	**BETTE DAVIS EYES**	*Kim Carnes*
544	**BETTER BE GOOD TO ME**	*Tina Turner*
2480	**BETTER BE HOME SOON**	*Crowded House*
760	**BETTER LOVE NEXT TIME**	*Dr. Hook*
3775	**BETTER THINGS**	*The Kinks*
3591	**(Between A) ROCK AND A HARD PLACE**	*Cutting Crew*
2632	**BEYOND**	*Herb Alpert*
2417	**THE BIG CRASH**	*Eddie Money*
1656	**BIG FUN**	*Kool & The Gang*
2775	**BIG IN JAPAN**	*Alphaville*
1306	**BIG LOG**	*Robert Plant*
697	**BIG LOVE**	*Fleetwood Mac*
3110	**BIG MISTAKE**	*Peter Cetera*
2182	**THE BIG MONEY**	*Rush*
4050	**BIG TALK**	*Warrant*
730	**BIG TIME**	*Peter Gabriel*
221	**BIGGEST PART OF ME**	*Ambrosia*
17	**BILLIE JEAN**	*Michael Jackson*
	Billie Jean...see also: DO IT AGAIN	
1935	**THE BIRD**	*The Time*
1928	**BIRTHDAY SUIT**	*Johnny Kemp*
3608	**BIT BY BIT**	*Stephanie Mills*
2355	**BLACK AND BLUE**	*Van Halen*
1966	**BLACK CARS**	*Gino Vannelli*
3588	**BLACK DOG**	*Newcity Rockers*
3804	**BLACK KISSES (Never Make You Blue)**	*Curtie And The Boombox*
4157	**BLACK LEATHER**	*Kings Of The Sun*
3020	**BLACK STATIONS/WHITE STATIONS**	*M+M*
2203	**BLAME IT ON LOVE**	*Smokey Robinson & Barbara Mitchell*
3747	**BLAME IT ON THE RADIO**	*John Parr*
168	**BLAME IT ON THE RAIN**	*Milli Vanilli*
3045	**BLAZE OF GLORY**	*Kenny Rogers*
2006	**BLESSED ARE THE BELIEVERS**	*Anne Murray*
970	**BLUE EYES**	*Elton John*
881	**BLUE JEAN**	*David Bowie*
3480	**BLUE KISS**	*Jane Wiedlin*
3039	**BLUE LIGHT**	*David Gilmour*
3105	**BLUE MONDAY 1988**	*New Order*
3295	**THE BLUE SIDE**	*Crystal Gayle*
3090	**BLUE WORLD**	*The Moody Blues*
2556	**THE BLUES**	*Randy Newman and Paul Simon*
3665	**BLUES POWER**	*Eric Clapton And His Band*
1216	**BOBBIE SUE**	*Oak Ridge Boys*
2701	**BODY**	*Jacksons*
1217	**BODY LANGUAGE**	*Queen*
2319	**BODY ROCK**	*Maria Vidal*
3147	**BODY TALK**	*The Deele*
2976	**BON BON VIE (Gimme The Good Life)**	*T.S. Monk*
3951	**BONGO BONGO**	*Steve Miller Band*
3426	**BOOGIE DOWN**	*Jarreau*
2138	**BOOM BOOM (Let's Go Back To My Room)**	*Paul Lekakis*
2590	**BOOM! THERE SHE WAS**	*Scritti Politti Featuring Roger*
1916	**BOP**	*Dan Seals*
1445	**BOP 'TIL YOU DROP**	*Rick Springfield*
1776	**THE BORDER**	*America*
638	**BORDERLINE**	*Madonna*
2196	**THE BORDERLINES**	*Jeffrey Osborne*
2383	**BORN IN EAST L.A.**	*Cheech & Chong* [P] see: BORN IN THE U.S.A.
823	**BORN IN THE U.S.A.**	*Bruce Springsteen*
454	**BORN TO BE MY BABY**	*Bon Jovi*
3196	**BORROWED TIME**	*Styx*
1249	**BOULEVARD**	*Jackson Browne*
3815	**BOUNCE, ROCK, SKATE, ROLL PART 1**	*Vaughan Mason And Crew*
2618	**BOUNCIN' OFF THE WALLS**	*Matthew Wilder*
3581	**BOY BLUE**	*Cyndi Lauper*
560	**BOY FROM NEW YORK CITY**	*The Manhattan Transfer*
1799	**BOY IN THE BOX**	*Corey Hart*
3938	**THE BOY IN THE BUBBLE**	*Paul Simon*
2195	**BOY, I'VE BEEN TOLD**	*Sa-Fire*
4150	**BOY TOY**	*Tia*
1887	**BOYS DO FALL IN LOVE**	*Robin Gibb*
1909	**BOYS NIGHT OUT**	*Timothy B. Schmit*
585	**THE BOYS OF SUMMER**	*Don Henley*
1168	**BRAND NEW LOVER**	*Dead Or Alive*
784	**BRASS IN POCKET (I'm Special)**	*The Pretenders*
2410	**BRASS MONKEY**	*Beastie Boys*
3356	**BREAD AND BUTTER**	*Robert John*
3501	**BREAK EVERY RULE**	*Tina Turner*
979	**BREAK IT TO ME GENTLY**	*Juice Newton*
1649	**BREAK IT UP**	*Foreigner*
351	**BREAK MY STRIDE**	*Matthew Wilder*
3613	**BREAK-A-WAY**	*Tracey Ullman*
2760	**BREAKAWAY**	*Big Pig*
722	**BREAKDANCE**	*Irene Cara*
1058	**BREAKDOWN DEAD AHEAD**	*Boz Scaggs*
2836	**BREAKFAST IN AMERICA**	*Supertramp*
2292	**BREAKIN' AWAY**	*Al Jarreau*
923	**BREAKIN'.. THERE'S NO STOPPING US**	*Ollie And Jerry*
1425	**BREAKING AWAY**	*Balance*
3420	**BREAKING UP IS HARD ON YOU (a/k/a Don't Take Ma Bell Away From Me)**	*The American Comedy Network*
1202	**BREAKING US IN TWO**	*Joe Jackson*
671	**BREAKOUT**	*Swing Out Sister*
3774	**THE BREAKS (Part 1)**	*Kurtis Blow*
966	**THE BREAKUP SONG (They Don't Write 'Em)**	*Greg Kihn Band*
746	**BRILLIANT DISGUISE**	*Bruce Springsteen*
2378	**BRING DOWN THE MOON**	*Boy Meets Girl*
3967	**BRING IT ALL BACK**	*Grayson Hugh*
2832	**BRINGIN' ON THE HEARTBREAK**	*Def Leppard*
3208	**BRITE EYES**	*Robbin Thompson Band*
4123	**BROKEN LAND**	*The Adventures*
118	**BROKEN WINGS**	*Mr. Mister*
3012	**BROOKLYN GIRLS**	*Robbie Dupree*
1787	**BRUCE**	*Rick Springfield*

Singles Alpha Index

428 BUFFALO STANCE *Neneh Cherry*	2373 CARS WITH THE BOOM *L'Trimm*	198 COLD HEARTED *Paula Abdul*
4060 BULLISH *Herb Alpert/Tijuana Brass*	713 CASANOVA *Levert*	1913 COLD LOVE *Donna Summer*
3514 BURN RUBBER (Why You Wanna Hurt Me) *The Gap Band*	3152 CASE OF YOU *Frank Stallone*	3312 COLORS *Ice-T*
1930 BURNIN' FOR YOU *Blue Öyster Cult*	1823 CASTLES IN THE AIR *Don McLean*	1393 THE COLOUR OF LOVE *Billy Ocean*
3187 BURNING DOWN ONE SIDE *Robert Plant*	3139 CAT PEOPLE (Putting Out Fire) *David Bowie*	1347 COME AS YOU ARE *Peter Wolf*
859 BURNING DOWN THE HOUSE *Talking Heads*	691 CATCH ME (I'm Falling) *Pretty Poison*	1769 COME BACK *The J. Geils Band*
3236 BURNING FLAME *Vitamin Z*	2136 CATCH ME I'M FALLING *Real Life*	1411 COME BACK AND STAY *Paul Young*
254 BURNING HEART *Survivor*	2475 CATCH MY FALL *Billy Idol*	644 COME DANCING *The Kinks*
1955 BURNING HEART *Vandenberg*	3254 CATCHING THE SUN *Spyro Gyra*	2718 COME GIVE YOUR LOVE TO ME *Janet Jackson*
3235 BURNING LIKE A FLAME *Dokken*	3552 CAUGHT IN THE GAME *Survivor*	626 COME GO WITH ME *Exposé*
465 BUST A MOVE *Young M.C.*	1819 CAUGHT UP IN THE RAPTURE *Anita Baker*	1379 COME GO WITH ME *The Beach Boys*
BUTT, DA' see...DA' BUTT	831 CAUGHT UP IN YOU *38 Special*	3165 COME HOME WITH ME BABY *Dead Or Alive*
2462 BUT YOU KNOW I LOVE YOU *Dolly Parton*	381 CAUSING A COMMOTION *Madonna*	126 COME ON EILEEN *Dexys Midnight Runners*
4101 BUTTON OFF MY SHIRT *Paul Carrack*	Cavalleria Rusticana...see: THEME FROM "RAGING BULL"	1669 COME ON, LET'S GO *Los Lobos*
	3632 CECILIA *Times Two*	3605 (Come On) SHOUT *Alex Brown*
# C	1912 CELEBRATE YOUTH *Rick Springfield*	3832 COME OUT FIGHTING *Easterhouse*
	84 CELEBRATION *Kool & The Gang*	3923 COME TO ME *Aretha Franklin*
1479 C-I-T-Y *John Cafferty And The Beaver Brown Band*	3859 THE CELTIC SOUL BROTHERS *Dexys Midnight Runners*	3153 COMIN' DOWN TONIGHT *Thirty Eight Special*
3629 CAFE AMORE *Spyro Gyra*	2209 CENTERFIELD *John Fogerty*	808 COMIN' IN AND OUT OF YOUR LIFE *Barbra Streisand*
2609 CALIFORNIA DREAMIN' *The Beach Boys*	20 CENTERFOLD *The J. Geils Band*	1398 COMING AROUND AGAIN *Carly Simon*
525 CALIFORNIA GIRLS *David Lee Roth*	1438 CENTIPEDE *Rebbie Jackson*	2189 COMING DOWN FROM LOVE *Bobby Caldwell*
1450 CALL IT LOVE *Poco*	3783 CENTURY'S END *Donald Fagen*	1465 COMING HOME *Cinderella*
6 CALL ME (Theme From...American Gigolo) *Blondie*	2870 A CERTAIN GIRL *Warren Zevon*	3996 COMING TO AMERICA (Part One) *The System*
1954 CALL ME *Skyy*	4139 CERTAIN THINGS ARE LIKELY *KTP*	2721 COMING UP CLOSE *'Til Tuesday*
2433 CALL ME *Go West*	245 C'EST LA VIE *Robbie Nevil*	22 COMING UP (Live At Glasgow) *Paul McCartney & Wings*
2522 CALL ME *Dennis DeYoung*	2907 CH CH CHERIE *Johnny Average Band*	3528 COMING UP YOU *The Cars*
3406 (Call Me) WHEN THE SPIRIT MOVES YOU *Touch*	3037 CHAIN REACTION (Special New Mix) *Diana Ross*	2185 COMMUNICATION *The Power Station*
1189 CALL TO THE HEART *Giuffria*	4121 --CHAIN REACTION *Diana Ross*	3002 COMMUNICATION *Spandau Ballet*
3070 CALLING ALL GIRLS *Queen*	1125 CHAINS OF LOVE *Erasure*	2601 COMPUTER GAME "THEME FROM THE CIRCUS" *Yellow Magic Orchestra*
1556 CALLING AMERICA *Electric Light Orchestra*	3566 A CHANCE FOR HEAVEN *Christopher Cross*	3040 CONCEALED WEAPONS *The J. Geils Band*
2761 CALLING IT LOVE *Animotion*	2629 CHANGE *John Waite*	707 CONGA *Miami Sound Machine*
2550 CAN WE STILL BE FRIENDS *Robert Palmer*	3337 CHANGE *Tears For Fears*	2738 CONGRATULATIONS *Vesta*
3576 CAN YOU FEEL IT *The Jacksons*	555 CHANGE OF HEART *Cyndi Lauper*	518 CONTROL *Janet Jackson*
2382 CAN YOU FEEL THE BEAT *Lisa Lisa and Cult Jam with Full Force*	1574 CHANGE OF HEART *Tom Petty and The Heartbreakers*	2871 CONTROVERSY *Prince*
2437 CAN YOU STAND THE RAIN *New Edition*	80 CHARIOTS OF FIRE - TITLES *Vangelis*	677 COOL CHANGE *Little River Band*
736 CANDLE IN THE WIND (Live) *Elton John*	3454 CHARM THE SNAKE *Christopher Cross*	420 COOL IT NOW *New Edition*
1537 CANDY *Cameo*	4029 CHEAP SUNGLASSES *ZZ Top*	1030 COOL LOVE *Pablo Cruise*
2237 CANDY GIRL *New Edition*	1286 CHECK IT OUT *John Cougar Mellencamp*	2896 COOL MAGIC *Steve Miller Band*
1898 CANNONBALL *Supertramp*	229 CHERISH *Kool & The Gang*	819 COOL NIGHT *Paul Davis*
103 CAN'T FIGHT THIS FEELING *REO Speedwagon*	453 CHERISH *Madonna*	3767 COOL (Part 1) *The Time*
3519 CAN'T GET STARTED *Peter Wolf*	3220 CHEROKEE *Europe*	2179 COOL PLACES *Sparks And Jane Wiedlin*
1709 CAN'T HELP FALLING IN LOVE *Corey Hart*	863 CHERRY BOMB *John Cougar Mellencamp*	2404 COULD I BE DREAMING *Pointer Sisters*
3868 CAN'T HOLD BACK (Your Loving) *Kano*	2555 CHINA *Red Rockers*	1798 COULD I HAVE THIS DANCE *Anne Murray*
3227 CAN'T LET GO *Stephen Stills Featuring Michael Finnigan*	852 CHINA GIRL *David Bowie*	2424 COULD IT BE LOVE *Jennifer Warnes*
3058 CAN'T PUT A PRICE ON LOVE *The Knack*	--Chip Off The Old Block...see: REAL PEOPLE	171 COULD'VE BEEN *Tiffany*
1642 CAN'T SHAKE LOOSE *Agnetha Faltskog*	1703 CHIQUITITA *ABBA*	2173 COUNT ME OUT *New Edition*
833 CAN'T STAY AWAY FROM YOU *Gloria Estefan and Miami Sound Machine*	2049 CHLOE *Elton John*	3972 COUNT ON ME *Gerard McMahon*
2669 CAN'T STOP *Rick James*	737 CHURCH OF THE POISON MIND *Culture Club*	3897 COUNT YOUR BLESSINGS *Ashford and Simpson*
3282 CAN'T WAIT ALL NIGHT *Juice Newton*	2461 CIRCLE *Edie Brickell & The New Bohemians*	504 COVER GIRL *New Kids On The Block*
2067 CAN'T WAIT ANOTHER MINUTE *Five Star*	955 CIRCLE IN THE SAND *Belinda Carlisle*	693 COVER ME *Bruce Springsteen*
662 CAN'T WE TRY *Dan Hill (Duet/Vonda Shepard)*	3003 CIRCLE OF LOVE *Steve Miller Band*	2055 COVER OF LOVE *Michael Damian*
2428 CAN'T WE TRY *Teddy Pendergrass*	2157 CIRCLES *Atlantic Starr*	167 COWARD OF THE COUNTY *Kenny Rogers*
1686 CAN'TCHA SAY (You Believe In Me)/ STILL IN LOVE *Boston*	1908 THE CLAPPING SONG *Pia Zadora*	3092 THE COWBOY AND THE LADY *John Denver*
4097 CANVAS OF LIFE *Minor Detail*	3537 CLEANIN' UP THE TOWN *The Bus Boys*	2041 CRAZY *Jesse Johnson (Featuring Sly Stone)*
1225 THE CAPTAIN OF HER HEART *Double*	2266 CLONES (We're All) *Alice Cooper*	1148 CRAZY *Icehouse*
2276 CARAVAN OF LOVE *Isley, Jasper, Isley*	2941 CLOSE ENOUGH TO PERFECT *Alabama*	3342 CRAZY *The Manhattans*
66 CARELESS WHISPER *Wham! Featuring George Michael*	752 CLOSE MY EYES FOREVER (remix) *Lita Ford (Duet With Ozzy Osbourne)*	3484 CRAZY *Kenny Rogers*
99 CARIBBEAN QUEEN (No More Love On The Run) *Billy Ocean*	2678 CLOSER THAN FRIENDS *Surface*	1090 CRAZY ABOUT HER *Rod Stewart*
511 CARRIE *Europe*	2676 CLOSER TO FINE *Indigo Girls*	3299 CRAZY CRAZY NIGHTS *KISS*
1964 CARRIE *Cliff Richard*	3222 CLOSER TO THE HEART *Rush*	75 CRAZY FOR YOU *Madonna*
3782 CARRIE'S GONE *Le Roux*	2224 THE CLOSER YOU GET *Alabama*	1367 CRAZY IN THE NIGHT (Barking At Airplanes) *Kim Carnes*
492 CARS *Gary Numan*	3228 (Closest Thing To) PERFECT *Jermaine Jackson*	2352 CRAZY (Keep On Falling) *The John Hall Band*
	3143 CLUB MICHELLE *Eddie Money*	18 CRAZY LITTLE THING CALLED LOVE *Queen*
	Cocaine...see: TULSA TIME	
	2051 COLD BLOODED *Rick James*	

Singles Alpha Index

3394 CRAZY WORLD *Big Trouble*	1215 DEAD GIVEAWAY *Shalamar*	1154 THE DOCTOR *The Doobie Brothers*
816 CRIMSON AND CLOVER	2584 THE DEAD HEART *Midnight Oil*	993 DOCTOR! DOCTOR! *Thompson Twins*
Joan Jett & The Blackhearts	4055 DEAR GOD *Midge Ure*	964 DR. FEELGOOD *Mötley Crüe*
2849 CRITICIZE *Alexander O'Neal*	3042 DEAR MR. JESUS *PowerSource*	1911 DR. HECKYLL & MR. JIVE *Men At Work*
858 CROSS MY BROKEN HEART *The Jets*	2935 DEDICATED TO THE ONE I LOVE	2650 DOCTORIN' THE TARDIS *The Timelords*
2255 CROSS MY HEART *Eighth Wonder*	*Bernadette Peters*	1835 DOES IT MAKE YOU REMEMBER *Kim Carnes*
3140 CROSS MY HEART *Lee Ritenour*	1540 DEEP INSIDE MY HEART *Randy Meisner*	889 DOING IT ALL FOR MY BABY
3976 CROSSROADS *Tracy Chapman*	3173 DEEP RIVER WOMAN *Lionel Richie*	*Huey Lewis And The News*
896 CRUEL SUMMER *Bananarama*	1053 DEJA VU *Dionne Warwick*	1606 DOMINO DANCING *Pet Shop Boys*
195 CRUISIN' *Smokey Robinson*	670 DELIRIOUS *Prince*	1308 DOMINOES *Robbie Nevil*
780 CRUMBLIN' DOWN *John Cougar Mellencamp*	474 DER KOMMISSAR *After The Fire*	2469 DON QUICHOTTE *Magazine 60*
477 CRUSH ON YOU *The Jets*	871 DESERT MOON *Dennis DeYoung*	1243 DON'T ANSWER ME *Alan Parsons Project*
1026 CRY *Waterfront*	286 DESIRE *Andy Gibb*	1322 DON'T ASK ME WHY *Billy Joel*
1251 CRY *Godley & Creme*	523 DESIRE *U2*	2577 DON'T ASK ME WHY *Eurythmics*
3659 CRY JUST A LITTLE *Paul Davis*	3257 DESIRE *Rockets*	3559 DON'T BE AFRAID OF THE DARK
2479 CRY LIKE A BABY *Kim Carnes*	3025 DESPERATE BUT NOT SERIOUS *Adam Ant*	*The Robert Cray Band*
2563 CRY WOLF *A-Ha*	2133 DESTINATION UNKNOWN *Missing Persons*	603 DON'T BE CRUEL *Cheap Trick*
1795 CRYIN' *Vixen*	3912 DESTROYER *The Kinks*	770 DON'T BE CRUEL *Bobby Brown*
433 CRYING *Don McLean*	3819 DEVIL IN A FAST CAR *Sheena Easton*	3895 DON'T BE MY ENEMY *Wang Chung*
1869 CUDDLY TOY (Feel For Me) *Roachford*	385 DEVIL INSIDE *INXS*	3362 DON'T BREAK MY HEART *Romeo's Daughter*
1246 CULT OF PERSONALITY *Living Colour*	3749 THE DEVIL MADE ME DO IT *Golden Earring*	3785 DON'T CHANGE *INXS*
447 CUM ON FEEL THE NOIZE *Quiet Riot*	1143 DIAL MY HEART *The Boys*	873 DON'T CLOSE YOUR EYES *Kix*
292 CUPID/I'VE LOVED YOU FOR A LONG TIME	702 DIAMONDS *Herb Alpert*	1149 DON'T COME AROUND HERE NO MORE
(Medley) *Spinners*	801 DID IT IN A MINUTE *Daryl Hall & John Oates*	*Tom Petty and The Heartbreakers*
1318 THE CURLY SHUFFLE *Jump 'N The Saddle*	991 DIDN'T I (Blow Your Mind) *New Kids On The Block*	882 DON'T CRY *Asia*
2756 CUTIE PIE *One Way*	2812 DIDN'T KNOW IT WAS LOVE *Survivor*	3270 DON'T CRY FOR ME ARGENTINA *Festival*
1250 CUTS LIKE A KNIFE *Bryan Adams*	218 DIDN'T WE ALMOST HAVE IT ALL	586 DON'T DISTURB THIS GROOVE *The System*
	Whitney Houston	4082 DON'T DO ME *Randy Bell*
D	842 A DIFFERENT CORNER *George Michael*	777 DON'T DO ME LIKE THAT
	2762 THE DIFFERENT STORY (World Of Lust And Crime)	*Tom Petty And The Heartbreakers*
2107 DA' BUTT *E.U.*	*Peter Schilling*	270 DON'T DREAM IT'S OVER *Crowded House*
1616 DADDY'S HOME *Cliff Richard*	2127 DIG THE GOLD *Joyce Cobb*	367 DON'T FALL IN LOVE WITH A DREAMER
2914 DANCE *Ratt*	1228 DIGGING YOUR SCENE *The Blow Monkeys*	*Kenny Rogers with Kim Carnes*
1050 DANCE HALL DAYS *Wang Chung*	1512 DIGITAL DISPLAY *Ready For The World*	1539 DON'T FIGHT IT *Kenny Loggins with Steve Perry*
2122 DANCE LITTLE SISTER (Part One)	2445 DINNER WITH GERSHWIN *Donna Summer*	406 DON'T FORGET ME (When I'm Gone) *Glass Tiger*
Terence Trent D'Arby	411 DIRTY DIANA *Michael Jackson*	1975 DON'T FORGET TO DANCE *The Kinks*
2877 DANCE WIT' ME - PART 1 *Rick James*	144 DIRTY LAUNDRY *Don Henley*	904 DON'T GET ME WRONG *The Pretenders*
3198 DANCIN' IN THE KEY OF LIFE *Steve Arrington*	4076 DIRTY LOOKS *Juice Newton*	3308 DON'T GIRLS GET LONELY *Glenn Shorrock*
3503 DANCIN' LIKE LOVERS *Mary MacGregor*	2498 DIRTY WATER *The Inmates*	1821 DON'T GIVE IT UP *Robbie Patton*
4053 DANCIN' WITH MY MIRROR *Corey Hart*	2825 DIRTY WATER *Rock And Hyde*	3393 DON'T GIVE UP *Peter Gabriel/Kate Bush*
3384 DANCING IN HEAVEN (Orbital Be-Bop) *Q-Feel*	3735 DISCIPLINE OF LOVE (Why Did You Do It)	Don't It Make You Feel Good…see: SHY BOY
3903 DANCING IN MY SLEEP *Secret Ties*	*Robert Palmer*	162 DON'T KNOW MUCH
78 DANCING IN THE DARK *Bruce Springsteen*	3706 DIVIDED HEARTS *Kim Carnes*	*Linda Ronstadt Featuring Aaron Neville*
3894 DANCING IN THE SHADOWS *After The Fire*	1060 DO I DO *Stevie Wonder*	3619 --ALL I NEED TO KNOW *Bette Midler*
1077 DANCING IN THE SHEETS *Shalamar*	2432 DO IT AGAIN *The Kinks*	3864 --DON'T KNOW MUCH *Bill Medley*
DANCING IN THE STREET	3504 DO IT AGAIN/BILLIE JEAN (Medley) *Club House*	1002 DON'T KNOW WHAT YOU GOT (Till It's Gone)
944 *Mick Jagger & David Bowie*	1784 DO IT FOR LOVE *Sheena Easton*	*Cinderella*
2082 *Van Halen*	2110 DO ME BABY *Meli'sa Morgan*	2269 DON'T LEAVE ME THIS WAY *Communards*
3167 --DANCIN' IN THE STREETS *Teri DeSario with K.C.*	1532 DO RIGHT *Paul Davis*	988 DON'T LET GO *Isaac Hayes*
248 DANCING ON THE CEILING *Lionel Richie*	11 DO THAT TO ME ONE MORE TIME	2171 DON'T LET GO *Wang Chung*
3062 DANCING UNDER A LATIN MOON *Candi*	*The Captain & Tennille*	3798 DON'T LET GO THE COAT *The Who*
4134 DANCING WITH THE MOUNTAINS *John Denver*	1595 DO THEY KNOW IT'S CHRISTMAS? *Band Aid*	1659 DON'T LET HIM GO *REO Speedwagon*
403 DANGER ZONE *Kenny Loggins*	925 DO WHAT YOU DO *Jermaine Jackson*	2295 DON'T LET HIM KNOW *Prism*
2534 DANGEROUS *Natalie Cole*	3899 DO YA DO YA (Wanna Please Me) *Samantha Fox*	584 DON'T LET IT END *Styx*
2965 DANGEROUS *Loverboy*	731 DO YOU BELIEVE IN LOVE	3181 DON'T LET ME IN *Sneaker*
995 DARE ME *Pointer Sisters*	*Huey Lewis And The News*	DON'T LOOK ANY FURTHER
3427 DARLIN' *Yipes!!*	3572 DO YOU BELIEVE IN SHAME? *Duran Duran*	3054 *The Kane Gang*
3778 DARLIN' *Frank Stallone*	3059 DO YOU COMPUTE? *Donnie Iris*	3363 *Dennis Edwards (Featuring Siedah Garrett)*
2842 DARLIN' DANIELLE DON'T *Henry Lee Summer*	DO YOU LOVE ME	1345 DON'T LOOK BACK *Fine Young Cannibals*
4056 DARLIN' I *Vanessa Williams*	1272 *The Contours*	2193 DON'T LOOK DOWN - THE SEQUEL *Go West*
1359 DAY BY DAY *Hooters*	3713 *Andy Fraser*	3770 DON'T LOSE ANY SLEEP *John Waite*
1646 DAY-IN DAY-OUT *David Bowie*	1721 DO YOU LOVE WHAT YOU FEEL *Rufus and Chaka*	571 DON'T LOSE MY NUMBER *Phil Collins*
DAYDREAM BELIEVER	93 DO YOU REALLY WANT TO HURT ME *Culture Club*	3032 DON'T MAKE A FOOL OF YOURSELF *Stacey Q*
792 *Anne Murray*	3521 DO YOU REMEMBER ME? *Jermaine Jackson*	3522 DON'T MAKE ME DO IT *Patrick Simmons*
3725 *The Monkees*	2247 DO YOU WANNA GET AWAY *Shannon*	DON'T MAKE ME OVER
3523 DAYS ARE NUMBERS (The Traveller)	3690 DO YOU WANNA HOLD ME? *Bow Wow Wow*	1278 *Sybil*
Alan Parsons Project	1424 DO YOU WANNA TOUCH ME (Oh Yeah)	3085 *Jennifer Warnes*
3663 DAYS GONE BY *Poco*	*Joan Jett & The Blackhearts*	1344 DON'T MAKE ME WAIT FOR LOVE
724 DE DO DO DO, DE DA DA DA *The Police*	2155 DO YOU WANT CRYING *Katrina And The Waves*	*Kenny G Vocal by Lenny Williams*

Singles Alpha Index

#	Title	Artist
530	DON'T MEAN NOTHING	Richard Marx
2600	DON'T MISUNDERSTAND ME	Rossington Collins Band
2270	DON'T NEED A GUN	Billy Idol
1735	DON'T PAY THE FERRYMAN	Chris DeBurgh
2400	DON'T PUSH IT DON'T FORCE IT	Leon Haywood
3557	DON'T RUN MY LIFE	Spys
426	DON'T RUSH ME	Taylor Dayne
2241	DON'T SAY GOODNIGHT (It's Time For Love) (Parts 1 And 2)	The Isley Brothers
3352	DON'T SAY NO	Billy Burnette
3193	DON'T SAY NO TONIGHT	Eugene Wilde
2409	DON'T SAY YOU LOVE ME	Billy Squier
802	DON'T SHED A TEAR	Paul Carrack
1157	DON'T SHUT ME OUT	Kevin Paige
829	DON'T STAND SO CLOSE TO ME ('81)	The Police
2588	--DON'T STAND SO CLOSE TO ME '86	The Police
2062	DON'T STOP	Jeffrey Osborne
824	DON'T STOP BELIEVIN'	Journey
2998	DON'T STOP ME BABY (I'm On Fire)	The Boys Band
1410	DON'T STOP THE MUSIC	Yarbrough & Peoples
3847	DON'T STOP TRYING	Rodway
3915	DON'T TALK	Larry Lee
64	DON'T TALK TO STRANGERS	Rick Springfield
1140	DON'T TELL ME LIES	Breathe
3575	DON'T TELL ME THE TIME	Martha Davis
2177	DON'T TELL ME YOU LOVE ME	Night Ranger
3265	DON'T TRY TO STOP IT	Roman Holliday
3935	DON'T WAIT FOR HEROES	Dennis DeYoung
1704	DON'T WALK AWAY	Rick Springfield
3318	DON'T WALK AWAY	Toni Childs
3988	DON'T WALK AWAY	Robert Tepper
277	DON'T WANNA LOSE YOU	Gloria Estefan
3524	DON'T WANT NO-BODY	J.D. Drews
2159	DON'T WANT TO WAIT ANYMORE	The Tubes
2272	DON'T WASTE YOUR TIME	Yarbrough & Peoples
265	DON'T WORRY BE HAPPY (Edit)	Bobby McFerrin
159	DON'T YOU (Forget About Me)	Simple Minds
1568	DON'T YOU GET SO MAD	Jeffrey Osborne
2922	DON'T YOU KNOW HOW MUCH I LOVE YOU	Ronnie Milsap
3300	DON'T YOU KNOW WHAT LOVE IS	Touch
793	DON'T YOU KNOW WHAT THE NIGHT CAN DO?	Steve Winwood
	Don't You Wanna Play This Game No More...see: SARTORIAL ELOQUENCE	
41	DON'T YOU WANT ME	The Human League
590	DON'T YOU WANT ME	Jody Watley
1490	DOUBLE DUTCH BUS	Frankie Smith
1697	DOWN BOYS	Warrant
2749	DOWN ON LOVE	Foreigner
32	DOWN UNDER	Men At Work
2052	DOWNTOWN	One 2 Many
3742	DOWNTOWN	Dolly Parton
2387	DOWNTOWN LIFE	Daryl Hall & John Oates
4152	DOWNTOWN TRAIN	Patty Smyth
1760	DRAW OF THE CARDS	Kim Carnes
1747	THE DREAM (Hold On To Your Dream)	Irene Cara
1253	DREAMER	Supertramp
3359	DREAMER	The Association
895	DREAMIN'	Vanessa Williams
2068	DREAMIN'	Will To Power
2675	DREAMIN'	John Schneider
1650	DREAMIN' IS EASY	Steel Breeze
3226	DREAMIN' OF LOVE	Stevie B
714	DREAMING	Cliff Richard
1355	DREAMING	Orchestral Manoeuvres In The Dark
1636	DREAMS	Van Halen
748	DREAMTIME	Daryl Hall
754	DRESS YOU UP	Madonna
1334	DRESSED FOR SUCCESS	Roxette
304	DRIVE	The Cars
2607	DRIVEN OUT	The Fixx
422	DRIVIN' MY LIFE AWAY	Eddie Rabbitt
3434	DROP THE PILOT	Joan Armatrading
1020	DUDE (Looks Like A Lady)	Aerosmith
	Dueling Bikes...see: ONE SUNNY DAY	
1132	DYNAMITE	Jermaine Jackson

E

#	Title	Artist
3926	EACH WORD'S A BEAT OF MY HEART	Mink DeVille
3268	EAGLES FLY	Sammy Hagar
	EARLY IN THE MORNING	
1422	Robert Palmer	
1612	The Gap Band	
1639	EARTH ANGEL	New Edition
2535	EASY FOR YOU TO SAY	Linda Ronstadt
2788	EASY LOVE	Dionne Warwick
151	EASY LOVER (Duet With Phil Collins)	Philip Bailey
1265	EAT IT	"Weird Al" Yankovic [P] see: BEAT IT
3494	EAT MY SHORTS	Rick Dees
3409	EATEN ALIVE	Diana Ross
12	EBONY AND IVORY	Paul McCartney
2048	EBONY EYES	Rick James featuring Smokey Robinson
1607	EDGE OF A BROKEN HEART	Vixen
3245	EDGE OF A DREAM	Joe Cocker
1158	THE EDGE OF HEAVEN	Wham!
1066	EDGE OF SEVENTEEN (Just Like The White Winged Dove)	Stevie Nicks
4057	EDIE (Ciao Baby)	The Cult
3464	EENIE MEENIE	Jeffrey Osborne
1958	853-5937	Squeeze
316	867-5309/JENNY	Tommy Tutone
615	18 AND LIFE	Skid Row
3506	8TH WONDER	Sugarhill Gang
642	ELECTION DAY	Arcadia
67	ELECTRIC AVENUE	Eddy Grant
771	ELECTRIC BLUE	Icehouse
3378	ELECTRIC KINGDOM	Twilight 22
1232	ELECTRIC YOUTH	Debbie Gibson
3539	ELECTRICLAND	Bad Company
376	ELVIRA	The Oak Ridge Boys
3067	THE ELVIS MEDLEY	Elvis Presley
1320	EMERGENCY	Kool & The Gang
3212	EMINENCE FRONT	The Who
3900	EMOTION	Barbra Streisand
1291	EMOTION IN MOTION	Ric Ocasek
217	EMOTIONAL RESCUE	The Rolling Stones
3387	EMOTIONS IN MOTION	Billy Squier
1297	EMPIRE STRIKES BACK (Medley)	Meco
1049	EMPTY GARDEN (Hey Hey Johnny)	Elton John
3177	ENCORE	Cheryl Lynn
938	THE END OF THE INNOCENCE	Don Henley
2912	END OF THE LINE	Traveling Wilburys
2	ENDLESS LOVE	Diana Ross & Lionel Richie
1429	ENDLESS NIGHTS	Eddie Money
378	ENDLESS SUMMER NIGHTS	Richard Marx
3881	ENGLISHMAN IN NEW YORK	Sting
2751	ENOUGH IS ENOUGH	April Wine
2604	ESCALATOR OF LIFE	Robert Hazard
291	ETERNAL FLAME	Bangles
3339	EUROPA AND THE PIRATE TWINS	Thomas Dolby
1810	EVEN IT UP	Heart
1100	EVEN NOW	Bob Seger & The Silver Bullet Band
390	EVEN THE NIGHTS ARE BETTER	Air Supply
3186	EVER SINCE THE WORLD BEGAN	Tommy Shaw
1068	EVERLASTING LOVE	Howard Jones
1886	EVERLASTING LOVE	Rex Smith/Rachel Sweet
3744	EVERY BEAT OF MY HEART	Rod Stewart
5	EVERY BREATH YOU TAKE	The Police
2984	EVERY HOME SHOULD HAVE ONE ('81)	Patti Austin
3213	--EVERY HOME SHOULD HAVE ONE (Remix) ('83)	Patti Austin
1266	EVERY LITTLE KISS ('87)	Bruce Hornsby And The Range
3036	--EVERY LITTLE KISS ('86)	Bruce Hornsby And The Range
466	EVERY LITTLE STEP	Bobby Brown
400	EVERY LITTLE THING SHE DOES IS MAGIC	The Police
3954	EVERY LOVE SONG	Greg Kihn Band
124	EVERY ROSE HAS ITS THORN	Poison
1755	EVERY STEP OF THE WAY	John Waite
343	EVERY WOMAN IN THE WORLD	Air Supply
1365	EVERYBODY DANCE	Ta Mara & The Seen
249	EVERYBODY HAVE FUN TONIGHT	Wang Chung
130	EVERYBODY WANTS TO RULE THE WORLD	Tears For Fears
1603	EVERYBODY WANTS YOU	Billy Squier
1104	EVERYBODY'S GOT TO LEARN SOMETIME	The Korgis
2591	EVERYDAY	James Taylor
1771	EVERYDAY I WRITE THE BOOK	Elvis Costello And The Attractions
2274	EVERYDAY PEOPLE	Joan Jett And The Blackhearts
2482	EVERYTHING I NEED	Men At Work
1696	EVERYTHING IN MY HEART	Corey Hart
3981	EVERYTHING IS ALRIGHT	Spider
2486	EVERYTHING MUST CHANGE	Paul Young
211	EVERYTHING SHE WANTS (Remix)	Wham!
2339	EVERYTHING WORKS IF YOU LET IT	Cheap Trick
600	EVERYTHING YOUR HEART DESIRES	Daryl Hall & John Oates
2850	EVERYTIME YOU CRY	The Outfield
191	EVERYTIME YOU GO AWAY	Paul Young
1185	EVERYWHERE	Fleetwood Mac
2748	EWOK CELEBRATION	Meco
371	EXPRESS YOURSELF	Madonna
183	EYE IN THE SKY	Alan Parsons Project
8	EYE OF THE TIGER	Survivor
3797	EYE OF THE ZOMBIE	John Fogerty
3089	EYE ON YOU	Billy Squier
3341	EYE TO EYE	Go West
3535	EYES THAT SEE IN THE DARK	Kenny Rogers
384	EYES WITHOUT A FACE	Billy Idol

F

#	Title	Artist
1578	FACE THE FACE	Pete Townshend
1789	FACTS OF LOVE	Jeff Lorber Featuring Karyn White
1619	FADE AWAY	Bruce Springsteen
3607	FADE AWAY	Loz Netto
2903	FADING AWAY	Will To Power
102	FAITH	George Michael
902	FAITHFULLY	Journey
1758	FAKE	Alexander O'Neal
2057	FAKE FRIENDS	Joan Jett & The Blackhearts
1145	FALL IN LOVE WITH ME	Earth, Wind & Fire
4107	FALL ON ME	R.E.M.
1298	FALLEN ANGEL	Poison
2552	FALLING IN LOVE	Balance

Singles Alpha Index

2908 FALLING IN LOVE AGAIN *Michael Stanley Band*	3834 THE FLYER *Saga*	2944 FRONT PAGE STORY *Neil Diamond*
1541 FALLING IN LOVE (Uh-Oh) *Miami Sound Machine*	3053 FOLLOW YOU *Glen Burtnick*	2413 FULL OF FIRE *Shalamar*
4032 FALLING OUT OF LOVE *Ivan Neville*	4038 FOLLOW YOUR HEART *Triumph*	533 FUNKY COLD MEDINA *Tone Loc*
357 FAME *Irene Cara*	2703 FOOL FOR A PRETTY FACE (Hurt By Love) *Humble Pie*	FUNKYTOWN
696 FAMILY MAN *Daryl Hall & John Oates*	3708 FOOL FOR YOUR LOVE *Jimmy Hall*	29 *Lipps, Inc.*
4042 FAMILY MAN *Fleetwood Mac*	1968 FOOL FOR YOUR LOVING ('89) *Whitesnake*	916 *Pseudo Echo*
2212 THE FANATIC *Felony*	2665 --FOOL FOR YOUR LOVING ('80) *Whitesnake*	2972 FUNNY HOW TIME SLIPS AWAY *Spinners*
2898 FANTASTIC VOYAGE *Lakeside*	1549 FOOL IN LOVE WITH YOU *Jim Photoglo*	1361 THE FUTURE'S SO BRIGHT, I GOTTA WEAR SHADES *Timbuk 3*
1577 FANTASY *Aldo Nova*	1408 FOOL IN THE RAIN *Led Zeppelin*	
2416 FANTASY GIRL *.38 Special*	2312 FOOL MOON FIRE *Walter Egan*	# G
894 FAR FROM OVER *Frank Stallone*	2217 FOOL THAT I AM *Rita Coolidge*	
2004 FAREWELL MY SUMMER LOVE *Michael Jackson*	1682 FOOLIN' *Def Leppard*	2504 GAMES *Phoebe Snow*
1456 FASCINATED *Company B*	3197 FOOLIN' YOURSELF *Aldo Nova*	1047 GAMES PEOPLE PLAY *Alan Parsons Project*
2412 FASCINATION STREET *The Cure*	332 FOOLISH BEAT *Debbie Gibson*	2456 GAMES WITHOUT FRONTIERS *Peter Gabriel*
2959 FASHION *David Bowie*	FOOLISH HEART	3336 THE GAP *Thompson Twins*
725 FAST CAR *Tracy Chapman*	1167 *Steve Perry*	3743 GARDEN PARTY *Herb Alpert*
4163 FAT *"Weird Al" Yankovic* [P] see: BAD	3928 *Sharon Bryant*	1777 GEE WHIZ *Bernadette Peters*
250 FATHER FIGURE *George Michael*	1998 FOOLISH PRIDE *Daryl Hall*	1121 GEMINI DREAM *The Moody Blues*
3718 FAVORITE WASTE OF TIME *Bette Midler*	3851 FOOL'S GAME *Michael Bolton*	1657 GENERAL HOSPI-TALE *The Afternoon Delights*
1947 FEEL IT AGAIN *Honeymoon Suite*	3776 FOOLS LIKE ME *Lorenzo Lamas*	1662 GENIUS OF LOVE *Tom Tom Club*
2608 FEEL LIKE A NUMBER *Bob Seger & The Silver Bullet Band*	43 FOOTLOOSE *Kenny Loggins*	2019 GET CLOSER *Linda Ronstadt*
3258 FEEL THE HEAT *Jean Beauvoir*	2468 FOR A ROCKER *Jackson Browne*	929 GET DOWN ON IT *Kool & The Gang*
2820 FEELINGS OF FOREVER *Tiffany*	1963 FOR AMERICA *Jackson Browne*	Get Fresh At The Weekend...see: SHOWING OUT
2036 FEELS SO GOOD *Van Halen*	3513 FOR THE LOVE OF MONEY *Bulletboys*	3687 GET IT *Stevie Wonder & Michael Jackson*
3415 FEELS SO REAL (Won't Let Go) *Patrice Rushen*	1525 FOR TONIGHT *Nancy Martinez*	3431 GET IT ON *Kingdom Come*
1210 FEELS SO RIGHT *Alabama*	362 FOR YOUR EYES ONLY *Sheena Easton*	967 GET IT ON *The Power Station*
3453 FEET DON'T FAIL ME NOW *Utopia*	1753 FOREVER *Kenny Loggins*	2840 GET IT RIGHT *Aretha Franklin*
3278 FEMALE INTUITION *Mai Tai*	2964 FOREVER *Little Steven And The Disciples Of Soul*	1208 GET ON YOUR FEET *Gloria Estefan*
2823 FIELDS OF FIRE *Big Country*	1471 (Forever) LIVE AND DIE *Orchestral Manoeuvres In The Dark*	222 GET OUTTA MY DREAMS, GET INTO MY CAR *Billy Ocean*
3001 FIGHT FIRE WITH FIRE *Kansas*	1773 FOREVER MAN *Eric Clapton*	2005 GET THAT LOVE *Thompson Twins*
936 THE FINAL COUNTDOWN *Europe*	1634 FOREVER MINE *The O'Jays*	2532 GET UP AND GO *Go-Go's*
2814 FINALLY *T.G. Sheppard*	2897 FOREVER MINE *The Motels*	1750 GETCHA BACK *The Beach Boys*
1715 FIND A WAY *Amy Grant*	1011 FOREVER YOUNG *Rod Stewart*	2732 THE GHOST IN YOU *Psychedelic Furs*
1439 FIND ANOTHER FOOL *Quarterflash*	2526 FOREVER YOUNG ('88) *Alphaville*	1818 (Ghost) RIDERS IN THE SKY *The Outlaws*
1847 FIND YOUR WAY BACK *Jefferson Starship*	4088 --FOREVER YOUNG ('85) *Alphaville*	1890 GHOST TOWN *Cheap Trick*
1467 A FINE FINE DAY *Tony Carey*	282 FOREVER YOUR GIRL *Paula Abdul*	49 GHOSTBUSTERS *Ray Parker Jr.*
728 THE FINER THINGS *Steve Winwood*	3668 FOREVER YOURS *Tony Terry*	2709 THE GIGOLO *O'Bryan*
2242 THE FINEST *The S.O.S. Band*	2418 FORGET ME NOT *Bad English*	2022 GIMME ALL YOUR LOVIN *ZZ Top*
1130 FINISH WHAT YA STARTED *Van Halen*	1483 FORGET ME NOTS *Patrice Rushen*	1312 GIMME SOME LOVIN' *Blues Brothers*
2619 FIRE (Live) *Bruce Springsteen & The E Street Band*	--Forgive Me, Girl...see WORKING MY WAY...	Gimme The Good Life...see: BON BON VIE
1327 FIRE AND ICE *Pat Benatar*	2356 FORGIVE ME FOR DREAMING *Elisa Fiorillo*	3685 GIMME YOUR GOOD LOVIN' *Diving For Pearls*
1728 FIRE IN THE MORNING *Melissa Manchester*	812 FORTRESS AROUND YOUR HEART *Sting*	1442 GIRL CAN'T HELP IT *Journey*
3672 FIRE IN THE SKY *The Dirt Band*	1570 FOUR IN THE MORNING... *Night Ranger*	2489 GIRL, DON'T LET IT GET YOU DOWN *The O'Jays*
568 FIRE LAKE *Bob Seger*	4002 FOUR LITTLE DIAMONDS *ELO*	2380 GIRL I AM SEARCHING FOR YOU *Stevie B*
3360 FIRE WITH FIRE *Wild Blue*	3351 FRANKIE *Sister Sledge*	196 GIRL I'M GONNA MISS YOU *Milli Vanilli*
2290 FIRE WOMAN *The Cult*	3927 FREAK-A-RISTIC *Atlantic Starr*	1957 A GIRL IN TROUBLE (Is A Temporary Thing) *Romeo Void*
3182 FIREFLIES *Fleetwood Mac*	2950 FREAK-A-ZOID *Midnight Star*	91 THE GIRL IS MINE *Michael Jackson/Paul McCartney*
2690 FIRST...BE A WOMAN *Lenore O'Malley*	3269 FREAKSHOW ON THE DANCE FLOOR *Bar-Kays*	2924 GIRL WITH THE HUNGRY EYES *Jefferson Starship*
2099 THE FIRST DAY OF SUMMER *Tony Carey*	Free Bird...see: BABY, I LOVE YOUR WAY	274 GIRL YOU KNOW IT'S TRUE *Milli Vanilli*
2679 FIRST NIGHT *Survivor*	2549 FREE ME *Roger Daltrey*	630 GIRLFRIEND *Pebbles*
2199 FIRST TIME LOVE *Livingston Taylor*	595 FREEDOM *Wham!*	2636 GIRLFRIEND *Bobby Brown*
1783 FISHNET *Morris Day*	2740 FREEDOM *Pointer Sisters*	1214 GIRLS *Dwight Twilley*
4100 500 MILES *Hooters*	1588 FREEDOM OVERSPILL *Steve Winwood*	2527 GIRLS AIN'T NOTHING BUT TROUBLE *D.J. Jazzy Jeff & The Fresh Prince*
Five Minutes Of Funk...see: FRIENDS	427 FREEWAY OF LOVE *Aretha Franklin*	2016 GIRLS ARE MORE FUN *Ray Parker Jr.*
256 THE FLAME *Cheap Trick*	356 FREEZE-FRAME *The J. Geils Band*	1719 GIRLS CAN GET IT *Dr. Hook*
1971 FLAMES OF PARADISE *Jennifer Rush (Duet With Elton John)*	2464 FRENCH KISS *Lil' Louis*	1102 GIRLS, GIRLS, GIRLS *Mötley Crüe*
7 FLASHDANCE...WHAT A FEELING *Irene Cara*	2599 FRENCH KISSIN *Debbie Harry*	117 GIRLS JUST WANT TO HAVE FUN *Cyndi Lauper*
3991 FLASHES *Tiggi Clay*	884 FRESH *Kool & The Gang*	2037 GIRLS WITH GUNS *Tommy Shaw*
2345 FLASH'S THEME AKA FLASH *Queen*	976 FRIENDS *Jody Watley with Eric B. & Rakim*	3216 GIVE *Missing Persons*
1839 FLESH FOR FANTASY *Billy Idol*	4020 FRIENDS//FIVE MINUTES OF FUNK [TSW] *Whodini*	2302 GIVE A LITTLE BIT MORE *Cliff Richard*
2183 FLIRTIN' WITH DISASTER *Molly Hatchet*	224 FRIENDS AND LOVERS *Gloria Loring & Carl Anderson*	1199 GIVE IT ALL YOU GOT *Chuck Mangione*
2288 FLY AWAY *Blackfoot*	2148 FRIENDS IN LOVE *Dionne Warwick And Johnny Mathis*	2175 GIVE IT TO ME BABY *Rick James*
FLY AWAY	4091 FRIGHT NIGHT *The J. Geils Band*	1070 GIVE IT UP *KC*
2783 *Peter Allen*		
3990 *Stevie Woods*		

Singles Alpha Index

2791 GIVE IT UP *Steve Miller Band*	4126 GRAVITY *James Brown*	709 HEART AND SOUL *Huey Lewis And The News*
2707 GIVE ME ALL NIGHT *Carly Simon*	2764 THE GREAT COMMANDMENT *Camouflage*	3908 HEART AND SOUL *The Monkees*
2385 GIVE ME ALL YOUR LOVE *Whitesnake*	2390 GREAT GOSH A'MIGHTY! *Little Richard*	215 HEART ATTACK *Olivia Newton-John*
2660 GIVE ME THE KEYS (And I'll Drive You Crazy) *Huey Lewis And The News*	3626 THE GREATEST GIFT OF ALL *Kenny Rogers & Dolly Parton*	2911 HEART DON'T FAIL ME NOW *Holly Knight*
352 GIVE ME THE NIGHT *George Benson*	146 GREATEST LOVE OF ALL *Whitney Houston*	2704 HEART DON'T LIE *La Toya Jackson*
2551 GIVE ME THE REASON *Luther Vandross*	237 GROOVY KIND OF LOVE *Phil Collins*	1571 HEART HOTELS *Dan Fogelberg*
2066 GIVE ME TONIGHT *Shannon*	303 GUILTY *Barbra Streisand & Barry Gibb*	2888 A HEART IN NEW YORK *Art Garfunkel*
1526 GIVE TO LIVE *Sammy Hagar*	3293 GUARANTEED FOR LIFE *Millions Like Us*	3390 THE HEART IS NOT SO SMART *El DeBarge with DeBarge*
633 GIVING IT UP FOR YOUR LOVE *Delbert McClinton*	1793 GUITAR MAN (Remix) *Elvis Presley*	1592 HEART LIKE A WHEEL *Steve Miller Band*
2370 GIVING UP ON LOVE *Rick Astley*	3753 GUNS FOR HIRE *AC/DC*	1996 HEART OF MINE *Boz Scaggs*
344 GIVING YOU THE BEST THAT I GOT *Anita Baker*	1163 GYPSY *Fleetwood Mac*	488 THE HEART OF ROCK & ROLL *Huey Lewis And The News*
542 THE GLAMOROUS LIFE *Sheila E.*	2932 GYPSY ROAD *Cinderella*	1407 HEART OF THE NIGHT *Juice Newton*
2097 GLAMOUR BOYS *Living Colour*	3751 GYPSY SPIRIT *Pendulum*	848 HEART TO HEART *Kenny Loggins*
2497 GLIDE *Pleasure*		2765 HEART TURNS TO STONE *Foreigner*
57 GLORIA *Laura Branigan*	# H	3495 HEARTACHE *Pepsi & Shirlie*
3096 GLORIA *The Doors*	2104 HAD A DREAM (Sleeping With The Enemy) *Roger Hodgson*	2772 HEARTACHE ALL OVER THE WORLD *Elton John*
602 GLORY DAYS *Bruce Springsteen*	3082 HALF MOON SILVER *Hotel*	2538 HEARTACHE AWAY *Don Johnson*
210 GLORY OF LOVE (Theme From The Karate Kid Part II) *Peter Cetera*	3447 HALLELUIAH MAN *Love and Money*	745 HEARTBEAT *Don Johnson*
2315 GO *Asia*	1161 HAND TO HOLD ON TO *John Cougar*	1722 HEARTBREAK BEAT *Psychedelic Furs*
3858 GO DOWN EASY *Dan Fogelberg*	2140 HANDLE WITH CARE *Traveling Wilburys*	1434 HEARTBREAK HOTEL *The Jacksons*
3103 GO FOR IT *Kim Wilde*	3112 HANDS ACROSS AMERICA *Voices Of America*	711 HEARTBREAKER *Dionne Warwick*
3429 GO FOR SODA *Kim Mitchell*	4070 HANDS ACROSS THE SEA *Modern English*	1224 HEARTBREAKER *Pat Benatar*
986 GO HOME *Stevie Wonder*	4026 HANDS ON THE RADIO *Henry Lee Summer*	482 HEARTLIGHT *Neil Diamond*
1433 GO INSANE *Lindsey Buckingham*	2213 HANDS TIED *Scandal Featuring Patty Smyth*	4094 HEARTLINE *Robin George*
3985 GO SEE THE DOCTOR *Kool Moe Dee*	309 HANDS TO HEAVEN *Breathe*	606 HEARTS *Marty Balin*
3174 GOIN' CRAZY! *David Lee Roth*	1671 HANG FIRE *The Rolling Stones*	4065 HEARTS AWAY *Night Ranger*
1223 GOIN' DOWN *Greg Guidry*	3700 HANG ON NOW *Kajagoogoo*	1332 HEARTS ON FIRE *Randy Meisner*
3872 GOIN' ON *The Beach Boys*	2452 HANGIN' ON A STRING (Contemplating) *Loose Ends*	1797 HEARTS ON FIRE *Bryan Adams*
2741 GOIN' TO THE BANK *Commodores*	347 HANGIN' TOUGH *New Kids On The Block*	2702 HEARTS ON FIRE *Steve Winwood*
1902 GOING BACK TO CALI *LL Cool J*	1885 HANGING ON A HEART ATTACK *Device*	3500 HEART'S ON FIRE *John Cafferty*
1807 GOING TO A GO-GO *The Rolling Stones*	1491 HAPPY *Surface*	266 THE HEAT IS ON *Glenn Frey*
1685 GOLD *Spandau Ballet*	2910 HAPPY ENDING *Joe Jackson*	2722 THE HEAT OF HEAT *Patti Austin*
1970 GOLDMINE *Pointer Sisters*	3430 HAPPY HOUR *Deodato*	373 HEAT OF THE MOMENT *Asia*
3707 GONE TOO FAR *Eddie Rabbitt*	3069 HAPPY MAN *Greg Kihn Band*	3021 HEAT OF THE MOMENT *After 7*
Gone, Gone, Gone...see: MY GIRL	2936 HAPPY TOGETHER (A Fantasy) *The Captain & Tennille*	740 HEAT OF THE NIGHT *Bryan Adams*
3350 GONNA MAKE IT *Sa-Fire*	3493 --HAPPY TOGETHER *The Nylons*	201 HEAVEN *Bryan Adams*
3907 GOOD FRIENDS *Joni Mitchell*	272 HARD HABIT TO BREAK *Chicago*	263 HEAVEN *Warrant*
3410 A GOOD HEART *Feargal Sharkey*	3518 HARD TIMES *James Taylor*	849 HEAVEN HELP ME *Deon Estus*
3044 GOOD LIFE *Inner City*	3237 HARD TIMES FOR LOVERS *Jennifer Holliday*	3818 HEAVEN IN YOUR ARMS *Dan Hartman*
3233 THE GOOD LORD LOVES YOU *Neil Diamond*	741 HARD TO SAY *Dan Fogelberg*	1110 HEAVEN IN YOUR EYES *Loverboy*
2774 GOOD MORNING GIRL/STAY AWHILE *Journey*	37 HARD TO SAY I'M SORRY *Chicago*	206 HEAVEN IS A PLACE ON EARTH *Belinda Carlisle*
3699 GOOD MUSIC *Joan Jett And The Blackhearts*	110 HARDEN MY HEART *Quarterflash*	4130 HEAVEN KNOWS *When In Rome*
Good Ol' Boys...see: THEME FROM "THE DUKES OF HAZZARD"	3794 THE HARDEST PART *Blondie*	3048 HEAVEN (Must Be There) *Eurogliders*
338 GOOD THING *Fine Young Cannibals*	732 HARLEM SHUFFLE *The Rolling Stones*	2514 HEAVEN'S ON FIRE *KISS*
2219 GOOD TIMES *INXS And Jimmy Barnes*	3038 HAS ANYONE EVER WRITTEN ANYTHING FOR YOU *Stevie Nicks*	1813 HEAVY METAL (Takin' A Ride) *Don Felder*
1275 GOODBYE *Night Ranger*	2947 HAVE YOU EVER LOVED SOMEBODY *Freddie Jackson*	1762 HE'LL NEVER LOVE YOU (Like I Do) *Freddie Jackson*
2201 GOODBYE IS FOREVER *Arcadia*	2265 HAVEN'T YOU HEARD *Patrice Rushen*	39 HELLO *Lionel Richie*
2852 GOODBYE TO YOU *Scandal*	Having A Party...see: BAD BOY	628 HELLO AGAIN *Neil Diamond*
2920 GOODNIGHT MY LOVE *Mike Pinera*	405 HAZY SHADE OF WINTER *Bangles*	1346 HELLO AGAIN *The Cars*
2818 GOODNIGHT SAIGON *Billy Joel*	1674 HE CAN'T LOVE YOU *Michael Stanley Band*	2757 HELP ME! *Marcy Levy & Robin Gibb*
782 GOODY TWO SHOES *Adam Ant*	3296 HE COULD BE THE ONE *Josie Cotton*	1078 HER TOWN TOO *James Taylor And J.D. Souther*
1152 THE GOONIES 'R' GOOD ENOUGH *Cyndi Lauper*	2989 HE GOT YOU *Ronnie Milsap*	2904 HERE COMES MY GIRL *Tom Petty And The Heartbreakers*
913 GOT A HOLD ON ME *Christine McVie*	3810 HE WANTS MY BODY *Starpoint*	363 HERE COMES THE RAIN AGAIN *Eurythmics*
3854 GOT A NEW LOVE *Good Question*	434 HEAD OVER HEELS *Tears For Fears*	437 HERE I AM (Just When I Thought I Was Over You) *Air Supply*
3252 GOT IT MADE *Crosby, Stills, Nash & Young*	900 HEAD OVER HEELS *Go-Go's*	180 HERE I GO AGAIN *Whitesnake*
220 GOT MY MIND SET ON YOU *George Harrison*	200 HEAD TO TOE *Lisa Lisa And Cult Jam*	3578 HERE SHE COMES *Bonnie Tyler*
3243 GOT TO BE THERE *Chaka Khan*	2233 HEADED FOR A FALL *Firefall*	3256 HERE TO LOVE YOU *The Doobie Brothers*
3276 GOT TO LOVE SOMEBODY *Sister Sledge*	1352 HEADED FOR A HEARTBREAK *Winger*	1383 HERE WITH ME *REO Speedwagon*
3621 GOT TO ROCK ON *Kansas*	2595 HEADED FOR THE FUTURE *Neil Diamond*	2686 HEROES *Commodores*
3542 GOTTA GET YOU HOME TONIGHT *Eugene Wilde*	3371 HEADLINES *Midnight Star*	2347 HE'S A LIAR *Bee Gees*
3884 GOTTA GIVE A LITTLE LOVE (Ten Years After) *Timmy Thomas*	1329 HEALING HANDS *Elton John*	3660 HE'S A PRETENDER *High Inergy*
2235 GOTTA HAVE MORE LOVE *Climax Blues Band*	496 HEART AND SOUL *T'Pau*	3623 HE'S MY GIRL *David Hallyday*
3509 GRACELAND *Paul Simon*		251 HE'S SO SHY *Pointer Sisters*

Singles Alpha Index

#	Title	Artist
1354	HEY BABY	Henry Lee Summer
	Hey Hey Johnny... see: EMPTY GARDEN	
2372	HEY LADIES	Beastie Boys
4108	HEY MAMBO	Barry Manilow with Kid Creole And The Coconuts
765	HEY NINETEEN	Steely Dan
1929	HEY THERE LONELY GIRL	Robert John
3017	HIDE YOUR HEART	KISS
3841	HIGH ENERGY	Evelyn Thomas
2033	HIGH ON EMOTION	Chris DeBurgh
956	HIGH ON YOU	Survivor
3438	HIGH ON YOUR LOVE	Debbie Jacobs
4080	HIGH SCHOOL NIGHTS	Dave Edmunds
2648	HIGH TIME	Styx
264	HIGHER LOVE	Steve Winwood
566	HIM	Rupert Holmes
479	HIP TO BE SQUARE	Huey Lewis And The News
2156	HIPPY HIPPY SHAKE	The Georgia Satellites
554	HIT ME WITH YOUR BEST SHOT	Pat Benatar
170	HOLD ME	Fleetwood Mac
1880	HOLD ME	Teddy Pendergrass
2698	HOLD ME	Menudo
2905	HOLD ME	Sheila E.
3787	HOLD ME	Laura Branigan
4165	HOLD ME	Colin James Hay
214	HOLD ME NOW	Thompson Twins
1792	HOLD ME 'TIL THE MORNIN' COMES	Paul Anka
1138	HOLD ON	Santana
2234	HOLD ON	Kansas
2962	HOLD ON	Badfinger
3274	HOLD ON	Donny Osmond
1576	HOLD ON LOOSELY	.38 Special
815	HOLD ON TIGHT	ELO
942	HOLD ON TO MY LOVE	Jimmy Ruffin
313	HOLD ON TO THE NIGHTS	Richard Marx
	Hold On To Your Dream...see: The DREAM	
4047	HOLD TIGHT	Change
2264	HOLDIN' ON	Tané Cain
3273	HOLDIN' ON FOR DEAR LOVE	Lobo
324	HOLDING BACK THE YEARS	Simply Red
1187	HOLDING ON	Steve Winwood
1865	HOLDING OUT FOR A HERO	Bonnie Tyler
2892	HOLE IN MY HEART (All The Way To China)	Cyndi Lauper
1035	HOLIDAY	Madonna
1726	HOLIDAY	The Other Ones
3253	HOLIDAY	Kool & The Gang
3948	HOLIDAY	Nazareth
3701	HOLIDAY ROAD	Lindsey Buckingham
3516	HOLLYWOOD	Shooting Star
3517	HOLYANNA	Toto
3857	HOME SWEET HOME	Mötley Crüe
1509	HONESTLY	Stryper
2525	HONEY, HONEY	David Hudson
1529	THE HONEYTHIEF	Hipsway
2597	HOOKED ON BIG BANDS	Frank Barber Orchestra
807	HOOKED ON CLASSICS	The Royal Philharmonic Orchestra
1997	HOOKED ON SWING	Larry Elgart And His Manhattan Swing Orchestra
	HOOKED ON YOU	
1548	Sweet Sensation ('89)	
2674	Sweet Sensation ('87)	
2113	HOPE YOU LOVE ME LIKE YOU SAY YOU DO	Huey Lewis And The News
	Hopelessly In Love...see: THE PARTY'S OVER	
2316	THE HORIZONTAL BOP	Bob Seger
2933	HOT FOR TEACHER	Van Halen
2931	HOT FUN IN THE SUMMERTIME	Dayton
924	HOT GIRLS IN LOVE	Loverboy
3108	HOT HOT HOT!!!	The Cure
2307	HOT HOT HOT (Radio Edit)	Buster Poindexter And His Banshees Of Blue
	HOT IN THE CITY	
1356	Billy Idol ('82)	
2515	Billy Idol ('87)	
920	HOT ROD HEARTS	Robbie Dupree
2961	HOT THING	Prince
3924	HOT WATER	Level 42
1274	HOURGLASS	Squeeze
3697	THE HOUSE OF THE RISING SUN	Dolly Parton
879	HOW AM I SUPPOSED TO LIVE WITHOUT YOU	Laura Branigan
887	HOW 'BOUT US	Champaign
412	HOW CAN I FALL?	Breathe
2967	HOW CAN I FORGET YOU	Elisa Fiorillo
3297	HOW CAN I LIVE WITHOUT HER	Christopher Atkins
2286	HOW CAN I REFUSE	Heart
3833	HOW CAN YOU LOVE ME	Ambrosia
726	HOW DO I MAKE YOU	Linda Ronstadt
1494	HOW DO I SURVIVE	Amy Holland
1711	HOW DO YOU KEEP THE MUSIC PLAYING (Theme From "Best Friends")	James Ingram And Patti Austin
1881	HOW DOES IT FEEL TO BE BACK	Daryl Hall & John Oates
2654	HOW LONG	Rod Stewart
1895	HOW MANY TIMES CAN WE SAY GOODBYE	Dionne Warwick And Luther Vandross
2688	HOW MUCH LOVE	Survivor
1510	(How To Be A) MILLIONAIRE	ABC
160	HOW WILL I KNOW	Whitney Houston
212	HUMAN	The Human League
699	HUMAN NATURE	Michael Jackson
1239	HUMAN TOUCH	Rick Springfield
3786	HUNGRY	Winger
478	HUNGRY EYES	Eric Carmen
345	HUNGRY HEART	Bruce Springsteen
149	HUNGRY LIKE THE WOLF	Duran Duran
3728	THE HUNTER	GTR
2787	HUNTERS OF THE NIGHT	Mr. Mister
3027	HURRY UP AND WAIT	The Isley Brothers
3777	HURT	Re-Flex
676	HURT SO BAD	Linda Ronstadt
26	HURTS SO GOOD	John Cougar
2578	HURTS TO BE IN LOVE	Gino Vannelli
2071	HYPERACTIVE	Robert Palmer
3005	HYPERACTIVE	Thomas Dolby
2154	HYPNOTIZE ME	Wang Chung
1120	HYSTERIA	Def Leppard

I

#	Title	Artist
856	I AIN'T GONNA STAND FOR IT	Stevie Wonder
	I Ain't Got Nobody...see: JUST A GIGOLO	
1536	I AM BY YOUR SIDE	Corey Hart
2420	I AM LOVE	Jennifer Holliday
1242	I BEG YOUR PARDON	Kon Kan
2142	I BELIEVE	Chilliwack
1325	I BELIEVE IN YOU	Don Williams
3941	I BELIEVE IN YOU	Stryper
3136	I CALL YOUR NAME	Switch
532	I CAN DREAM ABOUT YOU	Dan Hartman
4068	I CAN SURVIVE	Triumph
2223	I CAN TAKE CARE OF MYSELF	Billy & The Beaters
3737	I CANNOT BELIEVE IT'S TRUE	Phil Collins
1720	I CAN'T DRIVE 55	Sammy Hagar
3202	I CAN'T FACE THE FACT	Gina Go-Go
36	I CAN'T GO FOR THAT (No Can Do)	Daryl Hall & John Oates
1054	I CAN'T HELP IT	Andy Gibb & Olivia Newton-John
2227	I CAN'T HELP IT	Bananarama
2014	I CAN'T HELP MYSELF (Sugar Pie, Honey Bunch)	Bonnie Pointer
835	I CAN'T HOLD BACK	Survivor
1927	I CAN'T LET GO	Linda Ronstadt
4012	I CAN'T SAY GOODBYE TO YOU	Helen Reddy
818	I CAN'T STAND IT	Eric Clapton And His Band
2318	I CAN'T STAND STILL	Don Henley
3568	I CAN'T STOP THE FEELIN'	Pure Prairie League
609	I CAN'T TELL YOU WHY	Eagles
421	I CAN'T WAIT	Nu Shooz
1421	I CAN'T WAIT	Stevie Nicks
3119	I CAN'T WAIT	Deniece Williams
2866	I COULD BE GOOD FOR YOU	707
1203	I COULD NEVER MISS YOU (More Than I Do)	Lulu
1032	I COULD NEVER TAKE THE PLACE OF YOUR MAN	Prince
1926	I COULDN'T SAY NO	Robert Ellis Orrall With Carlene Carter
3113	I CRY JUST A LITTLE BIT	Shakin' Stevens
3610	I DID IT FOR LOVE	Night Ranger
3446	I DIDN'T MEAN TO STAY ALL NIGHT	Starship
	I DIDN'T MEAN TO TURN YOU ON	
415	Robert Palmer	
3381	Cherrelle	
1542	I DO (LIVE VERSION)	The J. Geils Band
1849	I DO'WANNA KNOW	REO Speedwagon
1808	I DO WHAT I DO (Theme For 9 1/2 Weeks)	John Taylor
1565	I DO YOU	The Jets
2144	I DON'T CARE ANYMORE	Phil Collins
3720	I DON'T KNOW	Michael Morales
2045	I DON'T KNOW WHERE TO START	Eddie Rabbitt
3423	I DON'T LIKE MONDAYS	The Boomtown Rats
1805	I DON'T MIND AT ALL	Bourgeois Tagg
299	I DON'T NEED YOU	Kenny Rogers
2979	I DON'T NEED YOU	Rupert Holmes
3760	I DON'T NEED YOU ANYMORE	Together? Featuring Jackie DeShannon
3171	I DON'T THINK THAT MAN SHOULD SLEEP ALONE	Ray Parker Jr.
2714	I DON'T WANNA DANCE	Eddy Grant
450	I DON'T WANNA GO ON WITH YOU LIKE THAT	Elton John
524	I DON'T WANNA LIVE WITHOUT YOUR LOVE	Chicago
3636	I DON'T WANT A LOVER	Texas
1961	I DON'T WANT TO BE A HERO	Johnny Hates Jazz
3721	I DON'T WANT TO BE LONELY	Dana Valery
3388	I DON'T WANT TO KNOW YOUR NAME	Glen Campbell
791	I DON'T WANT TO LIVE WITHOUT YOU	Foreigner
2406	I DON'T WANT TO TALK ABOUT IT	Rod Stewart
2013	I DON'T WANT TO WALK WITHOUT YOU	Barry Manilow
581	I DON'T WANT YOUR LOVE	Duran Duran
945	I DROVE ALL NIGHT	Cyndi Lauper
3087	I EAT CANNIBALS	Total Coelo
3507	I ENGINEER	Animotion
156	I FEEL FOR YOU	Chaka Khan
4008	I FEEL FREE	Belinda Carlisle
2098	I FEEL THE EARTH MOVE	Martika
3696	I FEEL THE MAGIC	Belinda Carlisle

Singles Alpha Index

1868 I FOUND SOMEBODY *Glenn Frey*	--I ONLY WANT TO BE WITH YOU	844 I WON'T HOLD YOU BACK *Toto*
I FOUND SOMEONE	2639 *Nicolette Larson*	1848 I WON'T STAND IN YOUR WAY *Stray Cats*
857 *Cher*	3755 *The Tourists*	963 I WOULD DIE 4 U *Prince And The Revolution*
3837 *Laura Branigan*	1169 I PLEDGE MY LOVE *Peaches & Herb*	3033 I WOULDN'T BEG FOR WATER *Sheena Easton*
1977 I GET EXCITED *Rick Springfield*	3071 I PREDICT *Sparks*	1285 I WOULDN'T HAVE MISSED IT FOR THE WORLD
3590 I GET OFF ON IT *Tony Joe White*	3481 I PRETEND *Kim Carnes*	*Ronnie Milsap*
491 I GET WEAK *Belinda Carlisle*	794 I RAN (SO FAR AWAY) *A Flock Of Seagulls*	4071 I WOULDN'T LIE *Yarbrough & Peoples*
2493 I GOT THE FEELIN' (It's Over) *Gregory Abbott*	2039 I REALLY DON'T NEED NO LIGHT *Jeffrey Osborne*	2646 I'D DO IT ALL AGAIN *Sam Harris*
2523 I GOT YOU *Split Enz*	862 I REMEMBER HOLDING YOU *Boys Club*	2143 I'D RATHER LEAVE WHILE I'M IN LOVE
1888 I GOT YOU BABE *UB40 With Chrissie Hynde*	1010 I SAW HIM STANDING THERE *Tiffany*	*Rita Coolidge*
2061 I GOTTA TRY *Michael McDonald*	3345 I SEND A MESSAGE *INXS*	1188 I'D STILL SAY YES *Klymaxx*
407 I GUESS THAT'S WHY THEY CALL IT THE BLUES	1883 I SHOULD BE SO LUCKY *Kylie Minogue*	3474 IF ANYBODY HAD A HEART *John Waite*
Elton John	3262 I SHOULDA LOVED YA *Narada Michael Walden*	1162 IF ANYONE FALLS *Stevie Nicks*
803 I HATE MYSELF FOR LOVING YOU	901 I STILL BELIEVE *Brenda K. Starr*	665 IF EVER YOU'RE IN MY ARMS AGAIN
Joan Jett And The Blackhearts	950 I STILL CAN'T GET OVER LOVING YOU	*Peabo Bryson*
3083 I HAVE THE SKILL *Sherbs*	*Ray Parker Jr.*	3380 IF I COULD GET YOU (Into My Life) *Gene Cotton*
3024 I HEAR YOU NOW *Jon & Vangelis*	219 I STILL HAVEN'T FOUND WHAT I'M LOOKING FOR	448 IF I COULD TURN BACK TIME *Cher*
529 I HEARD A RUMOUR *Bananarama*	*U2*	4018 IF I HAD A ROCKET LAUNCHER *Bruce Cockburn*
I HEARD IT THROUGH THE GRAPEVINE	3964 I STILL WANT YOU *The Del Fuegos*	2229 IF I HAD MY WISH TONIGHT *David Lasley*
3448 *Roger* (Part 1)	3158 I SURRENDER *Arlan Day*	2954 IF I SAY YES *Five Star*
3892 *The California Raisins*	1995 I THANK YOU *ZZ Top*	3408 IF I WAS YOUR GIRLFRIEND *Prince*
54 I JUST CALLED TO SAY I LOVE YOU *Stevie Wonder*	2988 I THINK I CAN BEAT MIKE TYSON	IF I WERE YOU
331 I JUST CAN'T STOP LOVING YOU *Michael Jackson*	*D.J. Jazzy Jeff & The Fresh Prince*	2210 *Lulu*
2991 I JUST CAN'T WALK AWAY *Four Tops*	1357 I THINK IT'S LOVE *Jermaine Jackson*	3458 *Toby Beau*
281 (I Just) DIED IN YOUR ARMS *Cutting Crew*	242 I THINK WE'RE ALONE NOW *Tiffany*	1261 IF I'D BEEN THE ONE *38 Special*
397 I KEEP FORGETTIN' *Michael McDonald*	4133 I THINK YOU'LL REMEMBER TONIGHT *Axe*	2337 IF IT AIN'T ONE THING... IT'S ANOTHER
3389 I KNEW THE BRIDE (When She Use To Rock And	1081 I WANNA BE A COWBOY *Boys Don't Cry*	*Richard "Dimples" Fields*
Roll) *Nick Lowe and His Cowboy Outfit*	3161 I WANNA BE LOVED *House Of Lords*	716 IF IT ISN'T LOVE *New Edition*
315 I KNEW YOU WERE WAITING (For Me)	1547 I WANNA BE THE ONE *Stevie B*	2508 IF LOOKS COULD KILL *Player*
Aretha Franklin & George Michael	935 I WANNA BE YOUR LOVER *Prince*	2855 IF LOOKS COULD KILL *Heart*
2017 I KNEW YOU WHEN *Linda Ronstadt*	115 I WANNA DANCE WITH SOMEBODY	3856 IF LOVE SHOULD GO *Streets*
3275 (I Know) I'M LOSING YOU *Uptown*	(Who Loves Me) *Whitney Houston*	2251 IF ONLY YOU KNEW *Patti LaBelle*
683 I KNOW THERE'S SOMETHING GOING ON *Frida*	I WANNA GO BACK	1790 IF SHE KNEW WHAT SHE WANTS *Bangles*
3150 I KNOW WHAT BOYS LIKE *The Waitresses*	1155 *Eddie Money*	1229 IF SHE WOULD HAVE BEEN FAITHFUL... *Chicago*
1076 I KNOW WHAT I LIKE *Huey Lewis And The News*	3604 *Billy Satellite*	1580 IF THE LOVE FITS WEAR IT *Leslie Pearl*
1919 I KNOW YOU'RE OUT THERE SOMEWHERE	899 I WANNA HAVE SOME FUN *Samantha Fox*	608 IF THIS IS IT *Huey Lewis And The News*
The Moody Blues	2129 I WANNA HEAR IT FROM YOUR LIPS *Eric Carmen*	2488 IF WE NEVER MEET AGAIN
3255 I LIKE *Guy*	3107 I WANNA ROCK *Twisted Sister*	*Tommy Conwell And The Young Rumblers*
3880 I LIKE *Men Without Hats*	538 I WANT A NEW DRUG *Huey Lewis And The News*	3554 IF YOU ASKED ME TO *Patti LaBelle*
634 I LIKE IT *Dino*	2642 I WANT ACTION *Poison*	329 IF YOU DON'T KNOW ME BY NOW *Simply Red*
1458 I LIKE IT *DeBarge*	2906 I WANT CANDY *Bow Wow Wow*	4111 IF YOU FEEL IT *Denise Lopez*
3962 I LIKE TO ROCK *April Wine*	700 I WANT HER *Keith Sweat*	556 IF YOU LEAVE *Orchestral Manoeuvres In The Dark*
2581 I LIKE YOU (Special Mix) *Phyllis Nelson*	2755 I WANT IT ALL *Queen*	3175 IF YOU LET ME STAY *Terence Trent D'Arby*
1901 I LIVE BY THE GROOVE *Paul Carrack*	3334 I WANT MY GIRL *Jesse Johnson's Revue*	382 IF YOU LOVE SOMEBODY SET THEM FREE *Sting*
1059 I LIVE FOR YOUR LOVE *Natalie Cole*	575 I WANT TO BE YOUR MAN *Roger*	1907 IF YOU SHOULD SAIL *Nielsen/Pearson*
3890 I LOST ON JEOPARDY *"Weird Al" Yankovic* [P]	3286 I WANT TO BE YOUR PROPERTY *Blue Mercedes*	3221 IF YOU WANNA GET BACK YOUR LADY
see also: JEOPARDY	2506 I WANT TO BREAK FREE *Queen*	*Pointer Sisters*
63 I LOVE A RAINY NIGHT *Eddie Rabbitt*	96 I WANT TO KNOW WHAT LOVE IS *Foreigner*	2287 IF YOU WANT MY LOVE *Cheap Trick*
4106 I LOVE MY TRUCK *Glen Campbell*	4146 I WANT TO MAKE THE WORLD TURN AROUND	3595 IF YOU WERE A WOMAN (And I Was A Man)
19 I LOVE ROCK 'N ROLL *Joan Jett & The*	*Steve Miller Band*	*Bonnie Tyler*
Blackhearts	3823 I WANT YOU *Animotion*	2592 IF YOUR HEART ISN'T IN IT *Atlantic Starr*
2881 I LOVE TO BASS *Bardeux*	1965 I WANT YOU, I NEED YOU *Chris Christian*	1641 I.G.Y. (What A Beautiful World) *Donald Fagen*
3977 I LOVE WOMEN *Jim Hurt*	2661 I WANT YOU SO BAD *Heart*	1173 IKO IKO *The Belle Stars*
749 I LOVE YOU *Climax Blues Band*	346 I WANT YOUR SEX *George Michael*	388 I'LL ALWAYS LOVE YOU *Taylor Dayne*
2000 I LOVED 'EM EVERY ONE *T.G. Sheppard*	3710 I WAS BORN TO LOVE YOU *Freddie Mercury*	1112 I'LL BE ALRIGHT WITHOUT YOU *Journey*
949 I MADE IT THROUGH THE RAIN *Barry Manilow*	3614 I WAS LOOKING FOR SOMEONE TO LOVE	3030 I'LL BE AROUND *What Is This*
3520 I MELT WITH YOU *Modern English*	*Leif Garrett*	2472 I'LL BE GOOD *René and Angela*
349 I MISS YOU *Klymaxx*	4110 I WASN'T THE ONE (Who Said Goodbye)	279 I'LL BE LOVING YOU (Forever)
1375 I MISSED AGAIN *Phil Collins*	*Agnetha Faltskog and Peter Cetera*	*New Kids On The Block*
2620 I MUST BE DREAMING *Giuffria*	2421 I WILL ALWAYS LOVE YOU *Dolly Parton*	951 I'LL BE OVER YOU *Toto*
2573 I NEED A MAN *Eurythmics*	2165 I WILL BE THERE *Glass Tiger*	4061 I'LL BE THERE *Kenny Loggins*
1388 I NEED LOVE *LL Cool J*	3577 I WILL FOLLOW *U2*	243 I'LL BE THERE FOR YOU *Bon Jovi*
2073 I NEED YOU *Paul Carrack*	1476 I WISH I HAD A GIRL *Henry Lee Summer*	2641 I'LL BE YOU *The Replacements*
2119 I NEED YOU *Pointer Sisters*	2439 I WISH I WAS EIGHTEEN AGAIN *George Burns*	2951 I'LL DRINK TO YOU *Duke Jupiter*
4154 I NEED YOU *Maurice White*	1530 I WONDER IF I TAKE YOU HOME	2340 I'LL FALL IN LOVE AGAIN *Sammy Hagar*
2112 I NEED YOU TONIGHT *Peter Wolf*	*Lisa Lisa and Cult Jam with Full Force*	2516 I'LL FIND MY WAY HOME *Jon & Vangelis*
1914 I NEED YOUR LOVIN' *Teena Marie*	1196 I WON'T BACK DOWN *Tom Petty*	1712 I'LL STILL BE LOVING YOU *Restless Heart*
2329 I NEED YOUR LOVING *The Human League*	3462 I WON'T BE HOME TONIGHT *Tony Carey*	2351 I'LL TRY SOMETHING NEW *A Taste Of Honey*
1976 I ONLY WANNA BE WITH YOU *Samantha Fox*	1153 I WON'T FORGET YOU *Poison*	753 I'LL TUMBLE 4 YA *Culture Club*

Singles Alpha Index

1113 I'LL WAIT *Van Halen*	3292 IN THE SHAPE OF A HEART *Jackson Browne*	3561 IT'S ALRIGHT (Baby's Coming Back) *Eurythmics*
2393 ILLEGAL ALIEN *Genesis*	483 IN TOO DEEP *Genesis*	2558 IT'S FOR YOU *Player*
2754 I'M A BELIEVER *Giant*	IN YOUR EYES	3766 IT'S GETTIN' LATE *The Beach Boys*
1302 I'M ALIVE *Electric Light Orchestra*	1734 Peter Gabriel ('86)	3748 IT'S GONNA BE SPECIAL *Patti Austin*
1939 I'M ALIVE *Neil Diamond*	2007 Peter Gabriel ('89)	941 IT'S GONNA TAKE A MIRACLE *Deniece Williams*
3060 I'M ALIVE *Gamma*	1553 IN YOUR LETTER *REO Speedwagon*	2367 IT'S HARD TO BE HUMBLE *Mac Davis*
1842 I'M ALMOST READY *Pure Prairie League*	661 IN YOUR ROOM *Bangles*	2101 IT'S INEVITABLE *Charlie*
569 I'M ALRIGHT *Kenny Loggins*	2403 IN YOUR SOUL *Corey Hart*	3966 IT'S JUST THE SUN *Don McLean*
3873 I'M BAD *LL Cool J*	2267 INDESTRUCTIBLE *Four Tops*	(It's Just) THE WAY THAT YOU LOVE ME
443 I'M COMING OUT *Diana Ross*	3653 INDUSTRIAL DISEASE *Dire Straits*	380 Paula Abdul ('89)
3980 I'M FOR REAL *Howard Hewett*	619 INFATUATION *Rod Stewart*	3862 Paula Abdul ('88)
1518 I'M FREE (Heaven Helps The Man) *Kenny Loggins*	4043 INFORMATION *Eric Martin*	3080 IT'S LIKE WE NEVER SAID GOODBYE *Crystal Gayle*
1131 I'M GOIN' DOWN *Bruce Springsteen*	4087 INJURED IN THE GAME OF LOVE *Donnie Iris*	2628 IT'S MONEY THAT MATTERS *Randy Newman*
3950 I'M GONNA LOVE HER FOR BOTH OF US *Meat Loaf*	3658 INNOCENT EYES *Graham Nash*	2837 IT'S MY JOB *Jimmy Buffett*
3780 I'M GONNA MISS YOU *Kenny Loggins*	847 AN INNOCENT MAN *Billy Joel*	1693 IT'S MY LIFE *Talk Talk*
1237 I'M GONNA TEAR YOUR PLAYHOUSE DOWN *Paul Young*	3052 INSIDE A DREAM *Jane Wiedlin*	3277 IT'S MY PARTY *Dave Stewart With Barbara Gaskin*
3248 I'M HAPPY JUST TO DANCE WITH YOU *Anne Murray*	2333 INSIDE LOVE (So Personal) *George Benson*	617 IT'S MY TURN *Diana Ross*
1845 I'M HAPPY THAT LOVE HAS FOUND YOU *Jimmy Hall*	3640 INSIDE OF YOU *Ray, Goodman & Brown*	880 IT'S NO CRIME *Babyface*
2034 I'M IN LOVE *Evelyn King*	2519 INSIDE OUTSIDE *The Cover Girls*	2100 IT'S NO SECRET *Kylie Minogue*
2529 I'M IN LOVE AGAIN *Pia Zadora*	4092 INTO MY LOVE *Greg Guidry*	2883 IT'S NOT A WONDER (Live) *Little River Band*
2880 I'M JUST TOO SHY *Jermaine Jackson*	4162 INTO MY SECRET *Alisha*	1195 IT'S NOT ENOUGH *Starship*
4017 I'M NEVER GONNA SAY GOODBYE *Billy Preston*	INTO THE NIGHT	1094 IT'S NOT OVER ('Til It's Over) *Starship*
2659 I'M NO ANGEL *Gregg Allman Band*	886 Benny Mardones ('80)	3992 IT'S NOT YOU, IT'S NOT ME *KBC Band*
3007 I'M NOT PERFECT (But I'm Perfect For You) *Grace Jones*	1448 Benny Mardones ('89)	1093 IT'S NOW OR NEVER *John Schneider*
3164 I'M NOT THE MAN I USED TO BE *Fine Young Cannibals*	2996 INTO YOU *Giant Steps*	1281 IT'S ONLY LOVE *Bryan Adams/Tina Turner*
2031 I'M NOT THE ONE *The Cars*	994 INVINCIBLE (Theme From The Legend Of Billie Jean) *Pat Benatar*	2693 IT'S ONLY LOVE *Simply Red*
3478 I'M NOT YOUR MAN *Tommy Conwell And The Young Rumblers*	1624 INVISIBLE *Alison Moyet*	933 IT'S RAINING AGAIN *Supertramp*
675 I'M ON FIRE *Bruce Springsteen*	2002 INVISIBLE HANDS *Kim Carnes*	2280 IT'S RAINING MEN *The Weather Girls*
I'M SO EXCITED	328 INVISIBLE TOUCH *Genesis*	34 IT'S STILL ROCK AND ROLL TO ME *Billy Joel*
787 Pointer Sisters ('84)	3382 INVITATION TO DANCE *Kim Carnes*	3086 IT'S THE END OF THE WORLD AS WE KNOW IT (And I Feel Fine) *R.E.M.*
1652 Pointer Sisters ('82)	2444 I.O.U. *Lee Greenwood*	2682 IT'S TRICKY *Run-D.M.C.*
4128 I'M SO GLAD I'M STANDING HERE TODAY *The Crusaders*	Irresistable Bitch...see: LET'S PRETEND WE'RE MARRIED	2708 IT'S YOU *Bob Seger & The Silver Bullet Band*
I'm Special...see: BRASS IN POCKET	1001 IS IT LOVE *Mr. Mister*	943 I'VE BEEN IN LOVE BEFORE *Cutting Crew*
3100 I'M STEPPING OUT *John Lennon*	4007 IS IT LOVE *J.J. Fad*	2750 I'VE BEEN WAITING FOR YOU ALL OF MY LIFE *Paul Anka*
2074 I'M STILL SEARCHING *Glass Tiger*	1287 IS IT YOU *Lee Ritenour*	663 I'VE DONE EVERYTHING FOR YOU *Rick Springfield*
918 I'M STILL STANDING *Elton John*	3137 IS THAT IT? *Katrina And The Waves*	1144 I'VE GOT A ROCK N' ROLL HEART *Eric Clapton*
1336 I'M THAT TYPE OF GUY *LL Cool J*	451 IS THERE SOMETHING I SHOULD KNOW *Duran Duran*	205 (I've Had) THE TIME OF MY LIFE *Bill Medley And Jennifer Warnes*
2208 I'M THE ONE *Roberta Flack*	310 IS THIS LOVE *Whitesnake*	3758 I'VE JUST BEGUN TO LOVE YOU *Dynasty*
3961 I'M THROUGH WITH LOVE *Eric Carmen*	762 IS THIS LOVE *Survivor*	I've Loved You For A Long Time...see: CUPID
489 I'M YOUR MAN *Wham!*	2758 IS THIS LOVE *Pat Travers Band*	377 I'VE NEVER BEEN TO ME *Charlene*
3843 I'M YOUR MAN *Barry Manilow*	3849 IS THIS THE END *New Edition*	
4104 I'M YOUR SUPERMAN *All Sports Band*	2150 ISLAND OF LOST SOULS *Blondie*	# J
3925 IMAGINATION *Belouis Some*	38 ISLANDS IN THE STREAM *Kenny Rogers Duet with Dolly Parton*	31 JACK & DIANE *John Cougar*
3914 IMAGINE *Tracie Spencer*	1254 IT AIN'T ENOUGH *Corey Hart*	4148 JACK THE LAD *3 Man Island*
1136 IN A BIG COUNTRY *Big Country*	2867 IT CAN HAPPEN *Yes*	2831 JACKIE *Blue Zone U.K.*
1022 IN AMERICA *Charlie Daniels Band*	2441 IT DIDN'T TAKE LONG *Spider*	2824 JACKIE BROWN *John Cougar Mellencamp*
3376 IN AND OUT OF LOVE *Bon Jovi*	3475 IT HURTS TO BE IN LOVE *Dan Hartman*	306 JACOB'S LADDER *Huey Lewis And The News*
4166 IN BETWEEN DAYS (Without You) *The Cure*	3803 IT HURTS TOO MUCH *Eric Carmen*	2341 JAM ON IT *Newcleus*
2250 IN GOD'S COUNTRY *U2*	2520 IT ISN'T, IT WASN'T, IT AIN'T NEVER GONNA BE *Aretha Franklin and Whitney Houston*	2008 JAM TONIGHT *Freddie Jackson*
3574 IN IT FOR LOVE *England Dan & John Ford Coley*	1326 IT MIGHT BE YOU (Theme From Tootsie) *Stephen Bishop*	3898 THE JAM WAS MOVING *Debbie Harry*
3267 IN LOVE WITH LOVE *Debbie Harry*	1816 IT MUST BE LOVE *Madness*	1118 JAMIE *Ray Parker Jr.*
1165 IN MY DREAMS *REO Speedwagon*	3729 IT TAKES TIME *The Marshall Tucker Band*	3860 JAMMIN *Teena Marie*
3512 IN MY DREAMS *Dokken*	1871 IT TAKES TWO *Rob Base & D.J. E-Z Rock*	1654 JAMMIN' ME *Tom Petty and The Heartbreakers*
1746 IN MY EYES *Stevie B*	1033 IT WOULD TAKE A STRONG STRONG MAN *Rick Astley*	Jammin'...see: MASTER BLASTER
625 IN MY HOUSE *Mary Jane Girls*	3543 IT'S A HARD LIFE *Queen*	1021 JANE *Jefferson Starship*
1894 IN NEON *Elton John*	1670 IT'S A LOVE THING *The Whispers*	3188 JANE'S GETTING SERIOUS *Jon Astley*
1380 IN THE AIR TONIGHT *Phil Collins*	1114 IT'S A MIRACLE *Culture Club*	3930 JANET *Commodores*
2135 IN THE DARK *Billy Squier*	621 IT'S A MISTAKE *Men At Work*	3731 JEALOUS GUY *John Lennon and the Plastic Ono Band (With The Flux Fiddlers)*
3492 IN THE DRIVER'S SEAT *John Schneider*	3399 IT'S A NIGHT FOR BEAUTIFUL GIRLS *The Fools*	188 JEOPARDY *Greg Kihn Band* see also: I LOST ON JEOPARDY
1877 IN THE MOOD *Robert Plant*	1008 IT'S A SIN *Pet Shop Boys*	821 JESSE *Carly Simon*
2830 IN THE NAME OF LOVE *Ralph MacDonald (With Bill Withers)*	2700 IT'S ALL I CAN DO *Anne Murray*	

Singles Alpha Index

2975 JESSE *Julian Lennon*
  48 JESSIE'S GIRL *Rick Springfield*
1960 JIMMY LEE *Aretha Franklin*
3844 JIMMY LOVES MARYANN *Josie Cotton*
3241 JIMMY MACK *Sheena Easton*
 247 JOANNA *Kool & The Gang*
2442 JODY *Jermaine Stewart*
2670 JOHNNY B *Hooters*
3821 JOHNNY B. GOODE *Peter Tosh*
2308 JOHNNY CAN'T READ *Don Henley*
3602 JOHNNY COME HOME *Fine Young Cannibals*
1211 JOJO *Boz Scaggs*
3391 JOLE BLON *Gary U.S. Bonds*
2317 JONES VS. JONES *Kool & The Gang*
3603 JOY *Teddy Pendergrass*
2605 JOY AND PAIN *Rob Base & D.J. E-Z Rock*
2637 JOYSTICK *Dazz Band*
2206 JUICY FRUIT *Mtume*
1878 JUKE BOX HERO *Foreigner*
  28 JUMP *Van Halen*
 278 JUMP (For My Love) *Pointer Sisters*
1207 JUMP START *Natalie Cole*
1826 JUMP TO IT *Aretha Franklin*
1824 JUMPIN' JACK FLASH *Aretha Franklin*
2539 JUNGLE BOY *John Eddie*
1117 JUNGLE LOVE *The Time*
1156 JUST A GIGOLO/I AIN'T GOT NOBODY
       David Lee Roth
3812 JUST ANOTHER DAY *Oingo Boingo*
2386 JUST ANOTHER DAY IN PARADISE *Bertie Higgins*
1137 JUST ANOTHER NIGHT *Mick Jagger*
1403 JUST AS I AM *Air Supply*
2305 JUST BE GOOD TO ME *The S.O.S. Band*
3335 JUST BE MY LADY *Larry Graham*
1222 JUST BECAUSE *Anita Baker*
1435 JUST BETWEEN YOU AND ME *April Wine*
3644 JUST CAN'T WAIT *The J. Geils Band*
2091 JUST CAN'T WIN 'EM ALL *Stevie Woods*
3790 JUST FOR THE MOMENT *Ray Kennedy*
1761 JUST GOT LUCKY *JoBoxers*
 989 JUST GOT PAID *Johnny Kemp*
1632 JUST LIKE HEAVEN *The Cure*
 773 JUST LIKE JESSE JAMES *Cher*
 864 JUST LIKE PARADISE *David Lee Roth*
3999 JUST LIKE PARADISE *Larry John McNally*
  24 (Just Like) STARTING OVER *John Lennon*
     Just Like The White Winged Dove...
     see: EDGE OF SEVENTEEN
 968 JUST ONCE *Quincy Jones Featuring James Ingram*
3011 JUST ONE MORE TIME *Headpins*
2568 JUST SO LONELY *Get Wet*
  88 JUST THE TWO OF US *Grover Washington, Jr.*
2680 JUST THE WAY YOU LIKE IT *The S.O.S. Band*
2684 JUST TO SATISFY YOU *Waylon & Willie*
 806 JUST TO SEE HER *Smokey Robinson*

K

4151 KAREN *B. E. Taylor Group*
  52 KARMA CHAMELEON *Culture Club*
3223 KAYLEIGH *Marillion*
 610 (Keep Feeling) FASCINATION *The Human League*
4084 KEEP IT CONFIDENTIAL *Nona Hendryx*
3637 KEEP IT TIGHT *Single Bullet Theory*
 121 KEEP ON LOVING YOU *REO Speedwagon*
 965 KEEP ON MOVIN' *Soul II Soul*
1938 KEEP THE FIRE *Kenny Loggins*
 593 KEEP THE FIRE BURNIN' *REO Speedwagon*

3281 KEEP THIS TRAIN A-ROLLIN' *The Doobie Brothers*
2533 KEEP YOUR EYE ON ME *Herb Alpert*
 417 KEEP YOUR HANDS TO YOURSELF
       Georgia Satellites
2459 KEEPING OUR LOVE ALIVE *Henry Paul Band*
1350 KEEPING THE FAITH *Billy Joel*
 507 KEY LARGO *Bertie Higgins*
3214 KICK THE WALL *Jimmy Davis & Junction*
2917 THE KID IS HOT TONITE *Loverboy*
1861 THE KID'S AMERICAN *Matthew Wilder*
1364 KIDS IN AMERICA *Kim Wilde*
1488 KILLIN' TIME *Fred Knoblock & Susan Anton*
3134 KILLING ME SOFTLY *Al B. Sure!*
2338 A KIND OF MAGIC *Queen*
 867 KING FOR A DAY *Thompson Twins*
 410 KING OF PAIN *The Police*
     see also: KING OF SUEDE
3131 KING OF SUEDE *"Weird Al" Yankovic* [P]
     see: KING OF PAIN
1825 KING OF THE HILL *Rick Pinette And Oak*
 154 KISS *Prince And The Revolution*
2088 KISS *The Art Of Noise featuring Tom Jones*
2075 KISS AND TELL *Bryan Ferry*
2799 KISS AND TELL *Breakfast Club*
2966 KISS AND TELL *Isley, Jasper, Isley*
1128 KISS HIM GOODBYE *The Nylons*
1038 KISS ME DEADLY *Lita Ford*
2114 KISS ME IN THE RAIN *Barbra Streisand*
 120 KISS ON MY LIST *Daryl Hall & John Oates*
1583 KISS THE BRIDE *Elton John*
3155 KISS YOU (When It's Dangerous) *Eight Seconds*
1170 KISSES ON THE WIND *Neneh Cherry*
4161 K.I.S.S.I.N.G. *Siedah Garrett*
 853 KISSING A FOOL *George Michael*
2258 KNOCKED OUT *Paula Abdul*
2956 KNOCKING AT YOUR BACK DOOR *Deep Purple*
 269 KOKOMO *The Beach Boys*
 161 KYRIE *Mr. Mister*

L

 133 LA BAMBA *Los Lobos*
3667 LA-DI-DA *Sad Café*
 519 LA ISLA BONITA *Madonna*
3386 LA LA MEANS I LOVE YOU *Tierra*
 521 LADIES NIGHT *Kool & The Gang*
  16 LADY *Kenny Rogers*
1836 LADY *The Whispers*
3417 LADY DOWN ON LOVE *Alabama*
 354 THE LADY IN RED *Chris DeBurgh*
4022 LADY, LADY, LADY *Joe "Bean" Esposito*
1664 LADY LOVE ME (One More Time) *George Benson*
3367 LADY OF MY HEART *Jack Wagner*
2411 LADY SOUL *The Temptations*
 570 LADY (You Bring Me Up) *Commodores*
2997 LAND OF A THOUSAND DANCES *The J. Geils Band*
 540 LAND OF CONFUSION *Genesis*
3986 LAND OF LA LA *Stevie Wonder*
2446 LANDLORD *Gladys Knight & The Pips*
2712 LANDSLIDE *Olivia Newton-John*
1092 THE LANGUAGE OF LOVE *Dan Fogelberg*
2363 THE LAST MILE *Cinderella*
3622 THE LAST SAFE PLACE ON EARTH *Le Roux*
2115 THE LAST TIME I MADE LOVE
       Joyce Kennedy & Jeffrey Osborne
2130 LAST TRAIN TO LONDON *Electric Light Orchestra*
1444 THE LAST WORTHLESS EVENING *Don Henley*
3008 LATE AT NIGHT *England Dan Seals*
 493 LATE IN THE EVENING *Paul Simon*

3010 LATELY *Stevie Wonder*
 948 LAWYERS IN LOVE *Jackson Browne*
3546 LAY ALL YOUR LOVE ON ME *Information Society*
2843 LAY DOWN YOUR ARMS *The Graces*
2321 LAY IT DOWN *Ratt*
 672 LAY YOUR HANDS ON ME *Thompson Twins*
 785 LAY YOUR HANDS ON ME *Bon Jovi*
3149 LAYIN' IT ON THE LINE *Jefferson Starship*
2725 LE BEL AGE *Pat Benatar*
3239 LEAD A DOUBLE LIFE *Loverboy*
4153 LEAD ME ON *Amy Grant*
 627 LEADER OF THE BAND *Dan Fogelberg*
2548 LEADER OF THE PACK *Twisted Sister*
 207 LEAN ON ME *Club Nouveau*
3194 LEARNING TO FLY *Pink Floyd*
 446 LEATHER AND LACE
       Stevie Nicks (With Don Henley)
1115 LEAVE A LIGHT ON *Belinda Carlisle*
1554 LEAVE A TENDER MOMENT ALONE *Billy Joel*
1498 LEAVE IT *Yes*
3370 LEAVING L.A. *Deliverance*
3306 THE LEBANON *The Human League*
2408 LEFT IN THE DARK *Barbra Streisand*
4025 LEFT TO MY OWN DEVICES *Pet Shop Boys*
3754 LEGAL TENDER *The B-52's*
1796 THE LEGEND OF WOOLEY SWAMP
       Charlie Daniels Band
 669 LEGS *ZZ Top*
3673 LEILA *ZZ Top*
3443 A LESSON IN LEAVIN' *Dottie West*
1057 LESSONS IN LOVE *Level 42*
2083 LET GO *Sharon Bryant*
2069 LET HIM GO *Animotion*
3711 LET IT ALL BLOW *Dazz Band*
2190 LET IT BE ME *Willie Nelson*
 404 LET IT WHIP *Dazz Band*
2477 LET ME BE *Korona*
1774 LET ME BE THE CLOCK *Smokey Robinson*
 759 LET ME BE THE ONE *Exposé*
2828 LET ME BE THE ONE *Five Star*
1181 LET ME BE YOUR ANGEL *Stacy Lattisaw*
3906 LET ME DOWN EASY *Roger Daltrey*
2285 LET ME GO *Ray Parker Jr.*
3477 LET ME GO *Heaven 17*
3569 LET ME GO *The Rings*
2063 LET ME GO, LOVE *Nicolette Larson*
3026 LET ME IN *Eddie Money*
2365 LET ME LOVE YOU ONCE *Greg Lake*
 809 LET ME LOVE YOU TONIGHT *Pure Prairie League*
3814 LET ME SLEEP ALONE *Cugini*
2369 LET ME TALK *Earth, Wind & Fire*
1404 LET ME TICKLE YOUR FANCY *Jermaine Jackson*
 798 LET MY LOVE OPEN THE DOOR *Pete Townshend*
2891 LET THE DAY BEGIN *The Call*
2248 LET THE FEELING FLOW *Peabo Bryson*
 623 LET THE MUSIC PLAY *Shannon*
2594 LET THE RIVER RUN *Carly Simon*
3121 LET'S BE LOVERS AGAIN
       Eddie Money with Valerie Carter
  53 LET'S DANCE *David Bowie*
3736 LET'S DANCE *Chris Rea*
3960 LET'S DANCE (Make Your Body Move)
       West Street Mob
3869 LET'S DO SOMETHING CHEAP AND SUPERFICIAL
       Burt Reynolds
2457 LET'S GET IT UP *AC/DC*
 629 LET'S GET SERIOUS *Jermaine Jackson*
 999 LET'S GO *Wang Chung*

Singles Alpha Index

640 LET'S GO ALL THE WAY *Sly Fox*	2613 LITTLE WALTER *Tony! Toni! Toné!*	4004 LOVE AND ROCK AND ROLL *Greg Kihn*
100 LET'S GO CRAZY *Prince And The Revolution*	2239 LIVE EVERY MINUTE *Ali Thomson*	267 LOVE BITES *Def Leppard*
1631 LET'S GO DANCIN' (Ooh La, La, La) *Kool & The Gang*	2120 LIVE EVERY MOMENT *REO Speedwagon*	981 A LOVE BIZARRE *Sheila E.*
4045 LET'S GO OUT TONIGHT *Nile Rodgers*	1690 LIVE IS LIFE *Opus*	1601 LOVE CHANGES (Everything) *Climie Fisher*
2695 LET'S GO 'ROUND AGAIN *Average White Band*	4051 LIVE IT UP *Gardner Cole*	4093 LOVE CHANGES EVERYTHING *Honeymoon Suite*
3412 LET'S GO UP *Diana Ross*	2257 LIVE MY LIFE *Boy George*	1397 LOVE COME DOWN *Evelyn King*
125 LET'S GROOVE *Earth, Wind & Fire*	193 LIVE TO TELL *Madonna*	2928 LOVE COMES QUICKLY *Pet Shop Boys*
2278 LET'S HANG ON *Barry Manilow*	1986 LIVIN' IN DESPERATE TIMES *Olivia Newton-John*	2463 LOVE CRIES *Stage Dolls*
71 LET'S HEAR IT FOR THE BOY *Deniece Williams*	111 LIVIN' ON A PRAYER *Bon Jovi*	4075 LOVE GRAMMAR *John Parr*
Let's Make Lots Of Money...see: OPPORUNITIES	4023 LIVIN' RIGHT *Glenn Frey*	3633 LOVE HAS A MIND OF ITS OWN *Donna Summer with Matthew Ward*
2576 LET'S PRETEND WE'RE MARRIED// IRRESISTABLE BITCH [TSW] *Prince*	2375 LIVING EYES *Bee Gees*	3816 LOVE HAS FINALLY COME AT LAST *Bobby Womack and Patti LaBelle*
3781 LET'S PUT THE FUN BACK IN ROCK N ROLL *Freddy Cannon & The Belmonts*	1469 LIVING IN A BOX *Living In A Box*	3043 LOVE HAS TAKEN ITS TOLL *Saraya*
4158 LET'S PUT THE X IN SEX *KISS*	2809 LIVING IN A DREAM *Pseudo Echo*	767 LOVE IN AN ELEVATOR *Aerosmith*
1496 LET'S STAY TOGETHER *Tina Turner*	1918 LIVING IN A FANTASY *Leo Sayer*	3959 LOVE IN SIBERIA *Laban*
2681 LET'S TALK ABOUT ME *Alan Parsons Project*	597 LIVING IN AMERICA *James Brown*	1524 LOVE IN STORE *Fleetwood Mac*
387 LET'S WAIT AWHILE *Janet Jackson*	888 LIVING IN SIN *Bon Jovi*	1017 LOVE IN THE FIRST DEGREE *Alabama*
2499 LET'S WORK *Mick Jagger*	3956 LIVING IN THE BACKGROUND *Baltimora*	2562 LOVE IN THE FIRST DEGREE *Bananarama*
2366 LICENCE TO CHILL *Billy Ocean*	480 LIVING INSIDE MYSELF *Gino Vannelli*	1567 THE LOVE IN YOUR EYES *Eddie Money*
2635 LICK IT UP *KISS*	3407 LIVING ON THE EDGE *Jim Capaldi*	360 LOVE IS A BATTLEFIELD *Pat Benatar*
1563 LIES *Thompson Twins*	2580 LIVING ON VIDEO *Trans-X*	3279 LOVE IS A HOUSE *Force M.D.'s*
1717 LIES *Jonathan Butler*	273 THE LIVING YEARS *Mike + The Mechanics*	1613 LOVE IS A STRANGER *Eurythmics*
4118 LIFE GETS BETTER *Graham Parker*	419 THE LOCO-MOTION *Kylie Minogue*	1374 LOVE IS ALRIGHT TONITE *Rick Springfield*
796 LIFE IN A NORTHERN TOWN *The Dream Academy*	3432 LOLA (Live Version) *The Kinks*	2767 LOVE IS CONTAGIOUS *Taja Sevelle*
1454 LIFE IN ONE DAY *Howard Jones*	2024 LONELY EYES *Robert John*	1180 LOVE IS FOREVER *Billy Ocean*
1985 A LIFE OF ILLUSION *Joe Walsh*	2857 LONELY IN LOVE *Giuffria*	832 LOVE IS IN CONTROL (Finger On The Trigger) *Donna Summer*
3982 LIFE'S WHAT YOU MAKE IT *Talk Talk*	3290 LONELY IS THE NIGHT *Air Supply*	2094 LOVE IS LIKE A ROCK *Donnie Iris*
2178 LIGHT OF DAY *The Barbusters (Joan Jett & The Blackhearts)*	3994 LONELY NIGHTS *Bryan Adams*	3739 LOVE IS THE HERO *Billy Squier*
1025 LIGHTS OUT *Peter Wolf*	656 LONELY OL' NIGHT *John Cougar Mellencamp*	3586 LOVE IS THE KEY *Maze Featuring Frankie Beverly*
3319 LIKE A CHILD *Noel*	2789 LONELY SCHOOL *Tommy Shaw*	1377 LOVE IS THE SEVENTH WAVE *Sting*
155 LIKE A PRAYER *Madonna*	2435 LONELY TOGETHER *Barry Manilow*	3280 LOVE KILLS *Freddie Mercury*
1273 LIKE A ROCK *Bob Seger & The Silver Bullet Band*	2942 LONELY WON'T LEAVE ME ALONE *Glenn Medeiros*	3909 LOVE LIGHT *Yutaka*
2694 LIKE A SURGEON *"Weird Al" Yankovic* [P] see: LIKE A VIRGIN	3009 LONG AND LASTING LOVE (Once In A Lifetime) *Glenn Medeiros*	1323 LOVE LIGHT IN FLIGHT *Stevie Wonder*
47 LIKE A VIRGIN *Madonna* see also: LIKE A SURGEON	650 THE LONG RUN *Eagles*	2776 LOVE ME AGAIN *The John Hall Band*
3801 LIKE FLAMES *Berlin*	3719 LONG TIME LOVIN' YOU *McGuffey Lane*	2320 LOVE ME IN A SPECIAL WAY *DeBarge*
1351 LIKE NO OTHER NIGHT *38 Special*	4167 LONG WAY TO LOVE *Britny Fox*	1451 LOVE ME TOMORROW *Chicago*
3229 LIKE THE WEATHER *10,000 Maniacs*	194 LONGER *Dan Fogelberg*	2194 LOVE MY WAY *Psychedelic Furs*
2564 LIKE TO GET TO KNOW YOU WELL *Howard Jones*	4034 THE LONGER YOU WAIT *Gino Vannelli*	2934 LOVE NEVER FAILS *Greg Kihn Band*
2739 LIMELIGHT *Rush*	975 THE LONGEST TIME *Billy Joel*	2503 LOVE OF A LIFETIME *Chaka Khan*
2672 LIPSTICK *Suzi Quatro*	225 THE LOOK *Roxette*	2354 LOVE OF THE COMMON PEOPLE *Paul Young*
2643 LISTEN LIKE THIEVES *INXS*	3413 LOOK AT THAT CADILLAC *Stray Cats*	2808 LOVE ON A SHOESTRING *The Captain & Tennille*
289 LISTEN TO YOUR HEART *Roxette*	141 LOOK AWAY *Chicago*	1489 LOVE ON A TWO WAY STREET *Stacy Lattisaw*
3791 A LITTLE BIT OF HEAVEN *Natalie Cole*	3411 LOOK MY WAY *The Vels*	3974 LOVE ON MY MIND TONIGHT *The Temptations*
3648 LITTLE BIT OF LOVE *Dwight Twilley*	845 THE LOOK OF LOVE (Part One) *ABC*	4021 LOVE ON THE AIRWAVES *Night*
2089 A LITTLE BIT OF LOVE (Is All It Takes) *New Edition*	2169 LOOK OUT ANY WINDOW *Bruce Hornsby & The Range*	4005 LOVE ON THE PHONE *Suzanne Fellini*
2092 LITTLE BY LITTLE *Robert Plant*	1029 LOOK WHAT YOU'VE DONE TO ME *Boz Scaggs*	85 LOVE ON THE ROCKS *Neil Diamond*
2505 LITTLE DARLIN' *Sheila*	514 LOOKIN' FOR LOVE *Johnny Lee*	2449 LOVE ON YOUR SIDE *Thompson Twins*
2282 LITTLE FIGHTER *White Lion*	Lookin' For The Lights... see: NOBODY SAID IT WAS EASY	2243 LOVE OR LET ME BE LONELY *Paul Davis*
3126 A LITTLE GOOD NEWS *Anne Murray*	202 LOOKING FOR A NEW LOVE *Jody Watley*	1401 LOVE OVERBOARD *Gladys Knight And The Pips*
978 A LITTLE IN LOVE *Cliff Richard*	2172 LOOKING FOR A STRANGER *Pat Benatar*	2141 THE LOVE PARADE *The Dream Academy*
3560 A LITTLE IS ENOUGH *Pete Townshend*	3344 LOOKING OVER MY SHOULDER *'til tuesday*	1661 LOVE PLUS ONE *Haircut One Hundred*
2125 LITTLE JACKIE WANTS TO BE A STAR *Lisa Lisa And Cult Jam*	2072 LOOKS LIKE LOVE AGAIN *Dann Rogers*	1171 LOVE POWER *Dionne Warwick & Jeffrey Osborne*
98 LITTLE JEANNIE *Elton John*	2596 LOOKS THAT KILL *Mötley Crüe*	3808 LOVE RESURRECTION *Alison Moyet*
3106 LITTLE LADY *Duke Jupiter*	1811 LOST HER IN THE SUN *John Stewart*	393 LOVE SHACK *The B-52's*
1267 LITTLE LIAR *Joan Jett And The Blackhearts*	312 LOST IN EMOTION *Lisa Lisa And Cult Jam*	559 LOVE SOMEBODY *Rick Springfield*
564 LITTLE LIES *Fleetwood Mac*	152 LOST IN LOVE *Air Supply*	578 LOVE SONG *The Cure*
2357 A LITTLE LOVE *Juice Newton*	1904 LOST IN LOVE *New Edition*	2502 A LOVE SONG *Kenny Rogers*
494 LITTLE RED CORVETTE *Prince*	1220 LOST IN YOU *Rod Stewart*	3327 MEDLEY: LOVE SONGS ARE BACK AGAIN *Band Of Gold*
1231 A LITTLE RESPECT *Erasure*	173 LOST IN YOUR EYES *Debbie Gibson*	2026 LOVE STINKS *The J. Geils Band*
3952 LITTLE SHEILA *Slade*	4052 LOUIE, LOUIE *Fat Boys*	3564 LOVE STRUCK *Jesse Johnson*
4074 LITTLE SUZI *Tesla*	3855 LOVE AGAIN *John Denver & Sylvie Vartan*	2559 LOVE THAT GOT AWAY *Firefall*
3313 LITTLE THING CALLED LOVE *Neil Young*	2401 LOVE ALL THE HURT AWAY *Aretha Franklin & George Benson*	1238 LOVE THE WORLD AWAY *Kenny Rogers*
1426 LITTLE TOO LATE *Pat Benatar*	2284 LOVE ALWAYS *El DeBarge*	Love Theme From Chances Are...see: AFTER ALL
	3538 LOVE AND LONELINESS *The Motors*	
	2546 LOVE AND PRIDE *King*	

Singles Alpha Index

Love Theme From Footloose...
see: ALMOST PARADISE
Love Theme From Goya...see: TILL I LOVED YOU
Love Theme From 'Leonard Part 6'...
see: WITHOUT YOU
3525 LOVE THEME FROM SHOGUN (Mariko's Theme) Meco
1107 LOVE THEME FROM ST. ELMO'S FIRE David Foster
Love Theme From TOP GUN...
see: TAKE MY BREATH AWAY
Love Theme From WHITE NIGHTS...
see: SEPARATE LIVES
2054 LOVE T.K.O. Teddy Pendergrass
735 LOVE TOUCH Rod Stewart
3234 LOVE TRAIN Holly Johnson
4036 LOVE, TRUTH AND HONESTY Bananarama
1470 LOVE WALKS IN Van Halen
850 LOVE WILL CONQUER ALL Lionel Richie
1677 LOVE WILL FIND A WAY Yes
1062 LOVE WILL SAVE THE DAY Whitney Houston
1946 LOVE WILL SHOW US HOW Christine McVie
939 LOVE WILL TURN YOU AROUND Kenny Rogers
2970 LOVE X LOVE George Benson
971 LOVE YOU DOWN Ready For The World
1648 LOVE YOU LIKE I NEVER LOVED BEFORE John O'Banion
1003 LOVE ZONE Billy Ocean
2807 LOVELINE Dr. Hook
1063 LOVELY ONE The Jacksons
3366 LOVER Michael Stanley Band
3404 LOVER COME BACK TO ME Dead Or Alive
460 THE LOVER IN ME Sheena Easton
322 LOVERBOY Billy Ocean
4040 LOVERBOY Karen Kamon
468 LOVERGIRL Teena Marie
2728 LOVERS AFTER ALL Melissa Manchester And Peabo Bryson
1876 A LOVER'S HOLIDAY Change
2528 LOVER'S LANE Georgio
4003 LOVE'S ABOUT TO CHANGE MY HEART Donna Summer
631 LOVE'S BEEN A LITTLE BIT HARD ON ME Juice Newton
2275 LOVE'S GOT A LINE ON YOU Scandal
3920 LOVE'S ONLY LOVE Engelbert
922 LOVIN' EVERY MINUTE OF IT Loverboy
2593 LOVIN' THE NIGHT AWAY The Dillman Band
3995 LOVING YOU Chris Rea
3232 LOVING YOU WITH MY EYES Starland Vocal Band
3428 LUANNE Foreigner
1950 LUCKY Greg Kihn
4028 LUCKY Eye To Eye
3093 A LUCKY GUY Rickie Lee Jones
2205 LUCKY IN LOVE Mick Jagger
2471 LUCKY ME Anne Murray
1423 THE LUCKY ONE Laura Branigan
522 LUCKY STAR Madonna
487 LUKA Suzanne Vega
3244 LULLABY The Cure
3211 LYING Peter Frampton

M

2884 MACHINERY Sheena Easton
414 MAD ABOUT YOU Belinda Carlisle
45 MAGIC Olivia Newton-John
960 MAGIC The Cars
3562 MAGIC CARPET RIDE Bardeux

3600 MAGIC MAN Herb Alpert
2495 MAGIC POWER Triumph
3505 MAGICAL John Parr
2874 MAGNETIC Earth, Wind & Fire
839 MAJOR TOM (Coming Home) Peter Schilling
1414 MAKE A LITTLE MAGIC The Dirt Band
587 MAKE A MOVE ON ME Olivia Newton-John
1841 MAKE BELIEVE Toto
3642 MAKE BELIEVE IT'S YOUR FIRST TIME Bobby Vinton
2794 MAKE IT BETTER (Forget About Me) Tom Petty and The Heartbreakers
2662 MAKE IT LAST FOREVER Keith Sweat (Duet With Jacci McGhee)
3958 MAKE IT MEAN SOMETHING Rob Jungklas
535 MAKE IT REAL The Jets
1628 MAKE LOVE STAY Dan Fogelberg
458 MAKE ME LOSE CONTROL Eric Carmen
3127 MAKE MY DAY T.G. Sheppard with Clint Eastwood
2491 MAKE NO MISTAKE, HE'S MINE Barbra Streisand (Duet With Kim Carnes)
2985 MAKE THAT MOVE Shalamar
3230 MAKE UP YOUR MIND Aurra
957 MAKING LOVE Roberta Flack
2046 MAKING LOVE IN THE RAIN Herb Alpert
107 MAKING LOVE OUT OF NOTHING AT ALL Air Supply
3160 MAMA Genesis
2028 MAMA USED TO SAY Junior
2350 MAMA WEER ALL CRAZEE NOW Quiet Riot
3827 MAN AGAINST THE WORLD Survivor
Man In Motion...see: ST. ELMO'S FIRE
227 MAN IN THE MIRROR Michael Jackson
2222 MAN ON THE CORNER Genesis
1065 MAN ON YOUR MIND Little River Band
1342 MAN SIZE LOVE Klymaxx
594 MANDOLIN RAIN Bruce Hornsby And The Range
42 MANEATER Daryl Hall & John Oates
61 MANIAC Michael Sembello
413 MANIC MONDAY Bangles
3533 MARY, MARY Run-D.M.C.
1437 MARY'S PRAYER Danny Wilson
3874 MASQUERADE Berlin
3953 MASTER AND SERVANT Depeche Mode
326 MASTER BLASTER (Jammin') Stevie Wonder
260 MATERIAL GIRL Madonna
3548 MATHEMATICS Melissa Manchester
1039 A MATTER OF TRUST Billy Joel
3848 MAYBE THIS DAY Kissing The Pink
3418 THE MAYOR OF SIMPLETON XTC
1840 ME MYSELF AND I De La Soul
1127 ME SO HORNY 2 Live Crew
2451 ME (Without You) Andy Gibb
3190 THE MEDICINE SONG Stephanie Mills
123 MEDLEY Stars On 45
3364 MEDLEY II (Medley) Stars On 45
3459 MEET EL PRESIDENTE Duran Duran
906 MEET ME HALF WAY Kenny Loggins
2691 MEETING IN THE LADIES ROOM Klymaxx
3176 MEGA FORCE 707
2990 MEMORIES Tierra
3330 MEMORIES OF DAYS GONE BY (Medley) Fred Parris & The Five Satins
2717 MEMORIZE YOUR NUMBER Leif Garrett
MEMORY
1979 Barry Manilow
2925 Barbra Streisand
3902 MEMPHIS Joe Jackson

3487 THE MEN ALL PAUSE Klymaxx
486 MERCEDES BOY Pebbles
2887 MERCY, MERCY, MERCY Phoebe Snow
3772 MERRY CHRISTMAS IN THE NFL Willis "The Guard" and Vigorish
3063 THE MESSAGE Grandmaster Flash & The Furious Five Featuring: Melle Mel and Duke Bootee
Metal Health...see: BANG YOUR HEAD
668 METHOD OF MODERN LOVE Daryl Hall & John Oates
2575 THE METRO Berlin
2731 MEXICAN RADIO Wall Of Voodoo
3115 MIAMI Bob Seger & The Silver Bullet Band
238 MIAMI VICE THEME Jan Hammer
69 MICKEY Toni Basil ...see also: RICKY
2161 MIDAS TOUCH Midnight Star
1260 MIDDLE OF THE ROAD The Pretenders
743 MIDNIGHT BLUE Lou Gramm
2077 MIDNIGHT BLUE Louise Tucker
3688 MIDNIGHT RAIN Poco
3625 MIDNIGHT RENDEZVOUS The Babys
1599 MIDNIGHT ROCKS Al Stewart
3289 MIDNITE MANIAC Krokus
3932 A MILLION MILES AWAY The Plimsouls
2040 MINIMUM LOVE Mac McAnally
1814 MIRACLES Stacy Lattisaw
4077 MIRAGE Eric Troyer
1855 MIRROR MAN The Human League
912 MIRROR, MIRROR Diana Ross
2252 MISFIT Curiosity Killed The Cat
817 MISLED Kool & The Gang
551 MISS ME BLIND Culture Club
1091 MISS SUN Boz Scaggs
738 MISS YOU LIKE CRAZY Natalie Cole
109 MISS YOU MUCH Janet Jackson
2087 MISSED OPPORTUNITY Daryl Hall & John Oates
72 MISSING YOU John Waite
733 MISSING YOU Diana Ross
1358 MISSING YOU Dan Fogelberg
1204 MISSIONARY MAN Eurythmics
1815 MISTAKE NO. 3 Culture Club
3104 MISTAKEN IDENTITY Kim Carnes
271 MR. ROBOTO Styx
2029 MISTER SANDMAN Emmylou Harris
1179 MR. TELEPHONE MAN New Edition
1004 MISUNDERSTANDING Genesis
861 MIXED EMOTIONS The Rolling Stones
1313 MODERN DAY DELILAH Van Stephenson
1259 MODERN GIRL Sheena Easton
1205 MODERN LOVE David Bowie
1197 MODERN WOMAN Billy Joel
2994 THE MOMENT OF TRUTH Survivor
2490 MONEY (That's What I Want) The Flying Lizards
1788 MONEY CHANGES EVERYTHING Cyndi Lauper
116 MONEY FOR NOTHING Dire Straits
1675 MONEY$ TOO TIGHT (To Mention) Simply Red
314 MONKEY George Michael
3422 THE MONKEY TIME The Tubes
4035 MONTEGO BAY Amazulu
246 MONY MONY "LIVE" Billy Idol
2902 MOONLIGHT ON WATER Kevin Raleigh
1763 MOONLIGHTING (Theme) Al Jarreau
3799 MORE BOUNCE TO THE OUNCE PART 1 Zapp
658 MORE LOVE Kim Carnes
2815 MORE STARS ON 45 (Medley) Stars On 45
101 MORE THAN I CAN SAY Leo Sayer
1741 MORE THAN JUST THE TWO OF US Sneaker
3647 MORE THAN PHYSICAL Bananarama

Singles Alpha Index

1395 MORE THAN YOU KNOW *Martika*	1027 NEVER BE THE SAME *Christopher Cross*	192 99 LUFTBALLONS *Nena*
2752 THE MORE YOU LIVE, THE MORE YOU LOVE *A Flock Of Seagulls*	1870 NEVER BEEN IN LOVE *Randy Meisner*	3195 96 TEARS *Garland Jeffreys*
1482 MORNIN' *Jarreau*	2647 NEVER CAN SAY GOODBYE *Communards*	Ninteen Eighty-Four...see: SEXCRIME
3013 MORNING DESIRE *Kenny Rogers*	3649 NEVER DIE YOUNG *James Taylor*	860 NITE AND DAY *Al B. Sure!*
3116 MORNING MAN *Rupert Holmes*	1303 NEVER ENDING STORY *Limahl*	A NITE AT THE APOLLO LIVE...see: THE WAY YOU DO THE THINGS YOU DO
148 MORNING TRAIN (Nine To Five) *Sheena Easton*	2790 NEVER ENOUGH *Patty Smyth*	3817 NO BIG DEAL *Love and Rockets*
2706 MOST OF ALL *Jody Watley*	2540 NEVER GIVE UP *Sammy Hagar*	1557 NO EASY WAY OUT *Robert Tepper*
1892 MOTHERS TALK *Tears For Fears*	2612 NEVER GIVE UP ON A GOOD THING *George Benson*	4001 NO FRILLS LOVE *Jennifer Holliday*
4019 THE MOTION OF LOVE *Gene Loves Jezebel*	184 NEVER GONNA GIVE YOU UP *Rick Astley*	2015 NO LOOKIN' BACK *Michael McDonald*
1862 MOTORTOWN *The Kane Gang*	234 NEVER GONNA LET YOU GO *Sergio Mendes*	598 NO MORE LONELY NIGHTS *Paul McCartney*
1803 MOUNTAINS *Prince And The Revolution*	3466 NEVER HAD A LOT TO LOSE *Cheap Trick*	No More Love On The Run... see: CARIBBEAN QUEEN
1226 MOVE AWAY *Culture Club*	1838 NEVER KNEW LOVE LIKE THIS *Alexander O'Neal featuring Cherrelle*	1493 NO MORE RHYME *Debbie Gibson*
2778 MOVE YOUR BOOGIE BODY *Bar-Kays*	490 NEVER KNEW LOVE LIKE THIS BEFORE *Stephanie Mills*	1441 NO MORE WORDS *Berlin*
2207 MURPHY'S LAW *Cheri*	1991 NEVER LET ME DOWN *David Bowie*	1464 NO NIGHT SO LONG *Dionne Warwick*
757 MUSCLES *Diana Ross*	2943 NEVER LET ME DOWN AGAIN *Depeche Mode*	3620 NO ONE CAN LOVE YOU MORE THAN ME *Melissa Manchester*
2296 MUSIC TIME *Styx*	2792 NEVER LET YOU GO *Sweet Sensation*	2093 NO ONE IN THE WORLD *Anita Baker*
2899 MUTUAL SURRENDER (What A Wonderful World) *Bourgeois Tagg*	3310 NEVER SAY DIE (Give A Little Bit More) *Cliff Richard*	558 NO ONE IS TO BLAME *Howard Jones*
3072 MY BABY *The Pretenders*	394 NEVER SURRENDER *Corey Hart*	3124 NO ONE LIKE YOU *Scorpions*
2011 MY BRAVE FACE *Paul McCartney*	826 NEVER TEAR US APART *INXS*	3401 NO PARKING (ON THE DANCE FLOOR) *Midnight Star*
1765 MY EVER CHANGING MOODS *The Style Council*	3943 NEVER TELL AN ANGEL (When Your Heart's On Fire) *The Stompers*	3328 NO PROMISES *Icehouse*
3074 MY FANTASY *Teddy Riley featuring Guy*	4122 NEVER THOUGHT I'D FALL IN LOVE *The Spinners*	1457 NO REPLY AT ALL *Genesis*
3294 MY FIRST NIGHT WITHOUT YOU *Cyndi Lauper*	1742 NEVER THOUGHT (That I Could Love) *Dan Hill*	4141 NO SOUVENIRS *Melissa Etheridge*
1602 MY GIRL *Suavé*	1629 NEVER TOO MUCH *Luther Vandross*	2105 NO TIME FOR TALK *Christopher Cross*
1820 MY GIRL *Donnie Iris*	1200 NEW ATTITUDE *Patti LaBelle*	3628 NO TIME TO LOSE *The Tarney/Spencer Band*
1247 MY GIRL (Gone, Gone, Gone) *Chilliwack*	2615 NEW DAY FOR YOU *Basia*	1370 NO WAY OUT *Jefferson Starship*
My Girl...see: WAY YOU DO THE THINGS YOU DO	3323 NEW FRONTIER *Donald Fagen*	1106 NOBODY *Sylvia (2)*
1627 MY GUY *Sister Sledge*	2987 NEW GIRL NOW *Honeymoon Suite*	1516 NOBODY SAID IT WAS EASY (Lookin' For The Lights) *Le Roux*
2923 MY GUY/MY GIRL *Amii Stewart & Johnny Bristol*	937 NEW MOON ON MONDAY *Duran Duran*	667 NOBODY TOLD ME *John Lennon*
528 MY HEART CAN'T TELL YOU NO *Rod Stewart*	1984 NEW ROMANCE (It's A Mystery) *Spider*	1618 NOBODY WINS *Elton John*
1962 MY HEART SKIPS A BEAT *The Cover Girls*	531 NEW SENSATION *INXS*	946 NOBODY'S FOOL *Kenny Loggins*
2232 MY HEROES HAVE ALWAYS BEEN COWBOYS *Willie Nelson*	1591 NEW SONG *Howard Jones*	1089 NOBODY'S FOOL *Cinderella*
865 MY HOMETOWN *Bruce Springsteen*	3144 NEW THING *Enuff Z'Nuff*	2862 NOBODY'S PERFECT *Mike + The Mechanics*
1806 MY KIND OF LADY *Supertramp*	1622 NEW WORLD MAN *Rush*	Not Another Lonely Night...see: WITHOUT YOU
2336 MY KINDA LOVER *Billy Squier*	2419 NEW YEAR'S DAY *U2*	1743 NOT ENOUGH LOVE IN THE WORLD *Don Henley*
596 MY LOVE *Lionel Richie*	3750 NEXT LOVE *Deniece Williams*	3455 NOT FADE AWAY *Eric Hine*
3761 MY LOVE *Julio Iglesias Featuring Stevie Wonder*	370 THE NEXT TIME I FALL *Peter Cetera w/Amy Grant*	1611 NOT JUST ANOTHER NIGHT *Ivan Neville*
3338 MY MISTAKE *The Kingbees*	3652 NEXT TIME YOU'LL KNOW *Sister Sledge*	1084 NOTHIN' AT ALL (Remix) *Heart*
1959 MY MOTHER'S EYES *Bette Midler*	4066 NIAGARA FALLS *Chicago*	868 NOTHIN' BUT A GOOD TIME *Poison*
3968 MY OBSESSION *Icehouse*	1897 NICE GIRLS *Eye To Eye*	3570 NOTHIN (That Compares 2 U) *Jacksons*
2095 MY OH MY *Slade*	2186 NICE GIRLS *Melissa Manchester*	2977 NOTHING EVER GOES AS PLANNED *Styx*
4131 MY ONE TEMPTATION *Mica Paris*	2683 NICE 'N' SLOW *Freddie Jackson*	2630 NOTHING IN COMMON *Thompson Twins*
3374 MY PARADISE *The Outfield*	1989 NICOLE *Point Blank*	Nothing Remains The Same... see: YESTERDAY ONCE MORE
2323 MY PRAYER *Ray, Goodman & Brown*	2277 THE NIGHT *The Animals*	841 NOTHING'S GONNA CHANGE MY LOVE FOR YOU *Glenn Medeiros*
137 MY PREROGATIVE *Bobby Brown*	2231 THE NIGHT IS STILL YOUNG *Billy Joel*	3643 NOTHING'S GONNA STOP ME NOW *Samantha Fox*
2481 MY TOOT TOOT *Jean Knight*	1610 NIGHT MOVES *Marilyn Martin*	134 NOTHING'S GONNA STOP US NOW *Starship*
2123 MY TOWN *Michael Stanley Band*	500 THE NIGHT OWLS *Little River Band*	284 NOTORIOUS *Duran Duran*
1572 MYSTERY LADY *Billy Ocean*	4062 NIGHT PULSE *Double Image*	2108 NOTORIOUS *Loverboy*
	2835 NIGHT SHIFT *Quarterflash*	3893 NOW AND FOREVER (You And Me) *Anne Murray*
## N	2328 A NIGHT TO REMEMBER *Shalamar*	3638 NOW IT'S MY TURN *Berlin*
2405 NAIL IT TO THE WALL *Stacy Lattisaw*	1791 NIGHTBIRD *Stevie Nicks (With Sandy Stewart)*	3199 NOW OR NEVER *Axe*
4142 NAME AND NUMBER *Big Noise*	2118 NIGHTIME *Pretty Poison*	4085 NOW YOU'RE IN HEAVEN *Julian Lennon*
463 NASTY *Janet Jackson*	1349 A NIGHTMARE ON MY STREET *DJ Jazzy Jeff & The Fresh Prince*	2876 NOWHERE TO RUN *Santana*
3207 NATURAL LOVE *Petula Clark*	4059 NIGHTRAIN *Guns N' Roses*	
3545 NATURE OF LOVE *Waterfront*	442 NIGHTSHIFT *Commodores*	## O
439 NAUGHTY GIRLS (Need Love Too) *Samantha Fox*	2361 NIGHTWALKER *Gino Vannelli*	
1396 NAUGHTY NAUGHTY *John Parr*	776 NIKITA *Elton John*	2572 THE OAK TREE *Morris Day*
2640 NEED A LITTLE TASTE OF LOVE *The Doobie Brothers*	59 9 TO 5 *Dolly Parton*	1235 OBJECT OF MY DESIRE *Starpoint*
223 NEED YOU TONIGHT *INXS*	1304 19 *Paul Hardcastle*	1827 OBSCENE PHONE CALLER *Rockwell*
2240 NEED YOUR LOVING TONIGHT *Queen*	804 1999 *Prince*	599 OBSESSION *Animotion*
2398 NEEDLES AND PINS *Tom Petty and The Heartbreakers with Stevie Nicks*	1360 99 *Toto*	2395 OFF ON YOUR OWN (Girl) *Al B. Sure!*
498 NEUTRON DANCE *Pointer Sisters*	3285 99 1/2 *Carol Lynn Townes*	874 OFF THE WALL *Michael Jackson*
462 NEVER *Heart*		
1688 NEVER AS GOOD AS THE FIRST TIME *Sade*		

Singles Alpha Index

3016 OH DADDY *Adrian Belew*
2170 OH GIRL *Boy Meets Girl*
2184 OH JULIE *Barry Manilow*
425 OH NO *Commodores*
2116 OH, PEOPLE *Patti LaBelle*
　　Oh Pretty Woman see: PRETTY WOMAN
337 OH SHEILA *Ready For The World*
335 OH SHERRIE *Steve Perry*
2557 OH YEAH *Yello*
1256 OLD-FASHION LOVE *Commodores*
2796 OLD FASHIONED LOVE *Smokey Robinson*
990 THE OLD MAN DOWN THE ROAD *John Fogerty*
1097 THE OLD SONGS *Barry Manilow*
2324 OLD TIME ROCK & ROLL
　　Bob Seger & The Silver Bullet Band
2960 OLYMPIA *Sergio Mendes*
2844 ON A CAROUSEL *Glass Moon*
3871 ON AND ON AND ON *ABBA*
94 ON MY OWN
　　Patti LaBelle And Michael McDonald
253 ON OUR OWN *Bobby Brown*
　　ON THE DARK SIDE
775 　*John Cafferty And The Beaver Brown Band*
2806 　*Eddie And The Cruisers*
3346 ON THE LINE *Tangier*
1419 ON THE LOOSE *Saga*
464 ON THE RADIO *Donna Summer*
2763 ON THE REBOUND *Russ Ballard*
1257 ON THE ROAD AGAIN *Willie Nelson*
2038 ON THE WAY TO THE SKY *Neil Diamond*
2465 ON THE WINGS OF A NIGHTINGALE
　　The Everly Brothers
1428 ON THE WINGS OF LOVE *Jeffrey Osborne*
4067 ONCE A NIGHT *Jackie English*
527 ONCE BITTEN TWICE SHY *Great White*
4024 ONCE IN A LIFETIME *Talking Heads*
1105 ONE *Bee Gees*
1936 ONE *Metallica*
2939 ONE DAY IN YOUR LIFE *Michael Jackson*
891 ONE FINE DAY *Carole King*
3703 ONE FOOT BACK IN YOUR DOOR *Roman Holliday*
2168 ONE FOR THE MOCKINGBIRD *Cutting Crew*
1924 ONE GOOD REASON *Paul Carrack*
647 ONE GOOD WOMAN *Peter Cetera*
1101 ONE HEARTBEAT *Smokey Robinson*
2020 ONE HIT (To The Body) *The Rolling Stones*
1109 ONE HUNDRED WAYS
　　Quincy Jones Featuring James Ingram
934 THE ONE I LOVE *R.E.M.*
1987 ONE IN A MILLION *The Romantics*
4027 ONE IN A MILLION *Eddie And The Tide*
921 ONE IN A MILLION YOU *Larry Graham*
4090 ONE LIFE TO LIVE *Wayne Massey*
1392 ONE LONELY NIGHT *REO Speedwagon*
2542 ONE LOVER AT A TIME *Atlantic Starr*
783 ONE MOMENT IN TIME *Whitney Houston*
3714 ONE MORE CHANCE *Diana Ross*
165 ONE MORE NIGHT *Phil Collins*
2720 ONE MORE NIGHT *Streek*
2537 ONE MORE TIME FOR LOVE
　　Billy Preston & Syreeta
136 ONE MORE TRY *George Michael*
　　ONE NIGHT IN BANGKOK
445 　*Murray Head*
3741 　*Robey*
1183 ONE NIGHT LOVE AFFAIR *Bryan Adams*
1288 ONE OF THE LIVING *Tina Turner*
539 ONE ON ONE *Daryl Hall & John Oates*

4089 ONE SIMPLE THING *Stabilizers*
1533 ONE STEP CLOSER *The Doobie Brothers*
1566 ONE STEP CLOSER TO YOU *Gavin Christopher*
1305 ONE STEP UP *Bruce Springsteen*
4138 ONE SUNNY DAY/DUELING BIKES FROM
　　QUICKSILVER *Ray Parker Jr. And Helen Terry*
3414 THE ONE THAT REALLY MATTERS *Survivor*
164 THE ONE THAT YOU LOVE *Air Supply*
1766 THE ONE THING *INXS*
399 ONE THING LEADS TO ANOTHER *The Fixx*
2364 ONE TO ONE *Carole King*
2298 ONE-TRICK PONY *Paul Simon*
579 1-2-3 *Gloria Estefan and Miami Sound Machine*
2699 ONE VISION *Queen*
3250 ONE WAY LOVE *TKA*
1192 THE ONE YOU LOVE *Glenn Frey*
2010 ONLY A LONELY HEART SEES *Felix Cavaliere*
4079 ONLY A MEMORY *The Smithereens*
2769 THE ONLY FLAME IN TOWN
　　Elvis Costello & The Atractions
2856 ONLY FOR LOVE *Limahl*
455 ONLY IN MY DREAMS *Debbie Gibson*
2839 ONLY LONELY *Bon Jovi*
3018 ONLY ONE YOU *T.G. Sheppard*
567 ONLY THE LONELY *The Motels*
4078 ONLY THE LONELY (Have A Reason To Be Sad)
　　La Flavour
1073 ONLY THE YOUNG *Journey*
1331 ONLY TIME WILL TELL *Asia*
4147 THE ONLY WAY IS UP
　　Yazz And The Plastic Population
3041 THE ONLY WAY OUT *Cliff Richard*
1882 ONLY WHEN YOU LEAVE *Spandau Ballet*
2225 ONLY YOU *Commodores*
3004 ONLY YOU *Yaz*
3170 OO-EE-DIDDLEY-BOP! *Peter Wolf*
2371 OOH OOH SONG *Pat Benatar*
3842 OOO LA LA LA *Teena Marie*
73 OPEN ARMS *Journey*
3796 OPEN LETTER (To A Landlord) *Living Colour*
294 OPEN YOUR HEART *Madonna*
1289 OPERATOR *Midnight Star*
1087 OPPORTUNITIES (Let's Make Lots Of Money)
　　Pet Shop Boys
3769 OPPOSITES DO ATTRACT *All Sports Band*
2893 ORIGINAL SIN *INXS*
1559 ORINOCO FLOW (Sail Away) *Enya*
763 THE OTHER GUY *Little River Band*
2841 THE OTHER SIDE OF LIFE *The Moody Blues*
365 THE OTHER WOMAN *Ray Parker Jr.*
657 OUR HOUSE *Madness*
911 OUR LIPS ARE SEALED *Go-Go's*
1191 OUT HERE ON MY OWN *Irene Cara*
2047 OUT OF MIND OUT OF SIGHT *Models*
497 OUT OF THE BLUE *Debbie Gibson*
92 OUT OF TOUCH *Daryl Hall & John Oates*
1394 OUT OF WORK *Gary U.S. Bonds*
4120 OUTLAW *War*
2687 OUTSIDE MY WINDOW *Stevie Wonder*
2565 OUTSTANDING *The Gap Band*
2388 OVER AND OVER *Pajama Party*
3550 OVER MY HEAD *Toni Basil*
3838 OVER THE LINE *Eddie Schwartz*
3826 OVER YOU *Roxy Music*
1768 OVERJOYED *Stevie Wonder*
236 OVERKILL *Men At Work*
2793 OWN THE NIGHT *Chaka Khan*
46 OWNER OF A LONELY HEART *Yes*

3645 OWWWW! *Chunky A*

P

786 PAC-MAN FEVER *Buckner & Garcia*
3789 PAINTED MOON *The Silencers*
3128 PAINTED PICTURE *Commodores*
1485 PAMELA *Toto*
1182 PANAMA *Van Halen*
178 PAPA DON'T PREACH *Madonna*
2723 PAPA WAS A ROLLIN' STONE *Wolf*
985 PAPER IN FIRE *John Cougar Mellencamp*
1833 PAPERLATE *Genesis*
1466 PARADISE *Sade*
3773 PARADISE *Change*
789 PARADISE CITY *Guns N' Roses*
2134 PARANOIMIA
　　The Art Of Noise with Max Headroom
1055 PARENTS JUST DON'T UNDERSTAND
　　D.J. Jazzy Jeff & The Fresh Prince
4140 PART OF ME THAT NEEDS YOU MOST *Jay Black*
177 PART-TIME LOVER *Stevie Wonder*
139 PARTY ALL THE TIME *Eddie Murphy*
1713 PARTYMAN *Prince*
1945 THE PARTY'S OVER (Hopelessly In Love) *Journey*
931 PASS THE DUTCHIE *Musical Youth*
396 PASSION *Rod Stewart*
613 PATIENCE *Guns N' Roses*
3311 PAY THE DEVIL (Ooo, Baby, Ooo) *The Knack*
3502 PAY YOU BACK WITH INTEREST *Gary O'*
3762 PEANUT BUTTER
　　Twennynine Featuring Lenny White
2279 PEEK-A-BOO *Siouxsie & The Banshees*
1299 A PENNY FOR YOUR THOUGHTS *Tavares*
712 PENNY LOVER *Lionel Richie*
1075 PEOPLE ARE PEOPLE *Depeche Mode*
2454 PEOPLE GET READY *Jeff Beck And Rod Stewart*
3661 PERFECT *Fairground Attraction*
3156 PERFECT COMBINATION
　　Stacy Lattisaw & Johnny Gill
827 PERFECT WAY *Scritti Politti*
534 PERFECT WORLD *Huey Lewis And The News*
3031 PERHAPS LOVE
　　Placido Domingo And John Denver
1119 PERSONALLY *Karla Bonoff*
2476 PETER GUNN
　　The Art Of Noise featuring Duane Eddy
996 PHOTOGRAPH *Def Leppard*
1 PHYSICAL *Olivia Newton-John*
814 PIANO IN THE DARK *Brenda Russell*
4039 PIECE BY PIECE *The Tubes*
3646 PIECE OF MY HEART *Sammy Hagar*
2044 PIECES OF ICE *Diana Ross*
927 PILOT OF THE AIRWAVES *Charlie Dore*
690 PINK CADILLAC *Natalie Cole*
885 PINK HOUSES *John Cougar Mellencamp*
2402 PLANET ROCK
　　Afrika Bambaataa And The Soul Sonic Force
3246 PLAY THAT FUNKY MUSIC *Roxanne*
2263 PLAY THE GAME *Queen*
1314 PLAY THE GAME TONIGHT *Kansas*
2805 PLAYING TO WIN *LRB*
3472 PLAYING WITH LIGHTNING *Shot In The Dark*
2713 PLAYING WITH THE BOYS *Kenny Loggins*
3073 PLEASE BE THE ONE *Karla Bonoff*
55 PLEASE DON'T GO *K.C. And The Sunshine Band*
932 PLEASE DON'T GO GIRL *New Kids On The Block*
3771 PLEASE MR. POSTMAN *Gentle Persuasion*
3264 PLEASURE AND PAIN *Divinyls*

Singles Alpha Index

1103 THE PLEASURE PRINCIPLE *Janet Jackson*
3476 PLEDGE PIN *Robert Plant*
723 POINT OF NO RETURN *Exposé*
1400 POINT OF NO RETURN *Nu Shooz*
878 POISON *Alice Cooper*
1447 POISON ARROW *ABC*
1333 THE POLITICS OF DANCING *Re-Flex*
1749 POOR MAN'S SON *Survivor*
2090 POP GOES THE MOVIES Part I *Meco*
1330 POP GOES THE WORLD *Men Without Hats*
947 POP LIFE *Prince & The Revolution*
1523 POP SINGER *John Cougar Mellencamp*
3936 POP SONG 89 *R.E.M.*
3224 POSSE' ON BROADWAY *Sir Mix-A-Lot*
1978 POSSESSION OBSESSION
    Daryl Hall & John Oates
323 POUR SOME SUGAR ON ME *Def Leppard*
2466 POWER *The Temptations*
3901 POWER *Kansas*
140 THE POWER OF LOVE *Huey Lewis And The News*
1500 POWER OF LOVE *Laura Branigan*
    THE POWER OF LOVE
2467   *Jennifer Rush*
3461   --(You Are My Lady) *Air Supply*
4149 POWER PLAY *Molly Hatchet*
3098 POWERFUL STUFF *The Fabulous Thunderbirds*
3209 PRAYING TO A NEW GOD *Wang Chung*
1495 PRECIOUS TO ME *Phil Seymour*
3585 PRESENCE OF LOVE *The Alarm*
1744 PRESS *Paul McCartney*
1363 PRESSURE *Billy Joel*
2622 PRETENDING *Eric Clapton*
4009 PRETTY BOYS AND PRETTY GIRLS *Book Of Love*
2299 PRETTY IN PINK *Psychedelic Furs*
3320 PRETTY MESS *Vanity*
1074 PRETTY WOMAN *Van Halen*
3132 PRIDE & PASSION
    John Cafferty And The Beaver Brown Band
1756 PRIDE (In The Name Of Love) *U2*
2096 PRIME TIME *Alan Parsons Project*
3309 PRIMITIVE LOVE RITES *Mondo Rock*
1949 THE PRISONER *Howard Jones*
810 PRIVATE DANCER *Tina Turner*
86 PRIVATE EYES *Daryl Hall & John Oates*
3396 PRIVATE IDAHO *The B-52's*
2494 PRIVATE NUMBER *The Jets*
855 THE PROMISE *When In Rome*
1757 PROMISE ME *The Cover Girls*
2623 PROMISES *Barbra Streisand*
2103 PROMISES IN THE DARK *Pat Benatar*
820 PROMISES, PROMISES *Naked Eyes*
2940 PROUD *Joe Chemay Band*
4114 PROVE ME WRONG *David Pack*
908 PROVE YOUR LOVE *Taylor Dayne*
2543 PSYCHOBABBLE *Alan Parsons Project*
1000 PUMP UP THE VOLUME *M/A/R/R/S*
228 PURPLE RAIN *Prince And The Revolution*
1086 PUSH IT *Salt-N-Pepa*
2963 PUSS N BOOTS/THESE BOOTS
    (Are Made For Walkin') *Kon Kan*
1042 PUT A LITTLE LOVE IN YOUR HEART
    Annie Lennox & Al Green
3526 PUT AWAY YOUR LOVE *Alessi*
1943 PUT IT IN A MAGAZINE *Sonny Charles*
3361 PUT THIS LOVE TO THE TEST *Jon Astley*
2021 PUT YOUR MOUTH ON ME *Eddie Murphy*
379 PUTTIN' ON THE RITZ *Taco*
977 P.Y.T. (Pretty Young Thing) *Michael Jackson*

Q

3231 QUE TE QUIERO *Katrina And The Waves*
185 QUEEN OF HEARTS *Juice Newton*
1937 QUEEN OF THE BROKEN HEARTS *Loverboy*

R

2742 RADAR LOVE *White Lion*
3587 RADIO FREE EUROPE *R.E.M.*
1412 RADIO GA-GA *Queen*
2496 RADIO ROMANCE *Tiffany*
1700 RADIOACTIVE *The Firm*
1386 RAG DOLL *Aerosmith*
3911 RAIN *Dragon*
983 THE RAIN *Oran "Juice" Jones*
2310 RAIN FOREST *Paul Hardcastle*
3508 RAIN IN MAY *Max Werner*
3162 RAIN IN THE SUMMERTIME *The Alarm*
1772 RAIN ON THE SCARECROW
    John Cougar Mellencamp
2668 RAINBOW'S END *Sergio Mendes*
4013 THE RAMBLER *Molly Hatchet*
2035 RAPPER'S DELIGHT *Sugarhill Gang*
3419 RAPPIN' RODNEY *Rodney Dangerfield*
127 RAPTURE *Blondie*
307 RASPBERRY BERET *Prince & The Revolution*
3792 REACH OUT *Giorgio Moroder*
2727 REACTION TO ACTION *Foreigner*
1252 READ 'EM AND WEEP *Barry Manilow*
4064 READY FOR LOVE *Silverado*
2553 READY OR NOT *Lou Gramm*
3828 THE REAL END *Rickie Lee Jones*
334 REAL LOVE *Jody Watley*
591 REAL LOVE *The Doobie Brothers*
3609 REAL LOVE *The Cretones*
4096 REAL LOVE *Dolly Parton (Duet w/Kenny Rogers)*
4124 REAL PEOPLE *Chic*
3870 REAL PEOPLE//CHIP OFF THE OLD BLOCK [TSW]
    Chic
3305 THE REAL THING *The Brothers Johnson*
3677 THE REAL THING *Jellybean Featuring Stephen Dante*
3712 REALISTIC *Shirley Lewis*
1263 REALLY WANNA KNOW YOU *Gary Wright*
3324 REAP THE WILD WIND *Ultravox*
2735 REASON TO LIVE *KISS*
4049 REASON TO TRY *Eric Carmen*
1994 REBEL YELL *Billy Idol*
3530 REBELS *Tom Petty and The Heartbreakers*
3120 REBELS ARE WE *Chic*
3452 RED HOT *Herb Alpert*
2327 RED LIGHT *Linda Clifford*
    RED RED WINE
258   *UB40 ('88)*
1680   *UB40 ('84)*
106 THE REFLEX *Duran Duran*
1064 REFUGEE *Tom Petty And The Heartbreakers*
    RELAX
1071   --(Mix) *Frankie Goes To Hollywood ('85)*
3051   *Frankie Goes To Hollywood ('84)*
1782 REMEMBER THE NIGHTS *The Motels*
3118 REMEMBER (Walking In The Sand) *Aerosmith*
3435 REMEMBER WHAT YOU LIKE *Jenny Burton*
3639 REMO'S THEME (What If) *Tommy Shaw*
3489 REMOTE CONTROL *The Reddings*
3490 REPETITION
    Information Society

3219 RESERVATIONS FOR TWO
    Dionne & Kashif
772 RESPECT YOURSELF *Bruce Willis*
2289 RESTLESS *Starpoint*
2889 RESTLESS HEART *John Waite*
3973 REV IT UP *Newcity Rockers*
673 RHYTHM IS GONNA GET YOU
    Gloria Estefan and Miami Sound Machine
2246 RHYTHM OF LOVE *Yes*
3563 RHYTHM OF LOVE *Scorpions*
342 RHYTHM OF THE NIGHT *DeBarge*
2833 RIBBON IN THE SKY *Stevie Wonder*
3852 RICH MAN *Terri Gibbs*
2921 RICKY *"Weird Al" Yankovic* [P] see: MICKEY
70 RIDE LIKE THE WIND *Christopher Cross*
3014 RIGHT AWAY *Hawks*
3240 RIGHT AWAY *Kansas*
3888 RIGHT BACK WHERE WE STARTED FROM *Sinitta*
2163 RIGHT BEFORE YOUR EYES *America*
2191 RIGHT BETWEEN THE EYES *Wax*
1856 RIGHT BY YOUR SIDE *Eurythmics*
2804 RIGHT HERE AND NOW *Bill Medley*
142 RIGHT HERE WAITING *Richard Marx*
2726 RIGHT KIND OF LOVE *Quarterflash*
3657 RIGHT NEXT DOOR (Because Of Me)
    The Robert Cray Band
2501 RIGHT NEXT TO ME *Whistle*
872 RIGHT ON TRACK *Breakfast Club*
2530 THE RIGHT STUFF *Vanessa Williams*
    Right Stuff...see: YOU GOT IT
3486 RIGHT THE FIRST TIME *Gamma*
1660 THE RIGHT THING *Simply Red*
1264 RIO *Duran Duran*
2322 RITUAL *Dan Reed Network*
4127 ROBERT De NIRO'S WAITING *Bananarama*
3845 ROCK-A-LOTT *Aretha Franklin*
1617 ROCK AND A HARD PLACE *The Rolling Stones*
    Rock And A Hard Place...see: (Between A)
1672 ROCK AND ROLL DREAMS COME THROUGH
    Jim Steinman
1630 ROCK AND ROLL GIRLS *John Fogerty*
3135 ROCK & ROLL STRATEGY *Thirty Eight Special*
429 R.O.C.K. IN THE U.S.A. (A Salute To 60's Rock)
    John Cougar Mellencamp
3192 ROCK IT *Lipps, Inc.*
2779 ROCK LOBSTER *The B-52's*
2631 ROCK ME *Great White*
190 ROCK ME AMADEUS *Falco*
1368 ROCK ME TONIGHT (For Old Times Sake)
    Freddie Jackson
1099 ROCK ME TONITE *Billy Squier*
1328 ROCK 'N' ROLL IS KING *ELO*
3314 ROCK 'N' ROLL TO THE RESCUE *The Beach Boys*
1219 ROCK OF AGES *Def Leppard*
1644 ROCK OF LIFE *Rick Springfield*
375 ROCK ON *Michael Damian*
727 ROCK STEADY *The Whispers*
562 ROCK THE CASBAH *The Clash*
1942 ROCK THE NIGHT *Europe*
611 ROCK THIS TOWN *Stray Cats*
897 ROCK WIT'CHA *Bobby Brown*
30 ROCK WITH YOU *Michael Jackson*
1522 ROCK YOU LIKE A HURRICANE *Scorpions*
1309 ROCKET *Def Leppard*
688 ROCKET 2 U *The Jets*
1867 ROCKIN' AT MIDNIGHT *The Honeydrippers*
2335 ROCKIN' INTO THE NIGHT *38-Special*
3034 ROCKIT *Herbie Hancock*
1860 ROLL ME AWAY
    Bob Seger And The Silver Bullet Band

Singles Alpha Index

#	Title	Artist
150	ROLL WITH IT	Steve Winwood
	Romance Is Not Over... see: WE'VE ONLY JUST BEGUN	
1503	ROMANCING THE STONE	Eddy Grant
875	ROMEO'S TUNE	Steve Forbert
645	RONI	Bobby Brown
4073	RONNIE'S RAPP	Ron And The D.C. Crew
1044	ROOM TO MOVE	Animotion
1497	ROOMS ON FIRE	Stevie Nicks
3076	ROOTY TOOT TOOT	John Cougar Mellencamp
51	ROSANNA	Toto
138	THE ROSE	Bette Midler
1684	ROTATION	Herb Alpert
1647	ROUGH BOY	ZZ Top
3887	ROUGH BOYS	Pete Townshend
2926	ROUND & ROUND	New Order
982	ROUND AND ROUND	Ratt
2106	ROUTE 101	Herb Alpert
3664	ROUTE 66	The Manhattan Transfer
2733	ROUTE 66/BEHIND THE WHEEL	Depeche Mode
3704	ROXANNE, ROXANNE	UTFO
2780	THE ROYAL MILE (Sweet Darlin')	Gerry Rafferty
1872	RUMBLESEAT	John Cougar Mellencamp
758	RUMORS	Timex Social Club
3218	THE RUMOUR	Olivia Newton-John
1385	RUN FOR THE ROSES	Dan Fogelberg
2974	RUN LIKE HELL	Pink Floyd
1382	RUN RUNAWAY	Slade
3321	RUN TO ME	Savoy Brown
3555	RUN TO PARADISE	Choirboys
604	RUN TO YOU	Bryan Adams
1951	RUNAWAY	Bon Jovi
3779	RUNAWAY	Luis Cardenas
3497	RUNAWAY RITA	Leif Garrett
1455	RUNNER	Manfred Mann's Earth Band
1699	RUNNIN' DOWN A DREAM	Tom Petty
3918	RUNNING	Chubby Checker
3682	RUNNING BACK	Eddie Money
3702	RUNNING BACK	Urgent
3886	RUNNING IN THE FAMILY	Level 42
2816	RUNNING SCARED	The Fools
1683	RUNNING UP THAT HILL	Kate Bush
576	RUNNING WITH THE NIGHT	Lionel Richie
1006	RUSH HOUR	Jane Wiedlin
1446	RUSSIANS	Sting
2737	RUTHLESS PEOPLE	Mick Jagger

S

#	Title	Artist
1234	SACRED EMOTION	Donny Osmond
4098	SAD GIRL	GQ
3863	SAD HEARTS	Four Tops
574	SAD SONGS (Say So Much)	Elton John
182	THE SAFETY DANCE	Men Without Hats
2743	SAIL AWAY	The Temptations
129	SAILING	Christopher Cross
1800	THE SALT IN MY TEARS	Martin Briley
3332	SAMANTHA (What You Gonna Do?) Cellarful Of Noise	
3879	THE SAME LOVE	The Jets
837	SAME OLD LANG SYNE	Dan Fogelberg
2145	SAME OLE LOVE (365 Days A Year)	Anita Baker
1248	SANCTIFY YOURSELF	Simple Minds
199	SARA	Starship
666	SARA	Fleetwood Mac
2913	SARA	Bill Champlin
2070	(Sartorial Eloquence) DON'T YA WANNA PLAY THIS GAME NO MORE	Elton John
2822	SATELLITE	Hooters
3527	SATISFACTION GUARANTEED	The Firm
436	SATISFIED	Richard Marx
3695	SATISFIED MAN	Molly Hatchet
3215	SATISFY ME	Billy Satellite
1645	SATURDAY LOVE	Cherrelle with Alexander O'Neal
1505	SAUSALITO SUMMERNIGHT	Diesel
1948	SAVANNAH NIGHTS	Tom Johnston
1399	SAVE A PRAYER	Duran Duran
3532	SAVE ME	Dave Mason
2078	SAVE THE LAST DANCE FOR ME	Dolly Parton
3592	SAVE THE NIGHT FOR ME	Maureen Steele
2919	SAVE THE OVERTIME (For Me)	Gladys Knight & The Pips
2554	SAVE YOUR LOVE	Great White
1284	SAVED BY ZERO	The Fixx
2656	SAVIN' MYSELF	Eria Fachin
233	SAVING ALL MY LOVE FOR YOU	Whitney Houston
1321	SAY GOODBYE TO HOLLYWOOD (Live)	Billy Joel
3955	SAY GOODBYE TO LITTLE JO	Steve Forbert
3326	SAY HELLO TO RONNIE	Janey Street
2043	SAY IT AGAIN	Jermaine Stewart
2230	SAY IT AGAIN	Santana
74	SAY IT ISN'T SO	Daryl Hall & John Oates
2853	SAY IT, SAY IT	E.G. Daily
2220	SAY IT'S GONNA RAIN	Will To Power
3	SAY SAY SAY	Paul McCartney and Michael Jackson
2056	SAY WHAT	Jesse Winchester
2795	SAY YOU REALLY WANT ME	Kim Wilde
56	SAY YOU, SAY ME	Lionel Richie
706	SAY YOU WILL	Foreigner
1502	SAY YOU'LL BE MINE	Christopher Cross
1626	SAY YOU'RE WRONG	Julian Lennon
1745	SAYIN' SORRY (Don't Make It Right)	Denise Lopez
4105	SCANDAL	RCR
4143	SCARLET FEVER	Kenny Rogers
3217	SCHOOL'S OUT	Krokus
3536	SCIENTIFIC LOVE	Midnight Star
3123	THE SCREAMS OF PASSION	The Family
1544	SE LA	Lionel Richie
	SEA OF LOVE	
321		The Honeydrippers
2027		Del Shannon
516	THE SEARCH IS OVER	Survivor
3722	SEASONS	Charles Fox
4137	SEASONS	Grace Slick
298	SEASONS CHANGE	Exposé
3867	SEASONS OF GOLD (Four Seasons Medley)	Gidea Park featuring Adrian Baker
3635	SEASONS OF THE HEART	John Denver
660	SECOND CHANCE	Thirty Eight Special
2164	SECOND NATURE	Dan Hartman
612	THE SECOND TIME AROUND	Shalamar
2570	SECRET	Orchestral Manoeuvres In The Dark
2603	SECRET JOURNEY	The Police
319	SECRET LOVERS	Atlantic Starr
2916	THE SECRET OF MY SUCCESS	Night Ranger
718	SECRET RENDEZVOUS	Karyn White
1593	SECRET SEPARATION	The Fixx
3444	SECRETS	Mac Davis
3473	SEDUCED	Leon Redbone
1689	THE SEDUCTION (Love Theme)	James Last Band
4063	SEE WHAT LOVE CAN DO	Eric Clapton
3263	SEEING IS BELIEVING	Mike + The Mechanics
353	SELF CONTROL	Laura Branigan
1508	SEND HER MY LOVE	Journey
1431	SEND ME AN ANGEL ('83)	Real Life
1575	--SEND ME AN ANGEL ('89)	Real Life
3904	SENDIN' ALL MY LOVE	The Jets
2909	THE SENSITIVE KIND	Santana
890	SENTIMENTAL STREET	Night Ranger
132	SEPARATE LIVES (Love Theme From White Nights)	Phil Collins and Marilyn Martin
513	SEPARATE WAYS (Worlds Apart)	Journey
1079	SEPTEMBER MORN'	Neil Diamond
1564	SEQUEL	Harry Chapin
1420	SERIOUS	Donna Allen
3266	SERIOUS KINDA GIRL	Christopher Max
1727	SET ME FREE	Utopia
3141	SET THE NIGHT ON FIRE	Oak
1517	SEVEN BRIDGES ROAD	Eagles
3993	777-9311	The Time
1531	SEVEN WONDERS	Fleetwood Mac
1416	SEVEN YEAR ACHE	Rosanne Cash
1874	17	Rick James
1620	SEVENTEEN	Winger
1732	SEX AS A WEAPON	Pat Benatar
2971	SEX (I'm A...)	Berlin
3612	SEX SHOOTER	Apollonia 6
2851	SEXAPPEAL	Georgio
3784	SEXCRIME (Nineteen Eighty-Four)	Eurythmics
147	SEXUAL HEALING	Marvin Gaye
386	SEXY EYES	Dr. Hook
1376	SEXY GIRL	Glenn Frey
2293	SHADDAP YOU FACE	Joe Dolce
1012	SHADOWS OF THE NIGHT	Pat Benatar
1851	SHAKE FOR THE SHEIK	The Escape Club
301	SHAKE IT UP	The Cars
3650	SHAKE IT UP	Bad Company
3168	SHAKE IT UP TONIGHT	Cheryl Lynn
189	SHAKE YOU DOWN	Gregory Abbott
432	SHAKE YOUR LOVE	Debbie Gibson
143	SHAKEDOWN	Bob Seger
2894	SHAKIN'	Eddie Money
1635	SHAME	The Motels
76	SHAME ON THE MOON	Bob Seger & The Silver Bullet Band
2423	SHANDI	KISS
1804	SHANGHAI BREEZES	John Denver
2930	SHANGRI-LA	Steve Miller Band
1080	SHARE YOUR LOVE WITH ME	Kenny Rogers
3987	SHARING THE LOVE	Rufus With Chaka Khan
2587	SHARP DRESSED MAN	ZZ Top
287	SHATTERED DREAMS	Johnny Hates Jazz
2541	SHATTERED GLASS	Laura Branigan
336	SHE BLINDED ME WITH SCIENCE	Thomas Dolby
261	SHE BOP	Cyndi Lauper
3251	SHE DID IT	Michael Damian
2436	SHE DON'T KNOW ME	Bon Jovi
3651	SHE DON'T LOOK BACK	Dan Fogelberg
285	SHE DRIVES ME CRAZY	Fine Young Cannibals
2882	SHE GOT THE GOLDMINE (I Got The Shaft)	Jerry Reed
3260	SHE LOOKS A LOT LIKE YOU	Clocks
3802	SHE LOVES MY CAR	Ronnie Milsap
843	SHE WANTS TO DANCE WITH ME	Rick Astley
2492	SHE WAS HOT	The Rolling Stones
2042	SHE WON'T TALK TO ME	Luther Vandross
169	SHE WORKS HARD FOR THE MONEY	Donna Summer
2271	SHELTER	Lone Justice
3969	SHELTER ME	Joe Cocker
3375	SHERRY	Robert John
1315	SHE'S A BAD MAMA JAMA (She's Built, She's Stacked)	Carl Carlton
869	SHE'S A BEAUTY	The Tubes

Singles Alpha Index

#	Title	Artist
3424	**SHE'S A RUNNER**	Billy Squier
	She's Built, She's Stacked... see: **SHE'S A BAD MAMA JAMA**	
3368	**SHE'S FLY**	Tony Terry
1513	**SHE'S GOT A WAY**	Billy Joel
2076	**SHE'S IN LOVE WITH YOU**	Suzi Quatro
416	**SHE'S LIKE THE WIND**	Patrick Swayze (Featuring Wendy Fraser)
1558	**SHE'S MINE**	Steve Perry
2431	**SHE'S ON THE LEFT**	Jeffrey Osborne
3905	**SHE'S ONLY 20**	Tami Show
761	**SHE'S OUT OF MY LIFE**	Michael Jackson
588	**(She's) SEXY + 17**	Stray Cats
1702	**SHE'S SO COLD**	The Rolling Stones
2342	**SHE'S STRANGE**	Cameo
3075	**SHE'S TIGHT**	Cheap Trick
3019	**SHE'S TROUBLE**	Musical Youth
1725	**SHINE ON**	L.T.D.
2484	**SHINE ON**	George Duke
2226	**SHINE SHINE**	Barry Gibb
348	**SHINING STAR**	The Manhattans
3465	**SHINY SHINY**	Haysi Fantayzee
3820	**SHIP OF FOOLS**	Robert Plant
1854	**SHIP OF FOOLS (Save Me From Tomorrow)**	World Party
3357	**SHIP TO SHORE**	Chris DeBurgh
3693	**SHIVER AND SHAKE**	The Silencers (2)
3957	**SHOCK**	The Motels
1381	**SHOCK THE MONKEY**	Peter Gabriel
2349	**SHOOT FOR THE MOON**	Poco
3876	**SHOOTING SHARK**	Blue Öyster Cult
3272	**SHOOTING STAR**	Dollar
3634	**SHOPPIN' FROM A TO Z**	Toni Basil
3078	**SHOT IN THE DARK**	Ozzy Osbourne
	SHOTGUN RIDER	
3271		Joe Sun
3442		Delbert McClinton
1133	**SHOULD I DO IT**	The Pointer Sisters
3615	**SHOULD I LOVE YOU**	Cee Farrow
1933	**SHOULD I SAY YES?**	Nu Shooz
2927	**SHOULD I SEE**	Frozen Ghost
2152	**SHOULD I STAY OR SHOULD I GO ('82)**	The Clash
2474	**--SHOULD I STAY OR SHOULD I GO ('83)**	The Clash
1582	**A SHOULDER TO CRY ON**	Tommy Page
485	**SHOULD'VE KNOWN BETTER**	Richard Marx
1150	**SHOULD'VE NEVER LET YOU GO**	Neil Sedaka & Dara Sedaka
122	**SHOUT**	Tears For Fears
1718	**SHOW ME**	The Pretenders
1779	**SHOW ME**	The Cover Girls
3061	**SHOW ME THE WAY**	Regina Belle
2204	**SHOW SOME RESPECT**	Tina Turner
695	**SHOWER ME WITH YOUR LOVE**	Surface
3529	**SHOWING OUT (Get Fresh At The Weekend)**	Mel & Kim
3763	**SHY BOY (Don't It Make You Feel Good)**	Bananarama
4109	**SHY BOYS**	Ana
3551	**SHY GIRL**	Stacey Q
3694	**SIDE BY SIDE**	Earth, Wind & Fire
1293	**SIDEWALK TALK**	Jellybean
517	**SIGN O' THE TIMES**	Prince
1706	**THE SIGN OF FIRE**	The Fixx
3022	**SIGN OF THE GYPSY QUEEN**	April Wine
3611	**SIGN OF THE TIMES**	The Belle Stars
572	**SIGN YOUR NAME**	Terence Trent D'Arby
1691	**SILENT MORNING**	Noel
705	**SILENT RUNNING (On Dangerous Ground)**	Mike + The Mechanics
1151	**SILHOUETTE**	Kenny G
2571	**SILLY**	Deniece Williams
4041	**SIMILAR FEATURES**	Melissa Etheridge
3416	**SIMPLE**	Johnny Mathis
311	**SIMPLY IRRESISTIBLE**	Robert Palmer
1637	**SINCE I DON'T HAVE YOU**	Don McLean
2447	**SINCE YOU'RE GONE**	The Cars
1738	**SINCE YOU'VE BEEN GONE**	The Outfield
1241	**SINCERELY YOURS**	Sweet Sensation
3913	**SING A SIMPLE SONG**	West Street Mob
2560	**SING ME AWAY**	Night Ranger
536	**SISTER CHRISTIAN**	Night Ranger
1461	**SISTERS ARE DOIN' IT FOR THEMSELVES**	Eurythmics And Aretha Franklin
3917	**SISTERS OF THE MOON**	Fleetwood Mac
	(Sittin' On) THE DOCK OF THE BAY	
1096		Michael Bolton
2734		The Reddings
1859	**SITTING AT THE WHEEL**	The Moody Blues
3111	**SITUATION**	Yazoo
510	**'65 LOVE AFFAIR**	Paul Davis
2753	**SKATEAWAY**	Dire Straits
1474	**SKELETONS**	Stevie Wonder
3738	**SKIN DEEP**	Cher
2394	**SKIN TRADE**	Duran Duran
244	**SLEDGEHAMMER**	Peter Gabriel
774	**SLEEPING BAG**	ZZ Top
3329	**SLEEPWALK**	Larry Carlton
3510	**SLIP AWAY**	Pablo Cruise
1817	**SLIPPING AWAY**	Dave Edmunds
3463	**SLIPSTREAM**	Allan Clarke
3829	**SLOW DANCIN'**	Peabo Bryson
128	**SLOW HAND**	Pointer Sisters
3949	**SMALL PARADISE**	John Cougar
601	**SMALL TOWN**	John Cougar Mellencamp
2802	**SMALL TOWN GIRL**	John Cafferty And The Beaver Brown Band
1931	**SMALL WORLD**	Huey Lewis And The News
1983	**SMALLTOWN BOY**	Bronski Beat
1723	**THE SMILE HAS LEFT YOUR EYES**	Asia
2392	**SMILING ISLANDS**	Robbie Patton
1317	**SMOKIN' IN THE BOYS ROOM**	Mötley Crüe
1695	**SMOKING GUN**	The Robert Cray Band
1282	**SMOKY MOUNTAIN RAIN**	Ronnie Milsap
893	**SMOOTH CRIMINAL**	Michael Jackson
546	**SMOOTH OPERATOR**	Sade
3095	**SMOOTH UP**	Bulletboys
1016	**SMUGGLER'S BLUES**	Glenn Frey
3395	**SNAKE EYES**	Alan Parsons Project
3878	**SNAP SHOT**	Slave
475	**SO ALIVE**	Love and Rockets
1463	**SO BAD**	Paul McCartney
3757	**SO. CENTRAL RAIN (I'm Sorry)**	R.E.M.
2109	**SO CLOSE**	Diana Ross
179	**SO EMOTIONAL**	Whitney Houston
1535	**SO FAR AWAY**	Dire Straits
2254	**SO FAR SO GOOD**	Sheena Easton
3680	**SO FINE**	Oak Ridge Boys
1581	**SO IN LOVE**	Orchestral Manoeuvres In The Dark
3946	**SO MUCH FOR LOVE**	The Venetians
2838	**SO MUCH IN LOVE**	Timothy B. Schmit
3840	**SO THE STORY GOES**	Living In A Box
1837	**SO WRONG**	Patrick Simmons
2847	**SO YOU RAN**	Orion The Hunter
2574	**SOLD ME DOWN THE RIVER**	The Alarm
435	**SOLDIER OF LOVE**	Donny Osmond
907	**SOLID**	Ashford & Simpson
3157	**SOLID ROCK**	Goanna
655	**SOLITAIRE**	Laura Branigan
2895	**SOLITAIRE**	Peter Mclan
4102	**SOLITUDE STANDING**	Suzanne Vega
3865	**SOLSBURY HILL**	Peter Gabriel
3249	**SOME CHANGES ARE FOR GOOD**	Dionne Warwick
1658	**SOME DAYS ARE DIAMONDS (Some Days Are Stone)**	John Denver
1018	**SOME GUYS HAVE ALL THE LUCK**	Rod Stewart
1511	**SOME KIND OF FRIEND**	Barry Manilow
1045	**SOME KIND OF LOVER**	Jody Watley
646	**SOME LIKE IT HOT**	The Power Station
2878	**SOME PEOPLE**	Paul Young
3373	**SOME PEOPLE**	Belouis Some
1475	**SOME THINGS ARE BETTER LEFT UNSAID**	Daryl Hall & John Oates
1037	**SOMEBODY**	Bryan Adams
3204	**SOMEBODY ELSE'S GUY**	Jocelyn Brown
2304	**SOMEBODY LIKE YOU**	38 Special
2915	**SOMEBODY LIKE YOU**	Robbie Nevil
3185	**SOMEBODY SAVE ME**	Cinderella
2875	**SOMEBODY SEND MY BABY HOME**	Lenny LeBlanc
3676	**SOMEBODY SOMEWHERE**	Platinum Blonde
3441	**SOMEBODY SPECIAL**	Rod Stewart
577	**SOMEBODY'S BABY**	Jackson Browne
4132	**SOMEBODY'S GONNA LOVE YOU**	Lee Greenwood
953	**SOMEBODY'S KNOCKIN'**	Terri Gibbs
1829	**SOMEBODY'S OUT THERE**	Triumph
104	**SOMEBODY'S WATCHING ME**	Rockwell
717	**SOMEDAY**	Glass Tiger
	SOMEDAY, SOMEWAY	
2147		Marshall Crenshaw
3674		Robert Gordon
3159	**SOMEONE**	El DeBarge
2651	**SOMEONE BELONGING TO SOMEONE**	Bee Gees
1175	**SOMEONE COULD LOSE A HEART TONIGHT**	Eddie Rabbitt
2949	**SOMEONE LIKE YOU**	Daryl Hall
3469	**SOMEONE LIKE YOU**	Michael Stanley Band
1218	**SOMEONE THAT I USED TO LOVE**	Natalie Cole
3331	**SOMEONE TO LOVE ME FOR ME**	Lisa Lisa & Cult Jam Featuring Full Force
2415	**SOMETHIN' 'BOUT YOU BABY I LIKE**	Glen Campbell and Rita Coolidge
632	**SOMETHING ABOUT YOU**	Level 42
3765	**SOMETHING IN MY HOUSE**	Dead Or Alive
3730	**SOMETHING JUST AIN'T RIGHT**	Keith Sweat
2032	**SOMETHING REAL (Inside Me/Inside You)**	Mr. Mister
766	**SOMETHING SO STRONG**	Crowded House
2473	**SOMETHING TO GRAB FOR**	Ric Ocasek
3541	**SOMETHING'S ON YOUR MIND**	"D" Train
2325	**SOMETIMES A FANTASY**	Billy Joel
1504	**SOMEWHERE DOWN THE ROAD**	Barry Manilow
2081	**SOMEWHERE (From "West Side Story")**	Barbra Streisand
2621	**SOMEWHERE IN AMERICA**	Survivor
430	**SOMEWHERE OUT THERE**	Linda Ronstadt And James Ingram
561	**SONGBIRD**	Kenny G
1853	**SOUL CITY**	Partland Brothers
1579	**SOUL KISS**	Olivia Newton-John
1409	**SOUL PROVIDER**	Michael Bolton
1427	**SOULS**	Rick Springfield
3666	**THE SOUND OF GOODBYE**	Crystal Gayle
4099	**SOUNDS OF YOUR VOICE**	Jon Butcher Axis

Singles Alpha Index

3449 SOUP FOR ONE Chic
1194 SOUTHERN CROSS Crosby, Stills & Nash
3433 SOUTHERN PACIFIC Neil Young & Crazy Horse
471 SOWING THE SEEDS OF LOVE Tears For Fears
1487 SPACE AGE LOVE SONG A Flock Of Seagulls
2667 SPACE AGE WHIZ KIDS Joe Walsh
2245 SPANISH EDDIE Laura Branigan
470 SPECIAL LADY Ray, Goodman & Brown
2860 SPECIAL WAY Kool & The Gang
3877 SPEND THE NIGHT IN LOVE The Four Seasons
1973 SPICE OF LIFE The Manhattan Transfer
854 SPIES LIKE US Paul McCartney
2810 SPIRIT IN THE SKY Doctor And The Medics
2582 THE SPIRIT OF RADIO Rush
1159 SPIRITS IN THE MATERIAL WORLD The Police
4164 SPRING LOVE The Cover Girls
1737 SPRING LOVE (Come Back To Me) Stevie B
1353 SPY IN THE HOUSE OF LOVE Was (Not Was)
2158 SQUARE BIZ Teena Marie
3746 SQUARE ROOMS Al Corley
163 ST. ELMO'S FIRE (Man In Motion) John Parr
3589 STACY Fortune
1625 STAGES ZZ Top
3937 STAIRWAY TO HEAVEN Far Corporation
742 STAND R.E.M.
481 STAND BACK Stevie Nicks
3671 STAND BACK The Fabulous Thunderbirds
2658 STAND BY Roman Holliday
    STAND BY ME
703   Ben E. King
1269   Mickey Gilley
2297   Maurice White
3322 STAND OR FALL The Fixx
3065 STAND UP David Lee Roth
3169 STAND UP Underworld
2973 STANDING ON THE TOP PART 1
    The Temptations Featuring Rick James
3091 STAR Earth, Wind & Fire
2918 STARGAZER Peter Brown
    STARS ON 45...see: MEDLEY, MEDLEY II
2086 STARS ON 45 III (A Tribute To Stevie Wonder)
    (Medley) Stars On
4145 START IT ALL OVER McGuffey Lane
81 START ME UP The Rolling Stones
2211 STARTING OVER AGAIN Dolly Parton
2348 STATE OF INDEPENDENCE Donna Summer
368 STATE OF SHOCK Jacksons
1552 STATE OF THE HEART Rick Springfield
3482 STATE OF THE NATION Industry
2869 STAY AWAKE Ronnie Laws
    Stay Awhile...see: GOOD MORNING GIRL
2817 STAY IN TIME Off Broadway USA
1178 STAY THE NIGHT Chicago
1373 STAY THE NIGHT Benjamin Orr
4116 STAY TRUE Sly Fox
1292 STAY WITH ME TONIGHT Jeffrey Osborne
1781 STAYING TOGETHER Debbie Gibson
2314 STAYING WITH IT Firefall
2945 STEADY Jules Shear
402 STEAL AWAY Robbie Dupree
1244 STEAL THE NIGHT Stevie Woods
438 STEP BY STEP Eddie Rabbitt
395 STEPPIN' OUT Joe Jackson
4046 STEPPIN' OUT Kool & The Gang
3800 STEREOTOMY Alan Parsons Project
1953 STICK AROUND Julian Lennon
3283 STILL John Schneider
2968 STILL A THRILL Jody Watley

4103 STILL CRUISIN' The Beach Boys
    Still In Love...see: CAN'TCHA SAY
    (You Believe In Me)
1585 STILL IN SAIGON Charlie Daniels Band
2448 STILL IN THE GAME Steve Winwood
3142 STILL LOVING YOU Scorpions
3405 STILL OF THE NIGHT Whitesnake
1707 STILL RIGHT HERE IN MY HEART
    Pure Prairie League
3479 STILL TAKING CHANCES Michael Martin Murphey
1316 STILL THEY RIDE Journey
4081 STIMULATION Wa Wa Nee
2111 STIR IT UP Patti LaBelle
684 STOMP! The Brothers Johnson
2151 STONE COLD Rainbow
1083 STONE LOVE Kool & The Gang
2845 STOP Sam Brown
4156 STOP! Erasure
2511 STOP DOGGIN' ME AROUND Klique
112 STOP DRAGGIN' MY HEART AROUND
    Stevie Nicks
    (With Tom Petty And The Heartbreakers)
1739 STOP IN THE NAME OF LOVE The Hollies
2566 STOP THIS GAME Cheap Trick
1134 STOP TO LOVE Luther Vandross
3288 STOP YOUR SOBBING The Pretenders
898 STRAIGHT FROM THE HEART Bryan Adams
2215 STRAIGHT FROM THE HEART
    The Allman Brothers Band
2798 STRAIGHT FROM THE HEART (Into Your Life)
    The Coyote Sisters
131 STRAIGHT UP Paula Abdul
1538 STRANGE BUT TRUE Times Two
2655 STRANGELOVE ('88) Depeche Mode
3483 --STRANGELOVE ('87) Depeche Mode
2547 STRANGER Jefferson Starship
2937 STRANGER Stephen Stills
3919 STRANGER IN MY HOME TOWN Foghat
1418 STRANGER IN MY HOUSE Ronnie Milsap
1666 STRANGER IN TOWN Toto
2801 STRANGERS IN A STRANGE WORLD
    Jenny Burton & Patrick Jude
3724 STRANGLEHOLD Paul McCartney
3929 STRAP ME IN The Cars
330 STRAY CAT STRUT Stray Cats
2664 STREET CORNER Ashford & Simpson
2652 STREET OF DREAMS Rainbow
2879 STRENGTH The Alarm
2221 STRIP Adam Ant
1122 THE STROKE Billy Squier
1834 STRONGER THAN BEFORE Carole Bayer Sager
2131 STRUNG OUT Steve Perry
607 STRUT Sheena Easton
300 STUCK ON YOU Lionel Richie
186 STUCK WITH YOU Huey Lewis And The News
3000 SUBURBIA Pet Shop Boys
4033 SUCH A SHAME Talk Talk
503 SUDDENLY Billy Ocean
1209 SUDDENLY Olivia Newton-John & Cliff Richard
698 SUDDENLY LAST SUMMER The Motels
2025 SUGAR DADDY Thompson Twins
1921 SUGAR DON'T BITE Sam Harris
2084 SUGAR FREE Wa Wa Nee
866 SUGAR WALLS Sheena Easton
231 SUKIYAKI A Taste Of Honey
3830 SUMMER '81 MEDLEY
    The Cantina Band
2890 SUMMER NIGHTS Survivor
639 SUMMER OF '69 Bryan Adams

2343 SUMMERGIRLS Dino
2685 SUMMERTIME GIRLS Y&T
3662 SUMMERTIME, SUMMERTIME Nocera
2826 THE SUN AIN'T GONNA SHINE ANYMORE
    Nielsen/Pearson
1430 THE SUN ALWAYS SHINES ON T.V. A-Ha
3468 THE SUN AND THE RAIN Madness
1917 SUN CITY Artists United Against Apartheid
583 SUNGLASSES AT NIGHT Corey Hart
1545 SUNSET GRILL Don Henley
1643 SUNSHINE Dino
3437 SUNSHINE IN THE SHADE
    The Fixx
2586 SUPER BOWL SHUFFLE
    The Chicago Bears Shufflin' Crew
1024 SUPER FREAK (Part I) Rick James
2374 SUPER TROUPER ABBA
3457 SUPERNATURAL LOVE Donna Summer
1521 SUPERSONIC J.J. Fad
3944 SUPERSTAR/UNTIL YOU COME BACK TO ME
    (That's What I'm Gonna Do)
    Luther Vandross
2003 SUPERSTITIOUS Europe
1061 SUPERWOMAN Karyn White
903 SURRENDER TO ME
    Ann Wilson and Robin Zander
3631 SURVIVE Jimmy Buffett
213 SUSSUDIO Phil Collins
1543 SUZANNE Journey
2747 SUZI Randy Vanwarmer
3531 SWEAR Sheena Easton
1227 SWEET BABY Stanley Clarke/George Duke
166 SWEET CHILD O' MINE Guns N' Roses
408 SWEET DREAMS Air Supply
44 SWEET DREAMS (Are Made Of This) Eurythmics
686 SWEET FREEDOM Michael McDonald
4135 SWEET LIES Robert Palmer
910 SWEET LOVE Anita Baker
3594 SWEET MERILEE Donnie Iris
3549 SWEET RACHEL Beau Coup
2900 SWEET SENSATION Stephanie Mills
1633 SWEET SIXTEEN Billy Idol
3580 SWEET, SWEET BABY (I'm Falling) Lone Justice
1651 SWEET TIME REO Speedwagon
620 THE SWEETEST TABOO Sade
444 THE SWEETEST THING (I've Ever Known)
    Juice Newton
800 SWEETHEART Franke And The Knockouts
2782 SWEETHEART LIKE YOU Bob Dylan
1391 SWEPT AWAY Diana Ross
1974 SWINGIN' John Anderson
2391 SWITCHIN' TO GLIDE The Kings
2585 SWITCHIN' TO GLIDE//
    THIS BEAT GOES ON [TSW] The Kings
1896 SYMPTOMS OF TRUE LOVE Tracie Spencer
1124 SYNCHRONICITY II The Police
2589 SYSTEM OF SURVIVAL Earth, Wind & Fire

T

547 TAINTED LOVE Soft Cell
1172 TAKE A LITTLE RHYTHM Ali Thomson
    Take A Look At Me Now...see: AGAINST ALL ODDS
2995 TAKE ANOTHER PICTURE Quarterflash
4015 TAKE AWAY Big Ric
755 TAKE IT AWAY Paul McCartney
1551 TAKE IT EASY Andy Taylor
739 TAKE IT EASY ON ME Little River Band
467 TAKE IT ON THE RUN REO Speedwagon

Singles Alpha Index

2561 TAKE IT WHILE IT'S HOT *Sweet Sensation*
2362 TAKE ME BACK *Bonnie Tyler*
1294 TAKE ME DOWN *Alabama*
 721 TAKE ME HOME *Phil Collins*
 580 TAKE ME HOME TONIGHT *Eddie Money*
3015 TAKE ME NOW *David Gates*
1040 TAKE ME TO HEART *Quarterflash*
1708 TAKE ME WITH U *Prince And The Revolution*
 297 TAKE MY BREATH AWAY
      (Love Theme From Top Gun) *Berlin*
1031 TAKE MY HEART (You Can Have It If You Want It)
      Kool & The Gang
3617 TAKE NO PRISONERS (In The Game Of Love)
      Peabo Bryson
1369 TAKE OFF *Bob & Doug McKenzie*
 230 TAKE ON ME *A-Ha*
2710 TAKE THE L. *The Motels*
1972 TAKE THE SHORT WAY HOME *Dionne Warwick*
3679 TAKE THE TIME *Michael Stanley Band*
3189 TAKE YOU TONIGHT *Ozark Mountain Daredevils*
 302 TAKE YOUR TIME (Do It Right) Part 1
      The S.O.S. Band
1764 TAKEN IN *Mike + The Mechanics*
4014 TAKIN' IT BACK *Breathless*
2430 TAKING IT ALL TOO HARD *Genesis*
1072 TALK DIRTY TO ME *Poison*
1453 TALK IT OVER *Grayson Hugh*
3579 TALK TALK *Talk Talk*
 520 TALK TO ME *Stevie Nicks*
1258 TALK TO ME *Chico DeBarge*
3122 TALK TO ME *Fiona*
3715 TALK TO ME *Quarterflash*
2001 TALK TO MYSELF *Christopher Williams*
3768 TALKIN' BOUT A REVOLUTION *Tracy Chapman*
2611 TALKING BACK TO THE NIGHT *Steve Winwood*
 153 TALKING IN YOUR SLEEP *The Romantics*
3138 TALKING OUT OF TURN *The Moody Blues*
1598 TALL COOL ONE *Robert Plant*
 962 TARZAN BOY *Baltimora*
2132 TASTY LOVE *Freddie Jackson*
3896 TAXI *J. Blackfoot*
2885 TAXI DANCING
      Rick Springfield & Randy Crawford
3183 TEACH ME TONIGHT *Al Jarreau*
2291 TEACHER TEACHER *Rockpile*
1852 TEACHER TEACHER *38 Special*
1980 TEARS *John Waite*
2368 TEARS ARE FALLING *KISS*
3154 TEARS RUN RINGS *Marc Almond*
 614 TELEFONE (Long Distance Love Affair)
      Sheena Easton
3403 TELL HER *Kenny Loggins*
 175 TELL HER ABOUT IT *Billy Joel*
1822 TELL HER NO *Juice Newton*
 751 TELL IT LIKE IT IS *Heart*
 618 TELL IT TO MY HEART *Taylor Dayne*
2616 TELL ME *White Lion*
2811 TELL ME IF YOU STILL CARE *The S.O.S. Band*
3006 TELL ME I'M NOT DREAMING *Robert Palmer*
2012 TELL ME TOMORROW - PART 1 *Smokey Robinson*
4010 TELL THAT GIRL TO SHUT UP *Transvision Vamp*
2353 TEMPTED *Squeeze*
1850 10-9-8 *Face To Face*
1486 TENDER IS THE NIGHT *Jackson Browne*
 992 TENDER LOVE *Force M.D.'s*
1759 TENDER YEARS ('84a)
      John Cafferty And The Beaver Brown Band
3681 --TENDER YEARS ('84b)
      John Cafferty And The Beaver Brown Band

1507 TENDERNESS *General Public*
3348 TEST OF TIME *The Romantics*
2617 TEXAS IN MY REAR VIEW MIRROR *Mac Davis*
1863 THANKS FOR MY CHILD *Cheryl "Pepsii" Riley*
1506 THAT AIN'T LOVE *REO Speedwagon*
3544 THAT DIDN'T HURT TOO BAD *Dr. Hook*
 333 THAT GIRL *Stevie Wonder*
1678 THAT GIRL COULD SING *Jackson Browne*
2719 THAT LOVIN' YOU FEELIN' AGAIN
      Roy Orbison & Emmylou Harris
1589 THAT OLD SONG *Ray Parker Jr. & Raydio*
4006 THAT WAS THEN BUT THIS IS NOW *ABC*
1586 THAT WAS THEN, THIS IS NOW
      Micky Dolenz And Peter Tork (Of The Monkees)
1212 THAT WAS YESTERDAY *Foreigner*
 573 THAT'S ALL *Genesis*
2999 THAT'S FREEDOM *Tom Kimmel*
3882 THAT'S LIFE *David Lee Roth*
1754 THAT'S LOVE *Jim Capaldi*
1663 THAT'S THE WAY *Katrina And The Waves*
  65 THAT'S WHAT FRIENDS ARE FOR
      Dionne & Friends
1174 THAT'S WHAT LOVE IS ALL ABOUT *Michael Bolton*
4168 THAT'S WHEN I THINK OF YOU *1927*
     The Morse Code Of Love...
      see: BABY COME BACK TO ME
     Theme From AMERICAN GIGOLO...see: CALL ME
     Theme From BEST FRIENDS...
      see: HOW DO YOU KEEP THE MUSIC PLAYING
     Theme from CATS...see: MEMORY
     Theme From Cheers...
      see: WHERE EVERYBODY KNOWS YOUR NAME
2992 THEME FROM "DOCTOR DETROIT" *Devo*
2649 THEME FROM DYNASTY *Bill Conti*
 828 THE THEME FROM HILL STREET BLUES
      Mike Post Featuring Larry Carlton
1460 THEME FROM MAGNUM P.I. *Mike Post*
1748 THEME FROM NEW YORK, NEW YORK
      Frank Sinatra
3889 THEME FROM "RAGING BULL"
      (Cavalleria Rusticana) *Joel Diamond*
3971 THEME FROM S-EXPRESS *S-Express*
3654 THEME FROM "TERMS OF ENDEARMENT"
      Michael Gore
     Theme From The Circus...see: COMPUTER GAME
1190 THEME FROM "THE DUKES OF HAZZARD"
      (Good Ol' Boys) *Waylon*
  97 THEME FROM "THE GREATEST AMERICAN HERO"
      (Believe It Or Not) *Joey Scarbury*
     Theme From The KARATE KID PART II...
      see: GLORY OF LOVE
     Theme From The LEGEND OF BILLIE JEAN...
      see: INVINCIBLE
     Theme From TOOTSIE...see: IT MIGHT BE YOU
2625 THEMES FROM E.T. (The Extra-Terrestrial)
      Walter Murphy
1499 THERE GOES MY BABY *Donna Summer*
1770 THERE MUST BE AN ANGEL
      (Playing With My Heart) *Eurythmics*
 203 THERE'LL BE SAD SONGS (To Make You Cry)
      Billy Ocean
2633 THERE'S NO EASY WAY *James Ingram*
 361 (There's) NO GETTIN' OVER ME *Ronnie Milsap*
2399 THERE'S NOTHING BETTER THAN LOVE
      Luther Vandross (Duet With Gregory Hines)
1085 THERE'S THE GIRL *Heart*
     These Boots Are Made For Walkin'...
      see: PUSS N BOOTS
 241 THESE DREAMS *Heart*
2407 THESE TIMES ARE HARD FOR LOVERS *John Waite*
 750 THEY DON'T KNOW *Tracey Ullman*

     They Don't Write 'Em...see: BREAKUP SONG
3764 THIEF OF HEARTS *Melissa Manchester*
3675 THIN LINE BETWEEN LOVE AND HATE
      The Pretenders
 553 THINGS CAN ONLY GET BETTER *Howard Jones*
1462 THINK ABOUT ME *Fleetwood Mac*
 997 THINK I'M IN LOVE *Eddie Money*
 840 THINK OF LAURA *Christopher Cross*
 928 THINKING OF YOU *Sa-Fire*
3049 THINKING OF YOU *Earth, Wind & Fire*
1290 THIRD TIME LUCKY (First Time I Was A Fool)
      Foghat
     This Beat Goes On...see: SWITCHIN' TO GLIDE
1015 THIS COULD BE THE NIGHT *Loverboy*
3064 THIS COULD BE THE RIGHT ONE *April Wine*
 637 THIS IS IT *Kenny Loggins*
2865 THIS IS MY NIGHT *Chaka Khan*
1899 THIS IS NOT AMERICA
      David Bowie/Pat Metheny Group
1371 THIS IS THE TIME *Billy Joel*
4095 THIS IS THE TIME *Dennis DeYoung*
3627 THIS IS THE WORLD CALLING *Bob Geldof*
 834 THIS LITTLE GIRL *Gary U.S. Bonds*
3836 THIS LOVE *Bad Company*
1843 THIS MAN IS MINE *Heart*
2864 THIS MUST BE THE PLACE (Naïve Melody)
      Talking Heads
4086 THIS ONE *Paul McCartney*
1459 THIS TIME *John Cougar*
1608 THIS TIME *Bryan Adams*
3540 THIS TIME *INXS*
3745 THIS TIME *Kiara (Duet With Shanice Williams)*
 876 THIS TIME I KNOW IT'S FOR REAL *Donna Summer*
1555 THIS WOMAN *Kenny Rogers*
3201 THORN IN MY SIDE *Eurythmics*
3125 THOSE GOOD OLD DREAMS *Carpenters*
1240 THREE TIMES IN LOVE *Tommy James*
 506 THRILLER *Michael Jackson*
2176 THROUGH THE FIRE *Chaka Khan*
1623 THROUGH THE STORM
      Aretha Franklin And Elton John
1135 THROUGH THE YEARS *Kenny Rogers*
3028 THROWAWAY *Mick Jagger*
 605 THROWING IT ALL AWAY *Genesis*
2786 THUNDER AND LIGHTNING *Chicago*
     Thunderdome...
      see: WE DON"T NEED ANOTHER HERO
2331 TI AMO *Laura Branigan*
  95 THE TIDE IS HIGH *Blondie*
2124 TIED UP *Olivia Newton-John*
1714 'TIL MY BABY COMES HOME *Luther Vandross*
3571 TIL YOU AND YOUR LOVER ARE LOVERS AGAIN
      Engelbert Humperdinck
1969 TILL I LOVED YOU (The Love Theme From Goya)
      Barbra Streisand And Don Johnson
 959 TIME *Alan Parsons Project*
  89 TIME AFTER TIME *Cyndi Lauper*
1546 TIME AND TIDE *Basia*
 172 TIME (Clock Of The Heart) *Culture Club*
3400 TIME FOR ME TO FLY *REO Speedwagon*
1139 TIME IS TIME *Andy Gibb*
1736 TIME OUT OF MIND *Steely Dan*
1036 TIME WILL REVEAL *DeBarge*
4159 TINA CHERRY *Georgio*
2724 TIP OF MY TONGUE *The Tubes*
3402 TIRED OF BEING BLONDE *Carly Simon*
 701 TIRED OF TOEIN' THE LINE *Rocky Burnette*
 508 TO ALL THE GIRLS I'VE LOVED BEFORE
      Julio Iglesias & Willie Nelson

Singles Alpha Index

#	Title	Artist
679	TO BE A LOVER	Billy Idol
3114	TO DREAM THE DREAM	Frankie Miller
1705	TO LIVE AND DIE IN L.A.	Wang Chung
3970	TOCCATA	Sky
3163	TODAY IS THE DAY	Bar-Kays
952	TOGETHER	Tierra
366	TOGETHER FOREVER	Rick Astley
2198	TOM SAWYER	Rush
1812	TOMORROW DOESN'T MATTER TONIGHT	Starship
2214	TOMORROW PEOPLE	Ziggy Marley And The Melody Makers
998	TONIGHT	Kool & The Gang
2569	TONIGHT	David Bowie
3824	TONIGHT	The Whispers
680	TONIGHT, I CELEBRATE MY LOVE	Peabo Bryson/Roberta Flack
1492	TONIGHT I'M YOURS (Don't Hurt Me)	Rod Stewart
3656	TONIGHT IS WHAT IT MEANS TO BE YOUNG	Fire Inc.
1992	TONIGHT IT'S YOU	Cheap Trick
781	TONIGHT SHE COMES	The Cars
2868	TONIGHT TONIGHT	Bill Champlin
557	TONIGHT, TONIGHT, TONIGHT	Genesis
3727	TOO GOOD TO TURN BACK NOW	Rick Bowles
461	TOO HOT	Kool & The Gang
3485	TOO LATE	Journey
622	TOO LATE FOR GOODBYES	Julian Lennon
4113	TOO MUCH AIN'T ENOUGH LOVE	Jimmy Barnes
3101	TOO MUCH LOVE TO HIDE	Crosby, Stills & Nash
769	TOO MUCH TIME ON MY HANDS	Styx
592	TOO SHY	Kajagoogoo
2283	TOO TIGHT	Con Funk Shun
2188	TOO YOUNG	Jack Wagner
4058	TOO YOUNG TO FALL IN LOVE	Mötley Crüe
1452	TORTURE	Jacksons
23	TOTAL ECLIPSE OF THE HEART	Bonnie Tyler
3583	TOUCH A FOUR LEAF CLOVER	Atlantic Starr
2065	TOUCH AND GO	The Cars
2981	TOUCH AND GO	Emerson, Lake & Powell
541	TOUCH ME (I Want Your Body)	Samantha Fox
3151	TOUCH ME TONIGHT	Shooting Star
1300	TOUCH ME WHEN WE'RE DANCING	Carpenters
1067	TOUCH OF GREY	Grateful Dead
3945	TOUCH THE FIRE	Icehouse
1477	TOUGH ALL OVER	John Cafferty And The Beaver Brown Band
2986	TOUGH WORLD	Donnie Iris
295	TOY SOLDIERS	Martika
1857	TRAGEDY	John Hunter
1501	TRAIN IN VAIN (Stand By Me)	The Clash
3565	TREASURE	The Brothers Johnson
2126	TREAT HER LIKE A LADY	The Temptations
1184	TREAT ME RIGHT	Pat Benatar
2545	TRIBUTE (Right On)	The Pasadenas
3534	A TRICK OF THE NIGHT	Bananarama
3317	TRICKLE TRICKLE	The Manhattan Transfer
635	TROUBLE	Lindsey Buckingham
1988	TROUBLE	Nia Peeples
3242	TROUBLE IN PARADISE	Jarreau
2381	TROUBLE ME	10,000 Maniacs
339	TRUE	Spandau Ballet
431	TRUE BLUE	Madonna
283	TRUE COLORS	Cyndi Lauper
1590	TRUE FAITH	New Order
1233	TRUE LOVE	Glenn Frey
3191	TRUE LOVE WAYS	Mickey Gilley
3304	TRUE TO YOU	Ric Ocasek
113	TRULY	Lionel Richie
3975	TRUST ME	Cindy Bullens
1255	TRY AGAIN	Champaign
543	TRYIN' TO LIVE MY LIFE WITHOUT YOU (Live)	Bob Seger
1046	TUFF ENUFF	The Fabulous Thunderbirds
2834	TUG OF WAR	Paul McCartney
3686	TULSA TIME	Eric Clapton And His Band
1701	TULSA TIME//COCAINE [TSW]	Eric Clapton And His Band
1095	TUNNEL OF LOVE	Bruce Springsteen
2117	TURN AND WALK AWAY	The Babys
3046	TURN AROUND	Neil Diamond
2813	TURN IT ON AGAIN	Genesis
1621	TURN ME LOOSE	Loverboy
3440	TURN OFF THE LIGHTS	The World Class Wreckin' Cru
2978	TURN ON YOUR RADAR	Prism
1830	TURN TO YOU	Go-Go's
1584	TURN UP THE RADIO	Autograph
449	TURN YOUR LOVE AROUND	George Benson
2085	TURNED AWAY	Chuckii Booker
1673	TURNING JAPANESE	The Vapors
2614	25 OR 6 TO 4	Chicago
2059	24/7	Dino
2294	20/20	George Benson
2384	TWILIGHT	ELO
1923	TWILIGHT WORLD	Swing Out Sister
641	TWILIGHT ZONE	Golden Earring
1802	TWILIGHT ZONE/TWILIGHT TONE	The Manhattan Transfer
1640	TWIST AND SHOUT	The Beatles
3793	TWIST MY ARM	Pointer Sisters
441	TWIST OF FATE	Olivia Newton-John
1440	THE TWIST (Yo, Twist!)	Fat Boys
3752	TWISTIN' THE NIGHT AWAY	Rod Stewart
235	TWO HEARTS	Phil Collins
2128	TWO HEARTS	Stephanie Mills featuring Teddy Pendergrass
1767	TWO LESS LONELY PEOPLE IN THE WORLD	Air Supply
1007	TWO OCCASIONS	The Deele
548	TWO OF HEARTS	Stacey Q
1915	TWO PEOPLE	Tina Turner
1614	TWO PLACES AT THE SAME TIME	Ray Parker Jr. & Raydio
2009	TWO SIDES OF LOVE	Sammy Hagar
1934	TWO TRIBES	Frankie Goes To Hollywood
341	TYPICAL MALE	Tina Turner

U

#	Title	Artist
418	U GOT THE LOOK	Prince
3931	UNCHAINED MELODY (Live)	Heart
2487	UNCONDITIONAL LOVE	Donna Summer
1569	UNDER PRESSURE	Queen & David Bowie
3068	UNDER THE BOARDWALK	Bruce Willis
3593	UNDER THE COVERS	Janis Ian
2438	UNDER THE GUN	Poco
3023	UNDER THE INFLUENCE	Vanity
1775	UNDER THE MILKY WAY	The Church
788	UNDERCOVER OF THE NIGHT	The Rolling Stones
3284	UNDERNEATH THE RADAR	Underworld
1276	UNDERSTANDING	Bob Seger & The Silver Bullet Band
4016	UNFAITHFULLY YOURS (One Love)	Stephen Bishop
239	UNION OF THE SNAKE	Duran Duran
2781	UNITED TOGETHER	Aretha Franklin
	Until You Come Back To Me...see: SUPERSTAR	
108	UP WHERE WE BELONG	Joe Cocker and Jennifer Warnes
14	UPSIDE DOWN	Diana Ross
90	UPTOWN GIRL	Billy Joel
325	URGENT	Foreigner
2429	US AND LOVE (We Go Together)	Kenny Nolan
2197	USED TO BE	Charlene & Stevie Wonder

V

#	Title	Artist
799	VACATION	Go-Go's
	VALERIE	
883		Steve Winwood ('87)
3471		Steve Winwood ('82)
1900	VALLEY GIRL	Frank Zappa
715	THE VALLEY ROAD	Bruce Hornsby & The Range
779	VALOTTE	Julian Lennon
3989	VANITY KILLS	ABC
2821	VARIETY TONIGHT	REO Speedwagon
2023	VELCRO FLY	ZZ Top
276	VENUS	Bananarama
1638	VERONICA	Elvis Costello
3740	THE VERY BEST IN YOU	Change
3709	VERY LAST TIME	Utopia
3922	VERY SPECIAL	Debra Laws
2060	VICTIM OF LOVE	Bryan Adams
909	VICTORY	Kool & The Gang
3349	VICTORY LINE	Limited Warranty
3910	VIDEO!	Jeff Lynne
1484	VIENNA CALLING (The New '86 Edit)	Falco
187	A VIEW TO A KILL	Duran Duran
3047	THE VISITORS	ABBA
2953	VITAMIN L	B. E. Taylor Group
1147	THE VOICE	The Moody Blues
2872	VOICE OF AMERICA'S SONS	John Cafferty And The Beaver Brown Band
3641	VOICE OF FREEDOM	Jim Kirk And The TM Singers
3298	VOICE ON THE RADIO	Conductor
1920	VOICES	Cheap Trick
764	VOICES CARRY	'til tuesday
1751	VOICES OF BABYLON	The Outfield
3077	VOLCANO	Jimmy Buffett
3445	VOO DOO	Rachel Sweet
1999	VOX HUMANA	Kenny Loggins
1893	VOYEUR	Kim Carnes

W

#	Title	Artist
917	WAIT	White Lion
1146	WAIT FOR ME	Daryl Hall & John Oates
3584	WAIT ON LOVE	Michael Bolton
1534	THE WAITING	Tom Petty And The Heartbreakers
10	WAITING FOR A GIRL LIKE YOU	Foreigner
440	WAITING FOR A STAR TO FALL	Boy Meets Girl
3379	WAITING FOR YOUR LOVE	Toto
3940	WAITING GAME	Swing Out Sister
1005	WAITING ON A FRIEND	The Rolling Stones
83	WAKE ME UP BEFORE YOU GO-GO	Wham!
1832	WAKE UP LITTLE SUSIE	Simon and Garfunkel
3056	WAKE UP MY LOVE	George Harrison
2146	WAKE UP (Next To You)	Graham Parker And The Shot
2139	WALK AWAY	Donna Summer
4144	WALK AWAY RENEE	Southside Johnny & The Jukes
2422	WALK LIKE A MAN	Mary Jane Girls
77	WALK LIKE AN EGYPTIAN	Bangles
659	WALK OF LIFE	Dire Straits
790	WALK ON WATER	Eddie Money

Singles Alpha Index

#	Title	Artist
3567	WALK RIGHT NOW	The Jacksons
972	WALK THE DINOSAUR	Was (Not Was)
654	WALK THIS WAY	Run-D.M.C.
3846	WALKIN' SHOES	Tora Tora
1009	WALKING AWAY	Information Society
1111	WALKING DOWN YOUR STREET	Bangles
3302	WALKING IN L.A.	Missing Persons
2771	WALKING IN MY SLEEP	Roger Daltrey
3655	WALKING INTO SUNSHINE	Central Line
1296	WALKING ON A THIN LINE	Huey Lewis And The News
813	WALKING ON SUNSHINE	Katrina And The Waves
2450	WALKING ON THE CHINESE WALL	Philip Bailey
2800	WALKING ON THIN ICE	Yoko Ono
3601	WALKING THROUGH WALLS	The Escape Club
1809	WALKS LIKE A LADY	Journey
3556	THE WALLS CAME DOWN	The Call
308	THE WANDERER	Donna Summer
3885	WANGO TANGO	Ted Nugent
501	WANNA BE STARTIN' SOMETHIN'	Michael Jackson
2873	WANNA BE WITH YOU	Earth, Wind & Fire
3079	(Want You) BACK IN MY LIFE AGAIN	Carpenters
3582	WANT YOU FOR MY GIRLFRIEND	4 By Four
689	WANTED DEAD OR ALIVE	Bon Jovi
3947	WANTED MAN	Ratt
954	WAR (Live)	Bruce Springsteen & The E Street Band
2332	WAR GAMES	Crosby, Stills & Nash
1514	THE WAR SONG	Culture Club
563	THE WARRIOR	Scandal Featuring Patty Smyth
1716	WASN'T THAT A PARTY	The Rovers
692	WASTED ON THE WAY	Crosby, Stills & Nash
3450	WATCHING OVER YOU	Glenn Medeiros
984	WATCHING THE WHEELS	John Lennon
3456	WATCHING YOU	Slave
2673	WATERFALL	Wendy & Lisa
3467	WAY COOL JR.	Ratt
1778	THE WAY HE MAKES ME FEEL	Barbra Streisand
216	THE WAY IT IS	Bruce Hornsby And The Range
2993	WAY OUT	J.J. Fad
2326	THE WAY TO YOUR HEART	Soulsister
1698	A NITE AT THE APOLLO LIVE: THE WAY YOU DO THE THINGS YOU DO/MY GIRL	Daryl Hall & John Oates Featuring David Ruffin & Eddie Kendricks
681	THE WAY YOU LOVE ME	Karyn White
364	THE WAY YOU MAKE ME FEEL	Michael Jackson
3439	WAYS TO BE WICKED	Lone Justice
1338	WE ALL SLEEP ALONE	Cher
79	WE ARE THE WORLD	USA for Africa
1473	WE ARE THE YOUNG	Dan Hartman
2638	WE ARE WHAT WE ARE	The Other Ones
476	WE BELONG	Pat Benatar
157	WE BUILT THIS CITY	Starship
3066	WE CAN GET TOGETHER	Icehouse
2426	WE CAN LAST FOREVER	Chicago
1903	WE CLOSE OUR EYES	Go West
1692	WE CONNECT	Stacey Q
3325	WE COULD BE TOGETHER	Debbie Gibson
145	WE DIDN'T START THE FIRE	Billy Joel
624	WE DON'T HAVE TO TAKE OUR CLOTHES OFF	Jermaine Stewart
372	WE DON'T NEED ANOTHER HERO (Thunderdome)	Tina Turner
549	WE DON'T TALK ANYMORE	Cliff Richard
119	WE GOT THE BEAT	Go-Go's
1615	WE LIVE FOR LOVE	Pat Benatar
4072	WE SHOULD BE SLEEPING	Eddie Money
1597	WE TWO	Little River Band
1731	WE WERE MEANT TO BE LOVERS	Photoglo
2948	WEAPONS OF LOVE	The Truth
4129	WEATHERMAN	Nick Jameson
3259	--WEATHERMAN SAYS	Jack Wagner
2453	WEIRD SCIENCE	Oingo Boingo
1573	WELCOME TO HEARTLIGHT	Kenny Loggins
3825	WELCOME TO PARADISE	John Waite
1952	WELCOME TO THE BOOMTOWN	David & David
744	WELCOME TO THE JUNGLE	Guns N' Roses
2627	WELCOME TO THE PLEASUREDOME (Trevor Horn Remix)	Frankie Goes To Hollywood
797	WE'LL BE TOGETHER	Sting
2228	WE'RE GOING ALL THE WAY	Jeffrey Osborne
914	WE'RE IN THIS LOVE TOGETHER	Al Jarreau
1481	WE'RE NOT GONNA TAKE IT	Twisted Sister
1098	WE'RE READY	Boston
2854	WEST COAST SUMMER NIGHTS	Tony Carey
181	WEST END GIRLS	Pet Shop Boys
2458	WET MY WHISTLE	Midnight Star
484	WE'VE GOT TONIGHT	Kenny Rogers And Sheena Easton
2730	WE'VE ONLY JUST BEGUN (The Romance Is Not Over)	Glenn Jones
2598	WE'VE SAVED THE BEST FOR LAST	Kenny G Vocal by Smokey Robinson
	What a Beautiful World...see: I.G.Y.	
	What A Feeling...see: FLASHDANCE	
2174	WHAT A WONDERFUL WORLD	Louis Armstrong
1801	WHAT ABOUT LOVE	'Til Tuesday
930	WHAT ABOUT LOVE?	Heart
1034	WHAT ABOUT ME ('82)	Moving Pictures
1932	--WHAT ABOUT ME ('89)	Moving Pictures
1270	WHAT ABOUT ME?	Kenny Rogers
1906	WHAT AM I GONNA DO (I'm So In Love With You)	Rod Stewart
915	WHAT ARE WE DOIN' IN LOVE	Dottie West
3238	WHAT CAN YOU GET A WOOKIEE FOR CHRISTMAS (When He Already Owns A Comb?)	Star Wars Intergalactic Droid Choir & Chorale
2716	WHAT CHA' GONNA DO FOR ME	Chaka Khan
2692	WHAT DO ALL THE PEOPLE KNOW	The Monroes
2121	WHAT DOES IT TAKE	Honeymoon Suite
423	WHAT HAVE I DONE TO DESERVE THIS?	Pet Shop Boys And Dusty Springfield
499	WHAT HAVE YOU DONE FOR ME LATELY	Janet Jackson
838	WHAT I AM	Edie Brickell & The New Bohemians
	WHAT I LIKE ABOUT YOU	
1879		Michael Morales
2579		The Romantics
3081	WHAT IF (I Said I Love You)	Unipop
	What If...see: REMO'S THEME	
2050	(What) IN THE NAME OF LOVE	Naked Eyes
1864	WHAT IS LOVE?	Howard Jones
958	WHAT KIND OF FOOL	Barbra Streisand & Barry Gibb
1436	WHAT KIND OF FOOL AM I	Rick Springfield
3145	WHAT LOVE IS	Marty Balin
3129	WHAT SHE DOES TO ME (The Diana Song)	The Producers
2396	WHAT THE BIG GIRLS DO	Van Stephenson
3369	WHAT YOU DO TO ME	Carl Wilson
940	WHAT YOU DON'T KNOW	Exposé
1340	WHAT YOU GET IS WHAT YOU SEE	Tina Turner
545	WHAT YOU NEED	INXS
1687	WHAT YOU SEE IS WHAT YOU GET	Brenda K. Starr
3689	WHAT YOU'RE MISSING	Chicago
1910	WHATCHA GONNA DO	Chilliwack
4125	WHATEVER HAPPENED TO OLD FASHIONED LOVE	B.J. Thomas
3717	WHATEVER YOU DECIDE	Randy Vanwarmer
1082	WHAT'S FOREVER FOR	Michael Murphey
1268	WHAT'S GOING ON	Cyndi Lauper
40	WHAT'S LOVE GOT TO DO WITH IT	Tina Turner
2079	WHAT'S NEW	Linda Ronstadt & The Nelson Riddle Orchestra
509	WHAT'S ON YOUR MIND (Pure Energy)	Information Society
3861	WHAT'S SHE GOT	Liquid Gold
3421	WHAT'S THE MATTER HERE?	10,000 Maniacs
3261	WHAT'S TOO MUCH	Smokey Robinson
3809	WHAT'S YOUR HURRY DARLIN'	Ironhorse
2053	WHEN A MAN LOVES A WOMAN	Bette Midler
1655	WHEN ALL IS SAID AND DONE	ABBA
27	WHEN DOVES CRY	Prince
1730	WHEN HE SHINES	Sheena Easton
4155	WHEN I FALL IN LOVE	Natalie Cole
1023	WHEN I LOOKED AT HIM	Exposé
257	WHEN I SEE YOU SMILE	Bad English
240	WHEN I THINK OF YOU	Janet Jackson
1366	WHEN I WANTED YOU	Barry Manilow
3618	WHEN I'M HOLDING YOU TIGHT	Michael Stanley Band
317	WHEN I'M WITH YOU ('88)	Sheriff
2946	--WHEN I'M WITH YOU ('83)	Sheriff
756	WHEN IT'S LOVE	Van Halen
1668	WHEN IT'S OVER	Loverboy
3133	WHEN LOVE COMES TO TOWN	U2 With B. B. King
2567	WHEN SHE DANCES	Joey Scarbury
846	WHEN SHE WAS MY GIRL	Four Tops
674	WHEN SMOKEY SINGS	ABC
565	WHEN THE CHILDREN CRY	White Lion
2521	WHEN THE FEELING COMES AROUND	Jennifer Warnes
383	WHEN THE GOING GETS TOUGH, THE TOUGH GET GOING	Billy Ocean
1198	WHEN THE HEART RULES THE MIND	GTR
3670	WHEN THE LADY SMILES	Golden Earring
1786	WHEN THE LIGHTS GO OUT	Naked Eyes
3499	WHEN THE RADIO IS ON	Paul Shaffer
2689	WHEN THE RAIN BEGINS TO FALL	Jermaine Jackson And Pia Zadora
3470	WHEN THE RAIN COMES DOWN	Andy Taylor
3963	WHEN THINGS GO WRONG	Robin Lane & The Chartbusters
3547	WHEN WE GET MARRIED	Larry Graham
2064	WHEN WE KISS	Bardeux
3200	WHEN WE MAKE LOVE	Alabama
1891	WHEN WE WAS FAB	George Harrison
3839	WHEN WILL I BE FAMOUS	Bros
1126	WHEN YOU CLOSE YOUR EYES	Night Ranger
4044	WHEN YOU WALK IN THE ROOM	Paul Carrack
3934	WHEN YOU WERE MINE	Mitch Ryder
1866	WHEN YOUR HEART IS WEAK	Cock Robin
974	WHERE ARE YOU NOW? ('89)	Jimmy Harnen w/Synch
3172	--WHERE ARE YOU NOW? ('86)	Synch
3942	WHERE DID WE GO WRONG	Frankie Valli Introducing Chris Forde
2666	WHERE DID YOUR HEART GO?	Wham!
296	WHERE DO BROKEN HEARTS GO	Whitney Houston
2180	WHERE DO THE CHILDREN GO	Hooters
2427	WHERE DOES THE LOVIN' GO	David Gates
3691	WHERE EVERYBODY KNOWS YOUR NAME (The Theme From "Cheers")	Gary Portnoy

Singles Alpha Index

#	Title	Artist
1301	WHERE THE STREETS HAVE NO NAME	U2
4000	WHERE'S YOUR ANGEL?	Lani Hall
3210	WHEREVER I LAY MY HAT (That's My Home)	Paul Young
636	WHILE YOU SEE A CHANCE	Steve Winwood
704	WHIP IT	Devo
1604	WHIRLY GIRL	Oxo
3148	WHISPER IN THE DARK	Dionne Warwick
1982	WHISPER TO A SCREAM (Birds Fly)	Icicle Works
1560	WHITE HORSE	Laid Back
2759	WHITE HOT	Red Rider
1780	WHITE WEDDING	Billy Idol
4115	WHITER SHADE OF PALE	Hagar, Schon, Aaronson, Shrieve
82	WHO CAN IT BE NOW?	Men At Work
1262	WHO DO YOU GIVE YOUR LOVE TO?	Michael Morales
2301	WHO DO YOU THINK YOU'RE FOOLIN'	Donna Summer
1372	WHO FOUND WHO	Jellybean (Elisa Fiorillo Vocals)
3616	WHO SAYS	Device
3247	WHO SHOT J.R.?	Gary Burbank with Band McNally
1341	WHO WEARS THESE SHOES?	Elton John
2770	WHO WERE YOU THINKIN' OF	The Doolittle Band
830	WHO WILL YOU RUN TO	Heart
1729	WHO'LL BE THE FOOL TONIGHT	Larsen-Feiten Band
2848	WHO'S BEHIND THE DOOR?	Zebra
369	WHO'S CRYING NOW	Journey
805	WHO'S HOLDING DONNA NOW	DeBarge
550	WHO'S JOHNNY (Short Circuit Theme)	El DeBarge
2162	WHO'S MAKING LOVE	Blues Brothers
305	WHO'S THAT GIRL	Madonna
1480	WHO'S THAT GIRL?	Eurythmics
729	WHO'S ZOOMIN' WHO	Aretha Franklin
3398	WHY	Carly Simon
2455	WHY CAN'T I BE YOU?	The Cure
1733	WHY CAN'T I HAVE YOU	The Cars
469	WHY CAN'T THIS BE LOVE	Van Halen
2663	WHY CAN'T THIS NIGHT GO ON FOREVER	Journey
552	WHY DO FOOLS FALL IN LOVE	Diana Ross
1665	WHY ME	Styx
2819	WHY ME?	Planet P
973	WHY ME?	Irene Cara
1362	WHY NOT ME	Fred Knoblock
2768	WHY SHOULD I CRY?	Nona Hendryx
2102	WHY YOU TREAT ME SO BAD	Club Nouveau
	Why You Wanna Hurt Me...see: BURN RUBBER	
3333	WHY YOU WANNA TRY ME	Commodores
3225	WILD AGAIN	Starship
2470	WILD AND CRAZY LOVE	Mary Jane Girls
114	THE WILD BOYS	Duran Duran
2259	WILD HORSES	Gino Vannelli
3166	THE WILD LIFE	Bananarama
350	WILD THING	Tone Loc
1443	WILD WILD LIFE	Talking Heads
197	WILD, WILD WEST	The Escape Club
2785	WILD, WILD WEST	Kool Moe Dee
1528	WILD WORLD	Maxi Priest
3599	WILL THE WOLF SURVIVE?	Los Lobos
512	WILL YOU STILL LOVE ME?	Chicago
2969	WILLIE AND THE HAND JIVE	George Thorogood & The Destroyers
	WIND BENEATH MY WINGS	
252		Bette Midler
3109		Lou Rawls
2901	WIND HIM UP	Saga
2938	WINDOWS	Missing Persons
2058	WINDS OF CHANGE	Jefferson Starship
537	THE WINNER TAKES IT ALL	ABBA
2440	WINNER TAKES IT ALL	Sammy Hagar
1245	WINNING	Santana
3984	WINTER GAMES	David Foster
1019	WIPEOUT	Fat Boys and The Beach Boys
3597	WIRED FOR SOUND	Cliff Richard
2982	WISE UP	Amy Grant
2929	WISHING I WAS LUCKY	Wet Wet Wet
1527	WISHING (If I Had A Photograph Of You)	A Flock Of Seagulls
262	WISHING WELL	Terence Trent D'Arby
505	WITH EVERY BEAT OF MY HEART	Taylor Dayne
158	WITH OR WITHOUT YOU	U2
2512	WITH YOU ALL THE WAY	New Edition
268	WITH YOU I'M BORN AGAIN	Billy Preston & Syreeta
3515	WITHOUT YOU	David Bowie
4037	WITHOUT YOU (Love Theme From 'Leonard Part 6')	Peabo Bryson & Regina Belle
1561	WITHOUT YOU (Not Another Lonely Night)	Franke And The Knockouts
1295	WITHOUT YOUR LOVE	Roger Daltrey
2216	WITHOUT YOUR LOVE	Toto
2773	WKRP IN CINCINNATI	Steve Carlisle
58	WOMAN	John Lennon
25	WOMAN IN LOVE	Barbra Streisand
3596	A WOMAN IN LOVE (It's Not Me)	Tom Petty And The Heartbreakers
1600	THE WOMAN IN ME	Donna Summer
3343	THE WOMAN IN ME	Crystal Gayle
1676	THE WOMAN IN YOU	Bee Gees
359	A WOMAN NEEDS LOVE (Just Like You Do)	Ray Parker Jr. & Raydio
2300	WOMEN	Foreigner
3734	WOMEN	Def Leppard
1324	WONDERING WHERE THE LIONS ARE	Bruce Cockburn
1520	WONDERLAND	Commodores
3998	WONDERLAND	Big Country
4031	WOOD BEEZ (Pray Like Aretha Franklin)	Scritti Politti
1653	A WORD IN SPANISH	Elton John
1941	THE WORD IS OUT	Jermaine Stewart
582	WORD UP	Cameo
2309	WORDS	Missing Persons
2777	WORDS	F.R. David
653	WORDS GET IN THE WAY	Miami Sound Machine
2626	WORK THAT BODY	Diana Ross
2334	WORKIN' FOR A LIVIN'	Huey Lewis And The News
3307	WORKING CLASS MAN	Jimmy Barnes
1390	WORKING FOR THE WEEKEND	Loverboy
2273	WORKING IN THE COAL MINE	Devo
135	WORKING MY WAY BACK TO YOU/ FORGIVE ME GIRL (Medley)	Spinners
3301	WORKING ON IT	Chris Rea
3933	WORLD SHUT YOUR MOUTH	Julian Cope
3117	WORLD WHERE YOU LIVE	Crowded House
2858	A WORLD WITHOUT HEROES	KISS
1166	WOT'S IT TO YA	Robbie Nevil
664	WOULD I LIE TO YOU?	Eurythmics
2187	WOULDN'T IT BE GOOD	Nik Kershaw
2149	WRACK MY BRAIN	Ringo Starr
1432	WRAP HER UP	Elton John
2513	WRAP IT UP	The Fabulous Thunderbirds
825	WRAPPED AROUND YOUR FINGER	The Police
3716	WYNKEN, BLYNKEN AND NOD	Doobie Brothers

#	Title	Artist
778	XANADU	Olivia Newton-John/Electric Light Orchestra

#	Title	Artist
1129	YAH MO B THERE	James Ingram (With Michael McDonald)
1384	YANKEE ROSE	David Lee Roth
1873	YEAH, YEAH, YEAH	Judson Spence
3088	YEARNING FOR YOUR LOVE	The Gap Band
1828	YEARS	Wayne Newton
2671	YEARS FROM NOW	Dr. Hook
2510	YES	Merry Clayton
176	YES, I'M READY	Teri DeSario
3916	YES OR NO	Go-Go's
2859	YESTERDAY ONCE MORE/ NOTHING REMAINS THE SAME (Medley)	Spinners
877	YESTERDAY'S SONGS	Neil Diamond
2509	YO' LITTLE BROTHER	Nolan Thomas
3099	YO NO SE'	Pajama Party
2244	YOU	Earth, Wind & Fire
3303	YOU AIN'T SEEN NOTHING YET	Figures On A Beach
401	YOU AND I	Eddie Rabbitt with Crystal Gayle
3726	YOU AND ME	Rockie Robbins
2517	YOU AND ME TONIGHT	Deja
424	YOU ARE	Lionel Richie
3035	YOU ARE FOREVER	Smokey Robinson
	YOU ARE IN MY SYSTEM	
2955		The System
3460		Robert Palmer
3795	YOU ARE MY EVERYTHING	Surface
2460	YOU ARE MY HEAVEN	Roberta Flack With Donny Hathaway
969	YOU ARE MY LADY	Freddie Jackson
1402	YOU ARE THE GIRL	The Cars
3939	YOU ARE THE ONE	TKA
1519	YOU BE ILLIN'	Run-D.M.C.
3788	YOU BELONG TO ME	The Doobie Brothers
290	YOU BELONG TO THE CITY	Glenn Frey
3179	YOU BETTER DANCE	The Jets
2236	YOU BETTER RUN	Pat Benatar
1335	YOU BETTER YOU BET	The Who
2443	YOU CAME	Kim Wilde
3598	YOU CAN	Madleen Kane
1193	YOU CAN CALL ME AL	Paul Simon
3883	YOU CAN CALL ME BLUE	Michael Johnson
515	YOU CAN DO MAGIC	America
2192	(You Can Still) ROCK IN AMERICA	Night Ranger
1186	YOU CAN'T GET WHAT YOU WANT (Till You Know What You Want)	Joe Jackson
651	YOU CAN'T HURRY LOVE	Phil Collins
2696	YOU CAN'T RUN FROM LOVE	Eddie Rabbitt
1013	YOU COULD HAVE BEEN WITH ME	Sheena Easton
1944	YOU COULD TAKE MY HEART AWAY	Silver Condor
2414	YOU DON'T BELIEVE	Alan Parsons Project
3498	YOU DON'T HAVE TO CRY	René and Angela
1468	YOU DON'T KNOW	Scarlett & Black
2531	YOU DON'T KNOW ME	Mickey Gilley
1176	YOU DON'T WANT ME ANYMORE	Steel Breeze
1752	YOU DROPPED A BOMB ON ME	The Gap Band

Singles Alpha Index

#	Title	Artist
457	YOU GIVE GOOD LOVE	Whitney Houston
275	YOU GIVE LOVE A BAD NAME	Bon Jovi
1028	YOU GOT IT	Roy Orbison
502	YOU GOT IT ALL	The Jets
526	YOU GOT IT (The Right Stuff)	New Kids On The Block
1160	YOU GOT LUCKY	Tom Petty and The Heartbreakers
3315	YOU GOT THE POWER	War
919	(You Gotta) FIGHT FOR YOUR RIGHT (To Party!)	Beastie Boys
3102	YOU HAVE PLACED A CHILL IN MY HEART	Eurythmics
280	YOU KEEP ME HANGIN' ON	Kim Wilde
2202	YOU KEEP RUNNIN' AWAY	38 Special
1201	YOU KNOW I LOVE YOU...DON'T YOU?	Howard Jones
1846	YOU KNOW THAT I LOVE YOU	Santana
3756	YOU KNOW WHAT TO DO	Carly Simon
2863	YOU LIKE ME DON'T YOU	Jermaine Jackson
2602	YOU LOOK MARVELOUS	Billy Crystal
3835	YOU MAKE ME WORK	Cameo
473	YOU MAKE MY DREAMS	Daryl Hall & John Oates
2736	YOU MAKE MY HEART BEAT FASTER (And That's All That Matters)	Kim Carnes
652	YOU MAY BE RIGHT	Billy Joel
3732	YOU, ME AND HE	Mtume
2784	YOU MIGHT NEED SOMEBODY	Turley Richards
649	YOU MIGHT THINK	The Cars
3678	YOU PUT THE BEAT IN MY HEART	Eddie Rabbitt
2262	YOU SAVED MY SOUL	Burton Cummings
3806	YOU SEND ME	The Manhattans
1681	YOU SHOOK ME ALL NIGHT LONG	AC/DC
1069	YOU SHOULD BE MINE (The Woo Woo Song)	Jeffrey Osborne
374	YOU SHOULD HEAR HOW SHE TALKS ABOUT YOU	Melissa Manchester
1108	YOU SPIN ME ROUND (Like A Record)	Dead Or Alive
2360	YOU TAKE ME UP	Thompson Twins
2379	YOU WEAR IT WELL	El DeBarge with DeBarge
3723	YOU WERE MADE FOR ME	Irene Cara
3669	YOU WILL KNOW	Stevie Wonder
3385	YOU WIN AGAIN	Bee Gees
1051	YOU'LL ACCOMP'NY ME	Bob Seger
3353	YOUNG BLOOD	Bruce Willis
2346	YOUNG LOVE	Air Supply
3097	YOUNG LOVE	Janet Jackson
3316	YOUNG THING, WILD DREAMS (Rock Me)	Red Rider
409	YOUNG TURKS	Rod Stewart
3807	YOUNGER DAYS	Joe Fagin
3180	YOUR DADDY DON'T KNOW	Toronto
1905	YOUR IMAGINATION	Daryl Hall & John Oates
648	YOUR LOVE	The Outfield
961	YOUR LOVE IS DRIVING ME CRAZY	Sammy Hagar
2485	YOUR LOVE IS KING	Sade
1177	YOUR MAMA DON'T DANCE	Poison
3978	YOUR PERSONAL TOUCH	Evelyn "Champagne" King
3205	YOUR SMILE	René and Angela
926	YOUR WILDEST DREAMS	The Moody Blues
1230	YOU'RE A FRIEND OF MINE	Clarence Clemons and Jackson Browne
3573	YOU'RE ALL I NEED	Mötley Crüe
	You're All I Need To Get By... see: AIN'T NOTHING LIKE THE REAL THING	
1990	YOU'RE DRIVING ME OUT OF MY MIND	Little River Band
3206	YOU'RE GONNA GET WHAT'S COMING	Bonnie Raitt
4083	YOU'RE IN LOVE	Ratt
2697	YOU'RE LOOKING LIKE LOVE TO ME	Peabo Bryson/Roberta Flack
3436	YOU'RE MINE TONIGHT	Pure Prairie League
3692	YOU'RE MY BLESSING	Lou Rawls
1889	YOU'RE MY GIRL	Franke And The Knockouts
2311	YOU'RE MY LATEST, MY GREATEST INSPIRATION	Teddy Pendergrass
1415	YOU'RE MY ONE AND ONLY (True Love)	Seduction
1141	YOU'RE NOT ALONE	Chicago
4117	YOU'RE NOT MY KIND OF GIRL	New Edition
980	YOU'RE ONLY HUMAN (Second Wind)	Billy Joel
4048	(You're Puttin') A RUSH ON ME	Stephanie Mills
3029	YOU'RE SO EASY TO LOVE	Tommy James
2524	(You're So Square) BABY, I DON'T CARE	Joni Mitchell
1858	YOU'RE SUPPOSED TO KEEP YOUR LOVE FOR ME	Jermaine Jackson
3553	YOU'RE THE BEST THING	The Style Council
355	YOU'RE THE INSPIRATION	Chicago
3831	YOU'RE THE ONLY LOVE	Paul Hyde And The Payolas
987	YOU'RE THE ONLY WOMAN (You And I)	Ambrosia
4160	YOUTH GONE WILD	Skid Row
3759	YOU'VE GOT A GOOD LOVE COMING	Van Stephenson
3184	YOU'VE GOT ANOTHER THING COMIN'	Judas Priest
3606	YOU'VE GOT WHAT I NEED	Shooting Star
747	YOU'VE LOST THAT LOVIN' FEELING	Daryl Hall & John Oates

The Singles Special Lists

Singles: 25 Weeks or More On Chart

Singles: 9 Weeks or More In Top 10

Singles: 16 Weeks or More In Top 40

Singles: Top 50 Each Year

Singles: Number 1s By Weeks

Singles: History of Number 1s

Singles: Weakest Follow Ups to Number 1s

Singles: Consecutive Number 1s

Singles: Highest Score Not Making Weekly Top 5

Singles: Lowest Scoring Number 1s

Singles: No Proximate Album

Singles: '80s Covers

Singles: 25 Weeks or More On Chart

TITLE Act	Wks
TAINTED LOVE Soft Cell	43
BUST A MOVE Young M.C.	39
GLORIA Laura Branigan	36
JESSIE'S GIRL Rick Springfield	32
BABY, COME TO ME Patti Austin (A duet with James Ingram)	32
ANOTHER ONE BITES THE DUST Queen	31
CELEBRATION Kool & The Gang	30
I'LL ALWAYS LOVE YOU Taylor Dayne	30
BORDERLINE Madonna	30
OUR LIPS ARE SEALED Go-Go's	30
ME SO HORNY 2 Live Crew	30
IN MY DREAMS REO Speedwagon	30
UPSIDE DOWN Diana Ross	29
TOTAL ECLIPSE OF THE HEART Bonnie Tyler	29
WIND BENEATH MY WINGS Bette Midler	29
WITH YOU I'M BORN AGAIN Billy Preston & Syreeta	29
HANDS TO HEAVEN Breathe	29
I MISS YOU Klymaxx	29
BREAK MY STRIDE Matthew Wilder	29
YOU AND I Eddie Rabbitt with Crystal Gayle	29
KEY LARGO Bertie Higgins	29
TONIGHT, I CELEBRATE MY LOVE Peabo Bryson/Roberta Flack	29
I KNOW THERE'S SOMETHING GOING ON Frida	29
YOU CAN CALL ME AL Paul Simon	29
HURTS SO GOOD John Cougar	28
WHAT'S LOVE GOT TO DO WITH IT Tina Turner	28
DON'T YOU WANT ME The Human League	28
I LOVE A RAINY NIGHT Eddie Rabbitt	28
CHARIOTS OF FIRE - TITLES Vangelis	28
KEEP ON LOVING YOU REO Speedwagon	28
HERE I GO AGAIN Whitesnake	28
KOKOMO The Beach Boys	28
BACK TO LIFE Soul II Soul	28
ONLY IN MY DREAMS Debbie Gibson	28
PLEASE DON'T GO GIRL New Kids On The Block	28
ENDLESS LOVE Diana Ross & Lionel Richie	27
DO THAT TO ME ONE MORE TIME The Captain & Tennille	27
ALL OUT OF LOVE Air Supply	27
MICKEY Toni Basil	27
WHO CAN IT BE NOW? Men At Work	27
QUEEN OF HEARTS Juice Newton	27
WILD, WILD WEST The Escape Club	27
TAKE ON ME A-Ha	27
THE FLAME Cheap Trick	27
867-5309/JENNY Tommy Tutone	27
A WOMAN NEEDS LOVE (Just Like You Do) Ray Parker Jr. & Raydio	27
LOVE SHACK The B-52's	27
STEPPIN' OUT Joe Jackson	27
THE LOCO-MOTION Kylie Minogue	27
NAUGHTY GIRLS (Need Love Too) Samantha Fox	27
HEART AND SOUL T'Pau	27
SOMETHING ABOUT YOU Level 42	27

TITLE Act	Wks
TWILIGHT ZONE Golden Earring	27
CONGA Miami Sound Machine	27
MISSING YOU Diana Ross	27
I LOVE YOU Climax Blues Band	27
1999 Prince	27
PHYSICAL Olivia Newton-John	26
BETTE DAVIS EYES Kim Carnes	26
SWEET DREAMS (Are Made Of This) Eurythmics	26
I JUST CALLED TO SAY I LOVE YOU Stevie Wonder	26
PLEASE DON'T GO K.C. And The Sunshine Band	26
9 TO 5 Dolly Parton	26
THE TIDE IS HIGH Blondie	26
THEME FROM "THE GREATEST AMERICAN HERO" (Believe It Or Not) Joey Scarbury	26
CARIBBEAN QUEEN (No More Love On The Run) Billy Ocean	26
TALKING IN YOUR SLEEP The Romantics	26
I FEEL FOR YOU Chaka Khan	26
DON'T KNOW MUCH Linda Ronstadt Featuring Aaron Neville	26
HE'S SO SHY Pointer Sisters	26
DON'T WORRY BE HAPPY (Edit) Bobby McFerrin	26
GIRL YOU KNOW IT'S TRUE Milli Vanilli	26
THE LADY IN RED Chris DeBurgh	26
FAME Irene Cara	26
YOU GOT IT ALL The Jets	26
YOU GOT IT (The Right Stuff) New Kids On The Block	26
ONCE BITTEN TWICE SHY Great White	26
THE WINNER TAKES IT ALL ABBA	26
THE GLAMOROUS LIFE Sheila E.	26
ALL CRIED OUT Lisa Lisa And Cult Jam With Full Force Featuring Paul Anthony And Bow Legged Lou	26
DON'T BE CRUEL Bobby Brown	26
I HATE MYSELF FOR LOVING YOU Joan Jett And The Blackhearts	26
I FOUND SOMEONE Cher	26
I STILL BELIEVE Brenda K. Starr	26
TARZAN BOY Baltimora	26
WHAT ABOUT ME ('82) Moving Pictures	26
CALL ME (Theme From...American Gigolo) Blondie	25
FLASHDANCE...WHAT A FEELING Irene Cara	25
EYE OF THE TIGER Survivor	25
ABRACADABRA Steve Miller Band	25
LADY Kenny Rogers	25
CENTERFOLD The J. Geils Band	25
ANOTHER BRICK IN THE WALL (Part II) Pink Floyd	25
DOWN UNDER Men At Work	25
BEAT IT Michael Jackson	25
ISLANDS IN THE STREAM Kenny Rogers Duet with Dolly Parton	25
BEING WITH YOU Smokey Robinson	25
DO YOU REALLY WANT TO HURT ME Culture Club	25

TITLE Act	Wks
MAKING LOVE OUT OF NOTHING AT ALL Air Supply	25
GIRLS JUST WANT TO HAVE FUN Cyndi Lauper	25
STRAIGHT UP Paula Abdul	25
WORKING MY WAY BACK TO YOU/FORGIVE ME GIRL (Medley) Spinners	25
THE ROSE Bette Midler	25
EYE IN THE SKY Alan Parsons Project	25
CRUISIN' Smokey Robinson	25
NEED YOU TONIGHT INXS	25
CHERISH Kool & The Gang	25
GROOVY KIND OF LOVE Phil Collins	25
RED RED WINE ('88) UB40	25
WISHING WELL Terence Trent D'Arby	25
HARD HABIT TO BREAK Chicago	25
SHINING STAR The Manhattans	25
WILD THING Tone Loc	25
SELF CONTROL Laura Branigan	25
FOR YOUR EYES ONLY Sheena Easton	25
YOU SHOULD HEAR HOW SHE TALKS ABOUT YOU Melissa Manchester	25
U GOT THE LOOK Prince	25
COOL IT NOW New Edition	25
DRIVIN' MY LIFE AWAY Eddie Rabbitt	25
WAITING FOR A STAR TO FALL Boy Meets Girl	25
ANGEL Aerosmith	25
THE LOVER IN ME Sheena Easton	25
HUNGRY EYES Eric Carmen	25
NEVER KNEW LOVE LIKE THIS BEFORE Stephanie Mills	25
CARS Gary Numan	25
WHAT'S ON YOUR MIND (Pure Energy) Information Society	25
MY HEART CAN'T TELL YOU NO Rod Stewart	25
I CAN DREAM ABOUT YOU Dan Hartman	25
STRUT Sheena Easton	25
TELL IT TO MY HEART Taylor Dayne	25
I LIKE IT Dino	25
LET'S GO ALL THE WAY Sly Fox	25
IF EVER YOU'RE IN MY ARMS AGAIN Peabo Bryson	25
THE WAY YOU LOVE ME Karyn White	25
WHIP IT Devo	25
CLOSE MY EYES FOREVER (remix) Lita Ford (duet with Ozzy Osbourne)	25
PIANO IN THE DARK Brenda Russell	25
PERFECT WAY Scritti Politti	25
NOTHING'S GONNA CHANGE MY LOVE FOR YOU Glenn Medeiros	25
THE LOOK OF LOVE (Part One) ABC	25
THE PROMISE When In Rome	25
MEET ME HALF WAY Kenny Loggins	25
PUSH IT Salt-N-Pepa	25
JUNGLE LOVE The Time	25
THAT'S WHAT LOVE IS ALL ABOUT Michael Bolton	25

Singles: 9 Weeks or More In Top 10

TITLE Act	Wks
HURTS SO GOOD *John Cougar*	16
ANOTHER ONE BITES THE DUST *Queen*	15
EYE OF THE TIGER *Survivor*	15
PHYSICAL *Olivia Newton-John*	15
WAITING FOR A GIRL LIKE YOU *Foreigner*	15
(Just Like) STARTING OVER *John Lennon*	14
ABRACADABRA *Steve Miller Band*	14
BETTE DAVIS EYES *Kim Carnes*	14
DO THAT TO ME ONE MORE TIME *The Captain & Tennille*	14
FLASHDANCE...WHAT A FEELING *Irene Cara*	14
UPSIDE DOWN *Diana Ross*	14
ALL NIGHT LONG (All Night) *Lionel Richie*	13
ENDLESS LOVE *Diana Ross & Lionel Richie*	13
EVERY BREATH YOU TAKE *The Police*	13
LADY *Kenny Rogers*	13
MANEATER *Daryl Hall & John Oates*	13
SAY SAY SAY *Paul McCartney and Michael Jackson*	13
ANOTHER BRICK IN THE WALL (Part II) *Pink Floyd*	12
ARTHUR'S THEME (Best That You Can Do) *Christopher Cross*	12
CALL ME (Theme From...American Gigolo) *Blondie*	12
CENTERFOLD *The J. Geils Band*	12
CRAZY LITTLE THING CALLED LOVE *Queen*	12
DON'T YOU WANT ME *The Human League*	12
EBONY AND IVORY *Paul McCartney*	12
HARD TO SAY I'M SORRY *Chicago*	12
HARDEN MY HEART *Quarterflash*	12
I CAN'T GO FOR THAT (No Can Do) *Daryl Hall & John Oates*	12
I LOVE ROCK 'N ROLL *Joan Jett & The Blackhearts*	12
ISLANDS IN THE STREAM *Kenny Rogers Duet with Dolly Parton*	12
JESSIE'S GIRL *Rick Springfield*	12
WOMAN *John Lennon*	12
BILLIE JEAN *Michael Jackson*	11
COMING UP (Live At Glasgow) *Paul McCartney & Wings*	11
DON'T TALK TO STRANGERS *Rick Springfield*	11
FOOTLOOSE *Kenny Loggins*	11
IT'S STILL ROCK AND ROLL TO ME *Billy Joel*	11
JUST THE TWO OF US *Grover Washington, Jr.*	11
LITTLE JEANNIE *Elton John*	11
PLEASE DON'T GO *K.C. And The Sunshine Band*	11
ROSANNA *Toto*	11
SLOW HAND *Pointer Sisters*	11
START ME UP *The Rolling Stones*	11
TOTAL ECLIPSE OF THE HEART *Bonnie Tyler*	11
WHEN DOVES CRY *Prince*	11
WOMAN IN LOVE *Barbra Streisand*	11
AGAINST ALL ODDS (Take A Look At Me Now) *Phil Collins*	10
ALL OUT OF LOVE *Air Supply*	10
ANOTHER DAY IN PARADISE *Phil Collins*	10
BEAT IT *Michael Jackson*	10
BEING WITH YOU *Smokey Robinson*	10
DIRTY LAUNDRY *Don Henley*	10
DOWN UNDER *Men At Work*	10
GHOSTBUSTERS *Ray Parker Jr.*	10
GLORIA *Laura Branigan*	10
HELLO *Lionel Richie*	10
HOLD ME *Fleetwood Mac*	10
I JUST CALLED TO SAY I LOVE YOU *Stevie Wonder*	10
JACK & DIANE *John Cougar*	10
JUMP *Van Halen*	10
LET'S DANCE *David Bowie*	10
LOVE ON THE ROCKS *Neil Diamond*	10
MICKEY *Toni Basil*	10
OPEN ARMS *Journey*	10
OWNER OF A LONELY HEART *Yes*	10
QUEEN OF HEARTS *Juice Newton*	10
SAY IT ISN'T SO *Daryl Hall & John Oates*	10
SEXUAL HEALING *Marvin Gaye*	10
STOP DRAGGIN' MY HEART AROUND *Stevie Nicks (with Tom Petty And The Heartbreakers)*	10
THAT'S WHAT FRIENDS ARE FOR *Dionne & Friends*	10
THE BEST OF TIMES *Styx*	10
THE GIRL IS MINE *Michael Jackson/Paul McCartney*	10
THE TIDE IS HIGH *Blondie*	10
THEME FROM "THE GREATEST AMERICAN HERO" (Believe It Or Not) *Joey Scarbury*	10
TRULY *Lionel Richie*	10
UPTOWN GIRL *Billy Joel*	10
WHAT'S LOVE GOT TO DO WITH IT *Tina Turner*	10
9 TO 5 *Dolly Parton*	9
BABY, COME TO ME *Patti Austin (A duet with James Ingram)*	9
BROKEN WINGS *Mr. Mister*	9
CARELESS WHISPER *Wham! Featuring George Michael*	9
CHARIOTS OF FIRE - TITLES *Vangelis*	9
CRAZY FOR YOU *Madonna*	9
DANCING IN THE DARK *Bruce Springsteen*	9
DO YOU REALLY WANT TO HURT ME *Culture Club*	9
FAITH *George Michael*	9
HUNGRY LIKE THE WOLF *Duran Duran*	9
I FEEL FOR YOU *Chaka Khan*	9
I LOVE A RAINY NIGHT *Eddie Rabbitt*	9
I WANNA DANCE WITH SOMEBODY (Who Loves Me) *Whitney Houston*	9
KARMA CHAMELEON *Culture Club*	9
KEEP ON LOVING YOU *REO Speedwagon*	9
LET'S GO CRAZY *Prince And The Revolution*	9
LET'S GROOVE *Earth, Wind & Fire*	9
LET'S HEAR IT FOR THE BOY *Deniece Williams*	9
LIKE A VIRGIN *Madonna*	9
FUNKYTOWN *Lipps, Inc.*	9
MAGIC *Olivia Newton-John*	9
MAKING LOVE OUT OF NOTHING AT ALL *Air Supply*	9
MANIAC *Michael Sembello*	9
MISSING YOU *John Waite*	9
MORE THAN I CAN SAY *Leo Sayer*	9
OUT OF TOUCH *Daryl Hall & John Oates*	9
PARTY ALL THE TIME *Eddie Murphy*	9
PRIVATE EYES *Daryl Hall & John Oates*	9
RIDE LIKE THE WIND *Christopher Cross*	9
ROCK WITH YOU *Michael Jackson*	9
SAY YOU, SAY ME *Lionel Richie*	9
SEPARATE LIVES (Love Theme From White Nights) *Phil Collins and Marilyn Martin*	9
SWEET DREAMS (Are Made Of This) *Eurythmics*	9
THAT GIRL *Stevie Wonder*	9
TIME (Clock Of The Heart) *Culture Club*	9
TIME AFTER TIME *Cyndi Lauper*	9
WE GOT THE BEAT *Go-Go's*	9
WHO CAN IT BE NOW? *Men At Work*	9

Singles: 16 Weeks or More In The Top 40

TITLE Act	Wks
DO THAT TO ME ONE MORE TIME The Captain & Tennille	22
GLORIA Laura Branigan	22
HURTS SO GOOD John Cougar	22
JESSIE'S GIRL Rick Springfield	22
ANOTHER ONE BITES THE DUST Queen	21
CELEBRATION Kool & The Gang	21
DON'T YOU WANT ME The Human League	21
PHYSICAL Olivia Newton-John	21
YOU AND I Eddie Rabbitt with Crystal Gayle	21
BETTE DAVIS EYES Kim Carnes	20
BUST A MOVE Young M.C.	20
CENTERFOLD The J. Geils Band	20
EVERY BREATH YOU TAKE The Police	20
FLASHDANCE...WHAT A FEELING Irene Cara	20
KEEP ON LOVING YOU REO Speedwagon	20
(Just Like) STARTING OVER John Lennon	19
ABRACADABRA Steve Miller Band	19
ANOTHER BRICK IN THE WALL (Part II) Pink Floyd	19
CALL ME (Theme From...American Gigolo) Blondie	19
DOWN UNDER Men At Work	19
ENDLESS LOVE Diana Ross & Lionel Richie	19
HARDEN MY HEART Quarterflash	19
IT'S STILL ROCK AND ROLL TO ME Billy Joel	19
LADY Kenny Rogers	19
QUEEN OF HEARTS Juice Newton	19
ROCK WITH YOU Michael Jackson	19
SHAME ON THE MOON Bob Seger & The Silver Bullet Band	19
START ME UP The Rolling Stones	19
WAITING FOR A GIRL LIKE YOU Foreigner	19
WOMAN IN LOVE Barbra Streisand	19
9 TO 5 Dolly Parton	18
BABY, COME TO ME Patti Austin (A duet with James Ingram)	18
BACK TO LIFE Soul II Soul	18
BEAT IT Michael Jackson	18
DO YOU REALLY WANT TO HURT ME Culture Club	18
EYE OF THE TIGER Survivor	18
HARD TO SAY I'M SORRY Chicago	18
I LOVE A RAINY NIGHT Eddie Rabbitt	18
ISLANDS IN THE STREAM Kenny Rogers Duet with Dolly Parton	18
MICKEY Toni Basil	18
PLEASE DON'T GO K.C. The Sunshine Band	18
ROSANNA Toto	18
SAY SAY SAY Paul McCartney and Michael Jackson	18
THE SWEETEST THING (I've Ever Known) Juice Newton	18
THEME FROM "THE GREATEST AMERICAN HERO" (Believe It Or Not) Joey Scarbury	18
TOTAL ECLIPSE OF THE HEART Bonnie Tyler	18
WHAT'S LOVE GOT TO DO WITH IT Tina Turner	18
ALL NIGHT LONG (All Night) Lionel Richie	17
ALL OUT OF LOVE Air Supply	17
ARTHUR'S THEME (Best That You Can Do) Christopher Cross	17
BILLIE JEAN Michael Jackson	17
CARELESS WHISPER Wham! Featuring George Michael	17
CARS Gary Numan	17
CRAZY LITTLE THING CALLED LOVE Queen	17
CRUISIN' Smokey Robinson	17
EVERY WOMAN IN THE WORLD Air Supply	17
EYE IN THE SKY Alan Parsons Project	17
HELLO Lionel Richie	17
HE'S SO SHY Pointer Sisters	17
I CAN'T GO FOR THAT (No Can Do) Daryl Hall & John Oates	17
I FEEL FOR YOU Chaka Khan	17
I LOVE YOU Climax Blues Band	17
I MISS YOU Klymaxx	17
JACK & DIANE John Cougar	17
KEY LARGO Bertie Higgins	17
KISS ON MY LIST Daryl Hall & John Oates	17
LITTLE JEANNIE Elton John	17
LOST IN LOVE Air Supply	17
LOVE ON THE ROCKS Neil Diamond	17
LOVE SHACK The B-52's	17
MAKING LOVE OUT OF NOTHING AT ALL Air Supply	17
MANEATER Daryl Hall & John Oates	17
NEED YOU TONIGHT INXS	17
OWNER OF A LONELY HEART Yes	17
PASSION Rod Stewart	17
PRIVATE EYES Daryl Hall & John Oates	17
RIDE LIKE THE WIND Christopher Cross	17
SHAKE IT UP The Cars	17
SHE WORKS HARD FOR THE MONEY Donna Summer	17
SWEET DREAMS (Are Made Of This) Eurythmics	17
THAT'S WHAT FRIENDS ARE FOR Dionne & Friends	17
THE TIDE IS HIGH Blondie	17
UPSIDE DOWN Diana Ross	17
URGENT Foreigner	17
WHO CAN IT BE NOW? Men At Work	17
WOMAN John Lennon	17
867-5309/JENNY Tommy Tutone	16
AFRICA Toto	16
AGAINST ALL ODDS (Take A Look At Me Now) Phil Collins	16
ALIVE & KICKING Simple Minds	16
ALLENTOWN Billy Joel	16
ANGEL OF THE MORNING Juice Newton	16
BEING WITH YOU Smokey Robinson	16
BURNING HEART Survivor	16
C'EST LA VIE Robbie Nevil	16
COMING UP (Live...) Paul McCartney & Wings	16
CONGA Miami Sound Machine	16
DON'T KNOW MUCH Linda Ronstadt Featuring Aaron Neville	16
DON'T TALK TO STRANGERS Rick Springfield	16
EASY LOVER (Duet With Phil Collins) Philip Bailey	16
FOOTLOOSE Kenny Loggins	16
HANDS TO HEAVEN Breathe	16
HEART AND SOUL T'Pau	16
HOW CAN I FALL? Breathe	16
HOW WILL I KNOW Whitney Houston	16
HUNGRY EYES Eric Carmen	16
HUNGRY LIKE THE WOLF Duran Duran	16
I CAN DREAM ABOUT YOU Dan Hartman	16
I LOVE ROCK 'N ROLL Joan Jett & The Blackhearts	16
I WANT TO KNOW WHAT LOVE IS Foreigner	16
I'LL ALWAYS LOVE YOU Taylor Dayne	16
JOANNA Kool & The Gang	16
JUST THE TWO OF US Grover Washington, Jr.	16
KARMA CHAMELEON Culture Club	16
LEADER OF THE BAND Dan Fogelberg	16
LET IT WHIP Dazz Band	16
LET'S GROOVE Earth, Wind & Fire	16
LOOK AWAY Chicago	16
MAGIC Olivia Newton-John	16
MANIAC Michael Sembello	16
MASTER BLASTER (Jammin') Stevie Wonder	16
MISSING YOU John Waite	16
MR. ROBOTO Styx	16
NEVER GONNA LET YOU GO Sergio Mendes	16
NEVER KNEW LOVE LIKE THIS BEFORE Stephanie Mills	16
ONLY IN MY DREAMS Debbie Gibson	16
OUT OF TOUCH Daryl Hall & John Oates	16
SAY YOU, SAY ME Lionel Richie	16
SEASONS CHANGE Exposé	16
SEPARATE LIVES (Love Theme From White Nights) Phil Collins and Marilyn Martin	16
SEPARATE WAYS (Worlds Apart) Journey	16
SHAKE YOU DOWN Gregory Abbott	16
SLOW HAND Pointer Sisters	16
STRAIGHT UP Paula Abdul	16
SUKIYAKI A Taste Of Honey	16
THE GLAMOROUS LIFE Sheila E.	16
THE ROSE Bette Midler	16
THE SAFETY DANCE Men Without Hats	16
THE WINNER TAKES IT ALL ABBA	16
THIS IS IT Kenny Loggins	16
TURN YOUR LOVE AROUND George Benson	16
UPTOWN GIRL Billy Joel	16
WAITING FOR A STAR TO FALL Boy Meets Girl	16
WHEN DOVES CRY Prince	16
WILD, WILD WEST The Escape Club	16
WORKING MY WAY BACK TO YOU/FORGIVE ME GIRL (Medley) Spinners	16
YES, I'M READY Teri DeSario	16
YOU ARE Lionel Richie	16
YOU CAN'T HURRY LOVE Phil Collins	16

Singles: Top 50 Each Year

1980 n=473

Yr Rank. TITLE Act [Decade Rank]

1. CALL ME (Theme From...American Gigolo) Blondie [6]
2. ANOTHER ONE BITES THE DUST Queen [9]
3. DO THAT TO ME ONE MORE TIME The Captain & Tennille [11]
4. UPSIDE DOWN Diana Ross [14]
5. LADY Kenny Rogers [16]
6. CRAZY LITTLE THING CALLED LOVE Queen [18]
7. ANOTHER BRICK IN THE WALL (Part II) Pink Floyd [21]
8. COMING UP (Live At Glasgow) Paul McCartney & Wings [22]
9. (Just Like) STARTING OVER John Lennon [24]
10. WOMAN IN LOVE Barbra Streisand [25]
11. FUNKYTOWN Lipps, Inc. [29]
12. ROCK WITH YOU Michael Jackson [30]
13. IT'S STILL ROCK AND ROLL TO ME Billy Joel [34]
14. MAGIC Olivia Newton-John [45]
15. PLEASE DON'T GO K.C. And The Sunshine Band [55]
16. ALL OUT OF LOVE Air Supply [60]
17. RIDE LIKE THE WIND
    Christopher Cross [70]
18. LITTLE JEANNIE
    Elton John [98]
19. MORE THAN I CAN SAY Leo Sayer [101]
20. SAILING Christopher Cross [129]
21. WORKING MY WAY BACK TO YOU/FORGIVE ME GIRL (Medley)
    Spinners [135]
22. THE ROSE Bette Midler [138]
23. LOST IN LOVE Air Supply [152]
24. COWARD OF THE COUNTY Kenny Rogers [167]
25. YES, I'M READY Teri DeSario [176]
26. LONGER Dan Fogelberg [194]
27. CRUISIN' Smokey Robinson [195]
28. EMOTIONAL RESCUE The Rolling Stones [217]
29. BIGGEST PART OF ME Ambrosia [221]
30. HE'S SO SHY Pointer Sisters [251]
31. WITH YOU I'M BORN AGAIN Billy Preston & Syreeta [268]
32. DESIRE Andy Gibb [286]
33. CUPID/I'VE LOVED YOU FOR A LONG TIME (Medley) Spinners [292]
34. TAKE YOUR TIME (Do It Right) Part 1 The S.O.S. Band [302]
35. THE WANDERER Donna Summer [308]
36. MASTER BLASTER (Jammin')
    Stevie Wonder [326]
37. HUNGRY HEART Bruce Springsteen [345]
38. SHINING STAR The Manhattans [348]
39. GIVE ME THE NIGHT George Benson [352]
40. FAME Irene Cara [357]
41. DON'T FALL IN LOVE WITH A DREAMER
    Kenny Rogers with Kim Carnes [367]
42. SEXY EYES Dr. Hook [386]
43. STEAL AWAY Robbie Dupree [402]
44. DRIVIN' MY LIFE AWAY Eddie Rabbitt [422]
45. I'M COMING OUT Diana Ross [443]
46. AGAINST THE WIND Bob Seger [452]
47. TOO HOT Kool & The Gang [461]
48. ON THE RADIO Donna Summer [464]
49. SPECIAL LADY Ray, Goodman & Brown [470]
50. NEVER KNEW LOVE LIKE THIS BEFORE Stephanie Mills [490]

1981 n=408

Yr Rank. TITLE Act [Decade Rank]

1. PHYSICAL Olivia Newton-John [1]
2. ENDLESS LOVE Diana Ross & Lionel Richie [2]
3. BETTE DAVIS EYES Kim Carnes [4]
4. WAITING FOR A GIRL LIKE YOU Foreigner [10]
5. JESSIE'S GIRL Rick Springfield [48]
6. ARTHUR'S THEME (Best That You Can Do) Christopher Cross [50]
7. WOMAN John Lennon [58]
8. 9 TO 5 Dolly Parton [59]
9. BEING WITH YOU Smokey Robinson [62]
10. I LOVE A RAINY NIGHT Eddie Rabbitt [63]
11. START ME UP The Rolling Stones [81]
12. CELEBRATION Kool & The Gang [84]
13. LOVE ON THE ROCKS Neil Diamond [85]
14. PRIVATE EYES Daryl Hall & John Oates [86]
15. JUST THE TWO OF US Grover Washington, Jr. [88]
16. THE TIDE IS HIGH Blondie [95]
17. THEME FROM "THE GREATEST AMERICAN HERO"
    (Believe It Or Not) Joey Scarbury [97]
18. STOP DRAGGIN' MY HEART AROUND
    Stevie Nicks (with Tom Petty And The Heartbreakers) [112]
19. KISS ON MY LIST Daryl Hall & John Oates [120]
20. KEEP ON LOVING YOU REO Speedwagon [121]
21. MEDLEY
    Stars On 45 [123]
22. LET'S GROOVE Earth, Wind & Fire [125]
23. RAPTURE Blondie [127]
24. SLOW HAND Pointer Sisters [128]
25. MORNING TRAIN (Nine To Five) Sheena Easton [148]
26. THE ONE THAT YOU LOVE Air Supply [164]
27. THE BEST OF TIMES Styx [174]
28. QUEEN OF HEARTS Juice Newton [185]
29. SUKIYAKI A Taste Of Honey [231]
30. ALL THOSE YEARS AGO George Harrison [288]
31. I DON'T NEED YOU Kenny Rogers [299]
32. GUILTY Barbra Streisand & Barry Gibb [303]
33. ANGEL OF THE MORNING Juice Newton [320]
34. URGENT Foreigner [325]
35. EVERY WOMAN IN THE WORLD Air Supply [343]
36. A WOMAN NEEDS LOVE (Just Like You Do)
    Ray Parker Jr. & Raydio [359]
37. (There's) NO GETTIN' OVER ME Ronnie Milsap [361]
38. FOR YOUR EYES ONLY Sheena Easton [362]
39. WHO'S CRYING NOW Journey [369]
40. ELVIRA The Oak Ridge Boys [376]
41. PASSION
    Rod Stewart [396]
42. EVERY LITTLE THING SHE DOES IS MAGIC The Police [400]
43. YOUNG TURKS Rod Stewart [409]
44. OH NO Commodores [425]
45. CRYING Don McLean [433]
46. HERE I AM (Just When I Thought I Was Over You) Air Supply [437]
47. STEP BY STEP Eddie Rabbitt [438]
48. TAKE IT ON THE RUN REO Speedwagon [467]
49. YOU MAKE MY DREAMS Daryl Hall & John Oates [473]
50. LIVING INSIDE MYSELF Gino Vannelli [480]

1982 n=424

Yr Rank. TITLE Act [Decade Rank]
1. EYE OF THE TIGER Survivor [8]
2. EBONY AND IVORY Paul McCartney [12]
3. ABRACADABRA Steve Miller Band [13]
4. I LOVE ROCK 'N ROLL Joan Jett & The Blackhearts [19]
5. CENTERFOLD The J. Geils Band [20]
6. HURTS SO GOOD John Cougar [26]
7. JACK & DIANE John Cougar [31]
8. I CAN'T GO FOR THAT (No Can Do) Daryl Hall & John Oates [36]
9. HARD TO SAY I'M SORRY Chicago [37]
10. DON'T YOU WANT ME The Human League [41]
11. MANEATER Daryl Hall & John Oates [42]
12. ROSANNA Toto [51]
13. GLORIA Laura Branigan [57]
14. DON'T TALK TO STRANGERS Rick Springfield [64]
15. MICKEY Toni Basil [69]
16. OPEN ARMS Journey [73]
17. CHARIOTS OF FIRE - TITLES Vangelis [80]
18. WHO CAN IT BE NOW? Men At Work [82]
19. UP WHERE WE BELONG Joe Cocker and Jennifer Warnes [108]
20. HARDEN MY HEART Quarterflash [110]
21. TRULY Lionel Richie [113]
22. WE GOT THE BEAT Go-Go's [119]
23. HOLD ME Fleetwood Mac [170]
24. EYE IN THE SKY Alan Parsons Project [183]
25. HEART ATTACK Olivia Newton-John [215]
26. SHAKE IT UP The Cars [301]
27. 867-5309/JENNY Tommy Tutone [316]
28. THAT GIRL Stevie Wonder [333]
29. FREEZE-FRAME The J. Geils Band [356]
30. THE OTHER WOMAN Ray Parker Jr. [365]
31. HEAT OF THE MOMENT Asia [373]
32. YOU SHOULD HEAR HOW SHE TALKS ABOUT YOU
    Melissa Manchester [374]
33. I'VE NEVER BEEN TO ME Charlene [377]
34. EVEN THE NIGHTS ARE BETTER Air Supply [390]
35. ALWAYS ON MY MIND Willie Nelson [391]
36. STEPPIN' OUT Joe Jackson [395]
37. I KEEP FORGETTIN' Michael McDonald [397]
38. LET IT WHIP Dazz Band [404]
39. SWEET DREAMS Air Supply [408]
40. THE SWEETEST THING (I've Ever Known) Juice Newton [444]
41. LEATHER AND LACE Stevie Nicks (with Don Henley) [446]
42. TURN YOUR LOVE AROUND George Benson [449]
43. HEARTLIGHT Neil Diamond [482]
44. KEY LARGO Bertie Higgins [507]
45. 65 LOVE AFFAIR Paul Davis [510]
46. YOU CAN DO MAGIC America [515]
47. TAINTED LOVE Soft Cell [547]
48. ONLY THE LONELY The Motels [567]
49. SOMEBODY'S BABY Jackson Browne [577]
50. MAKE A MOVE ON ME Olivia Newton-John [587]

1983 n=452

Yr Rank. TITLE Act [Decade Rank]
1. SAY SAY SAY Paul McCartney and Michael Jackson [3]
2. EVERY BREATH YOU TAKE The Police [5]
3. FLASHDANCE...WHAT A FEELING Irene Cara [7]
4. ALL NIGHT LONG (All Night) Lionel Richie [15]
5. BILLIE JEAN Michael Jackson [17]
6. TOTAL ECLIPSE OF THE HEART Bonnie Tyler [23]
7. DOWN UNDER Men At Work [32]
8. BEAT IT Michael Jackson [35]
9. ISLANDS IN THE STREAM Kenny Rogers Duet with Dolly Parton [38]
10. SWEET DREAMS (Are Made Of This) Eurythmics [44]
11. LET'S DANCE David Bowie [53]
12. MANIAC Michael Sembello [61]
13. ELECTRIC AVENUE Eddy Grant [67]
14. SAY IT ISN'T SO Daryl Hall & John Oates [74]
15. SHAME ON THE MOON Bob Seger & The Silver Bullet Band [76]
16. BABY, COME TO ME Patti Austin (A duet with James Ingram) [87]
17. UPTOWN GIRL Billy Joel [90]
18. THE GIRL IS MINE Michael Jackson/Paul McCartney [91]
19. DO YOU REALLY WANT TO HURT ME Culture Club [93]
20. MAKING LOVE OUT OF NOTHING AT ALL Air Supply [107]
21. COME ON EILEEN Dexys Midnight Runners [126]
22. DIRTY LAUNDRY Don Henley [144]
23. SEXUAL HEALING Marvin Gaye [147]
24. HUNGRY LIKE THE WOLF Duran Duran [149]
25. SHE WORKS HARD FOR THE MONEY Donna Summer [169]
26. TIME (Clock Of The Heart) Culture Club [172]
27. TELL HER ABOUT IT Billy Joel [175]
28. THE SAFETY DANCE Men Without Hats [182]
29. JEOPARDY Greg Kihn Band [188]
30. AFRICA Toto [204]
31. NEVER GONNA LET YOU GO Sergio Mendes [234]
32. OVERKILL
    Men At Work [236]
33. UNION OF THE SNAKE Duran Duran [239]
34. MR. ROBOTO Styx [271]
35. STRAY CAT STRUT Stray Cats [330]
36. SHE BLINDED ME WITH SCIENCE Thomas Dolby [336]
37. TRUE Spandau Ballet [339]
38. LOVE IS A BATTLEFIELD Pat Benatar [360]
39. PUTTIN' ON THE RITZ Taco [379]
40. ONE THING LEADS TO ANOTHER The Fixx [399]
41. YOU AND I Eddie Rabbitt with Crystal Gayle [401]
42. KING OF PAIN The Police [410]
43. YOU ARE Lionel Richie [424]
44. CUM ON FEEL THE NOIZE Quiet Riot [447]
45. IS THERE SOMETHING I SHOULD KNOW Duran Duran [451]
46. BACK ON THE CHAIN GANG The Pretenders [459]
47. DER KOMMISSAR After The Fire [474]
48. STAND BACK Stevie Nicks [481]
49. WE'VE GOT TONIGHT Kenny Rogers And Sheena Easton [484]
50. LITTLE RED CORVETTE Prince [494]

Singles: Special Lists

1984 n=432
Yr Rank. TITLE   Act [Decade Rank]
1. WHEN DOVES CRY Prince [27]
2. JUMP Van Halen [28]
3. AGAINST ALL ODDS (Take A Look At Me Now) Phil Collins [33]
4. HELLO Lionel Richie [39]
5. WHAT'S LOVE GOT TO DO WITH IT Tina Turner [40]
6. FOOTLOOSE Kenny Loggins [43]
7. OWNER OF A LONELY HEART Yes [46]
8. LIKE A VIRGIN Madonna [47]
9. GHOSTBUSTERS Ray Parker Jr. [49]
10. KARMA CHAMELEON Culture Club [52]
11. I JUST CALLED TO SAY I LOVE YOU   Stevie Wonder [54]
12. LET'S HEAR IT FOR THE BOY Deniece Williams [71]
13. MISSING YOU John Waite [72]
14. DANCING IN THE DARK Bruce Springsteen [78]
15. WAKE ME UP BEFORE YOU GO-GO Wham! [83]
16. TIME AFTER TIME Cyndi Lauper [89]
17. OUT OF TOUCH Daryl Hall & John Oates [92]
18. CARIBBEAN QUEEN (No More Love On The Run) Billy Ocean [99]
19. LET'S GO CRAZY Prince And The Revolution [100]
20. SOMEBODY'S WATCHING ME Rockwell [104]
21. THE REFLEX Duran Duran [106]
22. THE WILD BOYS Duran Duran [114]
23. GIRLS JUST WANT TO HAVE FUN Cyndi Lauper [117]
24. TALKING IN YOUR SLEEP The Romantics [153]
25. I FEEL FOR YOU Chaka Khan [156]
26. 99 LUFTBALLONS Nena [192]
27. HOLD ME NOW Thompson Twins [214]
28. PURPLE RAIN Prince And The Revolution [228]
29. JOANNA Kool & The Gang [247]
30. SHE BOP Cyndi Lauper [261]
31. HARD HABIT TO BREAK Chicago [272]
32. JUMP (For My Love) Pointer Sisters [278]
33. STUCK ON YOU Lionel Richie [300]
34. DRIVE The Cars [304]
35. OH SHERRIE Steve Perry [335]
36. BREAK MY STRIDE Matthew Wilder [351]
37. SELF CONTROL Laura Branigan [353]
38. HERE COMES THE RAIN AGAIN Eurythmics [363]
39. STATE OF SHOCK Jacksons [368]
40. EYES WITHOUT A FACE Billy Idol [384]
41. I GUESS THAT'S WHY THEY CALL IT THE BLUES Elton John [407]
42. TWIST OF FATE Olivia Newton-John [441]
43. AUTOMATIC Pointer Sisters [472]
44. THE HEART OF ROCK & ROLL Huey Lewis And The News [488]
45. THRILLER Michael Jackson [506]
46. TO ALL THE GIRLS I'VE LOVED BEFORE   Julio Iglesias & Willie Nelson [508]
47. LUCKY STAR Madonna [522]
48. I CAN DREAM ABOUT YOU Dan Hartman [532]
49. SISTER CHRISTIAN Night Ranger [536]
50. I WANT A NEW DRUG Huey Lewis And The News [538]

1985 n=406
Yr Rank. TITLE   Act [Decade Rank]
1. SAY YOU, SAY ME Lionel Richie [56]
2. CARELESS WHISPER Wham! Featuring George Michael [66]
3. CRAZY FOR YOU Madonna [75]
4. WE ARE THE WORLD USA for Africa [79]
5. I WANT TO KNOW WHAT LOVE IS Foreigner [96]
6. CAN'T FIGHT THIS FEELING REO Speedwagon [103]
7. MONEY FOR NOTHING Dire Straits [116]
8. BROKEN WINGS Mr. Mister [118]
9. SHOUT Tears For Fears [122]
10. EVERYBODY WANTS TO RULE THE WORLD Tears For Fears [130]
11. SEPARATE LIVES (Love Theme From White Nights)   Phil Collins and Marilyn Martin [132]
12. PARTY ALL THE TIME Eddie Murphy [139]
13. THE POWER OF LOVE Huey Lewis And The News [140]
14. EASY LOVER (Duet With Phil Collins) Philip Bailey [151]
15. WE BUILT THIS CITY Starship [157]
16. DON'T YOU (Forget About Me) Simple Minds [159]
17. ST. ELMO'S FIRE (Man In Motion) John Parr [163]
18. ONE MORE NIGHT Phil Collins [165]
19. PART-TIME LOVER Stevie Wonder [177]
20. A VIEW TO A KILL Duran Duran [187]
21. EVERYTIME YOU GO AWAY Paul Young [191]
22. HEAVEN Bryan Adams [201]
23. EVERYTHING SHE WANTS (Remix) Wham! [211]
24. SUSSUDIO Phil Collins [213]
25. CHERISH Kool & The Gang [229]
26. TAKE ON ME A-Ha [230]
27. SAVING ALL MY LOVE FOR YOU Whitney Houston [233]
28. MIAMI VICE THEME Jan Hammer [238]
29. MATERIAL GIRL Madonna [260]
30. THE HEAT IS ON Glenn Frey [266]
31. YOU BELONG TO THE CITY Glenn Frey [290]
32. ALIVE & KICKING Simple Minds [293]
33. RASPBERRY BERET Prince & The Revolution [307]
34. SEA OF LOVE The Honeydrippers [321]
35. LOVERBOY Billy Ocean [322]
36. OH SHEILA Ready For The World [337]
37. RHYTHM OF THE NIGHT DeBarge [342]
38. I MISS YOU Klymaxx [349]
39. YOU'RE THE INSPIRATION Chicago [355]
40. ALL I NEED Jack Wagner [358]
41. WE DON'T NEED ANOTHER HERO (Thunderdome) Tina Turner [372]
42. IF YOU LOVE SOMEBODY SET THEM FREE Sting [382]
43. AXEL F Harold Faltermeyer [392]
44. NEVER SURRENDER Corey Hart [394]
45. COOL IT NOW New Edition [420]
46. FREEWAY OF LOVE   Aretha Franklin [427]
47. HEAD OVER HEELS Tears For Fears [434]
48. NIGHTSHIFT Commodores [442]
49. ONE NIGHT IN BANGKOK Murray Head [445]
50. YOU GIVE GOOD LOVE Whitney Houston [457]

1986 n=397
Yr Rank. TITLE Act [Decade Rank]
1. THAT'S WHAT FRIENDS ARE FOR Dionne & Friends [65]
2. WALK LIKE AN EGYPTIAN Bangles [77]
3. ON MY OWN Patti LaBelle And Michael McDonald [94]
4. GREATEST LOVE OF ALL   Whitney Houston [146]
5. KISS Prince And The Revolution [154]
6. HOW WILL I KNOW Whitney Houston [160]
7. KYRIE Mr. Mister [161]
8. PAPA DON'T PREACH Madonna [178]
9. WEST END GIRLS Pet Shop Boys [181]
10. STUCK WITH YOU Huey Lewis And The News [186]
11. ROCK ME AMADEUS Falco [190]
12. LIVE TO TELL Madonna [193]
13. SARA   Starship [199]
14. THERE'LL BE SAD SONGS (To Make You Cry) Billy Ocean [203]
15. ADDICTED TO LOVE Robert Palmer [209]
16. GLORY OF LOVE (Theme From The Karate Kid Part II)   Peter Cetera [210]
17. HUMAN The Human League [212]
18. THE WAY IT IS Bruce Hornsby And The Range [216]
19. FRIENDS AND LOVERS Gloria Loring & Carl Anderson [224]
20. WHEN I THINK OF YOU Janet Jackson [240]
21. THESE DREAMS Heart [241]
22. SLEDGEHAMMER Peter Gabriel [244]
23. DANCING ON THE CEILING Lionel Richie [248]
24. EVERYBODY HAVE FUN TONIGHT Wang Chung [249]
25. BURNING HEART Survivor [254]
26. AMANDA Boston [255]
27. HIGHER LOVE Steve Winwood [264]
28. YOU GIVE LOVE A BAD NAME Bon Jovi [275]
29. VENUS Bananarama [276]
30. TRUE COLORS Cyndi Lauper [283]
31. TAKE MY BREATH AWAY (Love Theme From Top Gun) Berlin [297]
32. SECRET LOVERS   Atlantic Starr [319]
33. HOLDING BACK THE YEARS Simply Red [324]
34. INVISIBLE TOUCH Genesis [328]
35. TYPICAL MALE Tina Turner [341]
36. THE NEXT TIME I FALL Peter Cetera w/Amy Grant [370]
37. WHEN THE GOING GETS TOUGH, THE TOUGH GET GOING   Billy Ocean [383]
38. DANGER ZONE Kenny Loggins [403]
39. DON'T FORGET ME (When I'm Gone) Glass Tiger [406]
40. MANIC MONDAY Bangles [413]
41. MAD ABOUT YOU Belinda Carlisle [414]
42. I DIDN'T MEAN TO TURN YOU ON Robert Palmer [415]
43. I CAN'T WAIT Nu Shooz [421]
44. R.O.C.K. IN THE U.S.A. (A Salute To 60's Rock)   John Cougar Mellencamp [429]
45. TRUE BLUE Madonna [431]
46. NASTY Janet Jackson [463]
47. WHY CAN'T THIS BE LOVE Van Halen [469]
48. CRUSH ON YOU The Jets [477]
49. HIP TO BE SQUARE Huey Lewis And The News [479]
50. I'M YOUR MAN Wham! [489]

1987 n=398
Yr Rank. TITLE Act [Decade Rank]
1. FAITH George Michael [102]
2. ALONE Heart [105]
3. LIVIN' ON A PRAYER Bon Jovi [111]
4. I WANNA DANCE WITH SOMEBODY (Who Loves Me)   Whitney Houston [115]
5. LA BAMBA Los Lobos [133]
6. NOTHING'S GONNA STOP US NOW Starship [134]
7. SHAKEDOWN Bob Seger [143]
8. WITH OR WITHOUT YOU U2 [158]
9. HERE I GO AGAIN Whitesnake [180]
10. SHAKE YOU DOWN Gregory Abbott [189]
11. HEAD TO TOE Lisa Lisa And Cult Jam [200]
12. LOOKING FOR A NEW LOVE Jody Watley [202]
13. (I've Had) THE TIME OF MY LIFE   Bill Medley And Jennifer Warnes [205]
14. HEAVEN IS A PLACE ON EARTH Belinda Carlisle [206]
15. LEAN ON ME Club Nouveau [207]
16. ALWAYS   Atlantic Starr [208]
17. DIDN'T WE ALMOST HAVE IT ALL Whitney Houston [218]
18. I STILL HAVEN'T FOUND WHAT I'M LOOKING FOR U2 [219]
19. AT THIS MOMENT (Live) ('86) Billy Vera & The Beaters [232]
20. I THINK WE'RE ALONE NOW Tiffany [242]
21. C'EST LA VIE Robbie Nevil [245]
22. MONY MONY "LIVE" Billy Idol [246]
23. DON'T DREAM IT'S OVER Crowded House [270]
24. YOU KEEP ME HANGIN' ON Kim Wilde [280]
25. (I Just) DIED IN YOUR ARMS Cutting Crew [281]
26. NOTORIOUS Duran Duran [284]
27. OPEN YOUR HEART Madonna [294]
28. WHO'S THAT GIRL Madonna [305]
29. JACOB'S LADDER Huey Lewis And The News [306]
30. IS THIS LOVE Whitesnake [310]
31. LOST IN EMOTION Lisa Lisa And Cult Jam [312]
32. I KNEW YOU WERE WAITING (For Me)   Aretha Franklin & George Michael [315]
33. I JUST CAN'T STOP LOVING YOU Michael Jackson [331]
34. I WANT YOUR SEX George Michael [346]
35. THE LADY IN RED Chris DeBurgh [354]
36. CAUSING A COMMOTION Madonna [381]
37. LET'S WAIT AWHILE   Janet Jackson [387]
38. BAD Michael Jackson [398]
39. KEEP YOUR HANDS TO YOURSELF Georgia Satellites [417]
40. U GOT THE LOOK Prince [418]
41. SOMEWHERE OUT THERE Linda Ronstadt And James Ingram [430]
42. SHAKE YOUR LOVE Debbie Gibson [432]
43. ONLY IN MY DREAMS Debbie Gibson [455]
44. IN TOO DEEP   Genesis [483]
45. SHOULD'VE KNOWN BETTER Richard Marx [485]
46. LUKA Suzanne Vega [487]
47. HEART AND SOUL T'Pau [496]
48. YOU GOT IT ALL The Jets [502]
49. CARRIE Europe [511]
50. WILL YOU STILL LOVE ME? Chicago [512]

1988 n=387	
Yr Rank. TITLE Act [Decade Rank]	
1. EVERY ROSE HAS ITS THORN Poison [124]	
2. ONE MORE TRY George Michael [136]	
3. LOOK AWAY Chicago [141]	
4. ROLL WITH IT Steve Winwood [150]	
5. SWEET CHILD O' MINE Guns N' Roses [166]	
6. COULD'VE BEEN Tiffany [171]	
7. SO EMOTIONAL Whitney Houston [179]	
8. NEVER GONNA GIVE YOU UP Rick Astley [184]	
9. WILD, WILD WEST The Escape Club [197]	
10. GOT MY MIND SET ON YOU George Harrison [220]	
11. GET OUTTA MY DREAMS, GET INTO MY CAR Billy Ocean [222]	
12. NEED YOU TONIGHT INXS [223]	
13. ANYTHING FOR YOU Gloria Estefan and Miami Sound Machine [226]	
14. MAN IN THE MIRROR Michael Jackson [227]	
15. GROOVY KIND OF LOVE Phil Collins [237]	
16. FATHER FIGURE George Michael [250]	
17. THE FLAME Cheap Trick [256]	
18. RED RED WINE ('88) UB40 [258]	
19. (A) BABY, I LOVE YOUR WAY/(B) FREEBIRD MEDLEY (FREE BABY) Will To Power [259]	
20. WISHING WELL Terence Trent D'Arby [262]	
21. DON'T WORRY BE HAPPY (Edit) Bobby McFerrin [265]	
22. LOVE BITES Def Leppard [267]	
23. KOKOMO The Beach Boys [269]	
24. SHATTERED DREAMS Johnny Hates Jazz [287]	
25. WHERE DO BROKEN HEARTS GO Whitney Houston [296]	
26. SEASONS CHANGE Exposé [298]	
27. HANDS TO HEAVEN Breathe [309]	
28. SIMPLY IRRESISTIBLE Robert Palmer [311]	
29. HOLD ON TO THE NIGHTS Richard Marx [313]	
30. MONKEY George Michael [314]	
31. POUR SOME SUGAR ON ME Def Leppard [323]	
32. FOOLISH BEAT Debbie Gibson [332]	
33. BAD MEDICINE Bon Jovi [340]	
34. GIVING YOU THE BEST THAT I GOT Anita Baker [344]	
35. THE WAY YOU MAKE ME FEEL Michael Jackson [364]	
36. TOGETHER FOREVER Rick Astley [366]	
37. ENDLESS SUMMER NIGHTS Richard Marx [378]	
38. DEVIL INSIDE INXS [385]	
39. I'LL ALWAYS LOVE YOU Taylor Dayne [388]	
40. HAZY SHADE OF WINTER Bangles [405]	
41. DIRTY DIANA Michael Jackson [411]	
42. HOW CAN I FALL? Breathe [412]	
43. SHE'S LIKE THE WIND Patrick Swayze (featuring Wendy Fraser) [416]	
44. THE LOCO-MOTION Kylie Minogue [419]	
45. WHAT HAVE I DONE TO DESERVE THIS Pet Shop Boys And Dusty Springfield [423]	
46. NAUGHTY GIRLS (Need Love Too) Samantha Fox [439]	
47. WAITING FOR A STAR TO FALL Boy Meets Girl [440]	
48. I DON'T WANNA GO ON WITH YOU LIKE THAT Elton John [450]	
49. ANGEL Aerosmith [456]	
50. MAKE ME LOSE CONTROL Eric Carmen [458]	

1989 n=391
Yr Rank. TITLE Act [Decade Rank]
1. ANOTHER DAY IN PARADISE Phil Collins [68]
2. MISS YOU MUCH Janet Jackson [109]
3. STRAIGHT UP Paula Abdul [131]
4. MY PREROGATIVE Bobby Brown [137]
5. RIGHT HERE WAITING Richard Marx [142]
6. WE DIDN'T START THE FIRE Billy Joel [145]
7. LIKE A PRAYER Madonna [155]
8. DON'T KNOW MUCH Linda Ronstadt Featuring Aaron Neville [162]
9. BLAME IT ON THE RAIN Milli Vanilli [168]
10. LOST IN YOUR EYES Debbie Gibson [173]
11. GIRL I'M GONNA MISS YOU Milli Vanilli [196]
12. COLD HEARTED Paula Abdul [198]
13. THE LOOK Roxette [225]
14. TWO HEARTS Phil Collins [235]
15. I'LL BE THERE FOR YOU Bon Jovi [243]
16. WIND BENEATH MY WINGS Bette Midler [252]
17. ON OUR OWN Bobby Brown [253]
18. WHEN I SEE YOU SMILE Bad English [257]
19. HEAVEN Warrant [263]
20. THE LIVING YEARS Mike + The Mechanics [273]
21. GIRL YOU KNOW IT'S TRUE Milli Vanilli [274]
22. DON'T WANNA LOSE YOU Gloria Estefan [277]
23. I'LL BE LOVING YOU (Forever) New Kids On The Block [279]
24. FOREVER YOUR GIRL Paula Abdul [282]
25. SHE DRIVES ME CRAZY Fine Young Cannibals [285]
26. LISTEN TO YOUR HEART Roxette [289]
27. ETERNAL FLAME Bangles [291]
28. TOY SOLDIERS Martika [295]
29. WHEN I'M WITH YOU ('88) Sheriff [317]
30. BATDANCE Prince [318]
31. BABY DON'T FORGET MY NUMBER Milli Vanilli [327]
32. IF YOU DON'T KNOW ME BY NOW Simply Red [329]
33. REAL LOVE Jody Watley [334]
34. GOOD THING Fine Young Cannibals [338]
35. HANGIN' TOUGH New Kids On The Block [347]
36. WILD THING Tone Loc [350]
37. EXPRESS YOURSELF Madonna [371]
38. ROCK ON Michael Damian [375]
39. (It's Just) THE WAY THAT YOU LOVE ME ('89) Paula Abdul [380]
40. BACK TO LIFE Soul II Soul [389]
41. LOVE SHACK The B-52's [393]
42. DON'T RUSH ME Taylor Dayne [426]
43. BUFFALO STANCE Neneh Cherry [428]
44. SOLDIER OF LOVE Donny Osmond [435]
45. SATISFIED Richard Marx [436]
46. IF I COULD TURN BACK TIME Cher [448]
47. CHERISH Madonna [453]
48. BORN TO BE MY BABY Bon Jovi [454]
49. THE LOVER IN ME Sheena Easton [460]
50. BUST A MOVE Young M.C. [465]

Singles: Number 1s By Weeks

Entries by weeks, in order of declining score

TITLE Act	Wks
PHYSICAL Olivia Newton-John	10
ENDLESS LOVE Diana Ross & Lionel Richie	9
BETTE DAVIS EYES Kim Carnes	9
EVERY BREATH YOU TAKE The Police	8
EBONY AND IVORY Paul McCartney	7
BILLIE JEAN Michael Jackson	7
I LOVE ROCK 'N ROLL Joan Jett & The Blackhearts	7
SAY SAY SAY Paul McCartney and Michael Jackson	6
CALL ME (Theme From...American Gigolo) Blondie	6
FLASHDANCE...WHAT A FEELING Irene Cara	6
EYE OF THE TIGER Survivor	6
LADY Kenny Rogers	6
CENTERFOLD The J. Geils Band	6
LIKE A VIRGIN Madonna	6
(Just Like) STARTING OVER John Lennon	5
WHEN DOVES CRY Prince	5
JUMP Van Halen	5
UPSIDE DOWN Diana Ross	4
ALL NIGHT LONG (All Night) Lionel Richie	4
CRAZY LITTLE THING CALLED LOVE Queen	4
ANOTHER BRICK IN THE WALL (Part II) Pink Floyd	4
TOTAL ECLIPSE OF THE HEART Bonnie Tyler	4
FUNKYTOWN Lipps, Inc.	4
ROCK WITH YOU Michael Jackson	4
JACK & DIANE John Cougar	4
DOWN UNDER Men At Work	4
MANEATER Daryl Hall & John Oates	4
MAGIC Olivia Newton-John	4
SAY YOU, SAY ME Lionel Richie	4
THAT'S WHAT FRIENDS ARE FOR Dionne & Friends	4
ANOTHER DAY IN PARADISE Phil Collins	4
WALK LIKE AN EGYPTIAN Bangles	4
WE ARE THE WORLD USA for Africa	4
FAITH George Michael	4
MISS YOU MUCH Janet Jackson	4
LIVIN' ON A PRAYER Bon Jovi	4
ROLL WITH IT Steve Winwood	4
ANOTHER ONE BITES THE DUST Queen	3
COMING UP (Live At Glasgow) Paul McCartney & Wings	3
WOMAN IN LOVE Barbra Streisand	3
AGAINST ALL ODDS (Take A Look At Me Now) Phil Collins	3
BEAT IT Michael Jackson	3
WHAT'S LOVE GOT TO DO WITH IT Tina Turner	3
DON'T YOU WANT ME The Human League	3
FOOTLOOSE Kenny Loggins	3
GHOSTBUSTERS Ray Parker Jr.	3
ARTHUR'S THEME (Best That You Can Do) Christopher Cross	3
KARMA CHAMELEON Culture Club	3
I JUST CALLED TO SAY I LOVE YOU Stevie Wonder	3
CARELESS WHISPER Wham! Featuring George Michael	3
WAKE ME UP BEFORE YOU GO-GO Wham!	3

TITLE Act	Wks
ON MY OWN Patti LaBelle And Michael McDonald	3
CAN'T FIGHT THIS FEELING REO Speedwagon	3
ALONE Heart	3
UP WHERE WE BELONG Joe Cocker and Jennifer Warnes	3
MONEY FOR NOTHING Dire Straits	3
KISS ON MY LIST Daryl Hall & John Oates	3
SHOUT Tears For Fears	3
EVERY ROSE HAS ITS THORN Poison	3
STRAIGHT UP Paula Abdul	3
LA BAMBA Los Lobos	3
ONE MORE TRY George Michael	3
RIGHT HERE WAITING Richard Marx	3
GREATEST LOVE OF ALL Whitney Houston	3
LIKE A PRAYER Madonna	3
WITH OR WITHOUT YOU U2	3
LOST IN YOUR EYES Debbie Gibson	3
STUCK WITH YOU Huey Lewis And The News	3
ROCK ME AMADEUS Falco	3
ABRACADABRA Steve Miller Band	2
IT'S STILL ROCK AND ROLL TO ME Billy Joel	2
HARD TO SAY I'M SORRY Chicago	2
ISLANDS IN THE STREAM Kenny Rogers Duet with Dolly Parton	2
HELLO Lionel Richie	2
OWNER OF A LONELY HEART Yes	2
JESSIE'S GIRL Rick Springfield	2
9 TO 5 Dolly Parton	2
MANIAC Michael Sembello	2
I LOVE A RAINY NIGHT Eddie Rabbitt	2
LET'S HEAR IT FOR THE BOY Deniece Williams	2
CELEBRATION Kool & The Gang	2
PRIVATE EYES Daryl Hall & John Oates	2
BABY, COME TO ME Patti Austin (A duet with James Ingram)	2
TIME AFTER TIME Cyndi Lauper	2
OUT OF TOUCH Daryl Hall & John Oates	2
I WANT TO KNOW WHAT LOVE IS Foreigner	2
CARIBBEAN QUEEN (No More Love On The Run) Billy Ocean	2
LET'S GO CRAZY Prince And The Revolution	2
THE REFLEX Duran Duran	2
TRULY Lionel Richie	2
I WANNA DANCE WITH SOMEBODY (Who Loves Me) Whitney Houston	2
BROKEN WINGS Mr. Mister	2
RAPTURE Blondie	2
EVERYBODY WANTS TO RULE THE WORLD Tears For Fears	2
NOTHING'S GONNA STOP US NOW Starship	2
THE POWER OF LOVE Huey Lewis And The News	2
LOOK AWAY Chicago	2
WE DIDN'T START THE FIRE Billy Joel	2
MORNING TRAIN (Nine To Five) Sheena Easton	2
KISS Prince And The Revolution	2
WE BUILT THIS CITY Starship	2
HOW WILL I KNOW Whitney Houston	2
KYRIE Mr. Mister	2
ST. ELMO'S FIRE (Man In Motion) John Parr	2
ONE MORE NIGHT Phil Collins	2

TITLE Act	Wks
SWEET CHILD O' MINE Guns N' Roses	2
BLAME IT ON THE RAIN Milli Vanilli	2
COULD'VE BEEN Tiffany	2
PAPA DON'T PREACH Madonna	2
NEVER GONNA GIVE YOU UP Rick Astley	2
A VIEW TO A KILL Duran Duran	2
GIRL I'M GONNA MISS YOU Milli Vanilli	2
HEAVEN Bryan Adams	2
LEAN ON ME Club Nouveau	2
GLORY OF LOVE (Theme From The Karate Kid Part II) Peter Cetera	2
EVERYTHING SHE WANTS (Remix) Wham!	2
DIDN'T WE ALMOST HAVE IT ALL Whitney Houston	2
I STILL HAVEN'T FOUND WHAT I'M LOOKING FOR U2	2
GET OUTTA MY DREAMS, GET INTO MY CAR Billy Ocean	2
ANYTHING FOR YOU Gloria Estefan and Miami Sound Machine	2
MAN IN THE MIRROR Michael Jackson	2
AT THIS MOMENT (Live) ('86) Billy Vera & The Beaters	2
TWO HEARTS Phil Collins	2
GROOVY KIND OF LOVE Phil Collins	2
WHEN I THINK OF YOU Janet Jackson	2
I THINK WE'RE ALONE NOW Tiffany	2
FATHER FIGURE George Michael	2
AMANDA Boston	2
THE FLAME Cheap Trick	2
WHEN I SEE YOU SMILE Bad English	2
DON'T WORRY BE HAPPY (Edit) Bobby McFerrin	2
(I Just) DIED IN YOUR ARMS Cutting Crew	2
FOREVER YOUR GIRL Paula Abdul	2
TRUE COLORS Cyndi Lauper	2
TOY SOLDIERS Martika	2
WHERE DO BROKEN HEARTS GO Whitney Houston	2
MONKEY George Michael	2
I KNEW YOU WERE WAITING (For Me) Aretha Franklin & George Michael	2
BAD MEDICINE Bon Jovi	2
BAD Michael Jackson	2
DO THAT TO ME ONE MORE TIME The Captain & Tennille	1
I CAN'T GO FOR THAT (No Can Do) Daryl Hall & John Oates	1
SWEET DREAMS (Are Made Of This) Eurythmics	1
LET'S DANCE David Bowie	1
PLEASE DON'T GO K.C. & The Sunshine Band	1
MICKEY Toni Basil	1
MISSING YOU John Waite	1
CRAZY FOR YOU Madonna	1
CHARIOTS OF FIRE - TITLES Vangelis	1
WHO CAN IT BE NOW? Men At Work	1
THE TIDE IS HIGH Blondie	1
KEEP ON LOVING YOU REO Speedwagon	1
MEDLEY Stars On 45	1
COME ON EILEEN Dexys Midnight Runners	1
SAILING Christopher Cross	1

Singles: Special Lists

TITLE Act	Wks
SEPARATE LIVES	
(Love Theme From White Nights)	
Phil Collins and Marilyn Martin	1
MY PREROGATIVE Bobby Brown	1
SHAKEDOWN Bob Seger	1
DON'T YOU (Forget About Me) Simple Minds	1
THE ONE THAT YOU LOVE Air Supply	1
TELL HER ABOUT IT Billy Joel	1
PART-TIME LOVER Stevie Wonder	1
SO EMOTIONAL Whitney Houston	1
HERE I GO AGAIN Whitesnake	1
WEST END GIRLS Pet Shop Boys	1
SHAKE YOU DOWN Gregory Abbott	1
EVERYTIME YOU GO AWAY Paul Young	1
LIVE TO TELL Madonna	1
WILD, WILD WEST The Escape Club	1
COLD HEARTED Paula Abdul	1
SARA Starship	1
HEAD TO TOE Lisa Lisa And Cult Jam	1
THERE'LL BE SAD SONGS (To Make You Cry)	
Billy Ocean	1
AFRICA Toto	1
(I've Had) THE TIME OF MY LIFE	
Bill Medley And Jennifer Warnes	1
HEAVEN IS A PLACE ON EARTH	
Belinda Carlisle	1
ALWAYS Atlantic Starr	1
ADDICTED TO LOVE Robert Palmer	1
HUMAN The Human League	1
SUSSUDIO Phil Collins	1

TITLE Act	Wks
THE WAY IT IS Bruce Hornsby And The Range	1
GOT MY MIND SET ON YOU George Harrison	1
NEED YOU TONIGHT INXS	1
THE LOOK Roxette	1
TAKE ON ME A-Ha	1
SAVING ALL MY LOVE FOR YOU	
Whitney Houston	1
MIAMI VICE THEME Jan Hammer	1
THESE DREAMS Heart	1
I'LL BE THERE FOR YOU Bon Jovi	1
SLEDGEHAMMER Peter Gabriel	1
MONY MONY "LIVE" Billy Idol	1
WIND BENEATH MY WINGS Bette Midler	1
RED RED WINE ('88) UB40	1
A) BABY, I LOVE YOUR WAY/B) FREEBIRD	
MEDLEY (FREE BABY) Will To Power	1
WISHING WELL Terence Trent D'Arby	1
HIGHER LOVE Steve Winwood	1
LOVE BITES Def Leppard	1
KOKOMO The Beach Boys	1
YOU GIVE LOVE A BAD NAME Bon Jovi	1
THE LIVING YEARS Mike + The Mechanics	1
VENUS Bananarama	1
DON'T WANNA LOSE YOU Gloria Estefan	1
I'LL BE LOVING YOU (Forever)	
New Kids On The Block	1
YOU KEEP ME HANGIN' ON Kim Wilde	1
SHE DRIVES ME CRAZY Fine Young Cannibals	1
LISTEN TO YOUR HEART Roxette	1
ETERNAL FLAME Bangles	1

TITLE Act	Wks
OPEN YOUR HEART Madonna	1
TAKE MY BREATH AWAY	
(Love Theme From Top Gun) Berlin	1
SEASONS CHANGE Exposé	1
WHO'S THAT GIRL Madonna	1
JACOB'S LADDER Huey Lewis And The News	1
LOST IN EMOTION Lisa Lisa And Cult Jam	1
HOLD ON TO THE NIGHTS Richard Marx	1
WHEN I'M WITH YOU ('88) Sheriff	1
BATDANCE Prince	1
HOLDING BACK THE YEARS Simply Red	1
BABY DON'T FORGET MY NUMBER	
Milli Vanilli	1
INVISIBLE TOUCH Genesis	1
IF YOU DON'T KNOW ME BY NOW Simply Red	1
I JUST CAN'T STOP LOVING YOU	
Michael Jackson	1
FOOLISH BEAT Debbie Gibson	1
OH SHEILA Ready For The World	1
GOOD THING Fine Young Cannibals	1
HANGIN' TOUGH New Kids On The Block	1
THE WAY YOU MAKE ME FEEL	
Michael Jackson	1
TOGETHER FOREVER Rick Astley	1
THE NEXT TIME I FALL	
Peter Cetera w/Amy Grant	1
ROCK ON Michael Damian	1
DIRTY DIANA Michael Jackson	1
SATISFIED Richard Marx	1

Year, Chart Entries and Number 1s by year

Year	1980	1981	1982	1983	1984	1985	1986	1987	1988	1989
Entries	473	408	424	452	432	406	397	398	387	391
Number 1s	16	16	15	16	19	27	29	29	32	32

Singles: History of Number 1s

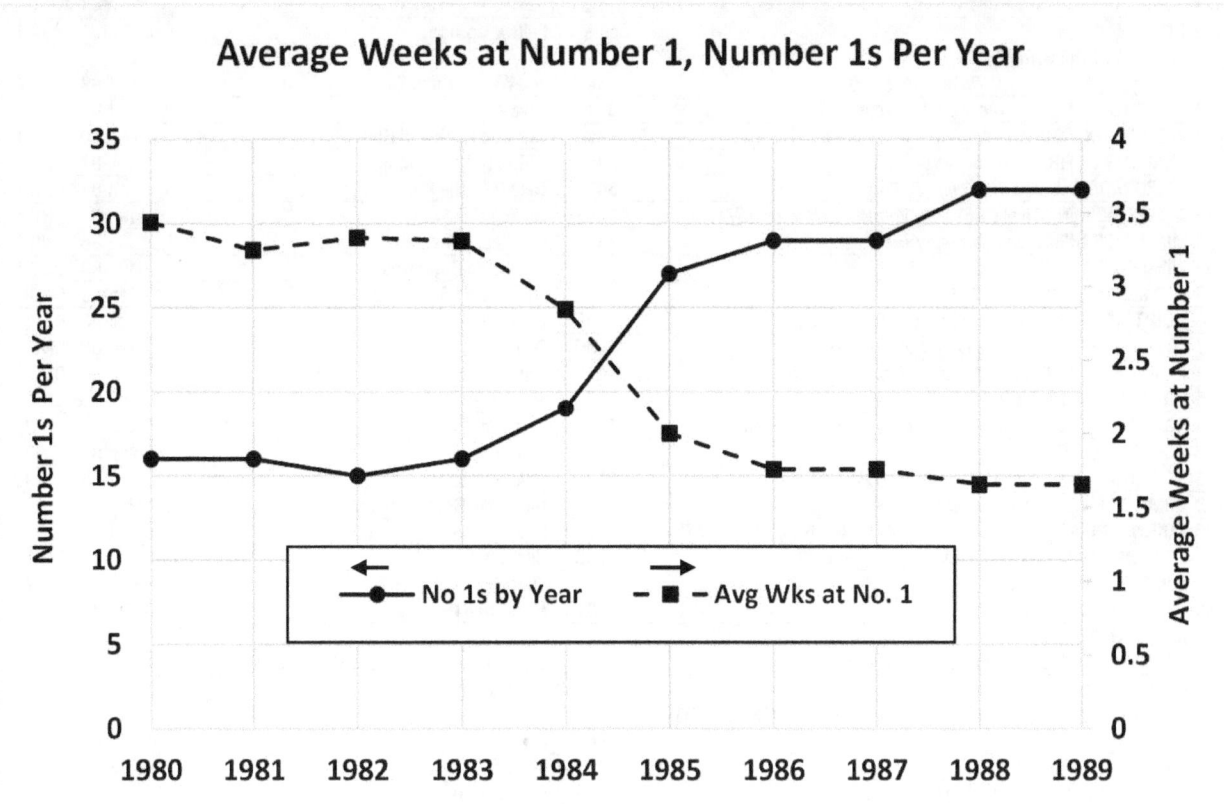

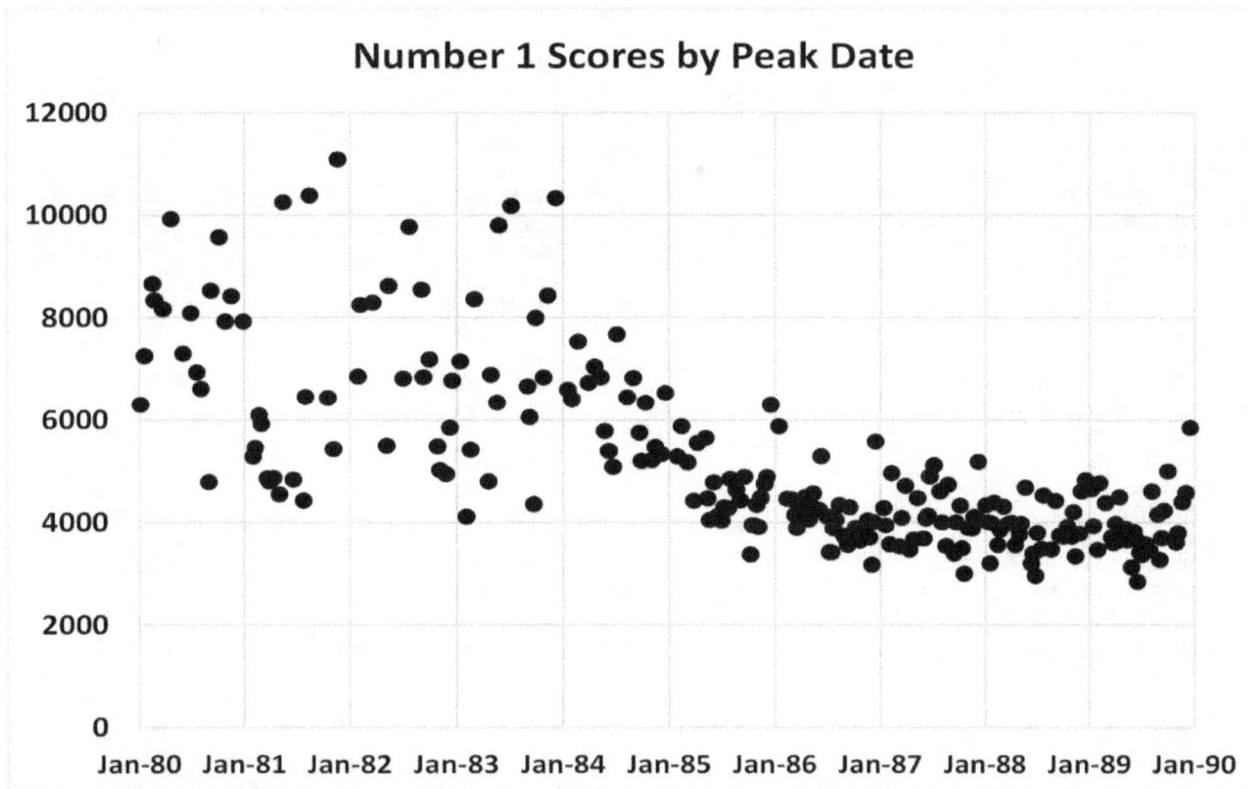

Clearly there was a transition in the charts in 1984. Note the greater opportunity for high outliers, in the first part of the decade. All of the highest charting singles of the decade came from the first half; no single from 1986 on ranked above number 69 for the decade. Generally average total duration on the chart for Number 1s was greater in the first part of the decade as well.

Singles: Weakest Follow Ups to Number 1s

Entry Date	TITLE Act (Yr)	Score	Follow-Up Title	Entry Date	Peak
27-Aug-83	ISLANDS IN THE STREAM *Kenny Rogers Duet with Dolly Parton (83)*	6841	You Were A Good Friend	Oct-83	DNC
6-Mar-82	DON'T YOU WANT ME *The Human League (82)*	6816	Love Action	Jul-82	DNC
25-Aug-79	PLEASE DON'T GO *K.C. And The Sunshine Band (80)*	6303	Let's Go Rock And Roll	Feb-80	DNC
12-Dec-81	CHARIOTS OF FIRE - TITLES *Vangelis (82)*	5496	To The Unknown Man	Nov-82	DNC
25-Oct-80	CELEBRATION *Kool & The Gang (81)*	5462	Take It To The Top	Feb-81	DNC
8-Nov-86	AT THIS MOMENT (Live) ('86) *Billy Vera & The Beaters (87)*	3948	Let You Get Away	Jan-87	DNC
7-Sep-85	MIAMI VICE THEME *Jan Hammer (85)*	3913	Crockett's Theme	Jan-88	DNC
4-Mar-89	WIND BENEATH MY WINGS *Bette Midler (89)*	3826	Under The Boardwalk	Jun-89	DNC
13-Aug-88	RED RED WINE ('88) *UB40 (88)*	3793	Where Did I Go Wrong	Jan-89	DNC
30-Jul-88	DON'T WORRY BE HAPPY (Edit) *Bobby McFerrin (88)*	3746	Good Lovin	Dec-88	DNC
1-Aug-87	LOST IN EMOTION *Lisa Lisa And Cult Jam (87)*	3495	Everything Will B-Fine	Jan-88	DNC
6-May-89	IF YOU DON'T KNOW ME BY NOW *Simply Red (89)*	3412	You've Got It	Sep-89	DNC
3-Aug-85	OH SHEILA *Ready For The World (85)*	3376	Slide Over	Oct-85	DNC
25-Feb-84	AGAINST ALL ODDS (Take A Look At Me Now) *Phil Collins (84)*	7041	In The Air Tonight	20-Oct-84	102
3-Sep-88	KOKOMO *The Beach Boys (88)*	3725	Still Cruisin'	26-Aug-89	93
22-Jun-85	ST. ELMO'S FIRE (Man In Motion) *John Parr (85)*	4434	Love Grammar	16-Nov-85	89
22-Jan-83	COME ON EILEEN *Dexys Midnight Runners (83)*	4804	The Celtic Soul Brothers	28-May-83	86
21-Jun-86	TAKE MY BREATH AWAY (Love Theme From Top Gun) *Berlin (86)*	3558	Like Flames	25-Oct-86	82
7-Apr-84	LET'S HEAR IT FOR THE BOY *Deniece Williams (84)*	5783	Next Love	11-Aug-84	81
4-Sep-82	MICKEY *Toni Basil (82)*	5851	Shoppin' From A To Z	26-Feb-83	77
28-Jun-86	VENUS *Bananarama (86)*	3701	More Than Physical	18-Oct-86	73
24-Apr-82	BABY, COME TO ME *Patti Austin (A duet with James Ingram) (83)*	5417	Every Home Should Have One (Remix)	19-Mar-83	69
11-Apr-81	MEDLEY *Stars On 45 (81)*	4837	Medley II (Medley)	18-Jul-81	67
10-Sep-88	(A) BABY, I LOVE YOUR WAY/(B) FREEBIRD MEDLEY (FREE BABY) *Will To Power (88)*	3786	Fading Away	4-Feb-89	65
29-Mar-80	FUNKYTOWN *Lipps, Inc. (80)*	7303	Rock It	2-Aug-80	64
7-Jan-89	THE LIVING YEARS *Mike + The Mechanics (89)*	3709	Seeing Is Believing	22-Apr-89	62
20-Sep-86	THE NEXT TIME I FALL *Peter Cetera w/Amy Grant (86)*	3174	Big Mistake	24-Jan-87	61
28-Mar-87	ALWAYS *Atlantic Starr (87)*	4081	One Lover At A Time	15-Aug-87	58
29-May-82	ABRACADABRA *Steve Miller Band (82)*	8547	Cool Magic	16-Oct-82	57
18-Oct-86	SHAKE YOU DOWN *Gregory Abbott (87)*	4286	I Got The Feelin' (It's Over)	21-Feb-87	56
20-Oct-79	DO THAT TO ME ONE MORE TIME *The Captain & Tennille (80)*	8662	Love On A Shoestring	8-Mar-80	55
19-Jan-80	ANOTHER BRICK IN THE WALL (Part II) *Pink Floyd (80)*	8171	Run Like Hell	10-May-80	53
6-Sep-80	WOMAN IN LOVE *Barbra Streisand (80)*	7930	Promises	23-May-81	48
5-Sep-87	MONY MONY "LIVE" *Billy Idol (87)*	3879	Hot In The City	12-Dec-87	48
16-Jul-83	TOTAL ECLIPSE OF THE HEART *Bonnie Tyler (83)*	7997	Take Me Back	3-Dec-83	46
16-Aug-80	ANOTHER ONE BITES THE DUST *Queen (80)*	9572	Need Your Loving Tonight	29-Nov-80	44
13-Sep-86	HUMAN *The Human League (86)*	4045	I Need Your Loving	6-Dec-86	44
28-Mar-87	YOU KEEP ME HANGIN' ON *Kim Wilde (87)*	3690	Say You Really Want Me	18-Jul-87	44
22-Dec-79	CRAZY LITTLE THING CALLED LOVE *Queen (80)*	8343	Play The Game	28-Jun-80	42
29-Nov-80	9 TO 5 *Dolly Parton (81)*	6099	But You Know I Love You	4-Apr-81	41
16-Feb-80	CALL ME (Theme From...American Gigolo) *Blondie (80)*	9931	Atomic	17-May-80	39
14-Feb-87	LEAN ON ME *Club Nouveau (87)*	4095	Why You Treat Me So Bad	30-May-87	39
7-Mar-87	(I Just) DIED IN YOUR ARMS *Cutting Crew (87)*	3669	One For The Mockingbird	6-Jun-87	38
23-Jun-84	MISSING YOU *John Waite (84)*	5752	Tears	20-Oct-84	37
31-Jan-81	RAPTURE *Blondie (81)*	4803	Island Of Lost Souls	29-May-82	37
4-Jun-83	MANIAC *Michael Sembello (83)*	6058	Automatic Man	24-Sep-83	34
8-Feb-86	ADDICTED TO LOVE *Robert Palmer (86)*	4061	Hyperactive	7-Jun-86	33
18-Mar-89	ROCK ON *Michael Damian (89)*	3129	Cover Of Love	17-Jun-89	31
4-Feb-89	ETERNAL FLAME *Bangles (89)*	3603	Be With You	6-May-89	30
28-Mar-81	BETTE DAVIS EYES *Kim Carnes (81)*	10255	Draw Of The Cards	8-Aug-81	28
20-Aug-88	WILD, WILD WEST *The Escape Club (88)*	4208	Shake For The Sheik	10-Dec-88	28
5-Apr-86	HOLDING BACK THE YEARS *Simply Red (86)*	3419	Money$ Too Tight (To Mention)	19-Jul-86	28

Singles: Consecutive Number 1s

Entry Date	Initial Number 1 (Yr)	Score	Following Number 1	Entry
15-Nov-80	THE TIDE IS HIGH *Blondie (81)*	5289	Rapture	31-Jan-81
6-Sep-86	YOU GIVE LOVE A BAD NAME *Bon Jovi (86)*	3706	Livin' On A Prayer	13-Dec-86
29-Aug-81	PRIVATE EYES *Daryl Hall & John Oates (81)*	5434	I Can't Go For That (No Can Do)	14-Nov-81
28-Jan-89	SHE DRIVES ME CRAZY *Fine Young Cannibals (89)*	3641	Good Thing	6-May-89
24-Oct-87	FAITH *George Michael (87)*	5187	Father Figure	16-Jan-88
16-Jan-88	FATHER FIGURE *George Michael (88)*	3850	One More Try	16-Apr-88
16-Apr-88	ONE MORE TRY *George Michael (88)*	4687	Monkey	9-Jul-88
29-Jun-85	THE POWER OF LOVE *Huey Lewis And The News (85)*	4636	Stuck With You	2-Aug-86
11-Apr-87	HEAD TO TOE *Lisa Lisa And Cult Jam (87)*	4139	Lost In Emotion	1-Aug-87
12-Apr-86	LIVE TO TELL *Madonna (86)*	4244	Papa Don't Preach	28-Jun-86
10-Jul-82	WHO CAN IT BE NOW? *Men At Work (82)*	5487	Down Under	6-Nov-82
22-Jan-83	BILLIE JEAN *Michael Jackson (83)*	8360	Beat It	26-Feb-83
8-Aug-87	I JUST CAN'T STOP LOVING YOU *Michael Jackson (87)*	3395	Bad	19-Sep-87
19-Sep-87	BAD *Michael Jackson (87)*	2998	The Way You Make Me Feel	21-Nov-87
21-Nov-87	THE WAY YOU MAKE ME FEEL *Michael Jackson (88)*	3201	Man In The Mirror	6-Feb-88
6-Feb-88	MAN IN THE MIRROR *Michael Jackson (88)*	3974	Dirty Diana	7-May-88
29-Apr-89	BABY DON'T FORGET MY NUMBER *Milli Vanilli (89)*	3415	Girl I'm Gonna Miss You	5-Aug-89
5-Aug-89	GIRL I'M GONNA MISS YOU *Milli Vanilli (89)*	4238	Blame It On The Rain	7-Oct-89
21-Sep-85	BROKEN WINGS *Mr. Mister (85)*	4890	Kyrie	21-Dec-85
1-Apr-89	I'LL BE LOVING YOU (Forever) *New Kids On The Block (89)*	3690	Hangin' Tough	15-Jul-89
24-May-80	MAGIC *Olivia Newton-John (80)*	6621	Physical	3-Oct-81
3-Dec-88	STRAIGHT UP *Paula Abdul (89)*	4776	Forever Your Girl	11-Mar-89
11-Mar-89	FOREVER YOUR GIRL *Paula Abdul (89)*	3654	Cold Hearted	24-Jun-89
24-Jun-89	COLD HEARTED *Paula Abdul (89)*	4158	Opposites Attract	16-Dec-89
7-Jun-86	GLORY OF LOVE (Theme From The Karate Kid Part II) *Peter Cetera (86)*	4056	The Next Time I Fall	20-Sep-86
9-Feb-85	ONE MORE NIGHT *Phil Collins (85)*	4427	Sussudio	11-May-85
3-Sep-88	GROOVY KIND OF LOVE *Phil Collins (88)*	3922	Two Hearts	19-Nov-88
19-Nov-88	TWO HEARTS *Phil Collins (89)*	3934	Another Day In Paradise	4-Nov-89
21-May-88	HOLD ON TO THE NIGHTS *Richard Marx (88)*	3490	Satisfied	6-May-89
6-May-89	SATISFIED *Richard Marx (89)*	2851	Right Here Waiting	8-Jul-89
19-Dec-87	NEVER GONNA GIVE YOU UP *Rick Astley (88)*	4313	Together Forever	16-Apr-88
7-Sep-85	WE BUILT THIS CITY *Starship (85)*	4486	Sara	28-Dec-85
16-Mar-85	EVERYBODY WANTS TO RULE THE WORLD *Tears For Fears (85)*	4785	Shout	15-Jun-85
29-Aug-87	I THINK WE'RE ALONE NOW *Tiffany (87)*	3894	Could've Been	28-Nov-87
21-Mar-87	WITH OR WITHOUT YOU *U2 (87)*	4480	I Still Haven't Found What I'm Looking For	13-Jun-87
8-Sep-84	WAKE ME UP BEFORE YOU GO-GO *Wham! (84)*	5473	Careless Whisper	22-Dec-84
22-Dec-84	CARELESS WHISPER *Wham! Featuring George Michael (85)*	5877	Everything She Wants (Remix)	23-Mar-85
17-Aug-85	SAVING ALL MY LOVE FOR YOU *Whitney Houston (85)*	3944	How Will I Know	7-Dec-85
7-Dec-85	HOW WILL I KNOW *Whitney Houston (86)*	4469	Greatest Love Of All	29-Mar-86
29-Mar-86	GREATEST LOVE OF ALL *Whitney Houston (86)*	4572	I Wanna Dance With Somebody (Who Loves Me)	16-May-87
16-May-87	I WANNA DANCE WITH SOMEBODY (Who Loves Me) *Whitney Houston (87)*	4899	Didn't We Almost Have It All	1-Aug-87
1-Aug-87	DIDN'T WE ALMOST HAVE IT ALL *Whitney Houston (87)*	4007	So Emotional	31-Oct-87
31-Oct-87	SO EMOTIONAL *Whitney Houston (88)*	4343	Where Do Broken Hearts Go	27-Feb-88

Act	Consecutive Number 1s
Whitney Houston	7
Michael Jackson	5
George Michael	4
Paula Abdul	4
Phil Collins	3
Richard Marx	3
Wham	3
Milli Vanilli	3

Singles: Highest Score Not Making Weekly Top 5

TITLE Act Peak (Yr)	Score
STEPPIN' OUT Joe Jackson Pk: 6 (82)	3003
YOU AND I Eddie Rabbitt with Crystal Gayle Pk: 7 (83)	2991
STEAL AWAY Robbie Dupree Pk: 6 (80)	2990
THE SWEETEST THING (I've Ever Known) Juice Newton Pk: 7 (82)	2829
LEATHER AND LACE Stevie Nicks (with Don Henley) Pk: 6 (82)	2825
BUST A MOVE Young M.C. Pk: 7 (89)	2743
LIVING INSIDE MYSELF Gino Vannelli Pk: 6 (81)	2640
WE'VE GOT TONIGHT Kenny Rogers And Sheena Easton Pk: 6 (83)	2632
THE HEART OF ROCK & ROLL Huey Lewis And The News Pk: 6 (84)	2607
NEVER KNEW LOVE LIKE THIS BEFORE Stephanie Mills Pk: 6 (80)	2602
CARS Gary Numan Pk: 9 (80)	2581
LATE IN THE EVENING Paul Simon Pk: 6 (80)	2579
LITTLE RED CORVETTE Prince Pk: 6 (83)	2573
NEUTRON DANCE Pointer Sisters Pk: 6 (85)	2532
THE NIGHT OWLS Little River Band Pk: 6 (81)	2525
KEY LARGO Bertie Higgins Pk: 8 (82)	2483
65 LOVE AFFAIR Paul Davis Pk: 6 (82)	2478
SEPARATE WAYS (Worlds Apart) Journey Pk: 8 (83)	2468
YOU CAN DO MAGIC America Pk: 8 (82)	2458
LADIES NIGHT Kool & The Gang Pk: 8 (80)	2426
I CAN DREAM ABOUT YOU Dan Hartman Pk: 6 (84)	2393
THE WINNER TAKES IT ALL ABBA Pk: 8 (81)	2383
I WANT A NEW DRUG Huey Lewis And The News Pk: 6 (84)	2382
ONE ON ONE Daryl Hall & John Oates Pk: 7 (83)	2378
THE GLAMOROUS LIFE Sheila E. Pk: 7 (84)	2352
TAINTED LOVE Soft Cell Pk: 8 (82)	2345
WE DON'T TALK ANYMORE Cliff Richard Pk: 7 (80)	2342
WHY DO FOOLS FALL IN LOVE Diana Ross Pk: 7 (81)	2330
HIT ME WITH YOUR BEST SHOT Pat Benatar Pk: 9 (80)	2328
BOY FROM NEW YORK CITY TManhattan Transfer Pk: 7 (81)	2291
ROCK THE CASBAH The Clash Pk: 8 (83)	2285
THE WARRIOR Scandal Featuring Patty Smyth Pk: 7 (84)	2284
HIM Rupert Holmes Pk: 6 (80)	2280
ONLY THE LONELY The Motels Pk: 9 (82)	2280
FIRE LAKE Bob Seger Pk: 6 (80)	2278
I'M ALRIGHT Kenny Loggins Pk: 7 (80)	2264
LADY (You Bring Me Up) Commodores Pk: 8 (81)	2260
THAT'S ALL Genesis Pk: 6 (84)	2253
RUNNING WITH THE NIGHT Lionel Richie Pk: 7 (84)	2232
SOMEBODY'S BABY Jackson Browne Pk: 7 (82)	2217
WORD UP Cameo Pk: 6 (86)	2212
SUNGLASSES AT NIGHT Corey Hart Pk: 7 (84)	2211
DON'T LET IT END Styx Pk: 6 (83)	2206
DON'T YOU WANT ME Jody Watley Pk: 6 (87)	2173
KEEP THE FIRE BURNIN' REO Speedwagon Pk: 7 (82)	2157
NO MORE LONELY NIGHTS Paul McCartney Pk: 6 (84)	2137
OBSESSION Animotion Pk: 6 (85)	2137
SMALL TOWN John Cougar Mellencamp Pk: 6 (85)	2127
RUN TO YOU Bryan Adams Pk: 6 (85)	2116
HEARTS Marty Balin Pk: 8 (81)	2113

Singles: Lowest Scoring Number 1s

TITLE Act (Year)	Score
SATISFIED Richard Marx (89)	2851
DIRTY DIANA Michael Jackson (88)	2958
BAD Michael Jackson (87)	2998
ROCK ON Michael Damian (89)	3129
THE NEXT TIME I FALL Peter Cetera w/Amy Grant (86)	3174
TOGETHER FOREVER Rick Astley (88)	3198
THE WAY YOU MAKE ME FEEL Michael Jackson (88)	3201
HANGIN' TOUGH New Kids On The Block (89)	3272
BAD MEDICINE Bon Jovi (88)	3342
GOOD THING Fine Young Cannibals (89)	3365
OH SHEILA Ready For The World (85)	3376
FOOLISH BEAT Debbie Gibson (88)	3393
I JUST CAN'T STOP LOVING YOU Michael Jackson (87)	3395
IF YOU DON'T KNOW ME BY NOW Simply Red (89)	3412
INVISIBLE TOUCH Genesis (86)	3412
BABY DON'T FORGET MY NUMBER Milli Vanilli (89)	3415
HOLDING BACK THE YEARS Simply Red (86)	3419
BATDANCE Prince (89)	3450
WHEN I'M WITH YOU ('88) Sheriff (89)	3465
I KNEW YOU WERE WAITING (For Me) Aretha Franklin & George Michael (87)	3467
MONKEY George Michael (88)	3473
HOLD ON TO THE NIGHTS Richard Marx (88)	3490
LOST IN EMOTION Lisa Lisa And Cult Jam (87)	3495
JACOB'S LADDER Huey Lewis And The News (87)	3536
WHO'S THAT GIRL Madonna (87)	3537
SEASONS CHANGE Exposé (88)	3558
TAKE MY BREATH AWAY (Love Theme From Top Gun) Berlin (86)	3558
WHERE DO BROKEN HEARTS GO Whitney Houston (88)	3563
TOY SOLDIERS Martika (89)	3573
OPEN YOUR HEART Madonna (87)	3575
ETERNAL FLAME Bangles (89)	3603
LISTEN TO YOUR HEART Roxette (89)	3612
SHE DRIVES ME CRAZY Fine Young Cannibals (89)	3641
TRUE COLORS Cyndi Lauper (86)	3650
FOREVER YOUR GIRL Paula Abdul (89)	3654
(I Just) DIED IN YOUR ARMS Cutting Crew (87)	3669
YOU KEEP ME HANGIN' ON Kim Wilde (87)	3690
I'LL BE LOVING YOU (Forever) New Kids On The Block (89)	3690
DON'T WANNA LOSE YOU Gloria Estefan (89)	3701
VENUS Bananarama (86)	3701
YOU GIVE LOVE A BAD NAME Bon Jovi (85)	3706
THE LIVING YEARS Mike + The Mechanics (89)	3709
KOKOMO The Beach Boys (88)	3725
LOVE BITES Def Leppard (88)	3728
DON'T WORRY BE HAPPY (Edit) Bobby McFerrin (88)	3746
HIGHER LOVE Steve Winwood (86)	3755
WISHING WELL Terence Trent D'Arby (88)	3764
(A) BABY, I LOVE YOUR WAY/ (B) FREEBIRD MEDLEY (FREE BABY) Will To Power (88)	3786
RED RED WINE ('88) UB40 (88)	3793
WHEN I SEE YOU SMILE Bad English ('88) UB40 (88)	3797

Singles: No Proximate Album

These are singles for which no album of the same time frame could be located as a contemporaneous source. In some cases, there was no album recorded by the Act, or the album followed the single by a significant period of time (*Make Me Lose Control*). In some cases, it was a benefit one-time collaboration (*Dancing In The Street*). *Twist And Shout* appeared in the movie Ferris Bueller's Day Off, but did not appear on the original Soundtrack.

TITLE - Act Peak (Year)
BREAKING UP IS HARD ON YOU
(a/k/a Don't Take Ma Bell Away From Me) -
The American Comedy Network Pk: 70 (84)
DREAMER - The Association Pk: 66 (81)
EVERY HOME SHOULD HAVE ONE - Patti Austin Pk: 69 (83)
DO THEY KNOW IT'S CHRISTMAS? - Band Aid Pk: 13 (84)
THE BEACH BOYS MEDLEY - The Beach Boys Pk: 12 (81)
THE BEATLES' MOVIE MEDLEY - The Beatles Pk: 12 (82)
TWIST AND SHOUT - The Beatles Pk: 23 (86)
DON'T DO ME - Randy Bell Pk: 90 (84)
NEVER GIVE UP ON A GOOD THING - George Benson Pk: 52 (82)
UNFAITHFULLY YOURS (One Love) - Stephen Bishop Pk: 87 (84)
PART OF ME THAT NEEDS YOU MOST - Jay Black Pk: 98 (80)
DANCING IN THE STREET - Mick Jagger & David Bowie Pk: 7 (85)
WHO SHOT J.R.? - Gary Burbank with Band McNally Pk: 67 (80)
LET'S PUT THE FUN BACK IN ROCK N ROLL -
Freddy Cannon & The Belmonts Pk: 81 (81)
MAKE ME LOSE CONTROL - Eric Carmen Pk: 3 (88)
SUPER BOWL SHUFFLE - Chicago Bears Shufflin' Crew Pk: 41 (86)
NATURAL LOVE - Petula Clark Pk: 66 (82)
DO IT AGAIN/BILLIE JEAN - Club House Pk: 75 (83)
DIG THE GOLD - Joyce Cobb Pk: 42 (79)
THEME FROM DYNASTY - Bill Conti Pk: 52 (82)
LET ME SLEEP ALONE - Cugini Pk: 88 (79)
DO YOU THINK I'M DISCO? -
Steve Dahl And Teenage Radiation Pk: 58 (79)
SHE DID IT - Michael Damian Pk: 69 (81)
TOM'S DINER (DNA Remix) - DNA Featuring Suzanne Vega Pk: 5 (90)
WHO WERE YOU THINKIN' OF - Doolittle Band Pk: 49 (80)
NIGHT PULSE - Double Image Pk: 92 (83)
GETTING AWAY WITH IT - Electronic Pk: 38 (90)
ONCE A NIGHT - Jackie English Pk: 94 (80)
ONLY WOMEN BLEED - Favorite Angel Pk: 69 (90)
PLEASE MR. POSTMAN - Gentle Persuasion Pk: 82 (83)
ALL I HAVE TO DO IS DREAM -
Andy Gibb And Victoria Principal Pk: 51 (81)
NOT FADE AWAY - Eric Hine Pk: 73 (81)
I LOVE WOMEN - Jim Hurt Pk: 90 (80)

TITLE - Act Peak (Year)
WHATCHA GONNA DO WITH MY LOVIN' - Inner City Pk: 76 (90)
YOU'RE SO EASY TO LOVE - Tommy James Pk: 58 (81)
BREAD AND BUTTER - Robert John Pk: 68 (83)
VOICE OF FREEDOM - Jim Kirk And The TM Singers Pk: 71 (80)
FOOLS LIKE ME - Lorenzo Lamas Pk: 85 (84)
WHAT'S SHE GOT - Liquid Gold Pk: 86 (83)
RUNNER - Manfred Mann's Earth Band Pk: 22 (84)
PUMP UP THE VOLUME - M/A/R/R/S Pk: 13 (87)
AMERICAN MEMORIES - Shamus M'Cool Pk: 80 (81)
LOVE THEME FROM SHOGUN (Mariko's Theme) - Meco Pk: 70 (80)
SUMMER '81 MEDLEY - The Cantina Band Pk: 81 (81)
YEARS - Wayne Newton Pk: 35 (80)
BREAKIN'.. THERE'S NO STOPPING US - Ollie And Jerry Pk: 9 (84)
WALKING ON THIN ICE - Yoko Ono Pk: 58 (81)
GYPSY SPIRIT - Pendulum Pk: 89 (80)
THEME FROM THE TV SHOW CHEERS
(Where Everybody Knows Your Name) -
Gary Portnoy Pk: 83 (83)
BACK ON THE CHAIN GANG - The Pretenders Pk: 5 (82)
RONNIE'S RAPP - Ron And The D.C. Crew Pk: 93 (87)
CHAIN REACTION - Diana Ross Pk: 66 (86)
I'M FREE - Soup Dragons Pk: 79 (90)
WHERE ARE YOU NOW? - Jimmy Harnen w/Synch Pk: 10 (89)
WHERE ARE YOU NOW? - Synch Pk: 77 (86)
JACK THE LAD - 3 Man Island Pk: 94 (88)
BOY TOY - Tia Pk: 97 (87)
DOCTORIN' THE TARDIS - The Timelords Pk: 66 (88)
MIRAGE - Eric Troyer Pk: 92 (80)
(I Know) I'M LOSING YOU - Uptown Pk: 80 (86)
I DON'T WANT TO BE LONELY - Dana Valery Pk: 87 (80)
THE LONGER YOU WAIT - Gino Vannelli Pk: 89 (82)
BODY ROCK - Maria Vidal Pk: 48 (84)
HANDS ACROSS AMERICA - Voices Of America Pk: 65 (86)
MERRY CHRISTMAS IN THE NFL -
Willis "The Guard" and Vigorish Pk: 82 (80)
THE ONLY WAY IS UP - Yazz And The Plastic Population Pk: 96 (88)
THERE'S A PARTY GOING ON - Yvonne Pk: 88 (90)

*The single *Back On The Chain Gang* entered the chart in December, 1982; the album *Learning To Crawl* entered in February, 1984.

Singles: Special Lists

Singles: '80s Covers

No self-covers, re-releases or parodies allowed. Original may also be an '80s single, but two must peak more than six months apart.

'80s Score	Rank. Title {Versions} Cover Act (Yr) Peak	First Version (Yr) Peak; Biggest Version (Yr) Peak
5195	1. More Than I Can Say {2} Leo SAYER (80) 2	F: Bobby VEE (61) 61; B: Leo SAYER (80) 2
4741	2. La Bamba {4} LOS LOBOS (87) 1	F: Ritchie VALENS (59) 22; B: LOS LOBOS (87) 1
4572	3. Greatest Love Of All {2} Whitney HOUSTON (86) 1	F: George BENSON (77) 24; B: Whitney HOUSTON (86) 1
4448	4. Don't Know Much {3} Linda RONSTADT (89) 2	F: Bill MEDLEY (81) 88; B: Linda RONSTADT (89) 2
4354	5. Yes, I'm Ready {2} Teri DeSARIO (80) 2	F: Barbara MASON (65) 5; B: Teri DeSARIO (80) 2
4095	6. Lean On Me {2} CLUB NOUVEAU (87) 1	F&B: Bill WITHERS (72) 1
3949	7. Sukiyaki {2} A TASTE OF HONEY (81) 3	F&B: Kyu SAKAMOTO (63) 1
3922	8. Groovy Kind Of Love {2} Phil COLLINS (88) 1	F&B: The MINDBENDERS (66) 2
3894	9. I Think We're Alone Now {3} TIFFANY (87) 1	F&B: Tommy JAMES And The SHONDELLS (67) 4
3879	10. Mony Mony {2} Billy IDOL (87) 1	F&B: Tommy JAMES And The SHONDELLS (68) 3
3826	11. Wind Beneath My Wings {2} Bette MIDLER (89) 1	F: Lou RAWLS (83) 65; B: Bette MIDLER (89) 1
3793	12. Red Red Wine {4} UB40 (88) 1	F: Neil DIAMOND (68) 62; B: UB40 (88) 1
3701	13. Venus {2} BANANARAMA (86) 1	F&B: SHOCKING BLUE (70) 1
3690	14. You Keep Me Hangin' On {5} Kim WILDE (87) 1	F&B: The SUPREMES (66) 1
3447	15. Angel Of The Morning {2} Juice NEWTON (81) 4	F: Merrilee RUSH (68) 7; B: Juice NEWTON (81) 4
3446	16. Sea Of Love {3} The HONEYDRIPPERS (85) 3	F&B: Phil PHILLIPS (59) 2
3412	17. If You Don't Know Me By Now {2} SIMPLY RED (89) 1	F&B: Harold MELVIN And The BLUE NOTES (72) 3
3129	18. Rock On {2} Michael DAMIAN (89) 1	F: David ESSEX (74) 5; B: Michael DAMIAN (89) 1
3123	19. Elvira {2} OAK RIDGE BOYS (81) 5	F: Dallas FRAZIER (66) 72; B: OAK RIDGE BOYS (81) 5
2999	20. I Keep Forgettin' (Every Time You're Near) {2} Michael McDONALD (82) 4	F: Chuck JACKSON (62) 55; B: Michael McDONALD (82) 4
2983	21. Hazy Shade Of Winter {2} The BANGLES (88) 2	F: SIMON & GARFUNKEL (66) 13; B: The BANGLES (88) 2
2941	22. I Didn't Mean To Turn You On {2} Robert PALMER (86) 2	F: CHERRELLE (84) 79; B: Robert PALMER (86) 2
2927	23. The Loco-Motion {3} Kylie MINOGUE (88) 3	F&B: LITTLE EVA (62) 1
2868	24. Crying {3} Don McLEAN (81) 5	F&B: Roy ORBISON (61) 2
2822	25. Cum On Feel The Noize {2} QUIET RIOT (83) 5	F: SLADE (73) 98; B: QUIET RIOT (83) 5
2632	26. We've Got Tonight {2} Kenny ROGERS & Sheena EASTON (83) 6	F: Bob SEGER & The SILVER BULLET BAND (79) 13; B: Kenny ROGERS & Sheena EASTON (83) 6
2418	27. California Girls {2} David Lee ROTH (85) 3	F&B: BEACH BOYS (65) 3
2330	28. Why Do Fools Fall In Love {6} Diana ROSS (81) 7	F&B: Frankie LYMON and the TEENAGERS (56) 7
2291	29. Boy From New York City {2} MANHATTAN TRANSFER (81) 7	F: AD LIBS (65) 8; B: MANHATTAN TRANSFER (81) 7
2120	30. Don't Be Cruel {4} CHEAP TRICK (88) 4	F&B: Elvis PRESLEY (56) 1
1974	31. You Can't Hurry Love {2} Phil COLLINS (83) 10	F&B: The SUPREMES (66) 1
1958	32. Walk This Way {2} RUN-D.M.C. (86) 4	F: AEROSMITH (77) 10; B: RUN-D.M.C. (86) 4
1950	33. More Love {2} Kim CARNES (80) 10	F: Smokey ROBINSON & The MIRACLES (67) 23; B: Kim CARNES (80) 10
1882	34. Hurt So Bad {3} Linda RONSTADT (80) 8	F: LITTLE ANTHONY And The IMPERIALS (65) 10; B: The LETTERMEN (69) 12
1852	35. Always Something There To Remind Me {5} NAKED EYES (83) 8	F: Lou JOHNSON (64) 49; B: NAKED EYES (83) 8
1802	36. Always On My Mind {2} PET SHOP BOYS (88) 4	F&B: Willie NELSON (82) 5
1773	37. Harlem Shuffle {3} ROLLING STONES (86) 5	F: BOB & EARL (64) 44; B: ROLLING STONES (86) 5
1734	38. You've Lost That Lovin' Feeling {5} Daryl HALL & John OATES (80) 12	F&B: RIGHTEOUS BROTHERS (65) 1
1716	39. Tell It Like It Is {3} HEART (81) 8	F&B: Aaron NEVILLE (67) 2
1683	40. Respect Yourself {2} Bruce WILLIS (87) 5	F&B: STAPLE SINGERS (71) 12
1648	41. Daydream Believer {2} Anne MURRAY (80) 12	F&B: The MONKEES (67) 1
1605	42. Crimson And Clover {2} Joan JETT & The BLACKHEARTS (82) 7	F&B: Tommy JAMES And The SHONDELLS (69) 1
1525	43. I Found Someone {2} CHER (88) 10	F: Laura BRANIGAN (86) 90; B: CHER (88) 10
1479	44. One Fine Day {4} Carole KING (80) 12	F&B: The CHIFFONS (63) 5
1431	45. Funkytown {2} PSEUDO ECHO (87) 6	F&B: LIPPS, INC. (80) 1
1384	46. It's Gonna Take A Miracle {2} Deniece WILLIAMS (82) 10	F: The ROYALETTES (65) 41; B: Deniece WILLIAMS (82) 10
1381	47. Dancing In The Street {6} Mick JAGGER And David BOWIE (85) 7	F&B: MARTHA & The VANDELLAS (64) 2
1369	48. Together {2} TIERRA (81) 18	F: The INTRUDERS (2) (67) 48; B: TIERRA (81) 18
1362	49. War {2} Bruce SPRINGSTEEN And The E STREET Band (86) 8	F&B: Edwin STARR (70) 1
1345	50. Get It On {2} POWER STATION (85) 9	F&B: T. REX (72) 10
1321	51. Break It To Me Gently {2} Juice NEWTON (82) 11	F&B: Brenda LEE (62) 4
1314	52. Don't Let Go {3} Isaac HAYES (80) 18	F: Roy HAMILTON (58) 13; B: Isaac HAYES (80) 18
1312	53. Didn't I (Blow Your Mind) {2} NEW KIDS ON THE BLOCK (89) 8	F&B: The DELFONICS (70) 10
1282	54. I Saw Him Standing There {2} TIFFANY (88) 7	F&B: The BEATLES (64) 14
1263	55. Some Guys Have All The Luck {2} Rod STEWART (84) 10	F: The PERSUADERS (73) 39; B: Rod STEWART (84) 10
1261	56. Wipeout {3} FAT BOYS (87) 12	F&B: The SURFARIS (63) 2
1205	57. Put A Little Love In Your Heart {2} Annie LENNOX Al GREEN (89) 9	F&B: Jackie DeSHANNON (69) 4
1163	58. (Oh) Pretty Woman {2} VAN HALEN (82) 12	F&B: Roy ORBISON (64) 1
1159	59. Share Your Love With Me {4} Kenny ROGERS (81) 14	F: Bobby BLAND (64) 42; B: Aretha FRANKLIN (69) 13
1152	60. Any Day Now {3} Ronnie MILSAP (82) 14	F: Chuck JACKSON (62) 23; B: Ronnie MILSAP (82) 14
1149	61. It's Now Or Never {2} John SCHNEIDER (81) 14	F&B: Elvis PRESLEY (60) 1
1144	62. (Sittin' On) The Dock Of The Bay {7} Michael BOLTON (88) 11	F&B: Otis REDDING (68) 1
1116	63. Kiss Him Goodbye {2} The NYLONS (87) 12	F&B: STEAM (69) 1
1077	64. I Wanna Go Back {2} Eddie MONEY (87) 14	F: BILLY SATELLITE (84) 78; B: Eddie MONEY (87) 14

Singles: Special Lists

'80s Score	Rank. Title {Versions} Cover Act (Yr) Peak	First Version (Yr) Peak; Biggest Version (Yr) Peak
1058	65. Iko Iko {3} BELLE STARS (89) 14	F: DIXIE CUPS (65) 20; B: BELLE STARS (89) 14
1054	66. Your Mama Don't Dance {2} POISON (89) 10	F&B: LOGGINS & MESSINA (73) 4
961	67. What's Going On {2} Cyndi LAUPER (87) 12	F&B: Marvin GAYE (71) 2
960	68. Stand By Me {6} Mickey GILLEY (80) 22	F&B: Ben E. King (61) 4
950	69. Don't Make Me Over {4} SYBIL (89) 20	F&B: Dionne WARWICK (63) 21
929	70. Touch Me When We're Dancing {2} CARPENTERS (81) 16	F: BAMA (79) 86; B: CARPENTERS (81) 16
911	71. Gimme Some Lovin' {3} BLUES BROTHERS (80) 18	F&B: Spencer DAVIS Group (67) 7
907	72. Smokin' In The Boys Room {2} MÖTLEY CRÜE (85) 16	F&B: BROWNSVILLE STATION (74) 3
843	73. Come Go With Me {3} BEACH BOYS (82) 18	F&B: DEL VIKINGS (57) 5
795	74. Early In The Morning {2} Robert PALMER (88) 19	F: GAP BAND (82) 24; B: Robert PALMER (88) 19
781	75. The Twist (Yo, Twist!) {5} FAT BOYS (88) 16	F: Hank BALLARD And The MIDNIGHTERS (60) 28; B: Chubby CHECKER (62) 1
739	76. Love On A Two Way Street {2} Stacy LATTISAW (81) 26	F&B: The MOMENTS (2) (70) 3
730	77. Let's Stay Together {3} Tina TURNER (84) 26	F&B: Al GREEN (72) 1
729	78. There Goes My Baby {2} Donna SUMMER (84) 21	F&B: The DRIFTERS (59) 2
727	79. Power Of Love {3} Laura BRANIGAN (88) 26	F: AIR SUPPLY (85) 68; B: Laura BRANIGAN (88) 26
708	80. Wild World {3} Maxi PRIEST (89) 25	F&B: Cat STEVENS (71) 11
696	81. I Do (Live Version) {2} J. GEILS Band (83) 24	F: The MARVELOWS (65) 37; B: J. GEILS Band (83) 24
655	82. My Girl {2} SUAVÉ (88) 20	F&B: The TEMPTATIONS (65) 1
642	83. Daddy's Home {4} Cliff RICHARD (82) 23	F&B: SHEP And The LIMELITES (61) 2
634	84. My Guy {3} SISTER SLEDGE (82) 23	F&B: Mary WELLS (64) 1
631	85. Since I Don't Have You {5} Don McLEAN (81) 23	F&B: The SKYLINERS (59) 12
629	86. Earth Angel {6} NEW EDITION (86) 21	F&B: The PENGUINS (55) 8
628	87. Twist And Shout {3} The BEATLES (86) 23	F: ISLEY BROTHERS (62) 17; B: The BEATLES (64) 2
610	88. Come On, Let's Go {3} LOS LOBOS (87) 21	F: Ritchie VALENS (58) 42; B: The McCOYS (66) 22
603	89. Red Red Wine {4} UB40 (84) 34	F: Neil DIAMOND (68) 62; B: UB40 (88) 1
585	90. Can't Help Falling In Love {4} Corey HART (87) 24	F&B: Elvis PRESLEY (62) 2
565	91. Stop In The Name Of Love {3} The HOLLIES (83) 29	F&B: The SUPREMES (65) 1
552	92. Gee Whiz {2} Bernadette PETERS (80) 31	F&B: Carla THOMAS (61) 10
541	93. Going To A Go-Go {2} ROLLING STONES (82) 25	F&B: The MIRACLES (66) 11
536	94. (Ghost) Riders In The Sky {4} The OUTLAWS (81) 31	F: The RAMRODS (61) 30; B: The OUTLAWS (81) 31
534	95. Tell Her No {2} Juice NEWTON (83) 27	F&B: The ZOMBIES (65) 6
534	96. Jumpin' Jack Flash {3} Aretha FRANKLIN (86) 21	F&B: ROLLING STONES (68) 3
532	97. Wake Up Little Susie {2} SIMON & GARFUNKEL (82) 27	F&B: EVERLY BROTHERS (57) 1
511	98. Rockin' At Midnight {2} The HONEYDRIPPERS (85) 25	F: Pat BOONE (59) 49; B: The HONEYDRIPPERS (85) 25
505	99. What I Like About You {2} Michael MORALES (89) 28	F: The ROMANTICS (80) 49; B: Michael MORALES (89) 28
501	100. Everlasting Love {3} Rex SMITH/Rachel SWEET (81) 32	F: Robert KNIGHT (67) 13; B: Carl CARLTON (74) 6
500	101. I Got You Babe {3} UB40 With Chrissie HYNDE (85) 28	F&B: SONNY & CHER (65) 1
491	102. The Clapping Song {2} Pia ZADORA (83) 36	F&B: Shirley ELLIS (65) 8
484	103. I Can't Let Go {2} Linda RONSTADT (80) 31	F&B: The HOLLIES (66) 42
483	104. Hey There Lonely Girl {3} Robert JOHN (80) 31	F: RUBY And The ROMANTICS (63) 27; B: Eddie HOLMAN (70) 2
463	105. I Only Wanna Be With You {5} Samantha FOX (89) 31	F: Dusty SPRINGFIELD (64) 12; B: BAY CITY ROLLERS (76) 12
463	106. Memory {2} Barry MANILOW (83) 39	F: Barbra STREISAND (82) 52; B: Barry MANILOW (83) 39
456	107. I Thank You {2} ZZ TOP (80) 34	F&B: SAM & DAVE (68) 9
451	108. I Don't Want To Walk Without You {2} Barry MANILOW (80) 36	F: Phyllis McGUIRE (64) 79; B: Barry MANILOW (80) 36
451	109. I Can't Help Myself (Sugar Pie, Honey Bunch) {3} Bonnie POINTER (80) 40	F&B: FOUR TOPS (65) 1
451	110. I Knew You When {2} Linda RONSTADT (83) 37	F&B: Billy Joe ROYAL (65) 14
447	111. Sea Of Love {3} Del SHANNON (82) 33	F&B: Phil PHILLIPS (59) 2
446	112. Mister Sandman {3} Emmylou HARRIS (81) 37	F&B: The CHORDETTES (54) 1
437	113. When A Man Loves A Woman {3} Bette MIDLER (80) 35	F&B: Percy SLEDGE (66) 1
427	114. Save The Last Dance For Me {4} Dolly PARTON (83) 45	F&B: The DRIFTERS (60) 1
425	115. Somewhere (From "West Side Story") {3} Barbra STREISAND (86) 43	F: P.J. PROBY (65) 91; B: Len BARRY (66) 26
425	116. Dancing In The Street {6} VAN HALEN (82) 38	F&B: MARTHA & The VANDELLAS (64) 2
423	117. Kiss {2} ART OF NOISE (89) 31	F&B: PRINCE And The REVOLUTION (86) 1
401	118. Someday, Someway {2} Marshall CRENSHAW (82) 36	F: Robert GORDON (81) 76; B: Marshall CRENSHAW (82) 36
398	119. Hippy Hippy Shake {2} GEORGIA SATELLITES (88) 45	F&B: SWINGING BLUE JEANS (64) 24
395	120. Who's Making Love {3} BLUES BROTHERS (81) 39	F&B: Johnnie TAYLOR (68) 5
385	121. Let It Be Me {6} Willie NELSON (82) 40	F: Jill COREY (57) 57; B: Betty EVERETT & Jerry BUTLER (64) 5
376	122. If I Were You {2} LULU (82) 44	F: TOBY BEAU (80) 70; B: LULU (82) 44
366	123. You Better Run {2} Pat BENATAR (80) 42	F&B: YOUNG RASCALS/RASCALS (66) 20
364	124. Love Or Let Me Be Lonely {2} Paul DAVIS (82) 40	F&B: FRIENDS OF DISTINCTION (70) 6
353	125. Don't Leave Me This Way {2} The COMMUNARDS (87) 40	F&B: Thelma HOUSTON (77) 1
352	126. Working In The Coal Mine {2} DEVO (81) 43	F&B: Lee DORSEY (66) 8
352	127. Everyday People {2} Joan JETT & The BLACKHEARTS (83) 37	F&B: SLY & The FAMILY STONE (69) 1
351	128. Let's Hang On {2} Barry MANILOW (82) 32	F&B: 4 SEASONS (65) 3
342	129. Stand By Me {6} Maurice WHITE (85) 50	F&B Ben E. King (61) 4
335	130. My Prayer {2} RAY, GOODMAN & BROWN (80) 47	F&B: The PLATTERS (56) 1
327	131. Mama Weer All Crazee Now {2} QUIET RIOT (84) 51	F: SLADE (73) 76; B: QUIET RIOT (84) 51
327	132. I'll Try Something New {3} A TASTE OF HONEY (82) 41	F: The MIRACLES (62) 39; B: Diana ROSS And The SUPREMES & The TEMPTATIONS (69) 25

Singles: Special Lists

'80s Score	Rank. Title {Versions} Cover Act (Yr) Peak	First Version (Yr) Peak; Biggest Version (Yr) Peak
326	133. Love Of The Common People {2} Paul YOUNG (84) 45	F&B: The WINSTONS (69) 54
315	134. Apache {4} SUGARHILL GANG (82) 53	F&B: Jorgen INGMANN (61) 2
314	135. Needles And Pins {4} Tom PETTY and The HEARTBREAKERS (86) 37	F: Jackie DeSHANNON (63) 84; B: The SEARCHERS (64) 13
304	136. Walk Like A Man {2} MARY JANE GIRLS (86) 41	F&B: 4 SEASONS (63) 1
294	137. People Get Ready {2} Jeff BECK And Rod STEWART (85) 48	F&B: The IMPRESSIONS (65) 14
292	138. But You Know I Love You {2} Dolly PARTON (81) 41	F&B: Kenny ROGERS And The FIRST EDITION (69) 19
290	139. The Power Of Love {3} Jennifer RUSH (86) 57	F: AIR SUPPLY (85) 68; B: Laura BRANIGAN (88) 26
287	140. Peter Gunn {4} ART OF NOISE (86) 50	F&B: Ray ANTHONY (59) 8
286	141. Cry Like A Baby {2} Kim CARNES (80) 44	F&B: BOX TOPS (68) 2
284	142. Money {4} FLYING LIZARDS (80) 50	F: Barrett STRONG (60) 23; B: The KINGSMEN (64) 16
282	143. Dirty Water {2} The INMATES (80) 51	F&B: The STANDELLS (66) 11
277	144. Stop Doggin' Me Around {2} KLIQUE (83) 50	F&B: Jackie WILSON (60) 15
277	145. Wrap It Up {2} FABULOUS THUNDERBIRDS (86) 50	F: Archie BELL & The DRELLS (70) 93; B: FABULOUS THUNDERBIRDS (86) 50
269	146. You Don't Know Me {5} Mickey GILLEY (81) 55	F: Jerry VALE (56) 14; B: Ray CHARLES (62) 2
266	147. Leader Of The Pack {2} TWISTED SISTER (86) 53	F&B: The SHANGRI-LAS (64) 1
266	148. Can We Still Be Friends {2} Robert PALMER (80) 52	F&B: Todd RUNDGREN (78) 29
253	149. Everyday {2} James TAYLOR (85) 61	F: John DENVER (72) 81; B: James TAYLOR (85) 61
248	150. California Dreamin' {4} BEACH BOYS (86) 57	F&B: MAMAS & The PAPAS (66) 4
246	151. Fire {2} Bruce SPRINGSTEEN And The E STREET Band (87) 46	F&B: POINTER SISTERS (79) 2
239	152. I Only Want To Be With You {5} Nicolette LARSON (82) 53	F: Dusty SPRINGFIELD (64) 12; B: BAY CITY ROLLERS (76) 12
238	153. All I Have To Do Is Dream {6} Andy GIBB And Victoria PRINCIPAL (81) 51	F&B: EVERLY BROTHERS (58) 1
237	154. Never Can Say Goodbye {4} The COMMUNARDS (88) 51	F&B: JACKSON 5 (71) 2
236	155. How Long {2} Rod STEWART (82) 49	F&B: ACE (75) 3
232	156. Dreamin' {2} John SCHNEIDER (82) 45	F&B: Johnny BURNETTE (60) 11
217	157. Papa Was A Rollin' Stone {3} WOLF (83) 55	F: UNDISPUTED TRUTH (72) 63; B: The TEMPTATIONS (72) 1
215	158. (Sittin' On) The Dock Of The Bay {7} The REDDINGS (82) 55	F&B: Otis REDDING (68) 1
213	159. Radar Love {2} WHITE LION (89) 59	F&B: GOLDEN EARRING (74) 13
195	160. Spirit In The Sky {3} DOCTOR And The MEDICS (86) 69	F&B: Norman GREENBAUM (70) 3
195	161. Running Scared {2} The FOOLS (81) 50	F&B: Roy ORBISON (61) 1
193	162. The Sun Ain't Gonna Shine Anymore {2} NIELSEN/PEARSON (81) 56	F&B: WALKER BROS. (66) 13
190	163. So Much In Love {2} Timothy B. SCHMIT (82) 59	F&B: The TYMES (63) 1
189	164. On A Carousel {2} GLASS MOON (82) 50	F&B: The HOLLIES (67) 11
182	165. A Certain Girl {2} Warren ZEVON (80) 57	F: Ernie K-DOE (61) 71; B: Warren ZEVON (80) 57
178	166. Mercy, Mercy, Mercy {2} Phoebe SNOW (81) 52	F&B: Don COVAY & The GOODTIMERS (64) 35

'80s Score	Rank. Title {Versions} Cover Act (Yr) Peak	First Version (Yr) Peak; Biggest Version (Yr) Peak
175	167. I Want Candy {2} BOW WOW WOW (82) 62	F&B: The STRANGELOVES (65) 11
170	168. Hot Fun In The Summertime {2} DAYTON (82) 58	F&B: SLY & The FAMILY STONE (69) 2
169	169. Dedicated To The One I Love {6} Bernadette PETERS (81) 65	F: The SHIRELLES (59) 83; B: The SHIRELLES (61) 3
169	170. Happy Together (A Fantasy) {3} CAPTAIN & TENNILLE (80) 53	F&B: The TURTLES (67) 1
162	171. Willie And The Hand Jive {4} George THOROGOOD & The DESTROYERS (85) 63	F&B: Johnny OTIS Show (58) 9
161	172. Funny How Time Slips Away {5} The SPINNERS (83) 67	F: Jimmy ELLEDGE (62) 22; B: Joe HINTON (64) 13
157	173. Land Of A Thousand Dances {6} J. GEILS Band (83) 60	F: Chris KENNER (63) 77; B: Wilson PICKETT (66) 6
150	174. I'll Be Around {2} WHAT IS THIS (85) 62	F&B: The SPINNERS (72) 3
145	175. Don't Look Any Further {2} KANE GANG (88) 64	F: Dennis EDWARDS (84) 72; B: KANE GANG (88) 64
143	176. Under The Boardwalk {3} Bruce WILLIS (87) 59	F&B: The DRIFTERS (64) 4
139	177. Don't Make Me Over {4} Jennifer WARNES (80) 67	F&B: Dionne WARWICK (63) 21
137	178. Gloria {4} The DOORS (84) 71	F: THEM (65) 93; B: SHADOWS OF KNIGHT (66) 10
133	179. Remember (Walking In The Sand) {3} AEROSMITH (80) 67	F&B: The SHANGRI-LAS (64) 5
130	180. Killing Me Softly {2} AL B. SURE! (88) 80	F&B: Roberta FLACK (73) 1
124	181. Dancin' In The Streets {6} Teri DeSARIO With KC (80) 66	F&B: MARTHA & The VANDELLAS (64) 2
123	182. Teach Me Tonight {5} Al JARREAU (82) 70	F&B: DeCASTRO SISTERS (54) 3
122	183. True Love Ways {2} Mickey GILLEY (80) 66	F&B: PETER And GORDON (65) 14
121	184. 96 Tears {3} Garland JEFFREYS (81) 66	F&B: ? (QUESTION MARK) & The MYSTERIANS (66) 1
118	185. School's Out {2} KROKUS (86) 67	F&B: ALICE COOPER (72) 7
113	186. Jimmy Mack {2} Sheena EASTON (86) 65	F&B: MARTHA & The VANDELLAS (67) 10
113	187. Got To Be There {2} Chaka KHAN (83) 67	F&B: Michael JACKSON (71) 4
113	188. Play That Funky Music {2} ROXANNE (88) 63	F&B: WILD CHERRY (76) 1
113	189. I'm Happy Just To Dance With You {2} Anne MURRAY (80) 64	F: The BEATLES (64) 95; B: Anne MURRAY (80) 64
112	190. She Did It {2} Michael DAMIAN (81) 69	F&B: Eric CARMEN (77) 23
112	191. Weatherman Says {2} Jack WAGNER (87) 67	F: Nick JAMESON (86) 95; B: Jack WAGNER (87) 67
110	192. (I Know) I'm Losing You {4} UPTOWN (87) 80	F: The TEMPTATIONS (66) 8; B: RARE EARTH (70) 7
109	193. It's My Party {2} Dave STEWART With Barbara GASKIN (82) 72	F&B: Lesley GORE (63) 1
109	194. Still {2} John SCHNEIDER (81) 69	F&B: The COMMODORES (79) 1
107	195. You Ain't Seen Nothing Yet {2} FIGURES ON A BEACH (89) 67	F&B: BACHMAN-TURNER OVERDRIVE (74) 2
103	196. Sleepwalk {2} Larry CARLTON (82) 74	F&B: SANTO & JOHNNY (59) 1
100	197. Young Blood {3} Bruce WILLIS (87) 68	F&B: The COASTERS (57) 8
99.5	198. All Right Now {4} PEPSI and SHIRLIE (88) 66	F&B: FREE (70) 4
99.2	199. Beast Of Burden {2} Bette MIDLER (84) 71	F&B: ROLLING STONES (78) 8
99.2	200. Bread And Butter {2} Robert JOHN (83) 68	F&B: The NEWBEATS (64) 2
97.0	201. Sherry {2} Robert JOHN (80) 70	F&B: 4 SEASONS (62) 1

Singles: Special Lists

'80s Score	Rank. Title {Versions} Cover Act (Yr) Peak	First Version (Yr) Peak; Biggest Version (Yr) Peak
94.8	202. La La Means I Love You {2} TIERRA (81) 72	F&B: The DELFONICS (68) 4
94.3	203. All Right Now {4} Rod STEWART (85) 72	F&B: FREE (70) 4
92.5	204. Tell Her {3} Kenny LOGGINS (89) 76	F&B: The EXCITERS (63) 4
90.6	205. Breaking Up Is Hard On You (a/k/a Don't Take Ma Bell Away From Me) {6} AMERICAN COMEDY NETWORK (84) 70	F&B: Neil SEDAKA (62) 1
90.1	206. The Monkey Time {2} The TUBES (83) 68	F&B: Major LANCE (63) 8
89.0	207. Darlin' {3} YIPES!! (80) 68	F&B: BEACH BOYS (68) 19
88.1	208. Lola (Live Version) {2} The KINKS (80) 81	F&B: The KINKS (70) 9
85.3	209. I Heard It Through The Grapevine (Part 1) {6} ROGER (81) 79	F: Gladys KNIGHT & The PIPS (67) 2; B: Marvin GAYE (68) 1
84.7	210. Beechwood 4-5789 {2} CARPENTERS (82) 74	F&B: The MARVELETTES (62) 17
84.3	211. Not Fade Away {3} Eric HINE (81) 73	F&B: ROLLING STONES (64) 48
82.2	212. It Hurts To Be In Love {2} Dan HARTMAN (81) 72	F&B: Gene PITNEY (64) 7
79.3	213. Happy Together {3} The NYLONS (87) 75	F&B: The TURTLES (67) 1
78.3	214. Pay You Back With Interest {2} GARY O' (81) 70	F&B: The HOLLIES (67) 28
77.2	215. For The Love Of Money {2} BULLETBOYS (89) 78	F&B: The O'JAYS (74) 9
72.6	216. When We Get Married {3} Larry GRAHAM (80) 76	F&B: The DREAMLOVERS (61) 10
68.0	217. Black Dog {2} NEWCITY ROCKERS (87) 80	F&B: LED ZEPPELIN (72) 15
64.1	218. All I Need To Know (Don't Know Much) {3} Bette MIDLER (83) 77	F: Bill MEDLEY (81) 88; B: Linda RONSTADT (89) 2
62.8	219. Cecilia {2} TIMES TWO (88) 79	F&B: SIMON & GARFUNKEL (70) 4
61.4	220. Piece Of My Heart {3} Sammy HAGAR (82) 73	F: Erma FRANKLIN (67) 62; B: BIG BROTHER And The HOLDING COMPANY (68) 12
58.5	221. Thin Line Between Love And Hate {2} The PRETENDERS (84) 83	F&B: The PERSUADERS (71) 15
57.9	222. So Fine {2} OAK RIDGE BOYS (82) 76	F&B: The FIESTAS (59) 11
56.0	223. The House Of The Rising Sun {4} Dolly PARTON (81) 77	F&B: The ANIMALS (64) 1
54.5	224. Do You Love Me {3} Andy FRASER (84) 82	F&B: The CONTOURS (62) 3
54.2	225. Wynken, Blynken And Nod {2} DOOBIE BROTHERS (81) 76	F&B: SIMON SISTERS (73) 73
51.0	226. Downtown {4} Dolly PARTON (84) 80	F&B: Petula CLARK (65) 1
49.7	227. Twistin' The Night Away {3} Rod STEWART (87) 80	F&B: Sam COOKE (62) 9
49.5	228. I Only Want To Be With You {5} The TOURISTS (80) 83	F: Dusty SPRINGFIELD (64) 12; B: BAY CITY ROLLERS (76) 12
47.6	229. Please Mr. Postman {3} GENTLE PERSUASION (83) 82	F&B: The MARVELETTES (61) 1
46.5	230. Runaway {5} Luis CARDENAS (86) 83	F&B: Del SHANNON (61) 1
45.5	231. You Belong To Me {2} DOOBIE BROTHERS (83) 79	F&B: Carly SIMON (78) 6
44.4	232. You Send Me {5} The MANHATTANS (85) 81	F&B: Sam COOKE (57) 1
43.2	233. Johnny B. Goode {4} Peter TOSH (83) 84	F&B: Chuck BERRY (58) 8
39.9	234. Jimmy Loves Maryann {2} Josie COTTON (84) 82	F&B: LOOKING GLASS (73) 33
35.7	235. That's Life {2} David Lee ROTH (86) 85	F&B: Frank SINATRA (66) 4
35.4	236. Right Back Where We Started From {2} SINITTA (89) 84	F&B: Maxine NIGHTINGALE (76) 2
35.0	237. I Heard It Through The Grapevine {6} CALIFORNIA RAISINS (88) 84	F: Gladys KNIGHT & The PIPS (67) 2; B: Marvin GAYE (68) 1
32.7	238. Sing A Simple Song {2} WEST STREET MOB (82) 89	F&B: SLY & The FAMILY STONE (69) 89
32.7	239. Imagine {2} Tracie SPENCER (89) 85	F&B: John LENNON/PLASTIC ONO BAND (71) 3
31.4	240. Foolish Heart {2} Sharon BRYANT (89) 90	F&B: Steve PERRY (85) 18
31.2	241. Unchained Melody (Live) {8} HEART (81) 83	F: Les BAXTER (55) 2; B: RIGHTEOUS BROTHERS (65) 4
25.6	242. Fly Away {2} Stevie WOODS (82) 84	F&B: Peter ALLEN (81) 55
20.5	243. Montego Bay {2} AMAZULU (87) 90	F&B: Bobby BLOOM (70) 8
19.7	244. When You Walk In The Room {3} Paul CARRACK (88) 90	F: Jackie DeSHANNON (64) 99; B: The SEARCHERS (64) 35
19.2	245. Louie, Louie {5} FAT BOYS (88) 89	F&B: The KINGSMEN (63) 2
17.5	246. Ain't Too Proud To Beg {3} Rick ASTLEY (89) 89	F&B: The TEMPTATIONS (66) 13
13.2	247. Sad Girl {2} GQ (82) 93	F&B: The INTRUDERS (2) (69) 47
13.0	248. 500 Miles {3} The HOOTERS (89) 97	F&B: Bobby BARE (63) 10
11.2	249. Whiter Shade Of Pale {4} HAGAR, SCHON, AARONSON, SHRIEVE (84) 94	F&B: PROCOL HARUM (67) 5
7.8	250. Walk Away Renee {3} SOUTHSIDE JOHNNY & The ASBURY JUKES (86) 98	F&B: LEFT BANKE (66) 5
4.7	251. When I Fall In Love {3} Natalie COLE (88) 95	F: Etta JONES (61) 65; B: The LETTERMEN (62) 7

Ranking the '80s:

The Albums

The Albums

Ranking the Albums

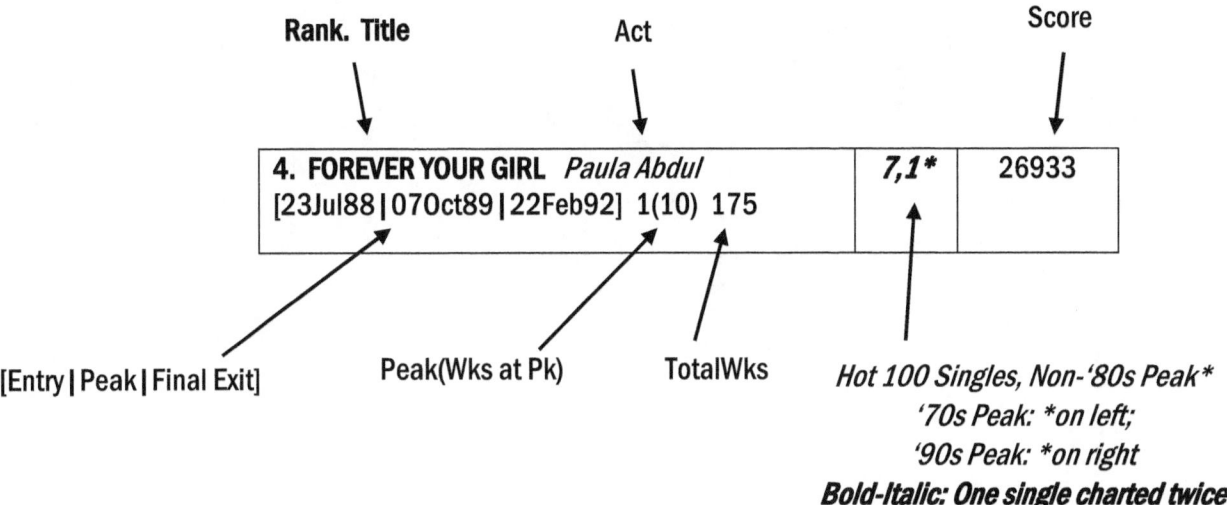

This Ranking covers albums that peaked in the '80s. By far, most of the Hot 100 singles derived from these albums also peaked in the '80s; however, because the life cycle of an album can be long, some derived singles peaked outside the decade. Albums with non-'80s peak singles are noted after the comma and with an asterisk to show when those singles peaked. In some cases, a single from an album was re-released and charted again and is counted twice.

Singles are only credited to the album from which they were originally drawn; thus, while Greatest Hits albums may contain a number of charted singles, they are not credited with singles that appeared on earlier albums.

Ranking the Albums

Rank. TITLE Act [Enter|Peak|Exit] Peak(Wks) TotWks	Hot 100	Score
1. THRILLER Michael Jackson [25Dec82|26Feb83|20Apr85] 1(37) 122	7	25933
2. BORN IN THE U.S.A. Bruce Springsteen [23Jun84|07Jul84|14Feb87] 1(7) 139	7	23827
3. HYSTERIA Def Leppard [22Aug87|23Jul88|03Mar90] 1(6) 133	7	19758
4. FOREVER YOUR GIRL Paula Abdul [23Jul88|07Oct89|22Feb92] 1(10) 175	7,1*	19418
5. WHITNEY HOUSTON Whitney Houston [30Mar85|08Mar86|14May88] 1(14) 162	4	18282
6. APPETITE FOR DESTRUCTION Guns N' Roses [29Aug87|06Aug88|30Jun90] 1(5) 147	4	17811
7. CAN'T SLOW DOWN Lionel Richie [12Nov83|03Dec83|14Feb87] 1(3) 160	5	17509
8. ESCAPE Journey [08Aug81|12Sep81|09Jun84] 1(1) 146	4	17166
9. DIRTY DANCING Soundtrack [19Sep87|14Nov87|15Jul89] 1(18) 96	4	16684
10. FAITH George Michael [21Nov87|16Jan88|15Jul89] 1(12) 87	6	16081
11. HI INFIDELITY REO Speedwagon [13Dec80|21Feb81|13Nov82] 1(15) 101	4	16076
12. 4 Foreigner [25Jul81|22Aug81|19Mar83] 1(10) 81	5	15590
13. DON'T BE CRUEL Bobby Brown [23Jul88|21Jan89|26May90] 1(6) 97	5	15398
14. SPORTS Huey Lewis & The News [08Oct83|30Jun84|16May87] 1(1) 158	5	15377
15. SLIPPERY WHEN WET Bon Jovi [13Sep86|25Oct86|25Jun88] 1(8) 94	3	14801
16. GIRL YOU KNOW IT'S TRUE Milli Vanilli [25Mar89|23Sep89|15Sep90] 1(7) 78	5,1*	14650
17. HANGIN' TOUGH New Kids On The Block [27Aug88|09Sep89|02Mar91] 1(2) 132	5	14550
18. NO JACKET REQUIRED Phil Collins [09Mar85|30Mar85|11Jul87] 1(7) 123	4	14141
19. JANET JACKSON'S RHYTHM NATION 1814 Janet Jackson [07Oct89|28Oct89|11Jan92] 1(4) 108	7,6*	13746
20. PURPLE RAIN (SOUNDTRACK) Prince And The Revolution [14Jul84|04Aug84|23Nov85] 1(24) 72	5	13717
21. SYNCHRONICITY The Police [02Jul83|23Jul83|01Dec84] 1(17) 75	4	13507
22. OFF THE WALL Michael Jackson [01Sep79|16Feb80|06Oct84] 3(3) 169	4,*1	13427
23. THE WALL Pink Floyd [15Dec79|19Jan80|29Sep90] 1(15) 123	2	13311
24. BAD Michael Jackson [26Sep87|26Sep87|20May89] 1(6) 87	7	13217
25. BUSINESS AS USUAL Men At Work [03Jul82|13Nov82|17Mar84] 1(15) 90	2	12719
26. CHRISTOPHER CROSS Christopher Cross [16Feb80|06Sep80|01May82] 6(3) 116	4	12682
27. BELLA DONNA Stevie Nicks [15Aug81|05Sep81|15Feb86] 1(1) 143	4	12532
28. CONTROL Janet Jackson [08Mar86|05Jul86|12Mar88] 1(2) 106	6	12415
29. THE JOSHUA TREE U2 [04Apr87|25Apr87|18Mar89] 1(9) 103	4	12139
30. KENNY ROGERS' GREATEST HITS Kenny Rogers [18Oct80|13Dec80|14Apr84] 1(2) 181	1	12108
31. BROTHERS IN ARMS Dire Straits [08Jun85|31Aug85|11Jun87] 1(9) 97	3	11915
32. WHITESNAKE Whitesnake [18Apr87|13Jun87|24Sep88] 2(10) 76	4	11814
33. PRIVATE DANCER Tina Turner [16Jun84|29Sep84|21Jun86] 3(11) 106	5	11671
34. GLASS HOUSES Billy Joel [22Mar80|14Jun80|14Nov81] 1(6) 73	4	11662
35. WHITNEY Whitney Houston [27Jun87|27Jun87|04Feb89] 1(11) 85	5	11594
36. ELIMINATOR ZZ Top [23Apr83|12Nov83|18Oct86] 9(1) 183	3	11592
37. AGAINST THE WIND Bob Seger & The Silver Bullet Band [15Mar80|03May80|09Apr83] 1(6) 110	4	11583
38. HEART Heart [13Jul85|21Dec85|19Sep87] 1(1) 92	5	11541
39. RECKLESS Bryan Adams [24Nov84|10Aug85|30May87] 1(2) 83	6	11440
40. SONGS FROM THE BIG CHAIR Tears For Fears [06Apr85|13Jul85|25Oct86] 1(5) 82	4	11182
41. PYROMANIA Def Leppard [05Feb83|14May83|25Feb89] 2(2) 116	3	11178
42. CRIMES OF PASSION Pat Benatar [23Aug80|17Jan81|29May82] 2(5) 93	3	11064
43. AN INNOCENT MAN Billy Joel [20Aug83|08Oct83|28Sep85] 4(5) 111	6	11007
44. PARADISE THEATER Styx [31Jan81|04Apr81|27Mar82] 1(3) 61	3	10999
45. LIKE A VIRGIN Madonna [01Dec84|09Feb85|19Sep87] 1(3) 108	4	10496
46. BACK IN BLACK AC/DC [23Aug80|20Dec80|23Mar91] 4(3) 131	2	10402
47. THE RAW & THE COOKED Fine Young Cannibals [11Mar89|03Jun89|19May90] 1(7) 63	5,1*	10227
48. KICK INXS [14Nov87|27Feb88|27May89] 3(4) 81	4	10126
49. TATTOO YOU The Rolling Stones [12Sep81|19Sep81|16Oct82] 1(9) 58	3	9921
50. FLASHDANCE Soundtrack [30Apr83|25Jun83|20Oct84] 1(2) 78	2	9902
51. GET LUCKY Loverboy [14Nov81|01May82|10Mar84] 7(2) 122	2	9866
52. LIONEL RICHIE Lionel Richie [23Oct82|27Nov82|22Jun85] 3(7) 140	3	9861
53. BEAUTY AND THE BEAT Go-Go's [01Aug81|06Mar82|11Dec82] 1(6) 72	2	9856
54. MAKE IT BIG Wham! [10Nov84|02Mar85|17May86] 1(3) 80	4	9851
55. ASIA Asia [03Apr82|15May82|29Oct83] 1(9) 64	2	9795
56. NEW JERSEY Bon Jovi [08Oct88|15Oct88|17Mar90] 1(4) 76	5	9744
57. ...BUT SERIOUSLY Phil Collins [02Dec89|30Dec89|07Sep91] 1(4) 90	5,4*	9734
58. AMERICAN FOOL John Cougar [08May82|11Sep82|10May86] 1(9) 106	3	9588
59. H2O Daryl Hall & John Oates [30Oct82|15Jan83|11Feb84] 3(15) 68	3	9556
60. RAPTURE Anita Baker [19Apr86|21Mar87|15Apr89] 11(2) 157	4	9452
61. DOUBLE FANTASY John Lennon & Yoko Ono [06Dec80|27Dec80|01May82] 1(8) 74	3	9397
62. OPEN UP AND SAY...AHH! Poison [21May88|18Jun88|16Sep89] 2(1) 70	4	9336
63. SHE'S SO UNUSUAL Cyndi Lauper [24Dec83|02Jun84|15Dec85] 4(4) 96	5	9304
64. GUILTY Barbra Streisand [11Oct80|25Oct80|12Sep81] 1(3) 49	4	9281
65. SCARECROW John Mellencamp [14Sep85|16Nov85|14Feb87] 2(3) 75	5	9110
66. ZENYATTA MONDATTA The Police [25Oct80|07Feb81|12May84] 5(6) 153	2	9097
67. INVISIBLE TOUCH Genesis [28Jun86|02Aug86|06Feb88] 3(2) 85	5	9040
68. BREAK OUT The Pointer Sisters [26Nov83|06Oct84|23Nov85] 8(2) 105	6	9038
69. FULL MOON FEVER Tom Petty [13May89|08Jul89|15Sep90] 3(2) 71	4,2*	9020
70. THE GAME Queen [19Jul80|20Sep80|09May81] 1(5) 43	4	9015
71. 1984 (MCMLXXXIV) Van Halen [28Jan84|17Mar84|13Jul85] 2(5) 77	4	9001
72. TRUE BLUE Madonna [19Jul86|16Aug86|06Feb88] 1(5) 82	5	8991
73. GHOST IN THE MACHINE The Police [24Oct81|05Dec81|12May84] 2(6) 109	3	8836
74. HEARTBEAT CITY The Cars [07Apr84|14Jul84|27Jul85] 3(1) 69	5	8802
75. PUMP Aerosmith [30Sep89|04Nov89|18Jan92] 5(3) 110	4,3*	8649
76. DR. FEELGOOD Mötley Crüe [23Sep89|14Oct89|19Oct91] 1(2) 109	5,4*	8619
77. 1999 Prince [20Nov82|28May83|19Oct85] 9(7) 153	4	8588
78. GRACELAND Paul Simon [13Sep86|04Apr87|16Jul88] 3(1) 97	3	8543
79. SKID ROW Skid Row [11Feb89|23Sep89|04Aug90] 6(1) 78	3,1*	8481
80. THE END OF THE INNOCENCE Don Henley [15Jul89|23Sep89|16May92] 8(1) 148	5,3*	8388
81. THE JAZZ SINGER (SOUNDTRACK) Neil Diamond [29Nov80|07Feb81|23Apr83] 3(7) 115	3	8275
82. FREEZE-FRAME The J. Geils Band [14Nov81|06Feb82|12Mar83] 1(4) 70	3	8234
83. TOP GUN Soundtrack [07Jun86|26Jul86|12Mar88] 1(5) 93	4	8187
84. FACE VALUE Phil Collins [14Mar81|04Jul81|25Apr87] 7(4) 164	2	8163
85. FEELS SO RIGHT Alabama [28Mar81|05Sep81|21Apr84] 16(2) 161	2	8135
86. LIKE A PRAYER Madonna [08Apr89|22Apr89|22Sep90] 1(6) 77	5,2*	8128
87. DON'T SAY NO Billy Squier [02May81|05Sep81|03Nov84] 5(3) 111	4	8125
88. DAMN THE TORPEDOES Tom Petty And The Heartbreakers [10Nov79|09Feb80|07Nov81] 2(7) 66	3	8108
89. BEACHES (SOUNDTRACK) Bette Midler [21Jan89|10Jun89|04Jul92] 2(3) 176	1	8076
90. FRONTIERS Journey [19Feb83|12Mar83|29Sep84] 2(9) 85	4	8070
91. COLOUR BY NUMBERS Culture Club [05Nov83|04Feb84|15Dec84] 2(6) 59	4	8038
92. BUILT FOR SPEED Stray Cats [03Jul82|27Nov82|26Nov83] 2(15) 74	2	7957
93. TRACY CHAPMAN Tracy Chapman [30Apr88|27Aug88|24Jun89] 1(1) 61	3	7949
94. FOOTLOOSE Soundtrack [18Feb84|21Apr84|18May85] 1(10) 61	4	7852
95. LICENSED TO ILL Beastie Boys [29Nov86|07Mar87|12Mar88] 1(7) 68	2	7721
96. REPEAT OFFENDER Richard Marx [20May89|02Sep89|18Aug90] 1(1) 66	5,2*	7671
97. TOTO IV Toto [24Apr82|10Jul82|12Nov83] 4(4) 82	5	7593
98. THE WAY IT IS Bruce Hornsby & The Range [21Jun86|07Mar87|07Nov87] 3(4) 73	4	7530
99. BACK IN THE HIGH LIFE Steve Winwood [19Jul86|06Sep86|05Mar88] 3(2) 86	4	7480
100. FORE! Huey Lewis & The News [13Sep86|18Oct86|07Nov87] 1(1) 61	5	7433
101. RAISING HELL Run-D.M.C. [14Jun86|20Sep86|17Oct87] 3(3) 71	3	7388

Ranking the Albums

Rank.	TITLE	Act	[Enter|Peak|Exit]	Peak(Wks)	TotWks	Hot 100	Score
102.	TIFFANY	Tiffany	[26Sep87|23Jan88|11Feb89]	1(2)	69	4	7298
103.	DIANA	Diana Ross	[14Jun80|04Oct80|06Jun81]	2(2)	52	2	7266
104.	STREET SONGS	Rick James	[02May81|01Aug81|25Sep82]	3(2)	74	2	7134
105.	SO	Peter Gabriel	[14Jun86|26Jul86|07Oct89]	2(3)	93	4	7120
106.	EMOTIONAL RESCUE	Rolling Stones	[19Jul80|26Jul80|28Nov81]	1(7)	51	2	7106
107.	THE DREAM OF THE BLUE TURTLES	Sting	[13Jul85|07Sep85|16Aug86]	2(6)	58	4	7101
108.	METAL HEALTH	Quiet Riot	[23Apr83|26Nov83|03Nov84]	1(1)	81	2	7033
109.	OU812	Van Halen	[18Jun88|25Jun88|13May89]	1(4)	48	4	6971
110.	MISTAKEN Identity	Kim Carnes	[02May81|27Jun81|24Apr82]	1(4)	52	3	6959
111.	LOOK WHAT THE CAT DRAGGED IN	Poison	[02Aug86|23May87|02Jul88]	3(2)	101	3	6911
112.	LET IT LOOSE	Gloria Estefan & Miami Sound Machine	[20Jun87|28May88|22Apr89]	6(2)	97	5	6903
113.	MADONNA	Madonna	[03Sep83|20Oct84|05Sep87]	8(3)	168	3	6893
114.	NIGHT SONGS	Cinderella	[19Jul86|07Feb87|14Nov87]	3(3)	70	2	6880
115.	LONG DISTANCE VOYAGER	The Moody Blues	[13Jun81|25Jul81|06Mar82]	1(3)	39	3	6873
116.	MOVING PICTURES	Rush	[07Mar81|28Mar81|19Mar83]	3(3)	68	2	6817
117.	RICHARD MARX	Richard Marx	[20Jun87|03Sep88|04Feb89]	8(2)	86	4	6799
118.	WORKING CLASS DOG	Rick Springfield	[14Mar81|05Sep81|31Jul82]	7(2)	73	3	6754
119.	ON THE RADIO: GREATEST HITS: VOLUMES I & II	Donna Summer	[03Nov79|05Jan80|26Jul80]	1(1)	39	1	6739
120.	SUDDENLY	Billy Ocean	[25Aug84|24Nov84|12Apr86]	9(2)	86	4	6704
121.	MIRAGE	Fleetwood Mac	[17Jul82|07Aug82|21May83]	1(5)	45	3	6695
122.	OUT OF THE BLUE	Debbie Gibson	[05Sep87|27Feb88|13May89]	7(2)	89	5	6673
123.	BEVERLY HILLS COP	Soundtrack	[12Jan85|22Jun85|15Mar86]	1(2)	62	4	6652
124.	5150	Van Halen	[12Apr86|26Apr86|27Sep87]	1(3)	64	3	6636
125.	PRECIOUS TIME	Pat Benatar	[25Jul81|15Aug81|31Jul82]	1(1)	54	2	6605
126.	THE RIVER	Bruce Springsteen	[01Nov80|08Nov80|15Feb86]	1(4)	107	2	6583
127.	CHICAGO 17	Chicago	[02Jun84|26Jan85|12Oct85]	4(1)	72	4	6572
128.	DIFFERENT LIGHT	Bangles	[01Feb86|31Jan87|22Aug87]	2(2)	82	4	6569
129.	DANCING ON THE CEILING	Lionel Richie	[30Aug86|27Sep86|03Oct87]	1(2)	58	6	6483
130.	ALWAYS ON MY MIND	Willie Nelson	[20Mar82|10Jul82|04Feb84]	2(4)	99	2	6455
131.	STORM FRONT	Billy Joel	[04Nov89|16Dec89|25May91]	1(1)	69	5,4*	6443
132.	G N' R LIES	Guns N' Roses	[17Dec88|13May89|27Jan90]	2(3)	53	1	6434
133.	HOLD OUT	Jackson Browne	[19Jul80|13Sep80|04Apr81]	1(1)	38	2	6411
134.	VOLUME 1	Traveling Wilburys	[12Nov88|28Jan89|24Mar90]	3(6)	53	2	6396
135.	PRIVATE EYES	Daryl Hall & John Oates	[26Sep81|13Feb82|20Nov82]	5(3)	61	4	6383
136.	GIVING YOU THE BEST THAT I GOT	Anita Baker	[05Nov88|24Dec88|19Aug89]	1(4)	42	2	6333
137.	COCKTAIL	Soundtrack	[13Aug88|07Jan89|07Oct89]	2(1)	61	4	6310
138.	VOICES	Daryl Hall & John Oates	[16Aug80|13Jun81|10Jul82]	17(1)	100	4	6297
139.	ELECTRIC YOUTH	Debbie Gibson	[11Feb89|11Mar89|27Jan90]	1(5)	51	4	6295
140.	THE DUDE	Quincy Jones	[04Apr81|24Apr82|09Oct82]	10(1)	80	3	6231
141.	THE LONESOME JUBILEE	John Mellencamp	[19Sep87|03Oct87|17Sep88]	6(7)	53	4	6213
142.	DUOTONES	Kenny G	[06Sep86|18Jul87|13Aug88]	6(2)	102	2	6209
143.	MIAMI VICE	TV Soundtrack	[12Oct85|02Nov85|31May86]	1(11)	34	3	6180
144.	BAD ANIMALS	Heart	[13Jun87|01Aug87|21May88]	2(3)	50	4	6177
145.	CHARIOTS OF FIRE (SOUNDTRACK)	Vangelis	[17Oct81|17Apr82|13Nov82]	1(4)	57	1	6161
146.	THIRD STAGE	Boston	[18Oct86|01Nov86|26Sep87]	1(4)	50	3	6153
147.	LET'S DANCE	David Bowie	[30Apr83|25Jun83|26Jan85]	4(1)	68	4	6138
148.	HOTTER THAN JULY	Stevie Wonder	[15Nov80|06Dec80|15Aug81]	3(7)	40	3	6054
149.	ARC OF A DIVER	Steve Winwood	[17Jan81|18Apr81|07Nov81]	3(6)	43	2	5990
150.	TANGO IN THE NIGHT	Fleetwood Mac	[02May87|23May87|28May88]	7(3)	57	5	5973
151.	PERMANENT VACATION	Aerosmith	[19Sep87|21Nov87|31Dec88]	11(4)	67	3	5898
152.	SEVEN AND THE RAGGED TIGER	Duran Duran	[10Dec83|11Feb84|23Feb85]	8(5)	64	3	5895
153.	RATTLE AND HUM (SOUNDTRACK)	U2	[29Oct88|12Nov88|15Jul89]	1(6)	38	4	5894
154.	PROMISE	Sade	[21Dec85|15Feb86|01Nov86]	1(2)	46	2	5873
155.	DIRTY DEEDS DONE DIRT CHEAP	AC/DC	[18Apr81|30May81|01May82]	3(6)	55	0	5829
156.	THE INNOCENT AGE	Dan Fogelberg	[12Sep81|10Oct81|13Nov82]	6(6)	62	4	5782
157.	WELCOME TO THE REAL WORLD	Mr. Mister	[31Aug85|01Mar86|04Oct86]	1(1)	58	3	5753
158.	ROLL WITH IT	Steve Winwood	[09Jul88|20Aug88|13May89]	1(1)	45	4	5737
159.	WHAT'S NEW	Linda Ronstadt & The Nelson Riddle Orchestra	[01Oct83|24Dec83|13Apr85]	3(5)	81	1	5736
160.	LONG COLD WINTER	Cinderella	[23Jul88|10Sep88|21Oct89]	10(5)	66	4	5711
161.	THE FINAL COUNTDOWN	Europe	[01Nov86|28Mar87|23Apr88]	8(2)	78	4	5668
162.	I LOVE ROCK 'N ROLL	Joan Jett & the Blackhearts	[19Dec81|10Apr82|29Jan83]	2(3)	59	2	5608
163.	MORE DIRTY DANCING	Soundtrack	[19Mar88|23Apr88|11Mar89]	3(5)	52	1	5578
164.	AROUND THE WORLD IN A DAY	Prince And The Revolution	[11May85|01Jun85|08Feb86]	1(3)	40	3	5561
165.	AFTERBURNER	ZZ Top	[16Nov85|07Dec85|28Mar87]	4(3)	70	4	5557
166.	IN THE HEAT OF THE NIGHT	Pat Benatar	[20Oct79|15Mar80|27Mar82]	12(2)	122	2	5547
167.	URBAN COWBOY	Soundtrack	[17May80|06Sep80|04Apr81]	3(2)	47	6	5546
168.	OUT OF ORDER	Rod Stewart	[04Jun88|01Apr89|14Oct89]	20(2)	72	4	5537
169.	PHYSICAL	Olivia Newton-John	[31Oct81|12Dec81|27Nov82]	6(7)	57	3	5534
170.	THE BIG CHILL	Soundtrack	[22Oct83|21Jan84|15Nov86]	17(2)	161	0	5525
171.	JUICE	Juice Newton	[07Mar81|20Feb82|27Nov82]	22(3)	86	3	5505
172.	KISSING TO BE CLEVER	Culture Club	[08Jan83|19Mar83|08Sep84]	14(10)	88	3	5493
173.	COMBAT ROCK	The Clash	[12Jun82|22Jan83|06Aug83]	7(5)	61	*3*	5485
174.	CENTERFIELD	John Fogerty	[26Jan85|23Mar85|11Jan86]	1(1)	51	3	5442
175.	THE BROADWAY ALBUM	Barbra Streisand	[23Nov85|25Jan86|11Jul87]	1(3)	50	1	5422
176.	PHOENIX	Dan Fogelberg	[08Dec79|08Mar80|30Aug80]	3(2)	39	2	5420
177.	EMERGENCY	Kool & The Gang	[15Dec84|12Oct85|10May86]	13(1)	74	4	5376
178.	KENNY	Kenny Rogers	[29Sep79|02Feb80|06Dec80]	5(2)	53	2,*1	5370
179.	WINELIGHT	Grover Washington Jr.	[15Nov80|11Apr81|07Nov81]	5(7)	52	1	5367
180.	TUNNEL OF LOVE	Bruce Springsteen	[24Oct87|07Nov87|27Aug88]	1(1)	45	3	5349
181.	HOOKED ON CLASSICS	Royal Philharmonic Orchestra	[14Nov81|16Jan82|26Feb83]	4(6)	68	1	5342
182.	UH-HUH	John Mellencamp	[05Nov83|28Jan84|24May86]	9(1)	66	3	5328
183.	JANE FONDA'S WORKOUT RECORD	Jane Fonda	[29May82|26Mar83|08Sep84]	15(4)	120	0	5318
184.	ABACAB	Genesis	[17Oct81|14Nov81|16Aug86]	7(1)	64	3	5305
185.	REBEL YELL	Billy Idol	[03Dec83|14Jul84|22Jun85]	6(3)	82	4	5290
186.	LIKE A ROCK	Bob Seger & The Silver Bullet Band	[19Apr86|24May86|20Jun87]	3(4)	62	4	5287
187.	CUTS BOTH WAYS	Gloria Estefan	[29Jul89|09Sep89|17Nov90]	8(2)	69	5,3*	5235
188.	INTRODUCING THE HARDLINE ACCORDING TO TERENCE TRENT D'ARBY	Terence Trent D'Arby	[24Oct87|07May88|10Dec88]	4(2)	60	4	5202
189.	A MOMENTARY LAPSE OF REASON	Pink Floyd	[26Sep87|24Oct87|15Oct88]	3(1)	56	1	5195
190.	RIO	Duran Duran	[05Jun82|12Mar83|02Feb85]	6(7)	129	2	5168
191.	WHENEVER YOU NEED SOMEBODY	Rick Astley	[23Jan88|12Mar88|11Mar89]	10(2)	60	3	5157
192.	JUST ONE NIGHT	Eric Clapton	[03May80|21Jun80|29Nov80]	2(6)	31	2	5135
193.	BREAKIN' AWAY	Al Jarreau	[22Aug81|10Oct81|06Aug83]	9(3)	103	3	5128
194.	LOC-ED AFTER DARK	Tone-Loc	[18Feb89|15Apr89|24Feb90]	1(1)	42	2	5062
195.	THE ONE THAT YOU LOVE	Air Supply	[13Jun81|01Aug81|31Jul82]	10(4)	60	3	5042

Ranking the Albums

Rank. TITLE *Act* [Enter|Peak|Exit] Peak(Wks) TotWks	Hot 100	Score
196. ...AND JUSTICE FOR ALL *Metallica* [24Sep88|08Oct88|21Apr90] 6(2) 83	1	4977
197. RIPTIDE *Robert Palmer* [23Nov85|17May86|08Aug87] 8(1) 90	4	4955
198. VIVID *Living Colour* [03Sep88|06May89|10Feb90] 6(2) 76	3	4950
199. DIRTY ROTTEN FILTHY STINKING RICH *Warrant* [04Mar89|16Sep89|26May90] 10(3) 65	4,1*	4909
200. NINE TONIGHT *Bob Seger & The Silver Bullet Band* [26Sep81|17Oct81|27Sep86] 3(4) 70	2	4899
201. JODY WATLEY *Jody Watley* [21Mar87|23May87|13Aug88] 10(1) 74	5	4882
202. MAD LOVE *Linda Ronstadt* [15Mar80|22Mar80|13Sep80] 3(4) 36	3	4864
203. BIG BAM BOOM *Daryl Hall & John Oates* [27Oct84|01Dec84|12Oct85] 5(2) 51	4	4851
204. EXPOSURE *Exposé* [21Feb87|20Feb88|16Jul88] 16(2) 74	4	4850
205. STEEL WHEELS *The Rolling Stones* [16Sep89|07Oct89|19May90] 3(4) 36	3,1*	4806
206. EMOTIONS IN MOTION *Billy Squier* [07Aug82|18Sep82|20Oct84] 5(8) 50	3	4779
207. CARGO *Men At Work* [07May83|21May83|07Apr84] 3(5) 49	3	4751
208. MIDNIGHT MADNESS *Night Ranger* [19Nov83|16Jun84|09Mar85] 15(10) 69	3	4751
209. GIRLS, GIRLS, GIRLS *Mötley Crüe* [13Jun87|27Jun87|23Apr88] 2(1) 46	2	4702
210. NOW AND ZEN *Robert Plant* [12Mar88|21May88|04Feb89] 6(1) 48	2	4702
211. SHOUT AT THE DEVIL *Mötley Crüe* [15Oct83|31Mar84|15Aug87] 17(2) 111	3	4698
212. MOUNTAIN MUSIC *Alabama* [13Mar82|08May82|12May84] 14(2) 114	2	4690
213. DIAMOND LIFE *Sade* [23Feb85|01Jun85|06Sep86] 5(2) 81	2	4672
214. SOMETHING SPECIAL *Kool & The Gang* [17Oct81|28Nov81|22Jan83] 12(2) 67	3	4669
215. LOVE ZONE *Billy Ocean* [17May86|28Jun86|11Apr87] 6(7) 48	3	4661
216. HE'S THE D.J., I'M THE RAPPER *D.J. Jazzy Jeff & The Fresh Prince* [23Apr88|24Sep88|06May89] 4(1) 55	2	4651
217. LOVE AT FIRST STING *Scorpions* [17Mar84|16Jun84|28Sep85] 6(2) 63	2	4637
218. SHOOTING RUBBERBANDS AT THE STARS *Edie Brickell & The New Bohemians* [24Sep88|18Feb89|30Sep89] 4(2) 54	2	4620
219. BATMAN (SOUNDTRACK) *Prince* [08Jul89|22Jul89|24Feb90] 1(6) 34	3	4618
220. RAISED ON RADIO *Journey* [10May86|31May86|15Aug87] 4(2) 67	5	4605
221. KEEP ON MOVIN' *Soul II Soul* [08Jul89|16Sep89|23Jun90] 14(4) 51	2	4558
222. IN SQUARE CIRCLE *Stevie Wonder* [19Oct85|23Nov85|27Sep86] 5(2) 50	4	4553
223. BIGGER AND DEFFER *LL Cool J* [20Jun87|29Aug87|18Jun88] 3(1) 53	2	4537
224. THE UNFORGETTABLE FIRE *U2* [20Oct84|24Nov84|19Sep87] 12(3) 132	1	4535
225. AUTOAMERICAN *Blondie* [13Dec80|21Feb81|01Aug81] 7(5) 34	3	4533
226. PRIDE *White Lion* [26Sep87|07May88|13May89] 11(1) 86	3	4513
227. WAR *U2* [19Mar83|07May83|19Sep87] 12(1) 179	1	4509
228. THE DOORS GREATEST HITS *Doors* [01Nov80|06Dec80|03Aug91] 17(2) 99	0	4507
229. NIGHT AND DAY *Joe Jackson* [17Jul82|27Nov82|13Aug83] 4(6) 57	2	4503
230. HELLO, I MUST BE GOING! *Phil Collins* [27Nov82|05Feb83|11Apr87] 8(5) 141	3	4492
231. BLIZZARD OF OZZ *Ozzy Osbourne* [18Apr81|08Aug81|09Aug91] 21(1) 104	0	4478
232. REACH THE BEACH *The Fixx* [28May83|15Oct83|20Oct84] 8(2) 54	3	4454
233. 90125 *Yes* [03Dec83|21Jan84|01Dec84] 5(4) 53	3	4449
234. STONE COLD RHYMIN' *Young M.C.* [23Sep89|09Dec89|18Aug90] 9(1) 48	3,2*	4442
235. GIVE ME THE NIGHT *George Benson* [09Aug80|04Oct80|25Apr81] 3(2) 38	3	4441
236. KNEE DEEP IN THE HOOPLA *Starship* [05Oct85|01Mar86|13Sep86] 7(3) 50	4	4436
237. MAKE IT LAST FOREVER *Keith Sweat* [09Jan88|30Apr88|15Jul89] 15(1) 67	3	4409
238. BEE GEES GREATEST *Bee Gees* [17Nov79|12Jan80|21Jun80] 1(1) 32	0	4406
239. THE WILD HEART *Stevie Nicks* [02Jul83|23Jul83|23Jun84] 5(7) 52	3	4401
240. HEART OF STONE *Cher* [22Jul89|14Oct89|21Jul90] 10(2) 53	3,1*	4398
241. FOR THOSE ABOUT TO ROCK (WE SALUTE YOU) *AC/DC* [12Dec81|26Dec81|03Jul82] 1(3) 30	1	4377
242. LOST IN LOVE *Air Supply* [17May80|27Sep80|05Jun82] 22(3) 104	3	4355
243. THE BRIDGE *Billy Joel* [16Aug86|20Sep86|04Jul87] 7(4) 47	4	4352
244. LOVERBOY *Loverboy* [31Jan81|23May81|26Mar83] 13(2) 105	2	4338
245. FAME *Soundtrack* [07Jun80|06Sep80|31Jul82] 7(2) 82	2	4319
246. OUT OF THE CELLAR *Ratt* [24Mar84|04Aug84|13Apr85] 7(4) 56	2	4307
247. HEART BREAK *New Edition* [09Jul88|29Oct88|17Jun89] 12(1) 50	3	4305
248. THE TURN OF A FRIENDLY CARD *Alan Parsons Project* [15Nov80|21Feb81|19Dec81] 13(2) 58	3	4304
249. TELL IT TO MY HEART *Taylor Dayne* [30Jan88|29Oct88|20May89] 21(1) 69	4	4302
250. SILHOUETTE *Kenny G* [22Oct88|03Dec88|18Nov89] 8(1) 57	2	4280
251. LA BAMBA (SOUNDTRACK) *Los Lobos* [25Jul87|12Sep87|21Aug88] 1(2) 44	2	4269
252. ROCK ME TONIGHT *Freddie Jackson* [25May85|07Dec85|26Jul86] 10(1) 62	3	4250
253. BUILDING THE PERFECT BEAST *Don Henley* [15Dec84|16Mar85|22Feb86] 13(5) 63	4	4238
254. HARD PROMISES *Tom Petty And The Heartbreakers* [23May81|25Jul81|19Dec81] 5(2) 31	2	4235
255. WINGER *Winger* [17Sep88|11Feb89|02Dec89] 21(3) 64	3	4224
256. SPANISH FLY *Lisa Lisa And Cult Jam* [09May87|20Jun87|02Apr88] 7(3) 48	3	4194
257. DIARY OF A MADMAN *Ozzy Osbourne* [21Nov81|19Dec81|09Apr83] 16(4) 73	0	4170
258. CELEBRATE! *Kool & The Gang* [18Oct80|07Mar81|15Aug81] 10(2) 44	2	4159
259. PLAY DEEP *The Outfield* [02Nov85|14Jun86|14Mar87] 9(1) 66	3	4156
260. PRETENDERS *The Pretenders* [26Jan80|07Jun80|18Jul81] 9(2) 78	2	4145
261. GET NERVOUS *Pat Benatar* [20Nov82|15Jan83|01Oct83] 4(5) 46	3	4128
262. GREATEST HITS VOL. I & II *Billy Joel* [20Jul85|28Sep85|09Feb91] 6(2) 65	2	4119
263. CUTS LIKE A KNIFE *Bryan Adams* [19Feb83|25Jun83|30Nov85] 8(3) 89	3	4109
264. XANADU (SOUNDTRACK) *Olivia Newton-John/Electric Light Orchestra* [12Jul80|04Oct80|14Mar81] 4(3) 36	5	4087
265. DARE *The Human League* [27Feb82|10Jul82|13Nov82] 3(3) 38	1	4081
266. HEAVEN ON EARTH *Belinda Carlisle* [24Oct87|27Feb88|08Oct88] 13(2) 51	4	4078
267. NERVOUS NIGHT *Hooters* [25May85|22Mar86|18Oct86] 12(1) 74	4	4059
268. EYE OF THE TIGER *Survivor* [26Jun82|14Aug82|02Apr83] 2(4) 41	3	4056
269. ZEBOP! *Santana* [18Apr81|13Jun81|21Nov81] 9(4) 32	2	4051
270. PRIMITIVE LOVE *Miami Sound Machine* [23Nov85|04Oct86|25Apr87] 21(1) 75	4	4048
271. NEW EDITION *New Edition* [13Oct84|23Feb85|19Oct85] 6(2) 54	3	4038
272. OLIVIA'S GREATEST HITS, VOL. 2 *Olivia Newton-John* [09Oct82|13Nov82|26May84] 16(4) 86	2	4032
273. THE PHANTOM OF THE OPERA *Original London Cast Recording* [23May87|19Mar88|03Jul93] 33(1) 255	0	4006
274. WINNER IN YOU *Patti LaBelle* [24May86|19Jul86|13Dec86] 1(1) 30	2	4001
275. KILROY WAS HERE *Styx* [19Mar83|30Apr83|05Nov83] 3(2) 34	3	3996
276. NO PARKING ON THE DANCE FLOOR *Midnight Star* [30Jul83|14Jan84|25May85] 27(2) 96	3	3986
277. SHAKE IT UP *The Cars* [28Nov81|19Dec81|04Sep82] 9(4) 41	3	3973
278. THE POWER STATION *The Power Station* [13Apr85|27Jul85|08Feb86] 6(2) 44	3	3972
279. EMPTY GLASS *Pete Townshend* [17May80|12Jul80|06Dec80] 5(3) 30	3	3960
280. SHARE YOUR LOVE *Kenny Rogers* [11Jul81|15Aug81|19Jun82] 6(1) 57	4	3955
281. QUARTERFLASH *Quarterflash* [31Oct81|27Feb82|23Oct82] 8(3) 52	3	3953
282. WHO'S ZOOMIN' WHO? *Aretha Franklin* [27Jul85|30Nov85|12Jul86] 13(1) 51	4	3951
283. TUG OF WAR *Paul McCartney* [15May82|29May82|27Nov82] 1(3) 29	3	3951
284. AGENT PROVOCATEUR *Foreigner* [05Jan85|27Apr85|09Nov85] 4(3) 45	4	3946
285. ALWAYS & FOREVER *Randy Travis* [30May87|11Jul87|13May89] 19(1) 103	0	3934
286. SUCCESS HASN'T SPOILED ME YET *Rick Springfield* [27Mar82|22May82|20Nov82] 2(3) 35	3	3922
287. ...TWICE SHY *Great White* [06May89|08Jul89|14Apr90] 9(1) 50	3,1*	3912
288. BRUCE SPRINGSTEEN & THE E STREET BAND LIVE 1975-1985 *Bruce Springsteen* [29Nov86|29Nov86|23May87] 1(7) 26	2	3897
289. THEATRE OF PAIN *Mötley Crüe* [13Jul85|17Aug85|29Aug87] 6(1) 72	2	3895
290. ...NOTHING LIKE THE SUN *Sting* [31Oct87|21Nov87|22Oct88] 9(4) 52	3	3861
291. THE DISTANCE *Bob Seger & The Silver Bullet Band* [15Jan83|19Feb83|08Oct83] 5(6) 39	3	3858
292. GENESIS *Genesis* [29Oct83|03Dec83|28Mar87] 9(1) 50	4	3843
293. GREEN *R.E.M.* [26Nov88|18Feb89|26Aug89] 12(3) 40	2	3841
294. MEMORIES *Barbra Streisand* [12Dec81|26Dec81|05Apr86] 10(6) 104	2	3838

Ranking the Albums

Rank. TITLE Act [Enter|Peak|Exit] Peak(Wks) TotWks	Hot 100	Score
295. LAP OF LUXURY *Cheap Trick* [07May88|24Sep88|25Mar89] 16(1) 47	4	3824
296. GREATEST HITS *Air Supply* [20Aug83|19Nov83|31Aug85] 7(1) 51	1	3820
297. THE ROSE (SOUNDTRACK) *Bette Midler* [22Dec79|23Feb80|25Oct80] 12(3) 45	2	3819
298. ABRACADABRA *The Steve Miller Band* [26Jun82|18Sep82|05Feb83] 3(6) 33	3	3819
299. EYES THAT SEE IN THE DARK *Kenny Rogers* [24Sep83|12Nov83|09Jun84] 6(4) 38	3	3792
300. KEEP THE FIRE *Kenny Loggins* [20Oct79|08Mar80|25Oct80] 16(2) 43	2	3776
301. THE NYLON CURTAIN *Billy Joel* [16Oct82|20Nov82|11Jun83] 7(4) 35	3	3773
302. ROCK 'N SOUL PART 1 *Daryl Hall & John Oates* [19Nov83|28Jan84|15Sep84] 7(2) 44	2	3750
303. UNDER A BLOOD RED SKY *U2* [10Dec83|28Jan84|19Sep87] 28(3) 180	1	3718
304. DAYLIGHT AGAIN *Crosby, Stills & Nash* [17Jul82|14Aug82|23Apr83] 8(5) 41	3	3714
305. MIDDLE MAN *Boz Scaggs* [19Apr80|14Jun80|29Nov80] 8(2) 33	2	3683
306. 7 WISHES *Night Ranger* [08Jun85|27Jul85|12Apr86] 10(3) 45	3	3665
307. WORD UP! *Cameo* [27Sep86|20Dec86|03Oct87] 8(1) 54	3	3662
308. DISINTEGRATION *The Cure* [20May89|10Jun89|21Jul90] 12(3) 55	4,1*	3656
309. RAISE! *Earth, Wind & Fire* [14Nov81|28Nov81|01May82] 5(8) 25	3	3652
310. THE EMPIRE STRIKES BACK [17May80|12Jul80|22Nov80] 4(4) 28	0	3643
311. GAUCHO *Steely Dan* [06Dec80|17Jan81|08Aug81] 9(3) 36	2	3629
312. WHEELS ARE TURNIN' *REO Speedwagon* [24Nov84|16Mar85|26Oct85] 7(1) 49	4	3592
313. ONE STEP CLOSER *Doobie Brothers* [11Oct80|25Oct80|18Apr81] 3(3) 28	3	3578
314. THE WOMAN IN RED (SOUNDTRACK) *Stevie Wonder* [22Sep84|10Nov84|22Jun85] 4(3) 40	2	3577
315. SIMPLE PLEASURES *Bobby McFerrin* [23Apr88|08Oct88|06May89] 5(3) 55	1	3562
316. THE WHISPERS *The Whispers* [05Jan80|12Apr80|30Aug80] 6(2) 35	2	3555
317. SPEAKING IN TONGUES *Talking Heads* [25Jun83|29Oct83|09Jun84] 15(2) 51	2	3554
318. EDDIE & THE CRUISERS *John Cafferty & The Beaver Brown Band* [15Oct83|06Oct84|13Jul85] 9(5) 62	4	3552
319. LEARNING TO CRAWL *The Pretenders* [04Feb84|25Feb84|17Nov84] 5(4) 42	3	3545
320. IN THE POCKET *Commodores* [11Jul81|29Aug81|10Apr82] 13(2) 40	3	3539
321. GIVE ME THE REASON *Luther Vandross* [18Oct86|21Feb87|17Oct87] 14(1) 53	3	3534
322. WILD-EYED SOUTHERN BOYS *.38 Special* [21Feb81|23May81|25Sep82] 18(2) 57	2	3509
323. TRUE COLORS *Cyndi Lauper* [04Oct86|15Nov86|01Aug87] 4(2) 44	4	3490
324. INTO THE GAP *Thompson Twins* [17Mar84|05May84|16Mar85] 10(2) 53	4	3487
325. STAY WITH ME TONIGHT *Jeffrey Osborne* [06Aug83|17Mar84|13Apr85] 25(1) 89	3	3485
326. EAT 'EM AND SMILE *David Lee Roth* [26Jul86|30Aug86|28Mar87] 4(2) 36	3	3485
327. STRONGER THAN PRIDE *Sade* [04Jun88|02Jul88|08Apr89] 7(2) 45	1	3456
328. SAVAGE AMUSEMENT *Scorpions* [07May88|04Jun88|25Feb89] 5(1) 43	1	3456
329. DEPARTURE *Journey* [22Mar80|26Apr80|12Dec81] 8(2) 57	3	3453
330. DIESEL AND DUST *Midnight Oil* [13Feb88|04Jun88|25Feb89] 21(3) 55	2	3447
331. NEW KIDS ON THE BLOCK *New Kids On The Block* [05Aug89|30Dec89|09Feb91] 25(3) 80	1	3437
332. DUKE *Genesis* [26Apr80|12Jul80|22Nov80] 11(2) 31	2	3426
333. A FLOCK OF SEAGULLS *A Flock Of Seagulls* [22May82|23Oct82|30Apr83] 10(3) 50	2	3417
334. CRUSHIN' *Fat Boys* [13Jun87|12Sep87|14May88] 8(3) 49	1	3405
335. ONCE UPON A TIME *Simple Minds* [09Nov85|01Mar86|23Aug86] 10(5) 42	3	3400
336. TUFF ENUFF *Fabulous Thunderbirds* [15Mar86|05Jul86|14Mar87] 13(3) 53	2	3398
337. DIVER DOWN *Van Halen* [08May82|12Jun82|09Jul83] 3(3) 65	2	3394
338. LITTLE CREATURES *Talking Heads* [06Jul85|27Jul85|20Dec86] 20(3) 77	1	3390
339. VITAL SIGNS *Survivor* [29Sep84|13Jul85|23Nov85] 16(3) 61	4	3378
340. PIRATES *Rickie Lee Jones* [08Aug81|26Sep81|20Feb82] 5(2) 29	1	3376
341. EYE IN THE SKY *Alan Parsons Project* [19Jun82|09Oct82|26Mar83] 7(6) 41	2	3372
342. LIVING IN OZ *Rick Springfield* [30Apr83|18Jun83|26May84] 12(3) 57	3	3368
343. IN EFFECT MODE *Al B. Sure!* [14May88|10Sep88|20May89] 20(1) 54	3	3363
344. BE YOURSELF TONIGHT *Eurythmics* [25May85|20Jul85|29Mar86] 9(1) 45	4	3357
345. KEEP IT UP *Loverboy* [02Jul83|23Jul83|24Mar84] 7(5) 39	2	3349
346. VALOTTE *Julian Lennon* [10Nov84|23Feb85|21Sep85] 17(2) 46	4	3339
347. 1100 BEL AIR PLACE *Julio Iglesias* [01Sep84|27Oct84|20Apr85] 5(2) 34	0	3337
348. SIGN 'O' THE TIMES *Prince* [18Apr87|09May87|23Aug88] 6(2) 54	5	3322
349. TOUCH *Eurythmics* [04Feb84|07Apr84|13Oct84] 7(3) 37	3	3314
350. FACE DANCES *The Who* [04Apr81|25Apr81|15Aug81] 4(4) 20	2	3313
351. LIGHT UP THE NIGHT *Brothers Johnson* [08Mar80|03May80|27Sep80] 5(2) 30	2	3294
352. DECEMBER *George Winston* [12Mar83|21Jan84|19Jan91] 54(1) 178	0	3285
353. LITA *Lita Ford* [20Feb88|28May88|16Sep89] 29(4) 62	4	3280
354. STAY HUNGRY *Twisted Sister* [07Jul84|15Sep84|10Aug85] 15(3) 51	2	3275
355. ICE CREAM CASTLE *The Time* [28Jul84|13Oct84|24Aug85] 24(3) 57	2	3275
356. THE CLOSER YOU GET *Alabama* [26Mar83|30Apr83|21Jul84] 10(1) 70	2	3270
357. THE JETS *The Jets* [05Apr86|19Jul86|05Sep87] 21(2) 70	5	3252
358. JOURNEY'S GREATEST HITS *Journey* [03Dec88|11Feb89|27Oct90] 10(2) 92	0	3252
359. WOMEN AND CHILDREN FIRST *Van Halen* [19Apr80|17May80|15Nov80] 6(5) 31	1	3216
360. STREET TALK *Steve Perry* [28Apr84|09Jun84|15Jun85] 12(4) 60	4	3214
361. SWEET SENSATION *Stephanie Mills* [03May80|21Jun80|28Feb81] 16(2) 44	2	3213
362. BEING WITH YOU *Smokey Robinson* [14Mar81|06Jun81|19Sep81] 10(2) 28	2	3208
363. BREAK EVERY RULE *Tina Turner* [27Sep86|08Nov86|19Sep87] 4(1) 52	4	3207
364. THE GREAT RADIO CONTROVERSY *Tesla* [18Feb89|11Mar89|30Jun90] 18(1) 67	2,2*	3201
365. CAPTURED *Journey* [21Feb81|14Mar81|12Jun82] 9(4) 69	1	3197
366. VOLUME ONE *The Honeydrippers* [20Oct84|01Dec84|18May85] 4(2) 31	2	3186
367. SPRING SESSION M *Missing Persons* [30Oct82|26Feb83|19May84] 17(6) 40	2	3181
368. STRONG PERSUADER *Robert Cray Band* [20Dec86|04Apr87|21Nov87] 13(2) 49	2	3174
369. EAGLES LIVE *Eagles* [29Nov80|20Dec80|23May81] 6(4) 26	1	3172
370. HORIZON *Eddie Rabbitt* [12Jul80|21Mar81|18Jul81] 19(2) 54	2	3151
371. ANNE MURRAY'S GREATEST HITS *Anne Murray* [04Oct80|06Dec80|21Jan84] 16(5) 64	0	3147
372. ALDO NOVA *Aldo Nova* [20Feb82|22May82|30Oct82] 8(6) 37	2	3135
373. LET'S GET SERIOUS *Jermaine Jackson* [12Apr80|12Jul80|25Oct80] 6(3) 29	2	3133
374. DREAM INTO ACTION *Howard Jones* [20Apr85|29Jun85|22Feb86] 10(1) 45	3	3115
375. LET'S GET IT STARTED *MC Hammer* [03Dec88|22Jul89|09Jun90] 30(1) 80	0	3114
376. SCENES FROM THE SOUTHSIDE *Bruce Hornsby & The Range* [21May88|09Jul88|19Nov88] 5(1) 27	2	3112
377. GUY *Guy* [30Jul88|20May89|25Nov89] 27(5) 70	1	3108
378. MOUTH TO MOUTH *Lipps Inc.* [19Apr80|31May80|11Oct80] 5(5) 26	2	3106
379. THE PRINCIPLE OF MOMENTS *Robert Plant* [30Jul83|08Oct83|28Apr84] 8(1) 40	2	3102
380. PARADE: MUSIC FROM THE MOTION PICTURE UNDER THE CHERRY MOON (SOUNDTRACK) *Prince And The Revolution* [19Apr86|03May86|25Oct86] 3(3) 28	3	3097
381. HIGH 'N' DRY *Def Leppard* [08Aug81|17Oct81|17Mar84] 38(1) 106	0	3096
382. BLACKOUT *Scorpions* [27Mar82|29May82|06Oct84] 10(2) 74	1	3091
383. INTO THE FIRE *Bryan Adams* [18Apr87|16May87|28Nov87] 7(1) 33	3	3091
384. THE OTHER SIDE OF LIFE *Moody Blues* [17May86|28Jun86|19Sep87] 9(4) 42	2	3087
385. CLOUD NINE *George Harrison* [21Nov87|23Jan88|18Jun88] 8(1) 31	2	3080
386. LIFE, LOVE & PAIN *Club Nouveau* [20Dec86|25Apr87|17Oct87] 6(1) 44	2	3075
387. CROWDED HOUSE *Crowded House* [30Aug86|16May87|16Jan88] 12(1) 58	3	3074
388. PERMANENT WAVES *Rush* [02Feb80|08Mar80|23May81] 4(3) 36	1	3066
389. PLEASE *Pet Shop Boys* [19Apr86|21Jun86|15Nov86] 7(1) 31	4	3061
390. FREEDOM AT POINT ZERO *Jefferson Starship* [01Dec79|02Feb80|07Jun80] 10(1) 28	2	3057
391. SWEET DREAMS (ARE MADE OF THIS) *Eurythmics* [28May83|27Aug83|07Jul84] 15(3) 59	2	3056
392. HEROES *Commodores* [28Jun80|12Jul80|07Feb81] 7(3) 33	2	3045

Ranking the Albums

Rank. TITLE Act [Enter|Peak|Exit] Peak(Wks) TotWks	Hot 100	Score
393. READY FOR THE WORLD *Ready For The World* [22Jun85|12Oct85|17May86] 17(5) 48	2	3043
394. CAN'T HOLD BACK *Eddie Money* [30Aug86|06Dec86|03Oct87] 20(3) 58	4	3034
395. HEAVY NOVA *Robert Palmer* [16Jul88|27Aug88|13May89] 13(3) 44	3	3027
396. MYSTERY GIRL *Roy Orbison* [18Feb89|08Apr89|19Aug89] 5(2) 27	1	3024
397. TRIUMPH *The Jacksons* [18Oct80|08Nov80|02May81] 10(4) 29	4	3024
398. FANCY FREE *The Oak Ridge Boys* [13Jun81|15Aug81|08May82] 14(1) 48	1	3024
399. FREEDOM OF CHOICE *Devo* [14Jun80|29Nov80|30May81] 22(2) 51	1	3023
400. LISTEN LIKE THIEVES *INXS* [02Nov85|12Apr86|15Nov86] 11(1) 55	3	3022
401. LONG AFTER DARK *Tom Petty And The Heartbreakers* [20Nov82|22Jan83|25Jun83] 9(3) 32	2	3020
402. SKYSCRAPER *David Lee Roth* [13Feb88|27Feb88|13Aug88] 6(6) 27	2	3013
403. LOVIN' EVERY MINUTE OF IT *Loverboy* [14Sep85|19Oct85|12Jul86] 13(5) 44	4	3011
404. DURAN DURAN *Duran Duran* [19Feb83|20Aug83|13Oct84] 10(1) 87	1	3007
405. TONIGHT I'M YOURS *Rod Stewart* [21Nov81|19Dec81|19Jun82] 11(4) 31	3	3006
406. PANORAMA *The Cars* [06Sep80|20Sep80|14Mar81] 5(4) 28	1	2985
407. PEBBLES *Pebbles* [13Feb88|11Jun88|29Oct88] 14(1) 38	2	2973
408. WHIPLASH SMILE *Billy Idol* [08Nov86|13Dec86|26Sep87] 6(1) 47	2	2968
409. TP *Teddy Pendergrass* [23Aug80|04Oct80|11Apr81] 14(3) 34	2	2966
410. SCREAMING FOR VENGEANCE *Judas Priest* [24Jul82|30Oct82|23Jul83] 17(8) 53	1	2965
411. 9 TO 5 AND ODD JOBS *Dolly Parton* [06Dec80|21Mar81|25Jul81] 11(2) 34	3	2959
412. PICTURES AT ELEVEN *Robert Plant* [17Jul82|07Aug82|17Mar84] 5(6) 53	2	2950
413. SOUTHERN ACCENTS *Tom Petty And The Heartbreakers* [13Apr85|11May85|16Nov85] 7(2) 32	3	2944
414. FASTER THAN THE SPEED OF NIGHT *Bonnie Tyler* [06Aug83|05Nov83|10Mar84] 4(1) 32	2	2933
415. HUNTING HIGH AND LOW *a-ha* [20Jul85|02Nov85|07Jun86] 15(2) 47	2	2928
416. IT TAKES TWO *Rob Base & D.J. E-Z Rock* [08Oct88|26Nov88|21Apr90] 31(1) 81	2	2923
417. WILLIE NELSON'S GREATEST HITS (& SOME THAT WILL BE) *Willie Nelson* [19Sep81|31Oct81|25Jun83] 27(1) 93	0	2922
418. KARYN WHITE *Karyn White* [15Oct88|11Mar89|21Oct89] 19(1) 54	3	2905
419. "...FAMOUS LAST WORDS..." *Supertramp* [13Nov82|27Nov82|21May83] 5(7) 28	1	2901
420. PRETTY IN PINK *Soundtrack* [01Mar86|03May86|30Aug86] 5(4) 27	2	2899
421. 7800 DEGREES FAHRENHEIT *Bon Jovi* [18May85|08Jun85|03Oct87] 37(3) 104	2	2896
422. THE GAP BAND III *The Gap Band* [27Dec80|14Mar81|05Sep81] 16(2) 37	2	2856
423. JUST LIKE THE FIRST TIME *Freddie Jackson* [15Nov86|14Mar87|31Oct87] 23(2) 51	3	2850
424. WHY DO FOOLS FALL IN LOVE *Diana Ross* [07Nov81|05Dec81|19Jun82] 15(6) 33	3	2846
425. THE ULTIMATE SIN *Ozzy Osbourne* [15Feb86|05Apr86|08Nov86] 6(2) 39	1	2840
426. BIG TYME *Heavy D & The Boyz* [01Jul89|09Sep89|16Jun90] 19(3) 51	0	2838
427. GAP BAND IV *The Gap Band* [12Jun82|11Sep82|04Jun83] 14(5) 52	3	2837
428. UP YOUR ALLEY *Joan Jett & the Blackhearts* [28May88|22Oct88|08Apr89] 19(1) 46	2	2833
429. NO BRAKES *John Waite* [14Jul84|29Sep84|11May85] 10(1) 43	3	2827
430. McCARTNEY II *Paul McCartney* [14Jun80|21Jun80|18Oct80] 3(5) 19	1	2823
431. EVERYTHING *Bangles* [05Nov88|22Apr89|19Aug89] 15(2) 42	3	2818
432. HEARTLIGHT *Neil Diamond* [16Oct82|27Nov82|04Jun83] 9(3) 34	3	2817
433. ANOTHER TICKET *Eric Clapton And His Band* [21Mar81|25Apr81|08Aug81] 7(2) 21	2	2816
434. ONE FOR THE ROAD *The Kinks* [28Jun80|30Aug80|07Feb81] 14(4) 33	1	2816
435. PICTURE THIS *Huey Lewis & The News* [27Feb82|12Jun82|23Feb85] 13(4) 59	3	2810
436. BAD ENGLISH *Bad English* [15Jul89|25Nov89|11Aug90] 21(2) 52	5,3*	2809
437. IN THE DARK *Grateful Dead* [25Jul87|22Aug87|12Mar88] 6(2) 34	1	2807
438. MEETING IN THE LADIES ROOM *Klymaxx* [02Feb85|22Feb86|10May86] 18(3) 67	3	2800
439. CHICAGO 16 *Chicago* [26Jun82|18Sep82|12Mar83] 9(5) 38	3	2792
440. SONIC TEMPLE *The Cult* [29Apr89|27May89|09Dec89] 10(6) 33	2	2792
441. LABOUR OF LOVE *UB40* [26Nov83|05Nov88|18Feb89] 14(1) 63	*2*	2783
442. MELISSA ETHERIDGE *Melissa Etheridge* [18Jun88|13May89|09Sep89] 22(1) 65	1	2781
443. WE ARE THE WORLD *USA For Africa* [20Apr85|27Apr85|14Sep85] 1(3) 22	1	2778
444. STRAIGHT OUTTA COMPTON *N.W.A.* [04Mar89|15Apr89|15Jul89] 37(3) 81	0	2777
445. ONCE BITTEN *Great White* [18Jul87|10Oct87|16Jan88] 23(2) 53	2	2769
446. FALCO 3 *Falco* [01Mar86|26Apr86|30Aug86] 3(1) 27	2	2764
447. SUPER TROUPER *ABBA* [13Dec80|28Feb81|29Aug81] 17(3) 38	3	2763
448. GEORGIA SATELLITES *Georgia Satellites* [01Nov86|28Feb87|15Aug87] 5(1) 42	2	2752
449. INVASION OF YOUR PRIVACY *Ratt* [29Jun85|27Jul85|12Apr86] 7(1) 42	2	2750
450. LOVE STINKS *The J. Geils Band* [09Feb80|19Apr80|27Mar82] 18(2) 42	3	2737
451. HONEYSUCKLE ROSE (SOUNDTRACK) *Willie Nelson & Family* [06Sep80|04Oct80|09May81] 11(3) 36	1	2735
452. TRASH *Alice Cooper* [12Aug89|18Nov89|02Jun90] 20(2) 45	3,2*	2735
453. GHOSTBUSTERS *Soundtrack* [07Jul84|18Aug84|23Feb85] 6(3) 34	2	2711
454. GOOD TROUBLE *REO Speedwagon* [10Jul82|07Aug82|18Dec82] 7(9) 24	2	2702
455. THE SECRET OF ASSOCIATION *Paul Young* [25May85|17Aug85|15Nov85] 19(6) 43	3	2699
456. MIDNIGHT LOVE *Marvin Gaye* [20Nov82|18Dec82|09Jun84] 7(5) 41	1	2691
457. DOCUMENT *R.E.M.* [26Sep87|07Nov87|07May88] 10(1) 33	2	2688
458. A PRIVATE HEAVEN *Sheena Easton* [20Oct84|02Feb85|15Jun85] 15(2) 35	3	2682
459. GOLD & PLATINUM *Lynyrd Skynyrd* [15Dec79|09Feb80|07Mar81] 12(2) 65	0	2678
460. ANYTIME, ANYPLACE, ANYWHERE *The Rossington Collins Band* [12Jul80|20Sep80|24Jan81] 13(3) 29	1	2669
461. EAZY-DUZ-IT *Eazy-E* [10Dec88|20May89|25Aug90] 41(1) 90	0	2665
462. A WOMAN NEEDS LOVE *Ray Parker Jr. & Raydio* [18Apr81|20Jun81|10Oct81] 13(2) 26	2	2655
463. FAIR WARNING *Van Halen* [30May81|13Jun81|31Oct81] 5(3) 23	0	2642
464. ARENA *Duran Duran* [01Dec84|05Jan85|08Jun85] 4(3) 28	2	2637
465. THE ALLNIGHTER *Glenn Frey* [14Jul84|29Jun85|25Jan86] 22(2) 65	3	2635
466. RHYTHM OF THE NIGHT *DeBarge* [23Mar85|01Jun85|15Feb86] 19(2) 48	4	2623
467. COMING AROUND AGAIN *Carly Simon* [25Apr87|19Sep87|11Jun88] 25(2) 60	3	2616
468. SLIP OF THE TONGUE *Whitesnake* [25Nov89|16Dec89|14Jul90] 10(1) 34	3,2*	2611
469. AMERICAN GIGOLO *Soundtrack* [01Mar80|26Apr80|16Aug80] 7(3) 25	2	2595
470. DAD LOVES HIS WORK *James Taylor* [21Mar81|02May81|22Aug81] 10(3) 23	2	2592
471. MIKE + THE MECHANICS *Mike + The Mechanics* [23Nov85|15Mar86|22Nov86] 26(6) 53	3	2589
472. VICTORY *The Jacksons* [21Jul84|04Aug84|09Feb85] 4(3) 30	3	2571
473. JERMAINE JACKSON *Jermaine Jackson* [19May84|16Jun84|20Apr85] 19(3) 49	2	2569
474. NEBRASKA *Bruce Springsteen* [09Oct82|30Oct82|16Feb85] 3(4) 29	0	2563
475. SONGS IN THE ATTIC *Billy Joel* [03Oct81|17Oct81|03Apr82] 8(3) 27	2	2554
476. MAKING MOVIES *Dire Straits* [15Nov80|27Dec80|13Jun81] 19(3) 31	1	2553
477. PICTURE BOOK *Simply Red* [19Apr86|12Jul86|06Jun87] 16(3) 60	2	2549
478. JULIO *Julio Iglesias* [02Apr83|21May83|08Dec84] 32(1) 89	2	2545
479. TRIO *Dolly Parton, Linda Ronstadt, Emmylou Harris* [28Mar87|02May87|05Mar88] 6(1) 48	0	2543
480. ROCK A LITTLE *Stevie Nicks* [14Dec85|08Feb86|09Aug86] 12(1) 35	3	2539
481. THE THIN RED LINE *Glass Tiger* [19Jul86|07Feb87|04Jul87] 27(2) 51	3	2531
482. CRAZY FROM THE HEAT *David Lee Roth* [23Feb85|16Mar85|05Oct85] 15(4) 33	2	2528
483. ANY LOVE *Luther Vandross* [22Oct88|19Nov88|03Jun89] 9(1) 33	2	2513
484. TIME AND TIDE *Basia* [20Feb88|26Nov88|05Aug89] 36(1) 77	2	2498
485. SURFING WITH THE ALIEN *Joe Satriani* [21Nov87|30Apr88|22Apr89] 29(4) 75	0	2477
486. THE LEXICON OF LOVE *ABC* [25Sep82|29Jan83|18Jun83] 24(10) 39	2	2473
487. HIGH ADVENTURE *Kenny Loggins* [25Sep82|23Oct82|23Jul83] 13(3) 44	3	2457
488. BEVERLY HILLS COP II *Soundtrack* [13Jun87|08Aug87|05Dec87] 8(1) 26	2	2457
489. AS THE BAND TURNS *Atlantic Starr* [18May85|29Mar86|30Aug86] 17(2) 68	3	2457

Ranking the Albums

Rank. TITLE *Act* [Enter | Peak | Exit] Peak(Wks) TotWks	Hot 100	Score
490. ALL FOUR ONE *The Motels* [24Apr82 | 14Aug82 | 29Jan83] 16(2) 41	3	2453
491. WORLD WIDE LIVE *Scorpions* [13Jul85 | 28Sep85 | 03May86] 14(1) 43	0	2451
492. ROLL ON *Alabama* [11Feb84 | 31Mar84 | 13Apr85] 21(2) 62	1	2447
493. THE CROSSING *Big Country* [24Sep83 | 05Nov83 | 07Jul84] 18(2) 42	2	2443
494. THE NIGHT I FELL IN LOVE *Luther Vandross* [06Apr85 | 15Jun85 | 26Apr86] 19(3) 56	1	2435
495. BEBE LE STRANGE *Heart* [08Mar80 | 22Mar80 | 02Aug80] 5(2) 22	1	2426
496. DIRTY WORK *The Rolling Stones* [12Apr86 | 03May86 | 27Sep86] 4(2) 25	2	2422
497. ACTUALLY *Pet Shop Boys* [03Oct87 | 21Nov87 | 06Aug88] 25(1) 45	2	2421
498. LIFE IS...TOO SHORT *Too Short* [25Feb89 | 20May89 | 18Aug90] 37(2) 78	0	2412
499. KENNY LOGGINS ALIVE *Kenny Loggins* [04Oct80 | 15Nov80 | 02May81] 11(2) 31	0	2402
500. GIVE THE PEOPLE WHAT THEY WANT *The Kinks* [12Sep81 | 17Oct81 | 15May82] 15(2) 36	2	2399
501. ONE HEARTBEAT *Smokey Robinson* [28Mar87 | 20Jun87 | 30Apr88] 26(2) 58	3	2397
502. NEVER TOO MUCH *Luther Vandross* [19Sep81 | 14Nov81 | 22May82] 19(2) 36	1	2394
503. BLIND MAN'S ZOO *10,000 Maniacs* [03Jun89 | 29Jul89 | 09Dec89] 13(1) 28	1	2375
504. WHERE THERE'S SMOKE *Smokey Robinson* [30Jun79 | 01Mar80 | 17May80] 17(2) 47	1	2370
505. GO ALL THE WAY *The Isley Brothers* [19Apr80 | 17May80 | 13Sep80] 8(4) 22	1	2369
506. NON-STOP EROTIC CABARET *Soft Cell* [30Jan82 | 07Aug82 | 06Nov82] 22(1) 41	1	2367
507. GOOD MORNING, VIETNAM *Soundtrack* [06Feb88 | 26Mar88 | 01Oct88] 10(2) 35	1	2363
508. THE SEEDS OF LOVE *Tears For Fears* [07Oct89 | 28Oct89 | 26May90] 8(3) 34	3,2*	2355
509. SPECIAL FORCES *38 Special* [29May82 | 17Jul82 | 12Mar83] 10(3) 42	2	2351
510. ALL THAT JAZZ *Breathe* [04Jun88 | 24Dec88 | 20May89] 34(3) 51	3	2350
511. HERE'S TO FUTURE DAYS *Thompson Twins* [19Oct85 | 22Feb86 | 14Jun86] 20(1) 35	2	2347
512. VITAL IDOL *Billy Idol* [10Oct87 | 14Nov87 | 23Apr88] 10(1) 29	2	2338
513. SLEEPING WITH THE PAST *Elton John* [16Sep89 | 07Oct89 | 15Sep90] 23(1) 53	3,2*	2337
514. STOP MAKING SENSE *Talking Heads* [22Sep84 | 13Oct84 | 20Dec86] 41(4) 118	1	2336
515. I FEEL FOR YOU *Chaka Khan* [20Oct84 | 01Dec84 | 21Sep85] 14(2) 49	3	2332
516. FOOLISH BEHAVIOUR *Rod Stewart* [06Dec80 | 27Dec80 | 25Apr81] 12(5) 21	2	2330
517. SOLITUDE STANDING *Suzanne Vega* [16May87 | 15Aug87 | 19Dec87] 11(2) 32	2	2315
518. NIGHTWALKER *Gino Vannelli* [11Apr81 | 30May81 | 03Oct81] 15(4) 26	2	2311
519. PIECE OF MIND *Iron Maiden* [11Jun83 | 20Aug83 | 17Nov84] 14(1) 45	0	2309
520. INSIDE INFORMATION *Foreigner* [26Dec87 | 27Feb88 | 03Sep88] 15(1) 37	3	2306
521. A NEW FLAME *Simply Red* [11Mar89 | 12Aug89 | 02Dec89] 22(1) 39	2	2305
522. REG STRIKES BACK *Elton John* [09Jul88 | 03Sep88 | 21Jan89] 16(1) 29	2	2304
523. SWEPT AWAY *Diana Ross* [29Sep84 | 10Nov84 | 03Aug85] 26(1) 45	4	2303
524. TIME EXPOSURE *Little River Band* [19Sep81 | 14Nov81 | 28Aug82] 21(1) 50	3	2288
525. ALLIED FORCES *Triumph* [19Sep81 | 21Nov81 | 20Apr85] 23(1) 59	1	2287
526. HOLD AN OLD FRIEND'S HAND *Tiffany* [10Dec88 | 21Jan89 | 24Jun89] 17(4) 29	2	2282
527. FULL MOON *The Charlie Daniels Band* [09Jul80 | 06Sep80 | 21Mar81] 11(2) 33	2	2276
528. BLACK & WHITE *The Pointer Sisters* [11Jul81 | 05Sep81 | 05Dec81] 12(2) 22	2	2271
529. TO HELL WITH THE DEVIL *Stryper* [22Nov86 | 07Feb87 | 16Apr88] 32(2) 74	1	2270
530. SELF CONTROL *Laura Branigan* [28Apr84 | 11Aug84 | 02Mar85] 23(2) 45	3	2270
531. TRIBUTE *Ozzy Osbourne/Randy Rhoads* [09May87 | 13Jun87 | 10Oct87] 6(2) 23	0	2266
532. BON JOVI *Bon Jovi* [25Feb84 | 28Apr84 | 03Oct87] 43(2) 86	2	2265
533. MODERN TIMES *Jefferson Starship* [18Apr81 | 06Jun81 | 28Nov81] 26(3) 33	2	2256
534. BOY IN THE BOX *Corey Hart* [20Jul85 | 24Aug85 | 29Mar86] 20(7) 37	3	2253
535. THE CONCERT IN CENTRAL PARK *Simon & Garfunkel* [13Mar82 | 17Apr82 | 01Oct83] 6(2) 34	1	2249
536. GREATEST HITS *The Cars* [23Nov85 | 21Dec85 | 16Aug86] 12(4) 39	1	2244
537. FANTASTIC VOYAGE *Lakeside* [29Nov80 | 21Feb81 | 25Jul81] 16(2) 35	1	2242
538. WALKING WITH A PANTHER *LL Cool J* [01Jul89 | 22Jul89 | 18Nov89] 6(1) 21	1	2231
539. ENDLESS LOVE *Soundtrack* [01Aug81 | 26Sep81 | 12Dec81] 9(2) 20	1	2231
540. STARS ON LONG PLAY *Stars On* [09May81 | 11Jul81 | 17Oct81] 9(4) 24	2	2230
541. FUTURE SHOCK *Herbie Hancock* [03Sep83 | 15Oct83 | 24Nov84] 43(2) 65	1	2229
542. CONSCIOUS PARTY *Ziggy Marley And The Melody Makers* [23Apr88 | 02Jul88 | 04Feb89] 23(1) 42	1	2220
543. SHE WORKS HARD FOR THE MONEY *Donna Summer* [16Jul83 | 27Aug83 | 18Feb84] 9(1) 32	3	2220
544. STRENGTH IN NUMBERS *38 Special* [17May86 | 28Jun86 | 13Dec86] 17(1) 31	2	2220
545. WHEN HARRY MET SALLY (SOUNDTRACK) *Harry Connick Jr.* [19Aug89 | 23Sep89 | 11Apr92] 42(2) 122	0	2220
546. PARIS *Supertramp* [11Oct80 | 01Nov80 | 04Apr81] 8(2) 26	2	2219
547. LAWYERS IN LOVE *Jackson Browne* [20Aug83 | 10Sep83 | 31Mar84] 8(3) 33	3	2216
548. THE PLEASURE PRINCIPLE *Gary Numan* [02Feb80 | 24May80 | 02Aug80] 16(2) 30	1	2216
549. ONE-TRICK PONY *Paul Simon* [06Sep80 | 18Oct80 | 28Feb81] 12(2) 26	2	2214
550. LEGEND *Bob Marley And The Wailers* [18Aug84 | 27Oct84 | 03Aug91] 54(2) 113	0	2209
551. THE GEORGE BENSON COLLECTION *George Benson* [21Nov81 | 30Jan82 | 15May82] 14(2) 26	0	2205
552. TOUR DE FORCE *38 Special* [03Dec83 | 04Feb84 | 25Aug84] 22(2) 39	2	2197
553. HEAVY METAL *Soundtrack* [08Aug81 | 03Oct81 | 13Feb82] 12(2) 28	1	2195
554. BACK FOR THE ATTACK *Dokken* [05Dec87 | 19Dec87 | 16Jul88] 13(1) 33	1	2188
555. KILLER ON THE RAMPAGE *Eddy Grant* [23Apr83 | 16Jul83 | 12Nov83] 10(3) 30	2	2188
556. ROCKY IV *Soundtrack* [16Nov85 | 22Feb86 | 07Jun86] 10(1) 30	2	2167
557. IN HEAT *The Romantics* [22Oct83 | 18Feb84 | 23Jun84] 14(1) 36	2	2167
558. IF THAT'S WHAT IT TAKES *Michael McDonald* [28Aug82 | 02Oct82 | 02Apr83] 6(6) 32	2	2160
559. FRIEND OR FOE *Adam Ant* [06Nov82 | 05Mar83 | 09Jul83] 16(3) 36	2	2159
560. KINGDOM COME *Kingdom Come* [19Mar88 | 30Apr88 | 01Oct88] 12(1) 29	1	2156
561. SOLITUDE/SOLITAIRE *Peter Cetera* [12Jul86 | 23Aug86 | 02May87] 23(1) 43	2	2156
562. IN MY TRIBE *10,000 Maniacs* [19Sep87 | 21May88 | 04Mar89] 37(1) 77	2	2156
563. THE B-52'S *The B-52s* [11Aug79 | 10May80 | 14Feb81] 59(2) 74	1	2154
564. LIVE FROM EARTH *Pat Benatar* [15Oct83 | 19Nov83 | 09Mar85] 13(2) 34	1	2147
565. JARREAU *Al Jarreau* [16Apr83 | 28May83 | 04Feb84] 13(2) 43	3	2144
566. CONTROVERSY *Prince* [07Nov81 | 21Nov81 | 04May85] 21(3) 64	1	2142
567. SOMEWHERE IN TIME *Iron Maiden* [11Oct86 | 22Nov86 | 14Feb87] 11(1) 39	0	2136
568. PETER GABRIEL (III) *Peter Gabriel* [21Jun80 | 23Aug80 | 03Jan81] 22(2) 29	1	2135
569. TILL I LOVED YOU *Barbra Streisand* [12Nov88 | 10Dec88 | 06May89] 10(1) 26	1	2133
570. I CAN'T STAND STILL *Don Henley* [04Sep82 | 04Dec82 | 30Apr83] 24(7) 35	3	2131
571. IT'S HARD *The Who* [25Sep82 | 23Oct82 | 30Apr83] 8(5) 32	2	2131
572. TALK SHOW *Go-Go's* [07Apr84 | 26May84 | 10Nov84] 18(3) 32	3	2131
573. BREAKIN' *Soundtrack* [02Jun84 | 21Jul84 | 03Nov84] 8(2) 23	0	2129
574. LARGER THAN LIFE *Jody Watley* [15Apr89 | 20May89 | 13Jan90] 16(2) 40	4,2*	2129
575. LIVING YEARS *Mike + The Mechanics* [19Nov88 | 08Apr89 | 29Jul89] 13(2) 37	3	2127
576. GIDEON *Kenny Rogers* [12Apr80 | 31May80 | 20Dec80] 12(2) 34	1	2123
577. THE NATURE OF THE BEAST *April Wine* [31Jan81 | 18Apr81 | 19Sep81] 26(1) 34	2	2119
578. SCARY MONSTERS *David Bowie* [04Oct80 | 15Nov80 | 04Apr81] 12(2) 27	1	2118
579. CAMOUFLAGE *Rod Stewart* [30Jun84 | 11Aug84 | 02Feb85] 18(2) 35	3	2113
580. GREATEST HITS/LIVE *Heart* [06Dec80 | 20Dec80 | 23May81] 13(3) 25	2	2112
581. STEVIE WONDER'S ORIGINAL MUSIQUARIUM I *Stevie Wonder* [29May82 | 12Jun82 | 04Dec82] 4(3) 28	3	2112
582. TOO LOW FOR ZERO *Elton John* [11Jun83 | 25Feb84 | 01Sep84] 25(1) 54	3	2107
583. GREATEST HITS *Queen* [14Nov81 | 05Dec81 | 29May82] 14(6) 26	0	2106
584. KEEP IT LIVE *Dazz Band* [03Apr82 | 10Jul82 | 20Nov82] 14(4) 34	1	2100
585. STAYING ALIVE (SOUNDTRACK) *Bee Gees* [16Jul83 | 27Aug83 | 14Jan84] 6(2) 27	2	2100
586. FUN AND GAMES *Chuck Mangione* [23Feb80 | 15Mar80 | 26Jul80] 8(4) 23	1	2095
587. REVENGE *Eurythmics* [09Aug86 | 04Oct86 | 21Mar87] 12(1) 33	2	2090
588. TOUGHER THAN LEATHER *Run-D.M.C.* [04Jun88 | 02Jul88 | 10Dec88] 9(1) 28	1	2081
589. ANIMALIZE *KISS* [06Oct84 | 01Dec84 | 22Jun85] 19(1) 38	1	2076

Ranking the Albums

Rank. TITLE Act [Enter|Peak|Exit] Peak(Wks) TotWks	Hot 100	Score
590. DREAMGIRLS Original Cast [22May82|21Aug82|04Dec82] 11(3) 29	1	2076
591. NIGHTSHIFT Commodores [16Feb85|27Apr85|26Oct85] 12(3) 37	3	2071
592. GTR GTR [17May86|12Jul86|08Nov86] 11(2) 26	2	2069
593. KISS ME, KISS ME, KISS ME The Cure [20Jun87|04Jul87|11Jun88] 35(4) 52	3	2063
594. DEGUELLO ZZ Top [24Nov79|16Feb80|10Mar84] 24(3) 43	3	2062
595. BILLY IDOL Billy Idol [31Jul82|16Jul83|08Dec84] 45(1) 104	2	2061
596. CRAZY NIGHTS KISS [10Oct87|31Oct87|28May88] 18(1) 34	2	2060
597. IF I SHOULD LOVE AGAIN Barry Manilow [17Oct81|14Nov81|03Apr82] 14(3) 25	3	2058
598. DAN FOGELBERG/GREATEST HITS Dan Fogelberg [13Nov82|18Dec82|09Jul83] 15(4) 35	2	2056
599. POWER WINDOWS Rush [09Nov85|14Dec85|17May86] 10(2) 28	1	2056
600. WARRIOR Scandal Featuring Patty Smyth [04Aug84|06Oct84|11May85] 17(3) 41	3	2050
601. BIG DADDY John Mellencamp [27May89|17Jun89|28Oct89] 7(1) 23	2	2048
602. SEE THE LIGHT The Jeff Healey Band [08Oct88|23Sep89|27Jan90] 22(3) 69	1	2041
603. MARTIKA Martika [04Feb89|19Aug89|02Dec89] 15(1) 39	3	2041
604. WHO'S THAT GIRL (SOUNDTRACK) Madonna [15Aug87|12Sep87|20Feb88] 7(2) 28	2	2038
605. SIGNS OF LIFE Billy Squier [04Aug84|01Sep84|16Feb85] 11(4) 29	3	2034
606. UNDERCOVER The Rolling Stones [26Nov83|10Dec83|28Apr84] 4(2) 23	2	2032
607. NOTHIN' MATTERS AND WHAT IF IT DID John Cougar [04Oct80|23May81|05Mar83] 37(2) 55	2	2032
608. PERFECT STRANGERS Deep Purple [01Dec84|09Feb85|06Jul85] 17(2) 32	1	2024
609. CROSSROADS Tracy Chapman [21Oct89|11Nov89|14Apr90] 9(2) 26	1	2020
610. EXIT...STAGE LEFT Rush [14Nov81|05Dec81|03Apr82] 10(3) 21	1	2016
611. NO CONTROL Eddie Money [10Jul82|23Oct82|07May83] 20(3) 44	2	2013
612. TEAR DOWN THESE WALLS Billy Ocean [19Mar88|07May88|15Oct88] 18(1) 31	2	2012
613. MECCA FOR MODERNS Manhattan Transfer [13Jun81|22Aug81|12Dec81] 22(4) 27	1	2009
614. BIG GENERATOR Yes [17Oct87|07Nov87|07May88] 15(3) 30	2	2009
615. MERRY, MERRY CHRISTMAS New Kids On The Block [14Oct89|23Dec89|19Jan91] 9(1) 28	1,1*	2001
616. NOTORIOUS Duran Duran [20Dec86|31Jan87|08Aug87] 12(1) 34	3	1997
617. SIGNALS Rush [02Oct82|27Nov82|30Jun84] 10(1) 33	1	1996
618. MAVERICK George Thorogood & The Destroyers [02Mar85|06Jul85|14Dec85] 32(1) 42	1	1995
619. LIVE RUST Neil Young With Crazy Horse [08Dec79|02Feb80|17May80] 15(2) 24	0	1985
620. GREAT WHITE NORTH Bob & Doug McKenzie [09Jan82|03Apr82|29May82] 8(2) 21	1	1984

Rank. TITLE Act [Enter|Peak|Exit] Peak(Wks) TotWks	Hot 100	Score
621. LOVE AND ROCKETS Love And Rockets [20May89|12Aug89|11Nov89] 14(1) 26	2	1984
622. THREE LOCK BOX Sammy Hagar [25Dec82|09Apr83|13Aug83] 17(2) 34	2	1980
623. BELINDA Belinda Carlisle [07Jun86|20Sep86|24Jan87] 13(1) 34	2	1971
624. ALL FOR LOVE New Edition [07Dec85|19Apr86|01Nov86] 32(1) 48	3	1967
625. THE FINAL CUT Pink Floyd [09Apr83|07May83|10Sep83] 6(2) 23	0	1966
626. CHARACTERS Stevie Wonder [05Dec87|19Dec87|02Jul88] 17(3) 31	3	1959
627. VACATION Go-Go's [14Aug82|18Sep82|19Feb83] 8(5) 28	2	1956
628. KEEP YOUR EYE ON ME Herb Alpert [21Mar87|13Jun87|17Oct87] 18(3) 31	3	1954
629. HOT, COOL AND VICIOUS Salt-N-Pepa [01Aug87|27Feb88|20Aug88] 26(4) 53	1	1947
630. FOREVER, FOR ALWAYS, FOR LOVE Luther Vandross [16Oct82|11Dec82|18Jun83] 20(5) 36	1	1945
631. FACES Earth, Wind & Fire [22Nov80|06Dec80|11Apr81] 10(2) 21	3	1942
632. ALL IN THE NAME OF LOVE Atlantic Starr [25Apr87|25Jul87|21Nov87] 18(1) 31	2	1929
633. FIRE OF UNKNOWN ORIGIN Blue Öyster Cult [11Jul81|19Sep81|06Feb82] 24(2) 31	1	1923
634. ONE EIGHTY Ambrosia [19Apr80|12Jul80|29Nov80] 25(1) 33	2	1910
635. GRACE UNDER PRESSURE Rush [05May84|19May84|03Nov84] 10(4) 27	0	1905
636. WATERMARK Enya [04Feb89|22Apr89|28Oct89] 25(4) 39	1	1899
637. MAGIC The Jets [07Nov87|09Jul88|15Oct88] 35(1) 50	5	1898
638. WARM THOUGHTS Smokey Robinson [15Mar80|24May80|02Aug80] 14(2) 21	1	1895
639. SKYY LINE Skyy [21Nov81|03Apr82|03Jul82] 18(2) 33	1	1892
640. RHYTHM OF YOUTH Men Without Hats [06Aug83|24Sep83|28Jan84] 13(3) 26	2	1887
641. THE FIRM The Firm [02Mar85|13Apr85|12Oct85] 17(2) 33	2	1884
642. ALPHA Asia [27Aug83|10Sep83|11Feb84] 6(4) 25	2	1881
643. SEVEN YEAR ACHE Rosanne Cash [28Mar81|01Aug81|31Oct81] 26(1) 32	1	1880
644. BIG FUN Shalamar [10Nov79|22Mar80|02Aug80] 23(2) 36	1	1879
645. QUIET LIES Juice Newton [29May82|17Jul82|09Apr83] 20(2) 46	3	1875
646. THE BLUES BROTHERS (SOUNDTRACK) Blues Brothers [28Jun80|09Aug80|01Nov80] 13(2) 19	1	1874
647. HOLD YOUR FIRE Rush [26Sep87|17Oct87|16Apr88] 13(2) 30	0	1873
648. HOW TO BE A...ZILLIONAIRE! ABC [05Oct85|16Nov85|12Jul86] 30(2) 41	3	1871
649. YENTL (SOUNDTRACK) Barbra Streisand [26Nov83|07Jan84|19May84] 9(2) 29	1	1871
650. BODY AND SOUL Joe Jackson [07Apr84|26May84|20Oct84] 20(4) 29	2	1868
651. HITS! Boz Scaggs [29Nov80|10Jan81|23May81] 24(6) 26	1	1868
652. S.O.S. The S.O.S. Band [28Jun80|09Aug80|08Nov80] 12(3) 20	1	1865
653. SUBSTANCE New Order [05Sep87|30Jan88|11Feb89] 36(1) 60	2	1864

Rank. TITLE Act [Enter|Peak|Exit] Peak(Wks) TotWks	Hot 100	Score
654. FLYING IN A BLUE DREAM Joe Satriani [18Nov89|02Dec89|11Aug90] 23(3) 39	0	1863
655. EVERY HOME SHOULD HAVE ONE Patti Austin [03Oct81|19Feb83|28May83] 36(8) 44	2	1862
656. SMALL WORLD Huey Lewis & The News [20Aug88|10Sep88|11Mar89] 11(2) 30	3	1851
657. PRINCE Prince [17Nov79|19Jan80|24May80] 22(2) 28	1	1850
658. THE OTHER WOMAN Ray Parker Jr. [24Apr82|12Jun82|23Oct82] 11(2) 27	2	1850
659. ONE VOICE Barbra Streisand [09May87|06Jun87|14Nov87] 9(2) 28	0	1848
660. STILL LIFE (AMERICAN CONCERT 1981) Rolling Stones [26Jun82|10Jul82|27Nov82] 5(4) 23	1	1847
661. OUTSIDE INSIDE The Tubes [02Apr83|02Jul83|19Nov83] 18(1) 34	3	1846
662. CHICAGO 19 Chicago [09Jul88|14Jan89|22Apr89] 37(2) 42	5,1*	1846
663. THE GLAMOROUS LIFE Sheila E. [07Jul84|06Oct84|18May85] 28(2) 46	2	1845
664. WELCOME TO THE PLEASURE DOME Frankie Goes To Hollywood [24Nov84|15Dec84|31Aug85] 33(5) 41	4	1841
665. WORLD MACHINE Level 42 [22Mar86|21Jun86|22Nov86] 18(2) 36	1	1841
666. ONLY FOUR YOU Mary Jane Girls [16Mar85|22Jun85|30Nov85] 18(2) 38	2	1841
667. SHAKE YOU DOWN Gregory Abbott [01Nov86|14Feb87|04Jul87] 22(1) 36	2	1838
668. EL LOCO ZZ Top [08Aug81|26Sep81|02Jan82] 17(2) 22	1	1836
669. FRIENDS Dionne Warwick [21Dec85|01Mar86|14Jun86] 12(2) 26	2	1836
670. STANDING HAMPTON Sammy Hagar [30Jan82|12Jun82|04Sep82] 28(2) 32	2	1836
671. SHEENA EASTON Sheena Easton [14Mar81|13Jun81|26Nov81] 24(1) 38	2	1828
672. RAY, GOODMAN & BROWN Ray, Goodman & Brown [26Jan80|05Apr80|28Jun80] 17(2) 23	2	1827
673. BLOW UP YOUR VIDEO AC/DC [05Mar88|26Mar88|13Aug88] 12(1) 24	0	1815
674. THEN & NOW...THE BEST OF THE MONKEES The Monkees [26Jul86|13Sep86|14Mar87] 17(2) 36	2	1814
675. TURBO Judas Priest [12Apr86|26Apr86|31Jan87] 17(2) 36	0	1811
676. CONTACT The Pointer Sisters [10Aug85|09Nov85|29Mar86] 24(1) 34	3	1809
677. NIGHT IN THE RUTS Aerosmith [01Dec79|19Jan80|05Apr80] 14(2) 19	1	1808
678. TRILOGY: PAST, PRESENT AND FUTURE Frank Sinatra [12Apr80|05Jul80|20Sep80] 17(2) 24	1	1804
679. THE GOLDEN AGE OF WIRELESS Thomas Dolby [19Mar83|18Jun83|24Sep83] 13(2) 28	2	1799
680. BROADCAST Cutting Crew [21Mar87|23May87|23Jan88] 16(2) 45	3	1795
681. LOVE APPROACH Tom Browne [26Jul80|01Nov80|17Jan81] 18(2) 26	0	1795
682. SEPTEMBER MORN Neil Diamond [12Jan80|16Feb80|24May80] 10(3) 20	2	1793
683. ST. ELMO'S FIRE Soundtrack [13Jul85|07Sep85|22Mar86] 21(2) 37	2	1793
684. EVERLASTING Natalie Cole [08Aug87|30Apr88|10Sep88] 42(2) 58	4	1792
685. ALL THIS LOVE DeBarge [11Sep82|18Jun83|25May85] 24(2) 48	2	1789

Ranking the Albums

Rank. TITLE Act [Enter | Peak | Exit] Peak(Wks) TotWks	Hot 100	Score
686. RECORDS Foreigner [25Dec82 | 12Feb83 | 11Jun83] 10(4) 25	0	1785
687. DEFENDERS OF THE FAITH Judas Priest [04Feb84 | 25Feb84 | 13Oct84] 18(4) 37	0	1785
688. BACK TO THE FUTURE Soundtrack [27Jul85 | 05Oct85 | 01Mar86] 12(2) 32	1	1783
689. TROPICO Pat Benatar [24Nov84 | 22Dec84 | 20Apr85] 14(4) 22	2	1780
690. BREAKING HEARTS Elton John [21Jul84 | 11Aug84 | 09Mar85] 20(5) 34	3	1779
691. EVERY BREATH YOU TAKE-THE SINGLES The Police [22Nov86 | 20Dec86 | 16May87] 7(1) 26	1	1778
692. THE TWO OF US Yarbrough & Peoples [27Dec80 | 28Mar81 | 06Jun81] 16(2) 24	1	1775
693. NO REST FOR THE WICKED Ozzy Osbourne [22Oct88 | 19Nov88 | 22Apr89] 13(1) 27	0	1774
694. ANGEL OF THE NIGHT Angela Bofill [03Nov79 | 08Mar80 | 14Jun80] 34(1) 33	0	1773
695. ON THE WAY TO THE SKY Neil Diamond [28Nov81 | 16Jan82 | 29May82] 17(2) 27	3	1772
696. LONDON CALLING The Clash [09Feb80 | 22Mar80 | 20Sep80] 27(3) 33	1	1770
697. REACH FOR THE SKY Ratt [19Nov88 | 17Dec88 | 20May89] 17(1) 27	1	1768
698. ANOTHER PAGE Christopher Cross [19Feb83 | 12Mar83 | 31Mar84] 11(5) 31	3	1765
699. THE NIGHTFLY Donald Fagen [30Oct82 | 27Nov82 | 30Apr83] 11(4) 27	2	1763
700. MUSIC FROM THE EDGE OF HEAVEN Wham! [19Jul86 | 23Aug86 | 24Jan87] 10(2) 28	4	1761
701. CAROL HENSEL'S EXERCISE AND DANCE PROGRAM Carol Hensel [21Mar81 | 23May81 | 19Jun82] 56(1) 55	0	1759
702. DELICATE SOUND OF THUNDER Pink Floyd [10Dec88 | 14Jan89 | 29Apr89] 11(1) 21	0	1759
703. SLIDE IT IN Whitesnake [19May84 | 25Aug84 | 12Mar88] 40(2) 85	0	1759
704. THREE SIDES LIVE Genesis [26Jun82 | 21Aug82 | 11Dec82] 10(3) 25	1	1758
705. BLOW MY FUSE Kix [15Oct88 | 02Dec89 | 24Mar90] 46(1) 60	1	1757
706. PICTURES FOR PLEASURE Charlie Sexton [30Nov85 | 22Mar86 | 19Jul86] 15(3) 34	1	1753
707. WHAT YOU DON'T KNOW Exposé [01Jul89 | 05Aug89 | 09Jun90] 33(2) 50	4,2*	1752
708. BORN TO LOVE Peabo Bryson & Roberta Flack [13Aug83 | 19Nov83 | 26May84] 25(1) 42	2	1749
709. GREATEST HITS Alabama [01Mar86 | 05Apr86 | 15Nov86] 24(3) 38	0	1740
710. STATE OF CONFUSION The Kinks [11Jun83 | 06Aug83 | 26Nov83] 12(1) 25	2	1739
711. JESSE JOHNSON'S REVUE Jesse Johnson's Revue [16Mar85 | 20Apr85 | 04Jan86] 43(5) 43	2	1738
712. BIG GAME White Lion [01Jul89 | 05Aug89 | 23Dec89] 19(2) 26	2	1730
713. HOOKED ON SWING Larry Elgart And His Manhattan Swing Orchestra [19Jun82 | 14Aug82 | 26Mar83] 24(5) 41	1	1729
714. PRETENDERS II The Pretenders [29Aug81 | 12Sep81 | 02Jan82] 10(3) 19	0	1727
715. CHER(2) Cher [05Dec87 | 07May88 | 10Sep88] 32(3) 41	3	1727
716. STEP BY STEP Eddie Rabbitt [22Aug81 | 17Oct81 | 10Apr82] 23(1) 34	3	1720
717. COLD BLOODED Rick James [27Aug83 | 29Oct83 | 10Mar84] 16(1) 29	2	1717
718. REMAIN IN LIGHT Talking Heads [01Nov80 | 06Dec80 | 02May81] 19(2) 27	0	1717
719. PARTNERS IN CRIME Rupert Holmes [10Nov79 | 15Mar80 | 07Jun80] 33(2) 31	4,*1	1717
720. WILD, WILD WEST The Escape Club [27Aug88 | 26Nov88 | 13May89] 27(3) 38	3	1716
721. IT'S BETTER TO TRAVEL Swing Out Sister [29Aug87 | 27Feb88 | 18Jun88] 40(1) 43	2	1716
722. TRUE STORIES: A FILM BY DAVID BYRNE, THE COMPLETE SOUNDTRACK Talking Heads [04Oct86 | 08Nov86 | 18Apr87] 17(3) 29	1	1713
723. VOICES CARRY 'Til Tuesday [20Apr85 | 27Jul85 | 16Nov85] 19(1) 31	2	1712
724. ...ALL THE RAGE General Public [27Oct84 | 16Feb85 | 20Jul85] 26(2) 39	1	1712
725. IN YOUR EYES George Benson [18Jun83 | 23Jul83 | 11Feb84] 27(1) 35	2	1711
726. JUST GETS BETTER WITH TIME Whispers [30May87 | 15Aug87 | 06Feb88] 22(2) 37	1	1710
727. UNDER LOCK AND KEY Dokken [21Dec85 | 15Feb86 | 28Mar87] 32(5) 67	1	1710
728. SHE'S THE BOSS Mick Jagger [16Mar85 | 20Apr85 | 28Sep85] 13(1) 29	2	1708
729. INDIGO GIRLS Indigo Girls [15Apr89 | 02Sep89 | 09Dec89] 22(3) 35	1	1704
730. MECHANICAL RESONANCE Tesla [31Jan87 | 04Apr87 | 26Mar88] 32(2) 61	1	1704
731. 21 AT 33 Elton John [31May80 | 19Jul80 | 18Oct80] 13(2) 21	2	1698
732. PIPES OF PEACE Paul McCartney [19Nov83 | 17Dec83 | 28Apr84] 15(1) 24	2	1698
733. NO FUN ALOUD Glenn Frey [26Jun82 | 14Aug82 | 12Mar83] 32(2) 38	3	1696
734. FLOWERS IN THE DIRT Paul McCartney [24Jun89 | 01Jul89 | 26May90] 21(3) 49	3,1*	1694
735. LIVE IN AUSTRALIA (WITH THE MELBOURNE SYMPHONY ORCHESTRA) Elton John [25Jul87 | 06Feb88 | 30Apr88] 24(2) 41	1	1685
736. AGAINST ALL ODDS Soundtrack [31Mar84 | 28Apr84 | 25Aug84] 12(4) 22	1	1684
737. AMMONIA AVENUE Alan Parsons Project [17Mar84 | 28Apr84 | 08Sep84] 15(1) 26	2	1684
738. BRAVE AND CRAZY Melissa Etheridge [07Oct89 | 11Nov89 | 10Nov90] 22(1) 58	1	1683
739. GREATEST HITS Fleetwood Mac [10Dec88 | 11Feb89 | 03Jun89] 14(2) 26	1	1681
740. IT MUST BE MAGIC Teena Marie [13Jun81 | 22Aug81 | 28Nov81] 23(1) 25	1	1680
741. IN THE HEART Kool & The Gang [10Dec83 | 03Mar84 | 18Aug84] 29(2) 37	2	1676
742. JUMP UP! Elton John [08May82 | 12Jun82 | 18Dec82] 17(3) 33	2	1675
743. WORLDS APART Saga [23Oct82 | 19Feb83 | 25Jun83] 29(5) 36	2	1669
744. FABLES OF THE RECONSTRUCTION R.E.M. [29Jun85 | 03Aug85 | 12Apr86] 28(2) 42	0	1668
745. AMERICAN DREAM Crosby, Stills, Nash & Young [03Dec88 | 07Jan89 | 29Apr89] 16(1) 22	1	1667
746. ROSES IN THE SNOW Emmylou Harris [24May80 | 12Jul80 | 10Jan81] 26(2) 34	0	1665
747. TIME Electric Light Orchestra [22Aug81 | 19Sep81 | 02Jan82] 16(3) 20	2	1665
748. IMAGINATION The Whispers [17Jan81 | 28Mar81 | 18Jul81] 23(2) 27	1	1662
749. CRASH AND BURN Pat Travers Band [05Apr80 | 17May80 | 20Sep80] 20(1) 25	1	1651
750. FOREVER Kool & The Gang [06Dec86 | 24Jan87 | 19Sep87] 25(1) 42	4	1651
751. ASYLUM KISS [05Oct85 | 16Nov85 | 19Apr86] 20(1) 29	1	1639
752. THE INNOCENTS Erasure [18Jun88 | 22Oct88 | 27May89] 49(1) 50	2	1638
753. I NEVER SAID GOODBYE Sammy Hagar [11Jul87 | 15Aug87 | 12Dec87] 14(1) 23	2	1636
754. ROCKIHNROLL Greg Kihn Band [11Apr81 | 29Aug81 | 14Nov81] 32(1) 32	1	1636
755. THIS TIME Al Jarreau [21Jun80 | 16Aug80 | 14Feb81] 27(2) 35	0	1635
756. TOM TOM CLUB Tom Tom Club [24Oct81 | 03Apr82 | 10Jul82] 23(1) 33	1	1635
757. INFORMATION SOCIETY Information Society [20Aug88 | 12Nov88 | 06May89] 25(1) 38	4	1635
758. SOMEBODY'S WATCHING ME Rockwell [11Feb84 | 31Mar84 | 01Sep84] 15(2) 30	2	1634
759. CHINESE WALL Philip Bailey [10Nov84 | 23Feb85 | 06Jul85] 22(3) 35	2	1633
760. SEVENTH SON OF A SEVENTH SON Iron Maiden [30Apr88 | 28May88 | 01Oct88] 12(1) 23	0	1631
761. ESCAPE Whodini [24Nov84 | 16Feb85 | 19Oct85] 35(1) 48	1	1624
762. CAMEOSIS Cameo [24May80 | 16Aug80 | 15Nov80] 25(3) 26	0	1618
763. WINDS OF CHANGE Jefferson Starship [30Oct82 | 11Dec82 | 28May83] 26(6) 31	2	1616
764. HEY RICKY Melissa Manchester [15May82 | 18Sep82 | 05Feb83] 19(5) 39	1	1614
765. RECKONING R.E.M. [05May84 | 02Jun84 | 04May85] 27(5) 53	1	1613
766. THE GREAT ADVENTURES OF SLICK RICK Slick Rick [21Jan89 | 10Jun89 | 21Oct89] 31(3) 40	0	1609
767. NEVER DIE YOUNG James Taylor [13Feb88 | 12Mar88 | 01Oct88] 25(2) 34	1	1606
768. ARETHA (II) Aretha Franklin [15Nov86 | 24Jan87 | 08Aug87] 32(2) 39	4	1604
769. HOW YA LIKE ME NOW Kool Moe Dee [28Nov87 | 21May88 | 05Nov88] 35(1) 50	1	1604
770. AFTER MIDNIGHT The Manhattans [19Apr80 | 12Jul80 | 11Oct80] 24(2) 26	1	1600
771. BARRY Barry Manilow [13Dec80 | 24Jan81 | 25Apr81] 15(2) 20	2	1599
772. PETER GABRIEL (SECURITY) Peter Gabriel [02Oct82 | 06Nov82 | 30Apr83] 28(9) 31	1	1598
773. WINDOWS AND WALLS Dan Fogelberg [18Feb84 | 10Mar84 | 18Aug84] 15(3) 27	2	1593
774. THE RETURN OF BRUNO Bruce Willis [14Feb87 | 07Mar87 | 29Aug87] 14(3) 29	3	1592
775. TOO-RYE-AY Dexys Midnight Runners [12Feb83 | 30Apr83 | 23Jul83] 14(2) 24	2	1589
776. MASTER OF PUPPETS Metallica [29Mar86 | 10May86 | 08Aug87] 29(2) 72	0	1587
777. WILD PLANET The B-52s [20Sep80 | 18Oct80 | 21Mar81] 18(1) 27	1	1586
778. YOU CAN DANCE Madonna [05Dec87 | 23Jan88 | 30Apr88] 14(1) 22	0	1585
779. STREET CALLED DESIRE René & Angela [06Jul85 | 15Mar86 | 01Nov86] 64(4) 70	3	1585
780. LIVES IN THE BALANCE Jackson Browne [22Mar86 | 05Apr86 | 18Oct86] 23(3) 31	2	1582

Ranking the Albums

Rank. TITLE Act [Enter|Peak|Exit] Peak(Wks) TotWks	Hot 100 Score
781. THREE FOR LOVE Shalamar [10Jan81|28Mar81|12Sep81] 40(2) 36	2 1577
782. VOA Sammy Hagar [11Aug84|15Dec84|13Apr85] 32(3) 36	2 1571
783. 24/7 Dino [25Mar89|02Sep89|17Feb90] 34(3) 48	5,1* 1569
784. SCREAM DREAM Ted Nugent [31May80|05Jul80|27Sep80] 13(2) 18	1 1565
785. LIFES RICH PAGEANT R.E.M. [23Aug86|11Oct86|28Mar87] 21(3) 32	1 1563
786. TOOTH AND NAIL Dokken [13Oct84|29Jun85|08Mar86] 49(2) 74	1 1562
787. THE OTHER SIDE OF THE MIRROR Stevie Nicks [10Jun89|08Jul89|28Oct89] 10(1) 21	1 1562
788. BLINDED BY SCIENCE Thomas Dolby [05Feb83|02Apr83|03Sep83] 20(4) 31	0 1559
789. SMASHES, THRASHES & HITS KISS [03Dec88|21Jan89|03Jun89] 21(2) 27	1 1558
790. KING OF ROCK Run-D.M.C. [23Feb85|02Mar85|15Mar86] 52(3) 56	0 1556
791. FIRST OFFENSE Corey Hart [14Jul84|15Sep84|16Mar85] 31(3) 36	2 1550
792. DONNA SUMMER Donna Summer [14Aug82|18Sep82|20(5) 37	3 1550
793. RANT 'N' RAVE WITH THE STRAY CATS Stray Cats [10Sep83|15Oct83|24Mar84] 14(2) 29	3 1548
794. TONIGHT David Bowie [20Oct84|24Nov84|30Mar85] 11(2) 24	2 1546
795. IN STEP Stevie Ray Vaughan And Double Trouble [01Jul89|26Aug89|10Nov90] 33(1) 47	0 1544
796. VISION QUEST Soundtrack [02Mar85|04May85|03Aug85] 11(1) 23	3 1542
797. LUSH LIFE Linda Ronstadt [08Dec84|05Jan85|01Jun85] 13(2) 26	0 1541
798. STAND BY ME Soundtrack [20Sep86|06Dec86|25Jul87] 31(6) 45	1 1541
799. HARD TO HOLD (SOUNDTRACK) Rick Springfield [07Apr84|05May84|12Jan85] 16(3) 36	4 1538
800. BUSY BODY Luther Vandross [24Dec83|11Feb84|29Sep84] 32(4) 41	3 1534
801. CATCHING THE SUN Spyro Gyra [22Mar80|03May80|04Oct80] 19(1) 29	1 1533
802. NO ONE CAN DO IT BETTER The D.O.C. [19Aug89|23Sep89|07Apr90] 20(2) 34	0 1531
803. BIG THING Duran Duran [05Nov88|26Nov88|29Apr89] 24(2) 26	3 1528
804. RUNNING IN THE FAMILY Level 42 [11Apr87|01Aug87|28Nov87] 23(1) 34	2 1525
805. KATRINA AND THE WAVES Katrina & The Waves [13Apr85|15Jun85|16Nov85] 25(2) 39	2 1525
806. FLEETWOOD MAC LIVE Fleetwood Mac [27Dec80|17Jan81|25Apr81] 14(3) 18	1 1524
807. TAO Rick Springfield [27Apr85|01Jun85|26Oct85] 21(2) 27	2 1524
808. TWENTY GREATEST HITS Kenny Rogers [12Nov83|07Jan84|02Jun84] 22(2) 30	0 1524
809. GAP BAND V - JAMMIN' The Gap Band [10Sep83|15Oct83|30Jun84] 28(1) 43	0 1519
810. FRANKE & THE KNOCKOUTS Franke & The Knockouts [28Mar81|20Jun81|26Sep81] 31(1) 27	2 1518
811. POWERLIGHT Earth, Wind & Fire [12Mar83|26Mar83|30Jul83] 12(4) 21	2 1518
812. GREATEST HITS Ronnie Milsap [25Oct80|28Feb81|01Aug81] 36(1) 41	1 1516
813. TRUE Spandau Ballet [14May83|29Oct83|31Mar84] 19(1) 37	3 1516
814. CRUSH Orchestral Manoeuvres In The Dark [27Jul85|23Nov85|26Jul86] 38(2) 53	2 1514
815. WHITE NIGHTS Soundtrack [02Nov85|01Feb86|26Apr86] 17(2) 26	1 1513
816. THE NUMBER OF THE BEAST Iron Maiden [10Apr82|12Jun82|10Nov84] 33(2) 65	0 1510
817. PONCHO & LEFTY Merle Haggard/Willie Nelson [12Feb83|20Aug83|11Feb84] 37(1) 53	0 1509
818. BARK AT THE MOON Ozzy Osbourne [10Dec83|28Jan84|23Jun84] 19(3) 29	0 1506
819. LIFE AS WE KNOW IT REO Speedwagon [28Feb87|04Apr87|23Jan88] 28(2) 48	3 1504
820. BULLETBOYS BulletBoys [29Oct88|11Feb89|16Sep89] 34(3) 47	2 1504
821. GREATEST HITS The Judds [27Aug88|10Sep88|23Jan93] 76(2) 97	0 1503
822. THE BREAKFAST CLUB Soundtrack [09Mar85|18May85|31Aug85] 17(2) 26	1 1501
823. THERE GOES THE NEIGHBORHOOD Joe Walsh [23May81|18Jul81|19Sep81] 20(1) 18	1 1499
824. TRUE CONFESSIONS Bananarama [16Aug86|20Sep86|21Feb87] 15(2) 28	3 1497
825. CODA Led Zeppelin [18Dec82|15Jan83|02Apr83] 6(3) 16	0 1495
826. MUSIC MAN Waylon Jennings [07Jun80|12Jul80|28Mar81] 36(2) 43	1 1492
827. HEARSAY Alexander O'Neal [22Aug87|26Sep87|21May88] 29(1) 40	3 1492
828. THE CLARKE/DUKE PROJECT Stanley Clarke & George Duke [09May81|04Jul81|10Oct81] 33(4) 23	1 1491
829. PHANTOMS The Fixx [08Sep84|20Oct84|23Mar85] 19(3) 29	2 1488
830. I'LL ALWAYS LOVE YOU Anne Murray [03Nov79|19Jan80|05Apr80] 24(2) 23	2,*1 1486
831. WHAT CHA' GONNA DO FOR ME Chaka Khan [09May81|20Jun81|05Sep81] 17(2) 18	1 1482
832. PRESTO Rush [02Dec89|16Dec89|02Jun90] 16(2) 27	0 1481
833. EVANGELINE Emmylou Harris [21Feb81|28Mar81|01Aug81] 22(3) 24	1 1480
834. SPECIAL BEAT SERVICE English Beat [13Nov82|09Apr83|10Sep83] 39(3) 44	0 1477
835. THE RIGHT STUFF Vanessa Williams [09Jul88|15Apr89|22Jul89] 38(1) 55	3 1476
836. MAKE YOUR MOVE Captain & Tennille [17Nov79|09Feb80|26Apr80] 23(2) 24	3 1476
837. THE LAST IN LINE Dio [21Jul84|25Aug84|16Mar85] 23(2) 35	0 1475
838. ONE IN A MILLION YOU Larry Graham [21Jun80|16Aug80|29Nov80] 26(3) 24	2 1475
839. WORD OF MOUTH Toni Basil [23Oct82|11Dec82|14May83] 22(7) 30	2 1472
840. HEADHUNTER Krokus [16Apr83|09Jul83|21Jan84] 25(2) 41	0 1468
841. FLICK OF THE SWITCH AC/DC [10Sep83|15Oct83|11Feb84] 15(2) 23	1 1466
842. EMOTION Barbra Streisand [27Oct84|10Nov84|04May85] 19(3) 28	3 1464
843. GET HAPPY!! Elvis Costello And The Attractions [22Mar80|12Apr80|28Jun80] 11(2) 15	0 1463
844. IT'S TIME FOR LOVE Teddy Pendergrass [03Oct81|31Oct81|03Apr82] 19(2) 27	1 1462
845. MUSIC FOR THE MASSES Depeche Mode [24Oct87|21Nov87|03Dec88] 35(2) 59	4 1461
846. ROBBIE NEVIL Robbie Nevil [29Nov86|07Feb87|10Oct87] 37(2) 46	3 1460
847. SOLID Ashford & Simpson [10Nov84|02Mar85|13Jul85] 29(2) 36	1 1459
848. THROWIN' DOWN Rick James [05Jun82|10Jul82|06Nov82] 13(4) 23	1 1459
849. DAWN PATROL Night Ranger [25Dec82|02Apr83|03Nov84] 38(4) 69	2 1458
850. GET CLOSE The Pretenders [15Nov86|20Dec86|30May87] 25(3) 29	2 1457
851. TOMMY TUTONE-2 Tommy Tutone [06Feb82|22May82|28Aug82] 20(3) 30	1 1453
852. VIXEN Vixen [01Oct88|15Apr89|01Jul89] 41(1) 40	2 1452
853. HEARTBREAKER Dionne Warwick [30Oct82|15Jan83|07May83] 25(2) 28	2 1452
854. GHOST RIDERS The Outlaws [13Dec80|21Feb81|06Jun81] 25(3) 26	1 1449
855. IN A SPECIAL WAY DeBarge [22Oct83|18Feb84|11May85] 36(1) 40	2 1448
856. THE WANDERER Donna Summer [08Nov80|22Nov80|07Mar81] 13(2) 18	3 1445
857. LICK IT UP KISS [15Oct83|12Nov83|05May84] 24(2) 30	1 1444
858. "WEIRD AL" YANKOVIC IN 3-D Weird Al" Yankovic [17Mar84|28Apr84|18Aug84] 17(3) 23	3 1444
859. COULDN'T STAND THE WEATHER Stevie Ray Vaughan And Double Trouble [23Jun84|14Jul84|09Mar85] 31(3) 38	0 1443
860. YOU COULD HAVE BEEN WITH ME Sheena Easton [28Nov81|03Apr82|26Nov83] 47(1) 53	2 1443
861. NOW AND FOREVER Air Supply [19Jun82|21Aug82|05Mar83] 25(3) 38	3 1443
862. NO PROTECTION Starship [25Jul87|15Aug87|09Jan88] 12(1) 25	3 1442
863. THE TRINITY SESSION Cowboy Junkies [28Jan89|01Apr89|12Aug89] 26(1) 29	0 1441
864. BRITNY FOX Britny Fox [23Jul88|08Oct88|01Apr89] 39(2) 37	1 1441
865. ESPECIALLY FOR YOU The Smithereens [16Aug86|28Feb87|25Jul87] 51(1) 50	0 1438
866. STORMS OF LIFE Randy Travis [19Jul86|30Aug86|11Jun88] 85(2) 100	0 1430
867. HEARTBEAT Don Johnson [13Sep86|18Oct86|14Mar87] 17(2) 27	2 1426
868. STARFISH The Church [12Mar88|25Jun88|12Nov88] 41(2) 36	1 1426
869. KIHNSPIRACY Greg Kihn Band [12Mar83|30Apr83|20Aug83] 15(2) 24	2 1424
870. DON'T LET GO Isaac Hayes [29Sep79|19Jan80|19Apr80] 39(1) 30	1 1423
871. WHO MADE WHO (SOUNDTRACK) AC/DC [21Jun86|23Aug86|04Apr87] 33(4) 42	0 1419
872. EMERSON, LAKE, & POWELL Emerson, Lake & Powell [14Jun86|26Jul86|06Dec86] 23(2) 26	1 1419
873. ROBERTA FLACK FEATURING DONNY HATHAWAY Roberta Flack Featuring Donny Hathaway [29Mar80|31May80|06Sep80] 25(2) 24	2 1419
874. NEVER SURRENDER Triumph [29Jan83|09Apr83|30Jul83] 26(2) 27	0 1416
875. CYCLES The Doobie Brothers [10Jun89|01Jul89|21Oct89] 17(1) 20	2 1414
876. BOOMTOWN David & David [16Aug86|06Dec86|02May87] 39(2) 38	2 1413

Ranking the Albums

Rank. TITLE Act [Enter|Peak|Exit] Peak(Wks) TotWks	Hot 100 Score
877. DON'T STOP *Jeffrey Osborne* [20Oct84|24Nov84|29Jun85] 39(4) 37	2 1412
878. JACKRABBIT SLIM *Steve Forbert* [10Nov79|01Mar80|03May80] 20(1) 26	2 1412
879. SHINE ON *L.T.D.* [06Sep80|01Nov80|14Mar81] 28(1) 28	1 1409
880. 3 FEET HIGH AND RISING *De La Soul* [01Apr89|10Jun89|14Oct89] 24(2) 29	1 1407
881. POINTS ON THE CURVE *Wang Chung* [25Feb84|14Jul84|03Nov84] 30(3) 37	3 1405
882. CONDITION CRITICAL *Quiet Riot* [04Aug84|18Aug84|09Feb85] 15(3) 28	1 1402
883. CHICAGO 18 *Chicago* [18Oct86|07Mar87|22Aug87] 35(2) 45	4 1402
884. LET ME UP (I'VE HAD ENOUGH) *Tom Petty And The Heartbreakers* [09May87|20Jun87|19Sep87] 20(2) 20	1 1401
885. FASCINATION! *The Human League* [18Jun83|20Aug83|31Dec83] 22(1) 29	2 1400
886. OUT OF THIS WORLD *Europe* [27Aug88|08Oct88|11Feb89] 19(1) 25	1 1396
887. RIT *Lee Ritenour* [09May81|18Jul81|10Oct81] 26(1) 23	1 1396
888. NO NIGHT SO LONG *Dionne Warwick* [09Aug80|20Sep80|24Jan81] 23(2) 25	2 1396
889. WHITE CITY - A NOVEL *Pete Townshend* [30Nov85|08Feb86|14Jun86] 26(3) 29	1 1391
890. GREATEST HITS, VOLUME 2 *Linda Ronstadt* [08Nov80|06Dec80|28Mar81] 26(3) 21	0 1391
891. UNGUARDED *Amy Grant* [15Jun85|07Sep85|01Mar86] 35(1) 38	2 1391
892. COME MORNING *Grover Washington Jr.* [12Dec81|13Feb82|12Jun82] 28(2) 27	1 1388
893. LISTEN *A Flock Of Seagulls* [28May83|16Jul83|29Oct83] 16(2) 23	1 1383
894. MILK AND HONEY *John Lennon & Yoko Ono* [11Feb84|10Mar84|16Jun84] 11(1) 19	2 1381
895. SPIKE *Elvis Costello* [25Feb89|22Apr89|12Aug89] 32(1) 25	1 1380
896. PERHAPS LOVE *Placido Domingo* [07Nov81|30Jan82|08May82] 18(2) 27	1 1375
897. DANCING UNDERCOVER *Ratt* [25Oct86|15Nov86|25Jul87] 26(2) 40	1 1372
898. SECONDS OF PLEASURE *Rockpile* [15Nov80|27Dec80|21Mar81] 27(3) 19	1 1371
899. THE DREAM ACADEMY *The Dream Academy* [09Nov85|08Mar86|19Jul86] 20(1) 37	2 1369
900. SOME DAYS ARE DIAMONDS *John Denver* [04Jul81|19Sep81|23Jan82] 32(2) 30	2 1369
901. HEAVEN AND HELL *Black Sabbath* [14Jun80|19Jul80|22Nov80] 28(3) 24	0 1367
902. AMADEUS (SOUNDTRACK) *Neville Marriner* [24Nov84|20Apr85|17May86] 56(2) 78	0 1365
903. HOLD ME IN YOUR ARMS *Rick Astley* [28Jan89|18Feb89|01Jul89] 19(3) 23	3 1365
904. CUT *Golden Earring* [11Dec82|09Apr83|02Jul83] 24(2) 30	2 1364
905. PELICAN WEST *Haircut One Hundred* [24Apr82|07Aug82|01Jan83] 31(3) 37	1 1363
906. LOVE OVER GOLD *Dire Straits* [16Oct82|13Nov82|21May83] 19(3) 32	1 1361
907. POOLSIDE *Nu Shooz* [31May86|12Jul86|03Jan87] 27(2) 32	2 1359
908. MAN OF COLOURS *Icehouse* [17Oct87|14May88|13Aug88] 43(1) 44	3 1358
909. THERE AND BACK *Jeff Beck* [12Jul80|09Aug80|22Nov80] 21(2) 20	0 1358
910. COME UPSTAIRS *Carly Simon* [12Jul80|15Nov80|14Feb81] 36(2) 32	1 1356
911. IF I WERE YOUR WOMAN *Stephanie Mills* [27Jun87|29Aug87|27Feb88] 30(1) 36	1 1353
912. AS ONE *Kool & The Gang* [09Oct82|18Dec82|19Mar83] 29(4) 24	2 1353
913. A VERY SPECIAL CHRISTMAS *Various Artists* [14Nov87|26Dec87|12Jan91] 20(3) 31	0 1351
914. BANANARAMA *Bananarama* [02Jun84|29Sep84|02Feb85] 30(5) 36	2 1349
915. RUN D.M.C. *Run-D.M.C.* [23Jun84|28Jul84|14Sep85] 53(2) 65	0 1346
916. THE LION AND THE COBRA *Sinead O'Connor* [06Feb88|23Apr88|16Jun90] 36(3) 38	0 1345
917. MAGIC TOUCH *Stanley Jordan* [25May85|27Jul85|23Aug86] 64(2) 66	0 1345
918. BANGIN' *The Outfield* [04Jul87|15Aug87|21Nov87] 18(1) 21	1 1345
919. HOT IN THE SHADE *KISS* [04Nov89|25Nov89|07Jul90] 29(3) 36	3,2* 1344
920. BUT THE LITTLE GIRLS UNDERSTAND *The Knack* [01Mar80|29Mar80|31May80] 15(1) 14	2 1344
921. STRAIGHT FROM THE HEART *Patrice Rushen* [01May82|26Jun82|06Nov82] 14(2) 28	1 1343
922. LOVESEXY *Prince* [28May88|11Jun88|22Oct88] 11(1) 21	1 1341
923. NAKED *Talking Heads* [02Apr88|07May88|20Aug88] 19(1) 21	0 1340
924. GHOSTBUSTERS II *Soundtrack* [01Jul89|29Jul89|04Nov89] 14(1) 19	0 1340
925. NAKED EYES *Naked Eyes* [16Apr83|18Jun83|28Jan84] 32(2) 42	3 1339
926. OLD 8 X 10 *Randy Travis* [30Jul88|03Sep88|20May89] 35(1) 43	0 1338
927. DEDICATION *Gary U.S. Bonds* [02May81|13Jun81|12Sep81] 27(2) 20	1 1338
928. HEADED FOR THE FUTURE *Neil Diamond* [24May86|19Jul86|25Oct86] 20(1) 23	1 1336
929. IRONS IN THE FIRE *Teena Marie* [13Sep80|22Nov80|28Mar81] 38(3) 29	1 1335
930. MESSAGES FROM THE BOYS *The Boys* [26Nov88|11Feb89|29Jul89] 33(2) 36	1 1335
931. ROCKY III *Soundtrack* [10Jul82|21Aug82|13Nov82] 15(5) 19	0 1334
932. SPEAK OF THE DEVIL *Ozzy Osbourne* [11Dec82|15Jan83|23Apr83] 14(4) 20	0 1334
933. PUNCH THE CLOCK *Elvis Costello And The Attractions* [13Aug83|01Oct83|21Jan84] 24(1) 24	1 1334
934. INFIDELS *Bob Dylan* [19Nov83|03Dec83|28Apr84] 20(5) 24	1 1332
935. DRAMA *Yes* [13Sep80|04Oct80|17Jan81] 18(2) 19	0 1330
936. JUMP TO IT *Aretha Franklin* [14Aug82|09Oct82|05Mar83] 23(4) 30	1 1328
937. WE'VE GOT TONIGHT *Kenny Rogers* [12Mar83|30Apr83|08Oct83] 18(1) 27	3 1326
938. EDDIE MURPHY: COMEDIAN *Eddie Murphy* [19Nov83|28Jan84|01Jun85] 35(1) 44	0 1323
939. NO NUKES/THE MUSE CONCERTS FOR A NON-NUCLEAR FUTURE *Various Artists* [22Dec79|09Feb80|19Apr80] 19(2) 18	0 1322
940. THE BEST OF BLONDIE *Blondie* [31Oct81|12Dec81|03Apr82] 30(2) 23	0 1318
941. STARCHILD *Teena Marie* [15Dec84|06Apr85|10Aug85] 31(2) 35	2 1315
942. BEAT STREET *Soundtrack* [02Jun84|21Jul84|20Oct84] 14(2) 21	2 1313
943. POWERSLAVE *Iron Maiden* [29Sep84|13Oct84|18May85] 21(3) 34	0 1313
944. MOTHER'S MILK *Red Hot Chili Peppers* [16Sep89|02Dec89|30Jun90] 52(1) 42	0 1311
945. AFTER EIGHT *Taco* [23Jul83|10Sep83|31Dec83] 23(2) 24	1 1311
946. NO HOLDIN' BACK *Randy Travis* [14Oct89|04Nov89|01Sep90] 33(1) 47	0 1311
947. STRENGTH *The Alarm* [09Nov85|15Feb86|12Jul86] 39(3) 36	1 1304
948. VISIONS *Gladys Knight & The Pips* [21May83|25Jun83|04Feb84] 34(2) 33	1 1303
949. LOST BOYS *Soundtrack* [01Aug87|03Oct87|02Jul88] 15(1) 39	1 1302
950. SOMEWHERE IN ENGLAND *George Harrison* [20Jun81|11Jul81|12Sep81] 11(2) 13	1 1295
951. OOH YEAH! *Daryl Hall & John Oates* [21May88|04Jun88|12Nov88] 24(2) 26	3 1295
952. AUDIO-VISIONS *Kansas* [04Oct80|08Nov80|21Feb81] 26(1) 21	2 1292
953. MONEY AND CIGARETTES *Eric Clapton* [19Feb83|26Mar83|25Jun83] 16(4) 19	1 1289
954. DON'T STOP *Billy Idol* [24Oct81|29Oct83|17Nov84] 71(1) 68	0 1288
955. ABOUT FACE *David Gilmour* [17Mar84|09Jun84|22Sep84] 32(3) 28	1 1288
956. SHANGO *Santana* [04Sep82|23Oct82|05Feb83] 22(3) 23	2 1288
957. STANDING ON A BEACH -- THE SINGLES *The Cure* [14Jun86|23Aug86|11Jul87] 48(2) 57	0 1287
958. HOW COULD IT BE *Eddie Murphy* [12Oct85|25Jan86|05Apr86] 26(1) 26	1 1285
959. BABYLON AND ON *Squeeze* [03Oct87|05Dec87|11Jun88] 36(2) 29	2 1284
960. BEST KEPT SECRET *Sheena Easton* [17Sep83|15Oct83|02Jun84] 33(2) 38	3 1284
961. LAW AND ORDER *Lindsey Buckingham* [07Nov81|12Dec81|17Apr82] 32(1) 24	1 1282
962. LISA LISA & CULT JAM WITH FULL FORCE *Lisa Lisa And Cult Jam With Full Force* [31Aug85|05Oct85|14Mar87] 52(2) 66	3 1280
963. EL DEBARGE *El DeBarge* [14Jun86|05Jul86|15Nov86] 24(3) 23	3 1280
964. SIGN IN PLEASE *Autograph* [05Jan85|30Mar85|20Jul85] 29(2) 29	1 1278
965. ELECTRIC *The Cult* [25Apr87|16May87|28Nov87] 38(3) 32	0 1278
966. EDDIE MURPHY *Eddie Murphy* [14Aug82|09Oct82|21Apr84] 52(5) 53	0 1275
967. LIVE *George Thorogood & The Destroyers* [23Aug86|20Sep86|06Jun87] 33(3) 42	0 1273
968. BACK IN BLACK *Whodini* [17May86|12Jul86|07Feb87] 35(2) 39	0 1272
969. GIUFFRIA *Giuffria* [08Dec84|02Mar85|22Jun85] 26(2) 29	2 1272
970. WIDE AWAKE IN DREAMLAND *Pat Benatar* [23Jul88|10Sep88|04Feb89] 28(2) 29	1 1272
971. THE MANY FACETS OF ROGER *Roger* [03Oct81|21Nov81|20Mar82] 26(1) 25	0 1271
972. THE GLOW OF LOVE *Change* [10May80|19Jul80|25Oct80] 29(2) 25	1 1270
973. GO TO HEAVEN *Grateful Dead* [17May80|28Jun80|04Oct80] 23(2) 21	1 1269

Rank. TITLE Act [Enter | Peak | Exit] Peak(Wks) TotWks	Hot 100	Score
974. OPERATION: MINDCRIME Queensryche [21May88 | 04Jun88 | 07Oct89] 50(2) 52	0	1266
975. IT'S YOUR NIGHT James Ingram [12Nov83 | 17Mar84 | 25Aug84] 46(1) 42	3	1262
976. BRANIGAN Laura Branigan [25Sep82 | 25Dec82 | 28May83] 34(5) 36	2	1260
977. THE HURTING Tears For Fears [07May83 | 16Jul83 | 08Mar86] 73(1) 69	1	1256
978. LITTLE ROBBERS The Motels [15Oct83 | 05Nov83 | 24Mar84] 22(4) 24	2	1254
979. LET THE MUSIC PLAY Shannon [11Feb84 | 24Mar84 | 20Oct84] 32(2) 37	2	1253
980. WE TOO ARE ONE Eurythmics [30Sep89 | 18Nov89 | 07Apr90] 34(2) 28	1	1252
981. PLEASURE VICTIM Berlin [19Feb83 | 07May83 | 08Oct83] 30(1) 34	3	1250
982. MIDNIGHT TO MIDNIGHT Psychedelic Furs [07Mar87 | 02May87 | 05Sep87] 29(2) 27	1	1246
983. ONE ON ONE Cheap Trick [29May82 | 09Oct82 | 27Nov82] 39(1) 27	2	1245
984. SERGIO MENDES(2) Sergio Mendes [07May83 | 30Jul83 | 05Nov83] 27(2) 27	2	1245
985. LIVE AFTER DEATH Iron Maiden [16Nov85 | 14Dec85 | 12Apr86] 19(1) 22	0	1244
986. CHAIN LIGHTNING Don McLean [14Feb81 | 11Apr81 | 04Jul81] 28(2) 21	3	1243
987. BY REQUEST (THE BEST OF BILLY VERA & THE BEATERS) Billy Vera & The Beaters [06Dec86 | 28Feb87 | 30May87] 15(1) 26	1	1239
988. BRILLIANCE Atlantic Starr [27Mar82 | 22May82 | 09Oct82] 18(2) 29	1	1239
989. CHRONICLES Steve Winwood [21Nov87 | 16Jan88 | 14May88] 26(1) 26	2	1238
990. GREATEST HITS Little River Band [04Dec82 | 12Feb83 | 25Jun83] 33(6) 30	1	1237
991. CRASH The Human League [04Oct86 | 22Nov86 | 21Mar87] 24(2) 25	2	1236
992. BORN TO BOOGIE Hank Williams Jr. [01Aug87 | 19Sep87 | 18Jun88] 28(1) 47	0	1235
993. A COLLECTION: GREATEST HITS...AND MORE Barbra Streisand [21Oct89 | 18Nov89 | 07Apr90] 26(3) 25	0	1234
994. THERE'S NO GETTIN' OVER ME Ronnie Milsap [05Sep81 | 17Oct81 | 03Apr82] 31(1) 31	2	1232
995. I WANNA HAVE SOME FUN Samantha Fox [26Nov88 | 18Feb89 | 15Jul89] 37(1) 34	2	1231
996. POINT OF ENTRY Judas Priest [04Apr81 | 23May81 | 19Sep81] 39(2) 25	0	1229
997. EXTENDED PLAY The Pretenders [18Apr81 | 16May81 | 31Oct81] 27(2) 29	0	1225
998. HOOKED ON CLASSICS II (CAN'T STOP THE CLASSICS) Royal Philharmonic Orchestra [28Aug82 | 09Oct82 | 04Jun83] 33(4) 41	0	1225
999. BAD LUCK STREAK IN DANCING SCHOOL Warren Zevon [08Mar80 | 05Apr80 | 21Jun80] 20(1) 16	1	1220
1000. CREST OF A KNAVE Jethro Tull [10Oct87 | 12Dec87 | 16Apr88] 32(1) 28	0	1219
1001. DOUBLE VISION Bob James/David Sanborn [14Jun86 | 19Jul86 | 29Aug86] 50(2) 64	0	1216
1002. BRANIGAN 2 Laura Branigan [09Apr83 | 11Jun83 | 17Nov84] 29(1) 37	2	1216
1003. BORN TO BE BAD George Thorogood & The Destroyers [06Feb88 | 05Mar88 | 16Jul88] 32(3) 24	0	1214
1004. BETWEEN THE SHEETS Isley Brothers [04Jun83 | 09Jul83 | 05Nov83] 19(2) 23	0	1214

Rank. TITLE Act [Enter | Peak | Exit] Peak(Wks) TotWks	Hot 100	Score
1005. FASTWAY Fastway [28May83 | 10Sep83 | 31Dec83] 31(1) 32	0	1213
1006. EVEN WORSE "Weird Al" Yankovic [07May88 | 02Jul88 | 29Oct88] 27(1) 26	1	1209
1007. 40 HOUR WEEK Alabama [23Feb85 | 30Mar85 | 23Nov85] 28(3) 40	0	1203
1008. TOUCH ME Samantha Fox [29Nov86 | 28Feb87 | 06Jun87] 24(2) 28	2	1202
1009. ANIMOTION Animotion [23Feb85 | 25May85 | 14Sep85] 28(1) 30	2	1200
1010. WHAT ABOUT ME? Kenny Rogers [22Sep84 | 17Nov84 | 20Apr85] 31(3) 31	2	1199
1011. LEGACY Poco [23Sep89 | 11Nov89 | 31Mar90] 40(2) 28	2,1*	1198
1012. ALBUM Joan Jett & the Blackhearts [16Jul83 | 13Aug83 | 26Nov83] 20(2) 20	2	1196
1013. AUGUST Eric Clapton [27Dec86 | 21Mar87 | 15Aug87] 37(2) 34	0	1193
1014. SOMEWHERE OVER THE RAINBOW Willie Nelson [21Mar81 | 11Apr81 | 22Aug81] 31(3) 23	0	1192
1015. PLANETARY INVASION Midnight Star [08Dec84 | 16Feb85 | 13Jul85] 32(2) 32	2	1191
1016. LOVE LIFE Berlin [31Mar84 | 09Jun84 | 20Oct84] 28(2) 30	2	1186
1017. I'M NO ANGEL Gregg Allman Band [07May87 | 09May87 | 12Sep87] 30(3) 28	1	1183
1018. ROBBIE ROBERTSON Robbie Robertson [14Nov87 | 26Dec87 | 02Jul88] 38(2) 34	0	1181
1019. LIVE IN NEW ORLEANS Maze Featuring Frankie Beverly [04Jul81 | 15Aug81 | 02Jan82] 34(2) 27	0	1181
1020. ZAPP Zapp [27Sep80 | 01Nov80 | 31Jan81] 19(2) 19	1	1180
1021. BROTHER WHERE YOU BOUND Supertramp [01Jun85 | 13Jul85 | 26Oct85] 21(2) 22	1	1176
1022. SILK ELECTRIC Diana Ross [23Oct82 | 04Dec82 | 02Apr83] 27(2) 24	2	1176
1023. BOBBIE SUE The Oak Ridge Boys [20Feb82 | 03Apr82 | 10Jul82] 20(3) 21	2	1175
1024. THUNDER SEVEN Triumph [08Dec84 | 02Mar85 | 29Jun85] 35(2) 30	1	1173
1025. STEPHANIE Stephanie Mills [16May81 | 27Jun81 | 17Oct81] 30(2) 23	1	1171
1026. SPECIAL THINGS The Pointer Sisters [30Aug80 | 29Nov80 | 07Feb81] 34(2) 24	2	1170
1027. SKYLARKIN' Grover Washington Jr. [08Mar80 | 19Apr80 | 02Aug80] 24(2) 22	0	1170
1028. NOW OR NEVER John Schneider [27Jun81 | 22Aug81 | 21Nov81] 37(1) 22	2	1167
1029. TECHNIQUE New Order [11Feb89 | 11Mar89 | 26Aug89] 32(1) 28	1	1165
1030. TWICE AS SWEET A Taste Of Honey [02Aug80 | 30May81 | 01Aug81] 36(2) 32	1	1161
1031. IT TAKES A NATION OF MILLIONS TO HOLD US BACK Public Enemy [23Jul88 | 13Aug88 | 19Aug89] 42(2) 51	0	1159
1032. TOUGH ALL OVER John Cafferty & The Beaver Brown Band [08Jun85 | 06Jul85 | 11Jan86] 40(3) 32	4	1156
1033. EMOTIONAL Jeffrey Osborne [28Jun86 | 23Aug86 | 20Dec86] 26(1) 26	2	1155
1034. BOY U2 [14Mar81 | 11Apr81 | 19Sep87] 63(2) 47	0	1154
1035. LIGHTS OUT Peter Wolf [11Aug84 | 15Sep84 | 02Feb85] 24(2) 26	3	1151
1036. TRUE COLOURS Split Enz [30Aug80 | 15Nov80 | 14Feb81] 40(2) 25	1	1146
1037. THE HUNGER Michael Bolton [10Oct87 | 02Apr88 | 16Jul88] 46(1) 41	3	1145

Rank. TITLE Act [Enter | Peak | Exit] Peak(Wks) TotWks	Hot 100	Score
1038. ME MYSELF I Joan Armatrading [07Jun80 | 09Aug80 | 08Nov80] 28(2) 23	0	1145
1039. NEW TRADITIONALISTS Devo [10Oct81 | 07Nov81 | 27Mar82] 23(2) 25	1	1144
1040. THE FOX Elton John [06Jun81 | 27Jun81 | 10Oct81] 21(2) 19	2	1144
1041. COMPUTER-WORLD Kraftwerk [06Jun81 | 22Aug81 | 20Mar82] 72(1) 42	0	1144
1042. RADIO ROMANCE Eddie Rabbitt [06Nov82 | 26Feb83 | 23Apr83] 31(2) 25	2	1143
1043. KNOWLEDGE IS KING Kool Moe Dee [17Jun89 | 15Jul89 | 18Nov89] 25(2) 23	0	1142
1044. QR III Quiet Riot [02Aug86 | 20Sep86 | 31Jan87] 31(2) 27	0	1142
1045. NUCLEAR FURNITURE Jefferson Starship [16Jun84 | 28Jul84 | 17Nov84] 28(4) 23	2	1140
1046. THE BLASTERS The Blasters [09Jan82 | 08May82 | 31Jul82] 36(3) 30	0	1135
1047. ALIVE SHE CRIED The Doors [05Nov83 | 03Dec83 | 17Mar84] 23(2) 20	1	1135
1048. ON THROUGH THE NIGHT Def Leppard [03May80 | 05Jul80 | 10Mar84] 51(1) 51	0	1134
1049. THE DISREGARD OF TIMEKEEPING Bonham [07Oct89 | 16Dec89 | 21Apr90] 38(1) 29	1,1*	1131
1050. ARETHA Aretha Franklin [25Oct80 | 06Dec80 | 16May81] 47(3) 30	2	1128
1051. INTO THE LIGHT Chris de Burgh [20Sep86 | 20Jun87 | 19Sep87] 25(2) 32	1	1127
1052. FREEDOM Neil Young [21Oct89 | 25Nov89 | 28Apr90] 35(2) 28	0	1127
1053. ALL THE GREAT HITS Diana Ross [24Oct81 | 28Nov81 | 29May82] 37(2) 32	0	1126
1054. ALL THE BEST COWBOYS HAVE CHINESE EYES Pete Townshend [10Jul82 | 11Sep82 | 01Jan83] 26(1) 26	0	1124
1055. MISSING PERSONS Missing Persons [15May82 | 25Sep82 | 02Apr83] 46(2) 47	2	1124
1056. KINGS OF THE WILD FRONTIER Adam And The Ants [28Feb81 | 13Jun81 | 24Oct81] 44(2) 35	0	1122
1057. JUICY FRUIT Mtume [28May83 | 09Jul83 | 22Oct83] 26(2) 22	1	1121
1058. WHAT TIME IS IT? The Time [25Sep82 | 30Oct82 | 07May83] 26(3) 33	1	1120
1059. ROD STEWART GREATEST HITS Rod Stewart [24Nov79 | 05Jan80 | 29Mar80] 22(2) 19	1	1118
1060. PAUL'S BOUTIQUE Beastie Boys [12Aug89 | 02Sep89 | 18Nov89] 14(2) 15	1	1118
1061. MOSAIC Wang Chung [01Nov86 | 06Dec86 | 04Jul87] 41(2) 36	3	1118
1062. THE $5.98 E.P.: GARAGE DAYS RE-REVISITED Metallica [12Sep87 | 10Oct87 | 02Apr88] 28(2) 30	0	1117
1063. THE COMPLETION BACKWARD PRINCIPLE The Tubes [30May81 | 25Jul81 | 28Nov81] 36(3) 27	1	1117
1064. SACRED HEART Dio [31Aug85 | 21Sep85 | 15Mar86] 29(4) 29	0	1116
1065. THE TIME The Time [12Sep81 | 31Oct81 | 17Apr82] 50(1) 32	1	1115
1066. ONE MORE SONG Randy Meisner [01Nov80 | 28Mar81 | 13Jun81] 50(1) 33	2	1114
1067. BLACK SEA XTC [22Nov80 | 21Feb81 | 02May81] 41(1) 24	0	1114
1068. ROCKIN' WITH THE RHYTHM The Judds [16Nov85 | 07Jun86 | 25Apr87] 66(1) 57	0	1113
1069. THE POET Bobby Womack [26Dec81 | 13Mar82 | 29May82] 29(2) 23	0	1109

Ranking the Albums

Rank. TITLE Act [Enter | Peak | Exit] Peak(Wks) TotWks	Hot 100	Score
1070. NIECY Deniece Williams [17Apr82 | 12Jun82 | 11Sep82] 20(3) 22	1	1107
1071. SOUL TO SOUL Stevie Ray Vaughan And Double Trouble [12Oct85 | 23Nov85 | 05Jul86] 34(2) 39	0	1106
1072. STAY WITH ME Regina Belle [16Sep89 | 18Nov89 | 04Aug90] 63(2) 44	2,1*	1106
1073. PACK UP THE PLANTATION - LIVE! Tom Petty And The Heartbreakers [14Dec85 | 01Mar86 | 07Jun86] 22(2) 26	1	1104
1074. ANNIE Soundtrack [29May82 | 14Aug82 | 29Jan83] 35(2) 31	0	1104
1075. THAT'S WHY I'M HERE James Taylor [23Nov85 | 21Dec85 | 14Jun86] 34(3) 30	1	1103
1076. AFTER DARK Andy Gibb [01Mar80 | 12Apr80 | 07Jun80] 21(1) 15	2	1099
1077. LOVE LANGUAGE Teddy Pendergrass [16Jun84 | 28Jul84 | 16Feb85] 38(4) 35	1	1095
1078. READY OR NOT Lou Gramm [28Feb87 | 25Apr87 | 22Aug87] 27(1) 26	2	1094
1079. UNION Toni Childs [25Jun88 | 15Oct88 | 29Apr89] 63(1) 45	1	1092
1080. RARITIES The Beatles [12Apr80 | 31May80 | 19Jul80] 21(1) 15	0	1091
1081. EAST SIDE STORY Squeeze [30May81 | 22Aug81 | 14Nov81] 44(1) 25	1	1091
1082. HOUNDS OF LOVE Kate Bush [26Oct85 | 14Dec85 | 26Apr86] 30(1) 27	1	1088
1083. THE WORKS Queen [17Mar84 | 28Apr84 | 28Jul84] 23(1) 20	3	1088
1084. SHE'S STRANGE Cameo [17Mar84 | 05May84 | 25Aug84] 27(1) 24	1	1083
1085. TOUCH THE WORLD Earth, Wind & Fire [21Nov87 | 19Dec87 | 28May88] 33(1) 28	2	1083
1086. RAW LIKE SUSHI Neneh Cherry [24Jun89 | 16Sep89 | 17Feb90] 40(1) 35	3,1*	1083
1087. 2ND WAVE Surface [26Nov88 | 09Sep89 | 02Dec89] 56(2) 39	3	1080
1088. LESS THAN ZERO Soundtrack [05Dec87 | 16Jan88 | 07May88] 31(2) 23	2	1080
1089. THE GAP BAND II The Gap Band [22Dec79 | 29Mar80 | 28Jun80] 42(1) 28	0	1079
1090. RE-AC-TOR Neil Young & Crazy Horse [21Nov81 | 19Dec81 | 13Mar82] 27(1) 17	1	1078
1091. TWO PLACES AT THE SAME TIME Ray Parker Jr. & Raydio [12Apr80 | 24May80 | 30Aug80] 33(2) 21	1	1077
1092. GET CLOSER Linda Ronstadt [16Oct82 | 13Nov82 | 23Apr83] 31(4) 28	3	1075
1093. FLASH GORDON (SOUNDTRACK) Queen [27Dec80 | 07Feb81 | 04Apr81] 23(2) 15	1	1073
1094. ARTHUR (THE ALBUM) Soundtrack [05Sep81 | 31Oct81 | 30Jan82] 32(3) 22	1	1071
1095. LIVING IN A FANTASY Leo Sayer [18Oct80 | 10Jan81 | 21Mar81] 36(2) 23	2	1070
1096. CANCIONES DE MI PADRE Linda Ronstadt [12Dec87 | 06Feb88 | 06Aug88] 42(2) 35	0	1069
1097. THE LOVER IN ME Sheena Easton [03Dec88 | 18Feb89 | 27May89] 44(4) 26	1	1067
1098. I'M THE MAN Anthrax [19Dec87 | 30Jan88 | 17Sep88] 53(1) 40	0	1066
1099. BEHIND THE SUN Eric Clapton [06Apr85 | 25May85 | 12Oct85] 34(1) 28	2	1065
1100. SUCKING IN THE SEVENTIES Rolling Stones [04Apr81 | 18Apr81 | 20Jun81] 15(2) 12	0	1065
1101. BEATIN' THE ODDS Molly Hatchet [20Sep80 | 11Oct80 | 07Feb81] 25(2) 21	1	1065
1102. WAKING UP WITH THE HOUSE ON FIRE Culture Club [24Nov84 | 01Dec84 | 06Apr85] 26(5) 20	2	1065
1103. LET ME BE YOUR ANGEL Stacy Lattisaw [05Jul80 | 18Oct80 | 10Jan81] 44(1) 28	1	1063
1104. BALIN Marty Balin [06Jun81 | 22Aug81 | 07Nov81] 35(2) 23	2	1063
1105. JOY AND PAIN Maze Featuring Frankie Beverly [02Aug80 | 04Oct80 | 24Jan81] 31(1) 23	0	1061
1106. WILLIE NELSON SINGS KRISTOFFERSON Willie Nelson [17Nov79 | 12Jan80 | 03May80] 42(2) 25	0	1061
1107. WHAT UP, DOG? Was (Not Was) [15Oct88 | 08Apr89 | 24Jun89] 43(2) 37	3	1057
1108. TWO OF A KIND (SOUNDTRACK) Soundtrack [03Dec83 | 04Feb84 | 14Apr84] 26(1) 20	2	1055
1109. BIG WORLD Joe Jackson [19Apr86 | 24May86 | 04Oct86] 34(2) 25	0	1053
1110. OUTRIDER Jimmy Page [09Jul88 | 30Jul88 | 19Nov88] 26(2) 20	0	1052
1111. FAHRENHEIT Toto [13Sep86 | 29Nov86 | 16May87] 40(2) 36	2	1052
1112. TEXAS FLOOD Stevie Ray Vaughan And Double Trouble [23Jul83 | 01Oct83 | 03Mar84] 38(1) 33	0	1049
1113. RADIO LL Cool J [11Jan86 | 15Mar86 | 27Sep86] 46(3) 38	0	1049
1114. LUXURY OF LIFE Five Star [21Sep85 | 08Mar86 | 09Aug86] 57(2) 47	2	1049
1115. EVERY GENERATION Ronnie Laws [16Feb80 | 24May80 | 21Jun80] 24(2) 19	0	1049
1116. JEFFREY OSBORNE Jeffrey Osborne [19Jun82 | 14Aug82 | 09Apr83] 49(3) 43	3	1048
1117. WILD HEART OF THE YOUNG Karla Bonoff [03Apr82 | 19Jun82 | 27Nov82] 49(1) 35	2	1046
1118. GET LOOSE Evelyn King [11Sep82 | 06Nov82 | 16Apr83] 27(3) 32	2	1044
1119. A SHIP ARRIVING TOO LATE TO SAVE A DROWNING WITCH Frank Zappa [12Jun82 | 21Aug82 | 06Nov82] 23(2) 22	1	1044
1120. E.T. - THE EXTRA-TERRESTRIAL Soundtrack [03Jul82 | 28Aug82 | 12Feb83] 37(2) 33	0	1043
1121. STATE OF EUPHORIA Anthrax [08Oct88 | 22Oct88 | 10Jun89] 30(2) 36	0	1043
1122. FLY ON THE WALL AC/DC [20Jul85 | 07Sep85 | 08Feb86] 32(2) 30	0	1041
1123. TALK IS CHEAP Keith Richards [22Oct88 | 19Nov88 | 25Mar89] 24(1) 23	0	1040
1124. RESTLESS Starpoint [05Oct85 | 01Feb86 | 23Aug86] 60(3) 47	2	1039
1125. SO RED THE ROSE Arcadia [21Dec85 | 18Jan86 | 12Apr86] 23(2) 17	2	1039
1126. DOOR TO DOOR The Cars [12Sep87 | 26Sep87 | 13Feb88] 26(3) 23	3	1037
1127. DOWN ON THE FARM Little Feat [08Dec79 | 05Jan80 | 26Apr80] 29(1) 21	0	1036
1128. NEVER LET ME DOWN David Bowie [23May87 | 30May87 | 14Nov87] 34(1) 26	2	1035
1129. ATF After The Fire [12Mar83 | 30Apr83 | 23Jul83] 25(3) 20	2	1034
1130. DON'T BE AFRAID OF THE DARK Robert Cray Band [27Aug88 | 24Sep88 | 01Apr89] 32(2) 32	1	1034
1131. SHOWTIME The J. Geils Band [04Dec82 | 29Jan83 | 09Apr83] 23(3) 19	2	1033
1132. SHAKEN 'N' STIRRED Robert Plant [15Jun85 | 13Jul85 | 19Oct85] 20(2) 19	1	1032
1133. GIVE MY REGARDS TO BROAD STREET Paul McCartney [10Nov84 | 24Nov84 | 09Mar85] 21(2) 18	1	1029
1134. POWER Ice-T [01Oct88 | 12Nov88 | 13May89] 35(1) 33	0	1029
1135. ROMANCE 1600 Sheila E. [21Sep85 | 08Mar86 | 03May86] 50(1) 33	1	1028
1136. ALL OUR LOVE Gladys Knight & The Pips [12Dec87 | 19Mar88 | 11Jun88] 39(1) 27	1	1027
1137. SO FAR, SO GOOD... SO WHAT! Megadeth [06Feb88 | 27Feb88 | 09Jul88] 28(1) 23	0	1027
1138. WE ARE ONE Maze Featuring Frankie Beverly [28May83 | 18Jun83 | 19Nov83] 25(3) 26	1	1026
1139. HOLY DIVER Dio [25Jun83 | 22Oct83 | 10Mar84] 56(1) 38	0	1026
1140. ADVENTURES IN UTOPIA Utopia [26Jan80 | 23Feb80 | 14Jun80] 32(2) 21	2	1025
1141. CHIPMUNK PUNK The Chipmunks [09Aug80 | 20Sep80 | 31Jan81] 34(2) 26	0	1024
1142. THE SENSUAL WORLD Kate Bush [04Nov89 | 25Nov89 | 28Apr90] 43(2) 26	0	1024
1143. A SALT WITH A DEADLY PEPA Salt-N-Pepa [13Aug88 | 24Sep88 | 11Mar89] 38(1) 31	0	1023
1144. JACKSONS LIVE The Jacksons [28Nov81 | 23Jan82 | 22Sep84] 30(2) 19	0	1022
1145. SANDINISTA! The Clash [07Feb81 | 07Mar81 | 20Jun81] 24(2) 20	0	1021
1146. VIEW FROM THE GROUND America [28Aug82 | 30Oct82 | 05Mar83] 41(8) 28	2	1020
1147. URBAN CHIPMUNK The Chipmunks [06Jun81 | 15Aug81 | 30Jan82] 56(1) 35	0	1020
1148. CHRISTINE MCVIE Christine McVie [18Feb84 | 17Mar84 | 21Jul84] 26(2) 23	2	1017
1149. SHOW ME The Cover Girls [15Aug87 | 27Feb88 | 05Nov88] 64(1) 61	5	1016
1150. THE BEST OF OMD Orchestral Manoeuvres In The Dark [26Mar88 | 14May88 | 08Oct88] 46(2) 29	1	1015
1151. VINYL CONFESSIONS Kansas [12Jun82 | 17Jul82 | 23Oct82] 16(2) 20	2	1015
1152. DREAMING #11 Joe Satriani [26Nov88 | 04Feb89 | 20May89] 42(1) 26	0	1012
1153. NAJEE'S THEME Najee [28Feb87 | 02May87 | 02Jan88] 56(2) 45	0	1012
1154. THE JEALOUS KIND Delbert McClinton [22Nov80 | 07Mar81 | 30May81] 34(1) 28	2	1011
1155. ZEBRA Zebra [14May83 | 13Aug83 | 19Nov83] 29(2) 28	1	1010
1156. THE PRESENT The Moody Blues [10Sep83 | 08Oct83 | 04Feb84] 26(3) 22	2	1010
1157. MEN AND WOMEN Simply Red [28Mar87 | 02May87 | 19Sep87] 31(2) 26	1	1006
1158. A FRESH AIRE CHRISTMAS Mannheim Steamroller [26Nov88 | 24Dec88 | 19Jan91] 36(3) 24	1	1006
1159. DESERT MOON Dennis DeYoung [06Oct84 | 17Nov84 | 23Mar85] 29(2) 25	2	1005
1160. SEVEN THE HARD WAY Pat Benatar [14Dec85 | 28Dec85 | 26Apr86] 26(4) 20	3	1004
1161. TOO FAST FOR LOVE Mötley Crüe [17Dec83 | 31Mar84 | 11Jul87] 77(1) 62	0	1004
1162. LIVE-STOMPIN' AT THE SAVOY Rufus And Chaka Khan [03Sep83 | 29Oct83 | 14Apr84] 50(1) 33	1	1003
1163. UNLIMITED! Roger [28Nov87 | 30Jan88 | 07May88] 35(1) 24	0	1000
1164. THE YOUTH OF TODAY Musical Youth [08Jan83 | 26Feb83 | 04Jun83] 23(4) 22	1	999
1165. DANA DANE WITH FAME Dana Dane [12Sep87 | 17Oct87 | 16Apr88] 46(2) 32	0	998

Ranking the Albums

Rank. Title *Act* [Enter|Peak|Exit] Peak(Wks) TotWks	Hot 100	Scr
1166. HERE COMES THE NIGHT *Barry Manilow* [18Dec82|22Jan83|18Jun83] 32(4) 27	2	998
1167. AIR SUPPLY *Air Supply* [29Jun85|20Jul85|16Nov85] 26(4) 21	2	997
1168. RADIANT *Atlantic Starr* [14Mar81|16May81|03Oct81] 47(1) 30	0	995
1169. ALL THAT JAZZ *Soundtrack* [22Mar80|19Apr80|23Aug80] 36(2) 23	0	995
1170. I'M IN LOVE *Evelyn King* [25Jul81|19Sep81|21Nov81] 28(1) 18	1	995
1171. LET IT ROLL *Little Feat* [20Aug88|08Oct88|01Apr89] 36(1) 33	0	994
1172. CIRCLE OF LOVE *The Steve Miller Band* [14Nov81|05Dec81|06Mar82] 26(2) 17	2	994
1173. GUITARS, CADILLACS, ETC., ETC. *Dwight Yoakam* [19Apr86|07Jun86|11Jul87] 61(2) 65	0	993
1174. SAM HARRIS *Sam Harris* [29Sep84|24Nov84|13Apr85] 35(2) 29	1	989
1175. COMING BACK HARD AGAIN *Fat Boys* [09Jul88|06Aug88|17Dec88] 33(2) 24	2	988
1176. TALKING BACK TO THE NIGHT *Steve Winwood* [21Aug82|18Sep82|05Feb83] 28(4) 25	2	988
1177. RUTHLESS PEOPLE *Soundtrack* [05Jul86|30Aug86|18Oct86] 20(1) 16	1	988
1178. ALLIGATOR WOMAN *Cameo* [10Apr82|15May82|18Sep82] 23(3) 24	0	987
1179. DIRTY MIND *Prince* [08Nov80|06Dec80|02Mar82] 45(1) 52	0	986
1180. ALL SHOOK UP *Cheap Trick* [15Nov80|06Dec80|21Feb81] 24(2) 15	1	985
1181. TOO TOUGH *Angela Bofill* [12Feb83|02Apr83|17Sep83] 40(3) 32	0	984
1182. I BELIEVE IN YOU *Don Williams* [04Oct80|20Dec80|02May81] 57(1) 31	1	981
1183. TRANS *Neil Young* [22Jan83|05Feb83|14May83] 19(6) 17	1	981
1184. LONG TIME COMING *Ready For The World* [06Dec86|21Feb87|30May87] 32(2) 26	1	980
1185. DO ME BABY *Meli'sa Morgan* [08Feb86|29Mar86|11Oct86] 41(3) 36	1	980
1186. BODY WISHES *Rod Stewart* [25Jun83|30Jul83|19Nov83] 30(1) 22	2	979
1187. THE BREAKFAST CLUB *The Breakfast Club* [28Mar87|06Jun87|17Oct87] 43(2) 30	2	978
1188. THE HEAD ON THE DOOR *The Cure* [05Oct85|07Dec85|06Sep86] 59(2) 49	1	978
1189. NIGHTCLUBBING *Grace Jones* [23May81|25Jul81|03Oct81] 32(2) 20	0	977
1190. IN VISIBLE SILENCE *The Art Of Noise* [03May86|28Jun86|22Nov86] 53(2) 30	2	977
1191. SCANDAL *Scandal* [29Jan83|11Jun83|03Sep83] 39(1) 32	2	974
1192. SWASS *Sir Mix-A-Lot* [22Oct88|21Jan89|25Nov89] 82(1) 58	1	973
1193. THE BROADSWORD AND THE BEAST *Jethro Tull* [01May82|19Jun82|21Aug82] 19(3) 17	0	973
1194. THE BLITZ *Krokus* [08Sep84|03Nov84|09Mar85] 31(1) 27	1	967
1195. JUST ANOTHER DAY IN PARADISE *Bertie Higgins* [27Feb82|19Jun82|14Aug82] 38(2) 25	2	966
1196. BAD TO THE BONE *George Thorogood & The Destroyers* [28Aug82|25Sep82|16Nov85] 43(4) 48	0	965
1197. I'M IN LOVE AGAIN *Patti LaBelle* [07Jan84|03Mar84|01Sep84] 40(2) 35	1	965
1198. GIPSY KINGS *Gipsy Kings* [17Dec88|15Apr89|30Sep89] 57(2) 42	0	965
1199. LOVE FOR LOVE *The Whispers* [02Apr83|21May83|15Oct83] 37(1) 29	1	964
1200. McVICAR (SOUNDTRACK) *Roger Daltrey* [16Aug80|13Sep80|22Nov80] 22(2) 15	2	962
1201. UPRISING *Bob Marley And The Wailers* [09Aug80|04Oct80|10Jan81] 45(1) 23	0	962
1202. WHAMMY! *The B-52s* [21May83|18Jun83|12Nov83] 29(2) 26	1	962
1203. CARNAVAL *Spyro Gyra* [01Nov80|13Dec80|23May81] 49(1) 30	1	959
1204. WILD THINGS RUN FAST *Joni Mitchell* [20Nov82|11Dec82|09Apr83] 25(4) 21	1	957
1205. MEAN BUSINESS *The Firm* [22Feb86|15Mar86|28Jun86] 22(3) 19	1	957
1206. DONE WITH MIRRORS *Aerosmith* [30Nov85|28Dec85|07Jun86] 36(3) 28	0	956
1207. RICHARD PRYOR LIVE ON THE SUNSET STRIP (SOUNDTRACK) *Richard Pryor* [17Apr82|29May82|07Aug82] 21(3) 17	0	956
1208. MY MELODY *Deniece Williams* [04Jul81|29Aug81|07Nov81] 74(1) 32	1	955
1209. REACH *Richard Simmons* [05Jun82|17Jul82|05Mar83] 44(2) 40	0	954
1210. SUPERSONIC--THE ALBUM *J.J. Fad* [23Jul88|13Aug88|11Feb89] 49(2) 30	3	953
1211. VOYEUR *David Sanborn* [18Apr81|23May81|12Sep81] 45(2) 22	0	953
1212. LOVE IS WHERE YOU FIND IT *The Whispers* [23Jan82|10Apr82|10Jul82] 35(3) 25	0	952
1213. PROGRESSIONS OF POWER *Triumph* [29Mar80|24May80|26Jul80] 32(1) 18	1	951
1214. L.A. GUNS *L.A. Guns* [06Feb88|09Apr88|03Dec88] 50(1) 33	0	951
1215. HUMAN'S LIB *Howard Jones* [24Mar84|30Jun84|05Oct85] 59(2) 43	2	951
1216. THIS WOMAN *K.T. Oslin* [24Sep88|05Nov88|16Sep89] 75(2) 52	0	951
1217. REFLECTOR *Pablo Cruise* [18Jul81|12Sep81|14Nov81] 34(2) 18	2	949
1218. POWER *Kansas* [15Nov86|24Jan87|16May87] 35(2) 27	2	949
1219. HOT BOX *Fatback* [19Apr80|14Jun80|18Oct80] 44(1) 27	0	946
1220. GRAND SLAM *The Isley Brothers* [21Mar81|25Apr81|11Jul81] 28(3) 17	1	946
1221. STONE JAM *Slave* [01Nov80|21Mar81|20Jun81] 53(2) 34	1	943
1222. RUNAWAY HORSES *Belinda Carlisle* [21Oct89|16Dec89|07Apr90] 37(1) 25	2,1*	943
1223. THE BIG THROWDOWN *Levert* [05Sep87|17Oct87|13Feb88] 32(2) 24	1	943
1224. HOW WILL THE WOLF SURVIVE *Los Lobos* [15Dec84|09Mar85|03Aug85] 47(3) 34	1	943
1225. IMPERIAL BEDROOM *Elvis Costello And The Attractions* [24Jul82|25Sep82|01Jan83] 30(3) 24	0	943
1226. WALKABOUT *The Fixx* [14Jun86|02Aug86|01Nov86] 30(2) 21	0	943
1227. FRIENDS *Shalamar* [20Feb82|22May82|07Aug82] 35(1) 25	1	943
1228. NEW YORK *Lou Reed* [28Jan89|08Apr89|24Jun89] 40(2) 22	0	942
1229. ONE WAY HOME *Hooters* [08Aug87|29Aug87|30Jan88] 27(3) 26	2	942
1230. CARL CARLTON *Carl Carlton* [08Aug81|10Oct81|12Dec81] 34(2) 19	1	941
1231. REACH FOR THE SKY *The Allman Brothers Band* [23Aug80|04Oct80|15Nov80] 27(2) 13	1	940
1232. VOX HUMANA *Kenny Loggins* [20Apr85|11May85|16Nov85] 41(1) 31	3	940
1233. FRANTIC ROMANTIC *Jermaine Stewart* [14Jun86|06Sep86|29Nov86] 32(2) 25	2	939
1234. KEEP YOUR HANDS OFF MY POWER SUPPLY *Slade* [05May84|30Jun84|06Oct84] 33(2) 23	2	937
1235. IN GOD WE TRUST *Stryper* [16Jul88|23Jul88|31Dec88] 32(3) 25	2	935
1236. PLAYING FOR KEEPS *Eddie Money* [09Aug80|04Oct80|29Nov80] 35(1) 17	2	934
1237. IN MY EYES *Stevie B* [11Mar89|24Jun89|14Apr90] 75(1) 46	4,1*	933
1238. FREETIME *Spyro Gyra* [29Aug81|03Oct81|27Feb82] 41(2) 27	0	930
1239. MAGIC *Tom Browne* [21Feb81|21Mar81|27Jun81] 37(2) 19	0	929
1240. SPIRIT OF LOVE *Con Funk Shun* [12Apr80|17May80|23Aug80] 30(3) 20	0	929
1241. LET'S GO ALL THE WAY *Sly Fox* [01Mar86|10May86|26Jul86] 31(2) 22	1	928
1242. GET HERE *Brenda Russell* [19Mar88|25Jun88|24Sep88] 49(2) 28	1	926
1243. ELO'S GREATEST HITS *Electric Light Orchestra* [08Dec79|05Jan80|15Mar80] 30(3) 15	0	925
1244. MARY JANE GIRLS *Mary Jane Girls* [14May83|17Sep83|18Feb84] 56(1) 41	0	924
1245. COPPERHEAD ROAD *Steve Earle* [12Nov88|28Jan89|20May89] 56(1) 28	0	924
1246. TWO OF A KIND *Earl Klugh & Bob James* [06Nov82|11Dec82|21May83] 44(4) 29	0	923
1247. PRIVATE REVOLUTION *World Party* [27Dec86|28Mar87|25Jul87] 39(3) 31	1	923
1248. KYLIE *Kylie Minogue* [10Sep88|07Jan89|18Mar89] 53(2) 28	3	922
1249. MOB RULES *Black Sabbath* [28Nov81|19Dec81|27Mar82] 29(3) 18	0	922
1250. ONE BRIGHT DAY *Ziggy Marley And The Melody Makers* [12Aug89|16Sep89|09Dec89] 26(2) 18	0	920
1251. ROUGH DIAMONDS *Bad Company* [04Sep82|09Oct82|01Jan83] 26(3) 18	1	920
1252. CITY NIGHTS *Tierra* [27Dec80|14Mar81|16May81] 38(3) 21	2	919
1253. GREATEST HITS-VOL. II *Barry Manilow* [03Dec83|14Jan84|07Apr84] 30(2) 19	1	919
1254. MADNESS *Madness* [30Apr83|16Jul83|12Nov83] 41(1) 29	3	918
1255. JONATHAN BUTLER *Jonathan Butler* [30May87|08Aug87|09Jan88] 50(2) 33	1	916
1256. UNION JACKS *The Babys* [19Jan80|15Mar80|14Jun80] 42(1) 22	2	916
1257. HANK WILLIAMS, JR.'S GREATEST HITS *Hank Williams Jr.* [13Nov82|22Jan83|10Mar84] 107(2) 70	0	915
1258. MURMUR *R.E.M.* [14May83|13Aug83|29Dec84] 36(1) 30	1	913
1259. RETURN OF THE JEDI *Soundtrack* [11Jun83|09Jul83|01Oct83] 20(2) 17	0	913
1260. NIGHTCRUISING *Bar-Kays* [14Nov81|16Jan82|29May82] 55(2) 29	0	912
1261. DANCIN' AND LOVIN' *Spinners* [19Jan80|05Apr80|31May80] 32(1) 20	1	910
1262. COCONUT TELEGRAPH *Jimmy Buffett* [21Feb81|04Apr81|20Jun81] 30(1) 18	1	909

Ranking the Albums

Rank. Title Act [Enter|Peak|Exit] Peak(Wks) TotWks	Hot 100	Scr
1263. ACTION REPLAY *Howard Jones* [03May86|12Jul86|11Oct86] 34(1) 24	0	909
1264. BETE NOIRE *Bryan Ferry* [21Nov87|23Apr88|18Jun88] 63(2) 31	1	907
1265. LOVE WILL TURN YOU AROUND *Kenny Rogers* [24Jul82|18Sep82|01Jan83] 34(2) 24	2	905
1266. COLOR OF SUCCESS *Morris Day* [19Oct85|23Nov85|17May86] 37(2) 31	1	905
1267. DANGEROUS AGE *Bad Company* [17Sep88|19Nov88|17Jun89] 58(2) 40	1	904
1268. FOREVER NOW *Psychedelic Furs* [13Nov82|02Apr83|18Jun83] 61(4) 32	1	902
1269. THE BIG PRIZE *Honeymoon Suite* [15Mar86|03May86|08Nov86] 61(2) 35	2	902
1270. FLESH + BLOOD *Roxy Music* [28Jun80|16Aug80|01Nov80] 35(2) 19	1	901
1271. HIGH CRIME *Al Jarreau* [24Nov84|08Dec84|20Jul85] 49(2) 35	1	901
1272. TONIGHT! *Four Tops* [12Sep81|21Nov81|30Jan82] 37(1) 21	1	900
1273. ON THE RISE *The S.O.S. Band* [27Aug83|05Nov83|10Mar84] 47(1) 29	2	900
1274. YOUTHQUAKE *Dead Or Alive* [13Jul85|31Aug85|23Nov85] 31(3) 20	2	898
1275. SOME GREAT REWARD *Depeche Mode* [19Jan85|17Aug85|30Nov85] 51(2) 42	2	896
1276. EXTENSIONS *The Manhattan Transfer* [08Dec79|07Jun80|16Aug80] 55(1) 37	2	895
1277. SEMINAR *Sir Mix-A-Lot* [18Nov89|25Nov89|25Aug90] 67(2) 41	0	893
1278. A COLLECTION OF GREAT DANCE SONGS *Pink Floyd* [12Dec81|09Jan82|27Mar82] 31(2) 16	0	892
1279. STRAIGHT BETWEEN THE EYES *Rainbow* [08May82|12Jun82|09Oct82] 30(3) 23	1	891
1280. FIRIN' UP *Pure Prairie League* [17May80|19Jul80|25Oct80] 37(2) 24	3	890
1281. THE SPORT OF KINGS *Triumph* [06Sep86|08Nov86|07Mar87] 33(2) 27	1	890
1282. GREEN THOUGHTS *The Smithereens* [09Apr88|07May88|05Nov88] 60(3) 31	1	890
1283. NEW SENSATIONS *Lou Reed* [16Jun84|20Oct84|19Jan85] 56(1) 32	0	889
1284. ADVENTURES IN THE LAND OF MUSIC *Dynasty* [02Aug80|04Oct80|20Dec80] 43(1) 21	1	888
1285. FOLLOW THE LEADER *Eric B. & Rakim* [13Aug88|03Sep88|26Nov88] 22(2) 16	0	887
1286. ANOTHER STEP *Kim Wilde* [04Apr87|13Jun87|26Sep87] 40(2) 26	2	887
1287. MANNHEIM STEAMROLLER CHRISTMAS *Mannheim Steamroller* [22Dec84|07Jan89|12Jan91] 50(1) 39	0	886
1288. DREAMS *Grace Slick* [05Apr80|03May80|19Jul80] 32(2) 16	1	884
1289. ALL I NEED *Jack Wagner* [22Sep84|09Feb85|06Apr85] 44(2) 29	2	884
1290. RHAPSODY AND BLUES *The Crusaders* [12Jul80|16Aug80|25Oct80] 29(2) 16	0	883
1291. GO WEST *Go West* [23Mar85|01Jun85|16Nov85] 60(1) 35	3	882
1292. WHAT BECOMES A SEMI-LEGEND MOST? *Joan Rivers* [23Apr83|28May83|10Sep83] 22(2) 21	0	880
1293. THREE HEARTS IN THE HAPPY ENDING MACHINE *Daryl Hall* [06Sep86|27Sep86|28Feb87] 29(3) 26	3	879
1294. TAKE ANOTHER PICTURE *Quarterflash* [09Jul83|27Aug83|26Nov83] 34(1) 21	2	878
1295. RAM IT DOWN *Judas Priest* [04Jun88|18Jun88|08Oct88] 31(3) 19	0	876
1296. BEATITUDE *Ric Ocasek* [29Jan83|05Mar83|14May83] 28(5) 16	1	875
1297. WINDOWS *The Charlie Daniels Band* [03Apr82|15May82|07Aug82] 26(3) 19	1	875
1298. HIGH PRIORITY *Cherrelle* [01Feb86|12Apr86|23Aug86] 36(2) 30	1	874
1299. HALL & OATES LIVE AT THE APOLLO WITH DAVID RUFFIN & EDDIE KENDRICK *Daryl Hall & John Oates* [28Sep85|26Oct85|25Jan86] 21(2) 18	1	873
1300. STREETS OF FIRE *Soundtrack* [16Jun84|04Aug84|03Nov84] 32(2) 21	0	871
1301. REEL MUSIC *The Beatles* [10Apr82|08May82|26Jun82] 19(3) 12	0	869
1302. I AM LOVE *Peabo Bryson* [28Nov81|27Feb82|08May82] 40(2) 24	1	868
1303. FLYING COWBOYS *Rickie Lee Jones* [14Oct89|18Nov89|31Mar90] 39(2) 25	0	868
1304. THIS SIDE OF PARADISE *Ric Ocasek* [11Oct86|15Nov86|14Mar87] 31(2) 23	2	868
1305. ANDERSON, BRUFORD, WAKEMAN, HOWE *Anderson, Bruford, Wakeman, Howe* [01Jul89|29Jul89|14Oct89] 30(1) 16	0	867
1306. OPEN SESAME *Whodini* [17Oct87|21Nov87|12Mar88] 30(2) 22	0	865
1307. 2 HYPE *Kid 'N Play* [17Dec88|21Jan89|04Nov89] 96(1) 47	0	865
1308. DANGEROUS TOYS *Dangerous Toys* [17Jun89|16Sep89|17Feb90] 65(3) 36	0	863
1309. SOUL SEARCHING *Glenn Frey* [03Sep88|05Nov88|07Jan89] 36(2) 19	2	862
1310. TRUST *Elvis Costello And The Attractions* [14Feb81|07Mar81|23May81] 28(2) 15	0	860
1311. DOWN TO THE MOON *Andreas Vollenweider* [02Aug86|27Sep86|25Apr87] 60(1) 39	0	859
1312. ICE ON FIRE *Elton John* [30Nov85|05Apr86|07Jun86] 48(2) 28	2	859
1313. HIGH COUNTRY SNOWS *Dan Fogelberg* [11May85|15Jun85|12Oct85] 30(2) 23	1	858
1314. INTROSPECTIVE *Pet Shop Boys* [05Nov88|03Dec88|01Apr89] 34(2) 22	3	857
1315. ALL THE GREAT HITS *Commodores* [04Dec82|22Jan83|14May83] 37(2) 24	1	854
1316. SUR LA MER *The Moody Blues* [25Jun88|23Jul88|29Oct88] 38(2) 19	1	852
1317. 2:00 A.M. PARADISE CAFÉ *Barry Manilow* [15Dec84|26Jan85|27Apr85] 28(2) 20	0	851
1318. FREHLEY'S COMET *Ace Frehley* [23May87|20Jun87|07Nov87] 43(2) 25	0	849
1319. LOVE SOMEBODY TODAY *Sister Sledge* [08May80|19Apr80|14Jun80] 31(2) 15	1	849
1320. WIDE RECEIVER *Michael Henderson* [30Aug80|01Nov80|10Jan81] 35(1) 18	0	849
1321. BENT OUT OF SHAPE *Rainbow* [01Oct83|19Nov83|18Feb84] 34(1) 21	1	848
1322. FOREIGN AFFAIR *Tina Turner* [07Oct89|28Oct89|24Feb90] 31(2) 21	2,1*	847
1323. STRAIGHT FROM THE HEART *Peabo Bryson* [16Jun84|11Aug84|08Dec84] 44(2) 26	2	847
1324. THE AGE OF CONSENT *Bronski Beat* [19Jan85|09Mar85|06Jul85] 36(2) 25	1	847
1325. HOT SPACE *Queen* [29May82|19Jun82|16Oct82] 22(3) 21	2	846
1326. ROD STEWART *Rod Stewart* [12Jul86|16Aug86|15Nov86] 28(2) 19	3	846
1327. ELVIS ARON PRESLEY *Elvis Presley* [23Aug80|20Sep80|22Nov80] 27(2) 14	0	845
1328. ZAPP II *Zapp* [14Aug82|18Sep82|18Dec82] 25(3) 19	0	845
1329. DON'T LOOK ANY FURTHER *Dennis Edwards* [03Mar84|02Jun84|01Sep84] 48(2) 27	1	845
1330. BRITISH STEEL *Judas Priest* [31May80|12Jul80|27Sep80] 34(2) 18	0	844
1331. PAC-MAN FEVER *Buckner & Garcia* [13Mar82|29May82|26Jun82] 24(2) 16	0	844
1332. THE VISITORS *ABBA* [09Jan82|13Feb82|01May82] 29(2) 17	2	844
1333. WHO? *Tony! Toni! Tone!* [28May88|02Jul88|08Apr89] 69(2) 46	1	844
1334. SIDE KICKS *Thompson Twins* [26Feb83|02Apr83|13Aug83] 34(3) 25	2	843
1335. B.L.T. *Jack Bruce/Bill Lordan/Robin Trower* [21Mar81|09May81|04Jul81] 37(1) 16	0	843
1336. PEACE SELLS...BUT WHO'S BUYING? *Megadeth* [25Oct86|29Nov86|12Sep87] 76(1) 47	0	843
1337. CROSSROADS *Eric Clapton* [07May88|14Jun88|29Oct88] 34(2) 26	0	842
1338. OPERA SAUVAGE *Vangelis* [13Dec86|11Apr87|05Sep87] 42(1) 39	0	839
1339. SOMETHING'S GOING ON *Frida* [13Nov82|09Apr83|21May83] 41(2) 28	1	839
1340. COMPUTER GAMES *George Clinton* [18Dec82|30Apr83|30Jul83] 40(1) 33	0	838
1341. TRULY FOR YOU *The Temptations* [17Nov84|09Mar85|06Jul85] 55(4) 34	1	835
1342. SEASONS OF THE HEART *John Denver* [20Mar82|22May82|30Oct82] 39(2) 33	2	835
1343. AEROSMITH'S GREATEST HITS *Aerosmith* [29Nov80|20Dec80|30Apr88] 53(3) 40	0	835
1344. A DECADE OF ROCK AND ROLL 1970 TO 1980 *REO Speedwagon* [19Apr80|31May80|20Jun81] 55(1) 34	0	835
1345. UB40 *UB40* [20Aug88|08Oct88|18Feb89] 44(2) 27	0	832
1346. T'PAU *T'Pau* [06Jun87|29Aug87|14Nov87] 31(1) 24	1	831
1347. THE HEART OF THE MATTER *Kenny Rogers* [19Oct85|28Dec85|26Apr86] 51(3) 28	1	827
1348. AMERICAN GARAGE *Pat Metheny Group* [24Nov79|02Feb80|03May80] 53(1) 24	0	826
1349. FEEL MY SOUL *Jennifer Holliday* [22Oct83|05Nov83|17Mar84] 31(2) 22	1	825
1350. FAT BOYS *Fat Boys* [05Jan85|09Feb85|05Oct85] 48(2) 40	0	824
1351. REEL LIFE *Boy Meets Girl* [22Oct88|14Jan89|15Apr89] 50(2) 26	2	824
1352. WAIATA *Split Enz* [23May81|20Jun81|26Sep81] 45(2) 19	0	824
1353. 80'S LADIES *K.T. Oslin* [12Dec87|26Mar88|16Jul88] 68(4) 32	0	823
1354. GREETINGS FROM TIMBUK 3 *Timbuk 3* [04Oct86|10Jan87|25Apr87] 50(1) 30	1	822
1355. COLORS *Soundtrack* [14May88|04Jun88|17Sep88] 31(2) 19	1	822
1356. SWEETS FROM A STRANGER *Squeeze* [29May82|10Jul82|18Dec82] 32(2) 30	0	822
1357. THE CALIFORNIA RAISINS *The California Raisins* [05Dec87|23Jan88|06Aug88] 60(2) 36	1	822
1358. SOMEBODY'S KNOCKIN' *Terri Gibbs* [14Feb81|11Apr81|01Aug81] 53(3) 25	2	822
1359. ROCK & ROLL STRATEGY *Thirty Eight Special* [22Oct88|20May89|29Jul89] 61(1) 41	3	821

Rank. Title *Act* [Enter|Peak|Exit] Peak(Wks) TotWks	Hot 100	Scr
1360. A SHOW OF HANDS *Rush* [28Jan89|18Feb89|06May89] 21(1) 15	0	820
1361. ALL AMERICAN GIRLS *Sister Sledge* [28Feb81|04Apr81|12Sep81] 42(1) 29	2	817
1362. MICKEY MOUSE DISCO *Various Artists* [12Apr80|07Jun80|11Oct80] 35(2) 27	0	815
1363. FEEL ME *Cameo* [06Dec80|17Jan81|28Mar81] 44(1) 17	0	814
1364. EYE OF THE ZOMBIE *John Fogerty* [11Oct86|01Nov86|14Feb87] 26(2) 19	1	814
1365. LIFE'S TOO GOOD *The Sugarcubes* [18Jun88|01Oct88|31Dec88] 54(2) 29	0	813
1366. EYES OF A STRANGER *The Deele* [27Feb88|30Apr88|13Aug88] 54(2) 25	1	813
1367. PSYCHO CAFE *Bang Tango* [01Jul89|28Oct89|24Mar90] 58(2) 39	0	813
1368. LITTLE BAGGARIDDIM *UB40* [17Aug85|05Oct85|01Feb86] 40(3) 25	1	811
1369. JERMAINE(2) *Jermaine Jackson* [06Dec80|21Feb81|06Jun81] 44(1) 23	1	810
1370. IN OUR LIFETIME *Marvin Gaye* [07Feb81|14Mar81|30May81] 32(2) 17	0	809
1371. BLUE *Double* [26Jul86|27Sep86|13Dec86] 30(1) 21	1	809
1372. 20/20 *George Benson* [26Jan85|16Feb85|31Aug85] 45(3) 32	1	809
1373. THE KARATE KID PART II *Soundtrack* [12Jul86|06Sep86|01Nov86] 30(2) 17	1	808
1374. PIZZAZZ *Patrice Rushen* [24Nov79|01Mar80|19Apr80] 39(2) 22	1	807
1375. WE'RE THE BEST OF FRIENDS *Natalie Cole & Peabo Bryson* [15Dec79|01Mar80|19Apr80] 44(1) 19	0	806
1376. VANITY 6 *Vanity 6* [02Oct82|11Dec82|30Apr83] 45(2) 31	0	805
1377. HOW 'BOUT US *Champaign* [21Mar81|16May81|01Aug81] 53(1) 20	1	805
1378. WOMAN OUT OF CONTROL *Ray Parker Jr.* [26Nov83|11Feb84|28Apr84] 45(2) 23	1	805
1379. KOOKOO *Debbie Harry* [29Aug81|26Sep81|14Nov81] 25(1) 12	2	803
1380. YOU BROKE MY HEART IN 17 PLACES *Tracey Ullman* [24Mar84|12May84|04Aug84] 34(2) 20	2	803
1381. BY THE LIGHT OF THE MOON *Los Lobos* [14Feb87|14Mar87|19Sep87] 47(2) 32	0	802
1382. HAWKS & DOVES *Neil Young* [22Nov80|20Dec80|07Mar81] 30(3) 16	0	800
1383. TAKE IT WHILE IT'S HOT *Sweet Sensation* [08Oct88|13May89|16Sep89] 63(1) 32	5	799
1384. SKYWAY *Skyy* [15Mar80|31May80|16Aug80] 61(1) 23	0	798
1385. MOVE SOMETHIN' *The 2 Live Crew* [04Jun88|02Jul88|18Mar89] 68(2) 42	0	798
1386. WHAT A FEELIN' *Irene Cara* [10Dec83|25Feb84|18Aug84] 77(1) 37	4	798
1387. AND IN THIS CORNER... *D.J. Jazzy Jeff & The Fresh Prince* [18Nov89|02Dec89|31Mar90] 39(4) 20	1	797
1388. SHUTTERED ROOM *The Fixx* [13Nov82|09Apr83|22Sep84] 106(1) 51	1	797
1389. IT'S A BIG DADDY THING *Big Daddy Kane* [07Oct89|21Oct89|28Apr90] 33(2) 30	0	796
1390. THE PROS & CONS OF HITCHHIKING *Roger Waters* [19May84|09Jun84|15Sep84] 31(3) 18	0	796
1391. ALF *Alison Moyet* [06Apr85|01Jun85|21Sep85] 45(2) 25	2	795
1392. AMONG THE LIVING *Anthrax* [11Apr87|30May87|12Dec87] 62(2) 36	0	795
1393. MIRACLES *Change* [18Apr81|13Jun81|12Sep81] 46(1) 22	2	793
1394. OH MERCY *Bob Dylan* [07Oct89|28Oct89|10Mar90] 30(2) 23	0	792
1395. MIRROR MOVES *Psychedelic Furs* [26May84|07Jul84|24Nov84] 43(2) 27	1	792
1396. TO LOVE AGAIN *Diana Ross* [14Mar81|04Apr81|13Jun81] 32(2) 14	1	791
1397. AUTOGRAPH *John Denver* [01Mar80|19Apr80|21Jun80] 39(2) 17	2	791
1398. NOTHING TO LOSE *Eddie Money* [22Oct88|03Dec88|06May89] 49(1) 29	3	791
1399. LOVE LIVES FOREVER *Minnie Riperton* [06Sep80|11Oct80|13Dec80] 35(2) 15	0	790
1400. SHORT SHARP SHOCKED *Michelle Shocked* [17Sep88|04Feb89|13May89] 73(1) 35	1	790
1401. A NIGHT TO REMEMBER *Cyndi Lauper* [27May89|24Jun89|14Oct89] 37(4) 21	2	788
1402. WASN'T TOMORROW WONDERFUL? *The Waitresses* [06Feb82|08May82|17Jul82] 41(1) 24	1	787
1403. LET'S HEAR IT FOR THE BOY *Deniece Williams* [09Jun84|14Jul84|13Oct84] 26(1) 19	2	786
1404. REAL PEOPLE *Chic* [26Jul80|30Aug80|01Nov80] 30(2) 15	3	784
1405. RAPPIN' RODNEY *Rodney Dangerfield* [12Nov83|24Dec83|24Mar84] 36(3) 20	1	783
1406. JUST SYLVIA *Sylvia (2)* [07Aug82|27Nov82|19Mar83] 56(1) 33	1	782
1407. VINNIE VINCENT INVASION *Vinnie Vincent Invasion* [20Sep86|15Nov86|04Apr87] 64(2) 29	0	781
1408. PAID IN FULL *Eric B. & Rakim* [12Sep87|24Oct87|28May88] 58(1) 38	0	780
1409. TOUCH *Con Funk Shun* [13Dec80|14Feb81|18Apr81] 51(1) 19	1	780
1410. THIS IS THE WAY *The Rossington Collins Band* [10Oct81|07Nov81|23Jan82] 24(1) 16	0	780
1411. MISPLACED CHILDHOOD *Marillion* [24Aug85|09Nov85|19Apr86] 47(2) 35	1	780
1412. COLLABORATION *George Benson/Earl Klugh* [11Jul87|29Aug87|06Feb88] 59(2) 31	0	779
1413. SHIRLEY MURDOCK! *Shirley Murdock* [14Feb87|28Mar87|04Apr87] 42(2) 26	1	777
1414. UNDER A RAGING MOON *Roger Daltrey* [12Oct85|23Nov85|05Apr86] 42(2) 26	2	777
1415. MARSHALL CRENSHAW *Marshall Crenshaw* [29May82|31Jul82|27Nov82] 50(2) 27	1	776
1416. CULTOSAURUS ERECTUS *Blue Öyster Cult* [12Jul80|23Aug80|25Oct80] 34(2) 16	0	775
1417. OUT OF AFRICA *Soundtrack* [01Feb86|26Apr86|28Jun86] 38(2) 22	0	773
1418. TEMPLE OF LOW MEN *Crowded House* [23Jul88|13Aug88|26Nov88] 40(2) 19	1	773
1419. SA-FIRE *SaFire* [08Oct88|13May89|19Aug89] 79(1) 46	3	773
1420. VICTIMS OF THE FURY *Robin Trower* [01Mar80|12Apr80|07Jun80] 34(2) 15	0	772
1421. ECHO & THE BUNNYMEN(2) *Echo & The Bunnymen* [08Aug87|17Oct87|16Apr88] 51(3) 37	0	771
1422. THE FRIENDS OF MR. CAIRO *Jon And Vangelis* [08Aug81|24Oct81|07Aug82] 64(1) 34	1	771
1423. ANIMAL MAGIC *The Blow Monkeys* [21Jun86|09Aug86|18Oct86] 35(2) 18	1	769
1424. LOVE AN ADVENTURE *Pseudo Echo* [21Mar87|11Jul87|19Sep87] 54(3) 27	2	769
1425. HIROSHIMA *Hiroshima* [22Dec79|08Mar80|21Jun80] 51(2) 27	0	768
1426. PRIMITIVE *Neil Diamond* [18Aug84|15Sep84|02Feb85] 35(3) 25	1	767
1427. DICE *Andrew Dice Clay* [29Apr89|21Oct89|17Mar90] 94(1) 47	0	767
1428. CUPID AND PSYCHE 85 *Scritti Politti* [03Aug85|18Jan86|12Apr86] 50(2) 28	2	767
1429. MALICE IN WONDERLAND *Nazareth* [16Feb80|15Mar80|21Jun80] 41(3) 19	1	766
1430. HEARTS AND BONES *Paul Simon* [19Nov83|07Jan84|17Mar84] 35(2) 18	1	766
1431. DON'T LET LOVE SLIP AWAY *Freddie Jackson* [13Aug88|27Aug88|04Mar89] 48(2) 30	1	762
1432. ESCAPE ARTIST *Garland Jeffreys* [21Mar81|09May81|18Jul81] 59(4) 18	0	762
1433. 12 GREATEST HITS VOL. II *Neil Diamond* [29May82|10Jul82|12Mar83] 48(2) 42	0	762
1434. SHABOOH SHOOBAH *INXS* [19Mar83|28May83|15Oct83] 46(3) 31	2	761
1435. RAIN MAN *Soundtrack* [11Mar89|29Apr89|24Jun89] 31(2) 16	1	760
1436. THE JOHN LENNON COLLECTION *John Lennon* [04Dec82|08Jan83|19Mar83] 33(4) 16	0	760
1437. GLOW *Rick James* [11May85|15Jun85|02Nov85] 50(2) 26	1	759
1438. IN THE EYE OF THE STORM *Roger Hodgson* [27Oct84|01Dec84|23Mar85] 46(5) 22	1	757
1439. THE HOUSE OF BLUE LIGHT *Deep Purple* [31Jan87|21Feb87|27Jun87] 34(3) 22	0	756
1440. WELCOME HOME *'Til Tuesday* [25Oct86|22Nov86|18Apr87] 49(3) 26	2	755
1441. ROBBIE DUPREE *Robbie Dupree* [14Jun80|23Aug80|22Nov80] 51(1) 24	2	755
1442. SOMEWHERE OVER CHINA *Jimmy Buffett* [23Jan82|13Feb82|01May82] 31(3) 15	0	755
1443. FOR SENTIMENTAL REASONS *Linda Ronstadt* [11Oct86|22Nov86|11Apr87] 46(1) 27	0	755
1444. AS FALLS WICHITA, SO FALLS WICHITA FALLS *Pat Metheny & Lyle Mays* [20Jun81|22Aug81|07Nov81] 50(1) 21	0	754
1445. JUNGLE *Dwight Twilley* [18Feb84|28Apr84|07Jul84] 39(1) 21	2	753
1446. TORCH *Carly Simon* [17Oct81|21Nov81|27Mar82] 50(1) 24	0	753
1447. CREATURES OF THE NIGHT *KISS* [20Nov82|29Jan83|26Mar83] 45(6) 19	0	752
1448. HAPPY TOGETHER *The Nylons* [23May87|15Aug87|31Oct87] 43(1) 24	2	752
1449. TIMES SQUARE *Soundtrack* [27Sep80|29Nov80|17Jan81] 37(2) 17	1	751
1450. WHEN SECONDS COUNT *Survivor* [08Nov86|27Dec86|18Apr87] 49(5) 24	3	750
1451. SAVED *Bob Dylan* [12Jul80|02Aug80|20Sep80] 24(2) 11	0	749
1452. TENTH *The Marshall Tucker Band* [22Mar80|26Apr80|28Jun80] 32(1) 15	1	749
1453. FROM LUXURY TO HEARTACHE *Culture Club* [26Apr86|07Jun86|16Aug86] 32(2) 17	1	748
1454. DIMPLES *Richard "Dimples" Fields* [25Jul81|12Sep81|14Nov81] 33(2) 17	0	748

Ranking the Albums

Rank. Title Act [Enter|Peak|Exit] Peak(Wks) TotWks	Hot 100	Scr
1455. JUST US Alabama [17Oct87|14Nov87|23Apr88] 55(1) 28	0	748
1456. 99 LUFTBALLONS Nena [24Mar84|14Apr84|23Jun84] 27(3) 14	1	747
1457. CAREFUL The Motels [12Jul80|16Aug80|22Nov80] 45(2) 20	0	746
1458. SURPRISE ATTACK Tora Tora [15Jul89|23Sep89|24Feb90] 47(1) 33	1	746
1459. BAD REPUTATION Joan Jett & the Blackhearts [14Mar81|09Oct82|08Jan83] 51(4) 21	1	746
1460. THE LOOK Shalamar [06Aug83|24Sep83|07Jan84] 38(1) 23	1	746
1461. THE HITS REO Speedwagon [25Jun88|10Sep88|19Nov88] 56(1) 22	1	745
1462. YOU KNOW HOW TO LOVE ME Phyllis Hyman [08Dec79|09Feb80|26Apr80] 50(1) 21	0	745
1463. DIVINE MADNESS (SOUNDTRACK) Bette Midler [29Nov80|20Dec80|28Feb81] 34(3) 14	1	745
1464. BACKSTREET David Sanborn [26Nov83|28Jan84|07Jul84] 81(1) 33	0	744
1465. HENRY LEE SUMMER Henry Lee Summer [12Mar88|07May88|13Aug88] 56(2) 23	3	744
1466. THE FLAT EARTH Thomas Dolby [17Mar84|14Apr84|14Jul84] 35(2) 18	1	743
1467. HOME Stephanie Mills [22Jul89|26Aug89|21Apr90] 82(3) 38	0	743
1468. TEDDY LIVE! COAST TO COAST Teddy Pendergrass [22Dec79|02Feb80|29Mar80] 33(3) 15	0	743
1469. LEATHER AND LACE Waylon Jennings & Jessi Colter [21Mar81|18Apr81|25Jul81] 43(2) 19	0	742
1470. TWO GQ [05Apr80|03May80|16Aug80] 46(2) 20	0	741
1471. WITHOUT A SONG Willie Nelson [26Nov83|14Jan84|14Jul84] 54(1) 34	0	740
1472. THE BIG CHILL: MORE SONGS FROM... Soundtrack [28Apr84|09Jun84|18May85] 85(1) 49	0	740
1473. THE ELECTRIC HORSEMAN (SOUNDTRACK) Willie Nelson/Dave Grusin [12Jan80|08Mar80|28Jun80] 52(2) 25	1	738
1474. JOY Teddy Pendergrass [28May88|16Jul88|05Nov88] 54(2) 24	1	737
1475. NO RESPECT Rodney Dangerfield [02Aug80|20Sep80|06Dec80] 48(1) 19	0	737
1476. CIMARRON Emmylou Harris [12Dec81|30Jan82|24Apr82] 46(1) 20	0	737
1477. ANOTHER PLACE Hiroshima [30Nov85|01Mar86|04Oct86] 79(2) 45	0	736
1478. BEYOND Herb Alpert [26Jul80|23Aug80|11Oct80] 28(2) 12	1	735
1479. THE SWING INXS [26May84|14Jul84|01Dec84] 52(2) 28	2	735
1480. POWER PLAY April Wine [10Jul82|14Aug82|20Nov82] 37(2) 20	1	735
1481. ROCK IN A HARD PLACE Aerosmith [25Sep82|16Oct82|29Jan83] 32(4) 19	0	734
1482. THE TOUCH Alabama [25Oct86|22Nov86|16May87] 42(1) 30	0	734
1483. RISING FORCE Yngwie Malmsteen [04May85|22Jun85|22Feb86] 60(3) 43	0	734
1484. WHITE FEATHERS Kajagoogoo [11Jun83|09Jul83|22Oct83] 38(2) 20	2	734
1485. COAL MINER'S DAUGHTER Soundtrack [29Mar80|17May80|09Aug80] 40(1) 20	0	733
1486. LOW RIDE Earl Klugh [07May83|11Jun83|15Oct83] 38(3) 24	0	733
1487. AN OFFICER AND A GENTLEMAN Soundtrack [30Oct82|11Dec82|02Apr83] 38(2) 23	1	732
1488. ORANGES AND LEMONS XTC [18Mar89|22Apr89|05Aug89] 44(2) 21	0	731
1489. BORDERLINE Ry Cooder [24Jan81|14Mar81|09May81] 43(2) 16	0	729
1490. IMAGINE: JOHN LENNON (SOUNDTRACK) John Lennon [22Oct88|19Nov88|18Feb89] 31(1) 18	1	729
1491. LOVE ALL THE HURT AWAY Aretha Franklin [29Aug81|03Oct81|19Dec81] 36(2) 17	1	729
1492. WOW! Bananarama [26Sep87|17Oct87|19Mar88] 44(2) 26	3	727
1493. AFL1-3603 Dave Davies [26Jul80|30Aug80|25Oct80] 42(2) 14	0	727
1494. SUN CITY Artists United Against Apartheid [23Nov85|14Dec85|22Mar86] 31(2) 18	1	726
1495. GREATEST HITS III Hank Williams Jr. [25Feb89|06May89|11Nov89] 61(1) 35	0	726
1496. VERY SPECIAL Debra Laws [11Apr81|29Aug81|10Oct81] 70(1) 27	1	726
1497. THE SECRET VALUE OF DAYDREAMING Julian Lennon [12Apr86|03May86|09Aug86] 32(2) 18	1	725
1498. I CAN DREAM ABOUT YOU Dan Hartman [03Nov84|22Dec84|11May85] 55(2) 28	3	725
1499. GIRLS WITH GUNS Tommy Shaw [20Oct84|24Nov84|06Apr85] 50(3) 25	2	725
1500. CLASSIC CRYSTAL Crystal Gayle [17Nov79|05Jan80|12Apr80] 62(3) 22	0	725
1501. I'M NO HERO Cliff Richard [11Oct80|13Dec80|30May81] 80(2) 34	3	723
1502. TURN BACK THE CLOCK Johnny Hates Jazz [16Apr88|11Jun88|01Oct88] 56(1) 25	2	723
1503. THE FIRST FAMILY RIDES AGAIN Rich Little [13Feb82|27Mar82|08May82] 29(2) 13	0	723
1504. TOMCATTIN' Blackfoot [21Jun80|23Aug80|01Nov80] 50(1) 20	0	722
1505. YES IT'S YOU LADY Smokey Robinson [20Feb82|20Mar82|12Jun82] 33(3) 17	2	722
1506. FIELDS OF FIRE Corey Hart [18Oct86|15Nov86|18Apr87] 55(2) 27	3	722
1507. MARCHING OUT Yngwie J. Malmsteen's Rising Force [07Sep85|26Oct85|15Mar86] 52(2) 28	0	721
1508. WILD-EYED DREAM Ricky Van Shelton [26Dec87|09Apr88|01Oct88] 76(1) 41	0	720
1509. BADLANDS Badlands [10Jun89|19Aug89|02Dec89] 57(2) 26	0	720
1510. KASHIF Kashif [09Apr83|11Jun83|19Nov83] 54(2) 33	0	720
1511. RECKONING Grateful Dead [18Apr81|23May81|01Aug81] 43(1) 16	0	720
1512. DANCING IN THE DRAGON'S JAWS Bruce Cockburn [23Feb80|28Jun80|02Aug80] 45(2) 24	1	720
1513. STEEL BREEZE Steel Breeze [18Sep82|06Nov82|26Mar83] 50(4) 28	2	719
1514. BUSTER Soundtrack [15Oct88|14Jan89|18Mar89] 54(2) 23	2	717
1515. KISS UNMASKED KISS [21Jun80|26Jul80|20Sep80] 35(2) 14	1	717
1516. AVALON SUNSET Van Morrison [01Jul89|05Aug89|24Mar90] 91(2) 39	0	716
1517. GOODBYE CRUEL WORLD Elvis Costello And The Attractions [07Jul84|28Jul84|24Nov84] 35(2) 21	1	715
1518. BEAST FROM THE EAST Dokken [03Dec88|17Dec88|25Mar89] 33(3) 17	0	715
1519. EXTRATERRESTRIAL LIVE Blue Öyster Cult [15May82|03Jul82|18Sep82] 29(2) 19	0	714
1520. NOW Patrice Rushen [16Jun84|28Jul84|01Dec84] 40(3) 25	1	714
1521. TEACHERS Soundtrack [27Oct84|01Dec84|09Feb85] 34(3) 16	3	714
1522. LIVE ALIVE Stevie Ray Vaughan And Double Trouble [20Dec86|24Jan87|06Jun87] 52(2) 25	0	713
1523. SHOW TIME Slave [10Oct81|21Nov81|13Mar82] 46(1) 23	1	712
1524. THE BEST IS YET TO COME Grover Washington Jr. [11Dec82|29Jan83|28May83] 50(3) 25	0	712
1525. EVITA Festival [09Feb80|29Mar80|07Jun80] 50(1) 18	0	711
1526. ABOUT LOVE Gladys Knight & The Pips [31May80|19Jul80|27Sep80] 48(1) 18	1	711
1527. THE SECRET POLICEMAN'S OTHER BALL Various Artists [20Mar82|15May82|03Jul82] 29(1) 16	0	710
1528. PRESS TO PLAY Paul McCartney [13Sep86|11Oct86|07Feb87] 30(2) 22	2	710
1529. ROCK THE HOUSE D.J. Jazzy Jeff & The Fresh Prince [25Apr87|17Dec88|25Mar89] 83(3) 35	1	709
1530. THE COLLECTION Amy Grant [20Sep86|07Feb87|02May87] 66(2) 33	0	709
1531. DECLARATION The Alarm [10Mar84|14Apr84|04Aug84] 50(2) 22	0	709
1532. NAUGHTY Chaka Khan [21Jun80|12Jul80|04Oct80] 43(4) 16	0	709
1533. EMPIRE BURLESQUE Bob Dylan [22Jun85|13Jul85|12Oct85] 33(2) 17	0	708
1534. LADY T Teena Marie [15Mar80|21Jun80|16Aug80] 45(1) 23	0	706
1535. THE MESSAGE Grandmaster Flash & The Furious Five [16Oct82|20Nov82|26Mar83] 53(4) 24	1	706
1536. BIOGRAPH Bob Dylan [07Dec85|11Jan86|03May86] 33(2) 22	0	705
1537. ONE GOOD REASON Paul Carrack [21Nov87|27Feb88|18Jun88] 67(1) 31	4	704
1538. BLACK ON BLACK Waylon Jennings [06Mar82|24Apr82|07Aug82] 39(2) 23	1	704
1539. STANDING ON THE EDGE Cheap Trick [17Aug85|12Oct85|14Dec85] 35(2) 18	1	704
1540. LONE JUSTICE Lone Justice [11May85|20Jul85|26Oct85] 56(1) 25	2	703
1541. BRENDA K. STARR Brenda K. Starr [21May88|16Jul88|29Oct88] 58(1) 24	2	700
1542. THE BEST OF THE ALAN PARSONS PROJECT The Alan Parsons Project [19Nov83|17Dec83|02Jun84] 53(4) 29	1	699
1543. SWEET DREAMS: THE LIFE AND TIMES OF PATSY CLINE (SOUNDTRACK) Patsy Cline [16Nov85|14Dec85|15Mar86] 29(2) 18	0	696
1544. BEST OF THE DOOBIES VOL. II Doobie Brothers [21Nov81|19Dec81|27Feb82] 39(1) 15	1	696
1545. OUT WHERE THE BRIGHT LIGHTS ARE GLOWING Ronnie Milsap [18Apr81|23May81|31Oct81] 89(2) 29	0	696

Ranking the Albums

Rank. Title *Act* [Enter|Peak|Exit] Peak(Wks) TotWks	Hot 100	Scr
1546. THE KEY *Joan Armatraiding* [30Apr83|11Jun83|24Sep83] 32(1) 22	1	696
1547. TA MARA & THE SEEN *Ta Mara & The Seen* [02Nov85|01Feb86|19Apr86] 54(3) 25	1	695
1548. CAN'T STOP THE LOVE *Maze Featuring Frankie Beverly* [30Mar85|04May85|19Oct85] 45(2) 30	1	695
1549. MEASURE FOR MEASURE *Icehouse* [24May86|09Aug86|01Nov86] 55(3) 24	1	694
1550. 20 GREATEST HITS *The Beatles* [13Nov82|15Jan83|12May84] 50(2) 28	0	693
1551. GHETTO MUSIC: THE BLUEPRINT OF HIP HOP *Boogie Down Productions* [22Jul89|12Aug89|11Nov89] 36(2) 17	0	693
1552. PRECIOUS MOMENTS *Jermaine Jackson* [22Mar86|03May86|16Aug86] 46(2) 22	2	693
1553. JOHN PARR *John Parr* [15Dec84|23Mar85|08Jun85] 48(2) 26	3	693
1554. ENUFF Z'NUFF *Enuff Z'Nuff* [30Sep89|25Nov89|19May90] 74(1) 34	2,1*	693
1555. A CHANGE OF HEART *David Sanborn* [14Feb87|28Mar87|16Jan88] 74(2) 37	0	692
1556. DON'T DISTURB THIS GROOVE *The System* [18Apr87|04Jul87|03Oct87] 62(5) 25	1	692
1557. JUST COOLIN' *Levert* [26Nov88|22Apr89|24Jun89] 79(1) 31	0	691
1558. Концерт (LIVE IN LENINGRAD) *Billy Joel* [07Nov87|05Dec87|26Mar88] 38(1) 18	0	691
1559. IT'S MY LIFE *Talk Talk* [07May84|16Jun84|01Sep84] 42(3) 22	2	690
1560. ANIMAL MAGNETISM *Scorpions* [17May80|05Jul80|04Oct80] 52(1) 21	0	690
1561. NEW CLEAR DAYS *The Vapors* [16Aug80|06Dec80|21Feb81] 62(1) 28	1	689
1562. I'VE NEVER BEEN TO ME *Charlene* [10Apr82|12Jun82|21Aug82] 36(2) 20	1	688
1563. JOHN COUGAR *John Cougar* [18Aug79|05Jan80|01Mar80] 64(2) 29	2,*1	687
1564. HANG ON FOR YOUR LIFE *Shooting Star* [19Sep81|05Dec81|10Apr82] 92(1) 30	1	686
1565. SOUL KISS *Olivia Newton-John* [02Nov85|23Nov85|15Feb86] 29(3) 16	1	685
1566. MISTRIAL *Lou Reed* [24May86|05Jul86|11Oct86] 47(2) 21	0	684
1567. SOLDIERS UNDER COMMAND *Stryper* [28Sep85|30Nov85|04Apr87] 84(2) 64	0	684
1568. CRAZY FOR YOU *Earl Klugh* [14Nov81|05Dec81|15May82] 53(2) 27	0	683
1569. PROPOSITIONS *Bar-Kays* [20Nov82|25Dec82|04Jun83] 51(3) 29	0	683
1570. GREEN LIGHT *Bonnie Raitt* [06Mar82|17Apr82|03Jul82] 38(2) 18	0	682
1571. FINE YOUNG CANNIBALS *Fine Young Cannibals* [25Jan86|24May86|02Aug86] 49(1) 28	1	682
1572. SOMETHING ABOUT YOU *Angela Bofill* [21Nov81|19Dec81|17Apr82] 61(4) 22	0	681
1573. THE YEAR 2000 *The O'Jays* [30Aug80|04Oct80|15Nov80] 36(2) 12	1	681
1574. THE ICEBERG (FREEDOM OF SPEECH...JUST WATCH WHAT YOU SAY) *Ice-T* [28Oct89|11Nov89|05May90] 37(2) 20	0	680
1575. RADIO ACTIVE *Pat Travers* [28Mar81|25Apr81|04Jul81] 37(2) 15	0	679
1576. HILLBILLY DELUXE *Dwight Yoakam* [16May87|13Jun87|21Nov87] 55(2) 28	0	679
1577. OUTLAW *War* [20Mar82|08May82|18Sep82] 48(2) 27	2	677
1578. BAD FOR GOOD *Jim Steinman* [16May81|11Jul81|05Sep81] 63(1) 17	1	676
1579. TIN MACHINE *Tin Machine* [10Jun89|01Jul89|30Sep89] 28(1) 17	0	676
1580. COOL NIGHT *Paul Davis* [19Dec81|22May82|03Jul82] 52(1) 29	3	676
1581. 1988 SUMMER OLYMPICS-ONE MOMENT IN TIME *Various Artists* [24Sep88|22Oct88|14Jan89] 31(2) 17	2	676
1582. PEARLS-SONGS OF GOFFIN AND KING *Carole King* [07Jun80|26Jul80|27Sep80] 44(1) 17	1	674
1583. SANDS OF TIME *The S.O.S. Band* [24May86|05Jul86|04Oct86] 44(2) 20	1	674
1584. ODYSSEY *Yngwie J. Malmsteen's Rising Force* [23Apr88|14May88|20Aug88] 40(2) 18	0	674
1585. AMERICAN MADE *The Oak Ridge Boys* [26Feb83|09Apr83|30Jul83] 51(3) 23	1	673
1586. MASK OF SMILES *John Waite* [31Aug85|12Oct85|14Dec85] 36(2) 16	2	672
1587. KEYED UP *Ronnie Milsap* [30Apr83|25Jun83|03Sep83] 36(1) 19	2	672
1588. SINGLE LIFE *Cameo* [13Jul85|17Aug85|11Jan86] 58(2) 27	0	671
1589. PASSIONWORKS *Heart* [17Sep83|22Oct83|04Feb84] 39(1) 21	2	671
1590. MANILOW *Barry Manilow* [30Nov85|28Dec85|09Aug86] 42(3) 24	1	671
1591. DRASTIC MEASURES *Kansas* [13Aug83|24Sep83|31Dec83] 41(2) 21	1	670
1592. ANDY GIBB'S GREATEST HITS *Andy Gibb* [06Dec80|31Jan81|04Apr81] 46(1) 18	2	670
1593. SILKY SOUL *Maze Featuring Frankie Beverly* [23Sep89|14Oct89|17Feb90] 37(2) 22	0	670
1594. INCOGNITO *Spyro Gyra* [23Oct82|20Nov82|02Apr83] 46(2) 24	0	669
1595. PRIMITIVE COOL *Mick Jagger* [03Oct87|17Oct87|13Feb88] 41(2) 20	2	668
1596. "A" *Jethro Tull* [13Sep80|11Oct80|29Nov80] 30(2) 12	0	667
1597. TOUGHER THAN LEATHER *Willie Nelson* [19Mar83|21May83|30Jul83] 39(1) 20	0	666
1598. WHITE WINDS *Andreas Vollenweider* [02Mar85|22Jun85|23Nov85] 76(2) 39	0	665
1599. HEARTLAND *Michael Stanley Band* [27Sep80|28Feb81|02May81] 86(2) 32	2	664
1600. UNDER THE BLUE MOON *New Edition* [20Dec86|07Feb87|23May87] 43(2) 23	1	663
1601. ISOLATION *Toto* [24Nov84|15Dec84|13Apr85] 42(4) 21	2	663
1602. ALPHABET CITY *ABC* [22Aug87|19Sep87|06Feb88] 48(3) 25	1	662
1603. CAUGHT IN THE ACT - LIVE *Styx* [21Apr84|19May84|28Jul84] 31(2) 15	1	662
1604. WHO'S FOOLIN' WHO *One Way* [03Apr82|12Jun82|04Sep82] 51(1) 23	1	662
1605. GLORYHALLASTOOPID (OR PIN THE TAIL ON THE FUNKY) *Parliament* [22Dec79|09Feb80|26Apr80] 44(2) 19	0	661
1606. WALK UNDER LADDERS *Joan Armatrading* [17Oct81|27Mar82|22May82] 88(2) 32	1	660
1607. MESOPOTAMIA *The B-52s* [20Feb82|13Mar82|19Jun82] 35(2) 18	0	660
1608. THE MAGAZINE *Rickie Lee Jones* [13Oct84|10Nov84|02Mar85] 44(2) 21	1	659
1609. AS FAR AS SIAM *Red Rider* [12Sep81|05Dec81|20Feb82] 65(2) 24	0	658
1610. SHADOWS AND LIGHT *Joni Mitchell* [04Oct80|25Oct80|17Jan81] 38(2) 16	0	657
1611. WILDSIDE *Loverboy* [12Sep87|17Oct87|30Jan88] 42(1) 21	1	657
1612. H *Bob James* [12Jul80|16Aug80|08Nov80] 47(3) 18	0	656
1613. STRANGEWAYS, HERE WE COME *The Smiths* [10Oct87|31Oct87|09Apr88] 55(2) 27	0	656
1614. DEAD SET *Grateful Dead* [19Sep81|10Oct81|28Nov81] 29(2) 11	0	656
1615. TAKING LIBERTIES *Elvis Costello* [11Oct80|08Nov80|10Jan81] 28(1) 14	0	654
1616. HOT HOUSE FLOWERS *Wynton Marsalis* [13Oct84|10Nov84|06Jul85] 90(2) 39	0	654
1617. EARTH - SUN - MOON *Love And Rockets* [31Oct87|30Jan88|07May88] 64(2) 28	0	653
1618. SAVAGE *Eurythmics* [26Dec87|06Feb88|30Apr88] 41(2) 19	2	653
1619. JOYSTICK *Dazz Band* [17Dec83|25Feb84|28Jul84] 73(2) 33	1	652
1620. FLASH *Jeff Beck* [20Jul85|17Aug85|16Nov85] 39(2) 18	1	652
1621. FLUSH THE FASHION *Alice Cooper* [24May80|05Jul80|13Sep80] 44(2) 17	1	651
1622. JUICE *Oran 'Juice' Jones* [20Sep86|15Nov86|14Feb87] 44(1) 22	1	651
1623. DIFFICULT TO CURE *Rainbow* [07Mar81|18Apr81|20Jun81] 50(2) 16	0	650
1624. SOME TOUGH CITY *Tony Carey* [31Mar84|19May84|08Sep84] 60(2) 24	2	649
1625. BETTER THAN HEAVEN *Stacey Q* [27Sep86|25Oct86|31Jan87] 59(2) 39	2	649
1626. THOSE OF YOU WITH OR WITHOUT CHILDREN, YOU'LL UNDERSTAND *Bill Cosby* [21Jun86|19Jul86|27Sep86] 26(2) 15	0	648
1627. VIVA HATE *Morrissey* [09Apr88|30Apr88|20Aug88] 48(2) 20	0	647
1628. TAKE NO PRISONERS *Molly Hatchet* [05Dec81|06Feb82|06Mar82] 36(1) 14	1	645
1629. THE PACIFIC AGE *Orchestral Manoeuvres In The Dark* [18Oct86|29Nov86|21Mar87] 47(2) 23	1	645
1630. DON'T SUPPOSE *Limahl* [27Apr85|22Jun85|07Sep85] 41(1) 20	2	642
1631. VOCALESE *The Manhattan Transfer* [10Aug85|12Oct85|10May86] 74(2) 40	0	641
1632. STREET OPERA *Ashford & Simpson* [29May82|03Jul82|09Oct82] 45(3) 20	1	641
1633. 25 #1 HITS FROM 25 YEARS *Various* [04Jun83|02Jul83|10Dec83] 42(2) 28	0	641
1634. THE BIZ NEVER SLEEPS *The Diabolical Biz Markie* [28Oct89|04Nov89|19May90] 66(2) 30	1,1*	640
1635. SMALL CHANGE *Prism* [06Feb82|03Apr82|19Jun82] 53(1) 20	2	640
1636. THE LAST COMMAND *W.A.S.P.* [23Nov85|14Dec85|26Apr86] 49(2) 23	0	639
1637. CLOSE-UP *David Sanborn* [16Jul88|23Jul88|21Jan89] 59(4) 28	0	638
1638. WANNA BE A STAR *Chilliwack* [03Oct81|19Dec81|24Apr82] 78(1) 30	2	638
1639. ELECTRIC UNIVERSE *Earth, Wind & Fire* [03Dec83|07Jan84|17Mar84] 40(2) 16	1	638
1640. SAN ANTONIO ROSE *Willie Nelson & Ray Price* [14Jun80|19Jul80|29Nov80] 70(2) 25	0	637
1641. AMERICAN EXCESS *Point Blank* [25Apr81|12Sep81|03Oct81] 80(1) 24	1	637
1642. 24 CARROTS *Al Stewart* [13Sep80|11Oct80|06Dec80] 37(2) 13	1	637

Ranking the Albums

Rank. Title *Act* [Enter|Peak|Exit] Peak(Wks) TotWks	Hot 100	Scr
1643. OFFRAMP *Pat Metheny Group* [22May82|19Jun82|27Nov82] 50(2) 28	0	637
1644. SOMETIMES YOU WIN *Dr. Hook* [24Nov79|12Jan80|05Jul80] 71(2) 32	3	635
1645. THE UP ESCALATOR *Graham Parker And The Rumour* [31May80|05Jul80|06Sep80] 40(1) 15	0	635
1646. THE KNIFE FEELS LIKE JUSTICE *Brian Setzer* [22Mar86|26Apr86|19Jul86] 45(3) 18	0	633
1647. PLANET P *Planet P* [26Mar83|28May83|27Aug83] 42(2) 23	1	633
1648. HEARTLAND *The Judds* [04Apr87|25Apr87|14Nov87] 52(2) 31	0	632
1649. CALL OF THE WEST *Wall Of Voodoo* [15Jan83|09Apr83|18Jun83] 45(2) 23	1	632
1650. THE GETAWAY *Chris de Burgh* [09Apr83|18Jun83|03Sep83] 43(2) 22	2	631
1651. GREATEST HITS, VOLUME TWO *George Strait* [26Sep87|07Nov87|23Apr88] 68(2) 31	0	631
1652. ROSS (II) *Diana Ross* [16Jul83|20Aug83|05Nov83] 32(1) 17	2	630
1653. RAGE FOR ORDER *Queensryche* [26Jul86|09Aug86|07Feb87] 47(5) 21	0	630
1654. AN AMERICAN TAIL *Soundtrack* [31Jan87|28Mar87|06Jun87] 42(2) 19	1	629
1655. CAROL HENSEL'S EXERCISE AND DANCE PROGRAM, VOLUME 2 *Carol Hensel* [19Dec81|06Mar82|26Jun82] 70(1) 28	0	628
1656. BIG LIFE *Night Ranger* [11Apr87|02May87|08Aug87] 28(2) 18	2	628
1657. OTHER ROADS *Boz Scaggs* [04Jun88|25Jun88|01Oct88] 47(4) 18	1	628
1658. LATE AT NIGHT *Billy Preston* [08Mar80|10May80|05Jul80] 49(2) 18	0	627
1659. CATHOLIC BOY *The Jim Carroll Band* [15Nov80|07Mar81|18Apr81] 73(2) 23	0	627
1660. SOMEWHERE IN AFRIKA *Manfred Mann's Earth Band* [28Jan84|24Mar84|16Jun84] 40(2) 21	0	627
1661. BUILT TO LAST *Grateful Dead* [18Nov89|02Dec89|24Feb90] 27(1) 15	0	626
1662. THE KINGS ARE HERE *The Kings* [16Aug80|11Oct80|07Feb81] 74(2) 26	2	626
1663. LIVE & MORE *Roberta Flack & Peabo Bryson* [20Dec80|14Feb81|25Apr81] 52(2) 19	0	626
1664. VULTURE CULTURE *Alan Parsons Project* [09Mar85|06Apr85|13Jul85] 46(3) 19	2	625
1665. SHELTER *Lone Justice* [29Nov86|28Mar87|20Jun87] 65(2) 30	1	625
1666. GO *Hiroshima* [15Aug87|12Sep87|19Mar88] 75(2) 32	0	624
1667. JANET JACKSON *Janet Jackson* [20Nov82|22Jan83|07May83] 63(1) 25	2	622
1668. RADIO K.A.O.S. *Roger Waters* [04Jul87|25Jul87|07Nov87] 50(2) 19	0	622
1669. BACK ON THE STREETS *Donnie Iris* [13Dec80|04Apr81|16May81] 57(1) 23	1	621
1670. GET IT RIGHT *Aretha Franklin* [30Jul83|03Sep83|26Nov83] 36(1) 18	1	621
1671. GREATEST HITS *Melissa Manchester* [26Feb83|16Apr83|16Jul83] 43(2) 21	0	620
1672. THE MIRACLE *Queen* [24Jun89|08Jul89|23Sep89] 24(1) 14	1	619
1673. A LITTLE LOVE *Aurra* [27Feb82|24Apr82|05Jun82] 38(2) 15	1	619
1674. ONE TO ONE *Howard Jones* [01Nov86|15Nov86|21Mar87] 56(4) 21	2	618
1675. MR. BIG *Mr. Big* [22Jul89|26Aug89|18Nov89] 46(2) 18	0	618
1676. WILD STREAK *Hank Williams Jr.* [16Jul88|13Aug88|19Nov88] 55(1) 19	0	617
1677. ANCIENT HEART *Tanita Tikaram* [11Feb89|13May89|15Jul89] 59(2) 23	0	616
1678. DANGER DANGER *Danger Danger* [19Aug89|11Nov89|15Sep90] 88(1) 42	1,1*	615
1679. INFORMATION *Dave Edmunds* [21May83|16Jul83|01Oct83] 51(2) 20	1	615
1680. MY HOME'S IN ALABAMA *Alabama* [19Jul80|30Aug80|06Dec80] 71(2) 21	0	614
1681. YOUR WISH IS MY COMMAND *Lakeside* [09Jan82|06Mar82|12Jun82] 58(1) 23	0	613
1682. BOX OF FROGS *Box Of Frogs* [07Jul84|25Aug84|17Nov84] 45(3) 20	0	613
1683. AVALON *Roxy Music* [19Jun82|17Jul82|18Dec82] 53(3) 27	0	613
1684. MODERN HEART *Champaign* [02Apr83|23Jul83|10Sep83] 64(1) 24	1	612
1685. SOLID GROUND *Ronnie Laws* [10Oct81|21Nov81|13Feb82] 51(2) 19	1	612
1686. EGO TRIP *Kurtis Blow* [13Oct84|16Mar85|22Jun85] 83(2) 37	1	612
1687. KIHNTINUED *Greg Kihn* [10Apr82|22May82|31Jul82] 33(3) 17	2	612
1688. HAVANA MOON *Carlos Santana* [23Apr83|28May83|13Aug83] 31(2) 17	0	611
1689. SAMANTHA FOX *Samantha Fox* [24Oct87|25Jun88|27Aug88] 51(1) 25	2	611
1690. THE QUEEN IS DEAD *The Smiths* [19Jul86|23Aug86|28Mar87] 70(1) 37	0	611
1691. OVER THE EDGE *Hurricane* [30Apr88|15Oct88|31Dec88] 92(2) 36	0	611
1692. CONCERTS FOR THE PEOPLE OF KAMPUCHEA *Various Artists* [18Apr81|09May81|04Jul81] 36(2) 12	0	610
1693. RHYME & REASON *Missing Persons* [31Mar84|12May84|14Jul84] 43(2) 16	1	609
1694. WILL TO POWER *Will To Power* [10Sep88|10Dec88|25Mar89] 68(2) 29	4	609
1695. UNDER THE GUN *Poco* [26Jul80|13Sep80|08Nov80] 46(1) 16	2	608
1696. DONNY OSMOND *Donny Osmond* [13May89|10Jun89|14Oct89] 54(3) 23	3	608
1697. THE FAT BOYS ARE BACK! *Fat Boys* [31Aug85|21Sep85|12Apr86] 63(2) 33	0	607
1698. DREAM COME TRUE *Earl Klugh* [19Apr80|17May80|23Aug80] 42(1) 19	0	607
1699. MAD, BAD AND DANGEROUS TO KNOW *Dead Or Alive* [27Dec86|14Mar87|13Jun87] 52(2) 25	2	607
1700. STEPHANIE MILLS *Stephanie Mills* [29Mar86|07Jun86|23Aug86] 47(2) 22	0	607
1701. TO BE CONTINUED... *The Temptations* [02Aug86|01Nov86|14Mar87] 74(2) 33	1	607
1702. BEYOND APPEARANCES *Santana* [23Mar85|20Apr85|10Aug85] 50(2) 21	1	605
1703. BEAT CRAZY *Joe Jackson* [08Nov80|13Dec80|21Feb81] 41(1) 16	0	605
1704. THROUGH THE FIRE *Hagar, Schon, Aaronson, Shrieve* [31Mar84|12May84|28Jul84] 42(2) 18	1	605
1705. THE POLITICS OF DANCING *Re-Flex* [24Dec83|24Mar84|30Jun84] 53(1) 28	2	604
1706. HEADLINES *Midnight Star* [14Jun86|05Jul86|13Dec86] 56(2) 27	2	604
1707. ICICLE WORKS *Icicle Works* [21Apr84|30Jun84|18Aug84] 40(2) 18	1	604
1708. HOY-HOY! *Little Feat* [22Aug81|26Sep81|14Nov81] 39(2) 13	0	604
1709. VOICES IN THE RAIN *Joe Sample* [31Jan81|14Mar81|13Jun81] 65(3) 20	0	603
1710. SEXAPPEAL *Georgio* [25Apr87|09Jan88|16Apr88] 117(1) 52	3	603
1711. DON'T TELL A SOUL *The Replacements* [18Feb89|11Mar89|24Jun89] 57(2) 19	1	602
1712. 8TH WONDER *Sugarhill Gang* [30Jan82|13Mar82|29May82] 50(1) 18	2	601
1713. KING OF AMERICA *Elvis Costello And The Attractions* [22Mar86|12Apr86|19Jul86] 39(2) 18	0	601
1714. DON'T LOOK BACK *Natalie Cole* [14Jun80|26Jul80|08Nov80] 77(1) 22	1	601
1715. WITH YOU *Stacy Lattisaw* [25Jul81|05Sep81|31Oct81] 46(1) 15	1	601
1716. POWER *The Temptations* [17May80|12Jul80|16Aug80] 45(1) 14	1	601
1717. A CHRISTMAS TOGETHER *John Denver & The Muppets* [10Nov79|05Jan80|26Jan80] 26(2) 12	0	600
1718. TANTALIZINGLY HOT *Stephanie Mills* [07Aug82|04Sep82|11Dec82] 48(6) 19	0	598
1719. LIVING EYES *Bee Gees* [21Nov81|19Dec81|06Feb82] 41(3) 12	2	598
1720. DAYDREAMING *Morris Day* [12Mar88|09Apr88|18Jun88] 41(2) 19	1	596
1721. NOTHING'S SHOCKING *Jane's Addiction* [17Sep88|25Feb89|13May89] 103(2) 35	0	595
1722. HAVE YOU SEEN ME LATELY? *Sam Kinison* [26Nov88|24Dec88|18Mar89] 43(3) 17	0	594
1723. FLASHBACK *38 Special* [22Aug87|26Sep87|12Dec87] 35(2) 17	1	591
1724. LIVING IN THE 20TH CENTURY *Steve Miller Band* [15Nov86|07Feb87|18Apr87] 65(1) 23	1	590
1725. SECRET MESSAGES *Electric Light Orchestra* [16Jul83|20Aug83|29Oct83] 36(1) 16	2	590
1726. SOMETIMES LATE AT NIGHT *Carole Bayer Sager* [16May81|11Jul81|10Oct81] 60(1) 22	1	590
1727. CHRISTMAS *Kenny Rogers* [21Nov81|09Jan82|15Jan83] 34(2) 13	0	589
1728. CHESS *Various Artists* [16Mar85|25May85|03Aug85] 47(2) 21	1	589
1729. ONE NIGHT WITH A STRANGER *Martin Briley* [07May83|30Jul83|01Oct83] 55(1) 22	1	589
1730. THE HISTORY MIX VOL. I *Godley & Creme* [17Aug85|05Oct85|23Nov85] 37(2) 15	1	589
1731. SHOCK *The Motels* [17Aug85|28Sep85|30Nov85] 36(1) 16	2	588
1732. LYLE LOVETT AND HIS LARGE BAND *Lyle Lovett* [18Feb89|22Apr89|08Jul89] 62(1) 21	0	588
1733. COCKER *Joe Cocker* [12Apr86|07Jun86|09Aug86] 50(2) 18	1	588
1734. TRILOGY *Yngwie Malmsteen* [11Oct86|01Nov86|14Mar87] 44(2) 23	0	587
1735. KING COOL *Donnie Iris* [26Sep81|07Nov81|26Jun82] 84(2) 31	3	587
1736. EB 84 *The Everly Brothers* [13Oct84|03Nov84|02Feb85] 38(2) 17	1	587
1737. KEEP YOUR DISTANCE *Curiosity Killed The Cat* [22Aug87|26Sep87|05Mar88] 55(1) 29	1	587
1738. WHERE DO YOU GO WHEN YOU DREAM *Anne Murray* [02May81|30May81|08Aug81] 55(2) 15	3	587
1739. STEREOTOMY *The Alan Parsons Project* [01Feb86|01Mar86|31May86] 43(2) 18	1	586

Ranking the Albums

Rank. Title *Act* [Enter|Peak|Exit] Peak(Wks) TotWks	Hot 100	Scr
1740. LANDING ON WATER *Neil Young* [16Aug86|27Sep86|29Nov86] 46(2) 16	0	586
1741. ARGYBARGY *Squeeze* [26Apr80|05Jul80|04Oct80] 71(1) 24	0	584
1742. BREAKING ALL THE RULES *Peter Frampton* [13Jun81|01Aug81|05Sep81] 43(1) 13	0	584
1743. DO YOU *Sheena Easton* [23Nov85|14Dec85|29Mar86] 40(2) 19	2	584
1744. YELLOW MOON *The Neville Brothers* [08Apr89|03Jun89|16Sep89] 66(3) 24	0	583
1745. PREMONITION *Survivor* [24Oct81|19Dec81|10Apr82] 82(3) 25	2	583
1746. YOUNGEST IN CHARGE *Special Ed* [03Jun89|22Jul89|09Dec89] 73(1) 28	0	583
1747. OCTOBER *U2* [07Nov81|28Nov81|19Sep87] 104(2) 38	0	583
1748. GAP BAND VI *The Gap Band* [19Jan85|09Feb85|22Jun85] 58(2) 23	0	583
1749. VOYEUR *Kim Carnes* [25Sep82|09Oct82|19Feb83] 49(3) 22	2	583
1750. GREATEST HITS *Ray Parker Jr.* [18Dec82|29Jan83|14May83] 51(3) 22	1	583
1751. ERROR IN THE SYSTEM *Peter Schilling* [08Oct83|21Jan84|10Mar84] 61(1) 23	1	582
1752. EXILES *Dan Fogelberg* [20Jun87|25Jul87|24Oct87] 48(3) 19	1	581
1753. HOUSE OF MUSIC *T.S. Monk* [31Jan81|11Apr81|27Jun81] 64(1) 22	1	581
1754. THE VISITOR *Mick Fleetwood* [18Jul81|29Aug81|17Oct81] 43(2) 14	0	581
1755. EXTREME *Extreme* [08Apr89|17Jun89|11Nov89] 80(1) 32	0	580
1756. THE NAME OF THIS BAND IS TALKING HEADS *Talking Heads* [17Apr82|22May82|17Jul82] 31(2) 14	0	579
1757. JOE JACKSON'S JUMPIN' JIVE *Joe Jackson* [01Aug81|05Sep81|24Oct81] 42(2) 13	0	579
1758. ROCK AWAY *Phoebe Snow* [04Apr81|23May81|01Aug81] 51(2) 18	2	579
1759. LET ME TOUCH YOU *The O'Jays* [10Oct87|05Dec87|26Mar88] 66(1) 25	0	577
1760. WORD OF MOUTH *The Kinks* [15Dec84|09Feb85|27Apr85] 57(3) 20	1	577
1761. BARBARA MANDRELL LIVE *Barbara Mandrell* [05Sep81|07Nov81|13Feb82] 86(2) 24	0	576
1762. GOOD TO BE BACK *Natalie Cole* [27May89|17Jun89|28Oct89] 59(2) 23	1	576
1763. SHADOWLAND *k.d. lang* [28May88|20Aug88|12Nov88] 73(2) 25	0	576
1764. MY EVER CHANGING MOODS *The Style Council* [07Apr84|30Jun84|01Sep84] 56(2) 22	2	576
1765. END OF THE CENTURY *The Ramones* [23Feb80|29Mar80|24May80] 44(1) 14	0	575
1766. EYE OF THE HURRICANE *The Alarm* [07Nov87|28Nov87|28May88] 77(2) 30	2	574
1767. BOYS AND GIRLS *Bryan Ferry* [29Jun85|20Jul85|14Dec85] 63(2) 25	0	574
1768. SARAYA *Saraya* [29Apr89|24Jun89|27Jan90] 79(3) 39	2	574
1769. LOVE *The Cult* [28Dec85|26Apr86|16Aug86] 87(3) 34	0	573
1770. STEADY NERVES *Graham Parker & The Shot* [20Apr85|15Jun85|07Sep85] 57(3) 21	1	573
1771. MADE IN AMERICA *Carpenters* [04Jul81|15Aug81|10Oct81] 52(1) 15	4	572
1772. CHANGE OF ADDRESS *Krokus* [03May86|28Jun86|23Aug86] 45(2) 17	1	571
1773. ELECTRIC LADY *Con Funk Shun* [18May85|22Jun85|09Nov85] 62(3) 26	0	571
1774. LAST MANGO IN PARIS *Jimmy Buffett* [06Jul85|24Aug85|16Nov85] 53(3) 20	0	568
1775. GEFFREY MORGAN... *UB40* [10Nov84|15Dec84|04May85] 60(1) 26	0	566
1776. PRIVATE AUDITION *Heart* [12Jun82|03Jul82|11Sep82] 25(2) 14	1	565
1777. ZAPP III *Zapp* [03Sep83|01Oct83|28Jan84] 39(1) 22	0	565
1778. RIDE THE LIGHTNING *Metallica* [29Sep84|09Mar85|28Jan86] 100(2) 50	0	564
1779. KNIGHTS OF THE SOUND TABLE *Cameo* [20Jun81|11Jul81|12Sep81] 44(2) 13	0	564
1780. THE TONIGHT SHOW BAND *The Tonight Show Band with Doc Severinsen* [01Nov86|27Dec86|25Apr87] 65(4) 26	0	563
1781. THE SEVENTH ONE *Toto* [19Mar88|30Apr88|16Jul88] 64(3) 18	1	562
1782. VOICES OF BABYLON *The Outfield* [15Apr89|06May89|16Sep89] 53(3) 23	2	562
1783. HIDEAWAY *David Sanborn* [08Mar80|17May80|12Jul80] 63(1) 19	0	561
1784. HOOLIGANS *The Who* [17Oct81|14Nov81|20Feb82] 52(2) 19	0	561
1785. WWII *Waylon Jennings & Willie Nelson* [30Oct82|20Nov82|26Mar83] 57(3) 22	0	561
1786. HIGHWAYMAN *Willie, Waylon, Johnny & Kris* [01Jun85|17Aug85|25Jan86] 92(1) 35	0	561
1787. CHALK MARK IN A RAIN STORM *Joni Mitchell* [09Apr88|30Apr88|23Jul88] 45(2) 16	0	560
1788. FAST TIMES AT RIDGEMONT HIGH *Soundtrack* [28Aug82|25Sep82|08Jan83] 54(3) 20	2	559
1789. POP GOES THE WORLD *Men Without Hats* [14Nov87|20Feb88|30Apr88] 73(2) 25	1	559
1790. ENGLISH SETTLEMENT *XTC* [20Mar82|22May82|31Jul82] 48(1) 20	0	559
1791. EXPRESS *Love And Rockets* [01Nov86|07Mar87|23May87] 72(2) 30	0	558
1792. JUST THE WAY YOU LIKE IT *The S.O.S. Band* [01Sep84|13Oct84|02Mar85] 60(2) 27	1	558
1793. STRICTLY BUSINESS *EPMD* [09Jul88|13Aug88|10Dec88] 80(1) 23	0	558
1794. SOUTHERN STAR *Alabama* [18Feb89|18Mar89|08Jul89] 62(2) 21	0	558
1795. REUNION *The Temptations* [01May82|05Jun82|28Aug82] 37(2) 18	1	557
1796. STILL CRUISIN' *The Beach Boys* [16Sep89|07Oct89|10Feb90] 46(1) 22	1	556
1797. OH NO! IT'S DEVO *Devo* [20Nov82|18Dec82|02Apr83] 47(3) 20	0	556
1798. LAST SAFE PLACE *Le Roux* [06Feb82|01May82|26Jun82] 64(2) 21	2	556
1799. YOU'VE GOT THE POWER *Third World* [20Mar82|05Jun82|18Sep82] 63(3) 27	0	555
1800. ON THE LINE *Gary U.S. Bonds* [26Jun82|14Aug82|16Oct82] 52(2) 17	1	555
1801. DANGEROUS *Bar-Kays* [21Apr84|16Jun84|15Sep84] 52(1) 22	1	555
1802. THE MAN WITH THE HORN *Miles Davis* [25Jul81|12Sep81|21Nov81] 53(2) 18	0	553
1803. ROMANCE DANCE *Kim Carnes* [05Jul80|16Aug80|25Oct80] 57(2) 17	2	553
1804. RENDEZ-VOUS *Jean Michel Jarre* [03May86|31May86|13Sep86] 52(4) 20	0	553
1805. SO EXCITED! *The Pointer Sisters* [17Jul82|04Sep82|22Jan83] 59(1) 28	3	552
1806. 101 *Depeche Mode* [01Apr89|15Apr89|05Aug89] 45(2) 19	0	551
1807. HEAVEN 17 *Heaven 17* [12Feb83|26Mar83|20Aug83] 68(3) 28	0	551
1808. PRIEST...LIVE *Judas Priest* [20Jun87|11Jul87|26Sep87] 38(2) 15	0	551
1809. WAITIN' FOR THE SUN TO SHINE *Ricky Skaggs* [12Jun82|10Jul82|01Jan83] 77(2) 30	0	551
1810. LET ME TICKLE YOUR FANCY *Jermaine Jackson* [21Aug82|09Oct82|04Dec82] 46(3) 16	1	551
1811. RAIN FOREST *Paul Hardcastle* [23Mar85|20Apr85|07Sep85] 63(2) 25	1	551
1812. PHANTOM, ROCKER & SLICK *Phantom, Rocker & Slick* [26Oct85|07Dec85|29Mar86] 61(2) 23	0	550
1813. THUNDER *Andy Taylor* [28Mar87|09May87|18Jul87] 46(4) 17	0	549
1814. VOICES OF THE HEART *Carpenters* [19Nov83|07Jan84|24Mar84] 46(2) 19	0	548
1815. FISHERMAN'S BLUES *The Waterboys* [10Dec88|11Feb89|03Jun89] 76(2) 26	0	548
1816. ROCK OF LIFE *Rick Springfield* [20Feb88|02Apr88|04Jun88] 55(1) 16	1	545
1817. SEND ME YOUR LOVE *Kashif* [21Jul84|08Sep84|08Dec84] 51(2) 21	0	545
1818. ONE PARTICULAR HARBOUR *Jimmy Buffett* [08Oct83|29Oct83|17Mar84] 59(3) 24	0	545
1819. THE BEST OF ERIC CARMEN *Eric Carmen* [11Jun88|27Aug88|22Oct88] 59(1) 20	0	543
1820. SINGLES 45'S AND UNDER *Squeeze* [08Jan83|26Feb83|28May83] 47(3) 21	0	543
1821. SURFACE *Surface* [30May87|04Jul87|03Oct87] 55(1) 19	1	542
1822. STRAIGHT TO THE HEART *David Sanborn* [09Feb85|02Mar85|14Sep85] 64(2) 32	0	542
1823. THE NET *Little River Band* [18Jun83|16Jul83|05Nov83] 61(1) 21	2	541
1824. NO MORE DIRTY DEALS *Johnny Van Zant Band* [06Sep80|08Nov80|13Dec80] 48(1) 15	0	540
1825. SWING STREET *Barry Manilow* [12Dec87|30Jan88|30Apr88] 70(1) 21	1	540
1826. RESERVATIONS FOR TWO *Dionne Warwick* [22Aug87|19Sep87|20Feb88] 56(2) 27	2	540
1827. FANTASTIC *Wham!* [20Aug83|08Oct83|16Nov85] 83(1) 44	1	540
1828. LIVING ALL ALONE *Phyllis Hyman* [11Oct86|11Apr87|25Jul87] 78(2) 41	0	537
1829. CROSS THAT LINE *Howard Jones* [15Apr89|06May89|09Sep89] 65(1) 22	2	537
1830. CATS *Original Broadway Cast Recording* [26Feb83|02Apr83|22Mar86] 113(3) 64	0	536
1831. WORLD IN MOTION *Jackson Browne* [24Jun89|08Jul89|07Oct89] 45(2) 16	0	536
1832. I'M THE ONE *Roberta Flack* [19Jun82|07Aug82|06Nov82] 59(2) 21	2	535
1833. BE MY LOVER *O'Bryan* [26May84|28Jul84|13Oct84] 64(1) 21	0	535
1834. GO INSANE *Lindsey Buckingham* [01Sep84|29Sep84|15Dec84] 45(4) 16	1	534
1835. DISCIPLINE *King Crimson* [31Oct81|28Nov81|20Feb82] 45(2) 17	0	534
1836. TRIUMPH AND AGONY *Warlock* [19Dec87|09Apr88|18Jun88] 80(1) 27	0	534
1837. HAPPY ANNIVERSARY, CHARLIE BROWN *Various Artists* [11Nov89|23Dec89|31Mar90] 65(3) 21	0	534

Ranking the Albums

Rank. Title *Act* [Enter|Peak|Exit] Peak(Wks) TotWks	Hot 100	Scr
1838. FOUND ALL THE PARTS *Cheap Trick* [05Jul80|19Jul80|20Sep80] 39(2) 12	0	533
1839. RIGHTEOUS ANGER *Van Stephenson* [02Jun84|11Aug84|13Oct84] 54(2) 20	2	533
1840. IN ROCK WE TRUST *Y&T* [18Aug84|15Sep84|08Dec84] 46(2) 17	0	533
1841. GREG LAKE *Greg Lake* [31Oct81|19Dec81|20Feb82] 62(4) 17	1	533
1842. A VIEW TO A KILL *Soundtrack* [29Jun85|27Jul85|05Oct85] 38(2) 15	1	533
1843. WISHES *Jon Butcher* [04Apr87|06Jun87|03Oct87] 77(2) 27	0	532
1844. FIRST CIRCLE *Pat Metheny Group* [13Oct84|03Nov84|08Jun85] 91(2) 35	0	532
1845. SECRET COMBINATION *Randy Crawford* [23May81|08Aug81|26Sep81] 71(1) 19	0	532
1846. ROCKIN' INTO THE NIGHT *38-Special* [05Jan80|29Mar80|10May80] 57(1) 19	1	531
1847. TWANGIN... *Dave Edmunds* [16May81|20Jun81|15Aug81] 48(2) 14	1	531
1848. SHOT OF LOVE *Bob Dylan* [05Sep81|26Sep81|31Oct81] 33(2) 9	0	531
1849. GIRL AT HER VOLCANO *Rickie Lee Jones* [02Jul83|06Aug83|15Oct83] 39(1) 16	0	531
1850. ACCESS ALL AREAS *Spyro Gyra* [14Jul84|25Aug84|17Nov84] 59(1) 19	0	531
1851. SKYLARKING *XTC* [24Jan87|06Jun87|08Aug87] 70(2) 29	0	531
1852. A MUSICAL AFFAIR *Ashford & Simpson* [23Aug80|13Sep80|08Nov80] 38(2) 12	0	530
1853. RUNNING SCARED *Soundtrack* [05Jul86|23Aug86|11Oct86] 43(2) 15	1	529
1854. BATMAN ORIGINAL MOTION PICTURE SCORE *Soundtrack* [26Aug89|16Sep89|11Nov89] 30(1) 12	0	528
1855. SNEAKIN' OUT *Stacy Lattisaw* [28Aug82|23Oct82|11Dec82] 55(4) 16	1	528
1856. UNCHAIN MY HEART *Joe Cocker* [14Nov87|16Jan88|14May88] 89(2) 27	1	528
1857. STRIP *Adam Ant* [10Dec83|24Dec83|02Jun84] 65(3) 26	1	527
1858. MAKE A LITTLE MAGIC *The Dirt Band* [19Jul80|23Aug80|01Nov80] 62(2) 16	1	525
1859. BROTHERS OF THE ROAD *Allman Brothers Band* [22Aug81|26Sep81|07Nov81] 44(1) 12	1	524
1860. PRIVATE PASSION *Jeff Lorber* [15Nov86|21Feb87|09May87] 68(2) 26	1	524
1861. MARAUDER *Blackfoot* [25Jul81|29Aug81|10Oct81] 48(2) 12	1	524
1862. SECRETS OF FLYING *Johnny Kemp* [11Jun88|30Jul88|15Oct88] 68(1) 19	1	521
1863. PEOPLE ARE PEOPLE *Depeche Mode* [28Jul84|24Aug85|26Oct85] 71(1) 30	0	521
1864. BOYS DON'T CRY *Boys Don't Cry* [21Jun86|12Jul86|25Oct86] 55(3) 19	1	520
1865. WHERE DO WE GO FROM HERE *Michael Damian* [17Jun89|15Jul89|03Mar90] 61(2) 27	3,1*	520
1866. TARANTELLA *Chuck Mangione* [16May81|06Jun81|22Aug81] 55(2) 15	0	519
1867. BLAH-BLAH-BLAH *Iggy Pop* [18Oct86|15Nov86|18Apr87] 75(2) 27	0	518
1868. COME OUT AND PLAY *Twisted Sister* [21Dec85|01Feb86|12Apr86] 53(2) 17	1	518
1869. ONE VICE AT A TIME *Krokus* [10Apr82|05Jun82|21Aug82] 53(2) 20	0	517
1870. EAGLES GREATEST HITS: VOLUME 2 *Eagles* [13Nov82|08Jan83|19Feb83] 52(3) 15	0	517
1871. STEALING FIRE *Bruce Cockburn* [25Aug84|23Feb85|13Apr85] 74(2) 31	1	517
1872. LET THE MUSIC DO THE TALKING *Joe Perry Project* [12Apr80|24May80|05Jul80] 47(2) 13	0	517
1873. RECONCILED *The Call* [08Mar86|24May86|27Sep86] 82(2) 30	0	517
1874. THE SPECIALS *The Specials* [26Jan80|15Mar80|14Jun80] 84(1) 21	0	516
1875. WELCOME TO THE WRECKING BALL *Grace Slick* [14Feb81|21Mar81|16May81] 48(1) 14	0	516
1876. TURN BACK *Toto* [07Feb81|07Mar81|11Apr81] 41(2) 10	0	516
1877. LET THE DAY BEGIN *The Call* [01Jul89|23Sep89|25Nov89] 64(1) 22	1	516
1878. BOYS IN HEAT *Britny Fox* [25Nov89|02Dec89|28Apr90] 79(3) 23	0	516
1879. SCOOP *Pete Townshend* [26Mar83|23Apr83|18Jun83] 35(2) 13	0	516
1880. BE YOURSELF *Patti LaBelle* [22Jul89|12Aug89|13Jan90] 86(1) 26	1	515
1881. MOUNTAIN DANCE *Dave Grusin* [21Mar81|23May81|18Jul81] 74(1) 18	0	515
1882. MY LIFE IN THE BUSH OF GHOSTS *Brian Eno - David Byrne* [21Mar81|11Apr81|13Jun81] 44(2) 13	0	514
1883. BEST OF FRIENDS *Twennynine Featuring Lenny White* [08Dec79|23Feb80|22Mar80] 54(2) 16	1	513
1884. THE CONCERT *Creedence Clearwater Revival* [20Dec80|14Mar81|02May81] 62(2) 20	0	512
1885. WISHFUL THINKING *Earl Klugh* [31Mar84|12May84|01Sep84] 69(2) 23	0	512
1886. ANOTHER PLACE AND TIME *Donna Summer* [20May89|24Jun89|30Sep89] 53(2) 20	2	512
1887. NEVER RUN NEVER HIDE *Benny Mardones* [07Jun80|27Sep80|08Jul89] 65(1) 24	1	511
1888. PETER GABRIEL/PLAYS LIVE *Peter Gabriel* [25Jun83|30Jul83|08Oct83] 44(1) 16	1	510
1889. CANDY GIRL *New Edition* [03Sep83|08Oct83|14Apr84] 90(2) 33	2	510
1890. MR. LOOK SO GOOD! *Richard "Dimples" Fields* [06Mar82|22May82|17Jul82] 63(1) 20	1	510
1891. CHILLIN' *Force M.D.'s* [22Feb86|03May86|09Aug86] 69(1) 25	1	509
1892. WHY NOT ME *The Judds* [01Dec84|02Mar85|25May85] 71(1) 26	0	509
1893. THIS IS BIG AUDIO DYNAMITE *Big Audio Dynamite* [23Nov85|15Feb86|23Aug86] 103(2) 35	0	507
1894. PEOPLE *Hothouse Flowers* [27Aug88|29Oct88|08Apr89] 88(2) 33	0	507
1895. STAGES *Triumph* [02Nov85|30Nov85|01Mar86] 50(2) 18	0	506
1896. KBC BAND *KBC Band* [08Nov86|14Feb87|18Apr87] 75(2) 24	1	506
1897. CLUES *Robert Palmer* [11Oct80|01Nov80|31Jan81] 59(2) 17	0	506
1898. CARAVAN OF LOVE *Isley Jasper Isley* [02Nov85|14Dec85|26Apr86] 77(2) 26	1	505
1899. BORN AGAIN *Black Sabbath* [22Oct83|12Nov83|04Feb84] 39(1) 16	0	503
1900. REFLECTIONS *Rick James* [25Aug84|22Sep84|29Dec84] 41(2) 19	1	503
1901. HEALING *Todd Rundgren* [21Feb81|07Mar81|16May81] 48(2) 13	0	503
1902. COUNT THREE AND PRAY *Berlin* [08Nov86|29Nov86|21Mar87] 61(2) 20	1	503
1903. HEART LAND *Real Life* [07Jan84|03Mar84|16Jun84] 58(1) 24	2	503
1904. MAD MAX BEYOND THUNDERDOME *Soundtrack* [24Aug85|21Sep85|16Nov85] 39(2) 13	2	503
1905. LITTLE SHOP OF HORRORS *Soundtrack* [17Jan87|28Feb87|09May87] 47(2) 17	0	502
1906. GAMMA 2 *Gamma* [13Sep80|29Nov80|17Jan81] 65(1) 19	0	502
1907. THE WARNING *Queensryche* [13Oct84|10Nov84|16Mar85] 61(3) 23	0	502
1908. PEEPSHOW *Siouxsie & The Banshees* [01Oct88|03Dec88|11Feb89] 68(2) 20	1	501
1909. WORKING GIRL *Soundtrack* [11Mar89|29Apr89|10Jun89] 45(1) 14	1	501
1910. CATS WITHOUT CLAWS *Donna Summer* [22Sep84|13Oct84|12Jan85] 40(2) 17	2	500
1911. THE PRESSURE IS ON *Hank Williams Jr.* [05Sep81|10Oct81|06Feb82] 76(2) 23	0	500
1912. A LITTLE SPICE *Loose Ends* [06Jul85|07Sep85|09Nov85] 46(2) 19	0	500
1913. DOG EAT DOG *Joni Mitchell* [23Nov85|14Dec85|29Mar86] 63(5) 19	1	499
1914. ALCHEMY-DIRE STRAITS LIVE *Dire Straits* [21Apr84|19May84|18Aug84] 46(2) 18	0	499
1915. FAREWELL MY SUMMER LOVE *Michael Jackson* [02Jun84|07Jul84|08Sep84] 46(2) 15	1	498
1916. SEQUEL *Harry Chapin* [01Nov80|20Dec80|07Feb81] 58(2) 15	1	498
1917. EATEN ALIVE *Diana Ross* [12Oct85|16Nov85|22Feb86] 45(2) 20	2	498
1918. MYSTICAL ADVENTURE *Jean-Luc Ponty* [13Feb82|17Apr82|15May82] 44(1) 14	0	497
1919. DON'T PLAY WITH FIRE *Peabo Bryson* [04Dec82|22Jan83|23Apr83] 55(3) 21	0	497
1920. THE SINGLES (THE FIRST TEN YEARS) *ABBA* [18Dec82|05Feb83|16Apr83] 62(6) 18	1	496
1921. CIVILIZED EVIL *Jean-Luc Ponty* [18Oct80|06Dec80|14Feb81] 73(1) 18	0	496
1922. CITY OF NEW ORLEANS *Willie Nelson* [04Aug84|15Sep84|26Jan85] 69(1) 26	0	495
1923. THE BEST YEARS OF OUR LIVES *Neil Diamond* [07Jan89|28Jan89|22Apr89] 46(2) 16	0	495
1924. JUST BE MY LADY *Larry Graham* [08Aug81|19Sep81|31Oct81] 46(1) 13	1	495
1925. FIRST OFFENCE *The Inmates* [01Dec79|09Feb80|22Mar80] 49(2) 17	1	495
1926. HEART'S HORIZON *Al Jarreau* [03Dec88|28Jan89|06May89] 75(2) 23	0	494
1927. THE DANCE OF LIFE *Narada Michael Walden* [05Jan80|01Mar80|10May80] 74(2) 19	1	491
1928. THE SON OF ROCK AND ROLL *Rocky Burnette* [21Jun80|23Aug80|20Sep80] 53(1) 14	1	491
1929. W.A.S.P. *W.A.S.P.* [06Oct84|03Nov84|04May85] 74(2) 31	0	490
1930. UNTOUCHABLES *Lakeside* [28May83|18Jun83|24Sep83] 42(2) 18	0	490
1931. DARE TO BE STUPID *"Weird Al" Yankovic* [13Jul85|10Aug85|26Oct85] 50(2) 16	1	490
1932. THIS NOTE'S FOR YOU *Neil Young & The Bluenotes* [30Apr88|21May88|27Aug88] 61(2) 18	0	490
1933. BODIES AND SOULS *Manhattan Transfer* [08Oct83|05Nov83|14Jul84] 52(2) 27	1	490

Ranking the Albums

Rank. Title *Act* [Enter|Peak|Exit] Peak(Wks) TotWks	Hot 100	Scr
1934. CASUAL GODS *Jerry Harrison: Casual Gods* [06Feb88|30Apr88|18Jun88] 78(1) 20	0	489
1935. NO BALLADS *Rockets* [02Feb80|08Mar80|10May80] 53(2) 15	1	489
1936. HOUSE OF LORDS *House Of Lords* [19Nov88|25Feb89|20May89] 78(1) 27	1	489
1937. AFTER THE SNOW *Modern English* [19Mar83|30Apr83|26May84] 70(1) 28	1	488
1938. OBJECTS OF DESIRE *Michael Franks* [30Jan82|27Feb82|01May82] 45(3) 14	0	488
1939. FIELD DAY *Marshall Crenshaw* [18Jun83|06Aug83|17Sep83] 52(1) 14	0	488
1940. SUBJECT: ALDO NOVA *Aldo Nova* [15Oct83|19Nov83|25Feb84] 56(2) 20	0	488
1941. DANCING WITH THE LION *Andreas Vollenweider* [15Apr89|06May89|19Aug89] 52(2) 19	0	488
1942. YESSHOWS *Yes* [20Dec80|31Jan81|07Mar81] 43(2) 12	0	488
1943. LOVING PROOF *Ricky Van Shelton* [29Oct88|12Nov88|08Apr89] 78(2) 24	0	487
1944. HOT WATER *Jimmy Buffett* [09Jul88|13Aug88|08Oct88] 46(2) 14	0	487
1945. PATTI *Patti LaBelle* [10Aug85|07Sep85|22Feb86] 72(3) 29	0	486
1946. WATTS IN A TANK *Diesel* [08Aug81|14Nov81|30Jan82] 68(2) 24	1	486
1947. EVERYBODY'S ROCKIN' *Neil and the Shocking Pinks* [20Aug83|24Sep83|26Nov83] 46(2) 15	0	486
1948. THE FINAL FRONTIER *Keel* [19Apr86|07Jun86|16Aug86] 53(2) 18	0	485
1949. LETTER FROM HOME *Pat Metheny Group* [22Jul89|26Aug89|18Nov89] 66(2) 18	0	485
1950. BALANCE OF POWER *Electric Light Orchestra* [01Mar86|05Apr86|07Jun86] 49(3) 15	1	484
1951. THE WHOLE STORY *Kate Bush* [20Dec86|07Feb87|20Jun87] 76(1) 27	0	484
1952. NON STOP *Julio Iglesias* [04Jun88|09Jul88|24Sep88] 52(2) 17	1	484
1953. KILLERS *Iron Maiden* [06Jun81|15Aug81|07Nov81] 78(2) 23	0	484
1954. LIVE EVIL *Black Sabbath* [05Feb83|19Feb83|23Apr83] 37(4) 12	0	483
1955. BEST SHOTS *Pat Benatar* [25Nov89|16Dec89|07Apr90] 67(1) 20	0	483
1956. ONE MORE STORY *Peter Cetera* [20Aug88|10Sep88|10Dec88] 58(3) 17	2	482
1957. THE SEER *Big Country* [19Jul86|09Aug86|08Nov86] 59(2) 17	0	482
1958. SPARKLE IN THE RAIN *Simple Minds* [18Feb84|07Apr84|28Jul84] 64(2) 24	0	481
1959. WINNERS *The Brothers Johnson* [18Jul81|15Aug81|10Oct81] 48(2) 13	1	481
1960. LADY *One Way* [26May84|07Jul84|06Oct84] 58(1) 20	0	481
1961. WILD! *Erasure* [11Nov89|25Nov89|14Apr90] 57(2) 23	0	481
1962. AS WE SPEAK *David Sanborn* [10Jul82|21Aug82|11Dec82] 70(4) 23	0	481
1963. GUITAR MAN *Elvis Presley* [14Feb81|28Mar81|02May81] 49(1) 12	1	480
1964. SOMETHING TO TALK ABOUT *Anne Murray* [15Feb86|29Mar86|19Jul86] 68(2) 23	1	479
1965. MISTER HEARTBREAK *Laurie Anderson* [17Mar84|21Apr84|21Jul84] 60(2) 19	0	478
1966. HOT TOGETHER *The Pointer Sisters* [29Nov86|20Dec86|28Mar87] 48(3) 18	2	478
1967. CHAKA KHAN *Chaka Khan* [18Dec82|12Feb83|16Apr83] 52(1) 18	1	477
1968. I'M SO PROUD *Deniece Williams* [04Jun83|25Jun83|08Oct83] 54(2) 19	0	477
1969. BY ALL MEANS NECESSARY *Boogie Down Productions* [30Apr88|28May88|01Oct88] 75(1) 23	0	477
1970. AS ONE *Bar-Kays* [13Dec80|24Jan81|28Mar81] 57(1) 16	0	476
1971. KEVIN PAIGE *Kevin Paige* [23Sep89|02Dec89|21Apr90] 107(2) 31	2,1*	476
1972. SHE SHOT ME DOWN *Frank Sinatra* [05Dec81|09Jan82|27Feb82] 52(2) 13	0	476
1973. THUNDER IN THE EAST *Loudness* [02Mar85|04May85|10Aug85] 74(2) 24	0	475
1974. WHEN IN ROME *When In Rome* [15Oct88|28Jan89|25Mar89] 84(1) 24	2	475
1975. DEAD RINGER *Meat Loaf* [19Sep81|10Oct81|28Nov81] 45(2) 11	1	474
1976. LOVE AMONG THE CANNIBALS *Starship* [19Aug89|30Sep89|16Dec89] 64(2) 18	2	473
1977. JUKEBOX *Dazz Band* [20Oct84|08Dec84|04May85] 83(1) 29	0	473
1978. A LITTLE GOOD NEWS *Anne Murray* [15Oct83|07Jan84|25Aug84] 72(1) 24	1	473
1979. VOICE OF AMERICA *Little Steven* [09Jun84|21Jul84|29Sep84] 55(2) 17	0	472
1980. POSH *Patrice Rushen* [29Nov80|07Feb81|28Mar81] 71(1) 18	0	472
1981. NO LOOKIN' BACK *Michael McDonald* [07Sep85|05Oct85|14Dec85] 45(2) 15	1	472
1982. WALK AWAY - COLLECTOR'S EDITION (THE BEST OF 1977-1980) *Donna Summer* [11Oct80|15Nov80|17Jan81] 50(2) 15	1	470
1983. TAKE 6 *Take 6* [11Mar89|13May89|15Jul89] 71(2) 19	0	470
1984. SENTIMENTAL HYGIENE *Warren Zevon* [27Jun87|25Jul87|24Oct87] 63(4) 18	0	469
1985. THE ROMANTICS *The Romantics* [02Feb80|29Mar80|10May80] 61(1) 15	1	468
1986. LOST IN SPACE *The Jonzun Crew* [14May83|23Jul83|24Sep83] 66(1) 20	0	467
1987. ALL AROUND THE TOWN *Bob James* [21Feb81|21Mar81|06Jun81] 66(1) 16	0	467
1988. ANOTHER GREY AREA *Graham Parker* [10Apr82|08May82|24Jul82] 51(2) 16	0	467
1989. CAN'T WE FALL IN LOVE AGAIN *Phyllis Hyman* [01Aug81|12Sep81|24Oct81] 57(1) 13	0	467
1990. MAMA AFRICA *Peter Tosh* [18Jun83|13Aug83|08Oct83] 59(1) 17	1	465
1991. PARADE *Spandau Ballet* [18Aug84|29Sep84|01Dec84] 50(3) 16	1	464
1992. SLOW TURNING *John Hiatt* [24Sep88|15Oct88|22Apr89] 98(3) 31	0	464
1993. CAMERON *Rafael Cameron* [02Aug80|04Oct80|29Nov80] 67(1) 18	0	464
1994. SHARP *Angela Winbush* [07Nov87|12Dec87|14May88] 81(1) 28	0	464
1995. DOOLITTLE *Pixies* [06May89|02Sep89|04Nov89] 98(1) 27	0	464
1996. LATE NIGHT GUITAR *Earl Klugh* [06Dec80|17Jan81|09May81] 98(3) 23	0	463
1997. LOUDER THAN BOMBS *The Smiths* [25Apr87|09May87|10Oct87] 62(2) 25	0	463
1998. FASTER PUSSYCAT *Faster Pussycat* [29Aug87|10Oct87|23Apr88] 97(2) 35	0	463
1999. HIPSWAY *Hipsway* [21Feb87|18Apr87|20Jun87] 55(2) 18	1	462
2000. LETHAL *UTFO* [03Oct87|14Nov87|13Feb88] 67(2) 20	0	462
2001. GREATEST HITS VOL.2 *ABBA* [22Dec79|19Jan80|22Mar80] 46(2) 14	0	462
2002. TOUCH *Laura Branigan* [01Aug87|26Dec87|05Mar88] 87(3) 28	2	462
2003. SO GOOD *The Whispers* [01Dec84|09Mar85|25May85] 88(1) 26	0	462
2004. CAT PEOPLE *Soundtrack* [17Apr82|05Jun82|17Jul82] 47(2) 14	0	461
2005. JAM ON REVENGE *Newcleus* [08Sep84|29Sep84|16Mar85] 74(2) 28	1	460
2006. FOR YOUR EYES ONLY *Soundtrack* [25Jul81|17Oct81|28Nov81] 84(2) 19	1	460
2007. FAMOUS BLUE RAINCOAT *Jennifer Warnes* [14Feb87|21Mar87|04Jul87] 72(2) 21	0	460
2008. WILD ANIMAL *Vanity* [22Sep84|03Nov84|23Feb85] 62(2) 23	1	460
2009. STAR TREK - THE MOTION PICTURE *Soundtrack* [05Jan80|16Feb80|15Mar80] 50(1) 11	0	460
2010. THE COLOUR OF SPRING *Talk Talk* [22Mar86|03May86|12Jul86] 58(2) 17	1	460
2011. CONSTRICTOR *Alice Cooper* [18Oct86|22Nov86|07Mar87] 59(1) 21	0	460
2012. BACKSTAGE PASS *Little River Band* [19Apr80|31May80|21Jun80] 44(1) 10	1	460
2013. THE RIGHT PLACE *Gary Wright* [27Jun81|19Sep81|31Oct81] 79(1) 19	1	460
2014. BALLS TO THE WALL *Accept* [04Feb84|17Mar84|28Jul84] 74(1) 26	0	459
2015. ALTERNATING CURRENTS *Spyro Gyra* [29Jun85|31Aug85|30Nov85] 66(2) 23	0	457
2016. CASINO LIGHTS *Various Artists* [13Nov82|11Dec82|19Mar83] 63(5) 19	0	457
2017. COCK ROBIN *Cock Robin* [13Jul85|05Oct85|16Nov85] 61(2) 19	1	457
2018. WHEELS *Restless Heart* [11Apr87|06Jun87|26Sep87] 73(2) 25	1	456
2019. LEGEND *Lynyrd Skynyrd* [10Oct87|31Oct87|30Jan88] 41(2) 17	0	456
2020. ALL THE BEST! *Paul McCartney* [19Dec87|09Jan88|09Apr88] 62(3) 17	0	456
2021. JEFF BECK'S GUITAR SHOP *Jeff Beck With Terry Bozzio & Tony Hymas* [21Oct89|28Oct89|17Feb90] 49(2) 18	0	455
2022. GLORIOUS RESULTS OF A MISSPENT YOUTH *Joan Jett & the Blackhearts* [27Oct84|15Dec84|16Mar85] 67(1) 21	0	454
2023. HEAVEN *BeBe & CeCe Winans* [04Mar89|03Jun89|19Aug89] 95(1) 25	0	453
2024. LOVE IS A SACRIFICE *Southside Johnny & The Asbury Jukes* [14Jun80|12Jul80|20Sep80] 67(2) 15	0	452
2025. THE FAMILY *The Family* [07Sep85|12Oct85|01Feb86] 62(2) 22	1	452
2026. ALL OVER THE PLACE *Bangles* [04Aug84|03Nov84|23Feb85] 80(2) 30	0	452
2027. SOUTH OF HEAVEN *Slayer* [06Aug88|20Aug88|10Dec88] 57(2) 19	0	452
2028. G FORCE *Kenny G* [24Mar84|12May84|11Aug84] 62(1) 21	0	451
2029. CONFETTI *Sergio Mendes* [19May84|18Aug84|13Oct84] 70(2) 22	2	451
2030. CATS *Original London Cast* [06Nov82|05Feb83|02Apr83] 86(2) 22	0	451
2031. HOT AUGUST NIGHT II *Neil Diamond* [21Nov87|09Jan88|12Mar88] 59(2) 17	0	450
2032. MEAT IS MURDER *The Smiths* [02Mar85|11May85|05Oct85] 110(2) 32	0	450
2033. BELOW THE BELT *Franke & The Knockouts* [10Apr82|12Jun82|07Aug82] 48(2) 18	1	449

Ranking the Albums

Rank. Title *Act* [Enter | Peak | Exit] Peak(Wks) TotWks	Hot 100	Scr
2034. BLACK CARS *Gino Vannelli* [29Jun85 | 20Jul85 | 14Dec85] 62(2) 25	2	449
2035. EL RAYO-X *David Lindley* [16May81 | 18Jul81 | 12Sep81] 83(1) 18	0	449
2036. AUTUMN *George Winston* [02Jun84 | 11Jan86 | 10Jan87] 139(1) 44	0	449
2037. DYLAN AND THE DEAD *Bob Dylan & The Grateful Dead* [18Feb89 | 04Mar89 | 29Apr89] 37(2) 11	0	449
2038. CITY LIFE *Boogie Boys* [31Aug85 | 02Nov85 | 21Dec85] 53(2) 17	0	448
2039. CAPTAIN SWING *Michelle Shocked* [11Nov89 | 18Nov89 | 05May90] 95(2) 26	0	448
2040. AND ONCE AGAIN *Isaac Hayes* [17May80 | 19Jul80 | 23Aug80] 59(1) 15	0	447
2041. COSI FAN TUTTI FRUTTI *Squeeze* [21Sep85 | 02Nov85 | 01Feb86] 57(2) 20	0	447
2042. LICENSE TO DREAM *Kleeer* [07Mar81 | 09May81 | 20Jun81] 81(2) 16	0	447
2043. NIGHT PASSAGE *Weather Report* [13Dec80 | 17Jan81 | 14Mar81] 57(1) 14	0	447
2044. SLAVE TO THE RHYTHM *Grace Jones* [23Nov85 | 21Dec85 | 05Apr86] 73(4) 20	0	447
2045. JOE'S GARAGE ACTS II + III *Frank Zappa* [15Dec79 | 26Jan80 | 01Mar80] 53(1) 12	0	446
2046. SKYYPORT *Skyy* [06Dec80 | 21Feb81 | 18Apr81] 85(1) 20	0	446
2047. DREAM OF A LIFETIME *Marvin Gaye* [08Jun85 | 29Jun85 | 14Sep85] 41(2) 15	0	446
2048. LIVING IN THE BACKGROUND *Baltimora* [18Jan86 | 15Mar86 | 10May86] 49(2) 17	2	446
2049. CALM ANIMALS *The Fixx* [11Feb89 | 01Apr89 | 10Jun89] 72(2) 18	1	446
2050. SIT DOWN AND TALK TO ME *Lou Rawls* [12Jan80 | 09Feb80 | 10May80] 81(1) 18	1	446
2051. CAN'T STOP THE MUSIC (SOUNDTRACK) *Village People* [21Jun80 | 26Jul80 | 06Sep80] 47(2) 12	0	445
2052. CENTIPEDE *Rebbie Jackson* [27Oct84 | 22Dec84 | 23Feb85] 63(3) 18	1	445
2053. MADE IN AMERICA *Blues Brothers* [27Dec80 | 14Feb81 | 14Mar81] 49(1) 12	1	445
2054. SYBIL *Sybil* [21Oct89 | 02Dec89 | 31Mar90] 75(2) 24	2,1*	445
2055. BLIND TO REASON *Grayson Hugh* [15Oct88 | 16Sep89 | 25Nov89] 71(2) 24	3,1*	445
2056. SUZANNE VEGA *Suzanne Vega* [15Jun85 | 31Aug85 | 14Jun86] 91(2) 31	0	445
2057. TIME AND TIDE *Split Enz* [08May82 | 26Jun82 | 18Sep82] 58(1) 20	0	445
2058. WAVES *Katrina & The Waves* [12Apr86 | 10May86 | 26Jul86] 49(2) 16	1	445
2059. A SENSE OF WONDER *Van Morrison* [09Mar85 | 06Apr85 | 29Jun85] 61(3) 17	0	444
2060. MUSIQUE/THE HIGH ROAD *Roxy Music* [09Apr83 | 11Jun83 | 03Sep83] 67(1) 22	0	443
2061. GREENPEACE/RAINBOW WARRIORS *Various Artists* [15Jul89 | 19Aug89 | 13Jan90] 68(2) 24	0	443
2062. I DON'T SPEAK THE LANGUAGE *Matthew Wilder* [07Jan84 | 03Mar84 | 21Apr84] 49(2) 16	2	443
2063. COME AS YOU ARE *Peter Wolf* [18Apr87 | 09May87 | 25Jul87] 53(2) 15	2	443
2064. MOONLIGHTING *TV Soundtrack* [08Aug87 | 29Aug87 | 07Nov87] 50(2) 14	1	442
2065. ROCK ISLAND *Jethro Tull* [30Sep89 | 14Oct89 | 27Jan90] 56(2) 18	0	442
2066. DREAM ON *George Duke* [06Mar82 | 10Apr82 | 22May82] 48(2) 12	1	442
2067. OPEN FIRE *Y&T* [20Jul85 | 07Sep85 | 09Nov85] 70(3) 17	1	442
2068. DEVO-LIVE *Devo* [18Apr81 | 09May81 | 04Jul81] 50(1) 12	0	442
2069. GREATEST HITS *Dolly Parton* [16Oct82 | 04Dec82 | 19Mar83] 77(2) 23	1	442
2070. CHANGESTWOBOWIE *David Bowie* [12Dec81 | 16Jan82 | 10Apr82] 68(1) 18	0	441
2071. HERO *Clarence Clemons* [23Nov85 | 18Jan86 | 22Mar86] 62(2) 18	1	441
2072. INSIDE YOU *The Isley Brothers* [31Oct81 | 21Nov81 | 23Jan82] 45(2) 13	0	440
2073. ALIENS ATE MY BUICK *Thomas Dolby* [07May88 | 02Jul88 | 10Sep88] 70(1) 19	0	440
2074. BIG & BEAUTIFUL *Fat Boys* [24May86 | 21Jun86 | 27Sep86] 62(2) 19	0	440
2075. WHILE THE CITY SLEEPS... *George Benson* [20Sep86 | 11Oct86 | 28Feb87] 77(1) 24	0	440
2076. REFLECTIONS *Gil Scott-Heron* [26Sep81 | 20Feb82 | 27Mar82] 106(1) 27	0	440
2077. BRASS CONSTRUCTION 5 *Brass Construction* [15Dec79 | 01Mar80 | 26Apr80] 89(1) 20	0	439
2078. HOT NUMBER *Fabulous Thunderbirds* [18Jul87 | 15Aug87 | 24Oct87] 49(1) 15	1	439
2079. LOVE TRIPPIN' *Spinners* [21Jun80 | 26Jul80 | 13Sep80] 53(2) 13	1	439
2080. SIGN OF THE TIMES *Bob James* [12Sep81 | 10Oct81 | 12Dec81] 56(1) 14	0	439
2081. YOU BOUGHT IT-YOU NAME IT *Joe Walsh* [09Jul83 | 20Aug83 | 08Oct83] 48(1) 14	1	438
2082. NICK THE KNIFE *Nick Lowe* [20Feb82 | 20Mar82 | 22May82] 50(2) 14	0	438
2083. IT'S HARD TO BE HUMBLE *Mac Davis* [24May80 | 12Jul80 | 30Aug80] 69(2) 15	1	437
2084. DAY BY DAY *Najee* [09Jul88 | 13Aug88 | 26Nov88] 76(2) 21	0	437
2085. DEEP SEA SKIVING *Bananarama* [16Apr83 | 11Jun83 | 20Aug83] 63(1) 19	1	437
2086. SINGLES COLLECTION - THE LONDON YEARS *The Rolling Stones* [09Sep89 | 28Oct89 | 03Feb90] 91(2) 22	0	436
2087. AT PEACE WITH WOMAN *The Jones Girls* [18Oct80 | 29Nov80 | 28Mar81] 96(1) 24	0	436
2088. THE CONFESSOR *Joe Walsh* [01Jun85 | 29Jun85 | 05Oct85] 65(2) 19	0	436
2089. QUEENSRYCHE *Queensryche* [17Sep83 | 05Nov83 | 11Feb84] 81(1) 22	0	435
2090. RUMBLE *Tommy Conwell And The Young Rumblers* [03Sep88 | 29Oct88 | 11Mar89] 103(1) 28	2	434
2091. O SOLE MIO - FAVORITE NEAPOLITAN SONGS *Luciano Pavarotti* [24Nov79 | 02Feb80 | 12Apr80] 77(1) 21	0	434
2092. WHERE'S THE PARTY? *Eddie Money* [05Nov83 | 03Dec83 | 10Mar84] 67(1) 19	2	434
2093. RIVER OF TIME *The Judds* [22Apr89 | 13May89 | 02Sep89] 51(2) 20	0	434
2094. QUARTET *Ultravox* [12Mar83 | 30Apr83 | 02Jul83] 61(1) 17	1	433
2095. NO PARLEZ *Paul Young* [14Apr84 | 09Jun84 | 05Oct85] 79(1) 23	3	432
2096. EPONYMOUS *R.E.M.* [22Oct88 | 12Nov88 | 25Feb89] 44(2) 19	0	432
2097. THE HUNTER *Blondie* [19Jun82 | 10Jul82 | 04Sep82] 33(2) 12	1	430
2098. ABOMINOG *Uriah Heep* [07Aug82 | 25Sep82 | 20Nov82] 56(3) 16	0	430
2099. THE BEACH BOYS *The Beach Boys* [29Jun85 | 20Jul85 | 28Sep85] 52(3) 14	2	429
2100. TELEVISION'S GREATEST HITS *Various* [09Nov85 | 08Feb86 | 28Jun86] 82(2) 34	0	429
2101. NEVER ENOUGH *Patty Smyth* [21Mar87 | 18Apr87 | 01Aug87] 66(2) 20	2	428
2102. ABSOLUTELY LIVE *Rod Stewart* [20Nov82 | 18Dec82 | 12Feb83] 46(3) 13	0	427
2103. NO PLACE TO RUN *UFO* [19Jan80 | 16Feb80 | 12Apr80] 51(1) 13	0	427
2104. YELLOW MAGIC ORCHESTRA *Yellow Magic Orchestra* [26Jan80 | 22Mar80 | 14Jun80] 81(1) 21	1	427
2105. RAT IN THE KITCHEN *UB40* [30Aug86 | 04Oct86 | 20Dec86] 53(2) 17	0	427
2106. BLACK CELEBRATION *Depeche Mode* [26Apr86 | 17May86 | 18Oct86] 90(2) 26	0	426
2107. DEDICATED *The Marshall Tucker Band* [23May81 | 06Jun81 | 08Aug81] 53(2) 12	0	426
2108. IGNITION *John Waite* [17Jul82 | 18Sep82 | 20Apr85] 68(3) 23	0	426
2109. PENETRATOR *Ted Nugent* [18Feb84 | 07Apr84 | 26May84] 56(2) 15	0	426
2110. BLUE MURDER *Blue Murder* [13May89 | 24Jun89 | 30Sep89] 69(2) 21	0	425
2111. UNFINISHED BUSINESS *EPMD* [19Aug89 | 09Sep89 | 18Nov89] 53(3) 14	0	424
2112. PARTY YOUR BODY *Stevie B* [23Jul88 | 24Sep88 | 10Dec88] 78(1) 21	2	424
2113. WELCOME TO THE CLUB *Ian Hunter* [26Apr80 | 24May80 | 16Aug80] 69(1) 17	0	423
2114. TOUCH THE SKY *Smokey Robinson* [29Jan83 | 26Feb83 | 21May83] 50(3) 17	0	423
2115. JUJU MUSIC *King Sunny Ade & His African Beats* [09Apr83 | 17Sep83 | 22Oct83] 111(1) 29	0	423
2116. UNSUNG HEROES *The Dregs* [18Apr81 | 23May81 | 18Jul81] 67(2) 14	0	422
2117. THE MONA LISA'S SISTER *Graham Parker* [28May88 | 02Jul88 | 01Oct88] 77(2) 19	0	422
2118. THINK OF ONE *Wynton Marsalis* [09Jul83 | 02Jun84 | 21Jul84] 102(1) 29	0	422
2119. BREAKOUT *Spyro Gyra* [12Jul86 | 27Sep86 | 15Nov86] 71(1) 19	0	421
2120. TODAY *Today* [14Jan89 | 11Feb89 | 10Jun89] 86(2) 22	0	421
2121. CITIZEN KIHN *Greg Kihn* [23Mar85 | 27Apr85 | 15Jun85] 51(2) 13	1	421
2122. CHILDREN OF TOMORROW *Frankie Smith* [08Aug81 | 05Sep81 | 10Oct81] 54(2) 10	1	420
2123. SEE YOU IN HELL *Grim Reaper* [25Aug84 | 27Oct84 | 23Feb85] 73(2) 27	0	420
2124. KALEIDOSCOPE WORLD *Swing Out Sister* [27May89 | 17Jun89 | 30Sep89] 61(2) 19	1	420
2125. BARKING AT AIRPLANES *Kim Carnes* [29Jun85 | 27Jul85 | 28Sep85] 48(2) 14	3	419
2126. YOU AND ME *Rockie Robbins* [07Jun80 | 02Aug80 | 20Sep80] 71(2) 16	0	419
2127. ASTRA *Asia* [07Dec85 | 28Dec85 | 29Mar86] 67(3) 17	1	418
2128. INSIDE THE ELECTRIC CIRCUS *W.A.S.P.* [08Nov86 | 22Nov86 | 14Mar87] 60(2) 19	0	418
2129. DIFFORD & TILBROOK *Difford & Tilbrook* [14Jul84 | 04Aug84 | 20Oct84] 55(3) 15	0	418
2130. THE DARK *Metal Church* [08Nov86 | 24Jan87 | 11Apr87] 92(1) 23	0	418
2131. INDIANA JONES AND THE TEMPLE OF DOOM (JOHN WILLIAMS) *Soundtrack* [16Jun84 | 14Jul84 | 18Aug84] 42(2) 10	0	416

Ranking the Albums

Rank. Title *Act* [Enter|Peak|Exit] Peak(Wks) TotWks	Hot 100	Scr
2132. TEASER *Angela Bofill* [26Nov83|11Feb84|14Apr84] 81(1) 21	0	416
2133. NO GUTS...NO GLORY *Molly Hatchet* [26Mar83|30Apr83|06Aug83] 59(1) 20	0	416
2134. CHICO DEBARGE *Chico DeBarge* [15Nov86|07Feb87|06Jun87] 90(2) 30	1	416
2135. THREE OF A PERFECT PAIR *King Crimson* [07Apr84|28Apr84|28Jul84] 58(2) 17	0	416
2136. TELEVISION THEME SONGS *Mike Post* [27Feb82|08May82|19Jun82] 70(2) 17	2	415
2137. ANYONE CAN SEE *Irene Cara* [30Jan82|13Mar82|22May82] 76(2) 17	1	415
2138. BRIAN WILSON *Brian Wilson* [30Jul88|13Aug88|22Oct88] 54(3) 13	0	415
2139. THROUGH THE STORM *Aretha Franklin* [20May89|10Jun89|16Sep89] 55(2) 18	2	415
2140. SKIN ON SKIN *Vanity* [22Mar86|10May86|02Aug86] 66(2) 20	1	415
2141. CHARTBUSTERS *Ray Parker Jr.* [15Dec84|26Jan85|23Mar85] 60(3) 15	1	414
2142. NORTH COAST *Michael Stanley Band* [01Aug81|10Oct81|07Nov81] 79(2) 15	1	414
2143. MACHISMO *Cameo* [12Nov88|26Nov88|18Mar89] 56(1) 19	1	414
2144. INSIDE MOVES *Grover Washington Jr.* [10Nov84|15Dec84|13Apr85] 79(1) 23	0	414
2145. LIVE *Alabama* [25Jun88|09Jul88|29Oct88] 76(3) 19	0	414
2146. FANDANGO *Herb Alpert* [29May82|04Sep82|20Nov82] 100(2) 26	1	414
2147. PERFECT TIMING *McAuley Schenker Group* [24Oct87|21Nov87|02Apr88] 95(2) 24	0	413
2148. TWISTING BY THE POOL *Dire Straits* [12Mar83|30Apr83|18Jun83] 53(2) 15	0	413
2149. IN YOUR FACE *Kingdom Come* [13May89|27May89|19Aug89] 49(2) 15	0	413
2150. KLYMAXX *Klymaxx* [06Dec86|01Aug87|19Sep87] 98(2) 31	2	413
2151. NEW GOLD DREAM (81-82-83-84) *Simple Minds* [19Feb83|09Apr83|25Jun83] 69(2) 19	0	413
2152. GENE CHANDLER '80 *Gene Chandler* [07Jun80|02Aug80|04Oct80] 87(2) 18	0	413
2153. ODORI *Hiroshima* [15Nov80|13Dec80|14Mar81] 72(1) 18	0	413
2154. WIDE AWAKE IN AMERICA *U2* [29Jun85|29Jun85|19Sep85] 37(2) 23	0	413
2155. NERUDA *Red Rider* [05Feb83|19Mar83|21May83] 66(4) 16	0	413
2156. THIS ONE'S FOR YOU *Teddy Pendergrass* [21Aug82|09Oct82|27Nov82] 59(1) 15	0	412
2157. LONG LIVE THE NEW FLESH *Flesh For Lulu* [12Dec87|12Mar88|21May88] 89(1) 24	0	412
2158. UPSTAIRS AT ERIC'S *Yazoo* [02Oct82|20Nov82|07May83] 92(1) 32	2	411
2159. KRUSH GROOVE *Soundtrack* [26Oct85|14Dec85|08Mar86] 79(2) 20	0	411
2160. LIVING MY LIFE *Grace Jones* [11Dec82|25Dec82|23Apr83] 86(4) 20	0	411
2161. SOMEBODY'S GONNA LOVE YOU *Lee Greenwood* [28May83|23Jul83|15Oct83] 73(1) 21	2	410
2162. NUGENT *Ted Nugent* [17Jul82|28Aug82|16Oct82] 51(2) 14	0	409
2163. ALLIES *Crosby, Stills & Nash* [02Jul83|06Aug83|17Sep83] 43(2) 12	1	409
2164. THE HEADLESS CHILDREN *W.A.S.P.* [22Apr89|29Apr89|15Jul89] 48(3) 13	0	409

Rank. Title *Act* [Enter|Peak|Exit] Peak(Wks) TotWks	Hot 100	Scr
2165. LOVE IS SUCH A FUNNY GAME *Michael Cooper* [16Jan88|12Mar88|02Jul88] 98(3) 25	0	409
2166. HEART BREAK *Shalamar* [08Dec84|05Jan85|18May85] 90(4) 24	2	409
2167. SOMEBODY'S WAITING *Anne Murray* [03May80|21Jun80|09Aug80] 88(4) 15	2	408
2168. HUMAN RACING *Nik Kershaw* [05May84|21Jul84|15Sep84] 70(2) 20	1	406
2169. STANDING TALL *The Crusaders* [10Oct81|31Oct81|23Jan82] 59(2) 16	1	404
2170. INTENSITIES IN 10 CITIES *Ted Nugent* [21Mar81|11Apr81|23May81] 51(2) 10	0	404
2171. HOW MANY TIMES CAN WE SAY GOODBYE *Dionne Warwick* [29Oct83|26Nov83|18Feb84] 57(2) 17	1	404
2172. BERRY GORDY'S THE LAST DRAGON *Soundtrack* [30Mar85|11May85|06Jul85] 58(2) 15	0	404
2173. STAND IN LINE *Impellitteri* [25Jun88|27Aug88|05Nov88] 91(2) 20	0	404
2174. WHAT IS BEAT? *The English Beat* [17Dec83|03Mar84|12May84] 87(1) 22	0	403
2175. STEELTOWN *Big Country* [24Nov84|15Dec84|16Mar85] 70(4) 17	0	403
2176. SMOOTH SAILIN' *The Isley Brothers* [20Jun87|18Jul87|10Oct87] 64(2) 17	0	403
2177. THE LACE *Benjamin Orr* [08Nov86|07Feb87|04Apr87] 86(2) 22	1	402
2178. LAST DATE *Emmylou Harris* [13Nov82|20Nov82|05Mar83] 65(4) 17	0	402
2179. ROWDY *Hank Williams Jr.* [21Feb81|14Mar81|30May81] 82(2) 15	0	402
2180. BLOWFLY'S PARTY [X-RATED] *Blowfly* [24May80|02Aug80|04Oct80] 82(2) 20	0	401
2181. PARTY MIX! *The B-52s* [08Aug81|22Aug81|17Oct81] 55(2) 11	0	401
2182. D.E. 7TH *Dave Edmunds* [01May82|12Jun82|31Jul82] 46(2) 14	0	401
2183. UTFO *UTFO* [15Jun85|29Jun85|26Oct85] 80(2) 20	1	399
2184. YOU CAN'T FIGHT FASHION *Michael Stanley Band* [24Sep83|26Nov83|14Jan84] 64(2) 17	2	399
2185. BEAUTIFUL VISION *Van Morrison* [06Mar82|03Apr82|15May82] 44(2) 11	0	399
2186. REACHING FOR TOMORROW *Switch* [12Apr80|17May80|12Jul80] 57(2) 14	0	398
2187. BLACK MARKET CLASH *The Clash* [22Nov80|27Dec80|07Mar81] 74(2) 16	0	398
2188. INSTINCTS *Romeo Void* [25Aug84|20Oct84|29Dec84] 68(4) 19	1	398
2189. TRAVELS *Pat Metheny Group* [25Jun83|23Jul83|15Oct83] 62(1) 17	0	398
2190. ALL SYSTEMS GO *Vinnie Vincent Invasion* [21May88|25Jun88|27Aug88] 64(2) 15	0	397
2191. SHOCKADELICA *Jesse Johnson* [18Oct86|15Nov86|28Feb87] 70(2) 20	1	397
2192. THE BLUE ALBUM *Harold Melvin And The Blue Notes* [22Mar80|19Apr80|02Aug80] 95(2) 20	0	397
2193. ALL FIRED UP *Fastway* [21Jul84|08Sep84|20Oct84] 92(2) 14	0	396
2194. ON TO VICTORY *Humble Pie* [12Apr80|31May80|12Jul80] 60(1) 14	1	396
2195. JEWEL OF THE NILE *Soundtrack* [28Dec85|01Mar86|19Apr86] 55(2) 17	1	395
2196. FLYING THE FLAG *Climax Blues Band* [25Apr81|04Jul81|08Aug81] 75(1) 16	2	395

Rank. Title *Act* [Enter|Peak|Exit] Peak(Wks) TotWks	Hot 100	Scr
2197. TO THE MAX *Con Funk Shun* [04Dec82|18Dec82|18Jun83] 115(1) 29	0	394
2198. HEAR & NOW *Billy Squier* [15Jul89|12Aug89|11Nov89] 64(1) 17	1	394
2199. CONFRONTATION *Bob Marley And The Wailers* [02Jul83|06Aug83|08Oct83] 55(1) 15	0	394
2200. NOTHIN' BUT TROUBLE *Nia Peeples* [14May88|25Jun88|01Oct88] 97(4) 21	1	393
2201. DEAD LETTER OFFICE *R.E.M.* [16May87|13Jun87|15Aug87] 52(1) 14	0	393
2202. HANK "LIVE" *Hank Williams Jr.* [14Feb87|25Apr87|25Jul87] 71(1) 24	0	392
2203. FRIENDSHIP *Ray Charles* [23Feb85|04May85|06Jul85] 75(1) 20	0	392
2204. ONE BAD HABIT *Michael Franks* [10May80|14Jun80|27Sep80] 83(1) 21	0	392
2205. SECRET SECRETS *Joan Armatrading* [30Mar85|22Jun85|03Aug85] 73(1) 19	0	392
2206. CHANCE *Manfred Mann's Earth Band* [24Jan81|28Mar81|09May81] 87(2) 16	0	391
2207. BEYOND THE BLUE NEON *George Strait* [04Mar89|18Mar89|12Aug89] 92(2) 24	0	391
2208. GOING FOR BROKE *Eddy Grant* [23Jun84|11Aug84|13Oct84] 64(2) 17	2	391
2209. GALAXIAN *Jeff Lorber Fusion* [18Apr81|30May81|25Jul81] 77(1) 15	0	390
2210. PERFECT *Soundtrack* [29Jun85|20Jul85|14Sep85] 45(2) 12	1	390
2211. DREGS OF THE EARTH *Dixie Dregs* [10May80|28Jun80|30Aug80] 81(1) 17	0	389
2212. THE BLIND LEADING THE NAKED *Violent Femmes* [15Feb86|29Mar86|26Jul86] 84(2) 24	0	389
2213. GO ON... *Mr. Mister* [26Sep87|03Oct87|16Jan88] 55(2) 17	1	389
2214. ELECTRIC RENDEZVOUS *Al Di Meola* [06Feb82|13Mar82|01May82] 55(2) 13	0	388
2215. RHYME PAYS *Ice-T* [15Aug87|14Nov87|13Feb88] 93(2) 27	0	388
2216. THE ALARM *The Alarm* [30Jul83|14Apr84|09Jun84] 126(1) 37	0	388
2217. ON THE ONE *Dazz Band* [12Feb83|19Mar83|28May83] 59(3) 16	0	387
2218. A KIND OF MAGIC *Queen* [19Jul86|09Aug86|11Oct86] 46(3) 13	1	387
2219. PAVAROTTI'S GREATEST HITS *Luciano Pavarotti* [07Jul80|26Jul80|04Oct80] 94(2) 18	0	387
2220. 9.9 *9.9* [14Sep85|02Nov85|08Feb86] 79(1) 22	1	387
2221. L IS FOR LOVER *Al Jarreau* [04Oct86|25Oct86|11Apr87] 81(2) 28	0	386
2222. SOMETHING IN THE NIGHT *Pure Prairie League* [02May81|06Jun81|08Aug81] 72(1) 15	2	386
2223. PLEASURES OF THE FLESH *Exodus* [28Nov87|23Jan88|09Apr88] 82(2) 20	0	386
2224. YOU AND I *O'Bryan* [12Mar83|30Apr83|10Sep83] 87(1) 27	0	386
2225. ORION THE HUNTER *Orion The Hunter* [19May84|07Jul84|18Aug84] 57(2) 14	1	386
2226. UP AND DOWN *Opus* [01Mar86|19Apr86|14Jun86] 64(2) 16	0	385
2227. DOLLY DOLLY DOLLY *Dolly Parton* [03May80|31May80|26Jul80] 71(1) 13	0	385
2228. BUENAS NOCHES FROM A LONELY ROOM *Dwight Yoakam* [20Aug88|17Sep88|26Nov88] 68(1) 15	0	384
2229. YOYO *Bourgeois Tagg* [24Oct87|19Dec87|12Mar88] 84(1) 21	1	384

Ranking the Albums

Rank. Title Act [Enter|Peak|Exit] Peak(Wks) TotWks	Hot 100	Scr
2230. DREAM EVIL Dio [15Aug87|29Aug87|24Oct87] 43(2) 11	0	384
2231. LAND OF DREAMS Randy Newman [15Oct88|10Dec88|18Feb89] 80(1) 19	1	384
2232. FANCY DANCER One Way [26Sep81|21Nov81|30Jan82] 79(2) 19	0	383
2233. PASSION Robin Trower [27Dec86|04Apr87|13Jun87] 100(2) 25	0	383
2234. A NICE PLACE TO BE George Howard [27Dec86|07Feb87|20Jun87] 109(3) 26	0	383
2235. SACRED SONGS Daryl Hall [29Mar80|26Apr80|14Jun80] 58(2) 12	0	383
2236. THE BEST SIDE OF GOODBYE Jane Olivor [23Feb80|29Mar80|10May80] 58(1) 12	0	382
2237. WEST SIDE STORY Leonard Bernstein [25May85|06Jul85|05Oct85] 70(2) 20	0	381
2238. HIGH 'N' DRY(2) Def Leppard [02Jun84|07Jul84|29Sep84] 72(2) 18	1	381
2239. YOURS FOREVER Atlantic Starr [19Nov83|21Jan84|26May84] 91(1) 28	1	381
2240. SHERIFF Sheriff [07Jan89|11Feb89|08Apr89] 60(2) 14	2	381
2241. ROCK YOU TO HELL Grim Reaper [01Aug87|07Nov87|19Dec87] 93(1) 21	0	381
2242. LIVE FROM NEW YORK Gilda Radner [01Dec79|19Jan80|16Feb80] 69(2) 12	0	380
2243. INDUSTRY STANDARD The Dregs [27Mar82|24Apr82|03Jul82] 56(2) 15	0	380
2244. SO GOOD Mica Paris [13May89|08Jul89|14Oct89] 86(2) 23	1	380
2245. FUR Jane Wiedlin [28May88|16Jul88|15Oct88] 105(2) 21	2	379
2246. CHEAT THE NIGHT Deborah Allen [03Dec83|03Mar84|14Apr84] 67(1) 20	1	379
2247. HONEYMOON SUITE Honeymoon Suite [25Aug84|27Oct84|15Dec84] 60(2) 17	1	378
2248. ONCE UPON A CHRISTMAS Kenny Rogers & Dolly Parton [08Dec84|05Jan85|26Jan85] 31(1) 8	1	378
2249. NO STRANGER TO LOVE Roy Ayers [15Dec79|01Mar80|12Apr80] 82(1) 18	0	378
2250. BONNIE POINTER (II) Bonnie Pointer [22Dec79|16Feb80|22Mar80] 63(2) 14	1	377
2251. ALMOST BLUE Elvis Costello And The Attractions [14Nov81|28Nov81|06Feb82] 50(2) 13	0	377
2252. TOMMY TUTONE Tommy Tutone [24May80|05Jul80|16Aug80] 68(2) 13	0	377
2253. MAN ON THE LINE Chris de Burgh [30Jun84|08Sep84|03Nov84] 69(2) 19	1	377
2254. PRINCE CHARMING Adam And The Ants [12Dec81|30Jan82|02Apr83] 94(2) 21	0	377
2255. TERENCE TRENT D'ARBY'S NEITHER FISH NOR FLESH Terence Trent D'Arby [25Nov89|02Dec89|03Mar90] 61(2) 15	0	377
2256. EXPOSED/A CHEAP PEEK AT TODAY'S PROVOCATIVE NEW ROCK Various Artists [27Jun81|18Jul81|22Aug81] 51(2) 9	0	376
2257. PHIL SEYMOUR Phil Seymour [21Feb81|04Apr81|06Jun81] 64(1) 16	1	376
2258. LIVE IN NEW YORK CITY John Lennon [22Mar86|05Apr86|31May86] 41(2) 11	0	376
2259. GHETTO BLASTER The Crusaders [21Apr84|19May84|15Sep84] 79(2) 22	0	375
2260. KOOL MOE DEE Kool Moe Dee [18Apr87|06Jun87|05Sep87] 83(2) 21	1	375
2261. FIVE-O Hank Williams Jr. [18May85|06Jul85|12Oct85] 72(2) 22	0	374
2262. TWO HEARTS Men At Work [22Jun85|13Jul85|14Sep85] 50(2) 13	1	374
2263. BLESSING IN DISGUISE Metal Church [11Mar89|08Apr89|17Jun89] 75(1) 15	0	374
2264. DEATH WISH II (SOUNDTRACK) Jimmy Page [03Apr82|24Apr82|05Jun82] 50(4) 10	0	374
2265. SCHOOL DAZE Soundtrack [19Mar88|07May88|23Jul88] 81(2) 17	1	374
2266. ADDICTIONS VOL. I Robert Palmer [25Nov89|30Dec89|17Mar90] 79(2) 17	0	373
2267. ON THE EDGE The Babys [15Nov80|13Dec80|21Feb81] 71(2) 15	1	373
2268. PLEASANT DREAMS The Ramones [08Aug81|19Sep81|17Oct81] 58(2) 11	0	372
2269. A LOT OF LOVE Melba Moore [23Aug86|04Apr87|18Jul87] 91(2) 29	0	372
2270. "JI" Junior [08May82|29May82|21Aug82] 71(3) 16	1	371
2271. BEAUTY STAB ABC [17Dec83|28Jan84|17Mar84] 69(2) 14	1	371
2272. FOR THE WORKING GIRL Melissa Manchester [13Sep80|11Oct80|22Nov80] 68(3) 11	1	371
2273. ICEHOUSE Icehouse [25Jul81|19Sep81|31Oct81] 82(2) 15	1	370
2274. UNDERTOW Firefall [12Apr80|10May80|19Jul80] 68(2) 15	2	370
2275. MAIDEN JAPAN Iron Maiden [31Oct81|28Nov81|10Nov84] 89(2) 30	0	370
2276. HEART OVER MIND Anne Murray [27Oct84|08Dec84|13Apr85] 92(1) 25	0	370
2277. SILK AND STEEL Five Star [04Oct86|01Nov86|21Mar87] 80(2) 25	2	369
2278. GAUDI The Alan Parsons Project [07Feb87|14Mar87|09May87] 57(2) 14	0	369
2279. LOVE JUNK The Pursuit Of Happiness [17Dec88|11Feb89|06May89] 93(1) 21	0	368
2280. OUT OF MIND OUT OF SIGHT Models [03May86|05Jul86|30Aug86] 84(1) 18	1	368
2281. THE POET II Bobby Womack [07Apr84|12May84|07Jul84] 60(1) 14	1	368
2282. MONEY FOR NOTHING Dire Straits [12Nov88|26Nov88|04Mar89] 62(2) 17	0	368
2283. TAKE IT TO THE LIMIT Willie Nelson With Waylon Jennings [21May83|25Jun83|03Sep83] 60(1) 16	0	368
2284. BRASIL The Manhattan Transfer [05Dec87|05Mar88|09Apr88] 96(1) 19	0	368
2285. DIRTY LOOKS Juice Newton [10Sep83|22Oct83|17Dec83] 52(1) 15	2	367
2286. BORN YESTERDAY The Everly Brothers [08Feb86|15Mar86|14Jun86] 83(4) 19	0	366
2287. THE LAST OF THE MOHICANS Bow Wow Wow [15May82|03Jul82|09Oct82] 67(1) 22	0	366
2288. OCEAN FRONT PROPERTY George Strait [14Feb87|21Mar87|22Aug87] 117(1) 28	0	365
2289. DREAM OF LIFE Patti Smith [30Jul88|06Aug88|05Nov88] 65(4) 15	0	365
2290. REACH UP AND TOUCH THE SKY Southside Johnny & The Asbury Jukes [09May81|23May81|25Jul81] 80(4) 12	0	364
2291. JAZZERCISE Judi Sheppard Missett [05Dec81|26Dec81|17Apr82] 117(3) 20	0	364
2292. TINSEL TOWN REBELLION Frank Zappa [30May81|04Jul81|08Aug81] 66(2) 11	0	364
2293. BREAKIN' 2 ELECTRIC BOOGALOO Soundtrack [12Jan85|16Feb85|06Apr85] 52(2) 13	0	363
2294. MAURICE WHITE Maurice White [05Oct85|09Nov85|08Feb86] 61(2) 19	2	363
2295. BLAZE OF GLORY Joe Jackson [06May89|27May89|23Sep89] 61(2) 21	0	363
2296. CITY KIDS Spyro Gyra [13Aug83|24Sep83|26Nov83] 66(1) 16	0	363
2297. MAGIC MAN Herb Alpert [22Aug81|26Sep81|24Oct81] 61(1) 10	1	362
2298. RAIDERS OF THE LOST ARK (SOUNDTRACK) London Symphony Orchestra/John Williams [04Jul81|08Aug81|26Sep81] 62(1) 13	0	362
2299. DANCIN' ON THE EDGE Lita Ford [04Aug84|29Sep84|17Nov84] 66(2) 16	0	361
2300. SAY ANYTHING Soundtrack [06May89|17Jun89|05Aug89] 62(2) 14	1	361
2301. WINDHAM HILL RECORDS SAMPLER '84 Various Artists [20Oct84|08Dec84|06Apr85] 108(4) 25	0	361
2302. MONSTER Herbie Hancock [19Apr80|12Jul80|16Aug80] 94(1) 18	0	361
2303. I'VE GOT EVERYTHING Henry Lee Summer [27May89|15Jul89|16Sep89] 78(1) 17	1	361
2304. CRUZADOS Cruzados [02Nov85|07Dec85|01Mar86] 76(2) 18	0	360
2305. LIVE AND UNCENSORED Millie Jackson [22Dec79|26Jan80|19Apr80] 94(2) 18	0	360
2306. HOT! LIVE AND OTHERWISE Dionne Warwick [13Jun81|18Jul81|12Sep81] 72(1) 14	1	359
2307. SO MANY RIVERS Bobby Womack [21Sep85|23Nov85|25Jan86] 66(2) 19	0	358
2308. OFFERING Axe [26Jun82|18Sep82|06Nov82] 81(3) 20	1	358
2309. SEASON OF GLASS Yoko Ono [27Jun81|11Jul81|22Aug81] 49(2) 9	0	358
2310. SHOTGUN MESSIAH Shotgun Messiah [21Oct89|02Dec89|24Mar90] 99(1) 23	0	358
2311. OFF TO SEE THE LIZARD Jimmy Buffett [15Jul89|05Aug89|07Oct89] 57(2) 13	0	358
2312. 10 1/2 The Dramatics [08Mar80|19Apr80|24May80] 61(2) 12	0	357
2313. THE BEST LITTLE WHOREHOUSE IN TEXAS Soundtrack [07Aug82|25Sep82|13Nov82] 63(1) 15	0	357
2314. NO FRILLS Bette Midler [27Aug83|08Oct83|19Nov83] 60(1) 13	3	357
2315. RADIOLAND Nicolette Larson [24Jan81|21Feb81|11Apr81] 62(1) 12	0	357
2316. A GOLDEN CELEBRATION Elvis Presley [17Nov84|26Jan85|23Mar85] 80(2) 19	0	356
2317. SHEFFIELD STEEL Joe Cocker [10Jul82|06Nov82|11Dec82] 105(2) 23	0	356
2318. HELLO BIG MAN Carly Simon [08Oct83|05Nov83|28Jan84] 69(2) 17	1	355
2319. THE IRON MAN (THE MUSICAL BY PETE TOWNSHEND) Pete Townshend [15Jul89|22Jul89|07Oct89] 58(4) 13	0	355
2320. GREATEST HITS The Oak Ridge Boys [22Nov80|13Dec80|11Apr81] 99(1) 21	0	355
2321. BEAT King Crimson [03Jul82|31Jul82|02Oct82] 52(2) 14	0	355
2322. VANDENBERG Vandenberg [08Jan83|12Mar83|07May83] 65(4) 18	1	355
2323. THE BOYS FROM DORAVILLE Atlanta Rhythm Section [16Aug80|13Sep80|25Oct80] 65(2) 11	0	354
2324. FIONA Fiona [30Mar85|01Jun85|27Jul85] 71(1) 18	1	354
2325. KNOCKED OUT LOADED Bob Dylan [02Aug86|23Aug86|25Oct86] 53(1) 13	0	353
2326. THE SINGLES The Pretenders [05Dec87|19Dec87|12Mar88] 69(3) 15	0	353
2327. SLEIGHT OF HAND Joan Armatrading [05Jul86|23Aug86|18Oct86] 68(1) 16	0	353

Ranking the Albums

Rank. Title *Act* [Enter|Peak|Exit] Peak(Wks) TotWks	Hot 100	Scr
2328. THIS IS MY DREAM *Switch* [15Nov80|24Jan81|07Mar81] 85(1) 17	0	353
2329. KIM WILDE *Kim Wilde* [05Jun82|07Aug82|30Oct82] 86(2) 22	1	352
2330. SOMETHING REAL *Phoebe Snow* [15Apr89|10Jun89|26Aug89] 75(1) 20	0	352
2331. LA CAGE AUX FOLLES *Original Cast* [24Sep83|29Oct83|31Dec83] 52(1) 15	0	352
2332. SHORT BACK 'N' SIDES *Ian Hunter* [29Aug81|03Oct81|07Nov81] 62(1) 11	0	352
2333. HOLD ME *Laura Branigan* [10Aug85|07Sep85|16Nov85] 71(3) 15	3	351
2334. QUINELLA *Atlanta Rhythm Section* [19Sep81|14Nov81|02Jan82] 70(1) 16	1	351
2335. MORE FUN IN THE NEW WORLD *X* [08Oct83|29Oct83|10Mar84] 86(2) 23	0	351
2336. FABULOUS DISASTER *Exodus* [25Feb89|06May89|17Jun89] 82(1) 17	0	351
2337. DAN REED NETWORK *Dan Reed Network* [02Apr88|07May88|06Aug88] 95(2) 19	1	351
2338. GORKY PARK *Gorky Park* [09Sep89|30Sep89|27Jan90] 80(2) 21	1,1*	350
2339. HARD LINE *The Blasters* [23Mar85|25May85|27Jul85] 86(2) 19	0	350
2340. JUST A TOUCH OF LOVE *Slave* [08Dec79|26Jan80|15Mar80] 92(2) 15	0	350
2341. SILK + STEEL *Giuffria* [24May86|21Jun86|23Aug86] 60(2) 14	1	349
2342. CHANGE. *The Alarm* [14Oct89|11Nov89|17Mar90] 75(1) 23	1	349
2343. THE FINE ART OF SURFACING *The Boomtown Rats* [01Dec79|16Feb80|15Mar80] 103(2) 16	1	349
2344. POETIC CHAMPIONS COMPOSE *Van Morrison* [10Oct87|31Oct87|05Mar88] 90(2) 22	0	348
2345. NON-STOP ECSTATIC DANCING *Soft Cell* [14Aug82|25Sep82|13Nov82] 57(2) 14	0	348
2346. APOLLONIA 6 *Apollonia 6* [27Oct84|17Nov84|16Feb85] 62(2) 17	1	348
2347. LINCOLN *They Might Be Giants* [24Dec88|25Feb89|29Apr89] 89(2) 19	0	346
2348. THE INVISIBLE MAN'S BAND *The Invisible Man's Band* [31May80|19Jul80|30Aug80] 90(1) 14	1	346
2349. BARRY MANILOW *Barry Manilow* [20May89|10Jun89|02Sep89] 64(2) 16	0	345
2350. HYSTERIA *The Human League* [16Jun84|21Jul84|08Sep84] 62(1) 13	1	345
2351. IF MY ANCESTORS COULD SEE ME NOW *Ivan Neville* [12Nov88|21Jan89|15Apr89] 107(1) 23	2	345
2352. OUTSIDE LOOKING IN *BoDeans* [10Oct87|21Nov87|20Feb88] 86(2) 20	0	345
2353. PASSION: MUSIC FOR THE LAST TEMPTATION OF CHRIST (SOUNDTRACK) *Peter Gabriel* [01Jul89|15Jul89|30Sep89] 60(2) 14	0	344
2354. THE FOOL CIRCLE *Nazareth* [14Feb81|14Mar81|09May81] 70(2) 13	0	344
2355. WILD & BLUE *John Anderson* [09Apr83|14May83|25Jun83] 58(1) 12	1	344
2356. PARADISE *Peabo Bryson* [03May80|24May80|16Aug80] 79(2) 16	0	344
2357. DO THE RIGHT THING *Soundtrack* [22Jul89|09Sep89|21Oct89] 68(2) 14	1	343
2358. GOOD AS GOLD *Red Rockers* [14May83|16Jul83|27Aug83] 71(2) 17	0	343
2359. THE WRESTLING ALBUM *Various Artists* [30Nov85|18Jan86|05Apr86] 84(2) 19	0	343
2360. UTOPIA *Utopia* [16Oct82|13Nov82|19Feb83] 84(2) 19	0	343
2361. TOUCH AND GO *Force M.D.'s* [15Aug87|03Oct87|28Nov87] 67(2) 16	1	343
2362. MY FAVORITE PERSON *The O'Jays* [15May82|05Jun82|07Aug82] 49(2) 13	0	343
2363. THE RAINMAKERS *The Rainmakers* [13Sep86|29Nov86|07Feb87] 85(1) 22	0	342
2364. BILL COSBY "HIMSELF" SOUNDTRACK) *Bill Cosby* [18Dec82|05Feb83|19Mar83] 64(2) 15	0	342
2365. LOW-LIFE *New Order* [08Jun85|29Jun85|02Nov85] 94(2) 22	0	342
2366. WATCH OUT! *Patrice Rushen* [28Mar87|09May87|01Aug87] 77(2) 19	0	342
2367. ALONG THE AXIS *Jon Butcher Axis* [12Oct85|07Dec85|01Feb86] 66(1) 17	1	340
2368. THE YELLOW AND BLACK ATTACK *Stryper* [23Aug86|13Sep86|04Apr87] 103(3) 30	0	339
2369. THE SAGA CONTINUES... *Roger* [02Jun84|14Jul84|01Sep84] 64(2) 14	0	339
2370. THE REAL CHUCKEEBOO *Loose Ends* [23Jul88|20Aug88|29Oct88] 80(3) 15	0	339
2371. 9 1/2 WEEKS *Soundtrack* [29Mar86|03May86|05Jul86] 59(2) 15	1	339
2372. 9 TO 5 *Soundtrack* [27Dec80|31Jan81|04Apr81] 77(1) 15	0	338
2373. WON'T BE BLUE ANYMORE *Dan Seals* [08Feb86|15Mar86|17May86] 59(1) 15	1	338
2374. ...KEEP SMILING *Laid Back* [31Mar84|05May84|07Jul84] 67(1) 15	1	338
2375. HARD TIMES IN THE LAND OF PLENTY *Omar And The Howlers* [27Jun87|08Aug87|31Oct87] 81(2) 19	0	338
2376. THE RIGHT TO ROCK *Keel* [09Mar85|25May85|27Jul85] 99(1) 21	0	338
2377. FOUR WINDS *Tangier* [29Jul89|23Sep89|18Nov89] 91(1) 17	1	338
2378. SOUND AFFECTS *The Jam* [07Feb81|14Mar81|18Apr81] 72(1) 11	0	337
2379. IT'S ABOUT TIME *John Denver* [15Oct83|12Nov83|21Jan84] 61(2) 15	0	337
2380. GREATEST HITS, VOL. 2 *The Oak Ridge Boys* [08Sep84|13Oct84|16Feb85] 71(2) 24	0	337
2381. CLUB NINJA *Blue Öyster Cult* [22Feb86|05Apr86|24May86] 63(2) 14	0	337
2382. TELL NO TALES *TNT* [23May87|13Jun87|10Oct87] 100(1) 21	0	335
2383. STREET BEAT *The Deele* [04Feb84|10Mar84|09Jun84] 78(2) 19	1	335
2384. THE SKILL *The Sherbs* [28Feb81|18Apr81|13Jun81] 100(2) 16	1	334
2385. DREAM STREET ROSE *Gordon Lightfoot* [05Apr80|03May80|14Jun80] 60(1) 11	0	334
2386. THE GREAT MUPPET CAPER *Soundtrack* [11Jul81|08Aug81|19Sep81] 66(2) 11	0	332
2387. THINK VISUAL *The Kinks* [20Dec86|14Feb87|04Apr87] 81(2) 16	0	332
2388. EVITA *Original Cast* [23Aug80|23Aug80|10Apr82] 105(2) 19	0	332
2389. ONE *Bee Gees* [19Aug89|30Sep89|11Nov89] 68(2) 13	1	331
2390. REAL LOVE *Ashford & Simpson* [06Sep86|04Oct86|03Jan87] 74(2) 18	1	331
2391. LIGHTNING STRIKES *Loudness* [31May86|12Jul86|13Sep86] 64(2) 16	0	331
2392. SWEET SIXTEEN *Reba McEntire* [03Jun89|24Jun89|30Sep89] 78(1) 18	0	331
2393. LIKE GANGBUSTERS *JoBoxers* [15Oct83|03Dec83|21Jan84] 70(2) 15	1	330
2394. BONK *Big Pig* [26Mar88|07May88|16Jul88] 93(2) 17	1	330
2395. DREAMLAND EXPRESS *John Denver* [06Jul85|07Sep85|09Nov85] 90(2) 19	0	330
2396. MINIMUM WAGE ROCK & ROLL *Bus Boys* [29Nov80|31Jan81|07Mar81] 85(1) 15	0	330
2397. MOSAIQUE *Gipsy Kings* [16Dec89|30Dec89|21Apr90] 95(2) 19	0	328
2398. STILL STANDING *Jason & The Scorchers* [22Nov86|14Feb87|28Mar87] 91(4) 19	0	328
2399. GYPSY BLOOD *Mason Ruffner* [13Jun87|08Aug87|26Sep87] 80(2) 16	0	327
2400. EAST *Hiroshima* [25Mar89|06May89|29Jul89] 105(2) 19	0	327
2401. STRONG STUFF *Hank Williams Jr.* [23Apr83|14May83|06Aug83] 64(2) 16	0	327
2402. THE ZAGORA *Loose Ends* [04Apr87|25Apr87|04Jul87] 59(2) 14	0	327
2403. MORE GEORGE THOROGOOD AND THE DESTROYERS *George Thorogood & The Destroyers* [08Nov80|06Dec80|24Jan81] 68(2) 12	0	327
2404. FLORIDAYS *Jimmy Buffett* [28Jun86|26Jul86|11Oct86] 66(1) 16	0	326
2405. THE REAL MACAW *Graham Parker* [20Aug83|15Oct83|19Nov83] 59(1) 14	1	326
2406. RHYTHM AND ROMANCE *Rosanne Cash* [22Jun85|07Sep85|09Nov85] 101(1) 21	0	325
2407. REI MOMO *David Byrne* [21Oct89|18Nov89|17Feb90] 71(2) 18	0	324
2408. ELTON JOHN'S GREATEST HITS, VOLUME III 1979-1987 *Elton John* [03Oct87|21Nov87|05Mar88] 84(2) 23	0	324
2409. BROKEN ENGLISH *Marianne Faithfull* [02Feb80|29Mar80|10May80] 82(1) 15	0	323
2410. WALKIN' THE RAZOR'S EDGE *Helix* [18Aug84|06Oct84|01Dec84] 69(2) 16	0	323
2411. PRIDE *Robert Palmer* [30Apr83|28May83|03Sep83] 112(2) 19	1	323
2412. MAJOR MOVES *Hank Williams Jr.* [09Jun84|21Jul84|13Oct84] 100(1) 19	0	322
2413. ENOUGH IS ENOUGH *Billy Squier* [18Oct86|08Nov86|31Jan87] 61(2) 16	1	322
2414. THE SISTERS *Sister Sledge* [13Feb82|13Mar82|15May82] 69(2) 14	1	322
2415. THE SWING OF DELIGHT *Devadip Carlos Santana* [06Sep80|04Oct80|08Nov80] 65(2) 10	0	321
2416. BETWEEN TWO FIRES *Paul Young* [22Nov86|27Dec86|14Mar87] 77(2) 17	1	321
2417. THE GIFT *The Jam* [27Mar82|22May82|10Jul82] 82(1) 16	0	320
2418. LEAD ME ON *Amy Grant* [23Jul88|06Aug88|15Oct88] 71(2) 13	1	319
2419. GOIN' OFF *Biz Markie* [19Mar88|23Apr88|16Jul88] 90(2) 18	0	319
2420. 2300 JACKSON ST. *The Jacksons* [17Jun89|01Jul89|26Aug89] 59(2) 11	1	319
2421. READY AN' WILLING *Whitesnake* [16Aug80|20Sep80|29Nov80] 90(1) 16	1	319
2422. AN AMERICAN DREAM *The Dirt Band* [26Jan80|22Mar80|26Apr80] 76(1) 14	1	319
2423. NONA *Nona Hendryx* [23Apr83|25Jun83|27Aug83] 83(2) 19	1	319
2424. 9 *Public Image Limited* [03Jun89|08Jul89|04Nov89] 106(2) 23	0	318
2425. FLOODLAND *Sisters Of Mercy* [06Feb88|12Mar88|21May88] 101(2) 16	0	317
2426. ROBBERY *Teena Marie* [26Nov83|04Feb84|05May84] 119(1) 24	0	317

Ranking the Albums

Rank. Title *Act* [Enter|Peak|Exit] Peak(Wks) TotWks	Hot 100	Scr
2427. IF I SHOULD FALL FROM GRACE WITH GOD *The Pogues* [27Feb88|02Apr88|11Jun88] 88(3) 16	0	316
2428. TROUBLE IN PARADISE *Randy Newman* [12Feb83|26Feb83|07May83] 64(4) 13	1	316
2429. SECRETS *Wilton Felder* [09Mar85|04May85|22Jun85] 81(1) 16	0	316
2430. ONE LOVE--ONE DREAM *Jeffrey Osborne* [27Aug88|01Oct88|10Dec88] 86(1) 16	1	315
2431. THE FALCON & THE SNOWMAN (SOUNDTRACK) *Pat Metheny Group* [09Mar85|06Apr85|11May85] 54(2) 10	1	315
2432. DAN HILL(2) *Dan Hill* [08Aug87|19Sep87|26Mar88] 90(1) 19	2	315
2433. CLOSER TO THE FLAME *Rob Jungklas* [14Jun86|23Aug86|14Mar87] 102(1) 22	1	315
2434. JULIA FORDHAM *Julia Fordham* [03Dec88|04Feb89|20May89] 118(2) 25	0	315
2435. BACK TO AVALON *Kenny Loggins* [20Aug88|03Sep88|19Nov88] 69(2) 14	3	314
2436. GREATEST HITS *Billy Ocean* [04Nov89|25Nov89|17Feb90] 77(2) 16	1	314
2437. PATTI AUSTIN *Patti Austin* [31Mar84|28Apr84|28Jul84] 87(1) 18	1	314
2438. THE REVOLUTION BY NIGHT *Blue Öyster Cult* [26Nov83|11Feb84|10Mar84] 93(1) 16	1	313
2439. LOVE MAGIC *L.T.D.* [28Nov81|09Jan82|13Feb82] 83(1) 12	0	313
2440. BIRTH, SCHOOL, WORK, DEATH *The Godfathers* [20Feb88|30Apr88|04Jun88] 91(2) 16	0	313
2441. SYREETA(2) *Syreeta* [17May80|12Jul80|23Aug80] 73(2) 15	0	313
2442. THESE DAYS *Crystal Gayle* [27Sep80|08Nov80|06Dec80] 79(1) 11	0	312
2443. DIAMOND SUN *Glass Tiger* [07May88|04Jun88|13Aug88] 82(2) 15	1	311
2444. STRAWBERRY MOON *Grover Washington Jr.* [29Aug87|19Sep87|12Dec87] 66(2) 16	0	311
2445. REQUIEM *Various Artists* [06Apr85|04May85|06Jul85] 77(2) 14	0	311
2446. 80/81 *Pat Metheny* [01Nov80|06Dec80|31Jan81] 89(2) 14	0	311
2447. NAKED TO THE WORLD *Teena Marie* [16Apr88|30Apr88|09Jul88] 65(2) 13	1	311
2448. TO LIVE AND DIE IN L.A. (SOUNDTRACK) *Wang Chung* [02Nov85|14Dec85|01Mar86] 85(2) 18	4	311
2449. HIGHWAYS AND HEARTACHES *Ricky Skaggs* [16Oct82|13Nov82|01Jan83] 61(3) 12	0	310
2450. SNAKES AND LADDERS *Gerry Rafferty* [14Jun80|12Jul80|09Aug80] 61(1) 9	1	310
2451. SOME KIND OF WONDERFUL *Soundtrack* [21Mar87|18Apr87|13Jun87] 57(2) 13	0	310
2452. ANSWERS TO NOTHING *Midge Ure* [11Feb89|08Apr89|27May89] 88(1) 16	1	310
2453. SCOUNDREL DAYS *a-ha* [01Nov86|15Nov86|14Mar87] 74(2) 20	1	308
2454. THE NEW ZAPP IV U *Zapp* [23Nov85|22Feb86|17May86] 110(2) 26	0	308
2455. TAKE A LITTLE RHYTHM *Ali Thomson* [05Jul80|09Aug80|11Oct80] 99(1) 15	2	308
2456. SOUTHSIDE *Texas* [19Aug89|07Oct89|02Dec89] 88(2) 16	0	308
2457. DESTINY *Chaka Khan* [23Aug86|13Sep86|08Nov86] 67(2) 12	1	308
2458. GUITAR TOWN *Steve Earle* [25Oct86|15Nov86|07Mar87] 89(2) 20	0	307
2459. ABOUT LAST NIGHT... *Soundtrack* [26Jul86|06Sep86|25Oct86] 72(1) 14	2	307
2460. OUTRAGEOUS *Lakeside* [28Jul84|18Aug84|03Nov84] 68(2) 15	0	307
2461. WINDHAM HILL RECORDS SAMPLER '86 *Various Artists* [29Mar86|24May86|26Jul86] 102(2) 18	0	306
2462. SAM-I-AM *Sam Harris* [15Feb86|29Mar86|17May86] 69(1) 14	1	305
2463. LOVE JONES *Johnny Guitar Watson* [05Jul80|26Jul80|04Oct80] 115(2) 14	0	305
2464. MICHAEL MARTIN MURPHEY *Michael Murphey* [04Sep82|13Nov82|18Dec82] 69(1) 16	2	305
2465. PARTY 'TIL YOU'RE BROKE *Rufus* [28Mar81|16May81|06Jun81] 73(1) 11	0	304
2466. GOOD LOVE *Meli'sa Morgan* [19Dec87|06Feb88|23Apr88] 108(1) 19	0	304
2467. BOSTON, MASS. *The Del Fuegos* [26Oct85|30Nov85|14Jun86] 132(1) 34	1	304
2468. IRON EAGLE *Soundtrack* [15Feb86|22Mar86|26Apr86] 54(2) 11	1	303
2469. TELEKON *Gary Numan* [04Oct80|01Nov80|06Dec80] 64(2) 10	0	303
2470. CADDYSHACK *Soundtrack* [23Aug80|04Oct80|08Nov80] 78(2) 12	1	303
2471. BUSTER POINDEXTER *Buster Poindexter And His Banshees Of Blue* [09Jan88|20Feb88|16Apr88] 90(1) 15	0	302
2472. YOU, ME AND HE *Mtume* [15Sep84|13Oct84|19Jan85] 77(2) 19	0	302
2473. SLEEPWALK *Larry Carlton* [30Jan82|27Mar82|15May82] 99(1) 16	1	302
2474. CON FUNK SHUN 7 *Con Funk Shun* [12Dec81|23Jan82|06Mar82] 82(1) 13	0	302
2475. STYLE *Cameo* [07May83|28May83|23Jul83] 53(1) 12	0	301
2476. JAM THE BOX! *Bill Summers & Summers Heat* [12Dec81|27Feb82|27Mar82] 92(1) 16	0	301
2477. ARE YOU GONNA BE THE ONE *Robert Gordon* [18Apr81|23May81|25Jul81] 117(4) 15	1	301
2478. CAMOUFLAGE *Rufus And Chaka Khan* [31Oct81|19Dec81|30Jan82] 98(1) 14	1	301
2479. CHRISTMAS WISHES *Anne Murray* [28Nov81|26Dec81|16Jan82] 54(2) 8	0	301
2480. REIGN IN BLOOD *Slayer* [15Nov86|20Dec86|14Mar87] 94(3) 18	0	301
2481. GLORIA LORING *Gloria Loring* [06Sep86|11Oct86|06Dec86] 61(2) 14	1	300
2482. DOWN IN THE GROOVE *Bob Dylan* [18Jun88|02Jul88|20Aug88] 61(3) 10	0	300
2483. SWEAT *The System* [12Mar83|07May83|13Aug83] 94(2) 23	1	300
2484. SKIN DIVE *Michael Franks* [15Jun85|29Jun85|14Dec85] 137(2) 27	0	300
2485. CONTAGIOUS *Y&T* [11Jul87|18Jul87|03Oct87] 78(2) 13	0	299
2486. INSIDE RONNIE MILSAP *Ronnie Milsap* [03Jul82|14Aug82|02Oct82] 66(2) 14	2	299
2487. EVERY SHADE OF LOVE *Jesse Johnson* [16Apr88|07May88|09Jul88] 79(2) 13	1	299
2488. MECHANIX *UFO* [20Feb82|10Apr82|22May82] 82(1) 14	0	298
2489. 1980 *Gil Scott-Heron & Brian Jackson* [08Mar80|05Apr80|24May80] 82(1) 12	0	298
2490. WEATHER REPORT(2) *Weather Report* [20Feb82|20Mar82|01May82] 68(2) 11	0	298
2491. PLAYING TO WIN *LRB* [09Feb85|09Mar85|11May85] 75(3) 14	1	298
2492. DIANA ROSS ANTHOLOGY *Diana Ross* [11Jun83|09Jul83|27Aug83] 63(1) 12	0	298
2493. THE HOUSE OF DOLLS *Gene Loves Jezebel* [14Nov87|12Mar88|23Apr88] 108(1) 22	1	298
2494. UNDER THE BIG BLACK SUN *X* [17Jul82|21Aug82|23Oct82] 76(3) 15	0	297
2495. PUCKER UP *Lipps Inc.* [11Oct80|22Nov80|06Dec80] 63(1) 9	0	297
2496. THE LUXURY GAP *Heaven 17* [04Jun83|16Jul83|27Aug83] 72(2) 13	1	297
2497. 'NARD *Bernard Wright* [14Mar81|23May81|13Jun81] 116(2) 14	0	297
2498. I ADVANCE MASKED *Andy Summers & Robert Fripp* [06Nov82|20Nov82|15Jan83] 60(3) 11	0	297
2499. MONTANA CAFE *Hank Williams Jr.* [19Jul86|23Aug86|15Nov86] 93(2) 18	0	296
2500. 22B3 *Device* [12Jul86|16Aug86|01Nov86] 73(2) 16	2	296
2501. LONG LIVE THE KANE *Big Daddy Kane* [16Jul88|24Sep88|19Nov88] 116(1) 19	0	295
2502. ABSOLUTE BEGINNERS *Soundtrack* [12Apr86|10May86|05Jul86] 62(2) 13	1	295
2503. LADIES OF THE EIGHTIES *A Taste Of Honey* [24Apr82|29May82|10Jul82] 73(4) 12	1	294
2504. CHICAGO XIV *Chicago* [09Aug80|13Sep80|04Oct80] 71(2) 9	1	294
2505. L.A. IS MY LADY *Frank Sinatra* [25Aug84|15Sep84|17Nov84] 58(2) 13	0	294
2506. EINZELHAFT *Falco* [07May83|28May83|30Jul83] 64(1) 13	0	293
2507. MISDEMEANOR *UFO* [05Apr86|03May86|09Aug86] 106(2) 19	0	293
2508. 2 LIVE CREW IS WHAT WE ARE *2 Live Crew* [11Apr87|09May87|20Feb88] 128(1) 33	0	293
2509. WONDERLAND *Big Country* [05May84|16Jun84|21Jul84] 65(2) 12	1	293
2510. BOI-NGO *Oingo Boingo* [21Mar87|18Apr87|04Jul87] 77(2) 16	0	293
2511. WHO'S LAST *The Who* [01Dec84|12Jan85|02Mar85] 81(2) 14	0	292
2512. MOONLIGHT MADNESS *Teri DeSario* [19Jan80|01Mar80|12Apr80] 80(1) 13	2	292
2513. MIKE'S MURDER (SOUNDTRACK) *Joe Jackson* [24Sep83|29Oct83|17Dec83] 64(1) 13	1	292
2514. INDIVIDUAL CHOICE *Jean-Luc Ponty* [27Aug83|01Oct83|03Dec83] 85(1) 15	0	292
2515. STAY HARD *Raven* [23Mar85|11May85|29Jun85] 81(3) 15	0	292
2516. TALK TALK TALK *Psychedelic Furs* [27Jun81|01Aug81|26Sep81] 89(2) 14	0	292
2517. WINTER INTO SPRING *George Winston* [12May84|02Jun84|04Apr87] 127(2) 32	0	291
2518. NASTY, NASTY *Black 'N Blue* [25Oct86|22Nov86|07Mar87] 110(1) 20	0	291
2519. ANIMAL GRACE *April Wine* [17Mar84|14Apr84|02Jun84] 62(1) 12	1	291
2520. YOURS TRULY *Tom Browne* [12Dec81|16Jan82|13Mar82] 97(1) 14	0	291
2521. TEXAS IN MY REAR VIEW MIRROR *Mac Davis* [18Oct80|15Nov80|13Dec80] 67(1) 9	2	291
2522. CONTINUATION *Philip Bailey* [10Sep83|22Oct83|10Dec83] 71(1) 14	0	290
2523. LITTLE MISS DANGEROUS *Ted Nugent* [22Mar86|19Apr86|21Jun86] 76(2) 14	0	290
2524. INTERMISSION *Dio* [28Jun86|12Jul86|11Oct86] 70(2) 16	0	290

Ranking the Albums

Rank. Title Act [Enter|Peak|Exit] Peak(Wks) TotWks	Hot 100	Scr
2525. MUSIC FROM THE ELDER KISS [05Dec81|23Jan82|13Feb82] 75(1) 11	1	290
2526. WE DON'T TALK ANYMORE Cliff Richard [08Dec79|02Feb80|15Mar80] 93(2) 15	2	290
2527. HANDS DOWN Bob James [17Jul82|14Aug82|06Nov82] 72(2) 17	0	289
2528. TURN THE HANDS OF TIME Peabo Bryson [28Feb81|11Apr81|09May81] 82(1) 11	0	289
2529. WANTED DREAD & ALIVE Peter Tosh [18Jul81|22Aug81|10Oct81] 91(1) 13	0	288
2530. THE LAST ONE TO KNOW Reba McEntire [10Oct87|07Nov87|20Feb88] 102(2) 20	0	288
2531. OPEN ALL NIGHT Georgia Satellites [02Jul88|16Jul88|24Sep88] 77(2) 13	0	288
2532. TOGETHER AGAIN The Temptations [24Oct87|21Nov87|12Mar88] 112(2) 21	0	288
2533. IN THE NIGHT Cheryl Lynn [11Jul81|22Aug81|03Oct81] 104(3) 13	1	288
2534. GAMMA 3 Gamma [20Mar82|08May82|05Jun82] 72(1) 12	1	288
2535. MENUDO Menudo [25May85|06Jul85|28Sep85] 100(2) 19	1	287
2536. STRANGE BEHAVIOR Animotion [15Mar86|12Apr86|14Jun86] 71(2) 14	2	287
2537. SKYYJAMMER Skyy [20Nov82|08Jan83|12Feb83] 81(2) 13	0	287
2538. KEEPER OF THE SEVEN KEYS - PART I Helloween [04Jul87|05Sep87|21Nov87] 104(1) 21	0	286
2539. BLUE AND GRAY Poco [25Jul81|15Aug81|26Sep81] 76(2) 10	0	286
2540. RAY, GOODMAN & BROWN II Ray, Goodman & Brown [04Oct80|22Nov80|20Dec80] 84(2) 12	1	286
2541. THE EVER POPULAR TORTURED ARTIST EFFECT Todd Rundgren [22Jan83|26Feb83|16Apr83] 66(3) 13	1	286
2542. MAHVELOUS Billy Crystal [21Sep85|19Oct85|14Dec85] 65(2) 13	1	286
2543. ARCADE Patrick Simmons [07May83|28May83|16Jul83] 52(1) 11	2	286
2544. LITTLE STEVIE ORBIT Steve Forbert [11Oct80|15Nov80|06Dec80] 70(1) 9	0	285
2545. TRY IT OUT Klique [08Oct83|19Nov83|07Jan84] 70(2) 14	1	285
2546. SAME GOES FOR YOU Leif Garrett [15Dec79|16Feb80|10May80] 129(1) 22	3,*1	285
2547. LOVE CHANGES Kashif [05Dec87|19Dec87|09Apr88] 118(2) 19	0	285
2548. THE FABULOUS BAKER BOYS (SOUNDTRACK) Dave Grusin [18Nov89|16Dec89|10Feb90] 74(2) 13	0	285
2549. FEVER Con Funk Shun [03Dec83|28Jan84|21Apr84] 105(1) 21	1	285
2550. DANGER ZONE Sammy Hagar [21Jun80|12Jul80|06Sep80] 85(1) 12	0	284
2551. GLENN JONES Glenn Jones [10Oct87|28Nov87|30Jan88] 94(1) 17	0	284
2552. RAINDANCING Alison Moyet [20Jun87|18Jul87|10Oct87] 94(5) 17	0	284
2553. SPELL Deon Estus [01Apr89|13May89|08Jul89] 89(1) 15	1	284
2554. SPONTANEOUS INVENTIONS Bobby McFerrin [21Mar87|02May87|25Jul87] 103(1) 19	0	284
2555. MR. JORDAN Julian Lennon [01Apr89|06May89|08Jul89] 87(1) 15	1	284
2556. KURTIS BLOW Kurtis Blow [18Oct80|22Nov80|20Dec80] 71(1) 10	1	283
2557. KILIMANJARO The Rippingtons Featuring Russ Freeman [07May88|04Jun88|13Aug88] 110(2) 15	0	283
2558. IT'S A FACT Jeff Lorber [27Mar82|15May82|19Jun82] 73(2) 13	0	283
2559. STRAIGHT AHEAD Larry Gatlin & The Gatlin Brothers Band [17Nov79|05Jan80|19Jul80] 102(1) 16	0	282
2560. MIRACLE The Kane Gang [21Nov87|26Dec87|02Apr88] 115(2) 20	2	282
2561. SPECIAL THINGS Pleasure [12Jul80|23Aug80|11Oct80] 97(2) 14	0	282
2562. ULTRA WAVE Bootsy [06Dec80|27Dec80|31Jan81] 70(3) 9	0	281
2563. COMMON ONE Van Morrison [20Sep80|18Oct80|22Nov80] 73(1) 10	0	281
2564. L.A. BOPPERS L.A. Boppers [15Mar80|12Apr80|24May80] 85(1) 11	0	281
2565. SOUND-SYSTEM Herbie Hancock [01Sep84|15Sep84|01Dec84] 71(2) 14	0	281
2566. INFECTED The The [14Feb87|21Mar87|13Jun87] 89(2) 18	0	281
2567. SONGS YOU KNOW BY HEART: JIMMY BUFFETT'S GREATEST HIT(S) Jimmy Buffett [16Nov85|14Dec85|29Sep90] 100(2) 24	0	280
2568. SPLENDIDO HOTEL Al Di Meola [12Jul80|09Aug80|11Oct80] 119(2) 14	0	280
2569. MEET DANNY WILSON Danny Wilson [18Jul87|12Sep87|31Oct87] 79(1) 16	1	280
2570. EYES ON THIS MC Lyte [21Oct89|18Nov89|03Mar90] 86(2) 20	0	280
2571. FACE TO FACE Evelyn "Champagne" King [24Dec83|18Feb84|05May84] 91(1) 20	0	280
2572. WORKIN' IT BACK Teddy Pendergrass [07Dec85|10May86|26Jul86] 96(2) 23	0	280
2573. ALL BY MYSELF Regina Belle [11Jul87|05Sep87|17Oct87] 85(1) 15	1	278
2574. ONE STEP CLOSER Gavin Christopher [05Jul86|09Aug86|11Oct86] 74(3) 15	1	278
2575. NEXT POSITION PLEASE Cheap Trick [10Sep83|15Oct83|19Nov83] 61(2) 11	0	278
2576. RED HOT RHYTHM & BLUES Diana Ross [30May87|20Jun87|29Aug87] 73(2) 14	0	278
2577. HOW CRUEL Joan Armatrading [08Dec79|19Jan80|05Apr80] 136(1) 18	0	278
2578. SHEILA E. Sheila E. [21Mar87|04Apr87|06Jun87] 56(2) 12	1	278
2579. A WINTER'S SOLSTICE Various Artists [07Dec85|11Jan86|17Jan87] 77(1) 19	0	278
2580. GLOBE OF FROGS Robyn Hitchcock And The Egyptians [05Mar88|23Apr88|11Jun88] 111(2) 15	0	278
2581. WHAT A LIFE! Divinyls [07Dec85|01Mar86|05Apr86] 91(2) 18	1	277
2582. STAND IN THE FIRE Warren Zevon [17Jan81|14Feb81|21Mar81] 80(3) 10	0	277
2583. GREAT BALLS OF FIRE (SOUNDTRACK) Jerry Lee Lewis [22Jul89|12Aug89|23Sep89] 62(1) 10	0	276
2584. A PLACE LIKE THIS Robbie Nevil [26Nov88|07Jan89|22Apr89] 118(2) 21	2	276
2585. AFTER DARK Cruzados [01Aug87|31Oct87|19Dec87] 106(2) 21	0	276
2586. CALL IT WHAT YOU WANT Bill Summers & Summers Heat [04Apr81|30May81|11Jul81] 129(1) 15	0	275
2587. YOU AND ME BOTH Yaz [13Aug83|10Sep83|05Nov83] 69(1) 13	0	275
2588. BEAUTIFUL FEELINGS Rick Springfield [08Dec84|19Jan85|02Mar85] 78(2) 13	1	275
2589. SEND YOUR LOVE Aurra [13Jun81|18Jul81|05Sep81] 103(1) 13	0	274
2590. PREMONITION Peter Frampton [08Feb86|15Mar86|10May86] 80(2) 14	1	274
2591. RAISE YOUR FIST AND YELL Alice Cooper [24Oct87|07Nov87|30Jan88] 73(2) 15	0	274
2592. SECOND SIGHTING Frehley's Comet [11Jun88|02Jul88|03Sep88] 81(3) 13	0	274
2593. CLOSE TO THE BONE Thompson Twins [25Apr87|23May87|25Jul87] 76(1) 14	1	274
2594. CLASSICAL GAS Mason Williams & Mannheim Steamroller [19Dec87|20Feb88|23Apr88] 118(1) 19	0	272
2595. CANDLES Heatwave [13Dec80|10Jan81|14Feb81] 71(2) 10	0	272
2596. MEN WITHOUT WOMEN Little Steven And The Disciples Of Soul [04Dec82|22Jan83|02Apr83] 118(3) 18	1	272
2597. MAGNETIC FIELDS Jean Michel Jarre [11Jul81|25Jul81|26Sep81] 98(2) 12	0	272
2598. NO GURU, NO METHOD, NO TEACHER Van Morrison [16Aug86|30Aug86|08Nov86] 70(2) 13	0	272
2599. THE BEST OF THE ART OF NOISE The Art Of Noise [17Dec88|28Jan89|18Mar89] 83(1) 14	1	271
2600. BERNADETTE PETERS Bernadette Peters [03May80|28Jun80|02Aug80] 114(1) 14	1	271
2601. BOYS CLUB Boys Club [26Nov88|28Jan89|11Mar89] 93(2) 16	1	271
2602. STEP BY STEP Jeff Lorber [09Mar85|27Apr85|22Jun85] 90(2) 16	0	271
2603. JOHN EDDIE John Eddie [21Jun86|02Aug86|27Sep86] 83(2) 15	1	271
2604. A CHIPMUNK CHRISTMAS Chipmunks [21Nov81|09Jan82|16Jan82] 72(2) 9	0	271
2605. INSIDE STORY Grace Jones [13Dec86|24Jan87|28Mar87] 81(2) 16	1	270
2606. THE GREAT PRETENDER Dolly Parton [18Feb84|31Mar84|19May84] 73(1) 14	2	270
2607. OBSESSION Bob James [22Nov86|07Feb87|23May87] 142(1) 27	0	270
2608. HOOKED ON BIG BANDS Frank Barber Orchestra [05Jun82|17Jul82|18Sep82] 94(3) 16	1	269
2609. GLENN MEDEIROS Glenn Medeiros [13Jun87|11Jul87|03Oct87] 83(2) 17	3	269
2610. THE WILD THE WILLING AND THE INNOCENT UFO [31Jan81|21Feb81|11Apr81] 77(2) 11	0	269
2611. MODERN ROMANS The Call [26Mar83|04Jun83|02Jul83] 84(1) 15	1	268
2612. ARK The Animals [10Sep83|08Oct83|12Nov83] 66(1) 10	1	268
2613. SUBURBAN VOODOO Paul Carrack [11Sep82|06Nov82|11Dec82] 78(1) 14	1	268
2614. LIVING LARGE... Heavy D & The Boyz [14Nov87|12Dec87|27Feb88] 92(2) 16	0	268
2615. ONE LORD, ONE FAITH, ONE BAPTISM Aretha Franklin [26Dec87|23Jan88|09Apr88] 106(3) 16	0	268
2616. '74 JAILBREAK AC/DC [17Nov84|08Dec84|16Feb85] 76(2) 14	0	268
2617. THE SECRET POLICEMAN'S BALL Various Artists [23May81|04Jul81|08Aug81] 106(1) 12	0	268
2618. MESSINA Jim Messina [20Jun81|25Jul81|29Aug81] 95(3) 11	0	267
2619. PRACTICE WHAT YOU PREACH Testament [02Sep89|23Sep89|18Nov89] 77(2) 12	0	267
2620. RICHARD PRYOR: HERE AND NOW (SOUNDTRACK) Richard Pryor [12Nov83|03Dec83|04Feb84] 71(1) 13	0	267
2621. KC TEN KC [04Feb84|14Apr84|02Jun84] 93(1) 18	1	267

Ranking the Albums

Rank. Title *Act* [Enter|Peak|Exit] Peak(Wks) TotWks	Hot 100	Scr
2622. I'VE GOT THE CURE *Stephanie Mills* [13Oct84|17Nov84|19Jan85] 73(2) 15	1	267
2623. UNDER THE VOLCANO *Rock And Hyde* [02May87|13Jun87|08Aug87] 94(1) 15	1	267
2624. BEELZEBUBBA *The Dead Milkmen* [24Dec88|22Apr89|03Jun89] 101(2) 23	0	267
2625. LOOKIN' FOR LOVE *Johnny Lee* [15Nov80|06Dec80|04Apr81] 132(1) 21	0	267
2626. MERCILESS *Stephanie Mills* [17Sep83|29Oct83|21Jan84] 104(1) 19	0	266
2627. SEAWIND(2) *Seawind* [25Oct80|29Nov80|03Jan81] 83(1) 11	0	266
2628. TOO LATE THE HERO *John Entwistle* [10Oct81|14Nov81|05Dec81] 71(2) 9	0	266
2629. KEEL *Keel* [27Jun87|08Aug87|19Sep87] 79(1) 13	0	265
2630. METAL HEART *Accept* [30Mar85|11May85|29Jun85] 94(2) 14	0	265
2631. DRUMS ALONG THE MOHAWK *Jean Beauvoir* [28Jun86|09Aug86|04Oct86] 93(2) 15	1	265
2632. NO. 10, UPPING STREET *Big Audio Dynamite* [01Nov86|22Nov86|04Apr87] 119(2) 23	0	265
2633. OLD WAYS *Neil Young* [07Sep85|05Oct85|23Nov85] 75(2) 12	0	264
2634. WHEN THE GOING GETS TOUGH, THE TOUGH GET GOING *Bow Wow Wow* [26Mar83|30Apr83|18Jun83] 82(1) 13	1	264
2635. DEFACE THE MUSIC *Utopia* [25Oct80|15Nov80|20Dec80] 65(2) 9	0	264
2636. LONG WAY TO HEAVEN *Helix* [29Jun85|17Aug85|19Oct85] 103(1) 17	0	264
2637. INTRODUCING...DAVID PEASTON *David Peaston* [05Aug89|23Sep89|02Dec89] 113(1) 18	0	264
2638. (WHO'S AFRAID OF?) THE ART OF NOISE! *The Art Of Noise* [14Jul84|04Aug84|06Oct84] 85(3) 13	0	264
2639. ROVER'S RETURN *John Waite* [11Jul87|08Aug87|26Sep87] 77(1) 12	2	263
2640. HARDWARE *Krokus* [04Apr81|23May81|20Jun81] 103(1) 12	0	263
2641. ZIGGY STARDUST-THE MOTION PICTURE (SOUNDTRACK) *David Bowie* [12Nov83|17Dec83|18Feb84] 89(1) 15	0	263
2642. HIGH VOLTAGE *AC/DC* [18Jul81|27Feb82|13Mar82] 146(1) 19	0	263
2643. THE MICHAEL SCHENKER GROUP *The Michael Schenker Group* [20Sep80|08Nov80|20Dec80] 100(1) 14	0	263
2644. CLOSER *Gino Soccio* [23May81|18Jul81|22Aug81] 96(2) 14	0	263
2645. SIOGO *Blackfoot* [11Jun83|13Aug83|03Sep83] 82(1) 13	0	263
2646. OBLIVION *Utopia* [11Feb84|24Mar84|28Apr84] 74(2) 12	0	263
2647. BARRY WHITE'S SHEET MUSIC *Barry White* [26Jul80|30Aug80|04Oct80] 85(1) 11	0	262
2648. IF YOU AIN'T LOVIN' (YOU AIN'T LIVIN') *George Strait* [19Mar88|09Apr88|18Jun88] 87(2) 14	0	262
2649. AIN'T LOVE GRAND *X* [17Aug85|14Sep85|16Nov85] 89(3) 14	0	262
2650. TINDERBOX *Siouxsie & The Banshees* [24May86|05Jul86|30Aug86] 88(1) 15	0	261
2651. YOU SHOULDN'T-NUF BIT FISH *George Clinton* [07Jan84|03Mar84|05May84] 102(1) 18	0	261
2652. ALEXANDER O'NEAL *Alexander O'Neal* [27Apr85|25May85|24Aug85] 92(2) 18	0	260
2653. BI-COASTAL *Peter Allen* [29Nov80|28Feb81|11Apr81] 123(1) 20	1	260
2654. GREASE 2 *Soundtrack* [19Jun82|07Aug82|11Sep82] 71(2) 13	1	260
2655. MEGATOP PHOENIX *Big Audio Dynamite* [23Sep89|11Nov89|16Dec89] 85(1) 13	0	259
2656. LIFE *Neil Young With Crazy Horse* [25Jul87|08Aug87|03Oct87] 75(2) 11	0	259
2657. BRING THE FAMILY *John Hiatt* [04Jul87|22Aug87|24Oct87] 107(2) 17	0	259
2658. EXIT 0 *Steve Earle And The Dukes* [13Jun87|01Aug87|12Sep87] 90(1) 14	0	259
2659. FERVOR *Jason & The Scorchers* [10Mar84|28Apr84|08Jun85] 116(1) 23	0	258
2660. PLATOON *Soundtrack* [04Apr87|25Apr87|27Jun87] 75(2) 13	0	258
2661. DEAD MAN'S PARTY *Oingo Boingo* [16Nov85|14Dec85|01Mar86] 98(2) 16	2	258
2662. KING OF STAGE *Bobby Brown* [13Dec86|24Jan87|04Apr87] 88(2) 17	1	258
2663. HONOR AMONG THIEVES *The Brandos* [26Sep87|24Oct87|30Jan88] 108(2) 19	0	258
2664. GREATEST HITS LIVE *Carly Simon* [27Aug88|10Sep88|19Nov88] 87(3) 13	0	258
2665. GRETCHEN GOES TO NEBRASKA *King's X* [05Aug89|14Oct89|02Dec89] 123(1) 18	0	257
2666. DO YOU WANNA GET AWAY *Shannon* [25May85|22Jun85|07Sep85] 92(2) 16	1	257
2667. CLOSE TO THE BONE *Tom Tom Club* [20Aug83|10Sep83|12Nov83] 73(1) 13	0	257
2668. TORNADO *The Rainmakers* [28Nov87|26Dec87|02Apr88] 116(2) 19	0	257
2669. STILL LIFE (TALKING) *Pat Metheny Group* [22Aug87|26Sep87|28Nov87] 86(2) 15	0	256
2670. IN DREAMS: THE GREATEST HITS *Roy Orbison* [07Jan89|18Feb89|15Apr89] 95(1) 15	0	256
2671. BLACK TIE *The Manhattans* [08May81|12Sep81|10Oct81] 86(3) 10	0	255
2672. IMAGINE THIS *Pieces Of A Dream* [25Feb84|31Mar84|02Jun84] 90(1) 15	0	255
2673. LOST & FOUND *Jason & The Scorchers* [30Mar85|01Jun85|06Jul85] 96(1) 15	0	255
2674. MAGIC MAN *Robert Winters And Fall* [09May81|23May81|27Jun81] 71(2) 8	0	255
2675. PRETTY PAPER *Willie Nelson* [01Dec79|19Jan80|19Jan80] 73(1) 8	0	255
2676. SODA FOUNTAIN SHUFFLE *Earl Klugh* [11May85|06Jul85|31Aug85] 110(1) 17	0	255
2677. THE COLOR PURPLE *Soundtrack* [08Mar86|05Apr86|31May86] 79(2) 13	0	254
2678. FRIDAY NIGHT IN SAN FRANCISCO *Al Di Meola/John McLaughlin/Paco De Lucia* [30May81|25Jul81|22Aug81] 97(1) 13	0	254
2679. A COUNTRY COLLECTION *Anne Murray* [09Feb80|01Mar80|05Apr80] 73(1) 9	0	254
2680. LOVE & HOPE & SEX & DREAMS *BoDeans* [07Jun86|09Aug86|11Oct86] 115(2) 19	0	254
2681. YO FRANKIE *Dion* [20May89|26Aug89|23Sep89] 130(1) 19	1	253
2682. SLINGSHOT *Michael Henderson* [19Sep81|10Oct81|28Nov81] 92(2) 11	0	253
2683. MAXI PRIEST *Maxi Priest* [03Dec88|11Feb89|25Mar89] 108(2) 17	1	253
2684. LIVE...IN THE RAW *W.A.S.P.* [10Oct87|07Nov87|09Jan88] 77(2) 14	0	253
2685. TOLD U SO *Nu Shooz* [23Apr88|28May88|23Jul88] 93(2) 14	1	253
2686. MICHAEL MORALES *Michael Morales* [17Jun89|22Jul89|28Oct89] 113(3) 20	3	253
2687. HOOKED ON CLASSICS III (JOURNEY THROUGH THE CLASSICS) *Royal Philharmonic Orchestra* [23Apr83|04Jun83|23Jul83] 89(1) 14	0	252
2688. PAT TRAVERS' BLACK PEARL *Pat Travers* [06Nov82|20Nov82|29Jan83] 74(3) 13	0	252
2689. STRAIGHT TO THE SKY *Lisa Lisa And Cult Jam* [13May89|27May89|12Aug89] 77(3) 13	1	252
2690. HAND TO MOUTH *General Public* [25Oct86|29Nov86|07Feb87] 83(1) 16	0	251
2691. DOIN' ALRIGHT *O'Bryan* [10Apr82|22May82|26Jun82] 80(1) 12	1	251
2692. MAGNUM CUM LOUDER *Hoodoo Gurus* [12Aug89|07Oct89|18Nov89] 101(1) 15	0	251
2693. THE COLOR OF MONEY *Soundtrack* [15Nov86|13Dec86|21Feb87] 81(2) 15	0	251
2694. NEGOTIATIONS AND LOVE SONGS (1971-1986) *Paul Simon* [12Nov88|07Jan89|11Feb89] 110(1) 14	0	251
2695. STOP AND SMELL THE ROSES *Ringo Starr* [14Nov81|12Dec81|30Jan82] 98(1) 12	1	251
2696. CLOUDS ACROSS THE SUN *Firefall* [10Jan81|14Feb81|04Apr81] 102(2) 13	1	250
2697. HOLLYWOOD, TENNESSEE *Crystal Gayle* [19Sep81|03Oct81|02Jan82] 99(1) 16	1	249
2698. VU *The Velvet Underground* [09Mar85|13Apr85|01Jun85] 85(1) 13	0	249
2699. OLD CREST ON A NEW WAVE *Dave Mason* [14Jun80|05Jul80|16Aug80] 74(2) 10	1	248
2700. FREIGHT TRAIN HEART *Jimmy Barnes* [11Jun88|09Jul88|17Sep88] 104(3) 15	1	248
2701. 'SNAZ *Nazareth* [10Oct81|07Nov81|05Dec81] 83(1) 9	0	248
2702. HIGH-RISE *Ashford & Simpson* [17Sep83|29Oct83|03Dec83] 84(1) 12	0	247
2703. BRIGHT LIGHTS, BIG CITY *Soundtrack* [02Apr88|30Apr88|11Jun88] 67(2) 11	1	247
2704. PIA & PHIL *Pia Zadora* [08Mar86|05Apr86|19Jul86] 113(2) 20	0	247
2705. STEVE ARRINGTON'S HALL OF FAME: I *Steve Arrington's Hall Of Fame* [12Mar83|21May83|02Jul83] 101(1) 17	0	247
2706. K-9 POSSE *K-9 Posse* [04May89|22Apr89|03Jun89] 98(3) 14	0	247
2707. SMOKING IN THE FIELDS *The Del Fuegos* [28Oct89|25Nov89|24Mar90] 139(1) 22	0	246
2708. SOUTHERN BY THE GRACE OF GOD/LYNYRD SKYNYRD TRIBUTE TOUR 1987 *Lynyrd Skynyrd* [16Apr88|30Apr88|25Jun88] 68(2) 11	0	246
2709. ...BEHIND THE GARDENS-BEHIND THE WALL-UNDER THE TREE... *Andreas Vollenweider* [01Dec84|12Jan85|30Mar85] 121(2) 18	0	246
2710. SEX AND THE SINGLE MAN *Ray Parker Jr.* [26Oct85|16Nov85|18Jan86] 65(2) 13	1	246
2711. YOUR MOVE *America* [02Jul83|03Sep83|01Oct83] 81(1) 14	1	246
2712. STREET FIGHTING YEARS *Simple Minds* [20May89|03Jun89|05Aug89] 70(2) 12	0	246
2713. IN OUTER SPACE *Sparks* [30Apr83|25Jun83|20Aug83] 88(1) 17	1	246
2714. GREATEST HITS VOL. 2 *Ronnie Milsap* [31Aug85|09Nov85|11Jan86] 102(2) 20	0	245
2715. BE A WINNER *Yarbrough & Peoples* [14Apr84|09Jun84|28Jul84] 90(2) 16	1	245

Ranking the Albums

Rank. Title *Act* [Enter|Peak|Exit] Peak(Wks) TotWks	Hot 100	Scr
2716. WENDY AND LISA *Wendy And Lisa* [19Sep87|07Nov87|12Dec87] 88(1) 13	1	245
2717. LOVE IS FOR SUCKERS *Twisted Sister* [01Aug87|22Aug87|10Oct87] 74(2) 11	0	245
2718. RIGHT BY YOU *Stephen Stills* [01Sep84|13Oct84|17Nov84] 75(2) 12	2	245
2719. SOMEWHERE IN THE STARS *Rosanne Cash* [10Jul82|24Jul82|25Sep82] 76(2) 12	0	244
2720. RUFF 'N' READY *Ready For The World* [15Oct88|29Oct88|17Dec88] 65(2) 10	0	244
2721. SHOT IN THE DARK *Great White* [16Aug86|11Oct86|08Nov86] 82(1) 13	0	244
2722. THAT'S THE STUFF *Autograph* [16Nov85|14Dec85|22Feb86] 92(2) 15	0	244
2723. SENSATIONAL *Starpoint* [21Mar87|25Apr87|20Jun87] 95(2) 14	1	244
2724. THE REAL DEAL *The Isley Brothers* [21Aug82|18Sep82|06Nov82] 87(3) 12	0	244
2725. YOU'LL NEVER KNOW *Rodney Franklin* [19Apr80|14Jun80|12Jul80] 104(1) 13	0	243
2726. A DECADE OF HITS *Charlie Daniels Band* [23Jul83|20Aug83|08Oct83] 84(1) 12	0	243
2727. NEW LIGHT THROUGH OLD WINDOWS *Chris Rea* [04Mar89|15Apr89|27May89] 92(4) 13	1	243
2728. STAR TREK II: THE WRATH OF KHAN *Soundtrack* [17Jul82|21Aug82|11Sep82] 61(2) 9	0	243
2729. RACING AFTER MIDNIGHT *Honeymoon Suite* [14May88|04Jun88|16Jul88] 86(2) 10	1	243
2730. HALL OF THE MOUNTAIN KING *Savatage* [10Oct87|30Jan88|12Mar88] 116(1) 23	0	242
2731. FREEDOM NO COMPROMISE *Little Steven* [13Jun87|04Jul87|29Aug87] 80(2) 12	0	242
2732. INSTANT LOVE *Cheryl Lynn* [17Jul82|28Aug82|27Nov82] 133(2) 20	0	242
2733. THE STORY OF A YOUNG HEART *A Flock Of Seagulls* [25Aug84|22Sep84|27Oct84] 66(1) 10	1	242
2734. CONCRETE BLONDE *Concrete Blonde* [21Feb87|18Apr87|06Jun87] 96(2) 16	0	242
2735. DOWNTOWN *Marshall Crenshaw* [12Oct85|07Dec85|08Feb86] 110(2) 18	0	242
2736. STORMS *Nanci Griffith* [16Sep89|11Nov89|16Dec89] 99(1) 14	0	242
2737. THE WORLD'S GREATEST ENTERTAINER *Doug E. Fresh & The Get Fresh Crew* [18Jun88|16Jul88|10Sep88] 88(2) 13	0	241
2738. HOME *BoDeans* [22Jul89|26Aug89|14Oct89] 94(2) 13	0	240
2739. CARNIVAL *Duran Duran* [02Oct82|13Nov82|08Jan83] 98(3) 15	0	240
2740. EVERYTHING *Climie Fisher* [28May88|16Jul88|10Sep88] 120(1) 16	1	240
2741. FEARGAL SHARKEY *Feargal Sharkey* [08Mar86|12Apr86|17May86] 75(2) 11	1	240
2742. LOOKIN' FOR TROUBLE *Joyce Kennedy* [08Sep84|13Oct84|01Dec84] 79(2) 13	0	240
2743. CATCHING UP WITH DEPECHE MODE *Depeche Mode* [07Dec85|01Feb86|05Apr86] 113(1) 18	0	240
2744. CAT IN THE HAT *Bobby Caldwell* [29Mar80|24May80|05Jul80] 113(2) 15	1	239
2745. POP GOES THE MOVIES *Meco* [03Apr82|24Apr82|29May82] 68(2) 9	1	239
2746. THE GENIE (THEMES & VARIATIONS FROM THE TV SERIES "TAXI") *Bob James* [04Jun83|23Jul83|13Aug83] 77(1) 11	0	239
2747. TOURIST IN PARADISE *The Rippingtons Featuring Russ Freeman* [10Jun89|08Jul89|26Aug89] 85(2) 12	0	239
2748. OUR BELOVED REVOLUTIONARY SWEETHEART *Camper van Beethoven* [18Jun88|27Aug88|08Oct88] 124(1) 17	0	238
2749. RIDDLES IN THE SAND *Jimmy Buffett* [29Sep84|03Nov84|29Dec84] 87(1) 14	0	238
2750. HEARTATTACK AND VINE *Tom Waits* [04Oct80|25Oct80|06Dec80] 96(2) 10	0	238
2751. TROMBIPULATION *Parliament* [10Jan81|24Jan81|21Feb81] 61(2) 7	0	238
2752. SMOKEY AND THE BANDIT 2 *Soundtrack* [06Sep80|04Oct80|15Nov80] 103(2) 11	1	238
2753. SERIOUS *The O'Jays* [27May89|15Jul89|16Sep89] 114(1) 17	0	238
2754. STAY AWAKE *Various Artists* [12Nov88|03Dec88|18Feb89] 119(4) 15	0	237
2755. UNDERNEATH THE RADAR *Underworld* [19Mar88|09Apr88|23Jul88] 139(2) 19	1	237
2756. THEM *King Diamond* [23Jul88|13Aug88|08Oct88] 89(2) 12	0	237
2757. THE CATHERINE WHEEL (ORIGINAL CAST) *David Byrne* [19Dec81|13Feb82|06Mar82] 104(1) 12	0	237
2758. 25TH ANNIVERSARY *Diana Ross & The Supremes* [17May86|05Jul86|06Sep86] 112(2) 17	0	237
2759. STREET FEVER *Moon Martin* [15Nov80|20Dec80|21Feb81] 138(3) 15	0	236
2760. BLUE BELL KNOLL *Cocteau Twins* [15Oct88|12Nov88|11Feb89] 109(2) 18	0	236
2761. HARDER THAN YOU *24-7 SPYZ* [17Jun89|12Aug89|30Sep89] 113(2) 16	0	236
2762. TREAT HER RIGHT *Treat Her Right* [09Apr88|18Jun88|06Aug88] 127(2) 18	0	236
2763. CLASSICS LIVE *Aerosmith* [26Apr86|14Jun86|12Jul86] 84(1) 12	0	236
2764. SERIOUS BUSINESS *Third World* [15Jul89|26Aug89|14Oct89] 107(1) 14	0	236
2765. ON *Off Broadway USA* [16Feb80|08Mar80|26Apr80] 101(2) 11	1	236
2766. FEAR NO EVIL *Grim Reaper* [06Jul85|24Aug85|05Oct85] 108(1) 14	0	236
2767. SEVENTH STAR *Black Sabbath Featuring Tony Iommi* [15Feb86|29Mar86|26Apr86] 78(2) 11	0	235
2768. MORE THAN FRIENDS *Jonathan Butler* [05Nov88|19Nov88|01Apr89] 113(2) 22	0	235
2769. CARL ANDERSON *Carl Anderson* [23Aug86|20Sep86|08Nov86] 87(2) 12	0	235
2770. FOLKWAYS: A VISION SHARED - A TRIBUTE TO WOODY GUTHRIE AND LEADBELLY *Various Artists* [17Sep88|08Oct88|19Nov88] 70(2) 10	0	235
2771. UNITED *Commodores* [22Nov86|20Dec86|28Feb87] 101(3) 15	1	235
2772. VOICES & IMAGES *Camouflage* [14Jan89|11Feb89|15Apr89] 100(3) 14	1	235
2773. EVERYTHING'S DIFFERENT NOW *'Til Tuesday* [19Nov88|17Dec88|25Mar89] 124(3) 19	1	234
2774. THE DUKES OF HAZZARD *TV Soundtrack* [17Apr82|12Jun82|17Jul82] 93(2) 14	0	234
2775. YOU WANNA DANCE WITH ME? *Jody Watley* [02Dec89|23Dec89|24Feb90] 86(3) 13	0	234
2776. INTRODUCING JONATHAN BUTLER *Jonathan Butler* [24May86|12Jul86|06Sep86] 101(2) 16	0	234
2777. I LOVE 'EM ALL *T.G. Sheppard* [25Apr81|20Jun81|11Jul81] 119(1) 12	1	234
2778. RECOVERY: LIVE! *Great White* [13Feb88|12Mar88|30Apr88] 99(2) 12	0	234
2779. RADIO ONE *The Jimi Hendrix Experience* [03Dec88|04Feb89|25Mar89] 119(2) 17	0	234
2780. GOD'S OWN MEDICINE *The Mission U.K.* [07Mar87|02May87|04Jul87] 108(1) 18	0	233
2781. SPEND THE NIGHT *The Isley Brothers Featuring Ronald Isley* [02Sep89|30Sep89|25Nov89] 89(1) 13	0	233
2782. ROCKS, PEBBLES AND SAND *Stanley Clarke* [28Jun80|12Jul80|06Sep80] 95(2) 11	0	233
2783. UNDER WRAPS *Jethro Tull* [27Oct84|24Nov84|12Jan85] 76(2) 12	0	233
2784. MANHATTANS GREATEST HITS *The Manhattans* [13Dec80|17Jan81|14Feb81] 87(2) 10	0	233
2785. 20 YEARS OF JETHRO TULL *Jethro Tull* [13Aug88|17Sep88|19Nov88] 97(2) 15	0	233
2786. MEANT FOR EACH OTHER *Barbara Mandrell & Lee Greenwood* [08Sep84|20Oct84|01Dec84] 89(1) 13	0	232
2787. SKY *Sky* [01Nov80|10Jan81|07Feb81] 125(1) 15	0	232
2788. SHINE *Average White Band* [31May80|12Jul80|16Aug80] 116(2) 12	1	232
2789. HEART ATTACK *Krokus* [07May88|02Jul88|16Jul88] 87(1) 11	0	232
2790. ANIMOTION(2) *Animotion* [25Mar89|06May89|15Jul89] 110(1) 17	2	232
2791. MIDNIGHT STAR *Midnight Star* [05Nov88|26Nov88|11Feb89] 96(2) 15	0	232
2792. BIG CITY *Merle Haggard* [07Nov81|21Aug82|30Oct82] 161(1) 28	0	232
2793. MARILYN MARTIN *Marilyn Martin* [22Feb86|22Mar86|03May86] 72(2) 11	1	232
2794. DELIRIOUS NOMAD *Armored Saint* [07Dec85|22Feb86|12Apr86] 108(2) 19	0	232
2795. EVERY GREAT MOTOWN HIT OF MARVIN GAYE *Marvin Gaye* [22Oct83|05May84|16Jun84] 80(1) 16	0	232
2796. STANDARDS, VOLUME 1 *Stanley Jordan* [14Feb87|14Mar87|13Jun87] 108(2) 18	0	231
2797. GAP GOLD/BEST OF THE GAP BAND *The Gap Band* [09Mar85|13Apr85|22Jun85] 103(2) 16	0	231
2798. ISLE OF MAN *Isle Of Man* [19Jul86|20Sep86|15Nov86] 110(2) 18	1	231
2799. TWICE THE LOVE *George Benson* [24Sep88|08Oct88|26Nov88] 76(2) 10	0	231
2800. MARVIN SEASE *Marvin Sease* [18Jul87|12Sep87|07Nov87] 114(2) 17	0	231
2801. ON A ROLL *Point Blank* [17Apr82|03Jul82|07Aug82] 119(1) 17	0	231
2802. LOUD AND CLEAR *Autograph* [11Apr87|23May87|18Jul87] 108(2) 15	0	231
2803. HUMANS *Bruce Cockburn* [18Oct80|15Nov80|13Dec80] 81(2) 9	0	231
2804. LATOYA JACKSON *LaToya Jackson* [18Oct80|06Dec80|10Jan81] 116(2) 13	0	230
2805. FAMILY *Hubert Laws* [08Nov80|13Dec80|31Jan81] 133(1) 13	0	230
2806. THE BOY GENIUS *Kwamé Featuring A New Beginning* [27May89|24Jun89|23Sep89] 114(2) 18	0	229
2807. PLEASED TO MEET ME *The Replacements* [30May87|22Aug87|03Oct87] 131(1) 19	0	229

Ranking the Albums

Rank. Title *Act* [Enter|Peak|Exit] Peak(Wks) TotWks	Hot 100	Scr
2808. RHYTHM ROMANCE *The Romantics* [21Sep85|26Oct85|30Nov85] 72(2) 11	1	229
2809. LIVE + 1 *Frehley's Comet* [27Feb88|12Mar88|30Apr88] 84(2) 10	0	229
2810. I'M REAL *James Brown* [18Jun88|16Jul88|17Sep88] 96(1) 14	0	229
2811. JON BUTCHER AXIS *Jon Butcher Axis* [26Mar83|04Jun83|13Aug83] 91(1) 13	0	228
2812. RAP'S GREATEST HITS *Various Artists* [08Nov86|31Jan87|28Feb87] 114(2) 17	0	228
2813. I AM WHAT I AM *George Jones* [13Jun81|05Sep81|12Sep81] 132(1) 14	0	228
2814. CITY STREETS *Carole King* [06May89|10Jun89|19Aug89] 111(1) 16	0	228
2815. RATT *Ratt* [30Jun84|11Aug84|03Nov84] 133(3) 19	0	228
2816. WA WA NEE *Wa Wa Nee* [07Nov87|19Dec87|27Feb88] 123(2) 17	2	228
2817. EYE TO EYE *Eye To Eye* [19Jun82|31Jul82|25Sep82] 99(2) 15	1	228
2818. SURVEILLANCE *Triumph* [28Nov87|12Dec87|20Feb88] 82(2) 13	0	227
2819. 1984 (FOR THE LOVE OF BIG BROTHER) (SOUNDTRACK) *Eurythmics* [05Jan85|23Feb85|06Apr85] 93(1) 14	1	227
2820. MARIA MCKEE *Maria McKee* [01Jul89|12Aug89|07Oct89] 120(2) 15	0	227
2821. BRIEF ENCOUNTER *Marillion* [22Mar86|12Apr86|24May86] 67(2) 10	0	227
2822. TO WHOM IT MAY CONCERN *The Pasadenas* [18Mar89|22Apr89|03Jun89] 89(2) 12	1	227
2823. SNAPSHOT *Sylvia (2)* [18Jun83|16Jul83|27Aug83] 77(1) 11	0	227
2824. STILL THE SAME OLE ME *George Jones* [28Nov81|23Jan82|27Feb82] 115(2) 14	0	227
2825. GOOD TO GO LOVER *Gwen Guthrie* [30Aug86|04Oct86|22Nov86] 89(2) 13	1	227
2826. ESPECIALLY FOR YOU *Don Williams* [25Jul81|19Sep81|03Oct81] 109(1) 11	0	227
2827. ONE WAY FEATURING AL HUDSON(2) *One Way Featuring Al Hudson* [02Aug80|27Sep80|18Oct80] 128(1) 12	0	226
2828. ROCKBIRD *Debbie Harry* [13Dec86|24Jan87|07Mar87] 97(2) 13	2	226
2829. FUN IN SPACE *Roger Taylor* [09May81|23May81|11Jul81] 121(1) 10	0	226
2830. BROTHERHOOD *New Order* [25Oct86|22Nov86|21Mar87] 117(1) 21	0	226
2831. GIRLS TO CHAT & BOYS TO BOUNCE *Foghat* [25Jul81|29Aug81|19Sep81] 92(1) 9	0	226
2832. LOVE HYSTERIA *Peter Murphy* [14May88|02Jul88|17Sep88] 135(2) 19	0	226
2833. HIGH NOTES *Hank Williams Jr.* [08May82|19Jun82|18Sep82] 123(1) 20	0	225
2834. GARDEN OF LOVE *Rick James* [23Aug80|13Sep80|25Oct80] 83(2) 10	0	225
2835. GO FOR IT *Shalamar* [24Oct81|07Nov81|30Jan82] 115(2) 15	0	225
2836. THE COMMUNARDS *The Communards* [20Dec86|28Feb87|04Apr87] 90(2) 16	1	225
2837. FEMALE TROUBLE *Nona Hendryx* [23May87|20Jun87|15Aug87] 96(2) 13	1	225
2838. EUGENE WILDE *Eugene Wilde* [26Jan85|16Mar85|04May85] 97(2) 15	0	225
2839. MR. CROWLEY *Ozzy Osbourne* [08May82|19Jun82|21Aug82] 120(1) 18	0	224
2840. SHADOWS *Gordon Lightfoot* [20Feb82|20Mar82|08May82] 87(2) 12	1	224
2841. BEST OF ELVIS COSTELLO/THE ATTRACTIONS *Elvis Costello And The Attractions* [30Nov85|18Jan86|15Mar86] 116(2) 16	0	224
2842. WITCH DOCTOR *Instant Funk* [08Dec79|09Feb80|01Mar80] 129(1) 13	0	224
2843. I LIKE YOUR STYLE *Jermaine Jackson* [26Sep81|24Oct81|28Nov81] 86(1) 10	1	224
2844. RIT/2 *Lee Ritenour* [04Dec82|18Dec82|05Mar83] 99(3) 14	1	223
2845. CAFE RACERS *Kim Carnes* [19Nov83|17Dec83|03Mar84] 97(1) 16	3	223
2846. THE BEST OF MANHATTAN TRANSFER *The Manhattan Transfer* [12Dec81|06Feb82|20Feb82] 103(2) 11	0	223
2847. EDEN ALLEY *Timbuk 3* [07May88|21May88|30Jul88] 107(5) 13	0	223
2848. SO HAPPY *Eddie Murphy* [26Aug89|16Sep89|21Oct89] 70(2) 9	1	223
2849. SUMMER HEAT *Brick* [05Sep81|03Oct81|07Nov81] 89(2) 10	0	222
2850. YOU'RE GETTIN' EVEN WHILE I'M GETTIN' ODD *The J. Geils Band* [24Nov84|22Dec84|26Jan85] 80(3) 14	1	222
2851. URBAN DAYDREAMS *David Benoit* [13May89|01Jul89|12Aug89] 101(2) 14	0	222
2852. DAYS OF INNOCENCE *Moving Pictures* [04Dec82|05Feb83|19Mar83] 101(3) 16	2	222
2853. DREAM BABIES GO TO HOLLYWOOD *John Stewart* [12Apr80|10May80|14Jun80] 85(2) 10	0	222
2854. SPREADING THE DISEASE *Anthrax* [21Dec85|29Mar86|19Apr86] 113(1) 18	0	221
2855. 3 *Violent Femmes* [04Feb89|18Feb89|29Apr89] 93(2) 13	0	221
2856. SONGS FROM LIQUID DAYS *Philip Glass* [12Apr86|10May86|05Jul86] 91(2) 13	0	221
2857. I JUST CAN'T STOP IT *The English Beat* [09Aug80|13Sep80|08Nov80] 142(1) 14	0	221
2858. PILEDRIVER -- THE WRESTLING ALBUM II *Various Artists* [17Oct87|31Oct87|27Feb88] 123(2) 20	0	221
2859. BARRY GOUDREAU *Barry Goudreau* [20Sep80|11Oct80|08Nov80] 88(3) 8	0	220
2860. THE DREAMS OF CHILDREN *Shadowfax* [17Nov84|08Dec84|30Mar85] 126(2) 20	0	220
2861. TRACIE SPENCER *Tracie Spencer* [25Jun88|16Jul88|14Jan89] 146(1) 21	2	220
2862. WILD WEST *Dottie West* [11Apr81|30May81|18Jul81] 126(1) 15	1	220
2863. ROAD HOUSE *Soundtrack* [03Jun89|01Jul89|05Aug89] 67(2) 10	0	220
2864. DUETS *Kenny Rogers* [05May84|23Jun84|14Jul84] 85(2) 11	0	220
2865. AFFAIR *Cherrelle* [19Nov88|24Dec88|25Feb89] 106(3) 15	0	220
2866. LOVE BYRD *Donald Byrd And 125th Street N.Y.C.* [03Oct81|31Oct81|05Dec81] 93(1) 10	0	220
2867. CUT THE CRAP *The Clash* [07Dec85|18Jan86|22Feb86] 88(2) 12	0	219
2868. NIGHT FADES AWAY *Nils Lofgren* [26Sep81|31Oct81|05Dec81] 99(1) 11	0	219
2869. BOTTOMS UP *The Chi-Lites* [04Jun83|02Jul83|20Aug83] 98(1) 12	0	219
2870. TWO TONS O' FUN *Two Tons Of Fun* [17May80|28Jun80|26Jul80] 91(2) 11	0	219
2871. JIMMY BARNES *Jimmy Barnes* [08Mar86|26Apr86|21Jun86] 109(2) 16	1	219
2872. AKIMBO ALOGO *Kim Mitchell* [18May85|06Jul85|24Aug85] 106(2) 15	1	218
2873. SAY IT AGAIN *Jermaine Stewart* [23Apr88|21May88|09Jul88] 98(2) 12	1	218
2874. AMERICAN ANTHEM *Soundtrack* [05Jul86|09Aug86|20Sep86] 91(3) 12	1	218
2875. RICCOCHET DAYS *Modern English* [24Mar84|21Apr84|09Jun84] 93(1) 12	1	217
2876. PONTIAC *Lyle Lovett* [20Feb88|02Apr88|28May88] 117(1) 14	0	217
2877. DELIVER *The Oak Ridge Boys* [19Nov83|17Dec83|18Feb84] 121(1) 14	0	217
2878. A CHORUS LINE-THE MOVIE *Soundtrack* [28Dec85|08Feb86|15Mar86] 77(2) 12	0	217
2879. 9012LIVE - THE SOLOS *Yes* [30Nov85|21Dec85|08Feb86] 81(3) 11	0	217
2880. CAMERON'S IN LOVE *Rafael Cameron* [18Jul81|12Sep81|03Oct81] 101(1) 12	0	217
2881. LITTLE AMERICA *Little America* [25Apr87|06Jun87|25Jul87] 102(2) 14	0	216
2882. WINNERS *Various Artists* [06Sep80|20Sep80|18Oct80] 69(2) 7	0	216
2883. R&B SKELETONS IN THE CLOSET *George Clinton* [24May86|28Jun86|09Aug86] 81(2) 12	0	216
2884. CLASS OF '55 *Carl Perkins, Jerry Lee Lewis, Roy Orbison, & Johnny Cash* [21Jun86|12Jul86|06Sep86] 87(2) 12	0	216
2885. THE RUMOUR *Olivia Newton-John* [03Sep88|17Sep88|29Oct88] 67(2) 9	1	216
2886. WINDHAM HILL RECORDS SAMPLER '88 *Various Artists* [27Feb88|12Mar88|11Jun88] 134(1) 16	0	216
2887. THE WORD IS OUT *Jermaine Stewart* [02Mar85|13Apr85|11May85] 90(2) 11	1	216
2888. THE HOTTEST NIGHT OF THE YEAR *Anne Murray* [28Aug82|09Oct82|13Nov82] 90(3) 12	0	215
2889. 4 OF A KIND *D.R.I.* [23Jul88|03Sep88|22Oct88] 116(2) 14	0	215
2890. LIVERPOOL *Frankie Goes To Hollywood* [15Nov86|06Dec86|07Feb87] 88(2) 13	0	215
2891. I'M YOUR PLAYMATE *Suave* [23Apr88|04Jun88|09Jul88] 101(1) 12	1	215
2892. TEQUILA SUNRISE *Soundtrack* [21Jan89|18Feb89|15Apr89] 101(1) 13	1	214
2893. COLONEL ABRAMS *Colonel Abrams* [19Apr86|24May86|12Jul86] 75(2) 11	0	214
2894. BOUNCING OFF THE SATELLITES *The B-52s* [04Oct86|11Oct86|10Jan87] 85(2) 15	0	214
2895. MEMORY IN THE MAKING *John Kilzer* [11Jun88|09Jul88|17Sep88] 110(2) 15	0	213
2896. EVERY MAN HAS A WOMAN *Various Artists* [13Oct84|10Nov84|15Dec84] 75(2) 10	0	213
2897. LEATHERWOLF *Leatherwolf* [05Mar88|23Apr88|21May88] 105(1) 12	0	213
2898. STRIKE LIKE LIGHTNING *Lonnie Mack* [15Jun85|20Jul85|02Nov85] 130(2) 21	0	213
2899. TOO FAR TO WHISPER *Shadowfax* [12Jul86|23Aug86|25Oct86] 114(1) 16	0	213
2900. INSIDE OUT *Philip Bailey* [24May86|21Jun86|02Aug86] 84(2) 11	0	213
2901. THIEF (SOUNDTRACK) *Tangerine Dream* [09May81|20Jun81|11Jul81] 115(1) 10	0	213
2902. WHAT'S NEXT *Frank Marino And Mahogany Rush* [08Mar80|29Mar80|03May80] 88(1) 9	0	213

Ranking the Albums

Rank. Title Act [Enter|Peak|Exit] Peak(Wks) TotWks	Hot 100	Scr
2903. HABITS OLD AND NEW Hank Williams Jr. [21Jun80|23Aug80|11Oct80] 154(1) 17	0	213
2904. EMERALD CITY Teena Marie [05Jul86|02Aug86|13Sep86] 81(2) 11	0	212
2905. "10" Soundtrack [05Jan80|26Jan80|01Mar80] 80(1) 9	0	212
2906. TIME PIECES -- THE BEST OF ERIC CLAPTON Eric Clapton [22May82|12Jun82|21Aug82] 101(2) 14	0	212
2907. SNEAKER Sneaker [12Dec81|27Feb82|03Apr82] 149(2) 17	2	211
2908. I WISH I WAS EIGHTEEN AGAIN George Burns [09Feb80|15Mar80|12Apr80] 93(2) 10	1	211
2909. MIAMI VICE II TV Soundtrack [06Dec86|10Jan87|21Feb87] 82(1) 12	1	211
2910. NO MUSS...NO FUSS Donnie Iris [16Mar85|18May85|22Jun85] 115(1) 15	1	211
2911. LES MISERABLES Original London Cast Recording [11Apr87|09May87|18Jul87] 106(2) 15	0	211
2912. WORKS Pink Floyd [18Jun83|09Jul83|13Aug83] 68(1) 9	0	211
2913. KEEPER OF THE SEVEN KEYS - PART II Helloween [29Oct88|19Nov88|11Feb89] 108(2) 16	0	211
2914. PEARL HARBOR & THE EXPLOSIONS Pearl Harbor And The Explosions [26Jan80|15Mar80|05Apr80] 107(1) 11	0	210
2915. THE COMPLETE STORY OF ROXANNE...THE ALBUM Dr. J.R. Kool & The Other Roxannes [20Jul85|07Sep85|12Oct85] 113(1) 13	0	210
2916. GREATEST HITS Reba McEntire [06Jun87|17Oct87|05Dec87] 139(2) 23	0	210
2917. FALSE ACCUSATIONS Robert Cray Band [05Apr86|26Apr86|16May87] 141(3) 21	0	210
2918. BOY MEETS GIRL Boy Meets Girl [04May85|08Jun85|13Jul85] 76(2) 11	1	210
2919. MAMMA Luciano Pavarotti [08Sep84|03Nov84|08Dec84] 103(2) 14	0	210
2920. WILD & LOOSE Oaktown's 3.5.7 [13May89|10Jun89|26Aug89] 126(1) 16	0	209
2921. HUSH John Klemmer [13Jun81|11Jul81|08Aug81] 99(1) 9	0	209
2922. WILL THE CIRCLE BE UNBROKEN, VOL.II Nitty Gritty Dirt Band [27May89|01Jul89|12Aug89] 95(2) 12	0	209
2923. TATTOOED BEAT MESSIAH Zodiac Mindwarp & The Love Reaction [26Mar88|16Apr88|02Jul88] 132(2) 15	0	209
2924. SPOILED GIRL Carly Simon [20Jul85|10Aug85|28Sep85] 88(2) 11	1	209
2925. WITH SYMPATHY Ministry [25Jun83|23Jul83|24Sep83] 96(1) 14	0	209
2926. NO EXIT Fates Warning [23Apr88|28May88|16Jul88] 111(3) 13	0	209
2927. FAREWELL SONG Janis Joplin [13Feb82|27Mar82|24Apr82] 104(1) 11	0	208
2928. BAD ATTITUDE Meat Loaf [18May85|22Jun85|20Jul85] 74(1) 10	0	208
2929. LIVE/INDIAN SUMMER Al Stewart [14Nov81|05Dec81|23Jan82] 110(2) 11	0	208
2930. GROOVE APPROVED Paul Carrack [11Nov89|09Dec89|10Mar90] 120(2) 18	1	207
2931. ALL DRESSED UP & NO PLACE TO GO Nicolette Larson [14Aug82|18Sep82|16Oct82] 75(1) 10	1	207
2932. IRISH HEARTBEAT Van Morrison & The Chieftains [23Jul88|13Aug88|15Oct88] 102(2) 13	0	207
2933. COUNTRY Soundtrack [01Dec84|12Jan85|09Mar85] 120(2) 15	0	207
2934. THE NEW ORDER Testament [25Jun88|13Aug88|24Sep88] 136(1) 14	0	206
2935. MADE IN U.S.A. The Beach Boys [26Jul86|30Aug86|11Oct86] 96(1) 12	2	206
2936. HOOKED ON SWING 2 Larry Elgart And His Manhattan Swing Orchestra [12Feb83|12Mar83|14May83] 89(2) 14	0	206
2937. LIBRA Julio Iglesias [24Aug85|05Oct85|09Nov85] 92(2) 12	0	204
2938. MTV'S ROCK 'N ROLL TO GO Various Artists [02Mar85|23Mar85|18May85] 91(2) 12	0	204
2939. TIGHT SHOES Foghat [21Jun80|12Jul80|23Aug80] 106(1) 10	1	204
2940. YOU WANT IT, YOU GOT IT Bryan Adams [30Jan82|20Mar82|24Apr82] 118(1) 13	1	204
2941. JUST LOOKIN' FOR A HIT Dwight Yoakam [14Oct89|04Nov89|16Dec89] 68(2) 10	0	204
2942. HE THINKS HE'S RAY STEVENS Ray Stevens [19Jan85|02Mar85|25May85] 118(2) 19	0	204
2943. THE OTHER SIDE OF THE RAINBOW Melba Moore [13Nov82|12Feb83|19Mar83] 152(1) 19	0	204
2944. THE BEST Quincy Jones [17Jul82|31Jul82|06Nov82] 122(2) 17	0	203
2945. TASTY JAM Fatback [20Jun81|18Jul81|08Aug81] 102(1) 8	0	203
2946. FOREVER YOURS Tony Terry [09Jan88|20Feb88|21May88] 151(1) 12	2	203
2947. ACROSS A CROWDED ROOM Richard Thompson [09Mar85|13Apr85|01Jun85] 102(2) 13	0	203
2948. I'VE GOT THE ROCK 'N' ROLLS AGAIN Joe Perry Project [04Jul81|08Aug81|05Sep81] 100(1) 10	0	202
2949. VICES Kick Axe [30Jun84|08Sep84|06Oct84] 126(2) 15	0	202
2950. IN THE POCKET Neil Sedaka [17May80|12Jul80|09Aug80] 135(1) 13	1	202
2951. 'ROUND MIDNIGHT Linda Ronstadt [11Oct86|01Nov86|31Jan87] 124(2) 17	0	202
2952. IT'S ALRIGHT (I SEE RAINBOWS) Yoko Ono [25Dec82|19Feb83|19Mar83] 98(2) 13	0	202
2953. CHANGE OF HEART Change [28Apr84|16Jun84|04Aug84] 102(2) 15	0	202
2954. THE SHOUTING STAGE Joan Armatrading [20Aug88|24Sep88|12Nov88] 100(1) 13	0	201
2955. THE HARD WAY Point Blank [31May80|19Jul80|23Aug80] 110(2) 13	0	201
2956. DON'T TAKE IT PERSONAL Jermaine Jackson [02Dec89|30Dec89|17Mar90] 115(2) 16	1,1*	201
2957. YEARS AGO The Statler Brothers [11Jul81|15Aug81|05Sep81] 103(1) 9	0	201
2958. SHOWDOWN! Albert Collins, Robert Cray, Johnny Copeland [15Feb86|12Apr86|14Jun86] 124(1) 18	0	201
2959. BULLISH Herb Alpert & The Tijuana Brass [25Aug84|29Sep84|27Oct84] 75(1) 10	1	201
2960. THE GOONIES (SOUNDTRACK) Dave Grusin [29Jun85|20Jul85|31Aug85] 73(2) 10	0	201
2961. THE MUSIC OF COSMOS TV Soundtrack [09May81|04Jul81|01Aug81] 136(2) 13	0	201
2962. FOR THE LONELY: A ROY ORBISON ANTHOLOGY, 1956-1965 Roy Orbison [07Jan89|18Feb89|01Apr89] 110(1) 13	0	200
2963. BOLD AS LOVE Bardeux [30Apr88|18Jun88|16Jul88] 104(2) 12	2	200
2964. ACT A FOOL King Tee [21Jan89|25Feb89|29Apr89] 125(1) 15	0	200
2965. LIVING PROOF Sylvester [24Nov79|19Jan80|09Feb80] 123(1) 12	0	200
2966. CHRISTMAS IN THE STARS/STAR WARS CHRISTMAS ALBUM Meco [13Dec80|27Dec80|17Jan81] 61(3) 6	1	200
2967. NEARLY HUMAN Todd Rundgren [17Jun89|08Jul89|26Aug89] 102(2) 11	0	200
2968. OCEAN RAIN Echo & The Bunnymen [09Jun84|14Jul84|18Aug84] 87(1) 11	0	199
2969. HARD MACHINE Stacey Q [05Mar88|02Apr88|14May88] 115(1) 11	1	199
2970. JOIN THE ARMY Suicidal Tendencies [23May87|11Jul87|15Aug87] 100(2) 13	0	199
2971. APPLE JUICE Tom Scott [11Jul81|08Aug81|19Sep81] 123(1) 11	0	199
2972. FOR MEN ONLY Millie Jackson [21Jun80|02Aug80|23Aug80] 100(1) 10	0	199
2973. NATURALLY Leon Haywood [17May80|28Jun80|19Jul80] 92(2) 10	1	199
2974. KEEP ON MOVING STRAIGHT AHEAD Lakeside [12Dec81|26Dec81|13Feb82] 109(4) 10	0	199
2975. THE REAL THING Angela Winbush [11Nov89|09Dec89|03Mar90] 113(2) 17	0	199
2976. FIERCE HEART Jim Capaldi [21May83|02Jul83|06Aug83] 91(1) 12	2	199
2977. THE FLAG Rick James [05Jul86|26Jul86|20Sep86] 95(2) 12	0	198
2978. LIVE 1980/86 Joe Jackson [21May88|04Jun88|06Aug88] 91(3) 12	0	198
2979. BADDEST Grover Washington Jr. [13Sep80|18Oct80|15Nov80] 96(1) 10	0	198
2980. CRACKERS INTERNATIONAL Erasure [13May89|20May89|15Jul89] 73(3) 10	1	198
2981. FIRE DOWN UNDER Riot [12Sep81|24Oct81|21Nov81] 99(1) 11	0	198
2982. LES PLUS GRANDS SUCCES DE CHIC - CHIC'S GREATEST HITS Chic [22Dec79|05Jan80|16Feb80] 88(2) 9	0	198
2983. MICHAEL BOLTON Michael Bolton [07May83|25Jun83|30Jul83] 89(1) 13	1	198
2984. BLOOD & CHOCOLATE Elvis Costello And The Attractions [11Oct86|01Nov86|20Dec86] 84(2) 11	0	197
2985. Y U I ORTA Ian Hunter/Mick Ronson [28Oct89|30Dec89|10Mar90] 157(3) 20	0	197
2986. WE ARE ONE Pieces Of A Dream [28Aug82|30Oct82|04Dec82] 114(3) 15	0	197
2987. RELEASED Patti LaBelle [12Apr80|31May80|05Jul80] 114(1) 13	0	197
2988. DUMB WAITERS The Korgis [08Nov80|06Dec80|24Jan81] 113(2) 12	1	197
2989. COMMON GROUND Rhythm Corps [13Aug88|24Sep88|12Nov88] 104(2) 14	0	197
2990. WHITE SHOES Emmylou Harris [19Nov83|14Jan84|11Feb84] 116(1) 13	0	197
2991. TAKE NO PRISONERS Peabo Bryson [06Jul85|10Aug85|28Sep85] 102(2) 13	1	196
2992. REWIND (1971-1984) The Rolling Stones [28Jul84|18Aug84|06Oct84] 86(1) 11	0	196
2993. SUPERCHARGED Tavares [08Mar80|29Mar80|19Apr80] 75(2) 7	1	196
2994. FRIENDS IN LOVE Dionne Warwick [22May82|12Jun82|07Aug82] 83(2) 12	1	196

Ranking the Albums

Rank. Title Act [Enter|Peak|Exit] Peak(Wks) TotWks	Hot 100	Scr
2995. DISCO Pet Shop Boys [27Dec86|24Jan87|14Mar87] 95(2) 12	0	196
2996. OH, JULIE! Barry Manilow [25Sep82|16Oct82|20Nov82] 69(2) 9	1	196
2997. TIGHTEN UP VOL. '88 Big Audio Dynamite [13Aug88|17Sep88|29Oct88] 102(1) 12	0	196
2998. LIGHT OF THE STABLE: THE CHRISTMAS ALBUM Emmylou Harris [29Nov80|27Dec80|24Jan81] 102(3) 9	0	196
2999. THE ENVOY Warren Zevon [14Aug82|18Sep82|06Nov82] 93(1) 13	0	195
3000. NIGHTSONGS Earl Klugh [27Oct84|24Nov84|16Feb85] 107(2) 17	0	195
3001. MAN OF STEEL Hank Williams Jr. [19Nov83|24Dec83|11Feb84] 116(2) 13	0	195
3002. MSG The Michael Schenker Group [24Oct81|21Nov81|12Dec81] 81(1) 8	0	195
3003. ECHOES OF AN ERA Various Artists [06Feb82|20Mar82|17Apr82] 105(1) 11	0	195
3004. SOUND + VISION David Bowie [14Oct89|04Nov89|27Jan90] 97(1) 16	0	195
3005. LIVE AND OUTRAGEOUS (RATED XXX) Millie Jackson [13Mar82|24Apr82|05Jun82] 113(1) 13	0	195
3006. 42ND STREET Original Cast [17Jan81|28Feb81|28Mar81] 120(1) 11	0	195
3007. SOULFORCE REVOLUTION 7 Seconds [04Nov89|30Dec89|10Mar90] 153(2) 19	0	194
3008. WILLIE & THE POOR BOYS Willie & The Poor Boys [25May85|22Jun85|10Aug85] 96(2) 12	0	194
3009. THE WINNING HAND Various Artists [15Jan83|19Mar83|16Apr83] 109(1) 14	0	194
3010. THIS IS ELVIS (SOUNDTRACK) Elvis Presley [25Apr81|16May81|27Jun81] 115(3) 10	0	194
3011. ORIGINAL STYLIN' Three Times Dope [22Apr89|20May89|09Sep89] 122(2) 18	0	194
3012. 5 TO 1 Tom Kimmel [04Jul87|08Aug87|10Oct87] 104(2) 15	1	194
3013. ALBUM Public Image Limited [08Mar86|12Apr86|21Jun86] 115(2) 16	0	194
3014. DAVE GRUSIN COLLECTION Dave Grusin [25Feb89|08Apr89|13May89] 110(3) 12	0	193
3015. STRAIGHT AHEAD Amy Grant [20Apr85|29Jun85|14Sep85] 133(2) 20	0	193
3016. LIVING IN A BOX Living In A Box [08Aug87|05Sep87|31Oct87] 89(2) 13	2	193
3017. FOOL IN LOVE WITH YOU Jim Photoglo [06Jun81|25Jul81|15Aug81] 119(1) 11	1	193
3018. THE OFFICIAL MUSIC OF THE XXIIIRD OLYMPIAD-LOS ANGELES 1984 Various Artists [14Jul84|01Sep84|06Oct84] 92(2) 13	2	192
3019. LOVE BOMB The Tubes [23Mar85|13Apr85|25May85] 87(2) 10	1	192
3020. LOVE NOTES Chuck Mangione [17Jul82|14Aug82|18Sep82] 83(2) 10	0	192
3021. WE GO A LONG WAY BACK Bloodstone [17Jul82|21Aug82|25Sep82] 95(2) 11	0	192
3022. FIFTH ANGEL Fifth Angel [16Apr88|11Jun88|09Jul88] 117(2) 13	0	192
3023. NO TELLIN' LIES Zebra [22Sep84|20Oct84|01Dec84] 84(2) 11	0	192
3024. LIGHT OF DAY Soundtrack [14Mar87|04Apr87|16May87] 82(2) 10	0	192
3025. TAKE ME TO YOUR HEAVEN Stevie Woods [05Dec81|16Jan82|22May82] 153(2) 25	3	192
3026. SILVER CONDOR Silver Condor [04Jul81|29Aug81|19Sep81] 141(4) 12	1	191
3027. CHRISTMAS Alabama [23Nov85|28Dec85|18Jan86] 75(2) 9	0	191
3028. ROUND TWO Johnny Van Zant Band [13Jun81|04Jul81|15Aug81] 119(2) 10	0	191
3029. EVERY STEP OF THE WAY David Benoit [04Jun88|25Jun88|03Sep88] 129(2) 14	0	191
3030. A CLASSIC CASE: THE MUSIC OF JETHRO TULL The London Symphony Orchestra/Ian Anderson [11Jan86|08Feb86|05Apr86] 93(2) 13	0	191
3031. DREAM A LITTLE DREAM Soundtrack [08Apr89|06May89|10Jun89] 94(2) 10	0	191
3032. SHORT STORIES Jon And Vangelis [31May80|04Oct80|25Oct80] 125(1) 15	1	191
3033. VOLUNTEER JAM VI Various Artists [19Jul80|02Aug80|13Sep80] 104(2) 9	0	191
3034. THIS KIND OF LOVIN' The Whispers [03Oct81|17Oct81|28Nov81] 100(3) 9	0	191
3035. GODDESS OF LOVE Phyllis Hyman [18Jun83|20Aug83|03Sep83] 112(1) 12	0	191
3036. FREEDOM Santana [07Mar87|28Mar87|16May87] 95(2) 11	0	191
3037. MAGNETS The Vapors [04Apr81|09May81|30May81] 109(2) 9	0	191
3038. LOOK OUT 20/20 [20Jun81|08Aug81|05Sep81] 127(2) 12	0	191
3039. HOT SPOT Dazz Band [17Aug85|28Sep85|02Nov85] 98(2) 12	0	190
3040. MASTER OF THE GAME George Duke [24Nov79|19Jan80|02Feb80] 125(1) 14	0	189
3041. FOREVER AND EVER Howard Hewett [16Apr88|07May88|02Jul88] 110(2) 12	0	189
3042. HOW WILL I LAUGH TOMORROW WHEN I CAN'T EVEN SMILE TODAY Suicidal Tendencies [01Oct88|29Oct88|17Dec88] 111(2) 12	0	189
3043. RED The Communards [06Feb88|27Feb88|02Apr88] 93(1) 9	1	189
3044. TAKE ME ALL THE WAY Stacy Lattisaw [11Oct86|06Dec86|07Mar87] 131(2) 22	1	189
3045. MR. NICE GUY Ronnie Laws [13Aug83|24Sep83|22Oct83] 98(1) 11	0	189
3046. COOL FROM THE WIRE Dirty Looks [21May88|09Jul88|20Aug88] 134(2) 14	0	188
3047. DANGEROUS MOMENTS Martin Briley [09Feb85|16Mar85|13Apr85] 85(2) 10	0	188
3048. JUNKYARD Junkyard [12Aug89|09Sep89|21Oct89] 105(2) 11	0	188
3049. INTUITION TNT [18Mar89|08Apr89|03Jun89] 115(1) 12	0	188
3050. CHANGE NO CHANGE Elliot Easton [09Mar85|06Apr85|18May85] 99(2) 11	0	188
3051. PRESSURE Bram Tchaikovsky [17May80|14Jun80|19Jul80] 108(1) 10	0	188
3052. SMOKE SIGNALS Smokey Robinson [15Feb86|22Mar86|10May86] 104(2) 13	0	188
3053. GRAVITY Kenny G [01Jun85|06Jul85|17Aug85] 97(2) 12	0	188
3054. POOL IT! The Monkees [19Sep87|03Oct87|14Nov87] 72(2) 9	1	188
3055. JEALOUSY The Dirt Band [05Sep81|03Oct81|31Oct81] 102(1) 9	1	187
3056. RANDY MEISNER Randy Meisner [21Aug82|18Sep82|30Oct82] 94(3) 11	1	187
3057. MADNESS, MONEY AND MUSIC Sheena Easton [16Oct82|30Oct82|01Jan83] 85(3) 12	2	187
3058. JUST VISITING THIS PLANET Jellybean [05Sep87|10Oct87|14Nov87] 101(2) 11	2	187
3059. WATERFRONT Waterfront [20May89|01Jul89|12Aug89] 103(1) 13	2	187
3060. METROPOLIS Soundtrack [25Aug84|20Oct84|17Nov84] 110(2) 13	2	187
3061. DOWN FOR THE COUNT Y&T [23Nov85|14Dec85|08Feb86] 91(2) 12	0	187
3062. TEASES AND DARES Kim Wilde [09Feb85|09Mar85|13Apr85] 84(2) 10	1	186
3063. WHO'S GREATEST HITS The Who [21May83|18Jun83|13Aug83] 94(1) 13	0	186
3064. I HEAR YOU ROCKIN' The Dave Edmunds Band [31Jan87|21Mar87|18Apr87] 106(1) 12	0	186
3065. MEAN STREAK Y&T [10Sep83|29Oct83|26Nov83] 103(1) 12	0	186
3066. HEARTBREAK EXPRESS Dolly Parton [24Apr82|22May82|10Jul82] 106(2) 12	0	185
3067. OUT OF CONTROL Brothers Johnson [04Aug84|15Sep84|13Oct84] 91(1) 11	0	185
3068. SHARING YOUR LOVE Change [15May82|12Jun82|10Jul82] 66(2) 9	1	185
3069. FROZEN GHOST Frozen Ghost [11Apr87|23May87|04Jul87] 107(2) 13	1	185
3070. PINK WORLD Planet P Project [01Dec84|02Feb85|02Mar85] 121(1) 14	0	185
3071. COLOR IN YOUR LIFE Missing Persons [09Aug86|06Sep86|18Oct86] 86(1) 11	0	184
3072. GRAB IT! L'Trimm [05Nov88|19Nov88|18Feb89] 132(2) 16	1	184
3073. POPEYE Soundtrack [27Dec80|07Feb81|28Feb81] 115(1) 9	0	184
3074. ONE SECOND Yello [26Sep87|10Oct87|28Nov87] 92(2) 10	1	184
3075. MAC BAND Mac Band Featuring The McCampbell Brothers [23Jul88|20Aug88|22Oct88] 109(2) 14	0	184
3076. INTO THE WOODS The Call [04Jul87|15Aug87|26Sep87] 123(2) 13	0	184
3077. FUEL FOR THE FIRE Naked Eyes [08Sep84|06Oct84|10Nov84] 83(3) 10	1	184
3078. MUSTA NOTTA GOTTA LOTTA Joe Ely [11Apr81|06Jun81|20Jun81] 135(1) 11	0	184
3079. THE ART OF FALLING APART Soft Cell [26Feb83|05Mar83|16Apr83] 84(4) 8	0	184
3080. AN IMITATION OF LOVE Millie Jackson [27Dec86|11Apr87|23May87] 119(2) 17	0	184
3081. TINA LIVE IN EUROPE Tina Turner [09Apr88|30Apr88|04Jun88] 86(2) 9	0	183
3082. DOWN AND OUT IN BEVERLY HILLS [05Apr86|26Apr86|17May86] 68(2) 7	2	183
3083. BLUE SKIES Kiri Te Kanawa/ Nelson Riddle And His Orchestra [07Dec85|08Feb86|22Mar86] 136(1) 16	0	183
3084. DISCOVERY Shanice Wilson [28Nov87|13Feb88|26Mar88] 149(2) 18	1	183
3085. SUBTERRANEAN JUNGLE The Ramones [26Mar83|09Apr83|21May83] 83(3) 9	0	183
3086. VICTORY Narada Michael Walden [18Oct80|22Nov80|06Dec80] 103(1) 8	0	182
3087. JOYRIDE Pieces Of A Dream [02Aug86|06Sep86|18Oct86] 102(2) 12	0	182
3088. NATURAL STATES David Lanz & Paul Speer [30Jan88|26Mar88|16Apr88] 125(2) 12	0	182
3089. GOOD MUSIC Joan Jett & the Blackhearts [25Oct86|22Nov86|07Feb87] 105(1) 16	2	182
3090. TALK TO YOUR DAUGHTER Robben Ford [06Aug88|10Sep88|29Oct88] 120(2) 13	0	182
3091. DROP DOWN AND GET ME Del Shannon [12Dec81|13Feb82|13Mar82] 123(1) 14	1	182

Ranking the Albums

Rank. Title *Act* [Enter|Peak|Exit] Peak(Wks) TotWks	Hot 100	Scr
3092. OH YES I CAN *David Crosby* [18Feb89|11Mar89|22Apr89] 104(2) 10	0	181
3093. ROBERT HAZARD *Robert Hazard* [26Mar83|30Apr83|04Jun83] 102(1) 11	1	181
3094. THE JIMI HENDRIX CONCERTS *Jimi Hendrix* [25Sep82|16Oct82|13Nov82] 79(2) 8	0	181
3095. NOW VOYAGER *Barry Gibb* [20Oct84|03Nov84|08Dec84] 72(2) 8	1	181
3096. BOINGO ALIVE *Oingo Boingo* [22Oct88|29Oct88|31Dec88] 90(3) 11	0	181
3097. BEHAVIOUR *Saga* [21Sep85|02Nov85|23Nov85] 87(2) 10	0	181
3098. THERE MUST BE A BETTER WORLD SOMEWHERE *B.B. King* [28Feb81|21Mar81|02May81] 131(1) 10	0	180
3099. THE MANILOW COLLECTION/TWENTY CLASSIC HITS *Barry Manilow* [29Jun85|20Jul85|14Sep85] 100(2) 12	0	180
3100. HEARTS IN MOTION *Air Supply* [06Sep86|04Oct86|01Nov86] 84(2) 9	1	180
3101. WILD AND FREE *Dazz Band* [30Aug86|27Sep86|08Nov86] 100(2) 11	0	180
3102. YOU'VE GOT A GOOD LOVE COMIN' *Lee Greenwood* [09Jun84|22Sep84|20Oct84] 150(2) 20	0	180
3103. THE INTRODUCTION *Steve Morse Band* [01Sep84|13Oct84|17Nov84] 101(1) 12	0	180
3104. WRAP YOUR ARMS AROUND ME *Agnetha Faltskog* [17Sep83|29Oct83|26Nov83] 102(1) 11	1	180
3105. COTTON CLUB *Soundtrack* [19Jan85|09Mar85|23Mar85] 93(1) 10	0	180
3106. SHADOWDANCE *Shadowfax* [19Nov83|18Feb84|24Mar84] 145(1) 19	0	180
3107. HEART PLAY *John Lennon & Yoko Ono* [14Jan84|25Feb84|31Mar84] 94(1) 12	0	179
3108. HURRY UP THIS WAY AGAIN *The Stylistics* [08Nov80|06Dec80|24Jan81] 127(1) 9	0	179
3109. LIVE *Stephane Grappelli/David Grisman* [06Jun81|01Aug81|08Aug81] 108(1) 10	0	179
3110. LABYRINTH (SOUNDTRACK) *David Bowie* [19Jul86|09Aug86|06Sep86] 68(2) 8	0	179
3111. FACE TO FACE *Face To Face* [16Jun84|25Aug84|29Sep84] 126(1) 16	1	179
3112. FINYL VINYL *Rainbow* [15Mar86|05Apr86|17May86] 87(2) 10	0	179
3113. SEE HOW WE ARE *X* [11Jul87|01Aug87|19Sep87] 107(2) 11	0	179
3114. BOSSA NOVA HOTEL *Michael Sembello* [08Oct83|12Nov83|10Dec83] 80(1) 10	1	178
3115. TOM COCHRANE & RED RIDER *Tom Cochrane & Red Rider* [02Aug86|23Aug86|18Oct86] 112(1) 12	0	178
3116. QUESTIONNAIRE *Chaz Jankel* [06Mar82|24Apr82|05Jun82] 126(1) 14	0	178
3117. TOUCH DANCE *Eurythmics* [07Jul84|08Sep84|15Sep84] 115(1) 11	0	177
3118. CHINATOWN *Thin Lizzy* [29Nov80|20Dec80|31Jan81] 120(1) 10	0	177
3119. SCUBA DIVERS *Dwight Twilley* [13Mar82|01May82|22May82] 109(2) 11	0	177
3120. FOXIE *Bob James* [08Oct83|29Oct83|31Dec83] 106(1) 13	0	177
3121. SAY YOU LOVE ME *Jennifer Holliday* [14Sep85|12Oct85|14Dec85] 110(2) 14	2	177
3122. RAISING FEAR *Armored Saint* [26Sep87|07Nov87|12Dec87] 114(2) 12	0	177
3123. KING'S RECORD SHOP *Rosanne Cash* [15Aug87|02Apr88|21May88] 138(1) 20	0	176
3124. BLOW YOUR COOL! *Hoodoo Gurus* [02May87|23May87|25Jul87] 120(4) 13	0	176
3125. STORIES WITHOUT WORDS *Spyro Gyra* [26Sep87|17Oct87|21Nov87] 84(1) 9	0	176
3126. SHE'S HAVING A BABY *Soundtrack* [12Mar88|09Apr88|30Apr88] 92(1) 8	0	175
3127. WHAT GOES AROUND *The Hollies* [09Jul83|06Aug83|03Sep83] 90(1) 9	1	175
3128. PROTECT THE INNOCENT *Rachel Sweet* [22Mar80|19Apr80|31May80] 123(1) 11	0	175
3129. CAROL HENSEL'S EXERCISE AND DANCE PROGRAM, VOLUME 3 *Carol Hensel* [22Jan83|19Mar83|09Apr83] 104(2) 12	0	175
3130. SYNCHRO SYSTEM *King Sunny Ade & His African Beats* [20Aug83|24Sep83|22Oct83] 91(1) 10	0	175
3131. TO THE POWER OF THREE *3* [19Mar88|09Apr88|21May88] 97(2) 10	0	175
3132. ST. JULIAN *Julian Cope* [04Apr87|18Apr87|20Jun87] 105(2) 12	1	175
3133. HERE TODAY, TOMORROW NEXT WEEK! *The Sugarcubes* [14Oct89|21Oct89|09Dec89] 70(2) 9	0	175
3134. STAR TREK III - THE SEARCH FOR SPOCK *Soundtrack* [23Jun84|21Jul84|11Aug84] 82(2) 8	0	175
3135. CROSSROADS (SOUNDTRACK) *Ry Cooder* [10May86|24May86|05Jul86] 85(3) 9	0	175
3136. RAGING SLAB *Raging Slab* [28Oct89|25Nov89|03Feb90] 113(1) 15	0	174
3137. INSTINCT *Iggy Pop* [23Jul88|06Aug88|08Oct88] 110(2) 12	0	174
3138. SEAMLESS *The Nylons* [29Mar86|24May86|12Jul86] 133(1) 16	0	174
3139. ROMEO'S ESCAPE *Dave Alvin* [26Sep87|24Oct87|19Dec87] 116(1) 13	0	174
3140. DR. HOOK/GREATEST HITS *Dr. Hook* [20Dec80|10Jan81|07Mar81] 142(3) 12	0	174
3141. TRUE LOVE *Crystal Gayle* [04Dec82|22Jan83|19Feb83] 120(1) 12	1	174
3142. ANNIE'S CHRISTMAS *No Artist* [20Nov82|25Dec82|15Jan83] 96(4) 9	0	174
3143. NO PAROLE FROM ROCK 'N' ROLL *Alcatrazz* [07Jan84|03Mar84|05May84] 128(1) 18	0	174
3144. I TOUCHED A DREAM *The Dells* [30Aug80|04Oct80|15Nov80] 137(1) 12	0	174
3145. UPLIFT MOFO PARTY PLAN *Red Hot Chili Peppers* [21Nov87|06Feb88|19Mar88] 148(1) 18	0	174
3146. CAN'T LOOK AWAY *Trevor Rabin* [19Aug89|16Sep89|21Oct89] 111(3) 10	0	174
3147. COMMODORES 13 *Commodores* [01Oct83|22Oct83|10Dec83] 103(1) 11	1	173
3148. WHITE KNUCKLE RIDE *Duke Jupiter* [02Jun84|28Jul84|18Aug84] 122(1) 12	1	173
3149. SONGS FROM THE FILM *Tommy Keene* [29Mar86|03May86|19Jul86] 148(2) 17	0	173
3150. MASK *Roger Glover* [16Jun84|11Aug84|01Sep84] 101(1) 12	0	173
3151. NICK LOWE AND HIS COWBOY OUTFIT *Nick Lowe And His Cowboy Outfit* [23Jun84|11Aug84|08Sep84] 113(1) 12	0	173
3152. WARM LEATHERETTE *Grace Jones* [21Jun80|02Aug80|23Aug80] 132(1) 10	0	173
3153. FLEX *Lene Lovich* [08Mar80|12Apr80|26Apr80] 94(1) 9	0	173
3154. CHARLIE SEXTON *Charlie Sexton* [18Feb89|25Mar89|15Apr89] 104(1) 9	0	173
3155. SCARLETT & BLACK *Scarlett & Black* [19Mar88|23Apr88|28May88] 107(2) 11	0	173
3156. MR. MUSIC HEAD *Adrian Belew* [22Jul89|26Aug89|30Sep89] 114(1) 11	1	173
3157. DANGEROUS ACQUAINTANCES *Marianne Faithfull* [17Oct81|21Nov81|12Dec81] 104(1) 9	0	172
3158. FLIP-FLOP *Guadalcanal Diary* [25Mar89|06May89|17Jun89] 132(1) 13	0	172
3159. BILLY & THE BEATERS *Billy Vera & The Beaters* [16May81|27Jun81|18Jul81] 118(1) 10	2	172
3160. THE KIDS FROM "FAME" LIVE! *The Kids From Fame* [26Mar83|30Apr83|04Jun83] 98(1) 11	0	172
3161. KINGS OF THE SUN *Kings Of The Sun* [30Apr88|21May88|13Aug88] 136(1) 16	1	172
3162. AMERICA'S GREATEST HERO *Joey Scarbury* [22Aug81|19Sep81|17Oct81] 104(2) 9	2	172
3163. LONE RHINO *Adrian Belew* [24Jul82|14Aug82|18Sep82] 82(2) 9	0	172
3164. WHAT PRICE PARADISE *China Crisis* [07Mar87|04Apr87|23May87] 114(2) 12	0	172
3165. GET OUT OF MY ROOM *Cheech & Chong* [12Oct85|02Nov85|21Dec85] 71(2) 11	1	171
3166. SIMPLE MINDS LIVE: IN THE CITY OF LIGHT *Simple Minds* [18Jul87|15Aug87|19Sep87] 96(2) 10	0	171
3167. 14 KARAT *Fatback* [01Nov80|06Dec80|13Dec80] 91(1) 7	0	171
3168. 25TH ANNIVERSARY *The Temptations* [17May86|28Jun86|30Aug86] 140(2) 16	0	171
3169. CLUTCHING AT STRAWS *Marillion* [11Jul87|08Aug87|19Sep87] 103(2) 11	0	171
3170. SOUL SURVIVOR *Al Green* [02May87|11Jul87|01Aug87] 131(1) 14	0	171
3171. SCISSORS CUT *Art Garfunkel* [12Sep81|03Oct81|31Oct81] 113(1) 8	1	171
3172. PROCESSION *Weather Report* [19Mar83|09Apr83|21May83] 96(2) 10	0	171
3173. HERE TO STAY *Neal Schon & Jan Hammer* [05Feb83|12Mar83|23Apr83] 122(4) 12	0	170
3174. THE FLYING LIZARDS *The Flying Lizards* [23Feb80|29Mar80|12Apr80] 99(1) 8	1	170
3175. POLICY *Martha Davis* [14Nov87|05Dec87|06Feb88] 127(1) 13	1	170
3176. STREET LIFE-20 GREAT HITS *Bryan Ferry/Roxy Music* [26Aug89|16Sep89|04Nov89] 100(1) 11	0	169
3177. YOU CAN'T STOP ROCK 'N' ROLL *Twisted Sister* [27Aug83|17Nov84|01Dec84] 130(1) 14	0	169
3178. THE LEAGUE OF GENTLEMEN *Robert Fripp* [04Apr81|18Apr81|16May81] 90(2) 7	0	169
3179. SUE SAAD AND THE NEXT *Sue Saad And The Next* [01Mar80|12Apr80|17May80] 131(1) 12	0	169
3180. ROUGH NIGHT IN JERICHO *Dreams So Real* [26Nov88|11Feb89|01Apr89] 150(1) 18	0	169
3181. AN OLD TIME CHRISTMAS *Randy Travis* [02Dec89|23Dec89|13Jan90] 70(1) 7	0	169
3182. THEY DON'T MAKE THEM LIKE THEY USED TO *Kenny Rogers* [13Dec86|07Feb87|21Mar87] 137(1) 15	0	169
3183. ZENO *Zeno* [10May86|07Jun86|12Jul86] 107(2) 10	0	168

Ranking the Albums

Rank. Title *Act* [Enter|Peak|Exit] Peak(Wks) TotWks	Hot 100	Scr
3184. LOVE CONFESSIONS *Miki Howard* [20Feb88|07May88|04Jun88] 145(2) 16	0	168
3185. WIZARD ISLAND *Jeff Lorber Fusion* [31May80|28Jun80|16Aug80] 123(1) 12	0	168
3186. A BLACK & WHITE NIGHT: LIVE (SOUNDTRACK) *Roy Orbison and Friends* [02Dec89|16Dec89|17Feb90] 123(4) 12	0	168
3187. FIGHT TO SURVIVE *White Lion* [16Apr88|23Apr88|16Jul88] 151(1) 14	0	168
3188. SHOCKER *Soundtrack* [18Nov89|02Dec89|03Feb90] 97(2) 12	1,1*	168
3189. QUIET RIOT *Quiet Riot* [19Nov88|24Dec88|28Jan89] 119(2) 11	0	168
3190. LOS HOMBRES MALO *The Outlaws* [01May82|22May82|26Jun82] 77(1) 9	0	168
3191. AFTER DARK *Ray Parker Jr.* [10Oct87|31Oct87|05Dec87] 86(2) 9	1	168
3192. THIN RED LINE *The Cretones* [29Mar80|10May80|31May80] 125(1) 10	1	167
3193. STAGE DOLLS *Stage Dolls* [19Aug89|23Sep89|04Nov89] 118(2) 12	0	167
3194. YO! BUM RUSH THE SHOW *Public Enemy* [23Jan88|27Feb88|09Apr88] 125(1) 12	0	167
3195. WHAT IF *Tommy Shaw* [26Oct85|23Nov85|21Dec85] 87(2) 9	1	167
3196. TURN OF THE SCREW *Dirty Looks* [19Aug89|30Sep89|28Oct89] 118(2) 11	0	167
3197. MOTHER WIT *Betty Wright* [23Apr88|04Jun88|16Jul88] 127(2) 13	0	167
3198. WORKBOOK *Bob Mould* [27May89|08Jul89|26Aug89] 127(1) 14	0	167
3199. NO FUEL LEFT FOR THE PILGRIMS *Disneyland After Dark* [30Sep89|25Nov89|09Dec89] 116(1) 11	0	167
3200. TIMOTHY B. *Timothy B. Schmit* [03Oct87|14Nov87|12Dec87] 106(2) 11	1	167
3201. LENA HORNE: THE LADY AND HER MUSIC (ORIGINAL CAST) *Lena Horne* [26Sep81|17Oct81|21Nov81] 112(2) 9	0	166
3202. GREATEST HITS *KC And The Sunshine Band* [22Mar80|19Apr80|31May80] 132(2) 11	0	166
3203. LOVE LIFE *Brenda Russell* [11Apr81|25Apr81|30May81] 107(2) 8	0	166
3204. ALIVE AND SCREAMIN' *Krokus* [22Nov86|13Dec86|07Feb87] 97(2) 12	0	166
3205. RANK *The Smiths* [01Oct88|08Oct88|19Nov88] 77(2) 8	0	166
3206. III WISHES *Shooting Star* [07Aug82|04Sep82|02Oct82] 82(2) 9	0	165
3207. AMY HOLLAND *Amy Holland* [30Aug80|11Oct80|29Nov80] 146(1) 14	1	165
3208. RODNEY CROWELL *Rodney Crowell* [03Oct81|07Nov81|21Nov81] 105(1) 8	0	165
3209. DREAMS *The Allman Brothers Band* [15Jul89|29Jul89|23Sep89] 103(2) 11	0	165
3210. AEROBIC DANCING *Barbara Ann Auer* [20Jun81|12Dec81|17Apr82] 145(2) 15	0	165
3211. THE DEED IS DONE *Molly Hatchet* [24Nov84|08Dec84|16Feb85] 117(2) 13	1	165
3212. WALL TO WALL *René & Angela* [22Aug81|26Sep81|10Oct81] 100(1) 8	0	164
3213. ROMEO KNIGHT *Boogie Boys* [19Mar88|09Apr88|28May88] 117(2) 11	0	164
3214. THE DOOBIE BROTHERS FAREWELL TOUR *Doobie Brothers* [23Jul83|20Aug83|17Sep83] 79(1) 9	1	164
3215. ME, MYSELF AND I *Cheryl Pepsii Riley* [12Nov88|03Dec88|21Jan89] 128(2) 11	1	164
3216. THE BROOKLYN, BRONX & QUEENS BAND *The Brooklyn, Bronx & Queens Band* [29Aug81|03Oct81|24Oct81] 109(1) 9	0	164
3217. ELECTRIC DREAMS *Soundtrack* [01Sep84|06Oct84|27Oct84] 94(2) 9	1	164
3218. MAGNETIC HEAVEN *Wax* [26Apr86|24May86|05Jul86] 101(2) 11	1	164
3219. SENSE OF PURPOSE *Third World* [13Apr85|11May85|22Jun85] 119(3) 11	0	163
3220. SIMPLICITY *Tim Curry* [29Aug81|26Sep81|17Oct81] 112(2) 8	0	163
3221. BREAKING THE CHAINS *Dokken* [15Oct83|17Dec83|07Jan84] 136(1) 13	0	163
3222. CONTAGIOUS *Bar-Kays* [07Nov87|28Nov87|06Feb88] 110(2) 14	0	163
3223. LIFE *Gladys Knight & The Pips* [23Mar85|11May85|08Jun85] 126(2) 12	0	162
3224. RITA COOLIDGE/GREATEST HITS *Rita Coolidge* [14Feb81|07Mar81|04Apr81] 107(2) 8	0	162
3225. ROACHFORD *Roachford* [20May89|08Jul89|05Aug89] 109(2) 12	1	162
3226. NEW YORK-LONDON-PARIS-MUNICH *M* [22Dec79|19Jan80|09Feb80] 79(2) 8	1,*1	162
3227. A TASTE OF YESTERDAY'S WINE *Merle Haggard/George Jones* [25Sep82|16Oct82|11Dec82] 123(4) 12	0	162
3228. NINE LIVES *Bonnie Raitt* [30Aug86|20Sep86|08Nov86] 115(1) 11	0	162
3229. WEIRD SCIENCE *Soundtrack* [31Aug85|12Oct85|09Nov85] 105(2) 11	0	162
3230. FINDER OF LOST LOVES *Dionne Warwick* [02Mar85|06Apr85|11May85] 106(2) 11	0	161
3231. DECISIONS *The Winans* [26Sep87|31Oct87|05Dec87] 109(1) 11	0	161
3232. YOU ARE WHAT YOU IS *Frank Zappa* [03Oct81|24Oct81|14Nov81] 93(1) 7	0	161
3233. TOUCH *Gladys Knight & The Pips* [05Sep81|10Oct81|24Oct81] 109(1) 8	0	161
3234. MAGNIFICENT MADNESS *John Klemmer* [09Aug80|30Aug80|18Oct80] 146(1) 11	0	160
3235. BETTER DAYS *The Blackbyrds* [17Jan81|07Mar81|28Mar81] 133(1) 11	0	160
3236. THE BEST OF EDDIE RABBITT *Eddie Rabbitt* [24Nov79|19Jan80|09Feb80] 151(1) 12	0	160
3237. REACHING OUT *Menudo* [10Mar84|07Apr84|26May84] 108(2) 12	0	160
3238. NOBODY'S PERFECT *Deep Purple* [23Jul88|13Aug88|17Sep88] 105(2) 9	0	160
3239. UNIVERSAL RHYTHM *Ralph MacDonald* [13Oct84|24Nov84|15Dec84] 108(2) 10	1	160
3240. FRANKS WILD YEARS *Tom Waits* [26Sep87|10Oct87|28Nov87] 115(2) 10	0	159
3241. THE KARATE KID *Soundtrack* [21Jul84|08Sep84|06Oct84] 114(2) 12	1	159
3242. SCROOGED *Soundtrack* [03Dec88|07Jan89|28Jan89] 93(2) 9	1	159
3243. TOO GOOD TO STOP NOW *John Schneider* [17Nov84|08Dec84|02Feb85] 111(2) 12	0	159
3244. MAN IN MOTION *Night Ranger* [22Oct88|19Nov88|10Dec88] 81(2) 8	1	159
3245. C.K. *Chaka Khan* [17Dec88|24Dec88|04Mar89] 125(3) 12	0	159
3246. OUT OF THE SHADOWS *Dave Grusin* [07Aug82|04Sep82|02Oct82] 88(2) 9	0	159
3247. 8 *Madhouse* [21Feb87|21Mar87|02May87] 107(2) 11	0	159
3248. AFTER THE WAR *Gary Moore* [25Mar89|29Apr89|20May89] 114(2) 9	0	158
3249. TAKE IT OFF *Chic* [19Dec81|23Jan82|13Feb82] 124(1) 9	0	158
3250. DON'T CLOSE YOUR EYES *Keith Whitley* [03Jun89|03Jun89|09Sep89] 121(2) 14	0	158
3251. HEADS OR TALES *Saga* [22Oct83|12Nov83|17Dec83] 92(1) 9	1	158
3252. BIG DREAMS IN A SMALL TOWN *Restless Heart* [27Aug88|10Sep88|05Nov88] 114(2) 11	0	158
3253. KEEPIN' THE SUMMER ALIVE *The Beach Boys* [12Apr80|26Apr80|17May80] 75(1) 6	1	157
3254. DROP THE BOMB *Trouble Funk* [08May82|26Jun82|07Aug82] 121(2) 14	0	157
3255. ATOMIC PLAYBOYS *Steve Stevens Atomic Playboys* [02Sep89|23Sep89|18Nov89] 119(1) 12	0	157
3256. JUST BEFORE THE BULLETS FLY *The Gregg Allman Band* [06Aug88|20Aug88|15Oct88] 117(2) 11	0	157
3257. MACALLA *Clannad* [22Mar86|19Apr86|07Jun86] 131(2) 12	0	157
3258. MILES *Miles Jaye* [12Dec87|16Jan88|27Feb88] 125(2) 12	0	157
3259. FLAUNT IT *Sigue Sigue Sputnik* [23Aug86|20Sep86|25Oct86] 96(2) 10	0	156
3260. A SPECIAL PART OF ME *Johnny Mathis* [10Mar84|31Mar84|25Aug84] 157(1) 19	1	156
3261. MARCH OF THE SAINT *Armored Saint* [22Dec84|09Mar85|06Apr85] 138(2) 16	0	156
3262. PERFECT TIMING *Donna Allen* [04Apr87|23May87|27Jun87] 133(2) 13	1	156
3263. SOUTHERN COMFORT *Conway Twitty* [13Feb82|06Mar82|22May82] 144(1) 15	0	156
3264. SYLVAIN SYLVAIN *Sylvain Sylvain* [16Feb80|08Mar80|05Apr80] 123(2) 8	0	156
3265. SOMETHING TO BELIEVE IN *Curtis Mayfield* [26Jul80|23Aug80|27Sep80] 128(2) 10	0	156
3266. SWING TO THE RIGHT *Utopia* [20Mar82|24Apr82|22May82] 102(1) 10	0	156
3267. CHRISTMAS *The Oak Ridge Boys* [04Dec82|08Jan83|15Jan83] 73(2) 7	0	156
3268. SPIDER *Spider* [17May80|05Jul80|19Jul80] 130(1) 10	2	156
3269. HIDING OUT *Soundtrack* [05Dec87|30Jan88|27Feb88] 146(2) 13	2	156
3270. ABOUT TIME *Ten Years After* [16Sep89|28Oct89|18Nov89] 120(2) 10	0	156
3271. LIVE IN LOS ANGELES *Maze Featuring Frankie Beverly* [20Sep86|04Oct86|29Nov86] 92(2) 11	0	155
3272. ABIGAIL *King Diamond* [11Jul87|19Sep87|03Oct87] 123(1) 13	0	155
3273. BILLY PRESTON & SYREETA *Billy Preston & Syreeta* [08Aug81|29Aug81|03Oct81] 127(4) 9	0	155
3274. FREE AS A BIRD *Supertramp* [31Oct87|21Nov87|09Jan88] 101(2) 11	0	155
3275. THE ROSE OF ENGLAND *Nick Lowe And His Cowboy Outfit* [21Sep85|19Oct85|07Dec85] 119(3) 12	1	155
3276. MILSAP MAGIC *Ronnie Milsap* [05Apr80|19Jul80|09Aug80] 137(2) 13	0	155
3277. FROM BRANCH TO BRANCH *Leon Redbone* [11Apr81|23May81|20Jun81] 152(1) 11	1	155
3278. DRIFTER *Sylvia (2)* [09May81|20Jun81|18Jul81] 139(1) 11	0	154
3279. LIVE IT UP *David Johansen* [03Jul82|21Aug82|09Oct82] 148(2) 15	0	154

Ranking the Albums

Rank. Title *Act* [Enter | Peak | Exit] Peak(Wks) TotWks	Hot 100	Scr
3280. WARM AND TENDER *Olivia Newton-John* [02Dec89 | 30Dec89 | 24Feb90] 124(2) 13	0	154
3281. SWEET, DELICIOUS & MARVELOUS *The California Raisins* [08Oct88 | 19Nov88 | 14Jan89] 140(1) 15	0	154
3282. DISTURBING THE PEACE *Alcatrazz* [20Apr85 | 25May85 | 03Aug85] 145(2) 16	0	154
3283. HIGH HAT *Boy George* [25Mar89 | 06May89 | 03Jun89] 126(1) 11	0	154
3284. UNDER THE INFLUENCE *Overkill* [30Jul88 | 10Sep88 | 22Oct88] 142(1) 13	0	154
3285. 8 FOR THE 80'S *Webster Lewis* [15Mar80 | 26Apr80 | 10May80] 114(1) 9	0	154
3286. N.E.W.S. *Golden Earring* [17Mar84 | 14Apr84 | 12May84] 107(1) 9	1	154
3287. SPELLBOUND *Joe Sample* [15Apr89 | 03Jun89 | 15Jul89] 129(1) 14	0	154
3288. A CHRISTMAS ALBUM *Barbra Streisand* [19Dec81 | 09Jan82 | 05Jan91] 108(2) 9	0	154
3289. TRUE DEMOCRACY *Steel Pulse* [17Jul82 | 21Aug82 | 09Oct82] 120(2) 13	0	154
3290. LONDON 0 HULL 4 *The Housemartins* [07Feb87 | 21Mar87 | 09May87] 124(2) 14	0	153
3291. LULU *Lulu* [26Sep81 | 07Nov81 | 28Nov81] 126(1) 10	2	153
3292. NEVER BUY TEXAS FROM A COWBOY *Brides Of Funkenstein* [16Feb80 | 01Mar80 | 29Mar80] 93(2) 7	0	153
3293. ONE TO ONE *Carole King* [03Apr82 | 08May82 | 12Jun82] 119(3) 11	1	153
3294. WILD FRONTIER *Gary Moore* [16May87 | 18Jul87 | 22Aug87] 139(1) 15	0	153
3295. WHISPER TAMES THE LION *Drivin' N' Cryin'* [02Apr88 | 14May88 | 18Jun88] 130(2) 12	0	153
3296. E-S-P *Bee Gees* [17Oct87 | 07Nov87 | 12Dec87] 96(2) 9	1	152
3297. GROWIN' UP TOO FAST *Billy Rankin* [24Mar84 | 05May84 | 02Jun84] 119(1) 11	1	152
3298. PARTING SHOULD BE PAINLESS *Roger Daltrey* [17Mar84 | 14Apr84 | 12May84] 102(2) 9	1	152
3299. ALL IS FORGIVEN *Red Siren* [08Apr89 | 13May89 | 24Jun89] 124(1) 12	0	152
3300. MONDO BONGO *The Boomtown Rats* [21Feb81 | 21Mar81 | 11Apr81] 116(2) 8	0	152
3301. GOLDEN YEARS *David Bowie* [27Aug83 | 17Sep83 | 22Oct83] 99(1) 9	0	152
3302. CARRY ON *Bobby Caldwell* [17Apr82 | 22May82 | 10Jul82] 133(1) 13	1	152
3303. IN FULL EFFECT *Mantronix* [09Apr88 | 30Apr88 | 28May88] 108(2) 8	0	152
3304. FORTUNE 410 *Donnie Iris* [02Jul83 | 06Aug83 | 17Sep83] 127(1) 12	1	151
3305. ARCHITECTURE AND MORALITY *Orchestral Manoeuvres In The Dark* [06Feb82 | 13Mar82 | 24Apr82] 144(4) 12	0	151
3306. ITALIAN X RAYS *The Steve Miller Band* [10Nov84 | 01Dec84 | 12Jan85] 101(2) 10	2	151
3307. NOEL *Noel* [22Oct88 | 26Nov88 | 14Jan89] 126(1) 13	2	151
3308. TEEVEE TOONS - THE COMMERCIALS *Various Artists* [10Jun89 | 12Aug89 | 07Oct89] 159(1) 18	0	150
3309. A WOMAN'S POINT OF VIEW *Shirley Murdock* [23Jul88 | 24Sep88 | 29Oct88] 137(1) 15	0	150
3310. BURLAP & SATIN *Dolly Parton* [04Jun83 | 09Jul83 | 13Aug83] 127(1) 11	0	150
3311. FREE *Concrete Blonde* [13May89 | 12Aug89 | 09Sep89] 148(1) 18	0	150
3312. A BRAZILIAN LOVE AFFAIR *George Duke* [31May80 | 05Jul80 | 26Jul80] 119(1) 9	0	150
3313. SHADES *J.J. Cale* [28Feb81 | 28Mar81 | 11Apr81] 110(1) 7	0	150
3314. SONG OF SEVEN *Jon Anderson* [06Dec80 | 17Jan81 | 14Feb81] 143(1) 11	0	149
3315. THE BITTEREST PILL (I EVER HAD TO SWALLOW) *The Jam* [27Nov82 | 11Dec82 | 26Feb83] 135(1) 14	0	149
3316. DISCOVER *Gene Loves Jezebel* [18Oct86 | 24Jan87 | 21Feb87] 155(2) 19	0	149
3317. LOST IN THE FIFTIES TONIGHT *Ronnie Milsap* [03May86 | 14Jun86 | 19Jul86] 121(2) 12	0	149
3318. TANE CAIN *Tane Cain* [11Sep82 | 16Oct82 | 13Nov82] 121(2) 10	1	149
3319. BORN 2B BLUE *Steve Miller* [08Oct88 | 22Oct88 | 10Dec88] 108(2) 9	0	149
3320. CHUCKII *Chuckii Booker* [22Jul89 | 02Sep89 | 23Sep89] 116(1) 10	1	148
3321. REBA *Reba McEntire* [21May88 | 11Jun88 | 23Jul88] 118(2) 10	0	148
3322. BENEFACTOR *Romeo Void* [04Sep82 | 25Sep82 | 27Nov82] 119(3) 13	0	148
3323. IT'S ONLY ROCK AND ROLL *Waylon Jennings* [30Apr83 | 14May83 | 09Jul83] 109(1) 11	0	148
3324. GEMS *Aerosmith* [10Dec88 | 07Jan89 | 18Feb89] 133(1) 11	0	148
3325. DOES FORT WORTH EVER CROSS YOUR MIND *George Strait* [10Nov84 | 08Dec84 | 23Feb85] 139(2) 16	0	148
3326. MY LIFE FOR A SONG *Placido Domingo* [09Apr83 | 04Jun83 | 18Jun83] 117(1) 11	0	148
3327. EVERYTHING'S KOOL & THE GANG: GREATEST HITS & MORE *Kool & The Gang* [20Aug88 | 03Sep88 | 29Oct88] 109(2) 11	0	148
3328. HORSESHOE IN THE GLOVE *So* [19Mar88 | 02Apr88 | 14May88] 124(3) 9	1	148
3329. LEATHER JACKETS *Elton John* [06Dec86 | 20Dec86 | 31Jan87] 91(3) 9	1	148
3330. ROCK FOR AMNESTY *Various Artists* [24Jan87 | 21Feb87 | 04Apr87] 121(2) 11	0	148
3331. LOVE FEVER *The O'Jays* [19Oct85 | 30Nov85 | 04Jan86] 121(2) 12	0	148
3332. TERMS OF ENDEARMENT *Soundtrack* [21Apr84 | 12May84 | 23Jun84] 111(1) 10	1	147
3333. RED 7 *Red 7* [25May85 | 06Jul85 | 27Jul85] 105(2) 10	0	147
3334. FRANK *Squeeze* [07Oct89 | 28Oct89 | 09Dec89] 113(3) 10	0	147
3335. SHOOTING STAR *Shooting Star* [15Mar80 | 12Apr80 | 06Mar82] 147(1) 14	1	147
3336. LOVELY *The Primitives* [10Sep88 | 08Oct88 | 05Nov88] 106(2) 9	0	147
3337. BIG SCIENCE *Laurie Anderson* [29May82 | 26Jun82 | 14Aug82] 124(1) 12	0	146
3338. CATCH ME, I'M FALLING *Pretty Poison* [30Apr88 | 07May88 | 18Jun88] 104(2) 8	2	146
3339. BOOM BOOM CHI BOOM BOOM *Tom Tom Club* [15Apr89 | 13May89 | 24Jun89] 114(2) 11	0	146
3340. UNTOLD PASSION *Neal Schon & Jan Hammer* [17Oct81 | 21Nov81 | 05Dec81] 115(2) 9	0	145
3341. THIS DAY AND AGE *D.L. Byron* [16Feb80 | 05Apr80 | 19Apr80] 133(1) 10	0	145
3342. 25 YEARS OF GRAMMY GREATS *Various Artists* [11Jun83 | 09Jul83 | 06Aug83] 107(1) 9	0	145
3343. THE IDOLMAKER *Soundtrack* [20Dec80 | 31Jan81 | 14Feb81] 130(1) 9	0	145
3344. MASSTERPIECE *Mass Production* [29Mar80 | 03May80 | 24May80] 133(1) 9	0	145
3345. NEW HOPE FOR THE WRETCHED *The Plasmatics* [21Feb81 | 21Mar81 | 25Apr81] 134(2) 10	0	145
3346. CORRIDORS OF POWER *Gary Moore* [23Apr83 | 04Jun83 | 16Jul83] 149(1) 13	0	144
3347. THE JUDDS *The Judds* [08Dec84 | 09Feb85 | 16Mar85] 153(1) 15	0	144
3348. THE TEMPTATIONS *The Temptations* [29Aug81 | 03Oct81 | 24Oct81] 119(1) 9	1	143
3349. VICTORY DAY *Tom Cochrane & Red Rider* [12Nov88 | 24Dec88 | 04Feb89] 144(2) 13	0	143
3350. BLACK 'N' BLUE *Black 'N Blue* [15Sep84 | 03Nov84 | 24Nov84] 116(2) 11	0	143
3351. THE HUNTER *Joe Sample* [16Apr83 | 14May83 | 16Jul83] 125(1) 14	0	143
3352. ALL OF THE ABOVE *The John Hall Band* [05Dec81 | 26Dec81 | 27Feb82] 158(3) 13	1	143
3353. THE MISSION *Soundtrack* [21Feb87 | 11Apr87 | 16May87] 132(1) 13	0	143
3354. BIG PLANS FOR EVERYBODY *Let's Active* [26Apr86 | 17May86 | 28Jun86] 111(2) 10	0	142
3355. A PLACE FOR MY STUFF! *George Carlin* [19Dec81 | 13Feb82 | 13Mar82] 145(1) 13	0	142
3356. NEVER SAY NEVER *Melba Moore* [24Dec83 | 07Apr84 | 21Apr84] 147(1) 14	0	142
3357. LOVE SEASON *Alex Bugnon* [01Apr89 | 29Apr89 | 10Jun89] 127(3) 11	0	142
3358. TELEVISION'S GREATEST HITS VOLUME II *Various Artists* [15Nov86 | 13Dec86 | 28Feb87] 149(2) 16	0	142
3359. UNDER THE BLADE *Twisted Sister* [06Jul85 | 10Aug85 | 14Sep85] 125(2) 11	0	142
3360. CONDITION OF THE HEART *Kashif* [21Dec85 | 08Feb86 | 22Mar86] 144(2) 14	0	142
3361. ZIGZAGGING THROUGH GHOSTLAND *The Radiators* [01Apr89 | 06May89 | 10Jun89] 122(2) 11	0	142
3362. SUPERMAN II *Soundtrack* [04Jul81 | 08Aug81 | 29Aug81] 133(1) 9	0	142
3363. KEEP MOVING *Madness* [17Mar84 | 14Apr84 | 05May84] 109(1) 8	0	141
3364. NUDE *Dead Or Alive* [22Jul89 | 05Aug89 | 16Sep89] 106(2) 9	1	141
3365. THE LEGEND OF JESSE JAMES *Various Artists* [06Dec80 | 27Dec80 | 28Feb81] 154(4) 13	0	141
3366. REAL LIVE *Bob Dylan* [05Jan85 | 19Jan85 | 02Mar85] 115(2) 9	0	141
3367. ONE STEP BEYOND *Madness* [08Mar80 | 12Apr80 | 03May80] 128(1) 9	0	141
3368. CAVERNA MAGICA (...UNDER THE TREE-IN THE CAVE...) *Andreas Vollenweider* [15Dec84 | 02Mar85 | 23Mar85] 149(1) 15	0	141
3369. INTERNATIONALISTS *The Style Council* [29Jun85 | 03Aug85 | 07Sep85] 123(1) 11	0	140
3370. YOU'RE UNDER ARREST *Miles Davis* [01Jun85 | 29Jun85 | 17Aug85] 111(2) 12	0	140
3371. HEAR 'N AID *Hear 'N Aid* [05Jul86 | 12Jul86 | 16Aug86] 80(2) 7	0	140
3372. FAME AND FORTUNE *Bad Company* [25Oct86 | 08Nov86 | 20Dec86] 106(2) 9	1	140
3373. CAN'T WAIT ALL NIGHT *Juice Newton* [14Jul84 | 01Sep84 | 15Sep84] 128(2) 10	2	140
3374. VOYAGER *Roger Whittaker* [09Feb80 | 29Mar80 | 26Apr80] 154(3) 12	0	140
3375. JEFFERSON AIRPLANE *Jefferson Airplane* [23Sep89 | 30Sep89 | 04Nov89] 85(2) 7	0	140

Ranking the Albums

Rank. Title Act [Enter | Peak | Exit] Peak(Wks) TotWks	Hot 100	Scr
3376. GUESS WHO'S COMIN' TO THE CRIB? *Full Force* [05Dec87 | 19Dec87 | 13Feb88] 126(3) 11	0	140
3377. MYSTERY STREET *John Brannen* [12Mar88 | 16Apr88 | 11Jun88] 156(1) 14	0	140
3378. LAW OF THE FISH *The Radiators* [19Dec87 | 06Feb88 | 02Apr88] 132(2) 16	0	139
3379. CAUGHT IN THE GAME *Survivor* [22Oct83 | 12Nov83 | 17Dec83] 82(1) 9	1	139
3380. MASTER OF DISGUISE *Lizzy Borden* [26Aug89 | 23Sep89 | 28Oct89] 133(1) 10	0	139
3381. I COULD RULE THE WORLD IF I COULD ONLY GET THE PARTS *The Waitresses* [18Dec82 | 22Jan83 | 19Feb83] 128(2) 10	0	139
3382. RUSS BALLARD *Russ Ballard* [09Jun84 | 11Aug84 | 01Sep84] 147(2) 13	0	139
3383. ONE DAY IN YOUR LIFE *Michael Jackson* [25Apr81 | 30May81 | 27Jun81] 144(1) 10	1	139
3384. NON FICTION *The Blasters* [14May83 | 04Jun83 | 02Jul83] 95(1) 8	0	138
3385. SURE SHOT *Crown Heights Affair* [29Mar80 | 31May80 | 14Jun80] 148(2) 12	0	138
3386. FOREVER BY YOUR SIDE *The Manhattans* [06Aug83 | 03Sep83 | 24Sep83] 104(1) 8	1	138
3387. INHERIT THE WIND *Wilton Felder* [08Nov80 | 06Dec80 | 31Jan81] 142(1) 13	0	138
3388. THE COST OF LOVING *The Style Council* [18Apr87 | 02May87 | 20Jun87] 122(2) 10	0	138
3389. FULL FORCE GET BUSY 1 TIME! *Full Force* [30Aug86 | 27Sep86 | 22Nov86] 141(3) 13	0	138
3390. 10 FROM 6 *Bad Company* [18Jan86 | 01Mar86 | 19Apr86] 137(1) 14	0	138
3391. PURE & NATURAL *T-Connection* [20Mar82 | 15May82 | 22May82] 123(1) 10	0	138
3392. PETE TOWNSHEND'S DEEP END LIVE! *Pete Townshend* [25Oct86 | 15Nov86 | 20Dec86] 98(2) 9	0	138
3393. FRIENDS *Larry Carlton* [18Jun83 | 16Jul83 | 27Aug83] 126(1) 11	0	138
3394. HERE I AM *Sharon Bryant* [09Sep89 | 14Oct89 | 02Dec89] 139(1) 13	2	138
3395. LET THE MUSIC PLAY *Dazz Band* [27Jun81 | 12Sep81 | 19Sep81] 154(2) 11	0	138
3396. CATS *Selections from Original Broadway Cast* [26Feb83 | 09Apr83 | 30Jul83] 131(2) 14	0	137
3397. TWENNYNINE WITH LENNY WHITE *Twennynine Featuring Lenny White* [01Nov80 | 15Nov80 | 20Dec80] 106(2) 8	0	137
3398. HEART OVER MIND *Jennifer Rush* [27Jun87 | 01Aug87 | 29Aug87] 118(3) 10	1	137
3399. EVERYTHING IS COOL *T-Connection* [21Mar81 | 02May81 | 09May81] 138(1) 7	0	137
3400. RUSSIAN ROULETTE *Accept* [17May86 | 07Jun86 | 12Jul86] 114(2) 9	0	137
3401. I'M NOT STRANGE I'M JUST LIKE YOU *Keith Sykes* [22Nov80 | 20Dec80 | 31Jan81] 147(3) 11	0	137
3402. HOMOSAPIEN *Pete Shelley* [26Jun82 | 31Jul82 | 28Aug82] 121(1) 10	0	136
3403. THE BEST PART OF THE FAT BOYS *Fat Boys* [03Oct87 | 10Oct87 | 05Dec87] 108(2) 10	0	136
3404. FINALLY! *T.G. Sheppard* [30Jan82 | 06Mar82 | 12Jun82] 152(2) 13	2	136
3405. THE BEST OF RITCHIE VALENS *Ritchie Valens* [29Aug87 | 26Sep87 | 31Oct87] 100(2) 10	0	136
3406. THE RIGHT NIGHT & BARRY WHITE *Barry White* [21Nov87 | 05Dec87 | 12Mar88] 159(2) 17	0	136
3407. KILL 'EM ALL(2) *Metallica* [13Feb88 | 27Feb88 | 02Apr88] 120(2) 8	0	136
3408. ROUND TRIP *The Knack* [07Nov81 | 21Nov81 | 12Dec81] 93(2) 6	1	136
3409. CYPRESS *Let's Active* [10Nov84 | 15Dec84 | 23Feb85] 138(1) 16	0	136
3410. AMERICA *Kurtis Blow* [02Nov85 | 21Dec85 | 08Feb86] 153(3) 15	0	136
3411. KEY LIME PIE *Camper van Beethoven* [07Oct89 | 21Oct89 | 23Dec89] 141(3) 12	0	135
3412. TRACK RECORD *Joan Armatrading* [21Jan84 | 03Mar84 | 24Mar84] 113(1) 10	0	135
3413. FEEL THE HEAT *Henry Paul Band* [02Aug80 | 16Aug80 | 20Sep80] 120(1) 8	0	135
3414. 2XS *Nazareth* [10Jul82 | 21Aug82 | 11Sep82] 122(2) 10	0	135
3415. OUT OF THE SILENT PLANET *King's X* [07May88 | 04Jun88 | 16Jul88] 144(1) 11	0	134
3416. THE BIG EASY *Soundtrack* [24Oct87 | 14Nov87 | 19Dec87] 107(2) 9	0	134
3417. IT'S MY TURN *Soundtrack* [22Nov80 | 13Dec80 | 31Jan81] 137(2) 11	1	134
3418. SCARS OF LOVE *TKA* [30Jan88 | 05Mar88 | 09Apr88] 135(2) 11	0	134
3419. BELIEVERS *Don McLean* [28Nov81 | 09Jan82 | 06Feb82] 156(1) 11	1	134
3420. THE SYMPHONY SESSIONS *David Foster* [20Feb88 | 12Mar88 | 09Apr88] 111(2) 8	1	134
3421. BALANCE *Balance* [29Aug81 | 03Oct81 | 14Nov81] 133(1) 12	2	134
3422. PROVISION *Scritti Politti* [16Jul88 | 30Jul88 | 03Sep88] 113(2) 8	1	133
3423. RITES OF SUMMER *Spyro Gyra* [16Jul88 | 06Aug88 | 03Sep88] 104(2) 8	0	133
3424. SHADES OF BLUE *Lou Rawls* [10Jan81 | 24Jan81 | 14Feb81] 110(2) 6	0	133
3425. MIND BOMB *The The* [22Jul89 | 05Aug89 | 07Oct89] 138(2) 12	0	133
3426. THE RIDDLE *Nik Kershaw* [27Apr85 | 18May85 | 29Jun85] 113(2) 10	0	133
3427. ROADIE *Soundtrack* [21Jun80 | 26Jul80 | 09Aug80] 125(2) 8	2	133
3428. X2 *Times Two* [30Apr88 | 11Jun88 | 09Jul88] 137(1) 11	2	133
3429. MORE THAN YOU KNOW *Toni Tennille* [09Jun84 | 28Jul84 | 18Aug84] 142(1) 11	0	132
3430. MIDNIGHT BLUE *Louise Tucker* [06Aug83 | 17Sep83 | 08Oct83] 127(1) 10	1	132
3431. PEACE & LOVE *The Pogues* [12Aug89 | 02Sep89 | 07Oct89] 118(1) 9	0	131
3432. STEALIN HORSES *Stealin Horses* [25Jun88 | 06Aug88 | 10Sep88] 146(2) 12	0	131
3433. HOT SHOT *Pat Travers* [05May84 | 02Jun84 | 23Jun84] 108(2) 8	0	131
3434. GOLD *Steely Dan* [03Jul82 | 07Aug82 | 28Aug82] 115(2) 9	0	131
3435. LIVE IN THE HEART OF THE CITY *Whitesnake* [27Dec80 | 07Feb81 | 14Mar81] 146(1) 12	0	131
3436. DISORDERLIES *Soundtrack* [12Sep87 | 26Sep87 | 31Oct87] 99(2) 8	0	131
3437. THE BEST OF THE STATLER BROS. RIDES AGAIN, VOL. II *The Statler Brothers* [02Feb80 | 08Mar80 | 12Apr80] 153(2) 11	0	130
3438. IN 'N' OUT *Stone City Band* [22Mar80 | 19Apr80 | 10May80] 122(2) 8	0	130
3439. SHOUT *Devo* [03Nov84 | 17Nov84 | 08Dec84] 83(2) 6	0	130
3440. TAKING IT HOME *Buckwheat Zydeco* [17Sep88 | 08Oct88 | 29Oct88] 104(1) 7	0	130
3441. TAKE WHAT YOU NEED *Robin Trower* [21May88 | 18Jun88 | 23Jul88] 133(2) 10	0	130
3442. DEEP IN THE HEART OF NOWHERE *Bob Geldof* [13Dec86 | 17Jan87 | 28Feb87] 130(1) 12	1	129
3443. VESTA 4 U *Vesta Williams* [02Sep89 | 14Oct89 | 04Nov89] 131(1) 10	1	129
3444. PICTURES FROM THE FRONT *Jon Butcher* [18Feb89 | 04Mar89 | 08Apr89] 121(3) 8	0	129
3445. MY NATION UNDERGROUND *Julian Cope* [10Dec88 | 18Feb89 | 04Mar89] 155(1) 13	0	129
3446. LIGHTING UP THE NIGHT *Jack Wagner* [19Oct85 | 16Nov85 | 25Jan86] 150(2) 15	1	129
3447. BOP DOO-WOP *Manhattan Transfer* [05Jan85 | 23Feb85 | 16Mar85] 127(2) 11	1	129
3448. BLACK CODES (FROM THE UNDERGROUND) *Wynton Marsalis* [19Oct85 | 09Nov85 | 21Dec85] 118(2) 10	0	128
3449. SECRET DREAMS & FORBIDDEN FIRE *Bonnie Tyler* [26Apr86 | 17May86 | 14Jun86] 106(2) 8	1	128
3450. LARSEN-FEITEN BAND *Larsen-Feiten Band* [13Sep80 | 25Oct80 | 15Nov80] 142(1) 10	1	128
3451. ALWAYS IN THE MOOD *Shirley Jones* [23Aug86 | 20Sep86 | 25Oct86] 128(1) 10	0	128
3452. GET AS MUCH LOVE AS YOU CAN *The Jones Girls* [05Dec81 | 06Feb82 | 13Mar82] 155(1) 15	0	128
3453. CHILDREN *The Mission U.K.* [30Apr88 | 07May88 | 02Jul88] 126(2) 10	0	128
3454. THE KINGBEES *The Kingbees* [31May80 | 28Jun80 | 16Aug80] 160(2) 12	1	127
3455. TROUBLEMAKER *Ian McLagan* [19Jan80 | 23Feb80 | 15Mar80] 125(1) 9	0	127
3456. POP ART *Transvision Vamp* [24Sep88 | 22Oct88 | 12Nov88] 115(1) 8	1	127
3457. HUMANESQUE *Jack Green* [18Oct80 | 29Nov80 | 06Dec80] 121(1) 8	0	127
3458. THE PACK IS BACK *Raven* [08Mar86 | 19Apr86 | 10May86] 121(2) 10	0	127
3459. THE MOTOWN STORY: THE FIRST 25 YEARS *Various Artists* [09Jul83 | 20Aug83 | 03Sep83] 114(1) 9	0	127
3460. WAREHOUSE: SONGS AND STORIES *Husker Du* [14Feb87 | 07Mar87 | 18Apr87] 117(2) 10	0	127
3461. BURNS LIKE A STAR *Stone Fury* [24Nov84 | 12Jan85 | 09Feb85] 144(2) 12	0	127
3462. OVER THERE (LIVE AT THE VENUE, LONDON) *The Blasters* [30Oct82 | 13Nov82 | 18Dec82] 117(4) 8	0	126
3463. GREATEST HITS - VOL. II *Eddie Rabbitt* [01Oct83 | 29Oct83 | 10Dec83] 131(3) 11	1	126
3464. REJOICING *Pat Metheny* [12May84 | 09Jun84 | 07Jul84] 116(2) 9	0	126
3465. FEEL THE SHAKE *Jetboy* [12Nov88 | 17Dec88 | 14Jan89] 135(1) 10	0	126
3466. HEADLESS CROSS *Black Sabbath* [13May89 | 10Jun89 | 01Jul89] 115(1) 8	0	126
3467. DANZIG *Danzig* [08Oct88 | 29Oct88 | 03Dec88] 125(2) 9	0	126
3468. CANYON *Paul Winter* [03May86 | 21Jun86 | 12Jul86] 138(1) 11	0	125
3469. GREAT WHITE *Great White* [24Mar84 | 28Apr84 | 09Jun84] 144(1) 12	0	125

Ranking the Albums

Rank. Title *Act* [Enter | Peak | Exit] Peak(Wks) TotWks	Hot 100	Scr
3470. A WINTER'S SOLSTICE II *Various Artists* [10Dec88 | 14Jan89 | 21Jan89] 108(1) 7	0	125
3471. LAMENT *Ultravox* [19May84 | 09Jun84 | 14Jul84] 115(2) 9	0	125
3472. LES MISERABLES *Original Broadway Cast Recording* [20Jun87 | 27Jun87 | 27Feb88] 117(2) 10	0	125
3473. THE CLARKE/DUKE PROJECT II *Stanley Clarke & George Duke* [26Nov83 | 07Jan84 | 28Jan84] 146(1) 10	0	125
3474. SHADAY *Ofra Haza* [21Jan89 | 25Feb89 | 18Mar89] 130(2) 9	0	124
3475. STAR FLEET PROJECT *Brian May And Friends* [19Nov83 | 17Dec83 | 14Jan84] 125(1) 9	0	124
3476. THIS IS SPINAL TAP (SOUNDTRACK) *Spinal Tap* [28Apr84 | 09Jun84 | 30Jun84] 121(2) 10	0	124
3477. ANTHOLOGY *Marvin Gaye* [21Apr84 | 05May84 | 09Jun84] 109(1) 8	0	124
3478. HIGH LAND, HARD RAIN *Aztec Camera* [10Sep83 | 22Oct83 | 12Nov83] 129(1) 10	0	124
3479. WAITING *Fun Boy Three* [30Jul83 | 27Aug83 | 10Sep83] 104(1) 7	0	124
3480. OPUS X *Chilliwack* [27Nov82 | 25Dec82 | 29Jan83] 112(2) 10	1	124
3481. STRIKES TWICE *Larry Carlton* [06Sep80 | 20Sep80 | 25Oct80] 138(1) 8	0	124
3482. CONSPIRACY *King Diamond* [30Sep89 | 14Oct89 | 18Nov89] 111(2) 8	0	124
3483. THE SONGSTRESS *Anita Baker* [29Oct83 | 03Dec83 | 07Jan84] 139(2) 11	0	123
3484. ALL OF THIS AND NOTHING *Psychedelic Furs* [24Sep88 | 15Oct88 | 12Nov88] 102(2) 8	0	123
3485. ROCKER *Elvis Presley* [08Dec84 | 12Jan85 | 02Mar85] 154(1) 13	0	123
3486. BLOW YOUR OWN HORN *Herb Alpert* [24Sep83 | 15Oct83 | 12Nov83] 120(1) 8	2	123
3487. SOLDIER *Iggy Pop* [08Mar80 | 29Mar80 | 19Apr80] 125(1) 7	0	123
3488. IN THE HEAT OF THE NIGHT *Jeff Lorber* [05May84 | 19May84 | 16Jun84] 106(1) 7	0	123
3489. RUN FOR THE ROSES *Jerry Garcia* [20Nov82 | 04Dec82 | 08Jan83] 100(2) 8	0	123
3490. I'LL PROVE IT TO YOU *Gregory Abbott* [04Jun88 | 09Jul88 | 30Jul88] 132(2) 9	0	123
3491. DANNY JOE BROWN AND THE DANNY JOE BROWN BAND *Danny Joe Brown And The Danny Joe Brown Band* [04Jul81 | 25Jul81 | 15Aug81] 120(2) 7	0	123
3492. WHA'PPEN *The English Beat* [27Jun81 | 25Jul81 | 01Aug81] 126(1) 6	0	123
3493. WINNERS *Kleeer* [26Apr80 | 31May80 | 28Jun80] 140(2) 10	0	123
3494. ANY WHICH WAY YOU CAN *Soundtrack* [17Jan81 | 07Mar81 | 14Mar81] 141(1) 9	0	123
3495. EMERGENCY *Melissa Manchester* [03Dec83 | 24Dec83 | 28Jan84] 135(3) 9	1	123
3496. GLASS MOON *Glass Moon* [10May80 | 28Jun80 | 05Jul80] 148(1) 9	0	123
3497. BACK TO THE WORLD *Dennis DeYoung* [29Mar86 | 19Apr86 | 17May86] 108(2) 8	2	123
3498. THE ROAD *The Kinks* [06Feb88 | 20Feb88 | 19Mar88] 110(2) 7	0	123
3499. AND YOU KNOW THAT! *Kirk Whalum* [19Mar88 | 23Apr88 | 21May88] 142(1) 10	0	122
3500. IN HEAT *Black 'N Blue* [23Apr88 | 14May88 | 18Jun88] 133(2) 9	0	122
3501. SHOGUN *TV Soundtrack* [04Oct80 | 25Oct80 | 08Nov80] 115(2) 6	0	122
3502. SUNDAY IN THE PARK WITH GEORGE *Original Broadway Cast Recording* [25Aug84 | 06Oct84 | 03Nov84] 149(3) 11	0	122
3503. MASTERPIECE *The Isley Brothers* [07Dec85 | 11Jan86 | 22Feb86] 140(2) 12	0	121
3504. LOVE WILL FOLLOW *George Howard* [19Apr86 | 17May86 | 28Jun86] 142(2) 11	0	121
3505. THE FIRST OF A MILLION KISSES *Fairground Attraction* [21Jan89 | 11Feb89 | 01Apr89] 137(2) 11	1	121
3506. A DIFFERENT KIND OF BLUES *Itzhak Perlman & Andre Previn* [14Mar81 | 02May81 | 09May81] 149(1) 9	0	121
3507. THROBBING PYTHON OF LOVE *Robin Williams* [02Apr83 | 07May83 | 28May83] 119(1) 9	0	121
3508. #7 *George Strait* [05Jul86 | 26Jul86 | 13Sep86] 126(1) 11	0	121
3509. ...AND THEN HE KISSED ME *Rachel Sweet* [05Sep81 | 19Sep81 | 17Oct81] 124(2) 7	1	121
3510. STEAMIN' HOT *The Reddings* [29May82 | 03Jul82 | 14Aug82] 153(3) 12	1	121
3511. SETTING SONS *The Jam* [16Feb80 | 08Mar80 | 05Apr80] 137(2) 8	0	121
3512. PLAYERS IN THE DARK *Dr. Hook* [03Apr82 | 24Apr82 | 15May82] 118(3) 7	2	120
3513. TRUCE *Jack Bruce & Robin Trower* [30Jan82 | 20Feb82 | 06Mar82] 109(2) 6	0	120
3514. GAP BAND VII *The Gap Band* [01Feb86 | 22Mar86 | 10May86] 159(1) 15	0	120
3515. CUT LOOSE *Paul Rodgers* [26Nov83 | 14Jan84 | 28Jan84] 135(1) 10	0	120
3516. SHRINER'S CONVENTION *Ray Stevens* [15Mar80 | 12Apr80 | 03May80] 132(1) 8	0	120
3517. CHARIOTS OF FIRE *Ernie Watts* [20Feb82 | 03Apr82 | 08May82] 161(2) 12	0	120
3518. CAPTURED *Rockwell* [23Feb85 | 16Mar85 | 20Apr85] 120(2) 9	0	120
3519. IF YOU CAN'T LICK 'EM...LICK 'EM *Ted Nugent* [05Mar88 | 12Mar88 | 16Apr88] 112(2) 7	0	120
3520. FRUIT AT THE BOTTOM *Wendy And Lisa* [08Apr89 | 29Apr89 | 27May89] 119(2) 8	0	120
3521. THE BEST OF EMERSON, LAKE AND PALMER *Emerson, Lake & Palmer* [29Nov80 | 06Dec80 | 10Jan81] 108(2) 7	0	119
3522. I WANT CANDY *Bow Wow Wow* [18Sep82 | 09Oct82 | 13Nov82] 123(1) 9	1	119
3523. CALL ON ME *Evelyn "Champagne" King* [11Oct80 | 25Oct80 | 22Nov80] 124(3) 7	0	119
3524. LINE OF FIRE *The Headpins* [21Jan84 | 03Mar84 | 17Mar84] 114(2) 9	1	119
3525. CARLIN ON CAMPUS *George Carlin* [04Aug84 | 06Oct84 | 13Oct84] 136(1) 11	0	119
3526. NEVER KICK A SLEEPING DOG *Mitch Ryder* [09Jul83 | 27Aug83 | 03Sep83] 120(2) 9	1	119
3527. BUSTIN' LOOSE *Roberta Flack* [27Jun81 | 25Jul81 | 05Sep81] 161(1) 11	0	119
3528. SEDUCTION *James Last Band* [28Jun80 | 02Aug80 | 16Aug80] 148(1) 8	0	119
3529. BEYOND THE VALLEY OF 1984 *The Plasmatics* [06Jun81 | 18Jul81 | 01Aug81] 142(1) 9	0	119
3530. READ MY LIPS *Melba Moore* [27Apr85 | 18May85 | 29Jun85] 130(2) 10	0	118
3531. TUCKERIZED *Marshall Tucker Band* [12Jun82 | 26Jun82 | 24Jul82] 95(2) 7	0	118
3532. BANGING THE WALL *Bar-Kays* [21Sep85 | 19Oct85 | 16Nov85] 115(2) 9	0	118
3533. PASSIONFRUIT *Michael Franks* [29Oct83 | 26Nov83 | 07Jan84] 141(1) 11	0	118
3534. KATHY SMITH'S AEROBIC FITNESS *Kathy Smith* [13Mar82 | 22May82 | 05Jun82] 144(2) 13	0	118
3535. CHANCES ARE *Bob Marley* [31Oct81 | 28Nov81 | 05Dec81] 117(1) 6	0	118
3536. BLAST OFF *Stray Cats* [29Apr89 | 13May89 | 24Jun89] 111(2) 9	0	118
3537. YOUNG MAN RUNNING *Corey Hart* [09Jul88 | 06Aug88 | 27Aug88] 121(2) 8	1	118
3538. ON GOLDEN POND *Soundtrack* [27Feb82 | 17Apr82 | 08May82] 147(1) 11	0	118
3539. RAGTIME *Soundtrack* [23Jan82 | 27Feb82 | 20Mar82] 134(2) 9	0	118
3540. THE PARTY'S OVER *Talk Talk* [18Sep82 | 20Nov82 | 01Jan83] 132(2) 16	1	118
3541. INARTICULATE SPEECH OF THE HEART *Van Morrison* [09Apr83 | 30Apr83 | 28May83] 116(1) 8	0	118
3542. COWBOYS & ENGLISHMEN *Poco* [20Feb82 | 13Mar82 | 10Apr82] 131(2) 8	0	118
3543. TIME EXPOSURE *Stanley Clarke* [28Apr84 | 09Jun84 | 04Aug84] 149(1) 12	0	118
3544. THE STARS WE ARE *Marc Almond* [28Jan89 | 18Feb89 | 08Apr89] 144(2) 11	1	118
3545. BORN TO LAUGH AT TORNADOES *Was (Not Was)* [15Oct83 | 19Nov83 | 10Dec83] 134(2) 9	0	118
3546. OVER THE TOP *Soundtrack* [07Mar87 | 28Mar87 | 25Apr87] 120(2) 8	2	117
3547. GOT ANY GUM? *Joe Walsh* [01Aug87 | 15Aug87 | 19Sep87] 113(2) 8	0	117
3548. SMALLCREEP'S DAY *Mike Rutherford* [05Apr80 | 17May80 | 14Jun80] 163(2) 11	0	117
3549. YOU'RE THE ONE FOR ME *"D" Train* [26Jun82 | 31Jul82 | 21Aug82] 128(2) 9	0	117
3550. SUNDAY MORNING SUITE *Frank Mills* [24Nov79 | 12Jan80 | 19Jan80] 149(1) 9	1,*1	117
3551. REFLECTIONS *George Howard* [18Jun88 | 16Jul88 | 06Aug88] 109(1) 8	0	117
3552. SURVIVE *Nuclear Assault* [13Aug88 | 08Oct88 | 22Oct88] 145(1) 11	0	117
3553. I COMMIT TO LOVE *Howard Hewett* [01Nov86 | 21Mar87 | 13Jun87] 159(2) 16	1	117
3554. HEYDEY *The Church* [21Jun86 | 09Aug86 | 30Aug86] 146(1) 11	0	116
3555. ALL OF THE GOOD ONES ARE TAKEN *Ian Hunter* [06Aug83 | 10Sep83 | 24Sep83] 125(1) 8	0	116
3556. TROUBLE WALKIN' *Ace Frehley* [11Nov89 | 25Nov89 | 06Jan90] 102(2) 9	0	116
3557. ESSAR *Smokey Robinson* [30Jun84 | 04Aug84 | 08Sep84] 141(2) 11	0	116
3558. DETROIT DIESEL *Alvin Lee* [23Aug86 | 27Sep86 | 18Oct86] 124(2) 9	0	116
3559. WORTH THE WAIT *Peaches & Herb* [11Oct80 | 01Nov80 | 15Nov80] 120(1) 6	0	116
3560. IN THE LAND OF SALVATION AND SIN *The Georgia Satellites* [11Nov89 | 25Nov89 | 03Feb90] 130(2) 13	0	116
3561. ISLANDS *Mike Oldfield* [27Feb88 | 19Mar88 | 16Apr88] 138(2) 8	0	116
3562. MORE SPECIALS *The Specials* [08Nov80 | 22Nov80 | 06Dec80] 98(1) 5	0	116
3563. STARS ON LONG PLAY II *Stars On* [31Oct81 | 14Nov81 | 05Dec81] 120(3) 6	1	116
3564. ON SOLID GROUND *Larry Carlton* [10Jun89 | 17Jun89 | 29Jul89] 126(2) 8	0	116
3565. ROMAN HOLLIDAY *Roman Holliday* [03Sep83 | 22Oct83 | 12Nov83] 142(1) 11	2	116

Ranking the Albums

Rank. Title *Act* [Enter | Peak | Exit] Peak(Wks) TotWks	Hot 100	Scr
3566. WISE GUY *Kid Creole & The Coconuts* [03Jul82 | 28Aug82 | 18Sep82] 145(2) 12	0	115
3567. IN LONDON *Al Jarreau* [21Sep85 | 19Oct85 | 16Nov85] 125(1) 9	0	115
3568. WEAPONS OF LOVE *The Truth* [30May87 | 13Jun87 | 18Jul87] 115(2) 8	1	115
3569. TWIN HYPE *Twin Hype* [26Aug89 | 16Sep89 | 04Nov89] 140(2) 11	0	115
3570. BURNIN' LOVE *Con Funk Shun* [19Jul86 | 09Aug86 | 27Sep86] 121(2) 11	0	115
3571. ELECTRIC CAFE *Kraftwerk* [29Nov86 | 13Dec86 | 28Feb87] 156(2) 14	0	115
3572. CIVILIZED MAN *Joe Cocker* [19May84 | 30Jun84 | 14Jul84] 133(1) 9	0	115
3573. CHRONICLE II *Creedence Clearwater Revival* [01Nov86 | 24Jan87 | 14Feb87] 165(2) 16	0	115
3574. MOVING TARGET *Gil Scott-Heron* [02Oct82 | 20Nov82 | 27Nov82] 123(1) 9	0	115
3575. HEAVEN ONLY KNOWS *Teddy Pendergrass* [07Jan84 | 11Feb84 | 03Mar84] 123(1) 9	0	115
3576. TOUCH *Sarah McLachlan* [29Apr89 | 03Jun89 | 15Jul89] 132(1) 12	0	115
3577. ARRIVE WITHOUT TRAVELLING *The Three O'Clock* [25May85 | 22Jun85 | 27Jul85] 125(2) 10	0	114
3578. LEGEND (SOUNDTRACK) *Tangerine Dream* [17May86 | 07Jun86 | 28Jun86] 96(2) 7	0	114
3579. SAWYER BROWN *Sawyer Brown* [23Feb85 | 30Mar85 | 11May85] 140(2) 11	0	114
3580. LISTEN TO THE MESSAGE *Club Nouveau* [18Jun88 | 25Jun88 | 23Jul88] 98(2) 6	0	114
3581. CURIOSITY *Regina* [04Oct86 | 18Oct86 | 22Nov86] 102(2) 8	1	114
3582. TROOP *Troop* [03Sep88 | 17Sep88 | 29Oct88] 133(2) 9	0	114
3583. VOICES IN THE SKY-BEST OF THE MOODY BLUES *The Moody Blues* [23Mar85 | 20Apr85 | 18May85] 132(2) 9	0	114
3584. STREET READY *Leatherwolf* [29Apr89 | 03Jun89 | 17Jun89] 123(1) 8	0	114
3585. KIHNTAGIOUS *Greg Kihn Band* [16Jun84 | 28Jul84 | 11Aug84] 121(1) 9	0	113
3586. IT'S TEE TIME *Sweet Tee* [25Feb89 | 11Mar89 | 20May89] 169(3) 13	0	113
3587. THE CAMERA NEVER LIES *Michael Franks* [01Aug87 | 29Aug87 | 10Oct87] 147(1) 11	0	113
3588. BELIEVE *The Jets* [02Sep89 | 23Sep89 | 14Oct89] 107(1) 7	2	113
3589. ROAD ISLAND *Ambrosia* [29May82 | 12Jun82 | 10Jul82] 115(2) 7	1	113
3590. BREAK OF HEARTS *Katrina & The Waves* [02Sep89 | 23Sep89 | 21Oct89] 122(2) 8	1	112
3591. POWERFUL STUFF *Fabulous Thunderbirds* [06May89 | 06May89 | 17Jun89] 118(3) 7	0	112
3592. MR. HANDS *Herbie Hancock* [29Nov80 | 13Dec80 | 03Jan81] 147(1) 6	0	112
3593. THREE TIMES IN LOVE *Tommy James* [22Mar80 | 19Apr80 | 03May80] 134(1) 7	1	112
3594. LIFE STORIES *Earl Klugh* [30Aug86 | 27Sep86 | 08Nov86] 143(2) 11	0	112
3595. GOD SAVE THE QUEEN/UNDER HEAVY MANNERS *Robert Fripp* [26Apr80 | 17May80 | 31May80] 110(2) 6	0	112
3596. BAD BOY *Robert Gordon* [02Feb80 | 08Mar80 | 29Mar80] 150(1) 9	0	112

Rank. Title *Act* [Enter | Peak | Exit] Peak(Wks) TotWks	Hot 100	Scr
3597. CITY *Roger McGuinn & Chris Hillman Featuring Gene Clark* [16Feb80 | 08Mar80 | 29Mar80] 136(2) 7	0	112
3598. IMAGINOS *Blue Öyster Cult* [20Aug88 | 10Sep88 | 08Oct88] 122(1) 8	0	112
3599. INTO THE NIGHT *Soundtrack* [13Apr85 | 11May85 | 01Jun85] 118(2) 8	0	112
3600. CCCP - LIVE IN MOSCOW *UB40* [29Aug87 | 26Sep87 | 17Oct87] 121(2) 8	0	112
3601. CONFLICTING EMOTIONS *Split Enz* [21Jul84 | 18Aug84 | 22Sep84] 137(1) 10	0	111
3602. ROCKIN' RADIO *Tom Browne* [03Dec83 | 07Jan84 | 18Feb84] 147(1) 12	0	111
3603. DOMINO THEORY *Weather Report* [24Mar84 | 05May84 | 12May84] 136(1) 8	0	111
3604. O.F.R. *Nitro* [12Aug89 | 09Sep89 | 07Oct89] 140(1) 9	0	111
3605. NURDS *The Roches* [22Nov80 | 13Dec80 | 03Jan81] 130(1) 7	0	111
3606. NEVER FELT SO GOOD *James Ingram* [13Sep86 | 27Sep86 | 08Nov86] 123(2) 9	0	110
3607. THE DOCTOR *Cheap Trick* [18Oct86 | 08Nov86 | 13Dec86] 115(2) 9	0	110
3608. WHEN A GUITAR PLAYS THE BLUES *Roy Buchanan* [03Aug85 | 21Sep85 | 26Oct85] 161(3) 13	0	110
3609. LABOR OF LOVE *Spinners* [04Apr81 | 11Apr81 | 09May81] 128(2) 6	1	110
3610. IN A SENTIMENTAL MOOD *Dr. John* [27May89 | 08Jul89 | 05Aug89] 142(2) 11	0	110
3611. QUEEN ELVIS *Robyn Hitchcock And The Egyptians* [01Apr89 | 06May89 | 27May89] 139(1) 9	0	110
3612. GREAT GONZOS! THE BEST OF TED NUGENT *Ted Nugent* [28Nov81 | 19Dec81 | 16Jan82] 140(1) 8	0	110
3613. THE MONROES *The Monroes* [19Jun82 | 17Jul82 | 14Aug82] 109(2) 9	1	110
3614. ON TARGET *Fastway* [22Apr89 | 13May89 | 24Jun89] 135(2) 10	0	110
3615. PORKY'S REVENGE *Soundtrack* [13Apr85 | 04May85 | 01Jun85] 122(2) 8	1	110
3616. THE ELVIS MEDLEY *Elvis Presley* [27Nov82 | 11Dec82 | 22Jan83] 133(2) 9	0	110
3617. TONGUE TWISTER *Shoes* [07Feb81 | 28Feb81 | 21Mar81] 140(3) 7	0	109
3618. STAY TUNED *Chet Atkins* [27Apr85 | 08Jun85 | 20Jul85] 145(2) 13	0	109
3619. THIRD GENERATION *Hiroshima* [20Aug83 | 08Oct83 | 15Oct83] 142(1) 9	0	109
3620. OLD ENOUGH *Lou Ann Barton* [24Apr82 | 12Jun82 | 19Jun82] 133(1) 9	0	109
3621. HOME OF THE BRAVE (SOUNDTRACK) *Laurie Anderson* [26Apr86 | 17May86 | 12Jul86] 145(2) 12	0	109
3622. THE SLIDE AREA *Ry Cooder* [12Jun82 | 26Jun82 | 24Jul82] 105(1) 7	0	109
3623. KEEP ON IT *Starpoint* [09May81 | 23May81 | 27Jun81] 138(2) 8	0	109
3624. TOOTSIE *Soundtrack* [26Feb83 | 26Mar83 | 14May83] 144(2) 12	1	109
3625. TOO LONG IN THE WASTELAND *James McMurtry* [14Oct89 | 18Nov89 | 09Dec89] 125(1) 9	0	109
3626. DIFFERENT STYLE! *Musical Youth* [17Dec83 | 18Feb84 | 03Mar84] 144(1) 12	1	109
3627. DEF, DUMB & BLONDE *Deborah Harry* [14Oct89 | 04Nov89 | 02Dec89] 123(2) 8	0	109
3628. WILL AND THE KILL *Will And The Kill* [09Apr88 | 07May88 | 28May88] 129(2) 8	0	109
3629. S.P.Y.S. *Spys* [14Aug82 | 18Sep82 | 16Oct82] 138(3) 10	1	109

Rank. Title *Act* [Enter | Peak | Exit] Peak(Wks) TotWks	Hot 100	Scr
3630. BET YOUR HEART ON ME *Johnny Lee* [24Oct81 | 21Nov81 | 12Dec81] 147(1) 8	1	109
3631. FLYING HOME *Stanley Jordan* [15Oct88 | 29Oct88 | 10Dec88] 131(2) 9	0	108
3632. DAVE GRUSIN AND THE GRP ALL-STARS/LIVE IN JAPAN *Dave Grusin and the GRP All-Stars* [18Jul81 | 15Aug81 | 29Aug81] 140(2) 7	0	108
3633. MY BEST *Kitaro* [10May86 | 24May86 | 12Jul86] 141(2) 10	0	108
3634. THE FUNK IS ON *Instant Funk* [18Oct80 | 08Nov80 | 22Nov80] 130(2) 6	0	108
3635. LOVE OR PHYSICAL *Ashford & Simpson* [18Mar89 | 22Apr89 | 06May89] 135(2) 8	0	108
3636. TWINS *Soundtrack* [21Jan89 | 04Feb89 | 08Apr89] 162(3) 12	0	108
3637. MAXIMUM SECURITY *Tony MacAlpine* [04Jul87 | 22Aug87 | 12Sep87] 146(2) 11	0	108
3638. FROLIC THROUGH THE PARK *Death Angel* [06Aug88 | 01Oct88 | 15Oct88] 143(1) 11	0	107
3639. GREATEST HITS OF THE OUTLAWS/HIGH TIDES FOREVER *The Outlaws* [27Nov82 | 15Jan83 | 22Jan83] 136(2) 9	0	107
3640. GO BANG! *Shriekback* [23Jul88 | 06Aug88 | 08Oct88] 169(2) 12	0	107
3641. EAT THE HEAT *Accept* [24Jun89 | 22Jul89 | 19Aug89] 139(2) 9	0	107
3642. URBAN COWBOY II *Soundtrack* [10Jan81 | 31Jan81 | 14Feb81] 134(1) 6	0	107
3643. SHE WAS ONLY A GROCER'S DAUGHTER *The Blow Monkeys* [25Apr87 | 23May87 | 13Jun87] 134(2) 9	0	107
3644. RICH AND POOR *Randy Crawford* [18Nov89 | 23Dec89 | 10Feb90] 159(3) 13	0	107
3645. NOW PLAYING *Bernadette Peters* [03Oct81 | 17Oct81 | 28Nov81] 151(1) 9	1	107
3646. MEGA FORCE *707* [03Jul82 | 31Jul82 | 28Aug82] 129(2) 9	1	107
3647. ALL RIGHT NOW *Pepsi & Shirlie* [27Feb88 | 26Mar88 | 23Apr88] 133(1) 9	2	107
3648. SOONER OR LATER *Larry Graham* [26Jun82 | 31Jul82 | 21Aug82] 142(1) 9	0	107
3649. KICK THE WALL *Jimmy Davis & Junction* [31Oct87 | 28Nov87 | 19Dec87] 122(2) 8	1	107
3650. THE LAST EMPEROR *Soundtrack* [27Feb88 | 14May88 | 21May88] 152(1) 10	0	107
3651. RESULTS *Liza Minnelli* [11Nov89 | 25Nov89 | 13Jan90] 128(1) 10	0	106
3652. LIVE SENTENCE *Alcatrazz* [09Jun84 | 07Jul84 | 11Aug84] 133(1) 10	0	106
3653. MIRAGE A TROIS *Yellowjackets* [28May83 | 25Jun83 | 30Jul83] 145(1) 10	0	106
3654. U.S. 1 *Head East* [08Nov80 | 06Dec80 | 13Dec80] 137(1) 6	0	106
3655. KEEPIN' LOVE NEW *Howard Johnson* [11Sep82 | 09Oct82 | 06Nov82] 122(1) 9	0	106
3656. RENEGADE *Thin Lizzy* [20Feb82 | 27Mar82 | 01May82] 157(1) 11	0	106
3657. THUNDERSTEEL *Riot* [14May88 | 28May88 | 16Jul88] 150(2) 10	0	106
3658. BLUE JEANS *Chocolate Milk* [12Dec81 | 09Jan82 | 13Feb82] 162(1) 10	0	106
3659. LET ME KNOW YOU *Stanley Clarke* [21Aug82 | 11Sep82 | 09Oct82] 114(2) 8	0	105
3660. LOOKING FOR JACK *Colin James Hay* [21Feb87 | 21Mar87 | 18Apr87] 126(2) 9	1	105
3661. PETER CETERA *Peter Cetera* [23Jan82 | 13Mar82 | 27Mar82] 143(2) 10	0	105
3662. THE AWAKENING *The Reddings* [20Dec80 | 27Dec80 | 07Mar81] 174(3) 12	1	105

Ranking the Albums

Rank. Title *Act* [Enter|Peak|Exit] Peak(Wks) TotWks	Hot 100	Scr
3663. AFOOT *Let's Active* [18Feb84|21Apr84|28Apr84] 154(1) 11	0	105
3664. A LETTER FROM ST. PAUL *The Silencers* [22Aug87|03Oct87|31Oct87] 147(2) 11	1	105
3665. COBRA *Soundtrack* [28Jun86|12Jul86|02Aug86] 100(2) 6	0	105
3666. IN HARMONY 2 *Various Artists* [21Nov81|05Dec81|23Jan82] 129(1) 10	0	105
3667. SALSA *Soundtrack* [25Jun88|02Jul88|30Jul88] 112(2) 6	0	105
3668. GLAMOUR *Dave Davies* [18Jul81|29Aug81|05Sep81] 152(1) 8	0	104
3669. WINDSONG *Randy Crawford* [26Jun82|07Aug82|28Aug82] 148(1) 10	0	104
3670. ANGEL EYES *Willie Nelson* [16Jun84|07Jul84|28Jul84] 116(2) 7	0	104
3671. CANDY APPLE GREY *Husker Du* [12Apr86|03May86|14Jun86] 140(2) 10	0	104
3672. CA$HFLOW *Ca$hflow* [03May86|14Jun86|16Aug86] 144(2) 11	0	104
3673. FORCE MAJEURE *Doro* [29Apr89|13May89|08Jul89] 154(2) 11	0	104
3674. BEAT STREET II *Soundtrack* [29Sep84|20Oct84|24Nov84] 137(2) 9	0	104
3675. I WONDER DO YOU THINK OF ME *Keith Whitley* [02Sep89|09Sep89|14Oct89] 115(2) 7	0	104
3676. PAST MASTERS - VOLUME 2 *The Beatles* [02Apr88|16Apr88|14May88] 121(2) 7	0	104
3677. NINE TO THE UNIVERSE *Jimi Hendrix* [26Apr80|24May80|07Jun80] 127(1) 7	0	104
3678. A.C. *Andre Cymone* [21Sep85|19Oct85|09Nov85] 121(2) 8	0	104
3679. POINT OF PLEASURE *Xavier* [24Apr82|22May82|05Jun82] 129(1) 7	0	104
3680. PERFECT COMBINATION *Stacy Lattisaw & Johnny Gill* [31Mar84|05May84|19May84] 139(1) 8	1	104
3681. LIVE! FOR LIFE *Various Artists* [07Jun86|28Jun86|05Jul86] 105(1) 7	0	104
3682. MUSIC FROM THE BILL COSBY SHOW-- A HOUSE FULL OF LOVE *TV Soundtrack* [01Mar86|05Apr86|12Apr86] 125(1) 7	0	103
3683. EVERYBODY LOVES THE PILOT (EXCEPT THE CREW) *Jon Astley* [01Aug87|22Aug87|03Oct87] 135(2) 10	1	103
3684. SELL MY SOUL *Sylvester* [27Sep80|11Oct80|15Nov80] 147(3) 8	0	103
3685. HERE'S THE WORLD FOR YA *Paul Hyde And The Payolas* [08Jun85|20Jul85|10Aug85] 144(2) 10	1	103
3686. REACT *The Fixx* [18Jul87|01Aug87|29Aug87] 110(2) 7	0	103
3687. PERFECT SYMMETRY *Fates Warning* [16Sep89|14Oct89|11Nov89] 141(1) 9	0	103
3688. TOO *The S.O.S. Band* [22Aug81|05Sep81|26Sep81] 117(2) 6	0	103
3689. POPPED IN SOULED OUT *Wet Wet Wet* [16Jul88|30Jul88|27Aug88] 123(3) 7	1	103
3690. THE TALE OF THE TAPE *Billy Squier* [07Jun80|05Jul80|22Aug81] 169(1) 12	0	103
3691. JULIAN COPE *Julian Cope* [21Feb87|14Mar87|28Mar87] 109(2) 6	0	102
3692. POSITIVE POWER *Steve Arrington's Hall Of Fame* [25Feb84|07Apr84|21Apr84] 141(2) 9	0	102
3693. VIENNA *Ultravox* [13Sep80|18Oct80|08Nov80] 164(1) 9	0	102
3694. RESCUE YOU *Joe Lynn Turner* [02Nov85|30Nov85|18Jan86] 143(2) 12	0	102
3695. DOUBLE TROUBLE LIVE *Molly Hatchet* [07Dec85|28Dec85|01Feb86] 130(3) 9	0	102
3696. A CHILD'S ADVENTURE *Marianne Faithfull* [26Mar83|09Apr83|07May83] 107(2) 7	0	102
3697. NEVER GONNA BE ANOTHER ONE *Thelma Houston* [30May81|27Jun81|04Jul81] 144(1) 6	0	102
3698. CHIPMUNK ROCK *The Chipmunks* [05Jun82|26Jun82|10Jul82] 109(1) 6	0	102
3699. 2400 FULTON ST. *Jefferson Airplane* [18Apr87|16May87|13Jun87] 138(2) 9	0	101
3700. THE FLAMINGO KID *Soundtrack* [16Feb85|23Mar85|06Apr85] 130(2) 8	0	101
3701. CLASS *The Reddings* [01Aug81|08Aug81|29Aug81] 106(2) 5	0	101
3702. THE KIDS FROM "FAME" *The Kids From Fame* [03Apr82|08May82|22May82] 146(2) 8	0	101
3703. DARKROOM *Angel City* [08Nov80|06Dec80|13Dec80] 133(1) 6	0	101
3704. SILK *Fuse One* [13Feb82|13Mar82|03Apr82] 139(1) 8	0	101
3705. EARTH & SKY *Graham Nash* [08Mar80|29Mar80|05Apr80] 117(1) 5	0	101
3706. ERIC CARMEN (II) *Eric Carmen* [09Feb85|02Mar85|13Apr85] 128(2) 10	2	101
3707. JANE FONDA'S WORKOUT RECORD FOR PREGNANCY, BIRTH AND RECOVERY *Jane Fonda* [21May83|18Jun83|02Jul83] 115(1) 7	0	101
3708. STANDING ON THE EDGE *Frankie Miller* [26Jun82|17Jul82|21Aug82] 135(2) 9	1	101
3709. THE BIG HEAT *Stan Ridgway* [12Apr86|17May86|07Jun86] 131(2) 9	0	101
3710. GONE TROPPO *George Harrison* [27Nov82|11Dec82|08Jan83] 108(2) 7	1	100
3711. LIVE NUDE GUITARS *Brian Setzer* [28May81|11Jun88|16Jul88] 140(2) 8	0	100
3712. SCHEMER-DREAMER *Steve Walsh* [16Feb80|15Mar80|22Mar80] 124(1) 6	0	100
3713. LIVE *Reba McEntire* [14Oct89|28Oct89|02Dec89] 124(3) 8	0	100
3714. IN THE SPIRIT OF THINGS *Kansas* [05Nov88|19Nov88|10Dec88] 114(2) 6	0	100
3715. BRONCO BILLY *Soundtrack* [05Jul80|19Jul80|09Aug80] 123(1) 6	0	99.9
3716. CONVERTIBLE MUSIC *Josie Cotton* [07Aug82|04Sep82|23Oct82] 147(2) 12	1	99.9
3717. THIS ISLAND *Eurogliders* [22Dec84|16Feb85|02Mar85] 140(2) 11	1	99.9
3718. OXO *Oxo* [30Apr83|28May83|11Jun83] 117(1) 7	1	99.8
3719. FUNLAND *Bram Tchaikovsky* [23May81|04Jul81|11Jul81] 158(1) 8	0	99.8
3720. IT'S BEGINNING TO AND BACK AGAIN *Wire* [08Jul89|12Aug89|09Sep89] 135(2) 10	0	99.8
3721. EARTH CRISIS *Steel Pulse* [31Mar84|12May84|16Jun84] 154(1) 12	0	99.7
3722. I WAS THE ONE *Elvis Presley* [21May83|11Jun83|25Jun83] 103(1) 6	0	99.2
3723. SWEAT BAND *Sweat Band* [13Dec80|10Jan81|31Jan81] 150(2) 8	0	99.1
3724. UK JIVE *The Kinks* [25Nov89|02Dec89|13Jan90] 122(2) 8	0	99.0
3725. RAIL *Rail* [25Aug84|13Oct84|27Oct84] 143(1) 10	0	98.9
3726. THE TOP TEN HITS *Elvis Presley* [15Aug87|12Sep87|03Oct87] 117(2) 8	0	98.8
3727. THE DREAMING *Kate Bush* [13Nov82|20Nov82|22Jan83] 157(2) 11	0	98.8
3728. THE ELECTRIC SPANKING OF WAR BABIES *Funkadelic* [29Aug81|12Sep81|19Sep81] 105(2) 4	0	98.7
3729. SPECIAL PAIN *Robert Ellis Orrall* [16Apr83|21May83|11Jun83] 146(2) 9	1	98.5
3730. ALONE/BUT NEVER ALONE *Larry Carlton* [28Jun86|19Jul86|06Sep86] 141(2) 11	0	98.4
3731. ...MILLIONS LIKE US *Millions Like Us* [19Dec87|06Feb88|05Mar88] 171(1) 12	1.0	98.4
3732. THE WORLD ACCORDING TO ME *Jackie Mason* [09Jan88|20Feb88|05Mar88] 146(1) 9	0	98.4
3733. ANTHOLOGY *The Babys* [07Nov81|28Nov81|19Dec81] 138(1) 7	0	98.3
3734. STARING AT THE SUN *Level 42* [29Oct88|19Nov88|10Dec88] 128(1) 7	0	98.2
3735. NEW DIRECTIONS *Tavares* [11Dec82|05Feb83|19Feb83] 137(1) 11	1	98.2
3736. THE STORY OF THE CLASH, VOLUME I *The Clash* [28May88|25Jun88|16Jul88] 142(2) 8	0	98.0
3737. UNLIMITED TOUCH *Unlimited Touch* [20Jun81|25Jul81|01Aug81] 142(1) 7	0	97.8
3738. SPECIAL FORCES *Alice Cooper* [19Sep81|03Oct81|17Oct81] 125(2) 5	0	97.8
3739. MAGIC *Four Tops* [29Jun85|10Aug85|24Aug85] 140(2) 9	0	97.8
3740. TOUCH ME *The Temptations* [25Jan86|15Feb86|29Mar86] 146(2) 10	0	97.7
3741. WILL POWER *Joe Jackson* [02May87|23May87|20Jun87] 131(2) 8	0	97.6
3742. ON THE NILE *The Egyptian Lover* [09Feb85|02Mar85|13Apr85] 146(2) 10	0	97.6
3743. THE PSYCHEDELIC FURS *Psychedelic Furs* [22Nov80|20Dec80|03Jan81] 140(1) 7	0	97.5
3744. SURVIVAL OF THE FRESHEST *Boogie Boys* [09Aug86|23Aug86|04Oct86] 124(2) 9	0	97.4
3745. THE SMITHS *The Smiths* [05May84|09Jun84|14Jul84] 150(2) 11	0	97.0
3746. BLAME IT ON LOVE AND ALL THE GREAT HITS *Smokey Robinson* [03Sep83|01Oct83|15Oct83] 124(1) 7	1	97.0
3747. DAVID GRISMAN - QUINTET "80" *David Grisman* [13Sep80|25Oct80|01Nov80] 152(1) 8	0	96.9
3748. HEAD FIRST *Uriah Heep* [04Jun83|09Jul83|06Aug83] 159(1) 10	0	96.9
3749. MENACE TO SOCIETY *Lizzy Borden* [01Nov86|29Nov86|03Jan87] 144(2) 10	0	96.9
3750. SOMETHING INSIDE SO STRONG *Kenny Rogers* [27May89|17Jun89|15Jul89] 141(2) 8	0	96.9
3751. TANTILLA *House Of Freaks* [06May89|03Jun89|08Jul89] 154(1) 10	0	96.8
3752. THE WARMER SIDE OF COOL *Wang Chung* [10Jun89|15Jul89|15Jul89] 123(1) 6	1	96.7
3753. THE PURSUIT OF HAPPINESS *The Beat Farmers* [05Sep87|17Oct87|24Oct87] 131(1) 8	0	96.7
3754. CENTRAL LINE *Central Line* [09Jan82|30Jan82|06Mar82] 145(2) 9	1	96.6
3755. PERFORMANCE *Ashford & Simpson* [17Oct81|07Nov81|21Nov81] 125(1) 6	0	96.5
3756. BRAIN DRAIN *The Ramones* [17Jun89|08Jul89|22Jul89] 122(2) 6	0	96.2

Ranking the Albums

Rank. Title Act [Enter|Peak|Exit] Peak(Wks) TotWks	Hot 100	Scr
3757. JANE FONDA'S WORKOUT RECORD NEW AND IMPROVED *Jane Fonda* [18Aug84|01Sep84|20Oct84] 135(2) 10	0	96.1
3758. TASTE THE MUSIC *Kleeer* [20Feb82|20Mar82|10Apr82] 139(1) 8	0	96.1
3759. PORCUPINE *Echo & The Bunnymen* [26Mar83|30Apr83|21May83] 137(1) 9	0	96.1
3760. GREATEST HITS VOLUME ONE *Elvis Presley* [19Dec81|09Jan82|30Jan82] 142(2) 7	0	96.1
3761. DISTANT LOVER *Alphonse Mouzon* [04Dec82|15Jan83|12Feb83] 146(3) 11	0	96.0
3762. THE ONE GIVETH, THE COUNT TAKETH AWAY *William "Bootsy" Collins* [29May82|26Jun82|17Jul82] 120(1) 8	0	96.0
3763. WORKIN' OVERTIME *Diana Ross* [24Jun89|01Jul89|29Jul89] 116(2) 6	0	96.0
3764. SUBSTANCE *Joy Division* [27Aug88|17Sep88|15Oct88] 146(2) 8	0	95.9
3765. WASN'T THAT A PARTY *The Rovers* [25Apr81|06Jun81|13Jun81] 157(1) 8	1	95.9
3766. KING SWAMP *King Swamp* [03Jun89|08Jul89|02Sep89] 159(1) 14	0	95.8
3767. LOVIN' THE NIGHT AWAY *The Dillman Band* [16May81|20Jun81|27Jun81] 145(2) 7	1	95.6
3768. BROADWAY'S CLOSER TO SUNSET BLVD. *Isley Jasper Isley* [09Feb85|23Mar85|13Apr85] 135(2) 10	1	95.5
3769. HAIRSPRAY *Soundtrack* [02Apr88|23Apr88|07May88] 114(1) 6	0	95.5
3770. 12 *Bob James* [27Oct84|24Nov84|29Dec84] 136(2) 10	0	95.4
3771. MONTY PYTHON'S CONTRACTUAL OBLIGATION ALBUM *Monty Python* [15Nov80|20Dec80|10Jan81] 164(1) 9	0	95.3
3772. LIGHTHOUSE *Kim Carnes* [14Jun86|05Jul86|26Jul86] 116(2) 7	1	94.9
3773. RUNNING FOR MY LIFE *Judy Collins* [03May80|17May80|07Jun80] 142(3) 6	0	94.9
3774. FACE TO FACE *GQ* [14Nov81|05Dec81|02Jan82] 140(2) 8	1	94.6
3775. ELECTRIC BREAKDANCE *Various Artists* [08Sep84|20Oct84|03Nov84] 147(2) 9	0	94.3
3776. JUDSON SPENCE *Judson Spence* [10Dec88|24Dec88|04Mar89] 168(4) 13	1	94.3
3777. ATTITUDES *Brass Construction* [22May82|29May82|10Jul82] 114(3) 8	0	94.2
3778. FRESH AIRE VI *Mannheim Steamroller* [13Dec86|10Jan87|14Mar87] 155(1) 14	0	94.1
3779. STORM WINDOWS *John Prine* [30Aug80|04Oct80|11Oct80] 144(1) 7	0	94.1
3780. "WEIRD AL" YANKOVIC *"Weird Al" Yankovic* [21May83|25Jun83|09Jul83] 139(1) 8	1	93.9
3781. GOLDEN DOWN *Willie Nile* [02May81|30May81|20Jun81] 158(1) 8	0	93.8
3782. SOUL MAN *Soundtrack* [15Nov86|13Dec86|10Jan87] 138(2) 9	0	93.5
3783. MIRRORS OF MY MIND *Roger Whittaker* [08Dec79|26Jan80|09Feb80] 157(2) 10	0	93.5
3784. BREAKING CURFEW *Red Rider* [23Jun84|28Jul84|11Aug84] 137(2) 8	1	93.3
3785. WILD ROMANCE *Herb Alpert* [24Aug85|14Sep85|26Oct85] 151(2) 10	0	93.3
3786. SOLD OUT *The Fools* [05Apr80|26Apr80|24May80] 151(1) 8	1	93.2
3787. STATE OF...EMERGENCY *Steel Pulse* [23Jul88|13Aug88|03Sep88] 127(2) 7	0	93.2
3788. PERSONAL ATTENTION *Stacy Lattisaw* [05Mar88|26Mar88|07May88] 153(2) 10	0	93.2
3789. BUT WHAT WILL THE NEIGHBORS THINK *Rodney Crowell* [26Apr80|24May80|12Jul80] 155(1) 10	1	93.2
3790. COOKIN' ON THE ROOF *Roman Holliday* [22Oct83|12Nov83|26Nov83] 116(1) 6	0	93.1
3791. ROCKAPELLA *The Nylons* [10Jun89|24Jun89|12Aug89] 136(2) 10	0	93.0
3792. THE TWO OF US *Ramsey Lewis & Nancy Wilson* [08Sep84|27Oct84|03Nov84] 144(1) 9	0	92.9
3793. GRASSHOPPER *J.J. Cale* [03Apr82|08May82|22May82] 149(1) 8	0	92.9
3794. IN NO SENSE? NONSENSE! *The Art Of Noise* [17Oct87|31Oct87|12Dec87] 134(1) 9	0	92.8
3795. IRON AGE *Mother's Finest* [23May81|11Jul81|11Jul81] 168(1) 8	0	92.8
3796. GROSS MISCONDUCT *M.O.D.* [11Mar89|08Apr89|29Apr89] 151(1) 8	0	92.7
3797. WHO'S MISSING *The Who* [28Dec85|01Feb86|01Feb86] 116(2) 8	0	92.6
3798. ORGASMATRON *Motorhead* [29Nov86|20Dec86|07Feb87] 157(3) 11	0	92.4
3799. BAD INFLUENCE *The Robert Cray Band* [07Mar87|11Apr87|23May87] 143(1) 11	0	92.4
3800. FRIENDS IN LOVE *Johnny Mathis* [08May82|19Jun82|03Jul82] 147(1) 9	1	92.2
3801. THE BAD C.C. *Carl Carlton* [23Oct82|20Nov82|04Dec82] 133(2) 7	0	92.1
3802. MAZARATI *Mazarati* [19Apr86|17May86|07Jun86] 133(1) 8	0	92.0
3803. CLASSICS *The Doors* [08Jun85|29Jun85|20Jul85] 124(2) 7	0	91.9
3804. LEFTY *Art Garfunkel* [16Apr88|30Apr88|04Jun88] 134(2) 8	0	91.8
3805. LOVE IS...ONE WAY *One Way* [07Mar81|28Mar81|16May81] 157(1) 8	0	91.7
3806. POWER & THE GLORY *Saxon* [18Jun83|13Aug83|20Aug83] 155(1) 10	0	91.7
3807. THE SECRET OF MY SUCCESS *Soundtrack* [13Jun87|04Jul87|01Aug87] 131(2) 8	0	91.6
3808. LIVE AT THE HOLLYWOOD BOWL *Doors* [11Jul87|01Aug87|19Sep87] 154(2) 11	0	91.5
3809. CHARLIE *Charlie* [23Jul83|03Sep83|17Sep83] 145(1) 9	1	91.4
3810. CLOSE *Kim Wilde* [01Oct88|22Oct88|05Nov88] 114(1) 6	1	91.2
3811. STATES OF EMERGENCY *Taxxi* [25Dec82|05Mar83|05Mar83] 161(1) 11	0	91.1
3812. INNOCENCE IS NO EXCUSE *Saxon* [02Nov85|07Dec85|21Dec85] 130(2) 8	0	91.0
3813. THE NUMBER ONE HITS *Elvis Presley* [08Aug87|05Sep87|03Oct87] 143(1) 9	0	90.9
3814. FRAGILE *Cherrelle* [08Sep84|20Oct84|27Oct84] 144(1) 8	1	90.8
3815. LOVE WARRIORS *Tuck & Patti* [24Jun89|15Jul89|02Sep89] 162(3) 11	0	90.5
3816. WORLD OF WONDERS *Bruce Cockburn* [26Jul86|23Aug86|13Sep86] 143(3) 8	0	90.3
3817. THEATER OF THE MIND *Mtume* [05Jul86|26Jul86|23Aug86] 135(2) 8	0	90.2
3818. BRASS CONSTRUCTION 6 *Brass Construction* [20Sep80|04Oct80|18Oct80] 121(2) 5	0	90.2
3819. PICTURES *Atlanta* [26May84|23Jun84|07Jul84] 140(2) 7	0	90.1
3820. UP *Le Roux* [23Aug80|13Sep80|27Sep80] 145(2) 6	0	90.0
3821. RAINBOW *Dolly Parton* [19Dec87|09Jan88|06Feb88] 153(1) 8	0	89.9
3822. OPEN MIND *Jean-Luc Ponty* [08Dec84|22Dec84|06Apr85] 171(3) 13	0	89.7
3823. CHRISTMAS IN AMERICA *Kenny Rogers* [16Dec89|30Dec89|20Jan90] 119(2) 6	0	89.7
3824. BANDED TOGETHER *Lee Ritenour* [23Jun84|21Jul84|11Aug84] 145(2) 8	0	89.7
3825. IN SEARCH OF THE RAINBOW SEEKERS *Mtume* [18Oct80|01Nov80|08Nov80] 119(1) 4	0	89.6
3826. DIANNE REEVES *Dianne Reeves* [23Apr88|07May88|16Jul88] 172(1) 12	0	89.5
3827. STEPS IN TIME *King* [17Aug85|28Sep85|12Oct85] 140(2) 9	1	89.5
3828. IN THE BEGINNING *Journey* [05Jan80|02Feb80|05Apr80] 152(1) 8	0	89.5
3829. ALL THE WAY STRONG *Third World* [01Oct83|22Oct83|12Nov83] 137(2) 7	0	89.4
3830. LET IT ROCK *Johnny And The Distractions* [20Feb82|27Mar82|17Apr82] 152(1) 9	0	89.2
3831. DEUCE *Kurtis Blow* [18Jul81|08Aug81|15Aug81] 137(1) 5	0	89.2
3832. I WANT OUT-LIVE *Helloween* [22Apr89|13May89|03Jun89] 123(2) 7	0	88.9
3833. CYCLE OF THE MOON *Prophet* [12Mar88|09Apr88|23Apr88] 137(1) 7	0	88.8
3834. BIKINI RED *The Screaming Blue Messiahs* [16Jan88|12Mar88|26Mar88] 172(1) 11	0	88.8
3835. ...IN A CHAMBER *Wire Train* [18Feb84|31Mar84|14Apr84] 150(1) 9	0	88.8
3836. REPLAY *Crosby, Stills & Nash* [10Jan81|17Jan81|07Feb81] 122(2) 5	0	88.8
3837. DECOY *Miles Davis* [30Jun84|28Jul84|08Sep84] 169(2) 11	0	88.6
3838. BEETLEJUICE *Soundtrack* [25Jun88|09Jul88|30Jul88] 118(1) 6	0	88.5
3839. ENDANGERED SPECIES *White Wolf* [21Jun86|19Jul86|09Aug86] 137(2) 8	0	88.5
3840. WILLIE NILE *Willie Nile* [12Apr80|03May80|17May80] 145(2) 6	0	88.5
3841. SOMETIMES WHEN WE TOUCH *Cleo Laine & James Galway* [26Jul80|16Aug80|30Aug80] 150(1) 6	0	88.4
3842. KILL 'EM ALL *Metallica* [05Apr86|26Apr86|07Jun86] 155(2) 10	0	88.4
3843. WHERE TO NOW *Charlie Dore* [26Apr80|24May80|07Jun80] 145(1) 7	1	88.0
3844. GLASSWORKS *Philip Glass* [10Apr82|24Apr82|15May82] 121(2) 6	0	88.0
3845. ANTHOLOGY *Grover Washington Jr.* [24Oct81|21Nov81|05Dec81] 149(1) 7	0	87.9
3846. ANYTIME *Henry Paul Band* [26Dec81|23Jan82|13Feb82] 158(1) 8	1	87.8
3847. POINT OF VIEW *Spyro Gyra* [08Jul89|15Jul89|12Aug89] 120(2) 6	0	87.8
3848. LAUGHING AT THE PIECES *Doctor And The Medics* [13Sep86|04Oct86|01Nov86] 125(1) 8	1	87.8
3849. STEVE MILLER BAND – LIVE *Steve Miller Band* [30Apr83|21May83|11Jun83] 125(1) 8	0	87.6
3850. FAMOUS AT NIGHT *John Hunter* [09Feb85|23Mar85|06Apr85] 148(2) 9	1	87.4
3851. LULLABY *Book Of Love* [23Jul88|06Aug88|01Oct88] 156(2) 10	0	87.3
3852. TRICK OR TREAT (SOUNDTRACK) *Fastway* [22Nov86|17Jan87|07Feb87] 156(1) 12	0	87.1

Ranking the Albums

Rank. Title *Act* [Enter|Peak|Exit] Peak(Wks) TotWks	Hot 100	Scr
3853. TUTU *Miles Davis* [25Oct86|15Nov86|24Jan87] 141(2) 10	0	87.1
3854. ROGUES GALLERY *Slade* [04May85|01Jun85|08Jun85] 132(2) 6	1	87.0
3855. DANGEROUS *Natalie Cole* [29Jun85|20Jul85|24Aug85] 140(2) 9	2	87.0
3856. WRABIT *Wrabit* [06Feb82|06Mar82|27Mar82] 157(1) 8	0	86.9
3857. 1ST *Streets* [03Dec83|28Jan84|11Feb84] 166(1) 11	1	86.8
3858. MYSTERY OF BULGARIAN VOICES *Bulgarian State Radio & T.V. Female Choir* [17Dec88|28Jan89|18Feb89] 165(1) 10	0	86.8
3859. ALL SYSTEMS GO *Donna Summer* [10Oct87|31Oct87|14Nov87] 122(2) 6	1	86.7
3860. DREAMBOY *Dreamboy* [14Jan84|17Mar84|24Mar84] 168(1) 11	0	86.6
3861. THE MINSTREL MAN *Willie Nelson* [01Aug81|15Aug81|12Sep81] 148(1) 7	0	86.5
3862. MECO PLAYS MUSIC FROM THE EMPIRE STRIKES BACK *Meco* [02Aug80|23Aug80|20Sep80] 140(1) 8	1	86.5
3863. AMERICAN WORKER *The Bus Boys* [21Aug82|11Sep82|02Oct82] 139(3) 7	0	86.3
3864. LOVE AND DANCING *The League Unlimited Orchestra* [18Sep82|09Oct82|30Oct82] 135(2) 7	0	86.1
3865. NO PLACE FOR DISGRACE *Flotsam And Jetsam* [18Jun88|02Jul88|06Aug88] 143(2) 8	0	86.1
3866. VAN GO *The Beat Farmers* [12Jul86|16Aug86|06Sep86] 135(2) 9	0	85.9
3867. STILL FEELS GOOD *Tom Johnston* [16May81|20Jun81|27Jun81] 158(1) 7	0	85.8
3868. STAR PEOPLE *Miles Davis* [21May83|18Jun83|02Jul83] 136(1) 7	0	85.8
3869. SHANGRI-LA *Bardeux* [14Oct89|11Nov89|25Nov89] 133(2) 7	1	85.8
3870. MUSICAL SHAPES *Carlene Carter* [04Oct81|01Nov80|08Nov80] 139(2) 6	0	85.7
3871. SAM COOKE LIVE AT THE HARLEM SQUARE CLUB *Sam Cooke* [22Jun85|20Jul85|10Aug85] 134(1) 8	0	85.5
3872. CLUB PARADISE (SOUNDTRACK) *Jimmy Cliff* [26Jul86|02Aug86|30Aug86] 122(2) 6	0	85.4
3873. GO FOR THE THROAT *Humble Pie* [09May81|06Jun81|13Jun81] 154(1) 6	0	85.4
3874. THE SPECKLESS SKY *Jane Siberry* [14Jun86|05Jul86|02Aug86] 149(2) 8	0	85.4
3875. NOTHING TO FEAR *Oingo Boingo* [04Sep82|25Sep82|30Oct82] 148(1) 9	0	85.3
3876. IRRESISTIBLE *Miles Jaye* [10Jun89|15Jul89|05Aug89] 160(2) 9	0	85.3
3877. SECOND TO NUNN *Bobby Nunn* [23Oct82|27Nov82|11Dec82] 148(1) 8	0	85.2
3878. STAY *Ray, Goodman & Brown* [09Jan82|13Feb82|20Feb82] 151(1) 7	0	85.2
3879. MAGIC WINDOWS *Herbie Hancock* [03Oct81|31Oct81|07Nov81] 140(1) 6	0	85.2
3880. DEFECTOR *Steve Hackett* [30Aug80|27Sep80|04Oct80] 144(1) 6	0	85.1
3881. BEAUTY AND THE BEAST: OF LOVE AND HOPE *TV Soundtrack* [17Jun89|26Aug89|09Sep89] 157(1) 10	0	85.1
3882. PRIMITIVE MAN *Icehouse* [09Oct82|06Nov82|13Nov82] 129(2) 6	0	85.1
3883. GREATEST HITS COLLECTION *Bananarama* [03Dec88|10Dec88|28Jan89] 151(2) 9	1	84.8
3884. THE GOLDEN CHILD *Soundtrack* [24Jan87|14Feb87|07Mar87] 126(2) 7	1	84.8
3885. COMMODORES ANTHOLOGY *Commodores* [11Jun83|09Jul83|23Jul83] 141(1) 7	0	84.5
3886. WE'RE MOVIN' UP *Atlantic Starr* [20May89|03Jun89|24Jun89] 125(2) 6	0	84.5
3887. THE BEST OF WAR...AND MORE *War* [30May87|04Jul87|01Aug87] 156(2) 10	0	84.5
3888. PERFECT VIEW *The Graces* [09Sep89|30Sep89|04Nov89] 147(1) 9	1	84.5
3889. PLAYING TO WIN *Rick Nelson* [21Feb81|21Mar81|28Mar81] 153(2) 6	0	84.4
3890. A CAPPELLA *Todd Rundgren* [12Oct85|09Nov85|30Nov85] 128(2) 8	0	84.4
3891. DIG THE NEW BREED *The Jam* [15Jan83|12Feb83|12Mar83] 131(1) 9	0	84.2
3892. AEROBIC DANCE HITS, VOLUME ONE *Carla Capuano* [13Mar82|10Apr82|01May82] 152(3) 8	0	84.0
3893. YESTERDAY ONCE MORE *Carpenters* [25May85|29Jun85|13Jul85] 144(2) 8	0	84.0
3894. THE BEST OF JOHNNY MATHIS 1975-1980 *Johnny Mathis* [27Dec80|24Jan81|07Feb81] 140(2) 7	0	83.9
3895. SERIOUS BUSINESS *Johnny Winter* [19Oct85|09Nov85|21Dec85] 156(2) 10	0	83.9
3896. THE SECOND ADVENTURE *Dynasty* [10Oct81|17Oct81|31Oct81] 119(2) 4	0	83.4
3897. T.R.A.S.H. (TUBES RARITIES AND SMASH HITS) *The Tubes* [29Aug81|26Sep81|03Oct81] 148(1) 6	0	83.2
3898. VIVE LE ROCK *Adam Ant* [19Oct85|16Nov85|30Nov85] 131(2) 7	0	83.1
3899. GET RHYTHM *Ry Cooder* [28Nov87|19Dec87|13Feb88] 177(1) 12	0	83.1
3900. THE ABOMINABLE SHOWMAN *Nick Lowe* [02Apr83|09Apr83|14May83] 129(3) 7	0	82.8
3901. THE UNDERTONES *The Undertones* [26Jan80|01Mar80|08Mar80] 154(1) 7	0	82.6
3902. BACK TO BASICS *The Temptations* [21Apr84|09Jun84|16Jun84] 152(1) 9	1	82.6
3903. IN CONCERT *Julio Iglesias* [25Aug84|15Sep84|20Oct84] 159(2) 9	0	82.5
3904. TALKIN' 'BOUT YOU *Diane Schuur* [12Nov88|26Nov88|14Jan89] 170(3) 10	0	82.2
3905. THE SEA OF LOVE *The Adventures* [16Apr88|21May88|11Jun88] 144(1) 9	1	82.1
3906. I BELIEVE IN LOVE *Rockie Robbins* [12Sep81|03Oct81|17Oct81] 147(2) 6	0	82.1
3907. FOREVER YOUNG *Alphaville* [22Dec84|12Jan85|30Mar85] 180(1) 15	3	82.1
3908. RAGE IN EDEN *Ultravox* [24Oct81|07Nov81|28Nov81] 144(2) 6	0	81.9
3909. EZO *EZO* [13Jun87|04Jul87|08Aug87] 150(2) 9	0	81.9
3910. NO RESPECT *Vain* [26Aug89|30Sep89|14Oct89] 154(2) 8	0	81.5
3911. 20/20 TWENTY NO.1 HITS FROM TWENTY YEARS AT MOTOWN *Various Artists* [12Apr80|26Apr80|17May80] 150(2) 6	0	81.5
3912. NO EASY WAY OUT *Robert Tepper* [19Apr86|10May86|07Jun86] 144(2) 8	2	81.5
3913. TWO B'S PLEASE *Robbin Thompson Band* [25Oct80|22Nov80|03Jan81] 168(1) 11	1	81.5
3914. SOLDIERS OF FORTUNE *The Outlaws* [08Nov86|22Nov86|10Jan87] 160(2) 10	0	81.4
3915. LUCIANO *Luciano Pavarotti* [24Apr82|15May82|05Jun82] 141(1) 7	0	81.4
3916. REBEL MUSIC *Bob Marley And The Wailers* [06Sep86|27Sep86|01Nov86] 140(1) 9	0	81.1
3917. INNOCENT EYES *Graham Nash* [26Apr86|24May86|07Jun86] 136(2) 7	1	81.1
3918. IN THE NAME OF LOVE *Thompson Twins* [26Jun82|24Jul82|14Aug82] 148(2) 8	0	81.0
3919. A WOMAN'S GOT THE POWER *The A's* [11Jul81|01Aug81|22Aug81] 146(1) 7	0	80.9
3920. DISGUISE *Chuck Mangione* [15Sep84|13Oct84|03Nov84] 148(2) 8	0	80.9
3921. WATCHING YOU, WATCHING ME *Bill Withers* [25May85|29Jun85|13Jul85] 143(2) 8	0	80.9
3922. ROUTES *Ramsey Lewis* [23Aug80|20Sep80|11Oct80] 173(1) 8	0	80.8
3923. EXPOSED II *Various Artists* [05Dec81|05Dec81|02Jan82] 124(2) 5	0	80.7
3924. HELP YOURSELF *Larry Gatlin & The Gatlin Brothers Band* [01Nov80|15Nov80|22Nov80] 118(1) 4	0	80.4
3925. CHRISTMAS RAP *Various Artists* [12Dec87|09Jan88|30Jan88] 130(1) 8	0	80.3
3926. JANE WIEDLIN *Jane Wiedlin* [26Oct85|16Nov85|30Nov85] 127(2) 6	1	80.3
3927. AMOUR *Richard Clayderman* [24Nov84|22Dec84|19Jan85] 160(3) 9	0	80.1
3928. EMOTION *DFX2* [20Aug83|24Sep83|08Oct83] 143(1) 8	0	80.0
3929. BALL ROOM *Sea Level* [23Aug80|13Sep80|27Sep80] 152(1) 6	0	79.9
3930. ROYAL JAM *The Crusaders With B.B. King And The Royal Philharmonic Orchestra* [17Jul82|07Aug82|28Aug82] 144(1) 7	0	79.8
3931. SKEEZER PLEEZER *UTFO* [09Aug86|06Sep86|27Sep86] 142(2) 8	0	79.6
3932. SIXTEEN *Stacy Lattisaw* [27Aug83|08Oct83|15Oct83] 160(1) 8	1	79.4
3933. UPTOWN *The Neville Brothers* [02May87|30May87|27Jun87] 155(1) 9	0	79.3
3934. BIG TRASH *Thompson Twins* [21Oct89|11Nov89|25Nov89] 143(2) 6	1	79.3
3935. EBONEE WEBB *Ebonee Webb* [12Sep81|03Oct81|24Oct81] 157(2) 7	0	79.0
3936. LIVIN' LARGE *E.U.* [22Apr89|03Jun89|17Jun89] 158(1) 9	0	78.6
3937. SWING, SWING, SWING *The Boston Pops Orchestra/John Williams* [17May86|07Jun86|05Jul86] 155(2) 8	0	78.6
3938. GOOD FOR YOUR SOUL *Oingo Boingo* [10Sep83|15Oct83|22Oct83] 144(1) 7	0	78.5
3939. I HEARD IT IN A LOVE SONG *McFadden & Whitehead* [04Oct80|25Oct80|08Nov80] 153(1) 6	0	78.4
3940. THE RHYTHMOTIST *Stewart Copeland* [07Sep85|12Oct85|26Oct85] 148(2) 8	0	78.4
3941. TRIAL BY FIRE: LIVE IN LENINGRAD *Yngwie Malmsteen* [11Nov89|11Nov89|06Jan90] 128(2) 8	0	78.4
3942. THE PASSENGER *Melvin James* [03Oct87|31Oct87|21Nov87] 146(2) 8	0	78.4
3943. ME & PAUL *Willie Nelson* [30Mar85|04May85|11May85] 152(1) 7	0	78.3
3944. CAPTURED LIVE *Peter Tosh* [22Sep84|20Oct84|10Nov84] 152(2) 8	0	78.1
3945. COME AN' GET IT *Whitesnake* [30May81|27Jun81|04Jul81] 151(2) 6	0	78.1
3946. FESTIVAL *Lee Ritenour* [21Jan89|25Feb89|11Mar89] 156(2) 8	0	77.9
3947. INTO THE WOODS *Original Cast Recording* [26Mar88|02Apr88|30Apr88] 126(2) 6	0	77.8
3948. SURFACE THRILLS *The Temptations* [19Mar83|09Apr83|14May83] 159(3) 9	1	77.8

Ranking the Albums

Rank. Title *Act* [Enter|Peak|Exit] Peak(Wks) TotWks	Hot 100	Scr
3949. GREMLINS *Soundtrack* [07Jul84|04Aug84|18Aug84] 143(1) 7	0	77.8
3950. PERFECT FIT *Jerry Knight* [11Apr81|25Apr81|16May81] 146(1) 6	0	77.8
3951. CHILL OUT *Black Uhuru* [24Jul82|21Aug82|04Sep82] 146(1) 7	0	77.7
3952. KINGS OF WEST COAST *L.A. Dream Team* [13Sep86|04Oct86|25Oct86] 138(1) 7	0	77.6
3953. ALIBI *America* [06Sep80|20Sep80|11Oct80] 142(2) 6	0	77.6
3954. SAY NO MORE *Badfinger* [28Mar81|18Apr81|02May81] 155(1) 6	1	77.3
3955. TAKE IT TO THE LIMIT *Norman Connors* [27Sep80|18Oct80|01Nov80] 145(1) 6	0	77.3
3956. STRANGE ANGELS *Laurie Anderson* [18Nov89|09Dec89|10Feb90] 171(1) 12	0	77.3
3957. SOUP FOR ONE *Soundtrack* [12Jun82|21Aug82|18Sep82] 168(1) 12	2	77.1
3958. KILIMANJARO *The Teardrop Explodes* [28Feb81|28Mar81|04Apr81] 156(2) 6	0	77.1
3959. FAVORITES *Crystal Gayle* [03May80|31May80|07Jun80] 149(1) 6	0	76.9
3960. RHINESTONE (SOUNDTRACK) *Dolly Parton* [21Jul84|11Aug84|01Sep84] 135(2) 7	0	76.5
3961. DON'T GIVE UP YOUR DAY JOB *Jack Wagner* [02May87|16May87|20Jun87] 151(2) 8	1	76.5
3962. LOOKS SO FINE *Instant Funk* [10Apr82|08May82|22May82] 147(2) 7	0	76.3
3963. GREATEST HITS *George Strait* [20Apr85|01Jun85|08Jun85] 157(1) 8	0	76.2
3964. MSB *Michael Stanley Band* [04Sep82|02Oct82|09Oct82] 136(2) 6	2	76.1
3965. TEN YEARS OF HARMONY (1970-1980) *The Beach Boys* [26Dec81|30Jan82|13Feb82] 156(1) 8	1	76.0
3966. MIGRATION *Dave Grusin* [21Oct89|11Nov89|09Dec89] 145(1) 8	0	75.4
3967. TWANG BAR KING *Adrian Belew* [01Oct83|22Oct83|12Nov83] 146(2) 7	0	75.4
3968. LOVE IS WHAT WE MAKE IT *Kenny Rogers* [20Apr85|11May85|01Jun85] 145(1) 7	0	74.9
3969. WHITE CHRISTMAS *John Schneider* [05Dec81|09Jan82|16Jan82] 155(1) 7	0	74.9
3970. WALK A FINE LINE *Paul Anka* [13Aug83|17Sep83|01Oct83] 156(1) 8	1	74.9
3971. THE PEABO BRYSON COLLECTION *Peabo Bryson* [14Jul84|04Aug84|15Sep84] 168(1) 10	0	74.9
3972. AGENT DOUBLE O SOUL *The Untouchables* [01Apr89|06May89|27May89] 162(2) 9	0	74.8
3973. TIME ODYSSEY *Vinnie Moore* [18Jun88|25Jun88|30Jul88] 147(2) 7	0	74.7
3974. SPIES OF LIFE *Player* [06Feb82|27Feb82|20Mar82] 152(1) 7	1	74.5
3975. FEARLESS *Nina Hagen* [28Jan84|03Mar84|17Mar84] 151(1) 8	0	74.5
3976. EDDIE & THE CRUISERS II (SOUNDTRACK) *John Cafferty & The Beaver Brown Band* [26Aug89|09Sep89|30Sep89] 121(2) 6	1	74.5
3977. MATERIAL THANGZ *The Deele* [06Jul85|03Aug85|24Aug85] 155(1) 8	0	74.5
3978. CHRISTIANE F. (SOUNDTRACK) *David Bowie* [03Apr82|10Apr82|15May82] 135(2) 7	0	74.4
3979. LISTEN TO THE RADIO *Don Williams* [01May82|15May82|19Jun82] 166(3) 8	0	74.4
3980. SPLASHDOWN *Breakwater* [07Jun80|28Jun80|05Jul80] 141(1) 5	0	74.3
3981. WAIT FOR NIGHT *Rick Springfield* [18Dec82|08Jan83|05Feb83] 159(2) 8	0	74.1
3982. THE CHRISTIANS *The Christians* [12Mar88|26Mar88|30Apr88] 158(2) 8	0	74.1
3983. HAPPY? *Public Image Limited* [24Oct87|07Nov87|20Feb88] 169(1) 10	0	74.0
3984. THE FLOWERS OF ROMANCE *Public Image Limited* [30May81|06Jun81|20Jun81] 114(2) 4	0	74.0
3985. GUARDIAN OF THE LIGHT *George Duke* [30Apr83|04Jun83|11Jun83] 147(1) 7	0	73.8
3986. FACE TO FACE *Angel City* [10May80|28Jun80|05Jul80] 152(2) 7	0	73.6
3987. WINDHAM HILL PIANO SAMPLER *Various Artists* [21Dec85|08Feb86|08Mar86] 167(2) 12	0	73.5
3988. THE SUN STILL SHINES *Sonny Charles* [25Dec82|29Jan83|05Feb83] 136(2) 7	1	73.4
3989. THE SECOND ALBUM *707* [07Feb81|14Mar81|14Mar81] 159(1) 6	0	73.2
3990. FLAG *Yello* [15Apr89|06May89|10Jun89] 152(2) 9	0	73.1
3991. RUN FOR COVER *Gary Moore* [15Mar86|12Apr86|26Apr86] 146(2) 7	0	73.1
3992. TENDERLY *George Benson* [05Aug89|19Aug89|09Sep89] 140(2) 6	0	73.1
3993. PRIVATE COLLECTION *Jon And Vangelis* [13Aug83|10Sep83|24Sep83] 148(1) 7	0	73.1
3994. CHANGE *Barry White* [02Oct82|23Oct82|06Nov82] 148(1) 6	0	73.0
3995. BOBBY & THE MIDNITES *Bobby And The Midnites* [21Nov81|12Dec81|02Jan82] 158(2) 7	0	72.8
3996. WONDERFUL *Rick James* [23Jul88|30Jul88|10Sep88] 148(2) 8	0	72.8
3997. FAME AND FASHION - DAVID BOWIE'S ALL TIME GREATEST HITS *David Bowie* [21Apr84|19May84|26May84] 147(1) 6	0	72.6
3998. BOURGEOIS TAGG *Bourgeois Tagg* [31May86|28Jun86|12Jul86] 139(2) 7	1	72.5
3999. OVER THE LINE *Greg Guidry* [17Apr82|22May82|29May82] 147(1) 7	2	72.4
4000. HEROES, ANGELS & FRIENDS *Janey Street* [03Nov84|24Nov84|08Dec84] 145(2) 6	1	72.0
4001. PASSION CRIMES *Darling Cruel* [09Sep89|30Sep89|28Oct89] 160(2) 8	0	72.0
4002. RADICAL DEPARTURE *Ranking Roger* [13Aug88|27Aug88|24Sep88] 151(2) 7	0	71.8
4003. THE MIND IS A TERRIBLE THING TO TASTE *Ministry* [09Dec89|23Dec89|10Feb90] 163(3) 10	0	71.8
4004. INDESTRUCTIBLE *Four Tops* [24Sep88|08Oct88|05Nov88] 149(2) 7	0	71.3
4005. COLD SPRING HARBOR *Billy Joel* [14Jan84|04Feb84|03Mar84] 158(1) 8	0	71.3
4006. CLASSIC YES *Yes* [09Jan82|23Jan82|06Feb82] 142(2) 5	0	71.1
4007. GOLDEN TOUCH *Rose Royce* [24Jan81|14Feb81|07Mar81] 160(2) 7	0	71.0
4008. DRAGNET *Soundtrack* [18Jul87|08Aug87|22Aug87] 137(2) 6	0	70.8
4009. REAL EYES *Gil Scott-Heron* [20Dec80|10Jan81|24Jan81] 159(2) 6	0	70.8
4010. SUNRISE *Jimmy Ruffin* [31May80|14Jun80|05Jul80] 152(2) 6	1	70.8
4011. GET IT ON CREDIT *Toronto* [04Sep82|09Oct82|06Nov82] 162(2) 10	1	70.8
4012. MARS NEEDS GUITARS! *Hoodoo Gurus* [10May86|31May86|21Jun86] 140(2) 7	0	70.5
4013. BIGGEST HITS *Marty Robbins* [22Jan83|12Mar83|19Mar83] 170(1) 9	0	70.5
4014. JOURNEY TO A RAINBOW *Chuck Mangione* [25Jun83|23Jul83|06Aug83] 154(1) 7	0	70.5
4015. HONI SOIT (O NEE SWA) *John Cale* [11Apr81|18Apr81|09May81] 154(2) 5	0	70.4
4016. WIRED FOR SOUND *Cliff Richard* [17Oct81|24Oct81|07Nov81] 132(2) 4	2	70.4
4017. NEVER SAY NEVER *Romeo Void* [06Mar82|03Apr82|10Apr82] 147(1) 6	0	70.4
4018. THE BEST OF DARK HORSE *George Harrison* [04Nov89|04Nov89|09Dec89] 132(2) 6	0	70.3
4019. BILLY SATELLITE *Billy Satellite* [01Sep84|15Sep84|06Oct84] 139(2) 6	2	70.3
4020. SONGS TO LEARN & SING *Echo & The Bunnymen* [11Jan86|22Feb86|08Mar86] 158(1) 9	0	70.3
4021. TONY CAREY [I WON'T BE HOME TONIGHT] *Tony Carey* [02Apr83|07May83|28May83] 167(1) 9	2	70.2
4022. VIVA SANTANA *Santana* [29Oct88|05Nov88|03Dec88] 142(2) 6	0	70.2
4023. PARTY OF ONE *Tim Weisberg* [02Aug80|09Aug80|13Sep80] 171(2) 7	0	70.1
4024. ACTING VERY STRANGE *Mike Rutherford* [09Oct82|30Oct82|13Nov82] 145(2) 6	0	70.1
4025. EVERY TURN OF THE WORLD *Christopher Cross* [30Nov85|14Dec85|04Jan86] 127(1) 6	1	70.1
4026. SHADOW MAN *Johnny Clegg & Savuka* [10Sep88|01Oct88|22Oct88] 155(2) 7	0	70.0
4027. THE OTHER ONES *The Other Ones* [16May87|06Jun87|20Jun87] 139(2) 6	2	70.0
4028. I HAD TO SAY IT *Millie Jackson* [07Feb81|21Feb81|28Feb81] 137(1) 4	0	69.8
4029. I'M A BLUES MAN *Z.Z. Hill* [07Jan84|11Feb84|03Mar84] 170(2) 9	0	69.5
4030. OASIS *Roberta Flack* [14Jan89|11Feb89|04Mar89] 159(1) 8	0	69.2
4031. SKELETONS IN THE CLOSET: THE BEST OF OINGO BOINGO *Oingo Boingo* [11Feb89|25Feb89|18Mar89] 150(2) 6	0	69.2
4032. THREE PIECE SUITE *Ramsey Lewis* [20Jun81|11Jul81|18Jul81] 152(2) 5	0	69.2
4033. HITS 1979-1989 *Rosanne Cash* [01Apr89|29Apr89|13May89] 152(1) 7	0	69.1
4034. ROCK THERAPY *Stray Cats* [27Sep86|11Oct86|25Oct86] 122(2) 5	0	69.0
4035. FUN & GAMES *The Connells* [06May89|03Jun89|08Jul89] 163(1) 10	0	68.9
4036. FIGHT FOR THE ROCK *Savatage* [21Jun86|26Jul86|02Aug86] 158(2) 7	0	68.7
4037. CONFIDENCE *Narada Michael Walden* [05Jun82|26Jun82|10Jul82] 135(1) 6	0	68.6
4038. ANOTHER PERFECT DAY *Motorhead* [23Jul83|03Sep83|03Sep83] 153(1) 7	0	68.5
4039. ASSAULT ATTACK *The Michael Schenker Group* [09Apr83|07May83|21May83] 151(1) 7	0	68.4
4040. MY ROAD OUR ROAD *Lee Oskar* [01Aug81|29Aug81|05Sep81] 162(2) 6	0	68.4
4041. FULL FORCE *Full Force* [15Feb86|15Mar86|05Apr86] 160(2) 8	0	68.4
4042. WRAP YOUR BODY *One Way* [10Aug85|31Aug85|05Oct85] 156(2) 9	0	68.1
4043. SCENARIO *Al Di Meola* [29Oct83|12Nov83|03Dec83] 128(1) 6	0	67.9

Ranking the Albums

Rank. Title *Act* [Enter|Peak|Exit] Peak(Wks) TotWks	Hot 100	Scr
4044. TENEMENT STEPS *The Motors* [12Apr80|26Apr80|31May80] 174(1) 8	1	67.6
4045. 4 BY FOUR *4 By Four* [27Jun87|18Jul87|08Aug87] 141(2) 7	1	67.6
4046. DRUMS AND WIRES *XTC* [26Jan80|16Feb80|15Mar80] 176(2) 8	0	67.4
4047. DURELL COLEMAN *Durell Coleman* [28Sep85|26Oct85|09Nov85] 155(2) 7	0	67.2
4048. VISUAL LIES *Lizzy Borden* [26Sep87|17Oct87|07Nov87] 146(1) 7	0	67.1
4049. ANY MAN'S HUNGER *Danny Wilde* [26Mar88|30Apr88|21May88] 176(1) 9	0	67.0
4050. ROOT HOG OR DIE *Mojo Nixon & Skid Roper* [06May89|27May89|17Jun89] 151(1) 7	0	66.9
4051. HAPPY LOVE *Natalie Cole* [26Sep81|03Oct81|17Oct81] 132(2) 4	0	66.8
4052. FEELING GOOD *Roy Ayers* [20Mar82|24Apr82|01May82] 160(1) 7	0	66.8
4053. TWIST OF SHADOWS *Xymox* [03Jun89|12Aug89|26Aug89] 165(1) 10	1,1*	66.8
4054. RIDIN' THE STORM OUT *REO Speedwagon* [12Jan74|21Feb81|14Mar81] 171(2) 8	0	66.7
4055. RENAISSANCE *Village People* [01Aug81|08Aug81|22Aug81] 138(2) 4	0	66.7
4056. MUSIC FOR THE KNEE PLAYS *David Byrne* [01Jun85|15Jun85|06Jul85] 141(2) 6	0	66.7
4057. COMPANY B *Company B* [18Jul87|08Aug87|22Aug87] 143(2) 6	0	66.7
4058. IN THE HEAT *Southside Johnny & The Jukes* [08Sep84|15Sep84|27Oct84] 164(2) 8	0	66.7
4059. SAD CAFÉ *Sad Café* [15Aug81|05Sep81|19Sep81] 160(2) 6	1	66.7
4060. DREAM STREET *Janet Jackson* [27Oct84|17Nov84|01Dec84] 147(2) 6	0	66.6
4061. THE STEVE MARTIN BROTHERS *Steve Martin* [14Nov81|21Nov81|05Dec81] 135(2) 4	0	66.6
4062. 21ST CENTURY MAN *Billy Thorpe* [08Nov80|29Nov80|06Dec80] 151(2) 5	0	66.5
4063. DISTANT SHORES *Robbie Patton* [15Aug81|12Sep81|19Sep81] 162(1) 6	1	66.5
4064. TURN UP THE MUSIC *Mass Production* [16May81|30May81|25Jul81] 166(2) 6	0	66.3
4065. STOP START *Modern English* [05Apr86|03May86|17May86] 154(2) 7	0	66.1
4066. STAY ON THESE ROADS *a-ha* [04Jun88|25Jun88|09Jul88] 148(2) 6	0	66.0
4067. DANCING ON THE EDGE *Roy Buchanan* [28Jun86|12Jul86|16Aug86] 153(2) 8	0	66.0
4068. FOOTSTEPS IN THE DARK: GREATEST HITS VOLUME 2 *Cat Stevens* [15Dec84|22Dec84|02Feb85] 165(3) 8	0	66.0
4069. IN CONCERT *Jane Olivor* [29May82|12Jun82|03Jul82] 144(2) 6	0	65.9
4070. READ MY LIPS *Fee Waybill* [10Nov84|01Dec84|15Dec84] 146(2) 6	0	65.9
4071. HEROES & ZEROS *Glen Burtnick* [24Oct87|31Oct87|28Nov87] 147(2) 6	1	65.9
4072. VAN-ZANT *Van-Zant* [04May85|08Jun85|22Jun85] 170(1) 8	0	65.8
4073. DARING ADVENTURES *Richard Thompson* [25Oct86|08Nov86|29Nov86] 142(2) 6	0	65.8
4074. JERRY KNIGHT *Jerry Knight* [24May80|14Jun80|05Jul80] 165(1) 7	0	65.7
4075. DON'T FIGHT IT *Red Rider* [26Apr80|17May80|24May80] 146(1) 5	1	65.7
4076. SUMMER LOVERS *Soundtrack* [28Aug82|25Sep82|09Oct82] 152(1) 7	0	65.7
4077. LIZA MINNELLI AT CARNEGIE HALL *Liza Minnelli* [14Nov87|28Nov87|02Jan88] 156(2) 8	0	65.5
4078. TOUCH ME TONIGHT-THE BEST OF SHOOTING STAR *Shooting Star* [04Nov89|02Dec89|16Dec89] 151(2) 7	1	65.5
4079. BARBEQUE KING *Jorma Kaukonen & Vital Parts* [14Feb81|07Mar81|21Mar81] 163(2) 6	0	65.5
4080. BIG NIGHT MUSIC *Shriekback* [21Feb87|14Mar87|28Mar87] 145(2) 6	0	65.2
4081. HYAENA *Siouxsie & The Banshees* [07Jul84|11Aug84|18Aug84] 157(1) 7	0	65.1
4082. IN HARMONY - A SESAME STREET RECORD *Various Artists* [10Jan81|24Jan81|07Feb81] 156(1) 5	1	65.0
4083. SURVIVOR *Survivor* [29Mar80|10May80|10May80] 169(1) 7	1	64.9
4084. WHAT'S MY NAME *Steady B* [31Oct87|31Oct87|23Jan88] 149(2) 7	0	64.9
4085. GRAND FUNK LIVES *Grand Funk Railroad* [17Oct81|31Oct81|14Nov81] 149(2) 5	0	64.8
4086. WE WANT MILES *Miles Davis* [29May82|19Jun82|10Jul82] 159(2) 7	0	64.8
4087. ME AND YOU *The Chi-Lites* [10Apr82|01May82|22May82] 162(2) 7	0	64.7
4088. SERIOUS SLAMMIN' *The Pointer Sisters* [19Mar88|09Apr88|23Apr88] 152(2) 6	0	64.7
4089. YOU DON'T KNOW ME *Mickey Gilley* [22Aug81|12Sep81|26Sep81] 170(2) 6	1	64.5
4090. LIGHTS IN THE NIGHT *Flash And The Pan* [31May80|14Jun80|05Jul80] 159(2) 6	0	64.4
4091. POWER & PASSION *Mama's Boys* [15Jun85|06Jul85|20Jul85] 151(2) 6	0	64.3
4092. HIGH ON YOUR LOVE *Debbie Jacobs* [09Feb80|23Feb80|05Apr80] 178(4) 7	0	64.2
4093. BUTT ROCKIN' *Fabulous Thunderbirds* [28Mar81|11Apr81|09May81] 176(1) 7	0	64.1
4094. EAST *Cold Chisel* [13Jun81|11Jul81|18Jul81] 171(2) 6	0	64.0
4095. TYRONE DAVIS *Tyrone Davis* [08Jan83|22Jan83|12Feb83] 137(2) 5	1	63.9
4096. SPIRIT OF ST. LOUIS *Ellen Foley* [04Apr81|11Apr81|25Apr81] 152(2) 4	0	63.8
4097. RISING *Dr. Hook* [06Dec80|20Dec80|24Jan81] 175(3) 8	2	63.8
4098. HOW TO BEAT THE HIGH COST OF LIVING (SOUNDTRACK) *Hubert Laws And Earl Klugh* [27Sep80|27Sep80|18Oct80] 134(2) 4	0	63.8
4099. TOGETHER *The Oak Ridge Boys* [29Mar80|19Apr80|03May80] 154(1) 6	0	63.8
4100. ELISA FIORILLO *Elisa Fiorillo* [20Feb88|05Mar88|09Apr88] 163(2) 8	2	63.7
4101. HEART DON'T LIE *LaToya Jackson* [09Jun84|30Jun84|14Jul84] 149(2) 6	1	63.7
4102. NOVO COMBO *Novo Combo* [10Oct81|07Nov81|14Nov81] 167(1) 6	0	63.5
4103. MONSTER *Fetchin Bones* [18Nov89|16Dec89|06Jan90] 175(1) 8	0	63.4
4104. BIG DREAMERS NEVER SLEEP *Gino Vannelli* [23May87|27Jun87|04Jul87] 160(2) 7	1	63.4
4105. OCTOPUSSY *Soundtrack* [16Jul83|30Jul83|13Aug83] 137(1) 5	1	63.3
4106. OLIVER & COMPANY *Soundtrack* [07Jan89|04Feb89|18Feb89] 170(2) 7	0	63.3
4107. TRUTH AND SOUL *Fishbone* [01Oct88|29Oct88|21Jan89] 153(1) 9	0	63.3
4108. ELOISE LAWS *Eloise Laws* [14Feb81|14Mar81|28Mar81] 175(1) 7	0	63.1
4109. VOLUNTEER JAM VII *Various Artists* [25Jul81|08Aug81|15Aug81] 149(2) 4	0	63.0
4110. FEMME FATALE *Femme Fatale* [28Jan89|11Feb89|25Feb89] 141(2) 5	0	63.0
4111. BET CHA SAY THAT TO ALL THE GIRLS *Sister Sledge* [04Jun83|02Jul83|23Jul83] 169(1) 8	0	62.8
4112. CONNECTIONS AND DISCONNECTIONS *Funkadelic(2)* [11Apr81|18Apr81|02May81] 151(2) 4	0	62.7
4113. WILD GIFT *X* [06Jun81|27Jun81|04Jul81] 165(1) 5	0	62.7
4114. MATHEMATICS *Melissa Manchester* [18May85|01Jun85|22Jun85] 144(2) 6	1	62.7
4115. DANNY DAVIS & WILLIE NELSON WITH THE NASHVILLE BRASS *Danny Davis And Willie Nelson With The Nashville Brass* [15Mar80|05Apr80|12Apr80] 150(1) 5	0	62.6
4116. THE SOURCE *Grandmaster Flash* [17May86|07Jun86|21Jun86] 145(2) 6	0	62.5
4117. DESIRE *Tom Scott* [25Sep82|23Oct82|06Nov82] 164(1) 7	0	62.4
4118. FUTURE WORLD *Pretty Maids* [20Jun87|18Jul87|08Aug87] 165(2) 8	0	62.4
4119. NEMESIS *Axe* [10Sep83|08Oct83|15Oct83] 156(1) 6	1	62.4
4120. SHARKY'S MACHINE *Soundtrack* [23Jan82|06Feb82|13Mar82] 171(1) 8	1	62.2
4121. THE ART OF EXCELLENCE *Tony Bennett* [21Jun86|12Jul86|09Aug86] 160(2) 8	0	62.2
4122. LIVE WITHOUT A NET *Angel* [23Feb80|08Mar80|15Mar80] 149(1) 4	0	62.2
4123. DANCING ON THE COUCH *Go West* [22Aug87|10Oct87|17Oct87] 172(1) 9	0	62.2
4124. BLAST! (THE LATEST AND THE GREATEST) *Brothers Johnson* [22Jan83|12Feb83|19Feb83] 138(2) 6	0	62.2
4125. CRYSTAL GAYLE'S GREATEST HITS *Crystal Gayle* [10Sep83|22Oct83|29Oct83] 169(1) 8	0	62.0
4126. MOST OF THE GIRLS LIKE TO DANCE BUT ONLY SOME OF THE BOYS LIKE TO *Don Dixon* [07Mar87|04Apr87|25Apr87] 162(2) 8	0	61.9
4127. THE RINGS *The Rings* [21Feb81|07Mar81|28Mar81] 164(2) 6	1	61.8
4128. WHEN WILL I SEE YOU AGAIN *The O'Jays* [13Aug83|03Sep83|10Sep83] 142(1) 5	0	61.7
4129. QUEST FOR FIRE *Soundtrack* [17Apr82|08May82|22May82] 154(2) 6	0	61.6
4130. TONIGHT YOU'RE MINE *Eric Carmen* [28Jun80|19Jul80|26Jul80] 160(1) 5	1	61.5
4131. LITE ME UP *Herbie Hancock* [29May82|19Jun82|03Jul82] 151(2) 6	0	61.5
4132. LEAVE SCARS *Dark Angel* [01Apr89|22Apr89|06May89] 159(1) 6	0	61.5
4133. ISLAND LIFE *Grace Jones* [18Jan86|15Feb86|01Mar86] 161(2) 7	0	61.5
4134. ROCK 'N' ROLL *Motorhead* [24Oct87|21Nov87|28Nov87] 150(2) 6	0	61.4
4135. DEKADANCE *INXS* [01Oct83|15Oct83|05Nov83] 148(2) 6	0	61.2
4136. TRUTHDARE DOUBLEDARE *Bronski Beat* [02Aug86|16Aug86|06Sep86] 147(2) 6	0	61.1
4137. PAST MASTERS - VOLUME 1 *The Beatles* [02Apr88|09Apr88|07May88] 149(2) 6	0	60.8

Ranking the Albums

Rank. Title *Act* [Enter|Peak|Exit] Peak(Wks) TotWks	Hot 100	Scr
4138. LI'L SUZY *Ozone* [04Sep82|02Oct82|09Oct82] 152(2) 6	0	60.6
4139. LEGENDARY HEARTS *Lou Reed* [09Apr83|30Apr83|21May83] 159(1) 7	0	60.6
4140. DRY DREAMS *The Jim Carroll Band* [22May82|12Jun82|03Jul82] 156(2) 7	0	60.6
4141. RIO *Lee Ritenour* [17Apr82|24Apr82|22May82] 163(3) 6	0	60.5
4142. THE WALK *The Cure* [13Aug83|16Jun84|16Jun84] 177(1) 9	0	60.5
4143. S.O.S. III *The S.O.S. Band* [25Dec82|12Feb83|12Feb83] 172(1) 8	0	60.5
4144. WHEN ALL THE PIECES FIT *Peter Frampton* [14Oct89|04Nov89|18Nov89] 152(2) 6	0	60.4
4145. ANIMAL BOY *The Ramones* [21Jun86|28Jun86|26Jul86] 143(2) 6	0	60.2
4146. THE HOUSE OF LOVE *The House Of Love* [17Sep88|15Oct88|29Oct88] 156(1) 7	0	60.2
4147. FASTER & LLOUDER *Foster & Lloyd* [13May89|03Jun89|17Jun89] 142(1) 6	0	60.0
4148. GUITAR SPEAK *Various Artists* [17Dec88|24Dec88|04Feb89] 171(3) 8	0	60.0
4149. PIECES OF A DREAM *Pieces Of A Dream* [31Oct81|28Nov81|05Dec81] 170(2) 6	0	59.7
4150. FITS LIKE A GLOVE *Howie Mandel* [21Jun86|12Jul86|26Jul86] 148(2) 6	0	59.6
4151. POSITIVE *Peabo Bryson* [13Feb88|27Feb88|19Mar88] 157(1) 6	0	59.6
4152. KATE BUSH *Kate Bush* [09Jul83|06Aug83|13Aug83] 148(1) 6	0	59.5
4153. THE BEST OF THE REST *Lynyrd Skynyrd* [20Nov82|20Nov82|01Jan83] 171(5) 7	0	59.5
4154. MAN OVERBOARD *Bob Welch* [11Oct80|01Nov80|08Nov80] 162(2) 5	0	59.5
4155. TRASH IT UP *Southside Johnny & The Jukes* [01Oct83|22Oct83|05Nov83] 154(2) 6	0	59.3
4156. THE JAM *The Jam* [19Dec81|26Dec81|30Jan82] 176(3) 7	0	59.3
4157. THIS IS YOUR TIME *Change* [02Apr83|30Apr83|14May83] 161(2) 7	0	59.3
4158. SEA HAGS *Sea Hags* [24Jun89|08Jul89|05Aug89] 163(2) 7	0	59.3
4159. MUMMER *XTC* [25Feb84|10Mar84|24Mar84] 145(2) 5	0	59.2
4160. DEEPEST PURPLE: THE VERY BEST OF DEEP PURPLE *Deep Purple* [01Nov80|22Nov80|22Nov80] 148(1) 4	0	59.0
4161. 1234 *Ronnie Wood* [19Sep81|10Oct81|17Oct81] 164(2) 5	0	58.9
4162. ABSOLUTELY *Madness* [22Nov80|06Dec80|13Dec80] 146(2) 4	0	58.9
4163. THE ETERNAL IDOL *Black Sabbath* [26Dec87|16Jan88|30Jan88] 168(1) 6	0	58.8
4164. BOTH SIDES OF LOVE *Paul Anka* [09May81|23May81|13Jun81] 171(2) 6	1	58.8
4165. HARMONY *Anne Murray* [20Jun87|04Jul87|25Jul87] 149(2) 6	0	58.8
4166. BIG TIME *Tom Waits* [08Oct88|29Oct88|12Nov88] 152(2) 6	0	58.8
4167. BACK TALK *Rockets* [08Aug81|05Sep81|05Sep81] 165(1) 5	0	58.6
4168. CHEQUERED PAST *Chequered Past* [15Sep84|06Oct84|20Oct84] 151(2) 6	0	58.5
4169. IN BLACK AND WHITE *Barbara Mandrell* [29May82|19Jun82|03Jul82] 153(2) 6	0	58.5
4170. ALL THE RIGHT MOVES *Soundtrack* [03Dec83|07Jan84|14Jan84] 165(1) 7	1	58.2
4171. THE METHOD TO OUR MADNESS *The Lords Of The New Church* [27Apr85|25May85|08Jun85] 158(1) 7	0	58.2
4172. STREET CORNER HEROES *Robbie Dupree* [13Jun81|11Jul81|11Jul81] 169(1) 5	1	58.0
4173. FOLK OF THE 80'S (PART III) *Men Without Hats* [06Oct84|13Oct84|27Oct84] 127(2) 4	0	57.9
4174. QUICKSILVER *Soundtrack* [01Mar86|15Mar86|29Mar86] 140(2) 5	1	57.7
4175. AFTER THE ROSES *Kenny Rankin* [28Jun80|19Jul80|02Aug80] 171(1) 6	0	57.7
4176. BIRDY (SOUNDTRACK) *Peter Gabriel* [20Apr85|18May85|01Jun85] 162(1) 7	0	57.6
4177. USED TO BE *Charlene* [27Nov82|18Dec82|08Jan83] 162(1) 7	1	57.5
4178. STORE AT THE SUN *Jon Butcher Axis* [31Mar84|28Apr84|05May84] 160(1) 6	0	57.5
4179. THE ART OF DEFENSE *Nona Hendryx* [05May84|26May84|16Jun84] 167(1) 7	0	57.4
4180. REMATCH *Sammy Hagar* [08Jan83|05Feb83|05Mar83] 171(2) 9	0	57.4
4181. EMPIRE OF THE SUN *Soundtrack* [13Feb88|20Feb88|12Mar88] 150(2) 5	0	57.2
4182. I'M YOURS *Linda Clifford* [04Oct80|25Oct80|08Nov80] 160(1) 6	1	57.1
4183. THE SCATTERING *Cutting Crew* [03Jun89|17Jun89|08Jul89] 150(2) 6	1	57.1
4184. STANDING ALONE *White Wolf* [16Feb85|16Mar85|30Mar85] 162(2) 7	0	56.8
4185. NOT FAKIN' IT *Michael Monroe* [07Oct89|28Oct89|25Nov89] 161(1) 8	0	56.8
4186. FLIP *Nils Lofgren* [22Jun85|06Jul85|20Jul85] 150(2) 5	0	56.8
4187. SWORDFISHTROMBONES *Tom Waits* [29Oct83|03Dec83|10Dec83] 167(1) 7	0	56.8
4188. GROWING UP IN PUBLIC *Lou Reed* [10May80|31May80|07Jun80] 158(2) 5	0	56.6
4189. DIFFERENT KINDA DIFFERENT *Johnny Mathis* [09Aug80|06Sep80|06Sep80] 164(1) 5	0	56.6
4190. GIGOLO *Fatback* [09Jan82|30Jan82|30Jan82] 148(1) 4	0	56.6
4191. RIGHT OR WRONG *George Strait* [03Mar84|31Mar84|14Apr84] 163(1) 7	0	56.5
4192. PLAIN FROM THE HEART *Delbert McClinton* [05Dec81|30Jan82|30Jan82] 181(1) 9	0	56.5
4193. SOLD *Boy George* [01Aug87|08Aug87|29Aug87] 145(2) 5	0	56.4
4194. GRAVITY *James Brown* [18Oct86|01Nov86|22Nov86] 156(2) 6	1	56.4
4195. THE SOUND OF MUSIC *The dB's* [28Nov87|05Dec87|16Jan88] 171(2) 6	0	56.4
4196. DAZZLE SHIPS *Orchestral Manoeuvres In The Dark* [23Apr83|21May83|28May83] 162(1) 6	0	56.2
4197. THE MAN & HIS MUSIC *Sam Cooke* [05Apr86|03May86|24May86] 175(1) 8	0	56.1
4198. JUST LIKE DREAMIN' *Twennynine Featuring Lenny White* [05Dec81|19Dec81|02Jan82] 162(3) 6	0	55.9
4199. LAST OF THE WILD ONES *Johnny Van Zant Band* [18Sep82|16Oct82|23Oct82] 159(1) 6	0	55.8
4200. HEAVENLY BODY *The Chi-Lites* [29Nov80|13Dec80|03Jan81] 179(3) 6	0	55.8
4201. DANCE & EXERCISE WITH THE HITS *Linda Fratianne* [20Feb82|03Apr82|03Apr82] 174(2) 7	0	55.7
4202. TOUR DE FORCE - "LIVE" *Al Di Meola* [25Dec82|29Jan83|05Feb83] 165(1) 7	0	55.6
4203. HEADING FOR A STORM *Vandenberg* [28Jan84|11Feb84|10Mar84] 169(2) 7	0	55.4
4204. THUNDER AND LIGHTNING *Thin Lizzy* [28May83|18Jun83|25Jun83] 159(2) 5	0	55.4
4205. FRANK ZAPPA MEETS THE MOTHERS OF PREVENTION *Frank Zappa* [18Jan86|01Feb86|22Feb86] 153(2) 6	0	55.4
4206. MUSIC FROM SONGWRITER (SOUNDTRACK) *Willie Nelson & Kris Kristofferson* [10Nov84|24Nov84|08Dec84] 152(2) 5	0	55.2
4207. ELECTRIC HONEY *Partland Brothers* [27Jun87|04Jul87|25Jul87] 146(2) 5	1	55.2
4208. GLASS HOUSE ROCK *Greg Kihn Band* [26Apr80|24May80|24May80] 167(1) 5	0	55.2
4209. OINGO BOINGO *Oingo Boingo* [25Oct80|08Nov80|22Nov80] 163(2) 5	0	55.1
4210. SCENES IN THE CITY *Branford Marsalis* [19May84|09Jun84|30Jun84] 164(2) 7	0	55.1
4211. BURNING *Shooting Star* [30Jul83|27Aug83|03Sep83] 162(1) 6	0	55.0
4212. CURRENT *Heatwave* [10Jul82|31Jul82|14Aug82] 156(2) 6	0	55.0
4213. COOL KIDS *Kix* [28May83|09Jul83|16Jul83] 177(1) 8	0	55.0
4214. NIGHT FLIGHT *Justin Hayward* [09Aug80|23Aug80|06Sep80] 166(2) 5	0	55.0
4215. MY GIFT TO YOU *Alexander O'Neal* [17Dec88|07Jan89|14Jan89] 149(1) 5	0	54.8
4216. DUNE (SOUNDTRACK) *Toto* [22Dec84|26Jan85|09Feb85] 168(2) 8	0	54.6
4217. ONLY A LAD *Oingo Boingo* [15Aug81|29Aug81|12Sep81] 172(1) 5	0	54.6
4218. GODDESS IN PROGRESS *Julie Brown* [02Feb85|16Feb85|16Mar85] 168(1) 7	0	54.4
4219. WHY LADY WHY *Gary Morris* [15Oct83|19Nov83|03Dec83] 174(1) 8	0	54.3
4220. DOMINGO-CON AMORE *Placido Domingo* [13Mar82|13Mar82|17Apr82] 164(2) 6	0	54.2
4221. THE ART OF CONTROL *Peter Frampton* [28Aug82|18Sep82|16Oct82] 174(2) 8	0	54.1
4222. I WEAR THE FACE *Mr. Mister* [14Apr84|26May84|26May84] 170(1) 7	1	54.0
4223. GREATEST HITS *Lee Greenwood* [18May85|25May85|06Jul85] 163(2) 8	0	54.0
4224. BEWITCHED *Andy Summers & Robert Fripp* [20Oct84|03Nov84|17Nov84] 155(2) 5	0	53.7
4225. NOTHING BUT THE TRUTH *Ruben Blades* [07May88|14May88|11Jun88] 156(2) 6	0	53.6
4226. SEEN ONE EARTH *Pete Bardens* [17Oct87|31Oct87|14Nov87] 148(2) 5	0	53.6
4227. DIFFERENT KIND OF TENSION *Buzzcocks* [23Feb80|08Mar80|29Mar80] 163(1) 6	0	53.4
4228. JOHN CONLEE'S GREATEST HITS *John Conlee* [11Jun83|09Jul83|16Jul83] 166(1) 6	0	53.4
4229. TOO HOT TO SLEEP *Sylvester* [11Jul81|01Dec81|01Aug81] 156(1) 4	0	53.1
4230. SWEET SOUND *Simon Townshend* [03Dec83|07Jan84|14Jan84] 169(1) 7	0	53.0
4231. SHINE ON ME *One Way* [20Aug83|17Sep83|24Sep83] 164(1) 6	0	53.0
4232. BACK INTO BLUE *Quarterflash* [05Oct85|19Oct85|02Nov85] 150(2) 5	1	53.0
4233. PEACE IN OUR TIME *Big Country* [29Oct88|19Nov88|03Dec88] 160(2) 6	0	52.9

Ranking the Albums

Rank. Title *Act* [Enter|Peak|Exit] Peak(Wks) TotWks	Hot 100	Scr
4234. THE PLIMSOULS *The Plimsouls* [04Apr81|18Apr81|25Apr81] 153(1) 4	0	52.9
4235. DOIN' IT! *UTFO* [10Jun89|17Jun89|01Jul89] 143(2) 4	0	52.8
4236. MR. BAD GUY *Freddie Mercury* [18May85|15Jun85|22Jun85] 159(1) 6	1	52.8
4237. BRUISEOLOGY *The Waitresses* [04Jun83|18Jun83|02Jul83] 155(2) 5	0	52.7
4238. NOW WE MAY BEGIN *Randy Crawford* [31May80|28Jun80|12Jul80] 180(1) 7	0	52.6
4239. BUCKY FELLINI *The Dead Milkmen* [01Aug87|12Sep87|12Sep87] 163(1) 7	0	52.6
4240. SLAM *Dan Reed Network* [21Oct89|11Nov89|25Nov89] 160(1) 6	0	52.6
4241. THE MYSTERY OF EDWIN DROOD *Original Broadway Cast Recording* [28Jun86|12Jul86|02Aug86] 150(2) 6	0	52.6
4242. MIDNIGHT MISSION *Textones* [24Nov84|12Jan85|12Jan85] 176(1) 8	0	52.5
4243. PARTY PARTY *Soundtrack* [05Feb83|19Feb83|12Mar83] 169(4) 6	0	52.3
4244. STEVE FORBERT *Steve Forbert* [24Jul82|14Aug82|28Aug82] 159(2) 6	0	52.3
4245. KISS THE SKY *Jimi Hendrix* [17Nov84|08Dec84|15Dec84] 148(1) 5	0	52.2
4246. TRON *Soundtrack* [31Jul82|14Aug82|28Aug82] 135(2) 5	0	52.0
4247. CONTROLLED BY HATRED/FEEL LIKE SHIT...DEJA VU *Suicidal Tendencies* [28Oct89|11Nov89|25Nov89] 150(2) 5	0	52.0
4248. AIRBORNE *Don Felder* [03Dec83|18Feb84|18Feb84] 178(1) 9	0	52.0
4249. STEP ON OUT *The Oak Ridge Boys* [20Apr85|18May85|18May85] 156(1) 5	0	51.9
4250. ROCK THE NATIONS *Saxon* [14Feb87|28Feb87|21Mar87] 149(2) 6	0	51.8
4251. THE MAN FROM UTOPIA *Frank Zappa* [16Apr83|07May83|14May83] 153(1) 5	0	51.7
4252. BIG DEAL *Killer Dwarfs* [28May88|18Jun88|02Jul88] 165(1) 6	0	51.6
4253. RIFF RAFF *Dave Edmunds* [13Oct84|20Oct84|03Nov84] 140(2) 4	0	51.5
4254. EAST OF MIDNIGHT *Gordon Lightfoot* [09Aug86|06Sep86|13Sep86] 165(1) 6	0	51.5
4255. START OF A ROMANCE *Skyy* [27May89|24Jun89|24Jun89] 155(1) 5	1,1*	51.5
4256. SEARCHPARTY *The John Hall Band* [05Mar83|12Mar83|02Apr83] 147(2) 5	1	51.3
4257. LUCKY *Marty Balin* [12Mar83|26Mar83|16Apr83] 156(1) 6	1	51.2
4258. STRANGE KIND OF LOVE *Love And Money* [25Mar89|08Apr89|06May89] 175(2) 7	1	51.2
4259. BULL DURHAM *Soundtrack* [06Aug88|20Aug88|10Sep88] 157(2) 6	0	51.2
4260. SCRIPT FOR A JESTER'S TEAR *Marillion* [25Jun83|09Jul83|06Aug83] 175(1) 7	0	51.2
4261. JEALOUS LOVER *Rainbow* [14Nov81|05Dec81|05Dec81] 147(1) 4	0	51.1
4262. SPEAK & SPELL *Depeche Mode* [26Dec81|20Feb82|20Feb82] 192(1) 9	0	51.1
4263. GIVE IT UP *Pleasure* [15May82|12Jun82|19Jun82] 164(1) 6	0	51.1
4264. 14 GREATEST HITS *Michael Jackson & The Jackson 5* [23Jun84|21Jul84|04Aug84] 168(1) 7	0	51.0
4265. GO FOR YOUR LIFE *Mountain* [27Apr85|18May85|01Jun85] 166(2) 6	0	51.0
4266. REUNION CONCERT *The Everly Brothers* [10Mar84|31Mar84|07Apr84] 162(2) 5	0	50.9
4267. CLASSICS THE EARLY YEARS *Neil Diamond* [25Jun83|09Jul83|06Aug83] 171(1) 7	0	50.8
4268. A GRP CHRISTMAS COLLECTION *Various Artists* [07Jan89|07Jan89|06Jan90] 140(1) 4	0	50.7
4269. ETERNAL NIGHTMARE *Vio-Lence* [20Aug88|10Sep88|24Sep88] 154(2) 6	0	50.7
4270. JOHN O'BANION *John O'Banion* [16May81|06Jun81|06Jun81] 164(1) 4	1	50.7
4271. MAYBE IT'S LIVE *Robert Palmer* [15May82|29May82|12Jun82] 148(1) 5	0	50.6
4272. XL-1 *Pete Shelley* [23Jul83|13Aug83|20Aug83] 151(1) 5	0	50.6
4273. HAI HAI *Roger Hodgson* [31Oct87|14Nov87|05Dec87] 163(2) 6	0	50.6
4274. THE SPIRIT'S IN IT *Patti LaBelle* [03Oct81|03Oct81|24Oct81] 156(3) 4	0	50.5
4275. STARS ON LONG PLAY III *Stars On* [08May82|22May82|12Jun82] 163(2) 6	1	50.5
4276. SPUN GOLD *Barbara Mandrell* [03Sep83|17Sep83|24Sep83] 140(1) 4	0	50.5
4277. PARTY *Iggy Pop* [19Sep81|03Oct81|17Oct81] 166(2) 5	0	50.4
4278. THIRTEEN *Emmylou Harris* [08Mar86|29Mar86|12Apr86] 157(2) 6	0	50.4
4279. WHISPERS AND PROMISES *Earl Klugh* [20May89|03Jun89|17Jun89] 150(2) 5	0	50.4
4280. MEAN *Montrose* [30May87|06Jun87|11Jul87] 165(2) 7	0	50.2
4281. SOME OF MY BEST JOKES ARE FRIENDS *George Clinton* [10Aug85|31Aug85|14Sep85] 163(2) 6	0	50.2
4282. COUP DE GRACE *Mink De Ville* [24Oct81|31Oct81|21Nov81] 161(2) 5	0	50.2
4283. POV *Utopia* [16Mar85|06Apr85|20Apr85] 161(2) 6	0	50.1
4284. REAL LIFE STORY *Terri Lyne Carrington* [29Apr89|13May89|10Jun89] 169(1) 7	0	49.9
4285. CAGE THE SONGBIRD *Crystal Gayle* [12Nov83|10Dec83|17Dec83] 171(1) 6	1	49.9
4286. SWING THE HEARTACHE - THE BBC SESSIONS *Bauhaus* [12Aug89|09Sep89|16Sep89] 169(1) 6	0	49.8
4287. MOONS OF JUPITER *Scruffy The Cat* [17Dec88|14Jan89|04Feb89] 177(1) 8	0	49.6
4288. MAMA'S BOYS *Mama's Boys* [11Aug84|01Sep84|29Sep84] 172(1) 8	0	49.6
4289. HEAVY MENTAL *The Fools* [28Mar81|11Apr81|18Apr81] 158(1) 4	1	49.5
4290. MAKING CONTACT *UFO* [30Apr83|14May83|28May83] 153(1) 5	0	49.4
4291. THE KING OF COMEDY *Soundtrack* [16Apr83|14May83|21May83] 162(1) 6	0	49.4
4292. BRUCE COCKBURN RESUME *Bruce Cockburn* [23May81|06Jun81|20Jun81] 174(2) 5	0	49.4
4293. DISTANT THUNDER *Aswad* [13Aug88|03Sep88|24Sep88] 173(3) 7	0	49.1
4294. IN CONTROL VOLUME I *Marley Marl* [08Oct88|22Oct88|05Nov88] 163(2) 5	0	49.0
4295. ADVENTURES IN MODERN RECORDING *The Buggles* [27Mar82|17Apr82|24Apr82] 161(2) 5	0	48.9
4296. MORE OF THE GOOD LIFE *T.S. Monk* [09Jan82|30Jan82|27Feb82] 176(1) 8	0	48.8
4297. THIS IS THE DAY...THIS IS THE HOUR...THIS IS THIS! *Pop Will Eat Itself* [26Aug89|16Sep89|30Sep89] 169(2) 6	0	48.7
4298. THE BEST OF DAVE EDMUNDS *Dave Edmunds* [09Jan82|30Jan82|06Feb82] 163(1) 5	0	48.6
4299. RISE AND SHINE *The Bears* [30Apr88|21May88|28May88] 159(1) 5	0	48.6
4300. INAMORATA *Poco* [19May84|09Jun84|23Jun84] 167(2) 6	1	48.6
4301. GREATEST HITS *Charley Pride* [21Nov81|05Dec81|02Jan82] 185(1) 7	0	48.5
4302. ROCK HARD *Suzi Quatro* [01Nov80|22Nov80|29Nov80] 165(1) 5	1	48.5
4303. I GOT THE MELODY *Odyssey* [18Jul81|08Aug81|15Aug81] 175(2) 5	0	48.5
4304. COME SHARE MY LOVE *Miki Howard* [14Mar87|28Mar87|23May87] 171(2) 6	0	48.5
4305. WYNTON MARSALIS *Wynton Marsalis* [06Mar82|03Apr82|03Apr82] 165(1) 5	0	48.4
4306. RAVEL BOLERO *Tomita* [09Feb80|23Feb80|08Mar80] 174(1) 5	0	48.2
4307. NATIONAL BREAKOUT *The Romantics* [06Dec80|13Dec80|17Jan81] 176(2) 7	0	48.1
4308. POPS IN SPACE *The Boston Pops Orchestra/John Williams* [20Dec80|24Jan81|24Jan81] 181(1) 6	0	48.0
4309. RUMBLE FISH (SOUNDTRACK) *Stewart Copeland* [17Dec83|14Jan84|14Jan84] 157(1) 5	0	47.8
4310. UHF/ORIGINAL MOTION PICTURE SOUNDTRACK AND OTHER STUFF *"Weird Al" Yankovic* [19Aug89|19Aug89|09Sep89] 146(2) 4	0	47.7
4311. CLOSER THAN CLOSE *Jean Carne* [06Sep86|20Sep86|11Oct86] 162(2) 6	0	47.6
4312. CHRISTMAS JOLLIES II *The Salsoul Orchestra* [19Dec81|09Jan82|16Jan82] 170(2) 5	0	47.3
4313. GENETIC WALK *Ahmad Jamal* [15Mar80|05Apr80|12Apr80] 173(1) 5	0	47.1
4314. IN THE MOOD FOR SOMETHING RUDE *Foghat* [13Nov82|04Dec82|11Dec82] 162(1) 5	0	47.0
4315. JAMMIN' IN MANHATTAN *Tyzik* [08Sep84|06Oct84|13Oct84] 172(1) 6	0	47.0
4316. PLAYIN' IT COOL *Timothy B. Schmit* [10Nov84|24Nov84|08Dec84] 160(2) 5	0	46.9
4317. SOMETHING HEAVY GOING DOWN - LIVE FROM THE TWILIGHT ZONE *Golden Earring* [24Nov84|08Dec84|29Dec84] 158(2) 6	0	46.8
4318. RICH AND FAMOUS *Blue Mercedes* [14May88|28May88|11Jun88] 165(2) 5	1	46.8
4319. LAUGHTER *Ian Dury And The Blockheads* [07Feb81|21Feb81|28Feb81] 159(1) 4	0	46.6
4320. HEARTBREAK RADIO *Rita Coolidge* [12Sep81|26Sep81|03Oct81] 160(1) 4	0	46.5
4321. THE PIRATE MOVIE *Soundtrack* [28Aug82|18Sep82|02Oct82] 166(2) 6	1	46.5
4322. MAX Q *Max Q* [07Oct89|07Oct89|25Nov89] 182(2) 8	0	46.4
4323. SOME COME RUNNING *Jim Capaldi* [17Dec88|04Feb89|11Feb89] 183(2) 8	0	46.4
4324. TOUGH *Kurtis Blow* [09Oct82|23Oct82|06Nov82] 167(2) 5	0	46.3
4325. IRON FIST *Motorhead* [22May82|05Jun82|26Jun82] 174(1) 6	0	46.2
4326. BABY TONIGHT *Marlon Jackson* [28Nov87|05Dec87|09Jan88] 175(2) 7	0	46.1
4327. LOVE IS FAIR *Barbara Mandrell* [27Sep80|11Oct80|01Nov80] 175(1) 6	0	46.0
4328. TOO HOT TO STOP IT *The Manhattans* [13Apr85|11May85|18May85] 171(2) 6	1	46.0
4329. CONVERSATIONS *Brass Construction* [11Jun83|09Jul83|16Jul83] 176(1) 6	0	45.8
4330. ESCAPADE *Tim Finn* [17Sep83|08Oct83|15Oct83] 161(1) 5	0	45.8

Ranking the Albums

Rank. Title *Act* [Enter|Peak|Exit] Peak(Wks) TotWks	Hot 100	Scr
4331. STAND UP *The Del Fuegos* [18Apr87|25Apr87|23May87] 167(2) 6	0	45.8
4332. EMPIRE JAZZ *Various Artists* [02Aug80|16Aug80|30Aug80] 168(1) 5	0	45.8
4333. ANGST IN MY PANTS *Sparks* [22May82|12Jun82|26Jun82] 173(2) 6	1	45.7
4334. NOT THE BOY NEXT DOOR *Peter Allen* [12Mar83|09Apr83|16Apr83] 170(2) 6	0	45.7
4335. BOOK OF DAYS *The Psychedelic Furs* [25Nov89|25Nov89|16Dec89] 138(2) 4	0	45.7
4336. JAMES BROWN...LIVE/HOT ON THE ONE *James Brown* [16Aug80|30Aug80|13Sep80] 170(2) 5	0	45.7
4337. A CURIOUS FEELING *Tony Banks* [15Dec79|05Jan80|12Jan80] 171(1) 5	0	45.6
4338. 10TH ANNIVERSARY *The Statler Brothers* [06Sep80|27Sep80|04Oct80] 169(1) 5	0	45.6
4339. 2 X 4 *Guadalcanal Diary* [16Jan88|30Jan88|27Feb88] 183(1) 7	0	45.5
4340. MARSALIS STANDARD TIME: VOL. 1 *Wynton Marsalis* [26Sep87|03Oct87|31Oct87] 153(1) 5	0	45.4
4341. ONE BIG DAY *Face To Face* [18Jun88|16Jul88|30Jul88] 176(1) 7	0	45.4
4342. MY MOVIE OF YOU *Leif Garrett* [12Dec81|16Jan82|23Jan82] 185(1) 7	1	45.4
4343. THE BEST OF THE BLUES BROTHERS *Blues Brothers* [09Jan82|16Jan82|23Jan82] 143(2) 3	0	45.4
4344. "NOW APPEARING" AT OLE' MISS *B.B. King* [26Apr80|10May80|17May80] 162(2) 4	0	45.3
4345. EVERYBODY WANTS SOME *Gucci Crew II* [23Sep89|30Sep89|28Oct89] 173(1) 6	0	45.3
4346. LOVE YOUR MAN *The Rossington Band* [16Jul88|30Jul88|06Aug88] 140(1) 4	0	45.1
4347. 1980 *B.T. Express* [31May80|14Jun80|21Jun80] 164(1) 4	0	45.1
4348. U.S.A. FOR M.O.D. *M.O.D.* [07Nov87|28Nov87|05Dec87] 153(2) 5	0	45.1
4349. D'YA LIKE SCRATCHIN' *Malcolm McLaren* [18Feb84|17Mar84|24Mar84] 173(2) 6	0	45.1
4350. THE BEST OF KANSAS *Kansas* [08Sep84|15Sep84|06Oct84] 154(2) 5	0	45.0
4351. COUNTERFEIT *Martin L. Gore* [12Aug89|19Aug89|09Sep89] 156(1) 5	0	44.9
4352. NIGHTS (FEEL LIKE GETTING DOWN) *Billy Ocean* [25Jul81|08Aug81|08Aug81] 152(1) 3	0	44.7
4353. TOO TOUGH TO DIE *The Ramones* [03Nov84|01Dec84|08Dec84] 171(1) 6	0	44.6
4354. MENLOVE AVENUE *John Lennon* [22Nov86|29Nov86|13Dec86] 127(2) 4	0	44.5
4355. VERY GREASY *David Lindley & El Rayo-X* [24Sep88|15Oct88|29Oct88] 174(1) 6	0	44.4
4356. DALLAS *Floyd Cramer* [24May80|14Jun80|21Jun80] 170(1) 5	0	44.1
4357. ANCIENT DREAMS *Candlemass* [21Jan89|04Feb89|25Feb89] 174(2) 6	0	44.1
4358. FARRENHEIT *Farrenheit* [09May87|06Jun87|20Jun87] 179(1) 7	0	43.8
4359. GRAND SLAM *Spinners* [08Jan83|22Jan83|12Feb83] 167(2) 6	1	43.8
4360. ROMAN GODS *The Fleshtones* [06Mar82|13Mar82|03Apr82] 174(3) 5	0	43.8
4361. YOUNGBLOOD *Soundtrack* [01Mar86|22Mar86|05Apr86] 166(3) 6	0	43.7
4362. CHICAGO - GREATEST HITS, VOLUME II *Chicago* [12Dec81|26Dec81|09Jan82] 171(2) 5	0	43.7
4363. RESTLESS EYES *Janis Ian* [04Jul81|18Jul81|18Jul81] 156(1) 3	1	43.5
4364. WHY NOT ME *Fred Knoblock* [04Oct80|18Oct80|01Nov80] 179(1) 5	1	43.5
4365. PROFILES *Nick Mason & Rick Fenn* [31Aug85|14Sep85|28Sep85] 154(2) 5	0	43.5
4366. ROCK & ROLL ADULT *Garland Jeffreys* [31Oct81|14Nov81|21Nov81] 163(2) 4	1	43.4
4367. WAITING ON YOU *Brick* [12Jul80|19Jul80|20Dec80] 179(2) 5	0	43.3
4368. GHOST ON THE BEACH *Insiders* [10Oct87|31Oct87|07Nov87] 167(1) 5	0	43.3
4369. FIVE MILES OUT *Mike Oldfield* [08May82|22May82|05Jun82] 164(2) 5	0	43.3
4370. OCEANLINER *Passport* [05Apr80|19Apr80|26Apr80] 163(1) 4	0	43.2
4371. A BROKEN FRAME *Depeche Mode* [04Dec82|18Dec82|22Jan83] 177(1) 8	0	43.2
4372. EVERLASTING LOVE *Rex Smith* [22Aug81|05Sep81|12Sep81] 167(2) 4	1	43.0
4373. CALLING *Noel Pointer* [16Aug80|30Aug80|06Sep80] 167(1) 5	0	43.0
4374. NIGHTHAWKS *The Nighthawks* [26Jul80|09Aug80|16Aug80] 166(1) 4	0	42.9
4375. GREATEST HITS, VOL. II *Hank Williams Jr.* [11Jan86|08Feb86|01Mar86] 183(1) 8	0	42.9
4376. MADAME X *Madame X* [10Oct87|17Oct87|07Nov87] 162(2) 5	0	42.9
4377. ALL THE PEOPLE ARE TALKIN' *John Anderson* [29Oct83|19Nov83|26Nov83] 163(1) 5	0	42.8
4378. LOVE FANTASY *Roy Ayers* [01Nov80|15Nov80|15Nov80] 157(1) 3	0	42.4
4379. FOLKSONGS FOR A NUCLEAR VILLAGE *Shadowfax* [14May88|04Jun88|11Jun88] 168(1) 5	0	42.3
4380. RAMONES MANIA *The Ramones* [25Jun88|02Jul88|23Jul88] 168(2) 5	0	42.2
4381. NIGHTLINE *Randy Crawford* [05Nov83|26Nov83|03Dec83] 164(1) 5	0	42.1
4382. INTRODUCING THE STYLE COUNCIL *The Style Council* [22Oct83|19Nov83|19Nov83] 172(2) 5	0	42.1
4383. THE STEVE HOWE ALBUM *Steve Howe* [16Feb80|23Feb80|08Mar80] 164(1) 4	0	42.0
4384. THE McCARTNEY INTERVIEW *Paul McCartney* [31Jan81|07Feb81|14Feb81] 158(2) 3	0	42.0
4385. IMITATION LIFE *Robin Lane & The Chartbusters* [25Apr81|02May81|16May81] 172(1) 4	0	41.9
4386. THE FIRST 25 YEARS-THE SILVER ANNIVERSARY ALBUM *Johnny Mathis* [25Jul81|08Aug81|15Aug81] 173(2) 4	0	41.6
4387. CONFESSIONS OF A POP GROUP *The Style Council* [13Aug88|10Sep88|17Sep88] 174(1) 6	0	41.5
4388. THE RHYTHM & THE BLUES *Z.Z. Hill* [05Feb83|05Mar83|05Mar83] 165(1) 5	0	41.4
4389. SUNSHINE DREAM *The Beach Boys* [03Jul82|17Jul82|07Aug82] 180(3) 6	0	41.4
4390. BORN IN AMERICA *Riot* [14Jan84|11Feb84|18Feb84] 175(1) 4	0	41.4
4391. DANCE *Gary Numan* [24Oct81|14Nov81|14Nov81] 167(1) 4	0	41.4
4392. THE ISLAND STORY, 1962-1987: THE 25TH ANNIVERSARY *Various Artists* [26Dec87|09Jan88|30Jan88] 180(2) 6	0	41.4
4393. THE BEST OF THOMPSON TWINS/GREATEST MIXES *Thompson Twins* [27Aug88|17Sep88|01Oct88] 175(2) 6	0	41.3
4394. PASSION, GRACE & FIRE *Al Di Meola/John McLaughlin/Paco De Lucia* [20Aug83|10Sep83|17Sep83] 171(2) 5	0	41.2
4395. POLTERGEIST *Soundtrack* [17Jul82|07Aug82|14Aug82] 168(1) 5	0	41.1
4396. ELVIS: THE FIRST LIVE RECORDINGS *Elvis Presley* [17Mar84|07Apr84|07Apr84] 163(1) 4	0	41.0
4397. ELECTRIC FOLKLORE LIVE *The Alarm* [29Oct88|12Nov88|26Nov88] 167(2) 5	0	40.8
4398. FICKLE *Michael Henderson* [04Jun83|25Jun83|02Jul83] 169(1) 5	0	40.8
4399. MUTUAL ATTRACTION *Sylvester* [14Feb87|21Feb87|14Mar87] 164(2) 5	0	40.7
4400. LIVE AT THE WHISKY A GO-GO ON THE FABULOUS SUNSET STRIP *X* [14May88|28May88|11Jun88] 175(2) 5	0	40.6
4401. JUNK CULTURE *Orchestral Manoeuvres In The Dark* [24Nov84|15Dec84|29Dec84] 182(3) 6	0	40.4
4402. VICTIMS OF THE FUTURE *Gary Moore* [09Jun84|07Jul84|07Jul84] 172(1) 5	0	40.3
4403. THE PEOPLE WHO GRINNED THEMSELVES TO DEATH *The Housemartins* [16Jan88|30Jan88|20Feb88] 177(1) 6	0	40.1
4404. WHEN THE NIGHT COMES *Lou Rawls* [14May83|04Jun83|04Jun83] 163(1) 4	1	40.0
4405. HEY JOE, HEY MOE *Moe Bandy & Joe Stampley* [11Apr81|25Apr81|02May81] 170(1) 4	0	40.0
4406. PUSH *Bros* [23Jul88|13Aug88|20Aug88] 171(1) 5	1	40.0
4407. CONAN THE BARBARIAN *Soundtrack* [12Jun82|26Jun82|10Jul82] 162(2) 5	0	39.9
4408. NEW AFFAIR *The Emotions* [26Sep81|03Oct81|17Oct81] 168(2) 4	0	39.9
4409. VISIONS OF THE LITE *Slave* [15Jan83|12Feb83|19Feb83] 177(1) 6	0	39.9
4410. MORE SONGS ABOUT LOVE & HATE *The Godfathers* [20May89|03Jun89|24Jun89] 174(1) 6	0	39.8
4411. LOVE LIGHT *Yutaka* [11Jul81|01Aug81|01Aug81] 174(1) 4	1	39.8
4412. WITCHDOCTOR *Sidewinders* [13May89|10Jun89|10Jun89] 169(1) 5	0	39.8
4413. KNIFE *Aztec Camera* [13Oct84|03Nov84|17Nov84] 175(2) 6	0	39.8
4414. WORD OF MOUTH *Jaco Pastorius* [15Aug81|22Aug81|29Aug81] 161(2) 3	0	39.7
4415. THE LAND OF RAPE AND HONEY *Ministry* [05Nov88|19Nov88|26Nov88] 164(2) 4	0	39.6
4416. FROM A CHILD TO A WOMAN *Julio Iglesias* [01Sep84|06Oct84|06Oct84] 181(1) 5	0	39.6
4417. BELO HORIZONTE *John McLaughlin* [12Dec81|19Dec81|02Jan82] 172(3) 4	0	39.6
4418. RAIN DOGS *Tom Waits* [16Nov85|30Nov85|08Feb86] 181(2) 7	0	39.5
4419. INTUITION *Linx* [20Jun81|11Jul81|11Jul81] 175(1) 4	0	39.5
4420. SUNDOWN *Rank And File* [07May83|28May83|04Jun83] 165(1) 5	0	39.3
4421. INSIDE LOOKIN' OUT *Junior* [23Jul83|13Aug83|27Aug83] 177(1) 6	0	39.3
4422. HERE COMES THE NIGHT *David Johansen* [11Jul81|25Jul81|25Jul81] 160(1) 3	0	39.3
4423. BIG CIRCUMSTANCE *Bruce Cockburn* [25Feb89|25Mar89|08Apr89] 182(1) 7	0	39.3
4424. IN LOVE *Cheryl Lynn* [19Jan80|02Feb80|09Feb80] 167(1) 4	0	39.3

Ranking the Albums

Rank. Title Act [Enter | Peak | Exit] Peak(Wks) TotWks	Hot 100	Scr
4425. ALL I NEED Sylvester [19Mar83 | 02Apr83 | 16Apr83] 168(2) 5	0	39.2
4426. BAD ENUFF Slave [22Oct83 | 12Nov83 | 19Nov83] 168(1) 5	0	39.2
4427. MUD WILL BE FLUNG TONIGHT! Bette Midler [21Dec85 | 11Jan86 | 25Jan86] 183(1) 6	0	39.1
4428. 3 SHIPS Jon Anderson [28Dec85 | 11Jan86 | 25Jan86] 166(2) 5	0	39.1
4429. TENDERNESS Ohio Players [11Apr81 | 18Apr81 | 25Apr81] 165(2) 3	0	39.0
4430. TOO MUCH PRESSURE The Selecter [03May80 | 24May80 | 24May80] 175(1) 4	0	39.0
4431. BILL & TED'S EXCELLENT ADVENTURE Soundtrack [08Apr89 | 22Apr89 | 29Apr89] 170(2) 4	0	38.9
4432. COME DANCING WITH THE KINKS/THE BEST OF THE KINKS 1977-1986 The Kinks [19Jul86 | 02Aug86 | 09Aug86] 159(1) 4	0	38.8
4433. IN LOVE Bunny Debarge [14Mar87 | 28Mar87 | 11Apr87] 172(2) 5	0	38.8
4434. AMAZON BEACH The Kings [26Sep81 | 10Oct81 | 17Oct81] 170(1) 4	0	38.8
4435. TAP Soundtrack [11Mar89 | 25Mar89 | 01Apr89] 166(2) 4	0	38.8
4436. WHITE NOISE Jay Ferguson [17Apr82 | 24Apr82 | 15May82] 178(3) 5	0	38.8
4437. RESCUE Clarence Clemons [05Nov83 | 26Nov83 | 03Dec83] 174(1) 5	0	38.6
4438. CRISTOFORI'S DREAM David Lanz [05Nov88 | 19Nov88 | 10Dec88] 180(1) 6	0	38.6
4439. V Zapp [07Oct89 | 14Oct89 | 28Oct89] 154(2) 4	0	38.5
4440. OZARK MOUNTAIN DAREDEVILS Ozark Mountain Daredevils [24May80 | 07Jun80 | 14Jun80] 170(1) 4	1	38.5
4441. LITTLE DREAMER Peter Green [25Oct80 | 15Nov80 | 22Nov80] 186(1) 5	0	38.4
4442. AFTERNOONS IN UTOPIA Alphaville [30Aug86 | 13Sep86 | 04Oct86] 174(1) 6	0	38.4
4443. VERTICAL SMILES Blackfoot [27Oct84 | 10Nov84 | 24Nov84] 176(3) 5	0	38.4
4444. DAVE GRUSIN AND THE NY/LA DREAM BAND Dave Grusin [16Apr83 | 07May83 | 21May83] 181(1) 6	0	38.2
4445. LE CHAT BLEU Mink De Ville [13Sep80 | 20Sep80 | 27Sep80] 163(2) 3	0	38.2
4446. NIGHT ATTACK Angel City [20Mar82 | 10Apr82 | 17Apr82] 174(1) 5	0	38.2
4447. THREE QUARTETS Chick Corea [01Aug81 | 15Aug81 | 22Aug81] 179(2) 4	0	38.2
4448. SIRIUS Clannad [05Mar88 | 19Mar88 | 02Apr88] 183(2) 5	0	38.0
4449. LIVE SHOTS Joe Ely [24Oct81 | 31Oct81 | 07Nov81] 159(2) 3	0	38.0
4450. SALUTE Gordon Lightfoot [13Aug83 | 03Sep83 | 10Sep83] 175(1) 5	0	38.0
4451. SWITCH V Switch [21Nov81 | 05Dec81 | 12Dec81] 174(1) 4	0	37.9
4452. BIG FUN Inner City [24Jun89 | 24Jun89 | 15Jul89] 162(2) 4	1	37.8
4453. AMANDLA Miles Davis [17Jun89 | 08Jul89 | 15Jul89] 177(2) 5	0	37.8
4454. NO EASY WALK TO FREEDOM Peter, Paul & Mary [14Mar87 | 04Apr87 | 11Apr87] 173(2) 5	0	37.8
4455. PAUL DAVIS Paul Davis [26Apr80 | 10May80 | 17May80] 173(2) 4	2	37.6
4456. THE ATLANTIC YEARS Roxy Music [21Jan84 | 18Feb84 | 25Feb84] 183(2) 6	0	37.6
4457. ONLY LIFE The Feelies [19Nov88 | 26Nov88 | 17Dec88] 173(2) 5	0	37.5
4458. ROCK AND ROLL DIARY 1967-1980 Lou Reed [20Dec80 | 27Dec80 | 10Jan81] 178(3) 4	0	37.4
4459. HEY! Julio Iglesias [01Sep84 | 29Sep84 | 06Oct84] 179(1) 6	0	37.4
4460. TIM The Replacements [01Feb86 | 15Feb86 | 15Mar86] 183(1) 7	0	37.3
4461. GREATEST HITS Dave & Sugar [07Mar81 | 14Mar81 | 28Mar81] 179(2) 4	0	37.1
4462. TENDER TOGETHERNESS Stanley Turrentine [10Oct81 | 17Oct81 | 24Oct81] 162(2) 3	0	37.1
4463. PREPPIE Cheryl Lynn [28Apr84 | 05May84 | 26May84] 161(1) 5	1	37.0
4464. TONGUE IN CHIC Chic [04Dec82 | 11Dec82 | 08Jan83] 173(2) 6	0	36.9
4465. VISAGE Visage [08Aug81 | 15Aug81 | 29Aug81] 178(2) 4	0	36.9
4466. LIVE AND LOWDOWN AT THE APOLLO VOL 1 James Brown [22Nov80 | 29Nov80 | 06Dec80] 163(2) 3	0	36.8
4467. PROFILE II: THE BEST OF EMMYLOU HARRIS Emmylou Harris [06Oct84 | 13Oct84 | 10Nov84] 176(1) 6	0	36.8
4468. MYSTERY WALK M + M [28Jul84 | 11Aug84 | 18Aug84] 163(2) 4	1	36.7
4469. NO FREE LUNCH Green On Red [03May86 | 07Jun86 | 07Jun86] 177(1) 6	0	36.6
4470. RAP'S GREATEST HITS, VOLUME 2 Various Artists [02May87 | 23May87 | 23May87] 167(1) 4	0	36.6
4471. RED SAILS IN THE SUNSET Midnight Oil [03Aug85 | 10Aug85 | 07Sep85] 177(2) 6	0	36.6
4472. THE EMPIRE STRIKES BACK/THE ADVENTURES OF LUKE SKYWALKER Soundtrack [06Sep80 | 13Sep80 | 27Sep80] 178(2) 4	0	36.5
4473. FABLES Jean-Luc Ponty [02Nov85 | 16Nov85 | 23Nov85] 166(2) 4	0	36.4
4474. CRUSADER Saxon [14Apr84 | 28Apr84 | 12May84] 174(2) 5	0	36.3
4475. DISCOVERY Larry Carlton [01Aug87 | 22Aug87 | 05Sep87] 180(1) 6	0	36.3
4476. WALKING WILD New England [18Jul81 | 01Aug81 | 08Aug81] 176(2) 4	0	36.1
4477. FOLLIES IN CONCERT Original Cast [25Jan86 | 08Feb86 | 01Mar86] 181(2) 6	0	36.0
4478. 2010 Soundtrack [02Feb85 | 23Feb85 | 02Mar85] 173(2) 6	0	36.0
4479. MY TOOT-TOOT Rockin' Sidney [24Aug85 | 31Aug85 | 14Sep85] 166(3) 4	0	35.7
4480. HOOKED ON ROCK CLASSICS London Symphony Orchestra [05Mar83 | 05Mar83 | 19Mar83] 145(2) 3	0	35.7
4481. CHRISTINE Soundtrack [21Jan84 | 18Feb84 | 18Feb84] 177(1) 5	0	35.6
4482. JUMPIN' JACK FLASH Soundtrack [22Nov86 | 29Nov86 | 13Dec86] 159(2) 4	0	35.3
4483. TOMMY PAGE Tommy Page [06May89 | 13May89 | 03Jun89] 166(2) 5	1	35.3
4484. DARKLANDS Jesus And Mary Chain [17Oct87 | 31Oct87 | 07Nov87] 161(2) 4	0	35.3
4485. LICENSE TO KILL Malice [11Apr87 | 25Apr87 | 16May87] 177(2) 6	0	35.3
4486. FLETCH Soundtrack [27Jul85 | 03Aug85 | 17Aug85] 160(1) 4	1	35.2
4487. THE FIRE STILL BURNS Russ Ballard [03Aug85 | 10Aug85 | 24Aug85] 166(2) 4	0	35.2
4488. SECOND TIME AROUND The Kinks [20Sep80 | 04Oct80 | 11Oct80] 177(2) 4	0	35.2
4489. ANIMATION Jon Anderson [03Jul82 | 17Jul82 | 31Jul82] 176(1) 5	0	35.2
4490. SURFIN' M.O.D. M.O.D. [17Sep88 | 17Sep88 | 22Oct88] 186(2) 6	0	35.1
4491. CHANGES The Monkees [08Nov86 | 22Nov86 | 29Nov86] 152(1) 4	0	34.9
4492. ROUND TRIP The Gap Band [09Dec89 | 30Dec89 | 10Feb90] 189(2) 7	0	34.9
4493. WILDER The Teardrop Explodes [06Feb82 | 13Feb82 | 27Feb82] 176(3) 4	0	34.9
4494. HARD Gang Of Four [08Oct83 | 22Oct83 | 29Oct83] 168(1) 4	0	34.9
4495. SPIRIT OF PLACE Goanna [25Jun83 | 23Jul83 | 23Jul83] 179(1) 5	1	34.9
4496. FIYO ON THE BAYOU Neville Brothers [29Aug81 | 12Sep81 | 12Sep81] 166(1) 3	0	34.8
4497. DANCING IN THE SUN George Howard [27Jul85 | 10Aug85 | 17Aug85] 169(2) 4	0	34.7
4498. THE BLUE MASK Lou Reed [27Feb82 | 13Mar82 | 20Mar82] 169(1) 4	0	34.6
4499. NICK MASON'S FICTITIOUS SPORTS Nick Mason [04Jul81 | 11Jul81 | 18Jul81] 170(2) 3	0	34.6
4500. TWO WHEELS GOOD Prefab Sprout [02Nov85 | 30Nov85 | 30Nov85] 178(1) 5	0	34.5
4501. ALL MIXED UP Alexander O'Neal [25Feb89 | 18Mar89 | 25Mar89] 185(2) 5	0	34.5
4502. HANG TOGETHER Odyssey [14Jun80 | 05Jul80 | 05Jul80] 181(1) 4	0	34.5
4503. THE BIG EXPRESS XTC [10Nov84 | 01Dec84 | 08Dec84] 178(1) 5	0	34.4
4504. SAYIN' SOMETHING! Peaches & Herb [12Sep81 | 19Sep81 | 26Sep81] 168(2) 3	0	34.2
4505. 1969 Soundtrack [17Dec88 | 21Jan89 | 21Jan89] 186(1) 6	0	34.0
4506. YES, GIORGIO Luciano Pavarotti [06Nov82 | 13Nov82 | 20Nov82] 158(2) 3	0	33.8
4507. SERIOUS Deja [05Dec87 | 09Jan88 | 09Jan88] 186(1) 6	1	33.7
4508. SHADDAP YOU FACE Joe Dolce [27Jun81 | 04Jul81 | 18Jul81] 181(2) 4	1	33.7
4509. FLAUNT THE IMPERFECTION China Crisis [01Jun85 | 22Jun85 | 22Jun85] 171(1) 4	0	33.7
4510. RUNAWAY Bill Champlin [06Feb82 | 13Feb82 | 27Feb82] 178(3) 4	2	33.5
4511. GIANT Woodentops [20Sep86 | 04Oct86 | 25Oct86] 185(1) 6	0	33.4
4512. THE BALLAD OF SALLY ROSE Emmylou Harris [25May85 | 08Jun85 | 15Jun85] 171(2) 4	0	33.3
4513. THE RIGHT COMBINATION Linda Clifford & Curtis Mayfield [19Jul80 | 26Jul80 | 09Aug80] 180(2) 4	0	33.3
4514. AMNESIA Richard Thompson [05Nov88 | 26Nov88 | 03Dec88] 182(2) 5	0	33.2
4515. BEST BITS Roger Daltrey [27Mar82 | 17Apr82 | 24Apr82] 185(2) 5	0	33.1
4516. THE BEST OF THE WHISPERS The Whispers [13Mar82 | 27Mar82 | 10Apr82] 180(1) 5	0	33.1
4517. CHRISTMAS WITH SLIM WHITMAN Slim Whitman [13Dec80 | 27Dec80 | 03Jan81] 184(2) 4	0	33.1
4518. URBAN BEACHES Cactus World News [09Aug86 | 06Sep86 | 06Sep86] 179(1) 5	0	33.0
4519. VICTORY Larry Graham [30Jul83 | 13Aug83 | 20Aug83] 173(1) 4	0	33.0
4520. ROCK & ROLL REBELS John Kay & Steppenwolf [26Sep87 | 10Oct87 | 17Oct87] 171(1) 4	0	32.9
4521. MEDICINE SHOW Dream Syndicate [04Aug84 | 25Aug84 | 25Aug84] 171(1) 4	0	32.8
4522. PIED PIPER Dave Valentin [08Aug81 | 08Aug81 | 29Aug81] 184(3) 4	0	32.8

Ranking the Albums

Rank. Title *Act* [Enter|Peak|Exit] Peak(Wks) TotWks	Hot 100	Scr
4523. ON A NIGHT LIKE THIS *Buckwheat Zydeco* [14Nov87|28Nov87|12Dec87] 172(1) 5	0	32.7
4524. JUST BETWEEN US *Gerald Albright* [27Feb88|12Mar88|26Mar88] 181(2) 5	0	32.7
4525. CASTLES IN THE SAND *David Allan Coe* [09Jul83|30Jul83|06Aug83] 179(1) 5	0	32.7
4526. DON'T FOLLOW ME, I'M LOST TOO *Pearl Harbor* [21Feb81|28Feb81|07Mar81] 170(2) 3	0	32.7
4527. THE NIGHT THE LIGHTS WENT OUT IN GEORGIA *Soundtrack* [22Aug81|12Sep81|19Sep81] 189(1) 5	1	32.7
4528. GREATEST HITS *Juice Newton* [21Jul84|04Aug84|18Aug84] 178(2) 5	0	32.6
4529. FOREVER, REX SMITH *Rex Smith* [12Jan80|19Jan80|26Jan80] 165(1) 3	0	32.5
4530. A COUNTRY CHRISTMAS *Various Artists* [25Dec82|08Jan83|15Jan83] 172(2) 4	0	32.3
4531. LIFE (IS SO STRANGE) *War* [23Jul83|06Aug83|13Aug83] 164(1) 4	0	32.3
4532. ESQUIRE *Esquire* [28Mar87|11Apr87|18Apr87] 165(2) 4	0	32.3
4533. TAP STEP *Chick Corea* [10May80|17May80|24May80] 170(2) 3	0	32.1
4534. STORIES *Gloria Gaynor* [24May80|07Jun80|14Jun80] 178(1) 4	0	31.9
4535. SHADES OF LACE *Lace* [23Jan88|06Feb88|06Feb88] 187(2) 5	0	31.9
4536. CURED *Steve Hackett* [24Oct81|31Oct81|07Nov81] 169(2) 3	0	31.7
4537. I PREFER THE MOONLIGHT *Kenny Rogers* [26Sep87|26Sep87|17Oct87] 163(2) 4	0	31.7
4538. OPEN ALL NIGHT *Leroi Bros.* [28Mar87|18Apr87|25Apr87] 181(2) 5	0	31.7
4539. A VALENTINE GIFT FOR YOU *Elvis Presley* [02Mar85|09Mar85|16Mar85] 154(1) 3	0	31.6
4540. STREET LANGUAGE *Rodney Crowell* [23Aug86|13Sep86|20Sep86] 177(1) 5	0	31.6
4541. ROCK 'N' ROLL WARRIORS *Savoy Brown* [25Jul81|15Aug81|15Aug81] 185(1) 4	0	31.6
4542. LA PISTOLA Y EL CORAZON *Los Lobos* [05Nov88|19Nov88|26Nov88] 179(1) 4	0	31.5
4543. SECOND EDITION *Public Image Limited* [10May80|17May80|24May80] 171(2) 3	0	31.5
4544. X-PERIMENT *The System* [31Mar84|21Apr84|28Apr84] 182(1) 5	0	31.4
4545. CACTUS AND A ROSE *Gary Stewart* [16Aug80|23Aug80|30Aug80] 165(1) 3	0	31.3
4546. THE BOOK OF PRIDE *Giant Steps* [12Nov88|03Dec88|10Dec88] 184(1) 5	2	31.3
4547. LOUDER THAN HELL *Sam Kinison* [08Nov86|29Nov86|06Dec86] 175(1) 5	0	31.2
4548. LOVE ME TENDER *B.B. King* [15May82|29May82|12Jun82] 179(1) 5	0	31.1
4549. WINDHAM HILL RECORDS SAMPLER '89 *Various Artists* [15Apr89|29Apr89|06May89] 176(2) 4	0	31.0
4550. HIT AND RUN *Girlschool* [22May82|29May82|19Jun82] 182(2) 5	0	31.0
4551. DREAMTIME *The Stranglers* [02May87|09May87|23May87] 172(2) 4	0	30.9
4552. LET'S MAKE A NEW DOPE DEAL *Cheech & Chong* [19Jul80|26Jul80|02Aug80] 173(1) 3	0	30.9
4553. BORDER WAVE *Sir Douglas Quintet* [14Feb81|21Feb81|07Mar81] 184(2) 4	0	30.9
4554. AQUA DREAM *McGuffey Lane* [23Jan82|13Feb82|27Feb82] 193(1) 6	1	30.9
4555. NO REFUGE *Eddie Schwartz* [06Feb82|27Feb82|13Mar82] 195(2) 6	2	30.9
4556. BAD TO THE BONE *L.A. Dream Team* [14Nov87|21Nov87|05Dec87] 162(2) 4	0	30.7
4557. FROM THE GREENHOUSE *Crack The Sky* [24Jun89|15Jul89|22Jul89] 186(1) 5	0	30.6
4558. BEAT SURRENDER *The Jam* [09Apr83|23Apr83|30Apr83] 171(1) 4	0	30.6
4559. STRAPHANGIN' *The Brecker Brothers* [20Jun81|27Jun81|04Jul81] 176(2) 3	0	30.6
4560. URGH! A MUSIC WAR *Various Artists* [26Sep81|03Oct81|10Oct81] 173(2) 3	0	30.5
4561. FANS *Malcolm McLaren* [02Feb85|16Feb85|09Mar85] 190(2) 6	0	30.5
4562. THIEF IN THE NIGHT *George Duke* [20Apr85|18May85|18May85] 183(1) 5	0	30.4
4563. SIMPLE THINGS *Richie Havens* [03Oct87|24Oct87|24Oct87] 173(1) 4	0	30.4
4564. COUNTRY BOY *Ricky Skaggs* [10Nov84|01Dec84|08Dec84] 180(1) 5	0	30.2
4565. THE BIGGEST PRIZE IN SPORT *999* [23Feb80|23Feb80|08Mar80] 177(2) 3	0	30.2
4566. QE 2 *Mike Oldfield* [04Jul81|18Jul81|18Jul81] 174(1) 3	0	30.1
4567. JAPANESE WHISPERS *The Cure* [25Feb84|10Mar84|24Mar84] 181(2) 5	0	30.1
4568. ANGEL BAND *Emmylou Harris* [01Aug87|15Aug87|22Aug87] 166(2) 4	0	30.1
4569. JOHNNY "GUITAR" WATSON AND THE FAMILY CLONE *Johnny Guitar Watson* [27Jun81|04Jul81|11Jul81] 177(2) 3	0	30.1
4570. LIVE IN CONCERT *Roger Whittaker* [13Jun81|27Jun81|27Jun81] 177(1) 3	0	30.1
4571. AEROBIC SHAPE UP II *Joanie Greggains* [18Jun83|02Jul83|09Jul83] 177(2) 4	0	30.0
4572. BURNING SENSATIONS *Burning Sensations* [30Jul83|20Aug83|20Aug83] 175(1) 4	0	29.9
4573. FIRE AND GASOLINE *Steve Jones* [21Oct89|28Oct89|11Nov89] 169(2) 4	0	29.9
4574. BLUES 'N JAZZ *B.B. King* [02Jul83|23Jul83|23Jul83] 172(1) 4	0	29.9
4575. TERROR RISING *Lizzy Borden* [02May87|16May87|06Jun87] 188(1) 6	0	29.8
4576. BLUE TATTOO *Passport* [29Aug81|05Sep81|12Sep81] 175(2) 4	0	29.7
4577. STEPPIN' OUT *George Howard* [01Sep84|15Sep84|22Sep84] 178(1) 4	0	29.7
4578. ABSTRACT EMOTIONS *Randy Crawford* [26Jul86|09Aug86|16Aug86] 178(2) 4	0	29.6
4579. DARLIN' *Tom Jones* [23May81|30May81|06Jun81] 179(1) 4	0	29.6
4580. HOW THE HELL DO YOU SPELL RYTHUM *Amazing Rhythm Aces* [04Oct80|18Oct80|18Oct80] 175(1) 3	0	29.6
4581. MY SPECIAL LOVE *LaToya Jackson* [12Sep81|19Sep81|26Sep81] 175(2) 3	0	29.5
4582. NAIVE ART *Red Flag* [23Sep89|07Oct89|14Oct89] 178(2) 4	0	29.4
4583. WHERE THE BEAT MEETS THE STREET *Bobby And The Midnites* [25Aug84|08Sep84|15Sep84] 166(1) 4	0	29.3
4584. BIRD *Soundtrack* [12Nov88|26Nov88|26Nov88] 169(1) 3	0	29.3
4585. SONGS I LOVE TO SING *Slim Whitman* [25Oct80|08Nov80|08Nov80] 175(1) 3	0	29.3
4586. SKY 3 *Sky* [02May81|16May81|16May81] 181(1) 3	0	29.2
4587. CLIMBING THE WALLS *Wrathchild America* [30Sep89|07Oct89|04Nov89] 190(2) 6	0	29.2
4588. THE RIGHT TO BE ITALIAN *Holly And The Italians* [11Jul81|18Jul81|25Jul81] 177(2) 3	0	29.0
4589. SURPRISE *Sylvia (2)* [28Apr84|12May84|19May84] 178(1) 4	0	28.9
4590. SWEET AND WONDERFUL *Jean Carn* [15Aug81|29Aug81|29Aug81] 176(1) 3	0	28.8
4591. ATLANTA BLUE *The Statler Brothers* [26May84|02Jun84|16Jun84] 177(2) 4	0	28.8
4592. VICTOR/VICTORIA *Soundtrack* [05Jun82|12Jun82|26Jun82] 174(2) 4	0	28.8
4593. THE STRIKERS *Strikers* [29Aug81|12Sep81|12Sep81] 174(1) 3	0	28.8
4594. FEELIN' RIGHT *Razzy Bailey* [27Feb82|13Mar82|20Mar82] 176(1) 4	0	28.7
4595. TIMES OF OUR LIVES *Judy Collins* [13Mar82|10Apr82|10Apr82] 190(1) 5	0	28.7
4596. IN BLACK AND WHITE *Jenny Burton* [24Mar84|07Apr84|14Apr84] 181(2) 5	1	28.7
4597. THAT'S ALL THAT MATTERS TO ME *Mickey Gilley* [30Aug80|06Sep80|13Sep80] 177(2) 3	1	28.5
4598. TONIGHT *France Joli* [28Jun80|12Jul80|12Jul80] 175(1) 3	0	28.5
4599. THE PIRATES OF PENZANCE *Original Cast* [06Jun81|20Jun81|20Jun81] 178(1) 3	0	28.4
4600. LOOKIN' FOR TROUBLE *Toronto* [30Aug80|20Sep80|20Sep80] 185(1) 4	0	28.4
4601. SCATTERLINGS *Juluka* [13Aug83|03Sep83|10Sep83] 186(1) 5	0	28.4
4602. IT'S A BEAUTIFUL THING *Maxine Nightingale* [08Jan83|22Jan83|29Jan83] 176(2) 4	0	28.3
4603. SAMURAI SAMBA *Yellowjackets* [13Apr85|04May85|04May85] 179(1) 4	0	28.3
4604. SONGS *The Kids From Fame* [15Jan83|15Jan83|05Feb83] 181(4) 4	0	28.2
4605. MY OWN WAY *Willie Nelson* [03Dec83|10Dec83|31Dec83] 182(1) 5	0	28.1
4606. I APPRECIATE *Alicia Myers* [08Dec84|22Dec84|05Jan85] 186(2) 5	0	28.1
4607. IT'S THE WORLD GONE CRAZY *Glen Campbell* [28Feb81|14Mar81|14Mar81] 178(1) 3	1	28.1
4608. THE FANATIC *Felony* [26Mar83|09Apr83|23Apr83] 185(2) 5	1	28.1
4609. DANCIN' IN THE KEY OF LIFE *Steve Arrington* [18May85|25May85|15Jun85] 185(2) 5	1	27.9
4610. NORTH OF A MIRACLE *Nick Heyward* [14Jan84|21Jan84|04Feb84] 178(2) 4	0	27.9
4611. PLANTATION HARBOR *Joe Vitale* [04Jul81|18Jul81|18Jul81] 181(1) 3	0	27.8
4612. BRAZIL CLASSICS 1: BELEZA TROPICAL *Various Artists* [22Apr89|06May89|13May89] 178(2) 4	0	27.8
4613. NIGHTHAWKS *Keith Emerson* [02May81|16May81|16May81] 183(1) 3	0	27.8
4614. POLKA PARTY! *"Weird Al" Yankovic* [15Nov86|29Nov86|06Dec86] 177(2) 4	0	27.6
4615. GUTS FOR LOVE *Garland Jeffreys* [26Feb83|05Mar83|19Mar83] 176(2) 4	0	27.6
4616. GANDHI *Soundtrack* [30Apr83|07May83|14May83] 168(1) 3	0	27.5
4617. PROOF THROUGH THE NIGHT *T-Bone Burnett* [01Oct83|15Oct83|29Oct83] 188(2) 5	0	27.3

Ranking the Albums

Rank. Title *Act* [Enter|Peak|Exit] Peak(Wks) TotWks	Hot 100	Scr
4618. INXS *INXS* [18Aug84|25Aug84|01Sep84] 164(1) 3	0	27.0
4619. MONDO MANDO *David Grisman* [24Oct81|07Nov81|07Nov81] 174(1) 3	0	27.0
4620. LETHAL WEAPON 2 *Soundtrack* [02Sep89|02Sep89|16Sep89] 164(2) 3	0	27.0
4621. THE BEST OF FIREFALL *Firefall* [26Dec81|09Jan82|16Jan82] 186(1) 4	0	26.9
4622. 10,9,8,7,6,5,4,3,2,1 *Midnight Oil* [04Feb84|11Feb84|03Mar84] 178(1) 5	0	26.9
4623. FEELING CAVALIER *Ebn-Ozn* [31Mar84|07Apr84|21Apr84] 185(2) 4	0	26.6
4624. MY TOOT TOOT *Jean Knight* [03Aug85|24Aug85|24Aug85] 180(1) 4	1	26.6
4625. SUPERMAN III *Soundtrack* [02Jul83|09Jul83|16Jul83] 163(1) 3	0	26.5
4626. THE PRODUCERS *The Producers* [06Jun81|13Jun81|13Jun81] 163(1) 2	1	26.5
4627. FAVORITE COUNTRY HITS *Ricky Skaggs* [09Mar85|23Mar85|30Mar85] 181(2) 4	0	26.4
4628. HAND OF KINDNESS *Richard Thompson* [30Jul83|13Aug83|27Aug83] 186(1) 5	0	26.3
4629. D.C. CAB *Soundtrack* [11Feb84|03Mar84|03Mar84] 181(1) 4	0	26.3
4630. WHAT MORE CAN I SAY? *Audio Two* [25Jun88|25Jun88|16Jul88] 185(2) 4	0	26.2
4631. HOLLYWOOD *Maynard Ferguson* [22May82|12Jun82|12Jun82] 185(1) 4	0	26.1
4632. THE TOP *The Cure* [23Jun84|14Jul84|14Jul84] 180(1) 4	0	26.0
4633. GREATEST HITS 1972-1978 *10cc* [22Dec79|05Jan80|12Jan80] 188(1) 4	0	25.9
4634. SHOULD I DO IT *Tanya Tucker* [01Aug81|15Aug81|15Aug81] 180(1) 3	0	25.9
4635. BEST OF SCORPIONS VOL. 2 *Scorpions* [04Aug84|11Aug84|25Aug84] 175(1) 4	0	25.8
4636. TRUTH IN DISGUISE *Denise Lopez* [26Nov88|17Dec88|17Dec88] 184(1) 4	2	25.8
4637. WALKING THROUGH FIRE *April Wine* [05Oct85|12Oct85|26Oct85] 174(2) 4	0	25.7
4638. THIEF OF HEARTS *Soundtrack* [22Dec84|05Jan85|12Jan85] 179(1) 4	1	25.4
4639. BLAST FROM THE BAYOU *Wayne Toups & Zydecajun* [18Mar89|01Apr89|08Apr89] 183(2) 4	0	25.4
4640. THE HIGH AND THE MIGHTY *Donnie Iris* [27Nov82|04Dec82|18Dec82] 180(1) 4	1	25.2
4641. SCOTT BAIO *Scott Baio* [04Sep82|11Sep82|25Sep82] 181(2) 4	0	25.2
4642. METAL PRIESTESS *The Plasmatics* [05Dec81|19Dec81|19Dec81] 177(1) 4	0	25.2
4643. CULTURE KILLED THE NATIVE *Victory* [06May89|20May89|03Jun89] 182(1) 5	0	25.2
4644. DREAMTIME *Tom Verlaine* [10Oct81|17Oct81|24Oct81] 177(1) 3	0	25.1
4645. DANSEPARC *Martha And The Muffins* [21May83|04Jun83|11Jun83] 184(1) 4	0	25.1
4646. BILL WITHERS GREATEST HITS *Bill Withers* [16May81|23May81|30May81] 183(1) 3	0	25.1
4647. TEN WOMEN *Wire Train* [02May87|16May87|23May87] 181(1) 4	0	25.0
4648. LOVE LETTERS *Force M.D.'s* [15Dec84|22Dec84|05Jan85] 185(3) 4	0	24.9
4649. TODAY *The Statler Brothers* [25Jun83|02Jul83|23Jul83] 193(1) 5	0	24.8
4650. LIVE FREE OR DIE *Balaam And The Angel* [30Apr88|14May88|14May88] 174(1) 3	0	24.6
4651. M.V.P. *Harvey Mason* [30May81|13Jun81|13Jun81] 186(1) 3	0	24.4
4652. MIDNIGHT CRAZY *Mac Davis* [16Jan82|23Jan82|30Jan82] 174(1) 3	0	24.3
4653. YOU GOTTA SAY YES TO ANOTHER EXCESS *Yello* [16Jul83|23Jul83|06Aug83] 184(2) 4	0	24.3
4654. GLORY ROAD *Gillan* [06Dec80|20Dec80|20Dec80] 183(1) 3	0	24.3
4655. SURVIVIN' IN THE 80'S *Andre Cymone* [15Oct83|29Oct83|05Nov83] 185(1) 4	0	24.2
4656. AFTER HERE THROUGH MIDLAND *Cock Robin* [05Sep87|26Sep87|03Oct87] 166(2) 3	0	24.2
4657. THE VELVETEEN RABBIT *Meryl Streep & George Winston* [20Apr85|27Apr85|11May85] 180(1) 4	0	24.2
4658. THE BEST OF EARTH, WIND & FIRE, VOL. II *Earth, Wind & Fire* [10Dec88|24Dec88|31Dec88] 190(2) 4	0	24.1
4659. NEW JERSEY *Joe Piscopo* [27Jul85|03Aug85|10Aug85] 168(1) 3	0	24.1
4660. POWER OF LOVE *Arlo Guthrie* [27Jun81|27Jun81|11Jul81] 184(2) 3	0	23.9
4661. THE BEST OF JERRY JEFF WALKER *Jerry Jeff Walker* [19Jul80|26Jul80|02Aug80] 185(1) 3	0	23.8
4662. UNITY *Shinehead* [05Nov88|03Dec88|03Dec88] 185(1) 4	0	23.8
4663. ROCK THE WORLD *Third World* [25Jul81|08Aug81|08Aug81] 186(1) 3	0	23.8
4664. METRO MUSIC *Martha And The Muffins* [13Sep80|27Sep80|27Sep80] 186(1) 3	0	23.8
4665. PATTERN DISRUPTIVE *Dickey Betts Band* [12Nov88|26Nov88|03Dec88] 187(2) 4	0	23.7
4666. SONGS OF THE FREE *Gang Of Four* [26Jun82|10Jul82|10Jul82] 175(1) 3	0	23.5
4667. PORTRAIT OF CARRIE *Carrie Lucas* [31Jan81|14Feb81|14Feb81] 185(1) 3	0	23.3
4668. NEW YORK CAKE *Kano* [09Jan82|23Jan82|30Jan82] 189(1) 4	1	23.3
4669. GUITAR SLINGER *Johnny Winter* [04Aug84|18Aug84|25Aug84] 183(1) 4	0	23.3
4670. WHEN THE SUN GOES DOWN *Red 7* [30May87|06Jun87|13Jun87] 175(2) 3	0	23.3
4671. POINTER SISTERS' GREATEST HITS *The Pointer Sisters* [13Nov82|20Nov82|27Nov82] 178(2) 3	0	23.1
4672. NIGHT CRUISER *Eumir Deodato* [27Sep80|04Oct80|11Oct80] 186(1) 3	0	23.0
4673. JUGGERNAUT *Frank Marino* [14Aug82|28Aug82|04Sep82] 185(1) 4	0	23.0
4674. REUNION *Jerry Jeff Walker* [20Jun81|20Jun81|04Jul81] 188(2) 3	0	22.8
4675. IN THE LONG GRASS *Boomtown Rats* [25May85|15Jun85|15Jun85] 188(1) 4	0	22.7
4676. THE BEST OF GINO VANNELLI *Gino Vannelli* [19Sep81|19Sep81|26Sep81] 172(2) 2	0	22.7
4677. GETTIN' AWAY WITH MURDER *Patti Austin* [09Nov85|23Nov85|30Nov85] 182(1) 4	1	22.7
4678. WHAT IS THIS *What Is This* [14Sep85|28Sep85|05Oct85] 187(1) 4	0	22.5
4679. LET'S BURN *Clarence Carter* [28Feb81|14Mar81|14Mar81] 189(1) 3	0	22.5
4680. UNITED STATES LIVE *Laurie Anderson* [26Jan85|02Feb85|23Feb85] 192(1) 5	0	22.4
4681. NO REST FOR THE WICKED *Helix* [22Oct83|05Nov83|12Nov83] 186(1) 4	0	22.1
4682. STRANGE LAND *Box Of Frogs* [14Jun86|21Jun86|28Jun86] 177(2) 3	0	22.0
4683. PLEASURE ONE *Heaven 17* [04Apr87|11Apr87|18Apr87] 177(2) 3	0	22.0
4684. SUPERBAD *Chris Jasper* [05Mar88|05Mar88|26Mar88] 182(2) 3	0	22.0
4685. HALFWAY TO SANITY *The Ramones* [10Oct87|17Oct87|24Oct87] 172(1) 3	0	21.9
4686. NATIONAL EMOTION *Tommy Tutone* [29Oct83|12Nov83|12Nov83] 179(1) 3	0	21.9
4687. NO MAN'S LAND *Lene Lovich* [15Jan83|05Feb83|05Feb83] 188(1) 4	0	21.8
4688. HOT ASH *Wishbone Ash* [23Jan82|13Feb82|13Feb82] 192(1) 4	0	21.7
4689. SMOKE SOME KILL *Schoolly D* [06Aug88|13Aug88|20Aug88] 180(1) 3	0	21.6
4690. FIRST THRILLS *Thrills* [27Jun81|18Jul81|25Jul81] 199(2) 4	0	21.6
4691. AZTEC CAMERA *Aztec Camera* [13Apr85|20Apr85|27Apr85] 181(2) 3	0	21.6
4692. PSYCHOCANDY *Jesus And Mary Chain* [22Feb86|01Mar86|15Mar86] 188(2) 4	0	21.5
4693. EVERYTIME TWO FOOLS COLLIDE *Kenny Rogers & Dottie West* [05Jan80|12Jan80|19Jan80] 186(2) 3	0	21.5
4694. TREACHEROUS: A HISTORY OF THE NEVILLE BROTHERS 1955-1985 *The Neville Brothers* [11Apr87|11Apr87|25Apr87] 178(2) 3	0	21.5
4695. REMEMBRANCE DAYS *The Dream Academy* [14Nov87|21Nov87|28Nov87] 181(2) 3	0	21.4
4696. EXTRA PLAY *Kaja* [27Apr85|11May85|18May85] 185(1) 4	0	21.3
4697. CAMERA CAMERA *Renaissance* [12Dec81|19Dec81|02Jan82] 196(3) 4	0	21.3
4698. WITH LOVE *Roger Whittaker* [29Nov80|29Nov80|06Dec80] 175(2) 2	0	21.3
4699. ON AND ON *Fat Boys* [28Oct89|04Nov89|11Nov89] 175(1) 3	0	21.2
4700. MERRY CHRISTMAS/HAPPY NEW YEAR'S *Montana Orchestra* [19Dec81|09Jan82|09Jan82] 195(1) 4	0	21.2
4701. FREE FALL *Alvin Lee* [20Dec80|10Jan81|10Jan81] 198(1) 4	0	21.1
4702. AWAKEN THE GUARDIAN *Fates Warning* [07Feb87|14Feb87|28Feb87] 191(3) 4	0	21.1
4703. EVERYWHERE AT ONCE *The Plimsouls* [23Jul83|06Aug83|13Aug83] 186(1) 4	1	21.0
4704. COCOON *Soundtrack* [27Jul85|10Aug85|17Aug85] 188(1) 4	0	21.0
4705. HALF NELSON *Willie Nelson* [12Oct85|19Oct85|26Oct85] 178(2) 3	0	20.9
4706. DIRECTIONS *Miles Davis* [11Apr81|11Apr81|18Apr81] 179(2) 2	0	20.9
4707. ART IN AMERICA *Art In America* [26Mar83|09Apr83|09Apr83] 176(1) 3	0	20.9
4708. STILL IN LOVE *Carrie Lucas* [11Sep82|25Sep82|25Sep82] 180(1) 3	0	20.9
4709. AT LEAST WE GOT SHOES *Southside Johnny & The Jukes* [21Jun86|05Jul86|12Jul86] 189(1) 4	1	20.7
4710. THE CINDERELLA THEORY *George Clinton* [02Sep89|16Sep89|23Sep89] 192(1) 4	0	20.7
4711. I'M READY *Natalie Cole* [17Sep83|01Oct83|01Oct83] 182(1) 3	0	20.6
4712. HIGH TENSION WIRES *Steve Morse* [24Jun89|24Jun89|08Jul89] 182(2) 3	0	20.6
4713. X-MULTIPLIES *Yellow Magic Orchestra* [20Sep80|20Sep80|27Sep80] 177(2) 2	0	20.6
4714. HURRICANE EYES *Loudness* [15Aug87|29Aug87|05Sep87] 190(1) 4	0	20.5

Ranking the Albums

Rank. Title *Act* [Enter|Peak|Exit] Peak(Wks) TotWks	Hot 100	Scr
4715. ONE MORE TRY FOR LOVE *Ronnie Milsap* [02Jun84|16Jun84|16Jun84] 180(1) 3	1	20.5
4716. MOTHER'S SPIRITUAL *Laura Nyro* [10Mar84|17Mar84|24Mar84] 182(1) 3	0	20.5
4717. SKYYLIGHT *Skyy* [06Aug83|20Aug83|20Aug83] 183(1) 3	0	20.4
4718. DO IT DEBBIE'S WAY *Debbie Reynolds* [26May84|02Jun84|09Jun84] 182(1) 3	0	20.4
4719. J MOOD *Wynton Marsalis* [01Nov86|08Nov86|22Nov86] 185(2) 4	0	20.3
4720. WOMAN IN FLAMES *Champaign* [10Nov84|17Nov84|24Nov84] 184(2) 3	0	20.2
4721. KEEP ON DOING *The Roches* [13Nov82|27Nov82|27Nov82] 183(1) 3	0	20.1
4722. SET MY LOVE IN MOTION *Syreeta* [30Jan82|30Jan82|13Feb82] 189(3) 3	0	19.9
4723. UP THE CREEK *Soundtrack* [12May84|26May84|26May84] 185(1) 3	0	19.9
4724. DA'KRASH *da'Krash* [16Apr88|16Apr88|30Apr88] 184(2) 3	0	19.7
4725. CAN'T SHAKE THIS FEELIN' *Spinners* [16Jan82|16Jan82|06Feb82] 196(2) 4	1	19.7
4726. A FINE MESS *Soundtrack* [23Aug86|06Sep86|06Sep86] 183(1) 3	1	19.6
4727. RIP IT TO SHREDS-THE ANIMALS GREATEST HITS LIVE *The Animals* [15Sep84|29Sep84|06Oct84] 193(2) 4	0	19.6
4728. RITES OF PASSAGE *Vitamin Z* [10Aug85|24Aug85|24Aug85] 183(1) 3	1	19.5
4729. PERSPECTIVE *America* [10Nov84|24Nov84|24Nov84] 185(1) 3	0	19.5
4730. MOMENTS *Julio Iglesias* [15Sep84|29Sep84|06Oct84] 191(1) 4	0	19.5
4731. MICK JONES *Mick Jones* [23Sep89|07Oct89|07Oct89] 184(1) 3	0	19.4
4732. JUST TESTING *Wishbone Ash* [29Mar80|05Apr80|05Apr80] 179(1) 2	0	19.2
4733. LIVE AT ST. DOUGLAS CONVENT *Father Guido Sarducci* [10May80|10May80|17May80] 179(2) 2	0	18.9
4734. NUNSEXMONKROCK *Nina Hagen* [05Jun82|12Jun82|19Jun82] 184(1) 3	0	18.9
4735. DARK CONTINENT *Wall Of Voodoo* [17Oct81|24Oct81|24Oct81] 177(1) 2	0	18.9
4736. FRESH FRUIT IN FOREIGN PLACES *Kid Creole & The Coconuts* [18Jul81|25Jul81|25Jul81] 180(1) 2	0	18.7
4737. MAKIN' FRIENDS *Razzy Bailey* [20Jun81|20Jun81|27Jun81] 183(2) 2	0	18.7
4738. 'LIFE' LIVE *Thin Lizzy* [28Jan84|11Feb84|11Feb84] 185(1) 3	0	18.6
4739. THE TWO RING CIRCUS *Erasure* [16Jan88|30Jan88|30Jan88] 186(1) 3	0	18.6
4740. THE BEST OF THE ALLMAN BROTHERS BAND *The Allman Brothers Band* [21Nov81|28Nov81|05Dec81] 189(3) 3	0	18.5
4741. TOTAL DEVO *Devo* [02Jul88|02Jul88|16Jul88] 189(2) 3	0	18.5
4742. WHOMP THAT SUCKER *Sparks* [15Aug81|15Aug81|22Aug81] 182(2) 2	0	18.3
4743. MARATHON *Rodney Franklin* [25Feb84|10Mar84|10Mar84] 187(2) 3	0	18.2
4744. LOOK HEAR? *10cc* [17May80|24May80|24May80] 180(1) 2	0	18.1
4745. THROUGH THE LOOKING GLASS *Siouxsie & The Banshees* [11Apr87|25Apr87|25Apr87] 188(1) 3	0	18.1
4746. VOLUME VIII *Average White Band* [20Sep80|27Sep80|27Sep80] 182(1) 2	0	18.1
4747. NOT JUST THE GIRL NEXT DOOR *Nancy Martinez* [21Feb87|28Feb87|07Mar87] 178(1) 3	1	17.9
4748. THE MANHATTAN TRANSFER LIVE *Manhattan Transfer* [30May87|06Jun87|13Jun87] 187(2) 3	0	17.9
4749. T.G. SHEPPARD'S GREATEST HITS *T.G. Sheppard* [11Jun83|18Jun83|25Jun83] 189(2) 3	0	17.8
4750. THE HEART NEVER LIES *Michael Martin Murphey* [29Oct83|12Nov83|12Nov83] 187(1) 3	0	17.7
4751. I'VE ALWAYS WANTED TO DO THIS *Jack Bruce And Friends* [13Dec80|20Dec80|20Dec80] 182(1) 3	0	17.7
4752. BARBED WIRE KISSES *Jesus And Mary Chain* [18Jun88|02Jul88|02Jul88] 192(1) 3	0	17.6
4753. ON THE STRENGTH *Grandmaster Flash & The Furious Five* [30Apr88|30Apr88|14May88] 189(2) 3	0	17.6
4754. HEARTBREAK HOTEL *Soundtrack* [29Oct88|29Oct88|05Nov88] 176(2) 2	0	17.6
4755. CARE *Shriekback* [25Jun83|09Jul83|09Jul83] 188(1) 3	0	17.5
4756. CARL WILSON *Carl Wilson* [02May81|09May81|09May81] 185(1) 2	0	17.4
4757. LANGUAGE *Gary Myrick* [06Aug83|20Aug83|20Aug83] 186(1) 3	0	17.4
4758. PHOTOGLO *Photoglo* [24May80|07Jun80|07Jun80] 194(1) 3	1	17.4
4759. BRUCE WOOLLEY & THE CAMERA CLUB *Bruce Woolley & The Camera Club* [08Mar80|15Mar80|15Mar80] 184(1) 2	0	17.2
4760. TO YOU HONEY, HONEY WITH LOVE *David Hudson* [23Aug80|30Aug80|30Aug80] 184(1) 2	1	17.2
4761. FLIRT *Evelyn "Champagne" King* [25Jun88|09Jul88|09Jul88] 192(1) 3	0	17.1
4762. TOUCH YOU *Jimmy Hall* [22Nov80|29Nov80|29Nov80] 183(1) 3	1	17.1
4763. BACK 2 BACK *Stargard* [04Jul81|04Jul81|11Jul81] 186(2) 2	0	17.1
4764. VELVET KISS, LICK OF THE LIME *Lions And Ghosts* [24Oct87|31Oct87|07Nov87] 187(2) 3	0	17.0
4765. LET'S DO IT TODAY *Lenny Williams* [15Nov80|15Nov80|22Nov80] 185(2) 2	0	16.9
4766. WAITING FOR SPRING *David Benoit* [11Nov89|25Nov89|25Nov89] 187(1) 3	0	16.9
4767. HEAVEN UP HERE *Echo & The Bunnymen* [25Jul81|01Aug81|01Aug81] 184(1) 2	0	16.9
4768. STRICTLY PERSONAL *The Romantics* [14Nov81|21Nov81|21Nov81] 182(1) 2	0	16.8
4769. THE LIVE ALBUM *Leon Russell And The New Grass Revival* [04Apr81|11Apr81|11Apr81] 187(1) 2	0	16.7
4770. LOVE *Aztec Camera* [19Dec87|19Dec87|02Jan88] 193(3) 3	0	16.6
4771. ALL OF YOU *Lillo Thomas* [06Oct84|13Oct84|20Oct84] 186(1) 3	0	16.5
4772. TALES OF THE NEW WEST *Beat Farmers* [08Jun85|15Jun85|22Jun85] 186(1) 3	0	16.5
4773. THIRD WORLD, PRISONER IN THE STREET (SOUNDTRACK) *Third World* [30Aug80|30Aug80|06Sep80] 186(2) 2	0	16.4
4774. PRIME TIME *Grey And Hanks* [23Feb80|23Feb80|08Mar80] 195(2) 3	0	16.4
4775. BETWEEN THE LINES *Spider* [11Jul81|18Jul81|18Jul81] 185(2) 2	1	16.3
4776. ROCK 'N' ROLL OUTLAW *Rose Tattoo* [29Nov80|13Dec80|13Dec80] 197(1) 3	0	16.1
4777. NOT GUILTY... *Larry Gatlin & The Gatlin Brothers Band* [17Oct81|24Oct81|24Oct81] 184(1) 2	0	16.0
4778. ECHO & THE BUNNYMEN *Echo & The Bunnymen* [11Feb84|18Feb84|25Feb84] 188(1) 3	0	16.0
4779. TALK MEMPHIS *Jesse Winchester* [27Jun81|04Jul81|04Jul81] 188(1) 2	1	15.7
4780. SOMEWHERE IN TIME *Soundtrack* [06Dec80|06Dec80|13Dec80] 187(2) 2	0	15.7
4781. NOTHING IN COMMON *Soundtrack* [20Sep86|27Sep86|04Oct86] 190(2) 3	1	15.7
4782. COMING TO AMERICA *Soundtrack* [30Jul88|06Aug88|06Aug88] 177(1) 2	1	15.5
4783. SPORTIN' LIFE *Weather Report* [27Apr85|11May85|11May85] 191(1) 3	0	15.5
4784. SEND ME AN ANGEL '89 *Real Life* [01Jul89|12Aug89|12Aug89] 191(1) 3	1	15.5
4785. HOMESICK HEROS *Charlie Daniels Band* [12Nov88|12Nov88|19Nov88] 181(2) 3	0	15.5
4786. IT'S MY TIME *Maynard Ferguson* [27Sep80|27Sep80|04Oct80] 188(2) 2	0	15.5
4787. JIMI PLAYS MONTEREY *Jimi Hendrix* [08Mar86|15Mar86|22Mar86] 192(2) 3	0	15.3
4788. WILD IN THE STREETS *Helix* [07Nov87|07Nov87|14Nov87] 179(2) 2	0	15.3
4789. THE CIRCUS *Erasure* [18Jul87|25Jul87|01Aug87] 190(1) 3	0	15.3
4790. RATHER BE ROCKIN' *Tantrum* [19Jan80|26Jan80|02Feb80] 199(2) 3	0	15.1
4791. THE CHRISTMAS ALBUM *Elvis Presley* [28Dec85|28Dec85|04Jan86] 178(2) 2	0	15.1
4792. SOLID GOLD *Gang Of Four* [06Jun81|06Jun81|13Jun81] 190(2) 2	0	15.1
4793. THE PRINCE'S TRUST 10TH ANNIVERSARY BIRTHDAY PARTY *Various Artists* [06Jun87|20Jun87|20Jun87] 194(1) 3	0	15.0
4794. LEARNING TO LOVE *Rodney Franklin* [22Jan83|22Jan83|05Feb83] 190(2) 3	0	14.9
4795. BARNET DOGS *Russ Ballard* [16Aug80|23Aug80|23Aug80] 187(1) 2	1	14.9
4796. RELIGHT MY FIRE *Dan Hartman* [15Mar80|22Mar80|22Mar80] 189(1) 2	0	14.8
4797. NEW LOOKS *B.J. Thomas* [21May83|04Jun83|04Jun83] 193(1) 3	1	14.5
4798. BLOODLINE *Levert* [25Oct86|08Nov86|08Nov86] 192(1) 3	0	14.5
4799. KING BEE *Muddy Waters* [16May81|16May81|23May81] 192(2) 2	0	14.5
4800. CONCRETE *999* [27Jun81|27Jun81|04Jul81] 192(2) 2	0	14.3
4801. TWITCH *Ministry* [05Apr86|05Apr86|19Apr86] 194(2) 3	0	14.3
4802. GHOST TOWN *Poco* [04Dec82|04Dec82|18Dec82] 195(2) 3	1	14.2
4803. THE SEARCHERS *The Searchers* [15Mar80|15Mar80|22Mar80] 191(2) 2	0	14.2
4804. FRAMED *Asleep At The Wheel* [06Sep80|13Sep80|13Sep80] 191(1) 2	0	13.9
4805. ME AND JOE *Rodney O & Joe Cooley* [04Mar89|04Mar89|11Mar89] 187(2) 2	0	13.8
4806. THE CHANGE HAS COME *Chubby Checker* [06Mar82|13Mar82|13Mar82] 186(1) 2	1	13.8
4807. FOUR FOR THE SHOW *The Statler Brothers* [09Aug86|09Aug86|16Aug86] 183(2) 2	0	13.7

Rank. Title *Act* [Enter|Peak|Exit] Peak(Wks) TotWks	Hot 100	Scr
4808. HIT AND RUN *TSOL* [04Jul87|11Jul87|11Jul87] 184(1) 2	0	13.7
4809. MARRIED TO THE MOB *Soundtrack* [01Oct88|01Oct88|15Oct88] 197(2) 3	0	13.7
4810. LOVE ACTION *Sniff 'N' The Tears* [19Sep81|19Sep81|26Sep81] 192(2) 2	0	13.7
4811. ROUND MIDNIGHT *Soundtrack* [24Jan87|07Feb87|07Feb87] 196(1) 3	0	13.6
4812. TOO HOT TO SLEEP *Survivor* [05Nov88|05Nov88|12Nov88] 187(2) 2	2	13.5
4813. SPECIAL *Jimmy Cliff* [14Aug82|14Aug82|21Aug82] 186(2) 2	0	13.2
4814. GRADUALLY GOING TORNADO *Bill Bruford* [29Mar80|05Apr80|05Apr80] 191(1) 2	0	13.2
4815. MY BABE *Roy Buchanan* [24Jan81|31Jan81|31Jan81] 193(1) 2	0	13.2
4816. DAVID FOSTER *David Foster* [19Jul86|19Jul86|02Aug86] 195(2) 3	1	13.2
4817. LAND OF THE THIRD EYE *Dave Valentin* [25Oct80|25Oct80|01Nov80] 194(2) 2	0	13.0
4818. WILD WEEKEND *NRBQ* [30Dec89|30Dec89|13Jan90] 198(3) 3	0	13.0
4819. PENNIES FROM HEAVEN *Soundtrack* [23Jan82|30Jan82|30Jan82] 195(2) 2	0	13.0
4820. THE UNFORGIVEN *The Unforgiven* [09Aug86|09Aug86|16Aug86] 185(2) 2	0	13.0
4821. WILD EXHIBITIONS *Walter Egan* [28May83|04Jun83|04Jun83] 187(1) 2	1	12.9
4822. CIRCUS OF POWER *Circus Of Power* [12Nov88|19Nov88|19Nov88] 185(1) 2	0	12.9
4823. SAVE YOUR PRAYERS *Waysted* [21Mar87|28Mar87|28Mar87] 185(1) 2	0	12.9
4824. SEE JUNGLE! SEE JUNGLE! GO JOIN YOUR GANG YEAH! CITY ALL OVER, GO APE CRAZY *Bow Wow Wow* [21Nov81|21Nov81|28Nov81] 192(2) 2	0	12.9
4825. OUTA HAND *Coney Hatch* [17Sep83|24Sep83|24Sep83] 186(1) 2	0	12.6
4826. GREATEST HITS *Steve Wariner* [17Oct87|17Oct87|24Oct87] 187(2) 2	0	12.6
4827. WOMAN OF THE YEAR *Original Cast* [27Jun81|27Jun81|04Jul81] 196(2) 2	0	12.4
4828. THE BEST OF ENGLAND DAN & JOHN FORD COLEY *England Dan & John Ford Coley* [05Jan80|05Jan80|12Jan80] 194(2) 2	1	12.3
4829. MINOR DETAIL *Minor Detail* [01Oct83|08Oct83|08Oct83] 187(1) 2	0	12.3
4830. ROMEO'S DAUGHTER *Romeo's Daughter* [19Nov88|19Nov88|26Nov88] 191(2) 2	1	11.9
4831. BACK TO THE BEACH *Soundtrack* [05Sep87|12Sep87|12Sep87] 188(1) 2	0	11.8
4832. GET BACK! *Ike & Tina Turner* [11May85|18May85|18May85] 189(1) 2	0	11.8
4833. BREAK OF DAWN *Firefall* [12Mar83|26Mar83|26Mar83] 199(1) 3	1	11.8
4834. BO-DAY-SHUS!!! *Mojo Nixon & Skid Roper* [10Oct87|17Oct87|17Oct87] 189(1) 2	0	11.7
4835. EXIT *Tangerine Dream* [21Nov81|21Nov81|28Nov81] 195(1) 2	0	11.6
4836. AFRICA, CENTER OF THE WORLD *Roy Ayers* [15Aug81|15Aug81|22Aug81] 197(2) 2	0	11.5
4837. SONGS FROM THE STAGE AND SCREEN *Michael Crawford* [30Jul88|30Jul88|06Aug88] 192(2) 2	0	11.2
4838. THE BELLE STARS *The Belle Stars* [28May83|04Jun83|04Jun83] 191(1) 2	1	11.2
4839. ICON *Icon* [09Jun84|16Jun84|16Jun84] 190(1) 2	0	11.2
4840. ZIG-ZAG WALK *Foghat* [25Jun83|02Jul83|02Jul83] 192(1) 2	0	11.0
4841. KWICK *Kwick* [07Jun80|14Jun80|14Jun80] 197(1) 2	0	10.9
4842. COVER GIRL *Phantom, Rocker & Slick* [18Oct86|25Oct86|25Oct86] 181(1) 2	0	10.9
4843. BACK IT UP *Robin Trower* [01Oct83|08Oct83|08Oct83] 191(1) 2	0	10.8
4844. SUCKER FOR A PRETTY FACE *Eric Martin Band* [24Sep83|01Oct83|01Oct83] 191(1) 2	0	10.8
4845. ANOTHER DAY/ANOTHER DOLLAR *Gang Of Four* [13Feb82|13Feb82|20Feb82] 195(2) 2	0	10.8
4846. BACHMAN TURNER OVERDRIVE(2) *Bachman-Turner Overdrive* [29Sep84|06Oct84|06Oct84] 191(1) 2	0	10.6
4847. HARD CORE *Paul Dean* [25Feb89|25Feb89|04Mar89] 195(1) 2	0	10.6
4848. SOUND ALARM *Michael Anderson* [13Aug88|13Aug88|20Aug88] 194(2) 2	0	10.3
4849. DREAMS OF TOMORROW *Lonnie Liston Smith* [30Jul83|06Aug83|06Aug83] 193(2) 2	0	10.3
4850. STRENGTH OF STEEL *Anvil* [18Jul87|25Jul87|25Jul87] 191(1) 2	0	10.2
4851. DIAMONDS IN THE RAW *The S.O.S. Band* [04Nov89|04Nov89|11Nov89] 194(2) 2	0	10.2
4852. RIP IT UP *Dead Or Alive* [30Jul88|30Jul88|06Aug88] 195(2) 2	0	10.2
4853. MOTOWN REMEMBERS MARVIN GAYE *Marvin Gaye* [03May86|10May86|10May86] 193(2) 2	0	10.1
4854. ASIA *Kitaro* [30Nov85|07Dec85|07Dec85] 191(2) 2	0	10.1
4855. MR. C. *Norman Connors* [05Dec81|12Dec81|12Dec81] 197(2) 2	0	10.0
4856. I HAVE A PONY *Steven Wright* [23Nov85|30Nov85|30Nov85] 192(1) 2	0	10.0
4857. DANGEROUS DREAMS *The Nails* [23Aug86|23Aug86|30Aug86] 194(2) 2	0	10.0
4858. IVORY COAST *Bob James* [10Sep88|10Sep88|17Sep88] 196(2) 2	0	10.0
4859. KASIM *Kasim Sulton* [27Feb82|27Feb82|06Mar82] 197(2) 2	0	10.0
4860. CHRIS ISAAK *Chris Isaak* [11Apr87|18Apr87|18Apr87] 194(1) 2	0	10.0
4861. THIS IS THIS *Weather Report* [23Aug86|23Aug86|30Aug86] 195(2) 2	0	9.7
4862. TRINERE AND FRIENDS GREATEST HITS *Trinere & Friends* [23Sep89|23Sep89|30Sep89] 196(2) 2	0	9.6
4863. ALL OF ME *Toni Tennille* [26Dec87|26Dec87|02Jan88] 198(2) 2	0	9.6
4864. BROTHER ARAB *Arabian Prince* [16Dec89|23Dec89|23Dec89] 193(2) 2	0	9.6
4865. EXCESS ALL AREAS *Shy* [27Jun87|11Jul87|11Jul87] 193(1) 2	0	9.5
4866. HARLEQUIN *Dave Grusin & Lee Ritenour* [05Oct85|12Oct85|12Oct85] 192(1) 2	0	9.5
4867. NO BLUE THING *Ray Lynch* [24Jun89|24Jun89|01Jul89] 197(2) 2	0	9.4
4868. RADIO SILENCE *Boris Grebenshikov* [26Aug89|26Aug89|02Sep89] 198(2) 2	0	9.2
4869. SHADES *Yellowjackets* [06Sep86|13Sep86|13Sep86] 195(1) 2	0	9.0
4870. THE VELVET UNDERGROUND *The Velvet Underground & Nico* [20Apr85|27Apr85|27Apr85] 197(1) 2	0	8.9
4871. THE HOLLYWOOD MUSICALS *Johnny Mathis & Henry Mancini* [10Jan87|10Jan87|17Jan87] 197(2) 2	0	8.9
4872. THE HOLE *Golden Earring* [12Jul86|12Jul86|19Jul86] 196(2) 2	0	8.8
4873. KINGDOM BLOW *Kurtis Blow* [13Dec86|13Dec86|20Dec86] 196(2) 2	0	8.6
4874. CANDLELAND *Ian McCulloch* [25Nov89|25Nov89|25Nov89] 179(1) 1	0	8.0
4875. THE PRINCESS BRIDE (SOUNDTRACK) *Mark Knopfler* [31Oct87|31Oct87|31Oct87] 180(1) 1	0	7.5
4876. TENKU *Kitaro* [04Apr87|04Apr87|04Apr87] 183(1) 1	0	7.0
4877. TIRAMI SU *Al Di Meola Project* [23Jan88|23Jan88|23Jan88] 190(1) 1	0	6.6
4878. BIG ONES *Loverboy* [23Dec89|23Dec89|23Dec89] 189(1) 1	1,1*	5.9
4879. TAKING OVER *Overkill* [11Apr87|11Apr87|11Apr87] 191(1) 1	0	5.6
4880. AN OLD FASHIONED CHRISTMAS *Carpenters* [05Jan85|05Jan85|05Jan85] 190(1) 1	0	5.5
4881. PAT MCLAUGHLIN *Pat McLaughlin* [16Apr88|16Apr88|16Apr88] 195(1) 1	0	5.4
4882. LET THE HUSTLERS PLAY *Steady B* [22Oct88|22Oct88|22Oct88] 193(1) 1	0	5.3
4883. BRIGHTER THAN A THOUSAND SUNS *Killing Joke* [11Apr87|11Apr87|11Apr87] 194(1) 1	0	5.1
4884. BLUES FOR SALVADOR *Carlos Santana* [07Nov87|07Nov87|07Nov87] 195(1) 1	0	4.8
4885. SLOW DANCE *Southside Johnny* [03Dec88|03Dec88|03Dec88] 198(1) 1	0	4.7
4886. SING *Soundtrack* [29Apr89|29Apr89|29Apr89] 196(1) 1	1	4.7
4887. BA-DOP-BOOM-BANG *Grandmaster Flash* [25Apr87|25Apr87|25Apr87] 197(1) 1	0	4.6
4888. WORKERS PLAYTIME *Billy Bragg* [05Nov88|05Nov88|05Nov88] 198(1) 1	0	4.6
4889. GOT TO BE TOUGH *M.C. Shy D* [27Jun87|27Jun87|27Jun87] 197(1) 1	0	4.5
4890. BLIND DATE *Soundtrack* [02May87|02May87|02May87] 198(1) 1	0	4.4
4891. ANOTHER SCOOP *Pete Townshend* [04Apr87|04Apr87|04Apr87] 198(1) 1	0	4.3
4892. ROACHES: THE BEGINNING *Bobby Jimmy & The Critters* [29Nov86|29Nov86|29Nov86] 200(1) 1	0	3.6

Ranking the Albums

Albums, Alphabetical Index with Rank

Album titles are alphabetized as though there were no spaces or punctuation. When the title contains a name it is alphabetized as a phrase and not as the act's last name. Caution is urged when looking for *Greatest Hits*, *Best Of* or *Live* albums: some start with the act's name, i.e., *Bill Withers' Greatest Hits* and some start with the category, *Greatest Hits*. Searching in the **Acts With Their Albums** section can help identify exact titles. Initial articles *A*, *An* and *The* are ignored.

Albums Alpha Index

A

#	Title	Artist
1596	"A"	Jethro Tull
184	ABACAB	Genesis
3272	ABIGAIL	King Diamond
3900	THE ABOMINABLE SHOWMAN	Nick Lowe
2098	ABOMINOG	Uriah Heep
955	ABOUT FACE	David Gilmour
2459	ABOUT LAST NIGHT...	Soundtrack
1526	ABOUT LOVE	Gladys Knight & The Pips
3270	ABOUT TIME	Ten Years After
298	ABRACADABRA	Steve Miller Band
2502	ABSOLUTE BEGINNERS	Soundtrack
4162	ABSOLUTELY	Madness
2102	ABSOLUTELY LIVE	Rod Stewart
4578	ABSTRACT EMOTIONS	Randy Crawford
3678	A.C.	Andre Cymone
3890	A CAPPELLA	Todd Rundgren
1850	ACCESS ALL AREAS	Spyro Gyra
2947	ACROSS A CROWDED ROOM	Richard Thompson
2964	ACT A FOOL	King Tee
4024	ACTING VERY STRANGE	Mike Rutherford
1263	ACTION REPLAY	Howard Jones
497	ACTUALLY	Pet Shop Boys
2266	ADDICTIONS VOL. I	Robert Palmer
4295	ADVENTURES IN MODERN RECORDING	The Buggles
1284	ADVENTURES IN THE LAND OF MUSIC	Dynasty
1140	ADVENTURES IN UTOPIA	Utopia
3892	AEROBIC DANCE HITS, VOLUME ONE	Carla Capuano
3210	AEROBIC DANCING	Barbara Ann Auer
4571	AEROBIC SHAPE UP II	Joanie Greggains
1343	AEROSMITH'S GREATEST HITS	Aerosmith
2865	AFFAIR	Cherrelle
1493	AFL1-3603	Dave Davies
3663	AFOOT	Let's Active
4836	AFRICA, CENTER OF THE WORLD	Roy Ayers
165	AFTERBURNER	ZZ Top
1076	AFTER DARK	Andy Gibb
2585	AFTER DARK	Cruzados
3191	AFTER DARK	Ray Parker Jr.
945	AFTER EIGHT	Taco
4656	AFTER HERE THROUGH MIDLAND	Cock Robin
770	AFTER MIDNIGHT	The Manhattans
4442	AFTERNOONS IN UTOPIA	Alphaville
4175	AFTER THE ROSES	Kenny Rankin
1937	AFTER THE SNOW	Modern English
3248	AFTER THE WAR	Gary Moore
736	AGAINST ALL ODDS	Soundtrack
37	AGAINST THE WIND	Bob Seger & The Silver Bullet Band
3972	AGENT DOUBLE O SOUL	The Untouchables
284	AGENT PROVOCATEUR	Foreigner
1324	THE AGE OF CONSENT	Bronski Beat
2649	AIN'T LOVE GRAND	X
4248	AIRBORNE	Don Felder
1167	AIR SUPPLY	Air Supply
2872	AKIMBO ALOGO	Kim Mitchell
2216	THE ALARM	The Alarm
1012	ALBUM	Joan Jett & the Blackhearts
3013	ALBUM	Public Image Limited
1914	ALCHEMY-DIRE STRAITS LIVE	Dire Straits
372	ALDO NOVA	Aldo Nova
2652	ALEXANDER O'NEAL	Alexander O'Neal
1391	ALF	Alison Moyet
3953	ALIBI	America
2073	ALIENS ATE MY BUICK	Thomas Dolby
3204	ALIVE AND SCREAMIN'	Krokus
1047	ALIVE SHE CRIED	The Doors
1361	ALL AMERICAN GIRLS	Sister Sledge
1987	ALL AROUND THE TOWN	Bob James
2573	ALL BY MYSELF	Regina Belle
2931	ALL DRESSED UP & NO PLACE TO GO	Nicolette Larson
2193	ALL FIRED UP	Fastway
624	ALL FOR LOVE	New Edition
490	ALL FOUR ONE	The Motels
525	ALLIED FORCES	Triumph
2163	ALLIES	Crosby, Stills & Nash
1178	ALLIGATOR WOMAN	Cameo
1289	ALL I NEED	Jack Wagner
4425	ALL I NEED	Sylvester
632	ALL IN THE NAME OF LOVE	Atlantic Starr
3299	ALL IS FORGIVEN	Red Siren
4501	ALL MIXED UP	Alexander O'Neal
465	THE ALLNIGHTER	Glenn Frey
4863	ALL OF ME	Toni Tennille
3352	ALL OF THE ABOVE	John Hall Band
3484	ALL OF THIS AND NOTHING	Psychedelic Furs
4771	ALL OF YOU	Lillo Thomas
1136	ALL OUR LOVE	Gladys Knight & The Pips
2026	ALL OVER THE PLACE	Bangles
3647	ALL RIGHT NOW	Pepsi & Shirlie
1180	ALL SHOOK UP	Cheap Trick
2190	ALL SYSTEMS GO	Vinnie Vincent Invasion
3859	ALL SYSTEMS GO	Donna Summer
510	ALL THAT JAZZ	Breathe
1169	ALL THAT JAZZ	Soundtrack
2020	ALL THE BEST!	Paul McCartney
1054	ALL THE BEST COWBOYS HAVE CHINESE EYES	Pete Townshend
3555	ALL OF THE GOOD ONES ARE TAKEN	Ian Hunter
1053	ALL THE GREAT HITS	Diana Ross
1315	ALL THE GREAT HITS	Commodores
4377	ALL THE PEOPLE ARE TALKIN'	John Anderson
724	...ALL THE RAGE	General Public
4170	ALL THE RIGHT MOVES	Soundtrack
3829	ALL THE WAY STRONG	Third World
685	ALL THIS LOVE	DeBarge
2251	ALMOST BLUE	Elvis Costello And The Attractions
3730	ALONE/BUT NEVER ALONE	Larry Carlton
2367	ALONG THE AXIS	Jon Butcher Axis
642	ALPHA	Asia
1602	ALPHABET CITY	ABC
2015	ALTERNATING CURRENTS	Spyro Gyra
285	ALWAYS & FOREVER	Randy Travis
3451	ALWAYS IN THE MOOD	Shirley Jones
130	ALWAYS ON MY MIND	Willie Nelson
902	AMADEUS (SOUNDTRACK)	Neville Marriner
4453	AMANDLA	Miles Davis
4434	AMAZON BEACH	The Kings
3410	AMERICA	Kurtis Blow
2874	AMERICAN ANTHEM	Soundtrack
745	AMERICAN DREAM	Crosby, Stills, Nash & Young
2422	AN AMERICAN DREAM	Dirt Band
1641	AMERICAN EXCESS	Point Blank
58	AMERICAN FOOL	John Cougar
1348	AMERICAN GARAGE	Pat Metheny Group
469	AMERICAN GIGOLO	Soundtrack
1585	AMERICAN MADE	Oak Ridge Boys
1654	AN AMERICAN TAIL	Soundtrack
3863	AMERICAN WORKER	Bus Boys
3162	AMERICA'S GREATEST HERO	Joey Scarbury
737	AMMONIA AVENUE	Alan Parsons Project
4514	AMNESIA	Richard Thompson
1392	AMONG THE LIVING	Anthrax
3927	AMOUR	Richard Clayderman
3207	AMY HOLLAND	Amy Holland
4357	ANCIENT DREAMS	Candlemass
1677	ANCIENT HEART	Tanita Tikaram
1305	ANDERSON, BRUFORD, WAKEMAN, HOWE	Anderson, Bruford, Wakeman, Howe
1387	AND IN THIS CORNER...	D.J. Jazzy Jeff & The Fresh Prince
196	...AND JUSTICE FOR ALL	Metallica
2040	AND ONCE AGAIN	Isaac Hayes
3509	...AND THEN HE KISSED ME	Rachel Sweet
1592	ANDY GIBB'S GREATEST HITS	Andy Gibb
3499	AND YOU KNOW THAT!	Kirk Whalum
4568	ANGEL BAND	Emmylou Harris
3670	ANGEL EYES	Willie Nelson
694	ANGEL OF THE NIGHT	Angela Bofill
4333	ANGST IN MY PANTS	Sparks
4145	ANIMAL BOY	The Ramones
2519	ANIMAL GRACE	April Wine
589	ANIMALIZE	KISS
1423	ANIMAL MAGIC	Blow Monkeys
1560	ANIMAL MAGNETISM	Scorpions
4489	ANIMATION	Jon Anderson
1009	ANIMOTION	Animotion
2790	ANIMOTION(2)	Animotion
1074	ANNIE	Soundtrack
3142	ANNIE'S CHRISTMAS	No Artist
371	ANNE MURRAY'S GREATEST HITS	Anne Murray
4845	ANOTHER DAY/ANOTHER DOLLAR	Gang Of Four
1988	ANOTHER GREY AREA	Graham Parker
698	ANOTHER PAGE	Christopher Cross
4038	ANOTHER PERFECT DAY	Motorhead
1477	ANOTHER PLACE	Hiroshima
1886	ANOTHER PLACE AND TIME	Donna Summer
4891	ANOTHER SCOOP	Pete Townshend
1286	ANOTHER STEP	Kim Wilde
433	ANOTHER TICKET	Eric Clapton And His Band
2452	ANSWERS TO NOTHING	Midge Ure
3477	ANTHOLOGY	Marvin Gaye
3733	ANTHOLOGY	The Babys
3845	ANTHOLOGY	Grover Washington Jr.
483	ANY LOVE	Luther Vandross
4049	ANY MAN'S HUNGER	Danny Wilde
2137	ANYONE CAN SEE	Irene Cara
3846	ANYTIME	Henry Paul Band
460	ANYTIME, ANYPLACE, ANYWHERE	Rossington Collins Band
3494	ANY WHICH WAY YOU CAN	Soundtrack
2346	APOLLONIA 6	Apollonia 6
6	APPETITE FOR DESTRUCTION	Guns N' Roses
2971	APPLE JUICE	Tom Scott
4554	AQUA DREAM	McGuffey Lane
2543	ARCADE	Patrick Simmons
3305	ARCHITECTURE AND MORALITY	Orchestral Manoeuvres In The Dark
149	ARC OF A DIVER	Steve Winwood
464	ARENA	Duran Duran
1050	ARETHA	Aretha Franklin
768	ARETHA (II)	Aretha Franklin
2477	ARE YOU GONNA BE THE ONE	Robert Gordon
1741	ARGYBARGY	Squeeze
2612	ARK	The Animals
164	AROUND THE WORLD IN A DAY	Prince And The Revolution
3577	ARRIVE WITHOUT TRAVELLING	The Three O'Clock
1094	ARTHUR (THE ALBUM)	Soundtrack
4707	ART IN AMERICA	Art In America
4221	THE ART OF CONTROL	Peter Frampton

Albums Alpha Index

4179 THE ART OF DEFENSE *Nona Hendryx*
4121 THE ART OF EXCELLENCE *Tony Bennett*
3079 THE ART OF FALLING APART *Soft Cell*
1444 AS FALLS WICHITA, SO FALLS WICHITA FALLS
       Pat Metheny & Lyle Mays
1609 AS FAR AS SIAM *Red Rider*
  55 ASIA *Asia*
4854 ASIA *Kitaro*
 912 AS ONE *Kool & The Gang*
1970 AS ONE *Bar-Kays*
4039 ASSAULT ATTACK *Michael Schenker Group*
 489 AS THE BAND TURNS *Atlantic Starr*
2127 ASTRA *Asia*
1962 AS WE SPEAK *David Sanborn*
 751 ASYLUM *KISS*
1129 ATF *After The Fire*
4591 ATLANTA BLUE *Statler Brothers*
4456 THE ATLANTIC YEARS *Roxy Music*
4709 AT LEAST WE GOT SHOES
       Southside Johnny And The Jukes
3255 ATOMIC PLAYBOYS
       Steve Stevens Atomic Playboys
2087 AT PEACE WITH WOMAN *Jones Girls*
3777 ATTITUDES *Brass Construction*
 952 AUDIO-VISIONS *Kansas*
1013 AUGUST *Eric Clapton*
 225 AUTOAMERICAN *Blondie*
1397 AUTOGRAPH *John Denver*
2036 AUTUMN *George Winston*
1683 AVALON *Roxy Music*
1516 AVALON SUNSET *Van Morrison*
3662 THE AWAKENING *The Reddings*
4702 AWAKEN THE GUARDIAN
       Fates Warning
4691 AZTEC CAMERA *Aztec Camera*

B

 959 BABYLON AND ON *Squeeze*
4326 BABY TONIGHT *Marlon Jackson*
4846 BACHMAN TURNER OVERDRIVE(2)
       Bachman-Turner Overdrive
 554 BACK FOR THE ATTACK *Dokken*
  46 BACK IN BLACK *AC/DC*
 968 BACK IN BLACK *Whodini*
  99 BACK IN THE HIGH LIFE *Steve Winwood*
4232 BACK INTO BLUE *Quarterflash*
4843 BACK IT UP *Robin Trower*
1669 BACK ON THE STREETS *Donnie Iris*
2012 BACKSTAGE PASS *Little River Band*
1464 BACKSTREET *David Sanborn*
4167 BACK TALK *Rockets*
2435 BACK TO AVALON *Kenny Loggins*
4763 BACK 2 BACK *Stargard*
3902 BACK TO BASICS *The Temptations*
4831 BACK TO THE BEACH *Soundtrack*
 688 BACK TO THE FUTURE *Soundtrack*
3497 BACK TO THE WORLD *Dennis DeYoung*
  24 BAD *Michael Jackson*
 144 BAD ANIMALS *Heart*
2928 BAD ATTITUDE *Meat Loaf*
3596 BAD BOY *Robert Gordon*
2979 BADDEST *Grover Washington Jr.*
3801 THE BAD C.C. *Carl Carlton*
 436 BAD ENGLISH *Bad English*
4426 BAD ENUFF *Slave*
1578 BAD FOR GOOD *Jim Steinman*
3799 BAD INFLUENCE *Robert Cray Band*
1509 BADLANDS *Badlands*

 999 BAD LUCK STREAK IN DANCING SCHOOL
       Warren Zevon
4887 BA-DOP-BOOM-BANG *Grandmaster Flash*
1459 BAD REPUTATION *Joan Jett & the Blackhearts*
1196 BAD TO THE BONE
       George Thorogood & The Destroyers
4556 BAD TO THE BONE *L.A. Dream Team*
3421 BALANCE *Balance*
1950 BALANCE OF POWER *Electric Light Orchestra*
1104 BALIN *Marty Balin*
4512 THE BALLAD OF SALLY ROSE *Emmylou Harris*
3929 BALL ROOM *Sea Level*
2014 BALLS TO THE WALL *Accept*
 914 BANANARAMA *Bananarama*
3824 BANDED TOGETHER *Lee Ritenour*
 918 BANGIN' *The Outfield*
3532 BANGING THE WALL *Bar-Kays*
1761 BARBARA MANDRELL LIVE *Barbara Mandrell*
4752 BARBED WIRE KISSES *Jesus And Mary Chain*
4079 BARBEQUE KING *Jorma Kaukonen & Vital Parts*
 818 BARK AT THE MOON *Ozzy Osbourne*
2125 BARKING AT AIRPLANES *Kim Carnes*
4795 BARNET DOGS *Russ Ballard*
 771 BARRY *Barry Manilow*
2859 BARRY GOUDREAU *Barry Goudreau*
2349 BARRY MANILOW *Barry Manilow*
2647 BARRY WHITE'S SHEET MUSIC *Barry White*
 219 BATMAN (SOUNDTRACK) *Prince*
1854 BATMAN ORIGINAL MOTION PICTURE SCORE
       Soundtrack
2099 THE BEACH BOYS *Beach Boys*
  89 BEACHES (SOUNDTRACK) *Bette Midler*
1518 BEAST FROM THE EAST *Dokken*
2321 BEAT *King Crimson*
1703 BEAT CRAZY *Joe Jackson*
1101 BEATIN' THE ODDS *Molly Hatchet*
1296 BEATITUDE *Ric Ocasek*
 942 BEAT STREET *Soundtrack*
3674 BEAT STREET II *Soundtrack*
4558 BEAT SURRENDER *The Jam*
2588 BEAUTIFUL FEELINGS *Rick Springfield*
2185 BEAUTIFUL VISION *Van Morrison*
3881 BEAUTY & THE BEAST: OF LOVE AND HOPE
       TV Soundtrack
  53 BEAUTY AND THE BEAT *Go-Go's*
2271 BEAUTY STAB *ABC*
2715 BE A WINNER *Yarbrough & Peoples*
 495 BEBE LE STRANGE *Heart*
 238 BEE GEES GREATEST *Bee Gees*
2624 BEELZEBUBBA *Dead Milkmen*
3838 BEETLEJUICE *Soundtrack*
3097 BEHAVIOUR *Saga*
2709 ...BEHIND THE GARDENS-BEHIND THE WALL-
       UNDER THE TREE... *Andreas Vollenweider*
1099 BEHIND THE SUN *Eric Clapton*
 362 BEING WITH YOU *Smokey Robinson*
3588 BELIEVE *The Jets*
3419 BELIEVERS *Don McLean*
 623 BELINDA *Belinda Carlisle*
  27 BELLA DONNA *Stevie Nicks*
4838 THE BELLE STARS *Belle Stars*
4417 BELO HORIZONTE *John McLaughlin*
2033 BELOW THE BELT *Franke & The Knockouts*
1833 BE MY LOVER *O'Bryan*
3322 BENEFACTOR *Romeo Void*
1321 BENT OUT OF SHAPE *Rainbow*
2600 BERNADETTE PETERS *Bernadette Peters*
2172 BERRY GORDY'S THE LAST DRAGON *Soundtrack*
2944 THE BEST *Quincy Jones*

4515 BEST BITS *Roger Daltrey*
1524 THE BEST IS YET TO COME *Grover Washington Jr.*
 960 BEST KEPT SECRET *Sheena Easton*
2313 THE BEST LITTLE WHOREHOUSE IN TEXAS
       Soundtrack
 940 THE BEST OF BLONDIE *Blondie*
4018 THE BEST OF DARK HORSE *George Harrison*
4298 THE BEST OF DAVE EDMUNDS *Dave Edmunds*
4658 THE BEST OF EARTH, WIND & FIRE, VOL. II
       Earth, Wind & Fire
3236 THE BEST OF EDDIE RABBITT *Eddie Rabbitt*
2841 BEST OF ELVIS COSTELLO/THE ATTRACTIONS
       Elvis Costello And The Attractions
3521 THE BEST OF EMERSON, LAKE AND PALMER
       Emerson, Lake & Palmer
4828 THE BEST OF ENGLAND DAN & JOHN FORD COLEY
       England Dan & John Ford Coley
1819 THE BEST OF ERIC CARMEN *Eric Carmen*
4621 THE BEST OF FIREFALL *Firefall*
1883 BEST OF FRIENDS
       Twennynine Featuring Lenny White
4676 THE BEST OF GINO VANNELLI *Gino Vannelli*
4661 THE BEST OF JERRY JEFF WALKER
       Jerry Jeff Walker
3894 THE BEST OF JOHNNY MATHIS 1975-1980
       Johnny Mathis
4350 THE BEST OF KANSAS *Kansas*
2846 THE BEST OF MANHATTAN TRANSFER
       Manhattan Transfer
1150 THE BEST OF OMD
       Orchestral Manoeuvres In The Dark
3405 THE BEST OF RITCHIE VALENS *Ritchie Valens*
4635 BEST OF SCORPIONS VOL. 2 *Scorpions*
1542 THE BEST OF THE ALAN PARSONS PROJECT
       Alan Parsons Project
4740 THE BEST OF THE ALLMAN BROTHERS BAND
       Allman Brothers Band
2599 THE BEST OF THE ART OF NOISE *Art Of Noise*
4343 THE BEST OF THE BLUES BROTHERS
       Blues Brothers
1544 BEST OF THE DOOBIES VOL. II *Doobie Brothers*
4153 BEST OF THE REST *Lynyrd Skynyrd*
3437 THE BEST OF THE STATLER BROS. RIDES AGAIN,
       VOL. II *Statler Brothers*
4516 THE BEST OF THE WHISPERS *The Whispers*
4393 THE BEST OF THOMPSON TWINS/
       GREATEST MIXES *Thompson Twins*
3887 THE BEST OF WAR...AND MORE *War*
3403 THE BEST PART OF THE FAT BOYS *Fat Boys*
1955 BEST SHOTS *Pat Benatar*
2236 THE BEST SIDE OF GOODBYE *Jane Olivor*
1923 THE BEST YEARS OF OUR LIVES *Neil Diamond*
4111 BET CHA SAY THAT TO ALL THE GIRLS *Sister Sledge*
1264 BETE NOIRE *Bryan Ferry*
3235 BETTER DAYS *The Blackbyrds*
1625 BETTER THAN HEAVEN *Stacey Q*
4775 BETWEEN THE LINES *Spider*
1004 BETWEEN THE SHEETS *Isley Brothers*
2416 BETWEEN TWO FIRES *Paul Young*
3630 BET YOUR HEART ON ME *Johnny Lee*
 123 BEVERLY HILLS COP *Soundtrack*
 488 BEVERLY HILLS COP II *Soundtrack*
4224 BEWITCHED *Andy Summers & Robert Fripp*
1478 BEYOND *Herb Alpert*
1702 BEYOND APPEARANCES *Santana*
2207 BEYOND THE BLUE NEON *George Strait*
3529 BEYOND THE VALLEY OF 1984 *The Plasmatics*
1880 BE YOURSELF *Patti LaBelle*
 344 BE YOURSELF TONIGHT *Eurythmics*
 563 THE B-52'S *The B-52s*
2653 BI-COASTAL *Peter Allen*

Albums Alpha Index

2074 BIG & BEAUTIFUL *Fat Boys*	1046 THE BLASTERS *The Blasters*	2869 BOTTOMS UP *The Chi-Lites*
203 BIG BAM BOOM *Daryl Hall & John Oates*	4639 BLAST FROM THE BAYOU *Wayne Toups & Zydecajun*	2894 BOUNCING OFF THE SATELLITES *The B-52s*
170 THE BIG CHILL *Soundtrack*	3536 BLAST OFF *Stray Cats*	3998 BOURGEOIS TAGG *Bourgeois Tagg*
1472 THE BIG CHILL: MORE SONGS FROM *Soundtrack*	2295 BLAZE OF GLORY *Joe Jackson*	1682 BOX OF FROGS *Box Of Frogs*
4423 BIG CIRCUMSTANCE *Bruce Cockburn*	2263 BLESSING IN DISGUISE *Metal Church*	1034 BOY *U2*
2792 BIG CITY *Merle Haggard*	4890 BLIND DATE *Soundtrack*	2806 THE BOY GENIUS *Kwamé Featuring A New Beginning*
601 BIG DADDY *John Mellencamp*	788 BLINDED BY SCIENCE *Thomas Dolby*	534 BOY IN THE BOX *Corey Hart*
4252 BIG DEAL *Killer Dwarfs*	2212 THE BLIND LEADING THE NAKED *Violent Femmes*	2918 BOY MEETS GIRL *Boy Meets Girl*
4104 BIG DREAMERS NEVER SLEEP *Gino Vannelli*	503 BLIND MAN'S ZOO *10,000 Maniacs*	1767 BOYS AND GIRLS *Bryan Ferry*
3252 BIG DREAMS IN A SMALL TOWN *Restless Heart*	2055 BLIND TO REASON *Grayson Hugh*	2601 BOYS CLUB *Boys Club*
3416 THE BIG EASY *Soundtrack*	1194 THE BLITZ *Krokus*	1864 BOYS DON'T CRY *Boys Don't Cry*
4503 THE BIG EXPRESS *XTC*	231 BLIZZARD OF OZZ *Ozzy Osbourne*	2323 THE BOYS FROM DORAVILLE *Atlanta Rhythm Section*
644 BIG FUN *Shalamar*	2984 BLOOD & CHOCOLATE *Elvis Costello And The Attractions*	1878 BOYS IN HEAT *Britny Fox*
4452 BIG FUN *Inner City*	4798 BLOODLINE *Levert*	3756 BRAIN DRAIN *The Ramones*
712 BIG GAME *White Lion*	2180 BLOWFLY'S PARTY [X-RATED] *Blowfly*	976 BRANIGAN *Laura Branigan*
614 BIG GENERATOR *Yes*	705 BLOW MY FUSE *Kix*	1002 BRANIGAN 2 *Laura Branigan*
223 BIGGER AND DEFFER *LL Cool J*	673 BLOW UP YOUR VIDEO *AC/DC*	2284 BRASIL *Manhattan Transfer*
4013 BIGGEST HITS *Marty Robbins*	3124 BLOW YOUR COOL! *Hoodoo Gurus*	2077 BRASS CONSTRUCTION 5 *Brass Construction*
4565 THE BIGGEST PRIZE IN SPORT *999*	3486 BLOW YOUR OWN HORN *Herb Alpert*	3818 BRASS CONSTRUCTION 6 *Brass Construction*
3709 THE BIG HEAT *Stan Ridgway*	1335 B.L.T. *Jack Bruce/Bill Lordan/Robin Trower*	738 BRAVE AND CRAZY *Melissa Etheridge*
1656 BIG LIFE *Night Ranger*	1371 BLUE *Double*	4612 BRAZIL CLASSICS 1: BELEZA TROPICAL *Various Artists*
4080 BIG NIGHT MUSIC *Shriekback*	2192 THE BLUE ALBUM *Harold Melvin And The Blue Notes*	3312 A BRAZILIAN LOVE AFFAIR *George Duke*
4878 BIG ONES *Loverboy*	2539 BLUE AND GRAY *Poco*	363 BREAK EVERY RULE *Tina Turner*
3354 BIG PLANS FOR EVERYBODY *Let's Active*	2760 BLUE BELL KNOLL *Cocteau Twins*	822 THE BREAKFAST CLUB *Soundtrack*
1269 THE BIG PRIZE *Honeymoon Suite*	3658 BLUE JEANS *Chocolate Milk*	1187 THE BREAKFAST CLUB *Breakfast Club*
3337 BIG SCIENCE *Laurie Anderson*	4498 THE BLUE MASK *Lou Reed*	573 BREAKIN' *Soundtrack*
803 BIG THING *Duran Duran*	2110 BLUE MURDER *Blue Murder*	2293 BREAKIN' 2 ELECTRIC BOOGALOO *Soundtrack*
1223 THE BIG THROWDOWN *Levert*	646 THE BLUES BROTHERS (SOUNDTRACK) *Blues Brothers*	193 BREAKIN' AWAY *Al Jarreau*
4166 BIG TIME *Tom Waits*	4884 BLUES FOR SALVADOR *Carlos Santana*	1742 BREAKING ALL THE RULES *Peter Frampton*
3934 BIG TRASH *Thompson Twins*	3083 BLUE SKIES *Kiri Te Kanawa/Nelson Riddle And His Orchestra*	3784 BREAKING CURFEW *Red Rider*
426 BIG TYME *Heavy D & The Boyz*	4574 BLUES 'N JAZZ *B.B. King*	690 BREAKING HEARTS *Elton John*
1109 BIG WORLD *Joe Jackson*	4576 BLUE TATTOO *Passport*	3221 BREAKING THE CHAINS *Dokken*
3834 BIKINI RED *Screaming Blue Messiahs*	1023 BOBBIE SUE *Oak Ridge Boys*	4833 BREAK OF DAWN *Firefall*
4431 BILL & TED'S EXCELLENT ADVENTURE *Soundtrack*	3995 BOBBY & THE MIDNITES *Bobby And The Midnites*	3590 BREAK OF HEARTS *Katrina & The Waves*
2364 BILL COSBY "HIMSELF" (SOUNDTRACK) *Bill Cosby*	4834 BO-DAY-SHUS!!! *Mojo Nixon & Skid Roper*	68 BREAK OUT *Pointer Sisters*
4646 BILL WITHERS GREATEST HITS *Bill Withers*	1933 BODIES AND SOULS *Manhattan Transfer*	2119 BREAKOUT *Spyro Gyra*
3159 BILLY & THE BEATERS *Billy Vera & The Beaters*	650 BODY AND SOUL *Joe Jackson*	1541 BRENDA K. STARR *Brenda K. Starr*
595 BILLY IDOL *Billy Idol*	1186 BODY WISHES *Rod Stewart*	2138 BRIAN WILSON *Brian Wilson*
3273 BILLY PRESTON & SYREETA *Billy Preston & Syreeta*	2510 BOI-NGO *Oingo Boingo*	243 THE BRIDGE *Billy Joel*
4019 BILLY SATELLITE *Billy Satellite*	3096 BOINGO ALIVE *Oingo Boingo*	2821 BRIEF ENCOUNTER *Marillion*
1536 BIOGRAPH *Bob Dylan*	2963 BOLD AS LOVE *Bardeux*	4883 BRIGHTER THAN A THOUSAND SUNS *Killing Joke*
4584 BIRD *Soundtrack*	532 BON JOVI *Bon Jovi*	2703 BRIGHT LIGHTS, BIG CITY *Soundtrack*
4176 BIRDY (SOUNDTRACK) *Peter Gabriel*	2394 BONK *Big Pig*	988 BRILLIANCE *Atlantic Starr*
2440 BIRTH, SCHOOL, WORK, DEATH *The Godfathers*	2250 BONNIE POINTER (II) *Bonnie Pointer*	2657 BRING THE FAMILY *John Hiatt*
3315 THE BITTEREST PILL (I EVER HAD TO SWALLOW) *The Jam*	4335 BOOK OF DAYS *Psychedelic Furs*	1330 BRITISH STEEL *Judas Priest*
1634 THE BIZ NEVER SLEEPS *Diabolical Biz Markie*	4546 THE BOOK OF PRIDE *Giant Steps*	864 BRITNY FOX *Britny Fox*
528 BLACK & WHITE *Pointer Sisters*	3339 BOOM BOOM CHI BOOM BOOM *Tom Tom Club*	680 BROADCAST *Cutting Crew*
3186 A BLACK & WHITE NIGHT: LIVE (SOUNDTRACK) *Roy Orbison and Friends*	876 BOOMTOWN *David & David*	1193 THE BROADSWORD AND THE BEAST *Jethro Tull*
2034 BLACK CARS *Gino Vannelli*	3447 BOP DOO-WOP *Manhattan Transfer*	175 THE BROADWAY ALBUM *Barbra Streisand*
2106 BLACK CELEBRATION *Depeche Mode*	1489 BORDERLINE *Ry Cooder*	3768 BROADWAY'S CLOSER TO SUNSET BLVD. *Isley Jasper Isley*
3448 BLACK CODES (FROM THE UNDERGROUND) *Wynton Marsalis*	4553 BORDER WAVE *Sir Douglas Quintet*	2409 BROKEN ENGLISH *Marianne Faithfull*
2187 BLACK MARKET CLASH *The Clash*	1899 BORN AGAIN *Black Sabbath*	4371 A BROKEN FRAME *Depeche Mode*
3350 BLACK 'N' BLUE *Black 'N Blue*	4390 BORN IN AMERICA *Riot*	3715 BRONCO BILLY *Soundtrack*
1538 BLACK ON BLACK *Waylon Jennings*	2 BORN IN THE U.S.A. *Bruce Springsteen*	3216 THE BROOKLYN, BRONX & QUEENS BAND *Brooklyn, Bronx & Queens Band*
382 BLACKOUT *Scorpions*	1003 BORN TO BE BAD *George Thorogood & The Destroyers*	4864 BROTHER ARAB *Arabian Prince*
1067 BLACK SEA *XTC*	3319 BORN 2B BLUE *Steve Miller*	2830 BROTHERHOOD *New Order*
2671 BLACK TIE *The Manhattans*	992 BORN TO BOOGIE *Hank Williams Jr.*	31 BROTHERS IN ARMS *Dire Straits*
1867 BLAH-BLAH-BLAH *Iggy Pop*	3545 BORN TO LAUGH AT TORNADOES *Was (Not Was)*	1859 BROTHERS OF THE ROAD *Allman Brothers Band*
3746 BLAME IT ON LOVE AND ALL THE GREAT HITS *Smokey Robinson*	708 BORN TO LOVE *Peabo Bryson & Roberta Flack*	1021 BROTHER WHERE YOU BOUND *Supertramp*
4124 BLAST! (THE LATEST AND THE GREATEST) *Brothers Johnson*	2286 BORN YESTERDAY *Everly Brothers*	4292 BRUCE COCKBURN RESUME *Bruce Cockburn*
	3114 BOSSA NOVA HOTEL *Michael Sembello*	288 BRUCE SPRINGSTEEN & THE E STREET BAND LIVE 1975-1985 *Bruce Springsteen*
	2467 BOSTON, MASS. *Del Fuegos*	4759 BRUCE WOOLLEY & THE CAMERA CLUB *Bruce Woolley & The Camera Club*
	4164 BOTH SIDES OF LOVE *Paul Anka*	4237 BRUISEOLOGY *The Waitresses*

Albums Alpha Index

4239 BUCKY FELLINI *Dead Milkmen*
2228 BUENAS NOCHES FROM A LONELY ROOM *Dwight Yoakam*
253 BUILDING THE PERFECT BEAST *Don Henley*
92 BUILT FOR SPEED *Stray Cats*
1661 BUILT TO LAST *Grateful Dead*
4259 BULL DURHAM *Soundtrack*
820 BULLETBOYS *BulletBoys*
2959 BULLISH *Herb Alpert & The Tijuana Brass*
3310 BURLAP & SATIN *Dolly Parton*
4211 BURNING *Shooting Star*
4572 BURNING SENSATIONS *Burning Sensations*
3570 BURNIN' LOVE *Con Funk Shun*
3461 BURNS LIKE A STAR *Stone Fury*
25 BUSINESS AS USUAL *Men At Work*
1514 BUSTER *Soundtrack*
2471 BUSTER POINDEXTER *Buster Poindexter And His Banshees Of Blue*
3527 BUSTIN' LOOSE *Roberta Flack*
800 BUSY BODY *Luther Vandross*
57 ...BUT SERIOUSLY *Phil Collins*
920 BUT THE LITTLE GIRLS UNDERSTAND *The Knack*
4093 BUTT ROCKIN' *Fabulous Thunderbirds*
3789 BUT WHAT WILL THE NEIGHBORS THINK *Rodney Crowell*
1969 BY ALL MEANS NECESSARY *Boogie Down Productions*
987 BY REQUEST (THE BEST OF BILLY VERA & THE BEATERS) *Billy Vera & The Beaters*
1381 BY THE LIGHT OF THE MOON *Los Lobos*

C

4545 CACTUS AND A ROSE *Gary Stewart*
2470 CADDYSHACK *Soundtrack*
2845 CAFE RACERS *Kim Carnes*
4285 CAGE THE SONGBIRD *Crystal Gayle*
1357 THE CALIFORNIA RAISINS *California Raisins*
4373 CALLING *Noel Pointer*
2586 CALL IT WHAT YOU WANT *Bill Summers & Summers Heat*
1649 CALL OF THE WEST *Wall Of Voodoo*
3523 CALL ON ME *Evelyn "Champagne" King*
2049 CALM ANIMALS *The Fixx*
762 CAMEOSIS *Cameo*
4697 CAMERA CAMERA *Renaissance*
3587 THE CAMERA NEVER LIES *Michael Franks*
1993 CAMERON *Rafael Cameron*
2880 CAMERON'S IN LOVE *Rafael Cameron*
579 CAMOUFLAGE *Rod Stewart*
2478 CAMOUFLAGE *Rufus And Chaka Khan*
1096 CANCIONES DE MI PADRE *Linda Ronstadt*
4874 CANDLELAND *Ian McCulloch*
2595 CANDLES *Heatwave*
3671 CANDY APPLE GREY *Husker Du*
1889 CANDY GIRL *New Edition*
394 CAN'T HOLD BACK *Eddie Money*
3146 CAN'T LOOK AWAY *Trevor Rabin*
4725 CAN'T SHAKE THIS FEELIN' *Spinners*
7 CAN'T SLOW DOWN *Lionel Richie*
1548 CAN'T STOP THE LOVE *Maze Featuring Frankie Beverly*
2051 CAN'T STOP THE MUSIC (SOUNDTRACK) *Village People*
3373 CAN'T WAIT ALL NIGHT *Juice Newton*
1989 CAN'T WE FALL IN LOVE AGAIN *Phyllis Hyman*
3468 CANYON *Paul Winter*
2039 CAPTAIN SWING *Michelle Shocked*
365 CAPTURED *Journey*

3518 CAPTURED *Rockwell*
3944 CAPTURED LIVE *Peter Tosh*
1898 CARAVAN OF LOVE *Isley Jasper Isley*
4755 CARE *Shriekback*
1457 CAREFUL *The Motels*
207 CARGO *Men At Work*
2769 CARL ANDERSON *Carl Anderson*
1230 CARL CARLTON *Carl Carlton*
3525 CARLIN ON CAMPUS *George Carlin*
4756 CARL WILSON *Carl Wilson*
1203 CARNAVAL *Spyro Gyra*
2739 CARNIVAL *Duran Duran*
701 CAROL HENSEL'S EXERCISE AND DANCE PROGRAM *Carol Hensel*
1655 CAROL HENSEL'S EXERCISE AND DANCE PROGRAM, VOLUME 2 *Carol Hensel*
3129 CAROL HENSEL'S EXERCISE AND DANCE PROGRAM, VOLUME 3 *Carol Hensel*
3302 CARRY ON *Bobby Caldwell*
3672 CA$HFLOW *Cashflow*
2016 CASINO LIGHTS *Various Artists*
4525 CASTLES IN THE SAND *David Allan Coe*
1934 CASUAL GODS *Jerry Harrison: Casual Gods*
801 CATCHING THE SUN *Spyro Gyra*
2743 CATCHING UP WITH DEPECHE MODE *Depeche Mode*
3338 CATCH ME, I'M FALLING *Pretty Poison*
2757 THE CATHERINE WHEEL (ORIGINAL CAST) *David Byrne*
1659 CATHOLIC BOY *Jim Carroll Band*
2744 CAT IN THE HAT *Bobby Caldwell*
2004 CAT PEOPLE *Soundtrack*
1830 CATS *Original Broadway Cast Recording*
2030 CATS *Original London Cast*
3396 CATS *Selections from Original Broadway Cast*
1910 CATS WITHOUT CLAWS *Donna Summer*
1603 CAUGHT IN THE ACT - LIVE *Styx*
3379 CAUGHT IN THE GAME *Survivor*
3368 CAVERNA MAGICA (...UNDER THE TREE-IN THE CAVE...) *Andreas Vollenweider*
3600 CCCP - LIVE IN MOSCOW *UB40*
258 CELEBRATE! *Kool & The Gang*
174 CENTERFIELD *John Fogerty*
2052 CENTIPEDE *Rebbie Jackson*
3754 CENTRAL LINE *Central Line*
986 CHAIN LIGHTNING *Don McLean*
1967 CHAKA KHAN *Chaka Khan*
1787 CHALK MARK IN A RAIN STORM *Joni Mitchell*
2206 CHANCE *Manfred Mann's Earth Band*
3535 CHANCES ARE *Bob Marley*
2342 CHANGE. *The Alarm*
3994 CHANGE *Barry White*
4806 THE CHANGE HAS COME *Chubby Checker*
3050 CHANGE NO CHANGE *Elliot Easton*
1772 CHANGE OF ADDRESS *Krokus*
1555 A CHANGE OF HEART *David Sanborn*
2953 CHANGE OF HEART *Change*
4491 CHANGES *The Monkees*
2070 CHANGESTWOBOWIE *David Bowie*
626 CHARACTERS *Stevie Wonder*
3517 CHARIOTS OF FIRE *Ernie Watts*
145 CHARIOTS OF FIRE (SOUNDTRACK) *Vangelis*
3809 CHARLIE *Charlie*
3154 CHARLIE SEXTON *Charlie Sexton*
2141 CHARTBUSTERS *Ray Parker Jr.*
2246 CHEAT THE NIGHT *Deborah Allen*
4168 CHEQUERED PAST *Chequered Past*
715 CHER(2) *Cher*
1728 CHESS *Various Artists*

2504 CHICAGO XIV *Chicago*
439 CHICAGO 16 *Chicago*
127 CHICAGO 17 *Chicago*
883 CHICAGO 18 *Chicago*
4362 CHICAGO - GREATEST HITS, VOLUME II *Chicago*
2134 CHICO DEBARGE *Chico DeBarge*
3453 CHILDREN *Mission U.K.*
2122 CHILDREN OF TOMORROW *Frankie Smith*
3696 A CHILD'S ADVENTURE *Marianne Faithfull*
1891 CHILLIN' *Force M.D.'s*
3951 CHILL OUT *Black Uhuru*
3118 CHINATOWN *Thin Lizzy*
759 CHINESE WALL *Philip Bailey*
2604 A CHIPMUNK CHRISTMAS *The Chipmunks*
1141 CHIPMUNK PUNK *The Chipmunks*
3698 CHIPMUNK ROCK *The Chipmunks*
2878 A CHORUS LINE-THE MOVIE *Soundtrack*
4860 CHRIS ISAAK *Chris Isaak*
3978 CHRISTIANE F. (SOUNDTRACK) *David Bowie*
3982 THE CHRISTIANS *The Christians*
4481 CHRISTINE *Soundtrack*
1148 CHRISTINE McVIE *Christine McVie*
1727 CHRISTMAS *Kenny Rogers*
3027 CHRISTMAS *Alabama*
3267 CHRISTMAS *Oak Ridge Boys*
3288 A CHRISTMAS ALBUM *Barbra Streisand*
4791 THE CHRISTMAS ALBUM *Elvis Presley*
3823 CHRISTMAS IN AMERICA *Kenny Rogers*
2966 CHRISTMAS IN THE STARS/STAR WARS CHRISTMAS ALBUM *Meco*
4312 CHRISTMAS JOLLIES II *Salsoul Orchestra*
3925 CHRISTMAS RAP *Various Artists*
1717 A CHRISTMAS TOGETHER *John Denver & The Muppets*
2479 CHRISTMAS WISHES *Anne Murray*
4517 CHRISTMAS WITH SLIM WHITMAN *Slim Whitman*
26 CHRISTOPHER CROSS *Christopher Cross*
3573 CHRONICLE II *Creedence Clearwater Revival*
989 CHRONICLES *Steve Winwood*
3320 CHUCKII *Chuckii Booker*
1476 CIMARRON *Emmylou Harris*
4710 THE CINDERELLA THEORY *George Clinton*
1172 CIRCLE OF LOVE *Steve Miller Band*
4789 THE CIRCUS *Erasure*
4822 CIRCUS OF POWER *Circus Of Power*
2121 CITIZEN KIHN *Greg Kihn*
3597 CITY *Roger McGuinn & Chris Hillman Ft. Gene Clark*
2296 CITY KIDS *Spyro Gyra*
2038 CITY LIFE *Boogie Boys*
1252 CITY NIGHTS *Tierra*
1922 CITY OF NEW ORLEANS *Willie Nelson*
2814 CITY STREETS *Carole King*
1921 CIVILIZED EVIL *Jean-Luc Ponty*
3572 CIVILIZED MAN *Joe Cocker*
3245 C.K. *Chaka Khan*
828 THE CLARKE/DUKE PROJECT *Stanley Clarke & George Duke*
3473 THE CLARKE/DUKE PROJECT II *Stanley Clarke & George Duke*
3701 CLASS *The Reddings*
2594 CLASSICAL GAS *Mason Williams & Mannheim Steamroller*
3030 A CLASSIC CASE: THE MUSIC OF JETHRO TULL *London Symphony Orchestra/Ian Anderson*
1500 CLASSIC CRYSTAL *Crystal Gayle*
3803 CLASSICS *The Doors*
2763 CLASSICS LIVE *Aerosmith*
4267 CLASSICS THE EARLY YEARS *Neil Diamond*

Albums Alpha Index

4006 CLASSIC YES Yes	535 THE CONCERT IN CENTRAL PARK Simon & Garfunkel	334 CRUSHIN' Fat Boys
2884 CLASS OF '55 Carl Perkins, Jerry Lee Lewis, Roy Orbison, & Johnny Cash	1692 CONCERTS FOR THE PEOPLE OF KAMPUCHEA Various Artists	2304 CRUZADOS Cruzados
4587 CLIMBING THE WALLS Wrathchild America	4800 CONCRETE 999	4125 CRYSTAL GAYLE'S GREATEST HITS Crystal Gayle
3810 CLOSE Kim Wilde	2734 CONCRETE BLONDE Concrete Blonde	1416 CULTOSAURUS ERECTUS Blue Öyster Cult
2644 CLOSER Gino Soccio	882 CONDITION CRITICAL Quiet Riot	4643 CULTURE KILLED THE NATIVE Victory
4311 CLOSER THAN CLOSE Jean Carne	3360 CONDITION OF THE HEART Kashif	1428 CUPID AND PSYCHE 85 Scritti Politti
2433 CLOSER TO THE FLAME Rob Jungklas	4387 CONFESSIONS OF A POP GROUP Style Council	4536 CURED Steve Hackett
356 THE CLOSER YOU GET Alabama	2088 THE CONFESSOR Joe Walsh	3581 CURIOSITY Regina
2593 CLOSE TO THE BONE Thompson Twins	2029 CONFETTI Sergio Mendes	4337 A CURIOUS FEELING Tony Banks
2667 CLOSE TO THE BONE Tom Tom Club	4037 CONFIDENCE Narada Michael Walden	4212 CURRENT Heatwave
1637 CLOSE-UP David Sanborn	3601 CONFLICTING EMOTIONS Split Enz	904 CUT Golden Earring
385 CLOUD NINE George Harrison	2199 CONFRONTATION Bob Marley And The Wailers	3515 CUT LOOSE Paul Rodgers
2696 CLOUDS ACROSS THE SUN Firefall	2474 CON FUNK SHUN 7 Con Funk Shun	187 CUTS BOTH WAYS Gloria Estefan
2381 CLUB NINJA Blue Öyster Cult	4112 CONNECTIONS AND DISCONNECTIONS Funkadelic(2)	263 CUTS LIKE A KNIFE Bryan Adams
3872 CLUB PARADISE (SOUNDTRACK) Jimmy Cliff	542 CONSCIOUS PARTY Ziggy Marley And The Melody Makers	2867 CUT THE CRAP The Clash
1897 CLUES Robert Palmer	3482 CONSPIRACY King Diamond	3833 CYCLE OF THE MOON Prophet
3169 CLUTCHING AT STRAWS Marillion	2011 CONSTRICTOR Alice Cooper	875 CYCLES The Doobie Brothers
1485 COAL MINER'S DAUGHTER Soundtrack	676 CONTACT Pointer Sisters	3409 CYPRESS Let's Active
3665 COBRA Soundtrack	2485 CONTAGIOUS Y&T	
1733 COCKER Joe Cocker	3222 CONTAGIOUS Bar-Kays	# D
2017 COCK ROBIN Cock Robin	2522 CONTINUATION Philip Bailey	
137 COCKTAIL Soundtrack	28 CONTROL Janet Jackson	470 DAD LOVES HIS WORK James Taylor
1262 COCONUT TELEGRAPH Jimmy Buffett	4247 CONTROLLED BY HATRED/FEEL LIKE SHIT... DEJA VU Suicidal Tendencies	4724 DA'KRASH da'Krash
4704 COCOON Soundtrack	566 CONTROVERSY Prince	4356 DALLAS Floyd Cramer
825 CODA Led Zeppelin	4329 CONVERSATIONS Brass Construction	88 DAMN THE TORPEDOES Tom Petty And The Heartbreakers
717 COLD BLOODED Rick James	3716 CONVERTIBLE MUSIC Josie Cotton	1165 DANA DANE WITH FAME Dana Dane
4005 COLD SPRING HARBOR Billy Joel	3790 COOKIN' ON THE ROOF Roman Holliday	4391 DANCE Gary Numan
1412 COLLABORATION George Benson/Earl Klugh	3046 COOL FROM THE WIRE Dirty Looks	4201 DANCE & EXERCISE WITH THE HITS Linda Fratianne
1530 THE COLLECTION Amy Grant	4213 COOL KIDS Kix	1927 THE DANCE OF LIFE Narada Michael Walden
993 A COLLECTION: GREATEST HITS...AND MORE Barbra Streisand	1580 COOL NIGHT Paul Davis	1261 DANCIN' AND LOVIN' Spinners
1278 A COLLECTION OF GREAT DANCE SONGS Pink Floyd	1245 COPPERHEAD ROAD Steve Earle	1512 DANCING IN THE DRAGON'S JAWS Bruce Cockburn
2893 COLONEL ABRAMS Colonel Abrams	3346 CORRIDORS OF POWER Gary Moore	4497 DANCING IN THE SUN George Howard
3071 COLOR IN YOUR LIFE Missing Persons	2041 COSI FAN TUTTI FRUTTI Squeeze	129 DANCING ON THE CEILING Lionel Richie
2693 THE COLOR OF MONEY Soundtrack	3388 THE COST OF LOVING Style Council	4123 DANCING ON THE COUCH Go West
1266 COLOR OF SUCCESS Morris Day	3105 COTTON CLUB Soundtrack	4067 DANCING ON THE EDGE Roy Buchanan
2677 THE COLOR PURPLE Soundtrack	859 COULDN'T STAND THE WEATHER Stevie Ray Vaughan And Double Trouble	897 DANCING UNDERCOVER Ratt
1355 COLORS Soundtrack	4351 COUNTERFEIT Martin L. Gore	1941 DANCING WITH THE LION Andreas Vollenweider
91 COLOUR BY NUMBERS Culture Club	2933 COUNTRY Soundtrack	4609 DANCIN' IN THE KEY OF LIFE Steve Arrington
2010 THE COLOUR OF SPRING Talk Talk	4564 COUNTRY BOY Ricky Skaggs	2299 DANCIN' ON THE EDGE Lita Ford
173 COMBAT ROCK The Clash	4530 A COUNTRY CHRISTMAS Various Artists	598 DAN FOGELBERG/GREATEST HITS Dan Fogelberg
3945 COME AN' GET IT Whitesnake	2679 A COUNTRY COLLECTION Anne Murray	1678 DANGER DANGER Danger Danger
2063 COME AS YOU ARE Peter Wolf	1902 COUNT THREE AND PRAY Berlin	1801 DANGEROUS Bar-Kays
4432 COME DANCING WITH THE KINKS/THE BEST OF THE KINKS 1977-1986 The Kinks	4282 COUP DE GRACE Mink De Ville	3855 DANGEROUS Natalie Cole
892 COME MORNING Grover Washington Jr.	4842 COVER GIRL Phantom, Rocker & Slick	3157 DANGEROUS ACQUAINTANCES Marianne Faithfull
1868 COME OUT AND PLAY Twisted Sister	3542 COWBOYS & ENGLISHMEN Poco	1267 DANGEROUS AGE Bad Company
4304 COME SHARE MY LOVE Miki Howard	2980 CRACKERS INTERNATIONAL Erasure	4857 DANGEROUS DREAMS The Nails
910 COME UPSTAIRS Carly Simon	991 CRASH Human League	3047 DANGEROUS MOMENTS Martin Briley
467 COMING AROUND AGAIN Carly Simon	749 CRASH AND BURN Pat Travers Band	1308 DANGEROUS TOYS Dangerous Toys
1175 COMING BACK HARD AGAIN Fat Boys	1568 CRAZY FOR YOU Earl Klugh	2550 DANGER ZONE Sammy Hagar
4782 COMING TO AMERICA Soundtrack	482 CRAZY FROM THE HEAT David Lee Roth	2432 DAN HILL(2) Dan Hill
3885 COMMODORES ANTHOLOGY Commodores	596 CRAZY NIGHTS KISS	4115 DANNY DAVIS & WILLIE NELSON WITH THE NASHVILLE BRASS Danny Davis And Willie Nelson With The Nashville Brass
3147 COMMODORES 13 Commodores	1447 CREATURES OF THE NIGHT KISS	3491 DANNY JOE BROWN AND THE DANNY JOE BROWN BAND Danny Joe Brown And The Danny Joe Brown Band
2989 COMMON GROUND Rhythm Corps	1000 CREST OF A KNAVE Jethro Tull	
2563 COMMON ONE Van Morrison	42 CRIMES OF PASSION Pat Benatar	
2836 THE COMMUNARDS The Communards	4438 CRISTOFORI'S DREAM David Lanz	2337 DAN REED NETWORK Dan Reed Network
4057 COMPANY B Company B	493 THE CROSSING Big Country	4645 DANSEPARC Martha And The Muffins
2915 THE COMPLETE STORY OF ROXANNE...THE ALBUM Dr. J.R. Kool & The Other Roxannes	609 CROSSROADS Tracy Chapman	3467 DANZIG Danzig
1063 THE COMPLETION BACKWARD PRINCIPLE The Tubes	1337 CROSSROADS Eric Clapton	265 DARE Human League
1340 COMPUTER GAMES George Clinton	3135 CROSSROADS (SOUNDTRACK) Ry Cooder	1931 DARE TO BE STUPID "Weird Al" Yankovic
1041 COMPUTER-WORLD Kraftwerk	1829 CROSS THAT LINE Howard Jones	4073 DARING ADVENTURES Richard Thompson
4407 CONAN THE BARBARIAN Soundtrack	387 CROWDED HOUSE Crowded House	2130 THE DARK Metal Church
1884 THE CONCERT Creedence Clearwater Revival	4474 CRUSADER Saxon	4735 DARK CONTINENT Wall Of Voodoo
	814 CRUSH Orchestral Manoeuvres In The Dark	4484 DARKLANDS Jesus And Mary Chain

Albums Alpha Index

3703 **DARKROOM** Angel City	2129 **DIFFORD & TILBROOK** Difford & Tilbrook	3695 **DOUBLE TROUBLE LIVE** Molly Hatchet
4579 **DARLIN'** Tom Jones	3891 **DIG THE NEW BREED** The Jam	1001 **DOUBLE VISION** Bob James/David Sanborn
3632 **DAVE GRUSIN AND THE GRP ALL-STARS/ LIVE IN JAPAN** Dave Grusin and the GRP All-Stars	1454 **DIMPLES** Richard "Dimples" Fields	3082 **DOWN AND OUT IN BEVERLY HILLS** Soundtrack
4444 **DAVE GRUSIN AND THE NY/LA DREAM BAND** Dave Grusin	4706 **DIRECTIONS** Miles Davis	3061 **DOWN FOR THE COUNT** Y&T
3014 **DAVE GRUSIN COLLECTION** Dave Grusin	9 **DIRTY DANCING** Soundtrack	2482 **DOWN IN THE GROOVE** Bob Dylan
4816 **DAVID FOSTER** David Foster	155 **DIRTY DEEDS DONE DIRT CHEAP** AC/DC	1127 **DOWN ON THE FARM** Little Feat
3747 **DAVID GRISMAN - QUINTET "80"** David Grisman	2285 **DIRTY LOOKS** Juice Newton	1311 **DOWN TO THE MOON** Andreas Vollenweider
849 **DAWN PATROL** Night Ranger	1179 **DIRTY MIND** Prince	2735 **DOWNTOWN** Marshall Crenshaw
2084 **DAY BY DAY** Najee	199 **DIRTY ROTTEN FILTHY STINKING RICH** Warrant	1743 **DO YOU** Sheena Easton
1720 **DAYDREAMING** Morris Day	496 **DIRTY WORK** Rolling Stones	2666 **DO YOU WANNA GET AWAY** Shannon
304 **DAYLIGHT AGAIN** Crosby, Stills & Nash	1835 **DISCIPLINE** King Crimson	--DR. see: DOCTOR
2852 **DAYS OF INNOCENCE** Moving Pictures	2995 **DISCO** Pet Shop Boys	4008 **DRAGNET** Soundtrack
4196 **DAZZLE SHIPS** Orchestral Manoeuvres In The Dark	3316 **DISCOVER** Gene Loves Jezebel	935 **DRAMA** Yes
4629 **D.C. CAB** Soundtrack	3084 **DISCOVERY** Shanice Wilson	1591 **DRASTIC MEASURES** Kansas
2201 **DEAD LETTER OFFICE** R.E.M.	4475 **DISCOVERY** Larry Carlton	899 **THE DREAM ACADEMY** Dream Academy
2661 **DEAD MAN'S PARTY** Oingo Boingo	3920 **DISGUISE** Chuck Mangione	3031 **DREAM A LITTLE DREAM** Soundtrack
1975 **DEAD RINGER** Meat Loaf	308 **DISINTEGRATION** The Cure	2853 **DREAM BABIES GO TO HOLLYWOOD** John Stewart
1614 **DEAD SET** Grateful Dead	3436 **DISORDERLIES** Soundtrack	3860 **DREAMBOY** Dreamboy
2264 **DEATH WISH II (SOUNDTRACK)** Jimmy Page	1049 **THE DISREGARD OF TIMEKEEPING** Bonham	1698 **DREAM COME TRUE** Earl Klugh
2726 **A DECADE OF HITS** Charlie Daniels Band	291 **THE DISTANCE** Bob Seger & The Silver Bullet Band	2230 **DREAM EVIL** Dio
1344 **A DECADE OF ROCK AND ROLL 1970 TO 1980** REO Speedwagon	3761 **DISTANT LOVER** Alphonse Mouzon	590 **DREAMGIRLS** Original Cast
352 **DECEMBER** George Winston	4063 **DISTANT SHORES** Robbie Patton	3727 **THE DREAMING** Kate Bush
3231 **DECISIONS** The Winans	4293 **DISTANT THUNDER** Aswad	1152 **DREAMING #11** Joe Satriani
1531 **DECLARATION** The Alarm	3282 **DISTURBING THE PEACE** Alcatrazz	374 **DREAM INTO ACTION** Howard Jones
3837 **DECOY** Miles Davis	337 **DIVER DOWN** Van Halen	2395 **DREAMLAND EXPRESS** John Denver
2107 **DEDICATED** Marshall Tucker Band	1463 **DIVINE MADNESS (SOUNDTRACK)** Bette Midler	2047 **DREAM OF A LIFETIME** Marvin Gaye
927 **DEDICATION** Gary U.S. Bonds	3607 **THE DOCTOR** Cheap Trick	2289 **DREAM OF LIFE** Patti Smith
3211 **THE DEED IS DONE** Molly Hatchet	76 **DR. FEELGOOD** Mötley Crüe	107 **THE DREAM OF THE BLUE TURTLES** Sting
4160 **DEEPEST PURPLE: THE VERY BEST OF DEEP PURPLE** Deep Purple	3140 **DR. HOOK/GREATEST HITS** Dr. Hook	2066 **DREAM ON** George Duke
3442 **DEEP IN THE HEART OF NOWHERE** Bob Geldof	457 **DOCUMENT** R.E.M.	1288 **DREAMS** Grace Slick
2085 **DEEP SEA SKIVING** Bananarama	3325 **DOES FORT WORTH EVER CROSS YOUR MIND** George Strait	3209 **DREAMS** Allman Brothers Band
2635 **DEFACE THE MUSIC** Utopia	1913 **DOG EAT DOG** Joni Mitchell	2860 **THE DREAMS OF CHILDREN** Shadowfax
3627 **DEF, DUMB & BLONDE** Deborah Harry	2691 **DOIN' ALRIGHT** O'Bryan	4849 **DREAMS OF TOMORROW** Lonnie Liston Smith
3880 **DEFECTOR** Steve Hackett	4235 **DOIN' IT!** UTFO	4060 **DREAM STREET** Janet Jackson
687 **DEFENDERS OF THE FAITH** Judas Priest	4718 **DO IT DEBBIE'S WAY** Debbie Reynolds	2385 **DREAM STREET ROSE** Gordon Lightfoot
594 **DEGUELLO** ZZ Top	2227 **DOLLY DOLLY DOLLY** Dolly Parton	4551 **DREAMTIME** The Stranglers
4135 **DEKADANCE** INXS	1185 **DO ME BABY** Meli'sa Morgan	4644 **DREAMTIME** Tom Verlaine
702 **DELICATE SOUND OF THUNDER** Pink Floyd	4220 **DOMINGO-CON AMORE** Placido Domingo	2211 **DREGS OF THE EARTH** Dixie Dregs
2794 **DELIRIOUS NOMAD** Armored Saint	3603 **DOMINO THEORY** Weather Report	3278 **DRIFTER** Sylvia (2)
2877 **DELIVER** Oak Ridge Boys	1206 **DONE WITH MIRRORS** Aerosmith	3091 **DROP DOWN AND GET ME** Del Shannon
329 **DEPARTURE** Journey	792 **DONNA SUMMER** Donna Summer	3254 **DROP THE BOMB** Trouble Funk
1159 **DESERT MOON** Dennis DeYoung	1696 **DONNY OSMOND** Donny Osmond	2631 **DRUMS ALONG THE MOHAWK** Jean Beauvoir
2182 **D.E. 7TH** Dave Edmunds	1130 **DON'T BE AFRAID OF THE DARK** Robert Cray Band	4046 **DRUMS AND WIRES** XTC
4117 **DESIRE** Tom Scott	13 **DON'T BE CRUEL** Bobby Brown	4140 **DRY DREAMS** Jim Carroll Band
2457 **DESTINY** Chaka Khan	3250 **DON'T CLOSE YOUR EYES** Keith Whitley	140 **THE DUDE** Quincy Jones
3558 **DETROIT DIESEL** Alvin Lee	1556 **DON'T DISTURB THIS GROOVE** The System	2864 **DUETS** Kenny Rogers
3831 **DEUCE** Kurtis Blow	4075 **DON'T FIGHT IT** Red Rider	332 **DUKE** Genesis
2068 **DEVO-LIVE** Devo	4526 **DON'T FOLLOW ME, I'M LOST TOO** Pearl Harbor	2774 **THE DUKES OF HAZZARD** TV Soundtrack
213 **DIAMOND LIFE** Sade	3961 **DON'T GIVE UP YOUR DAY JOB** Jack Wagner	2988 **DUMB WAITERS** The Korgis
4851 **DIAMONDS IN THE RAW** S.O.S. Band	870 **DON'T LET GO** Isaac Hayes	4216 **DUNE (SOUNDTRACK)** Toto
2443 **DIAMOND SUN** Glass Tiger	1431 **DON'T LET LOVE SLIP AWAY** Freddie Jackson	142 **DUOTONES** Kenny G
103 **DIANA** Diana Ross	1329 **DON'T LOOK ANY FURTHER** Dennis Edwards	404 **DURAN DURAN** Duran Duran
2492 **DIANA ROSS ANTHOLOGY** Diana Ross	1714 **DON'T LOOK BACK** Natalie Cole	4047 **DURELL COLEMAN** Durell Coleman
3826 **DIANNE REEVES** Dianne Reeves	1919 **DON'T PLAY WITH FIRE** Peabo Bryson	4349 **D'YA LIKE SCRATCHIN'** Malcolm McLaren
257 **DIARY OF A MADMAN** Ozzy Osbourne	87 **DON'T SAY NO** Billy Squier	2037 **DYLAN AND THE DEAD** Bob Dylan & The Grateful Dead
1427 **DICE** Andrew Dice Clay	877 **DON'T STOP** Jeffrey Osborne	
330 **DIESEL AND DUST** Midnight Oil	954 **DON'T STOP** Billy Idol	
4189 **DIFFERENT KINDA DIFFERENT** Johnny Mathis	1630 **DON'T SUPPOSE** Limahl	# E
3506 **A DIFFERENT KIND OF BLUES** Itzhak Perlman & Andre Previn	2956 **DON'T TAKE IT PERSONAL** Jermaine Jackson	1870 **EAGLES GREATEST HITS: VOLUME 2** Eagles
4227 **DIFFERENT KIND OF TENSION** Buzzcocks	1711 **DON'T TELL A SOUL** The Replacements	369 **EAGLES LIVE** Eagles
128 **DIFFERENT LIGHT** Bangles	3214 **THE DOOBIE BROTHERS FAREWELL TOUR** Doobie Brothers	3705 **EARTH & SKY** Graham Nash
3626 **DIFFERENT STYLE!** Musical Youth	1995 **DOOLITTLE** Pixies	3721 **EARTH CRISIS** Steel Pulse
1623 **DIFFICULT TO CURE** Rainbow	228 **THE DOORS GREATEST HITS** The Doors	1617 **EARTH - SUN - MOON** Love And Rockets
	1126 **DOOR TO DOOR** The Cars	2400 **EAST** Hiroshima
	2357 **DO THE RIGHT THING** Soundtrack	4094 **EAST** Cold Chisel
	61 **DOUBLE FANTASY** John Lennon & Yoko Ono	

Albums Alpha Index

#	Title	Artist
4254	EAST OF MIDNIGHT	Gordon Lightfoot
1081	EAST SIDE STORY	Squeeze
326	EAT 'EM AND SMILE	David Lee Roth
1917	EATEN ALIVE	Diana Ross
3641	EAT THE HEAT	Accept
461	EAZY-DUZ-IT	Eazy-E
1736	EB 84	Everly Brothers
3935	EBONEE WEBB	Ebonee Webb
4778	ECHO & THE BUNNYMEN	Echo & The Bunnymen
1421	ECHO & THE BUNNYMEN(2)	Echo & The Bunnymen
3003	ECHOES OF AN ERA	Various Artists
318	EDDIE & THE CRUISERS (SOUNDTRACK)	John Cafferty & The Beaver Brown Band
3976	EDDIE & THE CRUISERS II (SOUNDTRACK)	John Cafferty & The Beaver Brown Band
966	EDDIE MURPHY	Eddie Murphy
938	EDDIE MURPHY: COMEDIAN	Eddie Murphy
2847	EDEN ALLEY	Timbuk 3
1686	EGO TRIP	Kurtis Blow
3247	8	Madhouse
3285	8 FOR THE 80'S	Webster Lewis
1712	8TH WONDER	Sugarhill Gang
1353	80'S LADIES	K.T. Oslin
2446	80/81	Pat Metheny
2506	EINZELHAFT	Falco
963	EL DEBARGE	El DeBarge
965	ELECTRIC	The Cult
3775	ELECTRIC BREAKDANCE	Various Artists
3571	ELECTRIC CAFE	Kraftwerk
3217	ELECTRIC DREAMS	Soundtrack
4397	ELECTRIC FOLKLORE LIVE	The Alarm
4207	ELECTRIC HONEY	Partland Brothers
1473	THE ELECTRIC HORSEMAN (SOUNDTRACK)	Willie Nelson/Dave Grusin
1773	ELECTRIC LADY	Con Funk Shun
2214	ELECTRIC RENDEZVOUS	Al Di Meola
3728	THE ELECTRIC SPANKING OF WAR BABIES	Funkadelic
1639	ELECTRIC UNIVERSE	Earth, Wind & Fire
139	ELECTRIC YOUTH	Debbie Gibson
347	1100 BEL AIR PLACE	Julio Iglesias
36	ELIMINATOR	ZZ Top
4100	ELISA FIORILLO	Elisa Fiorillo
668	EL LOCO	ZZ Top
4108	ELOISE LAWS	Eloise Laws
1243	ELO'S GREATEST HITS	Electric Light Orchestra
2035	EL RAYO-X	David Lindley
2408	ELTON JOHN'S GREATEST HITS, VOLUME III 1979-1987	Elton John
4396	ELVIS: THE FIRST LIVE RECORDINGS	Elvis Presley
1327	ELVIS ARON PRESLEY	Elvis Presley
3616	THE ELVIS MEDLEY	Elvis Presley
2904	EMERALD CITY	Teena Marie
177	EMERGENCY	Kool & The Gang
3495	EMERGENCY	Melissa Manchester
872	EMERSON, LAKE, & POWELL	Emerson, Lake & Powell
842	EMOTION	Barbra Streisand
3928	EMOTION	DFX2
1033	EMOTIONAL	Jeffrey Osborne
106	EMOTIONAL RESCUE	Rolling Stones
206	EMOTIONS IN MOTION	Billy Squier
1533	EMPIRE BURLESQUE	Bob Dylan
4332	EMPIRE JAZZ	Various Artists
4181	EMPIRE OF THE SUN	Soundtrack
310	THE EMPIRE STRIKES BACK	Soundtrack
4472	THE EMPIRE STRIKES BACK/THE ADVENTURES OF LUKE SKYWALKER	Soundtrack
279	EMPTY GLASS	Pete Townshend
3839	ENDANGERED SPECIES	White Wolf
539	ENDLESS LOVE	Soundtrack
1765	END OF THE CENTURY	The Ramones
80	THE END OF THE INNOCENCE	Don Henley
1790	ENGLISH SETTLEMENT	XTC
2413	ENOUGH IS ENOUGH	Billy Squier
1554	ENUFF Z'NUFF	Enuff Z'Nuff
2999	THE ENVOY	Warren Zevon
2096	EPONYMOUS	R.E.M.
3706	ERIC CARMEN (II)	Eric Carmen
1751	ERROR IN THE SYSTEM	Peter Schilling
4330	ESCAPADE	Tim Finn
8	ESCAPE	Journey
761	ESCAPE	Whodini
1432	ESCAPE ARTIST	Garland Jeffreys
3296	E-S-P	Bee Gees
865	ESPECIALLY FOR YOU	The Smithereens
2826	ESPECIALLY FOR YOU	Don Williams
4532	ESQUIRE	Esquire
3557	ESSAR	Smokey Robinson
4163	THE ETERNAL IDOL	Black Sabbath
4269	ETERNAL NIGHTMARE	Vio-Lence
1120	E.T. - THE EXTRA-TERRESTRIAL	Soundtrack
2838	EUGENE WILDE	Eugene Wilde
833	EVANGELINE	Emmylou Harris
1006	EVEN WORSE	"Weird Al" Yankovic
684	EVERLASTING	Natalie Cole
4372	EVERLASTING LOVE	Rex Smith
2541	THE EVER POPULAR TORTURED ARTIST EFFECT	Todd Rundgren
3683	EVERYBODY LOVES THE PILOT (EXCEPT THE CREW)	Jon Astley
1947	EVERYBODY'S ROCKIN'	Neil and the Shocking Pinks
4345	EVERYBODY WANTS SOME	Gucci Crew II
691	EVERY BREATH YOU TAKE-THE SINGLES	The Police
1115	EVERY GENERATION	Ronnie Laws
2795	EVERY GREAT MOTOWN HIT OF MARVIN GAYE	Marvin Gaye
655	EVERY HOME SHOULD HAVE ONE	Patti Austin
2896	EVERY MAN HAS A WOMAN	Various Artists
2487	EVERY SHADE OF LOVE	Jesse Johnson
3029	EVERY STEP OF THE WAY	David Benoit
431	EVERYTHING	Bangles
2740	EVERYTHING	Climie Fisher
3399	EVERYTHING IS COOL	T-Connection
2773	EVERYTHING'S DIFFERENT NOW	'Til Tuesday
3327	EVERYTHING'S KOOL & THE GANG: GREATEST HITS & MORE	Kool & The Gang
4693	EVERYTIME TWO FOOLS COLLIDE	Kenny Rogers & Dottie West
4025	EVERY TURN OF THE WORLD	Christopher Cross
4703	EVERYWHERE AT ONCE	The Plimsouls
1525	EVITA	Festival
2388	EVITA	Original Cast
4865	EXCESS ALL AREAS	Shy
1752	EXILES	Dan Fogelberg
4835	EXIT	Tangerine Dream
610	EXIT...STAGE LEFT	Rush
2658	EXIT 0	Steve Earle And The Dukes
2256	EXPOSED/A CHEAP PEEK AT TODAY'S PROVOCATIVE NEW ROCK	Various Artists
3923	EXPOSED II	Various Artists
204	EXPOSURE	Exposé
1791	EXPRESS	Love And Rockets
997	EXTENDED PLAY	The Pretenders
1276	EXTENSIONS	Manhattan Transfer
4696	EXTRA PLAY	Kaja
1519	EXTRATERRESTRIAL LIVE	Blue Öyster Cult
1755	EXTREME	Extreme
341	EYE IN THE SKY	The Alan Parsons Project
1766	EYE OF THE HURRICANE	The Alarm
268	EYE OF THE TIGER	Survivor
1364	EYE OF THE ZOMBIE	John Fogerty
1366	EYES OF A STRANGER	The Deele
2570	EYES ON THIS	MC Lyte
299	EYES THAT SEE IN THE DARK	Kenny Rogers
2817	EYE TO EYE	Eye To Eye
3909	EZO	EZO

F

#	Title	Artist
4473	FABLES	Jean-Luc Ponty
744	FABLES OF THE RECONSTRUCTION	R.E.M.
2548	THE FABULOUS BAKER BOYS (SOUNDTRACK)	Dave Grusin
2336	FABULOUS DISASTER	Exodus
350	FACE DANCES	The Who
631	FACES	Earth, Wind & Fire
2571	FACE TO FACE	Evelyn "Champagne" King
3111	FACE TO FACE	Face To Face
3774	FACE TO FACE	GQ
3986	FACE TO FACE	Angel City
84	FACE VALUE	Phil Collins
1111	FAHRENHEIT	Toto
463	FAIR WARNING	Van Halen
10	FAITH	George Michael
446	FALCO 3	Falco
2431	THE FALCON & THE SNOWMAN (SOUNDTRACK)	Pat Metheny Group
2917	FALSE ACCUSATIONS	Robert Cray Band
245	FAME	Soundtrack
3372	FAME AND FORTUNE	Bad Company
3997	FAME AND FASHION - DAVID BOWIE'S ALL TIME GREATEST HITS	David Bowie
2025	THE FAMILY	The Family
2805	FAMILY	Hubert Laws
3850	FAMOUS AT NIGHT	John Hunter
2007	FAMOUS BLUE RAINCOAT	Jennifer Warnes
419	"...FAMOUS LAST WORDS..."	Supertramp
4608	THE FANATIC	Felony
2232	FANCY DANCER	One Way
398	FANCY FREE	Oak Ridge Boys
2146	FANDANGO	Herb Alpert
4561	FANS	Malcolm McLaren
1827	FANTASTIC	Wham!
537	FANTASTIC VOYAGE	Lakeside
1915	FAREWELL MY SUMMER LOVE	Michael Jackson
2927	FAREWELL SONG	Janis Joplin
4358	FARRENHEIT	Farrenheit
885	FASCINATION!	Human League
4147	FASTER & LLOUDER	Foster & Lloyd
1998	FASTER PUSSYCAT	Faster Pussycat
414	FASTER THAN THE SPEED OF NIGHT	Bonnie Tyler
1788	FAST TIMES AT RIDGEMONT HIGH	Soundtrack
1005	FASTWAY	Fastway
1350	FAT BOYS	Fat Boys
1697	THE FAT BOYS ARE BACK!	Fat Boys
4627	FAVORITE COUNTRY HITS	Ricky Skaggs
3959	FAVORITES	Crystal Gayle
2741	FEARGAL SHARKEY	Feargal Sharkey
3975	FEARLESS	Nina Hagen
2766	FEAR NO EVIL	Grim Reaper
4623	FEELING CAVALIER	Ebn-Ozn
4052	FEELING GOOD	Roy Ayers
4594	FEELIN' RIGHT	Razzy Bailey
1363	FEEL ME	Cameo

Albums Alpha Index

1349 FEEL MY SOUL *Jennifer Holliday*	333 A FLOCK OF SEAGULLS *A Flock Of Seagulls*	1052 FREEDOM *Neil Young*
85 FEELS SO RIGHT *Alabama*	2425 FLOODLAND *Sisters Of Mercy*	3036 FREEDOM *Santana*
3413 FEEL THE HEAT *Henry Paul Band*	2404 FLORIDAYS *Jimmy Buffett*	390 FREEDOM AT POINT ZERO *Jefferson Starship*
3465 FEEL THE SHAKE *Jetboy*	734 FLOWERS IN THE DIRT *Paul McCartney*	2731 FREEDOM NO COMPROMISE *Little Steven*
2837 FEMALE TROUBLE *Nona Hendryx*	3984 THE FLOWERS OF ROMANCE *Public Image Limited*	399 FREEDOM OF CHOICE *Devo*
4110 FEMME FATALE *Femme Fatale*	1621 FLUSH THE FASHION *Alice Cooper*	4701 FREE FALL *Alvin Lee*
2659 FERVOR *Jason & The Scorchers*	1303 FLYING COWBOYS *Rickie Lee Jones*	1238 FREETIME *Spyro Gyra*
3946 FESTIVAL *Lee Ritenour*	3631 FLYING HOME *Stanley Jordan*	82 FREEZE-FRAME *J. Geils Band*
2549 FEVER *Con Funk Shun*	654 FLYING IN A BLUE DREAM *Joe Satriani*	1318 FREHLEY'S COMET *Ace Frehley*
4398 FICKLE *Michael Henderson*	3174 THE FLYING LIZARDS *Flying Lizards*	2700 FREIGHT TRAIN HEART *Jimmy Barnes*
1939 FIELD DAY *Marshall Crenshaw*	2196 FLYING THE FLAG *Climax Blues Band*	1158 A FRESH AIRE CHRISTMAS *Mannheim Steamroller*
1506 FIELDS OF FIRE *Corey Hart*	1122 FLY ON THE WALL *AC/DC*	3778 FRESH AIRE VI *Mannheim Steamroller*
2976 FIERCE HEART *Jim Capaldi*	4173 FOLK OF THE 80'S (PART III) *Men Without Hats*	4736 FRESH FRUIT IN FOREIGN PLACES *Kid Creole & The Coconuts*
3022 FIFTH ANGEL *Fifth Angel*	4379 FOLKSONGS FOR A NUCLEAR VILLAGE *Shadowfax*	2678 FRIDAY NIGHT IN SAN FRANCISCO *Al Di Meola/John McLaughlin/Paco De Lucia*
124 5150 *Van Halen*	2770 FOLKWAYS: A VISION SHARED - A TRIBUTE TO WOODY GUTHRIE AND LEADBELLY *Various Artists*	559 FRIEND OR FOE *Adam Ant*
4036 FIGHT FOR THE ROCK *Savatage*	4477 FOLLIES IN CONCERT *Original Cast*	669 FRIENDS *Dionne Warwick*
3187 FIGHT TO SURVIVE *White Lion*	1285 FOLLOW THE LEADER *Eric B. & Rakim*	1227 FRIENDS *Shalamar*
161 THE FINAL COUNTDOWN *Europe*	2354 THE FOOL CIRCLE *Nazareth*	3393 FRIENDS *Larry Carlton*
625 THE FINAL CUT *Pink Floyd*	3017 FOOL IN LOVE WITH YOU *Jim Photoglo*	2203 FRIENDSHIP *Ray Charles*
1948 THE FINAL FRONTIER *Keel*	516 FOOLISH BEHAVIOUR *Rod Stewart*	2994 FRIENDS IN LOVE *Dionne Warwick*
3404 FINALLY! *T.G. Sheppard*	94 FOOTLOOSE *Soundtrack*	3800 FRIENDS IN LOVE *Johnny Mathis*
3230 FINDER OF LOST LOVES *Dionne Warwick*	4068 FOOTSTEPS IN THE DARK: GREATEST HITS VOLUME 2 *Cat Stevens*	1422 THE FRIENDS OF MR. CAIRO *Jon And Vangelis*
2343 THE FINE ART OF SURFACING *Boomtown Rats*	3673 FORCE MAJEURE *Doro*	3638 FROLIC THROUGH THE PARK *Death Angel*
4726 A FINE MESS *Soundtrack*	100 FORE! *Huey Lewis & The News*	4416 FROM A CHILD TO A WOMAN *Julio Iglesias*
1571 FINE YOUNG CANNIBALS *Fine Young Cannibals*	1322 FOREIGN AFFAIR *Tina Turner*	3277 FROM BRANCH TO BRANCH *Leon Redbone*
3112 FINYL VINYL *Rainbow*	750 FOREVER *Kool & The Gang*	1453 FROM LUXURY TO HEARTACHE *Culture Club*
2324 FIONA *Fiona*	3041 FOREVER AND EVER *Howard Hewett*	4557 FROM THE GREENHOUSE *Crack The Sky*
4573 FIRE AND GASOLINE *Steve Jones*	3386 FOREVER BY YOUR SIDE *The Manhattans*	90 FRONTIERS *Journey*
2981 FIRE DOWN UNDER *Riot*	630 FOREVER, FOR ALWAYS, FOR LOVE *Luther Vandross*	3069 FROZEN GHOST *Frozen Ghost*
633 FIRE OF UNKNOWN ORIGIN *Blue Öyster Cult*	1268 FOREVER NOW *Psychedelic Furs*	3520 FRUIT AT THE BOTTOM *Wendy And Lisa*
4487 THE FIRE STILL BURNS *Russ Ballard*	4529 FOREVER, REX SMITH *Rex Smith*	3077 FUEL FOR THE FIRE *Naked Eyes*
1280 FIRIN' UP *Pure Prairie League*	3907 FOREVER YOUNG *Alphaville*	4041 FULL FORCE *Full Force*
641 THE FIRM *The Firm*	4 FOREVER YOUR GIRL *Paula Abdul*	3389 FULL FORCE GET BUSY 1 TIME! *Full Force*
3857 1ST *Streets*	2946 FOREVER YOURS *Tony Terry*	527 FULL MOON *Charlie Daniels Band*
1844 FIRST CIRCLE *Pat Metheny Group*	2972 FOR MEN ONLY *Millie Jackson*	69 FULL MOON FEVER *Tom Petty*
1503 THE FIRST FAMILY RIDES AGAIN *Rich Little*	1443 FOR SENTIMENTAL REASONS *Linda Ronstadt*	4035 FUN & GAMES *The Connells*
3505 THE FIRST OF A MILLION KISSES *Fairground Attraction*	2962 FOR THE LONELY: A ROY ORBISON ANTHOLOGY, 1956-1965 *Roy Orbison*	586 FUN AND GAMES *Chuck Mangione*
1925 FIRST OFFENCE *The Inmates*	2272 FOR THE WORKING GIRL *Melissa Manchester*	2829 FUN IN SPACE *Roger Taylor*
791 FIRST OFFENSE *Corey Hart*	241 FOR THOSE ABOUT TO ROCK (WE SALUTE YOU) *AC/DC*	3634 THE FUNK IS ON *Instant Funk*
4690 FIRST THRILLS *Thrills*	3304 FORTUNE 410 *Donnie Iris*	3719 FUNLAND *Bram Tchaikovsky*
4386 THE FIRST 25 YEARS- THE SILVER ANNIVERSARY ALBUM *Johnny Mathis*	1007 40 HOUR WEEK *Alabama*	2245 FUR *Jane Wiedlin*
1815 FISHERMAN'S BLUES *The Waterboys*	3006 42ND STREET *Original Cast*	541 FUTURE SHOCK *Herbie Hancock*
4150 FITS LIKE A GLOVE *Howie Mandel*	2006 FOR YOUR EYES ONLY *Soundtrack*	4118 FUTURE WORLD *Pretty Maids*
4369 FIVE MILES OUT *Mike Oldfield*	1838 FOUND ALL THE PARTS *Cheap Trick*	4205 FRANK ZAPPA MEETS THE MOTHERS OF PREVENTION *Frank Zappa*
1062 THE $5.98 E.P.: GARAGE DAYS RE-REVISITED *Metallica*	12 4 *Foreigner*	
2261 FIVE-O *Hank Williams Jr.*	4045 4 BY FOUR *4 By Four*	# G
3012 5 TO 1 *Tom Kimmel*	4807 FOUR FOR THE SHOW *Statler Brothers*	2209 GALAXIAN *Jeff Lorber Fusion*
4496 FIYO ON THE BAYOU *Neville Brothers*	2889 4 OF A KIND *D.R.I.*	70 THE GAME *Queen*
2977 THE FLAG *Rick James*	4264 14 GREATEST HITS *Michael Jackson & The Jackson 5*	1906 GAMMA 2 *Gamma*
3990 FLAG *Yello*	3167 14 KARAT *Fatback*	2534 GAMMA 3 *Gamma*
3700 THE FLAMINGO KID *Soundtrack*	2377 FOUR WINDS *Tangier*	4616 GANDHI *Soundtrack*
1620 FLASH *Jeff Beck*	1040 THE FOX *Elton John*	1089 THE GAP BAND II *Gap Band*
1723 FLASHBACK *38 Special*	3120 FOXIE *Bob James*	422 THE GAP BAND III *Gap Band*
50 FLASHDANCE *Soundtrack*	3814 FRAGILE *Cherrelle*	427 GAP BAND IV *Gap Band*
1093 FLASH GORDON (SOUNDTRACK) *Queen*	4804 FRAMED *Asleep At The Wheel*	809 GAP BAND V- JAMMIN' *Gap Band*
1466 THE FLAT EARTH *Thomas Dolby*	3334 FRANK *Squeeze*	1748 GAP BAND VI *Gap Band*
3259 FLAUNT IT *Sigue Sigue Sputnik*	810 FRANKE & THE KNOCKOUTS *Franke & The Knockouts*	3514 GAP BAND VII *Gap Band*
4509 FLAUNT THE IMPERFECTION *China Crisis*	3240 FRANKS WILD YEARS *Tom Waits*	2797 GAP GOLD/BEST OF THE GAP BAND *Gap Band*
806 FLEETWOOD MAC LIVE *Fleetwood Mac*	1233 FRANTIC ROMANTIC *Jermaine Stewart*	2834 GARDEN OF LOVE *Rick James*
1270 FLESH + BLOOD *Roxy Music*	3311 FREE *Concrete Blonde*	311 GAUCHO *Steely Dan*
4486 FLETCH *Soundtrack*	3274 FREE AS A BIRD *Supertramp*	2278 GAUDI *Alan Parsons Project*
3153 FLEX *Lene Lovich*		1775 GEFFREY MORGAN... *UB40*
841 FLICK OF THE SWITCH *AC/DC*		
4186 FLIP *Nils Lofgren*		
3158 FLIP-FLOP *Guadalcanal Diary*		
4761 FLIRT *Evelyn "Champagne" King*		

Albums Alpha Index

#	Title	Artist
3324	GEMS	Aerosmith
2152	GENE CHANDLER '80	Gene Chandler
292	GENESIS	Genesis
4313	GENETIC WALK	Ahmad Jamal
2746	THE GENIE (THEMES & VARIATIONS FROM THE TV SERIES "TAXI")	Bob James
551	THE GEORGE BENSON COLLECTION	George Benson
448	GEORGIA SATELLITES	Georgia Satellites
3452	GET AS MUCH LOVE AS YOU CAN	Jones Girls
1650	THE GETAWAY	Chris de Burgh
4832	GET BACK!	Ike & Tina Turner
850	GET CLOSE	The Pretenders
1092	GET CLOSER	Linda Ronstadt
843	GET HAPPY!!	Elvis Costello And The Attractions
1242	GET HERE	Brenda Russell
4011	GET IT ON CREDIT	Toronto
1670	GET IT RIGHT	Aretha Franklin
1118	GET LOOSE	Evelyn King
51	GET LUCKY	Loverboy
261	GET NERVOUS	Pat Benatar
3165	GET OUT OF MY ROOM	Cheech & Chong
3899	GET RHYTHM	Ry Cooder
4677	GETTIN' AWAY WITH MURDER	Patti Austin
2028	G FORCE	Kenny G
2259	GHETTO BLASTER	The Crusaders
1551	GHETTO MUSIC: THE BLUEPRINT OF HIP HOP	Boogie Down Productions
453	GHOSTBUSTERS	Soundtrack
924	GHOSTBUSTERS II	Soundtrack
73	GHOST IN THE MACHINE	The Police
4368	GHOST ON THE BEACH	Insiders
854	GHOST RIDERS	The Outlaws
4802	GHOST TOWN	Poco
4511	GIANT	Woodentops
576	GIDEON	Kenny Rogers
2417	THE GIFT	The Jam
4190	GIGOLO	Fatback
1198	GIPSY KINGS	Gipsy Kings
1849	GIRL AT HER VOLCANO	Rickie Lee Jones
209	GIRLS, GIRLS, GIRLS	Mötley Crüe
2831	GIRLS TO CHAT & BOYS TO BOUNCE	Foghat
1499	GIRLS WITH GUNS	Tommy Shaw
16	GIRL YOU KNOW IT'S TRUE	Milli Vanilli
969	GIUFFRIA	Giuffria
4263	GIVE IT UP	Pleasure
235	GIVE ME THE NIGHT	George Benson
321	GIVE ME THE REASON	Luther Vandross
1133	GIVE MY REGARDS TO BROAD STREET	Paul McCartney
500	GIVE THE PEOPLE WHAT THEY WANT	The Kinks
136	GIVING YOU THE BEST THAT I GOT	Anita Baker
663	THE GLAMOROUS LIFE	Sheila E.
3668	GLAMOUR	Dave Davies
4208	GLASS HOUSE ROCK	Greg Kihn Band
34	GLASS HOUSES	Billy Joel
3496	GLASS MOON	Glass Moon
3844	GLASSWORKS	Philip Glass
2551	GLENN JONES	Glenn Jones
2609	GLENN MEDEIROS	Glenn Medeiros
2580	GLOBE OF FROGS	Robyn Hitchcock And The Egyptians
2481	GLORIA LORING	Gloria Loring
2022	GLORIOUS RESULTS OF A MISSPENT YOUTH	Joan Jett & the Blackhearts
1605	GLORYHALLASTOOPID (OR PIN THE TAIL ON THE FUNKY)	Parliament
4654	GLORY ROAD	Gillan
1437	GLOW	Rick James
972	THE GLOW OF LOVE	Change
132	G N' R LIES	Guns N' Roses
1666	GO	Hiroshima
505	GO ALL THE WAY	Isley Brothers
3640	GO BANG!	Shriekback
4218	GODDESS IN PROGRESS	Julie Brown
3035	GODDESS OF LOVE	Phyllis Hyman
3595	GOD SAVE THE QUEEN/UNDER HEAVY MANNERS	Robert Fripp
2780	GOD'S OWN MEDICINE	Mission U.K.
2835	GO FOR IT	Shalamar
3873	GO FOR THE THROAT	Humble Pie
4265	GO FOR YOUR LIFE	Mountain
2208	GOING FOR BROKE	Eddy Grant
2419	GOIN' OFF	Biz Markie
1834	GO INSANE	Lindsey Buckingham
3434	GOLD	Steely Dan
459	GOLD & PLATINUM	Lynyrd Skynyrd
679	THE GOLDEN AGE OF WIRELESS	Thomas Dolby
2316	A GOLDEN CELEBRATION	Elvis Presley
3884	THE GOLDEN CHILD	Soundtrack
3781	GOLDEN DOWN	Willie Nile
4007	GOLDEN TOUCH	Rose Royce
3301	GOLDEN YEARS	David Bowie
3710	GONE TROPPO	George Harrison
2358	GOOD AS GOLD	Red Rockers
1517	GOODBYE CRUEL WORLD	Elvis Costello And The Attractions
3938	GOOD FOR YOUR SOUL	Oingo Boingo
2466	GOOD LOVE	Meli'sa Morgan
507	GOOD MORNING, VIETNAM	Soundtrack
3089	GOOD MUSIC	Joan Jett & the Blackhearts
1762	GOOD TO BE BACK	Natalie Cole
2825	GOOD TO GO LOVER	Gwen Guthrie
454	GOOD TROUBLE	REO Speedwagon
2213	GO ON...	Mr. Mister
2960	THE GOONIES (SOUNDTRACK)	Dave Grusin
2338	GORKY PARK	Gorky Park
3547	GOT ANY GUM?	Joe Walsh
973	GO TO HEAVEN	Grateful Dead
4889	GOT TO BE TOUGH	M.C. Shy D
1291	GO WEST	Go West
3072	GRAB IT!	L'Trimm
78	GRACELAND	Paul Simon
635	GRACE UNDER PRESSURE	Rush
4814	GRADUALLY GOING TORNADO	Bill Bruford
4085	GRAND FUNK LIVES	Grand Funk Railroad
1220	GRAND SLAM	Isley Brothers
4359	GRAND SLAM	Spinners
3793	GRASSHOPPER	J.J. Cale
3053	GRAVITY	Kenny G
4194	GRAVITY	James Brown
2654	GREASE 2	Soundtrack
766	THE GREAT ADVENTURES OF SLICK RICK	Slick Rick
2583	GREAT BALLS OF FIRE (SOUNDTRACK)	Jerry Lee Lewis
296	GREATEST HITS	Air Supply
709	GREATEST HITS	Alabama
536	GREATEST HITS	The Cars
4461	GREATEST HITS	Dave & Sugar
739	GREATEST HITS	Fleetwood Mac
4223	GREATEST HITS	Lee Greenwood
821	GREATEST HITS	The Judds
3202	GREATEST HITS	KC And The Sunshine Band
990	GREATEST HITS	Little River Band
1671	GREATEST HITS	Melissa Manchester
2916	GREATEST HITS	Reba McEntire
812	GREATEST HITS	Ronnie Milsap
4528	GREATEST HITS	Juice Newton
2320	GREATEST HITS	Oak Ridge Boys
2436	GREATEST HITS	Billy Ocean
1750	GREATEST HITS	Ray Parker Jr.
2069	GREATEST HITS	Dolly Parton
4301	GREATEST HITS	Charley Pride
583	GREATEST HITS	Queen
3963	GREATEST HITS	George Strait
4826	GREATEST HITS	Steve Wariner
3883	GREATEST HITS COLLECTION	Bananarama
580	GREATEST HITS/LIVE	Heart
2664	GREATEST HITS LIVE	Carly Simon
4633	GREATEST HITS 1972-1978	10cc
3639	GREATEST HITS OF THE OUTLAWS/HIGH TIDES FOREVER	The Outlaws
3760	GREATEST HITS VOLUME ONE	Elvis Presley
262	GREATEST HITS VOL. I & II	Billy Joel
2001	GREATEST HITS VOL. 2	ABBA
1253	GREATEST HITS-VOL. II	Barry Manilow
2714	GREATEST HITS VOL. 2	Ronnie Milsap
2380	GREATEST HITS, VOL. 2	Oak Ridge Boys
3463	GREATEST HITS - VOL. II	Eddie Rabbitt
890	GREATEST HITS, VOLUME 2	Linda Ronstadt
1651	GREATEST HITS, VOLUME TWO	George Strait
4375	GREATEST HITS, VOL. II	Hank Williams Jr.
1495	GREATEST HITS III	Hank Williams Jr.
3612	GREAT GONZOS! THE BEST OF TED NUGENT	Ted Nugent
2386	THE GREAT MUPPET CAPER	Soundtrack
2606	THE GREAT PRETENDER	Dolly Parton
364	THE GREAT RADIO CONTROVERSY	Tesla
3469	GREAT WHITE	Great White
620	GREAT WHITE NORTH	Bob & Doug McKenzie
293	GREEN	R.E.M.
1570	GREEN LIGHT	Bonnie Raitt
2061	GREENPEACE/RAINBOW WARRIORS	Various Artists
1282	GREEN THOUGHTS	The Smithereens
1354	GREETINGS FROM TIMBUK 3	Timbuk 3
1841	GREG LAKE	Greg Lake
3949	GREMLINS	Soundtrack
2665	GRETCHEN GOES TO NEBRASKA	King's X
2930	GROOVE APPROVED	Paul Carrack
3796	GROSS MISCONDUCT	M.O.D.
4188	GROWING UP IN PUBLIC	Lou Reed
3297	GROWIN' UP TOO FAST	Billy Rankin
4268	A GRP CHRISTMAS COLLECTION	Various Artists
592	GTR	GTR
3985	GUARDIAN OF THE LIGHT	George Duke
3376	GUESS WHO'S COMIN' TO THE CRIB?	Full Force
64	GUILTY	Barbra Streisand
1963	GUITAR MAN	Elvis Presley
1173	GUITARS, CADILLACS, ETC., ETC.	Dwight Yoakam
4669	GUITAR SLINGER	Johnny Winter
4148	GUITAR SPEAK	Various Artists
2458	GUITAR TOWN	Steve Earle
4615	GUTS FOR LOVE	Garland Jeffreys
377	GUY	Guy
2399	GYPSY BLOOD	Mason Ruffner

H

#	Title	Artist
1612	H	Bob James
2903	HABITS OLD AND NEW	Hank Williams Jr.
4273	HAI HAI	Roger Hodgson

Albums Alpha Index

Cat#	Title	Artist
3769	HAIRSPRAY	Soundtrack
4705	HALF NELSON	Willie Nelson
4685	HALFWAY TO SANITY	The Ramones
1299	HALL & OATES LIVE AT THE APOLLO WITH DAVID RUFFIN & EDDIE KENDRICK	Daryl Hall & John Oates
2730	HALL OF THE MOUNTAIN KING	Savatage
4628	HAND OF KINDNESS	Richard Thompson
2527	HANDS DOWN	Bob James
2690	HAND TO MOUTH	General Public
17	HANGIN' TOUGH	New Kids On The Block
1564	HANG ON FOR YOUR LIFE	Shooting Star
4502	HANG TOGETHER	Odyssey
2202	HANK "LIVE"	Hank Williams Jr.
1257	HANK WILLIAMS, JR.'S GREATEST HITS	Hank Williams Jr.
3983	HAPPY?	Public Image Limited
1837	HAPPY ANNIVERSARY, CHARLIE BROWN	Various Artists
4051	HAPPY LOVE	Natalie Cole
1448	HAPPY TOGETHER	The Nylons
4494	HARD	Gang Of Four
4847	HARD CORE	Paul Dean
2761	HARDER THAN YOU	24-7 SPYZ
2339	HARD LINE	The Blasters
2969	HARD MACHINE	Stacey Q
254	HARD PROMISES	Tom Petty And The Heartbreakers
2375	HARD TIMES IN THE LAND OF PLENTY	Omar And The Howlers
799	HARD TO HOLD (SOUNDTRACK)	Rick Springfield
2640	HARDWARE	Krokus
2955	THE HARD WAY	Point Blank
4866	HARLEQUIN	Dave Grusin & Lee Ritenour
4165	HARMONY	Anne Murray
1688	HAVANA MOON	Carlos Santana
1722	HAVE YOU SEEN ME LATELY?	Sam Kinison
1382	HAWKS & DOVES	Neil Young
928	HEADED FOR THE FUTURE	Neil Diamond
3748	HEAD FIRST	Uriah Heep
840	HEADHUNTER	Krokus
4203	HEADING FOR A STORM	Vandenberg
2164	THE HEADLESS CHILDREN	W.A.S.P.
3466	HEADLESS CROSS	Black Sabbath
1706	HEADLINES	Midnight Star
1188	THE HEAD ON THE DOOR	The Cure
3251	HEADS OR TALES	Saga
1901	HEALING	Todd Rundgren
2198	HEAR & NOW	Billy Squier
3371	HEAR 'N AID	Hear 'N Aid
827	HEARSAY	Alexander O'Neal
38	HEART	Heart
1430	HEARTS AND BONES	Paul Simon
2789	HEART ATTACK	Krokus
2750	HEARTATTACK AND VINE	Tom Waits
867	HEARTBEAT	Don Johnson
74	HEARTBEAT CITY	The Cars
247	HEART BREAK	New Edition
2166	HEART BREAK	Shalamar
853	HEARTBREAKER	Dionne Warwick
3066	HEARTBREAK EXPRESS	Dolly Parton
4754	HEARTBREAK HOTEL	Soundtrack
4320	HEARTBREAK RADIO	Rita Coolidge
4101	HEART DON'T LIE	LaToya Jackson
1903	HEART LAND	Real Life
1599	HEARTLAND	Michael Stanley Band
1648	HEARTLAND	The Judds
432	HEARTLIGHT	Neil Diamond
4750	THE HEART NEVER LIES	Michael Martin Murphey
240	HEART OF STONE	Cher
1347	THE HEART OF THE MATTER	Kenny Rogers
2276	HEART OVER MIND	Anne Murray
3398	HEART OVER MIND	Jennifer Rush
3107	HEART PLAY	John Lennon & Yoko Ono
1926	HEART'S HORIZON	Al Jarreau
3100	HEARTS IN MOTION	Air Supply
2023	HEAVEN	BeBe & CeCe Winans
901	HEAVEN AND HELL	Black Sabbath
4200	HEAVENLY BODY	The Chi-Lites
266	HEAVEN ON EARTH	Belinda Carlisle
3575	HEAVEN ONLY KNOWS	Teddy Pendergrass
1807	HEAVEN 17	Heaven 17
4767	HEAVEN UP HERE	Echo & The Bunnymen
4289	HEAVY MENTAL	The Fools
553	HEAVY METAL	Soundtrack
395	HEAVY NOVA	Robert Palmer
2318	HELLO BIG MAN	Carly Simon
230	HELLO, I MUST BE GOING!	Phil Collins
3924	HELP YOURSELF	Larry Gatlin & The Gatlin Brothers Band
1465	HENRY LEE SUMMER	Henry Lee Summer
1166	HERE COMES THE NIGHT	Barry Manilow
4422	HERE COMES THE NIGHT	David Johansen
3394	HERE I AM	Sharon Bryant
3685	HERE'S THE WORLD FOR YA	Paul Hyde And The Payolas
511	HERE'S TO FUTURE DAYS	Thompson Twins
3133	HERE TODAY, TOMORROW NEXT WEEK!	The Sugarcubes
3173	HERE TO STAY	Neal Schon & Jan Hammer
2071	HERO	Clarence Clemons
392	HEROES	Commodores
4071	HEROES & ZEROS	Glen Burtnick
4000	HEROES, ANGELS & FRIENDS	Janey Street
216	HE'S THE D.J., I'M THE RAPPER	D.J. Jazzy Jeff & The Fresh Prince
2942	HE THINKS HE'S RAY STEVENS	Ray Stevens
4459	HEY!	Julio Iglesias
3554	HEYDEY	The Church
4405	HEY JOE, HEY MOE	Moe Bandy & Joe Stampley
764	HEY RICKY	Melissa Manchester
1783	HIDEAWAY	David Sanborn
3269	HIDING OUT	Soundtrack
487	HIGH ADVENTURE	Kenny Loggins
4640	THE HIGH AND THE MIGHTY	Donnie Iris
1313	HIGH COUNTRY SNOWS	Dan Fogelberg
1271	HIGH CRIME	Al Jarreau
3283	HIGH HAT	Boy George
3478	HIGH LAND, HARD RAIN	Aztec Camera
381	HIGH 'N' DRY	Def Leppard
2238	HIGH 'N' DRY(2)	Def Leppard
2833	HIGH NOTES	Hank Williams Jr.
4092	HIGH ON YOUR LOVE	Debbie Jacobs
1298	HIGH PRIORITY	Cherrelle
2702	HIGH-RISE	Ashford & Simpson
4712	HIGH TENSION WIRES	Steve Morse
2642	HIGH VOLTAGE	AC/DC
1786	HIGHWAYMAN	Willie, Waylon, Johnny & Kris
2449	HIGHWAYS AND HEARTACHES	Ricky Skaggs
11	HI INFIDELITY	REO Speedwagon
1576	HILLBILLY DELUXE	Dwight Yoakam
1999	HIPSWAY	Hipsway
1425	HIROSHIMA	Hiroshima
1730	THE HISTORY MIX VOL. I	Godley & Creme
4550	HIT AND RUN	Girlschool
4808	HIT AND RUN	TSOL
651	HITS!	Boz Scaggs
1461	THE HITS	REO Speedwagon
4033	HITS 1979-1989	Rosanne Cash
526	HOLD AN OLD FRIEND'S HAND	Tiffany
2333	HOLD ME	Laura Branigan
903	HOLD ME IN YOUR ARMS	Rick Astley
133	HOLD OUT	Jackson Browne
647	HOLD YOUR FIRE	Rush
4872	THE HOLE	Golden Earring
4631	HOLLYWOOD	Maynard Ferguson
4871	THE HOLLYWOOD MUSICALS	Johnny Mathis & Henry Mancini
2697	HOLLYWOOD, TENNESSEE	Crystal Gayle
1139	HOLY DIVER	Dio
1467	HOME	Stephanie Mills
2738	HOME	BoDeans
3621	HOME OF THE BRAVE (SOUNDTRACK)	Laurie Anderson
4785	HOMESICK HEROS	Charlie Daniels Band
3402	HOMOSAPIEN	Pete Shelley
2247	HONEYMOON SUITE	Honeymoon Suite
451	HONEYSUCKLE ROSE (SOUNDTRACK)	Willie Nelson & Family
4015	HONI SOIT (O NEE SWA)	John Cale
2663	HONOR AMONG THIEVES	The Brandos
2608	HOOKED ON BIG BANDS	Frank Barber Orchestra
181	HOOKED ON CLASSICS	Royal Philharmonic Orchestra
998	HOOKED ON CLASSICS II (CAN'T STOP THE CLASSICS)	Royal Philharmonic Orchestra
2687	HOOKED ON CLASSICS III (JOURNEY THROUGH THE CLASSICS)	Royal Philharmonic Orchestra
4480	HOOKED ON ROCK CLASSICS	London Symphony Orchestra
713	HOOKED ON SWING	Larry Elgart And His Manhattan Swing Orchestra
2936	HOOKED ON SWING 2	Larry Elgart And His Manhattan Swing Orchestra
1784	HOOLIGANS	The Who
370	HORIZON	Eddie Rabbitt
3328	HORSESHOE IN THE GLOVE	So
2306	HOT! LIVE AND OTHERWISE	Dionne Warwick
4688	HOT ASH	Wishbone Ash
2031	HOT AUGUST NIGHT II	Neil Diamond
1219	HOT BOX	Fatback
629	HOT, COOL AND VICIOUS	Salt-N-Pepa
1616	HOT HOUSE FLOWERS	Wynton Marsalis
919	HOT IN THE SHADE	KISS
2078	HOT NUMBER	Fabulous Thunderbirds
3433	HOT SHOT	Pat Travers
1325	HOT SPACE	Queen
3039	HOT SPOT	Dazz Band
148	HOTTER THAN JULY	Stevie Wonder
2888	THE HOTTEST NIGHT OF THE YEAR	Anne Murray
1966	HOT TOGETHER	Pointer Sisters
1944	HOT WATER	Jimmy Buffett
1082	HOUNDS OF LOVE	Kate Bush
1439	THE HOUSE OF BLUE LIGHT	Deep Purple
2493	THE HOUSE OF DOLLS	Gene Loves Jezebel
1936	HOUSE OF LORDS	House Of Lords
4146	THE HOUSE OF LOVE	House Of Love
1753	HOUSE OF MUSIC	T.S. Monk
1377	HOW 'BOUT US	Champaign
958	HOW COULD IT BE	Eddie Murphy
2577	HOW CRUEL	Joan Armatrading
2171	HOW MANY TIMES CAN WE SAY GOODBYE	Dionne Warwick
4580	HOW THE HELL DO YOU SPELL RYTHUM	Amazing Rhythm Aces

Albums Alpha Index

4098 HOW TO BEAT THE HIGH COST OF LIVING (SOUNDTRACK) *Hubert Laws And Earl Klugh*	2777 I LOVE 'EM ALL *T.G. Sheppard*	3438 IN 'N' OUT *Stone City Band*
648 HOW TO BE A...ZILLIONAIRE! *ABC*	162 I LOVE ROCK 'N ROLL *Joan Jett & the Blackhearts*	1370 IN OUR LIFETIME *Marvin Gaye*
3042 HOW WILL I LAUGH TOMORROW WHEN I CAN'T EVEN SMILE TODAY *Suicidal Tendencies*	4029 I'M A BLUES MAN *Z.Z. Hill*	2713 IN OUTER SPACE *Sparks*
1224 HOW WILL THE WOLF SURVIVE *Los Lobos*	748 IMAGINATION *The Whispers*	1840 IN ROCK WE TRUST *Y&T*
769 HOW YA LIKE ME NOW *Kool Moe Dee*	1490 IMAGINE: JOHN LENNON (SOUNDTRACK) *John Lennon*	3825 IN SEARCH OF THE RAINBOW SEEKERS *Mtume*
1708 HOY-HOY! *Little Feat*	2672 IMAGINE THIS *Pieces Of A Dream*	520 INSIDE INFORMATION *Foreigner*
59 H2O *Daryl Hall & John Oates*	3598 IMAGINOS *Blue Öyster Cult*	4421 INSIDE LOOKIN' OUT *Junior*
3457 HUMANESQUE *Jack Green*	1170 I'M IN LOVE *Evelyn King*	2144 INSIDE MOVES *Grover Washington Jr.*
2168 HUMAN RACING *Nik Kershaw*	1197 I'M IN LOVE AGAIN *Patti LaBelle*	2900 INSIDE OUT *Philip Bailey*
2803 HUMANS *Bruce Cockburn*	4385 IMITATION LIFE *Robin Lane & The Chartbusters*	2486 INSIDE RONNIE MILSAP *Ronnie Milsap*
1215 HUMAN'S LIB *Howard Jones*	3080 AN IMITATION OF LOVE *Millie Jackson*	2605 INSIDE STORY *Grace Jones*
1037 THE HUNGER *Michael Bolton*	1017 I'M NO ANGEL *Gregg Allman Band*	2128 INSIDE THE ELECTRIC CIRCUS *W.A.S.P.*
2097 THE HUNTER *Blondie*	1501 I'M NO HERO *Cliff Richard*	2072 INSIDE YOU *Isley Brothers*
3351 THE HUNTER *Joe Sample*	3401 I'M NOT STRANGE I'M JUST LIKE YOU *Keith Sykes*	222 IN SQUARE CIRCLE *Stevie Wonder*
415 HUNTING HIGH AND LOW *a-ha*	1225 IMPERIAL BEDROOM *Elvis Costello And The Attractions*	2732 INSTANT LOVE *Cheryl Lynn*
4714 HURRICANE EYES *Loudness*	4711 I'M READY *Natalie Cole*	795 IN STEP *Stevie Ray Vaughan And Double Trouble*
3108 HURRY UP THIS WAY AGAIN *The Stylistics*	2810 I'M REAL *James Brown*	3137 INSTINCT *Iggy Pop*
977 THE HURTING *Tears For Fears*	1968 I'M SO PROUD *Deniece Williams*	2188 INSTINCTS *Romeo Void*
2921 HUSH *John Klemmer*	1098 I'M THE MAN *Anthrax*	2170 INTENSITIES IN 10 CITIES *Ted Nugent*
4081 HYAENA *Siouxsie & The Banshees*	1832 I'M THE ONE *Roberta Flack*	2524 INTERMISSION *Dio*
3 HYSTERIA *Def Leppard*	2891 I'M YOUR PLAYMATE *Suave*	3369 INTERNATIONALISTS *Style Council*
2350 HYSTERIA *Human League*	4182 I'M YOURS *Linda Clifford*	3828 IN THE BEGINNING *Journey*
	3835 ...IN A CHAMBER *Wire Train*	437 IN THE DARK *Grateful Dead*
I	4300 INAMORATA *Poco*	1438 IN THE EYE OF THE STORM *Roger Hodgson*
2498 I ADVANCE MASKED *Andy Summers & Robert Fripp*	3541 INARTICULATE SPEECH OF THE HEART *Van Morrison*	741 IN THE HEART *Kool & The Gang*
1302 I AM LOVE *Peabo Bryson*	3610 IN A SENTIMENTAL MOOD *Dr. John*	4058 IN THE HEAT *Southside Johnny & The Jukes*
2813 I AM WHAT I AM *George Jones*	855 IN A SPECIAL WAY *DeBarge*	166 IN THE HEAT OF THE NIGHT *Pat Benatar*
4606 I APPRECIATE *Alicia Myers*	4169 IN BLACK AND WHITE *Barbara Mandrell*	3488 IN THE HEAT OF THE NIGHT *Jeff Lorber*
3906 I BELIEVE IN LOVE *Rockie Robbins*	4596 IN BLACK AND WHITE *Jenny Burton*	3560 IN THE LAND OF SALVATION AND SIN *Georgia Satellites*
1182 I BELIEVE IN YOU *Don Williams*	1594 INCOGNITO *Spyro Gyra*	4675 IN THE LONG GRASS *Boomtown Rats*
1498 I CAN DREAM ABOUT YOU *Dan Hartman*	3903 IN CONCERT *Julio Iglesias*	4314 IN THE MOOD FOR SOMETHING RUDE *Foghat*
570 I CAN'T STAND STILL *Don Henley*	4069 IN CONCERT *Jane Olivor*	3918 IN THE NAME OF LOVE *Thompson Twins*
1574 THE ICEBERG (FREEDOM OF SPEECH... JUST WATCH WHAT YOU SAY) *Ice-T*	4294 IN CONTROL VOLUME I *Marley Marl*	2533 IN THE NIGHT *Cheryl Lynn*
355 ICE CREAM CASTLE *The Time*	4004 INDESTRUCTIBLE *Four Tops*	320 IN THE POCKET *Commodores*
2273 ICEHOUSE *Icehouse*	2131 INDIANA JONES AND THE TEMPLE OF DOOM (John Williams) *Soundtrack*	2950 IN THE POCKET *Neil Sedaka*
1312 ICE ON FIRE *Elton John*	729 INDIGO GIRLS *Indigo Girls*	3714 IN THE SPIRIT OF THINGS *Kansas*
1707 ICICLE WORKS *Icicle Works*	2514 INDIVIDUAL CHOICE *Jean-Luc Ponty*	383 INTO THE FIRE *Bryan Adams*
3553 I COMMIT TO LOVE *Howard Hewett*	2670 IN DREAMS: THE GREATEST HITS *Roy Orbison*	324 INTO THE GAP *Thompson Twins*
4839 ICON *Icon*	2243 INDUSTRY STANDARD *The Dregs*	1051 INTO THE LIGHT *Chris de Burgh*
3381 I COULD RULE THE WORLD IF I COULD ONLY GET THE PARTS *The Waitresses*	343 IN EFFECT MODE *Al B. Sure!*	3599 INTO THE NIGHT *Soundtrack*
3343 THE IDOLMAKER *Soundtrack*	753 I NEVER SAID GOODBYE *Sammy Hagar*	3076 INTO THE WOODS *The Call*
2062 I DON'T SPEAK THE LANGUAGE *Matthew Wilder*	2566 INFECTED *The The*	3947 INTO THE WOODS *Original Cast Recording*
515 I FEEL FOR YOU *Chaka Khan*	934 INFIDELS *Bob Dylan*	2637 INTRODUCING...DAVID PEASTON *David Peaston*
2427 IF I SHOULD FALL FROM GRACE WITH GOD *The Pogues*	1679 INFORMATION *Dave Edmunds*	2776 INTRODUCING JONATHAN BUTLER *Jonathan Butler*
597 IF I SHOULD LOVE AGAIN *Barry Manilow*	757 INFORMATION SOCIETY *Information Society*	188 INTRODUCING THE HARDLINE ACCORDING TO TERENCE TRENT D'ARBY *Terence Trent D'Arby*
911 IF I WERE YOUR WOMAN *Stephanie Mills*	3303 IN FULL EFFECT *Mantronix*	4382 INTRODUCING THE STYLE COUNCIL *Style Council*
2351 IF MY ANCESTORS COULD SEE ME NOW *Ivan Neville*	1235 IN GOD WE TRUST *Stryper*	3103 THE INTRODUCTION *Steve Morse Band*
558 IF THAT'S WHAT IT TAKES *Michael McDonald*	3666 IN HARMONY 2 *Various Artists*	1314 INTROSPECTIVE *Pet Shop Boys*
2648 IF YOU AIN'T LOVIN' (YOU AIN'T LIVIN') *George Strait*	4082 IN HARMONY - A SESAME STREET RECORD *Various Artists*	3049 INTUITION *TNT*
3519 IF YOU CAN'T LICK 'EM...LICK 'EM *Ted Nugent*	557 IN HEAT *The Romantics*	4419 INTUITION *Linx*
2108 IGNITION *John Waite*	3500 IN HEAT *Black 'N Blue*	449 INVASION OF YOUR PRIVACY *Ratt*
4303 I GOT THE MELODY *Odyssey*	3387 INHERIT THE WIND *Wilton Felder*	2348 THE INVISIBLE MAN'S BAND *Invisible Man's Band*
4028 I HAD TO SAY IT *Millie Jackson*	3567 IN LONDON *Al Jarreau*	1190 IN VISIBLE SILENCE *Art Of Noise*
4856 I HAVE A PONY *Steven Wright*	4424 IN LOVE *Cheryl Lynn*	67 INVISIBLE TOUCH *Genesis*
3939 I HEARD IT IN A LOVE SONG *McFadden & Whitehead*	4433 IN LOVE *Bunny Debarge*	4618 INXS *INXS*
3064 I HEAR YOU ROCKIN' *Dave Edmunds Band*	1237 IN MY EYES *Stevie B*	725 IN YOUR EYES *George Benson*
2857 I JUST CAN'T STOP IT *English Beat*	562 IN MY TRIBE *10,000 Maniacs*	2149 IN YOUR FACE *Kingdom Come*
2843 I LIKE YOUR STYLE *Jermaine Jackson*	3812 INNOCENCE IS NO EXCUSE *Saxon*	4537 I PREFER THE MOONLIGHT *Kenny Rogers*
830 I'LL ALWAYS LOVE YOU *Anne Murray*	156 THE INNOCENT AGE *Dan Fogelberg*	2932 IRISH HEARTBEAT *Van Morrison & The Chieftains*
3490 I'LL PROVE IT TO YOU *Gregory Abbott*	3917 INNOCENT EYES *Graham Nash*	3795 IRON AGE *Mother's Finest*
	43 AN INNOCENT MAN *Billy Joel*	2468 IRON EAGLE *Soundtrack*
	752 THE INNOCENTS *Erasure*	4325 IRON FIST *Motorhead*
	3794 IN NO SENSE? NONSENSE! *Art Of Noise*	

Albums Alpha Index

2319 THE IRON MAN (THE MUSICAL BY PETE TOWNSHEND) *Pete Townshend*
929 IRONS IN THE FIRE *Teena Marie*
3876 IRRESISTIBLE *Miles Jaye*
4133 ISLAND LIFE *Grace Jones*
3561 ISLANDS *Mike Oldfield*
4392 THE ISLAND STORY, 1962-1987: THE 25TH ANNIVERSARY *Various Artists*
2798 ISLE OF MAN *Isle Of Man*
1601 ISOLATION *Toto*
3306 ITALIAN X RAYS *Steve Miller Band*
740 IT MUST BE MAGIC *Teena Marie*
3144 I TOUCHED A DREAM *The Dells*
4602 IT'S A BEAUTIFUL THING *Maxine Nightingale*
1389 IT'S A BIG DADDY THING *Big Daddy Kane*
2379 IT'S ABOUT TIME *John Denver*
2558 IT'S A FACT *Jeff Lorber*
2952 IT'S ALRIGHT (I SEE RAINBOWS) *Yoko Ono*
3720 IT'S BEGINNING TO AND BACK AGAIN *Wire*
721 IT'S BETTER TO TRAVEL *Swing Out Sister*
571 IT'S HARD *The Who*
2083 IT'S HARD TO BE HUMBLE *Mac Davis*
1559 IT'S MY LIFE *Talk Talk*
4786 IT'S MY TIME *Maynard Ferguson*
3417 IT'S MY TURN *Soundtrack*
3323 IT'S ONLY ROCK AND ROLL *Waylon Jennings*
3586 IT'S TEE TIME *Sweet Tee*
4607 IT'S THE WORLD GONE CRAZY *Glen Campbell*
844 IT'S TIME FOR LOVE *Teddy Pendergrass*
975 IT'S YOUR NIGHT *James Ingram*
1031 IT TAKES A NATION OF MILLIONS TO HOLD US BACK *Public Enemy*
416 IT TAKES TWO *Rob Base & D.J. E-Z Rock*
4751 I'VE ALWAYS WANTED TO DO THIS *Jack Bruce And Friends*
2303 I'VE GOT EVERYTHING *Henry Lee Summer*
2622 I'VE GOT THE CURE *Stephanie Mills*
2948 I'VE GOT THE ROCK 'N' ROLLS AGAIN *Joe Perry Project*
1562 I'VE NEVER BEEN TO ME *Charlene*
4858 IVORY COAST *Bob James*
995 I WANNA HAVE SOME FUN *Samantha Fox*
3522 I WANT CANDY *Bow Wow Wow*
3832 I WANT OUT-LIVE *Helloween*
3722 I WAS THE ONE *Elvis Presley*
4222 I WEAR THE FACE *Mr. Mister*
2908 I WISH I WAS EIGHTEEN AGAIN *George Burns*
3675 I WONDER DO YOU THINK OF ME *Keith Whitley*

J

878 JACKRABBIT SLIM *Steve Forbert*
1144 JACKSONS LIVE *The Jacksons*
4156 THE JAM *The Jam*
4336 JAMES BROWN...LIVE/HOT ON THE ONE *James Brown*
4315 JAMMIN' IN MANHATTAN *Tyzik*
2005 JAM ON REVENGE *Newcleus*
2476 JAM THE BOX! *Bill Summers & Summers Heat*
183 JANE FONDA'S WORKOUT RECORD *Jane Fonda*
3707 JANE FONDA'S WORKOUT RECORD FOR PREGNANCY, BIRTH AND RECOVERY *Jane Fonda*
3757 JANE FONDA'S WORKOUT RECORD NEW AND IMPROVED *Jane Fonda*
1667 JANET JACKSON *Janet Jackson*
19 JANET JACKSON'S RHYTHM NATION 1814 *Janet Jackson*
3926 JANE WIEDLIN *Jane Wiedlin*
4567 JAPANESE WHISPERS *The Cure*

565 JARREAU *Al Jarreau*
2291 JAZZERCISE *Judi Sheppard Missett*
81 THE JAZZ SINGER (SOUNDTRACK) *Neil Diamond*
1154 THE JEALOUS KIND *Delbert McClinton*
4261 JEALOUS LOVER *Rainbow*
3055 JEALOUSY *Dirt Band*
2021 JEFF BECK'S GUITAR SHOP *Jeff Beck With Terry Bozzio & Tony Hymas*
3375 JEFFERSON AIRPLANE *Jefferson Airplane*
1116 JEFFREY OSBORNE *Jeffrey Osborne*
1369 JERMAINE(2) *Jermaine Jackson*
473 JERMAINE JACKSON *Jermaine Jackson*
4074 JERRY KNIGHT *Jerry Knight*
711 JESSE JOHNSON'S REVUE *Jesse Johnson's Revue*
357 THE JETS *The Jets*
2195 JEWEL OF THE NILE *Soundtrack*
2270 "JI" *Junior*
3094 THE JIMI HENDRIX CONCERTS *Jimi Hendrix*
4787 JIMI PLAYS MONTEREY *Jimi Hendrix*
2871 JIMMY BARNES *Jimmy Barnes*
4719 J MOOD *Wynton Marsalis*
201 JODY WATLEY *Jody Watley*
1757 JOE JACKSON'S JUMPIN' JIVE *Joe Jackson*
2045 JOE'S GARAGE ACTS II + III *Frank Zappa*
4228 JOHN CONLEE'S GREATEST HITS *John Conlee*
1563 JOHN COUGAR *John Cougar*
2603 JOHN EDDIE *John Eddie*
1436 THE JOHN LENNON COLLECTION *John Lennon*
4569 JOHNNY "GUITAR" WATSON AND THE FAMILY CLONE *Johnny Guitar Watson*
4270 JOHN O'BANION *John O'Banion*
1553 JOHN PARR *John Parr*
2970 JOIN THE ARMY *Suicidal Tendencies*
1255 JONATHAN BUTLER *Jonathan Butler*
2811 JON BUTCHER AXIS *Jon Butcher Axis*
29 THE JOSHUA TREE *U2*
358 JOURNEY'S GREATEST HITS *Journey*
4014 JOURNEY TO A RAINBOW *Chuck Mangione*
1474 JOY *Teddy Pendergrass*
1105 JOY AND PAIN *Maze Featuring Frankie Beverly*
3087 JOYRIDE *Pieces Of A Dream*
1619 JOYSTICK *Dazz Band*
3347 THE JUDDS *The Judds*
3776 JUDSON SPENCE *Judson Spence*
4673 JUGGERNAUT *Frank Marino*
171 JUICE *Juice Newton*
1622 JUICE *Oran 'Juice' Jones*
1057 JUICY FRUIT *Mtume*
2115 JUJU MUSIC *King Sunny Ade & His African Beats*
1977 JUKEBOX *Dazz Band*
2434 JULIA FORDHAM *Julia Fordham*
3691 JULIAN COPE *Julian Cope*
478 JULIO *Julio Iglesias*
4482 JUMPIN' JACK FLASH *Soundtrack*
936 JUMP TO IT *Aretha Franklin*
742 JUMP UP! *Elton John*
1445 JUNGLE *Dwight Twilley*
4401 JUNK CULTURE *Orchestral Manoeuvres In The Dark*
3048 JUNKYARD *Junkyard*
1195 JUST ANOTHER DAY IN PARADISE *Bertie Higgins*
2340 JUST A TOUCH OF LOVE *Slave*
3256 JUST BEFORE THE BULLETS FLY *Gregg Allman Band*
1924 JUST BE MY LADY *Larry Graham*
4524 JUST BETWEEN US *Gerald Albright*
1557 JUST COOLIN' *Levert*
726 JUST GETS BETTER WITH TIME *The Whispers*

4198 JUST LIKE DREAMIN' *Twennynine Featuring Lenny White*
423 JUST LIKE THE FIRST TIME *Freddie Jackson*
2941 JUST LOOKIN' FOR A HIT *Dwight Yoakam*
192 JUST ONE NIGHT *Eric Clapton*
1406 JUST SYLVIA *Sylvia (2)*
4732 JUST TESTING *Wishbone Ash*
1792 JUST THE WAY YOU LIKE IT *S.O.S. Band*
1455 JUST US *Alabama*
3058 JUST VISITING THIS PLANET *Jellybean*

K

2124 KALEIDOSCOPE WORLD *Swing Out Sister*
3241 THE KARATE KID *Soundtrack*
1373 THE KARATE KID PART II *Soundtrack*
418 KARYN WHITE *Karyn White*
1510 KASHIF *Kashif*
4859 KASIM *Kasim Sulton*
4152 KATE BUSH *Kate Bush*
3534 KATHY SMITH'S AEROBIC FITNESS *Kathy Smith*
805 KATRINA AND THE WAVES *Katrina & The Waves*
2058 WAVES *Katrina & The Waves*
1896 KBC BAND *KBC Band*
2621 KC TEN *KC*
2629 KEEL *Keel*
2538 KEEPER OF THE SEVEN KEYS - PART I *Helloween*
2913 KEEPER OF THE SEVEN KEYS - PART II *Helloween*
3655 KEEPIN' LOVE NEW *Howard Johnson*
3253 KEEPIN' THE SUMMER ALIVE *Beach Boys*
584 KEEP IT LIVE *Dazz Band*
345 KEEP IT UP *Loverboy*
3363 KEEP MOVING *Madness*
4721 KEEP ON DOING *The Roches*
3623 KEEP ON IT *Starpoint*
221 KEEP ON MOVIN' *Soul II Soul*
2974 KEEP ON MOVING STRAIGHT AHEAD *Lakeside*
2374 ...KEEP SMILING *Laid Back*
300 KEEP THE FIRE *Kenny Loggins*
1737 KEEP YOUR DISTANCE *Curiosity Killed The Cat*
628 KEEP YOUR EYE ON ME *Herb Alpert*
1234 KEEP YOUR HANDS OFF MY POWER SUPPLY *Slade*
178 KENNY *Kenny Rogers*
499 KENNY LOGGINS ALIVE *Kenny Loggins*
30 KENNY ROGERS' GREATEST HITS *Kenny Rogers*
1971 KEVIN PAIGE *Kevin Paige*
1546 THE KEY *Joan Armatrading*
1587 KEYED UP *Ronnie Milsap*
3411 KEY LIME PIE *Camper van Beethoven*
48 KICK *INXS*
3649 KICK THE WALL *Jimmy Davis & Junction*
3702 THE KIDS FROM "FAME" *Kids From Fame*
3160 THE KIDS FROM "FAME" LIVE! *Kids From Fame*
869 KIHNSPIRACY *Greg Kihn Band*
3585 KIHNTAGIOUS *Greg Kihn Band*
1687 KIHNTINUED *Greg Kihn*
2557 KILIMANJARO *The Rippingtons Featuring Russ Freeman*
3958 KILIMANJARO *The Teardrop Explodes*
3407 KILL 'EM ALL(2) *Metallica*
3842 KILL 'EM ALL *Metallica*
555 KILLER ON THE RAMPAGE *Eddy Grant*
1953 KILLERS *Iron Maiden*
275 KILROY WAS HERE *Styx*
2329 KIM WILDE *Kim Wilde*
2218 A KIND OF MAGIC *Queen*
4799 KING BEE *Muddy Waters*

Albums Alpha Index

3454 THE KINGBEES *The Kingbees*	1798 LAST SAFE PLACE *Le Roux*	819 LIFE AS WE KNOW IT *REO Speedwagon*
1735 KING COOL *Donnie Iris*	1658 LATE AT NIGHT *Billy Preston*	4531 LIFE (IS SO STRANGE) *War*
4873 KINGDOM BLOW *Kurtis Blow*	1996 LATE NIGHT GUITAR *Earl Klugh*	498 LIFE IS...TOO SHORT *Too Short*
560 KINGDOM COME *Kingdom Come*	2804 LATOYA JACKSON *LaToya Jackson*	4738 'LIFE' LIVE *Thin Lizzy*
1713 KING OF AMERICA *Elvis Costello And The Attractions*	3848 LAUGHING AT THE PIECES *Doctor And The Medics*	386 LIFE, LOVE & PAIN *Club Nouveau*
4291 THE KING OF COMEDY *Soundtrack*	4319 LAUGHTER *Ian Dury And The Blockheads*	785 LIFES RICH PAGEANT *R.E.M.*
790 KING OF ROCK *Run-D.M.C.*	961 LAW AND ORDER *Lindsey Buckingham*	1365 LIFE'S TOO GOOD *The Sugarcubes*
2662 KING OF STAGE *Bobby Brown*	3378 LAW OF THE FISH *The Radiators*	3594 LIFE STORIES *Earl Klugh*
1662 THE KINGS ARE HERE *The Kings*	547 LAWYERS IN LOVE *Jackson Browne*	3772 LIGHTHOUSE *Kim Carnes*
3161 KINGS OF THE SUN *Kings Of The Sun*	2418 LEAD ME ON *Amy Grant*	3446 LIGHTING UP THE NIGHT *Jack Wagner*
1056 KINGS OF THE WILD FRONTIER *Adam And The Ants*	3178 THE LEAGUE OF GENTLEMEN *Robert Fripp*	2391 LIGHTNING STRIKES *Loudness*
3952 KINGS OF WEST COAST *L.A. Dream Team*	319 LEARNING TO CRAWL *The Pretenders*	3024 LIGHT OF DAY *Soundtrack*
3123 KING'S RECORD SHOP *Rosanne Cash*	4794 LEARNING TO LOVE *Rodney Franklin*	2998 LIGHT OF THE STABLE: THE CHRISTMAS ALBUM *Emmylou Harris*
3766 KING SWAMP *King Swamp*	1469 LEATHER AND LACE *Waylon Jennings & Jessi Colter*	4090 LIGHTS IN THE NIGHT *Flash And The Pan*
172 KISSING TO BE CLEVER *Culture Club*	3329 LEATHER JACKETS *Elton John*	1035 LIGHTS OUT *Peter Wolf*
593 KISS ME, KISS ME, KISS ME *The Cure*	2897 LEATHERWOLF *Leatherwolf*	351 LIGHT UP THE NIGHT *The Brothers Johnson*
4245 KISS THE SKY *Jimi Hendrix*	4132 LEAVE SCARS *Dark Angel*	86 LIKE A PRAYER *Madonna*
1515 KISS UNMASKED *KISS*	4445 LE CHAT BLEU *Mink De Ville*	186 LIKE A ROCK *Bob Seger & The Silver Bullet Band*
2150 KLYMAXX *Klymaxx*	3804 LEFTY *Art Garfunkel*	45 LIKE A VIRGIN *Madonna*
236 KNEE DEEP IN THE HOOPLA *Starship*	1011 LEGACY *Poco*	2393 LIKE GANGBUSTERS *JoBoxers*
4413 KNIFE *Aztec Camera*	2019 LEGEND *Bob Marley And The Wailers*	4138 LI'L SUZY *Ozone*
1646 THE KNIFE FEELS LIKE JUSTICE *Brian Setzer*	2019 LEGEND *Lynyrd Skynyrd*	2347 LINCOLN *They Might Be Giants*
1779 KNIGHTS OF THE SOUND TABLE *Cameo*	3578 LEGEND (SOUNDTRACK) *Tangerine Dream*	3524 LINE OF FIRE *The Headpins*
2706 K-9 POSSE *K-9 Posse*	4139 LEGENDARY HEARTS *Lou Reed*	916 THE LION AND THE COBRA *Sinead O'Connor*
2325 KNOCKED OUT LOADED *Bob Dylan*	3365 THE LEGEND OF JESSE JAMES *Various Artists*	52 LIONEL RICHIE *Lionel Richie*
1043 KNOWLEDGE IS KING *Kool Moe Dee*	3201 LENA HORNE: THE LADY AND HER MUSIC (ORIGINAL CAST) *Lena Horne*	962 LISA LISA & CULT JAM WITH FULL FORCE *Lisa Lisa And Cult Jam With Full Force*
1558 КОНЦЕРТ (LIVE IN LENINGRAD) *Billy Joel*	2911 LES MISERABLES *Original London Cast Recording*	2221 L IS FOR LOVER *Al Jarreau*
1379 KOOKOO *Debbie Harry*	3472 LES MISERABLES *Original Broadway Cast Recording*	893 LISTEN *A Flock Of Seagulls*
2260 KOOL MOE DEE *Kool Moe Dee*	2982 LES PLUS GRANDS SUCCES DE CHIC - CHIC'S GREATEST HITS *Chic*	400 LISTEN LIKE THIEVES *INXS*
2159 KRUSH GROOVE *Soundtrack*	1088 LESS THAN ZERO *Soundtrack*	3580 LISTEN TO THE MESSAGE *Club Nouveau*
2556 KURTIS BLOW *Kurtis Blow*	2000 LETHAL *UTFO*	3979 LISTEN TO THE RADIO *Don Williams*
4841 KWICK *Kwick*	4620 LETHAL WEAPON 2 *Soundtrack*	353 LITA *Lita Ford*
1248 KYLIE *Kylie Minogue*	112 LET IT LOOSE *Gloria Estefan & Miami Sound Machine*	4131 LITE ME UP *Herbie Hancock*
	3830 LET IT ROCK *Johnny And The Distractions*	2881 LITTLE AMERICA *Little America*
L	1171 LET IT ROLL *Little Feat*	1368 LITTLE BAGGARIDDIM *UB40*
251 LA BAMBA (SOUNDTRACK) *Los Lobos*	1103 LET ME BE YOUR ANGEL *Stacy Lattisaw*	338 LITTLE CREATURES *Talking Heads*
2564 L.A. BOPPERS *L.A. Boppers*	3659 LET ME KNOW YOU *Stanley Clarke*	4441 LITTLE DREAMER *Peter Green*
3609 LABOR OF LOVE *Spinners*	1810 LET ME TICKLE YOUR FANCY *Jermaine Jackson*	1978 A LITTLE GOOD NEWS *Anne Murray*
441 LABOUR OF LOVE *UB40*	1759 LET ME TOUCH YOU *The O'Jays*	1673 A LITTLE LOVE *Aurra*
3110 LABYRINTH (SOUNDTRACK) *David Bowie*	884 LET ME UP (I'VE HAD ENOUGH) *Tom Petty And The Heartbreakers*	2523 LITTLE MISS DANGEROUS *Ted Nugent*
2331 LA CAGE AUX FOLLES *Original Cast*	4679 LET'S BURN *Clarence Carter*	978 LITTLE ROBBERS *The Motels*
2177 THE LACE *Benjamin Orr*	147 LET'S DANCE *David Bowie*	1905 LITTLE SHOP OF HORRORS *Soundtrack*
2503 LADIES OF THE EIGHTIES *A Taste Of Honey*	4765 LET'S DO IT TODAY *Lenny Williams*	1912 A LITTLE SPICE *Loose Ends*
1960 LADY *One Way*	375 LET'S GET IT STARTED *MC Hammer*	2544 LITTLE STEVIE ORBIT *Steve Forbert*
1534 LADY T *Teena Marie*	373 LET'S GET SERIOUS *Jermaine Jackson*	967 LIVE *George Thorogood & The Destroyers*
1214 L.A. GUNS *L.A. Guns*	1241 LET'S GO ALL THE WAY *Sly Fox*	2145 LIVE *Alabama*
2505 L.A. IS MY LADY *Frank Sinatra*	1403 LET'S HEAR IT FOR THE BOY *Deniece Williams*	3109 LIVE *Stephane Grappelli/David Grisman*
3471 LAMENT *Ultravox*	4552 LET'S MAKE A NEW DOPE DEAL *Cheech & Chong*	3713 LIVE *Reba McEntire*
1740 LANDING ON WATER *Neil Young*	1949 LETTER FROM HOME *Pat Metheny Group*	1663 LIVE & MORE *Roberta Flack & Peabo Bryson*
2231 LAND OF DREAMS *Randy Newman*	3664 A LETTER FROM ST. PAUL *The Silencers*	985 LIVE AFTER DEATH *Iron Maiden*
4415 THE LAND OF RAPE AND HONEY *Ministry*	1877 LET THE DAY BEGIN *The Call*	4769 THE LIVE ALBUM *Leon Russell And The New Grass Revival*
4817 LAND OF THE THIRD EYE *Dave Valentin*	4882 LET THE HUSTLERS PLAY *Steady B*	1522 LIVE ALIVE *Stevie Ray Vaughan And Double Trouble*
4757 LANGUAGE *Gary Myrick*	1872 LET THE MUSIC DO THE TALKING *Joe Perry Project*	4466 LIVE AND LOWDOWN AT THE APOLLO VOL 1 *James Brown*
4542 LA PISTOLA Y EL CORAZON *Los Lobos*	979 LET THE MUSIC PLAY *Shannon*	3005 LIVE AND OUTRAGEOUS (RATED XXX) *Millie Jackson*
295 LAP OF LUXURY *Cheap Trick*	3395 LET THE MUSIC PLAY *Dazz Band*	2305 LIVE AND UNCENSORED *Millie Jackson*
574 LARGER THAN LIFE *Jody Watley*	486 THE LEXICON OF LOVE *ABC*	4733 LIVE AT ST. DOUGLAS CONVENT *Father Guido Sarducci*
3450 LARSEN-FEITEN BAND *Larsen-Feiten Band*	2937 LIBRA *Julio Iglesias*	3808 LIVE AT THE HOLLYWOOD BOWL *The Doors*
1636 THE LAST COMMAND *W.A.S.P.*	95 LICENSED TO ILL *Beastie Boys*	4400 LIVE AT THE WHISKY A GO-GO ON THE FABULOUS SUNSET STRIP *X*
2178 LAST DATE *Emmylou Harris*	2042 LICENSE TO DREAM *Kleeer*	1954 LIVE EVIL *Black Sabbath*
3650 THE LAST EMPEROR *Soundtrack*	4485 LICENSE TO KILL *Malice*	3681 LIVE! FOR LIFE *Various Artists*
837 THE LAST IN LINE *Dio*	857 LICK IT UP *KISS*	
1774 LAST MANGO IN PARIS *Jimmy Buffett*	2656 LIFE *Neil Young With Crazy Horse*	
2287 THE LAST OF THE MOHICANS *Bow Wow Wow*	3223 LIFE *Gladys Knight & The Pips*	
4199 LAST OF THE WILD ONES *Johnny Van Zant Band*		
2530 THE LAST ONE TO KNOW *Reba McEntire*		

Albums Alpha Index

4650	LIVE FREE OR DIE	*Balaam And The Angel*
564	LIVE FROM EARTH	*Pat Benatar*
2242	LIVE FROM NEW YORK	*Gilda Radner*
735	LIVE IN AUSTRALIA (WITH THE MELBOURNE SYMPHONY ORCHESTRA)	*Elton John*
4570	LIVE IN CONCERT	*Roger Whittaker*
2929	LIVE/INDIAN SUMMER	*Al Stewart*
3271	LIVE IN LOS ANGELES	*Maze Featuring Frankie Beverly*
1019	LIVE IN NEW ORLEANS	*Maze Featuring Frankie Beverly*
2258	LIVE IN NEW YORK CITY	*John Lennon*
3435	LIVE IN THE HEART OF THE CITY	*Whitesnake*
2684	LIVE...IN THE RAW	*W.A.S.P.*
3279	LIVE IT UP	*David Johansen*
2978	LIVE 1980/86	*Joe Jackson*
3711	LIVE NUDE GUITARS	*Brian Setzer*
2809	LIVE + 1	*Frehley's Comet*
2890	LIVERPOOL	*Frankie Goes To Hollywood*
619	LIVE RUST	*Neil Young With Crazy Horse*
3652	LIVE SENTENCE	*Alcatrazz*
4449	LIVE SHOTS	*Joe Ely*
780	LIVES IN THE BALANCE	*Jackson Browne*
1162	LIVE-STOMPIN' AT THE SAVOY	*Rufus And Chaka Khan*
4122	LIVE WITHOUT A NET	*Angel*
1828	LIVING ALL ALONE	*Phyllis Hyman*
1719	LIVING EYES	*Bee Gees*
3016	LIVING IN A BOX	*Living In A Box*
1095	LIVING IN A FANTASY	*Leo Sayer*
342	LIVING IN OZ	*Rick Springfield*
1724	LIVING IN THE 20TH CENTURY	*Steve Miller Band*
2048	LIVING IN THE BACKGROUND	*Baltimora*
2614	LIVING LARGE...	*Heavy D & The Boyz*
2160	LIVING MY LIFE	*Grace Jones*
2965	LIVING PROOF	*Sylvester*
575	LIVING YEARS	*Mike + The Mechanics*
3936	LIVIN' LARGE	*E.U.*
4077	LIZA MINNELLI AT CARNEGIE HALL	*Liza Minnelli*
194	LOC-ED AFTER DARK	*Tone-Loc*
696	LONDON CALLING	*The Clash*
3290	LONDON 0 HULL 4	*The Housemartins*
1540	LONE JUSTICE	*Lone Justice*
3163	LONE RHINO	*Adrian Belew*
141	THE LONESOME JUBILEE	*John Mellencamp*
401	LONG AFTER DARK	*Tom Petty And The Heartbreakers*
160	LONG COLD WINTER	*Cinderella*
115	LONG DISTANCE VOYAGER	*Moody Blues*
2501	LONG LIVE THE KANE	*Big Daddy Kane*
2157	LONG LIVE THE NEW FLESH	*Flesh For Lulu*
1184	LONG TIME COMING	*Ready For The World*
2636	LONG WAY TO HEAVEN	*Helix*
1460	THE LOOK	*Shalamar*
4744	LOOK HEAR?	*10cc*
2625	LOOKIN' FOR LOVE	*Johnny Lee*
2742	LOOKIN' FOR TROUBLE	*Joyce Kennedy*
4600	LOOKIN' FOR TROUBLE	*Toronto*
3660	LOOKING FOR JACK	*Colin James Hay*
3038	LOOK OUT	*20/20*
3962	LOOKS SO FINE	*Instant Funk*
111	LOOK WHAT THE CAT DRAGGED IN	*Poison*
3190	LOS HOMBRES MALO	*The Outlaws*
2673	LOST & FOUND	*Jason & The Scorchers*
949	LOST BOYS	*Soundtrack*
242	LOST IN LOVE	*Air Supply*
1986	LOST IN SPACE	*Jonzun Crew*
3317	LOST IN THE FIFTIES TONIGHT	*Ronnie Milsap*
2269	A LOT OF LOVE	*Melba Moore*

2802	LOUD AND CLEAR	*Autograph*
1997	LOUDER THAN BOMBS	*The Smiths*
4547	LOUDER THAN HELL	*Sam Kinison*
1769	LOVE	*The Cult*
4770	LOVE	*Aztec Camera*
2680	LOVE & HOPE & SEX & DREAMS	*BoDeans*
4810	LOVE ACTION	*Sniff 'N' The Tears*
1491	LOVE ALL THE HURT AWAY	*Aretha Franklin*
1976	LOVE AMONG THE CANNIBALS	*Starship*
1424	LOVE AN ADVENTURE	*Pseudo Echo*
3864	LOVE AND DANCING	*League Unlimited Orchestra*
621	LOVE AND ROCKETS	*Love And Rockets*
681	LOVE APPROACH	*Tom Browne*
217	LOVE AT FIRST STING	*Scorpions*
3019	LOVE BOMB	*The Tubes*
2866	LOVE BYRD	*Donald Byrd And 125th Street N.Y.C.*
2547	LOVE CHANGES	*Kashif*
3184	LOVE CONFESSIONS	*Miki Howard*
4378	LOVE FANTASY	*Roy Ayers*
3331	LOVE FEVER	*The O'Jays*
1199	LOVE FOR LOVE	*The Whispers*
2832	LOVE HYSTERIA	*Peter Murphy*
2024	LOVE IS A SACRIFICE	*Southside Johnny & The Asbury Jukes*
4327	LOVE IS FAIR	*Barbara Mandrell*
2717	LOVE IS FOR SUCKERS	*Twisted Sister*
3805	LOVE IS...ONE WAY	*One Way*
2165	LOVE IS SUCH A FUNNY GAME	*Michael Cooper*
3968	LOVE IS WHAT WE MAKE IT	*Kenny Rogers*
1212	LOVE IS WHERE YOU FIND IT	*The Whispers*
2463	LOVE JONES	*Johnny Guitar Watson*
2279	LOVE JUNK	*Pursuit Of Happiness*
1077	LOVE LANGUAGE	*Teddy Pendergrass*
4648	LOVE LETTERS	*Force M.D.'s*
1016	LOVE LIFE	*Berlin*
3203	LOVE LIFE	*Brenda Russell*
4411	LOVE LIGHT	*Yutaka*
1399	LOVE LIVES FOREVER	*Minnie Riperton*
3336	LOVELY	*The Primitives*
2439	LOVE MAGIC	*L.T.D.*
4548	LOVE ME TENDER	*B.B. King*
3020	LOVE NOTES	*Chuck Mangione*
3635	LOVE OR PHYSICAL	*Ashford & Simpson*
906	LOVE OVER GOLD	*Dire Straits*
244	LOVERBOY	*Loverboy*
1097	THE LOVER IN ME	*Sheena Easton*
3357	LOVE SEASON	*Alex Bugnon*
922	LOVESEXY	*Prince*
1319	LOVE SOMEBODY TODAY	*Sister Sledge*
450	LOVE STINKS	*J. Geils Band*
2079	LOVE TRIPPIN'	*Spinners*
3815	LOVE WARRIORS	*Tuck & Patti*
3504	LOVE WILL FOLLOW	*George Howard*
1265	LOVE WILL TURN YOU AROUND	*Kenny Rogers*
4346	LOVE YOUR MAN	*Rossington Band*
215	LOVE ZONE	*Billy Ocean*
403	LOVIN' EVERY MINUTE OF IT	*Loverboy*
1943	LOVING PROOF	*Ricky Van Shelton*
3767	LOVIN' THE NIGHT AWAY	*Dillman Band*
2365	LOW-LIFE	*New Order*
1486	LOW RIDE	*Earl Klugh*
3915	LUCIANO	*Luciano Pavarotti*
4257	LUCKY	*Marty Balin*
3851	LULLABY	*Book Of Love*
3291	LULU	*Lulu*
797	LUSH LIFE	*Linda Ronstadt*
2496	THE LUXURY GAP	*Heaven 17*
1114	LUXURY OF LIFE	*Five Star*

1732	LYLE LOVETT AND HIS LARGE BAND	*Lyle Lovett*

M

3257	MACALLA	*Clannad*
3075	MAC BAND	*Mac Band Feat. McCampbell Brothers*
2143	MACHISMO	*Cameo*
4376	MADAME X	*Madame X*
1699	MAD, BAD AND DANGEROUS TO KNOW	*Dead Or Alive*
1771	MADE IN AMERICA	*Carpenters*
2053	MADE IN AMERICA	*Blues Brothers*
2935	MADE IN U.S.A.	*Beach Boys*
202	MAD LOVE	*Linda Ronstadt*
1904	MAD MAX BEYOND THUNDERDOME	*Soundtrack*
1254	MADNESS	*Madness*
3057	MADNESS, MONEY AND MUSIC	*Sheena Easton*
113	MADONNA	*Madonna*
1608	THE MAGAZINE	*Rickie Lee Jones*
637	MAGIC	*The Jets*
1239	MAGIC	*Tom Browne*
3739	MAGIC	*Four Tops*
2297	MAGIC MAN	*Herb Alpert*
2674	MAGIC MAN	*Robert Winters And Fall*
917	MAGIC TOUCH	*Stanley Jordan*
3879	MAGIC WINDOWS	*Herbie Hancock*
2597	MAGNETIC FIELDS	*Jean Michel Jarre*
3218	MAGNETIC HEAVEN	*Wax*
3037	MAGNETS	*The Vapors*
3234	MAGNIFICENT MADNESS	*John Klemmer*
2692	MAGNUM CUM LOUDER	*Hoodoo Gurus*
2542	MAHVELOUS	*Billy Crystal*
2275	MAIDEN JAPAN	*Iron Maiden*
2412	MAJOR MOVES	*Hank Williams Jr.*
1858	MAKE A LITTLE MAGIC	*Dirt Band*
54	MAKE IT BIG	*Wham!*
237	MAKE IT LAST FOREVER	*Keith Sweat*
836	MAKE YOUR MOVE	*Captain & Tennille*
4737	MAKIN' FRIENDS	*Razzy Bailey*
4290	MAKING CONTACT	*UFO*
476	MAKING MOVIES	*Dire Straits*
1429	MALICE IN WONDERLAND	*Nazareth*
1990	MAMA AFRICA	*Peter Tosh*
4288	MAMA'S BOYS	*Mama's Boys*
2919	MAMMA	*Luciano Pavarotti*
4197	THE MAN & HIS MUSIC	*Sam Cooke*
4251	THE MAN FROM UTOPIA	*Frank Zappa*
2784	MANHATTANS GREATEST HITS	*The Manhattans*
4748	THE MANHATTAN TRANSFER LIVE	*Manhattan Transfer*
1590	MANILOW	*Barry Manilow*
3099	THE MANILOW COLLECTION/ TWENTY CLASSIC HITS	*Barry Manilow*
3244	MAN IN MOTION	*Night Ranger*
1287	MANNHEIM STEAMROLLER CHRISTMAS	*Mannheim Steamroller*
908	MAN OF COLOURS	*Icehouse*
3001	MAN OF STEEL	*Hank Williams Jr.*
2253	MAN ON THE LINE	*Chris de Burgh*
4154	MAN OVERBOARD	*Bob Welch*
1802	THE MAN WITH THE HORN	*Miles Davis*
971	THE MANY FACETS OF ROGER	*Roger*
4743	MARATHON	*Rodney Franklin*
1861	MARAUDER	*Blackfoot*
1507	MARCHING OUT	*Yngwie J. Malmsteen's Rising Force*
3261	MARCH OF THE SAINT	*Armored Saint*
2820	MARIA McKEE	*Maria McKee*
2793	MARILYN MARTIN	*Marilyn Martin*

Albums Alpha Index

4809 MARRIED TO THE MOB *Soundtrack*	2618 MESSINA *Jim Messina*	953 MONEY AND CIGARETTES *Eric Clapton*
4340 MARSALIS STANDARD TIME: VOL. 1 *Wynton Marsalis*	108 METAL HEALTH *Quiet Riot*	2282 MONEY FOR NOTHING *Dire Straits*
1415 MARSHALL CRENSHAW *Marshall Crenshaw*	2630 METAL HEART *Accept*	3613 THE MONROES *The Monroes*
4012 MARS NEEDS GUITARS! *Hoodoo Gurus*	4171 THE METHOD TO OUR MADNESS *Lords Of The New Church*	2302 MONSTER *Herbie Hancock*
603 MARTIKA *Martika*	4664 METRO MUSIC *Martha And The Muffins*	4103 MONSTER *Fetchin Bones*
2800 MARVIN SEASE *Marvin Sease*	3060 METROPOLIS *Soundtrack*	2499 MONTANA CAFE *Hank Williams Jr.*
1244 MARY JANE GIRLS *Mary Jane Girls*	143 MIAMI VICE *TV Soundtrack*	3771 MONTY PYTHON'S CONTRACTUAL OBLIGATION ALBUM *Monty Python*
3150 MASK *Roger Glover*	2909 MIAMI VICE II *TV Soundtrack*	2064 MOONLIGHTING *TV Soundtrack*
1586 MASK OF SMILES *John Waite*	2983 MICHAEL BOLTON *Michael Bolton*	2512 MOONLIGHT MADNESS *Teri DeSario*
3344 MASSTERPIECE *Mass Production*	2464 MICHAEL MARTIN MURPHEY *Michael Murphey*	4287 MOONS OF JUPITER *Scruffy The Cat*
3380 MASTER OF DISGUISE *Lizzy Borden*	2686 MICHAEL MORALES *Michael Morales*	163 MORE DIRTY DANCING *Soundtrack*
776 MASTER OF PUPPETS *Metallica*	2643 THE MICHAEL SCHENKER GROUP *Michael Schenker Group*	2335 MORE FUN IN THE NEW WORLD *X*
3040 MASTER OF THE GAME *George Duke*	1362 MICKEY MOUSE DISCO *Various Artists*	2403 MORE GEORGE THOROGOOD AND THE DESTROYERS *George Thorogood & The Destroyers*
3503 MASTERPIECE *Isley Brothers*	4731 MICK JONES *Mick Jones*	4296 MORE OF THE GOOD LIFE *T.S. Monk*
3977 MATERIAL THANGZ *The Deele*	305 MIDDLE MAN *Boz Scaggs*	4410 MORE SONGS ABOUT LOVE & HATE *The Godfathers*
4114 MATHEMATICS *Melissa Manchester*	3430 MIDNIGHT BLUE *Louise Tucker*	3562 MORE SPECIALS *The Specials*
2294 MAURICE WHITE *Maurice White*	4652 MIDNIGHT CRAZY *Mac Davis*	2768 MORE THAN FRIENDS *Jonathan Butler*
618 MAVERICK *George Thorogood & The Destroyers*	456 MIDNIGHT LOVE *Marvin Gaye*	3429 MORE THAN YOU KNOW *Toni Tennille*
3637 MAXIMUM SECURITY *Tony MacAlpine*	208 MIDNIGHT MADNESS *Night Ranger*	1061 MOSAIC *Wang Chung*
2683 MAXI PRIEST *Maxi Priest*	4242 MIDNIGHT MISSION *Textones*	2397 MOSAIQUE *Gipsy Kings*
4322 MAX Q *Max Q*	2791 MIDNIGHT STAR *Midnight Star*	4126 MOST OF THE GIRLS LIKE TO DANCE BUT ONLY SOME OF THE BOYS LIKE TO *Don Dixon*
4271 MAYBE IT'S LIVE *Robert Palmer*	982 MIDNIGHT TO MIDNIGHT *Psychedelic Furs*	944 MOTHER'S MILK *Red Hot Chili Peppers*
3802 MAZARATI *Mazarati*	3966 MIGRATION *Dave Grusin*	4716 MOTHER'S SPIRITUAL *Laura Nyro*
430 McCARTNEY II *Paul McCartney*	471 MIKE + THE MECHANICS *Mike + The Mechanics*	3197 MOTHER WIT *Betty Wright*
4384 THE McCARTNEY INTERVIEW *Paul McCartney*	2513 MIKE'S MURDER (SOUNDTRACK) *Joe Jackson*	4853 MOTOWN REMEMBERS MARVIN GAYE *Marvin Gaye*
1200 McVICAR (SOUNDTRACK) *Roger Daltrey*	3258 MILES *Miles Jaye*	3459 THE MOTOWN STORY: THE FIRST 25 YEARS *Various Artists*
3943 ME & PAUL *Willie Nelson*	894 MILK AND HONEY *John Lennon & Yoko Ono*	1881 MOUNTAIN DANCE *Dave Grusin*
4280 MEAN *Montrose*	3731 ...MILLIONS LIKE US *Millions Like Us*	212 MOUNTAIN MUSIC *Alabama*
1205 MEAN BUSINESS *The Firm*	3276 MILSAP MAGIC *Ronnie Milsap*	378 MOUTH TO MOUTH *Lipps Inc.*
4805 ME AND JOE *Rodney O & Joe Cooley*	3425 MIND BOMB *The The*	1385 MOVE SOMETHIN' *2 Live Crew*
4087 ME AND YOU *The Chi-Lites*	4003 THE MIND IS A TERRIBLE THING TO TASTE *Ministry*	116 MOVING PICTURES *Rush*
3065 MEAN STREAK *Y&T*	2396 MINIMUM WAGE ROCK & ROLL *Bus Boys*	3574 MOVING TARGET *Gil Scott-Heron*
2786 MEANT FOR EACH OTHER *Barbara Mandrell & Lee Greenwood*	4829 MINOR DETAIL *Minor Detail*	--MR. see: MISTER
1549 MEASURE FOR MEASURE *Icehouse*	3861 THE MINSTREL MAN *Willie Nelson*	3964 MSB *Michael Stanley Band*
2032 MEAT IS MURDER *The Smiths*	1672 THE MIRACLE *Queen*	3002 MSG *Michael Schenker Group*
613 MECCA FOR MODERNS *Manhattan Transfer*	2560 MIRACLE *Kane Gang*	2938 MTV'S ROCK 'N ROLL TO GO *Various Artists*
730 MECHANICAL RESONANCE *Tesla*	1393 MIRACLES *Change*	4427 MUD WILL BE FLUNG TONIGHT! *Bette Midler*
2488 MECHANIX *UFO*	121 MIRAGE *Fleetwood Mac*	4159 MUMMER *XTC*
3862 MECO PLAYS MUSIC FROM THE EMPIRE STRIKES BACK *Meco*	3653 MIRAGE A TROIS *Yellowjackets*	1258 MURMUR *R.E.M.*
4521 MEDICINE SHOW *Dream Syndicate*	1395 MIRROR MOVES *Psychedelic Furs*	1852 A MUSICAL AFFAIR *Ashford & Simpson*
2569 MEET DANNY WILSON *Danny Wilson*	3783 MIRRORS OF MY MIND *Roger Whittaker*	3870 MUSICAL SHAPES *Carlene Carter*
438 MEETING IN THE LADIES ROOM *Klymaxx*	2507 MISDEMEANOR *UFO*	4056 MUSIC FOR THE KNEE PLAYS *David Byrne*
3646 MEGA FORCE *707*	1411 MISPLACED CHILDHOOD *Marillion*	845 MUSIC FOR THE MASSES *Depeche Mode*
2655 MEGATOP PHOENIX *Big Audio Dynamite*	1055 MISSING PERSONS *Missing Persons*	4206 MUSIC FROM SONGWRITER (SOUNDTRACK) *Willie Nelson & Kris Kristofferson*
442 MELISSA ETHERIDGE *Melissa Etheridge*	3353 THE MISSION *Soundtrack*	3682 MUSIC FROM THE BILL COSBY SHOW-- A HOUSE FULL OF LOVE *TV Soundtrack*
294 MEMORIES *Barbra Streisand*	110 MISTAKEN Identity *Kim Carnes*	700 MUSIC FROM THE EDGE OF HEAVEN *Wham!*
2895 MEMORY IN THE MAKING *John Kilzer*	2839 MR. CROWLEY *Ozzy Osbourne*	2525 MUSIC FROM THE ELDER *KISS*
3215 ME, MYSELF AND I *Cheryl Pepsii Riley*	3592 MR. HANDS *Herbie Hancock*	826 MUSIC MAN *Waylon Jennings*
1038 ME MYSELF I *Joan Armatrading*	1965 MISTER HEARTBREAK *Laurie Anderson*	2961 THE MUSIC OF COSMOS *TV Soundtrack*
3749 MENACE TO SOCIETY *Lizzy Borden*	4236 MR. BAD GUY *Freddie Mercury*	2060 MUSIQUE/THE HIGH ROAD *Roxy Music*
1157 MEN AND WOMEN *Simply Red*	1675 MR. BIG *Mr. Big*	3078 MUSTA NOTTA GOTTA LOTTA *Joe Ely*
4354 MENLOVE AVENUE *John Lennon*	4855 MR. C. *Norman Connors*	4399 MUTUAL ATTRACTION *Sylvester*
4642 METAL PRIESTESS *The Plasmatics*	2555 MR. JORDAN *Julian Lennon*	4651 M.V.P. *Harvey Mason*
2535 MENUDO *Menudo*	1890 MR. LOOK SO GOOD! *Richard "Dimples" Fields*	4815 MY BABE *Roy Buchanan*
2596 MEN WITHOUT WOMEN *Little Steven And The Disciples Of Soul*	3156 MR. MUSIC HEAD *Adrian Belew*	3633 MY BEST *Kitaro*
2626 MERCILESS *Stephanie Mills*	3045 MR. NICE GUY *Ronnie Laws*	1764 MY EVER CHANGING MOODS *Style Council*
4700 MERRY CHRISTMAS/HAPPY NEW YEAR'S *Montana Orchestra*	1566 MISTRIAL *Lou Reed*	2362 MY FAVORITE PERSON *The O'Jays*
615 MERRY, MERRY CHRISTMAS *New Kids On The Block*	1249 MOB RULES *Black Sabbath*	4215 MY GIFT TO YOU *Alexander O'Neal*
1607 MESOPOTAMIA *The B-52s*	1684 MODERN HEART *Champaign*	1680 MY HOME'S IN ALABAMA *Alabama*
1535 THE MESSAGE *Grandmaster Flash & The Furious Five*	2611 MODERN ROMANS *The Call*	3326 MY LIFE FOR A SONG *Placido Domingo*
930 MESSAGES FROM THE BOYS *The Boys*	533 MODERN TIMES *Jefferson Starship*	
	189 A MOMENTARY LAPSE OF REASON *Pink Floyd*	
	4730 MOMENTS *Julio Iglesias*	
	2117 THE MONA LISA'S SISTER *Graham Parker*	
	3300 MONDO BONGO *Boomtown Rats*	
	4619 MONDO MANDO *David Grisman*	

Albums Alpha Index

#	Title	Artist
1882	MY LIFE IN THE BUSH OF GHOSTS	Brian Eno - David Byrne
1208	MY MELODY	Deniece Williams
4342	MY MOVIE OF YOU	Leif Garrett
3445	MY NATION UNDERGROUND	Julian Cope
4605	MY OWN WAY	Willie Nelson
4040	MY ROAD OUR ROAD	Lee Oskar
4581	MY SPECIAL LOVE	LaToya Jackson
396	MYSTERY GIRL	Roy Orbison
3858	MYSTERY OF BULGARIAN VOICES	Bulgarian State Radio & T.V. Female Choir
4241	THE MYSTERY OF EDWIN DROOD	Original Broadway Cast Recording
3377	MYSTERY STREET	John Brannen
4468	MYSTERY WALK	M + M
1918	MYSTICAL ADVENTURE	Jean-Luc Ponty
4479	MY TOOT-TOOT	Rockin' Sidney
4624	MY TOOT TOOT	Jean Knight

N

#	Title	Artist
4582	NAIVE ART	Red Flag
1153	NAJEE'S THEME	Najee
923	NAKED	Talking Heads
925	NAKED EYES	Naked Eyes
2447	NAKED TO THE WORLD	Teena Marie
1756	THE NAME OF THIS BAND IS TALKING HEADS	Talking Heads
2497	'NARD	Bernard Wright
2518	NASTY, NASTY	Black 'N Blue
4307	NATIONAL BREAKOUT	The Romantics
4686	NATIONAL EMOTION	Tommy Tutone
2973	NATURALLY	Leon Haywood
3088	NATURAL STATES	David Lanz & Paul Speer
577	THE NATURE OF THE BEAST	April Wine
1532	NAUGHTY	Chaka Khan
2967	NEARLY HUMAN	Todd Rundgren
474	NEBRASKA	Bruce Springsteen
2694	NEGOTIATIONS AND LOVE SONGS (1971-1986)	Paul Simon
4119	NEMESIS	Axe
2155	NERUDA	Red Rider
267	NERVOUS NIGHT	Hooters
1823	THE NET	Little River Band
3292	NEVER BUY TEXAS FROM A COWBOY	Brides Of Funkenstein
767	NEVER DIE YOUNG	James Taylor
2101	NEVER ENOUGH	Patty Smyth
3606	NEVER FELT SO GOOD	James Ingram
3697	NEVER GONNA BE ANOTHER ONE	Thelma Houston
3526	NEVER KICK A SLEEPING DOG	Mitch Ryder
1128	NEVER LET ME DOWN	David Bowie
1887	NEVER RUN NEVER HIDE	Benny Mardones
3356	NEVER SAY NEVER	Melba Moore
4017	NEVER SAY NEVER	Romeo Void
874	NEVER SURRENDER	Triumph
502	NEVER TOO MUCH	Luther Vandross
4408	NEW AFFAIR	The Emotions
1561	NEW CLEAR DAYS	The Vapors
3735	NEW DIRECTIONS	Tavares
271	NEW EDITION	New Edition
521	A NEW FLAME	Simply Red
2151	NEW GOLD DREAM (81-82-83-84)	Simple Minds
3345	NEW HOPE FOR THE WRETCHED	The Plasmatics
56	NEW JERSEY	Bon Jovi
4659	NEW JERSEY	Joe Piscopo
331	NEW KIDS ON THE BLOCK	New Kids On The Block
2727	NEW LIGHT THROUGH OLD WINDOWS	Chris Rea
4797	NEW LOOKS	B.J. Thomas
2934	THE NEW ORDER	Testament
3286	N.E.W.S.	Golden Earring
1283	NEW SENSATIONS	Lou Reed
1039	NEW TRADITIONALISTS	Devo
1228	NEW YORK	Lou Reed
4668	NEW YORK CAKE	Kano
3226	NEW YORK-LONDON-PARIS-MUNICH	M
2454	THE NEW ZAPP IV U	Zapp
2575	NEXT POSITION PLEASE	Cheap Trick
2234	A NICE PLACE TO BE	George Howard
3151	NICK LOWE AND HIS COWBOY OUTFIT	Nick Lowe And His Cowboy Outfit
4499	NICK MASON'S FICTITIOUS SPORTS	Nick Mason
2082	NICK THE KNIFE	Nick Lowe
1070	NIECY	Deniece Williams
229	NIGHT AND DAY	Joe Jackson
4446	NIGHT ATTACK	Angel City
1189	NIGHTCLUBBING	Grace Jones
4672	NIGHT CRUISER	Eumir Deodato
1260	NIGHTCRUISING	Bar-Kays
2868	NIGHT FADES AWAY	Nils Lofgren
4214	NIGHT FLIGHT	Justin Hayward
699	THE NIGHTFLY	Donald Fagen
4374	NIGHTHAWKS	The Nighthawks
4613	NIGHTHAWKS	Keith Emerson
494	THE NIGHT I FELL IN LOVE	Luther Vandross
677	NIGHT IN THE RUTS	Aerosmith
4381	NIGHTLINE	Randy Crawford
2043	NIGHT PASSAGE	Weather Report
4352	NIGHTS (FEEL LIKE GETTING DOWN)	Billy Ocean
591	NIGHTSHIFT	Commodores
114	NIGHT SONGS	Cinderella
3000	NIGHTSONGS	Earl Klugh
4527	THE NIGHT THE LIGHTS WENT OUT IN GEORGIA	Soundtrack
1401	A NIGHT TO REMEMBER	Cyndi Lauper
518	NIGHTWALKER	Gino Vannelli
2424	9	Public Image Limited
2371	9 1/2 WEEKS	Soundtrack
3228	NINE LIVES	Bonnie Raitt
233	90125	Yes
2879	9012LIVE - THE SOLOS	Yes
2220	9.9	9.9
662	CHICAGO 19	Chicago
2489	1980	Gil Scott-Heron & Brian Jackson
4347	1980	B.T. Express
1581	1988 SUMMER OLYMPICS-ONE MOMENT IN TIME	Various Artists
71	1984 (MCMLXXXIV)	Van Halen
2819	1984 (FOR THE LOVE OF BIG BROTHER) (SOUNDTRACK)	Eurythmics
77	1999	Prince
4505	1969	Soundtrack
2372	9 TO 5	Soundtrack
411	9 TO 5 AND ODD JOBS	Dolly Parton
200	NINE TONIGHT	Bob Seger & The Silver Bullet Band
3677	NINE TO THE UNIVERSE	Jimi Hendrix
1456	99 LUFTBALLONS	Nena
1935	NO BALLADS	Rockets
4867	NO BLUE THING	Ray Lynch
3238	NOBODY'S PERFECT	Deep Purple
429	NO BRAKES	John Waite
611	NO CONTROL	Eddie Money
4454	NO EASY WALK TO FREEDOM	Peter, Paul & Mary
3912	NO EASY WAY OUT	Robert Tepper
3307	NOEL	Noel
2926	NO EXIT	Fates Warning
4469	NO FREE LUNCH	Green On Red
2314	NO FRILLS	Bette Midler
3199	NO FUEL LEFT FOR THE PILGRIMS	Disneyland After Dark
733	NO FUN ALOUD	Glenn Frey
2598	NO GURU, NO METHOD, NO TEACHER	Van Morrison
2133	NO GUTS...NO GLORY	Molly Hatchet
946	NO HOLDIN' BACK	Randy Travis
18	NO JACKET REQUIRED	Phil Collins
1981	NO LOOKIN' BACK	Michael McDonald
4687	NO MAN'S LAND	Lene Lovich
1824	NO MORE DIRTY DEALS	Johnny Van Zant Band
2910	NO MUSS...NO FUSS	Donnie Iris
2423	NONA	Nona Hendryx
3384	NON FICTION	The Blasters
888	NO NIGHT SO LONG	Dionne Warwick
1952	NON STOP	Julio Iglesias
2345	NON-STOP ECSTATIC DANCING	Soft Cell
506	NON-STOP EROTIC CABARET	Soft Cell
939	NO NUKES/THE MUSE CONCERTS FOR A NON-NUCLEAR FUTURE	Various Artists
802	NO ONE CAN DO IT BETTER	The D.O.C.
276	NO PARKING ON THE DANCE FLOOR	Midnight Star
2095	NO PARLEZ	Paul Young
3143	NO PAROLE FROM ROCK 'N' ROLL	Alcatrazz
3865	NO PLACE FOR DISGRACE	Flotsam And Jetsam
2103	NO PLACE TO RUN	UFO
862	NO PROTECTION	Starship
4555	NO REFUGE	Eddie Schwartz
1475	NO RESPECT	Rodney Dangerfield
3910	NO RESPECT	Vain
693	NO REST FOR THE WICKED	Ozzy Osbourne
4681	NO REST FOR THE WICKED	Helix
2142	NORTH COAST	Michael Stanley Band
4610	NORTH OF A MIRACLE	Nick Heyward
2249	NO STRANGER TO LOVE	Roy Ayers
3023	NO TELLIN' LIES	Zebra
2632	NO. 10, UPPING STREET	Big Audio Dynamite
4185	NOT FAKIN' IT	Michael Monroe
4777	NOT GUILTY...	Larry Gatlin & The Gatlin Brothers Band
2200	NOTHIN' BUT TROUBLE	Nia Peeples
4225	NOTHING BUT THE TRUTH	Ruben Blades
4781	NOTHING IN COMMON	Soundtrack
290	...NOTHING LIKE THE SUN	Sting
1721	NOTHING'S SHOCKING	Jane's Addiction
3875	NOTHING TO FEAR	Oingo Boingo
1398	NOTHING TO LOSE	Eddie Money
607	NOTHIN' MATTERS AND WHAT IF IT DID	John Cougar
4747	NOT JUST THE GIRL NEXT DOOR	Nancy Martinez
616	NOTORIOUS	Duran Duran
4334	NOT THE BOY NEXT DOOR	Peter Allen
4102	NOVO COMBO	Novo Combo
1520	NOW	Patrice Rushen
861	NOW AND FOREVER	Air Supply
210	NOW AND ZEN	Robert Plant
4344	"NOW APPEARING" AT OLE' MISS	B.B. King
1028	NOW OR NEVER	John Schneider
3645	NOW PLAYING	Bernadette Peters
3095	NOW VOYAGER	Barry Gibb
4238	NOW WE MAY BEGIN	Randy Crawford
1045	NUCLEAR FURNITURE	Jefferson Starship
3364	NUDE	Dead Or Alive
2162	NUGENT	Ted Nugent
816	THE NUMBER OF THE BEAST	Iron Maiden
3813	THE NUMBER ONE HITS	Elvis Presley
3508	#7	George Strait

Albums Alpha Index

4734 **NUNSEXMONKROCK** *Nina Hagen*
3605 **NURDS** *The Roches*
301 **THE NYLON CURTAIN** *Billy Joel*

O

4030 **OASIS** *Roberta Flack*
1938 **OBJECTS OF DESIRE** *Michael Franks*
2646 **OBLIVION** *Utopia*
2607 **OBSESSION** *Bob James*
2288 **OCEAN FRONT PROPERTY** *George Strait*
4370 **OCEANLINER** *Passport*
2968 **OCEAN RAIN** *Echo & The Bunnymen*
1747 **OCTOBER** *U2*
4105 **OCTOPUSSY** *Soundtrack*
2153 **ODORI** *Hiroshima*
1584 **ODYSSEY** *Yngwie J. Malmsteen's Rising Force*
2308 **OFFERING** *Axe*
1487 **AN OFFICER AND A GENTLEMAN** *Soundtrack*
3018 **THE OFFICIAL MUSIC OF THE XXIIIRD OLYMPIAD-LOS ANGELES 1984** *Various Artists*
1643 **OFFRAMP** *Pat Metheny Group*
22 **OFF THE WALL** *Michael Jackson*
2311 **OFF TO SEE THE LIZARD** *Jimmy Buffett*
3604 **O.F.R.** *Nitro*
2996 **OH, JULIE!** *Barry Manilow*
1394 **OH MERCY** *Bob Dylan*
1797 **OH NO! IT'S DEVO** *Devo*
3092 **OH YES I CAN** *David Crosby*
4209 **OINGO BOINGO** *Oingo Boingo*
2699 **OLD CREST ON A NEW WAVE** *Dave Mason*
926 **OLD 8 X 10** *Randy Travis*
3620 **OLD ENOUGH** *Lou Ann Barton*
4880 **AN OLD FASHIONED CHRISTMAS** *Carpenters*
3181 **AN OLD TIME CHRISTMAS** *Randy Travis*
2633 **OLD WAYS** *Neil Young*
4106 **OLIVER & COMPANY** *Soundtrack*
272 **OLIVIA'S GREATEST HITS, VOL. 2** *Olivia Newton-John*
2765 **ON** *Off Broadway USA*
4699 **ON AND ON** *Fat Boys*
4523 **ON A NIGHT LIKE THIS** *Buckwheat Zydeco*
2801 **ON A ROLL** *Point Blank*
445 **ONCE BITTEN** *Great White*
2248 **ONCE UPON A CHRISTMAS** *Kenny Rogers & Dolly Parton*
335 **ONCE UPON A TIME** *Simple Minds*
2389 **ONE** *Bee Gees*
2204 **ONE BAD HABIT** *Michael Franks*
4341 **ONE BIG DAY** *Face To Face*
1250 **ONE BRIGHT DAY** *Ziggy Marley And The Melody Makers*
3383 **ONE DAY IN YOUR LIFE** *Michael Jackson*
634 **ONE EIGHTY** *Ambrosia*
434 **ONE FOR THE ROAD** *The Kinks*
3762 **THE ONE GIVETH, THE COUNT TAKETH AWAY** *William "Bootsy" Collins*
1537 **ONE GOOD REASON** *Paul Carrack*
501 **ONE HEARTBEAT** *Smokey Robinson*
838 **ONE IN A MILLION YOU** *Larry Graham*
2615 **ONE LORD, ONE FAITH, ONE BAPTISM** *Aretha Franklin*
2430 **ONE LOVE--ONE DREAM** *Jeffrey Osborne*
1066 **ONE MORE SONG** *Randy Meisner*
1956 **ONE MORE STORY** *Peter Cetera*
4715 **ONE MORE TRY FOR LOVE** *Ronnie Milsap*
1729 **ONE NIGHT WITH A STRANGER** *Martin Briley*
983 **ONE ON ONE** *Cheap Trick*
1806 **101** *Depeche Mode*

1818 **ONE PARTICULAR HARBOUR** *Jimmy Buffett*
3074 **ONE SECOND** *Yello*
3367 **ONE STEP BEYOND** *Madness*
313 **ONE STEP CLOSER** *Doobie Brothers*
2574 **ONE STEP CLOSER** *Gavin Christopher*
195 **THE ONE THAT YOU LOVE** *Air Supply*
1674 **ONE TO ONE** *Howard Jones*
3293 **ONE TO ONE** *Carole King*
549 **ONE-TRICK PONY** *Paul Simon*
4161 **1234** *Ronnie Wood*
1869 **ONE VICE AT A TIME** *Krokus*
659 **ONE VOICE** *Barbra Streisand*
2827 **ONE WAY FEATURING AL HUDSON(2)** *One Way Featuring Al Hudson*
1229 **ONE WAY HOME** *Hooters*
3538 **ON GOLDEN POND** *Soundtrack*
4217 **ONLY A LAD** *Oingo Boingo*
666 **ONLY FOUR YOU** *Mary Jane Girls*
4457 **ONLY LIFE** *The Feelies*
3564 **ON SOLID GROUND** *Larry Carlton*
3614 **ON TARGET** *Fastway*
2267 **ON THE EDGE** *The Babys*
1800 **ON THE LINE** *Gary U.S. Bonds*
3742 **ON THE NILE** *Egyptian Lover*
2217 **ON THE ONE** *Dazz Band*
119 **ON THE RADIO: GREATEST HITS: VOLUMES I & II** *Donna Summer*
1273 **ON THE RISE** *S.O.S. Band*
4753 **ON THE STRENGTH** *Grandmaster Flash & The Furious Five*
695 **ON THE WAY TO THE SKY** *Neil Diamond*
1048 **ON THROUGH THE NIGHT** *Def Leppard*
2194 **ON TO VICTORY** *Humble Pie*
951 **OOH YEAH!** *Daryl Hall & John Oates*
2531 **OPEN ALL NIGHT** *Georgia Satellites*
4538 **OPEN ALL NIGHT** *Leroi Bros.*
2067 **OPEN FIRE** *Y&T*
3822 **OPEN MIND** *Jean-Luc Ponty*
1306 **OPEN SESAME** *Whodini*
62 **OPEN UP AND SAY...AHH!** *Poison*
1338 **OPERA SAUVAGE** *Vangelis*
974 **OPERATION: MINDCRIME** *Queensryche*
3480 **OPUS X** *Chilliwack*
1488 **ORANGES AND LEMONS** *XTC*
3798 **ORGASMATRON** *Motorhead*
3011 **ORIGINAL STYLIN'** *Three Times Dope*
2225 **ORION THE HUNTER** *Orion The Hunter*
2091 **O SOLE MIO - FAVORITE NEAPOLITAN SONGS** *Luciano Pavarotti*
4027 **THE OTHER ONES** *Other Ones*
1657 **OTHER ROADS** *Boz Scaggs*
384 **THE OTHER SIDE OF LIFE** *Moody Blues*
787 **THE OTHER SIDE OF THE MIRROR** *Stevie Nicks*
2943 **THE OTHER SIDE OF THE RAINBOW** *Melba Moore*
658 **THE OTHER WOMAN** *Ray Parker Jr.*
109 **OU812** *Van Halen*
2748 **OUR BELOVED REVOLUTIONARY SWEETHEART** *Camper van Beethoven*
4825 **OUTA HAND** *Coney Hatch*
1577 **OUTLAW** *War*
1417 **OUT OF AFRICA** *Soundtrack*
3067 **OUT OF CONTROL** *Brothers Johnson*
2280 **OUT OF MIND OUT OF SIGHT** *Models*
168 **OUT OF ORDER** *Rod Stewart*
122 **OUT OF THE BLUE** *Debbie Gibson*
246 **OUT OF THE CELLAR** *Ratt*

3246 **OUT OF THE SHADOWS** *Dave Grusin*
3415 **OUT OF THE SILENT PLANET** *King's X*
886 **OUT OF THIS WORLD** *Europe*
2460 **OUTRAGEOUS** *Lakeside*
1110 **OUTRIDER** *Jimmy Page*
661 **OUTSIDE INSIDE** *The Tubes*
2352 **OUTSIDE LOOKING IN** *BoDeans*
1545 **OUT WHERE THE BRIGHT LIGHTS ARE GLOWING** *Ronnie Milsap*
1691 **OVER THE EDGE** *Hurricane*
3999 **OVER THE LINE** *Greg Guidry*
3462 **OVER THERE (LIVE AT THE VENUE, LONDON)** *The Blasters*
3546 **OVER THE TOP** *Soundtrack*
3718 **OXO** *Oxo*
4440 **OZARK MOUNTAIN DAREDEVILS** *Ozark Mountain Daredevils*

P

1629 **THE PACIFIC AGE** *Orchestral Manoeuvres In The Dark*
3458 **THE PACK IS BACK** *Raven*
1073 **PACK UP THE PLANTATION - LIVE!** *Tom Petty And The Heartbreakers*
1331 **PAC-MAN FEVER** *Buckner & Garcia*
1408 **PAID IN FULL** *Eric B. & Rakim*
406 **PANORAMA** *The Cars*
1991 **PARADE** *Spandau Ballet*
380 **PARADE: MUSIC FROM THE MOTION PICTURE UNDER THE CHERRY MOON (SOUNDTRACK)** *Prince And The Revolution*
2356 **PARADISE** *Peabo Bryson*
44 **PARADISE THEATER** *Styx*
546 **PARIS** *Supertramp*
3298 **PARTING SHOULD BE PAINLESS** *Roger Daltrey*
719 **PARTNERS IN CRIME** *Rupert Holmes*
4277 **PARTY** *Iggy Pop*
2181 **PARTY MIX!** *The B-52s*
4023 **PARTY OF ONE** *Tim Weisberg*
4243 **PARTY PARTY** *Soundtrack*
3540 **THE PARTY'S OVER** *Talk Talk*
2465 **PARTY 'TIL YOU'RE BROKE** *Rufus*
2112 **PARTY YOUR BODY** *Stevie B*
3942 **THE PASSENGER** *Melvin James*
2233 **PASSION** *Robin Trower*
2353 **PASSION: MUSIC FOR THE LAST TEMPTATION OF CHRIST (SOUNDTRACK)** *Peter Gabriel*
4001 **PASSION CRIMES** *Darling Cruel*
3533 **PASSIONFRUIT** *Michael Franks*
4394 **PASSION, GRACE & FIRE** *Al Di Meola/John McLaughlin/Paco De Lucia*
1589 **PASSIONWORKS** *Heart*
4137 **PAST MASTERS - VOLUME 1** *The Beatles*
3676 **PAST MASTERS - VOLUME 2** *The Beatles*
4881 **PAT McLAUGHLIN** *Pat McLaughlin*
4665 **PATTERN DISRUPTIVE** *Dickey Betts Band*
1945 **PATTI** *Patti LaBelle*
2437 **PATTI AUSTIN** *Patti Austin*
2688 **PAT TRAVERS' BLACK PEARL** *Pat Travers*
4455 **PAUL DAVIS** *Paul Davis*
1060 **PAUL'S BOUTIQUE** *Beastie Boys*
2219 **PAVAROTTI'S GREATEST HITS** *Luciano Pavarotti*
3971 **THE PEABO BRYSON COLLECTION** *Peabo Bryson*
3431 **PEACE & LOVE** *The Pogues*
4233 **PEACE IN OUR TIME** *Big Country*
1336 **PEACE SELLS...BUT WHO'S BUYING?** *Megadeth*
2914 **PEARL HARBOR & THE EXPLOSIONS** *Pearl Harbor And The Explosions*

Albums Alpha Index

1582 **PEARLS-SONGS OF GOFFIN AND KING** *Carole King*	2491 **PLAYING TO WIN** *LRB*	226 **PRIDE** *White Lion*
407 **PEBBLES** *Pebbles*	3889 **PLAYING TO WIN** *Rick Nelson*	2411 **PRIDE** *Robert Palmer*
1908 **PEEPSHOW** *Siouxsie & The Banshees*	4316 **PLAYIN' IT COOL** *Timothy B. Schmit*	1808 **PRIEST...LIVE** *Judas Priest*
905 **PELICAN WEST** *Haircut One Hundred*	2268 **PLEASANT DREAMS** *The Ramones*	4774 **PRIME TIME** *Grey And Hanks*
2109 **PENETRATOR** *Ted Nugent*	389 **PLEASE** *Pet Shop Boys*	1426 **PRIMITIVE** *Neil Diamond*
4819 **PENNIES FROM HEAVEN** *Soundtrack*	2807 **PLEASED TO MEET ME** *The Replacements*	1595 **PRIMITIVE COOL** *Mick Jagger*
1894 **PEOPLE** *Hothouse Flowers*	4683 **PLEASURE ONE** *Heaven 17*	270 **PRIMITIVE LOVE** *Miami Sound Machine*
1863 **PEOPLE ARE PEOPLE** *Depeche Mode*	548 **THE PLEASURE PRINCIPLE** *Gary Numan*	3882 **PRIMITIVE MAN** *Icehouse*
4403 **THE PEOPLE WHO GRINNED THEMSELVES TO DEATH** *The Housemartins*	2223 **PLEASURES OF THE FLESH** *Exodus*	657 **PRINCE** *Prince*
2210 **PERFECT** *Soundtrack*	981 **PLEASURE VICTIM** *Berlin*	2254 **PRINCE CHARMING** *Adam And The Ants*
3680 **PERFECT COMBINATION** *Stacy Lattisaw & Johnny Gill*	4234 **THE PLIMSOULS** *The Plimsouls*	4875 **THE PRINCESS BRIDE (SOUNDTRACK)** *Mark Knopfler*
3950 **PERFECT FIT** *Jerry Knight*	1069 **THE POET** *Bobby Womack*	4793 **THE PRINCE'S TRUST 10TH ANNIVERSARY BIRTHDAY PARTY** *Various Artists*
608 **PERFECT STRANGERS** *Deep Purple*	2344 **POETIC CHAMPIONS COMPOSE** *Van Morrison*	379 **THE PRINCIPLE OF MOMENTS** *Robert Plant*
3687 **PERFECT SYMMETRY** *Fates Warning*	2281 **THE POET II** *Bobby Womack*	1776 **PRIVATE AUDITION** *Heart*
2147 **PERFECT TIMING** *McAuley Schenker Group*	4671 **POINTER SISTERS' GREATEST HITS** *Pointer Sisters*	3993 **PRIVATE COLLECTION** *Jon And Vangelis*
3262 **PERFECT TIMING** *Donna Allen*	996 **POINT OF ENTRY** *Judas Priest*	33 **PRIVATE DANCER** *Tina Turner*
3888 **PERFECT VIEW** *The Graces*	3679 **POINT OF PLEASURE** *Xavier*	135 **PRIVATE EYES** *Daryl Hall & John Oates*
3755 **PERFORMANCE** *Ashford & Simpson*	3847 **POINT OF VIEW** *Spyro Gyra*	458 **A PRIVATE HEAVEN** *Sheena Easton*
896 **PERHAPS LOVE** *Placido Domingo*	881 **POINTS ON THE CURVE** *Wang Chung*	1860 **PRIVATE PASSION** *Jeff Lorber*
151 **PERMANENT VACATION** *Aerosmith*	3175 **POLICY** *Martha Davis*	1247 **PRIVATE REVOLUTION** *World Party*
388 **PERMANENT WAVES** *Rush*	1705 **THE POLITICS OF DANCING** *Re-Flex*	3172 **PROCESSION** *Weather Report*
3788 **PERSONAL ATTENTION** *Stacy Lattisaw*	4614 **POLKA PARTY!** *"Weird Al" Yankovic*	4626 **THE PRODUCERS** *The Producers*
4729 **PERSPECTIVE** *America*	4395 **POLTERGEIST** *Soundtrack*	4365 **PROFILES** *Nick Mason & Rick Fenn*
3392 **PETE TOWNSHEND'S DEEP END LIVE!** *Pete Townshend*	817 **PONCHO & LEFTY** *Merle Haggard/Willie Nelson*	4467 **PROFILE II: THE BEST OF EMMYLOU HARRIS** *Emmylou Harris*
3661 **PETER CETERA** *Peter Cetera*	2876 **PONTIAC** *Lyle Lovett*	1213 **PROGRESSIONS OF POWER** *Triumph*
568 **PETER GABRIEL (III)** *Peter Gabriel*	3054 **POOL IT!** *The Monkees*	154 **PROMISE** *Sade*
772 **PETER GABRIEL (SECURITY)** *Peter Gabriel*	907 **POOLSIDE** *Nu Shooz*	4617 **PROOF THROUGH THE NIGHT** *T-Bone Burnett*
1888 **PETER GABRIEL/PLAYS LIVE** *Peter Gabriel*	3456 **POP ART** *Transvision Vamp*	1569 **PROPOSITIONS** *Bar-Kays*
273 **THE PHANTOM OF THE OPERA** *Original London Cast Recording*	3073 **POPEYE** *Soundtrack*	1390 **THE PROS & CONS OF HITCHHIKING** *Roger Waters*
1812 **PHANTOM, ROCKER & SLICK** *Phantom, Rocker & Slick*	2745 **POP GOES THE MOVIES** *Meco*	3128 **PROTECT THE INNOCENT** *Rachel Sweet*
829 **PHANTOMS** *The Fixx*	1789 **POP GOES THE WORLD** *Men Without Hats*	3422 **PROVISION** *Scritti Politti*
2257 **PHIL SEYMOUR** *Phil Seymour*	3689 **POPPED IN SOULED OUT** *Wet Wet Wet*	3743 **THE PSYCHEDELIC FURS** *Psychedelic Furs*
176 **PHOENIX** *Dan Fogelberg*	4308 **POPS IN SPACE** *Boston Pops Orchestra/John Williams*	1367 **PSYCHO CAFE** *Bang Tango*
4758 **PHOTOGLO** *Photoglo*	3759 **PORCUPINE** *Echo & The Bunnymen*	4692 **PSYCHOCANDY** *Jesus And Mary Chain*
169 **PHYSICAL** *Olivia Newton-John*	3615 **PORKY'S REVENGE** *Soundtrack*	2495 **PUCKER UP** *Lipps Inc.*
2704 **PIA & PHIL** *Pia Zadora*	4667 **PORTRAIT OF CARRIE** *Carrie Lucas*	75 **PUMP** *Aerosmith*
477 **PICTURE BOOK** *Simply Red*	1980 **POSH** *Patrice Rushen*	933 **PUNCH THE CLOCK** *Elvis Costello And The Attractions*
3819 **PICTURES** *Atlanta*	4151 **POSITIVE** *Peabo Bryson*	3391 **PURE & NATURAL** *T-Connection*
412 **PICTURES AT ELEVEN** *Robert Plant*	3692 **POSITIVE POWER** *Steve Arrington's Hall Of Fame*	20 **PURPLE RAIN (SOUNDTRACK)** *Prince And The Revolution*
706 **PICTURES FOR PLEASURE** *Charlie Sexton*	4283 **POV** *Utopia*	3753 **THE PURSUIT OF HAPPINESS** *Beat Farmers*
3444 **PICTURES FROM THE FRONT** *Jon Butcher*	1134 **POWER** *Ice-T*	4406 **PUSH** *Bros*
435 **PICTURE THIS** *Huey Lewis & The News*	1218 **POWER** *Kansas*	41 **PYROMANIA** *Def Leppard*
519 **PIECE OF MIND** *Iron Maiden*	1716 **POWER** *The Temptations*	
4149 **PIECES OF A DREAM** *Pieces Of A Dream*	4091 **POWER & PASSION** *Mama's Boys*	# Q
4522 **PIED PIPER** *Dave Valentin*	3806 **POWER & THE GLORY** *Saxon*	
2858 **PILEDRIVER -- THE WRESTLING ALBUM II** *Various Artists*	3591 **POWERFUL STUFF** *Fabulous Thunderbirds*	4566 **QE 2** *Mike Oldfield*
3070 **PINK WORLD** *Planet P Project*	811 **POWERLIGHT** *Earth, Wind & Fire*	1044 **QR III** *Quiet Riot*
732 **PIPES OF PEACE** *Paul McCartney*	4660 **POWER OF LOVE** *Arlo Guthrie*	281 **QUARTERFLASH** *Quarterflash*
4321 **THE PIRATE MOVIE** *Soundtrack*	1480 **POWER PLAY** *April Wine*	2094 **QUARTET** *Ultravox*
340 **PIRATES** *Rickie Lee Jones*	943 **POWERSLAVE** *Iron Maiden*	3611 **QUEEN ELVIS** *Robyn Hitchcock And The Egyptians*
4599 **THE PIRATES OF PENZANCE** *Original Cast*	278 **THE POWER STATION** *Power Station*	1690 **THE QUEEN IS DEAD** *The Smiths*
1374 **PIZZAZZ** *Patrice Rushen*	599 **POWER WINDOWS** *Rush*	2089 **QUEENSRYCHE** *Queensryche*
3355 **A PLACE FOR MY STUFF!** *George Carlin*	2619 **PRACTICE WHAT YOU PREACH** *Testament*	4129 **QUEST FOR FIRE** *Soundtrack*
2584 **A PLACE LIKE THIS** *Robbie Nevil*	1552 **PRECIOUS MOMENTS** *Jermaine Jackson*	3116 **QUESTIONNAIRE** *Chaz Jankel*
4192 **PLAIN FROM THE HEART** *Delbert McClinton*	125 **PRECIOUS TIME** *Pat Benatar*	4174 **QUICKSILVER** *Soundtrack*
1015 **PLANETARY INVASION** *Midnight Star*	1745 **PREMONITION** *Survivor*	645 **QUIET LIES** *Juice Newton*
1647 **PLANET P** *Planet P*	2590 **PREMONITION** *Peter Frampton*	3189 **QUIET RIOT** *Quiet Riot*
4611 **PLANTATION HARBOR** *Joe Vitale*	4463 **PREPPIE** *Cheryl Lynn*	2334 **QUINELLA** *Atlanta Rhythm Section*
2660 **PLATOON** *Soundtrack*	1156 **THE PRESENT** *Moody Blues*	
259 **PLAY DEEP** *The Outfield*	1528 **PRESS TO PLAY** *Paul McCartney*	# R
3512 **PLAYERS IN THE DARK** *Dr. Hook*	3051 **PRESSURE** *Bram Tchaikovsky*	
1236 **PLAYING FOR KEEPS** *Eddie Money*	1911 **THE PRESSURE IS ON** *Hank Williams Jr.*	2883 **R&B SKELETONS IN THE CLOSET** *George Clinton*
	832 **PRESTO** *Rush*	
	260 **PRETENDERS** *The Pretenders*	
	714 **PRETENDERS II** *The Pretenders*	
	420 **PRETTY IN PINK** *Soundtrack*	
	2675 **PRETTY PAPER** *Willie Nelson*	

Albums Alpha Index

#	Title	Artist
2729	RACING AFTER MIDNIGHT	Honeymoon Suite
1168	RADIANT	Atlantic Starr
4002	RADICAL DEPARTURE	Ranking Roger
1113	RADIO	LL Cool J
1575	RADIO ACTIVE	Pat Travers
1668	RADIO K.A.O.S.	Roger Waters
2315	RADIOLAND	Nicolette Larson
2779	RADIO ONE	Jimi Hendrix Experience
1042	RADIO ROMANCE	Eddie Rabbitt
4868	RADIO SILENCE	Boris Grebenshikov
1653	RAGE FOR ORDER	Queensryche
3908	RAGE IN EDEN	Ultravox
3136	RAGING SLAB	Raging Slab
3539	RAGTIME	Soundtrack
2298	RAIDERS OF THE LOST ARK (SOUNDTRACK)	London Symphony Orchestra/John Williams
3725	RAIL	Rail
3821	RAINBOW	Dolly Parton
2552	RAINDANCING	Alison Moyet
4418	RAIN DOGS	Tom Waits
1811	RAIN FOREST	Paul Hardcastle
2363	THE RAINMAKERS	The Rainmakers
1435	RAIN MAN	Soundtrack
309	RAISE!	Earth, Wind & Fire
220	RAISED ON RADIO	Journey
2591	RAISE YOUR FIST AND YELL	Alice Cooper
3122	RAISING FEAR	Armored Saint
101	RAISING HELL	Run-D.M.C.
1295	RAM IT DOWN	Judas Priest
4380	RAMONES MANIA	The Ramones
3056	RANDY MEISNER	Randy Meisner
3205	RANK	The Smiths
793	RANT 'N' RAVE WITH THE STRAY CATS	Stray Cats
1405	RAPPIN' RODNEY	Rodney Dangerfield
2812	RAP'S GREATEST HITS	Various Artists
4470	RAP'S GREATEST HITS, VOLUME 2	Various Artists
60	RAPTURE	Anita Baker
1080	RARITIES	The Beatles
4790	RATHER BE ROCKIN'	Tantrum
2105	RAT IN THE KITCHEN	UB40
2815	RATT	Ratt
153	RATTLE AND HUM (SOUNDTRACK)	U2
4306	RAVEL BOLERO	Tomita
47	THE RAW & THE COOKED	Fine Young Cannibals
1086	RAW LIKE SUSHI	Neneh Cherry
672	RAY, GOODMAN & BROWN	Ray, Goodman & Brown
2540	RAY, GOODMAN & BROWN II	Ray, Goodman & Brown
1209	REACH	Richard Simmons
697	REACH FOR THE SKY	Ratt
1231	REACH FOR THE SKY	Allman Brothers Band
2186	REACHING FOR TOMORROW	Switch
3237	REACHING OUT	Menudo
232	REACH THE BEACH	The Fixx
2290	REACH UP AND TOUCH THE SKY	Southside Johnny & The Asbury Jukes
3686	REACT	The Fixx
1090	RE-AC-TOR	Neil Young & Crazy Horse
3530	READ MY LIPS	Melba Moore
4070	READ MY LIPS	Fee Waybill
2421	READY AN' WILLING	Whitesnake
393	READY FOR THE WORLD	Ready For The World
1078	READY OR NOT	Lou Gramm
2370	THE REAL CHUCKEEBOO	Loose Ends
2724	THE REAL DEAL	Isley Brothers
4009	REAL EYES	Gil Scott-Heron
4284	REAL LIFE STORY	Terri Lyne Carrington
3366	REAL LIVE	Bob Dylan
2390	REAL LOVE	Ashford & Simpson
2405	THE REAL MACAW	Graham Parker
1404	REAL PEOPLE	Chic
2975	THE REAL THING	Angela Winbush
3321	REBA	Reba McEntire
3916	REBEL MUSIC	Bob Marley And The Wailers
185	REBEL YELL	Billy Idol
39	RECKLESS	Bryan Adams
765	RECKONING	R.E.M.
1511	RECKONING	Grateful Dead
1873	RECONCILED	The Call
686	RECORDS	Foreigner
2778	RECOVERY: LIVE!	Great White
3043	RED	The Communards
2576	RED HOT RHYTHM & BLUES	Diana Ross
4471	RED SAILS IN THE SUNSET	Midnight Oil
3333	RED 7	Red 7
1351	REEL LIFE	Boy Meets Girl
1301	REEL MUSIC	The Beatles
1900	REFLECTIONS	Rick James
2076	REFLECTIONS	Gil Scott-Heron
3551	REFLECTIONS	George Howard
1217	REFLECTOR	Pablo Cruise
522	REG STRIKES BACK	Elton John
2480	REIGN IN BLOOD	Slayer
2407	REI MOMO	David Byrne
3464	REJOICING	Pat Metheny
2987	RELEASED	Patti LaBelle
4796	RELIGHT MY FIRE	Dan Hartman
718	REMAIN IN LIGHT	Talking Heads
4180	REMATCH	Sammy Hagar
4695	REMEMBRANCE DAYS	Dream Academy
4055	RENAISSANCE	Village People
1804	RENDEZ-VOUS	Jean Michel Jarre
3656	RENEGADE	Thin Lizzy
96	REPEAT OFFENDER	Richard Marx
3836	REPLAY	Crosby, Stills & Nash
2445	REQUIEM	Various Artists
4437	RESCUE	Clarence Clemons
3694	RESCUE YOU	Joe Lynn Turner
1826	RESERVATIONS FOR TWO	Dionne Warwick
1124	RESTLESS	Starpoint
4363	RESTLESS EYES	Janis Ian
3651	RESULTS	Liza Minnelli
774	THE RETURN OF BRUNO	Bruce Willis
1259	RETURN OF THE JEDI	Soundtrack
1795	REUNION	The Temptations
4674	REUNION	Jerry Jeff Walker
4266	REUNION CONCERT	Everly Brothers
587	REVENGE	Eurythmics
2438	THE REVOLUTION BY NIGHT	Blue Öyster Cult
2992	REWIND (1971-1984)	Rolling Stones
1290	RHAPSODY AND BLUES	The Crusaders
3960	RHINESTONE (SOUNDTRACK)	Dolly Parton
1693	RHYME & REASON	Missing Persons
2215	RHYME PAYS	Ice-T
2406	RHYTHM AND ROMANCE	Rosanne Cash
4388	THE RHYTHM & THE BLUES	Z.Z. Hill
466	RHYTHM OF THE NIGHT	DeBarge
640	RHYTHM OF YOUTH	Men Without Hats
3940	THE RHYTHMOTIST	Stewart Copeland
2808	RHYTHM ROMANCE	The Romantics
2875	RICCOCHET DAYS	Modern English
4318	RICH AND FAMOUS	Blue Mercedes
3644	RICH AND POOR	Randy Crawford
117	RICHARD MARX	Richard Marx
2620	RICHARD PRYOR: HERE AND NOW (SOUNDTRACK)	Richard Pryor
1207	RICHARD PRYOR LIVE ON THE SUNSET STRIP (SOUNDTRACK)	Richard Pryor
3426	THE RIDDLE	Nik Kershaw
2749	RIDDLES IN THE SAND	Jimmy Buffett
1778	RIDE THE LIGHTNING	Metallica
4054	RIDIN' THE STORM OUT	REO Speedwagon
4253	RIFF RAFF	Dave Edmunds
2718	RIGHT BY YOU	Stephen Stills
4513	THE RIGHT COMBINATION	Linda Clifford & Curtis Mayfield
1839	RIGHTEOUS ANGER	Van Stephenson
3406	THE RIGHT NIGHT & BARRY WHITE	Barry White
4191	RIGHT OR WRONG	George Strait
2013	THE RIGHT PLACE	Gary Wright
835	THE RIGHT STUFF	Vanessa Williams
4588	THE RIGHT TO BE ITALIAN	Holly And The Italians
2376	THE RIGHT TO ROCK	Keel
4127	THE RINGS	The Rings
190	RIO	Duran Duran
4141	RIO	Lee Ritenour
4727	RIP IT TO SHREDS- THE ANIMALS GREATEST HITS LIVE	The Animals
4852	RIP IT UP	Dead Or Alive
197	RIPTIDE	Robert Palmer
4299	RISE AND SHINE	The Bears
4097	RISING	Dr. Hook
1483	RISING FORCE	Yngwie Malmsteen
887	RIT	Lee Ritenour
2844	RIT/2	Lee Ritenour
3224	RITA COOLIDGE/GREATEST HITS	Rita Coolidge
4728	RITES OF PASSAGE	Vitamin Z
3423	RITES OF SUMMER	Spyro Gyra
126	THE RIVER	Bruce Springsteen
2093	RIVER OF TIME	The Judds
4892	ROACHES: THE BEGINNING	Bobby Jimmy & The Critters
3225	ROACHFORD	Roachford
3498	THE ROAD	The Kinks
2863	ROAD HOUSE	Soundtrack
3427	ROADIE	Soundtrack
3589	ROAD ISLAND	Ambrosia
2426	ROBBERY	Teena Marie
1441	ROBBIE DUPREE	Robbie Dupree
846	ROBBIE NEVIL	Robbie Nevil
1018	ROBBIE ROBERTSON	Robbie Robertson
873	ROBERTA FLACK FEATURING DONNY HATHAWAY	Roberta Flack Featuring Donny Hathaway
3093	ROBERT HAZARD	Robert Hazard
4520	ROCK & ROLL REBELS	John Kay & Steppenwolf
1359	ROCK & ROLL STRATEGY	Thirty Eight Special
480	ROCK A LITTLE	Stevie Nicks
4366	ROCK & ROLL ADULT	Garland Jeffreys
4458	ROCK AND ROLL DIARY 1967-1980	Lou Reed
3791	ROCKAPELLA	The Nylons
1758	ROCK AWAY	Phoebe Snow
2828	ROCKBIRD	Debbie Harry
3485	ROCKER	Elvis Presley
3330	ROCK FOR AMNESTY	Various Artists
4302	ROCK HARD	Suzi Quatro
754	ROCKIHNROLL	Greg Kihn Band
1481	ROCK IN A HARD PLACE	Aerosmith
1846	ROCKIN' INTO THE NIGHT	38-Special
3602	ROCKIN' RADIO	Tom Browne
1068	ROCKIN' WITH THE RHYTHM	The Judds
2065	ROCK ISLAND	Jethro Tull
252	ROCK ME TONIGHT	Freddie Jackson

Albums Alpha Index

4134 ROCK 'N' ROLL *Motorhead*	2369 THE SAGA CONTINUES... *Roger*	455 THE SECRET OF ASSOCIATION *Paul Young*
4776 ROCK 'N' ROLL OUTLAW *Rose Tattoo*	3667 SALSA *Soundtrack*	3807 THE SECRET OF MY SUCCESS *Soundtrack*
4541 ROCK 'N' ROLL WARRIORS *Savoy Brown*	1143 A SALT WITH A DEADLY PEPA *Salt-N-Pepa*	2617 THE SECRET POLICEMAN'S BALL *Various Artists*
302 ROCK 'N SOUL PART 1 *Daryl Hall & John Oates*	4450 SALUTE *Gordon Lightfoot*	1527 THE SECRET POLICEMAN'S OTHER BALL *Various Artists*
1816 ROCK OF LIFE *Rick Springfield*	1689 SAMANTHA FOX *Samantha Fox*	2429 SECRETS *Wilton Felder*
2782 ROCKS, PEBBLES AND SAND *Stanley Clarke*	3871 SAM COOKE LIVE AT THE HARLEM SQUARE CLUB *Sam Cooke*	2205 SECRET SECRETS *Joan Armatrading*
1529 ROCK THE HOUSE *D.J. Jazzy Jeff & The Fresh Prince*	2546 SAME GOES FOR YOU *Leif Garrett*	1862 SECRETS OF FLYING *Johnny Kemp*
4250 ROCK THE NATIONS *Saxon*	1174 SAM HARRIS *Sam Harris*	1497 THE SECRET VALUE OF DAYDREAMING *Julian Lennon*
4034 ROCK THERAPY *Stray Cats*	2462 SAM-I-AM *Sam Harris*	3528 SEDUCTION *James Last Band*
4663 ROCK THE WORLD *Third World*	4603 SAMURAI SAMBA *Yellowjackets*	508 THE SEEDS OF LOVE *Tears For Fears*
931 ROCKY III *Soundtrack*	1640 SAN ANTONIO ROSE *Willie Nelson & Ray Price*	3113 SEE HOW WE ARE *X*
556 ROCKY IV *Soundtrack*	1145 SANDINISTA! *The Clash*	4824 SEE JUNGLE! SEE JUNGLE! GO JOIN YOUR GANG YEAH! CITY ALL OVER, GO APE CRAZY *Bow Wow Wow*
2241 ROCK YOU TO HELL *Grim Reaper*	1583 SANDS OF TIME *S.O.S. Band*	4226 SEEN ONE EARTH *Pete Bardens*
3208 RODNEY CROWELL *Rodney Crowell*	1768 SARAYA *Saraya*	1957 THE SEER *Big Country*
1326 ROD STEWART *Rod Stewart*	1618 SAVAGE *Eurythmics*	602 SEE THE LIGHT *Jeff Healey Band*
1059 ROD STEWART GREATEST HITS *Rod Stewart*	328 SAVAGE AMUSEMENT *Scorpions*	2123 SEE YOU IN HELL *Grim Reaper*
3854 ROGUES GALLERY *Slade*	1451 SAVED *Bob Dylan*	530 SELF CONTROL *Laura Branigan*
492 ROLL ON *Alabama*	4823 SAVE YOUR PRAYERS *Waysted*	3684 SELL MY SOUL *Sylvester*
158 ROLL WITH IT *Steve Winwood*	3579 SAWYER BROWN *Sawyer Brown*	1277 SEMINAR *Sir Mix-A-Lot*
1803 ROMANCE DANCE *Kim Carnes*	2300 SAY ANYTHING *Soundtrack*	4784 SEND ME AN ANGEL '89 *Real Life*
1135 ROMANCE 1600 *Sheila E.*	4504 SAYIN' SOMETHING! *Peaches & Herb*	1817 SEND ME YOUR LOVE *Kashif*
4360 ROMAN GODS *The Fleshtones*	2873 SAY IT AGAIN *Jermaine Stewart*	2589 SEND YOUR LOVE *Aurra*
3565 ROMAN HOLLIDAY *Roman Holliday*	3954 SAY NO MORE *Badfinger*	2723 SENSATIONAL *Starpoint*
1985 THE ROMANTICS *The Romantics*	3121 SAY YOU LOVE ME *Jennifer Holliday*	3219 SENSE OF PURPOSE *Third World*
3213 ROMEO KNIGHT *Boogie Boys*	1191 SCANDAL *Scandal*	2059 A SENSE OF WONDER *Van Morrison*
4830 ROMEO'S DAUGHTER *Romeo's Daughter*	65 SCARECROW *John Mellencamp*	1142 THE SENSUAL WORLD *Kate Bush*
3139 ROMEO'S ESCAPE *Dave Alvin*	3155 SCARLETT & BLACK *Scarlett & Black*	1984 SENTIMENTAL HYGIENE *Warren Zevon*
4050 ROOT HOG OR DIE *Mojo Nixon & Skid Roper*	3418 SCARS OF LOVE *TKA*	682 SEPTEMBER MORN *Neil Diamond*
297 THE ROSE (SOUNDTRACK) *Bette Midler*	578 SCARY MONSTERS *David Bowie*	1916 SEQUEL *Harry Chapin*
3275 THE ROSE OF ENGLAND *Nick Lowe And His Cowboy Outfit*	4183 THE SCATTERING *Cutting Crew*	984 SERGIO MENDES(2) *Sergio Mendes*
746 ROSES IN THE SNOW *Emmylou Harris*	4601 SCATTERLINGS *Juluka*	2753 SERIOUS *The O'Jays*
1652 ROSS (II) *Diana Ross*	4043 SCENARIO *Al Di Meola*	4507 SERIOUS *Deja*
1251 ROUGH DIAMONDS *Bad Company*	376 SCENES FROM THE SOUTHSIDE *Bruce Hornsby & The Range*	2764 SERIOUS BUSINESS *Third World*
3180 ROUGH NIGHT IN JERICHO *Dreams So Real*	4210 SCENES IN THE CITY *Branford Marsalis*	3895 SERIOUS BUSINESS *Johnny Winter*
4811 ROUND MIDNIGHT *Soundtrack*	3712 SCHEMER-DREAMER *Steve Walsh*	4088 SERIOUS SLAMMIN' *Pointer Sisters*
2951 'ROUND MIDNIGHT *Linda Ronstadt*	2265 SCHOOL DAZE *Soundtrack*	4722 SET MY LOVE IN MOTION *Syreeta*
3408 ROUND TRIP *The Knack*	3171 SCISSORS CUT *Art Garfunkel*	3511 SETTING SONS *The Jam*
4492 ROUND TRIP *The Gap Band*	1879 SCOOP *Pete Townshend*	152 SEVEN AND THE RAGGED TIGER *Duran Duran*
3028 ROUND TWO *Johnny Van Zant Band*	4641 SCOTT BAIO *Scott Baio*	1160 SEVEN THE HARD WAY *Pat Benatar*
3922 ROUTES *Ramsey Lewis*	2453 SCOUNDREL DAYS *a-ha*	1781 THE SEVENTH ONE *Toto*
2639 ROVER'S RETURN *John Waite*	784 SCREAM DREAM *Ted Nugent*	760 SEVENTH SON OF A SEVENTH SON *Iron Maiden*
2179 ROWDY *Hank Williams Jr.*	410 SCREAMING FOR VENGEANCE *Judas Priest*	2767 SEVENTH STAR *Black Sabbath Featuring Tony Iommi*
3930 ROYAL JAM *The Crusaders With B.B. King And The Royal Philharmonic Orchestra*	4260 SCRIPT FOR A JESTER'S TEAR *Marillion*	421 7800 DEGREES FAHRENHEIT *Bon Jovi*
2720 RUFF 'N' READY *Ready For The World*	3242 SCROOGED *Soundtrack*	2616 '74 JAILBREAK *AC/DC*
2090 RUMBLE *Tommy Conwell And The Young Rumblers*	3119 SCUBA DIVERS *Dwight Twilley*	306 7 WISHES *Night Ranger*
4309 RUMBLE FISH (SOUNDTRACK) *Stewart Copeland*	4158 SEA HAGS *Sea Hags*	643 SEVEN YEAR ACHE *Rosanne Cash*
2885 THE RUMOUR *Olivia Newton-John*	3138 SEAMLESS *The Nylons*	2710 SEX AND THE SINGLE MAN *Ray Parker Jr.*
4510 RUNAWAY *Bill Champlin*	3905 THE SEA OF LOVE *The Adventures*	1710 SEXAPPEAL *Georgio*
1222 RUNAWAY HORSES *Belinda Carlisle*	4803 THE SEARCHERS *The Searchers*	1434 SHABOOH SHOOBAH *INXS*
915 RUN D.M.C. *Run-D.M.C.*	4256 SEARCHPARTY *John Hall Band*	3474 SHADAY *Ofra Haza*
3991 RUN FOR COVER *Gary Moore*	2309 SEASON OF GLASS *Yoko Ono*	4508 SHADDAP YOU FACE *Joe Dolce*
3489 RUN FOR THE ROSES *Jerry Garcia*	1342 SEASONS OF THE HEART *John Denver*	3313 SHADES *J.J. Cale*
3773 RUNNING FOR MY LIFE *Judy Collins*	2627 SEAWIND(2) *Seawind*	4869 SHADES *Yellowjackets*
804 RUNNING IN THE FAMILY *Level 42*	3896 THE SECOND ADVENTURE *Dynasty*	3424 SHADES OF BLUE *Lou Rawls*
1853 RUNNING SCARED *Soundtrack*	3989 THE SECOND ALBUM *707*	4535 SHADES OF LACE *Lace*
3382 RUSS BALLARD *Russ Ballard*	4543 SECOND EDITION *Public Image Limited*	3106 SHADOWDANCE *Shadowfax*
3400 RUSSIAN ROULETTE *Accept*	2592 SECOND SIGHTING *Frehley's Comet*	1763 SHADOWLAND *k.d. lang*
1177 RUTHLESS PEOPLE *Soundtrack*	898 SECONDS OF PLEASURE *Rockpile*	4026 SHADOW MAN *Johnny Clegg & Savuka*
	4488 SECOND TIME AROUND *The Kinks*	2840 SHADOWS *Gordon Lightfoot*
	3877 SECOND TO NUNN *Bobby Nunn*	1610 SHADOWS AND LIGHT *Joni Mitchell*
S	1087 2ND WAVE *Surface*	277 SHAKE IT UP *The Cars*
	1845 SECRET COMBINATION *Randy Crawford*	1132 SHAKEN 'N' STIRRED *Robert Plant*
1064 SACRED HEART *Dio*	3449 SECRET DREAMS & FORBIDDEN FIRE *Bonnie Tyler*	
2235 SACRED SONGS *Daryl Hall*	1725 SECRET MESSAGES *Electric Light Orchestra*	
4059 SAD CAFÉ *Sad Café*		
1419 SA-FIRE *SaFire*		

Albums Alpha Index

667 SHAKE YOU DOWN *Gregory Abbott*	3220 SIMPLICITY *Tim Curry*	3786 SOLD OUT *The Fools*
956 SHANGO *Santana*	4886 SING *Soundtrack*	847 SOLID *Ashford & Simpson*
3869 SHANGRI-LA *Bardeux*	1588 SINGLE LIFE *Cameo*	4792 SOLID GOLD *Gang Of Four*
280 SHARE YOUR LOVE *Kenny Rogers*	2326 THE SINGLES *The Pretenders*	1685 SOLID GROUND *Ronnie Laws*
3068 SHARING YOUR LOVE *Change*	1920 THE SINGLES (THE FIRST TEN YEARS) *ABBA*	561 SOLITUDE/SOLITAIRE *Peter Cetera*
4120 SHARKY'S MACHINE *Soundtrack*	2086 SINGLES COLLECTION - THE LONDON YEARS *Rolling Stones*	517 SOLITUDE STANDING *Suzanne Vega*
1994 SHARP *Angela Winbush*	1820 SINGLES 45'S AND UNDER *Squeeze*	2307 SO MANY RIVERS *Bobby Womack*
671 SHEENA EASTON *Sheena Easton*	2645 SIOGO *Blackfoot*	2161 SOMEBODY'S GONNA LOVE YOU *Lee Greenwood*
2317 SHEFFIELD STEEL *Joe Cocker*	4448 SIRIUS *Clannad*	1358 SOMEBODY'S KNOCKIN' *Terri Gibbs*
2578 SHEILA E. *Sheila E.*	2414 THE SISTERS *Sister Sledge*	2167 SOMEBODY'S WAITING *Anne Murray*
1665 SHELTER *Lone Justice*	2050 SIT DOWN AND TALK TO ME *Lou Rawls*	758 SOMEBODY'S WATCHING ME *Rockwell*
2240 SHERIFF *Sheriff*	3932 SIXTEEN *Stacy Lattisaw*	4323 SOME COME RUNNING *Jim Capaldi*
3126 SHE'S HAVING A BABY *Soundtrack*	3931 SKEEZER PLEEZER *UTFO*	900 SOME DAYS ARE DIAMONDS *John Denver*
1972 SHE SHOT ME DOWN *Frank Sinatra*	4031 SKELETONS IN THE CLOSET: THE BEST OF OINGO BOINGO *Oingo Boingo*	1275 SOME GREAT REWARD *Depeche Mode*
63 SHE'S SO UNUSUAL *Cyndi Lauper*	79 SKID ROW *Skid Row*	2451 SOME KIND OF WONDERFUL *Soundtrack*
1084 SHE'S STRANGE *Cameo*	2384 THE SKILL *The Sherbs*	4281 SOME OF MY BEST JOKES ARE FRIENDS *George Clinton*
728 SHE'S THE BOSS *Mick Jagger*	2484 SKIN DIVE *Michael Franks*	1572 SOMETHING ABOUT YOU *Angela Bofill*
3643 SHE WAS ONLY A GROCER'S DAUGHTER *Blow Monkeys*	2140 SKIN ON SKIN *Vanity*	4317 SOMETHING HEAVY GOING DOWN - LIVE FROM THE TWILIGHT ZONE *Golden Earring*
543 SHE WORKS HARD FOR THE MONEY *Donna Summer*	2787 SKY *Sky*	3750 SOMETHING INSIDE SO STRONG *Kenny Rogers*
2788 SHINE *Average White Band*	4586 SKY 3 *Sky*	2222 SOMETHING IN THE NIGHT *Pure Prairie League*
879 SHINE ON *L.T.D.*	1027 SKYLARKIN' *Grover Washington Jr.*	2330 SOMETHING REAL *Phoebe Snow*
4231 SHINE ON ME *One Way*	1851 SKYLARKING *XTC*	1339 SOMETHING'S GOING ON *Frida*
1119 A SHIP ARRIVING TOO LATE TO SAVE A DROWNING WITCH *Frank Zappa*	402 SKYSCRAPER *David Lee Roth*	214 SOMETHING SPECIAL *Kool & The Gang*
1413 SHIRLEY MURDOCK! *Shirley Murdock*	1384 SKYWAY *Skyy*	3265 SOMETHING TO BELIEVE IN *Curtis Mayfield*
1731 SHOCK *The Motels*	2537 SKYYJAMMER *Skyy*	1964 SOMETHING TO TALK ABOUT *Anne Murray*
2191 SHOCKADELICA *Jesse Johnson*	4717 SKYYLIGHT *Skyy*	1726 SOMETIMES LATE AT NIGHT *Carole Bayer Sager*
3188 SHOCKER *Soundtrack*	639 SKYY LINE *Skyy*	3841 SOMETIMES WHEN WE TOUCH *Cleo Laine & James Galway*
3501 SHOGUN *TV Soundtrack*	2046 SKYYPORT *Skyy*	1644 SOMETIMES YOU WIN *Dr. Hook*
218 SHOOTING RUBBERBANDS AT THE STARS *Edie Brickell & The New Bohemians*	4240 SLAM *Dan Reed Network*	1624 SOME TOUGH CITY *Tony Carey*
3335 SHOOTING STAR *Shooting Star*	2044 SLAVE TO THE RHYTHM *Grace Jones*	1660 SOMEWHERE IN AFRIKA *Manfred Mann's Earth Band*
2332 SHORT BACK 'N' SIDES *Ian Hunter*	513 SLEEPING WITH THE PAST *Elton John*	950 SOMEWHERE IN ENGLAND *George Harrison*
1400 SHORT SHARP SHOCKED *Michelle Shocked*	2473 SLEEPWALK *Larry Carlton*	2719 SOMEWHERE IN THE STARS *Rosanne Cash*
3032 SHORT STORIES *Jon And Vangelis*	2327 SLEIGHT OF HAND *Joan Armatrading*	567 SOMEWHERE IN TIME *Iron Maiden*
2310 SHOTGUN MESSIAH *Shotgun Messiah*	3622 THE SLIDE AREA *Ry Cooder*	4780 SOMEWHERE IN TIME *Soundtrack*
2721 SHOT IN THE DARK *Great White*	703 SLIDE IT IN *Whitesnake*	1442 SOMEWHERE OVER CHINA *Jimmy Buffett*
1848 SHOT OF LOVE *Bob Dylan*	2682 SLINGSHOT *Michael Henderson*	1014 SOMEWHERE OVER THE RAINBOW *Willie Nelson*
4634 SHOULD I DO IT *Tanya Tucker*	468 SLIP OF THE TONGUE *Whitesnake*	3314 SONG OF SEVEN *Jon Anderson*
3439 SHOUT *Devo*	15 SLIPPERY WHEN WET *Bon Jovi*	4604 SONGS *Kids From Fame*
211 SHOUT AT THE DEVIL *Mötley Crüe*	4885 SLOW DANCE *Southside Johnny*	2856 SONGS FROM LIQUID DAYS *Philip Glass*
2954 THE SHOUTING STAGE *Joan Armatrading*	1992 SLOW TURNING *John Hiatt*	40 SONGS FROM THE BIG CHAIR *Tears For Fears*
2958 SHOWDOWN! *Albert Collins, Robert Cray, Johnny Copeland*	1635 SMALL CHANGE *Prism*	3149 SONGS FROM THE FILM *Tommy Keene*
1149 SHOW ME *The Cover Girls*	3548 SMALLCREEP'S DAY *Mike Rutherford*	4837 SONGS FROM THE STAGE AND SCREEN *Michael Crawford*
1360 A SHOW OF HANDS *Rush*	656 SMALL WORLD *Huey Lewis & The News*	4585 SONGS I LOVE TO SING *Slim Whitman*
1523 SHOW TIME *Slave*	789 SMASHES, THRASHES & HITS *KISS*	475 SONGS IN THE ATTIC *Billy Joel*
1131 SHOWTIME *J. Geils Band*	3745 THE SMITHS *The Smiths*	4666 SONGS OF THE FREE *Gang Of Four*
3516 SHRINER'S CONVENTION *Ray Stevens*	3052 SMOKE SIGNALS *Smokey Robinson*	4020 SONGS TO LEARN & SING *Echo & The Bunnymen*
1388 SHUTTERED ROOM *The Fixx*	4689 SMOKE SOME KILL *Schoolly D*	3483 THE SONGSTRESS *Anita Baker*
1334 SIDE KICKS *Thompson Twins*	2752 SMOKEY AND THE BANDIT 2 *Soundtrack*	2567 SONGS YOU KNOW BY HEART: JIMMY BUFFETT'S GREATEST HIT(S) *Jimmy Buffett*
617 SIGNALS *Rush*	2707 SMOKING IN THE FIELDS *Del Fuegos*	440 SONIC TEMPLE *The Cult*
964 SIGN IN PLEASE *Autograph*	2176 SMOOTH SAILIN' *Isley Brothers*	1928 THE SON OF ROCK AND ROLL *Rocky Burnette*
2080 SIGN OF THE TIMES *Bob James*	2450 SNAKES AND LADDERS *Gerry Rafferty*	3648 SOONER OR LATER *Larry Graham*
348 SIGN 'O' THE TIMES *Prince*	2823 SNAPSHOT *Sylvia (2)*	1125 SO RED THE ROSE *Arcadia*
605 SIGNS OF LIFE *Billy Squier*	2701 'SNAZ *Nazareth*	652 S.O.S. *S.O.S. Band*
250 SILHOUETTE *Kenny G*	2907 SNEAKER *Sneaker*	4143 S.O.S. III *S.O.S. Band*
3704 SILK *Fuse One*	1855 SNEAKIN' OUT *Stacy Lattisaw*	3007 SOULFORCE REVOLUTION *7 Seconds*
2277 SILK AND STEEL *Five Star*	105 SO *Peter Gabriel*	1565 SOUL KISS *Olivia Newton-John*
2341 SILK + STEEL *Giuffria*	2676 SODA FOUNTAIN SHUFFLE *Earl Klugh*	3782 SOUL MAN *Soundtrack*
1022 SILK ELECTRIC *Diana Ross*	1805 SO EXCITED! *Pointer Sisters*	1309 SOUL SEARCHING *Glenn Frey*
1593 SILKY SOUL *Maze Featuring Frankie Beverly*	1137 SO FAR, SO GOOD... SO WHAT! *Megadeth*	3170 SOUL SURVIVOR *Al Green*
3026 SILVER CONDOR *Silver Condor*	2003 SO GOOD *The Whispers*	1071 SOUL TO SOUL *Stevie Ray Vaughan And Double Trouble*
3166 SIMPLE MINDS LIVE: IN THE CITY OF LIGHT *Simple Minds*	2244 SO GOOD *Mica Paris*	3004 SOUND + VISION *David Bowie*
315 SIMPLE PLEASURES *Bobby McFerrin*	2848 SO HAPPY *Eddie Murphy*	2378 SOUND AFFECTS *The Jam*
4563 SIMPLE THINGS *Richie Havens*	4193 SOLD *Boy George*	
	3487 SOLDIER *Iggy Pop*	
	3914 SOLDIERS OF FORTUNE *The Outlaws*	
	1567 SOLDIERS UNDER COMMAND *Stryper*	

Albums Alpha Index

4848 **SOUND ALARM** *Michael Anderson*	3868 **STAR PEOPLE** *Miles Davis*	2736 **STORMS** *Nanci Griffith*
4195 **THE SOUND OF MUSIC** *The dB's*	540 **STARS ON LONG PLAY** *Stars On*	866 **STORMS OF LIFE** *Randy Travis*
2565 **SOUND-SYSTEM** *Herbie Hancock*	3563 **STARS ON LONG PLAY II** *Stars On*	3779 **STORM WINDOWS** *John Prine*
3957 **SOUP FOR ONE** *Soundtrack*	4275 **STARS ON LONG PLAY III** *Stars On*	2733 **THE STORY OF A YOUNG HEART**
4116 **THE SOURCE** *Grandmaster Flash*	3544 **THE STARS WE ARE** *Marc Almond*	*A Flock Of Seagulls*
413 **SOUTHERN ACCENTS**	4255 **START OF A ROMANCE** *Skyy*	3736 **THE STORY OF THE CLASH, VOLUME I** *The Clash*
Tom Petty And The Heartbreakers	2728 **STAR TREK II: THE WRATH OF KHAN** *Soundtrack*	2559 **STRAIGHT AHEAD**
2708 **SOUTHERN BY THE GRACE OF GOD/**	3134 **STAR TREK III - THE SEARCH FOR SPOCK**	*Larry Gatlin & The Gatlin Brothers Band*
LYNYRD SKYNYRD TRIBUTE TOUR 1987	*Soundtrack*	3015 **STRAIGHT AHEAD** *Amy Grant*
Lynyrd Skynyrd	2009 **STAR TREK - THE MOTION PICTURE**	1279 **STRAIGHT BETWEEN THE EYES** *Rainbow*
3263 **SOUTHERN COMFORT** *Conway Twitty*	*Soundtrack*	921 **STRAIGHT FROM THE HEART** *Patrice Rushen*
1794 **SOUTHERN STAR** *Alabama*	710 **STATE OF CONFUSION** *The Kinks*	1323 **STRAIGHT FROM THE HEART** *Peabo Bryson*
2027 **SOUTH OF HEAVEN** *Slayer*	3787 **STATE OF...EMERGENCY** *Steel Pulse*	444 **STRAIGHT OUTTA COMPTON** *N.W.A.*
2456 **SOUTHSIDE** *Texas*	1121 **STATE OF EUPHORIA** *Anthrax*	1822 **STRAIGHT TO THE HEART** *David Sanborn*
256 **SPANISH FLY** *Lisa Lisa And Cult Jam*	3811 **STATES OF EMERGENCY** *Taxxi*	2689 **STRAIGHT TO THE SKY** *Lisa Lisa And Cult Jam*
1958 **SPARKLE IN THE RAIN** *Simple Minds*	3878 **STAY** *Ray, Goodman & Brown*	3956 **STRANGE ANGELS** *Laurie Anderson*
4262 **SPEAK & SPELL** *Depeche Mode*	2754 **STAY AWAKE** *Various Artists*	2536 **STRANGE BEHAVIOR** *Animotion*
317 **SPEAKING IN TONGUES** *Talking Heads*	2515 **STAY HARD** *Raven*	4258 **STRANGE KIND OF LOVE** *Love And Money*
932 **SPEAK OF THE DEVIL** *Ozzy Osbourne*	354 **STAY HUNGRY** *Twisted Sister*	4682 **STRANGE LAND** *Box Of Frogs*
4813 **SPECIAL** *Jimmy Cliff*	585 **STAYING ALIVE (SOUNDTRACK)** *Bee Gees*	1613 **STRANGEWAYS, HERE WE COME** *The Smiths*
834 **SPECIAL BEAT SERVICE** *English Beat*	4066 **STAY ON THESE ROADS** *a-ha*	4559 **STRAPHANGIN'** *Brecker Brothers*
509 **SPECIAL FORCES** *38 Special*	3618 **STAY TUNED** *Chet Atkins*	2444 **STRAWBERRY MOON** *Grover Washington Jr.*
3738 **SPECIAL FORCES** *Alice Cooper*	1072 **STAY WITH ME** *Regina Belle*	2383 **STREET BEAT** *The Deele*
3729 **SPECIAL PAIN** *Robert Ellis Orrall*	325 **STAY WITH ME TONIGHT** *Jeffrey Osborne*	779 **STREET CALLED DESIRE** *René & Angela*
3260 **A SPECIAL PART OF ME** *Johnny Mathis*	1770 **STEADY NERVES**	4172 **STREET CORNER HEROES** *Robbie Dupree*
1874 **THE SPECIALS** *The Specials*	*Graham Parker & The Shot*	2759 **STREET FEVER** *Moon Martin*
1026 **SPECIAL THINGS** *Pointer Sisters*	1871 **STEALING FIRE** *Bruce Cockburn*	2712 **STREET FIGHTING YEARS** *Simple Minds*
2561 **SPECIAL THINGS** *Pleasure*	3432 **STEALIN HORSES** *Stealin Horses*	4540 **STREET LANGUAGE** *Rodney Crowell*
3874 **THE SPECKLESS SKY** *Jane Siberry*	3510 **STEAMIN' HOT** *The Reddings*	3176 **STREET LIFE-20 GREAT HITS**
2553 **SPELL** *Deon Estus*	1513 **STEEL BREEZE** *Steel Breeze*	*Bryan Ferry/Roxy Music*
3287 **SPELLBOUND** *Joe Sample*	2175 **STEELTOWN** *Big Country*	1632 **STREET OPERA** *Ashford & Simpson*
2781 **SPEND THE NIGHT**	205 **STEEL WHEELS** *Rolling Stones*	3584 **STREET READY** *Leatherwolf*
Isley Brothers Featuring Ronald Isley	683 **ST. ELMO'S FIRE** *Soundtrack*	1300 **STREETS OF FIRE** *Soundtrack*
3268 **SPIDER** *Spider*	716 **STEP BY STEP** *Eddie Rabbitt*	104 **STREET SONGS** *Rick James*
3974 **SPIES OF LIFE** *Player*	2602 **STEP BY STEP** *Jeff Lorber*	360 **STREET TALK** *Steve Perry*
895 **SPIKE** *Elvis Costello*	1025 **STEPHANIE** *Stephanie Mills*	947 **STRENGTH** *The Alarm*
1240 **SPIRIT OF LOVE** *Con Funk Shun*	1700 **STEPHANIE MILLS** *Stephanie Mills*	544 **STRENGTH IN NUMBERS** *38 Special*
4495 **SPIRIT OF PLACE** *Goanna*	4249 **STEP ON OUT** *The Oak Ridge Boys*	4850 **STRENGTH OF STEEL** *Anvil*
4096 **SPIRIT OF ST. LOUIS** *Ellen Foley*	4577 **STEPPIN' OUT** *George Howard*	1793 **STRICTLY BUSINESS** *EPMD*
4274 **THE SPIRIT'S IN IT** *Patti LaBelle*	3827 **STEPS IN TIME** *King*	4768 **STRICTLY PERSONAL** *The Romantics*
3980 **SPLASHDOWN** *Breakwater*	1739 **STEREOTOMY**	2898 **STRIKE LIKE LIGHTNING** *Lonnie Mack*
2568 **SPLENDIDO HOTEL** *Al Di Meola*	*Alan Parsons Project*	4593 **THE STRIKERS** *Strikers*
2924 **SPOILED GIRL** *Carly Simon*	2705 **STEVE ARRINGTON'S HALL OF FAME: I**	3481 **STRIKES TWICE** *Larry Carlton*
2554 **SPONTANEOUS INVENTIONS** *Bobby McFerrin*	*Steve Arrington's Hall Of Fame*	1857 **STRIP** *Adam Ant*
4783 **SPORTIN' LIFE** *Weather Report*	4244 **STEVE FORBERT** *Steve Forbert*	327 **STRONGER THAN PRIDE** *Sade*
1281 **THE SPORT OF KINGS** *Triumph*	4383 **THE STEVE HOWE ALBUM** *Steve Howe*	368 **STRONG PERSUADER** *Robert Cray Band*
14 **SPORTS** *Huey Lewis & The News*	4061 **THE STEVE MARTIN BROTHERS** *Steve Martin*	2401 **STRONG STUFF** *Hank Williams Jr.*
2854 **SPREADING THE DISEASE** *Anthrax*	3849 **STEVE MILLER BAND - LIVE**	2475 **STYLE** *Cameo*
367 **SPRING SESSION M** *Missing Persons*	*The Steve Miller Band*	1940 **SUBJECT: ALDO NOVA** *Aldo Nova*
4276 **SPUN GOLD** *Barbara Mandrell*	581 **STEVIE WONDER'S ORIGINAL MUSIQUARIUM I**	653 **SUBSTANCE** *New Order*
3629 **S.P.Y.S.** *Spys*	*Stevie Wonder*	3764 **SUBSTANCE** *Joy Division*
3193 **STAGE DOLLS** *Stage Dolls*	1796 **STILL CRUISIN'** *Beach Boys*	3085 **SUBTERRANEAN JUNGLE** *The Ramones*
1895 **STAGES** *Triumph*	3867 **STILL FEELS GOOD** *Tom Johnston*	2613 **SUBURBAN VOODOO** *Paul Carrack*
2796 **STANDARDS, VOLUME 1** *Stanley Jordan*	4708 **STILL IN LOVE** *Carrie Lucas*	286 **SUCCESS HASN'T SPOILED ME YET**
798 **STAND BY ME** *Soundtrack*	660 **STILL LIFE (AMERICAN CONCERT 1981)**	*Rick Springfield*
4184 **STANDING ALONE** *White Wolf*	*Rolling Stones*	4844 **SUCKER FOR A PRETTY FACE** *Eric Martin Band*
670 **STANDING HAMPTON** *Sammy Hagar*	2669 **STILL LIFE (TALKING)** *Pat Metheny Group*	1100 **SUCKING IN THE SEVENTIES** *Rolling Stones*
957 **STANDING ON A BEACH -- THE SINGLES**	2398 **STILL STANDING** *Jason & The Scorchers*	120 **SUDDENLY** *Billy Ocean*
The Cure	2824 **STILL THE SAME OLE ME** *George Jones*	3179 **SUE SAAD AND THE NEXT**
1539 **STANDING ON THE EDGE** *Cheap Trick*	3132 **ST. JULIAN** *Julian Cope*	*Sue Saad And The Next*
3708 **STANDING ON THE EDGE** *Frankie Miller*	234 **STONE COLD RHYMIN'** *Young M.C.*	2849 **SUMMER HEAT** *Brick*
2169 **STANDING TALL** *The Crusaders*	1221 **STONE JAM** *Slave*	4076 **SUMMER LOVERS** *Soundtrack*
2173 **STAND IN LINE** *Impellitteri*	2695 **STOP AND SMELL THE ROSES** *Ringo Starr*	1494 **SUN CITY** *Artists United Against Apartheid*
2582 **STAND IN THE FIRE** *Warren Zevon*	514 **STOP MAKING SENSE** *Talking Heads*	3502 **SUNDAY IN THE PARK WITH GEORGE**
4331 **STAND UP** *Del Fuegos*	4065 **STOP START** *Modern English*	*Original Broadway Cast Recording*
941 **STARCHILD** *Teena Marie*	4178 **STORE AT THE SUN** *Jon Butcher Axis*	3550 **SUNDAY MORNING SUITE** *Frank Mills*
868 **STARFISH** *The Church*	4534 **STORIES** *Gloria Gaynor*	4420 **SUNDOWN** *Rank And File*
3475 **STAR FLEET PROJECT** *Brian May And Friends*	3125 **STORIES WITHOUT WORDS** *Spyro Gyra*	4010 **SUNRISE** *Jimmy Ruffin*
3734 **STARING AT THE SUN** *Level 42*	131 **STORM FRONT** *Billy Joel*	4389 **SUNSHINE DREAM** *Beach Boys*

Albums Alpha Index

#	Title	Artist
3988	THE SUN STILL SHINES	Sonny Charles
4684	SUPERBAD	Chris Jasper
2993	SUPERCHARGED	Tavares
3362	SUPERMAN II	Soundtrack
4625	SUPERMAN III	Soundtrack
1210	SUPERSONIC--THE ALBUM	J.J. Fad
447	SUPER TROUPER	ABBA
3385	SURE SHOT	Crown Heights Affair
1821	SURFACE	Surface
3948	SURFACE THRILLS	The Temptations
485	SURFING WITH THE ALIEN	Joe Satriani
4490	SURFIN' M.O.D.	M.O.D.
1316	SUR LA MER	Moody Blues
4589	SURPRISE	Sylvia (2)
1458	SURPRISE ATTACK	Tora Tora
2818	SURVEILLANCE	Triumph
3744	SURVIVAL OF THE FRESHEST	Boogie Boys
3552	SURVIVE	Nuclear Assault
4655	SURVIVIN' IN THE 80'S	Andre Cymone
4083	SURVIVOR	Survivor
2056	SUZANNE VEGA	Suzanne Vega
1192	SWASS	Sir Mix-A-Lot
2483	SWEAT	The System
3723	SWEAT BAND	Sweat Band
4590	SWEET AND WONDERFUL	Jean Carn
3281	SWEET, DELICIOUS & MARVELOUS	California Raisins
391	SWEET DREAMS (ARE MADE OF THIS)	Eurythmics
1543	SWEET DREAMS: THE LIFE AND TIMES OF PATSY CLINE (SOUNDTRACK)	Patsy Cline
361	SWEET SENSATION	Stephanie Mills
1356	SWEETS FROM A STRANGER	Squeeze
2392	SWEET SIXTEEN	Reba McEntire
4230	SWEET SOUND	Simon Townshend
523	SWEPT AWAY	Diana Ross
1479	THE SWING	INXS
2415	THE SWING OF DELIGHT	Devadip Carlos Santana
1825	SWING STREET	Barry Manilow
3937	SWING, SWING, SWING	Boston Pops Orchestra/John Williams
4286	SWING THE HEARTACHE - THE BBC SESSIONS	Bauhaus
3266	SWING TO THE RIGHT	Utopia
4451	SWITCH V	Switch
4187	SWORDFISHTROMBONES	Tom Waits
2054	SYBIL	Sybil
3264	SYLVAIN SYLVAIN	Sylvain Sylvain
3420	THE SYMPHONY SESSIONS	David Foster
21	SYNCHRONICITY	The Police
3130	SYNCHRO SYSTEM	King Sunny Ade & His African Beats
2441	SYREETA(2)	Syreeta

T

#	Title	Artist
2455	TAKE A LITTLE RHYTHM	Ali Thomson
1294	TAKE ANOTHER PICTURE	Quarterflash
3249	TAKE IT OFF	Chic
2283	TAKE IT TO THE LIMIT	Willie Nelson With Waylon Jennings
3955	TAKE IT TO THE LIMIT	Norman Connors
1383	TAKE IT WHILE IT'S HOT	Sweet Sensation
3044	TAKE ME ALL THE WAY	Stacy Lattisaw
3025	TAKE ME TO YOUR HEAVEN	Stevie Woods
1628	TAKE NO PRISONERS	Molly Hatchet
2991	TAKE NO PRISONERS	Peabo Bryson
1983	TAKE 6	Take 6
3441	TAKE WHAT YOU NEED	Robin Trower
3440	TAKING IT HOME	Buckwheat Zydeco
1615	TAKING LIBERTIES	Elvis Costello
4879	TAKING OVER	Overkill
3690	THE TALE OF THE TAPE	Billy Squier
4772	TALES OF THE NEW WEST	Beat Farmers
3904	TALKIN' 'BOUT YOU	Diane Schuur
1176	TALKING BACK TO THE NIGHT	Steve Winwood
1123	TALK IS CHEAP	Keith Richards
4779	TALK MEMPHIS	Jesse Winchester
572	TALK SHOW	Go-Go's
2516	TALK TALK TALK	Psychedelic Furs
3090	TALK TO YOUR DAUGHTER	Robben Ford
1547	TA MARA & THE SEEN	Ta Mara & The Seen
3318	TANE CAIN	Tane Cain
150	TANGO IN THE NIGHT	Fleetwood Mac
1718	TANTALIZINGLY HOT	Stephanie Mills
3751	TANTILLA	House Of Freaks
807	TAO	Rick Springfield
4435	TAP	Soundtrack
4533	TAP STEP	Chick Corea
1866	TARANTELLA	Chuck Mangione
3227	A TASTE OF YESTERDAY'S WINE	Merle Haggard/George Jones
3758	TASTE THE MUSIC	Kleeer
2945	TASTY JAM	Fatback
2923	TATTOOED BEAT MESSIAH	Zodiac Mindwarp & The Love Reaction
49	TATTOO YOU	Rolling Stones
1521	TEACHERS	Soundtrack
612	TEAR DOWN THESE WALLS	Billy Ocean
2132	TEASER	Angela Bofill
3062	TEASES AND DARES	Kim Wilde
1029	TECHNIQUE	New Order
1468	TEDDY LIVE! COAST TO COAST	Teddy Pendergrass
3308	TEEVEE TOONS - THE COMMERCIALS	Various Artists
2469	TELEKON	Gary Numan
2100	TELEVISION'S GREATEST HITS	Various Artists
3358	TELEVISION'S GREATEST HITS VOLUME II	Various Artists
2136	TELEVISION THEME SONGS	Mike Post
249	TELL IT TO MY HEART	Taylor Dayne
2382	TELL NO TALES	TNT
1418	TEMPLE OF LOW MEN	Crowded House
3348	THE TEMPTATIONS	The Temptations
2905	"10"	Soundtrack
2312	10 1/2	The Dramatics
3992	TENDERLY	George Benson
4429	TENDERNESS	Ohio Players
4462	TENDER TOGETHERNESS	Stanley Turrentine
4044	TENEMENT STEPS	The Motors
3390	10 FROM 6	Bad Company
4876	TENKU	Kitaro
4622	10,9,8,7,6,5,4,3,2,1	Midnight Oil
1452	TENTH	Marshall Tucker Band
4338	10TH ANNIVERSARY	Statler Brothers
4647	TEN WOMEN	Wire Train
3965	TEN YEARS OF HARMONY (1970-1980)	Beach Boys
2892	TEQUILA SUNRISE	Soundtrack
2255	TERENCE TRENT D'ARBY'S NEITHER FISH NOR FLESH	Terence Trent D'Arby
3332	TERMS OF ENDEARMENT	Soundtrack
4575	TERROR RISING	Lizzy Borden
1112	TEXAS FLOOD	Stevie Ray Vaughan/Double Trouble
2521	TEXAS IN MY REAR VIEW MIRROR	Mac Davis
4749	T.G. SHEPPARD'S GREATEST HITS	T.G. Sheppard
4597	THAT'S ALL THAT MATTERS TO ME	Mickey Gilley
2722	THAT'S THE STUFF	Autograph
1075	THAT'S WHY I'M HERE	James Taylor
3817	THEATER OF THE MIND	Mtume
289	THEATRE OF PAIN	Mötley Crüe
2756	THEM	King Diamond
674	THEN & NOW...THE BEST OF THE MONKEES	The Monkees
909	THERE AND BACK	Jeff Beck
823	THERE GOES THE NEIGHBORHOOD	Joe Walsh
3098	THERE MUST BE A BETTER WORLD SOMEWHERE	B.B. King
994	THERE'S NO GETTIN' OVER ME	Ronnie Milsap
2442	THESE DAYS	Crystal Gayle
3182	THEY DON'T MAKE THEM LIKE THEY USED TO	Kenny Rogers
2901	THIEF (SOUNDTRACK)	Tangerine Dream
4562	THIEF IN THE NIGHT	George Duke
4638	THIEF OF HEARTS	Soundtrack
2118	THINK OF ONE	Wynton Marsalis
2387	THINK VISUAL	The Kinks
481	THE THIN RED LINE	Glass Tiger
3192	THIN RED LINE	The Cretones
3619	THIRD GENERATION	Hiroshima
146	THIRD STAGE	Boston
4773	THIRD WORLD, PRISONER IN THE STREET (SOUNDTRACK)	Third World
4278	THIRTEEN	Emmylou Harris
3341	THIS DAY AND AGE	D.L. Byron
1893	THIS IS BIG AUDIO DYNAMITE	Big Audio Dynamite
3010	THIS IS ELVIS (SOUNDTRACK)	Elvis Presley
3717	THIS ISLAND	Eurogliders
2328	THIS IS MY DREAM	Switch
3476	THIS IS SPINAL TAP (SOUNDTRACK)	Spinal Tap
4297	THIS IS THE DAY...THIS IS THE HOUR...THIS IS THIS!	Pop Will Eat Itself
1410	THIS IS THE WAY	Rossington Collins Band
4861	THIS IS THIS	Weather Report
4157	THIS IS YOUR TIME	Change
3034	THIS KIND OF LOVIN'	The Whispers
1932	THIS NOTE'S FOR YOU	Neil Young & The Bluenotes
2156	THIS ONE'S FOR YOU	Teddy Pendergrass
1304	THIS SIDE OF PARADISE	Ric Ocasek
755	THIS TIME	Al Jarreau
1216	THIS WOMAN	K.T. Oslin
1626	THOSE OF YOU WITH OR WITHOUT CHILDREN, YOU'LL UNDERSTAND	Bill Cosby
2855	3	Violent Femmes
880	3 FEET HIGH AND RISING	De La Soul
781	THREE FOR LOVE	Shalamar
1293	THREE HEARTS IN THE HAPPY ENDING MACHINE	Daryl Hall
622	THREE LOCK BOX	Sammy Hagar
2135	THREE OF A PERFECT PAIR	King Crimson
4032	THREE PIECE SUITE	Ramsey Lewis
4447	THREE QUARTETS	Chick Corea
4428	3 SHIPS	Jon Anderson
704	THREE SIDES LIVE	Genesis
3593	THREE TIMES IN LOVE	Tommy James
3206	III WISHES	Shooting Star
1	THRILLER	Michael Jackson
3507	THROBBING PYTHON OF LOVE	Robin Williams
1704	THROUGH THE FIRE	Hagar, Schon, Aaronson, Shrieve
4745	THROUGH THE LOOKING GLASS	Siouxsie & The Banshees
2139	THROUGH THE STORM	Aretha Franklin
848	THROWIN' DOWN	Rick James
1813	THUNDER	Andy Taylor

Albums Alpha Index

4204 THUNDER AND LIGHTNING *Thin Lizzy*	4430 TOO MUCH PRESSURE *The Selecter*	2751 TROMBIPULATION *Parliament*
1973 THUNDER IN THE EAST *Loudness*	775 TOO-RYE-AY *Dexys Midnight Runners*	4246 TRON *Soundtrack*
1024 THUNDER SEVEN *Triumph*	786 TOOTH AND NAIL *Dokken*	3582 TROOP *Troop*
3657 THUNDERSTEEL *Riot*	1181 TOO TOUGH *Angela Bofill*	689 TROPICO *Pat Benatar*
102 TIFFANY *Tiffany*	4353 TOO TOUGH TO DIE *The Ramones*	2428 TROUBLE IN PARADISE *Randy Newman*
2997 TIGHTEN UP VOL. '88 *Big Audio Dynamite*	3624 TOOTSIE *Soundtrack*	3455 TROUBLEMAKER *Ian McLagan*
2939 TIGHT SHOES *Foghat*	4632 THE TOP *The Cure*	3556 TROUBLE WALKIN' *Ace Frehley*
569 TILL I LOVED YOU *Barbra Streisand*	83 TOP GUN *Soundtrack*	3513 TRUCE *Jack Bruce & Robin Trower*
4460 TIM *The Replacements*	3726 THE TOP TEN HITS *Elvis Presley*	813 TRUE *Spandau Ballet*
747 TIME *Electric Light Orchestra*	1446 TORCH *Carly Simon*	72 TRUE BLUE *Madonna*
1065 THE TIME *The Time*	2668 TORNADO *The Rainmakers*	323 TRUE COLORS *Cyndi Lauper*
484 TIME AND TIDE *Basia*	4741 TOTAL DEVO *Devo*	1036 TRUE COLOURS *Split Enz*
2057 TIME AND TIDE *Split Enz*	2197 TO THE MAX *Con Funk Shun*	824 TRUE CONFESSIONS *Bananarama*
524 TIME EXPOSURE *Little River Band*	3131 TO THE POWER OF THREE *3*	3289 TRUE DEMOCRACY *Steel Pulse*
3543 TIME EXPOSURE *Stanley Clarke*	97 TOTO IV *Toto*	3141 TRUE LOVE *Crystal Gayle*
3973 TIME ODYSSEY *Vinnie Moore*	349 TOUCH *Eurythmics*	722 TRUE STORIES: A FILM BY DAVID BYRNE, THE COMPLETE SOUNDTRACK *Talking Heads*
2906 TIME PIECES -- THE BEST OF ERIC CLAPTON *Eric Clapton*	1409 TOUCH *Con Funk Shun*	1341 TRULY FOR YOU *The Temptations*
4595 TIMES OF OUR LIVES *Judy Collins*	1482 THE TOUCH *Alabama*	1310 TRUST *Elvis Costello And The Attractions*
1449 TIMES SQUARE *Soundtrack*	2002 TOUCH *Laura Branigan*	4107 TRUTH AND SOUL *Fishbone*
3428 X2 *Times Two*	3233 TOUCH *Gladys Knight & The Pips*	4136 TRUTHDARE DOUBLEDARE *Bronski Beat*
3200 TIMOTHY B. *Timothy B. Schmit*	3576 TOUCH *Sarah McLachlan*	4636 TRUTH IN DISGUISE *Denise Lopez*
3081 TINA LIVE IN EUROPE *Tina Turner*	2361 TOUCH AND GO *Force M.D.'s*	2545 TRY IT OUT *Klique*
2650 TINDERBOX *Siouxsie & The Banshees*	3117 TOUCH DANCE *Eurythmics*	3531 TUCKERIZED *Marshall Tucker Band*
1579 TIN MACHINE *Tin Machine*	1008 TOUCH ME *Samantha Fox*	336 TUFF ENUFF *Fabulous Thunderbirds*
2292 TINSEL TOWN REBELLION *Frank Zappa*	3740 TOUCH ME *The Temptations*	283 TUG OF WAR *Paul McCartney*
4877 TIRAMI SU *Al Di Meola Project*	4078 TOUCH ME TONIGHT-THE BEST OF SHOOTING STAR *Shooting Star*	180 TUNNEL OF LOVE *Bruce Springsteen*
1701 TO BE CONTINUED... *The Temptations*	2114 TOUCH THE SKY *Smokey Robinson*	675 TURBO *Judas Priest*
2120 TODAY *Today*	1085 TOUCH THE WORLD *Earth, Wind & Fire*	1876 TURN BACK *Toto*
4649 TODAY *Statler Brothers*	4762 TOUCH YOU *Jimmy Hall*	1502 TURN BACK THE CLOCK *Johnny Hates Jazz*
3414 2XS *Nazareth*	4324 TOUGH *Kurtis Blow*	248 THE TURN OF A FRIENDLY CARD *Alan Parsons Project*
4099 TOGETHER *Oak Ridge Boys*	1032 TOUGH ALL OVER *John Cafferty & The Beaver Brown Band*	3196 TURN OF THE SCREW *Dirty Looks*
2532 TOGETHER AGAIN *The Temptations*	588 TOUGHER THAN LEATHER *Run-D.M.C.*	2528 TURN THE HANDS OF TIME *Peabo Bryson*
529 TO HELL WITH THE DEVIL *Stryper*	1597 TOUGHER THAN LEATHER *Willie Nelson*	4064 TURN UP THE MUSIC *Mass Production*
2685 TOLD U SO *Nu Shooz*	552 TOUR DE FORCE *38 Special*	3853 TUTU *Miles Davis*
2448 TO LIVE AND DIE IN L.A. (SOUNDTRACK) *Wang Chung*	4202 TOUR DE FORCE - "LIVE" *Al Di Meola*	3967 TWANG BAR KING *Adrian Belew*
1396 TO LOVE AGAIN *Diana Ross*	2747 TOURIST IN PARADISE *The Rippingtons Featuring Russ Freeman*	1847 TWANGIN... *Dave Edmunds*
1504 TOMCATTIN' *Blackfoot*	2822 TO WHOM IT MAY CONCERN *The Pasadenas*	3770 12 *Bob James*
3115 TOM COCHRANE & RED RIDER *Tom Cochrane & Red Rider*	4760 TO YOU HONEY, HONEY WITH LOVE *David Hudson*	1433 12 GREATEST HITS VOL. II *Neil Diamond*
4483 TOMMY PAGE *Tommy Page*	409 TP *Teddy Pendergrass*	3397 TWENNYNINE WITH LENNY WHITE *Twennynine Featuring Lenny White*
2252 TOMMY TUTONE *Tommy Tutone*	1346 T'PAU *T'Pau*	1372 20/20 *George Benson*
851 TOMMY TUTONE-2 *Tommy Tutone*	2861 TRACIE SPENCER *Tracie Spencer*	3911 20/20 TWENTY NO. 1 HITS FROM TWENTY YEARS AT MOTOWN *Various Artists*
756 TOM TOM CLUB *Tom Tom Club*	3412 TRACK RECORD *Joan Armatrading*	2758 25TH ANNIVERSARY *Diana Ross & The Supremes*
4464 TONGUE IN CHIC *Chic*	93 TRACY CHAPMAN *Tracy Chapman*	3168 25TH ANNIVERSARY *The Temptations*
3617 TONGUE TWISTER *Shoes*	1183 TRANS *Neil Young*	4062 21ST CENTURY MAN *Billy Thorpe*
794 TONIGHT *David Bowie*	452 TRASH *Alice Cooper*	1633 25 #1 HITS FROM 25 YEARS *Various Artists*
1272 TONIGHT! *Four Tops*	3897 T.R.A.S.H. (TUBES RARITIES AND SMASH HITS) *The Tubes*	3342 25 YEARS OF GRAMMY GREATS *Various Artists*
4598 TONIGHT *France Joli*	4155 TRASH IT UP *Southside Johnny & The Jukes*	1642 24 CARROTS *Al Stewart*
405 TONIGHT I'M YOURS *Rod Stewart*	2189 TRAVELS *Pat Metheny Group*	3699 2400 FULTON ST. *Jefferson Airplane*
1780 THE TONIGHT SHOW BAND *The Tonight Show Band with Doc Severinsen*	4694 TREACHEROUS: A HISTORY OF THE NEVILLE BROTHERS 1955-1985 *Neville Brothers*	783 24/7 *Dino*
4130 TONIGHT YOU'RE MINE *Eric Carmen*	2762 TREAT HER RIGHT *Treat Her Right*	808 TWENTY GREATEST HITS *Kenny Rogers*
4021 TONY CAREY [I WON'T BE HOME TONIGHT] *Tony Carey*	3941 TRIAL BY FIRE: LIVE IN LENINGRAD *Yngwie Malmsteen*	1550 20 GREATEST HITS *The Beatles*
3688 TOO *S.O.S. Band*	531 TRIBUTE *Ozzy Osbourne/Randy Rhoads*	731 21 AT 33 *Elton John*
2899 TOO FAR TO WHISPER *Shadowfax*	3852 TRICK OR TREAT (SOUNDTRACK) *Fastway*	2420 2300 JACKSON ST. *The Jacksons*
1161 TOO FAST FOR LOVE *Mötley Crüe*	1734 TRILOGY *Yngwie Malmsteen*	2500 22B3 *Device*
3243 TOO GOOD TO STOP NOW *John Schneider*	678 TRILOGY: PAST, PRESENT AND FUTURE *Frank Sinatra*	2785 20 YEARS OF JETHRO TULL *Jethro Tull*
4229 TOO HOT TO SLEEP *Sylvester*	4862 TRINERE AND FRIENDS GREATEST HITS *Trinere & Friends*	1030 TWICE AS SWEET *A Taste Of Honey*
4812 TOO HOT TO SLEEP *Survivor*	863 THE TRINITY SESSION *Cowboy Junkies*	287 ...TWICE SHY *Great White*
4328 TOO HOT TO STOP IT *The Manhattans*	479 TRIO *Dolly Parton, Linda Ronstadt, Emmylou Harris*	2799 TWICE THE LOVE *George Benson*
1307 2 HYPE *Kid 'N Play*	397 TRIUMPH *The Jacksons*	3569 TWIN HYPE *Twin Hype*
2628 TOO LATE THE HERO *John Entwistle*	1836 TRIUMPH AND AGONY *Warlock*	3636 TWINS *Soundtrack*
3625 TOO LONG IN THE WASTELAND *James McMurtry*		2148 TWISTING BY THE POOL *Dire Straits*
582 TOO LOW FOR ZERO *Elton John*		4053 TWIST OF SHADOWS *Xymox*
		397 TRIUMPH *The Jacksons*
		4801 TWITCH *Ministry*
		1470 TWO *GQ*

Albums Alpha Index

1317 **2:00 A.M. PARADISE CAFE** *Barry Manilow*	428 **UP YOUR ALLEY** *Joan Jett & the Blackhearts*	3583 **VOICES IN THE SKY-BEST OF THE MOODY BLUES** *Moody Blues*
3913 **TWO B'S PLEASE** *Robbin Thompson Band*	4518 **URBAN BEACHES** *Cactus World News*	1782 **VOICES OF BABYLON** *The Outfield*
4339 **2 X 4** *Guadalcanal Diary*	1147 **URBAN CHIPMUNK** *The Chipmunks*	1814 **VOICES OF THE HEART** *Carpenters*
2262 **TWO HEARTS** *Men At Work*	167 **URBAN COWBOY** *Soundtrack*	4746 **VOLUME VIII** *Average White Band*
2508 **2 LIVE CREW IS WHAT WE ARE** *2 Live Crew*	3642 **URBAN COWBOY II** *Soundtrack*	134 **VOLUME 1** *Traveling Wilburys*
1108 **TWO OF A KIND (SOUNDTRACK)** *Soundtrack*	2851 **URBAN DAYDREAMS** *David Benoit*	366 **VOLUME ONE** *The Honeydrippers*
1246 **TWO OF A KIND** *Earl Klugh & Bob James*	4560 **URGH! A MUSIC WAR** *Various Artists*	3033 **VOLUNTEER JAM VI** *Various Artists*
692 **THE TWO OF US** *Yarbrough & Peoples*	4348 **U.S.A. FOR M.O.D.** *M.O.D.*	4109 **VOLUNTEER JAM VII** *Various Artists*
3792 **THE TWO OF US** *Ramsey Lewis & Nancy Wilson*	4177 **USED TO BE** *Charlene*	1232 **VOX HUMANA** *Kenny Loggins*
1091 **TWO PLACES AT THE SAME TIME** *Ray Parker Jr. & Raydio*	3654 **U.S. 1** *Head East*	3374 **VOYAGER** *Roger Whittaker*
4739 **THE TWO RING CIRCUS** *Erasure*	2183 **UTFO** *UTFO*	1211 **VOYEUR** *David Sanborn*
4478 **2010** *Soundtrack*	2360 **UTOPIA** *Utopia*	1749 **VOYEUR** *Kim Carnes*
2870 **TWO TONS O' FUN** *Two Tons Of Fun*		2698 **VU** *Velvet Underground*
4500 **TWO WHEELS GOOD** *Prefab Sprout*	# V	1664 **VULTURE CULTURE** *The Alan Parsons Project*
4095 **TYRONE DAVIS** *Tyrone Davis*	4439 **V** *Zapp*	
	627 **VACATION** *Go-Go's*	# W
# U	4539 **A VALENTINE GIFT FOR YOU** *Elvis Presley*	1352 **WAIATA** *Split Enz*
303 **UNDER A BLOOD RED SKY** *U2*	346 **VALOTTE** *Julian Lennon*	3981 **WAIT FOR NIGHT** *Rick Springfield*
1345 **UB40** *UB40*	2322 **VANDENBERG** *Vandenberg*	1809 **WAITIN' FOR THE SUN TO SHINE** *Ricky Skaggs*
4310 **UHF/ORIGINAL MOTION PICTURE SOUNDTRACK AND OTHER STUFF** *"Weird Al" Yankovic*	3866 **VAN GO** *Beat Farmers*	3479 **WAITING** *Fun Boy Three*
182 **UH-HUH** *John Mellencamp*	1376 **VANITY 6** *Vanity 6*	4766 **WAITING FOR SPRING** *David Benoit*
3724 **UK JIVE** *The Kinks*	4072 **VAN-ZANT** *Van-Zant*	4367 **WAITING ON YOU** *Brick*
425 **THE ULTIMATE SIN** *Ozzy Osbourne*	4657 **THE VELVETEEN RABBIT** *Meryl Streep & George Winston*	1102 **WAKING UP WITH THE HOUSE ON FIRE** *Culture Club*
2562 **ULTRA WAVE** *Bootsy*	4764 **VELVET KISS, LICK OF THE LIME** *Lions And Ghosts*	4142 **THE WALK** *The Cure*
1856 **UNCHAIN MY HEART** *Joe Cocker*	4870 **THE VELVET UNDERGROUND** *Velvet Underground & Nico*	1226 **WALKABOUT** *The Fixx*
1414 **UNDER A RAGING MOON** *Roger Daltrey*	4443 **VERTICAL SMILES** *Blackfoot*	3970 **WALK A FINE LINE** *Paul Anka*
606 **UNDERCOVER** *Rolling Stones*	4355 **VERY GREASY** *David Lindley & El Rayo-X*	1982 **WALK AWAY - COLLECTOR'S EDITION (THE BEST OF 1977-1980)** *Donna Summer*
727 **UNDER LOCK AND KEY** *Dokken*	1496 **VERY SPECIAL** *Debra Laws*	2410 **WALKIN' THE RAZOR'S EDGE** *Helix*
2755 **UNDERNEATH THE RADAR** *Underworld*	913 **A VERY SPECIAL CHRISTMAS** *Various Artists*	4637 **WALKING THROUGH FIRE** *April Wine*
2494 **UNDER THE BIG BLACK SUN** *X*	3443 **VESTA 4 U** *Vesta Williams*	4476 **WALKING WILD** *New England*
3359 **UNDER THE BLADE** *Twisted Sister*	2949 **VICES** *Kick Axe*	538 **WALKING WITH A PANTHER** *LL Cool J*
1600 **UNDER THE BLUE MOON** *New Edition*	1420 **VICTIMS OF THE FURY** *Robin Trower*	1606 **WALK UNDER LADDERS** *Joan Armatrading*
1695 **UNDER THE GUN** *Poco*	4402 **VICTIMS OF THE FUTURE** *Gary Moore*	23 **THE WALL** *Pink Floyd*
3284 **UNDER THE INFLUENCE** *Overkill*	4592 **VICTOR/VICTORIA** *Soundtrack*	3212 **WALL TO WALL** *René & Angela*
2623 **UNDER THE VOLCANO** *Rock And Hyde*	472 **VICTORY** *The Jacksons*	856 **THE WANDERER** *Donna Summer*
3901 **THE UNDERTONES** *The Undertones*	3086 **VICTORY** *Narada Michael Walden*	1638 **WANNA BE A STAR** *Chilliwack*
2274 **UNDERTOW** *Firefall*	4519 **VICTORY** *Larry Graham*	2529 **WANTED DREAD & ALIVE** *Peter Tosh*
2783 **UNDER WRAPS** *Jethro Tull*	3349 **VICTORY DAY** *Tom Cochrane & Red Rider*	227 **WAR** *U2*
2111 **UNFINISHED BUSINESS** *EPMD*	3693 **VIENNA** *Ultravox*	3460 **WAREHOUSE: SONGS AND STORIES** *Husker Du*
224 **THE UNFORGETTABLE FIRE** *U2*	1146 **VIEW FROM THE GROUND** *America*	3280 **WARM AND TENDER** *Olivia Newton-John*
4820 **THE UNFORGIVEN** *The Unforgiven*	1842 **A VIEW TO A KILL** *Soundtrack*	3752 **THE WARMER SIDE OF COOL** *Wang Chung*
891 **UNGUARDED** *Amy Grant*	1407 **VINNIE VINCENT INVASION** *Vinnie Vincent Invasion*	3152 **WARM LEATHERETTE** *Grace Jones*
1079 **UNION** *Toni Childs*	1151 **VINYL CONFESSIONS** *Kansas*	638 **WARM THOUGHTS** *Smokey Robinson*
1256 **UNION JACKS** *The Babys*	4465 **VISAGE** *Visage*	1907 **THE WARNING** *Queensryche*
2771 **UNITED** *Commodores*	796 **VISION QUEST** *Soundtrack*	600 **WARRIOR** *Scandal Featuring Patty Smyth*
4680 **UNITED STATES LIVE** *Laurie Anderson*	948 **VISIONS** *Gladys Knight & The Pips*	3765 **WASN'T THAT A PARTY** *The Rovers*
4662 **UNITY** *Shinehead*	4409 **VISIONS OF THE LITE** *Slave*	1402 **WASN'T TOMORROW WONDERFUL?** *The Waitresses*
3239 **UNIVERSAL RHYTHM** *Ralph MacDonald*	1754 **THE VISITOR** *Mick Fleetwood*	1929 **W.A.S.P.** *W.A.S.P.*
1163 **UNLIMITED!** *Roger*	1332 **THE VISITORS** *ABBA*	3921 **WATCHING YOU, WATCHING ME** *Bill Withers*
3737 **UNLIMITED TOUCH** *Unlimited Touch*	4048 **VISUAL LIES** *Lizzy Borden*	2366 **WATCH OUT!** *Patrice Rushen*
2116 **UNSUNG HEROES** *The Dregs*	512 **VITAL IDOL** *Billy Idol*	3059 **WATERFRONT** *Waterfront*
3340 **UNTOLD PASSION** *Neal Schon & Jan Hammer*	339 **VITAL SIGNS** *Survivor*	636 **WATERMARK** *Enya*
1930 **UNTOUCHABLES** *Lakeside*	1627 **VIVA HATE** *Morrissey*	1946 **WATTS IN A TANK** *Diesel*
3820 **UP** *Le Roux*	4022 **VIVA SANTANA** *Santana*	2816 **WA WA NEE** *Wa Wa Nee*
2226 **UP AND DOWN** *Opus*	3898 **VIVE LE ROCK** *Adam Ant*	98 **THE WAY IT IS** *Bruce Hornsby & The Range*
1645 **THE UP ESCALATOR** *Graham Parker And The Rumour*	198 **VIVID** *Living Colour*	3568 **WEAPONS OF LOVE** *The Truth*
3145 **UPLIFT MOFO PARTY PLAN** *Red Hot Chili Peppers*	852 **VIXEN** *Vixen*	1138 **WE ARE ONE** *Maze Featuring Frankie Beverly*
1201 **UPRISING** *Bob Marley And The Wailers*	782 **VOA** *Sammy Hagar*	2986 **WE ARE ONE** *Pieces Of A Dream*
2158 **UPSTAIRS AT ERIC'S** *Yazoo*	1631 **VOCALESE** *Manhattan Transfer*	443 **WE ARE THE WORLD** *USA For Africa*
4723 **UP THE CREEK** *Soundtrack*	1979 **VOICE OF AMERICA** *Little Steven*	2490 **WEATHER REPORT(2)** *Weather Report*
3933 **UPTOWN** *Neville Brothers*	138 **VOICES** *Daryl Hall & John Oates*	2526 **WE DON'T TALK ANYMORE** *Cliff Richard*
	2772 **VOICES & IMAGES** *Camouflage*	3021 **WE GO A LONG WAY BACK** *Bloodstone*
	723 **VOICES CARRY** *'Til Tuesday*	3780 **"WEIRD AL" YANKOVIC** *"Weird Al" Yankovic*
	1709 **VOICES IN THE RAIN** *Joe Sample*	

Albums Alpha Index

858 "WEIRD AL" YANKOVIC IN 3-D "Weird Al" Yankovic	3148 WHITE KNUCKLE RIDE Duke Jupiter	4549 WINDHAM HILL RECORDS SAMPLER '89 Various Artists
3229 WEIRD SCIENCE Soundtrack	815 WHITE NIGHTS Soundtrack	1297 WINDOWS Charlie Daniels Band
1440 WELCOME HOME 'Til Tuesday	4436 WHITE NOISE Jay Ferguson	773 WINDOWS AND WALLS Dan Fogelberg
2113 WELCOME TO THE CLUB Ian Hunter	2990 WHITE SHOES Emmylou Harris	763 WINDS OF CHANGE Jefferson Starship
664 WELCOME TO THE PLEASURE DOME Frankie Goes To Hollywood	32 WHITESNAKE Whitesnake	3669 WINDSONG Randy Crawford
157 WELCOME TO THE REAL WORLD Mr. Mister	1598 WHITE WINDS Andreas Vollenweider	179 WINELIGHT Grover Washington Jr.
1875 WELCOME TO THE WRECKING BALL Grace Slick	35 WHITNEY Whitney Houston	255 WINGER Winger
2716 WENDY AND LISA Wendy And Lisa	5 WHITNEY HOUSTON Whitney Houston	274 WINNER IN YOU Patti LaBelle
3886 WE'RE MOVIN' UP Atlantic Starr	1333 WHO? Tony! Toni! Tone!	1959 WINNERS Brothers Johnson
1375 WE'RE THE BEST OF FRIENDS Natalie Cole & Peabo Bryson	1951 THE WHOLE STORY Kate Bush	2882 WINNERS Various Artists
2237 WEST SIDE STORY Leonard Bernstein	871 WHO MADE WHO (SOUNDTRACK) AC/DC	3493 WINNERS Kleeer
980 WE TOO ARE ONE Eurythmics	4742 WHOMP THAT SUCKER Sparks	3009 THE WINNING HAND Various Artists
937 WE'VE GOT TONIGHT Kenny Rogers	2638 (WHO'S AFRAID OF?) THE ART OF NOISE! Art Of Noise	2517 WINTER INTO SPRING George Winston
4086 WE WANT MILES Miles Davis	1604 WHO'S FOOLIN' WHO One Way	2579 A WINTER'S SOLSTICE Various Artists
1202 WHAMMY! The B-52s	3063 WHO'S GREATEST HITS The Who	3470 A WINTER'S SOLSTICE II Various Artists
3492 WHA'PPEN English Beat	2511 WHO'S LAST The Who	4016 WIRED FOR SOUND Cliff Richard
1010 WHAT ABOUT ME? Kenny Rogers	3797 WHO'S MISSING The Who	3566 WISE GUY Kid Creole & The Coconuts
1386 WHAT A FEELIN' Irene Cara	604 WHO'S THAT GIRL (SOUNDTRACK) Madonna	1843 WISHES Jon Butcher
2581 WHAT A LIFE! Divinyls	282 WHO'S ZOOMIN' WHO? Aretha Franklin	1885 WISHFUL THINKING Earl Klugh
1292 WHAT BECOMES A SEMI-LEGEND MOST? Joan Rivers	424 WHY DO FOOLS FALL IN LOVE Diana Ross	2842 WITCH DOCTOR Instant Funk
831 WHAT CHA' GONNA DO FOR ME Chaka Khan	4219 WHY LADY WHY Gary Morris	4412 WITCHDOCTOR Sidewinders
3127 WHAT GOES AROUND The Hollies	1892 WHY NOT ME The Judds	4698 WITH LOVE Roger Whittaker
3195 WHAT IF Tommy Shaw	4364 WHY NOT ME Fred Knoblock	1471 WITHOUT A SONG Willie Nelson
2174 WHAT IS BEAT? English Beat	2154 WIDE AWAKE IN AMERICA U2	2925 WITH SYMPATHY Ministry
4678 WHAT IS THIS What Is This	970 WIDE AWAKE IN DREAMLAND Pat Benatar	1715 WITH YOU Stacy Lattisaw
4630 WHAT MORE CAN I SAY? Audio Two	1320 WIDE RECEIVER Michael Henderson	3185 WIZARD ISLAND Jeff Lorber Fusion
3164 WHAT PRICE PARADISE China Crisis	1961 WILD! Erasure	4720 WOMAN IN FLAMES Champaign
4084 WHAT'S MY NAME Steady B	2355 WILD & BLUE John Anderson	314 THE WOMAN IN RED (SOUNDTRACK) Stevie Wonder
159 WHAT'S NEW Linda Ronstadt & The Nelson Riddle Orchestra	3101 WILD AND FREE Dazz Band	462 A WOMAN NEEDS LOVE Ray Parker Jr. & Raydio
2902 WHAT'S NEXT Frank Marino And Mahogany Rush	2920 WILD & LOOSE Oaktown's 3.5.7	4827 WOMAN OF THE YEAR Original Cast
1058 WHAT TIME IS IT? The Time	2008 WILD ANIMAL Vanity	1378 WOMAN OUT OF CONTROL Ray Parker Jr.
1107 WHAT UP, DOG? Was (Not Was)	4493 WILDER The Teardrop Explodes	3919 A WOMAN'S GOT THE POWER The A's
707 WHAT YOU DON'T KNOW Exposé	4821 WILD EXHIBITIONS Walter Egan	3309 A WOMAN'S POINT OF VIEW Shirley Murdock
2018 WHEELS Restless Heart	1508 WILD-EYED DREAM Ricky Van Shelton	359 WOMEN AND CHILDREN FIRST Van Halen
312 WHEELS ARE TURNIN' REO Speedwagon	322 WILD-EYED SOUTHERN BOYS .38 Special	3996 WONDERFUL Rick James
3608 WHEN A GUITAR PLAYS THE BLUES Roy Buchanan	3294 WILD FRONTIER Gary Moore	2509 WONDERLAND Big Country
4144 WHEN ALL THE PIECES FIT Peter Frampton	4113 WILD GIFT X	2373 WON'T BE BLUE ANYMORE Dan Seals
191 WHENEVER YOU NEED SOMEBODY Rick Astley	239 THE WILD HEART Stevie Nicks	2887 THE WORD IS OUT Jermaine Stewart
545 WHEN HARRY MET SALLY (SOUNDTRACK) Harry Connick Jr.	1117 WILD HEART OF THE YOUNG Karla Bonoff	839 WORD OF MOUTH Toni Basil
1974 WHEN IN ROME When In Rome	4788 WILD IN THE STREETS Helix	1760 WORD OF MOUTH The Kinks
1450 WHEN SECONDS COUNT Survivor	777 WILD PLANET The B-52s	4414 WORD OF MOUTH Jaco Pastorius
2634 WHEN THE GOING GETS TOUGH, THE TOUGH GET GOING Bow Wow Wow	3785 WILD ROMANCE Herb Alpert	307 WORD UP! Cameo
4404 WHEN THE NIGHT COMES Lou Rawls	1611 WILDSIDE Loverboy	3198 WORKBOOK Bob Mould
4670 WHEN THE SUN GOES DOWN Red 7	1676 WILD STREAK Hank Williams Jr.	4888 WORKERS PLAYTIME Billy Bragg
4128 WHEN WILL I SEE YOU AGAIN The O'Jays	2610 THE WILD THE WILLING AND THE INNOCENT UFO	118 WORKING CLASS DOG Rick Springfield
1865 WHERE DO WE GO FROM HERE Michael Damian	1204 WILD THINGS RUN FAST Joni Mitchell	1909 WORKING GIRL Soundtrack
1738 WHERE DO YOU GO WHEN YOU DREAM Anne Murray	4818 WILD WEEKEND NRBQ	2572 WORKIN' IT BACK Teddy Pendergrass
2092 WHERE'S THE PARTY? Eddie Money	2862 WILD WEST Dottie West	3763 WORKIN' OVERTIME Diana Ross
4583 WHERE THE BEAT MEETS THE STREET Bobby And The Midnites	720 WILD, WILD WEST Escape Club	1083 THE WORKS Queen
504 WHERE THERE'S SMOKE Smokey Robinson	3628 WILL AND THE KILL Will And The Kill	2912 WORKS Pink Floyd
3843 WHERE TO NOW Charlie Dore	3008 WILLIE & THE POOR BOYS Willie & The Poor Boys	3732 THE WORLD ACCORDING TO ME Jackie Mason
2075 WHILE THE CITY SLEEPS... George Benson	417 WILLIE NELSON'S GREATEST HITS (& SOME THAT WILL BE) Willie Nelson	1831 WORLD IN MOTION Jackson Browne
408 WHIPLASH SMILE Billy Idol	1106 WILLIE NELSON SINGS KRISTOFFERSON Willie Nelson	665 WORLD MACHINE Level 42
316 THE WHISPERS The Whispers	3840 WILLIE NILE Willie Nile	3816 WORLD OF WONDERS Bruce Cockburn
4279 WHISPERS AND PROMISES Earl Klugh	3741 WILL POWER Joe Jackson	743 WORLDS APART Saga
3295 WHISPER TAMES THE LION Drivin' N' Cryin'	2922 WILL THE CIRCLE BE UNBROKEN, VOL.II Nitty Gritty Dirt Band	2737 THE WORLD'S GREATEST ENTERTAINER Doug E. Fresh & The Get Fresh Crew
3969 WHITE CHRISTMAS John Schneider	1694 WILL TO POWER Will To Power	491 WORLD WIDE LIVE Scorpions
889 WHITE CITY - A NOVEL Pete Townshend	3987 WINDHAM HILL PIANO SAMPLER Various Artists	3559 WORTH THE WAIT Peaches & Herb
1484 WHITE FEATHERS Kajagoogoo	2301 WINDHAM HILL RECORDS SAMPLER '84 Various Artists	1492 WOW! Bananarama
	2461 WINDHAM HILL RECORDS SAMPLER '86 Various Artists	3856 WRABIT Wrabit
	2886 WINDHAM HILL RECORDS SAMPLER '88 Various Artists	3104 WRAP YOUR ARMS AROUND ME Agnetha Faltskog
		4042 WRAP YOUR BODY One Way

2359 THE WRESTLING ALBUM *Various Artists*
1785 WWII
   Waylon Jennings & Willie Nelson
4305 WYNTON MARSALIS
   Wynton Marsalis

 264 XANADU (SOUNDTRACK)
   Olivia Newton-John/Electric Light Orchestra
4272 XL-1 *Pete Shelley*
4713 X-MULTIPLIES *Yellow Magic Orchestra*
4544 X-PERIMENT *The System*

2957 YEARS AGO *Statler Brothers*
1573 THE YEAR 2000 *The O'Jays*
2368 THE YELLOW AND BLACK ATTACK *Stryper*
2104 YELLOW MAGIC ORCHESTRA
   Yellow Magic Orchestra
1744 YELLOW MOON *Neville Brothers*
 649 YENTL (SOUNDTRACK)
   Barbra Streisand
4506 YES, GIORGIO *Luciano Pavarotti*
1505 YES IT'S YOU LADY *Smokey Robinson*
1942 YESSHOWS *Yes*
3893 YESTERDAY ONCE MORE *Carpenters*
3194 YO! BUM RUSH THE SHOW *Public Enemy*
2681 YO FRANKIE *Dion*
2224 YOU AND I *O'Bryan*
2126 YOU AND ME *Rockie Robbins*
2587 YOU AND ME BOTH *Yaz*
3232 YOU ARE WHAT YOU IS *Frank Zappa*
2081 YOU BOUGHT IT-YOU NAME IT *Joe Walsh*
1380 YOU BROKE MY HEART IN 17 PLACES
   Tracey Ullman
 778 YOU CAN DANCE *Madonna*
2184 YOU CAN'T FIGHT FASHION *Michael Stanley Band*
3177 YOU CAN'T STOP ROCK 'N' ROLL *Twisted Sister*
 860 YOU COULD HAVE BEEN WITH ME *Sheena Easton*
4089 YOU DON'T KNOW ME *Mickey Gilley*
4653 YOU GOTTA SAY YES TO ANOTHER EXCESS
   Yello
1462 YOU KNOW HOW TO LOVE ME *Phyllis Hyman*
2725 YOU'LL NEVER KNOW *Rodney Franklin*
2472 YOU, ME AND HE *Mtume*
4361 YOUNGBLOOD *Soundtrack*
1746 YOUNGEST IN CHARGE *Special Ed*
3537 YOUNG MAN RUNNING *Corey Hart*
2850 YOU'RE GETTIN' EVEN WHILE I'M GETTIN' ODD
   J. Geils Band
3549 YOU'RE THE ONE FOR ME *"D" Train*
3370 YOU'RE UNDER ARREST *Miles Davis*
2711 YOUR MOVE *America*
2239 YOURS FOREVER *Atlantic Starr*
2520 YOURS TRULY *Tom Browne*
1681 YOUR WISH IS MY COMMAND *Lakeside*
2651 YOU SHOULDN'T-NUF BIT FISH
   George Clinton
1164 THE YOUTH OF TODAY *Musical Youth*
1274 YOUTHQUAKE *Dead Or Alive*
3102 YOU'VE GOT A GOOD LOVE COMIN'
   Lee Greenwood
1799 YOU'VE GOT THE POWER *Third World*
2775 YOU WANNA DANCE WITH ME? *Jody Watley*
2940 YOU WANT IT, YOU GOT IT *Bryan Adams*
2229 YOYO *Bourgeois Tagg*
2985 Y U I ORTA *Ian Hunter/Mick Ronson*

Z

2402 THE ZAGORA *Loose Ends*
1020 ZAPP *Zapp*
1328 ZAPP II *Zapp*
1777 ZAPP III *Zapp*
 269 ZEBOP! *Santana*
1155 ZEBRA *Zebra*
3183 ZENO *Zeno*
  66 ZENYATTA MONDATTA *The Police*
2641 ZIGGY STARDUST-THE MOTION PICTURE
   (SOUNDTRACK) *David Bowie*
3361 ZIGZAGGING THROUGH GHOSTLAND
   The Radiators
4840 ZIG-ZAG WALK *Foghat*

Albums Alpha Index

Albums: Special Lists

Albums: Special Lists

Albums: 75 Weeks or More On Chart

This table includes the entire lifecycle of albums peaking in the 1980s. In some cases this lifecycle extends outside the decade. Those albums are designated with an asterisk (*) in the title; to the left If the lifecycle begins before 1/1/80, to the right if beyond 12/31/89.

Title Act	Wks
THE PHANTOM OF THE OPERA*	
Original London Cast Recording	255
ELIMINATOR *ZZ Top*	183
KENNY ROGERS' GREATEST HITS *Kenny Rogers*	181
UNDER A BLOOD RED SKY *U2*	180
WAR *U2*	179
DECEMBER* *George Winston*	178
BEACHES (SOUNDTRACK)* *Bette Midler*	176
FOREVER YOUR GIRL* *Paula Abdul*	175
*OFF THE WALL *Michael Jackson*	169
MADONNA *Madonna*	168
FACE VALUE *Phil Collins*	164
WHITNEY HOUSTON *Whitney Houston*	162
FEELS SO RIGHT *Alabama*	161
THE BIG CHILL *Soundtrack*	161
CAN'T SLOW DOWN *Lionel Richie*	160
SPORTS *Huey Lewis & The News*	158
RAPTURE *Anita Baker*	157
ZENYATTA MONDATTA *The Police*	153
1999 *Prince*	153
THE END OF THE INNOCENCE* *Don Henley*	148
APPETITE FOR DESTRUCTION* *Guns N' Roses*	147
ESCAPE *Journey*	146
BELLA DONNA *Stevie Nicks*	143
HELLO, I MUST BE GOING! *Phil Collins*	141
LIONEL RICHIE *Lionel Richie*	140
BORN IN THE U.S.A. *Bruce Springsteen*	139
HYSTERIA* *Def Leppard*	133
HANGIN' TOUGH* *New Kids On The Block*	132
THE UNFORGETTABLE FIRE *U2*	132
BACK IN BLACK* *AC/DC*	131
RIO *Duran Duran*	129
NO JACKET REQUIRED *Phil Collins*	123
*THE WALL *Pink Floyd*	123
THRILLER *Michael Jackson*	122
GET LUCKY *Loverboy*	122
*IN THE HEAT OF THE NIGHT *Pat Benatar*	122
WHEN HARRY MET SALLY (SOUNDTRACK)*	
Harry Connick Jr.	122
JANE FONDA'S WORKOUT RECORD *Jane Fonda*	120
STOP MAKING SENSE *Talking Heads*	118
CHRISTOPHER CROSS *Christopher Cross*	116
PYROMANIA *Def Leppard*	116
THE JAZZ SINGER (SOUNDTRACK) *Neil Diamond*	115
MOUNTAIN MUSIC *Alabama*	114
LEGEND* *Bob Marley And The Wailers*	113
AN INNOCENT MAN *Billy Joel*	111
DON'T SAY NO *Billy Squier*	111
SHOUT AT THE DEVIL *Mötley Crüe*	111
AGAINST THE WIND	
Bob Seger & The Silver Bullet Band	110
PUMP* *Aerosmith*	110
GHOST IN THE MACHINE *The Police*	109
DR. FEELGOOD* *Mötley Crüe*	109

Title Act	Wks
JANET JACKSON'S RHYTHM NATION 1814*	
Janet Jackson	108
LIKE A VIRGIN *Madonna*	108
THE RIVER *Bruce Springsteen*	107
CONTROL *Janet Jackson*	106
PRIVATE DANCER *Tina Turner*	106
AMERICAN FOOL *John Cougar*	106
HIGH 'N' DRY *Def Leppard*	106
BREAK OUT *The Pointer Sisters*	105
LOVERBOY *Loverboy*	105
BLIZZARD OF OZZ *Ozzy Osbourne*	104
LOST IN LOVE *Air Supply*	104
MEMORIES *Barbra Streisand*	104
7800 DEGREES FAHRENHEIT *Bon Jovi*	104
BILLY IDOL *Billy Idol*	104
THE JOSHUA TREE *U2*	103
BREAKIN' AWAY *Al Jarreau*	103
ALWAYS & FOREVER *Randy Travis*	103
DUOTONES *Kenny G*	102
HI INFIDELITY *REO Speedwagon*	101
LOOK WHAT THE CAT DRAGGED IN *Poison*	101
VOICES *Daryl Hall & John Oates*	100
STORMS OF LIFE *Randy Travis*	100
ALWAYS ON MY MIND *Willie Nelson*	99
THE DOORS GREATEST HITS* *The Doors*	99
DON'T BE CRUEL* *Bobby Brown*	97
BROTHERS IN ARMS *Dire Straits*	97
GRACELAND *Paul Simon*	97
LET IT LOOSE	
Gloria Estefan & Miami Sound Machine	97
GREATEST HITS* *The Judds*	97
DIRTY DANCING *Soundtrack*	96
SHE'S SO UNUSUAL *Cyndi Lauper*	96
NO PARKING ON THE DANCE FLOOR	
Midnight Star	96
SLIPPERY WHEN WET *Bon Jovi*	94
CRIMES OF PASSION *Pat Benatar*	93
TOP GUN *Soundtrack*	93
SO *Peter Gabriel*	93
WILLIE NELSON'S GREATEST HITS	
(& SOME THAT WILL BE) *Willie Nelson*	93
HEART *Heart*	92
JOURNEY'S GREATEST HITS* *Journey*	92
BUSINESS AS USUAL *Men At Work*	90
...BUT SERIOUSLY* *Phil Collins*	90
RIPTIDE *Robert Palmer*	90
EAZY-DUZ-IT* *Eazy-E*	90
OUT OF THE BLUE *Debbie Gibson*	89
CUTS LIKE A KNIFE *Bryan Adams*	89
STAY WITH ME TONIGHT *Jeffrey Osborne*	89
JULIO *Julio Iglesias*	89
KISSING TO BE CLEVER *Culture Club*	88
FAITH *George Michael*	87
BAD *Michael Jackson*	87

Title Act	Wks
DURAN DURAN *Duran Duran*	87
BACK IN THE HIGH LIFE *Steve Winwood*	86
RICHARD MARX *Richard Marx*	86
SUDDENLY *Billy Ocean*	86
JUICE *Juice Newton*	86
PRIDE *White Lion*	86
OLIVIA'S GREATEST HITS, VOL. 2	
Olivia Newton-John	86
BON JOVI *Bon Jovi*	86
WHITNEY *Whitney Houston*	85
INVISIBLE TOUCH *Genesis*	85
FRONTIERS *Journey*	85
SLIDE IT IN *Whitesnake*	85
RECKLESS *Bryan Adams*	83
...AND JUSTICE FOR ALL* *Metallica*	83
SONGS FROM THE BIG CHAIR *Tears For Fears*	82
TRUE BLUE *Madonna*	82
TOTO IV *Toto*	82
DIFFERENT LIGHT *Bangles*	82
REBEL YELL *Billy Idol*	82
FAME *Soundtrack*	82
4 *Foreigner*	81
KICK *INXS*	81
METAL HEALTH *Quiet Riot*	81
WHAT'S NEW	
Linda Ronstadt & The Nelson Riddle Orchestra	81
DIAMOND LIFE *Sade*	81
IT TAKES TWO* *Rob Base & D.J. E-Z Rock*	81
STRAIGHT OUTTA COMPTON* *N.W.A.*	81
MAKE IT BIG *Wham!*	80
THE DUDE *Quincy Jones*	80
NEW KIDS ON THE BLOCK*	
New Kids On The Block	80
LET'S GET IT STARTED* *MC Hammer*	80
GIRL YOU KNOW IT'S TRUE* *Milli Vanilli*	78
FLASHDANCE *Soundtrack*	78
SKID ROW* *Skid Row*	78
THE FINAL COUNTDOWN *Europe*	78
PRETENDERS *The Pretenders*	78
LIFE IS...TOO SHORT* *Too Short*	78
AMADEUS (SOUNDTRACK) *Neville Marriner*	78
1984 (MCMLXXXIV) *Van Halen*	77
LIKE A PRAYER* *Madonna*	77
LITTLE CREATURES *Talking Heads*	77
TIME AND TIDE *Basia*	77
IN MY TRIBE *10,000 Maniacs*	77
WHITESNAKE *Whitesnake*	76
NEW JERSEY* *Bon Jovi*	76
VIVID* *Living Colour*	76
SYNCHRONICITY *The Police*	75
SCARECROW *John Mellencamp*	75
PRIMITIVE LOVE *Miami Sound Machine*	75
SURFING WITH THE ALIEN *Joe Satriani*	75

Albums: 13 Weeks or More In Top 10

This table includes the entire lifecycle of albums peaking in the 1980s. In some cases this lifecycle extends outside the decade. Those albums are designated with an asterisk (*) in the title; to the left if the lifecycle begins before 1/1/80, to the right if beyond 12/31/89.

Title Act	Wks
BORN IN THE U.S.A. Bruce Springsteen	84
THRILLER Michael Jackson	78
HYSTERIA* Def Leppard	78
FOREVER YOUR GIRL* Paula Abdul	64
CAN'T SLOW DOWN Lionel Richie	59
APPETITE FOR DESTRUCTION* Guns N' Roses	52
FAITH George Michael	51
DIRTY DANCING Soundtrack	48
WHITNEY HOUSTON Whitney Houston	46
SLIPPERY WHEN WET Bon Jovi	46
DON'T BE CRUEL* Bobby Brown	45
HANGIN' TOUGH* New Kids On The Block	45
SPORTS Huey Lewis & The News	42
GIRL YOU KNOW IT'S TRUE* Milli Vanilli	41
WHITESNAKE Whitesnake	41
SYNCHRONICITY The Police	40
RECKLESS Bryan Adams	40
BAD Michael Jackson	39
PRIVATE DANCER Tina Turner	39
ESCAPE Journey	38
PYROMANIA Def Leppard	38
CONTROL Janet Jackson	37
BROTHERS IN ARMS Dire Straits	37
HEART Heart	37
JANET JACKSON'S RHYTHM NATION 1814* Janet Jackson	35
THE JOSHUA TREE U2	35
4 Foreigner	34
FULL MOON FEVER* Tom Petty	34
LIKE A VIRGIN Madonna	33
H2O Daryl Hall & John Oates	33
PURPLE RAIN (SOUNDTRACK) Prince And The Revolution	32
SONGS FROM THE BIG CHAIR Tears For Fears	32
NO JACKET REQUIRED Phil Collins	31
BUSINESS AS USUAL Men At Work	31
WHITNEY Whitney Houston	31
HEARTBEAT CITY The Cars	31
HI INFIDELITY REO Speedwagon	30
AN INNOCENT MAN Billy Joel	30
PUMP* Aerosmith	30
COLOUR BY NUMBERS Culture Club	30
*OFF THE WALL Michael Jackson	29
CRIMES OF PASSION Pat Benatar	29
SCARECROW John Mellencamp	29
THE WALL Pink Floyd	27
PARADISE THEATER Styx	27
THE RAW & THE COOKED* Fine Young Cannibals	27
ASIA Asia	27
1984 (MCMLXXXIV) Van Halen	27
THE LONESOME JUBILEE John Mellencamp	27

Title Act	Wks
BELLA DONNA Stevie Nicks	26
OPEN UP AND SAY...AHH! Poison	26
INVISIBLE TOUCH Genesis	26
FORE! Huey Lewis & The News	26
GLASS HOUSES Billy Joel	25
FLASHDANCE Soundtrack	25
MAKE IT BIG Wham!	25
TRUE BLUE Madonna	25
GHOST IN THE MACHINE The Police	24
GRACELAND Paul Simon	24
BACK IN BLACK* AC/DC	23
THE WAY IT IS Bruce Hornsby & The Range	23
AGAINST THE WIND Bob Seger & The Silver Bullet Band	22
KICK INXS	22
TATTOO YOU The Rolling Stones	22
NEW JERSEY* Bon Jovi	22
AMERICAN FOOL John Cougar	22
DOUBLE FANTASY John Lennon & Yoko Ono	22
FRONTIERS Journey	22
TOTO IV Toto	22
TIFFANY Tiffany	22
BEVERLY HILLS COP Soundtrack	22
VOLUME 1* Traveling Wilburys	22
LIONEL RICHIE Lionel Richie	21
SHE'S SO UNUSUAL Cyndi Lauper	21
ZENYATTA MONDATTA The Police	21
THE GAME Queen	21
KENNY ROGERS' GREATEST HITS Kenny Rogers	20
...BUT SERIOUSLY* Phil Collins	20
TOP GUN Soundtrack	20
*DAMN THE TORPEDOES Tom Petty And The Heartbreakers	20
FOOTLOOSE Soundtrack	20
DANCING ON THE CEILING Lionel Richie	20
THIRD STAGE Boston	20
THE JAZZ SINGER (SOUNDTRACK) Neil Diamond	19
FREEZE-FRAME The J. Geils Band	19
LICENSED TO ILL Beastie Boys	19
THE DREAM OF THE BLUE TURTLES Sting	19
STORM FRONT* Billy Joel	19
COCKTAIL Soundtrack	19
GUILTY Barbra Streisand	18
BUILT FOR SPEED Stray Cats	18
TRACY CHAPMAN Tracy Chapman	18
DIANA Diana Ross	18
*ON THE RADIO: GREATEST HITS: VOLUMES I & II Donna Summer	18
MIRAGE Fleetwood Mac	18
G N' R LIES* Guns N' Roses	18
GIVING YOU THE BEST THAT I GOT Anita Baker	18
MIAMI VICE TV Soundtrack	18

Title Act	Wks
BAD ANIMALS Heart	18
THE BROADWAY ALBUM Barbra Streisand	18
METAL HEALTH Quiet Riot	17
LOOK WHAT THE CAT DRAGGED IN Poison	17
LIKE A PRAYER* Madonna	16
OU812 Van Halen	16
LET'S DANCE David Bowie	16
HOTTER THAN JULY Stevie Wonder	16
PROMISE Sade	16
KILROY WAS HERE Styx	16
CHRISTOPHER CROSS Christopher Cross	15
BEAUTY AND THE BEAT Go-Go's	15
NIGHT SONGS Cinderella	15
ALWAYS ON MY MIND Willie Nelson	15
WHAT'S NEW Linda Ronstadt & The Nelson Riddle Orchestra	15
AFTERBURNER ZZ Top	15
CENTERFIELD John Fogerty	15
LIKE A ROCK Bob Seger & The Silver Bullet Band	15
A MOMENTARY LAPSE OF REASON Pink Floyd	15
KNEE DEEP IN THE HOOPLA Starship	15
GET LUCKY Loverboy	14
BACK IN THE HIGH LIFE Steve Winwood	14
RAISING HELL Run-D.M.C.	14
EMOTIONAL RESCUE The Rolling Stones	14
MOVING PICTURES Rush	14
5150 Van Halen	14
PRECIOUS TIME Pat Benatar	14
CHARIOTS OF FIRE (SOUNDTRACK) Vangelis	14
SEVEN AND THE RAGGED TIGER Duran Duran	14
RATTLE AND HUM (SOUNDTRACK) U2	14
ROLL WITH IT Steve Winwood	14
AROUND THE WORLD IN A DAY Prince And The Revolution	14
URBAN COWBOY Soundtrack	14
PHYSICAL Olivia Newton-John	14
CARGO Men At Work	14
THE DISTANCE Bob Seger & The Silver Bullet Band	14
OUT OF THE BLUE Debbie Gibson	13
HOLD OUT Jackson Browne	13
PRIVATE EYES Daryl Hall & John Oates	13
ELECTRIC YOUTH* Debbie Gibson	13
ARC OF A DIVER Steve Winwood	13
MORE DIRTY DANCING Soundtrack	13
STEEL WHEELS* The Rolling Stones	13
LOVE AT FIRST STING Scorpions	13
IN SQUARE CIRCLE Stevie Wonder	13
BIGGER AND DEFFER LL Cool J	13
*BEE GEES GREATEST Bee Gees	13

Albums: 31 Weeks or More In The Top 40

This table includes the entire lifecycle of albums peaking in the 1980s. In some cases this lifecycle extends outside the decade. Those albums are designated with an asterisk in the title; to the left if the lifecycle begins before 1/1/80, to the right if beyond 12/31/89.

Title Act	Wks
BORN IN THE U.S.A. Bruce Springsteen	96
THRILLER Michael Jackson	96
HYSTERIA* Def Leppard	91
FOREVER YOUR GIRL* Paula Abdul	82
CAN'T SLOW DOWN Lionel Richie	81
APPETITE FOR DESTRUCTION* Guns N' Roses	78
FAITH George Michael	78
DIRTY DANCING Soundtrack	78
WHITNEY HOUSTON Whitney Houston	78
SLIPPERY WHEN WET Bon Jovi	77
DON'T BE CRUEL* Bobby Brown	77
HANGIN' TOUGH* New Kids On The Block	72
SPORTS Huey Lewis & The News	72
GIRL YOU KNOW IT'S TRUE* Milli Vanilli	71
WHITESNAKE Whitesnake	70
SYNCHRONICITY The Police	69
RECKLESS Bryan Adams	69
BAD Michael Jackson	68
PRIVATE DANCER Tina Turner	66
ESCAPE Journey	65
PYROMANIA Def Leppard	65
CONTROL Janet Jackson	63
BROTHERS IN ARMS Dire Straits	62
HEART Heart	62
JANET JACKSON'S RHYTHM NATION 1814* Janet Jackson	62
THE JOSHUA TREE U2	61
4 Foreigner	61
FULL MOON FEVER* Tom Petty	60
LIKE A VIRGIN Madonna	60
H2O Daryl Hall & John Oates	58
PURPLE RAIN (SOUNDTRACK) Prince And The Revolution	58
SONGS FROM THE BIG CHAIR Tears For Fears	58
NO JACKET REQUIRED Phil Collins	58
BUSINESS AS USUAL Men At Work	58
WHITNEY Whitney Houston	57
HEARTBEAT CITY The Cars	57
HI INFIDELITY REO Speedwagon	57
AN INNOCENT MAN Billy Joel	56
PUMP* Aerosmith	55
COLOUR BY NUMBERS Culture Club	55
*OFF THE WALL Michael Jackson	55
CRIMES OF PASSION Pat Benatar	54
SCARECROW John Mellencamp	54
THE WALL Pink Floyd	54
PARADISE THEATER Styx	54
THE RAW & THE COOKED* Fine Young Cannibals	53
ASIA Asia	52
1984 (MCMLXXXIV) Van Halen	52
THE LONESOME JUBILEE John Mellencamp	52
BELLA DONNA Stevie Nicks	52
OPEN UP AND SAY...AHH! Poison	52
INVISIBLE TOUCH Genesis	52
FORE! Huey Lewis & The News	52
GLASS HOUSES Billy Joel	52
FLASHDANCE Soundtrack	51
MAKE IT BIG Wham!	51
TRUE BLUE Madonna	51
GHOST IN THE MACHINE The Police	51
GRACELAND Paul Simon	51
BACK IN BLACK* AC/DC	50
THE WAY IT IS Bruce Hornsby & The Range	50
AGAINST THE WIND Bob Seger & The Silver Bullet Band	50
KICK INXS	49
TATTOO YOU The Rolling Stones	49
NEW JERSEY* Bon Jovi	48
AMERICAN FOOL John Cougar	48
DOUBLE FANTASY John Lennon & Yoko Ono	48
FRONTIERS Journey	48
TOTO IV Toto	47
TIFFANY Tiffany	47
BEVERLY HILLS COP Soundtrack	47
VOLUME 1* Traveling Wilburys	46
LIONEL RICHIE Lionel Richie	46
SHE'S SO UNUSUAL Cyndi Lauper	46
ZENYATTA MONDATTA The Police	45
THE GAME Queen	45
KENNY ROGERS' GREATEST HITS Kenny Rogers	45
...BUT SERIOUSLY* Phil Collins	45
TOP GUN Soundtrack	44
*DAMN THE TORPEDOES Tom Petty And The Heartbreakers	44
FOOTLOOSE Soundtrack	44
DANCING ON THE CEILING Lionel Richie	43
THIRD STAGE Boston	42
THE JAZZ SINGER (SOUNDTRACK) Neil Diamond	42
FREEZE-FRAME The J. Geils Band	42
LICENSED TO ILL Beastie Boys	42
THE DREAM OF THE BLUE TURTLES Sting	42
STORM FRONT* Billy Joel	42
COCKTAIL Soundtrack	41
GUILTY Barbra Streisand	41
BUILT FOR SPEED Stray Cats	41
TRACY CHAPMAN Tracy Chapman	41
DIANA Diana Ross	40
*ON THE RADIO: GREATEST HITS: VOLUMES I & II Donna Summer	40
MIRAGE Fleetwood Mac	39
G N' R LIES* Guns N' Roses	39
GIVING YOU THE BEST THAT I GOT Anita Baker	39
MIAMI VICE TV Soundtrack	38
BAD ANIMALS Heart	38
THE BROADWAY ALBUM Barbra Streisand	38
METAL HEALTH Quiet Riot	38
LOOK WHAT THE CAT DRAGGED IN Poison	38
LIKE A PRAYER* Madonna	38
OU812 Van Halen	38
LET'S DANCE David Bowie	38
HOTTER THAN JULY Stevie Wonder	37
PROMISE Sade	37
KILROY WAS HERE Styx	37
CHRISTOPHER CROSS Christopher Cross	37
BEAUTY AND THE BEAT Go-Go's	37
NIGHT SONGS Cinderella	36
ALWAYS ON MY MIND Willie Nelson	36
WHAT'S NEW Linda Ronstadt & The Nelson Riddle Orchestra	36
AFTERBURNER ZZ Top	36
CENTERFIELD John Fogerty	36
LIKE A ROCK Bob Seger & The Silver Bullet Band	36
A MOMENTARY LAPSE OF REASON Pink Floyd	36
KNEE DEEP IN THE HOOPLA Starship	36
GET LUCKY Loverboy	35
BACK IN THE HIGH LIFE Steve Winwood	35
RAISING HELL Run-D.M.C.	35
EMOTIONAL RESCUE The Rolling Stones	35
MOVING PICTURES Rush	35
5150 Van Halen	35
PRECIOUS TIME Pat Benatar	35
CHARIOTS OF FIRE (SOUNDTRACK) Vangelis	35
SEVEN AND THE RAGGED TIGER Duran Duran	35
RATTLE AND HUM (SOUNDTRACK) U2	35
ROLL WITH IT Steve Winwood	35
AROUND THE WORLD IN A DAY Prince And The Revolution	35
URBAN COWBOY Soundtrack	34
PHYSICAL Olivia Newton-John	34
CARGO Men At Work	34
THE DISTANCE Bob Seger & The Silver Bullet Band	34
OUT OF THE BLUE Debbie Gibson	34
HOLD OUT Jackson Browne	34
PRIVATE EYES Daryl Hall & John Oates	34
ELECTRIC YOUTH* Debbie Gibson	33
ARC OF A DIVER Steve Winwood	33
MORE DIRTY DANCING Soundtrack	33
STEEL WHEELS* The Rolling Stones	33
LOVE AT FIRST STING Scorpions	33
IN SQUARE CIRCLE Stevie Wonder	33
BIGGER AND DEFFER LL Cool J	33
*BEE GEES GREATEST Bee Gees	32
REPEAT OFFENDER* Richard Marx	32
STREET SONGS Rick James	32
SO Peter Gabriel	32
MISTAKEN IDENTITY Kim Carnes	32
LONG DISTANCE VOYAGER The Moody Blues	32
CHICAGO 17 Chicago	32
DIRTY DEEDS DONE DIRT CHEAP AC/DC	32
WELCOME TO THE REAL WORLD Mr. Mister	31
I LOVE ROCK 'N ROLL Joan Jett & the Blackhearts	31
LOC-ED AFTER DARK* Tone-Loc	31
MAD LOVE Linda Ronstadt	31
BIG BAM BOOM Daryl Hall & John Oates	31
GIRLS, GIRLS, GIRLS Mötley Crüe	31
LOVE ZONE Billy Ocean	31
90125 Yes	31

Albums: Top 50 By Year of Peak

1980 n=598

Yr Rank. TITLE Act [Decade Rank]
1. OFF THE WALL Michael Jackson [22]
2. THE WALL Pink Floyd [23]
3. CHRISTOPHER CROSS Christopher Cross [26]
4. KENNY ROGERS' GREATEST HITS Kenny Rogers [30]
5. GLASS HOUSES Billy Joel [34]
6. AGAINST THE WIND Bob Seger & The Silver Bullet Band [37]
7. BACK IN BLACK AC/DC [46]
8. DOUBLE FANTASY John Lennon & Yoko Ono [61]
9. GUILTY Barbra Streisand [64]
10. THE GAME Queen [70]
11. DAMN THE TORPEDOES Tom Petty And The Heartbreakers [88]
12. DIANA Diana Ross [103]
13. EMOTIONAL RESCUE The Rolling Stones [106]
14. ON THE RADIO: GREATEST HITS: VOLUMES I & II Donna Summer [119]
15. THE RIVER Bruce Springsteen [126]
16. HOLD OUT Jackson Browne [133]
17. HOTTER THAN JULY Stevie Wonder [148]
18. IN THE HEAT OF THE NIGHT Pat Benatar [166]
19. URBAN COWBOY Soundtrack [167]
20. PHOENIX Dan Fogelberg [176]
21. KENNY Kenny Rogers [178]
22. JUST ONE NIGHT Eric Clapton [192]
23. MAD LOVE Linda Ronstadt [202]
24. THE DOORS GREATEST HITS The Doors [228]
25. GIVE ME THE NIGHT George Benson [235]
26. BEE GEES GREATEST Bee Gees [238]
27. LOST IN LOVE Air Supply [242]
28. FAME Soundtrack [245]
29. PRETENDERS The Pretenders [260]
30. XANADU (SOUNDTRACK) Olivia Newton-John/Electric Light Orchestra [264]
31. EMPTY GLASS Pete Townshend [279]
32. THE ROSE (SOUNDTRACK) Bette Midler [297]
33. KEEP THE FIRE Kenny Loggins [300]
34. MIDDLE MAN Boz Scaggs [305]
35. THE EMPIRE STRIKES BACK Soundtrack [310]
36. ONE STEP CLOSER The Doobie Brothers [313]
37. THE WHISPERS The Whispers [316]
38. DEPARTURE Journey [329]
39. DUKE Genesis [332]
40. LIGHT UP THE NIGHT The Brothers Johnson [351]
41. WOMEN AND CHILDREN FIRST Van Halen [359]
42. SWEET SENSATION Stephanie Mills [361]
43. EAGLES LIVE Eagles [369]
44. ANNE MURRAY'S GREATEST HITS Anne Murray [371]
45. LET'S GET SERIOUS Jermaine Jackson [373]
46. MOUTH TO MOUTH Lipps Inc. [378]
47. PERMANENT WAVES Rush [388]
48. FREEDOM AT POINT ZERO Jefferson Starship [390]
49. HEROES Commodores [392]
50. TRIUMPH The Jacksons [397]

1981 n=597

Yr Rank. TITLE Act [Decade Rank]
1. ESCAPE Journey [8]
2. HI INFIDELITY REO Speedwagon [11]
3. 4 Foreigner [12]
4. BELLA DONNA Stevie Nicks [27]
5. CRIMES OF PASSION Pat Benatar [42]
6. PARADISE THEATER Styx [44]
7. TATTOO YOU The Rolling Stones [49]
8. ZENYATTA MONDATTA The Police [66]
9. GHOST IN THE MACHINE The Police [73]
10. THE JAZZ SINGER (SOUNDTRACK) Neil Diamond [81]
11. FACE VALUE Phil Collins [84]
12. FEELS SO RIGHT Alabama [85]
13. DON'T SAY NO Billy Squier [87]
14. STREET SONGS Rick James [104]
15. MISTAKEN IDENTITY Kim Carnes [110]
16. LONG DISTANCE VOYAGER The Moody Blues [115]
17. MOVING PICTURES Rush [116]
18. WORKING CLASS DOG Rick Springfield [118]
19. PRECIOUS TIME Pat Benatar [125]
20. VOICES Daryl Hall & John Oates [138]
21. ARC OF A DIVER Steve Winwood [149]
22. DIRTY DEEDS DONE DIRT CHEAP AC/DC [155]
23. THE INNOCENT AGE Dan Fogelberg [156]
24. PHYSICAL Olivia Newton-John [169]
25. WINELIGHT Grover Washington Jr. [179]
26. ABACAB Genesis [184]
27. BREAKIN' AWAY Al Jarreau [193]
28. THE ONE THAT YOU LOVE Air Supply [195]
29. NINE TONIGHT Bob Seger & The Silver Bullet Band [200]
30. SOMETHING SPECIAL Kool & The Gang [214]
31. AUTOAMERICAN Blondie [225]
32. BLIZZARD OF OZZ Ozzy Osbourne [231]
33. FOR THOSE ABOUT TO ROCK (WE SALUTE YOU) AC/DC [241]
34. LOVERBOY Loverboy [244]
35. THE TURN OF A FRIENDLY CARD The Alan Parsons Project [248]
36. HARD PROMISES Tom Petty And The Heartbreakers [254]
37. DIARY OF A MADMAN Ozzy Osbourne [257]
38. CELEBRATE! Kool & The Gang [258]
39. ZEBOP! Santana [269]
40. SHAKE IT UP The Cars [277]
41. SHARE YOUR LOVE Kenny Rogers [280]
42. MEMORIES Barbra Streisand [294]
43. RAISE! Earth, Wind & Fire [309]
44. GAUCHO Steely Dan [311]
45. IN THE POCKET Commodores [320]
46. WILD-EYED SOUTHERN BOYS .38 Special [322]
47. PIRATES Rickie Lee Jones [340]
48. FACE DANCES The Who [350]
49. BEING WITH YOU Smokey Robinson [362]
50. CAPTURED Journey [365]

1982 n=500	1983 n=479
Yr Rank. TITLE Act [Decade Rank]	Yr Rank. TITLE Act [Decade Rank]
1. BUSINESS AS USUAL Men At Work [25]	1. THRILLER Michael Jackson [1]
2. GET LUCKY Loverboy [51]	2. CAN'T SLOW DOWN Lionel Richie [7]
3. LIONEL RICHIE Lionel Richie [52]	3. SYNCHRONICITY The Police [21]
4. BEAUTY AND THE BEAT Go-Go's [53]	4. ELIMINATOR ZZ Top [36]
5. ASIA Asia [55]	5. PYROMANIA Def Leppard [41]
6. AMERICAN FOOL John Cougar [58]	6. AN INNOCENT MAN Billy Joel [43]
7. FREEZE-FRAME The J. Geils Band [82]	7. FLASHDANCE Soundtrack [50]
8. BUILT FOR SPEED Stray Cats [92]	8. H2O Daryl Hall & John Oates [59]
9. TOTO IV Toto [97]	9. 1999 Prince [77]
10. MIRAGE Fleetwood Mac [121]	10. FRONTIERS Journey [90]
11. ALWAYS ON MY MIND Willie Nelson [130]	11. METAL HEALTH Quiet Riot [108]
12. PRIVATE EYES Daryl Hall & John Oates [135]	12. LET'S DANCE David Bowie [147]
13. THE DUDE Quincy Jones [140]	13. WHAT'S NEW Linda Ronstadt & The Nelson Riddle Orchestra [159]
14. CHARIOTS OF FIRE (SOUNDTRACK) Vangelis [145]	14. KISSING TO BE CLEVER Culture Club [172]
15. I LOVE ROCK 'N ROLL Joan Jett & the Blackhearts [162]	15. COMBAT ROCK The Clash [173]
16. JUICE Juice Newton [171]	16. JANE FONDA'S WORKOUT RECORD Jane Fonda [183]
17. HOOKED ON CLASSICS Royal Philharmonic Orchestra Conducted By Louis Clark [181]	17. RIO Duran Duran [190]
18. EMOTIONS IN MOTION Billy Squier [206]	18. CARGO Men At Work [207]
19. MOUNTAIN MUSIC Alabama [212]	19. WAR U2 [227]
20. NIGHT AND DAY Joe Jackson [229]	20. HELLO, I MUST BE GOING! Phil Collins [230]
21. DARE The Human League [265]	21. REACH THE BEACH The Fixx [232]
22. EYE OF THE TIGER Survivor [268]	22. THE WILD HEART Stevie Nicks [239]
23. OLIVIA'S GREATEST HITS, VOL. 2 Olivia Newton-John [272]	23. GET NERVOUS Pat Benatar [261]
24. QUARTERFLASH Quarterflash [281]	24. CUTS LIKE A KNIFE Bryan Adams [263]
25. TUG OF WAR Paul McCartney [283]	25. KILROY WAS HERE Styx [275]
26. SUCCESS HASN'T SPOILED ME YET Rick Springfield [286]	26. THE DISTANCE Bob Seger & The Silver Bullet Band [291]
27. ABRACADABRA The Steve Miller Band [298]	27. GENESIS Genesis [292]
28. THE NYLON CURTAIN Billy Joel [301]	28. GREATEST HITS Air Supply [296]
29. DAYLIGHT AGAIN Crosby, Stills & Nash [304]	29. EYES THAT SEE IN THE DARK Kenny Rogers [299]
30. A FLOCK OF SEAGULLS A Flock Of Seagulls [333]	30. SPEAKING IN TONGUES Talking Heads [317]
31. DIVER DOWN Van Halen [337]	31. LIVING IN OZ Rick Springfield [342]
32. EYE IN THE SKY The Alan Parsons Project [341]	32. KEEP IT UP Loverboy [345]
33. ALDO NOVA Aldo Nova [372]	33. THE CLOSER YOU GET Alabama [356]
34. BLACKOUT Scorpions [382]	34. SPRING SESSION M Missing Persons [367]
35. SCREAMING FOR VENGEANCE Judas Priest [410]	35. THE PRINCIPLE OF MOMENTS Robert Plant [379]
36. PICTURES AT ELEVEN Robert Plant [412]	36. SWEET DREAMS (ARE MADE OF THIS) Eurythmics [391]
37. "...FAMOUS LAST WORDS..." Supertramp [419]	37. LONG AFTER DARK Tom Petty And The Heartbreakers [401]
38. GAP BAND IV The Gap Band [427]	38. DURAN DURAN Duran Duran [404]
39. HEARTLIGHT Neil Diamond [432]	39. FASTER THAN THE SPEED OF NIGHT Bonnie Tyler [414]
40. PICTURE THIS Huey Lewis & The News [435]	40. JULIO Julio Iglesias [478]
41. CHICAGO 16 Chicago [439]	41. THE LEXICON OF LOVE ABC [486]
42. GOOD TROUBLE REO Speedwagon [454]	42. THE CROSSING Big Country [493]
43. MIDNIGHT LOVE Marvin Gaye [456]	43. PIECE OF MIND Iron Maiden [519]
44. NEBRASKA Bruce Springsteen [474]	44. FUTURE SHOCK Herbie Hancock [541]
45. HIGH ADVENTURE Kenny Loggins [487]	45. SHE WORKS HARD FOR THE MONEY Donna Summer [543]
46. ALL FOUR ONE The Motels [490]	46. LAWYERS IN LOVE Jackson Browne [547]
47. NON-STOP EROTIC CABARET Soft Cell [506]	47. KILLER ON THE RAMPAGE Eddy Grant [555]
48. SPECIAL FORCES 38 Special [509]	48. FRIEND OR FOE Adam Ant [559]
49. THE CONCERT IN CENTRAL PARK Simon & Garfunkel [535]	49. LIVE FROM EARTH Pat Benatar [564]
50. THE GEORGE BENSON COLLECTION George Benson [551]	50. JARREAU Al Jarreau [565]

Albums: Special Lists

1984 n=456	
Yr Rank. TITLE *Act [Decade Rank]*	
1. BORN IN THE U.S.A. *Bruce Springsteen [2]*	
2. SPORTS *Huey Lewis & The News [14]*	
3. PURPLE RAIN (SOUNDTRACK) *Prince And The Revolution [20]*	
4. PRIVATE DANCER *Tina Turner [33]*	
5. SHE'S SO UNUSUAL *Cyndi Lauper [63]*	
6. BREAK OUT *The Pointer Sisters [68]*	
7. 1984 (MCMLXXXIV) *Van Halen [71]*	
8. HEARTBEAT CITY *The Cars [74]*	
9. COLOUR BY NUMBERS *Culture Club [91]*	
10. FOOTLOOSE *Soundtrack [94]*	
11. MADONNA *Madonna [113]*	
12. SUDDENLY *Billy Ocean [120]*	
13. SEVEN AND THE RAGGED TIGER *Duran Duran [152]*	
14. THE BIG CHILL *Soundtrack [170]*	
15. UH-HUH *John Mellencamp [182]*	
16. REBEL YELL *Billy Idol [185]*	
17. BIG BAM BOOM *Daryl Hall & John Oates [203]*	
18. MIDNIGHT MADNESS *Night Ranger [208]*	
19. SHOUT AT THE DEVIL *Mötley Crüe [211]*	
20. LOVE AT FIRST STING *Scorpions [217]*	
21. THE UNFORGETTABLE FIRE *U2 [224]*	
22. 90125 *Yes [233]*	
23. OUT OF THE CELLAR *Ratt [246]*	
24. NO PARKING ON THE DANCE FLOOR *Midnight Star [276]*	
25. ROCK 'N SOUL PART 1 *Daryl Hall & John Oates [302]*	
26. UNDER A BLOOD RED SKY *U2 [303]*	
27. THE WOMAN IN RED (SOUNDTRACK) *Stevie Wonder [314]*	
28. EDDIE & THE CRUISERS (SOUNDTRACK) *John Cafferty & The Beaver Brown Band [318]*	
29. LEARNING TO CRAWL *The Pretenders [319]*	
30. INTO THE GAP *Thompson Twins [324]*	
31. STAY WITH ME TONIGHT *Jeffrey Osborne [325]*	
32. 1100 BEL AIR PLACE *Julio Iglesias [347]*	
33. TOUCH *Eurythmics [349]*	
34. DECEMBER *George Winston [352]*	
35. STAY HUNGRY *Twisted Sister [354]*	
36. ICE CREAM CASTLE *The Time [355]*	
37. STREET TALK *Steve Perry [360]*	
38. VOLUME ONE *The Honeydrippers [366]*	
39. NO BRAKES *John Waite [429]*	
40. GHOSTBUSTERS *Soundtrack [453]*	
41. VICTORY *The Jacksons [472]*	
42. JERMAINE JACKSON *Jermaine Jackson [473]*	
43. ROLL ON *Alabama [492]*	
44. STOP MAKING SENSE *Talking Heads [514]*	
45. I FEEL FOR YOU *Chaka Khan [515]*	
46. SWEPT AWAY *Diana Ross [523]*	
47. SELF CONTROL *Laura Branigan [530]*	
48. BON JOVI *Bon Jovi [532]*	
49. LEGEND *Bob Marley And The Wailers [550]*	
50. TOUR DE FORCE *38 Special [552]*	

1985 n=412	
Yr Rank. TITLE *Act [Decade Rank]*	
1. NO JACKET REQUIRED *Phil Collins [18]*	
2. BROTHERS IN ARMS *Dire Straits [31]*	
3. HEART *Heart [38]*	
4. RECKLESS *Bryan Adams [39]*	
5. SONGS FROM THE BIG CHAIR *Tears For Fears [40]*	
6. LIKE A VIRGIN *Madonna [45]*	
7. MAKE IT BIG *Wham! [54]*	
8. SCARECROW *John Mellencamp [65]*	
9. THE DREAM OF THE BLUE TURTLES *Sting [107]*	
10. BEVERLY HILLS COP *Soundtrack [123]*	
11. CHICAGO 17 *Chicago [127]*	
12. MIAMI VICE *TV Soundtrack [143]*	
13. AROUND THE WORLD IN A DAY *Prince And The Revolution [164]*	
14. AFTERBURNER *ZZ Top [165]*	
15. CENTERFIELD *John Fogerty [174]*	
16. EMERGENCY *Kool & The Gang [177]*	
17. DIAMOND LIFE *Sade [213]*	
18. IN SQUARE CIRCLE *Stevie Wonder [222]*	
19. ROCK ME TONIGHT *Freddie Jackson [252]*	
20. BUILDING THE PERFECT BEAST *Don Henley [253]*	
21. GREATEST HITS VOL. I & II *Billy Joel [262]*	
22. NEW EDITION *New Edition [271]*	
23. THE POWER STATION *The Power Station [278]*	
24. WHO'S ZOOMIN' WHO? *Aretha Franklin [282]*	
25. AGENT PROVOCATEUR *Foreigner [284]*	
26. THEATRE OF PAIN *Mötley Crüe [289]*	
27. 7 WISHES *Night Ranger [306]*	
28. WHEELS ARE TURNIN' *REO Speedwagon [312]*	
29. LITTLE CREATURES *Talking Heads [338]*	
30. VITAL SIGNS *Survivor [339]*	
31. BE YOURSELF TONIGHT *Eurythmics [344]*	
32. VALOTTE *Julian Lennon [346]*	
33. DREAM INTO ACTION *Howard Jones [374]*	
34. READY FOR THE WORLD *Ready For The World [393]*	
35. LOVIN' EVERY MINUTE OF IT *Loverboy [403]*	
36. SOUTHERN ACCENTS *Tom Petty And The Heartbreakers [413]*	
37. HUNTING HIGH AND LOW *a-ha [415]*	
38. 7800 DEGREES FAHRENHEIT *Bon Jovi [421]*	
39. WE ARE THE WORLD *USA For Africa [443]*	
40. INVASION OF YOUR PRIVACY *Ratt [449]*	
41. THE SECRET OF ASSOCIATION *Paul Young [455]*	
42. A PRIVATE HEAVEN *Sheena Easton [458]*	
43. ARENA *Duran Duran [464]*	
44. THE ALLNIGHTER *Glenn Frey [465]*	
45. RHYTHM OF THE NIGHT *DeBarge [466]*	
46. CRAZY FROM THE HEAT *David Lee Roth [482]*	
47. WORLD WIDE LIVE *Scorpions [491]*	
48. THE NIGHT I FELL IN LOVE *Luther Vandross [494]*	
49. BOY IN THE BOX *Corey Hart [534]*	
50. GREATEST HITS *The Cars [536]*	

Albums: Special Lists

1986 n=438

Yr Rank. TITLE Act [Decade Rank]

1. WHITNEY HOUSTON Whitney Houston [5]
2. SLIPPERY WHEN WET Bon Jovi [15]
3. CONTROL Janet Jackson [28]
4. INVISIBLE TOUCH Genesis [67]
5. TRUE BLUE Madonna [72]
6. TOP GUN Soundtrack [83]
7. BACK IN THE HIGH LIFE Steve Winwood [99]
8. FORE! Huey Lewis & The News [100]
9. RAISING HELL Run-D.M.C. [101]
10. SO Peter Gabriel [105]
11. 5150 Van Halen [124]
12. DANCING ON THE CEILING Lionel Richie [129]
13. THIRD STAGE Boston [146]
14. PROMISE Sade [154]
15. WELCOME TO THE REAL WORLD Mr. Mister [157]
16. THE BROADWAY ALBUM Barbra Streisand [175]
17. LIKE A ROCK Bob Seger & The Silver Bullet Band [186]
18. RIPTIDE Robert Palmer [197]
19. LOVE ZONE Billy Ocean [215]
20. RAISED ON RADIO Journey [220]
21. KNEE DEEP IN THE HOOPLA Starship [236]
22. THE BRIDGE Billy Joel [243]
23. PLAY DEEP The Outfield [259]
24. NERVOUS NIGHT Hooters [267]
25. PRIMITIVE LOVE Miami Sound Machine [270]
26. WINNER IN YOU Patti LaBelle [274]
27. BRUCE SPRINGSTEEN & THE E STREET BAND LIVE 1975-1985 Bruce Springsteen [288]
28. WORD UP! Cameo [307]
29. TRUE COLORS Cyndi Lauper [323]
30. EAT 'EM AND SMILE David Lee Roth [326]
31. ONCE UPON A TIME Simple Minds [335]
32. TUFF ENUFF The Fabulous Thunderbirds [336]
33. THE JETS The Jets [357]
34. BREAK EVERY RULE Tina Turner [363]
35. PARADE: MUSIC FROM THE MOTION PICTURE UNDER THE CHERRY MOON (SOUNDTRACK) Prince And The Revolution [380]
36. THE OTHER SIDE OF LIFE The Moody Blues [384]
37. PLEASE Pet Shop Boys [389]
38. CAN'T HOLD BACK Eddie Money [394]
39. LISTEN LIKE THIEVES INXS [400]
40. WHIPLASH SMILE Billy Idol [408]
41. PRETTY IN PINK Soundtrack [420]
42. THE ULTIMATE SIN Ozzy Osbourne [425]
43. MEETING IN THE LADIES ROOM Klymaxx [438]
44. FALCO 3 Falco [446]
45. MIKE + THE MECHANICS Mike + The Mechanics [471]
46. PICTURE BOOK Simply Red [477]
47. ROCK A LITTLE Stevie Nicks [480]
48. AS THE BAND TURNS Atlantic Starr [489]
49. DIRTY WORK The Rolling Stones [496]
50. HERE'S TO FUTURE DAYS Thompson Twins [511]

1987 n=436

Yr Rank. TITLE Act [Decade Rank]

1. DIRTY DANCING Soundtrack [9]
2. BAD Michael Jackson [24]
3. THE JOSHUA TREE U2 [29]
4. WHITESNAKE Whitesnake [32]
5. WHITNEY Whitney Houston [35]
6. RAPTURE Anita Baker [60]
7. GRACELAND Paul Simon [78]
8. LICENSED TO ILL Beastie Boys [95]
9. THE WAY IT IS Bruce Hornsby & The Range [98]
10. LOOK WHAT THE CAT DRAGGED IN Poison [111]
11. NIGHT SONGS Cinderella [114]
12. DIFFERENT LIGHT Bangles [128]
13. THE LONESOME JUBILEE John Mellencamp [141]
14. DUOTONES Kenny G [142]
15. BAD ANIMALS Heart [144]
16. TANGO IN THE NIGHT Fleetwood Mac [150]
17. PERMANENT VACATION Aerosmith [151]
18. THE FINAL COUNTDOWN Europe [161]
19. TUNNEL OF LOVE Bruce Springsteen [180]
20. A MOMENTARY LAPSE OF REASON Pink Floyd [189]
21. JODY WATLEY Jody Watley [201]
22. GIRLS, GIRLS, GIRLS Mötley Crüe [209]
23. BIGGER AND DEFFER LL Cool J [223]
24. LA BAMBA (SOUNDTRACK) Los Lobos [251]
25. SPANISH FLY Lisa Lisa And Cult Jam [256]
26. ALWAYS & FOREVER Randy Travis [285]
27. ...NOTHING LIKE THE SUN Sting [290]
28. GIVE ME THE REASON Luther Vandross [321]
29. CRUSHIN' Fat Boys [334]
30. SIGN 'O' THE TIMES Prince [348]
31. STRONG PERSUADER The Robert Cray Band [368]
32. INTO THE FIRE Bryan Adams [383]
33. LIFE, LOVE & PAIN Club Nouveau [386]
34. CROWDED HOUSE Crowded House [387]
35. JUST LIKE THE FIRST TIME Freddie Jackson [423]
36. IN THE DARK Grateful Dead [437]
37. ONCE BITTEN Great White [445]
38. GEORGIA SATELLITES The Georgia Satellites [448]
39. DOCUMENT R.E.M. [457]
40. COMING AROUND AGAIN Carly Simon [467]
41. TRIO Dolly Parton, Linda Ronstadt, Emmylou Harris [479]
42. THE THIN RED LINE Glass Tiger [481]
43. BEVERLY HILLS COP II Soundtrack [488]
44. ACTUALLY Pet Shop Boys [497]
45. ONE HEARTBEAT Smokey Robinson [501]
46. VITAL IDOL Billy Idol [512]
47. SOLITUDE STANDING Suzanne Vega [517]
48. TO HELL WITH THE DEVIL Stryper [529]
49. TRIBUTE Ozzy Osbourne/Randy Rhoads [531]
50. BACK FOR THE ATTACK Dokken [554]

Albums: Special Lists

1988 n=477

Yr Rank. TITLE *Act [Decade Rank]*
1. HYSTERIA *Def Leppard [3]*
2. APPETITE FOR DESTRUCTION *Guns N' Roses [6]*
3. FAITH *George Michael [10]*
4. KICK *INXS [48]*
5. NEW JERSEY *Bon Jovi [56]*
6. OPEN UP AND SAY...AHH! *Poison [62]*
7. TRACY CHAPMAN *Tracy Chapman [93]*
8. TIFFANY *Tiffany [102]*
9. OU812 *Van Halen [109]*
10. LET IT LOOSE *Gloria Estefan & Miami Sound Machine [112]*
11. RICHARD MARX *Richard Marx [117]*
12. OUT OF THE BLUE *Debbie Gibson [122]*
13. GIVING YOU THE BEST THAT I GOT *Anita Baker [136]*
14. RATTLE AND HUM (SOUNDTRACK) *U2 [153]*
15. ROLL WITH IT *Steve Winwood [158]*
16. LONG COLD WINTER *Cinderella [160]*
17. MORE DIRTY DANCING *Soundtrack [163]*
18. INTRODUCING THE HARDLINE ACCORDING TO TERENCE TRENT D'ARBY *Terence Trent D'Arby [188]*
19. WHENEVER YOU NEED SOMEBODY *Rick Astley [191]*
20. ...AND JUSTICE FOR ALL *Metallica [196]*
21. EXPOSURE *Exposé [204]*
22. NOW AND ZEN *Robert Plant [210]*
23. HE'S THE D.J., I'M THE RAPPER *D.J. Jazzy Jeff & The Fresh Prince [216]*
24. PRIDE *White Lion [226]*
25. MAKE IT LAST FOREVER *Keith Sweat [237]*
26. HEART BREAK *New Edition [247]*
27. TELL IT TO MY HEART *Taylor Dayne [249]*
28. SILHOUETTE *Kenny G [250]*
29. HEAVEN ON EARTH *Belinda Carlisle [266]*
30. THE PHANTOM OF THE OPERA *Original London Cast Recording [273]*
31. LAP OF LUXURY *Cheap Trick [295]*
32. SIMPLE PLEASURES *Bobby McFerrin [315]*
33. STRONGER THAN PRIDE *Sade [327]*
34. SAVAGE AMUSEMENT *Scorpions [328]*
35. DIESEL AND DUST *Midnight Oil [330]*
36. IN EFFECT MODE *Al B. Sure! [343]*
37. LITA *Lita Ford [353]*
38. SCENES FROM THE SOUTHSIDE *Bruce Hornsby & The Range [376]*
39. CLOUD NINE *George Harrison [385]*
40. HEAVY NOVA *Robert Palmer [395]*
41. SKYSCRAPER *David Lee Roth [402]*
42. PEBBLES *Pebbles [407]*
43. IT TAKES TWO *Rob Base & D.J. E-Z Rock [416]*
44. UP YOUR ALLEY *Joan Jett & the Blackhearts [428]*
45. LABOUR OF LOVE *UB40 [441]*
46. ANY LOVE *Luther Vandross [483]*
47. TIME AND TIDE *Basia [484]*
48. SURFING WITH THE ALIEN *Joe Satriani [485]*
49. GOOD MORNING, VIETNAM *Soundtrack [507]*
50. ALL THAT JAZZ *Breathe [510]*

1989 n=499

Yr Rank. TITLE *Act [Decade Rank]*
1. FOREVER YOUR GIRL *Paula Abdul [4]*
2. DON'T BE CRUEL *Bobby Brown [13]*
3. GIRL YOU KNOW IT'S TRUE *Milli Vanilli [16]*
4. HANGIN' TOUGH *New Kids On The Block [17]*
5. JANET JACKSON'S RHYTHM NATION 1814 *Janet Jackson [19]*
6. THE RAW & THE COOKED *Fine Young Cannibals [47]*
7. ...BUT SERIOUSLY *Phil Collins [57]*
8. FULL MOON FEVER *Tom Petty [69]*
9. PUMP *Aerosmith [75]*
10. DR. FEELGOOD *Mötley Crüe [76]*
11. SKID ROW *Skid Row [79]*
12. THE END OF THE INNOCENCE *Don Henley [80]*
13. LIKE A PRAYER *Madonna [86]*
14. BEACHES (SOUNDTRACK) *Bette Midler [89]*
15. REPEAT OFFENDER *Richard Marx [96]*
16. STORM FRONT *Billy Joel [131]*
17. G N' R LIES *Guns N' Roses [132]*
18. VOLUME 1 *Traveling Wilburys [134]*
19. COCKTAIL *Soundtrack [137]*
20. ELECTRIC YOUTH *Debbie Gibson [139]*
21. OUT OF ORDER *Rod Stewart [168]*
22. CUTS BOTH WAYS *Gloria Estefan [187]*
23. LOC-ED AFTER DARK *Tone-Loc [194]*
24. VIVID *Living Colour [198]*
25. DIRTY ROTTEN FILTHY STINKING RICH *Warrant [199]*
26. STEEL WHEELS *The Rolling Stones [205]*
27. SHOOTING RUBBERBANDS AT THE STARS *Edie Brickell & The New Bohemians [218]*
28. BATMAN (SOUNDTRACK) *Prince [219]*
29. KEEP ON MOVIN' *Soul II Soul [221]*
30. STONE COLD RHYMIN' *Young M.C. [234]*
31. HEART OF STONE *Cher [240]*
32. WINGER *Winger [255]*
33. ...TWICE SHY *Great White [287]*
34. GREEN *R.E.M. [293]*
35. DISINTEGRATION *The Cure [308]*
36. NEW KIDS ON THE BLOCK *New Kids On The Block [331]*
37. JOURNEY'S GREATEST HITS *Journey [358]*
38. THE GREAT RADIO CONTROVERSY *Tesla [364]*
39. LET'S GET IT STARTED *MC Hammer [375]*
40. GUY *Guy [377]*
41. MYSTERY GIRL *Roy Orbison [396]*
42. KARYN WHITE *Karyn White [418]*
43. BIG TYME *Heavy D & The Boyz [426]*
44. EVERYTHING *Bangles [431]*
45. BAD ENGLISH *Bad English [436]*
46. SONIC TEMPLE *The Cult [440]*
47. MELISSA ETHERIDGE *Melissa Etheridge [442]*
48. STRAIGHT OUTTA COMPTON *N.W.A. [444]*
49. TRASH *Alice Cooper [452]*
50. EAZY-DUZ-IT *Eazy-E [461]*

Albums: Number 1s By Weeks

Entries by weeks, in order of declining score

Title Act	Peak Wks
THRILLER Michael Jackson	37
PURPLE RAIN (SOUNDTRACK) Prince And The Revolution	24
DIRTY DANCING Soundtrack	18
SYNCHRONICITY The Police	17
HI INFIDELITY REO Speedwagon	15
THE WALL Pink Floyd	15
BUSINESS AS USUAL Men At Work	15
WHITNEY HOUSTON Whitney Houston	14
FAITH George Michael	12
WHITNEY Whitney Houston	11
MIAMI VICE TV Soundtrack	11
FOREVER YOUR GIRL Paula Abdul	10
4 Foreigner	10
FOOTLOOSE Soundtrack	10
THE JOSHUA TREE U2	9
BROTHERS IN ARMS Dire Straits	9
TATTOO YOU The Rolling Stones	9
ASIA Asia	9
AMERICAN FOOL John Cougar	9
SLIPPERY WHEN WET Bon Jovi	8
DOUBLE FANTASY John Lennon & Yoko Ono	8
BORN IN THE U.S.A. Bruce Springsteen	7
GIRL YOU KNOW IT'S TRUE Milli Vanilli	7
NO JACKET REQUIRED Phil Collins	7
THE RAW & THE COOKED Fine Young Cannibals	7

Title Act	Peak Wks
LICENSED TO ILL Beastie Boys	7
EMOTIONAL RESCUE The Rolling Stones	7
BRUCE SPRINGSTEEN & THE E STREET BAND LIVE 1975-1985 Bruce Springsteen	7
HYSTERIA Def Leppard	6
DON'T BE CRUEL Bobby Brown	6
BAD Michael Jackson	6
GLASS HOUSES Billy Joel	6
AGAINST THE WIND Bob Seger & The Silver Bullet Band	6
BEAUTY AND THE BEAT Go-Go's	6
LIKE A PRAYER Madonna	6
RATTLE AND HUM (SOUNDTRACK) U2	6
BATMAN (SOUNDTRACK) Prince	6
APPETITE FOR DESTRUCTION Guns N' Roses	5
SONGS FROM THE BIG CHAIR Tears For Fears	5
THE GAME Queen	5
TRUE BLUE Madonna	5
TOP GUN Soundtrack	5
MIRAGE Fleetwood Mac	5
ELECTRIC YOUTH Debbie Gibson	5
JANET JACKSON'S RHYTHM NATION 1814 Janet Jackson	4
NEW JERSEY Bon Jovi	4
...BUT SERIOUSLY Phil Collins	4
FREEZE-FRAME The J. Geils Band	4
OU812 Van Halen	4

Title Act	Peak Wks
MISTAKEN IDENTITY Kim Carnes	4
THE RIVER Bruce Springsteen	4
GIVING YOU THE BEST THAT I GOT Anita Baker	4
CHARIOTS OF FIRE (SOUNDTRACK) Vangelis	4
THIRD STAGE Boston	4
CAN'T SLOW DOWN Lionel Richie	3
PARADISE THEATER Styx	3
LIKE A VIRGIN Madonna	3
MAKE IT BIG Wham!	3
GUILTY Barbra Streisand	3
LONG DISTANCE VOYAGER Moody Blues	3
5150 Van Halen	3
AROUND THE WORLD IN A DAY Prince And The Revolution	3
THE BROADWAY ALBUM Barbra Streisand	3
FOR THOSE ABOUT TO ROCK (WE SALUTE YOU) AC/DC	3
TUG OF WAR Paul McCartney	3
WE ARE THE WORLD USA For Africa	3
HANGIN' TOUGH New Kids On The Block	2
CONTROL Janet Jackson	2
KENNY ROGERS' GREATEST HITS Kenny Rogers	2
RECKLESS Bryan Adams	2
FLASHDANCE Soundtrack	2
DR. FEELGOOD Mötley Crüe	2
TIFFANY Tiffany	2

Title Act	Peak Wks
BEVERLY HILLS COP Soundtrack	2
DANCING ON THE CEILING Lionel Richie	2
PROMISE Sade	2
LA BAMBA (SOUNDTRACK) Los Lobos	2
ESCAPE Journey	1
SPORTS Huey Lewis & The News	1
BELLA DONNA Stevie Nicks	1
HEART Heart	1
TRACY CHAPMAN Tracy Chapman	1
REPEAT OFFENDER Richard Marx	1
FORE! Huey Lewis & The News	1
METAL HEALTH Quiet Riot	1
ON THE RADIO: GREATEST HITS: VOLUMES I & II Donna Summer	1
PRECIOUS TIME Pat Benatar	1
STORM FRONT Billy Joel	1
HOLD OUT Jackson Browne	1
WELCOME TO THE REAL WORLD Mr. Mister	1
ROLL WITH IT Steve Winwood	1
CENTERFIELD John Fogerty	1
TUNNEL OF LOVE Bruce Springsteen	1
LOC-ED AFTER DARK Tone-Loc	1
BEE GEES GREATEST Bee Gees	1
WINNER IN YOU Patti LaBelle	1

	1980	1981	1982	1983	1984	1985	1986	1987	1988	1989
Entries	598	597	500	479	456	412	438	436	477	499
Number 1s	12	10	8	5	4	12	14	7	10	14

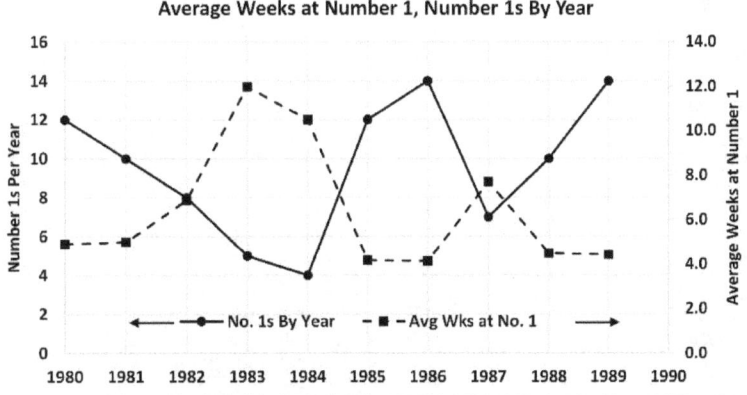

Average Weeks at Number 1, Number 1s By Year

The spike in average in 1983 and 1984 is driven by two exceptionally long-lived Number 1s: Thriller and Purple Rain.

Albums: Special Lists

Albums: Highest Scoring Missing The Weekly Top 5

Title - Act Peak (Year)	Score
CHRISTOPHER CROSS - Christopher Cross Pk: 6 (80)	12682
ELIMINATOR - ZZ Top Pk: 9 (83)	11592
GET LUCKY - Loverboy Pk: 7 (82)	9866
RAPTURE - Anita Baker Pk: 11 (87)	9452
BREAK OUT - The Pointer Sisters Pk: 8 (84)	9038
1999 - Prince Pk: 9 (83)	8588
SKID ROW - Skid Row Pk: 6 (89)	8481
THE END OF THE INNOCENCE - Don Henley Pk: 8 (89)	8388
FACE VALUE - Phil Collins Pk: 7 (81)	8163
FEELS SO RIGHT - Alabama Pk: 16 (81)	8135
LET IT LOOSE - Gloria Estefan & Miami Sound Machine Pk: 6 (88)	6903
MADONNA - Madonna Pk: 8 (84)	6893
RICHARD MARX - Richard Marx Pk: 8 (88)	6799
WORKING CLASS DOG - Rick Springfield Pk: 7 (81)	6754
SUDDENLY - Billy Ocean Pk: 9 (84)	6704
OUT OF THE BLUE - Debbie Gibson Pk: 7 (88)	6673
VOICES - Daryl Hall & John Oates Pk: 17 (81)	6297
THE DUDE - Quincy Jones Pk: 10 (82)	6231
THE LONESOME JUBILEE - John Mellencamp Pk: 6 (87)	6213
DUOTONES - Kenny G Pk: 6 (87)	6209
TANGO IN THE NIGHT - Fleetwood Mac Pk: 7 (87)	5973
PERMANENT VACATION - Aerosmith Pk: 11 (87)	5898
SEVEN AND THE RAGGED TIGER - Duran Duran Pk: 8 (84)	5895
THE INNOCENT AGE - Dan Fogelberg Pk: 6 (81)	5782
LONG COLD WINTER - Cinderella Pk: 10 (88)	5711

Albums: Lowest Scoring Number 1s

Title - Act Peak (Year)	Score
WE ARE THE WORLD - USA For Africa (85)	2778
BRUCE SPRINGSTEEN & THE E STREET BAND LIVE 1975-1985 - Bruce Springsteen (86)	3897
TUG OF WAR - Paul McCartney (82)	3951
WINNER IN YOU - Patti LaBelle (86)	4001
LA BAMBA (SOUNDTRACK) - Los Lobos (87)	4269
FOR THOSE ABOUT TO ROCK (WE SALUTE YOU) - AC/DC (81)	4377
BEE GEES GREATEST - Bee Gees (80)	4406
BATMAN (SOUNDTRACK) - Prince (89)	4618
LOC-ED AFTER DARK - Tone-Loc (89)	5062
TUNNEL OF LOVE - Bruce Springsteen (87)	5349
THE BROADWAY ALBUM - Barbra Streisand (86)	5422
CENTERFIELD - John Fogerty (85)	5442
AROUND THE WORLD IN A DAY - Prince And The Revolution (85)	5561
ROLL WITH IT - Steve Winwood (88)	5737
WELCOME TO THE REAL WORLD - Mr. Mister (86)	5753
PROMISE - Sade (86)	5873
RATTLE AND HUM (SOUNDTRACK) - U2 (88)	5894
THIRD STAGE - Boston (86)	6153
CHARIOTS OF FIRE (SOUNDTRACK) - Vangelis (82)	6161
MIAMI VICE - TV Soundtrack (85)	6180
ELECTRIC YOUTH - Debbie Gibson (89)	6295
GIVING YOU THE BEST THAT I GOT - Anita Baker (88)	6333
HOLD OUT - Jackson Browne (80)	6411
STORM FRONT - Billy Joel (89)	6443
DANCING ON THE CEILING - Lionel Richie (86)	6483

Albums: Highest Scoring with No Hot 100 Singles

Rank. Title - Act Peak(Peak Weeks) Total Weeks	Score
155. DIRTY DEEDS DONE DIRT CHEAP - AC/DC 3(6) 55	5829
170. THE BIG CHILL - Soundtrack 17(2) 161	5525
183. JANE FONDA'S WORKOUT RECORD - Jane Fonda 15(4) 120	5318
228. THE DOORS GREATEST HITS - The Doors 17(2) 99	4507
231. BLIZZARD OF OZZ - Ozzy Osbourne 21(1) 104	4478
238. BEE GEES GREATEST - Bee Gees 1(1) 32	4406
257. DIARY OF A MADMAN - Ozzy Osbourne 16(4) 73	4170
273. THE PHANTOM OF THE OPERA - Original London Cast Recording 33(1) 255	4006
285. ALWAYS & FOREVER - Randy Travis 19(1) 103	3934
310. THE EMPIRE STRIKES BACK - Soundtrack 4(4) 28	3643
347. 1100 BEL AIR PLACE - Julio Iglesias 5(2) 34	3337
352. DECEMBER - George Winston 54(1) 178	3285
358. JOURNEY'S GREATEST HITS - Journey 10(2) 92	3252
371. ANN MURRAY'S GREATEST HITS - Anne Murray 16(5) 64	3147
375. LET'S GET IT STARTED - MC Hammer 30(1) 80	3114
381. HIGH 'N' DRY - Def Leppard 38(1) 106	3096
417. WILLIE NELSON'S GREATEST HITS (& SOME THAT WILL BE) - Willie Nelson 27(1) 93	2922
426. BIG TYME - Heavy D & The Boyz 19(3) 51	2838
444. STRAIGHT OUTTA COMPTON - N.W.A. 37(3) 81	2777
459. GOLD & PLATINUM - Lynyrd Skynyrd 12(2) 65	2678
461. EAZY-DUZ-IT - Eazy-E 41(1) 90	2665
463. FAIR WARNING - Van Halen 5(3) 23	2642
474. NEBRASKA - Bruce Springsteen 3(4) 29	2563
478. JULIO - Julio Iglesias 32(1) 89	2545
479. TRIO - Dolly Parton, Linda Ronstadt, Emmylou Harris 6(1) 48	2543

Rank. Title - Act Peak(Peak Weeks) Total Weeks	Score
485. SURFING WITH THE ALIEN - Joe Satriani 29(4) 75	2477
491. WORLD WIDE LIVE - Scorpions 14(1) 43	2451
498. LIFE IS...TOO SHORT - Too Short 37(2) 78	2412
499. KENNY LOGGINS ALIVE - Kenny Loggins 11(2) 31	2402
519. PIECE OF MIND - Iron Maiden 14(1) 45	2309
531. TRIBUTE - Ozzy Osbourne/Randy Rhoads 6(2) 23	2266
545. WHEN HARRY MET SALLY (SOUNDTRACK) - Harry Connick Jr. 42(2) 122	2220
550. LEGEND - Bob Marley And The Wailers 54(2) 113	2209
551. THE GEORGE BENSON COLLECTION - George Benson 14(2) 26	2205
567. SOMEWHERE IN TIME - Iron Maiden 11(1) 39	2136
573. BREAKIN' - Soundtrack 8(2) 23	2129
583. GREATEST HITS - Queen 14(6) 26	2106
619. LIVE RUST - Neil Young With Crazy Horse 15(2) 24	1985
625. THE FINAL CUT - Pink Floyd 6(2) 23	1966
635. GRACE UNDER PRESSURE - Rush 10(4) 27	1905
647. HOLD YOUR FIRE - Rush 13(2) 30	1873
654. FLYING IN A BLUE DREAM - Joe Satriani 23(3) 39	1863
659. ONE VOICE - Barbra Streisand 9(2) 28	1848
673. BLOW UP YOUR VIDEO - AC/DC 12(1) 24	1815
675. TURBO - Judas Priest 17(2) 36	1811
681. LOVE APPROACH - Tom Browne 18(2) 26	1795
686. RECORDS - Foreigner 10(4) 25	1785
687. DEFENDERS OF THE FAITH - Judas Priest 18(4) 37	1785
693. NO REST FOR THE WICKED - Ozzy Osbourne 13(1) 27	1774
694. ANGEL OF THE NIGHT - Angela Bofill 34(1) 33	1773

Evaluating Albums Graphically Two Ways: Album Chart History vs. Score of Derived Singles

Albums: Special Lists

Albums: Evaluated by Album Chart or Derived Singles

This graph compares the ranking of albums based either on the album's chart history or the combined strength of singles derived from the album. Markers that lie close to the line, generally as Greek letters, designate albums that evaluate similarly either way. Upper case letters generally denote albums whose strength is the singles: the album could be thought of as a hits collection rather than a single work. Lower case letters denote those that evaluated more strongly as the album than the singles: the whole can be thought of as more than the parts.

There is, at the extremes, a real genre distinction. Hard rock or metal albums tend not to chart as many singles or those that chart strongly; pop acts are just the opposite.

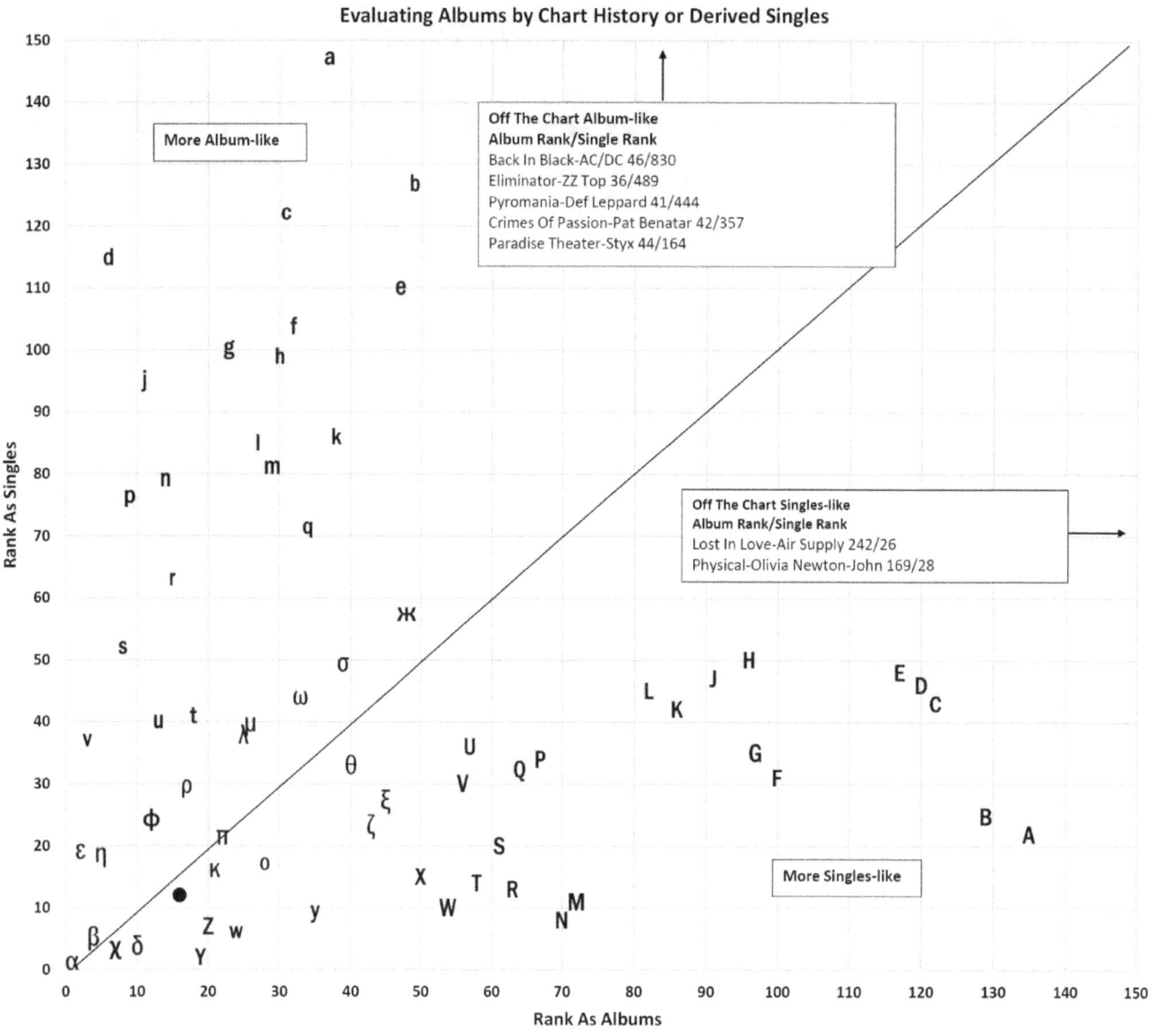

Albums: Special Lists

Legend: Rank As Singles vs. Rank As Albums

Title	Act	As Album	As Singles	Mark
THRILLER	Michael Jackson	1	1	α
BORN IN THE U.S.A.	Bruce Springsteen	2	19	ε
HYSTERIA	Def Leppard	3	37	v
FOREVER YOUR GIRL	Paula Abdul	4	5	β
WHITNEY HOUSTON	Whitney Houston	5	18	η
APPETITE FOR DESTRUCTION	Guns N' Roses	6	115	d
CAN'T SLOW DOWN	Lionel Richie	7	3	χ
ESCAPE	Journey	8	52	s
DIRTY DANCING	Soundtrack	9	76	p
FAITH	George Michael	10	4	δ
HI INFIDELITY	REO Speedwagon	11	95	j
4	Foreigner	12	24	φ
DON'T BE CRUEL	Bobby Brown	13	40	u
SPORTS	Huey Lewis & The News	14	79	n
SLIPPERY WHEN WET	Bon Jovi	15	63	r
GIRL YOU KNOW IT'S TRUE	Milli Vanilli	16	12	●
HANGIN' TOUGH	New Kids On The Block	17	29	ρ
NO JACKET REQUIRED	Phil Collins	18	41	t
JANET JACKSON'S RHYTHM NATION 1814	Janet Jackson	19	2	Y
PURPLE RAIN (SOUNDTRACK)	Prince And The Revolution	20	7	Z
SYNCHRONICITY	The Police	21	16	κ
OFF THE WALL	Michael Jackson	22	21	π
THE WALL	Pink Floyd	23	100	g
BAD	Michael Jackson	24	6	w
BUSINESS AS USUAL	Men At Work	25	38	λ
CHRISTOPHER CROSS	Christopher Cross	26	39	μ
BELLA DONNA	Stevie Nicks	27	85	l
CONTROL	Janet Jackson	28	17	o
CONTROL	Janet Jackson	28	17	o
THE JOSHUA TREE	U2	29	81	m
KENNY ROGERS' GREATEST HITS	Kenny Rogers	30	99	h
BROTHERS IN ARMS	Dire Straits	31	122	c
WHITESNAKE	Whitesnake	32	104	f
PRIVATE DANCER	Tina Turner	33	44	ω
GLASS HOUSES	Billy Joel	34	71	q
WHITNEY	Whitney Houston	35	9	y
ELIMINATOR	ZZ Top	36	489	OOR
AGAINST THE WIND	Bob Seger & The Silver Bullet Band	37	147	a
HEART	Heart	38	86	k
RECKLESS	Bryan Adams	39	49	σ
SONGS FROM THE BIG CHAIR	Tears For Fears	40	33	θ
PYROMANIA	Def Leppard	41	444	OOR
CRIMES OF PASSION	Pat Benatar	42	357	OOR
AN INNOCENT MAN	Billy Joel	43	23	ζ
PARADISE THEATER	Styx	44	164	OOR
LIKE A VIRGIN	Madonna	45	27	ξ
BACK IN BLACK	AC/DC	46	830	OOR
THE RAW & THE COOKED	Fine Young Cannibals	47	110	e
KICK	INXS	48	57	Ж
TATTOO YOU	The Rolling Stones	49	127	b
FLASHDANCE	Soundtrack	50	15	X
MAKE IT BIG	Wham!	54	10	W
NEW JERSEY	Bon Jovi	56	30	V
...BUT SERIOUSLY	Phil Collins	57	36	U
AMERICAN FOOL	John Cougar	58	14	T
DOUBLE FANTASY	John Lennon & Yoko Ono	61	20	S
SHE'S SO UNUSUAL	Cyndi Lauper	63	13	R
GUILTY	Barbra Streisand	64	32	Q
INVISIBLE TOUCH	Genesis	67	34	P
THE GAME	Queen	70	8	N
TRUE BLUE	Madonna	72	11	M
FREEZE-FRAME	The J. Geils Band	82	45	L
LIKE A PRAYER	Madonna	86	42	K
COLOUR BY NUMBERS	Culture Club	91	47	J
REPEAT OFFENDER	Richard Marx	96	50	H
TOTO IV	Toto	97	35	G
FORE!	Huey Lewis & The News	100	31	F
RICHARD MARX	Richard Marx	117	48	E
SUDDENLY	Billy Ocean	120	46	D
OUT OF THE BLUE	Debbie Gibson	122	43	C
DANCING ON THE CEILING	Lionel Richie	129	25	B
PRIVATE EYES	Daryl Hall & John Oates	135	22	A
PHYSICAL	Olivia Newton-John	169	28	OOR
LOST IN LOVE	Air Supply	242	26	OOR

OOR: Out of range of the axes of this graph

Albums: Ranked by Derived Singles

Rank by Singles. Title Act [Alb Rank] (Singles)	Score Singles
1. THRILLER Michael Jackson [1] (7)	28783
2. JANET JACKSON'S RHYTHM NATION 1814 Janet Jackson [19] (7)	24865
3. CAN'T SLOW DOWN Lionel Richie [7] (5)	22880
4. FAITH George Michael [10] (6)	22021
5. FOREVER YOUR GIRL Paula Abdul [4] (7)	20770
6. BAD Michael Jackson [24] (7)	18941
7. PURPLE RAIN (SOUNDTRACK) Prince And The Revolution [20] (5)	18772
8. THE GAME Queen [70] (4)	18635
9. WHITNEY Whitney Houston [35] (5)	17993
10. MAKE IT BIG Wham! [54] (4)	17554
11. TRUE BLUE Madonna [72] (5)	17492
12. GIRL YOU KNOW IT'S TRUE Milli Vanilli [16] (5)	17454
13. SHE'S SO UNUSUAL Cyndi Lauper [63] (5)	16788
14. AMERICAN FOOL John Cougar [58] (3)	16119
15. FLASHDANCE Soundtrack [50] (2)	15858
16. SYNCHRONICITY The Police [21] (4)	15854
17. CONTROL Janet Jackson [28] (6)	15805
18. WHITNEY HOUSTON Whitney Houston [5] (4)	15772
19. BORN IN THE U.S.A. Bruce Springsteen [2] (7)	15652
20. DOUBLE FANTASY John Lennon & Yoko Ono [61] (3)	15378
21. OFF THE WALL Michael Jackson [22] (4)	15301
22. PRIVATE EYES Daryl Hall & John Oates [135] (4)	14416
23. AN INNOCENT MAN Billy Joel [43] (6)	14151
24. 4 Foreigner [12] (5)	14125
25. DANCING ON THE CEILING Lionel Richie [129] (6)	13990
26. LOST IN LOVE Air Supply [242] (3)	13914
27. LIKE A VIRGIN Madonna [45] (4)	13897
28. PHYSICAL Olivia Newton-John [169] (3)	13503
29. HANGIN' TOUGH New Kids On The Block [17] (5)	13284
30. NEW JERSEY Bon Jovi [56] (3)	13169
31. FORE! Huey Lewis & The News [100] (5)	13126
32. GUILTY Barbra Streisand [64] (4)	13080
33. SONGS FROM THE BIG CHAIR Tears For Fears [40] (4)	12999
34. INVISIBLE TOUCH Genesis [67] (5)	12847
35. TOTO IV Toto [97] (5)	12717
36. ...BUT SERIOUSLY Phil Collins [57] (5)	12704
37. HYSTERIA Def Leppard [3] (7)	12696
38. BUSINESS AS USUAL Men At Work [25] (2)	12640
39. CHRISTOPHER CROSS Christopher Cross [26] (4)	12582
40. DON'T BE CRUEL Bobby Brown [13] (5)	12537
41. NO JACKET REQUIRED Phil Collins [18] (4)	12524
42. LIKE A PRAYER Madonna [86] (5)	12226
43. OUT OF THE BLUE Debbie Gibson [122] (5)	12152
44. PRIVATE DANCER Tina Turner [33] (5)	11905
45. FREEZE-FRAME The J. Geils Band [82] (3)	11879
46. SUDDENLY Billy Ocean [120] (4)	11852
47. COLOUR BY NUMBERS Culture Club [91] (4)	11630
48. RICHARD MARX Richard Marx [117] (4)	11628
49. RECKLESS Bryan Adams [39] (6)	11467
50. REPEAT OFFENDER Richard Marx [96] (5)	11466
51. XANADU (SOUNDTRACK) Olivia Newton-John/Electric Light Orchestra [264] (5)	11447
52. ESCAPE Journey [8] (4)	11420
53. KISSING TO BE CLEVER Culture Club [172] (3)	11399
54. DIANA Diana Ross [103] (2)	11359
55. BREAK OUT The Pointer Sisters [68] (6)	11316
56. PIPES OF PEACE Paul McCartney [732] (2)	11094
57. KICK INXS [48] (4)	11023
58. H2O Daryl Hall & John Oates [59] (3)	11004
59. MISTAKEN IDENTITY Kim Carnes [110] (3)	10949
60. EYE OF THE TIGER Survivor [268] (3)	10703
61. WELCOME TO THE REAL WORLD Mr. Mister [157] (3)	10646
62. JUICE Juice Newton [171] (3)	10585
63. SLIPPERY WHEN WET Bon Jovi [15] (3)	10533
64. TUG OF WAR Paul McCartney [283] (3)	10528
65. AMERICAN GIGOLO Soundtrack [469] (2)	10526
66. ENDLESS LOVE Soundtrack [539] (1)	10386
67. SEVEN AND THE RAGGED TIGER Duran Duran [152] (3)	10382
68. THE ONE THAT YOU LOVE Air Supply [195] (3)	10256
69. DIFFERENT LIGHT Bangles [128] (4)	10211
70. CHICAGO 19 Chicago [662] (5)	10155
71. GLASS HOUSES Billy Joel [34] (4)	10151
72. AUTOAMERICAN Blondie [225] (2)	10091
73. FOOTLOOSE Soundtrack [94] (4)	10054
74. LET IT LOOSE Gloria Estefan & Miami Sound Machine [112] (5)	10026
75. LIONEL RICHIE Lionel Richie [52] (3)	10005
76. DIRTY DANCING Soundtrack [9] (4)	9979
77. I LOVE ROCK 'N ROLL Joan Jett & the Blackhearts [162] (2)	9899
78. 1984 (MCMLXXXIV) Van Halen [71] (4)	9881
79. SPORTS Huey Lewis & The News [14] (5)	9860
80. VOICES Daryl Hall & John Oates [138] (4)	9779
81. THE JOSHUA TREE U2 [29] (4)	9774
82. PARTNERS IN CRIME Rupert Holmes [719] (4)	9762
83. TIFFANY Tiffany [102] (4)	9761
84. TELL IT TO MY HEART Taylor Dayne [249] (4)	9475
85. BELLA DONNA Stevie Nicks [27] (4)	9420
86. HEART Heart [38] (5)	9413
87. THE JAZZ SINGER (SOUNDTRACK) Neil Diamond [81] (3)	9362
88. KNEE DEEP IN THE HOOPLA Starship [236] (4)	9309
89. WORKING CLASS DOG Rick Springfield [118] (3)	9238
90. EXPOSURE Exposé [204] (4)	9103
91. MAKE YOUR MOVE Captain & Tennille [836] (3)	9027
92. CHICAGO 17 Chicago [127] (4)	9020
93. LET'S DANCE David Bowie [147] (4)	8979
94. ABRACADABRA The Steve Miller Band [298] (3)	8926
95. HI INFIDELITY REO Speedwagon [11] (4)	8906
96. HORIZON Eddie Rabbitt [370] (2)	8843
97. WHENEVER YOU NEED SOMEBODY Rick Astley [191] (3)	8732
98. BIG BAM BOOM Daryl Hall & John Oates [203] (4)	8465
99. KENNY ROGERS' GREATEST HITS Kenny Rogers [30] (1)	8423
100. THE WALL Pink Floyd [23] (2)	8331
101. HEARTBEAT CITY The Cars [74] (5)	8330
102. OPEN UP AND SAY...AHH! Poison [62] (4)	8327
103. FASTER THAN THE SPEED OF NIGHT Bonnie Tyler [414] (2)	8321
104. WHITESNAKE Whitesnake [32] (4)	8240
105. BEVERLY HILLS COP Soundtrack [123] (4)	8213
106. MCCARTNEY II Paul McCartney [430] (1)	8092
107. BAD ANIMALS Heart [144] (4)	8085
108. HEAVEN ON EARTH Belinda Carlisle [266] (4)	8076
109. SCARECROW John Mellencamp [65] (5)	8034
110. THE RAW & THE COOKED Fine Young Cannibals [47] (5)	8032
111. EMERGENCY Kool & The Gang [177] (4)	7941
112. TOP GUN Soundtrack [83] (4)	7896
113. JODY WATLEY Jody Watley [201] (5)	7889
114. BUSTER Soundtrack [1514] (2)	7856
115. APPETITE FOR DESTRUCTION Guns N' Roses [6] (3)	7852
116. SPANISH FLY Lisa Lisa And Cult Jam [256] (3)	7738
117. MIAMI VICE TV Soundtrack [143] (3)	7721
118. CHICAGO 16 Chicago [439] (3)	7669
119. EYES THAT SEE IN THE DARK Kenny Rogers [299] (3)	7603
120. ROCK 'N SOUL PART 1 Daryl Hall & John Oates [302] (2)	7557
121. ALL THAT JAZZ Breathe [510] (3)	7556
122. BROTHERS IN ARMS Dire Straits [31] (3)	7546
123. 90125 Yes [233] (3)	7513
124. RIPTIDE Robert Palmer [197] (4)	7482
125. ROLL WITH IT Steve Winwood [158] (4)	7450
126. MOUTH TO MOUTH Lipps Inc. [378] (2)	7424
127. TATTOO YOU The Rolling Stones [49] (3)	7388
128. SWEET DREAMS (ARE MADE OF THIS) Eurythmics [391] (2)	7310
129. BACK IN THE HIGH LIFE Steve Winwood [99] (4)	7294
130. THE WAY IT IS Bruce Hornsby & The Range [98] (4)	7279
131. THE DISTANCE Bob Seger & The Silver Bullet Band [291] (3)	7248
132. THE WOMAN IN RED (SOUNDTRACK) Stevie Wonder [314] (2)	7242
133. LARGER THAN LIFE Jody Watley [574] (4)	7217
134. URBAN COWBOY Soundtrack [167] (6)	7162
135. SUCCESS HASN'T SPOILED ME YET Rick Springfield [286] (3)	7132
136. VISION QUEST Soundtrack [796] (3)	7059
137. AGAINST ALL ODDS Soundtrack [736] (2)	7041
138. TRUE COLORS Cyndi Lauper [323] (4)	7006
139. CUTS BOTH WAYS Gloria Estefan [187] (5)	6966
140. WHEELS ARE TURNIN' REO Speedwagon [312] (4)	6937
141. SIGN 'O' THE TIMES Prince [348] (5)	6852
142. BAD ENGLISH Bad English [436] (5)	6838
143. DARE The Human League [265] (1)	6816
144. AGENT PROVOCATEUR Foreigner [284] (4)	6734
145. STORM FRONT Billy Joel [131] (5)	6697
146. WHO'S THAT GIRL (SOUNDTRACK) Madonna [604] (2)	6636
147. AGAINST THE WIND Bob Seger & The Silver Bullet Band [37] (4)	6612
148. INTRODUCING THE HARDLINE ACCORDING TO TERENCE TRENT D'ARBY Terence Trent D'Arby [188] (4)	6553
149. GHOSTBUSTERS Soundtrack [453] (2)	6519
150. LAP OF LUXURY Cheap Trick [295] (4)	6504

Albums: 3 or More Hot 100 Singles

Act name for single is shown if different from that of album. Singles released at virtually the same time from an Act's album and a Soundtrack are generally attributed to the Act's album. Some singles derived from these albums peaked outside the '80s. These are designated, e.g., 7,6*. The first digit is the total number of singles. The second is the number that peaked outside the decade. An asterisk prior to the number means peaked in the '70s; after the number means peaked in the '90s. 2117 '80s Albums charted Hot 100 singles; 1118 charted one and 570 charted two. Singles with two chart entries listed in italics.

Album Title - Act Pk [Rank] Entry Date
  Single Title Pk(WksOn) Peak Month

Seven Hot 100 Singles: 6 Albums

Forever Your Girl - Paula ABDUL 1 [4] Jul-88 7,1*
  Knocked Out 41(13) Aug-88
  (It's Just) The Way That You Love Me 88(5) Nov-88
  Straight Up 1(25) Feb-89
  Forever Your Girl 1(22) May-89
  Cold Hearted 1(21) Sep-89
  (It's Just) The Way That You Love Me 3(20) Dec-89
  Opposites Attract 1(23) Feb-90

Hysteria - DEF LEPPARD 1 [3] Aug-87
  Women 80(5) Sep-87
  Animal 19(19) Dec-87
  Hysteria 10(16) Mar-88
  Pour Some Sugar On Me 2(24) Jul-88
  Love Bites 1(23) Oct-88
  Armageddon It 3(18) Jan-89
  Rocket 12(13) Apr-89

**Janet Jackson's Rhythm Nation 1814 -
  Janet JACKSON 1 [19] Oct-89 7,6***
  Miss You Much 1(20) Oct-89
  Rhythm Nation 2(17) Jan-90
  Escapade 1(17) Mar-90
  Alright 4(16) Jun-90
  Come Back To Me 2(17) Aug-90
  Black Cat 1(16) Oct-90
  Love Will Never Do (Without You) 1(22) Jan-91

Thriller - Michael JACKSON 1 [1] Dec-82
  The Girl Is Mine - Michael Jackson/Paul McCartney
    2(18) Jan-83
  Billie Jean 1(24) Mar-83
  Beat It 1(25) Apr-83
  Wanna Be Startin' Somethin' 5(15) Jul-83
  Human Nature 7(14) Sep-83
  P.Y.T. (Pretty Young Thing) 10(16) Nov-83
  Thriller 4(14) Mar-84

Bad - Michael JACKSON 1 [24] Sep-87
  I Just Can't Stop Loving You 1(14) Sep-87
  Bad 1(14) Oct-87
  The Way You Make Me Feel 1(18) Jan-88
  Man In The Mirror 1(17) Mar-88
  Dirty Diana 1(14) Jul-88
  Another Part Of Me 11(13) Sep-88
  Smooth Criminal 7(15) Jan-89

**Born In The U.S.A. -
  Bruce SPRINGSTEEN 1 [2] Jun-84**
  Dancing In The Dark 2(21) Jun-84
  Cover Me 7(18) Oct-84
  Born In The U.S.A. 9(17) Jan-85
  I'm On Fire 6(20) Apr-85
  Glory Days 5(18) Aug-85
  I'm Goin' Down 9(13) Oct-85
  My Hometown 6(15) Jan-86

Six Hot 100 Singles: 7 Albums

Reckless - Bryan ADAMS 1 [39] Nov-84
  Run To You 6(19) Jan-85
  Somebody 11(17) Apr-85
  Heaven 1(19) Jun-85
  Summer Of '69 5(17) Aug-85
  One Night Love Affair 13(15) Nov-85
  It's Only Love -
    Bryan Adams/Tina Turner 15(14) Jan-86

Control - Janet JACKSON 1 [28] Mar-86
  What Have You Done For Me Lately? 4(21) May-86
  Nasty 3(19) Jul-86
  When I Think Of You 1(19) Oct-86
  Control 5(18) Jan-87
  Let's Wait Awhile 2(19) Mar-87
  The Pleasure Principle 14(18) Aug-87

An Innocent Man - Billy JOEL 4 [43] Aug-83
  Tell Her About It 1(18) Sep-83
  Uptown Girl 3(22) Nov-83
  An Innocent Man 10(18) Feb-84
  The Longest Time 14(18) May-84
  Leave A Tender Moment Alone 27(15) Aug-84
  Keeping The Faith 18(16) Mar-85

Faith - George MICHAEL 1 [10] Nov-87
  I Want Your Sex 2(20) Aug-87
  Faith 1(20) Dec-87
  Father Figure 1(17) Feb-88
  One More Try 1(18) May-88
  Monkey 1(16) Aug-88
  Kissing A Fool 5(15) Nov-88

Break Out - POINTER SISTERS 8 [68] Nov-83
  I Need You 48(15) Nov-83
  Automatic 5(20) Apr-84
  Jump (For My Love) 3(24) Jul-84
  I'm So Excited '84 9(24) Oct-84
  Neutron Dance 6(23) Feb-85
  Baby Come And Get It 44(11) Apr-85

**Dancing On The Ceiling -
  Lionel RICHIE 1 [129] Aug-86**
  Say You, Say Me 1(20) Dec-85
  Dancing On The Ceiling 2(17) Sep-86
  Love Will Conquer All 9(18) Nov-86
  Deep River Woman 71(8) Feb-87
  Ballerina Girl 7(18) Feb-87
  Se La 20(13) May-87

**Urban Cowboy -
  SOUNDTRACK-MOVIE 3 [167] May-80**
  Could I Have This Dance - Anne Murray 33(14) Nov-80
  Look What You've Done To Me - Boz Scaggs
    14(17) Oct-80
  All Night Long - Joe Walsh 19(16) Jul-80
  Lookin' For Love - Johnny Lee 5(21) Sep-80
  Love The World Away - Kenny Rogers 14(12) Aug-80
  Stand By Me - Mickey Gilley 22(18) Aug-80

Five Hot 100 Singles: 39 Albums

Bad English - BAD ENGLISH 21 [436] Jul-89 5,3*
  Forget Me Not 45(11) Aug-89
  When I See You Smile 1(22) Nov-89
  Price Of Love 5(19) Mar-90
  Heaven Is A 4 Letter Word 66(9) Apr-90
  Possession 21(17) Aug-90

New Jersey - BON JOVI 1 [56] Oct-88
  Bad Medicine 1(20) Nov-88
  Born To Be My Baby 3(20) Feb-89
  I'll Be There For You 1(22) May-89
  Lay Your Hands On Me 7(16) Jul-89
  Living In Sin 9(19) Dec-89

Don't Be Cruel - Bobby BROWN 1 [13] Jul-88
  Don't Be Cruel 8(26) Oct-88
  My Prerogative 1(24) Jan-89
  Roni 3(17) Mar-89
  Every Little Step 3(21) Jun-89
  Rock Wit'cha 7(21) Nov-89

Heartbeat City - The CARS 3 [74] Apr-84
  You Might Think 7(17) Apr-84
  Magic 12(17) Jul-84
  Drive 3(19) Sep-84
  Hello Again 20(15) Dec-84
  Why Can't I Have You 33(17) Mar-85

Chicago 19 - CHICAGO 37 [662] Jul-88 5,1*
  I Don't Wanna Live Without Your Love
    3(21) Aug-88
  Look Away 1(24) Dec-88
  You're Not Alone 10(17) Mar-89
  We Can Last Forever 55(12) Jun-89
  What Kind Of Man Would I Be? 5(18) Feb-90

...But Seriously - Phil COLLINS 1 [57] Dec-89 5,4*
  Another Day In Paradise 1(18) Dec-89
  I Wish It Would Rain Down 3(17) Mar-90
  Do You Remember? 4(19) Jun-90
  Something Happened On The Way To Heaven
    4(22) Oct-90
  Hang In Long Enough 23(13) Jan-91

Show Me - COVER GIRLS 64 [1149] Aug-87
  Show Me 44(18) May-87
  Spring Love 98(1) Aug-87
  Because Of You 27(20) Feb-88
  Promise Me 40(19) May-88
  Inside Outside 55(13) Aug-88

24/7 - DINO 34 [783] Mar-89 5,1*
  Summergirls 50(12) Aug-88
  24/7 42(14) Mar-89
  I Like It 7(25) Aug-89
  Sunshine 23(15) Nov-89
  Never 2 Much Of U 61(9) Jan-90

**Let It Loose - Gloria ESTEFAN/MIAMI SOUND
  MACHINE 6 [112] Jun-87**
  Rhythm Is Gonna Get You 5(17) Aug-87
  Betcha Say That - 36(11) Oct-87
  Can't Stay Away From You
    6(23) Mar-88
  Anything For You May-88
  1-2-3 3(19) Aug-88

Albums: Special Lists

Album Title - Act Pk [Rank] Entry Date
Single Title Pk(WksOn) Peak Date

Cuts Both Ways - Gloria ESTEFAN/
MIAMI SOUND MACHINE 8 [187] Jul-89 5,3*
- Don't Wanna Lose You - Gloria Estefan 1(18) Sep-89
- Get On Your Feet - Gloria Estefan 11(17) Nov-89
- Here We Are - Gloria Estefan 6(21) Mar-90
- Oye Mi Canto (Hear My Voice) -
  Gloria Estefan 48(7) May-90
- Cuts Both Ways - Gloria Estefan 44(14) Aug-90

The Raw & The Cooked -
FINE YOUNG CANNIBALS 1 [47] Mar-89 5,1*
- She Drives Me Crazy 1(23) Apr-89
- Good Thing 1(17) Jul-89
- Don't Look Back 11(12) Oct-89
- I'm Not The Man I Used To Be 54(6) Dec-89
- I'm Not Satisfied 90(3) Mar-90

Tango In The Night -
FLEETWOOD MAC 7 [150] May-87
- Big Love 5(16) May-87
- Seven Wonders 19(13) Aug-87
- Little Lies 4(21) Nov-87
- Everywhere 14(18) Feb-88
- Family Man 90(4) Apr-88

4 - FOREIGNER 1 [12] Jul-81
- Urgent 4(23) Sep-81
- Waiting For A Girl Like You 2(23) Nov-81
- Juke Box Hero 26(13) Apr-82
- Break It Up 26(13) Jun-82
- Luanne 75(6) Aug-82

Invisible Touch - GENESIS 3 [67] Jun-86
- Invisible Touch 1(17) Jul-86
- Throwing It All Away 4(16) Oct-86
- Land Of Confusion 4(21) Jan-87
- Tonight, Tonight, Tonight 3(15) Apr-87
- In Too Deep 3(17) Jun-87

Out Of The Blue - Debbie GIBSON 7 [122] Sep-87
- Only In My Dreams 4(28) Sep-87
- Shake Your Love 4(22) Dec-87
- Out Of The Blue 3(17) Apr-88
- Foolish Beat 1(20) Jun-88
- Staying Together 22(12) Sep-88

Heart - HEART 1 [38] Jul-85
- What About Love? 10(21) Aug-85
- Never 4(24) Dec-85
- These Dreams 1(20) Mar-86
- Nothin' At All 10(16) Jun-86
- If Looks Could Kill 54(9) Aug-86

The End Of The Innocence -
Don HENLEY 8 [80] Jul-89 5,3*
- The End Of The Innocence 8(18) Aug-89
- The Last Worthless Evening 21(18) Dec-89
- The Heart Of The Matter 21(21) May-90
- How Bad Do You Want It? 48(12) Aug-90
- New York Minute 48(16) Dec-90

Whitney - Whitney HOUSTON 1 [35] Jun-87
- I Wanna Dance With Somebody (Who Loves Me)
  1(18) Jun-87
- Didn't We Almost Have It All 1(17) Sep-87
- So Emotional 1(19) Jan-88
- Where Do Broken Hearts Go 1(18) Apr-88
- Love Will Save The Day 9(16) Aug-88

Magic - The JETS 35 [637] Nov-87
- Cross My Broken Heart 7(16) Aug-87
- I Do You 20(15) Dec-87
- Rocket 2 U 6(22) Apr-88
- Make It Real 4(20) Jun-88
- Sendin' All My Love 88(4) Aug-88

Storm Front - Billy JOEL 1 [131] Nov-89 5,4*
- We Didn't Start The Fire 1(19) Dec-89
- I Go To Extremes 6(16) Mar-90

Album Title - Act Pk [Rank] Entry Date
Single Title Pk(WksOn) Peak Date

- The Downeaster "Alexa" 57(8) Jun-90
- That's Not Her Style 77(6) Aug-90
- And So It Goes 37(12) Dec-90

Raised On Radio - JOURNEY 4 [220] May-86
- Be Good To Yourself 9(15) May-86
- Suzanne 17(13) Aug-86
- Girl Can't Help It 17(14) Nov-86
- I'll Be Alright Without You 14(21) Feb-87
- Why Can't This Night Go On Forever 60(12) May-87

She's So Unusual - Cyndi LAUPER 4 [63] Dec-83
- Girls Just Want To Have Fun 2(25) Mar-84
- Time After Time 1(20) Jun-84
- She Bop 3(18) Sep-84
- All Through The Night 5(19) Dec-84
- Money Changes Everything 27(13) Feb-85

Sports - Huey LEWIS And The NEWS 1 [14] Oct-83
- Heart And Soul 8(21) Nov-83
- I Want A New Drug 6(19) Mar-84
- The Heart Of Rock & Roll 6(20) Jun-84
- If This Is It 6(17) Sep-84
- Walking On A Thin Line 18(15) Dec-84

Fore! - Huey LEWIS And The NEWS 1 [100] Sep-86
- Stuck With You 1(19) Sep-86
- Hip To Be Square 3(16) Dec-86
- Jacob's Ladder 1(15) Mar-87
- I Know What I Like 9(14) May-87
- Doing It All For My Baby 6(16) Sep-87

True Blue - MADONNA 1 [72] Jul-86
- Live To Tell 1(18) Jun-86
- Papa Don't Preach 1(18) Aug-86
- True Blue 3(16) Nov-86
- Open Your Heart 1(18) Feb-87
- La Isla Bonita 4(17) May-87

Like A Prayer - MADONNA 1 [86] Apr-89 5,2*
- Like A Prayer 1(16) Apr-89
- Express Yourself 2(16) Jul-89
- Cherish 2(15) Oct-89
- Oh Father 20(13) Jan-90
- Keep It Together 8(13) Mar-90

Repeat Offender -
Richard MARX 1 [96] May-89 5,2*
- Satisfied 1(15) Jun-89
- Right Here Waiting 1(21) Aug-89
- Angelia 4(17) Dec-89
- Too Late To Say Goodbye 12(13) Mar-90
- Children Of The Night 13(15) Jun-90

Scarecrow - John MELLENCAMP 2 [65] Sep-85
- Lonely Ol' Night -
  John Cougar Mellencamp 6(20) Oct-85
- Small Town - John Cougar Mellencamp 6(18) Dec-85
- R.O.C.K. In The U.S.A. (A Salute To 60's Rock) -
  John Cougar Mellencamp 2(17) Apr-86
- Rain On The Scarecrow -
  John Cougar Mellencamp 21(12) Jun-86
- Rumbleseat - John Cougar Mellencamp 28(13) Aug-86

Girl You Know It's True -
MILLI VANILLI 1 [16] Mar-89 5,1*
- Girl You Know It's True 2(26) Apr-89
- Baby Don't Forget My Number 1(21) Jul-89
- Girl I'm Gonna Miss You 1(22) Sep-89
- Blame It On The Rain 1(23) Nov-89
- All Or Nothing 4(14) Feb-90

Dr. Feelgood -
MÖTLEY CRÜE 1 [76] Sep-89 5,4*
- Dr. Feelgood 6(16) Oct-89
- Kickstart My Heart 27(16) Jan-90
- Without You 8(17) Apr-90
- Don't Go Away Mad (Just Go Away) 19(16) Jul-90
- Same Ol' Situation (S.O.S.) 78(9) Sep-90

Album Title - Act Pk [Rank] Entry Date
Single Title Pk(WksOn) Peak Date

Hangin' Tough -
NEW KIDS ON THE BLOCK 1 [17] Aug-88
- Please Don't Go Girl 10(28) Oct-88
- You Got It (The Right Stuff) 3(26) Mar-89
- I'll Be Loving You (Forever) 1(21) Jun-89
- Hangin' Tough 1(17) Sep-89
- Cover Girl 2(18) Nov-89

Xanadu (Soundtrack) -
Olivia NEWTON-JOHN/
ELECTRIC LIGHT ORCHESTRA 4 [264] Jul-80
- I'm Alive - Electric Light Orchestra 16(15) Jul-80
- All Over The World -
  Electric Light Orchestra 13(16) Oct-80
- Xanadu 8(17) Oct-80
- Magic - Olivia Newton-John 1(23) Aug-80
- Suddenly - Olivia Newton-John &
  Cliff Richard 20(19) Jan-81

Purple Rain (Soundtrack) -
PRINCE and the REVOLUTION 1 [20] Jul-84
- When Doves Cry - Prince 1(21) Jul-84
- Let's Go Crazy 1(19) Sep-84
- Purple Rain 2(16) Nov-84
- I Would Die 4 U 8(15) Feb-85
- Take Me With U 25(12) Mar-85

Sign 'O' The Times - PRINCE 6 [348] Apr-87
- Sign O' The Times 3(14) Apr-87
- If I Was Your Girlfriend 67(6) Jun-87
- U Got The Look 2(25) Oct-87
- I Could Never Take The Place Of Your Man
  10(17) Feb-88
- Hot Thing 63(9) Feb-88

Can't Slow Down - Lionel RICHIE 1 [7] Nov-83
- All Night Long (All Night) 1(24) Nov-83
- Running With The Night 7(19) Feb-84
- Hello 1(24) May-84
- Stuck On You 3(19) Aug-84
- Penny Lover 8(18) Dec-84

Take It While It's Hot -
SWEET SENSATION 63 [1383] Oct-88
- Hooked On You 64(12) Feb-87
- Take It While It's Hot 57(13) May-88
- Never Let You Go 58(10) Oct-88
- Sincerely Yours 14(18) Apr-89
- Hooked On You 23(16) Aug-89

Toto IV - TOTO 4 [97] Apr-82
- Rosanna 2(23) Jul-82
- Make Believe 30(13) Sep-82
- Africa 1(21) Feb-83
- I Won't Hold You Back 10(17) May-83
- Waiting For Your Love 73(6) Jul-83

Private Dancer - Tina TURNER 3 [33] Jun-84
- Let's Stay Together 26(15) Mar-84
- What's Love Got To Do With It 1(28) Sep-84
- Better Be Good To Me 5(21) Nov-84
- Private Dancer 7(18) Mar-85
- Show Some Respect 37(10) Jun-85

Jody Watley - Jody WATLEY 10 [201] Mar-87
- Looking For A New Love 2(19) May-87
- Still A Thrill 56(7) Jul-87
- Don't You Want Me 6(23) Dec-87
- Some Kind Of Lover 10(17) Apr-88
- Most Of All 60(11) Jun-88

Albums: Special Lists

Four Hot 100 Singles: 110 Albums

Pump - AEROSMITH 5 [75] Sep-89 4,3*
- Love In An Elevator 5(16) Oct-89
- Janie's Got A Gun 4(18) Feb-90
- What It Takes 9(17) May-90
- The Other Side 22(15) Aug-90

Rapture - Anita BAKER 11 [60] Apr-86
- Sweet Love 8(22) Nov-86
- Caught Up In The Rapture 37(18) Feb-87
- Same Ole Love (365 Days A Year) 44(14) May-87
- No One In The World 44(17) Oct-87

Different Light - BANGLES 2 [128] Feb-86
- Manic Monday 2(20) Apr-86
- If She Knew What She Wants 29(14) Jul-86
- Walk Like An Egyptian 1(23) Dec-86
- Walking Down Your Street 11(16) Apr-87

Let's Dance - David BOWIE 4 [147] Apr-83
- Let's Dance 1(20) May-83
- China Girl 10(18) Aug-83
- Modern Love 14(13) Nov-83
- Without You 73(4) Mar-84

Eddie & The Cruisers (Soundtrack) - John CAFFERTY And The BEAVER BROWN BAND 9 [318] Oct-83
- On The Dark Side - Eddie And The Cruisers 64(9) Nov-83
- Tender Years 78(5) Feb-84
- On The Dark Side 7(18) Oct-84
- Tender Years 31(14) Jan-85

Tough All Over - John CAFFERTY And The BEAVER BROWN BAND 40 [1032] Jun-85
- Tough All Over 22(15) Jul-85
- C-I-T-Y 18(15) Oct-85
- Small Town Girl 64(10) Dec-85
- Voice Of America's Sons 62(8) Jul-86

What A Feelin' - Irene CARA 77 [1386] Dec-83
- Why Me? 13(15) Dec-83
- The Dream (Hold On To Your Dream) 37(14) Feb-84
- Breakdance 8(19) Jun-84
- You Were Made For Me 78(5) Aug-84

Heaven On Earth - Belinda CARLISLE 13 [266] Oct-87
- Heaven Is A Place On Earth 1(21) Dec-87
- I Get Weak 2(16) Mar-88
- Circle In The Sand 7(17) Jun-88
- I Feel Free 88(4) Aug-88

Made In America - CARPENTERS 52 [1771] Jul-81
- Touch Me When We're Dancing 16(14) Aug-81
- (Want You) Back In My Life Again 72(8) Oct-81
- Those Good Old Dreams 63(6) Jan-82
- Beechwood 4-5789 74(4) May-82

One Good Reason - Paul CARRACK 67 [1537] Nov-87
- Don't Shed A Tear 9(24) Feb-88
- One Good Reason 28(13) May-88
- When You Walk In The Room 90(3) Jun-88
- Button Off My Shirt 91(3) Sep-88

Lap Of Luxury - CHEAP TRICK 16 [295] May-88
- The Flame 1(27) Jul-88
- Don't Be Cruel 4(17) Oct-88
- Ghost Town 33(14) Dec-88
- Never Had A Lot To Lose 75(6) Mar-89

Chicago 17 - CHICAGO 4 [127] Jun-84
- Stay The Night 16(17) Jun-84
- Hard Habit To Break 3(25) Oct-84
- You're The Inspiration 3(22) Jan-85
- Along Comes A Woman 14(16) Apr-85

Chicago 18 - CHICAGO 35 [883] Oct-86
- 25 Or 6 To 4 48(8) Sep-86
- Will You Still Love Me? 3(23) Feb-87
- If She Would Have Been Faithful... 17(19) May-87
- Niagara Falls 91(3) Jul-87

Long Cold Winter - CINDERELLA 10 [160] Jul-88
- Don't Know What You Got (Till It's Gone) 12(22) Nov-88
- The Last Mile 36(10) Mar-89
- Coming Home 20(17) Jun-89
- Gypsy Road 51(7) Sep-89

Everlasting - Natalie COLE 42 [684] Aug-87
- Jump Start 13(18) Oct-87
- I Live For Your Love 13(22) Feb-88
- Pink Cadillac 5(17) May-88
- When I Fall In Love 95(1) Aug-88

No Jacket Required - Phil COLLINS 1 [18] Mar-85
- One More Night 1(18) Mar-85
- Sussudio 1(17) Jul-85
- Don't Lose My Number 4(18) Sep-85
- Take Me Home 7(16) May-86

Christopher Cross - Christopher CROSS 6 [26] Feb-80
- Ride Like The Wind 2(21) Apr-80
- Sailing 1(21) Aug-80
- Never Be The Same 15(19) Nov-80
- Say You'll Be Mine 20(14) May-81

Colour By Numbers - CULTURE CLUB 2 [91] Nov-83
- Church Of The Poison Mind 10(17) Dec-83
- Karma Chameleon 1(22) Feb-84
- Miss Me Blind 5(16) Apr-84
- It's A Miracle 13(13) Jun-84

Disintegration - The CURE 12 [308] May-89 4,1*
- Fascination Street 46(11) Jun-89
- Love Song 2(17) Oct-89
- Lullaby 74(8) Dec-89
- Pictures Of You 71(8) May-90

Introducing The Hardline According To Terence Trent D'Arby - Terence Trent D'ARBY 4 [188] Oct-87
- If You Let Me Stay 68(8) Nov-87
- Wishing Well 1(25) May-88
- Sign Your Name 4(21) Aug-88
- Dance Little Sister (Part One) 30(11) Oct-88

Tell It To My Heart - Taylor DAYNE 21 [249] Jan-88
- Tell It To My Heart 7(25) Jan-88
- Prove Your Love 7(18) May-88
- I'll Always Love You 3(30) Sep-88
- Don't Rush Me 2(20) Jan-89

Rhythm Of The Night - DeBARGE 19 [466] Mar-85
- Who's Holding Donna Now 6(19) Aug-85
- Rhythm Of The Night 3(22) Apr-85
- You Wear It Well - El DeBarge with DeBarge 46(10) Oct-85
- The Heart Is Not So Smart - El DeBarge with DeBarge 75(7) Jan-86

Music For The Masses - DEPECHE MODE 35 [845] Oct-87
- *Strangelove 76(6) Aug-87*
- Never Let Me Down Again 63(10) Feb-88
- Route 66/Behind The Wheel 61(11) May-88
- *Strangelove 50(9) Oct-88*

Primitive Love - MIAMI SOUND MACHINE 21 [270] Nov-85
- Conga 10(27) Feb-86
- Bad Boy 8(19) May-86
- Words Get In The Way 5(24) Sep-86
- Falling In Love (Uh-Oh) 25(16) Jan-87

The Final Countdown - EUROPE 8 [161] Nov-86
- The Final Countdown 8(18) Mar-87
- Rock The Night 30(13) Jun-87
- Carrie 3(19) Oct-87
- Cherokee 72(10) Dec-87

Be Yourself Tonight - EURYTHMICS 9 [344] May-85
- Would I Lie To You? 5(19) Jul-85
- There Must Be An Angel (Playing With My Heart) 22(11) Sep-85
- It's Alright (Baby's Coming Back) 78(6) Mar-86
- Sisters Are Doin' It For Themselves - Eurythmics and Aretha Franklin 18(15) Dec-85

Exposure - EXPOSÉ 16 [204] Feb-87
- Come Go With Me 5(19) Apr-87
- Point Of No Return 5(17) Jul-87
- Let Me Be The One 7(22) Oct-87
- Seasons Change 1(20) Feb-88

What You Don't Know - EXPOSÉ 33 [707] Jul-89 4,2*
- What You Don't Know 8(15) Jul-89
- When I Looked At Him 10(20) Oct-89
- Tell Me Why 9(15) Feb-90
- Your Baby Never Looked Good In Blue 17(16) May-90

The Innocent Age - Dan FOGELBERG 6 [156] Sep-81
- Same Old Lang Syne 9(18) Feb-81
- Hard To Say 7(19) Oct-81
- Leader Of The Band 9(20) Mar-82
- Run For The Roses 18(14) May-82

Agent Provocateur - FOREIGNER 4 [284] Jan-85
- I Want To Know What Love Is 1(21) Feb-85
- That Was Yesterday 12(15) May-85
- Reaction To Action 54(8) Jun-85
- Down On Love 54(8) Sep-85

Welcome To The Pleasure Dome - FRANKIE GOES TO HOLLYWOOD 33 [664] Nov-84
- Relax 67(7) May-84
- Two Tribes 43(15) Dec-84
- Relax (Mix) 10(16) Mar-85
- Welcome To The Pleasuredome (Trevor Horn Remix) 48(8) May-85

Who's Zoomin' Who? - Aretha FRANKLIN 13 [282] Jul-85
- Freeway Of Love 3(19) Aug-85
- Who's Zoomin' Who 7(19) Nov-85
- Another Night 22(14) Mar-86
- Sisters Are Doin' It For Themselves - Eurythmics And Aretha Franklin 18(15) Dec-85

Aretha (II) - Aretha FRANKLIN 32 [768] Nov-86
- Jumpin' Jack Flash 21(11) Nov-86
- Jimmy Lee 28(13) Feb-87
- I Knew You Were Waiting (For Me) - Aretha Franklin & George Michael 1(17) Apr-87
- Rock-A-Lott 82(4) Jul-87

So - Peter GABRIEL 2 [105] Jun-86
- Don't Give Up - Peter Gabriel/Kate Bush 72(6) Apr-87
- Sledgehammer 1(21) Jul-86
- In Your Eyes 26(14) Oct-86
- Big Time 8(23) Mar-87

Genesis - GENESIS 9 [292] Oct-83
- Mama 73(9) Oct-83
- That's All 6(20) Feb-84
- Illegal Alien 44(10) Apr-84
- Taking It All Too Hard 50(12) Jul-84

Electric Youth - Debbie GIBSON 1 [139] Feb-89
- Lost In Your Eyes 1(19) Mar-89
- Electric Youth 11(13) May-89
- No More Rhyme 17(14) Aug-89
- We Could Be Together 71(6) Sep-89

Albums: Special Lists

Album Title - Act Pk [Rank] Entry Date
Single Title Pk(WksOn) Peak Date

Appetite For Destruction -
GUNS N' ROSES 1 [6] Aug-87
- Sweet Child O' Mine 1(24) Sep-88
- Welcome To The Jungle 7(17) Dec-88
- Paradise City 5(17) Mar-89
- Nightrain 93(5) Aug-89

Voices - Daryl HALL & John OATES 17 [138] Aug-80
- How Does It Feel To Be Back 30(13) Sep-80
- You've Lost That Lovin' Feeling 12(20) Nov-80
- Kiss On My List 1(23) Apr-81
- You Make My Dreams 5(21) Jul-81

Private Eyes -
Daryl HALL & John OATES 5 [135] Sep-81
- Private Eyes 1(23) Nov-81
- I Can't Go For That (No Can Do) 1(21) Jan-82
- Did It In A Minute 9(16) May-82
- Your Imagination 33(11) Aug-82

Big Bam Boom -
Daryl HALL & John OATES 5 [203] Oct-84
- Out Of Touch 1(23) Dec-84
- Method Of Modern Love 5(19) Feb-85
- Some Things Are Better Left Unsaid 18(13) May-85
- Possession Obsession 30(12) Jul-85

Bad Animals - HEART 2 [144] Jun-87
- Alone 1(21) Jul-87
- Who Will You Run To 7(22) Oct-87
- There's The Girl 12(19) Jan-88
- I Want You So Bad 49(9) Mar-88

Building The Perfect Beast -
Don HENLEY 13 [253] Dec-84
- The Boys Of Summer 5(22) Feb-85
- All She Wants To Do Is Dance 9(19) May-85
- Not Enough Love In The World 34(17) Jul-85
- Sunset Grill 22(14) Oct-85

Partners In Crime -
Rupert HOLMES 33 [719] Nov-79 4,*1
- Him 6(17) Mar-80
- Answering Machine 32(11) Jun-80
- Morning Man 68(7) Dec-80
- Escape (The Pina Colada Song) 1(21) Dec-79

Nervous Night - The HOOTERS 12 [267] May-85
- All You Zombies - Hooters 58(11) Jun-85
- And We Danced - Hooters 21(20) Oct-85
- Day By Day - Hooters 18(18) Feb-86
- Where Do The Children Go - Hooters 38(12) May-86

The Way It Is -
Bruce HORNSBY And The RANGE 3 [98] Jun-86
- *Every Little Kiss 72(9) Aug-86*
- The Way It Is 1(22) Dec-86
- Mandolin Rain 4(18) Mar-87
- *Every Little Kiss 14(15) Jul-87*

Whitney Houston -
Whitney HOUSTON 1 [5] Mar-85
- You Give Good Love 3(21) Jul-85
- Saving All My Love For You 1(22) Oct-85
- How Will I Know 1(23) Feb-86
- Greatest Love Of All 1(18) May-86

Rebel Yell - Billy IDOL 6 [185] Dec-83
- Rebel Yell 46(14) Mar-84
- Eyes Without A Face 4(22) Jul-84
- Flesh For Fantasy 29(12) Oct-84
- Catch My Fall 50(11) Dec-84

Information Society -
INFORMATION SOCIETY 25 [757] Aug-88
- What's On Your Mind (Pure Energy) 3(25) Oct-88
- Walking Away 9(19) Feb-89
- Repetition 76(6) Apr-89
- Lay All Your Love On Me 83(8) Sep-89

Kick - INXS 3 [48] Nov-87
- Need You Tonight 1(25) Jan-88
- Devil Inside 2(17) Apr-88
- New Sensation 3(17) Jul-88
- Never Tear Us Apart 7(23) Nov-88

Off The Wall -
Michael JACKSON 3 [22] Sep-79 4,*1
- Rock With You 1(24) Jan-80
- Off The Wall 10(17) Apr-80
- She's Out Of My Life 10(16) Jun-80
- Don't Stop 'Til You Get Enough 1(21) Oct-79

Triumph - JACKSONS 10 [397] Oct-80
- Lovely One 12(18) Nov-80
- Heartbreak Hotel 22(16) Feb-81
- Can You Feel It 77(5) May-81
- Walk Right Now 73(4) Jul-81

Knee Deep In The Hoopla -
STARSHIP 7 [236] Oct-85
- We Built This City 1(24) Nov-85
- Sara 1(20) Mar-86
- Tomorrow Doesn't Matter Tonight 26(13) May-86
- Before I Go - Starship 68(7) Jul-86

Glass Houses - Billy JOEL 1 [34] Mar-80
- You May Be Right 7(15) May-80
- It's Still Rock And Roll To Me 1(21) Jul-80
- Don't Ask Me Why 19(15) Sep-80
- Sometimes A Fantasy 36(9) Nov-80

The Bridge - Billy JOEL 7 [243] Aug-86
- Modern Woman
  (From "Ruthless People") 10(15) Jul-86
- A Matter Of Trust 10(18) Oct-86
- This Is The Time 18(17) Jan-87
- Baby Grand 75(7) Apr-87

Escape - JOURNEY 1 [8] Aug-81
- Who's Crying Now 4(21) Oct-81
- Don't Stop Believin' 9(16) Dec-81
- Open Arms 2(18) Feb-82
- Still They Ride 19(14) Jul-82

Frontiers - JOURNEY 2 [90] Feb-83
- Separate Ways (Worlds Apart) 8(17) Mar-83
- Faithfully 12(16) Jun-83
- After The Fall 23(12) Aug-83
- Send Her My Love 23(15) Nov-83

Emergency - KOOL & The GANG 13 [177] Dec-84
- Misled 10(24) Mar-85
- Fresh 9(19) Jun-85
- Cherish 2(25) Sep-85
- Emergency 18(16) Dec-85

Forever - KOOL & The GANG 25 [750] Dec-86
- Victory 10(18) Jan-87
- Stone Love 10(18) May-87
- Holiday 66(7) Jul-87
- Special Way 72(14) Oct-87

True Colors - Cyndi LAUPER 4 [323] Oct-86
- True Colors 1(20) Oct-86
- Change Of Heart 3(17) Feb-87
- What's Going On 12(13) May-87
- Boy Blue 71(4) Jun-87

Valotte - Julian LENNON 17 [346] Nov-84
- Valotte 9(19) Jan-85
- Too Late For Goodbyes 5(17) Mar-85
- Say You're Wrong 21(12) Jun-85
- Jesse 54(6) Aug-85

Lovin' Every Minute Of It -
LOVERBOY 13 [403] Sep-85
- Lovin' Every Minute Of It 9(21) Nov-85
- Dangerous 65(9) Dec-85
- This Could Be The Night 10(18) Mar-86
- Lead A Double Life 68(7) May-86

Like A Virgin - MADONNA 1 [45] Dec-84
- Like A Virgin 1(19) Dec-84
- Material Girl 2(17) Mar-85
- Angel 5(17) Jun-85
- Dress You Up 5(16) Oct-85

Richard Marx - Richard MARX 8 [117] Jun-87
- Don't Mean Nothing 3(21) Aug-87
- Should've Known Better 3(21) Dec-87
- Endless Summer Nights 2(21) Mar-88
- Hold On To The Nights 1(21) Jul-88

The Lonesome Jubilee -
John Cougar MELLENCAMP 6 [141] Sep-87
- Paper In Fire 9(16) Oct-87
- Cherry Bomb 8(21) Jan-88
- Check It Out 14(15) Apr-88
- Rooty Toot Toot 61(8) Jun-88

Can't Hold Back - Eddie MONEY 20 [394] Aug-86
- Take Me Home Tonight 4(23) Nov-86
- I Wanna Go Back 14(21) Mar-87
- Endless Nights 21(19) Jun-87
- We Should Be Sleeping 90(3) Sep-87

Bella Donna - Stevie NICKS 1 [27] Aug-81
- Stop Draggin' My Heart Around - Stevie Nicks (with Tom Petty And The Heartbreakers) 3(21) Sep-81
- Leather And Lace - Stevie Nicks (with Don Henley) 6(19) Jan-82
- Edge Of Seventeen (Just Like The White Winged Dove) 11(14) Apr-82
- After The Glitter Fades 32(11) Jul-82

Suddenly - Billy OCEAN 9 [120] Aug-84
- Caribbean Queen (No More Love On The Run) 1(26) Nov-84
- Loverboy 2(21) Feb-85
- Suddenly 4(22) Jun-85
- Mystery Lady 24(15) Aug-85

Riptide - Robert PALMER 8 [197] Nov-85
- Discipline Of Love
  (Why Did You Do It) 82(5) Nov-85
- Addicted To Love 1(22) May-86
- Hyperactive 33(12) Jul-86
- I Didn't Mean To Turn You On 2(22) Nov-86

Street Talk - Steve PERRY 12 [360] Apr-84
- Oh Sherrie 3(20) Jun-84
- She's Mine 21(13) Aug-84
- Strung Out 40(13) Oct-84
- Foolish Heart 18(19) Feb-85

Please - PET SHOP BOYS 7 [389] Apr-86
- West End Girls 1(20) May-86
- Opportunities (Let's Make Lots Of Money) 10(16) Aug-86
- Love Comes Quickly 62(8) Oct-86
- Suburbia 70(10) Jan-87

Full Moon Fever -
Tom PETTY 3 [69] May-89 4,2*
- I Won't Back Down 12(15) Jul-89
- Runnin' Down A Dream 23(14) Sep-89
- Free Fallin' 7(21) Jan-90
- A Face In The Crowd 46(8) Mar-90

Open Up And Say...Ahh! -
POISON 2 [62] May-88
- Nothin' But A Good Time 6(19) Jul-88
- Fallen Angel 12(16) Oct-88
- Every Rose Has Its Thorn 1(21) Dec-88
- Your Mama Don't Dance 10(14) Apr-89

Synchronicity - The POLICE 1 [21] Jul-83
- Every Breath You Take 1(22) Jul-83
- King Of Pain 3(16) Oct-83
- Synchronicity II 16(14) Dec-83
- Wrapped Around Your Finger 8(16) Mar-84

Albums: Special Lists

Album Title - Act Pk [Rank] Entry Date
Single Title Pk(WksOn) Peak Date

1999 - PRINCE 9 [77] Nov-82
- Little Red Corvette 6(22) May-83
- 1999 12(21) Jul-83
- Delirious 8(18) Oct-83
- Let's Pretend We're Married/
  Irresistible Bitch 52(10) Jan-84

The Game - QUEEN 1 [70] Jul-80
- Crazy Little Thing Called Love 1(22) Feb-80
- Play The Game 42(9) Jul-80
- Another One Bites The Dust 1(31) Oct-80
- Need Your Loving Tonight 44(11) Dec-80

Hi Infidelity - REO SPEEDWAGON 1 [11] Dec-80
- Keep On Loving You 1(28) Mar-81
- Take It On The Run 5(20) May-81
- Don't Let Him Go 24(14) Aug-81
- In Your Letter 20(13) Sep-81

Wheels Are Turnin' REO SPEEDWAGON 7 [312] Nov-84
- I Do' Wanna Know 29(13) Dec-84
- Can't Fight This Feeling 1(18) Mar-85
- One Lonely Night 19(16) Jun-85
- Live Every Moment 34(11) Aug-85

Share Your Love - Kenny ROGERS 6 [280] Jul-81
- I Don't Need You 3(18) Aug-81
- Share Your Love With Me 14(15) Oct-81
- Blaze Of Glory 66(9) Dec-81
- Through The Years 13(15) Mar-82

Swept Away - Diana ROSS 26 [523] Sep-84
- Swept Away 19(14) Oct-84
- Missing You 10(27) Apr-85
- To All The Girls I've Loved Before -
  Julio Iglesias & Willie Nelson 5(21) May-84
- All Of You - Julio Iglesias & Diana Ross 19(16) Sep-84

Against The Wind - Bob SEGER &
The SILVER BULLET BAND 1 [37] Mar-80
- Fire Lake 6(16) May-80
- Against The Wind 5(17) Jun-80
- You'll Accomp'ny Me 14(16) Sep-80
- The Horizontal Bop 42(12) Dec-80

Like A Rock - Bob SEGER &
The SILVER BULLET BAND 3 [186] Apr-86
- American Storm 13(14) May-86
- Like A Rock 2(13) Jul-86
- It's You 52(9) Sep-86
- Miami 70(6) Nov-86

Footloose - SOUNDTRACK-MOVIE 1 [94] Feb-84
- Holding Out For A Hero - Bonnie Tyler 34(13) Apr-84
- Footloose - Kenny Loggins 1(23) Mar-84
- I'm Free (Heaven Helps The Man) -
  Kenny Loggins 22(14) Jul-84
- Almost Paradise...Love Theme From "Footloose" -
  Mike Reno And Ann Wilson 7(20) Jul-84

Beverly Hills Cop -
SOUNDTRACK-MOVIE 1 [123] Jan-85
- The Heat Is On - Glenn Frey 2(24) Mar-85
- Axel F - Harold Faltermeyer 3(19) Jun-85
- New Attitude - Patti LaBelle 17(21) May-85
- Stir It Up - Patti LaBelle 41(14) Aug-85

Top Gun - SOUNDTRACK-MOVIE 1 [83] Jun-86
- Take My Breath Away (Love Theme From "Top Gun") -
  Berlin 1(21) Sep-86
- Danger Zone - Kenny Loggins 2(21) Jul-86
- Playing With The Boys - Kenny Loggins 60(12) Sep-86
- Heaven In Your Eyes - Loverboy 12(17) Oct-86

Dirty Dancing - SOUNDTRACK-MOVIE 1 [9] Sep-87
- (I've Had) The Time Of My Life -
  Bill Medley And Jennifer Warnes 1(21) Nov-87
- Hungry Eyes - Eric Carmen 4(25) Feb-88
- Yes - Merry Clayton 45(11) Apr-88

Album Title - Act Pk [Rank] Entry Date
Single Title Pk(WksOn) Peak Date

- She's Like The Wind - Patrick Swayze 3(21) Feb-88

Cocktail - SOUNDTRACK-MOVIE 2 [137] Aug-88
- Kokomo - The Beach Boys 1(28) Nov-88
- Powerful Stuff - Fabulous Thunderbirds 65(7) Sep-88
- Hippy Hippy Shake - Georgia Satellites 45(14) Nov-88
- Wild Again - Starship 73(8) Jan-89

Hard To Hold (Soundtrack) -
Rick SPRINGFIELD 16 [799] Apr-84
- Taxi Dancing - Rick Springfield & Randy Crawford
  59(10) Dec-84
- Love Somebody 5(16) May-84
- Don't Walk Away 26(12) Jul-84
- Bop 'Til You Drop 20(15) Oct-84

In My Eyes - Stevie B 75 [1237] Mar-89 4,1*
- I Wanna Be The One 32(20) Apr-89
- In My Eyes 37(17) Jul-89
- Girl I Am Searching For You 56(15) Dec-89
- Love Me For Life 29(14) Mar-90

Out Of Order - Rod STEWART 20 [168] Jun-88
- Lost In You 12(18) Jul-88
- Forever Young 12(24) Oct-88
- My Heart Can't Tell You No 4(25) Apr-89
- Crazy About Her 11(17) Jul-89

The Dream Of The Blue Turtles - STING 2 [107] Jul-85
- If You Love Somebody Set Them Free 3(18) Aug-85
- Fortress Around Your Heart 8(20) Oct-85
- Love Is The Seventh Wave 17(13) Dec-85
- Russians 16(13) Mar-86

Guilty - Barbra STREISAND 1 [64] Oct-80
- Woman In Love 1(24) Oct-80
- Promises 48(9) Jun-81
- Guilty - Barbra Streisand & Barry Gibb 3(22) Jan-81
- What Kind Of Fool -
  Barbra Streisand & Barry Gibb 10(16) Mar-81

Vital Signs - SURVIVOR 16 [339] Sep-84
- I Can't Hold Back 13(23) Dec-84
- High On You 8(17) Mar-85
- The Search Is Over 4(21) Jul-85
- First Night 53(9) Sep-85

Songs From The Big Chair -
TEARS FOR FEARS 1 [40] Apr-85
- Everybody Wants To Rule The World 1(24) Jun-85
- Shout 1(19) Aug-85
- Head Over Heels 3(20) Nov-85
- Mothers Talk 27(12) May-86

Into The Gap - THOMPSON TWINS 10 [324] Mar-84
- Hold Me Now 3(21) May-84
- Doctor! Doctor! 11(16) Jul-84
- You Take Me Up 44(9) Oct-84
- The Gap 69(6) Dec-84

Tiffany - TIFFANY 1 [102] Sep-87
- I Think We're Alone Now 1(24) Nov-87
- Could've Been 1(20) Feb-88
- I Saw Him Standing There 7(14) Apr-88
- Feelings Of Forever 50(9) Jul-88

Break Every Rule - Tina TURNER 4 [363] Sep-86
- Typical Male 2(16) Oct-86
- Two People 30(12) Jan-87
- What You Get Is What You See 13(14) Apr-87
- Break Every Rule 74(5) May-87

The Joshua Tree - U2 1 [29] Apr-87
- With Or Without You 1(18) May-87
- I Still Haven't Found What I'm Looking For 1(17) Aug-87
- Where The Streets Have No Name 13(14) Nov-87
- In God's Country 44(12) Jan-88

Rattle And Hum (Soundtrack) - U2 1 [153] Oct-88
- When Love Comes To Town-U2/B. B. King 68(7) Apr-89
- Desire 3(17) Nov-88

Album Title - Act Pk [Rank] Entry Date
Single Title Pk(WksOn) Peak Date

- Angel Of Harlem 14(15) Feb-89
- All I Want Is You 83(4) Jul-89

1984 (MCMLXXXIV) - VAN HALEN 2 [71] Jan-84
- Jump 1(21) Feb-84
- I'll Wait 13(14) Jun-84
- Panama 13(15) Aug-84
- Hot For Teacher 56(7) Nov-84

OU812 - VAN HALEN 1 [109] Jun-88
- Black And Blue 34(10) Jun-88
- When It's Love 5(19) Sep-88
- Finish What Ya Started 13(20) Dec-88
- Feels So Good 35(14) Mar-89

Dirty Rotten Filthy Stinking Rich -
WARRANT 10 [199] Mar-89 4,1*
- Down Boys 27(16) Jul-89
- Heaven 2(19) Sep-89
- Big Talk 93(4) Nov-89
- Sometimes She Cries 20(17) Mar-90

Larger Than Life - Jody WATLEY 16 [574] Apr-89 4,2*
- Real Love 2(18) May-89
- Friends 9(18) Aug-89
- Everything 4(23) Jan-90
- Precious Love 87(3) Mar-90

Make It Big - WHAM! 1 [54] Nov-84
- Wake Me Up Before You Go-Go 1(24) Nov-84
- Careless Whisper - Wham! Featuring George Michael
  1(21) Feb-85
- Everything She Wants 1(20) May-85
- Freedom 3(18) Sep-85

Music From The Edge Of Heaven -
WHAM! 10 [700] Jul-86
- A Different Corner - George Michael 7(16) Jun-86
- I'm Your Man 3(18) Feb-86
- The Edge Of Heaven 10(13) Aug-86
- Where Did Your Heart Go? 50(8) Nov-86

Whitesnake - WHITESNAKE 2 [32] Apr-87
- Still Of The Night 79(7) Jul-87
- Here I Go Again 1(28) Oct-87
- Is This Love 2(19) Dec-87
- Give Me All Your Love 48(11) Mar-88

Will To Power - WILL TO POWER 68 [1694] Sep-88
- Dreamin' 50(16) Aug-87
- Say It's Gonna Rain 49(14) Jul-88
- (a)Baby, I Love Your Way/(b) FreeBird Medley
  (Free Baby) 1(24) Dec-88
- Fading Away 65(10) Feb-89

Back In The High Life -
Steve WINWOOD 3 [99] Jul-86
- Higher Love 1(22) Aug-86
- Freedom Overspill 20(15) Nov-86
- The Finer Things 8(23) Apr-87
- Back In The High Life Again 13(21) Aug-87

Roll With It - Steve WINWOOD 1 [158] Jul-88
- Roll With It 1(18) Jul-88
- Don't You Know What The Night Can Do? 6(17) Oct-88
- Holding On 11(17) Jan-89
- Hearts On Fire 53(9) Apr-89

In Square Circle - Stevie WONDER 5 [222] Oct-85
- Part-Time Lover 1(21) Nov-85
- Go Home 10(17) Feb-86
- Overjoyed 24(13) Apr-86
- Land Of La La 86(3) Jun-86

Afterburner - ZZ TOP 4 [165] Nov-85
- Sleeping Bag 8(17) Dec-85
- Stages 21(12) Mar-86
- Rough Boy 22(13) May-86
- Velcro Fly 35(12) Aug-86

Albums: Special Lists

Three Hot 100 Singles: 267 Albums

Super Trouper - ABBA 17 [447] Dec-80
The Winner Takes It All 8(26) Mar-81
Super Trouper 45(11) May-81
On And On And On 90(6) Jul-81

How To Be A...Zillionaire! - ABC 30 [648] Oct-85
Be Near Me 9(22) Nov-85
(How To Be A) Millionaire 20(14) Mar-86
Vanity Kills 91(4) May-86

Cuts Like A Knife - Bryan ADAMS 8 [263] Feb-83
Straight From The Heart 10(19) May-83
Cuts Like A Knife 15(14) Aug-83
This Time 24(12) Oct-83

Into The Fire - Bryan ADAMS 7 [383] Apr-87
Heat Of The Night 6(16) May-87
Hearts On Fire 26(13) Aug-87
Victim Of Love 32(12) Oct-87

Permanent Vacation - AEROSMITH 11 [151] Sep-87
Dude (Looks Like A Lady) 14(20) Dec-87
Angel 3(25) Apr-88
Rag Doll 17(17) Aug-88

Lost In Love - AIR SUPPLY 22 [242] May-80
Lost In Love 3(23) May-80
All Out Of Love 2(27) Sep-80
Every Woman In The World 5(22) Jan-81

**The One That You Love -
  AIR SUPPLY 10 [195] Jun-81**
The One That You Love 1(19) Jul-81
Here I Am (Just When I Thought I Was Over You) 5(20) Nov-81
Sweet Dreams 5(20) Mar-82

Now And Forever - AIR SUPPLY 25 [861] Jun-82
Even The Nights Are Better 5(18) Sep-82
Young Love 38(9) Oct-82
Two Less Lonely People In The World 38(14) Jan-83

In Effect Mode - AL B. SURE! 20 [343] May-88
Nite And Day 7(21) Jul-88
Off On Your Own (Girl) 45(12) Sep-88
Killing Me Softly 80(11) Nov-88

**Keep Your Eye On Me -
  Herb ALPERT 18 [628] Mar-87**
Keep Your Eye On Me 46(10) Apr-87
Diamonds 5(19) Jun-87
Making Love In The Rain 35(14) Sep-87

Forever Young - ALPHAVILLE 180 [3907] Dec-84
Big In Japan 66(10) Jan-85
Forever Young 93(4) Mar-85
Forever Young 65(14) Dec-88

**Whenever You Need Somebody -
  Rick ASTLEY 10 [191] Jan-88**
Never Gonna Give You Up 1(24) Mar-88
Together Forever 1(18) Jun-88
It Would Take A Strong Strong Man 10(16) Sep-88

Hold Me In Your Arms - Rick ASTLEY 19 [903] Jan-89
She Wants To Dance With Me 6(18) Feb-89
Giving Up On Love 38(10) May-89
Ain't Too Proud To Beg 89(3) Aug-89

**As The Band Turns -
  ATLANTIC STARR 17 [489] May-85**
Freak-A-Ristic 90(6) May-85
Secret Lovers 3(23) Mar-86
If Your Heart Isn't In It 57(12) May-86

True Confessions - BANANARAMA 15 [824] Aug-86
Venus 1(19) Sep-86
More Than Physical 73(5) Nov-86
A Trick Of The Night 76(7) Jan-87

Wow! - BANANARAMA 44 [1492] Sep-87
I Heard A Rumour 4(19) Sep-87
I Can't Help It 47(13) Jan-88
Love In The First Degree 48(10) Apr-88

Everything - BANGLES 15 [431] Nov-88
In Your Room 5(20) Jan-89
Eternal Flame 1(19) Apr-89
Be With You 30(12) Jun-89

Crimes Of Passion - Pat BENATAR 2 [42] Aug-80
You Better Run 42(11) Aug-80
Hit Me With Your Best Shot 9(24) Dec-80
Treat Me Right 18(18) Mar-81

Get Nervous - Pat BENATAR 4 [261] Nov-82
Shadows Of The Night 13(16) Dec-82
Little Too Late 20(14) Mar-83
Looking For A Stranger 39(10) May-83

Seven The Hard Way--Pat BENATAR 26 [1160] Dec-85
Invincible 10(17) Sep-85
Sex As A Weapon 28(13) Jan-86
Le Bel Age 54(8) Mar-86

Give Me The Night - George BENSON 3 [235] Aug-80
Give Me The Night 4(23) Sep-80
Love X Love 61(6) Nov-80
Turn Your Love Around 5(22) Feb-82

Pleasure Victim - BERLIN 30 [981] Feb-83
Sex (I'm A...) 62(7) Mar-83
The Metro 58(10) Jul-83
Masquerade 82(3) Oct-83

The Hunger - Michael BOLTON 46 [1037] Oct-87
That's What Love Is All About 19(25) Dec-87
(Sittin' On) The Dock Of The Bay 11(17) Mar-88
Wait On Love 79(6) Jun-88

Slippery When Wet - BON JOVI 1 [15] Sep-86
You Give Love A Bad Name 1(24) Nov-86
Livin' On A Prayer 1(21) Feb-87
Wanted Dead Or Alive 7(17) Jun-87

Third Stage - BOSTON 1 [146] Oct-86
Amanda 1(18) Nov-86
We're Ready 9(15) Feb-87
Can'tcha Say (You Believe In Me)/Still In Love 20(13) Apr-87

Self Control - Laura BRANIGAN 23 [530] Apr-84
Self Control 4(25) Jun-84
The Lucky One 20(15) Sep-84
Ti Amo 55(12) Dec-84

Hold Me - Laura BRANIGAN 71 [2333] Aug-85
Spanish Eddie 40(11) Sep-85
Hold Me 82(4) Oct-85
I Found Someone 90(6) Mar-86

All That Jazz - BREATHE 34 [510] Jun-88
Hands To Heaven 2(29) Aug-88
How Can I Fall? 3(22) Dec-88
Don't Tell Me Lies 10(16) Mar-89

**Lawyers In Love -
  Jackson BROWNE 8 [547] Aug-83**
Lawyers In Love 13(15) Sep-83
Tender Is The Night 25(17) Nov-83
For A Rocker 45(9) Feb-84

Word Up! - CAMEO 8 [307] Sep-86
Word Up 6(21) Nov-86
Candy 21(17) Mar-87
Back And Forth 50(8) May-87

**Make Your Move -
  CAPTAIN & TENNILLE 23 [836] Nov-79**
Do That To Me One More Time 1(27) Feb-80
Love On A Shoestring 55(7) Mar-80
Happy Together (A Fantasy) 53(6) Jun-80

Mistaken Identity - Kim CARNES 1 [110] May-81
Bette Davis Eyes 1(26) May-81
Draw Of The Cards 28(12) Sep-81
Mistaken Identity 60(6) Nov-81

Cafe Racers - Kim CARNES 97 [2845] Nov-83
Invisible Hands 40(13) Nov-83
You Make My Heart Beat Faster
  (And That's All That Matters) 54(8) Feb-84

I Pretend 74(5) Jun-84

Barking At Airplanes - Kim CARNES 48 [2125] Jun-85
Invitation To Dance 68(6) Feb-85
Crazy In The Night (Barking At Airplanes) 15(16) Jul-85
Abadabadango 67(4) Aug-85

Shake It Up - The CARS 9 [277] Nov-81
Shake It Up 4(22) Feb-82
Since You're Gone 41(9) May-82
I'm Not The One 32(11) Mar-86

Door To Door - The CARS 26 [1126] Sep-87
You Are The Girl 17(14) Oct-87
Strap Me In 85(4) Nov-87
Coming Up You 74(5) Feb-88

Tracy Chapman - Tracy CHAPMAN 1 [93] Apr-88
Fast Car 6(21) Aug-88
Talkin' Bout A Revolution 75(4) Oct-88
Baby Can I Hold You 48(12) Dec-88

Cher(2) - CHER 32 [715] Dec-87
I Found Someone 10(26) Mar-88
We All Sleep Alone 14(15) Jun-88
Skin Deep 79(4) Aug-88

Heart Of Stone - CHER 10 [240] Jul-89 3,1*
If I Could Turn Back Time 3(23) Sep-89
Just Like Jesse James 8(18) Dec-89
Heart Of Stone 20(14) Apr-90

**Raw Like Sushi -
  Neneh CHERRY 40 [1086] Jun-89 3,1***
Buffalo Stance 3(24) Jun-89
Kisses On The Wind 8(14) Sep-89
Heart 73(8) Jan-90

Real People - CHIC 30 [1404] Jul-80
Rebels Are We 61(6) Sep-80
Real People 89(1) Nov-80
Real People//Chip Off The Old Block 79(2) Nov-80

Chicago 16 - CHICAGO 9 [439] Jun-82
Hard To Say I'm Sorry 1(24) Sep-82
Love Me Tomorrow 22(15) Dec-82
What You're Missing 81(5) Jan-83

Just One Night - Eric CLAPTON 2 [192] May-80
Tulsa Time - Eric Clapton And His Band 64(2) Jun-80
Tulsa Time//Cocaine - Eric Clapton And His Band 30(12) Aug-80
Blues Power (Live) 76(5) Nov-80

Combat Rock - The CLASH 7 [173] Jun-82
Should I Stay Or Should I Go? 45(13) Sep-82
Rock The Casbah 8(24) Jan-83
Should I Stay Or Should I Go? 50(10) Mar-83

**Hello, I Must Be Going! -
  Phil COLLINS 8 [230] Nov-82**
You Can't Hurry Love 10(21) Feb-83
I Don't Care Anymore 39(11) Mar-83
I Cannot Believe It's True 79(4) May-83

**In The Pocket -
  The COMMODORES 13 [320] Jul-81**
Lady (You Bring Me Up) 8(22) Sep-81
Oh No 4(20) Dec-81
Why You Wanna Try Me 66(5) Feb-82

Nightshift - The COMMODORES 12 [591] Feb-85
Nightshift 3(22) Apr-85
Animal Instinct 43(9) Jun-85
Janet 87(4) Sep-85

Trash - ALICE COOPER 20 [452] Aug-89 3,2*
Poison 7(19) Nov-89
House Of Fire 56(9) Feb-90
Only My Heart Talkin' 89(3) May-90

**Daylight Again -
  CROSBY, STILLS & NASH 8 [304] Jul-82**
Wasted On The Way 9(15) Aug-82
Southern Cross 18(17) Nov-82
Too Much Love To Hide 69(6) Feb-83

248

Albums: Special Lists

Another Page - Christopher CROSS 11 [698] Feb-83
- All Right 12(16) Mar-83
- No Time For Talk 33(10) Jun-83
- Think Of Laura 9(17) Feb-84

**Crowded House -
CROWDED HOUSE 12 [387] Aug-86**
- Don't Dream It's Over 2(24) Apr-87
- Something So Strong 7(21) Jul-87
- World Where You Live 65(8) Sep-87

**Kissing To Be Clever -
CULTURE CLUB 14 [172] Jan-83**
- Do You Really Want To Hurt Me 2(25) Mar-83
- Time (Clock Of The Heart) 2(18) Jun-83
- I'll Tumble 4 Ya 9(16) Aug-83

**Kiss Me, Kiss Me, Kiss Me -
The CURE 35 [593] Jun-87**
- Why Can't I Be You? 54(12) Aug-87
- Just Like Heaven 40(19) Jan-88
- Hot Hot Hot!!! 65(7) Apr-88

Broadcast - CUTTING CREW 16 [680] Mar-87
- (I Just) Died In Your Arms 1(19) May-87
- One For The Mockingbird 38(11) Jul-87
- I've Been In Love Before 9(21) Nov-87

**Where Do We Go From Here -
Michael DAMIAN 61 [1865] Jun-89 3,1***
- Rock On 1(21) Jun-89
- Cover Of Love 31(12) Aug-89
- Was It Nothing At All 24(21) Feb-90

Cool Night - Paul DAVIS 52 [1580] Dec-81
- Cool Night 11(19) Feb-82
- '65 Love Affair 6(20) May-82
- Love Or Let Me Be Lonely 40(10) Aug-82

El DeBarge - El DeBARGE 24 [963] Jun-86
- Who's Johnny (Short Circuit Theme) 3(19) Jul-86
- Love Always 43(12) Sep-86
- Someone 70(9) Jan-87

Pyromania - DEF LEPPARD 2 [41] Feb-83
- Photograph 12(17) May-83
- Rock Of Ages 16(15) Aug-83
- Foolin' 28(14) Nov-83

**The Jazz Singer (Soundtrack) -
Neil DIAMOND 3 [81] Nov-80**
- Love On The Rocks 2(20) Jan-81
- Hello Again 6(16) Mar-81
- America 8(17) Jun-81

**On The Way To The Sky -
Neil DIAMOND 17 [695] Nov-81**
- Yesterday's Songs 11(15) Jan-82
- On The Way To The Sky 27(10) Mar-82
- Be Mine Tonight 35(11) Jul-82

Heartlight - Neil DIAMOND 9 [432] Oct-82
- Heartlight 5(19) Nov-82
- I'm Alive 35(12) Feb-83
- Front Page Story 65(8) May-83

Brothers In Arms - DIRE STRAITS 1 [31] Jun-85
- Money For Nothing 1(22) Sep-85
- Walk Of Life 7(21) Jan-86
- So Far Away 19(14) Apr-86

Sometimes You Win - DR. HOOK 71 [1644] Nov-79
- Better Love Next Time 12(19) Jan-80
- Sexy Eyes 5(21) May-80
- Years From Now 51(9) Aug-80

One Step Closer-DOOBIE BROTHERS 3 [313] Oct-80
- Real Love 5(16) Oct-80
- One Step Closer 24(14) Jan-81
- Keep This Train A-Rollin' 62(5) Mar-81

**Seven And The Ragged Tiger -
DURAN DURAN 8 [152] Dec-83**
- Union Of The Snake 3(17) Dec-83
- New Moon On Monday 10(16) Mar-84
- The Reflex 1(21) Jun-84

Notorious - DURAN DURAN 12 [616] Dec-86
- Notorious 2(17) Jan-87
- Skin Trade 39(9) Mar-87
- Meet El Presidente 70(5) May-87

Big Thing - DURAN DURAN 24 [803] Nov-88
- I Don't Want Your Love 4(16) Dec-88
- All She Wants Is 22(13) Feb-89
- Do You Believe In Shame? 72(5) Apr-89

Faces - EARTH, WIND & FIRE 10 [631] Nov-80
- Let Me Talk 44(9) Oct-80
- You 48(12) Dec-80
- And Love Goes On 59(7) Feb-81

Best Kept Secret - Sheena EASTON 33 [960] Sep-83
- Telefone (Long Distance Love Affair) 9(22) Oct-83
- Almost Over You 25(20) Mar-84
- Devil In A Fast Car 79(3) Apr-84

A Private Heaven - Sheena EASTON 15 [458] Oct-84
- Strut 7(25) Nov-84
- Sugar Walls 9(17) Mar-85
- Swear 80(6) Apr-85

Wild, Wild West - ESCAPE CLUB 27 [720] Aug-88
- Wild, Wild West 1(27) Nov-88
- Shake For The Sheik 28(14) Jan-89
- Walking Through Walls 81(6) Apr-89

Touch - EURYTHMICS 7 [349] Feb-84
- Here Comes The Rain Again 4(20) Mar-84
- Who's That Girl? 21(13) Jun-84
- Right By Your Side 29(12) Sep-84

Reach The Beach - The FIXX 8 [232] May-83
- Saved By Zero 20(16) Aug-83
- One Thing Leads To Another 4(19) Nov-83
- The Sign Of Fire 32(13) Jan-84

Mirage - FLEETWOOD MAC 1 [121] Jul-82
- Hold Me 4(17) Jul-82
- Gypsy 12(14) Oct-82
- Love In Store 22(14) Jan-83

Centerfield - John FOGERTY 1 [174] Jan-85
- The Old Man Down The Road 10(18) Mar-85
- Rock And Roll Girls 20(12) Apr-85
- Centerfield 44(13) Jun-85

**Inside Information -
FOREIGNER 15 [520] Dec-87**
- Say You Will 6(19) Feb-88
- I Don't Want To Live Without You 5(17) May-88
- Heart Turns To Stone 56(10) Aug-88

No Fun Aloud - Glenn FREY 32 [733] Jun-82
- I Found Somebody 31(13) Aug-82
- The One You Love 15(17) Nov-82
- All Those Lies 41(12) Jan-83

The Allnighter - Glenn FREY 22 [465] Jul-84
- Sexy Girl 20(15) Aug-84
- The Allnighter 54(6) Oct-84
- Smuggler's Blues 12(19) Jun-85

Gap Band IV - GAP BAND 14 [427] Jun-82
- Early In The Morning 24(14) Jul-82
- You Dropped A Bomb On Me 31(13) Sep-82
- Outstanding 51(8) Apr-83

**Same Goes For You -
Leif GARRETT 129 [2546] Dec-79 3,*1**
- Memorize Your Number 60(9) Jan-80
- I Was Looking For Someone To Love 78(5) May-80
- When I Think Of You 78(5) Nov-79

Love Stinks - J. GEILS Band 18 [450] Feb-80
- Come Back 32(12) Mar-80
- Love Stinks 38(12) May-80
- Just Can't Wait 78(5) Jul-80

Freeze-Frame - J. GEILS Band 1 [82] Nov-81
- Centerfold 1(25) Feb-82
- Freeze-Frame 4(16) Apr-82
- Angel In Blue 40(11) Jul-82

Abacab - GENESIS 7 [184] Oct-81
- No Reply At All 29(18) Nov-81
- Abacab 26(14) Feb-82
- Man On The Corner 40(11) May-82

Sexappeal - GEORGIO 117 [1710] Apr-87
- Sexappeal 58(9) Mar-87
- Tina Cherry 96(2) Aug-87
- Lover's Lane 59(12) Dec-87

**The Thin Red Line -
GLASS TIGER 27 [481] Jul-86**
- Don't Forget Me (When I'm Gone) 2(24) Oct-86
- Someday 7(21) Jan-87
- I Will Be There 34(11) Apr-87

Talk Show - The GO-GO'S 18 [572] Apr-84
- Head Over Heels 11(16) May-84
- Turn To You 32(14) Aug-84
- Yes Or No 84(3) Sep-84

Go West - GO WEST 60 [1291] Mar-85
- We Close Our Eyes 41(15) Apr-85
- Call Me 54(14) Jul-85
- Eye To Eye 73(7) Oct-85

...Twice Shy - GREAT WHITE 9 [287] May-89 3,1*
- Once Bitten Twice Shy 5(26) Aug-89
- The Angel Song 30(21) Dec-89
- House Of Broken Love 83(5) Mar-90

**Three Hearts In The Happy Ending Machine -
Daryl HALL 29 [1293] Sep-86**
- Dreamtime 5(15) Oct-86
- Foolish Pride 33(13) Dec-86
- Someone Like You 57(8) Feb-87

H2O - Daryl HALL & John OATES 3 [59] Oct-82
- Maneater 1(23) Dec-82
- One On One 7(18) Apr-83
- Family Man 6(16) Jun-83

**Ooh Yeah! -
Daryl HALL & John OATES 24 [951] May-88**
- Everything Your Heart Desires 3(16) Jun-88
- Missed Opportunity 29(11) Aug-88
- Downtown Life 31(9) Nov-88

Boy In The Box - Corey HART 20 [534] Jul-85
- Never Surrender 3(20) Aug-85
- Boy In The Box 26(12) Nov-85
- Everything In My Heart 30(15) Feb-86

Fields Of Fire - Corey HART 55 [1506] Oct-86
- I Am By Your Side 18(13) Nov-86
- Can't Help Falling In Love 24(14) Feb-87
- Dancin' With My Mirror 88(3) Apr-87

**I Can Dream About You -
Dan HARTMAN 55 [1498] Nov-84**
- I Can Dream About You 6(25) Aug-84
- We Are The Young 25(17) Dec-84
- Second Nature 39(12) Mar-85

**I Can't Stand Still -
Don HENLEY 24 [570] Sep-82**
- Johnny Can't Read 42(11) Oct-82
- Dirty Laundry 3(19) Jan-83
- I Can't Stand Still 48(11) Feb-83

**Blind To Reason -
Grayson HUGH 71 [2055] Oct-88 3,1***
- Talk It Over 19(18) Sep-89
- Bring It All Back 87(4) Nov-89
- How 'Bout Us -
Grayson Hugh and Betty Wright 67(7) Apr-90

Man Of Colours - ICEHOUSE 43 [908] Oct-87
- Crazy 14(21) Jan-88
- Electric Blue 7(21) May-88
- My Obsession 88(4) Jul-88

Whiplash Smile - Billy IDOL 6 [408] Nov-86
- To Be A Lover 6(18) Dec-86
- Don't Need A Gun 37(9) Mar-87
- Sweet Sixteen 20(14) Jun-87

Albums: Special Lists

It's Your Night -
  James INGRAM 46 [975] Nov-83
- Yah Mo B There 19(18) Mar-84
- There's No Easy Way 58(10) May-84
- How Do You Keep The Music Playing
  (Theme From "Best Friends") -
  James Ingram And Patti Austin 45(17) Jul-83

Listen Like Thieves - INXS 11 [400] Nov-85
- This Time 81(6) Dec-85
- What You Need 5(20) Apr-86
- Listen Like Thieves 54(9) Jun-86

King Cool - Donnie IRIS 84 [1735] Sep-81
- Sweet Merilee 80(6) Nov-81
- Love Is Like A Rock 37(14) Feb-82
- My Girl 25(14) May-82

Rock Me Tonight -
  Freddie JACKSON 10 [252] May-85
- Rock Me Tonight (For Old Times Sake) 18(19) Aug-85
- You Are My Lady 12(20) Nov-85
- He'll Never Love You (Like I Do) 25(15) Feb-86

Just Like The First Time -
  Freddie JACKSON 23 [423] Nov-86
- Tasty Love 41(12) Dec-86
- Have You Ever Loved Somebody 69(9) Mar-87
- Jam Tonight 32(13) Aug-87

Victory - JACKSONS 4 [472] Jul-84
- State Of Shock 3(15) Aug-84
- Torture 17(12) Sep-84
- Body 47(7) Nov-84

Breakin' Away - Al JARREAU 9 [193] Aug-81
- We're In This Love Together 15(24) Nov-81
- Breakin' Away 43(10) Jan-82
- Teach Me Tonight 70(7) Apr-82

Jarreau - Al JARREAU 13 [565] Apr-83
- Mornin' - Jarreau 21(15) May-83
- Boogie Down 77(6) Jul-83
- Trouble In Paradise 63(7) Oct-83

No Protection -
  STARSHIP 12 [862] Jul-87
- Nothing's Gonna Stop Us Now - Starship 1(22) Apr-87
- It's Not Over ('Til It's Over) - Starship 9(16) Aug-87
- Beat Patrol - Starship 46(10) Nov-87

The Jets - The JETS 21 [357] Apr-86
- Crush On You 3(20) Jun-86
- Private Number 47(11) Sep-86
- You Got It All 3(26) Mar-87

Supersonic--The Album - J.J. FAD 49 [1210] Jul-88
- Supersonic 30(22) Jun-88
- Way Out 61(9) Oct-88
- Is It Love 92(5) Dec-88

The Nylon Curtain - Billy JOEL 7 [301] Oct-82
- Pressure 20(17) Nov-82
- Allentown 17(22) Feb-83
- Goodnight Saigon 56(7) Apr-83

Too Low For Zero - Elton JOHN 25 [582] Jun-83
- I'm Still Standing 12(16) Jul-83
- Kiss The Bride 25(12) Oct-83
- I Guess That's Why They Call It The Blues 4(23) Jan-84

Breaking Hearts - Elton JOHN 20 [690] Jul-84
- Sad Songs (Say So Much) 5(19) Aug-84
- Who Wears These Shoes? 16(14) Nov-84
- In Neon 38(13) Jan-85

Sleeping With The Past -
  Elton JOHN 23 [513] Sep-89 3,2*
- Healing Hands 13(15) Oct-89
- Sacrifice 18(17) Mar-90
- Club At The End Of The Street 28(16) Jul-90

Dream Into Action - Howard JONES 10 [374] Apr-85
- Things Can Only Get Better 5(23) Jun-85
- Life In One Day 19(16) Sep-85
- No One Is To Blame 4(23) Jul-86

The Dude - Quincy JONES 10 [140] Apr-81
- Just Once - Quincy Jones Featuring James Ingram 17(23) Nov-81
- One Hundred Ways 14(21) Apr-82
- Ai No Corrida (I-No-Ko-ree-da) 28(12) May-81

Departure - JOURNEY 8 [329] Mar-80
- Any Way You Want It 23(15) Apr-80
- Walks Like A Lady 32(13) Jul-80
- Good Morning Girl/Stay Awhile 55(8) Sep-80

Katrina And The Waves -
  KATRINA & The WAVES 25 [805] Apr-85
- Walking On Sunshine 9(21) Jun-85
- Do You Want Crying 37(10) Sep-85
- Que Te Quiero 71(6) Oct-85

I Feel For You - Chaka KHAN 14 [515] Oct-84
- I Feel For You 3(26) Nov-84
- This Is My Night 60(9) Feb-85
- Through The Fire 60(19) May-85

Hot In The Shade - KISS 29 [919] Nov-89 3,2*
- Hide Your Heart 66(10) Dec-89
- Forever 8(17) Apr-90
- Rise To It 81(6) Jun-90

Meeting In The Ladies Room -
  KLYMAXX 18 [438] Feb-85
- Meeting In The Ladies Room 59(11) Jun-85
- I Miss You 5(29) Dec-85
- The Men All Pause 80(8) Mar-86

Something Special -
  KOOL & The GANG 12 [214] Oct-81
- Take My Heart (You Can Have It If You Want It) 17(17) Dec-81
- Steppin' Out 89(2) Feb-82
- Get Down On It 10(17) May-82

Double Fantasy -
  John LENNON & Yoko ONO 1 [61] Dec-80
- (Just Like) Starting Over - John Lennon 1(22) Dec-80
- Woman - John Lennon 2(20) Mar-81
- Watching The Wheels - John Lennon 10(17) May-81

Picture This - Huey LEWIS And The NEWS 13 [435] Feb-82
- Do You Believe In Love 7(17) Apr-82
- Hope You Love Me Like You Say You Do 36(11) Jun-82
- Workin' For A Livin' 41(9) Sep-82

Small World -
  Huey LEWIS And The NEWS 11 [656] Aug-88
- Perfect World 3(15) Sep-88
- Small World 25(11) Nov-88
- Give Me The Keys (And I'll Drive You Crazy) 47(8) Feb-89

Lisa Lisa & Cult Jam With Full Force - LISA LISA And CULT JAM 52 [962] Aug-85
- I Wonder If I Take You Home - Lisa Lisa and Cult Jam with Full Force 34(21) Aug-85
- Can You Feel The Beat 69(20) Dec-85
- All Cried Out - Lisa Lisa And Cult Jam/Full Force Ft. Paul Anthony And Bow Legged Lou 8(26) Oct-86

Spanish Fly -
  LISA LISA And CULT JAM 7 [256] May-87
- Head To Toe 1(20) Jun-87
- Lost In Emotion 1(20) Oct-87
- Someone To Love Me For Me 78(10) Dec-87

Time Exposure -
  LITTLE RIVER BAND 21 [524] Sep-81
- The Night Owls 6(21) Nov-81
- Take It Easy On Me 10(19) Mar-82
- Man On Your Mind 14(16) May-82

Vivid - LIVING COLOUR 6 [198] Sep-88
- Cult Of Personality 13(15) May-89
- Open Letter (To A Landlord) 82(5) Jul-89
- Glamour Boys 31(13) Oct-89

High Adventure -
  Kenny LOGGINS 13 [487] Sep-82
- Don't Fight It - Kenny Loggins with Steve Perry 17(12) Oct-82
- Heart To Heart 15(17) Jan-83
- Welcome To Heartlight 24(14) Apr-83

Vox Humana -
  Kenny LOGGINS 41 [1232] Apr-85
- Vox Humana 29(10) May-85
- Forever 40(22) Jul-85
- I'll Be There 88(2) Oct-85

Back To Avalon -
  Kenny LOGGINS 69 [2435] Aug-88
- Nobody's Fool 8(18) Sep-88
- I'm Gonna Miss You 82(5) Nov-88
- Tell Her 76(8) Feb-89

Madness - MADNESS 41 [1254] Apr-83
- Our House 7(19) Jul-83
- It Must Be Love 33(12) Oct-83
- The Sun And The Rain 72(5) Mar-84

Madonna - MADONNA 8 [113] Sep-83
- Holiday 16(21) Jan-84
- Borderline 10(30) Jun-84
- Lucky Star 4(16) Oct-84

If I Should Love Again -
  Barry MANILOW 14 [597] Oct-81
- The Old Songs 15(16) Nov-81
- Somewhere Down The Road 21(15) Feb-82
- Let's Hang On 32(10) May-82

Martika - MARTIKA 15 [603] Feb-89
- More Than You Know 18(20) Apr-89
- Toy Soldiers 1(20) Jul-89
- I Feel The Earth Move 25(11) Oct-89

Tug Of War - Paul McCARTNEY 1 [283] May-82
- Ebony And Ivory 1(19) May-82
- Take It Away 10(16) Aug-82
- Tug Of War 53(8) Oct-82

Flowers In The Dirt -
  Paul McCARTNEY 21 [734] Jun-89 3,1*
- My Brave Face 25(10) Jul-89
- This One 94(3) Sep-89
- Figure Of Eight 92(5) Jan-90

Chain Lightning - Don McLEAN 28 [986] Feb-81
- Crying 5(18) Mar-81
- Since I Don't Have You 23(14) May-81
- It's Just The Sun 83(2) Aug-81

Glenn Medeiros -
  Glenn MEDEIROS 83 [2609] Jun-87
- Nothing's Gonna Change My Love For You 12(25) Jun-87
- Watching Over You 80(8) Aug-87
- Lonely Won't Leave Me Alone 67(11) Jan-88

American Fool - John MELLENCAMP 1 [58] May-82
- Hurts So Good - John Cougar 2(28) Aug-82
- Jack & Diane - John Cougar 1(22) Oct-82
- Hand To Hold On To - John Cougar 19(18) Jan-83

Uh-Huh - John MELLENCAMP 9 [182] Nov-83
- Crumblin' Down - John Cougar Mellencamp 9(16) Nov-83
- Pink Houses - John Cougar Mellencamp 8(16) Feb-84
- The Authority Song - John Cougar Mellencamp 15(15) May-84

Cargo - MEN AT WORK 3 [207] May-83
- Overkill 3(16) Jun-83
- It's A Mistake 6(15) Aug-83
- Dr. Heckyll & Mr. Jive 28(11) Oct-83

No Frills -
  Bette MIDLER 60 [2314] Aug-83
- All I Need To Know (Don't Know Much) 77(4) Sep-83
- Favorite Waste Of Time 78(4) Nov-83
- Beast Of Burden 71(6) Mar-84

Albums: Special Lists

No Parking On The Dance Floor -
MIDNIGHT STAR 27 [276] Jul-83
- Freak-A-Zoid 66(8) Oct-83
- Wet My Whistle 61(11) Jan-84
- No Parking (On The Dance Floor) 81(8) Apr-84

Mike + The Mechanics -
MIKE + THE MECHANICS 26 [471] Nov-85
- Silent Running (On Dangerous Ground) 6(24) Mar-86
- All I Need Is A Miracle 5(19) Jun-86
- Taken In 32(15) Aug-86

Living Years -
MIKE + THE MECHANICS 13 [575] Nov-88
- Nobody's Perfect 63(11) Dec-88
- The Living Years 1(20) Mar-89
- Seeing Is Believing 62(6) May-89

Abracadabra - Steve MILLER Band 3 [298] Jun-82
- Abracadabra 1(25) Sep-82
- Cool Magic 57(8) Nov-82
- Give It Up 60(9) Jan-83

Kylie - Kylie MINOGUE 53 [1248] Sep-88
- I Should Be So Lucky 28(14) Jul-88
- The Loco-Motion 3(27) Nov-88
- It's No Secret 37(13) Feb-89

Welcome To The Real World -
MR. MISTER 1 [157] Aug-85
- Broken Wings 1(22) Dec-85
- Kyrie 1(20) Mar-86
- Is It Love 8(17) May-86

Nothing To Lose - Eddie MONEY 49 [1398] Oct-88
- Walk On Water 9(21) Dec-88
- The Love In Your Eyes 24(18) Mar-89
- Let Me In 60(7) May-89

Long Distance Voyager -
MOODY BLUES 1 [115] Jun-81
- Gemini Dream 12(15) Aug-81
- The Voice 15(17) Oct-81
- Talking Out Of Turn 65(7) Nov-81

Michael Morales -
Michael MORALES 113 [2686] Jun-89
- Who Do You Give Your Love To? 15(19) Jul-89
- What I Like About You 28(13) Oct-89
- I Don't Know 81(6) Dec-89

All Four One - The MOTELS 16 [490] Apr-82
- Only The Lonely 9(23) Jul-82
- Take The L. 52(9) Oct-82
- Forever Mine 60(8) Dec-82

Where Do You Go When You Dream -
Anne MURRAY 55 [1738] May-81
- Blessed Are The Believers 34(13) May-81
- It's All I Can Do 53(9) Oct-81
- Another Sleepless Night 44(9) Mar-82

Naked Eyes - NAKED EYES 32 [925] Apr-83
- Always Something There To Remind Me 8(22) Jun-83
- Promises, Promises 11(20) Oct-83
- When The Lights Go Out 37(14) Dec-83

Robbie Nevil - Robbie NEVIL 37 [846] Nov-86
- C'est La Vie 2(23) Jan-87
- Dominoes 14(16) Apr-87
- Wot's It To Ya 10(16) Aug-87

New Edition - NEW EDITION 6 [271] Oct-84
- Cool It Now 4(25) Jan-85
- Mr. Telephone Man 12(16) Feb-85
- Lost In Love 35(14) May-85

All For Love - NEW EDITION 32 [624] Dec-85
- Count Me Out 51(15) Dec-85
- A Little Bit Of Love (Is All It Takes) 38(15) Apr-86
- With You All The Way 51(11) Jul-86

Heart Break - NEW EDITION 12 [247] Jul-88
- If It Isn't Love 7(21) Sep-88
- You're Not My Kind Of Girl 95(4) Nov-88
- Can You Stand The Rain 44(13) Mar-89

Juice - Juice NEWTON 22 [171] Mar-81
- Angel Of The Morning 4(22) May-81
- Queen Of Hearts 2(27) Sep-81
- The Sweetest Thing (I've Ever Known) 7(24) Feb-82

Quiet Lies - Juice NEWTON 20 [645] May-82
- Love's Been A Little Bit Hard On Me 7(17) Jul-82
- Break It To Me Gently 11(17) Oct-82
- Heart Of The Night 25(16) Jan-83

Physical - Olivia NEWTON-JOHN 6 [169] Oct-81
- Physical 1(26) Nov-81
- Make A Move On Me 5(14) Apr-82
- Landslide 52(8) Jul-82

The Wild Heart - Stevie NICKS 5 [239] Jul-83
- Stand Back 5(19) Aug-83
- If Anyone Falls 14(14) Nov-83
- Nightbird 33(12) Jan-84

Rock A Little - Stevie NICKS 12 [480] Dec-85
- Talk To Me 4(18) Jan-86
- I Can't Wait 16(13) Apr-86
- Has Anyone Ever Written Anything For You 60(6) May-86

Midnight Madness -
NIGHT RANGER 15 [208] Nov-83
- (You Can Still) Rock In America 51(12) Jan-84
- Sister Christian 5(24) Jun-84
- When You Close Your Eyes 14(17) Sep-84

7 Wishes - NIGHT RANGER 10 [306] Jun-85
- Sentimental Street 8(17) Jul-85
- Four In The Morning
  (I Can't Take Any More) 19(13) Oct-85
- Goodbye 17(18) Feb-86

Love Zone - Billy OCEAN 6 [215] May-86
- There'll Be Sad Songs (To Make You Cry) 1(21) Jul-86
- Love Zone 10(16) Sep-86
- Love Is Forever 16(12) Dec-86

Hearsay - Alexander O'NEAL 29 [827] Aug-87
- Fake 25(15) Sep-87
- Criticize 70(11) Dec-87
- Never Knew Love Like This 28(14) Apr-88

Jeffrey Osborne -
Jeffrey OSBORNE 49 [1116] Jun-82
- I Really Don't Need No Light 39(15) Aug-82
- On The Wings Of Love 29(18) Dec-82
- Eenie Meenie 76(5) Apr-83

Stay With Me Tonight -
Jeffrey OSBORNE 25 [325] Aug-83
- Don't You Get So Mad 25(14) Sep-83
- Stay With Me Tonight 30(21) Jan-84
- We're Going All The Way 48(12) Apr-84

Donny Osmond -
Donny OSMOND 54 [1696] May-89
- Soldier Of Love 2(18) Jun-89
- Sacred Emotion 13(16) Aug-89
- Hold On 73(7) Oct-89

Play Deep - The OUTFIELD 9 [259] Nov-85
- Your Love 6(22) May-86
- All The Love In The World 19(16) Aug-86
- Everytime You Cry 66(10) Oct-86

Heavy Nova - Robert PALMER 13 [395] Jul-88
- Simply Irresistible 2(20) Sep-88
- Early In The Morning 19(15) Dec-88
- Tell Me I'm Not Dreaming 60(8) Aug-89

John Parr - John PARR 48 [1553] Dec-84
- Naughty Naughty 23(20) Mar-85
- Magical 73(5) Apr-85
- Love Grammar 89(2) Nov-85

The Turn Of A Friendly Card -
Alan PARSONS PROJECT 13 [248] Nov-80
- Games People Play 16(23) Mar-81
- Time 15(23) Aug-81

Snake Eyes 67(5) Nov-81

9 To 5 And Odd Jobs -
Dolly PARTON 11 [411] Dec-80
- 9 To 5 1(26) Feb-81
- But You Know I Love You 41(10) May-81
- The House Of The Rising Sun 77(4) Sep-81

Introspective - PET SHOP BOYS 34 [1314] Nov-88
- Always On My Mind 4(15) May-88
- Domino Dancing 18(14) Dec-88
- Left To My Own Devices 84(3) Feb-89

Damn The Torpedoes - Tom PETTY And The
HEARTBREAKERS 2 [88] Nov-79
- Don't Do Me Like That 10(18) Feb-80
- Refugee 15(14) Mar-80
- Here Comes My Girl 59(7) May-80

Southern Accents - Tom PETTY And The
HEARTBREAKERS 7 [413] Apr-85
- Don't Come Around Here No More 13(14) May-85
- Make It Better (Forget About Me) 54(8) Jul-85
- Rebels 74(5) Sep-85

So Excited! - POINTER SISTERS 59 [1805] Jul-82
- American Music 16(14) Aug-82
- I'm So Excited 30(16) Nov-82
- If You Wanna Get Back Your Lady 67(5) Apr-83

Contact - POINTER SISTERS 24 [676] Aug-85
- Dare Me 11(18) Sep-85
- Freedom 59(11) Nov-85
- Twist My Arm 83(5) Mar-86

Look What The Cat Dragged In -
POISON 3 [111] Aug-86
- Talk Dirty To Me 9(16) May-87
- I Want Action 50(10) Jul-87
- I Won't Forget You 13(21) Nov-87

Ghost In The Machine - The POLICE 2 [73] Oct-81
- Every Little Thing She Does Is Magic 3(19) Dec-81
- Spirits In The Material World 11(13) Mar-82
- Secret Journey 46(8) May-82

The Power Station -
POWER STATION 6 [278] Apr-85
- Some Like It Hot 6(18) May-85
- Get It On 9(15) Aug-85
- Communication 34(10) Oct-85

Learning To Crawl -
The PRETENDERS 5 [319] Feb-84
- Middle Of The Road 19(14) Feb-84
- Show Me 28(13) May-84
- Thin Line Between Love And Hate 83(5) Jul-84

Around The World In A Day -
PRINCE and the REVOLUTION 1 [164] May-85
- Raspberry Beret 2(17) Jul-85
- Pop Life 7(14) Sep-85
- America 46(7) Nov-85

Parade: Music From The Motion Picture Under The
Cherry Moon (Soundtrack) -
PRINCE and the REVOLUTION 3 [380] Apr-86
- Kiss 1(18) Apr-86
- Mountains 23(11) Jul-86
- Anotherloverholenyohead 63(10) Aug-86

Batman (Soundtrack) - PRINCE 1 [219] Jul-89
- Batdance 1(18) Aug-89
- Partyman 18(10) Oct-89
- The Arms Of Orion -
  Prince With Sheena Easton 36(14) Dec-89

Firin' Up - PURE PRAIRIE LEAGUE 37 [1280] May-80
- Let Me Love You Tonight 10(17) Jul-80
- I'm Almost Ready 34(13) Oct-80
- I Can't Stop This Feelin' 77(6) Dec-80

Quarterflash - QUARTERFLASH 8 [281] Oct-81
- Harden My Heart 3(24) Feb-82
- Find Another Fool 16(13) Apr-82
- Right Kind Of Love 56(8) Jul-82

Albums: Special Lists

The Works - QUEEN 23 [1083] Mar-84	You Be Illin' 29(18) Dec-86	In The Dark 35(12) Oct-81
Radio Ga-Ga 16(13) Apr-84	It's Tricky 57(10) Apr-87	My Kinda Lover 45(10) Jan-82
I Want To Break Free 45(8) May-84	**Sa-Fire - SA-FIRE 79 [1419] Oct-88**	**Emotions In Motion - Billy SQUIER 5 [206] Aug-82**
It's A Hard Life 72(4) Aug-84	Boy, I've Been Told 48(16) Nov-88	Emotions In Motion 68(6) Aug-82
Step By Step - Eddie RABBITT 23 [716] Aug-81	Thinking Of You 12(24) May-89	Everybody Wants You 32(17) Dec-82
Step By Step 5(22) Oct-81	Gonna Make It 71(7) Jul-89	She's A Runner 75(6) Feb-83
Someone Could Lose A Heart Tonight 15(15) Jan-82	**Warrior - SCANDAL/Patty SMYTH 17 [600] Aug-84**	**Signs Of Life - Billy SQUIER 11 [605] Aug-84**
I Don't Know Where To Start 35(13) Jun-82	The Warrior 7(21) Sep-84	Rock Me Tonite 15(16) Sep-84
Street Called Desire - RENÉ & ANGELA 64 [779] Jul-85	Hands Tied 41(13) Dec-84	All Night Long 75(3) Nov-84
I'll Be Good 47(10) Nov-85	Beat Of A Heart 41(14) Mar-85	Eye On You 71(8) Jan-85
Your Smile 62(6) Apr-86	**The Distance - Bob SEGER & The SILVER BULLET BAND 5 [291] Jan-83**	**Tonight I'm Yours - Rod STEWART 11 [405] Nov-81**
You Don't Have To Cry 75(7) Jul-86	Shame On The Moon 2(21) Feb-83	Young Turks 5(19) Dec-81
Life As We Know It - REO SPEEDWAGON 28 [819] Feb-87	Even Now 12(12) May-83	Tonight I'm Yours (Don't Hurt Me) 20(14) Mar-82
That Ain't Love 16(14) Apr-87	Roll Me Away 27(10) Jul-83	How Long 49(9) May-82
Variety Tonight 60(9) Jun-87	**Coming Around Again - Carly SIMON 25 [467] Apr-87**	**Camouflage - Rod STEWART 18 [579] Jun-84**
In My Dreams 19(30) Oct-87	Coming Around Again 18(17) Jan-87	Infatuation 6(18) Jul-84
I'm No Hero - Cliff RICHARD 80 [1501] Oct-80	Give Me All Night 61(12) Jun-87	Some Guys Have All The Luck 10(17) Oct-84
Dreaming 10(22) Nov-80	All I Want Is You 54(9) Feb-88	All Right Now 72(6) Jan-85
A Little In Love 17(22) Mar-81	**Graceland - Paul SIMON 3 [78] Sep-86**	**Rod Stewart - Rod STEWART 28 [1326] Jul-86**
Give A Little Bit More 41(11) Jun-81	Graceland 81(7) Jan-87	Love Touch 6(18) Aug-86
Lionel Richie - Lionel RICHIE 3 [52] Oct-82	The Boy In The Bubble 86(4) Mar-87	Another Heartache 52(9) Sep-86
Truly 1(18) Nov-82	You Can Call Me Al 23(29) May-87	Every Beat Of My Heart 83(6) Nov-86
You Are 4(18) Mar-83	**Once Upon A Time - SIMPLE MINDS 10 [335] Nov-85**	**...Nothing Like The Sun - STING 9 [290] Oct-87**
My Love 5(16) Jun-83	Alive And Kicking 3(20) Dec-85	We'll Be Together 7(18) Dec-87
One Heartbeat - Smokey ROBINSON 26 [501] Mar-87	Sanctify Yourself 14(14) Mar-86	Be Still My Beating Heart 15(14) Mar-88
Just To See Her 8(21) Jul-87	All The Things She Said 28(13) May-86	Englishman In New York 84(4) Apr-88
One Heartbeat 10(19) Oct-87	**Skid Row - SKID ROW 6 [79] Feb-89 3,1***	**Rant 'N' Rave With The Stray Cats - STRAY CATS 14 [793] Sep-83**
What's Too Much 79(10) Dec-87	Youth Gone Wild 99(2) Jun-89	(She's) Sexy + 17 5(15) Oct-83
We've Got Tonight - Kenny ROGERS 18 [937] Mar-83	18 And Life 4(20) Sep-89	I Won't Stand In Your Way 35(13) Dec-83
All My Life 37(11) Jun-83	I Remember You 6(20) Feb-90	Look At That Cadillac 68(5) Feb-84
Scarlet Fever 94(2) Aug-83	**Teachers - SOUNDTRACK-MOVIE 34 [1521] Oct-84**	**Emotion - Barbra STREISAND 19 [842] Oct-84**
We've Got Tonight - Kenny Rogers And Sheena Easton 6(18) Mar-83	Teacher, Teacher - 38 Special 25(12) Nov-84	Left In The Dark 50(12) Oct-84
	Understanding - Bob Seger & The Silver Bullet Band 17(15) Jan-85	Make No Mistake, He's Mine - Barbra Streisand (Duet With Kim Carnes) 51(10) Jan-85
Eyes That See In The Dark - Kenny ROGERS 6 [299] Sep-83	Edge Of A Dream - Joe Cocker 69(7) Nov-84	Emotion 79(2) Mar-85
Islands In The Stream - Kenny Rogers Duet with Dolly Parton 1(25) Oct-83	**Vision Quest - SOUNDTRACK-MOVIE 11 [796] Mar-85**	**Paradise Theater - STYX 1 [44] Jan-81**
This Woman 23(13) Mar-84	Change - John Waite 54(10) Apr-85	The Best Of Times 3(19) Mar-81
Eyes That See In The Dark 79(5) May-84	Only The Young - Journey 9(16) Mar-85	Too Much Time On My Hands 9(19) May-81
Tattoo You - ROLLING STONES 1 [49] Sep-81	Crazy For You - Madonna 1(21) May-85	Nothing Ever Goes As Planned 54(8) Aug-81
Start Me Up - The Rolling Stones 2(24) Oct-81	**Miami Vice - SOUNDTRACK-TV 1 [143] Oct-85**	**Kilroy Was Here - STYX 3 [275] Mar-83**
Waiting On A Friend - The Rolling Stones 13(15) Feb-82	Own The Night - Chaka Khan 57(9) Jan-86	Mr. Roboto 3(18) Apr-83
Hang Fire - The Rolling Stones 20(11) May-82	You Belong To The City - Glenn Frey 2(21) Nov-85	Don't Let It End 6(16) Jul-83
Steel Wheels - ROLLING STONES 3 [205] Sep-89 3,1*	Miami Vice Theme - Jan Hammer 1(22) Nov-85	High Time 48(7) Sep-83
Mixed Emotions 5(12) Oct-89	**True - SPANDAU BALLET 19 [813] May-83**	**The Wanderer - Donna SUMMER 13 [856] Nov-80**
Rock And A Hard Place 23(14) Dec-89	True 4(18) Oct-83	The Wanderer 3(20) Nov-80
Almost Hear You Sigh 50(9) Mar-90	Gold 29(12) Jan-84	Cold Love 33(12) Jan-81
Mad Love - Linda RONSTADT 3 [202] Mar-80	Communication 59(7) Apr-84	Who Do You Think You're Foolin' 40(11) Mar-81
How Do I Make You 10(16) Mar-80	**Working Class Dog - Rick SPRINGFIELD 7 [118] Mar-81**	**Donna Summer - Donna SUMMER 20 [792] Aug-82**
Hurt So Bad 8(14) May-80	Jessie's Girl 1(32) Aug-81	Love Is In Control (Finger On The Trigger) 10(18) Sep-82
I Can't Let Go 31(12) Aug-80	I've Done Everything For You 8(22) Nov-81	State Of Independence 41(10) Nov-82
Get Closer - Linda RONSTADT 31 [1092] Oct-82	Love Is Alright Tonite 20(16) Feb-82	The Woman In Me 33(12) Feb-83
Get Closer 29(12) Nov-82	**Success Hasn't Spoiled Me Yet - Rick SPRINGFIELD 2 [286] Mar-82**	**She Works Hard For The Money - Donna SUMMER 9 [543] Jul-83**
I Knew You When 37(12) Feb-83	Don't Talk To Strangers 2(21) May-82	She Works Hard For The Money 3(21) Aug-83
Easy For You To Say 54(10) May-83	What Kind Of Fool Am I 21(12) Jul-82	Unconditional Love 43(8) Oct-83
Why Do Fools Fall In Love - Diana ROSS 15 [424] Nov-81	I Get Excited 32(12) Oct-82	Love Has A Mind Of Its Own 70(4) Jan-84
Why Do Fools Fall In Love 7(20) Dec-81	**Living In Oz - Rick SPRINGFIELD 12 [342] Apr-83**	**Henry Lee Summer - Henry LEE SUMMER 56 [1465] Mar-88**
Mirror, Mirror 8(14) Mar-82	Affair Of The Heart 9(18) Jun-83	I Wish I Had A Girl 20(18) Apr-88
Work That Body 44(7) May-82	Human Touch 18(15) Sep-83	Darlin' Danielle Don't 57(8) Jun-88
Eat 'Em And Smile - David Lee ROTH 4 [326] Jul-86	Souls 23(15) Dec-83	Hands On The Radio 85(3) Sep-88
Yankee Rose 16(15) Aug-86	**Tunnel Of Love - Bruce SPRINGSTEEN 1 [180] Oct-87**	**2nd Wave - SURFACE 56 [1087] Nov-88**
Goin' Crazy! 66(7) Oct-86	Brilliant Disguise 5(16) Nov-87	Closer Than Friends 57(13) May-89
That's Life 85(4) Dec-86	Tunnel Of Love 9(16) Feb-88	Shower Me With Your Love 5(19) Sep-89
Raising Hell - RUN-D.M.C. 3 [101] Jun-86	One Step Up 13(15) Apr-88	You Are My Everything 84(5) Nov-89
Walk This Way 4(16) Sep-86	**Don't Say No - Billy SQUIER 5 [87] May-81**	**Eye Of The Tiger - SURVIVOR 2 [268] Jun-82**
	The Stroke 17(20) Aug-81	Eye Of The Tiger 1(25) Jul-82
		American Heartbeat 17(16) Nov-82
		The One That Really Matters 74(6) Feb-83

When Seconds Count - SURVIVOR 49 [1450] Nov-86
Is This Love 9(19) Jan-87
How Much Love 51(9) Mar-87
Man Against The World 86(5) May-87
Make It Last Forever - Keith SWEAT 15 [237] Jan-88
I Want Her 5(20) Apr-88
Something Just Ain't Right 79(5) May-88
Make It Last Forever 59(11) Aug-88
The Seeds Of Love -
  TEARS FOR FEARS 8 [508] Oct-89 3,2*
Sowing The Seeds Of Love 2(15) Oct-89
Woman In Chains 36(14) Feb-90
Advice For The Young At Heart 89(4) Mar-90
Rock & Roll Strategy -
  38 SPECIAL 61 [1359] Oct-88
Rock & Roll Strategy 67(8) Nov-88
Second Chance 6(21) May-89
Comin' Down Tonight 67(7) Jul-89
Empty Glass -
  Pete TOWNSHEND 5 [279] May-80
Let My Love Open The Door 9(19) Aug-80
A Little Is Enough 72(4) Oct-80
Rough Boys 89(4) Nov-80
Outside Inside - The TUBES 18 [661] Apr-83
She's A Beauty 10(20) Jul-83
Tip Of My Tongue 52(7) Aug-83
The Monkey Time 68(4) Oct-83
Give Me The Reason -
  Luther VANDROSS 14 [321] Oct-86
Give Me The Reason 57(11) Sep-86
Stop To Love 15(19) Feb-87
There's Nothing Better Than Love
  (Duet With Gregory Hines) 50(14) May-87
5150 - VAN HALEN 1 [124] Apr-86
Why Can't This Be Love 3(16) May-86
Dreams 22(14) Jul-86
Love Walks In 22(15) Oct-86
No Brakes - John WAITE 10 [429] Jul-84
Missing You 1(24) Sep-84
Tears 37(13) Nov-84
Restless Heart 59(8) Feb-85
Points On The Curve - WANG CHUNG 30 [881] Feb-84
Don't Let Go 38(11) Mar-84

Dance Hall Days 16(22) Jul-84
Don't Be My Enemy 86(3) Sep-84
Mosaic - WANG CHUNG 41 [1061] Nov-86
Everybody Have Fun Tonight 2(21) Dec-86
Let's Go 9(18) Apr-87
Hypnotize Me 36(12) Jul-87
What Up, Dog? - WAS (NOT WAS) 43 [1107] Oct-88
Spy In The House Of Love 16(17) Dec-88
Walk The Dinosaur 7(16) Apr-89
Anything Can Happen 75(6) May-89
Karyn White - Karyn WHITE 19 [418] Oct-88
The Way You Love Me 7(25) Feb-89
Superwoman 8(18) Apr-89
Secret Rendezvous 6(21) Aug-89
Pride - WHITE LION 11 [226] Sep-87
Wait 8(21) May-88
Tell Me 58(11) Aug-88
When The Children Cry 3(23) Feb-89
Slip Of The Tongue -
  WHITESNAKE 10 [468] Nov-89 3,2*
Fool For Your Loving 37(14) Dec-89
The Deeper The Love 28(14) Mar-90
Now You're Gone 96(2) Jun-90
The Right Stuff - Vanessa WILLIAMS 38 [835] Jul-88
The Right Stuff 44(10) Aug-88
Dreamin' 8(20) Apr-89
Darlin' I 88(4) Jun-89
The Return Of Bruno - Bruce WILLIS 14 [774] Feb-87
Respect Yourself 5(14) Mar-87
Young Blood 68(5) May-87
Under The Boardwalk 59(7) Jul-87
Winger - WINGER 21 [255] Sep-88
Seventeen 26(16) May-89
Headed For A Heartbreak 19(18) Aug-89
Hungry 85(6) Oct-89
Lights Out - Peter WOLF 24 [1035] Aug-84
Lights Out 12(14) Sep-84
I Need You Tonight 36(13) Nov-84
Oo-Ee-Diddley-Bop! 61(5) May-85
Hotter Than July - Stevie WONDER 3 [148] Nov-80
Master Blaster (Jammin') 5(23) Dec-80
I Ain't Gonna Stand For It 11(19) Mar-81
Lately 64(7) May-81

Stevie Wonder's Original Musiquarium I -
  Stevie WONDER 4 [581] May-82
That Girl 4(18) Mar-82
Do I Do 13(14) Jul-82
Ribbon In The Sky 54(7) Oct-82
Characters - Stevie WONDER 17 [626] Dec-87
Get It -
  Stevie Wonder & Michael Jackson 80(6) May-88
Skeletons 19(16) Dec-87
You Will Know 77(6) Feb-88
Take Me To Your Heaven -
  Stevie WOODS 153 [3025] Dec-81
Steal The Night 25(21) Dec-81
Just Can't Win 'Em All 38(12) Apr-82
Fly Away 84(2) May-82
"Weird Al" Yankovic In 3-D -
  "Weird Al" YANKOVIC 17 [858] Mar-84
Eat It 12(12) Apr-84
King Of Suede 62(6) May-84
I Lost On Jeopardy 81(3) Jul-84
90125 - YES 5 [233] Dec-83
Owner Of A Lonely Heart 1(23) Jan-84
Leave It 24(15) Apr-84
It Can Happen 51(7) Jul-84
No Parlez - Paul YOUNG 79 [2095] Apr-84
Wherever I Lay My Hat
  (That's My Home) 70(7) Oct-83
Come Back And Stay 22(15) Apr-84
Love Of The Common People 45(11) Jun-84
The Secret Of Association -
  Paul YOUNG 19 [455] May-85
Everytime You Go Away 1(23) Jul-85
I'm Gonna Tear Your Playhouse Down 13(14) Nov-85
Everything Must Change 56(11) Jan-86
Stone Cold Rhymin' -
  YOUNG M.C. 9 [234] Sep-89 3,2*
Bust A Move 7(39) Oct-89
Principal's Office 33(14) Jan-90
I Come Off 75(5) Apr-90
Eliminator - ZZ TOP 9 [36] Apr-83
Gimme All Your Lovin' 37(12) May-83
Sharp Dressed Man 56(9) Aug-83
Legs 8(19) Jul-84

Albums: Special Lists

Ranking the '80s:

The Acts

The Acts

Acts with Their Singles

Act Highest Peak [Top 10|Top 40|Total] Entries

Act Grouping   **Act Rank**

Act Total Score

Title (Year)-Act

Allocated credit for occasional collaborations*

Peak (Peak Wks)

Weeks [Top 10|Top 40|Total]

Single Score

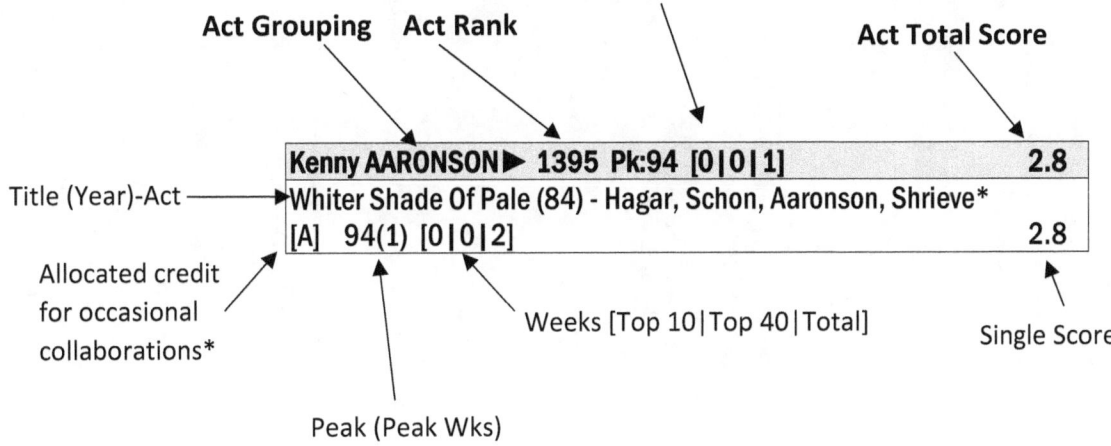

```
Kenny AARONSON ▶ 1395  Pk:94 [0|0|1]                    2.8
Whiter Shade Of Pale (84) - Hagar, Schon, Aaronson, Shrieve*
[A]   94(1)  [0|0|2]                                    2.8
```

Acts are listed alphabetically; Records are listed in order of chart strength.

*Act name is reproduced here if different from grouping name. Allocated credit is used for infrequent or one-off collaborations rather than creating a new Grouping. In this case, credit and score is divided among the four underlying Groupings. See "Methodology" in the appendix.

Acts With Singles

Acts With Singles

Act ▶ Rank Peak [Top10s|Top40s|Total] Title (Pk Yr) Peak(Wk) [Top10Wks|Top40Wks|TotalWks]	Score
A	
Kenny AARONSON▶ 1395 Pk:94 [0|0|1]	2.8
Whiter Shade Of Pale (84) - Hagar, Schon, Aaronson, Shrieve [A] 94(1) [0|0|2]	2.8
ABBA▶ 270 Pk:8 [1|3|6]	4088
The Winner Takes It All (81) 8(2) [4|16|26]	2383
When All Is Said And Done (82) 27(2) [0|8|14]	615
Chiquitita (80) 29(1) [0|6|12]	587
Super Trouper (81) 45(2) [0|0|11]	320
The Visitors (82) 63(2) [0|0|8]	146
On And On And On (81) 90(2) [0|0|6]	36.6
Gregory ABBOTT▶ 241 Pk:1 [1|1|2]	4569
Shake You Down (87) 1(1) [8|16|22]	4286
I Got The Feelin' (It's Over) (87) 56(2) [0|0|11]	283
ABC▶ 172 Pk:5 [2|5|7]	6629
When Smokey Sings (87) 5(1) [4|12|19]	1899
Be Near Me (85) 9(2) [3|11|22]	1646
The Look Of Love (Part One) (83) 18(3) [0|13|25]	1544
Poison Arrow (83) 25(3) [0|8|15]	772
(How To Be A) Millionaire (86) 20(1) [0|7|14]	717
Vanity Kills (86) 91(2) [0|0|4]	25.7
That Was Then But This Is Now (84) 89(1) [0|0|3]	24.0
Paula ABDUL▶ 65 Pk:1 [4|4|6]	16088
Straight Up (89) 1(3) [7|16|25]	4776
Cold Hearted (89) 1(1) [8|15|21]	4158
Forever Your Girl (89) 1(2) [6|14|22]	3654
(It's Just) The Way That You Love Me (89) 3(1) [6|14|20]	3106
Knocked Out (88) 41(1) [0|0|13]	357
(It's Just) The Way That You Love Me (88) 88(1) [0|0|5]	37.3
AC/DC▶ 495 Pk:35 [0|2|4]	1490
You Shook Me All Night Long (80) 35(1) [0|3|16]	603
Back In Black (81) 37(2) [0|5|15]	545
Let's Get It Up (82) 44(2) [0|0|9]	293
Guns For Hire (83) 84(1) [0|0|5]	49.6
Bryan ADAMS▶ 60 Pk:1 [5|12|13]	16851
Heaven (85) 1(2) [6|14|19]	4136
Run To You (85) 6(1) [4|12|19]	2116
Summer Of '69 (85) 5(2) [4|12|17]	2007
Heat Of The Night (87) 6(2) [4|10|16]	1759
Straight From The Heart (83) 10(2) [2|11|19]	1464
Somebody (85) 11(1) [0|10|17]	1214
One Night Love Affair (85) 13(1) [0|9|15]	1047
Cuts Like A Knife (83) 15(1) [0|8|14]	982
This Time (84) 24(1) [0|6|12]	649
Hearts On Fire (87) 26(1) [0|6|13]	544
It's Only Love (86) - Bryan Adams/Tina Turner [A] 15(1) [0|9|14]	474
Victim Of Love (87) 32(1) [0|5|12]	434
Lonely Nights (82) 84(1) [0|0|2]	24.8
The ADVENTURES▶ 1380 Pk:95 [0|0|1]	9.9
Broken Land (88) 95(1) [0|0|3]	9.9
AEROSMITH▶ 171 Pk:3 [2|4|5]	6717
Angel (88) 3(2) [5|15|25]	2787
Love In An Elevator (89) 5(1) [3|11|16]	1697
Dude (Looks Like A Lady) (87) 14(4) [0|10|20]	1261
Rag Doll (88) 17(1) [0|8|17]	840
Remember (Walking In The Sand) (80) 67(1) [0|0|6]	133
AFTER 7▶ 1036 Pk:74 [0|0|1]	151
Heat Of The Moment (89) 74(1) [0|0|12]	151
AFTERNOON DELIGHTS▶ 697 Pk:33 [0|1|1]	615
General Hospi-Tale (81) 33(2) [0|5|16]	615

Act ▶ Rank Peak [Top10s|Top40s|Total] Title (Pk Yr) Peak(Wk) [Top10Wks|Top40Wks|TotalWks]	Score
AFTER THE FIRE▶ 359 Pk:5 [1|1|2]	2703
Der Kommissar (83) 5(2) [5|14|21]	2668
Dancing In The Shadows (83) 85(1) [0|0|3]	34.9
A-HA▶ 223 Pk:1 [1|2|3]	5002
Take On Me (85) 1(1) [7|15|27]	3952
The Sun Always Shines On T.V. (86) 20(1) [0|8|17]	789
Cry Wolf (87) 50(1) [0|0|10]	262
AIR SUPPLY▶ 11 Pk:1 [8|11|13]	34148
All Out Of Love (80) 2(4) [10|17|27]	6080
Making Love Out Of Nothing At All (83) 2(3) [9|17|25]	5057
Lost In Love (80) 3(4) [6|17|23]	4516
The One That You Love (81) 1(1) [8|14|19]	4433
Every Woman In The World (81) 5(1) [8|17|22]	3319
Even The Nights Are Better (82) 5(2) [8|13|18]	3028
Sweet Dreams (82) 5(2) [7|15|20]	2974
Here I Am (Just When I Thought I Was Over You) (81) 5(3) [6|15|20]	2848
Just As I Am (85) 19(2) [0|10|15]	817
Two Less Lonely People In The World (83) 38(3) [0|5|14]	557
Young Love (82) 38(2) [0|2|9]	328
Lonely Is The Night (86) 76(1) [0|0|8]	108
The Power Of Love (You Are My Lady) (85) 68(1) [0|0|6]	83.8
ALABAMA▶ 279 Pk:15 [0|4|7]	3971
Love In The First Degree (82) 15(2) [0|10|21]	1264
Feels So Right (81) 20(2) [0|8|22]	1024
Take Me Down (82) 18(4) [0|8|13]	934
The Closer You Get (83) 38(2) [0|3|11]	369
Close Enough To Perfect (82) 65(2) [0|0|12]	168
When We Make Love (84) 72(1) [0|0|10]	120
Lady Down On Love (83) 76(1) [0|0|6]	91.0
The ALARM▶ 691 Pk:50 [0|0|4]	633
Sold Me Down The River (89) 50(1) [0|0|13]	258
Strength (86) 61(2) [0|0|10]	181
Rain In The Summertime (88) 71(2) [0|0|8]	126
Presence Of Love (88) 77(1) [0|0|6]	68.2
AL B. SURE!▶ 422 Pk:7 [1|1|3]	1960
Nite And Day (88) 7(1) [2|13|21]	1515
Off On Your Own (Girl) (88) 45(1) [0|0|12]	315
Killing Me Softly (88) 80(1) [0|0|11]	130
ALESSI▶ 1187 Pk:71 [0|0|1]	75.4
Put Away Your Love (82) 71(1) [0|0|4]	75.4
--Alice COOPER see: COOPER, Alice	
ALISHA▶ 894 Pk:68 [0|0|2]	278
Baby Talk (86) 68(3) [0|0|17]	275
Into My Secret (87) 97(1) [0|0|1]	3.1
Deborah ALLEN▶ 591 Pk:26 [0|1|1]	950
Baby I Lied (84) 26(3) [0|7|21]	950
Donna ALLEN▶ 621 Pk:21 [0|1|1]	797
Serious (87) 21(1) [0|9|18]	797
Peter ALLEN▶ 971 Pk:55 [0|0|1]	203
Fly Away (81) 55(1) [0|0|8]	203
Gregg ALLMAN Band▶ 933 Pk:49 [0|0|1]	235
I'm No Angel (87) 49(1) [0|0|10]	235
ALLMAN BROTHERS Band▶ 723 Pk:39 [0|1|2]	557
Straight From The Heart (81) 39(2) [0|2|11]	372
Angeline (80) 58(1) [0|0|8]	184
ALL SPORTS BAND▶ 1226 Pk:78 [0|0|2]	60.6
Opposites Do Attract (82) 78(2) [0|0|3]	47.8
I'm Your Superman (81) 93(1) [0|0|2]	12.8
Marc ALMOND▶ 1070 Pk:67 [0|0|1]	127
Tears Run Rings (89) 67(1) [0|0|6]	127
Herb ALPERT▶ 276 Pk:5 [1|4|9]	4008
Diamonds (87) 5(1) [4|12|19]	1839
Rotation (80) 30(2) [0|6|13]	602

Act ▶ Rank Peak [Top10s|Top40s|Total] Title (Pk Yr) Peak(Wk) [Top10Wks|Top40Wks|TotalWks]	Score
Making Love In The Rain (87) 35(1) [0|3|14]	439
Route 101 (82) 37(2) [0|4|10]	415
Keep Your Eye On Me (87) 46(1) [0|0|10]	269
Beyond (80) 50(1) [0|0|8]	241
Red Hot (83) 77(2) [0|0|5]	84.7
Magic Man (81) 79(1) [0|0|5]	66.4
Garden Party (83) 81(1) [0|0|4]	50.9
Herb ALPERT & The TIJUANA BRASS▶ 1359 Pk:90 [0|0|1]	18.2
Bullish (84) - Herb Alpert/Tijuana Brass 90(2) [0|0|2]	18.2
ALPHAVILLE▶ 758 Pk:65 [0|0|3]	492
Forever Young (88) 65(2) [0|0|14]	272
Big In Japan (85) 66(2) [0|0|10]	205
Forever Young (85) 93(1) [0|0|4]	14.4
AMAZULU▶ 1351 Pk:90 [0|0|1]	20.5
Montego Bay (87) 90(1) [0|0|4]	20.5
AMBROSIA▶ 209 Pk:3 [1|2|3]	5356
Biggest Part Of Me (80) 3(3) [8|14|19]	4001
You're The Only Woman (You And I) (80) 13(2) [0|10|18]	1314
How Can You Love Me (82) 86(2) [0|0|4]	41.2
AMERICA▶ 311 Pk:8 [1|2|3]	3406
You Can Do Magic (82) 8(5) [6|15|20]	2458
The Border (83) 33(3) [0|6|12]	552
Right Before Your Eyes (83) 45(2) [0|0|13]	395
AMERICAN COMEDY NETWORK▶ 1157 Pk:70 [0|0|1]	90.6
Breaking Up Is Hard On You (a/k/a Don't Take Ma Bell Away From Me) (84) 70(1) [0|0|5]	90.6
ANA▶ 1376 Pk:94 [0|0|1]	12.3
Shy Boys (87) 94(1) [0|0|3]	12.3
John ANDERSON▶ 772 Pk:43 [0|0|1]	464
Swingin' (83) 43(2) [0|0|13]	464
The ANIMALS▶ 839 Pk:48 [0|0|1]	351
The Night (83) 48(2) [0|0|10]	351
ANIMOTION▶ 269 Pk:6 [2|3|6]	4100
Obsession (85) 6(1) [5|14|24]	2137
Room To Move (89) 9(1) [1|11|18]	1205
Let Him Go (85) 39(1) [0|1|13]	429
Calling It Love (89) 53(1) [0|0|9]	208
I Engineer (86) 76(1) [0|0|6]	78.1
I Want You (86) 84(1) [0|0|4]	42.8
Paul ANKA▶ 636 Pk:40 [0|1|2]	759
Hold Me 'Til The Mornin' Comes (83) 40(2) [0|2|16]	548
I've Been Waiting For You All Of My Life (81) 48(1) [0|0|9]	211
Adam ANT▶ 400 Pk:12 [0|1|3]	2187
Goody Two Shoes (83) 12(3) [0|14|21]	1667
Strip (84) 42(1) [0|0|13]	370
Desperate But Not Serious (83) 66(3) [0|0|8]	150
Susan ANTON▶ 826 Pk:28 [0|1|1]	370
Killin' Time (81) - Fred Knoblock & Susan Anton [A] 28(1) [0|9|18]	370
APOLLONIA 6▶ 1212 Pk:85 [0|0|1]	65.1
Sex Shooter (84) 85(1) [0|0|6]	65.1
APRIL WINE▶ 522 Pk:21 [0|1|5]	1317
Just Between You And Me (81) 21(1) [0|7|16]	782
Enough Is Enough (82) 50(2) [0|0|8]	211
Sign Of The Gypsy Queen (81) 57(1) [0|0|8]	151
This Could Be The Right One (84) 58(1) [0|0|9]	144
I Like To Rock (80) 86(1) [0|0|3]	28.6
ARCADIA▶ 380 Pk:6 [1|2|2]	2371
Election Day (85) 6(2) [6|12|16]	1991
Goodbye Is Forever (86) 33(1) [0|3|10]	380

259

Acts With Singles

Act ▶ Rank Peak [Top10s|Top40s|Total] Title (Pk Yr) Peak(Wk) [Top10Wks|Top40Wks|TotalWks]	Score
Joan ARMATRADING ▶ 1165 Pk:78 [0|0|1]	**87.7**
Drop The Pilot (83) 78(1) [0|0|6]	87.7
Louis ARMSTRONG ▶ 817 Pk:32 [0|1|1]	**392**
What A Wonderful World (88) 32(1) [0|3|11]	392
Steve ARRINGTON ▶ 1079 Pk:68 [0|0|1]	**120**
Dancin' In The Key Of Life (85) 68(2) [0|0|6]	120
ARTISTS UNITED AGAINST APARTHEID ▶ 760 Pk:38 [0|1|1]	**488**
Sun City (85) 38(1) [0|3|13]	488
ART OF NOISE ▶ 559 Pk:31 [0|2|3]	**1116**
Kiss (89) - The Art Of Noise featuring Tom Jones 31(2) [0|6|11]	423
Paranoimia (86) - The Art Of Noise with Max Headroom 34(1) [0|4|12]	406
Peter Gunn (86) - The Art Of Noise featuring Duane Eddy 50(1) [0|0|11]	287
ASHFORD & SIMPSON ▶ 451 Pk:12 [0|1|3]	**1716**
Solid (85) 12(2) [0|11|24]	1448
Street Corner (82) 56(1) [0|0|10]	234
Count Your Blessings (86) 84(1) [0|0|4]	34.5
ASIA ▶ 178 Pk:4 [2|4|5]	**6447**
Heat Of The Moment (82) 4(3) [6|12|18]	3149
Don't Cry (83) 10(2) [2|11|13]	1486
Only Time Will Tell (82) 17(3) [0|8|14]	897
The Smile Has Left Your Eyes (83) 34(2) [0|5|13]	577
Go (86) 46(2) [0|0|11]	337
The ASSOCIATION ▶ 1140 Pk:66 [0|0|1]	**99.0**
Dreamer (81) 66(2) [0|0|5]	99.0
Jon ASTLEY ▶ 949 Pk:74 [0|0|2]	**221**
Jane's Getting Serious (87) 77(1) [0|0|10]	122
Put This Love To The Test (88) 74(1) [0|0|8]	98.8
Rick ASTLEY ▶ 106 Pk:1 [4|5|6]	**10618**
Never Gonna Give You Up (88) 1(2) [7|14|24]	4313
Together Forever (88) 1(1) [6|12|18]	3198
She Wants To Dance With Me (89) 6(1) [3|10|16]	1547
It Would Take A Strong Strong Man (88) 10(1) [1|10|16]	1222
Giving Up On Love (89) 38(1) [0|1|10]	321
Ain't Too Proud To Beg (89) 89(1) [0|0|3]	17.5
Christopher ATKINS ▶ 1118 Pk:71 [0|0|1]	**107**
How Can I Live Without Her (82) 71(2) [0|0|7]	107
ATLANTA RHYTHM SECTION ▶ 720 Pk:29 [0|1|1]	**564**
Alien (81) 29(2) [0|4|15]	564
ATLANTIC STARR ▶ 128 Pk:1 [2|3|7]	**8547**
Always (87) 1(1) [7|14|22]	4081
Secret Lovers (86) 3(2) [6|14|23]	3447
Circles (82) 38(2) [0|3|11]	398
One Lover At A Time (87) 58(1) [0|0|13]	267
If Your Heart Isn't In It (86) 57(1) [0|0|12]	253
Touch A Four Leaf Clover (83) 87(3) [0|0|7]	68.5
Freak-A-Ristic (85) 90(2) [0|0|6]	31.5
AURRA ▶ 1092 Pk:71 [0|0|1]	**116**
Make Up Your Mind (82) 71(1) [0|0|7]	116
Patti AUSTIN ▶ 182 Pk:1 [1|1|6]	**6253**
Baby, Come To Me (83) - Patti Austin (A duet with James Ingram) 1(2) [9|18|32]	5417
How Do You Keep The Music Playing (Theme From "Best Friends") (83) - James Ingram And Patti Austin [A] 5(1) [0|0|17]	292
The Heat Of Heat (86) 55(1) [0|0|9]	217
Every Home Should Have One (82) 62(2) [0|0|8]	158
Every Home Should Have One (Remix) (83) 69(1) [0|0|7]	119
It's Gonna Be Special (84) 82(1) [0|0|4]	50.1
AUTOGRAPH ▶ 675 Pk:29 [0|1|1]	**670**
Turn Up The Radio (85) 29(1) [0|5|19]	670
Johnny AVERAGE Band ▶ 1008 Pk:53 [0|0|1]	**174**
Ch Ch Cherie (81) 53(1) [0|0|7]	174

Act ▶ Rank Peak [Top10s|Top40s|Total] Title (Pk Yr) Peak(Wk) [Top10Wks|Top40Wks|TotalWks]	Score
AVERAGE WHITE BAND ▶ 944 Pk:53 [0|0|1]	**225**
Let's Go 'Round Again (80) 53(1) [0|0|8]	225
AXE ▶ 1066 Pk:64 [0|0|2]	**129**
Now Or Never (82) 64(2) [0|0|6]	120
I Think You'll Remember Tonight (83) 94(1) [0|0|2]	9.0

B

Act ▶ Rank Peak [Top10s|Top40s|Total] Title (Pk Yr) Peak(Wk) [Top10Wks|Top40Wks|TotalWks]	Score
BABYFACE ▶ 494 Pk:7 [1|1|1]	**1490**
It's No Crime (89) 7(1) [4|10|18]	1490
The BABYS ▶ 582 Pk:33 [0|1|3]	**979**
Back On My Feet Again (80) 33(1) [0|3|12]	502
Turn And Walk Away (80) 42(2) [0|0|12]	413
Midnight Rendezvous (80) 72(1) [0|0|4]	63.5
BAD COMPANY ▶ 1007 Pk:74 [0|0|3]	**175**
Electricland (82) 74(2) [0|0|4]	73.6
Shake It Up (89) 82(1) [0|0|8]	60.9
This Love (86) 85(2) [0|0|5]	40.5
BAD ENGLISH ▶ 268 Pk:1 [1|1|2]	**4102**
When I See You Smile (89) 1(2) [6|15|22]	3797
Forget Me Not (89) 45(1) [0|0|11]	305
BADFINGER ▶ 1023 Pk:56 [0|0|1]	**163**
Hold On (81) 56(2) [0|0|8]	163
Philip BAILEY ▶ 232 Pk:2 [1|1|2]	**4813**
Easy Lover (Duet With Phil Collins) (85) 2(2) [7|16|23]	4519
Walking On The Chinese Wall (85) 46(2) [0|0|12]	294
Anita BAKER ▶ 166 Pk:3 [2|4|6]	**7114**
Giving You The Best That I Got (88) 3(1) [7|15|22]	3304
Sweet Love (86) 8(1) [2|11|22]	1441
Just Because (89) 14(1) [0|11|16]	1009
Caught Up In The Rapture (87) 37(1) [0|2|18]	536
No One In The World (87) 44(1) [0|0|17]	422
Same Ole Love (365 Days A Year) (87) 44(2) [0|0|14]	403
BALANCE ▶ 569 Pk:22 [0|1|2]	**1058**
Breaking Away (81) 22(2) [0|9|17]	793
Falling In Love (82) 58(1) [0|0|11]	265
Marty BALIN ▶ 347 Pk:8 [1|2|3]	**2825**
Hearts (81) 8(2) [4|13|21]	2113
Atlanta Lady (Something About Your Love) (81) 27(3) [0|5|13]	584
What Love Is (83) 63(2) [0|0|6]	128
Russ BALLARD ▶ 962 Pk:58 [0|0|1]	**208**
On The Rebound (80) 58(1) [0|0|11]	208
BALTIMORA ▶ 515 Pk:13 [0|1|2]	**1379**
Tarzan Boy (86) 13(1) [0|10|26]	1350
Living In The Background (86) 87(1) [0|0|4]	29.2
Afrika BAMBAATAA and the SOUL SONIC FORCE ▶ 867 Pk:48 [0|0|1]	**312**
Planet Rock (82) 48(1) [0|0|11]	312
BANANARAMA ▶ 129 Pk:1 [3|3|11]	**8533**
Venus (86) 1(1) [7|12|19]	3701
I Heard A Rumour (87) 4(3) [6|12|19]	2398
Cruel Summer (84) 9(1) [2|11|18]	1465
I Can't Help It (88) 47(1) [0|0|13]	369
Love In The First Degree (88) 48(1) [0|0|10]	262
The Wild Life (84) 70(2) [0|0|6]	125
A Trick Of The Night (87) 76(1) [0|0|7]	74.5
More Than Physical (86) 73(1) [0|0|5]	61.1
Shy Boy (Don't It Make You Feel Good) (83) 83(1) [0|0|4]	48.7
Love, Truth And Honesty (88) 89(1) [0|0|3]	20.4
Robert DeNiro's Waiting (84) 95(1) [0|0|2]	9.4
BAND AID ▶ 680 Pk:13 [0|1|1]	**660**
Do They Know It's Christmas? (85) 13(1) [0|4|9]	660
BAND OF GOLD ▶ 1131 Pk:64 [0|0|1]	**103**
Medley: Love Songs Are Back Again (84) 64(1) [0|0|7]	103

Act ▶ Rank Peak [Top10s|Top40s|Total] Title (Pk Yr) Peak(Wk) [Top10Wks|Top40Wks|TotalWks]	Score
BANGLES ▶ 51 Pk:1 [5|8|8]	**19219**
Walk Like An Egyptian (86) 1(4) [8|15|23]	5581
Eternal Flame (89) 1(1) [6|14|19]	3603
Hazy Shade Of Winter (88) 2(1) [5|14|21]	2983
Manic Monday (86) 2(1) [5|14|20]	2950
In Your Room (89) 5(1) [4|12|20]	1943
Walking Down Your Street (87) 11(3) [0|9|16]	1131
If She Knew What She Wants (86) 29(1) [0|5|14]	548
Be With You (89) 30(1) [0|5|12]	478
Frank BARBER Orchestra ▶ 916 Pk:61 [0|0|1]	**250**
Hooked On Big Bands (82) 61(3) [0|0|12]	250
BARDEUX ▶ 667 Pk:36 [0|1|3]	**682**
When We Kiss (88) 36(1) [0|3|13]	431
I Love To Bass (89) 68(2) [0|0|10]	180
Magic Carpet Ride (88) 81(1) [0|0|7]	70.4
The BAR-KAYS ▶ 789 Pk:57 [0|0|3]	**441**
Move Your Boogie Body (80) 57(2) [0|0|7]	205
Today Is The Day (80) - Bar-Kays 60(1) [0|0|5]	125
Freakshow On The Dance Floor (84) 73(1) [0|0|8]	110
Jimmy BARNES ▶ 872 Pk:47 [0|0|3]	**303**
Good Times (87) - INXS And Jimmy Barnes [A] 47(1) [0|0|13]	186
Working Class Man (86) 74(1) [0|0|8]	106
Too Much Ain't Enough Love (88) 91(1) [0|0|2]	11.2
--BASE, Rob see ROB BASE	
BASIA ▶ 593 Pk:26 [0|1|2]	**940**
Time And Tide (88) 26(1) [0|7|20]	693
New Day For You (89) 53(1) [0|0|11]	247
Toni BASIL ▶ 186 Pk:1 [1|1|3]	**5985**
Mickey (82) 1(1) [10|18|27]	5851
Over My Head (84) 81(1) [0|0|6]	72.4
Shoppin' From A To Z (83) 77(2) [0|0|4]	62.3
BEACH BOYS ▶ 159 Pk:1 [1|5|10]	**7489**
Kokomo (88) 1(1) [5|15|28]	3725
The Beach Boys Medley (81) 12(2) [0|11|18]	1280
Come Go With Me (81) 18(2) [0|8|15]	843
Wipeout (87) - Fat Boys and The Beach Boys [A] 12(2) [0|11|19]	630
Getcha Back (85) 26(2) [0|7|12]	560
California Dreamin' (86) 57(1) [0|0|10]	248
Rock 'N' Roll To The Rescue (86) 68(1) [0|0|6]	105
It's Gettin' Late (85) 82(1) [0|0|5]	48.4
Goin' On (80) 83(1) [0|0|3]	36.6
Still Cruisin' (89) 93(1) [0|0|3]	12.8
BEASTIE BOYS ▶ 411 Pk:7 [1|2|3]	**2057**
(You Gotta) Fight For Your Right (To Party!) (87) 7(1) [2|10|18]	1427
Hey Ladies (89) 36(1) [0|2|10]	321
Brass Monkey (87) 48(1) [0|0|10]	309
The BEATLES ▶ 466 Pk:12 [0|2|2]	**1656**
The Beatles' Movie Medley (82) 12(3) [0|8|11]	1029
Twist And Shout (86) 23(1) [0|7|15]	628
--BEAU, Toby see: TOBY BEAU	
BEAU COUP ▶ 1194 Pk:80 [0|0|1]	**72.4**
Sweet Rachel (87) 80(1) [0|0|6]	72.4
Jean BEAUVOIR ▶ 1102 Pk:73 [0|0|1]	**112**
Feel The Heat (86) 73(1) [0|0|8]	112
Jeff BECK ▶ 1044 Pk:48 [0|0|1]	**147**
People Get Ready (85) - Jeff Beck And Rod Stewart [A] 48(1) [0|0|10]	147
BEE GEES ▶ 356 Pk:7 [1|3|6]	**2720**
One (89) 7(1) [1|10|14]	1136
The Woman In You (83) 24(3) [0|6|11]	605
He's A Liar (81) 30(1) [0|4|8]	328
Living Eyes (81) 45(2) [0|0|10]	320
Someone Belonging To Someone (83) 49(1) [0|0|6]	236
You Win Again (87) 75(1) [0|0|6]	95.0

260

Acts With Singles

Act ▶ Rank Peak [Top10s|Top40s|Total] Title (Pk Yr) Peak(Wk) [Top10Wks|Top40Wks|TotalWks]	Score
Adrian BELEW▶ 1034 Pk:58 [0|0|1]	**152**
Oh Daddy (89) 58(1) [0|0|8]	152
Randy BELL▶ 1368 Pk:90 [0|0|1]	**15.9**
Don't Do Me (84) 90(1) [0|0|3]	15.9
Regina BELLE▶ 855 Pk:60 [0|0|3]	**334**
Baby Come To Me (89) 60(1) [0|0|9]	179
Show Me The Way (87) 68(1) [0|0|9]	145
Without You (Love Theme From 'Leonard Part 6') (88) - Peabo Bryson & Regina Belle [A] 89(1) [0|0|3]	10.2
BELLE STARS▶ 557 Pk:14 [0|1|2]	**1123**
Iko Iko (89) 14(1) [0|10|18]	1058
Sign Of The Times (83) 75(2) [0|0|4]	65.1
Pat BENATAR▶ 56 Pk:5 [4|15|17]	**18270**
Love Is A Battlefield (83) 5(1) [7|14|22]	3213
We Belong (85) 5(2) [7|14|20]	2660
Hit Me With Your Best Shot (80) 9(3) [4|15|24]	2328
Invincible (85) 10(1) [1|11|17]	1310
Shadows Of The Night (82) 13(4) [0|10|16]	1280
Treat Me Right (81) 18(2) [0|10|18]	1046
Heartbreaker (80) 23(1) [0|10|18]	1008
Fire And Ice (81) 17(2) [0|9|15]	901
All Fired Up (88) 19(2) [0|8|17]	805
Little Too Late (83) 20(3) [0|7|14]	792
We Live For Love (80) 27(2) [0|6|14]	643
Sex As A Weapon (86) 28(1) [0|7|13]	571
Promises In The Dark (81) 38(2) [0|2|11]	416
Looking For A Stranger (83) 39(2) [0|3|10]	393
You Better Run (80) 42(2) [0|0|11]	366
Ooh Ooh Song (85) 36(2) [0|3|9]	321
Le Bel Age (The Best Years) (86) 54(2) [0|0|8]	217
George BENSON▶ 143 Pk:4 [2|3|8]	**7923**
Give Me The Night (80) 4(2) [7|14|23]	3251
Turn Your Love Around (82) 5(2) [6|16|22]	2821
Lady Love Me (One More Time) (83) 30(1) [0|6|13]	612
20/20 (85) 48(2) [0|0|13]	343
Inside Love (So Personal) (83) 43(1) [0|3|10]	331
Never Give Up On A Good Thing (82) 52(1) [0|0|9]	247
Love X Love (80) 61(3) [0|0|6]	162
Love All The Hurt Away (81) - Aretha Franklin & George Benson [A] 46(2) [0|0|10]	156
BERLIN▶ 227 Pk:1 [1|2|7]	**4898**
Take My Breath Away (Love Theme From Top Gun) (86) 1(1) [7|13|21]	3558
No More Words (84) 23(2) [0|8|17]	778
The Metro (83) 58(2) [0|0|10]	258
Sex (I'm A...) (83) 62(2) [0|0|7]	162
Now It's My Turn (84) 74(1) [0|0|4]	61.8
Like Flames (86) 82(1) [0|0|5]	44.6
Masquerade (83) 82(1) [0|0|3]	36.2
The B-52's▶ 316 Pk:3 [1|1|4]	**3355**
Love Shack (89) 3(2) [6|17|27]	3009
Rock Lobster (80) 56(1) [0|0|8]	204
Private Idaho (80) 74(1) [0|0|5]	93.3
Legal Tender (83) 81(1) [0|0|4]	49.5
BIG COUNTRY▶ 521 Pk:17 [0|1|3]	**1319**
In A Big Country (83) 17(2) [0|9|15]	1101
Fields Of Fire (84) 52(2) [0|0|6]	194
Wonderland (84) 86(1) [0|0|2]	24.6
BIG NOISE▶ 1387 Pk:97 [0|0|1]	**8.5**
Name And Number (89) 97(2) [0|0|3]	8.5
BIG PIG▶ 961 Pk:60 [0|0|1]	**208**
Breakaway (88) 60(1) [0|0|10]	208
BIG RIC▶ 1341 Pk:91 [0|0|1]	**23.2**
Take Away (83) 91(2) [0|0|3]	23.2
BIG TROUBLE▶ 1151 Pk:71 [0|0|1]	**93.9**
Crazy World (87) 71(1) [0|0|7]	93.9

Act ▶ Rank Peak [Top10s|Top40s|Total] Title (Pk Yr) Peak(Wk) [Top10Wks|Top40Wks|TotalWks]	Score
--BILLY & The BEATERS see: VERA, Billy & The BEATERS	
BILLY SATELLITE▶ 993 Pk:64 [0|0|2]	**184**
Satisfy Me (84) 64(1) [0|0|6]	118
I Wanna Go Back (84) 78(1) [0|0|5]	65.9
Stephen BISHOP▶ 595 Pk:25 [0|1|2]	**924**
It Might Be You (Theme From Tootsie) (83) 25(1) [0|8|20]	901
Unfaithfully Yours (One Love) (84) 87(1) [0|0|3]	23.0
Jay BLACK▶ 1386 Pk:98 [0|0|1]	**8.6**
Part Of Me That Needs You Most (80) 98(3) [0|0|4]	8.6
BLACKFOOT▶ 844 Pk:42 [0|0|1]	**345**
Fly Away (81) 42(1) [0|0|12]	345
J. BLACKFOOT▶ 1298 Pk:90 [0|0|1]	**34.6**
Taxi (84) 90(1) [0|0|5]	34.6
BLONDIE▶ 45 Pk:1 [3|5|6]	**20847**
Call Me (Theme from...American Gigolo) (80) 1(6) [12|19|25]	9931
The Tide Is High (81) 1(1) [10|17|26]	5289
Rapture (81) 1(2) [8|14|20]	4803
Island Of Lost Souls (82) 37(2) [0|3|10]	399
Atomic (80) 39(1) [0|3|9]	380
The Hardest Part (80) 84(2) [0|0|3]	44.9
Kurtis BLOW▶ 1039 Pk:71 [0|0|2]	**149**
Basketball (85) 71(2) [0|0|6]	102
The Breaks (Part 1) (80) 87(1) [0|0|6]	47.0
BLOW MONKEYS▶ 579 Pk:14 [0|1|1]	**1002**
Digging Your Scene (86) 14(1) [0|10|16]	1002
BLUE MERCEDES▶ 1115 Pk:66 [0|0|1]	**108**
I Want To Be Your Property (88) 66(1) [0|0|6]	108
BLUE ÖYSTER CULT▶ 742 Pk:40 [0|1|2]	**519**
Burnin' For You (81) 40(3) [0|3|14]	483
Shooting Shark (84) 83(1) [0|0|3]	36.0
BLUES BROTHERS▶ 526 Pk:18 [0|2|2]	**1306**
Gimme Some Lovin' (80) 18(2) [0|8|14]	911
Who's Making Love (81) 39(2) [0|2|11]	395
BLUE ZONE U.K.▶ 983 Pk:54 [0|0|1]	**192**
Jackie (88) 54(1) [0|0|9]	192
Michael BOLTON▶ 326 Pk:11 [0|3|5]	**3116**
(Sittin' On) The Dock Of The Bay (88) 11(2) [0|10|17]	1144
That's What Love Is All About (87) 19(2) [0|10|25]	1057
Soul Provider (89) 17(2) [0|7|17]	807
Wait On Love (88) 79(1) [0|0|6]	68.3
Fool's Game (83) 82(2) [0|0|3]	39.1
Gary (U.S.) BONDS▶ 372 Pk:11 [0|2|3]	**2492**
This Little Girl (81) 11(3) [0|13|18]	1565
Out Of Work (82) 21(2) [0|9|16]	833
Jole Blon (81) 65(1) [0|0|6]	94.5
BON JOVI▶ 27 Pk:1 [8|9|12]	**24759**
Livin' On A Prayer (87) 1(4) [7|13|21]	4968
I'll Be There For You (89) 1(1) [6|13|22]	3890
You Give Love A Bad Name (86) 1(1) [6|14|24]	3706
Bad Medicine (88) 1(2) [6|12|20]	3342
Born To Be My Baby (89) 3(1) [5|13|20]	2796
Wanted Dead Or Alive (87) 7(3) [5|12|17]	1860
Lay Your Hands On Me (89) 7(2) [4|11|16]	1661
Living In Sin (89) 9(1) [3|12|19]	1480
Runaway (84) 39(1) [0|1|13]	472
She Don't Know Me (84) 48(1) [0|0|11]	298
Only Lonely (85) 54(1) [0|0|8]	190
In And Out Of Love (85) 69(2) [0|0|6]	96.8
Karla BONOFF▶ 504 Pk:19 [0|1|3]	**1429**
Personally (82) 19(2) [0|12|18]	1122
Baby Don't Go (80) 69(1) [0|0|6]	164
Please Be The One (82) 63(2) [0|0|7]	143
Chuckii BOOKER▶ 799 Pk:42 [0|0|1]	**424**
Turned Away (89) 42(1) [0|0|14]	424

Act ▶ Rank Peak [Top10s|Top40s|Total] Title (Pk Yr) Peak(Wk) [Top10Wks|Top40Wks|TotalWks]	Score
BOOK OF LOVE▶ 1336 Pk:90 [0|0|1]	**23.8**
Pretty Boys And Pretty Girls (88) 90(2) [0|0|4]	23.8
BOOMTOWN RATS▶ 1158 Pk:73 [0|0|1]	**89.7**
I Don't Like Mondays (80) 73(1) [0|0|5]	89.7
BOSTON▶ 199 Pk:1 [2|3|3]	**5546**
Amanda (86) 1(2) [6|12|18]	3807
We're Ready (87) 9(1) [2|10|15]	1142
Can'tcha Say (You Believe In Me)/Still In Love (87) 20(2) [0|5|13]	598
BOURGEOIS TAGG▶ 653 Pk:38 [0|1|2]	**717**
I Don't Mind At All (87) 38(1) [0|2|17]	541
Mutual Surrender (What A Wonderful World) (86) 62(1) [0|0|10]	176
David BOWIE▶ 83 Pk:1 [4|9|14]	**13592**
Let's Dance (83) 1(1) [10|14|20]	6341
China Girl (83) 10(1) [1|11|18]	1531
Blue Jean (84) 8(2) [2|10|18]	1487
Modern Love (83) 14(1) [0|9|13]	1029
Dancing In The Street (85) - Mick Jagger & David Bowie [A] 7(1) [3|9|14]	691
Day-In Day-Out (87) 21(2) [0|7|12]	622
Never Let Me Down (87) 27(1) [0|5|11]	458
Under Pressure (82) - Queen & David Bowie [A] 29(2) [0|8|15]	339
Tonight (84) 53(3) [0|0|9]	259
This Is Not America (85) - David Bowie/ Pat Metheny Group [A] 32(2) [0|4|12]	248
Absolute Beginners (86) 53(1) [0|0|9]	216
Fashion (81) 70(1) [0|0|9]	164
Cat People (Putting Out Fire) (82) 67(1) [0|0|10]	129
Without You (84) 73(1) [0|0|4]	77.1
Rick BOWLES▶ 1250 Pk:77 [0|0|1]	**52.6**
Too Good To Turn Back Now (82) 77(2) [0|0|3]	52.6
BOW WOW WOW▶ 937 Pk:62 [0|0|2]	**231**
I Want Candy (82) 62(1) [0|0|7]	175
Do You Wanna Hold Me? (83) 77(1) [0|0|4]	56.2
Boy GEORGE▶ 835 Pk:40 [0|1|1]	**357**
Live My Life (88) 40(1) [0|1|12]	357
BOY MEETS GIRL▶ 305 Pk:5 [1|2|3]	**3546**
Waiting For A Star To Fall (88) 5(3) [6|16|25]	2834
Oh Girl (85) 39(1) [0|1|13]	393
Bring Down The Moon (89) 49(1) [0|0|11]	319
The BOYS▶ 561 Pk:13 [0|1|1]	**1097**
Dial My Heart (89) 13(1) [0|9|19]	1097
BOYS BAND▶ 1029 Pk:61 [0|0|1]	**157**
Don't Stop Me Baby (I'm On Fire) (82) 61(1) [0|0|8]	157
BOYS CLUB▶ 489 Pk:8 [1|1|1]	**1515**
I Remember Holding You (89) 8(1) [1|12|21]	1515
BOYS DON'T CRY▶ 548 Pk:12 [0|1|1]	**1158**
I Wanna Be A Cowboy (86) 12(1) [0|9|19]	1158
Laura BRANIGAN▶ 71 Pk:2 [3|7|12]	**15659**
Gloria (82) 2(3) [10|22|36]	6294
Self Control (84) 4(2) [6|15|25]	3250
Solitaire (83) 7(2) [4|13|17]	1953
How Am I Supposed To Live Without You (83) 12(1) [0|12|20]	1493
The Lucky One (84) 20(2) [0|8|15]	795
Power Of Love (88) 26(2) [0|9|18]	727
Spanish Eddie (85) 40(2) [0|2|11]	364
Ti Amo (84) 55(3) [0|0|12]	332
Shattered Glass (87) 48(1) [0|0|10]	268
All Night With Me (82) 69(1) [0|0|7]	97.6
Hold Me (85) 82(2) [0|0|4]	45.6
I Found Someone (86) 90(2) [0|0|6]	40.4
BREAKFAST CLUB▶ 457 Pk:7 [1|1|2]	**1700**
Right On Track (87) 7(1) [1|11|19]	1500
Kiss And Tell (87) 48(1) [0|0|9]	200

Acts With Singles

Act ▶ Rank Peak [Top10s|Top40s|Total] Title (Pk Yr) Peak(Wk) [Top10Wks|Top40Wks|TotalWks]	Score
BREATHE▶ 158 Pk:2 [3|3|3]	**7556**
Hands To Heaven (88) 2(2) [6|16|29]	3503
How Can I Fall? (88) 3(2) [5|16|22]	2953
Don't Tell Me Lies (89) 10(1) [1|10|16]	1099
BREATHLESS▶ 1340 Pk:92 [0|0|1]	**23.3**
Takin' It Back (80) 92(2) [0|0|4]	23.3
Edie BRICKELL & The NEW BOHEMIANS▶ 434 Pk:7 [1|1|2]	**1853**
What I Am (89) 7(1) [3|10|19]	1561
Circle (89) 48(2) [0|0|10]	292
Martin BRILEY▶ 727 Pk:36 [0|1|1]	**543**
The Salt In My Tears (83) 36(1) [0|3|15]	543
Britny FOX▶ 1399 Pk:100 [0|0|1]	**1.5**
Long Way To Love (88) 100(2) [0|0|2]	1.5
BRONSKI BEAT▶ 775 Pk:48 [0|0|1]	**461**
Smalltown Boy (85) 48(1) [0|0|16]	461
BROS▶ 1281 Pk:83 [0|0|1]	**40.2**
When Will I Be Famous (88) 83(1) [0|0|5]	40.2
BROTHERS JOHNSON▶ 414 Pk:7 [1|1|3]	**2045**
Stomp! (80) 7(2) [2|13|19]	1868
The Real Thing (81) 67(1) [0|0|6]	107
Treasure (80) 73(1) [0|0|4]	70.0
Alex BROWN▶ 1207 Pk:76 [0|0|1]	**65.9**
(Come On) Shout (85) 76(1) [0|0|6]	65.9
Bobby BROWN▶ 63 Pk:1 [6|6|7]	**16589**
My Prerogative (89) 1(1) [7|15|24]	4660
On Our Own (89) 2(3) [6|13|20]	3811
Every Little Step (89) 3(2) [6|13|21]	2738
Roni (89) 3(1) [4|11|17]	1982
Don't Be Cruel (88) 8(2) [2|14|26]	1692
Rock Wit'cha (89) 7(1) [3|11|21]	1465
Girlfriend (87) 57(2) [0|0|9]	240
James BROWN▶ 401 Pk:4 [1|1|2]	**2153**
Living In America (86) 4(1) [5|11|19]	2143
Gravity (86) 93(1) [0|0|2]	9.5
Jocelyn BROWN▶ 1082 Pk:75 [0|0|1]	**120**
Somebody Else's Guy (84) 75(2) [0|0|10]	120
Peter BROWN▶ 1010 Pk:59 [0|0|1]	**173**
Stargazer (80) 59(1) [0|0|8]	173
Sam BROWN▶ 988 Pk:65 [0|0|1]	**189**
Stop (89) 65(1) [0|0|10]	189
Jackson BROWNE▶ 170 Pk:7 [1|6|8]	**6785**
Somebody's Baby (82) 7(3) [6|12|19]	2217
Lawyers In Love (83) 13(2) [0|12|15]	1375
Boulevard (80) 19(2) [0|10|16]	984
Tender Is The Night (83) 25(1) [0|7|17]	740
That Girl Could Sing (80) 22(2) [0|5|13]	604
For America (86) 30(1) [0|5|12]	467
For A Rocker (84) 45(2) [0|0|9]	290
In The Shape Of A Heart (86) 70(1) [0|0|7]	108
Sharon BRYANT▶ 779 Pk:34 [0|1|2]	**457**
Let Go (89) 34(2) [0|4|13]	425
Foolish Heart (89) 90(1) [0|0|5]	31.4
Peabo BRYSON▶ 369 Pk:10 [1|1|6]	**2508**
If Ever You're In My Arms Again (84) 10(3) [3|13|25]	1922
Let The Feeling Flow (82) 42(1) [0|0|12]	362
Lovers After All (81) - Melissa Manchester And Peabo Bryson [A] 54(2) [0|0|9]	108
Take No Prisoners (In The Game Of Love) (85) 78(1) [0|0|6]	64.3
Slow Dancin' (84) 82(2) [0|0|4]	41.7
Without You (Love Theme From 'Leonard Part 6') (88) - Peabo Bryson & Regina Belle [A] 89(1) [0|0|3]	10.2
Peabo BRYSON/Roberta FLACK▶ 408 Pk:16 [0|1|2]	**2100**
Tonight, I Celebrate My Love (83) 16(2) [0|15|29]	1875
You're Looking Like Love To Me (84) 58(1) [0|0|11]	225

Act ▶ Rank Peak [Top10s|Top40s|Total] Title (Pk Yr) Peak(Wk) [Top10Wks|Top40Wks|TotalWks]	Score
Lindsey BUCKINGHAM▶ 344 Pk:9 [1|2|3]	**2850**
Trouble (82) 9(2) [5|14|19]	2011
Go Insane (84) 23(1) [0|9|16]	783
Holiday Road (83) 82(1) [0|0|5]	55.6
BUCKNER & GARCIA▶ 465 Pk:9 [1|1|1]	**1661**
Pac-Man Fever (82) 9(2) [3|14|19]	1661
Jimmy BUFFETT▶ 816 Pk:57 [0|0|3]	**395**
It's My Job (81) 57(1) [0|0|8]	190
Volcano (80) 66(2) [0|0|7]	141
Survive (80) 77(1) [0|0|5]	63.0
Cindy BULLENS▶ 1326 Pk:90 [0|0|1]	**27.0**
Trust Me (80) 90(1) [0|0|3]	27.0
BULLETBOYS▶ 956 Pk:71 [0|0|2]	**214**
Smooth Up (89) 71(1) [0|0|10]	137
For The Love Of Money (89) 78(1) [0|0|6]	77.2
Gary BURBANK with BAND McNALLY▶ 1099 Pk:67 [0|0|1]	**113**
Who Shot J.R.? (80) 67(1) [0|0|5]	113
Billy BURNETTE▶ 1139 Pk:68 [0|0|1]	**99.7**
Don't Say No (80) 68(2) [0|0|5]	99.7
Rocky BURNETTE▶ 437 Pk:8 [1|1|1]	**1840**
Tired Of Toein' The Line (80) 8(2) [2|12|19]	1840
George BURNS▶ 876 Pk:49 [0|0|1]	**297**
I Wish I Was Eighteen Again (80) 49(2) [0|0|10]	297
Glen BURTNICK▶ 1047 Pk:65 [0|0|1]	**146**
Follow You (87) 65(1) [0|0|8]	146
Jenny BURTON▶ 992 Pk:54 [0|0|2]	**187**
Strangers In A Strange World (84) - Jenny Burton & Patrick Jude [A] 54(2) [0|0|7]	99.8
Remember What You Like (84) 81(2) [0|0|6]	87.5
BUS BOYS▶ 1191 Pk:68 [0|0|1]	**74.0**
Cleanin' Up The Town (84) 68(1) [0|0|5]	74.0
Kate BUSH▶ 688 Pk:30 [0|1|2]	**649**
Running Up That Hill (85) 30(1) [0|4|20]	602
Don't Give Up (87) - Peter Gabriel/Kate Bush [A] 72(1) [0|0|6]	47.1
Jon BUTCHER AXIS▶ 1374 Pk:94 [0|0|1]	**13.2**
Sounds Of Your Voice (85) 94(2) [0|0|3]	13.2
Jonathan BUTLER▶ 711 Pk:27 [0|1|1]	**579**
Lies (87) 27(2) [0|5|14]	579

C

Act ▶ Rank Peak [Top10s|Top40s|Total] Title (Pk Yr) Peak(Wk) [Top10Wks|Top40Wks|TotalWks]	Score
John CAFFERTY▶ 1181 Pk:76 [0|0|1]	**78.4**
Heart's On Fire (86) 76(1) [0|0|6]	78.4
John CAFFERTY And The BEAVER BROWN BAND▶ 248 Pk:7 [1|4|9]	**4491**
On The Dark Side (84) 7(2) [3|11|18]	1673
Tough All Over (85) 22(1) [0|8|15]	747
C-I-T-Y (85) 18(1) [0|8|15]	745
Tender Years (85) 31(1) [0|7|14]	559
Small Town Girl (85) 64(1) [0|0|10]	200
On The Dark Side (83) - Eddie And The Cruisers 64(1) [0|0|9]	197
Voice Of America's Sons (86) 62(2) [0|0|8]	182
Pride & Passion (89) 66(1) [0|0|7]	131
Tender Years (84) 78(1) [0|0|5]	57.8
Tané CAIN▶ 838 Pk:37 [0|1|1]	**355**
Holdin' On (82) 37(3) [0|3|11]	355
Bobby CALDWELL▶ 763 Pk:42 [0|0|2]	**481**
Coming Down From Love (80) 42(2) [0|0|10]	385
All Of My Love (82) 77(3) [0|0|6]	95.9
CALIFORNIA RAISINS▶ 1297 Pk:84 [0|0|1]	**35.0**
I Heard It Through The Grapevine (88) 84(1) [0|0|4]	35.0
The CALL▶ 921 Pk:51 [0|0|2]	**249**
Let The Day Begin (89) 51(1) [0|0|9]	178
The Walls Came Down (83) 74(1) [0|0|5]	71.4
CAMEO▶ 308 Pk:6 [1|2|5]	**3497**
Word Up (86) 6(3) [5|14|21]	2212

Act ▶ Rank Peak [Top10s|Top40s|Total] Title (Pk Yr) Peak(Wk) [Top10Wks|Top40Wks|TotalWks]	Score
Candy (87) 21(1) [0|7|17]	702
She's Strange (84) 47(2) [0|0|11]	329
Back And Forth (87) 50(1) [0|0|8]	212
You Make Me Work (88) 85(1) [0|0|5]	41.1
CAMOUFLAGE▶ 963 Pk:59 [0|0|1]	**208**
The Great Commandment (89) 59(1) [0|0|12]	208
Glen CAMPBELL▶ 909 Pk:42 [0|0|3]	**261**
Somethin' 'Bout You Baby I Like (80) - Glen Campbell and Rita Coolidge [A] 42(2) [0|0|10]	153
I Don't Want To Know Your Name (81) 65(1) [0|0|5]	94.7
I Love My Truck (81) 94(1) [0|0|3]	12.6
CANDI▶ 1049 Pk:68 [0|0|1]	**145**
Dancing Under A Latin Moon (88) 68(1) [0|0|7]	145
Freddy CANNON & The BELMONTS▶ 1266 Pk:81 [0|0|1]	**46.3**
Let's Put The Fun Back In Rock N Roll (81) 81(2) [0|0|4]	46.3
Jim CAPALDI▶ 687 Pk:28 [0|1|2]	**651**
That's Love (83) 28(2) [0|5|13]	559
Living On The Edge (83) 75(2) [0|0|5]	91.6
CAPTAIN & TENNILLE▶ 119 Pk:1 [1|1|3]	**9027**
Do That To Me One More Time (80) 1(1) [14|22|28]	8662
Love On A Shoestring (80) 55(2) [0|0|7]	196
Happy Together (A Fantasy) (80) 53(1) [0|0|6]	169
Irene CARA▶ 55 Pk:1 [3|6|8]	**18325**
Flashdance...What A Feeling (83) 1(6) [14|20|25]	9800
Fame (80) 4(2) [6|12|26]	3221
Breakdance (84) 8(1) [3|11|19]	1793
Why Me? (83) 13(3) [0|10|13]	1330
Out Here On My Own (80) 19(2) [0|9|23]	1039
The Dream (Hold On To Your Dream) (84) 37(1) [0|3|14]	562
Anyone Can See (82) 42(2) [0|0|18]	527
You Were Made For Me (84) 78(1) [0|0|5]	53.4
Luis CARDENAS▶ 1265 Pk:83 [0|0|1]	**46.5**
Runaway (86) 83(1) [0|0|5]	46.5
Tony CAREY▶ 501 Pk:22 [0|2|4]	**1443**
A Fine Fine Day (84) 22(1) [0|8|15]	755
The First Day Of Summer (84) 33(1) [0|2|11]	417
West Coast Summer Nights (83) 64(1) [0|0|9]	187
I Won't Be Home Tonight (83) 79(2) [0|0|7]	83.8
Belinda CARLISLE▶ 93 Pk:1 [4|5|7]	**12201**
Heaven Is A Place On Earth (87) 1(1) [6|15|21]	4109
Mad About You (86) 3(2) [6|14|21]	2942
I Get Weak (88) 2(1) [5|13|16]	2582
Circle In The Sand (88) 7(1) [2|10|17]	1361
Leave A Light On (89) 11(1) [0|10|18]	1127
I Feel The Magic (86) 82(1) [0|0|5]	56.0
I Feel Free (88) 88(1) [0|4|4]	23.8
Steve CARLISLE▶ 967 Pk:65 [0|0|1]	**206**
WKRP In Cincinnati (81) 65(4) [0|0|10]	206
Carl CARLTON▶ 597 Pk:22 [0|1|1]	**909**
She's A Bad Mama Jama (She's Built, She's Stacked) (81) 22(2) [0|7|21]	909
Larry CARLTON▶ 601 Pk:10 [1|1|2]	**892**
The Theme From Hill Street Blues (81) - Mike Post Featuring Larry Carlton [A] 10(2) [2|10|22]	789
Sleepwalk (82) 74(1) [0|0|8]	103
Eric CARMEN▶ 188 Pk:3 [2|3|6]	**5934**
Make Me Lose Control (88) 3(1) [6|13|20]	2785
Hungry Eyes (88) 4(1) [5|16|25]	2648
I Wanna Hear It From Your Lips (85) 35(1) [0|4|11]	408
It Hurts Too Much (80) 75(1) [0|0|2]	44.6
I'm Through With Love (85) 87(2) [0|0|3]	28.6
Reason To Try (88) 87(1) [0|0|3]	19.4

Act ▶ Rank Peak [Top10s|Top40s|Total] Title (Pk Yr) Peak(Wk) [Top10Wks|Top40Wks|TotalWks]	Score
Kim CARNES▶ 66 Pk:1 [2|7|14]	16050
Bette Davis Eyes (81) 1(9) [14|20|26]	10255
More Love (80) 10(3) [3|15|19]	1950
Crazy In The Night (Barking At Airplanes) (85)	
15(1) [0|9|16]	855
Draw Of The Cards (81) 28(2) [0|6|12]	558
Does It Make You Remember (83) 36(2) [0|4|13]	530
Voyeur (82) 29(4) [0|6|12]	498
Invisible Hands (83) 40(2) [0|2|13]	455
Cry Like A Baby (80) 44(2) [0|0|8]	286
You Make My Heart Beat Faster	
(And That's All That Matters) (84) 54(1) [0|0|8]	214
Mistaken Identity (81) 60(1) [0|0|6]	136
Invitation To Dance (85) 68(1) [0|0|6]	95.9
I Pretend (84) 74(1) [0|0|5]	81.6
Abadabadango (85) 67(1) [0|0|4]	79.6
Divided Hearts (86) 79(2) [0|0|5]	55.4
CARPENTERS▶ 530 Pk:16 [0|1|4]	1286
Touch Me When We're Dancing (81)	
16(4) [0|8|14]	929
(Want You) Back In My Life Again (81)	
72(2) [0|0|8]	141
Those Good Old Dreams (82) 63(1) [0|0|6]	132
Beechwood 4-5789 (82) 74(2) [0|0|4]	84.7
Paul CARRACK▶ 330 Pk:9 [1|4|6]	3076
Don't Shed A Tear (88) 9(3) [3|13|24]	1635
I Live By The Groove (89) 31(1) [0|4|13]	494
One Good Reason (88) 28(2) [0|5|13]	486
I Need You (82) 37(2) [0|2|13]	428
When You Walk In The Room (88) 90(1) [0|0|3]	19.7
Button Off My Shirt (88) 91(1) [0|0|3]	13.0
The CARS▶ 72 Pk:3 [4|10|13]	15640
Shake It Up (82) 4(3) [7|17|22]	3549
Drive (84) 3(3) [7|14|19]	3543
You Might Think (84) 7(3) [4|11|17]	1977
Tonight She Comes (86) 7(1) [4|12|17]	1667
Magic (84) 12(3) [0|11|17]	1357
Hello Again (84) 20(2) [0|10|15]	882
You Are The Girl (87) 17(1) [0|9|14]	818
Why Can't I Have You (85) 33(1) [0|5|17]	571
I'm Not The One (86) 32(1) [0|4|11]	445
Touch And Go (80) 37(2) [0|3|11]	431
Since You're Gone (82) 41(1) [0|0|9]	294
Coming Up You (88) 74(1) [0|0|5]	75.3
Strap Me In (87) 85(1) [0|0|4]	31.3
Rosanne CASH▶ 620 Pk:22 [0|1|1]	802
Seven Year Ache (81) 22(2) [0|7|20]	802
Felix CAVALIERE▶ 784 Pk:36 [0|1|1]	452
Only A Lonely Heart Sees (80) 36(1) [0|3|11]	452
CELLARFUL OF NOISE▶ 1133 Pk:69 [0|0|1]	103
Samantha (What You Gonna Do?) (88)	
69(1) [0|0|7]	103
CENTRAL LINE▶ 1228 Pk:84 [0|0|1]	60.4
Walking Into Sunshine (81) 84(1) [0|0|6]	60.4
Peter CETERA▶ 108 Pk:1 [4|4|7]	10421
Glory Of Love (Theme From The Karate Kid Part II)	
(86) 1(2) [6|14|21]	4056
The Next Time I Fall (86) - Peter Cetera w/Amy Grant	
1(1) [6|15|21]	3174
One Good Woman (88) 4(1) [3|13|18]	1979
After All (Love Theme From "Chances Are") (89) -	
Cher and Peter Cetera [A] 6(1) [4|11|20]	883
Best Of Times (88) 59(1) [0|0|8]	189
Big Mistake (87) 61(2) [0|0|6]	134
I Wasn't The One (Who Said Goodbye) (88) -	
Agnetha Faltskog and Peter Cetera [A]	
93(1) [0|0|3]	6.1
CHAMPAIGN▶ 375 Pk:12 [0|2|2]	2461
How 'Bout Us (81) 12(1) [0|13|23]	1482
Try Again (83) 23(3) [0|9|20]	979

Act ▶ Rank Peak [Top10s|Top40s|Total] Title (Pk Yr) Peak(Wk) [Top10Wks|Top40Wks|TotalWks]	Score
Bill CHAMPLIN▶ 836 Pk:55 [0|0|2]	357
Tonight Tonight (82) 55(2) [0|0|8]	183
Sara (82) 61(1) [0|0|8]	173
CHANGE▶ 695 Pk:40 [0|1|4]	624
A Lover's Holiday (80) 40(1) [0|1|13]	506
The Very Best In You (82) 84(3) [0|0|5]	51.1
Paradise (81) 80(2) [0|0|4]	47.1
Hold Tight (81) 89(2) [0|0|2]	19.6
Harry CHAPIN▶ 669 Pk:23 [0|1|1]	679
Sequel (81) 23(2) [0|7|14]	679
Tracy CHAPMAN▶ 397 Pk:6 [1|1|4]	2229
Fast Car (88) 6(2) [4|12|21]	1788
Baby Can I Hold You (88) 48(3) [0|0|12]	365
Talkin' Bout A Revolution (88) 75(2) [0|0|4]	48.2
Crossroads (89) 90(2) [0|0|4]	26.9
CHARLENE▶ 320 Pk:3 [1|1|2]	3303
I've Never Been To Me (82) 3(3) [6|14|20]	3113
Used To Be (82) - Charlene & Stevie Wonder [A]	
46(3) [0|0|11]	190
Sonny CHARLES▶ 765 Pk:40 [0|1|1]	476
Put It In A Magazine (83) 40(2) [0|2|14]	476
CHARLIE▶ 800 Pk:38 [0|1|1]	416
It's Inevitable (83) 38(1) [0|2|11]	416
CHEAP TRICK▶ 130 Pk:1 [2|4|10]	8527
The Flame (88) 1(2) [6|14|27]	3801
Don't Be Cruel (88) 4(2) [5|12|17]	2120
Ghost Town (88) 33(3) [0|5|14]	499
Voices (80) 32(1) [0|3|11]	487
Tonight It's You (85) 44(1) [0|0|17]	457
If You Want My Love (82) 45(2) [0|0|11]	346
Everything Works If You Let It (80) 44(2) [0|0|10]	330
Stop This Game (80) 48(2) [0|0|12]	261
She's Tight (82) 65(3) [0|0|7]	142
Never Had A Lot To Lose (89) 75(1) [0|0|6]	83.4
Chubby CHECKER▶ 1305 Pk:91 [0|0|1]	32.3
Running (82) 91(3) [0|0|5]	32.3
CHEECH & CHONG▶ 863 Pk:48 [0|0|1]	317
Born In East L.A. (85) 48(2) [0|0|11]	317
Joe CHEMAY Band▶ 1018 Pk:68 [0|0|1]	168
Proud (81) 68(2) [0|0|8]	168
CHER▶ 147 Pk:3 [4|5|6]	7849
If I Could Turn Back Time (89) 3(2) [5|14|23]	2821
Just Like Jesse James (89) 8(4) [5|11|18]	1680
I Found Someone (88) 10(1) [1|12|26]	1525
We All Sleep Alone (88) 14(1) [0|9|15]	889
After All (Love Theme From Chances Are) (89) -	
Cher and Peter Cetera [A] 6(1) [4|11|20]	883
Skin Deep (88) 79(1) [0|0|4]	51.5
CHERI▶ 821 Pk:39 [0|1|1]	377
Murphy's Law (82) 39(2) [0|2|12]	377
CHERRELLE▶ 806 Pk:26 [0|1|2]	407
Saturday Love (86) - Cherrelle with	
Alexander O'Neal [A] 26(2) [0|6|17]	311
I Didn't Mean To Turn You On (84) 79(1) [0|0|9]	95.9
Neneh CHERRY▶ 282 Pk:3 [2|2|2]	3956
Buffalo Stance (89) 3(1) [6|14|24]	2898
Kisses On The Wind (89) 8(1) [1|8|14]	1059
CHIC▶ 907 Pk:61 [0|0|4]	264
Rebels Are We (80) 61(1) [0|0|6]	133
Soup For One (82) 80(2) [0|0|5]	85.0
Real People//Chip Off The Old Block (80)	
79(2) [0|0|2]	36.6
Real People (80) 89(1) [0|0|1]	9.9
CHICAGO▶ 15 Pk:1 [7|11|16]	29063
Hard To Say I'm Sorry (82) 1(2) [12|18|24]	6847
Look Away (88) 1(2) [8|16|24]	4616
Hard Habit To Break (84) 3(2) [6|15|25]	3712
You're The Inspiration (85) 3(2) [6|14|22]	3237
Will You Still Love Me? (87) 3(1) [4|13|23]	2470

Act ▶ Rank Peak [Top10s|Top40s|Total] Title (Pk Yr) Peak(Wk) [Top10Wks|Top40Wks|TotalWks]	Score
I Don't Wanna Live Without Your Love (88)	
3(1) [5|13|21]	2418
You're Not Alone (89) 10(1) [1|10|17]	1099
Stay The Night (84) 16(2) [0|10|17]	1052
Along Comes A Woman (85) 14(2) [0|10|16]	1018
If She Would Have Been Faithful... (87)	
17(1) [0|8|19]	1001
Love Me Tomorrow (82) 22(2) [0|8|15]	766
We Can Last Forever (89) 55(2) [0|0|12]	302
25 Or 6 To 4 (86) 48(3) [0|0|8]	247
Thunder And Lightning (80) 56(2) [0|0|9]	203
What You're Missing (83) 81(2) [0|0|5]	56.3
Niagara Falls (87) 91(1) [0|0|3]	17.7
CHICAGO BEARS SHUFFLIN' CREW▶ 914	
Pk:41 [0|0|1]	254
Super Bowl Shuffle (86) 41(2) [0|0|9]	254
Toni CHILDS▶ 1126 Pk:72 [0|0|1]	105
Don't Walk Away (88) 72(1) [0|0|7]	105
CHILLIWACK▶ 429 Pk:22 [0|2|3]	1884
My Girl (Gone, Gone, Gone) (81) 22(2) [0|11|19]	989
Whatcha Gonna Do (82) 41(3) [0|0|13]	491
I Believe (82) 33(2) [0|3|11]	403
CHOIRBOYS▶ 1195 Pk:80 [0|0|1]	71.6
Run To Paradise (89) 80(1) [0|0|7]	71.6
Chris CHRISTIAN▶ 755 Pk:37 [0|1|2]	499
I Want You, I Need You (81) 37(2) [0|3|14]	467
Ain't Nothing Like The Real Thing/	
You're All I Need To Get By (82) 88(3) [0|0|3]	32.2
Gavin CHRISTOPHER▶ 671 Pk:22 [0|1|1]	679
One Step Closer To You (86) 22(1) [0|7|17]	679
CHUNKY A▶ 1223 Pk:77 [0|0|1]	61.4
Owwww! (89) 77(1) [0|0|6]	61.4
The CHURCH▶ 725 Pk:24 [0|1|1]	552
Under The Milky Way (88) 24(1) [0|5|15]	552
CINDERELLA▶ 287 Pk:12 [0|4|6]	3818
Don't Know What You Got (Till It's Gone) (88)	
12(1) [0|11|22]	1292
Nobody's Fool (87) 13(1) [0|8|21]	1151
Coming Home (89) 20(1) [0|7|17]	758
The Last Mile (89) 36(1) [0|2|10]	324
Gypsy Road (89) 51(1) [0|0|7]	170
Somebody Save Me (87) 66(1) [0|0|7]	123
Eric CLAPTON▶ 262 Pk:10 [1|4|9]	4290
I Can't Stand It (81) - Eric Clapton And His Band	
10(2) [2|12|17]	1600
I've Got A Rock N' Roll Heart (83) 18(3) [0|10|16]	1097
Tulsa Time//Cocaine (80) -	
Eric Clapton And His Band 30(2) [0|5|12]	589
Forever Man (85) 26(1) [0|6|12]	553
Pretending (89) 55(1) [0|0|11]	245
Another Ticket (81) - Eric Clapton And His Band	
78(3) [0|0|11]	71.0
Blues Power (80) - Eric Clapton And His Band	
76(1) [0|0|5]	59.6
Tulsa Time (80) - Eric Clapton And His Band	
64(1) [0|0|2]	57.2
See What Love Can Do (85) 89(1) [0|0|2]	18.0
Petula CLARK▶ 1084 Pk:66 [0|0|1]	119
Natural Love (82) 66(1) [0|0|6]	119
Allan CLARKE▶ 1172 Pk:70 [0|0|1]	83.7
Slipstream (80) 70(2) [0|0|4]	83.7
Stanley CLARKE▶ 753 Pk:19 [0|1|1]	502
Sweet Baby (81) - Stanley Clarke/George Duke [A]	
19(2) [0|9|20]	502
The CLASH▶ 296 Pk:8 [1|2|4]	3696
Rock The Casbah (83) 8(4) [5|15|24]	2285
Train In Vain (Stand By Me) (80) 23(2) [0|7|14]	723
Should I Stay Or Should I Go (82) 45(2) [0|0|13]	399
Should I Stay Or Should I Go (83) 50(2) [0|0|10]	288
--CLAY, Tiggi see: TIGGI CLAY	

Acts With Singles

Act ▶ Rank Peak [Top10s|Top40s|Total] Title (Pk Yr) Peak(Wk) [Top10Wks|Top40Wks|TotalWks]	Score
Merry CLAYTON▶ 896 Pk:45 [0|0|1]	277
Yes (88) 45(1) [0|0|11]	277
Clarence CLEMONS▶ 580 Pk:18 [0|1|1]	1001
You're A Friend Of Mine (86) - Clarence Clemons and Jackson Browne 18(1) [0|12|19]	1001
Linda CLIFFORD▶ 857 Pk:41 [0|0|1]	333
Red Light (80) 41(1) [0|0|11]	333
CLIMAX BLUES BAND▶ 409 Pk:12 [0|1|2]	2088
I Love You (81) 12(1) [0|17|27]	1722
Gotta Have More Love (80) 47(2) [0|0|12]	366
CLIMIE FISHER▶ 683 Pk:23 [0|1|1]	655
Love Changes (Everything) (88) 23(1) [0|6|18]	655
CLOCKS▶ 1103 Pk:67 [0|0|1]	112
She Looks A Lot Like You (82) 67(2) [0|0|5]	112
CLUB HOUSE▶ 1184 Pk:75 [0|0|1]	78.2
Do It Again/Billie Jean (Medley) (83) 75(1) [0|0|5]	78.2
CLUB NOUVEAU▶ 247 Pk:1 [1|2|2]	4511
Lean On Me (87) 1(2) [6|12|17]	4095
Why You Treat Me So Bad (87) 39(1) [0|1|13]	416
Joyce COBB▶ 805 Pk:42 [0|0|1]	409
Dig The Gold (80) 42(1) [0|0|12]	409
Bruce COCKBURN▶ 594 Pk:21 [0|1|2]	924
Wondering Where The Lions Are (80) 21(2) [0|9|17]	901
If I Had A Rocket Launcher (85) 88(1) [0|0|3]	22.9
Joe COCKER▶ 361 Pk:1 [1|1|3]	2652
Up Where We Belong (82) - Joe Cocker and Jennifer Warnes [A] 1(3) [7|15|23]	2512
Edge Of A Dream (84) 69(1) [0|0|7]	113
Shelter Me (86) 91(2) [0|0|4]	27.6
COCK ROBIN▶ 749 Pk:35 [0|1|1]	511
When Your Heart Is Weak (85) 35(1) [0|3|16]	511
Gardner COLE▶ 1356 Pk:91 [0|0|1]	19.2
Live It Up (88) 91(1) [0|0|3]	19.2
Natalie COLE▶ 165 Pk:5 [2|5|8]	7168
Pink Cadillac (88) 5(2) [4|12|17]	1860
Miss You Like Crazy (89) 7(2) [4|13|19]	1760
I Live For Your Love (88) 13(1) [0|11|22]	1187
Jump Start (87) 13(2) [0|10|18]	1027
Someone That I Used To Love (80) 21(2) [0|9|21]	1016
Dangerous (85) 57(2) [0|0|10]	269
A Little Bit Of Heaven (85) 81(1) [0|0|6]	45.2
When I Fall In Love (88) 95(1) [0|0|1]	4.7
Phil COLLINS▶ 5 Pk:1 [10|13|14]	39768
Against All Odds (Take A Look At Me Now) (84) 1(3) [10|16|24]	7041
Another Day In Paradise (89) 1(4) [10|14|18]	5854
One More Night (85) 1(2) [6|12|18]	4427
Sussudio (85) 1(1) [6|14|17]	4040
Two Hearts (89) 1(2) [7|13|18]	3934
Groovy Kind Of Love (88) 1(2) [6|13|25]	3922

Separate Lives (Love Theme From White Nights) (85)
   - Phil Collins and Marilyn Martin [A]
   1(1) [9|16|21]
	2376

Don't Lose My Number (85) 4(1) [4|13|18]	2260
You Can't Hurry Love (83) 10(3) [3|16|21]	1974
Take Me Home (86) 7(3) [5|11|16]	1797
I Missed Again (81) 19(2) [0|9|16]	847
In The Air Tonight (81) 19(2) [0|8|17]	842
I Don't Care Anymore (83) 39(3) [0|3|11]	403
I Cannot Believe It's True (83) 79(1) [0|0|4]	51.6
The COMMODORES▶ 103 Pk:3 [3|5|12]	11046
Oh No (81) 4(3) [4|15|20]	2910
Nightshift (85) 3(1) [6|13|22]	2832
Lady (You Bring Me Up) (81) 8(3) [6|15|22]	2260
Old-Fashion Love (80) 20(2) [0|11|16]	973
Wonderland (80) 25(2) [0|6|15]	713
Only You (83) 54(2) [0|0|13]	369

Act ▶ Rank Peak [Top10s|Top40s|Total] Title (Pk Yr) Peak(Wk) [Top10Wks|Top40Wks|TotalWks]	Score
Animal Instinct (85) 43(1) [0|0|9]	281
Heroes (80) - Commodores 54(2) [0|0|9]	229
Goin' To The Bank (86) - 65(1) [0|0|12]	213
Painted Picture (83) 70(1) [0|0|6]	131
Why You Wanna Try Me (82) 66(2) [0|0|5]	103
Janet (85) 87(1) [0|0|4]	31.3
The COMMUNARDS▶ 706 Pk:40 [0|1|2]	590
Don't Leave Me This Way (87) 40(1) [0|1|13]	353
Never Can Say Goodbye (88) 51(1) [0|0|9]	237
COMPANY B▶ 634 Pk:21 [0|1|1]	763
Fascinated (87) 21(1) [0|8|18]	763
CONDUCTOR▶ 1119 Pk:63 [0|0|1]	107
Voice On The Radio (82) 63(1) [0|0|5]	107
CON FUNK SHUN▶ 807 Pk:40 [0|1|2]	407
Too Tight (81) 40(1) [0|1|10]	350
Baby I'm Hooked (Right Into Your Love) (84) 76(1) [0|0|5]	57.5
Bill CONTI▶ 930 Pk:52 [0|0|1]	237
Theme From Dynasty (82) 52(2) [0|0|9]	237
The CONTOURS▶ 589 Pk:11 [0|1|1]	958
Do You Love Me (88) 11(1) [0|8|16]	958
Tommy CONWELL And The YOUNG RUMBLERS▶ 829 Pk:48 [0|0|2]	366
If We Never Meet Again (89) 48(1) [0|0|11]	284
I'm Not Your Man (88) 74(1) [0|0|7]	81.9
Rita COOLIDGE▶ 496 Pk:36 [0|2|4]	1478
All Time High (83) 36(1) [0|4|13]	549
I'd Rather Leave While I'm In Love (80) 38(1) [0|2|10]	403
Fool That I Am (81) 46(2) [0|0|12]	372
Somethin' 'Bout You Baby I Like (80) - Glen Campbell and Rita Coolidge [A] 42(2) [0|0|10]	153
Alice COOPER▶ 435 Pk:7 [1|2|2]	1849
Poison (89) 7(1) [3|10|19]	1494
Clones (We're All) (80) 40(1) [0|1|9]	355
Julian COPE▶ 1309 Pk:84 [0|0|1]	31.2
World Shut Your Mouth (87) 84(1) [0|0|4]	31.2
Al CORLEY▶ 1255 Pk:80 [0|0|1]	50.3
Square Rooms (85) 80(1) [0|0|5]	50.3
Elvis COSTELLO▶ 693 Pk:19 [0|1|1]	630
Veronica (89) 19(1) [0|6|14]	630
Elvis COSTELLO & THE ATTRACTIONS▶ 635 Pk:36 [0|1|2]	760
Everyday I Write The Book (83) 36(1) [0|2|14]	554
The Only Flame In Town (84) 56(1) [0|0|9]	207
Gene COTTON▶ 1147 Pk:76 [0|0|1]	96.0
If I Could Get You (Into My Life) (82) 76(1) [0|0|8]	96.0
Josie COTTON▶ 1042 Pk:74 [0|0|2]	147
He Could Be The One (82) 74(1) [0|0|7]	108
Jimmy Loves Maryann (84) 82(1) [0|0|4]	39.9
COVER GIRLS▶ 360 Pk:27 [0|3|6]	2666
Because Of You (88) 27(2) [0|8|20]	811
Promise Me (88) 40(1) [0|1|19]	559
Show Me (87) 44(1) [0|0|18]	551
My Heart Skips A Beat (89) 38(1) [0|2|14]	469
Inside Outside (88) 55(2) [0|0|13]	275
Spring Love (87) 98(1) [0|0|1]	2.3
COYOTE SISTERS▶ 973 Pk:66 [0|0|1]	201
Straight From The Heart (Into Your Life) (84) 66(2) [0|0|10]	201
Randy CRAWFORD▶ 1159 Pk:59 [0|0|1]	89.4
Taxi Dancing (84) - Rick Springfield & Randy Crawford [A] 59(1) [0|0|10]	89.4
Robert CRAY Band▶ 648 Pk:22 [0|1|3]	722
Smoking Gun (87) 22(1) [0|6|14]	592
Don't Be Afraid Of The Dark (88) 74(1) [0|0|6]	70.6
Right Next Door (Because Of Me) (87) 80(1) [0|0|6]	59.9
Marshall CRENSHAW▶ 813 Pk:36 [0|1|1]	401
Someday, Someway (82) 36(2) [0|4|11]	401

Act ▶ Rank Peak [Top10s|Top40s|Total] Title (Pk Yr) Peak(Wk) [Top10Wks|Top40Wks|TotalWks]	Score
The CRETONES▶ 1210 Pk:79 [0|0|1]	65.3
Real Love (80) 79(1) [0|0|6]	65.3
CROSBY, STILLS & NASH▶ 315 Pk:9 [1|2|4]	3361
Wasted On The Way (82) 9(4) [5|12|15]	1857
Southern Cross (82) 18(3) [0|9|17]	1037
War Games (83) 45(1) [0|0|9]	331
Too Much Love To Hide (83) 69(3) [0|0|6]	136
CROSBY, STILLS, NASH & YOUNG▶ 1100 Pk:69 [0|0|1]	112
Got It Made (89) 69(1) [0|0|8]	112
Christopher CROSS▶ 34 Pk:1 [4|8|10]	22700
Arthur's Theme (Best That You Can Do) (81) 1(3) [12|17|24]	6437
Ride Like The Wind (80) 2(4) [9|17|21]	5822
Sailing (80) 1(1) [7|13|21]	4791
All Right (83) 12(3) [0|13|16]	1564
Think Of Laura (84) 9(2) [2|11|17]	1548
Never Be The Same (80) 15(3) [0|12|19]	1246
Say You'll Be Mine (81) 20(2) [0|7|14]	722
No Time For Talk (83) 33(1) [0|5|10]	416
Charm The Snake (85) 68(1) [0|0|5]	84.4
A Chance For Heaven (84) 76(1) [0|0|5]	70.0
CROWDED HOUSE▶ 190 Pk:2 [2|2|4]	5839
Don't Dream It's Over (87) 2(1) [7|15|24]	3721
Something So Strong (87) 7(1) [3|11|21]	1699
Better Be Home Soon (88) 42(1) [0|0|11]	286
World Where You Live (87) 65(1) [0|0|8]	133
Rodney CROWELL▶ 815 Pk:37 [0|1|1]	399
Ashes By Now (80) 37(1) [0|2|11]	399
The CRUSADERS▶ 1382 Pk:97 [0|0|1]	9.4
I'm So Glad I'm Standing Here Today (81) 97(2) [0|0|3]	9.4
Billy CRYSTAL▶ 922 Pk:58 [0|0|1]	249
You Look Marvelous (85) 58(1) [0|0|12]	249
CUGINI▶ 1274 Pk:88 [0|0|1]	43.6
Let Me Sleep Alone (80) 88(1) [0|0|4]	43.6
The CULT▶ 830 Pk:46 [0|0|2]	363
Fire Woman (89) 46(2) [0|0|11]	344
Edie (Ciao Baby) (89) 93(2) [0|0|4]	18.8
CULTURE CLUB▶ 24 Pk:1 [6|10|10]	25285
Karma Chameleon (84) 1(3) [9|16|22]	6404
Do You Really Want To Hurt Me (83) 2(3) [9|18|25]	5296
Time (Clock Of The Heart) (83) 2(2) [9|13|18]	4390
Miss Me Blind (84) 5(2) [6|12|16]	2336
Church Of The Poison Mind (83) 10(3) [3|12|17]	1763
I'll Tumble 4 Ya (83) 9(2) [4|12|16]	1713
It's A Miracle (84) 13(2) [0|8|13]	1127
Move Away (86) 12(2) [0|10|14]	1004
The War Song (84) 17(1) [0|7|13]	715
Mistake No. 3 (85) 33(2) [0|5|13]	537
Burton CUMMINGS▶ 837 Pk:37 [0|1|1]	356
You Saved My Soul (81) 37(2) [0|2|11]	356
The CURE▶ 295 Pk:2 [1|2|7]	3699
Love Song (89) 2(1) [4|12|17]	2215
Just Like Heaven (88) 40(1) [0|1|19]	632
Fascination Street (89) 46(2) [0|0|11]	309
Why Can't I Be You? (87) 54(1) [0|0|12]	294
Hot Hot Hot!!! (88) 65(1) [0|0|7]	135
Lullaby (89) 74(1) [0|0|8]	113
In Between Days (Without You) (86) 99(1) [0|0|1]	1.6
CURIOSITY KILLED THE CAT▶ 833 Pk:42 [0|0|1]	358
Misfit (87) 42(1) [0|0|13]	358
CURTIE And The BOOMBOX▶ 1271 Pk:81 [0|0|1]	44.5
Black Kisses (Never Make You Blue) (85) 81(1) [0|0|4]	44.5

Act ▶ Rank Peak [Top10s|Top40s|Total] Title (Pk Yr) Peak(Wk) [Top10Wks|Top40Wks|TotalWks]	Score
CUTTING CREW ▶ 202 Pk:1 [2|3|4]	5514
(I Just) Died In Your Arms (87) 1(2) [6|13|19]	3669
I've Been In Love Before (87) 9(2) [2|11|21]	1383
One For The Mockingbird (87) 38(1) [0|2|11]	395
(Between A) Rock And A Hard Place (89) 77(1) [0|0|5]	67.5

D

Act ▶ Rank Peak [Top10s|Top40s|Total] Title (Pk Yr) Peak(Wk) [Top10Wks|Top40Wks|TotalWks]	Score
E.G. DAILY ▶ 991 Pk:70 [0|0|1]	188
Say It, Say It (86) 70(3) [0|0|10]	188
Roger DALTREY ▶ 447 Pk:20 [0|1|5]	1771
Without Your Love (80) 20(2) [0|8|19]	933
After The Fire (85) 48(1) [0|0|13]	333
Free Me (80) 53(1) [0|0|10]	266
Walking In My Sleep (84) 62(1) [0|0|9]	206
Let Me Down Easy (86) 86(1) [0|0|4]	33.2
Michael DAMIAN ▶ 298 Pk:1 [1|2|3]	3678
Rock On (89) 1(1) [5|13|21]	3129
Cover Of Love (89) 31(1) [0|4|12]	436
She Did It (81) 69(1) [0|0|6]	112
Rodney DANGERFIELD ▶ 1156 Pk:83 [0|0|1]	90.9
Rappin' Rodney (84) 83(1) [0|0|3]	90.9
Charlie DANIELS Band ▶ 373 Pk:11 [0|3|3]	2468
In America (80) 11(2) [0|8|15]	1256
Still In Saigon (82) 22(2) [0|8|12]	668
The Legend Of Wooley Swamp (80) 31(1) [0|4|14]	544
DANNY WILSON ▶ 627 Pk:23 [0|1|1]	782
Mary's Prayer (87) 23(1) [0|8|20]	782
Terence Trent D'ARBY ▶ 173 Pk:1 [2|3|4]	6553
Wishing Well (88) 1(1) [6|15|25]	3764
Sign Your Name (88) 4(1) [5|13|21]	2254
Dance Little Sister (Part One) (88) 30(2) [0|5|11]	411
If You Let Me Stay (87) 68(1) [0|0|8]	124
DAVID & DAVID ▶ 654 Pk:37 [0|1|2]	716
Welcome To The Boomtown (86) 37(1) [0|3|16]	471
Ain't So Easy (87) 51(1) [0|0|11]	245
F.R. DAVID ▶ 968 Pk:62 [0|0|1]	205
Words (83) 62(1) [0|0|9]	205
Jimmy DAVIS & JUNCTION ▶ 1088 Pk:67 [0|0|1]	118
Kick The Wall (87) 67(2) [0|0|6]	118
Mac DAVIS ▶ 685 Pk:43 [0|0|3]	654
It's Hard To Be Humble (80) 43(1) [0|0|12]	323
Texas In My Rear View Mirror (80) 51(2) [0|0|9]	246
Secrets (81) 76(2) [0|0|6]	86.0
Martha DAVIS ▶ 1200 Pk:80 [0|0|1]	69.3
Don't Tell Me The Time (87) 80(1) [0|0|8]	69.3
Paul DAVIS ▶ 215 Pk:6 [1|4|5]	5202
'65 Love Affair (82) 6(2) [7|13|20]	2478
Cool Night (82) 11(2) [0|13|19]	1596
Do Right (80) 23(1) [0|6|14]	704
Love Or Let Me Be Lonely (82) 40(2) [0|2|10]	364
Cry Just A Little (80) 78(1) [0|0|4]	59.9
Tyrone DAVIS ▶ 1046 Pk:57 [0|0|1]	146
Are You Serious (83) 57(2) [0|0|6]	146
Arlan DAY ▶ 1073 Pk:71 [0|0|1]	126
I Surrender (81) 71(2) [0|0|7]	126
Morris DAY ▶ 617 Pk:23 [0|1|2]	809
Fishnet (88) 23(1) [0|6|13]	550
The Oak Tree (85) 65(1) [0|0|12]	259
Taylor DAYNE ▶ 95 Pk:2 [5|5|5]	11970
I'll Always Love You (88) 3(2) [6|16|30]	3037
Don't Rush Me (89) 2(1) [7|13|20]	2909
With Every Beat Of My Heart (89) 5(4) [6|13|18]	2496
Tell It To My Heart (88) 7(1) [4|14|25]	2084
Prove Your Love (88) 7(1) [3|11|18]	1445
DAYTON ▶ 1016 Pk:58 [0|0|1]	170
Hot Fun In The Summertime (82) 58(2) [0|0|7]	170

Act ▶ Rank Peak [Top10s|Top40s|Total] Title (Pk Yr) Peak(Wk) [Top10Wks|Top40Wks|TotalWks]	Score
DAZZ BAND ▶ 323 Pk:5 [1|1|3]	3280
Let It Whip (82) 5(2) [6|16|23]	2985
Joystick (84) 61(1) [0|0|11]	240
Let It All Blow (84) 84(1) [0|0|7]	54.7
DEAD OR ALIVE ▶ 374 Pk:11 [0|2|5]	2461
You Spin Me Round (Like A Record) (85) 11(1) [0|11|18]	1133
Brand New Lover (87) 15(1) [0|9|22]	1062
Come Home With Me Baby (89) 69(1) [0|0|8]	125
Lover Come Back To Me (85) 75(2) [0|0|7]	92.3
Something In My House (87) 85(2) [0|0|6]	48.5
DeBARGE ▶ 133 Pk:3 [2|5|6]	8457
Rhythm Of The Night (85) 3(2) [7|14|22]	3325
Who's Holding Donna Now (85) 6(1) [3|12|19]	1631
Time Will Reveal (84) 18(1) [0|11|21]	1215
All This Love (83) 17(3) [0|10|19]	1188
I Like It (83) 31(2) [0|6|17]	761
Love Me In A Special Way (84) 45(1) [0|0|11]	336
Chico DEBARGE ▶ 583 Pk:21 [0|1|1]	972
Talk To Me (87) 21(1) [0|11|20]	972
El DeBARGE ▶ 349 Pk:3 [1|1|3]	2816
Who's Johnny (Short Circuit Theme) (86) 3(1) [5|13|19]	2341
Love Always (86) 43(1) [0|0|12]	349
Someone (87) 70(1) [0|0|9]	126
El DeBARGE With DeBARGE ▶ 804 Pk:46 [0|0|2]	413
You Wear It Well (85) 46(1) [0|0|10]	319
The Heart Is Not So Smart (86) 75(1) [0|0|7]	94.5
Chris De BURGH ▶ 258 Pk:3 [1|2|4]	4354
The Lady In Red (87) 3(2) [6|14|26]	3242
Don't Pay The Ferryman (83) 34(1) [0|4|14]	569
High On Emotion (84) 44(2) [0|0|13]	444
Ship To Shore (83) 71(1) [0|0|5]	99.1
The DEELE ▶ 509 Pk:10 [1|1|2]	1413
Two Occasions (88) 10(1) [1|12|21]	1285
Body Talk (84) 77(2) [0|0|8]	128
DEEP PURPLE ▶ 1021 Pk:61 [0|0|1]	164
Knocking At Your Back Door (85) 61(2) [0|0|7]	164
Rick DEES ▶ 1178 Pk:75 [0|0|1]	79.3
Eat My Shorts (84) 75(2) [0|0|5]	79.3
DEF LEPPARD ▶ 67 Pk:1 [4|9|11]	15814
Love Bites (88) 1(1) [6|13|23]	3728
Pour Some Sugar On Me (88) 2(1) [7|15|24]	3419
Armageddon It (89) 3(2) [5|12|18]	2569
Photograph (83) 12(1) [0|9|17]	1309
Hysteria (88) 10(1) [1|10|16]	1121
Rock Of Ages (83) 16(1) [0|9|15]	1015
Rocket (89) 12(1) [0|9|13]	918
Animal (87) 19(2) [0|9|19]	890
Foolin' (83) 28(1) [0|5|14]	602
Bringin' On The Heartbreak (84) 61(2) [0|0|8]	192
Women (87) 80(1) [0|0|5]	51.9
DEJA ▶ 898 Pk:54 [0|0|1]	275
You And Me Tonight (87) 54(2) [0|0|12]	275
DE LA SOUL ▶ 738 Pk:34 [0|1|1]	528
Me Myself And I (89) 34(1) [0|3|17]	528
DEL FUEGOS ▶ 1321 Pk:87 [0|0|1]	28.2
I Still Want You (86) 87(2) [0|0|4]	28.2
DELIVERANCE ▶ 1145 Pk:71 [0|0|1]	97.7
Leaving L.A. (80) 71(2) [0|0|5]	97.7
John DENVER ▶ 445 Pk:31 [0|2|8]	1796
Some Days Are Diamonds (Some Days Are Stone) (81) 36(1) [0|4|20]	614
Shanghai Breezes (82) 31(1) [0|5|14]	541
Autograph (80) 52(2) [0|0|10]	338
The Cowboy And The Lady (81) 66(2) [0|0|8]	137
Perhaps Love (82) - Placido Domingo And John Denver [A] 59(1) [0|0|7]	74.8

Act ▶ Rank Peak [Top10s|Top40s|Total] Title (Pk Yr) Peak(Wk) [Top10Wks|Top40Wks|TotalWks]	Score
Seasons Of The Heart (82) 78(1) [0|0|5]	62.0
Love Again (84) - John Denver & Sylvie Vartan [A] 85(1) [0|0|4]	19.4
Dancing With The Mountains (80) 97(2) [0|0|3]	8.9
DEODATO ▶ 1162 Pk:70 [0|0|1]	88.8
Happy Hour (82) 70(2) [0|0|5]	88.8
DEPECHE MODE ▶ 427 Pk:13 [0|1|6]	1891
People Are People (85) 13(2) [0|10|18]	1163
Strangelove (88) 50(1) [0|0|9]	236
Route 66/Behind The Wheel (88) 61(2) [0|0|11]	215
Never Let Me Down Again (88) 63(1) [0|0|10]	168
Strangelove (87) 76(1) [0|0|6]	81.2
Master And Servant (85) 87(2) [0|0|3]	29.3
Teri DeSARIO ▶ 252 Pk:2 [1|1|2]	4416
Yes, I'm Ready (80) 2(2) [7|16|23]	4354
Dancin' In The Streets (80) - Teri DeSario with K.C. [A] 66(2) [0|0|6]	62.2
Jackie DeSHANNON ▶ 1259 Pk:86 [0|0|1]	49.1
I Don't Need You Anymore (80) - Together? Feat. Jackie DeShannon 86(1) [0|0|5]	49.1
DEVICE ▶ 718 Pk:35 [0|1|2]	566
Hanging On A Heart Attack (86) 35(3) [0|4|14]	502
Who Says (86) 79(1) [0|0|6]	64.4
--DeVILLE, Mink see MINK DeVILLE	
DEVO ▶ 383 Pk:14 [0|1|3]	2346
Whip It (80) 14(3) [0|15|25]	1837
Working In The Coal Mine (81) 43(1) [0|0|12]	352
Theme From "Doctor Detroit" (83) 59(1) [0|0|6]	157
DEXYS MIDNIGHT RUNNERS ▶ 230 Pk:1 [1|1|2]	4842
Come On Eileen (83) 1(1) [6|14|23]	4804
The Celtic Soul Brothers (83) 86(1) [0|0|4]	37.6
Dennis DeYOUNG ▶ 440 Pk:10 [1|1|4]	1818
Desert Moon (84) 10(1) [1|12|22]	1501
Call Me (86) 54(1) [0|0|11]	273
Don't Wait For Heroes (84) 83(1) [0|0|4]	30.9
This Is The Time (86) 93(1) [0|0|3]	13.6
Joel DIAMOND ▶ 1296 Pk:82 [0|0|1]	35.4
Theme From "Raging Bull" (Cavalleria Rusticana) (81) 82(1) [0|0|3]	35.4
Neil DIAMOND ▶ 62 Pk:2 [4|9|13]	16701
Love On The Rocks (81) 2(3) [10|17|20]	5450
Heartlight (82) 5(4) [6|11|19]	2634
Hello Again (81) 6(2) [5|12|16]	2040
America (81) 8(3) [4|13|17]	1871
Yesterday's Songs (81) 11(2) [0|12|15]	1494
September Morn' (80) 17(1) [0|10|16]	1159
I'm Alive (83) 35(4) [0|4|12]	479
Be Mine Tonight (82) 35(2) [0|4|11]	450
On The Way To The Sky (82) 27(2) [0|5|10]	443
Headed For The Future (86) 53(2) [0|0|10]	251
Front Page Story (83) 65(1) [0|0|8]	168
Turn Around (84) 62(1) [0|0|8]	146
The Good Lord Loves You (80) 67(1) [0|0|6]	115
DIESEL ▶ 651 Pk:25 [0|1|1]	719
Sausalito Summernight (81) 25(1) [0|6|18]	719
DILLMAN Band ▶ 915 Pk:45 [0|0|1]	253
Lovin' The Night Away (81) 45(1) [0|0|9]	253
DINO ▶ 312 Pk:7 [1|2|4]	3402
I Like It (89) 7(2) [4|14|25]	2013
Sunshine (89) 23(1) [0|6|15]	626
24/7 (89) 42(2) [0|0|14]	434
Summergirls (88) 50(3) [0|0|12]	329
DION ▶ 1238 Pk:75 [0|0|1]	57.6
And The Night Stood Still (89) 75(1) [0|0|5]	57.6
DIRE STRAITS ▶ 148 Pk:1 [2|3|5]	7818
Money For Nothing (85) 1(3) [8|13|22]	4894
Walk Of Life (86) 7(1) [4|15|21]	1949
So Far Away (86) 19(1) [0|7|14]	703

Acts With Singles

Act ▶ Rank Peak [Top10s|Top40s|Total] Title (Pk Yr) Peak(Wk) [Top10Wks|Top40Wks|TotalWks]	Score
DIRE STRAITS ▶ *Continued*	
Skateaway (81) 58(1) [0|0|10]	211
Industrial Disease (83) 75(1) [0|0|4]	60.5
DIVING FOR PEARLS ▶ 1239 Pk:84 [0|0|1]	**57.3**
Gimme Your Good Lovin' (89) 84(2) [0|0|6]	57.3
DIVINYLS ▶ 1105 Pk:76 [0|0|1]	**111**
Pleasure And Pain (86) 76(1) [0|0|7]	111
DJ JAZZY JEFF & THE FRESH PRINCE ▶ 371 Pk:12 [0|2|4]	**2496**
Parents Just Don't Understand (88) 12(2) [0|10|19]	1188
A Nightmare On My Street (88) 15(1) [0|9|16]	878
Girls Ain't Nothing But Trouble (88) 57(1) [0|0|12]	272
I Think I Can Beat Mike Tyson (89) 58(1) [0|0|9]	158
DOCTOR And The MEDICS ▶ 976 Pk:69 [0|0|1]	**195**
Spirit In The Sky (86) 69(2) [0|0|11]	195
DR. HOOK ▶ 179 Pk:5 [1|4|7]	**6424**
Sexy Eyes (80) 5(2) [6|15|21]	3047
Better Love Next Time (80) 12(2) [0|14|19]	1704
Baby Makes Her Blue Jeans Talk (82) 25(3) [0|6|12]	592
Girls Can Get It (80) 34(2) [0|6|14]	579
Years From Now (80) 51(2) [0|0|9]	232
Loveline (82) 60(2) [0|0|10]	197
That Didn't Hurt Too Bad (81) 69(1) [0|0|4]	72.9
DOKKEN ▶ 808 Pk:64 [0|0|3]	**404**
Alone Again (85) 64(2) [0|0|11]	213
Burning Like A Flame (88) 72(1) [0|0|8]	114
In My Dreams (86) 77(1) [0|0|7]	77.3
Thomas DOLBY ▶ 300 Pk:5 [1|1|3]	**3635**
She Blinded Me With Science (83) 5(4) [8|15|22]	3379
Hyperactive (84) 62(1) [0|0|7]	154
Europa And The Pirate Twins (83) 67(1) [0|0|5]	102
Joe DOLCE ▶ 846 Pk:53 [0|0|1]	**343**
Shaddap You Face (81) 53(1) [0|0|14]	343
Micky DOLENZ And Peter TORK ▶ 677 Pk:20 [0|1|1]	**667**
That Was Then, This Is Now (86) - Micky Dolenz And Peter Tork (Of The Monkees) 20(1) [0|7|14]	667
DOLLAR ▶ 1109 Pk:74 [0|0|1]	**110**
Shooting Star (80) 74(1) [0|0|6]	110
Placido DOMINGO ▶ 1190 Pk:59 [0|0|1]	**74.8**
Perhaps Love (82) - Placido Domingo And John Denver [A] 59(1) [0|0|7]	74.8
DOOBIE BROTHERS ▶ 245 Pk:5 [2|3|8]	**4515**
Real Love (80) 5(2) [5|11|16]	2173
The Doctor (89) 9(1) [1|9|14]	1079
One Step Closer (81) 24(2) [0|7|14]	704
Need A Little Taste Of Love (89) 45(1) [0|0|9]	239
Here To Love You (82) 65(2) [0|0|5]	112
Keep This Train A-Rollin' (81) 62(1) [0|0|5]	109
Wynken Blynken And Nod (81) 76(1) [0|0|4]	54.2
You Belong To Me (83) 79(1) [0|0|4]	45.5
DOOLITTLE Band ▶ 966 Pk:49 [0|0|1]	**206**
Who Were You Thinkin' Of (80) 49(2) [0|0|11]	206
The DOORS ▶ 1058 Pk:71 [0|0|1]	**137**
Gloria (84) 71(1) [0|0|7]	137
Charlie DORE ▶ 510 Pk:13 [1|1|1]	**1411**
Pilot Of The Airwaves (80) 13(2) [0|10|17]	1411
DOUBLE ▶ 578 Pk:16 [0|1|1]	**1006**
The Captain Of Her Heart (86) 16(2) [0|9|18]	1006
DOUBLE IMAGE ▶ 1360 Pk:92 [0|0|1]	**18.0**
Night Pulse (83) 92(1) [0|0|3]	18.0
DRAGON ▶ 1302 Pk:88 [0|0|1]	**33.1**
Rain (84) 88(2) [0|0|4]	33.1
DREAM ACADEMY ▶ 413 Pk:7 [1|2|2]	**2049**
Life In A Northern Town (86) 7(2) [3|11|21]	1645
The Love Parade (86) 36(2) [0|3|11]	404

Act ▶ Rank Peak [Top10s|Top40s|Total] Title (Pk Yr) Peak(Wk) [Top10Wks|Top40Wks|TotalWks]	Score
J.D. DREWS ▶ 1186 Pk:79 [0|0|1]	**75.6**
Don't Want No-Body (81) 79(1) [0|0|6]	75.6
"D" TRAIN ▶ 1193 Pk:79 [0|0|1]	**73.4**
Something's On Your Mind (84) 79(1) [0|0|6]	73.4
George DUKE ▶ 625 Pk:19 [0|1|2]	**787**
Sweet Baby (81) - Stanley Clarke/George Duke [A] 19(2) [0|9|20]	502
Shine On (82) 41(1) [0|0|9]	285
DUKE JUPITER ▶ 873 Pk:58 [0|0|2]	**301**
I'll Drink To You (82) 58(1) [0|0|7]	166
Little Lady (84) 68(2) [0|0|7]	135
Robbie DUPREE ▶ 242 Pk:6 [1|2|3]	**4569**
Steal Away (80) 6(2) [7|15|23]	2990
Hot Rod Hearts (80) 15(2) [0|12|18]	1426
Brooklyn Girls (81) 54(1) [0|0|7]	153
DURAN DURAN ▶ 7 Pk:1 [9|13|15]	**35594**
The Reflex (84) 1(2) [8|15|21]	5085
The Wild Boys (84) 2(4) [8|14|18]	4921
Hungry Like The Wolf (83) 3(3) [9|16|23]	4539
A View To A Kill (85) 1(2) [6|13|17]	4298
Union Of The Snake (83) 3(3) [6|12|17]	3910
Notorious (87) 2(1) [6|13|17]	3641
Is There Something I Should Know (83) 4(1) [6|12|17]	2818
I Don't Want Your Love (88) 4(2) [4|13|16]	2212
New Moon On Monday (84) 10(1) [1|10|16]	1387
Rio (83) 14(2) [0|9|13]	963
Save A Prayer (85) 16(2) [0|8|14]	820
All She Wants Is (89) 22(1) [0|6|13]	532
Skin Trade (87) 39(1) [0|1|9]	315
Meet El Presidente (87) 70(1) [0|0|5]	83.9
Do You Believe In Shame? (89) 72(1) [0|0|5]	69.6
Bob DYLAN ▶ 970 Pk:55 [0|0|1]	**204**
Sweetheart Like You (84) 55(1) [0|0|9]	204
DYNASTY ▶ 1258 Pk:87 [0|0|1]	**49.1**
I've Just Begun To Love You (80) 87(1) [0|0|6]	49.1

E

Act ▶ Rank Peak [Top10s|Top40s|Total] Title (Pk Yr) Peak(Wk) [Top10Wks|Top40Wks|TotalWks]	Score
EAGLES ▶ 233 Pk:8 [2|3|3]	**4797**
I Can't Tell You Why (80) 8(3) [4|12|16]	2108
The Long Run (80) 8(2) [3|12|15]	1975
Seven Bridges Road (81) 21(2) [0|7|14]	714
EARTH, WIND & FIRE ▶ 152 Pk:3 [1|2|11]	**7717**
Let's Groove (81) 3(5) [9|16|24]	4813
Fall In Love With Me (83) 17(3) [0|10|16]	1096
You (80) 48(3) [0|0|12]	364
Let Me Talk (80) 44(2) [0|0|9]	322
System Of Survival (87) 60(1) [0|0|13]	254
Wanna Be With You (82) 51(2) [0|0|6]	182
Magnetic (83) 57(2) [0|0|9]	182
And Love Goes On (81) 59(2) [0|0|7]	166
Thinking Of You (88) 67(2) [0|0|8]	146
Star (80) 64(1) [0|0|6]	137
Side By Side (83) 76(1) [0|0|4]	56.0
EASTERHOUSE ▶ 1280 Pk:82 [0|0|1]	**41.4**
Come Out Fighting (89) 82(1) [0|0|4]	41.4
Sheena EASTON ▶ 33 Pk:1 [7|12|18]	**22745**
Morning Train (Nine To Five) (81) 1(2) [6|15|21]	4553
For Your Eyes Only (81) 4(4) [5|14|25]	3209
The Lover In Me (89) 2(1) [5|14|25]	2769
Strut (84) 7(1) [3|15|25]	2111
Telefone (Long Distance Love Affair) (83) 9(2) [3|14|22]	2099
Sugar Walls (85) 9(2) [3|9|17]	1509
We've Got Tonight (83) - Kenny Rogers And Sheena Easton [A] 6(3) [7|15|18]	1316
You Could Have Been With Me (82) 15(2) [0|12|18]	1280

Act ▶ Rank Peak [Top10s|Top40s|Total] Title (Pk Yr) Peak(Wk) [Top10Wks|Top40Wks|TotalWks]	Score
Modern Girl (81) 18(2) [0|9|18]	971
Almost Over You (84) 25(2) [0|6|20]	889
When He Shines (82) 30(2) [0|6|15]	572
Do It For Love (85) 29(1) [0|4|14]	550
So Far So Good (86) 43(1) [0|0|12]	358
Machinery (82) 57(2) [0|0|7]	179
I Wouldn't Beg For Water (82) 64(3) [0|0|12]	149
Jimmy Mack (86) 65(2) [0|0|6]	113
Swear (85) 80(2) [0|0|6]	75.2
Devil In A Fast Car (84) 79(1) [0|0|3]	43.2
Clint EASTWOOD ▶ 1208 Pk:62 [0|0|1]	**65.8**
Make My Day (84) - T.G. Sheppard with Clint Eastwood [A] 62(1) [0|0|6]	65.8
John EDDIE ▶ 902 Pk:52 [0|0|1]	**268**
Jungle Boy (86) 52(3) [0|0|10]	268
EDDIE And The TIDE ▶ 1348 Pk:85 [0|0|1]	**21.7**
One In A Million (85) 85(1) [0|0|2]	21.7
Dave EDMUNDS ▶ 641 Pk:39 [0|1|3]	**745**
Slipping Away (83) 39(1) [0|1|15]	536
Almost Saturday Night (81) 54(2) [0|0|8]	193
High School Nights (85) 91(2) [0|0|2]	15.9
Dennis EDWARDS ▶ 1143 Pk:72 [0|0|1]	**98.6**
Don't Look Any Further (84) - Dennis Edwards (Featuring Siedah Garrett) 72(1) [0|0|6]	98.6
Walter EGAN ▶ 849 Pk:46 [0|0|1]	**338**
Fool Moon Fire (83) 46(1) [0|0|10]	338
EIGHTH WONDER ▶ 812 Pk:56 [0|0|2]	**401**
Cross My Heart (89) 56(2) [0|0|16]	358
Baby Baby (89) 84(1) [0|0|5]	43.7
EIGHT SECONDS ▶ 1071 Pk:72 [0|0|1]	**127**
Kiss You (When It's Dangerous) (87) 72(1) [0|0|8]	127
ELECTRIC LIGHT ORCHESTRA ▶ 168 Pk:8 [2|8|9]	**6914**
Hold On Tight (81) - ELO 10(2) [2|13|19]	1611
All Over The World (80) 13(2) [0|9|16]	1205
I'm Alive (80) 16(2) [0|8|15]	926
Rock 'N' Roll Is King (83) - ELO 19(2) [0|9|13]	901
Xanadu (80) - Olivia Newton-John/ Electric Light Orchestra [A] 8(2) [3|10|17]	834
Calling America (86) 18(1) [0|7|15]	687
Last Train To London (80) 39(1) [0|2|11]	408
Twilight (81) - ELO 38(2) [0|2|11]	316
Four Little Diamonds (83) - ELO 86(1) [0|0|5]	24.3
Larry ELGART And His MANHATTAN SWING Orchestra ▶ 780 Pk:31 [0|1|1]	**456**
Hooked On Swing (82) 31(2) [0|5|12]	456
EMERSON, LAKE & POWELL ▶ 1027 Pk:60 [0|0|1]	**159**
Touch And Go (86) 60(1) [0|0|8]	159
ENGLAND DAN & John Ford COLEY ▶ 1199 Pk:75 [0|0|1]	**69.5**
In It For Love (80) 75(1) [0|0|4]	69.5
Jackie ENGLISH ▶ 1362 Pk:94 [0|0|1]	**17.6**
Once A Night (81) 94(1) [0|0|4]	17.6
ENUFF Z'NUFF ▶ 1068 Pk:67 [0|0|1]	**128**
New Thing (89) 67(1) [0|0|7]	128
ENYA ▶ 665 Pk:24 [0|1|1]	**683**
Orinoco Flow (89) 24(1) [0|8|17]	683
ERASURE ▶ 403 Pk:12 [0|2|3]	**2123**
Chains Of Love (88) 12(1) [0|11|20]	1118
A Little Respect (89) 14(1) [0|9|17]	1000
Stop! (89) 97(1) [0|0|5]	4.6
ESCAPE CLUB ▶ 234 Pk:1 [1|2|3]	**4794**
Wild, Wild West (88) 1(1) [7|16|27]	4208
Shake For The Sheik (89) 28(3) [0|5|14]	520
Walking Through Walls (89) 81(1) [0|0|6]	66.1
Joe "Bean" ESPOSITO ▶ 1346 Pk:86 [0|0|1]	**22.5**
Lady, Lady, Lady (83) 86(1) [0|0|2]	22.5

Acts With Singles

Act ▶ Rank Peak [Top10s|Top40s|Total] Title (Pk Yr) Peak(Wk) [Top10Wks|Top40Wks|TotalWks]	Score
Gloria ESTEFAN/MIAMI SOUND MACHINE ▶ 44 Pk:1 [8|11|11]	**20947**
Anything For You (88) - Gloria Estefan and Miami Sound Machine 1(2) [7|14|23]	3981
Don't Wanna Lose You (89) - Gloria Estefan 1(1) [8|13|18]	3701
1-2-3 (88) - Gloria Estefan and Miami Sound Machine 3(1) [5|13|19]	2215
Words Get In The Way (86) - Miami Sound Machine 5(1) [3|13|24]	1973
Rhythm Is Gonna Get You (87) - Gloria Estefan and Miami Sound Machine 5(1) [5|12|17]	1904
Conga (86) - Miami Sound Machine 10(2) [2|16|27]	1829
Bad Boy (86) - Miami Sound Machine 8(3) [3|12|19]	1695
Can't Stay Away From You (88) - Gloria Estefan and Miami Sound Machine 6(1) [3|13|23]	1570
Get On Your Feet (89) - Gloria Estefan 11(1) [0|8|17]	1027
Falling In Love (Uh-Oh) (87) - Miami Sound Machine 25(1) [0|8|16]	697
Betcha Say That (87) - Gloria Estefan and Miami Sound Machine 36(2) [0|2|11]	356
Deon ESTUS ▶ 482 Pk:5 [1|1|1]	**1539**
Heaven Help Me (89) 5(1) [2|11|16]	1539
Melissa ETHERIDGE ▶ 1319 Pk:94 [0|0|2]	**28.6**
Similar Features (89) 94(3) [0|0|4]	20.1
No Souvenirs (89) 95(1) [0|0|3]	8.5
E.U. ▶ 802 Pk:35 [0|1|1]	**415**
Da' Butt (88) 35(2) [0|4|12]	415
EUROGLIDERS ▶ 1045 Pk:65 [0|0|1]	**146**
Heaven (Must Be There) (84) 65(1) [0|0|6]	146
EUROPE ▶ 225 Pk:3 [2|4|5]	**4908**
Carrie (87) 3(2) [5|12|19]	2472
The Final Countdown (87) 8(2) [2|9|18]	1387
Rock The Night (87) 30(1) [0|4|13]	477
Superstitious (88) 31(1) [0|4|13]	455
Cherokee (87) 72(1) [0|0|10]	118
EURYTHMICS ▶ 64 Pk:1 [3|10|15]	**16559**
Sweet Dreams (Are Made Of This) (83) 1(1) [9|17|26]	6665
Here Comes The Rain Again (84) 4(2) [7|14|20]	3206
Would I Lie To You? (85) 5(1) [4|13|19]	1926
Missionary Man (86) 14(1) [0|9|16]	1031
Who's That Girl? (84) 21(2) [0|7|13]	745
Love Is A Stranger (83) 23(1) [0|6|13]	646
There Must Be An Angel (Playing With My Heart) (85) 22(1) [0|7|11]	554
Right By Your Side (84) 29(1) [0|5|12]	518
Sisters Are Doin' It For Themselves (85) - Eurythmics And Aretha Franklin [A] 18(1) [0|8|15]	380
I Need A Man (88) 46(1) [0|0|10]	258
Don't Ask Me Why (89) 40(1) [0|1|9]	257
You Have Placed A Chill In My Heart (88) 64(1) [0|0|7]	136
Thorn In My Side (86) 68(2) [0|0|9]	120
It's Alright (Baby's Coming Back) (86) 78(1) [0|0|6]	70.4
Sexcrime (Nineteen Eighty-Four) (84) 81(1) [0|0|4]	46.1
EVERLY BROTHERS ▶ 883 Pk:50 [0|0|1]	**291**
On The Wings Of A Nightingale (84) 50(1) [0|0|12]	291
EXPOSÉ ▶ 96 Pk:1 [6|6|6]	**11741**
Seasons Change (88) 1(1) [6|16|20]	3558
Come Go With Me (87) 5(1) [5|12|19]	2047
Point Of No Return (87) 5(1) [4|11|17]	1791
Let Me Be The One (87) 7(1) [3|13|22]	1707

Act ▶ Rank Peak [Top10s|Top40s|Total] Title (Pk Yr) Peak(Wk) [Top10Wks|Top40Wks|TotalWks]	Score
What You Don't Know (89) 8(2) [3|11|15]	1384
When I Looked At Him (89) 10(1) [1|12|20]	1254
EYE TO EYE ▶ 743 Pk:37 [0|1|2]	**519**
Nice Girls (82) 37(2) [0|3|13]	497
Lucky (83) 88(1) [0|0|2]	21.6

F

Act ▶ Rank Peak [Top10s|Top40s|Total] Title (Pk Yr) Peak(Wk) [Top10Wks|Top40Wks|TotalWks]	Score
FABULOUS THUNDERBIRDS ▶ 463 Pk:10 [1|1|4]	**1676**
Tuff Enuff (86) 10(1) [1|10|19]	1204
Wrap It Up (86) 50(2) [0|0|10]	277
Powerful Stuff (88) 65(1) [0|0|7]	137
Stand Back (87) 76(1) [0|0|5]	58.8
FACE TO FACE ▶ 740 Pk:38 [0|1|1]	**521**
10-9-8 (84) 38(3) [0|3|15]	521
Erla FACHIN ▶ 932 Pk:50 [0|0|1]	**236**
Savin' Myself (88) 50(1) [0|0|10]	236
Donald FAGEN ▶ 629 Pk:26 [0|1|3]	**776**
I.G.Y. (What A Beautiful World) (82) 26(3) [0|7|14]	626
New Frontier (83) 70(1) [0|0|6]	104
Century's End (88) 83(2) [0|0|5]	46.1
Joe FAGIN ▶ 1272 Pk:80 [0|0|1]	**44.4**
Younger Days (82) 80(2) [0|0|3]	44.4
FAIRGROUND ATTRACTION ▶ 1232 Pk:80 [0|0|1]	**59.7**
Perfect (89) 80(1) [0|0|6]	59.7
FALCO ▶ 221 Pk:1 [1|2|2]	**5027**
Rock Me Amadeus (86) 1(3) [7|13|17]	4285
Vienna Calling (The New '86 Edit) (86) 18(1) [0|8|14]	742
Harold FALTERMEYER ▶ 331 Pk:3 [1|1|1]	**3024**
Axel F (85) 3(3) [5|12|19]	3024
Agnetha FALTSKOG ▶ 692 Pk:29 [0|1|2]	**632**
Can't Shake Loose (83) 29(1) [0|5|15]	626
I Wasn't The One (Who Said Goodbye) (88) - Agnetha Faltskog and Peter Cetera [A] 93(1) [0|0|3]	6.1
The FAMILY ▶ 1063 Pk:63 [0|0|1]	**132**
The Screams Of Passion (85) 63(1) [0|0|6]	132
FAR CORPORATION ▶ 1311 Pk:89 [0|0|1]	**30.7**
Stairway To Heaven (86) 89(1) [0|0|4]	30.7
Cee FARROW ▶ 1213 Pk:82 [0|0|1]	**64.5**
Should I Love You (83) 82(1) [0|0|6]	64.5
FAT BOYS ▶ 503 Pk:12 [0|2|3]	**1430**
The Twist (Yo, Twist!) (88) 16(2) [0|8|15]	781
Wipeout (87) - Fat Boys and The Beach Boys [A] 12(2) [0|11|19]	630
Louie, Louie (88) 89(1) [0|0|3]	19.2
Don FELDER ▶ 731 Pk:43 [0|0|1]	**539**
Heavy Metal (Takin' A Ride) (81) 43(1) [0|0|17]	539
Suzanne FELLINI ▶ 1335 Pk:87 [0|0|1]	**24.0**
Love On The Phone (80) 87(1) [0|0|2]	24.0
FELONY ▶ 822 Pk:42 [0|0|1]	**375**
The Fanatic (83) 42(2) [0|0|12]	375
Bryan FERRY ▶ 796 Pk:31 [0|1|1]	**427**
Kiss And Tell (88) 31(1) [0|3|13]	427
FESTIVAL ▶ 1107 Pk:72 [0|0|1]	**110**
Don't Cry For Me Argentina (80) 72(1) [0|0|8]	110
Richard "Dimples" FIELDS ▶ 858 Pk:47 [0|0|1]	**330**
If It Ain't One Thing... It's Another (82) 47(3) [0|0|10]	330
FIGURES ON A BEACH ▶ 1120 Pk:67 [0|0|1]	**107**
You Ain't Seen Nothing Yet (89) 67(1) [0|0|7]	107
FINE YOUNG CANNIBALS ▶ 141 Pk:1 [2|3|5]	**8080**
She Drives Me Crazy (89) 1(1) [7|14|23]	3641
Good Thing (89) 1(1) [6|13|17]	3365
Don't Look Back (89) 11(1) [0|8|12]	884

Act ▶ Rank Peak [Top10s|Top40s|Total] Title (Pk Yr) Peak(Wk) [Top10Wks|Top40Wks|TotalWks]	Score
I'm Not The Man I Used To Be (89) 54(1) [0|0|6]	125
Johnny Come Home (86) 76(1) [0|0|5]	66.1
FIONA ▶ 1062 Pk:64 [0|0|1]	**132**
Talk To Me (85) 64(1) [0|0|7]	132
Elisa FIORILLO ▶ 759 Pk:49 [0|0|2]	**488**
Forgive Me For Dreaming (88) 49(1) [0|0|12]	326
How Can I Forget You (88) 60(1) [0|0|8]	163
FIREFALL ▶ 531 Pk:35 [0|2|4]	**1286**
Headed For A Fall (80) 35(2) [0|3|9]	366
Staying With It (81) 37(2) [0|3|9]	337
Always (83) 59(5) [0|0|13]	320
Love That Got Away (80) 50(2) [0|0|9]	263
FIRE INC. ▶ 1229 Pk:80 [0|0|1]	**60.2**
Tonight Is What It Means To Be Young (84) 80(2) [0|0|5]	60.2
The FIRM ▶ 613 Pk:28 [0|1|3]	**829**
Radioactive (85) 28(2) [0|6|15]	590
All The Kings Horses (86) 61(2) [0|0|8]	164
Satisfaction Guaranteed (85) 73(2) [0|0|5]	75.4
FIVE SATINS ▶ 1132 Pk:71 [0|0|1]	**103**
Memories Of Days Gone By (Medley) (82) - Fred Parris & The Five Satins 71(1) [0|0|5]	103
FIVE STAR ▶ 577 Pk:41 [0|0|4]	**1009**
Can't Wait Another Minute (86) 41(2) [0|0|14]	430
All Fall Down (85) 65(3) [0|0|11]	220
Let Me Be The One (86) 59(2) [0|0|9]	193
If I Say Yes (87) 67(2) [0|0|11]	165
The FIXX ▶ 175 Pk:4 [1|5|8]	**6528**
One Thing Leads To Another (83) 4(1) [7|13|19]	2996
Saved By Zero (83) 20(1) [0|8|16]	941
Are We Ourselves? (84) 15(1) [0|8|15]	905
Secret Separation (86) 19(1) [0|6|14]	661
The Sign Of Fire (84) 32(2) [0|7|13]	586
Driven Out (89) 55(1) [0|0|10]	248
Stand Or Fall (82) 76(1) [0|0|8]	104
Sunshine In The Shade (84) 69(1) [0|0|5]	86.9
Roberta FLACK ▶ 394 Pk:13 [0|1|4]	**2251**
Making Love (82) 13(3) [0|11|21]	1361
I'm The One (82) 42(1) [0|0|11]	377
You Are My Heaven (80) - Roberta Flack With Donny Hathaway 47(1) [0|0|11]	292
Back Together Again (80) - Roberta Flack with Donny Hathaway 56(2) [0|0|8]	221
FLEETWOOD MAC ▶ 73 Pk:4 [4|9|13]	**15297**
Hold Me (82) 4(7) [10|15|17]	4397
Little Lies (87) 4(2) [5|13|21]	2284
Sara (80) 7(3) [5|11|14]	1920
Big Love (87) 5(1) [5|11|16]	1847
Gypsy (82) 12(3) [0|8|14]	1065
Everywhere (88) 14(2) [0|10|18]	1045
Think About Me (80) 20(2) [0|7|13]	759
Love In Store (83) 22(3) [0|8|14]	711
Seven Wonders (87) 19(1) [0|8|13]	704
As Long As You Follow (89) 43(1) [0|0|14]	389
Fireflies (81) 60(1) [0|0|6]	123
Sisters Of The Moon (80) 86(1) [0|0|3]	32.4
Family Man (88) 90(1) [0|0|4]	20.1
A FLOCK OF SEAGULLS ▶ 319 Pk:9 [1|3|4]	**3307**
I Ran (So Far Away) (82) 9(2) [4|10|22]	1647
Space Age Love Song (83) 30(2) [0|7|18]	740
Wishing (If I Had A Photograph Of You) (83) 26(2) [0|7|14]	709
The More You Live, The More You Love (84) 56(2) [0|0|9]	211
FLYING LIZARDS ▶ 889 Pk:50 [0|0|1]	**284**
Money (That's What I Want) (80) 50(1) [0|0|10]	284
Dan FOGELBERG ▶ 79 Pk:2 [4|9|12]	**14142**
Longer (80) 2(2) [7|13|22]	4243
Leader Of The Band (82) 9(2) [4|16|20]	2045
Hard To Say (81) 7(2) [4|10|19]	1758

Acts With Singles

Act ▶ Rank Peak [Top10s|Top40s|Total] Title (Pk Yr) Peak(Wk) [Top10Wks|Top40Wks|TotalWks]	Score
Dan FOGELBERG ▶ Continued	
Same Old Lang Syne (81) 9(2) [2|13|18]	1561
The Language Of Love (84) 13(1) [0|10|14]	1149
Missing You (82) 23(5) [0|9|16]	860
Run For The Roses (82) 18(1) [0|8|14]	840
Heart Hotels (80) 21(1) [0|6|13]	677
Make Love Stay (83) 29(3) [0|6|16]	633
Believe In Me (84) 48(1) [0|0|9]	277
She Don't Look Back (87) 84(1) [0|0|6]	60.6
Go Down Easy (85) 85(1) [0|0|4]	37.6
John FOGERTY ▶ 381 Pk:10 [1|2|4]	**2367**
The Old Man Down The Road (85) 10(1) [1|9|18]	1313
Rock And Roll Girls (85) 20(1) [0|6|12]	632
Centerfield (85) 44(2) [0|0|13]	377
Eye Of The Zombie (86) 81(1) [0|0|4]	44.8
FOGHAT ▶ 586 Pk:23 [0|1|2]	**971**
Third Time Lucky (First Time I Was A Fool) (80) 23(2) [0|10|15]	938
Stranger In My Home Town (80) 81(1) [0|0|3]	32.3
The FOOLS ▶ 886 Pk:50 [0|0|2]	**287**
Running Scared (81) 50(2) [0|0|7]	195
It's A Night For Beautiful Girls (80) 67(1) [0|0|4]	92.9
Steve FORBERT ▶ 485 Pk:11 [0|1|2]	**1527**
Romeo's Tune (80) 11(2) [0|12|19]	1497
Say Goodbye To Little Jo (80) 85(1) [0|0|3]	29.2
FORCE M.D.'S ▶ 508 Pk:10 [1|1|2]	**1421**
Tender Love (86) 10(2) [2|11|19]	1312
Love Is A House (87) 78(1) [0|0|9]	109
Lita FORD ▶ 336 Pk:8 [1|2|2]	**2928**
Close My Eyes Forever (remix) (89) - Lita Ford (duet with Ozzy Osbourne) 8(1) [4|12|25]	1716
Kiss Me Deadly (88) 12(2) [0|10|23]	1212
FOREIGNER ▶ 26 Pk:1 [5|8|13]	**24892**
Waiting For A Girl Like You (81) 2(10) [15|19|23]	9492
I Want To Know What Love Is (85) 1(2) [8|16|21]	5285
Urgent (81) 4(4) [7|17|23]	3417
Say You Will (88) 6(1) [4|12|19]	1833
I Don't Want To Live Without You (88) 5(1) [4|11|17]	1652
That Was Yesterday (85) 12(1) [0|10|15]	1021
Break It Up (82) 26(2) [0|6|13]	621
Juke Box Hero (82) 26(2) [0|6|13]	505
Women (80) 41(2) [0|0|9]	341
Reaction To Action (85) 54(1) [0|0|8]	216
Down On Love (85) 54(2) [0|0|8]	211
Heart Turns To Stone (88) 56(2) [0|0|10]	208
Luanne (82) 75(2) [0|0|6]	88.8
FORTUNE ▶ 1202 Pk:80 [0|0|1]	**67.7**
Stacy (86) 80(2) [0|0|6]	67.7
David FOSTER ▶ 540 Pk:15 [0|1|3]	**1198**
Love Theme From St. Elmo's Fire (85) 15(1) [0|10|22]	1134
The Best Of Me (86) - David Foster And Olivia Newton-John [A] 80(1) [0|0|8]	38.7
Winter Games (88) 85(1) [0|0|3]	25.9
4 BY FOUR ▶ 1201 Pk:79 [0|0|1]	**68.5**
Want You For My Girlfriend (87) 79(1) [0|0|6]	68.5
4 SEASONS ▶ 1291 Pk:91 [0|0|1]	**36.0**
Spend The Night In Love (80) - Four Seasons 91(3) [0|0|5]	36.0
FOUR TOPS ▶ 399 Pk:11 [0|2|5]	**2201**
When She Was My Girl (81) 11(2) [0|11|22]	1544
Indestructible (88) 35(1) [0|2|11]	354
I Just Can't Walk Away (83) 71(2) [0|0|9]	157
Back To School Again (82) 71(1) [0|0|7]	108
Sad Hearts (82) 84(2) [0|0|3]	37.2
Charles FOX ▶ 1248 Pk:75 [0|0|1]	**53.4**
Seasons (81) 75(1) [0|0|4]	53.4
Samantha FOX ▶ 164 Pk:3 [3|4|6]	**7224**
Naughty Girls (Need Love Too) (88) 3(1) [6|14|27]	2837

Act ▶ Rank Peak [Top10s|Top40s|Total] Title (Pk Yr) Peak(Wk) [Top10Wks|Top40Wks|TotalWks]	Score
Touch Me (I Want Your Body) (87) 4(1) [4|13|23]	2369
I Wanna Have Some Fun (89) 8(1) [1|12|23]	1459
I Only Wanna Be With You (89) 31(1) [0|4|13]	463
Nothing's Gonna Stop Me Now (87) 80(1) [0|0|5]	61.5
Do Ya Do Ya (Wanna Please Me) (87) 87(2) [0|0|5]	34.3
Peter FRAMPTON ▶ 1087 Pk:74 [0|0|1]	**119**
Lying (86) 74(2) [0|0|8]	119
FRANKE AND THE KNOCKOUTS ▶ 348 Pk:10 [1|3|3]	**2819**
Sweetheart (81) 10(2) [2|14|19]	1638
Without You (Not Another Lonely Night) (82) 24(2) [0|7|15]	681
You're My Girl (81) 27(2) [0|5|13]	500
FRANKIE GOES TO HOLLYWOOD ▶ 416 Pk:10 [1|1|4]	**2035**
Relax (Mix) (85) 10(2) [2|10|16]	1165
Two Tribes (84) 43(1) [0|0|15]	482
Welcome To The Pleasuredome (Trevor Horn Remix) (85) 48(2) [0|0|8]	242
Relax (84) 67(1) [0|0|7]	146
Aretha FRANKLIN ▶ 110 Pk:1 [3|9|15]	**10066**
Freeway Of Love (85) 3(1) [6|13|19]	2899
Who's Zoomin' Who (85) 7(1) [3|13|19]	1775
I Knew You Were Waiting (For Me) (87) - Aretha Franklin & George Michael [A] 1(2) [7|12|17]	1733
Another Night (86) 22(1) [0|7|14]	667
Jumpin' Jack Flash (86) 21(1) [0|6|11]	534
Jump To It (82) 24(2) [0|6|12]	533
Jimmy Lee (87) 28(1) [0|4|13]	469
Sisters Are Doin' It For Themselves (85) - Eurythmics And Aretha Franklin [A] 18(1) [0|8|13]	380
Through The Storm (89) - Aretha Franklin And Elton John [A] 16(1) [0|7|11]	318
United Together (81) 56(1) [0|0|8]	204
Get It Right (83) 61(1) [0|0|8]	190
Love All The Hurt Away (81) - Aretha Franklin & George Benson [A] 46(2) [0|0|10]	156
It Isn't, It Wasn't, It Ain't Never Gonna Be (89) - Aretha Franklin and Whitney Houston [A] 41(1) [0|0|8]	137
Rock-A-Lott (87) 82(1) [0|0|4]	39.8
Come To Me (81) 84(1) [0|0|3]	32.0
Andy FRASER ▶ 1246 Pk:82 [0|0|1]	**54.5**
Do You Love Me (84) 82(1) [0|0|5]	54.5
Glenn FREY ▶ 89 Pk:2 [2|7|10]	**12594**
The Heat Is On (85) 2(1) [6|13|24]	3741
You Belong To The City (85) 2(2) [7|13|21]	3607
Smuggler's Blues (85) 12(1) [0|11|19]	1274
The One You Love (82) 15(2) [0|11|17]	1039
True Love (88) 13(1) [0|9|15]	1000
Sexy Girl (84) 20(2) [0|9|15]	845
I Found Somebody (82) 31(2) [0|5|13]	510
All Those Lies (83) 41(3) [0|0|12]	397
The Allnighter (84) 54(1) [0|0|6]	159
Livin' Right (89) 90(1) [0|0|4]	22.4
FRIDA ▶ 430 Pk:13 [0|1|1]	**1869**
I Know There's Something Going On (83) 13(3) [0|12|29]	1869
FROZEN GHOST ▶ 1014 Pk:69 [0|0|1]	**171**
Should I See (87) 69(3) [0|0|10]	171

G

Act ▶ Rank Peak [Top10s|Top40s|Total] Title (Pk Yr) Peak(Wk) [Top10Wks|Top40Wks|TotalWks]	Score
Peter GABRIEL ▶ 145 Pk:1 [2|4|8]	**7899**
Sledgehammer (86) 1(1) [7|14|21]	3883
Big Time (87) 8(2) [4|11|23]	1775
Shock The Monkey (83) 29(2) [0|10|18]	841
In Your Eyes (86) 26(2) [0|7|14]	570
In Your Eyes (89) 41(1) [0|0|14]	453

Act ▶ Rank Peak [Top10s|Top40s|Total] Title (Pk Yr) Peak(Wk) [Top10Wks|Top40Wks|TotalWks]	Score
Games Without Frontiers (80) 48(2) [0|0|11]	293
Don't Give Up (87) - Peter Gabriel/Kate Bush [A] 72(1) [0|0|6]	47.1
Solsbury Hill (83) 84(1) [0|0|3]	36.8
GAMMA ▶ 943 Pk:60 [0|0|2]	**226**
I'm Alive (80) 60(1) [0|0|6]	145
Right The First Time (82) 77(2) [0|0|5]	80.8
GAP BAND ▶ 460 Pk:24 [0|2|5]	**1684**
Early In The Morning (82) 24(2) [0|6|14]	647
You Dropped A Bomb On Me (82) 31(5) [0|7|13]	559
Outstanding (83) 51(3) [0|0|8]	262
Yearning For Your Love (81) 60(1) [0|0|7]	138
Burn Rubber (Why You Wanna Hurt Me) (81) 84(1) [0|0|8]	77.2
Art GARFUNKEL ▶ 1002 Pk:66 [0|0|1]	**178**
A Heart In New York (81) 66(1) [0|0|9]	178
Leif GARRETT ▶ 831 Pk:60 [0|0|3]	**362**
Memorize Your Number (80) 60(1) [0|0|9]	218
Runaway Rita (81) 84(4) [0|0|6]	79.0
I Was Looking For Someone To Love (80) 78(1) [0|0|5]	64.9
Siedah GARRETT ▶ 1394 Pk:97 [0|0|1]	**3.1**
K.I.S.S.I.N.G. (88) 97(1) [0|0|1]	3.1
GARY O' ▶ 1182 Pk:70 [0|0|1]	**78.3**
Pay You Back With Interest (81) 70(2) [0|0|5]	78.3
David GATES ▶ 782 Pk:46 [0|0|2]	**454**
Where Does The Lovin' Go (80) 46(2) [0|0|8]	302
Take Me Now (81) 62(1) [0|0|7]	153
Marvin GAYE ▶ 240 Pk:3 [1|1|1]	**4570**
Sexual Healing (83) 3(3) [10|15|21]	4570
Crystal GAYLE ▶ 421 Pk:7 [1|1|6]	**1968**
You And I (83) - Eddie Rabbitt with Crystal Gayle [A] 7(4) [6|21|29]	1495
It's Like We Never Said Goodbye (80) 63(2) [0|0|6]	140
The Blue Side (80) 81(2) [0|0|8]	108
The Woman In Me (81) 76(2) [0|0|6]	101
Baby, What About You (83) 83(1) [0|0|5]	63.6
The Sound Of Goodbye (84) 84(2) [0|0|5]	59.5
J. GEILS Band ▶ 81 Pk:1 [2|6|10]	**13957**
Centerfold (82) 1(6) [12|20|25]	8247
Freeze-Frame (82) 4(4) [8|12|16]	3237
I Do (Live Version) (83) 24(3) [0|7|14]	696
Come Back (80) 32(2) [0|5|12]	555
Love Stinks (80) 38(2) [0|3|12]	447
Angel In Blue (82) 40(2) [0|2|11]	395
Land Of A Thousand Dances (83) 60(2) [0|0|6]	157
Concealed Weapons (84) 63(1) [0|0|7]	148
Just Can't Wait (80) 78(1) [0|0|6]	61.5
Fright Night (85) 91(1) [0|0|2]	14.1
Bob GELDOF ▶ 1216 Pk:82 [0|0|1]	**63.4**
This Is The World Calling (87) 82(1) [0|0|6]	63.4
GENE LOVES JEZEBEL ▶ 1343 Pk:87 [0|0|1]	**22.7**
The Motion Of Love (88) 87(1) [0|0|3]	22.7
GENERAL PUBLIC ▶ 652 Pk:27 [0|1|1]	**718**
Tenderness (85) 27(1) [0|5|18]	718
GENESIS ▶ 48 Pk:1 [6|11|15]	**19591**
Invisible Touch (86) 1(1) [6|12|17]	3412
In Too Deep (87) 3(1) [5|12|17]	2632
Land Of Confusion (87) 4(1) [4|15|21]	2370
Tonight, Tonight, Tonight (87) 3(1) [4|10|15]	2320
That's All (84) 6(2) [4|14|20]	2253
Throwing It All Away (86) 4(2) [4|12|16]	2113
Misunderstanding (80) 14(2) [0|11|18]	1289
No Reply At All (81) 29(2) [0|6|18]	763
Abacab (82) 26(2) [0|6|14]	603
Paperlate (82) 32(3) [0|5|14]	530

Acts With Singles

Act ▶ Rank Peak [Top10s|Top40s|Total] Title (Pk Yr) Peak(Wk) [Top10Wks|Top40Wks|TotalWks]	Score
GENESIS ▶ *Continued*	
Man On The Corner (82) 40(2) [0|2|11]	370
Illegal Alien (84) 44(1) [0|0|10]	315
Taking It All Too Hard (84) 50(1) [0|0|12]	300
Turn It On Again (80) 58(1) [0|0|8]	195
Mama (83) 73(1) [0|0|9]	126
GENTLE PERSUASION ▶ 1262 Pk:82 [0|0|1]	**47.6**
Please Mr. Postman (83) 82(2) [0|0|4]	47.6
Robin GEORGE ▶ 1371 Pk:92 [0|0|1]	**13.6**
Heartline (85) 92(1) [0|0|6]	13.6
GEORGIA SATELLITES ▶ 314 Pk:2 [1|1|3]	**3363**
Keep Your Hands To Yourself (87) 2(1) [5|14|20]	2929
Hippy Hippy Shake (88) 45(2) [0|0|14]	398
Battleship Chains (87) 86(1) [0|0|5]	35.1
GEORGIO ▶ 773 Pk:58 [0|0|3]	**464**
Lover's Lane (87) 59(4) [0|0|12]	271
Sexappeal (87) 58(1) [0|0|9]	188
Tina Cherry (87) 96(1) [0|0|2]	4.5
GET WET ▶ 910 Pk:39 [0|1|1]	**260**
Just So Lonely (81) 39(1) [0|2|9]	260
GIANT ▶ 958 Pk:56 [0|0|1]	**211**
I'm A Believer (89) 56(1) [0|0|10]	211
GIANT STEPS ▶ 516 Pk:13 [0|1|2]	**1361**
Another Lover (88) 13(1) [0|10|22]	1204
Into You (89) 58(1) [0|0|8]	157
Andy GIBB ▶ 191 Pk:4 [1|4|5]	**5742**
Desire (80) 4(4) [8|12|15]	3634
Time Is Time (81) 15(2) [0|11|17]	1100
I Can't Help It (80) - Andy Gibb & Olivia Newton-John [A] 12(2) [0|8|13]	595
Me (Without You) (81) 40(1) [0|1|8]	294
All I Have To Do Is Dream (81) - Andy Gibb And Victoria Principal [A] 51(2) [0|0|8]	119
Barry GIBB ▶ 827 Pk:37 [0|1|1]	**369**
Shine Shine (84) 37(2) [0|3|10]	369
Robin GIBB ▶ 703 Pk:37 [0|1|2]	**605**
Boys Do Fall In Love (84) 37(1) [0|4|12]	500
Help Me! (80) - Marcy Levy & Robin Gibb [A] 50(2) [0|0|10]	104
Terri GIBBS ▶ 512 Pk:13 [0|1|2]	**1406**
Somebody's Knockin' (81) 13(2) [0|12|22]	1367
Rich Man (81) 89(2) [0|0|5]	39.1
Debbie GIBSON ▶ 54 Pk:1 [5|8|9]	**18379**
Lost In Your Eyes (89) 1(3) [7|12|19]	4389
Foolish Beat (88) 1(1) [6|14|20]	3393
Shake Your Love (87) 4(1) [7|15|22]	2868
Only In My Dreams (87) 4(1) [5|16|28]	2792
Out Of The Blue (88) 3(1) [5|13|17]	2548
Electric Youth (89) 11(3) [0|8|13]	1000
No More Rhyme (89) 17(1) [0|7|14]	734
Staying Together (88) 22(1) [0|6|12]	551
We Could Be Together (89) 71(1) [0|0|6]	104
GIDEA PARK ▶ 1288 Pk:82 [0|0|1]	**36.7**
Seasons Of Gold (Four Seasons Medley) (82) - Gidea Park featuring Adrian Baker 82(1) [0|0|3]	36.7
Johnny GILL ▶ 1215 Pk:75 [0|0|1]	**63.5**
Perfect Combination (84) - Stacy Lattisaw & Johnny Gill [A] 75(1) [0|0|9]	63.5
Mickey GILLEY ▶ 518 Pk:22 [0|1|3]	**1351**
Stand By Me (80) 22(3) [0|9|18]	960
You Don't Know Me (81) 55(1) [0|0|12]	269
True Love Ways (80) 66(2) [0|0|7]	122
David GILMOUR ▶ 1040 Pk:62 [0|0|1]	**148**
Blue Light (84) 62(1) [0|0|7]	148
GINA GO-GO ▶ 1081 Pk:78 [0|0|1]	**120**
I Can't Face The Fact (89) 78(1) [0|0|11]	120
GIUFFRIA ▶ 497 Pk:15 [0|1|3]	**1472**
Call To The Heart (85) 15(2) [0|7|19]	1039
I Must Be Dreaming (86) 52(1) [0|0|10]	245

Act ▶ Rank Peak [Top10s|Top40s|Total] Title (Pk Yr) Peak(Wk) [Top10Wks|Top40Wks|TotalWks]	Score
Lonely In Love (85) 57(2) [0|0|8]	187
GLASS MOON ▶ 987 Pk:50 [0|0|1]	**189**
On A Carousel (82) 50(1) [0|0|7]	189
GLASS TIGER ▶ 196 Pk:2 [2|4|4]	**5611**
Don't Forget Me (When I'm Gone) (86) 2(1) [5|14|24]	2981
Someday (87) 7(2) [3|13|21]	1807
I'm Still Searching (88) 31(3) [0|5|11]	428
I Will Be There (87) 34(2) [0|5|11]	395
GOANNA ▶ 1072 Pk:71 [0|0|1]	**127**
Solid Rock (83) 71(2) [0|0|7]	127
GODLEY & CREME ▶ 581 Pk:16 [0|1|1]	**981**
Cry (85) 16(1) [0|10|17]	981
The GO-GO'S ▶ 109 Pk:2 [2|5|7]	**10243**
We Got The Beat (82) 2(3) [9|15|19]	4878
Vacation (82) 8(3) [4|9|14]	1638
Head Over Heels (84) 11(2) [0|10|16]	1455
Our Lips Are Sealed (81) 20(2) [0|13|30]	1438
Turn To You (84) 32(2) [0|5|14]	532
Get Up And Go (82) 50(2) [0|0|9]	269
Yes Or No (84) 84(1) [0|0|3]	32.4
GOLDEN EARRING ▶ 405 Pk:10 [1|1|3]	**2109**
Twilight Zone (83) 10(2) [2|15|27]	2000
When The Lady Smiles (84) 76(1) [0|0|4]	58.8
The Devil Made Me Do It (83) 79(2) [0|0|4]	50.1
GOOD QUESTION ▶ 1285 Pk:86 [0|0|1]	**38.9**
Got A New Love (88) 86(1) [0|0|5]	38.9
Robert GORDON ▶ 1236 Pk:76 [0|0|1]	**58.6**
Someday, Someway (81) 76(1) [0|0|4]	58.6
Michael GORE ▶ 1227 Pk:84 [0|0|1]	**60.5**
Theme From "Terms Of Endearment" (84) 84(1) [0|0|6]	60.5
GO WEST ▶ 533 Pk:39 [0|1|4]	**1278**
We Close Our Eyes (85) 41(1) [0|0|15]	493
Don't Look Down - The Sequel (87) 39(1) [0|2|13]	384
Call Me (85) 54(1) [0|0|14]	300
Eye To Eye (85) 73(1) [0|0|7]	102
GQ ▶ 1373 Pk:93 [0|0|1]	**13.2**
Sad Girl (82) 93(2) [0|0|2]	13.2
The GRACES ▶ 986 Pk:56 [0|0|1]	**189**
Lay Down Your Arms (89) 56(1) [0|0|9]	189
Larry GRAHAM ▶ 475 Pk:9 [1|1|3]	**1600**
One In A Million You (80) 9(2) [2|9|20]	1425
Just Be My Lady (81) 67(1) [0|0|5]	102
When We Get Married (80) 76(2) [0|0|4]	72.6
Lou GRAMM ▶ 417 Pk:5 [1|1|2]	**2017**
Midnight Blue (87) 5(1) [3|11|20]	1752
Ready Or Not (87) 54(2) [0|0|12]	265
GRANDMASTER FLASH And The FURIOUS FIVE ▶ **1050 Pk:62 [0|0|1]**	**144**
The Message (82) - Grandmaster Flash & The Furious Five Featuring: Melle Mel and Duke Bootee 62(2) [0|0|7]	144
GRANDMASTER MELLE MEL And **The FURIOUS FIVE ▶ 1338 Pk:86 [0|0|1]**	**23.5**
Beat Street Breakdown (84) - Grandmaster Melle Mel & The Furious Five 86(1) [0|0|2]	23.5
Amy GRANT ▶ 642 Pk:29 [0|1|3]	**744**
Find A Way (85) 29(1) [0|6|16]	581
Wise Up (85) 66(2) [0|0|9]	158
Lead Me On (88) 96(1) [0|0|2]	5.4
Eddy GRANT ▶ 169 Pk:2 [1|2|3]	**6805**
Electric Avenue (83) 2(5) [8|15|22]	5865
Romancing The Stone (84) 26(2) [0|6|17]	722
I Don't Wanna Dance (83) 53(1) [0|0|7]	219
GRATEFUL DEAD ▶ 528 Pk:9 [1|1|2]	**1291**
Touch Of Grey (87) 9(1) [2|9|15]	1171
Alabama Getaway (80) 68(1) [0|0|6]	120

Act ▶ Rank Peak [Top10s|Top40s|Total] Title (Pk Yr) Peak(Wk) [Top10Wks|Top40Wks|TotalWks]	Score
GREAT WHITE ▶ 304 Pk:5 [1|2|4]	**3554**
Once Bitten Twice Shy (89) 5(2) [5|14|26]	2400
The Angel Song (89) 30(1) [0|4|21]	648
Save Your Love (88) 57(1) [0|0|12]	265
Rock Me (87) 60(2) [0|0|14]	241
Lee GREENWOOD ▶ 871 Pk:53 [0|0|2]	**304**
I.O.U. (83) 53(2) [0|0|11]	295
Somebody's Gonna Love You (83) 96(2) [0|0|2]	9.0
GTR ▶ 564 Pk:14 [0|1|2]	**1087**
When The Heart Rules The Mind (86) 14(2) [0|10|16]	1035
The Hunter (86) 85(1) [0|0|6]	52.6
Greg GUIDRY ▶ 574 Pk:17 [0|1|2]	**1022**
Goin' Down (82) 17(3) [0|10|16]	1008
Into My Love (82) 92(1) [0|0|2]	13.8
GUNS N' ROSES ▶ 111 Pk:1 [4|4|5]	**9951**
Sweet Child O' Mine (88) 1(2) [7|14|24]	4427
Patience (89) 4(1) [5|10|18]	2099
Welcome To The Jungle (88) 7(2) [4|12|17]	1749
Paradise City (89) 5(2) [3|11|17]	1657
Nightrain (89) 93(1) [0|0|5]	18.5
Gwen GUTHRIE ▶ 834 Pk:42 [0|0|1]	**358**
Ain't Nothin' Goin' On But The Rent (86) 42(1) [0|0|13]	358
GUY ▶ 995 Pk:62 [0|0|2]	**183**
I Like (89) 70(1) [0|0|7]	112
My Fantasy (89) - Teddy Riley featuring Guy [A] 62(1) [0|0|6]	71.2

H

Act ▶ Rank Peak [Top10s|Top40s|Total] Title (Pk Yr) Peak(Wk) [Top10Wks|Top40Wks|TotalWks]	Score
Sammy HAGAR ▶ 266 Pk:13 [0|4|10]	**4163**
Your Love Is Driving Me Crazy (83) 13(2) [0|13|19]	1353
Give To Live (87) 23(1) [0|7|17]	710
I Can't Drive 55 (84) 26(1) [0|6|16]	578
Two Sides Of Love (84) 38(3) [0|3|12]	453
I'll Fall In Love Again (82) 43(2) [0|0|10]	329
Winner Takes It All (87) 54(1) [0|0|14]	297
Never Give Up (83) 46(1) [0|0|8]	268
Eagles Fly (87) 82(1) [0|0|13]	111
Piece Of My Heart (82) 73(1) [0|0|4]	61.4
Whiter Shade Of Pale (84) - Hagar, Schon, Aaronson, Shrieve [A] 94(1) [0|0|2]	2.8
HAIRCUT ONE HUNDRED ▶ 698 Pk:37 [0|1|1]	**614**
Love Plus One (82) 37(1) [0|4|17]	614
Daryl HALL ▶ 382 Pk:5 [1|2|3]	**2352**
Dreamtime (86) 5(1) [4|11|15]	1729
Foolish Pride (86) 33(1) [0|5|13]	456
Someone Like You (87) 57(1) [0|0|8]	166
Daryl HALL & John OATES ▶ 3 Pk:1 [13|22|22]	**55777**
I Can't Go For That (No Can Do) (82) 1(1) [12|17|21]	6855
Maneater (82) 1(4) [13|17|23]	6774
Say It Isn't So (83) 2(4) [10|15|18]	5735
Private Eyes (81) 1(2) [9|17|23]	5434
Out Of Touch (84) 1(2) [9|16|23]	5335
Kiss On My List (81) 1(3) [8|17|23]	4872
You Make My Dreams (81) 5(3) [6|14|21]	2669
One On One (83) 7(3) [6|15|18]	2378
Everything Your Heart Desires (88) 3(1) [5|11|16]	2134
Method Of Modern Love (85) 5(1) [4|11|19]	1918
Family Man (83) 6(1) [3|12|16]	1851
Adult Education (84) 8(1) [4|11|17]	1822
You've Lost That Lovin' Feeling (80) 12(3) [0|14|20]	1734
Did It In A Minute (82) 9(2) [4|11|16]	1635
Wait For Me (80) 18(1) [0|11|20]	1094
Some Things Are Better Left Unsaid (85) 18(1) [0|8|13]	748

Acts With Singles

Act ▶ Rank Peak [Top10s|Top40s|Total] Title (Pk Yr) Peak(Wk) [Top10Wks|Top40Wks|TotalWks]	Score
Daryl HALL & John OATES ▶ *Continued*	
A Nite At The Apollo Live:	
The Way You Do The Things You Do/My Girl (85) -	
Daryl Hall & John Oates Featuring David Ruffin &	
Eddie Kendricks 20(1) [0|7|11]	591
How Does It Feel To Be Back (80) 30(1) [0|4|13]	504
Your Imagination (82) 33(2) [0|5|11]	492
Possession Obsession (85) 30(1) [0|6|12]	463
Missed Opportunity (88) 29(1) [0|5|11]	423
Downtown Life (88) 31(1) [0|3|9]	316
Jimmy HALL ▶ 708 Pk:27 [0|1|2]	**581**
I'm Happy That Love Has Found You (80)	
27(1) [0|4|17]	526
Fool For Your Love (82) 77(2) [0|0|3]	55.3
John HALL Band ▶ 734 Pk:42 [0|0|2]	**532**
Crazy (Keep On Falling) (82) 42(1) [0|0|11]	327
Love Me Again (83) 64(4) [0|0|10]	205
Lani HALL ▶ 1334 Pk:88 [0|0|1]	**24.5**
Where's Your Angel? (81) 88(1) [0|0|3]	24.5
David HALLYDAY ▶ 1214 Pk:79 [0|0|1]	**63.6**
He's My Girl (87) 79(2) [0|0|6]	63.6
Jan HAMMER ▶ 284 Pk:1 [1|1|1]	**3913**
Miami Vice Theme (85) 1(1) [7|13|22]	3913
Herbie HANCOCK ▶ 1038 Pk:71 [0|0|1]	**149**
Rockit (83) 71(2) [0|0|9]	149
Paul HARDCASTLE ▶ 535 Pk:15 [0|1|2]	**1264**
19 (85) 15(2) [0|8|14]	925
Rain Forest (85) 57(1) [0|0|18]	339
Emmylou HARRIS ▶ 724 Pk:37 [0|1|2]	**555**
Mister Sandman (81) 37(1) [0|3|13]	446
That Lovin' You Feelin' Again (80) -	
Roy Orbison & Emmylou Harris [A] 55(2) [0|0|8]	109
Sam HARRIS ▶ 647 Pk:36 [0|1|2]	**724**
Sugar Don't Bite (84) 36(2) [0|3|14]	487
I'd Do It All Again (86) 52(1) [0|0|9]	237
George HARRISON ▶ 138 Pk:1 [2|3|4]	**8263**
Got My Mind Set On You (88) 1(1) [8|15|22]	4004
All Those Years Ago (81) 2(3) [6|11|16]	3615
When We Was Fab (88) 23(1) [0|6|11]	499
Wake Up My Love (82) 53(2) [0|0|5]	145
Debbie HARRY ▶ 640 Pk:43 [0|0|4]	**751**
Backfired (81) 43(1) [0|0|10]	357
French Kissin (87) 57(1) [0|0|11]	249
In Love With Love (87) 70(1) [0|0|7]	111
The Jam Was Moving (81) 82(1) [0|0|3]	34.4
Corey HART ▶ 121 Pk:3 [2|8|9]	**8952**
Never Surrender (85) 3(2) [6|14|20]	3008
Sunglasses At Night (84) 7(1) [5|15|23]	2211
It Ain't Enough (84) 17(2) [0|9|19]	980
I Am By Your Side (86) 18(1) [0|7|13]	703
Everything In My Heart (86) 30(1) [0|7|15]	591
Can't Help Falling In Love (87) 24(1) [0|5|14]	585
Boy In The Box (85) 26(1) [0|6|12]	544
In Your Soul (88) 38(1) [0|2|10]	311
Dancin' With My Mirror (87) 88(1) [0|0|3]	19.1
Dan HARTMAN ▶ 299 Pk:6 [1|3|5]	**3663**
I Can Dream About You (84)	
6(2) [4|16|25]	2393
We Are The Young (84) 25(1) [0|9|17]	749
Second Nature (85) 39(1) [0|2|12]	395
It Hurts To Be In Love (81) 72(1) [0|0|5]	82.2
Heaven In Your Arms (81) 86(2) [0|0|5]	43.2
--HATCHET, Molly see: MOLLY HATCHET	
HAWKS ▶ 1033 Pk:63 [0|0|1]	**153**
Right Away (81) 63(1) [0|0|7]	153
Colin James HAY ▶ 1398 Pk:99 [0|0|1]	**1.6**
Hold Me (87) 99(1) [0|0|1]	1.6
Isaac HAYES ▶ 524 Pk:18 [0|1|1]	**1314**
Don't Let Go (80) 18(2) [0|12|21]	1314

Act ▶ Rank Peak [Top10s|Top40s|Total] Title (Pk Yr) Peak(Wk) [Top10Wks|Top40Wks|TotalWks]	Score
HAYSI FANTAYZEE ▶ 1173 Pk:74 [0|0|1]	**83.5**
Shiny Shiny (83) 74(1) [0|0|5]	83.5
Leon HAYWOOD ▶ 866 Pk:49 [0|0|1]	**313**
Don't Push It Don't Force It (80) 49(1) [0|0|11]	313
Robert HAZARD ▶ 925 Pk:58 [0|0|1]	**248**
Escalator Of Life (83) 58(2) [0|0|9]	248
Murray HEAD ▶ 346 Pk:3 [1|1|1]	**2828**
One Night In Bangkok (85) 3(1) [6|13|20]	2828
HEADPINS ▶ 1032 Pk:70 [0|0|1]	**153**
Just One More Time (84) 70(1) [0|0|9]	153
Jeff HEALEY Band ▶ 432 Pk:5 [1|1|1]	**1866**
Angel Eyes (89) 5(1) [4|13|22]	1866
HEART ▶ 46 Pk:1 [7|10|15]	**20696**
Alone (87) 1(3) [8|15|21]	5121
These Dreams (86) 1(1) [6|13|20]	3896
Never (85) 4(1) [5|14|24]	2769
Tell It Like It Is (81) 8(2) [4|11|16]	1716
Who Will You Run To (87) 7(3) [3|11|22]	1573
What About Love? (85) 10(1) [1|12|21]	1405
Nothin' At All (Remix) (86) 10(1) [1|10|16]	1156
There's The Girl (88) 12(1) [0|11|19]	1156
Even It Up (80) 33(2) [0|4|12]	540
This Man Is Mine (82) 33(2) [0|4|13]	527
How Can I Refuse (83) 44(2) [0|0|11]	348
I Want You So Bad (88) 49(1) [0|0|9]	235
If Looks Could Kill (86) 54(1) [0|0|9]	187
Allies (83) 83(1) [0|0|4]	36.0
Unchained Melody (Live) (81) 83(1) [0|0|3]	31.2
HEAVEN 17 ▶ 1176 Pk:74 [0|0|1]	**81.9**
Let Me Go (83) 74(2) [0|0|5]	81.9
Nona HENDRYX ▶ 946 Pk:58 [0|0|2]	**222**
Why Should I Cry? (87) 58(1) [0|0|9]	207
Keep It Confidential (83) 91(1) [0|0|3]	15.2
Don HENLEY ▶ 90 Pk:3 [4|7|9]	**12425**
Dirty Laundry (83) 3(3) [10|14|19]	4595
The Boys Of Summer (85) 5(1) [4|14|22]	2201
All She Wants To Do Is Dance (85) 9(1) [2|11|19]	1534
The End Of The Innocence (89) 8(1) [2|12|18]	1386
The Last Worthless Evening (89) 21(1) [0|8|18]	775
Sunset Grill (85) 22(2) [0|8|14]	696
Not Enough Love In The World (85) 34(1) [0|5|17]	563
Johnny Can't Read (82) 42(2) [0|0|11]	339
I Can't Stand Still (83) 48(2) [0|0|11]	336
Howard HEWETT ▶ 1329 Pk:90 [0|0|1]	**26.3**
I'm For Real (86) 90(3) [0|0|3]	26.3
Bertie HIGGINS ▶ 351 Pk:8 [1|1|2]	**2799**
Key Largo (82) 8(2) [4|17|29]	2483
Just Another Day In Paradise (82) 46(2) [0|0|10]	316
HIGH INERGY ▶ 1231 Pk:82 [0|0|1]	**59.9**
He's A Pretender (83) 82(2) [0|0|5]	59.9
Dan HILL ▶ 370 Pk:6 [1|1|2]	**2503**
Can't We Try (w/Vonda Shepard) (87) 6(1) [4|13|24]	1939
Never Thought (That I Could Love) (88)	
43(1) [0|0|20]	563
Eric HINE ▶ 1170 Pk:73 [0|0|1]	**84.3**
Not Fade Away (81) 73(1) [0|0|5]	84.3
HIPSWAY ▶ 660 Pk:19 [0|1|1]	**708**
The Honeythief (87) 19(1) [0|6|15]	708
Roger HODGSON ▶ 801 Pk:48 [0|0|1]	**416**
Had A Dream (Sleeping With The Enemy) (84)	
48(2) [0|0|15]	416
Amy HOLLAND ▶ 644 Pk:22 [0|1|1]	**733**
How Do I Survive (80) 22(2) [0|6|16]	733
Jennifer HOLLIDAY ▶ 562 Pk:22 [0|1|4]	**1097**
And I Am Telling You I'm Not Going (82)	
22(3) [0|7|14]	654
I Am Love (83) 49(2) [0|0|11]	305
Hard Times For Lovers (85) 69(1) [0|0|7]	114
No Frills Love (86) 87(1) [0|0|3]	24.4

Act ▶ Rank Peak [Top10s|Top40s|Total] Title (Pk Yr) Peak(Wk) [Top10Wks|Top40Wks|TotalWks]	Score
The HOLLIES ▶ 719 Pk:29 [0|1|1]	**565**
Stop In The Name Of Love (83) 29(1) [0|6|12]	565
Rupert HOLMES ▶ 333 Pk:6 [1|2|4]	**3018**
Him (80) 6(2) [5|12|17]	2280
Answering Machine (80) 32(1) [0|3|11]	445
I Don't Need You (81) 56(1) [0|0|7]	159
Morning Man (80) 68(1) [0|0|7]	133
The HONEYDRIPPERS ▶ 281 Pk:3 [1|2|2]	**3957**
Sea Of Love (85) 3(1) [6|14|20]	3446
Rockin' At Midnight (85) 25(1) [0|6|11]	511
HONEYMOON SUITE ▶ 570 Pk:34 [0|1|4]	**1056**
Feel It Again (86) 34(1) [0|3|16]	472
What Does It Take (86) 52(1) [0|0|16]	412
New Girl Now (84) 57(1) [0|0|7]	158
Love Changes Everything (88) 91(1) [0|0|2]	13.7
--HOOK, Dr. see: D(OCTO)R HOOK	
The HOOTERS ▶ 345 Pk:18 [0|3|7]	**2834**
And We Danced (85) - Hooters 21(1) [0|8|20]	914
Day By Day (86) - Hooters 18(3) [0|7|18]	860
The HOOTERS ▶ 345 Pk:18 [0|3|7]	**2834**
Where Do The Children Go (86) 38(1) [0|1|12]	389
Johnny B (87) 61(1) [0|0|11]	233
All You Zombies (85) 58(1) [0|0|11]	232
Satellite (87) 61(1) [0|0|8]	194
500 Miles (89) 97(3) [0|0|5]	13.0
Bruce HORNSBY And The RANGE ▶ 116	
Pk:1 [3|5|6]	**9484**
The Way It Is (86) 1(1) [8|15|22]	4013
Mandolin Rain (87) 4(1) [5|12|18]	2155
The Valley Road (88) 5(1) [5|11|18]	1810
Every Little Kiss (87) 14(1) [0|9|15]	961
Look Out Any Window (88) - 35(1) [0|2|12]	395
Every Little Kiss (86) 72(2) [0|0|8]	149
HOT CHOCOLATE ▶ 1048 Pk:65 [0|0|1]	**145**
Are You Getting Enough Happiness (83)	
65(2) [0|0|7]	145
HOTEL ▶ 1054 Pk:72 [0|0|1]	**140**
Half Moon Silver (80) 72(2) [0|0|7]	140
HOUSE OF LORDS ▶ 1074 Pk:58 [0|0|1]	**126**
I Wanna Be Loved (89) 58(1) [0|0|7]	126
Whitney HOUSTON ▶ 8 Pk:1 [10|10|11]	**35563**
I Wanna Dance With Somebody (Who Loves Me) (87)	
1(2) [9|14|18]	4899
Greatest Love Of All (86) 1(3) [7|14|18]	4572
How Will I Know (86) 1(2) [6|16|23]	4469
So Emotional (88) 1(1) [8|14|19]	4343
Didn't We Almost Have It All (87) 1(2) [7|13|17]	4007
Saving All My Love For You (85) 1(1) [7|15|22]	3944
Where Do Broken Hearts Go (88) 1(2) [6|13|18]	3563
You Give Good Love (85) 3(1) [6|13|21]	2786
One Moment In Time (88) 5(1) [4|11|17]	1661
Love Will Save The Day (88) 9(1) [1|11|16]	1182
It Isn't, It Wasn't, It Ain't Never Gonna Be (89) -	
Aretha Franklin and Whitney Houston [A]	
41(1) [0|0|8]	137
David HUDSON ▶ 900 Pk:59 [0|0|1]	**272**
Honey, Honey (80) 59(1) [0|0|11]	272
Grayson HUGH ▶ 623 Pk:19 [0|1|2]	**793**
Talk It Over (89) 19(1) [0|8|18]	765
Bring It All Back (89) 87(1) [0|0|4]	27.7
HUGHES/THRALL ▶ 1179 Pk:79 [0|0|1]	**79.1**
Beg, Borrow Or Steal (83) 79(1) [0|0|5]	79.1
HUMAN LEAGUE ▶ 82 Pk:1 [3|4|6]	**13925**
Don't You Want Me (82) 1(3) [12|21|28]	6816
Human (86) 1(1) [7|15|20]	4045
(Keep Feeling) Fascination (83) 8(2) [3|13|20]	2106
Mirror Man (83) 30(1) [0|5|12]	518
I Need Your Loving (87) 44(1) [0|0|11]	333
The Lebanon (84) 64(2) [0|0|5]	107

Acts With Singles

Act ▶ Rank Peak [Top10s|Top40s|Total] Title (Pk Yr) Peak(Wk) [Top10Wks|Top40Wks|TotalWks]	Score
HUMBLE PIE▶ 947 Pk:52 [0|0|1]	222
Fool For A Pretty Face (Hurt By Love) (80) 52(2) [0|0|7]	222
Engelbert HUMPERDINCK▶ 1134 Pk:77 [0|0|2]	102
Til You And Your Lover Are Lovers Again (83) 77(1) [0|0|5]	69.9
Love's Only Love (80) - Engelbert 83(1) [0|0|2]	32.3
John HUNTER▶ 745 Pk:39 [0|1|1]	518
Tragedy (85) 39(2) [0|2|16]	518
Jim HURT▶ 1327 Pk:90 [0|0|1]	26.9
I Love Women (80) 90(2) [0|0|4]	26.9
Paul HYDE And The PAYOLAS▶ 1279 Pk:84 [0|0|1]	41.6
You're The Only Love (85) 84(1) [0|0|4]	41.6
Chrissie HYNDE▶ 918 Pk:28 [0|1|1]	250
I Got You Babe (85) - UB40 With Chrissie Hynde [A] 28(1) [0|4|14]	250

I

Act ▶ Rank Peak [Top10s|Top40s|Total]	Score
Janis IAN▶ 1205 Pk:71 [0|0|1]	67.4
Under The Covers (81) 71(2) [0|0|4]	67.4
ICEHOUSE▶ 329 Pk:7 [1|2|6]	3084
Electric Blue (88) 7(1) [3|13|23]	1689
Crazy (88) 14(1) [0|11|21]	1091
We Can Get Together (81) 62(2) [0|0|7]	144
No Promises (86) 79(1) [0|0|9]	103
Touch The Fire (89) 84(1) [0|0|4]	30.1
My Obsession (88) 88(1) [0|0|4]	27.7
ICE-T▶ 1123 Pk:70 [0|0|1]	105
Colors (88) 70(2) [0|0|7]	105
ICICLE WORKS▶ 774 Pk:37 [0|1|1]	461
Whisper To A Scream (Birds Fly) (84) 37(2) [0|4|12]	461
Billy IDOL▶ 87 Pk:1 [3|8|11]	12767
Mony Mony "Live" (87) 1(1) [6|12|22]	3879
Eyes Without A Face (84) 4(2) [6|14|22]	3061
To Be A Lover (86) 6(1) [3|13|18]	1879
Hot In The City (82) 23(4) [0|9|17]	864
Sweet Sixteen (87) 20(1) [0|7|14]	632
White Wedding (83) 36(1) [0|3|13]	551
Flesh For Fantasy (84) 29(2) [0|6|12]	528
Rebel Yell (84) 46(2) [0|0|14]	457
Don't Need A Gun (87) 37(1) [0|2|9]	353
Catch My Fall (84) 50(1) [0|0|11]	287
Hot In The City (88) 48(1) [0|0|10]	276
Julio IGLESIAS▶ 458 Pk:5 [1|2|3]	1688
To All The Girls I've Loved Before (84) - Julio Iglesias & Willie Nelson [A] 5(1) [6|12|21]	1241
All Of You (84) - Julio Iglesias & Diana Ross [A] 19(2) [0|8|16]	422
My Love (88) - Julio Iglesias Featuring Stevie Wonder [A] 80(1) [0|0|5]	24.3
INDIGO GIRLS▶ 936 Pk:52 [0|0|1]	232
Closer To Fine (89) 52(1) [0|0|9]	232
INDUSTRY▶ 1177 Pk:81 [0|0|1]	81.3
State Of The Nation (83) 81(2) [0|0|8]	81.3
INFORMATION SOCIETY▶ 283 Pk:3 [2|2|4]	3917
What's On Your Mind (Pure Energy) (88) 3(1) [6|14|25]	2481
Walking Away (89) 9(1) [3|10|19]	1284
Repetition (89) 76(1) [0|0|6]	79.6
Lay All Your Love On Me (89) 83(1) [0|0|8]	72.5
James INGRAM▶ 259 Pk:2 [1|4|6]	4330
Somewhere Out There (87) - Linda Ronstadt And James Ingram [A] 2(1) [5|12|22]	1445
Yah Mo B There (84) 19(1) [0|9|18]	1115
Just Once (81) - Quincy Jones Featuring James Ingram [A] 17(2) [0|10|23]	671

Act ▶ Rank Peak [Top10s|Top40s|Total]	Score
One Hundred Ways (82) - Quincy Jones Featuring James Ingram [A] 14(1) [0|11|21]	566
How Do You Keep The Music Playing (Theme From "Best Friends") (83) - James Ingram And Patti Austin [A] 45(1) [0|0|17]	292
There's No Easy Way (84) 58(1) [0|0|10]	241
The INMATES▶ 890 Pk:51 [0|0|1]	282
Dirty Water (80) 51(2) [0|0|10]	282
INNER CITY▶ 1043 Pk:73 [0|0|1]	147
Good Life (89) 73(1) [0|0|11]	147
INVISIBLE MAN'S BAND▶ 825 Pk:45 [0|0|1]	372
All Night Thing (80) 45(2) [0|0|10]	372
INXS▶ 76 Pk:1 [5|6|12]	14750
Need You Tonight (88) 1(1) [8|17|25]	3984
Devil Inside (88) 2(2) [5|12|17]	3056
New Sensation (88) 3(1) [5|12|17]	2395
What You Need (86) 5(1) [5|14|20]	2349
Never Tear Us Apart (88) 7(1) [3|11|23]	1587
The One Thing (83) 30(2) [0|5|14]	557
Listen Like Thieves (86) 54(2) [0|0|9]	238
Good Times (87) - INXS And Jimmy Barnes [A] 47(1) [0|0|13]	186
Original Sin (84) 58(1) [0|0|7]	177
I Send A Message (84) 77(2) [0|0|7]	101
This Time (85) 81(1) [0|0|6]	73.6
Don't Change (83) 80(1) [0|0|4]	46.0
Donnie IRIS▶ 419 Pk:25 [0|3|7]	2001
Ah! Leah! (81) 29(2) [0|6|18]	660
My Girl (82) 25(2) [0|6|14]	536
Love Is Like A Rock (82) 37(1) [0|2|14]	420
Tough World (82) 57(1) [0|0|6]	158
Do You Compute? (83) 64(1) [0|0|7]	145
Sweet Merilee (81) 80(1) [0|0|6]	67.1
Injured In The Game Of Love (85) 91(1) [0|0|2]	14.4
IRISH ROVERS▶ 710 Pk:37 [0|1|1]	579
Wasn't That A Party (81) - The Rovers 37(2) [0|4|17]	579
IRONHORSE▶ 1273 Pk:89 [0|0|1]	44.0
What's Your Hurry Darlin' (80) 89(1) [0|0|6]	44.0
ISLE OF MAN▶ 1332 Pk:90 [0|0|1]	24.6
Am I Forgiven (86) 90(2) [0|0|4]	24.6
ISLEY BROTHERS▶ 747 Pk:39 [0|1|2]	515
Don't Say Goodnight (It's Time For Love) (Parts 1 and 2) (80) 39(2) [0|2|9]	365
Hurry Up And Wait (81) 58(1) [0|0|7]	150
ISLEY, JASPER, ISLEY▶ 746 Pk:51 [0|0|2]	515
Caravan Of Love (86) 51(2) [0|0|14]	352
Kiss And Tell (85) 63(2) [0|0|7]	163

J

Act ▶ Rank Peak [Top10s|Top40s|Total]	Score
Freddie JACKSON▶ 275 Pk:12 [0|4|7]	4008
You Are My Lady (85) 12(2) [0|11|20]	1340
Rock Me Tonight (For Old Times Sake) (85) 18(1) [0|8|19]	854
He'll Never Love You (Like I Do) (86) 25(1) [0|6|15]	557
Jam Tonight (87) 32(1) [0|3|13]	453
Tasty Love (86) 41(1) [0|0|12]	407
Nice 'N' Slow (88) 61(1) [0|0|12]	229
Have You Ever Loved Somebody (87) 69(1) [0|0|9]	167
Janet JACKSON▶ 41 Pk:1 [6|7|9]	21166
Miss You Much (89) 1(4) [8|13|20]	5007
When I Think Of You (86) 1(2) [6|13|19]	3899
Let's Wait Awhile (87) 2(1) [5|11|19]	3037
Nasty (86) 3(1) [6|11|19]	2753
What Have You Done For Me Lately (86) 4(1) [6|13|21]	2532
Control (87) 5(1) [6|13|18]	2445
The Pleasure Principle (87) 14(1) [0|10|18]	1139

Act ▶ Rank Peak [Top10s|Top40s|Total]	Score
Come Give Your Love To Me (83) 58(3) [0|0|9]	218
Young Love (83) 64(2) [0|0|6]	137
Jermaine JACKSON▶ 161 Pk:9 [1|6|11]	7427
Let's Get Serious (80) 9(2) [3|14|23]	2039
Do What You Do (85) 13(1) [0|12|20]	1417
Dynamite (84) 15(2) [0|10|17]	1109
I Think It's Love (86) 16(1) [0|9|15]	861
Let Me Tickle Your Fancy (82) 18(2) [0|7|15]	813
You're Supposed To Keep Your Love For Me (80) 34(1) [0|4|13]	518
You Like Me Don't You (81) 50(1) [0|0|9]	184
I'm Just Too Shy (81) 60(1) [0|0|8]	181
(Closest Thing To) Perfect (85) 67(1) [0|0|7]	116
When The Rain Begins To Fall (85) - Jermaine Jackson And Pia Zadora [A] 54(2) [0|0|7]	114
Do You Remember Me? (86) 71(2) [0|0|5]	76.0
Joe JACKSON▶ 212 Pk:6 [1|3|5]	5290
Steppin' Out (82) 6(4) [8|15|27]	3003
You Can't Get What You Want (Till You Know What You Want) (84) 15(1) [0|9|16]	1045
Breaking Us In Two (83) 18(1) [0|10|16]	1034
Happy Ending (84) 57(2) [0|0|8]	174
Memphis (83) 85(1) [0|0|4]	34.2
La Toya JACKSON▶ 948 Pk:56 [0|0|1]	221
Heart Don't Lie (84) 56(1) [0|0|8]	221
Michael JACKSON▶ 1 Pk:1 [17|19|21]	61327
Billie Jean (83) 1(7) [11|17|24]	8360
Rock With You (80) 1(4) [9|19|24]	7256
Beat It (83) 1(3) [10|18|25]	6892
Say Say Say (83) - Paul McCartney and Michael Jackson [A] 1(6) [13|18|22]	5168
Man In The Mirror (88) 1(2) [7|13|17]	3974
I Just Can't Stop Loving You (87) 1(1) [6|11|14]	3395
The Way You Make Me Feel (88) 1(1) [6|13|18]	3201
Bad (87) 1(2) [5|11|14]	2998
Dirty Diana (88) 1(1) [5|11|14]	2958
The Girl Is Mine (83) - Michael Jackson/ Paul McCartney [A] 2(3) [10|14|18]	2673
Wanna Be Startin' Somethin' (83) 5(2) [6|11|15]	2524
Thriller (84) 4(2) [5|9|14]	2493
Human Nature (83) 7(2) [4|11|14]	1844
She's Out Of My Life (80) 10(2) [2|11|16]	1704
Off The Wall (80) 10(2) [2|11|17]	1498
Smooth Criminal (89) 7(2) [3|11|15]	1470
P.Y.T. (Pretty Young Thing) (83) 10(1) [1|9|16]	1323
Another Part Of Me (88) 11(1) [0|8|13]	945
Farewell My Summer Love (84) 38(2) [0|3|12]	454
One Day In Your Life (81) 55(1) [0|0|7]	168
Get It (88) - Stevie Wonder & Michael Jackson [A] 80(1) [0|0|6]	28.5
Rebbie JACKSON▶ 628 Pk:24 [0|1|1]	781
Centipede (84) 24(3) [0|8|19]	781
The JACKSONS▶ 180 Pk:3 [1|4|8]	6347
State Of Shock (84) 3(3) [6|11|15]	3191
Lovely One (80) 12(2) [0|8|18]	1177
Heartbreak Hotel (81) 22(2) [0|8|16]	782
Torture (84) 17(2) [0|8|12]	765
Body (84) 47(1) [0|0|7]	222
Walk Right Now (81) 73(2) [0|0|4]	70.0
Nothin (That Compares 2 U) (89) 77(1) [0|0|7]	69.9
Can You Feel It (81) 77(2) [0|0|5]	69.1
Debbie JACOBS▶ 1166 Pk:70 [0|0|1]	86.7
High On Your Love (80) 70(2) [0|0|4]	86.7
Mick JAGGER▶ 350 Pk:7 [1|4|6]	2814
Just Another Night (85) 12(2) [0|10|14]	1101
Dancing In The Street (85) - Mick Jagger & David Bowie [A] 7(1) [3|9|14]	691
Lucky In Love (85) 38(2) [0|3|11]	378
Let's Work (87) 39(1) [0|1|9]	281

271

Acts With Singles

Act ▶ Rank Peak [Top10s|Top40s|Total] Title (Pk Yr) Peak(Wk) [Top10Wks|Top40Wks|TotalWks]	Score
Mick JAGGER ▶ Continued	
Ruthless People (86) 51(2) [0|0|8]	214
Throwaway (87) 67(1) [0|0|9]	150
The JAGS ▶ 1322 Pk:84 [0|0|1]	**28.0**
Back Of My Hand (I've Got Your Number) (80)	
84(1) [0|0|2]	28.0
Rick JAMES ▶ 321 Pk:16 [0|4|8]	**3302**
Super Freak (Part I) (81) 16(2) [0|10|24]	1254
17 (84) 36(3) [0|3|14]	507
Cold Blooded (83) 40(1) [0|1|12]	437
Give It To Me Baby (81) 40(2) [0|2|14]	390
Can't Stop (85) 50(1) [0|0|8]	233
Ebony Eyes (84) - Rick James featuring Smokey Robinson [A] 43(1) [0|0|11]	219
Dance Wit' Me (Part 1) (82) 64(2) [0|0|9]	181
Standing On The Top Part 1 (82) - The Temptations Featuring Rick James [A] 66(1) [0|0|8]	80.4
Tommy JAMES ▶ 551 Pk:19 [0|1|2]	**1146**
Three Times In Love (80) 19(2) [0|9|16]	997
You're So Easy To Love (81) 58(2) [0|0|7]	150
Nick JAMESON ▶ 1383 Pk:95 [0|0|1]	**9.4**
Weatherman (86) 95(2) [0|0|2]	9.4
Al JARREAU ▶ 306 Pk:15 [0|3|8]	**3526**
We're In This Love Together (81) 15(2) [0|11|24]	1433
Mornin' (83) - Jarreau 21(2) [0|6|15]	743
Moonlighting (Theme) (87) 23(1) [0|5|13]	557
Breakin' Away (82) 43(1) [0|0|10]	343
After All (84) 69(1) [0|0|9]	124
Teach Me Tonight (82) 70(2) [0|0|7]	123
Trouble In Paradise (83) - Jarreau 63(1) [0|0|7]	113
Boogie Down (83) - Jarreau 77(1) [0|0|6]	89.2
JEFFERSON STARSHIP/STARSHIP ▶ 43 Pk:1 [4|11|18]	**20993**
Nothing's Gonna Stop Us Now (87) - Starship 1(2) [8|15|22]	4720
We Built This City (85) - Starship 1(2) [7|15|24]	4486
Sara (86) - Starship 1(1) [7|13|20]	4156
Jane (80) - Jefferson Starship 14(1) [0|10|15]	1259
It's Not Over ('Til It's Over) (87) - Starship 9(1) [1|10|16]	1147
It's Not Enough (89) - Starship 12(1) [0|9|16]	1036
No Way Out (84) - Jefferson Starship 23(1) [0|8|16]	851
Be My Lady (82) - Jefferson Starship 28(2) [0|6|16]	659
Tomorrow Doesn't Matter Tonight (86) - Starship 26(1) [0|6|13]	539
Find Your Way Back (81) - Jefferson Starship 29(2) [0|6|13]	523
Winds Of Change (83) - Jefferson Starship 38(2) [0|2|11]	434
Beat Patrol (87) - Starship 46(1) [0|0|10]	285
Stranger (81) - Jefferson Starship 48(2) [0|0|11]	266
Girl With The Hungry Eyes (80) - Jefferson Starship 55(1) [0|0|6]	172
Before I Go (86) - Starship 68(2) [0|0|6]	128
Layin' It On The Line (84) - Jefferson Starship 66(2) [0|0|6]	128
Wild Again (89) - Starship 73(2) [0|0|8]	117
I Didn't Mean To Stay All Night (89) - Starship 75(1) [0|0|8]	85.8
Garland JEFFREYS ▶ 1078 Pk:66 [0|0|1]	**121**
96 Tears (81) 66(2) [0|0|7]	121
JELLYBEAN ▶ 436 Pk:16 [0|2|3]	**1843**
Sidewalk Talk (86) 18(1) [0|9|18]	935
Who Found Who (87) - Jellybean (Elisa Fiorillo Vocals) 16(1) [0|8|15]	849
The Real Thing (87) - Jellybean Featuring Stephen Dante 82(1) [0|0|6]	58.4
Waylon JENNINGS ▶ 550 Pk:21 [0|1|2]	**1154**
Theme From "The Dukes Of Hazzard" (Good Ol' Boys) (80) - Waylon 21(2) [0|10|23]	1039

Act ▶ Rank Peak [Top10s|Top40s|Total] Title (Pk Yr) Peak(Wk) [Top10Wks|Top40Wks|TotalWks]	Score
Just To Satisfy You (82) - Waylon & Willie [A] 52(1) [0|0|9]	114
The JETS ▶ 94 Pk:3 [5|6|10]	**12108**
Crush On You (86) 3(2) [5|13|20]	2659
You Got It All (87) 3(1) [4|12|26]	2523
Make It Real (88) 4(2) [5|13|20]	2388
Rocket 2 U (88) 6(2) [4|13|22]	1861
Cross My Broken Heart (87) 7(1) [3|11|16]	1523
I Do You (87) 20(1) [0|6|15]	679
Private Number (86) 47(1) [0|0|11]	283
You Better Dance (89) 59(1) [0|0|7]	124
The Same Love (89) 87(1) [0|0|6]	35.9
Sendin' All My Love (88) 88(1) [0|0|4]	33.8
Joan JETT & The BLACKHEARTS ▶ 78 Pk:1 [3|8|9]	**14519**
I Love Rock 'N Roll (82) 1(7) [12|16|20]	8293
I Hate Myself For Loving You (88) 8(2) [3|12|26]	1633
Crimson And Clover (82) 7(2) [3|10|15]	1605
Little Liar (89) 19(1) [0|10|20]	961
Do You Wanna Touch Me (Oh Yeah) (82) 20(3) [0|7|14]	794
Fake Friends (83) 35(2) [0|4|10]	434
Light Of Day (87) - The Barbusters 33(1) [0|5|11]	390
Everyday People (83) 37(1) [0|2|9]	352
Good Music (86) 83(2) [0|0|6]	55.7
J.J. FAD ▶ 600 Pk:30 [0|1|3]	**893**
Supersonic (88) 30(1) [0|6|22]	712
Way Out (88) 61(2) [0|0|9]	157
Is It Love (88) 92(2) [0|0|5]	24.0
JOBOXERS ▶ 722 Pk:36 [0|1|1]	**558**
Just Got Lucky (83) 36(2) [0|4|15]	558
Billy JOEL ▶ 6 Pk:1 [9|20|22]	**38031**
It's Still Rock And Roll To Me (80) 1(2) [11|19|21]	6941
Uptown Girl (83) 3(5) [10|16|22]	5362
We Didn't Start The Fire (89) 1(2) [8|15|20]	4586
Tell Her About It (83) 1(1) [7|15|18]	4357
You May Be Right (80) 7(3) [4|11|15]	1973
Allentown (83) 17(6) [0|16|22]	1591
An Innocent Man (84) 10(1) [1|11|18]	1543
The Longest Time (84) 14(3) [0|11|18]	1325
You're Only Human (Second Wind) (85) 9(2) [2|11|16]	1320
A Matter Of Trust (86) 10(1) [1|10|18]	1210
Modern Woman (86) 10(1) [1|9|15]	1035
Say Goodbye To Hollywood (Live) (81) 17(1) [0|8|15]	904
Don't Ask Me Why (80) 19(3) [0|9|15]	903
Keeping The Faith (85) 18(1) [0|10|16]	876
Pressure (82) 20(3) [0|8|17]	859
This Is The Time (87) 18(1) [0|9|17]	850
She's Got A Way (82) 23(2) [0|9|14]	715
Leave A Tender Moment Alone (84) 27(2) [0|7|15]	688
The Night Is Still Young (85) 34(2) [0|3|10]	367
Sometimes A Fantasy (80) 36(2) [0|3|9]	334
Goodnight Saigon (83) 56(3) [0|0|7]	194
Baby Grand (87) - Billy Joel featuring Ray Charles 75(1) [0|0|7]	96.5
Elton JOHN ▶ 18 Pk:2 [6|19|20]	**27052**
Little Jeannie (80) 3(4) [11|17|21]	5217
I Guess That's Why They Call It The Blues (84) 4(1) [6|15|23]	2975
I Don't Wanna Go On With You Like That (88) 2(1) [6|13|18]	2818
Sad Songs (Say So Much) (84) 5(1) [5|13|19]	2244
Candle In The Wind (Live) (88) 6(1) [3|12|21]	1764
Nikita (86) 7(2) [4|11|13]	1673
I'm Still Standing (83) 12(1) [0|12|16]	1429
Blue Eyes (82) 12(3) [0|10|18]	1339

Act ▶ Rank Peak [Top10s|Top40s|Total] Title (Pk Yr) Peak(Wk) [Top10Wks|Top40Wks|TotalWks]	Score
Empty Garden (Hey Hey Johnny) (82) 13(2) [0|10|17]	1202
Healing Hands (89) 13(1) [0|9|15]	900
Who Wears These Shoes? (84) 16(1) [0|10|14]	887
Wrap Her Up (85) 20(2) [0|10|14]	786
Kiss The Bride (83) 25(1) [0|8|12]	670
Nobody Wins (81) 21(2) [0|6|13]	641
A Word In Spanish (88) 19(1) [0|6|13]	618
In Neon (85) 38(3) [0|3|13]	498
Chloe (81) 34(1) [0|3|13]	437
(Sartorial Eloquence) Don't Ya Wanna Play This Game No More (80) 39(1) [0|2|12]	429
Through The Storm (89) - Aretha Franklin And Elton John [A] 16(1) [0|7|11]	318
Heartache All Over The World (86) 55(1) [0|0|8]	206
Robert JOHN ▶ 556 Pk:31 [0|1|4]	**1127**
Hey There Lonely Girl (80) 31(1) [0|4|13]	483
Lonely Eyes (80) 41(1) [0|0|11]	447
Bread And Butter (83) 68(2) [0|0|4]	99.2
Sherry (80) 70(1) [0|0|5]	97.0
JOHNNY HATES JAZZ ▶ 271 Pk:2 [1|2|2]	**4084**
Shattered Dreams (88) 2(3) [6|13|19]	3615
I Don't Want To Be A Hero (88) 31(2) [0|5|12]	469
Don JOHNSON ▶ 395 Pk:5 [1|2|3]	**2243**
Heartbeat (86) 5(2) [3|10|15]	1742
Heartache Away (86) 56(3) [0|0|11]	268
Till I Loved You (The Love Theme From Goya) (88) - Barbra Streisand And Don Johnson [A] 25(1) [0|5|12]	232
Holly JOHNSON ▶ 1095 Pk:65 [0|0|1]	**115**
Love Train (89) 65(1) [0|0|6]	115
Jesse JOHNSON ▶ 609 Pk:53 [0|0|4]	**851**
Crazay (86) - Jesse Johnson (Featuring Sly Stone) 53(3) [0|0|16]	442
Be Your Man (85) - Jesse Johnson's Revue 61(1) [0|0|11]	236
I Want My Girl (85) - Jesse Johnson's Revue 76(1) [0|0|8]	103
Love Struck (88) 78(1) [0|0|8]	70.1
Michael JOHNSON ▶ 1292 Pk:86 [0|0|1]	**35.7**
You Can Call Me Blue (80) 86(1) [0|0|3]	35.7
Tom JOHNSTON ▶ 768 Pk:34 [0|1|1]	**472**
Savannah Nights (80) 34(2) [0|2|12]	472
JON & VANGELIS ▶ 798 Pk:51 [0|0|2]	**426**
I'll Find My Way Home (82) 51(2) [0|0|9]	275
I Hear You Now (80) 58(1) [0|0|6]	151
Glenn JONES ▶ 953 Pk:66 [0|0|1]	**216**
We've Only Just Begun (The Romance Is Not Over) (87) 66(1) [0|0|14]	216
Grace JONES ▶ 1031 Pk:69 [0|0|1]	**154**
I'm Not Perfect (But I'm Perfect For You) (87) 69(1) [0|0|9]	154
Howard JONES ▶ 114 Pk:4 [2|8|10]	**9582**
Things Can Only Get Better (85) 5(1) [6|14|23]	2329
No One Is To Blame (86) 4(1) [5|14|23]	2312
Everlasting Love (89) 12(1) [0|11|19]	1171
You Know I Love You...Don't You? (86) 17(3) [0|10|16]	1034
Life In One Day (85) 19(1) [0|8|16]	765
New Song (84) 27(1) [0|6|15]	661
What Is Love? (84) 33(3) [0|4|13]	514
The Prisoner (89) 30(1) [0|4|13]	472
Like To Get To Know You Well (85) 49(1) [0|0|9]	262
All I Want (87) 76(1) [0|0|5]	63.1
Oran "Juice" JONES ▶ 523 Pk:9 [1|1|1]	**1316**
The Rain (86) 9(1) [2|9|19]	1316
Quincy JONES ▶ 456 Pk:14 [0|3|3]	**1703**
Just Once (81) - Quincy Jones Featuring James Ingram [A] 17(2) [0|10|23]	671

Acts With Singles

Act ▶ Rank Peak [Top10s|Top40s|Total] Title (Pk Yr) Peak(Wk) [Top10Wks|Top40Wks|TotalWks]	Score
Quincy JONES ▶ Continued	
One Hundred Ways (82) - Quincy Jones Featuring James Ingram [A] 14(1) [0|11|21]	566
Ai No Corrida (I-No-Ko-ree-da) (81) 28(2) 0|5|12]	466
Rickie Lee JONES ▶ 1000 Pk:64 [0|0|2]	**179**
A Lucky Guy (81) 64(2) [0|0|7]	137
The Real End (84) 83(2) [0|0|4]	41.7
JOURNEY ▶ 29 Pk:2 [6|16|19]	**23887**
Open Arms (82) 2(6) [10|14|18]	5744
Who's Crying Now (81) 4(2) [8|14|21]	3180
Separate Ways (Worlds Apart) (83) 8(6) [7|16|17]	2468
Don't Stop Believin' (81) 9(3) [4|13|16]	1589
Faithfully (83) 12(3) [0|11|16]	1452
Only The Young (85) 9(1) [1|11|16]	1163
I'll Be Alright Without You (87) 14(2) [0|9|21]	1131
Be Good To Yourself (86) 9(1) [1|10|15]	1119
Still They Ride (82) 19(3) [0|9|14]	907
Girl Can't Help It (86) 17(1) [0|8|14]	777
Send Her My Love (83) 23(2) [0|7|15]	718
Any Way You Want It (80) 23(2) [0|6|15]	715
Suzanne (86) 17(1) [0|7|13]	696
After The Fall (83) 23(1) [0|8|12]	692
Walks Like A Lady (80) 32(2) [0|4|13]	540
The Party's Over (Hopelessly In Love) (81) 34(1) [0|4|13]	475
Why Can't This Night Go On Forever (87) 60(1) [0|0|12]	235
Good Morning Girl/Stay Awhile (80) 55(2) [0|0|8]	206
Too Late (80) 70(1) [0|0|4]	80.9
JUDAS PRIEST ▶ 1076 Pk:67 [0|0|1]	**123**
You've Got Another Thing Comin' (82) 67(2) [0|0|7]	123
Patrick JUDE ▶ 1138 Pk:54 [0|0|1]	**99.8**
Strangers In A Strange World (84) - Jenny Burton & Patrick Jude [A] 54(2) [0|0|7]	99.8
JUMP 'N THE SADDLE ▶ 598 Pk:15 [0|1|1]	**905**
The Curly Shuffle (84) 15(1) [0|7|14]	905
Rob JUNGKLAS ▶ 1317 Pk:86 [0|0|1]	**28.9**
Make It Mean Something (87) 86(1) [0|0|3]	28.9
JUNIOR ▶ 787 Pk:30 [0|1|1]	**446**
Mama Used To Say (82) 30(1) [0|3|13]	446

K

Act ▶ Rank Peak [Top10s|Top40s|Total]	Score
KAJAGOOGOO ▶ 398 Pk:5 [1|1|2]	**2216**
Too Shy (83) 5(1) [4|12|19]	2160
Hang On Now (83) 78(1) [0|0|4]	55.7
Karen KAMON ▶ 1352 Pk:88 [0|0|1]	**20.3**
Loverboy (84) 88(1) [0|0|2]	20.3
Madleen KANE ▶ 1206 Pk:77 [0|0|1]	**66.5**
You Can (82) 77(1) [0|0|5]	66.5
KANE GANG ▶ 679 Pk:36 [0|1|2]	**660**
Motortown (87) 36(1) [0|3|16]	515
Don't Look Any Further (88) 64(1) [0|0|8]	145
KANO ▶ 1289 Pk:89 [0|0|1]	**36.7**
Can't Hold Back (Your Loving) (82) 89(2) [0|0|5]	36.7
KANSAS ▶ 368 Pk:17 [0|3|7]	**2528**
Play The Game Tonight (82) 17(3) [0|9|15]	910
All I Wanted (87) 19(1) [0|10|18]	885
Hold On (80) 40(1) [0|1|11]	366
Fight Fire With Fire (83) 58(1) [0|0|10]	156
Right Away (82) 73(2) [0|0|6]	113
Got To Rock On (81) 76(1) [0|0|5]	64.0
Power (87) 84(1) [0|0|3]	34.3
KASHIF ▶ 1235 Pk:62 [0|0|1]	**58.8**
Reservations For Two – Dionne & Kashif [A] (87) 62(1) [0|0|7]	58.8
KATRINA & The WAVES ▶ 340 Pk:9 [1|3|5]	**2868**
Walking On Sunshine (85) 9(1) [3|13|21]	1612
That's The Way (89) 16(1) [0|6|12]	613
Do You Want Crying (85) 37(2) [0|2|10]	398

Act ▶ Rank Peak [Top10s|Top40s|Total]	Score
Is That It? (86) 70(1) [0|0|8]	130
Que Te Quiero (85) 71(2) [0|0|6]	116
KBC BAND ▶ 1331 Pk:89 [0|0|1]	**25.0**
It's Not You, It's Not Me (86) 89(1) [0|0|4]	25.0
KC ▶ 1218 Pk:66 [0|0|1]	**62.2**
Dancin' In The Streets (80) - Teri DeSario with K.C. [A] 66(2) [0|0|6]	62.2
KC & The SUNSHINE BAND ▶ 160 Pk:1 [1|2|2]	**7468**
Please Don't Go (80) - K.C. & The Sunshine Band 1(1) [11|18|26]	6303
Give It Up (84) - KC 18(1) [0|10|21]	1166
Johnny KEMP ▶ 444 Pk:10 [1|2|2]	**1797**
Just Got Paid (88) 10(1) [1|11|21]	1314
Birthday Suit (89) 36(2) [0|3|14]	483
Joyce KENNEDY ▶ 965 Pk:40 [0|1|1]	**206**
The Last Time I Made Love (84) - Joyce Kennedy & Jeffrey Osborne [A] 40(2) [0|2|12]	206
Ray KENNEDY ▶ 1268 Pk:82 [0|0|1]	**45.4**
Just For The Moment (80) 82(2) [0|0|3]	45.4
KENNY G ▶ 246 Pk:4 [1|3|4]	**4512**
Songbird (87) 4(1) [4|12|22]	2290
Silhouette (89) 13(1) [0|10|17]	1088
Don't Make Me Wait For Love (87) - Kenny G Vocal by Lenny Williams 15(1) [0|9|19]	884
We've Saved The Best For Last (89) - Kenny G Vocal by Smokey Robinson 47(1) [0|0|9]	250
Nik KERSHAW ▶ 818 Pk:46 [0|0|1]	**387**
Wouldn't It Be Good (84) 46(1) [0|0|13]	387
Chaka KHAN ▶ 189 Pk:3 [1|1|7]	**5874**
I Feel For You (84) 3(3) [9|17|26]	4486
Through The Fire (85) 60(1) [0|0|19]	390
Love Of A Lifetime (86) 53(1) [0|0|12]	280
What Cha' Gonna Do For Me (81) 53(1) [0|0|9]	219
Own The Night (86) 57(2) [0|0|9]	202
This Is My Night (85) 60(2) [0|0|9]	184
Got To Be There (83) 67(1) [0|0|5]	113
KIARA ▶ 1254 Pk:78 [0|0|1]	**50.6**
This Time (89) - Kiara (Duet with Shanice Wilson) 78(1) [0|0|5]	50.6
Greg KIHN ▶ 756 Pk:30 [0|1|2]	**496**
Lucky (85) 30(1) [0|4|12]	472
Love And Rock And Roll (86) 92(1) [0|0|5]	24.1
Greg KIHN Band ▶ 187 Pk:2 [1|2|5]	**5982**
Jeopardy (83) 2(1) [7|14|22]	4294
The Breakup Song (They Don't Write 'Em) (81) 15(2) [0|13|23]	1346
Love Never Fails (83) 59(2) [0|0|6]	169
Happy Man (82) 62(1) [0|0|7]	143
Every Love Song (82) 82(1) [0|0|5]	29.3
Tom KIMMEL ▶ 1030 Pk:64 [0|0|1]	**156**
That's Freedom (87) 64(2) [0|0|6]	156
KING ▶ 904 Pk:55 [0|0|1]	**266**
Love And Pride (85) 55(1) [0|0|11]	266
B.B. KING ▶ 1211 Pk:68 [0|0|1]	**65.2**
When Love Comes To Town (89) - U2 With B.B. King [A] 68(1) [0|0|7]	65.2
Ben E. KING ▶ 438 Pk:9 [1|1|1]	**1837**
Stand By Me (86) 9(3) [4|13|21]	1837
Carole KING ▶ 443 Pk:12 [0|1|2]	**1802**
One Fine Day (80) 12(2) [0|10|17]	1479
One To One (82) 45(2) [0|2|10]	323
Evelyn "Champagne" KING ▶ 453 Pk:17 [0|2|5]	**1707**
Love Come Down (82) - Evelyn King 17(2) [0|8|16]	822
I'm In Love (81) - Evelyn King 40(2) [0|2|14]	444
Betcha She Don't Love You (83) - Evelyn King 49(3) [0|0|11]	314
Action (84) 75(1) [0|0|7]	101
Your Personal Touch (86) 86(1) [0|0|4]	26.7

Act ▶ Rank Peak [Top10s|Top40s|Total]	Score
The KINGBEES ▶ 1135 Pk:81 [0|0|1]	**102**
My Mistake (80) 81(2) [0|0|8]	102
KINGDOM COME ▶ 1163 Pk:69 [0|0|1]	**88.5**
Get It On (88) 69(1) [0|0|6]	88.5
The KINGS ▶ 717 Pk:43 [0|0|2]	**570**
Switchin' To Glide (80) 56(2) [0|0|13]	315
Switchin' To Glide//This Beat Goes On (80) 43(1) [0|0|10]	255
KINGS OF THE SUN ▶ 1392 Pk:98 [0|0|1]	**4.6**
Black Leather (88) 98(2) [0|0|2]	4.6
The KINKS ▶ 338 Pk:6 [1|2|6]	**2921**
Come Dancing (83) 6(2) [3|12|17]	1990
Don't Forget To Dance (83) 29(1) [0|4|10]	464
Do It Again (85) 41(1) [0|0|10]	300
Lola (Live Version) (80) 81(2) [0|0|6]	88.1
Better Things (82) 92(2) [0|0|8]	46.9
Destroyer (81) 85(1) [0|0|4]	32.7
Jim KIRK & The TM Singers ▶ 1221 Pk:71 [0|0|1]	**61.5**
Voice Of Freedom (80) 71(1) [0|0|3]	61.5
KISS ▶ 442 Pk:47 [0|0|9]	**1808**
Tears Are Falling (85) 51(1) [0|0|13]	322
Shandi (80) 47(1) [0|0|10]	304
Heaven's On Fire (84) 49(1) [0|0|10]	276
Lick It Up (83) 66(3) [0|0|11]	240
Reason To Live (88) 64(1) [0|0|12]	215
A World Without Heroes (82) 56(2) [0|0|9]	187
Hide Your Heart (89) 66(1) [0|0|10]	152
Crazy Crazy Nights (87) 65(1) [0|0|7]	107
Let's Put The X In Sex (89) 97(1) [0|0|2]	4.6
KISSING THE PINK ▶ 1261 Pk:87 [0|0|2]	**48.0**
Maybe This Day (83) 87(2) [0|0|5]	39.5
Certain Things Are Likely (87) - KTP 97(1) [0|0|4]	8.6
KIX ▶ 492 Pk:11 [0|1|1]	**1498**
Don't Close Your Eyes (89) 11(1) [0|13|23]	1498
KLIQUE ▶ 897 Pk:50 [0|0|1]	**277**
Stop Doggin' Me Around (83) 50(1) [0|0|9]	277
KLYMAXX ▶ 204 Pk:5 [1|3|5]	**5495**
I Miss You (85) 5(4) [7|17|29]	3259
I'd Still Say Yes (87) 18(1) [0|9|20]	1042
Man Size Love (86) 15(1) [0|8|15]	887
Meeting In The Ladies Room (85) 59(1) [0|0|11]	227
The Men All Pause (86) 80(1) [0|0|8]	80.6
The KNACK ▶ 713 Pk:38 [0|1|3]	**576**
Baby Talks Dirty (80) 38(1) [0|2|8]	326
Can't Put A Price On Love (80) 62(2) [0|0|6]	145
Pay The Devil (Ooo, Baby, Ooo) (81) 67(2) [0|0|5]	106
Gladys KNIGHT & The PIPS ▶ 532 Pk:13 [0|1|3]	**1286**
Love Overboard (88) 13(1) [0|9|14]	818
Landlord (80) 46(1) [0|0|9]	295
Save The Overtime (For Me) (83) 66(1) [0|0|10]	173
Holly KNIGHT ▶ 1009 Pk:59 [0|0|1]	**173**
Heart Don't Fail Me Now (88) 59(1) [0|0|9]	173
Jean KNIGHT ▶ 888 Pk:50 [0|0|1]	**286**
My Toot Toot (85) 50(1) [0|0|15]	286
Fred KNOBLOCK ▶ 538 Pk:18 [0|2|2]	**1229**
Why Not Me (80) 18(2) [0|7|14]	859
Killin' Time (81) - Fred Knoblock & Susan Anton [A] 28(1) [0|9|18]	370
KON KAN ▶ 547 Pk:15 [0|1|2]	**1159**
I Beg Your Pardon (89) 15(2) [0|9|18]	995
Puss N Boots/These Boots (Are Made For Walkin') (89) 58(1) [0|0|8]	163
KOOL & The GANG ▶ 13 Pk:1 [10|16|19]	**30914**
Celebration (81) 1(2) [7|21|30]	5462
Cherish (85) 2(3) [7|15|25]	3953
Joanna (84) 2(1) [6|16|24]	3878
Too Hot (80) 5(2) [5|13|18]	2769
Ladies Night (80) 8(2) [5|14|24]	2426
Misled (85) 10(1) [1|13|24]	1600

Acts With Singles

Act ▶ Rank Peak [Top10s|Top40s|Total] Title (Pk Yr) Peak(Wk) [Top10Wks|Top40Wks|TotalWks]	Score
KOOL & The GANG▶ *Continued*	
Fresh (85) 9(1) [2|11|19]	1484
Victory (87) 10(1) [1|12|18]	1442
Get Down On It (82) 10(2) [2|9|17]	1405
Tonight (84) 13(2) [0|10|18]	1305
Take My Heart (You Can Have It If You Want It) (81) 17(3) [0|12|17]	1228
Stone Love (87) 10(1) [1|10|18]	1157
Emergency (85) 18(2) [0|8|16]	904
Let's Go Dancin' (Ooh La, La, La) (83) 30(2) [0|7|15]	632
Big Fun (82) 21(2) [0|7|11]	615
Jones Vs. Jones (81) 39(2) [0|2|11]	336
Special Way (87) 72(1) [0|0|14]	186
Holiday (87) 66(1) [0|0|7]	112
Steppin' Out (82) 89(2) [0|0|2]	19.7
KOOL MOE DEE▶ 938 Pk:62 [0|0|2]	229
Wild, Wild West (88) 62(2) [0|0|11]	203
Go See The Doctor (87) 89(1) [0|0|5]	25.9
The KORGIS▶ 554 Pk:18 [0|1|1]	1136
Everybody's Got To Learn Sometime (80) 18(2) [0|11|19]	1136
KORONA▶ 887 Pk:43 [0|0|1]	287
Let Me Be (80) 43(1) [0|0|8]	287
KROKUS▶ 942 Pk:67 [0|0|2]	226
School's Out (86) 67(1) [0|0|7]	118
Midnite Maniac (84) 71(1) [0|0|6]	108

L

Act ▶ Rank Peak [Top10s|Top40s|Total] Title (Pk Yr) Peak(Wk) [Top10Wks|Top40Wks|TotalWks]	Score
LABAN▶ 1318 Pk:88 [0|0|1]	28.7
Love In Siberia (86) 88(2) [0|0|4]	28.7
Patti LaBELLE▶ 224 Pk:1 [1|3|7]	4962
On My Own (86) - Patti LaBelle And Michael McDonald [A] 1(3) [7|15|23]	2647
New Attitude (85) 17(1) [0|9|21]	1035
Stir It Up (85) 41(1) [0|0|14]	414
Oh, People (86) 29(1) [0|3|12]	413
If Only You Knew (84) 46(1) [0|0|13]	360
If You Asked Me To (89) 79(2) [0|0|5]	71.8
Love Has Finally Come At Last (84) - Bobby Womack and Patti LaBelle [A] 88(2) [0|0|5]	21.6
La FLAVOUR▶ 1366 Pk:91 [0|0|1]	16.0
Only The Lonely (Have A Reason To Be Sad) (80) 91(1) [0|0|2]	16.0
LAID BACK▶ 666 Pk:26 [0|1|1]	683
White Horse (84) 26(2) [0|4|18]	683
Greg LAKE▶ 860 Pk:48 [0|0|1]	323
Let Me Love You Once (81) 48(3) [0|0|10]	323
LAKESIDE▶ 1004 Pk:55 [0|0|1]	176
Fantastic Voyage (81) 55(1) [0|0|8]	176
Lorenzo LAMAS▶ 1264 Pk:85 [0|0|1]	46.8
Fools Like Me (85) 85(1) [0|0|5]	46.8
Robin LANE & The CHARTBUSTERS▶ 1320 Pk:87 [0|0|1]	28.5
When Things Go Wrong (80) 87(2) [0|0|3]	28.5
LANIER AND CO.▶ 832 Pk:48 [0|0|1]	361
After I Cry Tonight (83) 48(2) [0|0|13]	361
LARSEN-FEITEN Band▶ 716 Pk:29 [0|1|1]	573
Who'll Be The Fool Tonight (80) 29(2) [0|6|14]	573
Nicolette LARSON▶ 672 Pk:35 [0|1|2]	671
Let Me Go, Love (80) 35(2) [0|3|11]	432
I Only Want To Be With You (82) 53(1) [0|0|9]	239
David LASLEY▶ 828 Pk:36 [0|1|1]	368
If I Had My Wish Tonight (82) 36(2) [0|3|10]	368
James LAST Band▶ 704 Pk:28 [0|1|1]	595
The Seduction (Love Theme) (80) 28(2) [0|6|13]	595
Stacy LATTISAW▶ 352 Pk:21 [0|3|6]	2787
Let Me Be Your Angel (80) 21(2) [0|10|24]	1047
Love On A Two Way Street (81) 26(2) [0|7|17]	739

Act ▶ Rank Peak [Top10s|Top40s|Total] Title (Pk Yr) Peak(Wk) [Top10Wks|Top40Wks|TotalWks]	Score
Miracles (83) 40(1) [0|1|16]	538
Nail It To The Wall (86) 48(1) [0|0|13]	311
Attack Of The Name Game (82) 70(2) [0|0|6]	89.2
Perfect Combination (84) - Stacy Lattisaw & Johnny Gill [A] 75(1) [0|0|9]	63.5
Cyndi LAUPER▶ 19 Pk:1 [8|10|13]	26545
Time After Time (84) 1(2) [9|14|20]	5392
Girls Just Want To Have Fun (84) 2(2) [8|14|25]	4890
She Bop (84) 3(3) [8|14|18]	3772
True Colors (86) 1(2) [6|12|20]	3650
Change Of Heart (87) 3(1) [5|13|17]	2326
All Through The Night (84) 5(1) [5|14|19]	2186
I Drove All Night (89) 6(1) [2|10|15]	1380
The Goonies 'R' Good Enough (85) 10(1) [1|9|15]	1086
What's Going On (87) 12(1) [0|10|13]	961
Money Changes Everything (85) 27(1) [0|6|13]	549
Hole In My Heart (All The Way To China) (88) 54(2) [0|0|8]	177
My First Night Without You (89) 62(1) [0|0|6]	108
Boy Blue (87) 71(1) [0|0|4]	68.5
Debra LAWS▶ 1306 Pk:90 [0|0|1]	32.1
Very Special (81) 90(1) [0|0|5]	32.1
Ronnie LAWS▶ 996 Pk:60 [0|0|1]	183
Stay Awake (81) 60(2) [0|0|9]	183
Lenny Le BLANC▶ 998 Pk:55 [0|0|1]	182
Somebody Send My Baby Home (81) 55(1) [0|0|7]	182
LED ZEPPELIN▶ 618 Pk:21 [0|1|1]	807
Fool In The Rain (80) 21(2) [0|8|13]	807
Johnny LEE▶ 357 Pk:5 [1|1|2]	2713
Lookin' For Love (80) 5(2) [5|13|21]	2465
Bet Your Heart On Me (81) 54(2) [0|0|9]	248
Larry LEE▶ 1304 Pk:81 [0|0|1]	32.6
Don't Talk (82) 81(1) [0|0|12]	32.6
Paul LEKAKIS▶ 809 Pk:43 [0|0|1]	404
Boom Boom (Let's Go Back To My Room) (87) 43(1) [0|0|13]	404
John LENNON▶ 59 Pk:1 [4|4|5]	17434
(Just Like) Starting Over (80) 1(5) [14|19|22]	7930
Woman (81) 2(3) [12|17|20]	6131
Nobody Told Me (84) 5(1) [4|11|14]	1920
Watching The Wheels (81) 10(2) [2|10|17]	1316
I'm Stepping Out (84) 55(1) [0|0|7]	136
John LENNON and the PLASTIC ONO BAND▶ 1252 Pk:80 [0|0|1]	52.1
Jealous Guy (88) - John Lennon and the Plastic Ono Band (With the Flux Fiddlers) 80(1) [0|0|4]	52.1
Julian LENNON▶ 222 Pk:5 [2|4|6]	5009
Too Late For Goodbyes (85) 5(1) [5|12|17]	2063
Valotte (85) 9(1) [4|12|19]	1668
Say You're Wrong (85) 21(2) [0|8|12]	634
Stick Around (85) 32(1) [0|4|13]	471
Jesse (85) 54(1) [0|0|6]	159
Now You're In Heaven (89) 93(1) [0|0|4]	15.0
Annie LENNOX & Al GREEN▶ 539 Pk:9 [1|1|1]	1205
Put A Little Love In Your Heart (89) 9(1) [2|10|17]	1205
Le ROUX▶ 615 Pk:18 [0|1|3]	824
Nobody Said It Was Easy (Lookin' For The Lights) (82) 18(2) [0|6|13]	714
The Last Safe Place On Earth (82) 77(1) [0|0|5]	63.9
Carrie's Gone (83) 81(2) [0|0|4]	46.2
LEVEL 42▶ 322 Pk:7 [1|2|4]	3284
Something About You (86) 7(2) [3|14|27]	2029
Lessons In Love (87) 12(1) [0|10|18]	1188
Running In The Family (87) 83(1) [0|0|4]	35.6
Hot Water (86) 87(2) [0|0|4]	31.9
LEVERT▶ 441 Pk:5 [1|1|1]	1814
Casanova (87) 5(1) [4|12|18]	1814

Act ▶ Rank Peak [Top10s|Top40s|Total] Title (Pk Yr) Peak(Wk) [Top10Wks|Top40Wks|TotalWks]	Score
Marcy LEVY▶ 1127 Pk:50 [0|0|1]	104
Help Me! (80) - Marcy Levy & Robin Gibb [A] 50(2) [0|0|10]	104
Huey LEWIS and the NEWS▶ 12 Pk:1 [12|15|17]	33246
The Power Of Love (85) 1(2) [8|15|19]	4636
Stuck With You (86) 1(3) [7|13|19]	4301
Jacob's Ladder (87) 1(1) [6|12|15]	3536
Hip To Be Square (86) 3(2) [5|12|16]	2647
The Heart Of Rock & Roll (84) 6(4) [7|14|20]	2607
Perfect World (88) 3(2) [5|12|15]	2388
I Want A New Drug (84) 6(2) [5|13|19]	2382
If This Is It (84) 6(2) [4|13|17]	2110
Heart And Soul (83) 8(2) [2|13|21]	1827
Do You Believe In Love (82) 7(3) [4|13|17]	1774
Doing It All For My Baby (87) 6(1) [3|11|16]	1480
I Know What I Like (87) 9(1) [1|10|14]	1162
Walking On A Thin Line (84) 18(2) [0|10|15]	933
Small World (88) 25(1) [0|6|11]	482
Hope You Love Me Like You Say You Do (82) 36(2) [0|4|11]	413
Workin' For A Livin' (82) 41(2) [0|0|9]	331
Give Me The Keys (And I'll Drive You Crazy) (89) 47(2) [0|0|8]	235
Shirley LEWIS▶ 1245 Pk:84 [0|0|1]	54.6
Realistic (89) 84(2) [0|0|6]	54.6
Gordon LIGHTFOOT▶ 926 Pk:50 [0|0|1]	248
Baby Step Back (82) 50(2) [0|0|8]	248
LIL LOUIS▶ 882 Pk:50 [0|0|1]	291
French Kiss (89) 50(1) [0|0|13]	291
LIMAHL▶ 560 Pk:17 [0|1|2]	1112
Never Ending Story (85) 17(1) [0|9|19]	925
Only For Love (85) 51(2) [0|0|7]	187
LIMITED WARRANTY▶ 1137 Pk:79 [0|0|1]	101
Victory Line (86) 79(2) [0|0|8]	101
LIPPS, INC.▶ 162 Pk:1 [1|1|2]	7424
Funkytown (80) 1(4) [9|15|23]	7303
Rock It (80) 64(1) [0|0|7]	121
LIQUID GOLD▶ 1287 Pk:86 [0|0|1]	37.4
What's She Got (83) 86(1) [0|0|4]	37.4
LISA LISA And CULT JAM▶ 142 Pk:1 [2|3|3]	8045
Head To Toe (87) 1(1) [7|14|20]	4139
Lost In Emotion (87) 1(1) [6|13|20]	3495
Little Jackie Wants To Be A Star (89) 29(1) [0|4|11]	410
LISA LISA And CULT JAM With FULL FORCE▶ 337 Pk:8 [1|2|4]	2926

All Cried Out (86) - Lisa Lisa And Cult Jam With
   Full Force Featuring Paul Anthony And
   Bow Legged Lou   8(1) [3|13|26]
	1799

I Wonder If I Take You Home (85) 34(1) [0|6|21]	707
Can You Feel The Beat (85) 69(1) [0|0|20]	317

Someone To Love Me For Me (87) -
   Lisa Lisa & Cult Jam Featuring Full Force
   78(1) [0|0|10]
	103

LITTLE RICHARD▶ 864 Pk:42 [0|0|1]	315
Great Gosh A'Mighty! (86) 42(2) [0|0|10]	315
LITTLE RIVER BAND▶ 107 Pk:6 [3|7|9]	10540
The Night Owls (81) 6(2) [6|14|21]	2525
Cool Change (80) 10(1) [1|13|18]	1881
Take It Easy On Me (82) 10(2) [2|15|19]	1760
The Other Guy (83) 11(3) [0|13|18]	1703
Man On Your Mind (82) 14(3) [0|8|16]	1174
We Two (83) 22(2) [0|6|12]	659
You're Driving Me Out Of My Mind (83) 35(1) [0|3|11]	459
Playing To Win (85) - LRB 60(1) [0|0|8]	199
It's Not A Wonder (Live) (80) 51(1) [0|0|6]	180

Acts With Singles

Act ▶ Rank Peak [Top10s|Top40s|Total] Title (Pk Yr) Peak(Wk) [Top 10Wks|Top40Wks|TotalWks]	Score
LITTLE STEVEN▶ 1024 Pk:63 [0|0|1]	**163**
Forever (83) - Little Steven And The Disciples Of Soul	
63(2) [0|0|9]	163
LIVING COLOUR▶ 499 Pk:13 [0|2|3]	**1454**
Cult Of Personality (89) 13(2) [0|9|15]	989
Glamour Boys (89) 31(1) [0|3|13]	420
Open Letter (To A Landlord) (89) 82(1) [0|0|5]	44.8
LIVING IN A BOX▶ 622 Pk:17 [0|1|2]	**794**
Living In A Box (87) 17(2) [0|7|15]	754
So The Story Goes (87) 81(1) [0|0|4]	40.1
LL COOL J▶ 391 Pk:14 [0|3|4]	**2259**
I'm That Type Of Guy (89) 15(2) [0|8|16]	890
I Need Love (87) 14(2) [0|8|13]	838
Going Back To Cali (88) 31(2) [0|5|14]	494
I'm Bad (87) 84(1) [0|0|4]	36.2
LOBO▶ 1110 Pk:75 [0|0|1]	**110**
Holdin' On For Dear Love (80) 75(2) [0|0|8]	110
Kenny LOGGINS▶ 36 Pk:1 [4|13|17]	**22325**
Footloose (84) 1(3) [11|16|23]	6733
Danger Zone (86) 2(1) [6|13|21]	2987
I'm Alright (80) 7(2) [5|12|22]	2264
This Is It (80) 11(2) [0|16|23]	2011
Heart To Heart (83) 15(5) [0|13|17]	1542
Meet Me Half Way (87) 11(2) [0|12|25]	1448
Nobody's Fool (88) 8(1) [3|11|18]	1379
I'm Free (Heaven Helps The Man) (84)	
22(2) [0|8|14]	714
Don't Fight It (82) -	
Kenny Loggins with Steve Perry 17(2) [0|6|12]	700
Welcome To Heartlight (83) 24(2) [0|7|14]	677
Forever (85) 40(1) [0|1|22]	559
Keep The Fire (80) 36(1) [0|2|13]	480
Vox Humana (85) 29(1) [0|4|10]	456
Playing With The Boys (86) 60(2) [0|0|12]	219
Tell Her (89) 76(1) [0|0|8]	92.5
I'm Gonna Miss You (88) 82(1) [0|0|5]	46.4
I'll Be There (85) 88(1) [0|0|2]	18.1
LONE JUSTICE▶ 751 Pk:47 [0|0|3]	**508**
Shelter (87) 47(2) [0|0|12]	353
Ways To Be Wicked (85) 71(1) [0|0|6]	86.7
Sweet, Sweet Baby (I'm Falling) (85)	
73(1) [0|0|5]	68.7
LOOSE ENDS▶ 878 Pk:43 [0|0|1]	**294**
Hangin' On A String (Contemplating) (85)	
43(1) [0|0|10]	294
Denise LOPEZ▶ 715 Pk:31 [0|1|2]	**575**
Sayin' Sorry (Don't Make It Right) (88)	
31(1) [0|5|17]	563
If You Feel It (88) 94(1) [0|0|3]	12.2
Jeff LORBER▶ 726 Pk:27 [0|1|1]	**548**
Facts Of Love (87) - Jeff Lorber Featuring	
Karyn White 27(1) [0|5|16]	548
Gloria LORING & Carl ANDERSON▶ 278	
Pk:2 [1|1|1]	**3983**
Friends And Lovers (86) 2(2) [7|14|21]	3983
LOS LOBOS▶ 207 Pk:1 [1|2|3]	**5418**
La Bamba (87) 1(3) [7|14|21]	4741
Come On, Let's Go (87) 21(2) [0|7|14]	610
Will The Wolf Survive? (85) 78(1) [0|0|5]	66.4
LOVE AND MONEY▶ 1169 Pk:75 [0|0|1]	**85.7**
Halleluiah Man (89) 75(1) [0|0|7]	85.7
LOVE AND ROCKETS▶ 358 Pk:3 [1|1|2]	**2706**
So Alive (89) 3(1) [6|12|20]	2663
No Big Deal (89) 81(1) [0|0|4]	43.2
LOVERBOY▶ 126 Pk:9 [2|9|12]	**8686**
Lovin' Every Minute Of It (85) 9(1) [1|11|21]	1425
Hot Girls In Love (83) 11(1) [0|11|16]	1420
This Could Be The Night (86) 10(1) [1|10|18]	1278
Heaven In Your Eyes (86) 12(2) [0|11|17]	1132
Working For The Weekend (82) 29(2) [0|8|20]	837

Act ▶ Rank Peak [Top10s|Top40s|Total] Title (Pk Yr) Peak(Wk) [Top 10Wks|Top40Wks|TotalWks]	Score
Turn Me Loose (81) 35(1) [0|6|17]	639
When It's Over (82) 26(2) [0|6|15]	611
Queen Of The Broken Hearts (83) 34(1) [0|3|12]	481
Notorious (87) 38(2) [0|3|14]	415
The Kid Is Hot Tonite (81) 55(1) [0|0|7]	173
Dangerous (85) 65(1) [0|0|9]	163
Lead A Double Life (86) 68(2) [0|0|7]	114
Nick LOWE And His COWBOY OUTFIT▶ 1150	
Pk:77 [0|0|1]	**94.5**
I Knew The Bride (When She Use To Rock And Roll)	
(86) 77(1) [0|0|9]	94.5
L.T.D.▶ 714 Pk:40 [0|1|1]	**576**
Shine On (81) 40(1) [0|1|16]	576
L'TRIMM▶ 861 Pk:54 [0|0|1]	**321**
Cars With The Boom (88) 54(1) [0|0|15]	321
LULU▶ 511 Pk:18 [0|1|2]	**1409**
I Could Never Miss You (More Than I Do) (81)	
18(2) [0|10|18]	1032
If I Were You (82) 44(1) [0|0|11]	376
Cheryl LYNN▶ 924 Pk:69 [0|0|2]	**248**
Shake It Up Tonight (81) 70(1) [0|0|7]	124
Encore (84) 69(1) [0|0|8]	124
Jeff LYNNE▶ 1303 Pk:85 [0|0|1]	**33.1**
Video! (84) 85(1) [0|0|3]	33.1

M

Act ▶ Rank Peak [Top10s|Top40s|Total] Title (Pk Yr) Peak(Wk) [Top 10Wks|Top40Wks|TotalWks]	Score
Ralph MacDONALD▶ 982 Pk:58 [0|0|1]	**192**
In The Name Of Love (84) - Ralph MacDonald	
(with Bill Withers) 58(1) [0|0|10]	192
Mary MacGREGOR▶ 1183 Pk:72 [0|0|1]	**78.3**
Dancin' Like Lovers (80) 72(2) [0|0|4]	78.3
MADNESS▶ 366 Pk:7 [1|2|3]	**2570**
Our House (83) 7(1) [4|13|19]	1950
It Must Be Love (83) 33(2) [0|5|12]	537
The Sun And The Rain (84) 72(1) [0|0|5]	83.0
MADONNA▶ 2 Pk:1 [17|18|18]	**59786**
Like A Virgin (84) 1(6) [9|14|19]	6531
Crazy For You (85) 1(1) [9|14|21]	5654
Like A Prayer (89) 1(3) [7|12|16]	4497
Papa Don't Preach (86) 1(2) [7|13|18]	4347
Live To Tell (86) 1(1) [6|13|18]	4244
Material Girl (85) 2(2) [6|12|17]	3774
Open Your Heart (87) 1(1) [6|14|18]	3575
Who's That Girl (87) 1(1) [6|11|16]	3537
Express Yourself (89) 2(2) [5|11|16]	3158
Causing A Commotion (87) 2(3) [5|11|18]	3098
True Blue (86) 3(3) [5|12|16]	2882
Cherish (89) 2(2) [5|12|15]	2798
La Isla Bonita (87) 4(3) [5|12|17]	2445
Lucky Star (84) 4(1) [5|12|16]	2425
Borderline (84) 10(1) [1|15|30]	2009
Angel (85) 5(1) [4|12|17]	1880
Dress You Up (85) 5(1) [4|11|16]	1711
Holiday (84) 16(2) [0|11|21]	1219
MAGAZINE 60▶ 884 Pk:56 [0|0|1]	**290**
Don Quichotte (86) 56(1) [0|0|11]	290
MAI TAI▶ 1113 Pk:71 [0|0|1]	**109**
Female Intuition (85) 71(1) [0|0|7]	109
Melissa MANCHESTER▶ 253 Pk:5 [1|2|7]	**4387**
You Should Hear How She Talks About You (82)	
5(3) [5|15|25]	3133
Fire In The Morning (80) 32(2) [0|5|13]	573
Nice Girls (83) 42(2) [0|0|11]	387
Lovers After All (81) - Melissa Manchester And	
Peabo Bryson [A] 54(1) [0|0|9]	108
Mathematics (85) 74(1) [0|0|7]	72.5
No One Can Love You More Than Me (83)	
78(1) [0|0|4]	64.0
Thief Of Hearts (84) 86(1) [0|0|6]	48.6

Act ▶ Rank Peak [Top10s|Top40s|Total] Title (Pk Yr) Peak(Wk) [Top 10Wks|Top40Wks|TotalWks]	Score
M + M▶ 1035 Pk:63 [0|0|1]	**151**
Black Stations/White Stations (84)	
63(1) [0|0|7]	151
MANFRED MANN'S EARTH BAND▶ 633	
Pk:22 [0|1|1]	**764**
Runner (84) 22(1) [0|8|15]	764
Chuck MANGIONE▶ 573 Pk:18 [0|1|1]	**1035**
Give It All You Got (80) 18(2) [0|9|16]	1035
The MANHATTANS▶ 310 Pk:5 [1|1|3]	**3415**
Shining Star (80) 5(3) [5|14|25]	3269
Crazy (83) 72(1) [0|0|6]	102
You Send Me (85) 81(1) [0|0|5]	44.4
MANHATTAN TRANSFER▶ 307 Pk:7 [1|3|6]	**3501**
Boy From New York City (81) 7(3) [6|13|21]	2291
Twilight Zone/Twilight Tone (80) 30(2) [0|4|12]	543
Spice Of Life (83) 40(2) [0|2|13]	464
Trickle Trickle (80) 73(1) [0|0|8]	105
Route 66 (82) 78(1) [0|0|5]	59.6
Baby Come Back To Me (The Morse Code Of Love)	
(85) 83(1) [0|0|3]	39.0
Barry MANILOW▶ 149 Pk:10 [1|10|13]	**7794**
I Made It Through The Rain (81) 10(1) [1|11|16]	1373
The Old Songs (81) 15(2) [0|10|16]	1143
Read 'Em And Weep (84) 18(2) [0|10|14]	981
When I Wanted You (80) 20(2) [0|7|16]	855
Somewhere Down The Road (82) 21(2) [0|7|15]	720
Some Kind Of Friend (83) 26(2) [0|7|16]	717
Memory (83) 39(2) [0|2|14]	463
I Don't Want To Walk Without You (80)	
36(3) [0|4|11]	451
Oh Julie (82) 38(2) [0|2|11]	388
Let's Hang On (82) 32(1) [0|3|10]	351
Lonely Together (81) 45(2) [0|0|10]	299
I'm Your Man (86) 86(2) [0|0|5]	40.0
Hey Mambo (88) - Barry Manilow with	
Kid Creole And The Coconuts 90(1) [0|0|2]	12.5
--MANN, Manfred see: MANFRED MANN	
Benny MARDONES▶ 393 Pk:11 [0|2|2]	**2254**
Into The Night (80) 11(2) [0|12|20]	1483
Into The Night (89) 20(1) [0|7|17]	771
Teena MARIE▶ 297 Pk:4 [1|2|5]	**3684**
Lovergirl (85) 4(1) [5|13|24]	2721
I Need Your Lovin' (81) 37(1) [0|3|14]	489
Square Biz (81) 50(1) [0|0|13]	397
Ooo La La La (88) 85(1) [0|0|5]	40.0
Jammin (85) 81(1) [0|0|3]	37.4
MARILLION▶ 1089 Pk:74 [0|0|1]	**117**
Kayleigh (85) 74(1) [0|0|8]	117
Ziggy MARLEY & The MELODY MAKERS▶ 824	
Pk:39 [0|1|1]	**373**
Tomorrow People (88) 39(1) [0|1|13]	373
M/A/R/R/S▶ 527 Pk:13 [0|1|1]	**1300**
Pump Up The Volume (88) 13(3) [0|11|23]	1300
MARSHALL TUCKER Band▶ 1251	
Pk:79 [0|0|1]	**52.4**
It Takes Time (80) 79(2) [0|0|3]	52.4
--MARTHA And The MUFFINS see: M + M	
MARTIKA▶ 231 Pk:1 [1|3|3]	**4821**
Toy Soldiers (89) 1(2) [6|13|20]	3573
More Than You Know (89) 18(1) [0|8|20]	832
I Feel The Earth Move (89) 25(1) [0|5|11]	417
Eric MARTIN▶ 1353 Pk:87 [0|0|1]	**19.8**
Information (85) 87(1) [0|0|2]	19.8
Marilyn MARTIN▶ 332 Pk:1 [1|2|2]	**3023**
Separate Lives (Love Theme From White Nights) (85)	
- Phil Collins and Marilyn Martin [A]	
1(1) [9|16|21]	2376
Night Moves (86) 28(1) [0|6|18]	647
Nancy MARTINEZ▶ 657 Pk:32 [0|1|1]	**711**
For Tonight (86) 32(3) [0|7|21]	711

Acts With Singles

Act ▶ Rank Peak [Top10s|Top40s|Total] Title (Pk Yr) Peak(Wk) [Top10Wks|Top40Wks|TotalWks]	Score
Richard MARX▶ 42 Pk:1 [7|7|7]	**21085**
Right Here Waiting (89) 1(3) [7|13|21]	4616
Hold On To The Nights (88) 1(1) [6|14|21]	3490
Endless Summer Nights (88) 2(2) [6|15|21]	3112
Satisfied (89) 1(1) [5|13|15]	2851
Should've Known Better (87) 3(1) [5|13|21]	2631
Don't Mean Nothing (87) 3(1) [4|12|21]	2396
Angelia (89) 4(1) [4|11|17]	1991
MARY JANE GIRLS▶ 362 Pk:7 [1|1|3]	**2643**
In My House (85) 7(3) [5|12|22]	2049
Walk Like A Man (86) 41(1) [0|0|10]	304
Wild And Crazy Love (85) 42(1) [0|0|10]	289
Dave MASON▶ 1189 Pk:71 [0|0|1]	**74.8**
Save Me (80) 71(1) [0|0|3]	74.8
Vaughan MASON And CREW▶ 1275 Pk:81 [0|0|1]	**43.6**
Bounce, Rock, Skate, Roll Pt. 1 (80) 81(1) [0|0|3]	43.6
Wayne MASSEY▶ 1370 Pk:92 [0|0|1]	**14.3**
One Life To Live (80) 92(1) [0|0|2]	14.3
Johnny MATHIS▶ 881 Pk:38 [0|1|2]	**291**
Friends In Love (82) - Dionne Warwick And Johnny Mathis [A] 38(2) [0|3|13]	200
Simple (84) 81(1) [0|0|8]	91.0
Christopher MAX▶ 1106 Pk:75 [0|0|1]	**111**
Serious Kinda Girl (89) 75(2) [0|0|8]	111
MAZE Featuring Frankie BEVERLY▶ 1124 Pk:80 [0|0|2]	**105**
Love Is The Key (83) 80(2) [0|0|5]	68.1
Back In Stride (85) 88(1) [0|0|6]	36.8
Mac McANALLY▶ 788 Pk:41 [0|0|1]	**442**
Minimum Love (83) 41(1) [0|0|12]	442
Paul McCARTNEY▶ 30 Pk:1 [6|9|12]	**23877**
Ebony And Ivory (82) 1(7) [12|15|19]	8625
Say Say Say (83) - Paul McCartney and Michael Jackson [A] 1(6) [13|18|22]	5168
The Girl Is Mine (83) - Michael Jackson/ Paul McCartney [A] 2(3) [10|14|18]	2673
No More Lonely Nights (84) 6(2) [5|14|18]	2137
Take It Away (82) 10(5) [5|11|16]	1711
Spies Like Us (86) 7(1) [3|11|17]	1530
So Bad (84) 23(2) [0|8|14]	759
Press (86) 21(2) [0|6|11]	563
My Brave Face (89) 25(1) [0|5|10]	452
Tug Of War (82) 53(2) [0|0|8]	191
Stranglehold (86) 81(2) [0|0|6]	53.4
This One (89) 94(1) [0|0|3]	14.9
Paul McCARTNEY & WINGS▶ 140 Pk:1 [1|1|1]	**8092**
Coming Up (Live At Glasgow) (80) 1(3) [11|16|21]	8092
Delbert McCLINTON▶ 404 Pk:8 [1|1|2]	**2115**
Giving It Up For Your Love (81) 8(3) [5|14|19]	2028
Shotgun Rider (81) 70(2) [0|0|6]	86.3
Michael McDONALD▶ 135 Pk:1 [3|4|5]	**8395**
I Keep Forgettin' (82) 4(3) [7|13|19]	2999
On My Own (86) - Patti LaBelle And Michael McDonald [A] 1(3) [7|15|23]	2647
Sweet Freedom (86) 7(3) [3|13|20]	1864
No Lookin' Back (85) 34(1) [0|4|12]	451
I Gotta Try (82) 44(4) [0|0|11]	433
Bobby McFERRIN▶ 292 Pk:1 [1|1|1]	**3746**
Don't Worry Be Happy (Edit) (88) 1(2) [6|13|26]	3746
McGUFFEY LANE▶ 1225 Pk:85 [0|0|2]	**61.0**
Long Time Lovin' You (81) 85(1) [0|0|7]	53.7
Start It All Over (82) 97(1) [0|0|3]	7.3
Peter McIAN▶ 1003 Pk:52 [0|0|1]	**177**
Solitaire (80) 52(1) [0|0|7]	177
Bob & Doug McKENZIE▶ 607 Pk:16 [0|1|1]	**854**
Take Off (82) 16(2) [0|9|14]	854
Don McLEAN▶ 273 Pk:5 [1|3|4]	**4061**
Crying (81) 5(3) [6|15|18]	2868

Act ▶ Rank Peak [Top10s|Top40s|Total] Title (Pk Yr) Peak(Wk) [Top10Wks|Top40Wks|TotalWks]	Score
Since I Don't Have You (81) 23(2) [0|6|14]	631
Castles In The Air (81) 36(3) [0|5|14]	534
It's Just The Sun (81) 83(1) [0|0|2]	27.8
Gerard McMAHON▶ 1325 Pk:85 [0|0|1]	**27.3**
Count On Me (83) 85(1) [0|0|3]	27.3
Larry John McNALLY▶ 1333 Pk:86 [0|0|1]	**24.6**
Just Like Paradise (81) 86(2) [0|0|2]	24.6
Shamus M'COOL▶ 1277 Pk:80 [0|0|1]	**43.1**
American Memories (81) 80(2) [0|0|3]	43.1
Christine McVIE▶ 426 Pk:10 [1|2|2]	**1909**
Got A Hold On Me (84) 10(1) [1|11|16]	1435
Love Will Show Us How (84) 30(1) [0|6|10]	474
MEAT LOAF▶ 1316 Pk:84 [0|0|1]	**29.6**
I'm Gonna Love Her For Both Of Us (81) 84(1) [0|0|3]	29.6
MECO▶ 459 Pk:18 [0|2|5]	**1684**
Empire Strikes Back (Medley) (80) 18(2) [0|8|14]	933
Pop Goes The Movies (Part I) (82) 35(2) [0|3|11]	422
Ewok Celebration (83) 60(1) [0|0|8]	212
Love Theme From Shogun (Mariko's Theme) (80) 70(2) [0|0|4]	75.5
Summer '81 Medley (81) - Cantina Band 81(2) [0|0|3]	41.7
Glenn MEDEIROS▶ 424 Pk:12 [0|1|4]	**1954**
Nothing's Gonna Change My Love For You (87) 12(1) [0|13|25]	1547
Lonely Won't Leave Me Alone (88) 67(1) [0|0|11]	168
Long And Lasting Love (Once In A Lifetime) (88) 68(1) [0|0|10]	154
Watching Over You (87) 80(1) [0|0|8]	85.0
Bill MEDLEY▶ 388 Pk:1 [1|1|3]	**2294**

(I've Had) The Time Of My Life (87) -
   Bill Medley And Jennifer Warnes [A]
   1(1) [6|15|21]
	2057

Right Here And Now (82) 58(3) [0|0|4]	199
Don't Know Much (81) 88(1) [0|0|4]	37.0
Randy MEISNER▶ 406 Pk:19 [0|3|3]	**2103**
Hearts On Fire (81) 19(3) [0|9|15]	895
Deep Inside My Heart (80) 22(2) [0|7|16]	699
Never Been In Love (82) 28(3) [0|6|11]	509
MEL & KIM▶ 1188 Pk:78 [0|0|1]	**75.3**
Showing Out (Get Fresh At The Weekend) (87) 78(1) [0|0|7]	75.3
John MELLENCAMP▶ 10 Pk:1 [9|17|20]	**35178**
Hurts So Good (82) - John Cougar 2(4) [16|22|28]	7863
Jack & Diane (82) - John Cougar 1(4) [10|17|22]	7190
R.O.C.K. In The U.S.A. (A Salute To 60's Rock) (86) - John Cougar Mellencamp 2(1) [5|11|17]	2895
Small Town (85) - John Cougar Mellencamp 6(4) [5|13|18]	2127
Lonely Ol' Night (85) - John Cougar Mellencamp 6(2) [5|13|20]	1951
Crumblin' Down (83) - John Cougar Mellencamp 9(3) [3|11|16]	1668
Cherry Bomb (88) - John Cougar Mellencamp 8(3) [3|12|21]	1512
Pink Houses (84) - John Cougar Mellencamp 8(2) [2|11|16]	1484
Paper In Fire (87) - John Cougar Mellencamp 9(3) [2|10|18]	1315
Ain't Even Done With The Night (81) - John Cougar 17(2) [0|12|21]	1205
Hand To Hold On To (83) - John Cougar 19(2) [0|11|18]	1066
The Authority Song (84) - John Cougar Mellencamp 15(1) [0|9|15]	1065
Check It Out (88) - John Cougar Mellencamp 14(1) [0|9|15]	941
This Time (80) - John Cougar 27(2) [0|7|17]	760

Act ▶ Rank Peak [Top10s|Top40s|Total] Title (Pk Yr) Peak(Wk) [Top10Wks|Top40Wks|TotalWks]	Score
Pop Singer (89) - John Cougar Mellencamp 15(1) [0|7|12]	712
Rain On The Scarecrow (86) - John Cougar Mellencamp 21(1) [0|6|12]	553
Rumbleseat (86) - John Cougar Mellencamp 28(1) [0|4|13]	508
Jackie Brown (89) - John Cougar Mellencamp 48(1) [0|0|8]	193
Rooty Toot Toot (88) - John Cougar Mellencamp 61(1) [0|0|8]	142
Small Paradise (80) - John Cougar 87(1) [0|0|3]	29.8
MEN AT WORK▶ 49 Pk:1 [4|5|6]	**19415**
Down Under (83) 1(4) [10|19|25]	7153
Who Can It Be Now? (82) 1(1) [9|17|27]	5487
Overkill (83) 3(1) [8|13|16]	3933
It's A Mistake (83) 6(2) [4|12|15]	2066
Dr. Heckyll & Mr. Jive (83) 28(2) [0|5|11]	490
Everything I Need (85) 47(1) [0|3|9]	286
Sergio MENDES▶ 217 Pk:4 [1|2|4]	**5147**
Never Gonna Let You Go (83) 4(4) [8|16|23]	3938
Alibis (84) 29(2) [0|7|19]	812
Rainbow's End (83) 52(2) [0|0|8]	234
Olympia (84) 58(1) [0|0|7]	164
MENUDO▶ 945 Pk:62 [0|0|1]	**224**
Hold Me (85) 62(1) [0|0|11]	224
MEN WITHOUT HATS▶ 213 Pk:3 [1|2|3]	**5256**
The Safety Dance (83) 3(4) [7|16|24]	4322
Pop Goes The World (88) 20(1) [0|10|21]	898
I Like (83) 84(1) [0|0|3]	35.9
Freddie MERCURY▶ 1022 Pk:69 [0|0|2]	**164**
Love Kills (84) 69(1) [0|0|6]	109
I Was Born To Love You (85) 76(1) [0|0|4]	55.0
METALLICA▶ 764 Pk:35 [0|1|1]	**481**
One (89) 35(1) [0|4|15]	481
Pat METHENY Group▶ 923 Pk:32 [0|1|1]	**248**
This Is Not America (85) - David Bowie/ Pat Metheny Group [A] 32(2) [0|4|12]	248
--George MICHAEL see also: WHAM	
George MICHAEL▶ 23 Pk:1 [8|8|8]	**25301**
Faith (87) 1(4) [9|15|20]	5187
One More Try (88) 1(3) [7|14|18]	4687
Father Figure (88) 1(2) [6|13|17]	3850
Monkey (88) 1(2) [6|12|16]	3473
I Want Your Sex (87) 2(1) [6|14|20]	3293

I Knew You Were Waiting (For Me) (87) -
   Aretha Franklin & George Michael [A]
   1(2) [7|12|17]
	1733

A Different Corner (86) 7(3) [4|10|16]	1547
Kissing A Fool (88) 5(1) [4|10|15]	1530
Bette MIDLER▶ 113 Pk:1 [2|4|7]	**9597**
The Rose (80) 3(3) [8|16|25]	4647
Wind Beneath My Wings (89) 1(1) [7|15|29]	3826
My Mother's Eyes (81) 39(2) [0|2|13]	470
When A Man Loves A Woman (80) 35(2) [0|3|10]	437
Beast Of Burden (84) 71(1) [0|0|6]	99.2
All I Need To Know (83) 77(1) [0|0|4]	64.1
Favorite Waste Of Time (83) 78(1) [0|0|4]	54.0
MIDNIGHT OIL▶ 545 Pk:17 [0|1|2]	**1176**
Beds Are Burning (88) 17(1) [0|9|22]	921
The Dead Heart (88) 53(1) [0|0|12]	255
MIDNIGHT STAR▶ 410 Pk:18 [0|1|7]	**2059**
Operator (85) 18(2) [0|8|17]	939
Midas Touch (86) 42(1) [0|0|14]	396
Wet My Whistle (84) 61(2) [0|0|11]	293
Freak-A-Zoid (83) 66(1) [0|0|8]	166
Headlines (86) 69(2) [0|0|7]	97.7
No Parking (On The Dance Floor) (84) 81(1) [0|0|8]	92.5
Scientific Love (85) 80(1) [0|0|3]	74.1

Acts With Singles

Act ▶ Rank Peak [Top10s|Top40s|Total] Title (Pk Yr) Peak(Wk) [Top10Wks|Top40Wks|TotalWks]	Score
MIKE + THE MECHANICS ▶ 139 Pk:1 [3|4|6]	**8260**
The Living Years (89) 1(1) [6|14|20]	3709
All I Need Is A Miracle (86) 5(1) [4|12|19]	1864
Silent Running (On Dangerous Ground) (86) 6(1) [4|11|24]	1835
Taken In (86) 32(4) [0|5|15]	557
Nobody's Perfect (88) 63(1) [0|0|11]	184
Seeing Is Believing (89) 62(1) [0|0|6]	111
Frankie MILLER ▶ 1061 Pk:62 [0|0|1]	**134**
To Dream The Dream (82) 62(1) [0|0|6]	134
Steve MILLER Band ▶ 112 Pk:1 [1|2|8]	**9948**
Abracadabra (82) 1(2) [14|19|25]	8547
Heart Like A Wheel (81) 24(2) [0|9|14]	661
Give It Up (83) 60(2) [0|0|9]	202
Cool Magic (82) 57(2) [0|0|8]	177
Shangri-La (84) 57(2) [0|0|6]	170
Circle Of Love (82) 55(2) [0|0|7]	155
Bongo Bongo (85) 84(1) [0|0|3]	29.6
I Want To Make The World Turn Around (86) 97(1) [0|0|3]	7.1
MILLIONS LIKE US ▶ 1117 Pk:69 [0|0|1]	**108**
Guaranteed For Life (87) 69(2) [0|0|9]	108
MILLI VANILLI ▶ 69 Pk:1 [4|4|4]	**15769**
Blame It On The Rain (89) 1(2) [6|14|23]	4409
Girl I'm Gonna Miss You (89) 1(2) [6|14|22]	4238
Girl You Know It's True (89) 2(1) [7|15|26]	3708
Baby Don't Forget My Number (89) 1(1) [6|14|21]	3415
Stephanie MILLS ▶ 313 Pk:6 [1|2|6]	**3393**
Never Knew Love Like This Before (80) 6(2) [5|16|25]	2602
Two Hearts (81) - Stephanie Mills featuring Teddy Pendergrass 40(2) [0|2|13]	409
Sweet Sensation (80) 52(2) [0|0|6]	176
The Medicine Song (84) 65(2) [0|0|9]	122
Bit By Bit (85) 78(1) [0|0|6]	65.3
(You're Puttin') A Rush On Me (87) 85(1) [0|0|2]	19.5
Ronnie MILSAP ▶ 163 Pk:5 [1|5|8]	**7422**
(There's) No Gettin' Over Me (81) 5(5) [8|15|20]	3210
Any Day Now (82) 14(2) [0|9|16]	1152
Smoky Mountain Rain (81) 24(2) [0|9|21]	945
I Wouldn't Have Missed It For The World (82) 20(2) [0|11|17]	941
Stranger In My House (83) 23(1) [0|8|16]	799
Don't You Know How Much I Love You (83) 58(1) [0|0|7]	172
He Got You (82) 59(2) [0|0|7]	158
She Loves My Car (84) 84(2) [0|0|4]	44.6
MINK DeVILLE ▶ 1307 Pk:89 [0|0|1]	**31.6**
Each Word's A Beat Of My Heart (84) 89(1) [0|0|4]	31.6
Kylie MINOGUE ▶ 286 Pk:3 [1|3|3]	**3847**
The Loco-Motion (88) 3(2) [6|13|27]	2927
I Should Be So Lucky (88) 28(1) [0|4|14]	503
It's No Secret (89) 37(1) [0|2|13]	417
MINOR DETAIL ▶ 1372 Pk:92 [0|0|1]	**13.2**
Canvas Of Life (83) 92(1) [0|0|2]	13.2
MISSING PERSONS ▶ 553 Pk:42 [0|0|5]	**1139**
Destination Unknown (82) 42(2) [0|0|14]	406
Words (82) 42(1) [0|0|11]	339
Windows (83) 63(2) [0|0|6]	169
Give (84) 67(1) [0|0|6]	118
Walking In L.A. (83) 70(2) [0|0|6]	107
MR. MISTER ▶ 101 Pk:1 [3|4|5]	**11293**
Broken Wings (85) 1(2) [9|15|22]	4890
Kyrie (86) 1(2) [7|13|20]	4458
Is It Love (86) 8(1) [1|11|17]	1297
Something Real (Inside Me/Inside You) (87) 29(1) [0|5|11]	445
Hunters Of The Night (84) 57(1) [0|0|8]	203

Act ▶ Rank Peak [Top10s|Top40s|Total] Title (Pk Yr) Peak(Wk) [Top10Wks|Top40Wks|TotalWks]	Score
Barbara MITCHELL ▶ 985 Pk:48 [0|0|1]	**189**
Blame It On Love (83) - Smokey Robinson & Barbara Mitchell [A] 48(1) [0|0|12]	189
Joni MITCHELL ▶ 869 Pk:47 [0|0|2]	**306**
(You're So Square) Baby, I Don't Care (82) 47(2) [0|0|9]	273
Good Friends (86) 85(1) [0|0|3]	33.2
Kim MITCHELL ▶ 1161 Pk:86 [0|0|1]	**88.8**
Go For Soda (85) 86(4) [0|0|9]	88.8
MODELS ▶ 791 Pk:37 [0|1|1]	**438**
Out Of Mind Out Of Sight (86) 37(1) [0|4|13]	438
MODERN ENGLISH ▶ 1152 Pk:78 [0|0|2]	**93.0**
I Melt With You (83) 78(1) [0|0|7]	76.1
Hands Across The Sea (84) 91(1) [0|0|3]	17.0
MOLLY HATCHET ▶ 767 Pk:42 [0|0|4]	**474**
Flirtin' With Disaster (80) 42(1) [0|0|10]	388
Satisfied Man (84) 81(1) [0|0|5]	56.0
The Rambler (81) 91(2) [0|0|3]	23.3
Power Play (82) 96(1) [0|0|2]	6.5
MONDO ROCK ▶ 1122 Pk:71 [0|0|1]	**106**
Primitive Love Rites (87) 71(1) [0|0|6]	106
Eddie MONEY ▶ 125 Pk:4 [2|6|13]	**8688**
Take Me Home Tonight (86) 4(1) [4|12|23]	2214
Walk On Water (88) 9(2) [4|13|21]	1653
Think I'm In Love (82) 16(3) [0|12|17]	1308
I Wanna Go Back (87) 14(1) [0|10|21]	1077
Endless Nights (87) 21(1) [0|7|19]	790
The Love In Your Eyes (89) 24(1) [0|7|18]	679
The Big Crash (84) 54(1) [0|0|11]	306
Shakin' (82) 63(1) [0|0|9]	177
Let Me In (89) 60(1) [0|0|7]	150
Let's Be Lovers Again (80) - Eddie Money with Valerie Carter 65(2) [0|0|6]	132
Club Michelle (84) 66(1) [0|0|7]	128
Running Back (80) 78(1) [0|0|4]	57.7
We Should Be Sleeping (87) 90(1) [0|0|3]	16.7
T.S. MONK ▶ 1026 Pk:63 [0|0|1]	**159**
Bon Bon Vie (Gimme The Good Life) (81) 63(1) [0|0|8]	159
The MONKEES ▶ 1168 Pk:79 [0|0|2]	**86.1**
Daydream Believer (86) 79(2) [0|0|4]	53.0
Heart And Soul (87) 87(1) [0|0|4]	33.2
The MONROES ▶ 941 Pk:59 [0|0|1]	**227**
What Do All The People Know (82) 59(1) [0|0|8]	227
MOODY BLUES ▶ 219 Pk:9 [1|5|8]	**5088**
Your Wildest Dreams (86) 9(2) [2|12|21]	1412
Gemini Dream (81) 12(1) [0|9|15]	1121
The Voice (81) 15(2) [0|11|17]	1093
Sitting At The Wheel (83) 27(1) [0|6|10]	518
I Know You're Out There Somewhere (88) 30(1) [0|4|16]	487
The Other Side Of Life (86) 58(1) [0|0|9]	189
Blue World (83) 62(2) [0|0|6]	138
Talking Out Of Turn (81) 65(2) [0|0|7]	130
Michael MORALES ▶ 486 Pk:15 [0|2|3]	**1523**
Who Do You Give Your Love To? (89) 15(1) [0|10|19]	965
What I Like About You (89) 28(1) [0|6|13]	505
I Don't Know (89) 81(1) [0|0|6]	53.4
Meli'sa MORGAN ▶ 803 Pk:46 [0|0|1]	**414**
Do Me Baby (86) 46(1) [0|0|14]	414
Giorgio MORODER ▶ 1269 Pk:81 [0|0|1]	**45.1**
Reach Out (84) 81(1) [0|0|4]	45.1
The MOTELS ▶ 192 Pk:9 [2|4|7]	**5733**
Only The Lonely (82) 9(4) [4|15|23]	2280
Suddenly Last Summer (83) 9(1) [2|13|20]	1846
Shame (85) 21(1) [0|7|13]	632
Remember The Nights (84) 36(1) [0|3|12]	551
Take The L. (82) 52(2) [0|0|9]	220

Act ▶ Rank Peak [Top10s|Top40s|Total] Title (Pk Yr) Peak(Wk) [Top10Wks|Top40Wks|TotalWks]	Score
Forever Mine (82) 60(2) [0|0|8]	177
Shock (85) 84(1) [0|0|3]	28.9
MÖTLEY CRÜE ▶ 290 Pk:6 [1|3|7]	**3773**
Dr. Feelgood (89) 6(1) [2|9|16]	1349
Girls, Girls, Girls (87) 12(1) [0|9|15]	1139
Smokin' In The Boys Room (85) 16(2) [0|9|15]	907
Looks That Kill (84) 54(2) [0|0|10]	251
You're All I Need (87) 83(1) [0|0|8]	69.5
Home Sweet Home (85) 89(1) [0|0|6]	38.3
Too Young To Fall In Love (84) 90(2) [0|0|2]	18.6
The MOTORS ▶ 1192 Pk:78 [0|0|1]	**73.8**
Love And Loneliness (80) 78(1) [0|0|5]	73.8
MOVING PICTURES ▶ 455 Pk:29 [0|1|2]	**1703**
What About Me (83) 29(1) [0|13|26]	1221
What About Me (89) 46(1) [0|0|17]	482
Alison MOYET ▶ 670 Pk:31 [0|1|2]	**679**
Invisible (85) 31(1) [0|6|17]	635
Love Resurrection (85) 82(1) [0|0|4]	44.1
--MR. MISTER see: MISTER, MR.	
MTUME ▶ 794 Pk:45 [0|0|2]	**429**
Juicy Fruit (83) 45(1) [0|0|12]	377
You, Me And He (84) 83(1) [0|0|5]	52.0
Shirley MURDOCK ▶ 631 Pk:23 [0|1|1]	**769**
As We Lay (87) 23(2) [0|7|18]	769
Michael Martin MURPHEY ▶ 537 Pk:19 [0|1|2]	**1239**
What's Forever For (82) - Michael Murphey 19(5) [0|11|20]	1157
Still Taking Chances (83) 76(2) [0|0|7]	81.8
Eddie MURPHY ▶ 218 Pk:2 [1|2|2]	**5095**
Party All The Time (85) 2(3) [9|14|22]	4645
Put Your Mouth On Me (89) 27(1) [0|4|13]	450
Walter MURPHY ▶ 928 Pk:47 [0|0|1]	**245**
Themes From E.T. (The Extra-Terrestrial) (82) 47(1) [0|0|9]	245
Anne MURRAY ▶ 293 Pk:12 [0|3|9]	**3740**
Daydream Believer (80) 12(3) [0|11|17]	1648
Could I Have This Dance (80) 33(2) [0|4|14]	544
Blessed Are The Believers (81) 34(2) [0|4|13]	454
Another Sleepless Night (82) 44(1) [0|0|9]	303
Lucky Me (80) 42(2) [0|0|8]	288
It's All I Can Do (81) 53(2) [0|0|9]	224
A Little Good News (83) 74(2) [0|0|9]	132
I'm Happy Just To Dance With You (80) 64(1) [0|0|6]	113
Now And Forever (You And Me) (86) 92(1) [0|0|6]	34.9
MUSICAL YOUTH ▶ 479 Pk:10 [1|1|2]	**1556**
Pass The Dutchie (83) 10(2) [2|10|18]	1404
She's Trouble (84) 65(1) [0|0|7]	151

N

NAKED EYES ▶ 250 Pk:8 [1|4|4]	**4434**
Always Something There To Remind Me (83) 8(2) [2|13|22]	1852
Promises, Promises (83) 11(1) [0|12|20]	1595
When The Lights Go Out (83) 37(2) [0|3|14]	549
(What) In The Name Of Love (84) 39(2) [0|2|12]	437
Graham NASH ▶ 1230 Pk:84 [0|0|1]	**59.9**
Innocent Eyes (86) 84(1) [0|0|7]	59.9
NAZARETH ▶ 1315 Pk:87 [0|0|1]	**29.9**
Holiday (80) 87(1) [0|0|3]	29.9
Phyllis NELSON ▶ 912 Pk:61 [0|0|1]	**256**
I Like You (Special Mix) (86) 61(2) [0|0|11]	256
Willie NELSON ▶ 184 Pk:5 [2|4|6]	**6107**
Always On My Mind (82) 5(3) [6|15|23]	3027
To All The Girls I've Loved Before (84) - Julio Iglesias & Willie Nelson [A] 5(1) [6|12|21]	1241
On The Road Again (80) 20(2) [0|10|20]	973
Let It Be Me (82) 40(3) [0|3|12]	385

Acts With Singles

Act ▶ Rank Peak [Top10s|Top40s|Total] Title (Pk Yr) Peak(Wk) [Top10Wks|Top40Wks|TotalWks]	Score
Willie NELSON ▶ *Continued*	
My Heroes Have Always Been Cowboys (80) 44(2) [0|0|10]	366
Just To Satisfy You (82) - Waylon & Willie [A] 52(1) [0|0|9]	114
NENA ▶ 264 Pk:2 [1|1|1]	**4276**
99 Luftballons (84) 2(1) [6|13|23]	4276
Loz NETTO ▶ 1209 Pk:82 [0|0|1]	**65.4**
Fade Away (83) 82(1) [0|0|6]	65.4
Robbie NEVIL ▶ 176 Pk:2 [2|4|5]	**6523**
C'est La Vie (87) 2(2) [7|16|23]	3882
Wot's It To Ya (87) 10(2) [2|9|16]	1064
Dominoes (87) 14(1) [0|9|16]	919
Back On Holiday (89) 34(1) [0|3|14]	485
Somebody Like You (89) 63(1) [0|0|11]	173
Ivan NEVILLE ▶ 676 Pk:26 [0|1|2]	**668**
Not Just Another Girl (88) 26(1) [0|6|19]	647
Falling Out Of Love (89) 91(2) [0|0|3]	21.0
NEWCITY ROCKERS ▶ 1148 Pk:80 [0|0|2]	**95.3**
Black Dog (87) 80(1) [0|0|6]	68.0
Rev It Up (87) 86(1) [0|0|4]	27.2
NEWCLEUS ▶ 859 Pk:56 [0|0|1]	**329**
Jam On It (84) 56(2) [0|0|15]	329
NEW EDITION ▶ 124 Pk:4 [2|6|12]	**8713**
Cool It Now (85) 4(1) [6|14|25]	2926
If It Isn't Love (88) 7(1) [4|13|21]	1807
Mr. Telephone Man (85) 12(2) [0|8|16]	1051
Earth Angel (86) 21(1) [0|6|14]	629
Lost In Love (85) 35(2) [0|4|14]	493
A Little Bit Of Love (Is All It Takes) (86) 38(2) [0|2|15]	423
Count Me Out (85) 51(1) [0|0|15]	392
Candy Girl (83) 46(1) [0|0|11]	366
Can You Stand The Rain (89) 44(1) [0|0|13]	298
With You All The Way (86) 51(1) [0|0|11]	277
Is This The End (83) 85(1) [0|0|4]	39.4
You're Not My Kind Of Girl (88) 95(2) [0|0|4]	10.7
NEW KIDS ON THE BLOCK ▶ 77 Pk:1 [6|6|6]	**14595**
I'll Be Loving You (Forever) (89) 1(1) [6|14|21]	3690
Hangin' Tough (89) 1(1) [6|12|17]	3272
Cover Girl (89) 2(1) [4|10|18]	2518
You Got It (The Right Stuff) (89) 3(1) [5|13|26]	2405
Please Don't Go Girl (88) 10(1) [1|12|28]	1400
Didn't I (Blow Your Mind) (89) 8(1) [2|10|19]	1312
Randy NEWMAN ▶ 823 Pk:51 [0|0|2]	**374**
It's Money That Matters (88) 60(2) [0|0|12]	242
The Blues (83) - Randy Newman and Paul Simon [A] 51(3) [0|0|8]	132
NEW ORDER ▶ 585 Pk:32 [0|1|3]	**971**
True Faith (87) 32(2) [0|8|18]	665
Round & Round (89) 64(2) [0|0|9]	171
Blue Monday 1988 (88) 68(1) [0|0|10]	135
Juice NEWTON ▶ 70 Pk:2 [4|7|10]	**15737**
Queen Of Hearts (81) 2(2) [10|19|27]	4309
Angel Of The Morning (81) 4(4) [6|16|22]	3447
The Sweetest Thing (I've Ever Known) (82) 7(2) [7|18|24]	2829
Love's Been A Little Bit Hard On Me (82) 7(2) [4|13|17]	2037
Break It To Me Gently (82) 11(3) [0|10|17]	1321
Heart Of The Night (83) 25(3) [0|10|16]	809
Tell Her No (83) 27(1) [0|5|11]	534
A Little Love (84) 44(2) [0|0|10]	326
Can't Wait All Night (84) 66(1) [0|0|6]	109
Dirty Looks (83) 90(1) [0|0|3]	16.2
Wayne NEWTON ▶ 733 Pk:35 [0|1|1]	**533**
Years (80) 35(1) [0|3|13]	533
Olivia NEWTON-JOHN ▶ 14 Pk:1 [6|11|18]	**30618**
Physical (81) 1(10) [15|21|26]	11089

Act ▶ Rank Peak [Top10s|Top40s|Total] Title (Pk Yr) Peak(Wk) [Top10Wks|Top40Wks|TotalWks]	Score
Magic (80) 1(4) [9|16|23]	6621
Heart Attack (82) 3(4) [7|13|21]	4023
Twist Of Fate (84) 5(2) [7|14|18]	2833
Make A Move On Me (82) 5(3) [5|10|14]	2194
Xanadu (80) - Olivia Newton-John/Electric Light Orchestra [A] 8(2) [3|10|17]	834
Soul Kiss (85) 20(1) [0|7|15]	671
I Can't Help It (80) - Andy Gibb & Olivia Newton-John [A] 12(2) [0|8|13]	595
Suddenly (81) - Olivia Newton-John & Cliff Richard [A] 20(1) [0|11|19]	513
Livin' In Desperate Times (84) 31(2) [0|5|10]	460
Tied Up (83) 38(3) [0|3|11]	410
Landslide (82) 52(2) [0|0|8]	220
The Rumour (88) 62(1) [0|0|6]	118
The Best Of Me (86) - David Foster And Olivia Newton-John [A] 80(1) [0|0|8]	38.7
Stevie NICKS ▶ 57 Pk:3 [4|11|12]	**17947**
Stop Draggin' My Heart Around (81) - Stevie Nicks (with Tom Petty And The Heartbreakers) 3(6) [10|15|21]	4960
Leather And Lace (82) - Stevie Nicks (with Don Henley) 6(3) [8|15|19]	2825
Stand Back (83) 5(1) [6|14|19]	2639
Talk To Me (86) 4(2) [5|13|18]	2443
Edge Of Seventeen (Just Like The White Winged Dove) (82) 11(2) [0|10|14]	1172
If Anyone Falls (83) 14(1) [0|9|14]	1066
I Can't Wait (86) 16(2) [0|8|13]	796
Rooms On Fire (89) 16(1) [0|7|14]	730
Nightbird (84) - Stevie Nicks (with Sandy Stewart) 33(2) [0|4|12]	548
After The Glitter Fades (82) 32(1) [0|4|11]	462
Needles And Pins (86) - Tom Petty and The Heartbreakers with Stevie Nicks [A] 37(2) [0|2|9]	157
Has Anyone Ever Written Anything For You (86) 60(2) [0|0|6]	149
NIELSEN/PEARSON ▶ 664 Pk:38 [0|1|2]	**684**
If You Should Sail (80) 38(1) [0|2|14]	491
The Sun Ain't Gonna Shine Anymore (81) 56(2) [0|0|8]	193
NIGHT ▶ 1345 Pk:87 [0|0|1]	**22.5**
Love On The Airwaves (81) 87(1) [0|0|3]	22.5
NIGHT RANGER ▶ 144 Pk:5 [2|6|11]	**7908**
Sister Christian (84) 5(2) [4|12|24]	2385
Sentimental Street (85) 8(2) [2|11|17]	1480
When You Close Your Eyes (84) 14(1) [0|11|17]	1118
Goodbye (86) 17(1) [0|10|18]	956
Four In The Morning (I Can't Take Any More) (85) 19(1) [0|6|13]	677
Don't Tell Me You Love Me (83) 40(3) [0|3|11]	390
(You Can Still) Rock In America (84) 51(1) [0|0|12]	384
Sing Me Away (83) 54(2) [0|0|9]	262
The Secret Of My Success (87) 64(3) [0|0|8]	173
I Did It For Love (88) 75(1) [0|0|5]	65.2
Hearts Away (87) 90(1) [0|0|3]	17.8
9.9 ▶ 842 Pk:51 [0|0|1]	**350**
All Of Me For All Of You (85) 51(1) [0|0|13]	350
1927 ▶ 1400 Pk:100 [0|0|1]	**0.8**
That's When I Think Of You (89) 100(1) [0|0|1]	0.8
NITTY GRITTY DIRT BAND ▶ 386 Pk:13 [0|2|3]	**2311**
An American Dream (80) - Dirt Band 13(2) [0|11|19]	1448
Make A Little Magic (80) - Dirt Band 25(3) [0|9|16]	804
Fire In The Sky (81) - Dirt Band 76(2) [0|0|4]	58.7
NOCERA ▶ 1233 Pk:84 [0|0|1]	**59.7**
Summertime, Summertime (87) 84(1) [0|0|7]	59.7

Act ▶ Rank Peak [Top10s|Top40s|Total] Title (Pk Yr) Peak(Wk) [Top10Wks|Top40Wks|TotalWks]	Score
NOEL ▶ 661 Pk:47 [0|0|2]	**699**
Silent Morning (87) 47(1) [0|0|22]	594
Like A Child (88) 67(1) [0|0|8]	104
Kenny NOLAN ▶ 874 Pk:44 [0|0|1]	**301**
Us And Love (We Go Together) (80) 44(1) [0|0|8]	301
Aldo NOVA ▶ 624 Pk:23 [0|1|2]	**793**
Fantasy (82) 23(2) [0|7|16]	672
Foolin' Yourself (82) 65(2) [0|0|6]	120
Ted NUGENT ▶ 1294 Pk:86 [0|0|1]	**35.6**
Wango Tango (80) 86(1) [0|0|4]	35.6
Gary NUMAN ▶ 364 Pk:9 [1|1|1]	**2581**
Cars (80) 9(3) [5|17|25]	2581
NU SHOOZ ▶ 265 Pk:3 [1|2|3]	**4223**
I Can't Wait (86) 3(1) [5|15|23]	2921
Point Of No Return (86) 28(1) [0|8|22]	820
Should I Say Yes? (88) 41(1) [0|0|16]	482
The NYLONS ▶ 541 Pk:12 [0|1|2]	**1195**
Kiss Him Goodbye (87) 12(1) [0|10|17]	1116
Happy Together (87) 75(1) [0|0|7]	79.3

O

Act ▶ Rank Peak [Top10s|Top40s|Total] Title (Pk Yr) Peak(Wk) [Top10Wks|Top40Wks|TotalWks]	Score
OAK ▶ 678 Pk:36 [0|1|2]	**663**
King Of The Hill (80) - Rick Pinette And Oak 36(2) [0|3|14]	534
Set The Night On Fire (81) 71(1) [0|0|6]	129
OAK RIDGE BOYS ▶ 261 Pk:5 [1|2|4]	**4291**
Elvira (81) 5(4) [8|14|22]	3123
Bobbie Sue (82) 12(2) [0|9|14]	1017
American Made (83) 72(2) [0|0|5]	93.3
So Fine (82) 76(1) [0|0|4]	57.9
John O'BANION ▶ 696 Pk:24 [0|1|1]	**622**
Love You Like I Never Loved Before (81) 24(2) [0|7|13]	622
O'BRYAN ▶ 950 Pk:57 [0|0|1]	**221**
The Gigolo (82) 57(2) [0|0|9]	221
Ric OCASEK ▶ 519 Pk:15 [0|1|3]	**1332**
Emotion In Motion (86) 15(1) [0|8|19]	937
Something To Grab For (83) 47(2) [0|0|9]	288
True To You (87) 75(1) [0|0|8]	107
Billy OCEAN ▶ 20 Pk:1 [7|11|11]	**26542**
Caribbean Queen (No More Love On The Run) (84) 1(2) [7|15|26]	5211
There'll Be Sad Songs (To Make You Cry) (86) 1(1) [7|14|21]	4121
Get Outta My Dreams, Get Into My Car (88) 1(2) [7|14|20]	3998
Loverboy (85) 2(1) [6|15|21]	3446
When The Going Gets Tough, The Tough Get Going (86) 2(1) [5|14|23]	3075
Suddenly (85) 4(2) [4|13|22]	2519
Love Zone (86) 10(1) [1|11|16]	1291
Love Is Forever (86) 16(3) [0|11|16]	1049
The Colour Of Love (88) 17(2) [0|8|16]	833
Mystery Lady (85) 24(1) [0|8|15]	677
Licence To Chill (89) 32(1) [0|2|9]	323
OFF BROADWAY (USA) ▶ 977 Pk:51 [0|0|1]	**194**
Stay In Time (80) 51(1) [0|0|7]	194
OINGO BOINGO ▶ 851 Pk:45 [0|0|2]	**338**
Weird Science (85) 45(2) [0|0|12]	294
Just Another Day (86) 85(1) [0|0|4]	43.9
The O'JAYS ▶ 596 Pk:28 [0|1|2]	**916**
Forever Mine (80) 28(1) [0|5|13]	632
Girl, Don't Let It Get You Down (80) 55(1) [0|0|11]	284
OLLIE And JERRY ▶ 507 Pk:9 [1|1|1]	**1422**
Breakin'.. There's No Stopping Us (84) 9(1) [1|11|18]	1422
Lenore O'MALLEY ▶ 940 Pk:53 [0|0|1]	**227**
First...Be A Woman (80) 53(2) [0|0|8]	227

Acts With Singles

Act ▶ Rank Peak [Top10s|Top40s|Total] Title (Pk Yr) Peak(Wk) [Top10Wks|Top40Wks|TotalWks]	Score
ONE 2 MANY ▶ 792 Pk:37 [0|1|1]	**437**
Downtown (89) 37(1) [0|4|13]	437
Alexander O'NEAL ▶ 476 Pk:25 [0|3|4]	**1586**
Fake (87) 25(1) [0|6|15]	559
Never Knew Love Like This (88) - Alexander O'Neal featuring Cherrelle 28(1) [0|6|14]	529
Saturday Love (86) - Cherrelle with Alexander O'Neal [A] 26(2) [0|6|17]	311
Criticize (87) 70(1) [0|0|11]	188
ONE TO ONE ▶ 1328 Pk:92 [0|0|1]	**26.7**
Angel In My Pocket (86) 92(2) [0|0|4]	26.7
ONE WAY ▶ 959 Pk:61 [0|0|1]	**210**
Cutie Pie (82) 61(1) [0|0|10]	210
Yoko ONO ▶ 974 Pk:58 [0|0|1]	**200**
Walking On Thin Ice (81) 58(1) [0|0|10]	200
OPUS ▶ 705 Pk:32 [0|1|1]	**594**
Live Is Life (86) 32(2) [0|5|16]	594
Roy ORBISON ▶ 517 Pk:9 [1|1|2]	**1354**
You Got It (89) 9(1) [1|11|18]	1245
That Lovin' You Feelin' Again (80) - Roy Orbison & Emmylou Harris [A] 55(2) [0|0|8]	109
ORCHESTRAL MANOEUVERS IN THE DARK ▶ 228 Pk:4 [1|4|5]	**4868**
If You Leave (86) 4(1) [5|13|20]	2321
Dreaming (88) 16(2) [0|9|17]	865
(Forever) Live And Die (86) 19(1) [0|7|17]	751
So In Love (85) 26(2) [0|7|17]	671
Secret (86) 63(2) [0|0|13]	259
ORION THE HUNTER ▶ 989 Pk:58 [0|0|1]	**188**
So You Ran (84) 58(2) [0|0|8]	188
Benjamin ORR ▶ 611 Pk:24 [0|1|1]	**849**
Stay The Night (87) 24(1) [0|6|20]	849
Robert Ellis ORRALL ▶ 762 Pk:32 [0|1|1]	**485**
I Couldn't Say No (83) - Robert Ellis Orrall With Carlene Carter 32(1) [0|3|12]	485
Jeffrey OSBORNE ▶ 181 Pk:12 [0|8|12]	**6316**
You Should Be Mine (The Woo Woo Song) (86) 13(1) [0|11|19]	1168
Stay With Me Tonight (84) 30(2) [0|8|21]	936
On The Wings Of Love (82) 29(3) [0|7|18]	790
Don't You Get So Mad (83) 25(1) [0|6|14]	678
Love Power (87) - Dionne Warwick & Jeffrey Osborne [A] 12(2) [0|9|14]	529
I Really Don't Need No Light (82) 39(2) [0|2|15]	443
Don't Stop (84) 44(1) [0|0|15]	433
The Borderlines (85) 38(1) [0|2|11]	382
We're Going All The Way (84) 48(1) [0|0|12]	368
She's On The Left (88) 48(3) [0|0|11]	300
The Last Time I Made Love (84) - Joyce Kennedy & Jeffrey Osborne [A] 40(2) [0|2|12]	206
Eenie Meenie (83) 76(2) [0|0|5]	83.7
Ozzy OSBOURNE ▶ 1052 Pk:68 [0|0|1]	**141**
Shot In The Dark (86) 68(1) [0|0|9]	141
Donny OSMOND ▶ 280 Pk:2 [1|2|3]	**3971**
Soldier Of Love (89) 2(1) [6|11|18]	2861
Sacred Emotion (89) 13(1) [0|9|16]	1000
Hold On (89) 73(1) [0|0|7]	110
OTHER ONES ▶ 616 Pk:29 [0|1|2]	**815**
Holiday (87) 29(1) [0|4|17]	575
We Are What We Are (87) 53(1) [0|0|9]	240
The OUTFIELD ▶ 267 Pk:6 [1|4|6]	**4136**
Your Love (86) 6(2) [4|12|22]	1977
All The Love In The World (86) 19(1) [0|7|16]	749
Since You've Been Gone (87) 31(1) [0|5|15]	565
Voices Of Babylon (89) 25(1) [0|6|14]	560
Everytime You Cry (86) 66(2) [0|0|10]	188
My Paradise (89) 72(2) [0|0|6]	97.4

Act ▶ Rank Peak [Top10s|Top40s|Total] Title (Pk Yr) Peak(Wk) [Top10Wks|Top40Wks|TotalWks]	Score
The OUTLAWS ▶ 732 Pk:31 [0|1|1]	**536**
(Ghost) Riders In The Sky (81) 31(1) [0|4|15]	536
OXO ▶ 686 Pk:28 [0|1|1]	**654**
Whirly Girl (83) 28(1) [0|6|14]	654
OZARK MOUNTAIN DAREDEVILS ▶ 1077 Pk:67 [0|0|1]	**122**
Take You Tonight (80) 67(2) [0|0|5]	122

P

Act ▶ Rank Peak [Top10s|Top40s|Total] Title (Pk Yr) Peak(Wk) [Top10Wks|Top40Wks|TotalWks]	Score
PABLO CRUISE ▶ 525 Pk:13 [0|1|2]	**1309**
Cool Love (81) 13(2) [0|11|17]	1232
Slip Away (81) 75(1) [0|0|5]	77.6
David PACK ▶ 1377 Pk:95 [0|0|1]	**11.2**
Prove Me Wrong (86) 95(1) [0|0|3]	11.2
Tommy PAGE ▶ 674 Pk:29 [0|1|1]	**670**
A Shoulder To Cry On (89) 29(1) [0|6|20]	670
Kevin PAIGE ▶ 566 Pk:18 [0|1|1]	**1076**
Don't Shut Me Out (89) 18(2) [0|10|24]	1076
PAJAMA PARTY ▶ 783 Pk:59 [0|0|2]	**452**
Over And Over (89) 59(1) [0|0|14]	316
Yo No Se' (89) 75(1) [0|0|10]	136
Robert PALMER ▶ 91 Pk:1 [3|5|10]	**12286**
Addicted To Love (86) 1(1) [7|14|22]	4061
Simply Irresistible (88) 2(2) [6|14|20]	3496
I Didn't Mean To Turn You On (86) 2(1) [5|13|22]	2941
Early In The Morning (88) 19(1) [0|9|15]	795
Hyperactive (86) 33(1) [0|5|12]	428
Can We Still Be Friends (80) 52(1) [0|0|9]	266
Tell Me I'm Not Dreaming (89) 60(1) [0|0|8]	154
You Are In My System (83) 78(1) [0|0|6]	83.9
Discipline Of Love (Why Did You Do It) (85) 82(2) [0|0|5]	51.7
Sweet Lies (88) 94(1) [0|0|2]	8.8
Mica PARIS ▶ 1384 Pk:97 [0|0|1]	**9.2**
My One Temptation (89) 97(2) [0|0|4]	9.2
Graham PARKER ▶ 1378 Pk:94 [0|0|1]	**10.6**
Life Gets Better (83) 94(1) [0|0|2]	10.6
Graham PARKER And The SHOT ▶ 810 Pk:39 [0|1|1]	**402**
Wake Up (Next To You) (85) 39(1) [0|3|12]	402
Ray PARKER Jr. ▶ 85 Pk:1 [2|7|9]	**13555**
Ghostbusters (84) 1(3) [10|14|21]	6445
The Other Woman (82) 4(2) [7|14|21]	3200
I Still Can't Get Over Loving You (84) 12(1) [0|11|19]	1373
Jamie (85) 14(1) [0|11|17]	1122
Bad Boy (83) 35(2) [0|4|12]	487
Girls Are More Fun (85) 34(1) [0|4|15]	451
Let Me Go (82) 38(2) [0|3|9]	349
I Don't Think That Man Should Sleep Alone (87) 68(1) [0|0|7]	124
One Sunny Day/Dueling Bikes From Quicksilver (86) - Ray Parker Jr. And Helen Terry [A] 96(2) [0|0|3]	4.3
Ray PARKER Jr. & RAYDIO ▶ 243 Pk:4 [1|3|3]	**4524**
A Woman Needs Love (Just Like You Do) (81) 4(2) [6|15|27]	3214
That Old Song (81) 21(2) [0|6|15]	665
Two Places At The Same Time (80) 30(2) [0|5|14]	645
John PARR ▶ 208 Pk:1 [1|2|5]	**5403**
St. Elmo's Fire (Man In Motion) (85) 1(2) [7|14|22]	4434
Naughty Naughty (85) 23(1) [0|8|20]	824
Magical (85) 73(1) [0|0|5]	78.1
Blame It On The Radio (86) 88(2) [0|0|6]	50.3
Love Grammar (85) 89(1) [0|0|2]	16.4
Alan PARSONS Project ▶ 117 Pk:3 [1|5|11]	**9317**
Eye In The Sky (82) 3(3) [8|17|25]	4321
Time (81) 15(3) [0|12|23]	1359
Games People Play (81) 16(2) [0|10|23]	1204
Don't Answer Me (84) 15(1) [0|8|15]	995

Act ▶ Rank Peak [Top10s|Top40s|Total] Title (Pk Yr) Peak(Wk) [Top10Wks|Top40Wks|TotalWks]	Score
Prime Time (84) 34(1) [0|3|11]	420
You Don't Believe (83) 54(3) [0|0|10]	308
Psychobabble (82) 57(4) [0|0|10]	267
Let's Talk About Me (85) 56(1) [0|0|10]	230
Snake Eyes (81) 67(1) [0|0|5]	93.6
Days Are Numbers (The Traveller) (85) 71(1) [0|0|5]	75.7
Stereotomy (86) 82(1) [0|0|4]	44.6
PARTLAND BROTHERS ▶ 741 Pk:27 [0|1|1]	**519**
Soul City (87) 27(1) [0|5|13]	519
Dolly PARTON ▶ 155 Pk:1 [1|2|9]	**7651**
9 To 5 (81) 1(2) [9|18|26]	6099
Save The Last Dance For Me (84) 45(1) [0|0|12]	427
Starting Over Again (80) 36(2) [0|3|10]	376
I Will Always Love You (82) 53(1) [0|0|14]	305
But You Know I Love You (81) 41(1) [0|0|10]	292
The House Of The Rising Sun (81) 77(2) [0|0|4]	56.0
Downtown (84) 80(1) [0|0|4]	51.0
The Greatest Gift Of All (85) - Kenny Rogers & Dolly Parton [A] 81(2) [0|0|4]	31.7
Real Love (85) - Dolly Parton (Duet With Kenny Rogers) 91(1) [0|0|3]	13.5
The PASADENAS ▶ 903 Pk:52 [0|0|1]	**266**
Tribute (Right On) (89) 52(1) [0|0|10]	266
Robbie PATTON ▶ 610 Pk:26 [0|1|2]	**851**
Don't Give It Up (81) 26(2) [0|6|13]	536
Smiling Islands (83) 52(1) [0|0|12]	315
Henry PAUL Band ▶ 879 Pk:50 [0|0|1]	**292**
Keeping Our Love Alive (82) 50(2) [0|0|10]	292
PEACHES & HERB ▶ 568 Pk:19 [0|1|1]	**1062**
I Pledge My Love (80) 19(1) [0|8|19]	1062
Leslie PEARL ▶ 673 Pk:28 [0|1|1]	**671**
If The Love Fits Wear It (82) 28(1) [0|7|16]	671
PEBBLES ▶ 237 Pk:2 [2|2|2]	**4654**
Mercedes Boy (88) 2(2) [5|11|18]	2616
Girlfriend (88) 5(1) [5|12|20]	2038
Nia PEEPLES ▶ 776 Pk:35 [0|1|1]	**460**
Trouble (88) 35(1) [0|3|15]	460
Teddy PENDERGRASS ▶ 468 Pk:43 [0|0|5]	**1647**
Hold Me (84) 46(2) [0|0|18]	505
Love T.K.O. (81) 44(1) [0|0|13]	437
You're My Latest, My Greatest Inspiration (82) 43(1) [0|0|11]	338
Can't We Try (80) 52(2) [0|0|12]	301
Joy (88) 77(2) [0|0|6]	66.0
PENDULUM ▶ 1256 Pk:89 [0|0|1]	**49.8**
Gypsy Spirit (80) 89(3) [0|0|7]	49.8
PEPSI and SHIRLIE ▶ 1001 Pk:66 [0|0|2]	**179**
All Right Now (88) 66(1) [0|0|6]	99.5
Heartache (87) 78(1) [0|0|8]	79.2
Steve PERRY ▶ 200 Pk:3 [1|4|4]	**5535**
Oh Sherrie (84) 3(1) [7|13|20]	3379
Foolish Heart (85) 18(1) [0|11|19]	1063
She's Mine (84) 21(2) [0|8|13]	686
Strung Out (84) 40(1) [0|1|13]	408
Bernadette PETERS ▶ 649 Pk:31 [0|1|2]	**721**
Gee Whiz (80) 31(2) [0|5|13]	552
Dedicated To The One I Love (81) 65(1) [0|0|8]	169
PET SHOP BOYS ▶ 104 Pk:1 [5|6|9]	**11027**
West End Girls (86) 1(1) [7|14|20]	4327
Always On My Mind (88) 4(1) [4|10|15]	1802
What Have I Done To Deserve This? (88) - Pet Shop Boys And Dusty Springfield [A] 2(2) [5|13|18]	1459
It's A Sin (87) 9(1) [2|10|19]	1284
Opportunities (Let's Make Lots Of Money) (86) 10(1) [1|9|16]	1154
Domino Dancing (88) 18(1) [0|6|14]	653
Love Comes Quickly (86) 62(1) [0|0|8]	170
Suburbia (87) 70(1) [0|0|10]	156
Left To My Own Devices (89) 84(1) [0|0|3]	22.3

Acts With Singles

Act ▶ Rank Peak [Top10s|Top40s|Total] Title (Pk Yr) Peak(Wk) [Top10Wks|Top40Wks|TotalWks]	Score
Tom PETTY▶ 472 Pk:12 [0|2|2]	**1625**
I Won't Back Down (89) 12(1) [0|9|15]	1035
Runnin' Down A Dream (89) 23(1) [0|7|14]	590
Tom PETTY And The HEARTBREAKERS▶ 153 **Pk:10 [1|8|12]**	**7678**
Don't Do Me Like That (80) 10(2) [2|13|18]	1671
Refugee (80) 15(2) [0|10|14]	1176
Don't Come Around Here No More (85) 13(1) [0|9|14]	1091
You Got Lucky (83) 20(3) [0|11|18]	1068
The Waiting (81) 19(2) [0|7|13]	703
Change Of Heart (83) 21(3) [0|7|11]	676
Jammin' Me (87) 18(1) [0|6|12]	617
Make It Better (Forget About Me) (85) 54(1) [0|0|8]	201
Here Comes My Girl (80) 59(1) [0|0|7]	175
Needles And Pins (86) - Tom Petty and The Heartbreakers with Stevie Nicks [A] 37(2) [0|2|9]	157
Rebels (85) 74(1) [0|0|5]	75.2
A Woman In Love (It's Not Me) (81) 79(1) [0|0|6]	66.8
PHOTOGLO▶ 536 Pk:25 [0|2|2]	**1263**
Fool In Love With You (81) - Jim Photoglo 25(2) [0|7|16]	692
We Were Meant To Be Lovers (80) 31(1) [0|4|12]	572
Mike PINERA▶ 1012 Pk:70 [0|0|1]	**172**
Goodnight My Love (80) 70(1) [0|0|8]	172
PINK FLOYD▶ 134 Pk:1 [1|1|3]	**8451**
Another Brick In The Wall (Part II) (80) 1(4) [12|19|25]	8171
Run Like Hell (80) 53(1) [0|0|6]	160
Learning To Fly (87) 70(1) [0|0|8]	121
PLANET P▶ 979 Pk:64 [0|0|1]	**194**
Why Me? (83) 64(1) [0|0|9]	194
Robert PLANT▶ 353 Pk:20 [0|4|7]	**2754**
Big Log (83) 20(2) [0|9|16]	921
Tall Cool One (88) 25(1) [0|7|18]	658
In The Mood (84) 39(2) [0|5|12]	506
Little By Little (85) 36(1) [0|4|11]	422
Burning Down One Side (82) 64(1) [0|0|6]	122
Pledge Pin (82) 74(2) [0|0|5]	81.9
Ship Of Fools (88) 84(1) [0|0|4]	43.2
PLATINUM BLONDE▶ 1237 Pk:82 [0|0|1]	**58.5**
Somebody Somewhere (86) 82(2) [0|0|5]	58.5
PLAYER▶ 728 Pk:46 [0|0|2]	**541**
If Looks Could Kill (82) 48(2) [0|0|9]	277
It's For You (80) 46(1) [0|0|8]	263
PLEASURE▶ 891 Pk:55 [0|0|1]	**282**
Glide (80) 55(2) [0|0|10]	282
The PLIMSOULS▶ 1308 Pk:82 [0|0|1]	**31.2**
A Million Miles Away (83) 82(1) [0|0|3]	31.2
POCO▶ 490 Pk:18 [0|1|5]	**1511**
Call It Love (89) 18(1) [0|8|16]	769
Shoot For The Moon (83) 50(1) [0|0|13]	327
Under The Gun (80) 48(1) [0|0|10]	298
Days Gone By (84) 80(1) [0|0|5]	59.6
Midnight Rain (80) 74(1) [0|0|4]	57.1
Buster POINDEXTER▶ 848 Pk:45 [0|0|1]	**339**
Hot Hot Hot (Radio Edit) (88) - Buster Poindexter And His Banshees Of Blue 45(1) [0|0|13]	339
POINT BLANK▶ 777 Pk:39 [0|1|1]	**459**
Nicole (81) 39(1) [0|2|14]	459
Bonnie POINTER▶ 785 Pk:40 [0|1|1]	**451**
I Can't Help Myself (Sugar Pie, Honey Bunch) (80) 40(2) [0|2|13]	451
POINTER SISTERS▶ 22 Pk:2 [6|11|19]	**25341**
Slow Hand (81) 2(3) [11|16|24]	4799
He's So Shy (80) 3(3) [5|17|26]	3840
Jump (For My Love) (84) 3(2) [8|15|24]	3698
Automatic (84) 5(1) [6|14|20]	2676

Act ▶ Rank Peak [Top10s|Top40s|Total] Title (Pk Yr) Peak(Wk) [Top10Wks|Top40Wks|TotalWks]	Score
Neutron Dance (85) 6(3) [6|14|23]	2532
I'm So Excited (84) 9(1) [2|12|24]	1658
Dare Me (85) 11(1) [0|13|18]	1309
Should I Do It (82) 13(2) [0|10|16]	1109
American Music (82) 16(3) [0|8|14]	952
I'm So Excited (82) 30(2) [0|6|16]	620
Goldmine (86) 33(1) [0|3|13]	465
I Need You (83) 48(2) [0|0|15]	412
Baby Come And Get It (85) 44(2) [0|0|11]	340
Could I Be Dreaming (80) 52(2) [0|0|11]	311
Be There (87) 42(1) [0|0|9]	236
Freedom (85) 59(2) [0|0|11]	214
If You Wanna Get Back Your Lady (83) 67(1) [0|0|5]	117
Twist My Arm (86) 83(2) [0|0|5]	45.0
All I Know Is The Way I Feel (87) 93(1) [0|0|2]	8.6
POISON▶ 105 Pk:1 [4|6|7]	**10815**
Every Rose Has Its Thorn (88) 1(3) [8|14|21]	4836
Nothin' But A Good Time (88) 6(1) [3|11|19]	1507
Talk Dirty To Me (87) 9(1) [2|9|16]	1164
I Won't Forget You (87) 13(1) [0|9|21]	1085
Your Mama Don't Dance (89) 10(1) [1|10|14]	1054
Fallen Angel (88) 12(1) [0|9|16]	930
I Want Action (87) 50(1) [0|0|10]	238
The POLICE▶ 31 Pk:1 [6|8|10]	**23786**
Every Breath You Take (83) 1(8) [13|20|22]	10182
Every Little Thing She Does Is Magic (81) 3(2) [4|15|19]	2994
King Of Pain (83) 3(2) [5|13|16]	2967
De Do Do Do, De Da Da Da (81) 10(2) [2|13|21]	1789
Wrapped Around Your Finger (84) 8(1) [2|10|16]	1588
Don't Stand So Close To Me (81) 10(3) [3|13|18]	1578
Synchronicity II (83) 16(4) [0|9|14]	1118
Spirits In The Material World (82) 11(2) [0|10|13]	1069
Don't Stand So Close To Me '86 (86) 46(1) [0|0|9]	254
Secret Journey (82) 46(2) [0|0|8]	249
Gary PORTNOY▶ 1240 Pk:83 [0|0|1]	**56.2**
Where Everybody Knows Your Name (The Theme From "Cheers') (83) 83(1) [0|0|4]	56.2
Mike POST▶ 481 Pk:10 [1|2|2]	**1549**
The Theme From Hill Street Blues (81) - Mike Post Featuring Larry Carlton [A] 10(2) [2|10|22]	789
Theme From Magnum P.I. (82) 25(2) [0|7|17]	760
POWERSOURCE▶ 1041 Pk:61 [0|0|1]	**148**
Dear Mr. Jesus (88) 61(1) [0|0|7]	148
POWER STATION▶ 294 Pk:6 [2|3|3]	**3712**
Some Like It Hot (85) 6(2) [5|12|18]	1980
Get It On (85) 9(2) [3|10|15]	1345
Communication (85) 34(2) [0|3|10]	387
Elvis PRESLEY▶ 662 Pk:28 [0|1|2]	**690**
Guitar Man (Remix) (81) 28(2) [0|5|14]	547
The Elvis Medley (82) 71(4) [0|0|7]	143
Billy PRESTON▶ 1342 Pk:88 [0|0|1]	**22.9**
I'm Never Gonna Say Goodbye (82) 88(1) [0|0|3]	22.9
Billy PRESTON & SYREETA▶ 277 Pk:4 [1|1|2]	**3996**
With You I'm Born Again (80) 4(4) [6|15|29]	3728
One More Time For Love (80) 52(2) [0|0|10]	268
The PRETENDERS▶ 151 Pk:5 [2|5|8]	**7742**
Back On The Chain Gang (83) 5(3) [5|14|24]	2775
Brass In Pocket (I'm Special) (80) 14(2) [0|12|22]	1661
Don't Get Me Wrong (86) 10(2) [2|12|18]	1450
Middle Of The Road (84) 19(2) [0|9|14]	968
Show Me (84) 28(1) [0|6|13]	579
My Baby (87) 64(1) [0|0|7]	143
Stop Your Sobbing (80) 65(2) [0|0|5]	108
Thin Line Between Love And Hate (84) 83(2) [0|0|5]	58.5
PRETTY POISON▶ 389 Pk:8 [1|2|2]	**2271**
Catch Me (I'm Falling) (87) 8(3) [3|14|23]	1859

Act ▶ Rank Peak [Top10s|Top40s|Total] Title (Pk Yr) Peak(Wk) [Top10Wks|Top40Wks|TotalWks]	Score
Nightime (88) 36(1) [0|4|12]	412
Maxi PRIEST▶ 659 Pk:25 [0|1|1]	**708**
Wild World (89) 25(1) [0|7|18]	708
PRINCE▶ 17 Pk:1 [8|12|16]	**28172**
When Doves Cry (84) 1(5) [11|16|21]	7679
Batdance (89) 1(1) [6|11|18]	3450
U Got The Look (87) 2(1) [6|13|25]	2927
Little Red Corvette (83) 6(2) [6|15|22]	2573
Sign O' The Times (87) 3(1) [5|11|14]	2446
Delirious (83) 8(4) [4|11|18]	1912
1999 (83) 12(2) [0|10|27]	1633
I Wanna Be Your Lover (80) 11(2) [0|12|16]	1389
I Could Never Take The Place Of Your Man (88) 10(1) [1|12|17]	1224
Alphabet St. (88) 8(1) [2|9|13]	1192
Partyman (89) 18(1) [0|7|10]	582
The Arms Of Orion (89) - Prince With Sheena Easton 36(1) [0|3|14]	470
Let's Pretend We're Married//Irresistable Bitch (84) 52(1) [0|0|10]	258
Controversy (81) 70(2) [0|0|11]	182
Hot Thing (88) 63(2) [0|0|9]	163
If I Was Your Girlfriend (87) 67(1) [0|0|6]	91.6
PRINCE And The REVOLUTION▶ 40 **Pk:1 [6|8|10]**	**21448**
Let's Go Crazy (84) 1(2) [9|14|19]	5200
Kiss (86) 1(2) [7|13|18]	4499
Purple Rain (84) 2(2) [7|11|16]	3959
Raspberry Beret (85) 2(1) [6|14|17]	3528
Pop Life (85) 7(1) [3|10|14]	1376
I Would Die 4 U (85) 8(1) [3|10|15]	1349
Take Me With U (85) 25(2) [0|6|12]	585
Mountains (86) 23(1) [0|6|11]	543
America (85) 46(2) [0|0|7]	208
Anotherloverholenyohead (86) 63(2) [0|0|10]	201
Victoria PRINCIPAL▶ 1086 Pk:51 [0|0|1]	**119**
All I Have To Do Is Dream (81) - Andy Gibb And Victoria Principal [A] 51(2) [0|0|8]	119
PRISM▶ 754 Pk:39 [0|1|2]	**502**
Don't Let Him Know (82) 39(2) [0|2|10]	343
Turn On Your Radar (82) 64(3) [0|0|7]	159
The PRODUCERS▶ 1064 Pk:61 [0|0|1]	**131**
What She Does To Me (The Diana Song) (81) 61(1) [0|0|6]	131
PSEUDO ECHO▶ 470 Pk:6 [1|1|2]	**1627**
Funkytown (87) 6(1) [3|10|15]	1431
Living In A Dream (87) 57(1) [0|0|9]	196
PSYCHEDELIC FURS▶ 488 Pk:26 [0|1|4]	**1517**
Heartbreak Beat (87) 26(1) [0|5|14]	577
Love My Way (83) 44(2) [0|0|10]	383
Pretty In Pink (86) 41(1) [0|0|11]	342
The Ghost In You (84) 59(1) [0|0|9]	215
PURE PRAIRIE LEAGUE▶ 339 Pk:10 [1|3|5]	**2895**
Let Me Love You Tonight (80) 10(2) [2|11|17]	1625
Still Right Here In My Heart (81) 28(2) [0|7|14]	586
I'm Almost Ready (80) 34(2) [0|4|13]	527
You're Mine Tonight (81) 68(2) [0|0|5]	87.4
I Can't Stop The Feelin' (80) 77(1) [0|0|6]	70.0

Q

	Score
Q-FEEL▶ 1149 Pk:75 [0|0|1]	**95.2**
Dancing In Heaven (Orbital Be-Bop) (89) 75(1) [0|0|7]	95.2
QUARTERFLASH▶ 157 Pk:3 [1|3|7]	**7578**
Harden My Heart (82) 3(2) [12|19|24]	4972
Take Me To Heart (83) 14(1) [0|11|16]	1206
Find Another Fool (82) 16(2) [0|7|13]	781
Right Kind Of Love (82) 56(1) [0|0|8]	216
Night Shift (82) 60(3) [0|0|8]	191

Acts With Singles

Act ▶ Rank Peak [Top10s|Top40s|Total] Title (Pk Yr) Peak(Wk) [Top10Wks|Top40Wks|TotalWks]	Score
QUARTERFLASH *Continued*	
Take Another Picture (83) 58(1) [0|0|6]	157
Talk To Me (85) 83(2) [0|0|6]	54.2
Suzi QUATRO ▶ 681 Pk:41 [0|0|2]	**660**
She's In Love With You (80) 41(2) [0|0|11]	427
Lipstick (81) 51(2) [0|0|9]	232
QUEEN ▶ 35 Pk:1 [2|5|14]	**22382**
Another One Bites The Dust (80) 1(3) [15|21|31]	9572
Crazy Little Thing Called Love (80) 1(4) [12|17|22]	8343
Body Language (82) 11(2) [0|8|14]	1016
Radio Ga-Ga (84) 16(1) [0|8|13]	805
Need Your Loving Tonight (80) 44(3) [0|0|11]	365
Play The Game (80) 42(2) [0|0|9]	355
Under Pressure (82) - Queen & David Bowie [A] 29(2) [0|8|15]	339
A Kind Of Magic (86) 42(1) [0|0|11]	330
Flash's Theme AKA Flash (81) 42(2) [0|0|10]	328
I Want To Break Free (84) 45(1) [0|0|8]	278
One Vision (86) 61(2) [0|0|10]	224
I Want It All (89) 50(1) [0|0|10]	210
Calling All Girls (82) 60(2) [0|0|6]	143
It's A Hard Life (84) 72(1) [0|0|4]	72.9
QUIET RIOT ▶ 303 Pk:5 [1|2|3]	**3606**
Cum On Feel The Noize (83) 5(2) [5|14|21]	2822
Bang Your Head (Metal Health) (84) 31(2) [0|4|12]	457
Mama Weer All Crazee Now (84) 51(2) [0|0|12]	327

R

Act ▶ Rank Peak [Top10s|Top40s|Total]	Score
Eddie RABBITT ▶ 74 Pk:1 [4|6|9]	**15013**
I Love A Rainy Night (81) 1(2) [9|18|28]	5924
Drivin' My Life Away (80) 5(2) [6|15|25]	2919
Step By Step (81) 5(2) [8|15|22]	2841
You And I (83) - Eddie Rabbitt with Crystal Gayle [A] 7(4) [6|21|29]	1495
Someone Could Lose A Heart Tonight (82) 15(2) [0|10|15]	1056
I Don't Know Where To Start (82) 35(1) [0|4|13]	440
You Can't Run From Love (83) 55(1) [0|0|8]	225
You Put The Beat In My Heart (83) 81(1) [0|0|5]	58.3
Gone Too Far (80) 82(1) [0|0|4]	55.3
Gerry RAFFERTY ▶ 969 Pk:54 [0|0|1]	**204**
The Royal Mile (Sweet Darlin') (80) 54(2) [0|0|8]	204
RAINBOW ▶ 690 Pk:40 [0|1|2]	**636**
Stone Cold (82) 40(1) [0|1|12]	399
Street Of Dreams (83) 60(2) [0|0|3]	236
Bonnie RAITT ▶ 1083 Pk:73 [0|0|1]	**120**
You're Gonna Get What's Coming (80) 73(1) [0|0|6]	120
Kevin RALEIGH ▶ 1006 Pk:60 [0|0|1]	**175**
Moonlight On Water (89) 60(1) [0|0|9]	175
Billy RANKIN ▶ 875 Pk:52 [0|0|1]	**300**
Baby Come Back (84) 52(1) [0|0|11]	300
RATT ▶ 423 Pk:12 [0|2|6]	**1954**
Round And Round (84) 12(2) [0|10|18]	1317
Lay It Down (85) 40(1) [0|1|11]	335
Dance (87) 59(2) [0|0|9]	173
Way Cool Jr. (89) 75(1) [0|0|7]	83.3
Wanted Man (84) 87(1) [0|0|4]	30.0
You're In Love (85) 89(1) [0|0|2]	15.7
Lou RAWLS ▶ 984 Pk:65 [0|0|2]	**191**
Wind Beneath My Wings (83) 65(2) [0|0|6]	135
You're My Blessing (80) 77(1) [0|0|3]	56.1
RAY, GOODMAN & BROWN ▶ 328 Pk:5 [1|1|3]	**3086**
Special Lady (80) 5(2) [4|14|18]	2690
My Prayer (80) 47(2) [0|0|10]	335
Inside Of You (80) 76(1) [0|0|4]	61.7
RCR ▶ 1375 Pk:94 [0|0|1]	**12.7**
Scandal (80) 94(2) [0|0|2]	12.7

Act ▶ Rank Peak [Top10s|Top40s|Total]	Score
Chris REA ▶ 994 Pk:73 [0|0|3]	**183**
Working On It (89) 73(2) [0|0|7]	107
Let's Dance (87) 81(2) [0|0|5]	51.6
Loving You (82) 88(2) [0|0|3]	24.7
READY FOR THE WORLD ▶ 206 Pk:1 [2|3|3]	**5433**
Oh Sheila (85) 1(1) [6|13|21]	3376
Love You Down (87) 9(1) [1|12|19]	1339
Digital Display (86) 21(1) [0|6|18]	717
REAL LIFE ▶ 431 Pk:26 [0|3|3]	**1869**
Send Me An Angel (84) 29(2) [0|6|19]	789
Send Me An Angel '89 (89) 26(1) [0|8|16]	675
Catch Me I'm Falling (84) 40(1) [0|1|11]	405
Leon REDBONE ▶ 1175 Pk:72 [0|0|1]	**82.5**
Seduced (81) 72(2) [0|0|6]	82.5
The REDDINGS ▶ 877 Pk:55 [0|0|2]	**295**
(Sittin' On) The Dock Of The Bay (82) 55(1) [0|0|9]	215
Remote Control (80) 89(2) [0|0|13]	79.9
Helen REDDY ▶ 1339 Pk:88 [0|0|1]	**23.3**
I Can't Say Goodbye To You (81) 88(1) [0|0|3]	23.3
RED RIDER ▶ 865 Pk:48 [0|0|2]	**313**
White Hot (80) 48(1) [0|0|17]	209
Young Thing, Wild Dreams (Rock Me) (84) 71(2) [0|0|6]	105
RED ROCKERS ▶ 906 Pk:53 [0|0|1]	**264**
China (83) 53(2) [0|0|10]	264
Dan REED Network ▶ 854 Pk:38 [0|1|1]	**335**
Ritual (88) 38(1) [0|2|11]	335
Jerry REED ▶ 999 Pk:57 [0|0|1]	**180**
She Got The Goldmine (I Got The Shaft) (82) 57(2) [0|0|9]	180
RE-FLEX ▶ 592 Pk:24 [0|1|2]	**940**
The Politics Of Dancing (84) 24(1) [0|5|21]	893
Hurt (84) 82(1) [0|0|4]	46.6
REGINA ▶ 491 Pk:10 [1|1|1]	**1504**
Baby Love (86) 10(2) [2|12|20]	1504
R.E.M. ▶ 309 Pk:6 [2|2|7]	**3447**
Stand (89) 6(3) [4|11|19]	1753
The One I Love (87) 9(1) [1|10|20]	1394
It's The End Of The World As We Know It (And I Feel Fine) (88) 69(1) [0|0|9]	139
Radio Free Europe (83) 78(1) [0|0|5]	68.0
So. Central Rain (I'm Sorry) (84) 85(1) [0|0|6]	49.3
Pop Song 89 (89) 86(1) [0|0|4]	30.8
Fall On Me (86) 94(1) [0|0|3]	12.6
RENÉ & ANGELA ▶ 761 Pk:47 [0|0|3]	**487**
I'll Be Good (85) 47(2) [0|0|10]	288
Your Smile (86) 62(2) [0|0|6]	120
You Don't Have To Cry (86) 75(1) [0|0|7]	79.0
Mike RENO ▶ 571 Pk:7 [1|1|1]	**1048**
Almost Paradise (84) - Mike Reno And Ann Wilson [A] 7(2) [4|13|20]	1048
REO SPEEDWAGON ▶ 39 Pk:1 [4|13|15]	**21530**
Can't Fight This Feeling (85) 1(3) [8|14|18]	5170
Keep On Loving You (81) 1(1) [9|20|28]	4866
Take It On The Run (81) 5(2) [6|15|20]	2737
Keep The Fire Burnin' (82) 7(3) [6|13|16]	2157
In My Dreams (87) 19(1) [0|8|30]	1064
Here With Me (88) 20(1) [0|9|19]	840
One Lonely Night (85) 19(1) [0|9|16]	834
That Ain't Love (87) 16(1) [0|7|14]	719
In Your Letter (81) 20(2) [0|7|13]	689
Sweet Time (82) 26(3) [0|6|14]	620
Don't Let Him Go (81) 24(2) [0|6|14]	614
I Do'wanna Know (84) 29(1) [0|5|13]	521
Live Every Moment (85) 34(1) [0|3|11]	412
Variety Tonight (87) 60(1) [0|0|9]	194
Time For Me To Fly (80) 77(1) [0|0|6]	92.7
The REPLACEMENTS ▶ 929 Pk:51 [0|0|1]	**239**
I'll Be You (89) 51(1) [0|0|10]	239

Act ▶ Rank Peak [Top10s|Top40s|Total]	Score
RESTLESS HEART ▶ 707 Pk:33 [0|1|1]	**583**
I'll Still Be Loving You (87) 33(2) [0|5|18]	583
Burt REYNOLDS ▶ 1290 Pk:88 [0|0|1]	**36.7**
Let's Do Something Cheap And Superficial (80) 88(1) [0|0|5]	36.7
Cliff RICHARD ▶ 150 Pk:7 [2|6|10]	**7757**
We Don't Talk Anymore (80) 7(2) [4|14|20]	2342
Dreaming (80) 10(3) [3|13|22]	1812
A Little In Love (81) 17(2) [0|11|22]	1321
Daddy's Home (82) 23(1) [0|8|13]	642
Suddenly (81) - Olivia Newton-John & Cliff Richard [A] 20(1) [0|11|19]	513
Carrie (80) 34(1) [0|3|11]	467
Give A Little Bit More (81) 41(2) [0|0|11]	340
The Only Way Out (82) 64(2) [0|0|7]	148
Never Say Die (Give A Little Bit More) (83) 73(1) [0|0|6]	106
Wired For Sound (81) 71(1) [0|0|4]	66.7
Turley RICHARDS ▶ 972 Pk:54 [0|0|1]	**203**
You Might Need Somebody (80) 54(2) [0|0|7]	203
Lionel RICHIE ▶ 4 Pk:1 [13|14|15]	**52069**
All Night Long (All Night) (83) 1(4) [13|17|24]	8437
Hello (84) 1(2) [10|17|24]	6840
Say You, Say Me (85) 1(4) [9|16|20]	6296
Endless Love (81) - Diana Ross & Lionel Richie [A] 1(9) [13|19|27]	5193
Truly (82) 1(2) [10|13|18]	4947
Dancing On The Ceiling (86) 2(2) [8|14|17]	3868
Stuck On You (84) 3(2) [7|14|19]	3553
You Are (83) 4(2) [4|16|18]	2911
Running With The Night (84) 7(2) [5|14|19]	2232
My Love (83) 5(1) [5|12|16]	2148
Penny Lover (84) 8(2) [4|13|18]	1819
Love Will Conquer All (86) 9(2) [3|10|18]	1534
Ballerina Girl (87) 7(1) [3|10|18]	1473
Se La (87) 20(2) [0|7|13]	696
Deep River Woman (87) 71(1) [0|0|8]	124
The Nelson RIDDLE Orchestra ▶ 957 Pk:53 [0|0|1]	**214**
What's New (83) - Linda Ronstadt & The Nelson Riddle Orchestra [A] 53(3) [0|0|14]	214
Cheryl Pepsii RILEY ▶ 748 Pk:32 [0|1|1]	**514**
Thanks For My Child (88) 32(2) [0|5|13]	514
Teddy RILEY ▶ 1197 Pk:62 [0|0|1]	**71.2**
My Fantasy (89) - Teddy Riley featuring Guy [A] 62(1) [0|0|6]	71.2
The RINGS ▶ 1198 Pk:75 [0|0|1]	**69.9**
Let Me Go (81) 75(2) [0|0|5]	69.9
Lee RITENOUR ▶ 567 Pk:15 [0|1|2]	**1070**
Is It You (81) 15(2) [0|9|16]	941
Cross My Heart (82) 69(3) [0|0|7]	129
ROACHFORD ▶ 750 Pk:25 [0|1|1]	**510**
Cuddly Toy (Feel For Me) (89) 25(1) [0|5|14]	510
ROB BASE ▶ 637 Pk:36 [0|1|2]	**757**
It Takes Two (88) - Rob Base & D.J. E-Z Rock 36(1) [0|3|16]	509
Joy And Pain (89) - Rob Base & D.J. E-Z Rock 58(1) [0|0|13]	248
Rockie ROBBINS ▶ 1249 Pk:80 [0|0|1]	**52.8**
You And Me (80) 80(2) [0|0|4]	52.8
ROBEY ▶ 1253 Pk:77 [0|0|1]	**51.0**
One Night In Bangkok (85) 77(1) [0|0|3]	51.0
Smokey ROBINSON ▶ 75 Pk:2 [4|6|11]	**14881**
Being With You (81) 2(3) [10|16|25]	5997
Cruisin' (80) 4(4) [7|17|25]	4240
Just To See Her (87) 8(1) [4|12|21]	1629
One Heartbeat (87) 10(1) [1|11|19]	1141
Let Me Be The Clock (80) 31(2) [0|4|14]	553
Tell Me Tomorrow - Part 1 (82) 33(1) [0|5|12]	451
Ebony Eyes (84) - Rick James featuring Smokey Robinson [A] 43(1) [0|0|11]	219

Acts With Singles

Act ▶ Rank Peak [Top10s|Top40s|Total] Title (Pk Yr) Peak(Wk) [Top10Wks|Top40Wks|TotalWks]	Score
Smokey ROBINSON ▶ Continued	
Old Fashioned Love (82) 60(1) [0|0|9]	201
Blame It On Love (83) - Smokey Robinson & Barbara Mitchell [A] 48(1) [0|0|12]	189
You Are Forever (81) 59(1) [0|0|7]	149
What's Too Much (87) 79(1) [0|0|10]	111
ROCK And HYDE ▶ 980 Pk:61 [0|0|1]	**193**
Dirty Water (87) 61(1) [0|0|13]	193
The ROCKETS ▶ 1101 Pk:70 [0|0|1]	**112**
Desire (80) - Rockets 70(2) [0|0|6]	112
ROCKPILE ▶ 845 Pk:51 [0|0|1]	**344**
Teacher Teacher (81) 51(1) [0|0|12]	344
ROCKWELL ▶ 194 Pk:2 [1|2|2]	**5674**
Somebody's Watching Me (84) 2(3) [8|14|19]	5142
Obscene Phone Caller (84) 35(1) [0|2|14]	533
Nile RODGERS ▶ 1354 Pk:88 [0|0|1]	**19.7**
Let's Go Out Tonight (85) 88(1) [0|0|3]	19.7
RODWAY ▶ 1284 Pk:83 [0|0|1]	**39.7**
Don't Stop Trying (82) 83(1) [0|0|5]	39.7
ROGER ▶ 385 Pk:3 [1|1|2]	**2327**
I Want To Be Your Man (88) 3(1) [5|13|21]	2242
I Heard It Through The Grapevine (Part 1) (81) 79(2) [0|0|7]	85.3
Dann ROGERS ▶ 795 Pk:41 [0|0|1]	**428**
Looks Like Love Again (80) 41(2) [0|0|11]	428
Kenny ROGERS ▶ 9 Pk:1 [6|13|20]	**35216**
Lady (80) 1(6) [13|19|25]	8423
Islands In The Stream (83) - Kenny Rogers Duet with Dolly Parton 1(2) [12|18|25]	6841
Coward Of The County (80) 3(4) [8|15|19]	4425
I Don't Need You (81) 3(2) [8|14|18]	3554
Don't Fall In Love With A Dreamer (80) - Kenny Rogers with Kim Carnes 4(3) [5|14|19]	3194
Love Will Turn You Around (82) 13(5) [0|10|17]	1386
We've Got Tonight (83) - Kenny Rogers And Sheena Easton [A] 6(3) [7|15|18]	1316
Share Your Love With Me (81) 14(2) [0|10|15]	1159
Through The Years (82) 13(2) [0|11|15]	1103
Love The World Away (80) 14(2) [0|8|12]	998
What About Me? (84) 15(1) [0|9|19]	959
This Woman (84) 23(2) [0|6|13]	688
All My Life (83) 37(2) [0|3|11]	395
A Love Song (82) 47(3) [0|0|10]	280
Morning Desire (85) 72(2) [0|0|9]	153
Blaze Of Glory (81) 66(2) [0|0|9]	147
Crazy (85) 79(1) [0|0|8]	81.2
Eyes That See In The Dark (84) 79(1) [0|0|5]	74.3
The Greatest Gift Of All (85) - Kenny Rogers & Dolly Parton [A] 81(2) [0|0|4]	31.7
Scarlet Fever (83) 94(1) [0|0|2]	8.2
ROLLING STONES ▶ 53 Pk:2 [5|11|12]	**18848**
Start Me Up (81) 2(3) [11|19|24]	5491
Emotional Rescue (80) 3(2) [8|14|19]	4012
Harlem Shuffle (86) 5(1) [5|10|13]	1773
Undercover Of The Night (83) 9(3) [3|10|14]	1657
Mixed Emotions (89) 5(2) [3|9|12]	1515
Waiting On A Friend (82) 13(3) [0|12|15]	1289
Rock And A Hard Place (89) 23(3) [0|8|14]	641
Hang Fire (82) 20(2) [0|6|11]	608
She's So Cold (80) 26(2) [0|5|13]	588
Going To A Go-Go (82) 25(3) [0|5|11]	541
One Hit (To The Body) (86) 28(1) [0|4|11]	450
She Was Hot (84) 44(2) [0|0|9]	283
ROMAN HOLLIDAY ▶ 811 Pk:54 [0|0|3]	**402**
Stand By (83) 54(1) [0|0|9]	235
Don't Try To Stop It (83) 68(1) [0|0|6]	111
One Foot Back In Your Door (85) 76(1) [0|0|5]	55.5
The ROMANTICS ▶ 210 Pk:3 [1|2|4]	**5318**
Talking In Your Sleep (84) 3(3) [7|15|26]	4501

Act ▶ Rank Peak [Top10s|Top40s|Total] Title (Pk Yr) Peak(Wk) [Top10Wks|Top40Wks|TotalWks]	Score
One In A Million (84) 37(2) [0|3|12]	460
What I Like About You (80) 49(2) [0|0|8]	256
Test Of Time (85) 71(1) [0|0|6]	101
ROMEO'S DAUGHTER ▶ 1142 Pk:73 [0|0|1]	**98.8**
Don't Break My Heart (88) 73(1) [0|0|7]	98.8
ROMEO VOID ▶ 771 Pk:35 [0|1|1]	**470**
A Girl In Trouble (Is A Temporary Thing) (84) 35(1) [0|2|13]	470
RON And The D.C. CREW ▶ 1363 Pk:93 [0|0|1]	**16.6**
Ronnie's Rapp (87) 93(1) [0|0|4]	16.6
Linda RONSTADT ▶ 99 Pk:2 [4|7|9]	**11425**
Don't Know Much (89) - Linda Ronstadt Featuring Aaron Neville 2(2) [8|16|26]	4448
Hurt So Bad (80) 8(3) [3|11|14]	1882
How Do I Make You (80) 10(3) [3|12|16]	1783
Somewhere Out There (87) - Linda Ronstadt And James Ingram [A] 2(1) [5|12|22]	1445
I Can't Let Go (80) 31(1) [0|4|12]	484
I Knew You When (83) 37(1) [0|3|12]	451
Get Closer (82) 29(2) [0|5|12]	450
Easy For You To Say (83) 54(1) [0|0|10]	269
What's New (83) - Linda Ronstadt & The Nelson Riddle Orchestra [A] 53(3) [0|0|14]	214
Diana ROSS ▶ 16 Pk:1 [8|12|18]	**28634**
Upside Down (80) 1(4) [14|17|29]	8529
Endless Love (81) - Diana Ross & Lionel Richie [A] 1(9) [13|19|27]	5193
I'm Coming Out (80) 5(3) [6|14|23]	2830
Why Do Fools Fall In Love (81) 7(3) [6|14|20]	2330
It's My Turn (81) 9(3) [3|15|21]	2085
Missing You (85) 10(2) [2|9|27]	1770
Muscles (82) 10(6) [6|10|17]	1707
Mirror, Mirror (82) 8(3) [3|10|14]	1437
Swept Away (84) 19(2) [0|8|14]	837
Pieces Of Ice (83) 31(1) [0|3|10]	441
All Of You (84) - Julio Iglesias & Diana Ross [A] 19(2) [0|8|16]	422
So Close (83) 40(2) [0|2|10]	415
Work That Body (82) 44(2) [0|0|7]	243
Chain Reaction (Spec New Mix) (86) 66(1) [0|0|8]	149
Eaten Alive (85) 77(1) [0|0|7]	91.5
Let's Go Up (84) 77(1) [0|0|6]	91.1
One More Chance (81) 79(1) [0|0|5]	54.4
Chain Reaction (85) 95(2) [0|0|3]	10.1
ROSSINGTON COLLINS BAND ▶ 919 Pk:55 [0|0|1]	**249**
Don't Misunderstand Me (80) 55(2) [0|0|9]	249
David Lee ROTH ▶ 183 Pk:3 [2|4|7]	**6148**
California Girls (85) 3(1) [5|11|16]	2418
Just Like Paradise (88) 6(1) [3|11|16]	1511
Just A Gigolo/I Ain't Got Nobody (85) 12(1) [0|10|17]	1077
Yankee Rose (86) 16(1) [0|8|15]	840
Stand Up (88) 64(1) [0|0|8]	144
Goin' Crazy! (86) 66(1) [0|0|7]	124
That's Life (86) 85(1) [0|0|4]	35.7
ROUGH TRADE ▶ 981 Pk:58 [0|0|1]	**193**
All Touch (83) 58(2) [0|0|7]	193
ROXANNE ▶ 1098 Pk:63 [0|0|1]	**113**
Play That Funky Music (88) 63(1) [0|0|7]	113
ROXETTE ▶ 131 Pk:1 [2|3|3]	**8484**
The Look (89) 1(1) [7|13|19]	3981
Listen To Your Heart (89) 1(1) [6|14|22]	3612
Dressed For Success (89) 14(1) [0|9|18]	891
ROXY MUSIC ▶ 1278 Pk:80 [0|0|1]	**42.0**
Over You (80) 80(1) [0|0|4]	42.0

Act ▶ Rank Peak [Top10s|Top40s|Total] Title (Pk Yr) Peak(Wk) [Top10Wks|Top40Wks|TotalWks]	Score
ROYAL PHILHARMONIC ORCHESTRA ▶ 471 Pk:10 [1|1|1]	**1627**
Hooked On Classics (82) 10(2) [2|12|20]	1627
RUBBER RODEO ▶ 1242 Pk:86 [0|0|1]	**55.9**
Anywhere With You (84) 86(2) [0|0|5]	55.9
Jimmy RUFFIN ▶ 514 Pk:10 [1|1|1]	**1383**
Hold On To My Love (80) 10(2) [2|9|14]	1383
RUFUS/Chaka KHAN ▶ 480 Pk:22 [0|2|3]	**1551**
Ain't Nobody (83) - Rufus And Chaka Khan 22(3) [0|8|19]	947
Do You Love What You Feel (80) - Rufus and Chaka 30(2) [0|4|15]	578
Sharing The Love (81) - Rufus With Chaka Khan 91(2) [0|0|5]	25.8
Todd RUNDGREN ▶ 1116 Pk:63 [0|0|1]	**108**
Bang The Drum All Day (83) 63(1) [0|0|5]	108
RUN-D.M.C. ▶ 334 Pk:4 [1|2|4]	**2975**
Walk This Way (86) 4(1) [5|10|16]	1958
You Be Illin' (86) 29(3) [0|7|18]	713
It's Tricky (87) 57(1) [0|0|10]	230
Mary, Mary (88) 75(1) [0|0|6]	74.6
RUSH ▶ 420 Pk:21 [0|1|6]	**1992**
New World Man (82) 21(3) [0|6|12]	636
The Big Money (86) 45(1) [0|0|14]	388
Tom Sawyer (81) 44(1) [0|0|13]	380
The Spirit Of Radio (80) 51(1) [0|0|8]	256
Limelight (81) 55(2) [0|0|7]	214
Closer To The Heart (82) 69(1) [0|0|7]	117
Jennifer RUSH ▶ 639 Pk:36 [0|1|2]	**754**
Flames Of Paradise (87) - Jennifer Rush (Duet With Elton John) 36(1) [0|3|13]	465
The Power Of Love (86) 57(1) [0|0|13]	290
Patrice RUSHEN ▶ 543 Pk:23 [0|1|3]	**1189**
Forget Me Nots (82) 23(3) [0|7|16]	743
Haven't You Heard (80) 42(2) [0|0|9]	355
Feels So Real (Won't Let Go) (84) 78(2) [0|0|6]	91.0
Brenda RUSSELL ▶ 474 Pk:6 [1|1|1]	**1611**
Piano In The Dark (88) 6(1) [2|13|25]	1611
Mitch RYDER ▶ 1310 Pk:87 [0|0|1]	**31.0**
When You Were Mine (83) 87(1) [0|0|4]	31.0

S

Act ▶ Rank Peak [Top10s|Top40s|Total] Title (Pk Yr) Peak(Wk) [Top10Wks|Top40Wks|TotalWks]	Score
SAD CAFÉ ▶ 1234 Pk:78 [0|0|1]	**59.2**
La-Di-Da (81) 78(1) [0|0|4]	59.2
SADE ▶ 185 Pk:5 [2|4|5]	**6062**
Smooth Operator (85) 5(2) [5|13|20]	2347
The Sweetest Taboo (86) 5(1) [4|13|22]	2075
Paradise (88) 16(1) [0|8|15]	758
Never As Good As The First Time (86) 20(1) [0|7|12]	596
Your Love Is King (85) 54(1) [0|0|11]	285
SA-FIRE ▶ 428 Pk:12 [0|1|3]	**1889**
Thinking Of You (89) 12(2) [0|12|24]	1406
Boy, I've Been Told (88) 48(2) [0|0|16]	382
Gonna Make It (89) 71(1) [0|0|7]	101
SAGA ▶ 575 Pk:26 [0|1|3]	**1015**
On The Loose (83) 26(3) [0|8|18]	798
Wind Him Up (83) 64(1) [0|0|8]	176
The Flyer (83) 79(1) [0|0|3]	41.2
Carole Bayer SAGER ▶ 736 Pk:30 [0|1|1]	**530**
Stronger Than Before (81) 30(2) [0|7|13]	530
SALT-N-PEPA ▶ 549 Pk:19 [0|1|1]	**1154**
Push It (88) 19(1) [0|13|25]	1154
SANTANA ▶ 317 Pk:15 [0|3|6]	**3342**
Hold On (82) 15(2) [0|10|14]	1100
Winning (81) 17(1) [0|11|18]	994
You Know That I Love You (80) 35(1) [0|3|13]	525
Say It Again (85) 46(1) [0|0|11]	367

Acts With Singles

Act ▶ Rank Peak [Top10s|Top40s|Total] Title (Pk Yr) Peak(Wk) [Top10Wks|Top40Wks|TotalWks]	Score
SANTANA▶ *Continued*	
Nowhere To Run (82) 66(2) [0|0|8]	182
The Sensitive Kind (81) 56(1) [0|0|8]	174
SARAYA▶ 870 Pk:63 [0|0|2]	306
Back To The Bullet (89) 63(1) [0|0|9]	158
Love Has Taken Its Toll (89) 64(1) [0|0|9]	147
SAVOY BROWN▶ 1128 Pk:68 [0|0|1]	104
Run To Me (81) 68(2) [0|0|5]	104
Leo SAYER▶ 193 Pk:2 [1|2|2]	5683
More Than I Can Say (80) 2(5) [9|15|23]	5195
Living In A Fantasy (81) 23(2) [0|6|12]	488
Boz SCAGGS▶ 220 Pk:14 [0|5|5]	5054
Look What You've Done To Me (80)	
14(2) [0|10|17]	1237
Breakdown Dead Ahead (80) 15(3) [0|9|14]	1188
Miss Sun (81) 14(2) [0|9|17]	1150
JoJo (80) 17(2) [0|9|17]	1024
Heart Of Mine (88) 35(2) [0|4|14]	456
SCANDAL/Patty SMYTH▶ 288 Pk:7 [1|1|7]	3811
The Warrior (84) -	
Scandal Featuring Patty Smyth 7(2) [5|15|21]	2284
Beat Of A Heart (85) -	
Scandal Featuring Patty Smyth 41(2) [0|0|14]	404
Hands Tied (84) -	
Scandal Featuring Patty Smyth 41(2) [0|0|13]	374
Love's Got A Line On You (83) -	
Scandal 59(2) [0|0|13]	352
Never Enough (87) - Patty Smyth 61(1) [0|0|11]	202
Goodbye To You (82) - Scandal 65(2) [0|0|11]	188
Downtown Train (87) - Patty Smyth 95(1) [0|0|2]	6.2
Joey SCARBURY▶ 201 Pk:2 [1|1|2]	5521
Theme From "The Greatest American Hero"	
(Believe It Or Not) (81) 2(2) [10|18|26]	5261
When She Dances (81) 49(1) [0|0|9]	260
SCARLETT & BLACK▶ 638 Pk:20 [0|1|1]	755
You Don't Know (88) 20(1) [0|8|18]	755
Peter SCHILLING▶ 448 Pk:14 [0|1|2]	1759
Major Tom (Coming Home) (83) 14(3) [0|10|22]	1551
The Different Story (World Of Lust And Crime) (89)	
61(1) [0|0|10]	208
Timothy B. SCHMIT▶ 668 Pk:25 [0|1|2]	681
Boys Night Out (87) 25(1) [0|5|13]	491
So Much In Love (82) 59(2) [0|0|8]	190
John SCHNEIDER▶ 477 Pk:14 [0|1|4]	1569
It's Now Or Never (81) 14(2) [0|11|19]	1149
Dreamin' (82) 45(2) [0|0|8]	232
Still (81) 69(2) [0|0|5]	109
In The Driver's Seat (82) 72(1) [0|0|6]	79.5
Neal SCHON▶ 1396 Pk:94 [0|0|1]	2.8
Whiter Shade Of Pale (84) - Hagar, Schon,	
Aaronson, Shrieve [A] 94(1) [0|0|2]	2.8
Eddie SCHWARTZ▶ 650 Pk:28 [0|1|2]	721
All Our Tomorrows (82) 28(2) [0|7|15]	680
Over The Line (82) 91(4) [0|0|5]	40.2
SCORPIONS▶ 572 Pk:25 [0|1|4]	1043
Rock You Like A Hurricane (84) 25(1) [0|7|16]	712
No One Like You (82) 65(2) [0|0|1]	132
Still Loving You (84) 64(2) [0|0|8]	128
Rhythm Of Love (88) 75(1) [0|0|6]	70.3
SCRITTI POLITTI▶ 433 Pk:11 [0|1|3]	1857
Perfect Way (85) 11(1) [0|13|25]	1583
Boom! There She Was (88) -	
Scritti Politti Featuring Roger 53(1) [0|0|11]	253
Wood Beez (pray like aretha franklin) (86)	
91(2) [0|0|4]	21.2
Dan SEALS▶ 689 Pk:42 [0|0|2]	642
Bop (86) 42(1) [0|0|15]	488
Late At Night (80) -	
England Dan Seals 57(2) [0|0|6]	154
SECRET TIES▶ 1299 Pk:91 [0|0|1]	34.0
Dancing In My Sleep (86) 91(2) [0|0|5]	34.0

Act ▶ Rank Peak [Top10s|Top40s|Total] Title (Pk Yr) Peak(Wk) [Top10Wks|Top40Wks|TotalWks]	Score
Neil SEDAKA & Dara SEDAKA▶ 563	
Pk:19 [0|1|1]	1089
Should've Never Let You Go (80) 19(2) [0|10|19]	1089
SEDUCTION▶ 619 Pk:23 [0|1|1]	804
You're My One And Only (True Love) (89)	
23(1) [0|8|21]	804
Bob SEGER▶ 84 Pk:1 [4|5|6]	13574
Shakedown (87) 1(1) [7|14|18]	4610
Against The Wind (80) 5(3) [6|11|17]	2801
Tryin' To Live My Life Without You (Live) (81)	
5(2) [5|12|19]	2351
Fire Lake (80) 6(2) [4|12|16]	2278
You'll Accomp'ny Me (80) 14(3) [0|9|16]	1197
The Horizontal Bop (80) 42(2) [0|0|12]	336
Bob SEGER & The SILVER BULLET BAND▶ 102	
Pk:2 [1|6|10]	11094
Shame On The Moon (83) 2(4) [8|19|21]	5590
Even Now (83) 12(1) [0|9|12]	1141
American Storm (86) 13(1) [0|9|14]	999
Like A Rock (86) 12(1) [0|9|13]	958
Understanding (85) 17(2) [0|8|15]	952
Roll Me Away (83) 27(2) [0|6|10]	516
Old Time Rock & Roll (83) 48(1) [0|0|11]	334
Feel Like A Number (82) 48(2) [0|0|8]	248
It's You (86) 52(1) [0|0|9]	221
Miami (86) 70(2) [0|0|9]	133
Michael SEMBELLO▶ 177 Pk:1 [1|2|2]	6484
Maniac (83) 1(2) [9|16|22]	6058
Automatic Man (83) 34(1) [0|2|10]	426
Taja SEVELLE▶ 964 Pk:62 [0|0|1]	207
Love Is Contagious (87) 62(2) [0|0|10]	207
707▶ 868 Pk:52 [0|0|2]	308
I Could Be Good For You (80) 52(2) [0|0|9]	184
Mega Force (82) 62(2) [0|0|6]	124
S-EXPRESS▶ 1324 Pk:91 [0|0|1]	27.3
Theme From S-Express (88) 91(1) [0|0|6]	27.3
Charlie SEXTON▶ 576 Pk:17 [0|1|1]	1010
Beat's So Lonely (86) 17(3) [0|10|20]	1010
Phil SEYMOUR▶ 645 Pk:22 [0|1|1]	731
Precious To Me (81) 22(2) [0|7|16]	731
Paul SHAFFER▶ 1180 Pk:81 [0|0|1]	78.7
When The Radio Is On (89) 81(1) [0|0|8]	78.7
SHALAMAR▶ 214 Pk:8 [1|3|7]	5226
The Second Time Around (80) 8(2) [4|13|23]	2101
Dancing In The Sheets (84) 17(2) [0|10|18]	1162
Dead Giveaway (83) 22(2) [0|10|20]	1018
A Night To Remember (82)	
44(2) [0|0|10]	333
Full Of Fire (81) 55(1) [0|0|12]	309
Make That Move (81) 60(1) [0|0|8]	158
Amnesia (84) 73(2) [0|0|9]	145
--SHANICE see: WILSON, Shanice	
SHANNON▶ 343 Pk:8 [1|1|3]	2852
Let The Music Play (84) 8(1) [3|12|24]	2058
Give Me Tonight (84) 46(2) [0|0|13]	430
Do You Wanna Get Away (85) 49(1) [0|0|15]	364
Del SHANNON▶ 786 Pk:33 [0|1|1]	447
Sea Of Love (82) 33(2) [0|4|12]	447
Feargal SHARKEY▶ 1153 Pk:74 [0|0|1]	91.3
A Good Heart (86) 74(2) [0|0|6]	91.3
Tommy SHAW▶ 614 Pk:33 [0|1|4]	829
Girls With Guns (84) 33(1) [0|3|12]	443
Lonely School (85) 60(2) [0|0|9]	202
Ever Since The World Began (88) 75(2) [0|0|9]	123
Remo's Theme (What If) (85) 81(2) [0|0|5]	61.8
Jules SHEAR▶ 1019 Pk:57 [0|0|1]	167
Steady (85) 57(1) [0|0|7]	167
SHEILA▶ 893 Pk:49 [0|0|1]	279
Little Darlin' (82) 49(1) [0|0|9]	279

Act ▶ Rank Peak [Top10s|Top40s|Total] Title (Pk Yr) Peak(Wk) [Top10Wks|Top40Wks|TotalWks]	Score
SHEILA E.▶ 251 Pk:7 [1|3|4]	4423
The Glamorous Life (84) 7(1) [5|16|26]	2352
A Love Bizarre (86) 11(1) [0|12|23]	1320
The Belle Of St. Mark (84) 34(4) [0|5|15]	576
Hold Me (87) 68(1) [0|0|10]	175
T.G. SHEPPARD▶ 604 Pk:37 [0|1|4]	868
I Loved 'Em Every One (81) 37(2) [0|2|14]	456
Finally (82) 58(2) [0|0|8]	195
Only One You (82) 68(1) [0|0|8]	152
Make My Day (84) -	
T.G. Sheppard with Clint Eastwood [A]	
62(1) [0|0|6]	65.8
SHERBS▶ 1055 Pk:61 [0|0|1]	140
I Have The Skill (81) 61(1) [0|0|7]	140
SHERIFF▶ 301 Pk:1 [1|1|2]	3632
When I'm With You (89) 1(1) [5|13|21]	3465
When I'm With You (83) 61(1) [0|0|7]	167
Michelle SHOCKED▶ 1057 Pk:66 [0|0|1]	137
Anchorage (89) 66(1) [0|0|8]	137
SHOOTING STAR▶ 901 Pk:67 [0|0|3]	270
Touch Me Tonight (89) 67(1) [0|0|9]	127
Hollywood (82) 70(1) [0|0|5]	76.9
You've Got What I Need (80) 76(1) [0|0|4]	65.8
Glenn SHORROCK▶ 1121 Pk:69 [0|0|1]	106
Don't Girls Get Lonely (83) 69(2) [0|0|6]	106
SHOT IN THE DARK▶ 1174 Pk:71 [0|0|1]	82.7
Playing With Lightning (81) 71(1) [0|0|5]	82.7
Michael SHRIEVE▶ 1397 Pk:94 [0|0|1]	2.8
Whiter Shade Of Pale (84) - Hagar, Schon,	
Aaronson, Shrieve [A] 94(1) [0|0|2]	2.8
The SILENCERS (2)▶ 1241 Pk:81 [0|0|1]	56.1
Shiver And Shake (80) 81(2) [0|0|5]	56.1
The SILENCERS▶ 1267 Pk:82 [0|0|1]	45.4
Painted Moon (87) 82(1) [0|0|6]	45.4
SILVERADO▶ 1361 Pk:92 [0|0|1]	17.9
Ready For Love (81) 92(1) [0|0|3]	17.9
SILVER CONDOR▶ 766 Pk:32 [0|1|1]	476
You Could Take My Heart Away (81)	
32(1) [0|4|13]	476
Patrick SIMMONS▶ 702 Pk:30 [0|1|2]	605
So Wrong (83) 30(1) [0|5|13]	529
Don't Make Me Do It (83) 75(1) [0|0|5]	75.8
Carly SIMON▶ 318 Pk:11 [0|2|8]	3336
Jesse (80) 11(2) [0|13|23]	1594
Coming Around Again (87) 18(1) [0|9|17]	821
Let The River Run (89) 49(2) [0|0|10]	252
Give Me All Night (87) 61(1) [0|0|12]	221
All I Want Is You (88) 54(2) [0|0|9]	212
Why (82) 74(2) [0|0|6]	92.9
Tired Of Being Blonde (85) 70(1) [0|0|5]	92.5
You Know What To Do (83) 83(1) [0|0|4]	49.4
Paul SIMON▶ 249 Pk:6 [1|3|7]	4468
Late In The Evening (80) 6(3) [7|12|16]	2579
You Can Call Me Al (87) 23(2) [0|7|29]	1039
One-Trick Pony (80) 40(2) [0|2|11]	342
Allergies (83) 44(1) [0|0|10]	269
The Blues (83) -	
Randy Newman and Paul Simon [A] 51(3) 0|0|8]	132
Graceland (87) 81(1) [0|0|7]	77.6
The Boy In The Bubble (87) 86(1) [0|0|4]	30.7
SIMON & GARFUNKEL▶ 735 Pk:27 [0|1|1]	532
Wake Up Little Susie (82) 27(2) [0|6|11]	532
SIMON F▶ 1379 Pk:91 [0|0|1]	10.5
American Dream (87) - Simon F. 91(1) [0|0|2]	10.5
SIMPLE MINDS▶ 115 Pk:1 [2|4|4]	9544
Don't You (Forget About Me) (85) 1(1) [8|14|22]	4475
Alive & Kicking (85) 3(2) [6|16|20]	3576
Sanctify Yourself (86) 14(3) [0|9|14]	986
All The Things She Said (86) 28(1) [0|6|13]	506

Acts With Singles

Act ▶ Rank Peak [Top10s|Top40s|Total] Title (Pk Yr) Peak(Wk) [Top10Wks|Top40Wks|TotalWks]	Score
SIMPLY RED ▶ 137 Pk:1 [2|4|5]	8277
Holding Back The Years (86) 1(1) [6|14|23]	3419
If You Don't Know Me By Now (89) 1(1) [6|15|22]	3412
The Right Thing (87) 27(2) [0|6|15]	614
Money$ Too Tight (To Mention) (86)	
28(1) [0|6|15]	607
It's Only Love (89) 57(1) [0|0|9]	226
Frank SINATRA ▶ 721 Pk:32 [0|1|1]	561
Theme From New York, New York (80)	
32(2) [0|6|12]	561
SINGLE BULLET THEORY ▶ 1220 Pk:78 [0|0|1]	61.8
Keep It Tight (83) 78(2) [0|0|4]	61.8
SINITTA ▶ 1295 Pk:84 [0|0|1]	35.4
Right Back Where We Started From (89)	
84(2) [0|0|4]	35.4
SIOUXSIE & The BANSHEES ▶ 840	
Pk:53 [0|0|1]	351
Peek-A-Boo (88) 53(1) [0|0|14]	351
SIR MIX-A-LOT ▶ 1090 Pk:70 [0|0|1]	117
Posse' On Broadway (89) 70(1) [0|0|9]	117
SISTER SLEDGE ▶ 588 Pk:23 [0|1|5]	960
My Guy (82) 23(2) [0|6|15]	634
Got To Love Somebody (80) 64(1) [0|0|5]	110
Frankie (85) 75(1) [0|0|8]	101
Next Time You'll Know (81) 82(3) [0|0|5]	60.5
All American Girls (81) 79(2) [0|0|5]	55.5
SKID ROW ▶ 407 Pk:4 [1|1|2]	2100
18 And Life (89) 4(1) [5|13|20]	2097
Youth Gone Wild (89) 99(2) [0|0|2]	3.2
SKY ▶ 1323 Pk:83 [0|0|1]	27.5
Toccata (81) 83(1) [0|0|2]	27.5
SKYY ▶ 769 Pk:26 [0|1|1]	471
Call Me (82) 26(2) [0|4|11]	471
SLADE ▶ 529 Pk:20 [0|2|3]	1290
Run Runaway (84) 20(1) [0|8|17]	840
My Oh My (84) 37(2) [0|3|11]	420
Little Sheila (85) 86(1) [0|0|3]	29.4
SLAVE ▶ 1080 Pk:78 [0|0|2]	120
Watching You (81) 78(1) [0|0|6]	84.2
Snap Shot (81) 91(2) [0|0|7]	35.9
Grace SLICK ▶ 1385 Pk:95 [0|0|1]	8.6
Seasons (80) 95(1) [0|0|2]	8.6
SLY FOX ▶ 418 Pk:7 [1|1|2]	2011
Let's Go All The Way (86) 7(2) [4|14|25]	2000
Stay True (86) 94(2) [0|0|2]	11.0
Frankie SMITH ▶ 643 Pk:30 [0|1|1]	738
Double Dutch Bus (81) 30(1) [0|7|19]	738
Rex SMITH ▶ 917 Pk:32 [0|1|1]	250
Everlasting Love (81) - Rex Smith/Rachel Sweet [A]	
32(2) [0|4|13]	250
The SMITHEREENS ▶ 1367 Pk:92 [0|0|1]	16.0
Only A Memory (88) 92(1) [0|0|4]	16.0
--SMYTH, Patti see: SCANDAL/Patti SMYTH	
SNEAKER ▶ 663 Pk:34 [0|1|2]	687
More Than Just The Two Of Us (82)	
34(1) [0|6|15]	564
Don't Let Me In (82) 63(2) [0|0|5]	123
Phoebe SNOW ▶ 778 Pk:46 [0|0|2]	458
Games (81) 46(2) [0|0|10]	280
Mercy, Mercy, Mercy (81) 52(2) [0|0|8]	178
SNUFF ▶ 1357 Pk:88 [0|0|1]	19.1
Bad, Bad Billy (83) 88(1) [0|0|2]	19.1
SO ▶ 862 Pk:41 [0|0|1]	319
Are You Sure (88) 41(1) [0|0|11]	319
SOFT CELL ▶ 384 Pk:8 [1|1|1]	2345
Tainted Love (82) 8(2) [3|15|43]	2345
Belouis SOME ▶ 1067 Pk:67 [0|0|2]	129
Some People (85) 67(1) [0|0|6]	97.5
Imagination (85) 88(1) [0|0|5]	31.7

Act ▶ Rank Peak [Top10s|Top40s|Total] Title (Pk Yr) Peak(Wk) [Top10Wks|Top40Wks|TotalWks]	Score
S.O.S. BAND ▶ 236 Pk:3 [1|1|5]	4674
Take Your Time (Do It Right) Part 1 (80)	
3(2) [7|14|21]	3545
The Finest (86) 44(1) [0|0|13]	364
Just Be Good To Me (83) 55(1) [0|0|14]	340
Just The Way You Like It (84) 64(1) [0|0|10]	230
Tell Me If You Still Care (84) 65(1) [0|0|11]	195
SOULSISTER ▶ 856 Pk:41 [0|0|1]	333
The Way To Your Heart (89) 41(2) [0|0|10]	333
SOUL II SOUL ▶ 255 Pk:4 [1|2|2]	4382
Back To Life (89) 4(1) [7|18|28]	3034
Keep On Movin' (89) 11(1) [0|10|20]	1348
J.D. SOUTHER ▶ 709 Pk:11 [0|1|1]	580
Her Town Too (81) -	
James Taylor And J.D. Souther [A] 11(2) [0|10|14]	580
SOUTHSIDE JOHNNY & The ASBURY JUKES ▶	
1388 Pk:98 [0|0|1]	7.8
Walk Away Renee (86) -	
Southside Johnny & The Jukes 98(2) [0|0|5]	7.8
SPANDAU BALLET ▶ 239 Pk:4 [1|3|4]	4610
True (83) 4(4) [6|13|18]	3350
Gold (84) 29(2) [0|6|12]	601
Only When You Leave (84) 34(2) [0|4|12]	503
Communication (84) 59(1) [0|0|7]	156
SPARKS ▶ 850 Pk:49 [0|0|2]	338
Cool Places (83) -	
Sparks And Jane Wiedlin [A] 49(2) [0|0|12]	195
I Predict (82) 60(1) [0|0|7]	143
Judson SPENCE ▶ 752 Pk:32 [0|1|1]	507
Yeah, Yeah, Yeah (88) 32(1) [0|4|14]	507
Tracie SPENCER ▶ 737 Pk:38 [0|1|2]	530
Symptoms Of True Love (88) 38(1) [0|3|16]	497
Imagine (89) 85(1) [0|0|4]	32.7
SPIDER ▶ 626 Pk:39 [0|1|3]	783
New Romance (It's A Mystery) (80) 39(2) [0|2|11]	461
It Didn't Take Long (81) 43(2) [0|0|10]	296
Everything Is Alright (80) 86(1) [0|0|3]	26.2
SPINNERS ▶ 127 Pk:2 [2|2|5]	8642
Working My Way Back To You/Forgive Me Girl	
(Medley) (80) 2(2) [8|16|25]	4689
Cupid/I've Loved You For A Long Time (Medley) (80)	
4(3) [7|14|19]	3595
Yesterday Once More/Nothing Remains The Same	
(Medley) (81) 52(2) [0|0|8]	186
Funny How Time Slips Away (83) 67(1) [0|0|8]	161
Never Thought I'd Fall In Love (82)	
95(2) [0|0|2]	10.0
SPLIT ENZ ▶ 899 Pk:53 [0|0|1]	273
I Got You (80) 53(1) [0|0|11]	273
Dusty SPRINGFIELD ▶ 498 Pk:2 [1|1|1]	1459
What Have I Done To Deserve This? (88) -	
Pet Shop Boys And Dusty Springfield [A]	
2(2) [5|13|18]	1459
Rick SPRINGFIELD ▶ 21 Pk:1 [5|16|17]	26085
Jessie's Girl (81) 1(2) [12|22|32]	6460
Don't Talk To Strangers (82) 2(4) [11|16|21]	5887
Love Somebody (84) 5(2) [5|12|16]	2297
I've Done Everything For You (81) 8(2) [4|12|22]	1929
Affair Of The Heart (83) 9(2) [3|13|18]	1827
Human Touch (83) 18(1) [0|11|15]	997
Love Is Alright Tonite (82) 20(2) [0|10|16]	848
Souls (83) 23(3) [0|6|15]	792
What Kind Of Fool Am I (82) 21(6) [0|9|12]	782
Bop 'Til You Drop (84) 20(1) [0|9|15]	775
State Of The Heart (85) 22(1) [0|8|15]	689
Rock Of Life (88) 22(1) [0|6|15]	624
Don't Walk Away (84) 26(2) [0|6|12]	587
Bruce (85) 27(1) [0|6|13]	549
Celebrate Youth (85) 26(1) [0|6|11]	490
I Get Excited (82) 32(2) [0|5|12]	463

Act ▶ Rank Peak [Top10s|Top40s|Total] Title (Pk Yr) Peak(Wk) [Top10Wks|Top40Wks|TotalWks]	Score
Taxi Dancing (84) - Rick Springfield &	
Randy Crawford [A] 59(1) [0|0|10]	89.4
Bruce SPRINGSTEEN ▶ 25 Pk:2 [11|13|14]	24999
Dancing In The Dark (84) 2(4) [9|15|21]	5565
Hungry Heart (80) 5(5) [8|14|18]	3294
Glory Days (85) 5(1) [5|13|18]	2124
I'm On Fire (85) 6(2) [4|12|20]	1895
Cover Me (84) 7(1) [4|13|18]	1853
Brilliant Disguise (87) 5(1) [5|11|16]	1737
Born In The U.S.A. (85) 9(1) [2|11|17]	1590
My Hometown (86) 6(2) [2|9|15]	1511
War (Live) (86) - Bruce Springsteen &	
The E Street Band 8(3) [3|9|12]	1362
Tunnel Of Love (88) 9(1) [1|11|16]	1146
I'm Goin' Down (85) 9(1) [1|9|13]	1114
One Step Up (88) 13(1) [0|8|15]	924
Fade Away (81) 20(2) [0|6|12]	640
Fire (Live) (87) - Bruce Springsteen &	
The E Street Band 46(2) [0|0|8]	246
SPYRO GYRA ▶ 1005 Pk:68 [0|0|2]	175
Catching The Sun (80) 68(1) [0|0|5]	112
Cafe Amore (81) 77(1) [0|0|5]	63.3
SPYS ▶ 1196 Pk:82 [0|0|1]	71.2
Don't Run My Life (82) 82(3) [0|0|5]	71.2
SQUEEZE ▶ 450 Pk:15 [0|2|3]	1754
Hourglass (87) 15(1) [0|10|19]	958
853-5937 (88) 32(1) [0|5|12]	470
Tempted (81) 49(2) [0|0|11]	327
Billy SQUIER ▶ 254 Pk:15 [0|4|11]	4386
Rock Me Tonite (84) 15(1) [0|12|16]	1141
The Stroke (81) 17(2) [0|11|20]	1121
Everybody Wants You (82) 32(3) [0|6|17]	654
In The Dark (81) 35(2) [0|3|12]	405
My Kinda Lover (82) 45(1) [0|0|10]	330
Don't Say You Love Me (89) 58(1) [0|0|15]	309
Eye On You (85) 71(1) [0|0|8]	138
Emotions In Motion (82) 68(2) [0|0|6]	94.7
She's A Runner (83) 75(2) [0|0|6]	89.5
All Night Long (84) 75(1) [0|0|3]	51.9
Love Is The Hero (86) 80(1) [0|0|5]	51.2
STABILIZERS ▶ 1369 Pk:93 [0|0|1]	14.4
One Simple Thing (87) 93(1) [0|0|3]	14.4
STACEY Q ▶ 325 Pk:3 [1|2|4]	3160
Two Of Hearts (86) 3(1) [4|13|22]	2345
We Connect (87) 35(2) [0|4|19]	594
Don't Make A Fool Of Yourself (88)	
66(1) [0|0|8]	150
Shy Girl (87) 89(1) [0|0|11]	72.2
STAGE DOLLS ▶ 880 Pk:46 [0|0|1]	291
Love Cries (89) 46(1) [0|0|13]	291
Frank STALLONE ▶ 469 Pk:10 [1|1|3]	1642
Far From Over (83) 10(2) [2|10|16]	1468
Case Of You (80) 67(2) [0|0|6]	127
Darlin' (84) 81(1) [0|0|4]	46.6
Michael STANLEY Band ▶ 493 Pk:33 [0|2|7]	1494
He Can't Love You (81) 33(2) [0|5|16]	607
My Town (83) 39(1) [0|1|10]	410
Falling In Love Again (81) 64(2) [0|0|8]	174
Lover (81) 68(2) [0|0|6]	97.8
Someone Like You (84) 75(1) [0|0|5]	82.8
When I'm Holding You Tight (82) 78(2) [0|0|4]	64.2
Take The Time (83) 81(2) [0|0|5]	58.0
STARLAND VOCAL BAND ▶ 1093	
Pk:71 [0|0|1]	115
Loving You With My Eyes (80) 71(1) [0|0|6]	115
STARPOINT ▶ 513 Pk:25 [0|1|3]	1388
Object Of My Desire (85) 25(1) [0|9|24]	1000
Restless (86) 46(2) [0|0|12]	344
He Wants My Body (87) 89(1) [0|0|7]	43.9

Acts With Singles

Act ► Rank Peak [Top10s|Top40s|Total] Title (Pk Yr) Peak(Wk) [Top10Wks|Top40Wks|TotalWks]	Score
Brenda K. STARR ► 412 Pk:13 [0|2|2]	2051
I Still Believe (88) 13(1) [0|12|26]	1454
What You See Is What You Get (88) 24(2) [0|7|13]	597
Ringo STARR ► 814 Pk:38 [0|1|1]	400
Wrack My Brain (81) 38(1) [0|2|11]	400
STARS ON ► 198 Pk:1 [1|2|4]	5553
Medley (81) - Stars On 45 1(1) [8|14|21]	4837
Stars On 45 III (A Tribute To Stevie Wonder) (Medley) (82) - Stars On 45 28(2) [0|5|10]	423
More Stars on 45 (Medley) (81) - Stars On 45 55(2) [0|0|7]	195
Medley II (Medley) (81) - Stars On 45 67(1) [0|0|6]	98.5
STAR WARS INTERGALACTIC DROID CHOIR & CHORALE ► 1097 Pk:69 [0|0|1]	114
What Can You Get A Wookiee For Christmas (When He Already Owns A Comb?) (80) 69(3) [0|0|6]	114
STEEL BREEZE ► 462 Pk:16 [0|2|2]	1677
You Don't Want Me Anymore (82) 16(1) [0|11|20]	1056
Dreamin' Is Easy (83) 30(3) [0|6|13]	621
Maureen STEELE ► 1204 Pk:77 [0|0|1]	67.4
Save The Night For Me (85) 77(1) [0|0|5]	67.4
STEELY DAN ► 390 Pk:10 [1|2|3]	2269
Hey Nineteen (81) 10(2) [2|13|19]	1701
Time Out Of Mind (81) 22(2) [0|7|11]	568
Jim STEINMAN ► 700 Pk:32 [0|1|1]	608
Rock And Roll Dreams Come Through (81) 32(1) [0|6|16]	608
Van STEPHENSON ► 534 Pk:22 [0|1|3]	1274
Modern Day Delilah (84) 22(2) [0|10|17]	910
What The Big Girls Do (84) 45(1) [0|0|9]	315
You've Got A Good Love Coming (81) 79(1) [0|0|4]	49.1
Shakin' STEVENS ► 1060 Pk:67 [0|0|1]	134
I Cry Just A Little Bit (84) 67(2) [0|0|6]	134
Stevie B ► 392 Pk:32 [0|2|5]	2256
I Wanna Be The One (89) 32(1) [0|5|20]	693
Spring Love (Come Back To Me) (88) 43(1) [0|0|18]	566
In My Eyes (89) 37(1) [0|2|17]	562
Girl I Am Searching For You (89) 56(1) [0|0|15]	318
Dreamin' Of Love (88) 80(2) [0|0|10]	117
Al STEWART ► 682 Pk:24 [0|1|1]	658
Midnight Rocks (80) 24(1) [0|6|13]	658
Amii STEWART & Johnny BRISTOL ► 1013 Pk:63 [0|0|1]	172
My Guy/My Girl (80) 63(2) [0|0|8]	172
Dave STEWART With Barbara GASKIN ► 1112 Pk:72 [0|0|1]	109
It's My Party (82) 72(2) [0|0|8]	109
Jermaine STEWART ► 324 Pk:5 [1|2|4]	3268
We Don't Have To Take Our Clothes Off (86) 5(2) [4|13|22]	2052
The Word Is Out (85) 41(1) [0|0|15]	478
Say It Again (88) 27(1) [0|5|12]	442
Jody (86) 42(1) [0|0|9]	296
John STEWART ► 730 Pk:34 [0|1|1]	539
Lost Her In The Sun (80) 34(1) [0|4|13]	539
Rod STEWART ► 47 Pk:4 [6|12|20]	20483
Passion (81) 5(2) [6|17|20]	3000
Young Turks (81) 5(4) [6|15|19]	2970
My Heart Can't Tell You No (89) 4(1) [5|13|25]	2400
Infatuation (84) 6(2) [4|13|18]	2076
Love Touch (86) 6(1) [4|12|18]	1765
Forever Young (88) 12(1) [0|10|24]	1281
Some Guys Have All The Luck (84) 10(1) [1|10|17]	1263
Crazy About Her (89) 11(1) [0|10|17]	1151
Baby Jane (83) 14(1) [0|9|14]	1124

Act ► Rank Peak [Top10s|Top40s|Total] Title (Pk Yr) Peak(Wk) [Top10Wks|Top40Wks|TotalWks]	Score
Lost In You (88) 12(1) [0|9|18]	1015
Tonight I'm Yours (Don't Hurt Me) (82) 20(2) [0|8|14]	735
What Am I Gonna Do (I'm So In Love With You) (83) 35(1) [0|3|12]	491
I Don't Want To Talk About It (80) 46(2) [0|0|11]	310
Another Heartache (86) 52(1) [0|0|9]	238
How Long (82) 49(2) [0|0|9]	236
People Get Ready (85) - Jeff Beck And Rod Stewart [A] 48(1) [0|0|10]	147
All Right Now (85) 72(1) [0|0|6]	94.3
Somebody Special (81) 71(2) [0|0|5]	86.5
Every Beat Of My Heart (86) 83(2) [0|0|6]	50.8
Twistin' The Night Away (87) 80(1) [0|0|4]	49.7
Stephen STILLS ► 1017 Pk:61 [0|0|1]	169
Stranger (84) 61(1) [0|0|8]	169
Stephen STILLS With Michael FINNIGAN ► 1091 Pk:67 [0|0|1]	116
Can't Let Go (84) - Stephen Stills Featuring Michael Finnigan 67(2) [0|0|6]	116
STING ► 123 Pk:3 [3|6|7]	8848
If You Love Somebody Set Them Free (85) 3(2) [6|14|18]	3098
We'll Be Together (87) 7(1) [4|12|18]	1644
Fortress Around Your Heart (85) 8(2) [3|11|20]	1614
Love Is The Seventh Wave (85) 17(3) [0|9|13]	844
Be Still My Beating Heart (88) 15(2) [0|8|14]	838
Russians (86) 16(2) [0|8|13]	774
Englishman In New York (88) 84(2) [0|0|4]	35.8
The STOMPERS ► 1313 Pk:88 [0|0|1]	30.3
Never Tell An Angel (When Your Heart's On Fire) (83) 88(1) [0|0|2]	30.3
STRAY CATS ► 136 Pk:3 [3|4|5]	8306
Stray Cat Strut (83) 3(3) [5|14|19]	3396
(She's) Sexy + 17 (83) 5(2) [4|12|15]	2192
Rock This Town (82) 9(5) [5|13|21]	2104
I Won't Stand In Your Way (83) 35(1) [0|3|13]	522
Look At That Cadillac (84) 68(1) [0|0|5]	91.1
STREEK ► 951 Pk:47 [0|0|1]	217
One More Night (81) 47(2) [0|0|7]	217
Janey STREET ► 1130 Pk:68 [0|0|1]	104
Say Hello To Ronnie (84) 68(2) [0|0|5]	104
STREETS ► 1286 Pk:87 [0|0|1]	38.7
If Love Should Go (83) 87(1) [0|0|5]	38.7
Barbra STREISAND ► 92 Pk:1 [1|5|11]	12222
Woman In Love (80) 1(3) [11|19|24]	7930
Comin' In And Out Of Your Life (82) 11(1) [0|11|16]	1626
The Way He Makes Me Feel (83) 40(2) [0|2|15]	551
Somewhere (From West Side Story) (86) 43(2) [0|0|14]	425
Kiss Me In The Rain (80) 37(2) [0|3|11]	413
Left In The Dark (84) 50(2) [0|0|12]	310
Make No Mistake, He's Mine (85) (Duet With Kim Carnes) 51(1) [0|0|10]	284
Promises (81) 48(2) [0|1|9]	245
Till I Loved You (The Love Theme From Goya) (88) - Barbra Streisand And Don Johnson [A] 25(1) [0|5|12]	232
Memory - The Theme From Andrew Lloyd Webber's Musical ''Cats'' (82) 52(1) [0|0|7]	172
Emotion (85) 79(1) [0|0|2]	34.3
Barbra STREISAND & Barry GIBB ► 226 Pk:3 [2|2|2]	4905
Guilty (81) 3(2) [7|15|22]	3545
What Kind Of Fool (81) 10(3) [3|10|16]	1360
STRYPER ► 602 Pk:23 [0|1|3]	879
Honestly (88) 23(1) [0|8|19]	718
Always There For You (88) 71(1) [0|0|8]	131
I Believe In You (88) 88(1) [0|0|5]	30.4

Act ► Rank Peak [Top10s|Top40s|Total] Title (Pk Yr) Peak(Wk) [Top10Wks|Top40Wks|TotalWks]	Score
STYLE COUNCIL ► 694 Pk:29 [0|1|2]	629
My Ever Changing Moods (84) 29(1) [0|6|14]	557
You're The Best Thing (84) 76(2) [0|0|5]	71.8
STYX ► 86 Pk:3 [4|6|9]	13467
The Best Of Times (81) 3(4) [10|15|19]	4383
Mr. Roboto (83) 3(2) [8|16|18]	3714
Don't Let It End (83) 6(1) [5|13|16]	2206
Too Much Time On My Hands (81) 9(2) [3|13|19]	1694
Why Me (80) 26(1) [0|5|13]	612
Music Time (84) 40(2) [0|2|9]	342
High Time (83) 48(1) [0|0|7]	237
Nothing Ever Goes As Planned (81) 54(2) [0|0|8]	159
Borrowed Time (80) 64(1) [0|0|6]	120
SUAVÉ ► 684 Pk:20 [0|1|1]	655
My Girl (88) 20(2) [0|7|15]	655
SUGARHILL GANG ► 612 Pk:36 [0|1|3]	837
Rapper's Delight (80) 36(1) [0|2|12]	444
Apache (82) - Sugar Hill Gang 53(2) [0|0|11]	315
8th Wonder (81) 82(1) [0|0|9]	78.1
Donna SUMMER ► 58 Pk:3 [5|10|16]	17446
She Works Hard For The Money (83) 3(3) [8|17|21]	4407
The Wanderer (80) 3(3) [6|13|20]	3526
On The Radio (80) 5(2) [6|12|17]	2750
Love Is In Control (Finger On The Trigger) (82) 10(1) [1|11|18]	1571
This Time I Know It's For Real (89) 7(2) [3|10|17]	1494
There Goes My Baby (84) 21(1) [0|8|14]	729
The Woman In Me (83) 33(3) [0|6|16]	657
Cold Love (81) 33(2) [0|3|12]	489
Walk Away (80) 36(2) [0|3|11]	404
Who Do You Think You're Foolin' (81) 40(2) [0|2|11]	341
State Of Independence (82) 41(2) [0|0|10]	328
Dinner With Gershwin (87) 48(1) [0|0|11]	295
Unconditional Love (83) 43(2) [0|0|8]	284
Supernatural Love (84) 75(2) [0|0|5]	84.1
Love Has A Mind Of Its Own (84) - Donna Summer with Matthew Ward 70(1) [0|0|4]	62.7
Love's About To Change My Heart (89) 85(1) [0|0|3]	24.1
Henry LEE SUMMER ► 439 Pk:18 [0|2|4]	1825
Hey Baby (89) 18(1) [0|8|18]	867
I Wish I Had A Girl (88) 20(1) [0|7|18]	747
Darlin' Danielle Don't (88) 57(1) [0|0|8]	189
Hands On The Radio (88) 85(1) [0|0|3]	21.9
Joe SUN ► 1108 Pk:71 [0|0|1]	110
Shotgun Rider (80) 71(1) [0|0|6]	110
SUPERTRAMP ► 302 Pk:11 [0|4|5]	3606
It's Raining Again (82) 11(4) [0|11|13]	1397
Dreamer (80) 15(2) [0|8|14]	981
My Kind Of Lady (83) 31(1) [0|5|12]	541
Cannonball (85) 28(2) [0|7|12]	497
Breakfast In America (80) 62(3) [0|0|8]	191
SURFACE ► 341 Pk:5 [1|2|4]	2866
Shower Me With Your Love (89) 5(1) [4|11|19]	1852
Happy (87) 20(2) [0|8|14]	738
Closer Than Friends (89) 57(1) [0|0|13]	232
You Are My Everything (89) 84(1) [0|0|5]	44.8
SURVIVOR ► 32 Pk:1 [5|8|18]	23583
Eye Of The Tiger (82) 1(6) [15|18|25]	9773
Burning Heart (86) 2(2) [6|16|22]	3808
The Search Is Over (85) 4(1) [5|14|21]	2456
Is This Love (87) 9(2) [3|13|19]	1703
I Can't Hold Back (84) 13(2) [0|13|23]	1564
High On You (85) 8(2) [2|11|17]	1361
American Heartbeat (82) 17(2) [0|7|16]	839
Poor Man's Son (81) 33(1) [0|4|14]	561
Somewhere In America (80) 70(2) [0|0|12]	245

Acts With Singles

Act ▶ Rank Peak [Top10s|Top40s|Total] Title (Pk Yr) Peak(Wk) [Top10Wks|Top40Wks|TotalWks]	Score
SURVIVOR ▶ *Continued*	
First Night (85) 53(1) [0|0|9]	231
How Much Love (87) 51(1) [0|0|9]	227
Didn't Know It Was Love (88) 61(1) [0|0|10]	195
Summer Nights (82) 62(2) [0|0|8]	178
The Moment Of Truth (84) 63(2) [0|0|7]	157
The One That Really Matters (83) 74(1) [0|0|6]	91.1
Across The Miles (89) 74(1) [0|0|6]	80.0
Caught In The Game (83) 77(2) [0|0|5]	72.0
Man Against The World (87) 86(1) [0|0|5]	42.0
Patrick SWAYZE ▶ 335 Pk:3 [1|1|1]	**2939**
She's Like The Wind (88) - Patrick Swayze (featuring Wendy Fraser) 3(3) [6|13|21]	2939
Keith SWEAT ▶ 402 Pk:5 [1|1|3]	**2129**
I Want Her (88) 5(1) [5|13|21]	1842
Make It Last Forever (88) - Keith Sweat (Duet with Jacci McGhee) 59(1) [0|0|11]	235
Something Just Ain't Right (88) 79(1) [0|0|5]	52.2
Rachel SWEET ▶ 852 Pk:32 [0|1|2]	**336**
Everlasting Love (81) - Rex Smith/Rachel Sweet [A] 32(2) [0|4|13]	250
Voo Doo (83) 72(2) [0|0|5]	86.0
SWEET SENSATION ▶ 379 Pk:14 [0|2|5]	**2384**
Sincerely Yours (89) 14(1) [0|8|18]	997
Hooked On You (89) 23(1) [0|7|16]	692
Take It While It's Hot (88) 57(1) [0|0|13]	262
Hooked On You (87) 64(1) [0|0|12]	232
Never Let You Go (88) 58(1) [0|0|10]	202
SWING OUT SISTER ▶ 376 Pk:6 [1|2|3]	**2428**
Breakout (87) 6(2) [4|11|23]	1911
Twilight World (88) 31(1) [0|3|15]	486
Waiting Game (89) 86(1) [0|0|4]	30.6
SWITCH ▶ 1065 Pk:83 [0|0|1]	**130**
I Call Your Name (80) 83(2) [0|0|15]	130
SYBIL ▶ 590 Pk:20 [0|1|1]	**950**
Don't Make Me Over (89) 20(1) [0|9|23]	950
SYLVIA ▶ 555 Pk:15 [0|1|1]	**1134**
Nobody (82) 15(3) [0|9|25]	1134
SYNCH ▶ 500 Pk:10 [1|1|2]	**1450**
Where Are You Now? (89) - Jimmy Harnen w/Synch 10(1) [1|11|24]	1326
Where Are You Now? (86) 77(1) [0|0|12]	124
The SYSTEM ▶ 378 Pk:4 [1|1|3]	**2389**
Don't Disturb This Groove (87) 4(1) [4|13|21]	2199
You Are In My System (83) 64(2) [0|0|8]	165
Coming To America (Part One) (88) 91(1) [0|0|5]	24.7

T

Act ▶ Rank Peak [Top10s|Top40s|Total]	Score
TACO ▶ 327 Pk:4 [1|1|1]	**3108**
Puttin' On The Ritz (83) 4(2) [6|14|21]	3108
TALKING HEADS ▶ 342 Pk:9 [1|2|5]	**2859**
Burning Down The House (83) 9(1) [1|11|20]	1522
Wild Wild Life (86) 25(1) [0|7|21]	777
And She Was (85) 54(1) [0|0|20]	353
This Must Be The Place (Naïve Melody) (83) 62(1) [0|0|8]	184
Once In A Lifetime (86) 91(1) [0|0|4]	22.3
TALK TALK ▶ 658 Pk:31 [0|1|4]	**709**
It's My Life (84) 31(2) [0|6|14]	593
Talk Talk (82) 75(1) [0|0|7]	68.8
Life's What You Make It (86) 90(2) [0|0|4]	26.1
Such A Shame (84) 89(1) [0|0|3]	20.9
TA MARA & The SEEN ▶ 606 Pk:24 [0|1|1]	**856**
Everybody Dance (86) 24(1) [0|10|21]	856
TAMI SHOW ▶ 1300 Pk:88 [0|0|1]	**33.5**
She's Only 20 (88) 88(1) [0|0|4]	33.5
TANGIER ▶ 1136 Pk:67 [0|0|1]	**101**
On The Line (89) 67(1) [0|0|1]	101

Act ▶ Rank Peak [Top10s|Top40s|Total]	Score
TARNEY/SPENCER Band ▶ 1217 Pk:74 [0|0|1]	**63.4**
No Time To Lose (81) 74(1) [0|0|4]	63.4
A TASTE OF HONEY ▶ 263 Pk:3 [1|1|2]	**4276**
Sukiyaki (81) 3(3) [8|16|24]	3949
I'll Try Something New (82) 41(3) [0|0|10]	327
TAVARES ▶ 544 Pk:33 [0|1|2]	**1185**
A Penny For Your Thoughts (82) 33(4) [0|9|21]	930
Bad Times (80) 47(2) [0|0|10]	256
Andy TAYLOR ▶ 630 Pk:24 [0|1|2]	**773**
Take It Easy (86) 24(1) [0|7|17]	690
When The Rain Comes Down (86) 73(1) [0|0|6]	82.8
B. E. TAYLOR Group ▶ 1011 Pk:66 [0|0|2]	**172**
Vitamin L (84) 66(1) [0|0|8]	166
Karen (86) 94(1) [0|0|2]	6.3
James TAYLOR ▶ 584 Pk:11 [0|1|4]	**971**
Her Town Too (81) - James Taylor And J.D. Souther [A] 11(2) [0|10|14]	580
Everyday (85) 61(1) [0|0|11]	253
Hard Times (81) 72(2) [0|0|5]	76.7
Never Die Young (88) 80(1) [0|0|5]	61.0
John TAYLOR ▶ 729 Pk:23 [0|1|1]	**540**
I Do What I Do (Theme For 9 1/2 Weeks) (86) 23(1) [0|6|12]	540
Livingston TAYLOR ▶ 820 Pk:38 [0|1|1]	**380**
First Time Love (80) 38(1) [0|2|10]	380
TEARS FOR FEARS ▶ 68 Pk:1 [4|5|6]	**15779**
Shout (85) 1(3) [7|13|19]	4854
Everybody Wants To Rule The World (85) 1(2) [8|14|24]	4785
Head Over Heels (85) 3(1) [6|12|20]	2862
Sowing The Seeds Of Love (89) 2(1) [5|12|15]	2677
Mothers Talk (86) 27(2) [0|6|12]	499
Change (83) 73(1) [0|0|6]	102
The TEMPTATIONS ▶ 505 Pk:43 [0|0|7]	**1426**
Treat Her Like A Lady (85) 48(2) [0|0|14]	409
Lady Soul (86) 47(2) [0|0|11]	309
Power (80) 43(1) [0|0|9]	290
Sail Away (84) 54(1) [0|0|8]	213
Aiming At Your Heart (81) 67(2) [0|0|5]	98.0
Standing On The Top Part 1 (82) - The Temptations Featuring Rick James [A] 66(1) [0|0|8]	80.4
Love On My Mind Tonight (83) 88(2) [0|0|3]	27.0
10,000 MANIACS ▶ 739 Pk:44 [0|0|3]	**524**
Trouble Me (89) 44(1) [0|0|12]	317
Like The Weather (88) 68(2) [0|0|11]	116
What's The Matter Here? (88) 80(1) [0|0|8]	90.6
Robert TEPPER ▶ 656 Pk:22 [0|1|2]	**713**
No Easy Way Out (86) 22(1) [0|7|16]	687
Don't Walk Away (86) 85(1) [0|0|3]	25.8
Helen TERRY ▶ 1393 Pk:96 [0|0|1]	**4.3**
One Sunny Day/Dueling Bikes From Quicksilver (86) - Ray Parker Jr. And Helen Terry [A] 96(2) [0|0|3]	4.3
Tony TERRY ▶ 1028 Pk:80 [0|0|2]	**157**
She's Fly (88) 80(1) [0|0|9]	97.8
Forever Yours (88) 80(1) [0|0|5]	59.1
TESLA ▶ 1364 Pk:91 [0|0|1]	**16.6**
Little Suzi (87) 91(1) [0|0|3]	16.6
TEXAS ▶ 1219 Pk:77 [0|0|1]	**61.9**
I Don't Want A Lover (89) 77(1) [0|0|6]	61.9
38 SPECIAL ▶ 118 Pk:6 [2|8|14]	**9241**
Second Chance (89) - Thirty Eight Special 6(1) [4|14|21]	1948
Caught Up In You (82) 10(3) [3|12|17]	1571
If I'd Been The One (84) 19(1) [0|9|16]	966
Like No Other Night (86) 14(1) [0|9|16]	874
Back Where You Belong (84) 20(2) [0|8|13]	746
Hold On Loosely (81) - .38 Special 27(2) [0|6|17]	675
Teacher, Teacher (84) 25(1) [0|5|12]	520
You Keep Runnin' Away (82) 38(2) [0|2|11]	379

Act ▶ Rank Peak [Top10s|Top40s|Total]	Score
Somebody Like You (86) 48(1) [0|0|12]	340
Rockin' Into The Night (80) - 43(1) [0|0|9]	330
Back To Paradise (87) 41(1) [0|0|11]	328
Fantasy Girl (81) - .38 Special 52(2) [0|0|10]	307
Rock & Roll Strategy (88) - Thirty Eight Special 67(1) [0|0|8]	130
Comin' Down Tonight (89) - Thirty Eight Special 67(1) [0|0|7]	127
B.J. THOMAS ▶ 1381 Pk:93 [0|0|1]	**9.8**
Whatever Happened To Old Fashioned Love (83) 93(1) [0|0|2]	9.8
Evelyn THOMAS ▶ 1282 Pk:85 [0|0|1]	**40.1**
High Energy (84) 85(1) [0|0|5]	40.1
Nolan THOMAS ▶ 895 Pk:57 [0|0|1]	**277**
Yo' Little Brother (85) 57(2) [0|0|13]	277
Timmy THOMAS ▶ 1293 Pk:80 [0|0|1]	**35.6**
Gotta Give A Little Love (Ten Years After) (84) 80(1) [0|0|3]	35.6
Chris THOMPSON ▶ 1347 Pk:85 [0|0|1]	**22.0**
All The Right Moves (83) - Jennifer Warnes/Chris Thompson [A] 85(1) [0|0|4]	22.0
ROBBIN THOMPSON Band ▶ 1085 Pk:66 [0|0|1]	**119**
Brite Eyes (80) 66(1) [0|0|9]	119
THOMPSON TWINS ▶ 100 Pk:3 [3|7|11]	**11301**
Hold Me Now (84) 3(2) [7|15|21]	4030
Lay Your Hands On Me (85) 6(2) [4|14|20]	1908
King For A Day (86) 8(1) [3|11|16]	1507
Doctor! Doctor! (84) 11(1) [0|9|16]	1310
Lies (83) 30(3) [0|5|16]	680
Get That Love (87) 31(1) [0|5|11]	454
Sugar Daddy (89) 28(1) [0|4|12]	447
You Take Me Up (84) 44(1) [0|0|9]	325
Love On Your Side (83) 45(1) [0|0|9]	294
Nothing In Common (86) 54(1) [0|0|10]	242
The Gap (84) 69(1) [0|0|6]	102
Ali THOMSON ▶ 506 Pk:15 [0|1|2]	**1424**
Take A Little Rhythm (80) 15(2) [0|9|17]	1058
Live Every Minute (80) 42(2) [0|0|11]	365
George THOROGOOD & The DESTROYERS ▶ 1025 Pk:63 [0|0|1]	**162**
Willie And The Hand Jive (85) 63(2) [0|0|8]	162
3 MAN ISLAND ▶ 1390 Pk:94 [0|0|1]	**6.7**
Jack The Lad (88) 94(1) [0|0|2]	6.7
TIA ▶ 1391 Pk:97 [0|0|1]	**6.3**
Boy Toy (87) 97(2) [0|0|2]	6.3
TIERRA ▶ 473 Pk:18 [0|1|3]	**1621**
Together (81) 18(2) [0|15|21]	1369
Memories (81) 62(2) [0|0|8]	158
La La Means I Love You (81) 72(1) [0|0|6]	94.8
TIFFANY ▶ 98 Pk:1 [4|5|6]	**11660**
Could've Been (88) 1(2) [6|14|20]	4391
I Think We're Alone Now (87) 1(2) [6|13|24]	3894
All This Time (89) 6(1) [3|14|21]	1617
I Saw Him Standing There (88) 7(1) [3|9|14]	1282
Radio Romance (89) 35(1) [0|1|9]	282
Feelings Of Forever (88) 50(1) [0|0|9]	194
TIGGI CLAY ▶ 1330 Pk:86 [0|0|1]	**25.2**
Flashes (84) 86(1) [0|0|3]	25.2
TIGHT FIT ▶ 1350 Pk:89 [0|0|1]	**21.4**
Back To The 60's (Medley) (81) 89(2) [0|0|3]	21.4
'TIL TUESDAY ▶ 365 Pk:8 [1|2|5]	**2575**
Voices Carry (85) 8(1) [2|13|21]	1701
What About Love (86) 26(1) [0|5|14]	543
Coming Up Close (87) 59(1) [0|0|10]	217
Looking Over My Shoulder (85) 61(1) [0|0|5]	101
(Believed You Were) Lucky (89) 95(2) [0|0|3]	11.4

Acts With Singles

Act ► Rank Peak [Top10s|Top40s|Total] Title (Pk Yr) Peak(Wk) [Top10Wks|Top40Wks|TotalWks]	Score
TIMBUK 3 ► 605 Pk:19 [0|1|1]	859
The Future's So Bright, I Gotta Wear Shades (86)	
19(2) [0|9|16]	859
The TIME ► 461 Pk:20 [0|2|4]	1677
Jungle Love (85) 20(1) [0|10|25]	1122
The Bird (85) 36(2) [0|2|13]	482
Cool (Part 1) (82) 90(1) [0|0|7]	48.4
777-9311 (83) 88(1) [0|0|3]	24.9
The TIMELORDS ► 931 Pk:66 [0|0|1]	237
Doctorin' The Tardis (89) 66(1) [0|0|13]	237
TIMES TWO ► 632 Pk:21 [0|1|2]	764
Strange But True (88) 21(2) [0|8|17]	702
Cecilia (88) 79(1) [0|0|6]	62.8
TIMEX SOCIAL CLUB ► 454 Pk:8 [1|1|1]	1707
Rumors (86) 8(2) [4|12|19]	1707
TKA ► 1051 Pk:75 [0|0|2]	143
One Way Love (86) 75(1) [0|0|9]	112
You Are The One (89) 91(1) [0|0|7]	30.6
TOBY BEAU ► 1171 Pk:70 [0|0|1]	84.0
If I Were You (80) 70(2) [0|0|4]	84.0
TOMMY TUTONE ► 291 Pk:4 [1|2|2]	3752
867-5309/Jenny (82) 4(3) [8|16|27]	3466
Angel Say No (80) 38(2) [0|2|8]	286
TOM TOM CLUB ► 699 Pk:31 [0|1|1]	614
Genius Of Love (82) 31(2) [0|4|17]	614
TONE LŌC ► 195 Pk:2 [2|2|2]	5651
Wild Thing (89) 2(1) [6|14|25]	3259
Funky Cold Medina (89) - Tone-Lōc	
3(1) [5|11|18]	2392
TONY! TONI! TONÉ! ► 927 Pk:47 [0|0|1]	247
Little Walter (88) 47(1) [0|0|10]	247
TORA TORA ► 1283 Pk:86 [0|0|1]	39.7
Walkin' Shoes (89) 86(1) [0|0|6]	39.7
TORONTO ► 1075 Pk:77 [0|0|1]	123
Your Daddy Don't Know (82) 77(1) [0|0|8]	123
Peter TOSH ► 1276 Pk:84 [0|0|1]	43.2
Johnny B. Goode (83) 84(1) [0|0|4]	43.2
TOTAL COELO ► 1056 Pk:66 [0|0|1]	138
I Eat Cannibals (83) 66(2) [0|0|6]	138
TOTO ► 61 Pk:1 [3|9|11]	16748
Rosanna (82) 2(5) [11|18|23]	6431
Africa (83) 1(1) [6|16|21]	4117
I Won't Hold You Back (83) 10(1) [1|12|17]	1544
I'll Be Over You (86) 11(1) [0|12|23]	1372
99 (80) 26(1) [0|8|17]	859
Pamela (88) 22(2) [0|8|19]	740
Stranger In Town (84) 30(2) [0|6|15]	611
Make Believe (82) 30(3) [0|5|13]	528
Without Your Love (87) 38(1) [0|2|11]	372
Waiting For Your Love (83) 73(1) [0|0|6]	96.4
Holyanna (85) 71(1) [0|0|5]	76.8
TOUCH ► 975 Pk:65 [0|0|2]	199
Don't You Know What Love Is (81)	
69(1) [0|0|6]	107
(Call Me) When The Spirit Moves You (80)	
65(1) [0|0|5]	91.7
The TOURISTS ► 1257 Pk:83 [0|0|1]	49.5
I Only Want To Be With You (80) 83(1) [0|0|4]	49.5
Carol Lynn TOWNES ► 1114 Pk:77 [0|0|1]	108
99 1/2 (84) 77(3) [0|0|9]	108
Pete TOWNSHEND ► 377 Pk:9 [1|2|4]	2421
Let My Love Open The Door (80) 9(3) [3|12|19]	1643
Face The Face (86) 26(1) [0|7|16]	672
A Little Is Enough (80) 72(2) [0|0|4]	70.5
Rough Boys (80) 89(2) [0|0|4]	35.4
T'PAU ► 367 Pk:4 [1|1|1]	2549
Heart And Soul (87) 4(1) [5|16|27]	2549
TRANSVISION VAMP ► 1337 Pk:87 [0|0|1]	23.7
Tell That Girl To Shut Up (88) 87(1) [0|0|3]	23.7

Act ► Rank Peak [Top10s|Top40s|Total] Title (Pk Yr) Peak(Wk) [Top10Wks|Top40Wks|TotalWks]	Score
TRANS-X ► 911 Pk:61 [0|0|1]	256
Living On Video (86) 61(1) [0|0|12]	256
TRAVELING WILBURYS ► 712 Pk:45 [0|0|2]	577
Handle With Care (88) 45(1) [0|0|14]	404
End Of The Line (89) 63(2) [0|0|9]	173
Pat TRAVERS ► 960 Pk:50 [0|0|1]	209
Is This Love (80) - Pat Travers Band 50(2) [0|0|7]	209
TRIUMPH ► 608 Pk:27 [0|1|4]	853
Somebody's Out There (86) 27(1) [0|5|15]	532
Magic Power (81) 51(1) [0|0|11]	282
Follow Your Heart (85) 88(1) [0|0|2]	20.4
I Can Survive (80) 91(2) [0|0|2]	17.5
Eric TROYER ► 1365 Pk:92 [0|0|1]	16.0
Mirage (80) 92(2) [0|0|2]	16.0
The TRUTH ► 1020 Pk:65 [0|0|1]	167
Weapons Of Love (87) 65(1) [0|0|9]	167
The TUBES ► 396 Pk:10 [1|2|5]	2230
She's A Beauty (83) 10(1) [1|12|20]	1506
Don't Want To Wait Anymore (81) 35(2) [0|3|12]	397
Tip Of My Tongue (83) 52(2) [0|0|7]	217
The Monkey Time (83) 68(1) [0|0|4]	90.1
Piece By Piece (85) 87(1) [0|0|2]	20.4
Louise TUCKER ► 797 Pk:46 [0|0|1]	427
Midnight Blue (83) 46(1) [0|0|13]	427
Tina TURNER ► 37 Pk:1 [5|12|13]	22054
What's Love Got To Do With It (84)	
1(3) [10|18|28]	6823
Typical Male (86) 2(3) [6|12|16]	3330
We Don't Need Another Hero (Thunderdome) (85)	
2(1) [6|12|18]	3151
Better Be Good To Me (84) 5(2) [5|13|21]	2351
Private Dancer (85) 7(2) [3|12|18]	1623
One Of The Living (85) 15(2) [0|10|18]	940
What You Get Is What You See (87)	
13(1) [0|7|14]	887
The Best (89) 15(1) [0|8|14]	800
Let's Stay Together (84) 26(1) [0|7|15]	730
Two People (87) 30(2) [0|5|12]	489
It's Only Love (86) - Bryan Adams/Tina Turner [A]	
15(1) [0|4|11]	474
Show Some Respect (85) 37(1) [0|3|10]	378
Break Every Rule (87) 74(2) [0|0|5]	78.4
TWENNYNINE ► 1260 Pk:83 [0|0|1]	48.7
Peanut Butter (80) -	
Twennynine Featuring Lenny White 83(1) [0|0|4]	48.7
TWILIGHT 22 ► 1146 Pk:79 [0|0|1]	96.5
Electric Kingdom (84) 79(1) [0|0|8]	96.5
Dwight TWILLEY ► 565 Pk:16 [0|1|2]	1079
Girls (84) 16(1) [0|10|16]	1018
Little Bit Of Love (84) 77(1) [0|0|4]	61.0
TWISTED SISTER ► 552 Pk:21 [0|1|3]	1146
We're Not Gonna Take It (84) 21(2) [0|7|15]	745
Leader Of The Pack (86) 53(1) [0|0|10]	266
I Wanna Rock (84) 68(2) [0|0|7]	135
2 LIVE CREW ► 558 Pk:26 [0|1|1]	1116
Me So Horny (89) 26(1) [0|9|30]	1116
Bonnie TYLER ► 120 Pk:1 [1|2|5]	8969
Total Eclipse Of The Heart (83)	
1(4) [11|18|29]	7997
Holding Out For A Hero (84) 34(3) [0|4|13]	512
Take Me Back (84) 46(1) [0|0|9]	324
Here She Comes (84) 76(1) [0|0|5]	69.0
If You Were A Woman (And I Was A Man) (86)	
77(1) [0|0|6]	66.9

U

Act ► Rank Peak [Top10s|Top40s|Total] Title (Pk Yr) Peak(Wk) [Top10Wks|Top40Wks|TotalWks]	Score
UB40 ► 238 Pk:1 [1|3|3]	4646
Red Red Wine (88) 1(1) [6|12|25]	3793
Red Red Wine (84) 34(1) [0|4|15]	603
I Got You Babe (85) -	
UB40 With Chrissie Hynde [A] 28(1) [0|4|14]	250
Tracey ULLMAN ► 446 Pk:8 [1|1|2]	1784
They Don't Know (84) 8(2) [3|11|17]	1719
Break-A-Way (84) 70(1) [0|0|4]	65.0
ULTRAVOX ► 1129 Pk:71 [0|0|1]	104
Reap The Wild Wind (83) 71(1) [0|0|5]	104
UNDERWORLD ► 934 Pk:67 [0|0|2]	233
Stand Up (89) 67(1) [0|0|8]	124
Underneath The Radar (88) 74(1) [0|0|8]	109
UNIPOP ► 1053 Pk:71 [0|0|1]	140
What If (I Said I Love You) (83) 71(3) [0|0|8]	140
UPTOWN ► 1111 Pk:80 [0|0|1]	110
(I Know) I'm Losing You (87) 80(2) [0|0|11]	110
Midge URE ► 1358 Pk:95 [0|0|1]	18.9
Dear God (89) 95(2) [0|0|5]	18.9
URGENT ► 1243 Pk:79 [0|0|1]	55.6
Running Back (85) 79(1) [0|0|5]	55.6
USA For AFRICA ► 197 Pk:1 [1|1|1]	5554
We Are The World (85) 1(4) [8|12|18]	5554
UTFO ► 1244 Pk:77 [0|0|1]	55.5
Roxanne, Roxanne (85) 77(1) [0|0|5]	55.5
UTOPIA ► 655 Pk:27 [0|1|3]	714
Set Me Free (80) 27(1) [0|5|12]	574
Feet Don't Fail Me Now (83) 82(3) [0|0|6]	84.5
The Very Last Time (80) 76(1) [0|0|3]	55.1
U2 ► 80 Pk:1 [3|6|11]	14112
With Or Without You (87) 1(3) [8|13|18]	4480
I Still Haven't Found What I'm Looking For (87)	
1(2) [7|13|17]	4006
Desire (88) 3(1) [5|13|17]	2423
Where The Streets Have No Name (87)	
13(2) [0|9|14]	927
Angel Of Harlem (89) 14(2) [0|8|15]	878
Pride (In The Name Of Love) (84) 33(1) [0|5|15]	559
In God's Country (88) 44(1) [0|0|12]	361
New Year's Day (83) 53(2) [0|0|12]	305
I Will Follow (84) 81(2) [0|0|5]	69.0
When Love Comes To Town (89) -	
U2 With B.B. King [A] 68(1) [0|0|7]	65.2
All I Want Is You (89) 83(1) [0|0|4]	39.2

V

Act ► Rank Peak [Top10s|Top40s|Total] Title (Pk Yr) Peak(Wk) [Top10Wks|Top40Wks|TotalWks]	Score
Dana VALERY ► 1247 Pk:87 [0|0|1]	53.4
I Don't Want To Be Lonely (80) 87(1) [0|0|5]	53.4
Frankie VALLI ► 1312 Pk:90 [0|0|1]	30.4
Where Did We Go Wrong (80) -	
Frankie Valli Introducing Chris Forde	
90(1) [0|0|4]	30.4
VANDENBERG ► 770 Pk:39 [0|1|1]	471
Burning Heart (83) 39(2) [0|2|14]	471
Luther VANDROSS ► 260 Pk:15 [0|5|10]	4300
Stop To Love (87) 15(1) [0|11|19]	1104
Never Too Much (81) 33(2) [0|4|15]	633
'Til My Baby Comes Home (85)	
29(1) [0|6|16]	582
She Won't Talk To Me (89) 30(1) [0|4|12]	442
Bad Boy/Having A Party (82) 55(5) [0|0|12]	357
Any Love (88) 44(1) [0|0|13]	326
There's Nothing Better Than Love (87) -	
Luther Vandross (Duet With Gregory Hines)	
50(1) [0|0|12]	313
Give Me The Reason (86) 57(3) [0|0|11]	265
How Many Times Can We Say Goodbye (83) -	
Dionne Warwick And Luther Vandross [A]	
27(2) [0|5|13]	249
Superstar/Until You Come Back To Me	
(That's What I'm Gonna Do) (84) 87(1) [0|0|4]	30.3
VANGELIS ► 203 Pk:1 [1|1|1]	5496
Chariots Of Fire - Titles (82) 1(1) [9|15|28]	5496

Acts With Singles

Act ▶ Rank Peak [Top10s|Top40s|Total] Title (Pk Yr) Peak(Wk) [Top10Wks|Top40Wks|TotalWks]	Score
VAN HALEN ▶ 50 Pk:1 [3|12|14]	**19350**
Jump (84) 1(5) [10|15|21]	7537
Why Can't This Be Love (86) 3(1) [5|11|16]	2706
When It's Love (88) 5(1) [3|12|19]	1708
Pretty Woman (82) 12(2) [0|9|16]	1163
I'll Wait (84) 13(2) [0|10|14]	1128
Finish What Ya Started (88) 13(1) [0|10|20]	1114
Panama (84) 13(1) [0|10|15]	1047
Love Walks In (86) 22(2) [0|9|15]	753
Dreams (86) 22(1) [0|7|14]	631
Feels So Good (89) 35(1) [0|3|14]	443
Dancing In The Street (82) 38(2) [0|3|11]	425
Black And Blue (88) 34(1) [0|2|10]	326
And The Cradle Will Rock... (80) 55(1) [0|0|7]	200
Hot For Teacher (84) 56(1) [0|0|7]	170
VANITY ▶ 913 Pk:56 [0|0|2]	**255**
Under The Influence (86) 56(1) [0|0|7]	151
Pretty Mess (84) 75(1) [0|0|7]	104
Gino VANNELLI ▶ 272 Pk:6 [1|1|6]	**4065**
Living Inside Myself (81) 6(3) [7|14|20]	2640
Black Cars (85) 42(1) [0|0|16]	467
Wild Horses (87) 55(2) [0|0|15]	357
Nightwalker (81) 41(2) [0|0|10]	324
Hurts To Be In Love (85) 57(1) [0|0|12]	257
The Longer You Wait (82) 89(1) [0|0|3]	20.8
Randy VANWARMER ▶ 905 Pk:55 [0|0|2]	**266**
Suzi (81) 55(1) [0|0|8]	212
Whatever You Decide (80) 77(2) [0|0|3]	54.1
The VAPORS ▶ 701 Pk:36 [0|1|1]	**608**
Turning Japanese (80) 36(1) [0|3|17]	608
Sylvie VARTAN ▶ 1355 Pk:85 [0|0|1]	**19.4**
Love Again (84) - John Denver & Sylvie Vartan [A] 85(1) [0|0|4]	19.4
Suzanne VEGA ▶ 363 Pk:3 [1|1|2]	**2627**
Luka (87) 3(1) [5|12|19]	2614
Solitude Standing (87) 94(1) [0|0|3]	12.9
The VELS ▶ 1154 Pk:72 [0|0|1]	**91.2**
Look My Way (85) 72(1) [0|0|6]	91.2
The VENETIANS ▶ 1314 Pk:88 [0|0|1]	**30.1**
So Much For Love (87) 88(1) [0|0|5]	30.1
Billy VERA & The BEATERS ▶ 257 Pk:1 [1|2|3]	**4362**
At This Moment (Live) (87) 1(2) [6|15|21]	3948
I Can Take Care Of Myself (81) - Billy & The Beaters 39(2) [0|2|11]	370
At This Moment (81) - Billy & The Beaters 79(1) [0|0|3]	44.5
VESTA ▶ 955 Pk:55 [0|0|1]	**214**
Congratulations (89) 55(1) [0|0|8]	214
Maria VIDAL ▶ 853 Pk:48 [0|0|1]	**336**
Body Rock (84) 48(2) [0|0|12]	336
Bobby VINTON ▶ 1222 Pk:78 [0|0|1]	**61.5**
Make Believe It's Your First Time (80) 78(2) [0|0|4]	61.5
VITAMIN Z ▶ 1096 Pk:73 [0|0|1]	**114**
Burning Flame (85) 73(1) [0|0|7]	114
VIXEN ▶ 542 Pk:22 [0|2|2]	**1195**
Edge Of A Broken Heart (88) 26(1) [0|5|21]	651
Cryin' (89) 22(1) [0|6|13]	544
VOICES OF AMERICA ▶ 1059 Pk:65 [0|0|1]	**134**
Hands Across America (86) 65(1) [0|0|8]	134

W

Act ▶ Rank Peak [Top10s|Top40s|Total] Title (Pk Yr) Peak(Wk) [Top10Wks|Top40Wks|TotalWks]	Score
Jack WAGNER ▶ 289 Pk:2 [1|1|4]	**3810**
All I Need (85) 2(2) [6|12|22]	3215
Too Young (85) 52(1) [0|0|14]	385
Weatherman Says (87) 67(1) [0|0|6]	112
Lady Of My Heart (85) 76(1) [0|0|8]	97.8
John WAITE ▶ 154 Pk:1 [1|3|9]	**7676**
Missing You (84) 1(1) [9|16|24]	5752

Act ▶ Rank Peak [Top10s|Top40s|Total] Title (Pk Yr) Peak(Wk) [Top10Wks|Top40Wks|TotalWks]	Score
Every Step Of The Way (85) 25(2) [0|6|12]	559
Tears (84) 37(2) [0|4|13]	462
These Times Are Hard For Lovers (87) 53(1) [0|0|16]	310
Change (85) 54(1) [0|0|10]	242
Restless Heart (85) 59(2) [0|0|8]	178
If Anybody Had A Heart (86) 76(1) [0|0|6]	82.4
Don't Lose Any Sleep (87) 81(1) [0|0|4]	47.8
Welcome To Paradise (85) 85(1) [0|0|4]	42.4
The WAITRESSES ▶ 1069 Pk:62 [0|0|1]	**128**
I Know What Boys Like (82) 62(2) [0|0|6]	128
Narada Michael WALDEN ▶ 1104 Pk:66 [0|0|1]	**111**
I Shoulda Loved Ya (80) 66(1) [0|0|6]	111
WALL OF VOODOO ▶ 954 Pk:58 [0|0|1]	**215**
Mexican Radio (83) 58(1) [0|0|9]	215
Joe WALSH ▶ 467 Pk:19 [0|2|3]	**1653**
All Night Long (80) 19(2) [0|8|16]	959
A Life Of Illusion (81) 34(2) [0|4|12]	460
Space Age Whiz Kids (83) 52(1) [0|0|8]	234
WANG CHUNG ▶ 146 Pk:2 [2|5|8]	**7893**
Everybody Have Fun Tonight (86) 2(2) [7|15|21]	3858
Let's Go (87) 9(1) [2|11|18]	1304
Dance Hall Days (84) 16(1) [0|10|22]	1199
To Live And Die In L.A. (85) 41(2) [0|0|18]	587
Hypnotize Me (87) 36(1) [0|2|12]	399
Don't Let Go (84) 38(2) [0|3|11]	393
Praying To A New God (89) 63(1) [0|0|7]	119
Don't Be My Enemy (84) 86(2) [0|0|3]	34.9
WAR ▶ 1094 Pk:66 [0|0|2]	**115**
You Got The Power (82) 66(2) [0|0|5]	105
Outlaw (82) 94(1) [0|0|3]	10.1
Jennifer WARNES ▶ 211 Pk:1 [2|2|6]	**5307**
Up Where We Belong (82) - Joe Cocker and Jennifer Warnes [A] 1(3) [7|15|23]	2512
(I've Had) The Time Of My Life (87) - Bill Medley And Jennifer Warnes [A] 1(1) [6|15|21]	2057
Could It Be Love (82) 47(2) [0|0|10]	304
When The Feeling Comes Around (80) 45(2) [0|0|8]	273
Don't Make Me Over (80) 67(2) [0|0|7]	139
All The Right Moves (83) - Jennifer Warnes/ Chris Thompson [A] 85(1) [0|0|4]	22.0
WARRANT ▶ 256 Pk:2 [1|2|3]	**4365**
Heaven (89) 2(2) [7|14|19]	3755
Down Boys (89) 27(1) [0|6|16]	591
Big Talk (89) 93(2) [0|0|4]	19.3
Dionne WARWICK ▶ 97 Pk:1 [2|7|13]	**11731**
That's What Friends Are For (86) - Dionne & Friends 1(4) [10|17|23]	5879
Heartbreaker (83) 10(2) [2|13|22]	1820
Deja Vu (80) 15(2) [0|11|19]	1191
No Night So Long (80) 23(2) [0|6|16]	758
Love Power (87) - Dionne Warwick & Jeffrey Osborne [A] 12(2) [0|9|14]	529
Take The Short Way Home (83) 41(2) [0|0|13]	464
How Many Times Can We Say Goodbye (83) - Dionne Warwick And Luther Vandross [A] 27(2) [0|5|13]	249
Easy Love (80) 62(2) [0|0|10]	202
Friends In Love (82) - Dionne Warwick And Johnny Mathis [A] 38(2) [0|3|13]	200
After You (80) 65(2) [0|0|6]	140
Whisper In The Dark (86) 72(1) [0|0|9]	128
Some Changes Are Good For You (81) 65(2) [0|0|6]	112
Reservations For Two - Dionne & Kashif [A] (87) 62(1) [0|0|7]	58.8
Grover WASHINGTON JR. ▶ 205 Pk:2 [1|2|2]	**5439**
Just The Two Of Us (81) 2(3) [11|16|24]	5413
Be Mine (Tonight) (82) 92(2) [0|0|4]	25.9

Act ▶ Rank Peak [Top10s|Top40s|Total] Title (Pk Yr) Peak(Wk) [Top10Wks|Top40Wks|TotalWks]	Score
WAS (NOT WAS) ▶ 387 Pk:7 [1|2|3]	**2301**
Walk The Dinosaur (89) 7(1) [3|9|16]	1333
Spy In The House Of Love (88) 16(1) [0|10|17]	869
Anything Can Happen (89) 75(2) [0|0|6]	99.0
WATERFRONT ▶ 520 Pk:10 [1|1|2]	**1322**
Cry (89) 10(2) [2|10|17]	1249
Nature Of Love (89) 70(1) [0|0|5]	72.6
Jody WATLEY ▶ 88 Pk:2 [5|5|7]	**12595**
Looking For A New Love (87) 2(4) [6|14|19]	4128
Real Love (89) 2(2) [6|12|18]	3382
Don't You Want Me (87) 6(3) [6|14|23]	2173
Friends (89) - Jody Watley with Eric B. & Rakim 9(1) [2|11|18]	1323
Some Kind Of Lover (88) 10(1) [1|10|17]	1205
Most Of All (88) 60(1) [0|0|11]	221
Still A Thrill (87) 56(1) [0|0|12]	162
WA WA NEE ▶ 790 Pk:35 [0|1|2]	**440**
Sugar Free (87) 35(1) [0|3|13]	424
Stimulation (88) 86(1) [0|0|2]	15.9
WAX ▶ 819 Pk:43 [0|0|1]	**384**
Right Between The Eyes (86) 43(2) [0|0|13]	384
WEATHER GIRLS ▶ 841 Pk:46 [0|0|1]	**350**
It's Raining Men (83) 46(3) [0|0|11]	350
WENDY And LISA ▶ 935 Pk:56 [0|0|1]	**232**
Waterfall (87) 56(1) [0|0|10]	232
Max WERNER ▶ 1185 Pk:74 [0|0|1]	**78.0**
Rain In May (81) 74(2) [0|0|4]	78.0
Dottie WEST ▶ 487 Pk:14 [0|1|2]	**1519**
What Are We Doin' In Love (81) 14(1) [0|12|20]	1433
A Lesson In Leavin' (80) 73(1) [0|0|5]	86.1
WEST STREET MOB ▶ 1224 Pk:88 [0|0|2]	**61.3**
Sing A Simple Song (82) 89(1) [0|0|4]	32.7
Let's Dance (Make Your Body Move) (81) 88(1) [0|0|3]	28.6
WET WET WET ▶ 1015 Pk:58 [0|0|1]	**170**
Wishing I Was Lucky (88) 58(2) [0|0|8]	170
WHAM! ▶ 38 Pk:1 [6|6|8]	**21683**
Careless Whisper (85) - Wham! Featuring George Michael 1(3) [9|17|21]	5877
Wake Me Up Before You Go-Go (84) 1(3) [8|14|24]	5473
Everything She Wants (Remix) (85) 1(2) [6|14|20]	4050
I'm Your Man (86) 3(2) [4|12|18]	2606
Freedom (85) 3(1) [4|12|18]	2153
The Edge Of Heaven (86) 10(2) [2|8|13]	1071
Where Did Your Heart Go? (86) 50(2) [0|0|8]	234
Bad Boys (83) - Wham! U.K. 60(2) [0|0|9]	219
WHAT IS THIS ▶ 1037 Pk:62 [0|0|1]	**150**
I'll Be Around (85) 62(1) [0|0|6]	150
WHEN IN ROME ▶ 483 Pk:11 [0|1|2]	**1538**
The Promise (88) 11(2) [0|13|25]	1528
Heaven Knows (89) 95(2) [0|0|2]	9.4
The WHISPERS ▶ 285 Pk:7 [1|4|5]	**3880**
Rock Steady (87) 7(1) [3|12|23]	1781
And The Beat Goes On (80) 19(2) [0|8|15]	917
It's A Love Thing (81) 28(2) [0|5|15]	610
Lady (80) 28(2) [0|4|11]	530
Tonight (83) 84(1) [0|0|4]	42.8
WHISTLE ▶ 892 Pk:60 [0|0|1]	**280**
Right Next To Me (89) 60(1) [0|0|13]	280
Karyn WHITE ▶ 229 Pk:6 [3|3|3]	**4863**
The Way You Love Me (89) 7(1) [4|14|25]	1874
Secret Rendezvous (89) 6(1) [3|12|21]	1805
Superwoman (89) 8(1) [1|10|18]	1183
Maurice WHITE ▶ 843 Pk:50 [0|0|2]	**346**
Stand By Me (85) 50(2) [0|0|13]	342
I Need You (86) 95(1) [0|0|1]	4.7

Acts With Singles

Act ▶ Rank Peak [Top10s|Top40s|Total] Title (Pk Yr) Peak(Wk) [Top10Wks|Top40Wks|TotalWks]	Score
Tony Joe WHITE▶ 1203 Pk:79 [0|0|1]	**67.7**
I Get Off On It (80) 79(1) [0|0|5]	67.7
WHITE LION▶ 244 Pk:3 [2|2|5]	**4523**
When The Children Cry (89)	
3(1) [4|12|23]	2283
Wait (88) 8(1) [2|11|21]	1431
Little Fighter (89) 52(1) [0|0|14]	350
Tell Me (88) 58(2) [0|0|11]	246
Radar Love (89) 59(1) [0|0|13]	213
WHITESNAKE▶ 122 Pk:1 [2|3|6]	**8939**
Here I Go Again (87) 1(1) [7|14|28]	4336
Is This Love (87) 2(1) [7|13|19]	3497
Fool For Your Loving (89) 37(2) [0|3|14]	465
Give Me All Your Love (88)	
48(1) [0|0|11]	316
Fool For Your Loving (80) 53(1) [0|0|8]	234
Still Of The Night (87) 79(1) [0|0|7]	91.9
The WHO▶ 464 Pk:18 [0|2|4]	**1665**
You Better You Bet (81) 18(3) [0|10|15]	891
Athena (82) 28(3) [0|6|14]	611
Eminence Front (83) 68(2) [0|0|6]	119
Don't Let Go The Coat (81) 84(1) [0|0|3]	44.8
WHODINI▶ 1344 Pk:87 [0|0|1]	**22.6**
Friends//Five Minutes Of Funk (85) 87(1) [0|0|3]	22.6
Jane WIEDLIN▶ 452 Pk:9 [1|1|4]	**1708**
Rush Hour (88) 9(1) [2|10|19]	1286
Cool Places (83) -	
Sparks And Jane Wiedlin [A] 49(2) [0|0|12]	195
Inside A Dream (88) 57(1) [0|0|7]	146
Blue Kiss (85) 77(1) [0|0|9]	81.6
WILD BLUE▶ 1141 Pk:71 [0|0|1]	**98.8**
Fire With Fire (86) 71(1) [0|0|6]	98.8
Eugene WILDE▶ 978 Pk:76 [0|0|2]	**194**
Don't Say No Tonight (86) 76(1) [0|0|10]	121
Gotta Get You Home Tonight (85) 83(1) [0|0|8]	73.3
Kim WILDE▶ 216 Pk:1 [1|2|5]	**5180**
You Keep Me Hangin' On (87) 1(1) [6|13|21]	3690
Kids In America (82) 25(4) [0|8|18]	858
You Came (88) 41(1) [0|0|10]	295
Say You Really Want Me (87) 44(1) [0|0|8]	201
Go For It (85) 65(1) [0|0|7]	136
Matthew WILDER▶ 274 Pk:5 [1|2|3]	**4019**
Break My Stride (84) 5(3) [7|14|29]	3257
The Kid's American (84) 33(2) [0|4|13]	516
Bouncin' Off The Walls (84) 52(1) [0|0|9]	246
Christopher WILLIAMS▶ 781 Pk:49 [0|0|1]	**455**
Talk To Myself (89) 49(2) [0|0|18]	455
Deniece WILLIAMS▶ 156 Pk:1 [2|2|5]	**7609**
Let's Hear It For The Boy (84) 1(2) [9|14|19]	5783
It's Gonna Take A Miracle (82) 10(2) [2|9|17]	1384
Silly (81) 53(2) [0|0|10]	259
I Can't Wait (88) 66(1) [0|0|8]	133
Next Love (84) 81(1) [0|0|4]	49.9
Don WILLIAMS▶ 599 Pk:24 [0|1|1]	**901**
I Believe In You (80) 24(3) [0|9|20]	901
Vanessa WILLIAMS▶ 449 Pk:8 [1|1|3]	**1756**
Dreamin' (89) 8(1) [2|11|20]	1466
The Right Stuff (88) 44(1) [0|0|10]	271
Darlin' I (89) 88(1) [0|0|4]	18.9
Bruce WILLIS▶ 425 Pk:5 [1|1|3]	**1926**
Respect Yourself (87) 5(1) [4|10|14]	1683
Under The Boardwalk (87) 59(1) [0|0|7]	143
Young Blood (87) 68(1) [0|0|5]	99.5
WILLIS "The GUARD" and VIGORISH▶ 1263	
Pk:82 [0|0|1]	**47.4**
Merry Christmas In The NFL (80) 82(3) [0|0|3]	47.4
WILL TO POWER▶ 235 Pk:1 [1|1|4]	**4761**
Baby, I Love Your Way/Freebird Medley (88)	
1(1) [6|15|24]	3786

Act ▶ Rank Peak [Top10s|Top40s|Total] Title (Pk Yr) Peak(Wk) [Top10Wks|Top40Wks|TotalWks]	Score
Dreamin' (87) 50(1) [0|0|16]	429
Say It's Gonna Rain (88) 49(1) [0|0|14]	371
Fading Away (89) 65(1) [0|0|10]	175
Ann WILSON▶ 415 Pk:6 [2|2|3]	**2040**
Almost Paradise (84) -	
Mike Reno And Ann Wilson [A] 7(2) [4|13|20]	1048
Surrender To Me (89) -	
Ann Wilson and Robin Zander [A] 6(1) [2|10|19]	726
The Best Man In The World (87) 61(2) [0|0|12]	267
Carl WILSON▶ 1144 Pk:72 [0|0|1]	**97.8**
What You Do To Me (83) 72(1) [0|0|6]	97.8
Shanice WILSON▶ 847 Pk:50 [0|0|1]	**339**
(Baby Tell Me) Can You Dance (87) 50(1) [0|0|13]	339
Jesse WINCHESTER▶ 793 Pk:32 [0|1|1]	**435**
Say What (81) 32(2) [0|5|12]	435
WINGER▶ 478 Pk:19 [0|2|3]	**1557**
Headed For A Heartbreak (89) 19(1) [0|9|18]	872
Seventeen (89) 26(1) [0|6|16]	640
Hungry (89) 85(1) [0|0|6]	45.6
--WINGS see: McCARTNEY, Paul & WINGS	
Steve WINWOOD▶ 52 Pk:1 [6|9|14]	**19105**
Roll With It (88) 1(4) [7|14|18]	4536
Higher Love (86) 1(1) [6|14|22]	3755
While You See A Chance (81) 7(2) [5|12|18]	2011
The Finer Things (87) 8(3) [4|12|23]	1776
Don't You Know What The Night Can Do? (88)	
6(1) [3|11|17]	1648
Valerie (87) 9(1) [1|12|20]	1485
Back In The High Life Again (87)	
13(1) [0|10|21]	1098
Holding On (89) 11(1) [0|11|17]	1045
Freedom Overspill (86) 20(2) [0|7|15]	666
Still In The Game (82) 47(2) [0|0|10]	294
Talking Back To The Night (88)	
57(1) [0|0|10]	247
Arc Of A Diver (81) 48(2) [0|0|9]	240
Hearts On Fire (89) 53(1) [0|0|9]	222
Valerie (82) 70(2) [0|0|4]	82.8
WOLF▶ 952 Pk:55 [0|0|1]	**217**
Papa Was A Rollin' Stone (83) 55(2) [0|0|9]	217
Peter WOLF▶ 354 Pk:12 [0|3|5]	**2744**
Lights Out (84) 12(2) [0|10|14]	1249
Come As You Are (87) 15(2) [0|9|15]	881
I Need You Tonight (84) 36(1) [0|2|13]	414
Oo-Ee-Diddley-Bop! (85) 61(1) [0|0|6]	124
Can't Get Started (87) 75(2) [0|0|5]	76.1
Bobby WOMACK▶ 1349 Pk:88 [0|0|1]	**21.6**
Love Has Finally Come At Last (84) - Bobby Womack	
and Patti LaBelle [A] 88(2) [0|0|3]	21.6
Stevie WONDER▶ 28 Pk:1 [5|10|18]	**24620**
I Just Called To Say I Love You (84)	
1(3) [10|15|26]	6339
Part-Time Lover (85) 1(1) [8|14|21]	4348
Master Blaster (Jammin') (80) 5(3) [8|16|23]	3415
That Girl (82) 4(3) [9|13|18]	3383
I Ain't Gonna Stand For It (81) 11(3) [0|11|19]	1526
Go Home (86) 10(1) [1|11|17]	1315
Do I Do (82) 13(3) [0|9|14]	1185
Love Light In Flight (85) 17(1) [0|10|16]	903
Skeletons (87) 19(1) [0|7|16]	748
Overjoyed (86) 24(2) [0|6|13]	556
Outside My Window (80) 52(2) [0|0|8]	228
Ribbon In The Sky (82) 54(2) [0|0|7]	192
Used To Be (82) - Charlene & Stevie Wonder [A]	
46(3) [0|0|11]	190
Lately (81) 64(2) [0|0|7]	154
You Will Know (88) 77(1) [0|0|6]	59.0
Get It (88) - Stevie Wonder & Michael Jackson [A]	
80(1) [0|0|6]	28.5
Land Of La La (86) 86(1) [0|0|3]	25.8

Act ▶ Rank Peak [Top10s|Top40s|Total] Title (Pk Yr) Peak(Wk) [Top10Wks|Top40Wks|TotalWks]	Score
My Love (88) - Julio Iglesias Featuring Stevie Wonder	
[A] 80(1) [0|0|5]	24.3
Stevie WOODS▶ 502 Pk:25 [0|2|3]	**1441**
Steal The Night (81) 25(3) [0|10|21]	994
Just Can't Win 'Em All (82) 38(1) [0|3|12]	422
Fly Away (82) 84(1) [0|0|5]	25.6
WORLD CLASS WRECKIN' CRU▶ 1167	
Pk:84 [0|0|1]	**86.5**
Turn Off The Lights (88) 84(1) [0|0|11]	86.5
WORLD PARTY▶ 744 Pk:27 [0|1|1]	**519**
Ship Of Fools (Save Me From Tomorrow) (87)	
27(1) [0|5|15]	519
Gary WRIGHT▶ 587 Pk:16 [0|1|1]	**964**
Really Wanna Know You (81) 16(2) [0|10|17]	964

X

XTC▶ 1155 Pk:72 [0|0|1]	**90.9**
The Mayor Of Simpleton (89) 72(1) [0|0|6]	90.9

Y

Y&T▶ 939 Pk:55 [0|0|1]	**229**
Summertime Girls (85) 55(1) [0|0|10]	229
"Weird Al" YANKOVIC▶ 484 Pk:12 [0|1|6]	**1529**
Eat It (84) 12(1) [0|7|12]	963
Like A Surgeon (85) 47(1) [0|0|8]	225
Ricky (83) 63(2) [0|0|8]	172
King Of Suede (84) 62(2) [0|0|6]	131
I Lost On Jeopardy (84) 81(1) [0|0|3]	35.1
Fat (88) 99(1) [0|0|2]	2.3
YARBROUGH & PEOPLES▶ 546 Pk:19 [0|1|3]	**1175**
Don't Stop The Music (81) 19(2) [0|7|16]	806
Don't Waste Your Time (84) 48(1) [0|0|12]	353
I Wouldn't Lie (86) 93(1) [0|0|4]	16.8
YAZ▶ 885 Pk:67 [0|0|2]	**289**
Only You (83) 67(3) [0|0|8]	155
Situation (82) - Yazoo 73(2) [0|0|8]	134
YAZZ And The PLASTIC POPULATION▶ 1389	
Pk:96 [0|0|1]	**6.8**
The Only Way Is Up (88) 96(1) [0|0|4]	6.8
YELLO▶ 908 Pk:51 [0|0|1]	**264**
Oh Yeah (87) 51(1) [0|0|11]	264
YELLOW MAGIC ORCHESTRA▶ 920	
Pk:60 [0|0|1]	**249**
Computer Game "Theme From The Circus" (80)	
60(2) [0|0|9]	249
YES▶ 132 Pk:1 [1|4|5]	**8482**
Owner Of A Lonely Heart (84) 1(2) [10|17|23]	6601
Leave It (84) 24(2) [0|7|15]	729
Love Will Find A Way (87) 30(1) [0|6|19]	605
Rhythm Of Love (88) 40(1) [0|1|12]	364
It Can Happen (84) 51(1) [0|0|7]	184
YIPES!!▶ 1160 Pk:68 [0|0|1]	**89.0**
Darlin' (80) 68(2) [0|0|5]	89.0
Neil YOUNG▶ 1125 Pk:71 [0|0|1]	**105**
Little Thing Called Love (83) 71(2) [0|0|6]	105
Neil YOUNG & CRAZY HORSE▶ 1164	
Pk:70 [0|0|1]	**88.0**
Southern Pacific (82) 70(1) [0|0|5]	88.0
Paul YOUNG▶ 167 Pk:1 [1|3|7]	**6998**
Everytime You Go Away (85) 1(1) [8|15|23]	4283
I'm Gonna Tear Your Playhouse Down (85)	
13(1) [0|10|17]	999
Come Back And Stay (84) 22(1) [0|8|15]	806
Love Of The Common People (84) 45(1) [0|0|11]	326
Everything Must Change (86) 56(2) [0|0|11]	285
Some People (86) 65(2) [0|0|10]	181
Wherever I Lay My Hat (That's My Home) (83)	
70(1) [0|0|7]	119

Acts With Singles

Act ▶ Rank Peak [Top10s|Top40s|Total] Title (Pk Yr) Peak(Wk) [Top10Wks|Top40Wks|TotalWks]	Score
YOUNG M.C. ▶ 355 Pk:7 [1|1|1]	2743
Bust A Move (89) 7(1) [4|20|39]	2743
YUTAKA ▶ 1301 Pk:81 [0|0|1]	33.1
Love Light (81) 81(1) [0|0|3]	33.1

Z

Act ▶ Rank Peak [Top10s|Top40s|Total] Title (Pk Yr) Peak(Wk) [Top10Wks|Top40Wks|TotalWks]	Score
Pia ZADORA ▶ 603 Pk:36 [0|1|3]	876
The Clapping Song (83) 36(2) [0|3|15]	491
I'm In Love Again (82) 45(1) [0|0|9]	271

When The Rain Begins To Fall (85) -
  Jermaine Jackson And Pia Zadora [A]
  54(2) [0|0|11]
	114

Robin ZANDER ▶ 646 Pk:6 [1|1|1]	726
Surrender To Me (89) - Ann Wilson and Robin Zander [A] 6(1) [2|10|19]	726
ZAPP ▶ 1270 Pk:86 [0|0|1]	44.7
More Bounce To The Ounce Part 1 (80) 86(1) [0|0|7]	44.7
Frank ZAPPA ▶ 757 Pk:32 [0|1|1]	496
Valley Girl (82) 32(2) [0|3|12]	496
ZEBRA ▶ 990 Pk:61 [0|0|1]	188
Who's Behind The Door? (83) 61(1) [0|0|8]	188
Warren ZEVON ▶ 997 Pk:57 [0|0|1]	182
A Certain Girl (80) 57(1) [0|0|7]	182
ZZ TOP ▶ 174 Pk:8 [2|7|10]	6533
Legs (84) 8(2) [4|12|19]	1913
Sleeping Bag (85) 8(1) [4|13|17]	1676
Stages (86) 21(1) [0|7|12]	634
Rough Boy (86) 22(1) [0|7|13]	622
I Thank You (80) 34(1) [0|3|11]	456
Gimme All Your Lovin (83) 37(1) [0|3|12]	449
Velcro Fly (86) 35(2) [0|3|12]	449
Sharp Dressed Man (83) 56(1) [0|0|9]	254
Leila (81) 77(2) [0|0|4]	58.7
Cheap Sunglasses (80) 89(2) [0|0|2]	21.4

Ranking The Singles Acts

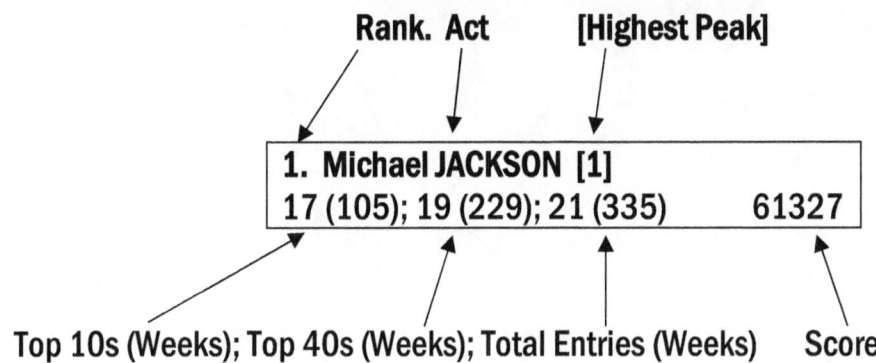

Singles Acts Ranked

Singles Acts Ranked

Rank. Act [Hi Peak] T10s (Wk); T40s (Wk); Ents (Wk)	Score
1. Michael JACKSON [1] 17 (105); 19 (229); 21 (335)	61327
2. MADONNA [1] 17 (95); 18 (222); 18 (325)	59786
3. Daryl HALL & John OATES [1] 13 (93); 22 (246); 22 (373)	55777
4. Lionel RICHIE [1] 13 (94); 14 (192); 15 (277)	52069
5. Phil COLLINS [1] 10 (66); 13 (158); 14 (244)	39768
6. Billy JOEL [1] 9 (45); 20 (210); 22 (344)	38031
7. DURAN DURAN [1] 9 (54); 13 (142); 15 (221)	35594
8. Whitney HOUSTON [1] 10 (61); 10 (134); 11 (197)	35563
9. Kenny ROGERS [1] 6 (53); 13 (152); 20 (273)	35216
10. John MELLENCAMP [1] 9 (52); 17 (185); 20 (316)	35178
11. AIR SUPPLY [1] 8 (62); 11 (142); 13 (226)	34148
12. Huey LEWIS and the NEWS [1] 12 (57); 15 (171); 17 (262)	33246
13. KOOL & The GANG [1] 10 (37); 16 (180); 19 (328)	30914
14. Olivia NEWTON-JOHN [1] 6 (46); 11 (118); 14 (209)	30618
15. CHICAGO [1] 7 (42); 11 (135); 16 (260)	29063
16. Diana ROSS [1] 8 (53); 12 (129); 18 (264)	28634
17. PRINCE [1] 8 (41); 12 (130); 16 (251)	28172
18. Elton JOHN [2] 6 (35); 19 (177); 20 (309)	27052
19. Cyndi LAUPER [1] 8 (44); 10 (116); 13 (193)	26545
20. Billy OCEAN [1] 7 (37); 11 (125); 11 (205)	26542
21. Rick SPRINGFIELD [1] 5 (35); 16 (157); 17 (270)	26085
22. POINTER SISTERS [2] 6 (38); 11 (128); 19 (287)	25341
23. George MICHAEL [1] 8 (49); 8 (100); 8 (139)	25301
24. CULTURE CLUB [1] 6 (40); 10 (113); 10 (167)	25285
25. Bruce SPRINGSTEEN [2] 11 (44); 13 (141); 14 (219)	24999
26. FOREIGNER [1] 5 (38); 8 (97); 13 (185)	24892
27. BON JOVI [1] 8 (42); 9 (101); 12 (197)	24759
28. Stevie WONDER [1] 5 (36); 10 (112); 18 (235)	24620
29. JOURNEY [2] 6 (31); 16 (151); 19 (273)	23887
30. Paul McCARTNEY [1] 6 (48); 9 (102); 12 (162)	23877
31. The POLICE [1] 6 (29); 8 (103); 10 (156)	23786
32. SURVIVOR [1] 5 (31); 8 (96); 18 (234)	23583
33. Sheena EASTON [1] 7 (32); 12 (133); 18 (279)	22745
34. Christopher CROSS [1] 4 (30); 8 (95); 10 (152)	22700
35. QUEEN [1] 2 (27); 5 (62); 14 (174)	22382
36. Kenny LOGGINS [1] 4 (25); 13 (121); 17 (261)	22325
37. Tina TURNER [1] 5 (30); 12 (116); 13 (203)	22054
38. WHAM! [1] 6 (33); 6 (77); 8 (131)	21683
39. REO SPEEDWAGON [1] 4 (29); 13 (122); 15 (241)	21530
40. PRINCE And The REVOLUTION [1] 6 (35); 8 (84); 10 (139)	21448
41. Janet JACKSON [1] 6 (37); 7 (84); 9 (149)	21166
42. Richard MARX [1] 7 (37); 7 (91); 7 (137)	21085
43. JEFFERSON STARSHIP/ STARSHIP [1] 4 (23); 11 (100); 18 (238)	20993
44. Gloria ESTEFAN/ MIAMI SOUND MACHINE [1] 8 (36); 11 (122); 11 (214)	20947
45. BLONDIE [1] 3 (30); 5 (56); 6 (93)	20847
46. HEART [1] 7 (28); 10 (105); 15 (220)	20696
47. Rod STEWART [4] 6 (26); 12 (129); 20 (276)	20483
48. GENESIS [1] 6 (27); 11 (105); 15 (220)	19591
49. MEN AT WORK [1] 4 (31); 5 (66); 6 (103)	19415
50. VAN HALEN [1] 3 (18); 12 (101); 14 (199)	19350
51. BANGLES [1] 5 (28); 8 (88); 8 (145)	19219
52. Steve WINWOOD [1] 6 (26); 9 (103); 14 (213)	19105
53. ROLLING STONES [2] 5 (30); 11 (102); 12 (166)	18848
54. Debbie GIBSON [1] 5 (30); 8 (91); 9 (151)	18379
55. Irene CARA [1] 3 (23); 6 (65); 8 (145)	18325
56. Pat BENATAR [5] 4 (19); 15 (129); 17 (257)	18270
57. Stevie NICKS [3] 4 (29); 11 (101); 12 (170)	17947
58. Donna SUMMER [3] 5 (24); 10 (85); 16 (198)	17446
59. John LENNON [1] 4 (32); 4 (57); 5 (79)	17434
60. Bryan ADAMS [1] 5 (20); 12 (112); 13 (189)	16851
61. TOTO [1] 3 (18); 9 (87); 11 (170)	16748
62. Neil DIAMOND [2] 4 (25); 9 (88); 13 (168)	16701
63. Bobby BROWN [1] 6 (28); 6 (77); 7 (138)	16589
64. EURYTHMICS [1] 3 (20); 10 (87); 15 (190)	16559
65. Paula ABDUL [1] 4 (27); 4 (59); 6 (106)	16088
66. Kim CARNES [1] 2 (17); 7 (62); 14 (153)	16050
67. DEF LEPPARD [1] 4 (19); 9 (91); 11 (172)	15814
68. TEARS FOR FEARS [1] 4 (26); 5 (57); 6 (96)	15779
69. MILLI VANILLI [1] 4 (25); 4 (57); 4 (92)	15769
70. Juice NEWTON [2] 4 (27); 7 (91); 10 (153)	15737
71. Laura BRANIGAN [2] 3 (20); 7 (81); 12 (181)	15659
72. The CARS [3] 4 (22); 10 (96); 13 (178)	15640
73. FLEETWOOD MAC [4] 4 (25); 9 (91); 13 (167)	15297
74. Eddie RABBITT [1] 4 (29); 6 (83); 9 (149)	15013
75. Smokey ROBINSON [2] 4 (22); 6 (65); 11 (165)	14881
76. INXS [1] 5 (26); 6 (71); 12 (162)	14750
77. NEW KIDS ON THE BLOCK [1] 6 (24); 6 (71); 6 (129)	14595
78. Joan JETT & The BLACKHEARTS [1] 3 (18); 8 (66); 9 (131)	14519
79. Dan FOGELBERG [2] 4 (17); 9 (91); 12 (171)	14142
80. U2 [1] 3 (20); 6 (61); 11 (136)	14112
81. J. GEILS Band [1] 2 (20); 6 (49); 10 (110)	13957
82. HUMAN LEAGUE [1] 3 (22); 4 (54); 6 (96)	13925
83. David BOWIE [1] 4 (16); 9 (77); 14 (174)	13592
84. Bob SEGER [1] 4 (22); 5 (58); 6 (98)	13574
85. Ray PARKER Jr. [1] 2 (17); 7 (61); 9 (124)	13555
86. STYX [3] 4 (26); 6 (64); 9 (115)	13467
87. Billy IDOL [1] 3 (15); 8 (66); 11 (162)	12767
88. Jody WATLEY [2] 5 (21); 5 (61); 7 (113)	12595
89. Glenn FREY [2] 2 (13); 7 (71); 10 (146)	12594
90. Don HENLEY [3] 4 (18); 7 (72); 9 (149)	12425
91. Robert PALMER [1] 3 (18); 5 (55); 10 (121)	12286
92. Barbra STREISAND [1] 1 (11); 5 (40); 11 (132)	12222
93. Belinda CARLISLE [1] 4 (19); 5 (62); 7 (102)	12201
94. The JETS [3] 5 (21); 6 (68); 10 (147)	12108
95. Taylor DAYNE [2] 5 (26); 5 (68); 5 (111)	11970
96. EXPOSÉ [1] 6 (22); 6 (75); 6 (113)	11741
97. Dionne WARWICK [1] 2 (12); 7 (64); 13 (171)	11731
98. TIFFANY [1] 4 (18); 5 (51); 6 (97)	11660
99. Linda RONSTADT [2] 4 (19); 7 (63); 9 (138)	11425
100. THOMPSON TWINS [3] 3 (14); 7 (63); 11 (146)	11301
101. MR. MISTER [1] 3 (17); 4 (44); 5 (78)	11293
102. Bob SEGER & The SILVER BULLET BAND [2] 1 (8); 6 (60); 10 (122)	11094
103. The COMMODORES [3] 3 (16); 5 (60); 12 (153)	11046
104. PET SHOP BOYS [1] 5 (19); 6 (62); 9 (123)	11027
105. POISON [1] 4 (14); 6 (62); 7 (117)	10815
106. Rick ASTLEY [1] 4 (17); 5 (47); 6 (89)	10618
107. LITTLE RIVER BAND [6] 3 (9); 7 (72); 9 (129)	10540
108. Peter CETERA [1] 4 (19); 4 (53); 7 (97)	10421
109. The GO-GO'S [2] 2 (13); 5 (52); 7 (105)	10243
110. Aretha FRANKLIN [1] 3 (16); 9 (76); 15 (172)	10066
111. GUNS N' ROSES [1] 4 (19); 4 (47); 5 (81)	9951
112. Steve MILLER Band [1] 1 (14); 2 (28); 8 (75)	9948
113. Bette MIDLER [1] 2 (15); 4 (36); 7 (91)	9597
114. Howard JONES [4] 2 (11); 8 (71); 10 (152)	9582
115. SIMPLE MINDS [1] 2 (14); 4 (45); 4 (69)	9544
116. Bruce HORNSBY And The RANGE [1] 3 (18); 5 (49); 6 (92)	9484
117. Alan PARSONS Project [3] 1 (8); 5 (50); 11 (141)	9317
118. 38 SPECIAL [6] 2 (7); 8 (65); 14 (180)	9241
119. CAPTAIN & TENNILLE [1] 1 (14); 1 (22); 3 (40)	9027
120. Bonnie TYLER [1] 1 (11); 2 (22); 5 (62)	8969
121. Corey HART [3] 2 (11); 8 (65); 9 (129)	8952
122. WHITESNAKE [1] 2 (14); 3 (30); 6 (87)	8939
123. STING [3] 3 (13); 6 (62); 7 (100)	8848
124. NEW EDITION [4] 2 (10); 6 (47); 12 (163)	8713
125. Eddie MONEY [4] 2 (8); 6 (61); 13 (166)	8688
126. LOVERBOY [9] 2 (2); 9 (69); 12 (173)	8686
127. SPINNERS [2] 2 (15); 2 (30); 5 (62)	8642
128. ATLANTIC STARR [1] 2 (13); 3 (31); 7 (94)	8547
129. BANANARAMA [1] 3 (15); 3 (35); 11 (108)	8533
130. CHEAP TRICK [1] 2 (11); 4 (34); 10 (132)	8527
131. ROXETTE [1] 2 (13); 3 (36); 3 (59)	8484
132. YES [1] 1 (10); 4 (31); 5 (76)	8482
133. DeBARGE [3] 2 (10); 5 (53); 6 (109)	8457
134. PINK FLOYD [1] 1 (12); 1 (19); 3 (39)	8451
135. Michael McDONALD [1] 3 (17); 4 (45); 5 (85)	8395
136. STRAY CATS [3] 3 (14); 4 (42); 5 (73)	8306
137. SIMPLY RED [1] 2 (12); 4 (41); 5 (84)	8277
138. George HARRISON [1] 2 (14); 3 (32); 4 (54)	8263

Singles Acts Ranked

Rank. Act [Hi Peak] T10s (Wk); T40s (Wk); Ents (Wk)	Score
139. MIKE + THE MECHANICS [1] 3 (14); 4 (42); 6 (95)	8260
140. Paul McCARTNEY & WINGS [1] 1 (11); 1 (16); 1 (21)	8092
141. FINE YOUNG CANNIBALS [1] 2 (13); 3 (35); 5 (63)	8080
142. LISA LISA And CULT JAM [1] 2 (13); 3 (31); 3 (51)	8045
143. George BENSON [4] 2 (13); 3 (36); 8 (106)	7923
144. NIGHT RANGER [5] 2 (6); 6 (53); 11 (137)	7908
145. Peter GABRIEL [1] 2 (11); 4 (42); 8 (110)	7899
146. WANG CHUNG [2] 2 (9); 5 (41); 8 (112)	7893
147. CHER [3] 4 (15); 5 (57); 6 (106)	7849
148. DIRE STRAITS [1] 2 (12); 3 (35); 5 (71)	7818
149. Barry MANILOW [10] 1 (1); 10 (63); 13 (156)	7794
150. Cliff RICHARD [7] 2 (7); 6 (60); 10 (136)	7757
151. The PRETENDERS [5] 2 (7); 5 (53); 8 (108)	7742
152. EARTH, WIND & FIRE [3] 1 (9); 2 (26); 11 (115)	7717
153. Tom PETTY And The HEARTBREAKERS [10] 1 (2); 8 (65); 12 (135)	7678
154. John WAITE [1] 1 (9); 3 (26); 9 (97)	7676
155. Dolly PARTON [1] 1 (9); 2 (21); 9 (87)	7651
156. Deniece WILLIAMS [1] 2 (11); 2 (23); 5 (58)	7609
157. QUARTERFLASH [3] 1 (12); 3 (37); 7 (81)	7578
158. BREATHE [2] 3 (12); 3 (42); 3 (67)	7556
159. BEACH BOYS [1] 1 (5); 5 (52); 10 (119)	7489
160. KC & The SUNSHINE BAND [1] 1 (11); 2 (28); 2 (47)	7468
161. Jermaine JACKSON [9] 1 (3); 6 (56); 11 (143)	7427
162. LIPPS, INC. [1] 1 (9); 1 (15); 2 (30)	7424
163. Ronnie MILSAP [5] 1 (8); 5 (52); 8 (108)	7422
164. Samantha FOX [3] 3 (11); 4 (43); 6 (96)	7224
165. Natalie COLE [5] 2 (8); 5 (55); 8 (114)	7168
166. Anita BAKER [3] 2 (9); 4 (39); 6 (109)	7114
167. Paul YOUNG [1] 1 (8); 3 (32); 7 (91)	6998
168. ELECTRIC LIGHT ORCHESTRA [8] 2 (5); 8 (60); 9 (119)	6914
169. Eddy GRANT [2] 1 (8); 2 (21); 3 (46)	6805
170. Jackson BROWNE [7] 1 (6); 6 (51); 8 (108)	6785
171. AEROSMITH [3] 2 (8); 4 (44); 5 (84)	6717
172. ABC [5] 2 (7); 5 (51); 7 (102)	6629
173. Terence Trent D'ARBY [1] 2 (11); 3 (33); 4 (65)	6553
174. ZZ TOP [8] 2 (8); 7 (48); 10 (111)	6533
175. The FIXX [4] 1 (7); 5 (42); 8 (100)	6528
176. Robbie NEVIL [2] 2 (9); 4 (37); 5 (80)	6523
177. Michael SEMBELLO [1] 1 (9); 2 (18); 2 (32)	6484
178. ASIA [4] 2 (8); 4 (36); 5 (69)	6447
179. DR. HOOK [5] 1 (6); 4 (41); 7 (89)	6424
180. The JACKSONS [3] 1 (6); 4 (36); 8 (84)	6347
181. Jeffrey OSBORNE [12] 0 (0); 8 (47); 12 (167)	6316
182. Patti AUSTIN [1] 1 (9); 1 (18); 6 (77)	6253
183. David Lee ROTH [3] 2 (8); 4 (40); 7 (83)	6148
184. Willie NELSON [5] 2 (12); 4 (40); 6 (95)	6107
185. SADE [5] 2 (9); 4 (41); 5 (80)	6062
186. Toni BASIL [1] 1 (10); 1 (18); 3 (37)	5985
187. Greg KIHN Band [2] 1 (7); 2 (27); 5 (60)	5982
188. Eric CARMEN [3] 2 (11); 3 (33); 6 (64)	5934
189. Chaka KHAN [3] 1 (9); 1 (17); 7 (91)	5874
190. CROWDED HOUSE [2] 2 (10); 2 (26); 4 (64)	5839
191. Andy GIBB [4] 1 (8); 4 (32); 5 (61)	5742
192. The MOTELS [9] 2 (6); 4 (38); 7 (88)	5733
193. Leo SAYER [2] 1 (9); 2 (21); 2 (35)	5683
194. ROCKWELL [2] 1 (8); 2 (16); 2 (33)	5674
195. TONE LŌC [2] 2 (11); 2 (25); 2 (43)	5651
196. GLASS TIGER [2] 2 (8); 4 (37); 4 (67)	5611
197. USA For AFRICA [1] 1 (8); 1 (12); 1 (18)	5554
198. STARS ON [1] 1 (8); 2 (19); 4 (44)	5553
199. BOSTON [1] 2 (8); 3 (27); 3 (46)	5546
200. Steve PERRY [3] 1 (7); 4 (33); 4 (65)	5535
201. Joey SCARBURY [2] 1 (10); 1 (18); 2 (35)	5521
202. CUTTING CREW [1] 2 (8); 3 (26); 4 (56)	5514
203. VANGELIS [1] 1 (9); 1 (15); 1 (28)	5496
204. KLYMAXX [5] 1 (7); 3 (34); 5 (83)	5495
205. Grover WASHINGTON JR. [2] 1 (11); 1 (16); 2 (28)	5439
206. READY FOR THE WORLD [1] 2 (7); 3 (31); 3 (58)	5433
207. LOS LOBOS [1] 1 (7); 2 (21); 3 (40)	5418
208. John PARR [1] 1 (7); 2 (22); 5 (55)	5403
209. AMBROSIA [3] 1 (8); 2 (24); 3 (41)	5356
210. The ROMANTICS [3] 1 (7); 2 (18); 4 (52)	5318
211. Jennifer WARNES [1] 2 (13); 2 (30); 6 (73)	5307
212. Joe JACKSON [6] 1 (8); 3 (34); 5 (71)	5290
213. MEN WITHOUT HATS [3] 1 (7); 2 (26); 3 (48)	5256
214. SHALAMAR [8] 1 (4); 3 (33); 7 (100)	5226
215. Paul DAVIS [6] 1 (7); 4 (34); 5 (67)	5202
216. Kim WILDE [1] 1 (6); 2 (21); 5 (64)	5180
217. Sergio MENDES [4] 1 (8); 2 (23); 4 (57)	5147
218. Eddie MURPHY [2] 1 (9); 2 (18); 2 (35)	5095
219. MOODY BLUES [9] 1 (2); 5 (42); 8 (101)	5088
220. Boz SCAGGS [14] 0 (0); 5 (41); 5 (79)	5054
221. FALCO [1] 1 (7); 2 (21); 2 (31)	5027
222. Julian LENNON [5] 2 (9); 4 (36); 6 (71)	5009
223. A-HA [1] 1 (7); 2 (23); 3 (54)	5002
224. Patti LaBELLE [1] 1 (7); 3 (27); 7 (93)	4962
225. EUROPE [3] 2 (7); 4 (29); 5 (73)	4908
226. Barbra STREISAND & Barry GIBB [3] 2 (10); 2 (25); 2 (38)	4905
227. BERLIN [1] 1 (7); 2 (21); 7 (67)	4898
228. ORCHESTRAL MANOEUVERS IN THE DARK [4] 1 (5); 4 (36); 5 (84)	4868
229. Karyn WHITE [6] 3 (8); 3 (36); 3 (64)	4863
230. DEXYS MIDNIGHT RUNNERS [1] 1 (6); 1 (14); 2 (27)	4842
231. MARTIKA [1] 1 (6); 3 (26); 3 (51)	4821
232. Philip BAILEY [2] 1 (7); 1 (16); 2 (35)	4813
233. EAGLES [8] 2 (7); 3 (31); 3 (45)	4797
234. ESCAPE CLUB [1] 1 (7); 2 (21); 3 (47)	4794
235. WILL TO POWER [1] 1 (6); 1 (15); 4 (64)	4761
236. S.O.S. BAND [3] 1 (7); 1 (14); 5 (69)	4674
237. PEBBLES [2] 2 (10); 2 (23); 2 (38)	4654
238. UB40 [1] 1 (6); 3 (20); 3 (54)	4646
239. SPANDAU BALLET [4] 1 (6); 3 (23); 4 (49)	4610
240. Marvin GAYE [3] 1 (10); 1 (15); 1 (21)	4570
241. Gregory ABBOTT [1] 1 (8); 1 (16); 2 (33)	4569
242. Robbie DUPREE [6] 1 (7); 2 (27); 3 (48)	4569
243. Ray PARKER Jr. & RAYDIO [4] 1 (6); 3 (26); 3 (56)	4524
244. WHITE LION [3] 2 (6); 2 (23); 5 (82)	4523
245. DOOBIE BROTHERS [5] 2 (6); 3 (27); 8 (71)	4515
246. KENNY G [4] 1 (4); 3 (31); 4 (67)	4512
247. CLUB NOUVEAU [1] 1 (6); 2 (13); 2 (30)	4511
248. John CAFFERTY And The BEAVER BROWN BAND [7] 1 (3); 4 (34); 9 (101)	4491
249. Paul SIMON [6] 1 (7); 3 (21); 7 (85)	4468
250. NAKED EYES [8] 1 (2); 4 (30); 4 (68)	4434
251. SHEILA E. [7] 1 (5); 3 (33); 4 (74)	4423
252. Teri DeSARIO [2] 1 (7); 1 (16); 2 (29)	4416
253. Melissa MANCHESTER [5] 1 (5); 2 (20); 7 (73)	4387
254. Billy SQUIER [15] 0 (0); 4 (32); 11 (118)	4386
255. SOUL II SOUL [4] 1 (7); 2 (28); 2 (48)	4382
256. WARRANT [2] 1 (7); 2 (20); 3 (39)	4365
257. Billy VERA & The BEATERS [1] 1 (6); 2 (17); 3 (35)	4362
258. Chris De BURGH [3] 1 (6); 2 (18); 4 (58)	4354
259. James INGRAM [2] 1 (5); 4 (42); 6 (111)	4330
260. Luther VANDROSS [15] 0 (0); 5 (30); 10 (129)	4300
261. OAK RIDGE BOYS [5] 1 (8); 2 (23); 4 (45)	4291
262. Eric CLAPTON [10] 1 (2); 4 (33); 9 (82)	4290
263. A TASTE OF HONEY [3] 1 (8); 1 (16); 2 (34)	4276
264. NENA [2] 1 (6); 1 (13); 1 (23)	4276
265. NU SHOOZ [3] 1 (5); 2 (23); 3 (61)	4223
266. Sammy HAGAR [13] 0 (0); 4 (29); 10 (115)	4163
267. The OUTFIELD [6] 1 (4); 4 (30); 6 (83)	4136
268. BAD ENGLISH [1] 1 (6); 1 (15); 2 (33)	4102
269. ANIMOTION [6] 2 (6); 3 (26); 6 (74)	4100
270. ABBA [8] 1 (4); 3 (30); 6 (77)	4088
271. JOHNNY HATES JAZZ [2] 1 (6); 2 (18); 2 (31)	4084
272. Gino VANNELLI [6] 1 (7); 1 (14); 6 (76)	4065
273. Don McLEAN [5] 1 (6); 3 (26); 4 (48)	4061
274. Matthew WILDER [5] 1 (7); 2 (18); 3 (51)	4019

Singles Acts Ranked

Rank. Act [Hi Peak] T10s (Wk); T40s (Wk); Ents (Wk)	Score
275. Freddie JACKSON [12] 0 (0); 4 (28); 7 (100)	4008
276. Herb ALPERT [5] 1 (4); 4 (25); 9 (88)	4008
277. Billy PRESTON & SYREETA [4] 1 (6); 1 (15); 2 (39)	3996
278. Gloria LORING & Carl ANDERSON [2] 1 (7); 1 (14); 1 (21)	3983
279. ALABAMA [15] 0 (0); 4 (29); 7 (95)	3971
280. Donny OSMOND [2] 1 (6); 2 (20); 3 (41)	3971
281. The HONEYDRIPPERS [3] 1 (6); 2 (20); 2 (31)	3957
282. Neneh CHERRY [3] 2 (7); 2 (22); 2 (38)	3956
283. INFORMATION SOCIETY [3] 2 (9); 2 (24); 4 (58)	3917
284. Jan HAMMER [1] 1 (7); 1 (13); 1 (22)	3913
285. The WHISPERS [7] 1 (3); 4 (29); 5 (68)	3880
286. Kylie MINOGUE [3] 1 (6); 3 (19); 3 (54)	3847
287. CINDERELLA [12] 0 (0); 4 (28); 6 (84)	3818
288. SCANDAL/Patty SMYTH [7] 1 (5); 1 (15); 7 (85)	3811
289. Jack WAGNER [2] 1 (6); 1 (12); 4 (50)	3810
290. MÖTLEY CRÜE [6] 1 (2); 3 (27); 7 (72)	3773
291. TOMMY TUTONE [4] 1 (8); 2 (18); 2 (35)	3752
292. Bobby McFERRIN [1] 1 (6); 1 (13); 1 (26)	3746
293. Anne MURRAY [12] 0 (0); 3 (19); 9 (91)	3740
294. POWER STATION [6] 2 (8); 3 (25); 3 (43)	3712
295. The CURE [2] 1 (4); 2 (13); 7 (75)	3699
296. The CLASH [8] 1 (5); 2 (22); 4 (61)	3696
297. Teena MARIE [4] 1 (5); 2 (15; 5 (59)	3684
298. Michael DAMIAN [1] 1 (5); 2 (17); 3 (39)	3678
299. Dan HARTMAN [6] 1 (4); 3 (27); 5 (64)	3663
300. Thomas DOLBY [5] 1 (8); 1 (15); 3 (34)	3635
301. SHERIFF [1] 1 (5); 1 (13); 2 (28)	3632
302. SUPERTRAMP [11] 0 (0); 4 (31); 5 (59)	3606
303. QUIET RIOT [5] 1 (5); 2 (18); 3 (45)	3606
304. GREAT WHITE [5] 1 (5); 2 (18); 4 (73)	3554
305. BOY MEETS GIRL [5] 1 (6); 2 (17); 3 (49)	3546
306. Al JARREAU [15] 0 (0); 3 (22); 8 (91)	3526
307. MANHATTAN TRANSFER [7] 1 (6); 3 (19); 6 (62)	3501
308. CAMEO [6] 1 (5); 2 (21); 5 (62)	3497
309. R.E.M. [6] 2 (5); 2 (21); 7 (66)	3447
310. The MANHATTANS [5] 1 (5); 1 (14); 3 (36)	3415
311. AMERICA [8] 1 (6); 2 (21); 3 (45)	3406
312. DINO [7] 1 (4); 2 (20); 4 (66)	3402
313. Stephanie MILLS [6] 1 (5); 2 (18); 6 (58)	3393
314. GEORGIA SATELLITES [2] 1 (5); 1 (14); 3 (39)	3363
315. CROSBY, STILLS & NASH [9] 1 (5); 2 (21); 4 (47)	3361
316. The B-52's [3] 1 (6); 1 (17); 4 (44)	3355
317. SANTANA [15] 0 (0); 3 (24); 6 (72)	3342
318. Carly SIMON [11] 0 (0); 2 (22); 8 (86)	3336
319. A FLOCK OF SEAGULLS [9] 1 (4); 3 (24); 4 (63)	3307
320. CHARLENE [3] 1 (6); 1 (14); 2 (31)	3303
321. Rick JAMES [16] 0 (0); 4 (16); 8 (100)	3302
322. LEVEL 42 [7] 1 (3); 2 (24); 4 (53)	3284
323. DAZZ BAND [5] 1 (4); 1 (16); 3 (41)	3280
324. Jermaine STEWART [5] 1 (4); 2 (18); 4 (58)	3268
325. STACEY Q [3] 1 (4); 2 (17); 4 (60)	3160
326. Michael BOLTON [11] 0 (0); 3 (27); 5 (68)	3116
327. TACO [4] 1 (6); 1 (14); 1 (21)	3108
328. RAY, GOODMAN & BROWN [5] 1 (4); 1 (14); 3 (32)	3086
329. ICEHOUSE [7] 1 (3); 2 (24); 6 (66)	3084
330. Paul CARRACK [9] 1 (3); 4 (24); 6 (69)	3076
331. Harold FALTERMEYER [3] 1 (5); 1 (12); 1 (19)	3024
332. Marilyn MARTIN [1] 1 (9); 2 (22); 2 (39)	3023
333. Rupert HOLMES [6] 1 (5); 2 (15); 4 (42)	3018
334. RUN-D.M.C. [4] 1 (5); 2 (17); 4 (50)	2975
335. Patrick SWAYZE [3] 1 (6); 1 (13); 1 (21)	2939
336. Lita FORD [8] 1 (4); 2 (22); 2 (48)	2928
337. LISA LISA And CULT JAM With FULL FORCE [8] 1 (3); 2 (19); 4 (77)	2926
338. The KINKS [6] 1 (3); 2 (16); 6 (55)	2921
339. PURE PRAIRIE LEAGUE [10] 1 (2); 3 (22); 5 (55)	2895
340. KATRINA & The WAVES [9] 1 (3); 3 (21); 5 (57)	2868
341. SURFACE [1] 1 (4); 2 (19); 4 (51)	2866
342. TALKING HEADS [9] 1 (1); 2 (18); 5 (73)	2859
343. SHANNON [8] 1 (3); 1 (12); 3 (52)	2852
344. Lindsey BUCKINGHAM [9] 1 (5); 2 (23); 3 (40)	2850
345. The HOOTERS [18] 0 (0); 3 (16); 7 (85)	2834
346. Murray HEAD [3] 1 (6); 1 (13); 1 (20)	2828
347. Marty BALIN [8] 1 (6); 2 (18); 3 (40)	2825
348. FRANKE AND THE KNOCKOUTS [10] 1 (2); 3 (26); 3 (47)	2819
349. El DeBARGE [3] 1 (5); 1 (13); 3 (40)	2816
350. Mick JAGGER [7] 1 (3); 4 (23); 6 (65)	2814
351. Bertie HIGGINS [8] 1 (4); 1 (17); 2 (39)	2799
352. Stacy LATTISAW [21] 0 (0); 3 (18); 6 (85)	2787
353. Robert PLANT [20] 0 (0); 4 (25); 7 (72)	2754
354. Peter WOLF [12] 0 (0); 3 (21); 5 (52)	2744
355. YOUNG M.C. [7] 1 (4); 1 (20); 1 (39)	2743
356. BEE GEES [7] 1 (1); 3 (20); 6 (55)	2720
357. Johnny LEE [5] 1 (5); 1 (13); 2 (30)	2713
358. LOVE AND ROCKETS [3] 1 (6); 1 (12); 2 (24)	2706
359. AFTER THE FIRE [5] 1 (5); 1 (14); 2 (24)	2703
360. COVER GIRLS [27] 0 (0); 3 (11); 6 (85)	2666
361. Joe COCKER [1] 1 (7); 1 (15); 3 (34)	2652
362. MARY JANE GIRLS [7] 1 (5); 1 (12); 3 (42)	2643
363. Suzanne VEGA [3] 1 (5); 1 (12); 2 (22)	2627
364. Gary NUMAN [9] 1 (5); 1 (17); 1 (25)	2581
365. 'TIL TUESDAY [8] 1 (2); 2 (18); 5 (53)	2575
366. MADNESS [7] 1 (4); 2 (18); 3 (36)	2570
367. T'PAU [4] 1 (5); 1 (16); 1 (27)	2549
368. KANSAS [17] 0 (0); 3 (20); 7 (66)	2528
369. Peabo BRYSON [10] 1 (3); 1 (13); 6 (59)	2508
370. Dan HILL [6] 1 (4); 1 (13); 2 (44)	2503
371. DJ JAZZY JEFF & THE FRESH PRINCE [12] 0 (0); 2 (19); 4 (56)	2496
372. Gary (U.S.) BONDS [11] 0 (0); 2 (22); 3 (40)	2492
373. Charlie DANIELS Band [11] 0 (0); 3 (20); 3 (41)	2468
374. DEAD OR ALIVE [11] 0 (0); 2 (20); 5 (61)	2461
375. CHAMPAIGN [12] 0 (0); 2 (22); 2 (43)	2461
376. SWING OUT SISTER [6] 1 (4); 2 (14); 3 (42)	2428
377. Pete TOWNSHEND [9] 1 (3); 2 (19); 4 (43)	2421
378. The SYSTEM [4] 1 (4); 1 (13); 3 (34)	2389
379. SWEET SENSATION [14] 0 (0); 2 (15); 5 (69)	2384
380. ARCADIA [6] 1 (6); 2 (15); 2 (26)	2371
381. John FOGERTY [10] 1 (1); 2 (15); 4 (47)	2367
382. Daryl HALL [5] 1 (4); 2 (16); 3 (36)	2352
383. DEVO [14] 0 (0); 1 (15); 3 (43)	2346
384. SOFT CELL [8] 1 (3); 1 (15); 1 (43)	2345
385. ROGER [3] 1 (5); 1 (13); 2 (28)	2327
386. NITTY GRITTY DIRT BAND [13] 0 (0); 2 (15); 2 (39)	2311
387. WAS (NOT WAS) [7] 1 (3); 2 (19); 3 (39)	2301
388. Bill MEDLEY [1] 1 (6); 1 (15); 3 (33)	2294
389. PRETTY POISON [8] 1 (3); 2 (18); 2 (35)	2271
390. STEELY DAN [10] 1 (2); 2 (20); 3 (30)	2269
391. LL COOL J [14] 0 (0); 3 (21); 4 (47)	2259
392. Stevie B [32] 0 (0); 2 (7); 5 (80)	2256
393. Benny MARDONES [11] 0 (0); 2 (19); 2 (37)	2254
394. Roberta FLACK [13] 0 (0); 1 (11); 4 (51)	2251
395. Don JOHNSON [5] 1 (5); 1 (15); 3 (38)	2243
396. The TUBES [10] 1 (1); 2 (15); 5 (45)	2230
397. Tracy CHAPMAN [6] 1 (4); 1 (12); 4 (41)	2229
398. KAJAGOOGOO [5] 1 (4); 1 (12); 2 (23)	2216
399. FOUR TOPS [11] 0 (0); 3 (13); 5 (52)	2201
400. Adam ANT [12] 0 (0); 1 (14); 3 (42)	2187
401. James BROWN [4] 1 (5); 1 (11); 2 (21)	2153
402. Keith SWEAT [5] 1 (5); 1 (13); 3 (36)	2129
403. ERASURE [12] 0 (0); 2 (20); 3 (39)	2123
404. Delbert McCLINTON [8] 1 (5); 1 (14); 2 (25)	2115
405. GOLDEN EARRING [10] 1 (2); 1 (15); 3 (35)	2109
406. Randy MEISNER [19] 0 (0); 3 (22); 3 (42)	2103
407. SKID ROW [4] 1 (5); 1 (13); 2 (22)	2100
408. Peabo BRYSON/Roberta FLACK [16] 0 (0); 1 (15); 2 (40)	2100
409. CLIMAX BLUES BAND [12] 0 (0); 1 (17); 2 (39)	2088
410. MIDNIGHT STAR [18] 0 (0); 1 (8); 7 (72)	2059
411. BEASTIE BOYS [7] 1 (2); 2 (12); 3 (38)	2057
412. Brenda K. STARR [13] 0 (0); 2 (19); 2 (39)	2051
413. DREAM ACADEMY [7] 1 (3); 2 (14); 2 (32)	2049
414. BROTHERS JOHNSON [7] 1 (2); 1 (13); 3 (29)	2045

Singles Acts Ranked

Rank. Act [Hi Peak] T10s (Wk); T40s (Wk); Ents (Wk)	Score
415. Ann WILSON [6] 2 (6); 2 (23); 3 (51)	2040
416. FRANKIE GOES TO HOLLYWOOD [10] 1 (2); 1 (10); 4 (46)	2035
417. Lou GRAMM [5] 1 (3); 1 (11); 2 (32)	2017
418. SLY FOX [7] 1 (4); 1 (14); 2 (27)	2011
419. Donnie IRIS [25] 0 (0); 3 (14); 7 (67)	2001
420. RUSH [21] 0 (0); 1 (6); 6 (63)	1992
421. Crystal GAYLE [7] 1 (6); 1 (21); 6 (59)	1968
422. AL B. SURE! [7] 1 (2); 1 (13); 3 (44)	1960
423. RATT [12] 0 (0); 2 (11); 6 (50)	1954
424. Glenn MEDEIROS [12] 0 (0); 1 (13); 4 (54)	1954
425. Bruce WILLIS [5] 1 (4); 1 (10); 3 (26)	1926
426. Christine McVIE [10] 1 (1); 2 (17); 2 (26)	1909
427. DEPECHE MODE [13] 0 (0); 1 (10); 6 (57)	1891
428. SA-FIRE [12] 0 (0); 1 (12); 3 (47)	1889
429. CHILLIWACK [22] 0 (0); 2 (14); 3 (43)	1884
430. FRIDA [13] 0 (0); 1 (12); 1 (29)	1869
431. REAL LIFE [26] 0 (0); 3 (15); 3 (46)	1869
432. Jeff HEALEY Band [5] 1 (4); 1 (13); 1 (22)	1866
433. SCRITTI POLITTI [11] 0 (0); 1 (13); 3 (40)	1857
434. Edie BRICKELL & The NEW BOHEMIANS [7] 1 (3); 1 (10); 2 (29)	1853
435. Alice COOPER [7] 1 (3); 2 (11); 2 (28)	1849
436. JELLYBEAN [16] 0 (0); 2 (17); 3 (39)	1843
437. Rocky BURNETTE [8] 1 (2); 1 (12); 1 (19)	1840
438. Ben E. KING [9] 1 (4); 1 (13); 1 (21)	1837
439. Henry LEE SUMMER [18] 0 (0); 2 (15); 4 (47)	1825
440. Dennis DeYOUNG [10] 1 (1); 1 (12); 4 (40)	1818
441. LEVERT [5] 1 (4); 1 (12); 1 (18)	1814
442. KISS [47] 0 (0); 0 (0); 9 (84)	1808
443. Carole KING [12] 0 (0); 1 (10); 2 (27)	1802
444. Johnny KEMP [10] 1 (1); 2 (14); 2 (35)	1797
445. John DENVER [31] 0 (0); 2 (9); 8 (70)	1796
446. Tracey ULLMAN [8] 1 (3); 1 (11); 2 (21)	1784
447. Roger DALTREY [20] 0 (0); 1 (8); 5 (55)	1771
448. Peter SCHILLING [14] 0 (0); 1 (10); 2 (32)	1759
449. Vanessa WILLIAMS [8] 1 (2); 1 (11); 3 (34)	1756

Rank. Act [Hi Peak] T10s (Wk); T40s (Wk); Ents (Wk)	Score
450. SQUEEZE [15] 0 (0); 2 (15); 3 (42)	1754
451. ASHFORD & SIMPSON [12] 0 (0); 1 (11); 3 (38)	1716
452. Jane WIEDLIN [9] 1 (2); 1 (10); 4 (47)	1708
453. Evelyn "Champagne" KING [17] 0 (0); 3 (10); 5 (52)	1707
454. TIMEX SOCIAL CLUB [8] 1 (4); 1 (12); 1 (19)	1707
455. MOVING PICTURES [29] 0 (0); 1 (13); 2 (43)	1703
456. Quincy JONES [14] 0 (0); 3 (26); 3 (56)	1703
457. BREAKFAST CLUB [7] 1 (1); 1 (11); 2 (28)	1700
458. Julio IGLESIAS [5] 1 (6); 2 (20); 3 (42)	1688
459. MECO [18] 0 (0); 2 (11); 5 (40)	1684
460. GAP BAND [24] 0 (0); 2 (13); 5 (50)	1684
461. The TIME [20] 0 (0); 2 (12); 4 (48)	1677
462. STEEL BREEZE [16] 0 (0); 2 (17); 2 (33)	1677
463. FABULOUS THUNDERBIRDS [10] 1 (1); 1 (10); 4 (41)	1676
464. The WHO [18] 0 (0); 2 (16); 4 (39)	1665
465. BUCKNER & GARCIA [9] 1 (3); 1 (14); 1 (19)	1661
466. The BEATLES [12] 0 (0); 2 (15); 2 (26)	1656
467. Joe WALSH [19] 0 (0); 2 (12); 3 (36)	1653
468. Teddy PENDERGRASS [43] 0 (0); 0 (0); 5 (60)	1647
469. Frank STALLONE [10] 1 (2); 1 (10); 3 (26)	1642
470. PSEUDO ECHO [6] 1 (3); 1 (10); 2 (24)	1627
471. ROYAL PHILHARMONIC ORCHESTRA [10] 1 (2); 1 (12); 1 (20)	1627
472. Tom PETTY [12] 0 (0); 2 (16); 2 (29)	1625
473. TIERRA [18] 0 (0); 1 (15); 3 (35)	1621
474. Brenda RUSSELL [6] 1 (2); 1 (13); 1 (25)	1611
475. Larry GRAHAM [9] 1 (1); 1 (9); 3 (29)	1600
476. Alexander O'NEAL [25] 0 (0); 3 (18); 4 (57)	1586
477. John SCHNEIDER [14] 0 (0); 1 (11); 4 (38)	1569
478. WINGER [19] 0 (0); 2 (15); 3 (40)	1557
479. MUSICAL YOUTH [10] 1 (2); 1 (10); 2 (25)	1556
480. RUFUS/ Chaka KHAN [22] 0 (0); 2 (12); 3 (39)	1551
481. Mike POST [10] 1 (2); 2 (17); 2 (39)	1549
482. Deon ESTUS [5] 1 (2); 1 (11); 1 (16)	1539
483. WHEN IN ROME [11] 0 (0); 1 (13); 2 (27)	1538

Rank. Act [Hi Peak] T10s (Wk); T40s (Wk); Ents (Wk)	Score
484. "Weird Al" YANKOVIC [12] 0 (0); 1 (7); 6 (39)	1529
485. Steve FORBERT [11] 0 (0); 1 (12); 2 (22)	1527
486. Michael MORALES [15] 0 (0); 2 (16); 3 (38)	1523
487. Dottie WEST [14] 0 (0); 1 (12); 2 (25)	1519
488. PSYCHEDELIC FURS [26] 0 (0); 1 (5); 4 (44)	1517
489. BOYS CLUB [8] 1 (1); 1 (12); 1 (21)	1515
490. POCO [18] 0 (0); 1 (8); 5 (48)	1511
491. REGINA [10] 1 (2); 1 (12); 1 (20)	1504
492. KIX [11] 0 (0); 1 (13); 1 (23)	1498
493. Michael STANLEY Band [33] 0 (0); 2 (6); 7 (54)	1494
494. BABYFACE [7] 1 (4); 1 (10); 1 (18)	1490
495. AC/DC [35] 0 (0); 2 (8); 4 (45)	1490
496. Rita COOLIDGE [36] 0 (0); 2 (6); 4 (45)	1478
497. GIUFFRIA [15] 0 (0); 1 (7); 3 (37)	1472
498. Dusty SPRINGFIELD [2] 1 (5); 1 (13); 1 (18)	1459
499. LIVING COLOUR [13] 0 (0); 2 (12); 3 (33)	1454
500. SYNCH [10] 1 (1); 1 (11); 2 (36)	1450
501. Tony CAREY [22] 0 (0); 2 (10); 4 (42)	1443
502. Stevie WOODS [25] 0 (0); 2 (13); 3 (35)	1441
503. FAT BOYS [12] 0 (0); 2 (19); 3 (37)	1430
504. Karla BONOFF [19] 0 (0); 1 (12); 3 (32)	1429
505. The TEMPTATIONS [43] 0 (0); 0 (0); 7 (58)	1426
506. Ali THOMSON [15] 0 (0); 1 (9); 2 (28)	1424
507. OLLIE And JERRY [9] 1 (1); 1 (11); 1 (18)	1422
508. FORCE M.D.'S [10] 1 (2); 1 (11); 2 (28)	1421
509. The DEELE [10] 1 (1); 1 (12); 2 (29)	1413
510. Charlie DORE [13] 0 (0); 1 (10); 1 (17)	1411
511. LULU [18] 0 (0); 1 (10); 2 (29)	1409
512. Terri GIBBS [13] 0 (0); 1 (12); 2 (27)	1406
513. STARPOINT [25] 0 (0); 1 (9); 3 (43)	1388
514. Jimmy RUFFIN [10] 1 (2); 1 (9); 1 (14)	1383
515. BALTIMORA [13] 0 (0); 1 (10); 2 (30)	1379
516. GIANT STEPS [13] 0 (0); 1 (10); 2 (30)	1361
517. Roy ORBISON [9] 1 (1); 1 (11); 2 (26)	1354
518. Mickey GILLEY [22] 0 (0); 1 (9); 3 (37)	1351
519. Ric OCASEK [15] 0 (0); 1 (8); 3 (36)	1332

Rank. Act [Hi Peak] T10s (Wk); T40s (Wk); Ents (Wk)	Score
520. WATERFRONT [10] 1 (2); 1 (10); 2 (22)	1322
521. BIG COUNTRY [17] 0 (0); 1 (9); 3 (23)	1319
522. APRIL WINE [21] 0 (0); 1 (7); 5 (41)	1317
523. Oran "Juice" JONES [9] 1 (2); 1 (9); 1 (19)	1316
524. Isaac HAYES [18] 0 (0); 1 (12); 1 (21)	1314
525. PABLO CRUISE [13] 0 (0); 1 (11); 2 (22)	1309
526. BLUES BROTHERS [18] 0 (0); 2 (10); 2 (25)	1306
527. M/A/R/R/S [13] 0 (0); 1 (11); 1 (23)	1300
528. GRATEFUL DEAD [9] 1 (2); 1 (9); 2 (21)	1291
529. SLADE [20] 0 (0); 2 (11); 3 (31)	1290
530. CARPENTERS [16] 0 (0); 1 (8); 4 (32)	1286
531. FIREFALL [35] 0 (0); 2 (6); 4 (40)	1286
532. Gladys KNIGHT & The PIPS [13] 0 (0); 1 (9); 3 (33)	1286
533. GO WEST [39] 0 (0); 1 (2); 4 (49)	1278
534. Van STEPHENSON [22] 0 (0); 1 (10); 3 (30)	1274
535. Paul HARDCASTLE [15] 0 (0); 1 (8); 2 (32)	1264
536. PHOTOGLO [25] 0 (0); 2 (11); 2 (30)	1263
537. Michael Martin MURPHEY [19] 0 (0); 1 (11); 2 (27)	1239
538. Fred KNOBLOCK [18] 0 (0); 2 (16); 2 (32)	1229
539. Annie LENNOX & Al GREEN [9] 1 (2); 1 (10); 1 (17)	1205
540. David FOSTER [15] 0 (0); 1 (10); 3 (33)	1198
541. The NYLONS [12] 0 (0); 1 (10); 2 (24)	1195
542. VIXEN [22] 0 (0); 2 (11); 2 (34)	1195
543. Patrice RUSHEN [23] 0 (0); 1 (7); 3 (31)	1189
544. TAVARES [33] 0 (0); 1 (9); 2 (31)	1185
545. MIDNIGHT OIL [17] 0 (0); 1 (9); 2 (32)	1176
546. YARBROUGH & PEOPLES [19] 0 (0); 1 (7); 3 (32)	1175
547. KON KAN [15] 0 (0); 1 (9); 2 (26)	1159
548. BOYS DON'T CRY [12] 0 (0); 1 (9); 1 (19)	1158
549. SALT-N-PEPA [19] 0 (0); 1 (13); 1 (25)	1154
550. Waylon JENNINGS [21] 0 (0); 1 (10); 2 (32)	1154
551. Tommy JAMES [19] 0 (0); 1 (9); 2 (23)	1146
552. TWISTED SISTER [21] 0 (0); 1 (7); 3 (32)	1146
553. MISSING PERSONS [42] 0 (0); 0 (0); 5 (45)	1139

Singles Acts Ranked

Rank. Act [Hi Peak] T10s (Wk); T40s (Wk); Ents (Wk)	Score
554. The KORGIS [18] 0 (0); 1 (11); 1 (19)	1136
555. SYLVIA [15] 0 (0); 1 (9); 1 (20)	1134
556. Robert JOHN [31] 0 (0); 1 (4); 4 (33)	1127
557. BELLE STARS [14] 0 (0); 1 (10); 2 (22)	1123
558. 2 LIVE CREW [26] 0 (0); 1 (9); 1 (30)	1116
559. ART OF NOISE [31] 0 (0); 2 (10); 3 (34)	1116
560. LIMAHL [17] 0 (0); 1 (9); 2 (26)	1112
561. The BOYS [13] 0 (0); 1 (9); 1 (19)	1097
562. Jennifer HOLLIDAY [22] 0 (0); 1 (7); 4 (35)	1097
563. Neil SEDAKA & Dara SEDAKA [19] 0 (0); 1 (10); 1 (19)	1089
564. GTR [14] 0 (0); 1 (10); 2 (22)	1087
565. Dwight TWILLEY [16] 0 (0); 1 (10); 2 (20)	1079
566. Kevin PAIGE [18] 0 (0); 1 (10); 1 (24)	1076
567. Lee RITENOUR [15] 0 (0); 1 (9); 2 (23)	1070
568. PEACHES & HERB [19] 0 (0); 1 (8); 1 (19)	1062
569. BALANCE [22] 0 (0); 1 (9); 2 (28)	1058
570. HONEYMOON SUITE [34] 0 (0); 1 (3); 4 (41)	1056
571. Mike RENO [7] 1 (4); 1 (13); 1 (20)	1048
572. SCORPIONS [25] 0 (0); 1 (7); 4 (35)	1043
573. Chuck MANGIONE [18] 0 (0); 1 (9); 1 (16)	1035
574. Greg GUIDRY [17] 0 (0); 1 (10); 2 (18)	1022
575. SAGA [26] 0 (0); 1 (8); 3 (29)	1015
576. Charlie SEXTON [17] 0 (0); 1 (10); 1 (20)	1010
577. FIVE STAR [41] 0 (0); 0 (0); 4 (45)	1009
578. DOUBLE [16] 0 (0); 1 (9); 1 (18)	1006
579. BLOW MONKEYS [14] 0 (0); 1 (10); 1 (19)	1002
580. Clarence CLEMONS [18] 0 (0); 1 (12); 1 (19)	1001
581. GODLEY & CREME [16] 0 (0); 1 (10); 1 (17)	981
582. The BABYS [33] 0 (0); 1 (3); 3 (28)	979
583. Chico DEBARGE [21] 0 (0); 1 (11); 1 (20)	972
584. James TAYLOR [11] 0 (0); 1 (10); 4 (35)	971
585. NEW ORDER [32] 0 (0); 1 (8); 3 (37)	971
586. FOGHAT [23] 0 (0); 1 (10); 2 (18)	971
587. Gary WRIGHT [16] 0 (0); 1 (10); 1 (17)	964
588. SISTER SLEDGE [23] 0 (0); 1 (6); 5 (38)	960
589. The CONTOURS [11] 0 (0); 1 (8); 1 (16)	958
590. SYBIL [20] 0 (0); 1 (9); 1 (23)	950
591. Deborah ALLEN [26] 0 (0); 1 (7); 1 (21)	950
592. RE-FLEX [24] 0 (0); 1 (5); 2 (25)	940
593. BASIA [26] 0 (0); 1 (7); 2 (31)	940
594. Bruce COCKBURN [21] 0 (0); 1 (9); 2 (20)	924
595. Stephen BISHOP [25] 0 (0); 1 (8); 2 (23)	924
596. The O'JAYS [28] 0 (0); 1 (5); 2 (24)	916
597. Carl CARLTON [22] 0 (0); 1 (7); 1 (21)	909
598. JUMP 'N THE SADDLE [15] 0 (0); 1 (7); 1 (14)	905
599. Don WILLIAMS [24] 0 (0); 1 (9); 1 (20)	901
600. J.J. FAD [30] 0 (0); 1 (4); 3 (36)	893
601. Larry CARLTON [10] 1 (2); 1 (10); 2 (30)	892
602. STRYPER [23] 0 (0); 1 (8); 3 (32)	879
603. Pia ZADORA [36] 0 (0); 1 (3); 3 (35)	876
604. T.G. SHEPPARD [37] 0 (0); 1 (2); 4 (36)	868
605. TIMBUK 3 [19] 0 (0); 1 (9); 1 (16)	859
606. TA MARA & The SEEN [24] 0 (0); 1 (10); 1 (21)	856
607. Bob & Doug McKENZIE [16] 0 (0); 1 (9); 1 (14)	854
608. TRIUMPH [27] 0 (0); 1 (5); 4 (30)	853
609. Jesse JOHNSON [53] 0 (0); 0 (0); 4 (43)	851
610. Robbie PATTON [26] 0 (0); 1 (6); 2 (25)	851
611. Benjamin ORR [24] 0 (0); 1 (6); 1 (20)	849
612. SUGARHILL GANG [36] 0 (0); 1 (3); 2 (32)	837
613. The FIRM [28] 0 (0); 1 (6); 3 (28)	829
614. Tommy SHAW [33] 0 (0); 1 (3); 4 (35)	829
615. Le ROUX [18] 0 (0); 1 (6); 3 (22)	824
616. OTHER ONES [29] 0 (0); 1 (4); 2 (26)	815
617. Morris DAY [23] 0 (0); 1 (6); 2 (25)	809
618. LED ZEPPELIN [21] 0 (0); 1 (8); 1 (13)	807
619. SEDUCTION [23] 0 (0); 1 (8); 1 (21)	804
620. Rosanne CASH [22] 0 (0); 1 (7); 1 (20)	802
621. Donna ALLEN [21] 0 (0); 1 (9); 1 (18)	797
622. LIVING IN A BOX [17] 0 (0); 1 (7); 2 (19)	794
623. Grayson HUGH [19] 0 (0); 1 (8); 2 (22)	793
624. Aldo NOVA [23] 0 (0); 1 (7); 2 (22)	793
625. George DUKE [19] 0 (0); 1 (9); 2 (29)	787
626. SPIDER [39] 0 (0); 1 (2); 3 (24)	783
627. DANNY WILSON [23] 0 (0); 1 (8); 1 (20)	782
628. Rebbie JACKSON [24] 0 (0); 1 (8); 1 (19)	781
629. Donald FAGEN [26] 0 (0); 1 (7); 3 (25)	776
630. Andy TAYLOR [24] 0 (0); 1 (7); 2 (23)	773
631. Shirley MURDOCK [23] 0 (0); 1 (7); 1 (18)	769
632. TIMES TWO [21] 0 (0); 1 (8); 2 (23)	764
633. MANFRED MANN'S EARTH BAND [22] 0 (0); 1 (8); 1 (15)	764
634. COMPANY B [21] 0 (0); 1 (8); 1 (18)	763
635. Elvis COSTELLO & THE ATTRACTIONS [36] 0 (0); 1 (2); 2 (23)	760
636. Paul ANKA [40] 0 (0); 1 (2); 2 (25)	759
637. ROB BASE [36] 0 (0); 1 (3); 2 (29)	757
638. SCARLETT & BLACK [20] 0 (0); 1 (8); 1 (18)	755
639. Jennifer RUSH [36] 0 (0); 1 (3); 2 (26)	754
640. Debbie HARRY [43] 0 (0); 0 (0); 4 (31)	751
641. Dave EDMUNDS [39] 0 (0); 1 (1); 3 (25)	745
642. Amy GRANT [29] 0 (0); 1 (6); 3 (27)	744
643. Frankie SMITH [30] 0 (0); 1 (7); 1 (19)	738
644. Amy HOLLAND [22] 0 (0); 1 (6); 1 (16)	733
645. Phil SEYMOUR [22] 0 (0); 1 (7); 1 (16)	731
646. Robin ZANDER [6] 1 (2); 1 (10); 1 (19)	726
647. Sam HARRIS [36] 0 (0); 1 (2); 3 (23)	724
648. Robert CRAY Band [22] 0 (0); 1 (6); 3 (26)	722
649. Bernadette PETERS [31] 0 (0); 1 (5); 2 (21)	721
650. Eddie SCHWARTZ [28] 0 (0); 1 (7); 2 (20)	721
651. DIESEL [25] 0 (0); 1 (6); 1 (18)	719
652. GENERAL PUBLIC [27] 0 (0); 1 (5); 1 (18)	718
653. BOURGEOIS TAGG [38] 0 (0); 1 (2); 2 (27)	717
654. DAVID & DAVID [37] 0 (0); 1 (3); 2 (27)	716
655. UTOPIA [27] 0 (0); 1 (5); 3 (21)	714
656. Robert TEPPER [22] 0 (0); 1 (7); 2 (19)	713
657. Nancy MARTINEZ [32] 0 (0); 1 (7); 1 (21)	711
658. TALK TALK [31] 0 (0); 1 (6); 4 (28)	709
659. Maxi PRIEST [25] 0 (0); 1 (7); 1 (18)	708
660. HIPSWAY [19] 0 (0); 1 (6); 1 (15)	708
661. NOEL [47] 0 (0); 0 (0); 2 (30)	699
662. Elvis PRESLEY [28] 0 (0); 1 (5); 2 (21)	690
663. SNEAKER [34] 0 (0); 1 (6); 2 (20)	687
664. NIELSEN/PEARSON [38] 0 (0); 1 (2); 2 (22)	684
665. ENYA [24] 0 (0); 1 (8); 1 (17)	683
666. LAID BACK [26] 0 (0); 1 (4); 1 (18)	683
667. BARDEUX [36] 0 (0); 1 (3); 3 (30)	682
668. Timothy B. SCHMIT [25] 0 (0); 1 (5); 2 (21)	681
669. Harry CHAPIN [23] 0 (0); 1 (7); 1 (14)	679
670. Alison MOYET [31] 0 (0); 1 (6); 2 (21)	679
671. Gavin CHRISTOPHER [22] 0 (0); 1 (7); 1 (17)	679
672. Nicolette LARSON [35] 0 (0); 1 (3); 2 (20)	671
673. Leslie PEARL [28] 0 (0); 1 (7); 1 (16)	671
674. Tommy PAGE [29] 0 (0); 1 (6); 1 (20)	670
675. AUTOGRAPH [29] 0 (0); 1 (5); 1 (19)	670
676. Ivan NEVILLE [26] 0 (0); 1 (6); 2 (22)	668
677. Micky DOLENZ And Peter TORK [20] 0 (0); 1 (7); 1 (14)	667
678. OAK [36] 0 (0); 1 (3); 2 (20)	663
679. KANE GANG [36] 0 (0); 1 (3); 2 (24)	660
680. BAND AID [13] 0 (0); 1 (4); 1 (9)	660
681. Suzi QUATRO [41] 0 (0); 0 (0); 2 (20)	660
682. Al STEWART [24] 0 (0); 1 (6); 1 (13)	658
683. CLIMIE FISHER [23] 0 (0); 1 (6); 1 (18)	655
684. SUAVÉ [20] 0 (0); 1 (7); 1 (15)	655
685. Mac DAVIS [43] 0 (0); 0 (0); 3 (27)	654
686. OXO [28] 0 (0); 1 (6); 1 (14)	654
687. Jim CAPALDI [28] 0 (0); 1 (5); 2 (18)	651
688. Kate BUSH [30] 0 (0); 1 (4); 2 (26)	649
689. Dan SEALS [42] 0 (0); 0 (0); 2 (21)	642
690. RAINBOW [40] 0 (0); 1 (1); 2 (22)	636
691. The ALARM [50] 0 (0); 0 (0); 4 (37)	633
692. Agnetha FALTSKOG [29] 0 (0); 1 (5); 2 (18)	632
693. Elvis COSTELLO [19] 0 (0); 1 (6); 1 (14)	630
694. STYLE COUNCIL [29] 0 (0); 1 (6); 2 (19)	629

Singles Acts Ranked

Rank. Act [Hi Peak] T10s (Wk); T40s (Wk); Ents (Wk)	Score
695. CHANGE [40] 0 (0); 1 (1); 4 (24)	624
696. John O'BANION [24] 0 (0); 1 (7); 1 (13)	622
697. AFTERNOON DELIGHTS [33] 0 (0); 1 (5); 1 (16)	615
698. HAIRCUT ONE HUNDRED [37] 0 (0); 1 (4); 1 (17)	614
699. TOM TOM CLUB [31] 0 (0); 1 (4); 1 (17)	614
700. Jim STEINMAN [32] 0 (0); 1 (6); 1 (16)	608
701. The VAPORS [36] 0 (0); 1 (3); 1 (15)	608
702. Patrick SIMMONS [30] 0 (0); 1 (5); 2 (18)	605
703. Robin GIBB [37] 0 (0); 1 (4); 2 (22)	605
704. James LAST Band [28] 0 (0); 1 (6); 1 (13)	595
705. OPUS [32] 0 (0); 1 (5); 1 (16)	594
706. The COMMUNARDS [40] 0 (0); 1 (1); 2 (22)	590
707. RESTLESS HEART [33] 0 (0); 1 (5); 1 (18)	583
708. Jimmy HALL [27] 0 (0); 1 (4); 2 (20)	581
709. J.D. SOUTHER [11] 0 (0); 1 (10); 1 (14)	580
710. IRISH ROVERS [37] 0 (0); 1 (4); 1 (17)	579
711. Jonathan BUTLER [27] 0 (0); 1 (5); 1 (14)	579
712. TRAVELING WILBURYS [45] 0 (0); 0 (0); 2 (23)	577
713. The KNACK [38] 0 (0); 1 (2); 3 (19)	576
714. L.T.D. [40] 0 (0); 1 (1); 1 (16)	576
715. Denise LOPEZ [31] 0 (0); 1 (5); 2 (20)	575
716. LARSEN-FEITEN Band [29] 0 (0); 1 (6); 1 (14)	573
717. The KINGS [43] 0 (0); 0 (0); 2 (23)	570
718. DEVICE [35] 0 (0); 1 (4); 2 (20)	566
719. The HOLLIES [29] 0 (0); 1 (6); 1 (12)	565
720. ATLANTA RHYTHM SECTION [29] 0 (0); 1 (4); 1 (15)	564
721. Frank SINATRA [32] 0 (0); 1 (6); 1 (12)	561
722. JOBOXERS [36] 0 (0); 1 (4); 1 (15)	558
723. ALLMAN BROTHERS Band [39] 0 (0); 1 (2); 2 (19)	557
724. Emmylou HARRIS [37] 0 (0); 1 (3); 2 (21)	555
725. The CHURCH [24] 0 (0); 1 (5); 1 (15)	552
726. Jeff LORBER [27] 0 (0); 1 (5); 1 (16)	548
727. Martin BRILEY [36] 0 (0); 1 (3); 1 (15)	543
728. PLAYER [46] 0 (0); 0 (0); 2 (17)	541
729. John TAYLOR [23] 0 (0); 1 (6); 1 (12)	540
730. John STEWART [34] 0 (0); 1 (4); 1 (13)	539
731. Don FELDER [43] 0 (0); 0 (0); 1 (17)	539
732. The OUTLAWS [31] 0 (0); 1 (4); 1 (15)	536
733. Wayne NEWTON [35] 0 (0); 1 (3); 1 (13)	533
734. John HALL Band [42] 0 (0); 0 (0); 2 (21)	532
735. SIMON & GARFUNKEL [27] 0 (0); 1 (6); 1 (11)	532
736. Carole Bayer SAGER [30] 0 (0); 1 (7); 1 (13)	530
737. Tracie SPENCER [38] 0 (0); 1 (3); 2 (20)	530
738. DE LA SOUL [34] 0 (0); 1 (3); 1 (17)	528
739. 10,000 MANIACS [44] 0 (0); 0 (0); 3 (28)	524
740. FACE TO FACE [38] 0 (0); 1 (3); 1 (15)	521
741. PARTLAND BROTHERS [27] 0 (0); 1 (5); 1 (13)	519
742. BLUE ÖYSTER CULT [40] 0 (0); 1 (3); 2 (17)	519
743. EYE TO EYE [37] 0 (0); 1 (3); 2 (15)	519
744. WORLD PARTY [27] 0 (0); 1 (5); 1 (15)	519
745. John HUNTER [39] 0 (0); 1 (2); 1 (16)	518
746. ISLEY, JASPER, ISLEY [51] 0 (0); 0 (0); 2 (21)	515
747. ISLEY BROTHERS [39] 0 (0); 1 (2); 2 (16)	515
748. Cheryl Pepsii RILEY [32] 0 (0); 1 (5); 1 (13)	514
749. COCK ROBIN [35] 0 (0); 1 (3); 1 (16)	511
750. ROACHFORD [25] 0 (0); 1 (5); 1 (14)	510
751. LONE JUSTICE [47] 0 (0); 0 (0); 3 (23)	508
752. Judson SPENCE [32] 0 (0); 1 (4); 1 (14)	507
753. Stanley CLARKE [19] 0 (0); 1 (9); 1 (20)	502
754. PRISM [39] 0 (0); 1 (2); 2 (17)	502
755. Chris CHRISTIAN [37] 0 (0); 1 (3); 2 (17)	499
756. Greg KIHN [30] 0 (0); 1 (4); 2 (17)	496
757. Frank ZAPPA [32] 0 (0); 1 (3); 1 (12)	496
758. ALPHAVILLE [65] 0 (0); 0 (0); 3 (28)	492
759. Elisa FIORILLO [49] 0 (0); 0 (0); 2 (20)	488
760. ARTISTS UNITED AGAINST APARTHEID [38] 0 (0); 1 (3); 1 (13)	488
761. RENÉ & ANGELA [47] 0 (0); 0 (0); 3 (23)	487
762. Robert Ellis ORRALL [32] 0 (0); 1 (3); 1 (12)	485
763. Bobby CALDWELL [42] 0 (0); 0 (0); 2 (16)	481
764. METALLICA [35] 0 (0); 1 (4); 1 (15)	481
765. Sonny CHARLES [40] 0 (0); 1 (2); 1 (14)	476
766. SILVER CONDOR [32] 0 (0); 1 (4); 1 (13)	476
767. MOLLY HATCHET [42] 0 (0); 0 (0); 4 (20)	474
768. Tom JOHNSTON [34] 0 (0); 1 (2); 1 (12)	472
769. SKYY [26] 0 (0); 1 (4); 1 (11)	471
770. VANDENBERG [39] 0 (0); 1 (2); 1 (14)	471
771. ROMEO VOID [35] 0 (0); 1 (2); 1 (13)	470
772. John ANDERSON [43] 0 (0); 0 (0); 1 (13)	464
773. GEORGIO [58] 0 (0); 0 (0); 3 (23)	464
774. ICICLE WORKS [37] 0 (0); 1 (4); 1 (12)	461
775. BRONSKI BEAT [48] 0 (0); 0 (0); 1 (16)	461
776. Nia PEEPLES [35] 0 (0); 1 (3); 1 (15)	460
777. POINT BLANK [39] 0 (0); 1 (2); 1 (14)	459
778. Phoebe SNOW [46] 0 (0); 0 (0); 2 (18)	458
779. Sharon BRYANT [34] 0 (0); 1 (4); 2 (18)	457
780. Larry ELGART And His MANHATTAN SWING Orchestra [31] 0 (0); 1 (5); 1 (12)	456
781. Christopher WILLIAMS [49] 0 (0); 0 (0); 1 (18)	455
782. David GATES [46] 0 (0); 0 (0); 2 (15)	454
783. PAJAMA PARTY [59] 0 (0); 0 (0); 2 (24)	452
784. Felix CAVALIERE [36] 0 (0); 1 (3); 1 (11)	452
785. Bonnie POINTER [40] 0 (0); 1 (2); 1 (13)	451
786. Del SHANNON [33] 0 (0); 1 (4); 1 (12)	447
787. JUNIOR [30] 0 (0); 1 (3); 1 (13)	446
788. Mac McANALLY [41] 0 (0); 0 (0); 1 (12)	442
789. The BAR-KAYS [57] 0 (0); 0 (0); 3 (20)	441
790. WA WA NEE [35] 0 (0); 1 (3); 2 (15)	440
791. MODELS [37] 0 (0); 1 (4); 1 (13)	438
792. ONE 2 MANY [37] 0 (0); 1 (4); 1 (13)	437
793. Jesse WINCHESTER [32] 0 (0); 1 (5); 1 (12)	435
794. MTUME [45] 0 (0); 0 (0); 2 (17)	429
795. Dann ROGERS [41] 0 (0); 1 (3); 1 (11)	428
796. Bryan FERRY [31] 0 (0); 1 (3); 1 (13)	427
797. Louise TUCKER [46] 0 (0); 0 (0); 1 (13)	427
798. JON & VANGELIS [51] 0 (0); 0 (0); 2 (15)	426
799. Chuckii BOOKER [42] 0 (0); 0 (0); 1 (14)	424
800. CHARLIE [38] 0 (0); 1 (2); 1 (11)	416
801. Roger HODGSON [48] 0 (0); 0 (0); 1 (15)	416
802. E.U. [35] 0 (0); 1 (4); 1 (12)	415
803. Meli'sa MORGAN [46] 0 (0); 0 (0); 1 (14)	414
804. El DeBARGE With DeBARGE [46] 0 (0); 0 (0); 2 (17)	413
805. Joyce COBB [42] 0 (0); 0 (0); 1 (12)	409
806. CHERRELLE [26] 0 (0); 1 (6); 2 (26)	407
807. CON FUNK SHUN [40] 0 (0); 1 (1); 2 (15)	407
808. DOKKEN [64] 0 (0); 0 (0); 3 (26)	404
809. Paul LEKAKIS [43] 0 (0); 0 (0); 1 (14)	404
810. Graham PARKER And The SHOT [39] 0 (0); 1 (3); 1 (12)	402
811. ROMAN HOLLIDAY [54] 0 (0); 0 (0); 3 (20)	402
812. EIGHTH WONDER [56] 0 (0); 0 (0); 2 (18)	401
813. Marshall CRENSHAW [36] 0 (0); 1 (4); 1 (11)	401
814. Ringo STARR [38] 0 (0); 1 (2); 1 (11)	400
815. Rodney CROWELL [37] 0 (0); 1 (2); 1 (11)	399
816. Jimmy BUFFETT [57] 0 (0); 0 (0); 3 (20)	395
817. Louis ARMSTRONG [32] 0 (0); 1 (3); 1 (11)	392
818. Nik KERSHAW [46] 0 (0); 0 (0); 1 (13)	387
819. WAX [43] 0 (0); 0 (0); 1 (13)	384
820. Livingston TAYLOR [38] 0 (0); 1 (2); 1 (10)	380
821. CHERI [39] 0 (0); 1 (2); 1 (12)	377
822. FELONY [42] 0 (0); 0 (0); 1 (12)	375
823. Randy NEWMAN [51] 0 (0); 0 (0); 2 (20)	374
824. Ziggy MARLEY & The MELODY MAKERS [39] 0 (0); 1 (1); 1 (13)	373
825. INVISIBLE MAN'S BAND [45] 0 (0); 0 (0); 1 (10)	372
826. Susan ANTON [28] 0 (0); 1 (9); 1 (18)	370
827. Barry GIBB [37] 0 (0); 1 (3); 1 (10)	369
828. David LASLEY [36] 0 (0); 1 (3); 1 (10)	368
829. Tommy CONWELL And The YOUNG RUMBLERS [48] 0 (0); 0 (0); 2 (18)	366
830. The CULT [46] 0 (0); 0 (0); 2 (15)	363

Singles Acts Ranked

Rank. Act [Hi Peak] T10s (Wk); T40s (Wk); Ents (Wk)	Score
831. Leif GARRETT [60] 0 (0); 0 (0); 3 (20)	362
832. LANIER AND CO. [48] 0 (0); 0 (0); 1 (13)	361
833. CURIOSITY KILLED THE CAT [42] 0 (0); 0 (0); 1 (13)	358
834. Gwen GUTHRIE [42] 0 (0); 0 (0); 1 (13)	358
835. Boy GEORGE [40] 0 (0); 1 (1); 1 (12)	357
836. Bill CHAMPLIN [55] 0 (0); 0 (0); 2 (16)	357
837. Burton CUMMINGS [37] 0 (0); 1 (2); 1 (11)	356
838. Tané CAIN [37] 0 (0); 1 (3); 1 (11)	355
839. The ANIMALS [48] 0 (0); 0 (0); 1 (10)	351
840. SIOUXSIE & The BANSHEES [53] 0 (0); 0 (0); 1 (14)	351
841. WEATHER GIRLS [46] 0 (0); 0 (0); 1 (11)	350
842. 9.9 [51] 0 (0); 0 (0); 1 (13)	350
843. Maurice WHITE [50] 0 (0); 0 (0); 2 (14)	346
844. BLACKFOOT [42] 0 (0); 0 (0); 1 (12)	345
845. ROCKPILE [51] 0 (0); 0 (0); 1 (12)	344
846. Joe DOLCE [53] 0 (0); 0 (0); 1 (14)	343
847. Shanice WILSON [50] 0 (0); 0 (0); 1 (13)	339
848. Buster POINDEXTER [45] 0 (0); 0 (0); 1 (13)	339
849. Walter EGAN [46] 0 (0); 0 (0); 1 (10)	338
850. SPARKS [49] 0 (0); 0 (0); 2 (19)	338
851. OINGO BOINGO [45] 0 (0); 0 (0); 2 (16)	338
852. Rachel SWEET [32] 0 (0); 1 (4); 2 (18)	336
853. Maria VIDAL [48] 0 (0); 0 (0); 1 (12)	336
854. Dan REED Network [38] 0 (0); 1 (2); 1 (11)	335
855. Regina BELLE [60] 0 (0); 0 (0); 3 (21)	334
856. SOULSISTER [41] 0 (0); 0 (0); 1 (10)	333
857. Linda CLIFFORD [41] 0 (0); 0 (0); 1 (11)	333
858. Richard "Dimples" FIELDS [47] 0 (0); 0 (0); 1 (10)	330
859. NEWCLEUS [56] 0 (0); 0 (0); 1 (15)	329
860. Greg LAKE [48] 0 (0); 0 (0); 1 (10)	323
861. L'TRIMM [54] 0 (0); 0 (0); 1 (15)	321
862. SO [41] 0 (0); 0 (0); 1 (11)	319
863. CHEECH & CHONG [48] 0 (0); 0 (0); 1 (11)	317
864. LITTLE RICHARD [42] 0 (0); 0 (0); 1 (10)	315
865. RED RIDER [48] 0 (0); 0 (0); 2 (13)	313
866. Leon HAYWOOD [49] 0 (0); 0 (0); 1 (11)	313
867. Afrika BAMBAATAA and the SOUL SONIC FORCE [48] 0 (0); 0 (0); 1 (11)	312
868. 707 [52] 0 (0); 0 (0); 2 (15)	308
869. Joni MITCHELL [47] 0 (0); 0 (0); 2 (12)	306
870. SARAYA [63] 0 (0); 0 (0); 2 (18)	306
871. Lee GREENWOOD [53] 0 (0); 0 (0); 2 (13)	304
872. Jimmy BARNES [47] 0 (0); 0 (0); 3 (23)	303
873. DUKE JUPITER [58] 0 (0); 0 (0); 2 (14)	301
874. Kenny NOLAN [44] 0 (0); 0 (0); 1 (8)	301
875. Billy RANKIN [52] 0 (0); 0 (0); 1 (11)	300
876. George BURNS [49] 0 (0); 0 (0); 1 (10)	297
877. The REDDINGS [55] 0 (0); 0 (0); 2 (22)	295
878. LOOSE ENDS [43] 0 (0); 0 (0); 1 (10)	294
879. Henry PAUL Band [50] 0 (0); 0 (0); 1 (10)	292
880. STAGE DOLLS [46] 0 (0); 0 (0); 1 (13)	291
881. Johnny MATHIS [38] 0 (0); 1 (3); 2 (21)	291
882. LIL LOUIS [50] 0 (0); 0 (0); 1 (13)	291
883. EVERLY BROTHERS [50] 0 (0); 0 (0); 1 (12)	291
884. MAGAZINE 60 [56] 0 (0); 0 (0); 1 (11)	290
885. YAZ [67] 0 (0); 0 (0); 2 (16)	289
886. The FOOLS [50] 0 (0); 0 (0); 2 (11)	287
887. KORONA [43] 0 (0); 0 (0); 1 (8)	287
888. Jean KNIGHT [50] 0 (0); 0 (0); 1 (15)	286
889. FLYING LIZARDS [50] 0 (0); 0 (0); 1 (10)	284
890. The INMATES [51] 0 (0); 0 (0); 1 (10)	282
891. PLEASURE [55] 0 (0); 0 (0); 1 (10)	282
892. WHISTLE [60] 0 (0); 0 (0); 1 (13)	280
893. SHEILA [49] 0 (0); 0 (0); 1 (9)	279
894. ALISHA [68] 0 (0); 0 (0); 2 (18)	278
895. Nolan THOMAS [57] 0 (0); 0 (0); 1 (13)	277
896. Merry CLAYTON [45] 0 (0); 0 (0); 1 (11)	277
897. KLIQUE [50] 0 (0); 0 (0); 1 (9)	277
898. DEJA [54] 0 (0); 0 (0); 1 (12)	275
899. SPLIT ENZ [53] 0 (0); 0 (0); 1 (11)	273
900. David HUDSON [59] 0 (0); 0 (0); 1 (11)	272
901. SHOOTING STAR [67] 0 (0); 0 (0); 3 (18)	270
902. John EDDIE [52] 0 (0); 0 (0); 1 (10)	268
903. The PASADENAS [52] 0 (0); 0 (0); 1 (10)	266
904. KING [55] 0 (0); 0 (0); 1 (11)	266
905. Randy VANWARMER [55] 0 (0); 0 (0); 2 (11)	266
906. RED ROCKERS [53] 0 (0); 0 (0); 1 (10)	264
907. CHIC [61] 0 (0); 0 (0); 4 (15)	264
908. YELLO [51] 0 (0); 0 (0); 1 (11)	264
909. Glen CAMPBELL [42] 0 (0); 0 (0); 3 (18)	261
910. GET WET [39] 0 (0); 1 (2); 1 (9)	260
911. TRANS-X [61] 0 (0); 0 (0); 1 (12)	256
912. Phyllis NELSON [61] 0 (0); 0 (0); 1 (11)	256
913. VANITY [56] 0 (0); 0 (0); 2 (14)	255
914. CHICAGO BEARS SHUFFLIN' CREW [41] 0 (0); 0 (0); 1 (9)	254
915. DILLMAN Band [45] 0 (0); 0 (0); 1 (9)	253
916. Frank BARBER Orchestra [61] 0 (0); 0 (0); 1 (12)	250
917. Rex SMITH [32] 0 (0); 1 (4); 1 (13)	250
918. Chrissie HYNDE [28] 0 (0); 1 (4); 1 (14)	250
919. ROSSINGTON COLLINS BAND [55] 0 (0); 0 (0); 1 (9)	249
920. YELLOW MAGIC ORCHESTRA [60] 0 (0); 0 (0); 1 (9)	249
921. The CALL [51] 0 (0); 0 (0); 2 (14)	249
922. Billy CRYSTAL [58] 0 (0); 0 (0); 1 (12)	249
923. Pat METHENY Group [32] 0 (0); 1 (4); 1 (12)	248
924. Cheryl LYNN [69] 0 (0); 0 (0); 2 (15)	248
925. Robert HAZARD [58] 0 (0); 0 (0); 1 (9)	248
926. Gordon LIGHTFOOT [50] 0 (0); 0 (0); 1 (8)	248
927. TONY! TONI! TONÉ! [47] 0 (0); 0 (0); 1 (10)	247
928. Walter MURPHY [47] 0 (0); 0 (0); 1 (10)	245
929. The REPLACEMENTS [51] 0 (0); 0 (0); 1 (11)	239
930. Bill CONTI [52] 0 (0); 0 (0); 1 (9)	237
931. The TIMELORDS [66] 0 (0); 0 (0); 1 (13)	237
932. Eria FACHIN [50] 0 (0); 0 (0); 1 (10)	236
933. Gregg ALLMAN Band [49] 0 (0); 0 (0); 1 (10)	235
934. UNDERWORLD [67] 0 (0); 0 (0); 2 (16)	233
935. WENDY And LISA [56] 0 (0); 0 (0); 1 (10)	232
936. INDIGO GIRLS [52] 0 (0); 0 (0); 1 (9)	232
937. BOW WOW WOW [62] 0 (0); 0 (0); 2 (11)	231
938. KOOL MOE DEE [62] 0 (0); 0 (0); 2 (16)	229
939. Y&T [55] 0 (0); 0 (0); 1 (10)	229
940. Lenore O'MALLEY [53] 0 (0); 0 (0); 1 (11)	227
941. The MONROES [59] 0 (0); 0 (0); 1 (8)	227
942. KROKUS [67] 0 (0); 0 (0); 2 (13)	226
943. GAMMA [60] 0 (0); 0 (0); 2 (11)	226
944. AVERAGE WHITE BAND/ AWB [53] 0 (0); 0 (0); 1 (8)	225
945. MENUDO [62] 0 (0); 0 (0); 1 (11)	224
946. Nona HENDRYX [58] 0 (0); 0 (0); 2 (12)	222
947. HUMBLE PIE [52] 0 (0); 0 (0); 1 (7)	222
948. La Toya JACKSON [56] 0 (0); 0 (0); 1 (8)	221
949. Jon ASTLEY [74] 0 (0); 0 (0); 2 (18)	221
950. O'BRYAN [57] 0 (0); 0 (0); 1 (9)	221
951. STREEK [47] 0 (0); 0 (0); 1 (7)	217
952. WOLF [55] 0 (0); 0 (0); 1 (9)	217
953. Glenn JONES [66] 0 (0); 0 (0); 1 (14)	216
954. WALL OF VOODOO [58] 0 (0); 0 (0); 1 (9)	215
955. VESTA [55] 0 (0); 0 (0); 1 (8)	214
956. BULLETBOYS [71] 0 (0); 0 (0); 2 (16)	214
957. The Nelson RIDDLE Orchestra [53] 0 (0); 0 (0); 1 (14)	214
958. GIANT [56] 0 (0); 0 (0); 1 (10)	211
959. ONE WAY [61] 0 (0); 0 (0); 1 (10)	210
960. Pat TRAVERS [50] 0 (0); 0 (0); 1 (7)	209
961. BIG PIG [60] 0 (0); 0 (0); 1 (10)	208
962. Russ BALLARD [58] 0 (0); 0 (0); 1 (8)	208
963. CAMOUFLAGE [59] 0 (0); 0 (0); 1 (12)	208
964. Taja SEVELLE [62] 0 (0); 0 (0); 1 (10)	207
965. Joyce KENNEDY [40] 0 (0); 1 (2); 1 (12)	206
966. DOOLITTLE Band [49] 0 (0); 0 (0); 1 (7)	206

Singles Acts Ranked

Rank. Act [Hi Peak] T10s (Wk); T40s (Wk); Ents (Wk)	Score
967. Steve CARLISLE [65] 0 (0); 0 (0); 1 (10)	206
968. F.R. DAVID [62] 0 (0); 0 (0); 1 (9)	205
969. Gerry RAFFERTY [54] 0 (0); 0 (0); 1 (8)	204
970. Bob DYLAN [55] 0 (0); 0 (0); 1 (9)	204
971. Peter ALLEN [55] 0 (0); 0 (0); 1 (8)	203
972. Turley RICHARDS [54] 0 (0); 0 (0); 1 (7)	203
973. COYOTE SISTERS [66] 0 (0); 0 (0); 1 (10)	201
974. Yoko ONO [58] 0 (0); 0 (0); 1 (10)	200
975. TOUCH [65] 0 (0); 0 (0); 2 (11)	199
976. DOCTOR And The MEDICS [69] 0 (0); 0 (0); 1 (11)	195
977. OFF BROADWAY (USA) [51] 0 (0); 0 (0); 1 (7)	194
978. Eugene WILDE [76] 0 (0); 0 (0); 2 (18)	194
979. PLANET P [64] 0 (0); 0 (0); 1 (9)	194
980. ROCK And HYDE [61] 0 (0); 0 (0); 1 (10)	193
981. ROUGH TRADE [58] 0 (0); 0 (0); 1 (7)	193
982. Ralph MacDONALD [58] 0 (0); 0 (0); 1 (10)	192
983. BLUE ZONE U.K. [54] 0 (0); 0 (0); 1 (9)	192
984. Lou RAWLS [65] 0 (0); 0 (0); 2 (13)	191
985. Barbara MITCHELL [48] 0 (0); 0 (0); 1 (12)	189
986. The GRACES [56] 0 (0); 0 (0); 1 (9)	189
987. GLASS MOON [50] 0 (0); 0 (0); 1 (7)	189
988. Sam BROWN [65] 0 (0); 0 (0); 1 (10)	189
989. ORION THE HUNTER [58] 0 (0); 0 (0); 1 (8)	188
990. ZEBRA [61] 0 (0); 0 (0); 1 (8)	188
991. E.G. DAILY [70] 0 (0); 0 (0); 1 (10)	188
992. Jenny BURTON [54] 0 (0); 0 (0); 2 (13)	187
993. BILLY SATELLITE [64] 0 (0); 0 (0); 2 (11)	184
994. Chris REA [73] 0 (0); 0 (0); 3 (15)	183
995. GUY [62] 0 (0); 0 (0); 2 (13)	183
996. Ronnie LAWS [60] 0 (0); 0 (0); 1 (9)	183
997. Warren ZEVON [57] 0 (0); 0 (0); 1 (7)	182
998. Lenny Le BLANC [55] 0 (0); 0 (0); 1 (7)	182
999. Jerry REED [57] 0 (0); 0 (0); 1 (9)	180
1000. Rickie Lee JONES [64] 0 (0); 0 (0); 2 (11)	179
1001. PEPSI and SHIRLIE [66] 0 (0); 0 (0); 2 (14)	179
1002. Art GARFUNKEL [66] 0 (0); 0 (0); 1 (9)	178
1003. Peter McIAN [52] 0 (0); 0 (0); 1 (7)	177
1004. LAKESIDE [55] 0 (0); 0 (0); 1 (8)	176
1005. SPYRO GYRA [68] 0 (0); 0 (0); 2 (10)	175
1006. Kevin RALEIGH [60] 0 (0); 0 (0); 1 (9)	175
1007. BAD COMPANY [74] 0 (0); 0 (0); 3 (17)	175
1008. Johnny AVERAGE Band [53] 0 (0); 0 (0); 1 (7)	174
1009. Holly KNIGHT [59] 0 (0); 0 (0); 1 (9)	173
1010. Peter BROWN [59] 0 (0); 0 (0); 1 (8)	173
1011. B. E. TAYLOR Group [66] 0 (0); 0 (0); 2 (10)	172
1012. Mike PINERA [70] 0 (0); 0 (0); 1 (8)	172
1013. Amii STEWART & Johnny BRISTOL [63] 0 (0); 0 (0); 1 (8)	172
1014. FROZEN GHOST [69] 0 (0); 0 (0); 1 (10)	171
1015. WET WET WET [58] 0 (0); 0 (0); 1 (8)	170
1016. DAYTON [58] 0 (0); 0 (0); 1 (7)	170
1017. Stephen STILLS [61] 0 (0); 0 (0); 1 (8)	169
1018. Joe CHEMAY Band [68] 0 (0); 0 (0); 1 (8)	168
1019. Jules SHEAR [57] 0 (0); 0 (0); 1 (7)	167
1020. The TRUTH [65] 0 (0); 0 (0); 1 (9)	167
1021. DEEP PURPLE [61] 0 (0); 0 (0); 1 (7)	164
1022. Freddie MERCURY [69] 0 (0); 0 (0); 2 (10)	164
1023. BADFINGER [56] 0 (0); 0 (0); 1 (8)	163
1024. LITTLE STEVEN [63] 0 (0); 0 (0); 1 (8)	163
1025. George THOROGOOD & The DESTROYERS [63] 0 (0); 0 (0); 1 (8)	162
1026. T.S. MONK [63] 0 (0); 0 (0); 1 (8)	159
1027. EMERSON, LAKE & POWELL [60] 0 (0); 0 (0); 1 (8)	159
1028. Tony TERRY [80] 0 (0); 0 (0); 2 (14)	157
1029. BOYS BAND [61] 0 (0); 0 (0); 1 (8)	157
1030. Tom KIMMEL [64] 0 (0); 0 (0); 1 (8)	156
1031. Grace JONES [69] 0 (0); 0 (0); 1 (8)	154
1032. HEADPINS [70] 0 (0); 0 (0); 1 (9)	153
1033. HAWKS [63] 0 (0); 0 (0); 1 (7)	153
1034. Adrian BELEW [58] 0 (0); 0 (0); 1 (8)	152
1035. M + M [63] 0 (0); 0 (0); 1 (7)	151
1036. AFTER 7 [74] 0 (0); 0 (0); 1 (12)	151
1037. WHAT IS THIS [62] 0 (0); 0 (0); 1 (6)	150
1038. Herbie HANCOCK [71] 0 (0); 0 (0); 1 (9)	149
1039. Kurtis BLOW [71] 0 (0); 0 (0); 2 (12)	149
1040. David GILMOUR [62] 0 (0); 0 (0); 1 (7)	148
1041. POWERSOURCE [61] 0 (0); 0 (0); 1 (7)	148
1042. Josie COTTON [74] 0 (0); 0 (0); 2 (11)	147
1043. INNER CITY [73] 0 (0); 0 (0); 1 (11)	147
1044. Jeff BECK [48] 0 (0); 0 (0); 1 (10)	147
1045. EUROGLIDERS [65] 0 (0); 0 (0); 1 (6)	146
1046. Tyrone DAVIS [57] 0 (0); 0 (0); 1 (6)	146
1047. Glen BURTNICK [65] 0 (0); 0 (0); 1 (8)	146
1048. HOT CHOCOLATE [65] 0 (0); 0 (0); 1 (7)	145
1049. CANDI [68] 0 (0); 0 (0); 1 (7)	145
1050. GRANDMASTER FLASH And The FURIOUS FIVE [62] 0 (0); 0 (0); 1 (7)	144
1051. TKA [75] 0 (0); 0 (0); 2 (16)	143
1052. Ozzy OSBOURNE [68] 0 (0); 0 (0); 1 (9)	141
1053. UNIPOP [71] 0 (0); 0 (0); 1 (8)	140
1054. HOTEL [72] 0 (0); 0 (0); 1 (7)	140
1055. SHERBS [61] 0 (0); 0 (0); 1 (7)	140
1056. TOTAL COELO [66] 0 (0); 0 (0); 1 (6)	138
1057. Michelle SHOCKED [66] 0 (0); 0 (0); 1 (8)	137
1058. The DOORS [71] 0 (0); 0 (0); 1 (7)	137
1059. VOICES OF AMERICA [65] 0 (0); 0 (0); 1 (8)	134
1060. Shakin' STEVENS [67] 0 (0); 0 (0); 1 (6)	134
1061. Frankie MILLER [62] 0 (0); 0 (0); 1 (6)	134
1062. FIONA [64] 0 (0); 0 (0); 1 (7)	132
1063. The FAMILY [63] 0 (0); 0 (0); 1 (6)	132
1064. The PRODUCERS [61] 0 (0); 0 (0); 1 (6)	131
1065. SWITCH [83] 0 (0); 0 (0); 1 (15)	130
1066. AXE [64] 0 (0); 0 (0); 2 (8)	129
1067. Belouis SOME [67] 0 (0); 0 (0); 2 (11)	129
1068. ENUFF Z'NUFF [67] 0 (0); 0 (0); 1 (7)	128
1069. The WAITRESSES [62] 0 (0); 0 (0); 1 (6)	128
1070. Marc ALMOND [67] 0 (0); 0 (0); 1 (8)	127
1071. EIGHT SECONDS [72] 0 (0); 0 (0); 1 (8)	127
1072. GOANNA [71] 0 (0); 0 (0); 1 (7)	127
1073. Arlan DAY [71] 0 (0); 0 (0); 1 (5)	126
1074. HOUSE OF LORDS [58] 0 (0); 0 (0); 1 (5)	126
1075. TORONTO [77] 0 (0); 0 (0); 1 (8)	123
1076. JUDAS PRIEST [67] 0 (0); 0 (0); 1 (7)	123
1077. OZARK MOUNTAIN DAREDEVILS [67] 0 (0); 0 (0); 1 (5)	122
1078. Garland JEFFREYS [66] 0 (0); 0 (0); 1 (7)	121
1079. Steve ARRINGTON [68] 0 (0); 0 (0); 1 (6)	120
1080. SLAVE [78] 0 (0); 0 (0); 2 (13)	120
1081. GINA GO-GO [78] 0 (0); 0 (0); 1 (11)	120
1082. Jocelyn BROWN [75] 0 (0); 0 (0); 1 (10)	120
1083. Bonnie RAITT [73] 0 (0); 0 (0); 1 (6)	120
1084. Petula CLARK [66] 0 (0); 0 (0); 1 (6)	119
1085. ROBBIN THOMPSON Band [66] 0 (0); 0 (0); 1 (9)	119
1086. Victoria PRINCIPAL [51] 0 (0); 0 (0); 1 (8)	119
1087. Peter FRAMPTON [74] 0 (0); 0 (0); 1 (8)	119
1088. Jimmy DAVIS & JUNCTION [67] 0 (0); 0 (0); 1 (6)	118
1089. MARILLION [74] 0 (0); 0 (0); 1 (8)	117
1090. SIR MIX-A-LOT [70] 0 (0); 0 (0); 1 (9)	117
1091. Stephen STILLS With Michael FINNIGAN [67] 0 (0); 0 (0); 1 (6)	116
1092. AURRA [71] 0 (0); 0 (0); 1 (7)	116
1093. STARLAND VOCAL BAND [71] 0 (0); 0 (0); 1 (6)	115
1094. WAR [66] 0 (0); 0 (0); 2 (9)	115
1095. Holly JOHNSON [65] 0 (0); 0 (0); 1 (8)	115
1096. VITAMIN Z [73] 0 (0); 0 (0); 1 (7)	114
1097. STAR WARS INTERGALACTIC DROID CHOIR & CHORALE [69] 0 (0); 0 (0); 1 (6)	114
1098. ROXANNE [63] 0 (0); 0 (0); 1 (7)	113
1099. Gary BURBANK with BAND McNALLY [67] 0 (0); 0 (0); 1 (5)	113
1100. CROSBY, STILLS, NASH & YOUNG [69] 0 (0); 0 (0); 1 (8)	112
1101. The ROCKETS [70] 0 (0); 0 (0); 1 (6)	112

Rank. Act [Hi Peak] T10s (Wk); T40s (Wk); Ents (Wk)	Score
1102. Jean BEAUVOIR [73] 0 (0); 0 (0); 1 (8)	112
1103. CLOCKS [67] 0 (0); 0 (0); 1 (5)	112
1104. Narada Michael WALDEN [66] 0 (0); 0 (0); 1 (6)	111
1105. DIVINYLS [76] 0 (0); 0 (0); 1 (7)	111
1106. Christopher MAX [75] 0 (0); 0 (0); 1 (8)	111
1107. FESTIVAL [72] 0 (0); 0 (0); 1 (8)	110
1108. Joe SUN [71] 0 (0); 0 (0); 1 (6)	110
1109. DOLLAR [74] 0 (0); 0 (0); 1 (6)	110
1110. LOBO [75] 0 (0); 0 (0); 1 (8)	110
1111. UPTOWN [80] 0 (0); 0 (0); 1 (11)	110
1112. Dave STEWART With Barbara GASKIN [72] 0 (0); 0 (0); 1 (7)	109
1113. MAI TAI [71] 0 (0); 0 (0); 1 (7)	109
1114. Carol Lynn TOWNES [77] 0 (0); 0 (0); 1 (9)	108
1115. BLUE MERCEDES [66] 0 (0); 0 (0); 1 (6)	108
1116. Todd RUNDGREN [63] 0 (0); 0 (0); 1 (5)	108
1117. MILLIONS LIKE US [69] 0 (0); 0 (0); 1 (6)	108
1118. Christopher ATKINS [71] 0 (0); 0 (0); 1 (7)	107
1119. CONDUCTOR [63] 0 (0); 0 (0); 1 (5)	107
1120. FIGURES ON A BEACH [67] 0 (0); 0 (0); 1 (7)	107
1121. Glenn SHORROCK [69] 0 (0); 0 (0); 1 (6)	106
1122. MONDO ROCK [71] 0 (0); 0 (0); 1 (6)	106
1123. ICE-T [70] 0 (0); 0 (0); 1 (7)	105
1124. MAZE Featuring Frankie BEVERLY [80] 0 (0); 0 (0); 2 (11)	105
1125. Neil YOUNG [71] 0 (0); 0 (0); 1 (6)	105
1126. Toni CHILDS [72] 0 (0); 0 (0); 1 (5)	105
1127. Marcy LEVY [50] 0 (0); 0 (0); 1 (10)	104
1128. SAVOY BROWN [68] 0 (0); 0 (0); 1 (5)	104
1129. ULTRAVOX [71] 0 (0); 0 (0); 1 (5)	104
1130. Janey STREET [68] 0 (0); 0 (0); 1 (5)	104
1131. BAND OF GOLD [64] 0 (0); 0 (0); 1 (7)	103
1132. FIVE SATINS [71] 0 (0); 0 (0); 1 (5)	103
1133. CELLARFUL OF NOISE [69] 0 (0); 0 (0); 1 (7)	103
1134. Engelbert HUMPERDINCK [77] 0 (0); 0 (0); 2 (7)	102
1135. The KINGBEES [81] 0 (0); 0 (0); 1 (8)	102

Rank. Act [Hi Peak] T10s (Wk); T40s (Wk); Ents (Wk)	Score
1136. TANGIER [67] 0 (0); 0 (0); 1 (7)	101
1137. LIMITED WARRANTY [79] 0 (0); 0 (0); 1 (8)	101
1138. Patrick JUDE [54] 0 (0); 0 (0); 1 (7)	99.8
1139. Billy BURNETTE [68] 0 (0); 0 (0); 1 (5)	99.7
1140. The ASSOCIATION [66] 0 (0); 0 (0); 1 (5)	99.0
1141. WILD BLUE [71] 0 (0); 0 (0); 1 (6)	98.8
1142. ROMEO'S DAUGHTER [73] 0 (0); 0 (0); 1 (7)	98.8
1143. Dennis EDWARDS [72] 0 (0); 0 (0); 1 (6)	98.6
1144. Carl WILSON [72] 0 (0); 0 (0); 1 (6)	97.8
1145. DELIVERANCE [71] 0 (0); 0 (0); 1 (5)	97.7
1146. TWILIGHT 22 [79] 0 (0); 0 (0); 1 (8)	96.5
1147. Gene COTTON [76] 0 (0); 0 (0); 1 (8)	96.0
1148. NEWCITY ROCKERS [80] 0 (0); 0 (0); 2 (10)	95.3
1149. Q-FEEL [75] 0 (0); 0 (0); 1 (7)	95.2
1150. Nick LOWE And His COWBOY OUTFIT [77] 0 (0); 0 (0); 1 (9)	94.5
1151. BIG TROUBLE [71] 0 (0); 0 (0); 1 (7)	93.9
1152. MODERN ENGLISH [78] 0 (0); 0 (0); 2 (10)	93.0
1153. Feargal SHARKEY [74] 0 (0); 0 (0); 1 (6)	91.3
1154. The VELS [72] 0 (0); 0 (0); 1 (6)	91.2
1155. XTC [72] 0 (0); 0 (0); 1 (6)	90.9
1156. Rodney DANGERFIELD [83] 0 (0); 0 (0); 1 (8)	90.9
1157. AMERICAN COMEDY NETWORK [70] 0 (0); 0 (0); 1 (5)	90.6
1158. BOOMTOWN RATS [73] 0 (0); 0 (0); 1 (5)	89.7
1159. Randy CRAWFORD [59] 0 (0); 0 (0); 1 (10)	89.4
1160. YIPES!! [68] 0 (0); 0 (0); 1 (5)	89.0
1161. Kim MITCHELL [86] 0 (0); 0 (0); 1 (9)	88.8
1162. DEODATO [70] 0 (0); 0 (0); 1 (5)	88.8
1163. KINGDOM COME [69] 0 (0); 0 (0); 1 (6)	88.5
1164. Neil YOUNG & CRAZY HORSE [70] 0 (0); 0 (0); 1 (5)	88.0
1165. Joan ARMATRADING [78] 0 (0); 0 (0); 1 (6)	87.7
1166. Debbie JACOBS [70] 0 (0); 0 (0); 1 (4)	86.7
1167. WORLD CLASS WRECKIN' CRU [84] 0 (0); 0 (0); 1 (11)	86.5
1168. The MONKEES [79] 0 (0); 0 (0); 2 (8)	86.1
1169. LOVE AND MONEY [75] 0 (0); 0 (0); 1 (7)	85.7

Rank. Act [Hi Peak] T10s (Wk); T40s (Wk); Ents (Wk)	Score
1170. Eric HINE [73] 0 (0); 0 (0); 1 (5)	84.3
1171. TOBY BEAU [70] 0 (0); 0 (0); 1 (4)	84.0
1172. Allan CLARKE [70] 0 (0); 0 (0); 1 (4)	83.7
1173. HAYSI FANTAYZEE [74] 0 (0); 0 (0); 1 (5)	83.5
1174. SHOT IN THE DARK [71] 0 (0); 0 (0); 1 (5)	82.7
1175. Leon REDBONE [72] 0 (0); 0 (0); 1 (6)	82.5
1176. HEAVEN 17 [74] 0 (0); 0 (0); 1 (6)	81.9
1177. INDUSTRY [81] 0 (0); 0 (0); 1 (8)	81.3
1178. Rick DEES [75] 0 (0); 0 (0); 1 (5)	79.3
1179. HUGHES/THRALL [79] 0 (0); 0 (0); 1 (5)	79.1
1180. Paul SHAFFER [81] 0 (0); 0 (0); 1 (8)	78.7
1181. John CAFFERTY [76] 0 (0); 0 (0); 1 (6)	78.4
1182. GARY O' [70] 0 (0); 0 (0); 1 (5)	78.3
1183. Mary MacGREGOR [72] 0 (0); 0 (0); 1 (4)	78.3
1184. CLUB HOUSE [75] 0 (0); 0 (0); 1 (5)	78.2
1185. Max WERNER [74] 0 (0); 0 (0); 1 (6)	78.0
1186. J.D. DREWS [79] 0 (0); 0 (0); 1 (6)	75.6
1187. ALESSI [71] 0 (0); 0 (0); 1 (4)	75.4
1188. MEL & KIM [78] 0 (0); 0 (0); 1 (7)	75.3
1189. Dave MASON [71] 0 (0); 0 (0); 1 (3)	74.8
1190. Placido DOMINGO [59] 0 (0); 0 (0); 1 (7)	74.8
1191. BUS BOYS [68] 0 (0); 0 (0); 1 (5)	74.0
1192. The MOTORS [78] 0 (0); 0 (0); 1 (5)	73.8
1193. "D" TRAIN [79] 0 (0); 0 (0); 1 (5)	73.4
1194. BEAU COUP [80] 0 (0); 0 (0); 1 (6)	72.4
1195. CHOIRBOYS [80] 0 (0); 0 (0); 1 (7)	71.6
1196. SPYS [82] 0 (0); 0 (0); 1 (5)	71.2
1197. Teddy RILEY [62] 0 (0); 0 (0); 1 (6)	71.2
1198. The RINGS [75] 0 (0); 0 (0); 1 (5)	69.9
1199. ENGLAND DAN & John Ford COLEY [75] 0 (0); 0 (0); 1 (4)	69.5
1200. Martha DAVIS [80] 0 (0); 0 (0); 1 (4)	69.3
1201. 4 BY FOUR [79] 0 (0); 0 (0); 1 (6)	68.5
1202. FORTUNE [80] 0 (0); 0 (0); 1 (6)	67.7
1203. Tony Joe WHITE [79] 0 (0); 0 (0); 1 (5)	67.7
1204. Maureen STEELE [77] 0 (0); 0 (0); 1 (5)	67.4

Rank. Act [Hi Peak] T10s (Wk); T40s (Wk); Ents (Wk)	Score
1205. Janis IAN [71] 0 (0); 0 (0); 1 (4)	67.4
1206. Madleen KANE [77] 0 (0); 0 (0); 1 (5)	66.5
1207. Alex BROWN [76] 0 (0); 0 (0); 1 (6)	65.9
1208. Clint EASTWOOD [62] 0 (0); 0 (0); 1 (6)	65.8
1209. Loz NETTO [82] 0 (0); 0 (0); 1 (6)	65.4
1210. The CRETONES [79] 0 (0); 0 (0); 1 (6)	65.3
1211. B.B. KING [68] 0 (0); 0 (0); 1 (7)	65.2
1212. APOLLONIA 6 [85] 0 (0); 0 (0); 1 (6)	65.1
1213. Cee FARROW [82] 0 (0); 0 (0); 1 (6)	64.5
1214. David HALLYDAY [79] 0 (0); 0 (0); 1 (6)	63.6
1215. Johnny GILL [75] 0 (0); 0 (0); 1 (9)	63.5
1216. Bob GELDOF [82] 0 (0); 0 (0); 1 (6)	63.4
1217. TARNEY/ SPENCER Band [74] 0 (0); 0 (0); 1 (4)	63.4
1218. KC [66] 0 (0); 0 (0); 1 (6)	62.2
1219. TEXAS [77] 0 (0); 0 (0); 1 (6)	61.9
1220. SINGLE BULLET THEORY [78] 0 (0); 0 (0); 1 (4)	61.8
1221. Jim KIRK And The TM Singers [71] 0 (0); 0 (0); 1 (3)	61.5
1222. Bobby VINTON [78] 0 (0); 0 (0); 1 (4)	61.5
1223. CHUNKY A [77] 0 (0); 0 (0); 1 (6)	61.4
1224. WEST STREET MOB [88] 0 (0); 0 (0); 2 (7)	61.3
1225. McGUFFEY LANE [85] 0 (0); 0 (0); 2 (10)	61.0
1226. ALL SPORTS BAND [78] 0 (0); 0 (0); 2 (5)	60.6
1227. Michael GORE [84] 0 (0); 0 (0); 1 (6)	60.5
1228. CENTRAL LINE [84] 0 (0); 0 (0); 1 (6)	60.4
1229. FIRE INC. [80] 0 (0); 0 (0); 1 (5)	60.2
1230. Graham NASH [84] 0 (0); 0 (0); 1 (7)	59.9
1231. HIGH INERGY [82] 0 (0); 0 (0); 1 (5)	59.9
1232. FAIRGROUND ATTRACTION [80] 0 (0); 0 (0); 1 (6)	59.7
1233. NOCERA [84] 0 (0); 0 (0); 1 (6)	59.7
1234. SAD CAFÉ [78] 0 (0); 0 (0); 1 (4)	59.2
1235. KASHIF [62] 0 (0); 0 (0); 1 (7)	58.8
1236. Robert GORDON [76] 0 (0); 0 (0); 1 (4)	58.6
1237. PLATINUM BLONDE [82] 0 (0); 0 (0); 1 (5)	58.5
1238. DION [75] 0 (0); 0 (0); 1 (5)	57.6

Singles Acts Ranked

Rank. Act [Hi Peak] T10s (Wk); T40s (Wk); Ents (Wk)	Score
1239. DIVING FOR PEARLS [84] 0 (0); 0 (0); 1 (6)	57.3
1240. Gary PORTNOY [83] 0 (0); 0 (0); 1 (4)	56.2
1241. The SILENCERS (2) [81] 0 (0); 0 (0); 1 (5)	56.1
1242. RUBBER RODEO [86] 0 (0); 0 (0); 1 (5)	55.9
1243. URGENT [79] 0 (0); 0 (0); 1 (5)	55.6
1244. UTFO [77] 0 (0); 0 (0); 1 (5)	55.5
1245. Shirley LEWIS [84] 0 (0); 0 (0); 1 (6)	54.6
1246. Andy FRASER [82] 0 (0); 0 (0); 1 (5)	54.5
1247. Dana VALERY [87] 0 (0); 0 (0); 1 (5)	53.4
1248. Charles FOX [75] 0 (0); 0 (0); 1 (4)	53.4
1249. Rockie ROBBINS [80] 0 (0); 0 (0); 1 (4)	52.8
1250. Rick BOWLES [77] 0 (0); 0 (0); 1 (3)	52.6
1251. MARSHALL TUCKER Band [79] 0 (0); 0 (0); 1 (3)	52.4
1252. John LENNON and the PLASTIC ONO BAND [80] 0 (0); 0 (0); 1 (4)	52.1
1253. ROBEY [77] 0 (0); 0 (0); 1 (3)	51.0
1254. KIARA [78] 0 (0); 0 (0); 1 (5)	50.6
1255. Al CORLEY [80] 0 (0); 0 (0); 1 (5)	50.3
1256. PENDULUM [89] 0 (0); 0 (0); 1 (7)	49.8
1257. The TOURISTS [83] 0 (0); 0 (0); 1 (4)	49.5
1258. DYNASTY [87] 0 (0); 0 (0); 1 (6)	49.1
1259. Jackie DeSHANNON [86] 0 (0); 0 (0); 1 (5)	49.1
1260. TWENNYNINE [83] 0 (0); 0 (0); 1 (4)	48.7
1261. KISSING THE PINK [87] 0 (0); 0 (0); 2 (9)	48.0
1262. GENTLE PERSUASION [82] 0 (0); 0 (0); 1 (4)	47.6
1263. WILLIS "The GUARD" and VIGORISH [82] 0 (0); 0 (0); 1 (3)	47.4
1264. Lorenzo LAMAS [85] 0 (0); 0 (0); 1 (5)	46.8
1265. Luis CARDENAS [83] 0 (0); 0 (0); 1 (5)	46.5
1266. Freddy CANNON & The BELMONTS [81] 0 (0); 0 (0); 1 (4)	46.3
1267. The SILENCERS [82] 0 (0); 0 (0); 1 (6)	45.4
1268. Ray KENNEDY [82] 0 (0); 0 (0); 1 (3)	45.4
1269. Giorgio MORODER [81] 0 (0); 0 (0); 1 (4)	45.1
1270. ZAPP [86] 0 (0); 0 (0); 1 (7)	44.7
1271. CURTIE And The BOOMBOX [81] 0 (0); 0 (0); 1 (4)	44.5
1272. Joe FAGIN [80] 0 (0); 0 (0); 1 (3)	44.4
1273. IRONHORSE [89] 0 (0); 0 (0); 1 (6)	44.0
1274. CUGINI [88] 0 (0); 0 (0); 1 (4)	43.6
1275. Vaughan MASON And CREW [81] 0 (0); 0 (0); 1 (4)	43.6
1276. Peter TOSH [84] 0 (0); 0 (0); 1 (4)	43.2
1277. Shamus M'COOL [80] 0 (0); 0 (0); 1 (3)	43.1
1278. ROXY MUSIC [80] 0 (0); 0 (0); 1 (4)	42.0
1279. Paul HYDE And The PAYOLAS [84] 0 (0); 0 (0); 1 (4)	41.6
1280. EASTERHOUSE [82] 0 (0); 0 (0); 1 (4)	41.4
1281. BROS [83] 0 (0); 0 (0); 1 (5)	40.2
1282. Evelyn THOMAS [85] 0 (0); 0 (0); 1 (4)	40.1
1283. TORA TORA [86] 0 (0); 0 (0); 1 (6)	39.7
1284. RODWAY [83] 0 (0); 0 (0); 1 (5)	39.7
1285. GOOD QUESTION [86] 0 (0); 0 (0); 1 (5)	38.9
1286. STREETS [87] 0 (0); 0 (0); 1 (5)	38.7
1287. LIQUID GOLD [86] 0 (0); 0 (0); 1 (4)	37.4
1288. GIDEA PARK [82] 0 (0); 0 (0); 1 (3)	36.7
1289. KANO [89] 0 (0); 0 (0); 1 (5)	36.7
1290. Burt REYNOLDS [88] 0 (0); 0 (0); 1 (4)	36.7
1291. 4 SEASONS [91] 0 (0); 0 (0); 1 (5)	36.0
1292. Michael JOHNSON [86] 0 (0); 0 (0); 1 (3)	35.7
1293. Timmy THOMAS [80] 0 (0); 0 (0); 1 (3)	35.6
1294. Ted NUGENT [86] 0 (0); 0 (0); 1 (4)	35.6
1295. SINITTA [84] 0 (0); 0 (0); 1 (4)	35.4
1296. Joel DIAMOND [82] 0 (0); 0 (0); 1 (3)	35.4
1297. CALIFORNIA RAISINS [84] 0 (0); 0 (0); 1 (4)	35.0
1298. J. BLACKFOOT [90] 0 (0); 0 (0); 1 (4)	34.6
1299. SECRET TIES [91] 0 (0); 0 (0); 1 (4)	34.0
1300. TAMI SHOW [88] 0 (0); 0 (0); 1 (4)	33.5
1301. YUTAKA [81] 0 (0); 0 (0); 1 (3)	33.1
1302. DRAGON [88] 0 (0); 0 (0); 1 (4)	33.1
1303. Jeff LYNNE [85] 0 (0); 0 (0); 1 (3)	33.1
1304. Larry LEE [81] 0 (0); 0 (0); 1 (2)	32.6
1305. Chubby CHECKER [91] 0 (0); 0 (0); 1 (5)	32.3
1306. Debra LAWS [90] 0 (0); 0 (0); 1 (5)	32.1
1307. MINK DeVILLE [89] 0 (0); 0 (0); 1 (3)	31.6
1308. The PLIMSOULS [82] 0 (0); 0 (0); 1 (3)	31.2
1309. Julian COPE [84] 0 (0); 0 (0); 1 (3)	31.2
1310. Mitch RYDER [87] 0 (0); 0 (0); 1 (4)	31.0
1311. FAR CORPORATION [89] 0 (0); 0 (0); 1 (4)	30.7
1312. Frankie VALLI [90] 0 (0); 0 (0); 1 (4)	30.4
1313. The STOMPERS [88] 0 (0); 0 (0); 1 (4)	30.3
1314. The VENETIANS [88] 0 (0); 0 (0); 1 (5)	30.1
1315. NAZARETH [87] 0 (0); 0 (0); 1 (3)	29.9
1316. MEAT LOAF [84] 0 (0); 0 (0); 1 (3)	29.6
1317. Rob JUNGKLAS [86] 0 (0); 0 (0); 1 (3)	28.9
1318. LABAN [88] 0 (0); 0 (0); 1 (4)	28.7
1319. Melissa ETHERIDGE [94] 0 (0); 0 (0); 2 (7)	28.6
1320. Robin LANE & The CHARTBUSTERS [87] 0 (0); 0 (0); 1 (4)	28.5
1321. DEL FUEGOS [87] 0 (0); 0 (0); 1 (4)	28.2
1322. The JAGS [84] 0 (0); 0 (0); 1 (2)	28.0
1323. SKY [83] 0 (0); 0 (0); 1 (2)	27.5
1324. S-EXPRESS [91] 0 (0); 0 (0); 1 (3)	27.3
1325. Gerard McMAHON [85] 0 (0); 0 (0); 1 (3)	27.3
1326. Cindy BULLENS [90] 0 (0); 0 (0); 1 (3)	27.0
1327. Jim HURT [90] 0 (0); 0 (0); 1 (4)	26.9
1328. ONE TO ONE [92] 0 (0); 0 (0); 1 (4)	26.7
1329. Howard HEWETT [90] 0 (0); 0 (0); 1 (3)	26.3
1330. TIGGI CLAY [86] 0 (0); 0 (0); 1 (3)	25.2
1331. KBC BAND [89] 0 (0); 0 (0); 1 (4)	25.0
1332. ISLE OF MAN [90] 0 (0); 0 (0); 1 (4)	24.6
1333. Larry John McNALLY [86] 0 (0); 0 (0); 1 (4)	24.6
1334. Lani HALL [88] 0 (0); 0 (0); 1 (3)	24.5
1335. Suzanne FELLINI [87] 0 (0); 0 (0); 1 (2)	24.0
1336. BOOK OF LOVE [90] 0 (0); 0 (0); 1 (2)	23.8
1337. TRANSVISION VAMP [87] 0 (0); 0 (0); 1 (3)	23.7
1338. GRANDMASTER MELLE MEL And The FURIOUS FIVE [86] 0 (0); 0 (0); 1 (2)	23.5
1339. Helen REDDY [88] 0 (0); 0 (0); 1 (3)	23.3
1340. BREATHLESS [92] 0 (0); 0 (0); 1 (4)	23.3
1341. BIG RIC [91] 0 (0); 0 (0); 1 (3)	23.2
1342. Billy PRESTON [88] 0 (0); 0 (0); 1 (3)	22.9
1343. GENE LOVES JEZEBEL [87] 0 (0); 0 (0); 1 (3)	22.7
1344. WHODINI [87] 0 (0); 0 (0); 1 (3)	22.6
1345. NIGHT [87] 0 (0); 0 (0); 1 (3)	22.5
1346. Joe "Bean" ESPOSITO [86] 0 (0); 0 (0); 1 (2)	22.5
1347. Chris THOMPSON [85] 0 (0); 0 (0); 1 (4)	22.0
1348. EDDIE And The TIDE [85] 0 (0); 0 (0); 1 (2)	21.7
1349. Bobby WOMACK [88] 0 (0); 0 (0); 1 (5)	21.6
1350. TIGHT FIT [89] 0 (0); 0 (0); 1 (3)	21.4
1351. AMAZULU [90] 0 (0); 0 (0); 1 (4)	20.5
1352. Karen KAMON [88] 0 (0); 0 (0); 1 (2)	20.3
1353. Eric MARTIN [87] 0 (0); 0 (0); 1 (2)	19.8
1354. Nile RODGERS [88] 0 (0); 0 (0); 1 (2)	19.7
1355. Sylvie VARTAN [85] 0 (0); 0 (0); 1 (4)	19.4
1356. Gardner COLE [91] 0 (0); 0 (0); 1 (3)	19.2
1357. SNUFF [88] 0 (0); 0 (0); 1 (2)	19.1
1358. Midge URE [95] 0 (0); 0 (0); 1 (5)	18.9
1359. Herb ALPERT & The TIJUANA BRASS [90] 0 (0); 0 (0); 1 (3)	18.2
1360. DOUBLE IMAGE [92] 0 (0); 0 (0); 1 (3)	18.0
1361. SILVERADO [92] 0 (0); 0 (0); 1 (3)	17.9
1362. Jackie ENGLISH [94] 0 (0); 0 (0); 1 (4)	17.6
1363. RON And The D.C. CREW [93] 0 (0); 0 (0); 1 (4)	16.6
1364. TESLA [91] 0 (0); 0 (0); 1 (3)	16.6
1365. Eric TROYER [92] 0 (0); 0 (0); 1 (2)	16.0
1366. La FLAVOUR [91] 0 (0); 0 (0); 1 (2)	16.0
1367. The SMITHEREENS [92] 0 (0); 0 (0); 1 (4)	16.0
1368. Randy BELL [90] 0 (0); 0 (0); 1 (3)	15.9
1369. STABILIZERS [93] 0 (0); 0 (0); 1 (3)	14.4
1370. Wayne MASSEY [92] 0 (0); 0 (0); 1 (3)	14.3
1371. Robin GEORGE [92] 0 (0); 0 (0); 1 (2)	13.6
1372. MINOR DETAIL [92] 0 (0); 0 (0); 1 (2)	13.2
1373. GQ [93] 0 (0); 0 (0); 1 (2)	13.2

Rank. Act [Hi Peak] T10s (Wk); T40s (Wk); Ents (Wk)	Score
1374. Jon BUTCHER AXIS [94] 0 (0); 0 (0); 1 (3)	13.2
1375. RCR [94] 0 (0); 0 (0); 1 (2)	12.7
1376. ANA [94] 0 (0); 0 (0); 1 (3)	12.3
1377. David PACK [95] 0 (0); 0 (0); 1 (3)	11.2
1378. Graham PARKER [94] 0 (0); 0 (0); 1 (2)	10.6
1379. SIMON F [91] 0 (0); 0 (0); 1 (2)	10.5
1380. The ADVENTURES [95] 0 (0); 0 (0); 1 (3)	9.9
1381. B.J. THOMAS [93] 0 (0); 0 (0); 1 (2)	9.8
1382. The CRUSADERS [97] 0 (0); 0 (0); 1 (3)	9.4
1383. Nick JAMESON [95] 0 (0); 0 (0); 1 (2)	9.4
1384. Mica PARIS [97] 0 (0); 0 (0); 1 (4)	9.2
1385. Grace SLICK [95] 0 (0); 0 (0); 1 (2)	8.6
1386. Jay BLACK [98] 0 (0); 0 (0); 1 (4)	8.6
1387. BIG NOISE [97] 0 (0); 0 (0); 1 (3)	8.5
1388. SOUTHSIDE JOHNNY & The ASBURY JUKES [98] 0 (0); 0 (0); 1 (5)	7.8
1389. YAZZ And The PLASTIC POPULATION [96] 0 (0); 0 (0); 1 (4)	6.8
1390. 3 MAN ISLAND [94] 0 (0); 0 (0); 1 (2)	6.7
1391. TIA [97] 0 (0); 0 (0); 1 (2)	6.3
1392. KINGS OF THE SUN [98] 0 (0); 0 (0); 1 (2)	4.6
1393. Helen TERRY [96] 0 (0); 0 (0); 1 (3)	4.3
1394. Siedah GARRETT [97] 0 (0); 0 (0); 1 (1)	3.1
1395. Kenny AARONSON [94] 0 (0); 0 (0); 1 (2)	2.8
1396. Neal SCHON [94] 0 (0); 0 (0); 1 (2)	2.8
1397. Michael SHRIEVE [94] 0 (0); 0 (0); 1 (2)	2.8
1398. Colin James HAY [99] 0 (0); 0 (0); 1 (1)	1.6
1399. Britny FOX [100] 0 (0); 0 (0); 1 (2)	1.5
1400. 1927 [100] 0 (0); 0 (0); 1 (1)	0.8

The Singles Acts Special Lists

50 Highest Charting Acts Within Each Year

Yearly Top 50s Through the Decade

Acts: 100 or Total Chart Weeks

Acts: 6 or More Chart Appearances

Acts: 3 or More Top 10 Entries

Acts: 4 or More Top 40 Entries

Acts: More than 40 Consecutive Weeks on Chart

Acts: 3 Singles on Chart Simultaneously

Acts: 2 Singles in Top 40 Simultaneously

Acts: Highest Average Record Score

Acts: Number 1 Records

Top 50 Acts and Their Writers

Top 50 Acts and Their Producers

Singles Acts: Top 50 Each Year

	1980	1981	1982	1983	1984
1	QUEEN	Daryl HALL & John OATES	John MELLENCAMP	Michael JACKSON	Cyndi LAUPER
2	Kenny ROGERS	John LENNON	J. GEILS Band	Lionel RICHIE	Lionel RICHIE
3	Diana ROSS	Kim CARNES	Daryl HALL & John OATES	The POLICE	DURAN DURAN
4	AIR SUPPLY	BLONDIE	Paul McCARTNEY	CULTURE CLUB	CULTURE CLUB
5	Christopher CROSS	AIR SUPPLY	Olivia NEWTON-JOHN	MEN AT WORK	Tina TURNER
6	BLONDIE	Sheena EASTON	Joan JETT & The BLACKHEARTS	Daryl HALL & John OATES	PRINCE And The REVOLUTION
7	Michael JACKSON	REO SPEEDWAGON	SURVIVOR	Irene CARA	VAN HALEN
8	Billy JOEL	Eddie RABBITT	Steve MILLER Band	Billy JOEL	Daryl HALL & John OATES
9	CAPTAIN & TENNILLE	Juice NEWTON	Rick SPRINGFIELD	DURAN DURAN	Bruce SPRINGSTEEN
10	Olivia NEWTON-JOHN	Diana ROSS	CHICAGO	David BOWIE	POINTER SISTERS
11	Barbra STREISAND	Rick SPRINGFIELD	TOTO	Kenny ROGERS	Huey LEWIS and the NEWS
12	SPINNERS	FOREIGNER	JOURNEY	Bonnie TYLER	PRINCE
13	PINK FLOYD	Neil DIAMOND	The GO-GO'S	Bob SEGER & The SILVER BULLET BAND	MADONNA
14	Paul McCARTNEY & WINGS	Christopher CROSS	HUMAN LEAGUE	EURYTHMICS	Ray PARKER Jr.
15	LIPPS, INC.	STYX	AIR SUPPLY	STRAY CATS	The CARS
16	Donna SUMMER	Dolly PARTON	MEN AT WORK	Michael SEMBELLO	Kenny LOGGINS
17	Bob SEGER	Olivia NEWTON-JOHN	FLEETWOOD MAC	STYX	Phil COLLINS
18	KC & The SUNSHINE BAND	Smokey ROBINSON	Juice NEWTON	Eddy GRANT	Stevie WONDER
19	KOOL & The GANG	Stevie NICKS	FOREIGNER	PRINCE	John WAITE
20	Elton JOHN	Kenny ROGERS	VANGELIS	JOURNEY	Deniece WILLIAMS
21	DR. HOOK	KOOL & The GANG	Laura BRANIGAN	Patti AUSTIN	THOMPSON TWINS
22	AMBROSIA	Joey SCARBURY	Stevie WONDER	AIR SUPPLY	Billy OCEAN
23	Bette MIDLER	ROLLING STONES	QUARTERFLASH	Donna SUMMER	YES
24	Dan FOGELBERG	Grover WASHINGTON JR.	Lionel RICHIE	TOTO	WHAM!
25	Cliff RICHARD	Lionel RICHIE	Alan PARSONS Project	Paul McCARTNEY	Elton JOHN
26	Smokey ROBINSON	STARS ON	Neil DIAMOND	DEXYS MIDNIGHT RUNNERS	ROCKWELL
27	Kenny LOGGINS	The POLICE	Diana ROSS	Laura BRANIGAN	CHICAGO
28	ROLLING STONES	POINTER SISTERS	ASIA	Greg KIHN Band	Steve PERRY
29	Andy GIBB	JOURNEY	Paul DAVIS	MEN WITHOUT HATS	KOOL & The GANG
30	Robbie DUPREE	The COMMODORES	Toni BASIL	Pat BENATAR	Michael JACKSON
31	Irene CARA	Ronnie MILSAP	Ray PARKER Jr.	Sergio MENDES	EURYTHMICS
32	Leo SAYER	Rod STEWART	TOMMY TUTONE	The FIXX	Laura BRANIGAN
33	Teri DeSARIO	A TASTE OF HONEY	Willie NELSON	NAKED EYES	The JACKSONS
34	POINTER SISTERS	Ray PARKER Jr. & RAYDIO	The CARS	Stevie NICKS	Billy IDOL
35	Linda RONSTADT	Don McLEAN	Stevie NICKS	SPANDAU BALLET	Chaka KHAN
36	EAGLES	George HARRISON	Dan FOGELBERG	Rick SPRINGFIELD	Rick SPRINGFIELD
37	Billy PRESTON & SYREETA	Barbra STREISAND & Barry GIBB	CHARLENE	Thomas DOLBY	Paul McCARTNEY
38	Boz SCAGGS	Dan FOGELBERG	Michael McDONALD	Sheena EASTON	NENA
39	S.O.S. BAND	OAK RIDGE BOYS	LITTLE RIVER BAND	Marvin GAYE	The ROMANTICS
40	Pat BENATAR	Pat BENATAR	Melissa MANCHESTER	Don HENLEY	Billy JOEL
41	Neil DIAMOND	Gino VANNELLI	DAZZ BAND	Bryan ADAMS	NIGHT RANGER
42	George BENSON	Marty BALIN	CROSBY, STILLS & NASH	TACO	Rod STEWART
43	ELECTRIC LIGHT ORCHESTRA	Alan PARSONS Project	KOOL & The GANG	The PRETENDERS	Corey HART
44	Eddie RABBITT	EARTH, WIND & FIRE	Kenny ROGERS	Elton JOHN	Dan HARTMAN
45	Stevie WONDER	LITTLE RIVER BAND	REO SPEEDWAGON	DEF LEPPARD	Sheena EASTON
46	The MANHATTANS	ABBA	POINTER SISTERS	AFTER THE FIRE	Matthew WILDER
47	Daryl HALL & John OATES	Bob SEGER	Jennifer WARNES	DeBARGE	SHEILA E.
48	John LENNON	MANHATTAN TRANSFER	AMERICA	HUMAN LEAGUE	SCANDAL/Patty SMYTH
49	Paul SIMON	MOODY BLUES	The MOTELS	QUIET RIOT	GENESIS
50	RAY, GOODMAN & BROWN	Bruce SPRINGSTEEN	Bertie HIGGINS	MADNESS	Irene CARA

Singles Acts: Top 50 Each Year

	1985	1986	1987	1988	1989
1	MADONNA	MADONNA	MADONNA	George MICHAEL	MILLI VANILLI
2	Phil COLLINS	Janet JACKSON	Whitney HOUSTON	Michael JACKSON	Paula ABDUL
3	TEARS FOR FEARS	Billy OCEAN	U2	INXS	NEW KIDS ON THE BLOCK
4	WHAM!	Lionel RICHIE	George MICHAEL	DEF LEPPARD	Bobby BROWN
5	Bryan ADAMS	Whitney HOUSTON	LISA LISA And CULT JAM	Whitney HOUSTON	MADONNA
6	Glenn FREY	Robert PALMER	Michael JACKSON	Rick ASTLEY	BON JOVI
7	KOOL & The GANG	Peter CETERA	HEART	Gloria ESTEFAN/ MIAMI SOUND MACHINE	Richard MARX
8	Whitney HOUSTON	Huey LEWIS and the NEWS	BON JOVI	Debbie GIBSON	Phil COLLINS
9	PRINCE And The REVOLUTION	BANGLES	WHITESNAKE	Steve WINWOOD	ROXETTE
10	DURAN DURAN	MR. MISTER	GENESIS	Richard MARX	FINE YOUNG CANNIBALS
11	Billy OCEAN	GENESIS	Huey LEWIS and the NEWS	CHICAGO	Debbie GIBSON
12	REO SPEEDWAGON	PET SHOP BOYS	JEFFERSON STARSHIP/STARSHIP	Taylor DAYNE	TONE LŌC
13	Tina TURNER	HEART	PRINCE	Terence Trent D'ARBY	CHER
14	FOREIGNER	Gloria ESTEFAN/ MIAMI SOUND MACHINE	EXPOSÉ	BREATHE	Janet JACKSON
15	SIMPLE MINDS	PRINCE And The REVOLUTION	Janet JACKSON	CHEAP TRICK	BANGLES
16	Bruce SPRINGSTEEN	John MELLENCAMP	Jody WATLEY	TIFFANY	MARTIKA
17	DIRE STRAITS	FALCO	CROWDED HOUSE	GUNS N' ROSES	Taylor DAYNE
18	Paul YOUNG	Dionne WARWICK	LOS LOBOS	POISON	Gloria ESTEFAN/ MIAMI SOUND MACHINE
19	USA For AFRICA	JEFFERSON STARSHIP/STARSHIP	CUTTING CREW	Eric CARMEN	Jody WATLEY
20	Stevie WONDER	Peter GABRIEL	Billy IDOL	Billy OCEAN	Billy JOEL
21	John PARR	Michael McDONALD	FLEETWOOD MAC	Elton JOHN	PRINCE
22	STING	Steve WINWOOD	Richard MARX	Phil COLLINS	Linda RONSTADT
23	SURVIVOR	MIKE + THE MECHANICS	Robbie NEVIL	PEBBLES	WARRANT
24	DeBARGE	VAN HALEN	Debbie GIBSON	The JETS	Karyn WHITE
25	Aretha FRANKLIN	SIMPLY RED	Bob SEGER	Belinda CARLISLE	SOUL II SOUL
26	Huey LEWIS and the NEWS	SURVIVOR	CLUB NOUVEAU	JOHNNY HATES JAZZ	GUNS N' ROSES
27	Don HENLEY	HUMAN LEAGUE	EUROPE	Robert PALMER	BAD ENGLISH
28	JEFFERSON STARSHIP/ STARSHIP	Gloria LORING & Carl ANDERSON	The JETS	ESCAPE CLUB	Neneh CHERRY
29	MR. MISTER	Tina TURNER	ATLANTIC STARR	AEROSMITH	Donny OSMOND
30	Philip BAILEY	BOSTON	TIFFANY	Bobby BROWN	MIKE + THE MECHANICS
31	A-HA	Cyndi LAUPER	Bruce HORNSBY And The RANGE	PET SHOP BOYS	Bette MIDLER
32	POINTER SISTERS	NU SHOOZ	BANGLES	WILL TO POWER	SIMPLY RED
33	Daryl HALL & John OATES	BANANARAMA	Steve WINWOOD	UB40	Michael DAMIAN
34	HEART	ATLANTIC STARR	Kim WILDE	Bobby McFERRIN	Rod STEWART
35	Jan HAMMER	WHAM!	Belinda CARLISLE	BANGLES	SHERIFF
36	Corey HART	BERLIN	Billy VERA & The BEATERS	FOREIGNER	DINO
37	POWER STATION	BON JOVI	Chris De BURGH	BON JOVI	GREAT WHITE
38	CHICAGO	GLASS TIGER	CHICAGO	BEACH BOYS	The B-52's
39	READY FOR THE WORLD	Bruce HORNSBY And The RANGE	KENNY G	Kylie MINOGUE	DEF LEPPARD
40	David Lee ROTH	ORCHESTRAL MANOEUVERS IN THE DARK	WANG CHUNG	EXPOSÉ	POISON
41	Julian LENNON	Kenny LOGGINS	Cyndi LAUPER	George HARRISON	LOVE AND ROCKETS
42	Howard JONES	The JETS	Gregory ABBOTT	Samantha FOX	YOUNG M.C.
43	The COMMODORES	SIMPLE MINDS	Smokey ROBINSON	VAN HALEN	TEARS FOR FEARS
44	Pat BENATAR	Howard JONES	Bruce SPRINGSTEEN	Daryl HALL & John OATES	EXPOSÉ
45	John MELLENCAMP	Patti LaBELLE	Gloria ESTEFAN/ MIAMI SOUND MACHINE	Patrick SWAYZE	The CURE
46	Harold FALTERMEYER	Belinda CARLISLE	GEORGIA SATELLITES	Huey LEWIS and the NEWS	Sheena EASTON
47	Lionel RICHIE	Stevie NICKS	Bryan ADAMS	Natalie COLE	WHITE LION
48	EURYTHMICS	The OUTFIELD	BANANARAMA	U2	Don HENLEY
49	Murray HEAD	El DeBARGE	Suzanne VEGA	INFORMATION SOCIETY	ROLLING STONES
50	SADE	JOURNEY	T'PAU	Anita BAKER	SURFACE

Singles Acts: All Top 50s Through The Decade

All acts appearing in any yearly Top 50 are shown here with their rank each year. 1-100 are listed numerically; 101-200 ="A"; 201-300="B" and greater than 301="C". Number of chart singles peaking in the 1970s or 1990s are shown.

Act	1980s Rank	1970s Charted	1980 n=347	1981 n=340	1982 n=330	1983 n=337	1984 n=317	1985 n=284	1986 n=307	1987 n=299	1988 n=301	1989 n=280	1990s Charted
ABBA	270	14	A	46	A								0
Gregory ABBOTT	241	0							91	42			0
Paula ABDUL	65	0									A	2	8
Bryan ADAMS	60	0			C	41	A	5	A	47			9
AEROSMITH	171	11	B							A	29	61	12
AFTER THE FIRE	359	0				46							0
A-HA	223	0						31	A	A			0
AIR SUPPLY	11	0	4	5	15	22	C	A	B				0
AMBROSIA	209	4	22		C								0
AMERICA	311	14			48	A							0
ASIA	178	0			28	58	C	B	A				1
Rick ASTLEY	106	0								B	6	57	3
ATLANTIC STARR	128	0			A	C	B	B	34	29			3
Patti AUSTIN	182	0		C	A	21	B		A				0
BAD ENGLISH	268	0										27	4
Philip BAILEY	232	0					A	30					0
Anita BAKER	166	0							85	87	50	58	4
Marty BALIN	347	0		42		B							0
BANANARAMA	129	0				C	71		33	48	A		0
BANGLES	51	0							9	32	35	15	0
Toni BASIL	186	0			30	62	B						0
BEACH BOYS	159	11	C	67	A		A	A	A	A	38	A	0
Pat BENATAR	56	0	40	40	99	30	58	44	A		A		0
George BENSON	143	6	42	A	54	A	B	A					0
BERLIN	227	0				A	A		36				0
The B-52's	316	0	A			B						38	2
BLONDIE	45	3	6	4	A								1
BON JOVI	27	0					A	A	37	8	37	6	8
BOSTON	199	6							30	74			1
David BOWIE	83	11	B	A	A	10	65	96	A	98			2
Laura BRANIGAN	71	0			21	27	32	A	B	A	A		1
BREATHE	158	0									14	86	2
Bobby BROWN	63	0							B	B	30	4	4
CAPTAIN & TENNILLE	119	11	9										0
Irene CARA	55	0	31	B	A	7	50						0
Belinda CARLISLE	93	0							46	35	25	94	2
Eric CARMEN	188	7	C					A		A	19		0
Kim CARNES	66	1	60	3	A	A	A	93	B				0
The CARS	72	5	A	A	34		15	74	92	A	B		0
Peter CETERA	108	0							7	A	59	A	4
CHARLENE	320	3			37	C							0
CHEAP TRICK	130	4	98	C	A			A			15	A	2
CHER	147	12								B	54	13	7
Neneh CHERRY	282	0										28	2
CHICAGO	15	28	B		10	B	27	38	A	38	11	53	3
CLUB NOUVEAU	247	0								26			0
Natalie COLE	165	7	A					A		84	47	56	3
Phil COLLINS	5	0		63	A	59	17	2	65		22	8	10
The COMMODORES	103	13	66	30	A	A		43	B	B			0
CROSBY, STILLS & NASH	315	2			42	A							0
Christopher CROSS	34	0	5	14	B	55	82	B					0

Singles Acts: Sp'l Lists

Act	1980s Rank	1970s Charted	1980 n=347	1981 n=340	1982 n=330	1983 n=337	1984 n=317	1985 n=284	1986 n=307	1987 n=299	1988 n=301	1989 n=280	1990s Charted
CROWDED HOUSE	190	0								17	A		1
CULTURE CLUB	24	0			B	4	4	A	A				0
The CURE	295	0							C	A	A	45	7
CUTTING CREW	202	0								19	B	B	0
Michael DAMIAN	298	0		B								33	2
Terence Trent D'ARBY	173	0								B	13		1
Paul DAVIS	215	10	A	A	29								0
Taylor DAYNE	95	0								A	12	17	5
DAZZ BAND	323	0			41		A	B					0
DeBARGE	133	0				47	A	24					0
El DeBARGE	349	0							49	B			0
Chris De BURGH	258	0				A	A			37			0
DEF LEPPARD	67	0				45	A			A	4	39	8
Teri DeSARIO	252	1	33										0
DEXYS MIDNIGHT RUNNERS	230	0				26							0
Neil DIAMOND	62	27	41	13	26	A	B		A				0
DINO	312	0								A		36	4
DIRE STRAITS	148	2	C	B		B		17	73				0
DR. HOOK	179	13	21	A	A								0
Thomas DOLBY	300	0			37	B							0
Robbie DUPREE	242	0	30	B									0
DURAN DURAN	7	0			C	9	3	10	83	53	63	A	5
EAGLES	233	16	36	A									1
EARTH, WIND & FIRE	152	19	A	44	53	86	C			B	A		1
Sheena EASTON	33	0		6	63	38	45	64	A		A	46	1
ELECTRIC LIGHT ORCHESTRA	168	15	43	58	C	A			A				0
ESCAPE CLUB	234	0								28	A		2
Gloria ESTEFAN/MIAMI SOUND MACHINE	44	0						A	14	45	7	18	0
EUROPE	225	0								27	A		0
EURYTHMICS	64	0				14	31	48	96	B	A	A	0
EXPOSÉ	96	0								14	40	44	6
FALCO	221	0							17				0
Harold FALTERMEYER	331	0						46					0
FINE YOUNG CANNIBALS	141	0							B			10	1
The FIXX	175	0			B	32	83		A			A	1
FLEETWOOD MAC	73	9	55	B	17	A				21	92	A	2
Dan FOGELBERG	79	1	24	38	36	A	80	B		B			0
FOREIGNER	26	8	A	12	19		A	14		B	36		1
Samantha FOX	164	0							A	59	42	63	0
Aretha FRANKLIN	110	26	C	A	A	B		25	87	56		A	4
Glenn FREY	89	0			71	A	92	6	B		94	B	2
Peter GABRIEL	145	1	A		A	A			20	73		A	2
Marvin GAYE	240	16			83	39							0
J. GEILS Band	81	8	A	A	2	A	B	B					0
GENESIS	48	2	78	A	74	A	49		11	10			5
GEORGIA SATELLITES	314	0							B	46	A	B	0
Andy GIBB	191	5	29	82									0
Debbie GIBSON	54	0								24	8	11	2
GLASS TIGER	196	0							38	71	A		0
The GO-GO'S	109	0		93	13		56						0
Eddy GRANT	169	0				18	A						0
GREAT WHITE	304	0								A	A	37	2
GUNS N' ROSES	111	0								17	26		7
Daryl HALL & John OATES	3	10	47	1	3	6	8	33		44			2
Jan HAMMER	284	0						35	B				0
George HARRISON	138	11		36	B					83	41		0
Corey HART	121	0					43	36	100	A	A		1
Dan HARTMAN	299	2		B			44	A					0
Murray HEAD	346	2						49					0

Singles Acts: Sp'l Lists

Act	1980s Rank	1970s Charted	1980 n=347	1981 n=340	1982 n=330	1983 n=337	1984 n=317	1985 n=284	1986 n=307	1987 n=299	1988 n=301	1989 n=280	1990s Charted
HEART	46	10	93	90	A	A		34	13	7	96		5
Don HENLEY	90	0			65	40	A	27				48	3
Bertie HIGGINS	351	0		B	50								0
Bruce HORNSBY And The RANGE	116	0							39	31	57		2
Whitney HOUSTON	8	0						8	5	2	5	B	16
HUMAN LEAGUE	82	0			14	48	B		27	A			2
Billy IDOL	87	0			A	A	34	B	86	20	A		2
INFORMATION SOCIETY	283	0									49	87	1
INXS	76	0				A	A	B	53	A	3	B	5
Janet JACKSON	41	0			B	A			2	15		14	16
Michael JACKSON	1	9	7	B	A	1	30			6	2	A	11
The JACKSONS	180	1	89	A			33					B	0
JEFFERSON STARSHIP/STARSHIP	43	19	84	A	A	A	A	28	19	12	B	90	1
The JETS	94	0							42	28	24	B	0
Joan JETT & The BLACKHEARTS	78	0			6	A			B	A	64	A	0
Billy JOEL	6	11	8	83	77	8	40	53	56	A		20	9
Elton JOHN	18	24	20	87	52	44	25	100	71	A	21	89	14
JOHNNY HATES JAZZ	271	0									26		0
Howard JONES	114	0					93	42	44	A		65	1
JOURNEY	29	5	77	29	12	20		87	50	86			2
KC & The SUNSHINE BAND	160	15	18			B	98						0
KENNY G	246	0								39	A	A	6
Chaka KHAN	189	1		A		B	35	98	A				1
Greg KIHN Band	187	0		72	B	28							0
KOOL & The GANG	13	13	19	21	43	A	29	7	A	55	B		0
Patti LaBELLE	224	0					A	78	45			B	1
Cyndi LAUPER	19	0				B	1	70	31	41	B	74	1
John LENNON	59	4	48	2			57						0
Julian LENNON	222	0					97	41	A			B	0
Huey LEWIS and the NEWS	12	0			51	70	11	26	8	11	46	A	4
LIPPS, INC.	162	0	15										0
LISA LISA And CULT JAM	142	0								5		A	1
LITTLE RIVER BAND	107	7	65	45	39	51		A					0
Kenny LOGGINS	36	3	27		A	61	16	95	41	80	78	B	1
Gloria LORING & Carl ANDERSON	278	0							28				0
LOS LOBOS	207	0						B		18			0
LOVE AND ROCKETS	358	0										41	0
MADNESS	366	0				50	B						0
MADONNA	2	0				A	13	1	1	1	B	5	25
Melissa MANCHESTER	253	8	A	B	40	A	B	B					0
The MANHATTANS	310	9	46			B		B					0
MANHATTAN TRANSFER	307	1	A	48	B	A		B					0
MARTIKA	231	0									B	16	2
Richard MARX	42	0								22	10	7	9
Paul McCARTNEY	30	2			4	25	37	A	79			A	3
Paul McCARTNEY & WINGS	140	23	14										0
Michael McDONALD	135	0			38	B		A	21				1
Bobby McFERRIN	292	0									34	B	0
Don McLEAN	273	6		35	B								0
John MELLENCAMP	10	1	A	73	1	53	51	45	16	57	70	A	7
MEN AT WORK	49	0			16	5	A						0
Sergio MENDES	217	0				31	A						0
MEN WITHOUT HATS	213	0				29				A	A		0
George MICHAEL	23	0							84	4	1	B	9
Bette MIDLER	113	8	23	A		B	B					31	3
MIKE + THE MECHANICS	139	0						B	23		B	30	1
Steve MILLER Band	112	7		A	8	B	A	B	C				1
MILLI VANILLI	69	0										1	1
Ronnie MILSAP	163	6	B	31	67	97	B						0

Singles Acts: Sp'l Lists

Act	1980s Rank	1970s Charted	1980 n=347	1981 n=340	1982 n=330	1983 n=337	1984 n=317	1985 n=284	1986 n=307	1987 n=299	1988 n=301	1989 n=280	1990s Charted
Kylie MINOGUE	286	0									39	A	0
MR. MISTER	101	0					A	29	10	A			0
MOODY BLUES	219	7		49		A			82		A		0
The MOTELS	192	0			49	57	A	A					0
NAKED EYES	250	0				33	A						0
Willie NELSON	184	3	86	C	33		87						0
NENA	264	0				B	38						0
Robbie NEVIL	176	0							A	23	A	A	2
NEW KIDS ON THE BLOCK	77	0									75	3	6
Olivia NEWTON-JOHN	14	20		9	18	89	A						2
Juice NEWTON	70	1	10	17	5	68	59	A	B		B		0
Stevie NICKS	57	0		19	35	34	A	A	47			A	2
NIGHT RANGER	144	0				A	41	56	A	B	B		0
NU SHOOZ	265	0							32	A			0
OAK RIDGE BOYS	261	0		39	92	B							0
Billy OCEAN	20	1				22	11	3	A	20	A		0
ORCHESTRAL MANOEUVERS IN THE DARK	228	0					A	40	B	100			0
Donny OSMOND	280	13									29		2
The OUTFIELD	267	0							48	A		A	2
Robert PALMER	91	3	B			B		B	6	B	27	A	2
Ray PARKER Jr.	85	0			31	A	14	88	C	B			0
Ray PARKER Jr. & RAYDIO	243	2	A	34									0
John PARR	208	0					B	21	B	B			0
Alan PARSONS Project	117	4	B	43	25	A	76	A	B				0
Dolly PARTON	155	8	A	16	A	B	A	B					0
PEBBLES	237	0									23		3
Steve PERRY	200	0					28	A					2
PET SHOP BOYS	104	0							12	78	31	B	3
PINK FLOYD	134	1	13							B			1
POINTER SISTERS	22	8	34	28	46	A	10	32	A	A			0
POISON	105	0							52	18	40		5
The POLICE	31	2	A	27	73	3	67		A				0
POWER STATION	294	0						37					0
Billy PRESTON & SYREETA	277	0	37										0
The PRETENDERS	151	0	71		B	43	78		A	A			0
PRINCE	17	1	87	B	A	19	12			13	55	21	14
PRINCE And The REVOLUTION	40	0					6	9	15				0
QUARTERFLASH	157	0		75	23	83		B					0
QUEEN	35	9	1	84	80		94	B	A			A	2
QUIET RIOT	303	0				49	A						0
Eddie RABBITT	74	5	44	8	78	77							0
RAY, GOODMAN & BROWN	328	0	50										0
READY FOR THE WORLD	206	0						39	A	96			0
REO SPEEDWAGON	39	3	B	7	45		A	12		65	99		1
Cliff RICHARD	150	3	25	55	A	B							0
Lionel RICHIE	4	0		25	24	2	2	47	4	54			2
Smokey ROBINSON	75	11	26	18	A	A	B			43	B		1
ROCKWELL	194	0					26						0
Kenny ROGERS	9	8	2	20	44	11	60	A	B				0
ROLLING STONES	53	16	28	23	57	93	A		63			49	5
The ROMANTICS	210	0	B			A	39	B					0
Linda RONSTADT	99	21	35		A	A	B		B	82		22	2
Diana ROSS	16	18	3	10	27	94	75	71	B				0
David Lee ROTH	183	0						40	A		72		0
ROXETTE	131	0										9	9
SADE	185	0						50	55		A		2
Leo SAYER	193	8	32	64									0
Boz SCAGGS	220	9	38	96						A			0
SCANDAL/Patty SMYTH	288	0			B	A	48	A		B			0

Act	1980s Rank	1970s Charted	1980 n=347	1981 n=340	1982 n=330	1983 n=337	1984 n=317	1985 n=284	1986 n=307	1987 n=299	1988 n=301	1989 n=280	1990s Charted
Joey SCARBURY	201	1		22									0
Bob SEGER	84	8	17	47	B					25			0
Bob SEGER & The SILVER BULLET BAND	102	4		B	A	13	A	A	60	B			1
Michael SEMBELLO	177	0				16							0
SHEILA E.	251	0					47	A	A	B			0
SHERIFF	301	0				B					A	35	0
Paul SIMON	249	9	49	C		A	C		A	A			1
SIMPLE MINDS	115	0						15	43				2
SIMPLY RED	137	0							25	A		32	2
S.O.S. BAND	236	0	39			A	A		A				0
SOUL II SOUL	255	0										25	2
SPANDAU BALLET	239	0				35	A						0
SPINNERS	127	0	12	B	B	B							0
Rick SPRINGFIELD	21	4		11	9	36	36	77			A		0
Bruce SPRINGSTEEN	25	4	73	50			9	16	74	44	68		5
STARS ON	198	0		26	A								0
Rod STEWART	47	19	A	32	61	74	42	A	67	B	52	34	12
STING	123	0						22	A	79	91		8
STRAY CATS	136	0			75	15	B						0
Barbra STREISAND	92	15	11	78	91	A	A	A	A		A	B	0
Barbra STREISAND & Barry GIBB	226	0	91	37									0
STYX	86	11	A	15		17	A						3
Donna SUMMER	58	14	16	A	62	23	A			A		68	2
SURFACE	341	0								A		50	2
SURVIVOR	32	0	B	A	7	B	73	23	26	89	A	B	0
Patrick SWAYZE	335	0								B	45		0
TACO	327	0				42							0
A TASTE OF HONEY	263	2		33	A								0
TEARS FOR FEARS	68	0				B		3	A			43	3
THOMPSON TWINS	100	0				98	21	69	78	A		A	0
TIFFANY	98	0								30	16	73	0
TOMMY TUTONE	291	0	A		32								0
TONE LŌC	195	0									B	12	1
TOTO	61	3	A		11	24	A	A	94	A	A		0
T'PAU	367	0								50			0
Tina TURNER	37	0					5	13	29	91		A	4
Bonnie TYLER	120	1				12	A		B				0
UB40	238	0					A	A			33	B	5
USA For AFRICA	197	0						19					0
U2	80	0				A	A	B		3	48	99	11
VANGELIS	203	0		C	20								0
VAN HALEN	50	4	B		72		7		24		43	A	4
Gino VANNELLI	272	4		41	C			A		A			0
Suzanne VEGA	363	0								49			1
Billy VERA & The BEATERS	257	0		A					A	36			0
John WAITE	154	0					19	94	B	A			1
WANG CHUNG	146	0					72	A	52	40		B	0
Jennifer WARNES	211	0	A	B	47	B				63	B		0
WARRANT	256	0										23	6
Dionne WARWICK	97	10	62	B	85	73		92	18	A			0
Grover WASHINGTON JR.	205	1		24	C								0
Jody WATLEY	88	0								16	60	19	6
WHAM!	38	0				B	24	4	35				0
Karyn WHITE	229	0									A	24	4
WHITE LION	244	0									65	47	0
WHITESNAKE	122	0	B							9	74	A	2
Kim WILDE	216	0			A			B		34	A		0
Matthew WILDER	274	0				92	46						0
Deniece WILLIAMS	156	2		A	79		20				B		0

Singles Acts: Sp'l Lists

Act	1980s Rank	1970s Charted	1980 n=347	1981 n=340	1982 n=330	1983 n=337	1984 n=317	1985 n=284	1986 n=307	1987 n=299	1988 n=301	1989 n=280	1990s Charted
WILL TO POWER	235	0								A	32	A	1
Steve WINWOOD	52	0		52	A				22	33	9	98	1
Stevie WONDER	28	20	45	54	22	C	18	20	81	A	A		2
YES	132	4				69	23			A	A		1
Paul YOUNG	167	0				B	96	18	A	B			2
YOUNG M.C.	355	0										42	3

Singles Acts: 100 Or More Total Chart Weeks

Act [Singles Rank]	Wks (Ent)
Daryl HALL & John OATES [3]	373 (22)
Billy JOEL [6]	344 (22)
Michael JACKSON [1]	335 (21)
KOOL & The GANG [13]	328 (19)
MADONNA [2]	325 (18)
John MELLENCAMP [10]	316 (20)
Elton JOHN [18]	309 (20)
POINTER SISTERS [22]	287 (19)
Sheena EASTON [33]	279 (18)
Lionel RICHIE [4]	277 (15)
Rod STEWART [47]	276 (20)
Kenny ROGERS [9]	273 (20)
JOURNEY [29]	273 (19)
Rick SPRINGFIELD [21]	270 (17)
Diana ROSS [16]	264 (18)
Huey LEWIS and The NEWS [12]	262 (17)
Kenny LOGGINS [36]	261 (17)
CHICAGO [15]	260 (16)
Pat BENATAR [56]	257 (17)
PRINCE [17]	251 (16)
Phil COLLINS [5]	244 (14)
REO SPEEDWAGON [39]	241 (15)
JEFFERSON STARSHIP/ STARSHIP [43]	238 (18)
Stevie WONDER [28]	235 (18)
SURVIVOR [32]	234 (18)
AIR SUPPLY [11]	226 (13)
DURAN DURAN [7]	221 (15)
HEART [46]	220 (15)
GENESIS [48]	220 (15)
Bruce SPRINGSTEEN [25]	219 (14)
Gloria ESTEFAN/ MIAMI SOUND MACHINE [44]	214 (11)
Steve WINWOOD [52]	213 (14)
Olivia NEWTON-JOHN [14]	209 (14)
Billy OCEAN [20]	205 (11)
Tina TURNER [37]	203 (13)
VAN HALEN [50]	199 (14)
Donna SUMMER [58]	198 (16)
Whitney HOUSTON [8]	197 (11)
BON JOVI [27]	197 (12)
Cyndi LAUPER [19]	193 (13)
EURYTHMICS [64]	190 (15)
Bryan ADAMS [60]	189 (13)
FOREIGNER [26]	185 (13)
Laura BRANIGAN [71]	181 (12)
38 SPECIAL [118]	180 (14)
The CARS [72]	178 (13)
QUEEN [35]	174 (14)
David BOWIE [83]	174 (14)
LOVERBOY [126]	173 (12)
DEF LEPPARD [67]	172 (11)
Aretha FRANKLIN [110]	172 (15)
Dan FOGELBERG [79]	171 (12)
Dionne WARWICK [97]	171 (13)
Stevie NICKS [57]	170 (12)
TOTO [61]	170 (11)
Neil DIAMOND [62]	168 (13)
CULTURE CLUB [24]	167 (10)
FLEETWOOD MAC [73]	167 (13)
Jeffrey OSBORNE [181]	167 (12)
ROLLING STONES [53]	166 (12)
Eddie MONEY [125]	166 (13)
Smokey ROBINSON [75]	165 (11)
NEW EDITION [124]	163 (12)
Paul McCARTNEY [30]	162 (12)
INXS [76]	162 (12)
Billy IDOL [87]	162 (11)
The POLICE [31]	156 (10)
Barry MANILOW [149]	156 (13)
Kim CARNES [66]	153 (14)
Juice NEWTON [70]	153 (10)
The COMMODORES [103]	153 (12)
Christopher CROSS [34]	152 (10)
Howard JONES [114]	152 (10)
Debbie GIBSON [54]	151 (9)
Janet JACKSON [41]	149 (9)
Eddie RABBITT [74]	149 (9)
Don HENLEY [90]	149 (9)
The JETS [94]	147 (10)
Glenn FREY [89]	146 (10)
THOMPSON TWINS [100]	146 (11)
BANGLES [51]	145 (8)
Irene CARA [55]	145 (8)
Jermaine JACKSON [161]	143 (11)
Alan PARSONS Project [117]	141 (11)
George MICHAEL [23]	139 (8)
PRINCE And The REVOLUTION [40]	139 (10)
Bobby BROWN [63]	138 (7)
Linda RONSTADT [99]	138 (9)
Richard MARX [42]	137 (7)
NIGHT RANGER [144]	137 (11)
U2 [80]	136 (11)
Cliff RICHARD [150]	136 (10)
Tom PETTY And The HEARTBREAKERS [153]	135 (12)
Barbra STREISAND [92]	132 (11)
CHEAP TRICK [130]	132 (10)
WHAM! [38]	131 (8)
Joan JETT & The BLACKHEARTS [78]	131 (9)
NEW KIDS ON THE BLOCK [77]	129 (6)
LITTLE RIVER BAND [107]	129 (9)
Corey HART [121]	129 (9)
Luther VANDROSS [260]	129 (10)
Ray PARKER Jr. [85]	124 (9)
PET SHOP BOYS [104]	123 (9)
Bob SEGER & The SILVER BULLET BAND [102]	122 (10)
Robert PALMER [91]	121 (10)
BEACH BOYS [159]	119 (10)
ELECTRIC LIGHT ORCHESTRA [168]	119 (9)
Billy SQUIER [254]	118 (11)
POISON [105]	117 (7)
STYX [86]	115 (9)
EARTH, WIND & FIRE [152]	115 (11)
Sammy HAGAR [266]	115 (10)
Natalie COLE [165]	114 (8)
Jody WATLEY [88]	113 (7)
EXPOSÉ [96]	113 (6)
WANG CHUNG [146]	112 (8)
Taylor DAYNE [95]	111 (5)
ZZ TOP [174]	111 (10)
James INGRAM [259]	111 (6)
J. GEILS Band [81]	110 (10)
Peter GABRIEL [145]	110 (8)
DeBARGE [133]	109 (6)
Anita BAKER [166]	109 (6)
BANANARAMA [129]	108 (11)
The PRETENDERS [151]	108 (8)
Ronnie MILSAP [163]	108 (8)
Jackson BROWNE [170]	108 (8)
Paula ABDUL [65]	106 (6)
George BENSON [143]	106 (8)
CHER [147]	106 (6)
The GO-GO'S [109]	105 (7)
MEN AT WORK [49]	103 (6)
Belinda CARLISLE [93]	102 (7)
ABC [172]	102 (7)
MOODY BLUES [219]	101 (8)
John CAFFERTY And The BEAVER BROWN BAND [248]	101 (9)
STING [123]	100 (7)
The FIXX [175]	100 (8)
SHALAMAR [214]	100 (7)
Freddie JACKSON [275]	100 (7)
Rick JAMES [321]	100 (8)

Singles Acts: 6 Or More Chart Entries

Act [Singles Rank]	Entries
Daryl HALL & John OATES [3]	22
Billy JOEL [6]	22
Michael JACKSON [1]	21
Kenny ROGERS [9]	20
John MELLENCAMP [10]	20
Elton JOHN [18]	20
Rod STEWART [47]	20
KOOL & The GANG [13]	19
POINTER SISTERS [22]	19
JOURNEY [29]	19
MADONNA [2]	18
Diana ROSS [16]	18
Stevie WONDER [28]	18
SURVIVOR [32]	18
Sheena EASTON [33]	18
JEFFERSON STARSHIP/ STARSHIP [43]	18
Huey LEWIS and the NEWS [12]	17
Rick SPRINGFIELD [21]	17
Kenny LOGGINS [36]	17
Pat BENATAR [56]	17
CHICAGO [15]	16
PRINCE [17]	16
Donna SUMMER [58]	16
Lionel RICHIE [4]	15
DURAN DURAN [7]	15
REO SPEEDWAGON [39]	15
HEART [46]	15
GENESIS [48]	15
EURYTHMICS [64]	15
Aretha FRANKLIN [110]	15
Phil COLLINS [5]	14
Olivia NEWTON-JOHN [14]	14
Bruce SPRINGSTEEN [25]	14
QUEEN [35]	14
VAN HALEN [50]	14
Steve WINWOOD [52]	14
Kim CARNES [66]	14
David BOWIE [83]	14
38 SPECIAL [118]	14
AIR SUPPLY [11]	13
Cyndi LAUPER [19]	13
FOREIGNER [26]	13
Tina TURNER [37]	13
Bryan ADAMS [60]	13
Neil DIAMOND [62]	13
The CARS [72]	13
FLEETWOOD MAC [73]	13
Dionne WARWICK [97]	13
Eddie MONEY [125]	13
Barry MANILOW [149]	13
BON JOVI [27]	12
Paul McCARTNEY [30]	12
ROLLING STONES [53]	12
Stevie NICKS [57]	12
Laura BRANIGAN [71]	12
INXS [76]	12
Dan FOGELBERG [79]	12
The COMMODORES [103]	12
NEW EDITION [124]	12
LOVERBOY [126]	12
Tom PETTY And The HEARTBREAKERS [153]	12
Jeffrey OSBORNE [181]	12
Whitney HOUSTON [8]	11
Billy OCEAN [20]	11
Gloria ESTEFAN/MIAMI SOUND MACHINE [44]	11
TOTO [61]	11
DEF LEPPARD [67]	11
Smokey ROBINSON [75]	11
U2 [80]	11
Billy IDOL [87]	11
Barbra STREISAND [92]	11
THOMPSON TWINS [100]	11
Alan PARSONS Project [117]	11
BANANARAMA [129]	11
NIGHT RANGER [144]	11
EARTH, WIND & FIRE [152]	11
Jermaine JACKSON [161]	11
Billy SQUIER [254]	11
CULTURE CLUB [24]	10
The POLICE [31]	10
Christopher CROSS [34]	10
PRINCE And The REVOLUTION [40]	10
Juice NEWTON [70]	10
J. GEILS Band [81]	10
Glenn FREY [89]	10
Robert PALMER [91]	10
The JETS [94]	10
Bob SEGER & The SILVER BULLET BAND [102]	10
Howard JONES [114]	10
CHEAP TRICK [130]	10
Cliff RICHARD [150]	10
BEACH BOYS [159]	10
ZZ TOP [174]	10
Luther VANDROSS [260]	10
Sammy HAGAR [266]	10
Janet JACKSON [41]	9
Debbie GIBSON [54]	9
Eddie RABBITT [74]	9
Joan JETT & The BLACKHEARTS [78]	9
Ray PARKER Jr. [85]	9
STYX [86]	9
Don HENLEY [90]	9
Linda RONSTADT [99]	9
PET SHOP BOYS [104]	9
LITTLE RIVER BAND [107]	9
Corey HART [121]	9
John WAITE [154]	9
Dolly PARTON [155]	9
ELECTRIC LIGHT ORCHESTRA [168]	9
John CAFFERTY/BEAVER BROWN BAND [248]	9
Eric CLAPTON [262]	9
Herb ALPERT [276]	9
Anne MURRAY [293]	9
KISS [442]	9
George MICHAEL [23]	8
WHAM! [38]	8
BANGLES [51]	8
Irene CARA [55]	8
Steve MILLER Band [112]	8
George BENSON [143]	8
Peter GABRIEL [145]	8
WANG CHUNG [146]	8
The PRETENDERS [151]	8
Ronnie MILSAP [163]	8
Natalie COLE [165]	8
Jackson BROWNE [170]	8
The FIXX [175]	8
The JACKSONS [180]	8
MOODY BLUES [219]	8
DOOBIE BROTHERS [245]	8
Al JARREAU [306]	8
Carly SIMON [318]	8
Rick JAMES [321]	8
John DENVER [445]	8
Richard MARX [42]	7
Bobby BROWN [63]	7
Jody WATLEY [88]	7
Belinda CARLISLE [93]	7
POISON [105]	7
Peter CETERA [108]	7
The GO-GO'S [109]	7
Bette MIDLER [113]	7
STING [123]	7
ATLANTIC STARR [128]	7
QUARTERFLASH [157]	7
Paul YOUNG [167]	7
ABC [172]	7
DR. HOOK [179]	7
David Lee ROTH [183]	7
Chaka KHAN [189]	7
The MOTELS [192]	7
SHALAMAR [214]	7
Patti LaBELLE [224]	7
BERLIN [227]	7
Paul SIMON [249]	7
Melissa MANCHESTER [253]	7
Freddie JACKSON [275]	7
ALABAMA [279]	7
SCANDAL/ Patty SMYTH [288]	7
MÖTLEY CRÜE [290]	7
The CURE [295]	7
R.E.M. [309]	7
The HOOTERS [345]	7
Robert PLANT [353]	7
KANSAS [368]	7
MIDNIGHT STAR [410]	7
Donnie IRIS [419]	7
Michael STANLEY Band [493]	7
The TEMPTATIONS [505]	7
BLONDIE [45]	6
MEN AT WORK [49]	6
Paula ABDUL [65]	6
TEARS FOR FEARS [68]	6
NEW KIDS ON THE BLOCK [77]	6
HUMAN LEAGUE [82]	6
Bob SEGER [84]	6
EXPOSÉ [96]	6
TIFFANY [98]	6
Rick ASTLEY [106]	6
Bruce HORNSBY And The RANGE [116]	6
WHITESNAKE [122]	6
DeBARGE [133]	6
MIKE + THE MECHANICS [139]	6
CHER [147]	6
Samantha FOX [164]	6
Anita BAKER [166]	6
Patti AUSTIN [182]	6
Willie NELSON [184]	6
Eric CARMEN [188]	6
Jennifer WARNES [211]	6
Julian LENNON [222]	6
James INGRAM [259]	6
The OUTFIELD [267]	6
ANIMOTION [269]	6
ABBA [270]	6
Gino VANNELLI [272]	6
CINDERELLA [287]	6
MANHATTAN TRANSFER [307]	6
Stephanie MILLS [313]	6
SANTANA [317]	6
ICEHOUSE [329]	6
Paul CARRACK [330]	6
The KINKS [338]	6
Mick JAGGER [350]	6
Stacy LATTISAW [352]	6
BEE GEES [356]	6
COVER GIRLS [360]	6
Peabo BRYSON [369]	6
RUSH [420]	6
Crystal GAYLE [421]	6
RATT [423]	6
DEPECHE MODE [427]	6
"Weird Al" YANKOVIC [484]	6

Singles Acts: 3 Or More Top 10 Entries

Act [Singles Rank]	T10s (Wks)
Michael JACKSON [1]	17 (105)
MADONNA [2]	17 (95)
Lionel RICHIE [4]	13 (94)
Daryl HALL & John OATES [3]	13 (93)
Huey LEWIS and the NEWS [12]	12 (57)
Bruce SPRINGSTEEN [25]	11 (44)
Phil COLLINS [5]	10 (66)
Whitney HOUSTON [8]	10 (61)
KOOL & The GANG [13]	10 (37)
DURAN DURAN [7]	9 (54)
John MELLENCAMP [10]	9 (52)
Billy JOEL [6]	9 (45)
AIR SUPPLY [11]	8 (62)
Diana ROSS [16]	8 (53)
George MICHAEL [23]	8 (49)
Cyndi LAUPER [19]	8 (44)
BON JOVI [27]	8 (42)
PRINCE [17]	8 (41)
Gloria ESTEFAN/ MIAMI SOUND MACHINE [44]	8 (36)
CHICAGO [15]	7 (42)
Billy OCEAN [20]	7 (37)
Richard MARX [42]	7 (37)
Sheena EASTON [33]	7 (32)
HEART [46]	7 (28)
Kenny ROGERS [9]	6 (53)
Paul McCARTNEY [30]	6 (48)
Olivia NEWTON-JOHN [14]	6 (46)
CULTURE CLUB [24]	6 (40)
POINTER SISTERS [22]	6 (38)
Janet JACKSON [41]	6 (37)
Elton JOHN [18]	6 (35)
PRINCE And The REVOLUTION [40]	6 (35)
WHAM! [38]	6 (33)
JOURNEY [29]	6 (31)
The POLICE [31]	6 (29)
Bobby BROWN [63]	6 (28)
GENESIS [48]	6 (27)
Rod STEWART [47]	6 (26)
Steve WINWOOD [52]	6 (26)
NEW KIDS ON THE BLOCK [77]	6 (24)
EXPOSÉ [96]	6 (22)
FOREIGNER [26]	5 (38)
Stevie WONDER [28]	5 (36)
Rick SPRINGFIELD [21]	5 (35)
SURVIVOR [32]	5 (31)
Tina TURNER [37]	5 (30)
ROLLING STONES [53]	5 (30)
Debbie GIBSON [54]	5 (30)
BANGLES [51]	5 (28)
INXS [76]	5 (26)
Taylor DAYNE [95]	5 (26)
Donna SUMMER [58]	5 (24)
The JETS [94]	5 (21)
Jody WATLEY [88]	5 (21)
Bryan ADAMS [60]	5 (20)
PET SHOP BOYS [104]	5 (19)
John LENNON [59]	4 (32)
MEN AT WORK [49]	4 (31)
Christopher CROSS [34]	4 (30)
REO SPEEDWAGON [39]	4 (29)
Stevie NICKS [57]	4 (29)
Eddie RABBITT [74]	4 (29)
Juice NEWTON [70]	4 (27)
Paula ABDUL [65]	4 (27)
STYX [86]	4 (26)
TEARS FOR FEARS [68]	4 (26)
Kenny LOGGINS [36]	4 (25)
Neil DIAMOND [62]	4 (25)
FLEETWOOD MAC [73]	4 (25)
MILLI VANILLI [69]	4 (25)
JEFFERSON STARSHIP/ STARSHIP [43]	4 (23)
The CARS [72]	4 (22)
Smokey ROBINSON [75]	4 (22)
Bob SEGER [84]	4 (22)
Pat BENATAR [56]	4 (19)
DEF LEPPARD [67]	4 (19)
Linda RONSTADT [99]	4 (19)
Belinda CARLISLE [93]	4 (19)
Peter CETERA [108]	4 (19)
GUNS N' ROSES [111]	4 (19)
Don HENLEY [90]	4 (18)
TIFFANY [98]	4 (18)
Dan FOGELBERG [79]	4 (17)
Rick ASTLEY [106]	4 (17)
David BOWIE [83]	4 (16)
CHER [147]	4 (15)
POISON [105]	4 (14)
BLONDIE [45]	3 (30)
Irene CARA [55]	3 (23)
HUMAN LEAGUE [82]	3 (22)
EURYTHMICS [64]	3 (20)
Laura BRANIGAN [71]	3 (20)
U2 [80]	3 (20)
VAN HALEN [50]	3 (18)
TOTO [61]	3 (18)
Joan JETT & The BLACKHEARTS [78]	3 (18)
Robert PALMER [91]	3 (18)
Bruce HORNSBY And The RANGE [116]	3 (18)
Michael McDONALD [135]	3 (17)
MR. MISTER [101]	3 (17)
Aretha FRANKLIN [110]	3 (16)
The COMMODORES [103]	3 (16)
Billy IDOL [87]	3 (15)
BANANARAMA [129]	3 (15)
THOMPSON TWINS [100]	3 (14)
MIKE + THE MECHANICS [139]	3 (14)
STRAY CATS [136]	3 (14)
STING [123]	3 (13)
BREATHE [158]	3 (12)
Samantha FOX [164]	3 (11)
LITTLE RIVER BAND [107]	3 (9)
Karyn WHITE [229]	3 (8)

Singles Acts: 4 Or More Top 40s

Act [Singles Rank]	T40s (Wks)
Daryl HALL & John OATES [3]	22 (246)
Billy JOEL [6]	20 (210)
Michael JACKSON [1]	19 (229)
Elton JOHN [18]	19 (177)
MADONNA [2]	18 (222)
John MELLENCAMP [10]	17 (185)
KOOL & The GANG [13]	16 (180)
Rick SPRINGFIELD [21]	16 (157)
JOURNEY [29]	16 (151)
Huey LEWIS and the NEWS [12]	15 (171)
Pat BENATAR [56]	15 (129)
Lionel RICHIE [4]	14 (192)
Phil COLLINS [5]	13 (158)
Kenny ROGERS [9]	13 (152)
DURAN DURAN [7]	13 (142)
Bruce SPRINGSTEEN [25]	13 (141)
REO SPEEDWAGON [39]	13 (122)
Kenny LOGGINS [36]	13 (121)
Sheena EASTON [33]	12 (133)
PRINCE [17]	12 (130)
Rod STEWART [47]	12 (129)
Diana ROSS [16]	12 (129)
Tina TURNER [37]	12 (116)
Bryan ADAMS [60]	12 (112)
VAN HALEN [50]	12 (101)
AIR SUPPLY [11]	11 (142)
CHICAGO [15]	11 (135)
POINTER SISTERS [22]	11 (128)
Billy OCEAN [20]	11 (125)
Gloria ESTEFAN/MIAMI SOUND MACHINE [44]	11 (122)
Olivia NEWTON-JOHN [14]	11 (118)
GENESIS [48]	11 (105)
ROLLING STONES [53]	11 (102)
Stevie NICKS [57]	11 (101)
JEFFERSON STARSHIP/STARSHIP [43]	11 (100)
Whitney HOUSTON [8]	10 (134)
Cyndi LAUPER [19]	10 (116)
CULTURE CLUB [24]	10 (113)
Stevie WONDER [28]	10 (112)
HEART [46]	10 (105)
The CARS [72]	10 (96)
EURYTHMICS [64]	10 (87)
Donna SUMMER [58]	10 (85)
Barry MANILOW [149]	10 (63)
Steve WINWOOD [52]	9 (103)
Paul McCARTNEY [30]	9 (102)
BON JOVI [27]	9 (101)
DEF LEPPARD [67]	9 (91)
Dan FOGELBERG [79]	9 (91)
FLEETWOOD MAC [73]	9 (91)
Neil DIAMOND [62]	9 (88)
TOTO [61]	9 (87)
David BOWIE [83]	9 (77)
Aretha FRANKLIN [110]	9 (76)
LOVERBOY [126]	9 (69)
The POLICE [31]	8 (103)
George MICHAEL [23]	8 (100)
FOREIGNER [26]	8 (97)
SURVIVOR [32]	8 (96)
Christopher CROSS [34]	8 (95)
Debbie GIBSON [54]	8 (91)
BANGLES [51]	8 (88)
PRINCE And The REVOLUTION [40]	8 (84)
Howard JONES [114]	8 (71)
Billy IDOL [87]	8 (66)
Joan JETT & The BLACKHEARTS [78]	8 (66)
38 SPECIAL [118]	8 (65)
Tom PETTY And The HEARTBREAKERS [153]	8 (65)
Corey HART [121]	8 (65)
ELECTRIC LIGHT ORCHESTRA [168]	8 (60)
Jeffrey OSBORNE [181]	8 (47)
Juice NEWTON [70]	7 (91)
Richard MARX [42]	7 (91)
Janet JACKSON [41]	7 (84)
Laura BRANIGAN [71]	7 (81)
Don HENLEY [90]	7 (72)
LITTLE RIVER BAND [107]	7 (72)
Glenn FREY [89]	7 (71)
Dionne WARWICK [97]	7 (64)
THOMPSON TWINS [100]	7 (63)
Linda RONSTADT [99]	7 (63)
Kim CARNES [66]	7 (62)
Ray PARKER Jr. [85]	7 (61)
ZZ TOP [174]	7 (48)
Eddie RABBITT [74]	6 (83)
Bobby BROWN [63]	6 (77)
WHAM! [38]	6 (77)
EXPOSÉ [96]	6 (75)
INXS [76]	6 (71)
NEW KIDS ON THE BLOCK [77]	6 (71)
The JETS [94]	6 (68)
Smokey ROBINSON [75]	6 (65)
Irene CARA [55]	6 (65)
STYX [86]	6 (64)
PET SHOP BOYS [104]	6 (62)
POISON [105]	6 (62)
STING [123]	6 (62)
Eddie MONEY [125]	6 (61)
U2 [80]	6 (61)
Cliff RICHARD [150]	6 (60)
Bob SEGER & The SILVER BULLET BAND [102]	6 (60)
Jermaine JACKSON [161]	6 (56)
NIGHT RANGER [144]	6 (53)
Jackson BROWNE [170]	6 (51)
J. GEILS Band [81]	6 (49)
NEW EDITION [124]	6 (47)
Taylor DAYNE [95]	5 (68)
MEN AT WORK [49]	5 (66)
QUEEN [35]	5 (62)
Belinda CARLISLE [93]	5 (62)
Jody WATLEY [88]	5 (61)
The COMMODORES [103]	5 (60)
Bob SEGER [84]	5 (58)
CHER [147]	5 (57)
TEARS FOR FEARS [68]	5 (57)
BLONDIE [45]	5 (56)
Robert PALMER [91]	5 (55)
Natalie COLE [165]	5 (55)
DeBARGE [133]	5 (53)
The PRETENDERS [151]	5 (53)
BEACH BOYS [159]	5 (52)
Ronnie MILSAP [163]	5 (52)
The GO-GO'S [109]	5 (52)
ABC [172]	5 (51)
TIFFANY [98]	5 (51)
Alan PARSONS Project [117]	5 (50)
Bruce HORNSBY And The RANGE [116]	5 (49)
Rick ASTLEY [106]	5 (47)
MOODY BLUES [219]	5 (42)
The FIXX [175]	5 (42)
WANG CHUNG [146]	5 (41)
Boz SCAGGS [220]	5 (41)
Barbra STREISAND [92]	5 (40)
Luther VANDROSS [260]	5 (30)
Paula ABDUL [65]	4 (59)
MILLI VANILLI [69]	4 (57)
John LENNON [59]	4 (57)
HUMAN LEAGUE [82]	4 (54)
Peter CETERA [108]	4 (53)
GUNS N' ROSES [111]	4 (47)
Michael McDONALD [135]	4 (45)
SIMPLE MINDS [115]	4 (45)
AEROSMITH [171]	4 (44)
MR. MISTER [101]	4 (44)
Samantha FOX [164]	4 (43)
James INGRAM [259]	4 (42)
Peter GABRIEL [145]	4 (42)
MIKE + THE MECHANICS [139]	4 (42)
STRAY CATS [136]	4 (42)
DR. HOOK [179]	4 (41)
SIMPLY RED [137]	4 (41)
SADE [185]	4 (41)
Willie NELSON [184]	4 (40)
David Lee ROTH [183]	4 (40)
Anita BAKER [166]	4 (39)
The MOTELS [192]	4 (38)
Robbie NEVIL [176]	4 (37)
GLASS TIGER [196]	4 (37)
Bette MIDLER [113]	4 (36)
The JACKSONS [180]	4 (36)
ORCHESTRAL MANOEUVERS IN THE DARK [228]	4 (36)
Julian LENNON [222]	4 (36)
ASIA [178]	4 (36)
CHEAP TRICK [130]	4 (34)
John CAFFERTY/BEAVER BROWN BAND [248]	4 (34)
Paul DAVIS [215]	4 (34)
Eric CLAPTON [262]	4 (33)
Steve PERRY [200]	4 (33)
Billy SQUIER [254]	4 (32)
Andy GIBB [191]	4 (32)
YES [132]	4 (31)
SUPERTRAMP [302]	4 (31)
The OUTFIELD [267]	4 (30)
NAKED EYES [250]	4 (30)
Sammy HAGAR [266]	4 (29)
ALABAMA [279]	4 (29)
EUROPE [225]	4 (29)
The WHISPERS [285]	4 (29)
Freddie JACKSON [275]	4 (28)
CINDERELLA [287]	4 (28)
Herb ALPERT [276]	4 (25)
Robert PLANT [353]	4 (25)
Paul CARRACK [330]	4 (24)
Mick JAGGER [350]	4 (23)
Rick JAMES [321]	4 (16)

Singles Acts: More Than 40 Consecutive Chart Weeks

*Denotes streak continuing into 1990

Act	Wks	Begin	End
Daryl HALL & John OATES	111	19-Jul-80	28-Aug-82
MADONNA	110	29-Oct-83	30-Nov-85
Bruce SPRINGSTEEN	95	26-May-84	15-Mar-86
CULTURE CLUB	88	4-Dec-82	4-Aug-84
DEF LEPPARD	86	10-Oct-87	27-May-89
Whitney HOUSTON	86	16-May-87	31-Dec-88
George MICHAEL	85	6-Jun-87	14-Jan-89
Janet JACKSON	83	22-Feb-86	19-Sep-87
WHAM!	82	8-Sep-84	29-Mar-86
NEW KIDS ON THE BLOCK	80*	25-Jun-88	30-Dec-89
Debbie GIBSON	77	9-May-87	22-Oct-88
Bobby BROWN	76*	23-Jul-88	30-Dec-89
Taylor DAYNE	76	10-Oct-87	18-Mar-89
Rick SPRINGFIELD	74	28-Mar-81	21-Aug-82
Sheena EASTON	74	14-Feb-81	10-Jul-82
Lionel RICHIE	73	17-Sep-83	2-Feb-85
Huey LEWIS and the NEWS	73	10-Sep-83	26-Jan-85
MEN AT WORK	73	10-Jul-82	26-Nov-83
Gloria ESTEFAN/MIAMI SOUND MACHINE	72	30-May-87	8-Oct-88
POINTER SISTERS	71	28-Jan-84	1-Jun-85
Laura BRANIGAN	71	10-Jul-82	12-Nov-83
Rod STEWART	70	7-May-88	2-Sep-89
Richard MARX	70	13-Jun-87	8-Oct-88
TOTO	69	17-Apr-82	6-Aug-83
HEART	68	1-Jun-85	13-Sep-86
JOURNEY	66	12-Apr-86	11-Jul-87
Cyndi LAUPER	66	17-Dec-83	16-Mar-85
INXS	65	24-Oct-87	14-Jan-89
EXPOSÉ	64	24-Jan-87	9-Apr-88
GENESIS	64	31-May-86	15-Aug-87
KOOL & The GANG	64	24-Nov-84	8-Feb-86
Billy JOEL	64	30-Jul-83	13-Oct-84
Michael JACKSON	63	8-Aug-87	15-Oct-88
Steve WINWOOD	63	7-Feb-87	16-Apr-88
Billy OCEAN	63	30-Nov-85	7-Feb-87
The CARS	63	10-Mar-84	18-May-85
GUNS N' ROSES	62	25-Jun-88	26-Aug-89
FOREIGNER	62	4-Jul-81	4-Sep-82
CHICAGO	61	4-Jun-88	29-Jul-89
Paula ABDUL	60*	12-Nov-88	30-Dec-89
Bryan ADAMS	60	3-Nov-84	21-Dec-85
AIR SUPPLY	59	9-Feb-80	21-Mar-81
Tina TURNER	58	19-May-84	22-Jun-85
Juice NEWTON	58	21-Feb-81	27-Mar-82
POISON	57	23-Apr-88	20-May-89
FLEETWOOD MAC	57	28-Mar-87	23-Apr-88
Bruce HORNSBY And The RANGE	57	26-Jul-86	22-Aug-87
John MELLENCAMP	57	24-Aug-85	20-Sep-86
SURVIVOR	57	15-Sep-84	12-Oct-85
Don HENLEY	56	10-Nov-84	30-Nov-85
BREATHE	55	16-Apr-88	29-Apr-89
EUROPE	54	24-Jan-87	30-Jan-88
Karyn WHITE	53	15-Oct-88	14-Oct-89
Eddie MONEY	53	16-Aug-86	15-Aug-87
Stevie NICKS	53	25-Jul-81	24-Jul-82
Christopher CROSS	53	16-Feb-80	14-Feb-81
MILLI VANILLI	52*	7-Jan-89	30-Dec-89
BON JOVI	52	24-Sep-88	16-Sep-89
AEROSMITH	52	3-Oct-87	24-Sep-88
Peter GABRIEL	52	10-May-86	2-May-87
Steve PERRY	52	7-Apr-84	30-Mar-85
PRINCE	52	26-Feb-83	18-Feb-84
Boz SCAGGS	52	29-Mar-80	21-Mar-81
Billy IDOL	51	28-Jan-84	12-Jan-85
VAN HALEN	50	21-May-88	29-Apr-89
STACEY Q	50	12-Jul-86	20-Jun-87
JEFFERSON STARSHIP/STARSHIP	50	7-Sep-85	16-Aug-86
Natalie COLE	49	25-Jul-87	25-Jun-88
U2	49	21-Mar-87	20-Feb-88
Robbie NEVIL	49	11-Oct-86	12-Sep-87
Robert PALMER	49	8-Feb-86	10-Jan-87
REO SPEEDWAGON	49	29-Nov-80	31-Oct-81
Bob SEGER	49	23-Feb-80	24-Jan-81
Belinda CARLISLE	48	26-Sep-87	20-Aug-88
Elton JOHN	48	7-May-83	31-Mar-84
LITTLE RIVER BAND	48	22-Aug-81	17-Jul-82
Eddie RABBITT	48	21-Jun-80	16-May-81
ROXETTE	47*	11-Feb-89	30-Dec-89
COVER GIRLS	47	28-Nov-87	15-Oct-88
The JETS	47	17-Oct-87	3-Sep-88
ORCHESTRAL MANOEUVERS IN THE DARK	47	31-Aug-85	19-Jul-86
DIRE STRAITS	47	13-Jul-85	31-May-86
The POLICE	47	4-Jun-83	21-Apr-84
DINO	46	4-Feb-89	16-Dec-89
Rick ASTLEY	46	19-Dec-87	29-Oct-88
WHITESNAKE	46	13-Jun-87	23-Apr-88
MIKE + THE MECHANICS	46	23-Nov-85	4-Oct-86
The HOOTERS	46	10-Aug-85	21-Jun-86
TEARS FOR FEARS	46	16-Mar-85	25-Jan-86
NAKED EYES	46	12-Mar-83	21-Jan-84
Paul DAVIS	46	7-Nov-81	18-Sep-82
Terence Trent D'ARBY	45	16-Jan-88	19-Nov-88
Anita BAKER	45	16-Aug-86	20-Jun-87
STING	45	8-Jun-85	12-Apr-86
DURAN DURAN	45	5-Nov-83	8-Sep-84
Dan FOGELBERG	45	29-Aug-81	3-Jul-82
CHEAP TRICK	44	9-Apr-88	4-Feb-89
GLASS TIGER	44	12-Jul-86	9-May-87
MR. MISTER	44	21-Sep-85	19-Jul-86
Aretha FRANKLIN	44	22-Jun-85	19-Apr-86
Freddie JACKSON	44	25-May-85	22-Mar-86
Jeffrey OSBORNE	44	16-Jul-83	12-May-84
INFORMATION SOCIETY	43	16-Jul-88	6-May-89
WILL TO POWER	43	18-Jun-88	8-Apr-89
Smokey ROBINSON	43	28-Mar-87	16-Jan-88
SOFT CELL	43	16-Jan-82	6-Nov-82
Kenny ROGERS	43	13-Jun-81	3-Apr-82
Cliff RICHARD	43	13-Sep-80	4-Jul-81
Pat BENATAR	43	26-Jul-80	16-May-81
ICEHOUSE	42	17-Oct-87	30-Jul-88
LOVERBOY	42	24-Aug-85	7-Jun-86
NIGHT RANGER	42	25-May-85	8-Mar-86
Alan PARSONS Project	42	6-Dec-80	19-Sep-81
Neil DIAMOND	42	1-Nov-80	15-Aug-81
BANGLES	41	15-Oct-88	22-Jul-89
Jody WATLEY	41	3-Oct-87	9-Jul-88
RUN-D.M.C.	41	26-Jul-86	2-May-87
The OUTFIELD	41	15-Feb-86	22-Nov-86
Phil COLLINS	41	9-Feb-85	16-Nov-85
NEW EDITION	41	22-Sep-84	29-Jun-85
The CLASH	41	17-Jul-82	23-Apr-83
The GO-GO'S	41	29-Aug-81	5-Jun-82
ROLLING STONES	41	22-Aug-81	29-May-82

Singles Acts: 3 Records on Chart Simultaneously

Note: Diana ROSS and NEW KIDS ON THE BLOCK also placed 3 in the Top 40 simultaneously. See their entries.

Act (Total Weeks)	Dates (Wks)
Diana ROSS (14)	10/25/80 - 1/24/81 (14)
	Top 40: 11/15 - 11/29/80
	Upside Down
	I'm Coming Out
	It's My Turn
MADONNA (10)	3/2/85 - 3/23/85 (4)
	Like A Virgin
	Material Girl
	Crazy For You
	4/27/85 - 6/1/85 (6)
	Material Girl
	Crazy For You
	Angel
Michael JACKSON (10)	5/28/83 - 7/2/83 (6)
	Billie Jean
	Beat It
	Wanna Be Startin' Somethin'
	7/23/83 - 8/13/83 (4)
	Beat It
	Wanna Be Startin' Somethin'
	Human Nature
NEW KIDS ON THE BLOCK (8)	9/16/89 - 11/4/89 (8)
	Top 40: 9/30 – 10/7/89
	Hangin' Tough
	Cover Girl
	Didn't I (Blow Your Mind)
Lionel RICHIE (4)	2/25/84 (1)
	All Night Long (All Night)
	Running With The Night
	Hello

Act (Total Weeks)	Dates (Wks)
Lionel RICHIE (cont'd)	1/17/87 - 1/31/87 (3)
	Love Will Conquer All
	Ballerina Girl
	Deep River Woman
QUEEN (4)	1/17/81 - 2/7/81 (4)
	Another One Bites The Dust
	Need Your Loving Tonight
	Flash's Theme AKA Flash
Bobby BROWN (4)	1/7/89 - 1/14/89 (2)
	Don't Be Cruel
	My Prerogative
	Roni
	3/25/89 - 4/1/89 (2)
	My Prerogative
	Roni
	Every Little Step
BILLY JOEL (1)	10/11/80 (1)
	It's Still Rock And Roll To Me
	Don't Ask Me Why
	Sometimes A Fantasy
John LENNON (1)	03/28/81 (1)
	(Just Like) Starting Over
	Woman
	Watching The Wheels
DOOBIE BROTHERS (1)	02/14/81 (1)
	One Step Closer
	Wynken Blynken And Nod
	Keep This Train A-Rollin'

Singles Acts: 2 Records In Top 10 Simultaneously

Act (Total Weeks)	Dates (Weeks)
John LENNON (2)	2/7/81 - 2/14/81
	(Just Like) Starting Over
	Woman
J. GEILS Band (1)	3/27/82
	Centerfold
	Freeze-Frame
NEW KIDS ON THE BLOCK (1)	11/11/89
	Cover Girl
	Didn't I (Blow Your Mind)
PRINCE And The REVOLUTION (2)	10/20/84 - 10/27/84
	Let's Go Crazy
	Purple Rain

Act (Total Weeks)	Dates (Weeks)
Diana ROSS (2)	11/1/80 - 11/8/80
	Upside Down
	I'm Coming Out
Michael JACKSON (4)	4/9/83 - 4/30/83
	Billie Jean
	Beat It
John MELLENCAMP (4)	9/11/82 - 10/2/82
	Hurts So Good
	Jack & Diane
MADONNA (3)	3/30/85 - 4/13/85
	Material Girl
	Crazy For You

Singles Acts: Highest Average Record Score

Five or more entries

Act [Singles Rank]	Entries	Avg Score
John LENNON [59]	5	3487
BLONDIE [45]	6	3474
Lionel RICHIE [4]	15	3471
MADONNA [2]	18	3321
MEN AT WORK [49]	6	3236
Whitney HOUSTON [8]	11	3233
George MICHAEL [23]	8	3163
Richard MARX [42]	7	3012
Michael JACKSON [1]	21	2920
Phil COLLINS [5]	14	2841
WHAM! [38]	8	2710
Paula ABDUL [65]	6	2681
TEARS FOR FEARS [68]	6	2630
AIR SUPPLY [11]	13	2627
Daryl HALL & John OATES [3]	22	2535
CULTURE CLUB [24]	10	2529
NEW KIDS ON THE BLOCK [77]	6	2433
Billy OCEAN [20]	11	2413
BANGLES [51]	8	2402
Taylor DAYNE [95]	5	2394
The POLICE [31]	10	2379
DURAN DURAN [7]	15	2373
Bobby BROWN [63]	7	2370
Janet JACKSON [41]	9	2352
HUMAN LEAGUE [82]	6	2321
Irene CARA [55]	8	2291
Christopher CROSS [34]	10	2270
Bob SEGER [84]	6	2262
MR. MISTER [101]	5	2259
Olivia NEWTON-JOHN [14]	14	2187
PRINCE And The REVOLUTION [40]	10	2145
BON JOVI [27]	12	2063
Debbie GIBSON [54]	9	2042
Cyndi LAUPER [19]	13	2042
GUNS N' ROSES [111]	5	1990
Paul McCARTNEY [30]	12	1990
EXPOSÉ [96]	6	1957
Huey LEWIS and the NEWS [12]	17	1956
TIFFANY [98]	6	1943
FOREIGNER [26]	13	1915
Gloria ESTEFAN/ MIAMI SOUND MACHINE [44]	11	1904
CHICAGO [15]	16	1816
Jody WATLEY [88]	7	1799
Bonnie TYLER [120]	5	1794
Bruce SPRINGSTEEN [25]	14	1786
Rick ASTLEY [106]	6	1770
Kenny ROGERS [9]	20	1761
PRINCE [17]	16	1761
John MELLENCAMP [10]	20	1759
Belinda CARLISLE [93]	7	1743
Billy JOEL [6]	22	1729
SPINNERS [127]	5	1728
Tina TURNER [37]	13	1696
YES [132]	5	1696
Michael McDONALD [135]	5	1679
Eddie RABBITT [74]	9	1668
STRAY CATS [136]	5	1661
SIMPLY RED [137]	5	1655
KOOL & The GANG [13]	19	1627
FINE YOUNG CANNIBALS [141]	5	1616
Joan JETT & The BLACKHEARTS [78]	9	1613
QUEEN [35]	14	1599
Diana ROSS [16]	18	1591
Bruce HORNSBY And The RANGE [116]	6	1581
Juice NEWTON [70]	10	1574
ROLLING STONES [53]	12	1571
DIRE STRAITS [148]	5	1564
POISON [105]	7	1545
Rick SPRINGFIELD [21]	17	1534
TOTO [61]	11	1523
Deniece WILLIAMS [156]	5	1522
Ray PARKER Jr. [85]	9	1506
STYX [86]	9	1496
Stevie NICKS [57]	12	1496
WHITESNAKE [122]	6	1490
Peter CETERA [108]	7	1489
The GO-GO'S [109]	7	1463
DEF LEPPARD [67]	11	1438
REO SPEEDWAGON [39]	15	1435
DeBARGE [133]	6	1410
J. GEILS Band [81]	10	1396
VAN HALEN [50]	14	1382
Don HENLEY [90]	9	1381
HEART [46]	15	1380
MIKE + THE MECHANICS [139]	6	1377
Bette MIDLER [113]	7	1371
Stevie WONDER [28]	18	1368
Steve WINWOOD [52]	14	1365
Smokey ROBINSON [75]	11	1353
Elton JOHN [18]	20	1353
AEROSMITH [171]	5	1343
POINTER SISTERS [22]	19	1334
Kenny LOGGINS [36]	17	1313
SURVIVOR [32]	18	1310
CHER [147]	6	1308
GENESIS [48]	15	1306
Laura BRANIGAN [71]	12	1305
Robbie NEVIL [176]	5	1305
Bryan ADAMS [60]	13	1296
ASIA [178]	5	1289
Neil DIAMOND [62]	13	1285
U2 [80]	11	1283
Linda RONSTADT [99]	9	1269
STING [123]	7	1264
Sheena EASTON [33]	18	1264
Glenn FREY [89]	10	1259
JOURNEY [29]	19	1257
Steve MILLER Band [112]	8	1244
INXS [76]	12	1229
Robert PALMER [91]	10	1229
PET SHOP BOYS [104]	9	1225
ATLANTIC STARR [128]	7	1221
SADE [185]	5	1212
The JETS [94]	10	1211
Samantha FOX [164]	6	1204
The CARS [72]	13	1203
Greg KIHN Band [187]	5	1196
Anita BAKER [166]	6	1186
Dan FOGELBERG [79]	12	1178
FLEETWOOD MAC [73]	13	1177
LITTLE RIVER BAND [107]	9	1171
JEFFERSON STARSHIP/STARSHIP [43]	18	1166
Billy IDOL [87]	11	1161
Andy GIBB [191]	5	1148
Kim CARNES [66]	14	1146
Barbra STREISAND [92]	11	1111
Bob SEGER & The SILVER BULLET BAND [102]	10	1109
EURYTHMICS [64]	15	1104
KLYMAXX [204]	5	1099
Donna SUMMER [58]	16	1090
QUARTERFLASH [157]	7	1083
John PARR [208]	5	1081
Pat BENATAR [56]	17	1075
Joe JACKSON [212]	5	1058
Patti AUSTIN [182]	6	1042
Paul DAVIS [215]	5	1040
Kim WILDE [216]	5	1036
THOMPSON TWINS [100]	11	1027
Rod STEWART [47]	20	1024
Willie NELSON [184]	6	1018
Boz SCAGGS [220]	5	1011
Paul YOUNG [167]	7	1000

Singles Acts: Number 1s

Alone or in collaboration

Act [Singles Rank]	No. 1s (Entries)
Michael JACKSON [1]	9 (21)
Whitney HOUSTON [8]	7 (11)
Phil COLLINS [5]	7 (14)
MADONNA [2]	7 (18)
George MICHAEL [23]	5 (8)
Lionel RICHIE [4]	5 (15)
Daryl HALL & John OATES [3]	5 (22)
BON JOVI [27]	4 (12)
MILLI VANILLI [69]	3 (4)
BLONDIE [45]	3 (6)
Paula ABDUL [65]	3 (6)
Richard MARX [42]	3 (7)
WHAM! [38]	3 (8)
Billy OCEAN [20]	3 (11)
Huey LEWIS and the NEWS [12]	3 (17)
JEFFERSON STARSHIP/STARSHIP [43]	3 (18)
Billy JOEL [6]	3 (22)
ROXETTE [131]	2 (3)
LISA LISA And CULT JAM [142]	2 (3)
MR. MISTER [101]	2 (5)
SIMPLY RED [137]	2 (5)
FINE YOUNG CANNIBALS [141]	2 (5)
MEN AT WORK [49]	2 (6)
TEARS FOR FEARS [68]	2 (6)
NEW KIDS ON THE BLOCK [77]	2 (6)
HUMAN LEAGUE [82]	2 (6)
TIFFANY [98]	2 (6)
Rick ASTLEY [106]	2 (6)
Jennifer WARNES [211]	2 (6)
Peter CETERA [108]	2 (7)
BANGLES [51]	2 (8)
Janet JACKSON [41]	2 (9)
Debbie GIBSON [54]	2 (9)
Christopher CROSS [34]	2 (10)
PRINCE And The REVOLUTION [40]	2 (10)
Gloria ESTEFAN/ MIAMI SOUND MACHINE [44]	2 (11)
U2 [80]	2 (11)
Paul McCARTNEY [30]	2 (12)
Cyndi LAUPER [19]	2 (13)
Olivia NEWTON-JOHN [14]	2 (14)
QUEEN [35]	2 (14)
Steve WINWOOD [52]	2 (14)
DURAN DURAN [7]	2 (15)
REO SPEEDWAGON [39]	2 (15)
HEART [46]	2 (15)
CHICAGO [15]	2 (16)
PRINCE [17]	2 (16)
Diana ROSS [16]	2 (18)
Stevie WONDER [28]	2 (18)
Kenny ROGERS [9]	2 (20)
Paul McCARTNEY & WINGS [140]	1 (1)
USA For AFRICA [197]	1 (1)
VANGELIS [203]	1 (1)
Jan HAMMER [284]	1 (1)
Bobby McFERRIN [292]	1 (1)
KC & The SUNSHINE BAND [160]	1 (2)
LIPPS, INC. [162]	1 (2)
Michael SEMBELLO [177]	1 (2)
FALCO [221]	1 (2)
DEXYS MIDNIGHT RUNNERS [230]	1 (2)
Gregory ABBOTT [241]	1 (2)
CLUB NOUVEAU [247]	1 (2)
BAD ENGLISH [268]	1 (2)
SHERIFF [301]	1 (2)
Marilyn MARTIN [332]	1 (2)
CAPTAIN & TENNILLE [119]	1 (3)
PINK FLOYD [134]	1 (3)
Toni BASIL [186]	1 (3)
BOSTON [199]	1 (3)
READY FOR THE WORLD [206]	1 (3)
LOS LOBOS [207]	1 (3)
A-HA [223]	1 (3)
MARTIKA [231]	1 (3)
ESCAPE CLUB [234]	1 (3)
UB40 [238]	1 (3)
Billy VERA & The BEATERS [257]	1 (3)
Michael DAMIAN [298]	1 (3)
Joe COCKER [361]	1 (3)
Bill MEDLEY [388]	1 (3)
SIMPLE MINDS [115]	1 (4)
George HARRISON [138]	1 (4)
Terence Trent D'ARBY [173]	1 (4)
STARS ON [198]	1 (4)
CUTTING CREW [202]	1 (4)
WILL TO POWER [235]	1 (4)
John LENNON [59]	1 (5)
GUNS N' ROSES [111]	1 (5)
Bonnie TYLER [120]	1 (5)
YES [132]	1 (5)
Michael McDONALD [135]	1 (5)
DIRE STRAITS [148]	1 (5)
Deniece WILLIAMS [156]	1 (5)
John PARR [208]	1 (5)
Kim WILDE [216]	1 (5)
Bob SEGER [84]	1 (6)
EXPOSÉ [96]	1 (6)
Bruce HORNSBY And The RANGE [116]	1 (6)
WHITESNAKE [122]	1 (6)
MIKE + THE MECHANICS [139]	1 (6)
Patti AUSTIN [182]	1 (6)
Bobby BROWN [63]	1 (7)
Belinda CARLISLE [93]	1 (7)
POISON [105]	1 (7)
Bette MIDLER [113]	1 (7)
ATLANTIC STARR [128]	1 (7)
Paul YOUNG [167]	1 (7)
Patti LaBELLE [224]	1 (7)
BERLIN [227]	1 (7)
Irene CARA [55]	1 (8)
Steve MILLER Band [112]	1 (8)
Peter GABRIEL [145]	1 (8)
Eddie RABBITT [74]	1 (9)
Joan JETT & The BLACKHEARTS [78]	1 (9)
Ray PARKER Jr. [85]	1 (9)
PET SHOP BOYS [104]	1 (9)
John WAITE [154]	1 (9)
Dolly PARTON [155]	1 (9)
CULTURE CLUB [24]	1 (10)
The POLICE [31]	1 (10)
J. GEILS Band [81]	1 (10)
Robert PALMER [91]	1 (10)
CHEAP TRICK [130]	1 (10)
BEACH BOYS [159]	1 (10)
TOTO [61]	1 (11)
DEF LEPPARD [67]	1 (11)
Billy IDOL [87]	1 (11)
Barbra STREISAND [92]	1 (11)
BANANARAMA [129]	1 (11)
INXS [76]	1 (12)
AIR SUPPLY [11]	1 (13)
FOREIGNER [26]	1 (13)
Tina TURNER [37]	1 (13)
Bryan ADAMS [60]	1 (13)
Dionne WARWICK [97]	1 (13)
VAN HALEN [50]	1 (14)
Kim CARNES [66]	1 (14)
David BOWIE [83]	1 (14)
GENESIS [48]	1 (15)
EURYTHMICS [64]	1 (15)
Aretha FRANKLIN [110]	1 (15)
Rick SPRINGFIELD [21]	1 (17)
Kenny LOGGINS [36]	1 (17)
SURVIVOR [32]	1 (18)
Sheena EASTON [33]	1 (18)
KOOL & The GANG [13]	1 (19)
John MELLENCAMP [10]	1 (20)

Acts with Their Albums

Acts are listed alphabetically; albums are listed in order of chart strength. Act scoring only includes points earned between 1/1/80 and 12/31/89; however, it includes all albums charted during that period. Act name* is reproduced here if different from grouping name. Allocated credit* is used for infrequent or one-off collaborations rather than creating a new Grouping. In this case, credit is divided among the four underlying Groupings. See "Methodology" in the appendix

Albums that charted only during the '80s:

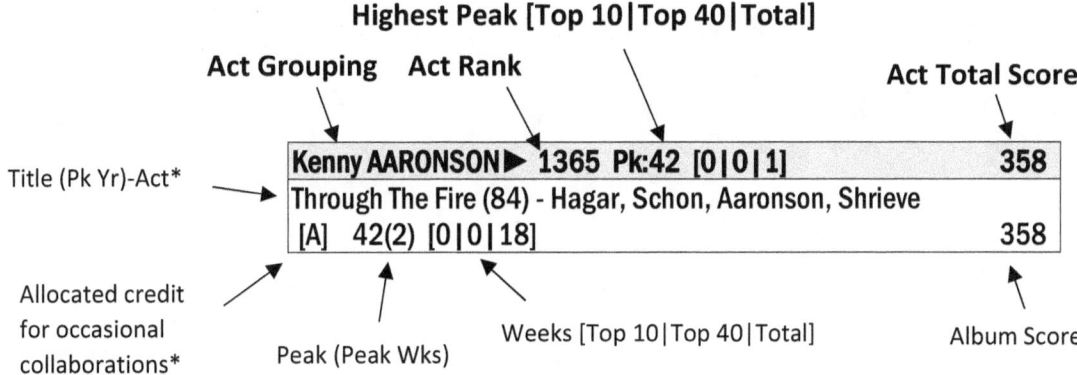

Albums that charted during the '80s but peaked in another decade:

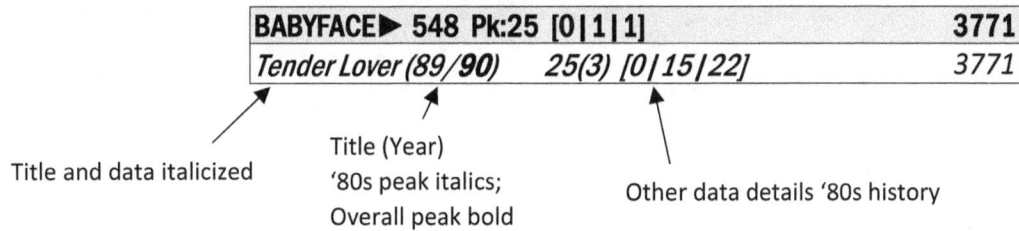

Albums that peaked during the '80s but also charted in another decade:

Paula ABDUL▶ 115 Pk:1 [1|1|1]	25251
Forever Your Girl (88) 1(1) [44|49|76]	*25251*

Only score is italicized.

Acts With Albums

Acts With Albums

Act ▶ Rank Peak [Top10s|Top40s|Total] Title (Yr) Pk(Wks) [T10 Wks|T40Wks|TotWks]	Score in '80s
A	
Kenny AARONSON ▶ 1365 Pk:42 [0|0|1]	358
Through The Fire (84)-Hagar, Schon, Aaronson, Shrieve [A] 42(2) [0|0|18]	358
ABBA ▶ 296 Pk:17 [0|2|5]	9596
Super Trouper (80) 17(3) [0|16|38]	5525
The Visitors (82) 29(2) [0|6|17]	1908
The Singles (The First Ten Years) (82) 62(6) [0|0|18]	1302
Greatest Hits Vol.2 (80) 46(2) [0|0|12]	851
Voulez-Vous (79/80) 200(1) [0|0|1]	*10.0*
Gregory ABBOTT ▶ 435 Pk:22 [0|1|2]	5290
Shake You Down (86) 22(1) [0|16|36]	5010
I'll Prove It To You (88) 132(2) [0|0|9]	280
ABC ▶ 220 Pk:24 [0|2|4]	13774
The Lexicon Of Love (82) 24(10) [0|20|39]	6041
How To Be A…Zillionaire! (85) 30(2) [0|15|41]	5108
Alphabet City (87) 48(3) [0|0|25]	1681
Beauty Stab (83) 69(2) [0|0|14]	944
Paula ABDUL ▶ 115 Pk:1 [1|1|1]	25251
Forever Your Girl (88) 1(1) [44|49|76]	*25251*
Colonel ABRAMS ▶ 1215 Pk:75 [0|0|1]	534
Colonel Abrams (86) 75(2) [0|0|11]	534
ACCEPT ▶ 690 Pk:74 [0|0|4]	2475
Balls To The Wall (84) 74(1) [0|0|26]	1193
Metal Heart (85) 94(2) [0|0|14]	674
Russian Roulette (86) 114(2) [0|0|9]	350
Eat The Heat (89) 139(2) [0|0|9]	259
AC/DC ▶ 24 Pk:1 [3|7|10]	58014
Back In Black (80) 4(3) [23|45|100]	*19962*
Dirty Deeds Done Dirt Cheap (81) 3(6) [12|19|55]	10890
For Those About To Rock (We Salute You) (81) 1(3) [12|16|30]	9505
Blow Up Your Video (88) 12(1) [0|12|24]	3961
Who Made Who (Soundtrack) (86) 33(4) [0|6|42]	3647
Flick Of The Switch (83) 15(2) [0|11|23]	3611
Highway To Hell (79/80) 63(2) [0|0|64]	*2699*
Fly On The Wall (85) 32(2) [0|5|30]	2548
'74 Jailbreak (84) 76(2) [0|0|14]	671
High Voltage (81) 146(1) [0|0|19]	521
Bryan ADAMS ▶ 37 Pk:1 [3|3|4]	47755
Reckless (84) 1(2) [40|66|83]	28951
Cuts Like A Knife (83) 8(3) [6|24|89]	10636
Into The Fire (87) 7(1) [5|23|33]	7709
You Want It, You Got It (82) 118(2) [0|0|13]	461
King Sunny ADE & His AFRICAN BEATS ▶ 874 Pk:91 [0|0|2]	1467
JuJu Music (83) 111(1) [0|0|29]	1039
Synchro System (83) 91(1) [0|0|10]	427
The ADVENTURES ▶ 1579 Pk:144 [0|0|1]	187
The Sea Of Love (88) 144(1) [0|0|9]	187
AEROSMITH ▶ 87 Pk:5 [1|5|8]	30948
Permanent Vacation (87) 11(4) [0|52|67]	15005
Pump (89) 5(3) [13|14|42]	*6345*
Night In The Ruts (80) 14(2) [0|9|14]	*2680*
Done With Mirrors (85) 36(3) [0|5|28]	2504
Rock In A Hard Place (82) 32(4) [0|6|19]	1794
Aerosmith's Greatest Hits (80) 53(3) [0|0|40]	1670
Classics Live (86) 84(1) [0|0|12]	605
Gems (88) 133(1) [0|0|11]	346
AFTER 7 ▶ 1379 Pk:132 [0|0|1]	344
After 7 (89/90) 132(2) [0|0|12]	*344*
AFTER THE FIRE ▶ 661 Pk:25 [0|1|1]	2685
ATF 25(3) [0|8|20]	2685
A-HA ▶ 329 Pk:15 [0|1|3]	8159
Hunting High And Low (85) 15(2) [0|25|47]	7168
Scoundrel Days (86) 74(2) [0|0|20]	841

Act ▶ Rank Peak [Top10s|Top40s|Total] Title (Yr) Pk(Wks) [T10 Wks|T40Wks|TotWks]	Score in '80s
Stay On These Roads (88) 148(2) [0|0|6]	150
AIR SUPPLY ▶ 72 Pk:7 [2|5|6]	34587
The One That You Love (81) 10(4) [4|25|60]	9643
Greatest Hits (83) 7(1) [8|26|51]	9329
Lost In Love (80) 22(3) [0|19|104]	8926
Now And Forever (82) 25(3) [0|8|38]	3618
Air Supply (85) 26(4) [0|8|21]	2585
Hearts In Motion (86) 84(2) [0|0|9]	485
ALABAMA ▶ 28 Pk:10 [1|6|12]	55776
Feels So Right (81) 16(2) [0|42|161]	14947
Mountain Music (82) 14(2) [0|20|114]	10820
The Closer You Get (83) 10(1) [1|15|70]	8506
Roll On (84) 21(2) [0|17|62]	6245
Greatest Hits (86) 24(3) [0|15|38]	4344
40 Hour Week (85) 28(3) [0|5|40]	3143
The Touch (86) 42(1) [0|0|30]	1983
Just Us (87) 55(1) [0|0|28]	1892
Southern Star (89) 62(2) [0|0|21]	1283
My Home's In Alabama (80) 71(1) [0|0|21]	1179
Live (88) 76(3) [0|0|19]	933
Christmas (85) 75(2) [0|0|9]	502
The ALARM ▶ 325 Pk:39 [0|1|6]	8331
Strength (85) 39(3) [0|3|36]	3422
Declaration (84) 50(2) [0|0|22]	1712
Eye Of The Hurricane (87) 77(2) [0|0|30]	1465
The Alarm (83) 126(1) [0|0|37]	963
Change. (89) 75(1) [0|0|12]	*672*
Electric Folklore Live (88) 167(2) [0|0|5]	96.6
Gerald ALBRIGHT ▶ 1798 Pk:181 [0|0|1]	71.3
Just Between Us (88) 181(2) [0|0|5]	71.3
AL B. SURE! ▶ 353 Pk:20 [0|1|1]	7619
In Effect Mode (88) 20(1) [0|25|54]	7619
ALCATRAZZ ▶ 970 Pk:128 [0|0|3]	1093
No Parole From Rock 'n' Roll (84) 128(1) [0|0|18]	450
Disturbing The Peace (85) 145(2) [0|0|16]	384
Live Sentence (84) 133(1) [0|0|10]	260
--Alice COOPER see: COOPER, Alice	
Deborah ALLEN ▶ 1016 Pk:67 [0|0|1]	953
Cheat The Night (83) 67(1) [0|0|20]	953
Donna ALLEN ▶ 1344 Pk:133 [0|0|1]	396
Perfect Timing (87) 133(1) [0|0|13]	396
Peter ALLEN ▶ 1145 Pk:123 [0|0|2]	639
Bi-Coastal (80) 123(1) [0|0|10]	521
Not The Boy Next Door (83) 170(2) [0|0|6]	119
Gregg ALLMAN Band ▶ 580 Pk:30 [0|1|2]	3421
I'm No Angel (87) 30(3) [0|6|28]	3037
Just Before The Bullets Fly (88) 117(2) [0|0|11]	384
ALLMAN BROTHERS Band ▶ 588 Pk:27 [0|1|4]	3342
Reach For The Sky (80) 27(2) [0|6|13]	1856
Brothers Of The Road (81) 44(1) [0|0|12]	1049
Dreams (89) 103(2) [0|0|11]	398
The Best Of The Allman Brothers Band (81) 189(1) [0|0|3]	39.1
Marc ALMOND ▶ 1469 Pk:144 [0|0|1]	271
The Stars We Are (89) 144(2) [0|0|11]	271
Herb ALPERT ▶ 251 Pk:18 [0|3|7]	11691
Keep Your Eye On Me (87) 18(3) [0|17|31]	5016
Rise (79/80) 19(1) [0|8|27]	*2986*
Beyond (80) 28(2) [0|4|12]	1380
Fandango (82) 100(2) [0|0|26]	1041
Magic Man (81) 61(1) [0|0|10]	725
Blow Your Own Horn (83) 120(1) [0|0|8]	301
Wild Romance (85) 151(2) [0|0|10]	242
Herb ALPERT & The TIJUANA BRASS ▶ 1227 Pk:75 [0|0|1]	520
Bullish (84) 75(1) [0|0|10]	520
ALPHAVILLE ▶ 1412 Pk:174 [0|0|2]	318
Forever Young (84) 180(1) [0|0|15]	217

Act ▶ Rank Peak [Top10s|Top40s|Total] Title (Yr) Pk(Wks) [T10 Wks|T40Wks|TotWks]	Score in '80s
Afternoons In Utopia (86) 174(1) [0|0|6]	101
Dave ALVIN ▶ 1303 Pk:116 [0|0|1]	434
Romeo's Escape (87) 116(1) [0|0|13]	434
AMAZING RHYTHM ACES ▶ 1826 Pk:175 [0|0|1]	58.9
How The Hell Do You Spell Rythum (80) 175(1) [0|0|3]	58.9
AMBROSIA ▶ 517 Pk:25 [0|1|2]	4102
One Eighty (80) 25(1) [0|7|33]	3819
Road Island (82) 115(2) [0|0|7]	283
AMERICA ▶ 586 Pk:41 [0|0|4]	3348
View From The Ground (82) 41(8) [0|0|28]	2540
Your Move (83) 81(1) [0|0|14]	605
Alibi (80) 142(2) [0|0|6]	154
Perspective (84) 185(1) [0|0|3]	48.5
Carl ANDERSON ▶ 1164 Pk:87 [0|0|1]	597
Carl Anderson (86) 87(2) [0|0|12]	597
John ANDERSON ▶ 1017 Pk:58 [0|0|2]	952
Wild & Blue (83) 58(1) [0|0|12]	846
All The People Are Talkin' (83) 163(1) [0|0|5]	106
Jon ANDERSON ▶ 1253 Pk:143 [0|0|3]	492
Song Of Seven (80) 143(1) [0|0|11]	302
3 Ships (85) 166(2) [0|0|5]	103
Animation (82) 176(1) [0|0|5]	86.6
Laurie ANDERSON ▶ 765 Pk:60 [0|0|5]	1980
Mister Heartbreak (84) 60(2) [0|0|19]	1151
Big Science (82) 124(1) [0|0|12]	368
Home Of The Brave (Soundtrack) (86) 145(2) [0|0|12]	280
Strange Angels (89) 171(1) [0|0|7]	*124*
United States Live (85) 192(1) [0|0|5]	58.2
Michael ANDERSON ▶ 1908 Pk:194 [0|0|1]	25.4
Sound Alarm (88) 194(2) [0|0|2]	25.4
ANGEL ▶ 1688 Pk:149 [0|0|1]	122
Live Without A Net (80) 149(1) [0|0|4]	122
ANGEL CITY ▶ 1293 Pk:133 [0|0|3]	441
Darkroom (80) 133(1) [0|0|6]	200
Face To Face (80) 152(2) [0|0|7]	152
Night Attack (82) 174(1) [0|0|5]	88.8
ANIMAL LOGIC ▶ 1746 Pk:147 [0|0|1]	90.1
Animal Logic (89/90) 147(1) [0|0|4]	*90.1*
The ANIMALS ▶ 1110 Pk:66 [0|0|2]	713
Ark (83) 66(1) [0|0|10]	661
Rip It To Shreds-The Animals Greatest Hits Live (84) 193(2) [0|0|4]	51.4
ANIMOTION ▶ 488 Pk:28 [0|1|3]	4412
Animotion (85) 28(1) [0|7|30]	3137
Strange Behavior (86) 71(2) [0|0|14]	726
Animotion(2) (89) 110(1) [0|0|17]	549
Paul ANKA ▶ 1444 Pk:156 [0|0|2]	290
Walk A Fine Line (83) 156(1) [0|0|8]	182
Both Sides Of Love (81) 171(2) [0|0|6]	108
Adam ANT ▶ 295 Pk:16 [0|1|5]	9635
Friend Or Foe (82) 16(3) [0|16|36]	5129
Kings Of The Wild Frontier (81)-Adam And The Ants 44(2) [0|0|35]	2132
Strip (83) 65(3) [0|0|16]	1337
Prince Charming (81)-Adam And The Ants 94(2) [0|0|21]	818
Vive Le Rock (85) 131(2) [0|0|7]	220
ANTHRAX ▶ 350 Pk:30 [0|1|4]	7679
State Of Euphoria (88) 30(2) [0|5|36]	2569
I'm The Man (87) 53(1) [0|0|40]	2541
Among The Living (87) 62(2) [0|0|36]	1995
Spreading The Disease (85) 113(1) [0|0|18]	575
ANVIL ▶ 1906 Pk:191 [0|0|1]	25.9
Strength Of Steel (87) 191(1) [0|0|2]	25.9
APOLLONIA 6 ▶ 1040 Pk:62 [0|0|1]	869
Apollonia 6 (84) 62(2) [0|0|17]	869

Acts With Albums

Act ▶ Rank Peak [Top10s|Top40s|Total] Title (Yr) Pk(Wks) [T10 Wks|T40Wks|TotWks]	Score in '80s
APRIL WINE ▶ 339 Pk:26 [0|2|5]	7858
The Nature Of The Beast (81) 26(1) [0|12|34]	4127
Power Play (82) 37(2) [0|3|20]	1837
Harder...Faster (79/80) 101(1) [0|0|32]	*1125*
Animal Grace (84) 62(1) [0|0|12]	700
Walking Through Fire (85) 174(2) [0|0|4]	70.2
ARABIAN PRINCE ▶ 1911 Pk:193 [0|0|1]	24.6
Brother Arab (89) 193(1) [0|0|2]	24.6
ARCADIA ▶ 659 Pk:23 [0|1|1]	2699
So Red The Rose (85) 23(2) [0|9|17]	2699
Joan ARMATRADING ▶ 314 Pk:28 [0|2|8]	8641
Me Myself I (80) 28(2) [0|5|23]	2296
The Key (83) 32(1) [0|3|22]	1726
Walk Under Ladders (81) 88(2) [0|0|32]	1380
Secret Secrets (85) 73(1) [0|0|19]	995
Sleight Of Hand (86) 68(1) [0|0|16]	944
The Shouting Stage (88) 100(1) [0|0|13]	498
How Cruel (80) 136(1) [0|0|14]	*457*
Track Record (84) 113(1) [0|0|10]	347
ARMORED SAINT ▶ 875 Pk:108 [0|0|3]	1467
Delirious Nomad (85) 108(2) [0|0|19]	614
Raising Fear (87) 114(2) [0|0|12]	440
March Of The Saint (84) 138(2) [0|0|16]	413
Steve ARRINGTON ▶ 1010 Pk:101 [0|0|3]	962
Steve Arrington's Hall Of Fame: I (83)- Steve Arrington's Hall Of Fame 101(1) [0|0|17]	641
Positive Power (84)- Steve Arrington's Hall Of Fame 141(1) [0|0|9]	250
Dancin' In The Key Of Life (85) 185(2) [0|0|5]	70.9
ART IN AMERICA ▶ 1838 Pk:176 [0|0|1]	54.3
Art In America (83) 176(1) [0|0|3]	54.3
ARTISTS UNITED AGAINST APARTHEID ▶ 783 Pk:31 [0|1|1]	1904
Sun City (85) 31(2) [0|5|18]	1904
ART OF NOISE ▶ 528 Pk:53 [0|0|4]	4018
In Visible Silence (86) 53(2) [0|0|30]	2504
(Who's Afraid Of?) The Art Of Noise! (84) 85(3) [0|0|13]	644
The Best Of The Art Of Noise (88) 83(1) [0|0|14]	636
In No Sense? Nonsense! (87) 134(1) [0|0|9]	235
The A's ▶ 1626 Pk:146 [0|0|1]	158
A Woman's Got The Power (81) 146(1) [0|0|7]	158
ASHFORD & SIMPSON ▶ 318 Pk:29 [0|2|8]	8420
Solid (84) 29(2) [0|10|36]	3634
Street Opera (82) 45(3) [0|0|20]	1613
A Musical Affair (80) 38(2) [0|2|12]	1046
Real Love (86) 74(2) [0|0|18]	890
High-Rise (83) 84(1) [0|0|12]	604
Love Or Physical (89) 135(2) [0|0|8]	257
Performance (81) 125(1) [0|0|6]	202
Stay Free (79/80) 127(1) [0|0|5]	*176*
ASIA ▶ 95 Pk:1 [2|2|3]	28736
Asia (82) 1(9) [27|35|64]	22988
Alpha (83) 6(4) [5|11|25]	4639
Astra (85) 67(3) [0|0|17]	1109
ASLEEP AT THE WHEEL ▶ 1896 Pk:191 [0|0|1]	27.7
Framed (80) 191(1) [0|0|2]	27.7
Jon ASTLEY ▶ 1479 Pk:135 [0|0|1]	266
Everybody Loves The Pilot (Except The Crew) (87) 135(2) [0|0|10]	266
Rick ASTLEY ▶ 205 Pk:10 [1|2|2]	14536
Whenever You Need Somebody (88) 10(2) [2|39|60]	11390
Hold Me In Your Arms (89) 19(3) [0|0|23]	3145
ASWAD ▶ 1694 Pk:173 [0|0|1]	121
Distant Thunder (88) 173(3) [0|0|7]	121
Chet ATKINS ▶ 1455 Pk:145 [0|0|1]	281
Stay Tuned (85) 145(2) [0|0|13]	281

Act ▶ Rank Peak [Top10s|Top40s|Total] Title (Yr) Pk(Wks) [T10 Wks|T40Wks|TotWks]	Score in '80s
ATLANTA ▶ 1540 Pk:140 [0|0|1]	219
Pictures (84) 140(2) [0|0|7]	219
ATLANTA RHYTHM SECTION ▶ 871 Pk:65 [0|0|3]	1482
The Boys From Doraville (80) 65(2) [0|0|11]	702
Quinella (81) 70(1) [0|0|16]	699
Are You Ready! (79/80) 164(1) [0|0|4]	*81.5*
ATLANTIC STARR ▶ 178 Pk:17 [0|3|6]	16908
As The Band Turns (85) 17(2) [0|14|68]	6234
All In The Name Of Love (87) 18(1) [0|15|31]	4738
Brilliance (82) 18(2) [0|7|29]	2923
Radiant (81) 47(1) [0|0|30]	1845
Yours Forever (83) 91(1) [0|0|28]	956
We're Movin' Up (89) 125(2) [0|0|6]	211
AUDIO TWO ▶ 1825 Pk:185 [0|0|1]	59.0
What More Can I Say? (88) 185(2) [0|0|4]	59.0
Barbara Ann AUER ▶ 1416 Pk:145 [0|0|1]	312
Aerobic Dancing (81) 145(2) [0|0|15]	312
AURRA ▶ 776 Pk:38 [0|1|2]	1941
A Little Love (82) 38(2) [0|3|15]	1416
Send Your Love (81) 103(1) [0|0|13]	525
Patti AUSTIN ▶ 470 Pk:36 [0|1|3]	4617
Every Home Should Have One (81) 36(8) [0|11|44]	3813
Patti Austin (84) 87(1) [0|0|18]	744
Gettin' Away With Murder (85) 182(1) [0|0|4]	59.5
AUTOGRAPH ▶ 472 Pk:29 [0|1|3]	4572
Sign In Please (85) 29(2) [0|12|29]	3353
That's The Stuff (85) 92(2) [0|0|15]	640
Loud And Clear (87) 108(2) [0|0|15]	579
AVERAGE WHITE BAND/AWB ▶ 1242 Pk:116 [0|0|2]	504
Shine (80)-Average White Band 116(2) [0|0|12]	469
Volume VIII (80)-Average White Band 182(1) [0|0|2]	35.8
AXE ▶ 982 Pk:81 [0|0|2]	1046
Offering (82) 81(3) [0|0|20]	892
Nemesis (83) 156(1) [0|0|6]	154
Roy AYERS ▶ 1009 Pk:82 [0|0|4]	964
No Stranger To Love (80) 82(2) [0|0|15]	702
Feeling Good (82) 160(1) [0|0|7]	155
Love Fantasy (80) 157(1) [0|0|3]	83.2
Africa, Center Of The World (81) 197(2) [0|0|2]	22.8
AZTEC CAMERA ▶ 1245 Pk:129 [0|0|4]	500
High Land, Hard Rain (83) 129(1) [0|0|10]	306
Knife (84) 175(2) [0|0|6]	101
Aztec Camera (85) 181(2) [0|0|3]	54.0
Love (87) 193(3) [0|0|3]	39.6

B

Act ▶ Rank Peak [Top10s|Top40s|Total] Title (Yr) Pk(Wks) [T10 Wks|T40Wks|TotWks]	Score in '80s
BABYFACE ▶ 548 Pk:25 [0|1|1]	3771
Tender Lover (89/90) 25(3) [0|15|22]	*3771*
BABYLON A.D. ▶ 1678 Pk:141 [0|0|1]	128
Babylon A.D. (89/90) 141(1) [0|0|5]	*128*
The BABYS ▶ 641 Pk:42 [0|0|3]	2833
Union Jacks (80) 42(1) [0|0|22]	1880
On The Edge (80) 71(2) [0|0|15]	740
Anthology (81) 138(1) [0|0|7]	214
BACHMAN-TURNER OVERDRIVE ▶ 1901 Pk:191 [0|0|1]	26.8
Bachman Turner Overdrive(2) (84) 191(1) [0|0|1]	26.8
BAD COMPANY ▶ 444 Pk:26 [0|1|4]	5159
Rough Diamonds (82) 26(3) [0|6|18]	2284
Dangerous Age (88) 58(2) [0|0|40]	2136
Fame And Fortune (86) 106(2) [0|0|9]	379
10 From 6 (86) 137(1) [0|0|14]	360
BAD ENGLISH ▶ 541 Pk:21 [0|1|1]	3825
Bad English (89) 21(2) [0|11|25]	*3825*

Act ▶ Rank Peak [Top10s|Top40s|Total] Title (Yr) Pk(Wks) [T10 Wks|T40Wks|TotWks]	Score in '80s
BADFINGER ▶ 1650 Pk:155 [0|0|1]	142
Say No More (81) 155(1) [0|0|6]	142
BADLANDS ▶ 809 Pk:57 [0|0|1]	1758
Badlands (89) 57(2) [0|0|26]	1758
Philip BAILEY ▶ 433 Pk:22 [0|1|3]	5327
Chinese Wall (84) 22(3) [0|11|35]	4067
Continuation (83) 71(1) [0|0|14]	715
Inside Out (86) 84(2) [0|0|11]	546
Razzy BAILEY ▶ 1726 Pk:176 [0|0|2]	101
Feelin' Right (82) 176(1) [0|0|4]	65.8
Makin' Friends (81) 183(2) [0|0|6]	35.4
Scott BAIO ▶ 1816 Pk:181 [0|0|1]	62.6
Scott Baio (82) 181(2) [0|0|4]	62.6
Anita BAKER ▶ 58 Pk:1 [1|2|3]	38866
Rapture (86) 11(2) [0|72|157]	23606
Giving You The Best That I Got (88) 1(4) [18|28|42]	14953
The Songstress (83) 139(2) [0|0|11]	307
BALAAM AND THE ANGEL ▶ 1832 Pk:174 [0|0|1]	56.6
Live Free Or Die (88) 174(1) [0|0|6]	56.6
BALANCE ▶ 1474 Pk:133 [0|0|1]	268
Balance (81) 133(1) [0|0|12]	268
Marty BALIN ▶ 736 Pk:35 [0|1|2]	2181
Balin (81) 35(2) [0|3|23]	2048
Lucky (83) 156(2) [0|0|6]	133
Russ BALLARD ▶ 1282 Pk:147 [0|0|3]	459
Russ Ballard (84) 147(2) [0|0|13]	338
The Fire Still Burns (85) 166(2) [0|0|4]	90.9
Barnet Dogs (80) 187(1) [0|0|2]	29.5
BALTIMORA ▶ 947 Pk:49 [0|0|1]	1165
Living In The Background (86) 49(2) [0|0|17]	1165
BANANARAMA ▶ 283 Pk:15 [0|2|5]	10253
True Confessions (86) 15(2) [0|11|28]	3851
Bananarama (84) 30(5) [0|7|36]	3307
Wow! (87) 44(2) [0|0|26]	1810
Deep Sea Skiving (83) 63(1) [0|0|19]	1088
Greatest Hits Collection (88) 151(2) [0|0|9]	197
Moe BANDY ▶ 1866 Pk:170 [0|0|1]	37.4
Hey Joe, Hey Moe (81)-Moe Bandy & Joe Stampley [A] 170(1) [0|0|4]	37.4
BANGLES ▶ 118 Pk:2 [1|2|3]	25004
Different Light (86) 2(2) [9|45|82]	17176
Everything (88) 15(2) [0|19|42]	6654
All Over the Place (84) 80(2) [0|0|30]	1174
BANG TANGO ▶ 844 Pk:58 [0|0|1]	1585
Psycho Cafe (89) 58(2) [0|0|27]	*1585*
Tony BANKS ▶ 1871 Pk:171 [0|0|1]	34.5
A Curious Feeling (80) 171(1) [0|0|2]	*34.5*
Frank BARBER Orchestra ▶ 1130 Pk:94 [0|0|1]	678
Hooked On Big Bands (82) 94(3) [0|0|16]	678
Pete BARDENS ▶ 1659 Pk:148 [0|0|1]	136
Seen One Earth (87) 148(2) [0|0|5]	136
BARDEUX ▶ 1126 Pk:104 [0|0|2]	687
Bold As Love (88) 104(2) [0|0|12]	460
Shangri-La (89) 133(2) [0|0|7]	227
The BAR-KAYS ▶ 342 Pk:35 [0|1|7]	7798
Nightcruising (81) 55(2) [0|0|29]	1944
Propositions (82) 51(3) [0|0|29]	1699
Dangerous (84) 52(1) [0|0|22]	1402
Injoy (79/80) 35(1) [0|1|16]	*1068*
As One (80) 57(1) [0|0|16]	952
Contagious (87) 110(2) [0|0|14]	415
Banging The Wall (85) 115(2) [0|0|9]	318
Jimmy BARNES ▶ 963 Pk:104 [0|0|2]	1114
Freight Train Heart (88) 104(3) [0|0|21]	564
Jimmy Barnes (86) 109(2) [0|0|16]	550
Lou Ann BARTON ▶ 1494 Pk:133 [0|0|1]	256
Old Enough (82) 133(1) [0|0|9]	256

Acts With Albums

Act ▶ Rank Peak [Top10s|Top40s|Total] Title (Yr) Pk(Wks) [T10 Wks|T40Wks|TotWks]	Score in '80s
--BASE, Rob see Rob BASE	
BASIA ▶ 422 Pk:36 [0|1|1]	**5587**
Time And Tide (88) 36(1) [0|3|77]	5587
Toni BASIL ▶ 569 Pk:22 [0|1|1]	**3523**
Word Of Mouth (82) 22(7) [0|9|30]	3523
BAUHAUS ▶ 1695 Pk:169 [0|0|1]	**121**
Swing The Heartache - The BBC Sessions (89) 169(1) [0|0|6]	121
BEACH BOYS ▶ 539 Pk:46 [0|0|7]	**3840**
Still Cruisin' (89) 46(1) [0|0|16]	1209
The Beach Boys (85) 52(3) [0|0|14]	1113
Made In U.S.A. (86) 96(1) [0|0|12]	522
*Endless Summer (**74**/81) 123(2) [0|0|16]*	*413*
Keepin' The Summer Alive (80) 75(1) [0|0|6]	317
Ten Years Of Harmony (1970-1980) (81) 156(1) [0|0|8]	164
Sunshine Dream (82) 180(3) [0|0|6]	102
The BEARS ▶ 1712 Pk:159 [0|0|1]	**112**
Rise And Shine (88) 159(1) [0|0|5]	112
BEASTIE BOYS ▶ 122 Pk:1 [1|2|2]	**24441**
Licensed To Ill (86) 1(7) [19|37|68]	21725
Paul's Boutique (89) 14(2) [0|8|15]	2716
BEAT FARMERS ▶ 1226 Pk:131 [0|0|3]	**520**
The Pursuit Of Happiness (87) 131(1) [0|0|8]	249
Van Go (86) 135(2) [0|0|9]	229
Tales Of The New West (85) 186(1) [0|0|3]	41.7
The BEATLES ▶ 200 Pk:19 [0|2|16]	**15084**
Rarities (80) 21(1) [0|9|15]	2202
Reel Music (82) 19(3) [0|8|12]	2034
*The Beatles 1967-1970 (**73**/81) 58(1) [0|0|37]*	*1762*
*The Beatles [White Album] (**68**/81) 67(1) [0|0|41]*	*1678*
20 Greatest Hits (82) 50(2) [0|0|28]	1675
*The Beatles 1962-1966 (**73**/81) 62(1) [0|0|38]*	*1647*
*Abbey Road (**69**/81) 69(2) [0|0|42]*	*1618*
*Sgt. Pepper's Lonely Hearts Club Band (**67**/81) 73(1) [0|0|27]*	*1058*
*Rubber Soul (**66**/81) 83(1) [0|0|8]*	*463*
Past Masters - Volume 2 (88) 121(2) [0|0|7]	238
*Love Songs (**77**/81) 149(1) [0|0|7]*	*193*
*Let It Be (Soundtrack) (**70**/87) 88(1) [0|0|4]*	*174*
*Magical Mystery Tour (Soundtrack) (**68**/84) 163(1) [0|0|9]*	*173*
Past Masters - Volume 1 (88) 149(2) [0|0|6]	139
*Yellow Submarine (Soundtrack) (**69**/87) 180(1) [0|0|1]*	*19.0*
*The Early Beatles (**65**/86) 197(1) [0|0|1]*	*11.4*
Jean BEAUVOIR ▶ 1121 Pk:93 [0|0|1]	**699**
Drums Along The Mohawk (86) 93(2) [0|0|15]	699
Jeff BECK ▶ 477 Pk:21 [0|2|3]	**4533**
There And Back (80) 21(2) [0|8|20]	2603
Flash (85) 39(2) [0|2|18]	1596
Jeff Beck's Guitar Shop (89)-Jeff Beck With Terry Bozzio & Tony Hymas [A] 49(2) [0|0|11]	*334*
BEE GEES ▶ 218 Pk:1 [2|2|7]	**13816**
Bee Gees Greatest (80) 1(1) [6|10|25]	5715
Staying Alive (Soundtrack) (83) 6(2) [6|14|27]	5217
Living Eyes (81) 41(3) [0|0|12]	1262
One (89) 68(2) [0|0|13]	802
E-S-P (87) 96(2) [0|0|5]	385
*Spirits Having Flown (**79**/80) 149(1) [0|0|9]*	*262*
*Saturday Night Fever (Soundtrack) (**78**/80) 163(2) [0|0|10]*	*171*
Adrian BELEW ▶ 986 Pk:82 [0|0|3]	**1030**
Lone Rhino (82) 82(2) [0|0|9]	435
Mr. Music Head (89) 114(2) [0|0|11]	408
Twang Bar King (83) 146(2) [0|0|7]	187
Regina BELLE ▶ 796 Pk:63 [0|0|2]	**1817**
Stay With Me (89) 63(2) [0|0|15]	1132
All By Myself (87) 85(1) [0|0|15]	685
BELLE STARS ▶ 1902 Pk:191 [0|0|1]	**26.4**
The Belle Stars (83) 191(1) [0|0|2]	26.4

Act ▶ Rank Peak [Top10s|Top40s|Total] Title (Yr) Pk(Wks) [T10 Wks|T40Wks|TotWks]	Score in '80s
Pat BENATAR ▶ 16 Pk:1 [3|8|9]	**71068**
Crimes Of Passion (80) 2(5) [29|38|93]	21849
Precious Time (81) 1(1) [14|30|54]	13084
Get Nervous (82) 4(5) [12|27|46]	10271
In The Heat Of The Night (80) 12(2) [0|15|111]	*9954*
Live From Earth (83) 13(2) [0|16|34]	5241
Tropico (84) 14(4) [0|15|22]	4504
Wide Awake In Dreamland (88) 28(2) [0|10|29]	3013
Seven The Hard Way (85) 26(4) [0|9|20]	2670
Best Shots (89) 67(1) [0|0|6]	*483*
Tony BENNETT ▶ 1622 Pk:160 [0|0|1]	**160**
The Art Of Excellence (86) 160(2) [0|0|8]	160
David BENOIT ▶ 985 Pk:101 [0|0|3]	**1035**
Urban Daydreams (89) 101(2) [0|0|14]	559
Every Step Of The Way (88) 129(2) [0|0|14]	435
Waiting For Spring (89) 187(1) [0|0|3]	41.8
George BENSON ▶ 139 Pk:3 [1|3|8]	**22156**
Give Me The Night (80) 3(2) [10|20|38]	8549
The George Benson Collection (81) 14(2) [0|14|26]	4657
In Your Eyes (83) 27(1) [0|11|35]	4014
20/20 (85) 45(3) [0|0|32]	2099
While The City Sleeps... (86) 77(1) [0|0|24]	1154
Collaboration (87)-George Benson/Earl Klugh [A] 59(2) [0|0|31]	*957*
Twice The Love (88) 76(2) [0|0|10]	550
Tenderly (89) 140(2) [0|0|6]	177
BERLIN ▶ 362 Pk:28 [0|2|3]	**7408**
Pleasure Victim (83) 30(1) [0|11|34]	3236
Love Life (84) 28(2) [0|4|30]	2811
Count Three And Pray (86) 61(2) [0|0|20]	1361
Leonard BERNSTEIN ▶ 1007 Pk:70 [0|0|1]	**975**
West Side Story (85) 70(2) [0|0|20]	975
Dickey BETTS ▶ 1830 Pk:187 [0|0|1]	**56.8**
Pattern Disruptive (88)-The Dickey Betts Band 187(2) [0|0|4]	56.8
The B-52's ▶ 170 Pk:6 [1|4|7]	**17533**
*Cosmic Thing (89/**90**) 6(1) [7|17|24]*	*6140*
Wild Planet (80) 18(1) [0|9|27]	3144
The B-52's (80) 59(2) [0|0|55]	*3066*
Whammy! (83) 29(2) [0|6|26]	2330
Mesopotamia (82) 35(2) [0|4|18]	1465
Party Mix! (81) 55(2) [0|0|11]	787
Bouncing Off The Satellites (86) 85(2) [0|0|15]	602
BIG AUDIO DYNAMITE ▶ 602 Pk:85 [0|0|4]	**3170**
This Is Big Audio Dynamite (85) 103(2) [0|0|35]	1330
No. 10, Upping Street (86) 119(2) [0|0|23]	722
Megatop Phoenix (89) 85(1) [0|0|13]	635
Tighten Up Vol. '88 (88) 102(1) [0|0|12]	483
BIG COUNTRY ▶ 303 Pk:18 [0|1|5]	**9098**
The Crossing (83) 18(2) [0|18|42]	5965
The Seer (86) 59(2) [0|0|17]	1267
Steeltown (84) 70(4) [0|0|17]	1019
Wonderland (84) 65(2) [0|0|12]	723
Peace In Our Time (88) 160(2) [0|0|6]	125
BIG PIG ▶ 1098 Pk:93 [0|0|1]	**740**
Bonk (88) 93(2) [0|0|17]	740
--Billy & The BEATERS see: VERA, Billy & The BEATERS	
BILLY SATELLITE ▶ 1586 Pk:139 [0|0|1]	**183**
Billy Satellite (84) 139(2) [0|0|6]	183
Biz MARKIE ▶ 917 Pk:66 [0|0|2]	**1308**
Goin' Off (88) 90(2) [0|0|18]	712
The Biz Never Sleeps (89)- The Diabolical Biz Markie 66(2) [0|0|10]	*596*
Clint BLACK ▶ 778 Pk:61 [0|0|1]	**1924**
*Killin' Time (89/**90**) 61(1) [0|0|30]*	*1924*
The BLACKBYRDS ▶ 1411 Pk:133 [0|0|1]	**320**
Better Days (81) 133(1) [0|0|11]	320

Act ▶ Rank Peak [Top10s|Top40s|Total] Title (Yr) Pk(Wks) [T10 Wks|T40Wks|TotWks]	Score in '80s
BLACKFOOT ▶ 565 Pk:48 [0|0|5]	**3570**
Tomcattin' (80) 50(1) [0|0|20]	1441
Marauder (81) 48(2) [0|0|12]	1038
Siogo (83) 82(1) [0|0|13]	624
*Strikes (79/**80**) 95(1) [0|0|7]*	*371*
Vertical Smiles (84) 176(3) [0|0|5]	95.9
BLACK 'N BLUE ▶ 880 Pk:110 [0|0|3]	**1443**
Nasty, Nasty (86) 110(1) [0|0|20]	786
Black 'N' Blue (84) 116(2) [0|0|11]	374
In Heat (88) 133(2) [0|0|9]	283
BLACK SABBATH ▶ 328 Pk:28 [0|4|7]	**8239**
Heaven And Hell (80) 28(3) [0|9|24]	2734
Mob Rules (81) 29(3) [0|8|18]	1992
Live Evil (83) 37(4) [0|4|12]	1230
Born Again (83) 39(1) [0|2|16]	1223
Seventh Star (86)-Black Sabbath Featuring Tony Iommi 78(2) [0|0|11]	611
Headless Cross (89) 115(1) [0|0|8]	316
The Eternal Idol (87) 168(1) [0|0|6]	134
BLACK UHURU ▶ 1565 Pk:146 [0|0|1]	**197**
Chill Out (82) 146(1) [0|0|7]	197
Ruben BLADES ▶ 1690 Pk:156 [0|0|1]	**122**
Nothing But The Truth (88) 156(2) [0|0|6]	122
The BLASTERS ▶ 519 Pk:36 [0|1|4]	**4094**
The Blasters (82) 36(3) [0|4|30]	2567
Hard Line (85) 86(2) [0|0|19]	889
Non Fiction (83) 95(1) [0|0|8]	336
Over There (Live At The Venue, London) (82) 117(4) [0|0|8]	303
BLONDIE ▶ 155 Pk:7 [1|4|5]	**19130**
Autoamerican (80) 7(5) [9|23|34]	9065
*Eat To The Beat (**79**/80) 29(1) [0|9|40]*	*4357*
The Best Of Blondie (81) 30(2) [0|11|23]	2819
*Parallel Lines (**79**/80) 81(1) [0|0|36]*	*1809*
The Hunter (82) 33(2) [0|4|12]	1080
BLOODSTONE ▶ 1263 Pk:95 [0|0|1]	**480**
We Go A Long Way Back (82) 95(2) [0|0|11]	480
Kurtis BLOW ▶ 649 Pk:71 [0|0|6]	**2769**
Ego Trip (84) 83(2) [0|0|37]	1551
Kurtis Blow (80) 71(1) [0|0|10]	556
America (85) 153(3) [0|0|15]	352
Deuce (81) 137(1) [0|0|5]	177
Tough (82) 167(2) [0|0|5]	110
Kingdom Blow (86) 196(2) [0|0|2]	23.6
BLOWFLY ▶ 1066 Pk:82 [0|0|1]	**807**
Blowfly's Party [X-Rated] (80) 82(2) [0|0|20]	807
BLOW MONKEYS ▶ 725 Pk:35 [0|1|2]	**2239**
Animal Magic (86) 35(2) [0|5|18]	1977
She Was Only A Grocer's Daughter (87) 134(2) [0|0|8]	263
BLUE MERCEDES ▶ 1720 Pk:165 [0|0|1]	**106**
Rich And Famous (88) 165(2) [0|0|5]	106
BLUE MURDER ▶ 977 Pk:69 [0|0|1]	**1070**
Blue Murder (89) 69(2) [0|0|21]	1070
BLUE ÖYSTER CULT ▶ 307 Pk:24 [0|3|6]	**8946**
Fire Of Unknown Origin (81) 24(2) [0|11|31]	3761
Extraterrestrial Live (82) 29(2) [0|5|19]	1792
Cultosaurus Erectus (80) 34(2) [0|3|16]	1486
Club Ninja (86) 63(2) [0|0|14]	859
The Revolution By Night (83) 93(1) [0|0|16]	772
Imaginos (88) 122(1) [0|0|8]	276
BLUES BROTHERS ▶ 467 Pk:13 [0|1|3]	**4638**
The Blues Brothers (Soundtrack) (80) 13(2) [0|12|19]	3638
Made In America (80) 49(1) [0|0|12]	897
The Best Of The Blues Brothers (82) 143(2) [0|0|3]	103
BOBBY And The MIDNITES ▶ 1531 Pk:158 [0|0|2]	**230**
Bobby & The Midnites (81) 158(2) [0|0|7]	154

327

Acts With Albums

Act ▶ Rank Peak [Top10s|Top40s|Total] Title (Yr) Pk(Wks) [T10 Wks|T40Wks|TotWks]	Score in '80s
BOBBY And The MIDNITES ▶ *Continued*	
Where The Beat Meets The Street (84) 166(1) [0|0|4]	76.0
BOBBY JIMMY & The CRITTERS ▶ 1927 Pk:200 [0|0|1]	10.0
Roaches: The Beginning (86) 200(1) [0|0|1]	10.0
BODEANS ▶ 751 Pk:86 [0|0|3]	2073
Outside Looking In (87) 86(2) [0|0|20]	856
Love & Hope & Sex & Dreams (86) 115(2) [0|0|19]	649
Home (89) 94(2) [0|0|13]	569
Angela BOFILL ▶ 359 Pk:34 [0|2|4]	7486
Too Tough (83) 40(3) [0|3|32]	2528
Angel Of The Night (80) 34(1) [0|6|24]	*2495*
Something About You (81) 61(4) [0|0|22]	1439
Teaser (83) 81(1) [0|0|21]	1024
Michael BOLTON ▶ 413 Pk:34 [0|1|3]	5895
The Hunger (87) 46(1) [0|0|41]	2839
Soul Provider (89/**90**) 34(1) [0|2|24]	*2567*
Michael Bolton (83) 89(1) [0|0|13]	489
Gary (U.S.) BONDS ▶ 531 Pk:27 [0|1|2]	3892
Dedication (81) 27(2) [0|7|20]	2510
On The Line (82) 52(2) [0|0|17]	1382
BONHAM ▶ 838 Pk:38 [0|1|1]	1621
The Disregard Of Timekeeping (89) 38(1) [0|2|14]	*1621*
BON JOVI ▶ 13 Pk:1 [2|3|4]	75821
Slippery When Wet (86) 1(8) [46|60|94]	39550
New Jersey (88) 1(4) [22|52|65]	*23386*
7800 Degrees Fahrenheit (85) 37(3) [0|5|104]	7350
Bon Jovi (84) 43(2) [0|0|86]	5535
Karla BONOFF ▶ 626 Pk:49 [0|0|2]	2952
Wild Heart Of The Young (82) 49(1) [0|0|35]	2455
Restless Nights (**79**/80) 114(1) [0|0|12]	*496*
BOOGIE BOYS ▶ 808 Pk:53 [0|0|3]	1759
City Life (85) 53(2) [0|0|17]	1141
Romeo Knight (88) 117(2) [0|0|11]	367
Survival Of The Freshest (86) 124(2) [0|0|5]	252
BOOGIE DOWN PRODUCTIONS ▶ 654 Pk:36 [0|1|2]	2736
Ghetto Music: The Blueprint Of Hip Hop (89) 36(2) [0|4|17]	1639
By All Means Necessary (88) 75(1) [0|0|23]	1097
Chuckii BOOKER ▶ 1373 Pk:116 [0|0|1]	351
Chuckii (89) 116(1) [0|0|10]	351
BOOK OF LOVE ▶ 1549 Pk:156 [0|0|1]	207
Lullaby (88) 156(2) [0|0|10]	207
BOOMTOWN RATS ▶ 1056 Pk:103 [0|0|3]	827
The Fine Art Of Surfacing (80) 103(2) [0|0|11]	*481*
Mondo Bongo (81) 116(2) [0|0|8]	289
In The Long Grass (85) 188(1) [0|0|4]	58.1
BOOTSY'S RUBBER BAND ▶ 1062 Pk:70 [0|0|2]	810
Ultra Wave (80)-Bootsy 70(3) [0|0|9]	569
The One Giveth, The Count Taketh Away (82)-William "Bootsy" Collins 120(1) [0|0|8]	242
BOSTON ▶ 160 Pk:1 [1|1|3]	18624
Third Stage (86) 1(4) [20|29|50]	17222
Boston (**76**/86) 98(1) [0|0|31]	*1163*
Don't Look Back (**78**/86) 146(2) [0|0|10]	*239*
BOSTON POPS Orchestra/John WILLIAMS ▶ 1433 Pk:155 [0|0|2]	297
Swing, Swing, Swing (86) 155(2) [0|0|8]	201
Pops In Space (80) 181(1) [0|0|6]	96.3
BOURGEOIS TAGG ▶ 948 Pk:84 [0|0|2]	1165
YoYo (87) 84(1) [0|0|11]	978
Bourgeois Tagg (86) 139(2) [0|0|7]	187
David BOWIE ▶ 94 Pk:4 [1|4|11]	29123
Let's Dance (83) 4(1) [16|35|68]	15224
Scary Monsters (80) 12(1) [0|12|27]	4216
Tonight (84) 11(2) [0|12|24]	3843

Act ▶ Rank Peak [Top10s|Top40s|Total] Title (Yr) Pk(Wks) [T10 Wks|T40Wks|TotWks]	Score in '80s
Never Let Me Down (87) 34(1) [0|4|26]	2568
Changestwobowie (81) 68(1) [0|0|18]	959
Ziggy Stardust-The Motion Picture (Soundtrack) (83) 89(1) [0|0|15]	650
Labyrinth (Soundtrack) (86) 68(2) [0|0|8]	472
Sound + Vision (89) 97(1) [0|0|12]	*458*
Golden Years (83) 99(1) [0|0|9]	375
Fame And Fashion - David Bowie's All Time Greatest Hits (84) 147(1) [0|0|6]	184
Christiane F. (Soundtrack) (82) 135(2) [0|0|7]	175
BOW WOW WOW ▶ 777 Pk:67 [0|0|4]	1926
The Last Of The Mohicans (82) 67(1) [0|0|22]	917
When The Going Gets Tough, The Tough Get Going (83) 82(1) [0|0|13]	688
I Want Candy (82) 123(1) [0|0|9]	294
See Jungle! See Jungle! Go Join Your Gang Yeah! City All Over, Go Ape Crazy (81) 192(2) [0|0|2]	27.2
BOX OF FROGS ▶ 845 Pk:45 [0|0|2]	1581
Box Of Frogs (84) 45(3) [0|0|20]	1525
Strange Land (86) 177(2) [0|0|3]	55.8
BOY GEORGE ▶ 1239 Pk:126 [0|0|2]	509
High Hat (89) 126(1) [0|0|11]	364
Sold (87) 145(2) [0|0|5]	145
BOY MEETS GIRL ▶ 675 Pk:50 [0|0|2]	2566
Reel Life (88) 50(2) [0|0|26]	2036
Boy Meets Girl (85) 76(2) [0|0|8]	531
The BOYS ▶ 603 Pk:33 [0|1|1]	3167
Messages From The Boys (88) 33(2) [0|7|36]	3167
BOYS CLUB ▶ 1142 Pk:93 [0|0|1]	643
Boys Club (88) 93(2) [0|0|16]	643
BOYS DON'T CRY ▶ 909 Pk:55 [0|0|1]	1337
Boys Don't Cry (86) 55(3) [0|0|19]	1337
Terry BOZZIO ▶ 1394 Pk:49 [0|0|1]	334
Jeff Beck's Guitar Shop (89)-Jeff Beck With Terry Bozzio & Tony Hymas [A] 49(2) [0|0|11]	*334*
Billy BRAGG ▶ 1925 Pk:198 [0|0|1]	10.9
Workers Playtime (88) 198(1) [0|0|1]	10.9
BRAM TCHAIKOVSKY ▶ 1186 Pk:108 [0|0|2]	574
Pressure (80) 108(1) [0|0|10]	385
Funland (81) 158(1) [0|0|8]	190
The BRANDOS ▶ 1143 Pk:108 [0|0|1]	641
Honor Among Thieves (87) 108(2) [0|0|19]	641
Laura BRANIGAN ▶ 215 Pk:23 [0|3|5]	13897
Self Control (84) 23(2) [0|19|45]	5728
Branigan (82) 34(5) [0|9|36]	3077
Branigan 2 (83) 29(1) [0|5|37]	2989
Touch (87) 87(3) [0|0|28]	1188
Hold Me (85) 71(3) [0|0|15]	915
John BRANNEN ▶ 1424 Pk:156 [0|0|1]	306
Mystery Street (88) 156(1) [0|0|14]	306
BRASS CONSTRUCTION ▶ 910 Pk:89 [0|0|4]	1336
Brass Construction 5 (80) 89(1) [0|0|17]	*811*
Attitudes (82) 114(3) [0|0|8]	237
Brass Construction 6 (80) 121(2) [0|0|5]	179
Conversations (83) 176(1) [0|0|6]	109
BREAKFAST CLUB ▶ 689 Pk:43 [0|0|1]	2481
The Breakfast Club (87) 43(2) [0|0|30]	2481
BREAKWATER ▶ 1641 Pk:141 [0|0|1]	149
Splashdown (80) 141(1) [0|0|5]	149
BREATHE ▶ 431 Pk:34 [0|1|1]	5343
All That Jazz (88) 34(3) [0|8|51]	5343
BRECKER BROTHERS ▶ 1827 Pk:176 [0|0|1]	57.9
Straphangin' (81) 176(2) [0|0|3]	57.9
BRICK ▶ 1221 Pk:89 [0|0|2]	528
Summer Heat (81) 89(2) [0|0|10]	445
Waiting On You (80) 179(2) [0|0|5]	83.0
Edie BRICKELL & The NEW BOHEMIANS ▶ 263 Pk:4 [1|1|1]	10990
Shooting Rubberbands At The Stars (88) 4(2) [8|28|54]	10990

Act ▶ Rank Peak [Top10s|Top40s|Total] Title (Yr) Pk(Wks) [T10 Wks|T40Wks|TotWks]	Score in '80s
BRIDES OF FUNKENSTEIN ▶ 1427 Pk:93 [0|0|1]	305
Never Buy Texas From A Cowboy (80) 93(2) [0|0|7]	305
Martin BRILEY ▶ 771 Pk:55 [0|0|2]	1952
One Night With A Stranger (83) 55(1) [0|0|22]	1457
Dangerous Moments (85) 85(2) [0|0|10]	495
BRITNY FOX ▶ 547 Pk:39 [0|1|2]	3785
Britny Fox (88) 39(2) [0|2|37]	3414
Boys In Heat (89) 79(3) [0|0|6]	*371*
BRONSKI BEAT ▶ 702 Pk:36 [0|1|2]	2394
The Age Of Consent (85) 36(2) [0|3|25]	2235
Truthdare Doubledare (86) 147(2) [0|0|6]	158
BROOKLYN, BRONX & QUEENS Band ▶ 1400 Pk:109 [0|0|1]	328
The Brooklyn, Bronx & Queens Band (81) 109(1) [0|0|9]	328
BROS ▶ 1736 Pk:171 [0|0|1]	94.7
Push (88) 171(1) [0|0|5]	94.7
BROTHERS JOHNSON ▶ 334 Pk:5 [1|1|4]	8089
Light Up The Night (80) 5(2) [7|16|30]	6493
Winners (81) 48(2) [0|0|13]	953
Out of Control (84) 91(1) [0|0|11]	481
Blast! (The Latest And The Greatest) (83) 138(2) [0|0|5]	163
Bobby BROWN ▶ 62 Pk:1 [1|2|3]	36950
Don't Be Cruel (88) 1(6) [45|69|76]	*35210*
Dance!...Ya Know It! (89/**90**) 18(1) [0|3|5]	1028
King Of Stage (86) 88(2) [0|0|17]	711
Danny Joe BROWN ▶ 1519 Pk:120 [0|0|1]	235
Danny Joe Brown And The Danny Joe Brown Band (81) 120(2) [0|0|7]	235
James BROWN ▶ 1045 Pk:96 [0|0|4]	842
I'm Real (88) 96(1) [0|0|14]	519
Gravity (86) 156(2) [0|0|6]	158
James Brown...Live/Hot On The One (80) 170(2) [0|0|5]	90.5
Live And Lowdown At The Apollo Vol 1 (80) 163(2) [0|0|3]	75.1
Julie BROWN ▶ 1651 Pk:168 [0|0|1]	141
Goddess In Progress (85) 168(1) [0|0|7]	141
Jackson BROWNE ▶ 133 Pk:1 [2|3|4]	22973
Hold Out (80) 1(1) [13|21|38]	12312
Lawyers In Love (83) 8(3) [4|12|33]	5412
Lives In The Balance (86) 23(3) [0|9|31]	3955
World In Motion (89) 45(2) [0|0|16]	1294
Tom BROWNE ▶ 409 Pk:18 [0|2|4]	6041
Love Approach (80) 18(2) [0|9|26]	3368
Magic (81) 37(2) [0|2|19]	1762
Yours Truly (81) 97(1) [0|0|14]	631
Rockin' Radio (83) 147(1) [0|0|12]	279
Jack BRUCE ▶ 1129 Pk:37 [0|1|3]	680
B.L.T. (81)-Jack Bruce/Bill Lordan/Robin Trower [A] 37(1) [0|3|16]	509
Truce (82)-Jack Bruce & Robin Trower [A] 109(1) [0|0|6]	136
I've Always Wanted To Do This (80)-Jack Bruce And Friends 182(1) [0|0|2]	35.3
Bill BRUFORD ▶ 1903 Pk:191 [0|0|1]	26.4
Gradually Going Tornado (80) 191(1) [0|0|1]	26.4
Sharon BRYANT ▶ 1391 Pk:139 [0|0|1]	334
Here I Am (89) 139(1) [0|0|13]	334
Peabo BRYSON ▶ 336 Pk:40 [0|1|9]	8053
Straight From The Heart (84) 44(2) [0|0|26]	2121
I Am Love (81) 40(2) [0|2|24]	1876
Don't Play With Fire (82) 55(3) [0|0|21]	1270
We're The Best Of Friends (80)-Natalie Cole & Peabo Bryson [A] 44(1) [0|0|16]	*729*
Paradise (80) 79(2) [0|0|16]	695
Turn The Hands Of Time (81) 82(1) [0|0|11]	548
Take No Prisoners (85) 102(2) [0|0|13]	500
The Peabo Bryson Collection (84) 168(1) [0|0|10]	183
Positive (88) 157(1) [0|0|6]	132

Acts With Albums

Act ▶ Rank Peak [Top10s|Top40s|Total] Title (Yr) Pk(Wks) [T10 Wks|T40Wks|TotWks]	Score in '80s
B.T. EXPRESS ▶ 1743 Pk:164 [0|0|1]	91.0
1980 (80) 164(1) [0|0|4]	91.0
Roy BUCHANAN ▶ 1259 Pk:153 [0|0|3]	485
When A Guitar Plays The Blues (85)	
161(1) [0|0|13]	285
Dancing On The Edge (86) 153(2) [0|0|8]	174
My Babe (81) 193(1) [0|0|2]	25.9
Lindsey BUCKINGHAM ▶ 506 Pk:32 [0|1|2]	4175
Law And Order (81) 32(1) [0|6|24]	2787
Go Insane (84) 45(4) [0|0|14]	1388
BUCKNER & GARCIA ▶ 773 Pk:24 [0|1|1]	1947
Pac-Man Fever (82) 24(2) [0|5|16]	1947
BUCKWHEAT ZYDECO ▶ 1350	
Pk:104 [0|0|2]	390
Taking It Home (88) 104(1) [0|0|7]	307
On A Night Like This (87) 172(1) [0|0|5]	83.4
Jimmy BUFFETT ▶ 270 Pk:30 [0|2|10]	10735
Coconut Telegraph (81) 30(1) [0|4|18]	1724
Somewhere Over China (82) 31(3) [0|6|15]	1711
Last Mango In Paris (85) 53(2) [0|0|20]	1445
One Particular Harbour (83) 59(3) [0|0|24]	1321
Hot Water (88) 46(2) [0|0|14]	1104
Off To See The Lizard (89) 57(2) [0|0|13]	862
Floridays (86) 66(1) [0|0|16]	860
Riddles In The Sand (84) 87(1) [0|0|14]	601
Songs You Know By Heart: Jimmy Buffett's	
Greatest Hit(s) (85) 100(2) [0|0|16]	600
Volcano **(79**/80) 100(1) *[0|0|12]*	*509*
The BUGGLES ▶ 1707 Pk:161 [0|0|1]	116
Adventures In Modern Recording (82)	
161(2) [0|0|5]	116
Alex BUGNON ▶ 1385 Pk:127 [0|0|1]	339
Love Season (89) 127(3) [0|0|11]	339
BULGARIAN STATE RADIO & T.V. FEMALE CHOIR	
▶ 1553 Pk:165 [0|0|1]	203
Mystery Of Bulgarian Voices (88)	
165(1) [0|0|10]	203
BULLETBOYS ▶ 566 Pk:34 [0|1|1]	3562
Bulletboys (88) 34(3) [0|5|47]	3562
T-Bone BURNETT ▶ 1809 Pk:188 [0|0|1]	67.6
Proof Through The Night (83) 188(2) [0|0|5]	67.6
Rocky BURNETTE ▶ 1003 Pk:53 [0|0|1]	979
The Son Of Rock And Roll (80) 53(1) [0|0|14]	979
BURNING SENSATIONS ▶ 1790	
Pk:175 [0|0|1]	74.4
Burning Sensations (83) 175(1) [0|0|4]	74.4
George BURNS ▶ 1311 Pk:93 [0|0|1]	425
I Wish I Was Eighteen Again (80) 93(2) [0|0|10]	425
Glen BURTNICK ▶ 1614 Pk:147 [0|0|1]	168
Heroes & Zeros (87) 147(2) [0|0|6]	168
Jenny BURTON ▶ 1805 Pk:181 [0|0|1]	69.4
In Black And White (84) 181(2) [0|0|4]	69.4
BUS BOYS ▶ 1039 Pk:85 [0|0|2]	870
Minimum Wage Rock & Roll (80) 85(1) [0|0|15]	659
American Worker (82) 139(3) [0|0|7]	211
Kate BUSH ▶ 418 Pk:30 [0|1|5]	5655
Hounds Of Love (85) 30(1) [0|6|27]	2871
The Whole Story (86) 76(1) [0|0|27]	1272
The Sensual World (89) 43(2) [0|0|9]	*1128*
The Dreaming (82) 157(2) [0|0|11]	239
Kate Bush (83) 148(1) [0|0|6]	146
Jon BUTCHER AXIS ▶ 592 Pk:66 [0|0|5]	3310
Wishes (87)-Jon Butcher 77(2) [0|0|27]	1349
Along The Axis (85) 66(1) [0|0|17]	933
Jon Butcher Axis (83) 91(1) [0|0|13]	594
Pictures From The Front (89)-Jon Butcher	
121(3) [0|0|8]	297
Store At The Sun (84) 160(1) [0|0|6]	136
Jonathan BUTLER ▶ 576 Pk:50 [0|0|3]	3459
Jonathan Butler (87) 50(2) [0|0|33]	2304

Act ▶ Rank Peak [Top10s|Top40s|Total] Title (Yr) Pk(Wks) [T10 Wks|T40Wks|TotWks]	Score in '80s
Introducing Jonathan Butler (86)	
101(2) [0|0|16]	600
More Than Friends (88) 113(2) [0|0|22]	555
BUZZCOCKS ▶ 1722 Pk:163 [0|0|1]	105
Different Kind Of Tension (80) 163(1) [0|0|6]	105
Donald BYRD And 125th STREET N.Y.C. ▶	
1285 Pk:93 [0|0|1]	450
Love Byrd (81) 93(1) [0|0|10]	450
David BYRNE ▶ 803 Pk:44 [0|0|4]	1805
Rei Momo (89) 71(2) [0|0|11]	*662*
The Catherine Wheel (Original Cast) (81)	
104(1) [0|0|12]	509
My Life In The Bush Of Ghosts (81)-Brian Eno -	
David Byrne [A] 44(2) [0|0|13]	466
Music For The Knee Plays (85) 141(2) [0|0|6]	169
D.L. BYRON ▶ 1447 Pk:133 [0|0|1]	289
This Day And Age (80) 133(1) [0|0|10]	289

C

Act ▶ Rank Peak [Top10s|Top40s|Total] Title (Yr) Pk(Wks) [T10 Wks|T40Wks|TotWks]	Score in '80s
CACTUS WORLD NEWS ▶ 1759	
Pk:179 [0|0|1]	85.2
Urban Beaches (86) 179(1) [0|0|5]	85.2
John CAFFERTY And The BEAVER BROWN BAND	
▶ 247 Pk:9 [1|2|3]	11772
Eddie & The Cruisers (Soundtrack) (83)	
9(5) [6|24|62]	8668
Tough All Over (85) 40(3) [0|3|32]	2928
Eddie & The Cruisers II (Soundtrack) (89)	
121(2) [0|0|6]	177
Tané CAIN ▶ 1359 Pk:121 [0|0|1]	372
Tane Cain (82)-Tane Cain 121(2) [0|0|10]	372
Bobby CALDWELL ▶ 1049 Pk:113 [0|0|2]	836
Cat In The Hat (80) 113(2) [0|0|15]	480
Carry On (82) 133(1) [0|0|13]	357
J.J. CALE ▶ 1244 Pk:110 [0|0|2]	502
Shades (81) 110(1) [0|0|7]	284
Grasshopper (82) 149(1) [0|0|8]	218
John CALE ▶ 1672 Pk:154 [0|0|1]	132
Honi Soit (o nee swa) (81) 154(2) [0|0|5]	132
CALIFORNIA RAISINS ▶ 708 Pk:60 [0|0|2]	2352
The California Raisins (87) 60(2) [0|0|36]	1972
Sweet, Delicious & Marvelous (88)	
140(1) [0|0|15]	380
The CALL ▶ 555 Pk:64 [0|0|4]	3678
Reconciled (86) 82(2) [0|0|30]	1300
Let The Day Begin (89) 64(1) [0|0|22]	1222
Modern Romans (83) 84(1) [0|0|15]	698
Into The Woods (87) 123(2) [0|0|13]	457
CAMEO ▶ 123 Pk:8 [1|4|9]	24359
Word Up! (86) 8(1) [3|31|54]	10085
Cameosis (80) 25(3) [0|10|26]	3256
She's Strange (84) 27(1) [0|9|24]	2605
Alligator Woman (82) 23(3) [0|6|24]	2312
Single Life (85) 58(2) [0|0|27]	1652
Feel Me (80) 44(1) [0|0|17]	1645
Knights Of The Sound Table (81) 44(2) [0|0|13]	1068
Machismo (88) 56(1) [0|0|19]	991
Style (83) 53(1) [0|0|12]	746
Rafael CAMERON ▶ 913 Pk:67 [0|0|2]	1319
Cameron (80) 67(1) [0|0|18]	891
Cameron's In Love (81) 101(1) [0|0|12]	429
CAMOUFLAGE ▶ 1212 Pk:100 [0|0|1]	540
Voices & Images (89) 100(3) [0|0|14]	540
Glen CAMPBELL ▶ 1839 Pk:178 [0|0|1]	53.4
It's The World Gone Crazy (81) 178(1) [0|0|3]	53.4
CAMPER VAN BEETHOVEN ▶ 1030	
Pk:124 [0|0|2]	898
Our Beloved Revolutionary Sweetheart (88)	
124(1) [0|0|17]	541
Key Lime Pie (89) 141(3) [0|0|12]	357

Act ▶ Rank Peak [Top10s|Top40s|Total] Title (Yr) Pk(Wks) [T10 Wks|T40Wks|TotWks]	Score in '80s
CANDLEMASS ▶ 1724 Pk:174 [0|0|1]	102
Ancient Dreams (89) 174(2) [0|0|6]	102
Jim CAPALDI ▶ 1171 Pk:91 [0|0|2]	590
Fierce Heart (83) 91(1) [0|0|12]	481
Some Come Running (88) 183(2) [0|0|8]	109
CAPTAIN & TENNILLE ▶ 693 Pk:23 [0|1|1]	2457
Make Your Move (80) 23(2) [0|9|17]	*2457*
Carla CAPUANO ▶ 1572 Pk:152 [0|0|1]	194
Aerobic Dance Hits, Volume One (82)	
152(3) [0|0|8]	194
Irene CARA ▶ 624 Pk:76 [0|0|2]	2960
What A Feelin' (83) 77(1) [0|0|37]	2023
Anyone Can See (82) 76(2) [0|0|17]	937
Tony CAREY ▶ 820 Pk:60 [0|0|2]	1714
Some Tough City (84) 60(2) [0|0|24]	1539
Tony Carey [I Won't Be Home Tonight] (83)	
167(1) [0|0|9]	175
George CARLIN ▶ 1157 Pk:136 [0|0|2]	615
Carlin on Campus (84) 136(1) [0|0|11]	309
A Place For My Stuff! (81) 145(1) [0|0|13]	306
Belinda CARLISLE ▶ 179 Pk:13 [0|3|3]	16829
Heaven On Earth (87) 13(2) [0|35|51]	10382
Belinda (86) 13(1) [0|16|34]	5044
Runaway Horses (89) 37(1) [0|2|11]	*1403*
Carl CARLTON ▶ 754 Pk:34 [0|1|2]	2067
Carl Carlton (81) 34(2) [0|3|19]	1846
The Bad C.C. (82) 133(2) [0|0|7]	221
Larry CARLTON ▶ 785 Pk:99 [0|0|6]	1886
Sleepwalk (82) 99(1) [0|0|16]	682
Friends (83) 126(1) [0|0|11]	323
On Solid Ground (89) 126(2) [0|0|8]	282
Alone/But Never Alone (86) 141(2) [0|0|11]	260
Strikes Twice (80) 138(1) [0|0|8]	246
Discovery (87) 180(1) [0|0|6]	93.4
Eric CARMEN ▶ 837 Pk:59 [0|0|3]	1622
The Best Of Eric Carmen (88) 59(1) [0|0|20]	1237
Eric Carmen (II) (85) 128(2) [0|0|10]	265
Tonight You're Mine (80) 160(1) [0|0|5]	119
Jean CARN ▶ 1582 Pk:162 [0|0|2]	185
Closer Than Close (86)-Jean Carne	
162(2) [0|0|3]	128
Sweet And Wonderful (81) 176(1) [0|0|3]	57.1
Kim CARNES ▶ 172 Pk:1 [1|1|6]	17433
Mistaken Identity (81) 1(4) [12|23|52]	13054
Voyeur (82) 49(3) [0|0|22]	1424
Barking At Airplanes (85) 48(2) [0|0|14]	1087
Romance Dance (80) 57(2) [0|0|17]	1068
Cafe Racers (83) 97(1) [0|0|16]	561
Lighthouse (86) 116(2) [0|0|7]	240
Mary Chapin CARPENTER ▶ 1834	
Pk:186 [0|0|1]	56.2
State Of The Heart **(89**/90) 186(2) *[0|0|4]*	*56.2*
CARPENTERS ▶ 636 Pk:46 [0|0|5]	2888
Voices Of The Heart (83) 46(2) [0|0|19]	1376
Made In America (81) 52(1) [0|0|15]	1091
Yesterday Once More (85) 144(2) [0|0|8]	215
The Singles 1969-1973 **(74**/83)	
118(2) *[0|0|5]*	*192*
An Old Fashioned Christmas (85)	
190(1) [0|0|1]	14.5
Paul CARRACK ▶ 651 Pk:67 [0|0|3]	2745
One Good Reason (87) 67(1) [0|0|31]	1788
Suburban Voodoo (82) 78(1) [0|0|14]	670
Groove Approved (89) 120(2) [0|0|8]	*286*
Terri Lyne CARRINGTON ▶ 1681	
Pk:169 [0|0|1]	127
Real Life Story (89) 169(1) [0|0|7]	127
Jim CARROLL Band ▶ 893 Pk:73 [0|0|2]	1396
Catholic Boy (80) 73(2) [0|0|23]	1244
Dry Dreams (82) 156(1) [0|0|7]	153

Acts With Albums

Act ▶ Rank Peak [Top10s|Top40s|Total] Title (Yr) Pk(Wks) [T10 Wks|T40Wks|TotWks]	Score in '80s
The CARS ▶ 36 Pk:3 [3|6|7]	**48668**
Heartbeat City (84) 3(1) [31|48|69]	21423
Shake It Up (81) 9(4) [7|24|41]	8580
Panorama (80) 5(4) [6|11|28]	5931
Greatest Hits (85) 12(4) [0|16|39]	5885
Door To Door (87) 26(3) [0|8|23]	2688
*The Cars (**79**/80) 82(1) [0|0|60]*	*2347*
*Candy-O (**79**/80) 38(2) [0|2|35]*	*1814*
Carlene CARTER ▶ 1606 Pk:139 [0|0|1]	**171**
Musical Shapes (80) 139(2) [0|0|6]	171
Clarence CARTER ▶ 1861 Pk:189 [0|0|1]	**42.7**
Let's Burn (81) 189(1) [0|0|3]	42.7
Johnny CASH ▶ 1655 Pk:87 [0|0|1]	**139**
Class Of '55 (86)-Carl Perkins, Jerry Lee Lewis, Roy Orbison, & Johnny Cash [A] 87(2) [0|0|12]	139
Rosanne CASH ▶ 424 Pk:26 [0|1|5]	**5519**
Seven Year Ache (81) 26(1) [0|9|32]	3455
Rhythm And Romance (85) 101(1) [0|0|21]	840
Somewhere In The Stars (82) 76(2) [0|0|12]	611
King's Record Shop (87) 138(1) [0|0|20]	448
Hits 1979-1989 (89) 152(1) [0|0|7]	165
CASHFLOW ▶ 1477 Pk:144 [0|0|1]	**266**
Ca$hflow (86)-Ca$hflow 144(2) [0|0|11]	266
CENTRAL LINE ▶ 1542 Pk:145 [0|0|1]	**219**
Central Line (82) 145(2) [0|0|9]	219
Peter CETERA ▶ 372 Pk:23 [0|1|3]	**7186**
Solitude/Solitaire (86) 23(1) [0|17|43]	5757
One More Story (88) 58(3) [0|0|17]	1190
Peter Cetera (82) 143(2) [0|0|10]	239
CHAMPAIGN ▶ 615 Pk:53 [0|0|3]	**3033**
Modern Heart (83) 64(1) [0|0|24]	1524
How 'Bout Us (81) 53(1) [0|0|20]	1458
Woman In Flames (84) 184(2) [0|0|3]	50.3
Bill CHAMPLIN ▶ 1784 Pk:178 [0|0|1]	**75.6**
Runaway (82) 178(3) [0|0|4]	75.6
Gene CHANDLER ▶ 1055 Pk:87 [0|0|1]	**828**
Gene Chandler '80 (80) 87(1) [0|0|18]	828
CHANGE ▶ 441 Pk:29 [0|1|5]	**5217**
The Glow Of Love (80) 29(2) [0|7|25]	2615
Miracles (81) 46(1) [0|0|22]	1483
Change Of Heart (84) 102(1) [0|0|15]	509
Sharing Your Love (82) 66(2) [0|0|9]	464
This Is Your Time (83) 161(1) [0|0|7]	148
Harry CHAPIN ▶ 1005 Pk:58 [0|0|1]	**977**
Sequel (80) 58(1) [0|0|15]	977
Tracy CHAPMAN ▶ 142 Pk:1 [2|2|2]	**21861**
Tracy Chapman (88) 1(1) [18|46|61]	18275
Crossroads (89) 9(2) [4|10|11]	*3586*
CHARLENE ▶ 810 Pk:36 [0|1|2]	**1757**
I've Never Been To Me (82) 36(2) [0|3|20]	1612
Used To Be (82) 162(1) [0|0|7]	145
Ray CHARLES ▶ 988 Pk:75 [0|0|1]	**1024**
Friendship (85) 75(1) [0|0|20]	1024
Sonny CHARLES ▶ 1574 Pk:136 [0|0|1]	**193**
The Sun Still Shines (82) 136(2) [0|0|7]	193
CHARLIE ▶ 1523 Pk:145 [0|0|1]	**233**
Charlie (83) 145(1) [0|0|9]	233
CHEAP TRICK ▶ 157 Pk:16 [0|5|9]	**18778**
Lap Of Luxury (88) 16(1) [0|28|47]	8713
One On One (82) 39(1) [0|10|27]	3133
All Shook Up (80) 24(2) [0|7|15]	1953
Standing On The Edge (85) 35(2) [0|4|18]	1809
Found All The Parts (80) 39(2) [0|2|12]	1030
*Dream Police (**79**/80) 45(2) [0|0|12]*	*976*
Next Position Please (83) 61(1) [0|0|11]	686
The Doctor (86) 115(2) [0|0|9]	309
*Cheap Trick At Budokan (**79**/80) 118(2) [0|0|8]*	*169*
Chubby CHECKER ▶ 1884 Pk:186 [0|0|1]	**31.3**
The Change Has Come (82) 186(1) [0|0|2]	31.3

Act ▶ Rank Peak [Top10s|Top40s|Total] Title (Yr) Pk(Wks) [T10 Wks|T40Wks|TotWks]	Score in '80s
CHEECH & CHONG ▶ 1218 Pk:71 [0|0|2]	**530**
Get Out Of My Room (85) 71(2) [0|0|11]	471
Let's Make A New Dope Deal (80) 173(1) [0|0|3]	59.4
CHEQUERED PAST ▶ 1632 Pk:151 [0|0|1]	**153**
Chequered Past (84) 151(2) [0|0|6]	153
CHER ▶ 277 Pk:10 [1|2|2]	**10470**
Heart Of Stone (89) 10(2) [2|19|24]	*6329*
Cher(2) (87) 32(3) [0|8|41]	4142
CHERRELLE ▶ 613 Pk:36 [0|1|3]	**3043**
High Priority (86) 36(2) [0|2|30]	2286
Affair (88) 106(3) [0|0|15]	521
Fragile (84) 180(1) [0|0|8]	236
Neneh CHERRY ▶ 688 Pk:40 [0|1|1]	**2483**
Raw Like Sushi (89) 40(1) [0|1|28]	*2483*
CHIC ▶ 732 Pk:30 [0|1|4]	**2207**
Real People (80) 30(2) [0|3|15]	1472
Take It Off (81) 124(1) [0|0|9]	340
Les Plus Grands Succes De Chic - Chic's Greatest Hits (80) 88(2) [0|0|7]	*300*
Tongue In Chic (82) 173(2) [0|0|6]	94.5
CHICAGO ▶ 82 Pk:4 [2|4|7]	**32083**
Chicago 17 (84) 4(1) [12|44|72]	16115
Chicago 16 (82) 9(5) [6|18|38]	6952
Chicago 19 (88) 37(2) [0|4|42]	4185
Chicago 18 (86) 35(2) [0|4|45]	3924
Chicago XIV (80) 71(2) [0|0|9]	567
*Greatest Hits 1982-1989 (89/**90**) 67(1) [0|0|4]*	*245*
Chicago - Greatest Hits, Volume II (81) 171(2) [0|0|5]	94.8
The CHIEFTAINS ▶ 1508 Pk:102 [0|0|1]	**245**
Irish Heartbeat (88)-Van Morrison & The Chieftains [A] 102(2) [0|0|13]	245
Toni CHILDS ▶ 691 Pk:63 [0|0|1]	**2463**
Union (88) 63(1) [0|0|45]	2463
The CHI-LITES ▶ 1081 Pk:98 [0|0|3]	**778**
Bottoms Up (83) 98(1) [0|0|12]	515
Me And You (82) 162(2) [0|0|7]	152
Heavenly Body (80) 179(3) [0|0|6]	112
CHILLIWACK ▶ 840 Pk:78 [0|0|2]	**1617**
Wanna Be A Star (81) 78(1) [0|0|30]	1306
Opus X (82) 112(2) [0|0|10]	311
CHINA CRISIS ▶ 1225 Pk:114 [0|0|2]	**526**
What Price Paradise (87) 114(2) [0|0|12]	441
Flaunt The Imperfection (85) 171(1) [0|0|4]	85.1
The CHIPMUNKS ▶ 461 Pk:34 [0|1|4]	**4763**
Chipmunk Punk (80) 34(2) [0|6|26]	1971
Urban Chipmunk (81) 56(1) [0|0|35]	1964
A Chipmunk Christmas (81) 72(2) [0|0|9]	572
Chipmunk Rock (82) 109(1) [0|0|6]	256
CHOCOLATE MILK ▶ 1533 Pk:162 [0|0|1]	**229**
Blue Jeans (81) 162(1) [0|0|10]	229
The CHRISTIANS ▶ 1620 Pk:158 [0|0|1]	**163**
The Christians (88) 158(2) [0|0|8]	163
Gavin CHRISTOPHER ▶ 1095 Pk:74 [0|0|1]	**745**
One Step Closer (86) 74(3) [0|0|15]	745
CHUNKY A ▶ 1554 Pk:77 [0|0|1]	**203**
*Large And In Charge (89/**90**) 77(1) [0|0|3]*	*203*
The CHURCH ▶ 579 Pk:41 [0|0|2]	**3426**
Starfish (88) 41(2) [0|0|36]	3126
Heydey (86) 146(1) [0|0|11]	299
CINDERELLA ▶ 83 Pk:3 [2|2|2]	**31632**
Night Songs (86) 3(3) [15|49|70]	18099
Long Cold Winter (88) 10(5) [5|32|66]	13533
CIRCUS OF POWER ▶ 1886 Pk:185 [0|0|1]	**30.9**
Circus Of Power (88) 185(1) [0|0|2]	30.9
CLANNAD ▶ 1266 Pk:131 [0|0|2]	**475**
Macalla (86) 131(2) [0|0|12]	392
Sirius (88) 183(2) [0|0|12]	83.0
Eric CLAPTON ▶ 96 Pk:2 [2|7|8]	**28460**
Just One Night (80) 2(6) [11|19|31]	10383

Act ▶ Rank Peak [Top10s|Top40s|Total] Title (Yr) Pk(Wks) [T10 Wks|T40Wks|TotWks]	Score in '80s
Another Ticket (81)-Eric Clapton And His Band 7(2) [8|13|21]	5103
Money And Cigarettes (83) 16(4) [0|10|19]	3335
August (86) 37(2) [0|5|34]	3120
Behind The Sun (85) 34(1) [0|7|28]	2685
Crossroads (88) 34(2) [0|3|26]	1918
*Journeyman (89/**90**) 17(1) [0|5|6]*	*1382*
Time Pieces -- The Best Of Eric Clapton (82) 101(2) [0|0|14]	533
Stanley CLARKE ▶ 995 Pk:95 [0|0|3]	**1007**
Rocks, Pebbles And Sand (80) 95(2) [0|0|11]	453
Time Exposure (84) 149(1) [0|0|12]	297
Let Me Know You (82) 114(2) [0|0|8]	258
Stanley CLARKE & George DUKE ▶ 612 Pk:33 [0|1|2]	**3053**
The Clarke/Duke Project (81) 33(4) [0|8|23]	2746
The Clarke/Duke Project II (83) 146(1) [0|0|10]	307
The CLASH ▶ 145 Pk:7 [1|3|6]	**20942**
Combat Rock (82) 7(5) [10|39|61]	13792
London Calling (80) 27(3) [0|19|33]	3559
Sandinista! (81) 24(2) [0|5|20]	1974
Black Market Clash (80) 74(2) [0|0|16]	813
Cut The Crap (85) 88(2) [0|0|12]	581
The Story Of The Clash, Volume I (88) 142(2) [0|0|8]	223
Andrew Dice CLAY ▶ 855 Pk:94 [0|0|1]	**1535**
Dice (89) 94(1) [0|0|36]	*1535*
Richard CLAYDERMAN ▶ 1556 Pk:160 [0|0|1]	**203**
Amour (84) 160(3) [0|0|9]	203
Johnny CLEGG & SAVUKA ▶ 1617 Pk:155 [0|0|1]	**166**
Shadow Man (88) 155(2) [0|0|7]	166
Clarence CLEMONS ▶ 930 Pk:62 [0|0|2]	**1253**
Hero (85) 62(2) [0|0|18]	1157
Rescue (83) 174(1) [0|0|5]	96.0
Jimmy CLIFF ▶ 1502 Pk:122 [0|0|2]	**249**
Club Paradise (Soundtrack) (86) 122(2) [0|0|6]	217
Special (82) 186(2) [0|0|6]	32.6
Linda CLIFFORD ▶ 1473 Pk:142 [0|0|3]	**268**
*Here's My Love (**79**/80) 142(1) [0|0|4]*	*123*
I'm Yours (80) 160(1) [0|0|6]	114
The Right Combination (80)-Linda Clifford & Curtis Mayfield [A] 180(2) [0|0|4]	32.0
CLIMAX BLUES Band ▶ 1102 Pk:75 [0|0|1]	**728**
Flying The Flag (81) 75(1) [0|0|16]	728
CLIMIE FISHER ▶ 1207 Pk:120 [0|0|1]	**546**
Everything (88) 120(1) [0|0|16]	546
Patsy CLINE ▶ 794 Pk:29 [0|1|1]	**1824**
Sweet Dreams: The Life And Times Of Patsy Cline (Soundtrack) (85) 29(2) [0|5|18]	1824
George CLINTON ▶ 563 Pk:40 [0|1|5]	**3609**
Computer Games (82) 40(1) [0|1|33]	2200
You Shouldn't-Nuf Bit Fish (84) 102(1) [0|0|18]	674
R&B Skeletons In The Closet (86) 81(2) [0|0|12]	555
Some Of My Best Jokes Are Friends (85) 163(2) [0|0|6]	131
The Cinderella Theory (89) 192(1) [0|0|4]	49.9
CLUB NOUVEAU ▶ 324 Pk:6 [1|1|2]	**8338**
Life, Love & Pain (86) 6(1) [7|18|44]	8080
Listen To The Message (88) 98(2) [0|0|6]	258
Tom COCHRANE/RED RIDER ▶ 567 Pk:65 [0|0|6]	**3544**
As Far As Siam (81)-Red Rider 65(2) [0|0|24]	1323
Neruda (83)-Red Rider 66(4) [0|0|16]	1051
Tom Cochrane & Red Rider (86)-Tom Cochrane & Red Rider 112(1) [0|0|12]	462
Victory Day (88)-Tom Cochrane & Red Rider 144(2) [0|0|13]	343
Breaking Curfew (84)-Red Rider 137(2) [0|0|8]	231
Don't Fight It (80)-Red Rider 146(1) [0|0|5]	134

330

Acts With Albums

Act ▶ Rank Peak [Top10s|Top40s|Total] Title (Yr) Pk(Wks) [T10 Wks|T40Wks|TotWks]	Score in '80s
Bruce COCKBURN ▶ 561 Pk:45 [0|0|6]	3623
Dancing In The Dragon's Jaws (80) 45(2) [0|0|24]	1417
Stealing Fire (84) 74(2) [0|0|31]	1339
Humans (80) 81(2) [0|0|9]	453
World Of Wonders (86) 143(3) [0|0|8]	229
Bruce Cockburn Resume (81) 174(2) [0|0|5]	93.7
Big Circumstance (89) 182(1) [0|0|7]	90.9
Joe COCKER ▶ 460 Pk:50 [0|0|5]	4862
Cocker (86) 50(2) [0|0|18]	1487
Unchain My Heart (87) 89(2) [0|0|27]	1344
Sheffield Steel (82) 105(2) [0|0|23]	890
One Night Of Sin (89/90) 72(1) [0|0|16]	*860*
Civilized Man (84) 133(1) [0|0|9]	282
COCK ROBIN ▶ 940 Pk:61 [0|0|2]	1186
Cock Robin (85) 61(2) [0|0|19]	1124
After Here Through Midland (87) 166(2) [0|0|3]	62.4
COCTEAU TWINS ▶ 1178 Pk:109 [0|0|1]	586
Blue Bell Knoll (88) 109(2) [0|0|14]	586
David Allan COE ▶ 1771 Pk:179 [0|0|1]	80.3
Castles In The Sand (83) 179(1) [0|0|5]	80.3
COLD CHISEL ▶ 1689 Pk:171 [0|0|1]	122
East (81) 171(2) [0|0|6]	122
Natalie COLE ▶ 320 Pk:42 [0|0|7]	8393
Everlasting (87) 42(2) [0|0|58]	4630
Good To Be Back (89) 59(2) [0|0|23]	1423
Don't Look Back (80) 77(1) [0|0|22]	1202
We're The Best Of Friends (80)-Natalie Cole & Peabo Bryson [A] 44(1) [0|0|16]	729
Dangerous (85) 140(2) [0|0|9]	226
Happy Love (81) 132(2) [0|0|4]	134
I'm Ready (83) 182(1) [0|0|3]	50.3
Durell COLEMAN ▶ 1589 Pk:155 [0|0|1]	180
Durell Coleman (85) 155(2) [0|0|7]	180
Albert COLLINS ▶ 1600 Pk:124 [0|0|1]	174
Showdown! (86)-Albert Collins, Robert Cray, Johnny Copeland [A] 124(1) [0|0|18]	174
Judy COLLINS ▶ 1490 Pk:142 [0|0|2]	258
Running For My Life (80) 142(3) [0|0|6]	192
Times Of Our Lives (82) 190(1) [0|0|5]	66.2
Phil COLLINS ▶ 19 Pk:1 [4|4|4]	64403
No Jacket Required (85) 1(7) [31|70|123]	35552
Face Value (81) 7(4) [10|26|164]	15139
Hello, I Must Be Going! (82) 8(5) [7|21|141]	11288
...But Seriously (89) 1(1) [3|4|5]	*2424*
Jessi COLTER ▶ 1133 Pk:43 [0|0|3]	672
Leather And Lace (81)-Waylon Jennings & Jessi Colter [A] 43(2) [0|0|19]	672
Shawn COLVIN ▶ 1810 Pk:165 [0|0|1]	64.3
Steady On (89/90) 165(1) [0|0|3]	*64.3*
The COMMODORES ▶ 116 Pk:7 [1|5|8]	25067
In The Pocket (81) 13(2) [0|19|40]	6921
Heroes (80) 7(3) [5|15|33]	5912
Nightshift (85) 12(3) [0|15|37]	5408
Midnight Magic (79/80) 18(4) [0|10|21]	*3376*
All The Great Hits (82) 37(3) [0|4|24]	2184
United (86) 101(3) [0|0|15]	635
Commodores 13 (83) 103(2) [0|0|11]	430
Commodores Anthology (83) 141(3) [0|0|7]	201
The COMMUNARDS ▶ 997 Pk:90 [0|0|2]	1004
The Communards (86) 90(2) [0|0|16]	590
Red (88) 93(1) [0|0|9]	413
COMPANY B ▶ 1612 Pk:143 [0|0|1]	169
Company B (87) 143(2) [0|0|5]	169
CONCRETE BLONDE ▶ 993 Pk:96 [0|0|2]	1012
Concrete Blonde (87) 96(2) [0|0|16]	635
Free (89) 148(1) [0|0|18]	378
CONEY HATCH ▶ 1888 Pk:186 [0|0|1]	30.8
Outa Hand (83) 186(1) [0|0|2]	30.8
CON FUNK SHUN ▶ 354 Pk:30 [0|1|7]	7565
Spirit Of Love (80) 30(3) [0|5|20]	1874
Touch (80) 51(1) [0|0|19]	1560
Electric Lady (85) 62(3) [0|0|26]	1448
To The Max (82) 115(1) [0|0|29]	1009
Fever (83) 105(1) [0|0|21]	715
Con Funk Shun 7 (81) 82(1) [0|0|13]	656
Burnin' Love (86) 121(2) [0|0|11]	303
John CONLEE ▶ 1680 Pk:166 [0|0|1]	127
John Conlee's Greatest Hits (83) 166(1) [0|0|6]	127
The CONNELLS ▶ 1599 Pk:163 [0|0|1]	174
Fun & Games (89) 163(1) [0|0|10]	174
Harry CONNICK Jr. ▶ 865 Pk:42 [0|0|1]	1498
When Harry Met Sally (Soundtrack) (89) 42(2) [0|0|20]	*1498*
Norman CONNORS ▶ 1597 Pk:145 [0|0|2]	176
Take It To The Limit (80) 145(1) [0|0|6]	154
Mr. C. (81) 197(1) [0|0|2]	21.9
Tommy CONWELL And The YOUNG RUMBLERS ▶ 981 Pk:103 [0|0|1]	1049
Rumble (88) 103(1) [0|0|28]	1049
Ry COODER ▶ 707 Pk:43 [0|0|4]	2354
Borderline (81) 43(2) [0|0|16]	1434
Crossroads (Soundtrack) (86) 85(3) [0|0|9]	441
The Slide Area (82) 105(1) [0|0|7]	274
Get Rhythm (87) 177(1) [0|0|12]	205
Sam COOKE ▶ 1363 Pk:134 [0|0|2]	362
Sam Cooke Live At The Harlem Square Club (85) 134(1) [0|0|8]	221
The Man & His Music (86) 175(1) [0|0|7]	141
Rita COOLIDGE ▶ 1231 Pk:107 [0|0|3]	518
Rita Coolidge/Greatest Hits (81) 107(2) [0|0|8]	313
Satisfied (79/80) 163(2) [0|0|5]	*111*
Heartbreak Radio (81) 160(1) [0|0|4]	93.6
Alice COOPER ▶ 346 Pk:20 [0|1|5]	7770
Trash (89) 20(2) [0|16|21]	*4279*
Flush The Fashion (80) 44(2) [0|0|17]	1311
Constrictor (86) 59(1) [0|0|21]	1287
Raise Your Fist And Yell (87) 73(2) [0|0|15]	698
Special Forces (81) 125(2) [0|0|5]	195
Michael COOPER ▶ 1028 Pk:98 [0|0|1]	901
Love Is Such A Funny Game (88) 98(3) [0|0|25]	901
Julian COPE ▶ 992 Pk:105 [0|0|3]	1013
St. Julian (87) 105(2) [0|0|12]	443
My Nation Underground (88) 155(1) [0|0|13]	301
Julian Cope (87) 109(2) [0|0|6]	269
Johnny COPELAND ▶ 1601 Pk:124 [0|0|1]	174
Showdown! (86)-Albert Collins, Robert Cray, Johnny Copeland [A] 124(1) [0|0|18]	174
Stewart COPELAND ▶ 1402 Pk:148 [0|0|2]	325
The Rhythmatist (85) 148(2) [0|0|8]	203
Rumble Fish (Soundtrack) (83) 157(1) [0|0|5]	122
Chick COREA ▶ 1648 Pk:170 [0|0|2]	142
Three Quartets (81) 179(2) [0|0|4]	76.2
Tap Step (80) 170(2) [0|0|3]	66.0
Bill COSBY ▶ 676 Pk:26 [0|1|2]	2564
Those Of You With Or Without Children, You'll Understand (86) 26(2) [0|4|15]	1666
Bill Cosby "Himself" Soundtrack (82) 64(2) [0|0|14]	898
Elvis COSTELLO ▶ 479 Pk:28 [0|2|6]	4496
Spike (89) 32(1) [0|11|25]	3195
Taking Liberties (80) 28(1) [0|3|14]	1301
Elvis COSTELLO & The ATTRACTIONS ▶ 194 Pk:11 [0|6|9]	15447
Punch The Clock (83) 24(1) [0|12|24]	3243
Get Happy!! (80) 11(2) [0|8|15]	2924
Imperial Bedroom (82) 30(3) [0|7|24]	2386
Goodbye Cruel World (84) 35(2) [0|3|21]	1779
Trust (81) 28(2) [0|4|15]	1664
King Of America (86) 39(2) [0|3|18]	1503
Almost Blue (81) 50(2) [0|0|13]	804
Best Of Elvis Costello/The Attractions (85) 116(2) [0|0|16]	586
Blood & Chocolate (86) 84(2) [0|0|11]	559
Josie COTTON ▶ 1503 Pk:147 [0|0|1]	249
Convertible Music (82) 147(2) [0|0|12]	249
COVER GIRLS ▶ 650 Pk:64 [0|0|2]	2753
Show Me (87) 64(1) [0|0|61]	2583
We Can't Go Wrong (89/90) 143(2) [0|0|6]	*170*
COWBOY JUNKIES ▶ 590 Pk:26 [0|1|1]	3320
The Trinity Session (89) 26(1) [0|9|29]	3320
CRACK THE SKY ▶ 1791 Pk:186 [0|0|1]	74.0
From The Greenhouse (89) 186(1) [0|0|5]	74.0
Floyd CRAMER ▶ 1751 Pk:170 [0|0|1]	88.8
Dallas (80) 170(1) [0|0|5]	88.8
Michael CRAWFORD ▶ 1897 Pk:192 [0|0|1]	27.2
Songs From The Stage And Screen (88) 192(2) [0|0|2]	27.2
Randy CRAWFORD ▶ 825 Pk:71 [0|0|6]	1695
Secret Combination (81) 71(1) [0|0|19]	1009
Windsong (82) 148(1) [0|0|10]	260
Rich And Poor (89) 159(2) [0|0|7]	*140*
Now We May Begin (80) 180(1) [0|0|7]	106
Nightline (83) 164(1) [0|0|5]	105
Abstract Emotions (86) 178(2) [0|0|4]	75.2
Robert CRAY Band ▶ 246 Pk:13 [0|2|5]	11790
Strong Persuader (86) 13(3) [0|26|49]	8341
Don't Be Afraid Of The Dark (88) 32(2) [0|7|32]	2512
False Accusations (86) 141(3) [0|0|21]	527
Bad Influence (87) 143(1) [0|0|11]	237
Showdown! (86)-Albert Collins, Robert Cray, Johnny Copeland [A] 124(1) [0|0|18]	174
CREEDENCE CLEARWATER REVIVAL ▶ 905 Pk:62 [0|0|2]	1342
The Concert (80) 62(1) [0|0|20]	1029
Chronicle II (86) 165(2) [0|0|16]	314
Marshall CRENSHAW ▶ 549 Pk:50 [0|0|3]	3761
Marshall Crenshaw (82) 50(2) [0|0|27]	1952
Field Day (83) 52(1) [0|0|14]	1145
Downtown (85) 110(2) [0|0|18]	663
The CRETONES ▶ 1389 Pk:125 [0|0|1]	335
Thin Red Line (80) 125(1) [0|0|10]	335
David CROSBY ▶ 1319 Pk:104 [0|0|1]	417
Oh Yes I Can (89) 104(2) [0|0|10]	417
CROSBY, STILLS & NASH ▶ 276 Pk:8 [1|1|3]	10475
Daylight Again (82) 8(5) [6|30|41]	9289
Allies (83) 43(2) [0|0|12]	1006
Replay (81) 122(2) [0|0|5]	180
CROSBY, STILLS, NASH & YOUNG ▶ 533 Pk:16 [0|1|1]	3862
American Dream (88) 16(1) [0|11|22]	3862
Christopher CROSS ▶ 90 Pk:6 [1|2|3]	29998
Christopher Cross (80) 6(3) [15|81|116]	25248
Another Page (83) 11(5) [0|11|31]	4567
Every Turn Of The World (85) 127(1) [0|0|6]	184
CROWDED HOUSE ▶ 291 Pk:12 [0|2|2]	9880
Crowded House (86) 12(1) [0|24|58]	8048
Temple Of Low Men (88) 40(2) [0|2|19]	1832
Rodney CROWELL ▶ 1159 Pk:105 [0|0|3]	609
Rodney Crowell (81) 105(1) [0|0|8]	339
But What Will The Neighbors Think (80) 155(1) [0|0|10]	190
Street Language (86) 177(1) [0|0|5]	80.3
CROWN HEIGHTS AFFAIR ▶ 1460 Pk:148 [0|0|1]	277
Sure Shot (80) 148(2) [0|0|12]	277
The CRUSADERS ▶ 529 Pk:29 [0|1|5]	3942
Rhapsody And Blues (80) 29(2) [0|4|16]	1694
Ghetto Blaster (84) 79(2) [0|0|22]	948

Acts With Albums

Act ▶ Rank Peak [Top10s|Top40s|Total] Title (Yr) Pk(Wks) [T10 Wks|T40Wks|TotWks]	Score in '80s
The CRUSADERS ▶ *Continued*	
Standing Tall (81) 59(2) [0|0|16]	838
Street Life (79/80) 135(1) [0|0|9]	*263*
Royal Jam (82)-The Crusaders With B.B. King And The Royal Philharmonic Orchestra 144(1) [0|0|7]	200
CRUZADOS ▶ 835 Pk:76 [0|0|2]	**1644**
Cruzados (85) 76(2) [0|0|18]	935
After Dark (87) 106(2) [0|0|21]	709
Billy CRYSTAL ▶ 1085 Pk:65 [0|0|1]	**767**
Mahvelous (85) 65(2) [0|0|13]	767
The CULT ▶ 250 Pk:10 [1|2|3]	**11733**
Sonic Temple (89) 10(6) [6|20|33]	7081
Electric (87) 38(3) [0|4|32]	3138
Love (85) 87(2) [0|0|34]	1515
CULTURE CLUB ▶ 56 Pk:2 [1|4|4]	**39249**
Colour By Numbers (83) 2(6) [30|38|59]	19976
Kissing To Be Clever (83) 14(10) [0|38|88]	14660
Waking Up With The House On Fire (84) 26(5) [0|9|20]	2695
From Luxury To Heartache (86) 32(2) [0|6|17]	1919
The CURE ▶ 152 Pk:12 [0|2|7]	**19658**
Disintegration (89) 12(3) [0|26|33]	*8220*
Kiss Me, Kiss Me, Kiss Me (87) 35(4) [0|8|52]	5226
Standing On A Beach -- The Singles (86) 48(2) [0|0|57]	3257
The Head On The Door (85) 59(2) [0|0|49]	2669
The Walk (83) 177(1) [0|0|9]	147
Japanese Whispers (84) 181(1) [0|0|5]	73.6
The Top (84) 180(1) [0|0|4]	64.4
CURIOSITY KILLED THE CAT ▶ 868 Pk:55 [0|0|1]	**1490**
Keep Your Distance (87) 55(1) [0|0|29]	1490
Tim CURRY ▶ 1185 Pk:77 [0|0|2]	**576**
Simplicity (81) 112(2) [0|0|8]	326
Fearless (79/80) 77(1) [0|0|7]	*250*
CUTTING CREW ▶ 463 Pk:16 [0|1|2]	**4752**
Broadcast (87) 16(2) [0|10|45]	4610
The Scattering (89) 150(2) [0|0|6]	142
Andre CYMONE ▶ 1386 Pk:121 [0|0|2]	**337**
A.C. (85) 121(2) [0|0|8]	278
Survivin' In The 80's (83) 185(1) [0|0|4]	59.1

D

Act ▶ Rank Peak [Top10s|Top40s|Total] Title (Yr) Pk(Wks) [T10 Wks|T40Wks|TotWks]	Score in '80s
DA'KRASH ▶ 1859 Pk:184 [0|0|1]	**44.9**
Da'Krash (88) 184(2) [0|0|3]	44.9
Roger DALTREY ▶ 481 Pk:22 [0|1|4]	**4485**
Under A Raging Moon (85) 42(2) [0|0|26]	2134
McVicar (Soundtrack) (80) 22(2) [0|6|15]	1907
Parting Should Be Painless (84) 102(2) [0|0|9]	366
Best Bits (82) 185(2) [0|0|5]	78.1
Michael DAMIAN ▶ 999 Pk:61 [0|0|1]	**1002**
Where Do We Go From Here (89) 61(2) [0|0|18]	*1002*
Dana DANE ▶ 669 Pk:46 [0|0|1]	**2587**
Dana Dane With Fame (87) 46(2) [0|0|32]	2587
DANGER DANGER ▶ 1038 Pk:88 [0|0|1]	**872**
Danger Danger (89) 88(1) [0|0|20]	*872*
Rodney DANGERFIELD ▶ 585 Pk:36 [0|1|2]	**3349**
Rappin' Rodney (83) 36(3) [0|3|20]	1934
No Respect (80) 48(1) [0|0|19]	1415
DANGEROUS TOYS ▶ 781 Pk:65 [0|0|1]	**1913**
Dangerous Toys (89) 65(3) [0|0|29]	*1913*
Charlie DANIELS Band ▶ 348 Pk:11 [0|2|6]	**7717**
Full Moon (80) 11(2) [0|9|33]	4381
Windows (82) 26(3) [0|6|19]	2053
A Decade Of Hits (83) 84(1) [0|0|12]	620
Million Mile Reflections (79/80) 106(2) [0|0|9]	*353*
Simple Man (89/90) 89(1) [0|0|6]	*273*
Homesick Heros (88) 181(2) [0|0|5]	37.2

Act ▶ Rank Peak [Top10s|Top40s|Total] Title (Yr) Pk(Wks) [T10 Wks|T40Wks|TotWks]	Score in '80s
DANNY WILSON ▶ 1114 Pk:79 [0|0|1]	**708**
Meet Danny Wilson (87) 79(1) [0|0|16]	708
DANZIG ▶ 1419 Pk:125 [0|0|1]	**309**
Danzig (88) 125(2) [0|0|9]	309
Terence Trent D'ARBY ▶ 222 Pk:4 [1|1|2]	**13724**
Introducing The Hardline According To Terence Trent D'Arby (87) 4(2) [8|32|60]	13244
Terence Trent D'Arby's Neither Fish Nor Flesh (89) 61(2) [0|0|6]	*480*
DARK ANGEL ▶ 1642 Pk:159 [0|0|1]	**147**
Leave Scars (89) 159(1) [0|0|6]	147
DARLING CRUEL ▶ 1598 Pk:160 [0|0|1]	**175**
Passion Crimes (89) 160(2) [0|0|8]	175
DAVE & SUGAR ▶ 1801 Pk:179 [0|0|1]	**70.3**
Greatest Hits (81) 179(2) [0|0|4]	70.3
DAVID & DAVID ▶ 560 Pk:39 [0|1|1]	**3635**
Boomtown (86) 39(2) [0|2|38]	3635
Dave DAVIES ▶ 847 Pk:42 [0|0|2]	**1571**
AFL1-3603 (80) 42(2) [0|0|14]	1364
Glamour (81) 152(1) [0|0|8]	207
Danny DAVIS ▶ 1819 Pk:150 [0|0|1]	**62.4**
Danny Davis & Willie Nelson With The Nashville Brass (80)-Danny Davis And Willie Nelson With The Nashville Brass [A] 150(1) [0|0|5]	62.4
Jimmy DAVIS & JUNCTION ▶ 1471 Pk:122 [0|0|1]	**269**
Kick The Wall (87) 122(2) [0|0|8]	269
Mac DAVIS ▶ 864 Pk:67 [0|0|3]	**1507**
It's Hard To Be Humble (80) 69(2) [0|0|15]	880
Texas In My Rear View Mirror (80) 67(1) [0|0|9]	571
Midnight Crazy (82) 174(1) [0|0|3]	55.7
Martha DAVIS ▶ 1304 Pk:127 [0|0|1]	**433**
Policy (87) 127(1) [0|0|13]	433
Miles DAVIS ▶ 699 Pk:53 [0|0|8]	**2407**
The Man With The Horn (81) 53(2) [0|0|18]	1096
You're Under Arrest (85) 111(2) [0|0|12]	355
Tutu (86) 141(2) [0|0|10]	235
Decoy (84) 169(2) [0|0|11]	220
Star People (83) 136(1) [0|0|7]	208
We Want Miles (82) 159(2) [0|0|5]	163
Amandla (89) 177(2) [0|0|5]	91.2
Directions (81) 179(2) [0|0|4]	39.0
Paul DAVIS ▶ 857 Pk:52 [0|0|2]	**1531**
Cool Night (81) 52(1) [0|0|29]	1454
Paul Davis (80) 173(2) [0|0|4]	76.6
Tyrone DAVIS ▶ 1608 Pk:137 [0|0|1]	**170**
Tyrone Davis (83) 137(2) [0|0|6]	170
Morris DAY ▶ 553 Pk:37 [0|1|2]	**3699**
Color Of Success (85) 37(2) [0|3|31]	2392
Daydreaming (88) 41(2) [0|0|15]	1308
Taylor DAYNE ▶ 281 Pk:21 [0|2|2]	**10269**
Tell It To My Heart (88) 21(2) [0|34|69]	9419
Can't Fight Fate (89/90) 39(1) [0|1|7]	*850*
DAZZ BAND ▶ 287 Pk:14 [0|1|7]	**9980**
Keep It Live (82) 14(4) [0|11|34]	4930
Joystick (83) 73(2) [0|0|33]	1659
Jukebox (84) 83(1) [0|0|29]	1175
On The One (83) 59(3) [0|0|16]	995
Hot Spot (85) 98(2) [0|0|12]	488
Wild And Free (86) 100(2) [0|0|11]	472
Let The Music Play (81) 154(2) [0|0|7]	261
DB'S ▶ 1654 Pk:171 [0|0|1]	**139**
The Sound Of Music (87) 171(2) [0|0|8]	139
DEAD MILKMEN ▶ 1091 Pk:101 [0|0|2]	**758**
Beelzebubba (88) 101(2) [0|0|23]	623
Bucky Fellini (87) 163(2) [0|0|7]	135
DEAD OR ALIVE ▶ 508 Pk:31 [0|1|4]	**4155**
Youthquake (85) 31(3) [0|6|20]	2209
Mad, Bad And Dangerous To Know (86) 52(2) [0|0|25]	1587

Act ▶ Rank Peak [Top10s|Top40s|Total] Title (Yr) Pk(Wks) [T10 Wks|T40Wks|TotWks]	Score in '80s
Nude (89) 106(2) [0|0|9]	334
Rip It Up (88) 195(2) [0|0|2]	24.6
Paul DEAN ▶ 1912 Pk:195 [0|0|1]	**24.6**
Hard Core (89) 195(2) [0|0|2]	24.6
DEATH ANGEL ▶ 1483 Pk:143 [0|0|1]	**263**
Frolic Through The Park (88) 143(1) [0|0|11]	263
DeBARGE ▶ 204 Pk:19 [0|3|3]	**14638**
Rhythm Of The Night (85) 19(2) [0|24|48]	6651
All This Love (82) 24(2) [0|7|48]	4471
In A Special Way (83) 36(2) [0|5|40]	3516
Bunny DeBARGE ▶ 1729 Pk:172 [0|0|1]	**98.7**
In Love (87) 172(2) [0|0|5]	98.7
Chico DeBARGE ▶ 958 Pk:90 [0|0|1]	**1126**
Chico DeBarge (86) 90(2) [0|0|30]	1126
El DeBARGE ▶ 598 Pk:24 [0|1|1]	**3241**
El DeBarge (86) 24(3) [0|11|23]	3241
Chris De BURGH ▶ 427 Pk:25 [0|1|3]	**5445**
Into The Light (86) 25(2) [0|8|32]	2958
The Getaway (83) 43(2) [0|0|22]	1551
Man On The Line (84) 69(1) [0|0|19]	936
The DEELE ▶ 642 Pk:54 [0|0|3]	**2830**
Eyes Of A Stranger (88) 54(2) [0|0|25]	1770
Street Beat (84) 78(2) [0|0|19]	870
Material Thangz (85) 155(1) [0|0|8]	190
DEEP PURPLE ▶ 349 Pk:17 [0|2|4]	**7692**
Perfect Strangers (84) 17(2) [0|18|32]	5236
The House Of Blue Light (87) 34(3) [0|4|22]	1961
Nobody's Perfect (88) 105(2) [0|0|9]	379
Deepest Purple: The Very Best Of Deep Purple (80) 148(1) [0|0|4]	116
DEF LEPPARD ▶ 6 Pk:1 [2|3|5]	**87705**
Hysteria (87) 1(6) [78|96|124]	*49929*
Pyromania (83) 2(2) [38|58|116]	28472
High 'N' Dry (81) 38(1) [0|18|106]	6077
On Through The Night (80) 51(1) [0|0|51]	2293
High 'N' Dry(2) (84) 72(2) [0|0|18]	934
DEJA ▶ 1769 Pk:186 [0|0|1]	**80.9**
Serious (87) 186(1) [0|0|6]	80.9
DE LA SOUL ▶ 584 Pk:24 [0|1|1]	**3354**
3 Feet High And Rising (89) 24(2) [0|10|29]	3354
DEL FUEGOS ▶ 937 Pk:132 [0|0|3]	**1199**
Boston, Mass. (85) 132(1) [0|0|34]	801
Smoking In The Fields (89) 139(1) [0|0|10]	*284*
Stand Up (87) 167(2) [0|0|8]	114
The DELLS ▶ 1377 Pk:137 [0|0|1]	**345**
I Touched A Dream (80) 137(1) [0|0|12]	345
John DENVER ▶ 344 Pk:32 [0|3|5]	**7778**
Some Days Are Diamonds (81) 32(2) [0|4|30]	2612
Seasons Of The Heart (82) 39(2) [0|2|33]	1942
Autograph (80) 39(2) [0|2|17]	1563
Dreamland Express (85) 90(2) [0|0|19]	839
It's About Time (83) 61(2) [0|0|15]	823
John DENVER & The MUPPETS ▶ 1200 Pk:26 [0|1|1]	**553**
A Christmas Together (80) 26(2) [0|2|4]	*553*
DEODATO ▶ 1856 Pk:186 [0|0|1]	**45.8**
Night Cruiser (80)-Eumir Deodato 186(1) [0|0|3]	45.8
DEPECHE MODE ▶ 272 Pk:35 [0|1|8]	**10696**
Music For The Masses (87) 35(2) [0|3|59]	3720
Some Great Reward (85) 51(2) [0|0|42]	2365
People Are People (84) 71(1) [0|0|30]	1348
101 (89) 45(2) [0|0|19]	1314
Black Celebration (86) 90(2) [0|0|26]	1093
Catching Up With Depeche Mode (85) 113(1) [0|0|18]	635
A Broken Frame (82) 177(1) [0|0|8]	110
Speak & Spell (81) 192(1) [0|0|9]	110
Teri DeSARIO ▶ 1163 Pk:80 [0|0|1]	**600**
Moonlight Madness (80) 80(1) [0|0|13]	600

Acts With Albums

Act ▶ Rank Peak [Top10s|Top40s|Total] Title (Yr) Pk(Wks) [T10 Wks|T40Wks|TotWks]	Score in '80s
DEVICE ▶ 1077 Pk:73 [0|0|1]	791
22B3 (86) 73(2) [0|0|16]	791
--DeVILLE, Mink see MINK DeVILLE	
DEVO ▶ 264 Pk:22 [0|2|6]	10989
Freedom Of Choice (80) 22(2) [0|16|51]	6045
New Traditionalists (81) 23(2) [0|6|25]	2371
Oh No! It's Devo (82) 47(3) [0|0|20]	1384
Devo-Live (81) 50(1) [0|0|12]	826
Shout (84) 83(2) [0|0|6]	322
Total Devo (88) 189(2) [0|0|3]	41.4
DEXYS MIDNIGHT RUNNERS ▶ 522 Pk:14 [0|1|1]	4083
Too-Rye-Ay (83) 14(2) [0|12|24]	4083
Dennis DeYOUNG ▶ 639 Pk:29 [0|1|2]	2845
Desert Moon (84) 29(3) [0|4|25]	2540
Back To The World (86) 108(2) [0|0|8]	305
DFX2 ▶ 1569 Pk:143 [0|0|1]	195
Emotion (83) 143(1) [0|0|8]	195
Neil DIAMOND ▶ 52 Pk:3 [3|6|10]	40629
The Jazz Singer (Soundtrack) (80) 3(7) [19|32|115]	16555
Heartlight (82) 9(3) [4|19|34]	6771
On The Way To The Sky (81) 17(2) [0|11|27]	3827
September Morn (80) 10(3) [3|10|20]	3723
Headed For The Future (86) 20(1) [0|11|23]	3428
Primitive (84) 35(3) [0|5|25]	1991
12 Greatest Hits Vol. II (82) 48(2) [0|0|42]	1917
The Best Years Of Our Lives (89) 46(2) [0|0|16]	1154
Hot August Night II (87) 59(2) [0|0|17]	1143
Classics The Early Years (83) 171(1) [0|0|7]	120
DIESEL ▶ 1015 Pk:68 [0|0|1]	954
Watts In A Tank (81) 68(2) [0|0|24]	954
DIFFORD & TILBROOK ▶ 989 Pk:55 [0|0|1]	1021
Difford & Tilbrook (84) 55(3) [0|0|15]	1021
DILLMAN Band ▶ 1588 Pk:145 [0|0|1]	180
Lovin' The Night Away (81) 145(2) [0|0|7]	180
Al DI MEOLA ▶ 815 Pk:55 [0|0|5]	1742
Electric Rendezvous (82) 55(2) [0|0|13]	875
Splendido Hotel (80) 119(2) [0|0|14]	537
Scenario (83) 128(1) [0|0|6]	169
Tour De Force - "Live" (82) 165(1) [0|0|7]	146
Tirami Su (88)-Al Di Meola Project 190(1) [0|0|1]	14.5
Al Di MEOLA/John McLAUGHLIN/ Paco De LUCIA ▶ 1180 Pk:97 [0|0|2]	583
Friday Night In San Francisco (81) 97(1) [0|0|13]	482
Passion, Grace & Fire (83) 171(2) [0|0|5]	101
DINO ▶ 581 Pk:34 [0|1|1]	3418
24/7 (89) 34(3) [0|6|41]	*3418*
DIO ▶ 268 Pk:23 [0|2|5]	10757
The Last In Line (84) 23(3) [0|10|35]	3757
Sacred Heart (85) 29(4) [0|10|29]	2843
Holy Diver (83) 56(1) [0|0|38]	2415
Dream Evil (87) 43(2) [0|0|11]	976
Intermission (86) 70(2) [0|0|16]	765
DION ▶ 1150 Pk:130 [0|0|1]	632
Yo Frankie (89) 130(1) [0|0|19]	632
DIRE STRAITS ▶ 49 Pk:1 [1|3|6]	41734
Brothers In Arms (85) 1(9) [37|55|97]	30184
Making Movies (80) 19(3) [0|17|31]	5062
Love Over Gold (82) 19(3) [0|8|32]	3273
Alchemy-Dire Straits Live (84) 46(2) [0|0|18]	1262
Twisting By The Pool (83) 53(1) [0|0|15]	1073
Money For Nothing (88) 62(2) [0|0|17]	881
DIRTY LOOKS ▶ 1053 Pk:118 [0|0|2]	829
Cool From The Wire (88) 134(2) [0|0|14]	425
Turn Of The Screw (89) 118(2) [0|0|11]	405
DISNEYLAND AFTER DARK ▶ 1299 Pk:116 [0|0|1]	436
No Fuel Left For The Pilgrims (89) 116(1) [0|0|11]	436

Act ▶ Rank Peak [Top10s|Top40s|Total] Title (Yr) Pk(Wks) [T10 Wks|T40Wks|TotWks]	Score in '80s
DIVINYLS ▶ 1100 Pk:91 [0|0|1]	734
What A Life! (85) 91(2) [0|0|18]	734
DIXIE DREGS ▶ 687 Pk:56 [0|0|3]	2487
Industry Standard (82)-The Dregs 56(2) [0|0|15]	897
Dregs Of The Earth (80) 81(1) [0|0|17]	801
Unsung Heroes (81)-The Dregs 67(2) [0|0|14]	789
Don DIXON ▶ 1623 Pk:162 [0|0|1]	159
Most Of The Girls Like To Dance But Only Some Of The Boys Like To (87) 162(1) [0|0|8]	159
DJ JAZZY JEFF & THE FRESH PRINCE ▶ 224 Pk:4 [1|2|3]	13511
He's The D.J., I'm The Rapper (88) 4(1) [10|23|55]	10762
Rock The House (87) 83(3) [0|0|35]	1742
And In This Corner... (89) 39(4) [0|5|7]	*1007*
The D.O.C. ▶ 601 Pk:20 [0|1|1]	3178
No One Can Do It Better (89) 20(2) [0|10|20]	*3178*
DOCTOR And The MEDICS ▶ 1520 Pk:125 [0|0|1]	235
Laughing At The Pieces (86) 125(1) [0|0|8]	235
DR. HOOK ▶ 811 Pk:71 [0|0|4]	1751
Sometimes You Win (80) 71(2) [0|0|26]	*990*
Dr. Hook/Greatest Hits (80) 142(3) [0|0|12]	349
Players In The Dark (82) 118(3) [0|0|7]	283
Rising (80) 175(3) [0|0|8]	129
DR. JOHN ▶ 1468 Pk:142 [0|0|1]	272
In A Sentimental Mood (89) 142(2) [0|0|11]	272
DR. J.R. KOOL & The OTHER ROXANNES ▶ 1234 Pk:113 [0|0|1]	515
The Complete Story Of Roxanne...The Album (85) 113(1) [0|0|13]	515
DOKKEN ▶ 191 Pk:13 [0|3|5]	15698
Back For The Attack (87) 13(1) [0|15|33]	5247
Under Lock And Key (85) 32(3) [0|5|67]	4442
Tooth And Nail (84) 49(2) [0|0|74]	3956
Beast From The East (88) 33(3) [0|4|17]	1655
Breaking The Chains (83) 136(1) [0|0|13]	397
Thomas DOLBY ▶ 256 Pk:13 [0|3|4]	11400
The Golden Age Of Wireless (83) 13(2) [0|13|28]	4639
Blinded By Science (83) 20(4) [0|11|31]	3971
The Flat Earth (84) 35(2) [0|3|18]	1787
Aliens Ate My Buick (88) 70(1) [0|0|19]	1003
Joe DOLCE ▶ 1812 Pk:181 [0|0|1]	64.0
Shaddap You Face (81) 181(2) [0|0|4]	64.0
Placido DOMINGO ▶ 574 Pk:18 [0|1|3]	3477
Perhaps Love (81) 18(2) [0|8|27]	2988
My Life For A Song (83) 117(1) [0|0|11]	364
Domingo-Con Amore (82) 164(2) [0|0|6]	125
DOOBIE BROTHERS ▶ 206 Pk:3 [1|3|6]	14298
One Step Closer (80) 3(3) [7|18|28]	7124
Cycles (89) 17(1) [0|11|20]	3452
Best Of The Doobies Vol. II (81) 39(1) [0|3|15]	1471
Minute By Minute (79/80) 84(1) [0|0|33]	*1061*
Best Of The Doobies (77/80) 149(3) [0|0|36]	*773*
Doobie Brothers Farewell Tour (83) 79(1) [0|0|9]	418
The DOORS ▶ 236 Pk:17 [0|2|6]	12484
The Doors Greatest Hits (80) 17(2) [0|17|80]	8188
Alive She Cried (83) 23(2) [0|9|20]	2820
*The Best Of The Doors(2) (87/**91**)* 127(2) [0|0|22]	*561*
The Doors (67/80) 128(1) [0|0|17]	*459*
Classics (85) 124(2) [0|0|7]	233
Live At The Hollywood Bowl (87) 154(2) [0|0|11]	225
Charlie DORE ▶ 1590 Pk:145 [0|0|1]	179
Where To Now (80) 145(1) [0|0|7]	179
DORO ▶ 1484 Pk:154 [0|0|1]	263
Force Majeure (89) 154(2) [0|0|11]	263
DOUBLE ▶ 757 Pk:30 [0|1|1]	2055
Blue (86) 30(1) [0|6|21]	2055
Doug E. FRESH & The GET FRESH CREW ▶ 1205 Pk:88 [0|0|1]	548
The World's Greatest Entertainer (88) 88(2) [0|0|13]	548

Act ▶ Rank Peak [Top10s|Top40s|Total] Title (Yr) Pk(Wks) [T10 Wks|T40 Wks|TotWks]	Score in '80s
The DRAMATICS ▶ 1119 Pk:61 [0|0|1]	704
10 1/2 (80) 61(2) [0|0|12]	704
DREAM ACADEMY ▶ 559 Pk:20 [0|1|2]	3648
The Dream Academy (85) 20(1) [0|9|37]	3594
Remembrance Days (87) 181(2) [0|0|3]	54.4
DREAMBOY ▶ 1538 Pk:168 [0|0|1]	221
Dreamboy (84) 168(1) [0|0|11]	221
DREAMS SO REAL ▶ 1339 Pk:150 [0|0|1]	400
Rough Night In Jericho (88) 150(1) [0|0|18]	400
DREAM SYNDICATE ▶ 1760 Pk:171 [0|0|1]	85.2
Medicine Show (84) 171(1) [0|0|4]	85.2
D.R.I. ▶ 1206 Pk:116 [0|0|2]	546
4 Of A Kind (88) 116(2) [0|0|14]	510
*Thrash Zone (89/**90**)* 180(1) [0|0|2]	*36.7*
DRIVIN' N' CRYIN' ▶ 1375 Pk:130 [0|0|1]	350
Whisper Tames The Lion (88) 130(2) [0|0|12]	350
"D" TRAIN ▶ 1440 Pk:128 [0|0|1]	291
You're The One For Me (82) 128(2) [0|0|9]	291
George DUKE ▶ 813 Pk:48 [0|0|5]	1745
Dream On (82) 48(2) [0|0|12]	1005
A Brazilian Love Affair (80) 119(1) [0|0|9]	302
Guardian Of The Light (83) 147(1) [0|0|7]	183
Master Of The Game (80) 125(1) [0|0|5]	*179*
Thief In The Night (85) 183(1) [0|0|5]	75.7
DUKE JUPITER ▶ 1309 Pk:122 [0|0|1]	425
White Knuckle Ride (84) 122(1) [0|0|12]	425
Robbie DUPREE ▶ 839 Pk:51 [0|0|2]	1620
Robbie Dupree (80) 51(1) [0|0|24]	1509
Street Corner Heroes (81) 169(1) [0|0|5]	111
DURAN DURAN ▶ 32 Pk:4 [4|6|8]	52241
Seven And The Ragged Tiger (83) 8(5) [14|41|64]	14952
Rio (82) 6(7) [11|21|129]	13006
Duran Duran (83) 10(1) [1|14|87]	7782
Arena (84) 4(3) [8|15|28]	6824
Notorious (86) 12(1) [0|14|34]	5249
Big Thing (88) 24(2) [0|14|26]	3607
Carnival (82) 98(3) [0|0|15]	585
*Decade (89/**90**)* 81(1) [0|0|4]	*237*
Ian DURY And The BLOCKHEADS ▶ 1745 Pk:159 [0|0|1]	90.1
Laughter (81) 159(1) [0|0|4]	90.1
Bob DYLAN ▶ 211 Pk:20 [0|7|11]	14175
Infidels (83) 20(5) [0|10|24]	3343
Biograph (85) 33(2) [0|2|22]	1867
Empire Burlesque (85) 33(3) [0|6|17]	1829
Oh Mercy (89) 30(2) [0|6|13]	*1699*
Saved (80) 24(2) [0|5|11]	1436
Shot Of Love (81) 33(2) [0|3|9]	1063
Knocked Out Loaded (86) 53(1) [0|0|13]	915
Down In The Groove (88) 61(3) [0|0|10]	681
Dylan And The Dead (89)-Bob Dylan & The Grateful Dead [A] 37(2) [0|3|11]	516
*Slow Train Coming (79/**80**)* 76(2) [0|0|9]	*455*
Real Live (85) 115(2) [0|0|9]	369
DYNASTY ▶ 786 Pk:43 [0|0|2]	1879
Adventures In The Land Of Music (80) 43(1) [0|0|21]	1706
The Second Adventure (81) 119(2) [0|0|4]	173

E

Act ▶ Rank Peak [Top10s|Top40s|Total] Title (Yr) Pk(Wks) [T10 Wks|T40 Wks|TotWks]	Score in '80s
EAGLES ▶ 154 Pk:2 [2|2|5]	19197
*The Long Run (79/**80**)* 2(2) [10|25|46]	*11279*
Eagles Live (80) 6(4) [7|16|26]	6346
Eagles Greatest Hits: Volume 2 (82) 52(3) [0|0|15]	1249
*Their Greatest Hits 1971-1975 (**76**/80)* 166(1) [0|0|10]	*186*
*Hotel California (**77**/80)* 157(1) [0|0|6]	*137*

333

Acts With Albums

Act ▶ Rank Peak [Top10s|Top40s|Total] Title (Yr) Pk(Wks) [T10 Wks|T40Wks|TotWks]	Score in '80s
Steve EARLE ▶ 614 Pk:56 [0|0|2]	3042
Copperhead Road (88) 56(1) [0|0|28]	2213
Guitar Town (86) 89(2) [0|0|20]	829
Steve EARLE And The DUKES ▶ 1138	
Pk:90 [0|0|1]	655
Exit 0 (87) 90(1) [0|0|14]	655
EARTH, WIND & FIRE ▶ 147 Pk:5 [2|5|8]	20527
Raise! (81) 5(8) [10|18|25]	7783
Faces (80) 10(2) [2|12|21]	3961
Powerlight (83) 12(4) [0|12|21]	3941
Touch The World (87) 33(1) [0|6|28]	2749
Electric Universe (83) 40(2) [0|2|16]	1602
I Am (79/80) 105(1) [0|0|9]	374
The Best Of Earth, Wind & Fire, Vol. I (79/80)	
172(1) [0|0|3]	61.1
The Best Of Earth, Wind & Fire, Vol. II (88)	
190(2) [0|0|4]	56.2
Elliot EASTON ▶ 1267 Pk:99 [0|0|1]	472
Change No Change (85) 99(2) [0|0|11]	472
Sheena EASTON ▶ 146 Pk:15 [0|4|7]	20756
A Private Heaven (84) 15(2) [0|22|35]	6665
Sheena Easton (81) 24(1) [0|8|38]	3390
Best Kept Secret (83) 33(2) [0|9|38]	3132
You Could Have Been With Me (81) 47(1) [0|0|53]	3116
The Lover In Me (88) 44(4) [0|0|26]	2472
Do You (85) 40(2) [0|2|19]	1531
Madness, Money And Music (82) 85(3) [0|0|12]	450
EAZY-E ▶ 447 Pk:41 [0|0|1]	5057
Eazy-Duz-It (88) 41(1) [0|0|56]	5057
EBN-OZN ▶ 1815 Pk:185 [0|0|1]	63.1
Feeling Cavalier (84) 185(2) [0|0|4]	63.1
EBONEE WEBB ▶ 1624 Pk:157 [0|0|1]	159
Ebonee Webb (81) 157(2) [0|0|7]	159
ECHO & The BUNNYMEN ▶ 622	
Pk:51 [0|0|6]	2988
Echo & The Bunnymen(2) (87) 51(1) [0|0|37]	1993
Ocean Rain (84) 87(1) [0|0|11]	486
Porcupine (83) 137(1) [0|0|9]	250
Songs To Learn & Sing (86) 158(1) [0|0|9]	184
Echo & The Bunnymen (84) 188(1) [0|0|3]	40.8
Heaven Up Here (81) 184(1) [0|0|4]	33.5
John EDDIE ▶ 1123 Pk:83 [0|0|1]	696
John Eddie (86) 83(2) [0|0|15]	696
Dave EDMUNDS ▶ 505 Pk:46 [0|0|6]	4183
Information (83) 51(2) [0|0|20]	1488
Twangin... (81) 48(2) [0|0|14]	1000
D.E. 7th (82) 46(2) [0|0|14]	971
I Hear You Rockin' (87)-The Dave Edmunds Band	
106(1) [0|0|12]	483
Riff Raff (84) 140(2) [0|0|4]	131
The Best Of Dave Edmunds (82) 163(1) [0|0|5]	110
Dennis EDWARDS ▶ 758 Pk:48 [0|0|1]	2053
Don't Look Any Further (84) 48(2) [0|0|27]	2053
Walter EGAN ▶ 1889 Pk:187 [0|0|1]	30.4
Wild Exhibitions (83) 187(1) [0|0|2]	30.4
EGYPTIAN LOVER ▶ 1492 Pk:146 [0|0|1]	256
On The Nile (85) 146(2) [0|0|10]	256
ELECTRIC LIGHT ORCHESTRA ▶ 255 Pk: 4 [1|4|6]	11414
Xanadu (Soundtrack) (80)-Olivia Newton-John/	
Electric Light Orchestra [A] 4(3) [8|15|36]	3918
Time (81) 16(3) [0|12|20]	3332
Secret Messages (83) 36(1) [0|1|16]	1466
ELO's Greatest Hits (80) 30(3) [0|5|11]	1250
Balance Of Power (86) 49(3) [0|0|15]	1209
Discovery (79/80) 129(1) [0|0|7]	240
Larry ELGART And His MANHATTAN SWING	
Orchestra ▶ 458 Pk:24 [0|1|2]	4865
Hooked On Swing (82) 24(5) [0|15|41]	4336
Hooked On Swing 2 (83) 89(2) [0|0|11]	528
Joe ELY ▶ 1312 Pk:135 [0|0|1]	424
Musta Notta Gotta Lotta (81) 135(1) [0|0|11]	343

Act ▶ Rank Peak [Top10s|Top40s|Total] Title (Yr) Pk(Wks) [T10 Wks|T40Wks|TotWks]	Score in '80s
Live Shots (81) 159(2) [0|0|3]	81.0
Keith EMERSON ▶ 1843 Pk:183 [0|0|1]	52.1
Nighthawks (81) 183(1) [0|0|3]	52.1
EMERSON, LAKE & PALMER ▶ 1318	
Pk:73 [0|0|2]	417
The Best Of Emerson, Lake And Palmer (80)	
108(2) [0|0|7]	239
Emerson, Lake & Palmer In Concert (79/80)	
73(1) [0|0|5]	179
EMERSON, LAKE & POWELL ▶ 564	
Pk:23 [0|1|1]	3593
Emerson, Lake, & Powell (86) 23(3) [0|12|26]	3593
The EMOTIONS ▶ 1436 Pk:96 [0|0|2]	295
Come Into Our World (79/80) 96(1) [0|0|6]	215
New Affair (81) 168(2) [0|0|4]	80.1
ENGLAND DAN & John Ford COLEY ▶ 1909	
Pk:194 [0|0|1]	25.4
The Best Of England Dan & John Ford Coley (80)	
194(2) [0|0|2]	25.4
ENGLISH BEAT ▶ 438 Pk:39 [0|1|4]	5250
Special Beat Service (82) 39(3) [0|3|44]	3566
What Is Beat? (83) 87(1) [0|0|22]	1026
I Just Can't Stop It (80) 142(1) [0|0|14]	425
Wha'ppen (81) 126(1) [0|0|6]	233
Brian ENO ▶ 1278 Pk:44 [0|0|1]	466
My Life In The Bush Of Ghosts (81)-	
Brian Eno - David Byrne [A] 44(2) [0|0|13]	466
John ENTWISTLE ▶ 1203 Pk:71 [0|0|1]	551
Too Late The Hero (81) 71(2) [0|0|9]	551
ENUFF Z'NUFF ▶ 1075 Pk:74 [0|0|1]	792
Enuff Z'Nuff (89) 74(1) [0|0|14]	792
ENYA ▶ 491 Pk:25 [0|1|1]	4356
Watermark (89) 25(4) [0|13|39]	4356
EPMD ▶ 714 Pk:53 [0|0|2]	2293
Strictly Business (88) 80(1) [0|0|23]	1265
Unfinished Business (89) 53(3) [0|0|14]	1028
ERASURE ▶ 454 Pk:49 [0|0|5]	4913
The Innocents (88) 49(1) [0|0|50]	3717
Wild! (89) 57(2) [0|0|8]	619
Crackers International (89) 73(3) [0|0|10]	499
The Two Ring Circus (88) 186(1) [0|0|3]	40.9
The Circus (87) 190(1) [0|0|3]	38.6
ERIC B. & RAKIM ▶ 499 Pk:22 [0|1|2]	4212
Follow The Leader (88) 22(2) [0|7|16]	2190
Paid In Full (87) 58(1) [0|0|38]	2022
ESCAPE CLUB ▶ 507 Pk:27 [0|1|1]	4170
Wild, Wild West (88) 27(3) [0|14|38]	4170
ESQUIRE ▶ 1767 Pk:165 [0|0|1]	82.0
Esquire (87) 165(2) [0|0|4]	82.0
Gloria ESTEFAN/MIAMI SOUND MACHINE ▶ 77	
Pk:6 [2|3|3]	34271
Let It Loose (87)-Gloria Estefan &	
Miami Sound Machine 6(2) [9|48|97]	17486
Primitive Love (85)-Miami Sound Machine	
21(1) [0|34|75]	10618
Cuts Both Ways (89)-Gloria Estefan 8(2) [4|22|23]	6167
Deon ESTUS ▶ 1132 Pk:89 [0|0|1]	677
Spell (89) 89(1) [0|0|15]	677
Melissa ETHERIDGE ▶ 308 Pk:22 [0|2|2]	8897
Melissa Etheridge (88) 22(1) [0|14|65]	6312
Brave And Crazy (89) 22(1) [0|10|13]	2585
E.U. ▶ 1559 Pk:158 [0|0|1]	199
Livin' Large (89) 158(1) [0|0|9]	199
EUROGLIDERS ▶ 1482 Pk:140 [0|0|1]	264
This Island (84) 140(2) [0|0|11]	264
EUROPE ▶ 156 Pk:8 [1|2|2]	18847
The Final Countdown (86) 8(2) [10|42|98]	15455
Out Of This World (88) 19(1) [0|11|25]	3392
EURYTHMICS ▶ 74 Pk:7 [2|5|8]	34372
Touch (84) 7(3) [6|23|37]	8611
Be Yourself Tonight (85) 9(1) [3|24|45]	8583

Act ▶ Rank Peak [Top10s|Top40s|Total] Title (Yr) Pk(Wks) [T10 Wks|T40Wks|TotWks]	Score in '80s
Sweet Dreams (Are Made Of This)	
(83) 15(2) [0|17|59]	7188
Revenge (86) 12(1) [0|15|33]	5399
We Too Are One (89) 34(2) [0|7|14]	2072
Savage (87) 41(2) [0|0|19]	1482
1984 (For The Love Of Big Brother) (Soundtrack)	
(85) 93(1) [0|0|14]	596
Touch Dance (84) 115(1) [0|0|11]	442
EVERLY BROTHERS ▶ 674 Pk:38 [0|1|3]	2568
EB 84 (84) 38(2) [0|3|17]	1486
Born Yesterday (86) 83(4) [0|0|19]	959
Reunion Concert (84) 162(2) [0|0|5]	123
EXODUS ▶ 807 Pk:82 [0|0|2]	1763
Pleasures Of The Flesh (87) 82(2) [0|0|20]	950
Fabulous Disaster (89) 82(1) [0|0|17]	813
EXPOSÉ ▶ 187 Pk:16 [0|2|2]	15860
Exposure (87) 16(2) [0|53|74]	12739
What You Don't Know (89) 33(2) [0|6|27]	3121
EXTREME ▶ 899 Pk:80 [0|0|1]	1377
Extreme (89) 80(1) [0|0|32]	1377
EYE TO EYE ▶ 1187 Pk:99 [0|0|1]	571
Eye To Eye (82) 99(2) [0|0|15]	571
EZO ▶ 1547 Pk:150 [0|0|1]	207
EZO (87) 150(2) [0|0|9]	207

F

Act ▶ Rank Peak [Top10s|Top40s|Total] Title (Yr) Pk(Wks) [T10 Wks|T40Wks|TotWks]	Score in '80s
FABULOUS THUNDERBIRDS ▶ 285	
Pk:13 [0|1|4]	10099
Tuff Enuff (86) 13(3) [0|25|53]	8587
Hot Number (87) 49(1) [0|0|15]	1111
Powerful Stuff (89) 118(3) [0|0|7]	284
Butt Rockin' (81) 176(1) [0|0|7]	118
FACE TO FACE ▶ 1202 Pk:126 [0|0|2]	551
Face To Face (84) 126(1) [0|0|16]	448
One Big Day (88) 176(1) [0|0|7]	103
Donald FAGEN ▶ 498 Pk:11 [0|1|1]	4219
The Nightfly (82) 11(4) [0|10|27]	4219
FAIRGROUND ATTRACTION ▶ 1454	
Pk:137 [0|0|1]	281
The First Of A Million Kisses (89)	
137(2) [0|0|11]	281
Marianne FAITHFULL ▶ 924 Pk:82 [0|0|3]	1275
Broken English (80) 82(1) [0|0|15]	650
Dangerous Acquaintances (81) 104(1) [0|0|9]	360
A Child's Adventure (83) 107(2) [0|0|7]	265
FALCO ▶ 352 Pk:3 [1|1|2]	7625
Falco 3 (86) 3(1) [6|18|27]	6898
Einzelhaft (83) 64(1) [0|0|13]	726
Agnetha FALTSKOG ▶ 1295 Pk:102 [0|0|1]	439
Wrap Your Arms Around Me (83)	
102(1) [0|0|11]	439
The FAMILY ▶ 946 Pk:62 [0|0|1]	1170
The Family (85) 62(2) [0|0|22]	1170
FARRENHEIT ▶ 1717 Pk:179 [0|0|1]	108
Farrenheit (87) 179(1) [0|0|7]	108
FASTER PUSSYCAT ▶ 763 Pk:60 [0|0|2]	2016
Faster Pussycat (87) 97(2) [0|0|35]	1167
Wake Me When It's Over (89/90)	
60(2) [0|0|15]	849
FASTWAY ▶ 489 Pk:31 [0|1|4]	4375
Fastway (83) 31(1) [0|6|32]	2853
All Fired Up (84) 59(2) [0|0|14]	1009
On Target (89) 135(2) [0|0|10]	277
Trick Or Treat (Soundtrack) (86)	
156(1) [0|0|12]	236
FATBACK ▶ 652 Pk:44 [0|0|4]	2740
Hot Box (80) 44(1) [0|0|27]	1892
Tasty Jam (81) 102(1) [0|0|11]	384
14 Karat (80) 91(1) [0|0|7]	335
Gigolo (82) 148(1) [0|0|4]	128

Acts With Albums

Act ▶ Rank Peak [Top10s|Top40s|Total] Title (Yr) Pk(Wks) [T10 Wks|T40Wks|TotWks]	Score in '80s
FAT BOYS ▶ 186 Pk:8 [1|2|7]	**16091**
Crushin' (87) 8(3) [4|21|49]	8621
Coming Back Hard Again (88) 33(2) [0|5|24]	2241
Fat Boys (85) 48(2) [0|0|40]	2162
The Fat Boys Are Back! (85) 63(2) [0|0|33]	1547
Big & Beautiful (86) 62(2) [0|0|19]	1129
The Best Part Of The Fat Boys (87) 108(2) [0|0|10]	337
On And On (89) 175(1) [0|0|3]	54.0
FATES WARNING ▶ 1078 Pk:111 [0|0|3]	**788**
No Exit (88) 111(3) [0|0|13]	483
Perfect Symmetry (89) 141(1) [0|0|9]	250
Awaken The Guardian (87) 191(3) [0|0|4]	54.6
The FEELIES ▶ 1752 Pk:173 [0|0|1]	**88.8**
Only Life (88) 173(2) [0|0|5]	88.8
Don FELDER ▶ 1674 Pk:178 [0|0|1]	**131**
Airborne (83) 178(1) [0|0|8]	131
Wilton FELDER ▶ 978 Pk:81 [0|0|2]	**1069**
Secrets (85) 81(1) [0|0|16]	795
Inherit The Wind (80) 142(1) [0|0|13]	274
FELONY ▶ 1792 Pk:185 [0|0|1]	**73.1**
The Fanatic (83) 185(2) [0|0|5]	73.1
FEMME FATALE ▶ 1644 Pk:141 [0|0|1]	**145**
Femme Fatale (89) 141(1) [0|0|5]	145
Rick FENN ▶ 1836 Pk:154 [0|0|1]	**55.4**
Profiles (85)-Nick Mason & Rick Fenn [A] 154(2) [0|0|5]	55.4
Jay FERGUSON ▶ 1744 Pk:178 [0|0|1]	**91.0**
White Noise (82) 178(3) [0|0|5]	91.0
Maynard FERGUSON ▶ 1733 Pk:185 [0|0|2]	**96.5**
Hollywood (82) 185(1) [0|0|4]	65.7
It's My Time (80) 188(2) [0|0|2]	30.8
Bryan FERRY ▶ 546 Pk:63 [0|0|2]	**3791**
Bete Noire (87) 63(2) [0|0|31]	2303
Boys And Girls (85) 63(2) [0|0|25]	1488
FESTIVAL ▶ 884 Pk:50 [0|0|1]	**1430**
Evita (80) 50(1) [0|0|18]	1430
FETCHIN BONES ▶ 1662 Pk:175 [0|0|1]	**135**
Monster (89) 175(1) [0|0|7]	135
Richard "Dimples" FIELDS ▶ 666 Pk:33 [0|1|2]	**2641**
Dimples (81) 33(2) [0|4|17]	1482
Mr. Look So Good! (82) 63(1) [0|0|20]	1159
FIFTH ANGEL ▶ 1297 Pk:117 [0|0|1]	**437**
Fifth Angel (88) 117(2) [0|0|13]	437
FINE YOUNG CANNIBALS ▶ 129 Pk:1 [1|1|2]	**23559**
The Raw & The Cooked (89) 1(7) [27|40|43]	21779
Fine Young Cannibals (86) 49(1) [0|0|28]	1780
Tim FINN ▶ 1713 Pk:161 [0|0|1]	**112**
Escapade (83) 161(1) [0|0|5]	112
FIONA ▶ 980 Pk:71 [0|0|2]	**1056**
Fiona (85) 71(1) [0|0|18]	898
Heart Like A Gun (89/90) 152(1) [0|0|6]	158
Elisa FIORILLO ▶ 1647 Pk:163 [0|0|1]	**143**
Elisa Fiorillo (88) 163(2) [0|0|8]	143
FIREFALL ▶ 904 Pk:68 [0|0|4]	**1344**
Undertow (80) 68(2) [0|0|15]	747
Clouds Across The Sun (81) 102(2) [0|0|13]	508
The Best Of Firefall (81) 186(1) [0|0|4]	58.0
Break Of Dawn (83) 199(1) [0|0|3]	30.5
The FIRM ▶ 370 Pk:17 [0|2|2]	**7259**
The Firm (85) 17(2) [0|16|33]	4817
Mean Business (86) 22(2) [0|7|19]	2442
FISHBONE ▶ 1630 Pk:153 [0|0|1]	**154**
Truth And Soul (88) 153(1) [0|0|9]	154
FIVE STAIRSTEPS ▶ 1722 Pk:90 [0|0|1]	**699**
The Invisible Man's Band (80)- The Invisible Man's Band 90(1) [0|0|14]	699
FIVE STAR ▶ 535 Pk:57 [0|0|2]	**3856**
Luxury Of Life (85) 57(2) [0|0|47]	2816

Act ▶ Rank Peak [Top10s|Top40s|Total] Title (Yr) Pk(Wks) [T10 Wks|T40Wks|TotWks]	Score in '80s
Silk And Steel (86) 80(2) [0|0|25]	1040
The FIXX ▶ 151 Pk:8 [1|3|6]	**19927**
Reach The Beach (83) 8(2) [10|28|54]	10476
Phantoms (84) 19(3) [0|10|29]	3858
Walkabout (86) 30(2) [0|8|21]	2387
Shuttered Room (82) 106(1) [0|0|51]	1925
Calm Animals (89) 72(2) [0|0|18]	1021
React (87) 110(2) [0|0|7]	260
Roberta FLACK ▶ 473 Pk:25 [0|1|4]	**4569**
Roberta Flack Featuring Donny Hathaway (80)- Roberta Flack Featuring Donny Hathaway 25(2) [0|10|24]	2842
I'm The One (82) 59(2) [0|0|21]	1342
Bustin' Loose (81) 161(1) [0|0|11]	225
Oasis (89) 159(1) [0|0|8]	159
Roberta FLACK & Peabo BRYSON ▶ 425 Pk:25 [0|1|2]	**5507**
Born To Love (83)- Peabo Bryson & Roberta Flack 25(1) [0|14|42]	4251
Live & More (80) 52(1) [0|0|19]	1256
FLASH & THE PAN ▶ 1675 Pk:159 [0|0|1]	**130**
Lights In The Night (80) 159(2) [0|0|6]	130
Mick FLEETWOOD ▶ 951 Pk:43 [0|0|1]	**1150**
The Visitor (81) 43(2) [0|0|14]	1150
FLEETWOOD MAC ▶ 42 Pk:1 [3|5|7]	**44721**
Mirage (82) 1(5) [18|21|45]	16743
Tango In The Night (87) 7(3) [6|44|57]	14772
Tusk (79/80) 8(3) [7|13|28]	5470
Greatest Hits (88) 14(1) [0|12|26]	3918
Fleetwood Mac Live (80) 14(3) [0|8|18]	3071
Rumours (77/80) 107(2) [0|0|16]	637
Fleetwood Mac (76/80) 158(1) [0|0|5]	110
FLESH FOR LULU ▶ 1002 Pk:89 [0|0|1]	**981**
Long Live The New Flesh (87) 89(1) [0|0|24]	981
The FLESHTONES ▶ 1728 Pk:174 [0|0|1]	**100**
Roman Gods (82) 174(3) [0|0|5]	99.5
A FLOCK OF SEAGULLS ▶ 237 Pk:10 [1|2|3]	**12484**
A Flock Of Seagulls (82) 10(3) [3|25|50]	8604
Listen (83) 16(2) [0|11|23]	3253
The Story Of A Young Heart (84) 66(1) [0|0|10]	627
FLOTSAM AND JETSAM ▶ 1568 Pk:143 [0|0|1]	**195**
No Place For Disgrace (88) 143(2) [0|0|8]	195
FLYING LIZARDS ▶ 1390 Pk:99 [0|0|1]	**335**
The Flying Lizards (80) 99(1) [0|0|8]	335
Dan FOGELBERG ▶ 78 Pk:3 [2|5|6]	**34168**
The Innocent Age (81) 6(6) [8|31|62]	11626
Phoenix (80) 3(2) [10|24|35]	10041
Dan Fogelberg/Greatest Hits (82) 15(4) [0|12|35]	4966
Windows And Walls (84) 15(3) [0|10|27]	3910
High Country Snows (85) 30(2) [0|5|23]	2151
Exiles (87) 48(3) [0|0|19]	1473
John FOGERTY ▶ 183 Pk:1 [1|2|2]	**16422**
Centerfield (85) 1(1) [15|28|51]	14119
Eye Of The Zombie (86) 26(2) [0|6|19]	2304
FOGHAT ▶ 911 Pk:92 [0|0|5]	**1329**
Girls To Chat & Boys To Bounce (81) 92(1) [0|0|9]	447
Tight Shoes (80) 106(1) [0|0|10]	408
Boogie Motel (79/80) 123(1) [0|0|9]	335
In The Mood For Something Rude (82) 162(1) [0|0|5]	114
Zig-Zag Walk (83) 192(1) [0|0|3]	25.9
Ellen FOLEY ▶ 1700 Pk:152 [0|0|1]	**118**
Spirit Of St. Louis (81) 152(2) [0|0|4]	118
Jane FONDA ▶ 217 Pk:15 [0|1|3]	**13874**
Jane Fonda's Workout Record (82) 15(4) [0|27|120]	13380
Jane Fonda's Workout Record New And Improved (84) 135(2) [0|0|10]	250

Act ▶ Rank Peak [Top10s|Top40s|Total] Title (Yr) Pk(Wks) [T10 Wks|T40Wks|TotWks]	Score in '80s
Jane Fonda's Workout Record For Pregnancy, Birth And Recovery (83) 115(1) [0|0|7]	244
The FOOLS ▶ 1457 Pk:151 [0|0|2]	**279**
Sold Out (80) 151(1) [0|0|8]	188
Heavy Mental (81) 158(1) [0|0|4]	90.9
Steve FORBERT ▶ 616 Pk:20 [0|1|3]	**3028**
Jackrabbit Slim (80) 20(1) [0|6|18]	2327
Little Stevie Orbit (80) 70(1) [0|0|9]	568
Steve Forbert (82) 159(2) [0|0|6]	132
FORCE M.D.'S ▶ 726 Pk:67 [0|0|3]	**2235**
Chillin' (86) 69(1) [0|0|25]	1298
Touch And Go (87) 67(2) [0|0|16]	872
Love Letters (84) 185(3) [0|0|4]	64.9
Lita FORD ▶ 326 Pk:29 [0|1|2]	**8274**
Lita (88) 29(4) [0|23|62]	7335
Dancin' On The Edge (84) 66(2) [0|0|16]	938
Robben FORD ▶ 1290 Pk:120 [0|0|1]	**445**
Talk To Your Daughter (88) 120(2) [0|0|13]	445
Julia FORDHAM ▶ 1101 Pk:118 [0|0|1]	**729**
Julia Fordham (88) 118(2) [0|0|25]	729
FOREIGNER ▶ 30 Pk:1 [3|5|7]	**54311**
4 (81) 1(10) [34|52|81]	30884
Agent Provocateur (85) 4(3) [12|24|45]	10350
Inside Information (87) 15(1) [0|14|37]	5234
Records (82) 10(4) [4|12|25]	4691
Head Games (79/80) 11(1) [0|8|27]	2747
Double Vision (78/81) 162(2) [0|0|15]	246
Foreigner (77/81) 132(1) [0|0|6]	160
David FOSTER ▶ 1393 Pk:111 [0|0|2]	**334**
The Symphony Sessions (88) 111(2) [0|0|8]	299
David Foster (86) 195(2) [0|0|3]	34.6
FOSTER & LLOYD ▶ 1636 Pk:142 [0|0|1]	**151**
Faster & Llouder (89) 142(1) [0|0|6]	151
4 BY FOUR ▶ 1607 Pk:141 [0|0|1]	**171**
4 By Four (87) 141(2) [0|0|7]	171
FOUR TOPS ▶ 727 Pk:37 [0|1|3]	**2233**
Tonight! (81) 37(1) [0|4|21]	1810
Magic (85) 140(2) [0|0|9]	254
Indestructible (88) 149(2) [0|0|7]	170
Samantha FOX ▶ 338 Pk:24 [0|2|3]	**7859**
Touch Me (86) 24(2) [0|11|28]	3381
I Wanna Have Some Fun (88) 37(1) [0|4|34]	2922
Samantha Fox (87) 51(1) [0|0|25]	1556
Peter FRAMPTON ▶ 743 Pk:43 [0|0|4]	**2130**
Breaking All The Rules (81) 43(1) [0|0|13]	1117
Premonition (86) 80(2) [0|0|14]	719
When All The Pieces Fit (89) 152(2) [0|0|6]	160
The Art Of Control (82) 174(2) [0|0|8]	135
FRANKE And The KNOCKOUTS ▶ 538 Pk:31 [0|1|2]	**3842**
Franke & The Knockouts (81) 31(1) [0|5|27]	2790
Below The Belt (82) 48(2) [0|0|18]	1052
FRANKIE GOES TO HOLLYWOOD ▶ 439 Pk:33 [0|1|2]	**5241**
Welcome To The Pleasure Dome (84) 33(5) [0|14|41]	4660
Liverpool (86) 88(2) [0|0|13]	581
Aretha FRANKLIN ▶ 119 Pk:13 [0|5|8]	**24552**
Who's Zoomin' Who? (85) 13(1) [0|35|51]	10090
Aretha (II) (86) 32(2) [0|8|39]	4345
Jump To It (82) 23(4) [0|9|30]	3280
Aretha (80) 47(3) [0|0|30]	2197
Get It Right (83) 36(1) [0|3|18]	1542
Love All The Hurt Away (81) 36(2) [0|3|17]	1456
Through The Storm (89) 55(2) [0|0|18]	1035
One Lord, One Faith, One Baptism (87) 106(3) [0|0|16]	608
Rodney FRANKLIN ▶ 1189 Pk:104 [0|0|3]	**570**
You'll Never Know (80) 104(1) [0|0|13]	486
Marathon (84) 187(1) [0|0|3]	44.5

Acts With Albums

Act ▶ Rank Peak [Top10s|Top40s|Total] Title (Yr) Pk(Wks) [T10 Wks|T40Wks|TotWks]	Score in '80s
Rodney FRANKLIN ▶ *Continued*	
Learning To Love (83) 190(2) [0|0|3]	39.0
Michael FRANKS ▶ 597 Pk:45 [0|0|5]	3254
Objects Of Desire (82) 45(3) [0|0|14]	1102
One Bad Habit (80) 83(1) [0|0|21]	807
Skin Dive (85) 137(2) [0|0|27]	761
Passionfruit (83) 141(1) [0|0|11]	294
The Camera Never Lies (87) 147(1) [0|0|11]	290
Linda FRATIANNE ▶ 1684 Pk:174 [0|0|1]	124
Dance & Exercise With The Hits (82) 174(1) [0|0|7]	124
Ace FREHLEY ▶ 703 Pk:43 [0|0|2]	2382
Frehley's Comet (87) 43(2) [0|0|25]	2108
Trouble Walkin' (89) 102(2) [0|0|8]	*275*
FREHLEY'S COMET ▶ 960 Pk:81 [0|0|2]	1123
Second Sighting (88) 81(3) [0|0|13]	625
Live + 1 (88) 84(2) [0|0|10]	498
Glenn FREY ▶ 231 Pk:22 [0|3|3]	12741
The Allnighter (84) 22(2) [0|14|65]	6438
No Fun Aloud (82) 32(3) [0|11|38]	4222
Soul Searching (88) 36(2) [0|7|19]	2082
FRIDA ▶ 761 Pk:41 [0|0|1]	2026
Something's Going On (82) 41(2) [0|0|28]	2026
Robert FRIPP ▶ 1210 Pk:90 [0|0|2]	541
The League Of Gentlemen (81) 90(2) [0|0|7]	312
God Save The Queen/Under Heavy Manners (80) 110(2) [0|0|6]	228
FROZEN GHOST ▶ 1280 Pk:107 [0|0|1]	464
Frozen Ghost (87) 107(2) [0|0|13]	464
FULL FORCE ▶ 1036 Pk:126 [0|0|3]	873
Full Force Get Busy 1 Time! (86) 141(3) [0|0|13]	361
Guess Who's Comin' To The Crib? (87) 126(3) [0|0|11]	335
Full Force (86) 160(2) [0|0|8]	177
FUN BOY THREE ▶ 1420 Pk:104 [0|0|1]	308
Waiting (83) 104(1) [0|0|7]	308
FUNKADELIC ▶ 1384 Pk:105 [0|0|2]	342
The Electric Spanking Of War Babies (81) 105(2) [0|0|4]	197
Uncle Jam Wants You (79/80) 141(1) [0|0|5]	*145*
FUNKADELIC(2) ▶ 1704 Pk:151 [0|0|1]	117
Connections And Disconnections (81) 151(1) [0|0|4]	117
FUSE ONE ▶ 1526 Pk:139 [0|0|1]	231
Silk (82) 139(1) [0|0|8]	231

G

Act ▶ Rank Peak [Top10s|Top40s|Total] Title (Yr) Pk(Wks) [T10 Wks|T40Wks|TotWks]	Score in '80s
Peter GABRIEL ▶ 98 Pk:2 [1|3|6]	28332
So (86) 2(3) [12|47|93]	18026
Peter Gabriel (III) (80) 22(2) [0|14|29]	4258
Peter Gabriel (Security) (82) 28(9) [0|12|31]	3887
Peter Gabriel/Plays Live (83) 44(1) [0|0|16]	1202
Passion: Music For The Last Temptation Of Christ (Soundtrack) (89) 60(2) [0|0|14]	815
Birdy (Soundtrack) (85) 162(1) [0|0|7]	144
James GALWAY ▶ 1764 Pk:150 [0|0|1]	83.0
Sometimes When We Touch (80)-Cleo Laine & James Galway [A] 150(1) [0|0|3]	83.0
GAMMA ▶ 789 Pk:65 [0|0|3]	1845
Gamma 2 (80) 65(1) [0|0|19]	993
Gamma 3 (82) 72(1) [0|0|12]	669
Gamma 1 (79/80) 144(1) [0|0|7]	*184*
GANG OF FOUR ▶ 1564 Pk:168 [0|0|4]	197
Hard (83) 168(1) [0|0|4]	84.6
Songs Of The Free (82) 175(1) [0|0|3]	58.4
Solid Gold (81) 190(1) [0|0|2]	29.0
Another Day/Another Dollar (82) 195(2) [0|0|2]	24.6
GAP BAND ▶ 143 Pk:14 [0|3|8]	21340
Gap Band IV (82) 14(5) [0|17|52]	7135

Act ▶ Rank Peak [Top10s|Top40s|Total] Title (Yr) Pk(Wks) [T10 Wks|T40Wks|TotWks]	Score in '80s
The Gap Band III (80) 16(2) [0|17|37]	5753
Gap Band V- Jammin' (83) 28(1) [0|7|43]	3744
The Gap Band II (80) 42(1) [0|0|26]	2224
Gap Band VI (85) 58(2) [0|0|23]	1539
Gap Gold/Best Of The Gap Band (85) 103(2) [0|0|16]	582
Gap Band VII (86) 159(1) [0|0|15]	314
Round Trip (89) 189(1) [0|0|4]	*49.5*
Jerry GARCIA ▶ 1422 Pk:100 [0|0|1]	306
Run For The Roses (82) 100(2) [0|0|10]	306
Art GARFUNKEL ▶ 1201 Pk:113 [0|0|2]	552
Scissors Cut (81) 113(1) [0|0|8]	343
Lefty (88) 134(2) [0|0|8]	209
Leif GARRETT ▶ 1156 Pk:129 [0|0|2]	619
Same Goes For You (80) 129(1) [0|0|19]	*521*
My Movie Of You (81) 185(1) [0|0|7]	98.6
Larry GATLIN/GATLIN BROTHERS Band ▶ 1268 Pk:102 [0|0|3]	472
Straight Ahead (80) 102(1) [0|0|9]	*280*
Help Yourself (80) 118(1) [0|0|4]	158
Not Guilty... (81) Band 184(1) [0|0|4]	33.5
Marvin GAYE ▶ 273 Pk:7 [1|2|9]	10693
Midnight Love (82) 7(5) [7|13|41]	6697
In Our Lifetime (81) 32(2) [0|4|17]	1565
Dream Of A Lifetime (85) 41(2) [0|0|15]	1129
Every Great Motown Hit Of Marvin Gaye (83) 80(1) [0|0|16]	562
Anthology (84) 109(1) [0|0|8]	314
Let's Get It On (73/84) 127(2) [0|0|7]	*240*
What's Going On (71/84) 154(1) [0|0|5]	*126*
I Want You (76/84) 183(1) [0|0|2]	*34.0*
Motown Remembers Marvin Gaye (86) 193(1) [0|0|2]	25.9
Crystal GAYLE ▶ 532 Pk:47 [0|0|8]	3867
Miss The Mississippi (79/80) 47(2) [0|0|14]	*1001*
Classic Crystal (80) 62(3) [0|0|15]	*873*
These Days (80) 79(1) [0|0|11]	620
Hollywood, Tennessee (81) 99(2) [0|0|16]	496
True Love (82) 120(1) [0|0|12]	445
Favorites (87) 149(1) [0|0|6]	156
Crystal Gayle's Greatest Hits (83) 169(1) [0|0|8]	153
Cage The Songbird (83) 171(1) [0|0|6]	123
Gloria GAYNOR ▶ 1811 Pk:178 [0|0|1]	64.3
Stories (80) 178(1) [0|0|4]	64.3
J. GEILS Band ▶ 111 Pk:1 [1|3|4]	26255
Freeze-Frame (81) 1(4) [19|29|70]	17546
Love Stinks (80) 18(2) [0|17|42]	5503
Showtime (82) 23(3) [0|10|19]	2643
You're Gettin' Even While I'm Gettin' Odd (84) 80(3) [0|0|10]	563
Bob GELDOF ▶ 1368 Pk:130 [0|0|1]	357
Deep In The Heart Of Nowhere (86) 130(1) [0|0|12]	357
GENE LOVES JEZEBEL ▶ 944 Pk:108 [0|0|2]	1175
The House Of Dolls (87) 108(1) [0|0|22]	758
Discover (86) 155(2) [0|0|19]	417
GENERAL PUBLIC ▶ 453 Pk:26 [0|1|2]	4955
...All The Rage (84) 26(2) [0|11|39]	4276
Hand To Mouth (86) 83(1) [0|0|16]	679
GENESIS ▶ 27 Pk:3 [4|5|6]	56117
Invisible Touch (86) 3(2) [26|61|85]	23858
Abacab (81) 7(1) [6|29|64]	11085
Genesis (83) 9(1) [3|27|50]	9559
Duke (80) 11(2) [0|21|31]	6975
Three Sides Live (82) 10(3) [3|11|25]	4377
...And Then There Were Three... (78/80) 93(3) [0|0|6]	*262*
GEORGIA SATELLITES ▶ 322 Pk:5 [1|1|3]	8374
Georgia Satellites (86) 5(1) [5|20|42]	7503
Open All Night (88) 77(2) [0|0|13]	646
In The Land Of Salvation And Sin (89) 130(2) [0|0|8]	*226*

Act ▶ Rank Peak [Top10s|Top40s|Total] Title (Yr) Pk(Wks) [T10 Wks|T40Wks|TotWks]	Score in '80s
GEORGIO ▶ 872 Pk:117 [0|0|1]	1481
Sexappeal (87) 117(1) [0|0|52]	1481
GIANT ▶ 1173 Pk:91 [0|0|1]	588
Last Of The Runaways (89/90) 91(2) [0|0|12]	*588*
GIANT STEPS ▶ 1787 Pk:184 [0|0|1]	74.9
The Book Of Pride (88) 184(1) [0|0|5]	74.9
Andy GIBB ▶ 570 Pk:21 [0|1|2]	3523
After Dark (80) 21(1) [0|7|15]	2171
Andy Gibb's Greatest Hits (80) 46(1) [0|0|18]	1353
Barry GIBB ▶ 1286 Pk:72 [0|0|1]	450
Now Voyager (84) 72(2) [0|0|8]	450
Terri GIBBS ▶ 843 Pk:53 [0|0|1]	1590
Somebody's Knockin' (81) 53(3) [0|0|25]	1590
Debbie GIBSON ▶ 84 Pk:1 [2|2|2]	31505
Out Of The Blue (87) 7(2) [13|45|89]	17204
Electric Youth (89) 1(5) [13|25|47]	*14300*
Johnny GILL ▶ 1686 Pk:139 [0|0|1]	123
Perfect Combination (84)-Stacy Lattisaw & Johnny Gill [A] 139(1) [0|0|8]	123
GILLAN ▶ 1850 Pk:183 [0|0|1]	49.0
Glory Road (80) 183(1) [0|0|3]	49.0
Mickey GILLEY ▶ 1581 Pk:170 [0|0|2]	186
You Don't Know Me (81) 170(2) [0|0|6]	129
That's All That Matters To Me (80) 177(2) [0|0|3]	56.7
David GILMOUR ▶ 610 Pk:32 [0|1|1]	3097
About Face (84) 32(3) [0|10|28]	3097
GIPSY KINGS ▶ 696 Pk:57 [0|0|2]	2423
Gipsy Kings (88) 57(2) [0|0|42]	2262
Mosaique (89) 95(1) [0|0|3]	*161*
GIRLSCHOOL ▶ 1777 Pk:182 [0|0|1]	78.0
Hit And Run (82) 182(2) [0|0|5]	78.0
GIUFFRIA ▶ 502 Pk:26 [0|1|2]	4198
Giuffria (84) 26(2) [0|7|29]	3302
Silk + Steel (86) 60(2) [0|0|14]	896
Philip GLASS ▶ 1087 Pk:91 [0|0|2]	765
Songs From Liquid Days (86) 91(2) [0|0|13]	559
Glassworks (82) 121(2) [0|0|6]	206
GLASS MOON ▶ 1499 Pk:148 [0|0|1]	253
Glass Moon (80) 148(1) [0|0|9]	253
GLASS TIGER ▶ 367 Pk:27 [0|1|2]	7367
The Thin Red Line (86) 27(2) [0|28|51]	6657
Diamond Sun (88) 82(2) [0|0|15]	710
Roger GLOVER ▶ 1302 Pk:101 [0|0|1]	434
Mask (84) 101(1) [0|0|12]	434
GOANNA ▶ 1766 Pk:179 [0|0|1]	82.1
Spirit Of Place (83) 179(1) [0|0|5]	82.1
The GODFATHERS ▶ 1070 Pk:91 [0|0|2]	799
Birth, School, Work, Death (88) 91(2) [0|0|16]	700
More Songs About Love & Hate (89) 174(1) [0|0|6]	*99.3*
GODLEY & CREME ▶ 863 Pk:37 [0|1|1]	1512
The History Mix Vol. I (85) 37(2) [0|3|15]	1512
The GO-GO'S ▶ 92 Pk:1 [2|3|3]	29694
Beauty And The Beat (81) 1(6) [15|38|72]	19679
Talk Show (84) 18(3) [0|18|32]	5186
Vacation (82) 8(5) [9|10|28]	4829
GOLDEN EARRING ▶ 527 Pk:24 [0|1|4]	4032
Cut (82) 24(2) [0|12|30]	3521
N.E.W.S. (84) 107(1) [0|0|9]	370
Something Heavy Going Down - Live From The Twilight Zone (84) 158(2) [0|0|6]	119
The Hole (86) 196(2) [0|0|2]	23.6
Robert GORDON ▶ 1079 Pk:117 [0|0|2]	787
Are You Gonna Be The One (81) 117(4) [0|0|15]	562
Bad Boy (80) 150(1) [0|0|9]	225
Martin L. GORE ▶ 1715 Pk:156 [0|0|1]	109
Counterfeit (89) 156(2) [0|0|5]	*109*
GORKY PARK ▶ 1072 Pk:80 [0|0|1]	794
Gorky Park (89) 80(2) [0|0|17]	*794*

336

Acts With Albums

Act ▶ Rank Peak [Top10s|Top40s|Total] Title (Yr) Pk(Wks) [T10 Wks|T40Wks|TotWks]	Score in '80s
Barry GOUDREAU ▶ 1296 Pk:88 [0|0|1]	**437**
Barry Goudreau (80) 88(3) [0|0|8]	437
GO WEST ▶ 701 Pk:60 [0|0|2]	**2394**
Go West (85) 60(1) [0|0|35]	2236
Dancing On The Couch (87) 172(1) [0|0|9]	158
GQ ▶ 826 Pk:46 [0|0|2]	**1694**
Two (80) 46(2) [0|0|20]	1493
Face To Face (81) 140(2) [0|0|8]	202
The GRACES ▶ 1550 Pk:147 [0|0|1]	**205**
Perfect View (89) 147(1) [0|0|9]	205
Larry GRAHAM ▶ 495 Pk:26 [0|1|4]	**4260**
One In A Million You (80) 26(3) [0|9|24]	2941
Just Be My Lady (81) 46(1) [0|0|13]	971
Sooner Or Later (82) 142(1) [0|0|9]	266
Victory (83) 173(1) [0|0|4]	81.9
Lou GRAMM ▶ 599 Pk:27 [0|1|2]	**3208**
Ready Or Not (87) 27(1) [0|7|26]	2887
*Long Hard Look (89/**90**) 122(2) [0|0|8]*	*321*
GRAND FUNK RAILROAD ▶ 1660 **Pk:149 [0|0|1]**	**135**
Grand Funk Lives (81) 149(2) [0|0|5]	135
GRANDMASTER FLASH ▶ 1605 **Pk:145 [0|0|2]**	**171**
The Source (86) 145(2) [0|0|6]	160
Ba-Dop-Boom-Bang (87) 197(1) [0|0|1]	11.4
GRANDMASTER FLASH & The FURIOUS FIVE ▶ **816 Pk:53 [0|0|2]**	**1738**
The Message (82) 53(4) [0|0|24]	1697
On The Strength (88) 189(2) [0|0|3]	40.5
Amy GRANT ▶ 392 Pk:35 [0|1|4]	**6630**
Unguarded (85) 35(1) [0|4|38]	3530
The Collection (86) 66(2) [0|0|33]	1861
Lead Me On (88) 71(2) [0|0|13]	756
Straight Ahead (85) 133(2) [0|0|20]	482
Eddy GRANT ▶ 397 Pk:10 [1|1|2]	**6409**
Killer On The Rampage (83) 10(3) [3|15|30]	5439
Going For Broke (84) 64(2) [0|0|17]	970
Stephane GRAPPELLI ▶ 1603 Pk:108 [0|0|1]	**173**
Live (81)-Stephane Grappelli/David Grisman [A] 108(1) [0|0|10]	173
GRATEFUL DEAD ▶ 221 Pk:6 [1|5|6]	**13772**
In The Dark (87) 6(2) [7|17|34]	6920
Go To Heaven (80) 23(2) [0|8|21]	2601
Reckoning (81) 43(1) [0|0|16]	1346
Dead Set (81) 29(2) [0|3|11]	1305
Built To Last (89) 27(1) [0|3|7]	*1085*
Dylan And The Dead (89)-Bob Dylan & The Grateful Dead [A] 37(2) [0|3|11]	516
GREAT WHITE ▶ 169 Pk:9 [1|2|5]	**17653**
...Twice Shy (89) 9(1) [4|26|35]	*9205*
Once Bitten (87) 23(2) [0|18|53]	7000
Shot In The Dark (86) 82(1) [0|0|13]	628
Recovery: Live! (88) 99(2) [0|0|12]	517
Great White (84) 144(1) [0|0|12]	304
Boris GREBENSHIKOV ▶ 1918 **Pk:198 [0|0|1]**	**21.8**
Radio Silence (89) 198(2) [0|0|2]	21.8
Al GREEN ▶ 1313 Pk:131 [0|0|1]	**422**
Soul Survivor (87) 131(1) [0|0|14]	422
Jack GREEN ▶ 1501 Pk:121 [0|0|1]	**250**
Humanesque (80) 121(1) [0|0|8]	250
Peter GREEN ▶ 1788 Pk:186 [0|0|1]	**74.9**
Little Dreamer (80) 186(1) [0|0|5]	74.9
GREEN ON RED ▶ 1737 Pk:177 [0|0|1]	**93.9**
No Free Lunch (86) 177(1) [0|0|6]	93.9
Lee GREENWOOD ▶ 790 Pk:73 [0|0|4]	**1842**
Somebody's Gonna Love You (83) 73(1) [0|0|21]	965
You've Got A Good Love Comin' (84) 150(1) [0|0|20]	439

Act ▶ Rank Peak [Top10s|Top40s|Total] Title (Yr) Pk(Wks) [T10 Wks|T40Wks|TotWks]	Score in '80s
Meant For Each Other (84)-Barbara Mandrell & Lee Greenwood [A] 89(1) [0|0|13]	301
Greatest Hits (85) 163(2) [0|0|8]	137
Joanie GREGGAINS ▶ 1802 Pk:177 [0|0|1]	**70.3**
Aerobic Shape Up II (83) 177(1) [0|0|4]	70.3
GREY And HANKS ▶ 1880 Pk:195 [0|0|1]	**32.3**
Prime Time (80) 195(1) [0|0|3]	32.3
Nanci GRIFFITH ▶ 1174 Pk:99 [0|0|1]	**588**
Storms (89) 99(1) [0|0|14]	588
GRIM REAPER ▶ 662 Pk:73 [0|0|3]	**2667**
See You In Hell (84) 73(2) [0|0|27]	1088
Rock You To Hell (87) 93(1) [0|0|21]	979
Fear No Evil (85) 108(1) [0|0|14]	600
David GRISMAN ▶ 1314 Pk:108 [0|0|3]	**422**
David Grisman - Quintet "80" (80) 152(1) [0|0|8]	192
Live (81)-Stephane Grappelli/David Grisman [A] 108(1) [0|0|10]	173
Mondo Mando (81) 174(1) [0|0|3]	57.5
Dave GRUSIN ▶ 524 Pk:52 [0|0|10]	**4056**
Mountain Dance (81) 74(1) [0|0|18]	933
The Electric Horseman (Soundtrack) (80)- Willie Nelson/Dave Grusin [A] 52(2) [0|0|25]	766
The Goonies (Soundtrack) (85) 73(2) [0|0|10]	520
The Fabulous Baker Boys (Soundtrack) (89) 74(2) [0|0|7]	*475*
Dave Grusin Collection (89) 110(3) [0|0|12]	448
Out Of The Shadows (82) 88(2) [0|0|9]	395
Dave Grusin and the GRP All-Stars/Live In Japan (81)-Dave Grusin and the GRP All-Stars 140(2) [0|0|7]	214
Migration (89) 145(1) [0|0|8]	198
Dave Grusin and the NY/LA Dream Band (83) 181(1) [0|0|6]	95.3
Harlequin (85)-Dave Grusin & Lee Ritenour [A] 192(1) [0|0|2]	13.0
GTR ▶ 436 Pk:11 [0|1|1]	**5286**
GTR (86) 11(2) [0|17|26]	5286
GUADALCANAL DIARY ▶ 1241 **Pk:132 [0|0|2]**	**507**
Flip-Flop (89) 132(2) [0|0|13]	407
2 X 4 (88) 183(1) [0|0|7]	100
GUCCI CREW II ▶ 1714 Pk:173 [0|0|1]	**111**
Everybody Wants Some (89) 173(1) [0|0|6]	111
Greg GUIDRY ▶ 1609 Pk:147 [0|0|1]	**170**
Over The Line (82) 147(1) [0|0|7]	170
GUNS N' ROSES ▶ 23 Pk:1 [2|2|2]	**59197**
Appetite For Destruction (87) 1(5) [52|78|123]	44164
G N' R Lies (88) 2(1) [18|33|50]	15034
Arlo GUTHRIE ▶ 1858 Pk:184 [0|0|1]	**45.3**
Power Of Love (81) 184(2) [0|0|3]	45.3
Gwen GUTHRIE ▶ 1167 Pk:89 [0|0|1]	**594**
Good To Go Lover (86) 89(2) [0|0|13]	594
GUY ▶ 357 Pk:27 [0|1|1]	**7517**
Guy (88) 27(5) [0|13|70]	7517

H

Act ▶ Rank Peak [Top10s|Top40s|Total] Title (Yr) Pk(Wks) [T10 Wks|T40Wks|TotWks]	Score in '80s
Steve HACKETT ▶ 1518 Pk:144 [0|0|2]	**237**
Defector (80) 144(1) [0|0|6]	169
Cured (81) 169(2) [0|0|3]	67.5
Sammy HAGAR ▶ 161 Pk:14 [0|4|7]	**18533**
Three Lock Box (82) 17(2) [0|14|34]	5205
Standing Hampton (82) 28(2) [0|14|32]	4143
VOA (84) 32(3) [0|9|36]	4083
I Never Said Goodbye (87) 14(1) [0|11|23]	4023
Danger Zone (80) 85(1) [0|0|12]	567
Through The Fire (84)-Hagar, Schon, Aaronson, Shrieve [A] 42(2) [0|0|18]	358
Rematch (83) 171(2) [0|0|9]	153

Act ▶ Rank Peak [Top10s|Top40s|Total] Title (Yr) Pk(Wks) [T10 Wks|T40Wks|TotWks]	Score in '80s
Nina HAGEN ▶ 1514 Pk:151 [0|0|2]	**239**
Fearless (84) 151(2) [0|0|8]	191
Nunsexmonkrock (82) 184(1) [0|0|3]	47.6
Merle HAGGARD ▶ 667 Pk:37 [0|1|3]	**2639**
Poncho & Lefty (83)-Merle Haggard/Willie Nelson [A] 37(1) [0|1|53]	1938
Big City (81) 161(1) [0|0|28]	504
A Taste Of Yesterday's Wine (82)-Merle Haggard/George Jones [A] 123(4) [0|0|12]	197
HAIRCUT ONE HUNDRED ▶ 600 **Pk:31 [0|1|1]**	**3195**
Pelican West (82) 31(3) [0|7|37]	3195
Daryl HALL ▶ 606 Pk:29 [0|1|2]	**3131**
Three Hearts In The Happy Ending Machine (86) 29(3) [0|6|26]	2365
Sacred Songs (80) 58(2) [0|0|12]	766
Daryl HALL & John OATES ▶ 12 Pk:3 [4|7|8]	**75922**
H2O (82) 3(15) [33|46|68]	22866
Private Eyes (81) 5(3) [13|32|61]	12806
Voices (80) 17(1) [0|35|100]	12479
Big Bam Boom (84) 5(2) [12|31|51]	12112
Rock 'N Soul Part 1 (83) 7(2) [11|28|44]	9415
Ooh Yeah! (88) 24(2) [0|7|26]	2919
Hall & Oates Live At The Apollo With David Ruffin & Eddie Kendrick (85) 21(2) [0|8|18]	2335
*X-Static (**79**/80) 48(2) [0|0|14]*	*991*
Jimmy HALL ▶ 1870 Pk:183 [0|0|1]	**34.9**
Touch You (80) 183(1) [0|0|2]	34.9
John HALL Band ▶ 1292 Pk:147 [0|0|2]	**443**
All Of The Above (81) 158(3) [0|0|13]	312
Searchparty (83) 147(2) [0|0|5]	131
Herbie HANCOCK ▶ 360 Pk:43 [0|0|6]	**7471**
Future Shock (83) 43(2) [0|0|65]	5468
Sound-System (84) 71(2) [0|0|14]	729
Monster (80) 94(1) [0|0|18]	721
Mr. Hands (80) 117(1) [0|0|6]	225
Magic Windows (81) 140(1) [0|0|6]	175
Lite Me Up (82) 151(2) [0|0|6]	155
Paul HARDCASTLE ▶ 894 Pk:63 [0|0|1]	**1396**
Rain Forest (85) 63(2) [0|0|25]	1396
Emmylou HARRIS ▶ 244 Pk:6 [1|3|11]	**12145**
Roses In The Snow (80) 26(2) [0|11|34]	3352
Evangeline (81) 22(3) [0|9|24]	2807
Trio (87)-Dolly Parton, Linda Ronstadt, Emmylou Harris [A] 6(1) [3|14|48]	2149
Cimarron (81) 46(1) [0|0|20]	1600
Last Date (82) 65(4) [0|0|17]	970
White Shoes (83) 116(1) [0|0|13]	494
Light Of The Stable: The Christmas Album (80) 102(3) [0|0|9]	391
Thirteen (86) 157(2) [0|0|6]	127
Profile II: Best Of Emmylou Harris (84) 176(1) [0|0|6]	93.0
The Ballad Of Sally Rose (85) 171(2) [0|0|4]	85.1
Angel Band (87) 166(2) [0|0|4]	77.4
Sam HARRIS ▶ 595 Pk:35 [0|1|2]	**3286**
Sam Harris (84) 35(2) [0|3|29]	2493
Sam-I-Am (86) 69(1) [0|0|14]	793
George HARRISON ▶ 271 Pk:8 [1|2|4]	**10697**
Cloud Nine (87) 8(1) [9|22|31]	7818
Somewhere In England (81) 11(2) [0|7|13]	2452
Gone Troppo (82) 108(2) [0|0|7]	252
The Best Of Dark Horse (89) 132(2) [0|0|6]	175
Jerry HARRISON: CASUAL GODS ▶ 976 **Pk:78 [0|0|1]**	**1071**
Casual Gods (88) 78(1) [0|0|20]	1071
Debbie HARRY ▶ 683 Pk:25 [0|1|3]	**2518**
KooKoo (81) 25(1) [0|4|12]	1606
Rockbird (86) 97(2) [0|0|13]	624
Def, Dumb & Blonde (89)-Deborah Harry 123(2) [0|0|8]	288

Acts With Albums

Act ▶ Rank Peak [Top10s|Top40s|Total] Title (Yr) Pk(Wks) [T10 Wks|T40Wks|TotWks]	Score in '80s
Corey HART ▶ 253 Pk:20 [0|2|4]	**11593**
Boy In The Box (85) 20(7) [0|15|37]	5516
First Offense (84) 31(3) [0|6|36]	3789
Fields Of Fire (86) 55(2) [0|0|27]	2021
Young Man Running (88) 121(2) [0|0|8]	268
Dan HARTMAN ▶ 792 Pk:55 [0|0|2]	**1828**
I Can Dream About You (84) 55(3) [0|0|28]	1799
Relight My Fire (80) 189(1) [0|0|2]	29.5
--HATCHET, Molly see: MOLLY HATCHET	
Richie HAVENS ▶ 1786 Pk:173 [0|0|1]	**75.2**
Simple Things (87) 173(1) [0|0|4]	75.2
Colin James HAY ▶ 1461 Pk:126 [0|0|1]	**277**
Looking For Jack (87) 126(2) [0|0|9]	277
Isaac HAYES ▶ 668 Pk:39 [0|1|3]	**2605**
Don't Let Go (80) 39(1) [0|1|16]	1584
And Once Again (80) 59(1) [0|0|15]	916
Royal Rappin's (79/80)-Millie Jackson & Isaac *Hayes [A] 122(1) [0|0|8]*	*105*
Justin HAYWARD ▶ 1721 Pk:166 [0|0|1]	**106**
Night Flight (80) 166(2) [0|0|5]	106
Leon HAYWOOD ▶ 1330 Pk:92 [0|0|2]	**407**
Naturally (80) 92(2) [0|0|10]	407
Ofra HAZA ▶ 1448 Pk:130 [0|0|1]	**289**
Shaday (89) 130(2) [0|0|9]	289
Robert HAZARD ▶ 1269 Pk:102 [0|0|1]	**472**
Robert Hazard (83) 102(1) [0|0|11]	472
The HEADBOYS ▶ 1585 Pk:143 [0|0|1]	**184**
The Headboys (79/80) 143(1) [0|0|7]	*184*
HEAD EAST ▶ 1223 Pk:119 [0|0|2]	**528**
A Different Kind Of Crazy (79/80) 119(1) [0|0|9]	*318*
U.S. 1 (80) 137(1) [0|0|6]	211
HEADPINS ▶ 1425 Pk:114 [0|0|1]	**306**
Line Of Fire (84) 114(1) [0|0|9]	306
Jeff HEALEY Band ▶ 452 Pk:22 [0|1|1]	**4977**
See The Light (88) 22(3) [0|8|65]	*4977*
HEAR 'N AID ▶ 1358 Pk:80 [0|0|1]	**375**
Hear 'N Aid (86) 80(2) [0|0|7]	375
HEART ▶ 26 Pk:1 [3|6|7]	**56224**
Heart (85) 1(1) [37|58|92]	28406
Bad Animals (87) 2(3) [18|36|50]	15638
Bebe Le Strange (80) 5(2) [4|13|22]	4781
Greatest Hits/Live (80) 13(3) [0|12|25]	4267
Passionworks (83) 39(1) [0|2|21]	1637
Private Audition (82) 25(2) [0|4|14]	1421
Dreamboat Annie (76/80) 177(1) [0|0|4]	*74.4*
HEATWAVE ▶ 1127 Pk:71 [0|0|2]	**682**
Candles (80) 71(2) [0|0|10]	544
Current (82) 156(2) [0|0|6]	138
HEAVEN 17 ▶ 738 Pk:68 [0|0|3]	**2171**
Heaven 17 (83) 68(3) [0|0|28]	1416
The Luxury Gap (83) 72(2) [0|0|13]	699
Pleasure One (87) 177(2) [0|0|3]	55.8
HEAVY D & The BOYZ ▶ 437 Pk:19 [0|1|2]	**5281**
Big Tyme (89) 19(3) [0|14|27]	*4598*
Living Large... (87) 92(2) [0|0|16]	683
HELIX ▶ 841 Pk:69 [0|0|4]	**1614**
Walkin' The Razor's Edge (84) 69(2) [0|0|16]	838
Long Way To Heaven (85) 103(1) [0|0|17]	684
No Rest For The Wicked (83) 186(1) [0|0|4]	53.6
Wild In The Streets (87) 179(2) [0|0|2]	39.0
HELLOWEEN ▶ 882 Pk:104 [0|0|3]	**1435**
Keeper Of The Seven Keys - Part I (87) 104(1) [0|0|21]	712
Keeper Of The Seven Keys - Part II (88) 108(2) [0|0|16]	499
I Want Out-Live (89) 123(2) [0|0|7]	225
Michael HENDERSON ▶ 719 Pk:35 [0|1|3]	**2286**
Wide Receiver (80) 35(1) [0|5|18]	1685
Slingshot (81) 86(2) [0|0|11]	504
Fickle (83) 169(1) [0|0|5]	96.0

Act ▶ Rank Peak [Top10s|Top40s|Total] Title (Yr) Pk(Wks) [T10 Wks|T40Wks|TotWks]	Score in '80s
Jimi HENDRIX ▶ 1060 Pk:79 [0|0|4]	**823**
The Jimi Hendrix Concerts (82) 79(2) [0|0|8]	443
Nine To The Universe (80) 127(1) [0|0|8]	211
Kiss The Sky (84) 148(1) [0|0|5]	131
Jimi Plays Monterey (86) 192(2) [0|0|3]	38.6
Jimi HENDRIX EXPERIENCE ▶ 1208 Pk:119 [0|0|1]	**541**
Radio One (88) 119(2) [0|0|17]	541
Nona HENDRYX ▶ 867 Pk:83 [0|0|3]	**1491**
Nona (83) 83(1) [0|0|19]	792
Female Trouble (87) 96(2) [0|0|18]	558
The Art Of Defense (84) 167(1) [0|0|7]	142
Don HENLEY ▶ 128 Pk:8 [1|3|3]	**23658**
Building The Perfect Beast (84) 13(5) [0|30|63]	11055
The End Of The Innocence (89) 8(7) [5|23|25]	*7311*
I Can't Stand Still (82) 24(7) [0|20|35]	5291
Carol HENSEL ▶ 450 Pk:56 [0|0|3]	**4997**
Carol Hensel's Exercise And Dance Program (81) 56(1) [0|0|55]	3188
Carol Hensel's Exercise And Dance Program, Volume 2 (81) 70(1) [0|0|28]	1351
Carol Hensel's Exercise And Dance Program, Volume 3 (83) 104(2) [0|0|12]	458
Howard HEWETT ▶ 1093 Pk:110 [0|0|2]	**749**
Forever And Ever (88) 110(2) [0|0|12]	431
I Commit To Love (86) 159(2) [0|0|16]	318
Nick HEYWARD ▶ 1799 Pk:178 [0|0|1]	**71.1**
North Of A Miracle (84) 178(2) [0|0|4]	71.1
John HIATT ▶ 812 Pk:98 [0|0|2]	**1747**
Slow Turning (88) 98(3) [0|0|31]	1104
Bring The Family (87) 107(2) [0|0|17]	643
Bertie HIGGINS ▶ 730 Pk:38 [0|1|1]	**2211**
Just Another Day In Paradise (82) 38(2) [0|3|25]	2211
Dan HILL ▶ 1061 Pk:90 [0|0|1]	**814**
Dan Hill(2) (87) 90(1) [0|0|19]	814
Z.Z. HILL ▶ 1451 Pk:165 [0|0|2]	**285**
I'm A Blues Man (84) 170(2) [0|0|9]	180
The Rhythm & The Blues (83) 165(1) [0|0|5]	106
HIPSWAY ▶ 935 Pk:55 [0|0|1]	**1215**
Hipsway (87) 55(2) [0|0|18]	1215
HIROSHIMA ▶ 379 Pk:51 [0|0|6]	**6962**
Another Place (85) 79(2) [0|0|45]	1928
Hiroshima (80) 51(2) [0|0|25]	*1589*
Go (87) 75(2) [0|0|32]	1587
Odori (80) 72(1) [0|0|18]	818
East (89) 105(2) [0|0|19]	773
Third Generation (83) 142(1) [0|0|9]	267
Robyn HITCHCOCK And The EGYPTIANS ▶ 1041 Pk:111 [0|0|2]	**868**
Globe Of Frogs (88) 111(1) [0|0|15]	606
Queen Elvis (89) 139(1) [0|0|9]	262
Roger HODGSON ▶ 762 Pk:46 [0|0|2]	**2017**
In The Eye Of The Storm (84) 46(5) [0|0|22]	1890
Hai Hai (87) 163(2) [0|0|6]	128
Amy HOLLAND ▶ 1399 Pk:146 [0|0|1]	**328**
Amy Holland (80) 146(1) [0|0|14]	328
Jennifer HOLLIDAY ▶ 694 Pk:31 [0|1|2]	**2453**
Feel My Soul (83) 31(1) [0|5|22]	2003
Say You Love Me (85) 110(2) [0|0|14]	450
The HOLLIES ▶ 1308 Pk:90 [0|0|1]	**430**
What Goes Around (83) 90(1) [0|0|9]	430
HOLLY And The ITALIANS ▶ 1831 Pk:177 [0|0|1]	**56.7**
The Right To Be Italian (81) 177(2) [0|0|3]	56.7
Rupert HOLMES ▶ 648 Pk:33 [0|1|1]	**2777**
Partners In Crime (80) 33(2) [0|12|23]	*2777*
The HONEYDRIPPERS ▶ 337 Pk:4 [1|1|1]	**7918**
Volume One (84) 4(2) [10|18|31]	7918
HONEYMOON SUITE ▶ 545 Pk:60 [0|0|3]	**3810**
The Big Prize (86) 61(2) [0|0|35]	2279

Act ▶ Rank Peak [Top10s|Top40s|Total] Title (Yr) Pk(Wks) [T10 Wks|T40Wks|TotWks]	Score in '80s
Honeymoon Suite (84) 60(2) [0|0|17]	981
Racing After Midnight (88) 86(2) [0|0|10]	550
HOODOO GURUS ▶ 934 Pk:101 [0|0|3]	**1223**
Magnum Cum Louder (89) 101(1) [0|0|15]	610
Blow Your Cool! (87) 120(4) [0|0|13]	435
Mars Needs Guitars! (86) 140(2) [0|0|7]	178
--HOOK, DR. see: DOCTOR HOOK	
John Lee HOOKER ▶ 1197 Pk:101 [0|0|1]	**558**
The Healer (89/90) 101(1) [0|0|13]	*558*
The HOOTERS ▶ 227 Pk:12 [0|2|3]	**12995**
Nervous Night (85)-Hooters 12(1) [0|30|74]	10378
One Way Home (87)-Hooters 27(3) [0|15|26]	2433
Zig Zag (89/90)-Hooters 120(2) [0|0|5]	*184*
Lena HORNE ▶ 1392 Pk:112 [0|0|1]	**334**
Lena Horne: The Lady And Her Music (Original Cast) (81) 112(2) [0|0|9]	334
Bruce HORNSBY And The RANGE ▶ 110 Pk:3 [2|2|2]	**26366**
The Way It Is (86) 3(4) [23|42|73]	19348
Scenes From The Southside (88) 5(1) [7|19|27]	7018
HOTHOUSE FLOWERS ▶ 932 Pk:88 [0|0|1]	**1231**
People (88) 88(2) [0|0|31]	1231
The HOUSEMARTINS ▶ 1255 Pk:124 [0|0|2]	**486**
London 0 Hull 4 (87) 124(2) [0|0|14]	398
The People Who Grinned Themselves To Death (88) 177(1) [0|0|6]	88.4
HOUSE OF FREAKS ▶ 1510 Pk:154 [0|0|1]	**245**
Tantilla (89) 154(1) [0|0|10]	245
HOUSE OF LORDS ▶ 950 Pk:78 [0|0|1]	**1158**
House Of Lords (88) 78(1) [0|0|27]	1158
HOUSE OF LOVE ▶ 1649 Pk:156 [0|0|1]	**142**
The House Of Love (88) 156(1) [0|0|7]	142
Thelma HOUSTON ▶ 1575 Pk:144 [0|0|1]	**193**
Never Gonna Be Another One (81) 144(1) [0|0|6]	193
Whitney HOUSTON ▶ 14 Pk:1 [2|2|2]	**75693**
Whitney Houston (85) 1(14) [46|78|162]	46434
Whitney (87) 1(11) [31|51|85]	29259
George HOWARD ▶ 817 Pk:109 [0|0|5]	**1735**
A Nice Place To Be (86) 109(3) [0|0|26]	1001
Love Will Follow (86) 142(2) [0|0|11]	303
Reflections (88) 109(1) [0|0|8]	265
Dancing In The Sun (85) 169(2) [0|0|4]	88.7
Steppin' Out (84) 178(1) [0|0|4]	77.0
Miki HOWARD ▶ 1248 Pk:145 [0|0|2]	**499**
Love Confessions (88) 145(2) [0|0|16]	376
Come Share My Love (87) 171(2) [0|0|6]	123
Steve HOWE ▶ 1762 Pk:164 [0|0|1]	**83.7**
The Steve Howe Album (80) 164(1) [0|0|4]	83.7
David HUDSON ▶ 1872 Pk:184 [0|0|1]	**34.0**
To You Honey, Honey With Love (80) 184(1) [0|0|2]	34.0
Grayson HUGH ▶ 967 Pk:71 [0|0|1]	**1103**
Blind To Reason (88) 71(2) [0|0|24]	1103
HUMAN LEAGUE ▶ 174 Pk:3 [1|3|5]	**17181**
Dare (82) 3(3) [9|20|38]	9340
Crash (86) 24(2) [0|11|25]	3479
Fascination! (83) 22(1) [0|8|29]	3285
Hysteria (84) 62(1) [0|0|13]	864
Love And Dancing (82)-The League Unlimited Orchestra 135(2) [0|0|7]	212
HUMBLE PIE ▶ 1013 Pk:60 [0|0|2]	**956**
On To Victory (80) 60(1) [0|0|14]	799
Go For The Throat (81) 154(1) [0|0|6]	157
Ian HUNTER ▶ 767 Pk:62 [0|0|4]	**1970**
Welcome To The Club (80) 69(1) [0|0|17]	862
Short Back 'N' Sides (81) 62(1) [0|0|11]	703
All Of The Good Ones Are Taken (83) 125(1) [0|0|8]	284
Y U I Orta (89)-Ian Hunter/Mick Ronson [A] *157(1) [0|0|10]*	*121*

338

Acts With Albums

Act ▶ Rank Peak [Top10s|Top40s|Total] Title (Yr) Pk(Wks) [T10 Wks|T40Wks|TotWks]	Score in '80s
John HUNTER ▶ 1530 Pk:148 [0|0|1]	230
Famous At Night (85) 148(2) [0|0|9]	230
HURRICANE ▶ 891 Pk:92 [0|0|1]	1404
Over The Edge (88) 92(2) [0|0|36]	1404
HUSKER DU ▶ 1168 Pk:117 [0|0|2]	593
Warehouse: Songs And Stories (87) 117(2) [0|0|10]	329
Candy Apple Grey (86) 140(2) [0|0|10]	264
Paul HYDE And The PAYOLAS ▶ 1488 Pk:144 [0|0|1]	261
Here's The World For Ya (85) 144(2) [0|0|10]	261
Phyllis HYMAN ▶ 510 Pk:50 [0|0|4]	4142
Living All Alone (86) 78(2) [0|0|41]	1521
You Know How To Love Me (80) 50(1) [0|0|17]	*1242*
Can't We Fall In Love Again (81) 57(1) [0|0|13]	932
Goddess Of Love (83) 112(1) [0|0|12]	448
Tony HYMAS ▶ 1395 Pk:49 [0|0|1]	334
Jeff Beck's Guitar Shop (89)-Jeff Beck With Terry Bozzio & Tony Hymas [A] 49(2) [0|0|11]	*334*

I

Act ▶ Rank Peak [Top10s|Top40s|Total] Title (Yr) Pk(Wks) [T10 Wks|T40Wks|TotWks]	Score in '80s
Janis IAN ▶ 1763 Pk:156 [0|0|1]	83.1
Restless Eyes (81) 156(1) [0|0|3]	83.1
ICEHOUSE ▶ 405 Pk:43 [0|0|4]	6150
Man Of Colours (87) 43(1) [0|0|44]	3434
Measure For Measure (86) 55(3) [0|0|24]	1781
Icehouse (81) 82(2) [0|0|15]	734
Primitive Man (82) 129(2) [0|0|6]	202
ICE-T ▶ 471 Pk:35 [0|2|3]	4580
Power (88) 35(1) [0|5|33]	2506
The Iceberg (Freedom Of Speech... Just Watch What You Say) (89) 37(2) [0|3|10]	*1088*
Rhyme Pays (87) 93(2) [0|0|27]	986
ICICLE WORKS ▶ 861 Pk:40 [0|1|1]	1526
Icicle Works (84) 40(2) [0|2|18]	1526
ICON ▶ 1898 Pk:190 [0|0|1]	27.2
Icon (84) 190(1) [0|0|2]	27.2
Billy IDOL ▶ 71 Pk:6 [3|3|5]	34986
Rebel Yell (83) 6(3) [5|38|82]	13293
Whiplash Smile (86) 6(1) [5|21|47]	8031
Vital Idol (87) 10(1) [1|18|29]	5795
Billy Idol (82) 45(1) [0|0|104]	5122
Don't Stop (81) 71(1) [0|0|68]	2746
Julio IGLESIAS ▶ 177 Pk:5 [1|2|8]	17095
1100 Bel Air Place (84) 5(2) [11|21|34]	8665
Julio (83) 32(1) [0|9|89]	6335
Non Stop (88) 52(2) [0|0|17]	1100
Libra (85) 92(2) [0|0|12]	531
In Concert (84) 159(2) [0|0|9]	214
From A Child To A Woman (84) 181(1) [0|0|6]	103
Hey! (84) 179(1) [0|0|6]	97.1
Moments (84) 191(1) [0|0|4]	51.0
IMPELLITTERI ▶ 1025 Pk:91 [0|0|1]	910
Stand In Line (88) 91(2) [0|0|20]	910
INDIGO GIRLS ▶ 492 Pk:22 [0|1|2]	4348
Indigo Girls (89) 22(3) [0|11|35]	4230
Strange Fire (89/*90*) 167(2) [0|0|3]	*118*
INFORMATION SOCIETY ▶ 525 Pk:25 [0|1|1]	4039
Information Society (88) 25(1) [0|9|38]	4039
James INGRAM ▶ 582 Pk:46 [0|0|2]	3412
It's Your Night (83) 46(1) [0|0|42]	3117
Never Felt So Good (86) 123(2) [0|0|9]	295
The INMATES ▶ 1058 Pk:49 [0|0|1]	825
First Offence (80) 49(2) [0|0|12]	*825*
INNER CITY ▶ 1740 Pk:162 [0|0|1]	91.4
Big Fun (89) 162(2) [0|0|4]	91.4
INSIDERS ▶ 1718 Pk:167 [0|0|1]	107
Ghost On The Beach (87) 167(1) [0|0|5]	107

Act ▶ Rank Peak [Top10s|Top40s|Total] Title (Yr) Pk(Wks) [T10 Wks|T40Wks|TotWks]	Score in '80s
INSTANT FUNK ▶ 1103 Pk:129 [0|0|3]	728
Witch Doctor (80) 129(1) [0|0|9]	*338*
The Funk Is On (80) 130(2) [0|0|6]	212
Looks So Fine (82) 147(2) [0|0|7]	179
INXS ▶ 60 Pk:3 [1|2|6]	37602
Kick (87) 3(4) [22|65|81]	25790
Listen Like Thieves (85) 11(1) [0|16|55]	7847
Shabooh Shoobah (83) 46(3) [0|0|31]	1962
The Swing (84) 52(2) [0|0|28]	1782
Dekadance (83) 148(2) [0|0|6]	152
INXS (84) 164(1) [0|0|3]	70.1
Donnie IRIS ▶ 583 Pk:57 [0|0|5]	3395
Back On The Streets (80) 57(1) [0|0|23]	1242
King Cool (81) 84(2) [0|0|31]	1178
No Muss...No Fuss (85) 115(1) [0|0|15]	539
Fortune 410 (83) 127(1) [0|0|12]	372
The High And The Mighty (82) 180(2) [0|0|4]	63.4
IRISH ROVERS ▶ 1594 Pk:157 [0|0|1]	177
Wasn't That A Party (81)-The Rovers 157(1) [0|0|8]	177
IRON MAIDEN ▶ 107 Pk:11 [0|6|8]	27106
Somewhere In Time (86) 11(1) [0|18|39]	6046
Piece Of Mind (83) 14(1) [0|14|45]	5482
Seventh Son Of A Seventh Son (88) 12(1) [0|11|23]	3750
The Number Of The Beast (82) 33(2) [0|5|65]	3535
Powerslave (84) 21(3) [0|7|34]	3308
Live After Death (85) 19(1) [0|10|22]	3261
Killers (81) 78(1) [0|0|23]	932
Maiden Japan (81) 89(2) [0|0|30]	791
Chris ISAAK ▶ 1498 Pk:149 [0|0|2]	253
Heart Shaped World (89/*91*) 149(2) [0|0|10]	*228*
Chris Isaak (87) 194(1) [0|0|2]	25.0
ISLE OF MAN ▶ 1160 Pk:110 [0|0|1]	608
Isle Of Man (86) 110(2) [0|0|18]	608
ISLEY BROTHERS ▶ 229 Pk:8 [1|3|8]	12749
Go All The Way (80) 8(4) [4|13|22]	4736
Between The Sheets (83) 19(1) [0|9|23]	2857
Grand Slam (81) 28(3) [0|5|17]	1715
Smooth Sailin' (87) 64(2) [0|0|17]	1020
Inside You (81) 45(1) [0|0|13]	942
The Real Deal (82) 87(3) [0|0|12]	596
Spend The Night (89)-The Isley Brothers Featuring Ronald Isley 89(1) [0|0|13]	563
Masterpiece (85) 140(2) [0|0|12]	322
ISLEY. JASPER. ISLEY ▶ 850 Pk:77 [0|0|2]	1562
Caravan Of Love (85) 77(2) [0|0|26]	1311
Broadway's Closer To Sunset Blvd. (85) 135(2) [0|0|10]	251

J

Act ▶ Rank Peak [Top10s|Top40s|Total] Title (Yr) Pk(Wks) [T10 Wks|T40Wks|TotWks]	Score in '80s
Freddie JACKSON ▶ 148 Pk:10 [1|2|3]	20465
Rock Me Tonight (85) 10(1) [1|36|62]	10867
Just Like The First Time (86) 23(2) [0|26|51]	7717
Don't Let Love Slip Away (88) 48(2) [0|0|30]	1882
Janet JACKSON ▶ 48 Pk:1 [2|2|4]	42313
Control (86) 1(2) [37|77|106]	31246
Janet Jackson's Rhythm Nation 1814 (89) 1(4) [12|13|13]	*9353*
Janet Jackson (82) 63(1) [0|0|25]	1548
Dream Street (84) 147(2) [0|0|6]	166
Jermaine JACKSON ▶ 166 Pk:6 [1|2|7]	17986
Let's Get Serious (80) 6(3) [5|17|29]	6323
Jermaine Jackson (84) 19(2) [0|16|49]	6294
Precious Moments (86) 46(2) [0|0|22]	1731
Jermaine(2) (80) 44(1) [0|0|23]	1636
Let Me Tickle Your Fancy (82) 46(3) [0|0|16]	1347
I Like Your Style (81) 86(1) [0|0|10]	449
Don't Take It Personal (89) 115(1) [0|0|5]	*206*

Act ▶ Rank Peak [Top10s|Top40s|Total] Title (Yr) Pk(Wks) [T10 Wks|T40Wks|TotWks]	Score in '80s
Joe JACKSON ▶ 124 Pk:4 [1|3|10]	24179
Night And Day (82) 4(6) [7|30|57]	11261
Body And Soul (84) 20(4) [0|16|29]	4547
Big World (86) 34(2) [0|8|25]	2630
Beat Crazy (80) 41(1) [0|0|16]	1198
Joe Jackson's Jumpin' Jive (81) 42(2) [0|0|13]	1156
I'm The Man (*79*/80) 46(1) [0|0|15]	*1067*
Blaze Of Glory (89) 61(2) [0|0|21]	918
Mike's Murder (Soundtrack) (83) 64(1) [0|0|13]	713
Live 1980/86 (88) 91(3) [0|0|12]	447
Will Power (87) 131(2) [0|0|8]	241
La Toya JACKSON ▶ 1135 Pk:116 [0|0|3]	667
LaToya Jackson (80) 116(2) [0|0|13]	452
Heart Don't Lie (84) 149(2) [0|0|6]	155
My Special Love (81) 175(2) [0|0|3]	59.4
Marlon JACKSON ▶ 1708 Pk:175 [0|0|1]	113
Baby Tonight (87) 175(2) [0|0|7]	113
Michael JACKSON ▶ 1 Pk:1 [3|3|6]	122517
Thriller (82) 1(37) [78|91|122]	68152
Bad (87) 1(6) [39|54|87]	32895
Off The Wall (80) 3(3) [21|35|151]	*19930*
Farewell My Summer Love (84) 46(2) [0|0|15]	1222
One Day In Your Life (81) 144(1) [0|0|10]	256
14 Greatest Hits (84)-Michael Jackson & The Jackson 5 [A] 168(1) [0|0|7]	63.3
Millie JACKSON ▶ 715 Pk:94 [0|0|6]	2289
Live And Uncensored (80) 94(2) [0|0|16]	*724*
An Imitation Of Love (86) 119(2) [0|0|17]	480
Live And Outrageous (Rated XXX) (82) 113(1) [0|0|13]	449
For Men Only (80) 100(1) [0|0|10]	396
I Had To Say It (81) 137(1) [0|0|4]	135
Royal Rappin's (*79*/80)-Millie Jackson & Isaac Hayes [A] 122(1) [0|0|8]	*105*
Rebbie JACKSON ▶ 964 Pk:63 [0|0|1]	1112
Centipede (84) 63(3) [0|0|18]	1112
JACKSON 5 ▶ 192 Pk:4 [2|3|5]	15530
Victory (84)-The Jacksons 4(3) [8|15|30]	6548
Triumph (80)-The Jacksons 10(4) [4|18|29]	5942
Jacksons Live (81)-The Jacksons 30(2) [0|8|19]	2208
2300 Jackson St. (89)-The Jacksons 59(2) [0|0|11]	770
14 Greatest Hits (84)-Michael Jackson & The Jackson 5 [A] 168(1) [0|0|7]	63.3
Debbie JACOBS ▶ 1676 Pk:178 [0|0|1]	129
High On Your Love (80) 178(4) [0|0|7]	129
Mick JAGGER ▶ 410 Pk:13 [0|1|2]	6023
She's The Boss (85) 13(1) [0|12|29]	4369
Primitive Cool (87) 41(2) [0|0|20]	1654
The JAM ▶ 695 Pk:72 [0|0|7]	2447
The Gift (82) 82(1) [0|0|16]	754
Sound Affects (81) 72(2) [0|0|11]	652
The Bitterest Pill (I Ever Had To Swallow) (82) 135(2) [0|0|14]	375
Setting Sons (80) 137(2) [0|0|8]	240
Dig The New Breed (83) 131(1) [0|0|9]	223
The Jam (81) 176(3) [0|0|7]	128
Beat Surrender (83) 171(1) [0|0|4]	75.3
Ahmad JAMAL Trio ▶ 1738 Pk:173 [0|0|1]	93.9
Genetic Walk (80)-Ahmad Jamal 173(1) [0|0|5]	93.9
Bob JAMES ▶ 368 Pk:47 [0|0|10]	7270
Double Vision (86)-Bob James/David Sanborn [A] 50(2) [0|0|64]	1539
H (80) 47(3) [0|0|18]	1258
All Around The Town (81) 66(1) [0|0|16]	886
Sign Of The Times (81) 56(1) [0|0|14]	882
Obsession (86) 142(1) [0|0|27]	729
Hands Down (82) 72(2) [0|0|12]	723
The Genie (Themes & Variations From The TV Series "Taxi") (83) 77(1) [0|0|11]	563
Foxie (83) 106(1) [0|0|13]	429

Acts With Albums

Act ▶ Rank Peak [Top10s|Top40s|Total] Title (Yr) Pk(Wks) [T10 Wks|T40Wks|TotWks]	Score in '80s
Bob JAMES ▶ *Continued*	
12 (84) 136(2) [0|0|10]	238
Ivory Coast (88) 196(2) [0|0|2]	23.6
Bob JAMES & Earl KLUGH ▶ 476	
Pk:30 [0|1|2]	4546
One On One (79/80)-Bob James/Earl Klugh	
30(1) [0|8|24]	*2354*
Two Of A Kind (82)-Earl Klugh & Bob James	
44(4) [0|0|29]	2192
Melvin JAMES ▶ 1571 Pk:146 [0|0|1]	194
The Passenger (87) 146(2) [0|0|8]	194
Rick JAMES ▶ 113 Pk:3 [1|3|9]	26007
Street Songs (81) 3(2) [12|27|74]	13383
Cold Blooded (83) 16(1) [0|12|29]	4237
Throwin' Down (82) 13(4) [0|10|23]	3671
Glow (85) 50(2) [0|0|26]	1903
Reflections (84) 41(2) [0|0|19]	1304
The Flag (86) 95(2) [0|0|12]	531
Garden Of Love (80) 83(2) [0|0|10]	445
Fire It Up (79/80) 115(2) [0|0|11]	*362*
Wonderful (88) 148(2) [0|0|8]	172
Tommy JAMES ▶ 1536 Pk:134 [0|0|1]	225
Three Times In Love (80) 134(1) [0|0|7]	225
JANE'S ADDICTION ▶ 890 Pk:103 [0|0|1]	1405
Nothing's Shocking (88)	
103(2) [0|0|35]	1405
Chaz JANKEL ▶ 1333 Pk:126 [0|0|1]	405
Questionnaire (82) 126(1) [0|0|14]	405
Jean Michel JARRE ▶ 772 Pk:52 [0|0|2]	1950
Rendez-Vous (86) 52(4) [0|0|20]	1418
Magnetic Fields (81) 98(2) [0|0|12]	532
Al JARREAU ▶ 127 Pk:9 [1|3|7]	23685
Breakin' Away (81) 9(3) [4|20|103]	10259
Jarreau (83) 13(2) [0|12|43]	5345
This Time (80) 27(2) [0|5|35]	3262
High Crime (84) 49(2) [0|0|35]	2280
Heart's Horizon (88) 75(2) [0|0|23]	1144
L Is For Lover (86) 81(2) [0|0|28]	1087
In London (85) 125(1) [0|0|9]	310
JASON & The SCORCHERS ▶ 741	
Pk:91 [0|0|3]	2158
Still Standing (86) 91(4) [0|0|19]	886
Lost & Found (85) 96(1) [0|0|15]	648
Fervor (84) 116(1) [0|0|23]	624
Chris JASPER ▶ 1851 Pk:182 [0|0|1]	48.0
Superbad (88) 182(2) [0|0|3]	48.0
Miles JAYE ▶ 1183 Pk:125 [0|0|2]	581
Miles (87) 125(2) [0|0|12]	373
Irresistible (89) 160(2) [0|0|9]	208
JEFFERSON AIRPLANE/STARSHIP ▶ 79	
Pk:7 [2|6|9]	33236
Knee Deep In The Hoopla (85)-	
Starship 7(3) [15|33|50]	12108
Freedom At Point Zero (80)-	
Jefferson Starship 10(1) [1|13|23]	*4897*
Modern Times (81)-	
Jefferson Starship 26(3) [0|16|33]	4215
Winds Of Change (82)-	
Jefferson Starship 26(6) [0|11|31]	3867
No Protection (87)-Starship 12(1) [0|9|25]	3555
Nuclear Furniture (84)-	
Jefferson Starship 28(4) [0|8|23]	2853
Love Among The Cannibals (89)-	
Starship 64(2) [0|0|18]	1146
Jefferson Airplane (89)-	
Jefferson Airplane 85(2) [0|0|7]	342
2400 Fulton St. (87)-	
Jefferson Airplane 138(2) [0|0|9]	253
Garland JEFFREYS ▶ 854 Pk:59 [0|0|3]	1545
Escape Artist (81) 59(4) [0|0|18]	1381
Rock & Roll Adult (81) 163(1) [0|0|4]	92.8
Guts For Love (83) 176(2) [0|0|4]	70.6

Act ▶ Rank Peak [Top10s|Top40s|Total] Title (Yr) Pk(Wks) [T10 Wks|T40Wks|TotWks]	Score in '80s
JELLYBEAN ▶ 1261 Pk:101 [0|0|1]	483
Just Visiting This Planet (87) 101(2) [0|0|11]	483
Waylon JENNINGS ▶ 259 Pk:36 [0|2|6]	11187
Greatest Hits (79/80) 69(1) [0|0|80]	*4434*
Music Man (80) 36(2) [0|3|43]	2993
Black On Black (82) 39(2) [0|2|23]	1601
What Goes Around Comes Around (79/80)	
54(1) [0|0|20]	*1119*
Leather And Lace (81)-Waylon Jennings &	
Jessi Colter [A] 43(2) [0|0|19]	672
It's Only Rock And Roll (83) 109(1) [0|0|11]	368
Waylon JENNINGS & Willie NELSON ▶ 906	
Pk:57 [0|0|1]	1342
WWII (82) 57(3) [0|0|22]	1342
JESUS AND MARY CHAIN ▶ 1452	
Pk:161 [0|0|4]	285
Automatic (89/90) 172(1) [0|0|6]	*101*
Darklands (87) 161(2) [0|0|4]	89.2
Psychocandy (86) 188(2) [0|0|4]	54.9
Barbed Wire Kisses (88) 192(1) [0|0|3]	40.0
JETBOY ▶ 1430 Pk:135 [0|0|1]	301
Feel The Shake (88) 135(1) [0|0|10]	301
JETHRO TULL ▶ 305 Pk:19 [0|3|7]	8962
Crest Of A Knave (87) 32(1) [0|8|28]	3021
The Broadsword And The Beast (82)	
19(3) [0|7|17]	2359
"A" (80) 30(2) [0|4|12]	1319
Rock Island (89) 56(2) [0|0|14]	*1042*
Under Wraps (84) 76(2) [0|0|12]	582
20 Years Of Jethro Tull (88) 97(2) [0|0|15]	575
Stormwatch (79/80) 174(1) [0|0|4]	*64.0*
The JETS ▶ 226 Pk:21 [0|2|3]	13279
The Jets (86) 21(2) [0|25|70]	8164
Magic (87) 35(2) [0|6|50]	4844
Believe (89) 107(1) [0|0|7]	272
Joan JETT & The BLACKHEARTS ▶ 121	
Pk:2 [1|3|6]	24484
I Love Rock 'N Roll (81) 2(3) [12|20|59]	12060
Up Your Alley (88) 19(1) [0|18|46]	6444
Album (83) 20(2) [0|10|20]	2971
Bad Reputation (81) 51(4) [0|0|21]	1383
Glorious Results Of A Misspent Youth (84)	
67(1) [0|0|21]	1134
Good Music (86) 105(1) [0|0|16]	491
JIVE BUNNY & The MASTERMIXERS ▶ 1868	
Pk:140 [0|0|1]	37.0
Jive Bunny - The Album (89/90) 140(1) [0|0|1]	*37.0*
J.J. FAD ▶ 722 Pk:49 [0|0|1]	2258
Supersonic--The Album (88) 49(2) [0|0|30]	2258
JOBOXERS ▶ 1067 Pk:70 [0|0|1]	806
Like Gangbusters (83) 70(2) [0|0|15]	806
Billy JOEL ▶ 3 Pk:1 [7|8|11]	94089
An Innocent Man (83) 4(5) [30|62|111]	26882
Glass Houses (80) 1(6) [25|35|73]	23314
The Bridge (86) 7(4) [11|29|47]	11197
Greatest Hits Vol. I & II (85) 6(2) [9|26|41]	*9576*
The Nylon Curtain (82) 7(7) [23|35]	9071
Storm Front (89) 1(1) [7|8|9]	*5481*
Songs In The Attic (81) 8(3) [5|10|27]	5231
Kohuept (Live In Leningrad) (87) 38(1) [0|3|18]	1763
The Stranger (77/80) 126(2) [0|0|31]	*862*
52nd Street (78/80) 154(1) [0|0|23]	*530*
Cold Spring Harbor (84) 158(1) [0|0|8]	182
David JOHANSEN ▶ 962 Pk:90 [0|0|3]	1119
Buster Poindexter (88)-Buster Poindexter And	
His Banshees Of Blue 90(1) [0|0|20]	663
Live It Up (82) 148(2) [0|0|15]	379
Here Comes The Night (81) 160(1) [0|0|5]	76.9
Elton JOHN ▶ 70 Pk:13 [0|8|12]	35198
Reg Strikes Back (88) 16(1) [0|19|29]	5224
Too Low For Zero (83) 25(1) [0|13|54]	5002

Act ▶ Rank Peak [Top10s|Top40s|Total] Title (Yr) Pk(Wks) [T10 Wks|T40Wks|TotWks]	Score in '80s
Breaking Hearts (84) 20(5) [0|14|34]	4531
Live In Australia (With The Melbourne Symphony	
Orchestra) (87) 24(2) [0|8|41]	4154
Jump Up! (82) 17(3) [0|8|33]	4151
21 At 33 (80) 13(2) [0|9|21]	3426
Sleeping With The Past (89) 23(1) [0|9|16]	*2638*
Ice On Fire (85) 48(2) [0|0|28]	2249
The Fox (81) 21(2) [0|5|19]	2204
Elton John's Greatest Hits, Volume III 1979-1987	
(87) 84(2) [0|0|23]	802
Leather Jackets (86) 91(3) [0|0|9]	410
Goodbye Yellow Brick Road (73/87)	
114(1) [0|0|12]	*407*
JOHNNY And The DISTRACTIONS ▶ 1561	
Pk:152 [0|0|1]	198
Let It Rock (82) 152(1) [0|0|9]	198
JOHNNY HATES JAZZ ▶ 834 Pk:56 [0|0|1]	1646
Turn Back The Clock (88) 56(1) [0|0|25]	1646
Don JOHNSON ▶ 544 Pk:17 [0|1|1]	3811
Heartbeat (86) 17(2) [0|11|27]	3811
Howard JOHNSON ▶ 1480 Pk:122 [0|0|1]	266
Keepin' Love New (82) 122(2) [0|0|9]	266
Jesse JOHNSON ▶ 401 Pk:43 [0|0|3]	6235
Jesse Johnson's Revue (85)-	
Jesse Johnson's Revue 43(5) [0|0|43]	4445
Shockadelica (86) 70(2) [0|0|20]	1111
Every Shade Of Love (88) 79(2) [0|0|13]	679
Tom JOHNSTON ▶ 1576 Pk:158 [0|0|2]	191
Still Feels Good (81) 158(1) [0|0|7]	162
Everything You've Heard Is True (79/80)	
188(1) [0|0|2]	*29.5*
France JOLI ▶ 1837 Pk:175 [0|0|1]	55.3
Tonight (80) 175(1) [0|0|3]	55.3
JON And VANGELIS ▶ 750 Pk:64 [0|0|3]	2075
The Friends Of Mr. Cairo (81) 64(1) [0|0|34]	1513
Short Stories (80) 125(1) [0|0|15]	385
Private Collection (83) 148(1) [0|0|7]	178
George JONES ▶ 959 Pk:115 [0|0|3]	1124
Still The Same Ole Me (81) 115(2) [0|0|14]	490
I Am What I Am (81) 132(1) [0|0|14]	436
A Taste Of Yesterday's Wine (82)-	
Merle Haggard/George Jones [A]	
123(4) [0|0|12]	197
Glenn JONES ▶ 1118 Pk:94 [0|0|1]	705
Glenn Jones (87) 94(1) [0|0|17]	705
Grace JONES ▶ 432 Pk:32 [0|1|6]	5337
Nightclubbing (81) 32(2) [0|4|20]	1854
Slave To The Rhythm (85) 73(4) [0|0|20]	1171
Living My Life (82) 86(4) [0|0|20]	1060
Inside Story (86) 81(2) [0|0|16]	746
Warm Leatherette (80) 132(1) [0|0|10]	345
Island Life (86) 161(2) [0|0|7]	161
Howard JONES ▶ 195 Pk:10 [1|2|5]	15417
Dream Into Action (85) 10(1) [1|21|45]	7765
Action Replay (85) 34(2) [0|6|24]	2330
Human's Lib (84) 59(2) [0|0|43]	2304
One To One (86) 56(4) [0|0|21]	1686
Cross That Line (89) 65(1) [0|0|22]	1333
Mick JONES ▶ 1852 Pk:184 [0|0|1]	47.6
Mick Jones (89) 184(1) [0|0|3]	47.6
Oran "Juice" JONES ▶ 821 Pk:44 [0|0|1]	1708
Juice (86) 44(1) [0|0|22]	1708
Quincy JONES ▶ 232 Pk:10 [1|2|3]	12676
The Dude (81) 10(1) [1|26|80]	11508
Back On The Block (89/90) 22(1) [0|2|4]	*660*
The Best (82) 122(2) [0|0|17]	508
Rickie Lee JONES ▶ 262 Pk:5 [1|3|4]	11065
Pirates (81) 5(2) [8|15|29]	6625
The Magazine (84) 44(2) [0|0|21]	1670
Flying Cowboys (89) 39(2) [0|2|12]	*1465*
Girl At Her Volcano (83) 39(1) [0|2|16]	1305

340

Acts With Albums

Act ▶ Rank Peak [Top10s|Top40s|Total] Title (Yr) Pk(Wks) [T10 Wks|T40Wks|TotWks]	Score in '80s
Shirley JONES ▶ 1403 Pk:128 [0|0|1]	325
Always In The Mood (86) 128(1) [0|0|10]	325
Steve JONES ▶ 1775 Pk:169 [0|0|1]	78.4
Fire And Gasoline (89) 169(2) [0|0|4]	78.4
Tom JONES ▶ 1835 Pk:179 [0|0|1]	56.2
Darlin' (81) 179(1) [0|0|3]	56.2
JONES GIRLS ▶ 955 Pk:96 [0|0|2]	1136
At Peace With Woman (80) 96(1) [0|0|24]	857
Get As Much Love As You Can (81) 155(1) [0|0|15]	279
JONZUN CREW ▶ 956 Pk:66 [0|0|1]	1135
Lost In Space (83) 66(1) [0|0|20]	1135
Janis JOPLIN ▶ 1265 Pk:104 [0|0|1]	476
Farewell Song (82) 104(1) [0|0|11]	476
Stanley JORDAN ▶ 493 Pk:64 [0|0|3]	4309
Magic Touch (85) 64(2) [0|0|66]	3439
Standards, Volume 1 (87) 116(2) [0|0|18]	602
Flying Home (88) 131(2) [0|0|9]	269
JOURNEY ▶ 5 Pk:1 [6|6|9]	89017
Escape (81) 1(1) [38|58|146]	33693
Frontiers (83) 2(9) [22|42|85]	20886
Raised On Radio (86) 4(2) [7|28|67]	11627
Departure (80) 8(2) [5|17|57]	6902
Journey's Greatest Hits (88) 10(2) [2|16|50]	6468
Captured (81) 9(4) [4|12|69]	6064
Evolution *(79/80)* 74(1) [0|0|58]	*1822*
Infinity *(78/80)* 117(3) [0|0|54]	*1370*
In The Beginning (80) 152(1) [0|0|8]	185
JOY DIVISION ▶ 1522 Pk:146 [0|0|1]	233
Substance (88) 146(2) [0|0|8]	233
JUDAS PRIEST ▶ 125 Pk:17 [0|7|7]	24080
Screaming For Vengeance (82) 17(8) [0|22|53]	7503
Defenders Of The Faith (84) 18(4) [0|12|37]	4637
Turbo (86) 17(2) [0|11|36]	4582
Point Of Entry (81) 39(2) [0|2|25]	2269
Ram It Down (88) 31(3) [0|6|19]	1991
British Steel (80) 34(2) [0|3|18]	1703
Priest...Live (87) 38(2) [0|3|15]	1396
The JUDDS ▶ 301 Pk:51 [0|0|6]	9261
Rockin' With The Rhythm (85) 66(1) [0|0|57]	2916
Greatest Hits (88) 76(2) [0|0|42]	*1959*
Heartland (87) 52(2) [0|0|31]	1601
Why Not Me (84) 71(1) [0|0|26]	1316
River Of Time (89) 51(2) [0|0|20]	1096
The Judds (84) 153(1) [0|0|15]	373
JULUKA ▶ 1806 Pk:186 [0|0|1]	69.0
Scatterlings (83) 186(1) [0|0|5]	69.0
Rob JUNGKLAS ▶ 1071 Pk:102 [0|0|1]	798
Closer To The Flame (86) 102(1) [0|0|22]	798
JUNIOR ▶ 990 Pk:71 [0|0|2]	1021
"Ji" (82) 71(3) [0|0|16]	920
Inside Lookin' Out (83) 177(1) [0|0|6]	100
JUNKYARD ▶ 1283 Pk:105 [0|0|1]	457
Junkyard (89) 105(2) [0|0|11]	457

K

Act ▶ Rank Peak [Top10s|Top40s|Total] Title (Yr) Pk(Wks) [T10 Wks|T40Wks|TotWks]	Score in '80s
KAJAGOOGOO ▶ 804 Pk:38 [0|1|2]	1796
White Feathers (83) 38(2) [0|2|20]	1742
Extra Play (85)-Kaja 185(1) [0|0|4]	54.9
Big Daddy KANE ▶ 753 Pk:33 [0|1|2]	2068
It's A Big Daddy Thing (89) 33(1) [0|4|13]	*1388*
Long Live The Kane (88) 116(1) [0|0|19]	680
KANE GANG ▶ 1105 Pk:115 [0|0|1]	717
Miracle (87) 115(2) [0|0|20]	717
KANO ▶ 1842 Pk:189 [0|0|1]	52.7
New York Cake (82) 189(1) [0|0|4]	52.7
KANSAS ▶ 293 Pk:16 [0|3|6]	9674
Audio-Visions (80) 26(1) [0|8|21]	2571
Power (86) 35(2) [0|5|27]	2569
Vinyl Confessions (82) 16(2) [0|6|20]	2553
Drastic Measures (83) 41(2) [0|0|21]	1629
In The Spirit Of Things (88) 114(2) [0|0|6]	236
The Best Of Kansas (84) 154(2) [0|0|5]	117
KASHIF ▶ 500 Pk:51 [0|0|4]	4210
Kashif (83) 54(2) [0|0|33]	1771
Send Me Your Love (84) 51(2) [0|0|21]	1387
Love Changes (87) 118(3) [0|0|19]	684
Condition Of The Heart (85) 144(2) [0|0|14]	369
KATRINA & The WAVES ▶ 442 Pk:25 [0|1|3]	5212
Katrina And The Waves (85) 25(2) [0|9|32]	3815
Waves (86) 49(2) [0|0|16]	1125
Break Of Hearts (89) 122(2) [0|0|8]	271
Jorma KAUKONEN & VITAL PARTS ▶ 1679 Pk:163 [0|0|1]	127
Barbeque King (81) 163(2) [0|0|6]	127
KBC BAND ▶ 900 Pk:75 [0|0|1]	1369
KBC Band (86) 75(2) [0|0|24]	1369
KC & The SUNSHINE BAND ▶ 793 Pk:53 [0|0|3]	1828
Do You Wanna Go Party *(79/80)* 53(2) [0|0|11]	*801*
KC Ten (84)-KC 93(1) [0|0|18]	694
Greatest Hits (80) 132(2) [0|0|11]	333
KEEL ▶ 655 Pk:53 [0|0|3]	2731
The Final Frontier (86) 53(2) [0|0|18]	1212
The Right To Rock (85) 99(1) [0|0|21]	849
Keel (87) 79(1) [0|0|13]	670
Tommy KEENE ▶ 1307 Pk:148 [0|0|1]	431
Songs From The Film (86) 148(2) [0|0|17]	431
Johnny KEMP ▶ 941 Pk:68 [0|0|1]	1186
Secrets Of Flying (88) 68(1) [0|0|19]	1186
Joyce KENNEDY ▶ 1153 Pk:79 [0|0|1]	622
Lookin' For Trouble (84) 79(2) [0|0|13]	622
KENNY G ▶ 93 Pk:6 [2|3|5]	29372
Duotones (86) 6(2) [10|41|102]	16700
Silhouette (88) 8(1) [9|27|57]	10573
G Force (84) 62(1) [0|0|21]	1093
Live (89/90) 29(1) [0|2|4]	*531*
Gravity (85) 97(2) [0|0|12]	474
KENTUCKY HEADHUNTERS ▶ 1677 Pk:104 [0|0|1]	128
Pickin' On Nashville *(89/90)* 104(1) [0|0|3]	*128*
Nik KERSHAW ▶ 903 Pk:70 [0|0|2]	1345
Human Racing (84) 70(2) [0|0|20]	1003
The Riddle (85) 113(2) [0|0|10]	342
Chaka KHAN ▶ 239 Pk:14 [0|2|6]	12347
I Feel For You (84) 14(3) [0|16|49]	5796
What Cha' Gonna Do For Me (81) 17(2) [0|9|18]	2730
Naughty (80) 43(4) [0|0|16]	1413
Chaka Khan (82) 52(1) [0|0|18]	1253
Destiny (86) 67(2) [0|0|12]	782
C.K. (88) 125(3) [0|0|12]	372
KICK AXE ▶ 1243 Pk:126 [0|0|1]	503
Vices (84) 126(2) [0|0|15]	503
KID CREOLE & The COCONUTS ▶ 1408 Pk:145 [0|0|2]	321
Wise Guy (82) 145(2) [0|0|12]	284
Fresh Fruit In Foreign Places (81) 180(1) [0|0|2]	37.1
KID 'N PLAY ▶ 760 Pk:96 [0|0|1]	2027
2 Hype (88) 96(1) [0|0|47]	2027
KIDS FROM FAME ▶ 1090 Pk:98 [0|0|3]	759
The Kids From "Fame" Live! (83) 98(1) [0|0|11]	448
The Kids From "Fame" (82) 146(2) [0|0|8]	237
Songs (83) 181(4) [0|0|4]	74.4
Greg KIHN Band ▶ 294 Pk:15 [0|3|6]	9646
Kihnspiracy (83) 15(1) [0|12|24]	3695
Rockihnroll (81) 32(1) [0|5|32]	3053
Kihntinued (82)-Greg Kihn 33(2) [0|4|17]	1433
Citizen Kihn (85)-Greg Kihn 51(2) [0|0|18]	1068
Kihntagious (84) 121(1) [0|0|9]	284

Act ▶ Rank Peak [Top10s|Top40s|Total] Title (Yr) Pk(Wks) [T10 Wks|T40Wks|TotWks]	Score in '80s
Glass House Rock (80) 167(1) [0|0|5]	112
KILLER DWARFS ▶ 1701 Pk:165 [0|0|1]	117
Big Deal (88) 165(1) [0|0|6]	117
KILLING JOKE ▶ 1921 Pk:194 [0|0|1]	12.7
Brighter Than A Thousand Suns (87) 194(1) [0|0|1]	12.7
John KILZER ▶ 1256 Pk:110 [0|0|1]	486
Memory In The Making (88) 110(2) [0|0|15]	486
Tom KIMMEL ▶ 1262 Pk:104 [0|0|1]	481
5 To 1 (87) 104(2) [0|0|15]	481
KING ▶ 1527 Pk:140 [0|0|1]	230
Steps In Time (85) 140(2) [0|0|9]	230
B.B. KING ▶ 1175 Pk:131 [0|0|4]	586
There Must Be A Better World Somewhere (81) 131(1) [0|0|10]	343
"Now Appearing" At Ole' Miss (80) 162(2) [0|0|4]	92.3
Love Me Tender (82) 179(1) [0|0|5]	78.0
Blues 'N Jazz (83) 172(1) [0|0|4]	73.4
Carole KING ▶ 716 Pk:44 [0|0|3]	2288
Pearls-Songs Of Goffin And King (80) 44(1) [0|0|17]	1352
City Streets (89) 111(1) [0|0|16]	577
One To One (82) 119(3) [0|0|11]	359
Evelyn "Champagne" KING ▶ 423 Pk:27 [0|2|5]	5579
Get Loose (82)-Evelyn King 27(3) [0|7|32]	2609
I'm In Love (81)-Evelyn King 28(1) [0|6|18]	1970
Face To Face (83) 91(1) [0|0|20]	724
Call On Me (80) 124(3) [0|0|7]	237
Flirt (88) 192(1) [0|0|3]	38.6
The KINGBEES ▶ 1491 Pk:160 [0|0|1]	257
The Kingbees (80) 160(2) [0|0|12]	257
KING CRIMSON ▶ 617 Pk:45 [0|0|3]	3027
Discipline (81) 45(2) [0|0|17]	1143
Three Of A Perfect Pair (84) 58(2) [0|0|17]	1012
Beat (82) 52(2) [0|0|14]	873
KING DIAMOND ▶ 928 Pk:89 [0|0|3]	1266
Them (88) 89(2) [0|0|12]	561
Abigail (87) 123(1) [0|0|13]	382
Conspiracy (89) 111(1) [0|0|8]	323
KINGDOM COME ▶ 414 Pk:12 [0|1|2]	5852
Kingdom Come (88) 12(1) [0|12|29]	4812
In Your Face (89) 49(2) [0|0|15]	1040
The KINGS ▶ 914 Pk:74 [0|0|2]	1319
The Kings Are Here (80) 74(2) [0|0|26]	1241
Amazon Beach (81) 170(1) [0|0|4]	77.8
KINGS OF THE SUN ▶ 1345 Pk:136 [0|0|1]	395
Kings Of The Sun (88) 136(1) [0|0|16]	395
KING SWAMP ▶ 1515 Pk:159 [0|0|1]	238
King Swamp (89) 159(1) [0|0|14]	238
KING'S X ▶ 1023 Pk:123 [0|0|2]	929
Gretchen Goes To Nebraska (89) 123(1) [0|0|18]	622
Out Of The Silent Planet (88) 144(1) [0|0|11]	306
KING TEE ▶ 1279 Pk:125 [0|0|1]	464
Act A Fool (89) 125(1) [0|0|15]	464
Sam KINISON ▶ 866 Pk:43 [0|0|2]	1494
Have You Seen Me Lately? (88) 43(3) [0|0|17]	1409
Louder Than Hell (86) 175(1) [0|0|5]	84.3
The KINKS ▶ 171 Pk:12 [0|3|9]	17414
One For The Road (80) 14(4) [0|14|33]	5467
Give The People What They Want (81) 15(2) [0|9|36]	4824
State Of Confusion (83) 12(1) [0|12|25]	4129
Word Of Mouth (84) 57(3) [0|0|20]	1505
Think Visual (86) 81(2) [0|0|16]	873
The Road (88) 110(2) [0|0|7]	269
UK Jive (89) 122(2) [0|0|6]	*207*
Come Dancing With The Kinks/The Best Of The Kinks 1977-1986 (86) 159(1) [0|0|4]	102
Second Time Around (80) 177(1) [0|0|4]	69.8

341

Acts With Albums

Act ▶ Rank Peak [Top10s|Top40s|Total] Title (Yr) Pk(Wks) [T10 Wks|T40Wks|TotWks]	Score in '80s
KISS ▶ 105 Pk:18 [0|7|9]	27343
Animalize (84) 19(1) [0|17|38]	5248
Crazy Nights (87) 18(1) [0|17|34]	5107
Asylum (85) 20(1) [0|16|29]	4475
Smashes, Thrashes & Hits (88) 21(2) [0|11|27]	3610
Lick It Up (83) 24(2) [0|7|30]	3525
Creatures Of The Night (82) 45(6) [0|0|19]	1872
Hot In The Shade (89) 29(1) [0|5|9]	*1445*
Kiss Unmasked (80) 35(2) [0|4|14]	1429
Music From The Elder (81) 75(1) [0|0|11]	632
KITARO ▶ 1413 Pk:141 [0|0|3]	317
My Best (86) 141(2) [0|0|10]	273
Asia (85) 191(1) [0|0|2]	26.4
Tenku (87) 183(1) [0|0|1]	17.7
KIX ▶ 554 Pk:46 [0|0|2]	3696
Blow My Fuse (88) 46(1) [0|0|48]	*3567*
Cool Kids (83) 177(1) [0|0|8]	129
KLEEER ▶ 915 Pk:81 [0|0|3]	1310
License To Dream (81) 81(2) [0|0|16]	846
Winners (80) 140(2) [0|0|10]	250
Taste The Music (82) 139(1) [0|0|8]	214
John KLEMMER ▶ 1112 Pk:99 [0|0|2]	709
Hush (81) 99(1) [0|0|9]	401
Magnificent Madness (80) 146(1) [0|0|11]	308
KLIQUE ▶ 1125 Pk:70 [0|0|1]	692
Try It Out (83) 70(2) [0|0|14]	692
Earl KLUGH ▶ 302 Pk:38 [0|1|11]	9202
Low Ride (83) 38(3) [0|3|24]	1815
Crazy For You (81) 53(2) [0|0|27]	1456
Wishful Thinking (84) 69(2) [0|0|23]	1215
Dream Come True (80) 42(1) [0|0|19]	1213
Collaboration (87)-George Benson/Earl Klugh [A] 59(2) [0|0|31]	957
Late Night Guitar (80) 98(3) [0|0|23]	936
Soda Fountain Shuffle (85) 110(1) [0|0|17]	638
Nightsongs (84) 107(2) [0|0|17]	488
Life Stories (86) 143(2) [0|0|11]	294
Whispers And Promises (89) 150(2) [0|0|5]	126
How To Beat The High Cost Of Living (Soundtrack) (80)-Hubert Laws And Earl Klugh [A] 134(2) [0|0|4]	63.5
KLYMAXX ▶ 319 Pk:18 [0|1|2]	8418
Meeting In The Ladies Room (85) 18(3) [0|15|67]	7273
Klymaxx (86) 98(2) [0|0|31]	1145
The KNACK ▶ 556 Pk:15 [0|1|3]	3664
But The Little Girls Understand (80) 15(1) [0|9|14]	2655
Get The Knack (79/80) 69(1) [0|0|13]	*714*
Round Trip (81) 93(2) [0|0|6]	295
Gladys KNIGHT & The PIPS ▶ 345 Pk:34 [0|2|5]	7771
Visions (83) 34(2) [0|6|33]	3156
All Our Love (87) 39(1) [0|3|27]	2448
About Love (80) 48(1) [0|0|18]	1434
Life (85) 126(2) [0|0|12]	412
Touch (81) 109(1) [0|0|8]	322
Jean KNIGHT ▶ 1808 Pk:180 [0|0|1]	68.5
My Toot Toot (85) 180(1) [0|0|4]	68.5
Jerry KNIGHT ▶ 1459 Pk:146 [0|0|2]	278
Perfect Fit (81) 146(1) [0|0|6]	145
Jerry Knight (80) 165(1) [0|0|7]	132
K-9 POSSE ▶ 1193 Pk:98 [0|0|1]	567
K-9 Posse (89) 98(3) [0|0|14]	567
Fred KNOBLOCK ▶ 1756 Pk:179 [0|0|1]	86.6
Why Not Me (80) 179(1) [0|0|5]	86.6
Mark KNOPFLER ▶ 1920 Pk:180 [0|0|1]	19.0
The Princess Bride (Soundtrack) (87) 180(1) [0|0|1]	19.0
KOOL & The GANG ▶ 35 Pk:10 [1|7|8]	49167
Emergency (84) 13(1) [0|51|74]	14022
Something Special (81) 12(2) [0|30|67]	9755
Celebrate! (80) 10(2) [2|24|44]	8171
Ladies' Night (79/80) 13(1) [0|17|30]	*4823*
Forever (86) 25(1) [0|9|42]	4575
In The Heart (83) 29(2) [0|11|37]	4251
As One (82) 29(4) [0|12|24]	3203
Everything's Kool & The Gang: Greatest Hits & More (88) 109(2) [0|0|11]	366
KOOL MOE DEE ▶ 351 Pk:25 [0|2|3]	7641
How Ya Like Me Now (87) 35(1) [0|4|50]	3950
Knowledge Is King (89) 25(2) [0|8|23]	2756
Kool Moe Dee (87) 83(2) [0|0|21]	935
The KORGIS ▶ 1351 Pk:113 [0|0|1]	390
Dumb Waiters (80) 113(2) [0|0|12]	390
KRAFTWERK ▶ 680 Pk:72 [0|0|2]	2527
Computer-World (81) 72(1) [0|0|42]	2203
Electric Cafe (86) 156(2) [0|0|14]	324
Lenny KRAVITZ ▶ 1534 Pk:96 [0|0|1]	227
Let Love Rule (89/90) 96(1) [0|0|6]	*227*
Kris KRISTOFFERSON ▶ 1807 Pk:152 [0|0|1]	68.7
Music From Songwriter (Soundtrack) (84)-Willie Nelson & Kris Kristofferson [A] 152(2) [0|0|5]	68.7
KROKUS ▶ 280 Pk:25 [0|2|7]	10306
Headhunter (83) 25(2) [0|8|41]	3660
The Blitz (84) 31(1) [0|6|27]	2508
Change Of Address (86) 45(2) [0|0|17]	1463
One Vice At A Time (82) 53(2) [0|0|20]	1211
Heart Attack (88) 87(1) [0|0|11]	529
Hardware (81) 103(1) [0|0|12]	486
Alive And Screamin' (86) 97(2) [0|0|12]	449
KWAMÉ & A NEW BEGINNING ▶ 1194 Pk:114 [0|0|1]	566
The Boy Genius (89)-Kwamé Featuring A New Beginning 114(2) [0|0|18]	566
KWICK ▶ 1917 Pk:197 [0|0|1]	21.9
Kwick (80) 197(1) [0|0|1]	21.9

L

Act ▶ Rank Peak [Top10s|Top40s|Total] Title (Yr) Pk(Wks) [T10 Wks|T40Wks|TotWks]	Score in '80s
Patti LaBELLE ▶ 189 Pk:1 [1|2|6]	15725
Winner In You (86) 1(1) [12|19|30]	10266
I'm In Love Again (84) 40(2) [0|2|35]	2496
Patti (85) 72(3) [0|0|29]	1267
Be Yourself (89) 86(1) [0|0|24]	*1194*
Released (80) 114(1) [0|0|13]	398
The Spirit's In It (81) 156(3) [0|0|4]	104
L.A. BOPPERS ▶ 1195 Pk:85 [0|0|1]	560
L.A. Boppers (80) 85(1) [0|0|11]	560
LACE ▶ 1800 Pk:187 [0|0|1]	70.5
Shades of Lace (88) 187(2) [0|0|5]	70.5
L.A. DREAM TEAM ▶ 1450 Pk:138 [0|0|2]	286
Kings Of West Coast (86) 138(1) [0|0|7]	207
Bad To The Bone (87) 162(2) [0|0|4]	78.3
L.A. GUNS ▶ 604 Pk:50 [0|0|2]	3137
L.A. Guns (88) 50(1) [0|0|33]	2084
Cocked & Loaded (89/90) 57(2) [0|0|16]	*1054*
LAID BACK ▶ 1068 Pk:67 [0|0|1]	801
...Keep Smiling (84) 67(1) [0|0|15]	801
Cleo LAINE ▶ 1765 Pk:150 [0|0|1]	83.0
Sometimes When We Touch (80)-Cleo Laine & James Galway [A] 150(1) [0|0|4]	83.0
Greg LAKE ▶ 954 Pk:62 [0|0|1]	1139
Greg Lake (81) 62(4) [0|0|17]	1139
LAKESIDE ▶ 317 Pk:16 [0|1|6]	8435
Fantastic Voyage (80) 16(2) [0|12|35]	4485
Your Wish Is My Command (82) 58(1) [0|0|23]	1386
Untouchables (83) 42(2) [0|0|18]	1153
Outrageous (84) 68(2) [0|0|15]	795
Keep On Moving Straight Ahead (81) 109(4) [0|0|10]	432
Rough Riders (79/80) 158(1) [0|0|9]	*185*
Robin LANE & The CHARTBUSTERS ▶ 1779 Pk:172 [0|0|1]	77.4
Imitation Life (81) 172(1) [0|0|4]	77.4
K.D. LANG ▶ 621 Pk:73 [0|0|2]	2991
Absolute Torch And Twang (89/90)-k.d. lang & The Reclines 76(2) [0|0|29]	*1682*
Shadowland (88) 73(2) [0|0|25]	1309
David LANZ ▶ 1443 Pk:125 [0|0|2]	290
Natural States (88)-David Lanz & Paul Speer [A] 125(1) [0|0|10]	199
Cristofori's Dream (88) 180(1) [0|0|6]	91.2
LARSEN-FEITEN BAND ▶ 1497 Pk:142 [0|0|1]	253
Larsen-Feiten Band (80) 142(1) [0|0|10]	253
Nicolette LARSON ▶ 842 Pk:62 [0|0|3]	1593
Radioland (81) 62(1) [0|0|12]	702
All Dressed Up & No Place To Go (82) 75(1) [0|0|10]	511
In The Nick Of Time (79/80) 132(1) [0|0|12]	*380*
James LAST Band ▶ 1528 Pk:148 [0|0|1]	230
Seduction (80) 148(1) [0|0|8]	230
Stacy LATTISAW ▶ 420 Pk:44 [0|0|7]	5613
Let Me Be Your Angel (80) 44(1) [0|0|28]	2052
Sneakin' Out (82) 55(4) [0|0|16]	1315
With You (81) 46(1) [0|0|15]	1190
Take Me All The Way (86) 131(2) [0|0|22]	534
Personal Attention (88) 153(2) [0|0|10]	203
Sixteen (83) 160(1) [0|0|8]	196
Perfect Combination (84)-Stacy Lattisaw & Johnny Gill [A] 139(1) [0|0|8]	123
Cyndi LAUPER ▶ 67 Pk:4 [2|3|3]	35855
She's So Unusual (83) 4(4) [21|62|96]	24082
True Colors (86) 4(2) [6|23|44]	9826
A Night To Remember (89) 37(4) [0|4|21]	1947
Debra LAWS ▶ 902 Pk:70 [0|0|1]	1355
Very Special (81) 70(1) [0|0|27]	1355
Eloise LAWS ▶ 1691 Pk:175 [0|0|1]	122
Eloise Laws (81) 175(1) [0|0|7]	122
Hubert LAWS ▶ 1229 Pk:133 [0|0|2]	519
Family (80) 133(1) [0|0|13]	455
How To Beat The High Cost Of Living (Soundtrack) (80)-Hubert Laws And Earl Klugh [A] 134(2) [0|0|4]	63.5
Ronnie LAWS ▶ 542 Pk:24 [0|1|3]	3816
Every Generation (80) 24(3) [0|6|19]	2088
Solid Ground (81) 51(2) [0|0|19]	1269
Mr. Nice Guy (83) 98(1) [0|0|11]	458
LEATHERWOLF ▶ 1092 Pk:105 [0|0|2]	753
Leatherwolf (88) 105(1) [0|0|12]	465
Street Ready (89) 123(1) [0|0|8]	288
LED ZEPPELIN ▶ 311 Pk:6 [2|2|3]	8746
In Through The Out Door (79/80) 6(1) [1|11|24]	*4296*
Coda (82) 6(3) [7|9|16]	3926
Led Zeppelin IV (71/80) 136(2) [0|0|26]	*524*
Alvin LEE ▶ 1387 Pk:124 [0|0|2]	337
Detroit Diesel (86) 124(2) [0|0|9]	294
Free Fall (80) 198(1) [0|0|4]	42.4
Johnny LEE ▶ 1089 Pk:132 [0|0|2]	760
Lookin' For Love (80) 132(1) [0|0|21]	529
Bet Your Heart On Me (81) 147(1) [0|0|8]	231
John LENNON ▶ 415 Pk:31 [0|2|6]	5743
The John Lennon Collection (82) 33(4) [0|8|16]	1944
Imagine: John Lennon (Soundtrack) (88) 31(1) [0|5|18]	1801
Live In New York City (86) 41(2) [0|0|11]	939
Mind Games (73/81) 86(2) [0|0|13]	*684*
Walls And Bridges (74/81) 130(3) [0|0|8]	*255*
Menlove Avenue (86) 127(2) [0|0|4]	120

Acts With Albums

Act ▶ Rank Peak [Top10s|Top40s|Total] Title (Yr) Pk(Wks) [T10 Wks|T40Wks|TotWks]	Score in '80s
John LENNON & Yoko ONO ▶ 134 Pk:1 [1|2|3]	22966
Double Fantasy (80) 1(8) [22|27|74]	18984
Milk And Honey (84) 11(1) [0|10|19]	3525
Heart Play (84) 94(1) [0|0|12]	458
John LENNON/PLASTIC ONO BAND ▶ 742 Pk:57 [0|0|3]	2155
Shaved Fish (75/81) 57(1) [0|0|18]	*1089*
Imagine (71/81) 63(1) [0|0|15]	*907*
John Lennon/Plastic Ono Band (71/81) 156(1) [0|0|6]	*159*
Julian LENNON ▶ 267 Pk:17 [0|2|3]	10827
Valotte (84) 17(2) [0|28|46]	8315
The Secret Value Of Daydreaming (86) 32(2) [0|5|18]	1836
Mr. Jordan (89) 87(1) [0|0|15]	676
LEROI BROS. ▶ 1772 Pk:181 [0|0|1]	80.3
Open All Night (87) 181(2) [0|0|5]	80.3
Le ROUX ▶ 883 Pk:64 [0|0|2]	1431
Last Safe Place (82) 64(2) [0|0|21]	1253
Up (80) 145(2) [0|0|6]	178
LET'S ACTIVE ▶ 1012 Pk:111 [0|0|3]	961
Big Plans For Everybody (86) 111(2) [0|0|10]	365
Cypress (84) 138(1) [0|0|16]	338
Afoot (84) 154(1) [0|0|11]	258
LEVEL 42 ▶ 313 Pk:18 [0|2|3]	8663
World Machine (86) 18(2) [0|13|36]	4602
Running In The Family (87) 23(1) [0|10|34]	3829
Staring At The Sun (88) 128(1) [0|0|7]	233
LEVERT ▶ 514 Pk:32 [0|1|3]	4112
The Big Throwdown (87) 32(2) [0|7|24]	2432
Just Coolin' (88) 79(1) [0|0|31]	1641
Bloodline (86) 192(1) [0|0|3]	39.1
Huey LEWIS And The NEWS ▶ 17 Pk:1 [2|4|4]	68157
Sports (83) 1(1) [42|72|158]	37290
Fore! (86) 1(1) [26|41|61]	19862
Picture This (82) 13(4) [0|14|59]	6431
Small World (88) 11(2) [0|12|30]	4573
Jerry Lee LEWIS ▶ 1076 Pk:62 [0|0|2]	792
Great Balls Of Fire (Soundtrack) (89) 62(1) [0|0|10]	653
Class Of '55 (86)-Carl Perkins, Jerry Lee Lewis, Roy Orbison, & Johnny Cash [A] 87(2) [0|0|12]	139
Ramsey LEWIS ▶ 1328 Pk:144 [0|0|3]	411
Routes (80) 173(1) [0|0|8]	160
Three Piece Suite (81) 152(2) [0|0|5]	131
The Two Of Us (84)-Ramsey Lewis & Nancy Wilson [A] 144(1) [0|0|9]	120
Webster LEWIS ▶ 1423 Pk:114 [0|0|1]	306
8 For The 80's (80) 114(1) [0|0|9]	306
Gordon LIGHTFOOT ▶ 895 Pk:60 [0|0|4]	1395
Dream Street Rose (80) 60(1) [0|0|11]	672
Shadows (82) 87(2) [0|0|12]	498
East Of Midnight (86) 165(1) [0|0|6]	133
Salute (83) 175(1) [0|0|5]	92.4
LIMAHL ▶ 831 Pk:41 [0|0|1]	1654
Don't Suppose (85) 41(1) [0|0|20]	1654
David LINDLEY ▶ 1019 Pk:83 [0|0|2]	951
El Rayo-X (81) 83(1) [0|0|18]	845
Very Greasy (88)-David Lindley & El Rayo-X 174(1) [0|0|6]	106
LINX ▶ 1789 Pk:175 [0|0|1]	74.7
Intuition (81) 175(1) [0|0|4]	74.7
LIONS AND GHOSTS ▶ 1860 Pk:187 [0|0|1]	43.2
Velvet Kiss, Lick Of The Lime (87) 187(2) [0|0|3]	43.2
LIPPS, INC. ▶ 387 Pk:5 [1|1|2]	6801
Mouth To Mouth (80) 5(5) [7|13|26]	6210
Pucker Up (80) 63(1) [0|0|9]	592
LISA LISA And CULT JAM ▶ 207 Pk:7 [1|1|3]	14244
Spanish Fly (87) 7(3) [7|29|48]	10349
Lisa Lisa & Cult Jam With Full Force (85)-Lisa Lisa And Cult Jam With Full Force 52(2) [0|0|66]	3261
Straight To The Sky (89) 77(3) [0|0|13]	634
Rich LITTLE ▶ 832 Pk:29 [0|1|1]	1651
The First Family Rides Again (82) 29(2) [0|6|13]	1651
LITTLE AMERICA ▶ 1217 Pk:102 [0|0|1]	532
Little America (87) 102(2) [0|0|14]	532
LITTLE FEAT ▶ 446 Pk:29 [0|3|3]	5124
Let It Roll (88) 36(1) [0|4|33]	2457
Down On The Farm (80) 29(2) [0|4|17]	*1459*
Hoy-Hoy! (81) 39(1) [0|2|13]	1208
LITTLE RIVER BAND ▶ 258 Pk:21 [0|2|6]	11295
Time Exposure (81) 21(1) [0|9|50]	4553
Greatest Hits (82) 33(6) [0|12|30]	3166
The Net (83) 61(1) [0|0|21]	1269
Backstage Pass (80) 44(1) [0|0|10]	919
Playing To Win (85)-LRB 75(3) [0|0|14]	783
First Under The Wire (79/80) 82(1) [0|0|11]	*606*
LITTLE STEVEN ▶ 692 Pk:55 [0|0|3]	2461
Voice Of America (84) 55(2) [0|0|17]	1152
Men Without Women (82)-Little Steven And The Disciples Of Soul 118(3) [0|0|18]	696
Freedom No Compromise (87) 80(2) [0|0|12]	613
LIVING COLOUR ▶ 249 Pk:6 [1|1|1]	11735
Vivid (88) 6(2) [8|28|70]	*11735*
LIVING IN A BOX ▶ 1247 Pk:89 [0|0|1]	499
Living In A Box (87) 89(2) [0|0|13]	499
LIZZY BORDEN ▶ 1050 Pk:133 [0|0|4]	835
Master Of Disguise (89) 133(1) [0|0|10]	330
Menace To Society (86) 144(2) [0|0|10]	264
Visual Lies (87) 146(1) [0|0|7]	167
Terror Rising (87) 188(1) [0|0|6]	73.7
LL COOL J ▶ 153 Pk:3 [2|2|3]	19530
Bigger And Deffer (87) 3(1) [13|23|53]	11492
Walking With A Panther (89) 6(1) [5|14|21]	5284
Radio (86) 46(3) [0|0|38]	2754
Nils LOFGREN ▶ 1177 Pk:99 [0|0|2]	586
Night Fades Away (81) 99(1) [0|0|11]	439
Flip (85) 150(2) [0|0|5]	147
Kenny LOGGINS ▶ 158 Pk:11 [0|3|5]	18744
High Adventure (82) 13(3) [0|9|44]	6003
Keep The Fire (80) 16(2) [0|14|32]	*4840*
Kenny Loggins Alive (80) 11(2) [0|11|31]	4781
Vox Humana (85) 41(1) [0|0|31]	2343
Back To Avalon (88) 69(2) [0|0|14]	777
LONDON SYMPHONY Orchestra ▶ 921 Pk:62 [0|0|3]	1284
Raiders Of The Lost Ark (Soundtrack) (81)-The London Symphony Orchestra/John Williams 62(1) [0|0|13]	691
A Classic Case: The Music Of Jethro Tull (86)-The London Symphony Orchestra/Ian Anderson 93(2) [0|0|12]	502
Hooked On Rock Classics (83) 145(2) [0|0|3]	91.3
LONE JUSTICE ▶ 571 Pk:56 [0|0|2]	3519
Lone Justice (85) 56(1) [0|0|25]	1762
Shelter (86) 65(2) [0|0|30]	1757
LOOSE ENDS ▶ 635 Pk:46 [0|0|3]	2904
A Little Spice (85) 46(2) [0|0|19]	1272
The Zagora (87) 59(2) [0|0|14]	828
The Real Chuckeeboo (88) 80(3) [0|0|15]	803
Denise LOPEZ ▶ 1822 Pk:184 [0|0|1]	61.2
Truth In Disguise (88) 184(1) [0|0|4]	61.2
Jeff LORBER ▶ 512 Pk:68 [0|0|6]	4139
Private Passion (86) 68(2) [0|0|17]	1419
Galaxian (81)-Jeff Lorber Fusion 77(1) [0|0|15]	729
Step By Step (85) 90(2) [0|0|16]	681
It's A Fact (82) 73(2) [0|0|13]	667
Wizard Island (80)-Jeff Lorber Fusion 123(1) [0|0|12]	339
In The Heat Of The Night (84) 106(1) [0|0|7]	304
Bill LORDAN ▶ 1238 Pk:37 [0|1|1]	509
B.L.T. (81)-Jack Bruce/Bill Lordan/Robin Trower [A] 37(1) [0|3|16]	509
LORDS Of The NEW CHURCH ▶ 1640 Pk:158 [0|0|1]	150
The Method To Our Madness (85) 158(1) [0|0|7]	150
Gloria LORING ▶ 1064 Pk:61 [0|0|1]	808
Gloria Loring (86) 61(2) [0|0|14]	808
Los LOBOS ▶ 198 Pk:1 [1|1|4]	15145
La Bamba (Soundtrack) (87) 1(2) [11|19|44]	10525
How Will The Wolf Survive (84) 47(3) [0|0|34]	2460
By The Light Of The Moon (87) 47(2) [0|0|32]	2085
La Pistola Y El Corazon (88) 179(2) [0|0|4]	74.4
LOUDNESS ▶ 746 Pk:64 [0|0|3]	2119
Thunder In The East (85) 74(2) [0|0|24]	1215
Lightning Strikes (86) 64(2) [0|0|16]	852
Hurricane Eyes (87) 190(1) [0|0|4]	52.2
LOVE AND MONEY ▶ 1696 Pk:175 [0|0|1]	121
Strange Kind Of Love (89) 175(2) [0|0|7]	121
LOVE AND ROCKETS ▶ 331 Pk:14 [0|1|3]	8122
Love And Rockets (89) 14(1) [0|17|26]	4950
Earth - Sun - Moon (87) 64(2) [0|0|28]	1649
Express (86) 72(2) [0|0|30]	1522
LOVERBOY ▶ 38 Pk:7 [2|4|6]	47081
Get Lucky (81) 7(2) [14|51|122]	21025
Loverboy (81) 13(2) [0|17|105]	8447
Keep It Up (83) 7(5) [7|22|39]	8232
Lovin' Every Minute Of It (85) 13(5) [0|22|44]	7660
Wildside (87) 42(1) [0|0|21]	1703
Big Ones (89) 189(1) [0|0|1]	15.0
Lyle LOVETT ▶ 791 Pk:62 [0|0|2]	1838
Lyle Lovett And His Large Band (89) 62(1) [0|0|21]	1352
Pontiac (88) 117(1) [0|0|14]	486
Lene LOVICH ▶ 1341 Pk:94 [0|0|2]	399
Flex (89) 94(1) [0|0|3]	341
No Man's Land (83) 188(1) [0|0|4]	57.6
Nick LOWE ▶ 942 Pk:50 [0|0|2]	1179
Nick The Knife (82) 50(1) [0|0|14]	973
The Abominable Showman (83) 129(2) [0|0|7]	206
Nick LOWE And His COWBOY OUTFIT ▶ 1044 Pk:113 [0|0|2]	845
Nick Lowe And His Cowboy Outfit (84) 113(1) [0|0|12]	429
The Rose Of England (85) 119(3) [0|0|7]	416
L.T.D. ▶ 575 Pk:28 [0|1|2]	3475
Shine On (80) 28(1) [0|8|28]	2798
Love Magic (81) 83(1) [0|0|7]	677
L'TRIMM ▶ 1300 Pk:132 [0|0|1]	435
Grab It! (88) 132(2) [0|0|16]	435
Carrie LUCAS ▶ 1730 Pk:180 [0|0|2]	97.5
Still In Love (82) 180(1) [0|0|3]	52.1
Portrait Of Carrie (81) 185(1) [0|0|3]	45.4
LULU ▶ 1421 Pk:126 [0|0|1]	308
Lulu (81) 126(1) [0|0|10]	308
Ray LYNCH ▶ 1915 Pk:197 [0|0|1]	22.8
No Blue Thing (89) 197(2) [0|0|2]	22.8
Cheryl LYNN ▶ 907 Pk:104 [0|0|4]	1342
Instant Love (82) 133(2) [0|0|20]	605
In The Night (81) 104(3) [0|0|13]	563
Preppie (84) 161(1) [0|0|5]	93.4
In Love (80) 167(1) [0|0|4]	80.6
LYNYRD SKYNYRD ▶ 361 Pk:12 [0|1|7]	7433
Gold & Platinum (80) 12(2) [0|11|62]	*5205*
Legend (87) 41(2) [0|0|17]	1130
Southern By The Grace Of God/ Lynyrd Skynyrd Tribute Tour 1987 (88) 68(2) [0|0|11]	560
One More From The Road (76/80) 108(1) [0|0|6]	*273*
The Best Of The Rest (82) 171(5) [0|0|7]	148
Street Survivors (77/80) 166(1) [0|0|3]	*73.3*
Lynyrd Skynyrd (pronounced leh-nerd skin-nerd) (75/80) 190(3) [0|0|3]	*43.5*

343

Acts With Albums

Act ▶ Rank Peak [Top10s|Top40s|Total] Title (Yr) Pk(Wks) [T10 Wks|T40Wks|TotWks]	Score in '80s

M

Act / Title	Score
M ▶ 1428 Pk:79 [0|0|1]	**304**
New York-London-Paris-Munich (80) 79(2) [0|0|6]	304
Tony MacALPINE ▶ 1475 Pk:146 [0|0|1]	**267**
Maximum Security (87) 146(2) [0|0|11]	267
MAC Band ▶ 1298 Pk:109 [0|0|1]	**436**
Mac Band (88)-Mac Band Featuring The McCampbell Brothers 109(2) [0|0|14]	436
Ralph MacDONALD ▶ 1336 Pk:108 [0|0|1]	**404**
Universal Rhythm (84) 108(2) [0|0|10]	404
Lonnie MACK ▶ 1209 Pk:130 [0|0|1]	**541**
Strike Like Lightning (85) 130(2) [0|0|21]	541
MADAME X ▶ 1719 Pk:162 [0|0|1]	**106**
Madame X (87) 162(2) [0|0|5]	106
MADHOUSE ▶ 1320 Pk:107 [0|0|1]	**417**
8 (87) 107(2) [0|0|11]	417
MADNESS ▶ 618 Pk:41 [0|0|4]	**3014**
Madness (83) 41(1) [0|0|29]	2277
Keep Moving (84) 109(1) [0|0|8]	340
One Step Beyond (80) 128(1) [0|0|9]	277
Absolutely (80) 146(2) [0|0|4]	120
MADONNA ▶ 4 Pk:1 [5|6|6]	**93686**
Like A Virgin (84) 1(3) [33|52|108]	27162
True Blue (86) 1(5) [25|52|82]	23653
Like A Prayer (89) 1(6) [16|31|39]	*16977*
Madonna (83) 8(3) [5|36|168]	16911
Who's That Girl (Soundtrack) (87) 7(2) [3|13|28]	5181
You Can Dance (87) 14(1) [0|12|22]	3802
MAHOGANY RUSH ▶ 1264 Pk:88 [0|0|2]	**476**
What's Next (80)-Frank Marino And Mahogany Rush 88(1) [0|0|4]	419
Juggernaut (82)-Frank Marino 185(1) [0|0|4]	56.8
MALICE ▶ 1753 Pk:177 [0|0|1]	**88.5**
License To Kill (87) 177(2) [0|0|6]	88.5
Yngwie MALMSTEEN ▶ 375 Pk:40 [0|1|5]	**7128**
Marching Out (85)-Yngwie J. Malmsteen's Rising Force 52(2) [0|0|28]	1866
Rising Force (85) 60(3) [0|0|43]	1857
Trilogy (86) 44(2) [0|0|23]	1663
Odyssey (88)- Yngwie J. Malmsteen's Rising Force 40(2) [0|2|18]	1559
Trial By Fire: Live In Leningrad (89) 128(2) [0|0|7]	*183*
MAMA'S BOYS ▶ 1438 Pk:151 [0|0|2]	**292**
Power & Passion (85) 151(2) [0|0|6]	163
Mama's Boys (84) 172(1) [0|0|8]	129
Melissa MANCHESTER ▶ 378 Pk:19 [0|1|6]	**7025**
Hey Ricky (82) 19(5) [0|8|39]	4048
Greatest Hits (83) 43(3) [0|0|21]	1587
For The Working Girl (80) 68(3) [0|0|11]	733
Emergency (83) 135(2) [0|0|9]	309
Melissa Manchester (*79/80*) 174(1) [0|0|12]	*189*
Mathematics (85) 144(2) [0|0|6]	159
Henry MANCINI and his Orchestra ▶ 1923 Pk:197 [0|0|1]	**11.4**
The Hollywood Musicals (87)-Johnny Mathis & Henry Mancini [A] 197(2) [0|0|2]	11.4
Howie MANDEL ▶ 1633 Pk:148 [0|0|1]	**153**
Fits Like A Glove (86) 148(2) [0|0|6]	153
Barbara MANDRELL ▶ 795 Pk:86 [0|0|5]	**1818**
Barbara Mandrell Live (81) 86(2) [0|0|24]	1154
Meant For Each Other (84)-Barbara Mandrell & Lee Greenwood [A] 89(1) [0|0|13]	301
In Black And White (82) 153(2) [0|0|11]	147
Spun Gold (83) 140(1) [0|0|4]	124
Love Is Fair (80) 175(1) [0|0|4]	91.6

Act / Title	Score
MANFRED MANN'S EARTH BAND ▶ 705 Pk:40 [0|1|2]	**2377**
Somewhere In Afrika (84) 40(2) [0|2|21]	1607
Chance (81) 87(2) [0|0|16]	771
Chuck MANGIONE ▶ 402 Pk:8 [1|1|6]	**6194**
Fun And Games (80) 8(4) [5|11|23]	4123
Tarantella (81) 55(2) [0|0|15]	977
Love Notes (82) 83(2) [0|0|10]	481
Feels So Good (*78/80*) 140(3) [0|0|8]	*236*
Disguise (84) 148(2) [0|0|8]	212
Journey To A Rainbow (83) 154(1) [0|0|7]	166
The MANHATTANS ▶ 469 Pk:24 [0|1|5]	**4619**
After Midnight (80) 24(2) [0|10|26]	3199
Black Tie (81) 86(3) [0|0|10]	501
Manhattans Greatest Hits (80) 87(2) [0|0|10]	466
Forever By Your Side (83) 104(1) [0|0|8]	338
Too Hot To Stop It (85) 171(2) [0|0|6]	115
MANHATTAN TRANSFER ▶ 284 Pk:22 [0|1|8]	**10209**
Mecca For Moderns (81) 22(4) [0|13|27]	3843
Extensions (80) 55(1) [0|0|33]	*1759*
Vocalese (85) 74(2) [0|0|40]	1671
Bodies And Souls (83) 52(2) [0|0|27]	1188
Brasil (87) 96(1) [0|0|19]	882
The Best Of Manhattan Transfer (81) 103(2) [0|0|11]	485
Bop Doo-Wop (85) 127(2) [0|0|11]	338
The Manhattan Transfer Live (87) 187(2) [0|0|3]	45.0
Barry MANILOW ▶ 141 Pk:14 [0|6|12]	**21986**
If I Should Love Again (81) 14(3) [0|11|25]	4301
Barry (80) 15(2) [0|9|20]	3197
Here Comes The Night (82) 32(4) [0|6|27]	2621
Greatest Hits-Vol. II (83) 30(2) [0|7|19]	2309
2:00 A.M. Paradise Cafe (84) 28(2) [0|6|20]	2220
Manilow (85) 42(3) [0|0|24]	1756
One Voice (*79/80*) 21(1) [0|4|14]	*1738*
Swing Street (87) 70(1) [0|0|21]	1287
Barry Manilow (89) 64(2) [0|0|16]	862
Greatest Hits (*79/80*) 131(1) [0|0|27]	*750*
Oh, Julie! (82) 69(2) [0|0|9]	478
The Manilow Collection/Twenty Classic Hits (85) 100(2) [0|0|12]	467
--**MANN, Manfred** see: **MANFRED MANN**	
MANNHEIM STEAMROLLER ▶ 540 Pk:36 [0|1|4]	**3829**
Mannheim Steamroller Christmas (84) 50(1) [0|0|29]	*1693*
A Fresh Aire Christmas (88) 36(3) [0|4|13]	*1551*
Classical Gas (87)-Mason Williams & Mannheim Steamroller [A] 118(1) [0|0|19]	325
Fresh Aire VI (86) 155(1) [0|0|14]	260
MANTRONIX ▶ 1382 Pk:108 [0|0|1]	**343**
In Full Effect (88) 108(2) [0|0|8]	343
Benny MARDONES ▶ 987 Pk:65 [0|0|1]	**1025**
Never Run Never Hide (80) 65(1) [0|0|24]	1025
Teena MARIE ▶ 230 Pk:23 [0|3|7]	**12745**
Starchild (84) 31(2) [0|9|35]	3429
It Must Be Magic (81) 23(1) [0|9|25]	3213
Irons In The Fire (80) 38(3) [0|3|29]	2642
Lady T (80) 45(1) [0|0|23]	1407
Robbery (83) 119(1) [0|0|24]	780
Naked To The World (88) 65(2) [0|0|13]	707
Emerald City (86) 81(2) [0|0|11]	567
MARILLION ▶ 605 Pk:47 [0|0|4]	**3134**
Misplaced Childhood (85) 47(2) [0|0|35]	2026
Brief Encounter (86) 67(2) [0|0|10]	568
Clutching At Straws (87) 103(2) [0|0|11]	420
Script For A Jester's Tear (83) 175(1) [0|0|7]	121
Bob MARLEY And The WAILERS ▶ 396 Pk:45 [0|0|6]	**6416**
Legend (84) 54(2) [0|0|59]	*2826*

Act / Title	Score
Uprising (80) 45(1) [0|0|23]	1852
Confrontation (83) 55(1) [0|0|15]	968
Survival (*79/80*) 85(1) [0|0|7]	*300*
Chances Are (81)-Bob Marley 117(1) [0|0|6]	253
Rebel Music (86) 140(1) [0|0|9]	218
Ziggy MARLEY & The MELODY MAKERS ▶ 365 Pk:23 [0|2|2]	**7373**
Conscious Party (88) 23(1) [0|16|42]	5137
One Bright Day (89) 26(2) [0|8|18]	2236
MARLEY MARL ▶ 1698 Pk:163 [0|0|1]	**121**
In Control Volume I (88) 163(2) [0|0|5]	121
Neville MARRINER ▶ 577 Pk:56 [0|0|1]	**3455**
Amadeus (Soundtrack) (84) 56(2) [0|0|78]	3455
Branford MARSALIS ▶ 1661 Pk:164 [0|0|1]	**135**
Scenes In The City (84) 164(2) [0|0|7]	135
Wynton MARSALIS ▶ 593 Pk:90 [0|0|6]	**3306**
Hot House Flowers (84) 90(2) [0|0|39]	1655
Think Of One (83) 102(1) [0|0|29]	1034
Black Codes (From The Underground) (85) 118(2) [0|0|10]	338
Marsalis Standard Time: Vol. 1 (87) 153(1) [0|0|5]	113
Wynton Marsalis (82) 165(1) [0|0|5]	110
J Mood (86) 185(2) [0|0|4]	55.4
MARSHALL TUCKER Band ▶ 657 Pk:32 [0|1|4]	**2719**
Tenth (80) 32(1) [0|5|15]	1497
Dedicated (81) 53(2) [0|0|12]	809
Tuckerized (82) 95(2) [0|0|7]	298
Greatest Hits (*78/81*) 158(1) [0|0|5]	*115*
--**MARTHA And The MUFFINS** see: **M + M**	
M + M ▶ 1555 Pk:163 [0|0|3]	**203**
Mystery Walk (84) 163(2) [0|0|7]	95.0
Danseparc (83)-Martha And The Muffins 184(1) [0|0|4]	60.8
Metro Music (80)-Martha And The Muffins 186(1) [0|0|3]	47.1
MARTIKA ▶ 465 Pk:15 [0|1|1]	**4681**
Martika (89) 15(1) [0|12|39]	4681
Eric MARTIN ▶ 1904 Pk:191 [0|0|1]	**26.4**
Sucker For A Pretty Face (83)-Eric Martin Band 191(1) [0|0|2]	26.4
Marilyn MARTIN ▶ 1170 Pk:72 [0|0|1]	**591**
Marilyn Martin (86) 72(2) [0|0|11]	591
Moon MARTIN ▶ 1272 Pk:138 [0|0|1]	**469**
Street Fever (80) 138(3) [0|0|15]	469
Steve MARTIN ▶ 1190 Pk:83 [0|0|2]	**569**
Comedy Is Not Pretty! (*79/80*) 83(1) [0|0|9]	*427*
The Steve Martin Brothers (81) 135(2) [0|0|4]	142
Nancy MARTINEZ ▶ 1853 Pk:178 [0|0|1]	**47.1**
Not Just The Girl Next Door (87) 178(1) [0|0|3]	47.1
Richard MARX ▶ 85 Pk:1 [2|2|2]	**31468**
Richard Marx (87) 8(2) [4|63|86]	17222
Repeat Offender (89) 1(1) [12|32|33]	*14247*
MARY JANE GIRLS ▶ 382 Pk:18 [0|1|2]	**6952**
Only Four You (85) 18(2) [0|11|38]	4708
Mary Jane Girls (83) 56(1) [0|0|41]	2244
Dave MASON ▶ 1252 Pk:74 [0|0|1]	**496**
Old Crest On A New Wave (80) 74(2) [0|0|10]	496
Harvey MASON ▶ 1854 Pk:186 [0|0|1]	**46.3**
M.V.P. (81) 186(1) [0|0|3]	46.3
Jackie MASON ▶ 1545 Pk:146 [0|0|1]	**216**
The World According To Me (88) 146(1) [0|0|9]	216
Nick MASON ▶ 1693 Pk:154 [0|0|2]	**121**
Nick Mason's Fictitious Sports (81) 170(2) [0|0|3]	66.0
Profiles (85)-Nick Mason & Rick Fenn [A] 154(2) [0|0|3]	55.4
MASS PRODUCTION ▶ 1323 Pk:133 [0|0|2]	**414**
Massterpiece (80) 133(1) [0|0|9]	290
Turn Up The Music (81) 166(2) [0|0|4]	125

Acts With Albums

Act ▶ Rank Peak [Top10s|Top40s|Total] Title (Yr) Pk(Wks) [T10 Wks|T40Wks|TotWks]	Score in '80s
Johnny MATHIS ▶ 1004 Pk:140 [0|0|6]	**978**
A Special Part Of Me (84) 157(1) [0|0|19]	377
Friends In Love (82) 147(1) [0|0|9]	228
The Best Of Johnny Mathis 1975-1980 (80) 140(2) [0|0|7]	169
Different Kinda Different (80) 164(1) [0|0|5]	109
The First 25 Years-The Silver Anniversary Album (81) 173(2) [0|0|4]	82.5
The Hollywood Musicals (87)-Johnny Mathis & Henry Mancini [A] 197(2) [0|0|2]	11.4
MAX Q ▶ 1687 Pk:182 [0|0|1]	**123**
Max Q (89) 182(2) [0|0|8]	123
Brian MAY ▶ 1415 Pk:125 [0|0|1]	**312**
Star Fleet Project (83)-Brian May And Friends 125(1) [0|0|9]	312
Curtis MAYFIELD ▶ 1404 Pk:128 [0|0|2]	**325**
Something To Believe In (80) 128(2) [0|0|10]	293
The Right Combination (80)-Linda Clifford & Curtis Mayfield [A] 180(2) [0|0|4]	32.0
Lyle MAYS ▶ 1107 Pk:50 [0|0|1]	**714**
As Falls Wichita, So Falls Wichita Falls (81)- Pat Metheny & Lyle Mays [A] 50(1) [0|0|21]	714
MAZARATI ▶ 1532 Pk:133 [0|0|1]	**230**
Mazarati (86) 133(1) [0|0|8]	230
MAZE Featuring Frankie BEVERLY ▶ 278 Pk:25 [0|4|6]	**10339**
We Are One (83) 25(3) [0|5|26]	2413
Live In New Orleans (81) 34(2) [0|4|27]	2254
Joy And Pain (80) 31(1) [0|4|23]	2038
Can't Stop The Love (85) 45(2) [0|0|30]	1765
Silky Soul (89) 37(2) [0|3|15]	1461
Live In Los Angeles (86) 92(2) [0|0|11]	408
MC HAMMER ▶ 398 Pk:30 [0|1|1]	**6372**
Let's Get It Started (88) 30(1) [0|23|57]	6372
MC LYTE ▶ 1240 Pk:86 [0|0|1]	**508**
Eyes On This (89) 86(2) [0|0|11]	508
MC SHY D ▶ 1924 Pk:197 [0|0|1]	**11.4**
Got To Be Tough (87)-M.C. Shy D 197(1) [0|0|1]	11.4
Paul McCARTNEY ▶ 99 Pk:1 [2|6|8]	**28257**
Tug Of War (82) 1(3) [9|18|29]	9912
McCartney II (80) 3(5) [7|12|19]	5644
Pipes Of Peace (83) 15(1) [0|13|24]	4263
Flowers In The Dirt (89) 21(3) [0|6|28]	2811
Give My Regards To Broad Street (84) 21(2) [0|10|18]	2563
Press To Play (86) 30(2) [0|5|22]	1897
All The Best! (87) 62(3) [0|0|17]	1086
The McCartney Interview (81) 158(2) [0|0|3]	81.8
Delbert McCLINTON ▶ 734 Pk:34 [0|1|2]	**2186**
The Jealous Kind (80) 34(1) [0|4|28]	2063
Plain From The Heart (81) 181(2) [0|0|9]	123
Ian McCULLOCH ▶ 1919 Pk:179 [0|0|1]	**19.5**
Candleland (89) 179(1) [0|0|1]	19.5
Michael McDONALD ▶ 393 Pk:6 [1|1|2]	**6598**
If That's What It Takes (82) 6(6) [6|11|32]	5377
No Lookin' Back (85) 45(2) [0|0|15]	1221
Reba McENTIRE ▶ 663 Pk:78 [0|0|5]	**2660**
Sweet Sixteen (89) 78(1) [0|0|18]	822
The Last One To Know (87) 102(2) [0|0|20]	715
Greatest Hits (87) 139(2) [0|0|23]	524
Reba (88) 118(2) [0|0|10]	335
Live (89) 124(3) [0|0|8]	265
McFADDEN & WHITEHEAD ▶ 1629 Pk:153 [0|0|1]	**156**
I Heard It In A Love Song (80) 153(1) [0|0|6]	156
Bobby McFERRIN ▶ 304 Pk:5 [1|1|2]	**8971**
Simple Pleasures (88) 5(3) [7|16|55]	8243
Spontaneous Inventions (87) 103(1) [0|0|19]	729
McGUFFEY LANE ▶ 1803 Pk:193 [0|0|1]	**70.0**
Aqua Dream (82) 193(1) [0|0|6]	70.0

Act ▶ Rank Peak [Top10s|Top40s|Total] Title (Yr) Pk(Wks) [T10 Wks|T40Wks|TotWks]	Score in '80s
McGUINN, CLARK & HILLMAN ▶ 1537 Pk:136 [0|0|1]	**223**
City (80)-McGuinn & Chris Hillman Featuring Gene Clark 136(2) [0|0|7]	223
Maria McKEE ▶ 1213 Pk:120 [0|0|1]	**538**
Maria McKee (89) 120(2) [0|0|15]	538
Bob & Doug McKENZIE ▶ 480 Pk:8 [1|1|1]	**4487**
Great White North (82) 8(2) [4|13|21]	4487
Sarah McLACHLAN ▶ 1442 Pk:132 [0|0|1]	**291**
Touch (89) 132(1) [0|0|23]	291
Ian McLAGAN ▶ 1487 Pk:125 [0|0|1]	**261**
Troublemaker (80) 125(1) [0|0|9]	261
Malcolm McLAREN ▶ 1577 Pk:173 [0|0|2]	**190**
D'ya Like Scratchin' (84) 173(2) [0|0|6]	111
Fans (85) 190(2) [0|0|6]	79.1
John McLAUGHLIN ▶ 1758 Pk:172 [0|0|1]	**85.9**
Belo Horizonte (81) 172(3) [0|0|4]	85.9
Pat McLAUGHLIN ▶ 1922 Pk:195 [0|0|1]	**12.3**
Pat McLaughlin (88) 195(1) [0|0|1]	12.3
Don McLEAN ▶ 660 Pk:28 [0|1|2]	**2694**
Chain Lightning (81) 28(2) [0|6|21]	2405
Believers (81) 156(1) [0|0|11]	290
James McMURTRY ▶ 1449 Pk:125 [0|0|1]	**289**
Too Long In The Wasteland (89) 125(1) [0|0|9]	289
Christine McVIE ▶ 684 Pk:26 [0|1|1]	**2497**
Christine McVie (84) 26(3) [0|7|23]	2497
MEAT LOAF ▶ 873 Pk:45 [0|0|2]	**1471**
Dead Ringer (81) 45(2) [0|0|11]	943
Bad Attitude (85) 74(1) [0|0|10]	528
MECO ▶ 957 Pk:61 [0|0|3]	**1127**
Pop Goes The Movies (82) 68(2) [0|0|9]	562
Christmas In The Stars/Star Wars Christmas Album (80) 61(3) [0|0|6]	399
Meco Plays Music From The Empire Strikes Back (80) 140(1) [0|0|8]	166
Glenn MEDEIROS ▶ 1128 Pk:83 [0|0|1]	**680**
Glenn Medeiros (87) 83(2) [0|0|17]	680
MEGADETH ▶ 478 Pk:28 [0|1|2]	**4527**
Peace Sells...But Who's Buying? (86) 76(1) [0|0|47]	2277
So Far, So Good... So What! (88) 28(1) [0|6|23]	2251
Randy MEISNER ▶ 665 Pk:50 [0|0|2]	**2642**
One More Song (80) 50(1) [0|0|33]	2184
Randy Meisner (82) 94(3) [0|0|11]	458
John MELLENCAMP ▶ 7 Pk:1 [5|6|7]	**85568**
American Fool (82)-John Cougar 1(9) [22|40|106]	23754
Scarecrow (85) 2(3) [29|48|75]	23177
The Lonesome Jubilee (87) 6(7) [27|37|53]	15806
Uh-Huh (83) 9(1) [2|36|66]	13240
Big Daddy (89) 7(1) [4|15|23]	5057
Nothin' Matters And What If It Did (80)- John Cougar 37(2) [0|3|55]	4043
John Cougar (80)-John Cougar 64(2) [0|0|9]	492
Harold MELVIN And The BLUE NOTES ▶ 1074 Pk:95 [0|0|1]	**793**
The Blue Album (80) 95(2) [0|0|20]	793
MEN AT WORK ▶ 43 Pk:1 [2|2|3]	**44027**
Business As Usual (82) 1(15) [31|48|90]	31295
Cargo (83) 3(5) [14|23|49]	11764
Two Hearts (85) 50(2) [0|0|13]	967
Sergio MENDES ▶ 504 Pk:27 [0|1|2]	**4188**
Sergio Mendes(2) (83) 27(2) [0|10|27]	3083
Confetti (84) 70(2) [0|0|22]	1105
MENUDO ▶ 961 Pk:100 [0|0|2]	**1121**
Menudo (85) 100(2) [0|0|19]	735
Reaching Out (84) 108(2) [0|0|12]	386
MEN WITHOUT HATS ▶ 403 Pk:13 [0|1|3]	**6183**
Rhythm Of Youth (83) 13(3) [0|14|26]	4614
Pop Goes The World (87) 73(2) [0|0|25]	1423
Folk Of The 80's (Part III) (84) 127(2) [0|0|4]	146

Act ▶ Rank Peak [Top10s|Top40s|Total] Title (Yr) Pk(Wks) [T10 Wks|T40Wks|TotWks]	Score in '80s
Freddie MERCURY ▶ 1664 Pk:159 [0|0|1]	**134**
Mr. Bad Guy (85) 159(1) [0|0|6]	134
Jim MESSINA ▶ 1151 Pk:95 [0|0|2]	**623**
Messina (81) 95(3) [0|0|11]	506
Oasis (79/80)-Jimmy Messina 119(1) [0|0|3]	117
METAL CHURCH ▶ 764 Pk:75 [0|0|2]	**1986**
The Dark (86) 92(1) [0|0|23]	1130
Blessing In Disguise (89) 75(1) [0|0|15]	856
METALLICA ▶ 149 Pk:6 [1|3|6]	**20203**
...And Justice For All (88) 6(2) [5|34|67]	11415
Master Of Puppets (86) 29(2) [0|7|72]	3948
The $5.98 E.P.: Garage Days Re-Revisited (87) 28(2) [0|8|30]	2896
Ride The Lightning (84) 100(2) [0|0|50]	1422
Kill 'Em All(2) (88) 120(2) [0|0|8]	300
Kill 'Em All (86) 155(2) [0|0|10]	222
Pat METHENY ▶ 1108 Pk:50 [0|0|1]	**714**
As Falls Wichita, So Falls Wichita Falls (81)- Pat Metheny & Lyle Mays [A] 50(1) [0|0|21]	714
PAT METHENY GROUP ▶ 312 Pk:50 [0|0|9]	**8680**
Offramp (82) 50(2) [0|0|28]	1603
First Circle (84) 91(2) [0|0|35]	1347
American Garage (80) 53(1) [0|0|18]	1287
Letter From Home (89) 66(2) [0|0|18]	1146
Travels (83) 62(1) [0|0|17]	937
The Falcon & The Snowman (Soundtrack) (85) 54(2) [0|0|10]	793
Still Life (Talking) (87) 86(2) [0|0|15]	650
80/81 (80)-Pat Metheny 89(1) [0|0|14]	609
Rejoicing (84)-Pat Metheny 116(2) [0|0|9]	308
--George MICHAEL see also: WHAM!	
George MICHAEL ▶ 51 Pk:1 [1|1|1]	**40818**
Faith (87) 1(12) [51|69|87]	40818
Bette MIDLER ▶ 131 Pk:2 [1|3|6]	**23242**
Beaches (Soundtrack) (89) 2(3) [11|27|50]	12751
The Rose (Soundtrack) (80) 12(3) [0|23|43]	7984
Divine Madness (Soundtrack) (80) 34(3) [0|4|14]	1490
No Frills (83) 60(1) [0|0|13]	880
Mud Will Be Flung Tonight! (85) 183(1) [0|0|6]	102
Thighs And Whispers (79/80) 169(1) [0|0|2]	35.8
MIDNIGHT OIL ▶ 343 Pk:21 [0|1|3]	**7790**
Diesel And Dust (88) 21(3) [0|26|55]	7626
Red Sails In The Sunset (85) 177(2) [0|0|6]	94.4
10,9,8,7,6,5,4,3,2,1 (84) 178(1) [0|0|5]	69.9
MIDNIGHT STAR ▶ 202 Pk:27 [0|2|4]	**15073**
No Parking On The Dance Floor (83) 27(2) [0|30|96]	9903
Planetary Invasion (84) 32(2) [0|7|32]	3093
Headlines (86) 56(2) [0|0|27]	1529
Midnight Star (88) 96(2) [0|0|15]	548
MIKE + THE MECHANICS ▶ 245 Pk:13 [0|2|2]	**11829**
Mike + The Mechanics (85) 26(6) [0|21|53]	6790
Living Years (88) 13(2) [0|13|37]	5040
Frankie MILLER ▶ 1500 Pk:135 [0|0|1]	**251**
Standing On The Edge (82) 135(2) [0|0|9]	251
Steve MILLER Band ▶ 210 Pk:3 [1|2|6]	**14184**
Abracadabra (82) 3(6) [12|18|33]	9507
Circle Of Love (81) 26(2) [0|5|17]	2118
Living In The 20th Century (86) 65(1) [0|0|23]	1599
Italian X Rays (84) 101(2) [0|0|10]	376
Born 2B Blue (88)-Steve Miller 108(2) [0|0|10]	367
Steve Miller Band - Live (83) 125(1) [0|0|7]	217
MILLIONS LIKE US ▶ 1521 Pk:171 [0|0|1]	**235**
...Millions Like Us (87) 171(1) [0|0|12]	235
MILLI VANILLI ▶ 132 Pk:1 [1|1|1]	**22975**
Girl You Know It's True (89) 1(6) [28|40|41]	22975
Frank MILLS ▶ 1776 Pk:149 [0|0|1]	**78.3**
Sunday Morning Suite (80) 149(1) [0|0|3]	78.3

Acts With Albums

Act ▶ Rank Peak [Top10s|Top40s|Total] Title (Yr) Pk(Wks) [T10 Wks|T40Wks|TotWks]	Score in '80s
Stephanie MILLS ▶ 168 Pk:16 [0|3|9]	**17834**
Sweet Sensation (80) 16(2) [0|22|44]	6497
If I Were Your Woman (87) 30(1) [0|9|36]	3415
Stephanie (81) 30(2) [0|5|23]	2204
Stephanie Mills (86) 47(2) [0|0|22]	1509
Tantalizingly Hot (82) 48(6) [0|0|19]	1489
Home (89) 82(3) [0|0|24]	*1383*
I've Got The Cure (84) 73(2) [0|0|15]	676
Merciless (83) 104(1) [0|0|19]	650
What Cha Gonna Do...With My Lovin'? *(79/80)* 197(1) [0|0|1]	11.4
Ronnie MILSAP ▶ 275 Pk:31 [0|3|9]	**10494**
Greatest Hits (80) 36(1) [0|4|41]	2954
There's No Gettin' Over Me (81) 31(1) [0|3|31]	2468
Keyed Up (83) 36(1) [0|3|19]	1667
Out Where The Bright Lights Are Glowing (81) 89(2) [0|0|29]	1300
Inside Ronnie Milsap (82) 66(2) [0|0|14]	736
Greatest Hits Vol. 2 (85) 102(2) [0|0|20]	625
Lost In The Fifties Tonight (86) 121(2) [0|0|12]	382
Milsap Magic (80) 137(2) [0|0|13]	312
One More Try For Love (84) 180(1) [0|0|3]	50.3
MINISTRY ▶ 1113 Pk:96 [0|0|4]	**708**
With Sympathy (83) 96(1) [0|0|14]	492
The Land Of Rape And Honey (88) 164(2) [0|0|4]	93.6
The Mind Is A Terrible Thing To Taste (89) 163(2) [0|0|4]	86.5
Twitch (86) 194(2) [0|0|3]	35.9
MINK DeVILLE ▶ 1587 Pk:161 [0|0|2]	**183**
Coup De Grace (81)-Mink DeVille 161(2) [0|0|5]	107
Le Chat Bleu (80) 163(2) [0|0|3]	75.6
Liza MINNELLI ▶ 1342 Pk:128 [0|0|2]	**397**
Results (89) 128(1) [0|0|8]	230
Liza Minnelli At Carnegie Hall (87) 156(2) [0|0|8]	167
Kylie MINOGUE ▶ 735 Pk:53 [0|0|1]	**2181**
Kylie (88) 53(2) [0|0|28]	2181
MINOR DETAIL ▶ 1890 Pk:187 [0|0|1]	**30.4**
Minor Detail (83) 187(1) [0|0|2]	30.4
Judi Sheppard MISSETT ▶ 1073 **Pk:117 [0|0|1]**	**794**
Jazzercise (81) 117(3) [0|0|20]	794
MISSING PERSONS ▶ 238 Pk:17 [0|1|4]	**12349**
Spring Session M (82) 17(6) [0|26|40]	7610
Missing Persons (82) 46(2) [0|0|47]	2819
Rhyme & Reason (84) 43(2) [0|0|16]	1444
Color In Your Life (86) 86(1) [0|0|11]	477
MISSION U.K. ▶ 1031 Pk:108 [0|0|2]	**893**
God's Own Medicine (87) 108(1) [0|0|18]	599
Children (88) 126(2) [0|0|10]	294
MR. BIG ▶ 877 Pk:46 [0|0|1]	**1461**
Mr. Big (89) 46(2) [0|0|18]	1461
MR. MISTER ▶ 188 Pk:1 [1|1|3]	**15754**
Welcome To The Real World (85) 1(1) [12|34|58]	14652
Go On... (87) 55(2) [0|0|17]	967
I Wear The Face (84) 170(1) [0|0|18]	135
Joni MITCHELL ▶ 400 Pk:25 [0|2|4]	**6269**
Wild Things Run Fast (82) 25(4) [0|5|21]	2382
Dog Eat Dog (85) 63(5) [0|0|19]	1310
Shadows And Light (80) 38(2) [0|3|16]	1308
Chalk Mark In A Rain Storm (88) 45(2) [0|0|16]	1269
Kim MITCHELL ▶ 1199 Pk:106 [0|0|1]	**553**
Akimbo Alogo (85) 106(2) [0|0|15]	553
M.O.D. ▶ 1329 Pk:151 [0|0|3]	**410**
Gross Misconduct (89) 151(1) [0|0|8]	212
U.S.A. For M.O.D. (87) 153(2) [0|0|5]	115
Surfin' M.O.D. (88) 186(2) [0|0|6]	83.0
MODELS ▶ 1021 Pk:84 [0|0|1]	**944**
Out Of Mind Out Of Sight (86) 84(1) [0|0|18]	944

Act ▶ Rank Peak [Top10s|Top40s|Total] Title (Yr) Pk(Wks) [T10 Wks|T40Wks|TotWks]	Score in '80s
MODERN ENGLISH ▶ 770 Pk:70 [0|0|3]	**1952**
After The Snow (83) 70(1) [0|0|28]	1260
Riccochet Days (84) 93(1) [0|0|12]	527
Stop Start (86) 154(2) [0|0|7]	166
MOLLY HATCHET ▶ 333 Pk:25 [0|3|7]	**8099**
Flirtin' With Disaster (79/80) 30(1) [0|8|34]	*2734*
Beatin' The Odds (80) 25(2) [0|6|21]	2112
Take No Prisoners (81) 36(1) [0|4|14]	1407
No Guts...No Glory (83) 59(1) [0|0|20]	1082
The Deed Is Done (84) 117(2) [0|0|13]	417
Double Trouble Live (85) 130(3) [0|0|9]	270
Molly Hatchet *(79/80)* 184(1) [0|0|6]	77.1
Eddie MONEY ▶ 165 Pk:20 [0|3|6]	**18044**
Can't Hold Back (86) 20(3) [0|22|58]	7944
No Control (82) 20(3) [0|13|44]	5035
Nothing To Lose (88) 49(1) [0|0|29]	1954
Playing For Keeps (80) 35(1) [0|5|17]	1798
Where's The Party? (83) 67(1) [0|0|19]	1078
Greatest Hits Sound Of Money *(89/90)* 84(1) [0|0|5]	236
T.S. MONK ▶ 931 Pk:64 [0|0|2]	**1242**
House Of Music (81) 64(1) [0|0|22]	1132
More Of The Good Life (82) 176(1) [0|0|8]	110
The MONKEES ▶ 298 Pk:21 [0|1|9]	**9415**
Then & Now...The Best Of The Monkees (86) 21(2) [0|12|34]	4605
The Monkees (66/86) 92(2) [0|0|24]	*1022*
More Of The Monkees (67/86) 96(2) [0|0|26]	*965*
The Monkees Greatest Hits(2) (76/86) 69(1) [0|0|14]	*823*
Headquarters (67/86) 121(2) [0|0|17]	*572*
Pisces, Aquarius, Capricorn, And Jones Ltd. (67/86) 124(2) [0|0|17]	*542*
Pool It! (87) 72(2) [0|0|9]	477
The Birds, The Bees & The Monkees (68/86) 145(1) [0|0|11]	*314*
Changes (86) 152(1) [0|0|4]	94.5
Michael MONROE ▶ 1639 Pk:161 [0|0|1]	**150**
Not Fakin' It (89) 161(1) [0|0|8]	150
The MONROES ▶ 1462 Pk:109 [0|0|1]	**275**
The Monroes (82) 109(2) [0|0|9]	275
MONTANA Orchestra ▶ 1857 Pk:195 [0|0|1]	**45.6**
Merry Christmas/Happy New Year's (81) 195(1) [0|0|4]	45.6
MONTROSE ▶ 1682 Pk:165 [0|0|1]	**126**
Mean (87) 165(2) [0|0|7]	126
MONTY PYTHON ▶ 1578 Pk:164 [0|0|1]	**189**
Monty Python's Contractual Obligation Album (80) 164(1) [0|0|9]	189
MOODY BLUES ▶ 114 Pk:1 [2|4|7]	**25949**
Long Distance Voyager (81) 1(3) [12|23|39]	13146
The Other Side Of Life (86) 9(4) [6|22|42]	7887
The Present (83) 26(3) [0|6|22]	2488
Sur La Mer (88) 38(2) [0|5|19]	1922
Voices In The Sky-Best Of The Moody Blues (85) 132(2) [0|0|8]	288
Greatest Hits (89/90) 127(1) [0|0|4]	*137*
Days Of Future Passed (72/86)-The Moody Blues With The London Festival Orchestra 154(1) [0|0|4]	*81.1*
Gary MOORE ▶ 896 Pk:114 [0|0|5]	**1395**
Wild Frontier (87) 139(1) [0|0|15]	379
After The War (89) 114(2) [0|0|9]	375
Corridors Of Power (83) 149(1) [0|0|13]	358
Run For Cover (86) 146(2) [0|0|12]	185
Victims Of The Future (84) 172(1) [0|0|5]	98.2
Melba MOORE ▶ 747 Pk:91 [0|0|4]	**2110**
A Lot Of Love (86) 91(2) [0|0|29]	946
The Other Side Of The Rainbow (82) 152(1) [0|0|19]	492
Never Say Never (83) 147(1) [0|0|14]	368

Act ▶ Rank Peak [Top10s|Top40s|Total] Title (Yr) Pk(Wks) [T10 Wks|T40Wks|TotWks]	Score in '80s
Read My Lips (85) 130(2) [0|0|10]	305
Vinnie MOORE ▶ 1611 Pk:147 [0|0|1]	**170**
Time Odyssey (88) 147(2) [0|0|7]	170
Michael MORALES ▶ 1158 Pk:113 [0|0|1]	**610**
Michael Morales (89) 113(3) [0|0|20]	610
Meli'sa MORGAN ▶ 594 Pk:41 [0|0|2]	**3293**
Do Me Baby (86) 41(1) [0|0|36]	2569
Good Love (87) 108(1) [0|0|19]	724
Gary MORRIS ▶ 1667 Pk:174 [0|0|1]	**132**
Why Lady Why (83) 174(1) [0|0|8]	132
Van MORRISON ▶ 411 Pk:44 [0|0|8]	**5956**
Avalon Sunset (89) 91(2) [0|0|27]	*1279*
A Sense Of Wonder (85) 61(3) [0|0|17]	1117
Beautiful Vision (82) 44(2) [0|0|11]	907
Poetic Champions Compose (87) 90(2) [0|0|22]	863
No Guru, No Method, No Teacher (86) 70(2) [0|0|13]	699
Common One (80) 73(1) [0|0|10]	558
Inarticulate Speech Of The Heart (83) 116(1) [0|0|8]	289
Irish Heartbeat (88)- Van Morrison & The Chieftains [A] 102(2) [0|0|13]	245
MORRISSEY ▶ 876 Pk:48 [0|0|1]	**1466**
Viva Hate (88) 48(2) [0|0|20]	1466
Steve MORSE Band ▶ 1232 Pk:101 [0|0|2]	**517**
The Introduction (84) 101(1) [0|0|12]	467
High Tension Wires (89)- Steve Morse 182(2) [0|0|3]	49.8
The MOTELS ▶ 248 Pk:16 [0|3|4]	**11752**
All Four One (82) 16(2) [0|15|41]	5751
Little Robbers (83) 22(4) [0|8|24]	3060
Shock (85) 36(1) [0|2|16]	1511
Careful (80) 45(2) [0|0|20]	1431
MOTHER'S FINEST ▶ 1596 Pk:168 [0|0|1]	**176**
Iron Age (81) 168(1) [0|0|8]	176
MÖTLEY CRÜE ▶ 47 Pk:1 [3|4|5]	**43313**
Girls, Girls, Girls (87) 2(1) [12|26|46]	11905
Shout At The Devil (83) 17(2) [0|28|111]	11465
Theatre Of Pain (85) 6(1) [9|18|72]	9587
Dr. Feelgood (89) 1(2) [11|15|15]	*7801*
Too Fast For Love (83) 77(1) [0|0|62]	2555
MOTORHEAD ▶ 1116 Pk:150 [0|0|4]	**707**
Orgasmatron (86) 157(3) [0|0|11]	260
Another Perfect Day (83) 153(1) [0|0|7]	175
Rock 'N' Roll (87) 150(2) [0|0|6]	156
Iron Fist (82) 174(1) [0|0|6]	116
The MOTORS ▶ 1658 Pk:174 [0|0|1]	**136**
Tenement Steps (80) 174(1) [0|0|8]	136
Bob MOULD ▶ 1325 Pk:127 [0|0|1]	**412**
Workbook (89) 127(1) [0|0|14]	412
MOUNTAIN ▶ 1673 Pk:166 [0|0|1]	**131**
Go For Your Life (85) 166(2) [0|0|6]	131
Alphonse MOUZON ▶ 1506 Pk:146 [0|0|1]	**246**
Distant Lover (82) 146(3) [0|0|11]	246
MOVING PICTURES ▶ 1192 Pk:101 [0|0|1]	**568**
Days Of Innocence (82) 101(3) [0|0|16]	568
Alison MOYET ▶ 656 Pk:45 [0|0|2]	**2723**
Alf 45(2) [0|0|25]	2004
Raindancing (87) 94(5) [0|0|17]	719
--Mr. MISTER see: MISTER, MR.	
MTUME ▶ 536 Pk:26 [0|1|4]	**3846**
Juicy Fruit (83) 26(2) [0|7|22]	2638
You, Me And He (84) 77(2) [0|0|19]	791
Theater Of The Mind (86) 135(2) [0|0|10]	241
In Search Of The Rainbow Seekers (80) 119(1) [0|0|8]	176
Shirley MURDOCK ▶ 704 Pk:44 [0|0|2]	**2379**
Shirley Murdock! (87) 44(2) [0|0|26]	2022
A Woman's Point Of View (88) 137(1) [0|0|15]	356

Acts With Albums

Act ▶ Rank Peak [Top10s|Top40s|Total] Title (Yr) Pk(Wks) [T10 Wks|T40Wks|TotWks]	Score in '80s
Michael Martin MURPHEY ▶ 1069 Pk:69 [0|0|2]	**801**
Michael Martin Murphey (82)-Michael Murphey 69(1) [0|0|16]	756
The Heart Never Lies (83) 187(1) [0|0|3]	44.1
Eddie MURPHY ▶ 274 Pk:26 [0|2|4]	**10526**
How Could It Be (85) 26(1) [0|10|26]	3527
Eddie Murphy: Comedian (83) 35(1) [0|3|44]	3322
Eddie Murphy (82) 52(5) [0|0|53]	3148
So Happy (89) 70(2) [0|0|2]	529
Peter MURPHY ▶ 1235 Pk:135 [0|0|1]	**511**
Love Hysteria (88) 135(2) [0|0|19]	511
Anne MURRAY ▶ 199 Pk:16 [0|2|11]	**15100**
Ann Murray's Greatest Hits (80) 16(5) [0|17|64]	6263
I'll Always Love You (80) 24(2) [0|5|14]	*1746*
Something To Talk About (86) 68(2) [0|0|23]	1244
A Little Good News (83) 72(1) [0|0|24]	1153
Where Do You Go When You Dream (81) 55(2) [0|0|15]	1100
Heart Over Mind (84) 92(1) [0|0|25]	923
Somebody's Waiting (80) 88(4) [0|0|15]	825
Christmas Wishes (81) 54(2) [0|0|8]	650
The Hottest Night Of The Year (82) 90(3) [0|0|12]	536
A Country Collection (80) 73(1) [0|0|9]	511
Harmony (87) 149(2) [0|0|6]	149
MUSICAL YOUTH ▶ 628 Pk:23 [0|1|2]	**2942**
The Youth Of Today (83) 23(4) [0|8|22]	2665
Different Style! (83) 144(1) [0|0|12]	277
Alicia MYERS ▶ 1793 Pk:186 [0|0|1]	**73.0**
I Appreciate (84) 186(2) [0|0|5]	73.0
Gary MYRICK ▶ 1862 Pk:186 [0|0|1]	**42.6**
Language (83) 186(1) [0|0|3]	42.6

N

Act ▶ Rank Peak [Top10s|Top40s|Total] Title (Yr) Pk(Wks) [T10 Wks|T40Wks|TotWks]	Score in '80s
The NAILS ▶ 1910 Pk:194 [0|0|1]	**25.4**
Dangerous Dreams (86) 194(2) [0|0|2]	25.4
NAJEE ▶ 558 Pk:56 [0|0|2]	**3661**
Najee's Theme (87) 56(2) [0|0|45]	2670
Day By Day (88) 76(2) [0|0|21]	991
NAKED EYES ▶ 543 Pk:32 [0|1|2]	**3813**
Naked Eyes (83) 32(2) [0|4|42]	3336
Fuel For The Fire (84) 83(3) [0|0|10]	477
Graham NASH ▶ 1331 Pk:117 [0|0|2]	**407**
Innocent Eyes (86) 136(2) [0|0|7]	208
Earth & Sky (80) 117(1) [0|0|5]	199
NAZARETH ▶ 611 Pk:41 [0|0|5]	**3068**
Malice In Wonderland (80) 41(3) [0|0|19]	1525
The Fool Circle (81) 70(2) [0|0|13]	666
'Snaz (81) 83(1) [0|0|9]	513
2XS (82) 122(2) [0|0|10]	337
Hair Of The Dog (76/81) 191(2) [0|0|2]	*28.2*
Rick NELSON ▶ 1621 Pk:153 [0|0|1]	**160**
Playing To Win (81) 153(2) [0|0|6]	160
Willie NELSON ▶ 46 Pk:2 [1|6|22]	**43579**
Always On My Mind (82) 2(4) [15|28|99]	15009
Willie Nelson's Greatest Hits (& Some That Will Be) (81) 27(1) [0|6|93]	5815
Honeysuckle Rose (Soundtrack) (80)- Willie Nelson & Family 11(3) [0|13|36]	5435
Stardust (78/80) 70(1) [0|0|68]	*3045*
Somewhere Over The Rainbow (81) 31(3) [0|6|23]	2161
Poncho & Lefty (83)-Merle Haggard/Willie Nelson [A] 37(1) [0|1|53]	1938
Without A Song (83) 54(1) [0|0|34]	1824
Tougher Than Leather (83) 39(1) [0|1|20]	1716
Willie Nelson Sings Kristofferson (80) 42(2) [0|0|18]	*1377*
City of New Orleans (84) 69(1) [0|0|26]	1286
Take It To The Limit (83)-Willie Nelson With Waylon Jennings 60(1) [0|0|16]	891
The Electric Horseman (Soundtrack) (80)- Willie Nelson/Dave Grusin [A] 52(2) [0|0|25]	766
San Antonio Rose (80)-Willie Nelson & Ray Price [A] 70(2) [0|0|25]	637
Willie Nelson And Family Live (79/80) *102(2) [0|0|14]*	*564*
Angel Eyes (84) 116(2) [0|0|7]	261
Pretty Paper (80) 73(1) [0|0|3]	*225*
Me & Paul (85) 152(1) [0|0|7]	199
The Minstrel Man (81) 148(1) [0|0|7]	173
My Own Way (83) 182(1) [0|0|5]	70.7
Music From Songwriter (Soundtrack) (84)-Willie Nelson & Kris Kristofferson [A] 152(2) [0|0|5]	68.7
Danny Davis & Willie Nelson With The Nashville Brass (80) [A] 150(1) [0|0|5]	62.4
Half Nelson (85) 178(2) [0|0|3]	57.5
NENA ▶ 799 Pk:27 [0|1|1]	**1810**
99 Luftballons (84) 27(3) [0|6|14]	1810
Robbie NEVIL ▶ 462 Pk:37 [0|1|2]	**4762**
Robbie Nevil (86) 37(2) [0|3|46]	4107
A Place Like This (88) 118(2) [0|0|21]	655
Ivan NEVILLE ▶ 1057 Pk:107 [0|0|1]	**827**
If My Ancestors Could See Me Now (88) 107(1) [0|0|23]	827
NEVILLE BROTHERS ▶ 823 Pk:66 [0|0|4]	**1704**
Yellow Moon (89) 66(3) [0|0|24]	1384
Uptown (87) 155(1) [0|0|9]	196
Fiyo On The Bayou (81) 166(1) [0|0|3]	69.6
Treacherous: A History Of The Neville Brothers 1955-1985 (87) 178(2) [0|0|3]	53.9
NEWCLEUS ▶ 939 Pk:74 [0|0|1]	**1193**
Jam On Revenge (84) 74(2) [0|0|28]	1193
NEW EDITION ▶ 101 Pk:6 [1|3|5]	**28195**
New Edition (84) 6(2) [6|29|54]	10226
Heart Break (88) 12(1) [0|37|50]	9762
All For Love (85) 32(1) [0|15|48]	5213
Under The Blue Moon (86) 43(2) [0|0|23]	1743
Candy Girl (83) 90(2) [0|0|33]	1252
NEW ENGLAND ▶ 1797 Pk:176 [0|0|1]	**71.5**
Walking Wild (81) 176(2) [0|0|4]	71.5
NEW KIDS ON THE BLOCK ▶ 80 Pk:1 [2|3|3]	**32612**
Hangin' Tough (88) 1(2) [41|47|71]	*26089*
Merry, Merry Christmas (89) 9(1) [2|11|12]	*3627*
New Kids On The Block (89) 25(1) [0|12|22]	*2896*
Randy NEWMAN ▶ 806 Pk:64 [0|0|2]	**1764**
Land Of Dreams (88) 80(1) [0|0|19]	952
Trouble In Paradise (83) 64(4) [0|0|13]	812
NEW ORDER ▶ 306 Pk:32 [0|2|4]	**8951**
Substance (87) 36(1) [0|7|60]	4806
Technique (89) 32(1) [0|6|28]	2669
Low-Life (85) 94(2) [0|0|22]	865
Brotherhood (86) 117(1) [0|0|21]	610
Juice NEWTON ▶ 181 Pk:20 [0|2|5]	**16477**
Juice (81) 22(3) [0|34|86]	10430
Quiet Lies (82) 20(2) [0|8|46]	4717
Dirty Looks (83) 52(1) [0|0|15]	905
Can't Wait All Night (84) 128(2) [0|0|10]	342
Greatest Hits (84) 178(2) [0|0|5]	83.0
Olivia NEWTON-JOHN ▶ 104 Pk:4 [2|4|6]	**27732**
Physical (81) 6(7) [14|27|57]	11836
Olivia's Greatest Hits, Vol. 2 (82) 16(4) [0|21|86]	9546
Xanadu (Soundtrack) (80)-Olivia Newton-John/ Electric Light Orchestra [A] 4(3) [8|15|36]	3918
Soul Kiss (85) 29(3) [0|5|16]	1779
The Rumour (88) 67(2) [0|0|9]	521
Warm And Tender (89) 124(1) [0|0|5]	*133*
Stevie NICKS ▶ 41 Pk:1 [3|4|4]	**46197**
Bella Donna (81) 1(1) [26|45|143]	24817

Act ▶ Rank Peak [Top10s|Top40s|Total] Title (Yr) Pk(Wks) [T10 Wks|T40Wks|TotWks]	Score in '80s
The Wild Heart (83) 5(7) [11|22|52]	10817
Rock A Little (85) 12(1) [0|17|35]	6751
The Other Side Of The Mirror (89) 10(1) [1|11|21]	3813
The NIGHTHAWKS ▶ 1770 Pk:166 [0|0|1]	**80.6**
Nighthawks (80) 166(1) [0|0|4]	80.6
Maxine NIGHTINGALE ▶ 1785 Pk:176 [0|0|1]	**75.6**
It's A Beautiful Thing (83) 176(2) [0|0|4]	75.6
NIGHT RANGER ▶ 108 Pk:10 [1|4|5]	**27014**
Midnight Madness (83) 15(10) [0|29|69]	11928
7 Wishes (85) 10(3) [3|25|45]	9284
Dawn Patrol (82) 38(4) [0|8|69]	3832
Big Life (87) 28(2) [0|3|18]	1577
Man In Motion (88) 81(2) [0|0|8]	392
Willie NILE ▶ 1371 Pk:145 [0|0|2]	**355**
Willie Nile (80) 145(2) [0|0|6]	179
Golden Down (81) 158(1) [0|0|8]	176
999 ▶ 1757 Pk:177 [0|0|2]	**86.6**
The Biggest Prize In Sport (80) 177(2) [0|0|3]	59.4
Concrete (81) 192(2) [0|0|2]	27.2
9.9 ▶ 1001 Pk:79 [0|0|1]	**984**
9.9 (85) 79(1) [0|0|22]	984
NITRO ▶ 1470 Pk:140 [0|0|1]	**270**
O.F.R. (89) 140(1) [0|0|9]	270
NITTY GRITTY DIRT BAND ▶ 678 Pk:62 [0|0|4]	**2542**
Make A Little Magic (80)- The Dirt Band 62(2) [0|0|16]	1008
An American Dream (80)- The Dirt Band 76(1) [0|0|14]	642
Will The Circle Be Unbroken, Vol.II (89) 95(2) [0|0|12]	517
Jealousy (81)-The Dirt Band 102(1) [0|0|9]	375
Mojo NIXON & Skid ROPER ▶ 1560 Pk:151 [0|0|2]	**198**
Root Hog Or Die (89) 151(1) [0|0|7]	169
Bo-Day-Shus!!! (87) 189(1) [0|0|2]	29.1
No Artist ▶ 1305 Pk:96 [0|0|1]	**433**
Annie's Christmas (82) 96(4) [0|0|9]	433
NOEL ▶ 1360 Pk:126 [0|0|1]	**372**
Noel (88) 126(1) [0|0|13]	372
Aldo NOVA ▶ 330 Pk:8 [1|1|2]	**8153**
Aldo Nova (82) 8(6) [8|16|37]	6961
Subject: Aldo Nova (83) 56(2) [0|0|20]	1191
NOVO COMBO ▶ 1670 Pk:167 [0|0|1]	**132**
Novo Combo (81) 167(1) [0|0|6]	132
NRBQ ▶ 1926 Pk:198 [0|0|1]	**10.9**
Wild Weekend (89) 198(1) [0|0|1]	*10.9*
NUCLEAR ASSAULT ▶ 1233 Pk:134 [0|0|2]	**515**
Survive (88) 145(1) [0|0|11]	288
Handle With Care (89/90) 134(1) [0|0|7]	*227*
Ted NUGENT ▶ 373 Pk:13 [0|1|7]	**7183**
Scream Dream (80) 13(2) [0|9|18]	3158
Penetrator (84) 56(2) [0|0|15]	1046
Nugent (82) 51(2) [0|0|14]	1024
Intensities In 10 Cities (81) 51(2) [0|0|10]	732
Little Miss Dangerous (86) 76(2) [0|0|14]	725
If You Can't Lick 'Em...Lick 'Em (88) 112(2) [0|0|7]	261
Great Gonzos! The Best Of Ted Nugent (81) 140(1) [0|0|8]	237
Gary NUMAN ▶ 445 Pk:16 [0|1|3]	**5141**
The Pleasure Principle (80) 16(2) [0|14|30]	4450
Telekon (80) 64(2) [0|0|10]	602
Dance (81) 167(1) [0|0|4]	88.2
Bobby NUNN ▶ 1551 Pk:148 [0|0|1]	**204**
Second To Nunn (83) 148(1) [0|0|8]	204
NU SHOOZ ▶ 521 Pk:27 [0|1|2]	**4086**
Poolside (86) 27(2) [0|8|32]	3501
Told U So (88) 93(2) [0|0|14]	585

Acts With Albums

Act ▶ Rank Peak [Top10s|Top40s|Total] Title (Yr) Pk(Wks) [T10 Wks|T40Wks|TotWks]	Score in '80s
N.W.A. ▶ 455 Pk:37 [0|1|1]	4901
Straight Outta Compton (89) 37(3) [0|9|44]	4901
The NYLONS ▶ 681 Pk:43 [0|0|3]	2527
Happy Together (87) 43(1) [0|0|24]	1866
Seamless (86) 133(1) [0|0|16]	434
Rockapella (89) 136(2) [0|0|10]	227
Laura NYRO ▶ 1848 Pk:182 [0|0|1]	49.4
Mother's Spiritual (84) 182(1) [0|0|3]	49.4

O

OAK RIDGE BOYS ▶ 228 Pk:14 [0|2|9]	12914
Fancy Free (81) 14(1) [0|10|48]	5784
Bobbie Sue (82) 20(3) [0|9|21]	2610
American Made (83) 51(3) [0|0|23]	1723
Greatest Hits, Vol. 2 (84) 71(2) [0|0|24]	873
Greatest Hits (80) 99(1) [0|0|21]	724
Deliver (83) 121(1) [0|0|14]	545
Christmas (82) 73(2) [0|0|7]	399
Step On Out (85) 156(1) [0|0|5]	129
Together (80) 154(1) [0|0|6]	128
OAKTOWN'S 3.5.7 ▶ 1224 Pk:126 [0|0|1]	527
Wild & Loose (89) 126(1) [0|0|16]	527
John O'BANION ▶ 1735 Pk:164 [0|0|1]	95.4
John O'Banion (81) 164(1) [0|0|4]	95.4
O'BRYAN ▶ 637 Pk:64 [0|0|3]	2887
Be My Lover (84) 64(1) [0|0|21]	1298
You And I (83) 87(1) [0|0|27]	1001
Doin' Alright (82) 80(1) [0|0|12]	589
Ric OCASEK ▶ 464 Pk:28 [0|2|2]	4705
This Side Of Paradise (86) 31(2) [0|5|23]	2456
Beatitude (83) 28(5) [0|9|16]	2249
Billy OCEAN ▶ 75 Pk:6 [2|3|5]	34370
Suddenly (84) 9(2) [2|57|86]	17373
Love Zone (86) 6(7) [12|30|48]	11907
Tear Down These Walls (88) 18(1) [0|11|31]	4491
Greatest Hits (89) 77(2) [0|0|9]	511
Nights (Feel Like Getting Down) (81) 152(1) [0|0|3]	88.5
Sinead O'CONNOR ▶ 653 Pk:36 [0|1|1]	2738
The Lion And The Cobra (88) 36(3) [0|5|28]	2738
ODYSSEY ▶ 1618 Pk:175 [0|0|2]	165
I Got The Melody (81) 175(2) [0|0|5]	96.0
Hang Together (80) 181(1) [0|0|4]	69.0
OFF BROADWAY USA ▶ 1271 Pk:101 [0|0|1]	469
On (80) 101(2) [0|0|11]	469
OHIO PLAYERS ▶ 1794 Pk:165 [0|0|1]	72.9
Tenderness (81) 165(2) [0|0|3]	72.9
OINGO BOINGO ▶ 664 Pk:77 [0|0|8]	2654
Boi-ngo (87) 77(2) [0|0|16]	751
Dead Man's Party (85) 98(2) [0|0|16]	676
Boingo Alive (88) 90(3) [0|0|11]	447
Nothing To Fear (82) 148(1) [0|0|6]	212
Good For Your Soul (83) 144(1) [0|0|7]	194
Skeletons In The Closet: The Best Of Oingo Boingo (89) 150(2) [0|0|6]	159
Only A Lad (81) 172(1) [0|0|5]	108
Oingo Boingo (80) 163(2) [0|0|5]	107
The O'JAYS ▶ 429 Pk:36 [0|1|7]	5419
Let Me Touch You (87) 66(1) [0|0|25]	1430
The Year 2000 (80) 36(2) [0|3|12]	1352
My Favorite Person (82) 49(2) [0|0|13]	860
Identify Yourself (79/80) 81(3) [0|0|14]	650
Serious (89) 114(1) [0|0|17]	587
Love Fever (85) 121(2) [0|0|12]	390
When Will I See You Again (83) 142(1) [0|0|5]	150
Mike OLDFIELD ▶ 1321 Pk:138 [0|0|3]	417
Islands (88) 138(2) [0|0|8]	252
Five Miles Out (82) 164(1) [0|0|5]	107
QE 2 (81) 174(1) [0|0|3]	57.5
Jane OLIVOR ▶ 1024 Pk:58 [0|0|2]	918
The Best Side Of Goodbye (80) 58(1) [0|0|12]	752
In Concert (82) 144(2) [0|0|6]	166
OMAR And The HOWLERS ▶ 1043 Pk:81 [0|0|1]	852
Hard Times In The Land Of Plenty (87) 81(2) [0|0|19]	852
Alexander O'NEAL ▶ 466 Pk:29 [0|1|4]	4667
Hearsay (87) 29(1) [0|8|40]	3788
Alexander O'Neal (85) 92(2) [0|0|18]	670
My Gift To You (88) 149(1) [0|0|5]	129
All Mixed Up (89) 185(2) [0|0|5]	79.9
ONE WAY ▶ 487 Pk:51 [0|0|8]	4434
Who's Foolin' Who (82) 51(1) [0|0|23]	1553
Lady (84) 58(1) [0|0|20]	1167
Fancy Dancer (81) 79(2) [0|0|19]	769
One Way Featuring Al Hudson(2) (80)-One Way Featuring Al Hudson 128(1) [0|0|12]	435
Wrap Your Body (85) 156(2) [0|0|7]	178
Love Is...One Way (81) 157(1) [0|0|8]	174
Shine On Me (83) 164(1) [0|0|6]	129
One Way Featuring Al Hudson (79/80)- One Way Featuring Al Hudson 188(2) [0|0|2]	30.8
Yoko ONO ▶ 936 Pk:49 [0|0|2]	1210
Season Of Glass (81) 49(2) [0|0|9]	680
It's Alright (I See Rainbows) (82) 98(2) [0|0|13]	530
OPUS ▶ 1011 Pk:64 [0|0|1]	961
Up And Down (86) 64(2) [0|0|16]	961
Roy ORBISON ▶ 321 Pk:5 [1|1|5]	8380
Mystery Girl (89) 5(2) [8|16|27]	6956
In Dreams: The Greatest Hits (89) 95(1) [0|0|15]	596
For The Lonely: A Roy Orbison Anthology, 1956- 1965 (89) 110(1) [0|0|13]	467
A Black & White Night: Live (Soundtrack) (89)- Roy Orbison and Friends 123(3) [0|0|5]	222
Class Of '55 (86)-Carl Perkins, Jerry Lee Lewis, Roy Orbison, & Johnny Cash [A] 87(2) [0|0|12]	139
ORCHESTRAL MANOEUVRES IN THE DARK ▶ 315 Pk:38 [0|1|6]	8528
Crush (85) 38(2) [0|5|53]	3866
The Best Of OMD (88) 46(2) [0|0|29]	2274
The Pacific Age (86) 47(2) [0|0|23]	1805
Architecture And Morality (82) 144(4) [0|0|12]	341
Dazzle Ships (83) 162(1) [0|0|6]	140
Junk Culture (84) 182(3) [0|0|6]	102
ORIGINAL CAST ▶ Pk:11 [0|2|17]	
Dreamgirls (82) 11(3) [0|15|29]	5227

The Phantom Of The Opera (87)-
   Original London Cast Recording
   33(1) [0|5|98]
	4623

Cats (83) 113(3) [0|0|64]	1374
Cats (82)-Original London Cast 86(2) [0|0|22]	1071
Annie (77/81) 105(2) [0|0|28]	915
La Cage Aux Folles (83) 52(1) [0|0|15]	859
Evita (80) 105(2) [0|0|19]	655
Les Miserables (87)-Original London Cast Recording 106(2) [0|0|15]	529
42nd Street (81) 120(1) [0|0|11]	388
Cats (83)-Selections from Original Broadway Cast 131(2) [0|0|14]	352
Les Miserables (87) 117(2) [0|0|10]	316
Sunday In The Park With George (84) 149(3) [0|0|11]	315
Into The Woods (88) 126(2) [0|0|6]	174
The Mystery Of Edwin Drood (86) 150(2) [0|0|6]	139
Follies In Concert (86) 181(2) [0|0|6]	93.9
The Pirates Of Penzance (81) 178(1) [0|0|3]	54.8
Woman Of The Year (81) 196(2) [0|0|2]	23.6
ORION THE HUNTER ▶ 1020 Pk:57 [0|0|1]	945
Orion The Hunter (84) 57(2) [0|0|14]	945
Benjamin ORR ▶ 972 Pk:86 [0|0|1]	1089
The Lace (86) 86(2) [0|0|22]	1089
Robert Ellis ORRALL ▶ 1507 Pk:146 [0|0|1]	245
Special Pain (83) 146(2) [0|0|9]	245
Jeffrey OSBORNE ▶ 162 Pk:25 [0|3|5]	18472
Stay With Me Tonight (83) 25(1) [0|23|89]	8519
Don't Stop (84) 39(4) [0|4|37]	3510
Emotional (86) 26(1) [0|11|26]	3047
Jeffrey Osborne (82) 49(3) [0|0|43]	2628
One Love--One Dream (88) 86(1) [0|0|16]	767
Ozzy OSBOURNE ▶ 55 Pk:6 [2|7|8]	39542
Diary Of A Madman (81) 16(4) [0|23|73]	8810
Blizzard Of Ozz (81) 21(1) [0|16|104]	8366
The Ultimate Sin (86) 6(2) [8|17|39]	7371
No Rest For The Wicked (88) 13(1) [0|13|27]	4382
Bark At The Moon (83) 19(3) [0|11|29]	3819
Speak Of The Devil (82) 14(4) [0|10|20]	3443
Tribute (87)-Ozzy Osbourne/Randy Rhoads [A] 6(2) [6|14|23]	2796
Mr. Crowley (82) 120(1) [0|0|18]	556
Lee OSKAR ▶ 1657 Pk:162 [0|0|1]	137
My Road Our Road (81) 162(2) [0|0|6]	137
K.T. OSLIN ▶ 497 Pk:68 [0|0|2]	4223
This Woman (88) 75(2) [0|0|52]	2261
80's Ladies (87) 68(4) [0|0|32]	1962
Donny OSMOND ▶ 858 Pk:54 [0|0|1]	1530
Donny Osmond (89) 54(3) [0|0|23]	1530
OTHER ONES ▶ 1602 Pk:139 [0|0|1]	173
The Other Ones (87) 139(2) [0|0|6]	173
The OUTFIELD ▶ 193 Pk:9 [1|2|3]	15529
Play Deep (85) 9(1) [6|29|66]	10792
Bangin' (87) 18(1) [0|11|21]	3342
Voices Of Babylon (89) 53(3) [0|0|23]	1395
The OUTLAWS ▶ 516 Pk:25 [0|1|5]	4106
Ghost Riders (80) 25(3) [0|7|26]	2897
Los Hombres Malo (82) 77(1) [0|0|9]	407
In The Eye Of The Storm (79/80) 98(1) [0|0|9]	312
Greatest Hits Of The Outlaws/High Tides Forever (82) 136(2) [0|0|9]	270
Soldiers Of Fortune (86) 160(2) [0|0|10]	220
OVERKILL ▶ 1326 Pk:142 [0|0|3]	412
Under The Influence (88) 142(1) [0|0|13]	372
The Years Of Decay (89/90) 194(2) [0|0|2]	25.4
Taking Over (87) 191(1) [0|0|1]	14.1
OXO ▶ 1504 Pk:117 [0|0|1]	248
Oxo (83) 117(1) [0|0|7]	248
OZARK MOUNTAIN DAREDEVILS ▶ 1780 Pk:170 [0|0|1]	77.4
Ozark Mountain Daredevils (80) 170(1) [0|0|4]	77.4
OZONE ▶ 1637 Pk:152 [0|0|1]	151
Li'l Suzy (82) 152(2) [0|0|6]	151

P

PABLO CRUISE ▶ 713 Pk:34 [0|1|2]	2300
Reflector (81) 34(2) [0|5|18]	1878
Part Of The Game (79/80) 103(2) [0|0|10]	422
Jimmy PAGE ▶ 596 Pk:26 [0|1|2]	3264
Outrider (88) 26(2) [0|6|20]	2386
Death Wish II (Soundtrack) (82) 50(4) [0|0|10]	878
Tommy PAGE ▶ 1749 Pk:166 [0|0|1]	89.2
Tommy Page (89) 166(2) [0|0|5]	89.2
Kevin PAIGE ▶ 1154 Pk:107 [0|0|1]	621
Kevin Paige (89) 107(2) [0|0|15]	621
Robert PALMER ▶ 138 Pk:8 [1|2|6]	22249
Riptide (85) 8(1) [4|35|90]	12995
Heavy Nova (88) 13(3) [0|20|44]	6978
Clues (80) 59(2) [0|0|17]	1007
Pride (83) 112(2) [0|0|19]	800
Addictions Vol. I (89) 79(1) [0|0|6]	342
Maybe It's Live (82) 148(1) [0|0|5]	127

Acts With Albums

Act ▶ Rank Peak [Top10s|Top40s|Total] Title (Yr) Pk(Wks) [T10 Wks|T40Wks|TotWks]	Score in '80s
Mica PARIS ▶ 1014 Pk:86 [0|0|1]	**956**
So Good (89) 86(2) [0|0|23]	956
Graham PARKER ▶ 513 Pk:40 [0|1|4]	**4128**
The Up Escalator (80)-Graham Parker And The Rumour 40(1) [0|1|15]	1281
Another Grey Area (82) 51(2) [0|0|16]	1093
The Mona Lisa's Sister (88) 77(2) [0|0|19]	959
The Real Macaw (83) 59(1) [0|0|14]	795
Graham PARKER And The SHOT ▶ 886 Pk:57 [0|0|1]	**1429**
Steady Nerves (85) 57(3) [0|0|21]	1429
Ray PARKER Jr. ▶ 286 Pk:	**9994**
The Other Woman (82) 11(2) [0|12|27]	4336
Woman Out Of Control (83) 45(2) [0|0|23]	1982
Greatest Hits (82) 51(3) [0|0|22]	1530
Chartbusters (84) 60(3) [0|0|15]	1081
Sex And The Single Man (85) 65(2) [0|0|13]	649
After Dark (87) 86(2) [0|0|9]	415
Ray PARKER Jr. & RAYDIO ▶ 374 Pk:11 [0|3|8]	**7133**
A Woman Needs Love (81) 13(2) [0|16|26]	4960
Two Places At The Same Time (80) 33(2) [0|5|21]	2173
PARLIAMENT ▶ 797 Pk:44 [0|0|2]	**1811**
Gloryhallastoopid (Or Pin The Tail On The Funky) (80) 44(2) [0|0|17]	1328
Trombipulation (81) 61(2) [0|0|7]	483
John PARR ▶ 800 Pk:48 [0|0|1]	**1807**
John Parr (84) 48(2) [0|0|26]	1807
Alan PARSONS PROJECT ▶ 100 Pk:7 [1|3|9]	**28199**
The Turn Of A Friendly Card (80) 13(2) [0|21|58]	8534
Eye In The Sky (82) 7(6) [7|21|41]	8459
Ammonia Avenue (84) 15(1) [0|13|26]	4049
The Best Of The Alan Parsons Project (83) 53(4) [0|0|29]	1756
Vulture Culture (85) 46(3) [0|0|19]	1572
Stereotomy (86) 43(2) [0|0|18]	1532
Gaudi (87) 57(2) [0|0|14]	956
I Robot (77/81) 86(1) [0|0|15]	*707*
Eve (79/80) 74(1) [0|0|11]	*633*
PARTLAND BROTHERS ▶ 1653 Pk:146 [0|0|1]	**139**
Electric Honey (87) 146(2) [0|0|5]	139
Dolly PARTON ▶ 240 Pk:6 [1|3|10]	**12318**
9 To 5 And Odd Jobs (80) 11(2) [0|15|34]	5977
Trio (87)-Dolly Parton, Linda Ronstadt, Emmylou Harris [A] 6(1) [3|14|48]	2149
Greatest Hits (82) 77(2) [0|0|23]	1062
Dolly Dolly Dolly (80) 71(1) [0|0|13]	778
The Great Pretender (84) 73(1) [0|0|14]	664
Once Upon A Christmas (84)- Kenny Rogers & Dolly Parton [A] 31(1) [0|4|8]	491
Heartbreak Express (82) 106(2) [0|0|12]	434
Burlap & Satin (83) 127(1) [0|0|11]	353
Rainbow (87) 153(1) [0|0|6]	214
Rhinestone (Soundtrack) (84) 135(2) [0|0|7]	195
The PASADENAS ▶ 1211 Pk:89 [0|0|1]	**541**
To Whom It May Concern (89) 89(2) [0|0|12]	541
PASSPORT ▶ 1643 Pk:163 [0|0|2]	**146**
Oceanliner (80) 163(1) [0|0|4]	87.0
Blue Tattoo (81) 175(2) [0|0|3]	59.4
Jaco PASTORIUS ▶ 1774 Pk:161 [0|0|1]	**78.7**
Word Of Mouth (81) 161(2) [0|0|3]	78.7
Robbie PATTON ▶ 1671 Pk:162 [0|0|1]	**132**
Distant Shores (81) 162(1) [0|0|6]	132
Henry PAUL Band ▶ 1288 Pk:120 [0|0|2]	**448**
Feel The Heat (80) 120(1) [0|0|7]	259
Anytime (81) 158(1) [0|0|8]	189
Luciano PAVAROTTI ▶ 717 Pk:77 [0|0|5]	**2286**
Pavarotti's Greatest Hits (80) 94(2) [0|0|18]	777
O Sole Mio - Favorite Neapolitan Songs (80) 77(1) [0|0|15]	*695*
Mamma (84) 103(2) [0|0|14]	544
Luciano (82) 141(1) [0|0|7]	191
Yes, Giorgio (82) 158(2) [0|0|3]	80.4
PEACHES & HERB ▶ 953 Pk:69 [0|0|3]	**1144**
Twice The Fire (79/80) 69(1) [0|0|22]	*845*
Worth The Wait (80) 120(1) [0|0|6]	231
Sayin' Something! (81) 168(2) [0|0|3]	68.7
PEARL HARBOR And The EXPLOSIONS ▶ 1257 Pk:107 [0|0|2]	**486**
Pearl Harbor & The Explosions (80) 107(1) [0|0|11]	424
Don't Follow Me, I'm Lost Too (81)- Pearl Harbor 170(2) [0|0|3]	62.0
David PEASTON ▶ 1146 Pk:113 [0|0|1]	**638**
Introducing...David Peaston (89) 113(1) [0|0|18]	638
PEBBLES ▶ 394 Pk:14 [0|1|1]	**6577**
Pebbles (88) 14(1) [0|19|38]	6577
Nia PEEPLES ▶ 1032 Pk:97 [0|0|1]	**890**
Nothin' But Trouble (88) 97(4) [0|0|21]	890
Teddy PENDERGRASS ▶ 180 Pk:14 [0|4|9]	**16799**
TP (80) 14(3) [0|16|34]	5857
It's Time For Love (81) 19(2) [0|7|27]	2995
Love Language (84) 38(4) [0|5|35]	2741
Joy (88) 54(2) [0|0|24]	1676
Teddy Live! Coast To Coast (80) 33(3) [0|4|24]	*1438*
This One's For You (82) 59(1) [0|0|15]	1008
Workin' It Back (85) 96(2) [0|0|23]	741
Heaven Only Knows (84) 123(1) [0|0|9]	297
Teddy (79/80) 165(1) [0|0|3]	*46.8*
Michael PENN ▶ 1516 Pk:98 [0|0|1]	**237**
March (89/90) 98(1) [0|0|6]	*237*
PEPSI and SHIRLIE ▶ 1524 Pk:133 [0|0|1]	**232**
All Right Now (88)-Pepsi & Shirlie 133(1) [0|0|17]	232
Carl PERKINS ▶ 1656 Pk:87 [0|0|1]	**139**
Class Of '55 (86)-Carl Perkins, Jerry Lee Lewis, Roy Orbison, & Johnny Cash [A] 87(2) [0|0|12]	139
Itzhak PERLMAN ▶ 1709 Pk:149 [0|0|1]	**112**
A Different Kind Of Blues (81)-Itzhak Perlman & Andre Previn [A] 149(1) [0|0|9]	112
Joe PERRY PROJECT ▶ 885 Pk:47 [0|0|2]	**1429**
Let The Music Do The Talking (80) 47(2) [0|0|13]	1043
I've Got The Rock 'N' Rolls Again (81) 100(1) [0|0|10]	386
Steve PERRY ▶ 332 Pk:12 [0|1|1]	**8110**
Street Talk (84) 12(4) [0|18|60]	8110
PETER, PAUL & MARY ▶ 1734 Pk:173 [0|0|1]	**96.1**
No Easy Walk To Freedom (87) 173(2) [0|0|5]	96.1
Bernadette PETERS ▶ 1086 Pk:114 [0|0|2]	**767**
Bernadette Peters (80) 114(1) [0|0|14]	548
Now Playing (81) 151(1) [0|0|9]	219
PET SHOP BOYS ▶ 185 Pk:7 [1|3|4]	**16179**
Please (86) 7(1) [6|21|31]	7645
Actually (87) 25(1) [0|22|45]	5998
Introspective (88) 34(2) [0|3|22]	2024
Disco (86) 95(2) [0|0|12]	512
Tom PETTY ▶ 201 Pk:3 [1|1|1]	**15075**
Full Moon Fever (89) 3(2) [25|33|34]	*15075*
Tom PETTY And The HEARTBREAKERS ▶ 45 Pk:2 [4|6|8]	**43688**
Damn The Torpedoes (80) 2(7) [16|22|58]	*14001*
Hard Promises (81) 5(2) [11|17|31]	8041
Long After Dark (82) 9(3) [4|22|32]	7515
Southern Accents (85) 7(2) [6|18|32]	7365
Let Me Up (I've Had Enough) (87) 20(2) [0|12|20]	3456
Pack Up The Plantation - Live! (85) 22(2) [0|9|26]	2935
Tom Petty And The Heartbreakers (78/80) 129(1) [0|0|12]	*375*
PHANTOM, ROCKER & SLICK ▶ 870 Pk:61 [0|0|2]	**1482**
Phantom, Rocker & Slick (85) 61(2) [0|0|23]	1452
Cover Girl (86) 181(1) [0|0|2]	30.4
PHOTOGLO ▶ 1332 Pk:119 [0|0|2]	**407**
Fool In Love With You (81)-Jim Photoglo 119(1) [0|0|11]	372
Photoglo (80) 194(1) [0|0|3]	35.0
PIECES OF A DREAM ▶ 819 Pk:90 [0|0|4]	**1714**
Imagine This (84) 90(1) [0|0|15]	624
We Are One (82) 114(3) [0|0|15]	491
Joyride (86) 102(2) [0|0|12]	472
Pieces Of A Dream (81) 170(2) [0|0|6]	128
PINK FLOYD ▶ 18 Pk:1 [3|5|8]	**65821**
The Wall (80) 1(15) [27|33|112]	*26497*
The Dark Side Of The Moon (73/80) 44(1) [0|0|450]	*14991*
A Momentary Lapse Of Reason (87) 3(1) [15|26|56]	12930
The Final Cut (83) 6(2) [5|12|23]	4833
Delicate Sound Of Thunder (88) 11(1) [0|10|21]	4100
A Collection Of Great Dance Songs (81) 31(2) [0|7|16]	1936
Works (83) 68(1) [0|0|9]	494
Meddle (71/80) 191(1) [0|0|3]	*40.0*
Joe PISCOPO ▶ 1821 Pk:168 [0|0|1]	**61.5**
New Jersey (85) 168(1) [0|0|3]	61.5
PIXIES ▶ 945 Pk:98 [0|0|1]	**1172**
Doolittle (89) 98(1) [0|0|27]	1172
PLANET P ▶ 744 Pk:42 [0|0|2]	**2124**
Planet P (83) 42(2) [0|0|23]	1647
Pink World (84)-Planet P Project 121(1) [0|0|14]	478
Robert PLANT ▶ 103 Pk:5 [3|4|4]	**28016**
Now And Zen (88) 6(1) [10|25|48]	10312
The Principle Of Moments (83) 8(1) [5|18|40]	7707
Pictures At Eleven (82) 5(6) [9|14|53]	7379
Shaken 'N' Stirred (85) 20(2) [0|8|19]	2618
The PLASMATICS ▶ 1198 Pk:134 [0|0|3]	**557**
New Hope For The Wretched (81) 134(2) [0|0|10]	274
Beyond The Valley Of 1984 (81) 142(1) [0|0|9]	228
Metal Priestess (81) 177(1) [0|0|3]	54.9
PLAYER ▶ 1613 Pk:152 [0|0|1]	**168**
Spies Of Life (82) 152(1) [0|0|7]	168
PLEASURE ▶ 1006 Pk:97 [0|0|3]	**975**
Special Things (80) 97(2) [0|0|14]	541
Future Now (79/80) 115(1) [0|0|8]	*306*
Give It Up (82) 164(1) [0|0|6]	128
The PLIMSOULS ▶ 1635 Pk:153 [0|0|2]	**151**
The Plimsouls (81) 153(1) [0|0|4]	97.7
Everywhere At Once (83) 186(1) [0|0|4]	53.6
POCO ▶ 523 Pk:40 [0|1|6]	**4067**
Legacy (89) 40(2) [0|2|15]	*1942*
Under The Gun (80) 46(1) [0|0|16]	1141
Blue And Gray (81) 76(2) [0|0|10]	567
Cowboys & Englishmen (82) 131(2) [0|0|8]	261
Inamorata (84) 167(2) [0|0|6]	119
Ghost Town (82) 195(2) [0|0|3]	36.4
The POGUES ▶ 994 Pk:88 [0|0|2]	**1009**
If I Should Fall From Grace With God (88) 88(3) [0|0|16]	689
Peace & Love (89) 118(1) [0|0|9]	319
POINT BLANK ▶ 745 Pk:80 [0|0|3]	**2123**
American Excess (81) 80(1) [0|0|24]	1175
On A Roll (82) 119(1) [0|0|17]	542
The Hard Way (80) 110(1) [0|0|13]	406

Acts With Albums

Act ▶ Rank Peak [Top10s|Top40s|Total] Title (Yr) Pk(Wks) [T10 Wks|T40Wks|TotWks]	Score in '80s
Bonnie POINTER ▶ 1097 Pk:63 [0|0|1]	742
Bonnie Pointer (II) (80) 63(2) [0|0|12]	*742*
Noel POINTER ▶ 1761 Pk:167 [0|0|1]	85.2
Calling (80) 167(1) [0|0|4]	85.2
POINTER SISTERS ▶ 64 Pk:8 [1|4|8]	36664
Break Out (83) 8(2) [6|65|105]	22261
Contact (85) 24(1) [0|17|34]	4714
Black & White (81) 12(2) [0|14|22]	4441
Special Things (80) 34(1) [0|5|24]	2324
So Excited! (82) 59(1) [0|0|28]	1381
Hot Together (86) 48(3) [0|0|18]	1344
Serious Slammin' (88) 152(2) [0|0|6]	144
Pointer Sisters' Greatest Hits (82) 178(2) [0|0|3]	55.7
POISON ▶ 57 Pk:2 [2|2|2]	38957
Open Up And Say...Ahh! (88) 2(1) [26|51|70]	21053
Look What The Cat Dragged In (86) 3(2) [17|47|101]	17904
The POLICE ▶ 10 Pk:1 [4|5|6]	79361
Synchronicity (83) 1(17) [40|50|75]	33196
Ghost In The Machine (81) 2(6) [24|30|109]	18830
Zenyatta Mondatta (80) 5(6) [21|31|153]	17724
Every Breath You Take-The Singles (86) 7(1) [4|13|26]	4811
Reggatta De Blanc (79/80) 36(1) [0|1|91]	*3741*
Outlandos D'Amour (79/83) 130(1) [0|0|36]	*1059*
Jean-Luc PONTY ▶ 562 Pk:44 [0|0|6]	3618
Mystical Adventure (82) 44(1) [0|0|14]	1135
Civilized Evil (80) 73(1) [0|0|18]	974
Individual Choice (83) 85(1) [0|0|15]	720
A Taste For Passion (79/80) *89(1) [0|0|11]*	*461*
Open Mind (84) 171(3) [0|0|13]	233
Fables (85) 166(2) [0|0|4]	94.5
Iggy POP ▶ 731 Pk:75 [0|0|4]	2208
Blah-Blah-Blah (86) 75(2) [0|0|27]	1451
Instinct (88) 110(2) [0|0|12]	413
Soldier (80) 125(1) [0|0|7]	243
Party (81) 166(2) [0|0|5]	100
POP WILL EAT ITSELF ▶ 1706 Pk:169 [0|0|1]	116
This Is The Day...This Is The Hour...This Is This! (89) 169(2) [0|0|6]	116
Mike POST ▶ 1018 Pk:70 [0|0|1]	951
Television Theme Songs (82) 70(2) [0|0|17]	951
POWER STATION ▶ 290 Pk:6 [1|1|1]	9938
The Power Station (85) 6(2) [11|25|44]	9938
PREFAB SPROUT ▶ 1747 Pk:178 [0|0|1]	89.7
Two Wheels Good (85) 178(1) [0|0|5]	89.7
Elvis PRESLEY ▶ 421 Pk:27 [0|1|13]	5593
Elvis Aron Presley (80) 27(2) [0|5|14]	1668
Guitar Man (81) 49(1) [0|0|12]	928
A Golden Celebration (84) 80(2) [0|0|19]	892
This Is Elvis (Soundtrack) (81) 115(3) [0|0|10]	358
Rocker (84) 154(1) [0|0|13]	320
The Elvis Medley (82) 133(2) [0|0|9]	275
The Top Ten Hits (87) 117(2) [0|0|8]	251
I Was The One (83) 103(1) [0|0|6]	240
The Number One Hits (87) 143(1) [0|0|9]	235
Greatest Hits Volume One (81) 142(1) [0|0|7]	207
Elvis: The First Live Recordings (84) 163(1) [0|0|4]	98.6
A Valentine Gift For You (85) 154(1) [0|0|3]	80.9
The Christmas Album (85) 178(2) [0|0|2]	39.8
Billy PRESTON ▶ 898 Pk:49 [0|0|2]	1389
Late At Night (80) 49(2) [0|0|18]	1237
Billy Preston & Syreeta (81)-Billy Preston & Syreeta [A] 127(4) [0|0|9]	152
The PRETENDERS ▶ 102 Pk:5 [3|5|6]	28086
Learning To Crawl (84) 5(4) [10|22|42]	9210
Pretenders (80) 9(2) [4|17|78]	8342
Get Close (86) 25(3) [0|14|29]	3946

Act ▶ Rank Peak [Top10s|Top40s|Total] Title (Yr) Pk(Wks) [T10 Wks|T40Wks|TotWks]	Score in '80s
Pretenders II (81) 10(3) [3|9|19]	3453
Extended Play (81) 27(2) [0|4|29]	2289
The Singles (87) 69(3) [0|0|15]	847
PRETTY MAIDS ▶ 1627 Pk:165 [0|0|1]	158
Future World (87) 165(2) [0|0|8]	158
PRETTY POISON ▶ 1388 Pk:104 [0|0|1]	336
Catch Me, I'm Falling (88) 104(2) [0|0|8]	336
Andre PREVIN ▶ 1710 Pk:149 [0|0|1]	112
A Different Kind Of Blues (81)-Itzhak Perlman & Andre Previn [A] 149(1) [0|0|7]	112
Ray PRICE ▶ 1147 Pk:70 [0|0|1]	637
San Antonio Rose (80)-Willie Nelson & Ray Price [A] 70(2) [0|0|25]	637
Charley PRIDE ▶ 1723 Pk:185 [0|0|1]	103
Greatest Hits (81) 185(1) [0|0|7]	103
Maxi PRIEST ▶ 1176 Pk:108 [0|0|1]	586
Maxi Priest (88) 108(1) [0|0|17]	586
The PRIMITIVES ▶ 1347 Pk:106 [0|0|2]	394
Lovely (88) 106(2) [0|0|19]	347
Pure (89/90) 170(2) [0|0|2]	*47.0*
PRINCE ▶ 31 Pk:1 [3|6|7]	52965
1999 (82) 9(7) [11|57|153]	21369
Batman (Soundtrack) (89) 1(6) [10|17|26]	*10949*
Sign 'O' The Times (87) 6(2) [4|12|54]	8285
Controversy (81) 21(3) [0|5|64]	4658
Lovesexy (88) 11(1) [0|9|21]	3049
Prince (80) 22(2) [0|7|21]	*2705*
Dirty Mind (80) 45(1) [0|0|52]	1952
PRINCE And The REVOLUTION ▶ 29 Pk:1 [3|3|3]	55192
Purple Rain (Soundtrack) (84) 1(24) [32|42|72]	33519
Around The World In A Day (85) 1(3) [14|27|40]	13937
Parade: Music From The Motion Picture Under The Cherry Moon (Soundtrack) (86) 3(3) [9|17|28]	7736
John PRINE ▶ 1580 Pk:144 [0|0|1]	187
Storm Windows (80) 144(1) [0|0|7]	187
PRISM ▶ 879 Pk:53 [0|0|1]	1443
Small Change (82) 53(1) [0|0|20]	1443
The PROCLAIMERS ▶ 1397 Pk:125 [0|0|1]	333
Sunshine On Leith (89/93) 125(2) [0|0|11]	*333*
The PRODUCERS ▶ 1846 Pk:163 [0|0|1]	51.1
The Producers (81) 163(1) [0|0|2]	51.1
PROPHET ▶ 1570 Pk:137 [0|0|1]	195
Cycle Of The Moon (88) 137(1) [0|0|7]	195
Richard PRYOR ▶ 634 Pk:21 [0|1|2]	2904
Richard Pryor Live On The Sunset Strip (Soundtrack) (82) 21(3) [7|7|17]	2244
Richard Pryor: Here And Now (Soundtrack) (83) 71(1) [0|0|13]	660
PSEUDO ECHO ▶ 766 Pk:54 [0|0|1]	1973
Love An Adventure (87) 54(3) [0|0|27]	1973
PSYCHEDELIC FURS ▶ 316 Pk:29 [0|1|7]	8455
Midnight To Midnight (87) 29(2) [0|11|27]	3199
Forever Now (82) 61(4) [0|0|32]	2178
Mirror Moves (84) 43(2) [0|0|27]	1920
Talk Talk Talk (81) 89(2) [0|0|14]	554
All Of This And Nothing (88) 102(2) [0|0|8]	293
The Psychedelic Furs (80) 140(1) [0|0|7]	199
Book Of Days (89) 138(2) [0|0|4]	112
PUBLIC ENEMY ▶ 608 Pk:42 [0|0|2]	3115
It Takes A Nation Of Millions To Hold Us Back (88) 42(2) [0|0|51]	2746
Yo! Bum Rush The Show (88) 125(1) [0|0|12]	369
PUBLIC IMAGE LIMITED ▶ 827 Pk:106 [0|0|5]	1672
9 (89) 106(2) [0|0|23]	791
Album (86) 115(2) [0|0|16]	487
Happy? (87) 169(1) [0|0|10]	189
The Flowers Of Romance (81) 114(2) [0|0|4]	140
Second Edition (80) 171(2) [0|0|3]	64.8

Act ▶ Rank Peak [Top10s|Top40s|Total] Title (Yr) Pk(Wks) [T10 Wks|T40Wks|TotWks]	Score in '80s
PURE PRAIRIE LEAGUE ▶ 677 Pk:37 [0|1|2]	2549
Firin' Up (80) 37(2) [0|3|24]	1825
Something In The Night (81) 72(1) [0|0|15]	724
PURSUIT OF HAPPINESS ▶ 1042 Pk:93 [0|0|1]	864
Love Junk (88) 93(1) [0|0|21]	864

Q

Act ▶ Rank Peak [Top10s|Top40s|Total] Title (Yr) Pk(Wks) [T10 Wks|T40Wks|TotWks]	Score in '80s
QUARTERFLASH ▶ 269 Pk:8 [1|2|3]	10752
Quarterflash (81) 8(3) [4|21|52]	8454
Take Another Picture (83) 34(1) [0|4|21]	2153
Back Into Blue (85) 150(2) [0|0|5]	145
Suzi QUATRO ▶ 1716 Pk:165 [0|0|2]	108
Rock Hard (80) 165(1) [0|0|5]	95.1
Suzi...And Other Four Letter Words (79/80) *193(1) [0|0|1]*	*13.2*
QUEEN ▶ 86 Pk:1 [1|6|7]	31224
The Game (80) 1(5) [21|31|43]	17315
Greatest Hits (81) 14(6) [0|13|26]	4489
The Works (84) 23(1) [0|9|20]	2616
Flash Gordon (Soundtrack) (80) 23(2) [0|7|15]	2162
Hot Space (82) 22(3) [0|5|21]	2129
The Miracle (89) 24(1) [0|4|14]	1496
A Kind Of Magic (86) 46(3) [0|0|13]	1018
Queen LATIFAH ▶ 1845 Pk:174 [0|0|1]	51.2
All Hail The Queen (89/90) 174(1) [0|0|3]	*51.2*
QUEENSRYCHE ▶ 388 Pk:47 [0|0|4]	6786
Operation: Mindcrime (88) 50(2) [0|0|52]	2855
Rage For Order (86) 47(5) [0|0|21]	1599
The Warning (84) 61(3) [0|0|23]	1270
Queensryche (83) 81(1) [0|0|22]	1062
QUIET RIOT ▶ 120 Pk:1 [1|3|4]	24485
Metal Health (83) 1(1) [17|36|81]	17487
Condition Critical (84) 15(3) [0|10|28]	3643
QR III (86) 31(2) [0|9|27]	2958
Quiet Riot (88) 119(2) [0|0|11]	398

R

Act ▶ Rank Peak [Top10s|Top40s|Total] Title (Yr) Pk(Wks) [T10 Wks|T40Wks|TotWks]	Score in '80s
Eddie RABBITT ▶ 233 Pk:19 [0|3|5]	12673
Horizon (80) 19(2) [0|15|54]	6040
Step By Step (81) 23(1) [0|8|34]	3442
Radio Romance (82) 31(2) [0|7|25]	2715
Greatest Hits - Vol. II (83) 131(3) [0|0|11]	313
The Best Of Eddie Rabbitt (80) 151(1) [0|0|6]	*162*
Trevor RABIN ▶ 1315 Pk:111 [0|0|1]	421
Can't Look Away (89) 111(3) [0|0|10]	421
The RADIATORS ▶ 1134 Pk:122 [0|0|2]	669
Zigzagging Through Ghostland (89) 122(2) [0|0|11]	338
Law Of The Fish (87) 139(2) [0|0|16]	332
Gilda RADNER ▶ 1276 Pk:69 [0|0|1]	467
Live From New York (80) 69(2) [0|0|7]	*467*
Gerry RAFFERTY ▶ 1155 Pk:61 [0|0|1]	620
Snakes And Ladders (80) 61(1) [0|0|9]	620
RAGING SLAB ▶ 1372 Pk:113 [0|0|1]	353
Raging Slab (89) 113(1) [0|0|10]	*353*
RAIL ▶ 1493 Pk:143 [0|0|1]	256
Rail (84) 143(1) [0|0|11]	256
RAINBOW ▶ 408 Pk:30 [0|2|5]	6101
Straight Between The Eyes (82) 30(3) [0|5|23]	2206
Bent Out Of Shape (83) 34(1) [0|4|21]	2102
Difficult To Cure (81) 50(2) [0|0|16]	1232
Finyl Vinyl (86) 87(2) [0|0|10]	452
Jealous Lover (81) 147(1) [0|0|4]	109
The RAINMAKERS ▶ 852 Pk:85 [0|0|2]	1546
The Rainmakers (86) 85(1) [0|0|21]	914
Tornado (87) 116(2) [0|0|19]	632

350

Acts With Albums

Act ▶ Rank Peak [Top10s|Top40s|Total] Title (Yr) Pk(Wks) [T10 Wks|T40Wks|TotWks]	Score in '80s
Bonnie RAITT ▶ 369 Pk:22 [0|2|4]	7267
Nick Of Time (89/90) 22(2) [0|14|38]	*4932*
Green Light (82) 38(1) [0|3|18]	1551
Nine Lives (86) 115(1) [0|0|11]	423
The Glow (79/80) 90(1) [0|0|9]	*361*
The RAMONES ▶ 623 Pk:44 [0|0|8]	2986
End Of The Century (80) 44(1) [0|0|14]	1133
Pleasant Dreams (81) 58(2) [0|0|11]	731
Subterranean Jungle (83) 83(3) [0|0|9]	475
Brain Drain (89) 122(2) [0|0|6]	232
Animal Boy (86) 143(2) [0|0|6]	155
Too Tough To Die (84) 171(1) [0|0|6]	111
Ramones Mania (88) 168(2) [0|0|5]	95.1
Halfway To Sanity (87) 172(1) [0|0|3]	54.3
RANK AND FILE ▶ 1731 Pk:165 [0|0|1]	97.4
Sundown (83) 165(1) [0|0|5]	97.4
Billy RANKIN ▶ 1361 Pk:119 [0|0|1]	369
Growin' Up Too Fast (84) 119(1) [0|0|11]	369
Kenny RANKIN ▶ 1711 Pk:171 [0|0|1]	112
After The Roses (80) 171(1) [0|0|6]	112
RANKING ROGER ▶ 1593 Pk:151 [0|0|1]	177
Radical Departure (88) 151(2) [0|0|7]	177
RATT ▶ 112 Pk:7 [2|4|5]	26024
Out Of The Cellar (84) 7(4) [9|26|56]	10432
Invasion Of Your Privacy (85) 7(1) [6|18|42]	7130
Reach For The Sky (86) 17(1) [0|13|27]	4190
Dancing Undercover (86) 26(2) [0|7|40]	3706
Ratt (84) 133(3) [0|0|19]	567
RAVEN ▶ 979 Pk:81 [0|0|2]	1059
Stay Hard (85) 81(3) [0|0|15]	740
The Pack Is Back (86) 121(2) [0|0|10]	320
Lou RAWLS ▶ 920 Pk:81 [0|0|3]	1293
Sit Down And Talk To Me (80) 81(1) [0|0|18]	925
Shades Of Blue (81) 110(2) [0|0|6]	270
When The Night Comes (83) 163(1) [0|0|4]	97.2
RAY, GOODMAN & BROWN ▶ 486 Pk:17 [0|1|3]	4438
Ray, Goodman & Brown (80) 17(2) [0|11|23]	3677
Ray, Goodman & Brown II (80) 84(2) [0|0|12]	569
Stay (82) 151(1) [0|0|7]	193
Chris REA ▶ 1196 Pk:92 [0|0|1]	559
New Light Through Old Windows (89) 92(4) [0|0|13]	559
READY FOR THE WORLD ▶ 260 Pk:17 [0|2|3]	11185
Ready For The World (85) 17(5) [0|30|48]	7863
Long Time Coming (86) 32(2) [0|5|26]	2717
Ruff 'N' Ready (88) 65(2) [0|0|10]	605
REAL LIFE ▶ 908 Pk:58 [0|0|2]	1337
Heart Land (84) 58(1) [0|0|24]	1301
Send Me An Angel '89 (89) 191(1) [0|0|3]	36.8
Leon REDBONE ▶ 1446 Pk:152 [0|0|1]	289
From Branch To Branch (81) 152(1) [0|0|11]	289
The REDDINGS ▶ 1106 Pk:106 [0|0|3]	716
Steamin' Hot (82) 153(2) [0|0|12]	304
The Awakening (80) 174(3) [0|0|12]	211
Class (81) 106(2) [0|0|5]	202
RED FLAG ▶ 1796 Pk:178 [0|0|1]	72.0
Naive Art (89) 178(2) [0|0|4]	72.0
RED HOT CHILI PEPPERS ▶ 759 Pk:52 [0|0|2]	2047
Mother's Milk (89) 52(1) [0|0|16]	*1606*
Uplift Mofo Party Plan (87) 148(1) [0|0|18]	441
RED ROCKERS ▶ 1051 Pk:71 [0|0|1]	834
Good As Gold (83) 71(2) [0|0|16]	834
RED 7 ▶ 1301 Pk:105 [0|0|2]	435
Red 7 (85) 105(2) [0|0|10]	377
When The Sun Goes Down (87) 175(2) [0|0|6]	58.5
RED SIREN ▶ 1364 Pk:124 [0|0|1]	361
All Is Forgiven (89) 124(1) [0|0|12]	361

Act ▶ Rank Peak [Top10s|Top40s|Total] Title (Yr) Pk(Wks) [T10 Wks|T40Wks|TotWks]	Score in '80s
Dan REED Network ▶ 1022 Pk:95 [0|0|2]	942
Dan Reed Network (88) 95(2) [0|0|19]	804
Slam (89) 160(1) [0|0|6]	138
Lou REED ▶ 395 Pk:40 [0|1|7]	6572
New Sensations (84) 56(1) [0|0|32]	2226
New York (89) 40(1) [0|1|22]	2170
Mistrial (86) 47(2) [0|0|21]	1756
Legendary Hearts (83) 159(1) [0|0|7]	149
Growing Up In Public (80) 158(2) [0|0|5]	117
The Blue Mask (82) 169(1) [0|0|4]	79.3
Rock And Roll Diary 1967-1980 (80) 178(3) [0|0|4]	75.1
Dianne REEVES ▶ 1548 Pk:172 [0|0|1]	207
Dianne Reeves (88) 172(1) [0|0|12]	207
RE-FLEX ▶ 849 Pk:53 [0|0|1]	1565
The Politics Of Dancing (83) 53(1) [0|0|28]	1565
REGINA ▶ 1410 Pk:102 [0|0|1]	320
Curiosity (86) 102(2) [0|0|8]	320
R.E.M. ▶ 81 Pk:10 [1|6|8]	32336
Green (88) 12(3) [0|27|40]	9115
Document (87) 10(1) [1|20|33]	6689
Fables Of The Reconstruction (85) 28(2) [0|14|42]	4324
Reckoning (84) 27(5) [0|6|53]	3983
Lifes Rich Pageant (86) 21(2) [0|12|32]	3972
Murmur (83) 36(1) [0|3|30]	2216
Eponymous (88) 44(2) [0|0|19]	1066
Dead Letter Office (87) 52(1) [0|0|14]	971
RENAISSANCE ▶ 1855 Pk:196 [0|0|1]	46.3
Camera Camera (81) 196(3) [0|0|4]	46.3
RENÉ & ANGELA ▶ 490 Pk:64 [0|0|2]	4362
Street Called Desire (85) 64(4) [0|7|70]	4033
Wall To Wall (81) 100(1) [0|0|8]	329
REO SPEEDWAGON ▶ 25 Pk:1 [3|4|10]	56817
Hi Infidelity (80) 1(15) [30|50|101]	32146
Wheels Are Turnin' (84) 7(1) [7|21|49]	9090
Good Trouble (82) 7(9) [10|16|24]	6757
Life As We Know It (87) 28(2) [0|7|48]	3969
The Hits (88) 56(1) [0|0|22]	1681
A Decade Of Rock And Roll 1970 To 1980 (80) 55(1) [0|0|34]	1669
You Can Tune A Piano But You Can't Tuna Fish (78/81) 100(1) [0|0|21]	*1029*
REO Speedwagon Live/You Get What You Play For (77/81) 117(2) [0|0|13]	*332*
Ridin' The Storm Out (81) 171(2) [0|0|8]	*81.1*
Nine Lives (79/81) 184(2) [0|0|4]	*63.0*
The REPLACEMENTS ▶ 755 Pk:57 [0|0|3]	2059
Don't Tell A Soul (89) 57(2) [0|0|19]	1385
Pleased To Meet Me (87) 131(1) [0|0|19]	577
Tim (86) 183(1) [0|0|7]	97.5
RESTLESS HEART ▶ 860 Pk:73 [0|0|2]	1528
Wheels (87) 73(2) [0|0|25]	1145
Big Dreams In A Small Town (88) 114(2) [0|0|11]	383
Debbie REYNOLDS ▶ 1849 Pk:182 [0|0|1]	49.4
Do It Debbie's Way (84) 182(1) [0|0|3]	49.4
Randy RHOADS ▶ 644 Pk:6 [1|1|1]	2796
Tribute (87)-Ozzy Osbourne/Randy Rhoads [A] 6(2) [6|14|23]	2796
RHYTHM CORPS ▶ 1258 Pk:104 [0|0|1]	485
Common Ground (88) 104(2) [0|0|14]	485
Cliff RICHARD ▶ 756 Pk:80 [0|0|3]	2055
I'm No Hero (80) 80(2) [0|0|34]	1440
We Don't Talk Anymore (80) 93(2) [0|0|11]	*468*
Wired For Sound (81) 132(2) [0|0|4]	147
Keith RICHARDS ▶ 672 Pk:24 [0|1|1]	2571
Talk Is Cheap (88) 24(1) [0|7|23]	2571
Lionel RICHIE ▶ 8 Pk:1 [3|3|3]	83826
Can't Slow Down (83) 1(3) [59|78|160]	43255
Lionel Richie (82) 3(7) [21|39|140]	23598

Act ▶ Rank Peak [Top10s|Top40s|Total] Title (Yr) Pk(Wks) [T10 Wks|T40Wks|TotWks]	Score in '80s
Dancing On The Ceiling (86) 1(2) [20|38|58]	16974
Stan RIDGWAY ▶ 1495 Pk:131 [0|0|1]	255
The Big Heat (86) 131(2) [0|0|9]	255
Cheryl Pepsii RILEY ▶ 1348 Pk:128 [0|0|1]	393
Me, Myself And I (88) 128(2) [0|0|11]	393
The RINGS ▶ 1702 Pk:164 [0|0|1]	117
The Rings (81) 164(2) [0|0|6]	117
RIOT ▶ 1096 Pk:99 [0|0|3]	744
Fire Down Under (81) 99(1) [0|0|11]	398
Thundersteel (88) 150(2) [0|0|10]	240
Born In America (84) 175(1) [0|0|6]	106
Minnie RIPERTON ▶ 848 Pk:35 [0|1|1]	1570
Love Lives Forever (80) 35(2) [0|4|15]	1570
The RIPPINGTONS ▶ 933 Pk:85 [0|0|2]	1227
Kilimanjaro (88)-The Rippingtons Featuring Russ Freeman 110(2) [0|0|15]	645
Tourist In Paradise (89)-The Rippingtons Featuring Russ Freeman 85(2) [0|0|12]	583
Lee RITENOUR ▶ 552 Pk:26 [0|1|6]	3701
Rit (81) 26(1) [0|7|23]	2571
Rit/2 (82) 99(3) [0|0|14]	572
Banded Together (84) 145(2) [0|0|8]	222
Festival (89) 156(2) [0|0|8]	181
Rio (82) 163(3) [0|0|6]	142
Harlequin (85)-Dave Grusin & Lee Ritenour [A] 192(1) [0|0|2]	13.0
Joan RIVERS ▶ 733 Pk:22 [0|1|1]	2189
What Becomes A Semi-Legend Most? (83) 22(2) [0|6|21]	2189
ROACHFORD ▶ 1337 Pk:109 [0|0|1]	404
Roachford (89) 109(2) [0|0|12]	404
ROB BASE ▶ 376 Pk:31 [0|1|2]	7071
It Takes Two (88)-Rob Base & D.J. E-Z Rock 31(1) [0|8|65]	*6697*
The Incredible Base (89/90) 56(2) [0|0|4]	*375*
Marty ROBBINS ▶ 1584 Pk:170 [0|0|1]	184
Biggest Hits (83) 170(1) [0|0|9]	184
Rockie ROBBINS ▶ 996 Pk:71 [0|0|2]	1006
You And Me (80) 71(2) [0|0|16]	841
I Believe In Love (81) 147(2) [0|0|6]	165
Robbie ROBERTSON ▶ 620 Pk:38 [0|1|1]	3009
Robbie Robertson (87) 38(2) [0|4|34]	3009
Smokey ROBINSON ▶ 135 Pk:10 [1|5|9]	22713
One Heartbeat (87) 26(2) [0|17|58]	6078
Being With You (81) 10(2) [2|17|28]	5950
Warm Thoughts (80) 14(2) [0|12|21]	3776
Where There's Smoke (80) 17(2) [0|11|20]	*3205*
Yes It's You Lady (82) 33(3) [0|4|17]	1604
Touch The Sky (83) 50(3) [0|0|17]	1087
Smoke Signals (86) 104(2) [0|0|13]	487
Essar (84) 141(2) [0|0|11]	288
Blame It On Love And All The Great Hits (83) 124(1) [0|0|7]	238
The ROCHES ▶ 1464 Pk:130 [0|0|2]	274
Nurds (80) 130(1) [0|0|7]	226
Keep On Doing (82) 183(1) [0|0|3]	48.5
ROCK And HYDE ▶ 1136 Pk:94 [0|0|1]	660
Under The Volcano (87) 94(1) [0|0|15]	660
The ROCKETS ▶ 969 Pk:53 [0|0|2]	1097
No Ballads (80) 53(2) [0|0|15]	982
Back Talk (81) 165(1) [0|0|5]	115
ROCKIN' SIDNEY ▶ 1739 Pk:166 [0|0|1]	92.7
My Toot-Toot (85) 166(3) [0|0|4]	92.7
ROCKPILE ▶ 658 Pk:27 [0|1|1]	2718
Seconds Of Pleasure (80) 27(3) [0|10|19]	2718
ROCKWELL ▶ 482 Pk:15 [0|1|2]	4482
Somebody's Watching Me (84) 15(2) [0|12|30]	4170
Captured (85) 120(2) [0|0|8]	313
Paul RODGERS ▶ 1434 Pk:135 [0|0|1]	296
Cut Loose (83) 135(1) [0|0|10]	296

Acts With Albums

Act ▶ Rank Peak [Top10s|Top40s|Total] Title (Yr) Pk(Wks) [T10 Wks|T40Wks|TotWks]	Score in '80s
RODNEY O & Joe COOLEY ▶ 1882 Pk:187 [0|0|1]	31.8
Me And Joe (89) 187(2) [0|0|2]	31.8
ROGER ▶ 412 Pk:26 [0|2|3]	5897
The Many Facets Of Roger (81) 26(1) [0|6|25]	2604
Unlimited! (87) 35(1) [0|5|24]	2461
The Saga Continues... (84) 64(2) [0|0|14]	832
Kenny ROGERS ▶ 15 Pk:1 [4|11|22]	74617
Kenny Rogers' Greatest Hits (80) 1(2) [20|36|181]	23789
Eyes That See In The Dark (83) 6(4) [8|24|38]	9258
Share Your Love (81) 6(2) [5|13|50]	7735
Kenny (80) 5(2) [10|13|39]	*6600*
Gideon (80) 12(2) [0|11|34]	4286
Twenty Greatest Hits (83) 22(2) [0|11|30]	3764
We've Got Tonight (83) 18(1) [0|12|27]	3443
The Gambler (**79**/80) 41(1) [0|0|57]	*3310*
What About Me? (84) 31(3) [0|8|31]	3157
Love Will Turn You Around (82) 34(2) [0|7|24]	2291
The Heart Of The Matter (85) 51(3) [0|0|28]	2185
Ten Years Of Gold (**78**/80) 83(2) [0|0|40]	*1346*
Christmas (81) 34(2) [0|4|13]	1245
Duets (84) 85(2) [0|0|11]	543
Once Upon A Christmas (84)- Kenny Rogers & Dolly Parton [A] 31(1) [0|4|8]	491
They Don't Make Them Like They Used To (86) 137(1) [0|0|15]	465
Something Inside So Strong (89) 141(2) [0|0|8]	239
Love Is What We Make It (85) 145(1) [0|0|7]	187
Christmas In America (89) 119(1) [0|0|3]	*129*
I Prefer The Moonlight (87) 163(2) [0|0|4]	78.8
Kenny Rogers (**77**/80) 187(2) [0|0|3]	*46.3*
Daytime Friends (**77**/80) 189(2) [0|0|2]	*30.0*
Kenny ROGERS & Dottie WEST ▶ 1741 Pk:170 [0|0|2]	91.4
Classics (**79**/80) 170(2) [0|0|2]	*47.0*
Everytime Two Fools Collide (80) 186(2) [0|0|3]	44.4
ROLLING STONES ▶ 20 Pk:1 [6|7|14]	64355
Tattoo You (81) 1(9) [22|30|58]	19948
Emotional Rescue (80) 1(7) [14|20|51]	13647
Steel Wheels (89) 3(4) [12|15|16]	*8057*
Dirty Work (86) 4(2) [6|15|25]	6129
Undercover (83) 4(2) [4|12|23]	5006
Still Life (American Concert 1981) (82) 5(4) [5|10|23]	4598
Hot Rocks 1964-1971 (**72**/80) 81(2) [0|0|71]	*2696*
Sucking In The Seventies (81) 15(2) [0|6|12]	1967
Singles Collection - The London Years (89) 91(1) [0|0|17]	*899*
Rewind (1971-1984) (84) 86(1) [0|0|11]	508
Some Girls (**78**/80) 124(1) [0|0|15]	*469*
Sticky Fingers (**71**/80) 133(1) [0|0|12]	*225*
Beggars Banquet (**69**/80) 169(1) [0|0|6]	*116*
Let It Bleed (**69**/80) 177(1) [0|0|5]	*90.1*
ROMAN HOLLIDAY ▶ 1237 Pk:116 [0|0|2]	510
Roman Holliday (83) 142(1) [0|0|11]	284
Cookin' On The Roof (83) 116(1) [0|0|6]	226
The ROMANTICS ▶ 381 Pk:14 [0|1|5]	6953
In Heat (83) 14(1) [0|15|36]	5264
The Romantics (80) 61(1) [0|0|15]	940
Rhythm Romance (85) 72(2) [0|0|11]	615
National Breakout (80) 176(2) [0|0|7]	97.1
Strictly Personal (81) 182(1) [0|0|5]	35.8
ROMEO'S DAUGHTER ▶ 1894 Pk:191 [0|0|1]	28.2
Romeo's Daughter (88) 191(2) [0|0|3]	28.2
ROMEO VOID ▶ 851 Pk:68 [0|0|3]	1561
Instincts (84) 68(4) [0|0|19]	1032
Benefactor (82) 119(3) [0|0|13]	369
Never Say Never (82) 147(1) [0|0|6]	160

Act ▶ Rank Peak [Top10s|Top40s|Total] Title (Yr) Pk(Wks) [T10 Wks|T40Wks|TotWks]	Score in '80s
Mick RONSON ▶ 1697 Pk:157 [0|0|1]	121
Y U I Orta (89)-Ian Hunter/Mick Ronson [A] 157(1) [0|0|10]	*121*
Linda RONSTADT ▶ 44 Pk:3 [4|7|10]	43801
What's New (83)- Linda Ronstadt & The Nelson Riddle Orchestra 3(5) [15|25|81]	14209
Mad Love (80) 3(4) [12|17|36]	9691
Lush Life (84) 13(2) [0|11|26]	4001
Cry Like A Rainstorm, Howl Like The Wind (**89**/90)- Linda Ronstadt (Featuring Aaron Neville) 9(1) [1|10|11]	*3156*
Greatest Hits, Volume 2 (80) 26(3) [0|10|21]	2755
Get Closer (82) 31(4) [0|6|28]	2584
Canciones de Mi Padre (87) 42(1) [0|0|35]	2548
Trio (87)-Dolly Parton, Linda Ronstadt, Emmylou Harris [A] 6(1) [3|14|48]	2149
For Sentimental Reasons (86) 46(1) [0|0|27]	2136
'Round Midnight (86) 124(2) [0|0|17]	571
ROSE ROYCE ▶ 1652 Pk:160 [0|0|1]	140
Golden Touch (81) 160(2) [0|0|7]	140
ROSE TATTOO ▶ 1881 Pk:197 [0|0|1]	32.3
Rock 'N' Roll Outlaw (80) 197(1) [0|0|3]	32.3
Diana ROSS ▶ 59 Pk:2 [1|7|12]	37923
Diana (80) 2(2) [18|34|52]	14528
Why Do Fools Fall In Love (81) 15(6) [0|20|33]	6188
Swept Away (84) 26(1) [0|18|45]	5805
Silk Electric (82) 27(2) [0|7|24]	2813
All The Great Hits (81) 37(2) [0|3|32]	2399
Ross (II) (83) 32(1) [0|4|17]	1565
To Love Again (81) 32(2) [0|4|14]	1468
Eaten Alive (85) 45(2) [0|0|20]	1366
Diana Ross Anthology (83) 63(1) [0|0|12]	707
Red Hot Rhythm & Blues (87) 73(2) [0|0|14]	700
Workin' Overtime (89) 116(2) [0|0|6]	*232*
The Boss (**79**/80) 150(1) [0|0|8]	*152*
ROSSINGTON COLLINS BAND ▶ 386 Pk:13 [0|2|3]	6836
Anytime, Anyplace, Anywhere (80) 13(3) [0|13|29]	5116
This Is The Way (81) 24(1) [0|5|16]	1616
Love Your Man (88)-The Rossington Band 140(1) [0|0|4]	104
David Lee ROTH ▶ 140 Pk:4 [2|3|3]	22118
Eat 'Em And Smile (86) 4(2) [8|21|36]	8849
Skyscraper (88) 6(6) [8|16|27]	6665
Crazy From The Heat (85) 15(4) [0|20|33]	6604
ROXETTE ▶ 494 Pk:28 [0|1|1]	4299
Look Sharp! (**89**/90) 28(3) [0|7|37]	*4299*
ROXY MUSIC ▶ 457 Pk:35 [0|1|5]	4875
Flesh + Blood (80) 35(2) [0|3|19]	1749
Avalon (82) 53(3) [0|0|27]	1537
Musique/The High Road (83) 67(1) [0|0|22]	1090
Street Life-20 Great Hits (89)- Bryan Ferry/Roxy Music 100(1) [0|0|11]	402
The Atlantic Years (84) 183(2) [0|0|6]	96.6
ROYAL PHILHARMONIC Orchestra ▶ 203 Pk:4 [1|2|3]	15060
Hooked On Classics (81)-Royal Philharmonic Orchestra Conducted By Louis Clark 4(6) [9|22|68]	11383
Hooked On Classics II (Can't Stop The Classics) (82) 33(4) [0|7|41]	3050
Hooked On Classics III (Journey Through The Classics) (83) 89(1) [0|0|14]	627
Jimmy RUFFIN ▶ 1646 Pk:152 [0|0|1]	143
Sunrise (80) 152(2) [0|0|6]	143
Mason RUFFNER ▶ 1054 Pk:80 [0|0|1]	829
Gypsy Blood (87) 80(2) [0|0|16]	829
RUFUS/Chaka KHAN ▶ 389 Pk:14 [0|1|4]	6766
Masterjam (**79**/80)-Rufus And Chaka Khan 14(1) [0|10|19]	*3103*

Act ▶ Rank Peak [Top10s|Top40s|Total] Title (Yr) Pk(Wks) [T10 Wks|T40Wks|TotWks]	Score in '80s
Live-Stompin' At The Savoy (83)-Rufus And Chaka Khan 50(1) [0|0|33]	2461
Camouflage (81)-Rufus And Chaka Khan 98(1) [0|0|14]	644
Party 'Til You're Broke (81)-Rufus 73(1) [0|0|9]	559
Todd RUNDGREN ▶ 697 Pk:48 [0|0|4]	2415
Healing (81) 48(2) [0|0|13]	954
The Ever Popular Tortured Artist Effect (83) 66(3) [0|0|13]	748
Nearly Human (89) 102(2) [0|0|11]	482
A Cappella (85) 128(2) [0|0|8]	232
RUN-D.M.C. ▶ 88 Pk:3 [2|2|4]	30840
Raising Hell (86) 3(3) [14|49|71]	18704
Tougher Than Leather (88) 9(1) [2|14|28]	4731
King Of Rock (85) 52(3) [0|0|56]	4066
Run D.M.C. (84) 53(2) [0|0|65]	3339
RUSH ▶ 40 Pk:3 [6|9|11]	46702
Moving Pictures (81) 3(3) [14|29|68]	12916
Permanent Waves (80) 4(3) [5|15|36]	6158
Power Windows (85) 10(2) [2|14|28]	5395
Signals (82) 10(1) [1|11|33]	4857
Grace Under Pressure (84) 10(4) [4|12|27]	4703
Hold Your Fire (87) 13(2) [0|14|30]	4660
Exit...Stage Left (81) 10(3) [3|14|21]	4296
A Show Of Hands (89) 21(1) [0|5|15]	1889
Presto (89) 16(2) [0|4|5]	*1236*
2112 (**76**/80) 142(2) [0|0|21]	*428*
All The World's A Stage: Recorded Live (**76**/80) 153(2) [0|0|7]	*166*
Jennifer RUSH ▶ 1376 Pk:118 [0|0|1]	346
Heart Over Mind (87) 118(3) [0|0|10]	346
Patrice RUSHEN ▶ 327 Pk:14 [0|3|5]	8255
Straight From The Heart (82) 14(2) [0|8|28]	3256
Now (84) 40(3) [0|3|25]	1787
Pizzazz (80) 39(2) [0|2|16]	*1402*
Posh (80) 71(1) [0|0|18]	945
Watch Out! (87) 77(2) [0|0|19]	866
Brenda RUSSELL ▶ 685 Pk:49 [0|0|3]	2496
Get Here (88) 49(2) [0|0|28]	2067
Love Life (81) 107(2) [0|0|8]	310
Brenda Russell (**79**/80) 137(1) [0|0|5]	*118*
Leon RUSSELL ▶ 1887 Pk:187 [0|0|1]	30.9
The Live Album (81)-Leon Russell And The New Grass Revival 187(1) [0|0|2]	30.9
Mike RUTHERFORD ▶ 1338 Pk:145 [0|0|2]	402
Smallcreep's Day (80) 163(2) [0|0|11]	236
Acting Very Strange (82) 145(2) [0|0|6]	166
Mitch RYDER ▶ 1441 Pk:120 [0|0|1]	291
Never Kick A Sleeping Dog (83) 120(1) [0|0|9]	291

S

Act ▶ Rank Peak [Top10s|Top40s|Total] Title (Yr) Pk(Wks) [T10 Wks|T40Wks|TotWks]	Score in '80s
Sue SAAD And The NEXT ▶ 1396 Pk:131 [0|0|1]	334
Sue Saad And The Next (80) 131(1) [0|0|12]	334
SAD CAFÉ ▶ 1668 Pk:160 [0|0|1]	132
Sad Café (81) 160(2) [0|0|6]	132
SADE ▶ 69 Pk:1 [3|3|3]	35322
Promise (85) 1(2) [16|27|46]	15256
Diamond Life (85) 5(2) [10|27|81]	12207
Stronger Than Pride (88) 7(2) [6|21|45]	7859
SA-FIRE ▶ 784 Pk:79 [0|0|1]	1904
Sa-Fire (88)-SaFire 79(1) [0|0|46]	1904
SAGA ▶ 459 Pk:29 [0|1|3]	4864
Worlds Apart (82) 29(5) [0|13|36]	3995
Behaviour (85) 87(2) [0|0|10]	485
Heads Of Tales (83) 92(1) [0|0|9]	384
Carole Bayer SAGER ▶ 965 Pk:60 [0|0|1]	1111
Sometimes Late At Night (81) 60(1) [0|0|22]	1111

Act ▶ Rank Peak [Top10s|Top40s|Total] Title (Yr) Pk(Wks) [T10 Wks|T40Wks|TotWks]	Score in '80s
SALSOUL Orchestra ▶ 1725 Pk:170 [0|0|1]	102
Christmas Jollies II (81) 170(2) [0|0|5]	102
SALT-N-PEPA ▶ 355 Pk:26 [0|2|2]	7535
Hot, Cool And Vicious (87) 26(4) [0|14|53]	5011
A Salt With A Deadly Pepa (88) 38(1) [0|4|31]	2524
Joe SAMPLE ▶ 782 Pk:65 [0|0|3]	1912
Voices In The Rain (81) 65(3) [0|0|20]	1174
Spellbound (89) 129(1) [0|0|14]	382
The Hunter (83) 125(1) [0|0|14]	356
David SANBORN ▶ 243 Pk:45 [0|0|8]	12156
Backstreet (83) 81(1) [0|0|33]	1833
A Change Of Heart (87) 74(2) [0|0|37]	1801
Voyeur (81) 45(2) [0|0|22]	1780
Double Vision (86)-Bob James/David Sanborn [A] 50(2) [0|0|64]	1539
Close-Up (88) 59(4) [0|0|28]	1471
Straight To The Heart (85) 64(2) [0|0|32]	1424
As We Speak (82) 70(4) [0|0|23]	1202
Hideaway (80) 63(1) [0|0|19]	1106
SANTANA ▶ 223 Pk:9 [1|2|6]	13684
Zebop! (81) 9(4) [7|21|32]	7569
Shango (82) 22(3) [0|0|10]	3198
Beyond Appearances (85) 50(2) [0|0|21]	1535
Marathon (79/80) 66(1) [0|0|11]	727
Freedom (87) 95(2) [0|0|11]	489
Viva Santana (88) 142(2) [0|0|6]	166
Carlos SANTANA ▶ 739 Pk:31 [0|1|3]	2171
Havana Moon (83) 31(2) [0|4|17]	1520
The Swing Of Delight (80)-Devadip Carlos Santana 65(2) [0|0|10]	639
Blues For Salvador (87) 195(1) [0|0|1]	12.3
SARAYA ▶ 892 Pk:79 [0|0|1]	1398
Saraya (89) 79(3) [0|0|35]	1398
Father Guido SARDUCCI ▶ 1865 Pk:179 [0|0|1]	39.0
Live At St. Douglas Convent (80) 179(2) [0|0|2]	39.0
Joe SATRIANI ▶ 282 Pk:23 [0|2|3]	10267
Surfing With The Alien (87) 29(4) [0|9|75]	6288
Dreaming #11 (88) 42(1) [0|0|26]	2402
Flying In A Blue Dream (89) 23(3) [0|6|7]	1577
SAVATAGE ▶ 1082 Pk:116 [0|0|2]	777
Hall Of The Mountain King (87) 116(1) [0|0|23]	601
Fight For The Rock (86) 158(2) [0|0|7]	176
SAVOY BROWN ▶ 1817 Pk:185 [0|0|1]	62.6
Rock 'N' Roll Warriors (81) 185(1) [0|0|4]	62.6
SAWYER BROWN ▶ 1432 Pk:140 [0|0|1]	297
Sawyer Brown (85) 140(2) [0|0|11]	297
SAXON ▶ 1131 Pk:130 [0|0|4]	677
Innocence Is No Excuse (85) 130(2) [0|0|8]	236
Power & The Glory (83) 155(1) [0|0|10]	215
Rock The Nations (87) 149(2) [0|0|6]	135
Crusader (84) 174(2) [0|0|5]	91.0
Leo SAYER ▶ 748 Pk:36 [0|1|1]	2102
Living In A Fantasy (80) 36(2) [0|6|23]	2102
Boz SCAGGS ▶ 235 Pk:8 [1|2|3]	12527
Middle Man (80) 8(2) [4|21|33]	7364
Hits! (80) 24(6) [0|14|26]	3737
Other Roads (88) 47(4) [0|0|18]	1427
SCANDAL/Patty SMYTH ▶ 341 Pk:17 [0|2|2]	7826
Warrior (84)-Scandal Featuring Patty Smyth 17(3) [0|14|41]	5324
Scandal (83)-Scandal 39(1) [0|3|32]	2502
Joey SCARBURY ▶ 1381 Pk:104 [0|0|1]	344
America's Greatest Hero (81) 104(2) [0|3|9]	344
SCARLETT & BLACK ▶ 1354 Pk:107 [0|0|1]	386
Scarlett & Black (88) 107(2) [0|0|11]	386
Michael SCHENKER Group ▶ 740 Pk: 81 [0|0|4]	2158
Perfect Timing (87)-McAuley Schenker Group 95(2) [0|0|24]	1052

Act ▶ Rank Peak [Top10s|Top40s|Total] Title (Yr) Pk(Wks) [T10 Wks|T40Wks|TotWks]	Score in '80s
The Michael Schenker Group (80) 100(1) [0|0|14]	522
MSG (81) 81(1) [0|0|8]	416
Assault Attack (83) 151(1) [0|0|7]	168
Peter SCHILLING ▶ 888 Pk:61 [0|0|1]	1412
Error In The System (83) 61(1) [0|0|23]	1412
Timothy B. SCHMIT ▶ 1219 Pk:106 [0|0|2]	529
Timothy B. (87) 106(2) [0|0|11]	413
Playin' It Cool (84) 160(2) [0|0|5]	117
John SCHNEIDER ▶ 646 Pk:37 [0|1|3]	2778
Now Or Never (81) 37(1) [0|3|22]	2217
Too Good To Stop Now (84) 111(2) [0|0|12]	398
White Christmas (81) 155(1) [0|0|7]	163
Neal SCHON ▶ 1366 Pk:42 [0|0|1]	358
Through The Fire (84)-Hagar, Schon, Aaronson, Shrieve [A] 42(2) [0|0|18]	358
Neal SCHON & Jan HAMMER ▶ 1099 Pk:115 [0|0|2]	738
Here To Stay (83) 122(4) [0|0|12]	434
Untold Passion (81) 115(2) [0|0|8]	304
SCHOOLLY D ▶ 1841 Pk:180 [0|0|1]	53.0
Smoke Some Kill (88) 180(1) [0|0|3]	53.0
Diane SCHUUR ▶ 1563 Pk:170 [0|0|1]	197
Talkin' 'Bout You (88) 170(3) [0|0|10]	197
Eddie SCHWARTZ ▶ 1804 Pk:195 [0|0|1]	69.6
No Refuge (82) 195(2) [0|0|6]	69.6
SCORPIONS ▶ 76 Pk:5 [3|4|7]	34310
Love At First Sting (84) 6(2) [13|27|63]	11150
Savage Amusement (88) 5(1) [7|22|43]	7874
Blackout (82) 10(2) [2|17|74]	7294
World Wide Live (85) 14(1) [0|16|43]	6031
Animal Magnetism (80) 52(1) [0|0|21]	1414
Best Of Rockers N' Ballads (89/90) 51(1) [0|0|5]	479
Best Of Scorpions Vol. 2 (84) 175(1) [0|0|4]	67.1
Tom SCOTT ▶ 1165 Pk:123 [0|0|3]	596
Apple Juice (81) 123(1) [0|0|11]	389
Desire (82) 164(1) [0|0|7]	153
Street Beat (79/80) 162(1) [0|0|3]	54.3
Gil SCOTT-HERON ▶ 918 Pk:106 [0|0|3]	1304
Reflections (81) 106(1) [0|0|27]	882
Moving Target (82) 123(1) [0|0|9]	280
Real Eyes (80) 159(2) [0|0|6]	142
Gil SCOTT-HERON & Brian JACKSON ▶ 1172 Pk:82 [0|0|1]	588
1980 (80) 82(1) [0|0|12]	588
SCREAMING BLUE MESSIAHS ▶ 1567 Pk:172 [0|0|1]	196
Bikini Red (88) 172(1) [0|0|11]	196
SCRITTI POLITTI ▶ 718 Pk:50 [0|0|2]	2286
Cupid And Psyche 85 (85) 50(2) [0|0|28]	1979
Provision (88) 113(2) [0|0|8]	308
SCRUFFY THE CAT ▶ 1705 Pk:177 [0|0|1]	116
Moons Of Jupiter (88) 177(1) [0|0|8]	116
SEA HAGS ▶ 1645 Pk:163 [0|0|1]	143
Sea Hags (89) 163(2) [0|0|7]	143
SEA LEVEL ▶ 1628 Pk:152 [0|0|1]	158
Ball Room (80) 152(1) [0|0|6]	158
Dan SEALS ▶ 1034 Pk:59 [0|0|1]	886
Won't Be Blue Anymore (86) 59(2) [0|0|15]	886
The SEARCHERS ▶ 1895 Pk:191 [0|0|1]	28.2
The Searchers (80) 191(2) [0|0|4]	28.2
Marvin SEASE ▶ 1179 Pk:114 [0|0|1]	584
Marvin Sease (87) 114(2) [0|0|17]	584
SEAWIND ▶ 1230 Pk:83 [0|0|1]	518
Seawind(2) (80) 83(1) [0|0|11]	518
Neil SEDAKA ▶ 1324 Pk:135 [0|0|1]	414
In The Pocket (80) 135(1) [0|0|13]	414
SEDUCTION ▶ 1184 Pk:78 [0|0|1]	581
Nothing Matters Without Love (89) 78(1) [0|0|10]	581

Act ▶ Rank Peak [Top10s|Top40s|Total] Title (Yr) Pk(Wks) [T10 Wks|T40Wks|TotWks]	Score in '80s
Bob SEGER And The SILVER BULLET Band ▶ 22 Pk:1 [4|4|7]	60040
Against The Wind (80) 1(6) [22|43|110]	23078
Like A Rock (86) 3(4) [15|28|62]	13204
The Distance (83) 5(6) [14|23|39]	10197
Nine Tonight (81) 3(4) [10|21|70]	9829
'Live' Bullet (76/80) 103(1) [0|0|62]	2015
Stranger In Town (78/80) 111(1) [0|0|29]	986
Night Moves (77/80) 113(1) [0|0|23]	731
The SELECTER ▶ 1773 Pk:175 [0|0|1]	78.8
Too Much Pressure (80) 175(1) [0|0|5]	78.8
Michael SEMBELLO ▶ 1306 Pk:80 [0|0|1]	433
Bossa Nova Hotel (83) 80(1) [0|0|10]	433
Brian SETZER ▶ 798 Pk:45 [0|0|2]	1810
The Knife Feels Like Justice (86) 45(3) [0|0|18]	1582
Live Nude Guitars (88) 140(2) [0|0|8]	228
707 ▶ 1335 Pk:129 [0|0|2]	404
Mega Force (82) 129(2) [0|0|9]	263
The Second Album (81) 159(1) [0|0|6]	142
7 SECONDS ▶ 1529 Pk:153 [0|0|1]	230
Soulforce Revolution (89) 153(1) [0|0|9]	230
Doc SEVERINSEN And The TONIGHT SHOW Orchestra ▶ 856 Pk:65 [0|0|1]	1535
The Tonight Show Band (86)-The Tonight Show Band with Doc Severinsen 65(4) [0|0|26]	1535
Charlie SEXTON ▶ 451 Pk:15 [0|1|2]	4989
Pictures For Pleasure (85) 15(3) [0|13|34]	4592
Charlie Sexton (89) 104(1) [0|0|9]	398
Phil SEYMOUR ▶ 1109 Pk:64 [0|0|1]	713
Phil Seymour (81) 64(1) [0|10|16]	713
SHADOWFAX ▶ 829 Pk:114 [0|0|4]	1667
Too Far To Whisper (86) 114(1) [0|0|16]	569
The Dreams Of Children (84) 126(2) [0|0|20]	552
Shadowdance (83) 145(1) [0|0|19]	451
Folksongs For A Nuclear Village (88) 168(2) [0|0|5]	95.9
SHALAMAR ▶ 241 Pk:23 [0|4|6]	12309
Big Fun (80) 23(2) [0|11|31]	3652
Three For Love (81) 40(2) [0|2|25]	3200
Friends (82) 35(1) [0|3|25]	2093
The Look (83) 38(1) [0|2|23]	1823
Heart Break (84) 90(4) [0|0|24]	1061
Go For It (81) 115(2) [0|0|15]	479
--SHANICE see: WILSON, SHANICE	
SHANNON ▶ 534 Pk:32 [0|1|2]	3856
Let The Music Play (84) 32(2) [0|4|37]	3199
Do You Wanna Get Away (85) 92(2) [0|0|16]	657
Del SHANNON ▶ 1346 Pk:123 [0|0|1]	395
Drop Down And Get Me (81) 123(1) [0|0|14]	395
Feargal SHARKEY ▶ 1162 Pk:75 [0|0|1]	604
Feargal Sharkey (86) 75(2) [0|0|11]	604
Tommy SHAW ▶ 724 Pk:50 [0|0|2]	2242
Girls With Guns (84) 50(3) [0|0|25]	1802
What If (85) 87(2) [0|0|9]	441
SHEILA E. ▶ 335 Pk:28 [0|1|3]	8061
The Glamorous Life (84) 28(2) [0|9|46]	4590
Romance 1600 (85) 50(1) [0|0|33]	2758
Sheila E. (87) 56(2) [0|0|12]	713
Pete SHELLEY ▶ 1274 Pk:121 [0|0|2]	468
Homosapien (82) 121(1) [0|0|10]	339
XL-1 (83) 151(1) [0|0|5]	129
Ricky Van SHELTON ▶ 645 Pk:76 [0|0|2]	2789
Wild-Eyed Dream (87) 76(1) [0|0|41]	1635
Loving Proof (88) 78(2) [0|0|24]	1154
T.G. SHEPPARD ▶ 1080 Pk:119 [0|0|3]	781
I Love 'Em All (81) 119(1) [0|0|12]	431
Finally! (82) 152(2) [0|0|13]	307
T.G. Sheppard's Greatest Hits (83) 189(2) [0|0|3]	42.3

Acts With Albums

Act ▶ Rank Peak [Top10s|Top40s|Total] Title (Yr) Pk(Wks) [T10 Wks|T40Wks|TotWks]	Score in '80s
SHERBS ▶ 1148 Pk:100 [0|0|1]	**635**
The Skill (81)-The Sherbs 100(2) [0|0|16]	635
SHERIFF ▶ 1033 Pk:60 [0|0|1]	**888**
Sheriff (89) 60(2) [0|0|14]	888
SHINEHEAD ▶ 1833 Pk:185 [0|0|1]	**56.3**
Unity (88) 185(1) [0|0|4]	56.3
Michelle SHOCKED ▶ 728 Pk:73 [0|0|2]	**2231**
Short Sharp Shocked (88) 73(1) [0|0|35]	1866
Captain Swing (89) 95(2) [0|0|8]	*365*
SHOES ▶ 1546 Pk:140 [0|0|1]	**211**
Tongue Twister (81) 140(3) [0|0|7]	211
SHOOTING STAR ▶ 706 Pk:82 [0|0|5]	**2369**
Hang On For Your Life (81) 92(1) [0|0|30]	1365
III Wishes (82) 82(2) [0|0|9]	412
Shooting Star (80) 147(1) [0|0|14]	293
Touch Me Tonight-The Best Of Shooting Star (89) 151(2) [0|0|7]	163
Burning (83) 162(1) [0|0|6]	137
SHOTGUN MESSIAH ▶ 1222 Pk:99 [0|0|1]	**528**
Shotgun Messiah (89) 99(1) [0|0|11]	*528*
SHRIEKBACK ▶ 1277 Pk:145 [0|0|3]	**467**
Go Bang! (88) 169(2) [0|0|12]	254
Big Night Music (87) 145(2) [0|0|6]	171
Care (83) 188(1) [0|0|3]	41.3
Michael SHRIEVE ▶ 1367 Pk:42 [0|0|1]	**358**
Through The Fire (84)-Hagar, Schon, Aaronson, Shrieve [A] 42(2) [0|0|18]	358
SHY ▶ 1913 Pk:193 [0|0|1]	**24.1**
Excess All Areas (87) 193(1) [0|0|2]	24.1
Jane SIBERRY ▶ 1544 Pk:149 [0|0|1]	**216**
The Speckless Sky (86) 149(2) [0|0|8]	216
SIDEWINDERS ▶ 1727 Pk:169 [0|0|1]	**100**
Witchdoctor (89) 169(1) [0|0|5]	100
SIGUE SIGUE SPUTNIK ▶ 1343 Pk:96 [0|0|1]	**397**
Flaunt It (86) 96(2) [0|0|10]	397
The SILENCERS ▶ 1476 Pk:147 [0|0|1]	**267**
A Letter From St. Paul (87) 147(2) [0|0|11]	267
SILVER CONDOR ▶ 1362 Pk:141 [0|0|1]	**365**
Silver Condor (81) 141(4) [0|0|12]	365
Patrick SIMMONS ▶ 1117 Pk:52 [0|0|1]	**707**
Arcade (83) 52(1) [0|0|11]	707
Richard SIMMONS ▶ 700 Pk:44 [0|0|1]	**2402**
Reach (82) 44(2) [0|0|40]	2402
Carly SIMON ▶ 234 Pk:25 [0|2|6]	**12598**
Coming Around Again (87) 25(2) [0|10|60]	6425
Come Upstairs (80) 36(2) [0|4|32]	2600
Torch (81) 50(1) [0|0|24]	1573
Hello Big Man (83) 69(2) [0|0|17]	862
Greatest Hits Live (88) 87(3) [0|0|13]	626
Spoiled Girl (85) 88(2) [0|0|11]	512
Paul SIMON ▶ 91 Pk:3 [1|3|4]	**29749**
Graceland (86) 3(1) [24|52|97]	22828
One-Trick Pony (80) 12(2) [0|13|26]	4398
Hearts And Bones (83) 35(2) [0|8|18]	1923
Negotiations And Love Songs (1971-1986) (88) 110(1) [0|0|14]	600
SIMON & GARFUNKEL ▶ 443 Pk:6 [1|1|1]	**5189**
The Concert In Central Park (82) 6(2) [5|11|34]	5189
SIMPLE MINDS ▶ 242 Pk:10 [1|1|5]	**12219**
Once Upon A Time (85) 10(5) [5|27|42]	8923
Sparkle In The Rain (84) 64(2) [0|0|24]	1181
New Gold Dream (81-82-83-84) (83) 69(2) [0|0|19]	1069
Street Fighting Years (89) 70(2) [0|0|12]	614
Simple Minds Live: In The City Of Light (87) 96(2) [0|0|10]	432
SIMPLY RED ▶ 209 Pk:16 [0|3|3]	**14189**
Picture Book (86) 16(3) [0|18|60]	6365
A New Flame (89) 22(1) [0|14|39]	5272

Act ▶ Rank Peak [Top10s|Top40s|Total] Title (Yr) Pk(Wks) [T10 Wks|T40Wks|TotWks]	Score in '80s
Men And Women (87) 31(2) [0|5|26]	2551
Frank SINATRA ▶ 428 Pk:17 [0|1|3]	**5440**
Trilogy: Past, Present And Future (80) 17(2) [0|12|24]	3641
She Shot Me Down (81) 52(2) [0|0|13]	1038
L.A. Is My Lady (84) 58(2) [0|0|13]	761
SIOUXSIE & The BANSHEES ▶ 749 Pk:68 [0|0|4]	**2097**
Peepshow (88) 68(1) [0|0|20]	1220
Tinderbox (86) 88(1) [0|0|15]	669
Hyaena (84) 157(1) [0|0|7]	162
Through The Looking Glass (87) 188(1) [0|0|3]	45.4
SIR DOUGLAS Quintet ▶ 1824 Pk:184 [0|0|1]	**59.8**
Border Wave (81) 184(2) [0|0|4]	59.8
Sir MIX-A-LOT ▶ 629 Pk:67 [0|0|2]	**2922**
Swass (88) 82(1) [0|0|58]	2405
Seminar (89) 67(2) [0|0|7]	*517*
SISTER SLEDGE ▶ 515 Pk:31 [0|1|4]	**4108**
Love Somebody Today (80) 31(2) [0|5|15]	1673
All American Girls (81) 42(1) [0|0|29]	1553
The Sisters (82) 69(2) [0|0|14]	735
Bet Cha Say That To All The Girls (83) 169(1) [0|0|8]	148
SISTERS OF MERCY ▶ 1124 Pk:101 [0|0|2]	**695**
Floodland (88) 101(2) [0|0|16]	695
Ricky SKAGGS ▶ 720 Pk:61 [0|0|4]	**2273**
Waitin' For The Sun To Shine (82) 77(2) [0|0|30]	1386
Highways And Heartaches (82) 61(3) [0|0|12]	746
Country Boy (84) 180(1) [0|0|5]	75.3
Favorite Country Hits (85) 181(2) [0|0|4]	66.3
SKID ROW ▶ 208 Pk:6 [1|1|1]	**14202**
Skid Row (89) 6(1) [11|42|47]	*14202*
SKY ▶ 1236 Pk:125 [0|0|2]	**510**
Sky (80) 125(1) [0|0|15]	455
Sky 3 (81) 181(1) [0|0|3]	54.8
SKYY ▶ 364 Pk:18 [0|1|6]	**7381**
Skyy Line (81) 18(2) [0|11|33]	3998
Skyway (80) 61(3) [0|0|23]	1590
Skyyport (80) 85(1) [0|0|20]	901
Skyyjammer (82) 81(2) [0|0|13]	715
Start Of A Romance (89) 155(1) [0|0|5]	127
Skyylight (83) 183(1) [0|0|3]	49.9
SLADE ▶ 679 Pk:33 [0|1|2]	**2534**
Keep Your Hands Off My Power Supply (84) 33(2) [0|4|23]	2314
Rogues Gallery (85) 132(2) [0|0|6]	220
SLAVE ▶ 526 Pk:46 [0|0|5]	**4038**
Stone Jam (80) 53(2) [0|0|34]	1849
Show Time (81) 46(1) [0|0|23]	1476
Just A Touch Of Love (80) 92(2) [0|0|11]	*512*
Visions Of The Lite (83) 177(1) [0|0|6]	106
Bad Enuff (83) 168(1) [0|0|5]	95.1
SLAYER ▶ 779 Pk:57 [0|0|2]	**1921**
South Of Heaven (88) 57(2) [0|0|19]	1107
Reign In Blood (86) 94(3) [0|0|18]	815
Grace SLICK ▶ 647 Pk:32 [0|1|2]	**2778**
Dreams (80) 32(2) [0|4|16]	1779
Welcome To The Wrecking Ball (81) 48(1) [0|0|14]	998
SLICK RICK ▶ 551 Pk:31 [0|1|1]	**3734**
The Great Adventures Of Slick Rick (89) 31(3) [0|5|40]	3734
SLY FOX ▶ 710 Pk:31 [0|1|1]	**2317**
Let's Go All The Way (86) 31(2) [0|4|22]	2317
Frankie SMITH ▶ 1059 Pk:54 [0|0|1]	**825**
Children Of Tomorrow (81) 54(2) [0|0|10]	825
Kathy SMITH ▶ 1467 Pk:144 [0|0|1]	**273**
Kathy Smith's Aerobic Fitness (82) 144(2) [0|0|13]	273

Act ▶ Rank Peak [Top10s|Top40s|Total] Title (Yr) Pk(Wks) [T10 Wks|T40Wks|TotWks]	Score in '80s
Lonnie Liston SMITH ▶ 1907 Pk:193 [0|0|1]	**25.5**
Dreams Of Tomorrow (83) 193(1) [0|0|2]	25.5
Patti SMITH ▶ 1035 Pk:65 [0|0|1]	**882**
Dream Of Life (88) 65(4) [0|0|15]	882
Rex SMITH ▶ 1631 Pk:165 [0|0|2]	**154**
Everlasting Love (81) 167(2) [0|0|4]	86.1
Forever, Rex Smith (80) 165(1) [0|0|3]	67.5
The SMITHEREENS ▶ 407 Pk:51 [0|0|3]	**6122**
Especially For You (86) 51(1) [0|0|50]	3701
Green Thoughts (88) 60(3) [0|0|31]	2015
11 (89/90) 84(2) [0|0|7]	*406*
The SMITHS ▶ 404 Pk:55 [0|0|6]	**6163**
Strangeways, Here We Come (87) 55(2) [0|0|27]	1626
The Queen Is Dead (86) 70(1) [0|0|37]	1607
Meat Is Murder (85) 110(2) [0|0|32]	1150
Louder Than Bombs (87) 62(2) [0|0|25]	1137
Rank (88) 77(2) [0|0|8]	403
The Smiths (84) 150(2) [0|0|11]	240
--SMYTH, Patti see also: SCANDAL/Patti SMYTH	
Patty SMYTH ▶ 968 Pk:66 [0|0|1]	**1098**
Never Enough (87) 66(2) [0|0|20]	1098
SNEAKER ▶ 1281 Pk:149 [0|0|1]	**459**
Sneaker (81) 149(2) [0|0|17]	459
SNIFF 'N' The TEARS ▶ 1899 Pk:192 [0|0|1]	**27.2**
Love Action (81) 192(2) [0|0|1]	27.2
Phoebe SNOW ▶ 775 Pk:51 [0|0|2]	**1943**
Rock Away (81) 51(1) [0|0|18]	1069
Something Real (89) 75(1) [0|0|20]	874
SO ▶ 1398 Pk:124 [0|0|1]	**330**
Horseshoe In The Glove (88) 124(3) [0|0|9]	330
Gino SOCCIO ▶ 1246 Pk:96 [0|0|1]	**499**
Closer (81) 96(2) [0|0|14]	499
SOFT CELL ▶ 391 Pk:22 [0|1|3]	**6671**
Non-Stop Erotic Cabaret (82) 22(1) [0|18|41]	5341
Non-Stop Ecstatic Dancing (82) 57(2) [0|0|14]	859
The Art Of Falling Apart (83) 84(4) [0|0|8]	471
S.O.S. BAND ▶ 299 Pk:12 [0|1|7]	**9408**
S.O.S. (80) 12(3) [0|11|20]	3621
On The Rise (83) 47(1) [0|0|29]	2220
Sands Of Time (86) 44(2) [0|0|20]	1728
Just The Way You Like It (84) 60(2) [0|0|27]	1449
Too (81) 117(2) [0|0|6]	206
S.O.S. III (82) 172(1) [0|0|8]	159
Diamonds In The Raw (89) 194(2) [0|0|2]	25.4
SOUL II SOUL ▶ 380 Pk:14 [0|1|1]	**6953**
Keep On Movin' (89) 14(4) [0|23|26]	*6953*
SOUNDTRACK-MOVIE ▶ Pk:1 [18|51|185]	
Dirty Dancing (87) 1(18) [48|68|96]	42442
Flashdance (83) 1(2) [25|54|78]	24557
Top Gun (86) 1(5) [20|33|93]	20959
Footloose (84) 1(10) [20|27|61]	19277
Beverly Hills Cop (85) 1(2) [22|32|62]	17526
Cocktail (88) 2(1) [19|28|61]	15573
The Big Chill (83) 17(2) [0|19|161]	13420
More Dirty Dancing (88) 3(5) [13|27|52]	12448
Urban Cowboy (80) 3(2) [14|23|47]	11366
Fame (80) 7(2) [5|16|82]	8664
The Empire Strikes Back (80) 4(4) [9|17|28]	7465
Pretty In Pink (86) 5(4) [9|17|27]	7236
Ghostbusters (84) 6(3) [8|17|34]	6744
Beverly Hills Cop II (87) 8(1) [6|17|26]	6220
Rocky IV (85) 10(1) [1|16|30]	5679
Breakin' (84) 8(2) [4|14|23]	5221
Good Morning, Vietnam (88) 10(2) [2|14|35]	5176
American Gigolo (80) 7(3) [5|15|25]	5126
Back To The Future (85) 12(2) [0|14|32]	4553
Endless Love (81) 9(2) [2|14|20]	4454
St. Elmo's Fire (85) 21(2) [0|11|37]	4413

Acts With Albums

Act ▶ Rank Peak [Top10s|Top40s|Total] Title (Yr) Pk(Wks) [T10 Wks|T40Wks|TotWks]	Score in '80s
SOUNDTRACK-MOVIE ▶ *Continued*	
Heavy Metal (81) 12(2) [0|13|28]	4307
Stand By Me (86) 31(6) [0|10|45]	4043
Against All Odds (84) 12(4) [0|12|22]	3992
Vision Quest (85) 11(1) [0|12|23]	3942
White Nights (85) 17(2) [0|12|26]	3930
The Breakfast Club (85) 17(2) [0|13|26]	3774
Lost Boys (87) 15(1) [0|9|39]	3351
Rocky III (82) 15(5) [0|10|19]	3337
Beat Street (84) 14(2) [0|9|21]	3220
Ghostbusters II (89) 14(1) [0|10|19]	3172
Annie (82) 35(2) [0|7|31]	2777
Two Of A Kind (Soundtrack) (83) 26(1) [0|8|20]	2651
Ruthless People (86) 20(1) [0|9|16]	2643
Less Than Zero (87) 31(2) [0|7|23]	2591
E.T. - The Extra-Terrestrial (82) 37(2) [0|5|33]	2566
Streets Of Fire (84) 32(2) [0|6|21]	2181
Return Of The Jedi (83) 20(2) [0|6|17]	2167
The Karate Kid Part II (86) 30(2) [0|9|17]	2159
Arthur (The Album) (81) 32(3) [0|7|22]	2144
Out Of Africa (86) 38(2) [0|2|22]	2022
All That Jazz (80) 36(2) [0|3|23]	1988
The Big Chill: More Songs From The Original Soundtrack (84) 85(1) [0|0|49]	1867
Colors (88) 31(2) [0|5|19]	1863
Teachers (84) 34(3) [0|4|16]	1782
Buster (88) 54(2) [0|0|23]	1778
An Officer And A Gentleman (82) 38(2) [0|3|23]	1752
Rain Man (89) 31(2) [0|5|16]	1739
An American Tail (87) 42(2) [0|0|19]	1633
Times Square (80) 37(2) [0|4|17]	1495
Coal Miner's Daughter (80) 40(1) [0|1|20]	1469
Running Scared (86) 43(2) [0|0|15]	1416
Fast Times At Ridgemont High (82) 54(3) [0|0|20]	1391
A View To A Kill (85) 38(2) [0|4|15]	1381
Mad Max Beyond Thunderdome (85) 39(2) [0|2|13]	1307
Little Shop Of Horrors (87) 47(2) [0|0|17]	1298
Batman Original Motion Picture Score (89) 30(1) [0|4|12]	1254
Working Girl (89) 45(2) [0|0|14]	1146
Krush Groove (85) 79(2) [0|0|20]	1084
Cat People (82) 47(2) [0|0|14]	1082
Jewel Of The Nile (85) 55(2) [0|0|17]	1045
Indiana Jones And The Temple Of Doom (John Williams) (84) 42(2) [0|0|10]	1043
Berry Gordy's The Last Dragon (85) 58(2) [0|0|15]	1025
Perfect (85) 45(2) [0|0|12]	1011
Breakin' 2 Electric Boogaloo (85) 52(2) [0|0|13]	958
Star Trek - The Motion Picture (80) 50(1) [0|0|11]	950
Say Anything (89) 62(2) [0|0|14]	913
For Your Eyes Only (81) 84(2) [0|0|19]	912
The Best Little Whorehouse In Texas (82) 63(1) [0|0|15]	888
9 1/2 Weeks (86) 59(2) [0|0|15]	842
School Daze (88) 81(2) [0|0|17]	835
Do The Right Thing (89) 68(2) [0|0|14]	812
Some Kind Of Wonderful (87) 57(2) [0|0|13]	796
Iron Eagle (86) 54(2) [0|0|11]	787
About Last Night... (86) 72(2) [0|0|14]	780
Absolute Beginners (86) 62(2) [0|0|13]	747
9 To 5 (80) 77(2) [0|0|15]	681
The Color Of Money (86) 81(2) [0|0|15]	679
The Muppet Movie: Original Soundtrack Recording (79/80) 55(1) [0|0|10]	*658*
Platoon (87) 75(2) [0|0|13]	655
Grease 2 (82) 71(2) [0|0|13]	651
The Great Muppet Caper (81) 66(2) [0|0|11]	650
The Color Purple (86) 79(2) [0|0|13]	640
Star Trek II: The Wrath Of Khan (82) 61(2) [0|0|9]	608
Caddyshack (80) 78(2) [0|0|12]	597
American Anthem (86) 91(3) [0|0|12]	582
A Chorus Line-The Movie (85) 77(2) [0|0|12]	573
Bright Lights, Big City (88) 67(2) [0|0|11]	567
Road House (89) 67(2) [0|0|10]	547
Country (84) 120(2) [0|0|15]	535
Tequila Sunrise (89) 101(1) [0|0|13]	497
Light Of Day (87) 82(2) [0|0|10]	488
Metropolis (84) 110(2) [0|0|13]	485
Cotton Club (85) 93(2) [0|0|10]	475
Smokey And The Bandit 2 (80) 103(2) [0|0|11]	473
Down And Out In Beverly Hills (86) 68(2) [0|0|7]	460
Dream A Little Dream (89) 94(2) [0|0|10]	453
"10" (80) 80(1) [0|0|9]	438
Star Trek III - The Search For Spock (84) 82(2) [0|0|8]	433
Electric Dreams (84) 94(2) [0|0|9]	426
Weird Science (85) 105(2) [0|0|11]	411
The Karate Kid (84) 114(2) [0|0|12]	406
She's Having A Baby (88) 92(1) [0|0|8]	385
The Mission (87) 132(1) [0|0|13]	375
Hiding Out (87) 146(2) [0|0|13]	373
Terms Of Endearment (84) 111(1) [0|0|10]	373
Popeye (80) 115(1) [0|0|10]	371
Scrooged (88) 93(1) [0|0|9]	368
The Big Easy (87) 107(2) [0|0|9]	342
Disorderlies (87) 99(2) [0|0|8]	338
Shocker (89) 97(2) [0|0|7]	*314*
Over The Top (87) 120(2) [0|0|8]	302
The Idolmaker (80) 130(1) [0|0|9]	291
Into The Night (85) 118(2) [0|0|8]	280
Tootsie (83) 144(3) [0|0|12]	279
Cobra (86) 100(2) [0|0|6]	277
It's My Turn (80) 137(2) [0|0|11]	274
Porky's Revenge (85) 122(2) [0|0|8]	274
Superman II (81) 133(1) [0|0|9]	270
On Golden Pond (82) 147(1) [0|0|11]	270
Ragtime (82) 134(2) [0|0|9]	267
Roadie (80) 125(2) [0|0|8]	265
The Flamingo Kid (85) 130(2) [0|0|8]	264
Beat Street II (84) 137(2) [0|0|9]	261
Soul Man (86) 138(2) [0|0|9]	253
Twins (89) 162(3) [0|0|12]	250
Any Which Way You Can (81) 141(1) [0|0|9]	245
Salsa (88) 112(2) [0|0|6]	236
The Last Emperor (88) 152(1) [0|0|10]	232
The Secret Of My Success (87) 131(2) [0|0|8]	232
The Little Mermaid (89/90) 61(1) [0|0|3]	*229*
Hairspray (88) 114(1) [0|0|6]	219
Urban Cowboy (81) 134(1) [0|0|6]	217
The Golden Child (87) 126(2) [0|0|7]	217
Beetlejuice (88) 118(1) [0|0|6]	200
Soup For One (82) 168(1) [0|0|12]	194
Gremlins (84) 143(1) [0|0|7]	194
Bronco Billy (80) 123(1) [0|0|6]	193
Dragnet (87) 137(2) [0|0|6]	179
Summer Lovers (82) 152(1) [0|0|7]	164
Octopussy (83) 137(1) [0|0|5]	157
Oliver & Company (89) 170(2) [0|0|7]	148
All The Right Moves (83) 165(1) [0|0|7]	146
Quest For Fire (82) 154(2) [0|0|7]	145
Quicksilver (86) 140(2) [0|0|5]	144
Sharky's Machine (82) 171(1) [0|0|7]	141
Party Party (83) 169(4) [0|0|11]	133
Tron (82) 135(2) [0|0|6]	129
Empire Of The Sun (88) 150(2) [0|0|5]	127
Bull Durham (88) 157(1) [0|0|6]	125
The King Of Comedy (83) 162(1) [0|0|6]	123
The Pirate Movie (82) 166(2) [0|0|6]	116
Youngblood (86) 166(2) [0|0|6]	109
Poltergeist (82) 168(1) [0|0|5]	103
Conan The Barbarian (82) 162(2) [0|0|5]	100
Jumpin' Jack Flash (86) 159(2) [0|0|4]	95.5
2010 (85) 173(2) [0|0|5]	93.4
Bill & Ted's Excellent Adventure (89) 170(2) [0|0|4]	92.2
Christine (84) 177(1) [0|0|5]	91.5
Fletch (85) 160(1) [0|0|4]	90.0
Tap (89) 166(2) [0|0|4]	88.7
1969 (88) 186(1) [0|0|6]	79.8
Grease (78/80) 186(1) [0|0|5]	*72.7*
The Empire Strikes Back/The Adventures Of Luke Skywalker (80) 178(2) [0|0|4]	72.5
Victor/Victoria (82) 174(2) [0|0|4]	72.5
Bird (88) 169(1) [0|0|3]	70.2
Gandhi (83) 168(1) [0|0|3]	68.3
Thief Of Hearts (84) 179(1) [0|0|3]	67.2
D.C. Cab (84) 181(1) [0|0|4]	67.1
The Night The Lights Went Out In Georgia (81) 189(1) [0|0|5]	65.4
Superman III (83) 163(1) [0|0|3]	65.2
Lethal Weapon 2 (89) 164(2) [0|0|3]	65.1
Cocoon (85) 188(1) [0|0|4]	53.6
A Fine Mess (86) 183(1) [0|0|3]	49.9
Up The Creek (84) 185(1) [0|0|3]	48.5
Heartbreak Hotel (88) 176(2) [0|0|2]	41.6
Nothing In Common (86) 190(2) [0|0|3]	41.3
Coming To America (88) 177(1) [0|0|2]	37.6
Round Midnight (87) 196(1) [0|0|3]	34.6
Married To The Mob (88) 197(2) [0|0|3]	33.3
Somewhere In Time (80) 187(2) [0|0|3]	31.8
Back To The Beach (87) 188(1) [0|0|2]	30.4
Pennies From Heaven (82) 188(1) [0|0|2]	29.5
Sing (89) 196(1) [0|0|1]	11.8
Blind Date (87) 198(1) [0|0|1]	10.9
SOUNDTRACK-TV ▶ Pk:1 [1|1|8]	
Miami Vice (85) 1(11) [18|22|34]	16966
Moonlighting (87) 50(2) [0|0|14]	1143
Miami Vice II (86) 82(1) [0|0|12]	585
The Dukes Of Hazzard (82) 93(2) [0|0|14]	550
The Music Of Cosmos (81) 136(2) [0|0|13]	370
Music From The Bill Cosby Show-- A House Full Of Love (86) 125(1) [0|0|7]	258
Shogun (80) 115(2) [0|0|6]	242
Beauty & The Beast: Of Love And Hope (89) 157(1) [0|0|10]	205
J.D. SOUTHER ▶ 1485 Pk:124 [0|0|1]	262
You're Only Lonely (79/80)-John David Souther 124(2) [0|0|7]	*262*
SOUTHSIDE JOHNNY & The ASBURY JUKES ▶ 769 Pk:67 [0|0|6]	1960
Love Is A Sacrifice (80) 67(2) [0|0|15]	904
Reach Up And Touch The Sky (81) 80(4) [0|0|12]	672
In The Heat (84)-Southside Johnny & The Jukes 164(2) [0|0|8]	173
Trash It Up (83)-Southside Johnny & The Jukes 154(2) [0|0|6]	147
At Least We Got Shoes (86)-Southside Johnny & The Jukes 189(1) [0|0|4]	53.2
Slow Dance (88)-Southside Johnny 198(1) [0|0|1]	10.9
SPANDAU BALLET ▶ 456 Pk:19 [0|1|2]	4885
True (83) 19(1) [0|8|37]	3680
Parade (84) 50(3) [0|0|16]	1205
SPARKS ▶ 1088 Pk:88 [0|0|3]	761
In Outer Space (83) 88(1) [0|0|17]	610
Angst In My Pants (82) 173(2) [0|0|6]	115
Whomp That Sucker (81) 182(2) [0|0|2]	36.2

Acts With Albums

Act ▶ Rank Peak [Top10s|Top40s|Total] Title (Yr) Pk(Wks) [T10 Wks|T40Wks|TotWks]	Score in '80s
SPECIAL ED ▶ 878 Pk:73 [0|0|1]	**1450**
Youngest In Charge (89) 73(1) [0|0|28]	1450
The SPECIALS ▶ 927 Pk:84 [0|0|2]	**1268**
The Specials (80) 84(1) [0|0|21]	1039
More Specials (80) 98(1) [0|0|5]	229
Paul SPEER ▶ 1557 Pk:125 [0|0|1]	**199**
Natural States (88)-David Lanz & Paul Speer [A] 125(1) [0|0|12]	199
Judson SPENCE ▶ 1539 Pk:168 [0|0|1]	**220**
Judson Spence (88) 168(4) [0|0|13]	220
Tracie SPENCER ▶ 1250 Pk:146 [0|0|1]	**497**
Tracie Spencer (88) 146(1) [0|0|21]	497
SPIDER ▶ 1374 Pk:130 [0|0|2]	**351**
Spider (80) 130(1) [0|0|10]	319
Between The Lines (81) 185(1) [0|0|2]	31.8
SPINAL TAP ▶ 1414 Pk:121 [0|0|1]	**314**
This Is Spinal Tap (Soundtrack) (84) 121(2) [0|0|10]	314
SPINNERS ▶ 609 Pk:32 [0|1|5]	**3109**
Dancin' And Lovin' (80) 32(1) [0|4|20]	1868
Love Trippin' (80) 53(2) [0|0|13]	875
Labor Of Love (81) 128(2) [0|0|6]	204
Grand Slam (83) 167(2) [0|0|6]	117
Can't Shake This Feelin' (82) 196(2) [0|0|4]	45.0
SPLIT ENZ ▶ 440 Pk:40 [0|1|4]	**5226**
True Colours (80) 40(2) [0|2|25]	2276
Waiata (81) 45(2) [0|0|19]	1564
Time And Tide (82) 58(1) [0|0|20]	1102
Conflicting Emotions (84) 137(1) [0|0|10]	284
Rick SPRINGFIELD ▶ 54 Pk:2 [2|5|8]	**39927**
Working Class Dog (81) 7(2) [5|38|73]	12526
Success Hasn't Spoiled Me Yet (82) 2(3) [12|15|35]	9256
Living In Oz (83) 12(2) [0|24|57]	8353
Tao (85) 21(2) [0|11|27]	3922
Hard To Hold (Soundtrack) (84) 16(3) [0|9|36]	3744
Rock Of Life (88) 55(1) [0|0|16]	1219
Beautiful Feelings (84) 78(2) [0|0|13]	713
Wait For Night (84) 159(2) [0|0|8]	195
Bruce SPRINGSTEEN ▶ 2 Pk:1 [5|5|9]	**107845**
Born In The U.S.A. (84) 1(7) [84|96|139]	59099
Tunnel Of Love (87) 1(1) [10|33|45]	13619
The River (80) 1(4) [8|22|107]	12905
Bruce Springsteen & The E Street Band Live 1975-1985 (86) 1(7) [11|15|26]	10964
Nebraska (82) 3(4) [7|11|29]	6068
Born To Run (75/80) 66(5) [0|0|81]	*3552*
Darkness On The Edge Of Town (78/80) *105(1) [0|0|42]*	*1088*
Greetings From Asbury Park, N.J. (75/80) *135(1) [0|0|16]*	*358*
The Wild, The Innocent And The E Street Shuffle *(75/80) 111(1) [0|0|5]*	*192*
SPYRO GYRA ▶ 216 Pk:19 [0|1|12]	**13892**
Catching The Sun (80) 19(1) [0|8|29]	3064
Carnaval (80) 49(1) [0|0|30]	1880
Freetime (81) 41(2) [0|0|27]	1859
Incognito (82) 46(2) [0|0|24]	1600
Access All Areas (84) 59(1) [0|0|19]	1297
Alternating Currents (85) 66(2) [0|0|23]	1185
Breakout (86) 71(1) [0|0|19]	1125
City Kids (83) 66(1) [0|0|16]	883
Stories Without Words (87) 84(1) [0|0|9]	438
Rites Of Summer (88) 104(2) [0|0|8]	307
Point Of View (89) 120(2) [0|0|6]	215
Morning Dance (79/80) 166(1) [0|0|2]	*38.5*
SPYS ▶ 1472 Pk:138 [0|0|1]	**268**
S.P.Y.S. (82) 138(3) [0|0|10]	268
SQUEEZE ▶ 254 Pk:32 [0|2|7]	**11546**
Babylon And On (87) 36(2) [0|7|29]	3180
East Side Story (81) 44(1) [0|0|25]	2071

Act ▶ Rank Peak [Top10s|Top40s|Total] Title (Yr) Pk(Wks) [T10 Wks|T40Wks|TotWks]	Score in '80s
Sweets From A Stranger (82) 32(2) [0|4|30]	2069
Singles 45's And Under (83) 47(3) [0|0|21]	1450
Cosi Fan Tutti Frutti (85) 57(2) [0|0|20]	1200
Argybargy (80) 71(1) [0|0|24]	1190
Frank (89) 113(3) [0|0|10]	388
Billy SQUIER ▶ 73 Pk:5 [2|3|6]	**34473**
Don't Say No (81) 5(3) [7|38|111]	15241
Emotions In Motion (82) 5(8) [9|29|50]	11894
Signs Of Life (84) 11(4) [0|14|29]	5282
Hear & Now (89) 64(1) [0|0|17]	950
Enough Is Enough (86) 61(2) [0|0|16]	902
The Tale Of The Tape (80) 169(1) [0|0|12]	206
STACEY Q ▶ 729 Pk:59 [0|0|2]	**2222**
Better Than Heaven (86) 59(2) [0|0|39]	1787
Hard Machine (88) 115(1) [0|0|11]	435
STAGE DOLLS ▶ 1334 Pk:118 [0|0|1]	**405**
Stage Dolls (89) 118(1) [0|0|12]	405
Joe STAMPLEY ▶ 1867 Pk:170 [0|0|1]	**37.4**
Hey Joe, Hey Moe (81)-Moe Bandy & Joe Stampley [A] 170(1) [0|0|4]	37.4
Michael STANLEY Band ▶ 591 Pk:64 [0|0|4]	**3312**
Heartland (80) 86(2) [0|0|32]	1322
You Can't Fight Fashion (83) 64(2) [0|0|17]	974
North Coast (81) 79(2) [0|0|15]	827
MSB (82) 136(2) [0|0|6]	189
STARGARD ▶ 1879 Pk:186 [0|0|1]	**32.6**
Back 2 Back (81) 186(2) [0|0|2]	32.6
STARPOINT ▶ 557 Pk:60 [0|0|3]	**3663**
Restless (85) 60(3) [0|0|47]	2837
Sensational (87) 95(2) [0|0|14]	626
Keep On It (81) 138(2) [0|0|8]	201
Brenda K. STARR ▶ 846 Pk:58 [0|0|1]	**1579**
Brenda K. Starr (88) 58(1) [0|0|24]	1579
Ringo STARR ▶ 1216 Pk:98 [0|0|1]	**534**
Stop And Smell The Roses (81) 98(1) [0|0|12]	534
STARS ON ▶ 483 Pk:9 [1|1|3]	**4481**
Stars On Long Play (81) 9(4) [5|8|24]	4108
Stars On Long Play II (81) 120(3) [0|0|6]	247
Stars On Long Play III (82) 163(2) [0|0|6]	125
STATLER BROTHERS ▶ 1026 Pk:103 [0|0|6]	**909**
Years Ago (81) 103(1) [0|0|9]	393
The Best Of The Statler Bros. Rides Again, Vol. II (80) 153(2) [0|0|11]	262
10th Anniversary (80) 169(1) [0|0|5]	90.6
Atlanta Blue (84) 177(2) [0|0|4]	69.9
Today (83) 193(1) [0|0|5]	58.3
Four For The Show (86) 183(2) [0|0|5]	35.4
STEADY B ▶ 1595 Pk:149 [0|0|2]	**177**
What's My Name (87) 149(2) [0|0|7]	164
Let The Hustlers Play (88) 193(1) [0|0|3]	13.2
STEALIN HORSES ▶ 1435 Pk:146 [0|0|1]	**296**
Stealin Horses (88) 146(2) [0|0|12]	296
STEEL BREEZE ▶ 805 Pk:50 [0|0|1]	**1774**
Steel Breeze (82) 50(4) [0|0|28]	1774
STEEL PULSE ▶ 1046 Pk:120 [0|0|3]	**841**
True Democracy (82) 120(2) [0|0|13]	384
Earth Crisis (84) 154(1) [0|0|12]	236
State Of...Emergency (88) 127(2) [0|0|7]	221
STEELY DAN ▶ 347 Pk:9 [1|1|4]	**7743**
Gaucho (80) 9(3) [6|19|36]	7332
Gold (82) 115(1) [0|0|9]	322
Katy Lied (75/81) 172(2) [0|0|2]	*45.2*
The Royal Scam (76/81) 174(2) [0|0|2]	*43.4*
Jim STEINMAN ▶ 926 Pk:63 [0|0|1]	**1273**
Bad For Good (81) 63(1) [0|0|17]	1273
Van STEPHENSON ▶ 916 Pk:54 [0|0|1]	**1308**
Righteous Anger (84) 54(2) [0|0|20]	1308
STEPPENWOLF ▶ 1768 Pk:171 [0|0|1]	**82.0**
Rock & Roll Rebels (87)-John Kay & Steppenwolf 171(1) [0|0|4]	82.0

Act ▶ Rank Peak [Top10s|Top40s|Total] Title (Yr) Pk(Wks) [T10 Wks|T40Wks|TotWks]	Score in '80s
Cat STEVENS ▶ 1604 Pk:165 [0|0|1]	**172**
Footsteps In The Dark: Greatest Hits Volume 2 (84) 165(3) [0|0|8]	172
Ray STEVENS ▶ 1083 Pk:118 [0|0|2]	**777**
He Thinks He's Ray Stevens (85) 118(2) [0|0|19]	538
Shriner's Convention (80) 132(1) [0|0|8]	239
Steve STEVENS ATOMIC PLAYBOYS ▶ 1355 Pk:119 [0|0|1]	**379**
Atomic Playboys (89) 119(1) [0|0|12]	379
Stevie B ▶ 619 Pk:75 [0|0|2]	**3013**
In My Eyes (89) 75(1) [0|0|38]	*2009*
Party Your Body (88) 78(1) [0|0|21]	1004
Al STEWART ▶ 824 Pk:37 [0|1|2]	**1703**
24 Carrots (80) 37(2) [0|4|13]	1260
Live/Indian Summer (81) 110(2) [0|0|11]	443
Gary STEWART ▶ 1820 Pk:165 [0|0|1]	**62.0**
Cactus And A Rose (80) 165(1) [0|0|3]	62.0
Jermaine STEWART ▶ 578 Pk:32 [0|1|3]	**3433**
Frantic Romantic (86) 32(2) [0|5|25]	2378
The Word Is Out (85) 90(2) [0|0|11]	551
Say It Again (88) 98(2) [0|0|12]	504
John STEWART ▶ 1289 Pk:85 [0|0|1]	**447**
Dream Babies Go To Hollywood (80) 85(2) [0|0|10]	447
Rod STEWART ▶ 66 Pk:11 [0|7|9]	**35979**
Out Of Order (88) 20(2) [0|46|72]	12590
Tonight I'm Yours (81) 11(4) [0|20|31]	6349
Camouflage (84) 18(2) [0|19|35]	5251
Foolish Behaviour (80) 12(5) [0|15|21]	4708
Body Wishes (83) 30(1) [0|4|22]	2305
Rod Stewart (86) 28(2) [0|7|19]	2258
Rod Stewart Greatest Hits (80) 22(2) [0|2|13]	*1123*
Absolutely Live (82) 46(3) [0|0|13]	1063
Storyteller/Complete Anthology: 1964-1990 *(89/**90**) 63(1) [0|0|5]*	*333*
Stephen STILLS ▶ 1149 Pk:75 [0|0|1]	**635**
Right By You (84) 75(2) [0|0|12]	635
STING ▶ 106 Pk:2 [2|2|2]	**27228**
The Dream Of The Blue Turtles (85) 2(6) [19|37|58]	17478
...Nothing Like The Sun (87) 9(4) [4|26|52]	9750
STONE CITY BAND ▶ 1489 Pk:122 [0|0|1]	**260**
In 'N' Out (80) 122(2) [0|0|8]	260
STONE FURY ▶ 1409 Pk:144 [0|0|1]	**320**
Burns Like A Star (84) 144(2) [0|0|12]	320
George STRAIT ▶ 448 Pk:68 [0|0|8]	**5023**
Greatest Hits, Volume Two (87) 68(2) [0|0|31]	1570
Ocean Front Property (87) 117(1) [0|0|28]	949
Beyond The Blue Neon (89) 92(2) [0|0|24]	899
If You Ain't Lovin' (You Ain't Livin') (88) 87(2) [0|0|14]	585
Does Fort Worth Ever Cross Your Mind (84) 139(2) [0|0|11]	369
#7 (86) 126(1) [0|0|11]	324
Greatest Hits (85) 157(1) [0|0|8]	190
Right Or Wrong (84) 163(1) [0|0|7]	137
The STRANGLERS ▶ 1782 Pk:172 [0|0|1]	**76.5**
Dreamtime (87) 172(2) [0|0|4]	76.5
STRAY CATS ▶ 126 Pk:2 [1|2|4]	**23881**
Built For Speed (82) 2(15) [18|37|74]	19578
Rant 'N' Rave With The Stray Cats (83) 14(2) [0|10|29]	3814
Blast Off (89) 111(2) [0|0|9]	300
Rock Therapy (86) 122(2) [0|0|5]	190
Meryl STREEP ▶ 1891 Pk:180 [0|0|1]	**30.2**
The Velveteen Rabbit (85)-Meryl Streep & George Winston [A] 180(1) [0|0|4]	30.2
Janey STREET ▶ 1591 Pk:145 [0|0|1]	**179**
Heroes, Angels & Friends (84) 145(2) [0|0|6]	179

Acts With Albums

Act ▶ Rank Peak [Top10s|Top40s|Total] Title (Yr) Pk(Wks) [T10 Wks|T40Wks|TotWks]	Score in '80s
STREETS ▶ 1543 Pk:166 [0|0|1]	218
1st (83) 166(1) [0|0|11]	218
Barbra STREISAND ▶ 21 Pk:1 [6|9|10]	63929
Guilty (80) 1(3) [18|33|49]	18478
The Broadway Album (85) 1(3) [18|24|50]	14221
Memories (81) 10(6) [6|15|104]	8335
Till I Loved You (88) 10(1) [1|14|26]	5109
Yentl (Soundtrack) (83) 9(2) [2|13|26]	4609
One Voice (87) 9(2) [3|12|28]	4561
Emotion (84) 19(3) [0|11|28]	3655
Wet (79/80) 17(2) [0|6|17]	*2665*
A Collection: Greatest Hits...And More (89) 26(3) [0|10|11]	2059
A Christmas Album (81) 108(2) [0|0|5]	239
The STRIKERS ▶ 1828 Pk:174 [0|0|1]	57.5
The Strikers (81)-Strikers 174(1) [0|0|3]	57.5
STRYPER ▶ 265 Pk:32 [0|2|4]	10988
To Hell With The Devil (86) 32(2) [0|12|74]	6142
In God We Trust (88) 32(3) [0|5|25]	2155
Soldiers Under Command (85) 84(2) [0|0|64]	1829
The Yellow And Black Attack (86) 103(3) [0|0|30]	862
STYLE COUNCIL ▶ 711 Pk:56 [0|0|5]	2314
My Ever Changing Moods (84) 56(2) [0|0|22]	1401
Internationalists (85) 123(1) [0|0|11]	364
The Cost Of Loving (87) 122(2) [0|0|10]	344
Confessions Of A Pop Group (88) 174(1) [0|0|6]	102
Introducing The Style Council (83) 172(1) [0|0|5]	102
The STYLISTICS ▶ 1370 Pk:127 [0|0|1]	355
Hurry Up This Way Again (80) 127(1) [0|0|12]	355
STYX ▶ 50 Pk:1 [3|4|7]	41640
Paradise Theater (81) 1(3) [27|35|61]	21414
Kilroy Was Here (83) 3(2) [16|22|34]	10304
Cornerstone (79/80) 3(1) [7|14|48]	*6500*
Caught In The Act - Live (84) 31(2) [0|6|15]	1672
Pieces Of Eight (78/80) 98(1) [0|0|26]	*939*
The Grand Illusion (78/80) 126(2) [0|0|25]	*621*
Equinox (76/80) 153(1) [0|0|8]	*190*
SUAVE ▶ 1251 Pk:101 [0|0|1]	497
I'm Your Playmate (88) 101(1) [0|0|12]	497
The SUGARCUBES ▶ 712 Pk:54 [0|0|2]	2309
Life's Too Good (88) 54(2) [0|0|29]	1846
Here Today, Tomorrow Next Week! (89) 70(2) [0|0|9]	463
SUGARHILL GANG ▶ 901 Pk:50 [0|0|1]	1357
8th Wonder (82) 50(1) [0|0|18]	1357
SUICIDAL TENDENCIES ▶ 973 Pk:100 [0|0|3]	1087
Join The Army (87) 100(2) [0|0|13]	494
How Will I Laugh Tomorrow When I Can't Even Smile Today (88) 111(2) [0|0|12]	460
Controlled By Hatred/Feel Like Shit...Deja Vu (89) 150(2) [0|0|5]	133
Kasim SULTON ▶ 1916 Pk:197 [0|0|1]	22.8
Kasim (82) 197(2) [0|0|2]	22.8
Donna SUMMER ▶ 117 Pk:1 [2|5|10]	25039
On The Radio: Greatest Hits: Volumes I & II (80) 1(1) [10|15|30]	*8126*
She Works Hard For The Money (83) 9(1) [1|15|32]	5514
Donna Summer (82) 20(5) [0|9|37]	3827
The Wanderer (80) 13(3) [0|9|18]	2862
Cats Without Claws (84) 40(2) [0|2|17]	1317
Another Place And Time (89) 53(2) [0|0|20]	1277
Walk Away - Collector's Edition (The Best Of 1977-1980) (80) 50(2) [0|0|15]	936
Bad Girls (79/80) 82(2) [0|0|15]	*746*
Live And More (78/80) 133(1) [0|0|7]	*218*
All Systems Go (87) 122(2) [0|0|6]	215
Henry LEE SUMMER ▶ 682 Pk:56 [0|0|2]	2522
Henry Lee Summer (88) 56(2) [0|0|23]	1632
I've Got Everything (89) 78(1) [0|0|17]	890
Andy SUMMERS & Robert FRIPP ▶ 1048 Pk:60 [0|0|2]	838
I Advance Masked (82) 60(3) [0|0|11]	704
Bewitched (84) 155(2) [0|0|5]	133
Bill SUMMERS & SUMMERS HEAT ▶ 949 Pk:92 [0|0|2]	1162
Jam The Box! (81) 92(1) [0|0|16]	654
Call It What You Want (81) 129(1) [0|0|15]	508
SUPERTRAMP ▶ 163 Pk:5 [2|4|7]	18335
"...Famous Last Words..." (82) 5(7) [7|17|28]	7007
Paris (80) 8(2) [3|11|26]	4419
Breakfast In America (79/80) 28(1) [0|9|48]	*3303*
Brother Where You Bound (85) 21(2) [0|10|22]	2974
Free As A Bird (87) 101(2) [0|0|11]	391
Crime Of The Century (75/80) 151(1) [0|0|8]	*172*
Even In The Quietest Moments... (77/80) *182(1) [0|0|5]*	*69.9*
The SUPREMES/Diana ROSS And The SUPREMES ▶ 1161 Pk:112 [0|0|1]	605
25th Anniversary (86)-Diana Ross & The Supremes 112(2) [0|0|17]	605
SURFACE ▶ 530 Pk:55 [0|0|2]	3928
2nd Wave (88) 56(2) [0|0|39]	2564
Surface (87) 55(1) [0|0|19]	1364
SURVIVOR ▶ 137 Pk:2 [1|2|7]	22386
Eye Of The Tiger (82) 2(4) [12|19|41]	10099
Vital Signs (84) 16(3) [0|27|61]	8514
When Seconds Count (86) 49(5) [0|0|24]	2030
Premonition (81) 82(3) [0|0|25]	1243
Caught In The Game (83) 82(1) [0|0|9]	338
Survivor (80) 169(1) [0|0|7]	130
Too Hot To Sleep (88) 187(2) [0|0|2]	31.8
Keith SWEAT ▶ 292 Pk:15 [0|1|1]	9675
Make It Last Forever (88) 15(1) [0|27|67]	9675
SWEAT BAND ▶ 1562 Pk:150 [0|0|1]	198
Sweat Band (80) 150(2) [0|0|8]	198
Rachel SWEET ▶ 1169 Pk:123 [0|0|2]	592
Protect The Innocent (80) 123(1) [0|0|11]	351
...And Then He Kissed Me (81) 124(2) [0|0|7]	242
SWEET SENSATION ▶ 768 Pk:63 [0|0|1]	1968
Take It While It's Hot (88) 63(1) [0|0|32]	1968
SWEET TEE ▶ 1486 Pk:169 [0|0|1]	261
It's Tee Time (89) 169(3) [0|0|13]	261
SWING OUT SISTER ▶ 430 Pk:40 [0|1|2]	5367
It's Better To Travel (87) 40(1) [0|1|43]	4331
Kaleidoscope World (89) 61(2) [0|0|19]	1036
SWITCH ▶ 830 Pk:57 [0|0|4]	1661
Reaching For Tomorrow (80) 57(1) [0|0|14]	804
This Is My Dream (80) 85(1) [0|0|17]	699
Switch V (81) 174(2) [0|0|4]	80.1
Switch II (79/80) 180(1) [0|0|5]	*76.7*
SYBIL ▶ 1137 Pk:75 [0|0|1]	655
Sybil (89) 75(2) [0|0|11]	*655*
Keith SYKES ▶ 1458 Pk:147 [0|0|1]	279
I'm Not Strange I'm Just Like You (80) 147(3) [0|0|11]	279
Sylvain SYLVAIN ▶ 1418 Pk:123 [0|0|1]	311
Sylvain Sylvain (80) 123(2) [0|0|11]	311
SYLVESTER ▶ 1104 Pk:123 [0|0|5]	727
Living Proof (80) 123(1) [0|0|6]	*211*
Sell My Soul (80) 147(3) [0|0|8]	205
Mutual Attraction (87) 164(2) [0|0|5]	106
Too Hot To Sleep (81) 156(1) [0|0|4]	104
All I Need (83) 168(2) [0|0|5]	101
SYLVIA ▶ 640 Pk:56 [0|0|4]	2836
Just Sylvia (82) 56(1) [0|0|33]	1946
Snapshot (83) 77(1) [0|0|11]	533
Drifter (81) 139(1) [0|0|11]	284
Surprise (84) 178(1) [0|0|4]	72.9
SYREETA ▶ 1047 Pk:73 [0|0|3]	838
Syreeta(2) (80) 73(2) [0|0|15]	641
Billy Preston & Syreeta (81)- Billy Preston & Syreeta [A] 127(4) [0|0|9]	152
Set My Love In Motion (82) 189(3) [0|0|3]	45.0
The SYSTEM ▶ 671 Pk:62 [0|0|3]	2579
Don't Disturb This Groove (87) 62(5) [0|0|25]	1726
Sweat (83) 94(1) [0|0|23]	779
X-Periment (84) 182(1) [0|0|5]	74.4

T

Act ▶ Rank Peak [Top10s|Top40s|Total] Title (Yr) Pk(Wks) [T10 Wks|T40Wks|TotWks]	Score in '80s
TACO ▶ 589 Pk:23 [0|1|1]	3339
After Eight (83) 23(2) [0|11|24]	3339
TAKE 6 ▶ 974 Pk:71 [0|0|1]	1074
Take 6 (89) 71(2) [0|0|17]	1074
TALKING HEADS ▶ 65 Pk:15 [0|6|8]	36295
Little Creatures (85) 20(3) [0|24|77]	8626
Speaking In Tongues (83) 15(2) [0|25|51]	8368
Stop Making Sense (84) 41(4) [0|0|118]	6150
True Stories: A Film By David Byrne, The Complete Soundtrack (86) 17(3) [0|15|29]	4822
Remain In Light (80) 19(2) [0|9|27]	3366
Naked (88) 19(1) [0|10|21]	3073
The Name Of This Band Is Talking Heads (82) 31(2) [0|4|14]	1360
Fear Of Music (79/80) 95(2) [0|0|12]	*530*
TALK TALK ▶ 607 Pk:42 [0|0|3]	3120
It's My Life (84) 42(3) [0|0|22]	1680
The Colour Of Spring (86) 58(2) [0|0|17]	1149
The Party's Over (82) 132(2) [0|0|16]	290
TA MARA & The SEEN ▶ 802 Pk:54 [0|0|1]	1806
Ta Mara & The Seen (85) 54(3) [0|0|25]	1806
TANGERINE DREAM ▶ 1115 Pk:96 [0|0|3]	708
Thief (Soundtrack) (81) 115(1) [0|0|10]	392
Legend (Soundtrack) (86) 96(2) [0|0|7]	291
Exit (81) 195(2) [0|0|2]	24.6
TANGIER ▶ 1065 Pk:91 [0|0|1]	808
Four Winds (89) 91(1) [0|0|17]	808
TANTRUM ▶ 1885 Pk:199 [0|0|1]	31.0
Rather Be Rockin' (80) 199(2) [0|0|3]	31.0
A TASTE OF HONEY ▶ 630 Pk:36 [0|1|2]	2921
Twice As Sweet (80) 36(2) [0|4|32]	2230
Ladies Of The Eighties (82) 73(4) [0|0|12]	690
TAVARES ▶ 1144 Pk:75 [0|0|2]	640
Supercharged (80) 75(2) [0|0|7]	386
New Directions (82) 137(1) [0|0|14]	253
TAXXI ▶ 1513 Pk:161 [0|0|1]	239
States Of Emergency (82) 161(1) [0|0|11]	239
Andy TAYLOR ▶ 897 Pk:46 [0|0|1]	1391
Thunder (87) 46(4) [0|0|17]	1391
James TAYLOR ▶ 261 Pk:10 [1|3|3]	11142
Dad Loves His Work (81) 10(3) [3|12|23]	4696
Never Die Young (88) 25(2) [0|11|34]	3552
That's Why I'm Here (85) 34(3) [0|8|30]	2894
Roger TAYLOR ▶ 1322 Pk:121 [0|0|1]	416
Fun In Space (81) 121(1) [0|0|10]	416
T-CONNECTION ▶ 1191 Pk:123 [0|0|2]	568
Pure & Natural (81) 123(1) [0|0|8]	320
Everything Is Cool (81) 138(1) [0|0|8]	248
TEARDROP EXPLODES ▶ 1535 Pk:156 [0|0|2]	225
Kilimanjaro (81) 156(2) [0|0|6]	147
Wilder (82) 176(3) [0|0|4]	78.7
TEARS FOR FEARS ▶ 68 Pk:1 [2|2|3]	35324
Songs From The Big Chair (85) 1(5) [32|55|82]	28185
The Seeds Of Love (89) 8(3) [3|12|13]	*4028*
The Hurting (83) 73(1) [0|0|69]	3111

Acts With Albums

Act ▶ Rank Peak [Top10s|Top40s|Total] Title (Yr) Pk(Wks) [T10 Wks|T40Wks|TotWks]	Score in '80s
TECHNOTRONIC ▶ 1634 Pk:70 [0|0|1]	**153**
Pump Up The Jam - The Album (89/**90**) 70(1) [0|0|2]	153
Kiri TE KANAWA ▶ 1260 Pk:136 [0|0|1]	**485**
Blue Skies (85)-Kiri Te Kanawa/Nelson Riddle And His Orchestra 136(1) [0|0|16]	485
The TEMPTATIONS ▶ 323 Pk:37 [0|1|10]	**8366**
Truly For You (84) 55(4) [0|0|34]	2093
To Be Continued... (86) 74(2) [0|0|33]	1571
Reunion (82) 37(2) [0|2|18]	1351
Power (80) 45(1) [0|0|14]	1231
Together Again (87) 112(2) [0|0|21]	733
25th Anniversary (86) 140(1) [0|0|16]	437
The Temptations (81) 119(1) [0|0|9]	286
Touch Me (86) 146(2) [0|0|10]	255
Back To Basics (84) 152(1) [0|0|9]	209
Surface Thrills (83) 159(3) [0|0|9]	201
10cc ▶ 1818 Pk:180 [0|0|2]	**62.5**
Look Hear? (80) 180(1) [0|0|2]	37.1
Greatest Hits 1972-1978 (80) 188(1) [0|0|2]	25.4
Toni TENNILLE ▶ 1380 Pk:142 [0|0|2]	**344**
More Than You Know (84) 142(1) [0|0|11]	322
All Of Me (87) 198(2) [0|0|2]	21.8
10,000 MANIACS ▶ 257 Pk:13 [0|2|2]	**11387**
Blind Man's Zoo (89) 13(1) [0|19|28]	5903
In My Tribe (87) 37(1) [0|2|77]	5484
TEN YEARS AFTER ▶ 1356 Pk:120 [0|0|1]	**378**
About Time (89) 120(2) [0|0|10]	378
Robert TEPPER ▶ 1552 Pk:144 [0|0|1]	**204**
No Easy Way Out (86) 144(2) [0|0|8]	204
Tony TERRY ▶ 1291 Pk:151 [0|0|1]	**445**
Forever Yours (88) 151(1) [0|0|20]	445
TESLA ▶ 309 Pk:18 [0|2|2]	**8876**
The Great Radio Controversy (89) 18(1) [0|10|41]	4455
Mechanical Resonance (87) 32(2) [0|8|61]	4421
TESTAMENT ▶ 966 Pk:77 [0|0|2]	**1110**
Practice What You Preach (89) 77(2) [0|0|12]	645
The New Order (88) 136(1) [0|0|14]	465
TEXAS ▶ 1094 Pk:88 [0|0|1]	**747**
Southside (89) 88(2) [0|0|16]	747
TEXTONES ▶ 1666 Pk:176 [0|0|1]	**133**
Midnight Mission (84) 176(1) [0|0|8]	133
The THE ▶ 983 Pk:89 [0|0|2]	**1045**
Infected (87) 89(2) [0|0|18]	730
Mind Bomb (89) 138(2) [0|0|12]	315
THEY MIGHT BE GIANTS ▶ 1063 **Pk:89 [0|0|1]**	**809**
Lincoln (88) 89(2) [0|0|19]	809
THIN LIZZY ▶ 1084 Pk:120 [0|0|4]	**769**
Chinatown (80) 120(1) [0|0|10]	355
Renegade (82) 157(1) [0|0|11]	236
Thunder And Lightning (83) 159(2) [0|0|8]	130
'Life' Live (84) 185(1) [0|0|3]	47.7
3rd BASS ▶ 1401 Pk:76 [0|0|1]	**325**
The Cactus Album (89/**90**) 76(3) [0|0|5]	325
THIRD WORLD ▶ 673 Pk:63 [0|0|6]	**2569**
You've Got The Power (82) 63(1) [0|0|27]	1291
Serious Business (89) 107(1) [0|0|14]	568
Sense Of Purpose (85) 119(3) [0|0|11]	409
All The Way Strong (83) 137(1) [0|0|7]	221
Rock The World (81) 186(1) [0|0|3]	47.2
Third World, Prisoner in The Street (Soundtrack) (80) 186(2) [0|0|2]	32.6
38 SPECIAL ▶ 97 Pk:10 [1|5|7]	**28387**
Wild-Eyed Southern Boys (81)-.38 Special 18(2) [0|18|57]	6654
Special Forces (82) 10(3) [3|11|42]	5916
Strength In Numbers (86) 17(1) [0|20|31]	5670
Tour De Force (83) 22(2) [0|20|39]	5520

Act ▶ Rank Peak [Top10s|Top40s|Total] Title (Yr) Pk(Wks) [T10 Wks|T40Wks|TotWks]	Score in '80s
Rock & Roll Strategy (88)-Thirty Eight Special 61(1) [0|0|41]	2030
Flashback (87) 35(2) [0|3|17]	1499
Rockin' Into The Night (80)-38-Special 57(1) [0|0|19]	1098
B.J. THOMAS ▶ 1869 Pk:193 [0|0|1]	**35.1**
New Looks (83) 193(1) [0|0|3]	35.1
Lillo THOMAS ▶ 1863 Pk:186 [0|0|1]	**41.8**
All Of You (84) 186(1) [0|0|3]	41.8
Richard THOMPSON ▶ 1052 Pk:102 [0|0|4]	**831**
Across A Crowded Room (85) 102(2) [0|0|13]	510
Daring Adventures (86) 142(2) [0|0|6]	178
Amnesia (88) 182(2) [0|0|5]	78.4
Hand Of Kindness (83) 186(1) [0|0|5]	65.4
Robbin THOMPSON Band ▶ 1625 **Pk:168 [0|0|1]**	**159**
Two B's Please (80) 168(1) [0|0|11]	159
THOMPSON TWINS ▶ 167 Pk:10 [1|3|7]	**17929**
Into The Gap (84) 10(2) [2|24|53]	8385
Here's To Future Days (85) 20(1) [0|24|35]	6201
Side Kicks (83) 34(3) [0|4|25]	2160
Close To The Bone (87) 76(1) [0|0|14]	673
Big Trash (89) 143(2) [0|0|6]	208
In The Name Of Love (82) 148(2) [0|0|8]	202
The Best Of Thompson Twins/Greatest Mixes (88) 175(2) [0|0|6]	100
Ali THOMSON ▶ 1166 Pk:99 [0|0|1]	**595**
Take A Little Rhythm (80) 99(1) [0|0|15]	595
George THOROGOOD & The DESTROYERS ▶ **214 Pk:32 [0|3|5]**	**14044**
Maverick (85) 32(1) [0|15|42]	5100
Live (86) 33(3) [0|4|42]	3234
Born To Be Bad (88) 32(3) [0|10|24]	2661
Bad To The Bone (82) 43(4) [0|0|48]	2404
More George Thorogood And The Destroyers (80) 68(2) [0|0|13]	647
Billy THORPE ▶ 1669 Pk:151 [0|0|1]	**132**
21st Century Man (80) 151(2) [0|0|7]	132
3 ▶ 1349 Pk:97 [0|0|1]	**390**
To The Power Of Three (88) 97(2) [0|0|10]	390
The THREE O'CLOCK ▶ 1439 Pk:125 [0|0|1]	**292**
Arrive Without Travelling (85) 125(2) [0|0|10]	292
THREE TIMES DOPE ▶ 1254 Pk:122 [0|0|1]	**490**
Original Stylin' (89) 122(2) [0|0|18]	490
THRILLS ▶ 1864 Pk:199 [0|0|1]	**41.0**
First Thrills (81) 199(2) [0|0|4]	41.0
TIERRA ▶ 788 Pk:38 [0|1|1]	**1852**
City Nights (80) 38(3) [0|4|21]	1852
TIFFANY ▶ 130 Pk:1 [1|2|2]	**23481**
Tiffany (87) 1(2) [22|35|69]	18162
Hold An Old Friend's Hand (88) 17(4) [0|19|29]	5319
Tanita TIKARAM ▶ 889 Pk:59 [0|0|1]	**1412**
Ancient Heart (89) 59(1) [0|0|23]	1412
'TIL TUESDAY ▶ 384 Pk:19 [0|1|3]	**6863**
Voices Carry (85) 19(1) [0|12|31]	4268
Welcome Home (86) 49(3) [0|0|26]	2039
Everything's Different Now (88) 124(3) [0|0|19]	555
TIMBUK 3 ▶ 643 Pk:50 [0|0|2]	**2824**
Greetings From Timbuk 3 (86) 50(1) [0|0|30]	2316
Eden Alley (88) 107(5) [0|0|13]	508
The TIME ▶ 225 Pk:24 [0|2|3]	**13456**
Ice Cream Castle (84) 24(3) [0|34|57]	8479
What Time Is It? (82) 26(3) [0|8|33]	2735
The Time (81) 50(1) [0|0|32]	2242
TIMES TWO ▶ 1426 Pk:137 [0|0|1]	**305**
X2 (88) 137(1) [0|0|11]	305
TIN MACHINE ▶ 833 Pk:28 [0|1|1]	**1651**
Tin Machine (89) 28(1) [0|5|17]	1651
TKA ▶ 1437 Pk:135 [0|0|1]	**294**
Scars Of Love (88) 135(2) [0|0|11]	294

Act ▶ Rank Peak [Top10s|Top40s|Total] Title (Yr) Pk(Wks) [T10 Wks|T40Wks|TotWks]	Score in '80s
TNT ▶ 922 Pk:100 [0|0|2]	**1281**
Tell No Tales (87) 100(1) [0|0|21]	833
Intuition (89) 115(1) [0|0|12]	448
TODAY ▶ 1008 Pk:86 [0|0|1]	**970**
Today (89) 86(2) [0|0|22]	970
Isao TOMITA ▶ 1732 Pk:174 [0|0|1]	**96.9**
Ravel Bolero (80)-Tomita 174(1) [0|0|5]	96.9
TOMMY TUTONE ▶ 520 Pk:20 [0|1|3]	**4090**
Tommy Tutone-2 (82) 20(3) [0|8|30]	3277
Tommy Tutone (80) 68(2) [0|0|13]	759
National Emotion (83) 179(1) [0|0|3]	54.4
TOM TOM CLUB ▶ 485 Pk:23 [0|1|3]	**4474**
Tom Tom Club (81) 23(1) [0|10|33]	3485
Close To The Bone (83) 73(1) [0|0|13]	627
Boom Boom Chi Boom Boom (89) 114(2) [0|0|11]	363
TONE LŌC ▶ 252 Pk:1 [1|1|1]	**11607**
Loc-ed After Dark (89)- Tone-Lōc 1(1) [1|12|23|40]	11607
TONY! TONI! TONÉ! ▶ 780 Pk:69 [0|0|1]	**1918**
Who? (88)-Tony! Toni! Toné! 69(2) [0|0|46]	1918
TOO SHORT ▶ 474 Pk:37 [0|1|1]	**4568**
Life Is...Too Short (89) 37(2) [0|8|45]	4568
TORA TORA ▶ 836 Pk:47 [0|0|1]	**1634**
Surprise Attack (89) 47(1) [0|0|25]	1634
TORONTO ▶ 1525 Pk:162 [0|0|2]	**232**
Get It On Credit (82) 162(2) [0|0|10]	176
Lookin' For Trouble (80) 185(1) [0|0|4]	56.4
Peter TOSH ▶ 787 Pk:59 [0|0|2]	**1868**
Mama Africa (83) 59(1) [0|0|17]	1092
Wanted Dread & Alive (81) 91(1) [0|0|13]	571
Captured Live (84) 152(2) [0|0|8]	206
TOTO ▶ 109 Pk:4 [1|2|7]	**26649**
Toto IV (82) 4(4) [22|42|82]	17803
Fahrenheit (86) 40(2) [0|2|36]	2810
Hydra (79/**80**) 41(1) [0|0|22]	1962
Isolation (84) 42(4) [0|0|21]	1678
The Seventh One (88) 64(3) [0|0|18]	1255
Turn Back (81) 41(2) [0|0|10]	998
Dune (Soundtrack) (84) 168(2) [0|0|8]	144
Wayne TOUPS & ZYDECAJUN ▶ 1823 **Pk:183 [0|0|1]**	**60.4**
Blast From The Bayou (89) 183(2) [0|0|4]	60.4
Pete TOWNSHEND ▶ 176 Pk:5 [1|4|7]	**17148**
Empty Glass (80) 5(3) [9|19|30]	8115
White City - A Novel (85) 26(3) [0|11|29]	3644
All The Best Cowboys Have Chinese Eyes (82) 26(1) [0|9|26]	2810
Scoop (83) 35(2) [0|4|13]	1341
The Iron Man (The Musical By Pete Townshend) (89) 58(4) [0|0|13]	855
Pete Townshend's Deep End Live! (86) 98(2) [0|0|13]	372
Another Scoop (87) 198(1) [0|0|1]	10.9
Simon TOWNSHEND ▶ 1665 Pk:169 [0|0|1]	**133**
Sweet Sound (83) 169(1) [0|0|7]	133
T'PAU ▶ 752 Pk:31 [0|1|1]	**2072**
T'Pau (87) 31(1) [0|3|24]	2072
TRANSVISION VAMP ▶ 1429 Pk:115 [0|0|1]	**303**
Pop Art (88) 115(1) [0|0|8]	303
TRAVELING WILBURYS ▶ 197 Pk:3 [1|1|1]	**15180**
Volume 1 (88) 3(6) [22|32|47]	15180
Pat TRAVERS ▶ 426 Pk:20 [0|2|4]	**5494**
Crash And Burn (80)- Pat Travers Band 20(1) [0|12|25]	3324
Radio Active (81) 37(2) [0|2|15]	1248
Pat Travers' Black Pearl (82) 74(3) [0|0|13]	599
Hot Shot (84) 108(2) [0|0|8]	324

Acts With Albums

Act ▶ Rank Peak [Top10s|Top40s|Total] Title (Yr) Pk(Wks) [T10 Wks|T40Wks|TotWks]	Score in '80s
Randy TRAVIS ▶ 159 Pk:19 [0|3|5]	18717
Always & Forever (87) 19(1) [0|21|103]	9897
Storms Of Life (86) 85(2) [0|0|100]	3761
Old 8 x 10 (88) 35(1) [0|6|43]	3237
No Holdin' Back (89) 33(1) [0|3|12]	1515
An Old Time Christmas (89) 70(1) [0|0|5]	308
TREAT HER RIGHT ▶ 1214 Pk:127 [0|0|1]	535
Treat Her Right (88) 127(2) [0|0|18]	535
TRINERE & FRIENDS ▶ 1914 Pk:196 [0|0|1]	23.6
Trinere & Friends Greatest Hits (89) 196(2) [0|0|2]	23.6
TRIUMPH ▶ 173 Pk:23 [0|5|7]	17409
Allied Forces (81) 23(1) [0|9|59]	4551
Never Surrender (83) 26(2) [0|12|27]	3638
Thunder Seven (84) 35(2) [0|7|30]	3046
The Sport Of Kings (86) 33(2) [0|5|27]	2395
Progressions Of Power (80) 32(1) [0|6|18]	1905
Stages (85) 50(2) [0|0|18]	1315
Surveillance (87) 82(2) [0|0|13]	560
TROOP ▶ 1463 Pk:133 [0|0|1]	275
Troop (88) 133(2) [0|0|9]	275
TROUBLE FUNK ▶ 1352 Pk:121 [0|0|1]	389
Drop The Bomb (82) 121(2) [0|0|14]	389
Robin TROWER ▶ 573 Pk:34 [0|2|6]	3490
Victims Of The Fury (80) 34(1) [0|3|15]	1524
Passion (86) 100(2) [0|0|25]	1002
B.L.T. (81)-Jack Bruce/Bill Lordan/Robin Trower [A] 37(1) [0|3|16]	509
Take What You Need (88) 133(2) [0|0|10]	292
Truce (82)-Jack Bruce & Robin Trower [A] 109(2) [0|0|6]	136
Back It Up (83) 191(1) [0|0|2]	26.8
The TRUTH ▶ 1445 Pk:115 [0|0|1]	290
Weapons Of Love (87) 115(2) [0|0|8]	290
TSOL ▶ 1873 Pk:184 [0|0|1]	34.0
Hit And Run (87) 184(1) [0|0|2]	34.0
The TUBES ▶ 366 Pk:18 [0|2|4]	7369
Outside Inside (83) 18(1) [0|14|34]	4595
The Completion Backward Principle (81) 36(3) [0|4|27]	2121
Love Bomb (85) 87(2) [0|0|10]	488
T.R.A.S.H. (Tubes Rarities And Smash Hits) (81) 148(1) [0|0|6]	166
TUCK & PATTI ▶ 1541 Pk:162 [0|0|1]	219
Love Warriors (89) 162(3) [0|0|11]	219
Louise TUCKER ▶ 1406 Pk:127 [0|0|1]	322
Midnight Blue (83) 127(1) [0|0|10]	322
Tanya TUCKER ▶ 1610 Pk:121 [0|0|2]	170
Tear Me Apart (79/80) 121(1) [0|0|3]	118
Should I Do It (81) 180(1) [0|0|3]	51.7
Ike & Tina TURNER ▶ 1893 Pk:189 [0|0|1]	29.5
Get Back! (85) 189(1) [0|0|2]	29.5
Joe Lynn TURNER ▶ 1481 Pk:143 [0|0|1]	265
Rescue You (85) 143(2) [0|0|12]	265
Tina TURNER ▶ 53 Pk:3 [2|3|4]	40259
Private Dancer (84) 3(11) [39|62|106]	29221
Break Every Rule (86) 4(1) [6|17|52]	8832
Foreign Affair (89) 31(2) [0|5|13]	1790
Tina Live In Europe (88) 86(2) [0|0|9]	415
Stanley TURRENTINE ▶ 1781 Pk:162 [0|0|1]	76.8
Tender Togetherness (81) 162(2) [0|0|3]	76.8
TWENNYNINE ▶ 912 Pk:54 [0|0|3]	1321
Best Of Friends (80)-Twennynine Featuring Lenny White 54(2) [0|0|12]	931
Twennynine With Lenny White (80)-Twennynine Featuring Lenny White 106(2) [0|0|8]	269
Just Like Dreamin' (81)-Twennynine Featuring Lenny White 162(3) [0|0|5]	122
24-7 SPYZ ▶ 1188 Pk:113 [0|0|1]	570
Harder Than You (89) 113(2) [0|0|8]	570
20/20 ▶ 1294 Pk:127 [0|0|2]	439
Look Out (81) 127(2) [0|0|12]	361
20/20 (79/80) 158(1) [0|0|4]	78.7
Dwight TWILLEY ▶ 723 Pk:39 [0|1|2]	2256
Jungle (84) 39(1) [0|3|21]	1848
Scuba Divers (82) 109(2) [0|0|11]	408
TWIN HYPE ▶ 1465 Pk:140 [0|0|1]	274
Twin Hype (89) 140(2) [0|0|11]	274
TWISTED SISTER ▶ 266 Pk:15 [0|1|5]	10900
Stay Hungry (84) 15(3) [0|26|51]	8147
Come Out And Play (85) 53(2) [0|0|17]	1345
Love Is For Suckers (87) 74(2) [0|0|11]	630
You Can't Stop Rock 'N' Roll (83) 130(1) [0|0|14]	417
Under The Blade (85) 125(2) [0|0|11]	361
Conway TWITTY ▶ 1369 Pk:144 [0|0|1]	357
Southern Comfort (82) 144(1) [0|0|15]	357
2 LIVE CREW ▶ 399 Pk:30 [0|1|3]	6344
As Nasty As They Wanna Be (89/90) 30(1) [0|17|23]	3795
Move Somethin' (88) 68(2) [0|0|42]	1814
2 Live Crew Is What We Are (87) 128(1) [0|0|33]	735
Bonnie TYLER ▶ 358 Pk:4 [1|1|2]	7498
Faster Than The Speed Of Night (83) 4(1) [7|18|32]	7170
Secret Dreams & Forbidden Fire (86) 106(2) [0|0|8]	328
TYZIK ▶ 1692 Pk:172 [0|0|1]	122
Jammin' In Manhattan (84) 172(1) [0|0|6]	122

U

Act ▶ Rank Peak [Top10s|Top40s|Total] Title (Yr) Pk(Wks) [T10 Wks|T40Wks|TotWks]	Score in '80s
UB40 ▶ 219 Pk:14 [0|2|6]	13804
Labour Of Love (83) 14(1) [0|15|63]	6854
Little Baggariddim (85) 40(3) [0|3|25]	2085
UB40 (88) 44(2) [0|0|27]	2057
Geffrey Morgan... (84) 60(1) [0|0|26]	1410
Rat In The Kitchen (86) 53(2) [0|0|17]	1117
CCCP - Live In Moscow (87) 121(2) [0|0|8]	282
UFO ▶ 631 Pk:51 [0|0|5]	2920
No Place To Run (80) 51(1) [0|0|13]	876
Misdemeanor (86) 106(2) [0|0|19]	736
Mechanix (82) 82(1) [0|0|14]	663
The Wild The Willing And The Innocent (81) 77(2) [0|0|11]	523
Making Contact (83) 153(1) [0|0|5]	123
Tracey ULLMAN ▶ 774 Pk:34 [0|1|1]	1944
You Broke My Heart In 17 Places (84) 34(2) [0|5|20]	1944
ULTRAVOX ▶ 801 Pk:61 [0|0|4]	1806
Quartet (83) 61(1) [0|0|17]	1123
Lament (84) 115(2) [0|0|9]	307
Vienna (80) 164(1) [0|0|9]	202
Rage In Eden (81) 144(2) [0|0|6]	175
The UNDERTONES ▶ 1615 Pk:154 [0|0|1]	166
The Undertones (80) 154(1) [0|0|7]	166
UNDERWORLD ▶ 1220 Pk:139 [0|0|1]	529
Underneath The Radar (88) 139(2) [0|0|19]	529
The UNFORGIVEN ▶ 1875 Pk:185 [0|0|1]	33.6
The Unforgiven (86) 185(2) [0|0|8]	33.6
UNLIMITED TOUCH ▶ 1583 Pk:142 [0|0|1]	185
Unlimited Touch (81) 142(1) [0|0|8]	185
The UNTOUCHABLES ▶ 1592 Pk:162 [0|0|1]	178
Agent Double O Soul (89) 162(2) [0|0|9]	178
Midge URE ▶ 1111 Pk:88 [0|0|1]	710
Answers To Nothing (89) 88(1) [0|0|16]	710
URIAH HEEP ▶ 919 Pk:56 [0|0|2]	1297
Abominog (82) 56(3) [0|0|16]	1069
Head First (83) 159(1) [0|0|7]	228
USA For AFRICA ▶ 383 Pk:1 [1|1|1]	6924
We Are The World (85) 1(3) [7|11|22]	6924
UTFO ▶ 686 Pk:67 [0|0|3]	2492
Lethal (87) 67(2) [0|0|20]	1144
UTFO (85) 80(2) [0|0|20]	1014
Skeezer Pleezer (86) 142(2) [0|0|8]	206
Doin' It! (89) 143(2) [0|0|4]	129
UTOPIA ▶ 475 Pk:32 [0|1|6]	4562
Adventures In Utopia (80) 32(2) [0|5|21]	2062
Utopia (82) 84(2) [0|0|19]	825
Oblivion (84) 74(2) [0|0|12]	670
Deface The Music (80) 65(2) [0|0|9]	515
Swing To The Right (82) 102(1) [0|0|10]	363
POV (85) 161(2) [0|0|6]	128
U2 ▶ 9 Pk:1 [2|6|8]	81520
The Joshua Tree (87) 1(9) [35|58|103]	30759
Rattle And Hum (Soundtrack) (88) 1(6) [14|23|38]	13957
War (83) 12(1) [0|16|179]	11627
The Unforgettable Fire (84) 12(3) [0|22|132]	11271
Under A Blood Red Sky (83) 28(3) [0|13|180]	9430
Boy (81) 63(2) [0|0|47]	2140
October (81) 104(2) [0|0|38]	1268
Wide Awake In America (85) 37(1) [0|1|23]	1070

V

Act ▶ Rank Peak [Top10s|Top40s|Total] Title (Yr) Pk(Wks) [T10 Wks|T40Wks|TotWks]	Score in '80s
VAIN ▶ 1573 Pk:154 [0|0|1]	194
No Respect (89) 154(2) [0|0|8]	194
Ritchie VALENS ▶ 1383 Pk:100 [0|0|1]	343
The Best Of Ritchie Valens (87) 100(2) [0|0|10]	343
Dave VALENTIN ▶ 1748 Pk:184 [0|0|2]	89.7
Pied Piper (81) 184(3) [0|0|4]	64.3
Land Of The Third Eye (80) 194(2) [0|0|2]	25.4
VANDENBERG ▶ 971 Pk:65 [0|0|2]	1089
Vandenberg (83) 65(4) [0|0|18]	946
Heading For A Storm (84) 169(2) [0|0|7]	142
Luther VANDROSS ▶ 61 Pk:9 [1|7|7]	37178
Give Me The Reason (86) 14(1) [0|32|53]	9891
Any Love (88) 9(1) [3|15|33]	6209
The Night I Fell In Love (85) 19(3) [0|15|56]	6138
Never Too Much (81) 19(2) [0|10|36]	4764
Forever, For Always, For Love (82) 20(5) [0|12|36]	4676
Busy Body (83) 32(4) [0|9|41]	3971
The Best Of Luther Vandross...The Best Of Love (89/90) 28(1) [0|6|9]	1529
VANGELIS ▶ 196 Pk:1 [1|1|2]	15187
Chariots Of Fire (Soundtrack) (81) 1(4) [14|20|57]	12872
Opera Sauvage (86) 42(1) [0|0|39]	2316
VAN HALEN ▶ 11 Pk:1 [6|6|8]	78155
1984 (MCMLXXXIV) (84) 2(5) [27|52|77]	23083
5150 (86) 1(3) [14|32|64]	16974
OU812 (88) 1(4) [16|35|48]	15820
Diver Down (82) 3(3) [9|16|65]	8409
Women And Children First (80) 6(5) [9|13|31]	6428
Fair Warning (81) 5(3) [5|12|23]	5015
Van Halen (78/80) 117(1) [0|0|76]	2378
Van Halen II (79/80) 137(1) [0|0|9]	228
VANITY ▶ 501 Pk:45 [0|0|3]	4206
Vanity 6 (82)-Vanity 6 45(2) [0|0|31]	1959
Wild Animal (84) 62(2) [0|0|23]	1211
Skin On Skin (86) 66(2) [0|0|20]	1037
Gino VANNELLI ▶ 417 Pk:15 [0|1|4]	5682
Nightwalker (81) 15(4) [0|15|26]	4315
Black Cars (85) 62(2) [0|0|25]	1165
Big Dreamers Never Sleep (87) 160(2) [0|0|7]	157
The Best Of Gino Vannelli (81) 172(2) [0|0|2]	45.2
Johnny Van ZANT Band ▶ 814 Pk:48 [0|0|4]	1744
No More Dirty Deals (80) 48(1) [0|0|15]	1074
Round Two (81) 119(2) [0|0|10]	366
Van-Zant (85)-Van-Zant 170(1) [0|0|8]	167
Last Of The Wild Ones (82) 159(1) [0|0|6]	138

Acts With Albums

Act ▶ Rank Peak [Top10s|Top40s|Total] Title (Yr) Pk(Wks) [T10 Wks|T40Wks|TotWks]	Score in '80s
The VAPORS ▶ 818 Pk:62 [0|0|2]	1717
New Clear Days (80) 62(1) [0|0|28]	1365
Magnets (81) 109(1) [0|0|9]	352
VARIOUS ARTISTS ▶ Pk:19 [0|6|59]	
A Very Special Christmas (87) 20(3) [0|7|22]	2867
No Nukes/The MUSE Concerts For A Non-Nuclear Future (80) 19(2) [0|9|16]	2668
The Secret Policeman's Other Ball (82) 29(1) [0|5|16]	1651
Mickey Mouse Disco (80) 35(2) [0|4|27]	1645
1988 Summer Olympics-One Moment In Time (88) 31(2) [0|3|17]	1607
25 #1 Hits From 25 Years (83) 42(2) [0|0|28]	1509
Chess (85) 47(2) [0|0|21]	1506
Concerts For The People Of Kampuchea (81) 36(1) [0|3|12]	1139
Television's Greatest Hits (85) 82(2) [0|0|34]	1126
Casino Lights (82) 63(5) [0|0|19]	1104
Greenpeace/Rainbow Warriors (89) 68(2) [0|0|22]	1018
The Wrestling Album (85) 84(2) [0|0|19]	899
Windham Hill Records Sampler '84 (84) 108(4) [0|0|15]	896
Requiem (85) 77(2) [0|0|14]	784
Windham Hill Records Sampler '86 (86) 102(2) [0|0|18]	761
A Winter's Solstice (85) 77(1) [0|0|19]	736
Exposed/A Cheap Peek At Today's Provocative New Rock (81) 51(2) [0|0|9]	715
Rap's Greatest Hits (86) 114(2) [0|0|17]	618
Happy Anniversary, Charlie Brown (89) 65(2) [0|0|8]	608
Stay Awake (88) 119(4) [0|0|15]	568
Piledriver -- The Wrestling Album II (87) 123(2) [0|0|20]	558
Folkways: A Vision Shared - A Tribute To Woody Guthrie And Leadbelly (88) 70(2) [0|0|10]	555
Every Man Has A Woman (84) 75(2) [0|0|10]	540
MTV's Rock 'N Roll To Go (85) 91(2) [0|0|12]	522
The Winning Hand (83) 109(1) [0|0|14]	513
The Secret Policeman's Ball (81) 106(1) [0|0|12]	508
The Official Music Of The XXIIIrd Olympiad-Los Angeles 1984 (84) 92(2) [0|0|13]	470
Windham Hill Records Sampler '88 (88) 134(1) [0|0|16]	470
Echoes Of An Era (82) 105(1) [0|0|11]	440
Winners (80) 69(2) [0|0|7]	430
Television's Greatest Hits Volume II (86) 149(2) [0|0|16]	384
Rock For Amnesty (87) 121(2) [0|0|11]	377
TeeVee Toons - The Commercials (89) 159(1) [0|0|18]	367
Volunteer Jam VI (80) 104(2) [0|0|9]	366
25 Years Of Grammy Greats (83) 107(1) [0|0|9]	344
The Motown Story: The First 25 Years (83) 114(1) [0|0|9]	311
A Winter's Solstice II (88) 108(1) [0|0|7]	292
The Legend Of Jesse James (80) 154(4) [0|0|13]	285
Live! For Life (86) 105(2) [0|0|7]	265
Electric Breakdance (84) 147(2) [0|0|9]	245
In Harmony 2 (81) 129(2) [0|0|10]	221
Christmas Rap (87) 130(2) [0|0|8]	191
Windham Hill Piano Sampler (85) 167(2) [0|0|5]	191
Exposed II (81) 124(2) [0|0|5]	176
20/20 Twenty No.1 Hits From Twenty Years At Motown (80) 150(2) [0|0|6]	165
Make A Difference Foundation: Stairway To Heaven/Highway To Hell (89/90) 100(2) [0|0|3]	*155*
Guitar Speak (88) 171(2) [0|0|8]	141
In Harmony - A Sesame Street Record (81) 156(1) [0|0|5]	132
Volunteer Jam VII (81) 149(2) [0|0|4]	125
The Island Story, 1962-1987: The 25th Anniversary (87) 180(2) [0|0|6]	93.9
A GRP Christmas Collection (89) 140(1) [0|0|3]	*91.2*
Rap's Greatest Hits, Volume 2 (87) 167(1) [0|0|4]	90.5
Empire Jazz (80) 168(1) [0|0|5]	87.9
A Country Christmas (82) 172(2) [0|0|4]	85.0
Windham Hill Records Sampler '89 (89) 176(2) [0|0|4]	77.0
Brazil Classics 1: Beleza Tropical (89) 178(2) [0|0|4]	70.2
Urgh! A Music War (81) 173(2) [0|0|3]	61.2
The Prince's Trust 10th Anniversary Birthday Party (87) 194(1) [0|0|3]	37.3
A Night At Studio 54 (79/80) 167(1) [0|0|2]	*37.2*
Stevie Ray VAUGHAN And DOUBLE TROUBLE ▶ 212 Pk:31 [0|4|5]	14073
Couldn't Stand The Weather (84) 31(3) [0|8|38]	3580
Soul To Soul (85) 34(2) [0|6|39]	3037
In Step (89) 33(1) [0|6|27]	*2907*
Texas Flood (83) 38(1) [0|3|33]	2674
Live Alive (86) 52(2) [0|0|25]	1875
Suzanne VEGA ▶ 385 Pk:11 [0|1|2]	6858
Solitude Standing (87) 11(2) [0|16|31]	5728
Suzanne Vega (85) 91(2) [0|0|31]	1129
VELVET UNDERGROUND ▶ 1139 Pk:85 [0|0|2]	647
VU (85) 85(2) [0|0|13]	625
The Velvet Underground (85)-The Velvet Underground & Nico 197(1) [0|0|2]	22.3
Billy VERA & The BEATERS ▶ 550 Pk:15 [0|1|2]	3757
By Request (The Best of Billy Vera & The Beaters) (86) 15(1) [0|10|26]	3433
Billy & The Beaters (81) 118(1) [0|0|10]	324
Tom VERLAINE ▶ 1844 Pk:177 [0|0|1]	52.1
Dreamtime (81) 177(1) [0|0|3]	52.1
VESTA ▶ 1417 Pk:131 [0|0|1]	312
Vesta 4 U (89)-Vesta Williams 131(1) [0|0|10]	312
VICTORY ▶ 1813 Pk:182 [0|0|1]	63.6
Culture Killed The Native (89) 182(1) [0|0|5]	63.6
VILLAGE PEOPLE ▶ 859 Pk:47 [0|0|3]	1529
Can't Stop The Music (Soundtrack) (80) 47(2) [0|0|12]	888
Live And Sleazy (79/80) 58(1) [0|0|9]	*507*
Renaissance (81) 138(2) [0|0|4]	133
Vinnie VINCENT INVASION ▶ 627 Pk:64 [0|0|2]	2946
Vinnie Vincent Invasion (86) 64(2) [0|0|29]	2050
All Systems Go (88) 64(2) [0|0|15]	896
VIO-LENCE ▶ 1683 Pk:154 [0|0|1]	125
Eternal Nightmare (88) 154(2) [0|0|6]	125
VIOLENT FEMMES ▶ 862 Pk:84 [0|0|2]	1515
The Blind Leading The Naked (86) 84(2) [0|0|24]	1009
3 (89) 93(3) [0|0|13]	507
VISAGE ▶ 1795 Pk:178 [0|0|1]	72.4
Visage (81) 178(2) [0|0|4]	72.4
Joe VITALE ▶ 1840 Pk:181 [0|0|1]	53.1
Plantation Harbor (81) 181(1) [0|0|3]	53.1
VITAMIN Z ▶ 1847 Pk:183 [0|0|1]	50.8
Rites Of Passage (85) 183(1) [0|0|3]	50.8
VIXEN ▶ 568 Pk:41 [0|0|1]	3537
Vixen (88) 41(1) [0|0|40]	3537
VOIVOD ▶ 1814 Pk:168 [0|0|1]	63.3
Nothingface (89/90) 168(2) [0|0|3]	*63.3*
Andreas VOLLENWEIDER ▶ 406 Pk:52 [0|0|5]	6141
Down To The Moon (86) 60(1) [0|0|39]	2226
White Winds (85) 76(2) [0|0|39]	1700
Dancing With The Lion (89) 52(2) [0|0|19]	1211
...Behind The Gardens-Behind The Wall-Under The Tree... (84) 121(2) [0|0|18]	637
Caverna Magica (...Under The Tree-In The Cave...) (84) 149(1) [0|0|15]	367

W

Act ▶ Rank Peak [Top10s|Top40s|Total] Title (Yr) Pk(Wks) [T10 Wks|T40Wks|TotWks]	Score in '80s
Jack WAGNER ▶ 638 Pk:44 [0|0|3]	2856
All I Need (84) 44(2) [0|0|29]	2326
Lighting Up The Night (85) 150(2) [0|0|15]	341
Don't Give Up Your Day Job (87) 151(2) [0|0|8]	189
John WAITE ▶ 279 Pk:10 [1|2|4]	10334
No Brakes (84) 10(1) [1|17|43]	6908
Mask Of Smiles (85) 36(2) [0|3|16]	1713
Ignition (82) 68(3) [0|0|23]	1065
Rover's Return (87) 77(1) [0|0|12]	648
The WAITRESSES ▶ 721 Pk:41 [0|0|3]	2264
Wasn't Tomorrow Wonderful? (82) 41(1) [0|0|24]	1775
I Could Rule The World If I Could Only Get The Parts (82) 128(2) [0|0|10]	365
Bruiseology (83) 155(2) [0|0|5]	124
Tom WAITS ▶ 929 Pk:96 [0|0|5]	1261
Heartattack And Vine (80) 96(2) [0|0|10]	474
Franks Wild Years (87) 115(2) [0|0|10]	397
Big Time (88) 152(2) [0|0|6]	145
Swordfishtrombones (83) 167(1) [0|0|7]	141
Rain Dogs (85) 181(2) [0|0|7]	104
Narada Michael WALDEN ▶ 853 Pk:74 [0|0|3]	1546
The Dance Of Life (80) 74(2) [0|0|19]	1015
Victory (80) 103(1) [0|0|8]	358
Confidence (82) 135(1) [0|0|6]	173
Jerry Jeff WALKER ▶ 1750 Pk:185 [0|0|2]	88.9
The Best Of Jerry Jeff Walker (80) 185(1) [0|0|3]	45.8
Reunion (81) 188(2) [0|0|3]	43.1
WALL OF VOODOO ▶ 822 Pk:45 [0|0|2]	1707
Call Of The West (83) 45(2) [0|0|23]	1668
Dark Continent (81) 177(1) [0|0|2]	39.4
Joe WALSH ▶ 434 Pk:20 [0|1|4]	5324
There Goes The Neighborhood (81) 20(1) [0|9|24]	2846
The Confessor (85) 65(2) [0|0|19]	1101
You Bought It-You Name It (83) 48(1) [0|0|14]	1075
Got Any Gum? (87) 113(2) [0|0|8]	302
Steve WALSH ▶ 1558 Pk:124 [0|0|1]	199
Schemer-Dreamer (80) 124(1) [0|0|5]	199
WANG CHUNG ▶ 356 Pk:30 [0|1|4]	7524
Points On The Curve (84) 30(3) [0|6|37]	3434
Mosaic (86) 41(2) [0|0|36]	3047
To Live And Die In L.A. (Soundtrack) (85) 85(2) [0|0|18]	806
The Warmer Side Of Cool (89) 123(1) [0|0|6]	236
WAR ▶ 737 Pk:48 [0|0|4]	2178
Outlaw (82) 48(2) [0|0|27]	1574
The Music Band 2 (79/80) 111(1) [0|0|9]	*310*
The Best Of War...And More (87) 156(2) [0|0|10]	213
Life (Is So Strange) (83) 164(1) [0|0|4]	82.4
Steve WARINER ▶ 1883 Pk:187 [0|0|1]	31.8
Greatest Hits (87) 187(2) [0|0|1]	31.8
WARLOCK ▶ 925 Pk:80 [0|0|1]	1273
Triumph And Agony (87) 80(1) [0|0|27]	1273
Jennifer WARNES ▶ 938 Pk:72 [0|0|1]	1197
Famous Blue Raincoat (87) 72(2) [0|0|21]	1197
WARRANT ▶ 300 Pk:10 [1|1|1]	9325
Dirty Rotten Filthy Stinking Rich (89) 10(3) [3|33|44]	*9325*

Acts With Albums

Act ▶ Rank Peak [Top10s|Top40s|Total] Title (Yr) Pk(Wks) [T10 Wks|T40Wks|TotWks]	Score in '80s
Dionne WARWICK ▶ 182 Pk:12 [0|3|10]	**16457**
Friends (85) 12(2) [0|13|26]	4770
Heartbreaker (82) 25(3) [0|10|28]	3473
No Night So Long (80) 23(2) [0|6|25]	2687
Dionne(2) 79/80 55(2) [0|0|24]	*1536*
Reservations For Two (87) 56(2) [0|0|27]	1371
How Many Times Can We Say Goodbye (83) 57(2) [0|0|17]	1004
Hot! Live And Otherwise (81) 72(1) [0|0|14]	686
Friends In Love (82) 83(2) [0|0|12]	494
Finder Of Lost Loves (85) 106(2) [0|0|11]	412
Greatest Hits 1979-1990 (89/90) 196(2) [0|0|2]	*23.6*
Grover WASHINGTON Jr. ▶ 150 Pk:5 [1|3|8]	**20194**
Winelight (80) 5(7) [8|27|52]	10642
Come Morning (81) 28(2) [0|5|27]	3014
Skylarkin' (80) 24(2) [0|6|22]	2305
The Best Is Yet To Come (82) 50(3) [0|0|25]	1838
Inside Moves (84) 79(1) [0|0|23]	1030
Strawberry Moon (87) 66(2) [0|0|16]	785
Baddest (80) 96(1) [0|0|10]	392
Anthology (81) 149(1) [0|0|7]	187
WAS (NOT WAS) ▶ 633 Pk:43 [0|0|2]	**2906**
What Up, Dog? (88) 43(2) [0|0|37]	2619
Born To Laugh At Tornadoes (83) 134(2) [0|0|9]	287
W.A.S.P. ▶ 416 Pk:48 [0|0|5]	**5707**
The Last Command (85) 49(2) [0|0|23]	1676
W.A.S.P. (84) 74(2) [0|0|31]	1240
Inside The Electric Circus (86) 60(2) [0|0|19]	1132
The Headless Children (89) 48(3) [0|0|13]	1033
Live...In The Raw (87) 77(2) [0|0|14]	627
The WATERBOYS ▶ 923 Pk:76 [0|0|1]	**1277**
Fisherman's Blues (88) 76(2) [0|0|26]	1277
WATERFRONT ▶ 1275 Pk:103 [0|0|1]	**467**
Waterfront (89) 103(1) [0|0|13]	467
Muddy WATERS ▶ 1900 Pk:192 [0|0|1]	**27.2**
King Bee (81) 192(2) [0|0|2]	27.2
Roger WATERS ▶ 572 Pk:31 [0|1|2]	**3495**
The Pros & Cons Of Hitchhiking (84) 31(3) [0|7|18]	1950
Radio K.A.O.S. (87) 50(2) [0|0|19]	1545
Jody WATLEY ▶ 164 Pk:10 [1|2|3]	**18078**
Jody Watley (87) 10(1) [1|44|74]	12532
Larger Than Life (89) 16(2) [0|17|38]	*5259*
You Wanna Dance With Me? (89) 86(2) [0|0|5]	*287*
Johnny Guitar WATSON ▶ 1140 **Pk:115 [0|0|2]**	**646**
Love Jones (80) 115(2) [0|0|14]	589
Johnny "Guitar" Watson And The Family Clone (81) 177(2) [0|0|3]	57.1
Ernie WATTS ▶ 1478 Pk:161 [0|0|1]	**266**
Chariots Of Fire (82) 161(2) [0|0|12]	266
WA WA NEE ▶ 1182 Pk:123 [0|0|1]	**582**
Wa Wa Nee (87) 123(1) [0|0|17]	582
WAX ▶ 1316 Pk:101 [0|0|1]	**419**
Magnetic Heaven (86) 101(2) [0|0|11]	419
Fee WAYBILL ▶ 1619 Pk:146 [0|0|1]	**164**
Read My Lips (84) 146(2) [0|0|6]	164
WAYSTED ▶ 1877 Pk:185 [0|0|1]	**33.1**
Save Your Prayers (87) 185(1) [0|0|2]	33.1
WEATHER GIRLS ▶ 1287 Pk:91 [0|0|1]	**448**
Two Tons O' Fun (80)-Two Tons Of Fun 91(2) [0|0|11]	448
WEATHER REPORT ▶ 709 Pk:57 [0|0|6]	**2328**
Night Passage (80) 57(1) [0|0|14]	893
Weather Report(2) (82) 68(2) [0|0|11]	661
Procession (83) 96(2) [0|0|10]	440
Domino Theory (84) 136(1) [0|0|8]	269
Sportin' Life (85) 191(1) [0|0|3]	40.0
This Is This (86) 195(2) [0|0|2]	24.6

Act ▶ Rank Peak [Top10s|Top40s|Total] Title (Yr) Pk(Wks) [T10 Wks|T40Wks|TotWks]	Score in '80s
Tim WEISBERG ▶ 1663 Pk:171 [0|0|1]	**135**
Party Of One (80) 171(2) [0|0|7]	135
Bob WELCH ▶ 1512 Pk:105 [0|0|2]	**241**
The Other One (79/80) 105(1) [0|0|3]	*122*
Man Overboard (80) 162(2) [0|0|5]	118
WENDY and LISA ▶ 1027 Pk:88 [0|0|2]	**907**
Wendy And Lisa (87) 88(1) [0|0|13]	624
Fruit At The Bottom (89) 119(2) [0|0|8]	284
Dottie WEST ▶ 1327 Pk:126 [0|0|1]	**411**
Wild West (81) 126(1) [0|0|15]	411
WET WET WET ▶ 1517 Pk:123 [0|0|1]	**237**
Popped In Souled Out (88) 123(3) [0|0|7]	237
Kirk WHALUM ▶ 1466 Pk:142 [0|0|1]	**273**
And You Know That! (88) 142(1) [0|0|10]	273
WHAM! ▶ 89 Pk:1 [2|2|3]	**30487**
Make It Big (84) 1(3) [25|56|80]	24537
Music From The Edge Of Heaven (86) 10(2) [2|11|28]	4632
Fantastic (83) 83(1) [0|0|44]	1319
WHAT IS THIS ▶ 1829 Pk:187 [0|0|1]	**57.3**
What Is This (85) 187(1) [0|0|4]	57.3
WHEN IN ROME ▶ 943 Pk:84 [0|0|1]	**1178**
When In Rome (88) 84(1) [0|0|24]	1178
The WHISPERS ▶ 144 Pk:6 [1|5|8]	**21183**
The Whispers (80) 6(2) [9|17|35]	7346
Just Gets Better With Time (87) 22(2) [0|12|37]	4302
Imagination (81) 23(2) [0|11|27]	3317
Love For Love (83) 37(1) [0|3|29]	2399
Love Is Where You Find It (82) 35(3) [0|4|25]	2157
So Good (84) 88(1) [0|0|26]	1194
This Kind Of Lovin' (81) 100(2) [0|0|9]	391
The Best Of The Whispers (82) 180(1) [0|0|5]	76.3
Barry WHITE ▶ 991 Pk:85 [0|0|3]	**1015**
Barry White's Sheet Music (80) 85(1) [0|0|11]	492
The Right Night & Barry White (87) 159(2) [0|0|17]	345
Change (82) 148(1) [0|0|6]	178
Karyn WHITE ▶ 371 Pk:19 [0|1|1]	**7200**
Karyn White (88) 19(1) [0|17|54]	7200
Maurice WHITE ▶ 1000 Pk:61 [0|0|1]	**991**
Maurice White (85) 61(2) [0|0|19]	991
WHITE LION ▶ 190 Pk:11 [0|2|3]	**15711**
Pride (87) 11(1) [0|31|86]	11231
Big Game (89) 19(2) [0|13|26]	4098
Fight To Survive (88) 151(1) [0|0|14]	382
WHITESNAKE ▶ 63 Pk:2 [2|3|6]	**36780**
Whitesnake (87) 2(10) [41|54|76]	29467
Slide It In (84) 40(2) [0|2|85]	4309
Slip Of The Tongue (89) 10(1) [1|6|6]	*1960*
Ready An' Willing (80) 90(1) [0|0|16]	632
Live In The Heart Of The City (80) 146(1) [0|0|12]	264
Come An' Get It (81) 151(2) [0|0|6]	148
WHITE WOLF ▶ 1357 Pk:137 [0|0|2]	**376**
Endangered Species (86) 137(2) [0|0|8]	227
Standing Alone (85) 162(2) [0|0|7]	148
Keith WHITLEY ▶ 1141 Pk:115 [0|0|2]	**643**
Don't Close Your Eyes (89) 121(2) [0|0|14]	393
I Wonder Do You Think Of Me (89) 115(2) [0|0|7]	250
Slim WHITMAN ▶ 1685 Pk:175 [0|0|2]	**123**
Christmas With Slim Whitman (80) 184(2) [0|0|4]	66.1
Songs I Love To Sing (80) 175(2) [0|0|3]	57.1
Roger WHITTAKER ▶ 1228 Pk:154 [0|0|4]	**520**
Voyager (80) 154(3) [0|0|12]	281
Mirrors Of My Mind (80) 157(2) [0|0|6]	*138*
Live In Concert (81) 177(1) [0|0|3]	57.5
With Love (80) 175(2) [0|0|5]	42.6

Act ▶ Rank Peak [Top10s|Top40s|Total] Title (Yr) Pk(Wks) [T10 Wks|T40Wks|TotWks]	Score in '80s
The WHO ▶ 213 Pk:4 [2|2|7]	**14070**
Face Dances (81) 4(4) [8|14|20]	6119
It's Hard (82) 8(5) [7|10|32]	5205
Hooligans (81) 52(2) [0|0|19]	1172
Who's Last (84) 81(2) [0|0|14]	756
Who's Greatest Hits (83) 94(1) [0|0|13]	450
Who's Missing (85) 116(2) [0|0|8]	245
Quadrophenia (Soundtrack) (79/80) *148(3) [0|0|4]*	*122*
WHODINI ▶ 297 Pk:30 [0|3|3]	**9547**
Escape (84) 35(1) [0|5|48]	4110
Back In Black (86) 35(2) [0|4|39]	3250
Open Sesame (87) 30(2) [0|6|22]	2188
Jane WIEDLIN ▶ 975 Pk:105 [0|0|2]	**1074**
Fur (88) 105(2) [0|0|21]	862
Jane Wiedlin (85) 127(2) [0|0|6]	212
Danny WILDE ▶ 1638 Pk:176 [0|0|1]	**150**
Any Man's Hunger (88) 176(1) [0|0|9]	150
Eugene WILDE ▶ 1181 Pk:97 [0|0|1]	**583**
Eugene Wilde (85) 97(2) [0|0|15]	583
Kim WILDE ▶ 537 Pk:40 [0|1|4]	**3845**
Another Step (87) 40(2) [0|2|26]	2247
Kim Wilde (82) 86(2) [0|0|22]	887
Teases And Dares (85) 84(2) [0|0|10]	489
Close (88) 114(1) [0|0|6]	222
Matthew WILDER ▶ 952 Pk:49 [0|0|1]	**1145**
I Don't Speak The Language (84) 49(2) [0|0|22]	1145
WILL And The KILL ▶ 1505 Pk:129 [0|0|1]	**246**
Will And The Kill (88) 129(2) [0|0|8]	246
Deniece WILLIAMS ▶ 363 Pk:20 [0|2|4]	**7400**
Niecy (82) 20(3) [0|7|22]	2599
Let's Hear It For The Boy (84) 26(1) [0|6|19]	1915
My Melody (81) 74(1) [0|0|32]	1764
I'm So Proud (83) 54(2) [0|0|19]	1123
Don WILLIAMS ▶ 670 Pk:57 [0|0|3]	**2583**
I Believe In You (80) 57(1) [0|0|31]	1953
Especially For You (81) 109(1) [0|0|11]	449
Listen To The Radio (82) 166(3) [0|0|8]	180
Hank WILLIAMS Jr. ▶ 184 Pk:28 [0|1|15]	**16188**
Born To Boogie (87) 28(1) [0|3|47]	3178
Hank Williams, Jr.'s Greatest Hits (82) 107(2) [0|0|70]	2209
Greatest Hits III (89) 61(1) [0|0|35]	1680
Wild Streak (88) 55(1) [0|0|19]	1423
Hank "Live" (87) 71(1) [0|0|24]	1020
The Pressure Is On (81) 76(2) [0|0|23]	1002
Five-O (85) 72(2) [0|0|22]	950
Strong Stuff (83) 64(2) [0|0|16]	813
Major Moves (84) 100(1) [0|0|19]	786
Montana Cafe (86) 93(2) [0|0|18]	779
Rowdy (81) 82(2) [0|0|15]	762
High Notes (82) 123(1) [0|0|20]	558
Man Of Steel (83) 116(2) [0|0|13]	490
Habits Old And New (80) 154(1) [0|0|17]	424
Greatest Hits, Vol. II (86) 183(1) [0|0|8]	113
Lenny WILLIAMS ▶ 1876 Pk:185 [0|0|1]	**33.6**
Let's Do It Today (80) 185(2) [0|0|2]	33.6
Mason WILLIAMS ▶ 1405 Pk:118 [0|0|1]	**325**
Classical Gas (87)-Mason Williams & Mannheim Steamroller [A] 118(1) [0|0|19]	325
Robin WILLIAMS ▶ 1431 Pk:119 [0|0|1]	**301**
Throbbing Python Of Love (83) 119(1) [0|0|9]	301
Vanessa WILLIAMS ▶ 587 Pk:38 [0|1|1]	**3347**
The Right Stuff (88) 38(1) [0|4|55]	3347
WILLIE & The POOR BOYS ▶ 1249 **Pk:96 [0|0|1]**	**497**
Willie & The Poor Boys (85) 96(2) [0|0|12]	497

Acts With Albums

Act ▶ Rank Peak [Top10s|Top40s|Total] Title (Yr) Pk(Wks) [T10 Wks|T40Wks|TotWks]	Score in '80s
WILLIE, WAYLON, JOHNNY & KRIS ▶ 887 Pk:92 [0|0|1]	**1417**
Highwayman (85) 92(1) [0|0|35]	1417
Bruce WILLIS ▶ 511 Pk:14 [0|1|1]	**4142**
The Return Of Bruno (87) 14(3) [0|11|29]	4142
WILL TO POWER ▶ 881 Pk:68 [0|0|1]	**1440**
Will To Power (88) 68(1) [0|0|29]	1440
Shanice WILSON ▶ 1284 Pk:149 [0|0|1]	**450**
Discovery (87) 149(2) [0|0|18]	450
Brian WILSON ▶ 998 Pk:54 [0|0|1]	**1003**
Brian Wilson (88) 54(3) [0|0|13]	1003
Carl WILSON ▶ 1878 Pk:185 [0|0|1]	**32.7**
Carl Wilson (81) 185(1) [0|0|2]	32.7
Nancy WILSON ▶ 1699 Pk:144 [0|0|1]	**120**
The Two Of Us (84)-Ramsey Lewis & Nancy Wilson [A] 144(1) [0|0|9]	120
The WINANS ▶ 1340 Pk:109 [0|0|1]	**400**
Decisions (87) 109(1) [0|0|11]	400
BeBe & CeCe WINANS ▶ 984 Pk:95 [0|0|1]	**1042**
Heaven (89) 95(1) [0|0|25]	1042
Angela WINBUSH ▶ 869 Pk:81 [0|0|2]	**1489**
Sharp (87) 81(1) [0|0|28]	1183
The Real Thing (89) 113(2) [0|0|8]	*305*
Jesse WINCHESTER ▶ 1892 Pk:188 [0|0|1]	**29.9**
Talk Memphis (81) 188(1) [0|0|2]	29.9
WINGER ▶ 288 Pk:21 [0|1|1]	**9978**
Winger (88) 21(3) [0|35|64]	9978
George WINSTON ▶ 289 Pk:54 [0|0|4]	**9973**
December (83) 54(1) [0|0|168]	*8132*
Autumn (84) 139(1) [0|0|44]	1101
Winter Into Spring (84) 127(2) [0|0|32]	711
The Velveteen Rabbit (85)-Meryl Streep & George Winston [A] 180(1) [0|0|4]	30.2
Johnny WINTER ▶ 1453 Pk:156 [0|0|2]	**282**
Serious Business (85) 156(2) [0|0|10]	222
Guitar Slinger (84) 183(1) [0|0|4]	60.4
Paul WINTER ▶ 1407 Pk:138 [0|0|1]	**321**
Canyon (86) 138(1) [0|0|11]	321
Robert WINTERS and FALL ▶ 1270 Pk:71 [0|0|1]	**470**
Magic Man (81) 71(2) [0|0|8]	470
Steve WINWOOD ▶ 34 Pk:1 [3|5|5]	**50199**
Back In The High Life (86) 3(2) [14|60|86]	19677
Roll With It (88) 1(1) [14|31|45]	13007
Arc Of A Diver (81) 3(6) [13|26|43]	11956
Chronicles (87) 26(1) [0|10|26]	3142
Talking Back To The Night (82) 28(4) [0|6|25]	2417
WIRE ▶ 1509 Pk:135 [0|0|1]	**245**
It's Beginning To And Back Again (89) 135(2) [0|0|10]	245
WIRE TRAIN ▶ 1456 Pk:150 [0|0|2]	**280**
...In A Chamber (84) 150(1) [0|0|9]	218
Ten Women (87) 181(1) [0|0|4]	61.8
WISHBONE ASH ▶ 1755 Pk:179 [0|0|2]	**87.6**
Hot Ash (82) 192(1) [0|0|4]	49.1
Just Testing (80) 179(1) [0|0|2]	38.5
Bill WITHERS ▶ 1496 Pk:143 [0|0|2]	**254**
Watching You, Watching Me (85) 143(2) [0|0|8]	207
Bill Withers Greatest Hits (81) 183(1) [0|0|3]	47.2
Peter WOLF ▶ 518 Pk:24 [0|1|2]	**4095**
Lights Out (84) 24(2) [0|8|26]	2991
Come As You Are (87) 53(2) [0|0|15]	1104
Bobby WOMACK ▶ 496 Pk:29 [0|1|3]	**4246**
The Poet (81) 29(2) [0|6|23]	2388
So Many Rivers (85) 66(2) [0|0|19]	962
The Poet II (84) 60(1) [0|0|14]	896
Stevie WONDER ▶ 39 Pk:3 [5|6|8]	**47016**
In Square Circle (85) 5(2) [13|29|50]	12031
Hotter Than July (80) 3(7) [16|25|40]	12004

Act ▶ Rank Peak [Top10s|Top40s|Total] Title (Yr) Pk(Wks) [T10 Wks|T40Wks|TotWks]	Score in '80s
The Woman In Red (Soundtrack) (84) 4(3) [11|21|40]	9416
Stevie Wonder's Original Musiquarium I (82) 4(3) [6|8|28]	5313
Characters (87) 17(3) [0|13|31]	4698
Journey Through The Secret Life Of Plants (*79/80*) 4(1) [3|9|16]	*3461*
Innervisions (*73/81*) 165(2) [0|0|2]	*51.6*
Songs In The Key Of Life (*76/81*) 192(3) [0|0|3]	*40.8*
Ronnie WOOD ▶ 1703 Pk:164 [0|0|1]	**117**
1234 (81) 164(2) [0|0|5]	117
The WOODENTOPS ▶ 1754 Pk:185 [0|0|1]	**87.6**
Giant (86)-Woodentops 185(1) [0|0|6]	87.6
Stevie WOODS ▶ 1317 Pk:153 [0|0|1]	**418**
Take Me To Your Heaven (81) 153(2) [0|0|25]	418
Bruce WOOLLEY & The CAMERA CLUB ▶ 1874 Pk:184 [0|0|1]	**34.0**
Bruce Woolley & The Camera Club (80) 184(1) [0|0|2]	34.0
WORLD PARTY ▶ 698 Pk:39 [0|1|1]	**2413**
Private Revolution (86) 39(3) [0|3|31]	2413
WRABIT ▶ 1566 Pk:157 [0|0|1]	**196**
Wrabit (82) 157(1) [0|0|8]	196
WRATHCHILD AMERICA ▶ 1783 Pk:190 [0|0|1]	**76.3**
Climbing The Walls (89) 190(2) [0|0|6]	76.3
Bernard WRIGHT ▶ 1204 Pk:116 [0|0|1]	**551**
'Nard (81) 116(2) [0|0|14]	551
Betty WRIGHT ▶ 1353 Pk:127 [0|0|1]	**386**
Mother Wit (88) 127(2) [0|0|13]	386
Gary WRIGHT ▶ 1037 Pk:79 [0|0|1]	**873**
The Right Place (81) 79(1) [0|0|19]	873
Steven WRIGHT ▶ 1905 Pk:192 [0|0|1]	**26.3**
I Have A Pony (85) 192(1) [0|0|2]	26.3

X

Act ▶ Rank Peak [Top10s|Top40s|Total] Title (Yr) Pk(Wks) [T10 Wks|T40Wks|TotWks]	Score in '80s
X ▶ 632 Pk:76 [0|0|6]	**2920**
More Fun In The New World (83) 86(2) [0|0|23]	852
Under The Big Black Sun (82) 76(3) [0|0|15]	744
Ain't Love Grand (85) 89(3) [0|0|14]	673
See How We Are (87) 107(2) [0|0|11]	439
Wild Gift (81) 165(1) [0|0|5]	121
Live At The Whisky A Go-Go On The Fabulous Sunset Strip (88) 175(2) [0|0|5]	92.0
XAVIER ▶ 1511 Pk:129 [0|0|1]	**243**
Point Of Pleasure (82) 129(1) [0|0|7]	243
XTC ▶ 377 Pk:41 [0|0|7]	**7031**
Black Sea (80) 41(1) [0|0|24]	2272
Oranges And Lemons (89) 44(3) [0|0|21]	1740
Skylarking (87) 70(2) [0|0|29]	1355
English Settlement (82) 48(1) [0|0|20]	1299
Mummer (84) 145(2) [0|0|5]	145
Drums And Wires (80) 176(2) [0|0|8]	136
The Big Express (84) 178(1) [0|0|5]	85.7
XYMOX ▶ 1616 Pk:165 [0|0|1]	**166**
Twist Of Shadows (89) 165(1) [0|0|10]	166
XYZ ▶ 1742 Pk:144 [0|0|1]	**91.2**
XYZ (*89/90*) 144(1) [0|0|3]	*91.2*

Y

Act ▶ Rank Peak [Top10s|Top40s|Total] Title (Yr) Pk(Wks) [T10 Wks|T40Wks|TotWks]	Score in '80s
Y&T ▶ 509 Pk:46 [0|0|5]	**4147**
In Rock We Trust (84) 46(2) [0|0|17]	1383
Open Fire (85) 70(3) [0|0|17]	1082
Contagious (87) 78(2) [0|0|13]	736
Down For The Count (85) 91(2) [0|0|12]	489
Mean Streak (83) 103(1) [0|0|12]	457
"Weird Al" YANKOVIC ▶ 340 Pk:17 [0|2|6]	**7852**
"Weird Al" Yankovic In 3-D (84) 17(3) [0|11|23]	3473
Even Worse (88) 27(1) [0|9|26]	2755

Act ▶ Rank Peak [Top10s|Top40s|Total] Title (Yr) Pk(Wks) [T10 Wks|T40Wks|TotWks]	Score in '80s
Dare To Be Stupid (85) 50(2) [0|0|16]	1207
"Weird Al" Yankovic (83) 139(1) [0|0|8]	227
UHF/Original Motion Picture Soundtrack And Other Stuff (89) 146(2) [0|0|6]	116
Polka Party! (86) 177(2) [0|0|4]	74.8
YARBROUGH & PEOPLES ▶ 503 Pk:16 [0|1|2]	**4189**
The Two Of Us (80) 16(2) [0|11|24]	3575
Be A Winner (84) 90(2) [0|0|16]	615
YAZ ▶ 828 Pk:69 [0|0|2]	**1668**
Upstairs At Eric's (82)-Yazoo 92(1) [0|0|32]	999
You And Me Both (83) 69(1) [0|0|13]	668
YELLO ▶ 1120 Pk:92 [0|0|3]	**700**
One Second (87) 92(2) [0|0|10]	458
Flag (89) 152(2) [0|0|9]	182
You Gotta Say Yes To Another Excess (83) 184(2) [0|0|4]	60.3
YELLOWJACKETS ▶ 1378 Pk:145 [0|0|3]	**345**
Mirage A Trois (83) 145(1) [0|0|10]	250
Samurai Samba (85) 179(1) [0|0|4]	70.7
Shades (86) 195(1) [0|0|2]	24.1
YELLOW MAGIC ORCHESTRA ▶ 1029 Pk:81 [0|0|2]	**900**
Yellow Magic Orchestra (80) 81(1) [0|0|21]	859
X-Multiplies (80) 177(2) [0|0|2]	40.8
YES ▶ 136 Pk:5 [1|4|8]	**22680**
90125 (83) 5(4) [12|28|53]	11179
Big Generator (87) 15(3) [0|18|30]	5079
Drama (80) 18(2) [0|7|19]	2632
Anderson, Bruford, Wakeman, Howe (89) 30(1) [0|7|16]	2053
Yesshows (80) 43(2) [0|0|12]	979
9012Live - The Solos (85) 81(3) [0|0|11]	568
Classic Yes (82) 142(2) [0|0|5]	161
Fragile (*72/84*) 188(1) [0|0|2]	*29.5*
Dwight YOAKAM ▶ 419 Pk:55 [0|0|4]	**5650**
Guitars, Cadillacs, Etc., Etc. (86) 61(2) [0|0|65]	2481
Hillbilly Deluxe (87) 55(2) [0|0|28]	1680
Buenas Noches From A Lonely Room (88) 68(1) [0|0|28]	950
Just Lookin' For A Hit (89) 68(2) [0|0|10]	540
Neil YOUNG ▶ 175 Pk:15 [0|5|11]	**17171**
Live Rust (80)- Neil Young With Crazy Horse 15(2) [0|10|20]	*3212*
Trans (83) 19(6) [0|7|17]	2566
Re-ac-tor (81)- Neil Young & Crazy Horse 27(3) [0|9|17]	2277
Hawks & Doves (80) 30(3) [0|6|16]	1633
Freedom (89) 35(2) [0|4|11]	*1543*
Landing On Water (86) 46(2) [0|0|16]	1508
Everybody's Rockin' (83)- Neil and the Shocking Pinks 46(2) [0|0|15]	1186
This Note's For You (88)- Neil Young & The Bluenotes 61(2) [0|0|18]	1127
Rust Never Sleeps (*79/80*)- Neil Young & Crazy Horse 87(2) [0|0|15]	*798*
Old Ways (85) 75(2) [0|0|12]	685
Life (87)- Neil Young With Crazy Horse 75(2) [0|0|11]	638
Paul YOUNG ▶ 310 Pk:19 [0|1|3]	**8853**
The Secret Of Association (85) 19(6) [0|24|43]	6901
No Parlez (84) 79(1) [0|0|23]	1083
Between Two Fires (86) 77(2) [0|0|17]	869
YOUNG M.C. ▶ 484 Pk:9 [1|1|1]	**4475**
Stone Cold Rhymin' (89) 9(1) [2|13|15]	4475
YUTAKA ▶ 1778 Pk:174 [0|0|1]	**77.8**
Love Light (81) 174(1) [0|0|4]	77.8

Z

Act ▶ Rank Peak [Top10s|Top40s|Total] Title (Yr) Pk(Wks) [T10 Wks|T40Wks|TotWks]	Score in '80s
Pia ZADORA ▶ 1152 Pk:113 [0|0|1]	**622**
Pia & Phil (86) 113(2) [0|0|20]	622

Acts With Albums

Act ▶ Rank Peak [Top10s|Top40s|Total] Title (Yr) Pk(Wks) [T10 Wks|T40Wks|TotWks]	Score in '80s
ZAPP ▶ 390 Pk:19 [0|3|5]	**6730**
Zapp (80) 19(2) [0|7|19]	2348
Zapp II (82) 25(3) [0|6|19]	2086
Zapp III (83) 39(1) [0|2|22]	1385
The New Zapp IV U (85) 110(2) [0|0|26]	809
V (89) 154(2) [0|0|4]	102
Frank ZAPPA ▶ 449 Pk:23 [0|1|7]	**5001**
A Ship Arriving Too Late To Save A Drowning Witch (82) 23(2) [0|7|22]	2625
Joe's Garage Acts II + III (80) 53(1) [0|0|9]	*691*
Tinsel Town Rebellion (81) 66(2) [0|0|11]	690
*Joe's Garage Act I (**79**/80) 108(1) [0|0|10]*	*393*
You Are What You Is (81) 93(1) [0|0|7]	329
Frank Zappa Meets The Mothers Of Prevention (86) 153(2) [0|0|6]	145
The Man From Utopia (83) 153(1) [0|0|5]	129
ZEBRA ▶ 625 Pk:29 [0|1|2]	**2957**
Zebra (83) 29(2) [0|7|28]	2452
No Tellin' Lies (84) 84(2) [0|0|11]	505
ZENO ▶ 1310 Pk:107 [0|0|1]	**425**
Zeno (86) 107(2) [0|0|10]	425
Warren ZEVON ▶ 468 Pk:20 [0|1|4]	**4624**
Bad Luck Streak In Dancing School (80) 20(1) [0|8|16]	2405
Sentimental Hygiene (87) 63(4) [0|0|18]	1184
Stand In The Fire (81) 80(3) [0|0|10]	553
The Envoy (82) 93(1) [0|0|13]	483
ZODIAC MINDWARP & The LOVE REACTION ▶ 1273 Pk:132 [0|0|1]	**469**
Tattooed Beat Messiah (88) 132(2) [0|0|15]	469
ZZ TOP ▶ 33 Pk:4 [2|4|5]	**50774**
Eliminator (83) 9(1) [4|82|183]	28823
Afterburner (85) 4(3) [15|36|70]	14561
Deguello (80) 24(3) [0|12|37]	*3635*
El Loco (81) 17(2) [0|12|22]	3605
*The Best Of ZZ Top (**78**/83) 182(1) [0|0|11]*	*150*

Acts With Albums

Ranking The Album Acts

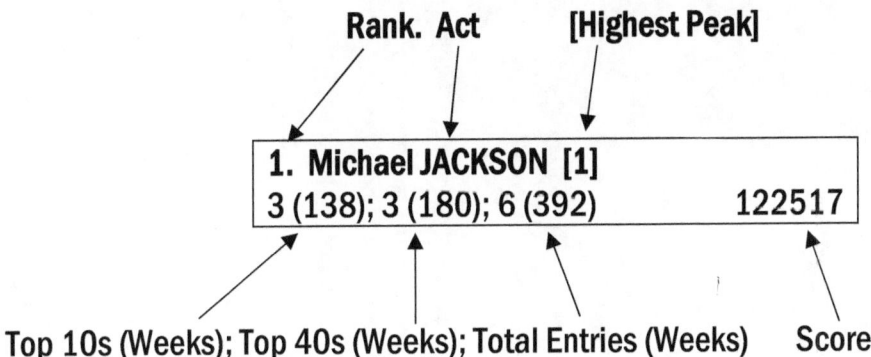

Act scoring only includes points earned between 1/1/80 and 12/31/89; however, it includes all albums charted during that period regardless of date of peak. Full lifecycle statistics for all albums peaking in the '80s can be found in the Ranking the Albums section. See Methodology in the Appendix.

Album Acts Ranked

Album Acts Ranked

Rank. Act [Hi Peak] T10 (Wk); T40 (Wk); Ent (Wk)	Score In '80s
1. Michael JACKSON [1] 3 (138); 3 (180); 6 (392)	122517
2. Bruce SPRINGSTEEN [1] 5 (120); 5 (177); 9 (490)	107845
3. Billy JOEL [1] 7 (94); 8 (196); 11 (423)	94089
4. MADONNA [1] 5 (82); 6 (196); 6 (447)	93686
5. JOURNEY [1] 6 (78); 6 (173); 9 (594)	89017
6. DEF LEPPARD [1] 2 (116); 3 (157); 5 (415)	87705
7. John MELLENCAMP [1] 5 (84); 6 (179); 7 (387)	85568
8. Lionel RICHIE [1] 3 (100); 3 (155); 3 (358)	83826
9. U2 [1] 2 (49); 6 (133); 8 (740)	81520
10. The POLICE [1] 4 (89); 5 (125); 6 (490)	79361
11. VAN HALEN [1] 6 (80); 6 (160); 8 (393)	78155
12. Daryl HALL & John OATES [3] 4 (69); 7 (187); 8 (382)	75922
13. BON JOVI [1] 2 (68); 3 (117); 4 (349)	75821
14. Whitney HOUSTON [1] 2 (77); 2 (129); 2 (247)	75693
15. Kenny ROGERS [1] 4 (43); 11 (143); 22 (653)	74617
16. Pat BENATAR [1] 3 (55); 8 (160); 9 (415)	71068
17. Huey LEWIS And The NEWS [1] 2 (68); 4 (139); 4 (308)	68157
18. PINK FLOYD [1] 3 (47); 5 (88); 8 (690)	65821
19. Phil COLLINS [1] 4 (51); 4 (121); 4 (433)	64403
20. ROLLING STONES [1] 6 (63); 7 (108); 14 (345)	64355
21. Barbra STREISAND [1] 6 (48); 9 (138); 10 (344)	63929
22. Bob SEGER And The SILVER BULLET BAND [1] 4 (61); 4 (115); 7 (395)	60040
23. GUNS N' ROSES [1] 2 (70); 2 (111); 2 (173)	59197
24. AC/DC [1] 3 (47); 7 (114); 10 (401)	58014
25. REO SPEEDWAGON [1] 3 (47); 4 (94); 10 (320)	56817
26. HEART [1] 3 (59); 6 (125); 7 (228)	56224
27. GENESIS [3] 4 (38); 5 (149); 6 (261)	56117
28. ALABAMA [10] 1 (1); 6 (114); 12 (613)	55776
29. PRINCE And The REVOLUTION [1] 3 (55); 3 (86); 3 (140)	55192
30. FOREIGNER [1] 3 (50); 5 (110); 7 (236)	54311
31. PRINCE [1] 3 (25); 6 (107); 7 (391)	52965
32. DURAN DURAN [4] 4 (34); 6 (119); 8 (387)	52241
33. ZZ TOP [4] 2 (19); 4 (142); 5 (323)	50774
34. Steve WINWOOD [1] 3 (41); 5 (133); 5 (225)	50199
35. KOOL & The GANG [10] 1 (2); 7 (154); 8 (329)	49167
36. The CARS [3] 3 (44); 6 (109); 7 (295)	48668
37. Bryan ADAMS [1] 3 (51); 3 (113); 4 (218)	47755
38. LOVERBOY [7] 2 (21); 4 (112); 6 (332)	47081
39. Stevie WONDER [3] 5 (49); 6 (105); 8 (210)	47016
40. RUSH [3] 6 (29); 9 (118); 11 (291)	46702
41. Stevie NICKS [1] 3 (38); 4 (95); 4 (251)	46197
42. FLEETWOOD MAC [1] 3 (31); 5 (98); 7 (195)	44721
43. MEN AT WORK [1] 2 (45); 2 (71); 3 (152)	44027
44. Linda RONSTADT [3] 4 (31); 7 (93); 10 (330)	43801
45. Tom PETTY And The HEARTBREAKERS [2] 4 (37); 6 (100); 7 (211)	43688
46. Willie NELSON [2] 1 (15); 6 (55); 22 (592)	43579
47. MÖTLEY CRÜE [1] 3 (32); 4 (87); 5 (306)	43313
48. Janet JACKSON [1] 2 (49); 2 (90); 4 (150)	42313
49. DIRE STRAITS [1] 1 (37); 3 (80); 6 (210)	41734
50. STYX [1] 3 (50); 4 (77); 7 (217)	41640
51. George MICHAEL [1] 1 (51); 1 (69); 1 (87)	40818
52. Neil DIAMOND [3] 3 (26); 6 (88); 10 (326)	40629
53. Tina TURNER [3] 2 (45); 3 (84); 4 (180)	40259
54. Rick SPRINGFIELD [2] 2 (17); 5 (97); 8 (265)	39927
55. Ozzy OSBOURNE [6] 2 (14); 7 (104); 8 (333)	39542
56. CULTURE CLUB [2] 1 (30); 4 (91); 4 (184)	39249
57. POISON [2] 2 (43); 2 (98); 2 (171)	38957
58. Anita BAKER [1] 1 (18); 2 (100); 3 (210)	38866
59. Diana ROSS [2] 1 (18); 7 (90); 12 (277)	37923
60. INXS [3] 1 (22); 2 (81); 6 (204)	37602
61. Luther VANDROSS [9] 1 (3); 7 (99); 7 (264)	37178
62. Bobby BROWN [1] 1 (45); 2 (72); 3 (98)	36950
63. WHITESNAKE [2] 2 (42); 3 (62); 6 (201)	36780
64. POINTER SISTERS [8] 1 (6); 4 (101); 8 (240)	36664
65. TALKING HEADS [15] 0 (0); 6 (87); 8 (349)	36295
66. Rod STEWART [11] 0 (0); 7 (113); 9 (231)	35979
67. Cyndi LAUPER [4] 2 (27); 3 (89); 3 (161)	35855
68. TEARS FOR FEARS [1] 2 (35); 2 (67); 3 (164)	35324
69. SADE [1] 3 (32); 3 (75); 3 (172)	35322
70. Elton JOHN [13] 0 (0); 8 (85); 12 (319)	35198
71. Billy IDOL [6] 3 (11); 3 (77); 5 (330)	34986
72. AIR SUPPLY [7] 2 (12); 5 (86); 6 (283)	34587
73. Billy SQUIER [5] 2 (16); 3 (81); 6 (235)	34473
74. EURYTHMICS [7] 2 (9); 5 (86); 8 (232)	34372
75. Billy OCEAN [6] 2 (14); 3 (98); 5 (177)	34370
76. SCORPIONS [5] 3 (22); 4 (82); 7 (253)	34310
77. Gloria ESTEFAN/MIAMI SOUND MACHINE [6] 2 (13); 3 (104); 3 (195)	34271
78. Dan FOGELBERG [3] 2 (18); 5 (82); 6 (201)	34168
79. JEFFERSON AIRPLANE/ STARSHIP [7] 2 (16); 6 (90); 9 (219)	33236
80. NEW KIDS ON THE BLOCK [1] 2 (43); 3 (70); 3 (105)	32612
81. R.E.M. [10] 1 (1); 6 (82); 8 (263)	32336
82. CHICAGO [4] 2 (18); 4 (70); 7 (215)	32083
83. CINDERELLA [3] 2 (20); 2 (81); 2 (136)	31632
84. Debbie GIBSON [1] 2 (26); 2 (70); 2 (136)	31505
85. Richard MARX [1] 2 (16); 2 (95); 2 (119)	31468
86. QUEEN [1] 1 (21); 6 (69); 7 (152)	31224
87. AEROSMITH [5] 1 (13); 5 (86); 8 (205)	30948
88. RUN-D.M.C. [3] 2 (16); 2 (63); 4 (220)	30840
89. WHAM! [1] 2 (27); 2 (67); 3 (152)	30487
90. Christopher CROSS [6] 1 (15); 2 (92); 3 (153)	29998
91. Paul SIMON [1] 1 (24); 3 (73); 4 (155)	29749
92. The GO-GO'S [1] 2 (24); 3 (66); 3 (132)	29694
93. KENNY G [6] 2 (19); 3 (70); 5 (196)	29372
94. David BOWIE [4] 1 (16); 4 (63); 11 (220)	29123
95. ASIA [1] 2 (32); 2 (46); 3 (106)	28736
96. Eric CLAPTON [2] 2 (19); 7 (62); 8 (179)	28460
97. 38 SPECIAL [10] 1 (3); 5 (72); 7 (246)	28387
98. Peter GABRIEL [2] 1 (12); 3 (73); 6 (190)	28332
99. Paul McCARTNEY [1] 2 (16); 6 (64); 8 (160)	28257
100. Alan PARSONS PROJECT [7] 1 (7); 3 (55); 9 (231)	28199
101. NEW EDITION [6] 1 (6); 3 (81); 5 (208)	28195
102. The PRETENDERS [5] 3 (17); 5 (66); 6 (212)	28086
103. Robert PLANT [5] 3 (24); 4 (65); 4 (160)	28016
104. Olivia NEWTON-JOHN [4] 2 (22); 4 (68); 6 (209)	27732
105. KISS [18] 0 (0); 7 (77); 9 (211)	27343
106. STING [2] 2 (23); 2 (63); 2 (110)	27228
107. IRON MAIDEN [11] 0 (0); 6 (65); 8 (281)	27106
108. NIGHT RANGER [10] 1 (3); 4 (65); 5 (209)	27014
109. TOTO [4] 1 (22); 2 (44); 7 (197)	26649
110. Bruce HORNSBY And The RANGE [3] 2 (30); 2 (61); 2 (100)	26366
111. J. GEILS Band [1] 1 (19); 3 (56); 4 (141)	26255
112. RATT [7] 2 (15); 4 (64); 5 (184)	26024
113. Rick JAMES [3] 1 (12); 3 (49); 9 (212)	26007
114. MOODY BLUES [1] 2 (18); 4 (56); 7 (139)	25949
115. Paula ABDUL [1] 1 (44); 1 (49); 1 (76)	25251
116. The COMMODORES [7] 1 (5); 5 (63); 8 (188)	25067
117. Donna SUMMER [1] 2 (11); 5 (50); 10 (197)	25039
118. BANGLES [2] 1 (9); 2 (64); 3 (154)	25004
119. Aretha FRANKLIN [13] 0 (0); 5 (58); 8 (219)	24552
120. QUIET RIOT [1] 1 (17); 3 (55); 4 (147)	24485
121. Joan JETT & The BLACKHEARTS [2] 1 (12); 3 (48); 6 (183)	24484
122. BEASTIE BOYS [1] 1 (19); 2 (45); 2 (83)	24441
123. CAMEO [8] 1 (3); 4 (56); 9 (216)	24359
124. Joe JACKSON [4] 1 (7); 3 (54); 10 (209)	24179
125. JUDAS PRIEST [17] 0 (0); 5 (79); 7 (203)	24080
126. STRAY CATS [2] 1 (18); 2 (47); 4 (117)	23881
127. Al JARREAU [9] 1 (4); 3 (37); 7 (276)	23685
128. Don HENLEY [8] 1 (5); 3 (73); 3 (123)	23658
129. FINE YOUNG CANNIBALS [1] 1 (27); 1 (40); 2 (71)	23559
130. TIFFANY [1] 1 (22); 2 (54); 2 (98)	23481
131. Bette MIDLER [2] 1 (11); 3 (54); 6 (128)	23242
132. MILLI VANILLI [1] 1 (28); 1 (40); 1 (41)	22975
133. Jackson BROWNE [1] 2 (17); 3 (42); 4 (118)	22973
134. John LENNON/Yoko ONO [1] 1 (22); 2 (37); 3 (105)	22966
135. Smokey ROBINSON [10] 1 (2); 5 (61); 9 (192)	22713
136. YES [5] 1 (12); 4 (60); 8 (148)	22680

Album Acts Ranked

Rank. Act [Hi Peak] T10 (Wk); T40 (Wk); Ent (Wk)	Score In '80s
137. SURVIVOR [2] 1 (12); 2 (46); 7 (169)	22386
138. Robert PALMER [8] 1 (4); 2 (55); 6 (181)	22249
139. George BENSON [3] 1 (10); 3 (45); 8 (202)	22156
140. David Lee ROTH [4] 2 (16); 3 (57); 3 (96)	22118
141. Barry MANILOW [14] 0 (0); 6 (43); 12 (234)	21986
142. Tracy CHAPMAN [1] 2 (22); 2 (56); 2 (72)	21861
143. GAP BAND [14] 0 (0); 3 (41); 8 (216)	21340
144. The WHISPERS [6] 1 (9); 5 (47); 8 (193)	21183
145. The CLASH [7] 1 (10); 3 (53); 6 (150)	20942
146. Sheena EASTON [15] 0 (0); 4 (41); 7 (221)	20756
147. EARTH, WIND & FIRE [5] 2 (12); 5 (50); 8 (127)	20527
148. Freddie JACKSON [10] 1 (1); 2 (62); 3 (143)	20465
149. METALLICA [6] 1 (5); 3 (49); 6 (237)	20203
150. Grover WASHINGTON Jr. [5] 1 (8); 3 (38); 8 (182)	20194
151. The FIXX [8] 1 (10); 3 (46); 6 (180)	19927
152. The CURE [12] 0 (0); 2 (34); 7 (209)	19658
153. LL COOL J [3] 2 (18); 2 (37); 3 (112)	19530
154. EAGLES [2] 2 (17); 2 (41); 5 (103)	19197
155. BLONDIE [7] 1 (9); 4 (47); 5 (145)	19130
156. EUROPE [8] 1 (10); 2 (53); 2 (103)	18847
157. CHEAP TRICK [16] 0 (0); 5 (51); 9 (159)	18778
158. Kenny LOGGINS [11] 0 (0); 3 (34); 5 (152)	18744
159. Randy TRAVIS [19] 0 (0); 3 (30); 5 (263)	18717
160. BOSTON [1] 1 (20); 1 (29); 3 (91)	18624
161. Sammy HAGAR [14] 0 (0); 4 (48); 7 (164)	18533
162. Jeffrey OSBORNE [25] 0 (0); 3 (38); 5 (211)	18472
163. SUPERTRAMP [5] 2 (10); 4 (47); 7 (148)	18335
164. Jody WATLEY [10] 1 (1); 2 (61); 3 (107)	18078
165. Eddie MONEY [20] 0 (0); 3 (40); 6 (172)	18044
166. Jermaine JACKSON [6] 1 (5); 2 (33); 7 (154)	17986
167. THOMPSON TWINS [10] 1 (2); 3 (52); 7 (147)	17929
168. Stephanie MILLS [16] 0 (0); 3 (36); 9 (203)	17834
169. GREAT WHITE [9] 1 (4); 2 (44); 5 (125)	17653
170. The B-52's [6] 1 (7); 4 (36); 7 (176)	17533
171. The KINKS [12] 0 (0); 3 (35); 9 (151)	17444
172. Kim CARNES [1] 1 (12); 1 (23); 6 (128)	17433

Rank. Act [Hi Peak] T10 (Wk); T40 (Wk); Ent (Wk)	Score In '80s
173. TRIUMPH [23] 0 (0); 5 (39); 7 (192)	17409
174. HUMAN LEAGUE [3] 1 (9); 3 (39); 5 (112)	17181
175. Neil YOUNG [15] 0 (0); 5 (36); 11 (168)	17171
176. Pete TOWNSHEND [5] 1 (9); 4 (43); 7 (121)	17148
177. Julio IGLESIAS [5] 1 (11); 2 (30); 8 (177)	17095
178. ATLANTIC STARR [17] 0 (0); 3 (36); 6 (192)	16908
179. Belinda CARLISLE [13] 0 (0); 3 (53); 3 (96)	16829
180. Teddy PENDERGRASS [14] 0 (0); 3 (42); 9 (183)	16799
181. Juice NEWTON [20] 0 (0); 2 (42); 5 (162)	16477
182. Dionne WARWICK [12] 0 (0); 3 (29); 10 (186)	16457
183. John FOGERTY [1] 1 (15); 2 (34); 2 (70)	16422
184. Hank WILLIAMS Jr. [28] 0 (0); 3 (15); 13 (366)	16188
185. PET SHOP BOYS [7] 1 (6); 3 (46); 4 (110)	16179
186. FAT BOYS [8] 1 (4); 2 (26); 7 (178)	16091
187. EXPOSÉ [16] 0 (0); 2 (59); 2 (101)	15860
188. MR. MISTER [1] 1 (12); 1 (34); 3 (82)	15754
189. Patti LaBELLE [1] 1 (12); 2 (21); 6 (135)	15725
190. WHITE LION [11] 0 (0); 2 (44); 3 (126)	15711
191. DOKKEN [13] 0 (0); 3 (24); 5 (204)	15698
192. JACKSON 5 [4] 2 (12); 3 (41); 5 (96)	15530
193. The OUTFIELD [9] 1 (6); 2 (40); 3 (110)	15529
194. Elvis COSTELLO & The ATTRACTIONS [11] 0 (0); 6 (37); 9 (157)	15447
195. Howard JONES [10] 1 (1); 2 (27); 5 (155)	15417
196. VANGELIS [1] 1 (14); 1 (20); 2 (96)	15187
197. TRAVELING WILBURYS [3] 1 (22); 1 (32); 1 (47)	15180
198. Los LOBOS [1] 1 (11); 1 (19); 4 (114)	15145
199. Anne MURRAY [16] 0 (0); 2 (22); 11 (215)	15100
200. The BEATLES [19] 0 (0); 2 (17); 16 (283)	15084
201. Tom PETTY [3] 1 (25); 1 (33); 1 (34)	15075
202. MIDNIGHT STAR [27] 0 (0); 2 (37); 4 (170)	15073
203. ROYAL PHILHARMONIC Orchestra [4] 1 (9); 2 (29); 3 (123)	15060
204. DeBARGE [19] 0 (0); 3 (36); 3 (136)	14638
205. Rick ASTLEY [10] 1 (2); 2 (49); 2 (83)	14536
206. DOOBIE BROTHERS [3] 1 (7); 3 (32); 6 (141)	14298
207. LISA LISA And CULT JAM [7] 1 (7); 1 (29); 3 (127)	14244

Rank. Act [Hi Peak] T10 (Wk); T40 (Wk); Ent (Wk)	Score In '80s
208. SKID ROW [6] 1 (11); 1 (42); 1 (47)	14202
209. SIMPLY RED [16] 0 (0); 3 (37); 3 (125)	14189
210. Steve MILLER Band [3] 1 (12); 2 (23); 6 (100)	14184
211. Bob DYLAN [20] 0 (0); 7 (35); 11 (148)	14175
212. Stevie Ray VAUGHAN And DOUBLE TROUBLE [31] 0 (0); 4 (23); 5 (162)	14073
213. The WHO [4] 2 (15); 2 (24); 7 (110)	14070
214. George THOROGOOD & The DESTROYERS [32] 0 (0); 3 (29); 5 (168)	14044
215. Laura BRANIGAN [23] 0 (0); 3 (33); 5 (161)	13897
216. SPYRO GYRA [19] 0 (0); 1 (8); 12 (212)	13892
217. Jane FONDA [15] 0 (0); 1 (27); 3 (137)	13874
218. BEE GEES [1] 2 (12); 2 (24); 7 (105)	13816
219. UB40 [14] 0 (0); 2 (18); 6 (166)	13804
220. ABC [24] 0 (0); 2 (35); 4 (119)	13774
221. GRATEFUL DEAD [6] 1 (7); 5 (34); 6 (100)	13772
222. Terence Trent D'ARBY [4] 1 (8); 1 (32); 2 (66)	13724
223. SANTANA [9] 1 (7); 2 (31); 6 (104)	13684
224. DJ JAZZY JEFF & THE FRESH PRINCE [4] 1 (10); 2 (28); 3 (97)	13511
225. The TIME [24] 0 (0); 2 (42); 3 (122)	13456
226. The JETS [21] 0 (0); 2 (31); 3 (127)	13279
227. The HOOTERS [12] 0 (0); 2 (35); 3 (105)	12995
228. OAK RIDGE BOYS [14] 0 (0); 2 (19); 9 (169)	12914
229. ISLEY BROTHERS [8] 1 (4); 3 (27); 8 (129)	12749
230. Teena MARIE [23] 0 (0); 3 (21); 7 (160)	12745
231. Glenn FREY [22] 0 (0); 3 (32); 3 (122)	12741
232. Quincy JONES [10] 1 (1); 2 (28); 3 (101)	12676
233. Eddie RABBITT [19] 0 (0); 3 (30); 5 (130)	12673
234. Carly SIMON [25] 0 (0); 2 (14); 6 (157)	12598
235. Boz SCAGGS [8] 1 (4); 2 (35); 3 (77)	12527
236. The DOORS [17] 0 (0); 2 (26); 6 (157)	12484
237. A FLOCK OF SEAGULLS [10] 1 (3); 2 (36); 3 (83)	12484
238. MISSING PERSONS [17] 0 (0); 1 (26); 4 (114)	12349
239. Chaka KHAN [14] 0 (0); 2 (25); 6 (125)	12347
240. Dolly PARTON [6] 1 (3); 3 (33); 10 (178)	12318
241. SHALAMAR [23] 0 (0); 4 (18); 6 (154)	12309

Rank. Act [Hi Peak] T10 (Wk); T40 (Wk); Ent (Wk)	Score In '80s
242. SIMPLE MINDS [10] 1 (5); 1 (27); 5 (107)	12219
243. David SANBORN [45] 0 (0); 0 (0); 8 (258)	12156
244. Emmylou HARRIS [6] 1 (3); 3 (34); 11 (185)	12145
245. MIKE + THE MECHANICS [13] 0 (0); 2 (34); 2 (90)	11829
246. Robert CRAY Band [13] 0 (0); 2 (33); 5 (131)	11790
247. John CAFFERTY And The BEAVER BROWN BAND [9] 1 (6); 2 (27); 3 (100)	11772
248. The MOTELS [16] 0 (0); 3 (25); 4 (101)	11752
249. LIVING COLOUR [6] 1 (8); 1 (28); 1 (70)	11735
250. The CULT [10] 1 (6); 2 (24); 3 (99)	11733
251. Herb ALPERT [18] 0 (0); 3 (29); 7 (124)	11691
252. TONE LŌC [1] 1 (12); 1 (23); 1 (40)	11607
253. Corey HART [20] 0 (0); 2 (21); 4 (108)	11593
254. SQUEEZE [32] 0 (0); 2 (11); 7 (159)	11546
255. ELECTRIC LIGHT ORCHESTRA [4] 1 (8); 4 (33); 6 (105)	11414
256. Thomas DOLBY [13] 0 (0); 3 (27); 4 (96)	11400
257. 10,000 MANIACS [13] 0 (0); 2 (21); 2 (105)	11387
258. LITTLE RIVER BAND [21] 0 (0); 2 (21); 6 (136)	11295
259. Waylon JENNINGS [36] 0 (0); 2 (5); 6 (196)	11187
260. READY FOR THE WORLD [17] 0 (0); 2 (35); 3 (84)	11185
261. James TAYLOR [10] 1 (3); 3 (31); 3 (87)	11142
262. Rickie Lee JONES [5] 1 (8); 3 (19); 4 (78)	11065
263. Edie BRICKELL & The NEW BOHEMIANS [4] 1 (8); 1 (28); 1 (54)	10990
264. DEVO [22] 0 (0); 2 (22); 6 (117)	10989
265. STRYPER [32] 0 (0); 2 (17); 4 (193)	10988
266. TWISTED SISTER [15] 0 (0); 1 (26); 5 (104)	10900
267. Julian LENNON [17] 0 (0); 2 (33); 3 (79)	10827
268. DIO [23] 0 (0); 2 (20); 5 (129)	10757
269. QUARTERFLASH [8] 1 (4); 2 (25); 3 (78)	10752
270. Jimmy BUFFETT [30] 0 (0); 2 (10); 10 (162)	10735
271. George HARRISON [8] 1 (9); 2 (29); 4 (57)	10697
272. DEPECHE MODE [35] 0 (0); 1 (3); 8 (211)	10696
273. Marvin GAYE [7] 1 (7); 2 (17); 9 (113)	10693
274. Eddie MURPHY [26] 0 (0); 2 (13); 4 (132)	10526
275. Ronnie MILSAP [31] 0 (0); 3 (10); 9 (182)	10494

Album Acts Ranked

Rank. Act [Hi Peak] T10 (Wk); T40 (Wk); Ent (Wk)	Score In '80s
276. CROSBY, STILLS & NASH [8] 1 (6); 1 (30); 3 (58)	10475
277. CHER [10] 1 (2); 2 (27); 2 (65)	10470
278. MAZE Featuring Frankie BEVERLY [25] 0 (0); 4 (16); 6 (132)	10339
279. John WAITE [10] 1 (1); 2 (20); 4 (94)	10334
280. KROKUS [25] 0 (0); 2 (14); 7 (140)	10306
281. Taylor DAYNE [21] 0 (0); 2 (35); 2 (76)	10269
282. Joe SATRIANI [23] 0 (0); 2 (15); 3 (108)	10267
283. BANANARAMA [15] 0 (0); 2 (18); 5 (118)	10253
284. MANHATTAN TRANSFER [22] 0 (0); 1 (13); 8 (171)	10209
285. FABULOUS THUNDERBIRDS [13] 0 (0); 1 (25); 4 (82)	10099
286. Ray PARKER Jr. [11] 0 (0); 1 (12); 6 (109)	9994
287. DAZZ BAND [14] 0 (0); 1 (11); 7 (146)	9980
288. WINGER [21] 0 (0); 1 (35); 1 (64)	9978
289. George WINSTON [54] 0 (0); 0 (0); 4 (248)	9973
290. POWER STATION [6] 1 (11); 1 (25); 1 (44)	9938
291. CROWDED HOUSE [12] 0 (0); 2 (26); 2 (77)	9880
292. Keith SWEAT [15] 0 (0); 1 (27); 1 (67)	9675
293. KANSAS [16] 0 (0); 3 (19); 6 (100)	9674
294. Greg KIHN Band [15] 0 (0); 3 (21); 6 (100)	9646
295. Adam ANT [16] 0 (0); 1 (16); 5 (125)	9635
296. ABBA [17] 0 (0); 2 (22); 5 (86)	9596
297. WHODINI [30] 0 (0); 3 (15); 3 (109)	9547
298. The MONKEES [21] 0 (0); 1 (12); 9 (156)	9415
299. S.O.S. BAND [12] 0 (0); 1 (11); 7 (112)	9408
300. WARRANT [10] 1 (3); 1 (33); 1 (44)	9325
301. The JUDDS [51] 0 (0); 0 (0); 6 (191)	9261
302. Earl KLUGH [38] 0 (0); 1 (3); 11 (201)	9202
303. BIG COUNTRY [18] 0 (0); 1 (18); 5 (94)	9098
304. Bobby McFERRIN [5] 1 (7); 1 (16); 2 (74)	8971
305. JETHRO TULL [19] 0 (0); 3 (19); 7 (102)	8962
306. NEW ORDER [32] 0 (0); 2 (13); 4 (131)	8951
307. BLUE ÖYSTER CULT [24] 0 (0); 3 (19); 6 (104)	8946
308. Melissa ETHERIDGE [22] 0 (0); 2 (24); 2 (78)	8897
309. TESLA [18] 0 (0); 2 (18); 2 (102)	8876
310. Paul YOUNG [19] 0 (0); 1 (24); 3 (83)	8853
311. LED ZEPPELIN [6] 2 (8); 2 (20); 3 (66)	8746
312. PAT METHENY GROUP [50] 0 (0); 0 (0); 9 (164)	8680
313. LEVEL 42 [18] 0 (0); 2 (23); 3 (77)	8663
314. Joan ARMATRADING [28] 0 (0); 2 (8); 8 (149)	8641
315. ORCHESTRAL MANOEUVRES IN THE DARK [38] 0 (0); 1 (5); 6 (129)	8528
316. PSYCHEDELIC FURS [29] 0 (0); 1 (11); 7 (119)	8455
317. LAKESIDE [16] 0 (0); 1 (12); 6 (110)	8435
318. ASHFORD & SIMPSON [29] 0 (0); 2 (12); 8 (117)	8420
319. KLYMAXX [18] 0 (0); 1 (15); 2 (98)	8418
320. Natalie COLE [42] 0 (0); 0 (0); 7 (135)	8393
321. Roy ORBISON [5] 1 (8); 1 (16); 5 (72)	8380
322. GEORGIA SATELLITES [5] 1 (5); 1 (20); 3 (63)	8374
323. The TEMPTATIONS [37] 0 (0); 1 (2); 10 (173)	8366
324. CLUB NOUVEAU [6] 1 (7); 1 (18); 2 (50)	8338
325. The ALARM [39] 0 (0); 1 (3); 6 (142)	8331
326. Lita FORD [29] 0 (0); 1 (23); 2 (78)	8274
327. Patrice RUSHEN [14] 0 (0); 3 (13); 5 (106)	8255
328. BLACK SABBATH [28] 0 (0); 4 (23); 7 (95)	8239
329. A-HA [15] 0 (0); 1 (25); 3 (73)	8159
330. Aldo NOVA [8] 1 (8); 1 (16); 2 (57)	8153
331. LOVE AND ROCKETS [14] 0 (0); 1 (17); 3 (84)	8122
332. Steve PERRY [12] 0 (0); 1 (18); 1 (60)	8110
333. MOLLY HATCHET [25] 0 (0); 3 (18); 7 (117)	8099
334. BROTHERS JOHNSON [5] 1 (7); 1 (16); 4 (59)	8089
335. SHEILA E. [28] 0 (0); 1 (9); 3 (91)	8061
336. Peabo BRYSON [40] 0 (0); 1 (2); 9 (143)	8053
337. The HONEYDRIPPERS [4] 1 (10); 1 (18); 1 (31)	7918
338. Samantha FOX [24] 0 (0); 2 (15); 3 (87)	7859
339. APRIL WINE [26] 0 (0); 2 (15); 5 (102)	7858
340. "Weird Al" YANKOVIC [17] 0 (0); 2 (20); 6 (81)	7852
341. SCANDAL/ Patty SMYTH [17] 0 (0); 2 (17); 2 (73)	7826
342. The BAR-KAYS [35] 0 (0); 1 (1); 7 (135)	7798
343. MIDNIGHT OIL [21] 0 (0); 1 (26); 3 (66)	7790
344. John DENVER [32] 0 (0); 3 (8); 5 (114)	7778
345. Gladys KNIGHT & The PIPS [34] 0 (0); 2 (9); 5 (98)	7771
346. Alice COOPER [20] 0 (0); 1 (16); 5 (79)	7770
347. STEELY DAN [9] 1 (6); 1 (19); 4 (49)	7743
348. Charlie DANIELS Band [11] 0 (0); 2 (15); 6 (81)	7717
349. DEEP PURPLE [17] 0 (0); 2 (22); 4 (67)	7692
350. ANTHRAX [30] 0 (0); 1 (5); 4 (130)	7679
351. KOOL MOE DEE [25] 0 (0); 2 (12); 3 (94)	7641
352. FALCO [3] 1 (6); 1 (18); 2 (40)	7625
353. AL B. SURE! [20] 0 (0); 1 (25); 1 (54)	7619
354. CON FUNK SHUN [30] 0 (0); 1 (5); 7 (139)	7565
355. SALT-N-PEPA [26] 0 (0); 2 (18); 2 (84)	7535
356. WANG CHUNG [30] 0 (0); 1 (6); 4 (97)	7524
357. GUY [27] 0 (0); 1 (13); 1 (70)	7517
358. Bonnie TYLER [4] 1 (7); 1 (18); 2 (40)	7498
359. Angela BOFILL [34] 0 (0); 2 (9); 4 (99)	7486
360. Herbie HANCOCK [43] 0 (0); 0 (0); 6 (115)	7471
361. LYNYRD SKYNYRD [12] 0 (0); 1 (11); 7 (109)	7433
362. BERLIN [28] 0 (0); 2 (15); 3 (84)	7408
363. Deniece WILLIAMS [20] 0 (0); 2 (13); 4 (92)	7400
364. SKYY [18] 0 (0); 1 (11); 6 (97)	7381
365. Ziggy MARLEY & The MELODY MAKERS [23] 0 (0); 2 (24); 2 (60)	7373
366. The TUBES [18] 0 (0); 2 (18); 4 (77)	7369
367. GLASS TIGER [27] 0 (0); 1 (28); 2 (66)	7367
368. Bob JAMES [47] 0 (0); 0 (0); 10 (192)	7270
369. Bonnie RAITT [22] 0 (0); 2 (17); 4 (76)	7267
370. The FIRM [17] 0 (0); 2 (23); 2 (52)	7259
371. Karyn WHITE [19] 0 (0); 1 (17); 1 (54)	7200
372. Peter CETERA [23] 0 (0); 1 (17); 3 (70)	7186
373. Ted NUGENT [13] 0 (0); 1 (9); 7 (86)	7183
374. Ray PARKER Jr. & RAYDIO [13] 0 (0); 2 (21); 2 (47)	7133
375. Yngwie MALMSTEEN [40] 0 (0); 1 (2); 5 (119)	7128
376. Rob BASE [31] 0 (0); 1 (8); 2 (69)	7071
377. XTC [41] 0 (0); 0 (0); 7 (112)	7031
378. Melissa MANCHESTER [19] 0 (0); 1 (8); 6 (98)	7025
379. HIROSHIMA [51] 0 (0); 0 (0); 6 (148)	6962
380. SOUL II SOUL [14] 0 (0); 1 (23); 1 (26)	6953
381. The ROMANTICS [14] 0 (0); 1 (15); 5 (71)	6953
382. MARY JANE GIRLS [18] 0 (0); 1 (11); 2 (79)	6952
383. USA For AFRICA [1] 1 (7); 1 (11); 1 (22)	6924
384. 'TIL TUESDAY [19] 0 (0); 1 (12); 3 (76)	6863
385. Suzanne VEGA [11] 0 (0); 1 (16); 2 (63)	6858
386. ROSSINGTON COLLINS BAND [13] 0 (0); 2 (18); 3 (49)	6836
387. LIPPS, INC. [5] 1 (7); 1 (13); 2 (35)	6801
388. QUEENSRYCHE [47] 0 (0); 0 (0); 4 (118)	6786
389. RUFUS/Chaka KHAN [14] 0 (0); 1 (0); 4 (77)	6766
390. ZAPP [19] 0 (0); 3 (15); 5 (90)	6730
391. SOFT CELL [22] 0 (0); 1 (18); 3 (63)	6671
392. Amy GRANT [35] 0 (0); 1 (4); 4 (104)	6630
393. Michael McDONALD [6] 1 (6); 1 (17); 2 (47)	6598
394. PEBBLES [14] 0 (0); 1 (19); 1 (38)	6577
395. Lou REED [40] 0 (0); 1 (1); 7 (95)	6572
396. Bob MARLEY And The WAILERS [45] 0 (0); 0 (0); 6 (119)	6416
397. Eddy GRANT [23] 1 (3); 1 (15); 2 (47)	6409
398. MC HAMMER [30] 0 (0); 1 (23); 1 (57)	6372
399. 2 LIVE CREW [30] 0 (0); 1 (17); 3 (98)	6344
400. Joni MITCHELL [25] 0 (0); 2 (8); 4 (72)	6269
401. Jesse JOHNSON [43] 0 (0); 0 (0); 3 (76)	6235
402. Chuck MANGIONE [8] 1 (5); 1 (11); 6 (71)	6194
403. MEN WITHOUT HATS [13] 0 (0); 1 (14); 3 (55)	6183
404. The SMITHS [55] 0 (0); 0 (0); 6 (140)	6163
405. ICEHOUSE [43] 0 (0); 0 (0); 4 (89)	6150
406. Andreas VOLLENWEIDER [52] 0 (0); 0 (0); 5 (130)	6141
407. The SMITHEREENS [51] 0 (0); 0 (0); 3 (88)	6122

Album Acts Ranked

Rank. Act [Hi Peak] T10 (Wk); T40 (Wk); Ent (Wk)	Score In '80s
408. RAINBOW [30] 0 (0); 2 (9); 5 (74)	6101
409. Tom BROWNE [18] 0 (0); 2 (11); 4 (71)	6041
410. Mick JAGGER [13] 0 (0); 1 (12); 2 (49)	6023
411. Van MORRISON [44] 0 (0); 0 (0); 8 (121)	5956
412. ROGER [26] 0 (0); 2 (11); 3 (63)	5897
413. Michael BOLTON [34] 0 (0); 1 (2); 3 (78)	5895
414. KINGDOM COME [12] 0 (0); 1 (12); 2 (44)	5852
415. John LENNON [31] 0 (0); 2 (13); 6 (70)	5743
416. W.A.S.P. [48] 0 (0); 0 (0); 5 (100)	5707
417. Gino VANNELLI [15] 0 (0); 1 (15); 4 (60)	5682
418. Kate BUSH [30] 0 (0); 1 (6); 5 (80)	5655
419. Dwight YOAKAM [55] 0 (0); 0 (0); 4 (118)	5650
420. Stacy LATTISAW [44] 0 (0); 0 (0); 7 (107)	5613
421. Elvis PRESLEY [27] 0 (0); 1 (5); 13 (116)	5593
422. BASIA [36] 0 (0); 1 (3); 1 (77)	5587
423. Evelyn "Champagne" KING [27] 0 (0); 2 (13); 5 (80)	5579
424. Rosanne CASH [26] 0 (0); 1 (9); 5 (92)	5519
425. Roberta FLACK & Peabo BRYSON [25] 0 (0); 1 (14); 2 (61)	5507
426. Pat TRAVERS [20] 0 (0); 2 (14); 4 (61)	5494
427. Chris De BURGH [25] 0 (0); 1 (8); 3 (73)	5445
428. Frank SINATRA [17] 0 (0); 1 (12); 3 (50)	5440
429. The O'JAYS [36] 0 (0); 1 (3); 7 (98)	5419
430. SWING OUT SISTER [40] 0 (0); 1 (1); 2 (62)	5367
431. BREATHE [34] 0 (0); 1 (8); 1 (51)	5343
432. Grace JONES [32] 0 (0); 1 (4); 6 (93)	5337
433. Philip BAILEY [22] 0 (0); 1 (11); 3 (60)	5327
434. Joe WALSH [20] 0 (0); 1 (9); 4 (59)	5324
435. Gregory ABBOTT [22] 0 (0); 1 (16); 2 (45)	5290
436. GTR [11] 0 (0); 1 (17); 1 (26)	5286
437. HEAVY D & The BOYZ [19] 0 (0); 1 (14); 2 (43)	5281
438. ENGLISH BEAT [39] 0 (0); 1 (3); 4 (86)	5250
439. FRANKIE GOES TO HOLLYWOOD [33] 0 (0); 1 (14); 2 (54)	5241
440. SPLIT ENZ [40] 0 (0); 1 (2); 4 (74)	5226
441. CHANGE [29] 0 (0); 1 (7); 5 (78)	5217

Rank. Act [Hi Peak] T10 (Wk); T40 (Wk); Ent (Wk)	Score In '80s
442. KATRINA & The WAVES [25] 0 (0); 1 (9); 3 (56)	5212
443. SIMON & GARFUNKEL [6] 1 (5); 1 (11); 1 (34)	5189
444. BAD COMPANY [26] 0 (0); 1 (6); 4 (81)	5159
445. Gary NUMAN [16] 0 (0); 1 (14); 3 (44)	5141
446. LITTLE FEAT [29] 0 (0); 3 (10); 3 (63)	5124
447. EAZY-E [41] 0 (0); 0 (0); 1 (56)	5057
448. George STRAIT [68] 0 (0); 0 (0); 8 (139)	5023
449. Frank ZAPPA [23] 0 (0); 1 (7); 7 (70)	5001
450. Carol HENSEL [56] 0 (0); 0 (0); 3 (95)	4997
451. Charlie SEXTON [15] 0 (0); 1 (13); 2 (43)	4989
452. Jeff HEALEY Band [22] 0 (0); 1 (8); 1 (65)	4977
453. GENERAL PUBLIC [26] 0 (0); 1 (11); 2 (55)	4955
454. ERASURE [49] 0 (0); 0 (0); 5 (74)	4913
455. N.W.A. [37] 0 (0); 1 (9); 1 (44)	4901
456. SPANDAU BALLET [19] 0 (0); 1 (8); 2 (53)	4885
457. ROXY MUSIC [35] 0 (0); 1 (3); 5 (85)	4875
458. Larry ELGART And His MANHATTAN SWING Orchestra [24] 0 (0); 1 (15); 2 (55)	4865
459. SAGA [29] 0 (0); 1 (13); 3 (55)	4864
460. Joe COCKER [50] 0 (0); 0 (0); 5 (93)	4862
461. The CHIPMUNKS [34] 0 (0); 1 (6); 4 (76)	4763
462. Robbie NEVIL [37] 0 (0); 1 (3); 2 (67)	4762
463. CUTTING CREW [16] 0 (0); 1 (10); 2 (51)	4752
464. Ric OCASEK [28] 0 (0); 2 (14); 2 (39)	4705
465. MARTIKA [15] 0 (0); 1 (12); 1 (39)	4681
466. Alexander O'NEAL [29] 0 (0); 1 (8); 4 (68)	4667
467. BLUES BROTHERS [13] 0 (0); 1 (12); 3 (34)	4638
468. Warren ZEVON [20] 0 (0); 1 (8); 4 (57)	4624
469. The MANHATTANS [24] 0 (0); 1 (10); 5 (60)	4619
470. Patti AUSTIN [36] 0 (0); 1 (11); 3 (66)	4617
471. ICE-T [35] 0 (0); 2 (8); 3 (70)	4580
472. AUTOGRAPH [29] 0 (0); 1 (12); 3 (59)	4572
473. Roberta FLACK [25] 0 (0); 1 (10); 4 (64)	4569
474. TOO SHORT [37] 0 (0); 1 (8); 1 (45)	4568
475. UTOPIA [32] 0 (0); 1 (5); 6 (77)	4562

Rank. Act [Hi Peak] T10 (Wk); T40 (Wk); Ent (Wk)	Score In '80s
476. Bob JAMES & Earl KLUGH [30] 0 (0); 1 (8); 2 (53)	4546
477. Jeff BECK [21] 0 (0); 2 (10); 3 (49)	4533
478. MEGADETH [28] 0 (0); 1 (6); 2 (70)	4527
479. Elvis COSTELLO [28] 0 (0); 2 (14); 2 (39)	4496
480. Bob & Doug McKENZIE [8] 1 (4); 1 (13); 1 (21)	4487
481. Roger DALTREY [22] 0 (0); 1 (6); 4 (55)	4485
482. ROCKWELL [15] 0 (0); 1 (12); 2 (39)	4482
483. STARS ON [9] 1 (5); 1 (8); 3 (36)	4481
484. YOUNG M.C. [9] 1 (2); 1 (13); 1 (15)	4475
485. TOM TOM CLUB [23] 0 (0); 1 (10); 3 (57)	4474
486. RAY, GOODMAN & BROWN [17] 0 (0); 1 (11); 3 (42)	4438
487. ONE WAY [51] 0 (0); 0 (0); 8 (99)	4434
488. ANIMOTION [28] 0 (0); 1 (7); 3 (61)	4412
489. FASTWAY [31] 0 (0); 1 (6); 4 (68)	4375
490. RENÉ & ANGELA [64] 0 (0); 1 (3); 2 (78)	4362
491. ENYA [25] 0 (0); 1 (13); 1 (39)	4356
492. INDIGO GIRLS [22] 0 (0); 1 (11); 2 (41)	4348
493. Stanley JORDAN [64] 0 (0); 0 (0); 3 (93)	4309
494. ROXETTE [28] 0 (0); 1 (7); 1 (37)	4299
495. Larry GRAHAM [26] 0 (0); 1 (9); 4 (50)	4260
496. Bobby WOMACK [29] 0 (0); 1 (6); 3 (56)	4246
497. K.T. OSLIN [68] 0 (0); 0 (0); 2 (84)	4223
498. Donald FAGEN [11] 0 (0); 1 (11); 1 (27)	4219
499. ERIC B. & RAKIM [22] 0 (0); 1 (7); 2 (54)	4212
500. KASHIF [51] 0 (0); 0 (0); 4 (87)	4210
501. VANITY [45] 0 (0); 0 (0); 3 (74)	4206
502. GIUFFRIA [26] 0 (0); 1 (7); 2 (43)	4198
503. YARBROUGH & PEOPLES [16] 0 (0); 1 (11); 2 (40)	4189
504. Sergio MENDES [27] 0 (0); 1 (10); 2 (49)	4188
505. Dave EDMUNDS [46] 0 (0); 0 (0); 6 (69)	4183
506. Lindsey BUCKINGHAM [32] 0 (0); 1 (6); 2 (40)	4175
507. ESCAPE CLUB [27] 0 (0); 1 (14); 1 (38)	4170
508. DEAD OR ALIVE [31] 0 (0); 1 (6); 4 (56)	4155

Rank. Act [Hi Peak] T10 (Wk); T40 (Wk); Ent (Wk)	Score In '80s
509. Y&T [46] 0 (0); 0 (0); 5 (71)	4147
510. Phyllis HYMAN [50] 0 (0); 0 (0); 4 (83)	4142
511. Bruce WILLIS [14] 0 (0); 1 (11); 1 (29)	4142
512. Jeff LORBER [68] 0 (0); 0 (0); 6 (89)	4139
513. Graham PARKER [40] 0 (0); 1 (1); 4 (64)	4128
514. LEVERT [32] 0 (0); 1 (7); 3 (58)	4112
515. SISTER SLEDGE [31] 0 (0); 1 (5); 4 (66)	4108
516. The OUTLAWS [25] 0 (0); 1 (7); 5 (63)	4106
517. AMBROSIA [25] 0 (0); 1 (7); 2 (40)	4102
518. Peter WOLF [24] 0 (0); 1 (8); 2 (41)	4095
519. The BLASTERS [36] 0 (0); 1 (4); 4 (65)	3791
520. TOMMY TUTONE [20] 0 (0); 1 (8); 3 (46)	4090
521. NU SHOOZ [27] 0 (0); 1 (8); 2 (46)	4086
522. DEXYS MIDNIGHT RUNNERS [14] 0 (0); 1 (12); 1 (24)	4083
523. POCO [40] 0 (0); 1 (2); 6 (58)	4067
524. Dave GRUSIN [52] 0 (0); 0 (0); 10 (104)	4056
525. INFORMATION SOCIETY [25] 0 (0); 1 (9); 1 (38)	4039
526. SLAVE [46] 0 (0); 0 (0); 5 (79)	4038
527. GOLDEN EARRING [24] 0 (0); 1 (12); 4 (47)	4032
528. ART OF NOISE [53] 0 (0); 0 (0); 4 (66)	4018
529. The CRUSADERS [29] 0 (0); 1 (4); 5 (70)	3942
530. SURFACE [55] 0 (0); 0 (0); 2 (58)	3928
531. Gary (U.S.) BONDS [27] 0 (0); 1 (7); 2 (37)	3892
532. Crystal GAYLE [47] 0 (0); 0 (0); 8 (88)	3867
533. CROSBY, STILLS, NASH & YOUNG [16] 0 (0); 1 (11); 1 (22)	3862
534. SHANNON [32] 0 (0); 1 (4); 2 (53)	3856
535. FIVE STAR [57] 0 (0); 0 (0); 2 (72)	3856
536. MTUME [26] 0 (0); 1 (7); 4 (53)	3846
537. Kim WILDE [40] 0 (0); 1 (2); 4 (64)	3845
538. FRANKE And The KNOCKOUTS [31] 0 (0); 1 (5); 2 (45)	3842
539. BEACH BOYS [46] 0 (0); 0 (0); 7 (78)	3840
540. MANNHEIM STEAMROLLER [36] 0 (0); 1 (4); 4 (75)	3829
541. BAD ENGLISH [21] 0 (0); 1 (11); 1 (25)	3825

Album Acts Ranked

Rank. Act [Hi Peak] T10 (Wk); T40 (Wk); Ent (Wk)	Score In '80s
542. Ronnie LAWS [24] 0 (0); 1 (6); 3 (49)	3816
543. NAKED EYES [32] 0 (0); 1 (4); 2 (52)	3813
544. Don JOHNSON [17] 0 (0); 1 (11); 1 (27)	3811
545. HONEYMOON SUITE [60] 0 (0); 0 (0); 3 (62)	3810
546. Bryan FERRY [63] 0 (0); 0 (0); 2 (56)	3791
547. BRITNY FOX [39] 0 (0); 1 (2); 2 (43)	3785
548. BABYFACE [25] 0 (0); 1 (15); 1 (22)	3771
549. Marshall CRENSHAW [50] 0 (0); 0 (0); 3 (59)	3761
550. Billy VERA & The BEATERS [15] 0 (0); 1 (10); 2 (36)	3757
551. SLICK RICK [31] 0 (0); 1 (5); 1 (40)	3734
552. Lee RITENOUR [26] 0 (0); 1 (7); 6 (61)	3701
553. Morris DAY [37] 0 (0); 1 (3); 2 (46)	3699
554. KIX [46] 0 (0); 0 (0); 2 (56)	3696
555. The CALL [64] 0 (0); 0 (0); 4 (80)	3678
556. The KNACK [15] 0 (0); 1 (9); 3 (33)	3664
557. STARPOINT [60] 0 (0); 0 (0); 3 (69)	3663
558. NAJEE [56] 0 (0); 0 (0); 2 (66)	3661
559. DREAM ACADEMY [20] 0 (0); 1 (9); 2 (40)	3648
560. DAVID & DAVID [39] 0 (0); 1 (2); 1 (38)	3635
561. Bruce COCKBURN [45] 0 (0); 0 (0); 6 (84)	3623
562. Jean-Luc PONTY [44] 0 (0); 0 (0); 6 (75)	3618
563. George CLINTON [40] 0 (0); 1 (1); 5 (73)	3609
564. EMERSON, LAKE & POWELL [23] 0 (0); 1 (12); 1 (26)	3593
565. BLACKFOOT [48] 0 (0); 0 (0); 5 (57)	3570
566. BULLETBOYS [34] 0 (0); 1 (5); 1 (47)	3562
567. Tom COCHRANE/ RED RIDER [65] 0 (0); 0 (0); 6 (78)	3544
568. VIXEN [41] 0 (0); 0 (0); 1 (40)	3537
569. Toni BASIL [22] 0 (0); 1 (9); 1 (30)	3523
570. Andy GIBB [21] 0 (0); 1 (7); 2 (33)	3523
571. LONE JUSTICE [56] 0 (0); 0 (0); 2 (55)	3519
572. Roger WATERS [31] 0 (0); 1 (7); 2 (37)	3495
573. Robin TROWER [34] 0 (0); 2 (6); 6 (74)	3490
574. Placido DOMINGO [18] 0 (0); 1 (8); 3 (44)	3477
575. L.T.D. [28] 0 (0); 1 (8); 2 (40)	3475
576. Jonathan BUTLER [50] 0 (0); 0 (0); 3 (71)	3459
577. Neville MARRINER [56] 0 (0); 0 (0); 1 (78)	3455
578. Jermaine STEWART [32] 0 (0); 1 (5); 3 (48)	3433
579. The CHURCH [41] 0 (0); 0 (0); 2 (47)	3426
580. Gregg ALLMAN Band [30] 0 (0); 1 (6); 2 (39)	3421
581. DINO [34] 0 (0); 1 (6); 1 (41)	3418
582. James INGRAM [46] 0 (0); 0 (0); 2 (51)	3412
583. Donnie IRIS [57] 0 (0); 0 (0); 5 (85)	3395
584. DE LA SOUL [24] 0 (0); 1 (10); 1 (29)	3354
585. Rodney DANGERFIELD [36] 0 (0); 1 (3); 2 (39)	3349
586. AMERICA [41] 0 (0); 0 (0); 4 (51)	3348
587. Vanessa WILLIAMS [38] 0 (0); 1 (4); 1 (55)	3347
588. ALLMAN BROTHERS Band [27] 0 (0); 1 (6); 4 (39)	3342
589. TACO [23] 0 (0); 1 (11); 1 (24)	3339
590. COWBOY JUNKIES [26] 0 (0); 1 (9); 1 (29)	3320
591. Michael STANLEY Band [64] 0 (0); 0 (0); 4 (70)	3312
592. Jon BUTCHER AXIS [66] 0 (0); 0 (0); 5 (71)	3310
593. Wynton MARSALIS [90] 0 (0); 0 (0); 6 (92)	3306
594. Meli'sa MORGAN [41] 0 (0); 0 (0); 2 (55)	3293
595. Sam HARRIS [35] 0 (0); 1 (3); 2 (43)	3286
596. Jimmy PAGE [26] 0 (0); 1 (6); 2 (30)	3264
597. Michael FRANKS [45] 0 (0); 0 (0); 5 (84)	3254
598. El DeBARGE [24] 0 (0); 1 (11); 1 (23)	3241
599. Lou GRAMM [27] 0 (0); 1 (7); 2 (34)	3208
600. HAIRCUT ONE HUNDRED [31] 0 (0); 1 (7); 1 (37)	3195
601. The D.O.C. [20] 0 (0); 1 (10); 1 (20)	3178
602. BIG AUDIO DYNAMITE [85] 0 (0); 0 (0); 4 (83)	3170
603. The BOYS [33] 0 (0); 1 (7); 1 (36)	3167
604. L.A. GUNS [50] 0 (0); 0 (0); 2 (49)	3137
605. MARILLION [47] 0 (0); 0 (0); 4 (63)	3134
606. Daryl HALL [29] 0 (0); 1 (6); 2 (38)	3131
607. TALK TALK [42] 0 (0); 0 (0); 3 (55)	3120
608. PUBLIC ENEMY [42] 0 (0); 0 (0); 2 (63)	3115
609. SPINNERS [32] 0 (0); 1 (4); 5 (49)	3109
610. David GILMOUR [32] 0 (0); 1 (10); 1 (28)	3097
611. NAZARETH [41] 0 (0); 0 (0); 5 (53)	3068
612. Stanley CLARKE & George DUKE [33] 0 (0); 1 (8); 2 (33)	3053
613. CHERRELLE [36] 0 (0); 1 (2); 3 (53)	3043
614. Steve EARLE [56] 0 (0); 0 (0); 2 (48)	3042
615. CHAMPAIGN [53] 0 (0); 0 (0); 3 (47)	3033
616. Steve FORBERT [20] 0 (0); 1 (6); 3 (33)	3028
617. KING CRIMSON [45] 0 (0); 0 (0); 3 (48)	3027
618. MADNESS [41] 0 (0); 0 (0); 4 (50)	3014
619. Stevie B [75] 0 (0); 0 (0); 2 (59)	3013
620. Robbie ROBERTSON [38] 0 (0); 1 (4); 1 (34)	3009
621. K.D. LANG [73] 0 (0); 0 (0); 2 (54)	2991
622. ECHO & The BUNNYMEN [51] 0 (0); 0 (0); 6 (71)	2988
623. The RAMONES [44] 0 (0); 0 (0); 8 (60)	2986
624. Irene CARA [76] 0 (0); 0 (0); 2 (54)	2960
625. ZEBRA [29] 0 (0); 1 (7); 2 (39)	2957
626. Karla BONOFF [49] 0 (0); 0 (0); 2 (47)	2952
627. Vinnie VINCENT INVASION [64] 0 (0); 0 (0); 2 (44)	2946
628. MUSICAL YOUTH [23] 0 (0); 1 (8); 2 (34)	2942
629. Sir MIX-A-LOT [67] 0 (0); 0 (0); 2 (65)	2922
630. A TASTE OF HONEY [36] 0 (0); 1 (4); 2 (44)	2921
631. UFO [51] 0 (0); 0 (0); 5 (62)	2920
632. X [76] 0 (0); 0 (0); 6 (73)	2920
633. WAS (NOT WAS) [43] 0 (0); 0 (0); 2 (46)	2906
634. Richard PRYOR [21] 0 (0); 1 (7); 2 (30)	2904
635. LOOSE ENDS [46] 0 (0); 0 (0); 3 (48)	2904
636. CARPENTERS [46] 0 (0); 0 (0); 5 (48)	2888
637. O'BRYAN [64] 0 (0); 0 (0); 3 (60)	2887
638. Jack WAGNER [44] 0 (0); 0 (0); 3 (52)	2856
639. Dennis DeYOUNG [29] 0 (0); 1 (4); 2 (33)	2845
640. SYLVIA [56] 0 (0); 0 (0); 4 (59)	2836
641. The BABYS [42] 0 (0); 0 (0); 3 (44)	2833
642. The DEELE [54] 0 (0); 0 (0); 3 (52)	2830
643. TIMBUK 3 [50] 0 (0); 0 (0); 2 (43)	2824
644. Randy RHOADS [6] 1 (6); 1 (14); 1 (23)	2796
645. Ricky Van SHELTON [76] 0 (0); 0 (0); 2 (65)	2789
646. John SCHNEIDER [37] 0 (0); 1 (3); 3 (41)	2778
647. Grace SLICK [32] 0 (0); 1 (4); 2 (30)	2778
648. Rupert HOLMES [33] 0 (0); 1 (12); 1 (23)	2777
649. Kurtis BLOW [71] 0 (0); 0 (0); 6 (74)	2769
650. COVER GIRLS [64] 0 (0); 0 (0); 2 (67)	2753
651. Paul CARRACK [67] 0 (0); 0 (0); 3 (53)	2745
652. FATBACK [44] 0 (0); 0 (0); 4 (46)	2740
653. Sinead O'CONNOR [36] 0 (0); 1 (5); 1 (28)	2738
654. BOOGIE DOWN PRODUCTIONS [36] 0 (0); 1 (4); 2 (40)	2736
655. KEEL [53] 0 (0); 0 (0); 3 (52)	2731
656. Alison MOYET [45] 0 (0); 0 (0); 2 (42)	2723
657. MARSHALL TUCKER Band [32] 0 (0); 1 (5); 4 (39)	2719
658. ROCKPILE [27] 0 (0); 1 (10); 1 (19)	2718
659. ARCADIA [23] 0 (0); 1 (9); 1 (17)	2699
660. Don McLEAN [28] 0 (0); 1 (6); 2 (32)	2694
661. AFTER THE FIRE [25] 0 (0); 1 (8); 1 (20)	2685
662. GRIM REAPER [73] 0 (0); 0 (0); 3 (62)	2667
663. Reba McENTIRE [78] 0 (0); 0 (0); 5 (79)	2660
664. OINGO BOINGO [77] 0 (0); 0 (0); 8 (75)	2654
665. Randy MEISNER [50] 0 (0); 0 (0); 2 (44)	2642
666. Richard "Dimples" FIELDS [33] 0 (0); 1 (4); 2 (37)	2641
667. Merle HAGGARD [37] 0 (0); 1 (1); 3 (93)	2639
668. Isaac HAYES [39] 0 (0); 1 (1); 3 (39)	2605
669. Dana DANE [46] 0 (0); 0 (0); 1 (32)	2587
670. Don WILLIAMS [57] 0 (0); 0 (0); 3 (50)	2583
671. The SYSTEM [62] 0 (0); 0 (0); 3 (53)	2579
672. Keith RICHARDS [24] 0 (0); 1 (7); 1 (23)	2571
673. THIRD WORLD [63] 0 (0); 0 (0); 6 (64)	2569
674. EVERLY BROTHERS [38] 0 (0); 1 (3); 3 (41)	2568
675. BOY MEETS GIRL [50] 0 (0); 0 (0); 2 (37)	2566
676. Bill COSBY [26] 0 (0); 1 (4); 2 (29)	2564
677. PURE PRAIRIE LEAGUE [37] 0 (0); 1 (3); 2 (39)	2549

Album Acts Ranked

Rank. Act [Hi Peak] T10 (Wk); T40 (Wk); Ent (Wk)	Score In '80s
678. NITTY GRITTY DIRT BAND [62] 0 (0); 0 (0); 4 (51)	2542
679. SLADE [33] 0 (0); 1 (4); 2 (29)	2534
680. KRAFTWERK [72] 0 (0); 0 (0); 2 (56)	2527
681. The NYLONS [43] 0 (0); 0 (0); 3 (50)	2527
682. Henry LEE SUMMER [56] 0 (0); 0 (0); 2 (40)	2522
683. Debbie HARRY [25] 0 (0); 1 (4); 3 (33)	2518
684. Christine McVIE [26] 0 (0); 1 (7); 1 (23)	2497
685. Brenda RUSSELL [49] 0 (0); 0 (0); 3 (41)	2496
686. UTFO [67] 0 (0); 0 (0); 4 (52)	2492
687. DIXIE DREGS [56] 0 (0); 0 (0); 3 (46)	2487
688. Neneh CHERRY [40] 0 (0); 1 (1); 1 (28)	2483
689. BREAKFAST CLUB [43] 0 (0); 0 (0); 1 (30)	2481
690. ACCEPT [74] 0 (0); 0 (0); 4 (58)	2475
691. Toni CHILDS [63] 0 (0); 0 (0); 1 (45)	2463
692. LITTLE STEVEN [55] 0 (0); 0 (0); 3 (47)	2461
693. CAPTAIN & TENNILLE [23] 0 (0); 1 (9); 1 (17)	2457
694. Jennifer HOLLIDAY [31] 0 (0); 1 (5); 2 (36)	2453
695. The JAM [72] 0 (0); 0 (0); 7 (69)	2447
696. GIPSY KINGS [57] 0 (0); 0 (0); 2 (45)	2423
697. Todd RUNDGREN [48] 0 (0); 0 (0); 4 (45)	2415
698. WORLD PARTY [39] 0 (0); 1 (3); 1 (31)	2413
699. Miles DAVIS [53] 0 (0); 0 (0); 8 (72)	2407
700. Richard SIMMONS [44] 0 (0); 0 (0); 1 (40)	2402
701. GO WEST [60] 0 (0); 0 (0); 2 (44)	2394
702. BRONSKI BEAT [36] 0 (0); 1 (3); 2 (31)	2394
703. Ace FREHLEY [43] 0 (0); 0 (0); 2 (33)	2382
704. Shirley MURDOCK [44] 0 (0); 0 (0); 2 (41)	2379
705. MANFRED MANN'S EARTH BAND [40] 0 (0); 1 (2); 2 (37)	2377
706. SHOOTING STAR [82] 0 (0); 0 (0); 5 (66)	2369
707. Ry COODER [43] 0 (0); 0 (0); 4 (44)	2354
708. CALIFORNIA RAISINS [60] 0 (0); 0 (0); 2 (51)	2352
709. WEATHER REPORT [57] 0 (0); 0 (0); 6 (48)	2328
710. SLY FOX [31] 0 (0); 1 (4); 1 (22)	2317
711. STYLE COUNCIL [56] 0 (0); 0 (0); 5 (54)	2314
712. The SUGARCUBES [54] 0 (0); 0 (0); 2 (38)	2309
713. PABLO CRUISE [34] 0 (0); 1 (5); 2 (28)	2300
714. EPMD [53] 0 (0); 0 (0); 2 (37)	2293
715. Millie JACKSON [94] 0 (0); 0 (0); 6 (68)	2289
716. Carole KING [44] 0 (0); 0 (0); 3 (44)	2288
717. Luciano PAVAROTTI [77] 0 (0); 0 (0); 5 (57)	2286
718. SCRITTI POLITTI [50] 0 (0); 0 (0); 2 (36)	2286
719. Michael HENDERSON [35] 0 (0); 1 (5); 3 (34)	2286
720. Ricky SKAGGS [61] 0 (0); 0 (0); 4 (51)	2273
721. The WAITRESSES [41] 0 (0); 0 (0); 3 (39)	2264
722. J.J. FAD [49] 0 (0); 0 (0); 1 (30)	2258
723. Dwight TWILLEY [39] 0 (0); 1 (3); 2 (32)	2256
724. Tommy SHAW [50] 0 (0); 0 (0); 2 (34)	2242
725. BLOW MONKEYS [35] 0 (0); 1 (5); 2 (26)	2239
726. FORCE M.D.'S [67] 0 (0); 0 (0); 3 (45)	2235
727. FOUR TOPS [37] 0 (0); 1 (4); 3 (37)	2233
728. Michelle SHOCKED [73] 0 (0); 0 (0); 2 (43)	2231
729. STACEY Q [59] 0 (0); 0 (0); 2 (50)	2222
730. Bertie HIGGINS [38] 0 (0); 0 (0); 1 (25)	2211
731. Iggy POP [75] 0 (0); 0 (0); 4 (51)	2208
732. CHIC [30] 0 (0); 1 (3); 4 (37)	2207
733. Joan RIVERS [22] 0 (0); 1 (6); 1 (21)	2189
734. Delbert McCLINTON [34] 0 (0); 1 (4); 2 (37)	2186
735. Kylie MINOGUE [53] 0 (0); 0 (0); 1 (28)	2181
736. Marty BALIN [35] 0 (0); 1 (3); 2 (29)	2181
737. WAR [48] 0 (0); 0 (0); 4 (50)	2178
738. HEAVEN 17 [68] 0 (0); 0 (0); 3 (44)	2171
739. Carlos SANTANA [31] 0 (0); 1 (4); 3 (28)	2171
740. Michael SCHENKER Group [81] 0 (0); 0 (0); 4 (53)	2158
741. JASON & The SCORCHERS [91] 0 (0); 0 (0); 3 (57)	2158
742. John LENNON/ PLASTIC ONO BAND [57] 0 (0); 0 (0); 3 (39)	2155
743. Peter FRAMPTON [43] 0 (0); 0 (0); 4 (41)	2130
744. PLANET P [42] 0 (0); 0 (0); 2 (37)	2124
745. POINT BLANK [80] 0 (0); 0 (0); 3 (54)	2123
746. LOUDNESS [64] 0 (0); 0 (0); 3 (44)	2119
747. Melba MOORE [91] 0 (0); 0 (0); 4 (72)	2110
748. Leo SAYER [36] 0 (0); 1 (6); 1 (23)	2102
749. SIOUXSIE & The BANSHEES [68] 0 (0); 0 (0); 3 (47)	2097
750. JON And VANGELIS [64] 0 (0); 0 (0); 3 (56)	2075
751. BODEANS [86] 0 (0); 0 (0); 3 (52)	2073
752. T'PAU [31] 0 (0); 1 (3); 1 (24)	2072
753. Big Daddy KANE [33] 0 (0); 1 (4); 2 (32)	2068
754. Carl CARLTON [34] 0 (0); 1 (3); 2 (26)	2067
755. The REPLACEMENTS [57] 0 (0); 0 (0); 3 (45)	2059
756. Cliff RICHARD [80] 0 (0); 0 (0); 3 (49)	2055
757. DOUBLE [30] 0 (0); 1 (6); 1 (21)	2055
758. Dennis EDWARDS [48] 0 (0); 0 (0); 1 (27)	2053
759. RED HOT CHILI PEPPERS [52] 0 (0); 0 (0); 2 (34)	2047
760. KID 'N PLAY [96] 0 (0); 0 (0); 1 (47)	2027
761. FRIDA [41] 0 (0); 0 (0); 1 (28)	2026
762. Roger HODGSON [46] 0 (0); 0 (0); 2 (28)	2017
763. FASTER PUSSYCAT [60] 0 (0); 0 (0); 2 (50)	2016
764. METAL CHURCH [75] 0 (0); 0 (0); 2 (38)	1986
765. Laurie ANDERSON [60] 0 (0); 0 (0); 5 (55)	1980
766. PSEUDO ECHO [54] 0 (0); 0 (0); 1 (27)	1973
767. Ian HUNTER [62] 0 (0); 0 (0); 4 (46)	1970
768. SWEET SENSATION [63] 0 (0); 0 (0); 1 (32)	1968
769. SOUTHSIDE JOHNNY & The ASBURY JUKES [67] 0 (0); 0 (0); 6 (46)	1960
770. MODERN ENGLISH [70] 0 (0); 0 (0); 3 (47)	1952
771. Martin BRILEY [55] 0 (0); 0 (0); 2 (32)	1952
772. Jean Michel JARRE [52] 0 (0); 0 (0); 2 (32)	1950
773. BUCKNER & GARCIA [24] 0 (0); 1 (5); 1 (16)	1947
774. Tracey ULLMAN [34] 0 (0); 1 (5); 1 (20)	1944
775. Phoebe SNOW [51] 0 (0); 0 (0); 2 (38)	1943
776. AURRA [38] 0 (0); 0 (0); 2 (28)	1941
777. BOW WOW WOW [67] 0 (0); 0 (0); 4 (46)	1926
778. Clint BLACK [61] 0 (0); 0 (0); 1 (30)	1924
779. SLAYER [57] 0 (0); 0 (0); 2 (37)	1921
780. TONY! TONI! TONÉ! [69] 0 (0); 0 (0); 1 (46)	1918
781. DANGEROUS TOYS [65] 0 (0); 0 (0); 1 (29)	1913
782. Joe SAMPLE [65] 0 (0); 0 (0); 3 (48)	1912
783. ARTISTS UNITED AGAINST APARTHEID [31] 0 (0); 1 (5); 1 (18)	1904
784. SA-FIRE [79] 0 (0); 0 (0); 1 (46)	1904
785. Larry CARLTON [99] 0 (0); 0 (0); 6 (60)	1886
786. DYNASTY [43] 0 (0); 0 (0); 2 (25)	1879
787. Peter TOSH [59] 0 (0); 0 (0); 3 (38)	1868
788. TIERRA [38] 0 (0); 1 (4); 1 (21)	1852
789. GAMMA [65] 0 (0); 0 (0); 3 (38)	1845
790. Lee GREENWOOD [73] 0 (0); 0 (0); 4 (62)	1842
791. Lyle LOVETT [62] 0 (0); 0 (0); 2 (35)	1838
792. Dan HARTMAN [55] 0 (0); 0 (0); 2 (30)	1828
793. KC & The SUNSHINE BAND [53] 0 (0); 0 (0); 3 (40)	1828
794. Patsy CLINE [29] 0 (0); 1 (5); 1 (18)	1824
795. Barbara MANDRELL [86] 0 (0); 0 (0); 5 (53)	1818
796. Regina BELLE [63] 0 (0); 0 (0); 2 (31)	1817
797. PARLIAMENT [44] 0 (0); 0 (0); 2 (24)	1811
798. Brian SETZER [45] 0 (0); 0 (0); 2 (26)	1810
799. NENA [27] 0 (0); 1 (6); 1 (14)	1810
800. John PARR [48] 0 (0); 0 (0); 1 (26)	1807
801. ULTRAVOX [61] 0 (0); 0 (0); 4 (41)	1806
802. TA MARA & The SEEN [54] 0 (0); 0 (0); 1 (25)	1806
803. David BYRNE [44] 0 (0); 0 (0); 4 (42)	1805
804. KAJAGOOGOO [38] 0 (0); 1 (2); 2 (24)	1796
805. STEEL BREEZE [50] 0 (0); 0 (0); 1 (28)	1774
806. Randy NEWMAN [64] 0 (0); 0 (0); 2 (32)	1764
807. EXODUS [82] 0 (0); 0 (0); 2 (37)	1763
808. BOOGIE BOYS [53] 0 (0); 0 (0); 3 (37)	1759
809. BADLANDS [57] 0 (0); 0 (0); 1 (26)	1758
810. CHARLENE [36] 0 (0); 1 (3); 2 (27)	1757
811. DR. HOOK [71] 0 (0); 0 (0); 4 (53)	1751
812. John HIATT [98] 0 (0); 0 (0); 2 (48)	1747
813. George DUKE [48] 0 (0); 0 (0); 5 (38)	1745

Album Acts Ranked

Rank. Act [Hi Peak] T10 (Wk); T40 (Wk); Ent (Wk)	Score In '80s
814. Johnny Van ZANT Band [48] 0 (0); 0 (0); 4 (39)	1744
815. Al Di MEOLA [55] 0 (0); 0 (0); 5 (41)	1742
816. GRANDMASTER FLASH & The FURIOUS FIVE [53] 0 (0); 0 (0); 2 (27)	1738
817. George HOWARD [109] 0 (0); 0 (0); 5 (53)	1735
818. The VAPORS [62] 0 (0); 0 (0); 2 (37)	1717
819. PIECES OF A DREAM [90] 0 (0); 0 (0); 4 (48)	1714
820. Tony CAREY [60] 0 (0); 0 (0); 2 (33)	1714
821. Oran "Juice" JONES [44] 0 (0); 0 (0); 1 (22)	1708
822. WALL OF VOODOO [45] 0 (0); 0 (0); 2 (25)	1707
823. NEVILLE BROTHERS [66] 0 (0); 0 (0); 4 (39)	1704
824. Al STEWART [37] 0 (0); 1 (4); 2 (24)	1703
825. Randy CRAWFORD [71] 0 (0); 0 (0); 6 (52)	1695
826. GQ [46] 0 (0); 0 (0); 2 (28)	1694
827. PUBLIC IMAGE LIMITED [106] 0 (0); 0 (0); 5 (56)	1672
828. YAZ [69] 0 (0); 0 (0); 2 (45)	1668
829. SHADOWFAX [114] 0 (0); 0 (0); 4 (60)	1667
830. SWITCH [57] 0 (0); 0 (0); 4 (40)	1661
831. LIMAHL [41] 0 (0); 0 (0); 1 (20)	1654
832. Rich LITTLE [29] 0 (0); 1 (6); 1 (13)	1651
833. TIN MACHINE [28] 0 (0); 1 (5); 1 (17)	1651
834. JOHNNY HATES JAZZ [56] 0 (0); 0 (0); 1 (25)	1646
835. CRUZADOS [76] 0 (0); 0 (0); 2 (39)	1644
836. TORA TORA [47] 0 (0); 0 (0); 1 (25)	1634
837. Eric CARMEN [59] 0 (0); 0 (0); 3 (35)	1622
838. BONHAM [38] 0 (0); 1 (2); 1 (13)	1621
839. Robbie DUPREE [51] 0 (0); 0 (0); 2 (29)	1620
840. CHILLIWACK [78] 0 (0); 0 (0); 2 (40)	1617
841. HELIX [69] 0 (0); 0 (0); 4 (39)	1614
842. Nicolette LARSON [62] 0 (0); 0 (0); 3 (34)	1593
843. Terri GIBBS [53] 0 (0); 0 (0); 1 (25)	1590
844. BANG TANGO [58] 0 (0); 0 (0); 1 (27)	1585
845. BOX OF FROGS [45] 0 (0); 0 (0); 2 (23)	1581
846. Brenda K. STARR [58] 0 (0); 0 (0); 1 (24)	1579
847. Dave DAVIES [42] 0 (0); 0 (0); 2 (22)	1571
848. Minnie RIPERTON [35] 0 (0); 1 (4); 1 (15)	1570
849. RE-FLEX [53] 0 (0); 0 (0); 1 (28)	1565
850. ISLEY. JASPER. ISLEY [77] 0 (0); 0 (0); 2 (36)	1562
851. ROMEO VOID [68] 0 (0); 0 (0); 3 (38)	1561
852. The RAINMAKERS [85] 0 (0); 0 (0); 2 (41)	1546
853. Narada Michael WALDEN [74] 0 (0); 0 (0); 3 (33)	1546
854. Garland JEFFREYS [59] 0 (0); 0 (0); 3 (26)	1545
855. Andrew Dice CLAY [94] 0 (0); 0 (0); 1 (36)	1535
856. Doc SEVERINSEN And The TONIGHT SHOW Orchestra [65] 0 (0); 0 (0); 1 (26)	1535
857. Paul DAVIS [52] 0 (0); 0 (0); 2 (33)	1531
858. Donny OSMOND [54] 0 (0); 0 (0); 1 (23)	1530
859. VILLAGE PEOPLE [47] 0 (0); 0 (0); 3 (25)	1529
860. RESTLESS HEART [73] 0 (0); 0 (0); 2 (36)	1528
861. ICICLE WORKS [40] 0 (0); 1 (2); 1 (18)	1526
862. VIOLENT FEMMES [84] 0 (0); 0 (0); 2 (37)	1515
863. GODLEY & CREME [37] 0 (0); 1 (3); 1 (15)	1512
864. Mac DAVIS [67] 0 (0); 0 (0); 3 (27)	1507
865. Harry CONNICK Jr. [42] 0 (0); 0 (0); 1 (20)	1498
866. Sam KINISON [43] 0 (0); 0 (0); 2 (22)	1494
867. Nona HENDRYX [83] 0 (0); 0 (0); 3 (39)	1491
868. CURIOSITY KILLED THE CAT [55] 0 (0); 0 (0); 1 (29)	1490
869. Angela WINBUSH [81] 0 (0); 0 (0); 2 (36)	1489
870. PHANTOM, ROCKER & SLICK [61] 0 (0); 0 (0); 2 (25)	1482
871. ATLANTA RHYTHM SECTION [65] 0 (0); 0 (0); 3 (31)	1482
872. GEORGIO [117] 0 (0); 0 (0); 1 (52)	1481
873. MEAT LOAF [45] 0 (0); 0 (0); 2 (21)	1471
874. King Sunny ADE & His AFRICAN BEATS [91] 0 (0); 0 (0); 2 (39)	1467
875. ARMORED SAINT [108] 0 (0); 0 (0); 3 (47)	1467
876. MORRISSEY [48] 0 (0); 0 (0); 1 (20)	1466
877. MR. BIG [46] 0 (0); 0 (0); 1 (18)	1461
878. SPECIAL ED [73] 0 (0); 0 (0); 1 (28)	1450
879. PRISM [53] 0 (0); 0 (0); 1 (20)	1443
880. BLACK 'N BLUE [110] 0 (0); 0 (0); 3 (40)	1443
881. WILL TO POWER [68] 0 (0); 0 (0); 1 (29)	1440
882. HELLOWEEN [104] 0 (0); 0 (0); 3 (44)	1435
883. Le ROUX [64] 0 (0); 0 (0); 2 (27)	1431
884. FESTIVAL [50] 0 (0); 0 (0); 1 (18)	1430
885. Joe PERRY PROJECT [47] 0 (0); 0 (0); 2 (23)	1429
886. Graham PARKER And The SHOT [57] 0 (0); 0 (0); 1 (21)	1429
887. WILLIE, WAYLON, JOHNNY & KRIS [92] 0 (0); 0 (0); 1 (35)	1417
888. Peter SCHILLING [61] 0 (0); 0 (0); 1 (23)	1412
889. Tanita TIKARAM [59] 0 (0); 0 (0); 1 (23)	1412
890. JANE'S ADDICTION [103] 0 (0); 0 (0); 1 (35)	1405
891. HURRICANE [92] 0 (0); 0 (0); 1 (36)	1404
892. SARAYA [79] 0 (0); 0 (0); 1 (35)	1398
893. Jim CARROLL Band [73] 0 (0); 0 (0); 2 (30)	1396
894. Paul HARDCASTLE [63] 0 (0); 0 (0); 1 (25)	1396
895. Gordon LIGHTFOOT [60] 0 (0); 0 (0); 4 (34)	1395
896. Gary MOORE [114] 0 (0); 0 (0); 5 (49)	1395
897. Andy TAYLOR [46] 0 (0); 0 (0); 1 (17)	1391
898. Billy PRESTON [49] 0 (0); 0 (0); 2 (27)	1389
899. EXTREME [80] 0 (0); 0 (0); 1 (32)	1377
900. KBC BAND [75] 0 (0); 0 (0); 1 (24)	1369
901. SUGARHILL GANG [50] 0 (0); 0 (0); 1 (18)	1357
902. Debra LAWS [70] 0 (0); 0 (0); 1 (27)	1355
903. Nik KERSHAW [70] 0 (0); 0 (0); 2 (30)	1345
904. FIREFALL [68] 0 (0); 0 (0); 4 (35)	1344
905. CREEDENCE CLEARWATER REVIVAL [62] 0 (0); 0 (0); 2 (36)	1342
906. Waylon JENNINGS & Willie NELSON [57] 0 (0); 0 (0); 1 (22)	1342
907. Cheryl LYNN [104] 0 (0); 0 (0); 4 (42)	1342
908. REAL LIFE [58] 0 (0); 0 (0); 2 (27)	1337
909. BOYS DON'T CRY [55] 0 (0); 0 (0); 1 (19)	1337
910. BRASS CONSTRUCTION [89] 0 (0); 0 (0); 4 (36)	1336
911. FOGHAT [92] 0 (0); 0 (0); 5 (35)	1329
912. TWENNYNINE [54] 0 (0); 0 (0); 3 (25)	1321
913. Rafael CAMERON [67] 0 (0); 0 (0); 2 (30)	1319
914. The KINGS [74] 0 (0); 0 (0); 2 (30)	1319
915. KLEEER [81] 0 (0); 0 (0); 3 (34)	1310
916. Van STEPHENSON [54] 0 (0); 0 (0); 1 (20)	1308
917. Biz MARKIE [66] 0 (0); 0 (0); 2 (28)	1308
918. Gil SCOTT-HERON [106] 0 (0); 0 (0); 3 (42)	1304
919. URIAH HEEP [56] 0 (0); 0 (0); 2 (26)	1297
920. Lou RAWLS [81] 0 (0); 0 (0); 3 (28)	1293
921. LONDON SYMPHONY Orchestra [62] 0 (0); 0 (0); 3 (29)	1284
922. TNT [100] 0 (0); 0 (0); 2 (33)	1281
923. The WATERBOYS [76] 0 (0); 0 (0); 1 (26)	1277
924. Marianne FAITHFULL [82] 0 (0); 0 (0); 3 (31)	1275
925. WARLOCK [80] 0 (0); 0 (0); 1 (27)	1273
926. Jim STEINMAN [63] 0 (0); 0 (0); 1 (17)	1273
927. The SPECIALS [84] 0 (0); 0 (0); 2 (26)	1268
928. KING DIAMOND [89] 0 (0); 0 (0); 3 (33)	1266
929. Tom WAITS [96] 0 (0); 0 (0); 5 (40)	1261
930. Clarence CLEMONS [62] 0 (0); 0 (0); 2 (23)	1253
931. T.S. MONK [64] 0 (0); 0 (0); 2 (30)	1242
932. HOTHOUSE FLOWERS [88] 0 (0); 0 (0); 1 (33)	1231
933. The RIPPINGTONS [85] 0 (0); 0 (0); 2 (27)	1227
934. HOODOO GURUS [101] 0 (0); 0 (0); 3 (35)	1223
935. HIPSWAY [55] 0 (0); 0 (0); 1 (18)	1215
936. Yoko ONO [49] 0 (0); 0 (0); 2 (22)	1210
937. DEL FUEGOS [132] 0 (0); 0 (0); 3 (50)	1199
938. Jennifer WARNES [72] 0 (0); 0 (0); 1 (21)	1197
939. NEWCLEUS [74] 0 (0); 0 (0); 1 (28)	1193
940. COCK ROBIN [61] 0 (0); 0 (0); 2 (22)	1186
941. Johnny KEMP [68] 0 (0); 0 (0); 1 (19)	1186
942. Nick LOWE [50] 0 (0); 0 (0); 2 (21)	1179
943. WHEN IN ROME [84] 0 (0); 0 (0); 1 (24)	1178
944. GENE LOVES JEZEBEL [108] 0 (0); 0 (0); 2 (41)	1175
945. PIXIES [98] 0 (0); 0 (0); 1 (27)	1172
946. The FAMILY [62] 0 (0); 0 (0); 1 (22)	1170
947. BALTIMORA [49] 0 (0); 0 (0); 1 (17)	1165

Album Acts Ranked

Rank. Act [Hi Peak] T10 (Wk); T40 (Wk); Ent (Wk)	Score In '80s
948. BOURGEOIS TAGG [84] 0 (0); 0 (0); 2 (28)	1165
949. Bill SUMMERS & SUMMERS HEAT [92] 0 (0); 0 (0); 2 (31)	1162
950. HOUSE OF LORDS [78] 0 (0); 0 (0); 1 (27)	1158
951. Mick FLEETWOOD [43] 0 (0); 0 (0); 1 (14)	1150
952. Matthew WILDER [49] 0 (0); 0 (0); 1 (16)	1145
953. PEACHES & HERB [69] 0 (0); 0 (0); 3 (31)	1144
954. Greg LAKE [62] 0 (0); 0 (0); 1 (17)	1139
955. JONES GIRLS [96] 0 (0); 0 (0); 2 (39)	1136
956. JONZUN CREW [66] 0 (0); 0 (0); 1 (20)	1135
957. MECO [61] 0 (0); 0 (0); 3 (23)	1127
958. Chico DeBARGE [90] 0 (0); 0 (0); 1 (30)	1126
959. George JONES [115] 0 (0); 0 (0); 3 (40)	1124
960. FREHLEY'S COMET [81] 0 (0); 0 (0); 2 (23)	1123
961. MENUDO [100] 0 (0); 0 (0); 2 (31)	1121
962. David JOHANSEN [90] 0 (0); 0 (0); 3 (33)	1119
963. Jimmy BARNES [104] 0 (0); 0 (0); 2 (31)	1114
964. Rebbie JACKSON [63] 0 (0); 0 (0); 1 (18)	1112
965. Carole Bayer SAGER [60] 0 (0); 0 (0); 1 (22)	1111
966. TESTAMENT [77] 0 (0); 0 (0); 2 (26)	1110
967. Grayson HUGH [71] 0 (0); 0 (0); 1 (24)	1103
968. Patty SMYTH [66] 0 (0); 0 (0); 1 (20)	1098
969. The ROCKETS [53] 0 (0); 0 (0); 2 (20)	1097
970. ALCATRAZZ [128] 0 (0); 0 (0); 3 (44)	1093
971. VANDENBERG [65] 0 (0); 0 (0); 2 (25)	1089
972. Benjamin ORR [86] 0 (0); 0 (0); 1 (22)	1089
973. SUICIDAL TENDENCIES [100] 0 (0); 0 (0); 3 (30)	1087
974. TAKE 6 [71] 0 (0); 0 (0); 1 (19)	1074
975. Jane WIEDLIN [105] 0 (0); 0 (0); 2 (27)	1074
976. Jerry HARRISON: CASUAL GODS [78] 0 (0); 0 (0); 1 (20)	1071
977. BLUE MURDER [69] 0 (0); 0 (0); 1 (21)	1070
978. Wilton FELDER [81] 0 (0); 0 (0); 2 (29)	1069
979. RAVEN [81] 0 (0); 0 (0); 2 (25)	1059
980. FIONA [71] 0 (0); 0 (0); 2 (24)	1056
981. Tommy CONWELL/ YOUNG RUMBLERS [103] 0 (0); 0 (0); 1 (28)	1049

Rank. Act [Hi Peak] T10 (Wk); T40 (Wk); Ent (Wk)	Score In '80s
982. AXE [81] 0 (0); 0 (0); 2 (26)	1046
983. The THE [89] 0 (0); 0 (0); 2 (30)	1045
984. BeBe & CeCe WINANS [95] 0 (0); 0 (0); 1 (25)	1042
985. David BENOIT [101] 0 (0); 0 (0); 3 (31)	1035
986. Adrian BELEW [82] 0 (0); 0 (0); 3 (27)	1030
987. Benny MARDONES [65] 0 (0); 0 (0); 1 (24)	1025
988. Ray CHARLES [75] 0 (0); 0 (0); 1 (20)	1024
989. DIFFORD & TILBROOK [55] 0 (0); 0 (0); 1 (15)	1021
990. JUNIOR [71] 0 (0); 0 (0); 2 (22)	1021
991. Barry WHITE [85] 0 (0); 0 (0); 3 (34)	1015
992. Julian COPE [105] 0 (0); 0 (0); 3 (31)	1013
993. CONCRETE BLONDE [96] 0 (0); 0 (0); 2 (34)	1012
994. The POGUES [88] 0 (0); 0 (0); 2 (25)	1009
995. Stanley CLARKE [95] 0 (0); 0 (0); 3 (31)	1007
996. Rockie ROBBINS [71] 0 (0); 0 (0); 2 (22)	1006
997. The COMMUNARDS [90] 0 (0); 0 (0); 2 (25)	1004
998. Brian WILSON [54] 0 (0); 0 (0); 1 (13)	1003
999. Michael DAMIAN [61] 0 (0); 0 (0); 1 (18)	1002
1000. Maurice WHITE [61] 0 (0); 0 (0); 1 (19)	991
1001. 9.9 [79] 0 (0); 0 (0); 1 (22)	984
1002. FLESH FOR LULU [89] 0 (0); 0 (0); 1 (24)	981
1003. Rocky BURNETTE [53] 0 (0); 0 (0); 1 (14)	979
1004. Johnny MATHIS [140] 0 (0); 0 (0); 6 (46)	978
1005. Harry CHAPIN [58] 0 (0); 0 (0); 1 (15)	977
1006. PLEASURE [97] 0 (0); 0 (0); 3 (28)	975
1007. Leonard BERNSTEIN [70] 0 (0); 0 (0); 1 (20)	975
1008. TODAY [86] 0 (0); 0 (0); 1 (22)	970
1009. Roy AYERS [82] 0 (0); 0 (0); 4 (27)	964
1010. Steve ARRINGTON [101] 0 (0); 0 (0); 3 (31)	962
1011. OPUS [64] 0 (0); 0 (0); 1 (16)	961
1012. LET'S ACTIVE [111] 0 (0); 0 (0); 3 (37)	961
1013. HUMBLE PIE [60] 0 (0); 0 (0); 2 (20)	956
1014. Mica PARIS [86] 0 (0); 0 (0); 1 (23)	956
1015. DIESEL [68] 0 (0); 0 (0); 1 (24)	954
1016. Deborah ALLEN [67] 0 (0); 0 (0); 1 (20)	953
1017. John ANDERSON [58] 0 (0); 0 (0); 2 (17)	952

Rank. Act [Hi Peak] T10 (Wk); T40 (Wk); Ent (Wk)	Score In '80s
1018. Mike POST [70] 0 (0); 0 (0); 1 (17)	951
1019. David LINDLEY [83] 0 (0); 0 (0); 2 (24)	951
1020. ORION THE HUNTER [57] 0 (0); 0 (0); 1 (14)	945
1021. MODELS [84] 0 (0); 0 (0); 2 (18)	944
1022. Dan REED Network [95] 0 (0); 0 (0); 2 (25)	942
1023. KING'S X [123] 0 (0); 0 (0); 2 (29)	929
1024. Jane OLIVOR [58] 0 (0); 0 (0); 2 (18)	918
1025. IMPELLITTERI [91] 0 (0); 0 (0); 1 (20)	910
1026. STATLER BROTHERS [103] 0 (0); 0 (0); 6 (36)	909
1027. WENDY and LISA [88] 0 (0); 0 (0); 2 (21)	907
1028. Michael COOPER [98] 0 (0); 0 (0); 1 (25)	901
1029. YELLOW MAGIC ORCHESTRA [81] 0 (0); 0 (0); 2 (23)	900
1030. CAMPER VAN BEETHOVEN [124] 0 (0); 0 (0); 2 (29)	898
1031. MISSION U.K. [108] 0 (0); 0 (0); 2 (28)	893
1032. Nia PEEPLES [97] 0 (0); 0 (0); 1 (21)	890
1033. SHERIFF [60] 0 (0); 0 (0); 1 (14)	888
1034. Dan SEALS [59] 0 (0); 0 (0); 1 (15)	886
1035. Patti SMITH [65] 0 (0); 0 (0); 1 (15)	882
1036. FULL FORCE [126] 0 (0); 0 (0); 3 (32)	873
1037. Gary WRIGHT [79] 0 (0); 0 (0); 1 (19)	873
1038. DANGER DANGER [88] 0 (0); 0 (0); 1 (20)	872
1039. BUS BOYS [85] 0 (0); 0 (0); 2 (22)	870
1040. APOLLONIA 6 [62] 0 (0); 0 (0); 1 (17)	869
1041. Robyn HITCHCOCK And The EGYPTIANS [111] 0 (0); 0 (0); 2 (24)	868
1042. PURSUIT OF HAPPINESS [93] 0 (0); 0 (0); 1 (21)	864
1043. OMAR And The HOWLERS [81] 0 (0); 0 (0); 1 (19)	852
1044. Nick LOWE And His COWBOY OUTFIT [113] 0 (0); 0 (0); 2 (24)	845
1045. James BROWN [96] 0 (0); 0 (0); 4 (28)	842
1046. STEEL PULSE [120] 0 (0); 0 (0); 3 (32)	841
1047. SYREETA [73] 0 (0); 0 (0); 3 (27)	838
1048. Andy SUMMERS & Robert FRIPP [60] 0 (0); 0 (0); 2 (16)	838
1049. Bobby CALDWELL [113] 0 (0); 0 (0); 2 (28)	836

Rank. Act [Hi Peak] T10 (Wk); T40 (Wk); Ent (Wk)	Score In '80s
1050. LIZZY BORDEN [133] 0 (0); 0 (0); 4 (33)	835
1051. RED ROCKERS [71] 0 (0); 0 (0); 1 (16)	834
1052. Richard THOMPSON [102] 0 (0); 0 (0); 4 (29)	831
1053. DIRTY LOOKS [118] 0 (0); 0 (0); 2 (25)	829
1054. Mason RUFFNER [80] 0 (0); 0 (0); 2 (24)	829
1055. Gene CHANDLER [87] 0 (0); 0 (0); 1 (18)	828
1056. BOOMTOWN RATS [103] 0 (0); 0 (0); 3 (23)	827
1057. Ivan NEVILLE [107] 0 (0); 0 (0); 1 (23)	827
1058. The INMATES [49] 0 (0); 0 (0); 1 (12)	825
1059. Frankie SMITH [54] 0 (0); 0 (0); 1 (10)	825
1060. Jimi HENDRIX [79] 0 (0); 0 (0); 4 (23)	823
1061. Dan HILL [90] 0 (0); 0 (0); 1 (19)	814
1062. BOOTSY'S RUBBER BAND [70] 0 (0); 0 (0); 2 (17)	810
1063. THEY MIGHT BE GIANTS [89] 0 (0); 0 (0); 1 (19)	809
1064. Gloria LORING [61] 0 (0); 0 (0); 1 (14)	808
1065. TANGIER [91] 0 (0); 0 (0); 1 (17)	808
1066. BLOWFLY [82] 0 (0); 0 (0); 1 (20)	807
1067. JOBOXERS [70] 0 (0); 0 (0); 1 (15)	806
1068. LAID BACK [67] 0 (0); 0 (0); 1 (15)	801
1069. Michael Martin MURPHEY [69] 0 (0); 0 (0); 2 (19)	801
1070. The GODFATHERS [91] 0 (0); 0 (0); 2 (22)	799
1071. Rob JUNGKLAS [102] 0 (0); 0 (0); 1 (22)	798
1072. GORKY PARK [80] 0 (0); 0 (0); 1 (17)	794
1073. Judi Sheppard MISSETT [117] 0 (0); 0 (0); 1 (20)	794
1074. Harold MELVIN And The BLUE NOTES [95] 0 (0); 0 (0); 1 (20)	793
1075. ENUFF Z'NUFF [74] 0 (0); 0 (0); 1 (14)	792
1076. Jerry Lee LEWIS [62] 0 (0); 0 (0); 2 (22)	792
1077. DEVICE [73] 0 (0); 0 (0); 1 (16)	791
1078. FATES WARNING [111] 0 (0); 0 (0); 3 (26)	788
1079. Robert GORDON [117] 0 (0); 0 (0); 2 (24)	787
1080. T.G. SHEPPARD [119] 0 (0); 0 (0); 3 (28)	781
1081. The CHI-LITES [98] 0 (0); 0 (0); 3 (25)	778
1082. SAVATAGE [116] 0 (0); 0 (0); 2 (30)	777

Album Acts Ranked

Rank. Act [Hi Peak] T10 (Wk); T40 (Wk); Ent (Wk)	Score In '80s
1083. Ray STEVENS [118] 0 (0); 0 (0); 2 (27)	777
1084. THIN LIZZY [120] 0 (0); 0 (0); 4 (29)	769
1085. Billy CRYSTAL [65] 0 (0); 0 (0); 1 (13)	767
1086. Bernadette PETERS [114] 0 (0); 0 (0); 2 (23)	767
1087. Philip GLASS [91] 0 (0); 0 (0); 2 (19)	765
1088. SPARKS [88] 0 (0); 0 (0); 3 (25)	761
1089. Johnny LEE [132] 0 (0); 0 (0); 2 (29)	760
1090. KIDS FROM FAME [98] 0 (0); 0 (0); 3 (23)	759
1091. DEAD MILKMEN [101] 0 (0); 0 (0); 2 (30)	758
1092. LEATHERWOLF [105] 0 (0); 0 (0); 2 (20)	753
1093. Howard HEWETT [110] 0 (0); 0 (0); 2 (28)	749
1094. TEXAS [88] 0 (0); 0 (0); 1 (16)	747
1095. Gavin CHRISTOPHER [74] 0 (0); 0 (0); 1 (15)	745
1096. RIOT [99] 0 (0); 0 (0); 3 (27)	744
1097. Bonnie POINTER [63] 0 (0); 0 (0); 1 (12)	742
1098. BIG PIG [93] 0 (0); 0 (0); 1 (17)	740
1099. Neal SCHON & Jan HAMMER [115] 0 (0); 0 (0); 2 (20)	738
1100. DIVINYLS [91] 0 (0); 0 (0); 1 (18)	734
1101. Julia FORDHAM [118] 0 (0); 0 (0); 1 (25)	729
1102. CLIMAX BLUES Band [75] 0 (0); 0 (0); 1 (16)	728
1103. INSTANT FUNK [129] 0 (0); 0 (0); 3 (22)	728
1104. SYLVESTER [123] 0 (0); 0 (0); 5 (28)	727
1105. KANE GANG [115] 0 (0); 0 (0); 1 (20)	717
1106. The REDDINGS [106] 0 (0); 0 (0); 3 (29)	716
1107. Lyle MAYS [50] 0 (0); 0 (0); 1 (21)	714
1108. Pat METHENY [50] 0 (0); 0 (0); 1 (21)	714
1109. Phil SEYMOUR [64] 0 (0); 0 (0); 1 (16)	713
1110. The ANIMALS [66] 0 (0); 0 (0); 2 (14)	713
1111. Midge URE [88] 0 (0); 0 (0); 1 (16)	710
1112. John KLEMMER [99] 0 (0); 0 (0); 2 (20)	709
1113. MINISTRY [96] 0 (0); 0 (0); 4 (25)	708
1114. DANNY WILSON [79] 0 (0); 0 (0); 1 (16)	708
1115. TANGERINE DREAM [96] 0 (0); 0 (0); 3 (19)	708
1116. MOTORHEAD [150] 0 (0); 0 (0); 4 (30)	707
1117. Patrick SIMMONS [52] 0 (0); 0 (0); 1 (11)	707
1118. Glenn JONES [94] 0 (0); 0 (0); 1 (17)	705
1119. The DRAMATICS [61] 0 (0); 0 (0); 1 (12)	704
1120. YELLO [92] 0 (0); 0 (0); 3 (23)	700
1121. Jean BEAUVOIR [93] 0 (0); 0 (0); 1 (15)	699
1122. FIVE STAIRSTEPS [90] 0 (0); 0 (0); 1 (14)	699
1123. John EDDIE [83] 0 (0); 0 (0); 1 (15)	696
1124. SISTERS OF MERCY [101] 0 (0); 0 (0); 1 (16)	695
1125. KLIQUE [70] 0 (0); 0 (0); 1 (14)	692
1126. BARDEUX [104] 0 (0); 0 (0); 2 (19)	687
1127. HEATWAVE [71] 0 (0); 0 (0); 2 (16)	682
1128. Glenn MEDEIROS [83] 0 (0); 0 (0); 1 (17)	680
1129. Jack BRUCE [37] 0 (0); 1 (3); 3 (24)	680
1130. Frank BARBER Orchestra [94] 0 (0); 0 (0); 1 (16)	678
1131. SAXON [130] 0 (0); 0 (0); 4 (29)	677
1132. Deon ESTUS [89] 0 (0); 0 (0); 1 (15)	677
1133. Jessi COLTER [43] 0 (0); 0 (0); 1 (19)	672
1134. The RADIATORS [122] 0 (0); 0 (0); 2 (27)	669
1135. La Toya JACKSON [116] 0 (0); 0 (0); 3 (22)	667
1136. ROCK And HYDE [94] 0 (0); 0 (0); 1 (15)	660
1137. SYBIL [75] 0 (0); 0 (0); 1 (11)	655
1138. Steve EARLE And The DUKES [90] 0 (0); 0 (0); 1 (14)	655
1139. VELVET UNDERGROUND [85] 0 (0); 0 (0); 2 (15)	647
1140. Johnny Guitar WATSON [115] 0 (0); 0 (0); 2 (17)	646
1141. Keith WHITLEY [115] 0 (0); 0 (0); 2 (21)	643
1142. BOYS CLUB [93] 0 (0); 0 (0); 1 (16)	643
1143. The BRANDOS [108] 0 (0); 0 (0); 1 (19)	641
1144. TAVARES [75] 0 (0); 0 (0); 2 (18)	640
1145. Peter ALLEN [123] 0 (0); 0 (0); 2 (26)	639
1146. David PEASTON [113] 0 (0); 0 (0); 1 (18)	638
1147. Ray PRICE [70] 0 (0); 0 (0); 1 (25)	637
1148. SHERBS [100] 0 (0); 0 (0); 1 (16)	635
1149. Stephen STILLS [75] 0 (0); 0 (0); 1 (12)	635
1150. DION [130] 0 (0); 0 (0); 1 (19)	632
1151. Jim MESSINA [95] 0 (0); 0 (0); 2 (14)	623
1152. Pia ZADORA [113] 0 (0); 0 (0); 1 (20)	622
1153. Joyce KENNEDY [79] 0 (0); 0 (0); 1 (13)	622
1154. Kevin PAIGE [107] 0 (0); 0 (0); 1 (15)	621
1155. Gerry RAFFERTY [61] 0 (0); 0 (0); 1 (9)	620
1156. Leif GARRETT [129] 0 (0); 0 (0); 2 (26)	619
1157. George CARLIN [136] 0 (0); 0 (0); 2 (24)	615
1158. Michael MORALES [113] 0 (0); 0 (0); 1 (20)	610
1159. Rodney CROWELL [105] 0 (0); 0 (0); 3 (23)	609
1160. ISLE OF MAN [110] 0 (0); 0 (0); 1 (18)	608
1161. The SUPREMES/ Diana ROSS And The SUPREMES [112] 0 (0); 0 (0); 1 (17)	605
1162. Feargal SHARKEY [75] 0 (0); 0 (0); 1 (11)	604
1163. Teri DeSARIO [80] 0 (0); 0 (0); 1 (13)	600
1164. Carl ANDERSON [87] 0 (0); 0 (0); 1 (12)	597
1165. Tom SCOTT [123] 0 (0); 0 (0); 3 (21)	596
1166. Ali THOMSON [99] 0 (0); 0 (0); 1 (15)	595
1167. Gwen GUTHRIE [89] 0 (0); 0 (0); 1 (13)	594
1168. HUSKER DU [117] 0 (0); 0 (0); 2 (20)	593
1169. Rachel SWEET [123] 0 (0); 0 (0); 2 (18)	592
1170. Marilyn MARTIN [72] 0 (0); 0 (0); 1 (11)	591
1171. Jim CAPALDI [91] 0 (0); 0 (0); 2 (20)	590
1172. Gil SCOTT-HERON & Brian JACKSON [82] 0 (0); 0 (0); 1 (12)	588
1173. GIANT [91] 0 (0); 0 (0); 1 (12)	588
1174. Nanci GRIFFITH [99] 0 (0); 0 (0); 1 (14)	588
1175. B.B. KING [131] 0 (0); 0 (0); 4 (23)	586
1176. Maxi PRIEST [108] 0 (0); 0 (0); 1 (17)	586
1177. Nils LOFGREN [99] 0 (0); 0 (0); 2 (16)	586
1178. COCTEAU TWINS [109] 0 (0); 0 (0); 1 (18)	586
1179. Marvin SEASE [114] 0 (0); 0 (0); 1 (17)	584
1180. Al DI MEOLA/ John McLAUGHLIN/ Paco De LUCIA [97] 0 (0); 0 (0); 2 (18)	583
1181. Eugene WILDE [97] 0 (0); 0 (0); 1 (15)	583
1182. WA WA NEE [123] 0 (0); 0 (0); 1 (17)	582
1183. Miles JAYE [125] 0 (0); 0 (0); 2 (21)	581
1184. SEDUCTION [78] 0 (0); 0 (0); 1 (10)	581
1185. Tim CURRY [77] 0 (0); 0 (0); 2 (15)	576
1186. BRAM TCHAIKOVSKY [108] 0 (0); 0 (0); 2 (18)	574
1187. EYE TO EYE [99] 0 (0); 0 (0); 1 (15)	571
1188. 24-7 SPYZ [113] 0 (0); 0 (0); 1 (16)	570
1189. Rodney FRANKLIN [104] 0 (0); 0 (0); 3 (19)	570
1190. Steve MARTIN [83] 0 (0); 0 (0); 2 (13)	569
1191. T-CONNECTION [123] 0 (0); 0 (0); 2 (18)	568
1192. MOVING PICTURES [101] 0 (0); 0 (0); 1 (16)	568
1193. K-9 POSSE [98] 0 (0); 0 (0); 1 (14)	567
1194. KWAMÉ & A NEW BEGINNING [114] 0 (0); 0 (0); 1 (18)	566
1195. L.A. BOPPERS [85] 0 (0); 0 (0); 1 (11)	560
1196. Chris REA [92] 0 (0); 0 (0); 1 (13)	559
1197. John Lee HOOKER [101] 0 (0); 0 (0); 1 (13)	558
1198. The PLASMATICS [134] 0 (0); 0 (0); 3 (22)	557
1199. Kim MITCHELL [106] 0 (0); 0 (0); 1 (15)	553
1200. John DENVER & The MUPPETS [26] 0 (0); 1 (2); 1 (4)	553
1201. Art GARFUNKEL [113] 0 (0); 0 (0); 2 (16)	552
1202. FACE TO FACE [126] 0 (0); 0 (0); 2 (23)	551
1203. John ENTWISTLE [71] 0 (0); 0 (0); 1 (16)	551
1204. Bernard WRIGHT [116] 0 (0); 0 (0); 1 (14)	551
1205. Doug E. FRESH & The GET FRESH CREW [88] 0 (0); 0 (0); 1 (13)	548
1206. D.R.I. [116] 0 (0); 0 (0); 2 (16)	546
1207. CLIMIE FISHER [120] 0 (0); 0 (0); 1 (16)	546
1208. Jimi HENDRIX EXPERIENCE [119] 0 (0); 0 (0); 1 (17)	541
1209. Lonnie MACK [130] 0 (0); 0 (0); 1 (21)	541
1210. Robert FRIPP [90] 0 (0); 0 (0); 2 (13)	541
1211. The PASADENAS [89] 0 (0); 0 (0); 1 (12)	541
1212. CAMOUFLAGE [100] 0 (0); 0 (0); 1 (14)	540
1213. Maria McKEE [120] 0 (0); 0 (0); 1 (15)	538
1214. TREAT HER RIGHT [127] 0 (0); 0 (0); 1 (18)	535
1215. Colonel ABRAMS [75] 0 (0); 0 (0); 1 (11)	534
1216. Ringo STARR [98] 0 (0); 0 (0); 1 (12)	534
1217. LITTLE AMERICA [102] 0 (0); 0 (0); 1 (14)	532

Album Acts Ranked

Rank. Act [Hi Peak] / T10 (Wk); T40 (Wk); Ent (Wk)	Score In '80s
1218. CHEECH & CHONG [71] / 0 (0); 0 (0); 2 (14)	530
1219. Timothy B. SCHMIT [106] / 0 (0); 0 (0); 2 (16)	529
1220. UNDERWORLD [139] / 0 (0); 0 (0); 1 (19)	529
1221. BRICK [89] / 0 (0); 0 (0); 2 (15)	528
1222. SHOTGUN MESSIAH [99] / 0 (0); 0 (0); 1 (11)	528
1223. HEAD EAST [119] / 0 (0); 0 (0); 2 (15)	528
1224. OAKTOWN'S 3.5.7 [126] / 0 (0); 0 (0); 1 (16)	527
1225. CHINA CRISIS [114] / 0 (0); 0 (0); 2 (16)	526
1226. BEAT FARMERS [131] / 0 (0); 0 (0); 3 (20)	520
1227. Herb ALPERT & The TIJUANA BRASS [75] / 0 (0); 0 (0); 1 (10)	520
1228. Roger WHITTAKER [154] / 0 (0); 0 (0); 4 (23)	520
1229. Hubert LAWS [133] / 0 (0); 0 (0); 2 (17)	519
1230. SEAWIND [83] / 0 (0); 0 (0); 1 (11)	518
1231. Rita COOLIDGE [107] / 0 (0); 0 (0); 3 (17)	518
1232. Steve MORSE Band [101] / 0 (0); 0 (0); 2 (15)	517
1233. NUCLEAR ASSAULT [134] / 0 (0); 0 (0); 2 (18)	515
1234. DR. J.R. KOOL & The OTHER ROXANNES [113] / 0 (0); 0 (0); 1 (13)	515
1235. Peter MURPHY [135] / 0 (0); 0 (0); 1 (19)	511
1236. SKY [125] / 0 (0); 0 (0); 2 (18)	510
1237. ROMAN HOLLIDAY [116] / 0 (0); 0 (0); 2 (17)	510
1238. Bill LORDAN [37] / 0 (0); 1 (3); 1 (16)	509
1239. BOY GEORGE [126] / 0 (0); 0 (0); 2 (16)	509
1240. MC LYTE [86] / 0 (0); 0 (0); 1 (11)	508
1241. GUADALCANAL DIARY [132] / 0 (0); 0 (0); 2 (20)	507
1242. AVERAGE WHITE BAND/AWB [116] / 0 (0); 0 (0); 2 (14)	504
1243. KICK AXE [126] / 0 (0); 0 (0); 1 (15)	503
1244. J.J. CALE [110] / 0 (0); 0 (0); 2 (15)	502
1245. AZTEC CAMERA [129] / 0 (0); 0 (0); 4 (22)	500
1246. Gino SOCCIO [96] / 0 (0); 0 (0); 1 (14)	499
1247. LIVING IN A BOX [89] / 0 (0); 0 (0); 1 (13)	499
1248. Miki HOWARD [145] / 0 (0); 0 (0); 2 (22)	499
1249. WILLIE & The POOR BOYS [96] / 0 (0); 0 (0); 1 (12)	497
1250. Tracie SPENCER [146] / 0 (0); 0 (0); 1 (21)	497
1251. SUAVE [101] / 0 (0); 0 (0); 1 (12)	497
1252. Dave MASON [74] / 0 (0); 0 (0); 1 (10)	496
1253. Jon ANDERSON [143] / 0 (0); 0 (0); 3 (21)	492
1254. THREE TIMES DOPE [122] / 0 (0); 0 (0); 1 (18)	490
1255. The HOUSEMARTINS [124] / 0 (0); 0 (0); 2 (20)	486
1256. John KILZER [110] / 0 (0); 0 (0); 1 (15)	486
1257. PEARL HARBOR And The EXPLOSIONS [107] / 0 (0); 0 (0); 2 (14)	486
1258. RHYTHM CORPS [104] / 0 (0); 0 (0); 1 (14)	485
1259. Roy BUCHANAN [153] / 0 (0); 0 (0); 3 (23)	485
1260. Kiri TE KANAWA [136] / 0 (0); 0 (0); 1 (16)	485
1261. JELLYBEAN [101] / 0 (0); 0 (0); 1 (11)	483
1262. Tom KIMMEL [104] / 0 (0); 0 (0); 1 (15)	481
1263. BLOODSTONE [95] / 0 (0); 0 (0); 1 (11)	480
1264. MAHOGANY RUSH [88] / 0 (0); 0 (0); 2 (13)	476
1265. Janis JOPLIN [104] / 0 (0); 0 (0); 1 (14)	476
1266. CLANNAD [131] / 0 (0); 0 (0); 2 (17)	475
1267. Elliot EASTON [99] / 0 (0); 0 (0); 1 (11)	472
1268. Larry GATLIN/GATLIN BROTHERS Band [102] / 0 (0); 0 (0); 3 (15)	472
1269. Robert HAZARD [102] / 0 (0); 0 (0); 1 (11)	472
1270. Robert WINTERS and FALL [71] / 0 (0); 0 (0); 1 (8)	470
1271. OFF BROADWAY USA [101] / 0 (0); 0 (0); 1 (11)	469
1272. Moon MARTIN [138] / 0 (0); 0 (0); 1 (15)	469
1273. ZODIAC MINDWARP & The LOVE REACTION [132] / 0 (0); 0 (0); 1 (15)	469
1274. Pete SHELLEY [121] / 0 (0); 0 (0); 2 (15)	468
1275. WATERFRONT [103] / 0 (0); 0 (0); 1 (13)	467
1276. Gilda RADNER [69] / 0 (0); 0 (0); 1 (7)	467
1277. SHRIEKBACK [145] / 0 (0); 0 (0); 3 (21)	467
1278. Brian ENO [44] / 0 (0); 0 (0); 1 (13)	466
1279. KING TEE [125] / 0 (0); 0 (0); 1 (15)	464
1280. FROZEN GHOST [107] / 0 (0); 0 (0); 1 (13)	464
1281. SNEAKER [149] / 0 (0); 0 (0); 1 (17)	459
1282. Russ BALLARD [147] / 0 (0); 0 (0); 3 (19)	459
1283. JUNKYARD [105] / 0 (0); 0 (0); 1 (11)	457
1284. Shanice WILSON [149] / 0 (0); 0 (0); 1 (18)	450
1285. Donald BYRD And 125th STREET N.Y.C. [93] / 0 (0); 0 (0); 1 (10)	450
1286. Barry GIBB [72] / 0 (0); 0 (0); 1 (8)	450
1287. WEATHER GIRLS [91] / 0 (0); 0 (0); 1 (11)	448
1288. Henry PAUL Band [120] / 0 (0); 0 (0); 2 (16)	448
1289. John STEWART [85] / 0 (0); 0 (0); 1 (10)	447
1290. Robben FORD [120] / 0 (0); 0 (0); 1 (13)	445
1291. Tony TERRY [151] / 0 (0); 0 (0); 1 (20)	445
1292. John HALL Band [147] / 0 (0); 0 (0); 2 (18)	443
1293. ANGEL CITY [133] / 0 (0); 0 (0); 3 (18)	441
1294. 20/20 [127] / 0 (0); 0 (0); 2 (16)	439
1295. Agnetha FALTSKOG [102] / 0 (0); 0 (0); 1 (11)	439
1296. Barry GOUDREAU [88] / 0 (0); 0 (0); 1 (8)	437
1297. FIFTH ANGEL [117] / 0 (0); 0 (0); 1 (13)	437
1298. MAC Band [109] / 0 (0); 0 (0); 1 (14)	436
1299. DISNEYLAND AFTER DARK [116] / 0 (0); 0 (0); 1 (11)	436
1300. L'TRIMM [132] / 0 (0); 0 (0); 1 (16)	435
1301. RED 7 [105] / 0 (0); 0 (0); 2 (13)	435
1302. Roger GLOVER [101] / 0 (0); 0 (0); 1 (12)	434
1303. Dave ALVIN [116] / 0 (0); 0 (0); 1 (13)	434
1304. Martha DAVIS [127] / 0 (0); 0 (0); 1 (13)	433
1305. No Artist [96] / 0 (0); 0 (0); 1 (9)	433
1306. Michael SEMBELLO [80] / 0 (0); 0 (0); 1 (10)	433
1307. Tommy KEENE [148] / 0 (0); 0 (0); 1 (17)	431
1308. The HOLLIES [90] / 0 (0); 0 (0); 1 (9)	430
1309. DUKE JUPITER [122] / 0 (0); 0 (0); 1 (12)	425
1310. ZENO [107] / 0 (0); 0 (0); 1 (10)	425
1311. George BURNS [93] / 0 (0); 0 (0); 1 (10)	425
1312. Joe ELY [135] / 0 (0); 0 (0); 2 (14)	424
1313. Al GREEN [131] / 0 (0); 0 (0); 1 (14)	422
1314. David GRISMAN [108] / 0 (0); 0 (0); 3 (21)	422
1315. Trevor RABIN [111] / 0 (0); 0 (0); 1 (10)	421
1316. WAX [101] / 0 (0); 0 (0); 1 (11)	419
1317. Stevie WOODS [153] / 0 (0); 0 (0); 1 (25)	418
1318. EMERSON, LAKE & PALMER [73] / 0 (0); 0 (0); 2 (12)	417
1319. David CROSBY [104] / 0 (0); 0 (0); 1 (11)	417
1320. MADHOUSE [107] / 0 (0); 0 (0); 1 (11)	417
1321. Mike OLDFIELD [138] / 0 (0); 0 (0); 3 (16)	417
1322. Roger TAYLOR [121] / 0 (0); 0 (0); 1 (10)	416
1323. MASS PRODUCTION [133] / 0 (0); 0 (0); 2 (15)	414
1324. Neil SEDAKA [135] / 0 (0); 0 (0); 1 (13)	414
1325. Bob MOULD [127] / 0 (0); 0 (0); 1 (14)	412
1326. OVERKILL [142] / 0 (0); 0 (0); 3 (16)	412
1327. Dottie WEST [126] / 0 (0); 0 (0); 1 (15)	411
1328. Ramsey LEWIS [144] / 0 (0); 0 (0); 3 (22)	411
1329. M.O.D. [151] / 0 (0); 0 (0); 3 (19)	410
1330. Leon HAYWOOD [92] / 0 (0); 0 (0); 1 (10)	407
1331. Graham NASH [117] / 0 (0); 0 (0); 2 (12)	407
1332. PHOTOGLO [119] / 0 (0); 0 (0); 2 (14)	407
1333. Chaz JANKEL [126] / 0 (0); 0 (0); 1 (11)	405
1334. STAGE DOLLS [118] / 0 (0); 0 (0); 1 (12)	405
1335. 707 [129] / 0 (0); 0 (0); 2 (15)	404
1336. Ralph MacDONALD [108] / 0 (0); 0 (0); 1 (10)	404
1337. ROACHFORD [109] / 0 (0); 0 (0); 1 (12)	404
1338. Mike RUTHERFORD [145] / 0 (0); 0 (0); 2 (17)	402
1339. DREAMS SO REAL [150] / 0 (0); 0 (0); 1 (18)	400
1340. The WINANS [109] / 0 (0); 0 (0); 1 (11)	400
1341. Lene LOVICH [94] / 0 (0); 0 (0); 2 (12)	399
1342. Liza MINNELLI [128] / 0 (0); 0 (0); 2 (16)	397
1343. SIGUE SIGUE SPUTNIK [96] / 0 (0); 0 (0); 1 (10)	397
1344. Donna ALLEN [133] / 0 (0); 0 (0); 2 (14)	396
1345. KINGS OF THE SUN [136] / 0 (0); 0 (0); 1 (16)	395
1346. Del SHANNON [123] / 0 (0); 0 (0); 1 (14)	395
1347. The PRIMITIVES [106] / 0 (0); 0 (0); 2 (11)	394
1348. Cheryl Pepsii RILEY [128] / 0 (0); 0 (0); 1 (11)	393
1349. 3 [97] / 0 (0); 0 (0); 1 (10)	390
1350. BUCKWHEAT ZYDECO [104] / 0 (0); 0 (0); 2 (12)	390
1351. The KORGIS [113] / 0 (0); 0 (0); 1 (12)	390

Album Acts Ranked

Rank. Act [Hi Peak] T10 (Wk); T40 (Wk); Ent (Wk)	Score In '80s
1352. TROUBLE FUNK [121] 0 (0); 0 (0); 1 (14)	389
1353. Betty WRIGHT [127] 0 (0); 0 (0); 1 (13)	386
1354. SCARLETT & BLACK [107] 0 (0); 0 (0); 1 (11)	386
1355. Steve STEVENS ATOMIC PLAYBOYS [119] 0 (0); 0 (0); 1 (12)	379
1356. TEN YEARS AFTER [120] 0 (0); 0 (0); 1 (10)	378
1357. WHITE WOLF [137] 0 (0); 0 (0); 2 (15)	376
1358. HEAR 'N AID [80] 0 (0); 0 (0); 1 (7)	375
1359. Tané CAIN [121] 0 (0); 0 (0); 1 (10)	372
1360. NOEL [126] 0 (0); 0 (0); 1 (13)	372
1361. Billy RANKIN [119] 0 (0); 0 (0); 1 (11)	369
1362. SILVER CONDOR [141] 0 (0); 0 (0); 1 (12)	365
1363. Sam COOKE [134] 0 (0); 0 (0); 2 (16)	362
1364. RED SIREN [124] 0 (0); 0 (0); 1 (12)	361
1365. Kenny AARONSON [42] 0 (0); 0 (0); 1 (18)	358
1366. Neal SCHON [42] 0 (0); 0 (0); 1 (18)	358
1367. Michael SHRIEVE [42] 0 (0); 0 (0); 1 (18)	358
1368. Bob GELDOF [130] 0 (0); 0 (0); 1 (12)	357
1369. Conway TWITTY [144] 0 (0); 0 (0); 1 (15)	357
1370. The STYLISTICS [127] 0 (0); 0 (0); 1 (12)	355
1371. Willie NILE [145] 0 (0); 0 (0); 2 (14)	355
1372. RAGING SLAB [113] 0 (0); 0 (0); 1 (10)	353
1373. Chuckii BOOKER [116] 0 (0); 0 (0); 1 (10)	351
1374. SPIDER [130] 0 (0); 0 (0); 2 (12)	351
1375. DRIVIN' N' CRYIN' [130] 0 (0); 0 (0); 1 (10)	350
1376. Jennifer RUSH [118] 0 (0); 0 (0); 1 (10)	346
1377. The DELLS [137] 0 (0); 0 (0); 1 (12)	345
1378. YELLOWJACKETS [145] 0 (0); 0 (0); 3 (16)	345
1379. AFTER 7 [132] 0 (0); 0 (0); 1 (12)	344
1380. Toni TENNILLE [142] 0 (0); 0 (0); 2 (13)	344
1381. Joey SCARBURY [104] 0 (0); 0 (0); 1 (9)	344
1382. MANTRONIX [108] 0 (0); 0 (0); 1 (8)	343
1383. Ritchie VALENS [100] 0 (0); 0 (0); 1 (11)	343
1384. FUNKADELIC [105] 0 (0); 0 (0); 2 (9)	342
1385. Alex BUGNON [127] 0 (0); 0 (0); 1 (11)	339
1386. Andre CYMONE [121] 0 (0); 0 (0); 2 (12)	337
1387. Alvin LEE [124] 0 (0); 0 (0); 1 (10)	337
1388. PRETTY POISON [104] 0 (0); 0 (0); 1 (8)	336
1389. The CRETONES [125] 0 (0); 0 (0); 1 (10)	335
1390. FLYING LIZARDS [99] 0 (0); 0 (0); 1 (8)	335
1391. Sharon BRYANT [139] 0 (0); 0 (0); 1 (13)	334
1392. Lena HORNE [112] 0 (0); 0 (0); 1 (9)	334
1393. David FOSTER [111] 0 (0); 0 (0); 2 (11)	334
1394. Terry BOZZIO [49] 0 (0); 0 (0); 1 (11)	334
1395. Tony HYMAS [49] 0 (0); 0 (0); 1 (11)	334
1396. Sue SAAD And The NEXT [131] 0 (0); 0 (0); 1 (12)	334
1397. The PROCLAIMERS [125] 0 (0); 0 (0); 1 (11)	333
1398. SO [124] 0 (0); 0 (0); 1 (9)	330
1399. Amy HOLLAND [146] 0 (0); 0 (0); 1 (14)	328
1400. BROOKLYN, BRONX & QUEENS Band [109] 0 (0); 0 (0); 1 (9)	328
1401. 3rd BASS [76] 0 (0); 0 (0); 1 (5)	325
1402. Stewart COPELAND [148] 0 (0); 0 (0); 2 (13)	325
1403. Shirley JONES [128] 0 (0); 0 (0); 1 (10)	325
1404. Curtis MAYFIELD [128] 0 (0); 0 (0); 2 (14)	325
1405. Mason WILLIAMS [118] 0 (0); 0 (0); 1 (19)	325
1406. Louise TUCKER [127] 0 (0); 0 (0); 1 (10)	322
1407. Paul WINTER [138] 0 (0); 0 (0); 1 (11)	321
1408. KID CREOLE & The COCONUTS [145] 0 (0); 0 (0); 2 (14)	321
1409. STONE FURY [144] 0 (0); 0 (0); 1 (12)	320
1410. REGINA [102] 0 (0); 0 (0); 1 (8)	320
1411. The BLACKBYRDS [133] 0 (0); 0 (0); 1 (11)	320
1412. ALPHAVILLE [174] 0 (0); 0 (0); 2 (21)	318
1413. KITARO [141] 0 (0); 0 (0); 3 (13)	317
1414. SPINAL TAP [121] 0 (0); 0 (0); 1 (10)	314
1415. Brian MAY [125] 0 (0); 0 (0); 1 (9)	312
1416. Barbara Ann AUER [145] 0 (0); 0 (0); 1 (15)	312
1417. VESTA [131] 0 (0); 0 (0); 1 (10)	312
1418. Sylvain SYLVAIN [123] 0 (0); 0 (0); 1 (8)	311
1419. DANZIG [125] 0 (0); 0 (0); 1 (9)	309
1420. FUN BOY THREE [104] 0 (0); 0 (0); 1 (7)	308
1421. LULU [126] 0 (0); 0 (0); 1 (10)	308
1422. Jerry GARCIA [100] 0 (0); 0 (0); 1 (8)	306
1423. Webster LEWIS [114] 0 (0); 0 (0); 1 (9)	306
1424. John BRANNEN [156] 0 (0); 0 (0); 1 (14)	306
1425. HEADPINS [114] 0 (0); 0 (0); 1 (9)	306
1426. TIMES TWO [137] 0 (0); 0 (0); 1 (11)	305
1427. BRIDES OF FUNKENSTEIN [93] 0 (0); 0 (0); 1 (7)	305
1428. M [79] 0 (0); 0 (0); 1 (6)	304
1429. TRANSVISION VAMP [115] 0 (0); 0 (0); 1 (8)	303
1430. JETBOY [135] 0 (0); 0 (0); 1 (10)	301
1431. Robin WILLIAMS [119] 0 (0); 0 (0); 1 (9)	301
1432. SAWYER ROWN [140] 0 (0); 0 (0); 1 (11)	297
1433. BOSTON POPS Orchestra/ John WILLIAMS [155] 0 (0); 0 (0); 2 (14)	297
1434. Paul RODGERS [135] 0 (0); 0 (0); 1 (11)	296
1435. STEALIN HORSES [146] 0 (0); 0 (0); 1 (12)	296
1436. The EMOTIONS [96] 0 (0); 0 (0); 2 (14)	295
1437. TKA [135] 0 (0); 0 (0); 1 (11)	294
1438. MAMA'S BOYS [151] 0 (0); 0 (0); 2 (14)	292
1439. The THREE O'CLOCK [125] 0 (0); 0 (0); 2 (14)	292
1440. "D" TRAIN [128] 0 (0); 0 (0); 1 (9)	291
1441. Mitch RYDER [120] 0 (0); 0 (0); 1 (9)	291
1442. Sarah McLACHLAN [132] 0 (0); 0 (0); 1 (12)	291
1443. David LANZ [125] 0 (0); 0 (0); 2 (18)	290
1444. Paul ANKA [156] 0 (0); 0 (0); 2 (14)	290
1445. The TRUTH [115] 0 (0); 0 (0); 1 (8)	290
1446. Leon REDBONE [152] 0 (0); 0 (0); 1 (11)	289
1447. D.L. BYRON [133] 0 (0); 0 (0); 1 (10)	289
1448. Ofra HAZA [130] 0 (0); 0 (0); 1 (9)	289
1449. James McMURTRY [125] 0 (0); 0 (0); 1 (9)	289
1450. L.A. DREAM TEAM [138] 0 (0); 0 (0); 2 (11)	286
1451. Z.Z. HILL [165] 0 (0); 0 (0); 2 (14)	285
1452. JESUS AND MARY CHAIN [161] 0 (0); 0 (0); 4 (17)	285
1453. Johnny WINTER [156] 0 (0); 0 (0); 2 (14)	282
1454. FAIRGROUND ATTRACTION [137] 0 (0); 0 (0); 1 (11)	281
1455. Chet ATKINS [145] 0 (0); 0 (0); 1 (13)	281
1456. WIRE TRAIN [150] 0 (0); 0 (0); 2 (13)	280
1457. The FOOLS [151] 0 (0); 0 (0); 2 (12)	279
1458. Keith SYKES [147] 0 (0); 0 (0); 1 (11)	279
1459. Jerry KNIGHT [146] 0 (0); 0 (0); 2 (13)	278
1460. CROWN HEIGHTS AFFAIR [148] 0 (0); 0 (0); 1 (12)	277
1461. Colin James HAY [126] 0 (0); 0 (0); 1 (9)	277
1462. The MONROES [109] 0 (0); 0 (0); 1 (9)	275
1463. TROOP [133] 0 (0); 0 (0); 1 (9)	275
1464. The ROCHES [130] 0 (0); 0 (0); 2 (10)	274
1465. TWIN HYPE [140] 0 (0); 0 (0); 1 (11)	274
1466. Kirk WHALUM [142] 0 (0); 0 (0); 1 (11)	273
1467. Kathy SMITH [144] 0 (0); 0 (0); 1 (13)	273
1468. DR. JOHN [142] 0 (0); 0 (0); 1 (11)	272
1469. Marc ALMOND [144] 0 (0); 0 (0); 1 (11)	271
1470. NITRO [140] 0 (0); 0 (0); 1 (9)	270
1471. Jimmy DAVIS & JUNCTION [122] 0 (0); 0 (0); 1 (8)	269
1472. SPYS [138] 0 (0); 0 (0); 1 (10)	268
1473. Linda CLIFFORD [142] 0 (0); 0 (0); 3 (14)	268
1474. BALANCE [133] 0 (0); 0 (0); 1 (12)	268
1475. Tony MacALPINE [146] 0 (0); 0 (0); 1 (11)	267
1476. The SILENCERS [147] 0 (0); 0 (0); 1 (11)	267
1477. CASHFLOW [144] 0 (0); 0 (0); 1 (11)	266
1478. Ernie WATTS [161] 0 (0); 0 (0); 1 (12)	266
1479. Jon ASTLEY [135] 0 (0); 0 (0); 1 (10)	266
1480. Howard JOHNSON [122] 0 (0); 0 (0); 1 (9)	266
1481. Joe Lynn TURNER [143] 0 (0); 0 (0); 1 (12)	265
1482. EUROGLIDERS [140] 0 (0); 0 (0); 1 (11)	264
1483. DEATH ANGEL [143] 0 (0); 0 (0); 1 (11)	263
1484. DORO [154] 0 (0); 0 (0); 1 (11)	263
1485. J.D. SOUTHER [124] 0 (0); 0 (0); 1 (7)	262
1486. SWEET TEE [169] 0 (0); 0 (0); 1 (13)	261
1487. Ian McLAGAN [125] 0 (0); 0 (0); 1 (9)	261
1488. Paul HYDE And The PAYOLAS [144] 0 (0); 0 (0); 1 (10)	261

Album Acts Ranked

Rank. Act [Hi Peak] T10 (Wk); T40 (Wk); Ent (Wk)	Score In '80s
1489. STONE CITY BAND [122] 0 (0); 0 (0); 1 (8)	260
1490. Judy COLLINS [142] 0 (0); 0 (0); 2 (11)	258
1491. The KINGBEES [160] 0 (0); 0 (0); 1 (12)	257
1492. EGYPTIAN LOVER [146] 0 (0); 0 (0); 1 (10)	256
1493. RAIL [143] 0 (0); 0 (0); 1 (10)	256
1494. Lou Ann BARTON [133] 0 (0); 0 (0); 1 (9)	256
1495. Stan RIDGWAY [131] 0 (0); 0 (0); 1 (9)	255
1496. Bill WITHERS [143] 0 (0); 0 (0); 2 (11)	254
1497. LARSEN-FEITEN BAND [142] 0 (0); 0 (0); 1 (10)	253
1498. Chris ISAAK [149] 0 (0); 0 (0); 2 (12)	253
1499. GLASS MOON [148] 0 (0); 0 (0); 1 (9)	253
1500. Frankie MILLER [135] 0 (0); 0 (0); 1 (9)	251
1501. Jack GREEN [121] 0 (0); 0 (0); 1 (8)	250
1502. Jimmy CLIFF [122] 0 (0); 0 (0); 2 (8)	249
1503. Josie COTTON [147] 0 (0); 0 (0); 1 (12)	249
1504. OXO [117] 0 (0); 0 (0); 1 (7)	248
1505. WILL And The KILL [129] 0 (0); 0 (0); 1 (8)	246
1506. Alphonse MOUZON [146] 0 (0); 0 (0); 1 (11)	246
1507. Robert Ellis ORRALL [146] 0 (0); 0 (0); 1 (9)	245
1508. The CHIEFTAINS [102] 0 (0); 0 (0); 1 (13)	245
1509. WIRE [135] 0 (0); 0 (0); 1 (10)	245
1510. HOUSE OF FREAKS [154] 0 (0); 0 (0); 1 (10)	245
1511. XAVIER [129] 0 (0); 0 (0); 1 (7)	243
1512. Bob WELCH [105] 0 (0); 0 (0); 2 (8)	241
1513. TAXXI [161] 0 (0); 0 (0); 1 (11)	239
1514. Nina HAGEN [151] 0 (0); 0 (0); 2 (11)	239
1515. KING SWAMP [159] 0 (0); 0 (0); 1 (14)	238
1516. Michael PENN [98] 0 (0); 0 (0); 1 (6)	237
1517. WET WET WET [123] 0 (0); 0 (0); 1 (7)	237
1518. Steve HACKETT [144] 0 (0); 0 (0); 2 (9)	237
1519. Danny Joe BROWN [120] 0 (0); 0 (0); 1 (7)	235
1520. DOCTOR And The MEDICS [125] 0 (0); 0 (0); 1 (8)	235
1521. MILLIONS LIKE US [171] 0 (0); 0 (0); 1 (12)	235
1522. JOY DIVISION [146] 0 (0); 0 (0); 1 (8)	233
1523. CHARLIE [145] 0 (0); 0 (0); 1 (9)	233
1524. PEPSI and SHIRLIE [133] 0 (0); 0 (0); 1 (9)	232
1525. TORONTO [162] 0 (0); 0 (0); 2 (14)	232
1526. FUSE ONE [139] 0 (0); 0 (0); 1 (9)	231
1527. KING [140] 0 (0); 0 (0); 1 (9)	230
1528. James LAST Band [148] 0 (0); 0 (0); 1 (9)	230
1529. 7 SECONDS [153] 0 (0); 0 (0); 1 (9)	230
1530. John HUNTER [148] 0 (0); 0 (0); 1 (9)	230
1531. BOBBY And The MIDNITES [158] 0 (0); 0 (0); 2 (11)	230
1532. MAZARATI [133] 0 (0); 0 (0); 1 (8)	230
1533. CHOCOLATE MILK [162] 0 (0); 0 (0); 1 (10)	229
1534. Lenny KRAVITZ [96] 0 (0); 0 (0); 1 (6)	227
1535. TEARDROP EXPLODES [156] 0 (0); 0 (0); 2 (10)	225
1536. Tommy JAMES [134] 0 (0); 0 (0); 1 (7)	225
1537. McGUINN, CLARK & HILLMAN [136] 0 (0); 0 (0); 1 (7)	223
1538. DREAMBOY [168] 0 (0); 0 (0); 1 (11)	221
1539. Judson SPENCE [168] 0 (0); 0 (0); 1 (13)	220
1540. ATLANTA [140] 0 (0); 0 (0); 1 (7)	219
1541. TUCK & PATTI [162] 0 (0); 0 (0); 1 (11)	219
1542. CENTRAL LINE [145] 0 (0); 0 (0); 1 (9)	219
1543. STREETS [166] 0 (0); 0 (0); 1 (11)	218
1544. Jane SIBERRY [149] 0 (0); 0 (0); 1 (8)	216
1545. Jackie MASON [146] 0 (0); 0 (0); 1 (9)	216
1546. SHOES [140] 0 (0); 0 (0); 1 (7)	211
1547. EZO [150] 0 (0); 0 (0); 1 (9)	207
1548. Dianne REEVES [172] 0 (0); 0 (0); 1 (12)	207
1549. BOOK OF LOVE [156] 0 (0); 0 (0); 1 (10)	207
1550. The GRACES [147] 0 (0); 0 (0); 1 (9)	205
1551. Bobby NUNN [148] 0 (0); 0 (0); 1 (8)	204
1552. Robert TEPPER [144] 0 (0); 0 (0); 1 (8)	204
1553. BULGARIAN STATE RADIO & T.V. FEMALE CHOIR [165] 0 (0); 0 (0); 1 (10)	203
1554. CHUNKY A [77] 0 (0); 0 (0); 1 (3)	203
1555. M + M [163] 0 (0); 0 (0); 3 (11)	203
1556. Richard CLAYDERMAN [160] 0 (0); 0 (0); 1 (9)	203
1557. Paul SPEER [125] 0 (0); 0 (0); 1 (12)	199
1558. Steve WALSH [124] 0 (0); 0 (0); 1 (6)	199
1559. E.U. [158] 0 (0); 0 (0); 1 (9)	199
1560. Mojo NIXON & Skid ROPER [151] 0 (0); 0 (0); 2 (9)	198
1561. JOHNNY And The DISTRACTIONS [152] 0 (0); 0 (0); 1 (9)	198
1562. SWEAT BAND [150] 0 (0); 0 (0); 1 (8)	198
1563. Diane SCHUUR [170] 0 (0); 0 (0); 1 (10)	197
1564. GANG OF FOUR [168] 0 (0); 0 (0); 4 (11)	197
1565. BLACK UHURU [146] 0 (0); 0 (0); 1 (7)	197
1566. WRABIT [157] 0 (0); 0 (0); 1 (8)	196
1567. SCREAMING BLUE MESSIAHS [172] 0 (0); 0 (0); 1 (11)	196
1568. FLOTSAM AND JETSAM [143] 0 (0); 0 (0); 1 (8)	195
1569. DFX2 [143] 0 (0); 0 (0); 1 (8)	195
1570. PROPHET [137] 0 (0); 0 (0); 1 (7)	195
1571. Melvin JAMES [146] 0 (0); 0 (0); 1 (8)	194
1572. Carla CAPUANO [152] 0 (0); 0 (0); 1 (8)	194
1573. VAIN [154] 0 (0); 0 (0); 1 (8)	194
1574. Sonny CHARLES [136] 0 (0); 0 (0); 1 (7)	193
1575. Thelma OUSTON [144] 0 (0); 0 (0); 1 (6)	193
1576. Tom JOHNSTON [158] 0 (0); 0 (0); 2 (9)	191
1577. Malcolm McLAREN [173] 0 (0); 0 (0); 2 (12)	190
1578. MONTY PYTHON [164] 0 (0); 0 (0); 1 (9)	189
1579. The ADVENTURES [144] 0 (0); 0 (0); 1 (7)	187
1580. John PRINE [144] 0 (0); 0 (0); 1 (7)	187
1581. Mickey GILLEY [170] 0 (0); 0 (0); 2 (9)	186
1582. Jean CARN [162] 0 (0); 0 (0); 2 (9)	185
1583. UNLIMITED TOUCH [142] 0 (0); 0 (0); 1 (7)	185
1584. Marty ROBBINS [170] 0 (0); 0 (0); 1 (9)	184
1585. The HEADBOYS [143] 0 (0); 0 (0); 1 (7)	184
1586. BILLY SATELLITE [139] 0 (0); 0 (0); 1 (7)	183
1587. MINK DeVILLE [161] 0 (0); 0 (0); 2 (8)	183
1588. DILLMAN Band [145] 0 (0); 0 (0); 1 (7)	180
1589. Durell COLEMAN [155] 0 (0); 0 (0); 1 (7)	180
1590. Charlie DORE [145] 0 (0); 0 (0); 1 (7)	179
1591. Janey STREET [145] 0 (0); 0 (0); 1 (6)	179
1592. The UNTOUCHABLES [162] 0 (0); 0 (0); 1 (9)	178
1593. RANKING ROGER [151] 0 (0); 0 (0); 1 (7)	177
1594. IRISH ROVERS [157] 0 (0); 0 (0); 1 (7)	177
1595. STEADY B [149] 0 (0); 0 (0); 2 (8)	177
1596. MOTHER'S FINEST [168] 0 (0); 0 (0); 1 (8)	176
1597. Norman CONNORS [145] 0 (0); 0 (0); 2 (8)	176
1598. DARLING CRUEL [160] 0 (0); 0 (0); 1 (8)	175
1599. The CONNELLS [163] 0 (0); 0 (0); 1 (10)	174
1600. Albert COLLINS [124] 0 (0); 0 (0); 1 (18)	174
1601. Johnny COPELAND [124] 0 (0); 0 (0); 1 (18)	174
1602. OTHER ONES [139] 0 (0); 0 (0); 1 (7)	173
1603. Stephane GRAPELLI [108] 0 (0); 0 (0); 1 (10)	173
1604. Cat STEVENS [165] 0 (0); 0 (0); 1 (8)	172
1605. GRANDMASTER FLASH [145] 0 (0); 0 (0); 2 (7)	171
1606. Carlene CARTER [139] 0 (0); 0 (0); 1 (6)	171
1607. 4 BY FOUR [141] 0 (0); 0 (0); 1 (7)	171
1608. Tyrone DAVIS [137] 0 (0); 0 (0); 1 (6)	170
1609. Greg GUIDRY [147] 0 (0); 0 (0); 1 (7)	170
1610. Tanya TUCKER [121] 0 (0); 0 (0); 2 (6)	170
1611. Vinnie MOORE [147] 0 (0); 0 (0); 1 (7)	170
1612. COMPANY B [143] 0 (0); 0 (0); 1 (6)	169
1613. PLAYER [152] 0 (0); 0 (0); 1 (7)	168
1614. Glen BURTNICK [147] 0 (0); 0 (0); 1 (6)	168
1615. The UNDERTONES [154] 0 (0); 0 (0); 1 (7)	166
1616. XYMOX [165] 0 (0); 0 (0); 1 (10)	166
1617. Johnny CLEGG & SAVUKA [155] 0 (0); 0 (0); 1 (7)	166
1618. ODYSSEY [175] 0 (0); 0 (0); 2 (9)	165
1619. Fee WAYBILL [146] 0 (0); 0 (0); 1 (6)	164
1620. The CHRISTIANS [158] 0 (0); 0 (0); 1 (8)	163
1621. Rick NELSON [153] 0 (0); 0 (0); 1 (6)	160
1622. Tony BENNETT [160] 0 (0); 0 (0); 1 (8)	160
1623. Don DIXON [162] 0 (0); 0 (0); 1 (8)	159

Album Acts Ranked

Rank. Act [Hi Peak] T10 (Wk); T40 (Wk); Ent (Wk)	Score In '80s
1624. EBONEE WEBB [157] 0 (0); 0 (0); 1 (7)	159
1625. Robbin THOMPSON Band [168] 0 (0); 0 (0); 1 (11)	159
1626. The A's [146] 0 (0); 0 (0); 1 (7)	158
1627. PRETTY MAIDS [165] 0 (0); 0 (0); 1 (8)	158
1628. SEA LEVEL [152] 0 (0); 0 (0); 1 (6)	158
1629. McFADDEN & WHITEHEAD [153] 0 (0); 0 (0); 1 (6)	156
1630. FISHBONE [153] 0 (0); 0 (0); 1 (9)	154
1631. Rex SMITH [165] 0 (0); 0 (0); 2 (7)	154
1632. CHEQUERED PAST [151] 0 (0); 0 (0); 1 (6)	153
1633. Howie MANDEL [148] 0 (0); 0 (0); 1 (6)	153
1634. TECHNOTRONIC [70] 0 (0); 0 (0); 1 (2)	153
1635. The PLIMSOULS [153] 0 (0); 0 (0); 2 (8)	151
1636. FOSTER & LLOYD [142] 0 (0); 0 (0); 1 (6)	151
1637. OZONE [152] 0 (0); 0 (0); 1 (6)	151
1638. Danny WILDE [176] 0 (0); 0 (0); 1 (9)	150
1639. Michael MONROE [161] 0 (0); 0 (0); 1 (8)	150
1640. LORDS Of The NEW CHURCH [158] 0 (0); 0 (0); 1 (7)	150
1641. BREAKWATER [141] 0 (0); 0 (0); 1 (5)	149
1642. DARK ANGEL [159] 0 (0); 0 (0); 1 (6)	147
1643. PASSPORT [163] 0 (0); 0 (0); 2 (7)	146
1644. FEMME FATALE [141] 0 (0); 0 (0); 1 (5)	145
1645. SEA HAGS [163] 0 (0); 0 (0); 1 (7)	143
1646. Jimmy RUFFIN [152] 0 (0); 0 (0); 1 (6)	143
1647. Elisa FIORILLO [163] 0 (0); 0 (0); 1 (8)	143
1648. Chick COREA [170] 0 (0); 0 (0); 2 (7)	142
1649. HOUSE OF LOVE [156] 0 (0); 0 (0); 1 (7)	142
1650. BADFINGER [155] 0 (0); 0 (0); 1 (6)	142
1651. Julie BROWN [168] 0 (0); 0 (0); 1 (7)	141
1652. ROSE ROYCE [160] 0 (0); 0 (0); 1 (7)	140
1653. PARTLAND BROTHERS [146] 0 (0); 0 (0); 1 (5)	139
1654. DB'S [171] 0 (0); 0 (0); 1 (8)	139
1655. Johnny CASH [87] 0 (0); 0 (0); 1 (12)	139
1656. Carl PERKINS [87] 0 (0); 0 (0); 1 (12)	139
1657. Lee OSKAR [162] 0 (0); 0 (0); 1 (6)	137
1658. The MOTORS [174] 0 (0); 0 (0); 1 (8)	136
1659. Pete BARDENS [148] 0 (0); 0 (0); 1 (5)	136
1660. GRAND FUNK RAILROAD [149] 0 (0); 0 (0); 1 (5)	135
1661. Branford MARSALIS [164] 0 (0); 0 (0); 1 (7)	135
1662. FETCHIN BONES [175] 0 (0); 0 (0); 1 (6)	135
1663. Tim WEISBERG [171] 0 (0); 0 (0); 1 (7)	135
1664. Freddie MERCURY [159] 0 (0); 0 (0); 1 (6)	134
1665. Simon TOWNSHEND [169] 0 (0); 0 (0); 1 (7)	133
1666. TEXTONES [176] 0 (0); 0 (0); 1 (7)	133
1667. Gary MORRIS [174] 0 (0); 0 (0); 1 (8)	132
1668. SAD CAFÉ [160] 0 (0); 0 (0); 1 (6)	132
1669. Billy THORPE [151] 0 (0); 0 (0); 1 (5)	132
1670. NOVO COMBO [167] 0 (0); 0 (0); 1 (6)	132
1671. Robbie PATTON [162] 0 (0); 0 (0); 1 (6)	132
1672. John CALE [154] 0 (0); 0 (0); 1 (5)	132
1673. MOUNTAIN [166] 0 (0); 0 (0); 1 (5)	131
1674. Don FELDER [178] 0 (0); 0 (0); 1 (8)	131
1675. FLASH & THE PAN [159] 0 (0); 0 (0); 1 (6)	130
1676. Debbie JACOBS [178] 0 (0); 0 (0); 1 (7)	129
1677. KENTUCKY HEADHUNTERS [104] 0 (0); 0 (0); 1 (3)	128
1678. BABYLON A.D. [141] 0 (0); 0 (0); 1 (5)	128
1679. Jorma KAUKONEN & VITAL PARTS [163] 0 (0); 0 (0); 1 (6)	127
1680. John CONLEE [166] 0 (0); 0 (0); 1 (6)	127
1681. Terri Lyne CARRINGTON [169] 0 (0); 0 (0); 1 (7)	127
1682. MONTROSE [165] 0 (0); 0 (0); 1 (7)	126
1683. VIO-LENCE [154] 0 (0); 0 (0); 1 (6)	125
1684. Linda FRATIANNE [174] 0 (0); 0 (0); 1 (7)	124
1685. Slim WHITMAN [175] 0 (0); 0 (0); 2 (7)	123
1686. Johnny GILL [139] 0 (0); 0 (0); 1 (8)	123
1687. MAX Q [182] 0 (0); 0 (0); 1 (8)	123
1688. ANGEL [149] 0 (0); 0 (0); 1 (4)	122
1689. COLD CHISEL [171] 0 (0); 0 (0); 1 (6)	122
1690. Ruben BLADES [156] 0 (0); 0 (0); 1 (6)	122
1691. Eloise LAWS [175] 0 (0); 0 (0); 1 (7)	122
1692. TYZIK [172] 0 (0); 0 (0); 1 (6)	122
1693. Nick MASON [154] 0 (0); 0 (0); 2 (8)	121
1694. ASWAD [173] 0 (0); 0 (0); 1 (7)	121
1695. BAUHAUS [169] 0 (0); 0 (0); 1 (6)	121
1696. LOVE AND MONEY [175] 0 (0); 0 (0); 1 (7)	121
1697. Mick RONSON [157] 0 (0); 0 (0); 1 (10)	121
1698. MARLEY MARL [163] 0 (0); 0 (0); 1 (5)	121
1699. Nancy WILSON [144] 0 (0); 0 (0); 1 (9)	120
1700. Ellen FOLEY [152] 0 (0); 0 (0); 1 (4)	118
1701. KILLER DWARFS [165] 0 (0); 0 (0); 1 (6)	117
1702. The RINGS [164] 0 (0); 0 (0); 1 (6)	117
1703. Ronnie WOOD [164] 0 (0); 0 (0); 1 (5)	117
1704. FUNKADELIC(2) [151] 0 (0); 0 (0); 1 (4)	117
1705. SCRUFFY THE CAT [177] 0 (0); 0 (0); 1 (8)	116
1706. POP WILL EAT ITSELF [169] 0 (0); 0 (0); 1 (6)	116
1707. The BUGGLES [161] 0 (0); 0 (0); 1 (5)	116
1708. Marlon JACKSON [175] 0 (0); 0 (0); 1 (7)	113
1709. Itzhak PERLMAN [149] 0 (0); 0 (0); 1 (9)	112
1710. Andre PREVIN [149] 0 (0); 0 (0); 1 (9)	112
1711. Kenny RANKIN [171] 0 (0); 0 (0); 1 (5)	112
1712. The BEARS [159] 0 (0); 0 (0); 1 (5)	112
1713. Tim FINN [161] 0 (0); 0 (0); 1 (5)	112
1714. GUCCI CREW II [173] 0 (0); 0 (0); 1 (6)	111
1715. Martin L. GORE [156] 0 (0); 0 (0); 1 (5)	109
1716. Suzi QUATRO [165] 0 (0); 0 (0); 2 (6)	108
1717. FARRENHEIT [179] 0 (0); 0 (0); 1 (7)	108
1718. INSIDERS [167] 0 (0); 0 (0); 1 (5)	107
1719. MADAME X [162] 0 (0); 0 (0); 1 (5)	106
1720. BLUE MERCEDES [165] 0 (0); 0 (0); 1 (5)	106
1721. Justin HAYWARD [166] 0 (0); 0 (0); 1 (5)	106
1722. BUZZCOCKS [163] 0 (0); 0 (0); 1 (6)	105
1723. Charley PRIDE [185] 0 (0); 0 (0); 1 (7)	103
1724. CANDLEMASS [174] 0 (0); 0 (0); 1 (6)	102
1725. SALSOUL Orchestra [170] 0 (0); 0 (0); 1 (5)	102
1726. Razzy BAILEY [176] 0 (0); 0 (0); 2 (6)	101
1727. SIDEWINDERS [169] 0 (0); 0 (0); 1 (5)	100
1728. The FLESHTONES [174] 0 (0); 0 (0); 1 (5)	99.5
1729. Bunny DeBARGE [172] 0 (0); 0 (0); 1 (5)	98.7
1730. Carrie LUCAS [180] 0 (0); 0 (0); 2 (6)	97.5
1731. RANK AND FILE [165] 0 (0); 0 (0); 1 (5)	97.4
1732. Isao TOMITA [174] 0 (0); 0 (0); 1 (5)	96.9
1733. Maynard FERGUSON [185] 0 (0); 0 (0); 2 (6)	96.5
1734. PETER, PAUL & MARY [173] 0 (0); 0 (0); 1 (5)	96.1
1735. John O'BANION [164] 0 (0); 0 (0); 1 (4)	95.4
1736. BROS [171] 0 (0); 0 (0); 1 (5)	94.7
1737. GREEN ON RED [177] 0 (0); 0 (0); 1 (6)	93.9
1738. Ahmad JAMAL Trio [173] 0 (0); 0 (0); 1 (5)	93.9
1739. ROCKIN' SIDNEY [166] 0 (0); 0 (0); 1 (4)	92.7
1740. INNER CITY [162] 0 (0); 0 (0); 1 (4)	91.4
1741. Kenny ROGERS & Dottie WEST [170] 0 (0); 0 (0); 2 (5)	91.4
1742. XYZ [144] 0 (0); 0 (0); 1 (3)	91.2
1743. B.T. EXPRESS [164] 0 (0); 0 (0); 1 (4)	91.0
1744. Jay FERGUSON [178] 0 (0); 0 (0); 1 (5)	91.0
1745. Ian DURY And The BLOCKHEADS [159] 0 (0); 0 (0); 1 (5)	90.1
1746. ANIMAL LOGIC [147] 0 (0); 0 (0); 1 (4)	90.1
1747. PREFAB SPROUT [178] 0 (0); 0 (0); 1 (5)	89.7
1748. Dave VALENTIN [184] 0 (0); 0 (0); 2 (6)	89.7
1749. Tommy PAGE [166] 0 (0); 0 (0); 1 (5)	89.2
1750. Jerry Jeff WALKER [185] 0 (0); 0 (0); 2 (6)	88.9
1751. Floyd CRAMER [170] 0 (0); 0 (0); 1 (5)	88.8
1752. The FEELIES [173] 0 (0); 0 (0); 1 (5)	88.8
1753. MALICE [177] 0 (0); 0 (0); 1 (6)	88.5
1754. The WOODENTOPS [185] 0 (0); 0 (0); 1 (6)	87.6
1755. WISHBONE ASH [179] 0 (0); 0 (0); 2 (6)	87.6
1756. Fred KNOBLOCK [179] 0 (0); 0 (0); 1 (5)	86.6
1757. 999 [177] 0 (0); 0 (0); 2 (5)	86.6
1758. John McLAUGHLIN [172] 0 (0); 0 (0); 1 (4)	85.9
1759. CACTUS WORLD NEWS [179] 0 (0); 0 (0); 1 (5)	85.2

Album Acts Ranked

Rank. Act [Hi Peak] T10 (Wk); T40 (Wk); Ent (Wk)	Score In '80s
1760. DREAM SYNDICATE [171] 0 (0); 0 (0); 1 (4)	85.2
1761. Noel POINTER [167] 0 (0); 0 (0); 1 (4)	85.2
1762. Steve HOWE [164] 0 (0); 0 (0); 1 (4)	83.7
1763. Janis IAN [156] 0 (0); 0 (0); 1 (3)	83.1
1764. James GALWAY [150] 0 (0); 0 (0); 1 (6)	83.0
1765. Cleo LAINE [150] 0 (0); 0 (0); 1 (6)	83.0
1766. GOANNA [179] 0 (0); 0 (0); 1 (5)	82.1
1767. ESQUIRE [165] 0 (0); 0 (0); 1 (4)	82.0
1768. STEPPENWOLF [171] 0 (0); 0 (0); 1 (4)	82.0
1769. DEJA [186] 0 (0); 0 (0); 1 (6)	80.9
1770. The NIGHTHAWKS [166] 0 (0); 0 (0); 1 (4)	80.6
1771. David Allan COE [179] 0 (0); 0 (0); 1 (5)	80.3
1772. LEROI BROS. [181] 0 (0); 0 (0); 1 (5)	80.3
1773. The SELECTER [175] 0 (0); 0 (0); 1 (4)	78.8
1774. Jaco PASTORIUS [161] 0 (0); 0 (0); 1 (3)	78.7
1775. Steve JONES [169] 0 (0); 0 (0); 1 (4)	78.4
1776. Frank MILLS [149] 0 (0); 0 (0); 1 (3)	78.3
1777. GIRLSCHOOL [182] 0 (0); 0 (0); 1 (5)	78.0
1778. YUTAKA [174] 0 (0); 0 (0); 1 (4)	77.8
1779. Robin LANE & The CHARTBUSTERS [172] 0 (0); 0 (0); 1 (4)	77.4
1780. OZARK MOUNTAIN DAREDEVILS [170] 0 (0); 0 (0); 1 (4)	77.4
1781. Stanley TURRENTINE [162] 0 (0); 0 (0); 1 (3)	76.8
1782. The STRANGLERS [172] 0 (0); 0 (0); 1 (4)	76.5
1783. WRATHCHILD AMERICA [190] 0 (0); 0 (0); 1 (6)	76.3
1784. Bill CHAMPLIN [178] 0 (0); 0 (0); 1 (4)	75.6
1785. Maxine NIGHTINGALE [176] 0 (0); 0 (0); 1 (4)	75.6
1786. Richie HAVENS [173] 0 (0); 0 (0); 1 (4)	75.2
1787. GIANT STEPS [184] 0 (0); 0 (0); 1 (5)	74.9
1788. Peter GREEN [186] 0 (0); 0 (0); 1 (5)	74.9
1789. LINX [175] 0 (0); 0 (0); 1 (4)	74.7
1790. BURNING SENSATIONS [175] 0 (0); 0 (0); 1 (4)	74.4
1791. CRACK THE SKY [186] 0 (0); 0 (0); 1 (5)	74.0
1792. FELONY [185] 0 (0); 0 (0); 1 (5)	73.1
1793. Alicia MYERS [186] 0 (0); 0 (0); 1 (5)	73.0
1794. OHIO PLAYERS [165] 0 (0); 0 (0); 1 (3)	72.9
1795. VISAGE [178] 0 (0); 0 (0); 1 (4)	72.4
1796. RED FLAG [178] 0 (0); 0 (0); 1 (4)	72.0
1797. NEW ENGLAND [176] 0 (0); 0 (0); 1 (4)	71.5
1798. Gerald ALBRIGHT [181] 0 (0); 0 (0); 1 (5)	71.3
1799. Nick HEYWARD [178] 0 (0); 0 (0); 1 (4)	71.1
1800. LACE [187] 0 (0); 0 (0); 1 (5)	70.5
1801. DAVE & SUGAR [179] 0 (0); 0 (0); 1 (4)	70.3
1802. Joanie GREGGAINS [177] 0 (0); 0 (0); 1 (4)	70.3
1803. McGUFFEY LANE [193] 0 (0); 0 (0); 1 (6)	70.0
1804. Eddie SCHWARTZ [195] 0 (0); 0 (0); 1 (6)	69.6
1805. Jenny BURTON [181] 0 (0); 0 (0); 1 (4)	69.4
1806. JULUKA [186] 0 (0); 0 (0); 1 (5)	69.0
1807. Kris KRISTOFFERSON [152] 0 (0); 0 (0); 1 (5)	68.7
1808. Jean KNIGHT [180] 0 (0); 0 (0); 1 (4)	68.5
1809. T-Bone BURNETT [188] 0 (0); 0 (0); 1 (5)	67.6
1810. Shawn COLVIN [165] 0 (0); 0 (0); 1 (3)	64.3
1811. Gloria GAYNOR [178] 0 (0); 0 (0); 1 (4)	64.3
1812. Joe DOLCE [181] 0 (0); 0 (0); 1 (4)	64.0
1813. VICTORY [182] 0 (0); 0 (0); 1 (5)	63.6
1814. VOIVOD [168] 0 (0); 0 (0); 1 (3)	63.3
1815. EBN-OZN [185] 0 (0); 0 (0); 1 (4)	63.1
1816. Scott BAIO [181] 0 (0); 0 (0); 1 (4)	62.6
1817. SAVOY BROWN [185] 0 (0); 0 (0); 1 (4)	62.6
1818. 10cc [180] 0 (0); 0 (0); 2 (4)	62.5
1819. Danny DAVIS [150] 0 (0); 0 (0); 1 (5)	62.4
1820. Gary STEWART [165] 0 (0); 0 (0); 1 (4)	62.0
1821. Joe PISCOPO [168] 0 (0); 0 (0); 1 (3)	61.5
1822. Denise LOPEZ [184] 0 (0); 0 (0); 1 (4)	61.2
1823. Wayne TOUPS & ZYDECAJUN [183] 0 (0); 0 (0); 1 (4)	60.4
1824. SIR DOUGLAS Quintet [184] 0 (0); 0 (0); 1 (4)	59.8
1825. AUDIO TWO [185] 0 (0); 0 (0); 1 (4)	59.0
1826. AMAZING RHYTHM ACES [175] 0 (0); 0 (0); 1 (3)	58.9
1827. BRECKER BROTHERS [176] 0 (0); 0 (0); 1 (3)	57.9
1828. The STRIKERS [174] 0 (0); 0 (0); 1 (3)	0.0
1829. WHAT IS THIS [187] 0 (0); 0 (0); 1 (4)	57.3
1830. Dickey BETTS [187] 0 (0); 0 (0); 1 (4)	56.8
1831. HOLLY And The ITALIANS [177] 0 (0); 0 (0); 1 (3)	56.7
1832. BALAAM AND THE ANGEL [174] 0 (0); 0 (0); 1 (3)	56.6
1833. SHINEHEAD [185] 0 (0); 0 (0); 1 (4)	56.3
1834. Mary Chapin CARPENTER [186] 0 (0); 0 (0); 1 (4)	56.2
1835. Tom JONES [179] 0 (0); 0 (0); 1 (3)	56.2
1836. Rick FENN [154] 0 (0); 0 (0); 1 (3)	55.4
1837. France JOLI [175] 0 (0); 0 (0); 1 (4)	55.3
1838. ART IN AMERICA [176] 0 (0); 0 (0); 1 (3)	54.3
1839. Glen CAMPBELL [178] 0 (0); 0 (0); 1 (3)	53.4
1840. Joe VITALE [181] 0 (0); 0 (0); 1 (4)	53.1
1841. SCHOOLLY D [180] 0 (0); 0 (0); 1 (4)	53.0
1842. KANO [189] 0 (0); 0 (0); 1 (4)	52.7
1843. Keith EMERSON [183] 0 (0); 0 (0); 1 (3)	52.1
1844. Tom VERLAINE [177] 0 (0); 0 (0); 1 (3)	52.1
1845. Queen LATIFAH [174] 0 (0); 0 (0); 1 (3)	51.2
1846. The PRODUCERS [163] 0 (0); 0 (0); 1 (2)	51.1
1847. VITAMIN Z [183] 0 (0); 0 (0); 1 (3)	50.8
1848. Laura NYRO [182] 0 (0); 0 (0); 1 (3)	49.4
1849. Debbie REYNOLDS [182] 0 (0); 0 (0); 1 (3)	49.4
1850. GILLAN [183] 0 (0); 0 (0); 1 (3)	49.0
1851. Chris JASPER [182] 0 (0); 0 (0); 1 (3)	48.0
1852. Mick JONES [184] 0 (0); 0 (0); 1 (3)	47.6
1853. Nancy MARTINEZ [178] 0 (0); 0 (0); 1 (3)	47.1
1854. Harvey MASON [186] 0 (0); 0 (0); 1 (3)	46.3
1855. RENAISSANCE [196] 0 (0); 0 (0); 1 (4)	46.3
1856. DEODATO [186] 0 (0); 0 (0); 1 (3)	45.8
1857. MONTANA Orchestra [195] 0 (0); 0 (0); 1 (3)	45.6
1858. Arlo GUTHRIE [184] 0 (0); 0 (0); 1 (3)	45.3
1859. DA'KRASH [184] 0 (0); 0 (0); 1 (3)	44.9
1860. LIONS AND GHOSTS [187] 0 (0); 0 (0); 1 (3)	43.2
1861. Clarence CARTER [189] 0 (0); 0 (0); 1 (3)	42.7
1862. Gary MYRICK [186] 0 (0); 0 (0); 1 (3)	42.6
1863. Lillo THOMAS [186] 0 (0); 0 (0); 1 (3)	41.8
1864. THRILLS [199] 0 (0); 0 (0); 1 (4)	41.0
1865. Father Guido SARDUCCI [179] 0 (0); 0 (0); 1 (2)	39.0
1866. Moe BANDY [170] 0 (0); 0 (0); 1 (4)	37.4
1867. Joe STAMPLEY [170] 0 (0); 0 (0); 1 (4)	37.4
1868. JIVE BUNNY & The MASTERMIXERS [140] 0 (0); 0 (0); 1 (1)	37.0
1869. B.J. THOMAS [193] 0 (0); 0 (0); 1 (3)	35.1
1870. Jimmy HALL [183] 0 (0); 0 (0); 1 (2)	34.9
1871. Tony BANKS [171] 0 (0); 0 (0); 1 (2)	34.5
1872. David HUDSON [184] 0 (0); 0 (0); 1 (2)	34.0
1873. TSOL [184] 0 (0); 0 (0); 1 (2)	34.0
1874. Bruce WOOLLEY & The CAMERA CLUB [184] 0 (0); 0 (0); 1 (2)	34.0
1875. The UNFORGIVEN [185] 0 (0); 0 (0); 1 (2)	33.6
1876. Lenny WILLIAMS [185] 0 (0); 0 (0); 1 (2)	33.6
1877. WAYSTED [185] 0 (0); 0 (0); 1 (2)	33.1
1878. Carl WILSON [185] 0 (0); 0 (0); 1 (2)	32.7
1879. STARGARD [186] 0 (0); 0 (0); 1 (2)	32.6
1880. GREY And HANKS [195] 0 (0); 0 (0); 1 (3)	32.3
1881. ROSE TATTOO [197] 0 (0); 0 (0); 1 (3)	32.3
1882. RODNEY O & Joe COOLEY [187] 0 (0); 0 (0); 1 (2)	31.8
1883. Steve WARINER [187] 0 (0); 0 (0); 1 (2)	31.8
1884. Chubby CHECKER [186] 0 (0); 0 (0); 1 (2)	31.3
1885. TANTRUM [199] 0 (0); 0 (0); 1 (3)	31.0
1886. CIRCUS OF POWER [185] 0 (0); 0 (0); 1 (2)	30.9
1887. Leon RUSSELL [187] 0 (0); 0 (0); 1 (2)	30.9
1888. CONEY HATCH [186] 0 (0); 0 (0); 1 (2)	30.8
1889. Walter EGAN [187] 0 (0); 0 (0); 1 (2)	30.4
1890. MINOR DETAIL [187] 0 (0); 0 (0); 1 (2)	30.4
1891. Meryl STREEP [180] 0 (0); 0 (0); 1 (4)	30.2
1892. Jesse WINCHESTER [188] 0 (0); 0 (0); 1 (2)	29.9
1893. Ike & Tina TURNER [189] 0 (0); 0 (0); 1 (2)	29.5

Album Acts Ranked

Rank. Act [Hi Peak] T10 (Wk); T40 (Wk); Ent (Wk)	Score In '80s
1894. ROMEO'S DAUGHTER [191] 0 (0); 0 (0); 1 (2)	28.2
1895. The SEARCHERS [191] 0 (0); 0 (0); 1 (2)	28.2
1896. ASLEEP AT THE WHEEL [191] 0 (0); 0 (0); 1 (2)	27.7
1897. Michael CRAWFORD [192] 0 (0); 0 (0); 1 (2)	27.2
1898. ICON [190] 0 (0); 0 (0); 1 (2)	27.2
1899. SNIFF 'N' The TEARS [192] 0 (0); 0 (0); 1 (2)	27.2
1900. Muddy WATERS [192] 0 (0); 0 (0); 1 (2)	27.2
1901. BACHMAN-TURNER OVERDRIVE [191] 0 (0); 0 (0); 1 (2)	26.8
1902. BELLE STARS [191] 0 (0); 0 (0); 1 (2)	26.4
1903. Bill BRUFORD [191] 0 (0); 0 (0); 1 (2)	26.4
1904. Eric MARTIN [191] 0 (0); 0 (0); 1 (2)	26.4
1905. Steven WRIGHT [192] 0 (0); 0 (0); 1 (2)	26.3
1906. ANVIL [191] 0 (0); 0 (0); 1 (2)	25.9
1907. Lonnie Liston SMITH [193] 0 (0); 0 (0); 1 (2)	25.5
1908. Michael ANDERSON [194] 0 (0); 0 (0); 1 (2)	25.4
1909. ENGLAND DAN & John Ford COLEY [194] 0 (0); 0 (0); 1 (2)	25.4
1910. The NAILS [194] 0 (0); 0 (0); 1 (2)	25.4
1911. ARABIAN PRINCE [193] 0 (0); 0 (0); 1 (2)	24.6
1912. Paul DEAN [195] 0 (0); 0 (0); 1 (2)	24.6
1913. SHY [193] 0 (0); 0 (0); 1 (2)	24.1
1914. TRINERE & FRIENDS [196] 0 (0); 0 (0); 1 (2)	23.6
1915. Ray LYNCH [197] 0 (0); 0 (0); 1 (2)	22.8
1916. Kasim SULTON [197] 0 (0); 0 (0); 1 (2)	22.8
1917. KWICK [197] 0 (0); 0 (0); 1 (2)	21.9
1918. Boris GREBENSHIKOV [198] 0 (0); 0 (0); 1 (2)	21.8
1919. Ian McCULLOCH [179] 0 (0); 0 (0); 1 (1)	19.5
1920. Mark KNOPFLER [180] 0 (0); 0 (0); 1 (1)	19.0
1921. KILLING JOKE [194] 0 (0); 0 (0); 1 (1)	12.7
1922. Pat McLAUGHLIN [195] 0 (0); 0 (0); 1 (1)	12.3
1923. Henry MANCINI and his Orchestra [197] 0 (0); 0 (0); 1 (2)	11.4
1924. MC SHY D [197] 0 (0); 0 (0); 1 (1)	11.4
1925. Billy BRAGG [198] 0 (0); 0 (0); 1 (1)	10.9
1926. NRBQ [198] 0 (0); 0 (0); 1 (1)	10.9
1927. BOBBY JIMMY & The CRITTERS [200] 0 (0); 0 (0); 1 (1)	10.0

Album Acts Ranked

Album Acts Special Lists

Acts: Top 50 Each Year

Acts: Yearly Top 50s Through the Decade

Acts: 150 or Total Chart Weeks

Acts: 6 or More Chart Appearances

Acts: Top 10 Entries

Acts: 3 or More Top 40 Entries

Acts: More than 50 Consecutive Weeks on Chart

Acts: Most Albums on Chart Simultaneously

Acts: 2 Albums in Top 10 Simultaneously

Acts: 3 Albums on Chart Simultaneously

Acts: Number 1 Albums

Album Acts: Top 50 Each Year

	1980	1981	1982	1983	1984
1	PINK FLOYD	REO SPEEDWAGON	ASIA	Michael JACKSON	Lionel RICHIE
2	Bob SEGER And The SILVER BULLET Band	Pat BENATAR	The GO-GO'S	MEN AT WORK	Michael JACKSON
3	Billy JOEL	AC/DC	LOVERBOY	DEF LEPPARD	Huey LEWIS And The NEWS
4	Kenny ROGERS	STYX	JOURNEY	The POLICE	PRINCE And The REVOLUTION
5	Pat BENATAR	JOURNEY	John MELLENCAMP	Daryl HALL & John OATES	VAN HALEN
6	Michael JACKSON	Kenny ROGERS	J. GEILS Band	JOURNEY	Bruce SPRINGSTEEN
7	QUEEN	The POLICE	FLEETWOOD MAC	Lionel RICHIE	CULTURE CLUB
8	Tom PETTY And The HEARTBREAKERS	ROLLING STONES	Willie NELSON	DURAN DURAN	DURAN DURAN
9	ROLLING STONES	FOREIGNER	FOREIGNER	STRAY CATS	The CARS
10	Christopher CROSS	John LENNON & Yoko ONO	Joan JETT & The BLACKHEARTS	CULTURE CLUB	Cyndi LAUPER
11	EAGLES	RUSH	Rick SPRINGFIELD	Billy JOEL	Billy JOEL
12	Diana ROSS	Neil DIAMOND	The POLICE	David BOWIE	Billy IDOL
13	Barbra STREISAND	Kim CARNES	Daryl HALL & John OATES	PRINCE	ZZ TOP
14	Donna SUMMER	MOODY BLUES	Billy SQUIER	Kenny ROGERS	Tina TURNER
15	Linda RONSTADT	Daryl HALL & John OATES	ROLLING STONES	QUIET RIOT	POINTER SISTERS
16	Jackson BROWNE	Steve WINWOOD	VANGELIS	ALABAMA	MADONNA
17	Eric CLAPTON	Stevie NICKS	ROYAL PHILHARMONIC Orchestra	Pat BENATAR	The POLICE
18	Dan FOGELBERG	Rick JAMES	ALABAMA	Stevie NICKS	NIGHT RANGER
19	Willie NELSON	Grover WASHINGTON Jr.	Olivia NEWTON-JOHN	LOVERBOY	SCORPIONS
20	AC/DC	Christopher CROSS	MEN AT WORK	The FIXX	Daryl HALL & John OATES
21	The CARS	AIR SUPPLY	GENESIS	ZZ TOP	John MELLENCAMP
22	The COMMODORES	Barbra STREISAND	Stevie NICKS	Bob SEGER	Julio IGLESIAS
23	Bette MIDLER	Bob SEGER And The SILVER BULLET Band	Ozzy OSBOURNE	STYX	MÖTLEY CRÜE
24	JOURNEY	Billy SQUIER	AC/DC	TOTO	EURYTHMICS
25	Boz SCAGGS	Phil COLLINS	SURVIVOR	Bryan ADAMS	RATT
26	Kenny LOGGINS	BLONDIE	Steve MILLER Band	Willie NELSON	YES
27	Pete TOWNSHEND	KOOL & The GANG	Paul McCARTNEY	Thomas DOLBY	Kenny ROGERS
28	Bruce SPRINGSTEEN	Rick SPRINGFIELD	HUMAN LEAGUE	PINK FLOYD	The PRETENDERS
29	Stevie WONDER	Tom PETTY And The HEARTBREAKERS	Neil DIAMOND	Jane FONDA	U2
30	The PRETENDERS	Alan PARSONS PROJECT	REO SPEEDWAGON	Phil COLLINS	Linda RONSTADT
31	George BENSON	Stevie WONDER	KOOL & The GANG	U2	QUIET RIOT
32	VAN HALEN	ALABAMA	TOTO	Robert PLANT	ALABAMA
33	The WHISPERS	SANTANA	The CLASH	Rick SPRINGFIELD	THOMPSON TWINS
34	GENESIS	The WHO	Juice NEWTON	TALKING HEADS	CHICAGO
35	STYX	Diana ROSS	STRAY CATS	AIR SUPPLY	Elton JOHN
36	BLONDIE	QUEEN	Alan PARSONS PROJECT	John MELLENCAMP	TWISTED SISTER
37	DOOBIE BROTHERS	LOVERBOY	Dan FOGELBERG	Linda RONSTADT	DEF LEPPARD
38	Smokey ROBINSON	Ozzy OSBOURNE	VAN HALEN	Bonnie TYLER	JACKSON 5
39	LIPPS, INC.	Willie NELSON	QUARTERFLASH	MISSING PERSONS	Jeffrey OSBORNE
40	SUPERTRAMP	Juice NEWTON	AIR SUPPLY	Al JARREAU	Steve PERRY
41	KOOL & The GANG	Quincy JONES	Kenny ROGERS	IRON MAIDEN	Stevie WONDER
42	Jermaine JACKSON	The PRETENDERS	A FLOCK OF SEAGULLS	Tom PETTY And The HEARTBREAKERS	Barbra STREISAND
43	Waylon JENNINGS	Rickie Lee JONES	CROSBY, STILLS & NASH	Donna SUMMER	GENESIS
44	BROTHERS JOHNSON	Billy JOEL	Aldo NOVA	Joe JACKSON	MIDNIGHT STAR
45	RUSH	STEELY DAN	Diana ROSS	EURYTHMICS	PRINCE
46	FLEETWOOD MAC	Eddie RABBITT	The CARS	The CLASH	John CAFFERTY And The BEAVER BROWN BAND
47	Teddy PENDERGRASS	38 SPECIAL	RUSH	Eddy GRANT	John WAITE
48	BEE GEES	Bruce SPRINGSTEEN	CHICAGO	ASIA	Laura BRANIGAN
49	Stephanie MILLS	Smokey ROBINSON	Robert PLANT	Olivia NEWTON-JOHN	Billy SQUIER
50	HEART	The COMMODORES	Joe JACKSON	Sammy HAGAR	Rick SPRINGFIELD

Album Acts: Top 50 Each Year

	1985	1986	1987	1988	1989
1	Bruce SPRINGSTEEN	Whitney HOUSTON	BON JOVI	George MICHAEL	NEW KIDS ON THE BLOCK
2	Phil COLLINS	Janet JACKSON	U2	DEF LEPPARD	GUNS N' ROSES
3	MADONNA	SADE	Whitney HOUSTON	GUNS N' ROSES	Bobby BROWN
4	Bryan ADAMS	HEART	WHITESNAKE	INXS	Paula ABDUL
5	PRINCE And The REVOLUTION	MADONNA	BEASTIE BOYS	Michael JACKSON	MILLI VANILLI
6	TEARS FOR FEARS	VAN HALEN	Paul SIMON	TIFFANY	FINE YOUNG CANNIBALS
7	WHAM!	John MELLENCAMP	MADONNA	POISON	MADONNA
8	DIRE STRAITS	Bob SEGER And The SILVER BULLET Band	POISON	U2	Debbie GIBSON
9	Tina TURNER	Bruce SPRINGSTEEN	Bruce HORNSBY And The RANGE	Debbie GIBSON	Tom PETTY
10	John FOGERTY	ZZ TOP	EUROPE	Tracy CHAPMAN	BON JOVI
11	Whitney HOUSTON	RUN-D.M.C.	HEART	VAN HALEN	Richard MARX
12	STING	BON JOVI	Janet JACKSON	Steve WINWOOD	SKID ROW
13	POINTER SISTERS	DIRE STRAITS	Anita BAKER	Terence Trent D'ARBY	Bette MIDLER
14	Billy OCEAN	Barbra STREISAND	KENNY G	Gloria ESTEFAN/ MIAMI SOUND MACHINE	TRAVELING WILBURYS
15	KOOL & The GANG	MR. MISTER	GENESIS	Richard MARX	TONE LŌC
16	HEART	Peter GABRIEL	Bruce SPRINGSTEEN	Rick ASTLEY	LIVING COLOUR
17	Billy JOEL	Phil COLLINS	CINDERELLA	AEROSMITH	PRINCE
18	Don HENLEY	GENESIS	Michael JACKSON	BON JOVI	DEF LEPPARD
19	FOREIGNER	Billy OCEAN	FLEETWOOD MAC	DJ JAZZY JEFF & THE FRESH PRINCE	Janet JACKSON
20	SADE	Lionel RICHIE	Los LOBOS	Whitney HOUSTON	WARRANT
21	Daryl HALL & John OATES	Robert PALMER	Steve WINWOOD	Robert PLANT	GREAT WHITE
22	POWER STATION	Patti LaBELLE	MÖTLEY CRÜE	Anita BAKER	Anita BAKER
23	U2	The OUTFIELD	Huey LEWIS And The NEWS	John MELLENCAMP	ROLLING STONES
24	MÖTLEY CRÜE	Huey LEWIS And The NEWS	LL COOL J	Keith SWEAT	Edie BRICKELL & The NEW BOHEMIANS
25	NEW EDITION	Steve WINWOOD	LISA LISA And CULT JAM	Elton JOHN	WINGER
26	John MELLENCAMP	Gloria ESTEFAN/ MIAMI SOUND MACHINE	BOSTON	CINDERELLA	Roy ORBISON
27	TALKING HEADS	BOSTON	DEF LEPPARD	WHITE LION	The CURE
28	CHICAGO	JEFFERSON AIRPLANE/STARSHIP	BANGLES	PINK FLOYD	MÖTLEY CRÜE
29	EURYTHMICS	Billy JOEL	Robert CRAY Band	Bobby BROWN	POISON
30	REO SPEEDWAGON	JOURNEY	Billy IDOL	CHEAP TRICK	Tracy CHAPMAN
31	Stevie WONDER	FABULOUS THUNDERBIRDS	EXPOSÉ	Belinda CARLISLE	R.E.M.
32	NIGHT RANGER	The MONKEES	PINK FLOYD	SCORPIONS	Don HENLEY
33	Howard JONES	BANGLES	CLUB NOUVEAU	WHITESNAKE	KENNY G
34	Lionel RICHIE	David Lee ROTH	CROWDED HOUSE	Bruce SPRINGSTEEN	Melissa ETHERIDGE
35	Tom PETTY And The HEARTBREAKERS	PRINCE And The REVOLUTION	FAT BOYS	Randy TRAVIS	The CULT
36	SURVIVOR	MOODY BLUES	Bryan ADAMS	MIDNIGHT OIL	SOUL II SOUL
37	RATT	PET SHOP BOYS	Jody WATLEY	SADE	WHITE LION
38	Freddie JACKSON	TALKING HEADS	Lionel RICHIE	Bobby McFERRIN	Rod STEWART
39	USA For AFRICA	Ozzy OSBOURNE	Luther VANDROSS	Bruce HORNSBY And The RANGE	Gloria ESTEFAN/ MIAMI SOUND MACHINE
40	DeBARGE	SIMPLE MINDS	Randy TRAVIS	Taylor DAYNE	AEROSMITH
41	David Lee ROTH	Stevie WONDER	RUN-D.M.C.	NEW EDITION	METALLICA
42	Paul YOUNG	FALCO	PRINCE	KENNY G	CHER
43	Julian LENNON	INXS	John MELLENCAMP	David Lee ROTH	Karyn WHITE
44	Cyndi LAUPER	MIKE + THE MECHANICS	GEORGIA SATELLITES	STING	MC HAMMER
45	Huey LEWIS And The NEWS	Anita BAKER	Freddie JACKSON	AL B. SURE!	The B-52's
46	Aretha FRANKLIN	TEARS FOR FEARS	GRATEFUL DEAD	PEBBLES	10,000 MANIACS
47	George THOROGOOD & The DESTROYERS	Tina TURNER	Peter GABRIEL	SALT-N-PEPA	U2
48	SCORPIONS	ROLLING STONES	CAMEO	George HARRISON	GUY
49	ZZ TOP	The HOOTERS	Gloria ESTEFAN/ MIAMI SOUND MACHINE	Rod STEWART	Jody WATLEY
50	Luther VANDROSS	Stevie NICKS	R.E.M.	Joe SATRIANI	BANGLES

Album Acts: All Yearly Top 50s Through The Decade

All acts appearing in any yearly Top 50 are shown here with their rank each year.
1-100 are listed numerically; 101-200 = "A"; 201-300= "B" and greater than 301= "C".
1980s Charted albums includes recurrent pre-1980 releases. The number of charted acts in the year is represented by "n".

Act	'80s Rank	1980s Charted	1980 n=465	1981 n=605	1982 n=522	1983 n=508	1984 n=481	1985 n=467	1986 n=505	1987 n=531	1988 n=574	1989 n=580	1990s Releases
Paula ABDUL	115	1									B	4	3
AC/DC	24	10	20	3	24	88	B	A	95	C	70		1
Bryan ADAMS	37	4			C	25	A	4	98	36			5
AEROSMITH	87	8	98	B	A	C		B	A	58	17	40	3
AIR SUPPLY	72	6	71	21	40	35	A	A	C				0
ALABAMA	28	12	B	32	18	16	32	74	55	A	A	A	3
AL B. SURE!	353	1									45	B	2
ASIA	95	3			1	48	C	C	B				0
Rick ASTLEY	205	2									16	86	2
Anita BAKER	58	3				C	C		45	13	22	22	2
BANGLES	118	3					B	C	33	28	A	50	1
BEASTIE BOYS	122	2							B	5	C	A	7
BEE GEES	218	7	48	B	C	51	C			C		B	1
Pat BENATAR	16	9	5	2	54	17	79	88	A		92	C	3
George BENSON	139	8	31	A	87	73	C	A	B	A	B	C	1
The B-52's	170	7	53	A	A	A			B	C		45	3
BLONDIE	155	5	36	26	A								1
BON JOVI	13	4					88	78	12	1	18	10	5
BOSTON	160	3							27	26			2
David BOWIE	94	11	100	A	B	12	54	B	C	A		C	4
Laura BRANIGAN	215	5			A	62	48	A		B	C		1
Edie BRICKELL & The NEW BOHEMIANS	263	1									A	24	2
BROTHERS JOHNSON	334	4	44	B		C	B						0
Bobby BROWN	62	3							C	B	29	3	2
Jackson BROWNE	133	4	16	B		55	B		77			A	3
John CAFFERTY And The BEAVER BROWN BAND	247	3				B	46	60	C			C	0
CAMEO	123	9	80	A	A	B	A	A	78	48	B	C	1
Belinda CARLISLE	179	3							62	A	31	A	0
Kim CARNES	172	6	B	13	A	B	C	B	C				0
The CARS	36	7	21	A	46		9	54	76	A	C		0
Tracy CHAPMAN	142	2									10	30	2
CHEAP TRICK	157	9	88	B	98	B		A	C		30	B	4
CHER	277	2								C	73	42	4
CHICAGO	82	7	B	C	48	B	34	28	A	A	100	A	5
CINDERELLA	83	2							51	17	26	55	2
Eric CLAPTON	96	8	17	62	B	91		A	C	94	A	A	9
The CLASH	145	6	78	A	33	46		C	C		C		1
CLUB NOUVEAU	324	2							C	33	C		0
Phil COLLINS	19	4		25	A	30	B	2	17	A		A	4
The COMMODORES	116	8	22	50	A	A		51	C	C			0
Robert CRAY Band	246	5							B	29	A	C	6
CROSBY, STILLS & NASH	276	3		C	43	96							3
Christopher CROSS	90	3	10	20	B	72	B	C	C				0
CROWDED HOUSE	291	2								C	34	A	2
The CULT	250	3						C	A	92		35	2
CULTURE CLUB	56	4				10	7	A	A				2
The CURE	152	7				C	C	B	73	68	A	27	6
Terence Trent D'ARBY	222	2								B	13	C	2
Taylor DAYNE	281	2									40	85	1
DeBARGE	204	3			C	52	A	40	C				0
DEF LEPPARD	6	5	B	A	B	3	37			27	2	18	5
Neil DIAMOND	52	10	55	12	29	66	A	C	88	C	B	B	5
DIRE STRAITS	49	6	A	85	A	A	A	8	13	C	B	C	2
DJ JAZZY JEFF & THE FRESH PRINCE	224	3								B	19	A	3
Thomas DOLBY	256	4			27	A					B		0
DOOBIE BROTHERS	206	6	37	99	B	C						87	1

387

Album Acts: Sp'l Lists

Act	'80s Rank	1980s Charted	1980 n=465	1981 n=605	1982 n=522	1983 n=508	1984 n=481	1985 n=467	1986 n=505	1987 n=531	1988 n=574	1989 n=580	1990s Releases
DURAN DURAN	32	8			A	8	8	57	C	61	A	A	5
EAGLES	154	5	11	75	B	B							1
Gloria ESTEFAN/MIAMI SOUND MACHINE	77	3						C	26	49	14	39	8
Melissa ETHERIDGE	308	2									A	34	3
EUROPE	156	2							C	10	65	C	0
EURYTHMICS	74	8				45	24	29	57	C	A	A	2
EXPOSÉ	187	2								31	67	96	1
FABULOUS THUNDERBIRDS	285	4			C				31	A		C	0
FALCO	352	2				B			42				0
FAT BOYS	186	7						94	A	35	83	C	0
FINE YOUNG CANNIBALS	129	2						A				6	0
The FIXX	151	6			C	20	60	B	A	C		B	1
FLEETWOOD MAC	42	7	46	96	7	B				19	74	89	2
A FLOCK OF SEAGULLS	237	3			42	53	B						0
Dan FOGELBERG	78	6	18	52	37	93	72	A		A			1
John FOGERTY	183	2						10	A	C			1
Jane FONDA	217	3			88	29	A						0
FOREIGNER	30	7	A	9	9	54		19		C	55		3
Aretha FRANKLIN	119	8	B	A	A	A		46	64	83	B	B	3
Peter GABRIEL	98	6	74	C	A	A		C	16	47		A	3
J. GEILS Band	111	4	57	A	6	A	C	C					0
GENESIS	27	6	34	78	21	85	43		18	15	C		4
GEORGIA SATELLITES	322	3							B	44	B	C	0
Debbie GIBSON	84	2								A	9	8	2
The GO-GO'S	92	3		76	2	C	51						1
Eddy GRANT	397	2				47	B						0
GRATEFUL DEAD	221	6	A	A					46	C	A		8
GREAT WHITE	169	5					C		B	73	78	21	4
GUNS N' ROSES	23	2								A	3	2	4
GUY	357	1									A	48	1
Sammy HAGAR	161	7	B		79	50	81	B		70			3
Daryl HALL & John OATES	12	8	66	15	13	5	20	21	C		93		2
George HARRISON	271	4		A	C	C				A	48	C	0
HEART	26	7	50	94	A	A	C	16	4	11	A		5
Don HENLEY	128	3			97	A	B	18	C			32	1
The HOOTERS	227	3						70	49	A	C	C	0
Bruce HORNSBY And The RANGE	110	2							56	9	39		4
Whitney HOUSTON	14	2						11	1	3	20	C	4
HUMAN LEAGUE	174	5			28	92	B		A	B			0
Billy IDOL	71	5		C	B	77	12	A	A	30	A		2
Julio IGLESIAS	177	8				A	22	A			B		2
INXS	60	6				A	A	B	43	A	4	A	6
IRON MAIDEN	107	8		A	A	41	83	A	61	A	75		6
Freddie JACKSON	148	3						38	66	45	A	C	4
Janet JACKSON	48	4			C	B	C		2	12	C	19	3
Jermaine JACKSON	166	7	42	A	A		53	A	A			C	0
Joe JACKSON	124	10	A	A	50	44	61		A	C	C	B	1
Michael JACKSON	1	6	6	B	C	1	2	B		18	5	A	3
JACKSON 5	192	5	81	95	A		38	C				B	0
Rick JAMES	113	9	B	18	55	75	A	A	C		C		1
Al JARREAU	127	7	A	51	84	40	B	A	B	C	C	B	2
JEFFERSON AIRPLANE/STARSHIP	79	9	62	73	A	A	96	93	28	76	C	A	0
Waylon JENNINGS	259	6	43	A	A	C							2
Joan JETT & The BLACKHEARTS	121	6		C	10	A	B	C	C	C	57	A	1
Billy JOEL	3	11	3	44	58	11	11	17	29	56	B	51	2
Elton JOHN	70	12	91	A	81	A	35	B	A	A	25	A	6
Howard JONES	195	5					A	33	90	B		A	0
Quincy JONES	232	3		41	60							B	2
Rickie Lee JONES	262	4		43	C	B	A	C				A	4
JOURNEY	5	9	24	5	4	6	A		30	85	B	52	2
KENNY G	93	5					A	B	B	14	42	33	6
KOOL & The GANG	35	8	41	27	31	A	68	15	A	71	C		0
Patti LaBELLE	189	6	C	C			A	B	22			B	5

Album Acts: Sp'l Lists

Act	'80s Rank	1980s Charted	1980 n=465	1981 n=605	1982 n=522	1983 n=508	1984 n=481	1985 n=467	1986 n=505	1987 n=531	1988 n=574	1989 n=580	1990s Releases
Cyndi LAUPER	67	3				C	10	44	60	62		A	3
John LENNON & Yoko ONO	134	3	A	10	C		71						0
Julian LENNON	267	3					A	43	A			B	0
Huey LEWIS And The NEWS	17	4			57	A	3	45	24	23	69	C	3
LIPPS, INC.	387	2	39										0
LISA LISA And CULT JAM	207	3						A	A	25	B	B	1
LIVING COLOUR	249	1									B	16	3
LL COOL J	153	3							A	24	A	53	5
Kenny LOGGINS	158	5	26	A	95	A		A			B		6
Los LOBOS	198	4					C	A		20	A		5
LOVERBOY	38	6		37	3	19	B	67	92	A	C	C	0
MADONNA	4	6				B	16	3	5	7	82	7	6
Richard MARX	85	2								53	15	11	4
MC HAMMER	398	1									C	44	4
Paul McCARTNEY	99	8	52	C	27	A	70	B	A	C	B	A	10
Bobby McFERRIN	304	2								B	38	B	2
John MELLENCAMP	7	7	A	A	5	36	21	26	7	43	23	57	7
MEN AT WORK	43	3			20	2	A	B					0
METALLICA	149	6					C	B	80	97	52	41	6
George MICHAEL	51	1								A	1	84	4
Bette MIDLER	131	6	23	B		B		C	C			13	6
MIDNIGHT OIL	343	3					C	C			36	C	4
MIDNIGHT STAR	202	4				82	44	90	A		C	C	0
MIKE + THE MECHANICS	245	2						C	44		C	62	1
Steve MILLER Band	210	6		A	26	C	C	C	C	B	C		1
MILLI VANILLI	132	1										5	1
Stephanie MILLS	168	9	49	A	A	B	B	C	A	91	C	A	0
MISSING PERSONS	238	4			78	39	A		C				0
MR. MISTER	188	3					C	A	15	B	C		0
The MONKEES	298	9							32	A			0
MOODY BLUES	114	7		14	B	A	C	C	36	C	A	C	3
MÖTLEY CRÜE	47	5				A	23	24	A	22	A	28	4
Willie NELSON	46	22	19	39	8	26	87	C					8
NEW EDITION	101	5				B	A	25	63	A	41	98	2
NEW KIDS ON THE BLOCK	80	3									A	1	3
Juice NEWTON	181	5		40	34	A	C						0
Olivia NEWTON-JOHN	104	6	86	83	19	49	A	A	C		C	C	2
Stevie NICKS	41	4		17	22	18	A	B	50			81	3
NIGHT RANGER	108	5			C	86	18	32	A	A	C		0
Aldo NOVA	330	2			44	B	C						1
Billy OCEAN	75	5		C			57	14	19	A	66	C	0
Roy ORBISON	321	5							C			26	2
Jeffrey OSBORNE	162	5			A	71	39	A	A		B		1
Ozzy OSBOURNE	55	8		38	23	57	93		39	A	89	A	6
The OUTFIELD	193	3						C	23	78		A	1
Robert PALMER	138	6	B	C	C	B		C	21	A	51	A	2
Alan PARSONS PROJECT	100	9	A	30	36	A	52	A	A	B			1
PEBBLES	394	1									46		1
Teddy PENDERGRASS	180	9	47	90	A		95	C	B		A		3
Steve PERRY	332	1					40	A					1
PET SHOP BOYS	185	4							37	A	58	B	7
Tom PETTY	201	1										9	1
Tom PETTY And The HEARTBREAKERS	45	7	8	29	A	42		35	A	82			4
PINK FLOYD	18	8	1	A	74	28	A	A	A	32	28	100	2
Robert PLANT	103	4			49	32	A	A			21	C	4
POINTER SISTERS	64	8	A	67	B	C	15	13	B	B	C		0
POISON	57	2							B	8	7	29	3
The POLICE	10	6	63	7	12	4	17		A	98			2
POWER STATION	290	1						22	C				0
The PRETENDERS	102	6	30	42	C		28		A	A	B		4
PRINCE	31	7	87	A	91	13	45	A		42	68	17	15
PRINCE And The REVOLUTION	29	3					4	5	35				0
QUARTERFLASH	269	3		B	39	A		C					0

Album Acts: Sp'l Lists

Act	'80s Rank	1980s Charted	1980 n=465	1981 n=605	1982 n=522	1983 n=508	1984 n=481	1985 n=467	1986 n=505	1987 n=531	1988 n=574	1989 n=580	1990s Releases
QUEEN	86	7	7	36	72		A		B			A	6
QUIET RIOT	120	4				15	31	C	A	C	C	C	0
Eddie RABBITT	233	5	A	46	A	A							0
RATT	112	5					25	37	A	A	A	A	3
R.E.M.	81	8				A	76	68	79	50	54	31	6
REO SPEEDWAGON	25	10	B	1	30		B	30		74	A		1
Lionel RICHIE	8	3				66	7	1	34	20	38		3
Smokey ROBINSON	135	9	38	49	A	A	C		C	52	B		2
Kenny ROGERS	15	22	4	6	41	14	27	A	A	C		C	5
ROLLING STONES	20	14	9	8	15	A	99		48			23	6
Linda RONSTADT	44	10	15	B	A	37	30	84	A	88	A	94	7
Diana ROSS	59	12	12	35	45	97	A	62	C	B		C	4
David Lee ROTH	140	3						41	34	B	43		4
ROYAL PHILHARMONIC Orchestra	203	3		B	17	A							0
RUN-D.M.C.	88	4					A	56	11	41	63		3
RUSH	40	11	45	11	47	B	56	A	A	80	A	95	5
SADE	69	3						20	3		37	B		2
SALT-N-PEPA	355	2								B	47	C	3
SANTANA	223	6	B	33	A	C		A		C	C		4
Joe SATRIANI	282	3								C	50	77	5
Boz SCAGGS	235	3	25	98							A		2
SCORPIONS	76	7	A		51	C	19	48	B		32	B	3
Bob SEGER And The SILVER BULLET Band	22	7	2	23	A	22			8	B			2
Paul SIMON	91	4	75	C		B	B		58	6	A	C	4
SIMPLE MINDS	242	5				B	A	A	40	C		B	2
SKID ROW	208	1										12	3
SOUL II SOUL	380	1										36	2
Rick SPRINGFIELD	54	8		28	11	33	50	64			A		1
Bruce SPRINGSTEEN	2	9	28	48	63	B	6	1	9	16	34		7
Billy SQUIER	73	6	C	24	14	58	49	C	B	C		B	1
STEELY DAN	347	4	B	45	C								1
Rod STEWART	66	9	A	58	64	A	55	C	A		49	38	6
STING	106	2						12	71	93	44		7
STRAY CATS	126	4			35	9	B		C			C	0
Barbra STREISAND	21	10	13	22	69	80	42	86	14	57	A	67	6
STYX	50	7	35	4	C	23	A						4
Donna SUMMER	117	10	14	B	94	43	A	C		C		A	0
SUPERTRAMP	163	7	40	B	99	67		100		C	C		0
SURVIVOR	137	7	C	C	25	B	A	36	B	A	C		0
Keith SWEAT	292	1									24	B	5
TALKING HEADS	65	8	A	A	A	34	94	27	38	A	88		1
TEARS FOR FEARS	68	3				A		6	46			78	3
10,000 MANIACS	257	2								C	60	46	4
38 SPECIAL	97	7	B	47	53	B	58		54	A	B	A	1
THOMPSON TWINS	167	7			C	A	33	A	81	B	C	C	0
George THOROGOOD & The DESTROYERS	214	5	C	C	A	C		47	A	B	A		3
TIFFANY	130	2								90	6	66	0
TONE LŌC	252	1										15	0
TOTO	109	7	A	B	32	24	B	B	A	B	A		1
Pete TOWNSHEND	176	7	27		A	A		B	89	C		B	1
TRAVELING WILBURYS	197	1									94	14	1
Randy TRAVIS	159	5							B	40	35	99	9
Tina TURNER	53	4					14	9	47	81	C	A	3
TWISTED SISTER	266	5				C	36	A	B	B			0
Bonnie TYLER	358	2				38	B		C				0
USA For AFRICA	383	1						39					0
U2	9	8		A	C	31	29	23	91	2	8	47	5
Luther VANDROSS	61	7		A	71	A	73	50	A	39	79	75	7
VANGELIS	196	2		C	16				C	A			0
VAN HALEN	11	8	32	63	38	C	5	A	6	A	11	A	5
John WAITE	279	4			B		47	A		B			0
WARRANT	300	1										20	2
Grover WASHINGTON Jr.	150	8	95	19	A	A	B	B		B			2

Act	'80s Rank	1980s Charted	1980 n=465	1981 n=605	1982 n=522	1983 n=508	1984 n=481	1985 n=467	1986 n=505	1987 n=531	1988 n=574	1989 n=580	1990s Releases
Jody WATLEY	164	3								37	59	49	2
WHAM!	89	3				B	A	7	53	C			0
The WHISPERS	144	8	33	80	A	A	C	B		67	C		12
Karyn WHITE	371	1									B	43	2
WHITE LION	190	3								B	27	37	1
WHITESNAKE	63	6	B	C			A	C		4	33	A	1
The WHO	213	7	C	34	70	A	C	B	C				3
WINGER	288	1									A	25	2
Steve WINWOOD	34	5		16	A	C			25	21	12	A	2
Stevie WONDER	39	8	29	31	62		41	31	41	B	76		2
YES	136	8	A	B	C	A	26	C	C	A	A	A	6
Paul YOUNG	310	3					A	42	B	C			1
ZZ TOP	33	5	89	86	C	21	13	49	10	C			5

Album Acts: 150 Or More Total Chart Weeks

Act [Rank]	Wks (Ents)
U2 [9]	740 (8)
PINK FLOYD [18]	690 (8)
Kenny ROGERS [15]	653 (22)
ALABAMA [28]	613 (12)
JOURNEY [5]	594 (9)
Willie NELSON [46]	592 (22)
The POLICE [10]	490 (6)
Bruce SPRINGSTEEN [2]	490 (9)
MADONNA [4]	447 (6)
Phil COLLINS [19]	433 (4)
Billy JOEL [3]	423 (11)
DEF LEPPARD [6]	415 (5)
Pat BENATAR [16]	415 (9)
AC/DC [24]	401 (10)
Bob SEGER/SILVER BULLET Band [22]	395 (7)
VAN HALEN [11]	393 (8)
Michael JACKSON [1]	392 (6)
PRINCE [31]	391 (7)
John MELLENCAMP [7]	387 (7)
DURAN DURAN [32]	387 (8)
Daryl HALL & John OATES [12]	382 (8)
Hank WILLIAMS Jr. [184]	366 (15)
Lionel RICHIE [8]	358 (3)
BON JOVI [13]	349 (4)
TALKING HEADS [65]	349 (8)
ROLLING STONES [20]	345 (14)
Barbra STREISAND [21]	344 (10)
Ozzy OSBOURNE [55]	333 (8)
LOVERBOY [38]	332 (6)
Billy IDOL [71]	330 (5)
Linda RONSTADT [44]	330 (10)
KOOL & The GANG [35]	329 (8)
Neil DIAMOND [52]	326 (10)
ZZ TOP [33]	323 (5)
REO SPEEDWAGON [25]	320 (10)
Elton JOHN [70]	319 (12)
Huey LEWIS And The NEWS [17]	308 (4)
MÖTLEY CRÜE [47]	306 (5)
The CARS [36]	295 (7)
RUSH [40]	291 (11)
AIR SUPPLY [72]	283 (6)
The BEATLES [200]	283 (16)
IRON MAIDEN [107]	281 (8)
Diana ROSS [59]	277 (12)
Al JARREAU [127]	276 (7)
Rick SPRINGFIELD [54]	265 (8)
Luther VANDROSS [61]	264 (7)
Randy TRAVIS [159]	263 (5)
R.E.M. [81]	263 (8)
GENESIS [27]	261 (6)
David SANBORN [243]	258 (8)
SCORPIONS [76]	253 (7)
Stevie NICKS [41]	251 (4)
George WINSTON [289]	248 (4)
Whitney HOUSTON [14]	247 (2)
38 SPECIAL [97]	246 (7)
POINTER SISTERS [64]	240 (8)
METALLICA [149]	237 (6)
FOREIGNER [30]	236 (7)

Act [Rank]	Wks (Ents)
Billy SQUIER [73]	235 (6)
Barry MANILOW [141]	234 (12)
EURYTHMICS [74]	232 (8)
Rod STEWART [66]	231 (9)
Alan PARSONS PROJECT [100]	231 (9)
HEART [26]	228 (7)
Steve WINWOOD [34]	225 (5)
Sheena EASTON [146]	221 (7)
RUN-D.M.C. [88]	220 (4)
David BOWIE [94]	220 (11)
Aretha FRANKLIN [119]	219 (8)
JEFFERSON AIRPLANE/STARSHIP [79]	219 (9)
Bryan ADAMS [37]	218 (4)
STYX [50]	217 (7)
GAP BAND [143]	216 (8)
CAMEO [123]	216 (9)
CHICAGO [82]	215 (7)
Anne MURRAY [199]	215 (11)
The PRETENDERS [102]	212 (6)
Rick JAMES [113]	212 (9)
SPYRO GYRA [216]	212 (12)
Jeffrey OSBORNE [162]	211 (5)
Tom PETTY/HEARTBREAKERS [45]	211 (7)
DEPECHE MODE [272]	211 (8)
KISS [105]	211 (9)
Anita BAKER [58]	210 (3)
DIRE STRAITS [49]	210 (6)
Stevie WONDER [39]	210 (8)
NIGHT RANGER [108]	209 (5)
Olivia NEWTON-JOHN [104]	209 (6)
The CURE [152]	209 (7)
Joe JACKSON [124]	209 (10)
NEW EDITION [101]	208 (5)
AEROSMITH [87]	205 (8)
DOKKEN [191]	204 (5)
INXS [60]	204 (6)
JUDAS PRIEST [125]	203 (7)
Stephanie MILLS [168]	203 (9)
George BENSON [139]	202 (8)
WHITESNAKE [63]	201 (6)
Dan FOGELBERG [78]	201 (6)
Earl KLUGH [302]	201 (11)
TOTO [109]	197 (7)
Donna SUMMER [117]	197 (10)
KENNY G [93]	196 (5)
Waylon JENNINGS [259]	196 (6)
Gloria ESTEFAN/ MIAMI SOUND MACHINE [77]	195 (3)
FLEETWOOD MAC [42]	195 (7)
STRYPER [265]	193 (4)
The WHISPERS [144]	193 (8)
ATLANTIC STARR [178]	192 (6)
TRIUMPH [173]	192 (7)
Smokey ROBINSON [135]	192 (9)
Bob JAMES [368]	192 (10)
The JUDDS [301]	191 (6)
Peter GABRIEL [98]	190 (6)
The COMMODORES [116]	188 (8)
Dionne WARWICK [182]	186 (10)

Act [Rank]	Wks (Ents)
Emmylou HARRIS [244]	185 (11)
CULTURE CLUB [56]	184 (4)
RATT [112]	184 (5)
Joan JETT & The BLACKHEARTS [121]	183 (6)
Teddy PENDERGRASS [180]	183 (9)
Grover WASHINGTON Jr. [150]	182 (8)
Ronnie MILSAP [275]	182 (9)
Robert PALMER [138]	181 (6)
Tina TURNER [53]	180 (4)
The FIXX [151]	180 (6)
Eric CLAPTON [96]	179 (8)
FAT BOYS [186]	178 (7)
Dolly PARTON [240]	178 (10)
Billy OCEAN [75]	177 (5)
Julio IGLESIAS [177]	177 (8)
The B-52's [170]	176 (7)
GUNS N' ROSES [23]	173 (2)
The TEMPTATIONS [323]	173 (10)
SADE [69]	172 (3)
Eddie MONEY [165]	172 (6)
POISON [57]	171 (2)
MANHATTAN TRANSFER [284]	171 (8)
MIDNIGHT STAR [202]	170 (4)
SURVIVOR [137]	169 (7)
OAK RIDGE BOYS [228]	169 (9)
George THOROGOOD & The DESTROYERS [214]	168 (5)
Neil YOUNG [175]	168 (11)
UB40 [219]	166 (6)
TEARS FOR FEARS [68]	164 (3)
Sammy HAGAR [161]	164 (7)
PAT METHENY GROUP [312]	164 (9)
Juice NEWTON [181]	162 (5)
Stevie Ray VAUGHAN And DOUBLE TROUBLE [212]	162 (5)
Jimmy BUFFETT [270]	162 (10)
Cyndi LAUPER [67]	161 (3)
Laura BRANIGAN [215]	161 (5)
Robert PLANT [103]	160 (4)
Teena MARIE [230]	160 (7)
Paul McCARTNEY [99]	160 (8)
SQUEEZE [254]	159 (7)
CHEAP TRICK [157]	159 (9)
Carly SIMON [234]	157 (6)
The DOORS [236]	157 (6)
Elvis COSTELLO/ATTRACTIONS [194]	157 (9)
The MONKEES [298]	156 (9)
Paul SIMON [91]	155 (4)
Howard JONES [195]	155 (5)
BANGLES [118]	154 (3)
SHALAMAR [241]	154 (6)
Jermaine JACKSON [166]	154 (7)
Christopher CROSS [90]	153 (3)
MEN AT WORK [43]	152 (3)
WHAM! [89]	152 (4)
Kenny LOGGINS [158]	152 (5)
QUEEN [86]	152 (7)
The KINKS [171]	151 (9)
Janet JACKSON [48]	150 (4)
The CLASH [145]	150 (6)

Album Acts: 6 Or More Chart Appearances

Act [Rank]	Ent
Kenny ROGERS [15]	22
Willie NELSON [46]	22
The BEATLES [200]	16
Hank WILLIAMS Jr. [184]	15
ROLLING STONES [20]	14
Elvis PRESLEY [421]	13
ALABAMA [28]	12
Diana ROSS [59]	12
Elton JOHN [70]	12
Barry MANILOW [141]	12
SPYRO GYRA [216]	12
Billy JOEL [3]	11
RUSH [40]	11
David BOWIE [94]	11
Neil YOUNG [175]	11
Anne MURRAY [199]	11
Bob DYLAN [211]	11
Emmylou HARRIS [244]	11
Earl KLUGH [302]	11
Barbra STREISAND [21]	10
AC/DC [24]	10
REO SPEEDWAGON [25]	10
Linda RONSTADT [44]	10
Neil DIAMOND [52]	10
Donna SUMMER [117]	10
Joe JACKSON [124]	10
Dionne WARWICK [182]	10
Dolly PARTON [240]	10
Jimmy BUFFETT [270]	10
The TEMPTATIONS [323]	10
Bob JAMES [368]	10
Dave GRUSIN [524]	10
Bruce SPRINGSTEEN [2]	9
JOURNEY [5]	9
Pat BENATAR [16]	9
Rod STEWART [66]	9
JEFFERSON AIRPLANE/ STARSHIP [79]	9
Alan PARSONS PROJECT [100]	9
KISS [105]	9
Rick JAMES [113]	9
CAMEO [123]	9
Smokey ROBINSON [135]	9
CHEAP TRICK [157]	9
Stephanie MILLS [168]	9
The KINKS [171]	9
Teddy PENDERGRASS [180]	9
Elvis COSTELLO & The ATTRACTIONS [194]	9
OAK RIDGE BOYS [228]	9
Marvin GAYE [273]	9
Ronnie MILSAP [275]	9
The MONKEES [298]	9
PAT METHENY GROUP [312]	9
Peabo BRYSON [336]	9
U2 [9]	8
VAN HALEN [11]	8
Daryl HALL & John OATES [12]	8
PINK FLOYD [18]	8

Act [Rank]	Ent
DURAN DURAN [32]	8
KOOL & The GANG [35]	8
Stevie WONDER [39]	8
Rick SPRINGFIELD [54]	8
Ozzy OSBOURNE [55]	8
POINTER SISTERS [64]	8
TALKING HEADS [65]	8
EURYTHMICS [74]	8
R.E.M. [81]	8
AEROSMITH [87]	8
Eric CLAPTON [96]	8
Paul McCARTNEY [99]	8
IRON MAIDEN [107]	8
The COMMODORES [116]	8
Aretha FRANKLIN [119]	8
YES [136]	8
George BENSON [139]	8
GAP BAND [143]	8
The WHISPERS [144]	8
EARTH, WIND & FIRE [147]	8
Grover WASHINGTON Jr. [150]	8
Julio IGLESIAS [177]	8
ISLEY BROTHERS [229]	8
David SANBORN [243]	8
DEPECHE MODE [272]	8
MANHATTAN TRANSFER [284]	8
Joan ARMATRADING [314]	8
ASHFORD & SIMPSON [318]	8
Van MORRISON [411]	8
George STRAIT [448]	8
ONE WAY [487]	8
Crystal GAYLE [532]	8
The RAMONES [623]	8
OINGO BOINGO [664]	8
Miles DAVIS [699]	8
John MELLENCAMP [7]	7
Bob SEGER And The SILVER BULLET Band [22]	7
HEART [26]	7
FOREIGNER [30]	7
PRINCE [31]	7
The CARS [36]	7
FLEETWOOD MAC [42]	7
Tom PETTY And The HEARTBREAKERS [45]	7
STYX [50]	7
Luther VANDROSS [61]	7
SCORPIONS [76]	7
CHICAGO [82]	7
QUEEN [86]	7
38 SPECIAL [97]	7
TOTO [109]	7
MOODY BLUES [114]	7
JUDAS PRIEST [125]	7
Al JARREAU [127]	7
SURVIVOR [137]	7
Sheena EASTON [146]	7
The CURE [152]	7

Act [Rank]	Ent
Sammy HAGAR [161]	7
SUPERTRAMP [163]	7
Jermaine JACKSON [166]	7
THOMPSON TWINS [167]	7
The B-52's [170]	7
TRIUMPH [173]	7
Pete TOWNSHEND [176]	7
FAT BOYS [186]	7
The WHO [213]	7
BEE GEES [218]	7
Teena MARIE [230]	7
Herb ALPERT [251]	7
SQUEEZE [254]	7
KROKUS [280]	7
DAZZ BAND [287]	7
S.O.S. BAND [299]	7
JETHRO TULL [305]	7
PSYCHEDELIC FURS [316]	7
Natalie COLE [320]	7
BLACK SABBATH [328]	7
MOLLY HATCHET [333]	7
The BAR-KAYS [342]	7
CON FUNK SHUN [354]	7
LYNYRD SKYNYRD [361]	7
Ted NUGENT [373]	7
XTC [377]	7
Lou REED [395]	7
Stacy LATTISAW [420]	7
The O'JAYS [429]	7
Frank ZAPPA [449]	7
BEACH BOYS [539]	7
The JAM [695]	7
Michael JACKSON [1]	6
MADONNA [4]	6
The POLICE [10]	6
GENESIS [27]	6
LOVERBOY [38]	6
DIRE STRAITS [49]	6
INXS [60]	6
WHITESNAKE [63]	6
AIR SUPPLY [72]	6
Billy SQUIER [73]	6
Dan FOGELBERG [78]	6
Peter GABRIEL [98]	6
The PRETENDERS [102]	6
Olivia NEWTON-JOHN [104]	6
Joan JETT & The BLACKHEARTS [121]	6
Bette MIDLER [131]	6
Robert PALMER [138]	6
The CLASH [145]	6
METALLICA [149]	6
The FIXX [151]	6
Eddie MONEY [165]	6
Kim CARNES [172]	6
ATLANTIC STARR [178]	6
Patti LaBELLE [189]	6
DOOBIE BROTHERS [206]	6

Act [Rank]	Ent
Steve MILLER Band [210]	6
UB40 [219]	6
GRATEFUL DEAD [221]	6
SANTANA [223]	6
Carly SIMON [234]	6
The DOORS [236]	6
Chaka KHAN [239]	6
SHALAMAR [241]	6
ELECTRIC LIGHT ORCHESTRA [255]	6
LITTLE RIVER BAND [258]	6
Waylon JENNINGS [259]	6
DEVO [264]	6
MAZE Featuring Frankie BEVERLY [278]	6
Ray PARKER Jr. [286]	6
KANSAS [293]	6
Greg KIHN Band [294]	6
The JUDDS [301]	6
BLUE ÖYSTER CULT [307]	6
ORCHESTRAL MANOEUVRES IN THE DARK [315]	6
LAKESIDE [317]	6
The ALARM [325]	6
"Weird Al" YANKOVIC [340]	6
Charlie DANIELS Band [348]	6
Herbie HANCOCK [360]	6
SKYY [364]	6
Melissa MANCHESTER [378]	6
HIROSHIMA [379]	6
Bob MARLEY And The WAILERS [396]	6
Chuck MANGIONE [402]	6
The SMITHS [404]	6
John LENNON [415]	6
Grace JONES [432]	6
UTOPIA [475]	6
Dave EDMUNDS [505]	6
Jeff LORBER [512]	6
POCO [523]	6
Lee RITENOUR [552]	6
Bruce COCKBURN [561]	6
Jean-Luc PONTY [562]	6
Tom COCHRANE/RED RIDER [567]	6
Robin TROWER [573]	6
Wynton MARSALIS [593]	6
ECHO & The BUNNYMEN [622]	6
X [632]	6
Kurtis BLOW [649]	6
THIRD WORLD [673]	6
WEATHER REPORT [709]	6
Millie JACKSON [715]	6
SOUTHSIDE JOHNNY & The ASBURY JUKES [769]	6
Larry CARLTON [785]	6
Randy CRAWFORD [825]	6
Johnny MATHIS [1004]	6
STATLER BROTHERS [1026]	6

Album Acts: Top 10 Entries

Act [Rank]	No.(Wk)
Billy JOEL [3]	7 (94)
VAN HALEN [11]	6 (80)
JOURNEY [5]	6 (78)
ROLLING STONES [20]	6 (63)
Barbra STREISAND [21]	6 (48)
RUSH [40]	6 (29)
Bruce SPRINGSTEEN [2]	5 (120)
John MELLENCAMP [7]	5 (84)
MADONNA [4]	5 (82)
Stevie WONDER [39]	5 (49)
The POLICE [10]	4 (89)
Daryl HALL & John OATES [12]	4 (69)
Bob SEGER [22]	4 (61)
Phil COLLINS [19]	4 (51)
Kenny ROGERS [15]	4 (43)
GENESIS [27]	4 (38)
Tom PETTY And The HEARTBREAKERS [45]	4 (37)
DURAN DURAN [32]	4 (34)
Linda RONSTADT [44]	4 (31)
Michael JACKSON [1]	3 (138)
Lionel RICHIE [8]	3 (100)
HEART [26]	3 (59)
Pat BENATAR [16]	3 (55)
PRINCE And The REVOLUTION [29]	3 (55)
Bryan ADAMS [37]	3 (51)
FOREIGNER [30]	3 (50)
STYX [50]	3 (50)
PINK FLOYD [18]	3 (47)
AC/DC [24]	3 (47)
REO SPEEDWAGON [25]	3 (47)
The CARS [36]	3 (44)
Steve WINWOOD [34]	3 (41)
Stevie NICKS [41]	3 (38)
MÖTLEY CRÜE [47]	3 (32)
SADE [69]	3 (32)
FLEETWOOD MAC [42]	3 (31)
Neil DIAMOND [52]	3 (26)
PRINCE [31]	3 (25)
Robert PLANT [103]	3 (24)
SCORPIONS [76]	3 (22)
The PRETENDERS [102]	3 (17)
Billy IDOL [71]	3 (11)
DEF LEPPARD [6]	2 (116)
Whitney HOUSTON [14]	2 (77)
GUNS N' ROSES [23]	2 (70)
BON JOVI [13]	2 (68)
Huey LEWIS And The NEWS [17]	2 (68)
U2 [9]	2 (49)
Janet JACKSON [48]	2 (49)
MEN AT WORK [43]	2 (45)
Tina TURNER [53]	2 (45)
POISON [57]	2 (43)
NEW KIDS ON THE BLOCK [80]	2 (43)
WHITESNAKE [63]	2 (42)
TEARS FOR FEARS [68]	2 (35)
ASIA [95]	2 (32)
Bruce HORNSBY And The RANGE [110]	2 (30)
Cyndi LAUPER [67]	2 (27)
WHAM! [89]	2 (27)
Debbie GIBSON [84]	2 (26)
The GO-GO'S [92]	2 (24)
STING [106]	2 (23)
Olivia NEWTON-JOHN [104]	2 (22)
Tracy CHAPMAN [142]	2 (22)
LOVERBOY [38]	2 (21)
CINDERELLA [83]	2 (20)
ZZ TOP [33]	2 (19)
KENNY G [93]	2 (19)
Eric CLAPTON [96]	2 (19)
Dan FOGELBERG [78]	2 (18)
CHICAGO [82]	2 (18)
MOODY BLUES [114]	2 (18)
LL COOL J [153]	2 (18)
Rick SPRINGFIELD [54]	2 (17)
Jackson BROWNE [133]	2 (17)
EAGLES [154]	2 (17)
Billy SQUIER [73]	2 (16)
JEFFERSON AIRPLANE/STARSHIP [79]	2 (16)
Richard MARX [85]	2 (16)
RUN-D.M.C. [88]	2 (16)
Paul McCARTNEY [99]	2 (16)
David Lee ROTH [140]	2 (16)
RATT [112]	2 (15)
The WHO [213]	2 (15)
Ozzy OSBOURNE [55]	2 (14)
Billy OCEAN [75]	2 (14)
Gloria ESTEFAN/MIAMI SOUND MACHINE [77]	2 (13)
AIR SUPPLY [72]	2 (12)
EARTH, WIND & FIRE [147]	2 (12)
JACKSON 5 [192]	2 (12)
BEE GEES [218]	2 (12)
Donna SUMMER [117]	2 (11)
SUPERTRAMP [163]	2 (10)
EURYTHMICS [74]	2 (9)
LED ZEPPELIN [311]	2 (8)
George MICHAEL [51]	1 (51)
Bobby BROWN [62]	1 (45)
Paula ABDUL [115]	1 (44)
DIRE STRAITS [49]	1 (37)
CULTURE CLUB [56]	1 (30)
MILLI VANILLI [132]	1 (28)
FINE YOUNG CANNIBALS [129]	1 (27)
Tom PETTY [201]	1 (25)
Paul SIMON [91]	1 (24)
INXS [60]	1 (22)
TOTO [109]	1 (22)
TIFFANY [130]	1 (22)
John LENNON & Yoko ONO [134]	1 (22)
TRAVELING WILBURYS [197]	1 (22)
QUEEN [86]	1 (21)
BOSTON [160]	1 (20)
J. GEILS Band [111]	1 (19)
BEASTIE BOYS [122]	1 (19)
Anita BAKER [58]	1 (18)
Diana ROSS [59]	1 (18)
STRAY CATS [126]	1 (18)
QUIET RIOT [120]	1 (17)
David BOWIE [94]	1 (16)
Willie NELSON [46]	1 (15)
Christopher CROSS [90]	1 (15)
John FOGERTY [183]	1 (15)
VANGELIS [196]	1 (14)
AEROSMITH [87]	1 (13)
Peter GABRIEL [98]	1 (12)
Rick JAMES [113]	1 (12)
Joan JETT & The BLACKHEARTS [121]	1 (12)
YES [136]	1 (12)
SURVIVOR [137]	1 (12)
Kim CARNES [172]	1 (12)
MR. MISTER [188]	1 (12)
Patti LaBELLE [189]	1 (12)
Steve MILLER Band [210]	1 (12)
TONE LŌC [252]	1 (12)
Bette MIDLER [131]	1 (11)
Julio IGLESIAS [177]	1 (11)
Los LOBOS [198]	1 (11)
SKID ROW [208]	1 (11)
POWER STATION [290]	1 (11)
George BENSON [139]	1 (10)
The CLASH [145]	1 (10)
The FIXX [151]	1 (10)
EUROPE [156]	1 (10)
DJ JAZZY JEFF & THE FRESH PRINCE [224]	1 (10)
The HONEYDRIPPERS [337]	1 (10)
BANGLES [118]	1 (9)
The WHISPERS [144]	1 (9)
BLONDIE [155]	1 (9)
HUMAN LEAGUE [174]	1 (9)
Pete TOWNSHEND [176]	1 (9)
ROYAL PHILHARMONIC Orchestra [203]	1 (9)
George HARRISON [271]	1 (9)
Grover WASHINGTON Jr. [150]	1 (8)
Terence Trent D'ARBY [222]	1 (8)
LIVING COLOUR [249]	1 (8)
ELECTRIC LIGHT ORCHESTRA [255]	1 (8)
Rickie Lee JONES [262]	1 (8)
Edie BRICKELL & The NEW BOHEMIANS [263]	1 (8)
Roy ORBISON [321]	1 (8)
Aldo NOVA [330]	1 (8)
Alan PARSONS PROJECT [100]	1 (7)
Joe JACKSON [124]	1 (7)
The B-52's [170]	1 (7)
DOOBIE BROTHERS [206]	1 (7)
LISA LISA And CULT JAM [207]	1 (7)
GRATEFUL DEAD [221]	1 (7)
SANTANA [223]	1 (7)
Marvin GAYE [273]	1 (7)
Bobby McFERRIN [304]	1 (7)
CLUB NOUVEAU [324]	1 (7)
BROTHERS JOHNSON [334]	1 (7)
Bonnie TYLER [358]	1 (7)
USA For AFRICA [383]	1 (7)
LIPPS, INC. [387]	1 (7)
POINTER SISTERS [64]	1 (6)
NEW EDITION [101]	1 (6)
PET SHOP BOYS [185]	1 (6)
The OUTFIELD [193]	1 (6)
John CAFFERTY And The BEAVER BROWN BAND [247]	1 (6)
The CULT [250]	1 (6)
CROSBY, STILLS & NASH [276]	1 (6)
STEELY DAN [347]	1 (6)
FALCO [352]	1 (6)
Michael McDONALD [393]	1 (6)
Randy RHOADS [644]	1 (6)
The COMMODORES [116]	1 (5)
Don HENLEY [128]	1 (5)
METALLICA [149]	1 (5)
Jermaine JACKSON [166]	1 (5)
SIMPLE MINDS [242]	1 (5)
GEORGIA SATELLITES [322]	1 (5)
Chuck MANGIONE [402]	1 (5)
SIMON & GARFUNKEL [443]	1 (5)
STARS ON [483]	1 (5)
Al JARREAU [127]	1 (4)
Robert PALMER [138]	1 (4)
GREAT WHITE [169]	1 (4)
FAT BOYS [186]	1 (4)
ISLEY BROTHERS [229]	1 (4)
Boz SCAGGS [235]	1 (4)
QUARTERFLASH [269]	1 (4)
Bob & Doug McKENZIE [480]	1 (4)
Luther VANDROSS [61]	1 (3)
38 SPECIAL [97]	1 (3)
NIGHT RANGER [108]	1 (3)
CAMEO [123]	1 (3)
A FLOCK OF SEAGULLS [237]	1 (3)
Dolly PARTON [240]	1 (3)
Emmylou HARRIS [244]	1 (3)
James TAYLOR [261]	1 (3)
WARRANT [300]	1 (3)
Eddy GRANT [397]	1 (3)
KOOL & The GANG [35]	1 (2)
Smokey ROBINSON [135]	1 (2)
THOMPSON TWINS [167]	1 (2)
Rick ASTLEY [205]	1 (2)
CHER [277]	1 (2)
YOUNG M.C. [484]	1 (2)
ALABAMA [28]	1 (1)
R.E.M. [81]	1 (1)
Freddie JACKSON [148]	1 (1)
Jody WATLEY [164]	1 (1)
Howard JONES [195]	1 (1)
Quincy JONES [232]	1 (1)
John WAITE [279]	1 (1)

Albums Acts: 3 Or More Top 40s

Act [Rank]	No. (Wk)
Kenny ROGERS [15]	11 (143)
Barbra STREISAND [21]	9 (138)
RUSH [40]	9 (118)
Billy JOEL [3]	8 (196)
Pat BENATAR [16]	8 (160)
Elton JOHN [70]	8 (85)
Daryl HALL & John OATES [12]	7 (187)
KOOL & The GANG [35]	7 (154)
AC/DC [24]	7 (114)
Rod STEWART [66]	7 (113)
ROLLING STONES [20]	7 (108)
Ozzy OSBOURNE [55]	7 (104)
Luther VANDROSS [61]	7 (99)
Linda RONSTADT [44]	7 (93)
Diana ROSS [59]	7 (90)
KISS [105]	7 (77)
Eric CLAPTON [96]	7 (62)
JUDAS PRIEST [125]	7 (59)
Bob DYLAN [211]	7 (35)
MADONNA [4]	6 (196)
John MELLENCAMP [7]	6 (179)
JOURNEY [5]	6 (173)
VAN HALEN [11]	6 (160)
U2 [9]	6 (133)
HEART [26]	6 (125)
DURAN DURAN [32]	6 (119)
ALABAMA [28]	6 (114)
The CARS [36]	6 (109)
PRINCE [31]	6 (107)
Stevie WONDER [39]	6 (105)
Tom PETTY & HEARTBREAKERS [45]	6 (100)
JEFFERSON AIRPLANE/ STARSHIP [79]	6 (90)
Neil DIAMOND [52]	6 (88)
TALKING HEADS [65]	6 (87)
R.E.M. [81]	6 (82)
QUEEN [86]	6 (69)
IRON MAIDEN [107]	6 (65)
Paul McCARTNEY [99]	6 (64)
Willie NELSON [46]	6 (55)
Barry MANILOW [141]	6 (43)
Elvis COSTELLO & The ATTRACTIONS [194]	6 (37)
Bruce SPRINGSTEEN [2]	5 (177)
GENESIS [27]	5 (149)
Steve WINWOOD [34]	5 (133)
The POLICE [10]	5 (125)
FOREIGNER [30]	5 (110)
FLEETWOOD MAC [42]	5 (98)
Rick SPRINGFIELD [54]	5 (97)
PINK FLOYD [18]	5 (88)
AIR SUPPLY [72]	5 (86)
EURYTHMICS [74]	5 (86)
AEROSMITH [87]	5 (86)
Dan FOGELBERG [78]	5 (82)
38 SPECIAL [97]	5 (72)
The PRETENDERS [102]	5 (66)
The COMMODORES [116]	5 (63)
Smokey ROBINSON [135]	5 (61)
Aretha FRANKLIN [119]	5 (58)
CHEAP TRICK [157]	5 (51)

Act [Rank]	No. (Wk)
Donna SUMMER [117]	5 (50)
EARTH, WIND & FIRE [147]	5 (50)
The WHISPERS [144]	5 (47)
TRIUMPH [173]	5 (39)
Neil YOUNG [175]	5 (36)
GRATEFUL DEAD [221]	5 (34)
ZZ TOP [33]	4 (142)
Huey LEWIS And The NEWS [17]	4 (139)
Phil COLLINS [19]	4 (121)
Bob SEGER/SILVER BULLET Band [22]	4 (115)
LOVERBOY [38]	4 (112)
POINTER SISTERS [64]	4 (101)
Stevie NICKS [41]	4 (95)
REO SPEEDWAGON [25]	4 (94)
CULTURE CLUB [56]	4 (91)
MÖTLEY CRÜE [47]	4 (87)
SCORPIONS [76]	4 (82)
STYX [50]	4 (77)
CHICAGO [82]	4 (70)
Olivia NEWTON-JOHN [104]	4 (68)
Robert PLANT [103]	4 (65)
NIGHT RANGER [108]	4 (65)
RATT [112]	4 (64)
David BOWIE [94]	4 (63)
YES [136]	4 (60)
MOODY BLUES [114]	4 (56)
CAMEO [123]	4 (56)
Sammy HAGAR [161]	4 (48)
BLONDIE [155]	4 (47)
SUPERTRAMP [163]	4 (47)
Pete TOWNSHEND [176]	4 (43)
Sheena EASTON [146]	4 (41)
The B-52's [170]	4 (36)
ELECTRIC LIGHT ORCHESTRA [255]	4 (33)
Teddy PENDERGRASS [180]	4 (32)
Stevie Ray VAUGHAN And DOUBLE TROUBLE [212]	4 (23)
BLACK SABBATH [328]	4 (23)
SHALAMAR [241]	4 (18)
MAZE Featuring Frankie BEVERLY [278]	4 (16)
Michael JACKSON [1]	3 (180)
DEF LEPPARD [6]	3 (157)
Lionel RICHIE [8]	3 (155)
BON JOVI [13]	3 (117)
Bryan ADAMS [37]	3 (113)
Gloria ESTEFAN/ MIAMI SOUND MACHINE [77]	3 (104)
Billy OCEAN [75]	3 (98)
Cyndi LAUPER [67]	3 (89)
PRINCE And The REVOLUTION [29]	3 (86)
Tina TURNER [53]	3 (84)
Billy SQUIER [73]	3 (81)
NEW EDITION [101]	3 (81)
DIRE STRAITS [49]	3 (80)
Billy IDOL [71]	3 (77)
SADE [69]	3 (75)
Paul SIMON [91]	3 (73)
Peter GABRIEL [98]	3 (73)
Don HENLEY [128]	3 (73)
NEW KIDS ON THE BLOCK [80]	3 (70)

Act [Rank]	No. (Wk)
KENNY G [93]	3 (70)
The GO-GO'S [92]	3 (66)
WHITESNAKE [63]	3 (62)
David Lee ROTH [140]	3 (57)
J. GEILS Band [111]	3 (56)
Alan PARSONS PROJECT [100]	3 (55)
QUIET RIOT [120]	3 (55)
Joe JACKSON [124]	3 (54)
Bette MIDLER [131]	3 (54)
The CLASH [145]	3 (53)
Belinda CARLISLE [179]	3 (53)
THOMPSON TWINS [167]	3 (52)
Rick JAMES [113]	3 (49)
METALLICA [149]	3 (49)
Joan JETT & The BLACKHEARTS [121]	3 (48)
The FIXX [151]	3 (46)
PET SHOP BOYS [185]	3 (46)
George BENSON [139]	3 (45)
Jackson BROWNE [133]	3 (42)
GAP BAND [143]	3 (41)
JACKSON 5 [192]	3 (41)
Eddie MONEY [165]	3 (40)
HUMAN LEAGUE [174]	3 (39)
Grover WASHINGTON Jr. [150]	3 (38)
Jeffrey OSBORNE [162]	3 (38)
Al JARREAU [127]	3 (37)
SIMPLY RED [209]	3 (37)
Stephanie MILLS [168]	3 (36)
ATLANTIC STARR [178]	3 (36)
DeBARGE [204]	3 (36)
The KINKS [171]	3 (35)
Kenny LOGGINS [158]	3 (34)
Emmylou HARRIS [244]	3 (34)
Laura BRANIGAN [215]	3 (33)
Dolly PARTON [240]	3 (33)
DOOBIE BROTHERS [206]	3 (32)
Glenn FREY [231]	3 (32)
James TAYLOR [261]	3 (31)
Randy TRAVIS [159]	3 (30)
Eddie RABBITT [233]	3 (30)
Dionne WARWICK [182]	3 (29)
George THOROGOOD & The DESTROYERS [214]	3 (29)
Herb ALPERT [251]	3 (29)
ISLEY BROTHERS [229]	3 (27)
Thomas DOLBY [256]	3 (27)
The MOTELS [248]	3 (25)
DOKKEN [191]	3 (24)
Teena MARIE [230]	3 (21)
Greg KIHN Band [294]	3 (21)
Rickie Lee JONES [262]	3 (19)
KANSAS [293]	3 (19)
JETHRO TULL [305]	3 (19)
BLUE ÖYSTER CULT [307]	3 (19)
MOLLY HATCHET [333]	3 (18)
WHODINI [297]	3 (15)
ZAPP [390]	3 (15)
Patrice RUSHEN [327]	3 (13)
Ronnie MILSAP [275]	3 (10)
LITTLE FEAT [446]	3 (10)
John DENVER [344]	3 (8)

Album Acts: More Than 70 Consecutive Chart Weeks

Leading asterisk denotes streak in progress 1/1/80; trailing asterisk denotes continuing streak after 12/31/89

Act	Wks	Begin	End
Lionel RICHIE [8]	259	23-Oct-82	3-Oct-87
ALABAMA [28]	252	28-Mar-81	18-Jan-86
MADONNA [4]	244	3-Sep-83	30-Apr-88
Willie NELSON [46]	242	14-Jun-80	26-Jan-85
Daryl HALL & John OATES [12]	214	16-Aug-80	15-Sep-84
Whitney HOUSTON [14]	202	30-Mar-85	4-Feb-89
ZZ TOP [33]	202	23-Apr-83	28-Feb-87
U2 [9]	195	19-Mar-83	6-Dec-86
JOURNEY [5]	189	21-Feb-81	29-Sep-84
Anita BAKER [58]	175	19-Apr-86	19-Aug-89
POISON [57]	164	2-Aug-86	16-Sep-89
DURAN DURAN [32]	158	5-Jun-82	8-Jun-85
Bruce SPRINGSTEEN [2]	153	23-Jun-84	23-May-87
PRINCE [31]	153	20-Nov-82	19-Oct-85
Stevie NICKS [41]	150	15-Aug-81	23-Jun-84
Randy TRAVIS [159]	149	19-Jul-86	20-May-89
Phil COLLINS [19]	143	20-Oct-84	11-Jul-87
The CURE [152]	141	5-Oct-85	11-Jun-88
Billy JOEL [3]	141	20-Aug-83	26-Apr-86
TALKING HEADS [65]	135	22-Sep-84	18-Apr-87
Olivia NEWTON-JOHN [104]	135	31-Oct-81	26-May-84
Pat BENATAR [16]	*135	5-Jan-80	31-Jul-82
Neil DIAMOND [52]	132	29-Nov-80	4-Jun-83
Kenny ROGERS [15]	*132	5-Jan-80	10-Jul-82
DOKKEN [191]	129	13-Oct-84	28-Mar-87
Al JARREAU [127]	129	22-Aug-81	4-Feb-84
Huey LEWIS And The NEWS [17]	128	8-Oct-83	15-Mar-86
Jane FONDA [217]	126	29-May-82	20-Oct-84
DEF LEPPARD [6]	124*	22-Aug-87	30-Dec-89
LOVERBOY [38]	124	14-Nov-81	24-Mar-84
GUNS N' ROSES [23]	123*	29-Aug-87	30-Dec-89
POINTER SISTERS [64]	123	26-Nov-83	29-Mar-86
Debbie GIBSON [84]	122*	5-Sep-87	30-Dec-89
Michael JACKSON [1]	122	25-Dec-82	20-Apr-85
The POLICE [10]	121	25-Oct-80	12-Feb-83
CULTURE CLUB [56]	118	8-Jan-83	6-Apr-85
Christopher CROSS [90]	116	16-Feb-80	1-May-82
Bob SEGER [22]	115	15-Mar-80	22-May-82
Billy SQUIER [73]	111	2-May-81	11-Jun-83
Billy IDOL [71]	109	28-May-83	22-Jun-85
Julio IGLESIAS [177]	108	2-Apr-83	20-Apr-85
Janet JACKSON [48]	106	8-Mar-86	12-Mar-88
Tina TURNER [53]	106	16-Jun-84	21-Jun-86
Ozzy OSBOURNE [55]	106	18-Apr-81	23-Apr-83
REO SPEEDWAGON [25]	106	13-Dec-80	18-Dec-82
MIDNIGHT STAR [202]	103	30-Jul-83	13-Jul-85
KENNY G [93]	102	6-Sep-86	13-Aug-88
The PRETENDERS [102]	102	26-Jan-80	2-Jan-82
Jeffrey OSBORNE [162]	100	6-Aug-83	29-Jun-85
AC/DC [24]	98	23-Aug-80	3-Jul-82
Gloria ESTEFAN/MIAMI SOUND MACHINE [77]	97	20-Jun-87	22-Apr-89
Paul SIMON [91]	97	13-Sep-86	16-Jul-88
DIRE STRAITS [49]	97	8-Jun-85	11-Apr-87
Steve WINWOOD [34]	96	19-Jul-86	14-May-88
Cyndi LAUPER [67]	96	24-Dec-83	19-Oct-85
Dan FOGELBERG [78]	96	12-Sep-81	9-Jul-83
QUIET RIOT [120]	95	23-Apr-83	9-Feb-85
BON JOVI [13]	94	13-Sep-86	25-Jun-88
MEN AT WORK [43]	93	3-Jul-82	7-Apr-84
Grover WASHINGTON Jr. [150]	92	13-Sep-80	12-Jun-82
RUN-D.M.C. [88]	91	23-Jun-84	15-Mar-86
STRAY CATS [126]	91	3-Jul-82	24-Mar-84
AIR SUPPLY [72]	91	13-Jun-81	5-Mar-83
STRYPER [265]	90	2-Aug-86	16-Apr-88
Robert PALMER [138]	90	23-Nov-85	8-Aug-87
SADE [69]	89	23-Feb-85	1-Nov-86
ROYAL PHILHARMONIC Orchestra [203]	89	14-Nov-81	23-Jul-83
Rick SPRINGFIELD [54]	89	14-Mar-81	20-Nov-82
Linda RONSTADT [44]	88	1-Oct-83	1-Jun-85
George MICHAEL [51]	87	21-Nov-87	15-Jul-89
WHITE LION [190]	86	26-Sep-87	13-May-89
Richard MARX [85]	86	20-Jun-87	4-Feb-89
Billy OCEAN [75]	86	25-Aug-84	12-Apr-86
GENESIS [27]	85	28-Jun-86	6-Feb-88
Dwight YOAKAM [419]	84	19-Apr-86	21-Nov-87
HEART [26]	84	13-Jul-85	14-Feb-87
MÖTLEY CRÜE [47]	84	15-Oct-83	18-May-85
Quincy JONES [232]	84	4-Apr-81	6-Nov-82
PRINCE And The REVOLUTION [29]	83	14-Jul-84	8-Feb-86
BANGLES [118]	82	1-Feb-86	22-Aug-87
TEARS FOR FEARS [68]	82	6-Apr-85	25-Oct-86
Bryan ADAMS [37]	82	24-Nov-84	14-Jun-86
TOTO [109]	82	24-Apr-82	12-Nov-83
The GO-GO'S [92]	82	1-Aug-81	19-Feb-83
INXS [60]	81	14-Nov-87	27-May-89
SALT-N-PEPA [355]	81	29-Aug-87	11-Mar-89
WHAM! [89]	80	10-Nov-84	17-May-86
Rick JAMES [113]	80	2-May-81	6-Nov-82
The DOORS [236]	80	1-Nov-80	8-May-82
Joe SATRIANI [282]	79	21-Nov-87	20-May-89
EUROPE [156]	78	1-Nov-86	23-Apr-88
Neville MARRINER [577]	78	24-Nov-84	17-May-86
BASIA [422]	77	20-Feb-88	5-Aug-89
10,000 MANIACS [257]	77	19-Sep-87	4-Mar-89
VAN HALEN [11]	77	28-Jan-84	13-Jul-85
Bobby BROWN [62]	76*	23-Jul-88	30-Dec-89
Paula ABDUL [115]	76*	23-Jul-88	30-Dec-89
Rick ASTLEY [205]	76	23-Jan-88	1-Jul-89
WHITESNAKE [63]	76	18-Apr-87	24-Sep-88
ANTHRAX [350]	76	11-Apr-87	17-Sep-88
Peter GABRIEL [98]	76	14-Jun-86	21-Nov-87
Ronnie MILSAP [275]	76	25-Oct-80	3-Apr-82
AEROSMITH [87]	75	19-Sep-87	18-Feb-89
SIMPLY RED [209]	75	19-Sep-87	19-Sep-87
John MELLENCAMP [7]	75	14-Sep-85	14-Feb-87
John CAFFERTY/BEAVER BROWN BAND [247]	75	11-Aug-84	11-Jan-86
The FIXX [151]	75	13-Nov-82	14-Apr-84
KOOL & The GANG [35]	75	17-Oct-81	19-Mar-83
Jody WATLEY [164]	74	21-Mar-87	13-Aug-88
EXPOSÉ [187]	74	21-Feb-87	16-Jul-88
The HOOTERS [227]	74	25-May-85	18-Oct-86
J. GEILS Band [111]	74	14-Nov-81	9-Apr-83
John LENNON & Yoko ONO [134]	74	6-Dec-80	1-May-82
Bruce HORNSBY And The RANGE [110]	73	21-Jun-86	7-Nov-87
METALLICA [149]	73	22-Mar-86	8-Aug-87
EURYTHMICS [74]	73	28-May-83	13-Oct-84
Rod STEWART [66]	72	4-Jun-88	14-Oct-89
CHICAGO [82]	72	2-Jun-84	12-Oct-85
FOREIGNER [30]	72	25-Jul-81	4-Dec-82
NEW KIDS ON THE BLOCK [80]	71*	27-Aug-88	30-Dec-89
Hank WILLIAMS Jr. [184]	71	14-Feb-87	18-Jun-88

Album Acts: Most Albums on Chart Simultaneously

Act (Total Weeks)	Dates
U2 (22)	**Seven Albums** 4/25/87-9/19/87 Boy October War Under A Blood Red Sky The Unforgettable Fire Wide Awake In America The Joshua Tree
Julio IGLESIAS (4)	**Six Albums** 9/15/84-10/06/84 Julio In Concert 1100 Bel Air Place From A Child To A Woman Hey! Moments
Michael JACKSON (7)	**Four Albums** 6/23/84-8/04/84 Off The Wall Thriller Farewell My Summer Love 14 Greatest Hits- Michael Jackson/Jackson 5
MADONNA (3)	8/22/87-9/05/87 Madonna Like A Virgin True Blue Who's That Girl (Soundtrack)

Act (Total Weeks)	Dates
AC/DC (13)	**Four Albums** 12/19/81-3/13/82 Back In Black Dirty Deeds Done Dirt Cheap High Voltage For Those About To Rock (We Salute You)
ALABAMA (11)	2/11/84-4/21/84 Feels So Right Mountain Music The Closer You Get Roll On

Willie NELSON (6)	5/21/83-6/25/83
	Willie Nelson's Greatest Hits
	   (& Some That Will Be)
	Always On My Mind
	Tougher Than Leather
	Take It To The Limit-
	   Willie Nelson with Waylon Jennings

MÖTLEY CRÜE (3)	6/27/87-7/11/87 Shout At The Devil Too Fast For Love Theatre Of Pain Girls, Girls, Girls
Ozzy OSBOURNE (2)	1/08/83-1/15/83 Blizzard Of Ozz Diary Of A Madman Mr. Crowley Speak Of The Devil
IRON MAIDEN (2)	11/03/84-11/10/84 Maiden Japan The Number Of The Beast Piece Of Mind Powerslave

Album Acts: 2 Albums In Top 10 Simultaneously

Act (Total Weeks)	Dates
GUNS N' ROSES (9)	1/21/89-3/18/89 Appetite For Destruction G N' R Lies
MEN AT WORK (2)	5/14/83-5/21/83 Business As Usual Cargo

Act (Total Weeks)	Dates
NEW KIDS ON THE BLOCK (2)	12/23/89-12/30/89 Hangin' Tough Merry, Merry Christmas

Album Acts: 3 Albums In Top 40 Simultaneously

Act (Total Weeks)	Dates
NEW KIDS ON THE BLOCK (11)	10/21/89-12/30/89* Hangin' Tough New Kids On The Block Merry, Merry Christmas

*Streak continued two weeks into 1990.

Acts: Number 1 Albums

Alone or in collaboration

Act [Rank]	No. (Total Wks)
Bruce SPRINGSTEEN [2]	4 (19)
MADONNA [4]	3 (14)
Michael JACKSON [1]	2 (43)
PRINCE And The REVOLUTION [29]	2 (27)
Whitney HOUSTON [14]	2 (25)
ROLLING STONES [20]	2 (16)
U2 [9]	2 (15)
BON JOVI [13]	2 (12)
Phil COLLINS [19]	2 (8)
Billy JOEL [3]	2 (7)
VAN HALEN [11]	2 (7)
Barbra STREISAND [21]	2 (6)
Janet JACKSON [48]	2 (6)
Lionel RICHIE [8]	2 (5)
Huey LEWIS And The NEWS [17]	2 (2)
The POLICE [10]	1 (17)
PINK FLOYD [18]	1 (15)
REO SPEEDWAGON [25]	1 (15)
MEN AT WORK [43]	1 (15)
George MICHAEL [51]	1 (12)
FOREIGNER [30]	1 (10)
John MELLENCAMP [7]	1 (9)
DIRE STRAITS [49]	1 (9)
ASIA [95]	1 (9)
John LENNON & Yoko ONO [134]	1 (8)
BEASTIE BOYS [122]	1 (7)
FINE YOUNG CANNIBALS [129]	1 (7)
DEF LEPPARD [6]	1 (6)
Bob SEGER/SILVER BULLET Band [22]	1 (6)
PRINCE [31]	1 (6)
Bobby BROWN [62]	1 (6)
The GO-GO'S [92]	1 (6)
MILLI VANILLI [132]	1 (6)
GUNS N' ROSES [23]	1 (5)
FLEETWOOD MAC [42]	1 (5)
TEARS FOR FEARS [68]	1 (5)

Act [Rank]	No. (Total Wks)
Debbie GIBSON [84]	1 (5)
QUEEN [86]	1 (5)
Anita BAKER [58]	1 (4)
J. GEILS Band [111]	1 (4)
BOSTON [160]	1 (4)
Kim CARNES [172]	1 (4)
VANGELIS [196]	1 (4)
AC/DC [24]	1 (3)
STYX [50]	1 (3)
WHAM! [89]	1 (3)
Paul McCARTNEY [99]	1 (3)
MOODY BLUES [114]	1 (3)
USA For AFRICA [383]	1 (3)
Kenny ROGERS [15]	1 (2)
Bryan ADAMS [37]	1 (2)
MÖTLEY CRÜE [47]	1 (2)
SADE [69]	1 (2)
NEW KIDS ON THE BLOCK [80]	1 (2)
TIFFANY [130]	1 (2)
Los LOBOS [198]	1 (2)
JOURNEY [5]	1 (1)
Pat BENATAR [16]	1 (1)
HEART [26]	1 (1)
Steve WINWOOD [34]	1 (1)
Stevie NICKS [41]	1 (1)
Richard MARX [85]	1 (1)
Paula ABDUL [115]	1 (1)
Donna SUMMER [117]	1 (1)
QUIET RIOT [120]	1 (1)
Jackson BROWNE [133]	1 (1)
Tracy CHAPMAN [142]	1 (1)
John FOGERTY [183]	1 (1)
MR. MISTER [188]	1 (1)
Patti LaBELLE [189]	1 (1)
BEE GEES [218]	1 (1)
TONE LŌC [252]	1 (1)

Alphabetical Index of Acts:
Singles and Albums Rank

Acts Alpha Index

Act	Sing Rank	Alb Rank
A		
Kenny AARONSON	1394	1365
ABBA	270	296
Gregory ABBOTT	241	435
ABC	172	220
Paula ABDUL	65	115
Colonel ABRAMS		1215
ACCEPT		690
AC/DC	495	24
Bryan ADAMS	60	37
King Sunny ADE & His AFRICAN BEATS		874
The ADVENTURES	1379	1579
AEROSMITH	171	87
AFTER 7	1036	1379
AFTERNOON DELIGHTS	697	
AFTER THE FIRE	359	661
A-HA	223	329
AIR SUPPLY	11	72
ALABAMA	279	28
The ALARM	691	325
Gerald ALBRIGHT		1798
AL B. SURE!	422	353
ALCATRAZZ		970
ALESSI	1187	
ALISHA	894	
Deborah ALLEN	591	1016
Donna ALLEN	621	1344
Peter ALLEN	971	1145
Gregg ALLMAN Band	933	580
ALLMAN BROTHERS Band	723	588
ALL SPORTS BAND	1226	
Marc ALMOND	1070	1469
Herb ALPERT	276	251
Herb ALPERT & The TIJUANA BRASS	1358	1227
ALPHAVILLE	758	1412
Dave ALVIN		1303
AMAZING RHYTHM ACES		1826
AMAZULU	1350	
AMBROSIA	209	517
AMERICA	311	586
AMERICAN COMEDY NETWORK	1157	
ANA	1375	
Carl ANDERSON		1164
John ANDERSON	772	1017
Jon ANDERSON		1253
Laurie ANDERSON		765
Michael ANDERSON		1908
ANGEL		1688
ANGEL CITY		1293
ANIMAL LOGIC		1746
The ANIMALS	839	1110
ANIMOTION	269	488
Paul ANKA	636	1444
Adam ANT	400	295
ANTHRAX		350
Susan ANTON	826	
ANVIL		1906
APOLLONIA 6	1212	1040
APRIL WINE	522	339
ARABIAN PRINCE		1911
ARCADIA	380	659
Joan ARMATRADING	1165	314
ARMORED SAINT		875
Louis ARMSTRONG	817	
Steve ARRINGTON	1079	1010
ART IN AMERICA		1838
ARTISTS UNITED AGAINST APARTHEID	760	783
ART OF NOISE	559	528
ASHFORD & SIMPSON	451	318
ASIA	178	95
ASLEEP AT THE WHEEL		1896
The ASSOCIATION	1140	
The A's		1626
Jon ASTLEY	949	1479
Rick ASTLEY	106	205
ASWAD		1694
Chet ATKINS		1455
Christopher ATKINS	1118	
ATLANTA		1540
ATLANTA RHYTHM SECTION	720	871
ATLANTIC STARR	128	178
AUDIO TWO		1825
Barbara Ann AUER		1416
AURRA	1092	776
Patti AUSTIN	182	470
AUTOGRAPH	675	472
Johnny AVERAGE Band	1008	
AVERAGE WHITE BAND/ AWB	944	1242
AXE	1066	982
Roy AYERS		1009
AZTEC CAMERA		1245
B		
BABYFACE	494	548
BABYLON A.D.		1678
The BABYS	582	641
BACHMAN-TURNER OVERDRIVE		1901
BAD COMPANY	1007	444
BAD ENGLISH	268	541
BADFINGER	1023	1650
BADLANDS		809
Philip BAILEY	232	433
Razzy BAILEY		1726
Scott BAIO		1816
Anita BAKER	166	58
BALAAM AND THE ANGEL		1832
BALANCE	569	1474
Marty BALIN	347	736
Russ BALLARD	962	1282
BALTIMORA	515	947
Afrika BAMBAATAA and the SOUL SONIC FORCE	867	
BANANARAMA	129	283
BAND AID	680	
BAND OF GOLD	1131	
Moe BANDY		1866
BANGLES	51	118
BANG TANGO		844
Tony BANKS		1871
Frank BARBER Orchestra	916	1130
Pete BARDENS		1659
BARDEUX	667	1126
The BAR-KAYS	789	342
Jimmy BARNES	872	963
Lou Ann BARTON		1494
BASIA	593	422
Toni BASIL	186	569
BAUHAUS		1695
BEACH BOYS	159	539
The BEARS		1712
BEASTIE BOYS	411	122
BEAT FARMERS		1226
The BEATLES	466	200
BEAU COUP	1194	
Jean BEAUVOIR	1102	1121
Jeff BECK	1044	477
BEE GEES	356	218
Adrian BELEW	1034	986
Randy BELL	1367	
Regina BELLE	855	796
BELLE STARS	557	1902
Pat BENATAR	56	16
Tony BENNETT		1622
David BENOIT		985
George BENSON	143	139
BERLIN	227	362
Leonard BERNSTEIN		1007
Dickey BETTS		1830
The B-52's	316	170
BIG AUDIO DYNAMITE		602
BIG COUNTRY	521	303
BIG NOISE	1386	
BIG PIG	961	1098
BIG RIC	1340	
BIG TROUBLE	1151	
BILLY SATELLITE	993	1586
Stephen BISHOP	595	
Biz MARKIE		917
Clint BLACK		778
Jay BLACK	1385	
The BLACKBYRDS		1411
BLACKFOOT	844	565
J. BLACKFOOT	1297	
BLACK 'N BLUE		880
BLACK SABBATH		328
BLACK UHURU		1565
Ruben BLADES		1690
The BLASTERS		519
BLONDIE	45	155
BLOODSTONE		1263
Kurtis BLOW	1039	649
BLOWFLY		1066
BLOW MONKEYS	579	725
BLUE MERCEDES	1115	1720
BLUE MURDER		977
BLUE ÖYSTER CULT	742	307
BLUES BROTHERS	526	467
BLUE ZONE U.K.	983	
BOBBY And The MIDNITES		1531
BOBBY JIMMY & The CRITTERS		1927
BODEANS		751
Angela BOFILL		359
Michael BOLTON	326	413
Gary (U.S.) BONDS	372	531
BONHAM		838
BON JOVI	27	13
Karla BONOFF	504	626
BOOGIE BOYS		808
BOOGIE DOWN PRODUCTIONS		654
Chuckii BOOKER	799	1373
BOOK OF LOVE	1335	1549
BOOMTOWN RATS	1158	1056
BOOTSY'S RUBBER BAND		1062
BOSTON	199	160
BOSTON POPS Orchestra/ John WILLIAMS		1433
BOURGEOIS TAGG	653	948
David BOWIE	83	94
Rick BOWLES	1249	
BOW WOW WOW	937	777
BOX OF FROGS		845
Boy GEORGE	835	1239
BOY MEETS GIRL	305	675
The BOYS	561	603
BOYS BAND	1029	
BOYS CLUB	489	1142
BOYS DON'T CRY	548	909
Terry BOZZIO		1394
Billy BRAGG		1925
BRAM TCHAIKOVSKY		1186
The BRANDOS		1143
Laura BRANIGAN	71	215
John BRANNEN		1424
BRASS CONSTRUCTION		910
BREAKFAST CLUB	457	689
BREAKWATER		1641
BREATHE	158	431
BREATHLESS	1339	
BRECKER BROTHERS		1827
BRICK		1221
Edie BRICKELL & The NEW BOHEMIANS	434	263
BRIDES OF FUNKENSTEIN		1427
Martin BRILEY	727	771
Britny FOX	1398	547
BRONSKI BEAT	775	702
BROOKLYN, BRONX & QUEENS Band		1400
BROS	1280	1736
BROTHERS JOHNSON	414	334
Alex BROWN	1207	
Bobby BROWN	63	62
Danny Joe BROWN		1519
James BROWN	401	1045
Jocelyn BROWN	1082	
Julie BROWN		1651
Peter BROWN	1010	
Sam BROWN	988	
Jackson BROWNE	170	133
Tom BROWNE		409
Jack BRUCE		1129
Bill BRUFORD		1903
Sharon BRYANT	779	1391
Peabo BRYSON	369	336
Peabo BRYSON/ Roberta FLACK	408	*425*
--also see Roberta FLACK and Peabo BRYSON		
B.T. EXPRESS		1743
Roy BUCHANAN		1259
Lindsey BUCKINGHAM	344	506
BUCKNER & GARCIA	465	773
BUCKWHEAT ZYDECO		1350
Jimmy BUFFETT	816	270
The BUGGLES		1707
Alex BUGNON		1385
BULGARIAN STATE RADIO & T.V. FEMALE CHOIR		1553
Cindy BULLENS	1325	
BULLETBOYS	956	566
Gary BURBANK with BAND McNALLY	1099	
T-Bone BURNETT		1809
Billy BURNETTE	1139	
Rocky BURNETTE	437	1003
BURNING SENSATIONS		1790
George BURNS	876	1311
Glen BURTNICK	1047	1614
Jenny BURTON	992	1805

Acts Alpha Index

Act	Sing Rank	Alb Rank
BUS BOYS	1191	1039
Kate BUSH	688	418
Jon BUTCHER AXIS	1373	592
Jonathan BUTLER	711	576
BUZZCOCKS		1722
Donald BYRD And 125th STREET N.Y.C.		1285
David BYRNE		803
D.L. BYRON		1447

C

Act	Sing Rank	Alb Rank
CACTUS WORLD NEWS		1759
John CAFFERTY	1181	
John CAFFERTY And The BEAVER BROWN BAND	248	247
Tané CAIN	838	1359
Bobby CALDWELL	763	1049
J.J. CALE		1244
John CALE		1672
CALIFORNIA RAISINS	1296	708
The CALL	921	555
CAMEO	308	123
Rafael CAMERON		913
CAMOUFLAGE	963	1212
Glen CAMPBELL	909	1839
CAMPER VAN BEETHOVEN		1030
CANDI	1049	
CANDLEMASS		1724
Freddy CANNON & The BELMONTS	1265	
Jim CAPALDI	687	1171
CAPTAIN & TENNILLE	119	693
Carla CAPUANO		1572
Irene CARA	55	624
Luis CARDENAS	1264	
Tony CAREY	501	820
George CARLIN		1157
Belinda CARLISLE	93	179
Steve CARLISLE	967	
Carl CARLTON	597	754
Larry CARLTON	601	785
Eric CARMEN	188	837
Jean CARN		1582
Kim CARNES	66	172
Mary Chapin CARPENTER		1834
CARPENTERS	530	636
Paul CARRACK	330	651
Terri Lyne CARRINGTON		1681
Jim CARROLL Band		893
The CARS	72	36
Carlene CARTER		1606
Clarence CARTER		1861
Johnny CASH		1655
Rosanne CASH	620	424
CASHFLOW		1477
Felix CAVALIERE	784	
CELLARFUL OF NOISE	1133	
CENTRAL LINE	1228	1542
Peter CETERA	108	372
CHAMPAIGN	375	615
Bill CHAMPLIN	836	1784
Gene CHANDLER		1055
CHANGE	695	441
Harry CHAPIN	669	1005
Tracy CHAPMAN	397	142
CHARLENE	320	810
Ray CHARLES		988
Sonny CHARLES	765	1574
CHARLIE	800	1523

Act	Sing Rank	Alb Rank
CHEAP TRICK	130	157
Chubby CHECKER	1304	1884
CHEECH & CHONG	863	1218
Joe CHEMAY Band	1018	
CHEQUERED PAST		1632
CHER	147	277
CHERI	821	
CHERRELLE	806	613
Neneh CHERRY	282	688
CHIC	907	732
CHICAGO	15	82
CHICAGO BEARS SHUFFLIN' CREW	914	
The CHIEFTAINS		1508
Toni CHILDS	1126	691
The CHI-LITES		1081
CHILLIWACK	429	840
CHINA CRISIS		1225
The CHIPMUNKS		461
CHOCOLATE MILK		1533
CHOIRBOYS	1195	
Chris CHRISTIAN	755	
The CHRISTIANS		1620
Gavin CHRISTOPHER	671	1095
CHUNKY A	1223	1554
The CHURCH	725	579
CINDERELLA	287	83
CIRCUS OF POWER		1886
CLANNAD		1266
Eric CLAPTON	262	96
Petula CLARK	1084	
Allan CLARKE	1172	
Stanley CLARKE	753	995
Stanley CLARKE & George DUKE		612
The CLASH	296	145
Andrew Dice CLAY		855
Richard CLAYDERMAN		1556
Merry CLAYTON	896	
Johnny CLEGG & SAVUKA		1617
Clarence CLEMONS	580	930
Jimmy CLIFF		1502
Linda CLIFFORD	857	1473
CLIMAX BLUES BAND	409	1102
CLIMIE FISHER	683	1207
Patsy CLINE		794
George CLINTON		563
CLOCKS	1103	
CLUB HOUSE	1184	
CLUB NOUVEAU	247	324
Joyce COBB	805	
Tom COCHRANE/ RED RIDER		567
Bruce COCKBURN	594	561
Joe COCKER	361	460
COCK ROBIN	749	940
COCTEAU TWINS		1178
David Allan COE		1771
COLD CHISEL		1689
Gardner COLE	1355	
Natalie COLE	165	320
Durell COLEMAN		1589
Albert COLLINS		1600
Judy COLLINS		1490
Phil COLLINS.	5	19
Jessi COLTER		1133
Shawn COLVIN		1810
The COMMODORES	103	116
The COMMUNARDS	706	997
COMPANY B	634	1612

Act	Sing Rank	Alb Rank
CONCRETE BLONDE		993
CONDUCTOR	1119	
CONEY HATCH		1888
CON FUNK SHUN	807	354
John CONLEE		1680
The CONNELLS		1599
Harry CONNICK Jr.		865
Norman CONNORS		1597
Bill CONTI	930	
The CONTOURS	589	
Tommy CONWELL And The YOUNG RUMBLERS	829	981
Ry COODER		707
Sam COOKE		1363
Rita COOLIDGE	496	1231
Alice COOPER	435	346
Michael COOPER		1028
Julian COPE	1308	992
Johnny COPELAND		1601
Stewart COPELAND		1402
Chick COREA		1648
Al CORLEY	1254	
Bill COSBY		676
Elvis COSTELLO	693	479
Elvis COSTELLO And The ATTRACTIONS	635	194
Gene COTTON	1147	
Josie COTTON	1042	1503
COVER GIRLS	360	650
COWBOY JUNKIES		590
COYOTE SISTERS	973	
CRACK THE SKY		1791
Floyd CRAMER		1751
Michael CRAWFORD		1897
Randy CRAWFORD	1159	825
Robert CRAY Band	648	246
CREEDENCE CLEARWATER REVIVAL		905
Marshall CRENSHAW	813	549
The CRETONES	1210	1389
David CROSBY		1319
CROSBY, STILLS & NASH	315	276
CROSBY, STILLS, NASH & YOUNG	1100	533
Christopher CROSS	34	90
CROWDED HOUSE	190	291
Rodney CROWELL	815	1159
CROWN HEIGHTS AFFAIR		1460
The CRUSADERS	1381	529
CRUZADOS		835
Billy CRYSTAL	922	1085
CUGINI	1273	
The CULT	830	250
CULTURE CLUB	24	56
Burton CUMMINGS	837	
The CURE	295	152
CURIOSITY KILLED THE CAT	833	868
Tim CURRY		1185
CURTIE And The BOOMBOX	1270	
CUTTING CREW	202	463
Andre CYMONE		1386

D

Act	Sing Rank	Alb Rank
E.G. DAILY	991	
DA'KRASH		1859
Roger DALTREY	447	481
Michael DAMIAN	298	999
Dana DANE		669
DANGER DANGER		1038

Act	Sing Rank	Alb Rank
Rodney DANGERFIELD	1156	585
DANGEROUS TOYS		781
Charlie DANIELS Band	373	348
DANNY WILSON	627	1114
DANZIG		1419
Terence Trent D'ARBY	173	222
DARK ANGEL		1642
DARLING CRUEL		1598
DAVE & SUGAR		1801
DAVID & DAVID	654	560
F.R. DAVID	968	
Dave DAVIES		847
Danny DAVIS		1819
Jimmy DAVIS & JUNCTION	1088	1471
Mac DAVIS	685	864
Martha DAVIS	1200	1304
Miles DAVIS		699
Paul DAVIS	215	857
Tyrone DAVIS	1046	1608
Arlan DAY	1073	
Morris DAY	617	553
Taylor DAYNE	95	281
DAYTON	1016	
DAZZ BAND	323	287
DB'S		1654
DEAD MILKMEN		1091
DEAD OR ALIVE	374	508
Paul DEAN		1912
DEATH ANGEL		1483
DeBARGE	133	204
Bunny DeBARGE		1729
Chico DEBARGE	583	958
El DeBARGE	349	598
El DeBARGE With DeBARGE	804	
Chris De BURGH	258	427
The DEELE	509	642
DEEP PURPLE	1021	349
Rick DEES	1178	
DEF LEPPARD	67	6
DEJA	898	1769
DE LA SOUL	738	584
DEL FUEGOS	1320	937
DELIVERANCE	1145	
The DELLS		1377
John DENVER	445	344
John DENVER & The MUPPETS		1200
DEODATO	1162	1856
DEPECHE MODE	427	272
Teri DeSARIO	252	1163
Jackie DeSHANNON	1258	
DEVICE	718	1077
DEVO	383	264
DEXYS MIDNIGHT RUNNERS	230	522
Dennis DeYOUNG	440	639
DFX2		1569
Joel DIAMOND	1295	
Neil DIAMOND	62	52
DIESEL	651	1015
DIFFORD & TILBROOK		989
DILLMAN Band	915	1588
Al Di MEOLA		815
Al Di MEOLA/ John McLAUGHLIN/ Paco De LUCIA		1180
DINO	312	581
DIO		268
DION	1237	1150
DIRE STRAITS	148	49

Acts Alpha Index

Act	Sing Rank	Alb Rank
DIRTY LOOKS		1053
DISNEYLAND AFTER DARK		1299
DIVING FOR PEARLS	1238	
DIVINYLS	1105	1100
DIXIE DREGS		687
Don DIXON		1623
DJ JAZZY JEFF & The FRESH PRINCE	371	224
The D.O.C.		601
DOCTOR And The MEDICS	976	1520
DR. HOOK	179	811
DR. JOHN		1468
DR. J.R. KOOL & The OTHER ROXANNES		1234
DOKKEN	808	191
Thomas DOLBY	300	256
Joe DOLCE	846	1812
Micky DOLENZ And Peter TORK	677	
DOLLAR	1109	
Placido DOMINGO	1190	574
DOOBIE BROTHERS	245	206
DOOLITTLE Band	966	
The DOORS	1058	236
Charlie DORE	510	1590
DORO		1484
DOUBLE	578	757
DOUBLE IMAGE	1359	
Doug E. FRESH & The GET FRESH CREW		1205
DRAGON	1301	
The DRAMATICS		1119
DREAM ACADEMY	413	559
DREAMBOY		1538
DREAMS SO REAL		1339
DREAM SYNDICATE		1760
J.D. DREWS	1186	
D.R.I.		1206
DRIVIN' N' CRYIN'		1375
"D" TRAIN	1193	1440
George DUKE	625	813
DUKE JUPITER	873	1309
Robbie DUPREE	242	839
DURAN DURAN	7	32
Ian DURY And The BLOCKHEADS		1745
Bob DYLAN	970	211
DYNASTY	1257	786

E

Act	Sing Rank	Alb Rank
EAGLES	233	154
Steve EARLE		614
Steve EARLE And The DUKES		1138
EARTH, WIND & FIRE	152	147
EASTERHOUSE	1279	
Elliot EASTON		1267
Sheena EASTON	33	146
Clint EASTWOOD	1208	
EAZY-E		447
EBN-OZN		1815
EBONEE WEBB		1624
ECHO & The BUNNYMEN		622
John EDDIE	902	1123
EDDIE And The TIDE	1347	
Dave EDMUNDS	641	505
Dennis EDWARDS	1143	758
Walter EGAN	849	1889
EGYPTIAN LOVER		1492
EIGHTH WONDER	812	
EIGHT SECONDS	1071	
ELECTRIC LIGHT ORCHESTRA	168	255
Larry ELGART And His MANHATTAN SWING Orchestra	780	458
Joe ELY		1312
Keith EMERSON		1843
EMERSON, LAKE & PALMER		1318
EMERSON, LAKE & POWELL	1027	564
The EMOTIONS		1436
ENGLAND DAN & John Ford COLEY	1199	1909
Jackie ENGLISH	1361	
ENGLISH BEAT		438
Brian ENO		1278
John ENTWISTLE		1203
ENUFF Z'NUFF	1068	1075
ENYA	665	491
EPMD		714
ERASURE	403	454
ERIC B. & RAKIM		499
ESCAPE CLUB	234	507
Joe "Bean" ESPOSITO	1345	
ESQUIRE		1767
Gloria ESTEFAN/ MIAMI SOUND MACHINE	44	77
Deon ESTUS	482	1132
Melissa ETHERIDGE	1318	308
E.U.	802	1559
EUROGLIDERS	1045	1482
EUROPE	225	156
EURYTHMICS	64	74
EVERLY BROTHERS	883	674
EXODUS		807
EXPOSÉ	96	187
EXTREME		899
EYE TO EYE	743	1187
EZO		1547

F

Act	Sing Rank	Alb Rank
FABULOUS THUNDERBIRDS	463	285
FACE TO FACE	740	1202
Eria FACHIN	932	
Donald FAGEN	629	498
Joe FAGIN	1271	
FAIRGROUND ATTRACTION	1232	1454
Marianne FAITHFULL		924
FALCO	221	352
Harold FALTERMEYER	331	
Agnetha FALTSKOG	692	1295
The FAMILY	1063	946
FAR CORPORATION	1310	
FARRENHEIT		1717
Cee FARROW	1213	
FASTER PUSSYCAT		763
FASTWAY		489
FATBACK		652
FAT BOYS	503	186
FATES WARNING		1078
The FEELIES		1752
Don FELDER	731	1674
Wilton FELDER		978
Suzanne FELLINI	1334	
FELONY	822	1792
FEMME FATALE		1644
Rick FENN		1836
Jay FERGUSON		1744
Maynard FERGUSON		1733
Bryan FERRY	796	546
FESTIVAL	1107	884
FETCHIN BONES		1662
Richard "Dimples" FIELDS	858	666
FIFTH ANGEL		1297
FIGURES ON A BEACH	1120	
FINE YOUNG CANNIBALS	141	129
Tim FINN		1713
FIONA	1062	980
Elisa FIORILLO	759	1647
FIREFALL	531	904
FIRE INC.	1229	
The FIRM	613	370
FISHBONE		1630
FIVE SATINS	1132	
FIVE STAIRSTEPS		1122
FIVE STAR	577	535
The FIXX	175	151
Roberta FLACK	394	473
Roberta FLACK & Peabo BRYSON	408	425
--also see Peabo BRYSON/ Roberta FLACK		
FLASH & THE PAN		1675
Mick FLEETWOOD		951
FLEETWOOD MAC	73	42
FLESH FOR LULU		1002
The FLESHTONES		1728
A FLOCK OF SEAGULLS	319	237
FLOTSAM AND JETSAM		1568
FLYING LIZARDS	889	1390
Dan FOGELBERG	79	78
John FOGERTY	381	183
FOGHAT	586	911
Ellen FOLEY		1700
Jane FONDA		217
The FOOLS	886	1457
Steve FORBERT	485	616
FORCE M.D.'S	508	726
Lita FORD	336	326
Robben FORD		1290
Julia FORDHAM		1101
FOREIGNER	26	30
FORTUNE	1202	
David FOSTER	540	1393
FOSTER & LLOYD		1636
4 BY FOUR	1201	1607
4 SEASONS	1290	
FOUR TOPS	399	727
Charles FOX	1247	
Samantha FOX	164	338
Peter FRAMPTON	1087	743
FRANKE AND THE KNOCKOUTS	348	538
FRANKIE GOES TO HOLLYWOOD	416	439
Aretha FRANKLIN	110	119
Rodney FRANKLIN		1189
Michael FRANKS		597
Andy FRASER	1245	
Linda FRATIANNE		1684
Ace FREHLEY		703
FREHLEY'S COMET		960
Glenn FREY	89	231
FRIDA	430	761
Robert FRIPP		1210
FROZEN GHOST	1014	1280
FULL FORCE		1036
FUN BOY THREE		1420
FUNKADELIC		1384
FUNKADELIC(2)		1704
FUSE ONE		1526

G

Act	Sing Rank	Alb Rank
Peter GABRIEL	145	98
James GALWAY		1764
GAMMA	943	789
GANG OF FOUR		1564
GAP BAND	460	143
Jerry GARCIA		1422
Art GARFUNKEL	1002	1201
Leif GARRETT	831	1156
Siedah GARRETT	1393	
GARY O'	1182	
David GATES	782	
Larry GATLIN/ GATLIN BROTHERS Band		1268
Marvin GAYE	240	273
Crystal GAYLE	421	532
Gloria GAYNOR		1811
J. GEILS Band	81	111
Bob GELDOF	1216	1368
GENE LOVES JEZEBEL	1342	944
GENERAL PUBLIC	652	453
GENESIS	48	27
GENTLE PERSUASION	1261	
Robin GEORGE	1370	
GEORGIA SATELLITES	314	322
GEORGIO	773	872
GET WET	910	
GIANT	958	1173
GIANT STEPS	516	1787
Andy GIBB	191	570
Barry GIBB	827	1286
Robin GIBB	703	
Terri GIBBS	512	843
Debbie GIBSON	54	84
GIDEA PARK	1287	
Johnny GILL	1215	1686
GILLAN		1850
Mickey GILLEY	518	1581
David GILMOUR	1040	610
GINA GO-GO	1081	
GIPSY KINGS		696
GIRLSCHOOL		1777
GIUFFRIA	497	502
Philip GLASS		1087
GLASS MOON	987	1499
GLASS TIGER	196	367
Roger GLOVER		1302
GOANNA	1072	1766
The GODFATHERS		1070
GODLEY & CREME	581	863
The GO-GO's	109	92
GOLDEN EARRING	405	527
GOOD QUESTION	1284	
Robert GORDON	1235	1079
Martin L. GORE		1715
Michael GORE	1227	
GORKY PARK		1072
Barry GOUDREAU		1296
GO WEST	533	701
GQ	1372	826
The GRACES	986	1550
Larry GRAHAM	475	495
Lou GRAMM	417	599
GRAND FUNK RAILROAD		1660
GRANDMASTER FLASH		1605

403

Acts Alpha Index

Act	Sing Rank	Alb Rank
GRANDMASTER FLASH And The FURIOUS FIVE	1050	816
GRANDMASTER MELLE MEL And The FURIOUS FIVE	1337	
Amy GRANT	642	392
Eddy GRANT	169	397
Stephane GRAPELLI		1603
GRATEFUL DEAD	528	221
GREAT WHITE	304	169
Boris GREBENSHIKOV		1918
Al GREEN		1313
Jack GREEN		1501
Peter GREEN		1788
GREEN ON RED		1737
Lee GREENWOOD	871	790
Joanie GREGGAINS		1802
GREY And HANKS		1880
Nanci GRIFFITH		1174
GRIM REAPER		662
David GRISMAN		1314
Dave GRUSIN		524
GTR	564	436
GUADALCANAL DIARY		1241
GUCCI CREW II		1714
Greg GUIDRY	574	1609
GUNS N' ROSES	111	23
Arlo GUTHRIE		1858
Gwen GUTHRIE	834	1167
GUY	995	357

H

Act	Sing Rank	Alb Rank
Steve HACKETT		1518
Sammy HAGAR	266	161
Nina HAGEN		1514
Merle HAGGARD		667
HAIRCUT ONE HUNDRED	698	600
Daryl HALL	382	606
Daryl HALL & John OATES	3	12
Jimmy HALL	708	1870
John HALL Band	734	1292
Lani HALL	1333	
David HALLYDAY	1214	
Jan HAMMER	284	
Herbie HANCOCK	1038	360
Paul HARDCASTLE	535	894
Emmylou HARRIS	724	244
Sam HARRIS	647	595
George HARRISON	138	271
Jerry HARRISON: CASUAL GODS		976
Debbie HARRY	640	683
Corey HART	121	253
Dan HARTMAN	299	792
Richie HAVENS		1786
HAWKS	1033	
Colin James HAY	1397	1461
Isaac HAYES	524	668
HAYSI FANTAYZEE	1173	
Justin HAYWARD		1721
Leon HAYWOOD	866	1330
Ofra HAZA		1448
Robert HAZARD	925	1269
Murray HEAD	346	
The HEADBOYS		1585
HEAD EAST		1223
HEADPINS	1032	1425
Jeff HEALEY Band	432	452
HEAR 'N AID		1358
HEART	46	26

Act	Sing Rank	Alb Rank
HEATWAVE		1127
HEAVEN 17	1176	738
HEAVY D & The BOYZ		437
HELIX		841
HELLOWEEN		882
Michael HENDERSON		719
Jimi HENDRIX		1060
Jimi HENDRIX EXPERIENCE		1208
Nona HENDRYX	946	867
Don HENLEY	90	128
Carol HENSEL		450
Howard HEWETT	1328	1093
Nick HEYWARD		1799
John HIATT		812
Bertie HIGGINS	351	730
HIGH INERGY	1231	
Dan HILL	370	1061
Z.Z. HILL		1451
Eric HINE	1170	
HIPSWAY	660	935
HIROSHIMA		379
Robyn HITCHCOCK And The EGYPTIANS		1041
Roger HODGSON	801	762
Amy HOLLAND	644	1399
Jennifer HOLLIDAY	562	694
The HOLLIES	719	1308
HOLLY And The ITALIANS		1831
Rupert HOLMES	333	648
The HONEYDRIPPERS	281	337
HONEYMOON SUITE	570	545
HOODOO GURUS		934
John Lee HOOKER		1197
The HOOTERS	345	227
Lena HORNE		1392
Bruce HORNSBY And The RANGE	116	110
HOT CHOCOLATE	1048	
HOTEL	1054	
HOTHOUSE FLOWERS		932
The HOUSEMARTINS		1255
HOUSE OF FREAKS		1510
HOUSE OF LORDS	1074	950
HOUSE OF LOVE		1649
Thelma HOUSTON		1575
Whitney HOUSTON	8	14
George HOWARD		817
Miki HOWARD		1248
Steve HOWE		1762
David HUDSON	900	1872
Grayson HUGH	623	967
HUGHES/THRALL	1179	
HUMAN LEAGUE	82	174
HUMBLE PIE	947	1013
Engelbert HUMPERDINCK	1134	
Ian HUNTER		767
John HUNTER	745	1530
HURRICANE		891
Jim HURT	1326	
HUSKER DU		1168
Paul HYDE And The PAYOLAS	1278	1488
Phyllis HYMAN		510
Tony HYMAS		1395
Chrissie HYNDE	918	

I

Act	Sing Rank	Alb Rank
Janis IAN	1205	1763
ICEHOUSE	329	405

Act	Sing Rank	Alb Rank
ICE-T	1123	471
ICICLE WORKS	774	861
ICON		1898
Billy IDOL	87	71
Julio IGLESIAS	458	177
IMPELLITTERI		1025
INDIGO GIRLS	936	492
INDUSTRY	1177	
INFORMATION SOCIETY	283	525
James INGRAM	259	582
The INMATES	890	1058
INNER CITY	1043	1740
INSIDERS		1718
INSTANT FUNK		1103
INVISIBLE MAN'S BAND	825	
INXS	76	60
Donnie IRIS	419	583
IRISH ROVERS	710	1594
IRONHORSE	1272	
IRON MAIDEN		107
Chris ISAAK		1498
ISLE OF MAN	1331	1160
ISLEY BROTHERS	747	229
ISLEY, JASPER, ISLEY	746	850

J

Act	Sing Rank	Alb Rank
Freddie JACKSON	275	148
Janet JACKSON	41	48
Jermaine JACKSON	161	166
Joe JACKSON	212	124
La Toya JACKSON	948	1135
Marlon JACKSON		1708
Michael JACKSON	1	1
Millie JACKSON		715
Rebbie JACKSON	628	964
JACKSON 5		192
The JACKSONS	180	
Debbie JACOBS	1166	1676
Mick JAGGER	350	410
The JAGS	1321	
The JAM		695
Ahmad JAMAL Trio		1738
Bob JAMES		368
Bob JAMES & Earl KLUGH		476
Melvin JAMES		1571
Rick JAMES	321	113
Tommy JAMES	551	1536
Nick JAMESON	1382	
JANE'S ADDICTION		890
Chaz JANKEL		1333
Jean Michel JARRE		772
Al JARREAU	306	127
JASON & The SCORCHERS		741
Chris JASPER		1851
Miles JAYE		1183
JEFFERSON AIRPLANE/ STARSHIP	43	79
Garland JEFFREYS	1078	854
JELLYBEAN	436	1261
Waylon JENNINGS	550	259
Waylon JENNINGS & Willie NELSON		906
JESUS AND MARY CHAIN		1452
JETBOY		1430
JETHRO TULL		305
The JETS	94	226
Joan JETT/BLACKHEARTS	78	121
JIVE BUNNY & The MASTERMIXERS		1868

Act	Sing Rank	Alb Rank
J.J. FAD	600	722
JOBOXERS	722	1067
Billy JOEL	6	3
David JOHANSEN		962
Elton JOHN	18	70
Robert JOHN	556	
JOHNNY And The DISTRACTIONS		1561
JOHNNY HATES JAZZ	271	834
Don JOHNSON	395	544
Holly JOHNSON	1095	
Howard JOHNSON		1480
Jesse JOHNSON	609	401
Michael JOHNSON	1291	
Tom JOHNSTON	768	1576
France JOLI		1837
JON & VANGELIS	798	750
George JONES		959
Glenn JONES	953	1118
Grace JONES	1031	432
Howard JONES	114	195
Mick JONES		1852
Oran "Juice" JONES	523	821
Quincy JONES	456	232
Rickie Lee JONES	1000	262
Shirley JONES		1403
Steve JONES		1775
Tom JONES		1835
JONES GIRLS		955
JONZUN CREW		956
Janis JOPLIN		1265
Stanley JORDAN		493
JOURNEY	29	5
JOY DIVISION		1522
JUDAS PRIEST	1076	125
The JUDDS		301
Patrick JUDE	1138	
JULUKA		1806
JUMP 'N THE SADDLE	598	
Rob JUNGKLAS	1316	1071
JUNIOR	787	990
JUNKYARD		1283

K

Act	Sing Rank	Alb Rank
KAJAGOOGOO	398	804
Karen KAMON	1351	
Big Daddy KANE		753
Madleen KANE	1206	
KANE GANG	679	1105
KANO	1288	1842
KANSAS	368	293
KASHIF	1235	500
KATRINA & The WAVES	340	442
Jorma KAUKONEN & VITAL PARTS		1679
KBC BAND	1330	900
KC	1218	
KC & The SUNSHINE BAND	160	793
KEEL		655
Tommy KEENE		1307
Johnny KEMP	444	941
Joyce KENNEDY	965	1153
Ray KENNEDY	1267	
KENNY G	246	93
KENTUCKY HEADHUNTERS		1677
Nik KERSHAW	818	903
Chaka KHAN	189	239
KIARA	1253	
KICK AXE		1243

Acts Alpha Index

Act	Sing Rank	Alb Rank
KID CREOLE & The COCONUTS		1408
KID 'N PLAY		760
KIDS FROM FAME		1090
Greg KIHN	756	
Greg KIHN Band	187	294
KILLER DWARFS		1701
KILLING JOKE		1921
John KILZER		1256
Tom KIMMEL	1030	1262
KING	904	1527
B.B. KING	1211	1175
Ben E. KING	438	
Carole KING	443	716
Evelyn "Champagne" KING	453	423
The KINGBEES	1135	1491
KING CRIMSON		617
KING DIAMOND		928
KINGDOM COME	1163	414
The KINGS	717	914
KINGS OF THE SUN	1391	1345
KING SWAMP		1515
KING'S X		1023
KING TEE		1279
Sam KINISON		866
The KINKS	338	171
Jim KIRK And The TM Singers	1221	
KISS	442	105
KISSING THE PINK	1260	
KITARO		1413
KIX	492	554
KLEEER		915
John KLEMMER		1112
KLIQUE	897	1125
Earl KLUGH		302
KLYMAXX	204	319
The KNACK	713	556
Gladys KNIGHT & The PIPS	532	345
Holly KNIGHT	1009	
Jean KNIGHT	888	1808
Jerry KNIGHT		1459
K-9 POSSE		1193
Fred KNOBLOCK	538	1756
Mark KNOPFLER		1920
KON KAN	547	
KOOL & The GANG	13	35
KOOL MOE DEE	938	351
The KORGIS	554	1351
KORONA	887	
KRAFTWERK		680
Lenny KRAVITZ		1534
Kris KRISTOFFERSON		1807
KROKUS	942	280
KWAMÉ & A NEW BEGINNING		1194
KWICK		1917

L

Act	Sing Rank	Alb Rank
LABAN	1317	
Patti LaBELLE	224	189
L.A. BOPPERS		1195
LACE		1800
L.A. DREAM TEAM		1450
La FLAVOUR	1365	
L.A. GUNS		604
LAID BACK	666	1068
Cleo LAINE		1765
Greg LAKE	860	954
LAKESIDE	1004	317
Lorenzo LAMAS	1263	
Robin LANE & The CHARTBUSTERS	1319	1779
K.D. LANG		621
LANIER AND CO.	832	
David LANZ		1443
LARSEN-FEITEN Band	716	1497
Nicolette LARSON	672	842
David LASLEY	828	
James LAST Band	704	1528
Stacy LATTISAW	352	420
Cyndi LAUPER	19	67
Debra LAWS	1305	902
Eloise LAWS		1691
Hubert LAWS		1229
Ronnie LAWS	996	542
LEATHERWOLF		1092
Lenny Le BLANC	998	
LED ZEPPELIN	618	311
Alvin LEE		1387
Johnny LEE	357	1089
Larry LEE	1303	
Paul LEKAKIS	809	
John LENNON	59	415
John LENNON & Yoko ONO		134
John LENNON and the PLASTIC ONO BAND	1251	742
Julian LENNON	222	267
Annie LENNOX & Al GREEN	539	
LEROI BROS.		1772
Le ROUX	615	883
LET'S ACTIVE		1012
LEVEL 42	322	313
LEVERT	441	514
Marcy LEVY	1127	
Huey LEWIS and the NEWS	12	17
Jerry Lee LEWIS		1076
Ramsey LEWIS		1328
Shirley LEWIS	1244	
Webster LEWIS		1423
Gordon LIGHTFOOT	926	895
LIL LOUIS	882	
LIMAHL	560	831
LIMITED WARRANTY	1137	
David LINDLEY		1019
LINX		1789
LIONS AND GHOSTS		1860
LIPPS, Inc.	162	387
LIQUID GOLD	1286	
LISA LISA And CULT JAM	142	207
LISA LISA And CULT JAM With FULL FORCE	337	
Rich LITTLE		832
LITTLE AMERICA		1217
LITTLE FEAT		446
LITTLE RICHARD	864	
LITTLE RIVER BAND	107	258
LITTLE STEVEN	1024	692
LIVING COLOUR	499	249
LIVING IN A BOX	622	1247
LIZZY BORDEN		1050
LL COOL J	391	153
LOBO	1110	
Nils LOFGREN		1177
Kenny LOGGINS	36	158
LONDON SYMPHONY Orchestra		921
LONE JUSTICE	751	571
LOOSE ENDS	878	635

Act	Sing Rank	Alb Rank
Denise LOPEZ	715	1822
Jeff LORBER	726	512
Bill LORDAN		1238
LORDS Of The NEW CHURCH		1640
Gloria LORING		1064
Gloria LORING & Carl ANDERSON	278	
LOS LOBOS	207	198
LOUDNESS		746
LOVE AND MONEY	1169	1696
LOVE AND ROCKETS	358	331
LOVERBOY	126	38
Lyle LOVETT		791
Lene LOVICH		1341
Nick LOWE		942
Nick LOWE And His COWBOY OUTFIT	1150	1044
L.T.D.	714	575
L'TRIMM	861	1300
Carrie LUCAS		1730
LULU	511	1421
Ray LYNCH		1915
Cheryl LYNN	924	907
Jeff LYNNE	1302	
LYNYRD SKYNYRD		361

M

Act	Sing Rank	Alb Rank
M		1428
Tony MacALPINE		1475
MAC Band		1298
Ralph MacDONALD	982	1336
Mary MacGREGOR	1183	
Lonnie MACK		1209
MADAME X		1719
MADHOUSE		1320
MADNESS	366	618
MADONNA	2	4
MAGAZINE 60	884	
MAHOGANY RUSH		1264
MAI TAI	1113	
MALICE		1753
Yngwie MALMSTEEN		375
MAMA'S BOYS		1438
Melissa MANCHESTER	253	378
Henry MANCINI and his Orchestra		1923
Howie MANDEL		1633
M + M	1035	1555
Barbara MANDRELL		795
MANFRED MANN'S EARTH BAND	633	705
Chuck MANGIONE	573	402
The MANHATTANS	310	469
MANHATTAN TRANSFER	307	284
Barry MANILOW	149	141
MANNHEIM STEAMROLLER		540
MANTRONIX		1382
Benny MARDONES	393	987
Teena MARIE	297	230
MARILLION	1089	605
Bob MARLEY And The WAILERS		396
Ziggy MARLEY & The MELODY MAKERS	824	365
MARLEY MARL		1698
Neville MARRINER		577
M/A/R/R/S	527	
Branford MARSALIS		1661
Wynton MARSALIS		593

Act	Sing Rank	Alb Rank
MARSHALL TUCKER Band	1250	657
MARTIKA	231	465
Eric MARTIN	1352	1904
Marilyn MARTIN	332	1170
Moon MARTIN		1272
Steve MARTIN		1190
Nancy MARTINEZ	657	1853
Richard MARX	42	85
MARY JANE GIRLS	362	382
Dave MASON	1189	1252
Harvey MASON		1854
Jackie MASON		1545
Nick MASON		1693
Vaughan MASON And CREW	1274	
Wayne MASSEY	1369	
MASS PRODUCTION		1323
Johnny MATHIS	881	1004
Christopher MAX	1106	
MAX Q		1687
Brian MAY		1415
Curtis MAYFIELD		1404
Lyle MAYS		1107
MAZARATI		1532
MAZE Featuring Frankie BEVERLY	1124	278
MC HAMMER		398
MC LYTE		1240
MC SHY D		1924
Mac McANALLY	788	
Paul McCARTNEY	30	99
Paul McCARTNEY & WINGS	140	
Delbert McCLINTON	404	734
Ian McCULLOCH		1919
Michael McDONALD	135	393
Reba McENTIRE		663
McFADDEN & WHITEHEAD		1629
Bobby McFERRIN	292	304
McGUFFEY LANE	1225	1803
McGUINN, CLARK & HILLMAN		1537
Peter McIAN	1003	
Maria McKEE		1213
Bob & Doug McKENZIE	607	480
Sarah McLACHLAN		1442
Ian McLAGAN		1487
Malcolm McLAREN		1577
John McLAUGHLIN		1758
Pat McLAUGHLIN		1922
Don McLEAN	273	660
Gerard McMAHON	1324	
James McMURTRY		1449
Larry John McNALLY	1332	
Shamus M'COOL	1276	
Christine McVIE	426	684
MEAT LOAF	1315	873
MECO	459	957
Glenn MEDEIROS	424	1128
Bill MEDLEY	388	
MEGADETH		478
Randy MEISNER	406	665
MEL & KIM	1188	
John MELLENCAMP	10	7
Harold MELVIN And The BLUE NOTES		1074
MEN AT WORK	49	43
Sergio MENDES	217	504
MENUDO	945	961
MEN WITHOUT HATS	213	403
Freddie MERCURY	1022	1664
Jim MESSINA		1151

Acts Alpha Index

Act	Sing Rank	Alb Rank
METAL CHURCH		764
METALLICA	764	149
Pat METHENY		1108
Pat METHENY Group	923	312
George MICHAEL	23	51
Bette MIDLER	113	131
MIDNIGHT OIL	545	343
MIDNIGHT STAR	410	202
MIKE + THE MECHANICS	139	245
Frankie MILLER	1061	1500
Steve MILLER Band	112	210
MILLIONS LIKE US	1117	1521
MILLI VANILLI	69	132
Frank MILLS		1776
Stephanie MILLS	313	168
Ronnie MILSAP	163	275
MINISTRY		1113
MINK DeVILLE	1306	1587
Liza MINNELLI		1342
Kylie MINOGUE	286	735
MINOR DETAIL	1371	1890
Judi Sheppard MISSETT		1073
MISSING PERSONS	553	238
MISSION U.K.		1031
MR. BIG		877
MR. MISTER	101	188
Barbara MITCHELL	985	
Joni MITCHELL	869	400
Kim MITCHELL	1161	1199
M.O.D.		1329
MODELS	791	1021
MODERN ENGLISH	1152	770
MOLLY HATCHET	767	333
MONDO ROCK	1122	
Eddie MONEY	125	165
T.S. MONK	1026	931
The MONKEES	1168	298
Michael MONROE		1639
The MONROES	941	1462
MONTANA Orchestra		1857
MONTROSE		1682
MONTY PYTHON		1578
MOODY BLUES	219	114
Gary MOORE		896
Melba MOORE		747
Vinnie MOORE		1611
Michael MORALES	486	1158
Meli'sa MORGAN	803	594
Giorgio MORODER	1268	
Gary MORRIS		1667
Van MORRISON		411
MORRISSEY		876
Steve MORSE Band		1232
The MOTELS	192	248
MOTHER'S FINEST		1596
MÖTLEY CRÜE	290	47
MOTORHEAD		1116
The MOTORS	1192	1658
Bob MOULD		1325
MOUNTAIN		1673
Alphonse MOUZON		1506
MOVING PICTURES	455	1192
Alison MOYET	670	656
MTUME	794	536
Shirley MURDOCK	631	704
Michael Martin MURPHEY	537	1069
Eddie MURPHY	218	274
Peter MURPHY		1235
Walter MURPHY	928	

Act	Sing Rank	Alb Rank
Anne MURRAY	293	199
MUSICAL YOUTH	479	628
Alicia MYERS		1793
Gary MYRICK		1862

N

Act	Sing Rank	Alb Rank
The NAILS		1910
NAJEE		558
NAKED EYES	250	543
Graham NASH	1230	1331
NAZARETH	1314	611
Phyllis NELSON	912	
Rick NELSON		1621
Willie NELSON	184	46
NENA	264	799
Loz NETTO	1209	
Robbie NEVIL	176	462
Ivan NEVILLE	676	1057
NEVILLE BROTHERS		823
NEWCITY ROCKERS	1148	
NEWCLEUS	859	939
NEW EDITION	124	101
NEW ENGLAND		1797
NEW KIDS ON THE BLOCK	77	80
Randy NEWMAN	823	806
NEW ORDER	585	306
Juice NEWTON	70	181
Wayne NEWTON	733	
Olivia NEWTON-JOHN	14	104
Stevie NICKS	57	41
NIELSEN/PEARSON	664	
NIGHT	1344	
The NIGHTHAWKS		1770
Maxine NIGHTINGALE		1785
NIGHT RANGER	144	108
Willie NILE		1371
999		1757
9.9	842	1001
1927		1399
NITRO		1470
NITTY GRITTY DIRT BAND	386	678
Mojo NIXON & Skid ROPER		1560
No Artist		1305
NOCERA	1233	
NOEL	661	1360
Kenny NOLAN	874	
Aldo NOVA	624	330
NOVO COMBO		1670
NRBQ		1926
NUCLEAR ASSAULT		1233
Ted NUGENT	1293	373
Gary NUMAN	364	445
Bobby NUNN		1551
NU SHOOZ	265	521
N.W.A.		455
The NYLONS	541	681
Laura NYRO		1848

O

Act	Sing Rank	Alb Rank
OAK	678	
OAK RIDGE BOYS	261	228
OAKTOWN'S 3.5.7		1224
John O'BANION	696	1735
O'BRYAN	950	637
Ric OCASEK	519	464
Billy OCEAN	20	75
Sinead O'CONNOR		653

Act	Sing Rank	Alb Rank
ODYSSEY		1618
OFF BROADWAY (USA)	977	1271
OHIO PLAYERS		1794
OINGO BOINGO	851	664
The O'JAYS	596	429
Mike OLDFIELD		1321
Jane OLIVOR		1024
OLLIE And JERRY	507	
Lenore O'MALLEY	940	
OMAR And The HOWLERS		1043
ONE 2 MANY	792	
Alexander O'NEAL	476	466
ONE TO ONE	1327	
ONE WAY	959	487
Yoko ONO	974	936
OPUS	705	1011
Roy ORBISON	517	321
ORCHESTRAL MANOEUVERS IN THE DARK	228	315
ORION THE HUNTER	989	1020
Benjamin ORR	611	972
Robert Ellis ORRALL	762	1507
Jeffrey OSBORNE	181	162
Ozzy OSBOURNE	1052	55
Lee OSKAR		1657
K.T. OSLIN		497
Donny OSMOND	280	858
OTHER ONES	616	1602
The OUTFIELD	267	193
The OUTLAWS	732	516
OVERKILL		1326
OXO	686	1504
OZARK MOUNTAIN DAREDEVILS	1077	1780
OZONE		1637

P

Act	Sing Rank	Alb Rank
PABLO CRUISE	525	713
David PACK	1376	
Jimmy PAGE		596
Tommy PAGE	674	1749
Kevin PAIGE	566	1154
PAJAMA PARTY	783	
Robert PALMER	91	138
Mica PARIS	1383	1014
Graham PARKER	1377	513
Graham PARKER And The SHOT	810	886
Ray PARKER Jr.	85	286
Ray PARKER Jr. & RAYDIO	243	374
PARLIAMENT		797
John PARR	208	800
Alan PARSONS Project	117	100
PARTLAND BROTHERS	741	1653
Dolly PARTON	155	240
The PASADENAS	903	1211
PASSPORT		1643
Jaco PASTORIUS		1774
Robbie PATTON	610	1671
Henry PAUL Band	879	1288
Luciano PAVAROTTI		717
PEACHES & HERB	568	953
Leslie PEARL	673	
PEARL HARBOR And The EXPLOSIONS		1257
David PEASTON		1146
PEBBLES	237	394
Nia PEEPLES	776	1032
Teddy PENDERGRASS	468	180

Act	Sing Rank	Alb Rank
PENDULUM	1255	
Michael PENN		1516
PEPSI and SHIRLIE	1001	1524
Carl PERKINS		1656
Itzhak PERLMAN		1709
Joe PERRY PROJECT		885
Steve PERRY	200	332
PETER, PAUL & MARY		1734
Bernadette PETERS	649	1086
PET SHOP BOYS	104	185
Tom PETTY	472	201
Tom PETTY And The HEARTBREAKERS	153	45
PHANTOM, ROCKER & SLICK		870
PHOTOGLO	536	1332
PIECES OF A DREAM		819
Mike PINERA	1012	
PINK FLOYD	134	18
Joe PISCOPO		1821
PIXIES		945
PLANET P	979	744
Robert PLANT	353	103
The PLASMATICS		1198
PLATINUM BLONDE	1236	
PLAYER	728	1613
PLEASURE	891	1006
The PLIMSOULS	1307	1635
POCO	490	523
The POGUES		994
Buster POINDEXTER		848
POINT BLANK	777	745
Bonnie POINTER	785	1097
Noel POINTER		1761
POINTER SISTERS	22	64
POISON	105	57
The POLICE	31	10
Jean-Luc PONTY		562
Iggy POP		731
POP WILL EAT ITSELF		1706
Gary PORTNOY	1239	
Mike POST	481	1018
POWERSOURCE	1041	
POWER STATION	294	290
PREFAB SPROUT		1747
Elvis PRESLEY	662	421
Billy PRESTON	1341	898
Billy PRESTON & SYREETA	277	
The PRETENDERS	151	102
PRETTY MAIDS		1627
PRETTY POISON	389	1388
Andre PREVIN		1710
Ray PRICE		1147
Charley PRIDE		1723
Maxi PRIEST	659	1176
The PRIMITIVES		1347
PRINCE	17	31
PRINCE And The REVOLUTION	40	29
Victoria PRINCIPAL	1086	
John PRINE		1580
PRISM	754	879
The PROCLAIMERS		1397
The PRODUCERS	1064	1846
PROPHET		1570
Richard PRYOR		634
PSEUDO ECHO	470	766
PSYCHEDELIC FURS	488	316
PUBLIC ENEMY		608

Acts Alpha Index

Act	Sing Rank	Alb Rank
PUBLIC IMAGE LIMITED		827
PURE PRAIRIE LEAGUE	339	677
PURSUIT OF HAPPINESS		1042

Q

Act	Sing Rank	Alb Rank
Q-FEEL	1149	
QUARTERFLASH	157	269
Suzi QUATRO	681	1716
QUEEN	35	86
Queen LATIFAH		1845
QUEENSRYCHE		388
QUIET RIOT	303	120

R

Act	Sing Rank	Alb Rank
Eddie RABBITT	74	233
Trevor RABIN		1315
The RADIATORS		1134
Gilda RADNER		1276
Gerry RAFFERTY	969	1155
RAGING SLAB		1372
RAIL		1493
RAINBOW	690	408
The RAINMAKERS		852
Bonnie RAITT	1083	369
Kevin RALEIGH	1006	
The RAMONES		623
RANK AND FILE		1731
Billy RANKIN	875	1361
Kenny RANKIN		1711
RANKING ROGER		1593
RATT	423	112
RAVEN		979
Lou RAWLS	984	920
RAY, GOODMAN & BROWN	328	486
RCR	1374	
Chris REA	994	1196
READY FOR THE WORLD	206	260
REAL LIFE	431	908
Leon REDBONE	1175	1446
The REDDINGS	877	1106
Helen REDDY	1338	
RED FLAG		1796
RED HOT CHILI PEPPERS		759
RED RIDER	865	
RED ROCKERS	906	1051
RED 7		1301
RED SIREN		1364
Dan REED Network	854	1022
Jerry REED	999	
Lou REED		395
Dianne REEVES		1548
RE-FLEX	592	849
REGINA	491	1410
R.E.M.	309	81
RENAISSANCE		1855
RENÉ & ANGELA	761	490
Mike RENO	571	
REO SPEEDWAGON	39	25
The REPLACEMENTS	929	755
RESTLESS HEART	707	860
Burt REYNOLDS	1289	
Debbie REYNOLDS		1849
Randy RHOADS		644
RHYTHM CORPS		1258
Cliff RICHARD	150	756
Keith RICHARDS		672
Turley RICHARDS	972	
Lionel RICHIE	4	8
The Nelson RIDDLE Orchestra	957	
Stan RIDGWAY		1495
Cheryl Pepsii RILEY	748	1348
Teddy RILEY	1197	
The RINGS	1198	1702
RIOT		1096
Minnie RIPERTON		848
The RIPPINGTONS		933
Lee RITENOUR	567	552
Joan RIVERS		733
ROACHFORD	750	1337
ROB BASE	637	376
Marty ROBBINS		1584
Rockie ROBBINS	1248	996
Robbie ROBERTSON		620
ROBEY	1252	
Smokey ROBINSON	75	135
The ROCHES		1464
ROCK And HYDE	980	1136
The ROCKETS	1101	969
ROCKIN' SIDNEY		1739
ROCKPILE	845	658
ROCKWELL	194	482
Nile RODGERS	1353	
Paul RODGERS		1434
RODNEY O & Joe COOLEY		1882
RODWAY	1283	
ROGER	385	412
Dann ROGERS	795	
Kenny ROGERS	9	15
Kenny ROGERS & Dottie WEST		1741
ROLLING STONES	53	20
ROMAN HOLLIDAY	811	1237
The ROMANTICS	210	381
ROMEO'S DAUGHTER	1142	1894
ROMEO VOID	771	851
RON And The D.C. CREW	1362	
Mick RONSON		1697
Linda RONSTADT	99	44
ROSE ROYCE		1652
ROSE TATTOO		1881
Diana ROSS	16	59
ROSSINGTON COLLINS BAND	919	386
David Lee ROTH	183	140
ROUGH TRADE	981	
ROXANNE	1098	
ROXETTE	131	494
ROXY MUSIC	1277	457
ROYAL PHILHARMONIC ORCHESTRA	471	203
RUBBER RODEO	1241	
Jimmy RUFFIN	514	1646
Mason RUFFNER		1054
RUFUS/ Chaka KHAN	480	389
Todd RUNDGREN	1116	697
RUN-D.M.C.	334	88
RUSH	420	40
Jennifer RUSH	639	1376
Patrice RUSHEN	543	327
Brenda RUSSELL	474	685
Leon RUSSELL		1887
Mike RUTHERFORD		1338
Mitch RYDER	1309	1441

S

Act	Sing Rank	Alb Rank
Sue SAAD And The NEXT		1396
SAD CAFÉ	1234	1668
SADE	185	69
SA-FIRE	428	784
SAGA	575	459
Carole Bayer SAGER	736	965
SALSOUL Orchestra		1725
SALT-N-PEPA	549	355
Joe SAMPLE		782
David SANBORN		243
SANTANA	317	223
Carlos SANTANA		739
SARAYA	870	892
Father Guido SARDUCCI		1865
Joe SATRIANI		282
SAVATAGE		1082
SAVOY BROWN	1128	1817
SAWYER BROWN		1432
SAXON		1131
Leo SAYER	193	748
Boz SCAGGS	220	235
SCANDAL	288	341
Joey SCARBURY	201	1381
SCARLETT & BLACK	638	1354
Michael SCHENKER Group		740
Peter SCHILLING	448	888
Timothy B. SCHMIT	668	1219
John SCHNEIDER	477	646
Neal SCHON	1395	1366
Neal SCHON & Jan HAMMER		1099
SCHOOLLY D		1841
Diane SCHUUR		1563
Eddie SCHWARTZ	650	1804
SCORPIONS	572	76
Tom SCOTT		1165
Gil SCOTT-HERON		918
Gil SCOTT-HERON & Brian JACKSON		1172
SCREAMING BLUE MESSIAHS		1567
SCRITTI POLITTI	433	718
SCRUFFY THE CAT		1705
SEA HAGS		1645
SEA LEVEL		1628
Dan SEALS	689	1034
The SEARCHERS		1895
Marvin SEASE		1179
SEAWIND		1230
SECRET TIES	1298	
Neil SEDAKA		1324
Neil SEDAKA & Dara SEDAKA	563	
SEDUCTION	619	1184
Bob SEGER	84	
Bob SEGER & The SILVER BULLET BAND	102	22
The SELECTER		1773
Michael SEMBELLO	177	1306
Brian SETZER		798
Taja SEVELLE	964	
707	868	1335
7 SECONDS		1529
Doc SEVERINSEN And The TONIGHT SHOW Orchestra		856
S-EXPRESS	1323	
Charlie SEXTON	576	451
Phil SEYMOUR	645	1109
SHADOWFAX		829
Paul SHAFFER	1180	
SHALAMAR	214	241
SHANNON	343	534
Del SHANNON	786	1346
Feargal SHARKEY	1153	1162
Tommy SHAW	614	724
Jules SHEAR	1019	
SHEILA	893	
SHEILA E.	251	335
Pete SHELLEY		1274
Ricky Van SHELTON		645
T.G. SHEPPARD	604	1080
SHERBS	1055	1148
SHERIFF	301	1033
SHINEHEAD		1833
Michelle SHOCKED	1057	728
SHOES		1546
SHOOTING STAR	901	706
Glenn SHORROCK	1121	
SHOTGUN MESSIAH		1222
SHOT IN THE DARK	1174	
SHRIEKBACK		1277
Michael SHRIEVE	1396	1367
SHY		1913
Jane SIBERRY		1544
SIDEWINDERS		1727
SIGUE SIGUE SPUTNIK		1343
The SILENCERS (2)	1240	
The SILENCERS	1266	1476
SILVERADO	1360	
SILVER CONDOR	766	1362
Patrick SIMMONS	702	1117
Richard SIMMONS		700
Carly SIMON	318	234
Paul SIMON	249	91
SIMON & GARFUNKEL	735	443
SIMON F	1378	
SIMPLE MINDS	115	242
SIMPLY RED	137	209
Frank SINATRA	721	428
SINGLE BULLET THEORY	1220	
SINITTA	1294	
SIOUXSIE & The BANSHEES	840	749
SIR DOUGLAS Quintet		1824
SIR MIX-A-LOT	1090	629
SISTER SLEDGE	588	515
SISTERS OF MERCY		1124
Ricky SKAGGS		720
SKID ROW	407	208
SKY	1322	1236
SKYY	769	364
SLADE	529	679
SLAVE	1080	526
SLAYER		779
Grace SLICK	1384	647
SLICK RICK		551
SLY FOX	418	710
Frankie SMITH	643	1059
Kathy SMITH		1467
Lonnie Liston SMITH		1907
Patti SMITH		1035
Rex SMITH	917	1631
The SMITHEREENS	1366	407
The SMITHS		404
Patty SMYTH		968
SNEAKER	663	1281
SNIFF 'N' The TEARS		1899

407

Acts Alpha Index

Act	Sing Rank	Alb Rank
Phoebe SNOW	778	775
SNUFF	1356	
SO	862	1398
Gino SOCCIO		1246
SOFT CELL	384	391
Belouis SOME	1067	
S.O.S. BAND	236	299
SOULSISTER	856	
SOUL II SOUL	255	380
J.D. SOUTHER	709	1485
SOUTHSIDE JOHNNY & The ASBURY JUKES	1387	769
SPANDAU BALLET	239	456
SPARKS	850	1088
SPECIAL ED		878
The SPECIALS		927
Paul SPEER		1557
Judson SPENCE	752	1539
Tracie SPENCER	737	1250
SPIDER	626	1374
SPINAL TAP		1414
SPINNERS	127	609
SPLIT ENZ	899	440
Dusty SPRINGFIELD	498	
Rick SPRINGFIELD	21	54
Bruce SPRINGSTEEN	25	2
SPYRO GYRA	1005	216
SPYS	1196	1472
SQUEEZE	450	254
Billy SQUIER	254	73
STABILIZERS	1368	
STACEY Q	325	729
STAGE DOLLS	880	1334
Frank STALLONE	469	
Joe STAMPLEY		1867
Michael STANLEY Band	493	591
STARGARD		1879
STARLAND VOCAL BAND	1093	
STARPOINT	513	557
Brenda K. STARR	412	846
Ringo STARR	814	1216
STARS ON	198	483
STAR WARS INTERGALACTIC DROID CHOIR & CHORALE	1097	
STATLER BROTHERS		1026
STEADY B		1595
STEALIN HORSES		1435
STEEL BREEZE	462	805
Maureen STEELE	1204	
STEEL PULSE		1046
STEELY DAN	390	347
Jim STEINMAN	700	926
Van STEPHENSON	534	916
STEPPENWOLF		1768
Cat STEVENS		1604
Ray STEVENS		1083
Shakin' STEVENS	1060	
Steve STEVENS ATOMIC PLAYBOYS		1355
Stevie B	392	619
Al STEWART	682	824
Amii STEWART & Johnny BRISTOL	1013	
David A. STEWART With Barbara GASKIN	1112	
Gary STEWART		1820
Jermaine STEWART	324	578
John STEWART	730	1289
Rod STEWART	47	66
Stephen STILLS	1017	1149

Act	Sing Rank	Alb Rank
Stephen STILLS With Michael FINNIGAN		1091
STING	123	106
STOMPERS	1312	
STONE CITY BAND		1489
STONE FURY		1409
George STRAIT		448
The STRANGLERS		1782
STRAY CATS	136	126
STREEK	951	
Meryl STREEP		1891
Janey STREET	1130	1591
STREETS	1285	1543
Barbra STREISAND	92	21
Barbra STREISAND & Barry GIBB	226	
The STRIKERS		1828
STRYPER	602	265
STYLE COUNCIL	694	711
The STYLISTICS		1370
STYX	86	50
SUAVE		1251
SUAVÉ	684	
The SUGARCUBES		712
SUGARHILL GANG	612	901
SUICIDAL TENDENCIES		973
Kasim SULTON		1916
Donna SUMMER	58	117
Henry LEE SUMMER	439	682
Andy SUMMERS & Robert FRIPP		1048
Bill SUMMERS & SUMMERS HEAT		949
Joe SUN	1108	
SUPERTRAMP	302	163
The SUPREMES/Diana ROSS And The SUPREMES		1161
SURFACE	341	530
SURVIVOR	32	137
Patrick SWAYZE	335	
Keith SWEAT	402	292
SWEAT BAND		1562
Rachel SWEET	852	1169
SWEET SENSATION	379	768
SWEET TEE		1486
SWING OUT SISTER	376	430
SWITCH	1065	830
SYBIL	590	1137
Keith SYKES		1458
Sylvain SYLVAIN		1418
SYLVESTER		1104
SYLVIA	555	640
SYNCH	500	
SYREETA		1047
The SYSTEM	378	671

T

Act	Sing Rank	Alb Rank
TACO	327	589
TAKE 6		974
TALKING HEADS	342	65
TALK TALK	658	607
TA MARA & The SEEN	606	802
TAMI SHOW	1299	
TANGERINE DREAM		1115
TANGIER	1136	1065
TANTRUM		1885
TARNEY/SPENCER Band	1217	
A TASTE OF HONEY	263	630
TAVARES	544	1144

Act	Sing Rank	Alb Rank
TAXXI	1513	
Andy TAYLOR	630	897
B. E. TAYLOR Group	1011	
James TAYLOR	584	261
John TAYLOR	729	
Livingston TAYLOR	820	
Roger TAYLOR		1322
T-CONNECTION		1191
TEARDROP EXPLODES		1535
TEARS FOR FEARS	68	68
TECHNOTRONIC		1634
Kiri TE KANAWA		1260
The TEMPTATIONS	505	323
10cc		1818
Toni TENNILLE		1380
10,000 MANIACS	739	257
TEN YEARS AFTER		1356
Robert TEPPER	656	1552
Helen TERRY	1392	
Tony TERRY	1028	1291
TESLA	1363	309
TESTAMENT		966
TEXAS	1219	1094
TEXTONES		1666
The THE		983
THEY MIGHT BE GIANTS		1063
THIN LIZZY		1084
3rd BASS		1401
THIRD WORLD		673
38 SPECIAL	118	97
B.J. THOMAS	1380	1869
Evelyn THOMAS	1281	
Lillo THOMAS		1863
Nolan THOMAS	895	
Timmy THOMAS	1292	
Chris THOMPSON	1346	
Richard THOMPSON		1052
Robbin THOMPSON Band	1085	1625
THOMPSON TWINS	100	167
Ali THOMSON	506	1166
George THOROGOOD & The DESTROYERS	1025	214
Billy THORPE		1669
3		1349
3 MAN ISLAND	1389	
The THREE O'CLOCK		1439
THREE TIMES DOPE		1254
THRILLS		1864
TIA	1390	
TIERRA	473	788
TIFFANY	98	130
TIGGI CLAY	1329	
TIGHT FIT	1349	
Tanita TIKARAM		889
'TIL TUESDAY	365	384
TIMBUK 3	605	643
The TIME	461	225
The TIMELORDS	931	
TIMES TWO	632	1426
TIMEX SOCIAL CLUB	454	
TIN MACHINE		833
TKA	1051	1437
TNT		922
TOBY BEAU	1171	
TODAY		1008
Isao TOMITA		1732
TOMMY TUTONE	291	520
TOM TOM CLUB	699	485
TONE LŌC	195	252

Act	Sing Rank	Alb Rank
TONY! TONI! TONÉ!	927	780
TOO SHORT		474
TORA TORA	1282	836
TORONTO	1075	1525
Peter TOSH	1275	787
TOTAL COELO	1056	
TOTO	61	109
TOUCH	975	
Wayne TOUPS & ZYDECAJUN		1823
The TOURISTS	1256	
Carol Lynn TOWNES	1114	
Pete TOWNSHEND	377	176
Simon TOWNSHEND		1665
T'PAU	367	752
TRANSVISION VAMP	1336	1429
TRANS-X	911	
TRAVELING WILBURYS	712	197
Pat TRAVERS	960	426
Randy TRAVIS		159
TREAT HER RIGHT		1214
TRINERE & FRIENDS		1914
TRIUMPH	608	173
TROOP		1463
TROUBLE FUNK		1352
Robin TROWER		573
Eric TROYER	1364	
The TRUTH	1020	1445
TSOL		1873
The TUBES	396	366
TUCK & PATTI		1541
Louise TUCKER	797	1406
Tanya TUCKER		1610
Ike & Tina TURNER		1893
Joe Lynn TURNER		1481
Tina TURNER	37	53
Stanley TURRENTINE		1781
TWENNYNINE	1259	912
24-7 SPYZ		1188
20/20		1294
TWILIGHT 22	1146	
Dwight TWILLEY	565	723
TWIN HYPE		1465
TWISTED SISTER	552	266
Conway TWITTY		1369
2 LIVE CREW	558	399
Bonnie TYLER	120	358
TYZIK		1692

U

Act	Sing Rank	Alb Rank
UB40	238	219
UFO		631
Tracey ULLMAN	446	774
ULTRAVOX	1129	801
The UNDERTONES		1615
UNDERWORLD	934	1220
The UNFORGIVEN		1875
UNIPOP	1053	
UNLIMITED TOUCH		1583
The UNTOUCHABLES		1592
UPTOWN	1111	
Midge URE	1357	1111
URGENT	1242	
URIAH HEEP		919
USA For AFRICA	197	383
UTFO	1243	686
UTOPIA	655	475
U2	80	9

Acts Alpha Index

Act	Sing Rank	Alb Rank
V		
VAIN		1573
Ritchie VALENS		1383
Dave VALENTIN		1748
Dana VALERY	1246	
Frankie VALLI	1311	
VANDENBERG	770	971
Luther VANDROSS	260	61
VANGELIS	203	196
VAN HALEN	50	11
VANITY	913	501
Gino VANNELLI	272	417
Randy VANWARMER	905	
Johnny Van ZANT Band		814
The VAPORS	701	818
Sylvie VARTAN	1354	
Stevie Ray VAUGHAN And DOUBLE TROUBLE		212
Suzanne VEGA	363	385
The VELS	1154	
VELVET UNDERGROUND		1139
The VENETIANS	1313	
Billy VERA & The BEATERS	257	550
Tom VERLAINE		1844
VESTA	955	1417
VICTORY		1813
Maria VIDAL	853	
VILLAGE PEOPLE		859
Vinnie VINCENT INVASION		627
Bobby VINTON	1222	
VIO-LENCE		1683
VIOLENT FEMMES		862
VISAGE		1795
Joe VITALE		1840
VITAMIN Z	1096	1847
VIXEN	542	568
VOICES OF AMERICA	1059	
VOIVOD		1814
Andreas VOLLENWEIDER		406
W		
Jack WAGNER	289	638
John WAITE	154	279
The WAITRESSES	1069	721
Tom WAITS		929
Narada Michael WALDEN	1104	853
Jerry Jeff WALKER		1750
WALL OF VOODOO	954	822
Joe WALSH	467	434
Steve WALSH		1558
WANG CHUNG	146	356
WAR	1094	737
Steve WARINER		1883
WARLOCK		925
Jennifer WARNES	211	938
WARRANT	256	300
Dionne WARWICK	97	182
Grover WASHINGTON Jr.	205	150
WAS (NOT WAS)	387	633
W.A.S.P.		416
The WATERBOYS		923
WATERFRONT	520	1275
Muddy WATERS		1900
Roger WATERS		572
Jody WATLEY	88	164
Johnny Guitar WATSON		1140
Ernie WATTS		1478
WA WA NEE	790	1182
WAX	819	1316
Fee WAYBILL		1619
WAYSTED		1877
WEATHER GIRLS	841	1287
WEATHER REPORT		709
Tim WEISBERG		1663
Bob WELCH		1512
WENDY And LISA	935	1027
Max WERNER	1185	
Dottie WEST	487	1327
WEST STREET MOB	1224	
WET WET WET	1015	1517
Kirk WHALUM		1466
WHAM!	38	89
WHAT IS THIS	1037	1829
WHEN IN ROME	483	943
The WHISPERS	285	144
WHISTLE	892	
Barry WHITE		991
Karyn WHITE	229	371
Maurice WHITE	843	1000
Tony Joe WHITE	1203	
WHITE LION	244	190
WHITESNAKE	122	63
WHITE WOLF		1357
Keith WHITLEY		1141
Slim WHITMAN		1685
Roger WHITTAKER		1228
The WHO	464	213
WHODINI	1343	297
Jane WIEDLIN	452	975
WILD BLUE	1141	
Danny WILDE		1638
Eugene WILDE	978	1181
Kim WILDE	216	537
Matthew WILDER	274	952
WILL And The KILL		1505
Christopher WILLIAMS	781	
Deniece WILLIAMS	156	363
Don Williams	599	670
Hank WILLIAMS Jr.		184
Lenny WILLIAMS		1876
Mason WILLIAMS		1405
Robin WILLIAMS		1431
Vanessa WILLIAMS	449	587
WILLIE & The POOR BOYS		1249
WILLIE, WAYLON, JOHNNY & KRIS		887
Bruce WILLIS	425	511
WILLIS "The GUARD" and VIGORISH	1262	
WILL TO POWER	235	881
Ann WILSON	415	
Brian WILSON		998
Carl WILSON	1144	1878
Nancy WILSON		1699
Shanice WILSON	847	1284
The WINANS		1340
BeBe & CeCe WINANS		984
Angela WINBUSH		869
Jesse WINCHESTER	793	1892
WINGER	478	288
George WINSTON		289
Johnny WINTER		1453
Paul WINTER		1407
Robert WINTERS and FALL		1270
Steve WINWOOD	52	34
WIRE		1509
WIRE TRAIN		1456
WISHBONE ASH		1755
Bill WITHERS		1496
WOLF	952	
Peter WOLF	354	518
Bobby WOMACK	1348	496
Stevie WONDER	28	39
Ronnie WOOD		1703
The WOODENTOPS		1754
Stevie WOODS	502	1317
Bruce WOOLLEY & The CAMERA CLUB		1874
WORLD CLASS WRECKIN' CRU	1167	
WORLD PARTY	744	698
WRABIT		1566
WRATHCHILD AMERICA		1783
Bernard WRIGHT		1204
Betty WRIGHT		1353
Gary WRIGHT	587	1037
Steven WRIGHT		1905
X		
X		632
XAVIER		1511
XTC	1155	377
XYMOX		1616
XYZ		1742
Y		
Y&T	939	509
"Weird Al" YANKOVIC	484	340
YARBROUGH & PEOPLES	546	503
YAZ	885	828
YAZZ And The PLASTIC POPULATION	1388	
YELLO	908	1120
YELLOWJACKETS		1378
YELLOW MAGIC ORCHESTRA	920	1029
YES	132	136
YIPES!!	1160	
Dwight YOAKAM		419
Neil YOUNG	1125	175
Neil YOUNG & CRAZY HORSE	1164	
Paul YOUNG	167	310
YOUNG M.C.	355	484
YUTAKA	1300	1778
Z		
Pia ZADORA	603	1152
Robin ZANDER	646	
ZAPP	1269	390
Frank ZAPPA	757	449
ZEBRA	990	625
ZENO		1310
Warren ZEVON	997	468
ZODIAC MINDWARP & The LOVE REACTION		1273
ZZ TOP	174	33

Acts Alpha Index

Evaluating Acts Two Ways: Singles-Centric vs. Album-Centric Acts

Chart Strengths of the Top 20 Singles Acts. Michael Jackson, Madonna, Hall & Oates and Lionel Richie stand head and shoulders above the rest of the '80s singles acts. After that, the separations are nowhere near as profound. It's worth asking the question: are the acts ranked 5 and 6 or for that matter 19 and 20 really different? To put that in perspective, since the "average" single scores 1000 points by definition*, Phil Collins and Billy Joel are separated by more than one average single; probably significant. The rest of the acts are much closer together. Acts number 7 through 10 are separated by fewer than 500 points; however, number 8 Whitney Houston achieved that ranking with 11 singles rather than 20 for number 9 Kenny Rogers and number 10 John Mellencamp. Finally, number 19 Cyndi Lauper and number 20 Billy Ocean are separated by only 3 points. Only two of the top 20 singles acts have higher album rank than singles rank: Billy Joel and John Mellencamp.

*A single peaking around 20 with about a ten week chart life would score about 1000 points and be an "average" single

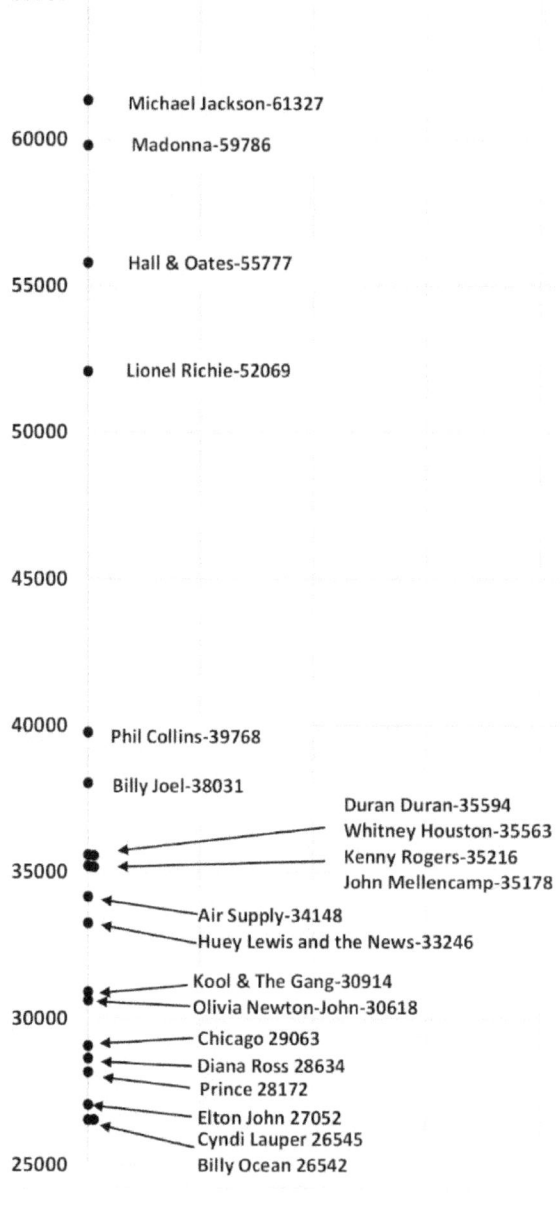

Top 20 Singles Acts by Total Singles Points

- Michael Jackson-61327
- Madonna-59786
- Hall & Oates-55777
- Lionel Richie-52069
- Phil Collins-39768
- Billy Joel-38031
- Duran Duran-35594
- Whitney Houston-35563
- Kenny Rogers-35216
- John Mellencamp-35178
- Air Supply-34148
- Huey Lewis and the News-33246
- Kool & The Gang-30914
- Olivia Newton-John-30618
- Chicago 29063
- Diana Ross 28634
- Prince 28172
- Elton John 27052
- Cyndi Lauper 26545
- Billy Ocean 26542

Chart Strength of the Top 20 Album Acts. Once again, Michael Jackson leads the list but Bruce Springsteen, the number 32 singles act, places second. Seven of the top 20 album acts ranked higher in singles performance than albums. Some acts earned points in the '80s from albums released in earlier decades that returned to the chart in the '80s but those points are generally small and only 2 are larger than 5% of the act's total. Pink Floyd had the highest percentage of non-'80s-release points (23%), most of which are derived from the continued chart presence of *The Dark Side Of The Moon*.

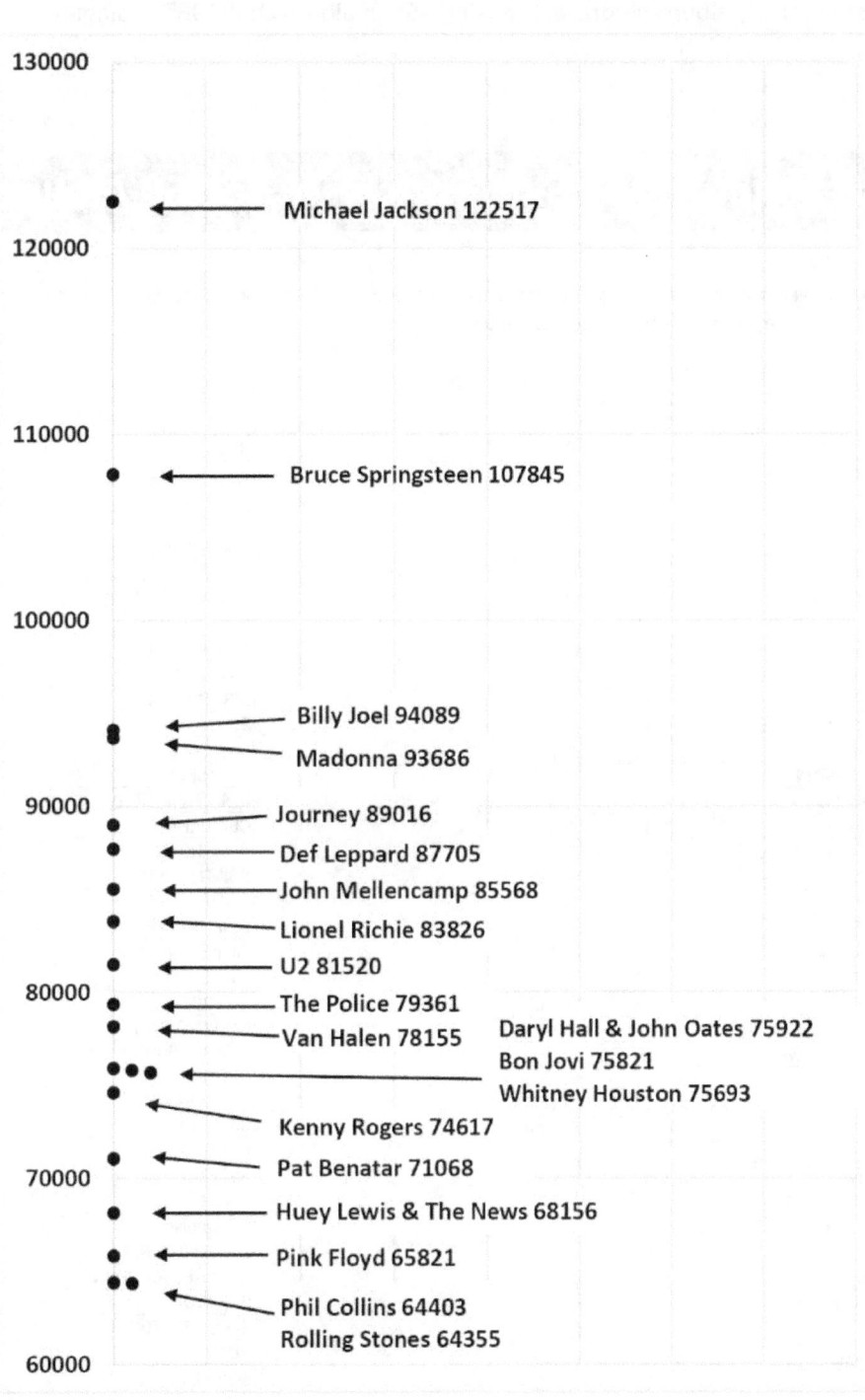

Top 20 Album Acts by Total Album Points

Acts: Single vs Album

Album Acts vs Singles Acts. On the graph on the next page are shown the top 50 acts based on singles points and the top 50 acts based on album points. Those acts and their coordinates are shown in the table on the page following.

The statistic used is rank since the scoring systems for singles and albums are different. The diagonal line indicates an equal rank for both media. Above the line are acts whose singles rank is higher than albums rank; below the line are acts whose album performance is stronger.

Acts that lie close to the line rank comparably in the singles and albums realm; however, most acts tend to rank significantly higher in one vs. the other. The boxes show top 50 acts whose other rank is greater than 100, and thus off the graph. Olivia Newton-John is the most relatively singles-centric act, ranking 14th in singles but 126th in albums. AC/DC is the most relatively album-centric act ranking 25th in albums but 496th in singles.

Legend: Top 50 Singles Acts vs. Top 50 Album Acts

Acts that lie close to the line, generally as Greek letters, designate acts that show relatively similar strength in both media. Upper case letters denote more singles-centric acts. Lower case letters denote more album-centric acts

Act	Albums Rank	Singles Rank
Michael JACKSON	1	1 α
Bruce SPRINGSTEEN	2	25 z
Billy JOEL	3	6 β
MADONNA	4	2 γ
JOURNEY	5	29 y
DEF LEPPARD	6	67 h
John MELLENCAMP	7	10 v
Lionel RICHIE	8	4 δ
U2	9	80 e
The POLICE	10	31 x
VAN HALEN	11	50 u
Daryl HALL & John OATES	12	3 ε
BON JOVI	13	27 p
Whitney HOUSTON	14	8 ζ
Kenny ROGERS	15	9 η
Pat BENATAR	16	56 s
Huey LEWIS and the NEWS	17	12 λ
PINK FLOYD	18	134 OOR
Phil COLLINS	19	5 L
ROLLING STONES	20	53 t
Barbra STREISAND	21	92 b
Bob SEGER	22	84 d
GUNS N' ROSES	23	111 OOR
AC/DC	24	495 OOR
REO SPEEDWAGON	25	39 n
HEART	26	46 w
GENESIS	27	48 v
ALABAMA	28	279 OOR
PRINCE And The REVOLUTION	29	40 m
FOREIGNER	30	26 μ
PRINCE	31	17 P
DURAN DURAN	32	7 Q
ZZ TOP	33	174 OOR
Steve WINWOOD	34	52 r
KOOL & The GANG	35	13 N
The CARS	36	72 g
Bryan ADAMS	37	60 j
LOVERBOY	38	126 OOR
Stevie WONDER	39	28 M
RUSH	40	420 OOR
Stevie NICKS	41	57 k
FLEETWOOD MAC	42	73 f
MEN AT WORK	43	49 π
Linda RONSTADT	44	99 a
Tom PETTY And The HEARTBREAKERS	45	153 OOR
Willie NELSON	46	184 OOR
MÖTLEY CRÜE	47	290 OOR
Janet JACKSON	48	41 σ
DIRE STRAITS	49	148 OOR
STYX	50	86 c
George MICHAEL	51	23 S
Tina TURNER	53	37 R
Rick SPRINGFIELD	54	21 X
CULTURE CLUB	56	24 T
Diana ROSS	59	16 Y
POINTER SISTERS	64	22 U
Rod STEWART	66	47 K
Cyndi LAUPER	67	19 V
Elton JOHN	70	18 W
AIR SUPPLY	72	11 Z
Billy OCEAN	75	20 G
Gloria ESTEFAN/ MIAMI SOUND MACHINE	77	44 J
JEFFERSON AIRPLANE/STARSHIP	79	43 H
CHICAGO	82	15 F
Richard MARX	85	42 E
QUEEN	86	35 D
WHAM!	89	38 C
Christopher CROSS	90	34 B
Paul McCARTNEY	99	30 A
Olivia NEWTON-JOHN	104	14 OOR
SURVIVOR	137	32 OOR
Sheena EASTON	146	33 OOR
BLONDIE	155	45 OOR
Kenny LOGGINS	158	36 OOR

OOR = Out Of Range of Graph Axis

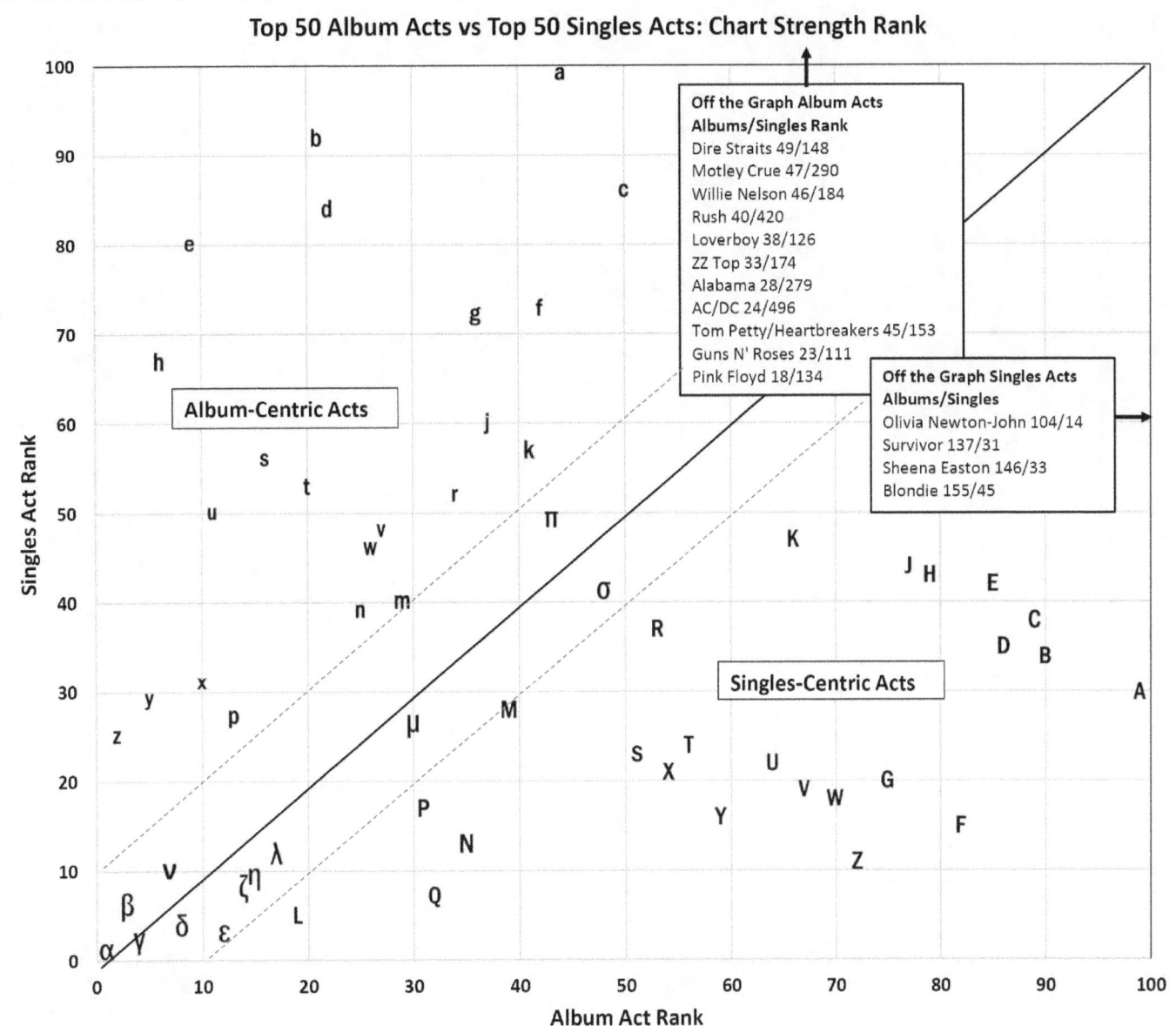

Acts: Single vs Album

Chronologies

The purpose of this Section is to provide a succinct visual snapshot of the decadal performance of the top 20 singles acts and top 20 album acts through the chart history of their singles, albums and interactions of the two.

Act Headers. Act scoring for singles is the sum of the entire lifecycle of singles peaking in the '80s. Act scoring for albums includes only actual album scores earned week-to-week between 1/1/80 and 12/31/89 and may include recurrent albums from the '60s or '70s. This is discussed in the Acts With Their Albums section and the Methodology section of the Appendix.

The Header for each act includes singles and albums performance.

MICHAEL JACKSON ▶ Singles Rank: 1 (21) 61327; Albums Rank: 1 (5/5) 122754 100%

- Number of singles charted
- Sum of scores of all '80s singles
- '80s Albums/ All albums charting
- Sum of all weekly album scores earned
- Pct. of points earned by '80s albums

Singles Graphs. All Hot 100 singles peaking in the '80s are included. Explanation of scoring for singles is detailed in the Appendix, but incorporates the full week-to-week chart lifecycle in what is called the area under the curve method. The singles graph shows the score of each of the act's singles plotted vs. that single's peak date. The marker on the graph denotes the underlying album from which the singles are drawn. Singles for which credit is divided between two artists in a one-off collaboration are shown with an upward arrow to denote that total score is higher, and titles are italicized. The total points for the single are also listed.

Singles are labeled as follows:

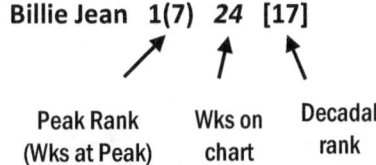

Billie Jean 1(7) 24 [17]

- Peak Rank (Wks at Peak)
- Wks on chart
- Decadal rank

Singles peaking in the weekly top 5 include Peak Rank, Weeks at Peak and Weeks on Chart; those peaking in the weekly top 10 include Peak, Weeks at Peak and Weeks on chart, others just Peak and Weeks at Peak.

Album Graphs. Albums that peaked on the Billboard 200 in the '80s or yielded '80s singles are shown on the graph, although statistics are only shown for the former. The graph shows the chart position of each album week to week. Individual album scoring includes the entire lifecycle of the album and is normalized. Album lifecycles are characterized in the legend:

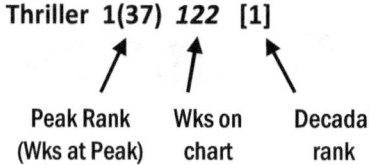

Top 20 Singles Acts

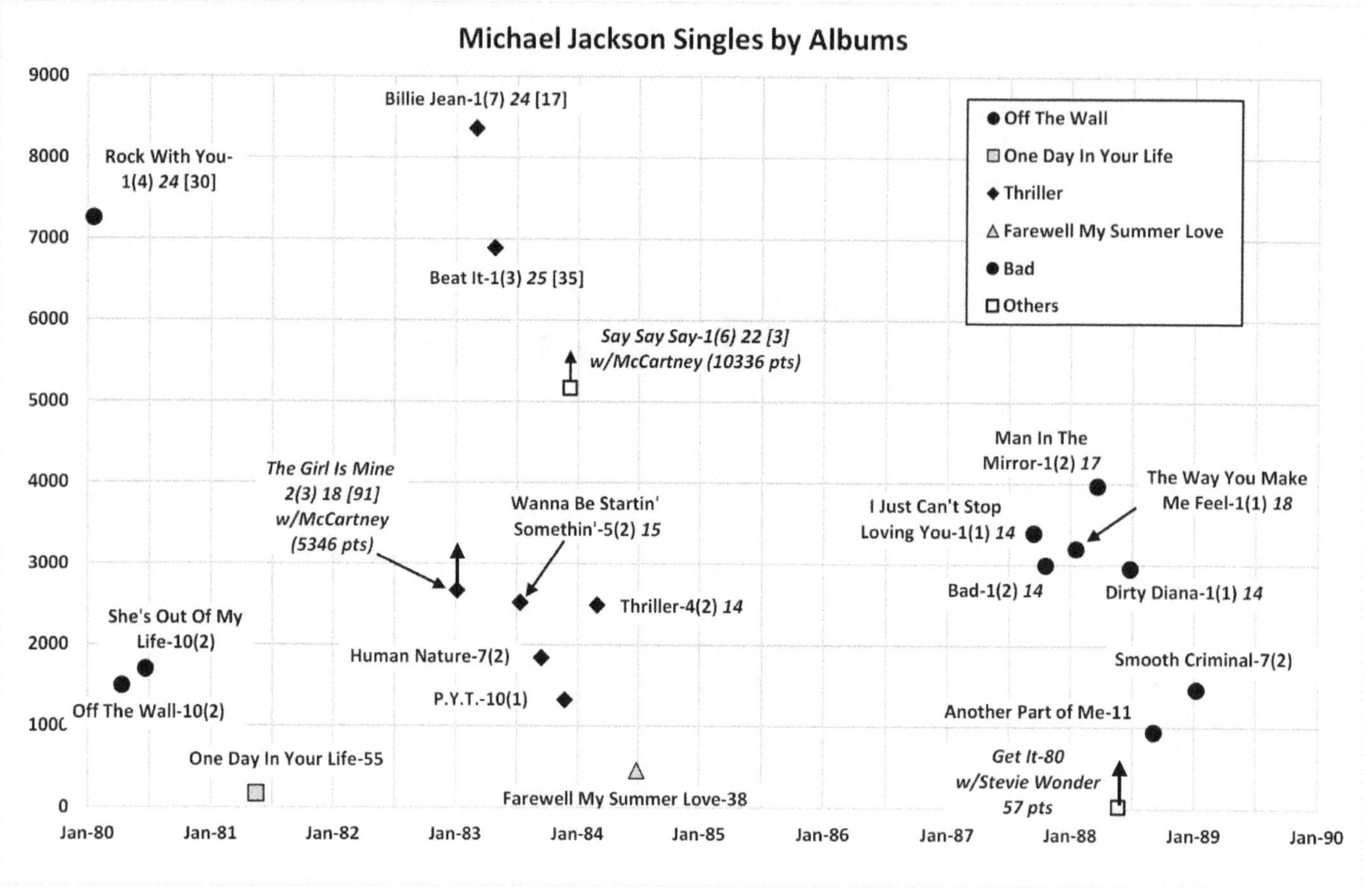
Michael Jackson Singles by Albums
9000
8000
7000
6000
5000
4000
3000
2000
1000
0
Billie Jean-1(7) 24 [17]
Rock With You-1(4) 24 [30]
Beat It-1(3) 25 [35]
Say Say Say-1(6) 22 [3] w/McCartney (10336 pts)
The Girl Is Mine 2(3) 18 [91] w/McCartney (5346 pts)
Wanna Be Startin' Somethin'-5(2) 15
Thriller-4(2) 14
Human Nature-7(2)
P.Y.T.-10(1)
She's Out Of My Life-10(2)
Off The Wall-10(2)
One Day In Your Life-55
Farewell My Summer Love-38
Man In The Mirror-1(2) 17
The Way You Make Me Feel-1(1) 18
I Just Can't Stop Loving You-1(1) 14
Bad-1(2) 14
Dirty Diana-1(1) 14
Smooth Criminal-7(2)
Another Part of Me-11
Get It-80 w/Stevie Wonder 57 pts
Off The Wall
One Day In Your Life
Thriller
Farewell My Summer Love
Bad
Others
Jan-80
Jan-81
Jan-82
Jan-83
Jan-84
Jan-85
Jan-86
Jan-87
Jan-88
Jan-89
Jan-90

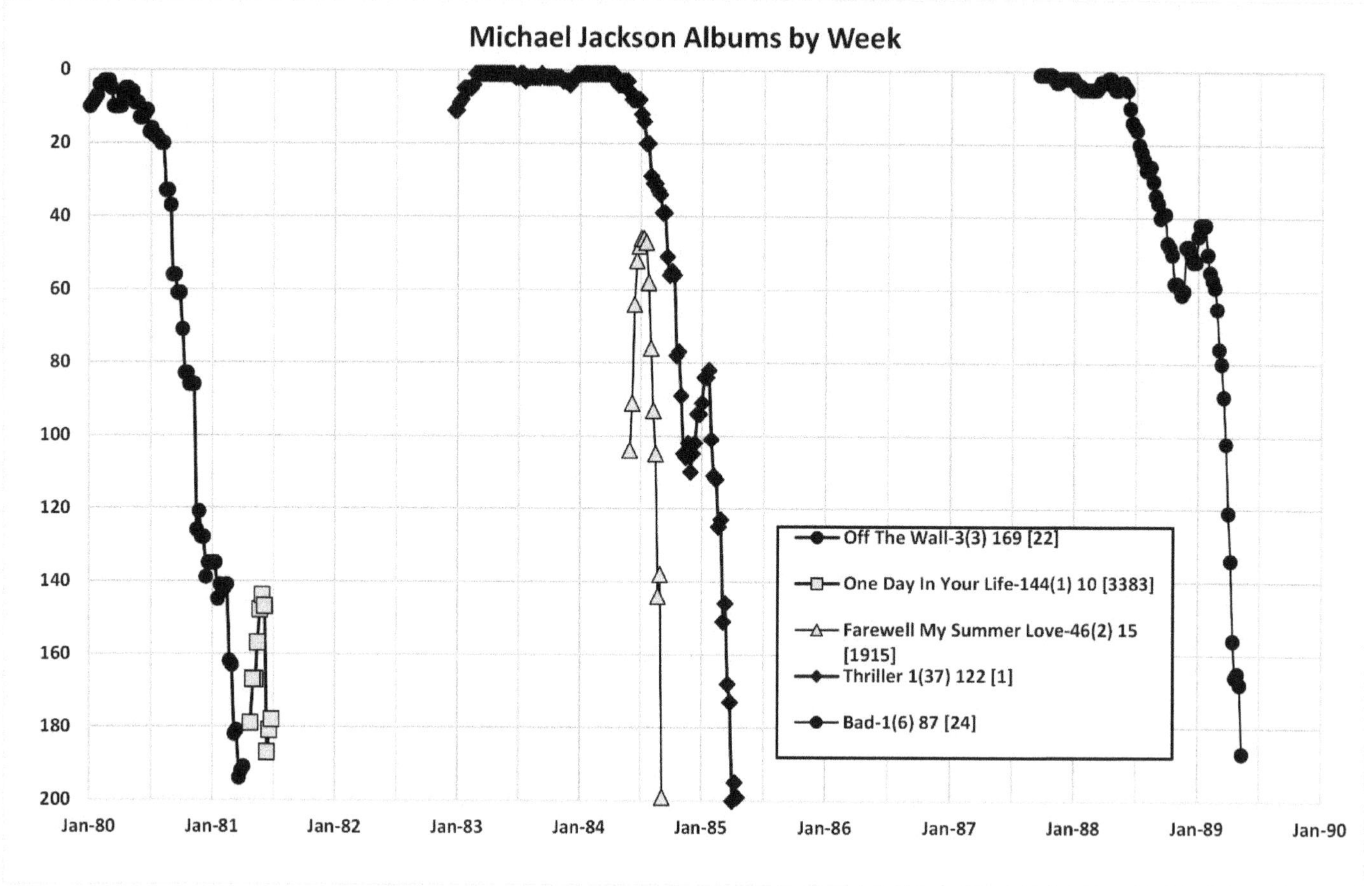
Michael Jackson Albums by Week
0
20
40
60
80
100
120
140
160
180
200
Off The Wall-3(3) 169 [22]
One Day In Your Life-144(1) 10 [3383]
Farewell My Summer Love-46(2) 15 [1915]
Thriller 1(37) 122 [1]
Bad-1(6) 87 [24]
Jan-80
Jan-81
Jan-82
Jan-83
Jan-84
Jan-85
Jan-86
Jan-87
Jan-88
Jan-89
Jan-90

Note: 14 Greatest Hits-Michael Jackson and the Jackson Five not shown

MICHAEL JACKSON ▶ Singles Rank: 1 (21) 61327; Albums Rank: 1 (5/5) 122517 100%

The King of Pop was the number 1 singles act in this decade, but not by much, besting Madonna by only one decent performing record. Album performance was similarly strong, also finishing number 1 in the decade, and well clear of the rest. Five singles in which he appeared finished in the top 100 of the decade; nine of his singles were number 1's, and three more were top 5. All three albums of new material (*One Day In Your Life* was a compilation from the Motown era and *Farewell My Summer Love* consisted of previously "lost" early '70s tracks) were number 1's and ranked in the top 25 for the decade.

The singles from the three albums performed differently. *Off The Wall* had a clear lead cut but only three singles total (however, all three were top 10's); Both *Thriller* and *Bad* went seven singles deep and only one of those 14 failed to make the top 10, although the singles from *Bad* did not score as strongly. *Thriller* was certified 20x Platinum by RIAA before the end of 1984; *Bad* and *Off The Wall* were certified 6x Platinum by the end of 1988. When scored by singles, *Thriller* ranked number 1, and *Bad* ranked number 4, up from number 24 scored as an album.

How the singles supported the sales of the album differs between *Thriller* and *Bad* as well. *Thriller's* position only started to decline well after the final single, *Thriller,* had run out of gas. On the other hand, *Bad* started to fall after the fourth single, *The Way You Make Me Feel*, and even the number 1 single *Dirty Diana* could only partially buoy the performance. Clearly, *Bad* was a singles album.

There are unexplained spikes in the lifecycles of both albums, both coming around the end of the year in 1984 and 1988. The spike in *Bad* corresponds to the chart life of the final single of the album, *Smooth Criminal*, which only peaked at number 7. There is no obvious reason for the spike in *Thriller*.

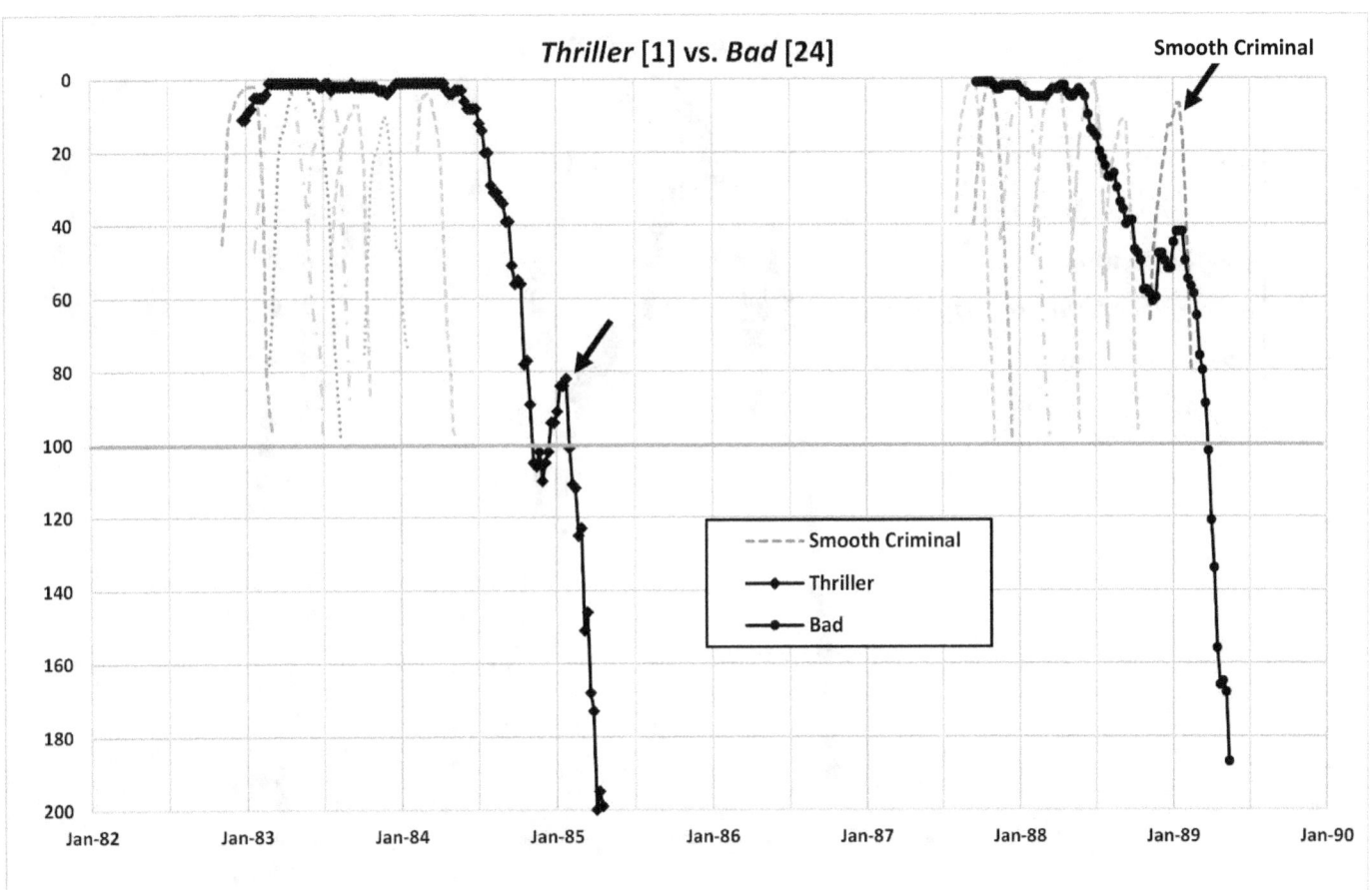

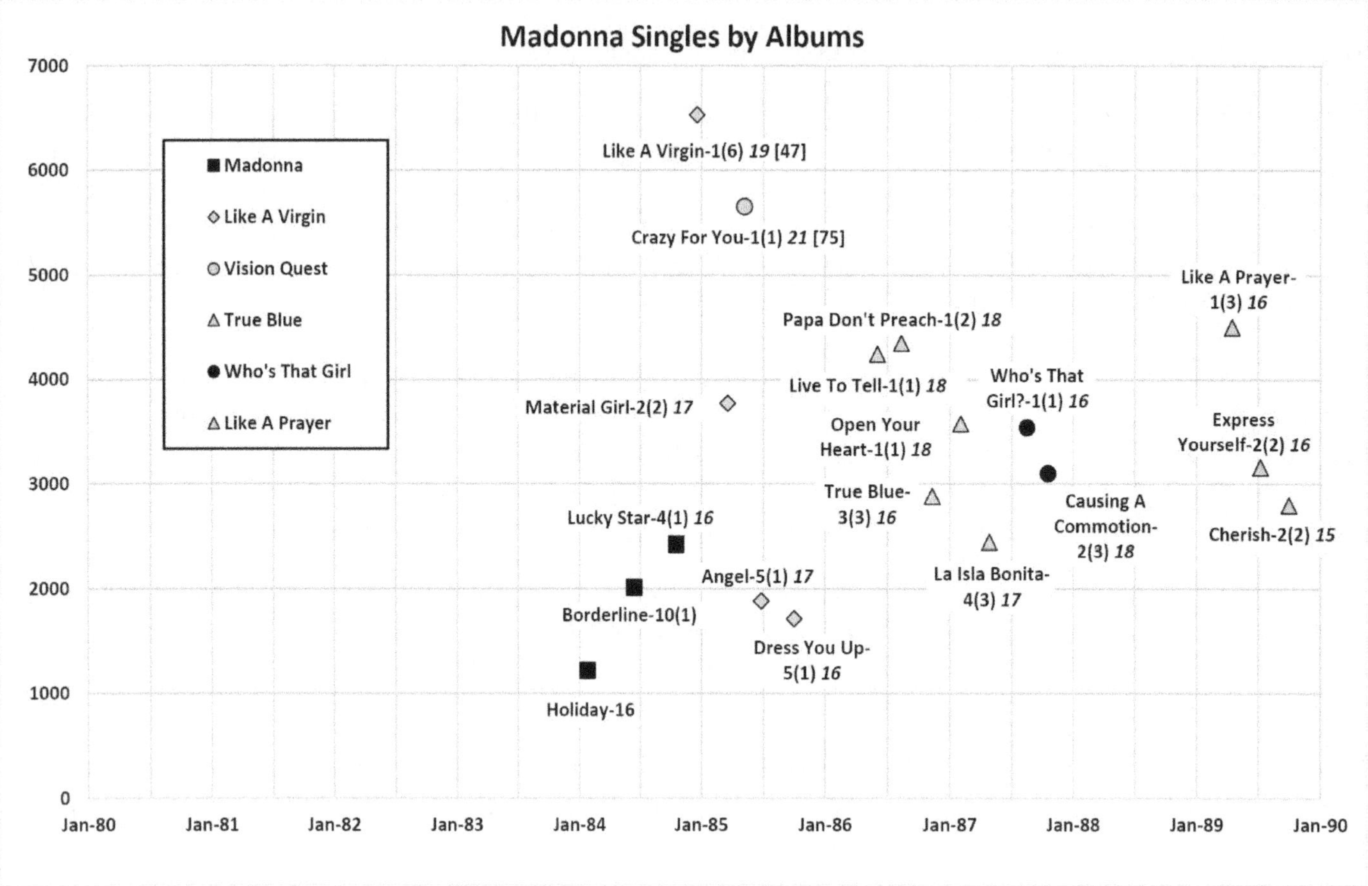

Chronologies

Madonna Singles by Albums

Legend:
- ■ Madonna
- ◇ Like A Virgin
- ○ Vision Quest
- △ True Blue
- ● Who's That Girl
- △ Like A Prayer

Data points:
- Holiday-16
- Borderline-10(1)
- Lucky Star-4(1) *16*
- Like A Virgin-1(6) *19* [47]
- Material Girl-2(2) *17*
- Crazy For You-1(1) *21* [75]
- Angel-5(1) *17*
- Dress You Up-5(1) *16*
- Live To Tell-1(1) *18*
- Papa Don't Preach-1(2) *18*
- Open Your Heart-1(1) *18*
- True Blue-3(3) *16*
- La Isla Bonita-4(3) *17*
- Who's That Girl?-1(1) *16*
- Causing A Commotion-2(3) *18*
- Like A Prayer-1(3) *16*
- Express Yourself-2(2) *16*
- Cherish-2(2) *15*

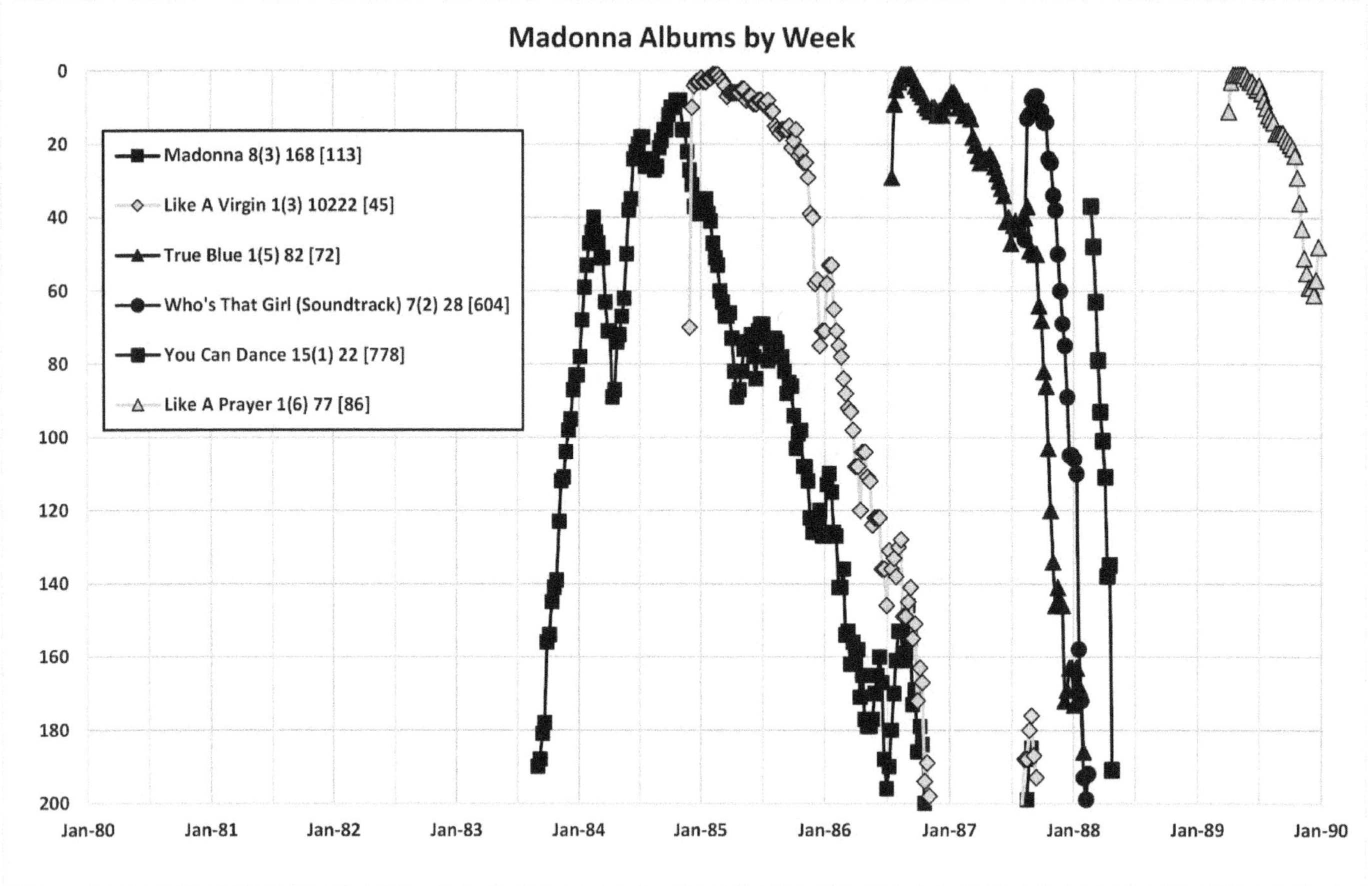

Madonna Albums by Week

Legend:
- ■ Madonna 8(3) 168 [113]
- ◇ Like A Virgin 1(3) 10222 [45]
- ▲ True Blue 1(5) 82 [72]
- ● Who's That Girl (Soundtrack) 7(2) 28 [604]
- ■ You Can Dance 15(1) 22 [778]
- △ Like A Prayer 1(6) 77 [86]

MADONNA ▶ Singles Rank: 2 (18) 59786; Album Rank: 4 (6/6) 93686 100%

Prior to *Holiday,* Madonna had a Bubbling Under record *Everybody*, which peaked at 107. She charted two of the top 100 singles of the decade, *Like A Virgin* and *Crazy For You*. Seven singles peaked at number 1 and four more at number 2. Her overall singles performance was consistently high, averaging over 3300 points per record, second best of the decade.

Neglecting soundtracks and the remix album *You Can Dance,* three of her four albums of original material charted in the top 100 of the decade, *Like A Virgin* (number 45 and 7x Platinum by the end of 1987), *True Blue* (72 and 5x Platinum by the end of 1987) and *Like A Prayer* (86 and 3x Platinum by the end of 1989). Her eponymous debut, *Madonna*, ranked 113. When scored by singles, *True Blue* rose to number 9 and *Like A Virgin* rose to number 25. She was the number 4 album act of the decade.

The record *Live To Tell* is listed as the first single from *True Blue*, but as a single preceded the album by about three months. Its Wikipedia article notes that it was originally composed as an instrumental by longtime Madonna collaborator Patrick Leonard. Leonard showed it to Madonna who liked it and proposed it for the score of then-husband Sean Penn's upcoming movie "At Close Range."

The question is: should it be considered to have originated with *True Blue*? The soundtrack of "At Close Range" retreads a number of previously recorded songs: *Miss You* by the Rolling Stones and *Boogie Oogie Oogie* by A Taste Of Honey. No album for the soundtrack can be found on Discogs, and it did not chart. As a result, despite its first manifestation in the movie, I have attributed it to *True Blue*, in line with convention.

Madonna also defies the trend of the decade. In general, there were fewer decade-leading singles in the later years. Madonna's trendline is clearly upward headed into 1990.

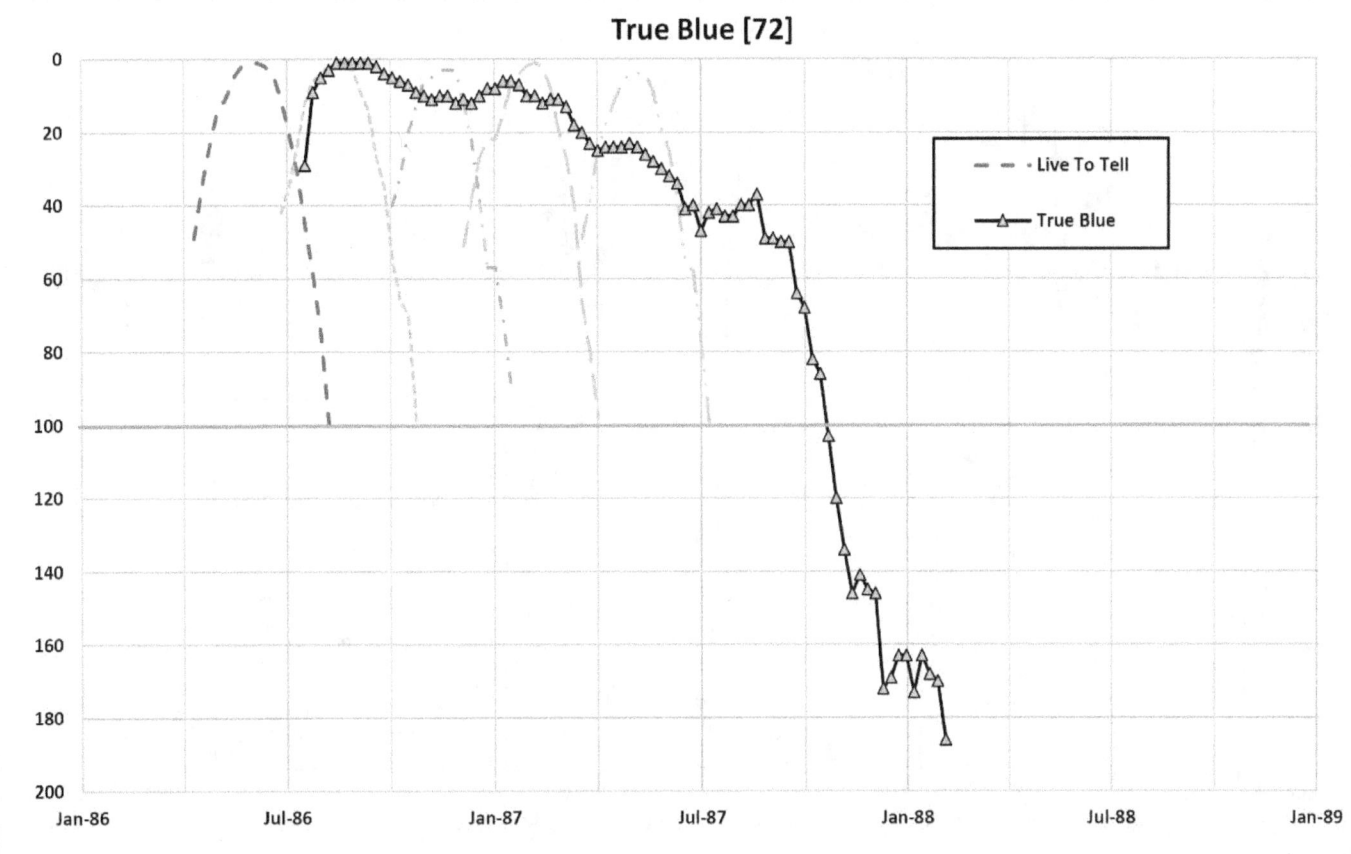

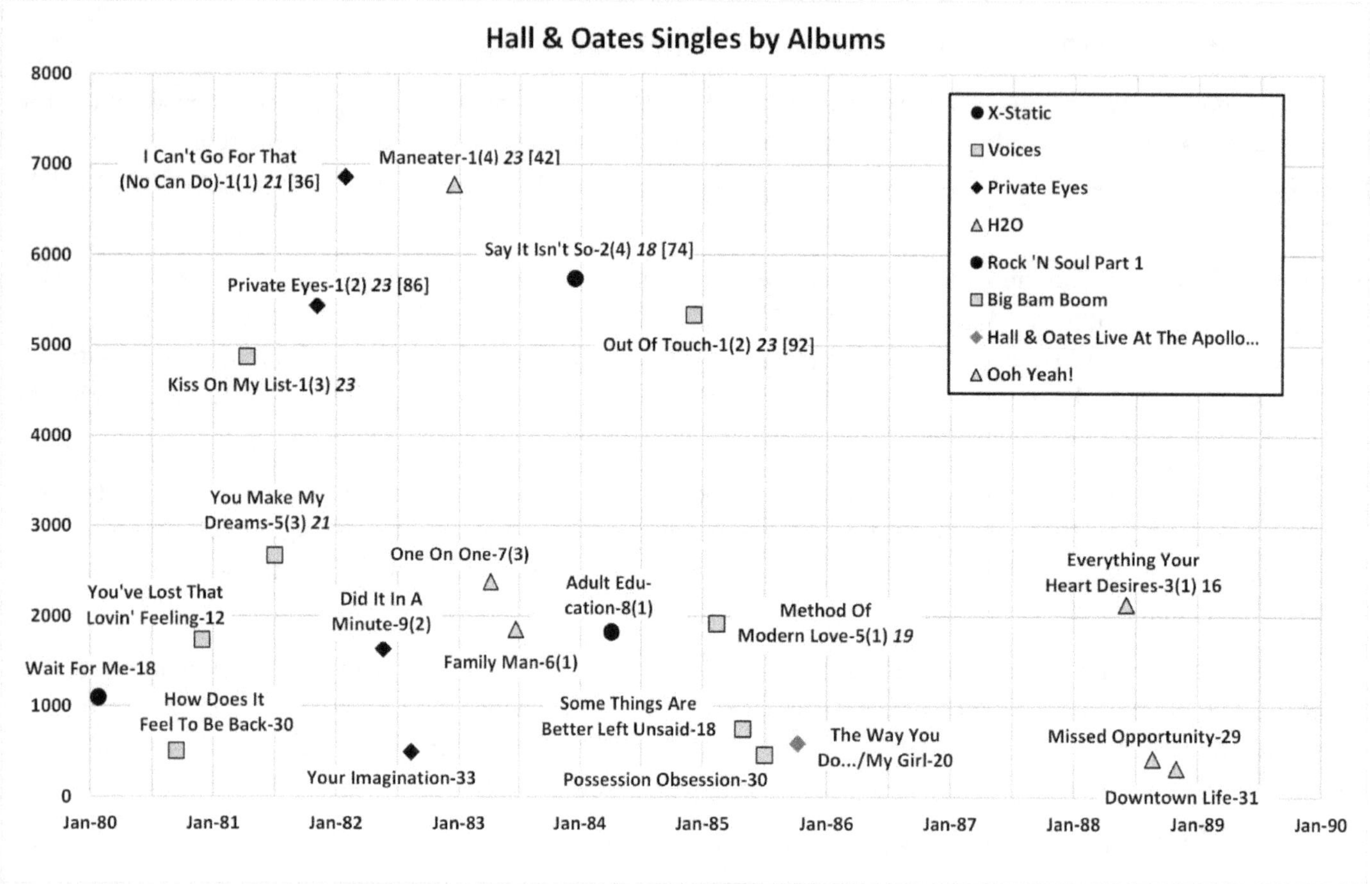

Chronologies

Hall & Oates Singles by Albums

Legend:
- ● X-Static
- □ Voices
- ♦ Private Eyes
- △ H2O
- ● Rock 'N Soul Part 1
- □ Big Bam Boom
- ♦ Hall & Oates Live At The Apollo...
- △ Ooh Yeah!

Data points:
- Wait For Me-18
- You've Lost That Lovin' Feeling-12
- How Does It Feel To Be Back-30
- Kiss On My List-1(3) 23
- You Make My Dreams-5(3) 21
- I Can't Go For That (No Can Do)-1(1) 21 [36]
- Private Eyes-1(2) 23 [86]
- Did It In A Minute-9(2)
- Your Imagination-33
- Maneater-1(4) 23 [42]
- One On One-7(3)
- Family Man-6(1)
- Say It Isn't So-2(4) 18 [74]
- Adult Education-8(1)
- Possession Obsession-30
- Out Of Touch-1(2) 23 [92]
- Method Of Modern Love-5(1) 19
- Some Things Are Better Left Unsaid-18
- The Way You Do.../My Girl-20
- Everything Your Heart Desires-3(1) 16
- Missed Opportunity-29
- Downtown Life-31

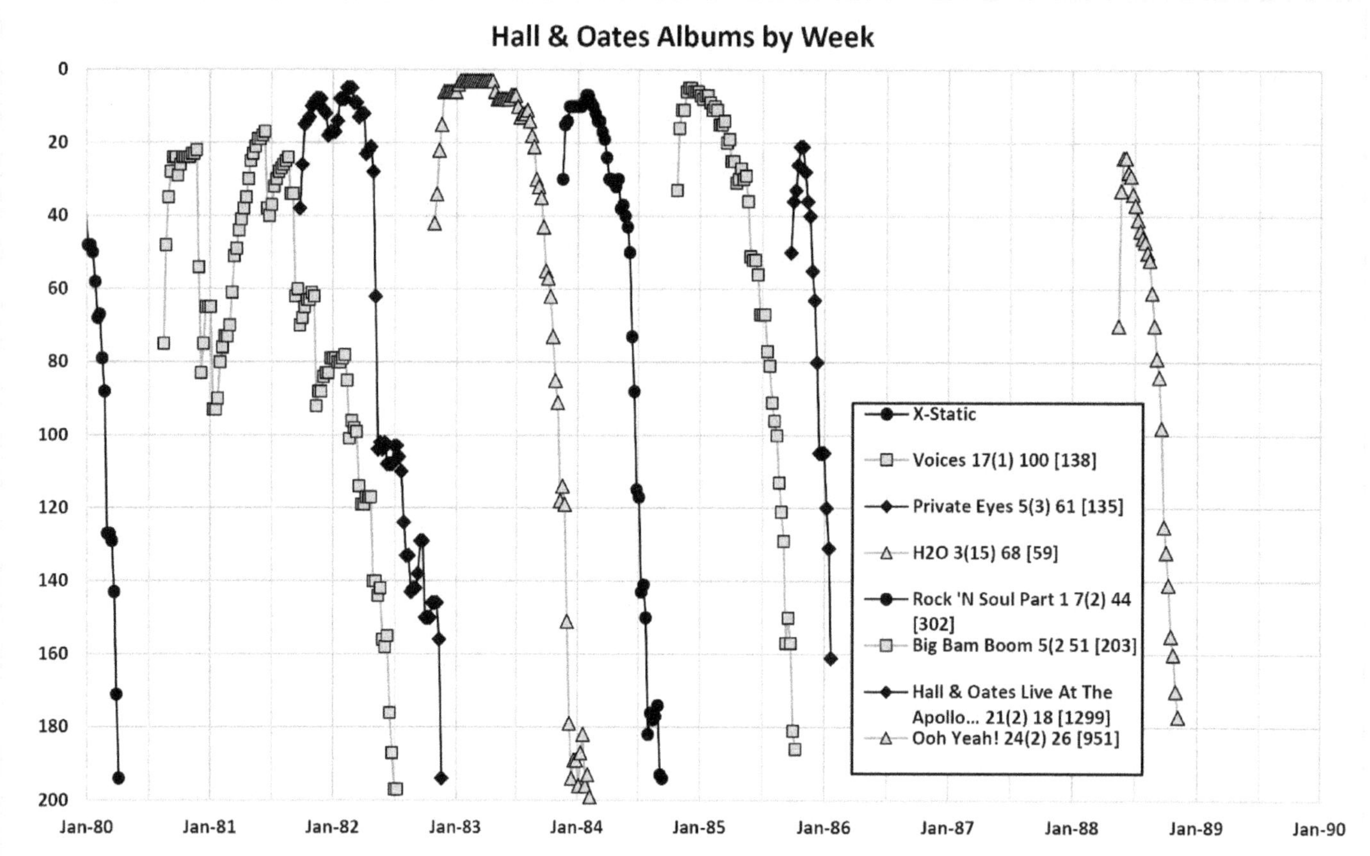

Hall & Oates Albums by Week

Legend:
- ● X-Static
- □ Voices 17(1) 100 [138]
- ♦ Private Eyes 5(3) 61 [135]
- △ H2O 3(15) 68 [59]
- ● Rock 'N Soul Part 1 7(2) 44 [302]
- □ Big Bam Boom 5(2 51 [203]
- ♦ Hall & Oates Live At The Apollo... 21(2) 18 [1299]
- △ Ooh Yeah! 24(2) 26 [951]

HALL & OATES ▶ Singles Rank: 3 (22) 55777; Albums Rank: 12 (7/8) 74922 98.7%

Hall and Oates was a top 100 act for the '70s even though they were only truly active on the charts in the second half. That second half of the '70s set the stage for explosive performance in the first half of the '80s. Their singles started modestly with *How Does It Feel To Be Back* peaking at 30 and *You've Lost That Lovin' Feeling* peaking at 12, but 12 of the next 13 singles went to the Top 10, including five of the top 100 singles of the decade. However, after *Method Of Modern Love* there would be only one more Top 10—*Everything Your Heart Desires*.

Album-wise, *X-Static* ended the '70s and was their second biggest album of that decade—points-wise nearly the twin of *Ooh Yeah!*, their final album of the '80s. The album apex is *H2O*, album 59 of the decade, although when scored by singles, *Private Eyes* exceeds it by rising from 135 to 20 on the strength of *I Can't Go For That* and *Private Eyes*. The album trajectories as the '80s progressed had a similar high entry, quick peak and decreasing time at apex before a fairly steep fall.

Voices and *Private Eyes* are interesting chart cases to compare. These two albums had virtually the same score and rank but very different life cycles: longer life and lower peak vs. shorter life and higher peak.

It's unusual for the third single from an album to be a breakout—in this case, *Kiss On My List*--going to Number 1 for three weeks after the first two peaked at 30 and 12 respectively. The singles performance had a dramatic effect on *Voices*; its chart path had stalled in the low 20s, and it had fallen 70 positions with the exit of *You've Lost That Lovin' Feeling* until the success of *Kiss On My List* revived it. As *Kiss* was exiting, *Voices* fell again, only to be revived by Top 10 *You Make My Dreams*. After the release of *Private Eyes*, lesser effects on *Voices* can be seen from the new album's singles. *Private Eyes* itself saw no such drama. The two number 1 singles *Private Eyes* and *I Can't Go For That* drove it to the Top 10 and Top 5 of albums respectively, but the lesser performance of *Did It In A Minute* and *Your Imagination* did not stem the decline.

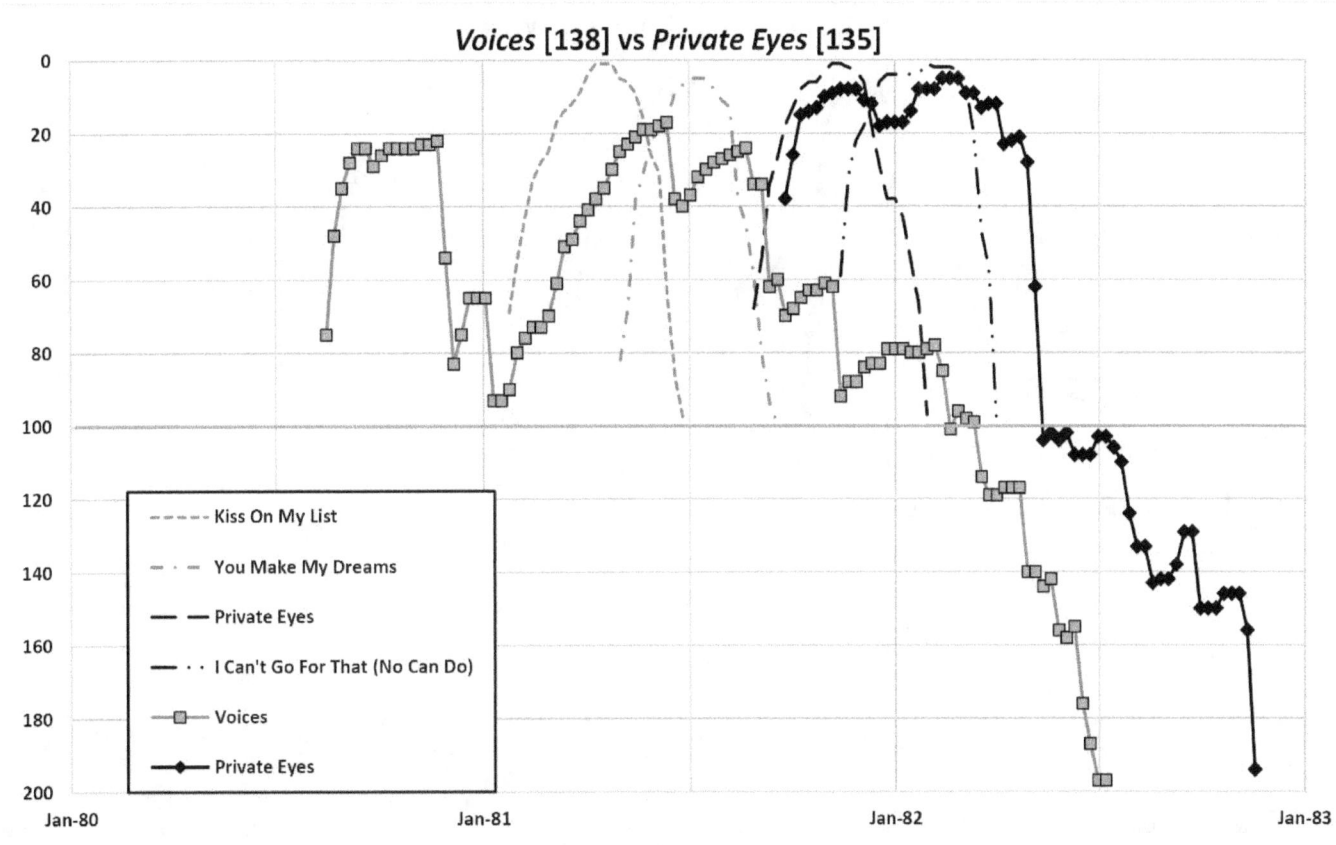

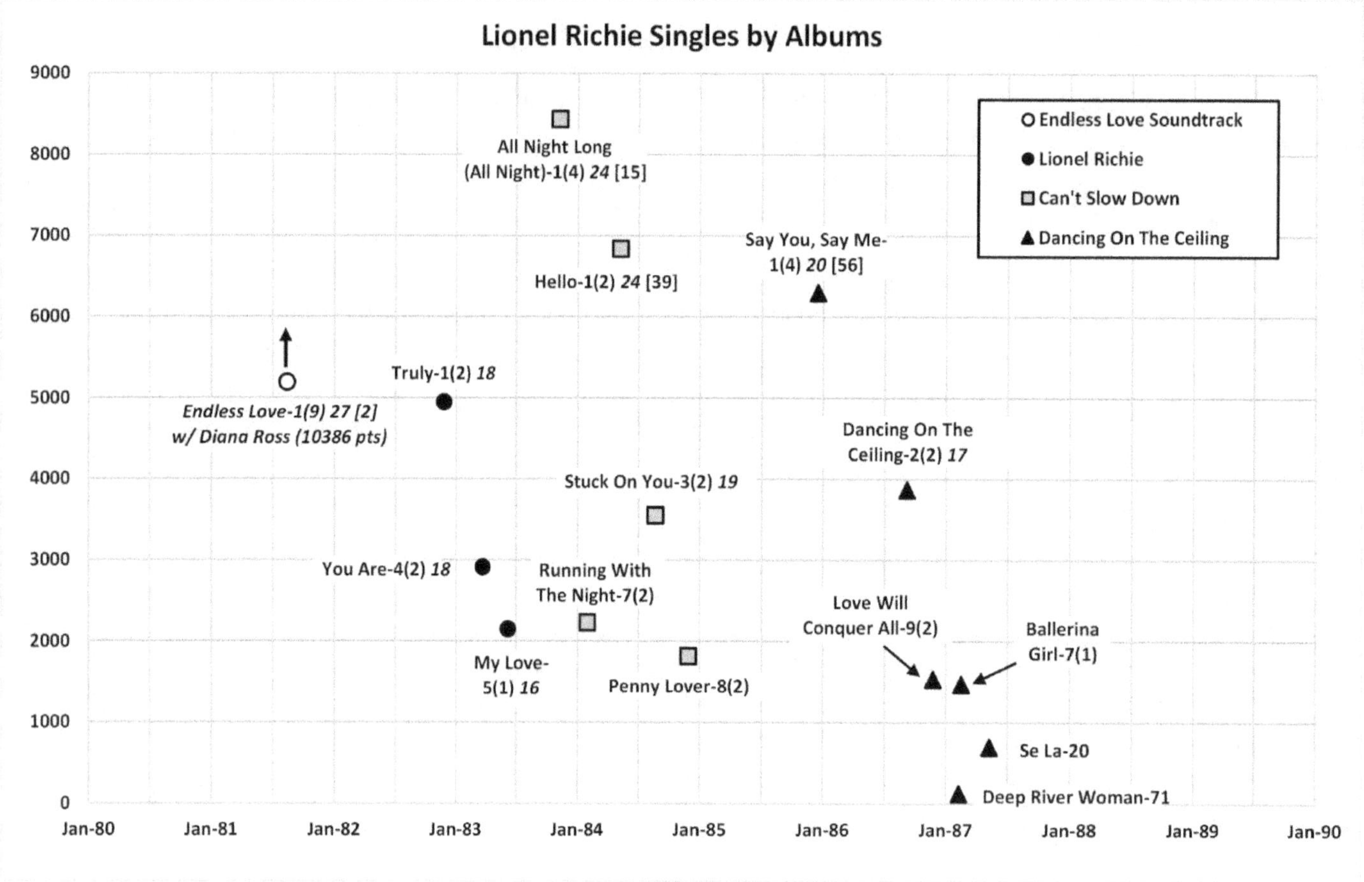

Chronologies

Lionel Richie Singles by Albums

Legend:
- ○ Endless Love Soundtrack
- ● Lionel Richie
- □ Can't Slow Down
- ▲ Dancing On The Ceiling

Data points:
- Endless Love-1(9) 27 [2] w/ Diana Ross (10386 pts)
- Truly-1(2) 18
- All Night Long (All Night)-1(4) 24 [15]
- Hello-1(2) 24 [39]
- Say You, Say Me-1(4) 20 [56]
- Dancing On The Ceiling-2(2) 17
- Stuck On You-3(2) 19
- You Are-4(2) 18
- Running With The Night-7(2)
- Love Will Conquer All-9(2)
- Ballerina Girl-7(1)
- My Love-5(1) 16
- Penny Lover-8(2)
- Se La-20
- Deep River Woman-71

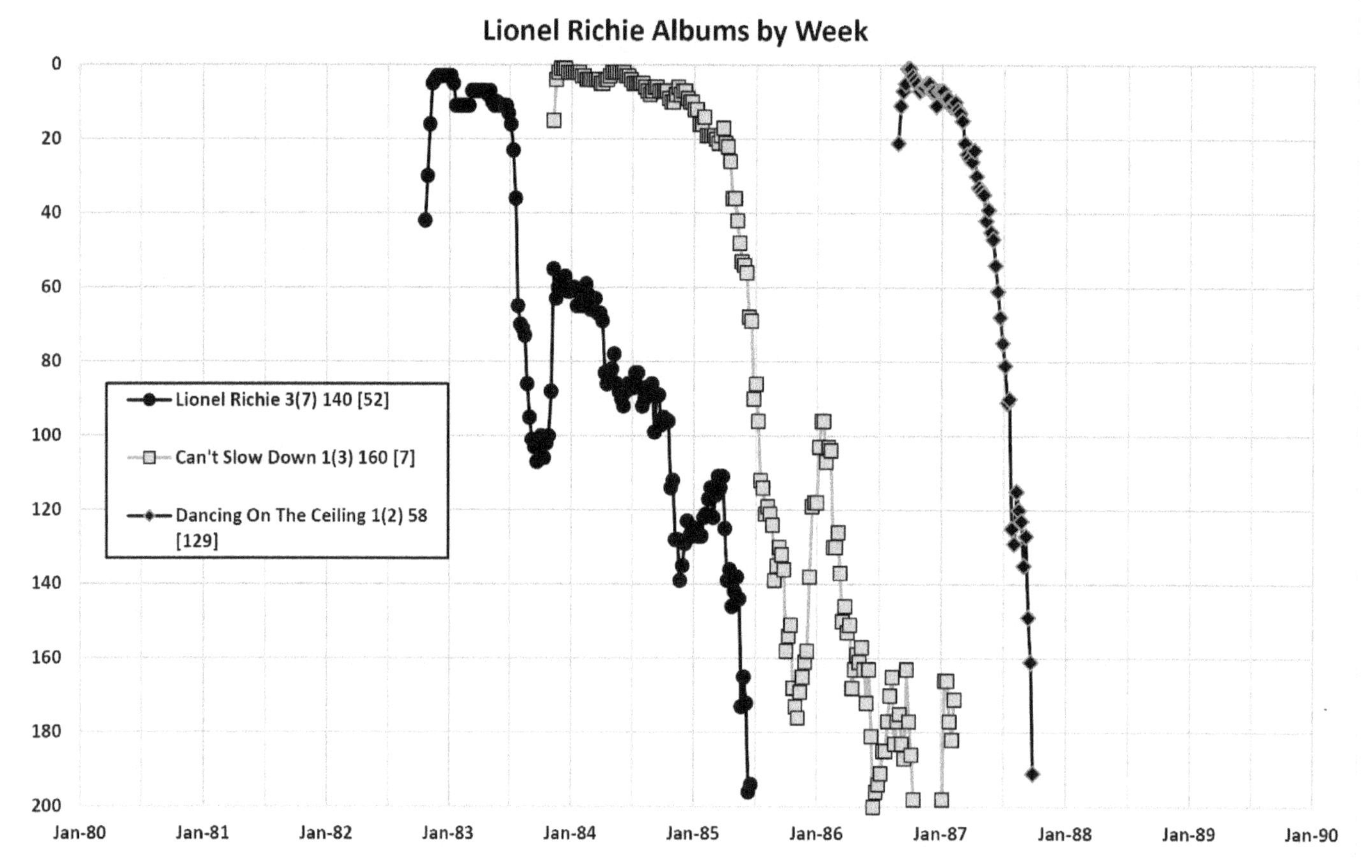

Lionel Richie Albums by Week

Legend:
- ● Lionel Richie 3(7) 140 [52]
- □ Can't Slow Down 1(3) 160 [7]
- ♦ Dancing On The Ceiling 1(2) 58 [129]

426

LIONEL RICHIE ▶ Singles Rank: 4 (15) 52069; Albums Rank: 8 (3/3) 83826 100%

Lionel Richie stepped out on his own in 1982, but his duet with Diana Ross, *Endless Love*, predated his first solo records. Because the credit for *Endless Love*--the number 2 single of the period--is divided, the singles graph doesn't visually do it justice—it scored over 10,000 points and spent nine weeks at number 1.

The releases from his three solo albums in the '80s followed similar trajectories: by far, the strongest single was the first, with generally declining subsequent performance. The albums themselves moved quickly into the top 5, with both *Lionel Richie* [52, 4x Platinum by 1984] and *Can't Slow Down* [7, 10x Platinum by 1985] having substantial chart time. Those two albums rose to 23 and 2 respectively when scored by singles. *Dancing On The Ceiling* screamed to the top of the chart, but fell relatively quickly by comparison.

His singles catalog includes four number 1's and four other top 5's; the 15 singles averaged 3471 points, good for third place. Despite having two of his three albums in the top 100 of the decade, because of the small number he was only the number 8 album act of the decade.

It is interesting to see how the singles propped up the album chart performance, particularly for *Can't Slow Down*. The release of *Can't Slow Down* and its singles added 50 places to *Lionel Richie* virtually immediately. The number 56 single of the period, *Say You, Say Me* boosted *Can't Slow Down* by nearly 100 chart ranks, even though it was not on that album—and its story is a bit complicated.

Say You, Say Me was in the movie White Nights, but Motown did not want it on the soundtrack. It was added as a bonus track later, on a reissued soundtrack CD. It was recorded for and led the release of *Dancing On the Ceiling* by nearly a year, apparently fueling the demand for the then-current album, *Can't Slow Down*. Viewing his whole singles catalog, it is impossible to miss a general downward slope of Richie's score vs. time over the decade.

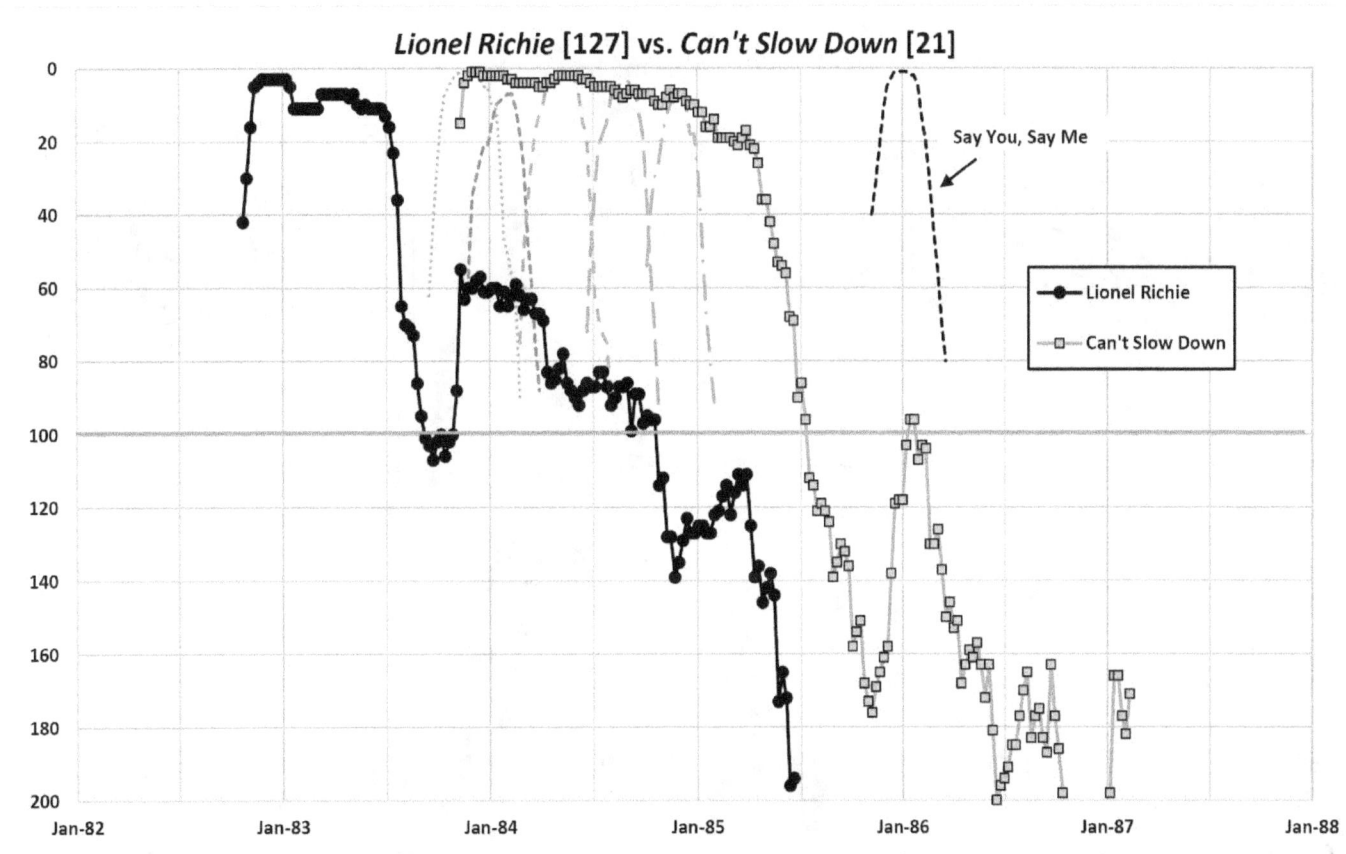

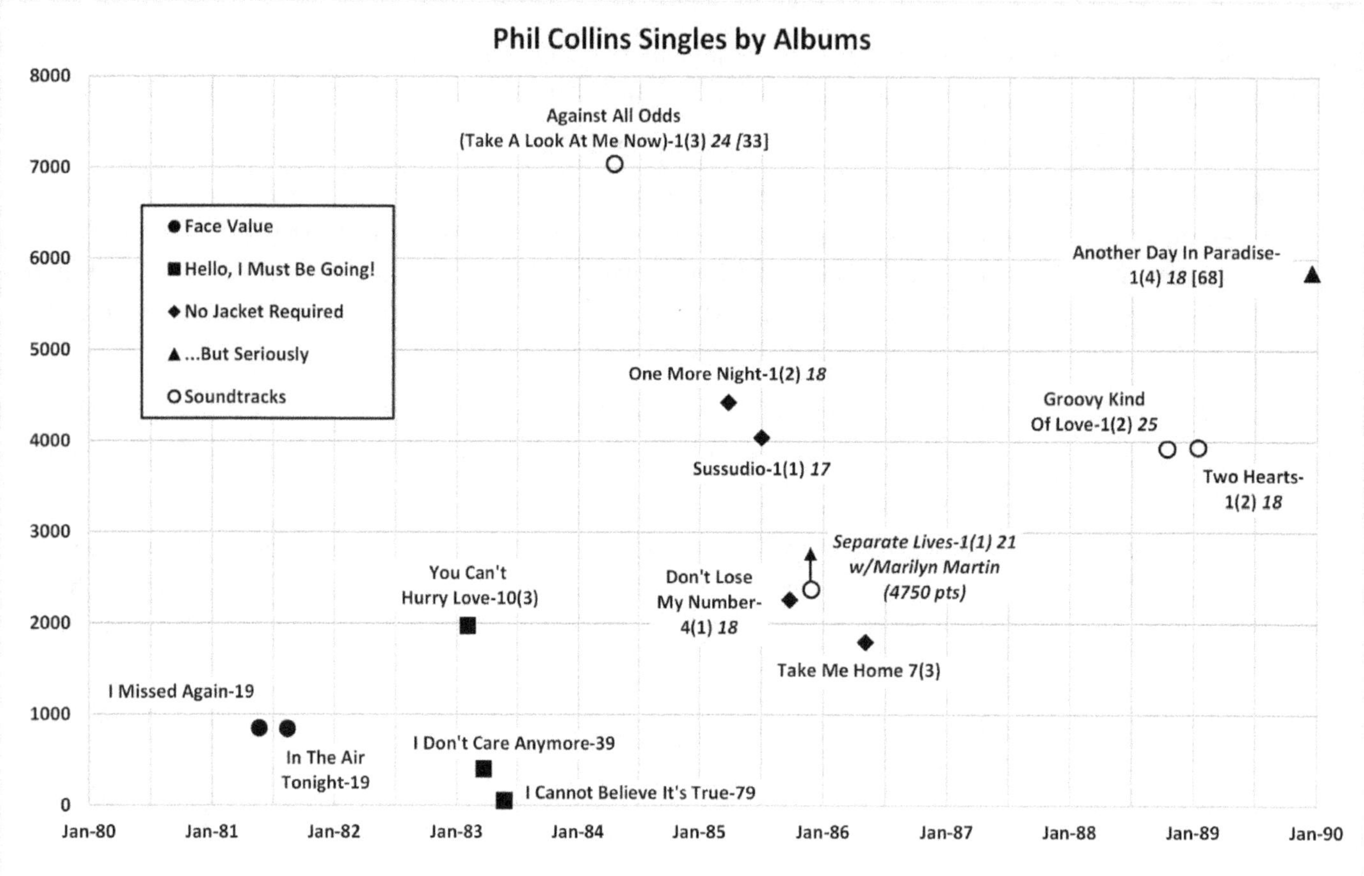

Chronologies

Phil Collins Singles by Albums

- ● Face Value
- ■ Hello, I Must Be Going!
- ◆ No Jacket Required
- ▲ ...But Seriously
- ○ Soundtracks

- Against All Odds (Take A Look At Me Now)-1(3) *24 [33]*
- Another Day In Paradise-1(4) *18 [68]*
- One More Night-1(2) *18*
- Sussudio-1(1) *17*
- Groovy Kind Of Love-1(2) *25*
- Two Hearts-1(2) *18*
- You Can't Hurry Love-10(3)
- Don't Lose My Number-4(1) *18*
- *Separate Lives-1(1) 21 w/Marilyn Martin (4750 pts)*
- Take Me Home 7(3)
- I Missed Again-19
- In The Air Tonight-19
- I Don't Care Anymore-39
- I Cannot Believe It's True-79

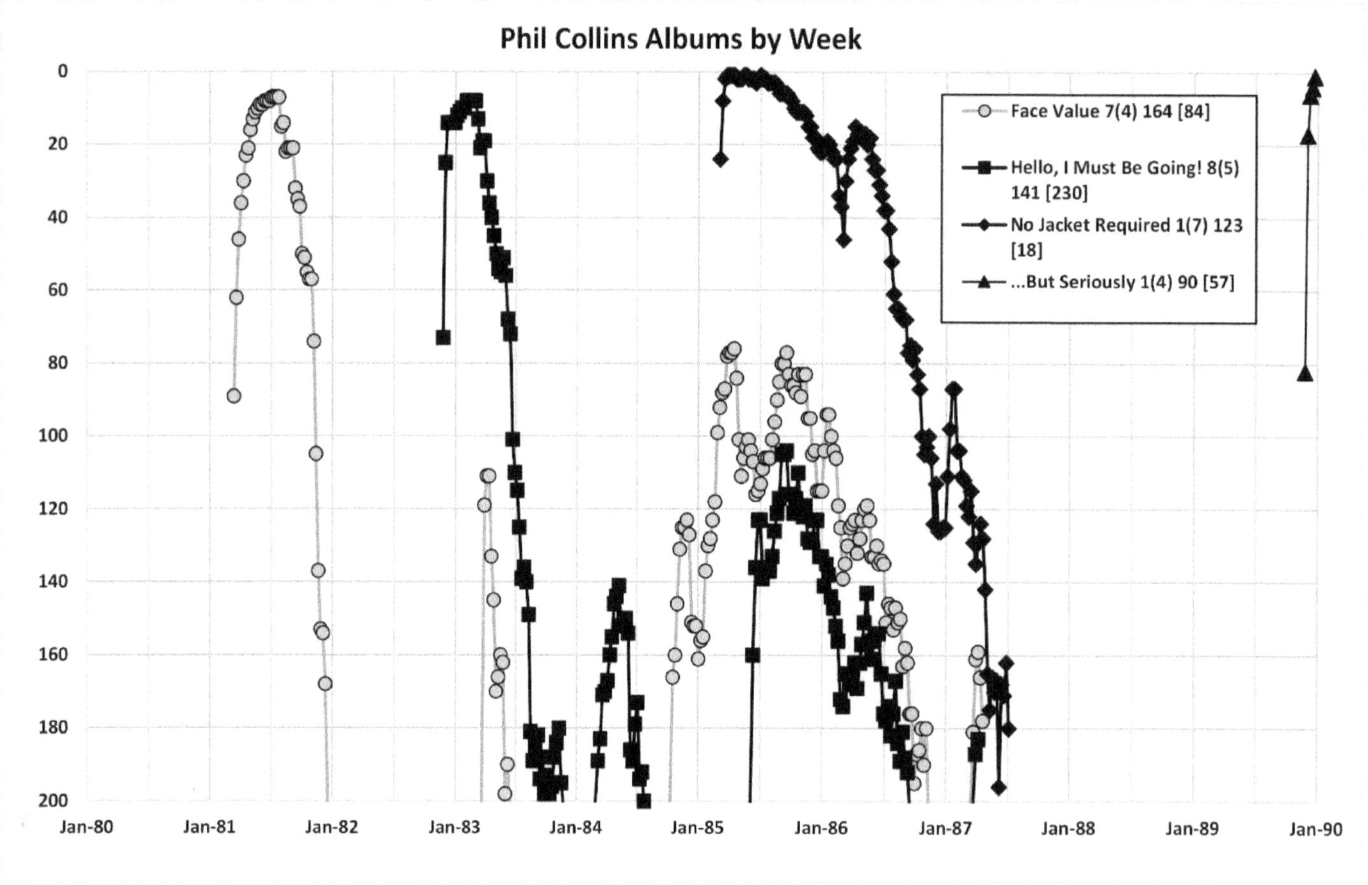

Phil Collins Albums by Week

- ─○─ Face Value 7(4) 164 [84]
- ─■─ Hello, I Must Be Going! 8(5) 141 [230]
- ─◆─ No Jacket Required 1(7) 123 [18]
- ─▲─ ...But Seriously 1(4) 90 [57]

428

PHIL COLLINS ▶ Singles Rank: 5 (14) 39768; Albums Rank: 19 (4/4) 64403 100%

The three remaining members of Genesis started their extracurricular work in 1979: Tony Banks and Mike Rutherford worked on solo albums and Phil Collins played and toured with Brand X. Genesis came back together for *Duke* in 1980 and after that Collins began work on *Face Value*, which was released in February, 1981.

Overall, Phil Collins had seven number 1 singles and three other top 10 singles out of 14 entries. He averaged 2841 points, good for tenth. Three of his four albums were in the top 100: *No Jacket Required* spent seven weeks at number 1 and was 6x Platinum by the end of 1989. When scored by its five singles, however, the whole was more than the sum of the parts: it fell from number 18 to number 39.

Face Value only charted two singles, both of which peaked at number 19. When scored by singles it ranked 654th. The life cycle of *Face Value* appeared to be over by the end of 1981. But the release of *Hello, I Must Be Going* and the three singles it spawned appeared to breathe new life into *Face Value* as an album entity: it had a recurrent spike as *Hello, I Must Be Going* peaked, which eventually vaulted it to the number 84 album spot.

Face Value had another resurrection and a new life much longer than the first with the debut of the TV series Miami Vice. Collins' single *In The Air Tonight* figured strongly in the ethos of the show and while it failed to chart a second time as a single, it appears to have renewed interest in *Face Value* and dragged that album back on to the chart; the release of *No Jacket Required* energized both the previous albums and extended their life cycles to the very end of that of *No Jacket Required*.

There is one other curiosity. For all the times *Face Value* got up off the mat, the single *Against All Odds* only seemed to revitalize *Hello, I Must Be Going*. There is a distinct coincident peak for that album at the same time as the single from the movie soundtrack. From the singles chart through the decade, it is clear that by 1990 Phil Collins' directional arrow is pointed up.

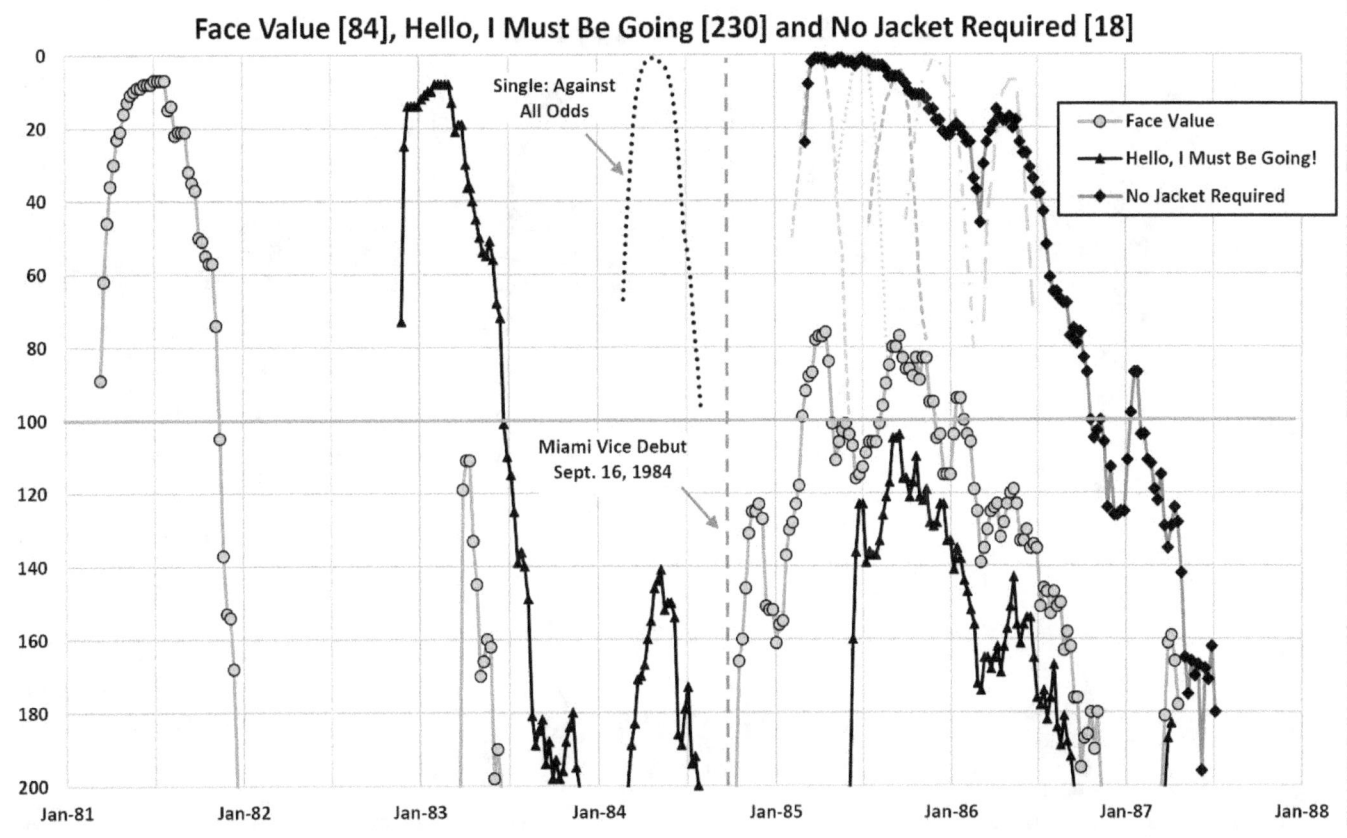

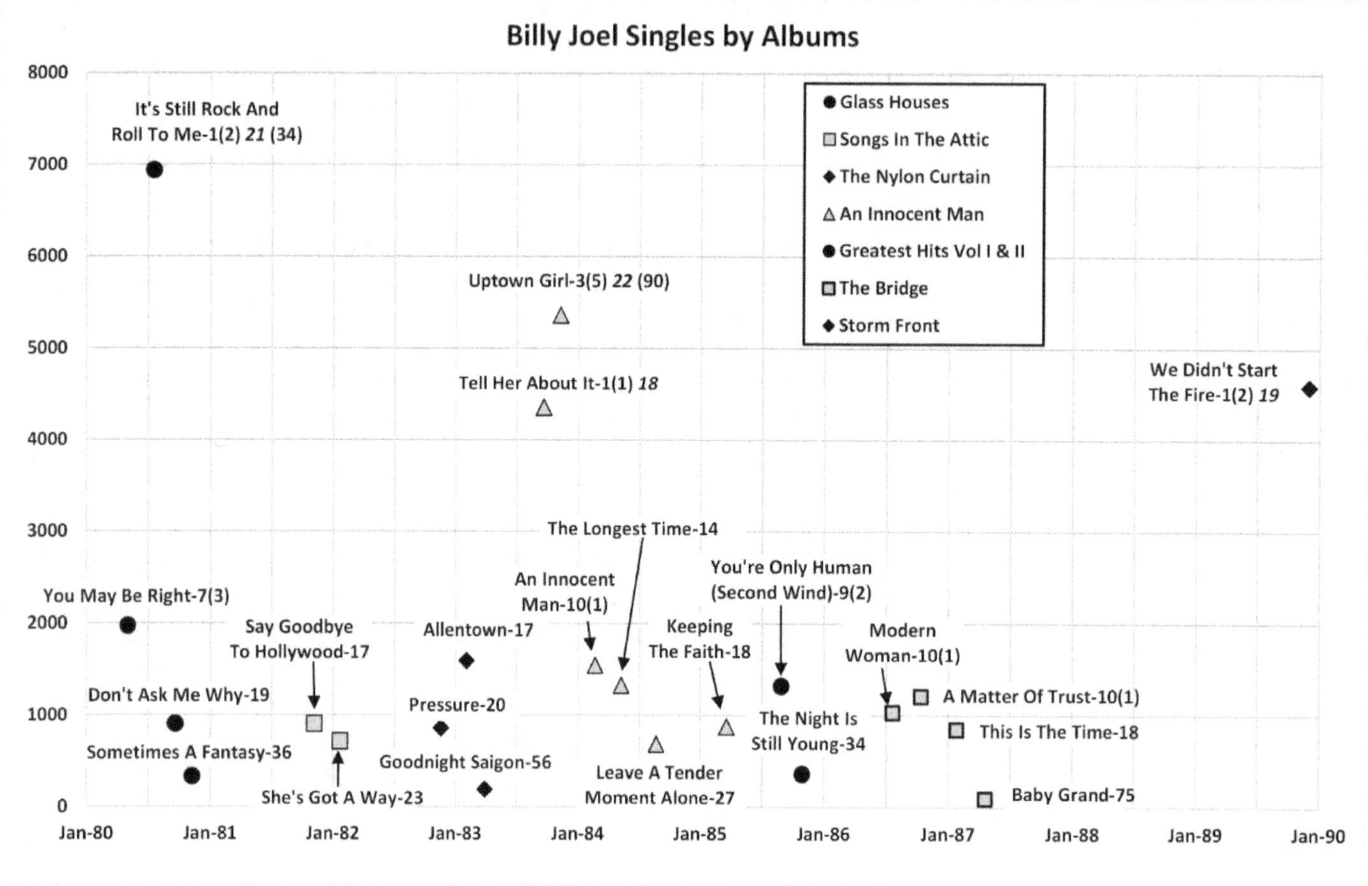

Chronologies

Billy Joel Singles by Albums

Legend:
- ● Glass Houses
- ☐ Songs In The Attic
- ◆ The Nylon Curtain
- △ An Innocent Man
- ● Greatest Hits Vol I & II
- ☐ The Bridge
- ◆ Storm Front

Data points:
- It's Still Rock And Roll To Me-1(2) *21* (34)
- Uptown Girl-3(5) *22* (90)
- Tell Her About It-1(1) *18*
- We Didn't Start The Fire-1(2) *19*
- The Longest Time-14
- An Innocent Man-10(1)
- You're Only Human (Second Wind)-9(2)
- You May Be Right-7(3)
- Say Goodbye To Hollywood-17
- Allentown-17
- Keeping The Faith-18
- Modern Woman-10(1)
- Don't Ask Me Why-19
- Pressure-20
- A Matter Of Trust-10(1)
- Sometimes A Fantasy-36
- Goodnight Saigon-56
- The Night Is Still Young-34
- This Is The Time-18
- She's Got A Way-23
- Leave A Tender Moment Alone-27
- Baby Grand-75

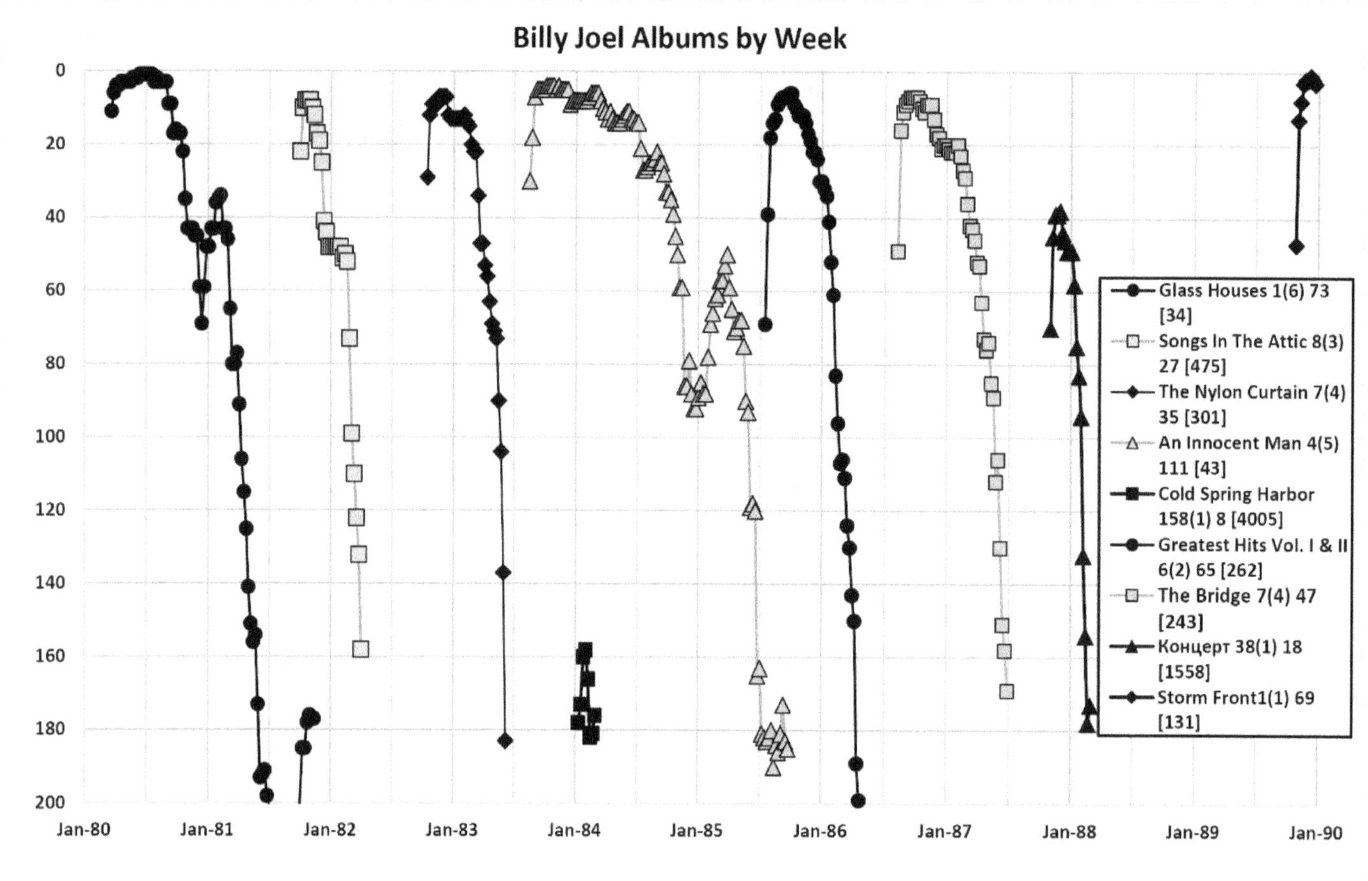

Billy Joel Albums by Week

Legend:
- ● Glass Houses 1(6) 73 [34]
- ☐ Songs In The Attic 8(3) 27 [475]
- ◆ The Nylon Curtain 7(4) 35 [301]
- △ An Innocent Man 4(5) 111 [43]
- ■ Cold Spring Harbor 158(1) 8 [4005]
- ● Greatest Hits Vol. I & II 6(2) 65 [262]
- ☐ The Bridge 7(4) 47 [243]
- ▲ Концерт 38(1) 18 [1558]
- ◆ Storm Front 1(1) 69 [131]

BILLY JOEL ▶ Singles Rank: 6 (22) 38031; Albums Rank: 3 (9/11) 94089 98.5%

Following the success of *The Stranger* and *52nd Street* in the late 1970s, Billy Joel's trajectory was truly headed up. The number one chart performance of *Glass Houses* and its first single *It's Still Rock And Roll To Me* did not disappoint. But looking through the decade was a different story. He certainly was productive—his 22 charted singles tied for most in the '80s.

Those 22 singles included three number 1s plus *Uptown Girl* which only peaked at number 3 for five weeks but was the 90th ranked single of the decade (It was blocked by *Say Say Say* and *All Night Long (All Night)*, numbers 3 and 15 of the decade respectively). But only five of the other 18 cracked the top ten, the highest of which was 7. He ranked 51st in average score per single. His trip to the number 6-ranked singles act and number 3-ranked album act was fueled by volume.

Seven of his nine albums made the top ten—largest number in the decade; two went to number 1. *Glass Houses*, the strongest, ranked 34th and was 5x Platinum by the end of 1984; however, its rank dropped to 67 when scored by singles. *Storm Front* [131] was the other number 1 album, driven by *We Didn't Start The Fire*—a number 1 single.

His second biggest album of the decade, *An Innocent Man*, is an interesting case. It ranks number 43 for the decade, 4x Platinum by the end of 1984 and went six singles deep, including the aforementioned *Uptown Girl*, and *Tell Her About It*, his second '80s number 1. And yet the album itself only peaked at number four for five weeks. The first five singles peaked progressively lower, and after *Leave A Tender Moment Alone*, the album appeared to be on its way out. However, the sixth single, *Keeping The Faith* peaked higher and appeared to drag the album performance up with it, perhaps extending its chart life another 20 weeks beyond what it would have been otherwise. When scored by its singles, *An Innocent Man* rose from 43 to 21.

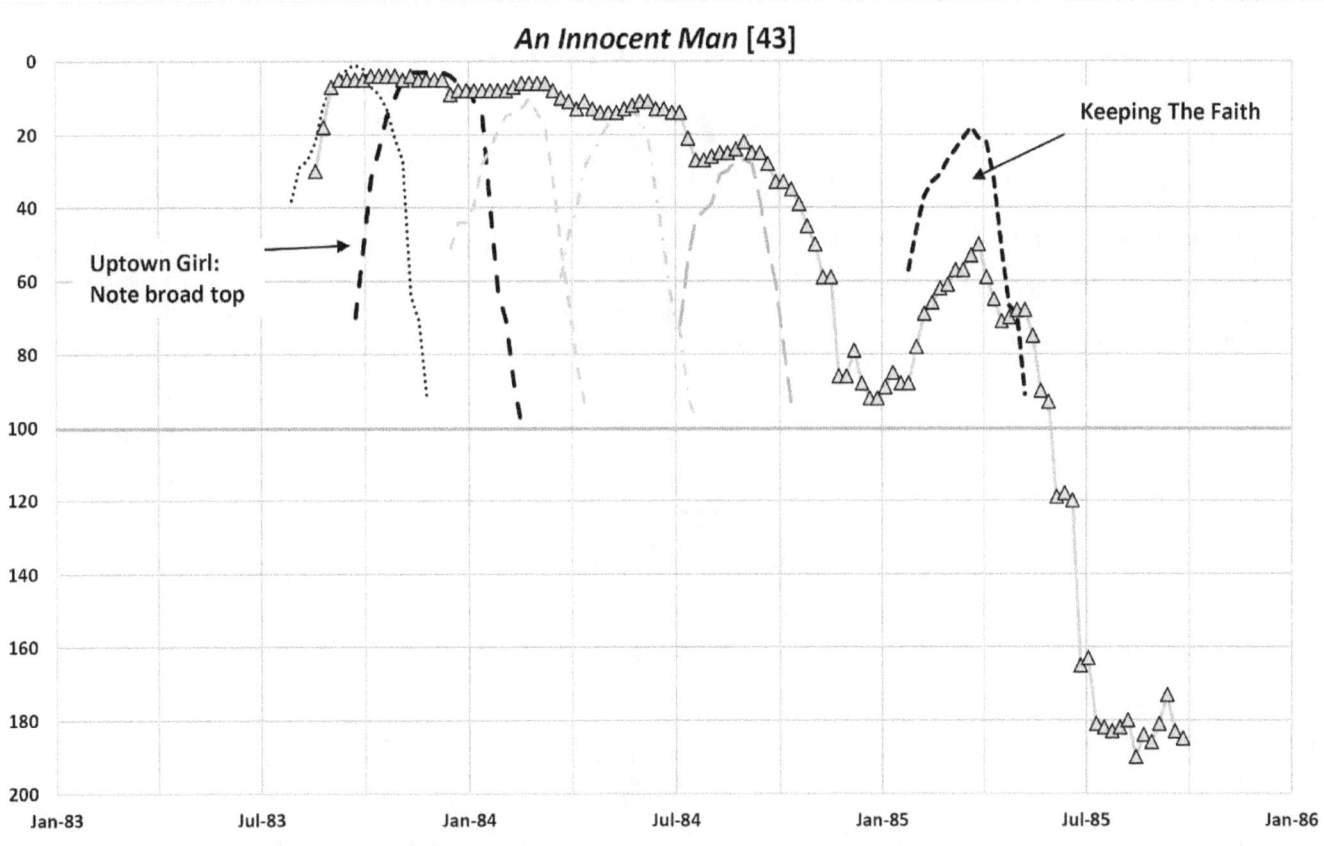

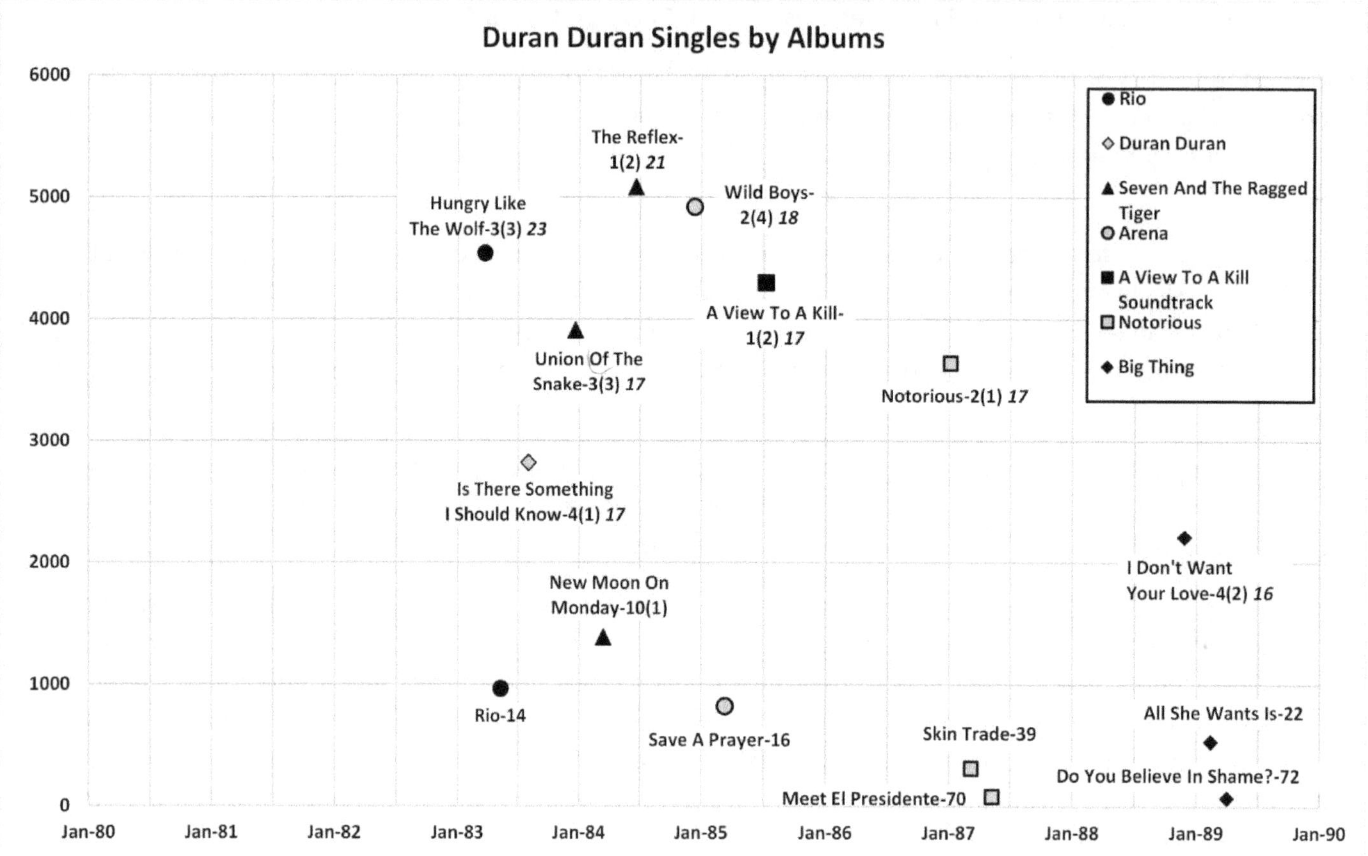

Chronologies

Duran Duran Singles by Albums

Legend:
- ● Rio
- ◇ Duran Duran
- ▲ Seven And The Ragged Tiger
- ○ Arena
- ■ A View To A Kill Soundtrack
- □ Notorious
- ◆ Big Thing

Data points:
- Hungry Like The Wolf-3(3) 23
- The Reflex-1(2) 21
- Wild Boys-2(4) 18
- A View To A Kill-1(2) 17
- Union Of The Snake-3(3) 17
- Notorious-2(1) 17
- Is There Something I Should Know-4(1) 17
- New Moon On Monday-10(1)
- I Don't Want Your Love-4(2) 16
- Rio-14
- Save A Prayer-16
- Skin Trade-39
- All She Wants Is-22
- Meet El Presidente-70
- Do You Believe In Shame?-72

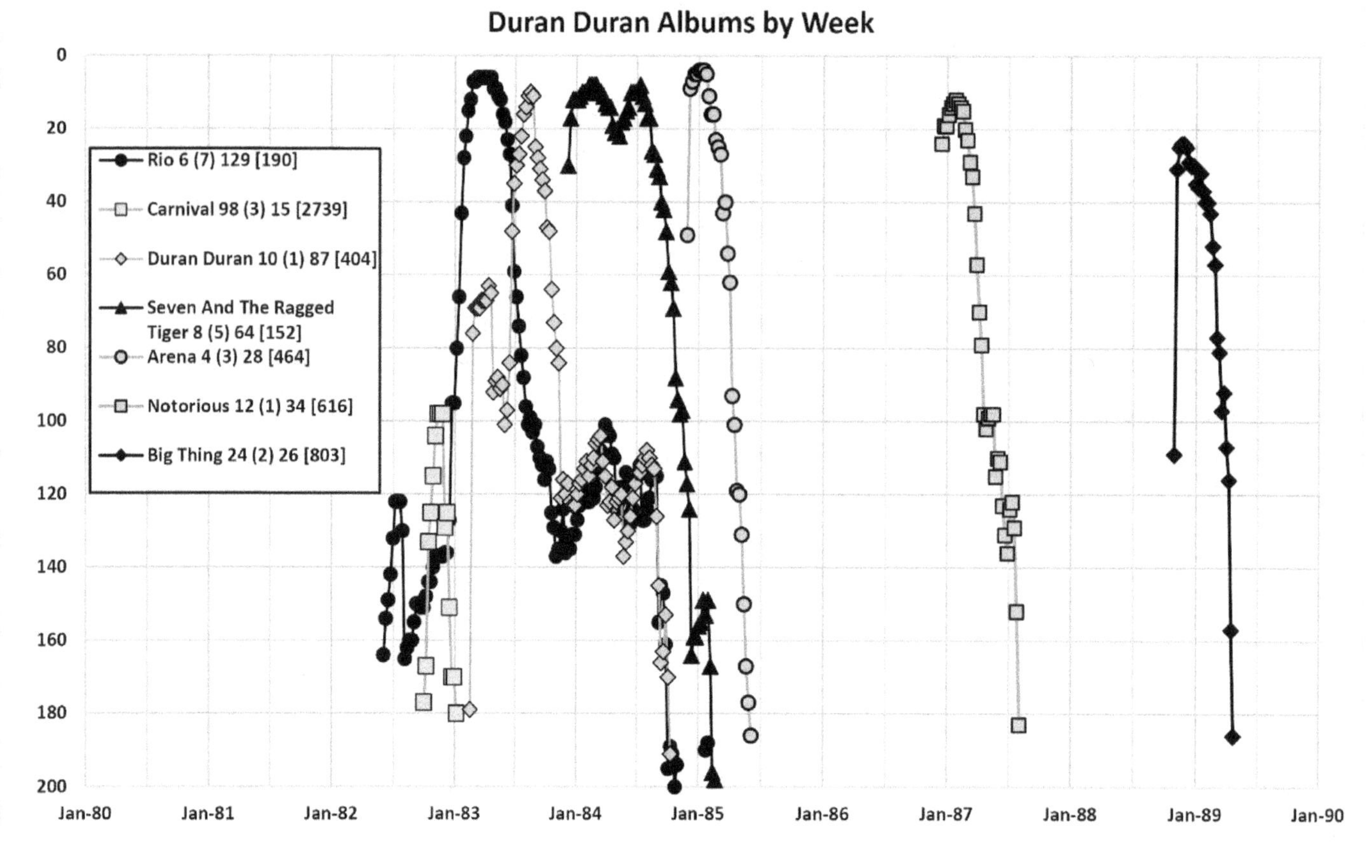

Duran Duran Albums by Week

Legend:
- ● Rio 6 (7) 129 [190]
- □ Carnival 98 (3) 15 [2739]
- ◇ Duran Duran 10 (1) 87 [404]
- ▲ Seven And The Ragged Tiger 8 (5) 64 [152]
- ○ Arena 4 (3) 28 [464]
- □ Notorious 12 (1) 34 [616]
- ◆ Big Thing 24 (2) 26 [803]

DURAN DURAN ▶ Singles Rank: 7 (15) 35594; Albums Rank : 33 (8/8) 52241 100%

Your first thought might be, "How did Duran Duran make it to number 7 with only fifteen singles, none of which placed in the decade top 100 and a mid-decade group hiatus and personnel change?" The answer is eight of those fifteen peaked in the top 4, if briefly.

Duran Duran had a hard time getting started in the US. Singles, and the eponymous album released in 1981 made no impression on the charts. *Rio* was released in May, 1982, and the single *Hungry Like The Wolf* followed in July. The single helped drive album sales a bit, but it didn't make the Hot 100.

Rio was remixed and rereleased in November 1982, and repositioned as a dance album. This time both the single and the album took off in a bigger way. *Carnival*, an EP of dance remixes, charted with the albums, and *Duran Duran* got new life as *Rio* found its legs.

The band took a hiatus for rest and side projects starting after the Live Aid Concert, with *A View To A Kill*—the first Bond theme to go to number 1--on the chart in 1985. Andy Taylor and Roger Taylor left and Simon LeBon, Nick Rhodes and John Taylor generated *Notorious* in 1986 with outside support.

Rio's album lifecycle shows the slow start with the minor boost from the original *Hungry Like The Wolf*. Clearly the album's two singles kept it at its apex, and the release of *Seven And The Ragged Tiger* with its singles slowed the decline. *Rio* spent 129 weeks on the album chart—well more than the 64 weeks by *Seven And The Ragged Tiger*—but most of those weeks were at lower echelons.

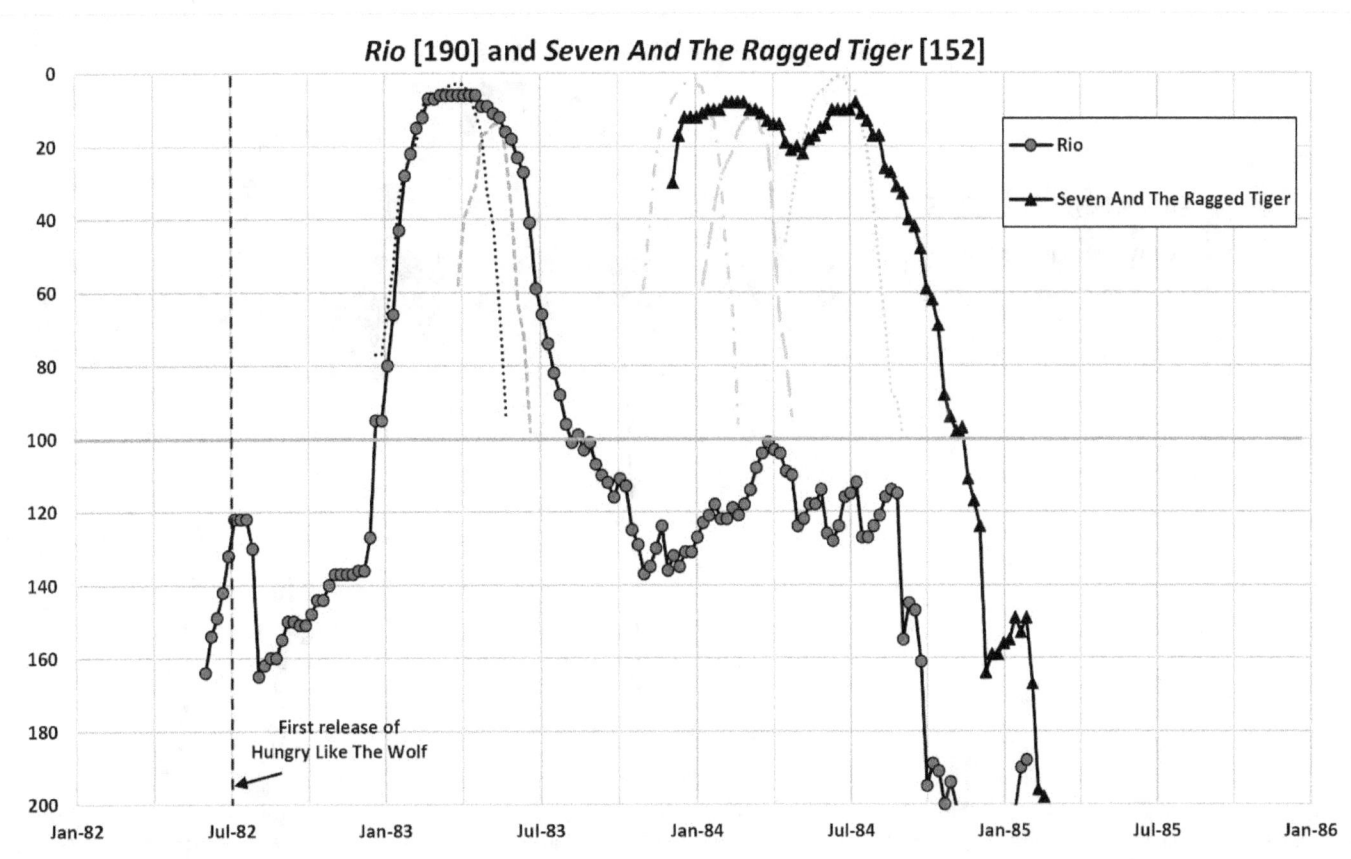

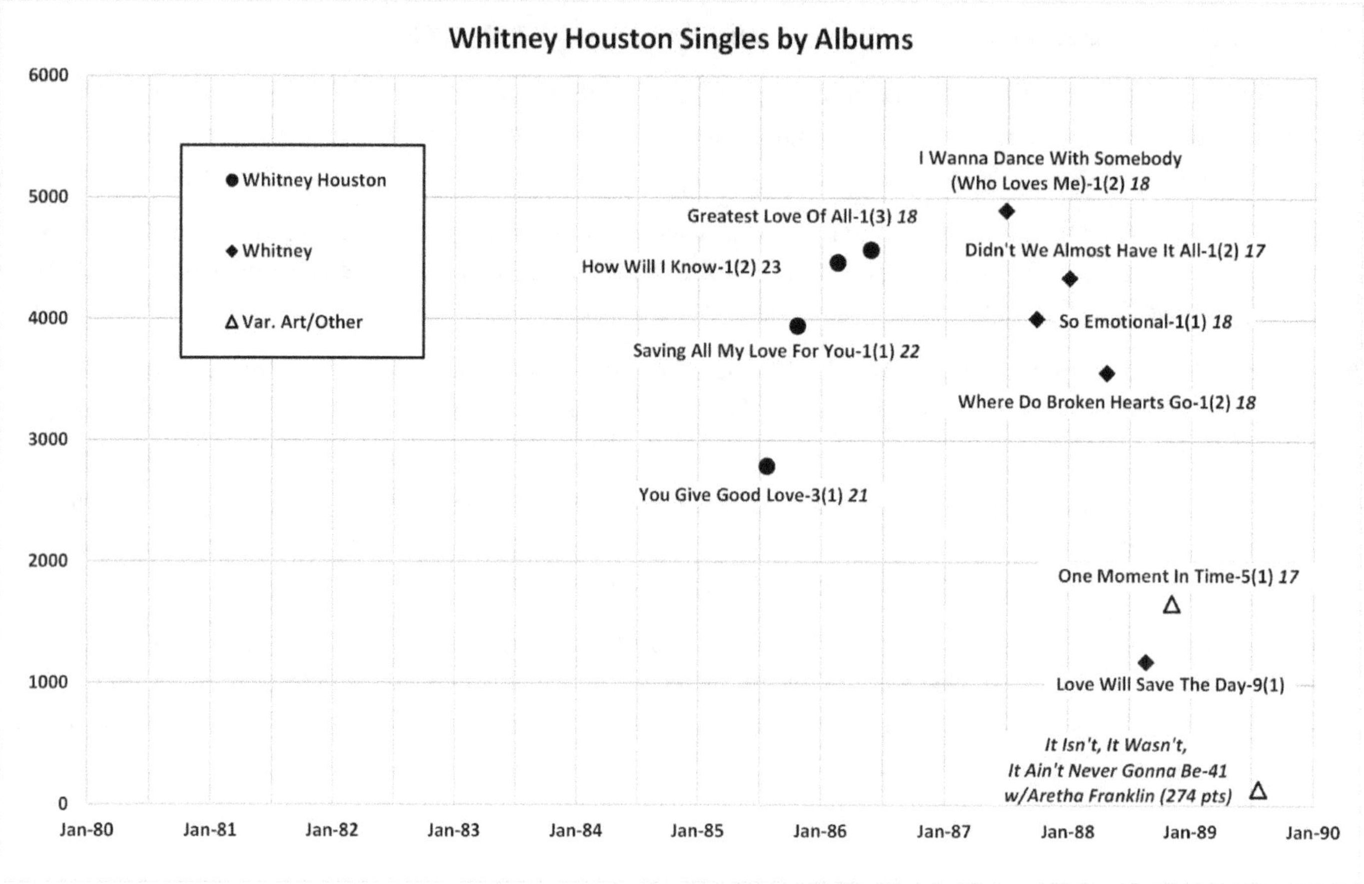
Whitney Houston Singles by Albums
6000
5000
4000
3000
2000
1000
0
Whitney Houston
Whitney
Var. Art/Other
I Wanna Dance With Somebody (Who Loves Me)-1(2) *18*
Greatest Love Of All-1(3) *18*
How Will I Know-1(2) 23
Didn't We Almost Have It All-1(2) *17*
So Emotional-1(1) *18*
Saving All My Love For You-1(1) *22*
Where Do Broken Hearts Go-1(2) *18*
You Give Good Love-3(1) *21*
One Moment In Time-5(1) *17*
Love Will Save The Day-9(1)
It Isn't, It Wasn't, It Ain't Never Gonna Be-41 w/Aretha Franklin (274 pts)
Jan-80
Jan-81
Jan-82
Jan-83
Jan-84
Jan-85
Jan-86
Jan-87
Jan-88
Jan-89
Jan-90

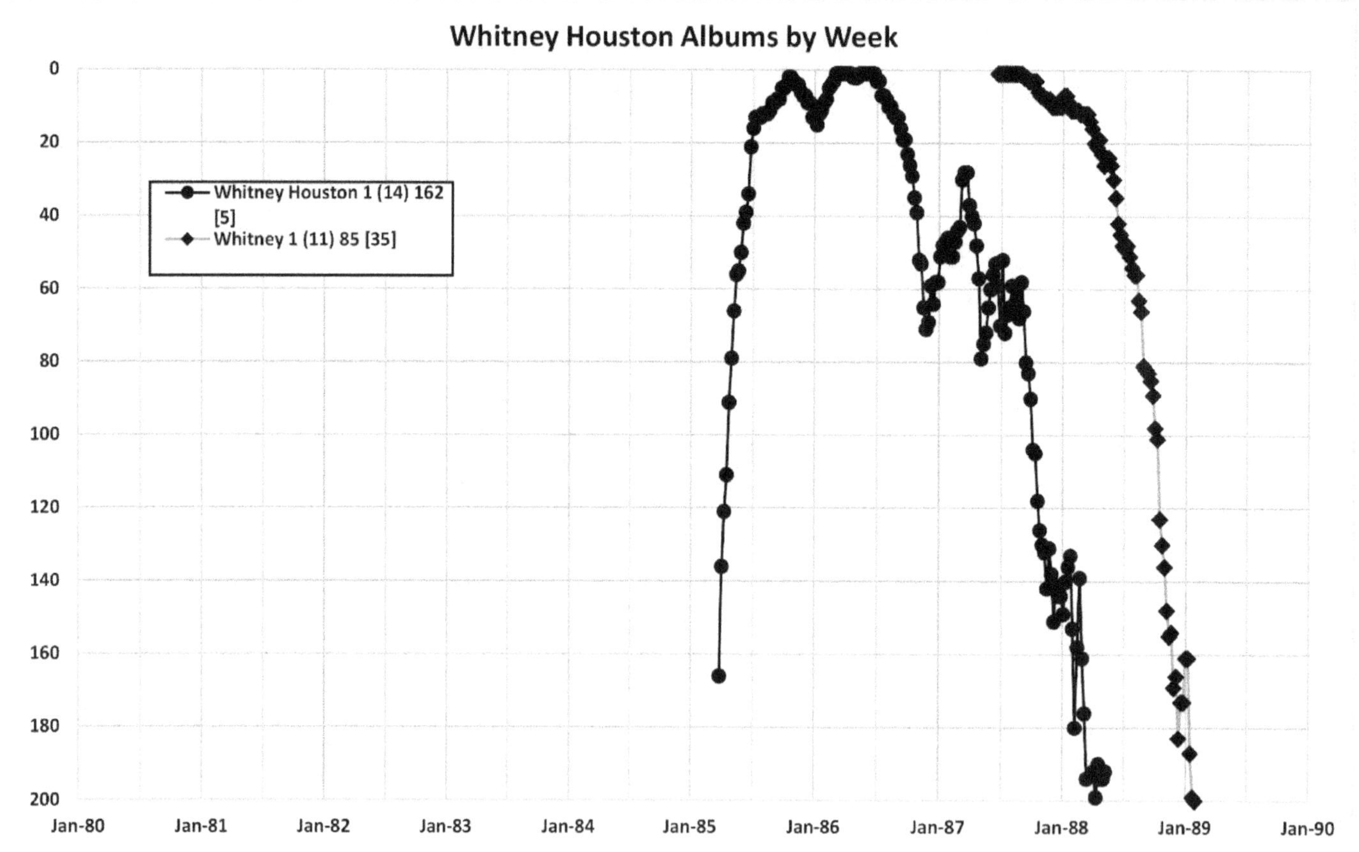
Whitney Houston Albums by Week
0
20
40
60
80
100
120
140
160
180
200
Whitney Houston 1 (14) 162 [5]
Whitney 1 (11) 85 [35]
Jan-80
Jan-81
Jan-82
Jan-83
Jan-84
Jan-85
Jan-86
Jan-87
Jan-88
Jan-89
Jan-90

WHITNEY HOUSTON ▶ Singles Rank: 8 (11) 35563; Albums Rank: 14 (2/2) 75693 100%

No other act in the '80s had seven consecutive releases go to number 1. The question for Whitney Houston is the opposite of that of Duran Duran: How can such an act score as low as number 8 for the decade? The reason is there were very few transcendent chart performances by singles in the last half of the decade compared to the first. Generally, that means more turnover at number 1, fewer weeks of very high score, and thus fewer singles of very high score—in other words, number 1s were worth less in 1988 than 1982. Only one of Whitney's seven consecutive number 1 singles lasted as long as three weeks at number 1, and yet she still ranked sixth in average single score.

Compare those weeks at the top to the champion, *Physical* by Olivia Newton-John, with 10. The difference is stark: the number 1 record of the year 1988, Poison's *Every Rose Has Its Thorn*, ranked 124th in the decade. No record charting between 1986 and 1990 ranked higher than 65th of the decade.

In the last half of the decade there were between 26 and 32 different number 1s in the year—nearly a different one every week. However, in the first half of the decade there were about 15 per year. What the charts are saying is either there are lots of truly great songs at the same time and people can't decide what to listen to, or there are very few with no agreement on greatness—the latter is generally the case, I think--but in either event there is no consensus that creates high scoring records of long duration at the top.

The performance of the two albums *Whitney Houston* and *Whitney* is also very different. Most important is that both made the top 50 for the decade. *Whitney Houston* went four singles deep but stayed in the top 20 for well over a year and lingered long after that. That glide path was probably somewhat lengthened by the popularity of *Whitney*, but it was formidable on its own. *Whitney* went five singles deep and debuted at number 1, but had nowhere near the longevity, 85 vs 162 weeks.

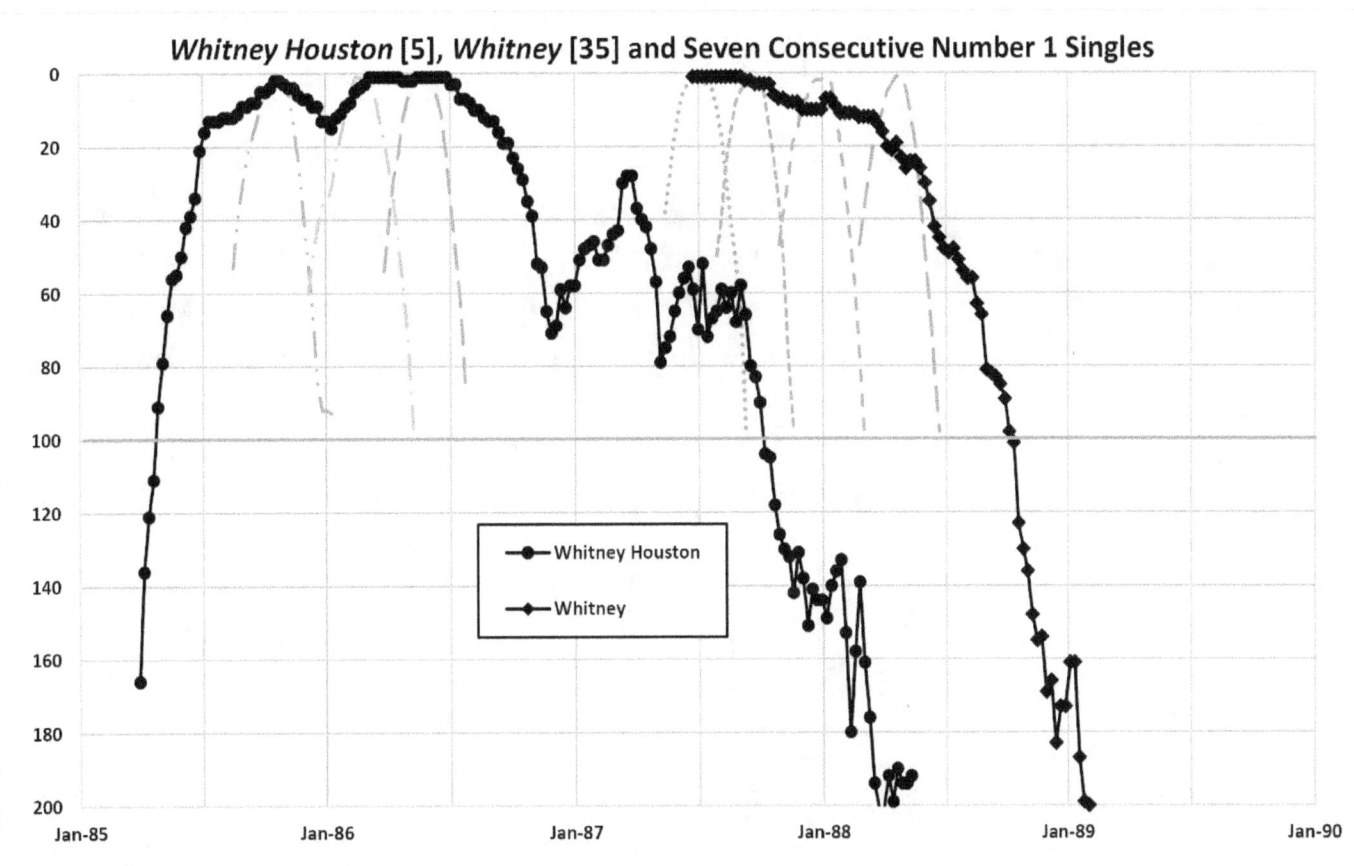

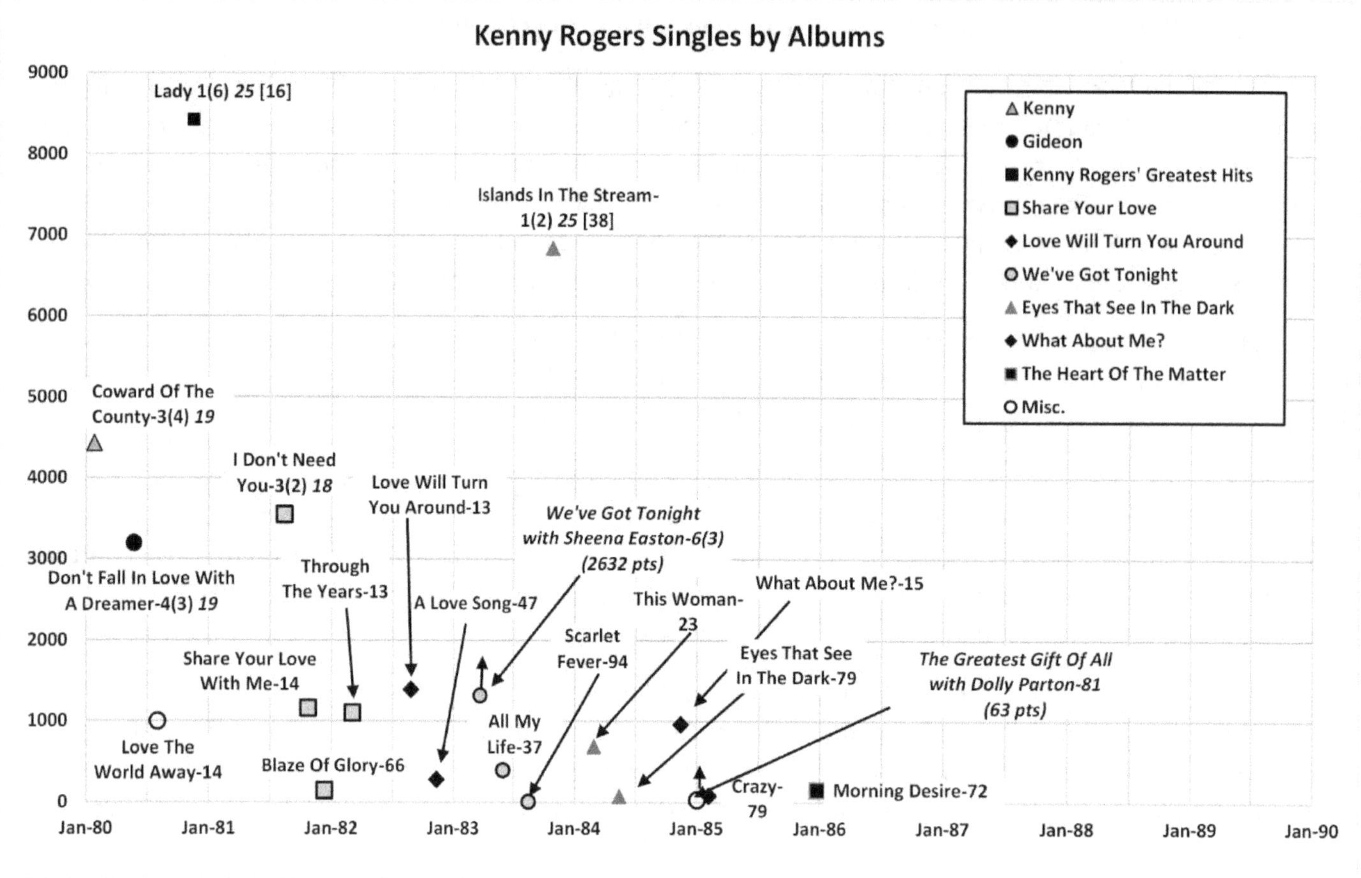

Chronologies

Kenny Rogers Singles by Albums

Legend:
- △ Kenny
- ● Gideon
- ■ Kenny Rogers' Greatest Hits
- ☐ Share Your Love
- ◆ Love Will Turn You Around
- ○ We've Got Tonight
- ▲ Eyes That See In The Dark
- ♦ What About Me?
- ■ The Heart Of The Matter
- ○ Misc.

Data points labeled:
- Lady 1(6) 25 [16]
- Islands In The Stream-1(2) 25 [38]
- Coward Of The County-3(4) 19
- I Don't Need You-3(2) 18
- Don't Fall In Love With A Dreamer-4(3) 19
- Love Will Turn You Around-13
- We've Got Tonight with Sheena Easton-6(3) (2632 pts)
- Through The Years-13
- A Love Song-47
- This Woman-23
- What About Me?-15
- Share Your Love With Me-14
- Scarlet Fever-94
- Eyes That See In The Dark-79
- The Greatest Gift Of All with Dolly Parton-81 (63 pts)
- Love The World Away-14
- Blaze Of Glory-66
- All My Life-37
- Crazy-79
- Morning Desire-72

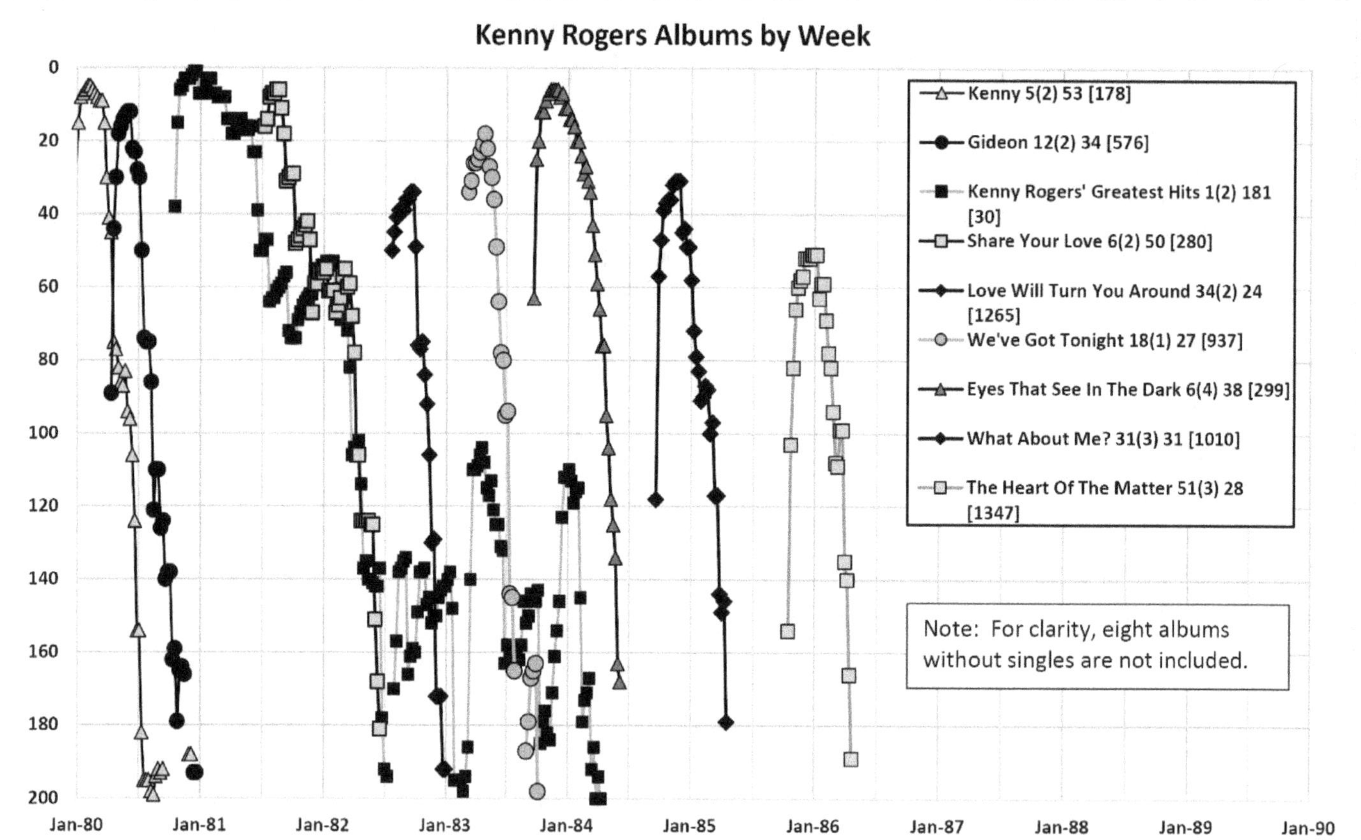

Kenny Rogers Albums by Week

Legend:
- △ Kenny 5(2) 53 [178]
- ● Gideon 12(2) 34 [576]
- ■ Kenny Rogers' Greatest Hits 1(2) 181 [30]
- ☐ Share Your Love 6(2) 50 [280]
- ◆ Love Will Turn You Around 34(2) 24 [1265]
- ○ We've Got Tonight 18(1) 27 [937]
- ▲ Eyes That See In The Dark 6(4) 38 [299]
- ♦ What About Me? 31(3) 31 [1010]
- ☐ The Heart Of The Matter 51(3) 28 [1347]

Note: For clarity, eight albums without singles are not included.

KENNY ROGERS ▶ Singles Rank: 9 (20) 35216; Albums Rank: 15 (18/22) 74617 93.7%

Over 40 percent of the points that propelled Kenny Rogers to the number 9 singles spot came from two records: *Lady* and *Islands In The Stream*. Both placed in the top 40 for the decade each with 25 weeks on the chart. Three more made the weekly top 5 early in the decade, and *We've Got Tonight*, with points shared with Sheena Easton, completes his list of top 10 singles.

The album *Kenny Rogers' Greatest Hits* is mostly drawn from the '70s, although *Don't Fall In Love With A Dreamer*, *Love The World Away* and *Lady* (which first appeared on this album) were recorded in 1980. Two First Edition recordings were also included.

This album is the highest ranked Greatest Hits album of the decade, and its longevity and resilience are remarkable. After its release it went quickly to number 1 for two weeks. Every subsequent new album release boosted its chart presence until it exited finally in 1984 after nearly three years on the chart. It was certified 12x Platinum in 1997.

He was the 16th ranked album artist of the decade, largely due to *Greatest Hits* and the sheer volume of albums charted (17). He was 47th in average single score.

As the '80s progressed and genres partitioned, Rogers continued to have hits on the country chart but less success on the Hot 100. *Islands In The Stream* was the last country song to peak at number 1 on the Hot 100 for the next sixteen years (the next was *Amazed* by Lonestar in 2000).

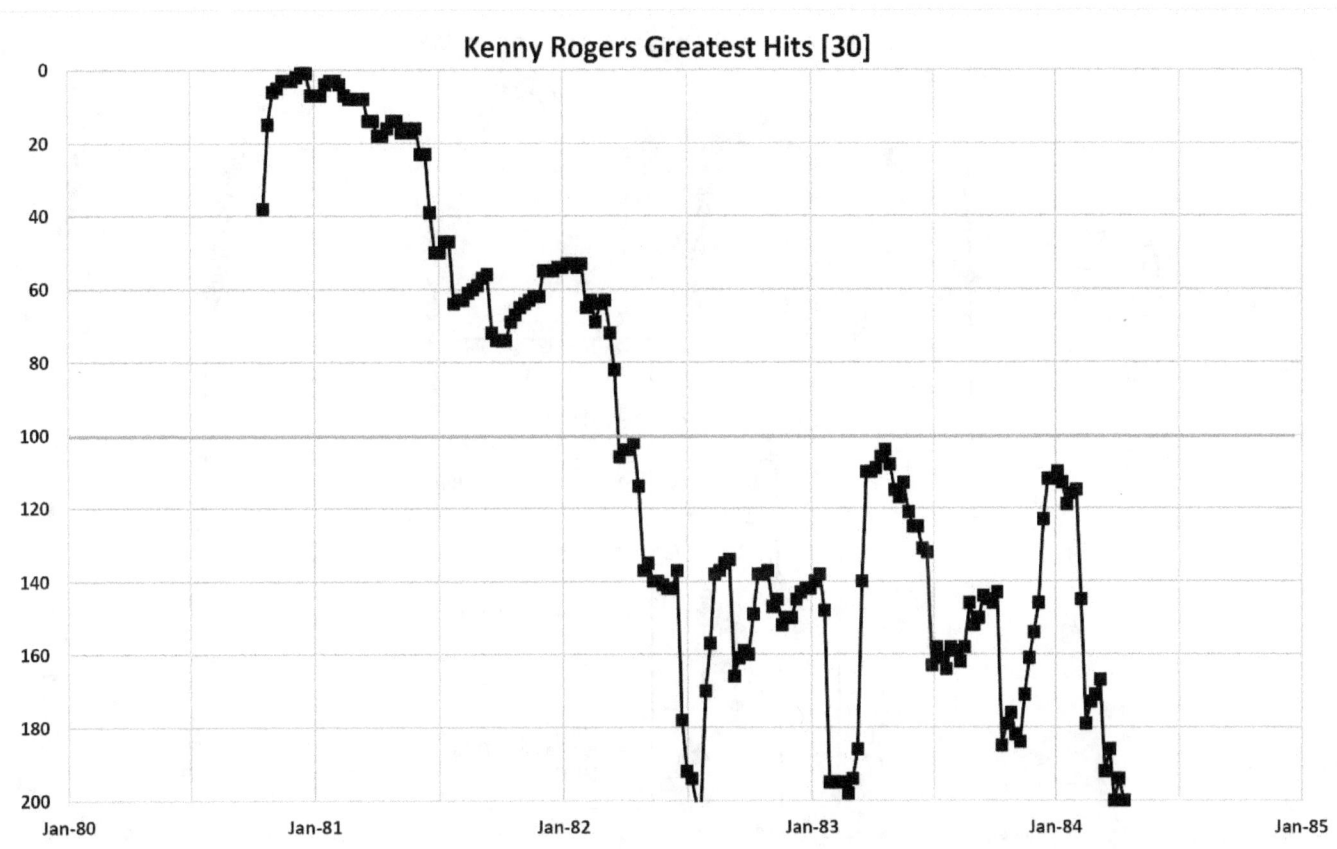

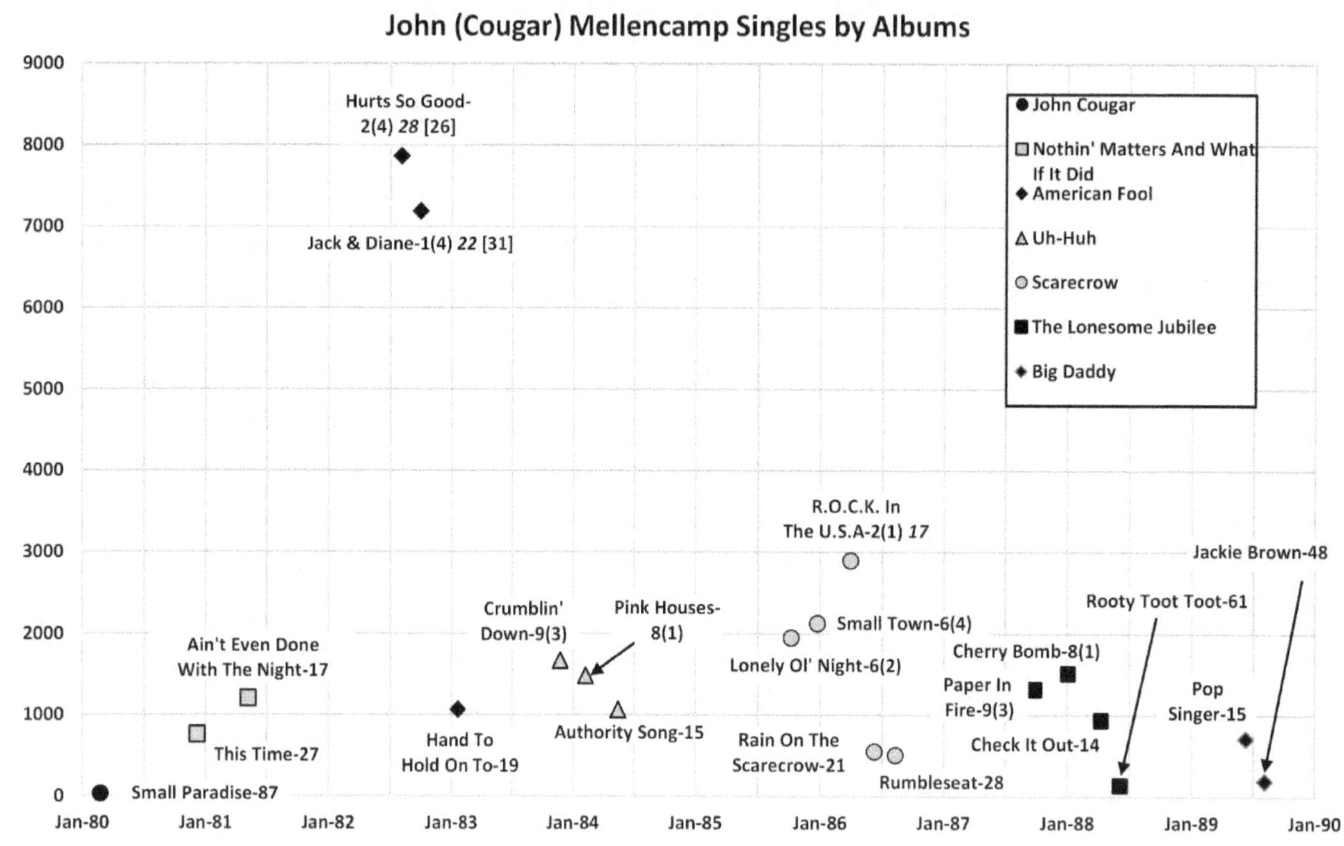
John (Cougar) Mellencamp Singles by Albums
9000
8000
7000
6000
5000
4000
3000
2000
1000
0
Jan-80
Jan-81
Jan-82
Jan-83
Jan-84
Jan-85
Jan-86
Jan-87
Jan-88
Jan-89
Jan-90
John Cougar
Nothin' Matters And What If It Did
American Fool
Uh-Huh
Scarecrow
The Lonesome Jubilee
Big Daddy
Hurts So Good-2(4) *28* [26]
Jack & Diane-1(4) *22* [31]
R.O.C.K. In The U.S.A-2(1) *17*
Jackie Brown-48
Rooty Toot Toot-61
Crumblin' Down-9(3)
Pink Houses-8(1)
Small Town-6(4)
Lonely Ol' Night-6(2)
Cherry Bomb-8(1)
Ain't Even Done With The Night-17
Paper In Fire-9(3)
Pop Singer-15
Hand To Hold On To-19
Authority Song-15
Rain On The Scarecrow-21
Check It Out-14
This Time-27
Rumbleseat-28
Small Paradise-87

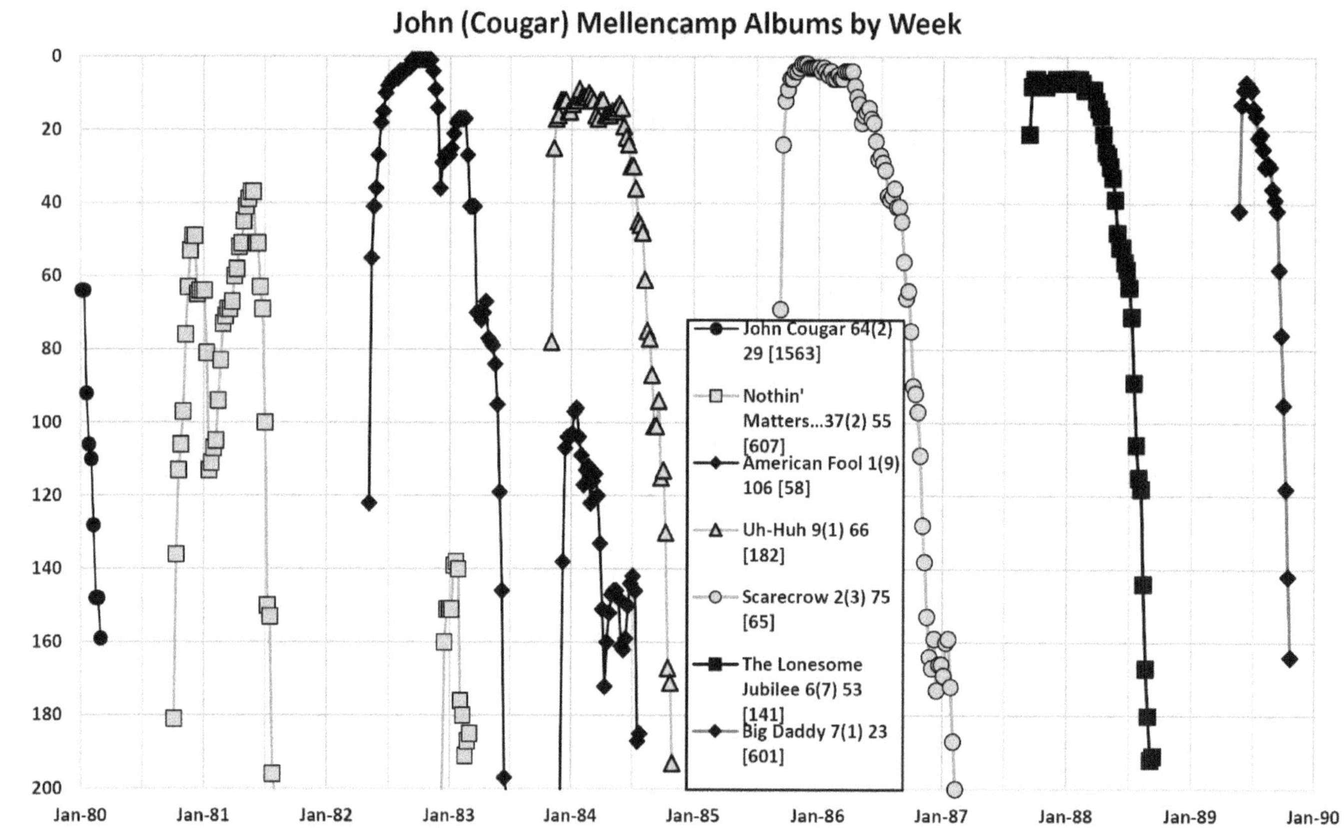
John (Cougar) Mellencamp Albums by Week
0
20
40
60
80
100
120
140
160
180
200
Jan-80
Jan-81
Jan-82
Jan-83
Jan-84
Jan-85
Jan-86
Jan-87
Jan-88
Jan-89
Jan-90
John Cougar 64(2) 29 [1563]
Nothin' Matters...37(2) 55 [607]
American Fool 1(9) 106 [58]
Uh-Huh 9(1) 66 [182]
Scarecrow 2(3) 75 [65]
The Lonesome Jubilee 6(7) 53 [141]
Big Daddy 7(1) 23 [601]

JOHN MELLENCAMP ▶ Singles Rank: 10 (20) 35178; Albums Rank: 7 (7/7) 85568 100%

Like Kenny Rogers, John Mellencamp's '80s singles portfolio consisted of two skyrockets that accounted for over 40% of his total points, and a handful of other top 10s in a 20-record catalog. There was really nothing to suggest the wild success of *Hurts So Good* and *Jack & Diane* from the earlier singles or albums, since none had cracked the top 10. He was even still known as John Cougar—and those two were the only gold records of his singles career.

Other than those two records, there was really no pattern to his '80s trajectory, although the singles from *Big Daddy* were disappointing performers compared to the earlier albums. It is a bit surprising that some well-known songs of his—*Pink Houses, Authority Song, Lonely Ol' Night* and *Small Town*—did not score better. They may have fallen victim to the proliferation of radio genres and ultimately poorer longevity despite a top 10 peak.

Only two of his albums placed in the top 100 for the decade but he was the number 7 album act on the volume strength of his seven albums. Like Kenny Rogers, his average single score was well down the list at 49th.

His second, third and fourth albums, and the singles derived from them, show interesting interactive behavior. On a relative basis, the singles *This Time* and *Ain't Even Done With The Night* showed more strength than the underlying album, *Nothin' Matters*.... *Hurts So Good* and *Jack & Diane* were driven by exceptional longevity at 28 and 22 weeks respectively; those two and *Hand To Hold On To* drove *American Fool*, which fell off rather quickly after them, but not before dragging *Nothin' Matters*... back to the charts.

The pattern for *Uh-Huh* was similar, falling off after three singles, but helping *American Fool* back to the charts during their run. None of the other albums showed marked longevity beyond the singles.

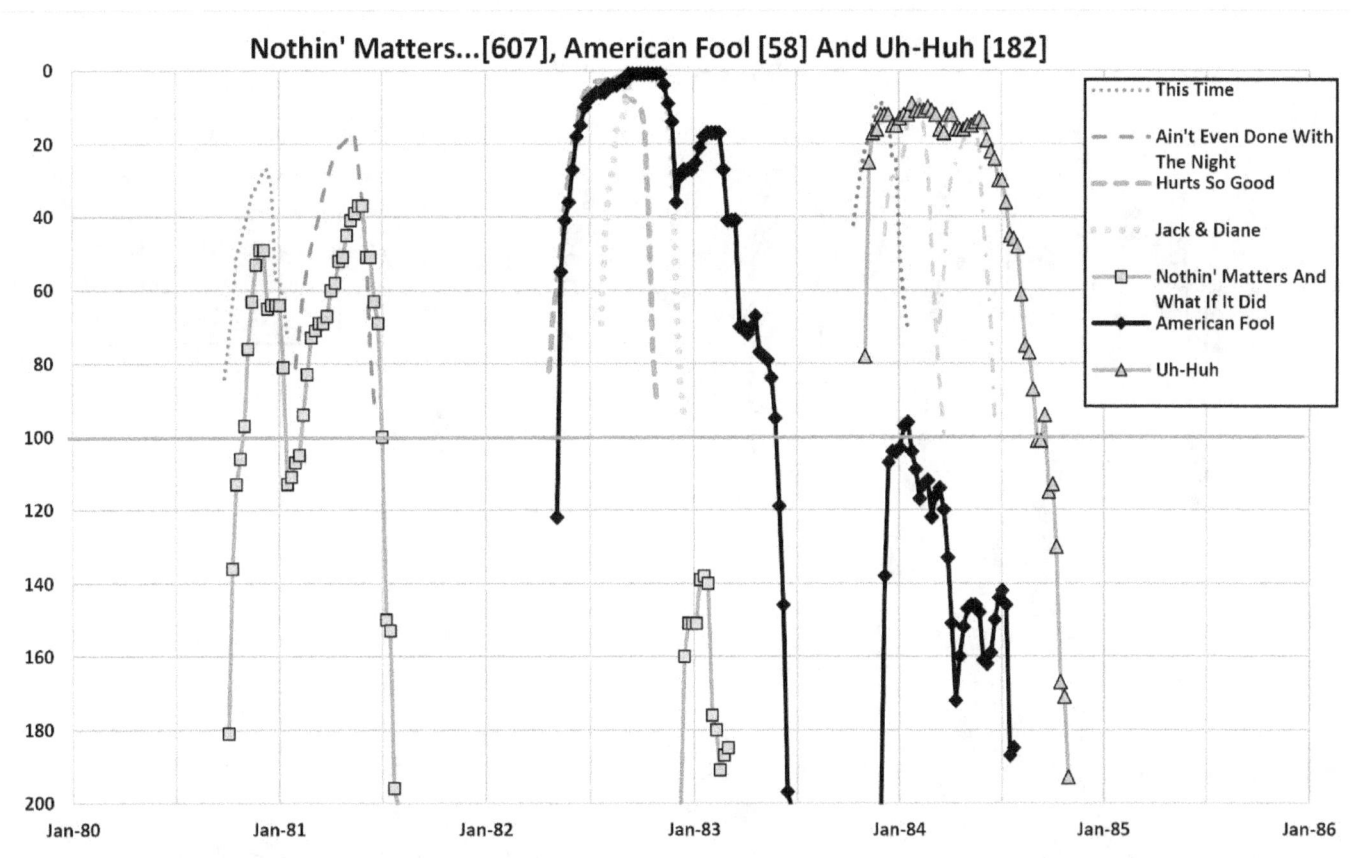

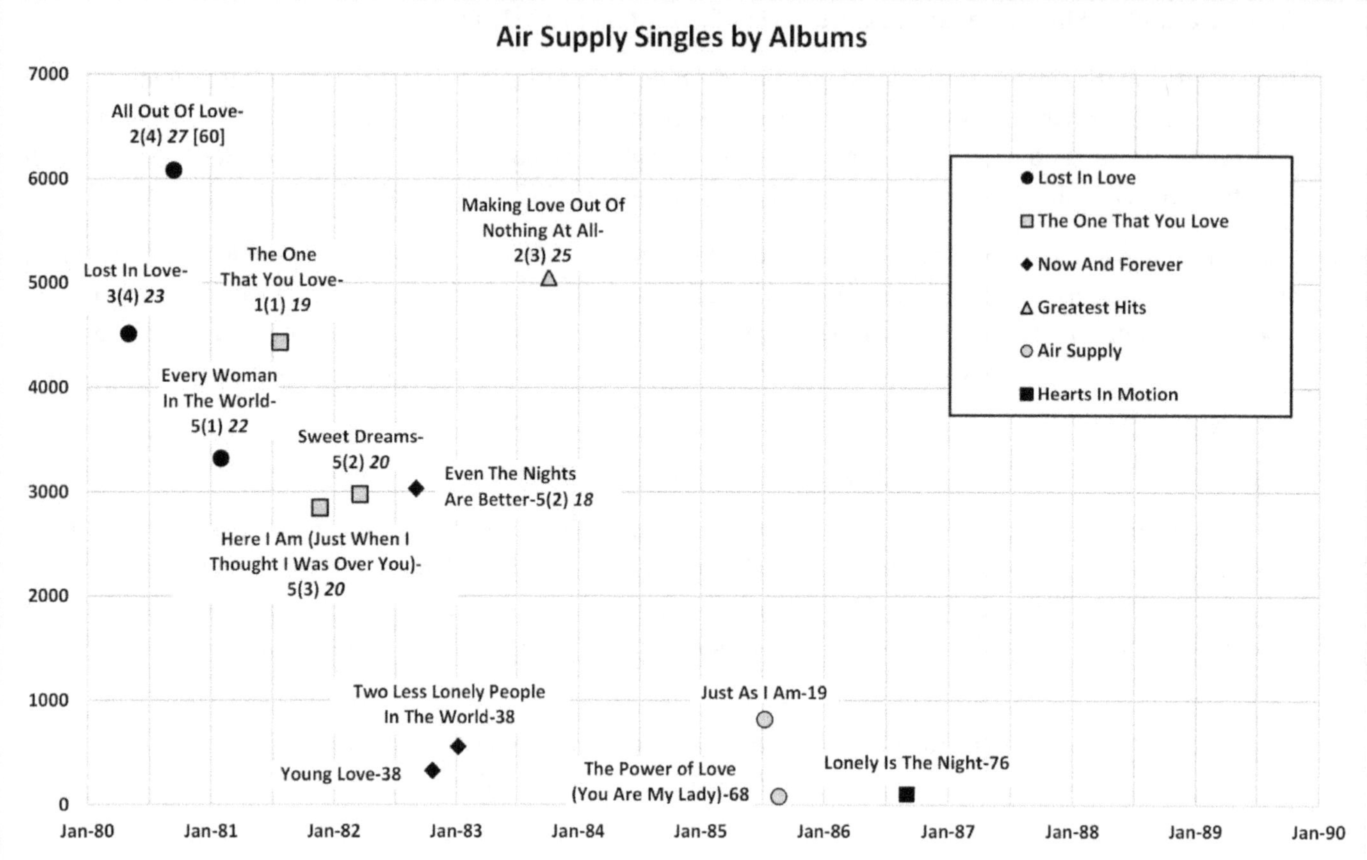
Air Supply Singles by Albums
7000
6000
5000
4000
3000
2000
1000
0
All Out Of Love-
2(4) 27 [60]
Lost In Love-
3(4) 23
The One
That You Love-
1(1) 19
Making Love Out Of
Nothing At All-
2(3) 25
Every Woman
In The World-
5(1) 22
Sweet Dreams-
5(2) 20
Even The Nights
Are Better-5(2) 18
Here I Am (Just When I
Thought I Was Over You)-
5(3) 20
Two Less Lonely People
In The World-38
Young Love-38
Just As I Am-19
The Power of Love
(You Are My Lady)-68
Lonely Is The Night-76
Lost In Love
The One That You Love
Now And Forever
Greatest Hits
Air Supply
Hearts In Motion
Jan-80
Jan-81
Jan-82
Jan-83
Jan-84
Jan-85
Jan-86
Jan-87
Jan-88
Jan-89
Jan-90

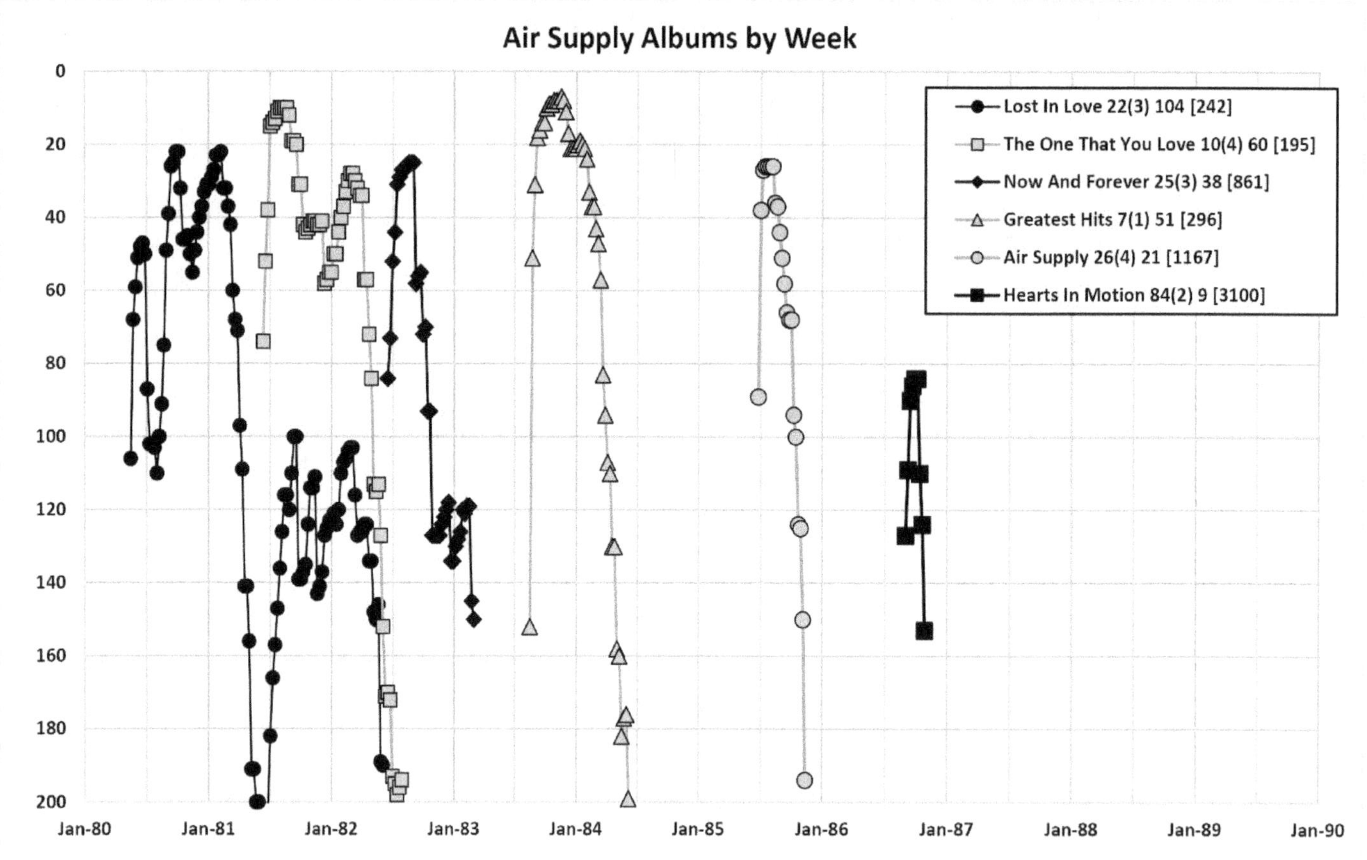
Air Supply Albums by Week
0
20
40
60
80
100
120
140
160
180
200
Lost In Love 22(3) 104 [242]
The One That You Love 10(4) 60 [195]
Now And Forever 25(3) 38 [861]
Greatest Hits 7(1) 51 [296]
Air Supply 26(4) 21 [1167]
Hearts In Motion 84(2) 9 [3100]
Jan-80
Jan-81
Jan-82
Jan-83
Jan-84
Jan-85
Jan-86
Jan-87
Jan-88
Jan-89
Jan-90

AIR SUPPLY ▶ Singles Rank: 11 (13) 34148; Albums Rank: 72 (6/6) 34587 100%

The core of Air Supply is a duo: Graham Russell and Russell Hitchcock. Graham Russell is English, but the band formed in Australia in the mid-'70s. Clive Davis heard and liked their original version of *Lost In Love* and had it rerecorded and remixed. This single was released in January 1980.

There's no denying that their 13-single catalog was pretty potent. Their first seven entries went to the top 5; however, only one of their last six cracked the top 10—*Making Love Out Of Nothing At All*, which spent three weeks at number 2 and was their second biggest hit. They averaged over 2600 points per single, good for 14th. But it's difficult to miss the downward trajectory of their singles over the decade. A revolving door of personnel by mid-decade may have contributed to the decline.

Albums were not their long suit. None made the weekly top 5; they were simply a vehicle for singles. *Lost In Love* ranked 242nd scored as an album, but 24th when scored by singles; of the six albums, the strongest, *The One That You Love*, ranked 195 in the decade (although it rose to 65 when scored by singles). In fact, all six of their albums scored higher by singles than as an album. They ranked 73rd as an album act.

There is an interesting interaction among the first albums and their singles that is illustrative of the band history. Note that all six singles from the first two albums charted higher than the albums themselves. The impact of each single can be seen on the lifecycle of its underlying album; however, the singles from *The One That You Love* also dragged *Lost In Love* back to the chart. The other four albums had simple rise and fall trajectories.

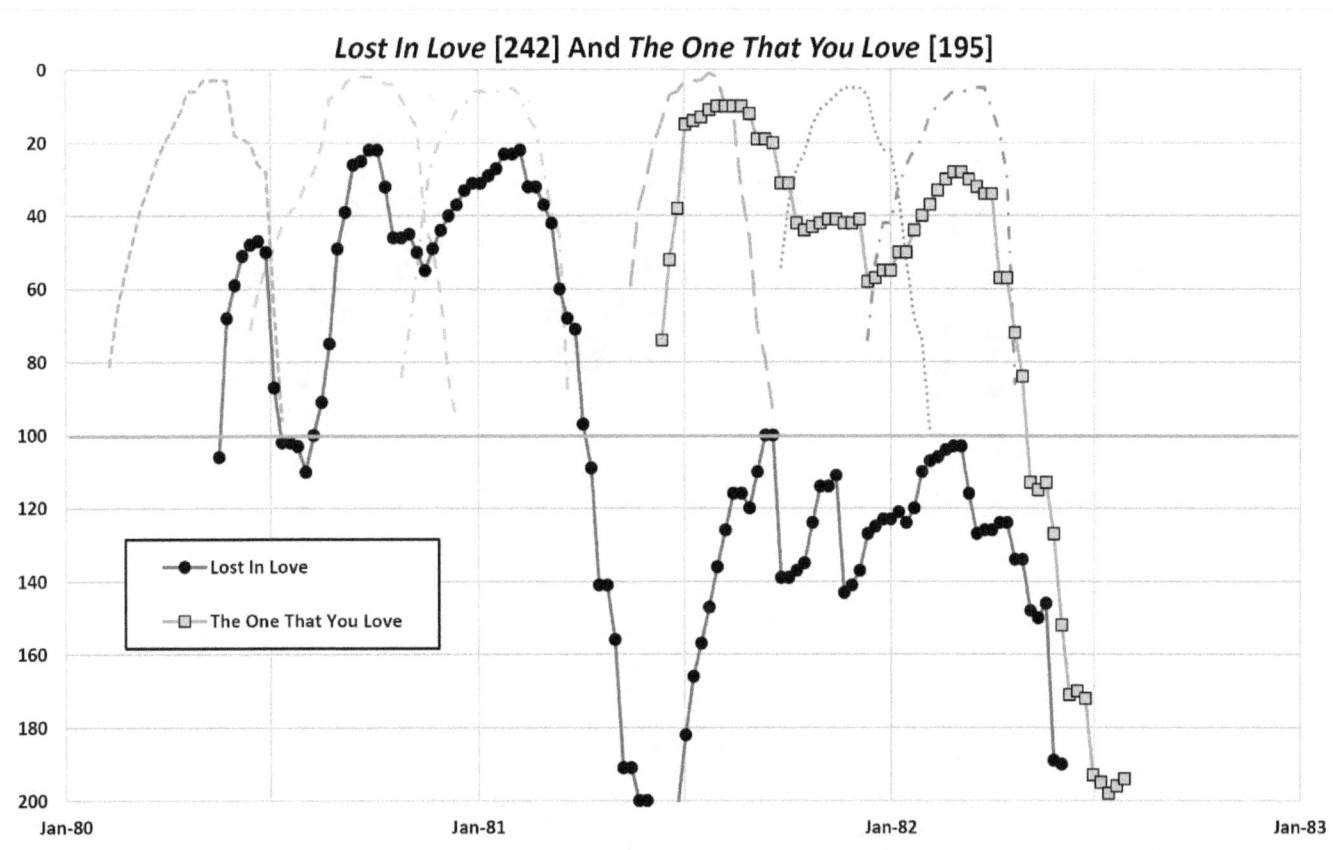

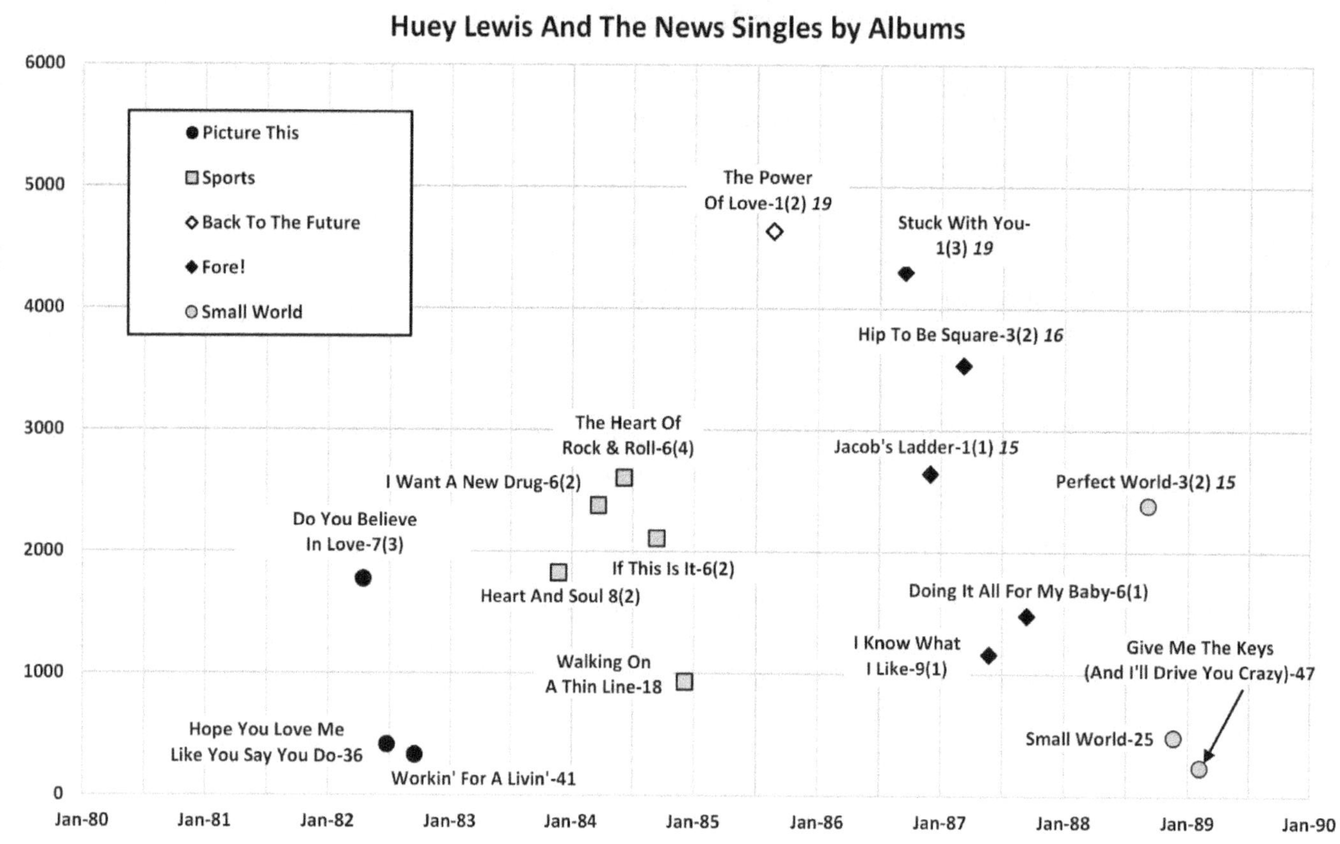
Huey Lewis And The News Singles by Albums
Picture This
Sports
Back To The Future
Fore!
Small World
The Power Of Love-1(2) 19
Stuck With You-1(3) 19
Hip To Be Square-3(2) 16
The Heart Of Rock & Roll-6(4)
Jacob's Ladder-1(1) 15
I Want A New Drug-6(2)
Perfect World-3(2) 15
Do You Believe In Love-7(3)
If This Is It-6(2)
Heart And Soul 8(2)
Doing It All For My Baby-6(1)
I Know What I Like-9(1)
Give Me The Keys (And I'll Drive You Crazy)-47
Walking On A Thin Line-18
Hope You Love Me Like You Say You Do-36
Small World-25
Workin' For A Livin'-41
0
1000
2000
3000
4000
5000
6000
Jan-80
Jan-81
Jan-82
Jan-83
Jan-84
Jan-85
Jan-86
Jan-87
Jan-88
Jan-89
Jan-90

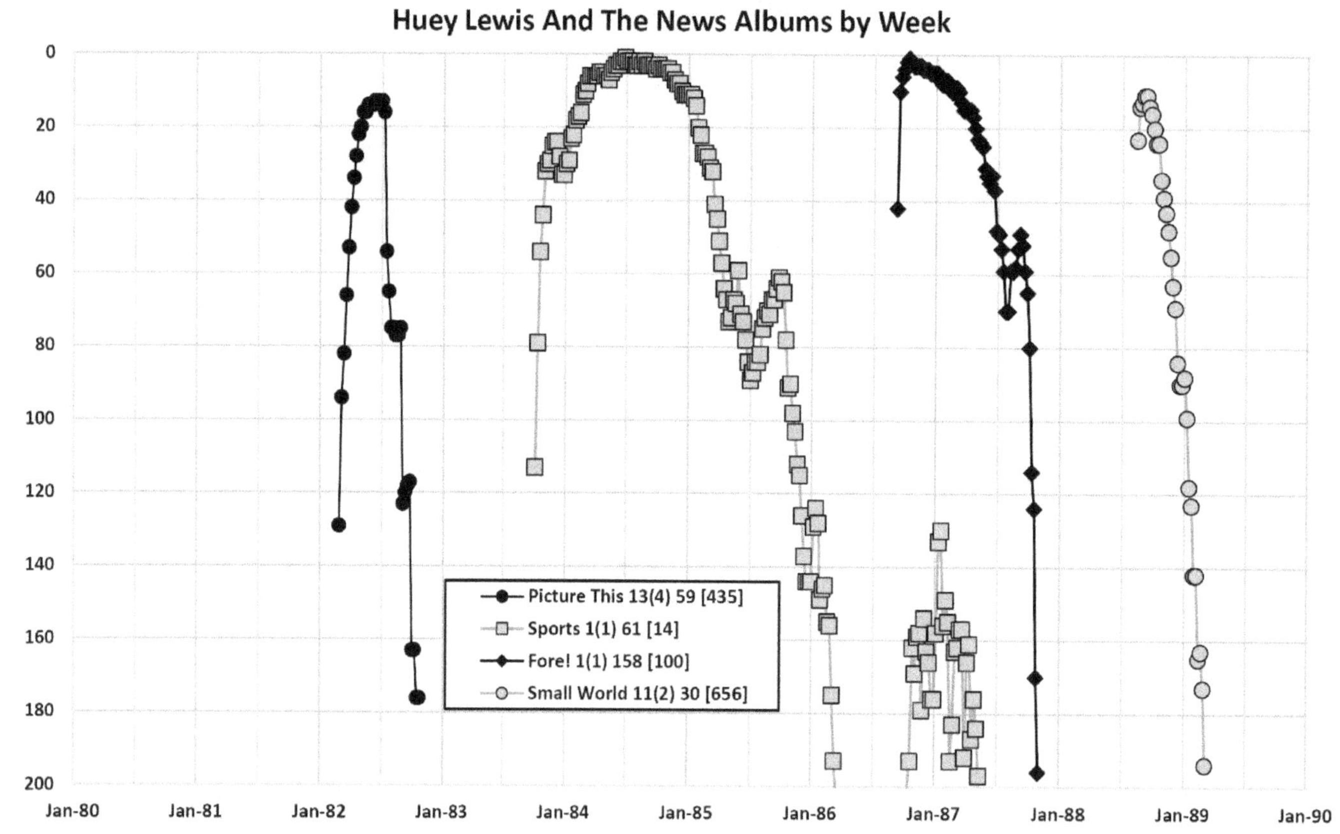
Huey Lewis And The News Albums by Week
Picture This 13(4) 59 [435]
Sports 1(1) 61 [14]
Fore! 1(1) 158 [100]
Small World 11(2) 30 [656]
0
20
40
60
80
100
120
140
160
180
200
Jan-80
Jan-81
Jan-82
Jan-83
Jan-84
Jan-85
Jan-86
Jan-87
Jan-88
Jan-89
Jan-90

HUEY LEWIS AND THE NEWS ▶ Singles Rank: 12 (17) 33246; Albums Rank: 17 (4/4) 68157 100%

Huey Lewis And The News' eponymous first album was released in 1980, and neither it nor the two singles from it charted. *Picture This*, released in 1982 broke through on the basis of three singles, including *Do You Believe In Love*, which made the top 10. Twelve of their seventeen '80s entries made the top 10, including five top 5s of which three went to number 1. None of those records placed in the top 100 for the decade, although *Stuck With You* ranked number 10 for the year 1986.

To the naked eye, their singles trajectory through the decade seems to be upward, but it is more properly viewed as rising to a peak with *The Power Of Love* and *Stuck With You*, but then in decline for the remainder of the decade and beyond. *Perfect World* was their final top 10 record.

As an album act, they ranked 17th, with two very big albums, *Sports* and *Fore!*, both of which ranked in the top 100 for the decade. As an album, *Sports* had by far the best chart history [14]; although scored by singles it dropped to 74th. On the other hand, *Fore!* ranked number 100 but when scored by singles rose to number 30—much higher than the stronger album, *Sports*. In terms of sales, *Sports* was certified 7x Platinum by the end of the decade; *Fore!* only 3x Platinum at the same time. *Picture This* and *Small World* peaked in the top 20 but were well behind.

The interaction among the singles and those two albums is interesting. Even after the fifth single, *Sports* declined in chart rank only slowly. It was rejuvenated twenty places by *The Power Of Love*, even though that song was not on that album. The singles of *Fore!* also helped drag *Sports* back to the charts, although *Fore!* itself declined rather quickly after the final single *Doing It All For My Baby* left the chart.

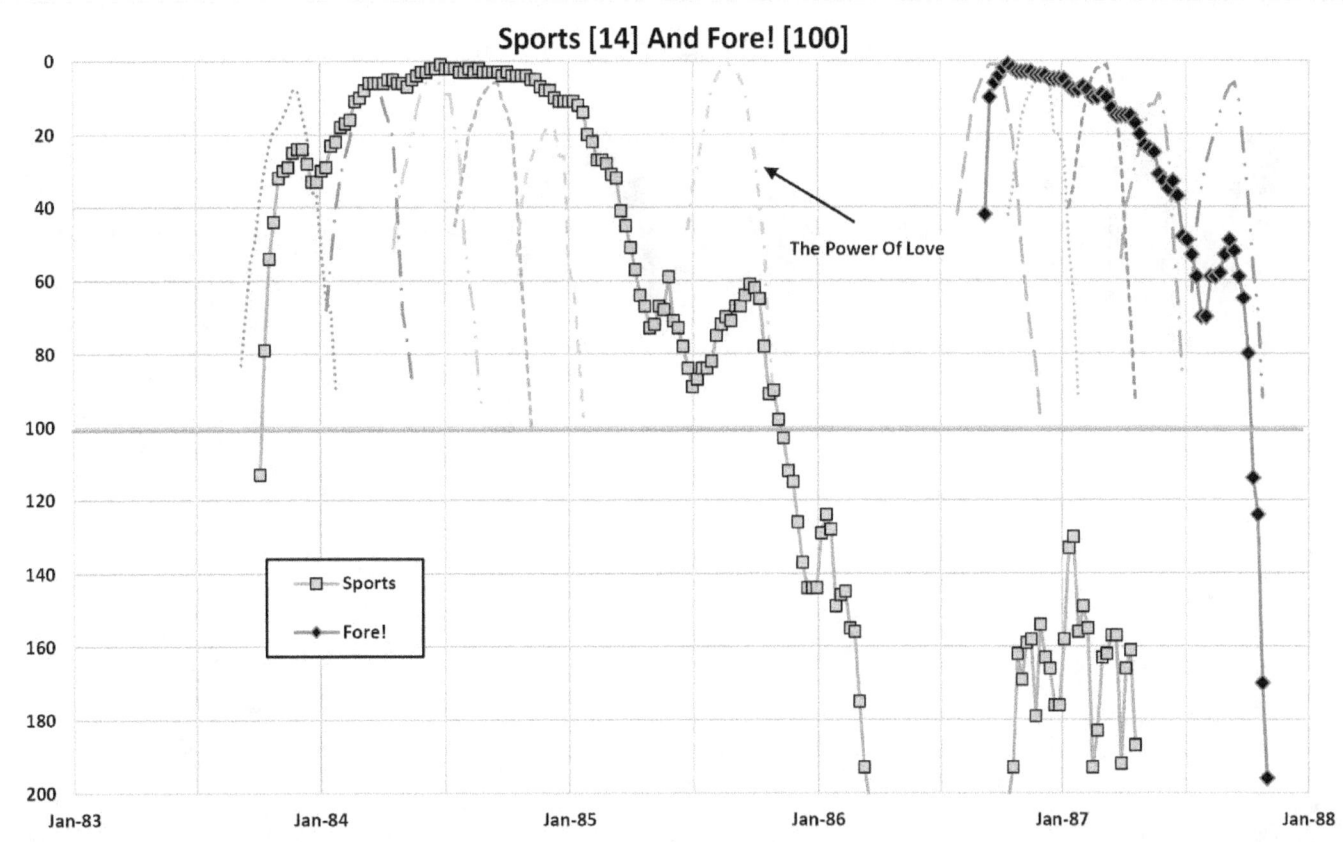

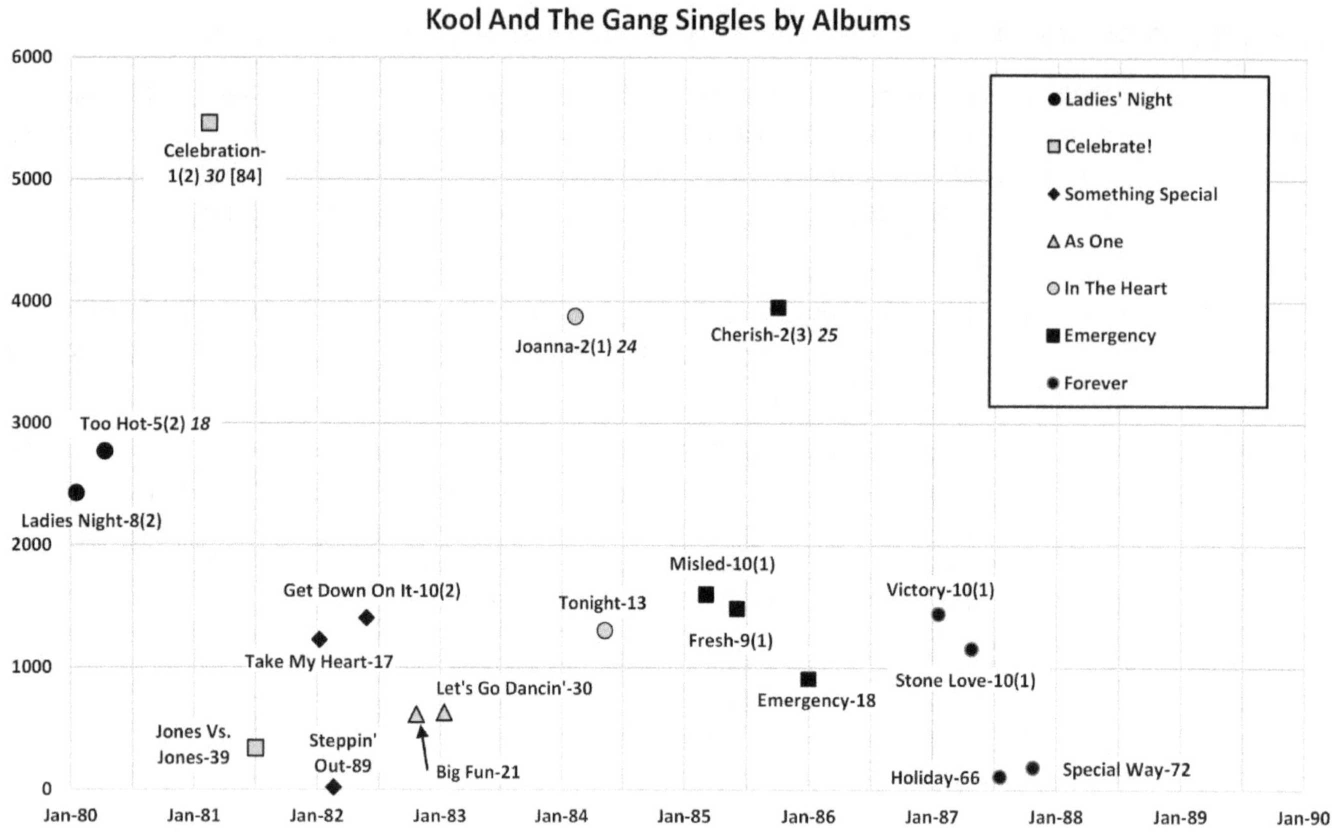
Kool And The Gang Singles by Albums
Ladies' Night
Celebrate!
Something Special
As One
In The Heart
Emergency
Forever
Celebration-1(2) 30 [84]
Joanna-2(1) 24
Cherish-2(3) 25
Too Hot-5(2) 18
Ladies Night-8(2)
Misled-10(1)
Get Down On It-10(2)
Tonight-13
Victory-10(1)
Fresh-9(1)
Take My Heart-17
Stone Love-10(1)
Let's Go Dancin'-30
Emergency-18
Jones Vs. Jones-39
Steppin' Out-89
Big Fun-21
Holiday-66
Special Way-72
6000
5000
4000
3000
2000
1000
0
Jan-80
Jan-81
Jan-82
Jan-83
Jan-84
Jan-85
Jan-86
Jan-87
Jan-88
Jan-89
Jan-90

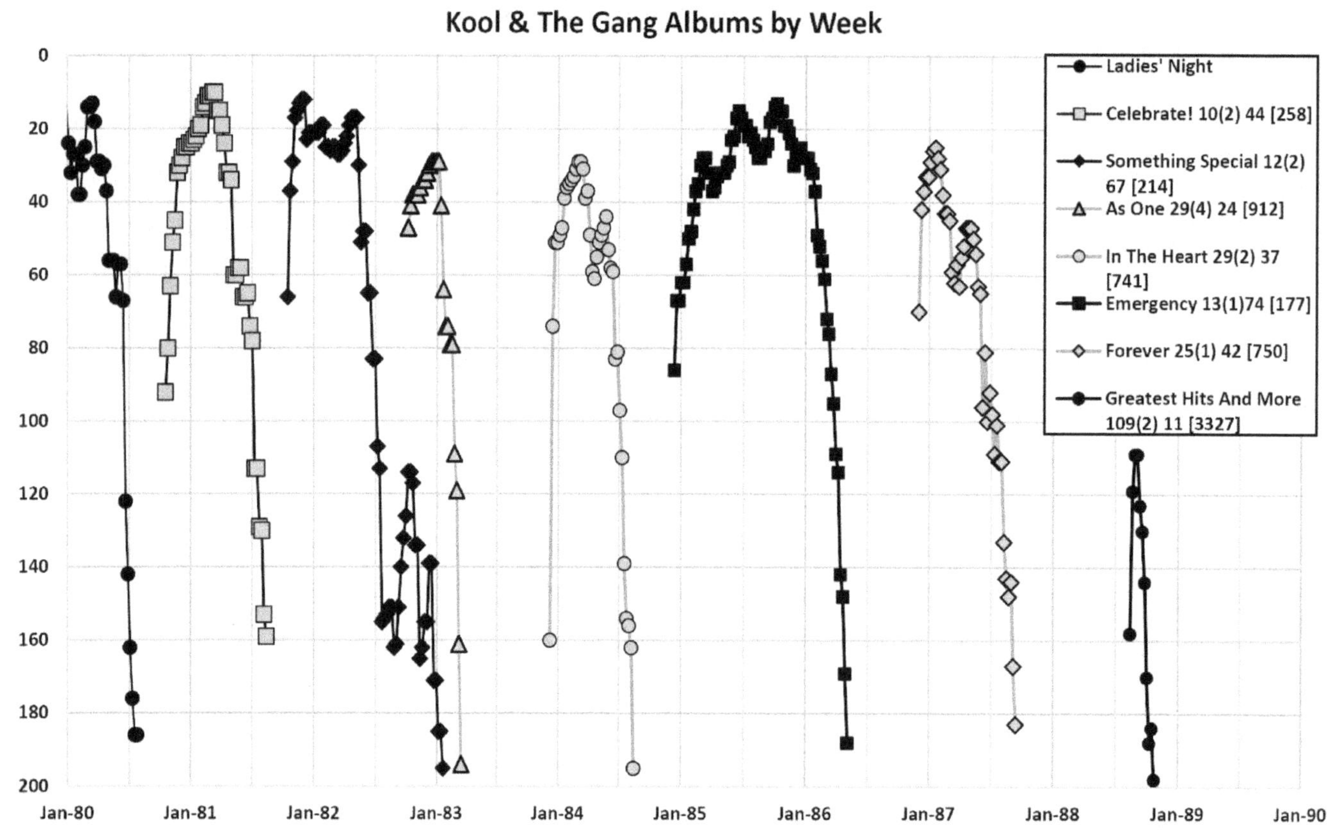
Kool & The Gang Albums by Week
Ladies' Night
Celebrate! 10(2) 44 [258]
Something Special 12(2) 67 [214]
As One 29(4) 24 [912]
In The Heart 29(2) 37 [741]
Emergency 13(1)74 [177]
Forever 25(1) 42 [750]
Greatest Hits And More 109(2) 11 [3327]
0
20
40
60
80
100
120
140
160
180
200
Jan-80
Jan-81
Jan-82
Jan-83
Jan-84
Jan-85
Jan-86
Jan-87
Jan-88
Jan-89
Jan-90

KOOL & THE GANG ▶ Singles Rank: 13 (19) 30914; Albums Rank: 35 (7/8) 49167 90.2%

Kool & The Gang had an unlikely start in the mid-1960s as a teen-age jazz band. By the 1980s they were veterans with a large catalog of albums and singles already to their credit. What they weren't very good at was disco—when they tried, they alienated their old jazz-boogie fan base and didn't attract new.

Two personnel changes sparked an '80s resurgence: vocalist J.T. Taylor and producer Eumir Deodato. Their relationship with the latter lasted only four albums, the first two of which included three of their biggest five singles. Starting with *In The Heart,* they produced themselves in association with Jim Bonneford.

Kool & The Gang were a singles band; ten of their nineteen charted singles went to the top ten, although only *Celebration* went to number 1 and made the top 100 of the decade. That consistency with volume accounted for their 13th rank as an act.

Their singles' success didn't translate as well to albums: only one '80s album cracked the top 10. They ranked 35th as an album act in the '80s, which includes the '80s portion of *Ladies' Night*.

Both *Emergency* and *Forever* yielded four singles each but the interaction between singles and albums is quite different between the two. The impact of the four singles on the lifecycle of *Emergency* can be seen clearly, and each single achieved a higher rank than the album. Curiously, the biggest single, *Cherish*, was the third release. *Emergency* stayed on the chart for nearly a year and a half. Even though it ranked 177th as an album, it rose to 108th when scored by the singles.

Forever was a different story. Never as popular as an album, its rank was exceeded by the first two singles, it exited soon after the third, and did not chart at all during the chart time of the fourth single, *Special Way*.

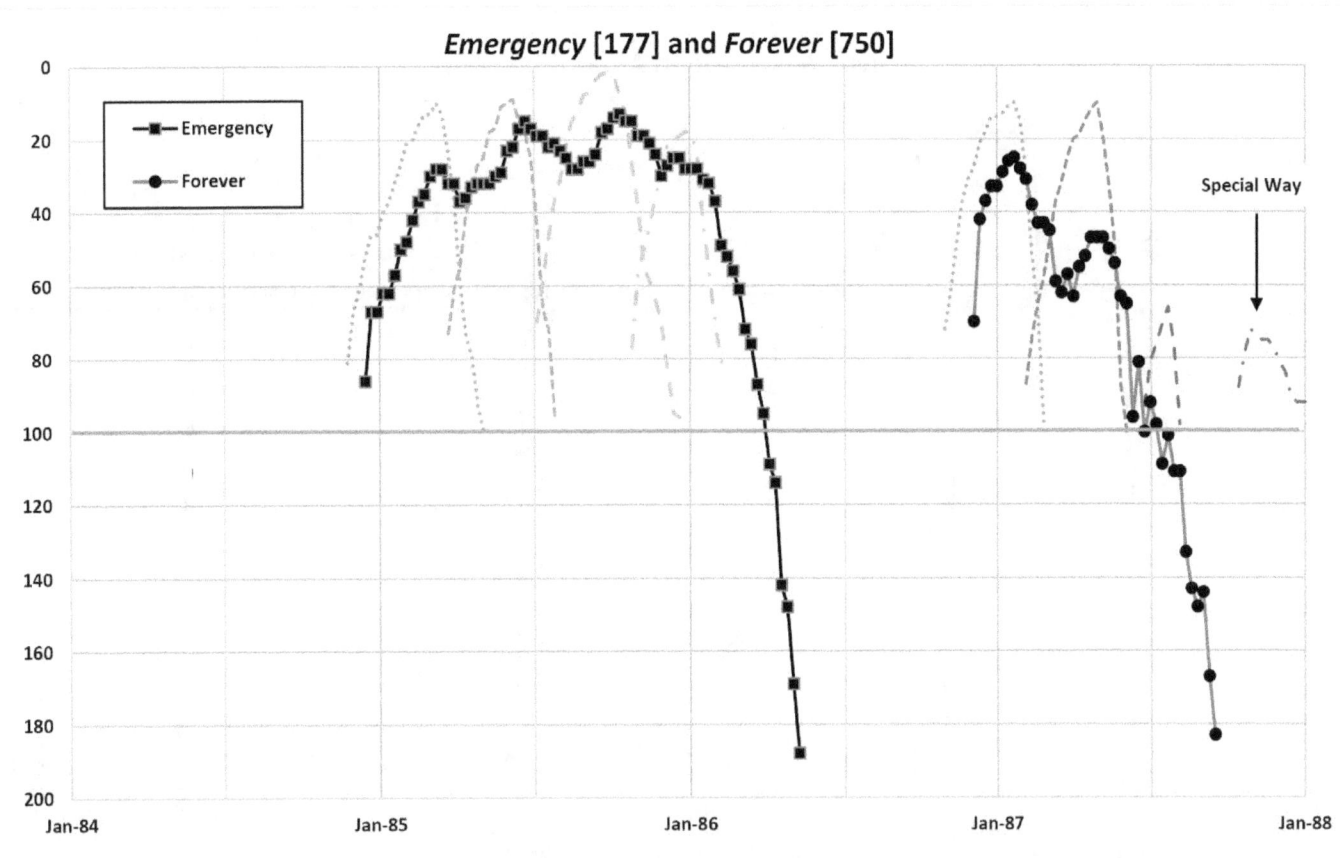

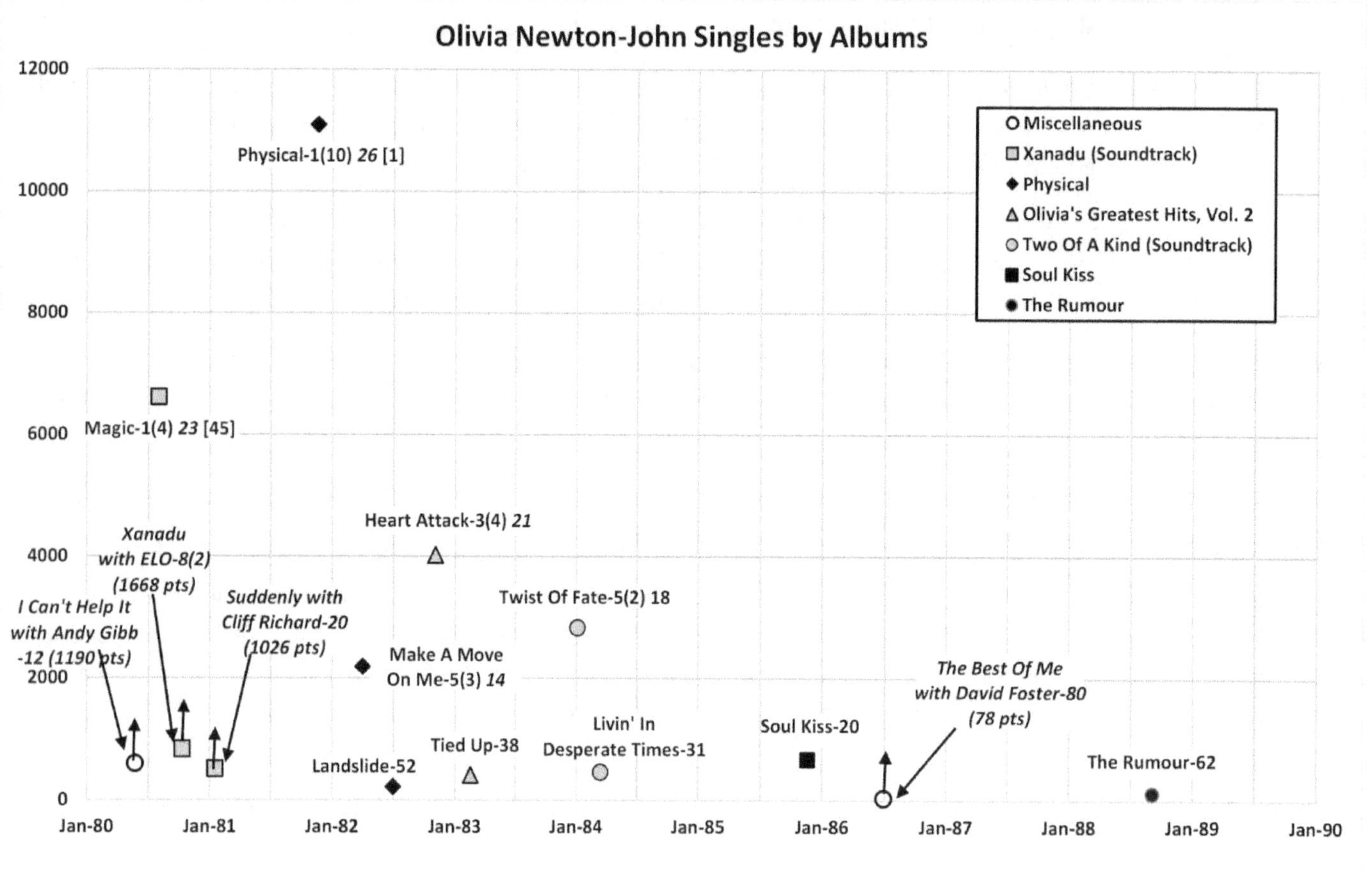

Chronologies

Olivia Newton-John Singles by Albums

Legend:
- ○ Miscellaneous
- □ Xanadu (Soundtrack)
- ◆ Physical
- △ Olivia's Greatest Hits, Vol. 2
- ○ Two Of A Kind (Soundtrack)
- ■ Soul Kiss
- ● The Rumour

Data points:
- Physical-1(10) 26 [1]
- Magic-1(4) 23 [45]
- Xanadu with ELO-8(2) (1668 pts)
- I Can't Help It with Andy Gibb -12 (1190 pts)
- Suddenly with Cliff Richard-20 (1026 pts)
- Heart Attack-3(4) 21
- Twist Of Fate-5(2) 18
- Make A Move On Me-5(3) 14
- Landslide-52
- Tied Up-38
- Livin' In Desperate Times-31
- Soul Kiss-20
- The Best Of Me with David Foster-80 (78 pts)
- The Rumour-62

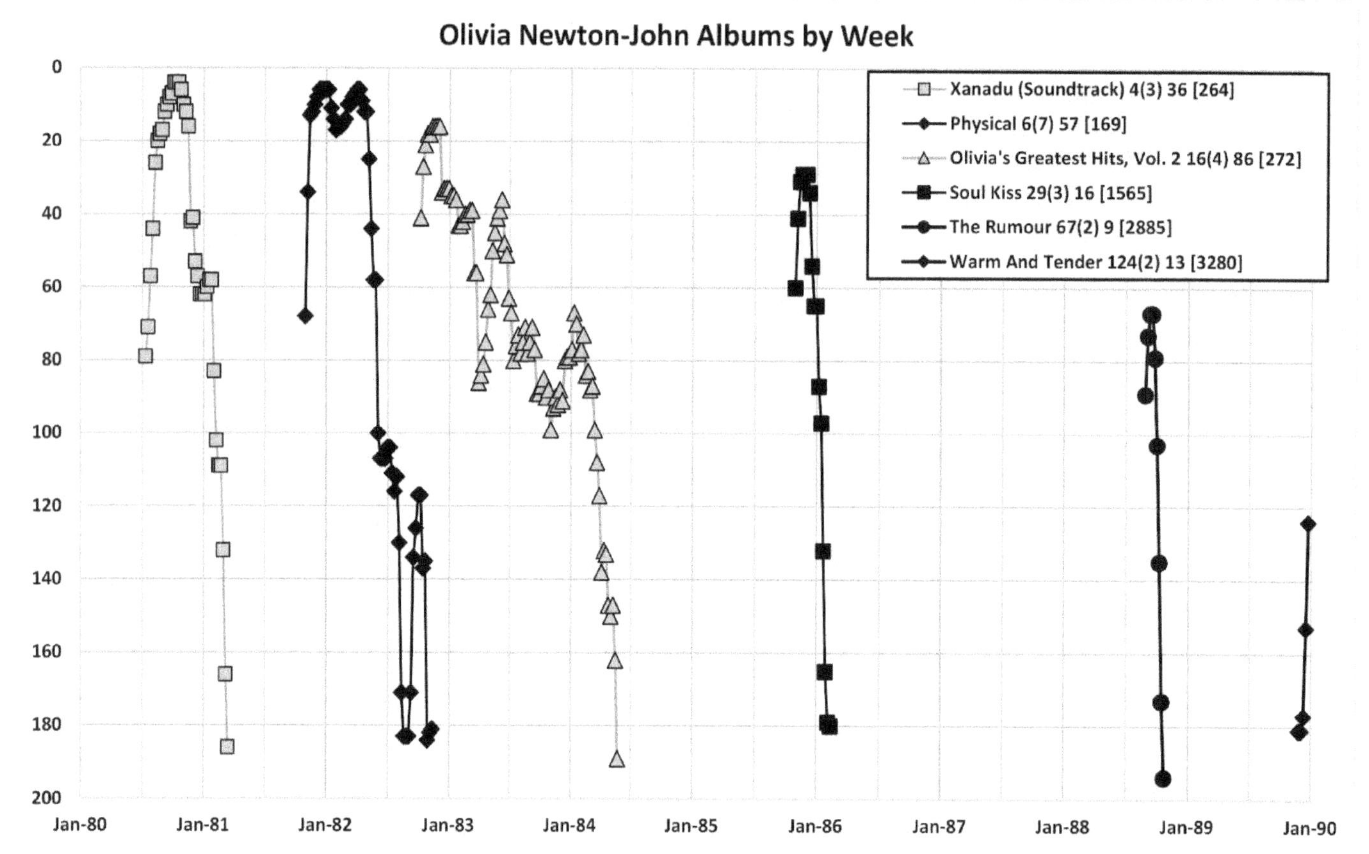

Olivia Newton-John Albums by Week

Legend:
- □ Xanadu (Soundtrack) 4(3) 36 [264]
- ◆ Physical 6(7) 57 [169]
- △ Olivia's Greatest Hits, Vol. 2 16(4) 86 [272]
- ■ Soul Kiss 29(3) 16 [1565]
- ● The Rumour 67(2) 9 [2885]
- ◆ Warm And Tender 124(2) 13 [3280]

446

OLIVIA NEWTON-JOHN ▶ Singles Rank: 14 (14) 30618; Albums Rank: 104 (5/5) 27732 100%

Olivia Newton-John was already a superstar by 1980, having had numerous country-flavored and soft rock hits (including the top single of 1974, *I Honestly Love You*). In addition, she starred in *Grease* with John Travolta. After *Grease*, she was ready for a bit more aggressive persona, which led to the 1978 album *Totally Hot*, and ultimately to *Physical*, her second-highest charting album.

She was clearly a singles artist, but the catalog was uneven. Together, the four top 5 singles of her 14 total entries accounted for 82% of her total points. As an album artist she ranked below number 100 even including half the *Xanadu* Soundtrack to her credit.

The album *Physical* ranked 160th and peaked at number 6. Because of the monster strength of the single of the same name, it rose to the number 27 album when measured by its singles success.

It's easy to see the impact of the singles on two early '80s albums, *Physical* and *Olivia's Greatest Hits, Vol 2*. All three singles from the first album drove inflections in the album curve. Apparently, the single *Heart Attack*, and perhaps *Greatest Hits* itself rejuvenated *Physical*. *Tied Up* and *Twist Of Fate* from the *Two Of A Kind* Soundtrack impacted *Greatest Hits* sales, but there is no apparent reason for its 40 position rise in March-June 1983 (arrow).

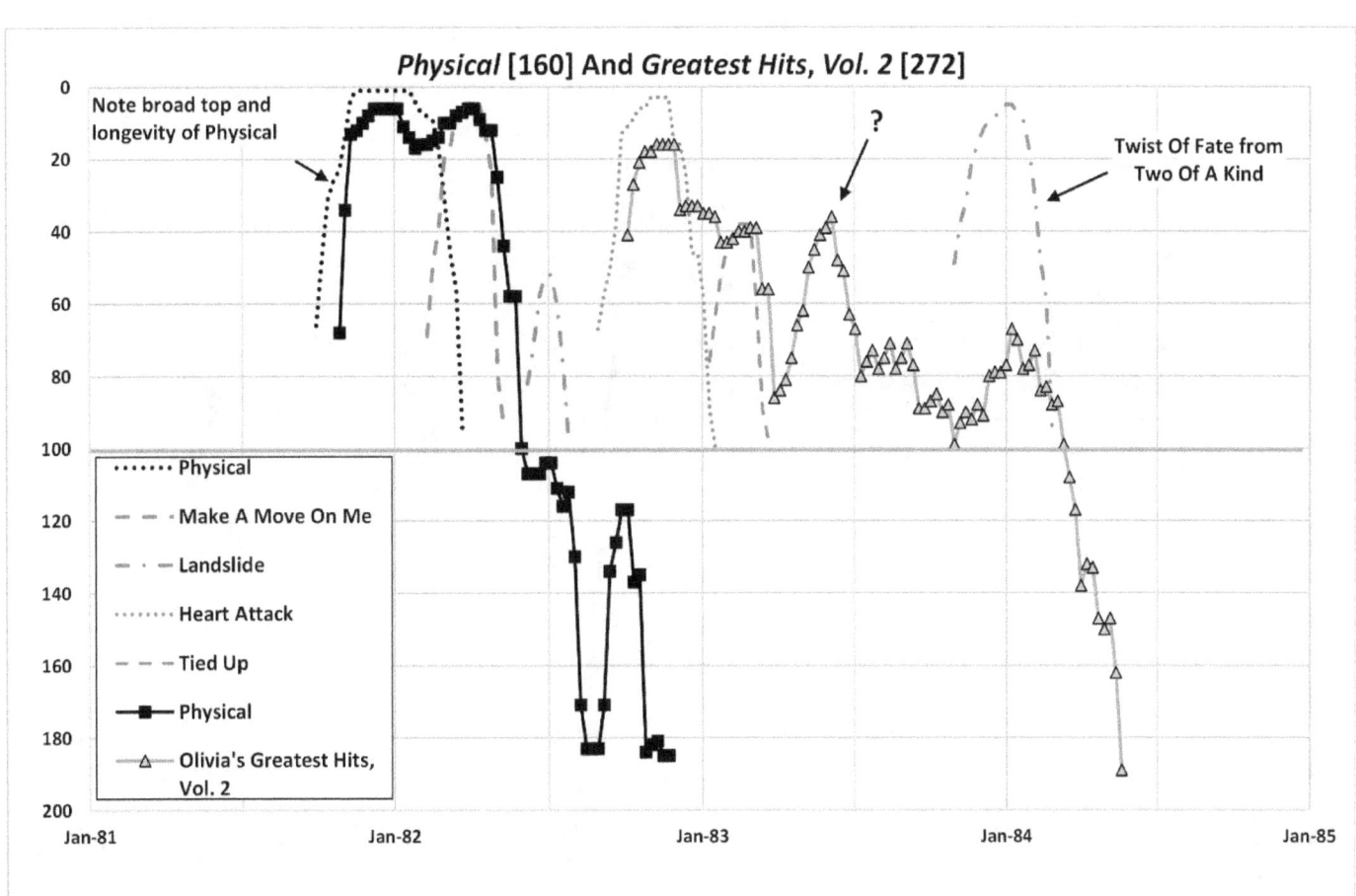

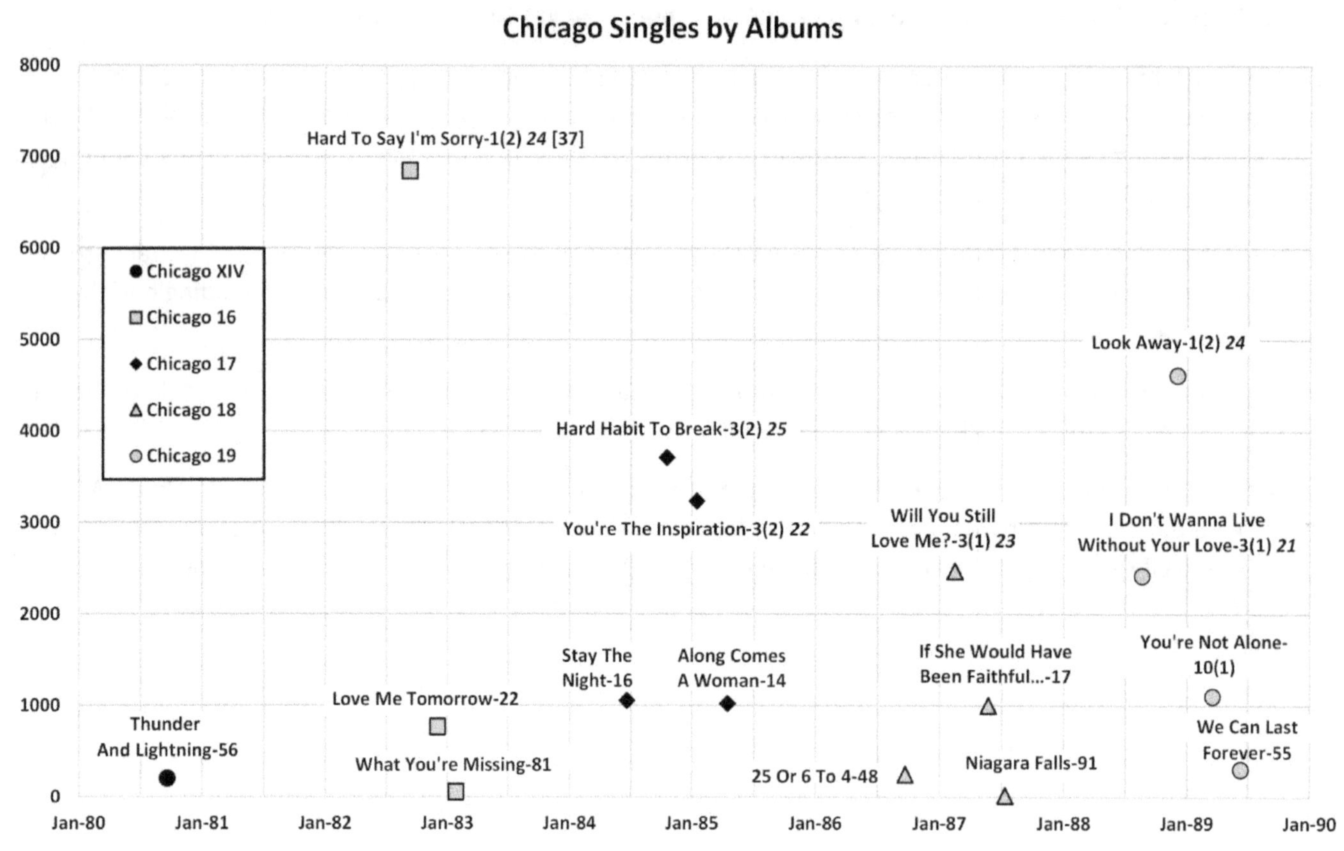

Chronologies

Chicago Singles by Albums

- ● Chicago XIV
- ☐ Chicago 16
- ◆ Chicago 17
- △ Chicago 18
- ○ Chicago 19

Hard To Say I'm Sorry-1(2) *24* [37]
Look Away-1(2) *24*
Hard Habit To Break-3(2) *25*
You're The Inspiration-3(2) *22*
Will You Still Love Me?-3(1) *23*
I Don't Wanna Live Without Your Love-3(1) *21*
Stay The Night-16
Along Comes A Woman-14
If She Would Have Been Faithful...-17
You're Not Alone-10(1)
Love Me Tomorrow-22
Thunder And Lightning-56
What You're Missing-81
25 Or 6 To 4-48
Niagara Falls-91
We Can Last Forever-55

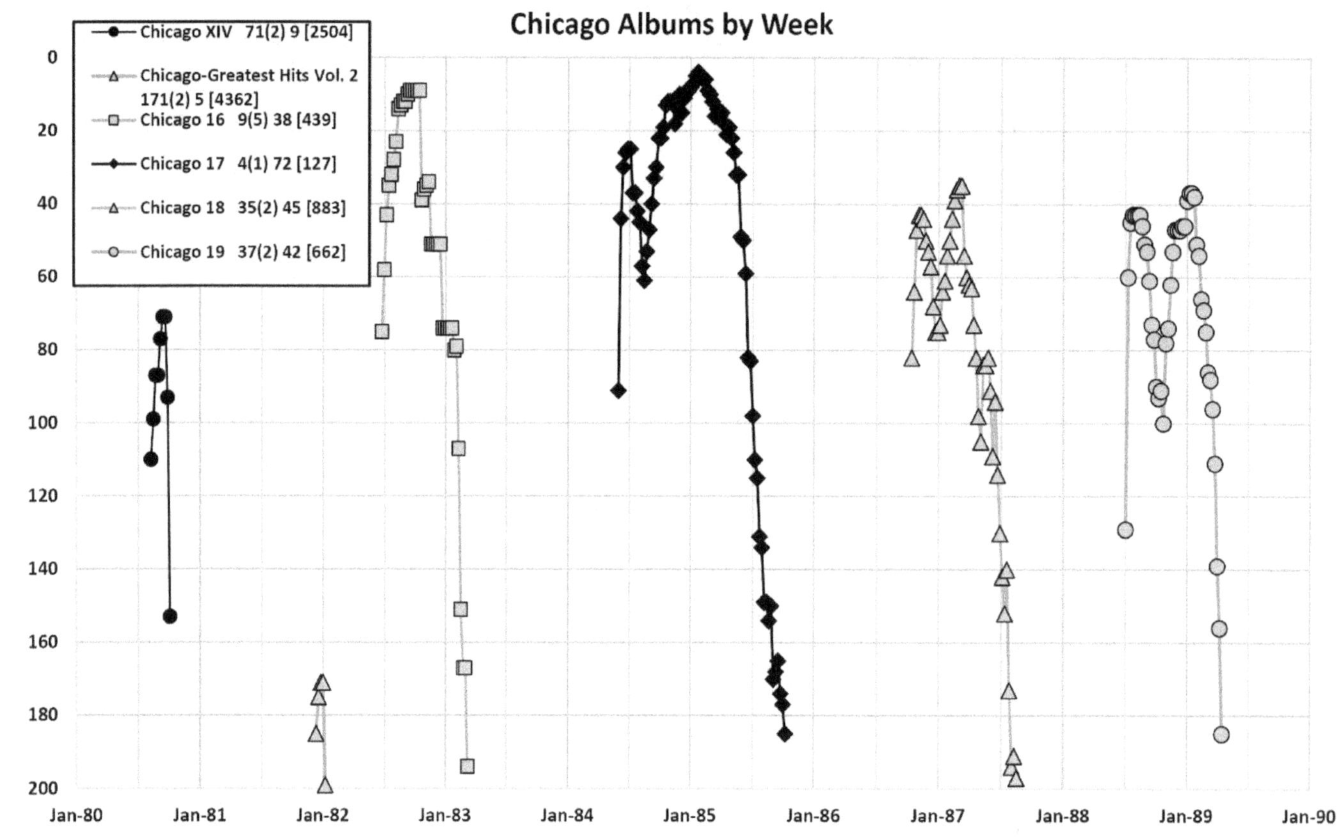

Chicago Albums by Week

- ●— Chicago XIV   71(2) 9 [2504]
- △— Chicago-Greatest Hits Vol. 2   171(2) 5 [4362]
- ☐— Chicago 16   9(5) 38 [439]
- ◆— Chicago 17   4(1) 72 [127]
- △— Chicago 18   35(2) 45 [883]
- ○— Chicago 19   37(2) 42 [662]

CHICAGO ▶ Singles Rank: 15 (16) 29063; Albums Rank: 82 (6/7) 32083 99.2%

By 1980, Chicago was a band with problems. The tragic death of Terry Kath was a blow but even more importantly, music was changing. Chicago was squeezed between disco and punk rock, and horn bands were out of style.

Their first album release of the '80s, *Chicago XIV*, yielded only one charted single and was their worst charting album to date, a record it held only until *Chicago's Greatest Hits Volume II*, the 15th album in the series, had an even more underwhelming chart life. At that point, no one was happy, and Columbia bought them out of their record deal.

In the fall of 1981, they asked Bill Champlin to join the group, and he introduced them to David Foster, who would produce their next three albums and co-write many of the songs. Peter Cetera left the band for a solo career after *Chicago 17*.

With all that turmoil, how did Chicago get to be the number 15 singles act? For better or worse, the new configuration on *Chicago 16* produced a huge hit, *Hard To Say I'm Sorry*, the number 9 record of 1982 and number 37 of the decade. *Chicago 17* produced two top 5 and two more top 20 singles, *Chicago 18* yielded one top 5 and one top 20 single and *Chicago 19* generated their final number 1, *Look Away*, as well as two more top 10s.

The irony is: Chicago, a band that helped launch album-oriented radio, was now a singles act, ranking only number 83 as an album act in the '80s. Their entire '80s catalog combined had fewer chart points than *Chicago II*. To show how much the album market had changed, *Chicago 17*, 127th album of the decade, was their largest selling album of all time (4x Platinum by the end of the 80s), but on the basis of score ranked toward the middle of their catalog. On the basis of its singles, however, the album moved up to number 89.

The three David Foster albums are shown below. Clearly, *Chicago 16* is driven by *Hard To Say I'm Sorry*. The two top 5 singles in *Chicago 17* sustained most of its top, and by *Chicago 18*, singles were clearly outrunning the album.

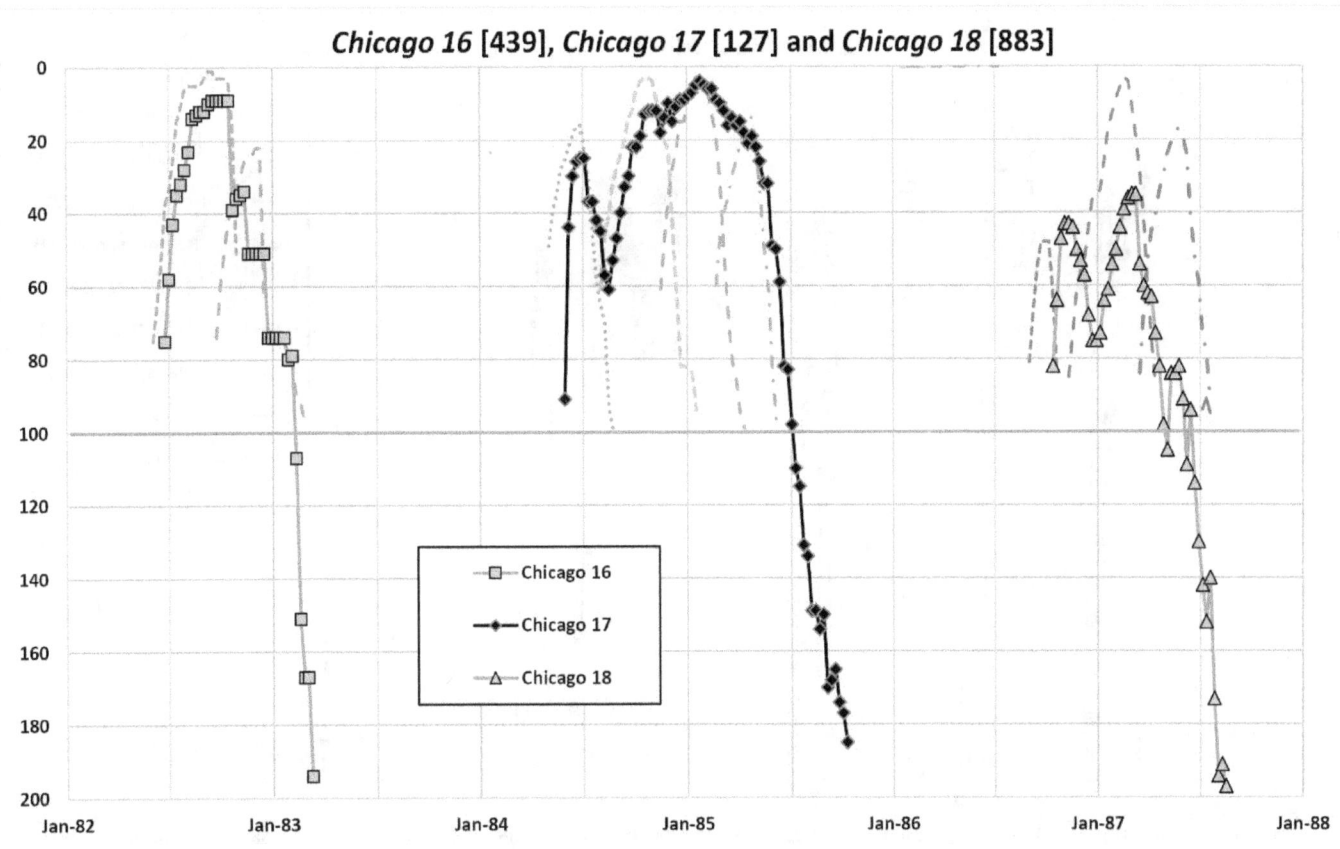

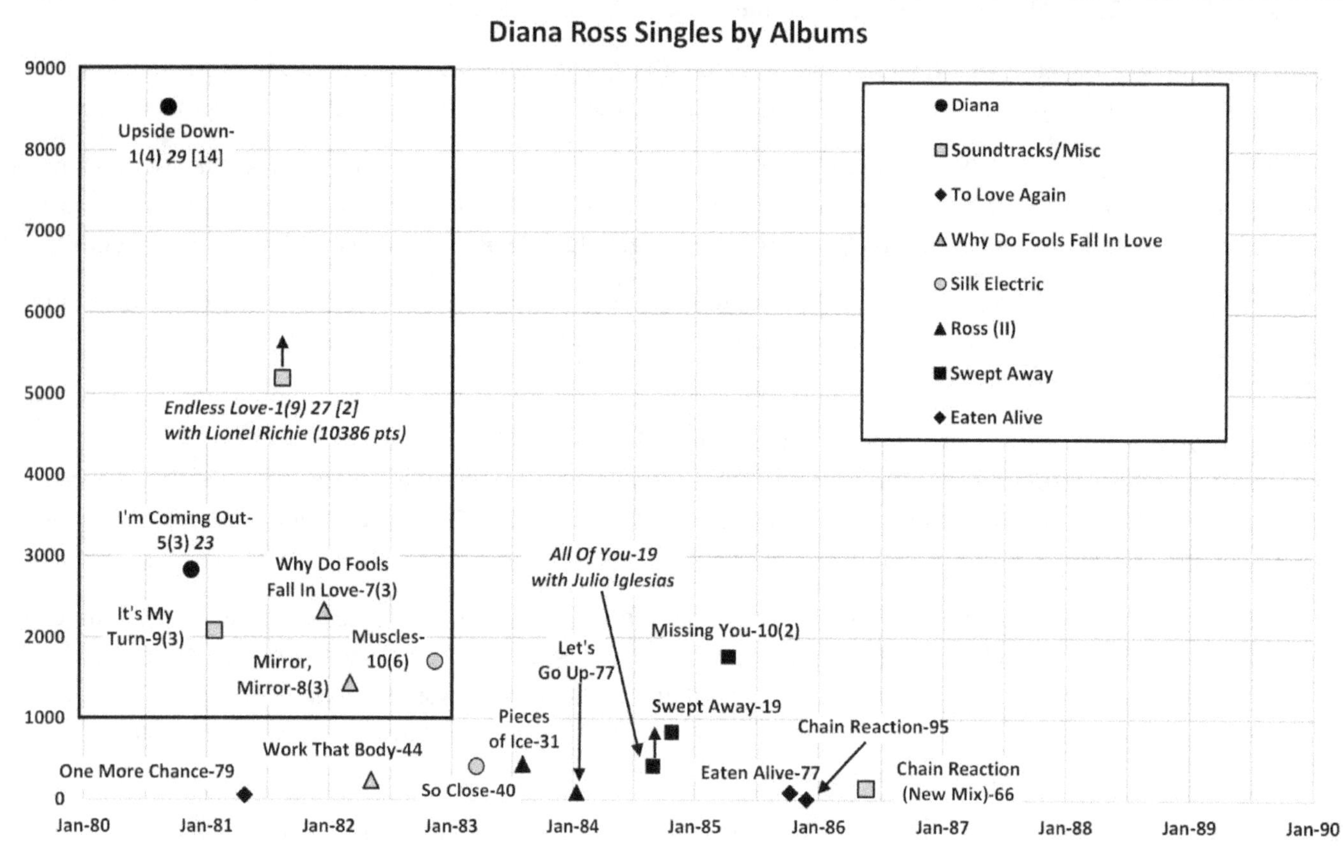

Chronologies

Diana Ross Singles by Albums

Legend:
- ● Diana
- □ Soundtracks/Misc
- ♦ To Love Again
- △ Why Do Fools Fall In Love
- ○ Silk Electric
- ▲ Ross (II)
- ■ Swept Away
- ♦ Eaten Alive

Data points:
- Upside Down-1(4) *29* [14]
- Endless Love-1(9) 27 [2] with Lionel Richie (10386 pts)
- I'm Coming Out-5(3) *23*
- It's My Turn-9(3)
- Why Do Fools Fall In Love-7(3)
- Mirror, Mirror-8(3)
- Muscles-10(6)
- One More Chance-79
- Work That Body-44
- Pieces of Ice-31
- So Close-40
- Let's Go Up-77
- All Of You-19 with Julio Iglesias
- Missing You-10(2)
- Swept Away-19
- Eaten Alive-77
- Chain Reaction-95
- Chain Reaction (New Mix)-66

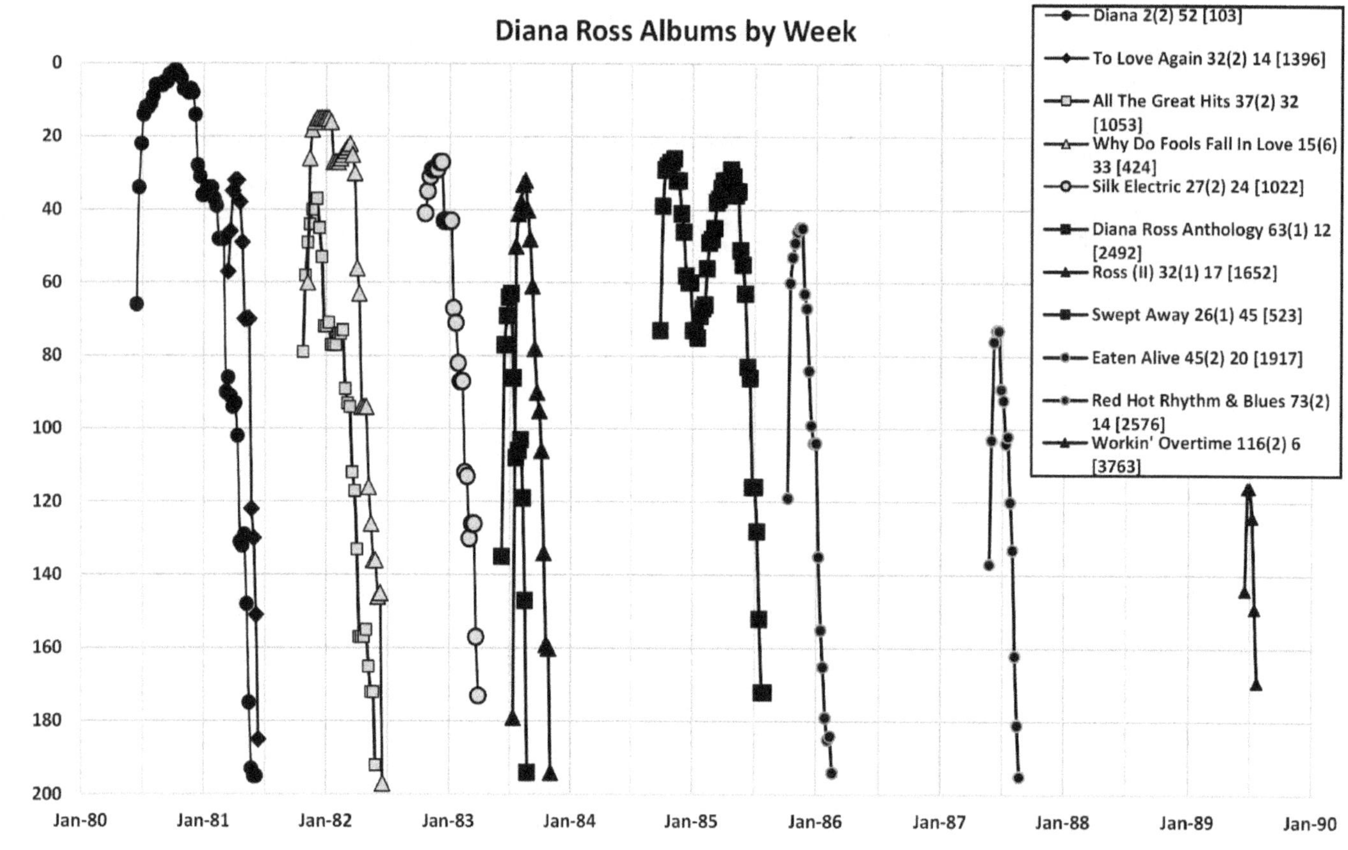

Diana Ross Albums by Week

Legend:
- ● Diana 2(2) 52 [103]
- ♦ To Love Again 32(2) 14 [1396]
- □ All The Great Hits 37(2) 32 [1053]
- △ Why Do Fools Fall In Love 15(6) 33 [424]
- ○ Silk Electric 27(2) 24 [1022]
- ■ Diana Ross Anthology 63(1) 12 [2492]
- ▲ Ross (II) 32(1) 17 [1652]
- ■ Swept Away 26(1) 45 [523]
- ● Eaten Alive 45(2) 20 [1917]
- ● Red Hot Rhythm & Blues 73(2) 14 [2576]
- ▲ Workin' Overtime 116(2) 6 [3763]

DIANA ROSS ▶ Singles Rank: 16 (18) 28634; Albums Rank: 59 (11/12) 37923 99.6%

Diana Ross was known as a diva and entertainer, but her chart record as a solo artist after the Supremes is spotty. Simple inspection of the '80s album and single history shows a steep decline over the decade; however, the first part of the decade was quite good for her.

Upside Down with its incredible 29-week run was the number 4 record of 1980 and 14th overall for the decade; *Endless Love*, which charted for 27 weeks, and for which she split credits with Lionel Richie, was the number 2 song of 1981 and number 2 in the decade as well. *I'm Coming Out*, the second single from the *Diana* album was also a top 5 record. During that time, she was one of only two acts in the decade to have three singles simultaneously in the top 40. She had four other lesser but still important top 10 singles before the end of 1982. Combined, her point production during the first two years of the decade accounted for nearly 85% of her '80s chart points.

By comparison, in the 1970s she had four number 1s including Ain't No Mountain High Enough; but fourteen other singles that didn't crack the top 10. The years 1980 to 1982 marked her longest continuous run of singles chart success.

As an album act, she hit her apex with the first album of the decade, *Diana*, which placed 103rd. *Diana* was driven by *Upside Down* and when scored by singles, rose to 50th. Beyond that, her success was in contributions to soundtracks such as *Endless Love* and *It's My Turn*. She ranked 57th as an album act.

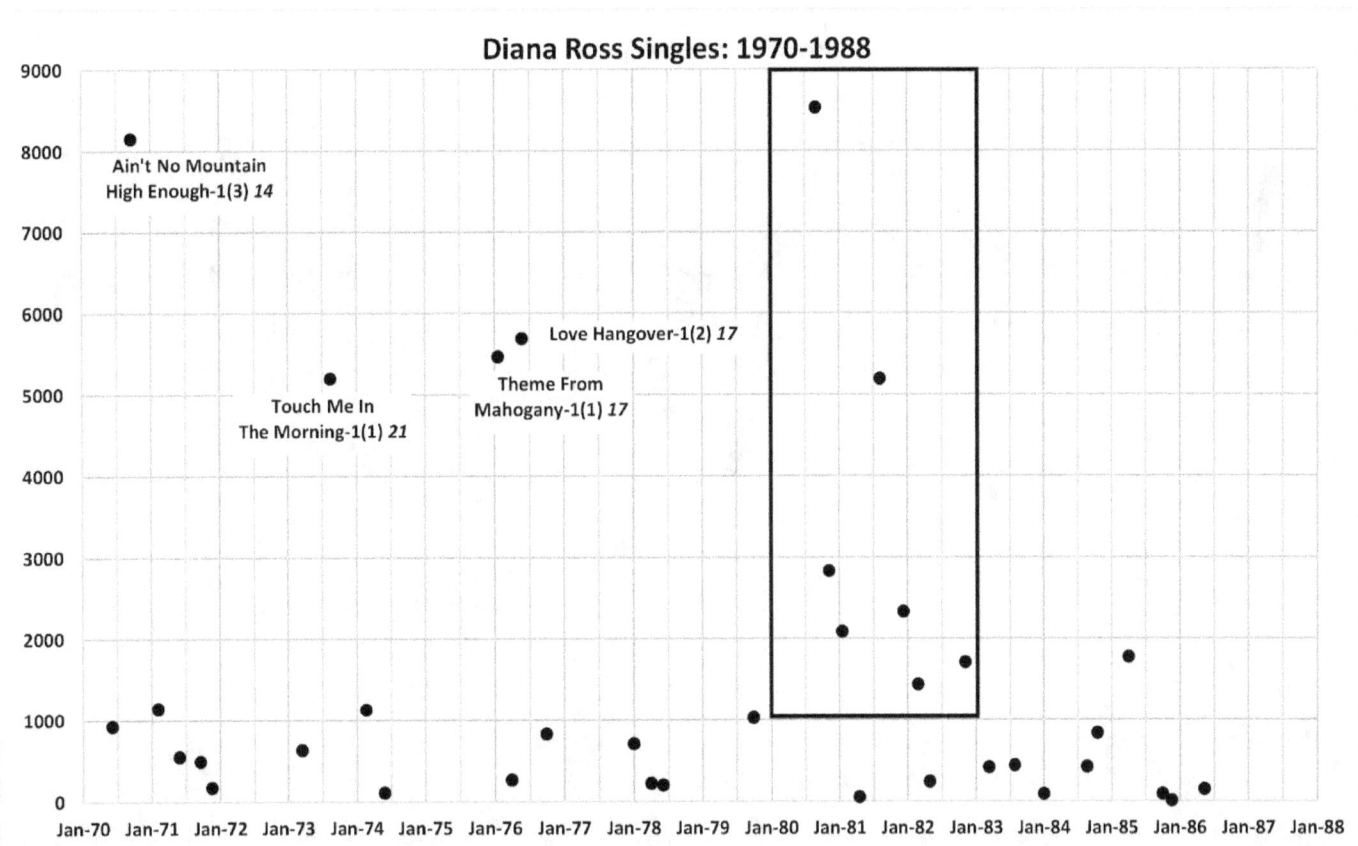

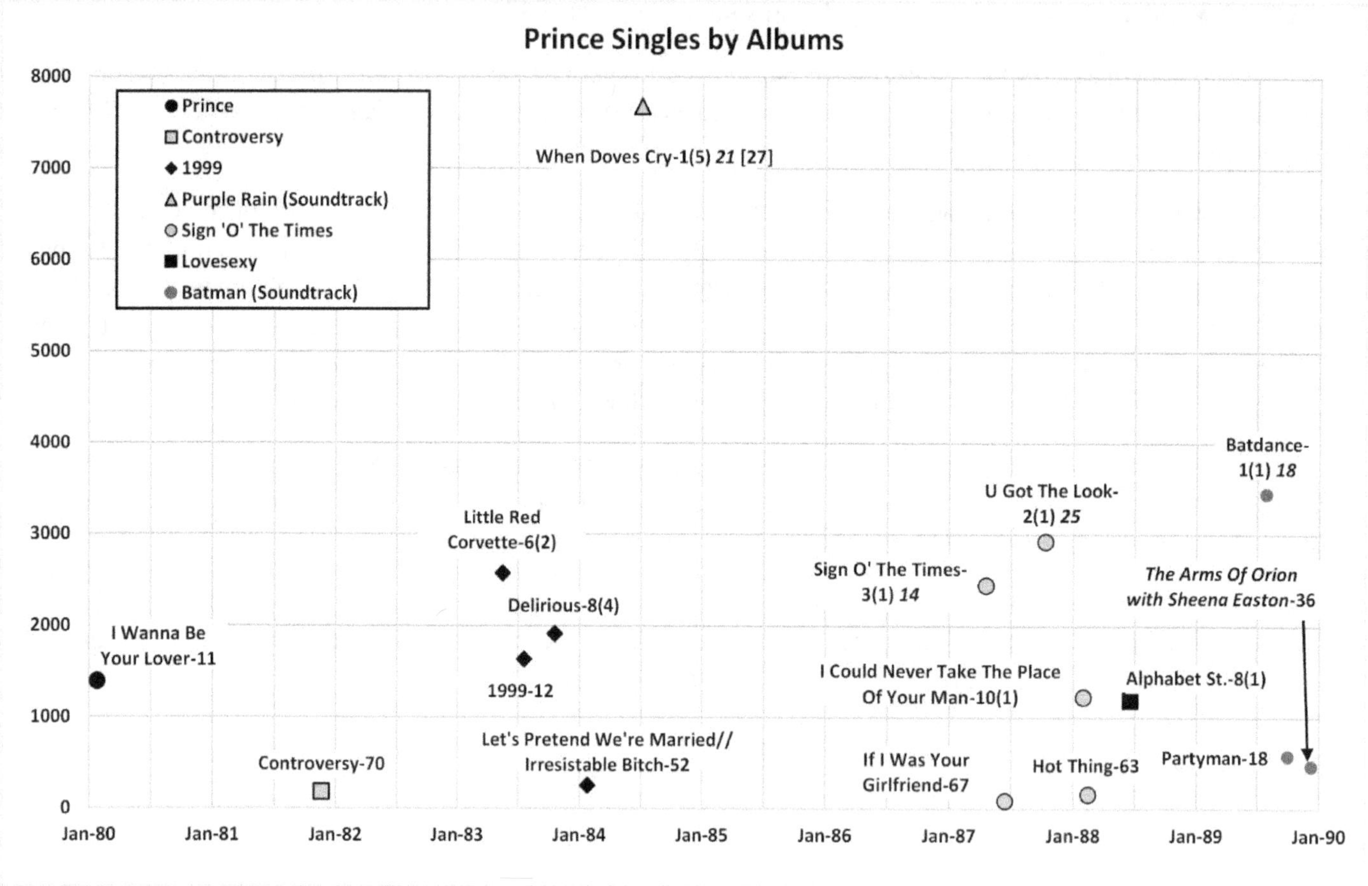

Chronologies

Prince Singles by Albums

- ● Prince
- ☐ Controversy
- ◆ 1999
- △ Purple Rain (Soundtrack)
- ○ Sign 'O' The Times
- ■ Lovesexy
- ● Batman (Soundtrack)

When Doves Cry-1(5) *21* [27]

Batdance-1(1) *18*

U Got The Look-2(1) *25*

Little Red Corvette-6(2)

Sign O' The Times-3(1) *14*

The Arms Of Orion with Sheena Easton-36

Delirious-8(4)

I Wanna Be Your Lover-11

1999-12

I Could Never Take The Place Of Your Man-10(1)

Alphabet St.-8(1)

Controversy-70

Let's Pretend We're Married// Irresistable Bitch-52

If I Was Your Girlfriend-67

Hot Thing-63

Partyman-18

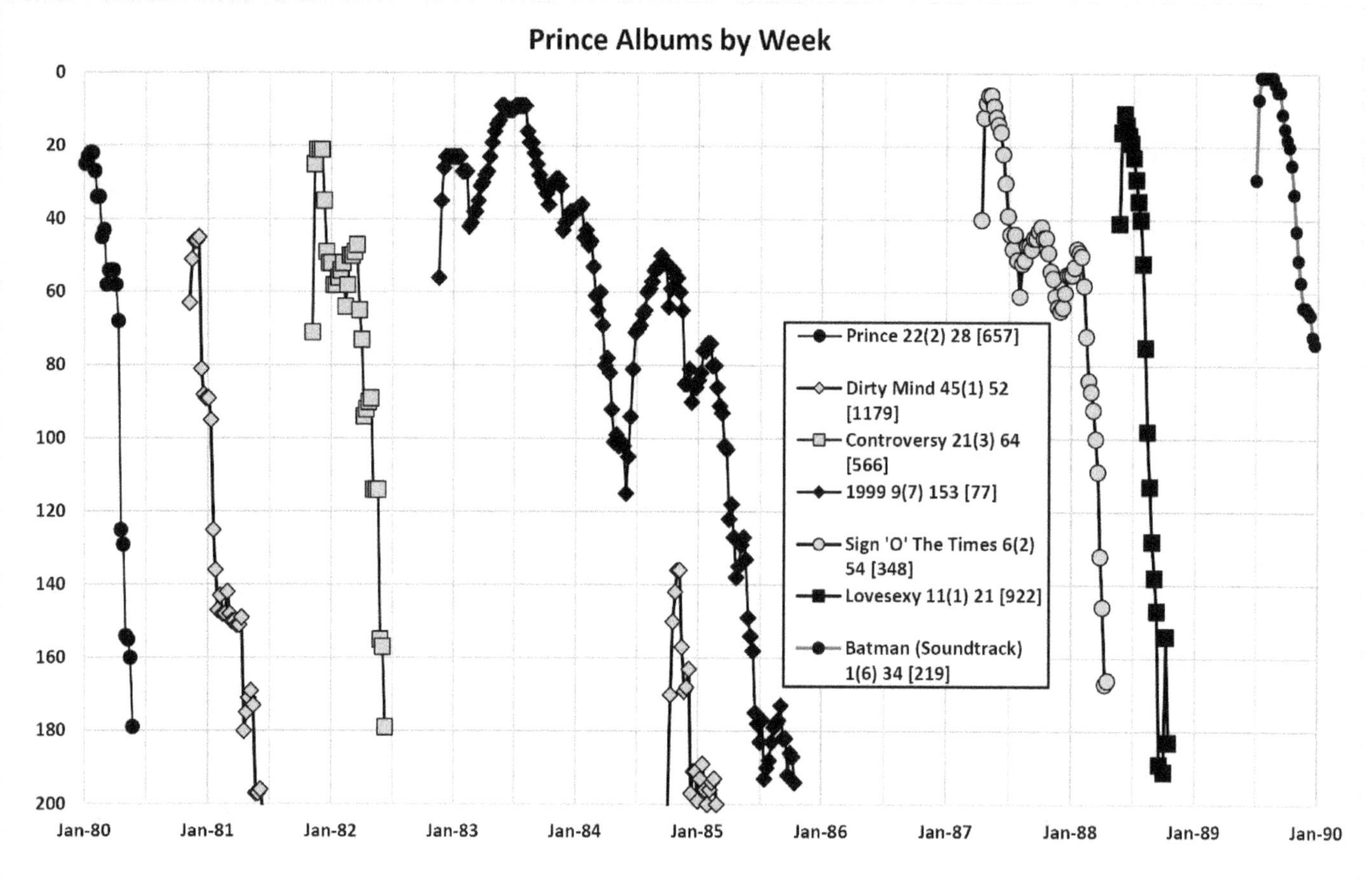

Prince Albums by Week

- —●— Prince 22(2) 28 [657]
- —◇— Dirty Mind 45(1) 52 [1179]
- —☐— Controversy 21(3) 64 [566]
- —◆— 1999 9(7) 153 [77]
- —○— Sign 'O' The Times 6(2) 54 [348]
- —■— Lovesexy 11(1) 21 [922]
- —●— Batman (Soundtrack) 1(6) 34 [219]

PRINCE ▶ Singles Rank: 17 (16) 28172; Albums Rank: 31 (7/7) 52965 100%

In this analysis, Prince is separated from Prince And The Revolution and the *Purple Rain* album—the number 20 album of the decade--marks the transition from the former to the latter. *When Doves Cry,* the first single from *Purple Rain* and the number 27 single of the '80s is credited to Prince on the 45; the album and the rest of the singles are credited to Prince And The Revolution. Wikipedia treats The Revolution as a separate reference, formed by Prince and disbanded ca. 1986. *Sign 'O' The Times* in 1987 was once again attributed just to Prince.

Eight of his sixteen singles hit the top 10, and though albums generally tended to peak higher through the decade, chart longevity declined. *1999* and *Sign 'O' The Times* yielded four and five singles respectively, and when scored on those singles, *1999* chart strength declined from 77th to 151st but *Sign 'O' The Times* increased from 348th to 134th, indicating that 1999 was a stronger album, but *Sign* was stronger for its singles. As an album act, Prince ranked 32nd.

Taken as a separate entity, Prince And The Revolution ranks 40th for singles with 21448 points and 10 entries, and as an album act, 29th for the three albums. Those singles included three number 1s and a number 2, one of which was the number 100 single of the decade, *Let's Go Crazy,* and the 13x Platinum album *Purple Rain* . If the two acts were combined, they would rank 5th for singles score and 2nd for albums.

Although shown on a different scale below, it's clear that The Revolution albums were declining in longevity as the decade progressed, with *Parade/Under The Cherry Moon* Soundtrack only lasting 28 weeks compared to *Purple Rain's* 72.

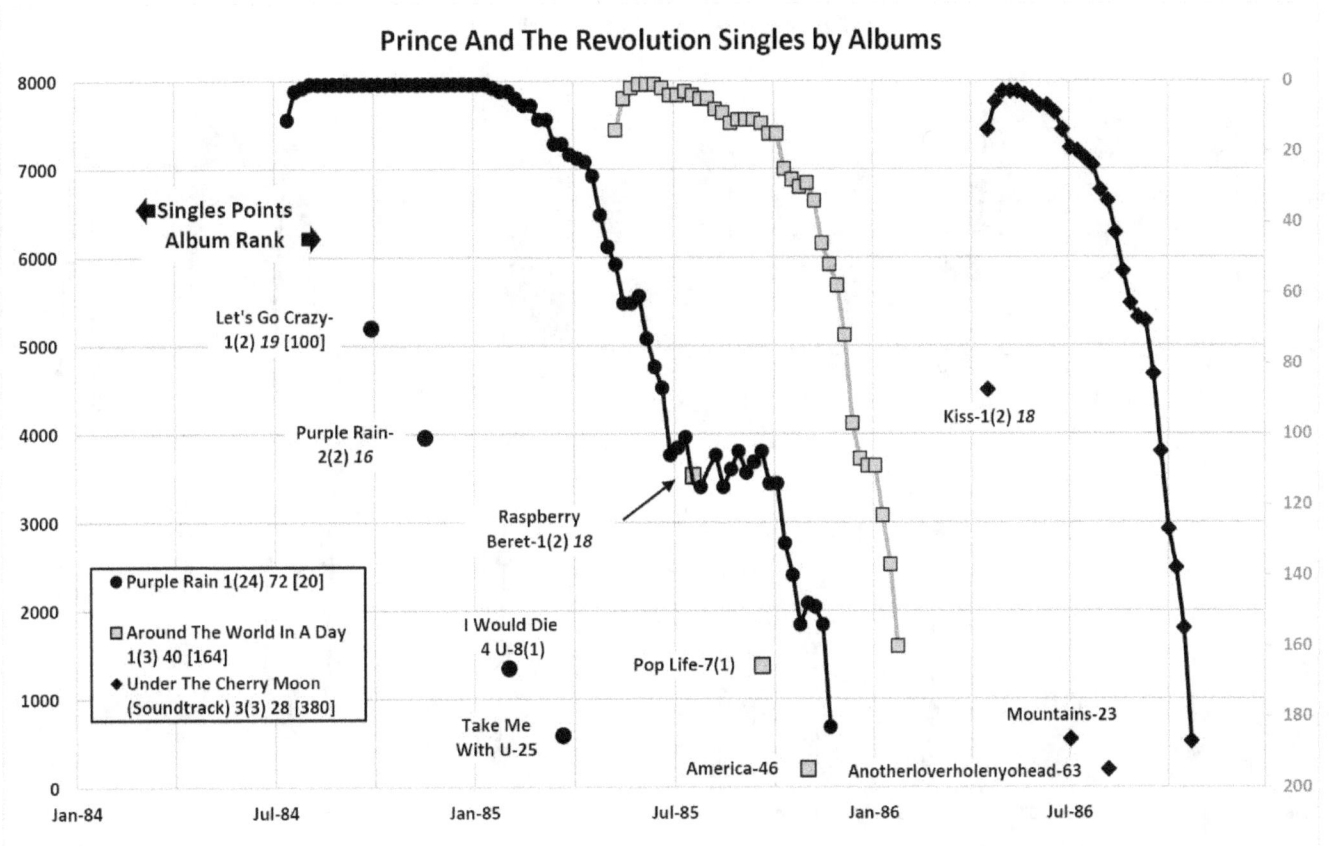

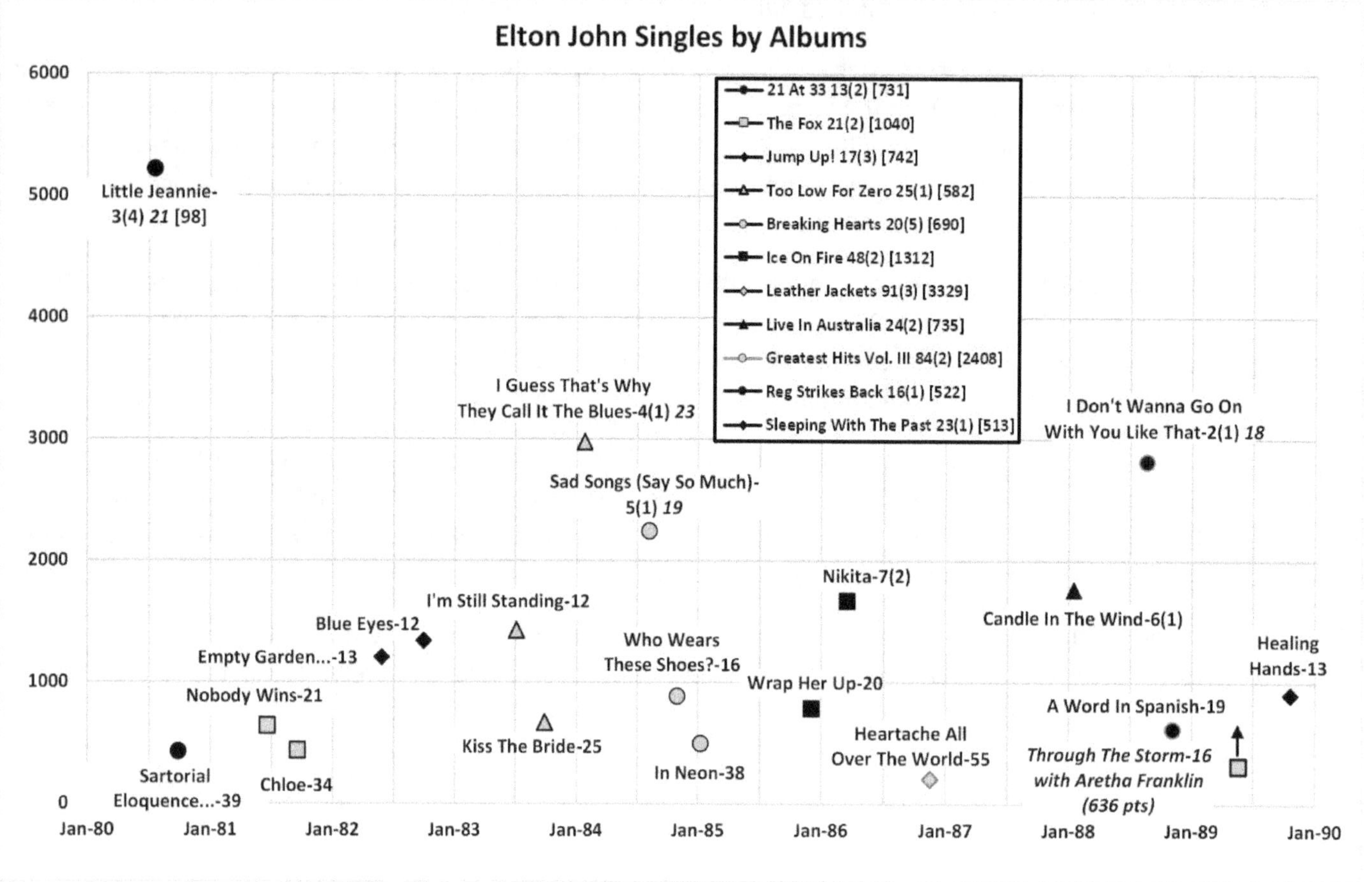
Elton John Singles by Albums
21 At 33 13(2) [731]
The Fox 21(2) [1040]
Jump Up! 17(3) [742]
Too Low For Zero 25(1) [582]
Breaking Hearts 20(5) [690]
Ice On Fire 48(2) [1312]
Leather Jackets 91(3) [3329]
Live In Australia 24(2) [735]
Greatest Hits Vol. III 84(2) [2408]
Reg Strikes Back 16(1) [522]
Sleeping With The Past 23(1) [513]
Little Jeannie-3(4) *21* [98]
I Guess That's Why They Call It The Blues-4(1) *23*
I Don't Wanna Go On With You Like That-2(1) *18*
Sad Songs (Say So Much)-5(1) *19*
Nikita-7(2)
I'm Still Standing-12
Blue Eyes-12
Candle In The Wind-6(1)
Who Wears These Shoes?-16
Empty Garden...-13
Healing Hands-13
Wrap Her Up-20
Nobody Wins-21
A Word In Spanish-19
Heartache All Over The World-55
Kiss The Bride-25
Through The Storm-16 with Aretha Franklin (636 pts)
Sartorial Eloquence...-39
Chloe-34
In Neon-38
6000
5000
4000
3000
2000
1000
0
Jan-80
Jan-81
Jan-82
Jan-83
Jan-84
Jan-85
Jan-86
Jan-87
Jan-88
Jan-89
Jan-90

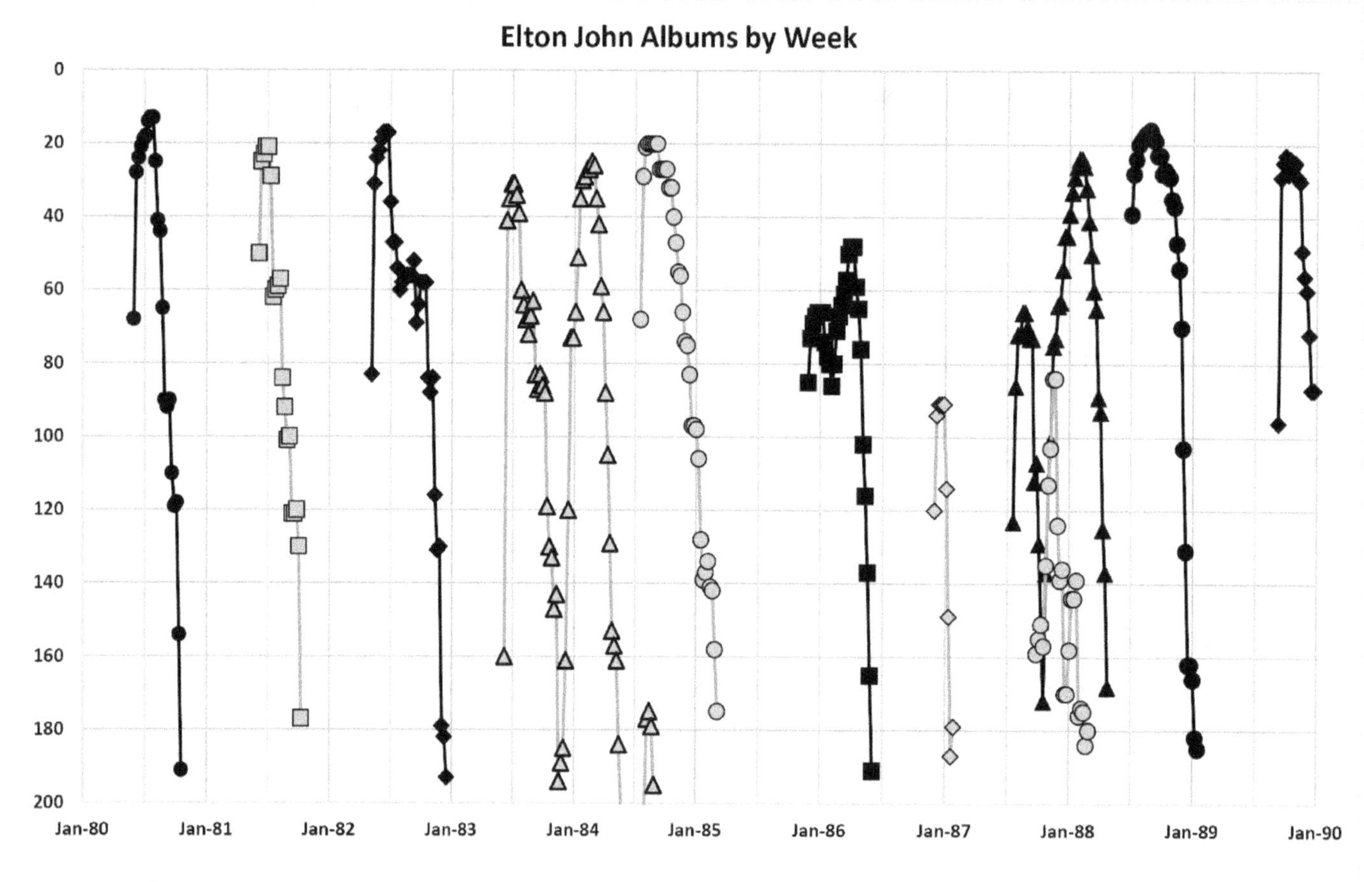
Elton John Albums by Week
0
20
40
60
80
100
120
140
160
180
200
Jan-80
Jan-81
Jan-82
Jan-83
Jan-84
Jan-85
Jan-86
Jan-87
Jan-88
Jan-89
Jan-90

ELTON JOHN ▶ Singles Rank: 18 (16) 27052; Albums Rank: 70 (11/12) 35198 98.8%

The first half of the '70s for Elton John included a string of number 1 singles and albums. By the second half, his chart performance had cooled, and as the number 18 singles act of the '80, the decade could hardly be called a failure, unless, perhaps, you're Elton John.

Of his 16 singles, four made the top 5, including the number 98 record of the decade, *Little Jeannie*, and two more made the top 10. But those six accounted for 62% of his points. His singles averaged 1353 points, good for 90th. And yet, it was a better singles decade than album decade for him.

Leaving *Little Jeannie* out of it, the first three years of the '80s were kind of bleak; however, things perked up a bit when John and Bernie Taupin reunited to write all the songs on *Too Low For Zero*; original band members Davey Johnstone, Nigel Olssen and Tony Murray also returned to the band. *I Guess That's Why They Call It The Blues* marked a return to the top 5.

He was the 69th ranked album act, charting 11 albums in the '80s, none of which made the weekly top 10 or the top 500 for the decade. In the middle of the decade the albums *Leather Jackets* and *Greatest Hits Volume 3* barely broke into the weekly top 100; those two albums were his final efforts for Geffen before moving back to MCA. And yet eight of the 11 albums peaked between 10 and 25; fairly consistent if not spectacular performance.

In comparing *Too Low For Zero*, *Breaking Hearts* and *Ice On Fire*, two things are clear. First, the singles peak higher than the albums that spawned them. Second, the album moves on the chart only after the single has moved, and none of these albums have sustained chart presence without a single to drive them.

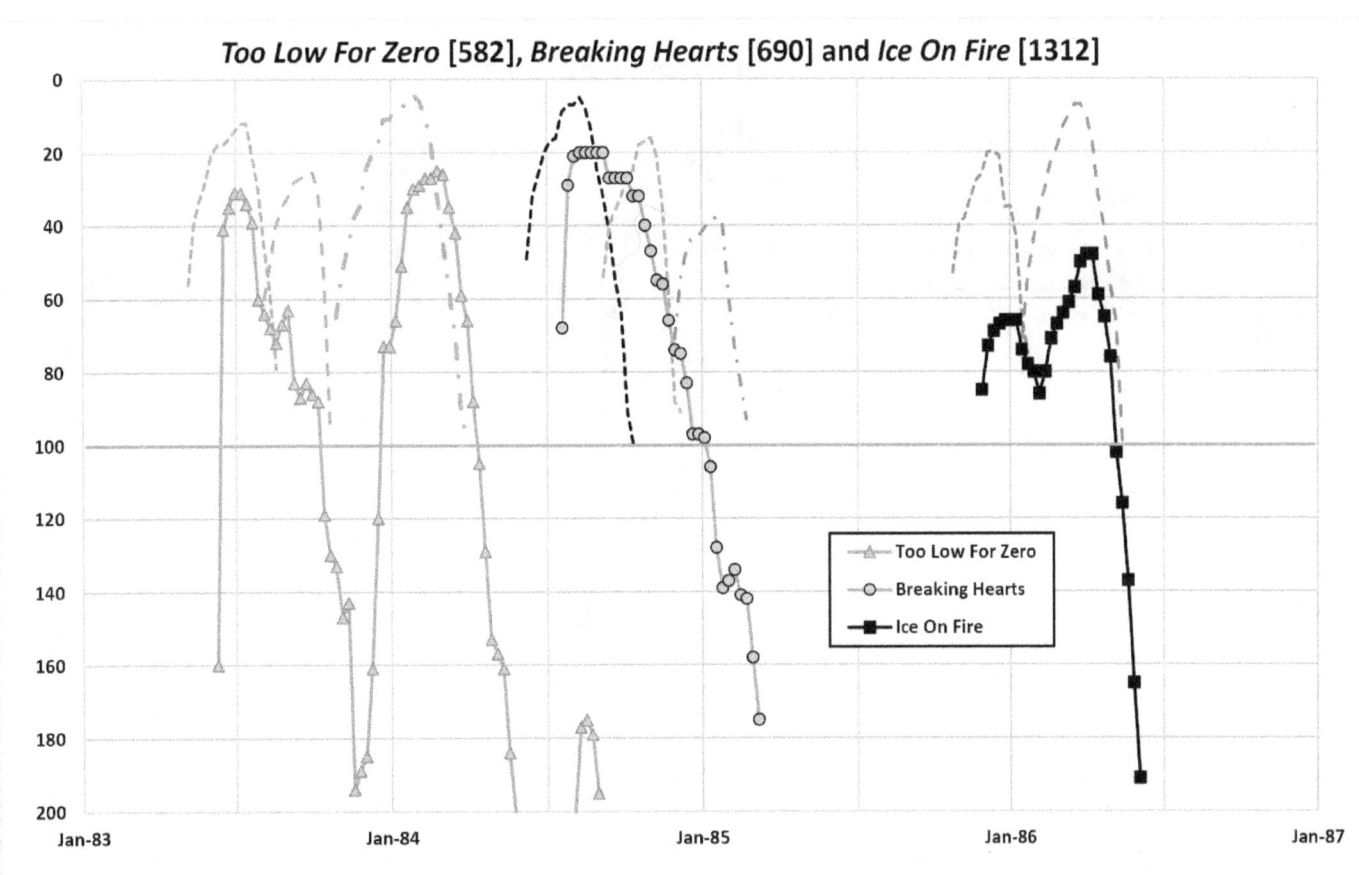

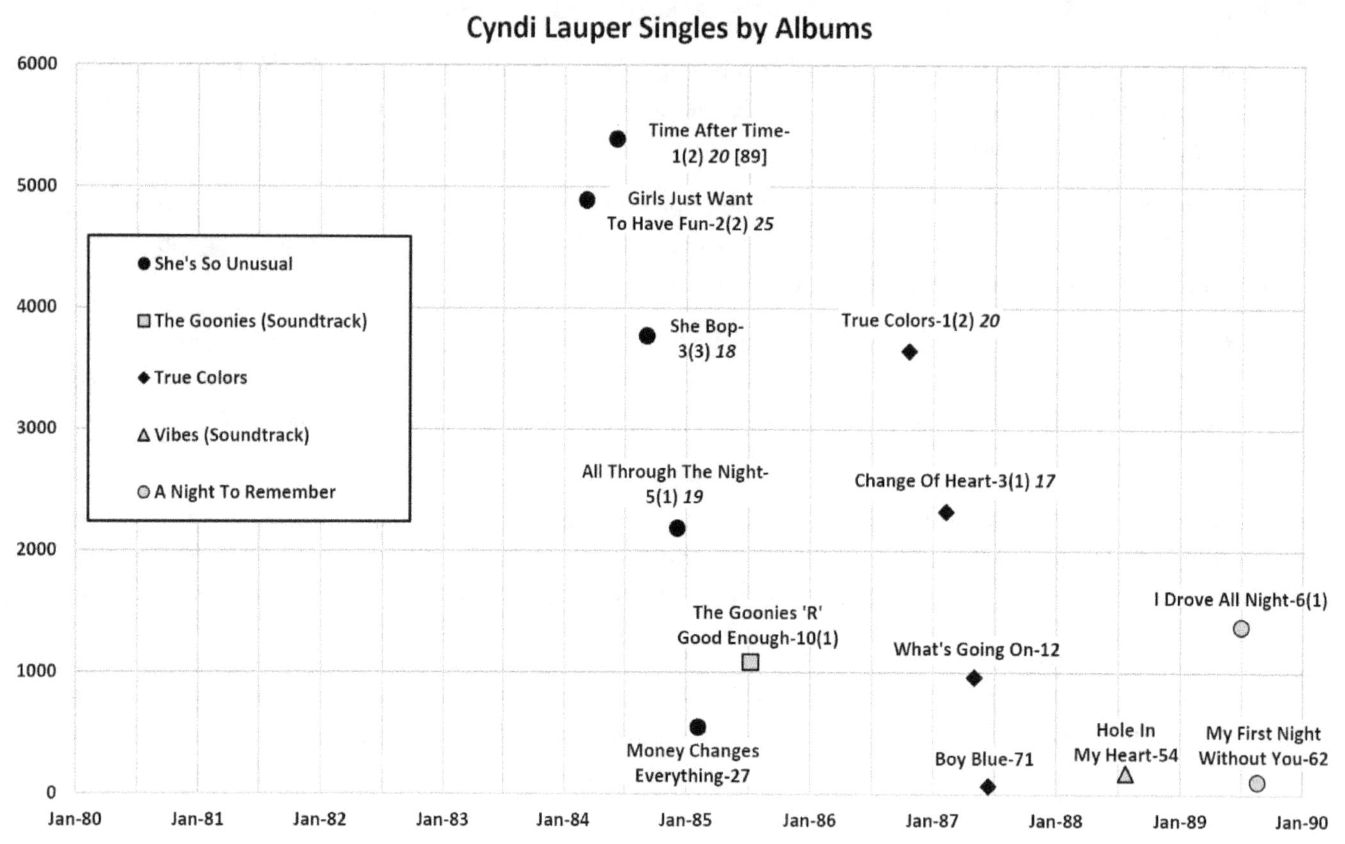

Cyndi Lauper Singles by Albums
6000
5000
4000
3000
2000
1000
0
She's So Unusual
The Goonies (Soundtrack)
True Colors
Vibes (Soundtrack)
A Night To Remember
Time After Time-1(2) *20* [89]
Girls Just Want To Have Fun-2(2) *25*
She Bop-3(3) *18*
True Colors-1(2) *20*
All Through The Night-5(1) *19*
Change Of Heart-3(1) *17*
I Drove All Night-6(1)
The Goonies 'R' Good Enough-10(1)
What's Going On-12
Money Changes Everything-27
Boy Blue-71
Hole In My Heart-54
My First Night Without You-62
Jan-80
Jan-81
Jan-82
Jan-83
Jan-84
Jan-85
Jan-86
Jan-87
Jan-88
Jan-89
Jan-90

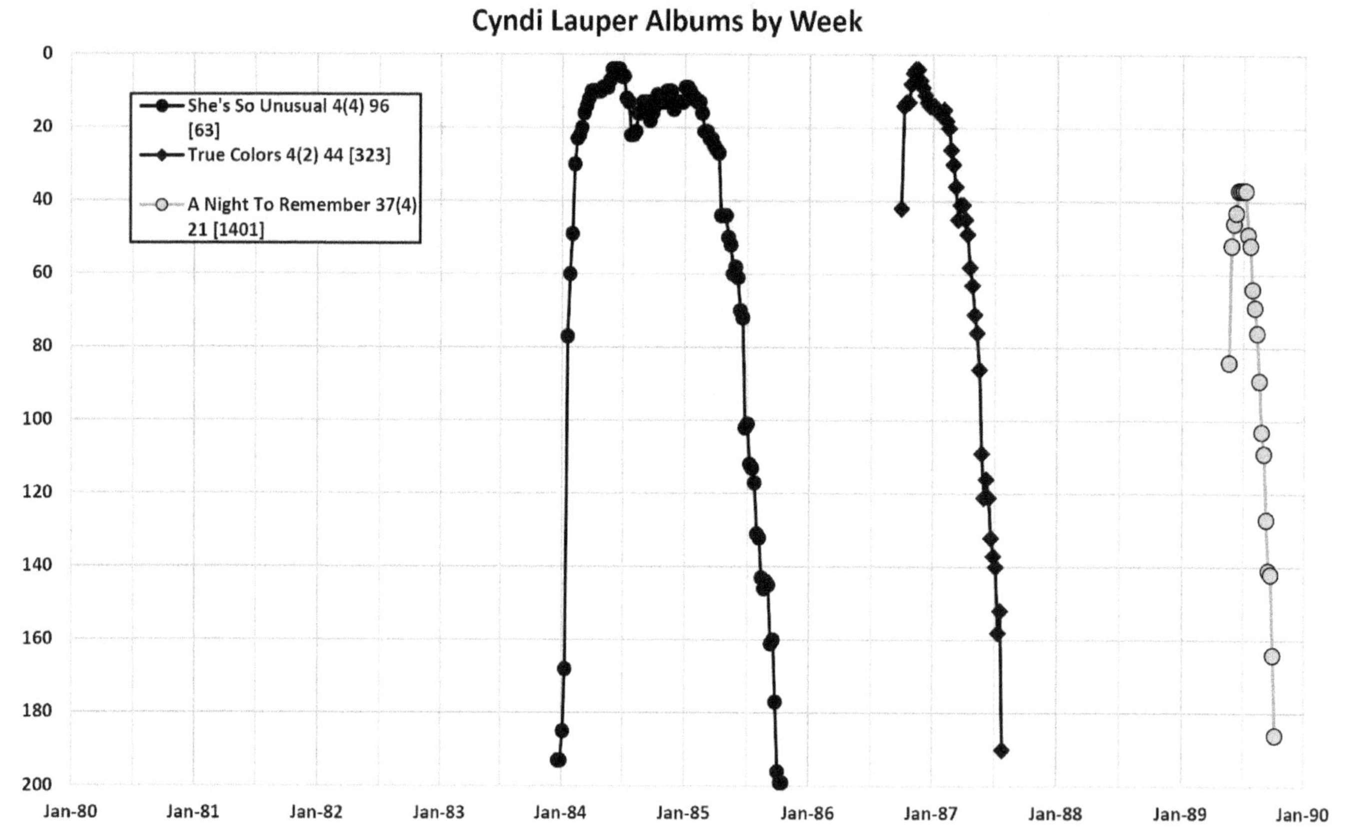

Cyndi Lauper Albums by Week
0
20
40
60
80
100
120
140
160
180
200
She's So Unusual 4(4) 96 [63]
True Colors 4(2) 44 [323]
A Night To Remember 37(4) 21 [1401]
Jan-80
Jan-81
Jan-82
Jan-83
Jan-84
Jan-85
Jan-86
Jan-87
Jan-88
Jan-89
Jan-90

CYNDI LAUPER ▶ Singles Rank: 19 (13) 26545; Albums Rank: 67 (3/3) 35855 100%

Cyndi Lauper became visible right at the seam between the early '80s of blockbuster hits and the late '80s of genre diversification. Her quirky voice, looks and personality—along with her association with the World Wrestling Federation fit well with MTV, as it morphed into a Top 40-type outlet about the time of *Girls Just Want To Have Fun*.

She also worked closely with two other '80s acts—the Hooters and Jules Shear. The Hooters were a Philadelphia band whose signature sound was derived from a melodica—a sort of keyboard harmonica. Eric Bazillian and Rob Hyman played on the songs of *She's So Unusual* and Hyman co-wrote *Time After Time*. Jules Shear wrote *All Through The Night* (he also contributed *If She Knew What She Wants* for Bangles).

Despite her first album ranking number 63 in the decade, Cyndi Lauper was really a singles artist. For her singles scores, she ranked 19th in the decade, but as an album artist only 66th. While only charting 13 singles in the '80s, eight peaked in the top 10, with her first four releases peaking in the top 5. *Time After Time* ranked number 89 for the decade and her average singles score for the 13 entries was 2042, good for 33rd.

Her albums charted more strongly as collections of singles than as an album entity. All the singles from *She's So Unusual* outran the album itself, which, if scored by the singles derived from it, rises from 63rd to number 10; on the same basis, True Colors rises from 323rd to 132nd.

Her discography resembles that of Prince and the Revolution: First album is big and yields multiple singles. Second album peaks strongly, also produces multiple singles but without the longevity of the first. Third album not as strong.

She's So Unusual only peaked at number 4 but logged 21 weeks in the top 10 and over a year in the top 40; it was 4x Platinum by the end of 1985. Even after the decline of the final single *Money Changes Everything* the album declined at only a modest rate, that decline possibly slowed by the number 10 *The Goonies 'R' Good Enough*.

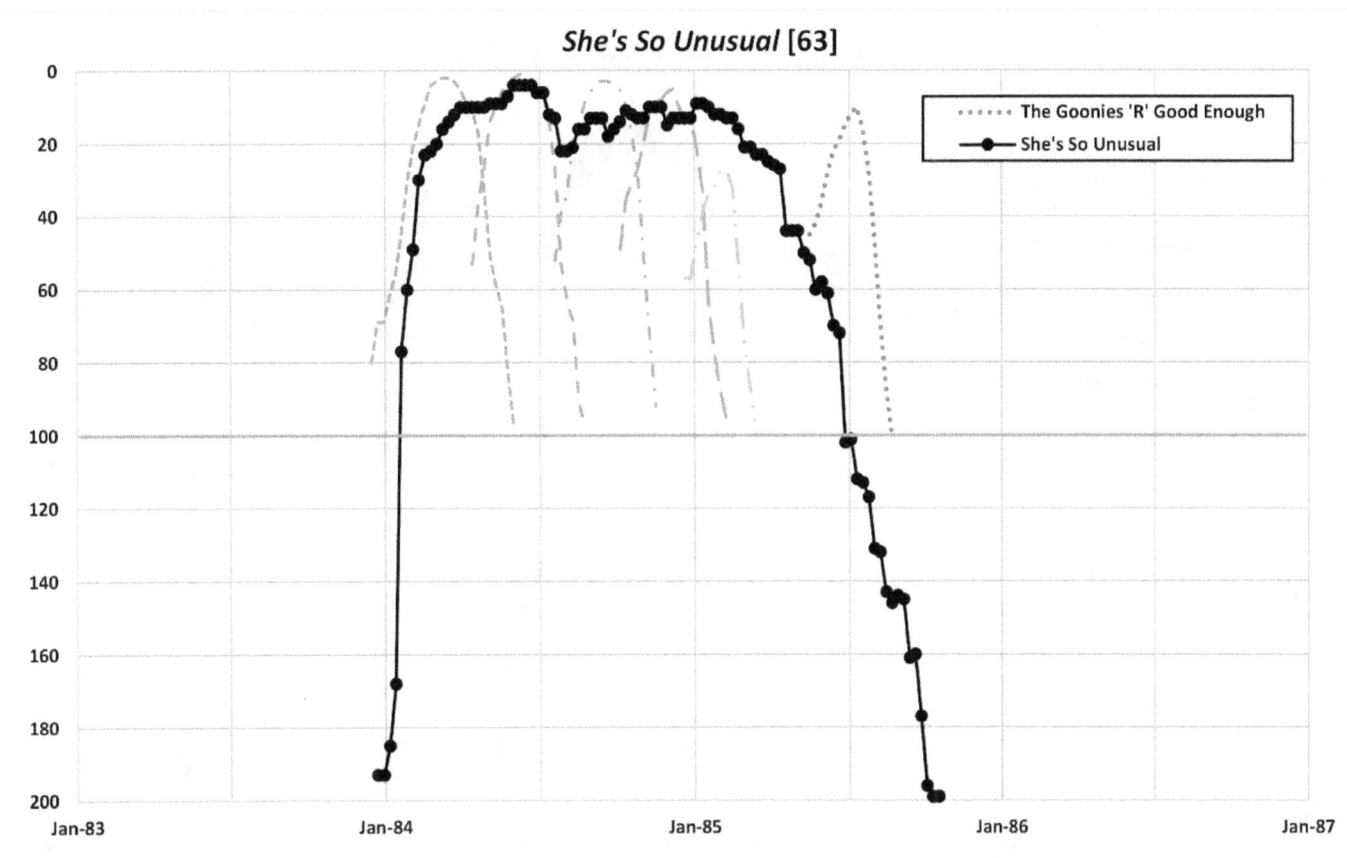

Billy Ocean (20) Singles by Albums
Suddenly
Jewel Of The Nile Soundtrack
Love Zone
Tear Down These Walls
Greatest Hits
Caribbean Queen-1(2) 26 [99]
There'll Be Sad Songs
(To Make You Cry)- 1(1) 21
I Get Outta My Dreams,
Get Into My Car-1(2) 20
Loverboy-2(1) 21
When The Going Gets Tough,
The Tough Get Going-2(1) 23
Suddenly-4(2) 22
Love Zone-10(1)
The Colour Of Love-17
Mystery Lady-24
Love Is Forever-16
License To Chill-32
6000
5000
4000
3000
2000
1000
0
Jan-80
Jan-81
Jan-82
Jan-83
Jan-84
Jan-85
Jan-86
Jan-87
Jan-88
Jan-89
Jan-90

Billy Ocean (20) Albums by Week
Suddenly 9(2) [120]
Love Zone 6(7) [215]
Tear Down These Walls 18(1) [612]
Greatest Hits 77(2) [2436]
0
20
40
60
80
100
120
140
160
180
200
Jan-80
Jan-81
Jan-82
Jan-83
Jan-84
Jan-85
Jan-86
Jan-87
Jan-88
Jan-89
Jan-90

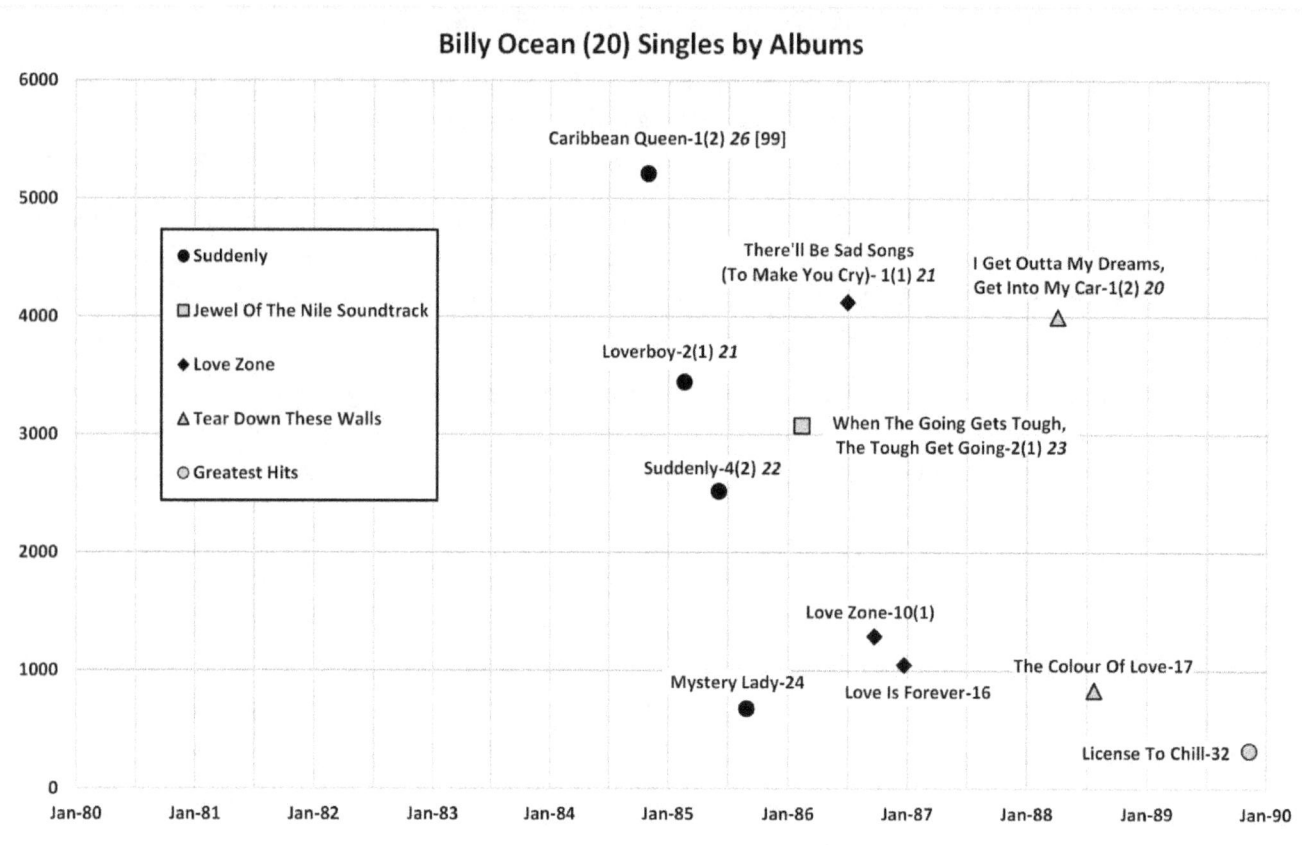

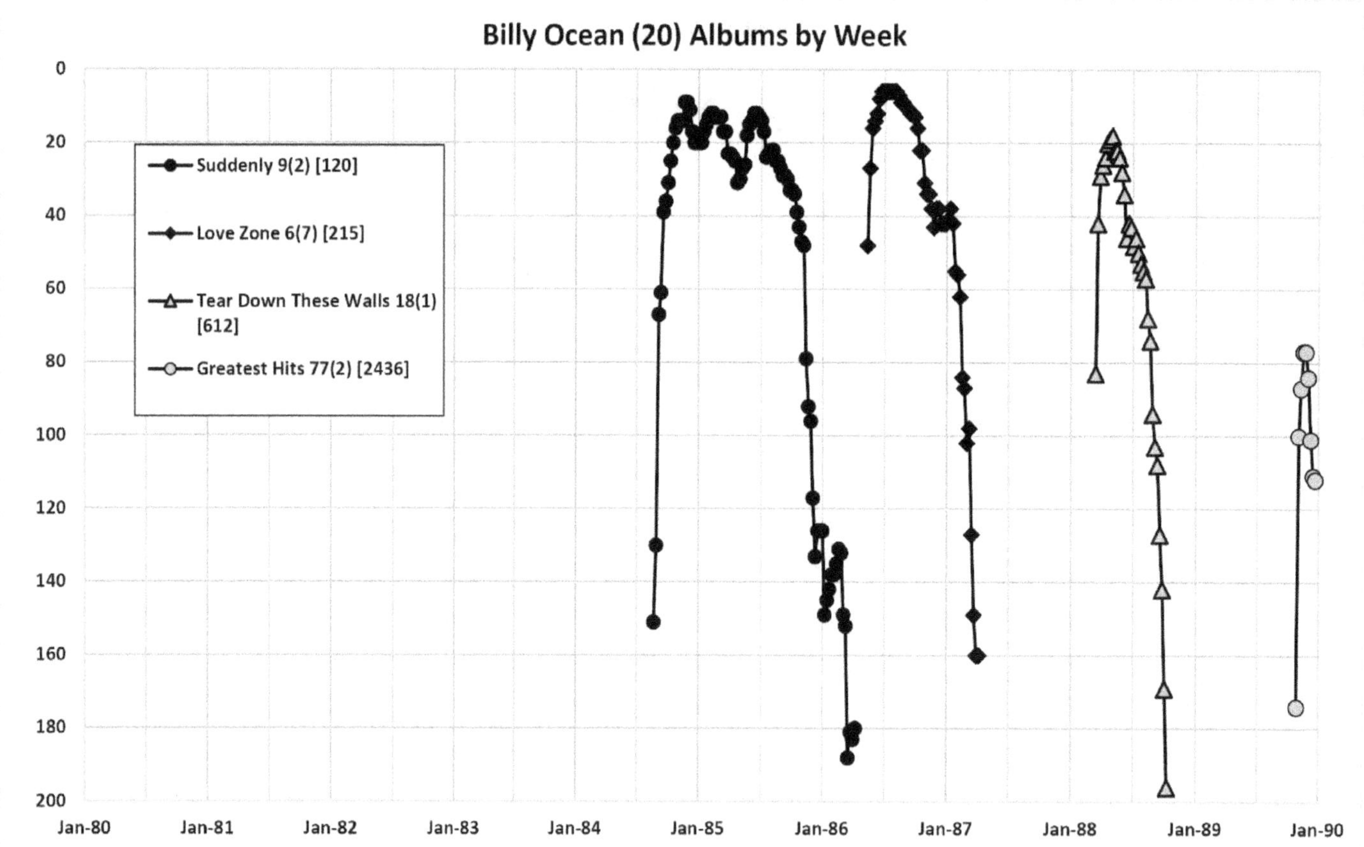

BILLY OCEAN ▶ Singles Rank: 20 (11) 26542; Albums Rank: 75 (5/5) 34370

Like Madonna, Whitney Houston and Cyndi Lauper, Billy Ocean was a phenomenon of the last half of the decade. Also similar to Lauper and Prince and the Revolution, his first charted album *Suddenly* produced multiple singles, and well over a year in longevity, not matched by subsequent albums. Unlike Madonna and Whitney Houston, the charts would not be kind to him in the '90s.

But there was a before-time for Leslie Sebastian Charles, with one Billboard Top 40 hit in 1976, *Love Really Hurts Without You*. He was more successful in the UK during the late '70s and early '80s until reappearing on the US charts in 1984.

His '80s catalog of 11 singles all peaked in the top 40; this included three number 1s (one of which was the number 99 single, *Caribbean Queen*), two number 2s and one number 4. His average score per single was 2418, good for 18th.

His four albums charting in the '80s showed a progression of declining album score, even though the second album, *Love Zone*, had a higher weekly peak (6) than his first and stronger entry, *Suddenly* (9), both of which are 2x Platinum. From each album, the first single release was strongest, and following singles scored lower.

Billy Ocean ranked 76th as an album act, clearly making him a stronger singles act. Both *Suddenly* (43 vs 120) and *Love Zone* (146 vs 215) outperformed when scored by their singles. The four singles from *Suddenly*, with three top 5 peaks, clearly buoyed the chart history of the album, whose decline was softened a bit by the chart success of *When The Going Gets Tough The Tough Get Going* from the *Romancing The Stone* soundtrack.

Top 20 Albums Acts Not Already Covered

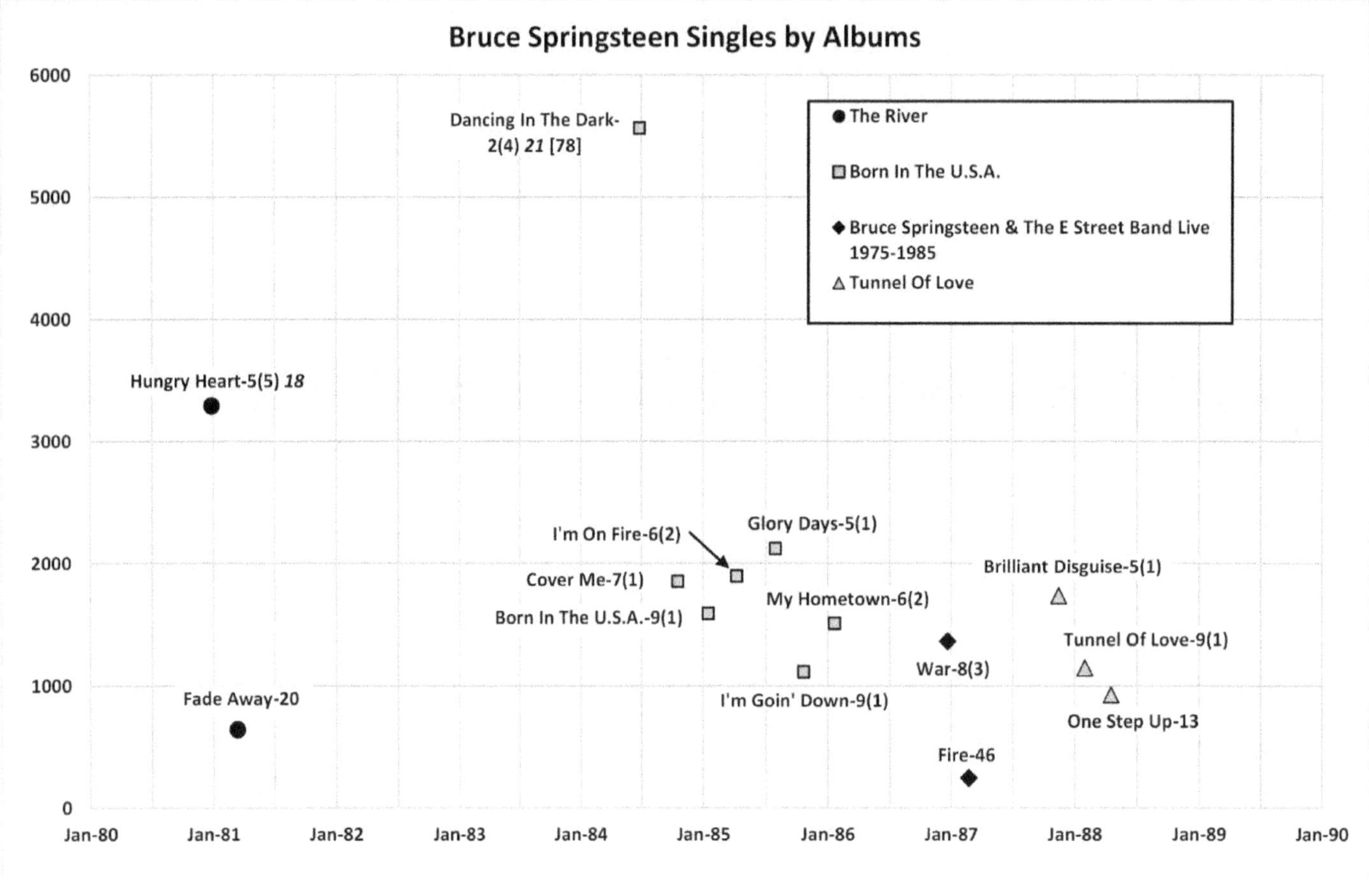
Bruce Springsteen Singles by Albums
6000
5000
4000
3000
2000
1000
0
Jan-80
Jan-81
Jan-82
Jan-83
Jan-84
Jan-85
Jan-86
Jan-87
Jan-88
Jan-89
Jan-90
The River
Born In The U.S.A.
Bruce Springsteen & The E Street Band Live 1975-1985
Tunnel Of Love
Dancing In The Dark-2(4) *21* [78]
Hungry Heart-5(5) *18*
Fade Away-20
I'm On Fire-6(2)
Glory Days-5(1)
Cover Me-7(1)
Born In The U.S.A.-9(1)
My Hometown-6(2)
I'm Goin' Down-9(1)
War-8(3)
Fire-46
Brilliant Disguise-5(1)
Tunnel Of Love-9(1)
One Step Up-13

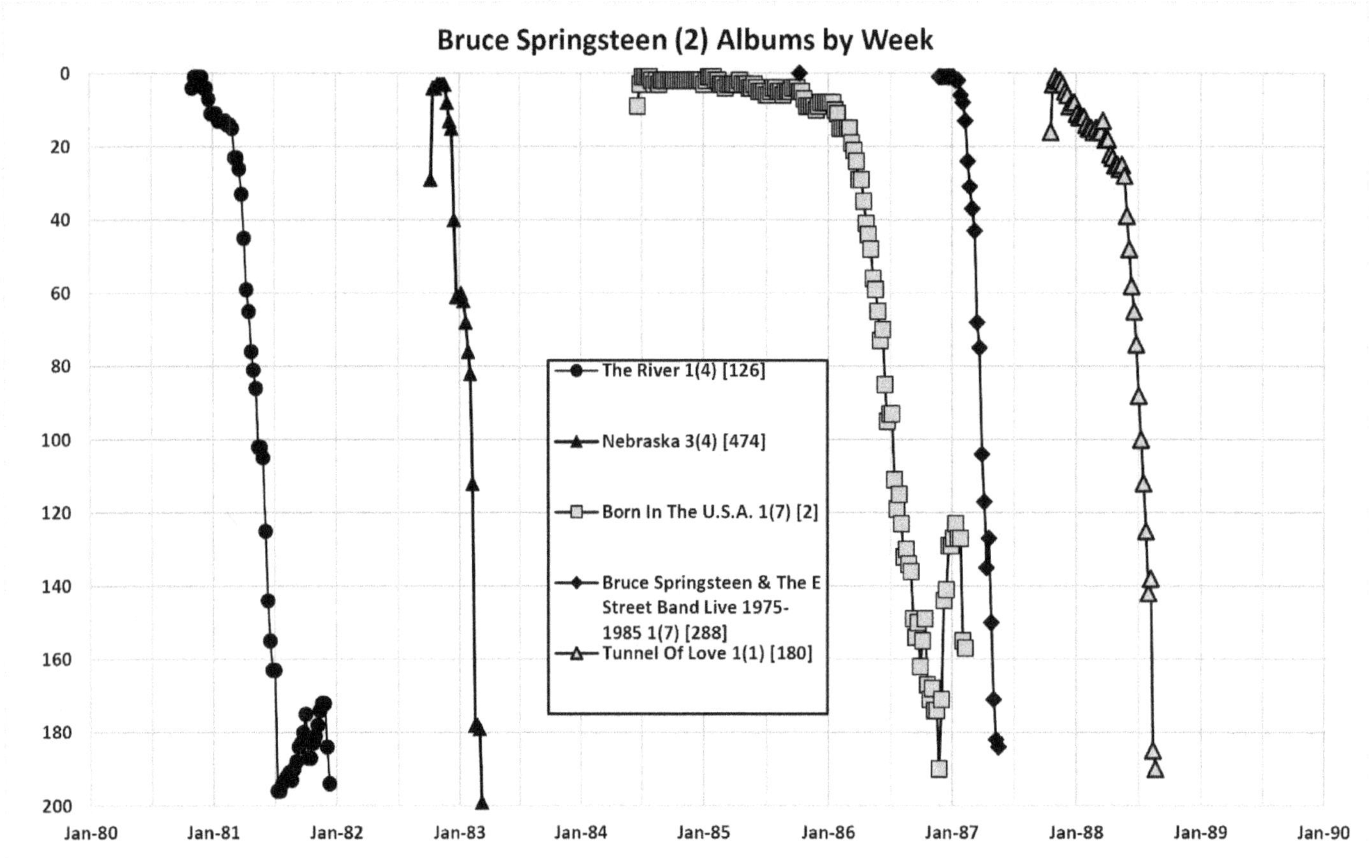
Bruce Springsteen (2) Albums by Week
0
20
40
60
80
100
120
140
160
180
200
Jan-80
Jan-81
Jan-82
Jan-83
Jan-84
Jan-85
Jan-86
Jan-87
Jan-88
Jan-89
Jan-90
The River 1(4) [126]
Nebraska 3(4) [474]
Born In The U.S.A. 1(7) [2]
Bruce Springsteen & The E Street Band Live 1975-1985 1(7) [288]
Tunnel Of Love 1(1) [180]

BRUCE SPRINGSTEEN ▶ Singles Rank: 25 (14) 24999; Albums Rank: 2 (5/9) 107845 95.2%

All five albums released in the '80s by Bruce Springsteen ranked in the top 500—approximately the top 10%--and 11 of his 14 singles made the top 10; however, late in the decade a top 10 peak did not necessarily guarantee a score much above average. His average single scored just under 2000 points.

Of course, there are two real standouts: *Born In The U.S.A.*, the number 2 album of the decade, and its lead single, *Dancing In The Dark*, ranked 78. It's curious that a single ranked so high would not have spent some time at number 1 but its duration of 21 weeks along with four weeks at number 2 combined for its overall strength (Duran Duran's *The Reflex* [106] and Prince's *When Doves Cry* [27] blocked it).

This relative difference in strength of singles and albums distinguishes Springsteen as an '80s act. Four of the five albums peaked at number 1. *Nebraska*, the only one without a charted single, still peaked at a relatively strong number 3.

Born In The U.S.A. is one of four albums in the '80s to yield seven charted singles in the '80s—in fact, all seven were top 10 singles—a feat only matched by Michael Jackson's *Thriller*. *Born In The U.S.A.*'s lifecycle shows continued strength through all of those singles, with only a gradual decline after. Most surprising is its rejuvenation by the release of the *Live* album and the single *War*. By that time it had already been certified 10x Platinum.

As an album, the whole was still greater than the sum of its parts. *Born In The U.S.A.* dropped from 2 to 18 when scored on the basis of its seven top 10 singles.

Four of his previous albums also charted in the '80s, comprising about 5% of his point total.

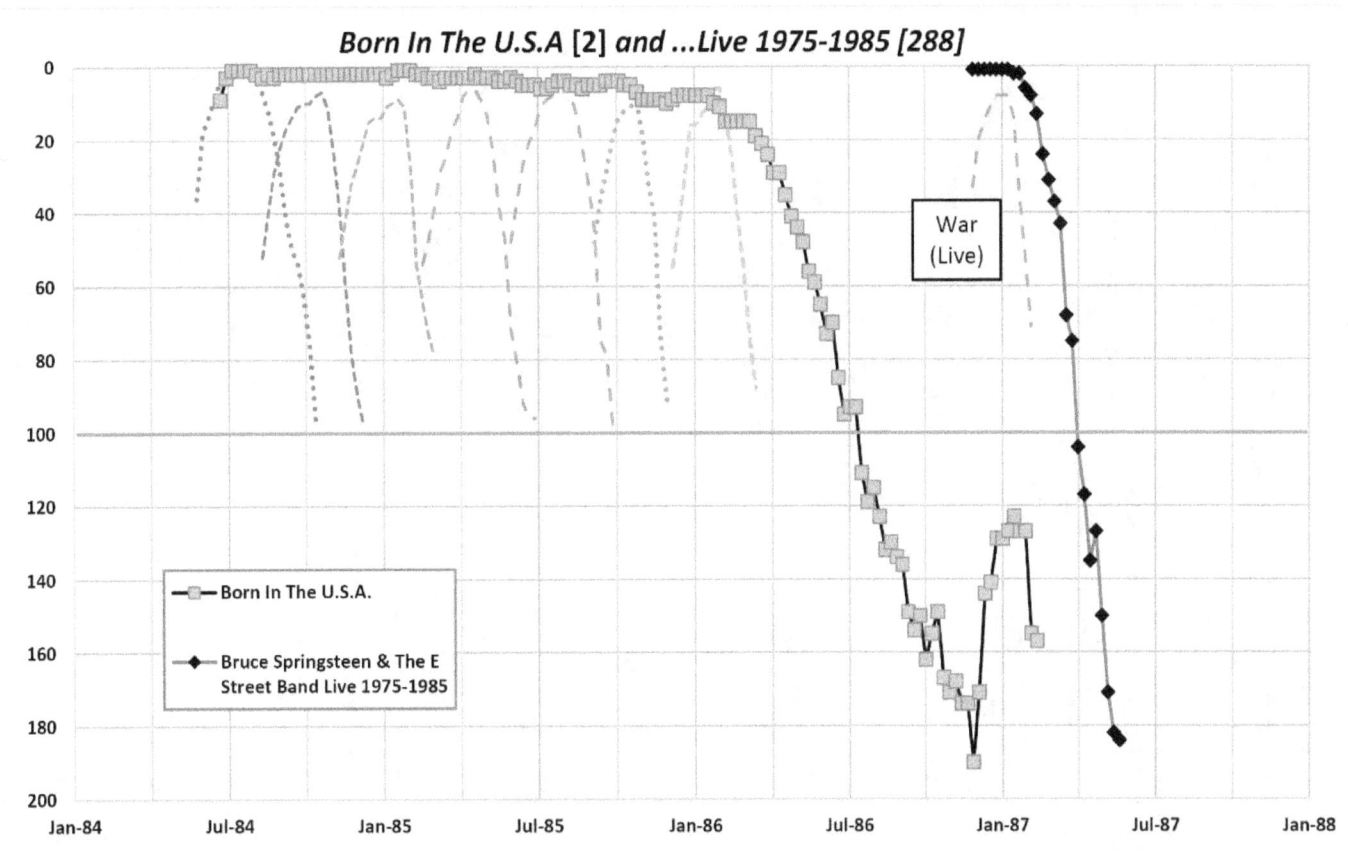

Chronologies

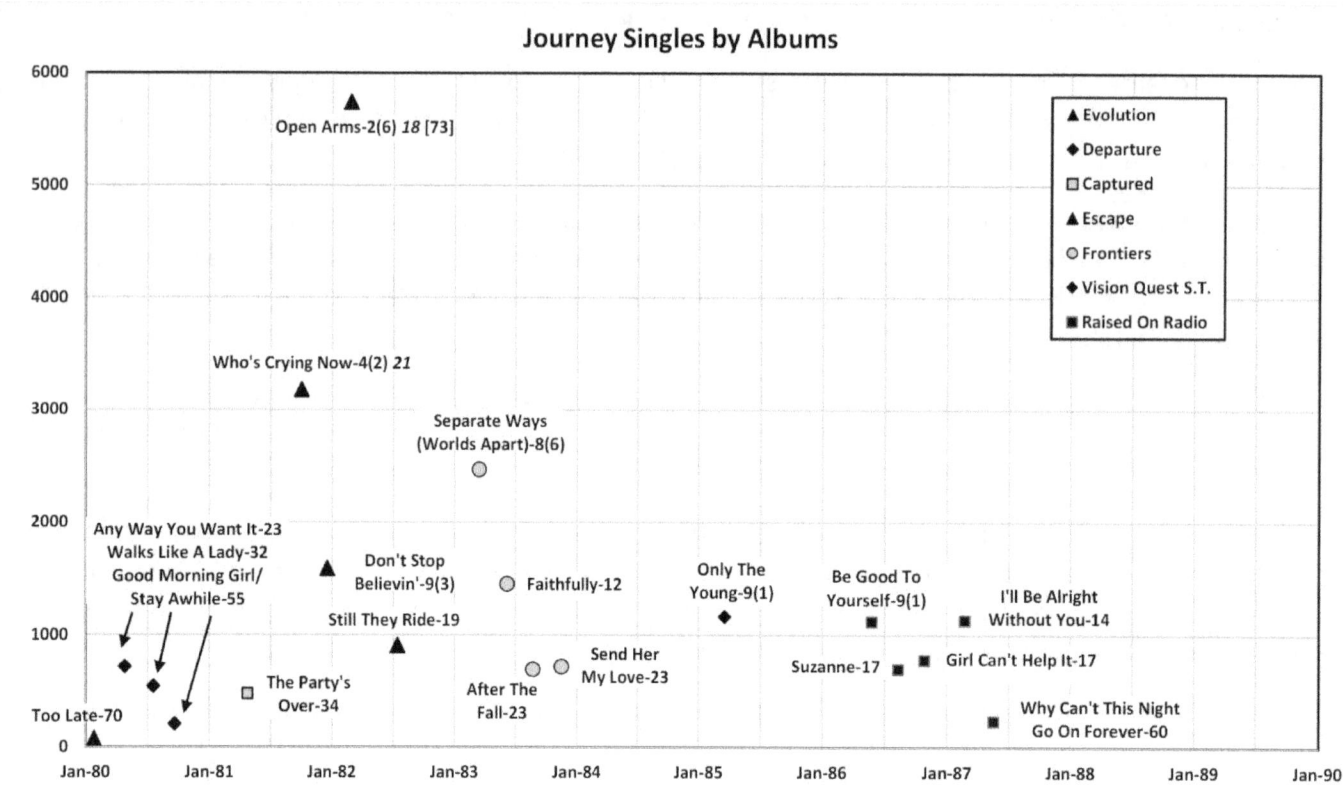

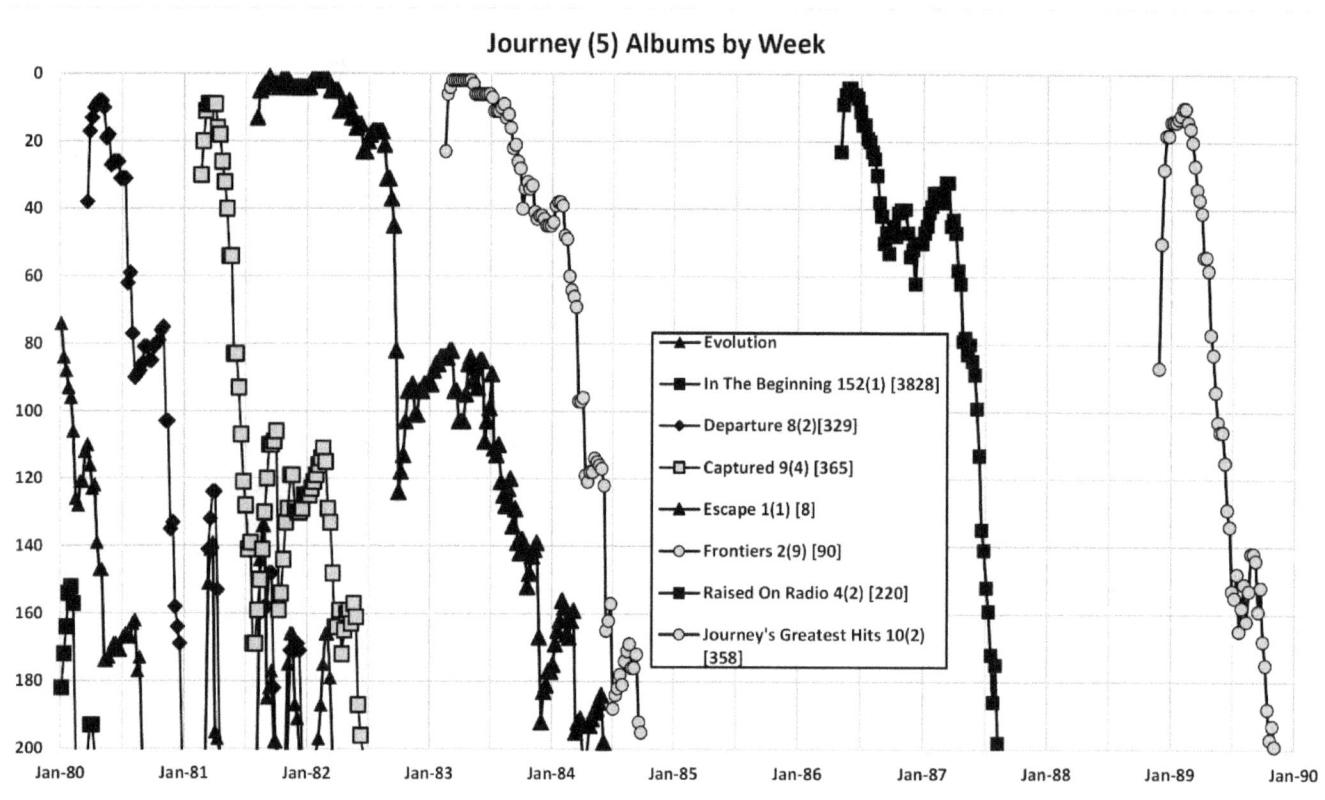

JOURNEY ▶ Singles Rank: 28 (19) 23887; Albums Rank: 5 (7/9) 89016 96.4%

Although the group formed in 1973, it only started to show its potential after the addition of Steve Perry in 1977 and Jonathan Cain in 1981, with the release of its two highest charting albums, *Escape* and *Frontiers*. As the album chart shows, however, the band was on pause between *Frontiers* and *Raised On Radio*, then a more lengthy hiatus after the latter album. There were numerous personnel changes along the way.

From *Departure* on, all their albums peaked in the top 10, and ranked in the top 10% of '80s albums, led by *Escape*, number 8 for the decade. On the other hand, only 6 of their 19 singles made the weekly top 10.

Escape had the strongest charting singles with two top 5s, a top 10 and a top 20. The two top 5s, *Open Arms* and *Who's Crying Now* made the top 10% of singles, with *Open Arms* ranking 73 for the decade. The anthemic *Don't Stop Believin'* was surprisingly less popular than those two. Journey's singles averaged 1257 points, only good for 197th.

When scored by its singles, *Escape* fell from rank 8 as an album to rank 48 when scored by the singles—the whole greater than the sum of its parts. It was certified 6x Platinum by RIAA by the end of 1984.

The Album chart for Journey looks hopelessly complicated when looking at the entire decade. On closer inspection, we find that *Evolution*, *Departure* and *Captured* were all dragged back to the charts by the success of *Open Arms* and *Escape*. This seems common among album acts; earlier albums seem to get a second look when a later album has success.

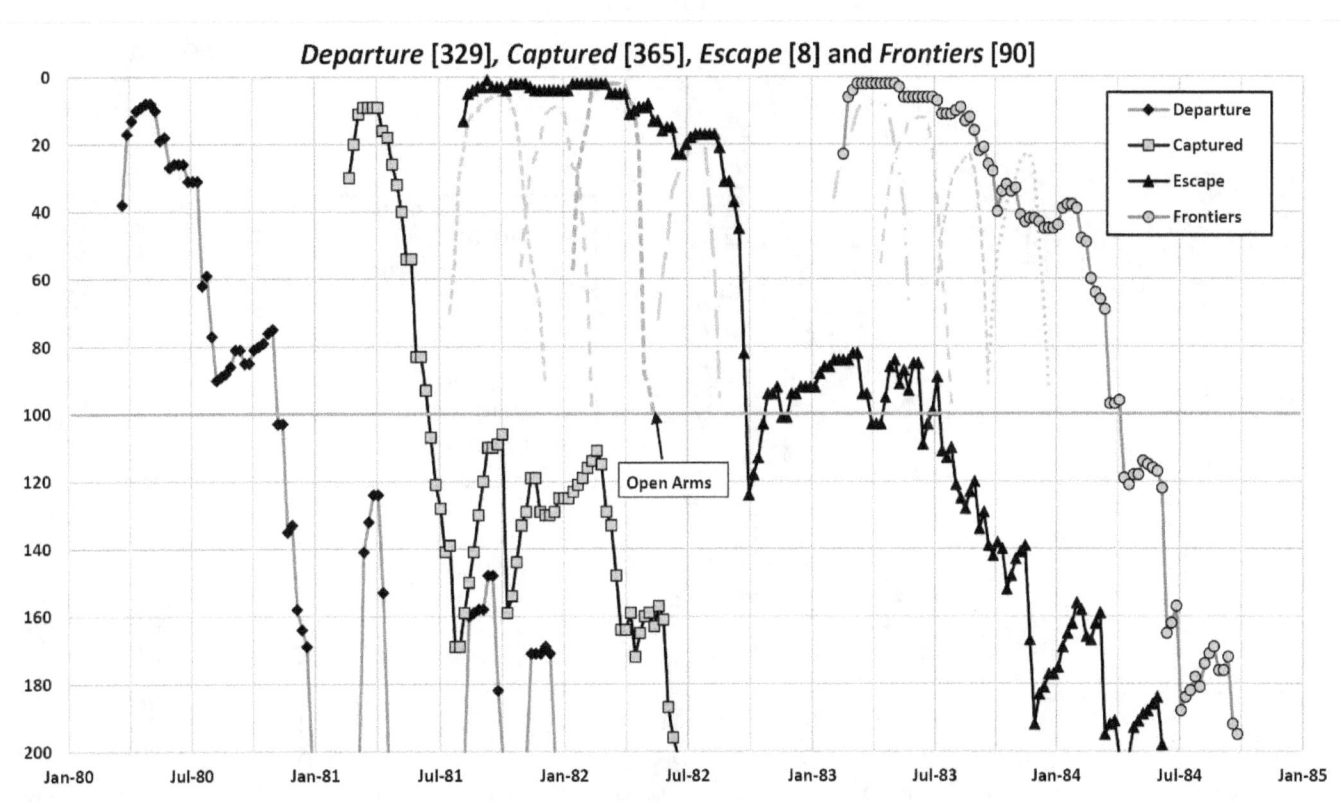

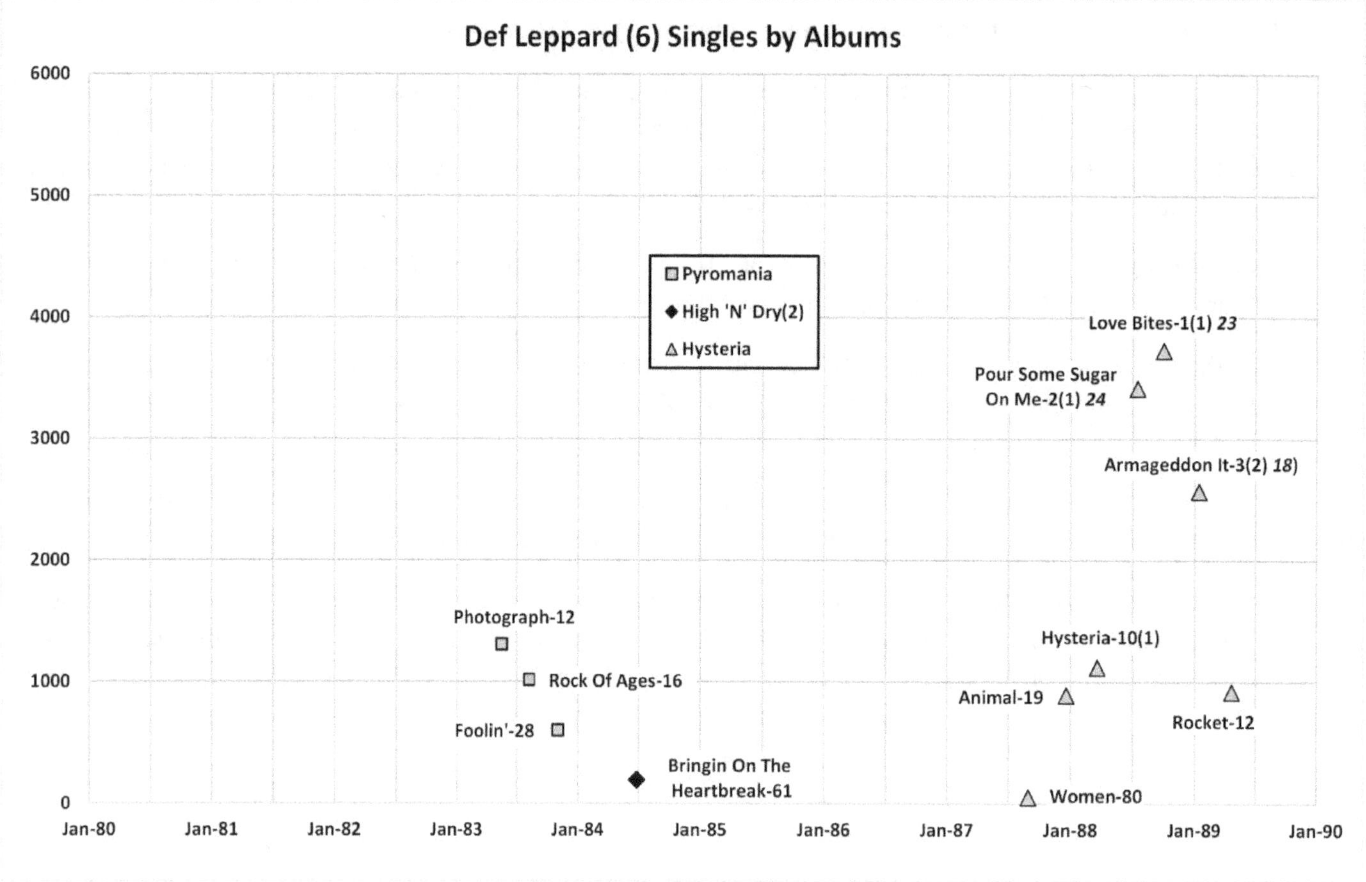

Chronologies

Def Leppard (6) Singles by Albums

Legend:
- □ Pyromania
- ◆ High 'N' Dry(2)
- △ Hysteria

Data points:
- Love Bites-1(1) *23*
- Pour Some Sugar On Me-2(1) *24*
- Armageddon It-3(2) *18*
- Photograph-12
- Rock Of Ages-16
- Hysteria-10(1)
- Animal-19
- Rocket-12
- Foolin'-28
- Bringin On The Heartbreak-61
- Women-80

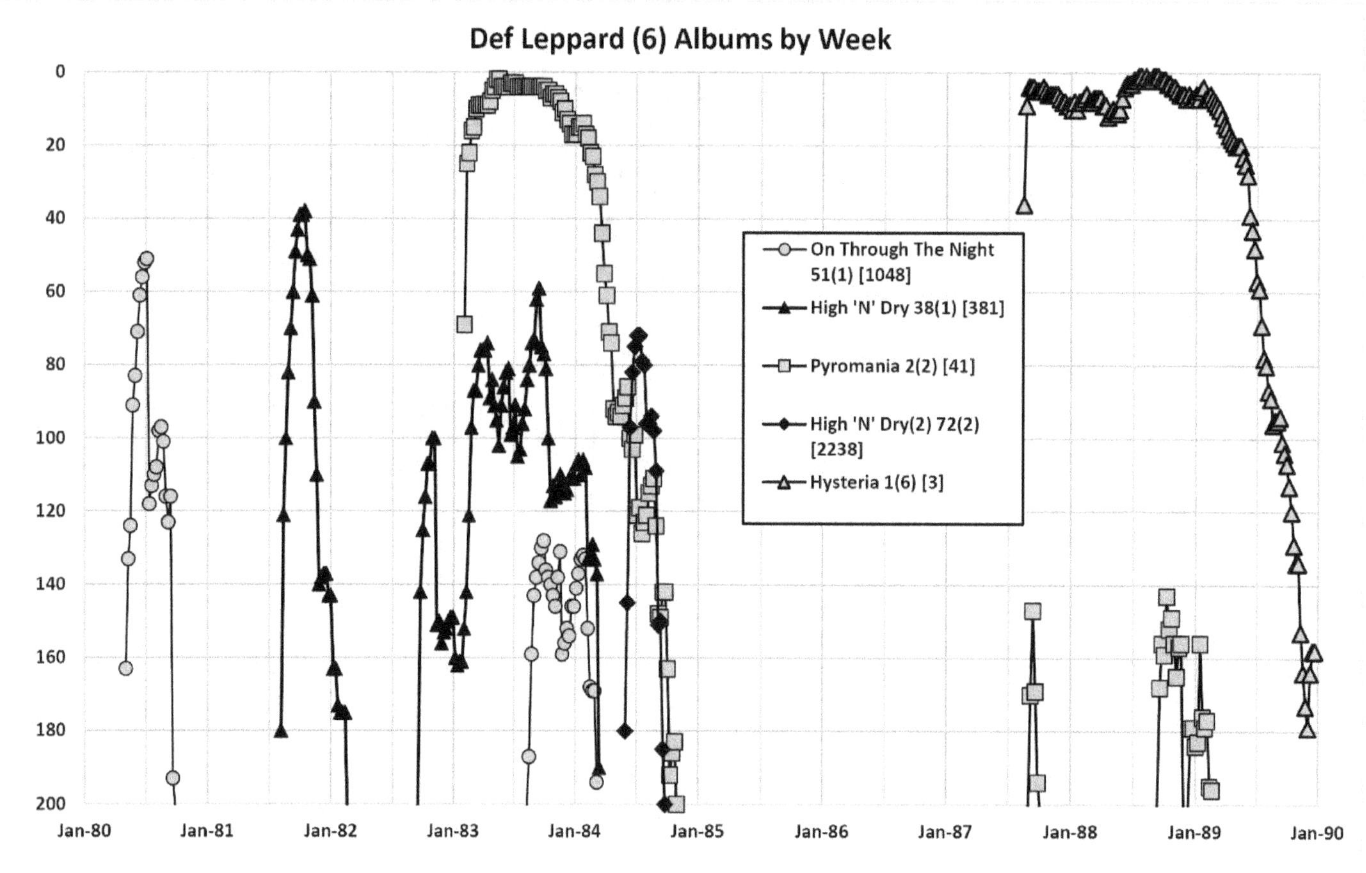

Def Leppard (6) Albums by Week

Legend:
- ─○─ On Through The Night 51(1) [1048]
- ─▲─ High 'N' Dry 38(1) [381]
- ─□─ Pyromania 2(2) [41]
- ─◆─ High 'N' Dry(2) 72(2) [2238]
- ─△─ Hysteria 1(6) [3]

DEF LEPPARD ▶ Singles Rank: 67 (11) 15814; Albums Rank: 6 (5/5) 87705 100%

Def Leppard didn't even chart a single until their third album even though their first two albums had reasonable success. The third album, *Pyromania* charted three singles, two of which made the weekly top 20; but the combined effect of those singles and a top 5 album pulled those two previous albums back to the chart, and even triggered a re-issue of *High 'N' Dry* with two bonus tracks. One of those tracks charted as their fourth single.

Their fifth album, *Hysteria*, charted seven '80s singles, one of only four '80s albums to do so. Of those seven, three were top 5 and one top 10; none placed in the top 100 for the decade.

On the other hand, three of their five albums placed in the top 10% of the decade, and two placed in the top 50. Although *Pyromania* only peaked at number 2 it spent 38 weeks of its 116-week life cycle in the top 10 and finished number 41 for the decade. *Hysteria* held down the number 1 position on the chart for six weeks and 78 weeks in the top 10. It was the number 3 album for the decade. By the end of the '80s, *Pyromania* was certified 7x Platinum, and *Hysteria* 9x Platinum.

It will probably come as no surprise to know that when scored by their singles, *Pyromania* and *Hysteria* fall to numbers 432 and 36 respectively. The whole greater than the parts is a characteristic of album-centric acts.

Just as *Pyromania* dragged *On Through The Night* and *High 'N' Dry* back to the chart, *Hysteria* and its more successful singles had the same effect on *Pyromania*.

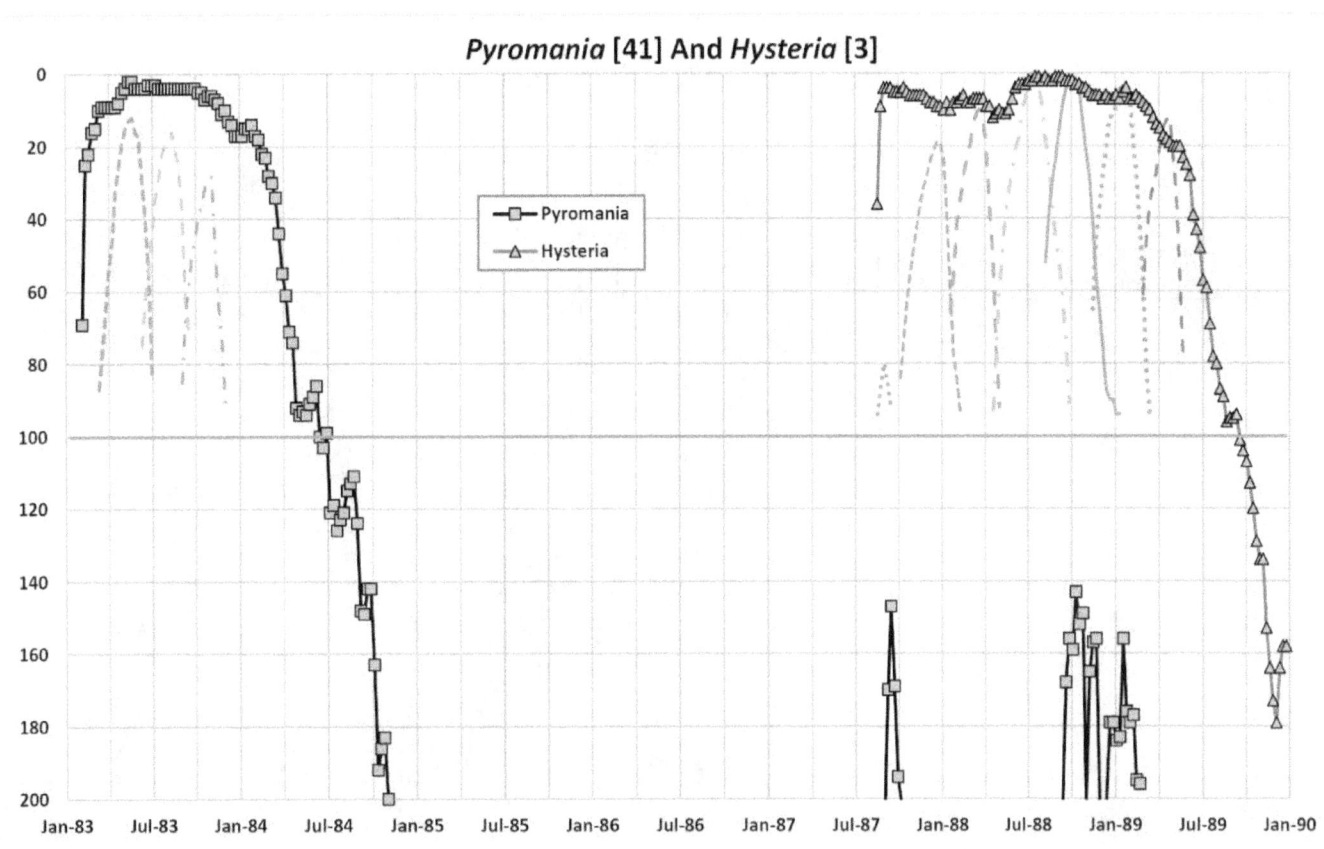

Chronologies

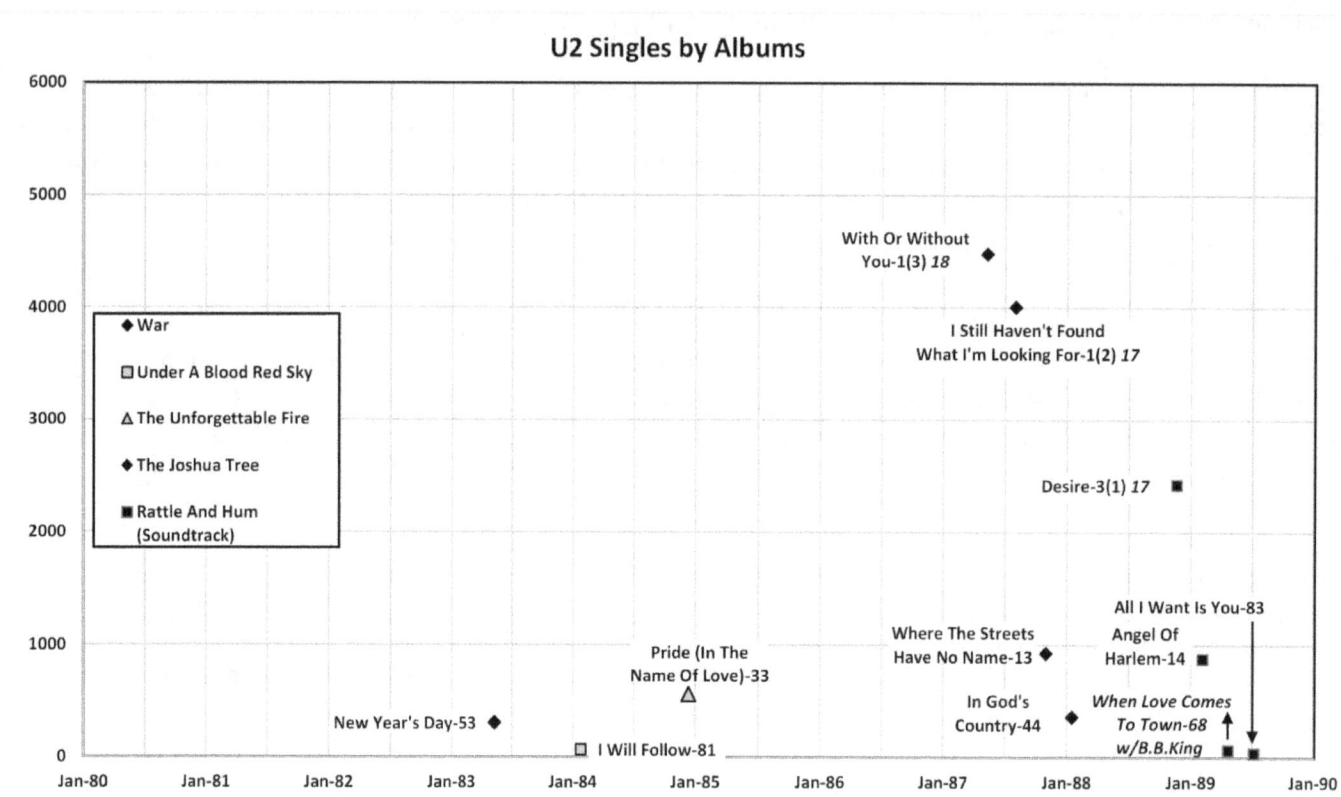

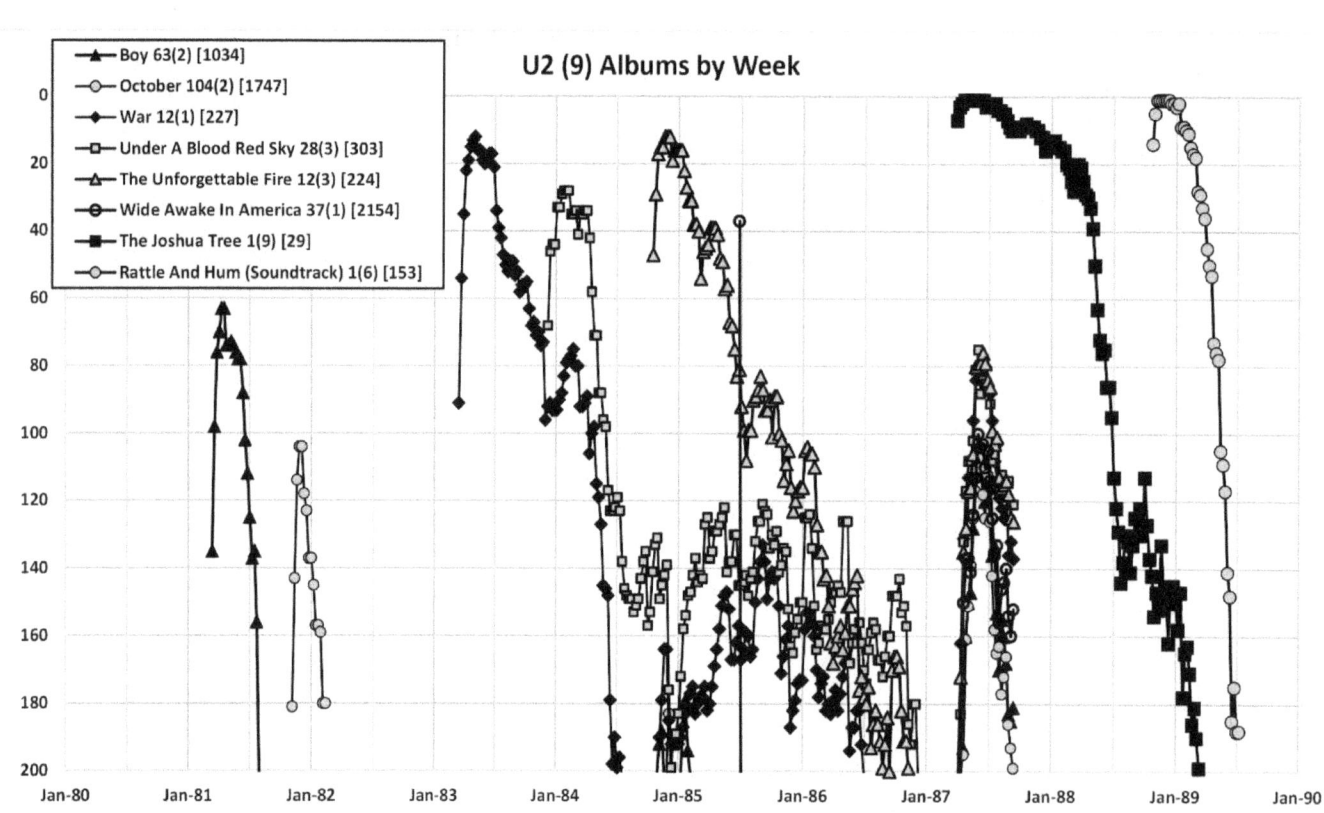

U2 ▶ Singles Rank: 80 (11) 14112; Albums: 9 (8/8) 81520 100%

U2's chart record with its eleven '80s singles points to good success for a very small number and mediocrity for the rest. Of the four singles from Joshua Tree, two went to number 1; one single from Rattle And Hum peaked at number 3, and none of the other eight made the top 10. This explains why a legendary band like U2 is only the number 80 singles act for the decade.

Their chart record with albums is much more substantial. Their first two albums, *Boy* and *October*, were at best modest successes in their first trip to the chart. U2's trajectory upward began with *War*; while only peaking at number 12, it spent 179 weeks on the chart and yielded the single *New Year's Day*. *Under A Blood Red Sky* was a bit less successful, but still ranked in the top 10% of albums in the '80s. *The Unforgettable Fire* became its most successful record to date and spawned its most successful single to date, *Pride (In The Name Of Love)*.

Perhaps the strangest chart history for a record of this time is that of the EP *Wide Awake In America* (which charted with the albums). When it was released, it entered the chart at 37, stayed there one week, and dropped off. Nearly two years later it and the entire earlier catalog was dragged back to the charts for 22 weeks by *The Joshua Tree*.

Stranger yet, that entire earlier catalog of six albums arrived back on the charts within one week of one another in April of '87 and every one of them dropped off on September 19, 1987—one plummeting from 121. There was no explanation in the magazine for this unusual event.

The Joshua Tree was a landmark that peaked at number 1 for nine weeks, 35 weeks in the top 10, and 103 weeks total on the chart. It was certified 5x Platinum by the end of the '80s Its rejuvenation in August 1988 seems to have been due to the release of *Rattle And Hum*, and the single *Desire*. *Rattle And Hum* itself spent six weeks at number 1 but was neither as critically acclaimed nor as successful as *The Joshua Tree*.

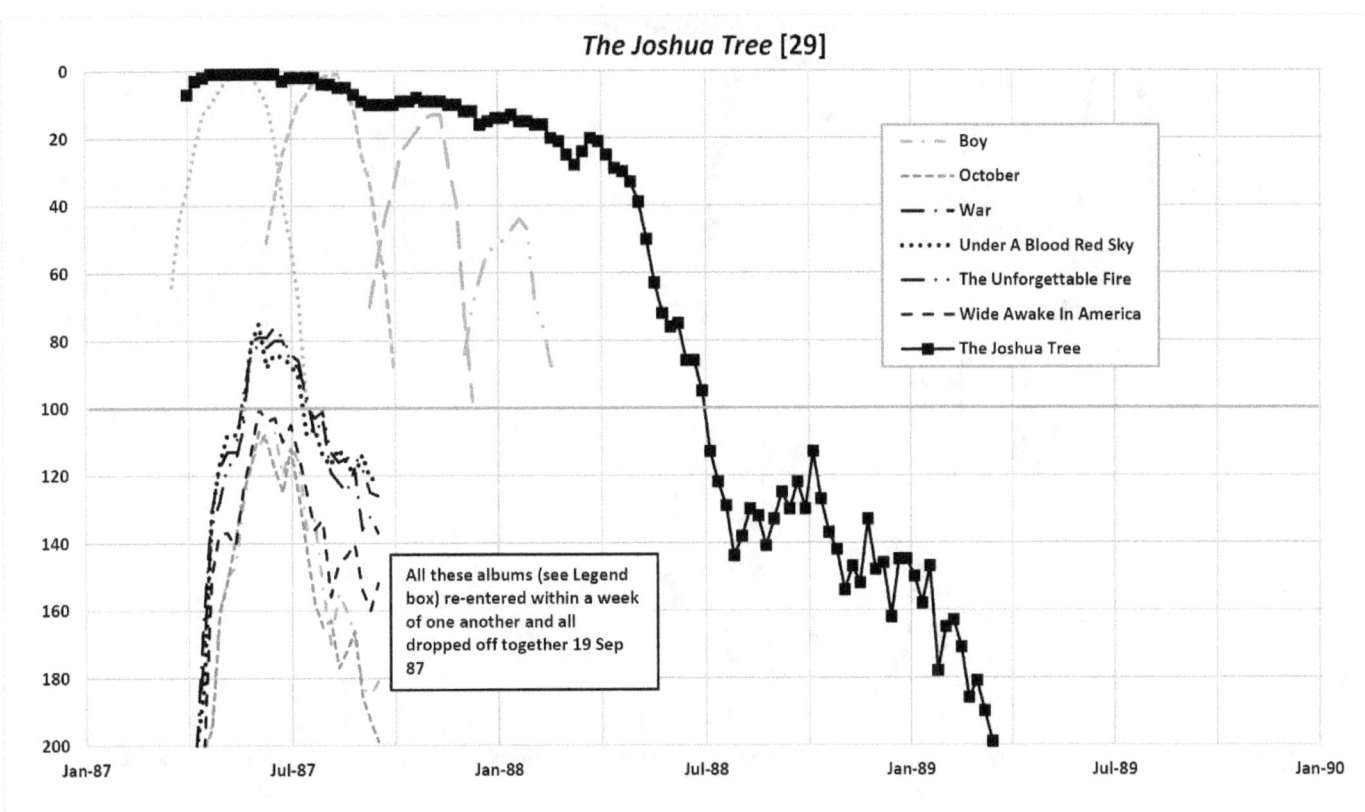

Chronologies

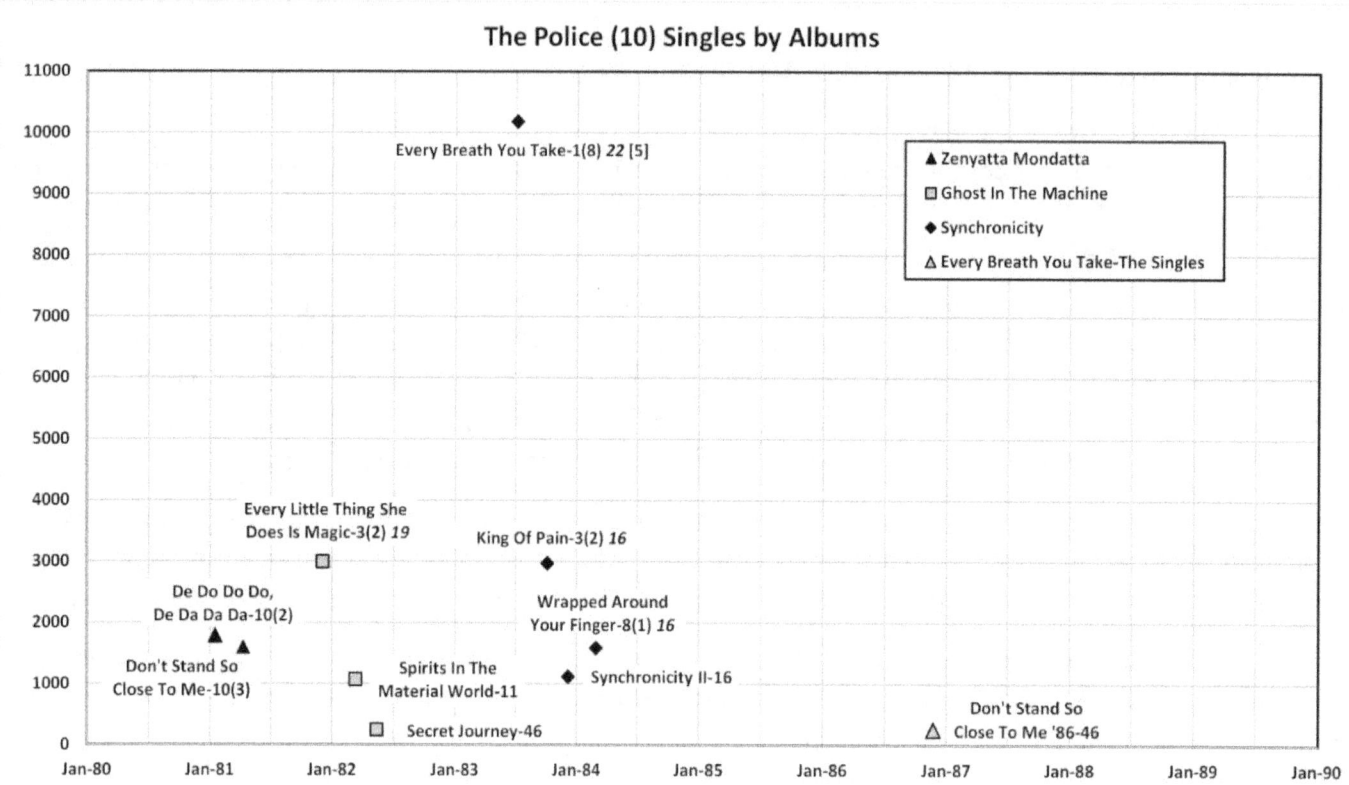

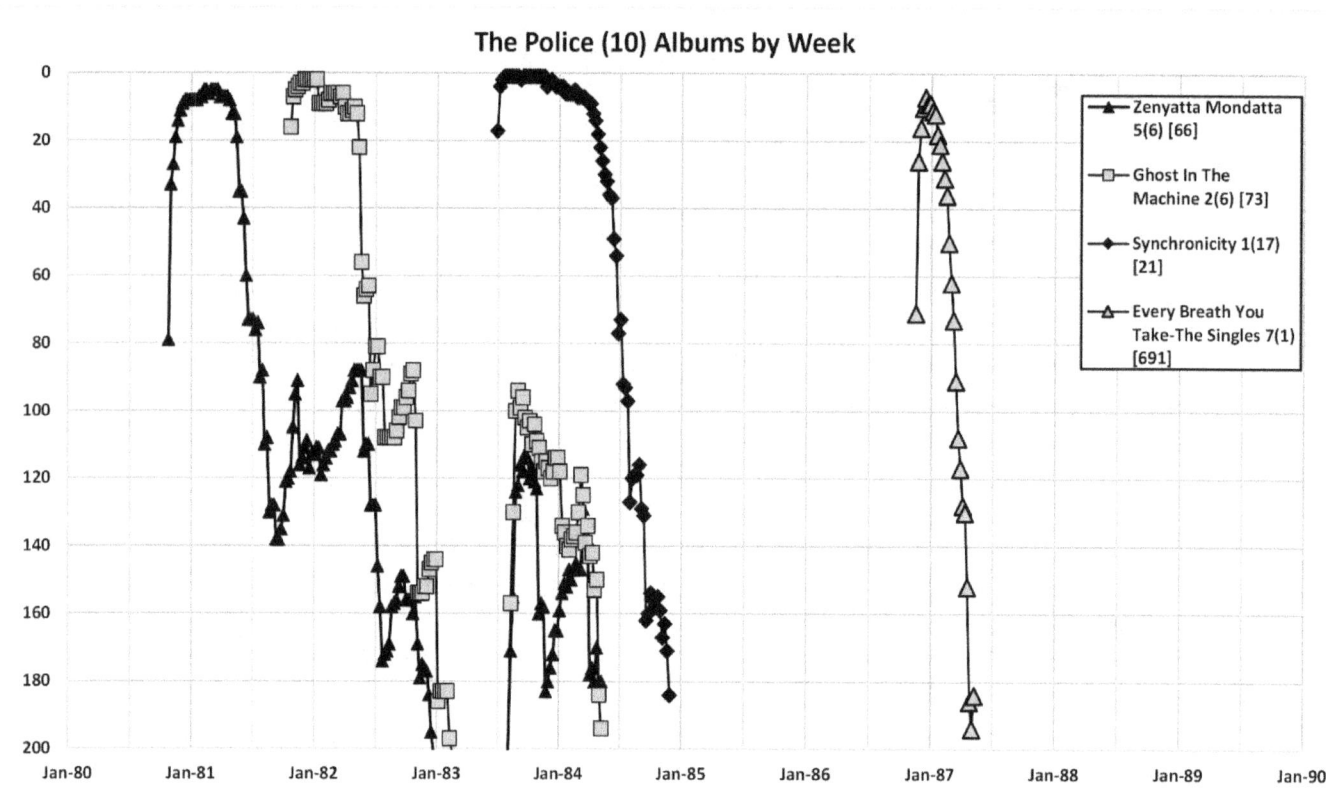

THE POLICE ▶ Singles Rank: 30 (10) 23786; Albums Rank: 10 (4/6) 79361 94.0%

The singles chart history for the Police is somewhat remarkable: it effectively only happened in the first third of the decade, as their '70s entries didn't crack the top 30 and their recording career for new material ended in 1983 with the album *Synchronicity*. During the '80s, they charted ten singles; six made the top 10 and one more peaked at number 11. It's important to notice that the point scale is expanded to show *Every Breath You Take*, the number 5 single of the decade, which logged seven weeks at number 1. Thus, their other top 5s—Every Little Thing She Does Is Magic, and King Of Pain—seem less consequential due to the compressed scale; however, they were both among the top 10% of singles for the decade.

With such a record, then, how could they be called an album act? The answer is volume. With only ten singles they simply did not have the brute force to crack the decadal top 20 for singles; on the other hand, their average per single was nearly 2400 points, good for 22nd, and higher than many of the top 20 singles acts.

Their career as an album act was also mainly in the '80s—their two '70s albums also did not make the top 20, but their three '80s albums of original work all ranked in the top 100 of the decade. *Synchronicity* led their '80s catalog with 17 weeks at number 1, and was certified 4x Platinum in 1984 only about a year after its release. *Zenyatta Mondata* and *Ghost In The Machine* were also certified Platinum in the '80s. Extrapolating their chart record to two more original albums with three singles apiece would place them securely in the top 10 singles acts and challenging for the top album spot.

Synchronicity and its singles make for an interesting chart history. *Every Breath You Take* preceded the album to the charts, but took it to number 1 in four weeks. Note its broad top comprising seven weeks at number 1. Additionally, *Synchronicity* dragged the two previous albums back to the chart for about a year each. Finally, its decline was quite gradual, taking nearly a year to leave the chart after leaving the top 40. Scored by singles, its rank rose from 21 to 14.

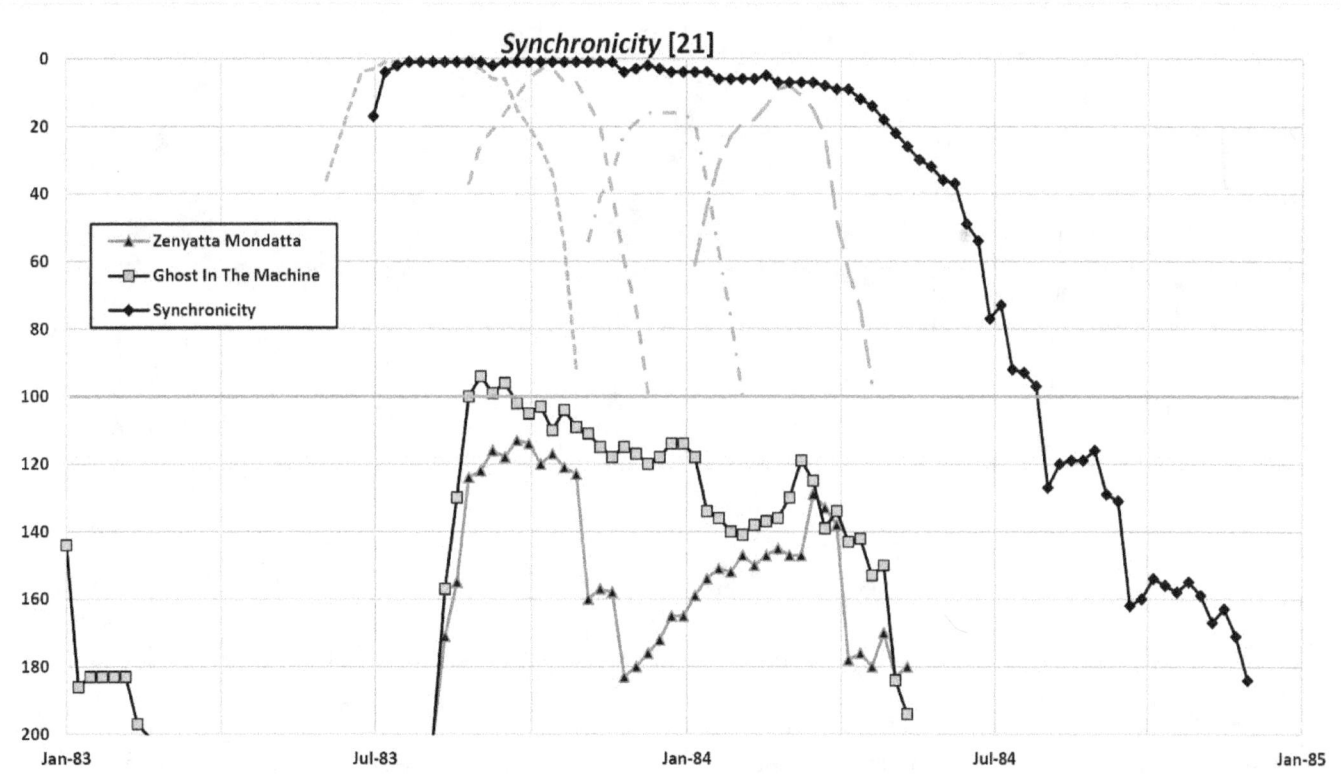

Chronologies

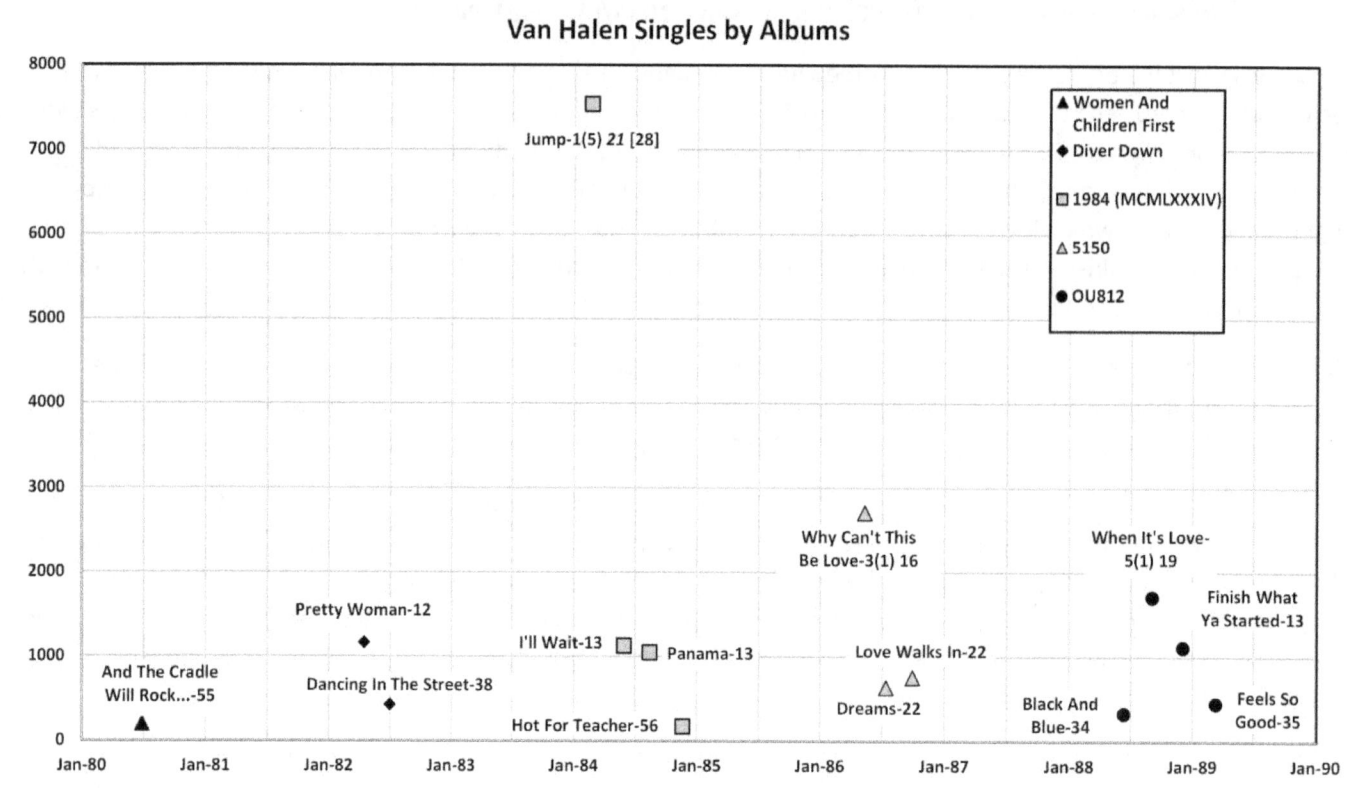

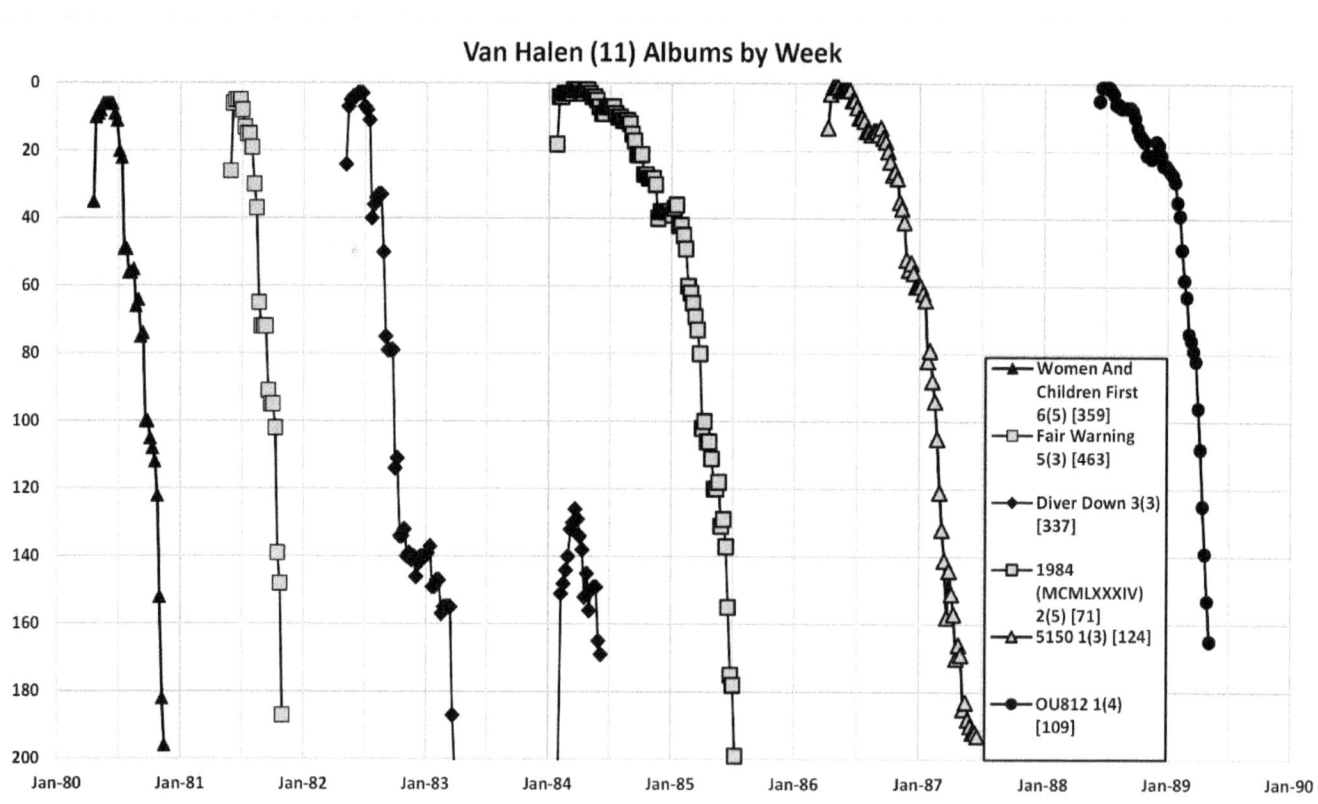

VAN HALEN ▶ Singles Rank: 50 (14) 19350; Albums Rank: 11 (6/8) 78154 96.7%

Van Halen's singles chart record is reminiscent of other album-centric acts: *Jump*, which spent 5 weeks at number 1 and was the number 28 single of the decade leads the pack, and like *Every Breath You Take*, has forced an expanded points scale. Beyond that, two top 5's and a number of reasonable but not remarkable chart performances adds up to finishing as the 50th ranked singles act.

Album-wise, it's a different story. Five of the six charted albums finished in the top 10%, three of which were in the top 125 for the decade, peaking at either number 1 or 2 for multiple weeks, with duration of a year to a year and a half. Curiously, in only one case did two of their albums appear on the chart simultaneously. The success of *1984* dragged *Diver Down* back to the chart for an additional four months.

1984 was certified 6x Platinum by the end of the '80s. It charted four singles: a big number 1, two number 13s and a number 56. These don't seem strong enough to hold an album on the charts by themselves, and in fact, the album was in decline, but that decline was quite gradual, showing its strength. Scored by its singles, *1984* ranked 73—nearly exactly where it ranked as an album—but the singles score was driven by that one skyrocket, *Jump*.

Synchronicity and *1984* had very similar life cycles. Both peaked at number 1, both yielded four singles, one of which was extremely successful, both had about 75 weeks total on the chart and both drew previous albums back on. *Synchronicity* ranks higher because it had more weeks at higher ranks in the top 40, even though *1984* outperformed in the last six months of gradually declining rank.

Between *1984* and *5150*, Van Halen changed its front man from David Lee Roth to Sammy Hagar. Although *5150* and *OU812* did not score quite as well as *1984*, both of the latter two charted multiple weeks at number 1. With Hagar, the lineup would remain stable through the first half of the '90s.

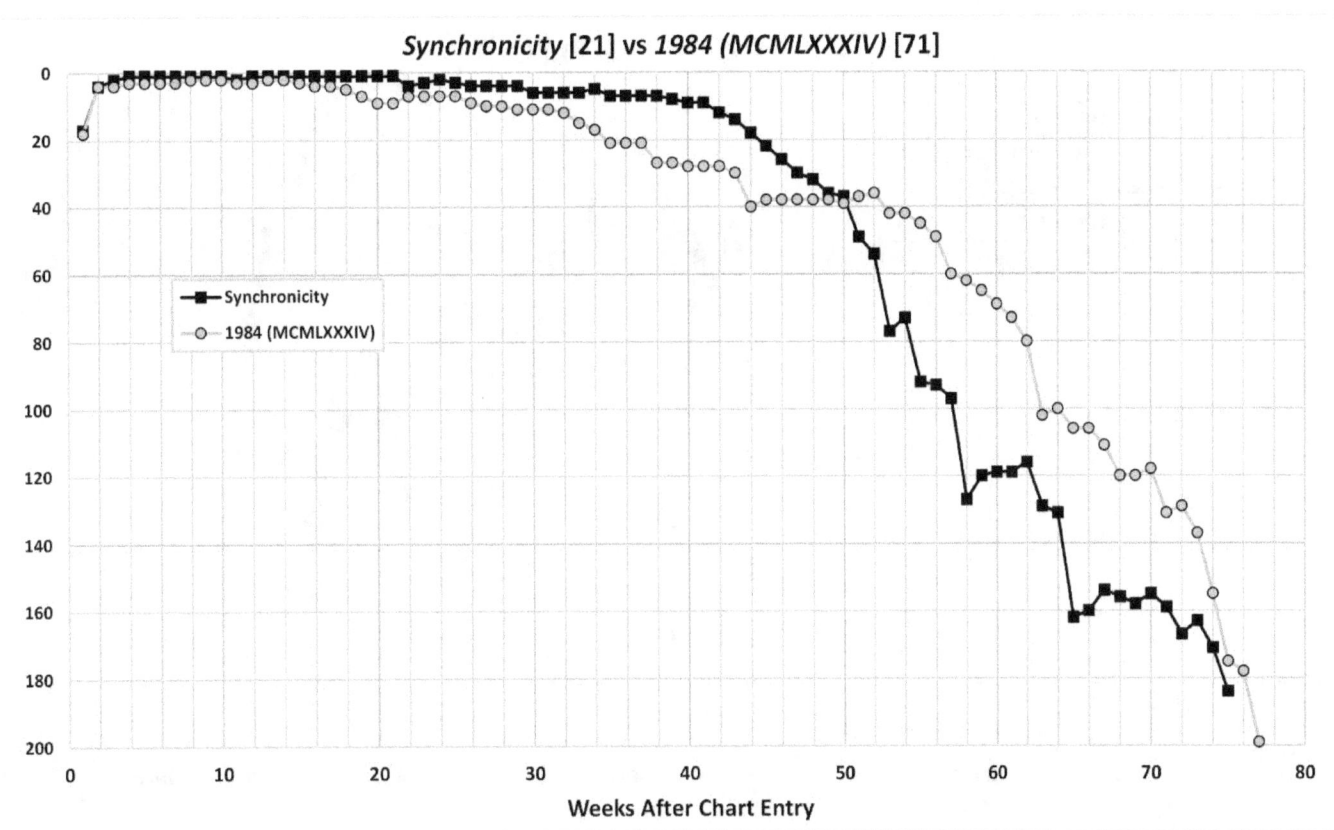

Chronologies

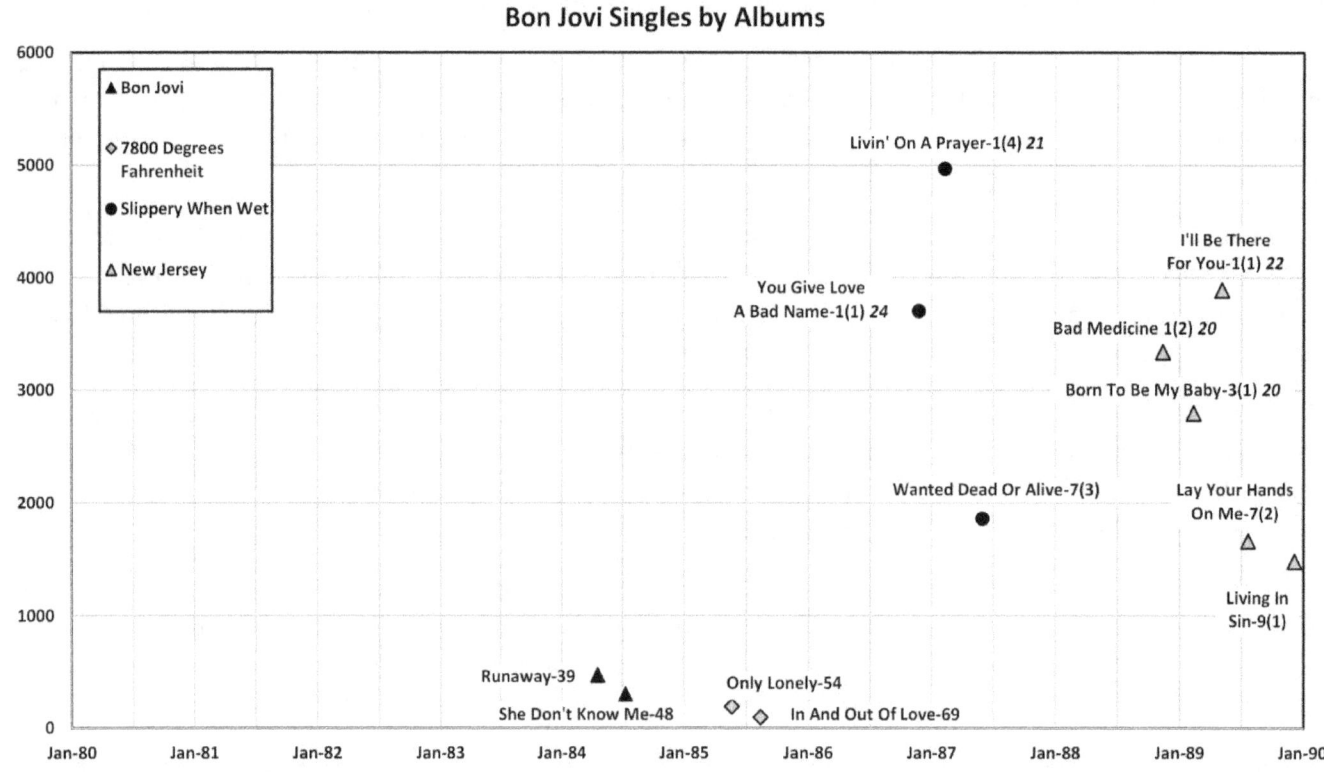

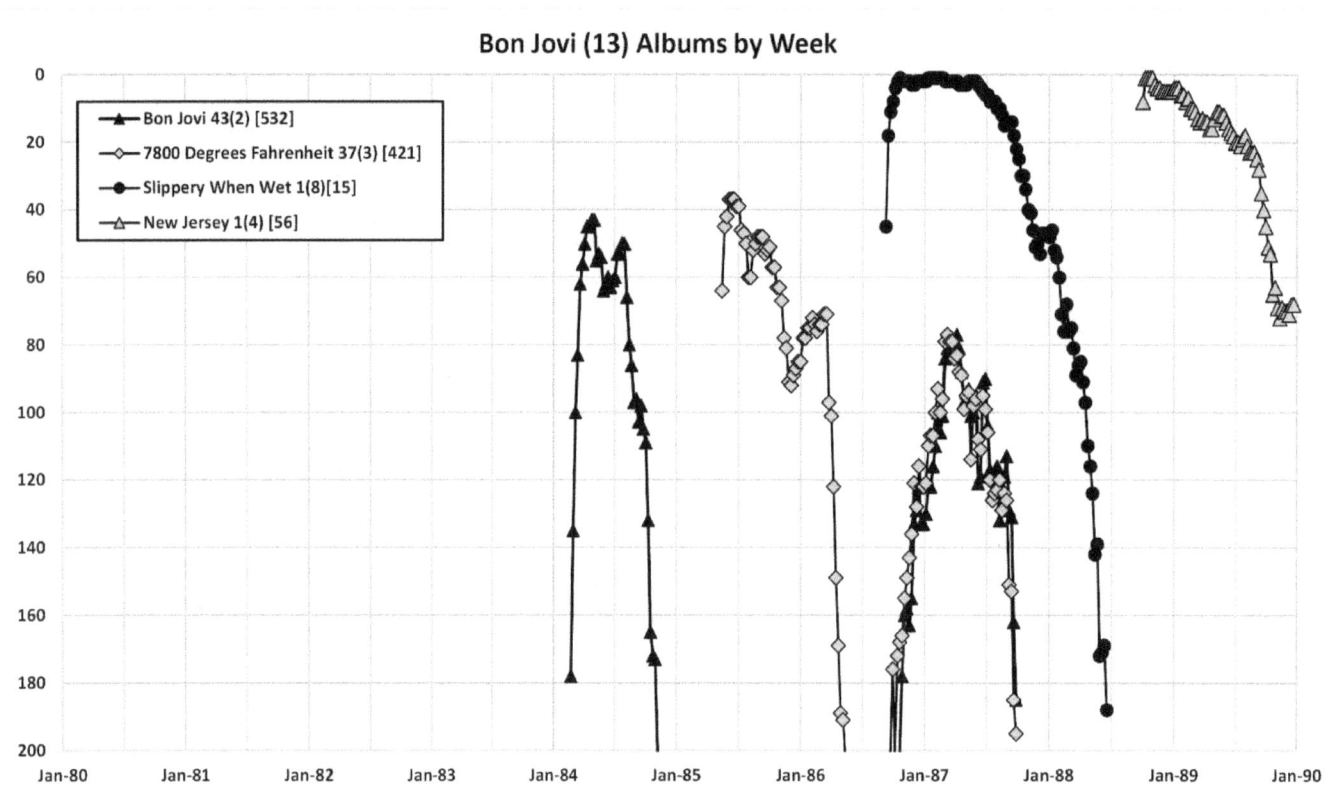

BON JOVI ▶ Singles Rank: 26 (12) 24759; Albums Rank: 13 (4/4) 75820 100%

Bon Jovi and The Police are near complements in the 1980s. The last new Police material would have only overlapped the first Bon Jovi material in 1984. After recording *New Jersey* and touring it, the band took a hiatus until 1992.

Bon Jovi charted 12 singles in the '80s, and the trendline was generally up in terms of number of singles per album and singles score. The first two albums, *Bon Jovi* and *7800 Degrees Fahrenheit*, charted two singles each, all of which peaked in the 40-70 range. After these two albums, personnel changed in a significant way. Desmond Child became a writing collaborator and Bruce Fairbairn produced. *Slippery When Wet* took three months to record, and the effort paid huge dividends. Three singles charted from this album: two went to number 1 and the third peaked at number 7.

The fourth album, *New Jersey*, charted five singles: two went to number 1 and the other three made the weekly top 10, although none stayed at the top for long enough to make the top 100 singles of the decade.

The first two albums were modest successes themselves, peaking around number 40, but *Slippery When Wet* was a huge increment upward. It spent eight weeks at number 1 and 94 weeks on the chart, ending as the number 15 album of the decade. It had sufficient resurrection power to drag both *Bon Jovi* and *7800 Degrees Fahrenheit* into the top 100 simultaneously, giving them each nearly a year's second chart life. Additionally, it maintained its own chart presence after the final single, declining only gradually.

New Jersey's third, fourth and fifth singles outran the album's popularity at that time, and its chart rank dropped rather quickly after the fifth single. The importance of these two observations can be seen when scoring these albums by their singles, *Slippery When Wet* dropped from 15 to 59 when scored by singles; *New Jersey* rose from 56 to 29 when scored by singles.

* New Jersey and Living In Sin both peaked in the '80s and were scored in full despite a few weeks carryover into 1990.

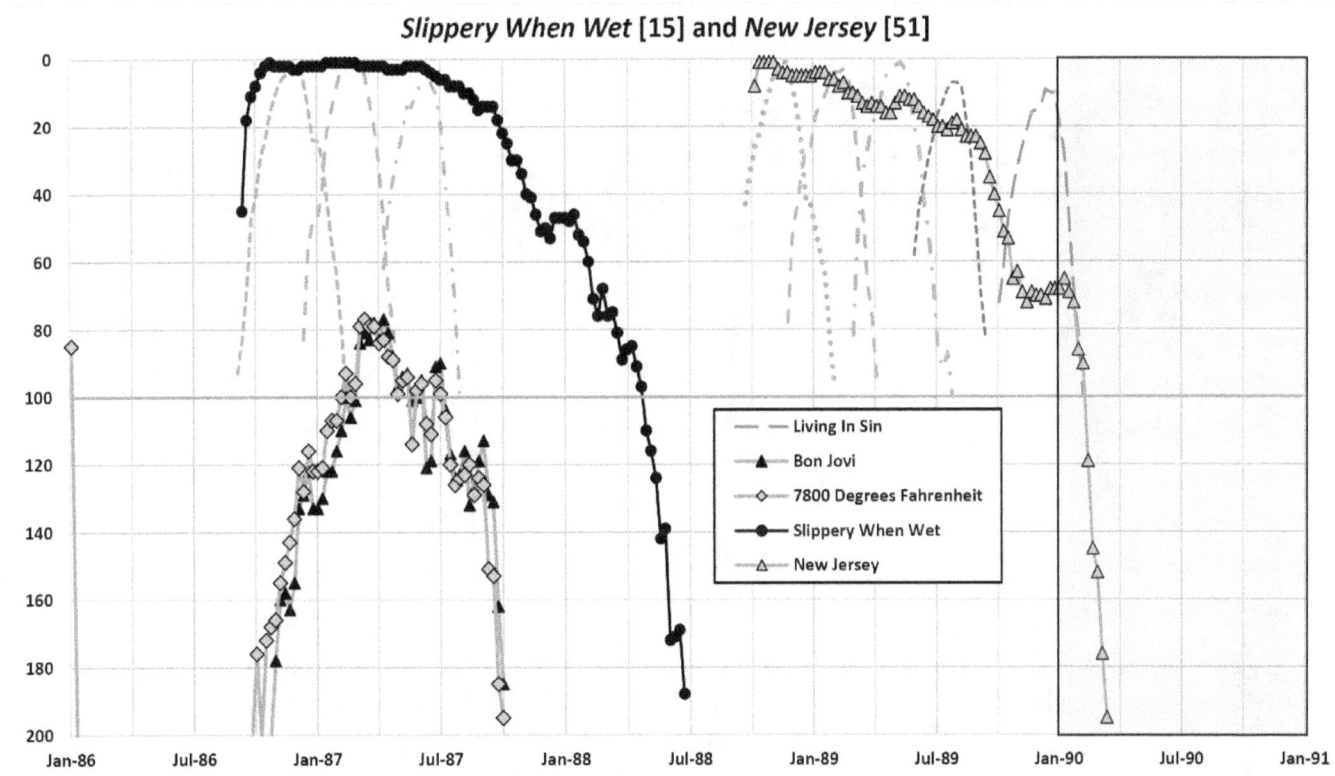

Slippery When Wet [15] and *New Jersey* [51]

Chronologies

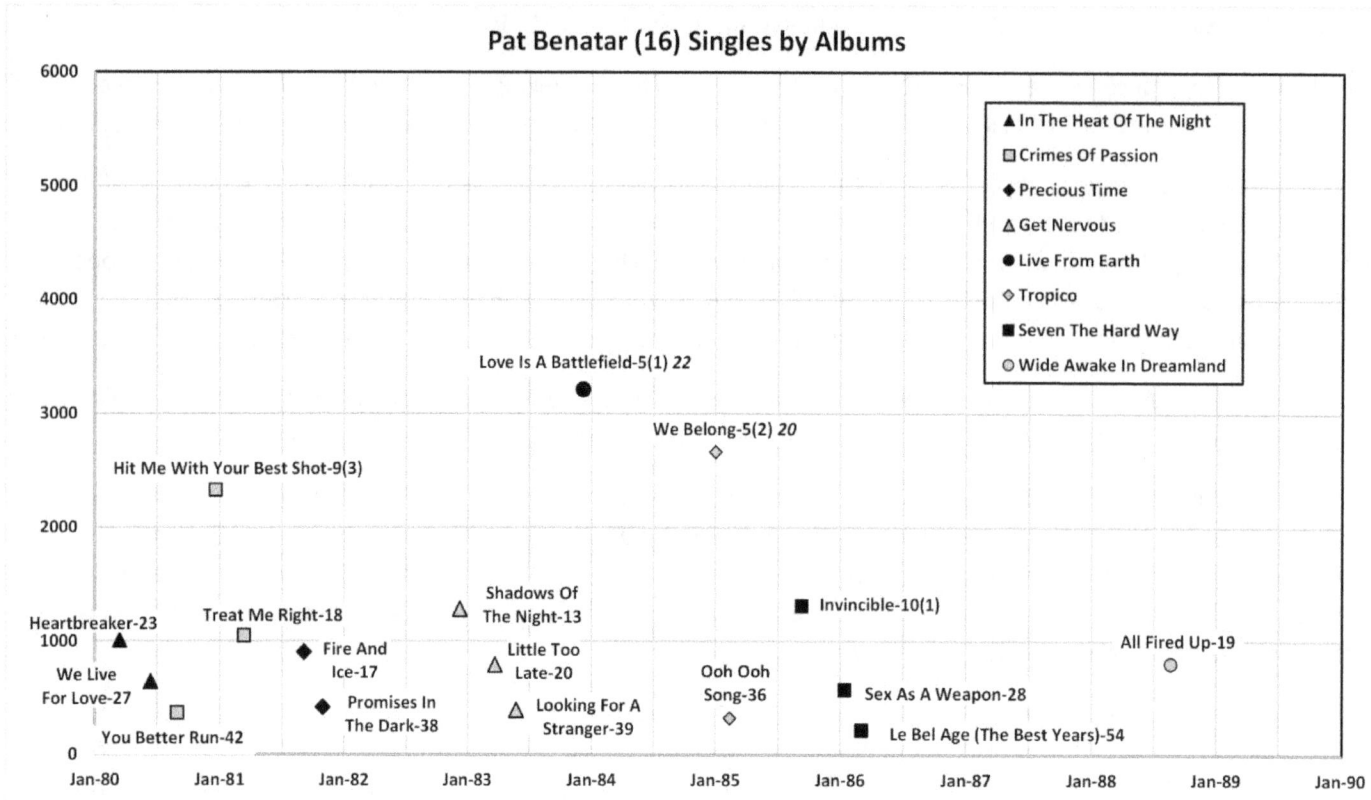

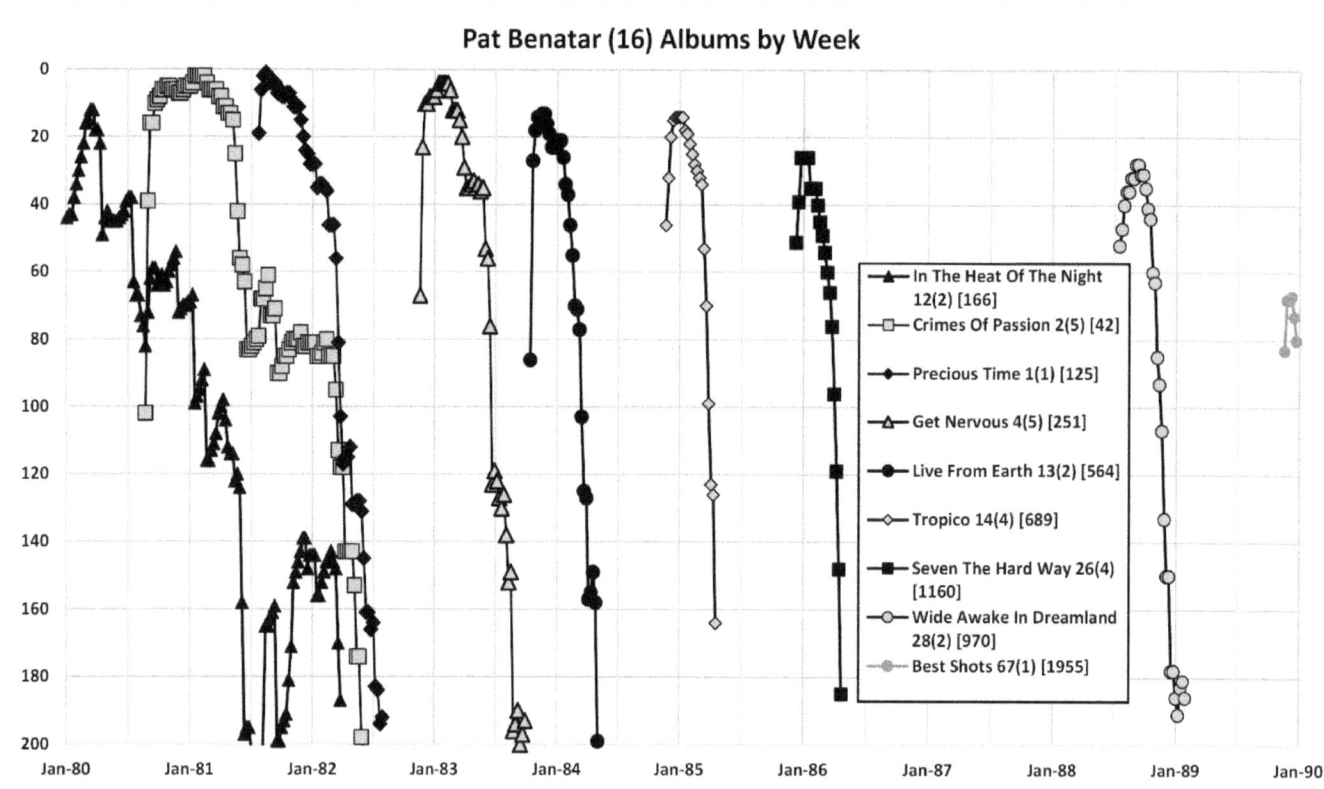

PAT BENATAR ▶ Singles Rank: 56 (17) 18270; Albums Rank: 16 (9/9) 71068 100%

Overall, Pat Benatar charted 17 singles in the '80s, four of which peaked in the top 10, the highest at number 5. Her singles score is as high as it is because of the sheer volume of material. On average her singles scored 1075, only good for 133rd, and not much different than an average record. However, the high points of her singles catalog are anthemic: *Hit Me With Your Best Shot* and *Love Is A Battlefield*.

Her album catalog is also large with nine albums charted during the '80s. Her greatest success was in the first part of the decade, during which time her first four albums placed in the top 10%. *Crimes Of Passion,* which peaked at number 2 for five weeks, ranked 42nd for the decade. Still, after the high point, the overall trajectory for her albums was downward from *Crimes Of Passion*.

At one point in 1981 and 1982 her first three albums were on the chart simultaneously, and it is interesting to see how their success and the success of the underlying singles reinforced one another. *In The Heat Of The Night*, driven by *Heartbreaker*, peaked at number 12 and was on a slow trajectory downward when it was briefly refreshed by the entrance of *Crimes Of Passion* and *Hit Me With Your Best Shot*. *Crimes Of Passion* logged 29 weeks in the top 10 and had begun its decline when *Precious Time* was released. *Crimes Of Passion's* decline was slowed, apparently by the popularity of *Precious Time*. All three began their final fade at the same time.

Crimes Of Passion was certified 4x Platinum by the end of 1984. By the same time, *Precious Time* was 2x Platinum. When scored by singles, it is no surprise that number 42 *Crimes of Passion* dropped to 346. Pat Benatar's albums were clearly stronger than her singles.

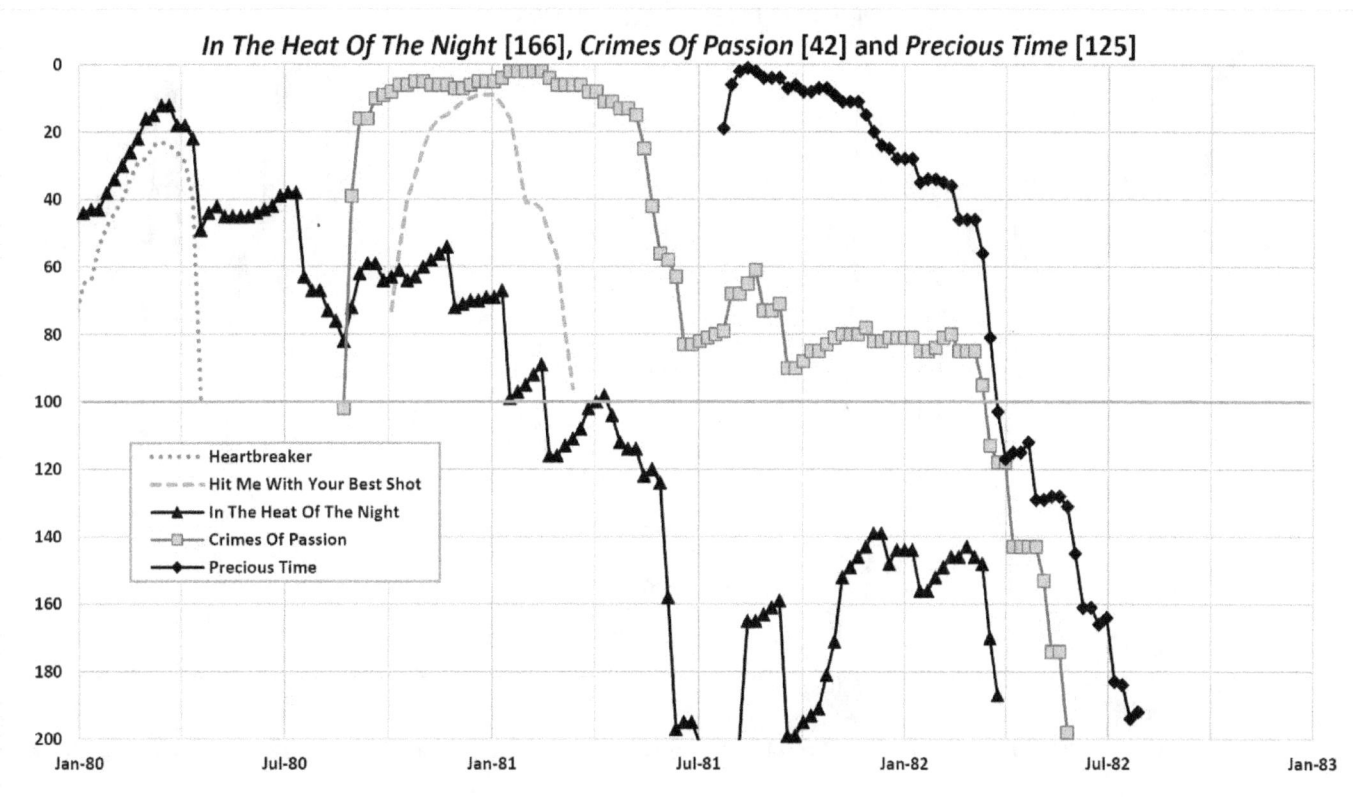

Chronologies

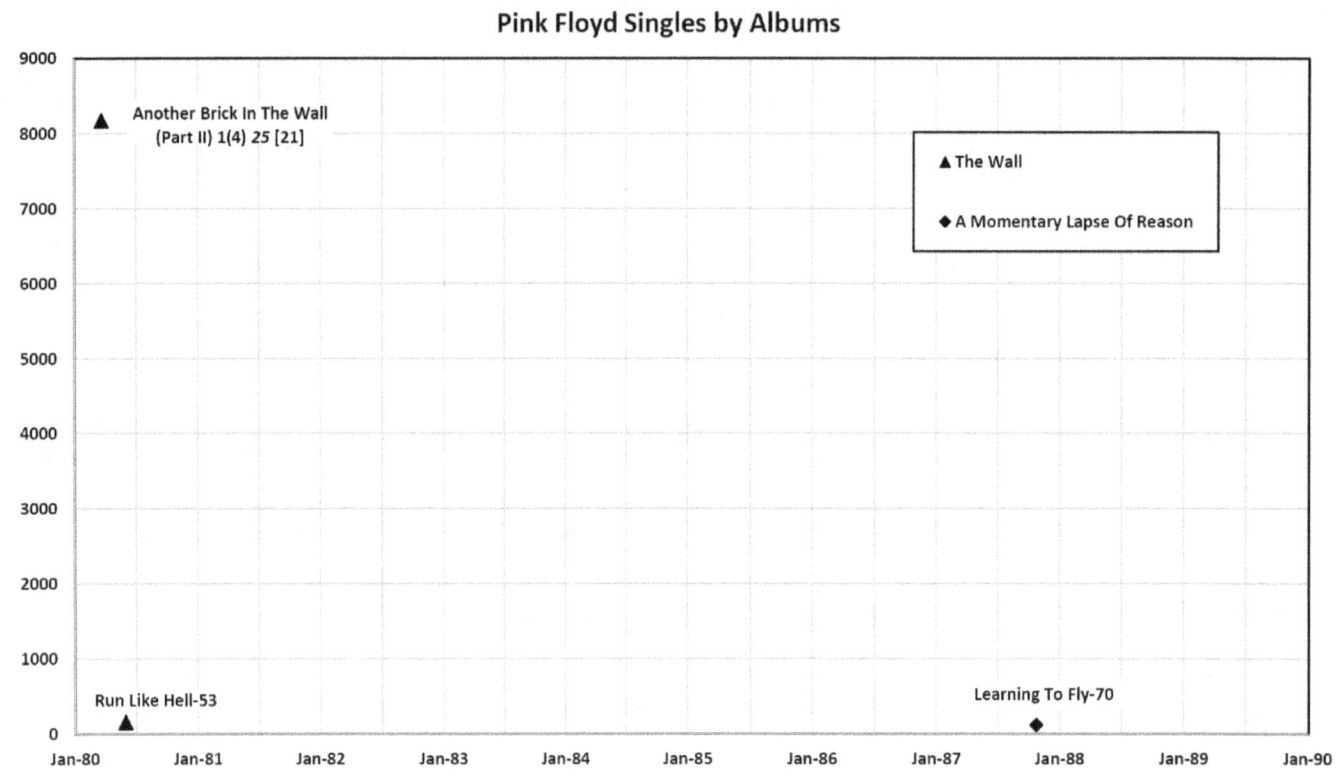

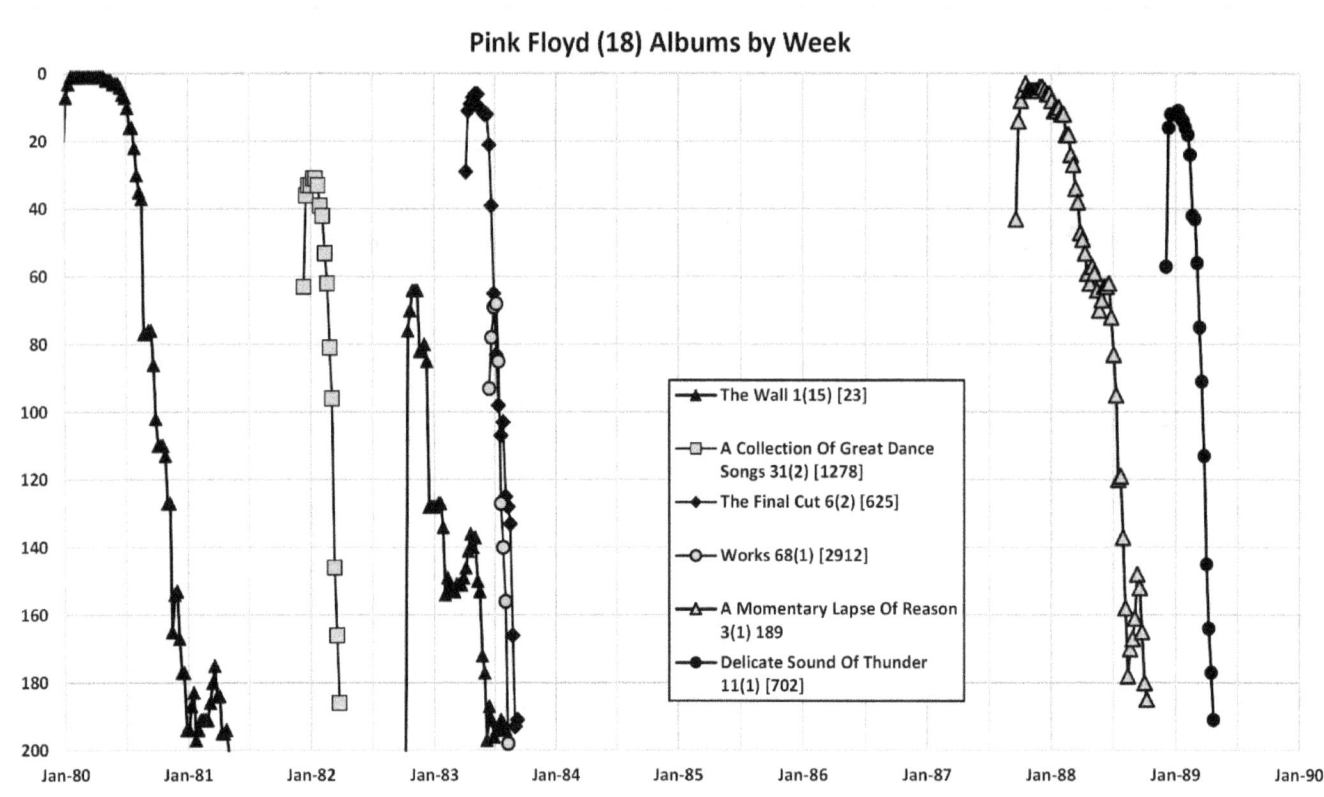

PINK FLOYD ▶ Singles Rank: 134 (3) 8451; Albums Rank: 18 (6/8) 65821 77.2%

Pink Floyd is at once the simplest and the most complex act in this section. It doesn't take much commentary to point out the salient features of their singles graph. Only one entry matters—*Another Brick In The Wall (Part II)*—and it's at least arguable that, big as it was, it didn't matter either. Because Pink Floyd is such an album-centric act, *The Wall* might well have had similar success if the single had not charted. With that said, *Another Brick In The Wall* (Part II) spent four weeks at number 1 and is ranked number 21 for the decade.

As a result, Pink Floyd is the poster child for an album-centric act, but a complicated one at that. Only two of its six '80s albums reached the top 10%, but *The Wall* ranked 23rd for the decade with 15 weeks at number 1. It will come as no surprise that when scored by its singles, *The Wall* drops from 23 to 97, with two entries, one of which was huge on the chart.

But what actually powered Pink Floyd into the top 20 album acts is a holdover from the '70s, *The Dark Side Of The Moon*. Released in March, 1973, this concept album had been trialed and honed by touring before final recording. In addition to critical acclaim, it assumed an extraordinary chart presence: while it only spent one week at number 1, it spent 741 weeks—over fourteen years—on the Billboard Top 200 chart between 1973 and 2000, and another 200+ weeks on various other mainline Billboard album charts. It ranked number 7 of the Billboard albums from 1963-1989.

The Dark Side of the Moon provided nearly a quarter of Pink Floyd's album points in the 1980s. It spiked up into the mid-40s in 1980, boosted by the success of *The Wall*, and for most of the rest of the decade bounced around between 100 and 200. It was the highest scoring '60s or '70s holdover album in the '80s by a wide margin.

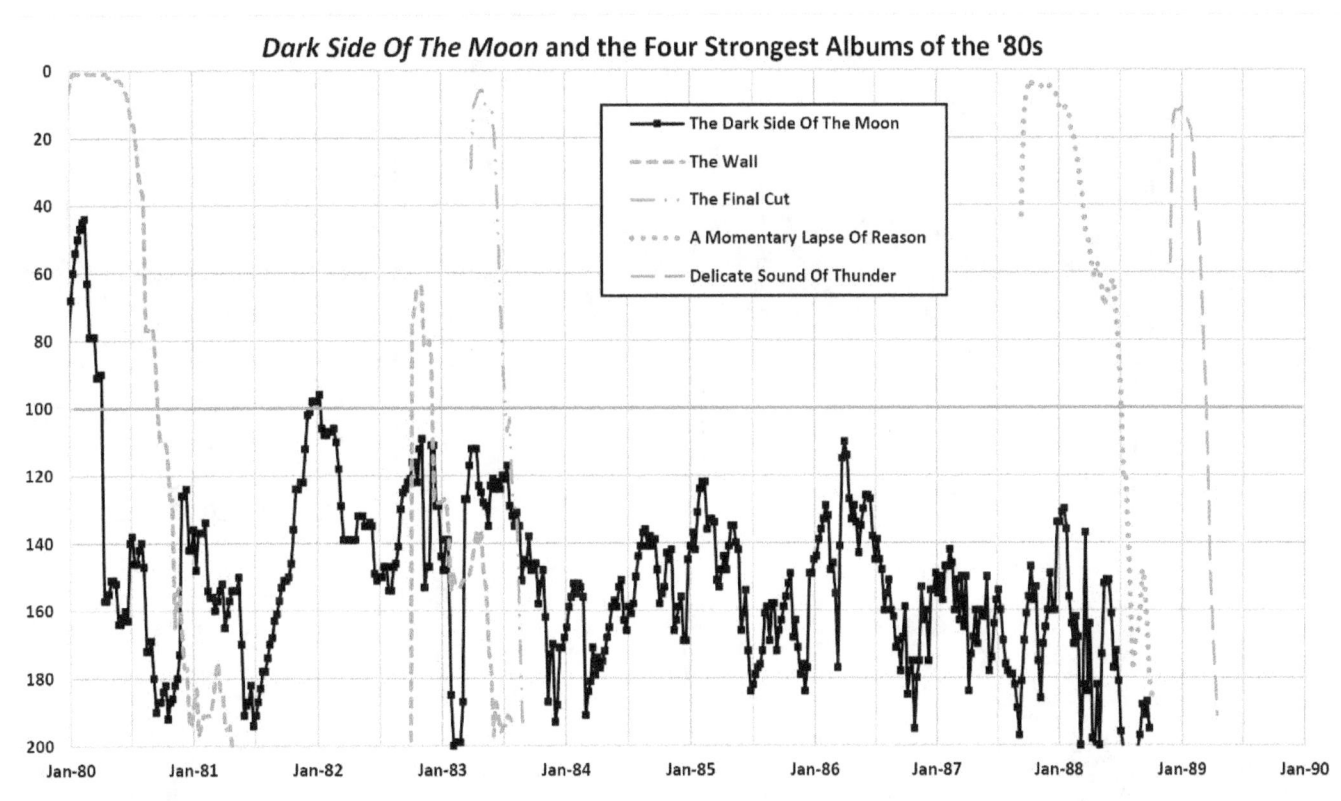

Chronologies

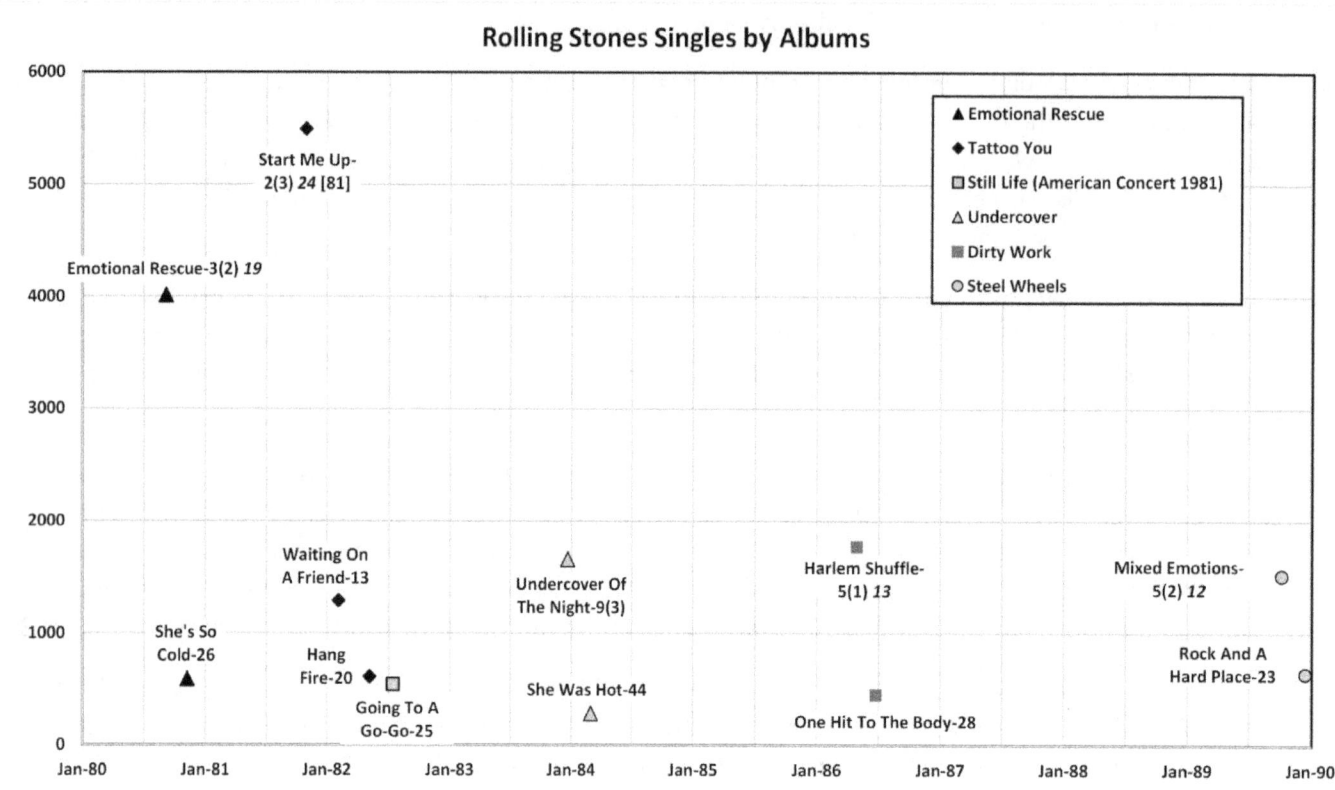

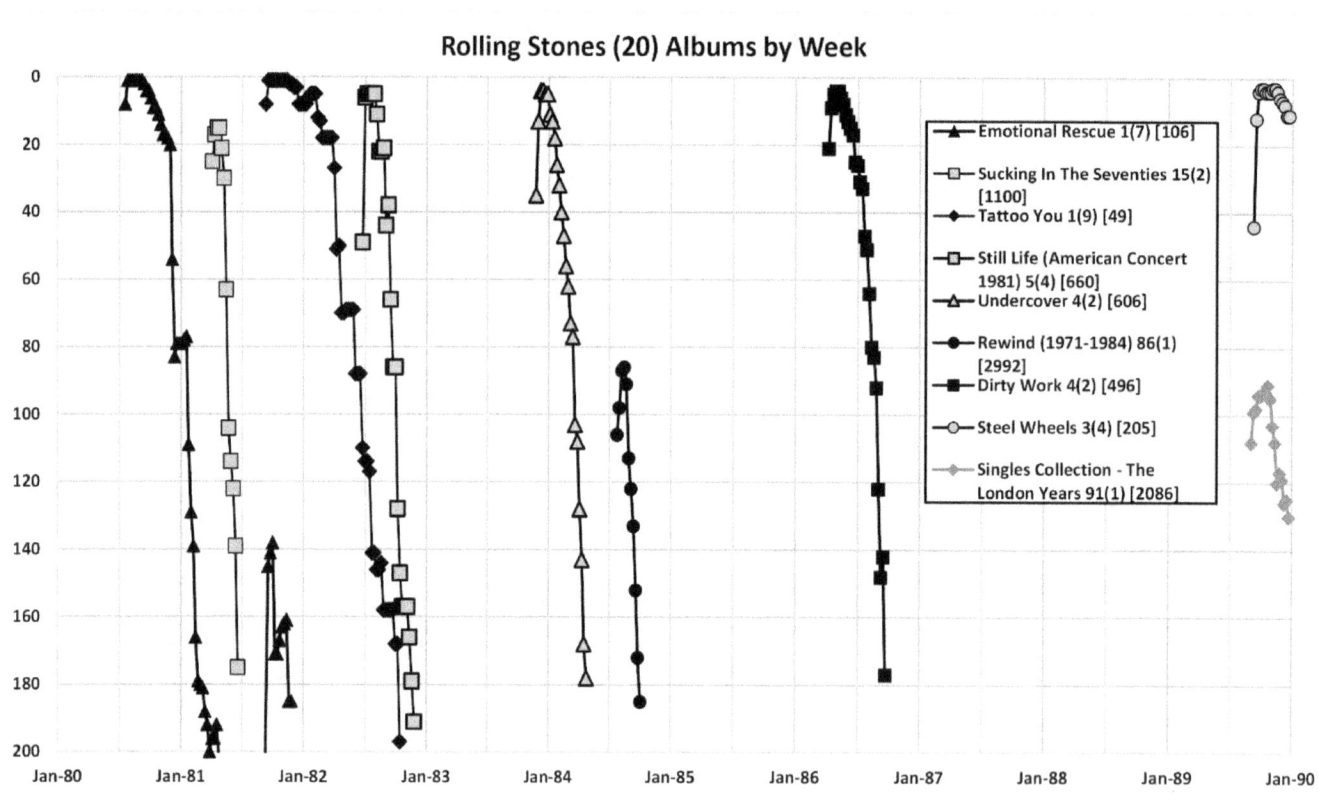

ROLLING STONES ▶ Singles Rank: 53 (12) 18848; Albums Rank: 20 (9/14) 64354 94.4%

There's probably not a lot left unsaid about The Rolling Stones in the nearly 60 years they've been around, but it's worth repeating that they were an exceptionally consequential act on the charts for three decades, which is remarkable.

Of their twelve singles in the '80s, four made the top 5, one more made the top 10 and eleven made the top 30. *Start Me Up*, the strongest, ranked number 81 for the decade. The poorest performance of the twelve was *She Was Hot*, which peaked at number 44. Remembering that there are nearly 1400 acts listed in this book, finishing 53rd in the singles list is still a pretty good record. Generally, they released two singles per album, and the first single always charted higher.

Of their nine albums, five consisted of new material, three were anthologies and one was live. Additionally, four earlier albums from the '60s and '70s charted briefly; however, 94% of their points earned in the '80s were from albums released in the '80s.

Emotional Rescue and *Tattoo You*—the number 49 album of the decade—both went to number 1, completing a string of eight consecutive number 1 albums of new material, dating back to *Sticky Fingers*. The other three '80s releases peaked in the top 5.

Tattoo You is their second highest charting album from the '60s through the '80s. Of their nine number 1 albums, it logged the most time there—nine weeks. That period, powered by the single *Start Me Up*, also drew *Emotional Rescue* back to the chart. However, their albums peaked fast and had relatively steep declines. Of the five albums by various acts in the '80s that spent nine weeks at number 1, *Tattoo You* spent the fewest weeks in the top 40 and the fewest total weeks on the chart. When scored by singles it dropped from 49 to 124.

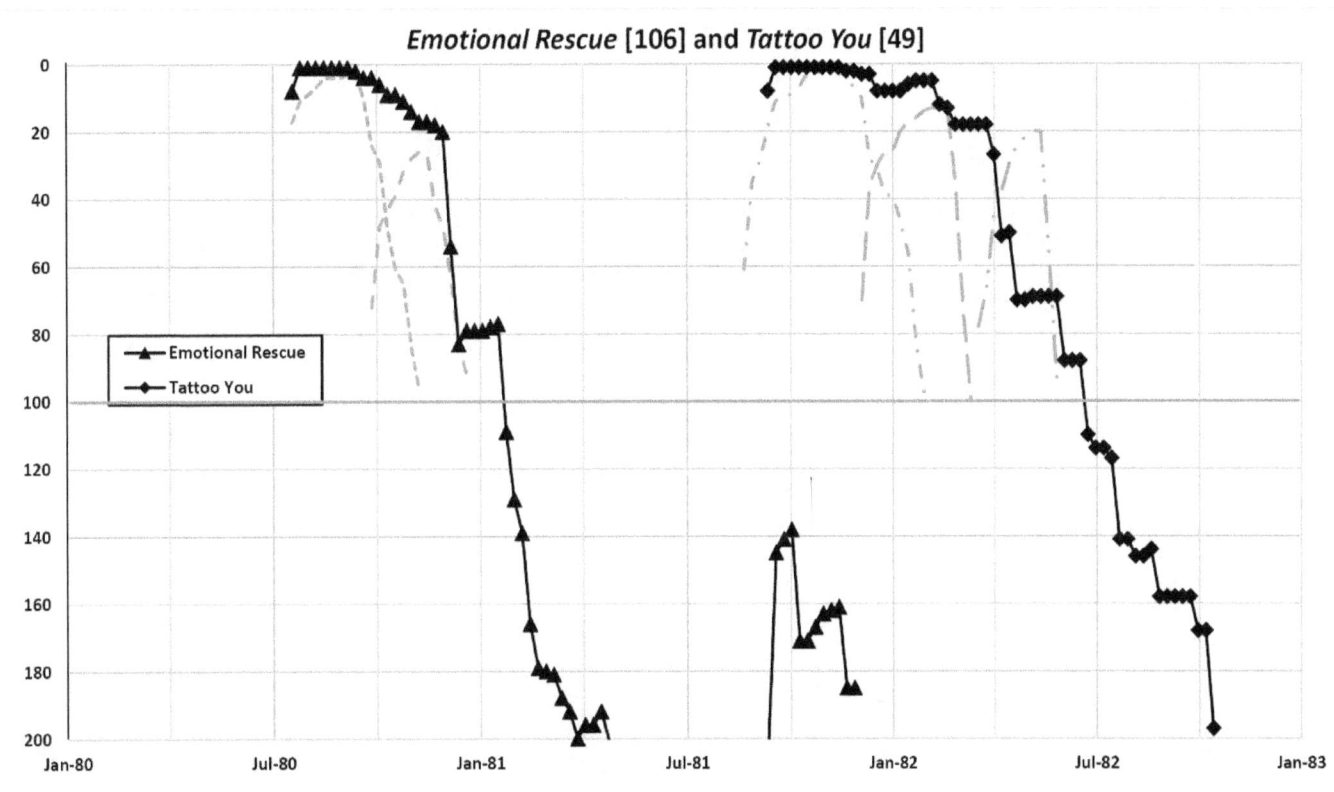

Chronologies

Ranking the '80s:

The Writers
The Producers

Ranking the Writers and Writer Teams

The difference between writers and writer teams is the allocation of score. As writers, credit for each single is divided equally among the writers of record, and a decadal score is calculated by summing the individual pieces. As writer teams, the entire score for a single is allocated to that team. Thus a single writer such as Prince can be a writer or a writer team. As the former, he may score points as an allocation from a collaboration. As the latter, he only scores points for solo writing efforts.

Top 200 Writers

Top 150 Writer Teams

Top 50 Writer Teams With Their Acts And Singles

Top 50 Singles Acts With Their Writers

Writers: 50 or More Consecutive Chart Weeks

Writers: Four or More Songs Charting Simultaneously

Writers Ranked

Writers: The Top 200

Rank. Writer (Writer Credits)	Score
1. Lionel RICHIE (22)	65988
2. PRINCE (42)	64433
3. Michael JACKSON (24)	52137
4. George MICHAEL (15)	42848
5. Phil COLLINS (27)	38672
6. Billy JOEL (22)	38031
7. STING (19)	34759
8. Paul McCARTNEY (56)	34644
9. Daryl HALL (24)	33221
10. John MELLENCAMP (20)	30136
11. Diane WARREN (21)	29685
12. Bruce SPRINGSTEEN (18)	28390
13. Stevie WONDER (43)	26922
14. Rod TEMPERTON (13)	24399
15. Rick SPRINGFIELD (16)	22568
16. John LENNON (49)	22514
17. Richard MARX (12)	21324
18. Madonna L. CICCONE (13)	20588
19. Ray PARKER Jr. (12)	19126
20. Debbie GIBSON (9)	18379
21. Michael Leslie JONES (13)	18104
22. Christopher CROSS (10)	17784
23. David FOSTER (38)	17575
24. James Samuel HARRIS (27)	17435
25. Terry LEWIS (27)	17435
26. Ric OCASEK (16)	16972
27. Giorgio MORODER (21)	16422
28. Kevin CRONIN (11)	16123
29. Bob SEGER (14)	15950
30. Colin James HAY (7)	15840
31. Jim STEINMAN (8)	15299
32. Graham RUSSELL (5)	15291
33. Dean PITCHFORD (15)	14802
34. Billy STEINBERG (13)	14754
35. Dan FOGELBERG (11)	14104
36. Neil DIAMOND (16)	14089
37. Thomas F. KELLY (15)	13807
38. Dennis DeYOUNG (12)	13531
39. Barry GIBB (22)	13514
40. Maurice STARR (7)	13486
41. Carole Bayer SAGER (20)	13286
42. Gloria ESTEFAN (6)	13284
43. Smokey ROBINSON (20)	13076
44. Jim PETERIK (24)	13040
45. Steve PERRY (27)	12817
46. David PAICH (11)	12467
47. Elton JOHN (19)	12348
48. Frankie SULLIVAN (18)	12102
49. Will JENNINGS (17)	12020
50. Tom SNOW (17)	11984
51. Lewis MARTINEÉ (6)	11741
52. Bernie TAUPIN (18)	11732
53. Steve KIPNER (9)	11693
54. Cynthia WEIL (21)	11614
55. Burt BACHARACH (17)	11612
56. Jonathan CAIN (24)	11594
57. David BOWIE (13)	11426
58. Robin GIBB (17)	11373
59. Jim VALLANCE (27)	11293
60. Michael MASSER (13)	11163
61. Kenneth EDMONDS (18)	10952
62. Jon BON JOVI (13)	10930
63. Seth JUSTMAN (7)	10866
64. Stevie NICKS (14)	10784
65. Kenny LOGGINS (16)	10734
66. Billy OCEAN (11)	10530
67. Keith FORSEY (10)	10513
68. John DEACON (6)	10460
69. John OATES (10)	10431
70. Holly KNIGHT (23)	10373
71. Desmond CHILD (18)	10306
72. Mick JAGGER (19)	10163
73. Freddie MERCURY (9)	10141
74. Mike RUTHERFORD (19)	10108
75. Mutt LANGE (19)	10072
76. Rick JAMES (11)	9814
77. Jeff LYNNE (17)	9812
78. Bruce HORNSBY (8)	9765
79. Peter CETERA (10)	9706
80. Steve MILLER (7)	9672
81. Patrick Raymond LEONARD (8)	9548
82. Bryan ADAMS (21)	9453
83. David A. STEWART (19)	9327
84. Howard JONES (10)	9326
85. John FARRAR (5)	9167
86. Huey LEWIS (11)	9096
87. Richie SAMBORA (9)	9022
88. Tom PETTY (18)	8988
89. Elliot WOLFF (2)	8934
90. Steve GREENBERG (3)	8855
91. Nigel John TAYLOR (18)	8813
92. Sara ALLEN (10)	8736
93. Simon Le BON (17)	8666
94. Nick RHODES (17)	8666
95. Toni TENNILLE (1)	8662
96. Robert PALMER (7)	8579
97. Keith RICHARDS (14)	8512
98. Antonio REID (15)	8478
99. Michael ZAGER (6)	8471
100. Annie LENNOX (15)	8469
101. Corey HART (8)	8368
102. Steve WINWOOD (15)	8329
103. John BETTIS (12)	8316
104. Brian SETZER (5)	8306
105. Roger WATERS (2)	8251
106. Mike CHAPMAN (16)	8176
107. Terry BRITTEN (12)	7968
108. Peter GABRIEL (8)	7946
109. Roger TAYLOR (17)	7917
110. Lou GRAMM (10)	7796
111. Stephen BRAY (9)	7601
112. Graham LYLE (8)	7458
113. Rodney CROWELL (3)	7437
114. Marv ROSS (7)	7424
115. Paul SIMON (10)	7393
116. Sammy HAGAR (14)	7384
117. Andy FARRISS (11)	7264
118. Andy TAYLOR (14)	7255
119. Gerry GOFFIN (10)	7248
120. Michael HUTCHENCE (10)	7191
121. Roland ORZABAL (6)	7143
122. Graham GOBLE (7)	7084
123. Mark KNOPFLER (6)	6994
124. Russ BALLARD (9)	6943
125. Harold FALTERMEYER (7)	6845
126. Eddy GRANT (3)	6805
127. Oliver LEIBER (3)	6797
128. Don HENLEY (10)	6797
129. Chrissie HYNDE (6)	6745
130. Maurice GIBB (16)	6680
131. Per GESSLE (3)	6678
132. Christine McVIE (8)	6649
133. Charlotte CAFFEY (7)	6643
134. Barry MANN (13)	6555
135. Narada Michael WALDEN (17)	6511
136. David PACK (5)	6413
137. Dolly PARTON (2)	6404
138. George M. BROWN (19)	6395
139. Nile RODGERS (13)	6259
140. Bernard EDWARDS (10)	6172
141. Peter F. WOLF (6)	6154
142. Albert HAMMOND (7)	6150
143. Glenn FREY (11)	6117
144. Donna SUMMER (8)	6105
145. Daryl SIMMONS (10)	6093
146. Keith DIAMOND (7)	6023
147. Danny KORTCHMAR (9)	5984
148. Tony BANKS (13)	5977
149. Bill WITHERS (3)	5963
150. Michael CRAIG (10)	5895
151. Roy HAY (10)	5895
152. Jon MOSS (10)	5895
153. George O'DOWD (10)	5895
154. VANGELIS (4)	5872
155. Lamont DOZIER (14)	5863
156. Chris HAYES (7)	5777
157. George MERRILL (5)	5712
158. Shannon RUBICAM (5)	5712
159. Kennedy GORDY (2)	5674
160. Jody WATLEY (7)	5597
161. Jackie DeSHANNON (6)	5587
162. Liam STERNBERG (1)	5581
163. Terry SHADDICK (1)	5545
164. Rupert HOLMES (5)	5541
165. Nick VAN EEDE (4)	5480
166. Tom SCHOLZ (6)	5447
167. Donna WEISS (3)	5385
168. Martha DAVIS (8)	5366
169. Randy GOODRUM (13)	5320
170. Cyndi LAUPER (7)	5306
171. Janna ALLEN (4)	5298
172. Joe JACKSON (5)	5290
173. Mike CAMPBELL (9)	5269
174. Neil FINN (5)	5262
175. Ivan DOROSCHUK (3)	5256
176. Deborah HARRY (6)	5241
177. Clement A. BOZEWSKI (5)	5186
178. Jackson BROWNE (8)	5183
179. Dennis W. MORGAN (8)	5173
180. Robert HAZARD (2)	5138
181. Paul DAVIS (5)	5082
182. George GREEN (3)	5042
183. Nicky CHINN (8)	5029
184. TRADITIONAL (3)	5008
185. Billy IDOL (9)	4985
186. Barry EASTMOND (9)	4947
187. Alex CALL (3)	4914
188. Alan TARNEY (6)	4878
189. Ritchie CORDELL (4)	4865
190. Florrie PALMER (2)	4839
191. Chris LOWE (8)	4827
192. Neil TENNANT (8)	4827
193. Irene CARA (6)	4785
194. Stephen BISHOP (2)	4774
195. André CYMONE (5)	4769
196. David MALLOY (9)	4756
197. Even STEVENS (8)	4752
198. Lindsey BUCKINGHAM (5)	4707
199. Terence Trent D'ARBY (4)	4671
200. Alan PARSONS (11)	4658

Writers: The Top 150 Teams

Rank. Writer Team (Team Credits)	Score
1. PRINCE (34)	61364
2. Lionel RICHIE (13)	57417
3. Michael JACKSON (14)	41438
4. George MICHAEL (13)	39140
5. Billy JOEL (22)	38031
6. James Samuel HARRIS/Terry LEWIS (24)	32873
7. STING (16)	31860
8. Phil COLLINS (12)	29221
9. Bruce SPRINGSTEEN (17)	28284
10. Stevie WONDER (24)	25391
11. John MELLENCAMP (17)	25094
12. Simon Le BON/Nick RHODES/Andy TAYLOR/Nigel John TAYLOR/Roger TAYLOR (8)	24443
13. Paul McCARTNEY (10)	23913
14. Diane WARREN (12)	23431
15. Jim PETERIK/Frankie SULLIVAN (15)	23171
16. Rod TEMPERTON (9)	22898
17. Thomas F. KELLY/Billy STEINBERG (8)	22023
18. Rick SPRINGFIELD (14)	21729
19. Ray PARKER Jr. (12)	19126
20. Sara ALLEN/Daryl HALL/John OATES (6)	19007
21. Barry GIBB/Maurice GIBB/Robin GIBB (12)	18718
22. Richard MARX (6)	18689
23. Debbie GIBSON (9)	18379
24. Tony BANKS/Phil COLLINS/Mike RUTHERFORD (13)	17932
25. John LENNON (7)	17519
26. Daryl HALL (9)	17486
27. Bryan ADAMS/Jim VALLANCE (17)	17031
28. Ric OCASEK (16)	16972
29. Annie LENNOX/David A. STEWART (15)	16938
30. Michael CRAIG/Roy HAY/Jon MOSS/George O'DOWD (7)	16751
31. Mick JAGGER/Keith RICHARDS (12)	16723
32. Christopher CROSS (7)	16109
33. Kenneth EDMONDS/Antonio REID/Daryl SIMMONS (9)	16081
34. Peter CETERA/David FOSTER (5)	15958
35. Jon BON JOVI/Desmond CHILD/Richie SAMBORA (5)	15700
36. Curtis BEDEAU/Gerard CHARLES/Hugh CLARKE/Brian GEORGE/Lucien GEORGE/ Paul GEORGE (9)	15678
37. Kevin CRONIN (9)	15171
38. Madonna L. CICCONE/Patrick Raymond LEONARD (4)	15076
39. Jim STEINMAN (7)	15043
40. Elton JOHN/Bernie TAUPIN (13)	15028
41. Terry BRITTEN/Graham LYLE (5)	14680
42. Bob SEGER (13)	14414
43. Andy FARRISS/Michael HUTCHENCE (8)	14207
44. Adam CLAYTON/Dave EVANS/Paul HEWSON/Larry MULLEN (11)	14177
45. Dan FOGELBERG (11)	14104
46. Jonathan CAIN/Steve PERRY (7)	13732
47. Lou GRAMM/Michael Leslie JONES (8)	13575
48. Dennis DeYOUNG (11)	13471
49. Will JENNINGS/Steve WINWOOD (10)	13441
50. Maurice STARR (5)	13284
51. David MALLOY/Eddie RABBITT/Even STEVENS (7)	13216
52. Rick ALLEN/Steve CLARK/Phil COLLEN/Joe ELLIOTT/Mutt LANGE/Rick SAVAGE (6)	12645
53. Colin James HAY (6)	12263
54. Graham RUSSELL (4)	12251
55. Burt BACHARACH/Carole Bayer SAGER (3)	12231
56. Bernard EDWARDS/Nile RODGERS (10)	12217
57. Lewis MARTINEÉ (6)	11741
58. Michael Leslie JONES (5)	11317
59. Gloria ESTEFAN (4)	11224
60. Irene CARA/Keith FORSEY/Giorgio MORODER (2)	11130
61. Steve KIPNER/Terry SHADDICK (1)	11089
62. Stephen BRAY/Madonna L. CICCONE (4)	11019
63. Dean PITCHFORD/Tom SNOW (3)	10682
64. Michael JACKSON/Paul McCARTNEY (1)	10336
65. Jackie DeSHANNON/Donna WEISS (1)	10255
66. John DEACON (3)	10216
67. Tom BAILEY/Alannah CURRIE/Joe LEEWAY (8)	10157
68. George GREEN/John MELLENCAMP (3)	10084
69. Michael ANTHONY/David Lee ROTH/Alex VAN HALEN/Eddie VAN HALEN (5)	10080
70. John LENNON/Paul McCARTNEY (22)	9990
71. Steven ADLER/Saul HUDSON/Duff McKAGAN/Axl ROSE/Izzy STRADLIN (5)	9951
72. David BOWIE (7)	9936
73. Clement A. BOZEWSKI/Deborah HARRY/Giorgio MORODER (1)	9931
74. Freddie MERCURY (5)	9842
75. Matt AITKEN/Mike STOCK/Pete WATERMAN (8)	9813
76. Steve GEORGE/John LANG/Richard PAGE (3)	9793
77. Smokey ROBINSON (7)	9780
78. Bobby DALL/C.C. DeVILLE/Bret MICHAELS/Rikki ROCKETT (6)	9761
79. Steve MILLER (5)	9572
80. David PAICH (5)	9340
81. Alan PARSONS/Eric WOOLFSON (11)	9317
82. Howard JONES (9)	9069
83. Rick JAMES (9)	9042
84. Elliot WOLFF (2)	8934
85. André CYMONE/Jody WATLEY (4)	8877
86. Steve GREENBERG (3)	8855
87. Glenn FREY/Jack TEMPCHIN (9)	8853
88. Toni TENNILLE (1)	8662
89. Seth JUSTMAN (2)	8642
90. Gerry GOFFIN/Michael MASSER (6)	8591
91. Michael CRAIG/Roy HAY/Jon MOSS/George O'DOWD/Phil PICKETT (3)	8535
92. Michael ZAGER (3)	8471
93. George MERRILL/Shannon RUBICAM (4)	8445
94. Giorgio MORODER/Tom WHITLOCK (5)	8383
95. Corey HART (8)	8368
96. Mike CHAPMAN/Nicky CHINN (5)	8346
97. Chris HAYES/Huey LEWIS (4)	8327
98. Jeff LYNNE (11)	8317
99. Brian SETZER (5)	8306
100. Jake HOOKER/Alan MERRILL (1)	8293
101. Neil DIAMOND (6)	8237
102. Barry GIBB/Robin GIBB (2)	8175
103. Roger WATERS (1)	8171
104. Albert HAMMOND/Diane WARREN (5)	8156
105. Roland GIFT/David STEELE (5)	8080
106. Peter GABRIEL (4)	7946
107. Bruce HORNSBY/John HORNSBY (4)	7895
108. John FARRAR (3)	7866
109. Keith DIAMOND/Billy OCEAN (2)	7730
110. Michael ANTHONY/Sammy HAGAR/Alex VAN HALEN/Eddie VAN HALEN (7)	7681
111. Robert PALMER (3)	7676
112. Stevie NICKS (7)	7601
113. Chris LOWE/Neil TENNANT (6)	7595
114. Michael DELAHUNTY/David GLASPER/Marcus LILLINGTON/Ian SPICE (3)	7556
115. Mike CAMPBELL/Tom PETTY (6)	7533
116. Dennis MATKOSKY/Michael SEMBELLO (2)	7496
117. Kenny LOGGINS/Dean PITCHFORD (2)	7446
118. Rodney CROWELL (3)	7437
119. Jon BON JOVI/Richie SAMBORA (3)	7411
120. Paul SIMON (8)	7378
121. Ina WOLF/Peter F. WOLF (3)	7347
122. Marv ROSS (5)	7333
123. Colin James HAY/Ron STRYKERT (1)	7153
124. Lamont DOZIER/Brian HOLLAND/Eddie HOLLAND (8)	7002
125. Elton JOHN/Gary OSBORNE (3)	6994
126. Graham GOBLE (5)	6826
127. John William CALLIS/Philip OAKEY/Philip WRIGHT (1)	6816
128. Eddy GRANT (3)	6805
129. Oliver LEIBER (3)	6797
130. Janna ALLEN/Daryl HALL (2)	6790
131. Simon Le BON/Nick RHODES/Nigel John TAYLOR (2)	6784
132. Jonathan CAIN/Steve PERRY/Neal SCHON (6)	6687
133. Russ BALLARD (8)	6667
134. Gilbert BÉCAUD/Neil DIAMOND (2)	6609
135. Jon ANDERSON/Trevor HORN/Trevor RABIN/Chris SQUIRE (1)	6601
136. Rick NOWELS/Ellen SHIPLEY (3)	6597
137. Wayne BRATHWAITE/Barry EASTMOND/Billy OCEAN (3)	6461
138. Peter ALLEN/Burt BACHARACH/Christopher CROSS/Carole Bayer SAGER (1)	6437
139. Dolly PARTON (2)	6404
140. John BETTIS/Michael CLARK (4)	6383
141. Giancarlo BIGAZZI/Umberto TOZZI/Trevor VEITCH (1)	6294
142. Giorgio MORODER/Donna SUMMER (2)	6277
143. Kenneth GAMBLE/Leon HUFF (6)	6091
144. Clive DAVIS/Graham RUSSELL (1)	6080
145. Frank BEARD/Billy GIBBONS/Dusty HILL (9)	6077
146. Barry MANN/Cynthia WEIL (5)	6011
147. Chrissie HYNDE (5)	5914
148. George MICHAEL/Andrew RIDGELEY (1)	5877
149. Geoffrey DOWNES/John WETTON (4)	5870
150. Simon CLIMIE/Dennis W. MORGAN (2)	5866

Writers: Top 50 Writer Teams, Their Acts And Singles

Rank. Writer Team (Entries) / Act (Entries) / Title	Score
1. PRINCE (34)	**61364**
PRINCE (15)	27702
When Doves Cry	7679
Batdance	3450
U Got The Look	2927
Little Red Corvette	2573
Sign O' The Times	2446
Delirious	1912
1999	1633
I Wanna Be Your Lover	1389
I Could Never Take The Place Of Your Man	1224
Alphabet St.	1192
Partyman	582
Let's Pretend We're Married// Irresistible Bitch	258
Controversy	182
Hot Thing	163
If I Was Your Girlfriend	91.6
PRINCE And The REVOLUTION (8)	20698
Let's Go Crazy	5200
Kiss	4499
Purple Rain	3959
Raspberry Beret	3528
Pop Life	1376
I Would Die 4 U	1349
Take Me With U	585
Anotherloverholenyohead	201
Chaka KHAN (1)	4486
I Feel For You	4486
BANGLES (1)	2950
Manic Monday	2950
SHEILA E. (2)	2928
The Glamorous Life	2352
The Belle Of St. Mark	576
Sheena EASTON (1)	1509
Sugar Walls	1509
ART OF NOISE (1)	423
Kiss	423
Meli'sa MORGAN (1)	414
Do Me Baby	414
The FAMILY (1)	132
The Screams Of Passion	132
APOLLONIA 6 (1)	65.1
Sex Shooter	65.1
Mitch RYDER (1)	31.0
When You Were Mine	31.0
The TIME (1)	24.9
777-9311	24.9
2. Lionel RICHIE (13)	**57417**
Lionel RICHIE (9)	44204
Endless Love	10386
All Night Long (All Night)	8437
Hello	6840
Say You, Say Me	6296
Truly	4947
Stuck On You	3553
My Love	2148
Ballerina Girl	1473
Deep River Woman	124
Kenny ROGERS (1)	8423
Lady	8423
The COMMODORES (1)	2910
Oh No	2910
Diana ROSS (1)	1770
Missing You	1770
John SCHNEIDER (1)	109
Still	109
3. Michael JACKSON (13)	**41359**
Michael JACKSON (10)	38089
Billie Jean	8360
Beat It	6892
The Girl Is Mine	5346
I Just Can't Stop Loving You	3395
The Way You Make Me Feel	3201
Bad	2998
Dirty Diana	2958
Wanna Be Startin' Somethin'	2524
Smooth Criminal	1470
Another Part Of Me	945
Diana ROSS (1)	1707
Muscles	1707
The JACKSONS (1)	782
Heartbreak Hotel	782
Rebbie JACKSON (1)	781
Centipede	781
4. George MICHAEL (13)	**39140**
George MICHAEL (7)	23568
Faith	5187
One More Try	4687
Father Figure	3850
Monkey	3473
I Want Your Sex	3293
A Different Corner	1547
Kissing A Fool	1530
WHAM! (6)	15572
Wake Me Up Before You Go-Go	5473
Everything She Wants (Remix)	4050
I'm Your Man	2606
Freedom	2153
The Edge Of Heaven	1071
Bad Boys	219
5. Billy JOEL (22)	**38031**
Billy JOEL (22)	38031
It's Still Rock And Roll To Me	6941
Uptown Girl	5362
We Didn't Start The Fire	4586
Tell Her About It	4357
You May Be Right	1973
Allentown	1591
An Innocent Man	1543
The Longest Time	1325
You're Only Human (Second Wind)	1320
A Matter Of Trust	1210
Modern Woman (From "Ruthless People")	1035
Say Goodbye To Hollywood (Live)	904
Don't Ask Me Why	903
Keeping The Faith	876
Pressure	859
This Is The Time	850
She's Got A Way	715
Leave A Tender Moment Alone	688
The Night Is Still Young	367
Sometimes A Fantasy	334
Goodnight Saigon	194
Baby Grand	96.5
6. James Samuel HARRIS/Terry LEWIS (24)	**32873**
Janet JACKSON (5)	16636
Miss You Much	5007
When I Think Of You	3899
Nasty	2753
What Have You Done For Me Lately	2532
Control	2445
HUMAN LEAGUE (1)	4045
Human	4045
Robert PALMER (1)	2941
I Didn't Mean To Turn You On	2941
Herb ALPERT (3)	2548
Diamonds	1839
Making Love In The Rain	439
Keep Your Eye On Me	269
NEW EDITION (3)	2116
If It Isn't Love	1807
Can You Stand The Rain	298
You're Not My Kind Of Girl	10.7
FORCE M.D.'S (1)	1312
Tender Love	1312
S.O.S. BAND (4)	1129
The Finest	364
Just Be Good To Me	340
Just The Way You Like It	230
Tell Me If You Still Care	195
Fake	559
Never Knew Love Like This	529
CHERRELLE (2)	719
Saturday Love	623
I Didn't Mean To Turn You On	95.9
Patti AUSTIN (1)	217
The Heat Of Heat	217
Cheryl LYNN (1)	124
Encore	124
Alexander O'NEAL (2)	1087
7. STING (16)	**31860**
The POLICE (10)	23786
Every Breath You Take	10182
Every Little Thing She Does Is Magic	2994
King Of Pain	2967
De Do Do Do, De Da Da Da	1789
Wrapped Around Your Finger	1588
Don't Stand So Close To Me	1578
Synchronicity II	1118
Spirits In The Material World	1069
Don't Stand So Close To Me '86	254
Secret Journey	249
STING (6)	8073
If You Love Somebody Set Them Free	3098
We'll Be Together	1644
Fortress Around Your Heart	1614
Love Is The Seventh Wave	844
Be Still My Beating Heart	838
Englishman In New York	35.8
8. Phil COLLINS (12)	**29221**
Phil COLLINS (10)	27562
Against All Odds (Take A Look At Me Now)	7041
Another Day In Paradise	5854
One More Night	4427
Sussudio	4040
Don't Lose My Number	2260
Take Me Home	1797

Writers Ranked

Rank. Writer Team (Entries) / Act (Entries) / Title	Score
Phil COLLINS Continued	
I Missed Again	847
In The Air Tonight	842
I Don't Care Anymore	403
I Cannot Believe It's True	51.6
GENESIS (2)	1659
Misunderstanding	1289
Man On The Corner	370
9. Bruce SPRINGSTEEN (17)	**28284**
Bruce SPRINGSTEEN (13)	23637
Dancing In The Dark	5565
Hungry Heart	3294
Glory Days	2124
I'm On Fire	1895
Cover Me	1853
Brilliant Disguise	1737
Born In The U.S.A.	1590
My Hometown	1511
Tunnel Of Love	1146
I'm Goin' Down	1114
One Step Up	924
Fade Away	640
Fire (Live)	246
Gary (U.S.) BONDS (2)	2397
This Little Girl	1565
Out Of Work	833
Natalie COLE (1)	1860
Pink Cadillac	1860
Joan JETT & The BLACKHEARTS (1)	390
Light Of Day	390
10. Stevie WONDER (18)	**25001**
Stevie WONDER (15)	24377
I Just Called To Say I Love You	6339
Part-Time Lover	4348
Master Blaster (Jammin')	3415
That Girl	3383
I Ain't Gonna Stand For It	1526
Go Home	1315
Do I Do	1185
Love Light In Flight	903
Skeletons	748
Overjoyed	556
Outside My Window	228
Ribbon In The Sky	192
Lately	154
You Will Know	59.0
Land Of La La	25.8
Jermaine JACKSON (1)	518
You're Supposed To Keep Your Love For Me	518
Michael JACKSON (1)	57.1
Get It	57.1
Julio IGLESIAS (1)	48.7
My Love	48.7
11. John MELLENCAMP (17)	**25094**
John MELLENCAMP (17)	25094
Jack & Diane	7190
R.O.C.K. In The U.S.A. (A Salute To 60's Rock)	2895
Small Town	2127
Lonely Ol' Night	1951
Cherry Bomb	1512
Pink Houses	1484
Paper In Fire	1315
Ain't Even Done With The Night	1205
Hand To Hold On To	1066
Authority Song	1065
Check It Out	941
This Time	760
Pop Singer	712

Rank. Writer Team (Entries) / Act (Entries) / Title	Score
Rumbleseat	508
Jackie Brown	193
Rooty Toot Toot	142
Small Paradise	29.8
12. Simon Le BON/Nick RHODES/Andy TAYLOR/ Nigel John TAYLOR/Roger TAYLOR (8)	**24443**
DURAN DURAN (8)	24443
The Reflex	5085
The Wild Boys	4921
Hungry Like The Wolf	4539
Union Of The Snake	3910
Is There Something I Should Know	2818
New Moon On Monday	1387
Rio	963
Save A Prayer	820
13. Paul McCARTNEY (10)	**23913**
Paul McCARTNEY (8)	15531
Ebony And Ivory with vocals by Stevie Wonder	8625
No More Lonely Nights	2137
Take It Away	1711
Spies Like Us	1530
So Bad	759
Press	563
Tug Of War	191
This One	14.9
Paul McCARTNEY & WINGS (1)	8092
Coming Up	8092
EVERLY BROTHERS (1)	291
On The Wings Of A Nightingale	291
14. Diane WARREN (12)	**23431**
CHICAGO (1)	4616
Look Away	4616
MILLI VANILLI (1)	4409
Blame It On The Rain	4409
BAD ENGLISH (1)	3797
When I See You Smile	3797
DeBARGE (1)	3325
Rhythm Of The Night	3325
CHER (1)	2821
If I Could Turn Back Time	2821
Belinda CARLISLE (1)	2582
I Get Weak	2582
HEART (1)	1573
Who Will You Run To	1573
El DeBARGE With DeBARGE (1)	94.5
The Heart Is Not So Smart	94.5
Patti LaBELLE (1)	71.8
If You Asked Me To	71.8
DION (1)	57.6
And The Night Stood Still	57.6
John WAITE (1)	47.8
Don't Lose Any Sleep	47.8
The JETS (1)	35.9
The Same Love	35.9
15. Jim PETERIK/Frankie SULLIVAN (15)	**23171**
SURVIVOR (14)	23048
Eye Of The Tiger	9773
Burning Heart	3808
The Search Is Over	2456
Is This Love	1703
I Can't Hold Back	1564
High On You	1361
American Heartbeat	839
Poor Man's Son	561
First Night	231
How Much Love	227
Didn't Know It Was Love	195
Summer Nights	178

Rank. Writer Team (Entries) / Act (Entries) / Title	Score
Across The Miles	80.0
Caught In The Game	72.0
Tommy SHAW (1)	123
Ever Since The World Began	123
16. Rod TEMPERTON (9)	**22898**
Michael JACKSON (3)	11247
Rock With You	7256
Thriller	2493
Off The Wall	1498
Patti AUSTIN (1)	5417
Baby, Come To Me	5417
George BENSON (2)	3413
Give Me The Night	3251
Love X Love	162
Michael McDONALD (1)	1864
Sweet Freedom	1864
KLYMAXX (1)	887
Man Size Love	887
BROTHERS JOHNSON (1)	70.0
Treasure	70.0
17. Thomas F. KELLY/Billy STEINBERG (8)	**22023**
MADONNA (1)	6531
Like A Virgin	6531
HEART (2)	5356
Alone	5121
I Want You So Bad	235
Cyndi LAUPER (2)	5030
True Colors	3650
I Drove All Night	1380
Whitney HOUSTON (1)	4343
So Emotional	4343
Pat BENATAR (1)	571
Sex As A Weapon	571
BLUE ZONE U.K. (1)	192
Jackie	192
18. Rick SPRINGFIELD (14)	**21729**
Rick SPRINGFIELD (13)	21550
Jessie's Girl	6460
Don't Talk To Strangers	5887
Love Somebody	2297
Human Touch	997
Love Is Alright Tonite	848
Souls	792
What Kind Of Fool Am I	782
Bop 'Til You Drop	775
Rock Of Life	624
Don't Walk Away	587
Bruce	549
Celebrate Youth	490
I Get Excited	463
Randy CRAWFORD (1)	179
Taxi Dancing	179
19. Ray PARKER Jr. (12)	**19126**
Ray PARKER Jr. (8)	13550
Ghostbusters	6445
The Other Woman	3200
I Still Can't Get Over Loving You	1373
Jamie	1122
Bad Boy	487
Girls Are More Fun	451
Let Me Go	349
I Don't Think That Man Should Sleep Alone	124
Ray PARKER Jr. & RAYDIO (3)	4524
A Woman Needs Love (Just Like You Do)	3214
That Old Song	665
Two Places At The Same Time	645
NEW EDITION (1)	1051
Mr. Telephone Man	1051

Writers Ranked

Rank. Writer Team (Entries) Act (Entries) Title	Score
20. Sara ALLEN/Daryl HALL/John OATES (6)	**19007**
Daryl HALL & John OATES (6)	19007
I Can't Go For That (No Can Do)	6855
Maneater	6774
You Make My Dreams	2669
Adult Education	1822
Possession Obsession	463
Missed Opportunity	423
21. Barry GIBB/Maurice GIBB/Robin GIBB (12)	**18718**
Kenny ROGERS (1)	6841
Islands In The Stream	6841
Barbra STREISAND & Barry GIBB (1)	3545
Guilty	3545
BEE GEES (6)	2720
One	1136
The Woman In You	605
He's A Liar	328
Living Eyes	320
Someone Belonging To Someone	236
You Win Again	95.0
Dionne WARWICK (1)	1820
Heartbreaker	1820
Diana ROSS (2)	159
Chain Reaction (Special New Mix)	149
Chain Reaction	10.1
Andy GIBB (1)	3634
Desire	3634
22. Richard MARX (6)	**18689**
Richard MARX (6)	18689
Right Here Waiting	4616
Hold On To The Nights	3490
Endless Summer Nights	3112
Satisfied	2851
Should've Known Better	2631
Angelia	1991
23. Debbie GIBSON (9)	**18379**
Debbie GIBSON (9)	18379
Lost In Your Eyes	4389
Foolish Beat	3393
Shake Your Love	2868
Only In My Dreams	2792
Out Of The Blue	2548
Electric Youth	1000
No More Rhyme	734
Staying Together	551
We Could Be Together	104
24. Tony BANKS/Phil COLLINS/Mike RUTHERFORD (13)	**17932**
GENESIS (13)	17932
Invisible Touch	3412
In Too Deep	2632
Land Of Confusion	2370
Tonight, Tonight, Tonight	2320
That's All	2253
Throwing It All Away	2113
No Reply At All	763
Abacab	603
Paperlate	530
Illegal Alien	315
Taking It All Too Hard	300
Turn It On Again	195
Mama	126
25. John LENNON (7)	**17519**
John LENNON (5)	17434
(Just Like) Starting Over	7930
Woman	6131
Nobody Told Me	1920
Watching The Wheels	1316
I'm Stepping Out	136
John LENNON and the PLASTIC ONO BAND (1)	52.1
Jealous Guy	52.1
Tracie SPENCER (1)	32.7
Imagine	32.7
26. Daryl HALL (9)	**17486**
Daryl HALL & John OATES (6)	12581
Say It Isn't So	5735
One On One	2378
Everything Your Heart Desires	2134
Wait For Me	1094
Some Things Are Better Left Unsaid	748
Your Imagination	492
Paul YOUNG (1)	4283
Everytime You Go Away	4283
Daryl HALL (2)	623
Foolish Pride	456
Someone Like You	166
27. Bryan ADAMS/Jim VALLANCE (17)	**17031**
Bryan ADAMS (12)	15860
Heaven	4136
Run To You	2116
Summer Of '69	2007
Heat Of The Night	1759
Somebody	1214
One Night Love Affair	1047
Cuts Like A Knife	982
It's Only Love	947
This Time	649
Hearts On Fire	544
Victim Of Love	434
Lonely Nights	24.8
38 SPECIAL (1)	520
Teacher, Teacher	520
PRISM (1)	343
Don't Let Him Know	343
LOVERBOY (1)	163
Dangerous	163
Joe COCKER (1)	113
Edge Of A Dream	113
Roger DALTREY (1)	33.2
Let Me Down Easy	33.2
28. Ric OCASEK (16)	**16972**
The CARS (13)	15640
Shake It Up	3549
Drive	3543
You Might Think	1977
Tonight She Comes	1667
Magic	1357
Hello Again	882
You Are The Girl	818
Why Can't I Have You	571
I'm Not The One	445
Touch And Go	431
Since You're Gone	294
Coming Up You	75.3
Strap Me In	31.3
Ric OCASEK (3)	1332
Emotion In Motion	937
Something To Grab For	288
True To You	107
29. Annie LENNOX/David A. STEWART (15)	**16938**
EURYTHMICS (15)	16938
Sweet Dreams (Are Made Of This)	6665
Here Comes The Rain Again	3206
Would I Lie To You?	1926
Missionary Man	1031
Sisters Are Doin' It For Themselves	759
Who's That Girl?	745
Love Is A Stranger	646
There Must Be An Angel (Playing With My Heart)	554
Right By Your Side	518
I Need A Man	258
Don't Ask Me Why	257
You Have Placed A Chill In My Heart	136
Thorn In My Side	120
It's Alright (Baby's Coming Back)	70.4
Sexcrime (Nineteen Eighty-Four)	46.1
30. Michael CRAIG/Roy HAY/Jon MOSS/George O'DOWD (7)	**16751**
CULTURE CLUB (7)	16751
Do You Really Want To Hurt Me	5296
Time (Clock Of The Heart)	4390
Miss Me Blind	2336
Church Of The Poison Mind	1763
I'll Tumble 4 Ya	1713
The War Song	715
Mistake No. 3	537
31. Mick JAGGER/Keith RICHARDS (11)	**16718**
ROLLING STONES (9)	16085
Start Me Up	5491
Emotional Rescue	4012
Undercover Of The Night	1657
Mixed Emotions	1515
Waiting On A Friend	1289
Rock And A Hard Place	641
Hang Fire	608
She's So Cold	588
She Was Hot	283
Aretha FRANKLIN (1)	534
Jumpin' Jack Flash	534
Bette MIDLER (1)	99.2
Beast Of Burden	99.2
32. Christopher CROSS (7)	**16109**
Christopher CROSS (7)	16109
Ride Like The Wind	5822
Sailing	4791
All Right	1564
Think Of Laura	1548
Never Be The Same	1246
Say You'll Be Mine	722
No Time For Talk	416
33. Kenneth EDMONDS/Antonio REID/Daryl SIMMONS (9)	**16081**
Bobby BROWN (2)	5504
On Our Own	3811
Don't Be Cruel	1692
Karyn WHITE (3)	4863
The Way You Love Me	1874
Secret Rendezvous	1805
Superwoman	1183
Sheena EASTON (1)	2769
The Lover In Me	2769
BABYFACE (1)	1490
It's No Crime	1490
The BOYS (1)	1097
Dial My Heart	1097
Paula ABDUL (1)	357
Knocked Out	357
34. Peter CETERA/David FOSTER (5)	**15958**
CHICAGO (4)	11902
Hard To Say I'm Sorry	6847
You're The Inspiration	3237
Stay The Night	1052
Love Me Tomorrow	766

Writers Ranked

Rank. Writer Team (Entries) Act (Entries) Title	Score
34. Peter CETERA/David FOSTER *Continued*	
Peter CETERA (1)	4056
Glory Of Love (Theme From The Karate Kid Part II)	4056
35. Jon BON JOVI/Desmond CHILD/ Richie SAMBORA (5)	**15700**
BON JOVI (4)	14811
Livin' On A Prayer	4968
You Give Love A Bad Name	3706
Bad Medicine	3342
Born To Be My Baby	2796
CHER (1)	889
We All Sleep Alone	889
36. Curtis BEDEAU/Gerard CHARLES/Hugh CLARKE/Brian GEORGE/Lucien GEORGE/ Paul GEORGE (9)	**15678**
LISA LISA And CULT JAM (3)	8045
Head To Toe	4139
Lost In Emotion	3495
Little Jackie Wants To Be A Star	410
Samantha FOX (2)	4296
Naughty Girls (Need Love Too)	2837
I Wanna Have Some Fun	1459
LISA LISA And CULT JAM With FULL FORCE (3)	2823
All Cried Out	1799
I Wonder If I Take You Home	707
Can You Feel The Beat	317
Cheryl Pepsii RILEY (1)	514
Thanks For My Child	514
37. Kevin CRONIN (9)	**15171**
REO SPEEDWAGON (9)	15171
Can't Fight This Feeling	5170
Keep On Loving You	4866
Keep The Fire Burnin'	2157
That Ain't Love	719
Sweet Time	620
Don't Let Him Go	614
I Do 'wanna Know	521
Live Every Moment	412
Time For Me To Fly	92.7
38. Madonna L. CICCONE/ Patrick Raymond LEONARD (4)	**15076**
MADONNA (4)	15076
Like A Prayer	4497
Live To Tell	4244
Who's That Girl	3537
Cherish	2798
39. Jim STEINMAN (7)	**15043**
Bonnie TYLER (1)	7997
Total Eclipse Of The Heart	7997
AIR SUPPLY (1)	5057
Making Love Out Of Nothing At All	5057
Barry MANILOW (1)	981
Read 'Em And Weep	981
Jim STEINMAN (1)	608
Rock And Roll Dreams Come Through	608
Barbra STREISAND (1)	310
Left In The Dark	310
FIRE INC. (1)	60.2
Tonight Is What It Means To Be Young	60.2
MEAT LOAF (1)	29.6
I'm Gonna Love Her For Both Of Us	29.6
40. Elton JOHN/Bernie TAUPIN (13)	**15028**
Elton JOHN (12)	14910
I Don't Wanna Go On With You Like That	2818

Rank. Writer Team (Entries) Act (Entries) Title	Score
Sad Songs (Say So Much)	2244
Candle In The Wind (Live)	1764
Nikita	1673
I'm Still Standing	1429
Empty Garden (Hey Hey Johnny)	1202
Healing Hands	900
Who Wears These Shoes?	887
Kiss The Bride	670
A Word In Spanish	618
In Neon	498
Heartache All Over The World	206
Olivia NEWTON-JOHN (1)	118
The Rumour	118
41. Terry BRITTEN/Graham LYLE (5)	**14680**
Tina TURNER (5)	14680
What's Love Got To Do With It	6823
Typical Male	3330
We Don't Need Another Hero (Thunderdome)	3151
What You Get Is What You See	887
Two People	489
42. Bob SEGER (13)	**14414**
Bob SEGER (4)	6612
Against The Wind	2801
Fire Lake	2278
You'll Accomp'ny Me	1197
The Horizontal Bop	336
Bob SEGER & The SILVER BULLET BAND (8)	5169
Even Now	1141
American Storm	999
Like A Rock	958
Understanding	952
Roll Me Away	516
Feel Like A Number	248
It's You	221
Miami	133
Sheena EASTON (1)	2632
We've Got Tonight	2632
43. Andy FARRISS/Michael HUTCHENCE (8)	**14207**
INXS (8)	14207
Need You Tonight	3984
Devil Inside	3056
New Sensation	2395
What You Need	2349
Never Tear Us Apart	1587
The One Thing	557
Original Sin	177
I Send A Message	101
44. Adam CLAYTON/Dave EVANS/Paul HEWSON/Larry MULLEN (11)	**14177**
U2 (10)	14047
With Or Without You	4480
I Still Haven't Found What I'm Looking For	4006
Desire	2423
Where The Streets Have No Name	927
Angel Of Harlem	878
Pride (In The Name Of Love)	559
In God's Country	361
New Year's Day	305
I Will Follow	69.0
All I Want Is You	39.2
B.B. KING (1)	130
When Love Comes To Town	130
45. Dan FOGELBERG (11)	**14104**
Dan FOGELBERG (11)	14104

Rank. Writer Team (Entries) Act (Entries) Title	Score
Longer	4243
Leader Of The Band	2045
Hard To Say	1758
Same Old Lang Syne	1561
The Language Of Love	1149
Missing You	860
Run For The Roses	840
Heart Hotels	677
Make Love Stay	633
Believe In Me	277
She Don't Look Back	60.6
46. Jonathan CAIN/Steve PERRY (7)	**13732**
JOURNEY (7)	13732
Open Arms	5744
Who's Crying Now	3180
Separate Ways (Worlds Apart)	2468
Send Her My Love	718
Suzanne	696
After The Fall	692
Why Can't This Night Go On Forever	235
47. Lou GRAMM/Michael Leslie JONES (8)	**13575**
FOREIGNER (8)	13575
Waiting For A Girl Like You	9492
Say You Will	1833
That Was Yesterday	1021
Juke Box Hero	505
Reaction To Action	216
Down On Love	211
Heart Turns To Stone	208
Luanne	88.8
48. Dennis DeYOUNG (11)	**13471**
STYX (7)	11653
The Best Of Times	4383
Mr. Roboto	3714
Don't Let It End	2206
Why Me	612
Music Time	342
High Time	237
Nothing Ever Goes As Planned	159
Dennis DeYOUNG (4)	1818
Desert Moon	1501
Call Me	273
Don't Wait For Heroes	30.9
This Is The Time	13.6
49. Will JENNINGS/Steve WINWOOD (10)	**13441**
Steve WINWOOD (10)	13441
Higher Love	3755
While You See A Chance	2011
The Finer Things	1776
Don't You Know What The Night Can Do?	1648
Valerie	1485
Back In The High Life Again	1098
Holding On	1045
Still In The Game	294
Talking Back To The Night	247
Valerie	82.8
50. Maurice STARR (5)	**13284**
NEW KIDS ON THE BLOCK (5)	13284
I'll Be Loving You (Forever)	3690
Hangin' Tough	3272
Cover Girl	2518
You Got It (The Right Stuff)	2405
Please Don't Go Girl	1400

Top 50 Singles Acts and Their Writers

Writers Ranked

Rank. Act (Chart Entries) Writer (Writer Credits)	Score
1. Michael JACKSON (21)	**61327**
Michael Jackson (11)	37999
Rod Temperton (3)	11247
Paul McCartney (1)	2584
Glen Ballard (1)	1987
Siedah Garrett (1)	1987
Tom Bahler (1)	1704
John Bettis (1)	922
Steve Porcaro (1)	922
James Ingram (1)	662
Quincy Jones (1)	662
Keni St. Lewis (1)	454
Renée Armand (1)	84.2
Samuel F. Brown III (1)	84.2
Stevie Wonder (1)	28.5
2. MADONNA (18)	**59786**
Madonna L. Ciccone (12)	19652
Patrick Raymond Leonard (5)	8353
Stephen Bray (4)	5509
Thomas F. Kelly (1)	3265
Billy Steinberg (1)	3265
John Bettis (1)	2827
Jon Lind (1)	2827
Brian Elliot (1)	2174
Reggie Lucas (1)	2009
Peter H. Brown (1)	1887
Robert Rans (1)	1887
Gardner Cole (1)	1192
Peter Rafelson (1)	1192
Andrea LaRusso (1)	856
Peggy Stanziale (1)	856
Bruce Gaitsch (1)	815
Curtis Hudson (1)	610
Lisa Stevens (1)	610
3. Daryl HALL & John OATES (22)	**55777**
Daryl Hall (18)	26961
John Oates (9)	9586
Sara Allen (9)	8318
Janna Allen (4)	5298
Warren Pash (1)	1358
Barry Mann (1)	578
Phil Spector (1)	578
Cynthia Weil (1)	578
Tim Cross (1)	309
Richard Fenn (1)	309
Mike Frye (1)	309
Mike Oldfield (1)	309
Morris Pert (1)	309
Maggie Reilly (1)	309
Smokey Robinson (2)	295
Robert Rogers (1)	148
Ronnie White (1)	148
Rick Iantosca (1)	78.9
4. Lionel RICHIE (15)	**52069**
Lionel Richie (15)	44640
Brenda Richie (2)	2365
Cynthia Weil (2)	1627
Michael Frenchik (1)	1289
Carlos Rios (1)	1289
Greg Phillinganes (2)	859
5. Phil COLLINS (14)	**39768**
Phil Collins (11)	29529
Lamont Dozier (2)	2625
Stephen Bishop (1)	2376
Carole Bayer Sager (1)	1961
Toni Wine (1)	1961
Brian Holland (1)	658
Eddie Holland (1)	658
6. Billy JOEL (22)	**38031**
Billy Joel (22)	38031
7. DURAN DURAN (15)	**35594**
Simon Le Bon (15)	7876
Nick Rhodes (15)	7876
Nigel John Taylor (15)	7876
Andy Taylor (9)	5605
Roger Taylor (9)	5605
John Barry (1)	716
Eleanor Broadwater (1)	9.9
Robert Chaisson (1)	9.9
Dale Hawkins (1)	9.9
Stanley J. Lewis (1)	9.9
8. Whitney HOUSTON (11)	**35563**
Michael Masser (3)	6262
George Merrill (2)	3939
Shannon Rubicam (2)	3939
LaForrest Cope (1)	2786
Linda Creed (1)	2286
Thomas F. Kelly (1)	2171
Billy Steinberg (1)	2171
Will Jennings (1)	2003
Gerry Goffin (1)	1972
Charles Henry Jackson Jr. (1)	1781
Frank Wildhorn (1)	1781
Narada Michael Walden (1)	1490
Toni Colandreo (1)	1182
Albert Hammond (2)	899
John Bettis (1)	831
Diane Warren (1)	68.3
9. Kenny ROGERS (20)	**35216**
Lionel Richie (1)	8423
Ricky Lynn Christian (1)	3554
Barry Gibb (3)	2661
Maurice Gibb (2)	2317
Robin Gibb (1)	2280
Roger Bowling (1)	2213
Billy Edd Wheeler (1)	2213
Kim Carnes (1)	1597
Dave Ellingson (1)	1597
Bob Seger (1)	1316
Kenny Rogers (3)	707
Al Braggs (1)	579
Deadric Malone (1)	579
Steve Dorff (1)	552
Marty Panzer (1)	552
Bob Morrison (1)	499
Johnny Wilson (1)	499
Richard Marx (2)	360
David Malloy (1)	347
Thom Schuyler (1)	347
Even Stevens (1)	347
Albhy Galuten (1)	344
David Foster (1)	320
Lee Greenwood (1)	280
Dave Loggins (1)	153
Charles David Robbins (1)	132
Jeff Silbar (1)	132
Van Stephenson (1)	132
Larry Keith (1)	48.9
Danny Morrison (1)	48.9
Johnny Slate (1)	48.9
John Jarvis (1)	31.7
Mike Dekle (1)	8.2
10. John MELLENCAMP (20)	**35178**
John Mellencamp (20)	30136
George Green (3)	5042
11. AIR SUPPLY (13)	**34148**
Graham Russell (5)	15291
Jim Steinman (1)	5057
Clive Davis (1)	3040
Norman Sallitt (1)	2848
Dominic Bugatti (1)	1660
Frank Musker (1)	1660
Kenneth Edward Bell (1)	1009
Terry Skinner (1)	1009
J.L. Wallace (1)	1009
Rob Hegel (1)	409
Dick Wagner (1)	409
Howard Greenfield (1)	278
Ken Hirsch (1)	278
Albert Hammond (1)	54.0
Diane Warren (1)	54.0
Mary Applegate (1)	20.9
Candy DeRouge (1)	20.9
Gunther Mende (1)	20.9
Jennifer Rush (1)	20.9
12. Huey LEWIS and the NEWS (17)	**33246**
Huey Lewis (10)	9084
Chris Hayes (6)	5764
Johnny Colla (4)	3959
Alex Call (1)	2388
Mutt Lange (1)	1774
Bruce Hornsby (1)	1768
John Hornsby (1)	1768
Mike Duke (2)	1153
Bill Gibson (3)	1016
Sean Hopper (2)	938
Mike Chapman (1)	914
Nicky Chinn (1)	914
Phil Cody (1)	740
André Pessis (1)	467
Kevin Wells (1)	467
Steve Lewis (1)	78.4
Mario Cipollina (1)	55.1
13. KOOL & The GANG (19)	**30914**
George M. Brown (19)	6395
Ronald Bell (18)	3627
J.T. Taylor (18)	3627
Robert Bell (17)	3589
Claydes Smith (17)	3589
Curtis Fitzgerald Williams (11)	2234
James Bonneford (6)	1691
Dennis Thomas (5)	1195
Eumir Deodato (7)	1171
Robert Mickens (7)	1049
Earl Toon (1)	948
Clifford Adams (5)	831
Michael Ray (4)	346
Meekaaeel Muhammed (1)	303
Sandy Linzer (1)	185
Amir Bayyan (1)	57.5
Dwania Kyles (1)	37.3
Kendal Stubbs (1)	37.3
Amos Guider (1)	2.5
14. Olivia NEWTON-JOHN (14)	**30618**
Steve Kipner (3)	8972
John Farrar (5)	8655
Terry Shaddick (1)	5545
Paul Steven Bliss (1)	2011
Peter Beckett (1)	1416
Tom Snow (2)	1327
Jeff Lynne (1)	834
Mark Goldenberg (1)	671
Barry Gibb (1)	595
Barry Alfonso (1)	230
Lee Ritenour (1)	205
Elton John (1)	58.9
Bernie Taupin (1)	58.9
David Foster (1)	12.9
Jeremy Lubbock (1)	12.9
Richard Marx (1)	12.9
15. CHICAGO (16)	**29063**
David Foster (5)	6774
Peter Cetera (6)	6528
Diane Warren (2)	5825
Steve Kipner (3)	2366
John Lewis Parker (1)	1856
Albert Hammond (1)	1209
James Scott (1)	1099
Richard Baskin (1)	823
Tom Keane (1)	823
Mark Goldenberg (1)	509
Randy Goodrum (1)	501
Robert Lamm (2)	314
John Dexter (1)	151
Jason Scheff (1)	151
Danny Seraphine (1)	67.7
Jay Gruska (1)	28.2
Joseph Amos Williams (1)	28.2
Bobby Caldwell (1)	8.9
16. Diana ROSS (18)	**28634**
Lionel Richie (2)	6964
Bernard Edwards (2)	5680
Nile Rodgers (2)	5680
Michael Jackson (2)	1738
Michael Masser (2)	1070

Writers Ranked

Rank. Act (Chart Entries) Writer (Writer Credits)	Score
16. Diana ROSS *Continued*	
Carole Bayer Sager (1)	1042
Dennis Matkosky (1)	719
Michael Sembello (1)	719
George Goldner (1)	466
Morris Levy (1)	466
Frankie Lymon (1)	466
Jimmy Merchant (1)	466
Herman Santiago (1)	466
Sara Allen (1)	418
Daryl Hall (1)	418
John Capek (1)	220
Marc Jordan (1)	220
Diana Ross (2)	219
Julio Iglesias (1)	141
Tony Renis (1)	141
Cynthia Weil (1)	141
Rob Mounsey (1)	138
Bill Wray (1)	138
Barry Gibb (3)	83.4
Maurice Gibb (3)	83.4
Ray Chew (1)	80.9
Paul Jabara (1)	80.9
Robin Gibb (2)	52.9
Franne Golde (1)	45.6
Peter Ivers (1)	45.6
Gerry Goffin (1)	27.2
17. PRINCE (16)	**28172**
Prince (16)	27937
Sheena Easton (1)	235
18. Elton JOHN (20)	**27052**
Elton John (18)	12289
Bernie Taupin (14)	8578
Gary Osborne (4)	3817
Davey Johnstone (2)	1123
Jean-Paul Dreau (1)	320
Tom Robinson (1)	214
Albert Hammond (1)	159
Diane Warren (1)	159
Fred Mandel (1)	131
Charlie Morgan (1)	131
Paul Westwood (1)	131
19. Cyndi LAUPER (13)	**26545**
Cyndi Lauper (6)	5223
Robert Hazard (1)	4890
Rob Hyman (1)	2696
Thomas F. Kelly (3)	2551
Billy Steinberg (3)	2551
Jules Shear (1)	2186
Stephen Lunt (3)	1328
Essra Mohawk (1)	1163
Rick Chertoff (1)	943
Gary Corbett (1)	943
Thomas Gray (1)	549
Arthur Stead (1)	362
Renaldo Benson (1)	320
Al Cleveland (1)	320
Marvin Gaye (1)	320
Richard Orange (1)	177
Jeff Bova (1)	22.8
20. Billy OCEAN (11)	**26542**
Billy Ocean (11)	10530
Keith Diamond (4)	5239
Mutt Lange (4)	4077

Rank. Act (Chart Entries) Writer (Writer Credits)	Score
Wayne Brathwaite (5)	3131
Barry Eastmond (5)	3131
James Woodley (1)	226
Jolyon Skinner (1)	208
21. Rick SPRINGFIELD (17)	**26085**
Rick Springfield (16)	22478
Sammy Hagar (1)	1929
Danny Tate (1)	609
Blaise Tosti (1)	609
Eric McCusker (1)	230
Tim Pierce (1)	230
22. POINTER SISTERS (19)	**25341**
John Bettis (1)	2400
Michael Clark (1)	2400
Cynthia Weil (2)	2034
Tom Snow (1)	1920
Allee Willis (2)	1384
Mark Goldenberg (1)	1338
Brock Walsh (1)	1338
Danny Sembello (1)	1266
Stephen Mitchell (1)	1233
Marti Sharron (1)	1233
Gary Skardina (1)	1233
Layng Martine Jr. (1)	1109
Parker McGee (1)	952
Trevor Lawrence (3)	673
Anita Pointer (3)	673
David Innis (1)	655
Sam Lorber (1)	655
June Pointer (2)	569
Ruth Pointer (2)	569
Andy Goldmark (2)	255
Bruce Roberts (2)	255
David McHugh (1)	214
John Alden Black (1)	137
Richard Feldman (1)	137
Nan O'Byrne (1)	137
Franne Golde (1)	118
James Ingram (1)	113
Barry Mann (1)	113
Marlo Henderson (1)	104
John Lewis Parker (1)	58.7
Brian Potter (1)	58.7
Estelle Levitt (1)	4.3
Jerry Ragovoy (1)	4.3
23. George MICHAEL (8)	**25301**
George Michael (7)	23568
Simon Climie (1)	867
Dennis W. Morgan (1)	867
24. CULTURE CLUB (10)	**25285**
Michael Craig (10)	5895
Roy Hay (10)	5895
Jon Moss (10)	5895
George O'Dowd (10)	5895
Phil Pickett (3)	1707
25. Bruce SPRINGSTEEN (14)	**24999**
Bruce Springsteen (13)	23637
Barrett Strong (1)	681
Norman Whitfield (1)	681
26. FOREIGNER (13)	**24892**
Michael Leslie Jones (13)	18104
Lou Gramm (8)	6787
27. BON JOVI (12)	**24759**
Jon Bon Jovi (11)	10551

Rank. Act (Chart Entries) Writer (Writer Credits)	Score
Richie Sambora (7)	8642
Desmond Child (4)	4937
Mark Avsec (1)	298
George Karak (1)	236
Dave Bryan (1)	95.0
28. Stevie WONDER (18)	**24620**
Stevie Wonder (17)	24430
Ken Hirsch (1)	95.1
Ronald Norman Miller (1)	95.1
29. JOURNEY (19)	**23887**
Steve Perry (19)	10610
Jonathan Cain (14)	10547
Neal Schon (10)	2730
30. Paul McCARTNEY (12)	**23877**
Paul McCartney (11)	18367
Michael Jackson (2)	5257
Elvis Costello (1)	226
Eric Stewart (1)	26.7
31. The POLICE (10)	**23786**
Sting (10)	23786
32. SURVIVOR (18)	**23583**
Jim Peterik (17)	11874
Frankie Sullivan (15)	11538
Peter Beckett (1)	52.3
Bill Conti (1)	52.3
Dennis Lambert (1)	52.3
Jimi Jamison (1)	14.0
33. Sheena EASTON (18)	**22745**
Florrie Palmer (2)	4839
Mike Leeson (1)	1679
Bill Conti (1)	1605
Prince (1)	1509
Bob Seger (1)	1316
Lea Maalfrid (1)	1280
Greg Mathieson (2)	1071
Trevor Veitch (2)	1071
Charlie Dore (1)	1055
Julian Littman (1)	1055
Kenneth Edmonds (1)	923
Antonio Reid (1)	923
Daryl Simmons (1)	923
Dominic Bugatti (2)	771
Frank Musker (1)	485
Jennifer Kimball (1)	444
Cindy Richardson (1)	444
Adele Bertei (1)	275
Mary Alice Kessler (1)	275
Julia Downes (1)	179
Tom Snow (1)	179
Cynthia Weil (1)	179
Tim Scott (1)	75.2
Peter Vale (1)	74.6
Lamont Dozier (1)	37.8
Brian Holland (1)	37.8
Eddie Holland (1)	37.8
34. Christopher CROSS (10)	**22700**
Christopher Cross (10)	17784
Burt Bacharach (2)	1633
Carole Bayer Sager (2)	1633
Peter Allen (1)	1609
Michael Omartian (1)	42.2
35. QUEEN (14)	**22382**
John Deacon (6)	10392
Freddie Mercury (7)	9964

Rank. Act (Chart Entries) Writer (Writer Credits)	Score
Roger Taylor (6)	1454
Brian May (4)	504
David Bowie (1)	67.8
36. Kenny LOGGINS (17)	**22325**
Kenny Loggins (13)	9840
Dean Pitchford (3)	3957
Giorgio Moroder (2)	2217
Tom Whitlock (2)	2217
Michael McDonald (2)	1519
David Foster (3)	707
Mike Towers (1)	690
E. Ein Loggins (4)	660
Steve Perry (1)	233
Bert Berns (1)	92.5
Ina Wolf (1)	73.1
Peter F. Wolf (1)	73.1
Jeff Pescetto (1)	15.5
Pam Reswick (1)	15.5
Steve Werfel (1)	15.5
37. Tina TURNER (13)	**22054**
Terry Britten (6)	7529
Graham Lyle (5)	7340
Holly Knight (3)	2123
Mark Knopfler (1)	1623
Mike Chapman (2)	1184
Nicky Chinn (1)	784
Al Green (1)	243
Al Jackson (1)	243
Willie Mitchell (1)	243
Bryan Adams (1)	237
Jim Vallance (1)	237
Sue Shifrin (1)	189
Rupert Hine (1)	39.2
Jeannette Obstoj (1)	39.2
38. WHAM! (8)	**21683**
George Michael (7)	18511
Andrew Ridgeley (1)	2939
David Was (1)	117
Don Was (1)	117
39. REO SPEEDWAGON (15)	**21530**
Kevin Cronin (11)	16123
Gary Richrath (2)	3426
Neal Doughty (2)	1028
Thomas F. Kelly (1)	532
Rick Braun (1)	420
40. PRINCE And The REVOLUTION (10)	**21448**
Prince (10)	20913
Lisa Coleman (2)	216
Wendy Melvoin (2)	216
Mark Alton Brown (1)	34.6
Matt Fink (1)	34.6
Robert Rivkin (1)	34.6
41. Janet JACKSON (9)	**21166**
James Samuel Harris (6)	9077
Terry Lewis (6)	9077
Monte Moir (1)	1139
Melanie Andrews (1)	759
Janet Jackson (1)	759
Glen Barbee (1)	109
Charmaine Sylvers (1)	109
René Moore (1)	68.3
Angela Winbush (1)	68.3

Rank. Act (Chart Entries) Writer (Writer Credits)	Score
42. Richard MARX (7)	**21085**
Richard Marx (7)	19887
Bruce Gaitsch (1)	1198
43. JEFFERSON AIRPLANE/ STARSHIP (18)	**20993**
Peter F. Wolf (3)	3625
Ina Wolf (2)	2503
Albert Hammond (1)	2360
Diane Warren (1)	2360
Martin Page (2)	1639
Dennis Lambert (1)	1122
Bernie Taupin (1)	1122
Jeannette Sears (3)	680
Pete Sears (3)	680
Craig Chaquico (3)	640
Tommy Funderburk (1)	518
Paul Kantner (2)	487
Phil Galdston (1)	382
Robbie Nevil (1)	382
John Van Tongeren (1)	382
David Freiberg (1)	315
Jim McPherson (1)	315
Johnny Warman (1)	285
Steven Cristol (1)	269
Robin Randall (1)	269
Thomas Borsdorf (1)	262
David Scott Roberts (1)	128
Mutt Lange (1)	85.8
Mickey Thomas (1)	63.9
John Bettis (1)	58.6
Michael Clark (1)	58.6

Rank. Act (Chart Entries) Writer (Writer Credits)	Score
44. Gloria ESTEFAN/ MIAMI SOUND MACHINE (11)	**20947**
Gloria Estefan (6)	13284
Enrique Garcia (3)	3888
Larry Dermer (3)	916
Joe Galdo (3)	916
Rafael Vigil (3)	916
Jorge Casas (1)	342
John DeFaria (1)	342
Clay Ostwald (1)	342
45. BLONDIE (6)	**20847**
Clement A. Bozewski (5)	5186
Deborah Harry (5)	5186
Giorgio Moroder (1)	3310
John Holt (1)	2644
Duke Reid (1)	2644
Christopher Stein (3)	1749
Jimmy Destri (1)	127
46. HEART (15)	**20696**
Thomas F. Kelly (2)	2678
Billy Steinberg (2)	2678
Martin Page (1)	1948
Bernie Taupin (1)	1948
Nancy Wilson (5)	1684
Diane Warren (1)	1573
Holly Knight (2)	1270
Mark Mueller (1)	1156
Ann Wilson (4)	1106
George Richard Davis Jr. (1)	858
Lee Diamond (1)	858
Gene Black (1)	692

Rank. Act (Chart Entries) Writer (Writer Credits)	Score
Brian Allen (1)	468
Sheron Alton (1)	468
Jim Vallance (1)	468
Sue Ennis (3)	413
Jack Conrad (1)	93.7
Bob Garrett (1)	93.7
Mark Andes (1)	57.9
Denny Carmassi (1)	57.9
Howard Leese (1)	57.9
Jonathan Cain (1)	36.0
Alex North (1)	15.6
Hy Zaret (1)	15.6
47. Rod STEWART (20)	**20483**
Rod Stewart (12)	4316
Kevin Savigar (6)	1948
Duane Hitchings (3)	1818
Jim Cregan (5)	1563
Jeff Fortgang (1)	1263
Simon Climie (1)	1200
Dennis W. Morgan (1)	1200
Carmine Appice (1)	742
Jay Davis (2)	726
Roland Robinson (1)	692
Phil Chen (2)	614
Gary Grainger (2)	614
Gene Black (1)	588
Mike Chapman (1)	588
Holly Knight (1)	588
Andy Taylor (1)	507
Bob Dylan (1)	320

Rank. Act (Chart Entries) Writer (Writer Credits)	Score
Danny Whitten (1)	310
Paul Carrack (1)	236
Tony Brock (1)	164
Curtis Mayfield (1)	147
Bryan Adams (1)	59.4
Jim Vallance (1)	59.4
Randy Wayne (1)	59.4
Sam Cooke (1)	49.7
Andy Fraser (1)	47.1
Paul Rodgers (1)	47.1
Steve Harley (1)	14.4
48. GENESIS (15)	**19591**
Phil Collins (15)	7637
Tony Banks (13)	5977
Mike Rutherford (13)	5977
49. MEN AT WORK (6)	**19415**
Colin James Hay (6)	15838
Ron Strykert (1)	3577
50. VAN HALEN (14)	**19350**
Michael Anthony (12)	4440
Alex Van Halen (12)	4440
Eddie Van Halen (12)	4440
David Lee Roth (5)	2520
Sammy Hagar (7)	1920
William Dees (1)	582
Roy Orbison (1)	582
Marvin Gaye (1)	142
Ivy Jo Hunter (1)	142
William Stevenson (1)	142

Writers: 50 or More Consecutive Weeks on the Hot 100

Writer	Weeks	Begin	End
Richard Marx	134	13-Jun-87	30-Dec-89*
Prince	121	02-Jun-84	20-Sep-86
Antonio Reid	101	30-Jan-88	30-Dec-89*
Kenneth Edmonds	101	30-Jan-88	30-Dec-89*
Lionel Richie	97	17-Sep-83	20-Jul-85
Keith Diamond	96	11-Aug-84	07-Jun-86
Bruce Springsteen	95	26-May-84	15-Mar-86
Desmond Child	95	20-Jun-87	08-Apr-89
Holly Knight	94	30-Jun-84	12-Apr-86
James Samuel Harris	92	01-Feb-86	31-Oct-87
Madonna L. Ciccone	92	12-Apr-86	09-Jan-88
Terry Lewis	92	01-Feb-86	31-Oct-87
Mike Rutherford	91	23-Nov-85	15-Aug-87
George O'Dowd	88	04-Dec-82	04-Aug-84
Jon Moss	88	04-Dec-82	04-Aug-84
Michael Craig	88	04-Dec-82	04-Aug-84
Roy Hay	88	04-Dec-82	04-Aug-84
Joe Elliott	86	10-Oct-87	27-May-89
Mutt Lange	86	10-Oct-87	27-May-89
Phil Collen	86	10-Oct-87	27-May-89
Rick Allen	86	10-Oct-87	27-May-89
Rick Savage	86	10-Oct-87	27-May-89
Steve Clark	86	10-Oct-87	27-May-89
George Michael	85	06-Jun-87	14-Jan-89
Daryl Hall	84	24-Jan-81	28-Aug-82
Barry Eastmond	83	07-Sep-85	04-Apr-87
Daryl Simmons	81	18-Jun-88	30-Dec-89*
Maurice Starr	80	25-Jun-88	30-Dec-89*
Jim Vallance	80	29-Sep-84	05-Apr-86
Jonathan Cain	78	18-Jan-86	11-Jul-87
David Foster	77	17-Apr-82	01-Oct-83
Debbie Gibson	77	09-May-87	22-Oct-88
Peter F. Wolf	77	07-Sep-85	21-Feb-87
Phil Collins	77	21-Mar-81	04-Sep-82
Bryan Adams	74	29-Sep-84	22-Feb-86
Narada Michael Walden	74	22-Jun-85	15-Nov-86
Stevie Wonder	74	29-Dec-79	23-May-81
Colin James Hay	73	10-Jul-82	26-Nov-83
Steve Perry	70	18-Jul-81	13-Nov-82
Matt Aitken	68	18-Jul-87	29-Oct-88
Mike Stock	68	18-Jul-87	29-Oct-88
Pete Waterman	68	18-Jul-87	29-Oct-88
Andy Farriss	65	24-Oct-87	14-Jan-89
Michael Hutchence	65	24-Oct-87	14-Jan-89
Billy Joel	64	30-Jul-83	13-Oct-84
Claydes Smith	64	24-Nov-84	08-Feb-86
Curtis Fitzgerald Williams	64	24-Nov-84	08-Feb-86
George M. Brown	64	24-Nov-84	08-Feb-86
J.T. Taylor	64	24-Nov-84	08-Feb-86
James Bonneford	64	24-Nov-84	08-Feb-86
Lewis Martineé	64	24-Jan-87	09-Apr-88
Robert Bell	64	24-Nov-84	08-Feb-86
Ronald Bell	64	24-Nov-84	08-Feb-86
Tony Banks	64	31-May-86	15-Aug-87
Billy Ocean	63	30-Nov-85	07-Feb-87
Ric Ocasek	63	10-Mar-84	18-May-85
Steve Winwood	63	07-Feb-87	16-Apr-88
Wayne Brathwaite	63	30-Nov-85	07-Feb-87
Will Jennings	63	07-Feb-87	16-Apr-88
Jeff Lynne	62	29-Oct-88	30-Dec-89*
Tom Petty	62	29-Oct-88	30-Dec-89*
Axl Rose	62	25-Jun-88	26-Aug-89
Duff McKagan	62	25-Jun-88	26-Aug-89
Izzy Stradlin	62	25-Jun-88	26-Aug-89
Michael Leslie Jones	62	04-Jul-81	04-Sep-82
Sara Allen	62	02-May-81	03-Jul-82
Saul Hudson	62	25-Jun-88	26-Aug-89
Steven Adler	62	25-Jun-88	26-Aug-89
Billy Steinberg	57	13-Aug-88	09-Sep-89
Bruce Hornsby	57	26-Jul-86	22-Aug-87
Frankie Sullivan	57	15-Sep-84	12-Oct-85
Jim Peterik	57	15-Sep-84	12-Oct-85
John Mellencamp	57	24-Aug-85	20-Sep-86
Thomas F. Kelly	57	13-Aug-88	09-Sep-89
David Glasper	55	16-Apr-88	29-Apr-89
Ian Spice	55	16-Apr-88	29-Apr-89
Marcus Lillington	55	16-Apr-88	29-Apr-89
Michael Delahunty	55	16-Apr-88	29-Apr-89
Andy Taylor	54	03-Nov-84	09-Nov-85
Joey Tempest	54	24-Jan-87	30-Jan-88
Nigel John Taylor	54	03-Nov-84	09-Nov-85
Christopher Cross	53	16-Feb-80	14-Feb-81
Peter Gabriel	53	10-May-86	09-May-87
Robbie Nevil	53	11-Oct-86	10-Oct-87
Jon Bon Jovi	52	24-Sep-88	16-Sep-89
Richie Sambora	52	24-Sep-88	16-Sep-89
Steven Tyler	52	03-Oct-87	24-Sep-88
Billy Idol	51	28-Jan-84	12-Jan-85
Burt Bacharach	51	09-Nov-85	25-Oct-86
Carole Bayer Sager	51	09-Nov-85	25-Oct-86
Alex Van Halen	50	21-May-88	29-Apr-89
Cynthia Weil	50	26-Nov-83	03-Nov-84
Eddie Van Halen	50	21-May-88	29-Apr-89
Michael Anthony	50	21-May-88	29-Apr-89
Rod Temperton	50	29-Dec-79	06-Dec-80
Sammy Hagar	50	21-May-88	29-Apr-89

*Designates streak that continued into 1990

Writers: 4 or More Singles Simultaneously on Hot 100

Seven

Diane Warren: 28-Oct to 04-Nov-89
If I Could Turn Back Time - Cher (Diane Warren)
When I See You Smile - Bad English (Diane Warren)
Blame It On The Rain - Milli Vanilli (Diane Warren)
If You Asked Me To - Patti LaBelle (Diane Warren)
Just Like Jesse James - Cher (Desmond Child/Diane Warren)
The Same Love - The Jets (Diane Warren)
When The Night Comes - Joe Cocker (Bryan Adams/Jim Vallance/
   Diane Warren)

Prince: 27-Oct to 24-Nov-84
The Glamorous Life - Sheila E. (Prince)
Let's Go Crazy - Prince And The Revolution (Prince)
I Feel For You - Chaka Khan (Prince)
Purple Rain - Prince And The Revolution (Prince)
Sex Shooter - Apollonia 6 (Prince)
The Belle Of St. Mark - Sheila E. (Prince)
Jungle Love - The Time (Morris Day/Jesse Woods Johnson/Prince)

Six

James Samuel Harris/Terry Lewis: 17-May to 07-Jun-86
Tender Love - Force M.D.'s (James Samuel Harris/Terry Lewis)
Saturday Love - Cherrelle with Alexander O'Neal
   (James Samuel Harris/Terry Lewis)
What Have You Done For Me Lately? - Janet Jackson
   (James Samuel Harris/Terry Lewis)
The Heat Of Heat - Patti Austin (James Samuel Harris/Terry Lewis)
The Finest - The S.O.S. Band (James Samuel Harris/Terry Lewis)
Nasty - Janet Jackson (James Samuel Harris/Terry Lewis)

Kenneth Edmonds: 25-Mar to 01-Apr-89
The Way You Love Me - Karyn White
   (Kenneth Edmonds/Antonio Reid/Daryl Simmons)
The Lover In Me - Sheena Easton
   (Kenneth Edmonds/Antonio Reid/Daryl Simmons)
Dial My Heart - The Boys (Kenneth Edmonds/Antonio Reid/Daryl Simmons)
Roni - Bobby Brown (Darnell Bristol/Kenneth Edmonds)
Superwoman - Karyn White
   (Kenneth Edmonds/Antonio Reid/Daryl Simmons)
Every Little Step - Bobby Brown (Kenneth Edmonds/Antonio Reid)

Five

Antonio Reid: 25-Mar to 01-Apr-89
The Way You Love Me - Karyn White
   (Kenneth Edmonds/Antonio Reid/Daryl Simmons)
The Lover In Me - Sheena Easton
   (Kenneth Edmonds/Antonio Reid/Daryl Simmons)
Dial My Heart - The Boys (Kenneth Edmonds/Antonio Reid/Daryl Simmons)
Superwoman - Karyn White
   (Kenneth Edmonds/Antonio Reid/Daryl Simmons)
Every Little Step - Bobby Brown (Kenneth Edmonds/Antonio Reid)

Dean Pitchford: 12-May to 19-May-84
Footloose - Kenny Loggins (Kenny Loggins/Dean Pitchford)
Holding Out For Hero - Bonnie Tyler (Dean Pitchford/Jim Steinman)
Dancing In The Sheets - Shalamar (Dean Pitchford/Bill Wolfer)
Let's Hear It For The Boy - Deniece Williams (Dean Pitchford/Tom Snow)
Almost Paradise...Love Theme From "Footloose" -
   Mike Reno And Ann Wilson (Eric Carmen/Dean Pitchford)

Dean Pitchford: 16-Jun to 30-Jun-84
Footloose - Kenny Loggins (Kenny Loggins/Dean Pitchford)
Dancing In The Sheets - Shalamar (Dean Pitchford/Bill Wolfer)
Let's Hear It For The Boy - Deniece Williams (Dean Pitchford/Tom Snow)
Almost Paradise...Love Theme From "Footloose" -
   Mike Reno And Ann Wilson (Eric Carmen/Dean Pitchford)
I'm Free (Heaven Helps The Man) - Kenny Loggins
   (Kenny Loggins/Dean Pitchford)

David Foster: 01-Jun to 08-Jun-85
Through The Fire - Chaka Khan (David Foster/Tom Keane/Cynthia Weil)
You're The Only Love - Paul Hyde And The Payolas
   (David Foster/Paul Hyde/Miriam Nelson/Bob Rock)
Forever - Kenny Loggins (David Foster/E. Ein Loggins/Kenny Loggins)
Lady Of My Heart - Jack Wagner (Glen Ballard/David Foster/Jay Graydon)
Who's Holding Donna Now - DeBarge
   (David Foster/Randy Goodrum/Jay Graydon)

David Foster: 22-Jun to 13-Jul-85
Through The Fire - Chaka Khan (David Foster/Tom Keane/Cynthia Weil)
Forever - Kenny Loggins (David Foster/E. Ein Loggins/Kenny Loggins)
Lady Of My Heart - Jack Wagner (Glen Ballard/David Foster/Jay Graydon)
Who's Holding Donna Now - DeBarge
   (David Foster/Randy Goodrum/Jay Graydon)
St. Elmo's Fire (Man In Motion) - John Parr (David Foster/John Parr)

Four

Matt Aitken, Mike Stock, Pete Waterman: 07-May to 21-May-88
Never Gonna Give You Up - Rick Astley
   (Matt Aitken/Mike Stock/Pete Waterman)
Love In The First Degree - Bananarama (Matt Aitken/Sarah Dallin/
   Siobhan Fahey/Mike Stock/Pete Waterman/Keren Woodward)
Together Forever - Rick Astley (Matt Aitken/Mike Stock/Pete Waterman)
I Should Be So Lucky - Kylie Minogue
   (Matt Aitken/Mike Stock/Pete Waterman)

Barry Gibb: 31-Jan to 14-Feb-81
Woman In Love - Barbra Streisand (Barry Gibb/Robin Gibb)
Guilty - Barbra Streisand & Barry Gibb
   (Barry Gibb/Maurice Gibb/Robin Gibb)
Time Is Time - Andy Gibb (Andy Gibb/Barry Gibb)
What Kind Of Fool - Barbra Streisand & Barry Gibb
   (Albhy Galuten/Barry Gibb)

Bernie Taupin: 18-Jan to 25-Jan-86
These Dreams - Heart (Martin Page/Bernie Taupin)
Nikita - Elton John (Elton John/Bernie Taupin)
We Built This City - Starship
   (Dennis Lambert/Martin Page/Bernie Taupin/Peter F. Wolf)
Wrap Her Up -
   Elton John (Elton John/Davey Johnstone/Fred Mandel/
   Charlie Morgan/Bernie Taupin/Paul Westwood)

Will Jennings: 10-Oct to 17-Oct-87
Back In The High Life Again - Steve Winwood
   (Will Jennings/Steve Winwood)
Didn't We Almost Have It All - Whitney Houston
   (Will Jennings/Michael Masser)
Boys Night Out - Timothy B. Schmit
   (Bruce Gaitsch/Will Jennings/Timothy B. Schmit)
Valerie - Steve Winwood
   (Will Jennings/Steve Winwood)

Writers Ranked

Desmond Child: 26-Nov to 17-Dec-88
I Hate Myself For Loving You - Joan Jett And The Blackhearts
    (Desmond Child/Joan Jett)
Bad Medicine - Bon Jovi (Jon Bon Jovi/Desmond Child/Richie Sambora)
Little Liar - Joan Jett And The Blackhearts (Desmond Child/Joan Jett)
Born To Be My Baby - Bon Jovi
    (Jon Bon Jovi/Desmond Child/Richie Sambora)

Desmond Child: 14-Jan to 21-Jan-89
Bad Medicine - Bon Jovi
    (Jon Bon Jovi/Desmond Child/Richie Sambora)
Little Liar - Joan Jett And The Blackhearts
    (Desmond Child/Joan Jett)
Born To Be My Baby - Bon Jovi
    (Jon Bon Jovi/Desmond Child/Richie Sambora)
Let's Put The X In Sex - KISS (Desmond Child/Paul Stanley)

Daryl Simmons: 10-Dec to 14-Jan-89
The Way You Love Me - Karyn White
    (Kenneth Edmonds/Antonio Reid/Daryl Simmons)
The Lover In Me - Sheena Easton
    (Kenneth Edmonds/Antonio Reid/Daryl Simmons)
Don't Be Cruel - Bobby Brown
    (Kenneth Edmonds/Antonio Reid/Daryl Simmons)
Dial My Heart - The Boys (Kenneth Edmonds/Antonio Reid/Daryl Simmons)

Daryl Simmons: 28-Jan to 01-Apr-89
The Way You Love Me - Karyn White
    (Kenneth Edmonds/Antonio Reid/Daryl Simmons)
The Lover In Me - Sheena Easton
    (Kenneth Edmonds/Antonio Reid/Daryl Simmons)
Dial My Heart - The Boys (Kenneth Edmonds/Antonio Reid/Daryl Simmons)
Superwoman - Karyn White
    (Kenneth Edmonds/Antonio Reid/Daryl Simmons)

Ranking the Producers

Top 150 Producer Teams

Top 50 Producer Teams with Their Acts and Singles

Producers

Ranking The Producer Teams 1-150

Singles are scored, attributed to a producer team and the scores are summed.

Rank. Producer Team (Entries)	Score
1. Quincy Jones (35)	82259
2. James Anthony Carmichael/ Lionel Richie (15)	48647
3. George Michael (14)	43098
4. Narada Michael Walden (28)	41107
5. Daryl Hall/John Oates (12)	40933
6. Michael Omartian (24)	40118
7. Phil Ramone (29)	38706
8. Chris Thomas (27)	37727
9. Ron Nevison (31)	36276
10. Giorgio Moroder (16)	35510
11. Keith Olsen (30)	33986
12. David Foster (31)	33723
13. Prince (18)	33480
14. Huey Lewis And The News (17)	33246
15. Richard Perry (25)	33023
16. Bruce Fairbairn (20)	32876
17. Don Gehman/John Mellencamp (15)	32277
18. George Martin (15)	31721
19. Albhy Galuten/Barry Gibb/ Karl Richardson (15)	29343
20. Jimmy Jam Harris/Terry Lewis (21)	28893
21. Kenneth Edmonds/L.A. Reid (17)	27592
22. Stevie Wonder (18)	26991
23. John Farrar (8)	26252
24. Steve Levine (12)	24944
25. Lionel Richie (6)	24772
26. Arif Mardin (18)	24622
27. Mike Chapman (23)	23444
28. Matt Aitken/Mike Stock/ Pete Waterman (22)	22997
29. Mutt Lange (17)	22932
30. Jimmy Iovine (19)	22567
31. George Tobin (18)	22052
32. Michael Masser (12)	22007
33. Ted Templeman (23)	21455
34. Patrick Leonard/Madonna (6)	21096
35. Phil Collins/Hugh Padgham (7)	21060
36. Hugh Padgham/Police (7)	20166
37. Rick Chertoff (14)	19940
38. Val Garay (19)	19840
39. Peter Mclan (7)	19509
40. Ray Parker Jr. (14)	19259
41. Queen (4)	18635
42. Christopher Neil (18)	18381
43. Kevin Elson/Mike Stone (9)	17913
44. Richard Landis (15)	17724
45. David A. Stewart (16)	17441
46. Bryan Adams/Bob Clearmountain (13)	17324
47. Peter Asher (16)	16729
48. David Malloy (10)	16522
49. Prince And The Revolution (8)	16140
50. Rupert Hine (23)	16127
51. Genesis/Hugh Padgham (9)	15841
52. Frank Farian (5)	15800
53. Full Force (10)	15781
54. Jack Douglas/John Lennon/ Yoko Ono (4)	15578
55. Keith Forsey (14)	15518
56. Nile Rodgers (10)	15491
57. David Cole/Richard Marx (5)	15342
58. Jim Steinman (8)	15308
59. Terry Britten (8)	15271
60. Jim Peterik/Frankie Sullivan (6)	15250
61. Tom Werman (23)	15236
62. Toto (9)	15149
63. Eumir Deodato (10)	15068
64. Alan Tarney (10)	14629
65. George Duke (14)	14458
66. Seth Justman (12)	14317
67. Harry Maslin (7)	14213
68. Mick Jones/Mutt Lange (5)	14125
69. Jon Landau/Chuck Plotkin/ Bruce Springsteen/ Steven Van Zandt (6)	14061
70. Burt Bacharach/Carole Bayer Sager (5)	13941
71. Keith Diamond (6)	13909
72. Stewart Levine (9)	13625
73. Stephen Hague (12)	13539
74. Stephen Bray/Madonna (4)	13486
75. Chris Hughes (5)	13471
76. Styx (9)	13467
77. Bernard Edwards (8)	13367
78. Ronald Bell/Jim Bonneford/ Kool & The Gang (6)	13123
79. Richie Zito (14)	13073
80. John Boylan (14)	13005
81. Glimmer Twins (6)	12529
82. Lewis A. Martinée (7)	12394
83. Bernard Edwards/Nile Rodgers (10)	12217
84. Ric Wake (5)	11970
85. Jack White (9)	11870
86. Trevor Horn (9)	11865
87. Bob Gaudio (11)	11620
88. Michael Lloyd (9)	11316
89. Jay Graydon (16)	11228
90. Michael Jackson (3)	5679
91. David Kahne (6)	10946
92. Rick James (12)	10889
93. Phil Collins (8)	10800
94. Larry Butler (10)	10770
95. Ritchie Cordell/Kenny Laguna (3)	10693
96. Paul DeVilliers/Mr. Mister (3)	10646
97. Mitchell Froom (6)	10609
98. Mike Clink (6)	10483
99. Bob Clearmountain/Daryl Hall/ John Oates (5)	10414
100. Dave Edmunds (12)	10403
101. Jellybean Benitez (8)	10396
102. Wayne Brathwaite/Barry Eastmond (5)	10369
103. Brian Eno/Daniel Lanois (5)	10333
104. Cyndi Lauper/Lennie Petze (9)	10305
105. Emilio Estefan (5)	10175
106. Clive Langer/Alan Winstanley (8)	9600
107. Jeff Lynne (12)	9563
108. John Ryan (16)	9495
109. Steve Perry (9)	9493
110. Bill Drescher/Rick Springfield (10)	9481
111. Human League/Martin Rushent (3)	9440
112. Jimmy Iovine/Tom Petty (7)	9428
113. Roy Thomas Baker (11)	9387
114. Alex Sadkin (5)	9381
115. Rick Nowels (6)	9352
116. Alan Parsons (11)	9317
117. Jeremy Smith/Peter Wolf (4)	9309
118. Dan Fogelberg/Marty Lewis (9)	9161
119. Peter Bunetta/Rick Chudacoff (10)	9062
120. Daryl Dragon (3)	9027
121. David Bowie/Nile Rodgers (4)	8979
122. Duran Duran/Nile Rodgers (4)	8961
123. Elliot Wolff (3)	8942
124. Gary Mallaber/Steve Miller (3)	8926
125. Bill Szymczyk (8)	8923
126. Kevin Beamish/Kevin Cronin/ Gary Richrath (4)	8906
127. Harold Faltermeyer/Keith Forsey (3)	8765
128. Peter Coleman/Neil Geraldo (6)	8659
129. Peter Wolf (8)	8572
130. Tom Dowd (11)	8566
131. Maurice Starr (4)	8501
132. Clarence Ofwerman (3)	8484
133. Michael Zager (3)	8471
134. Matt Dike/Michael Ross (3)	8394
135. Bob Ezrin/David Gilmour/ Roger Waters (2)	8331
136. Mike Post (6)	8206
137. Bob Sargeant (4)	8170
138. Maurice White (12)	8111
139. Paul McCartney (2)	8107
140. David Lewis/Wayne Lewis (5)	8081
141. Don Henley/Danny Kortchmar/ Greg Ladanyi (6)	8062
142. Lindsey Buckingham/ Richard Dashut (7)	7968
143. Dennis Lambert (8)	7946
144. Beau Hill (12)	7877
145. Debbie Gibson (2)	7782
146. Tom Lord-Alge/Steve Winwood (6)	7781
147. Rodney Mills (12)	7654
148. Pat Glasser (8)	7652
149. Neil Dorfsman/Mark Knopfler (3)	7546
150. Harry Wayne Casey/Richard Finch (2)	7468

Top 50 Producer Teams, Their Acts and Their Singles

Rank. Producer Team (Credits) Act (Entries) Title	Score
1. Quincy Jones (35)	**82259**
Michael JACKSON (16)	52835
Billie Jean	8360
Rock With You	7256
Beat It	6892
Man In The Mirror	3974
I Just Can't Stop Loving You	3395
The Way You Make Me Feel	3201
Bad	2998
Dirty Diana	2958
Wanna Be Startin' Somethin'	2524
Thriller	2493
Human Nature	1844
She's Out Of My Life	1704
Off The Wall	1498
Smooth Criminal	1470
P.Y.T. (Pretty Young Thing)	1323
Another Part Of Me	945
Patti AUSTIN (4)	5744
Baby, Come To Me	5417
Every Home Should Have One ('81)	158
Every Home Should Have One ('83)	119
It's Gonna Be Special	50
USA For AFRICA (1)	5554
We Are The World	5554
Michael JACKSON/ Paul McCARTNEY (1)	5346
The Girl Is Mine	5346
George BENSON (2)	3413
Give Me The Night	3251
Love X Love	162
Donna SUMMER (3)	2555
Love Is In Control (Finger On The Trigger)	1571
The Woman In Me	657
State Of Independence	328
Quincy JONES Featuring James INGRAM (2)	2475
Just Once	1343
One Hundred Ways	1132
BROTHERS JOHNSON (2)	1938
Stomp!	1868
Treasure	70
James INGRAM (2)	1356
Yah Mo B There	1115
There's No Easy Way	241
RUFUS/Chaka KHAN (1)	578
Do You Love What You Feel	578
Quincy JONES (1)	466
Ai No Corrida (I-No-Ko-ree-da)	466
2. James Anthony Carmichael/ Lionel Richie (15)	**48647**
Lionel RICHIE (14)	46876
All Night Long (All Night)	8437
Hello	6840
Say You, Say Me	6296
Truly	4947
Dancing On The Ceiling	3868
Stuck On You	3553
You Are	2911
Running With The Night	2232
My Love	2148
Penny Lover	1819
Love Will Conquer All	1534
Ballerina Girl	1473
Se La	696
Deep River Woman	124
Diana ROSS (1)	1770
Missing You	1770
3. George Michael (14)	**43098**
WHAM! (7)	21464
Careless Whisper	5877
Wake Me Up Before You Go-Go	5473
Everything She Wants	4050
I'm Your Man	2606
Freedom	2153
The Edge Of Heaven	1071
Where Did Your Heart Go?	234
George MICHAEL (6)	20095
Faith	5187
One More Try	4687
Father Figure	3850
I Want Your Sex	3293
A Different Corner	1547
Kissing A Fool	1530
Deon ESTUS (1)	1539
Heaven Help Me	1539
4. Narada Michael Walden (28)	**41107**
Whitney HOUSTON (5)	18935
I Wanna Dance With Somebody (Who Loves Me)	4899
How Will I Know	4469
So Emotional	4343
Where Do Broken Hearts Go	3563
One Moment In Time	1661
Aretha FRANKLIN (5)	5850
Freeway Of Love	2899
Who's Zoomin' Who	1775
Another Night	667
Jimmy Lee	469
Rock-A-Lott	40
JEFFERSON STARSHIP/ STARSHIP (1)	4720
Nothing's Gonna Stop Us Now	4720
Aretha FRANKLIN and George Michael (1)	3467
I Knew You Were Waiting (For Me)	3467
Stacy LATTISAW (4)	2413
Let Me Be Your Angel	1047
Love On A Two Way Street	739
Miracles	538
Attack Of The Name Game	89
Jermaine STEWART (2)	2348
We Don't Have To Take Our Clothes Off	2052
Jody	296
Clarence CLEMONS (1)	1001
You're A Friend Of Mine	1001
Aretha FRANKLIN and Elton JOHN (1)	636
Through The Storm	636
Eddie MURPHY (1)	450
Put Your Mouth On Me	450
Sheena EASTON (1)	358
So Far So Good	358

Rank. Producer Team (Credits) Act (Entries) Title	Score
Aretha FRANKLIN and Whitney HOUSTON (1)	273
It Isn't, It Wasn't, It Ain't Never Gonna Be	273
POINTER SISTERS (1)	236
Be There	236
Regina BELLE (1)	179
Baby Come To Me	179
Stacy LATTISAW & Johnny GILL (1)	127
Perfect Combination	127
SISTER SLEDGE (2)	116
Next Time You'll Know	61
All American Girls	55
5. Daryl Hall/John Oates (12)	**40933**
Daryl HALL & John OATES (12)	40933
I Can't Go For That (No Can Do)	6855
Maneater	6774
Say It Isn't So	5735
Private Eyes	5434
Kiss On My List	4872
You Make My Dreams	2669
One On One	2378
Family Man	1851
You've Lost That Lovin' Feeling	1734
Did It In A Minute	1635
How Does It Feel To Be Back	504
Your Imagination	492
6. Michael Omartian (24)	**40118**
Christopher CROSS (10)	22700
Arthur's Theme (Best That You Can Do)	6437
Ride Like The Wind	5822
Sailing	4791
All Right	1564
Think Of Laura	1548
Never Be The Same	1246
Say You'll Be Mine	722
No Time For Talk	416
Charm The Snake	84
A Chance For Heaven	70
Peter CETERA (3)	7364
Glory Of Love (Theme From The Karate Kid Part II)	4056
The Next Time I Fall	3174
Big Mistake	134
Donna SUMMER (5)	5567
She Works Hard For The Money	4407
There Goes My Baby	729
Unconditional Love	284
Supernatural Love	84
Love Has A Mind Of Its Own	63
Rod STEWART (3)	3433
Infatuation	2076
Some Guys Have All The Luck	1263
All Right Now	94
Jermaine JACKSON (3)	1053
I Think It's Love	861
(Closest Thing To) Perfect	116
Do You Remember Me?	76
7. Phil Ramone (29)	**38706**
Billy JOEL (20)	32541
It's Still Rock And Roll To Me	6941
Uptown Girl	5362
Tell Her About It	4357
You May Be Right	1973
Allentown	1591
An Innocent Man	1543
The Longest Time	1325
You're Only Human (Second Wind)	1320
A Matter Of Trust	1210
Modern Woman (From "Ruthless People")	1035
Don't Ask Me Why	903
Keeping The Faith	876
Pressure	859
This Is The Time	850
She's Got A Way	715
Leave A Tender Moment Alone	688
The Night Is Still Young	367
Sometimes A Fantasy	334
Goodnight Saigon	194
Baby Grand	97
Julian LENNON (5)	4994
Too Late For Goodbyes	2063
Valotte	1668
Say You're Wrong	634
Stick Around	471
Jesse	159
Barbra STREISAND and Don JOHNSON (1)	465
Till I Loved You (The Love Theme From Goya)	465
Michael SEMBELLO (1)	426
Automatic Man	426
GET WET (1)	260
Just So Lonely	260
Karen KAMON (1)	20
Loverboy	20
8. Chris Thomas (27)	**37727**
Elton JOHN (11)	15581
I Guess That's Why They Call It The Blues	2975
I Don't Wanna Go On With You Like That	2818
Sad Songs (Say So Much)	2244
I'm Still Standing	1429
Blue Eyes	1339
Empty Garden (Hey Hey Johnny)	1202
Healing Hands	900
Who Wears These Shoes?	887
Kiss The Bride	670
A Word In Spanish	618
In Neon	498
INXS (7)	13683
Need You Tonight	3984
Devil Inside	3056
New Sensation	2395
What You Need	2349
Never Tear Us Apart	1587
Listen Like Thieves	238
This Time	74
The PRETENDERS (5)	6041
Back On The Chain Gang	2775
Brass In Pocket (I'm Special)	1661
Middle Of The Road	968
Show Me	579
Thin Line Between Love And Hate	58

Producers

Rank. Producer Team (Credits) / Act (Entries) / Title	Score
8. Chris Thomas *Continued*	
Pete TOWNSHEND (4)	2421
Let My Love Open The Door	1643
Face The Face	672
A Little Is Enough	70
Rough Boys	35
9. Ron Nevison (31)	**36276**
HEART (9)	17498
Alone	5121
These Dreams	3896
Never	2769
Who Will You Run To	1573
What About Love?	1405
Nothin' At All	1156
There's The Girl	1156
I Want You So Bad	235
If Looks Could Kill	187
CHICAGO (4)	8436
Look Away	4616
I Don't Wanna Live Without Your Love	2418
You're Not Alone	1099
We Can Last Forever	302
SURVIVOR (5)	5769
The Search Is Over	2456
I Can't Hold Back	1564
High On You	1361
First Night	231
The Moment Of Truth	157
JEFFERSON STARSHIP/ STARSHIP (6)	3199
Jane	1259
No Way Out	851
Find Your Way Back	523
Stranger	266
Girl With The Hungry Eyes	172
Layin' It On The Line	128
EUROPE (1)	455
Superstitious	455
KISS (2)	322
Reason To Live	215
Crazy Crazy Nights	107
Ann WILSON (1)	267
The Best Man In The World	267
Eddie MONEY (2)	190
Let's Be Lovers Again	132
Running Back	58
Ozzy OSBOURNE (1)	141
Shot In The Dark	141
10. Giorgio Moroder (16)	**35510**
Irene CARA (4)	13485
Flashdance... What A Feeling	9800
Breakdance	1793
Why Me?	1330
The Dream (Hold On To Your Dream)	562
BLONDIE (1)	9931
Call Me	9931
Kenny LOGGINS (2)	4435
Danger Zone	2987
Meet Me Half Way	1448
BERLIN (1)	3558
Take My Breath Away (Love Theme From "Top Gun")	3558
Donna SUMMER (1)	2750
On The Radio	2750
LIMAHL (1)	925
Never Ending Story	925
David BOWIE (1)	129
Cat People (Putting Out Fire)	129
BIG TROUBLE (1)	94
Crazy World	94
Bonnie TYLER (1)	69
Here She Comes	69
Madleen KANE (1)	66
You Can	66
Giorgio MORODER (1)	45
Reach Out	45
Joe "Bean" ESPOSITO (1)	22
Lady, Lady, Lady	22
11. Keith Olsen (30)	**33986**
Rick SPRINGFIELD (5)	15522
Jessie's Girl	6460
Don't Talk To Strangers	5887
I've Done Everything For You	1929
What Kind Of Fool Am I	782
I Get Excited	463
WHITESNAKE (1)	4336
Here I Go Again	4336
Pat BENATAR (3)	3739
Hit Me With Your Best Shot	2328
Treat Me Right	1046
You Better Run	366
Mike RENO And Ann WILSON (1)	2095
Almost Paradise... Love Theme From "Footloose"	2095
Sammy HAGAR (4)	2011
Your Love Is Driving Me Crazy	1353
I'll Fall In Love Again	329
Never Give Up	268
Piece Of My Heart	61
38 SPECIAL (2)	1214
Like No Other Night	874
Somebody Like You	340
JEFFERSON STARSHIP/ STARSHIP (1)	1147
It's Not Over ('Til It's Over)	1147
The BABYS (3)	979
Back On My Feet Again	502
Turn And Walk Away	413
Midnight Rendezvous	63
REO SPEEDWAGON (1)	840
Here With Me	840
Kim CARNES (3)	751
Invisible Hands	455
You Make My Heart Beat Faster (And That's All That Matters)	214
I Pretend	82
SANTANA (1)	525
You Know That I Love You	525
HEART (2)	384
How Can I Refuse	348
Allies	36
SHEILA (1)	279
Little Darlin'	279
707 (1)	124
Mega Force	124
BAD COMPANY (1)	40
This Love	40
12. David Foster (31)	**33723**
CHICAGO (11)	20425
Hard To Say I'm Sorry	6847
Hard Habit To Break	3712
You're The Inspiration	3237
Will You Still Love Me?	2470
Stay The Night	1052
Along Comes A Woman	1018
If She Would Have Been Faithful...	1001
Love Me Tomorrow	766
25 Or 6 To 4	247
What You're Missing	56
Niagara Falls	18
John PARR (1)	4434
St. Elmo's Fire (Man In Motion)	4434
Olivia NEWTON-JOHN (2)	3293
Twist Of Fate	2833
Livin' In Desperate Times	460
The TUBES (4)	2210
She's A Beauty	1506
Don't Want To Wait Anymore	397
Tip Of My Tongue	217
The Monkey Time	90
Daryl HALL & John OATES (1)	1094
Wait For Me	1094
Jennifer HOLLIDAY (1)	654
And I Am Telling You I'm Not Going	654
Barbra STREISAND (1)	425
Somewhere (From "West Side Story")	425
Bill CHAMPLIN (2)	357
Tonight Tonight	183
Sara	173
AVERAGE WHITE BAND/AWB (1)	225
Let's Go 'Round Again	225
Peter ALLEN (1)	203
Fly Away	203
NIGHT RANGER (1)	173
The Secret Of My Success	173
Kenny ROGERS (1)	81
Crazy	81
Ray KENNEDY (1)	45
Just For The Moment	45
Paul HYDE And The PAYOLAS (1)	42
You're The Only Love	42
Anne MURRAY (1)	35
Now And Forever (You And Me)	35
David FOSTER (1)	26
Winter Games	26
13. Prince (18)	**33480**
PRINCE (16)	28172
When Doves Cry	7679
Batdance	3450
U Got The Look	2927
Little Red Corvette	2573
Sign O' The Times	2446
Delirious	1912
1999	1633
I Wanna Be Your Lover	1389
I Could Never Take The Place Of Your Man	1224
Alphabet St.	1192
Partyman	582
The Arms Of Orion	470
Let's Pretend We're Married/ Irresistible Bitch	258
Controversy	182
Hot Thing	163
If I Was Your Girlfriend	92
PRINCE And The REVOLUTION (2)	5308
Purple Rain	3959
I Would Die 4 U	1349
14. Huey Lewis And The News (17)	**33246**
Huey LEWIS and the NEWS (17)	33246
The Power Of Love	4636
Stuck With You	4301
Jacob's Ladder	3536
Hip To Be Square	2647
The Heart Of Rock & Roll	2607
Perfect World	2388
I Want A New Drug	2382
If This Is It	2110
Heart And Soul	1827
Do You Believe In Love	1774
Doing It All For My Baby	1480
I Know What I Like	1162
Walking On A Thin Line	933
Small World	482
Hope You Love Me Like You Say You Do	413
Workin' For A Livin'	331
Give Me The Keys (And I'll Drive You Crazy)	235
15. Richard Perry (25)	**33023**
POINTER SISTERS (18)	25106
Slow Hand	4799
He's So Shy	3840
Jump (For My Love)	3698
Automatic	2676
Neutron Dance	2532
I'm So Excited '84	1658
Dare Me	1309
Should I Do It	1109
American Music	952
I'm So Excited	620
Goldmine	465
I Need You	412
Baby Come And Get It	340
Could I Be Dreaming	311
Freedom	214
If You Wanna Get Back Your Lady	117
Twist My Arm	45
All I Know Is The Way I Feel	9
DeBARGE (1)	3325
Rhythm Of The Night	3325
Julio IGLESIAS & Willie NELSON (1)	2482
To All The Girls I've Loved Before	2482
Jeffrey OSBORNE (1)	1168
You Should Be Mine (The Woo Woo Song)	1168
Patti LaBELLE (1)	413
Oh, People	413
Donna SUMMER (1)	295
Dinner With Gershwin	295
Bill MEDLEY (1)	199
Right Here And Now	199
Barbra STREISAND (1)	34
Emotion	34
16. Bruce Fairbairn (20)	**32876**
BON JOVI (8)	23702
Livin' On A Prayer	4968
I'll Be There For You	3890
You Give Love A Bad Name	3706
Bad Medicine	3342
Born To Be My Baby	2796
Wanted Dead Or Alive	1860
Lay Your Hands On Me	1661
Living In Sin	1480

Producers

Rank. Producer Team (Credits) / Act (Entries) / Title	Score
16. Bruce Fairbairn *Continued*	
AEROSMITH (4)	6584
Angel	2787
Love In An Elevator	1697
Dude (Looks Like A Lady)	1261
Rag Doll	840
LOVERBOY (3)	1227
Turn Me Loose	639
Notorious	415
The Kid Is Hot Tonite	173
HONEYMOON SUITE (2)	884
Feel It Again	472
What Does It Take	412
Dan REED Network (1)	335
Ritual	335
KROKUS (1)	108
Midnite Maniac	108
BLUE ÖYSTER CULT (1)	36
Shooting Shark	36
17. Don Gehman / John Mellencamp (15)	**32277**
John MELLENCAMP (15)	32277
Hurts So Good	7863
Jack & Diane	7190
R.O.C.K. In The U.S.A. (A Salute To 60's Rock)	2895
Small Town	2127
Lonely Ol' Night	1951
Crumblin' Down	1668
Cherry Bomb	1512
Pink Houses	1484
Paper In Fire	1315
Hand To Hold On To	1066
Authority Song	1065
Check It Out	941
Rain On The Scarecrow	553
Rumbleseat	508
Rooty Toot Toot	142
18. George Martin (15)	**31721**
Paul McCARTNEY (5)	13424
Ebony And Ivory	8625
No More Lonely Nights	2137
Take It Away	1711
So Bad	759
Tug Of War	191
Paul McCARTNEY And Michael JACKSON (1)	10336
Say Say Say	10336
LITTLE RIVER BAND (3)	5458
The Night Owls	2525
Take It Easy On Me	1760
Man On Your Mind	1174
The BEATLES (2)	1656
The Beatles' Movie Medley	1029
Twist And Shout	628
CHEAP TRICK (2)	590
Everything Works If You Let It	330
Stop This Game	261
Kenny ROGERS (1)	153
Morning Desire	153
ULTRAVOX (1)	104
Reap The Wild Wind	104
19. Albhy Galuten/Barry Gibb/ Karl Richardson (15)	**29343**
Barbra STREISAND (2)	8175
Woman In Love	7930
Promises	245
Kenny ROGERS (3)	7603
Islands In The Stream	6841
This Woman	688
Eyes That See In The Dark	74
Andy GIBB (3)	5028
Desire	3634
Time Is Time	1100
Me (Without You)	294
Barbra STREISAND & Barry GIBB (2)	4905
Guilty	3545
What Kind Of Fool	1360
Dionne WARWICK (2)	2283
Heartbreaker	1820
Take The Short Way Home	464
Andy GIBB & Olivia NEWTON-JOHN (1)	1190
I Can't Help It	1190
Diana ROSS (2)	159
Chain Reaction (Special New Mix)	149
Chain Reaction	10
20. Jimmy Jam Harris/ Terry Lewis (21)	**28893**
Janet JACKSON (5)	16636
Miss You Much	5007
When I Think Of You	3899
Nasty	2753
What Have You Done For Me Lately?	2532
Control	2445
HUMAN LEAGUE (2)	4378
Human	4045
I Need Your Loving	333
Herb ALPERT (2)	2279
Diamonds	1839
Making Love In The Rain	439
NEW EDITION (3)	2116
If It Isn't Love	1807
Can You Stand The Rain	298
You're Not My Kind Of Girl	11
S.O.S. BAND (4)	1129
The Finest	364
Just Be Good To Me	340
Just The Way You Like It	230
Tell Me If You Still Care	195
Alexander O'NEAL (2)	1087
Fake	559
Never Knew Love Like This	529
CHERRELLE with Alexander O'NEAL (1)	623
Saturday Love	623
Morris DAY (1)	550
Fishnet	550
CHERRELLE (1)	96
I Didn't Mean To Turn You On	96
21. Kenneth Edmonds/ L.A. Reid (17)	**27592**
Bobby BROWN (5)	11689
On Our Own	3811
Every Little Step	2738
Roni	1982
Don't Be Cruel	1692
Rock Wit'cha	1465
Karyn WHITE (3)	4863
The Way You Love Me	1874
Secret Rendezvous	1805
Superwoman	1183
Sheena EASTON (1)	2769
The Lover In Me	2769
PEBBLES (1)	2038
Girlfriend	2038
The WHISPERS (1)	1781
Rock Steady	1781
BABYFACE (1)	1490
It's No Crime	1490
The DEELE (1)	1285
Two Occasions	1285
The BOYS (1)	1097
Dial My Heart	1097
Paula ABDUL (1)	357
Knocked Out	357
AFTER 7 (1)	151
Heat Of The Moment	151
The JACKSONS (1)	70
Nothin (that compares 2 U)	70
22. Stevie Wonder (18)	**26991**
Stevie WONDER (15)	24377
I Just Called To Say I Love You	6339
Part-Time Lover	4348
Master Blaster (Jammin')	3415
That Girl	3383
I Ain't Gonna Stand For It	1526
Go Home	1315
Do I Do	1185
Love Light In Flight	903
Skeletons	748
Overjoyed	556
Outside My Window	228
Ribbon In The Sky	192
Lately	154
You Will Know	59
Land Of La La	26
Jermaine JACKSON (2)	2557
Let's Get Serious	2039
You're Supposed To Keep Your Love For Me	518
Stevie WONDER & Michael JACKSON (1)	57
Get It	57
23. John Farrar (8)	**26252**
Olivia NEWTON-JOHN (7)	25227
Physical	11089
Magic	6621
Heart Attack	4023
Make A Move On Me	2194
Soul Kiss	671
Tied Up	410
Landslide	220
Olivia NEWTON-JOHN & Cliff RICHARD (1)	1025
Suddenly	1025
24. Steve Levine (12)	**24944**
CULTURE CLUB (9)	24281
Karma Chameleon	6404
Do You Really Want To Hurt Me	5296
Time (Clock Of The Heart)	4390
Miss Me Blind	2336
Church Of The Poison Mind	1763
I'll Tumble 4 Ya	1713
It's A Miracle	1127
The War Song	715
Mistake No. 3	537
BEACH BOYS (2)	608
Getcha Back	560
It's Gettin' Late	48
QUARTERFLASH (1)	54
Talk To Me	54
25. Lionel Richie (6)	**24772**
Kenny ROGERS (5)	14386
Lady	8423
I Don't Need You	3554
Share Your Love With Me	1159
Through The Years	1103
Blaze Of Glory	147
Diana ROSS & Lionel RICHIE (1)	10386
Endless Love	10386
26. Arif Mardin (18)	**24622**
Phil COLLINS (1)	7041
Against All Odds (Take A Look At Me Now)	7041
Chaka KHAN (3)	4783
I Feel For You	4486
This Is My Night	184
Got To Be There	113
Bette MIDLER (1)	3826
Wind Beneath My Wings	3826
Melissa MANCHESTER (3)	3584
You Should Hear How She Talks About You	3133
Nice Girls	387
No One Can Love You More Than Me	64
BOY MEETS GIRL (2)	3153
Waiting For A Star To Fall	2834
Bring Down The Moon	319
Howard JONES (2)	1097
You Know I Love You...Don't You?	1034
All I Want	63
George BENSON (1)	612
Lady Love Me (One More Time)	612
Aretha FRANKLIN & George BENSON (1)	312
Love All The Hurt Away	312
BEE GEES (1)	95
You Win Again	95
Peabo BRYSON (1)	64
Take No Prisoners (In The Game Of Love)	64
Aretha FRANKLIN (1)	32
Come To Me	32
SCRITTI POLITTI (1)	21
Wood Beez (pray like aretha franklin)	21
27. Mike Chapman (23)	**23444**
BLONDIE (5)	10915
The Tide Is High	5289
Rapture	4803
Island Of Lost Souls	399
Atomic	380
The Hardest Part	45
SCANDAL/Patti SMYTH (3)	3063
The Warrior	2284
Beat Of A Heart	404
Hands Tied	374
Lita FORD (2)	2928
Close My Eyes Forever (remix)	1716
Kiss Me Deadly	1212
Rod STEWART (1)	1765
Love Touch	1765
Pat BENATAR (1)	1310
Invincible	1310
Tina TURNER (1)	940
One Of The Living	940
Suzi QUATRO (2)	660
She's In Love With You	427
Lipstick	232

Producers

Rank. Producer Team (Credits) Act (Entries) Title	Score
27. Mike Chapman *Continued*	
Agnetha FALTSKOG (1)	626
Can't Shake Loose	626
DEVICE (2)	566
Hanging On A Heart Attack	502
Who Says	64
The KNACK (2)	471
Baby Talks Dirty	326
Can't Put A Price On Love	145
DIVINYLS (1)	111
Pleasure And Pain	111
BOW WOW WOW (1)	56
Do You Wanna Hold Me?	56
TAMI SHOW (1)	33
She's Only 20	33
28. Matt Aitken/Mike Stock/ Pete Waterman (22)	**22997**
Rick ASTLEY (4)	8750
Never Gonna Give You Up	4313
Together Forever	3198
It Would Take A Strong Strong Man	1222
Ain't Too Proud To Beg	17
BANANARAMA (6)	6811
Venus	3701
I Heard A Rumour	2398
I Can't Help It	369
Love In The First Degree	262
More Than Physical	61
Love, Truth And Honesty	20
Kylie MINOGUE (3)	3847
The Loco-Motion	2927
I Should Be So Lucky	503
It's No Secret	417
Donna SUMMER (2)	1519
This Time I Know It's For Real	1494
Love's About To Change My Heart	24
DEAD OR ALIVE (3)	1203
Brand New Lover	1062
Lover Come Back To Me	92
Something In My House	48
Samantha FOX (2)	525
I Only Wanna Be With You	463
Nothing's Gonna Stop Me Now	61
Laura BRANIGAN (1)	268
Shattered Glass	268
MEL & KIM (1)	75
Showing Out (Get Fresh At The Weekend)	75
29. Mutt Lange (17)	**22932**
DEF LEPPARD (11)	15814
Love Bites	3728
Pour Some Sugar On Me	3419
Armageddon It	2569
Photograph	1309
Hysteria	1121
Rock Of Ages	1015
Rocket	918
Animal	890
Foolin'	602
Bringin' On The Heartbreak	192
Women	52
Billy OCEAN (2)	4320
Get Outta My Dreams, Get Into My Car	3998
Licence To Chill	323
AC/DC (3)	1440
You Shook Me All Night Long	603

Rank. Producer Team (Credits) Act (Entries) Title	Score
Back In Black	545
Let's Get It Up	293
The CARS (1)	1357
Magic	1357
30. Jimmy Iovine (19)	**22567**
Stevie NICKS (6)	8712
Leather And Lace	2825
Stand Back	2639
Edge Of Seventeen (Just Like The White Winged Dove)	1172
If Anyone Falls	1066
Nightbird	548
After The Glitter Fades	462
Bob SEGER & The SILVER BULLET BAND (3)	7248
Shame On The Moon	5590
Even Now	1141
Roll Me Away	516
U2 (3)	3340
Desire	2423
Angel Of Harlem	878
All I Want Is You	39
BREAKFAST CLUB (1)	1500
Right On Track	1500
Tom PETTY and The HEARTBREAKERS (1)	1068
You Got Lucky	1068
Joan JETT & The BLACKHEARTS (1)	390
Light Of Day	390
LONE JUSTICE (2)	155
Ways To Be Wicked	87
Sweet, Sweet Baby (I'm Falling)	69
U2 With B. B. KING (1)	130
When Love Comes To Town	130
GENE LOVES JEZEBEL (1)	23
The Motion Of Love	23
31. George Tobin (18)	**22052**
TIFFANY (6)	11660
Could've Been	4391
I Think We're Alone Now	3894
All This Time	1617
I Saw Him Standing There	1282
Radio Romance	282
Feelings Of Forever	194
Smokey ROBINSON (4)	6798
Being With You	5997
Tell Me Tomorrow - Part 1	451
Old Fashioned Love	201
You Are Forever	149
Kim CARNES (2)	2235
More Love	1950
Cry Like A Baby	286
Robert JOHN (3)	644
Lonely Eyes	447
Bread And Butter	99
Sherry	97
Smokey ROBINSON & Barbara MITCHELL (1)	379
Blame It On Love	379
NEW EDITION (1)	277
With You All The Way	277
HIGH INERGY (1)	60
He's A Pretender	60
32. Michael Masser (12)	**22007**
Whitney HOUSTON (3)	12523
Greatest Love Of All	4572
Didn't We Almost Have It All	4007
Saving All My Love For You	3944

Rank. Producer Team (Credits) Act (Entries) Title	Score
Natalie COLE (2)	2776
Miss You Like Crazy	1760
Someone That I Used To Love	1016
Diana ROSS (2)	2139
It's My Turn	2085
One More Chance	54
Peabo BRYSON (1)	1922
If Ever You're In My Arms Again	1922
Peabo BRYSON/ Roberta FLACK (1)	1875
Tonight, I Celebrate My Love	1875
Teddy PENDERGRASS (1)	505
Hold Me	505
Glenn MEDEIROS (1)	154
Long And Lasting Love (Once In A Lifetime)	154
Dionne WARWICK (1)	112
Some Changes Are For Good	112
33. Ted Templeman (23)	**21455**
VAN HALEN (7)	11669
Jump	7537
(Oh) Pretty Woman	1163
I'll Wait	1128
Panama	1047
Dancing In The Street	425
And The Cradle Will Rock...	200
Hot For Teacher	170
David Lee ROTH (5)	4494
California Girls	2418
Just A Gigolo/I Ain't Got Nobody	1077
Yankee Rose	840
Goin' Crazy!	124
That's Life	36
DOOBIE BROTHERS (5)	3143
Real Love	2173
One Step Closer	704
Here To Love You	112
Keep This Train A-Rollin'	109
You Belong To Me	45
Sammy HAGAR (2)	1031
I Can't Drive 55	578
Two Sides Of Love	453
Tom JOHNSTON (1)	472
Savannah Nights	472
Nicolette LARSON (1)	432
Let Me Go, Love	432
BULLETBOYS (2)	214
Smooth Up	137
For The Love Of Money	77
34. Patrick Leonard/ Madonna (6)	**21096**
MADONNA (6)	21096
Like A Prayer	4497
Live To Tell	4244
Open Your Heart	3575
Who's That Girl	3537
Cherish	2798
La Isla Bonita	2445
35. Phil Collins/ Hugh Padgham (7)	**21060**
Phil COLLINS (5)	18378
Another Day In Paradise	5854
One More Night	4427
Sussudio	4040
Don't Lose My Number	2260
Take Me Home	1797
Howard JONES (1)	2312
No One Is To Blame	2312

Rank. Producer Team (Credits) Act (Entries) Title	Score
Adam ANT (1)	370
Strip	370
36. Hugh Padgham/ Police (7)	**20166**
The POLICE (7)	20166
Every Breath You Take	10182
Every Little Thing She Does Is Magic	2994
King Of Pain	2967
Wrapped Around Your Finger	1588
Synchronicity II	1118
Spirits In The Material World	1069
Secret Journey	249
37. Rick Chertoff (14)	**19940**
Cyndi LAUPER (4)	16240
Time After Time	5392
Girls Just Want To Have Fun	4890
She Bop	3772
All Through The Night	2186
The HOOTERS (7)	2834
And We Danced	914
Day By Day	860
Where Do The Children Go	389
Johnny B	233
All You Zombies	232
Satellite	194
500 Miles	13
Rex SMITH/Rachel SWEET (1)	501
Everlasting Love	501
Tommy CONWELL And The YOUNG RUMBLERS (2)	366
If We Never Meet Again	284
I'm Not Your Man	82
38. Val Garay (19)	**19840**
Kim CARNES (6)	12033
Bette Davis Eyes	10255
Draw Of The Cards	558
Does It Make You Remember	530
Voyeur	498
Mistaken Identity	136
Divided Hearts	55
The MOTELS (5)	5073
Only The Lonely	2280
Suddenly Last Summer	1846
Remember The Nights	551
Take The L	220
Forever Mine	177
Randy MEISNER (2)	1594
Hearts On Fire	895
Deep Inside My Heart	699
Dolly PARTON (2)	478
Save The Last Dance For Me	427
Downtown	51
SANTANA (1)	367
Say It Again	367
Marty BALIN (1)	128
What Love Is	128
Joan ARMATRADING (1)	88
Drop The Pilot	88
The NYLONS (1)	79
Happy Together	79
39. Peter Mclan (7)	**19509**
MEN AT WORK (5)	19129
Down Under	7153
Who Can It Be Now?	5487
Overkill	3933
It's A Mistake	2066
Dr. Heckyll & Mr. Jive	490

Producers

Rank. Producer Team (Credits) Act (Entries) Title	Score
39. Peter McIan *Continued*	
MR. MISTER (1)	203
Hunters Of The Night	203
Peter McIAN (1)	177
Solitaire	177
40. Ray Parker Jr. (14)	**19259**
Ray PARKER Jr. (8)	13550
Ghostbusters	6445
The Other Woman	3200
I Still Can't Get Over Loving You	1373
Jamie	1122
Bad Boy	487
Girls Are More Fun	451
Let Me Go	349
I Don't Think That Man Should Sleep Alone	124
Ray PARKER Jr./RAYDIO (3)	4524
A Woman Needs Love (Just Like You Do)	3214
That Old Song	665
Two Places At The Same Time	645
NEW EDITION (1)	1051
Mr. Telephone Man	1051
Cheryl LYNN (1)	124
Shake It Up Tonight	124
Ray PARKER Jr. And Helen TERRY (1)	9
One Sunny Day/Dueling Bikes From "Quicksilver"	9
41. Queen (4)	**18635**
QUEEN (4)	18635
Another One Bites The Dust	9572
Crazy Little Thing Called Love	8343
Need Your Loving Tonight	365
Play The Game	355
42. Christopher Neil (18)	**18381**
Sheena EASTON (7)	10913
Morning Train (Nine To Five)	4553
For Your Eyes Only	3209
You Could Have Been With Me	1280
Modern Girl	971
When He Shines	572
Machinery	179
I Wouldn't Beg For Water	149
MIKE + THE MECHANICS (3)	4255
All I Need Is A Miracle	1864
Silent Running (On Dangerous Ground)	1835
Taken In	557
Paul CARRACK (4)	2154
Don't Shed A Tear	1635
One Good Reason	486
When You Walk In The Room	20
Button Off My Shirt	13
OTHER ONES (2)	815
Holiday	575
We Are What We Are	240

Rank. Producer Team (Credits) Act (Entries) Title	Score
Shakin' STEVENS (1)	134
I Cry Just A Little Bit	134
DOLLAR (1)	110
Shooting Star	110
43. Kevin Elson/ Mike Stone (9)	**17913**
JOURNEY (9)	17913
Open Arms	5744
Who's Crying Now	3180
Separate Ways (Worlds Apart)	2468
Don't Stop Believin'	1589
Faithfully	1452
Only The Young	1163
Still They Ride	907
Send Her My Love	718
After The Fall	692
44. Richard Landis (15)	**17724**
Juice NEWTON (10)	15737
Queen Of Hearts	4309
Angel Of The Morning	3447
The Sweetest Thing (I've Ever Known)	2829
Love's Been A Little Bit Hard On Me	2037
Break It To Me Gently	1321
Heart Of The Night	809
Tell Her No	534
A Little Love	326
Can't Wait All Night	109
Dirty Looks	16
Van STEPHENSON (2)	1225
Modern Day Delilah	910
What The Big Girls Do	315
NIELSEN/PEARSON (2)	684
If You Should Sail	491
The Sun Ain't Gonna Shine Anymore	193
GARY O' (1)	78
Pay You Back With Interest	78
45. David A. Stewart (16)	**17332**
EURYTHMICS (12)	15276
Sweet Dreams (Are Made Of This)	6665
Here Comes The Rain Again	3206
Would I Lie To You?	1926
Missionary Man	1031
Who's That Girl?	745
There Must Be An Angel (Playing With My Heart)	554
Right By Your Side	518
I Need A Man	258
You Have Placed A Chill In My Heart	136
Thorn In My Side	120
It's Alright (Baby's Coming Back)	70
Sexcrime (Nineteen Eighty-Four)	46
Annie LENNOX & Al GREEN (1)	1205
Put A Little Love In Your Heart	1205

Rank. Producer Team (Credits) Act (Entries) Title	Score
EURYTHMICS and Aretha FRANKLIN (1)	759
Sisters Are Doin' It For Themselves	759
Feargal SHARKEY (1)	91
A Good Heart	91
46. Bryan Adams/ Bob Clearmountain (13)	**17324**
Bryan ADAMS (12)	16377
Heaven	4136
Run To You	2116
Summer Of '69	2007
Heat Of The Night	1759
Straight From The Heart	1464
Somebody	1214
One Night Love Affair	1047
Cuts Like A Knife	982
This Time	649
Hearts On Fire	544
Victim Of Love	434
Lonely Nights	25
Bryan ADAMS/Tina TURNER (1)	947
It's Only Love	947
47. Peter Asher (16)	**16729**
Linda RONSTADT (7)	9766
Don't Know Much	4448
Hurt So Bad	1882
How Do I Make You	1783
I Can't Let Go	484
I Knew You When	451
Get Closer	450
Easy For You To Say	269
Linda RONSTADT And James INGRAM (1)	2890
Somewhere Out There	2890
CHER and Peter CETERA (1)	1766
After All (Love Theme From "Chances Are")	1766
James TAYLOR And J.D. SOUTHER (1)	1160
Her Town Too	1160
10,000 MANIACS (3)	524
Trouble Me	317
Like The Weather	116
What's The Matter Here?	91
Linda RONSTADT & The Nelson RIDDLE Orchestra (1)	427
What's New	427
Bonnie RAITT (1)	120
You're Gonna Get What's Coming	120
James TAYLOR (1)	77
Hard Times	77
48. David Malloy (10)	**16522**
Eddie RABBITT (8)	13518
I Love A Rainy Night	5924
Drivin' My Life Away	2919
Step By Step	2841

Rank. Producer Team (Credits) Act (Entries) Title	Score
Someone Could Lose A Heart Tonight	1056
I Don't Know Where To Start	440
You Can't Run From Love	225
You Put The Beat In My Heart	58
Gone Too Far	55
Eddie RABBITT with Crystal GAYLE (1)	2991
You And I	2991
Dolly PARTON (1)	13
Real Love	13
49. Prince And The Revolution (8)	**16140**
PRINCE And The REVOLUTION (8)	16140
Let's Go Crazy	5200
Kiss	4499
Raspberry Beret	3528
Pop Life	1376
Take Me With U	585
Mountains	543
America	208
Anotherloverholenyohead	201
50. Rupert Hine (23)	**16127**
The FIXX (7)	6280
One Thing Leads To Another	2996
Saved By Zero	941
Are We Ourselves?	905
Secret Separation	661
The Sign Of Fire	586
Stand Or Fall	104
Sunshine In The Shade	87
Howard JONES (4)	3870
Things Can Only Get Better	2329
Life In One Day	765
What Is Love?	514
Like To Get To Know You Well	262
Tina TURNER (2)	2430
Better Be Good To Me	2351
Break Every Rule	78
Chris De BURGH (3)	1112
Don't Pay The Ferryman	569
High On Emotion	444
Ship To Shore	99
SAGA (3)	1015
On The Loose	798
Wind Him Up	176
The Flyer	41
Stevie NICKS (1)	730
Rooms On Fire	730
THOMPSON TWINS (1)	454
Get That Love	454
EIGHT SECONDS (1)	127
Kiss You (When It's Dangerous)	127
UNDERWORLD (1)	109
Underneath The Radar	109

Appendix

Methodology

Singles with Allocated Credit

Answers to Trivia Questions

Acknowledgements

About the Author

Appendix

Appendix

Methodology

Many notes on methodology have been relocated to the preambles of the various chapters for reader convenience. *Ranking the '80s* covers the period January 1, 1980 to December 31, 1989 for the Billboard Hot 100 singles chart and Top 200 album chart. Information for titles and names for acts are drawn from that magazine and facsimiles of labels. Songwriter information was drawn from multiple sources including the royalty rights organizations such as ASCAP, BMI and SESAC, copyright data from volumes compiled by the US Copyright Office and the labels of the 45s themselves. Writer scores are based only on Billboard in this book; however, more information on the writers' methodology can be found in *Ranking the Rock Writers*. Producer information is transcribed from the labels of the 45s themselves.

Singles are attributed to their "proximate" album; that is, the album that is most likely to be the source for the single. In many cases this is straightforward; in some cases it is not. Singles that first appeared in a soundtrack album are attributed to it even if a later album by the act itself contains it; on the other hand, singles that originated elsewhere and are included in a soundtrack are attributed to the original source. In cases where the soundtrack and the act's album appeared about the same time, preference is given to the act's album

In three cases a single was attributed to two albums. The act for each of those singles was an occasional collaboration and there was no common album, only albums for the two collaborating acts.

Scoring. The basics of "area under the curve" scoring have been discussed in the previous works[1,2,3,4] and can be found at the website **http://ranking.rocks/methodology**. The system is similar for singles and albums in that it starts with the lifecycle of a record week to week. Each chart position carries a value of 1-1000 points, with the values increasing with higher positions on the charts. The weekly points values are summed to obtain a raw score. Then those raw scores are adjusted to account for differences in chart construction over the period studied. Usually the differences involve large changes in the number of records appearing on the charts in a year. It is easy to show mathematically that years with more chart entries tend to have records with lower average scores and time on the chart. To compare rationally across those years, the differences must be taken into account.

The method for doing so is called normalization. Normalizing shifts from comparing actual record scores to a comparison of each record with those that charted around it with the idea of isolating those that truly stand out. While the exact process differs between singles and albums due to the longer life cycle of albums, in both cases it involves dividing a record's raw score by the average score of contemporaneous records. An average record, thus, has a score of 1 and that score is multiplied by 1000 to avoid decimals.

The lifecycle of an "average" record scores 1000 points; thus, a record scoring 10,000 points has ten times the chart popularity of an average record. 10,000 points generally places a record at or near the top 1% of the distribution.

---

[1] Carroll, William F., Jr. "Not So Lonely At The Top: Billboard #1s and a New Methodology for Comparing Records, 1958-1975." Popular Music And Society 38, 586-610 (2015) DOI: 10.1080/03007766.2014.991188. Published Online 20 Dec 2014. **http://dx.doi.org/10.1080/03007766.2014.991188**.

[2] Dann Isbell and Bill Carroll "Ranking the '70s: A Complete Compilation of the Chart Songs and Acts from Pop's Eclectic Decade" © 2015 Jefrian Books, Dallas, TX. http://ranking.rocks

[3] Bill Carroll "Ranking the Rock Writers" © 2018 Carroll Applied Science, LLC Dallas, TX http://ranking.rocks

[4] Bill Carroll "Ranking the Albums – The Stereo LP Era: 1963-1989" © 2020 Carroll Applied Science, LLC Dallas, TX http://ranking.rocks

Appendix

<u>Rational comparisons.</u> A study of a given decade, a constrained period of time, requires some definitions and ground rules. *Ranking the '80s* includes all singles that peaked between 1/1/1980 and 12/31/1989, and when singles are compared, it is on the basis of their entire lifecycle. Fewer than 1% of points earned by singles peaking in the '80s were earned in weeks outside the '80s. Similarly, comparing singles acts is simplified by adding the scores of their singles that peaked in the '80s.

Because producer and writer scoring is based on singles performance, their scores are found by summing the performance of their singles that peaked in the '80s.

Even though album lifecycles are longer, and albums have a much greater chance of re-entering the chart having fallen off, when albums are compared, it is on their total lifecycle as well. On the other hand, album acts are compared on the basis of the *total points earned only in the '80s*. This takes into account all the activity by all the albums, whether released in the '80s or not but does not credit activity outside the '80s.

<u>Proximate Album.</u> This is an attempt to link the charted singles with their most likely album root. While it was common in the '50s and '60s for singles to be recorded and released, with an album product generated afterwards if needed, by the '80s albums—vinyl, cassette or CD—were outselling singles by 10 to 1 and more in units. It was common to buy the album for one single—at least I did that. In the same way, in the '80s, the album usually came first. This also changed the dynamic of releasing singles, whereupon it became much more common to release three or more singles from an album.

This linkage becomes difficult because of the potential for the root album of singles being a soundtrack. I've made a genuine effort to find those, and the ground rules for deciding whether the soundtrack or an act's contemporaneous album is proximate for a single is documented at the beginning of the Ranking the Singles section.

<u>Grouping Acts for Scoring and Allocated Credit and an All-Purpose Confession.</u> Generally, the Billboard reference and label attributions are used to determine the name of the act. In cases where two nominal acts are mentioned joined with "featuring" or "with," if the second act is listed in smaller type than the first act it is not credited. Acts joined by "and" or listed in equal font size are generally both credited.

There are cases of occasional equal-to-equal collaborations between two well-known acts for which there is no continuing entity created (Kenny Rogers and Sheena Easton). In a case where the two artists have records of their own and are charted, points from a one-off are divided and allocated to the underlying acts. If the occasional collaboration is ongoing, a new entity is created. A list of allocated credits is included in Table 1 following.

There are numerous and well-known cases where an act's name changes but the personnel largely do not. In those cases, the differently named acts are scored under one name. In some cases, a solo act appears with a band but uses the same name (Eric Clapton and His Band; Bruce Springsteen and the E Street Band). For those cases, where the entity is substantially the same, they are grouped with the solo act.

In some cases, the band has changed altogether; thus, it is grouped as a separate entity (Prince vs. Prince and the Revolution; Ray Parker Jr. vs Ray Parker Jr. and Raydio. Admittedly, these are judgment calls and in some cases are ambiguous. The author remains vexed by Jefferson Airplane/Jefferson Starship/Starship and the Miami Sound Machine continuum to Gloria Estefan and apologizes for any inconsistency or violence done to common sense.

And a mea culpa for acts that drop an initial "The" from their names mid-career. I'm pretty sure a couple of those got past me as well. I have also not been scrupulous about D.J. vs DJ, & vs and, (Part 1, I or One) vs. Part 1, I or One. Yes, I do realize there is a special level in hell for me. Fortunately, there are also **www.45cat.com** and **www.discogs.com** where actual label facsimiles can be seen for those to whom it really matters and whose OCD is worse than mine.

Appendix

Table 1. Singles with Credit Allocated to the Underlying Acts.

Endless Love - Diana Ross & Lionel Richie
Say Say Say - Paul McCartney and Michael Jackson
The Girl Is Mine - Michael Jackson/Paul McCartney
On My Own - Patti LaBelle And Michael McDonald
Up Where We Belong -   Joe Cocker and Jennifer Warnes
Separate Lives (Love Theme From White Nights) -   Phil Collins and Marilyn Martin
(I've Had) The Time Of My Life -   Bill Medley And Jennifer Warnes
I Knew You Were Waiting (For Me) -   Aretha Franklin & George Michael
You And I - Eddie Rabbitt with Crystal Gayle
What Have I Done To Deserve This? -   Pet Shop Boys And Dusty Springfield
Somewhere Out There -   Linda Ronstadt And James Ingram
We've Got Tonight - Kenny Rogers And Sheena Easton
To All The Girls I've Loved Before -   Julio Iglesias & Willie Nelson
Almost Paradise - Mike Reno And Ann Wilson
After All (Love Theme From CHANCES ARE) -   Cher and Peter Cetera
Xanadu - Olivia Newton-John/Electric Light Orchestra
Surrender To Me - Ann Wilson and Robin Zander
Dancing In The Street - Mick Jagger & David Bowie
Just Once - Quincy Jones Featuring James Ingram
I Can't Help It - Andy Gibb & Olivia Newton-John
Her Town Too - James Taylor And J.D. Souther
Love Power - Dionne Warwick & Jeffrey Osborne
Suddenly - Olivia Newton-John & Cliff Richard
Sweet Baby - Stanley Clarke/George Duke
It's Only Love - Bryan Adams/Tina Turner
All Of You - Julio Iglesias & Diana Ross
Sisters Are Doin' It For Themselves -   Eurythmics And Aretha Franklin
Killin' Time - Fred Knoblock & Susan Anton
Under Pressure - Queen & David Bowie
Through The Storm - Aretha Franklin And Elton John
Saturday Love - Cherrelle with Alexander O'Neal
How Do You Keep The Music Playing (Theme From "Best Friends") - James Ingram And Patti Austin
Everlasting Love - Rex Smith/Rachel Sweet
I Got You Babe - UB40 With Chrissie Hynde
How Many Times Can We Say Goodbye -   Dionne Warwick And Luther Vandross
This Is Not America -   David Bowie/Pat Metheny Group
Till I Loved You (The Love Theme From Goya) -   Barbra Streisand And Don Johnson
Ebony Eyes - Rick James featuring Smokey Robinson

What's New -   Linda Ronstadt & The Nelson Riddle Orchestra
The Last Time I Made Love -   Joyce Kennedy & Jeffrey Osborne
Friends In Love - Dionne Warwick And Johnny Mathis
Cool Places - Sparks And Jane Wiedlin
Used To Be - Charlene & Stevie Wonder
Blame It On Love -   Smokey Robinson & Barbara Mitchell
Good Times - INXS And Jimmy Barnes
Somethin' 'Bout You Baby I Like -   Glen Campbell and Rita Coolidge
People Get Ready - Jeff Beck And Rod Stewart
It Isn't, It Wasn't, It Ain't Never Gonna Be -   Aretha Franklin and Whitney Houston
The Blues - Randy Newman and Paul Simon
All I Have To Do Is Dream -   Andy Gibb And Victoria Principal
When The Rain Begins To Fall -   Jermaine Jackson And Pia Zadora
That Lovin' You Feelin' Again -   Roy Orbison & Emmylou Harris
Lovers After All -   Melissa Manchester And Peabo Bryson
Help Me! - Marcy Levy & Robin Gibb
Strangers In A Strange World -   Jenny Burton & Patrick Jude
Taxi Dancing - Rick Springfield & Randy Crawford
Perhaps Love - Placido Domingo And John Denver
My Fantasy - Teddy Riley featuring Guy
Make My Day - T.G. Sheppard with Clint Eastwood
When Love Comes To Town - U2 With B.B. King
Perfect Combination - Stacy Lattisaw & Johnny Gill
Don't Give Up - Peter Gabriel/Kate Bush
The Best Of Me -   David Foster And Olivia Newton-John
Reservations For Two – Dionne Warwick And Kashif
The Greatest Gift Of All - Kenny Rogers & Dolly Parton
Get It - Stevie Wonder & Michael Jackson
My Love - Julio Iglesias Featuring Stevie Wonder
All The Right Moves -   Jennifer Warnes/Chris Thompson
Love Has Finally Come At Last -   Bobby Womack and Patti LaBelle
Whiter Shade Of Pale -   Hagar, Schon, Aaronson, Shrieve

Answers to Ranking the '80s: Twenty (Trivia) Questions and Other Challenges

1. Which four albums that peaked in the '80s also charted seven singles on the Hot 100 in the '80s?
   Thriller - Michael Jackson; Bad – Michael Jackson; Born In The USA – Bruce Springsteen; Hysteria – Def Leppard
2. Which single spent the most weeks at number 2 but never made number 1?
   Waiting For A Girl Like You – Foreigner 10 weeks
3. Which two acts charted three singles in top 40 simultaneously?
   Diana Ross, 1980: Upside Down, I'm Coming Out, It's My Turn
   New Kids On The Block, 1989: Hangin' Tough, Cover Girl, Didn't I (Blow Your Mind)
4. This 1983 hit was the last country record to top the Hot 100 for nearly 17 years. Name it.
   Islands In The Stream - Kenny Rogers and Dolly Parton
5. Name the three James Bond themes that charted in the '80s. Which was the first to go to number 1?
   For Your Eyes Only-Sheena Easton - #4; All Time High (Octopussy) - #36; View To A Kill-Duran Duran - #1
6. Which act had seven consecutive released singles go to number 1?
   Whitney Houston
7. Which act scored its eighth consecutive number 1 album of new material in the '80s?
   Rolling Stones
8. Name the only '80s charted single and the only charting singles act beginning with the letter X.
   Single: Xanadu – Olivia Newton-John/Electric Light Orchestra; Act: XTC
9. Name the highest ranking male and female solo singles acts having only one name.
   Prince; Madonna
10. What was the highest-ranking singles duo of the '80s? What was the highest ranking singles group?
    Hall and Oates; Duran Duran
11. Which act had seven albums simultaneously on the Top 200 chart?
    U2
12. What was the highest charting album with no Hot 100 singles?
    Dirty Deeds Done Dirt Cheap – AC/DC
13. What was the highest ranked '80s cover of a previous Hot 100 single?
    More Than I Can Say – Leo Sayer (Bobby Vee original)
14. Name the three covers of previous number 1 records that also went to number 1 in the '80s.
    Venus – Bananarama (Shocking Blue original); Lean On Me – Club Nouveau (Bill Withers original); You Keep Me Hanging On – Kim Wilde (Supremes original)
15. Name the three writers to have songs simultaneously occupy in number 1 and 2 on the Hot 100.
    Jim Steinman: Total Eclipse Of The Heart 1, Making Love Out Of Nothing At All 2 (also produced both);
    Prince: Kiss 1, Manic Monday 2;
    Diane Warren: When I See You Smile-Bad English 1, Blame It On The Rain-Milli Vanilli 2
16. Who was the highest charting producer of the '80s measured by singles chart scores?
    Quincy Jones
17. Which act charted the most albums (alone or in collaboration) that peaked in the '80s?
    Willie Nelson – 20 (Kenny Rogers – 17 without collaborations)
18. Which act charted the most albums (alone or in collaboration) including those holdovers or recurrent from previous decades?
    Kenny Rogers - 22
19. Which writer, alone or in a team, charted the most songs on the Hot 100 in the '80s?
    Paul McCartney - Writer – 56; Prince-Writer Team-34
20. What are the top three soundtracks of the '80s?
    Dirty Dancing - #9; Purple Rain - #20; Flashdance - #50

Bonus 1: What was the highest charting writing team, who were not primarily performers?
Jimmy Jam Harris & Terry Lewis

Bonus 2: The titles of The Police's two biggest hits both started with "Every." Name them.
Every Breath You Take;   Every Little Thing She Does Is Magic

Bonus 3: One act ranked number 1 for three consecutive years: 1985, 1986, 1987. Name that act.
Madonna

Bonus 4: Three solo male acts, three solo female acts and three duos each having one name, charted number 1 singles in the '80s. Name them.
Males: Prince (2), Falco (1), Vangelis (1)
Females: Madonna (7), Tiffany (2), Martika (1)
Duos: Wham! (3), Roxette (2), Eurythmics (1)

Bonus 5: These three '80s acts, which were not one-off star collaborations (i.e. USA For Africa)—had one single ever in the Hot 100 and it peaked at number 1. Two were instrumental and one was a cappella. Name them.
Vangelis (Chariots Of Fire Theme); Jan Hammer (Miami Vice Theme); Bobby McFerrin (Don't Worry Be Happy).

Bonus Bonus: One other non-US-based '80s act peaked at number 2 and never made the Hot 100 again. Who was it?
Nena (99 Luftballoons/99 Red Balloons)

Who was bigger:

Go-Gos (109, 10243) or **Belinda Carlisle (93, 12201)**
Genesis (48, 19591) or **Phil Collins (5, 39768)**
George Michael (23, 25301) or Wham! (38, 21683)
Stevie Nicks (57, 17947) or Fleetwood Mac (73, 15297)
Peter Cetera (108, 10421) or **Chicago (15, 29063)**
Bobby Brown (63, 16588) or New Edition (124, 8712)
Don Henley (90, 12425) or **Glenn Frey (89, 12594)**
Neil Diamond (62, 16701) or Barbra Streisand (92, 12222)
Rick Astley (106, 10618) or Bananarama (129, 8533)
The Fixx (175, 6528) or A-Ha (223, 5002)
The Pretenders (151, 7742) or The Motels (192, 5733)
Pat Benatar (56, 18270) or **Debbie Gibson (54, 18379)**
Billy Joel (6, 38031) or Elton John (18, 27052)

ACKNOWLEDGEMENTS

Thanks to Doug Heatherly, Ph.D. for tech support on cover art and file uploading. Doug has been a part of the Ranking team from the beginning, although we may be running him out of splashy color combinations.

It's important to mention the chart hounds who have bought copies of the books in this series, and particularly those who took time to share compliments and critiques. The audience is small, but passionate.

Thanks to my writing colleague Dann Isbell—the founder of the *Ranking* franchise and author of the Foreword for this book--who I met online nearly the first day I started this line of work in July, 2013. Through *Ranking the '70s*, he taught me how to create a book like this one. He is a marvelous sounding board, proofreader, ground wire and friend.

Lou Simon of Sirius XM is an unwavering source of encouragement to this enterprise. Lou maintains an oasis on satellite radio called The Diner, where music fans of all ages congregate once a week to talk about the music they love. Out of that common interest has grown a virtual community of people who have also found ways to meet and care for one another in the real world. Few people could have created such a community out of the void using nothing but warmth and personality. Lou did.

The now ubiquitous use of videoconferencing has allowed members of that community, the Diner Friends, to regularly gather and swap tales. I've tried out concepts and trivia questions used in this book on them, and I appreciate their warm input.

Thanks to Rich Appel, professor and radio entrepreneur, a source of information, support and friendship since I started chart work in earnest. In that time, he has harpooned his white whale—bringing *That Thing with Rich Appel* to life on over 100 stations, where he is heard weekly, literally around the world.

Finally, and saving the best for last, I want to thank my wife Mary, daughter Allison and sons Will and Quin. They have been kind enough to only roll their eyes when they thought I wasn't looking, enduring the gestation and birth of now four books over the past seven years. They didn't scream when I hijacked interesting conversations on other topics to describe this work to poor souls who were forced to chew a leg off to escape the trap. Mary observes that writing these books has kept me quiet and occupied during the Covid-19 pandemic and lockdown, thus preventing us from driving each other crazy.

My late brother Jim provided the first impetus to start the work by bequeathing me his collection of chart books. My sister Mary Ellen's evaluation of '80s pop as "metal machine music" still reverberates in my head. I can't hear "Dancing Queen" without a fond remembrance of my late sister Kate.

It is truly odd that after a career in science and science policy, my experiments now are dry, done *in silico* and have nothing to do with chemistry, but, I'll argue, are appropriately Weird Science nonetheless.

March 2021

About The Author

BILL CARROLL, Ph.D. is an Adjunct Professor of Chemistry at Indiana University, Bloomington, IN and lives with his wife Mary in Dallas, TX. They have three children, Allison, Will and Quin.

Bill has been President and Chair of the Board of the American Chemical Society, the world's largest single-discipline scientific professional organization with over 150,000 members.

In 2015, Bill retired after a 37-year career in industry and now spends his time as a globally recognized industrial chemistry consultant through his company, Carroll Applied Science, LLC. His first love for 60 years, however, has been the pop charts. As an author of peer-reviewed articles, as well as *Ranking the Albums*, the massive *Ranking the Rock Writers* and co-author with Dann Isbell of the highly-rated *Ranking the '70s*, he is fulfilling his dream of applying science-based data analytics to popular music.

ca. 1980

www.ingramcontent.com/pod-product-compliance
Lightning Source LLC
Chambersburg PA
CBHW081206230426
43666CB00015B/2666